2

3

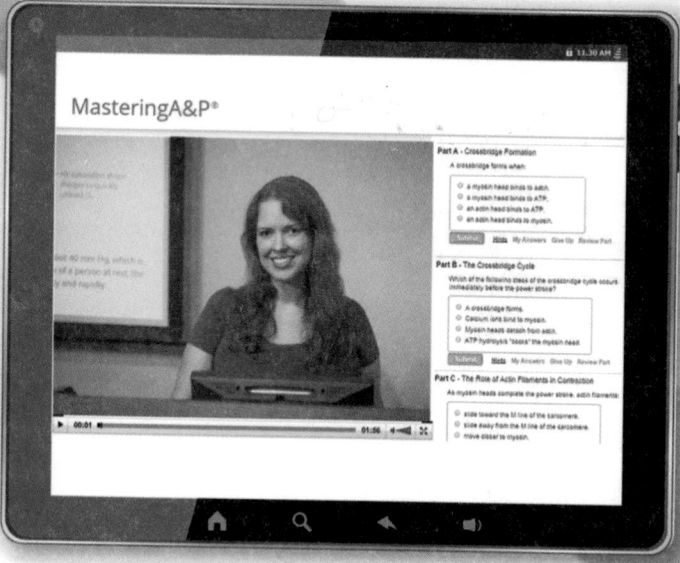

Big Picture figures with Animations

- Scan the QR code and watch **Big Picture figures** come alive as Amerman narrates physiological events. With these **mobile-ready Big Picture Animations,** you can study A&P on-the-go, anywhere, any time.
- Play with interactive quizzes for each animation, where you can draw, predict, apply and more.

If you don't have a QR code reader on your smart phone, go to your phone's app store and download one of the free QR reader apps.

MasteringA&P®

- Access **MasteringA&P** assignments for Concept Boost Video Tutors and Big Picture Animations, Practice Quizzes, *Practicing A&P: Active-Learning Worksheets,* **NEW!** Interactive Physiology 2.0, and more.

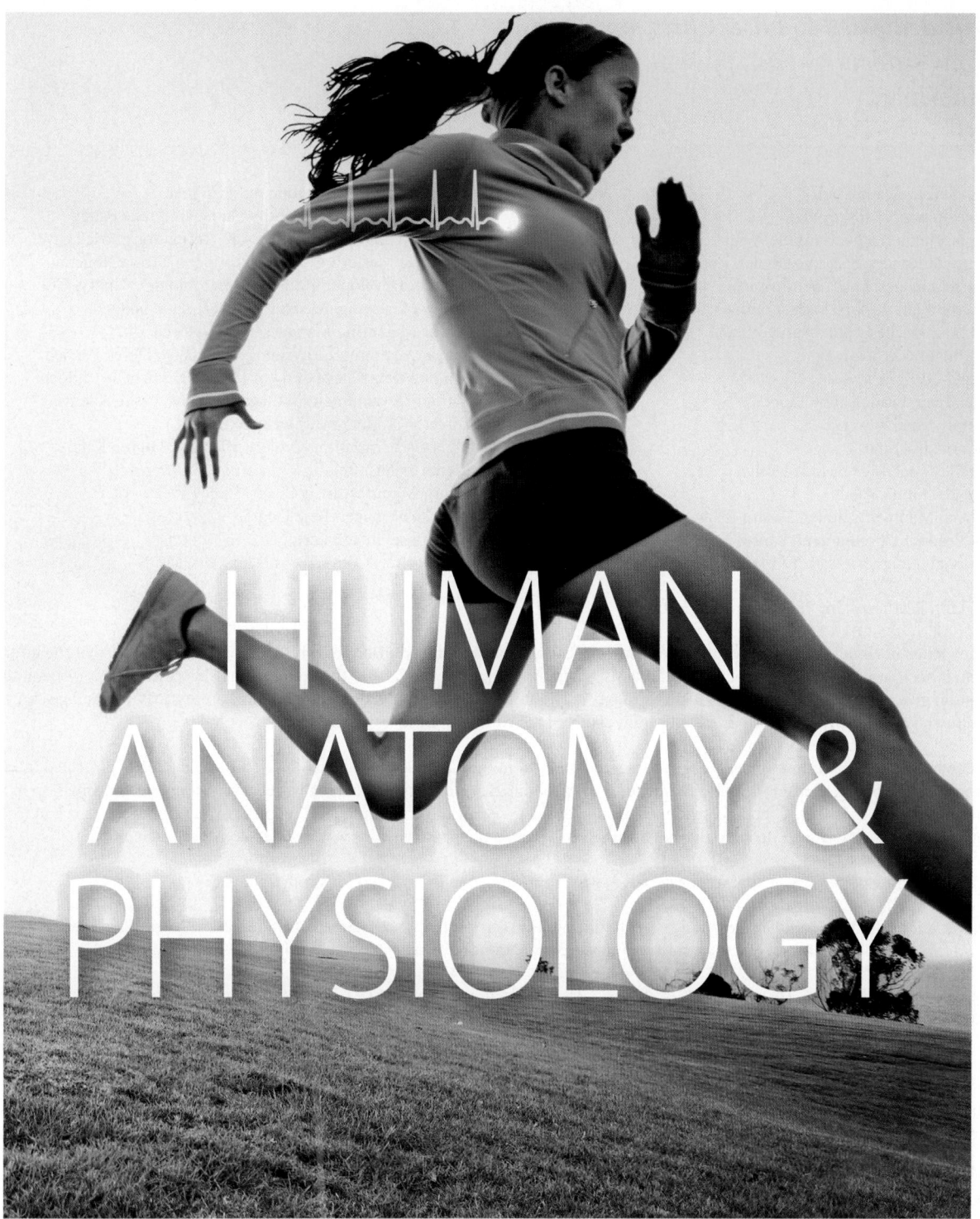

HUMAN ANATOMY & PHYSIOLOGY

Erin C. Amerman

Florida State College at Jacksonville

PEARSON

This book is dedicated to Elise, Chris, my mom,
and all of my students—past, present, and future.
—Erin Amerman

Editor-in-Chief: Serina Beauparlant
Director of Development: Barbara Yien
Project Managers: Nicole Taché, Caroline Ayres
Senior Development Editors: Suzanne Olivier, Alice Fugate
Art Development Manager: Laura Southworth
Senior Art Development Editors: Laura Southworth, Elisheva Marcus
Editorial Assistants: Arielle Grant, Daniel Wikey
Director of Production: Erin Gregg
Program Manager Team Lead: Mike Early
Project Manager Team Lead: Nancy Tabor
Production Editor: Jane Hoover
Copyeditor: Bonnie Boehme
Text and Cover Designer: Side by Side Design
Photo Image Lead: Donna Kalal
Photo Researchers: Maureen Spuhler, Kristin Piljay
VP, Director of Product Development: Lauren Fogel
Senior Instructional Designer: Sarah Young-Dualan

Executive Content Producer: Liz Winer
Content Producers: Timothy Hainley, Natalie Pettry
Development Editor, Big Picture Animations: Eric Leaver
Associate Mastering® Media Producer: Nicola Houston
Content Developer, A&P and Microbiology: Cheryl Chi
Senior Mastering® Media Producer: Katie Foley
Senior Marketing Manager: Allison Rona
Senior Anatomy & Physiology Specialist: Derek Perrigo
Senior Market Development Manager: Michelle Cadden
Market Development Manager, Science: Leslie Allen
Director of Marketing: Christy Lesko
Rights & Permissions Project Manager: Donna Kalal
Rights & Permissions Management: Lumina
Senior Manufacturing Buyer: Stacey Weinberger
IRDVD Manager: Kyle Doctor
Art House: Imagineering
Compositor: Cenveo® Publisher Services

Cover Photo Credit: Peathegee, Inc./Blend Images/Corbis

Library of Congress Cataloging-in-Publication Data
Amerman, Erin C., 1977-
  Human anatomy & physiology / Erin C. Amerman, Florida State College at Jacksonville.
    pages cm
  Includes index.
  ISBN 978-0-8053-8295-2 (student edition)—ISBN 978-0-321-99061-7 (instructor's review copy)
  1. Human physiology—Textbooks. 2. Human physiology—Study and teaching (Higher)
  3. Human anatomy—Textbooks. 4. Human anatomy—Study
and teaching (Higher) I. Title. II. Title: Human anatomy and physiology.
  QP34.5.A44 2014
  612—dc23
                    2014043310

www.pearsonhighered.com

2 3 4 5 6 7 8 9 10—V057—19 18 17 16 15
Student Edition: ISBN 10: 0-8053-8295-X; ISBN 13: 978-0-8053-8295-2
Instructor's Review Copy: ISBN 10: 0-321-99061-7; ISBN 13: 978-0-321-99061-7

About the Author

Erin C. Amerman

Erin Amerman teaches anatomy and physiology at Florida State College at Jacksonville; she has been involved in anatomy and physiology education for more than 14 years as an author and professor. She received a B.S. in cellular and molecular biology from the University of West Florida and a doctorate in podiatric medicine from Des Moines University. She is also the author of the best-selling *Exploring Anatomy and Physiology in the Laboratory*, now in its second edition, with Morton Publishing Company. Erin is deeply committed to helping her students succeed in the A&P course and to generating curiosity and excitement about the material and its application in their future health careers. She is a member of the Human Anatomy and Physiology Society (HAPS) and enjoys attending the annual HAPS conferences, especially when they are in locations that have ample hiking opportunities and many different species of snake.

When not writing or teaching, Erin enjoys spending time with her family and her menagerie of rescued cats, dogs, pet rats, snakes, tetras, and a turtle. She also practices karate and kobudo (she recently earned her second degree black belt) and loves photographing the local wildlife around her home in rural northern Florida. She can often be found hiking around the prairie with a camera and snake hook in hand.

About the Media Author

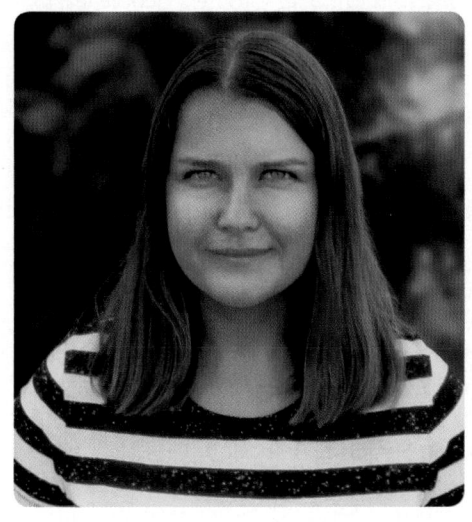

Virginia Irintcheva

Virginia Irintcheva is the author of the Big Picture Animations and interactive figures and assessments in MasteringA&P®. She also served as an editorial advisor and reviewer of key concepts in this book. She is an associate professor at Black Hawk College in Moline, Illinois, where she has been teaching courses in anatomy and physiology, medical terminology, and biology for more than 8 years. Virginia was born and raised in Bulgaria. She earned a B.S. in biology and Spanish from St. Louis University, under whose auspices she studied in Madrid, Spain, as well as in Missouri. She received her Ph.D. in pharmacological and physiological sciences in 2006 from that school. Her primary research interests are cell signaling and, specifically, the effects of the coagulation protein thrombin and the family of Ras proteins on cell growth and proliferation. As a teacher, her fundamental goal is to create an environment of effective communication by promoting mutual respect and successful learning. Virginia strives to teach her students not only what to learn but also how to learn and why they are learning.

Outside the classroom, Virginia likes to travel, visit family in Europe, snowboard, hike with her dog, and ride her horse.

Preface

To the Student

Welcome to the fascinating study of the human body! Though you and I might never meet in person, I consider you and every other student who uses this textbook to be "my" student. Just as I want to ensure the success of the students in my classroom, I am similarly invested in your success. For this reason, this book was designed with *you* in mind—every feature, study tool, and media presentation is intended to help you achieve your goals.

This book was written not only for you, but also *about* you. The great thing about human A&P is that no matter what your goals are, it is relevant to your life. Human A&P is you; it's also me, your family, your friends, and indeed every human who ever lived or will live. There's nothing in the study of A&P that is irrelevant or esoteric, because every single detail revolves around you and your life. How many other courses can make that claim?

So dive right in and begin to explore the science of you. I sincerely hope that you enjoy your study of human A&P and find it as fascinating and wondrous as I do.

—Dr. Amerman
erin.amerman2014@gmail.com

To the Instructor

Why a New A&P Textbook?

I get the question "Why did you write this book?" quite regularly. The short answer to this is that writing and teaching are just in my DNA somewhere. For the long answer, we have to look back in time and start with my 5-year-old self.

When I was in kindergarten, I was placed with another kid, Kyle, into a separate group for reading time because we were the only two kids in the class who could already read. It struck my 5-year-old brain as inconceivable that so many of my classmates couldn't read. Reading was so *easy;* anyone could do it! Maybe, I reasoned, they just needed a book to teach them how to read. So I gathered up some construction paper and crayons and got to writing. And thus my first "textbook" was born: *The Bird and Mr. Bear.*

Fast forward a few years to my medical school education. While in medical school, I co-taught a human physiology course, and during my first class I had one of those "aha" moments: Teaching somehow just "felt right." I connected with my students, and they connected with me. This feeling only grew over the next two semesters. But still, I was in medical school, and who would be crazy enough to go through the pain of medical school, graduate, and then not ever practice as a physician?

Well, it turns out that *I* was crazy enough to do just that. I was lucky enough to find a full-time position teaching anatomy and physiology. And while I loved teaching, there were far more challenges than I had anticipated. My students were different

from my former classmates. The difference wasn't in intelligence—my students were smart. But, this new generation of students seemed to be ill-prepared for the rigors of a college science course. They lacked study skills, they had little to no background in science, and—alarmingly—they couldn't read or understand their textbooks. For these reasons, so many bright, motivated students struggled with the course.

As a teacher, this was the last thing I wanted to see. So I did the same thing I did in kindergarten: grabbed some paper and started writing. First came my own lab exercises, which were followed by lecture outlines and notes. As I wrote, I "Amermanized" the content (a term coined by a student) with concise prose, simple diagrams, stories/analogies, and active learning exercises. My students' responses were enthusiastic; indeed, many asked if they could return their textbooks and just use my notes instead.

A vision for a new textbook began to form in my mind: one for today's students. It would:

- be written at a level my students could understand and, at the same time, still provide the information they need;
- anticipate where they need help with the science and provide the necessary in-the-moment coaching; and
- reduce cognitive overload and present information—in both text *and* art—in manageable chunks that are more easily digestible.

Eight years later, my vision finally became reality with the publication of *Human Anatomy & Physiology.* I am thrilled to be able to offer this text and its ancillary materials and I sincerely hope it has a positive impact on students' lives and education. This is what I have wanted since *The Bird and Mr. Bear*—to help people learn.

Key Features

Many of the key features found in this textbook, the companion workbook, and media came directly from my experience teaching and working with a range of students and seeing what helps them learn. These features include the following:

- **How to Succeed in A&P** in Chapter 1 introduces students to core study skills, including how to manage time, how to take notes, and how to study for an A&P exam. I also guide students through how to use the textbook, workbook, and online tools.
- Recurring **Core Principles** icons appear throughout the book and remind students to recall and apply four core principles introduced in Chapter 1: Structure-Function, Feedback Loops, Gradients, and Cell-Cell Communication.
- Over 50 **Concept Boosts** and **Study Boosts** coach students on key A&P concepts that are often difficult or tricky. Additional emphasis is placed on explaining challenging topics, often incorporating familiar analogies and simple illustrations, giving students a boost in fully understanding the content.
- **Concept Boost Video Tutors** walk students through selected Concept Boost topics that are particularly tough to understand.

These Video Tutors are assignable in MasteringA&P® and are also available in the Study Area of MasteringA&P.

- **One-concept-at-a-time art** focuses on teaching one concept per figure so that a student can instantly grasp the key idea without being distracted by a sea of details. For key physiology concepts, unique sequence figures unpack information systematically so that each scene contains only the most important information, again making it easier for today's students to focus on key details.
- **In-the-moment visual reminders** of key foundational science concepts appear as needed to help students recall and apply given concepts.
- **Big Picture figures** visually summarize key physiological processes and anatomy concepts, highlighting only what is most important.
- Mobile-ready **Big Picture Animations** with interactive quizzes bring the Big Picture figures to life and help reinforce students' understanding of each step in a key process. These animations are assignable in MasteringA&P and are also available in the Study Area of MasteringA&P.
- **HAPS-based Learning Outcomes** begin each module within a chapter. Additionally, the assessments in MasteringA&P are organized by these Learning Outcomes.
- **Pronunciations** use phonetic sounds (instead of traditional symbols) to help students learn correct pronunciations.
- **Flashback** questions encourage students to think about previously learned concepts they will need to apply in order to understand upcoming discussions.
- **Quick Check** questions appear throughout each module to test students' basic understanding of the material. Answers to Quick Check questions are available in the Study Area of MasteringA&P.
- **Apply What You Learned** questions at the end of each module ask students to think critically and apply what they've just learned to a real-world scenario.
- **A&P in the Real World** features highlight clinical conditions and disorders that illustrate and reinforce key A&P concepts discussed in the chapter.
- **Chapter running case studies** with assessments challenge students to apply their knowledge of key A&P concepts to a real-world clinical scenario, while allowing instructors to "flip" the classroom and incorporate critical thinking and/or group activities. These cases can be found in the Instructor's Guide in MasteringA&P and are also assignable in MasteringA&P.
- *Practicing A&P: Active-Learning Workbook* helps engage the kinesthetic learner with labeling, drawing, and build-your-own summary-table exercises that students can complete as they read the textbook. This workbook is available as a print supplement and also as downloadable PDF worksheets in the Study Area of MasteringA&P.

The Development Story

Making this book, including the text, the art, and the media, was an intensive, collaborative process. Each draft of the manuscript was sent to A&P instructors and other content experts to evaluate the breadth, depth, and accuracy of coverage; the art

program; and the overall pedagogical effectiveness. Our team worked closely together to analyze the feedback and determine which changes were necessary to improve each chapter. Each subsequent draft took into account the reviewers' feedback to make sure that we identified the key challenging concepts that students struggle with and that we included a way to help students better understand those concepts (such as giving them closer attention via a Concept Boost, or "unpacking" a complex topic gradually over a series of well-paced figures).

Accuracy and Currency

During the production process, when art was rendered and each book page was laid out, we went through another extensive reviewing process to ensure that accuracy and clarity were maintained in all materials. We also reviewed the pages with various focus groups to be certain that the final presentation delivers an effective, reader-friendly experience for students and instructors alike.

One of our goals was ensuring that this book is not only accurate but also as up-to-date as possible. Key topics throughout the book were researched, and in some cases we updated our presentation to reflect current understanding. Examples include up-to-date discussions of endocytosis and calcium homeostasis, of new evidence suggesting that PTH is the main regulatory hormone in humans and calcitonin is not as relevant as we once thought, of learning and memory in the central nervous system chapter, and of increased public health concern about children not being vaccinated because parents believe vaccines are linked to autism.

Class-Tested and Approved

Anatomy and physiology students across the country also contributed to the development of *Human Anatomy & Physiology*. Over 4000 students provided feedback through extensive class testing prior to publication. We asked students to use the chapters in place of, or alongside, their current A&P textbook during their course. We then asked them to evaluate numerous aspects of the text, including how clear and understandable the writing style is, how well it explains and coaches on difficult topics, whether the text helps them see the "big picture" of key physiology processes and anatomy concepts, whether the figures are easier to read and understand, and how well the Big Picture Animations help them to visualize and grasp physiology topics. Through these student reviews, the strengths of *Human Anatomy & Physiology* were put to the test, and it passed. Overwhelmingly, the majority of the students who class tested it would prefer to use this textbook over their current textbook.

In addition, our market development team interviewed over 75 A&P instructors, gathering feedback on how well the text builds conceptual understanding, how well the author coaching helps demystify tough or tricky concepts, and how well the *A&P in the Real World* boxes illustrate key concepts and cover key clinical topics. Instructors also reported on the accuracy and depth of the content overall. All comments, suggestions, and corrections were analyzed and addressed by the author and editorial team prior to publication.

Are your students overwhelmed by the amount of information in the course?

One-concept-at-a-time art

Drawing from her experience in the classroom and the latest research in cognitive science, Amerman reduces cognitive overload by visually unpacking key information using one-concept-at-a-time art and Big Picture figure visual summaries.

This figure shows the first concept: how synaptic transmission occurs at a chemical synapse.

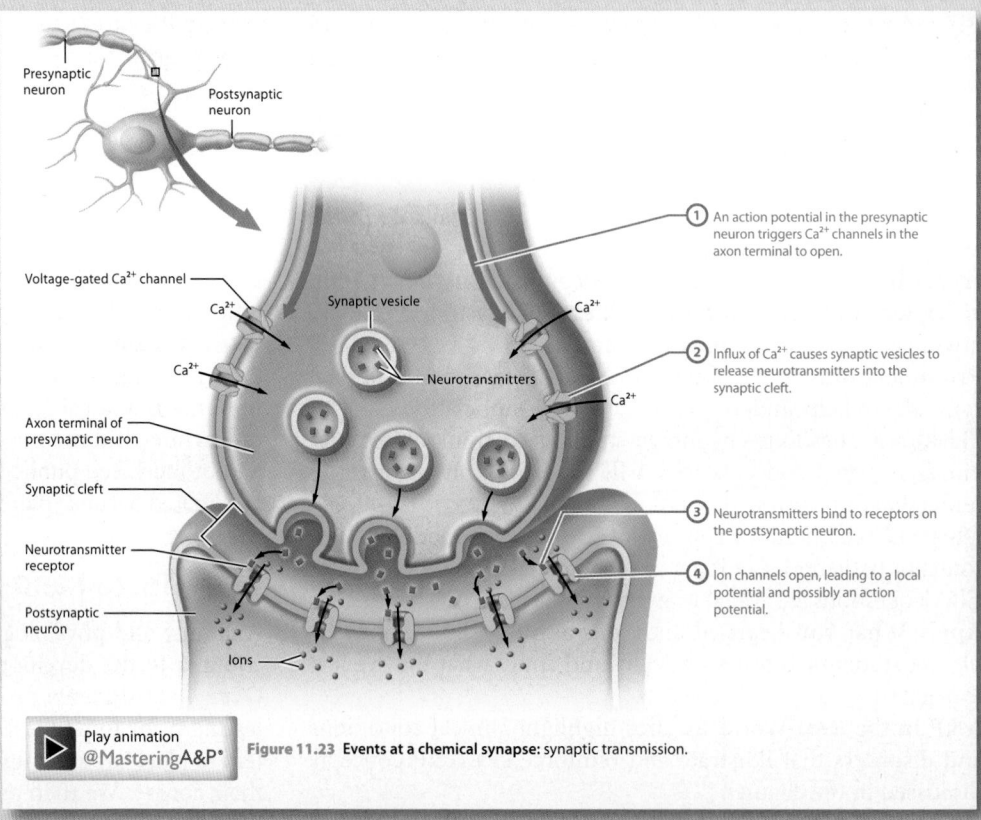

① An action potential in the presynaptic neuron triggers Ca²⁺ channels in the axon terminal to open.

② Influx of Ca²⁺ causes synaptic vesicles to release neurotransmitters into the synaptic cleft.

③ Neurotransmitters bind to receptors on the postsynaptic neuron.

④ Ion channels open, leading to a local potential and possibly an action potential.

Play animation @MasteringA&P®

Figure 11.23 Events at a chemical synapse: synaptic transmission.

This figure shows the second concept: how postsynaptic potentials are triggered.

Play animation @MasteringA&P®

Figure 11.24 Postsynaptic potentials.

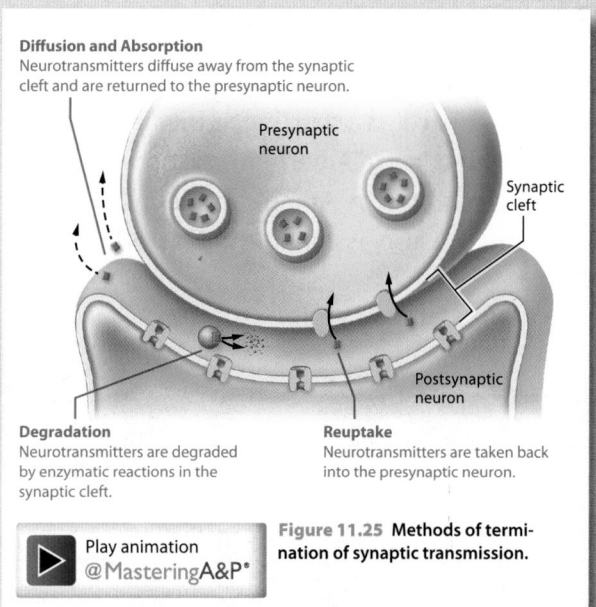

Diffusion and Absorption
Neurotransmitters diffuse away from the synaptic cleft and are returned to the presynaptic neuron.

Presynaptic neuron

Synaptic cleft

Postsynaptic neuron

Degradation
Neurotransmitters are degraded by enzymatic reactions in the synaptic cleft.

Reuptake
Neurotransmitters are taken back into the presynaptic neuron.

Play animation
@MasteringA&P®

Figure 11.25 Methods of termination of synaptic transmission.

The next figure shows the third concept: how synaptic transmission is terminated.

The Big Picture of Chemical Synaptic Transmission
Figure 11.26

Play animation
@MasteringA&P®

① **Action potential:** An action potential reaches the axon terminal of the presynaptic neuron.

Action potential

Axon terminal of presynaptic neuron

Ca²⁺

Neurotransmitters

Ca²⁺

Synaptic cleft

Open receptor/ ion channel

Postsynaptic neuron

② **Synaptic transmission:** Ca²⁺ channels open in the presynaptic neuron; neurotransmitters are released from synaptic vesicles and bind to receptors on the postsynaptic neuron (see Figure 11.23).

OR

EPSP   Threshold   IPSP   Threshold

③ **Postsynaptic potentials:** Neurotransmitters trigger an EPSP or IPSP, moving the membrane potential of the postsynaptic neuron either closer to or farther from threshold (see Figure 11.24).

Diffusion

Reuptake

Degradation   Closed receptor/ ion channel

④ **Termination of synaptic transmission:** Neurotransmitter concentration in the synaptic cleft decreases and synaptic transmission is terminated (see Figure 11.25).

The Big Picture figure summarizes the key events of chemical synaptic transmission.

vii

Would your students benefit from another mode of learning A&P concepts?

Big Picture Animations

With Amerman's mobile-ready Big Picture Animations, students can study and review A&P topics anywhere, any time. These animations help students visualize events occurring at the molecular level and reflect the same terminology and explanations found in the Amerman textbook to ensure a consistent learning experience. Each animation is narrated by Amerman and includes interactive quizzes with instant feedback. These animations are also assignable in MasteringA&P® and are available in the Study Area of MasteringA&P.

Scan to watch Figure 24.24.

Concept Boost Video Tutors

Students get just-in-time coaching on-the-go with Amerman's mobile-ready Concept Boost Video Tutors. Select Video Tutors feature the author talking directly to students and coaching them through tough-to-understand or tricky topics. These Video Tutors are assignable in MasteringA&P® and are also available in the Study Area of Mastering A&P.

"The Big Picture Animations are AWESOME! I truly love the way they are presented—plus, the quizzes that follow are sure to engage my students and give them the immediate feedback they expect."
—**William Huber, St. Louis Community College at Forest Park**

Are there concepts you know in advance your students will struggle with?

Concept Boosts

In her classroom, Amerman gives her students extra coaching in advance of those tough-to-understand concepts, *right when they need it,* and she has built that same strategy into her textbook. She anticipates where students will need extra help and then provides just-in-time coaching via Concept Boosts. Each Concept Boost focuses on tough-to-understand or tricky concepts.

> Amerman is first and foremost a great teacher and it shows. She anticipates where students get tripped up and provides the extra help. Her writing style is simple and straight to the point. She writes much like how an instructor teaches a lecture, which I love.
>
> —Richard Gonzalez Diaz, Seminole State

ConceptBOOST »»»

"But It All Looks Pink!" Part 2

In Part 1 of this Concept Boost, we discussed how to orient yourself to the different components of a tissue section. Now we'll take it a step further and identify the tissue from which a section was taken. Let's try it with these examples:

| Example A | Example B | Example C |

This isn't as difficult as it appears, particularly if you use a methodical, step-by-step approach:

1. **Identify the cells and the ECM.** Start with the basics you were given in the first part of this Concept Boost (on p. 128) and label the cells you see in Examples A, B, and C. Now move on to the ECM. Remember, the ECM consists only of ground substance and protein fibers. The ground substance will generally stain a uniform color (or simply appear clear). The protein fibers can take on various forms in different tissues, but they will generally stain darker than the ground substance, and they will always lack nuclei. Identify the protein fibers and ground substance in Examples A, B, and C.
2. **Notice how the cells are shaped and arranged.** Are the cells packed tightly together, or are they widely spaced? Do they form a continuous sheet, as in epithelial tissue—or do they seem to be surrounded by ECM, as in connective tissue? Are the cells all identical, or are there clearly different types? Do the cells have "arms" extending from a central body? Explain how the cells are shaped and arranged in Examples A, B, and C.
3. **Notice how the ECM is arranged.** Is the ECM confined to one specific part of the tissue, or is it spaced evenly between the cells? Does ground substance predominate, or are protein fibers the main elements? What types of protein fibers can you see? Explain how the ECM is arranged in Examples A, B and C.
4. **Determine the class of tissue.** Using your analysis in the preceding steps, now you are ready to identify the class of tissue. Determine the type of tissue in Examples A, B, and C. ■

Answers: A: loose connective tissue, B: skeletal muscle tissue, C: simple columnar epithelium

ConceptBOOST »»»

How Changes in Arteriolar Diameter Influence the GFR

As we discussed earlier, filtration will occur only when a net pressure gradient in the glomerulus drives fluid out of the blood and into the capsular space. The size of this gradient determines how much filtration takes place—a small gradient will lead to only minimal filtration, whereas a large gradient leads to heavy filtration. Several factors determine the size of the pressure gradient in the glomerulus, but one of the most easily adjustable factors is the diameter of the afferent (entering) and efferent (leaving) arterioles. When either arteriole constricts or dilates, this changes the glomerular hydrostatic pressure (GHP), and therefore the entire net pressure gradient also changes.

You can think of blood flowing in and out of the glomerulus as being similar to water flowing in and out of a sink, where the afferent arteriole is the faucet, the basin is the glomerulus, and the efferent arteriole is the drainpipe. Keep this analogy in mind as we explore how this mechanism works:

- Vasoconstriction of the afferent arteriole "turns down the faucet." This allows less blood to flow into the glomerulus, which decreases the GHP and the GFR:

- Vasoconstriction of the efferent arteriole "clogs the drain." This causes blood to back up within the glomerulus, which increases the GHP and thus increases the GFR:

Vasodilation has the opposite effects:

- Vasodilation of the afferent arteriole "turns up the faucet." This increases the GHP and the GFR:

- Vasodilation of the efferent arteriole "unclogs the drain," allowing increased flow out of the glomerulus. This decreases the GHP and the GFR:

In the upcoming sections, you'll see that the main mechanisms the body uses to control or maintain GFR work by causing one or more of these physical changes. ∎

Are your students prepared for the rigor of A&P?

MyReadinessTest™

MyReadinessTest for A&P prepares students *before* their A&P course begins.

Students can get free online access the moment they register for your A&P course. MyReadinessTest assesses students' proficiency in study skills and foundation concepts in science and math, and tutors them in core areas where they need additional practice and review, before they even set foot in an A&P classroom. It offers:

- **Student online access** upon registration for their A&P course

- **Diagnostic Test and Cumulative Test** based on learning outcomes from a widely used primer, *Get Ready for A&P* by Lori Garrett

- **Personalized Study Plan** based on students' test results that includes practice questions with tutorials

- **Flexible Testing** that allows instructors to edit the Diagnostic Test or implement their own placement test or exit exam

- **Gradebook** that automatically records students' results

MyReadinessTest™   Help ▾

4.1 Basic Chemistry   Overview

Objective: Differentiate between ionic, covalent, and hydrogen bonding.

◀◀ ◀ ① ② ③ ④ ⑤ ⑥ ⑦ ⑧ ▶ ▶▶

View the lesson about chemical bonding, and then answer the following question.

During chemical bonding, which specific subatomic particles form the bond?

○ A. protons   ○ B. neutrons

○ C. electrons in the firs

✓ **Excellent!**

Excellent! Only the oute
chemical bonding.

CC
is off

nucleus

▶   00:43 ━━━━━━━━━━   05:13   ◀••◖◗◖   ⛶

How to Succeed in A&P

Amerman includes a special section, "How to Succeed in A&P," in Chapter 1, with discussion of how to manage time, how to take notes, how to study for an A&P exam, and how to use the textbook, the companion *Practicing A&P: Active-Learning Workbook,* and online tools.

Core Principles

In Chapter 1, Amerman introduces **four core principles** and then highlights them throughout the textbook to remind students of the overall theme of human anatomy and physiology—homeostasis—and show how the core principles revolve around maintaining it.

ing that we have only about 3 liters of plasma. Therefore, your entire plasma volume is filtered by your kidneys about 60 times per day.

The kidneys are able to filter blood so efficiently in part because the glomerular capillaries are remarkably permeable.
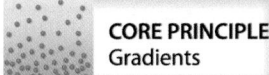
CORE PRINCIPLE
Gradients
However, even with fenestrated capillaries, filtration will happen only if a pressure gradient is present to push water and solutes through the filtration membrane (an example of the Gradients Core Principle, p. 26). In this section we discuss the forces that allow this process to occur.

Filtration Pressures

Let's first review the two forces that drive fluid movement in a typical capillary bed:

- **Hydrostatic pressure.** *Hydrostatic pressure* is the force of a fluid on the wall of its container. In the case of blood cap-

CORE PRINCIPLE
Feedback Loops

CORE PRINCIPLE
Structure-Function

CORE PRINCIPLE
Gradients

CORE PRINCIPLE
Cell-Cell Communication

In-the-Moment Visual Reminders

In addition to coaching via Concept Boosts, Amerman provides in-the-moment visual reminders of key foundational science concepts students need to recall and apply to understand a given concept.

increases. These changes are easy to visualize with a syringe, like the ones you've seen in a doctor's office. When you depress the plunger, you decrease the syringe's volume and increase the pressure. When you pull the plunger out, you increase the syringe's volume and decrease the pressure:

Do your students understand A&P concepts the first time they encounter them?

MasteringA&P® Assignable Content

With MasteringA&P, students get it by doing self-paced tutorials that reflect your course objectives, provide personalized coaching, and respond to each student's progress. Coaching activities include immediate specific wrong-answer feedback and hints that emulate the office-hour experience to focus the students' learning and keep them on track.

MasteringA&P helps instructors maximize their class time with easy-to-assign, customizable, and automatically graded assignments that motivate students to learn outside of class and arrive better prepared for lecture and lab. Pre-built MasteringA&P courses and assignments are available to get you started easily and quickly.

Assignable Coaching Activities include:

Big Picture Animation Activities

The Big Picture Animation Activities help students visualize physiological events occurring at the molecular level and reinforce understanding of key concepts learned in the chapter. These mobile-ready animations are narrated by the author and reflect the same terminology and explanations found in the Amerman textbook. Activity assignments include multiple-choice questions with hints and specific wrong-answer feedback, interactive ranking and sorting exercises, and labeling activities.

Concept Boost Video Tutor Activities

These videos and assignments feature the author talking directly to students and coaching them through tough-to-understand or tricky topics.

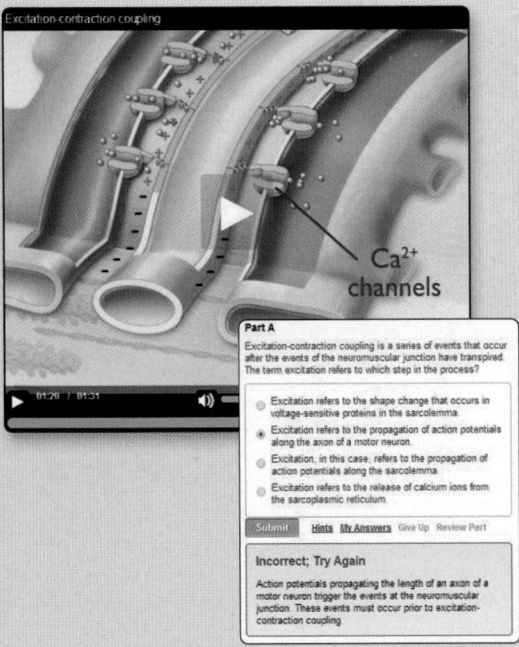

Interactive Physiology 2.0 Activities

This award-winning tutorial program helps students advance beyond memorization to a genuine understanding of complex physiological processes. I.P. 2.0 features brand-new graphics, quicker navigation, and more robust interactivity, so students can explore, draw, predict, and more. Includes activities for 2.0 and 1.0.

A&P Flix Activities

These 3D movie-quality animations of key physiological processes include coaching activity assignments that use a variety of question types and levels.

Learning Catalytics

This classroom lecture tool is a "bring your own device" engagement, assessment, and classroom intelligence system. With Learning Catalytics, instructors can flip the classroom and assess students in real time using open-ended tasks to probe their understanding. Students use their smartphone, tablet, or laptop to respond to questions in class.

Also Assignable in MasteringA&P

- **NEW!** *Get Ready for A&P* Learning Styles Assessment
- *Get Ready for A&P* Diagnostic Test, Chapter Tests, and Cumulative Test
- *Get Ready for A&P* Video Tutors
- Chemistry Review Activities
- Reading Quiz questions
- Art-labeling and Sequence-the-steps Activities
- Art-based questions
- Chapter Running Case for each chapter in Amerman
- Chapter Test questions
- Test Bank Questions
- Clinical Case Study Coaching Activities
- PAL 3.0 Test Bank with images
- **NEW!** Bone and Dissection Videos
- PhysioEx 9.1 with assessments

Do your students have the tools to suit their different learning styles?

MasteringA&P® Study Area

The Study Area includes a wide selection of study and practice tools to engage students and meet the needs of different learning styles and study strategies.

▶ Big Picture Animations

Narrated by the author, these animations help students visualize key physiological processes and reinforce student understanding via interactive self-quizzing with immediate feedback.

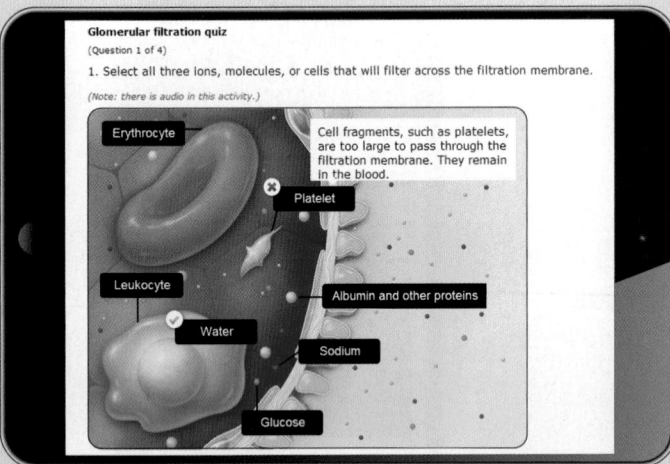

▶ Concept Boost Video Tutors

These videos of select Concept Boosts highlighted in the textbook feature author Amerman teaching directly to students and coaching them through tough-to-understand or tricky topics.

Practice quizzes, art-labeling, and sequence-the-step activities

A rich variety of practice quizzes and activities gives visual and kinesthetic learners varied practice learning key structures and processes.

Practicing A&P: Active-Learning Worksheets

These worksheets written by Amerman help engage the kinesthetic learner with labeling, drawing, and build-your-own-summary-table exercises that students can complete *as they read the textbook*. Available in downloadable PDFs in the Study Area of MasteringA&P®. They can also be packaged as a print supplement with the Amerman textbook at no additional cost.

---

52   Practicing A&P Active Learning Exercises

Key Concept: What are the three main types of ion channels and how do they differ?

_____

_____

_____

Draw It: Ion Gradients and Ion Movements

Below is an illustration of a plasma membrane with sodium and potassium ion channels. Draw the sodium and potassium ions in the correct distribution for a resting cell. Then, draw arrows to show in which directions the ions will diffuse when the channels open.

K^+ leak channel Na^+ leak channel

Key Concept: How does the movement of positive ions lead to a negative resting membrane potential?

_____

_____

_____

Describe the Steps: Changing the Membrane Potential

Fill in the blanks to describe what happens to the membrane potential of a neuron when ion channels open and close.

A small, local change in the membrane potential of the neuron is called a/an _____ _____. It

may have one of two effects: It may make the membrane potential more positive, a change called _____,

or it may make the membrane potential more negative, a change called _____. If a positive change in

membrane potential reaches a value called _____, _____ channels open, initiating an event

known as a/an _____ _____. This event has three phases in a neuron: (1) _____,

mediated by the influx of _____ ions; (2) _____, mediated by the outflow of _____ ions;

and (3) _____, mediated by the continued outflow of _____ ions.

Author podcasts

Narrated by the author, these podcasts give on-the-go students and auditory learners the opportunity to review the tough topics in A&P. Available as downloadable MP3 audio files in the Study Area of MasteringA&P.

PAL | practice anatomy lab™ **Practice Anatomy Lab™** PAL is a virtual anatomy study tool that gives students 24/7 access to the most widely used lab specimens including human cadaver, anatomical models, histology, cat, and fetal pig. PAL 3.0™ includes randomized multiple-choice quizzes and fill-in-the-blank lab practical questions.

 Dynamic Study Modules

Designed to enable students to study effectively on their own, Dynamic Study Modules help them quickly access and learn the concepts they need to be more successful on quizzes and exams. These flashcard-style questions adapt to a student's performance and include art and explanations from the Amerman textbook to cement the student's understanding. Modules can also be assigned in MasteringA&P.

Also Available in the Study Area

- *Get Ready for A&P*
- Student Lecture Outlines
- Answers to Chapter Quick Check questions
- Chapter Running Cases
- Clinical case studies
- eText

- Interactive Physiology 2.0 and 1.0
- A&P Flix 3D animations
- PAL 3.0 app
- Bone and Dissection Videos
- PhysioEx 9.1

Instructor Supplements

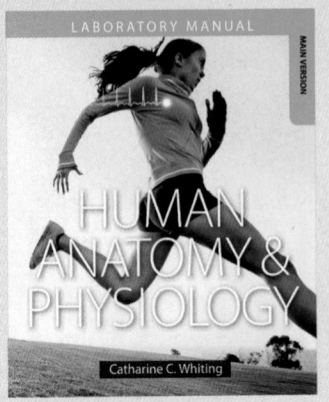

Human Anatomy & Physiology Laboratory Manual: Making Connections

by Catharine C. Whiting

0-13-397855-9 / 978-0-13-397855-1

Whiting's brand-new lab manual takes an active-learning approach that uses a rich variety of activities along with guided questions to engage students and help them apply concepts learned in lecture to the lab. Students can also better prepare for lab and assess their understanding of material with pre-lab and post-lab assignments. Additional pre-lab and post-lab assignments are available in MasteringA&P®. The Whiting Lab Manual is available in three versions: Cat, Main, and Pig.

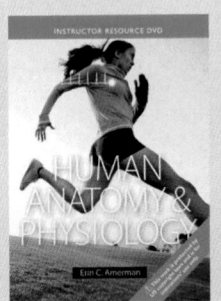

Instructor Resource DVD with PowerPoint® Lecture Outlines

by Suzanne Pundt

0-321-99062-5 / 978-0-321-99062-4

- All illustrations and photos from the text as presentation-ready labeled and unlabeled JPEG files
- All illustrations and photos from the text as presentation-ready editable PowerPoint® files
- Customizable PowerPoint® lecture presentations including embedded Big Picture Animations, Concept Boost Video Tutors, Interactive Physiology, and A&P Flix
- *Instructor Guide with Chapter Running Case Studies* in Microsoft Word®
- *Student Lecture Outlines* in Microsoft Word®
- Test Bank in Microsoft Word®
- TestGen test generation and management software

Instructor Guide with Chapter Running Case Studies for Human Anatomy & Physiology

by Christopher Amerman

0-321-99064-1 / 978-0-321-99064-8

Includes chapter learning outcomes, suggested lecture outlines, and chapter running case studies with questions that are tied directly to the chapters in the Amerman textbook. Available via MasteringA&P and the Instructor Resource DVD.

Student Lecture Outlines for Human Anatomy & Physiology

by Patty Bostwick Taylor

Fill-in-the-blank chapter outlines ask students to fill in key details, allowing them to stay engaged during lecture. Available as customizable Word files via MasteringA&P and the Instructor Resource DVD.

Printed Test Bank for Human Anatomy & Physiology

by Patty Bostwick Taylor

0-321-99063-3 / 978-0-321-99063-1

Contains thousands of test questions including multiple-choice, matching, true/false, short answer, and essay. Available as a print supplement and also accessible via MasteringA&P and the Instructor Resource DVD.

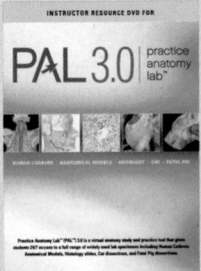

PAL 3.0 Instructor Resource DVD with Test Bank

by Nora Hebert, Ruth E. Heisler, Jett Chinn, Karen M. Krabbenhoft, Olga Malakhova

0-321-74963-4 / 978-0-321-74963-5

Includes everything an instructor needs to present and assess PAL 3.0 in lecture and lab. The DVD includes images in PowerPoint with editable labels and leader lines, labeled and unlabeled images in JPEG and PowerPoint® format, and more.

Student Supplements

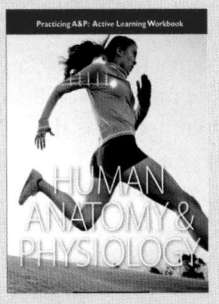

Practicing A&P: Active-Learning Workbook

by Erin C. Amerman
0-321-94989-7 / 978-0321-94989-9
Active learning exercises help engage the kinesthetic learner with labeling, drawing, and build-your-own-summary-table exercises that students can complete as they read the text. Available as downloadable PDF files in the Study Area or packaged as a print supplement with the Amerman textbook at no extra cost.

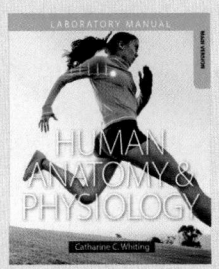

Human Anatomy & Phsyiology Laboratory Manual: Making Connections

by Catharine C. Whiting
0-13-397855-9 /
978-0-13-397855-1
See the description of this manual on p. xviii.

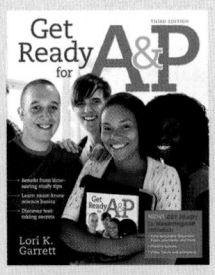

Get Ready for A&P

by Lori K. Garrett
0-321-81336-7 / 978-0-321-81336-7
Hands-on workbook that quickly gets students up to speed on basic study skills, math skills, anatomical terminology, basic chemistry, cell biology, and other basics of the human body. Video Tutors feature author Lori Garrett walking students through basic concepts in chemistry and cell biology.

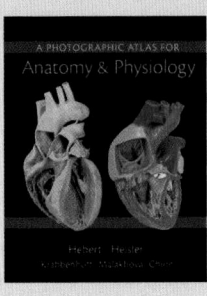

A Photographic Atlas for A&P

by Nora Hebert, Ruth E. Heisler, Jett Chinn, Karen M. Krabbenhoft, Olga Malakhova
0-321-86925-7 / 978-0-321-86925-8
NEW! Visual lab study tool that helps students learn and identify key anatomical structures. Featuring photos from Practice Anatomy Lab™ 3.0 and other sources, the Atlas includes over 250 cadaver dissection photos, histology photomicrographs, and cat dissection photos plus over 50 photos of anatomical models.

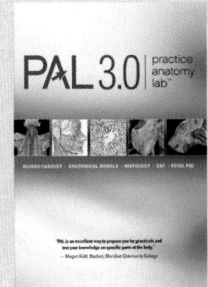

Practice Anatomy Lab™ (PAL™) 3.0 DVD

by Nora Hebert, Ruth E. Heisler, Jett Chinn, Karen M. Krabbenhoft, Olga Malakhova
0-321-68211-4 / 978-0-321-68211-6
See the description of this study tool on p. xvii.

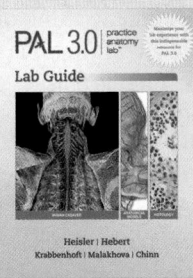

PAL 3.0™ Lab Guide

0-321-84025-9 / 978-0-321-84025-7
Written by the PAL authors to accompany PAL 3.0, Practice Anatomy Lab 3.0 Lab Guide contains lab exercises that direct students to select images and features in PAL 3.0 and then assess students' understanding using labeling, matching, short-answer, and fill-in-the-blank questions. Exercises cover key lab specimens in PAL 3.0—human cadaver, anatomical models, and histology.

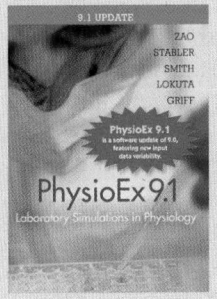

PhysioEx™ 9.1 Laboratory Simulations in Physiology

by Peter Zao, Timothy Stabler, Lori A. Smith, Andrew Lokuta, and Edwin Griff
0-321-92964-0 / 978-0-321-92964-8
This easy-to-use laboratory simulation software and lab manual consists of 12 exercises containing 63 physiology lab activities that can supplement or substitute for wet labs safely and cost-effectively. Now with input data variability.

Acknowledgments

Believe it or not, this book you are now holding has been nine and a half years in the making. When I first started writing it, my daughter wasn't even 2 years old; now she is halfway through fifth grade. But I was certainly not alone on this journey, as a huge number of people were involved in bringing this book to life. Saying a simple "thank you" in the acknowledgments seems so insufficient given the quality and quantity of their contributions, but these thanks are genuine and heartfelt.

I will start with my family because they lived for nine and a half years with the day-to-day stresses that come with a project of this magnitude. Were it not for the help and understanding of my husband Chris Amerman, my daughter Elise, my mother Cathy Young, and my dear friend David Ferguson, this book would have never been completed. They served as a source of unwavering support, encouragement, and ideas. Elise was also incredibly patient and understanding with how much I had to work, and I am so thankful for that. I should also thank my dogs for making sure to bark and howl each and every time I was in a phone meeting, and my cats for never failing to lie exactly in the middle of whatever I was trying to do.

Next is the core team of the book, which I've come to think of as parts of the brain, each performing absolutely vital functions that maintained homeostasis of the whole book. First is Serina Beauparlant, who, as editor-in-chief, is our brainstem. She has tirelessly performed all of those critical behind-the-scenes functions, ranging from wrangling budgets and securing administrative support to running focus groups and analyzing reviewer feedback. It has been Serina's driving force that kept the book alive over these long years. Simply put, without her, there would be no book.

Our team's cerebral hemispheres are our two brilliant developmental editors: Suzanne Olivier and Laura Southworth. As our text development editor, Suzanne is the left cerebral hemisphere. Her ability to logically and patiently approach a chapter from a "big picture" perspective ensured our chapters maintained a consistent narrative flow. It's impossible to overstate her role—not only did Suzanne always manage to find a chapter's sticking points, but she also always proposed solutions to these problems that made the chapter better. The readability, logical flow, and text-art coordination of this book are largely due to Suzanne's efforts.

Laura Southworth, as our art development editor, is the right cerebral hemisphere. Laura not only is a very talented artist but also has an incredible ability to analyze a figure and work magic to make it teach better. This is in part due to her amazing skill for visual-spatial layout (a skill I absolutely lack), which is arguably the most important part of a figure. No matter what we gave her or how rough our ideas or sketches, Laura turned it into gold. This is why "Let's ask Laura" became our mantra when Suzanne and I were working on a chapter. Any time we were perplexed by a figure, Laura unfailingly found a solution.

The role of team thalamus was played by project editor/manager Nicole Tache (neé Graziano). This is a high compliment, as without a functional thalamus, absolutely nothing can get done! Just as the thalamus processes and routes information into and out of the cerebral hemispheres, Nicole processes, edits, and sorts all material for the chapters and supplements for this project. Basically, without Nicole, we would have all been utterly lost.

Rounding out the team is Barbara Yien, our cerebellum. Barbara has been involved with this project from the very start, first as a project editor and now as the Director of Development. Her even-keeled approach has helped troubleshoot scheduling, budgeting, and our marketing efforts. Whenever we come to a sticking point, we look to Barbara, who always manages to find a way to correct the "motor error" and keep everything balanced and on track.

(Now that I've written this, I'm wondering exactly what part of the brain I represent on the team. The basal nuclei? Maybe the hypothalamus? Hopefully not the pineal gland, as I don't want to make my students sleepy. . . .)

Every member of this core "brain" team deserves the highest praise for their skills, dedication, and willingness to persistently climb the mountain that was this book. I am beyond grateful to them for this, and I am also deeply thankful for their friendship.

Assisting the core team was a group of incredibly talented people without whom the book could not have happened: art development editor Elisheva Marcus and text development editor Alice Fugate; our indomitable marketing team of Allison Rona, Derek Perrigo, Michelle Cadden, Leslie Allen, Jessica Moro, Mansour Bethoney, Patrice Jones, Tim Galligan, Christy Lesko, and Yez Alayan; design director Mark Ong; copyeditor Bonnie Boehme; photo researcher Maureen Spuhler; media development editor Eric Leaver; the Pearson media team, including Lauren Fogel, Stacy Treco, Laura Tommasi, Caroline Power, Katie Foley, Cheryl Chi, Kristen Sanchez, Aimee Pavy, Sarah Young-Dualan, and Kyle Doctor; our production team of Nancy Tabor, Jane Hoover, Caroline Ayres, Dorothy Cox, and Kirsten Forsberg; our manufacturing buyer Stacey Weinberger; Animated Biomedical Productions; and editorial assistant Arielle Grant.

Next I want to thank and acknowledge everyone who contributed to the book, including Virginia Irintcheva, who authored the script and storyboards for the book's animations and interactive figures and assessments in MasteringA&P; Chris Amerman, who contributed Chapter 8 and authored the Instructor Guide; Bert Atsma and Mark Seifert, who contributed Chapter 9; Sheri Boyce, who contributed Chapter 15; Karen Keller, who contributed Chapters 26 and 27; Patty Bostwick Taylor, who authored the Test Bank, Student Lecture Outlines, and Dynamic Study Modules; Suzi Pundt, who authored the PowerPoint lecture slides; Jeff Engel, who provided answers to the Quick Check questions; Betsy Brantley, who reviewed the book's clinical coverage; Winston Charles Poulton, who helped produce the Concept Boost Video Tutors; photographer John Wilson White and model Luiza Silva, who contributed the body movements photos in Chapter 8; and William Karkow, Nina Zanetti, Patricia Wilhelm, and Tom Appleton, who carefully reviewed/contributed light micrographs for select chapters. Each of these people devoted a huge amount of time and effort to this project—as I'm sure they will tell you, authoring materials is hard work! I am so grateful that each of them was willing to share his or her talents and play a role in the success of this project. I am also grateful to the Editorial Consultants who provided invaluable feedback on teaching ideas and carefully accuracy-checked pages and to all of the many academic reviewers, class testers, and focus group attendees who have shared their time, expertise, and ideas with us.

I would also like to sincerely thank Lauren Harp, Executive Marketing Manager. I met Lauren in 2005 when she was the marketing manager for natural sciences. She passed my name along to Serina as a potential author after we had a two-hour-long conversation in my office about what I would like to see in a textbook. Had she not done this, *Human Anatomy & Physiology* likely wouldn't exist.

Finally, none of this would have been possible without the unwavering support of Vice President and Editorial Director of Life Sciences Frank Ruggirello, Managing Director of Pearson Science Paul Corey, Editorial Director of Pearson Science Adam Jaworski, and Finance Director of Pearson Science Hogan Nymberg. All have supported this project from the beginning, and it was only because of their continued encouragement and belief in our team that you are holding this book right now. They have my eternal gratitude for allowing us to bring our vision to life.

Editorial Consultants

Emily Allen, *Rowan College of Gloucester County*
Chris Amerman, *Santa Fe College*
Bert Atsma, *Union County College*
Mary Lou Bareither, *University of Illinois at Chicago*
Patty Bostwick Taylor, *Florence-Darlington Technical College*
Sheri L. Boyce, *Messiah College*
Robert G. Carroll, *East Carolina University, Brody School of Medicine*
Linda Costanzo, *Virginia Commonwealth University School of Medicine*
Sharon S. Ellerton, *Queensborough Community College*
Jeff E. Engel, *Western Illinois University*
Karen L. Keller, *Frostburg State University*
Naomi Machell, *Delaware County Community College*
Ken Malachowsky, *Florence-Darlington Technical College*
Suzanne Pundt, *University of Texas at Tyler*
Mark Seifert, *Indiana University–Purdue University, Indianapolis, School of Medicine*
Cindy L. Stanfield, *University of South Alabama*
Cynthia Surmacz, *Bloomsburg University*

Text and Art Reviewers

Joslyn Ahlgren, *University of Florida–Gainesville*
Anitha Akkal, *Joliet Junior College*
Ticiano Alegre, *North Lake College*
Ayanna Alexander-Street, *Lehman College*
Emily Allen, *Rowan College of Gloucester County*
Matt Allen, *Indiana University–Purdue University, Indianapolis, School of Medicine*
Beth Altschafl, *University of Wisconsin–Madison*
Teresa Alvarez, *St. Louis Community College–Forest Park*
Kathy Pace Ames, *Illinois Central College*
Heather Evans Anderson, *Winthrop University*
March Ard, *University of Mississippi Medical Center/Hinds Community College*
Bert Atsma, *Union County College*
Vince Austin, *Bluegrass Community & Technical College*
Stephanie Baiyasi, *Delta College*
Tim Ballard, *University of North Carolina–Wilmington*
Stephen Bambas, *University of South Dakota*
Michelle Baragona, *Northeast Mississippi Community College*
Mary Lou Bareither, *University of Illinois–Chicago*
Marianne Baricevic, *Raritan Valley Community College*
Verona Barr, *Heartland Community College*
David Barton, *Indian Hills Community College*
Jerry Barton, *Tarrant County College–South*
Thomas Bell, *Rutgers University at Camden*
Dena Berg, *Tarrant County College–Northwest*
Tom Betsy, *Bergen County Community College*
J. Gordon Betts, *Tyler Junior College*
Laura Bianco, *Delaware Technical Community College of Wilmington*
Heather Billings, *West Virginia School of Medicine*
Ruth Birch, *St. Louis Community College–Florissant Valley*
Bonnie Blazer-Yost, *Indiana University–Purdue University, Indianapolis*
Rob Blum, *Lehigh Carbon Community College*
Sue Bodine, *University of California, Davis*
Franklyn F. Bolander, *University of South Carolina*
Patty Bostwick Taylor, *Florence-Darlington Technical College*
Sherry Bowen, *Indian River State College*
Sheri Boyce, *Messiah College*
Laura Branagan, *Foothill College*
Eldon Braun, *University of Arizona*
Carol Britson, *University of Mississippi*
David Brown, *East Carolina University, Brody School of Medicine*
Jack Brown, *Paris Junior College*
Kristen Bruzzini, *Maryville University*
Diep Burbridge, *Long Beach City College*
Warren Burggren, *University of North Texas*
Ed Burke, *Truckee Meadows Community College*
Steve Burnett, *Clayton State University*
Rebecca Burton, *Alverno College*
Beth Campbell, *Itawamba Community College*
Jamie Campbell, *Truckee Meadows Community College*
Geralyn Caplan, *Owensboro Community & Technical College*

Steven J. Carlisle, *John Tyler Community College*
Tamyra Carmona, *Cosumnes River College*
Robert Carroll, *East Carolina University, Brody School of Medicine*
Jana Causey, *Pearl River Community College*
Karen Chooljian, *California State University, Fresno*
Robert Clark, *Ozarks Technical Community College*
Pamela Cole, *Shelton State Community College*
Francisco Coro, *Miami Dade College*
Ron Cortright, *East Carolina University, Brody School of Medicine*
Linda Costanzo, *Virginia Commonwealth University*
Ken Crane, *Texarkana College*
Marianne Crocker, *Ozarks Technical Community College*
Robert Crocker, *SUNY-Farmingdale State College*
James Crowder, *Brookdale Community College*
Michael Cryder, *Riverside Community College*
Paul Currie, *Hazard Community & Technical College*
Judith D'Aleo, *Plymouth State University*
Lynnette Danzl-Tauer, *Rock Valley College*
Mary Beth Dawson, *Kingsborough Community College*
Danielle Desroches, *William Patterson University*
Gary Diffee, *University of Wisconsin-Madison*
Josh Drouin, *Lock Haven University*
Joseph D'Silva, *Norfolk State University*
Sondra Dubowsky, *McLennan Community College*
Kathryn Durham, *Lorain County Community College*
Idrissi Abdeslem El, *College of Staten Island*
Sharon Ellerton, *Queensborough Community College*
Kurt Elliott, *Northwest Vista College*
Jeff Engel, *Western Illinois University*
Greg Erianne, *County College of Morris*
Martha Eshleman, *Pulaski Technical College*
Marirose Ethington, *Genesee Community College*
David Evans, *Pennsylvania College of Technology*
Brian Feige, *Mott Community College*
Michael Ferrari, *University of Michigan, School of Medicine*
Linda Flora, *Delaware County Community College*
Maria Florez, *Lone Star College-CyFair*
Cliff Fontenot, *Southeastern Louisiana University*
Barbara Fritz, *Rochester Community & Technical College*
Larry Frolich, *Miami Dade College-Wolfson*
Van Fronhofer, *Hudson Valley Community College*
Chris Gan, *Highline Community College*
Joseph Gar, *West Kentucky Community & Technical College*
Esther Gardner, *New York University, Langone Medical Center*
Lynn Gargan, *Tarrant County College-Northeast*
Lori Garrett, *Parkland College*
Michelle K. Gaston, *Northern Virginia Community College*
Jane Gavin, *University of South Dakota, Sanford School of Medicine*
Michelle Gibson, *Montcalm Community College*
Mike Gilbert, *Fresno City College*
Theresa Gillian, *Virginia Tech*
Lauren Gollahon, *Texas Tech University*
Matthew Gosses, *Owens Community College*
Margaret Grant, *Hudson Valley Community College*

Bruce Gray, *Simmons College*
Melissa Greene, *Northwest Mississippi Community College*
Patricia Halpin, *University of New Hampshire-Manchester*
Chris Harendza, *Montgomery County Community College*
Rebecca Harris, *Pitt Community College*
Carol Haspel, *LaGuardia Community College*
Nora Hebert, *Red Rocks Community College*
Gary Heisermann, *Salem State University*
DJ Hennager, *Kirkwood Community College*
Kristin Hensley, *Rowan University*
Brent Hill, *University of Central Arkansas*
Karen Hlinka, *West Kentucky Community & Technical College*
Elizabeth Hoffman, *Baker College-Clinton*
Rodney Holmes, *Waubonsee Community College*
Mark Hubley, *Prince George's Community College*
Julie Huggins, *Arkansas State University*
Jason Hunt, *Brigham Young University-Idaho*
Peggy Hunter, *Camosun College-Victoria, British Columbia*
Jim Hutchins, *Weber State University*
Sandra Hutchinson, *Sinclair Community College*
Alexander Ibe, *Weatherford College*
Virginia Irintcheva, *Black Hawk College*
Bruce Johnson, *Cornell University*
Corey S. Johnson, *University of North Carolina-Chapel Hill*
Eddie Johnson, *Central Oregon Community College*
Cindy Jones, *Colorado Community College Online*
Margaret Kauffman, *Ohlone College*
Karen Keller, *Frostburg State University*
Suzanne Keller, *Indian Hills Community College*
Will Kleinelp, *Middlesex County College*
Michael Klemsz, *Indiana University-Purdue University, Indianapolis, School of Medicine*
Chad Knight, *Northern Virginia Community College-Alexandria Campus*
Karen Krabbenhoft, *University of Wisconsin-Madison, School of Medicine*
Chris Kule, *Pennsylvania College of Technology*
Edward LaBelle, *Rowan College of Gloucester County*
Jason LaPres, *Lone Star College-University Park*
Barbara Lax, *Community College of Allegheny*
Steven Leadon, *Durham Technical Community College*
Marian Leal, *Sacred Heart University*
Lisa Lee, *University of Colorado, Denver, School of Medicine*
Peggy LePage, *North Hennepin Community College*
John Lepri, *University of North Carolina-Greensboro*
Michael Levitzky, *Louisiana State University Health Sciences Center*
Robert Logan, *North Shore Community College-Lynn*
Shawn Macauley, *Muskegon Community College*
Naomi Machell, *Delaware County Community College*
Erin MacKenzie, *Howard College*
Ken Malachowsky, *Florence-Darlington Technical College*
Bruce Maring, *Daytona State College-West/Deland*
Patricia Marquardt, *Wayne County Community College*
Karen Martin, *Fulton-Montgomery Community College*
Theresa Martin, *College of San Mateo*

Alice McAfee, *University of Toledo*
Jameson McCann, *Guilford Technical Community College*
Jenny McFarland, *Edmonds Community College*
Cherie McKeever, *Montana State University–Great Falls College*
Karen McLellan, *Indiana University–Purdue University, Ft. Wayne*
Mark Meade, *Jacksonville State University*
Jaime Mergliano, *John Tyler Community College*
Anthony Mescher, *Indiana University School of Medicine-Bloomington*
Steve Meyer, *Florida State College at Jacksonville*
Claire Miller, *Community College of Denver*
Justin Moore, *American River College*
Erin Morrey, *Georgia Perimeter College–Dunwoody*
Susan Moss, *Imperial Valley College*
Howard Motoike, *LaGuardia Community College*
Jen Musa, *Broome Community College*
Barbara Musolf, *Clayton State University*
Cheryl Neudauer, *Minneapolis Community & Technical College*
Chad Newton, *Bakersfield College*
Mary Jane Niles, *University of South Florida*
Maria Oehler, *Florida State College at Jacksonville*
Justicia Opoku-Edusei, *University of Maryland*
David Osborne, *Texas Tech Health Science Center*
Betsy Ott, *Tyler Junior College*
Ellen Ott-Reeves, *Blinn College*
Anthony Paganini, *Michigan State University*
Russ Palmeri, *Asheville-Buncombe Technical Community College*
Michele Paradies, *SUNY–Orange County Community College*
Diane Pelletier, *Green River Community College*
Chris Picken, *SUNY–Suffolk County Community College*
Rebecca Pierce, *Delta Community College*
Melissa Piliang, *Cleveland Clinic*
John Placyk, *University of Texas at Tyler*
Brandon Poe, *Springfield Technical Community College*
Peter Porter, *Moraine Valley Community College*
Frank Powell, *University of California, San Diego*
Renee Prenitzer, *Greenville Technical College*
Cynthia Prentice-Craver, *Chemeketa Community College*
Steven Price, *Virginia Commonwealth University*
Suzanne Pundt, *University of Texas at Tyler*
David Quadagno, *Florida State University*
Saeed Rahmanian, *Roane State Community College*
Scott Rahschulte, *Ivy Tech Community College–Lawrenceburg*
Elizabeth Randolph, *Front Range Community College*
Wendy Rappazzo, *Harford Community College*
Terrence Ravine, *University of South Alabama*
Laura Ritt, *Burlington County College*
Dawn Roberts, *Pellissippi State Community College*
Mark Robertson, *Delta College*
Robin Robison, *Northwest Mississippi Community College*
Alex Robling, *Indiana University School of Medicine*
Susan Rohde, *Triton College*
Amanda Rosenzweig, *Delgado Community College*
Kyla Ross, *Georgia State University*

John Rowe, *Florida Gateway College*
Stephen Sarikas, *Lasell College*
Leif Saul, *University of Colorado-Boulder*
Lou Scala, *Passaic County Community College*
Connie Scanga, *University of Pennsylvania School of Nursing*
Sharon Schapel, *Mott Community College*
Steve Schenk, *College of Central Florida-Citrus*
Catherine Scholz, *Gwinnett Technical College*
Mark Seifert, *Indiana University–Purdue University, Indianapolis, School of Medicine*
Donald Shaw, *University of Tennessee at Martin*
Matthew "Doc" Sheehan, *Mt. Wachusett Community College*
Gidi Shemer, *University of North Carolina-Chapel Hill*
Michael Shipley, *Midwestern State University*
Brian Shmaefsky, *Lone Star College-Kirkwood*
Marilyn Shopper, *Johnson County Community College*
Pam Siergiej, *Roane State Community College*
Lynnda Skidmore, *Wayne County Community College*
Patricia M. Smeltz, *John Tyler Community College*
Debra Socci, *Seminole State College of Florida*
Ashley Spring, *Brevard Community College*
Maria Squire, *University of Scranton*
Claudia Stanescu, *University of Arizona*
Cindy Stanfield, *University of South Alabama*
George Steer, *Jefferson College of Health Sciences*
Jill Stein, *Essex County College*
Nora Stevens, *Portland Community College-Cascade*
Leo Stouder, *Broward College*
Diana Sturges, *Georgia Southern University*
Carole Subotich, *Rowan College of Gloucester County*
Eric Sun, *Middle Georgia State College*
Cynthia Surmacz, *Bloomsburg University*
Robert Swatski, *Harrisburg Area Community College-York*
Yong Tang, *Front Range Community College*
George Tanner, *Indiana University School of Medicine*
Shyra Tedesco, *Ivy Tech Community College-Madison*
Alvin Telser, *Northwestern University, Feinberg School of Medicine*
Deb Temperly, *Delta College*
Terry Thompson, *Wor-Wic Community College*
Maureen Tubbiola, *St. Cloud State University*
Katherine "Kate" Van de Wal, *Community College of Baltimore County*
Padmaja Vedartham, *Lone Star College-CyFair*
Heather Walker, *Clemson University*
Michael Walls, *Ivy Tech Community College-Evansville*
Lynn Wandrey, *Mott Community College*
Delon Washo-Krupps, *Arizona State University*
Amy Way, *Lock Haven University*
Chad Wayne, *University of Houston*
Mary Weis, *Collin College*
Lisa Welch, *Weatherford College-Decatur*
Corrie Whisner, *Cornell University*
Shirley Whitescarver, *Bluegrass Community & Technical College*
Sheila Wicks, *Malcolm X College*
Michael Wiley, *University of Toronto*

Samia Williams, *Santa Fe College*
Peggie Williamson, *Central Texas College*
Larry Wilmore, *Lamar State University–Orange*
Colleen Winters, *Towson University*
Diane Wood, *Southeast Missouri State University*
Amber Wyman, *Finger Lakes Community College*
Jim Yount, *Brevard Community College*
Anne Marie Yunker, *Cuyahoga Community College*
Nina Zanetti, *Siena College*
Scott Zimmerman, *Missouri State University*

Media Reviewers

Emily Allen, *Rowan College of Gloucester County*
Vince Austin, *Bluegrass Community & Technical College*
Marianne Baricevic, *Raritan Valley Community College*
Jerry Barton, *Tarrant County College*
C. Audra Bassett-Touchell, *Asheville-Buncombe Technical Community College*
David Bastedo, *San Bernardino Valley College*
Carol Britson, *University of Mississippi*
Jack Brown, *Paris Junior College*
Carolyn Bunde, *Idaho State University*
C. Steven Cahill, *West Kentucky Community & Technical College*
Maria Carles, *Northern Essex Community College*
Brendan K. Chastain, *West Kentucky Community & Technical College*
Ken Crane, *Texarkana College*
Sharon Ellerton, *Queensborough Community College*
Theresa Gillian, *Virginia Tech*
Lauren Gollahon, *Texas Tech University*
Mark Hollier, *Georgia Perimeter College–Clarkston*
William F. Huber, *St. Louis Community College*
Julie Huggins, *Arkansas State University*
Tom Jordan, *Pima Community College–NW*
Michelle Klein, *Prince George's Community College*
Jodi Long, *Santa Fe College*
Bruce Maring, *Daytona State College*
Howard Motoike, *LaGuardia Community College*
Maria Oehler, *Florida State College at Jacksonville*
John Patillo, *Middle Georgia State College*
Suzanne Pundt, *University of Texas at Tyler*
Elizabeth Randolph, *Front Range Community College*
Rozanne Redlinski, *Erie Community College–South Campus*
Ann Riedl, *Front Range Community College*
Sharon Schapel, *Mott Community College*
Jason Shaw, *Brigham Young University–Idaho*
Mark Slivkoff, *Collin College*

Instructor Focus Groups
Media (2014, Jacksonville, Florida)

Gordon Betts, *Tyler Junior College*
Patty Bostwick Taylor, *Florence-Darlington Technical College*

Ken Malachowsky, *Florence-Darlington Technical College*
Suzanne Pundt, *University of Texas at Tyler*
Kathy Warren, *Daytona State College*
Catharine Whiting, *University of North Georgia*
Peggie Williamson, *Central Texas College*

Mastering and Media (2013, San Francisco, California)

Willie Asobayire, *Essex County Community College*
Susan Burgoon, *Amarillo College*
Steve Cahill, *West Kentucky Technical & Community College*
Rita Moyes, *Texas A&M University*

Text, Art, and Media (2012, Tiburon, California)

Emily Allen, *Rowan College of Gloucester County*
Verona Barr, *Heartland Community College*
Heather Billings, *West Virginia University*
Sheri Boyce, *Messiah College*
Sharon Ellerton, *Queensborough Community College*
Jeff Engel, *Western Illinois University*
Jane Gavin, *University of South Dakota*
Virginia Irintcheva, *Black Hawk College*
Suzanne Pundt, *University of Texas at Tyler*

Art and Media (2012, Tulsa, Oklahoma)

Robert Crocker, *SUNY–Farmingdale State College*
Lynn Gargan, *Tarrant County College–Northeast*
Adewale Laditan, *Baltimore City Community College*
Howard Motoike, *LaGuardia Community College*
Patricia Visser, *Jackson Community College*
Pete Wickley, *Cuyahoga Community College*
Peggie Williamson, *Central Texas College*

Media (2011, Victoria, British Columbia)

Brian Feige, *Mott Community College*
Pamela Gregory, *Tyler Junior College*
Betsy Ott, *Tyler Junior College*
Sharon Schapel, *Mott Community College*
Joanne Settel, *Baltimore City Community College*
Catharine Whiting, *University of North Georgia*

Student Focus Groups
College of San Mateo

Professor Carlene Tonini-Boutacoff
Reyna Cabrera, Andrew Denys, Julian-Vincent Garcia, Sebastian Lopez, Christian Thompson, Wes Turner

Cosumnes River College

Professor Tamyra Carmona
Tiffany Atinan, Mariya Brushchenko, Sara Catania, Kim Ford, Kathy Hua, Hienvy Huynh, Hoa Huynh, Edith Madrigal, Kevin Nguyen, Galina Oleksiyenko

Des Moines Area Community College

Professor Qian Moss
Chasidy Atkinson Bennett, Staci D. Braue, Kristin Cox, Benjamin J. Dorin, Rob J. Fernandez, Anthony Finney, Lauren Hester, Jackie Maxwell, Sarah Thomas, Taylor Williams

Frostburg State University

Professor Karen Keller
Abena A. Abu, Gillian D. Hasslinger, Megan C. Kenney, Courtney A. Strubin

Truckee Meadows Community College

Professor Steve Schenk
Cambria Breithaupt, Danielle Eaton, Tiarra Ellis, Kenzie Espalin, Levi Evans, Adriana Gonzalez, Christine Graf, Rosana Hill-Thomas, Patrick Hurbain, Kimberly Jamieson, Amanda Jones, Kevin Jorde, Simerpreet Kaur, Carol Ledford, David Lloyd, Lisa McDonald, Aaron McGarry, Kelsey McKechnie, Sarah Morgan, Jan Overbay, Claudia Pfau, Ashley Percival, Erica Pratt, Jenna Sampson, Amrinder Singh, Jagpreet Singh, Erin Smith, Rachel Smith-Tutson, Claire Studebaker, Traci Taylor, Brian Van Dyne, Tina Wall, Nate Weigl

Waubonsee Community College

Professor Rodney Holmes
Diane Barry, Loredana Battisti, Jessica Cobo, Maria Davis, Jeff Hamilton, Carrie Hyde, Kristin Jury, Mia Magracia, Rana Moussa, Maria C. Puello, Bonnie Ritter, Alanna Russo, Parupkar Sidhu, Kim Voigt, Amy Wegner

Brief Contents

Contents

4 Histology   123

Unit 2 Body Coverings and Movement

5 The Integumentary System   160

Unit 3 Integration, Control, and Maintenance of Homeostasis

11 Introduction to the Nervous System and Nervous Tissue 381

12 The Central Nervous System 424

Unit 4 Transport and Immunity

17 The Cardiovascular System I: The Heart — 631

18 The Cardiovascular System II: The Blood Vessels — 669

Unit 5 Regulation of the Body's Intake and Output

22 The Digestive System 846

23 Metabolism and Nutrition 897

24 The Urinary System 941

25 Fluid, Electrolyte, and Acid-Base Homeostasis 985

Unit 6 Continuity of Life

26 The Reproductive System 1013

27 Development and Heredity 1057

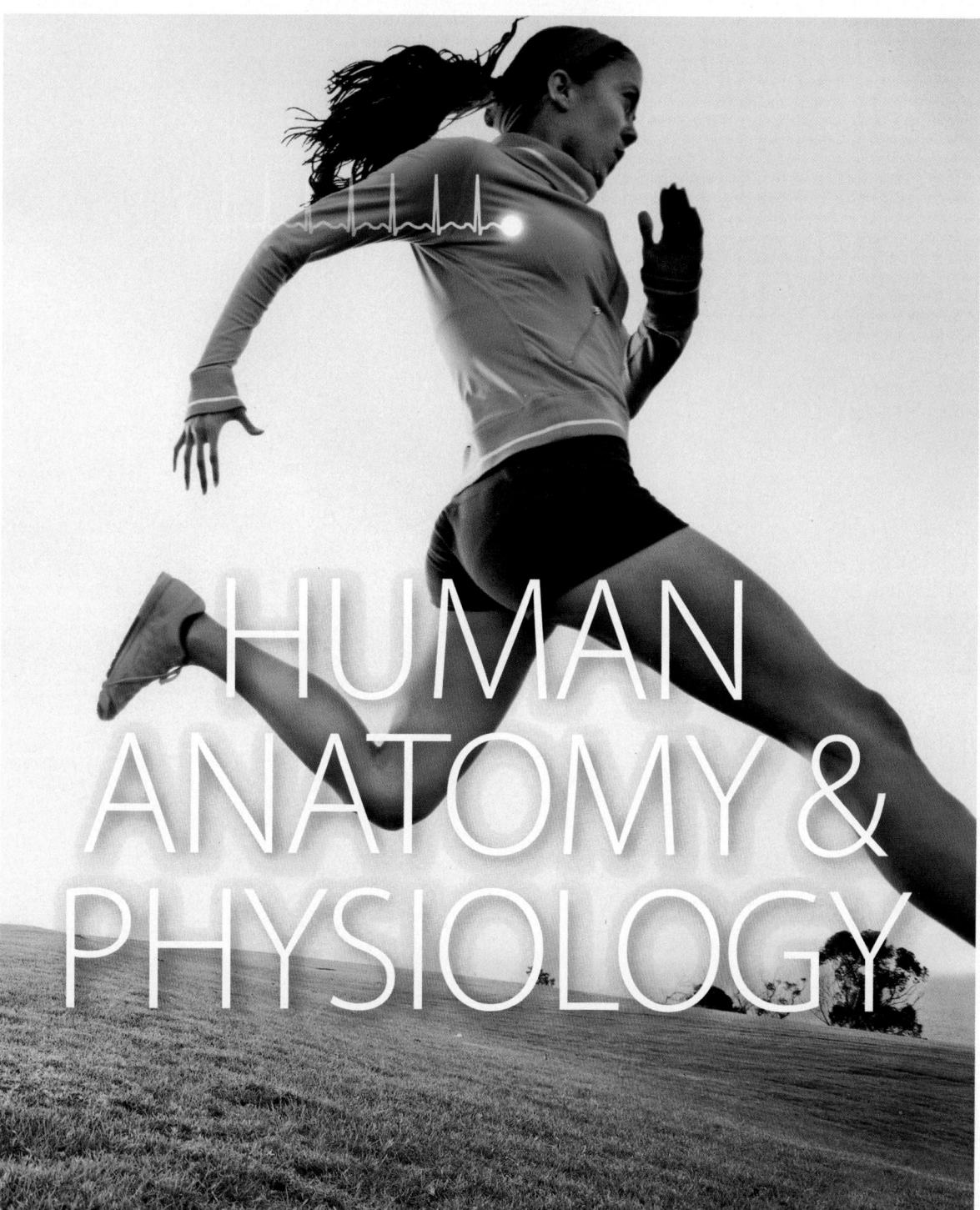

HUMAN
ANATOMY &
PHYSIOLOGY

1

Introduction to Anatomy and Physiology

The body has fascinated humankind since ancient times. Many of the earliest ideas about the form and function of the human body were wildly inaccurate. For example, gods and demons were believed to cause most diseases, the heart was perceived as the seat of intelligence, and the brain was viewed as merely a useless mass of tissue. However, through centuries of observation, experimentation, and imagination, we have dramatically expanded our knowledge of the human body. Much of this has resulted from the application of *science*, a way of observing and measuring natural phenomena in order to try to explain them. But so much remains to learn about the way our bodies are built and how they function. Studying the human body is like a treasure hunt, with fascinating "gems" still to discover.

This chapter introduces you to the world of anatomy and physiology. **Human anatomy** is the study of the structure or form of the human body, whereas **human physiology** (fiz′-ee-AWL-oh-gee) is the study of the body's functions. Although we define the two terms separately, the body's structure and function are closely related, a fact that we explore in depth throughout this text. The common name of your course, anatomy and physiology (A&P), reflects this connection.

The topics we discuss in this chapter form the foundation for the remaining chapters. We begin with a special section on how to succeed in an A&P course. An overview of the field and a discussion of the properties of all living organisms follow. After examining the language of A&P and the basic organization of the human body, the chapter concludes with recurring themes in A&P—those ideas that form the core of the field and the basis of our study in the rest of this book.

MODULE **1.1**

How to Succeed in Your Anatomy and Physiology Course

Learning Outcomes

1. Describe how to determine your learning style, read a textbook, budget your time, and study for quizzes and exams.
2. Explain how to make the best use of class and laboratory time.
3. Describe how to use this book and its associated materials.

Photo: Colored magnetic resonance imaging (MRI) whole body scans of a man (left), a woman (right), and a nine-year-old boy.

The world of anatomy and physiology often feels like a foreign and intimidating place to new students. And in a way it is—A&P is new territory for most people, and comes with a language all its own. Mastery of this material requires not only attendance at all lecture and laboratory sessions but also investment in a great deal of study time outside class. But what is the best way for you to study? And how should you make the most of your time in class and in lab? Furthermore, how should you use this book and its associated resources? This introductory module answers these questions and more. Read on, and find out how you can develop the skills needed for success in your A&P course.

How to Develop Study Skills

"I don't know why I don't get good grades. I study so much, and yet it doesn't seem to help me do well on the exams. How can I improve my test scores?" Sentences like these have been uttered by students everywhere. Many students put in a great deal of effort for little reward. This can be demoralizing, making a person feel just not "smart enough" to excel in tough courses. But this is usually not even remotely the case; instead, the problem centers on simply not knowing how to study and make effective use of time. The next subsections offer help and guidance with these tasks.

What's Your Learning Style?

It might surprise you that people learn in different ways. For example, you might be able to understand a process just from hearing it described, another person might need to see the process diagrammed, and yet another might have to manipulate a three-dimensional model to truly understand the process. These different ways of learning are known as **learning styles.** They are also called *learning preferences,* since they are not absolute.

We can classify learning styles in several ways. The most common way, and the one we use here, categorizes learners according to four basic styles:

- **Visual/Verbal.** Individuals classified as **visual/verbal** learners fare best when reading written material such as notes and textbooks and looking at diagrams, illustrations, and visual multimedia presentations (such as animations).
- **Visual/Nonverbal.** A **visual/nonverbal** learner usually best understands concepts through the use of diagrams, illustrations, and other visual media without text.
- **Auditory/Verbal.** As implied by the name, an **auditory/verbal** learner does well when listening to lectures, presentations, or when discussing material with a group. These individuals tend to prefer text-based materials rather than visuals.
- **Tactile/Kinesthetic.** The **tactile/kinesthetic** learner thrives in an environment where he or she can physically manipulate a specimen. These individuals excel in practical settings, including the anatomy and physiology lab.

You may be able to determine your learning style just from reading the descriptions, or you may wish to take an online test to determine it. Note, though, that very few people have strictly one type of learning style. Realistically, most people use a combination of more

than one. The key is figuring out which one or which combination fits you the best. For example, you might find that you retain and understand information better in class if you listen rather than take notes, a feature of an auditory/verbal learner. However, you might also need to make notes and draw diagrams to help you study, which characterizes a visual/verbal learner. If this is the case, try recording your lectures so that you can listen during class, and then taking notes *after* class from the recorded lecture. Experiment with different strategies to find what works best for you.

How to Read a Textbook

Reading your textbook is a must, but what is the best way to go about it? The first mistake many students make is trying to read a textbook as if it were a novel, starting at the first page and reading through it sequentially until the end. However, it's pretty obvious that textbooks aren't novels, and they therefore shouldn't be read as such. Instead, try an approach known as the SQ3R method, which involves the following steps: *survey, question, read, recite,* and *review.* Let's look at each of these steps in more detail.

- **Survey.** Before reading any chapter, you should do an initial survey, during which you skim the chapter and its figures. Begin by reading the headings and the Learning Outcomes at the beginning of each section. As you skim, take note of the key terms that are in boldface as well as the figures and tables. After finishing the initial survey, read the chapter summary. This initial survey will give an idea of what you will be focusing on and will make you more familiar with the content as you read.
- **Question.** The next step is to form questions about the chapter's content that you should answer as you read. These questions can be fairly broad and simple. For example, if you're reading about the heart, ask the following questions: "What do I know about the heart?" "How is the heart structured?" and "How does the heart's structure relate to its function?" If you find yourself struggling to come up with questions, a good place to start is the Learning Outcomes at the beginning of each section. Take a look back at the Learning Outcomes on the first page of this section, and notice how easily each outcome can be turned into a question.
- **Read—actively.** Now you are ready to go on and read the chapter. As you read, it is most important to read *actively,* as active reading—the process of reading while engaging in some sort of activity with the text—promotes active learning. In other words, to learn most effectively, students must do more than just read: They must also listen, write, discuss, and be engaged in solving problems. A novel is fine to read passively because you are interested in the plot, which keeps you turning pages. In addition, you don't necessarily have to remember the details of a novel—they are more like icing on the cake of the plot. But with a textbook, there is no overarching plot, no matter how interesting you might find the material, and the details are often critical. Active reading is, therefore, a must. Reading can be made active in many ways, the first of which is to take notes and make diagrams as you read. In addition, this book has many other

resources to help you read actively, which are detailed in a later section of this module.

- **Recite.** Part of reading actively is reciting the material—speaking aloud—as you read through it. This might seem strange, but speaking while reading engages more parts of your brain and helps you retain more of what you've read. It is particularly important in sciences such as A&P, which have a great number of new, often foreign-sounding words. We've provided pronunciations after many new terms to help with this.
- **Review.** The final step is reviewing what you have read. This can be done in many ways, including answering quiz questions found in the book, writing summaries, or discussing things aloud with study partners.

Note that you should read your textbook in this manner *before* going to the class at which the topic is discussed. We explain why in a later section.

Managing Your Time

Whatever your approach, one fact remains: Studying takes time, and we don't have a way to add extra hours to the day. Perhaps the number one mistake made by students is not allotting enough time for regular studying every week, which eventually necessitates "cramming" for the exam. It goes (almost) without saying that cramming does not lead to success, in either the short or the long term.

So, how can you avoid the need for cramming? The answer is surprisingly simple: make a schedule and budget your time. Plan out your time at least a week before your exam, and budget several hours each day for studying. Allow more hours than you are likely to need in case something unexpected comes up and disrupts your schedule. You might want to schedule specific activities each day; for example, Monday might be for reviewing notes, Tuesday for meeting with a study group, and Wednesday for a different activity. To maximize study time, think about *all* of your time—you might be able to study, for instance, with a study group or during your commute to and from school or a job (**Figure 1.1**).

How to Study for an Exam

Now that you know how to read the book, you are probably wondering how to apply this in studying for quizzes and exams. The answer to that is, of course, complicated, as no two people arrive at identical study strategies. However, the following methods have been found to work for many people:

- **Find out as much as you can about the exam.** Your professor isn't in the business of tricking you and will generally answer questions about the exam well before it is administered. Ask about the format of the exam—are the questions multiple choice, short answer, matching, or a mixture of many different types? Will the questions be fact-based or application-based? In addition, find out if your professor has made available sample exam questions or old exams for students to practice on. If you are unable to glean much information, use your first exam in the class for future reference.
- **Take advantage of the resources available to you.** A variety of resources for this book are available to help you study for exams, including the companion workbook and several features within the text. These resources also include many online practice tools. (See the section later in this module for details.) The more questions you answer, the better prepared you will be for the exam, because you will find out what you do know and, more importantly, what you *don't* yet know. Your college or university might also have resources for help, including tutors, computer labs, and the A&P lab.
- **Form a study group.** People with certain learning styles, particularly auditory/verbal and tactile/kinesthetic, do quite well in study groups. Find a group of classmates at the beginning of the semester with learning styles similar to yours, and meet regularly to go over tricky concepts, take practice tests, and offer one another support. Note that visual/verbal and visual/nonverbal learners may experience more success in studying alone, at least until they are ready to practice with the material.
- **Use whatever study techniques help you.** Depending on your learning style, a variety of techniques can help you master the material, including concept maps, coloring and

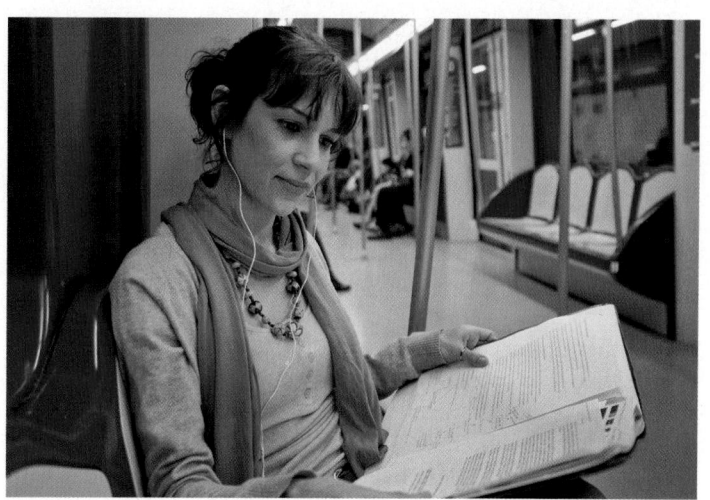

Figure 1.1 Some ways to maximize your study time.

drawing exercises, flashcards, models, mnemonics, and other tools that you can find online. Some materials may prove more useful in preparing for lab quizzes, whereas others may work better for lecture exams.

- **Take care of yourself and manage your stress.** This bit of advice might sound obvious, but some students tend to neglect their basic needs, including nutrition, health, and sleep. Multiple studies have demonstrated that a lack of these three necessities, particularly sleep, interferes with the ability to learn and retain information. As part of taking care of yourself, do what works best to alleviate stress, perhaps going for a run or singing to a favorite piece of music. Stress can be particularly problematic in the week before an exam, when you might favor an extra hour of studying over an extra hour of sleep. However, in this case, another hour of sleep might actually be more helpful to your grade than an additional hour of studying.

Quick Check

☐ 1. What are learning styles? Which learning style(s) best describe(s) you?

☐ 2. How should you approach reading a textbook, including this one?

☐ 3. What are some study strategies to improve your chances of success?

How to Make the Best Use of Class and Lab Time

To this point, we have been discussing what you do *outside* class to maximize your chances of success. But what you do *in* class is every bit as important to your success. It is doubly important in A&P, because you will likely have a laboratory period as well. The next two sections discuss how you can make the best use of time in class and in lab.

Come Prepared

Perhaps the most important step you can take to enhance your classroom and laboratory experience is to come prepared. You are assigned reading to help you become familiar with the material *before* you come to class or lab. Why is this important? There are several reasons. The first is simple: Attending a lecture without first learning about the material is like visiting a foreign country without knowing one word of its language. It is as difficult to find your way through unfamiliar material as it is to navigate foreign roads whose signs are in a language you don't speak. In other words, it's very easy to wind up lost.

Another important reason to read the lecture material before class is that your professor prepares lectures with the assumption that you have already read the assigned material. He or she uses the lecture to discuss details of the reading and explain its concepts, not start at the very beginning. If you are unfamiliar with the material in general, you won't benefit from a discussion of its details.

A final reason has to do with the way the brain processes information. Consider this example: A person is driving past a horse field and says to his companion, "Hey, look, my friend told me that there's a zebra in that field." The companion looks at the field blankly and sees only horses. "What zebra?" the companion asks. The driver points and says, "It's right there, in front of you." After a moment of staring, the companion finally realizes that there is, indeed, a zebra standing in the field, surrounded by horses. Why couldn't the companion see the zebra, when it was there all along? The answer is that her brain did not expect a zebra to be present in the middle of a horse field, and so her brain simply didn't perceive the zebra's presence. The driver knew the zebra was there and so his brain did perceive its presence. This is a true story—the companion who had trouble seeing the zebra was yours truly, the author of this book.

In the same way that someone might not perceive a zebra that he or she doesn't expect to be there, your brain will not perceive concepts and information that it didn't expect in class. If the information in the lecture is foreign to you, your brain simply isn't likely to register it. However, if you read the material before lecture, you will be able to "see the zebra" in your classroom, and do better on your exams and quizzes as a result.

How to Take Good Notes

Many professors present their lectures using PowerPoint®, and many also make these presentations available online for students to read before class. You might think that if you print out your professor's presentation and take it to lecture, your notes for class are as good as taken. This couldn't be further from the truth. Although it's not a bad idea to look over your professor's presentations before class, these are by no means a substitute for class notes.

The best way to take notes during class is to bring along the notes you took while reading. Leave several lines between each main point in your notes so that you can add as needed while your instructor lectures. Come to class equipped with a highlighter so that you can identify points emphasized by your instructor.

As with other topics we have discussed, note taking should be individualized according to what works best for you. If you have an auditory/verbal learning style, ask your professor if you can record his or her lectures. If you are a visual/nonverbal learner, you might do best if you take notes in the form of diagrams. Experiment with different strategies to find out what works best for you.

Quick Check

☐ 4. Why is it important to read the material before you come to class and lab?

☐ 5. What are some strategies for taking good notes in class?

How to Use This Book and Its Associated Materials

The study tips and strategies we have discussed so far are fairly universal and could be applied to any course you are taking, including A&P. Let's now move on to some hints and tips for this course, and in particular, this book and its associated materials. You should know above all that this book was written specifically with *you* in mind. Students today are different from those who took A&P 20 years ago. Today's students are busier; many work in addition to taking a full-time class load, some take care of families, and some might not have any background in the sciences. Every single feature of this book was carefully crafted to help meet the needs of today's student—*your* needs. The hallmark of the approach in this book is to provide coaching that guides you through the material and gives additional explanation of the most critical and difficult topics. Let's tour the features found in this book, and discover how to make the best use of them.

Modules

Each chapter is divided into sections that we call *modules*. A module is a block of text with definitive starting and stopping points that covers one core concept and its related principles. The text is organized in this manner to divide the material into more manageable chunks and to help you make the most efficient use of limited time. It usually works best to master the material in one module before moving on to the next one. Every module contains several features to help you master the material, which we discuss next.

Learning Outcomes

As you likely noticed at the start of this module, every module starts with Learning Outcomes, or a list containing the core concepts and relating principles you should come to understand in each module. Students have a tendency to breeze past the Learning Outcomes, but, as with all of the other features of this book, they are there for one main reason: to help you. Use them to your advantage—read them prior to starting the module as part of the "survey" portion of the SQ3R method, and then revisit them after finishing the module. Take a look at the Learning Outcomes for this module; you'll notice they are written in a way that makes it easy to adapt them into a quiz for yourself and/or your study group.

Try an experiment for this chapter: Take the end-of-chapter quiz (called Assess What You Learned) after reading the chapter, without first looking at the Learning Outcomes. Then, work with the Learning Outcomes and write down the answers to each one of them. Now, take the quiz a second time to see how much better you can do with the help of the Learning Outcomes.

Concept Boosts and Study Boosts

Without question, certain concepts of A&P are more challenging than others. Such concepts often require a bit of coaching, perhaps more explanation, or another way of looking at them. This is provided for you in the form of short sections called Concept Boosts (an example is shown in **Figure 1.2**), which are intended

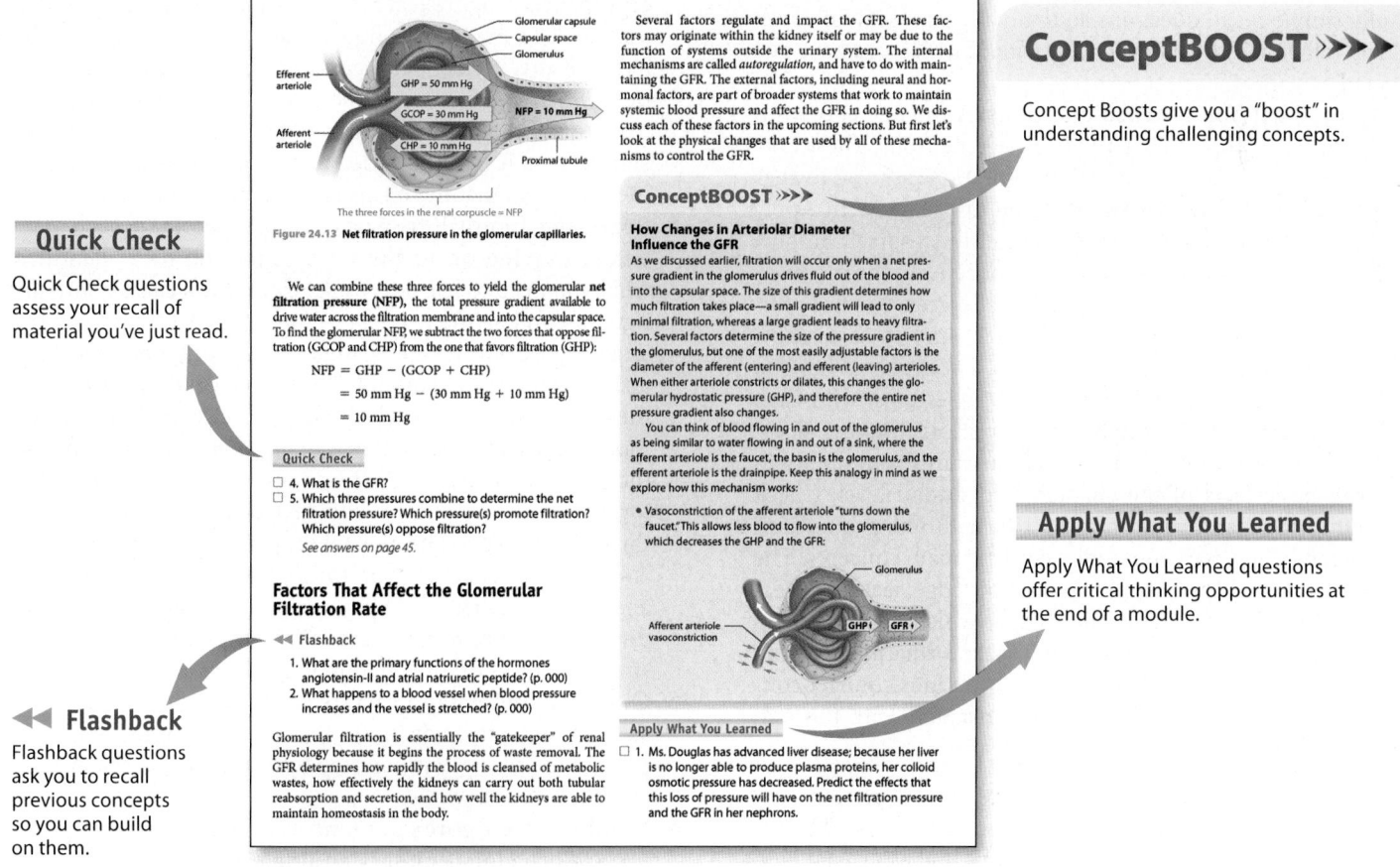

Figure 1.2 Selected features of this textbook.

to give you a "boost" in understanding a concept. You shouldn't treat these sections as optional; in fact, it's a good idea to give them as much attention as the regular text.

In addition to the Concept Boosts, you'll find Study Boosts in some chapters. Study Boosts give you study tips for particularly challenging material. They often include mnemonic devices to help you memorize anatomical structures or the steps of physiological processes. Like Concept Boosts, Study Boosts are there for your benefit, and it's to your advantage to pay close attention to them.

Questions

You will find many different opportunities for self-assessment throughout each chapter, several examples of which are shown in Figure 1.2. Each type of question serves a different purpose, although all are intended to help you discover how well you met the Learning Outcomes. The following list provides an overview of the categories of questions you will find in this book:

- **Flashback questions.** Many sections start with one or more Flashback questions that ask you to recall material from previous modules or chapters. Each Flashback question has an associated page number to help you find the concept in the book. Flashbacks are intended to help you remember previous terms and concepts so that you can build on them and better understand the current material.
- **Quick Check questions.** Quick Check questions are found after each major section within a module. These are generally simple recall questions so that you can assess how well you remember the basic concepts about which you have just read.
- **Apply What You Learned questions.** A module ends with two or more Apply What You Learned questions. These critical thinking and problem-solving questions ask you to analyze and apply the material that you just read. With Apply What You Learned questions, you can test yourself on how well you understand and recall the concepts in the module.
- **Assess What You Learned quiz.** Finally, once you have finished the chapter, you have another chance to assess how well you met the Learning Outcomes. The Assess What You Learned quiz contains three levels of questions, with each level requiring increasingly greater critical thinking skills. The first level, Check Your Recall, measures your ability to recall basic facts of the chapter. The second level, Check Your Understanding, contains critical thinking questions that target how well you understand the main concepts. The third and final level, Apply Your Knowledge, features questions that are often clinically based, which allows you to apply and analyze situations using the information you read about in the chapter. Some of these questions connect concepts you've discovered in previous chapters with those in the current chapter.

Like the other features in the book, these questions are intended for your benefit; use them to your advantage. The in-module questions, including the Flashback, Quick Check, and Apply What You Learned questions, are generally best answered as you read. This enables reading to be active rather than passive, which will increase the ability to retain and understand the material. Experiment with the other question types. You might want to try combining them with the Learning Outcomes to build quizzes for yourself and/or your study group.

Figures

In an A&P course, the art is every bit as important as the text. Indeed, in some cases, the figures are actually *more* important than the text in terms of developing an understanding of the concepts. The figures have been developed to coach you through these concepts. Following are some tips for using the art in this book as a valuable learning tool.

- **Examine the figures as you do your initial chapter survey.** As you read earlier in this module, it's a good idea to start with a chapter survey before you fully read the chapter, as part of the SQ3R method. This chapter survey should include a perusal of the figures as well as the text. The figures of a chapter will tell a story in and of themselves, even if removed from the text. Examining the figures as part of your initial chapter survey will give you a preview of the story told by the art and will make you better prepared to understand its details as you read later.
- **Identify the concept that the figure teaches first.** Once you start reading the chapter and delving into the figures, begin your interpretation by first identifying the exact concept that each figure teaches. Every effort has been made in this book to make each figure teach a single concept. The name of this concept can generally be found in the figure title, which you can see in the example shown in **Figure 1.3**. Note that if the concept illustrated in the figure sounds foreign to you, it is a good idea to re-read the text section on the subject so that the terms become more familiar.
- **Break the figure into parts, and understand each part before moving on to the next.** After you have identified what you need to learn from the figure, you can explore the figure's content more deeply. The content of some of the larger figures in this book might look intimidating at first, just in terms of the amount of information they contain. In these cases, it is a good idea to break the figure into smaller, more manageable parts. Often in physiology figures, such as the example in Figure 1.3, this is already done for you in the form of numbered steps with text boxes, which walk you through a process. The blue text explaining these steps is the author's voice coaching you through a concept. However, in anatomical figures, it may be helpful to imagine lines that divide the figure into sections. This gives you "boxes" with smaller chunks of information to study.
- **Once you understand each part of the figure, examine it as a whole.** After you have mastered and understood the parts of the figure, you can step back and look at the process or structure as a whole.
- **Combine the figure's content with that of other figures for a more global understanding.** For more complicated anatomical structures or physiology concepts, the content

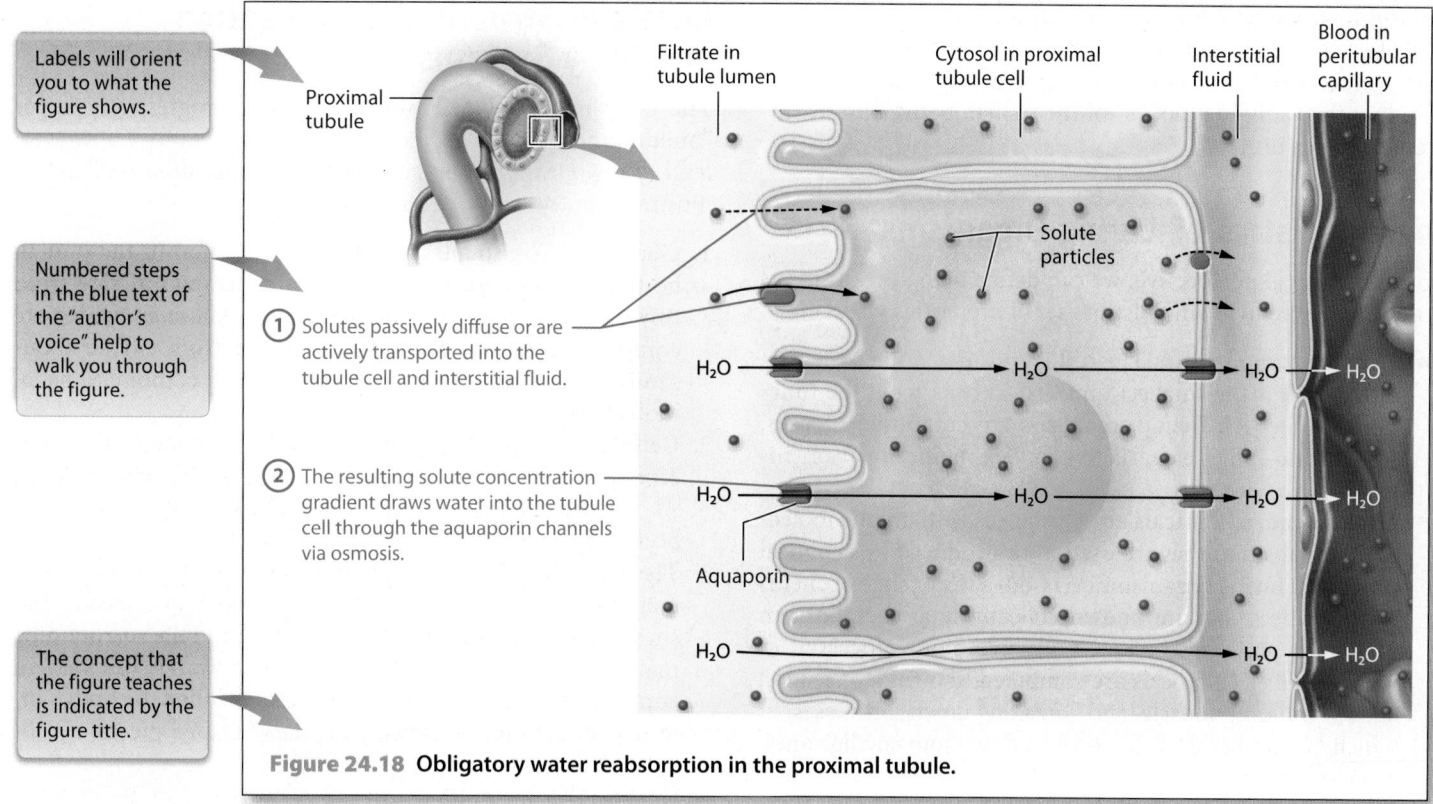

Labels will orient you to what the figure shows.

Numbered steps in the blue text of the "author's voice" help to walk you through the figure.

The concept that the figure teaches is indicated by the figure title.

Proximal tubule

Filtrate in tubule lumen

Cytosol in proximal tubule cell

Interstitial fluid

Blood in peritubular capillary

Solute particles

1 Solutes passively diffuse or are actively transported into the tubule cell and interstitial fluid.

2 The resulting solute concentration gradient draws water into the tubule cell through the aquaporin channels via osmosis.

H_2O

Aquaporin

Figure 24.18 Obligatory water reabsorption in the proximal tubule.

Figure 1.3 How to approach a physiology figure.

might actually be spread out over several figures. In such cases, follow the same steps outlined previously for each figure. After you have interpreted each one, put them all together so that you can see how all the parts combine to make the whole. Sometimes this is done for you in sections and figures called *Putting It All Together: The Big Picture.*

Companion Workbook

Another active learning feature of this text is the companion workbook, *Practicing A&P.* The workbook is not a study guide that you use after reading the chapter; instead it is intended to be used *as* you read. It contains multiple features: sequencing-the-steps exercises; labeling, coloring, and drawing activities; and group exercises. All of these are designed to guide you through the reading, to provide many opportunities to assess your understanding, and to help you practice active reading.

Online Practice Tools

Online practice tools for this textbook can be found on MasteringA&P®. They include chapter practice tests, reading quizzes, art-labeling and sequencing-the-steps exercises (🔲 icon), and author-narrated podcasts (🔊 icon). Throughout the book, you will see animation icons (▶), which signal you to view animations. These animations allow you to see an entire physiological process and interact with the textbook figures in an online setting. Additional media found on MasteringA&P include Interactive Physiology® tutorials, PhysioEx™ 9.1 Laboratory Simulations, and Practice Anatomy Lab™ (PAL™) 3.0 virtual anatomy lab study tool.

Quick Check

☐ 6. How can you use the features found in each chapter?
☐ 7. How should you approach the study of figures in this book?

Apply What You Learned

☐ 1. Design a study schedule for yourself up to the first exam.
☐ 2. Determine several study strategies that will help you succeed in your course.

See answers in Appendix A.

MODULE **1.2**

Overview of Anatomy and Physiology

Learning Outcomes

1. List the characteristics of life and the processes carried out by living organisms.
2. Describe the major structural levels of organization in the human body.
3. Define the types of anatomy and physiology.
4. Describe the organ systems of the human body and their major components.
5. Explain the major functions of each organ system.

We begin our study of human anatomy and physiology with the characteristics common to all living organisms. Then, discussion turns to the introductory principles of anatomy and physiology, including the "building blocks" of the human body, and the different subfields of A&P.

Characteristics of Living Organisms

From a biological perspective, we can define living organisms as those sharing the following distinct set of properties:

- **Cellular composition.** The **cell** is considered the basic unit of life, as it is the smallest unit that can carry out the functions of life. All living organisms are composed of cells, from single-celled bacteria to complex multicellular organisms such as humans.
- **Metabolism. Chemicals** are substances with unique molecular composition that are used in or produced by chemical processes. Living organisms carry out a wide range of chemical processes that are known collectively as **metabolism.** Metabolic processes include both "building" processes, in which smaller chemicals are combined to form larger ones, such as building muscle, and "breaking down" processes, in which larger chemicals are broken down into smaller ones, as in digestion.
- **Growth.** When the "building" metabolic processes outweigh the "breaking down" processes, the result is **growth.** Growth may come in two forms: an increase in the size of individual cells and/or an increase in the number of cells.
- **Excretion.** Metabolic processes can yield chemicals called *waste products* that the organism cannot use for any purpose. Waste products are toxic if they accumulate, and so the organism must have a way to separate the wastes and remove them. The process by which this occurs is known as **excretion.**
- **Responsiveness.** Living organisms sense and react to changes in their environment known as *stimuli;* this property is known as **responsiveness** or **irritability.** Humans and other animals respond to stimuli perceived through the senses, including sight, smell, hearing, touch, and pain.
- **Movement.** Another key property of life is **movement.** Including this may seem strange at first, because clearly plants don't get up and take a walk. However, plants do exhibit movement inside and between their cells. Other forms of movement include motion of one or more cells within the organism and movement of the organism itself.
- **Reproduction.** The final property common to life is the ability to carry out **reproduction.** Reproduction takes two forms in multicellular organisms: (1) individual cells reproduce within the organism during growth and to replace damaged or old cells, and (2) the organism itself reproduces to yield similar offspring.

Quick Check

☐ 1. List the properties common to all living organisms.

Levels of Structural Organization and Body Systems

The body is constructed from a series of progressively larger "building blocks." Each type of block is known as a *structural level of organization.* The six levels of organization we'll cover, illustrated in **Figure 1.4**, are as follows:

1. **Chemical.** The smallest level of organization in the human body is the **chemical level.** Chemicals range in size from tiny *atoms* to complex structures called *molecules,* which are composed of atoms ranging in number from two to thousands. All other levels are made up of combinations of molecules.
2. **Cellular.** Groups of several different types of molecules combine in specific ways to form structures at the **cellular level.** As you will discover in upcoming chapters, the cells in the body vary widely in size, shape, and function.
3. **Tissue.** At the **tissue level,** two or more cell types that cooperate to perform a common function make up a **tissue.** Tissues consist of two components: the cells and the surrounding material known as the *extracellular matrix.* Tissues vary in appearance from the thin, broad sheets that line the body's internal surfaces to the short, irregularly shaped pieces of rubbery cartilage that make up the nose.
4. **Organ.** The **organ level** consists of two or more tissue types combined to form a structure called an **organ,** which has a recognizable shape and performs a specialized task. Examples of organs include the skin, the heart, and the bones of the skeleton.
5. **Organ system.** At the **organ system level,** the body's organs are grouped into **organ systems.** An organ system consists of two or more organs that together carry out a broad function in the body. For example, the organs of the *cardiovascular system,* the heart and blood vessels, work together to transport and deliver blood through the body, and the organs of the *digestive system* work together to ingest food, absorb nutrients, and eliminate wastes.
6. **Organism.** The organ systems function together to make up the working human body, an **organism,** at the **organism level** of organization.

Throughout this book we will examine all of these levels of organization, starting at the chemical, moving to the cellular and tissue, and then to organs and individual organ systems. The human body has 11 organ systems; the components and major functions of each organ system are shown in **Figure 1.5.** As we examine organs and organ systems, never lose sight of the fact that all of these organ systems work together to ensure the survival of the organism as a whole.

Quick Check

☐ 2. What are the six levels of organization of the human body?
☐ 3. List the 11 organ systems in the body.

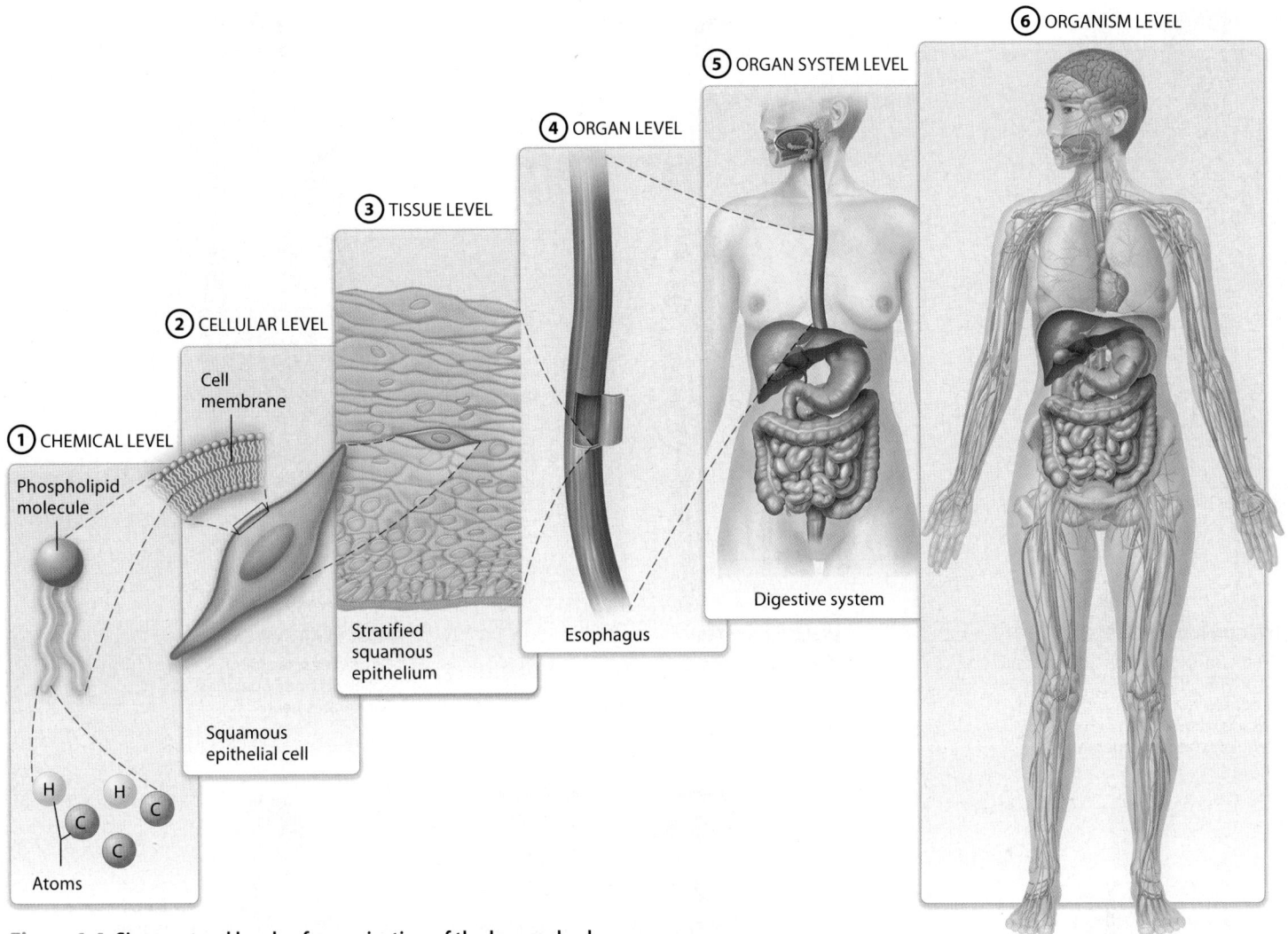

① CHEMICAL LEVEL

Phospholipid
molecule

Atoms

② CELLULAR LEVEL

Cell
membrane

Squamous
epithelial cell

③ TISSUE LEVEL

Stratified
squamous
epithelium

④ ORGAN LEVEL

Esophagus

⑤ ORGAN SYSTEM LEVEL

Digestive system

⑥ ORGANISM LEVEL

Figure 1.4 Six structural levels of organization of the human body.

Types of Anatomy and Physiology

We can approach the study of anatomy in several ways. In this book, we examine the human body primarily by looking at individual organ systems, which is known as **systemic anatomy.** In another approach, the body can be divided into regions such as the back or head and neck; this is **regional anatomy.** Finally, we can study the surface markings of the body, which is called **surface anatomy.**

Within the broad field of anatomy are numerous subfields for more specialized study. Anatomical subfields are generally classified by the structural level of organization being studied. For example, the field of **gross anatomy** examines structures, including organs and organ systems, that can be seen with the unaided eye. In contrast, the structures studied in **microscopic anatomy** require the aid of a microscope. Topics within the field of microscopic anatomy include **histology** (hiss-TOL-uh-jee; *histo-* = "tissue," *-logy* = "study of"), the study of tissues, and **cytology** (sy-TOL-uh-jee; *cyto-* = "cell"), the study of cells.

Physiology also has numerous subfields. Typically, physiological specializations are classified according to the organ or organ system being studied. For example, *neurophysiology* studies the brain and nerves (*neuro-* = "nerve"), and *cardiovascular physiology* studies the heart and blood vessels. Physiologists may also specialize in levels of organization other than the systemic—some physiologists study the body's chemical and cellular processes, and others study specific tissues or organs.

Quick Check

☐ 4. How do gross anatomy and microscopic anatomy differ?
☐ 5. How are physiological specializations classified?

Apply What You Learned

☐ 1. The condition *hypothyroidism* is characterized by a decrease in the synthesis and secretion of the chemical thyroid hormone from the thyroid gland, an organ in the neck that is part of the endocrine system. Explain how this condition involves all levels of organization in the body.

INTEGUMENTARY SYSTEM
- Protects the body from the external environment
- Produces vitamin D
- Retains water
- Regulates body temperature

SKELETAL SYSTEM
- Supports the body
- Protects internal organs
- Provides leverage for movement
- Produces blood cells
- Stores calcium salts

MUSCULAR SYSTEM
- Produces movement
- Controls body openings
- Generates heat

NERVOUS SYSTEM
- Regulates body functions
- Provides for sensation, movement, automatic functions, and higher mental functions via nerve impulses

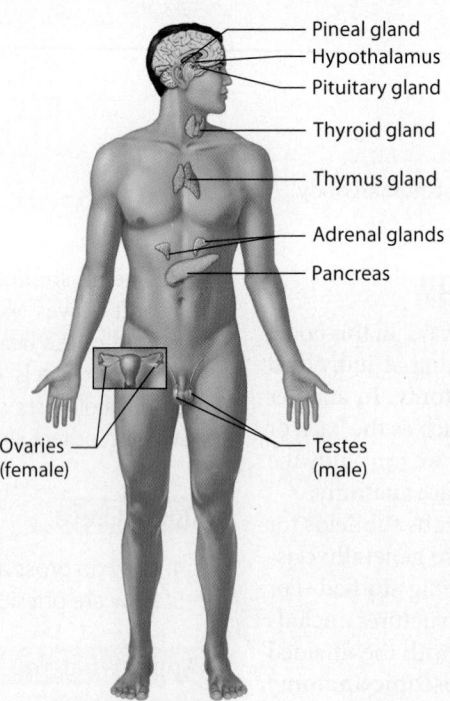

ENDOCRINE SYSTEM
- Regulates body functions
- Regulates the functions of muscles, glands, and other tissues through the secretion of chemicals called hormones

CARDIOVASCULAR SYSTEM
- Pumps and delivers oxygen-poor blood to the lungs and oxygen-rich blood to the tissues
- Removes wastes from the tissues
- Transports cells, nutrients, and other substances

Figure 1.5 The 11 organ systems of the human body.

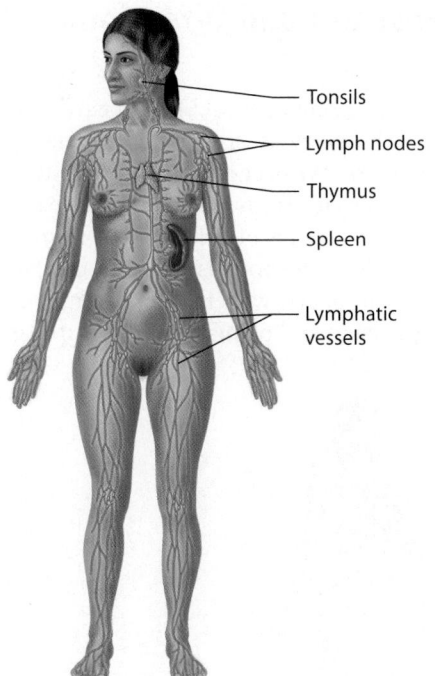

LYMPHATIC SYSTEM
- Returns excess tissue fluid to the cardiovascular system
- Provides immunity (protection against disease)

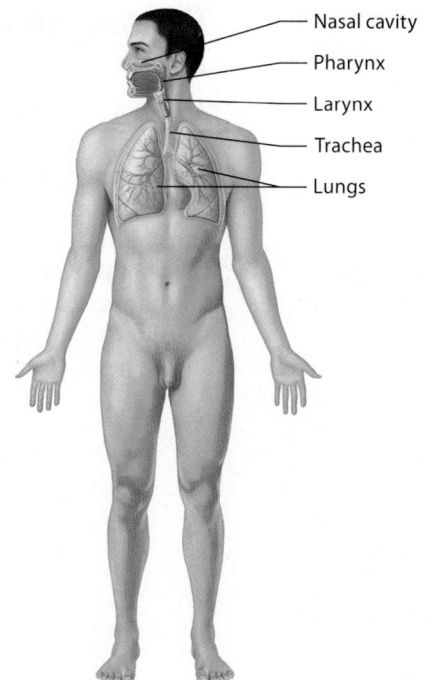

RESPIRATORY SYSTEM
- Delivers oxygen to the blood
- Removes carbon dioxide from the body
- Maintains the acid-base balance of the blood

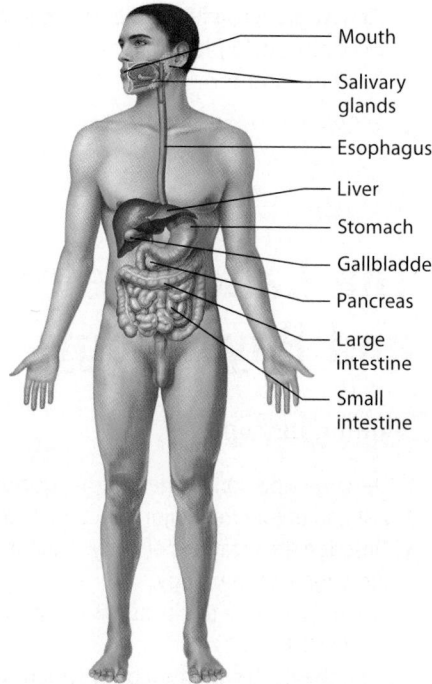

DIGESTIVE SYSTEM
- Digests food
- Absorbs nutrients into the blood
- Removes food waste
- Regulates fluid, electrolyte, and acid-base balance

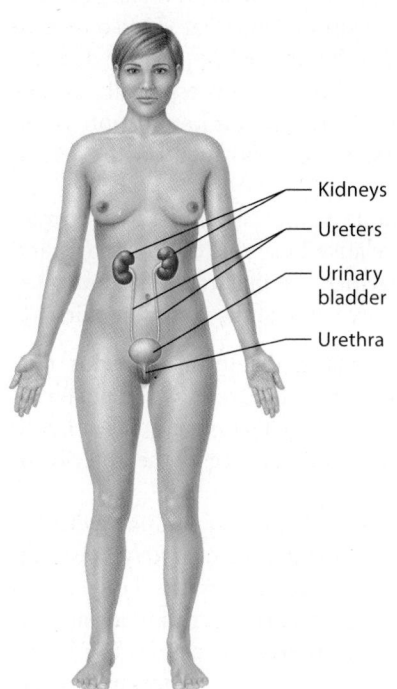

URINARY SYSTEM
- Removes metabolic wastes from the blood
- Regulates fluid, electrolyte, and acid-base balance
- Stimulates blood cell production

REPRODUCTIVE SYSTEM: MALE
- Produces and transports sperm
- Secretes hormones
- Sexual function

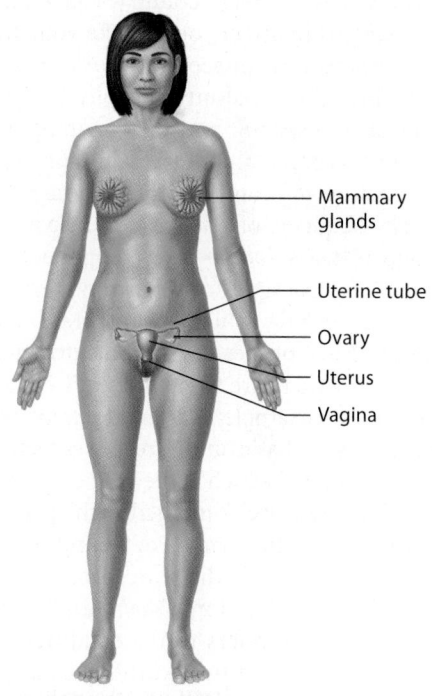

REPRODUCTIVE SYSTEM: FEMALE
- Produces and transports eggs
- Site of fetal development, fetal nourishment, childbirth, and lactation
- Secretes hormones
- Sexual function

Figure 1.5 *(continued)*

☐ 2. At first glance, a human and a rose plant seem to have little in common. Explain what these two organisms do actually have in common.

See answers in Appendix A.

MODULE **1.3**

The Language of Anatomy and Physiology

Learning Outcomes

1. Describe a person in anatomical position.
2. List and define the major directional terms used in anatomy.
3. Describe the locations of body structures using appropriate directional terminology.
4. List and describe the locations of the major anatomical regions of the body.
5. Describe the locations of body structures using regional terminology.
6. Identify the various planes in which a body or body part might be dissected, and describe the appearance of a body presented along each of those planes.

If you visited a foreign country, you would first begin to learn its language so that you could make your way. The same is true in the perhaps foreign-seeming world of science. All fields of science, including anatomy and physiology, come with their own language—a specific way of describing things. And as with your trip to that foreign country, you must learn the language of science before you can hope to fully grasp it.

The language of science, and in particular A&P, is built on a group of *word roots*—core components of words with specific meanings. Examples of word roots include *card-*, which means heart, *encephal-*, which means brain, and *myo-*, which means muscle. Most of these word roots originated in Latin and Greek; this is why there are sometimes more than one root for the same concept, for example, *uter-* and *hystero-*, which are the word roots for uterus in Latin and Greek, respectively (think of the term "hysterectomy," which is the surgical removal of the uterus).

Word roots are combined with specific prefixes and suffixes to yield scientific terms. For example, we can combine the prefix *an-* ("without"), the word root *encephal-*, and the suffix *–ic* ("condition of") to form the word "*anencephalic*" (an'-en-seh-FAL-ik), which refers to the condition of lacking a part of the brain. To combine two word roots, add an "o" between them, as in the term for the heart and blood vessels, *cardiovascular* (*vascul-* = "vessel"). Building anatomical terms from word roots, prefixes, and suffixes is discussed on the inside back cover of the textbook.

This module serves as an introduction to the language of A&P. We'll talk about the accepted terms used by all scientists as well as health and medical professionals to communicate about the human body.

The Anatomical Position and Directional Terms

Accurate communication among scientists and health care professionals when discussing the body is crucial to preventing experimental and medical errors. We accomplish this by establishing a common frame of reference from which all body parts and regions are described. This frame of reference is termed the **anatomical position,** in which the body is standing upright and the feet are shoulder width apart, with the upper limbs at the sides of the trunk and the head and palms facing forward, as shown here:

We always refer to the body as if it were in anatomical position, even if it's actually in another position. Note that we also refer to "right" and "left" as the right and left of the body we are describing, not our own right and left. This system is especially important in the health care field, where miscommunication could result, for example, in surgery being performed on the wrong side of the body.

Another way to ensure accurate communication is to use **directional terms** to describe the relative locations of body parts and markings. Directional terms, which are illustrated in **Figure 1.6**, generally come in pairs. These are some of the common pairs:

- **Anterior/Posterior.** In humans, the terms **anterior** (an-TEER-ee-ur) and **ventral** refer to the front, while **posterior** and **dorsal** refer to the back. These terms can refer either to the body as a whole or to a body part (e.g., the anterior or posterior surface of the heart).
- **Superior/Inferior.** You are likely already familiar with the terms superior and inferior from your everyday language. They have a more specific meaning in the language of A&P—**superior,** also known as **cranial,** means toward the head (*crani-* = "head"), and **inferior** or **caudal** (KAW-d'l; *caud-* = "tail") means away from the head or toward the tail. We use these terms to refer to positions only on the head, neck, and trunk; these terms are not applied to the upper and lower limbs.
- **Proximal/Distal.** Instead of using superior and inferior for the limbs, we use the terms proximal and distal. The term **proximal** (PROKS-ih-mul) refers to something being closer to the point of origin. You know this from the everyday term *approximate,* which refers to closeness.

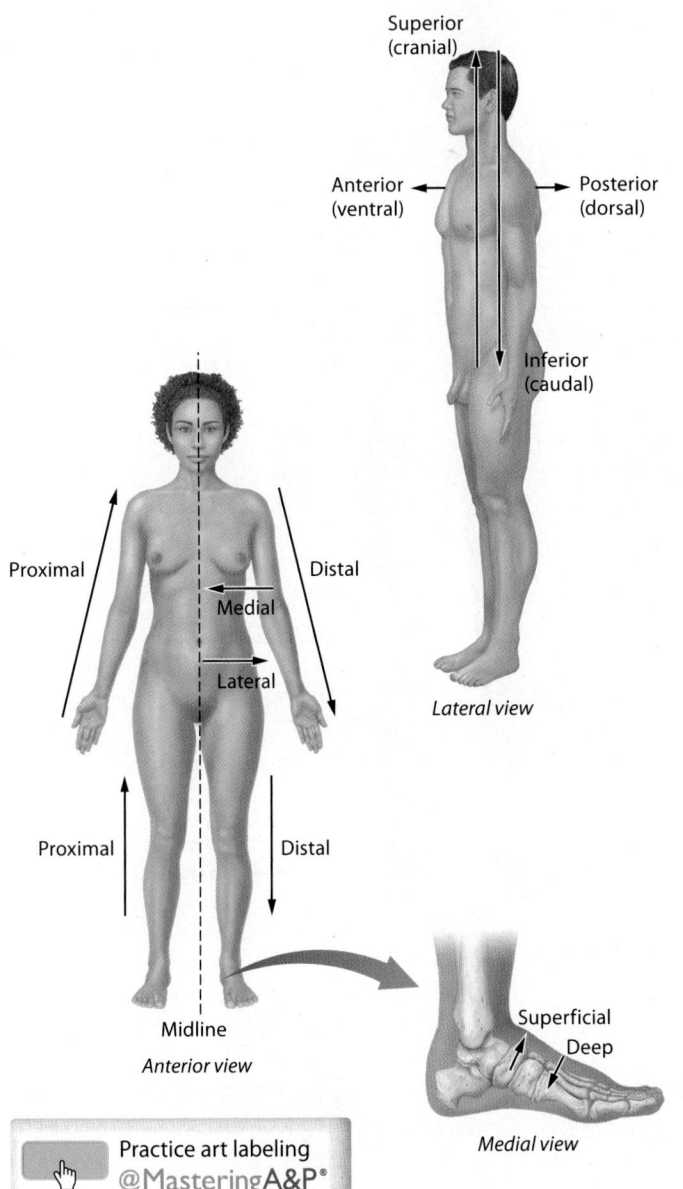

Superior
(cranial)

Anterior
(ventral)

Posterior
(dorsal)

Inferior
(caudal)

Lateral view

Proximal

Distal

Medial

Lateral

Proximal

Distal

Midline

Anterior view

Superficial

Deep

Medial view

Practice art labeling
@MasteringA&P®

Figure 1.6 Directional terms.

TERM	DEFINITION	EXAMPLES
Anterior (ventral)	Toward the front	• The palms are on the anterior side of the body. • The esophagus is anterior to the spinal cord.
Posterior (dorsal)	Toward the back	• The occipital bone is on the posterior cranium (skull). • The spinal cord is posterior to the esophagus.
Superior (cranial)	Toward the head	• The nose is superior to the mouth. • The neck is superior to the chest.
Inferior (caudal)	Toward the tail	• The nose is inferior to the forehead. • The umbilicus (belly button) is inferior to the chest.
Proximal	Closer to the point of origin (generally the trunk)	• The knee is proximal to the ankle. • The shoulder is proximal to the elbow.
Distal	Farther away from the point of origin (generally the trunk)	• The foot is distal to the hip. • The wrist is distal to the elbow.
Medial	Closer to the midline of the body or a body part; on the inner side of	• The ear is medial to the shoulder. • The index finger is medial to the thumb.
Lateral	Farther away from the midline of the body or a body part; on the outer side of	• The shoulder is lateral to the chest. • The thumb is lateral to the index finger.
Superficial	Closer to the surface	• The skin is superficial to the muscle. • Muscle is superficial to bone.
Deep	Farther below the surface	• Bone is deep to the skin. • Bone is deep to muscle.

Distal refers to the opposite condition—being farther away or *distant* from the point of origin. In the case of the upper and lower limbs, the point of origin is the trunk. So structures nearer the trunk are proximal, and those farther away are distal.

• **Medial/Lateral.** The body's *midline* is an imaginary line that runs down the middle of the body on the inside (see Figure 1.6). The term **medial** refers to a position that is closer to this midline, whereas **lateral** refers to a position that is farther away from the midline. This almost always refers to the body's midline, not the midline of an individual body part.

• **Superficial/Deep.** This is another pair of terms that you probably use as part of your everyday language. The term **superficial** refers to structures that are closer to the surface of the body, whereas **deep** refers to those farther below.

Quick Check

☐ 1. Define anatomical position.
☐ 2. Fill in the blanks:
   a. The nose is _____ to the mouth.
   b. The ankle is _____ to the knee.
   c. The ring finger is _____ to the index finger.
   d. The skin is _____ to the bone.
   e. The trachea is _____ to the spine.
   f. The shoulder is _____ to the wrist.

Regional Terms

Broadly, we can divide the body into two regions: the **axial region** (AKS-ee-ul), which includes the head, neck, and trunk, and the **appendicular region** (ap'-en-DIK-yoo-lar), which

Cephalic
Occipital
Cervical

Thoracic
Dorsal
Sternal
Arm
Vertebral

Upper limb
Lumbar
Abdominal
Forearm
Sacral
Pelvic
Gluteal

Manual
Palmar
Thigh

Inguinal
Lower limb
Popliteal
Pubic

Sural
Leg

Pedal
Plantar

Anterior
Posterior

(a) Anterior and posterior views of the full body

Cranial
Coxal

Frontal
Acromial

Ocular
Axillary
Otic
Femoral
Nasal
Buccal
Brachial
Oral
Mental
Antecubital
Patellar

Antebrachial

(b) Head, anterior view
Crural

Carpal
Metacarpal
Pollex
Tarsal
Metatarsal
Digital
Hallux
Digital

(c) Upper limb, anterior view

Figure 1.7 Regional terms.

(d) Lower limb, anterior view

includes the upper and lower limbs. The main components of the axial and appendicular regions are shown in **Figure 1.7**.

Notice in Figure 1.7 that each broad region can be divided into several smaller regions. These may be named as nouns, such as *brachium* ("arm"; BRAY-kee-um), *crus* ("shin"; KROOS), and *thorax* ("trunk"), or we can add the suffix *-al* or *-ic* to the terms and turn them into adjectives. To use these adjectives, we must pair them with the word "region," which gives us terms such as the *brachial region, crural region,* and *thoracic region* (thoh-RAH-sik).

A summary of regional terms, including the adjectival forms, is given in the companion workbook. Some of the terms listed there may already be familiar to you. Try to start with terms you already know, and use the word roots to work out the meanings.

ConceptBOOST »»»

Putting Anatomical Terms Together

To put together the directional and regional terms we encountered in the past two sections (see Figures 1.6 and 1.7), let's use these terms to describe the surgical incision shown here:

The first thing to do is name the region, which in this case is the cervical region. Now we need to add some descriptive directional terms to tell where, exactly, the incision is located in the cervical region. We can say that the incision is on the anterior side, it is just lateral to the midline, and it begins inferior to the mental region and ends superior to the thoracic region. We could also describe the depth of the incision, which is deep to the skin and muscle but superficial to the underlying larynx. So let's put this all together:

The incision was made on the anterior cervical region just lateral to the midline. It extended vertically from 1 centimeter inferior to the mental region to 2 centimeters superior to the thoracic region, and was deep to the skin and muscle but superficial to the larynx.

Let's do one more example using the following illustration in which a person has a wound from a pellet gun:

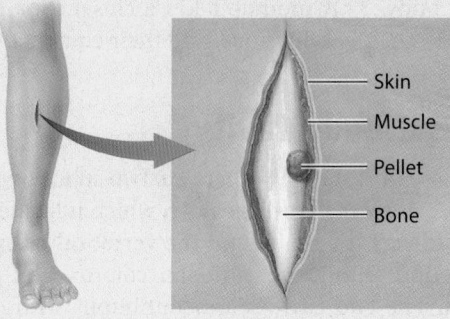

Start with the region, the left crural region, and add descriptive directional terms (don't forget that we need to use proximal and distal here, since we are describing the limbs): anterior, on the medial side, proximal to the tarsal region and distal to the patellar region, and deep to the skin and muscle but superficial to the bone. Now we put it all together:

The wound is on the left anteromedial crural region, 6 centimeters proximal to the tarsal region and 10 centimeters distal to the patellar region. The pellet is lodged deep to the skin and muscle but superficial to the bone. ■

Quick Check

☐ 3. Fill in the blanks:
   a. The wrist is also known as the _____ region.
   b. The arm is also known as the _____ region.
   c. The armpit is also known as the _____ region.
   d. The neck is also known as the _____ region.

Planes of Section

When studying a body or body part, we can learn only so much by looking at it from the outside. To gain a better understanding of its form and function, we need to look inside. This is accomplished through the use of **planes of section**—standard ways of dividing a body or body part to examine its internal structure. The three primary planes of section, shown in **Figure 1.8**, are as follows:

- **Sagittal plane.** The **sagittal plane** (SAJ-ih-tul) divides the body or body part into right and left sections (see Figure 1.8a). There are two variations of the sagittal plane: (1) the **midsagittal plane**, also known as the **median plane** (not to be confused with medial), which divides the body or body part into equal right and left parts, and (2) the **parasagittal plane** (pair'-ah-SAJ-ih-tul; *para-* = "near"), which divides the body or body part into unequal right and left sections.
- **Frontal plane.** The **frontal plane,** also known as the **coronal plane,** divides the body or body part into anterior and posterior sections (see Figure 1.8b). The frontal plane is easy to remember because the *frontal* plane gives a *front* section and a back section.
- **Transverse plane.** The third commonly used plane is the **transverse plane,** also known as the **horizontal plane** or a **cross section** (see Figure 1.8c). The transverse plane divides the body or body part into a superior and inferior part. Note that it also divides the appendicular region into proximal and distal parts.

Another type of section that is less standardized is the **oblique plane.** An oblique section is taken at an angle and is useful for examining structures, such as the knee joint, that aren't easy to study with standard planes of section. However, these sections can be difficult to interpret and are used less frequently than the others we have discussed.

Parasagittal section — Midsagittal section

Brain

Midsagittal section of brain

(a) Sagittal plane

Brain — Frontal plane

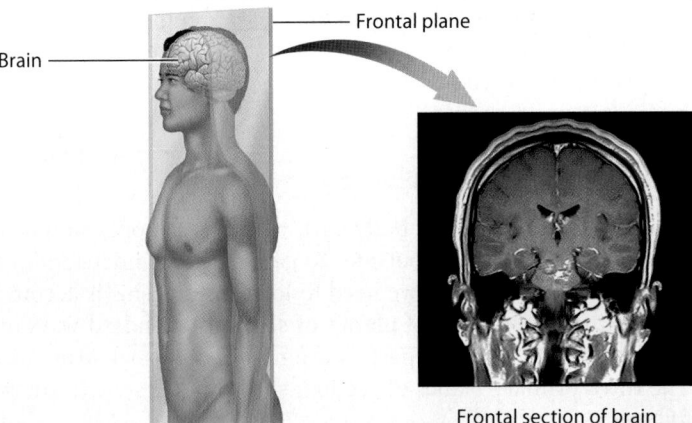

Frontal section of brain

(b) Frontal plane

Brain —

Transverse planes

Transverse section of brain

(c) Transverse plane

Figure 1.8 Planes of section.

Practice art labeling
@MasteringA&P®

Note that planes of section are *not* the same as directional terms. You cannot say, "The bullet is located on the brachial region along the parasagittal plane." You need to describe the location of the bullet using directional terms such as medial/lateral, anterior/posterior, and so on. However, as a surgeon you can make a cut along a parasagittal plane of the arm to locate and remove the bullet.

Quick Check

☐ 4. Explain how the three main planes of section differ.

Apply What You Learned

☐ 1. Locate a mole, scar, tattoo, or other mark on your skin and describe its location using regional and directional terms.
☐ 2. You need to examine the internal anatomy of both lungs at the same time but can make only one cut. Which plane or planes of section would allow you to do this?

See answers in Appendix A.

MODULE **1.4**

The Organization of the Human Body

Learning Outcomes

1. Describe the location of the body cavities, and identify the major organs found in each cavity.
2. Describe the location of the four quadrants and nine regions of the abdominopelvic cavity, and list the major organs located in each.
3. List and describe the serous membranes that line the body cavities.

Internally, the axial region of the body—the head, neck, and thorax—is divided into several cavities. A *cavity* is any fluid-filled space within the body. Body cavities both protect internal organs and allow them to move and expand as necessary. The two major cavities are the **dorsal body cavity,** which is largely located on the posterior (dorsal) side of the body, and the **ventral body cavity,** which is largely located on the anterior (ventral) side of the body. This module takes a closer look at these two body cavities, their subdivisions, and their contents.

The Dorsal Body Cavity

As you can see in **Figure 1.9a,** there are two subcavities within the dorsal body cavity: the **cranial cavity,** which is located within the skull and protects the brain, and the **vertebral** or **spinal cavity,** which is found within the vertebral column and protects the spinal cord. The two cavities are continuous with one another,

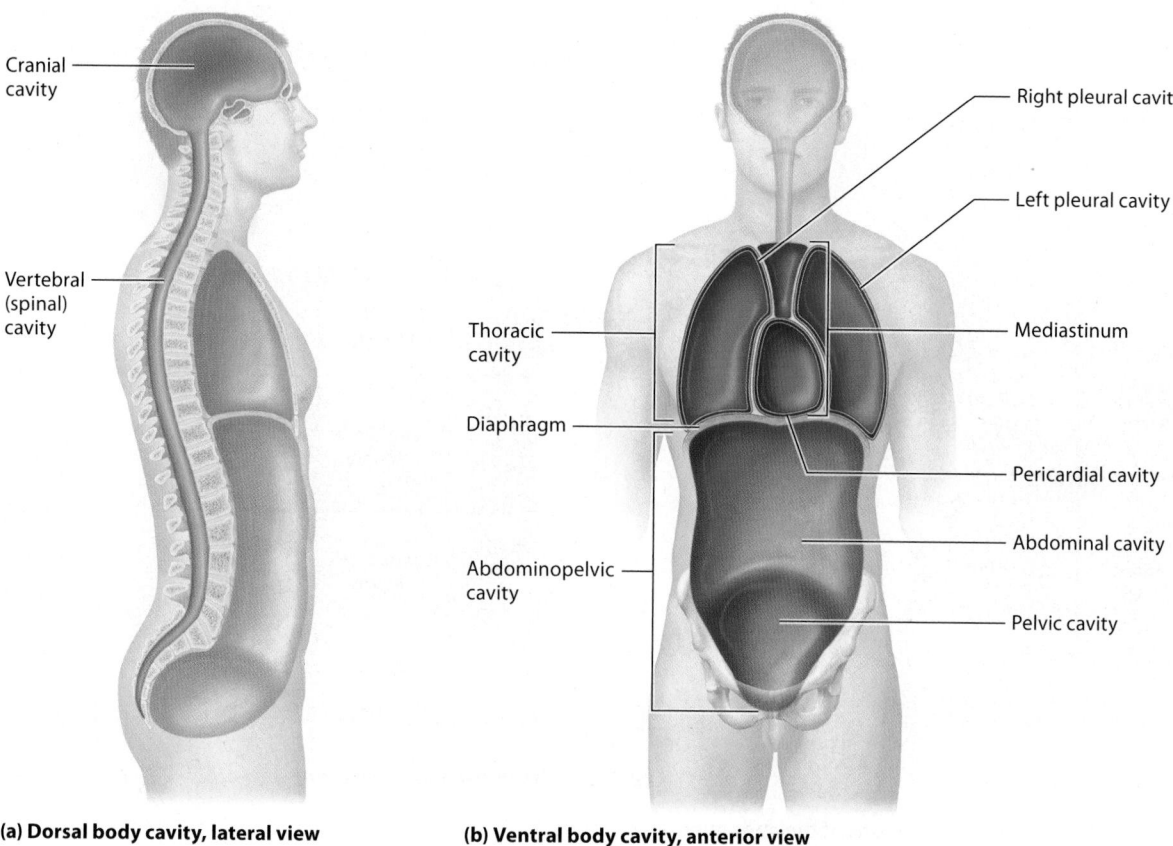

Cranial cavity

Vertebral (spinal) cavity

Thoracic cavity

Diaphragm

Abdominopelvic cavity

Right pleural cavity

Left pleural cavity

Mediastinum

Pericardial cavity

Abdominal cavity

Pelvic cavity

(a) Dorsal body cavity, lateral view

(b) Ventral body cavity, anterior view

Figure 1.9 The dorsal and ventral body cavities.

Practice art labeling @MasteringA&P®

which reflects the fact that the brain and spinal cord are actually a single, continuous structure. Both cavities are filled with a fluid called *cerebrospinal fluid* (sehr′-eh-broh-SPY-nul; *cerebro-* = "head"), or *CSF,* which bathes both organs and keeps the brain buoyant within the skull.

Quick Check

☐ 1. What are the two subcavities of the dorsal body cavity?

The Ventral Body Cavity

The ventral body cavity has two main divisions that are separated by a dome-shaped muscle called the **diaphragm** (DY-uh-fram), which functions in breathing (**Figure 1.9b**). The cavity superior to the diaphragm encompasses the area of the thorax, which is why it is referred to as the **thoracic cavity.** The cavity inferior to the diaphragm encompasses the abdomen and pelvis, and is therefore known as the **abdominopelvic cavity** (ab-dom′-ih-noh-PEL-vik). These cavities are described next, along with the smaller cavities within them. Note that some of these smaller cavities are formed by thin sheets of tissue termed *serous membranes* (SEER-us), and so the discussion concludes with a look at the structure and function of this tissue.

The Thoracic Cavity

If you look at Figure 1.9b, you can see that the thoracic cavity houses three smaller cavities. These cavities are as follows:

● **Pleural cavities.** There are right and left **pleural cavities** (PLOO-ral), each of which surrounds one lung. These cavities are located within serous membranes.
● **Mediastinum.** Between the pleural cavities we find the **mediastinum** (meh′-dee-ass-TY-num; *med-* = "middle"). The mediastinum houses the heart, great blood vessels, trachea (windpipe), and esophagus. The mediastinum is not located within a serous membrane.
● **Pericardial cavity.** Within the mediastinum we find another cavity within a serous membrane called the **pericardial cavity** (*peri-* = "around") that surrounds the heart, as its name implies.

The Abdominopelvic Cavity

There are two divisions within the abdominopelvic cavity: the **abdominal cavity,** which is the area from the diaphragm to the bony pelvis, and the **pelvic cavity,** which is the area within the bony pelvis. The abdominal cavity contains some of the organs of multiple systems, including the digestive, lymphatic, and urinary systems. Some of these organs are located in another subcavity within serous membranes called the **peritoneal cavity** (pear′-ih-ton-EE-ul; note that you cannot see the peritoneal cavity in Figure 1.9). We'll revisit

(a) The four abdominopelvic quadrants

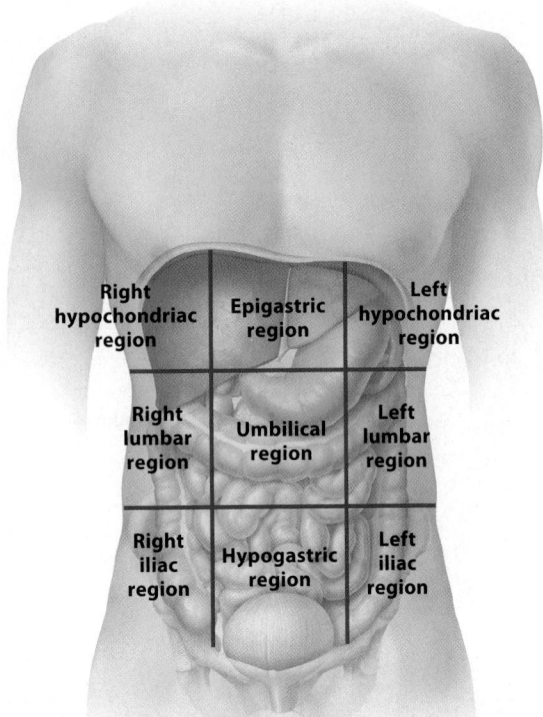

(b) The nine abdominopelvic regions

Figure 1.10 **The four quadrants and nine regions of the abdominopelvic cavity.**

the peritoneal cavity shortly. The pelvic cavity also contains the organs of several systems, including the reproductive, digestive, and urinary systems.

Observe in **Figure 1.10a** that we can divide the abdominopelvic cavity into different segments by drawing imaginary lines through its surface. The simplest way to divide the abdominopelvic cavity is to use two lines through the umbilical (belly button) region, one in the transverse plane and one in the midsagittal plane. This gives us four quadrants: the **right upper quadrant (RUQ),** the **left upper quadrant (LUQ),** the **right lower quadrant (RLQ),** and the **left lower quadrant (LLQ).**

A second system uses four lines—two in the parasagittal plane and two in the transverse plane—to divide the abdominopelvic cavity into nine regions. These regions, shown in **Figure 1.10b,** are as follows:

- The **right** and **left hypochondriac regions** (hy′-poh-KON-dree-ak) are the right and left superior divisions, respectively. They are named for the fact that they lie below the cartilage of the ribs (*hypo-* = "below," *chondro-* = "cartilage").
- The **epigastric region** (ep-ih-GAS-trik) is the middle, superior region between the right and left hypochondriac regions that lies over the stomach (*epi-* = "above," *gastr-* = "stomach").
- The right and left middle regions are the **right** and **left lumbar regions,** so named because they are located in approximately the same region as the lumbar vertebrae.
- Between the right and left lumbar regions is the **umbilical region,** which lies over the umbilicus.

- The right and left inferior regions are the **right** and **left iliac** (EE-lee-ak) or **inguinal regions** (ING-gwih-nul), respectively. Like other regions, these are named for the anatomical regions over which they lie.
- Between the two iliac regions we find the **hypogastric region,** so named because it lies below the stomach.

The nine-region system is preferred by anatomists, whereas the four-quadrant system is widely used in the medical field. *A&P in the Real World: Abdominal Pain,* on page 20, discusses the main organs that can be found in each quadrant.

Serous Membranes

As you've just read, certain cavities within the ventral body cavity are formed by thin sheets of tissue called **serous membranes.** When sectioned, serous membranes appear to be two membranes, but they actually consist of a single, continuous layer of tissue that folds over on itself to enclose a single space, much like a bean bag chair is one continuous layer of fabric that encloses stuffing. Within the cavity between the two layers (instead of stuffing), we find an extremely thin layer of fluid called **serous fluid.** Serous fluid, which is produced by the cells of the membrane, is a watery, slippery liquid that lubricates organs just like oil lubricates the pistons in a car—it prevents friction as an organ moves against adjacent structures.

In the body, serous membranes surround certain organs, including the heart, the lungs, and many abdominal organs.

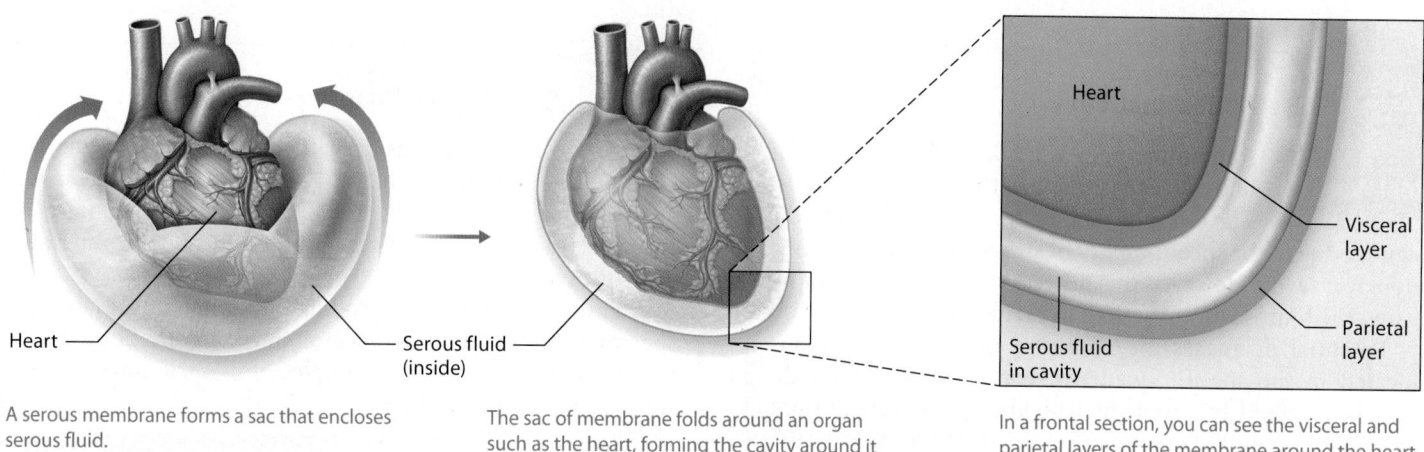

Heart

Serous fluid (inside)

Heart

Serous fluid in cavity

Visceral layer

Parietal layer

A serous membrane forms a sac that encloses serous fluid.

The sac of membrane folds around an organ such as the heart, forming the cavity around it (in this case, the pericardial cavity).

In a frontal section, you can see the visceral and parietal layers of the membrane around the heart.

Figure 1.11 How a serous membrane envelops the heart.

Figure 1.11 shows how a serous membrane folds around and envelops the heart, forming the pericardial cavity. Part of the membrane comes into contact with the heart itself, and part has contact with the structures surrounding the heart. Although it is all one continuous membrane, we call these two parts of the same membrane different names: The inner layer that contacts the organ is known as its **visceral layer** (VISS-er-ul; *viscer-* = "organ"), and the outer layer that attaches to surrounding structures is called the **parietal layer** (pah-RY-eh-tal).

We can envision the relationship between these two layers of membrane by going back to our bean bag chair example. Imagine you are sitting in such a chair. You represent an organ—the part of the chair that is in contact with your body is akin to the visceral layer of a serous membrane, and the part in contact with the floor is like its parietal layer.

The body has three main serous membranes, within which are three serous body cavities. These membranes are illustrated in **Figure 1.12** and discussed next.

Inferior view

Parietal pericardium

Pericardial cavity

Visceral pericardium

Anterior

Mediastinum

Heart

Right lung

Left lung

Parietal pleura

Pleural cavity

Visceral pleura

Posterior

(a) Transverse section through the pleural and pericardial cavities. Notice that you are looking at an inferior view.

Organs surrounded by visceral peritoneum (intraperitoneal organs)

Anterior

Parietal peritoneum

Visceral peritoneum

Peritoneal cavity containing serous fluid

Retroperitoneal organs

Posterior

(b) Midsagittal section through the peritoneal cavity

Figure 1.12 The serous membranes of the ventral body cavities.

Practice art labeling @MasteringA&P®

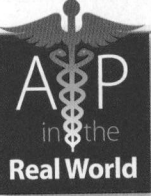

Abdominal Pain

Abdominal pain is a common complaint of individuals visiting health care professionals. The cause of such pain can be difficult to diagnose due to the sheer number of structures in the abdominopelvic cavity. However, this process can be made easier by using the four-quadrant system to divide the cavity.

Potential diagnoses are narrowed down by the organs located in the specific quadrant. For example, RLQ pain could be caused by problems with the appendix, the right ovary (in females), the first part of the large intestine, or the last portion of the small intestine. Similarly, pain in the LUQ could result from problems with the stomach, the spleen, the pancreas, and parts of the large intestine. Although more information is needed to arrive at a conclusive diagnosis, knowing the quadrant in which the pain is located is a helpful first step in finding the cause of the pain.

- The **pleural membranes** around the lungs consist of the *parietal pleura,* which you can see in Figure 1.12a as the outer layer curving along the thoracic wall, and the *visceral pleura,* which is shown as the inner layer running along the surfaces of the lungs. The thin space enclosed by the pleural membranes forms the pleural cavities.
- The **pericardial membranes** around the heart consist of the *parietal pericardium,* which is the outer layer encircling the heart in the mediastinum, and the *visceral pericardium,* which is attached to the heart muscle itself (see Figure 1.12a). Between the two we find the pericardial cavity.
- The **peritoneal membranes** around some abdominal organs consist of the *parietal peritoneum* and the *visceral peritoneum.* The peritoneal cavity is the space between the parietal and visceral peritoneal membranes. As you can see in Figure 1.12b, the peritoneal cavity is much more extensive than either the pleural cavities or the pericardial cavity. Notice also that there are several abdominal organs that are not located within the peritoneal cavity; instead, these organs are actually behind the parietal peritoneum (it covers their anterior surfaces). Such organs, including the kidneys, are said to be *retroperitoneal* (*retro-* = "behind").

See *A&P in the Real World: Medical Imaging* to read about the modern imaging techniques that can be used to show these membranes and the cavities they form. We explore the structure and function of these membranes in greater detail in later chapters.

Quick Check

☐ 2. List the subdivisions of the thoracic and abdominopelvic cavities.

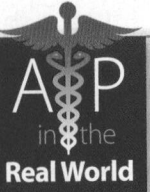

Medical Imaging

To look inside patients without resorting to surgery, medical professionals rely on *medical imaging.* These techniques use different forms of radiation to form images of the body's internal structures.

Many medical images produce sectional images along specific planes so that we can look into different body cavities. Following are several examples of medical images that show sections of body planes.

- **X-Ray.** An x-ray uses ionizing radiation to produce an image of internal body structures. The image to the right is a chest x-ray film, which shows the thoracic cavity and all of the smaller cavities contained within it.

- **Computed tomography scan.** A computed tomography (CT) scan also uses ionizing radiation to gather data, but these data are fed into a computer to produce a three-dimensional image. The image to the right is a CT scan in a transverse section of the abdominopelvic and peritoneal cavities.
- **Magnetic resonance image.** A magnetic resonance image, or MRI, is produced by placing the body within a magnetic field. As with a CT scan, a computer compiles the data from an MRI scan and produces a three-dimensional image. The following two MRI scans show inside the brain and cranial cavity (left) and inside the vertebral cavity (right), both in a midsagittal section.

☐ 3. What are serous membranes, and what are their functions?

☐ 4. Explain how serous membranes form certain ventral body cavities.

Apply What You Learned

☐ 1. In an industrial accident, a piece of shattered equipment enters a man's body near the right anterior pelvis and travels through the organs until it rests inside the left lung. List all of the body cavities and serous membranes through which the fragment passed on its way through this man's body.

☐ 2. Certain conditions lead to inadequate serous fluid in the pleural cavities. Predict the effects of these conditions.

☐ 3. Inflammation of serous membranes may cause the cavities to fill with excess serous fluid. Predict how this will affect the function of the organs they surround.

See answers in Appendix A.

MODULE **1.5**

Core Principles in Anatomy and Physiology

Learning Outcomes

1. Describe the principle of homeostasis.
2. Describe the components of a feedback loop, and explain the function of each component.
3. Compare and contrast negative and positive feedback in terms of the relationship between stimulus and response, and explain why negative feedback is the most commonly used mechanism to maintain homeostasis in the body.
4. Describe how structure and function are related.
5. Define the term gradient, and give examples of the types of gradients that drive processes in the body.
6. Describe how cells communicate with one another, and state why such communication is necessary in a multicellular organism.

As noted in the introduction, this chapter forms the foundation for the rest of this book, and for your continued studies of the human body. In this module, we complete our foundation with an introduction to a set of basic concepts that we revisit repeatedly in this text. We call these concepts the "core principles" of anatomy and physiology. However, there is one overarching theme in anatomy and physiology that all of these core principles relate to—*homeostasis.* Let's explore this overall theme and the four related principles: (1) feedback loops, (2) the relationship of structure and function, (3) gradients, and (4) cell-cell communication.

Overall Theme: Physiological Processes Operate to Maintain the Body's Homeostasis

In the upcoming chapters, you will discover the incredible number of physiological processes that occur in the trillions of cells of your body every second. Yet in spite of this complexity, your body continues to function; this requires the creation and maintenance of a relatively stable internal environment in which these cells can function. This maintenance of the body's internal environment is called **homeostasis** (hoh´-mee-oh-STAY-sis). Even when your body is subject to harsh conditions, such as being outside on a cold day without a coat, your internal environment doesn't change very much. This maintenance of homeostasis in the face of changing conditions requires the body to expend an immense amount of energy.

Homeostasis can be likened to a scale that we are using to try to balance a quantity in the internal environment, for example sugar. Adding too much sugar to one side causes the scale to tip to that side. Similarly, removing sugar from one side also causes the scale to lose its balance and tip. This occurs in the body as well—having too little sugar in the blood causes confusion, dizziness, weakness, nausea, and even coma. Conversely, having too much sugar in the blood on a long-term basis damages nerves, blood vessels, the kidneys, and the eyes. Both conditions are disturbances in homeostasis, or *homeostatic imbalances,* which can result in disease or even death if uncorrected. The body must avoid tipping the scale to one side or the other in order to prevent a homeostatic imbalance.

The body's internal environment is the result of a wide range of coordinated processes or *variables,* including the temperature, the chemical composition of blood and other body fluids, and many more. To prevent homeostatic imbalances, most variables in the internal environment are controlled so they stay close to a particular normal value; these variables are known as *regulated variables.*

Quick Check

☐ 1. What is homeostasis, and why is it important?

☐ 2. What is a homeostatic imbalance?

Core Principle One: Feedback Loops Are a Key Mechanism Used to Maintain Homeostasis

A major function of the body's organ systems is to maintain homeostasis to ensure survival. They often do this through homeostatic control mechanisms called **feedback loops,** in which a change in a regulated variable causes effects that *feed back* and in turn affect that same variable. Feedback loops are made up of a series of events that lead to an effect or *output* of some sort. As the loops continue, this output then influences the events of the loops themselves.

This feedback can be either negative (opposing the initial change and reducing the output) or positive (reinforcing the initial change and increasing the output). These are the two types of feedback loops.

Negative Feedback Loops

Maintaining homeostasis is based on a relatively simple principle: If one of the body's regulated variables shifts too far from its normal value, then physiological processes increase or decrease their activity to return it to normal. When a change in a regulated variable is detected, actions are triggered that will bring the regulated variable back toward normal. This type of regulation, in which a change in a regulated variable in one direction results in actions that cause changes in the variable in the opposite direction, is called a **negative feedback loop** (or *mechanism*), and this is one of the primary means for maintaining homeostasis in the body.

Each regulated variable has its own **set point,** or normal value. For example, normal body temperature is 37° C (98.6° F), and normal blood pressure is 110/70 mm Hg (millimeters of mercury, a way of measuring pressure). However, in reality, it's not an exact value that is maintained, but rather a **normal range** around that set point, as shown here:

The ranges differ for individual variables, with some having wider ranges and others, narrower ones. Note also that normal ranges for regulated variables can vary between individuals.

A regulated variable has to move outside its normal range in order to trigger a negative feedback mechanism. For example, if your body temperature on a hot day was 37.2° C, your body would probably not initiate a negative feedback loop, as this is usually within the normal range. However, if your body temperature reached 37.4° C, those mechanisms would probably kick in, and your body would start sweating to lose heat and so return your temperature to within the normal range.

Let's see how a negative feedback loop works in general. When a regulated variable is outside its normal range, this information (called a **stimulus**) is picked up by a cellular structure called a **receptor, or sensor.** The receptor can then send this information to a **control center** by a variety of mechanisms, usually requiring the use of cells of the nervous or endocrine system. The control center is often cells in the brain or an endocrine organ called a gland. The control center compares the current value of the variable to its set point, and determines that it's out of the normal range. The control center then sends signals to cells or organs, called **effectors,** that cause physiological **responses** that return the variable to the normal homeostatic range.

As the variable returns to normal, there is no longer a stimulus for the negative feedback loop, and so the control center allows the activity of the effectors to return to normal levels. The activity of the effectors will continue at normal levels until another change in a variable is detected, at which point the negative feedback loop will again be activated.

Negative feedback loops don't just occur in the human body; they can also be found in mechanical devices. A common example of a mechanical negative feedback loop is the control of room temperature (**Figure 1.13a**). Let's say you set the thermostat in your house to 21° C (70° F), the set point. In the winter, cold temperatures outside cause the temperature inside the house to drop below this set point. The receptor, in this case the thermometer, detects the room's current temperature and sends this information via electrical wires to the control center, which is probably a computer chip in the thermostat. The thermostat compares this current value to the set point. When it determines that the room temperature is too far from the set point, it sends commands via wires to turn on the effector, which in this case is the heater. The heater then produces the response, which is heat to warm the house; this moves the room temperature back toward the set point. Once the temperature has returned to the set point, the thermostat turns the heater off.

Now we'll compare an example in the body. **Figure 1.13b** shows how body temperature is controlled through a negative feedback loop. Say you go outside without a coat on that cold day, and your body temperature drops below the normal range. This stimulus is detected by receptors in cells of the brain. The brain cells, which in this case also act as the control center, then compare the information about body temperature with the set point, and find that it's too far outside its normal range. Therefore, the brain cells activate other nerve cells that send signals to skeletal muscle cells, the effectors, which start shivering, This shivering produces heat, causing the body temperature to rise. When the body temperature returns to the normal range, the stimulus ends, and the brain cells eventually stop their signals to the muscles. There are also different negative feedback loops that are initiated if body temperature increases above the normal range.

Positive Feedback Loops

Although less common, **positive feedback loops** also occur in the body and contribute to homeostasis. The key difference between negative and positive feedback loops is that in a negative feedback loop, the effector activity opposes the initial stimulus and shuts off when conditions return to the normal range. However, in a positive feedback loop, the effector's activity actually *increases*—positive feedback reinforces the initial stimulus using a loop of increasing output that amplifies the response. A positive feedback loop therefore causes a rapid change in a variable.

You might be wondering how this process can contribute to homeostasis. Positive feedback loops are often found within a

STIMULUS

Room temperature decreases below set point of 70° F.

Set point

RECEPTOR

Thermometer detects change in temperature.

Information sent

CONTROL CENTER

Thermostat receives information and determines current temperature is below set point.

Signal sent

RETURN TO SET POINT

As room temperature returns to set point, feedback turns off heater.

Set point

EFFECTOR/RESPONSE

Heater turns on, produces heat, room warms.

(a) Control of room temperature by a negative feedback loop

STIMULUS

Body temperature decreases below normal range.

Normal range

RECEPTOR

Receptors in brain cells detect change in temperature.

Information sent

CONTROL CENTER

Brain cells receive information and determine body temperature is below normal range.

Signal sent

IN HOMEOSTATIC RANGE

As body temperature returns to normal, feedback stops shivering of skeletal muscles.

Normal range

EFFECTOR/RESPONSE

Nerve cells activate skeletal muscle cells to start shivering, producing heat.

(b) Control of body temperature by a negative feedback loop

Figure 1.13 Comparison of how negative feedback mechanisms control room and body temperature.

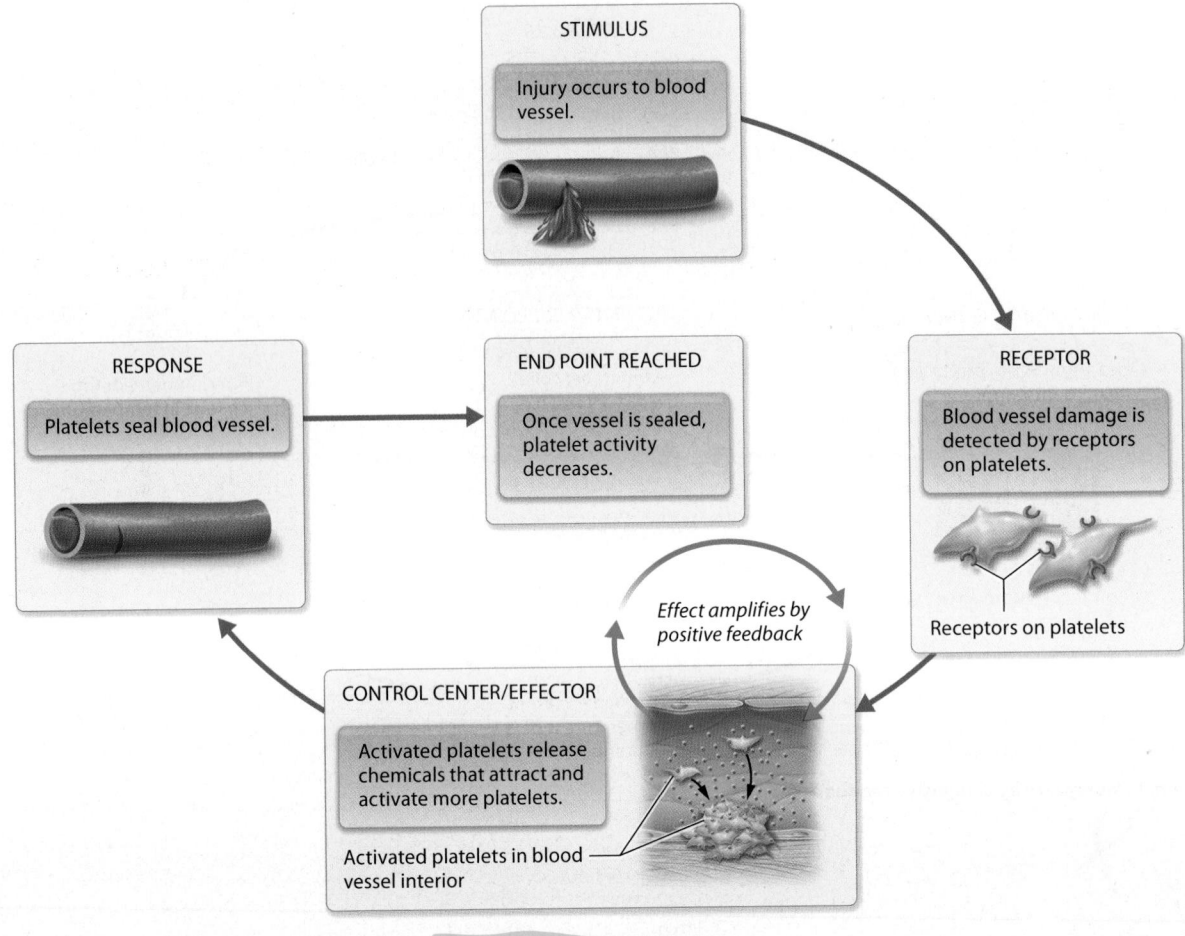

Figure 1.14 **Control of blood clotting by a positive feedback mechanism.**

negative feedback loop to produce a quicker response. A common example is the formation of a blood clot (Figure 1.14). As you can see, this loop begins with an injury to a blood vessel that causes blood loss. Damage to the blood vessel wall activates receptors on tiny cellular fragments in the blood called *platelets*. The activated platelets act as both control centers and effectors, releasing chemicals that attract and activate more platelets to aggregate and plug the wound, and these additional platelets attract and activate more platelets, and so on, amplifying the effect. This continues until the response is complete, when the vessel is sealed by a blood clot and the blood loss stops. In this example, positive feedback occurred when platelets stimulated other platelets to form the clot. The overall effect, however, is negative feedback, as the stimulus—the injury and the bleeding it caused—was stopped.

Although the effects in a positive feedback loop are amplified, they do not continue indefinitely. Every correctly functioning positive feedback loop eventually reaches a point at which positive feedback ceases in response to an external stimulus (an event that is not part of the loop). In the case of blood clotting, the system shuts down once the torn vessel is sealed, thereby removing the stimulus that started the positive feedback loop. Other positive feedback loops operate in a similar manner and shut down when some end point is reached or some outside event intervenes. Read about the

positive feedback loop that results in the birth of a baby in *A&P in the Real World: Childbirth, Pitocin, and Positive Feedback Loops* on page 26.

ConceptBOOST »»»

Debunking Some Common Misconceptions about Homeostasis*

As you've read, homeostasis is the overall theme of human anatomy and physiology. With such a broad concept, misconceptions are bound to happen. Here are a few common misconceptions and the truth behind them:

- **Misconception 1: Negative feedback is bad for the body; positive feedback is good.** When most of us hear the term "negative," we think of bad consequences. Likewise, when we hear "positive," good consequences come to mind. However, remember that the terms negative feedback and positive feedback refer simply to the direction of the output in response to an initial change. Under normal conditions, both negative feedback and positive feedback promote homeostasis.
- **Misconception 2: Maintaining homeostasis means that the body's internal environment is static or unchanging.**

Though we talk about the maintenance of "set points" or "normal ranges," it doesn't mean that the body's internal environment is unchanging. Think about what happens in your body when you sit down to eat dinner. The levels of sugar and other nutrients in your blood increase, your water balance increases, the levels of various other chemicals in your blood change, your digestive system undergoes multiple physiological processes to digest the food, and more. That's hardly a static environment, and these sorts of changes are occurring constantly.

- **Misconception 3: Regulatory mechanisms and feedback loops are either "on" or "off," like a switch.** Since the body's internal environment is a dynamic place, feedback loops are constantly engaged in some degree of activity. A good example is blood sugar regulation. The level of sugar in the blood is constantly changing, and the negative feedback loops that regulate blood sugar therefore must always be in operation.

- **Misconception 4: Any physiological variable can be controlled.** A physiological variable can be controlled through feedback loops only if cells with receptors exist to detect changes in a set point. In addition, changes in an unregulated variable can disrupt the homeostasis for a regulated variable. For example, vitamin D is an important chemical involved in calcium ion homeostasis. Receptors exist for the level of calcium ions, but not for vitamin D. So if the vitamin D level falls too low, the body has no mechanism to return the level to normal. A lower vitamin D level, in turn, impacts the level of calcium ions in the body. ■

*Jenny McFarland, Joel Michael, Mary Pat Wenderoth, Harold Modell, Ann Wright and William Cliff, "Conceptual Framework and Misconceptions Associated with the Core Principles of Homeostasis," Human Anatomy & Physiology Society workshop (2012).

Quick Check

☐ 3. How do negative feedback loops maintain homeostasis?
☐ 4. Explain how positive and negative feedback loops differ.

Core Principle Two: Structure and Function Are Related at All Levels of Organization

One of the most basic principles in anatomy and physiology is known as the **principle of complementarity of structure and function.** This principle may be stated more simply as *form follows function:* The form of a structure is always such that it best suits its function. This is easy to see at the organ level of organization—blood vessels are hollow tubes that can transport blood through the body, hard and strong bones support and frame the body, and the hollow and muscular urinary bladder stores and expels urine. However, this principle applies at all levels of organization, even down to the chemical level.

Take a look at **Figure 1.15**, which shows this structure-function relationship in the tissues of the lung. You can see that the lung tissue is extremely thin. This form follows its function of allowing gases such as oxygen and carbon dioxide to cross. If the lung tissue were thick, gases would take too long to cross and our cells would suffocate from lack of oxygen. We will revisit this principle and encounter additional examples in each chapter of this book.

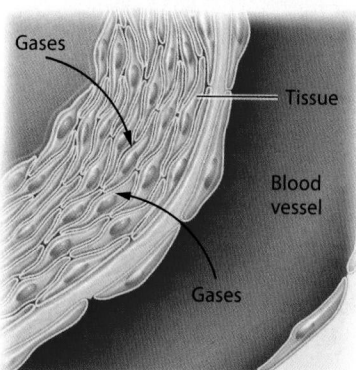

⊘ The tissues in the lungs are thin, allowing gases to cross rapidly.

⊗ If the tissues were thick, gases would have difficulty crossing rapidly enough to maintain homeostasis.

Figure 1.15 **The relationship between structure and function. This figure explores structure and function using the example of what would happen if the tissues in the lung were thick instead of thin.**

☐ 5. How are structure and function related? To what levels of organization does this relationship apply?

A&P in the Real World

Childbirth, Pitocin, and Positive Feedback Loops

Pregnancy generally lasts about 40 weeks, after which the process of childbirth begins with *labor*. Labor occurs by a positive feedback loop, whereby several stimuli, including the baby's head stretching the cervix, cause the control center in the brain to send signals to the uterus. The uterus responds by producing receptors for the hormone *oxytocin* (ok-sih-TOH-sin), which stimulates contractions of the uterus. Uterine contractions, in turn, move the baby's head, causing more stretching of the cervix, which stimulates the release of more oxytocin, and the effect continues to be amplified until the baby is born.

Sometimes it becomes necessary to initiate the labor process artificially, which is known as labor induction. Induction is carried out with the help of the drug Pitocin (pih-TOH-sin), the synthetic version of oxytocin. Pitocin stimulates uterine contractions, which then initiate the positive feedback loop that leads to the release of more oxytocin, progressively stronger contractions, and eventual childbirth.

Core Principle Three: Gradients Drive Many Physiological Processes

A **gradient** is present any time more of something exists in one area than another and the two areas are connected. **Figure 1.16** shows three examples of gradients; these same types of gradients exist in the human body. Figure 1.16a shows a *temperature gradient,* in which heat is concentrated in a room closest to a heater and the temperature gradually decreases as we move away from it. Figure 1.16b shows a *concentration gradient,* in which a pill is dissolving in a beaker of water. A concentration gradient is present when the amount of a substance changes over a range. In Figure 1.16b, the area around the pill has a high concentration of the medication, and the concentration gradually decreases as we move away from the pill. Finally, Figure 1.16c shows a *pressure gradient,* in which we have a syringe with a plunger pushing down. The contents of the syringe are being compressed, and are under high pressure. The air outside the syringe is not as compressed and is therefore under lower pressure. This pressure difference is the pressure gradient.

Gradients form the cornerstone of much of our physiology, as they drive many of our physiological processes. Everything from respiration to nutrient exchange in the blood vessels to the formation of urine is driven by gradients. We explore the roles of different types of gradients in nearly every chapter in which physiology is prominent.

☐ 6. Define gradient, and give examples of different types of gradients.

Temperature gradient
80° 75° 70°

Heater

Heat is concentrated closer to the heater, showing a temperature gradient.

(a) Temperature gradient

Concentration gradient

Pill

The dissolving pill is more concentrated in the area around the pill and decreases away from it, showing a concentration gradient.

(b) Concentration gradient

Pressure gradient

Higher pressure   Lower pressure

The higher air pressure inside the compressed syringe and the lower air pressure outside show a pressure gradient.

(c) Pressure gradient

Figure 1.16 Examples of gradients.

Core Principle Four: Cell-Cell Communication Is Required to Coordinate Body Functions

Cells in the body must work together in a coordinated fashion to ensure that homeostasis of the entire organism is maintained. Just as you and a friend can coordinate a study date only if you communicate in some way, cells can work together only if they communicate with one another. Such communication generally comes in the form of electrical signals or chemical messengers, in which one cell triggers a response from another cell. Electrical signals are typically transmitted directly between neighboring cells. Chemical messengers, however, may be released from one cell directly onto another cell or into the fluid surrounding another cell, or they may reach another cell after traveling through the blood. In this way, cells can communicate with one another irrespective of their location in the body and the distance between them.

An example of cell-cell communication is shown in Figure 1.17, in which a nerve cell is stimulating a muscle cell. The nerve cell releases chemical messengers into the space near the muscle cell; these messengers trigger changes that lead to a muscle contraction. This type of cell-cell communication is the main way in which the cells of the nervous system communicate with other cells of the body. We will see many other types of cell-cell communication throughout this book.

Each of the four core principles is highlighted throughout the book with a core principles icon, which you can see in Figure 1.18. Keep on the lookout for these icons to remind yourself of the overall theme of human anatomy and physiology—homeostasis—and the core principles that revolve around maintaining it.

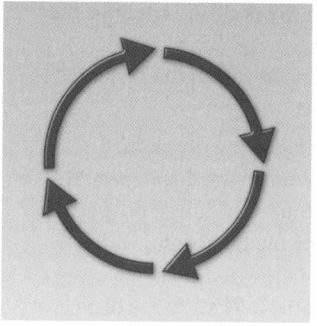

Figure 1.17 **Communication between a nerve cell and a muscle cell.**

Nerve cell

The nerve cell releases chemical messengers into the space near the muscle cell.

Muscle cell

The chemical messengers trigger a response in the muscle cell.

Quick Check

☐ 7. Why is cell-cell communication important?
☐ 8. What are the two major methods by which cells communicate?

Apply What You Learned

☐ 1. The cells of cancerous tumors undergo changes that cause them to bear little to no resemblance to their original cell type. Predict how this will affect the ability of these cells to perform their functions. (*Hint:* Consider the structure-function relationship.)
☐ 2. Explain what would happen if body temperature increased above its normal range, but, rather than being regulated by a negative feedback loop, it was instead regulated by a positive feedback loop.
☐ 3. You plug in an electric air freshener and notice that you can smell the air freshener near the electrical outlet but not on the other side of the room. Explain how this represents a gradient.

See answers in Appendix A.

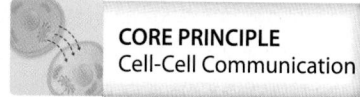

CORE PRINCIPLE Feedback Loops

CORE PRINCIPLE Structure-Function

CORE PRINCIPLE Gradients

CORE PRINCIPLE Cell-Cell Communication

Figure 1.18 **Core principles icons.**

CHAPTER SUMMARY

Take full advantage of the invaluable online practice and study tools for this chapter by going to MasteringA&P®. There you will find:

- Practice tests, module quizzes, art-labeling and sequence-the-steps activities
- Animated figures with quizzes
- *Practicing A&P* worksheets
- Author-narrated podcast
- Dynamic Study Modules
- iP *Interactive Physiology*

MODULE 1.1
How to Succeed in Your Anatomy and Physiology Course 1

- The four general learning styles are **visual/verbal, visual/nonverbal, auditory/verbal,** and **tactile/kinesthetic.** Each learning style is characterized by specific ways in which an individual learns best.
- One of the best ways to approach reading a textbook is to use the SQ3R method: do an initial *survey,* form *questions, read* the chapter actively, *recite* the material, and *review* the material.
- Make a schedule so that you can budget your time effectively and give yourself adequate time to study for an exam or quiz. (Figure 1.1, p. 3)
- In studying for an exam, first find out all you can about the exam, take advantage of the resources available to you through this book and your school, form a study group, use a variety of study techniques that suit your learning style, and take care of your physical needs.
- You can make the best use of class and laboratory time by coming prepared.
- This book contains several key features to assist you in mastery of anatomy and physiology. (Figure 1.2, p. 5)
- The best way to interpret the figures is as follows: examine them during your initial chapter survey, identify the concept that each figure teaches, break the figure into parts and interpret each part, then put the figure together and interpret it as a whole. The figure may be combined with other figures to gain a more global understanding of a structure or process—see Big Picture figures. (Figure 1.3, p. 7)

MODULE 1.2
Overview of Anatomy and Physiology 7

- Human **anatomy** is the study of the form of the human body; human **physiology** is the study of its functions.
- Characteristics common to all forms of life are cellular composition, metabolism, growth, excretion, responsiveness, movement, and reproduction.
- The six levels of organization of the body are the **chemical, cellular, tissue, organ, organ system,** and **organism** levels. There are 11 major organ systems in the body. (Figure 1.4, p. 9; Figure 1.5, p. 10)

- Anatomy may be studied from a systemic point of view or from a regional point of view. Subfields of anatomy include **gross anatomy** and **microscopic anatomy.** Microscopic anatomy can be divided into **cytology** and **histology.**

MODULE 1.3
The Language of Anatomy and Physiology 12

- The common frame of reference from which all body positions are described is the **anatomical position.**
- Directional terms are used to describe body parts and body markings. The common pairs include: **anterior** and **posterior** (front and back); **superior** and **inferior** (toward the head and away from the head); **proximal** and **distal** (closer to the point of origin and farther away from the point of origin); **medial** and **lateral** (closer to the midline and farther away from the midline); and **superficial** and **deep** (closer to the body's surface and farther below the body's surface). (Figure 1.6, p. 13)

  - Practice art-labeling exercise: Figure 1.6.

- The body is divided into two main regions: the **axial region,** which includes the head, neck, and trunk, and the **appendicular region,** which includes the upper and lower limbs. The body is also divided into more specific regions. (Figure 1.7, p. 14)

  - Practice art-labeling exercise: Figure 1.7.

- There are three main planes of section: the **sagittal plane,** the **frontal plane,** and the **transverse plane.** (Figure 1.8, p. 16)

  - Practice art-labeling exercise: Figure 1.8.

MODULE 1.4
The Organization of the Human Body 16

- The **dorsal body cavity** has two divisions, which are continuous with one another and contain *cerebrospinal fluid (CSF).* (Figure 1.9, p. 17)

  - Practice art-labeling exercise: Figure 1.9.

  - The **cranial cavity** is located within the skull and contains the brain.
  - The **vertebral** or **spinal cavity** is located within the vertebral column and contains the spinal cord.
- The **ventral body cavity** contains the following divisions:
  - The **thoracic cavity** is the area superior to the diaphragm. It contains the two **pleural cavities** around the lungs, the **pericardial cavity** around the heart, and the **mediastinum,** which is between the two pleural cavities.
  - The **abdominopelvic cavity** is the area inferior to the diaphragm. It consists of the **abdominal cavity** and the **pelvic cavity.** Certain organs within the abdominal cavity are located within the **peritoneal cavity;** some are retroperitoneal. The abdominopelvic cavity may also be divided into four quadrants or nine regions. (Figure 1.10, p. 18)
- The pleural, pericardial, and peritoneal body cavities are formed by **serous membranes,** continuous sheets of tissue that enclose

a single cavity. The layer of the serous membrane attached to the organ is the **visceral layer**, and the layer attached to surrounding structures is the **parietal layer**. Between the two layers is **serous fluid,** which is a thin, watery liquid that lubricates the membranes so adjacent organs can move in the cavity without friction. (Figure 1.11, p. 19; Figure 1.12, p. 19)

Practice art-labeling exercise: Figure 1.12.

MODULE 1.5
Core Principles in Anatomy and Physiology 21

- Overall Theme: Homeostasis. The body's processes work to keep the regulated variables of the internal environment within set ranges, a condition known as **homeostasis.**

- Core Principle One: Feedback Loops. **Feedback loops** are a series of events that function to restore homeostasis. Each process or variable in the body is maintained within a **normal range** of a **set point.** Stimuli that cause deviations from normal ranges are monitored by **receptors** or **sensors,** which send information to a **control center.** This center mediates changes that work to restore conditions to the normal range through an **effector** that initiates some sort of output. The output feeds back to influence the process itself and restore homeostasis.
  - ○ **Negative feedback loops** are the more common type of feedback loop. This kind of process returns a regulated variable to within the normal range, and the activity of the effector then declines. (Figure 1.13, p. 23)
  - ○ **Positive feedback loops** feature an escalating response to a deviation from a normal range. When the effector response is triggered, it triggers additional effectors, which amplifies the response. This cycle is ended by an outside event. (Figure 1.14, p. 24)

- Core Principle Two: Structure and Function. Structure and function are related at all levels of organization, a concept known as the **principle of complementarity of structure and function.** (Figure 1.15, p. 25)

- Core Principle Three: Gradients. We find a **gradient** any time there is more of something in one area than in another, including heat, a substance, or pressure. Gradients drive many of our physiological processes. (Figure 1.16, p. 26)

- Core Principle Four: Cell-to-Cell Communication. Cells communicate to coordinate their functions and maintain homeostasis via chemical messengers and/or electrical signals. (Figure 1.17, p. 27)

- Figure 1.18, p. 27, introduces the icons that will be used to point out instances where these four core principles apply throughout this text.

ASSESS WHAT YOU LEARNED

See answers in Appendix A.

LEVEL 1 Check Your Recall

1. Fill in the blanks: The study of the form of the body is _____; the study of its functions is _____.

2. Mark the following statements as true or false. If a statement is false, correct it to make a true statement.
   a. Auditory/verbal learners tend to learn best from diagrams and figures.
   b. You should budget your time for studying at least two days before an exam.
   c. Textbooks are best read using the survey, question, read, recite, and review method.
   d. It is better to forgo sleep in favor of studying before an exam.

3. Explain why it is important to come to class prepared.

4. Which of the following resources does the book and its associated materials offer to help you succeed?
   a. Concept Boosts
   b. Multiple opportunities for assessment
   c. Companion workbook
   d. Both a and b are correct.
   e. All of the above are correct.

5. Groups of many cells working together to perform a common function are known as a(n):
   a. cell.
   b. organ.
   c. organ system.
   d. tissue.

6. Which of the following correctly describes the functions of the endocrine system?
   a. Transport of blood through the body and through the lungs
   b. Regulation of body functions through hormone secretion
   c. Regulation of body functions through nerve impulses
   d. Immunity and returning extra tissue fluid to the blood vessels

7. Mark the following statements as true or false. If a statement is false, correct it to make a true statement.
   a. Histology is the division of microscopic anatomy that studies the cellular level of organization.
   b. All living organisms are composed of one or more cells.
   c. Living organisms react to changes in their environment known as stimuli.
   d. The anatomical position features the person facing forward, feet shoulder width apart, and palms facing posteriorly.

8. Match the following terms with the correct definition.
   _____ Anterior   a. Toward the back
   _____ Lateral   b. Closer to the midline
   _____ Proximal   c. Farther away from the point of origin
   _____ Posterior   d. Toward the front
   _____ Inferior   e. Closer to the point of origin
   _____ Distal   f. Toward the head
   _____ Medial   g. Away from the body's midline
   _____ Superior   h. Toward the tail

9. The upper and lower limbs are known broadly as the _____ region.
   a. appendicular
   b. axial
   c. crural
   d. dorsal

10. The arm is known as the _____ region; the neck is known as the _____ region.
    a. otic; axillary
    b. antebrachial; dorsal
    c. brachial; cervical
    d. sural; crural

11. A parasagittal section divides the body or body part into:
    a. equal right and left parts.
    b. front and back parts.
    c. superior/proximal and inferior/distal parts.
    d. unequal right and left parts.

12. Fill in the blanks: The two divisions of the dorsal body cavity are the _____ and _____ body cavities.

13. Fill in the blanks: The two main divisions of the ventral body cavity are the _____ and _____ body cavities.

14. In which of the following cavities do serous membranes envelop the organs? (Circle all that apply.)
    a. Mediastinum
    b. Pericardial cavity
    c. Peritoneal cavity
    d. Abdominal cavity
    e. Pelvic cavity
    f. Pleural cavity

15. Serous fluid functions in:
    a. providing temperature stability.
    b. lubricating serous membranes as organs move in the cavity.
    c. protecting the organs from mechanical trauma.
    d. decreasing the weight of the organs and keeping them buoyant.

16. Which organs would you expect to find in the left upper quadrant?
    a. Spleen, liver, appendix
    b. Liver, ovary, spleen
    c. Spleen, stomach, pancreas
    d. Gallbladder, liver, large intestine

17. Mark the following statements as true or false. If a statement is false, correct it to make a true statement.
    a. Structure and function are closely related at the organ level only.
    b. Homeostasis is the condition of having a dynamic equilibrium of the internal environment.
    c. Positive feedback loops are triggered by a deviation from a homeostatic set point and are shut down when conditions return to the set point.

    d. The effects of negative feedback loops are amplified to create an escalating response.
    e. Cell-cell communication is required to coordinate the activities of the whole body.

18. A gradient exists:
    a. when heat is concentrated in one area of the body.
    b. when more of a substance is located in one place than in another.
    c. when there is higher pressure in one place than in another.
    d. All of the above are correct.
    e. None of the above are correct.

LEVEL 2 Check Your Understanding

1. Examine the structure of the skull, and predict its likely functions based on its form.

2. Use the correct regional and directional terms to describe the location of each of the following organs in the body. You may use Figure 1.7 for reference.
   a. Esophagus
   b. Brain
   c. Urinary bladder (in a female)

LEVEL 3 Apply Your Knowledge

1. Ms. Curtis presents to the clinic with left upper quadrant pain. Predict the organs and cavities that may be involved in causing her pain.

2. During a procedure on Ms. Curtis's pancreas, a surgeon makes the initial incision in the left anterior hypochondriac region. List all of the organs, serous membranes, and body cavities that the surgeon will encounter as she moves through the body to get to the pancreas. (*Hint*: Refer to *A&P in the Real World: Abdominal Pain* on page 20 for help.)

3. Later that same day, the surgeon performs a procedure on Ms. Curtis's right kidney. She makes the incision in the right posterior lumbar region. Will she cut through the same serous membrane(s) and cavities as in the previous procedure? Why or why not? How would this change if the incision were made on the anterior lumbar region?

4. The *baroreceptor reflex* causes blood pressure to drop when it rises dangerously high. Predict whether this is a positive or negative feedback loop. Explain your reasoning.

2

The Chemistry of Life

As you begin this chapter, you may wonder why we are starting with chemistry instead of something like bones or digestion or respiration—topics that seem to fit better with the study of the human body. In fact, why study chemistry at all in a human anatomy and physiology course? Well, consider the following examples:

- Bones are strong because they are composed of *minerals* and the *protein collagen*.
- The food we eat is broken down by *chemical reactions* with the help of *acid* and *enzymes*.
- We breathe to take in *oxygen* so our cells can convert the *chemical energy* in food to a form the cells can use.

Even if some of the words in these statements seem foreign to you, one common thread should be clear: We are talking about chemistry (the words in italics) with each statement.

In the first chapter of this book, you learned that the chemical level of organization is the most basic one in the body (see Figure 1.4). What this means is that we are made of chemicals, and no modern study of anatomy and physiology could be carried out without discussing them. Chemistry is the foundation not only of our bodies but also for all of the concepts in this course.

MODULE **2.1**

Atoms and Elements

Learning Outcomes

1. Describe the charge, mass, and relative location of electrons, protons, and neutrons.
2. Compare and contrast the terms atoms and elements.
3. Identify the four major elements found in the human body.
4. Distinguish atomic number from mass number.
5. Compare and contrast the terms isotopes and radioisotopes.
6. Explain how isotopes are produced by changing the relative number of specific subatomic particles.

Matter is defined as anything that has mass and occupies space. Matter can exist in three states, as a *solid*, a *liquid*, or a *gas*. **Chemistry** is the study of matter and its interactions, and so we begin our exploration of the world of chemistry with matter. In this module, we introduce the components and properties of matter and discuss how they apply to the human body.

Computer-generated image: All of the structures that make up our bodies, such as the strand of DNA shown in this image, are composed of chemicals.

Atoms and Atomic Structure

An **atom** is defined as the smallest unit of matter that still retains its original properties. Atoms certainly are tiny—an average human cell has roughly 100 trillion atoms. They are themselves composed of even smaller parts called **subatomic particles.** Let's examine the subatomic particles and how they are organized within an atom.

Subatomic Particles

There are three basic subatomic particles, which are illustrated in **Figure 2.1**. These particles are as follows:

- **Protons.** The positively charged **protons** (p^+) reside in the central core of the atom, called the **atomic nucleus** (NEW-klee-us).
- **Neutrons.** Notice in Figure 2.1 that protons are not alone in the atomic nucleus—they are joined by a second type of subatomic particle: the **neutrons** (n^0; NEW-tronz). Neutrons are just slightly larger than protons and, as their name implies, have no charge.
- **Electrons.** Surrounding the atomic nucleus are the tiny, negatively charged **electrons** (e^-).

All atoms are electrically neutral, meaning an atom as a whole has no charge. The number of positively charged protons in an atom equals the number of negatively charged electrons. However, the number of protons and the number of neutrons in an atom do not have to be equal.

The bulk of an atom's mass comes from its protons and neutrons. Electrons are so small that they contribute practically no mass to an atom, and 99.95% or more of an atom's mass is in the nucleus. Note, however, that the nucleus makes up only a tiny fraction of an atom's total volume, and most of the atom is simply empty space. As an analogy, think of a football stadium: The entire stadium is the atom, a marble at its center is its nucleus, and the specks of pollen floating around in the stadium are its electrons. Everything except the marble and the pollen grains is just empty space.

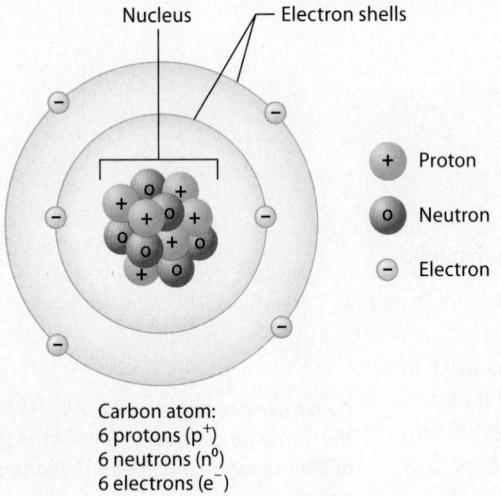

Carbon atom:
6 protons (p^+)
6 neutrons (n^0)
6 electrons (e^-)

Figure 2.1 Structure of a representative atom.

Electron Shells

You may be familiar with the classical atomic model in which the atom looks like a miniature solar system, with the electrons traveling in circular orbits around the nucleus. We often still use that model to draw atoms for the sake of simplicity, but it is not entirely correct. Although we can determine an electron's distance from the nucleus, we can't determine precisely where the electron is going to be found or the path that it is going to take. However, we can predict the likelihood or *probability* of finding electrons in a given area, and loosely define regions where they are likely to be found. These regions, shown in Figure 2.1, are termed **electron shells.**

Each shell can hold a certain maximum number of electrons:

- the first shell, which is closest to the nucleus, can hold 2 electrons;
- the second shell can hold 8 electrons; and
- the third shell can hold 18 electrons (but is "satisfied" with only 8).

The atom shown in Figure 2.1 has two shells. The number of shells an atom has depends on its number of electrons—the electrons fill the shells closest to the nucleus first. Many atoms have more than three shells, but for simplicity we will restrict our discussion in this chapter to those with three or fewer shells.

Quick Check

- [ ] 1. What is matter?
- [ ] 2. What are atoms?
- [ ] 3. How do the three types of subatomic particles differ?

Elements in the Periodic Table and the Human Body

The number of protons that an atom contains is called its **atomic number.** An atom's atomic number defines its particular **element,** which is a substance that cannot be broken down into simpler substances by chemical means. Each element is made of atoms with the same atomic number; therefore, the atoms of different elements have different numbers of protons. For example, every atom of the element oxygen has 8 protons and every atom of the element iron has 26 protons. There are 92 elements that occur naturally.

Elements are listed in order of increasing atomic number in the **periodic table of the elements,** which is shown in abbreviated form in **Figure 2.2**. This way of organizing the elements shows that certain properties of the elements repeat in a regular way, or *periodically.* Elements are listed on the periodic table by an abbreviation called a *chemical symbol.* Generally, the chemical symbol is the first letter or two of the element's name, such as C for carbon or Ca for calcium. However, you'll find that some symbols are based on the element's Latin or German name, such as sodium's symbol, Na, which is from the Latin word *natrium.*

Very few elements are found in significant quantities in the human body (see Figure 2.2). In fact, four *major elements*

These are the rare earth elements, which are not found in the human body.

Figure 2.2 Elements in the human body and their positions in the periodic table.

account for 96% of the body's mass: Oxygen (O) makes up 65%, carbon (C) 18%, hydrogen (H) 10%, and nitrogen (N) 3%. The seven *mineral elements* account for less than 4% of the total mass of the body; these elements are sodium (Na), potassium (K), calcium (Ca), chlorine (Cl), magnesium (Mg), phosphorus (P), and sulfur (S). The 13 *trace elements,* which include iron (Fe), copper (Cu), iodine (I), and zinc (Zn), are found in extremely small amounts in the body.

Notice in Figure 2.2 that a red dividing line that looks like a staircase cuts across the right side of the periodic table. This "staircase" divides the periodic table into two sections: The section to the left contains metals, and the section to the right contains nonmetals. *Metals* are shiny substances that conduct electricity, whereas *nonmetals* exist as gases or brittle solids and are usually poor conductors of electricity. Metals and nonmetals differ in how they interact with other elements, which we discuss in Module 2.2.

Note that hydrogen is unusual in that it behaves differently than you would expect from its properties. Hydrogen's properties make it like the metals, and it is placed with them in the periodic table. However, it doesn't experience the right conditions here on Earth to act like a metal. Elemental hydrogen requires extreme pressures and temperatures to exist as a metal; in fact, you would have to go to the core of Jupiter to find conditions extreme enough to produce metallic hydrogen. For this reason, hydrogen on Earth acts like a nonmetal when it interacts with other elements.

Quick Check

☐ 4. What is an element?
☐ 5. How are elements arranged in the periodic table?

Isotopes and Radioactivity

Recall that the neutrons in the nucleus of an atom do not contribute to the atom's overall charge or the placement of the element in the periodic table, but they do contribute to the atom's mass. The mass of an atom is represented by a number called the **mass number,** which is equal to the number of neutrons plus the number of protons in the atom. However, atoms can gain or lose neutrons. When this happens, an atom's mass number changes but its atomic number remains the same—in other words, it is still the same element. Atoms with the same atomic numbers but different mass numbers are called **isotopes** (EYE-soh-towpz; *iso-* = "same"). For example, hydrogen has three isotopes: Regular hydrogen has no neutrons, *deuterium* (doo-TEER-ee-um) has one neutron, and *tritium* (TRIT-ee-um) has two neutrons. The three isotopes of hydrogen are written this way:

$$\begin{array}{ccc} {}_{1}^{1}\text{H} & {}_{1}^{2}\text{H} & {}_{1}^{3}\text{H} \\ \textit{Hydrogen} & \textit{Deuterium} & \textit{Tritium} \end{array}$$

The upper number is hydrogen's mass number, which varies depending on the number of neutrons. The lower number is hydrogen's atomic number, which doesn't change. Another way to represent an isotope is to write the name of the element followed by the isotope's mass number. For example, $_{53}^{131}\text{I}$ may also be written as iodine-131.

Certain isotopes have very high energy, which makes them unstable. Such isotopes will release this energy in the form of *radiation*. This energy release, called *radioactive decay,* allows the isotope to assume a more stable form. Isotopes that do this are called **radioisotopes,** and they form the basis for the field of *nuclear medicine,* the use of radioactive materials in medical diagnosis and treatment. See *A&P in the Real World: Nuclear Medicine* on page 34 for more information.

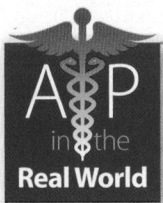

Nuclear Medicine

Some of the more common applications of radioisotopes in nuclear medicine are as follows:

- **Cancer radiation therapy.** The radiation from certain radioisotopes damages the structural features of cancer cells and interferes with their functions.
- **Radiotracers.** Many radioisotopes are used as *radiotracers*. Radiotracers are injected into a patient, and the radiation they produce is detected by a camera and analyzed by a computer. The resulting picture shows the size, shape, and activity of certain organs and the cells in the blood. For example, shown here is a colored gamma scan of a healthy human skeleton seen from the front (left) and back (right):

- **Treatment of thyroid disorders.** High doses of the radioisotope iodine-131 are used to treat overactive thyroid and thyroid cancers. The isotope accumulates in the thyroid gland and damages the cells, interfering with their functions.

Quick Check

☐ 6. What are isotopes?

Apply What You Learned

☐ 1. You have collected two equal-size samples of carbon, both of which are pure (i.e., they contain no other elements). However, when you measure them, you find that one has a greater mass than the other. How do you explain this?

☐ 2. The element lithium has an atomic number of 3 and a mass number of 7. How many protons, neutrons, and electrons does a lithium atom have?

See answers in Appendix A.

MODULE **2.2**

Matter Combined: Mixtures and Chemical Bonds

Learning Outcomes

1. Distinguish among the terms solution, solute, solvent, colloid, and suspension.
2. Relate the number of electrons in an atom's valence shell to its chemical stability and ability to form chemical bonds.
3. Define the terms molecules and ions.
4. Explain the mechanisms of ionic and covalent bonds, and discuss their relative strengths.
5. Explain the difference between polar and nonpolar covalent bonds.
6. Describe how hydrogen bonds form and how they give water the property of surface tension.

Matter may be combined in two ways: by being physically intermixed or chemically combined. When atoms of two or more elements are physically intermixed, producing a **mixture,** the chemical nature of those atoms does not change. The atoms retain their original properties and may be physically separated. In contrast, when atoms of two or more elements are combined by forming *chemical bonds*, the atoms are changed chemically. Two or more atoms combined by chemical bonds form a **molecule** (MOL-eh-kyul). Molecules have different properties from those of their original atoms, and can be separated only by chemical means. Let's take a look at these two ways of combining matter and how they apply to the human body.

Mixtures

There are three basic types of mixtures: *suspensions, colloids,* and *solutions* (**Figure 2.3**). All three are found in the human body, so let's examine each more closely.

Suspensions

A **suspension** is a mixture that generally consists of a liquid mixed with a solid. The solid particles in a suspension are usually visible (making the suspension appear cloudy or opaque); if left alone, these particles tend to settle out of the mixture. An example of a suspension in the human body is blood, in which the blood cells are "suspended" in the plasma, the fluid portion of blood (Figure 2.3a). If a blood sample is allowed to sit, the blood

(a) **Suspension:** Particles large and usually visible; settle out.

(b) **Colloid:** Two distinct components; particles small and not visible; do not settle out.

(c) **Solution:** Particles extremely small and not visible; do not settle out; one component dissolves in the other component.

Figure 2.3 **The three types of mixtures.**

cells will settle to the bottom of the container. This is one reason why we need our heart to keep our blood moving—when it stops beating, the blood cells stop moving, meaning that they can no longer deliver oxygen to our cells.

Colloids

Like suspensions, **colloids** (KOLL-oydz) consist of two distinct components. Colloids appear opaque; however, the solid particles in a colloid are quite small and are not visible with the naked eye (or even with a microscope). Because of their small size, the particles in a colloid remain dispersed and do not settle out. An example of a colloid in the human body is milk (Figure 2.3b). Milk contains numerous proteins (PROH-teenz; large molecules that we'll discuss later in this chapter) that are too small to settle out, but large enough to be considered a separate component. Other examples of commonly encountered colloids include gelatin, aerosols, and cytosol, the fluid material inside our cells.

Solutions

A **solution** in a living organism generally consists of a solid, liquid, or gas mixed with a liquid (usually water). Solutions generally appear clear. A solution is usually described by saying that one substance *dissolves* in another. The substance that dissolves is called the **solute** (SOL-yoot), and the substance in which the solute dissolves is called the **solvent.** For example, in Figure 2.3c, glucose is the solute and water is the solvent. The degree to which a solute dissolves in a solvent is called its *solubility* (sol′-yoo-BIL-ih-tee). Water is the most important solvent in the human body—it makes up most of our blood plasma and the fluid in and around our cells. Numerous substances dissolve in our plasma and other body fluids, and a chemical's solubility in water is a major factor in determining how it is transported in the body.

The amount of solute present in a solution is called the solution's *concentration.* We can express a solution's concentration in several ways; the simplest is by stating the percentage of a solution that is solute. For example, if a solution has 10 grams (g) of salt and 90 milliliters (ml; equal to 90 g) of water, we call it a 10/100 = 10% salt solution; if it has 40 g of salt and 60 ml of water,

we call it a 40/100 = 40% salt solution. Solutions are often compared with one another by their concentrations, and the solution with the higher concentration of solute is called the *more concentrated* solution. In our example, the 40% salt solution is the more concentrated solution.

The individual solute particles in a solution are extremely small and not individually visible, even with a microscope. Due to their small size, the solute particles will not settle out if the solution is left to stand. However, remember that a solution is still a mixture, and that neither the solvent nor the solute changes chemically when they form a solution. So, if we want to separate the salt from the water in our 40% salt solution, we can do so by simple physical means, such as evaporating the water, which leaves the salt behind.

Quick Check

☐ 1. What is a mixture?
☐ 2. How do the three types of mixtures differ?

Chemical Bonds

Elemental sodium (Na) is a soft metal that reacts violently with water, and elemental chlorine (Cl) is a poisonous green gas. But when you take a small piece of sodium metal and drop it into a beaker containing chlorine gas, something interesting happens: A small explosion occurs, the sodium bursts into flames, and when it finishes burning, a white powder is left behind. The white powder that forms, believe it or not, is something that you sprinkle on your food—sodium chloride (NaCl), or table salt. How exactly did two highly toxic elements become something we can eat? The answer is that when the two elements were combined, each one was altered *chemically* by a **chemical bond** formed between them. A bond is not a physical structure but rather an energy relationship between atoms, that is, an attractive force. In this section we cover how bonds form and the different types of bonds that hold molecules together.

Molecules, Compounds, and Molecular Formulas

As we discussed in the module introduction, a molecule forms when two or more atoms are chemically bonded. We use three different terms to refer to molecules of differing size and composition. Two or more atoms of the same element that are chemically bonded are called a *molecule of an element*. For example, O_2 is a molecule of oxygen, two oxygen atoms bonded together. Two or more atoms of different elements that are chemically bonded are called a molecule of a **compound.** When we combined sodium metal and chlorine gas, we formed molecules of the compound NaCl, sodium chloride. Very large molecules composed of many atoms, such as those that make up our hair and nails, are known as **macromolecules.**

In this discussion, we've seen examples of *molecular formulas,* ways to represent molecules symbolically with letters and numbers. Such formulas show the kinds and numbers of atoms in a molecule. For example, in CH_4, the formula for a methane molecule, the use of the element symbols next to each other tells us that this molecule is made up of atoms of carbon and hydrogen, and the subscript number only on H tells us that four atoms of hydrogen—but just one atom of carbon—are found in each molecule. Similarly, in O_2, the subscript 2 tells us that two oxygen atoms are bound to one another in the molecule. All molecules are formed by joining atoms with chemical bonds, but how exactly are these bonds formed?

Valence Electrons and Chemical Bonding

Chemical bonds are formed when the electrons located in the outermost electron shells of atoms interact. This electron shell is called the **valence shell** (VAY-lents), and the electrons in this shell are called the **valence electrons.** These electrons determine how an atom interacts with other atoms and whether it will form bonds with a specific atom. In general, the atoms that are relevant to physiology follow a rule called the **octet rule,** which states that an atom is most stable when it has eight electrons in its valence shell. Atoms that have filled valence shells, such as helium and neon (He and Ne; see the rightmost column of the periodic table), are nonreactive, or *inert*. Atoms that do not meet the octet rule are said to be *reactive*—that is, they are unstable and so will interact with other atoms and fill their valence shells. Note that an exception to this rule exists for atoms with five or fewer electrons; such atoms are most stable with only their first electron shell filled (because the first electron shell holds only two electrons, this exception is called the *duet rule*).

As you could see with the formation of NaCl, a chemical bond can change an element's properties drastically. This is because a chemical bond alters the valence electrons of an atom. The bond allows reactive atoms to obey the octet or duet rule, and the atoms become more stable. Two things may happen to valence electrons when a chemical bond is formed: (1) The electrons may be transferred from one atom to another, which is called an *ionic bond* (EYE-on-ik); or (2) the electrons may be shared between two or more atoms, which is called a *covalent bond*. The upcoming sections examine these two types of chemical bonds.

Quick Check

☐ 3. What are valence electrons, and why are they important?

Ions and Ionic Bonds

An **ionic bond** results when electrons are transferred between a metal atom and a nonmetal atom. Let's continue to use our example of elemental sodium and chlorine to explore how and why these bonds form.

Here you see the basic structure of a sodium atom (Na; note that we've left off the neutrons for simplicity):

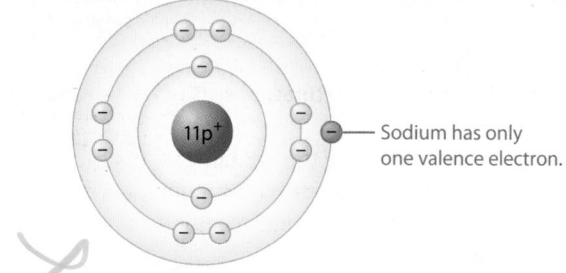

Sodium has only one valence electron.

You can see that a sodium atom's valence shell has only one electron—it does not obey the octet rule, and for this reason, elemental sodium is violently reactive. The sodium can become stable in one of two ways: (1) The sodium atom could give up its lone valence electron, leaving behind its full second shell to become the valence shell, or (2) the sodium atom could steal seven electrons from another atom. The second option would require so much energy that it wouldn't happen naturally. For this reason, we are left with the first option: The sodium atom gives up its lone outer electron.

When this occurs, notice that the sodium still has 11 protons, but now has only 10 electrons. Sodium has lost a negative charge and switched from an electrically neutral atom to a particle with a +1 charge. Such a charged particle is called an **ion** (EYE-on), or more specifically, a positive ion called a **cation** (KAT-eye-on). The positive charge on the sodium ion is represented by adding a superscript plus sign, as Na^+. Note that other metals behave in the same way as sodium—they donate electrons to become cations.

Now let's move on to the chlorine atom (Cl), shown here:

Chlorine has seven valence electrons.

As you can see, the chlorine atom's valence shell has seven electrons—it needs one more electron to obey the octet rule. Therefore, as step ① in **Figure 2.4** shows, the chlorine atom accepts an electron from another atom (in this case sodium),

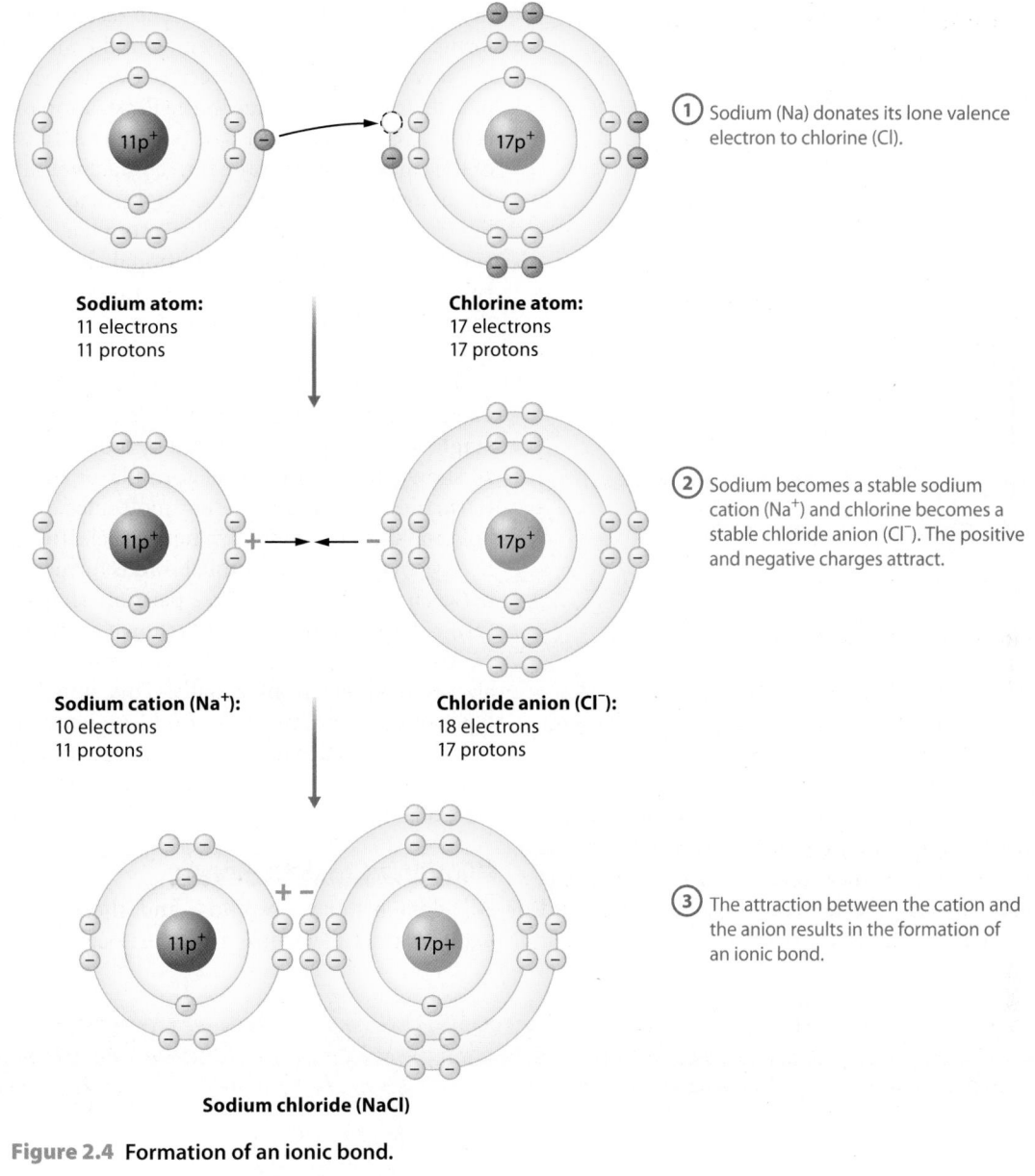

1. Sodium (Na) donates its lone valence electron to chlorine (Cl).

Sodium atom:
11 electrons
11 protons

Chlorine atom:
17 electrons
17 protons

2. Sodium becomes a stable sodium cation (Na$^+$) and chlorine becomes a stable chloride anion (Cl$^-$). The positive and negative charges attract.

Sodium cation (Na$^+$):
10 electrons
11 protons

Chloride anion (Cl$^-$):
18 electrons
17 protons

3. The attraction between the cation and the anion results in the formation of an ionic bond.

Sodium chloride (NaCl)

Figure 2.4 **Formation of an ionic bond.**

giving it eight electrons in its valence shell. The number of protons has not changed, but the extra negative charge turns the chlorine atom into a negative ion, or **anion** (AN-eye-on), with a −1 charge (note that in this state it is called *chloride* and is represented by adding a superscript minus sign, as Cl$^-$). Other nonmetals behave the same way as chlorine, also accepting electrons to become anions.

Now that we have two charged particles, we can get to the actual ionic bond. Stated simply, particles of opposite charge attract one another. In our example, the positively charged sodium ion and the negatively charged chloride ion attract each other (Figure 2.4, step ②). This attraction holds, or "bonds," the two ions together, and they form a compound known as a *salt* (Figure 2.4, step ③; salts are discussed in Module 2.4).

Ionic bonds are easy to identify because metals will *only* bond with a nonmetal via an ionic bond. So, if a metal is in

the molecule, the bond must be ionic. Remember, though, that hydrogen is an exception; it behaves like a nonmetal instead of a metal (unless you are in the core of Jupiter).

Quick Check

☐ 4. What is an ionic bond?
☐ 5. How is an ionic bond formed?

Covalent Bonds

Whereas ionic bonding involves a transfer of electrons from a metal to a nonmetal, **covalent bonding** involves two or more nonmetals *sharing* electrons, which makes this the strongest type of chemical bond. An atom will share electrons until its outer

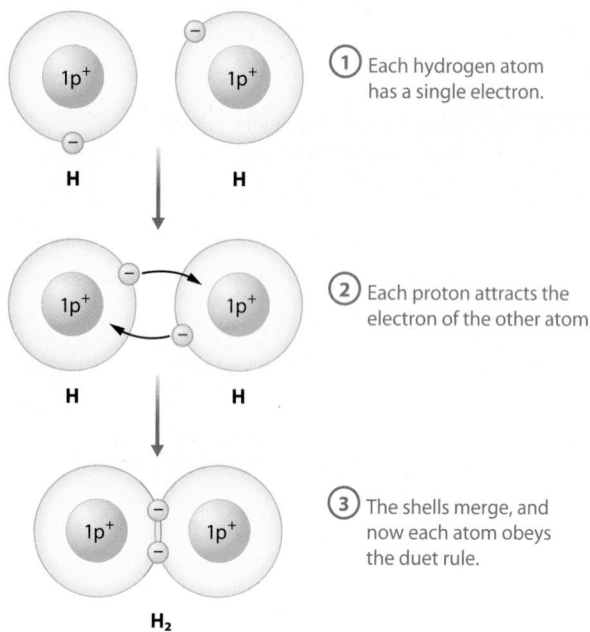

① Each hydrogen atom has a single electron.

② Each proton attracts the electron of the other atom.

③ The shells merge, and now each atom obeys the duet rule.

Figure 2.5 Formation of a covalent bond.

shell is full. **Figure 2.5** shows the formation of a simple covalent bond. In step ① of this figure we have two hydrogen atoms, each with one electron. To become stable, each atom needs another electron to satisfy the duet rule. In step ②, the hydrogen atoms approach one another and the proton (with a positive charge) in each atom's nucleus attracts the other atom's electron (with a negative charge). Finally, in step ③, the electron shells merge and the electrons are shared between the two hydrogen atoms, which

now both obey the duet rule. The result is a covalent bond that forms a molecule of hydrogen, with the molecular formula H_2.

Now let's apply this pattern of events to a slightly more complicated example. Notice in the illustration of the methane molecule (CH_4) in the first row of **Table 2.1** how the valence shells of the carbon atoms come into contact with the valence shells of the hydrogen atoms. Why does this happen? The answer again lies in the octet rule: A carbon atom has four valence electrons and therefore needs four additional electrons to obey the octet rule. So, when a carbon atom bonds with four hydrogen atoms, they share their electrons and each atom obeys the octet or duet rule.

Two atoms can share one or more pairs of electrons (see Table 2.1). When only one pair is shared, the bond is called a **single bond** and is represented by a line between the two atoms, as shown in the structural formula of the methane molecule. (*Structural formulas* represent molecules by using element symbols and lines for bonds.) When two electron pairs are shared, as in oxygen gas (O_2), a **double bond** is formed (shown as two lines between atoms). When three pairs are shared, as in nitrogen gas (N_2), a **triple bond** is formed (shown as three lines).

All covalent bonds involve electron sharing, but atoms don't always share electrons equally. This situation gives rise to two types of covalent bonds: A bond is *nonpolar covalent* if the electrons are shared equally or *polar covalent* if the electrons are shared unequally (i.e., they are attracted more to one pole, or side, of a molecule).

Nonpolar Covalent Bonds

All elements have protons, and those protons can to some extent attract electrons, a property called *electronegativity*. An

TABLE 2.1 ELECTRON SHARING IN COVALENT BONDS

Number of Electron Pairs Shared	Molecular Structure	Structural Formula	Molecular Formula
One (single bond)		H H—C—H H	CH_4, methane
Two (double bond)		O=O	O_2, oxygen gas
Three (triple bond)		N≡N	N_2, nitrogen gas

element's electronegativity generally increases from the bottom left to the upper right of the periodic table—fluorine (F) is the most electronegative element (see Figure 2.2). The more electronegative an element is, the more strongly it attracts electrons. Stated another way, elements with high electronegativity are "greedy" for electrons, and pull them away from less electronegative elements.

When two nonmetals in a molecule have identical or nearly identical electronegativities, they both tug on the electrons with the same force and the electrons are shared equally. This kind of bond is called a **nonpolar covalent bond.** Molecules in which nonpolar bonds predominate are referred to as **nonpolar molecules.** Nonpolar molecules occur in three situations:

- **The atoms sharing the electrons are of the same element.** Identical atoms also have identical electronegativities, and an atom can't pull on electrons any more strongly than its identical twin. You can see an example of a nonpolar covalent molecule in **Figure 2.6a**, which shows H_2. Notice that the yellow electron cloud is evenly distributed around the two atoms. If you look again at Table 2.1, you'll see two other examples: O_2 (a molecule of oxygen) and N_2 (a molecule of nitrogen).

- **The arrangement of the atoms makes one atom unable to pull more strongly than another atom.** Electronegativity isn't the only factor that influences how electrons are shared—the arrangement of the atoms is equally important. Consider, for example, the molecule CO_2 (carbon dioxide):

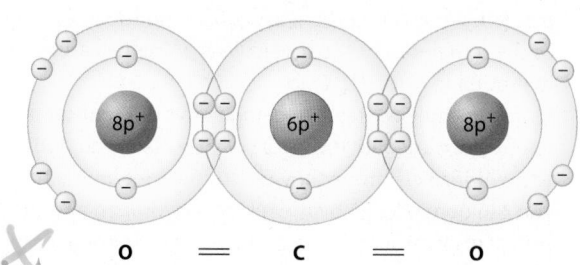

$$O = C = O$$

Here the carbon atom is double-bonded to two very electronegative oxygen atoms in two polar double bonds, yet CO_2 is *nonpolar*. This seemingly contradictory fact is due to the arrangement of the oxygen atoms. Notice how the molecule is straight, and the oxygen atoms are on either side of the carbon atom. Both oxygen atoms pull the electrons away from the carbon atom in a game of molecular tug-of-war. But the two oxygen atoms are pulling the electrons with equal strength, and as a result, neither "wins."

- **The bond is between carbon and hydrogen.** Technically speaking, carbon has a slightly higher electronegativity than hydrogen. However, this difference is so slight that any bond between carbon and hydrogen atoms is functionally nonpolar.

Note that the electronegativity of most metals is so low that they usually lose electrons (which we just saw in our discussion of ionic bonds). This is why metals cannot participate in covalent bonds: Electronegative nonmetals tug at the valence electrons

(a) Nonpolar covalent bond—H_2 (hydrogen molecule): Electrons spend equal time around the two hydrogen atoms.

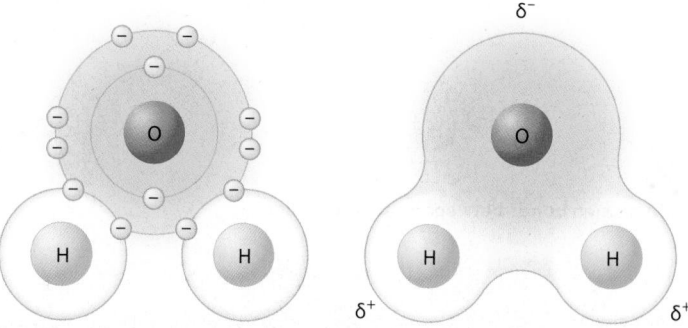

(b) Polar covalent bond—H_2O (water): Electrons spend more time around the more electronegative oxygen atom.

Figure 2.6 Nonpolar vs. polar covalent bonds.

of a metal so strongly that they are just pulled away rather than shared.

Polar Covalent Bonds and Hydrogen Bonds

When nonmetals of different electronegativities interact, a tug-of-war between the two nonmetals occurs and the stronger (more electronegative) nonmetal "wins." However, this is only a partial victory, as the winning atom is not electronegative enough to steal the electrons completely. Take a look at **Figure 2.6b**, which shows a water molecule (H_2O) with its two hydrogen atoms and its single oxygen atom. The oxygen atom is much more electronegative than the two hydrogen atoms, so the oxygen atom attracts the electrons more strongly. This is indicated by the dark yellow color of the electron cloud of the oxygen atom, which tells us that the electrons spend more time around it. This gives the oxygen atom a partial negative charge (represented by the delta symbol and a superscript minus sign, δ^-). You can see that the electron clouds of the hydrogen atoms are nearly white, which indicates that the electrons spend little time around them. The hydrogen atoms therefore have a partial positive charge (represented by the symbol δ^+).

Molecules with such partially positive and negative ends have what is known as a **dipole** (DY-pol; "two sides"), or a **polar covalent bond.** All "polar" really means is that there are two opposite sides or "poles." In the case of a polar bond, we have a partially positive pole and a partially negative pole. Molecules that contain primarily polar bonds are known as **polar molecules.** Keep in mind that, unlike an ion, a polar molecule still has an equal number of protons and electrons, but how the electrons are shared produces an unequal distribution of charges.

When polar molecules are mixed together, their partially positive and partially negative ends attract one another. The polar water molecules align so that the partially positive hydrogen atoms line up next to the partially negative oxygen atoms, which results in weak attractions called **hydrogen bonds** (indicated in

(a) Hydrogen bonds between water molecules

Hydrogen bonds

(b) Hydrogen bonds between water molecules create surface tension that causes blood to form droplets.

Figure 2.7 Hydrogen bonding and surface tension between water molecules.

Figure 2.7a by dashed lines). Note that a hydrogen bond can form between a partially positive hydrogen atom and any partially negative nonmetal. Nonpolar covalent molecules cannot form hydrogen bonds because they lack positive and negative poles.

Hydrogen bonds differ from ionic and covalent bonds in two key ways. First, they are not true bonds because electrons are neither transferred nor shared in a hydrogen bond. Instead, they are simply weak ionic attractions between two polar molecules. A second key difference is that hydrogen bonds are generally *between* molecules, whereas covalent and ionic bonds generally form *within* the molecules themselves.

Hydrogen bonding between water molecules is responsible for one of the key properties of water: *surface tension*. Where air and water meet, the polar water molecules are more strongly attracted to one another than they are to the nonpolar air

molecules. This mutual attraction causes the water molecules to cluster together and form hydrogen bonds; each water molecule can form up to four such bonds. The resulting "web" of molecular connections produces **surface tension,** a visible "film" on top of the water. Surface tension is what causes rain to fall in drops, water to form puddles, and a blood sample to form a droplet, as shown in **Figure 2.7b**. Hydrogen bonds play extremely important biological roles. For example, they help to form the structure of our proteins and of our genetic material.

The types of chemical bonds are summarized in **Table 2.2**.

ConceptBOOST »»»

Determining the Type of Bonds in a Molecule

If you're having difficulty determining if a molecule contains mostly polar or nonpolar bonds, here are some basic "rules" you can keep in mind:

- If the compound contains both a metal and a nonmetal, the bond is *ionic*.
- If the molecule contains two or more nonmetals, the bond is *covalent*. Don't forget that hydrogen behaves like a nonmetal.
  - If the molecule contains two identical nonmetals, it is *nonpolar covalent* (e.g., O_2).
  - If the molecule contains only or primarily carbon and hydrogen, it is *nonpolar covalent* (e.g., CH_4). Note that a compound can contain a couple of strongly electronegative elements, such as oxygen, and still be nonpolar as long as the remainder of the molecule is primarily carbon and hydrogen (e.g., $C_{16}H_{32}O_2$).
  - If the molecule contains two nonmetals of significantly different electronegativities, it is *polar covalent*. You don't need to memorize electronegativity values to determine this—you can generally go by the rule that hydrogen and carbon have low electronegativities, whereas elements like oxygen, nitrogen, and phosphorus have high electronegativities. So, molecules with a proportionally high number of C—O, C—N, P—O, H—N, and H—O bonds are polar covalent (e.g., $C_5H_{10}O_5$, CN^-, PO_4^{-3}).

Remember that these "rules" don't apply to every situation and that the arrangement of the atoms plays a role in determining whether or not a molecule is polar or nonpolar covalent. But, in general, they are a good place to start when figuring out the types of bonds in a molecule. ∎

TABLE 2.2 CHEMICAL BONDS		
Type of Bond	**Definition**	**Example**
Ionic Bond	Bond resulting from the transfer of electrons from a metal to a nonmetal	NaCl
Nonpolar Covalent Bond	Bond resulting from the equal sharing of electrons between two nonmetals	H_2
Polar Covalent Bond	Bond resulting from the unequal sharing of electrons between two nonmetals; the electrons spend more time around the more electronegative atom, which creates a dipole	H_2O

☐ 6. Explain how polar and nonpolar covalent bonds differ.

☐ 7. What are hydrogen bonds? Why do hydrogen bonds form only between polar covalent molecules?

Apply What You Learned

☐ 1. Which would be more reactive—an atom of fluorine or an atom of neon? Why? (*Hint:* Look at the valence electrons of each element.)

☐ 2. Would a molecule of hydrogen (H_2) form hydrogen bonds? Why or why not?

☐ 3. Explain why the molecule Na_2 does not exist in nature.

See answers in Appendix A.

MODULE **2.3**

Chemical Reactions

Learning Outcomes

1. Explain what happens during a chemical reaction.
2. Describe the forms and types of energy, and apply the principles of energy to chemical bonds and reactions.
3. Differentiate between endergonic and exergonic reactions.
4. Describe and explain the differences between the three types of chemical reactions.
5. Describe the factors that influence reaction rates.
6. Explain the properties, actions, and importance of enzymes.

In the previous module, we opened our discussion of chemical bonds with the example of combining explosive elemental sodium with poisonous chlorine gas to yield table salt. This is an example of a **chemical reaction.** Any time chemical bonds are formed, broken, or rearranged, or electrons are transferred between two or more atoms (or molecules), a chemical reaction has taken place. Although the reaction of sodium metal with chlorine gas is not one that occurs in the human body, literally trillions of other chemical reactions are occurring in our bodies at any given time. In this module, we explore the types of reactions that take place in our cells. We also introduce the concept of energy and discuss how it applies to chemical reactions.

Chemical Notation

Before we discuss chemical reactions, we must first address how we represent them. **Chemical notation** is a series of symbols and abbreviations that is used to demonstrate what occurs in a reaction. You have already been introduced to chemical symbols— the letter or letters that represent an element, as well as the subscript numbers in a molecular formula that tell you how many atoms of each element make up a molecule.

The other basic form of chemical notation is the **chemical equation.** Let's use the chemical equation that shows the conversion of carbon dioxide and water into carbonic acid as an example:

$$CO_2 \quad + \quad H_2O \quad \rightleftarrows \quad H_2CO_3$$

Carbon dioxide   *Water*   *Carbonic acid*

As you can see, this chemical equation has two parts:

- **Reactants.** Both carbon dioxide and water are **reactants**— the substances we are starting with that will undergo a reaction. The plus sign (+) is read as "reacts with." By convention, the reactants are placed on the left-hand side of the equation.
- **Products.** On the right-hand side of the equation, we have the **products**—in this case carbonic acid—the substance or substances produced in the reaction.

In standard chemical notation, an arrow between the reactants and products signifies that the reactants on the left have interacted to form the products on the right. Notice that in this equation, there are two arrows pointing in opposite directions. This indicates that the reaction is a **reversible reaction:** It can progress in both the forward and the reverse directions—that is, the carbonic acid can also split to form carbon dioxide and water.

Energy and Chemical Reactions

Matter and energy are two of the most basic concepts of the sciences, and the two are inseparable. Broadly, **energy** is defined as the capacity to do work. This means that energy can put matter into motion—it is energy that fuels our chemical reactions, causes the cells in our hearts to contract, drives our digestion and respiration, and allows you to read this book. Let's take a look at the different forms of energy in the human body and how they apply to chemical reactions.

Potential and Kinetic Energy

There are two general types of energy:

- **Potential energy. Potential energy** is energy that is stored, ready to be released and used to do work. The classic example of potential energy is a ball sitting atop a hill. The ball is stationary, but it has the *potential* to roll down the hill.
- **Kinetic energy.** Potential energy becomes **kinetic energy** (*kinet-* = "movement") when it is used. Kinetic energy is energy in motion, which means that work of some sort is being done. When the ball rolls down the hill, its energy has changed from potential to kinetic:

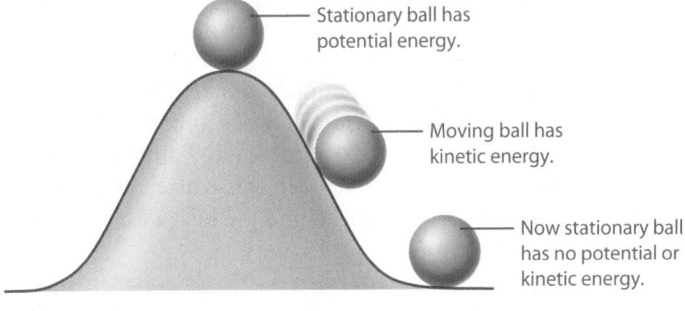

Stationary ball has potential energy.

Moving ball has kinetic energy.

Now stationary ball has no potential or kinetic energy.

All atoms have kinetic energy because atoms are constantly in motion. The faster the atoms move, the higher their kinetic energy.

Note that kinetic and potential energy are interconvertible—each may be turned into the other. In our ball and hill analogy, kinetic energy is required to get the ball to the top of the hill. Once the ball is at the top, that kinetic energy has been converted to potential energy. When the ball begins to roll down the hill, its potential energy has been reconverted to kinetic energy. When the ball is at the bottom of the hill and has stopped rolling, however, it has neither potential nor kinetic energy.

Forms of Energy in the Human Body

Potential and kinetic energy may be found in different forms. In the human body, energy takes on three forms: *chemical energy, electrical energy,* and *mechanical energy,* which may each be potential or kinetic, depending on the location or process. The form of energy that drives nearly all of our cellular processes is **chemical energy,** which is the energy inherent in chemical bonds. This might seem abstract—how can energy be stored in a shared or transferred pair of electrons? To understand this, first remember that electrons are always in motion and therefore have kinetic energy. Second, consider that it takes *work* to transfer and share electrons (i.e., protons must attract electrons). Since we can't do work without energy, energy must be present in all chemical bonds.

Certain cells in the body also harness **electrical energy.** Electrical energy is generated by the movement of charged particles. For example, the electricity that powers your house is generated by the flow of electrons through metal wiring. In our cells, however, electrical energy results from the flow of ions. This form of energy allows nerve cells to communicate and plays a role in heart and skeletal muscle contractions.

The third form of energy that drives processes in the human body is **mechanical energy.** Mechanical energy is best explained as energy that has been directly transferred from one object to another. For example, when you swing a bat to hit a baseball, you are transferring your mechanical energy first to the bat, and then to the ball. Similarly, when the tiny proteins in skeletal muscles move past one another and generate a muscle contraction, they transfer their mechanical energy first to the muscle cell, then to the whole muscle, and finally to the body part that moves.

Note that these types of energy are interconvertible (with a few exceptions). For example, electrical energy coming into a blender is converted into mechanical energy that spins its blades. Most of the time, such conversions occur easily but they are rather inefficient, and much of the energy is lost as heat. This is why an appliance or engine feels hot to the touch after it has been running for a while. This heat production is useful in the human body, though, as it helps us to maintain our relatively high body temperature. For example, we shiver when our body temperature drops, which converts chemical energy into mechanical energy that moves our muscle cells. In this conversion, much of the chemical energy is lost as heat, and the heat raises our body temperature.

Endergonic and Exergonic Reactions

As we have already discussed, energy is inherent in all chemical bonds. This means that some amount of energy must be invested any time a chemical reaction occurs. Some reactions require an input of energy that is *greater* than the energy of the reactants. These reactions, called **endergonic reactions** (en-der-GON-ik; *ender-* = "within"), require chemical, electrical, or mechanical energy from another source to proceed. The products of an endergonic reaction contain *more* energy than the reactants because energy was invested in the reaction.

The reverse situation is also true—often the energy of the reactants exceeds the amount of energy required for the reaction to proceed; such a reaction is called an **exergonic reaction** (eks-er-GON-ik; *exer-* = "without"). In this case, the excess energy stored in the reactants is released, leaving the products of the reaction with *less* energy than the reactants. An example of an exergonic reaction is the one between elemental sodium and chlorine gas. Recall that when the two elements combine, a small explosion occurs. This explosion is due to the release of energy as sodium and chlorine are converted from their unstable elemental forms into stable table salt.

Quick Check

☐ 1. How do kinetic energy and potential energy differ?
☐ 2. Explain the differences between the three types of energy that drive processes in the human body.
☐ 3. How do endergonic and exergonic reactions differ?

Homeostasis and Types of Chemical Reactions

Before you consider the kinds of reactions commonly carried out in the body, think first about what the body has to do to maintain *homeostasis,* or a relatively constant internal environment (see Chapter 1). For one, you need nutrients in the form of food, so you must have a way to break down the food you eat. Then, you have to take the broken-down food and convert the energy in its chemical bonds into a form that the body can use. Finally, you use that energy to repair or replace damaged cells and for normal growth and maintenance. So we have three fundamental processes that occur in the body: breaking down molecules (e.g., food), converting the energy in food into a usable form, and building new molecules. Each of these fundamental processes is carried out by one of three basic types of chemical reactions: (1) *catabolic reactions,* which break chemical bonds; (2) *exchange reactions,* which transfer atoms or electrons between reactants; and (3) *anabolic reactions,* which form new chemical bonds.

Catabolic Reactions

During a **catabolic reaction** (kat-ah-BOL-ik; *cato-* = "downward"), also called a *decomposition reaction,* larger substances are broken down into smaller ones. These reactions have the general form AB → A + B. Catabolic reactions are required to break down the nutrients in the foods we eat, to break apart and recycle old or damaged cells, to defend the body against invading organisms, and more. Catabolic reactions are generally exergonic because chemical bonds are broken.

Exchange Reactions, Including Oxidation-Reduction Reactions

During an **exchange reaction,** one or more atoms from the reactants are exchanged for another. In this type of reaction, bonds are both broken and formed, as indicated by the general form: AB + CD → AD + BC. A simple example is shown here—the conversion of hydrochloric acid (HCl) and sodium hydroxide (NaOH) to water and sodium chloride.

HCl + NaOH → H_2O + NaCl
Hydrochloric acid   Sodium hydroxide   Water   Sodium chloride

A special kind of exchange reaction is the **oxidation-reduction reaction** (sometimes shortened to *redox reaction*). During an oxidation-reduction reaction, electrons are exchanged instead of atoms. The reactant that loses electrons is said to be *oxidized,* and the reactant that gains electrons is said to be *reduced.* The transferred electrons carry energy with them. Oxidation-reduction reactions are generally exergonic and can release large amounts of energy. In fact, one of the fundamental energy-liberating reactions carried out by our cells is the oxidation of the sugar glucose. This topic is given full coverage in the metabolism chapter (see Chapter 23).

Anabolic Reactions

Our bodies are obviously not made of solitary atoms floating around, but rather of large and often complex molecules. These molecules are built from simpler subunits united by chemical bonds. Reactions that create new chemical bonds are called **anabolic reactions** (an-ah-BOL-ik; *ana-* = "upward") or **synthesis reactions;** they take the general form A + B → AB. Anabolic reactions in the human body are endergonic, and they are fueled by chemical energy.

Quick Check

☐ 4. What are the three main types of chemical reactions that take place in the human body? What are the differences between these types of reactions?

Reaction Rates and Enzymes

The time it takes for a chemical reaction to finish can vary widely. For example, explosions are chemical reactions that occur in a fraction of a second, but the chemical reactions of metal corrosion (such as rusting) can take years to complete. To understand why reaction rates vary so much, first consider that for a reaction to take place, molecules must come into physical contact with one another. Because all atoms are in motion, they sometimes collide with one another. However, the electrons of atoms also repel one another (just as opposite charges attract, like charges repel). So for a reaction to occur, atoms must collide with enough force to overcome the repulsion of their electrons. The faster these collisions occur, the faster the reaction proceeds.

All chemical reactions require the input of some amount of energy to overcome the repulsion of the atoms' electrons, and to allow adequately strong collisions to take place. This energy is called the **activation energy (E_a),** and all reactions, even those that are exergonic, must overcome this energy barrier in order to proceed. Consider the example in **Figure 2.8**, where you have a ball that you want to get to the other side of a hill. To do this, you must first ① expend energy to roll the ball up the hill. Once the ball is on top of the hill ②, it has an equal probability of rolling backward or forward, and has reached a point called the *transition state.* When the ball has reached this state ③, it can roll down the other side of the hill using its own kinetic energy. This analogy can be applied to a chemical reaction—the activation energy must be supplied so that the reactants reach their transition states (i.e., get to the top of the energy "hill") in order to react and form the products (i.e., roll down the hill).

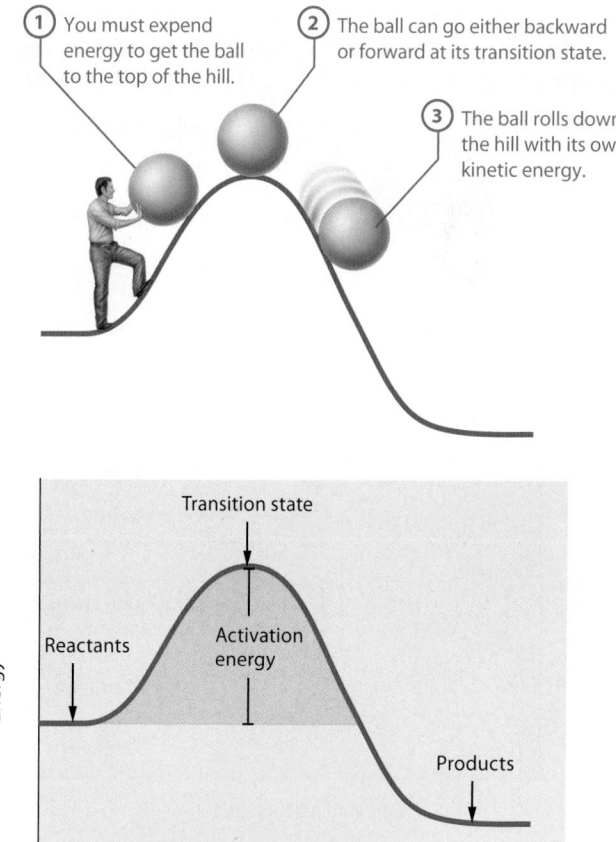

① You must expend energy to get the ball to the top of the hill.

② The ball can go either backward or forward at its transition state.

③ The ball rolls down the hill with its own kinetic energy.

Figure 2.8 Activation energy.

Several factors increase the rate of a reaction by either reducing the activation energy or increasing the likelihood of strong collisions between atoms. These factors include the following:

- **Concentration.** When the reactant concentration increases, more reactant particles are present, which increases the chances for collisions between the particles.
- **Temperature.** Raising the temperature of the reactants increases the kinetic energy of their atoms, resulting in more forceful and effective collisions. Note that for biological reactions, however, this works only up to a point, because many molecules in the human body are destroyed at high temperatures.
- **Properties of the reactants.** Both the size of the particles and the phase (solid, liquid, or gas) of the reactants influence reaction rates. Typically, smaller particles move faster and therefore have more energy than larger particles. Also, reactant particles that are in the gaseous phase have higher kinetic energy than those in the liquid or solid phase. For this reason, reactions with two gases generally proceed slightly faster than those with two liquids, and a great deal faster than those with two solids.
- **Presence or absence of a catalyst.** A *catalyst* is a substance that increases a reaction rate by lowering the activation energy but is not consumed or altered in the reaction. Biological catalysts are called **enzymes.**

You may have heard it said that an enzyme "breaks something down," but this is not actually correct. Enzymes are involved in essentially all of our catabolic reactions, but they are also involved in all of our exchange and anabolic reactions. Furthermore, the enzyme itself doesn't actually chemically alter or "break" the bonds in the reactants. What an enzyme does instead is to increase the rate at which a reaction takes place.

Each cell has roughly 4000 different kinds of enzymes. Most enzymes are named by adding the suffix "-ase" to their root words; for example, the enzyme that catalyzes the reaction breaking down the sugar in milk is called *lactase* (*lact-* = "milk"). Nearly all enzymes are macromolecules called proteins, with the following properties:

- **Enzymes speed up reactions by lowering the activation energy.** Take another look at Figure 2.8, the figure with our ball and hill analogy. You learned that to get the ball to the other side of the hill, you must first supply the necessary activation energy to move the ball to the hilltop. Notice in **Figure 2.9** what the enzyme does—it lowers the height of the hill. With a smaller hill, less activation energy is required to get the ball to the top. The lowered activation energy also allows the reaction to proceed faster.
- **Enzymes are highly specific for individual substrates and reactions.** Each enzyme contains a unique region called the **active site** that binds to specific molecules. A substance that binds to the active site of an enzyme is called that enzyme's **substrate.** Not only is an enzyme specific for its substrate, but it is also specific for an individual reaction. One substrate may bind two different enzymes and produce two different products, depending on the conditions in the cell.

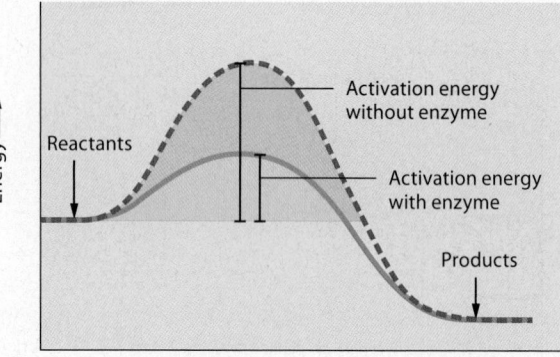

Figure 2.9 The effect of enzymes on activation energy.

Practice art labeling
@MasteringA&P®

A&P
in the
Real World

Enzyme Deficiencies

Given the important role of enzymes in the body's homeostasis, deficiencies of certain enzymes can be extremely damaging. The following are some examples of common enzyme deficiencies:

- *Tay-Sachs disease* is a deficiency of the enzyme hexosaminidase-A (hek'-soh-sah-MIN-ih-dayz), which results in the buildup of chemicals called *gangliosides* (GANG-glee-oh-sydz) around the nerve cells in the brain. The disease usually causes death by age three.
- *Severe combined immunodeficiency syndrome,* or SCIDS, may be caused by a deficiency of the enzyme adenosine deaminase (ah-DEE-noh-seen dee-AM-ih-nays). This disease causes a nearly complete absence of the immune system, and affected patients must live in a sterile "bubble."
- *Phenylketonuria* (fee'-nul-kee-tohn-YUR-ee-ah) is a deficiency of the enzyme phenylalanine hydroxylase (fen'-el-AL-ah-neen hy-DROK-syl-ays), which converts the amino acid (a building block of proteins) called phenylalanine into a different amino acid called tyrosine. The symptoms of the disease include seizures and mental retardation; however, these symptoms can be prevented with dietary modifications.

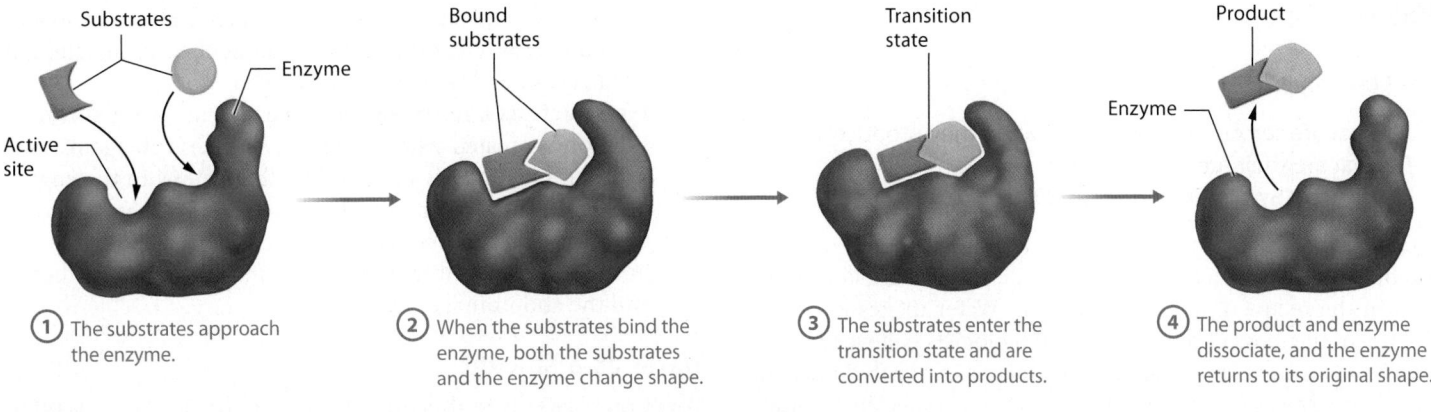

① The substrates approach the enzyme.

② When the substrates bind the enzyme, both the substrates and the enzyme change shape.

③ The substrates enter the transition state and are converted into products.

④ The product and enzyme dissociate, and the enzyme returns to its original shape.

Figure 2.10 Enzyme-substrate interaction.

Sequence the steps
@MasteringA&P®

- **Enzymes do not alter the reactants or products.** The nature of a chemical reaction is the same whether or not an enzyme is present—the enzyme does not chemically change the reactants or the products of a reaction.
- **Enzymes are not permanently altered in the reaction.** Although the chemical bonds of the reactants change during a reaction, those of the enzyme do not change. The structure of the enzyme does alter somewhat during the reaction, but reverts to its original form when the reaction is complete. The enzyme isn't "used up" in the reaction, and so can go on to bind another substrate molecule and catalyze another reaction.

An enzyme's mechanism of action becomes clear when you look at its interaction with its substrate(s). Such enzyme-substrate interaction, illustrated in Figure 2.10, is currently thought to occur by an "induced fit" mechanism, in which the enzyme and substrate both change shape slightly when they bind, making a tighter "fit" between them ②. Notice that this shape change brings the substrates to their transition states. This significantly reduces the required activation energy and increases the rates of reactions between 10 million and 100 trillion times.

The importance of enzymes in the human body cannot be overemphasized. Although many chemical reactions can be carried out in the laboratory without the help of a catalyst, that isn't true for the chemical reactions in the body. Most biological molecules have low kinetic energy and are pretty stable—otherwise, we would risk exploding. Because the kinetic energy of biological molecules is low, the activation energy for biological reactions is high. Without enzymes, the chemical reactions required to carry out all of our basic processes (to maintain homeostasis) would occur far too slowly for us to remain alive. For this reason, enzymes are absolutely vital to our survival; see *A&P in the Real World: Enzyme Deficiencies* for some examples of diseases that result when enzymes are missing or faulty.

Quick Check

☐ 5. What factors can influence the rate of a chemical reaction?
☐ 6. What is an enzyme, and what does an enzyme do?

Apply What You Learned

☐ 1. Why do you feel hot during and immediately after exercise? (*Hint:* Think about what happens when you convert one form of energy to another.)
☐ 2. Explain why most biological molecules are stable and nonreactive.
☐ 3. Many naturally occurring poisons function by inhibiting a cell's enzymes. Why might this effect be lethal to the cell?

See answers in Appendix A.

MODULE **2.4**

Inorganic Compounds: Water, Acids, Bases, and Salts

Learning Outcomes

1. Discuss the physiologically important properties of water.
2. Explain why certain compounds will dissolve in water, whereas others will not.
3. Define and describe the properties of acids and bases with respect to hydrogen ions.
4. Explain what the pH scale measures, and state acidic, neutral, and basic pH values.
5. Explain the function of a buffer.
6. Define and describe the terms salt and electrolyte, and give examples of their physiological roles.

In the world of **biochemistry,** the chemistry of life, compounds are grouped into one of two types: inorganic and organic compounds. **Inorganic compounds** are those that generally do not contain carbon bonded to hydrogen. Important inorganic compounds in the human body include water, acids, bases, and salts. **Organic compounds** are broadly defined as those that contain carbon bonded to hydrogen. This module and the next explore the properties of these chemicals.

Water

◀◀ **Flashback**

1. What are serous membranes, what do they produce, and what are their functions? (p. 18)
2. What are hydrogen bonds? (p. 39)

One of the first things astronomers search for in their quest to find life on other planets is liquid **water**, H_2O, as it is a key ingredient in the recipe of life as we know it. Water makes up about 60–80% of the mass of our bodies, and biologists believe that it is the most abundant chemical in all organisms on Earth. Let's look at the properties of water that make it unique, especially its ability to act as a solvent.

Properties of Water

Water has a number of key properties that make it vital to living organisms; many of these properties are related to its ability to form hydrogen bonds. We've already mentioned the important property of surface tension. Other properties include the following:

- **Water absorbs heat without changing significantly in temperature itself.** If you have ever boiled water, you have witnessed water's high *heat capacity*, which means that a large amount of heat must be applied to disrupt enough hydrogen bonds to change water from a liquid to a gas. This is why water must be heated to 100°C (212°F) before it boils, and it is also why you can walk in the desert in the summer, take a hot shower, or sit in a sauna without damaging your cells. Even as the air around you increases in temperature, the water in and around your cells doesn't change temperature appreciably.

- **Water carries heat with it when it changes from a liquid to a gas.** If you are outside exercising in the heat of summer, before long you will begin sweating. The sweat will begin to evaporate when enough heat is available to disrupt the hydrogen bonds between the water molecules. This heat is provided by your body, which means in essence that the water pulls heat out of your body when it turns into water vapor. The overall effect is that your body cools down. When the humidity in the air is very high, water evaporates much less efficiently, which explains why humid heat often feels hotter than dry heat.

- **Water cushions and protects the body's structures.** The cushioning ability of water has to do with its relatively high *density*. A substance's density relates its mass to its volume. Stated simply, the more of a substance that you can fit into a particular volume, the higher its density. Water's density is high because the hydrogen bonds pull water molecules closely together so that they occupy less space. Such a property is useful because anything that is less dense than water will float in it to some degree (this explains why ice cubes float in water—the molecules in ice form a crystalline structure and are therefore farther apart than they are in liquid water). Since most of the body's organs are less dense than water, they are able to remain slightly suspended in their watery surroundings. This is particularly important for organs such as the brain.

The watery fluid that surrounds the brain reduces its weight by about 97%. Were the brain not suspended in this fluid, it would crush itself under its own weight.

- **Water acts as a lubricant between two adjacent surfaces.** Friction is created when two solid surfaces rub against one another. Like the oil surrounding the pistons in the engine of a car, a thin layer of water between body surfaces acts as a lubricant that reduces friction. Water has this function in the serous membranes surrounding the lungs, the heart, and the abdominal organs.

Water as a Solvent

Water serves as the body's primary solvent. In fact, water is often called the *universal solvent* because so many solutes will dissolve in it either fully or at least partially. Water's ability to dissolve substances has to do with both its molecular shape and its polar covalent bonds. Recall that the water molecule is bent—that is, the two hydrogen atoms sit at an angle to one another (see

(a) Ionic compounds are hydrophilic.

(b) Polar covalent compounds are hydrophilic.

(c) Nonpolar covalent compounds are hydrophobic.

Figure 2.11 The behavior of hydrophilic and hydrophobic molecules in water.

Figure 2.6). Remember also the partial charges on the oxygen and hydrogen atoms that result from its polar covalent bonds.

These two factors combine to make water a polar molecule; this gives it "hands" with which it can "grab" certain solutes and pull them apart. Water molecules can then surround these solutes and keep them apart. Water is only able to dissolve solutes that are known as **hydrophilic** (hy-droh-FIL-ik; "water loving"). Notice in **Figure 2.11a and b** that hydrophilic solutes have fully or partially charged ends. These charged ends attract the oppositely charged ends of water molecules, due to the attraction between negative and positive charges. Recall from Module 2.2 that ionic and polar covalent molecules have fully and partially charged ends, respectively, and these compounds are therefore hydrophilic. Solutes that lack charged ends do not have "handles" for water to grab and will not dissolve in water. These solutes are referred to as **hydrophobic** (hy-droh-FOH-bik; "water hating") and include uncharged nonpolar covalent molecules such as oils and fats (**Figure 2.11c**).

A solute's solubility in water follows the general rule of thumb "*like dissolves like.*" This is a simple way of saying that partially or fully charged molecules, such as those with polar covalent and ionic bonds, will interact with one another but will not interact with uncharged molecules. However, uncharged nonpolar covalent molecules will not interact with charged molecules but will dissolve readily in nonpolar covalent solvents, such as paint thinner.

Many solutes vital to our homeostasis are hydrophilic, so our bodies are able to carry them in blood plasma and transport them safely and efficiently from one place to another. However, certain hydrophobic substances, such as fats, oxygen, and carbon dioxide, also require transport through the body but have very little ability to dissolve in water. As we progress through this book, we will see several examples of how our bodies are adapted to meet this challenge.

Quick Check

☐ 1. What are four properties of water that make it a good solvent for living organisms?

☐ 2. Which molecules are likely to be hydrophilic? Which are likely to be hydrophobic?

(a) Some water molecules dissociate to H⁺ and OH⁻. In pure water, the numbers of H⁺ and OH⁻ are equal.

(b) An acid (HCl) releases H⁺ and so increases the H⁺ concentration of the solution.

(c) A base (NaHCO₃) binds free H⁺ and so decreases the H⁺ concentration of the solution.

Figure 2.12 The behavior of acids and bases in water.

Acids and Bases

The study of acids and bases is really the study of the **hydrogen ion (H⁺).** Earlier you learned that hydrogen behaves as a nonmetal most of the time: It forms covalent rather than ionic bonds, and elemental hydrogen exists on Earth as a gas, not a metal. However, when certain hydrogen-containing compounds are placed in water, hydrogen can switch roles and start behaving like a metal—it donates its electron to produce a hydrogen ion. In fact, this even happens to the hydrogen in water to a small extent. As shown in **Figure 2.12a** and the following reaction, a water molecule in a solution may break apart, or *dissociate*, to form a hydrogen ion and an anion called the **hydroxide ion (OH⁻):**

$$\text{H}_2\text{O} \rightleftharpoons \text{H}^+ + \text{OH}^-$$

$$\textit{Water} \qquad \textit{Hydrogen} \qquad \textit{Hydroxide}$$
$$\textit{ion} \qquad \qquad \textit{ion}$$

(Sometimes anions can consist of more than one atom, but they act as single anions.)

Acids and bases are defined according to their behavior with respect to hydrogen ions:

- **An acid is a hydrogen ion donor.** As shown in **Figure 2.12b** and the following equation, when you place an **acid** in water, it splits apart into a hydrogen cation (H^+) and an anion. Let's look at what happens with hydrochloric acid (HCl):

$$In\ H_2O$$

$$\underset{\substack{Hydrochloric \\ acid}}{HCl} \quad \rightleftharpoons \quad \underset{\substack{Hydrogen \\ ion}}{H^+} \quad + \quad \underset{\substack{Chloride \\ ion}}{Cl^-}$$

In this process, HCl "donates" its hydrogen ion to the solution, and the number of hydrogen ions in the solution increases. After an acid dissociates, the anion that forms can bond with or "accept" the hydrogen ion back if the concentration of hydrogen ions rises again.

Examples of acids found in the human body include HCl in the stomach and H_2CO_3 (carbonic acid) in the blood. Note that acids are sometimes called *proton donors* rather than hydrogen ion donors. This is because a hydrogen ion has only one subatomic particle—the lone proton in its nucleus.

- **A base is a hydrogen ion acceptor.** When a **base** (also called an **alkali;** AL-kah-lye) is mixed with water, the number of hydrogen ions in the solution decreases. Notice in **Figure 2.12c** that when the base sodium bicarbonate ($NaHCO_3$) is placed in water, it dissociates into sodium cations (Na^+) and bicarbonate anions (HCO_3^-). The bicarbonate ions then accept the free hydrogen ions in the solution. Most bases used in the chemistry laboratory have $-OH$ (hydroxide) in their formulas. However, the main bases in the human body, such as bicarbonate ions (HCO_3^-), do not.

The pH Scale

Concentrations of acids and bases affect many of the chemical reactions that occur in our bodies. A simple way to represent the hydrogen ion concentration of a solution is via the **pH scale.** The term "pH" has widely, and incorrectly, been reported to mean "potential of hydrogen." Actually, *p* is a mathematical abbreviation for the $-\log$ (read as "negative logarithm") of a number, and *H* stands for the concentration of hydrogen ions in the solution. So, the term "pH" means $-\log[H^+]$, or the "negative logarithm of the solution's hydrogen ion concentration" (brackets indicate concentration).

The pH scale, shown in **Figure 2.13**, goes from 0 to 14. Notice in the column on the left that the more hydrogen ions present in a solution, the lower its pH. As we move up the scale, we find progressively fewer hydrogen ions in the solution, and the pH increases. However, the column to the right, which represents the number of base ions in the solution, shows the opposite trend. At a lower pH, there are fewer base ions, and as we move up the scale, we have progressively more base ions. Notice now what you see in both columns when the pH is 7—the numbers of hydrogen ions and base ions are equal. At this pH a solution is called **neutral.** Pure water at 25°C has a pH of 7, but few other solutions do—nearly all of them fall into the category of either *acidic* or *basic*.

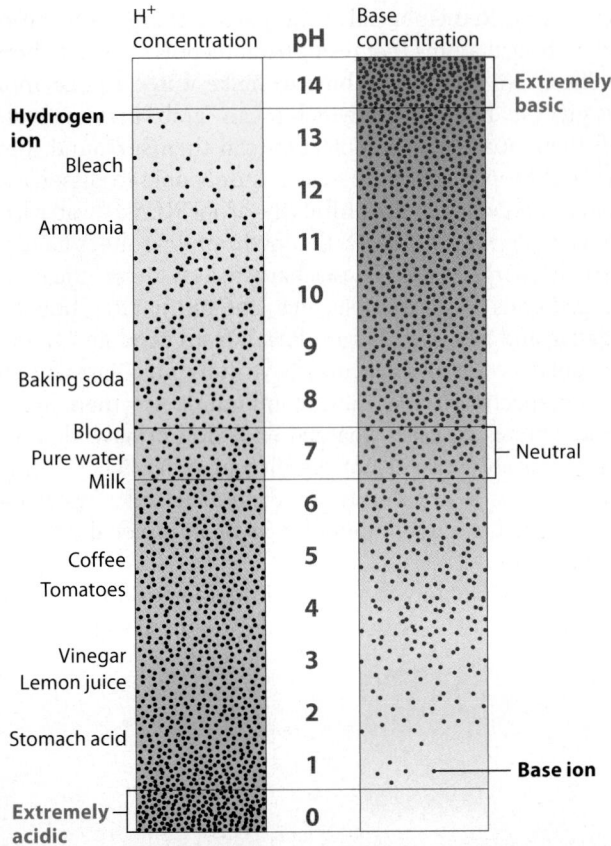

Figure 2.13 The pH scale.

- **A solution with a pH less than 7 is called acidic.** As you can see in Figure 2.13, **acidic** solutions have more hydrogen ions than base ions. A strongly acidic solution, called a *strong acid,* with a pH near 0 has essentially no base ions and an extremely high concentration of hydrogen ions. A *weak acid* with a pH just below 7 has only slightly more hydrogen ions than base ions.
- **A solution with a pH greater than 7 is called basic or alkaline. Basic** solutions have the reverse conditions of acidic solutions—they have fewer hydrogen ions and more base ions. A strongly basic solution, a *strong base,* with a pH near 14 has essentially no hydrogen ions and a high concentration of base ions, whereas a *weak base* has a pH just above 7 and has only slightly more base ions than hydrogen ions.

The pH of most fluids in the human body hovers just above the neutral mark at a slightly basic pH. For example, the pH of the blood averages about 7.35–7.45, and the pH inside our cells is around 7.2. Examples of other common solutions and their corresponding pH values are also shown in Figure 2.13.

Something important to note about the pH scale is that we are dealing with *base 10* logarithms. What this means is that each single-digit change in the pH number corresponds to a 10-fold change in hydrogen ion concentration. So when the pH decreases from 7 to 6, the hydrogen ion concentration increases 10 times. When the pH decreases from 7 to 5, the hydrogen ion concentration increases 100 times, and it increases 1000 times with a pH decrease from 7 to 4. The same is true in reverse—a pH increase from 7 to 8 represents a 10-fold decrease in the solution's hydrogen ion concentration.

✓

ConceptBOOST »»»

Making Sense of the pH Scale

Why does the pH *decrease* if the solution has *more* hydrogen ions? It all goes back to that –log concept that we talked about, which is not nearly as difficult as it sounds. Grab your calculator, and you'll see. On your calculator, enter the following hydrogen ion concentrations that decrease in 10-fold increments and calculate their –log:

$-\log(0.01)$ = _____

$-\log(0.001)$ = _____

$-\log(0.0001)$ = _____

$-\log(0.00001)$ = _____

Okay, that should be enough for you to notice the general trend here: *The smaller the number, the bigger its negative log.* Notice also that each single-digit change in the negative logarithm (e.g., from 2 to 3) accompanies a 10-fold change in the hydrogen ion concentration (e.g., from 0.01 to 0.001).

Let's do one quick example. You have two solutions. Solution A has a hydrogen ion concentration of 0.015 *M* (the *M* here stands for molar; *molarity* is a measure of concentration). Solution B has a hydrogen ion concentration of 0.0003 *M*.

• Which solution has more hydrogen ions? _____
• What is the –log of solution A? _____
• What is the –log of solution B? _____

So, as you can see, the solution with the *higher* hydrogen ion concentration has the *lower* –log. For this reason, the more acidic a solution, the lower its pH. ■

Answers: 2, 3, 4, 5; solution A, –log of solution A = 1.82, –log of solution B = 3.52

Buffers

The reason that the pH range in our blood and cells is so narrow is that the function of many important molecules in the body, such as enzymes and other proteins, can be disrupted by even slight changes in hydrogen ion concentration. A pH decrease of only about 0.3 reflects a doubling of the hydrogen ion concentration, and as you just learned, a pH decrease of 1 reflects a 10-fold increase in the hydrogen ion concentration. For this reason, it is vital to have mechanisms in place to keep the pH of the body's fluids within their normal ranges.

This task is accomplished by the body's **buffers,** chemical systems that resist changes in pH and prevent large swings in the pH when acid or base is added to a solution. Typically, a buffer consists of a weak acid and its corresponding anion. One of the major buffers in the body is the *carbonic acid–bicarbonate buffer*

system. In this system, the weak acid is carbonic acid (H_2CO_3), and its anion is bicarbonate (HCO_3^-), as shown here:

$$H_2CO_3 \quad \rightleftarrows \quad H^+ \quad + \quad HCO_3^-$$

Carbonic acid   *Hydrogen ion*   *Bicarbonate ion*

This buffer system works in a simple way: When the pH of the blood rises (i.e., the blood becomes too basic), the carbonic acid releases hydrogen ions into the blood to offset the increase in pH. When the pH of the blood falls (i.e., the blood becomes too acidic), the bicarbonate binds to hydrogen ions in the blood, taking them out of solution and offsetting the decrease in pH.

This and the other buffer systems in the body are generally very effective at keeping the blood pH in the normal range. However, these systems can be overwhelmed. One condition that may result is *acidosis* (ah-sih-DOH-sis), which is a blood pH less than 7.35 that is caused by an increase in the total body concentration of hydrogen ions. The other condition is *alkalosis* (al-kah-LOH-sis), which is a blood pH greater than 7.45 that is caused by a loss of hydrogen ions or a gain of basic ions.

The two main organ systems in the body that work to correct these pH imbalances are the respiratory and urinary systems. The specific mechanisms by which these body systems compensate for pH disturbances are discussed fully in the chapters that discuss those systems (see Chapters 21 and 24, respectively).

Quick Check

☐ 3. Define the terms acid and base.

☐ 4. What is the pH scale? Which pH values are considered acidic, basic, and neutral?

☐ 5. What is the effect of a buffer on a solution?

Salts and Electrolytes

◀◀ **Flashback**

1. What type of bond forms between a metal and a nonmetal? (p. 36)
2. Are compounds with this type of bond hydrophilic or hydrophobic? Why? (p. 46)

You are likely familiar with the term "salt" from common table salt, sodium chloride (NaCl). Technically, however, the term **salt** refers to any metal cation and nonmetal anion held together by ionic bonds. As you learned earlier, ionic compounds such as salts dissolve when placed in water. The resulting cations and anions in the solution are called **electrolytes** (ee-LEK-troh-lytz). Electrolytes are named for the fact that they will conduct an electric current in water. This is why you risk electrocution if an electric appliance comes into contact with tap water with you in it—tap water contains electrolytes, and the electrolytes conduct the electricity from the electrical device into you. All salts dissociate into electrolytes, as do some acids and bases.

Salts play a wide range of vital roles in the body. Solid calcium phosphate ($Ca_3(PO_4)_2$) salts are the major component of bones and teeth. Other salts dissolve into their component electrolytes in the body's fluids. Dissolved sodium and potassium ions are required for muscles to contract and for our nerve cells to function. Calcium ions are also required for muscles to contract, and are essential for the heart muscle to keep a normal rhythm and pace. We will encounter these salts and their electrolytes repeatedly in our study of anatomy and physiology.

Quick Check

☐ 6. What is a salt?
☐ 7. What does an electrolyte do in a solution?

Apply What You Learned

☐ 1. Explain how the water surrounding a fetus in the womb protects the fetus as it develops.
☐ 2. Why might a person who is severely dehydrated feel hot, even if he or she is not in a hot environment?

See answers in Appendix A.

MODULE **2.5**

Organic Compounds: Carbohydrates, Lipids, Proteins, and Nucleotides

Learning Outcomes

1. Explain the relationship between monomers and polymers.
2. Define and give examples of dehydration synthesis and hydrolysis reactions.
3. Compare and contrast the general molecular structures of carbohydrates, lipids, proteins, and nucleic acids.
4. For carbohydrates, lipids, proteins, and nucleic acids, identify their monomers and polymers, provide examples, and discuss their physiological and structural roles in the human body.
5. Describe the four levels of protein structure and discuss the importance of protein shape for protein function.
6. Describe the generalized reversible reaction for release of energy from ATP and explain the role of ATP in the cell.

We now begin our exploration of the world of organic compounds. These chemicals are produced by living organisms and so make up much of the human body and other living things. They are generally much larger and more complex than inorganic compounds.

Certain organic compounds, called *hydrocarbons,* contain only hydrogen and carbon. As shown in the next column, they can exist in two forms—chains and rings.

Hydrocarbon chain = Another way to show a hydrocarbon chain; the C atoms are at each point, and the H atoms are not shown.

Hydrocarbon ring = Another way to show a hydrocarbon ring; the C atoms are at each corner, and the H atoms are not shown.

This carbon-hydrogen backbone forms the basis for all of our organic compounds, with the addition of groups of atoms called *functional groups.* They have this name because each different group, when bonded to a molecule, somehow modifies the function of that molecule. We'll see examples of functional groups throughout this chapter.

The organic compounds covered in this chapter are *carbohydrates, proteins, lipids* (fats), and *nucleic acids* (including *nucleotides*). You have likely heard of the nutritional functions of many of these molecules; however, their structural and functional roles in the body extend far beyond this. We explore these roles as well as the structure of each of these molecules in this module.

Monomers and Polymers

Monomers (MON-oh-merz; "one part") are single subunits that can be combined to build larger structures. **Polymers** (POL-ih-merz; "many parts") are the larger structures that consist of many monomers linked together. Each type of organic molecule we discuss here has its own monomer and a corresponding polymer built from those subunits.

Polymers are built by an anabolic reaction called **dehydration synthesis,** in which two monomers are linked by a covalent bond and a water molecule is formed. The reverse reaction can also take place by a process called **hydrolysis** (hy-DROL-uh-sis; "water splitting"). In this process, a molecule of water is added to the polymer, the atoms in the water molecule are split apart, and the monomers are separated. You will encounter several examples of dehydration synthesis and hydrolysis in the upcoming sections.

Quick Check

☐ 1. How do polymers and monomers differ?

Carbohydrates

◄◄ Flashback

1. What types of chemical bonds form between hydrogen and oxygen atoms? (p. 9)

(a) Pentoses: five-carbon sugars   **(b) Hexoses: six-carbon sugars**

Deoxyribose ($C_5H_{10}O_4$)   Ribose ($C_5H_{10}O_5$)   Glucose ($C_6H_{12}O_6$)   Fructose ($C_6H_{12}O_6$)   Galactose ($C_6H_{12}O_6$)

Isomers of glucose

Figure 2.14 Carbohydrates: structure of monosaccharides.

2. Are compounds with these types of bonds hydrophilic or hydrophobic? Why? (p. 46)

The elements that make up the group of organic molecules called **carbohydrates** are easy to figure out just from the name: *carbo-* means "carbon," and *-hydrate* refers to the elements in water—hydrogen and oxygen. Indeed, most carbohydrate monomers are composed of carbon, hydrogen, and oxygen in the ratio of 1C:2H:1O (there are two hydrogen atoms for every carbon and oxygen atom). Because carbohydrates contain several –OH groups (hydroxyl groups, an example of a functional group), the molecules on the whole are polar and therefore hydrophilic.

Carbohydrates account for about 1% of the body's mass. They have only a few structural roles in the body—some carbohydrates are scattered throughout the membranes that surround our cells, and others are found in our genetic material. The main role of carbohydrates in the body is that of fuel. Through catabolic and oxidation-reduction reactions, our bodies liberate the energy stored in carbohydrates and use it to drive our cellular processes.

The monomer of carbohydrates is the so-called *simple sugar,* more formally called the *monosaccharide* (mon'-oh-SAK-ah-ryd; *mono-* = "one," *sacchar-* = "sugar"). Carbohydrates are classified according to how many of these subunits are present. Monosaccharides, of course, have just one subunit; *disaccharides* (dy-SAK-ah-rydz; *di-* = "two") have two subunits; and *polysaccharides* (pol'-ee-SAK-ah-rydz; *poly-* = "many") have many subunits.

Monosaccharides and Disaccharides

Monosaccharides have from three to seven carbon atoms, and all but the smallest generally take on a ring structure. The majority of carbohydrates in the body have either five carbons (the *pentoses*) or six carbons (the *hexoses*). The two primary pentose sugars are *deoxyribose* (dee'-oks-ee-RY-bohs) and *ribose* (RY-bohs; **Figure 2.14a**). Neither pentose is used directly for fuel, but both are found in our genetic material (discussed later in this module).

The main hexose in the body is the sugar **glucose** (GLUW-kohs), shown in **Figure 2.14b**. Glucose is the body's primary source of fuel, and the term "blood sugar" refers to the concentration of glucose dissolved in the blood plasma. Two other hexoses are found to a large extent in our diets: *fructose* (FRUWK-tohs), found in fruits and certain vegetables, and *galactose* (gah-LAK-tohs), found in dairy products. Notice in Figure 2.14b that both fructose and galactose have the same molecular formula as glucose: $C_6H_{12}O_6$. However, the arrangement of the atoms is different in all three molecules. Molecules like these with the same molecular formula but different structures are called **isomers** (EYE-soh-merz; "same parts"). In spite of similarities among the three hexoses, however, glucose is the only sugar that the body can use directly in its catabolic reactions.

A **disaccharide** is a molecule with two monosaccharides joined by a polar covalent bond. As shown in **Figure 2.15**, disaccharides are formed by dehydration synthesis—a hydrogen atom is removed from fructose and a hydroxyl group (−OH) is removed from glucose, and a molecule of water is formed. Disaccharides can be broken down into monosaccharides by hydrolysis, also shown in Figure 2.15.

The two most prevalent disaccharides are sucrose and lactose. Sucrose is the sweetest of the disaccharides, which is why we use it for sweetening our food and call it "table sugar." It consists of glucose bound to fructose and comes primarily from

Glucose (monosaccharide)   +   Fructose (monosaccharide) H_2O Dehydration synthesis   Hydrolysis H_2O Sucrose (disaccharide)

Figure 2.15 Carbohydrates: formation and breakdown of disaccharides.

sugar cane. The disaccharide lactose, which consists of glucose and galactose, is found in milk and other dairy products. Adults commonly lack the enzyme required to digest lactose, which, as you learned, is lactase, causing the digestive problems associated with *lactose intolerance.*

Polysaccharides

The polymers known as **polysaccharides** are the largest and least soluble of the carbohydrates. Polysaccharides are composed of long, branching chains of monosaccharides joined by covalent bonds and formed from dehydration synthesis. This structure causes their low solubility, as molecules that large are difficult for water molecules to separate from one another, regardless of the types of bonds in the molecule.

Because of their structure and low solubility, polysaccharides represent a way for plants and animals to store some of their glucose for later use. Plants store their glucose in a form called *starch,* and animals store their glucose in a form called **glycogen** (GLY-koh-jen), found primarily in the liver and skeletal muscles. As you can see in **Figure 2.16**, glycogen's structure is heavily branched, which allows the glucose in glycogen to be stored in a compact form. This form is a distinct advantage, as glucose molecules are released via enzymatic reactions only from the ends of a polysaccharide. The highly branched structure of glycogen gives the enzymes many ends on which to work, so the hydrolysis of glycogen stores can quickly increase the blood glucose concentration when necessary. However, these glycogen stores don't last that long—the supply in the liver and the muscle is exhausted in about two hours during strenuous exercise, after which the liver and muscle cells must take in new glucose molecules and "restock" their glycogen supplies.

Some polysaccharides are found attached by covalent bonds to proteins and lipids, yielding compounds called **glycoproteins** and **glycolipids,** respectively. In this form, they help maintain the structural integrity of our cells and play various roles depending on where they are located in the body.

Quick Check

- [ ] 2. What three elements compose carbohydrates? In what ratio are they found?
- [ ] 3. How do monosaccharides, disaccharides, and polysaccharides differ?
- [ ] 4. How are two monosaccharides linked to form a disaccharide? How is a disaccharide split into two monosaccharides?

Lipids

◄◄ **Flashback**

1. What types of chemical bonds form between carbon and hydrogen atoms? (p. 9)
2. Are molecules in which these bonds predominate hydrophilic or hydrophobic? Why? (p. 46)

Like carbohydrates, **lipids** (LIP-idz), which include fats and oils, contain only carbon, hydrogen, and oxygen atoms. However, when you compare the molecular formulas of glucose and a typical lipid, you can see how different they are. Whereas glucose ($C_6H_{12}O_6$) has an equal number of carbon and oxygen atoms and many hydroxyl groups (−OH), the lipid has the molecular formula $C_{15}H_{31}COOH$—carbon and hydrogen are its main components, and oxygen is very much in the minority. The predominance of carbon and hydrogen atoms in the lipid molecule makes it overall very nonpolar and therefore hydrophobic. However, lipids will dissolve in and can act as a solvent for other nonpolar compounds, which are often said to be *lipid-soluble.* Lipids take on many forms in the body, which we examine in the following sections.

Fatty Acids and Triglycerides

A basic lipid monomer is a molecule called a **fatty acid** (**Figure 2.17**). A fatty acid consists of a *carboxylic acid* group (−COOH, another functional group; car-bok-SIL-ik) bound to a hydrocarbon chain

Figure 2.16 Carbohydrates: the polysaccharide glycogen.

Glucose molecules

that can have 4 to 20 or more carbon atoms. The hydrocarbon chain may have one or more double bonds between its carbon atoms, which provides a convenient way to classify fatty acids. The three classes are as follows:

- **Saturated fatty acids.** A **saturated fatty acid** has no double bonds between carbon atoms in its hydrocarbon chain, and so its carbon atoms are "saturated" with hydrogen atoms. Saturated fatty acids are found predominantly in animal

fats, and they are mostly solid at room temperature. The fatty acid shown in Figure 2.17a, palmitic acid, is saturated.
- **Monounsaturated fatty acids.** A **monounsaturated fatty acid** has one double bond between two carbons in its hydrocarbon chain, which causes the chain to bend, as you can see in the three-dimensional model of the molecule. This means that these hydrocarbons can't pack together as tightly. These fats have a lower melting point than saturated fats, so they are generally liquid at room temperature. Plant oils such as olive, grapeseed, and peanut oils have high concentrations of monounsaturated fats. The example shown in Figure 2.17b is oleic acid (oh-LAY-ik), the predominant fatty acid in olive oil.
- **Polyunsaturated fatty acids.** A **polyunsaturated fatty acid** has two or more double bonds between its carbon atoms. The melting points of polyunsaturated fatty acids are lower than those of both saturated and monounsaturated fatty acids, and they are liquid at room temperature. The example in Figure 2.17c is linoleic acid (lin-oh-LAY-ik).

Our bodies cannot package fatty acids into long chains as they can monosaccharides. Instead, fatty acids are stored by linking three of them via dehydration synthesis to a modified three-carbon sugar called **glycerol** (GLISS-er-ahl). These reactions produce a polymer called a **neutral fat,** or **triglyceride** (try-GLISS-er-iyd; Figure 2.18).

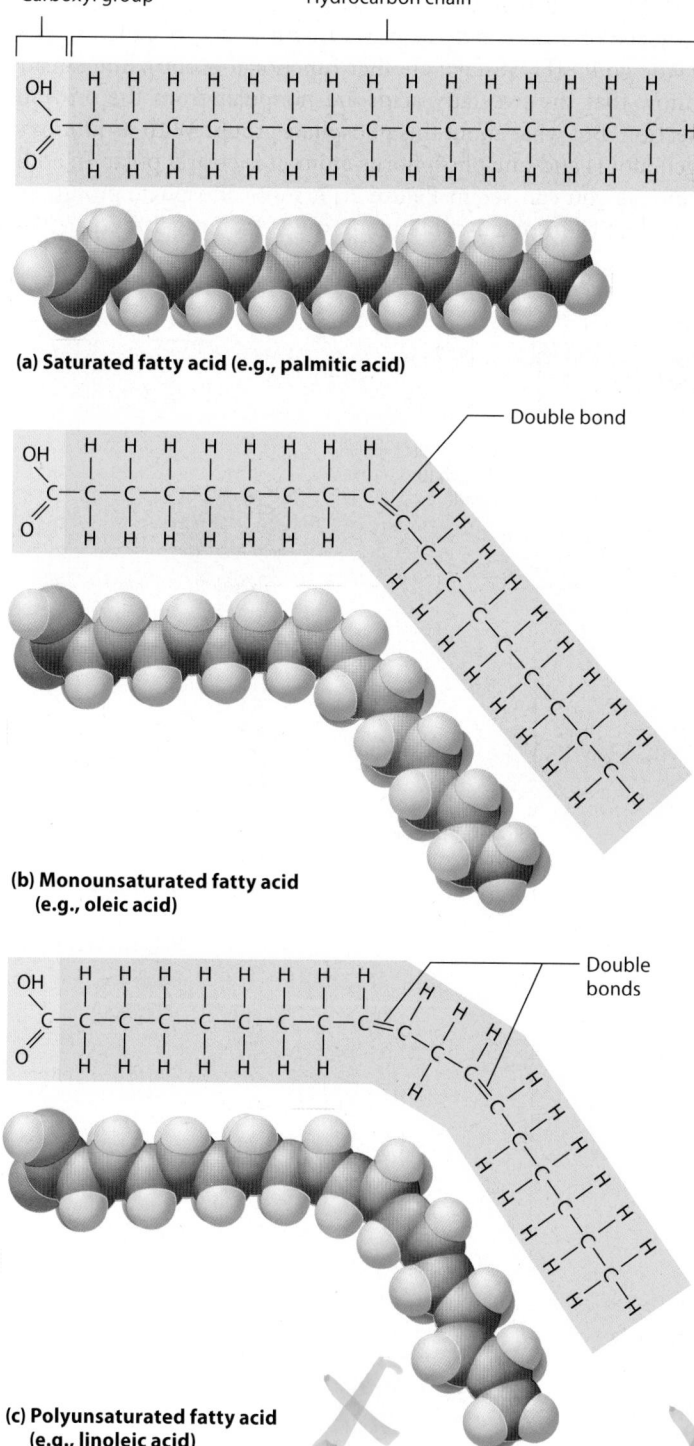

(a) **Saturated fatty acid (e.g., palmitic acid)**

(b) **Monounsaturated fatty acid (e.g., oleic acid)**

(c) **Polyunsaturated fatty acid (e.g., linoleic acid)**

Figure 2.17 Lipids: structure of fatty acids.

Figure 2.18 Lipids: structure and formation of triglycerides.

Although all triglycerides share the same basic structure, their properties vary because of the different fatty acids that may be attached to the glycerol. It is partly this variation that determines if a triglyceride is solid (a fat) or liquid (an oil) at room temperature. As you can see in the linoleic and oleic acid structures shown in Figure 2.17, the long hydrocarbon chains of unsaturated fatty acids are kinked. This prevents them from being packed closely together and solidifying. For this reason, a triglyceride that contains one or more unsaturated fatty acids is likely to be liquid. The long, straight chains of the saturated fatty acids allow them to be packaged together tightly, and triglycerides with three saturated fatty acids are solid. See *A&P in the Real World* on page 56 for information on how these properties could affect your health.

Triglycerides are stored in fat cells, where they help to insulate our internal organs against physical shock and regulate body temperature. In addition, triglycerides are a ready supply of fuel. When they are released to be catabolized, their fatty acids are removed from the glycerol backbone. These fatty acids are then broken apart into smaller two- and three-carbon subunits that are oxidized to convert their chemical energy into a form that is usable by the cell. Each triglyceride can yield multiple smaller subunits due to its long fatty acid chains, which makes triglycerides a very efficient source of energy.

Phospholipids

Take a look at the **phospholipid** (foss-foh-LIP-id) molecule in **Figure 2.19**. Part of the molecule looks familiar, with a glycerol backbone and two fatty acids as in the triglycerides. However, in place of a third fatty acid, we find a structure called a *phosphate group* (FOSS-fayt; another functional group). You already know that the two fatty acids are nonpolar from the previous section. But what about this phosphate group? With its four oxygen atoms and one phosphorus atom, it is clearly polar. In addition, as you can see in Figure 2.19, most phosphate groups are attached to nitrogen-containing groups that are also polar. This

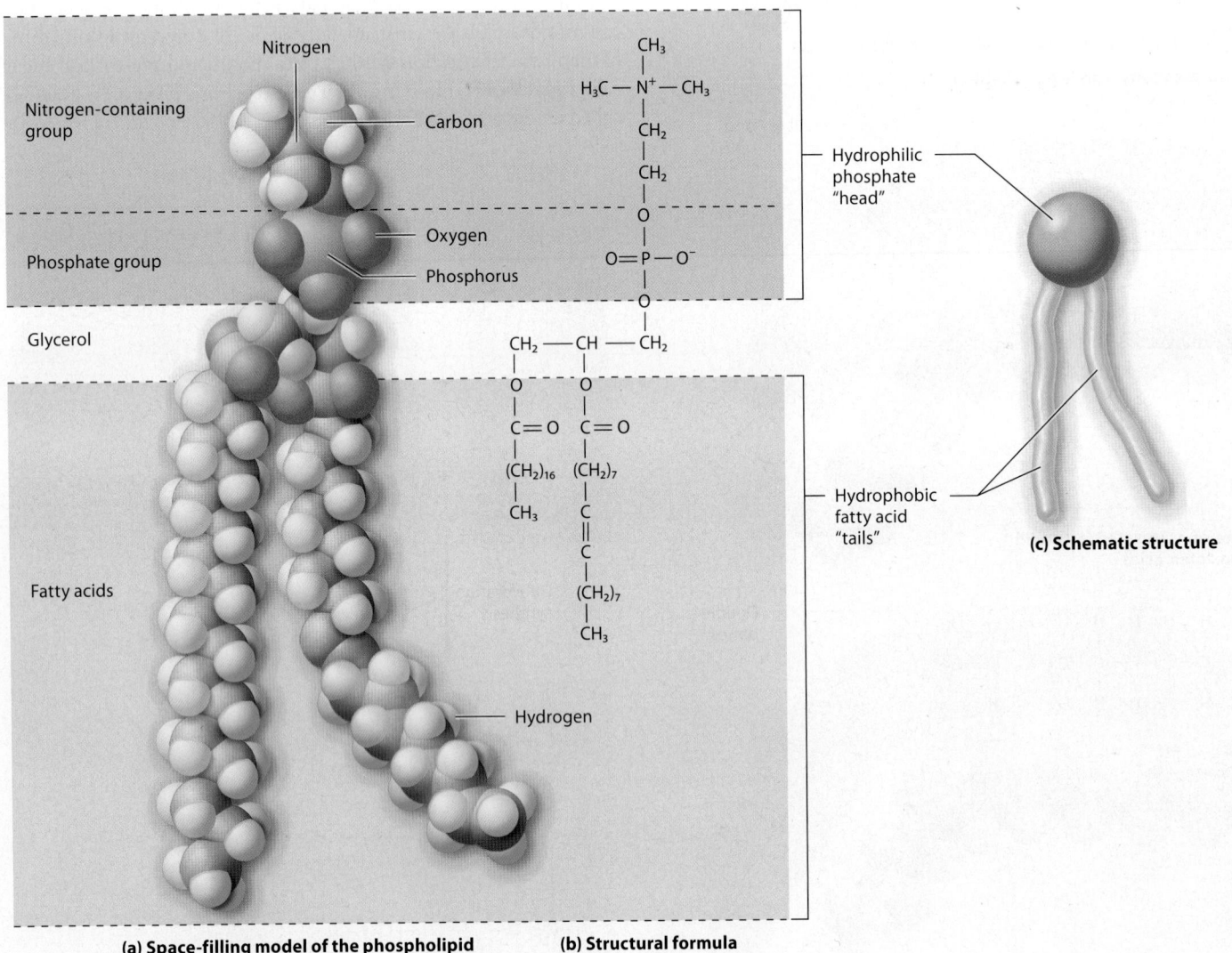

(a) Space-filling model of the phospholipid phosphatidylcholine

(b) Structural formula

(c) Schematic structure

Figure 2.19 Lipids: structure of phospholipids.

Practice art labeling
@MasteringA&P®

(a) Steroid nucleus found in
all steroids

(b) Cholesterol

Figure 2.20 Lipids: structure of steroids.

leaves us with a two-sided molecule: One side, called the *phosphate head,* is very polar and hydrophilic, and the other, called the *fatty acid tails,* is very nonpolar and hydrophobic. Molecules such as phospholipids with strongly polar and strongly nonpolar parts are known as *amphiphilic* (am-fih-FIL-ik; "likes both") molecules. This amphiphilic nature gives phospholipids a number of important properties that allow them to serve as the main structural components of *cell membranes,* the barrier between our cells and their environment. The schematic structure shown in Figure 2.19 is the one we'll use to represent phospholipids throughout this text. We explore phospholipids in more detail in the cell chapter (see Chapter 3).

Steroids

The class of lipids known as **steroids** all share a four-ring hydrocarbon structure, shown in **Figure 2.20**, called the *steroid nucleus.* The steroid *cholesterol* forms the basis for the body's other steroids, including bile acids that function in digestion and the sex hormones estrogen and testosterone (look ahead to Table 2.3 on page 61). The hydrocarbon rings of the steroid nucleus are nonpolar, and most molecules derived from it are nonpolar as well.

Quick Check

☐ 5. What atoms make up lipids? Are these molecules polar or nonpolar?
☐ 6. How do phospholipids and triglycerides differ?
☐ 7. What are steroids?

Proteins

You may not be aware of it, but every time you look at your reflection in the mirror, you are seeing mostly proteins. Proteins make up your hair, nails, and the outer part of your skin; they color your skin, eyes, and hair; they compose the whites of your eyes; they give your bones their structure; and so on. The human body produces from 50,000 to 100,000 different types of proteins, and they constitute about 20% of the body's mass. **Proteins** are macromolecules with numerous functions: They act as enzymes, play vital structural roles, function in the body's defenses, allow cells to communicate, make muscles contract, can be oxidized as fuel, and much more. In this section about proteins, let's first take a look at the monomers that make up proteins and protein structure.

Amino Acids

The monomer of all proteins is the **amino acid,** illustrated in **Figure 2.21**. Each type of protein in the human body is made up of a unique series of the 20 different amino acids. As you can see in Figure 2.21a, all amino acids share the same core structure and contain carbon, hydrogen, oxygen, and nitrogen atoms. In the core structure, each has a central carbon atom bound to (1) a hydrogen atom, (2) a nitrogen-containing amino group ($-NH_2$; another functional group), and (3) a carboxylic acid group ($-COOH$). The name "amino acid" is derived from these *amino* and carboxylic *acid* groups. Each amino acid is made unique by the "R" group also attached to its central carbon. (The "R" in the name "R group" is just a generic abbreviation that stands for "residue.") Note in Figure 2.21b that the R groups of amino acids vary widely in terms of their complexity and properties, from the hydrogen atom of the simplest amino acid glycine to the large and complicated ring structure of the amino acid tryptophan. R groups also vary in their polarity and whether they act as acids

(a) **Amino acid structure**

(b) **Examples of amino acids**

Figure 2.21 Proteins: structure of amino acids.

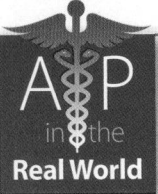

The Good, the Bad, and the Ugly of Fatty Acids

When it comes to the world of lipids, not all fatty acids were created equally. Some fatty acids are good for you, others are bad for you in excess, and still others are downright ugly:

- *The Good: Omega-3 Fats.* Omega-3 fatty acids, found in items such as flaxseed oil and fish oil, are a family of polyunsaturated fatty acids that the human body cannot manufacture, so they must be supplied in the diet. Research has indicated that omega-3 fatty acids have positive effects on cardiovascular health.
- *The Bad: Saturated Fats.* Saturated fatty acids are found in animal fats as well as certain oils such as palm and coconut. Hundreds of scientific studies have demonstrated a link between overconsumption of saturated fats and increased risk for certain diseases, especially heart disease.
- *The Ugly: Trans Fats.* Trans fatty acids are produced by adding hydrogen atoms to certain unsaturated plant oils (producing "partially hydrogenated oils"). There is no dietary requirement for trans fats, and in fact there is no safe level of trans fat consumption period, as they significantly increase the risk for heart disease.

or bases. The behavior of a protein in water is largely determined by the properties of the R groups in its amino acids.

Peptides and Peptide Bonds

Just as fatty acids can be linked to glycerol and monosaccharides to one another by dehydration synthesis, amino acid monomers may also be linked by dehydration synthesis to form polymers. Notice in the example in **Figure 2.22** that during this process the amino group of glycine loses a hydrogen atom and the carboxylic acid group ($-COOH$) of alanine loses a hydroxyl group ($-OH$). The result is the formation of a molecule known as a **peptide,** in which two amino acids are joined by a polar covalent bond called a **peptide bond,** and a water molecule. (Note, however, that proteins may have both polar and nonpolar sections.)

In the same way as other organic polymers, peptides may also be degraded by hydrolysis. Peptides are named for the number of amino acids they contain—*dipeptides* consist of 2, *tripeptides* have 3, and **polypeptides** contain 10 or more amino acids. **Proteins** consist of one or more polypeptide chains that are folded into a distinct structure. A protein must maintain this structure in order to be functional.

Protein Structure

There are two basic types of proteins that may be classified according to their structure: fibrous and globular. The *fibrous proteins* are long protein strands composed of mostly nonpolar amino acids. They are found in structures such as hair, nails, tendons, and bone, and they are tough, durable proteins that give a high degree of strength to a tissue. Fibrous proteins resemble ropes, and indeed they are, in essence, the "ropes" of the body. Just as a rope ties two structures together, fibrous proteins help to connect or "tie" the body's structures together, and allow our tissues to resist stretching and twisting forces. In *globular proteins,* the polypeptide chains assemble into a "globe" or roughly spherical shape. Most globular proteins are polar and function as enzymes, hormones, and other cell messengers.

The structures of fibrous and globular proteins are often highly complex. Many are even composed of multiple polypeptide chains that interact to form the protein as a whole. Due to this complexity, we divide a protein's structure into four levels (**Figure 2.23**):

1. **Primary structure.** A protein's **primary structure** is simply the amino acid sequence of its polypeptide, which includes all of the amino acids held together by covalent peptide bonds (Figure 2.23a).
2. **Secondary structure.** One or more segments of the polypeptide fold over on themselves in characteristic ways, and these folds are called the polypeptide's **secondary structure.** Two of the most common folding patterns are shown in Figure 2.23b: the **alpha helix,** which resembles a spring, and the **beta-pleated sheet,** which resembles Venetian blinds. Each of these structures is stabilized by hydrogen bonds between the amino acids. Note that a given protein may have several regions of different secondary structures.
3. **Tertiary structure.** A protein's **tertiary structure** (Figure 2.23c) refers to the final three-dimensional shape that a polypeptide chain assumes. It consists of the twists, folds, and coils that form between and around the alpha helices and

Figure 2.22 Proteins: formation and breakdown of dipeptides.

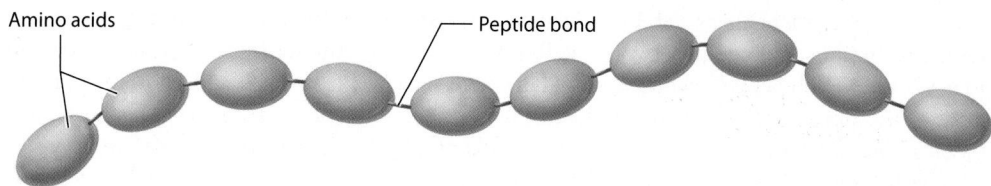

(a) Primary structure:
The amino acid sequence of the polypeptide.

Amino acids

Peptide bond

(b) Secondary structure:
Amino acids of the protein interact via H bonds to form an alpha helix or a beta-pleated sheet.

Beta-pleated sheet

Alpha helix

Amino acids

Hydrogen bonds

Beta-pleated sheets

Alpha helices

(c) Tertiary structure:
Interactions between R groups help to determine the shape the protein assumes after it is folded.

Globular protein

Fibrous protein

(d) Quaternary structure:
The assembly of two or more polypeptide chains into the functional protein. Note that some proteins lack a quaternary structure.

Practice art labeling
@MasteringA&P®

Figure 2.23 Levels of protein structure.

beta-pleated sheets. This shape is stabilized by hydrogen bonds and ionic interactions that occur between the R groups of the amino acids in the polypeptide chain. Notice that the tertiary structure of the fibrous protein is nearly identical to its secondary structure. This is because most fibrous proteins simply consist of very long alpha helices or beta-pleated sheets.

4. **Quaternary structure.** Many proteins consist of more than one polypeptide chain, and the manner in which these polypeptide chains assemble makes up the **quaternary structure** of the protein (Figure 2.23d). Each polypeptide chain that contributes to the protein's quaternary structure has its own primary, secondary, and tertiary structure.

You learned previously about the Structure-Function Core Principle—the way structures in the body are shaped allows them to best carry out their functions (see page 25). This is true even at the molecular level: A protein's shape determines its function. Each polypeptide and protein contains a unique number and sequence of amino acids, and even minor changes to its primary structure can alter all of its subsequent levels of structure and render it nonfunctional. A classic example is found in the disease *sickle cell anemia,* in which the substitution of a single amino acid in the oxygen-carrying protein hemoglobin causes the protein to form chains inside blood cells. The chains of hemoglobin distort the blood cells and cause them to assume the characteristic "sickle" shape, which produces the symptoms of the disease (see Figure 19.7).

CORE PRINCIPLE
Structure-Function

Protein Denaturation

A protein that loses its shape cannot function properly. For example, enzymes are highly specific for their individual substrates. When an enzyme loses its shape, it is no longer able to bind to its substrate and carry out its functions. The process that destroys a protein's shape is called **denaturation.** Many things can denature proteins, including heat, pH changes, and chemicals such as alcohol, as they disrupt hydrogen bonds and ionic interactions within the protein that stabilize its structure. The human body cannot tolerate extensive temperature and pH swings because they denature our proteins.

Quick Check

☐ 8. What is an amino acid? What four components does each amino acid contain?
☐ 9. How are amino acids linked to form peptides and proteins? What are the bonds called?
☐ 10. What are the four levels of structural organization of a protein?

Nucleotides and Nucleic Acids

◀◀ Flashback

1. What is chemical energy? How is energy stored in chemical bonds? (p. 12)
2. What is an exergonic reaction? (p. 12)

The last class of organic molecules in the body we'll cover is the **nucleotides** (NEW-klee-oh-tydz). The name "nucleotide" comes from the fact that many of these compounds are found in the central core of the cell, the *nucleus*. Nucleotides are the monomers that form our genetic material, the **nucleic acids** *deoxyribonucleic acid* (*DNA;* dee-ok-see-RY-boh-new-klay'-ik) and *ribonucleic acid* (*RNA*), which are vital to many cellular processes. Unlike the other organic molecules, nucleic acids are not oxidized by the cell for fuel. However, nucleotides do play a key role in the energy systems of the body—the nucleotide *adenosine triphosphate* (*ATP*) is our main source of chemical energy, the molecule that drives nearly all of our cellular processes. In this section, we explore the structure and function of these molecules in greater depth.

Nucleotide Structure

Notice in **Figure 2.24a** that a nucleotide is composed of three parts: (1) a nitrogenous base, (2) a five-carbon (pentose) sugar, and (3) a phosphate group. The first part, the nitrogenous **base,** is a nitrogen-containing molecule with a hydrocarbon ring structure. (Be careful not to confuse the term "base" here with the bases we discussed earlier with reference to pH.) There are two types of nitrogenous bases: purines and pyrimidines (**Figure 2.24b**). *Purines*

Nitrogenous base — | Phosphate group
Pentose sugar

(a) Structure of a nucleotide

Purines

Adenine (A)   Guanine (G)

Pyrimidines

Cytosine (C)   Thymine (T)   Uracil (U)

(b) The nitrogenous bases can be either purines or pyrimidines.

Figure 2.24 Structure of nucleotides.

(PYUR-eenz) are double-ringed molecules and include the bases **adenine** and **guanine.** *Pyrimidines* (pir-IM-ih-deenz) are single-ringed molecules and include the bases **cytosine, thymine,** and **uracil.** Each nitrogenous base is abbreviated by the first letter of its name, so the five bases are usually written simply as A, G, C, T, and U, respectively.

Attached to the nitrogenous base is the second component of a nucleotide: a pentose sugar, which may be the sugar ribose or deoxyribose (see Figure 2.14a). Covalently bound to the pentose sugar we find the third component, a group that you saw previously in phospholipids: the phosphate group. Nucleotides may contain one, two, or three phosphate groups.

Adenosine Triphosphate (ATP)

As we discussed in Module 2.2, the body's main source of energy is *chemical energy,* or the energy present in chemical bonds. The primary source of chemical energy in the body comes in the form of a nucleotide called **adenosine triphosphate, or ATP** (ah-DEN-ah-seen; **Figure 2.25a**). ATP contains the nitrogenous base adenine attached to ribose—these two together are known as *adenosine*—and three phosphate groups. Adenosine with two phosphates is *adenosine diphosphate* (*ADP*), and with only one phosphate is *adenosine monophosphate* (*AMP*). Notice that each phosphate group is negatively charged.

ATP is generally synthesized by adding a third phosphate to ADP (**Figure 2.25b**). Just as opposite charges attract, the negatively charged phosphates, being like charges, repel each other, so a great deal of energy is required to add a third phosphate group to ADP. When this bond is broken by hydrolysis, resulting in ADP and the inorganic phosphate group (abbreviated P_i), the chemical energy in the bond is released as the molecule assumes a more stable form. Note that this bond is sometimes referred to as a "high-energy bond," but in truth it contains no more energy than any other molecular bond. A large amount of energy is released when the bond is *broken,* but energy is not stored within the bond itself.

The cell makes ATP through the catabolism of organic molecules, primarily the monosaccharide glucose. Glucose undergoes a series of reactions in the cell during which its chemical bonds are broken, and the chemical energy released drives the cellular processes that add a phosphate group to ADP. The processes by which this happens are discussed in the metabolism chapter (see Chapter 23).

A molecule of ATP lasts maybe 60 seconds in the cell before it is hydrolyzed, as nearly every process in the cell is driven by the energy released from ATP hydrolysis. ATP is not stored in any significant amount by our cells, and our entire supply of ATP is exhausted in only 60–90 seconds. For this reason, a cell must continually replenish its ATP supply or risk failure of essentially all of its functions. This becomes critically obvious when a person stops breathing: Oxygen is required for most ATP synthesis, and, in fact, the need for ATP is the main reason why we breathe oxygen. When breathing stops, cells can no longer make sufficient ATP, and death occurs within only three to four minutes.

The Nucleic Acids: DNA and RNA

The nucleic acids account for only about 2% of the body's mass, but they are some of the most important molecules in the body. The two main nucleic acids are **deoxyribonucleic acid, or DNA,** and **ribonucleic acid, or RNA.** Together they are responsible for the storage and execution of the genetic code.

(a) The structure of ATP

(b) The formation of ATP from ADP requires energy; the reverse reaction liberates energy.

Figure 2.25 Nucleotides: structure and formation of ATP.

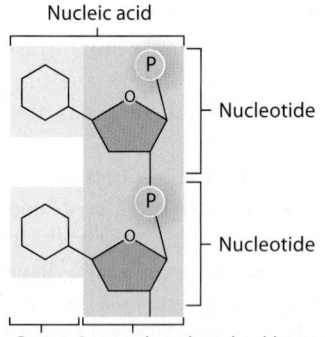

Nucleic acid

Nucleotide

Nucleotide

Bases Sugar-phosphate backbone

Nucleic acids are polymers of linked nucleotides. The nucleotides in a nucleic acid are linked by dehydration synthesis, which joins the phosphate group of one nucleotide to the sugar of the next by a polar covalent bond. Notice in the illustration of part of a nucleic acid, to the left, that the linked sugar and phosphate groups form the molecule's "backbone"; the nitrogenous bases extend out to the side and do not participate in the linkage of the nucleotide strand.

Let's examine the role of DNA first. In the previous section, we introduced proteins and their various roles in the body. Recall that proteins are long chains of amino acids joined by covalent peptide bonds, and that each type of protein has a unique amino acid

Sugar-phosphate backbone

Hydrogen bonds

Sugar-phosphate backbone

Base uracil in place of thymine

(a) Double-helical DNA

(b) Single-stranded RNA

Sugar-phosphate backbone:

Deoxyribose

Phosphate group

Hydrogen bonds

Complementary base pairing

Key:

A = Adenine

G = Guanine

T = Thymine

C = Cytosine

U = Uracil (RNA)

Figure 2.26 Structure of the nucleic acids DNA and RNA.

Practice art labeling
@MasteringA&P®

sequence. Each cell produces thousands of different proteins, so how does a cell know which amino acids to join to make a specific protein? The answer is in the cell's DNA.

DNA is an extremely large molecule found in the nucleus of the cell. As you can see in **Figure 2.26a**, DNA is composed of two long chains or strands of nucleotides that twist around one another to form a **double helix.** Other structural features of DNA include the following:

- **DNA contains the sugar deoxyribose.** Recall from our discussion of monosaccharides that deoxyribose is a five-carbon sugar. It is called *deoxy*ribose because one of its carbon atoms lacks an oxygen-containing hydroxyl group (–OH) and has just a hydrogen atom instead.
- **DNA contains the bases adenine, guanine, cytosine, and thymine.** Only the bases A, G, C, and T are found in DNA.
- **The two strands are held together by hydrogen bonds between the bases.** The nucleotide bases form hydrogen bonds with bases on the other strand. These hydrogen bonds are the major force that keeps the two strands of DNA together.
- **DNA exhibits complementary base pairing.** Notice in the blow-up from Figure 2.26a that the two strands of DNA are not identical. However, a pattern is followed throughout the entire molecule: *The purine A always bonds with the pyrimidine T, and the purine G always bonds with the pyrimidine C.* This type of pairing is called **complementary base pairing.** The bases that can pair together are determined by the number of hydrogen bonds that each base can form. Both the bases A and T can form only two hydrogen bonds, whereas the bases C and G can form three hydrogen bonds. As a result, these bases are complementary, and pair up together. For the bases to bond correctly, the two strands of

DNA must run in opposite directions; as you can see in the blow-up in Figure 2.26a, one strand is "upside down" relative to the other.

Within the DNA we find the *genetic code,* which means that it contains the "recipe" for each and every protein in our bodies. The recipe for one individual protein is found within a segment of DNA called a *gene.* The sequence of the nucleotides within each gene ultimately specifies the amino acid sequence of a protein; the cell can read this sequence and assemble the encoded protein via a process called *protein synthesis* (see Chapter 3).

Protein synthesis requires more than just DNA. In fact, DNA is not directly involved in protein assembly at all. The task of assembling the correct amino acid sequence for a protein falls to RNA (**Figure 2.26b**). Unlike DNA, RNA consists only of a single strand of nucleotides, and it is not confined to the nucleus of the cell. In addition, RNA contains the sugar ribose and the nitrogenous base uracil instead of thymine. Uracil binds to adenine in the same manner as thymine, and when an RNA molecule binds to another nucleotide strand, U pairs with A and C with G.

RNA plays a crucial role in protein synthesis. Remember that DNA contains genes, and one gene has the "recipe" for one protein. But the DNA is in the nucleus, which it can't leave, and the "kitchen," with all of the ingredients to make this protein, is outside the nucleus. This means that to make a protein, the cell must first copy the recipe and take it into the cellular kitchen, where it can then put the ingredients together. These two processes, called *transcription* and *translation,* are carried out with the help of RNA molecules. Both transcription and translation are discussed in the cell chapter (see Chapter 3).

Table 2.3 provides a summary of the information we've covered on organic molecules.

TABLE 2.3 ORGANIC MOLECULES			
Type of Molecule	**Structure**	**General Functions**	**Examples/Location**
Carbohydrates			
Monosaccharides Glucose	Carbon, hydrogen and oxygen in 1C:2H:1O ratio; can take on a ring form	• Energy: glucose is the main fuel for all cells • Structure: ribose and deoxyribose are found in nucleic acids	Ribose, deoxyribose, glucose, galactose, fructose/Found in almost all cells
Disaccharides Sucrose	Contain two monosaccharide subunits joined by a polar covalent bond	• Energy: disaccharides are broken down into monosaccharides, which are then used for fuel	Lactose, sucrose, and maltose

(continued)

TABLE 2.3 ORGANIC MOLECULES (continued)

Type of Molecule	Structure	General Functions	Examples/Location
Carbohydrates (continued)			
Polysaccharides Glycogen	Highly branched molecule with hundreds of monosaccharide subunits	• Energy storage: when needed, stored polysaccharides are broken down into monosaccharides and released to be used as fuel	Glycogen is the main storage polysaccharide/Stored in liver and skeletal muscle cells
Lipids			
Triglycerides 	Three fatty acids bonded to a glycerol backbone	• Energy storage: triglycerides are released as needed • Protection and insulation: cushioning and insulation of organs	Multiple combinations of saturated and unsaturated fatty acids are possible/Stored in adipose (fat) cells
Phospholipids 	Two fatty acids and a phosphate group bound to a glycerol backbone	• Structure: main component of cell membranes	Phosphatidylcholine/Found in all membranes
Steroids Testosterone	Four-ringed hydrocarbon with modifications for different molecules	• Regulation: steroid hormones control many physiological processes • Structure: cholesterol forms part of barrier between cells and their environment	Hormones such as testosterone, estrogen, aldosterone, and cortisol; other molecules include cholesterol and bile salts/Made in glands such as testes and ovaries

TABLE 2.3 *(continued)*

Type of Molecule	Structure	General Functions	Examples/Location
Proteins			
Di- and polypeptides	Two or 10–50 amino acids, respectively, joined by a peptide bond or bonds	• Regulation: most function as hormones or chemical messengers in the brain	Substance P, endorphins, glucagon, calcitonin, and insulin

Peptide bond

Dipeptide

Proteins	Long strands of more than 50 amino acids that fold into a characteristic three-dimensional shape	• Structure: primary structural molecule in the body • Movement: involved in cell and muscle movement • Catalysis: function as enzymes • Transport: transport substances through the body • Defense: function in body's defenses	Collagen, keratin, amylase, albumin, hemoglobin, and many others/Found throughout the body

Globular   Fibrous

Nucleotides and Nucleic Acids			
ATP	The nitrogenous base adenine bound to ribose and three phosphate groups	• Energy: main source of chemical energy for the body—drives cellular work	ATP

Nucleic acids	Strings of nucleotides arranged in either a double helix (DNA) or a single strand (RNA)	• Information storage: DNA contains the instructions for building every protein in the body • Information retrieval: RNA carries out the instructions to build proteins	DNA, RNA/Found in nuclei and cytosol of cells

DNA   RNA

Quick Check

☐ 11. What are the three components of a nucleotide?
☐ 12. How do DNA and RNA differ?
☐ 13. What are the components and roles of ATP?

Apply What You Learned

☐ 1. Which molecule would be the most soluble in water: glucose, a triglyceride, or a large protein? Explain.

☐ 2. In Module 2.3, you learned that increasing temperature will also increase reaction rates. Why is there a limit on how high the body's temperature can be raised? (*Hint:* What happens to proteins at high temperatures?)

☐ 3. How could a defect in a gene lead to a malfunctioning enzyme?

See answers in Appendix A.

CHAPTER SUMMARY

Take full advantage of the invaluable online practice and study tools for this chapter by going to MasteringA&P®. There you will find:

 Practice tests, module quizzes, art-labeling and sequence-the-steps activities

▷ Animated figures with quizzes

 Practicing A&P worksheets

))) Author-narrated podcast

🔥 Dynamic Study Modules

iP *Interactive Physiology*

MODULE 2.1
Atoms and Elements 31

- **Matter** is defined as anything that has mass and takes up space. It can exist as a solid, a liquid, or a gas.

- An **atom** is the smallest particle of matter that still retains its properties. It is composed of **subatomic particles:** (Figure 2.1, p. 32)
  ○ **Protons** have a positive charge and reside in the **atomic nucleus.**
  ○ **Neutrons** have no charge and also reside in the atomic nucleus.
  ○ **Electrons** have a negative charge and reside in regions around the atomic nucleus called **electron shells.**

- An **element** is a substance that cannot be broken down into simpler substances by chemical means. It is defined by its number of protons, called the **atomic number.**

- Elements are arranged on the **periodic table** in order of increasing atomic number.

- Four major elements are found in the human body: hydrogen, oxygen, carbon, and nitrogen. The body also contains 7 mineral elements and 13 trace elements. (Figure 2.2, p. 33)

- The number of protons plus the number of neutrons is an atom's **mass number.** Isotopes are atoms with the same atomic number but different mass numbers.

MODULE 2.2
Matter Combined: Mixtures and Chemical Bonds 34

- Matter can be combined physically or chemically.

- Physical combinations of matter are called **mixtures.** There are three types of mixtures: (Figure 2.3, p. 35)
  ○ **Suspensions** contain two or more components with large, unevenly distributed particles that will settle out when left to stand.
  ○ **Colloids** contain two or more components with small, evenly distributed particles that will not settle out.
  ○ **Solutions** contain two or more components with extremely small, evenly distributed particles that will not settle out. A solution contains a **solvent** that dissolves a **solute.** The amount of solute present in a solution is the solution's *concentration*.

- Atoms are combined chemically by **chemical bonds.** Two or more atoms chemically bonded are known as a **molecule.** When the atoms are of different elements, it is a molecule of a **compound.**

- The electrons in an atom's outermost shell are called the **valence electrons.** Atoms with six or more electrons are most stable with eight electrons in the **valence shell** and obey the **octet rule.** Atoms with five or fewer electrons are most stable with two electrons in their first shell and obey the *duet rule.*

- An **ionic bond** forms when a metal and nonmetal transfer electrons so that both atoms obey the octet rule. The charged particles that form when electrons are transferred are called **ions.** (Figure 2.4, p. 37)
  ○ The metal loses one or more electrons to become a positively charged **cation.**
  ○ The nonmetal accepts one or more electrons to become a negatively charged **anion.**
  ○ The ionic bond results from the attraction of the metal cation and the nonmetal anion.

- A **covalent bond** forms when two or more atoms share electrons so that each atom obeys the octet rule. (Figure 2.5 and Table 2.1, p. 38)
  ○ **Nonpolar covalent bonds** result from the equal sharing of electrons between two nonmetals. They form between atoms of identical (or nearly identical) *electronegativities*, or ability to attract electrons. Molecules in which nonpolar covalent bonds predominate are **nonpolar molecules.** (Figure 2.6, p. 39)
  ○ **Polar covalent bonds** result from the unequal sharing of electrons between two nonmetals. The electrons are pulled more strongly by the more electronegative atom, and a **dipole** results: The more electronegative atom has a partially negative charge, and the less electronegative atom has a partially positive charge. Molecules in which polar covalent bonds predominate are called **polar molecules.**
  ○ **Hydrogen bonds** form between polar covalent molecules. They result from the weak attractions between the partially positive

hydrogen atoms and a fully or partially negative nonmetal. Hydrogen bonds give water the property of **surface tension** and many of its other important biological properties. (Figure 2.7, p. 40)

- The types of chemical bonds are summarized in Table 2.2. (p. 40)

MODULE **2.3**

Chemical Reactions 41

- Any time chemical bonds are formed, broken, or rearranged, or electrons are transferred between atoms, a **chemical reaction** has occurred.
- The basic form of **chemical notation** is the **chemical equation,** which generally shows the **reactants** on the left and the **products** on the right.
- **Energy** is defined as the capacity to do work. There are two types of energy: **potential energy,** which is energy that is stored, and **kinetic energy,** which is energy in motion. Three forms of potential and kinetic energy are found in the human body:
  - **Chemical energy** is energy inherent in chemical bonds.
  - **Electrical energy** is energy from the movement of charged particles.
  - **Mechanical energy** is energy that has been directly transferred from one object to another.
- A reaction that releases energy is an **exergonic reaction;** a reaction that consumes energy is an **endergonic reaction.**
- There are three types of reactions in the human body:
  - **Catabolic** or *decomposition* **reactions** break larger molecules into smaller ones, and are generally exergonic.
  - During an **exchange reaction,** one or more atoms from the reactants are exchanged for another. In an **oxidation-reduction (redox) reaction,** electrons are transferred between reactants.
  - **Anabolic** or **synthesis reactions** form larger molecules from smaller ones, and are generally endergonic.
- The rate of a reaction is determined by the concentration of the reactants, temperature, size and phase of the reactants, and the presence or absence of a *catalyst*. It is also determined by the **activation energy,** which is the amount of energy that must be put into any reaction for it to proceed to completion. (Figure 2.8, p. 43)
- **Enzymes** are biological catalysts that increase the speed of a reaction by stabilizing the *transition state,* which lowers a reaction's activation energy. Enzymes do not change the products or reactants, are not consumed during the reaction, and are highly specific for individual **substrates.** (Figure 2.9, p. 44; Figure 2.10, p. 45)

  Practice art-labeling exercise: Figure 2.9.

  Practice sequencing the steps: Figure 2.10.

MODULE **2.4**

Inorganic Compounds: Water, Acids, Bases, and Salts 45

- **Water** has a high *heat capacity,* takes a great deal of heat with it when it evaporates, and has a high *density,* properties that make it a useful solvent in the human body.
- Water acts as a universal *solvent*. Substances with polar covalent and ionic bonds will interact with and dissolve in water and so are **hydrophilic,** whereas those with nonpolar covalent bonds will not dissolve in water and are **hydrophobic.** (Figure 2.11, p. 46)

- An **acid** is a hydrogen ion donor, and a **base** is a hydrogen ion acceptor. (Figure 2.12, p. 47)
- The **pH scale** is a logarithmic scale that measures the hydrogen ion concentration of a solution. (Figure 2.13, p. 48)
  - A pH less than 7 is **acidic;** as the pH decreases, the hydrogen ion concentration increases and the solution becomes more acidic.
  - A pH of 7 is called **neutral.**
  - A pH greater than 7 is **basic** (alkaline); as the pH increases, the hydrogen ion concentration decreases and the solution becomes more basic.
- A **buffer** is a system of chemicals that resist a change in pH. Buffers help to maintain the pH of the blood between 7.35 and 7.45.
- A **salt** is a metal cation bonded ionically to a nonmetal anion. Salts are **electrolytes** because they dissociate in water and conduct an electric current. Some acids and bases can also be electrolytes.

  ▶ **Play Interactive Physiology tutorial on Fluid, Electrolyte and Acid/Base Balance: Introduction to Body Fluids.**

MODULE **2.5**

Organic Compounds: Carbohydrates, Lipids, Proteins, and Nucleotides 50

- Each type of organic compound has its own **monomer** and the corresponding **polymer** built from those subunits.
- **Carbohydrates** are composed of carbon, hydrogen, and oxygen. They function primarily as fuels in the body but also play structural roles.
  - **Monosaccharides** have from three to seven carbons, and are the monomers from which all carbohydrates are made. The most abundant monosaccharides in the body are **glucose,** *fructose, galactose, ribose,* and *deoxyribose.* (Figure 2.14, p. 51)
  - **Disaccharides** are formed from the union of two monosaccharides by **dehydration synthesis.** (Figure 2.15, p. 51)
  - **Polysaccharides** consist of many monosaccharides joined by dehydration synthesis. **Glycogen** is the body's main polysaccharide. (Figure 2.16, p. 52)
- **Lipids** are nonpolar, hydrophobic molecules composed mostly of carbon and hydrogen.
  - A monomer of fats is the **fatty acid.** There are saturated and unsaturated fatty acids. Three fatty acids are joined to a molecule of glycerol to form a **triglyceride.** (Figure 2.17, Figure 2.18, p. 53)
  - A **phospholipid** contains two fatty acids and a *phosphate group* bound to glycerol. It has both polar and nonpolar parts, and so is an *amphiphilic* molecule. (Figure 2.19, p. 54)

  Practice art-labeling exercise: Figure 2.19.

  - **Steroids** are a class of lipids based upon a four-ring hydrocarbon *steroid nucleus.* The steroid *cholesterol* forms the basis for the body's other steroids. (Figure 2.20, p. 55)
- **Proteins** function as enzymes, play structural roles, are involved in movement, function in the body's defenses, can be used as fuel, and more.

○ Each protein is composed of a unique sequence of 50 or more **amino acid** monomers joined together into a **polypeptide** by **peptide bonds.** (Figure 2.21, p. 55; Figure 2.22, p. 56)

○ Proteins have four levels of organization to their structure: The **primary structure** is the amino acid sequence, the **secondary structure** is the repeating ways in which the amino acid strand folds on itself (stabilized by H bonds), the **tertiary structure** is the final three-dimensional shape the protein assumes, and the **quaternary structure** is the arrangement of multiple polypeptide chains within a functional protein. (Figure 2.23, p. 57)

🖑 **Practice art-labeling exercise: Figure 2.23.**

○ *Fibrous proteins* consist of long strands of **alpha helices** or **beta-pleated sheets.** They are found in tissues requiring a great degree of resistance to tension.

○ *Globular proteins* are roughly spherical in shape and function as regulatory proteins or enzymes.

○ Heat, pH changes, and certain chemicals cause a protein to permanently lose its shape and therefore its function. This is called protein **denaturation.**

• A **nucleotide** is a monomer composed of three parts: a nitrogenous base, a sugar, and a phosphate group.

• There are two types of nitrogenous bases:
  ○ The *purines* have a double ring structure and include **adenine** (A) and **guanine** (G).
  ○ The *pyrimidines* have a single ring structure and include **cytosine** (C), **thymine** (T), and **uracil** (U).

• **Adenosine triphosphate (ATP)** is the body's chief source of chemical energy. ATP is synthesized from adenosine diphosphate (ADP) and a phosphate group using the energy from the oxidation of fuels such as glucose. The production of large amounts of ATP requires oxygen, which is the main reason why we breathe oxygen. (Figure 2.25, p. 59)

• The **nucleic acids** DNA and RNA are polymers of nucleotides linked by dehydration synthesis. (Figure 2.26, p. 60)

🖑 **Practice art-labeling exercise: Figure 2.26.**

• **Deoxyribonucleic acid,** or **DNA,** is a large molecule confined to the nucleus of the cell. It consists of two strands of nucleotides that are joined by hydrogen bonds and twist around one another to form a **double helix.**
  ○ DNA contains the sugar deoxyribose and the nitrogenous bases C, T, A, and G.
  ○ The two strands of nucleotides exhibit **complementary base pairing:** C always binds with G, and T always binds with A.
  ○ DNA contains *genes* that provide the "recipe" or code for every protein in the body.

• **Ribonucleic acid,** or **RNA,** is a single strand of nucleotides that can move between the cell's nucleus and the cytosol.
  ○ RNA contains the sugar ribose and the nitrogenous bases A, C, G, and U.
  ○ RNA is a copy of the protein "recipe" from DNA that is made via transcription. It helps to assemble amino acids into a protein outside the nucleus via translation.

• The types of organic molecules are summarized in Table 2.3 (p. 61).

ASSESS WHAT YOU LEARNED

See answers in Appendix A.

LEVEL 1 Check Your Recall

1. Mark the following statements as true or false. If a statement is false, correct it to make a true statement.
   a. The mass number of an atom is the sum of its neutrons and protons.
   b. Electrons travel in orderly, circular orbits around the nucleus.
   c. Protons and neutrons have a positive charge and electrons have a negative charge.
   d. Valence electrons are the electrons in the outermost shell of an atom.
   e. Every element has a characteristic number of protons, which is called the element's mass number.

2. Fill in the blanks: Isotopes are atoms with the same _____ number but different _____ numbers.

3. Which of the following statements correctly describes a solution?
   a. In a solution the solute is chemically dissolved by the solvent.
   b. Solutions involve large particles suspended in another component.
   c. The particles in a solution will settle out if left to sit.
   d. The amount of solute in a solution is expressed as the solution's concentration.

4. Explain the difference between an ionic and a covalent bond.

5. Identify each of the following molecules as ionic, polar covalent, or nonpolar covalent.
   a. H_2O _____
   b. LiI _____
   c. $C_{10}H_{22}$ _____
   d. N_2 _____
   e. $MgBr_2$ _____
   f. H_2S _____

6. What are hydrogen bonds, and how do they form?

7. Describe what takes place during a chemical reaction.

8. Match the following terms with the correct definition:
   _____ Endergonic reaction
   _____ Potential energy
   _____ Electrical energy
   _____ Anabolic reaction
   _____ Oxidation-reduction reaction
   _____ Chemical energy
   _____ Catabolic reaction
   _____ Kinetic energy

   a. Energy in motion
   b. Energy stored in chemical bonds
   c. Reaction that consumes energy
   d. A decomposition reaction
   e. The energy of moving charged particles
   f. Stored energy
   g. Synthesis reaction
   h. Reaction where electrons are transferred between reactants

9. Which of the following would *not* result in an increase in reaction rate?
   a. Increase in temperature
   b. Particles in the solid phase
   c. Presence of a catalyst
   d. Smaller particle size

10. An enzyme is a:

    a. biological catalyst that increases the concentration of the products.

    b. biological catalyst that increases the concentration of the reactants.

    c. chemical that is used in a reaction, after which it can no longer be used by the cell.

    d. biological catalyst that works by bringing its substrates closer to their transition states.

11. Which of the following is *not* a property of water?

    a. Water has a high heat capacity.

    b. Water doesn't take heat with it when it evaporates.

    c. Water is a polar solvent in which many solutes will dissolve.

    d. Water serves as a cushion and lubricant in the body.

12. With respect to their solubility in water, substances with non-polar covalent bonds are _____, while substances with polar covalent and ionic bonds are _____.

13. Mark the following statements as true or false. If a statement is false, correct it to make a true statement.

    a. Acids dissociate in water to give hydrogen ions and an anion.

    b. Bases are defined as hydrogen ion donors.

    c. As a solution's pH rises, it becomes more acidic.

    d. A buffer is a system of chemicals that resists a change in pH.

    e. A solution with a pH of 10 has more hydrogen ions than a solution with a pH of 2.

14. A salt is:

    a. a metal cation bound to a nonmetal anion.

    b. a hydrogen ion donor.

    c. an electrolyte that conducts electricity in water.

    d. Both a and c are correct.

    e. Both b and c are correct.

15. Compare and contrast the structures of phospholipids and triglycerides.

16. Mark the following statements as true or false. If a statement is false, correct it to make a true statement.

    a. Carbohydrates are the main source of fuel for the human body.

    b. The main storage form of glucose in the human body is starch.

    c. Lipids contain oxygen and are therefore polar covalent molecules.

    d. Steroids, phospholipids, and triglycerides are all types of lipids.

    e. Proteins are composed of strings of fatty acids linked by glycosidic bonds.

    f. Amino acids have a carboxylic group on one end and an amino group on their other end.

17. Why is it important for a protein to maintain its structure? What is it called when a protein loses its structure?

18. Which of the following is *not* part of a nucleotide?

    a. An amino acid

    b. A nitrogenous base

    c. One or more phosphate groups

    d. A sugar

19. Mark the following properties as belonging to either DNA or RNA.

    a. Contains the nucleotides A, G, C, and U _____

    b. Found only in the nucleus of the cell _____

    c. Can move in and out of the cell's nucleus _____

    d. Contains two strands of nucleotides linked by hydrogen bonds _____

    e. Contains the code for every protein in the body _____

    f. Contains the sugar ribose _____

20. Which of the following statements is/are true regarding ATP?

    a. It is the body's primary source of chemical energy.

    b. Its production is the main reason humans breathe oxygen.

    c. It is produced through the oxidation of molecular fuels like glucose.

    d. All of the above statements are true.

LEVEL 2 Check Your Understanding

1. In certain types of radioactive decay, the isotope releases a particle called an alpha particle, which contains two protons and two neutrons. When this happens, is the product still the same element? Why or why not?

2. Considering that water is a main component of the juices in the stomach and intestines, explain why digestion of lipids is more complicated than digestion of carbohydrates and proteins.

3. Explain why monosaccharides are polar and fatty acids are nonpolar even though they both contain the same atoms.

LEVEL 3 Apply Your Knowledge

PART A: Application and Analysis

1. The polysaccharide cellulose is not digestible by humans, as we lack the enzyme *cellulase*, which is required to break it down. Certain dietary supplements contain the enzyme cellulase, and claim that being able to break down cellulose will help a person lose weight. But what do you think would happen if we could digest the cellulose we ate?

2. Some claim that the pH of your blood can be affected by eating acidic foods such as citrus. Do you believe this to be true? Explain your answer. (*Hint:* What happens when extra hydrogen ions are added to the blood?)

3. Many drugs and poisons exert their effects by blocking one or more enzymes. How could blocking an enzyme lead to the death of a cell?

4. You have just dropped some phospholipids into water. Predict how the hydrophilic and hydrophobic ends of these molecules will align themselves. (*Hint:* The molecules in air are nonpolar.)

PART B: Make the Connection

5. Explain how buffer systems in the body work if the pH of body fluids increases. Is this an example of a negative or a positive feedback loop? Explain. (*Connects to Chapter 1*)

3

The Cell

Our **cells** are the smallest units that can carry out the functions of life (see Chapter 1). Humans are composed of trillions of cells that work together to maintain homeostasis for the whole organism. In this chapter, we take a tour of a typical human cell, exploring the structure and functions of its various parts, the ways in which it interacts with its environment, and how it reproduces itself.

MODULE **3.1**

Introduction to Cells

Learning Outcomes

1. Identify the three main parts of a cell, and list the general functions of each.
2. Describe the location and components of intracellular and extracellular fluid.
3. Explain how cytoplasm and cytosol are different.
4. Define the term organelle, and describe the basic functions of organelles.

Within any cell many chemical reactions and cellular processes are carried out continuously in order for the cell to survive. In addition to these basic processes, cells also perform a host of specialized functions that are unique to each cell type. In this module, we examine these basic processes and the common cellular components that perform them. We conclude the module with a look at how cell types differ in structure, and how this relates to a cell type's specialized functions.

Basic Processes of Cells

Several basic processes are common to all cell types. These processes include cell metabolism, transport of substances, communication, and cell reproduction.

- **Cell metabolism.** The chemical reactions that a cell carries out to maintain life are known collectively as **cell metabolism.** Metabolic reactions include all of the different types of reactions that you read about earlier (see Chapter 2), including:
  ○ *Anabolic reactions.* Anabolic, or building, reactions lead to the production of macromolecules. This includes synthesis of polysaccharides, lipids, proteins, and nucleic acids.
  ○ *Catabolic reactions.* In other metabolic reactions, cells use catabolic reactions to break down macromolecules into smaller compounds. Some catabolic reactions are required to degrade old and worn-out cellular components, whereas others break down nutrients such as carbohydrates and lipids.
  ○ *Oxidation-reduction reactions.* Some of the cell's oxidation-reduction reactions convert the energy in the chemical bonds of nutrients such as glucose into a

Computer-generated image: The series of events from cell formation to cell division, shown here, is called the cell cycle.

form of chemical energy that the cell can use to fuel its processes, namely, *adenosine triphosphate*, or ATP.

- **Transport of substances.** A cell must transport molecules that it has produced or ingested to a variety of destinations. Some molecules are transported for release from the cell, whereas others are simply transported to a different location within the cell.
- **Communication.** A cell communicates with itself, with its surrounding environment, and with other cells in the body. This communication occurs through a variety of means, including chemical and electrical signals.
- **Cell reproduction.** Almost all cells in the body can reproduce themselves through a process called *cell division*. Cell division is necessary for growth and development, and for replacement of old and damaged cells.

Quick Check

☐ 1. What are the general functions of cells?

Overview of Cell Structure

Most animal cells have three basic components: the plasma membrane, cytoplasm, and nucleus (**Figure 3.1**). We introduce the basic structure and functions of each of these three components here, and we examine each in greater depth in later modules.

Plasma Membrane

Surrounding each cell is a "fence" called the **plasma membrane.** The plasma membrane serves several important functions, including structural support, communication with other cells, and cell identification. And, of course, one of the plasma membrane's primary functions is the same as that of any fence—to physically isolate the cell from its surroundings.

But what is the plasma membrane separating the cell from? The body contains two fluid compartments: (1) the *intracellular space*, which is the space within our cells that contains the *cytosol* (SY-tow-sol) or *intracellular fluid* (*intra* = "within"), and (2) the *extracellular space*, which is the space outside our cells that contains fluid called **extracellular fluid** (*extra* = "outside"), or simply **ECF.** The plasma membrane forms a barrier between the ECF and the cytosol, and keeps the two fluid compartments separate. We will see in later modules that the plasma membrane regulates the exchange of substances between the cell and its external environment.

Cytoplasm

Inside the plasma membrane is a cell's **cytoplasm.** The cytoplasm consists of both the fluid cytosol and the structures embedded within it—the *organelles* and the *cytoskeleton.*

Cytosol The **cytosol,** or **intracellular fluid (ICF),** is the fluid portion of the cytoplasm that makes up about half of the cell's total volume. It consists mostly of water with many proteins and dissolved solutes. The cytosol has several primary functions, including storing materials and providing a site for critical reactions, including protein synthesis. Certain cell types contain clusters of identical storage molecules called **inclusions** in the cytosol. These inclusions store various necessary materials, such as nutrients from the ECF and proteins made by the cell. For example, the cells in the liver and skeletal muscle have inclusions that contain glycogen, the storage form of glucose.

The cytosol is the location of many of our most important biochemical reactions. For example, the series of enzyme-catalyzed reactions known collectively as *glycolytic catabolism* or *glycolysis* (gly-KOL-uh-sis; *glyco* = "sugar," *lysis* = "splitting") takes place in the cytosol. As implied by their name, these reactions split a molecule of glucose and in the process generate ATP, the main source of chemical energy for the cell. We will revisit glycolysis in later chapters (see Chapters 10 and 23).

Organelles Suspended in the cytosol are the **organelles** (or-gan-ELZ), which are like molecular machines that perform specific functions within the cell. Organelles are vital to a cell because they compartmentalize the cell's functions. Compartmentalization is essential for efficiency—it enables the cell to keep related enzymes and proteins together in one place so that they are not scattered throughout the cell. Compartmentalization is also critical for survival because many of the organelles contain enzymes and chemicals that would be toxic if they were free in the cytosol. This compartmentalization is achieved for many organelles by enclosure in a phospholipid bilayer membrane. We'll learn more about these organelles later (see Module 3.4).

Cytoskeleton Another component within the cytoplasm is the **cytoskeleton,** which consists of a network of protein filaments. The cytoskeleton supports the cell, helps create and maintain its shape, and holds the organelles in place. The protein filaments of the cytoskeleton also function as the cell's molecular "highways"

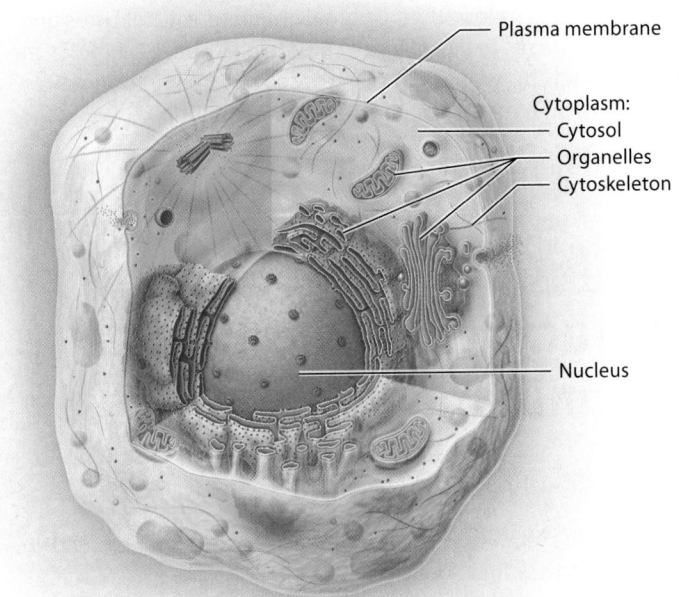

Figure 3.1 The basic components of a generalized cell.

Plasma membrane

Cytoplasm:
Cytosol
Organelles
Cytoskeleton

Nucleus

that allow substances and even organelles to be transported within the cell. Finally, the cytoskeleton is involved in cellular movement and cell division.

Nucleus

Most cells contain a single roughly spherical organelle called the **nucleus** (NEW-klee-us; plural, *nuclei,* NEW-klee-eye). The nucleus is surrounded by a phospholipid bilayer similar to the plasma membrane, known as the *nuclear envelope.* The nucleus contains most of the cell's *deoxyribonucleic acid* (*DNA*) and is the primary location for producing much of the cell's *ribonucleic acid* (*RNA*). The DNA and RNA control all of the more specific organelle functions by coding for and creating proteins.

Quick Check

☐ 2. What three components make up most animal cells, and what are their functions?
☐ 3. Where are intracellular and extracellular fluids located?

Cell Size and Diversity

◄◄ **Flashback**

1. How are structure and function related? (p. 25)

The cell shown in Figure 3.1 and in later figures in this chapter is a "generic" or *generalized* animal cell. However, most types of cells do not precisely fit this model. The cells in the human body vary greatly in appearance and size from the tiny (6–8 micrometers [μm; one millionth of a meter]), concave red blood cells that lack nuclei and most organelles to the long (1 millimeter [mm] to over a meter), branching nerve cells

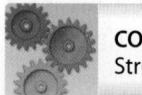

CORE PRINCIPLE
Structure-Function

(Figure 3.2). Note that this structural variation is an example of the Structure-Function Core Principle (p. 25). Red blood cells carry oxygen through the blood, so they must be small and flexible enough to fit through the smallest blood vessels. Nerve cells, however, contact and interact with many other cells, sometimes in different parts of the body, and so must have long extensions to reach those cells.

The composition of the cytoplasm also varies among cell types, which allows for cells to be *specialized* and so perform functions unique to their particular cell type. For example, cells of the liver serve critical detoxification functions (the ability to make dangerous chemicals harmless), and so they contain a large number of organelles that carry out detoxification reactions. Another example is the cells of skeletal muscle, whose cytoplasm is filled with contractile proteins that enable the cells to contract and generate movement. A unique composition of organelles, proteins, and other structures allows each cell type to be specialized. This specialization is vital. Without it, all of our organs and tissues would be identical. Cells of the heart would be indistinguishable from cells of the bone or

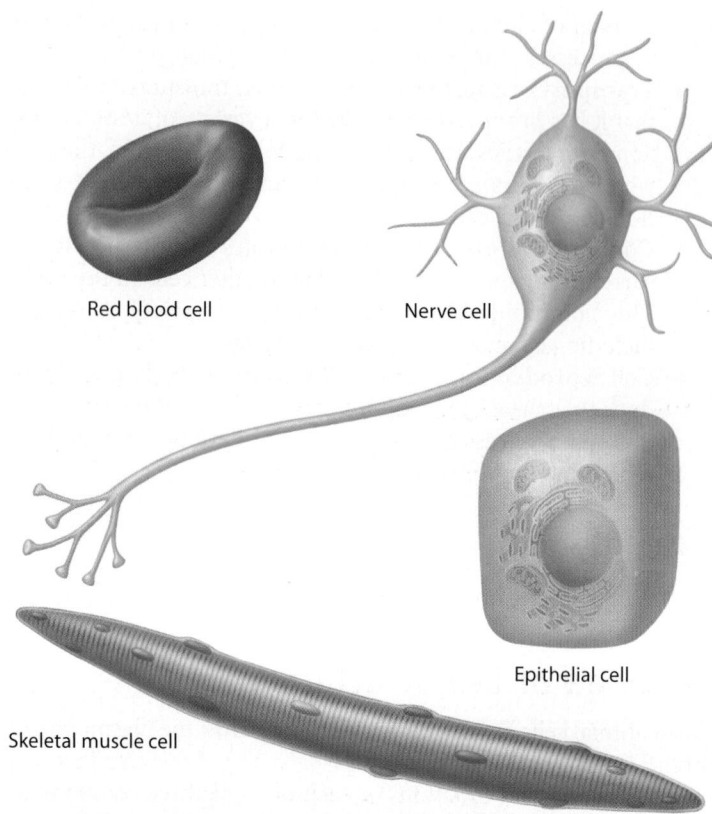

Red blood cell   Nerve cell

Epithelial cell

Skeletal muscle cell

Figure 3.2 Cell diversity. Note that cells are not drawn to same scale.

kidneys, making these organs unable to carry out their unique functions.

Apply What You Learned

☐ 1. Bacterial cells lack nuclei and most cytoplasmic organelles. Explain how these cells must differ from our cells in structure and function.
☐ 2. Cancer cells lose their original characteristic features and become more like the cell shown in Figure 3.1. Predict the impact this would have on these cells' ability to function. (*Hint:* Remember that form follows function.)

See answers in Appendix A.

MODULE **3.2**

Structure of the Plasma Membrane

Learning Outcomes

1. Describe how lipids are distributed in a cell membrane, and explain their functions.
2. Describe how carbohydrates and proteins are distributed in a cell membrane, and explain their functions.

3. Explain the overall structure of the plasma membrane according to the fluid mosaic model.

Let's now begin our exploration of the cell with its outermost component: the plasma membrane. Our main focus in this module is the plasma membrane's structure, but keep in mind that structure and function are tightly correlated. In the next module, we will discuss how the plasma membrane's structure relates to its many functional roles in maintaining homeostasis of the cell and the body as a whole.

The Phospholipid Bilayer

◀◀ Flashback

1. How do compounds with ionic, polar covalent, and nonpolar covalent bonds differ? (p. 40)
2. How are the terms hydrophilic, hydrophobic, and amphiphilic defined? (pp. 47 and 55)
3. What is a phospholipid? (p. 54)

In order for the plasma membrane to form an effective barrier between the ECF and cytosol, the molecules that make up the membrane must have two key properties. First, the molecules must have parts that can interact with the water in both of these fluid compartments. We've discussed that compounds with ionic bonds and compounds with polar covalent bonds are both *hydrophilic,* or able to interact with water (see Chapter 2). Yet ionic compounds wouldn't make good material for our barrier, because they dissociate in water. Obviously such a barrier would fall apart easily (imagine building your house out of salt—it would dissolve the first time it rained!). Compounds with polar covalent bonds, however, are hydrophilic but their individual atoms do not dissociate in water. So the membrane's hydrophilic regions that interact with water must have polar covalent bonds, whose polar ends can form hydrogen bonds with surrounding water molecules.

Second, the molecules must also have parts that repel water to keep the water in the ECF and cytosol separated. If the whole plasma membrane were polar and hydrophilic, then water, ions, and many organic molecules could move in and out of our cells freely and the membrane would not function as a barrier. So part of the barrier also has to be *hydrophobic,* or water-repelling. Because the major hydrophobic compounds in the body have nonpolar covalent bonds, the membrane's hydrophobic regions that repel water must contain nonpolar covalent bonds.

As you can see in **Figure 3.3a**, the molecules that cells use to build the plasma membrane must have both polar and nonpolar parts. Certain molecules fit this bill precisely: the phospholipids (see Chapter 2). Recall that phospholipids are a special class of lipid that is *amphiphilic*—they have both polar and nonpolar parts. Their phosphate-containing "heads" are relatively large, highly polar regions, and their fatty acid "tails" are large, nonpolar regions.

Notice in **Figure 3.3b** that phospholipids do something interesting when placed on the surface of water: The polar phosphate heads line up so that they face the water, whereas the nonpolar fatty acid tails point away from the water and toward the air (which consists chiefly of nonpolar gases). The result is a single layer, or *monolayer,* of phospholipids. Remember (from Chapter 2) that *like interacts with like*—the polar heads interact with the polar water molecules, and the nonpolar tails interact with the nonpolar gas molecules in air.

However, see what happens when we add another layer of water to the tube in **Figure 3.3c**. The phospholipids rearrange themselves into two layers, so that their fatty acid tails still face away from the water molecules, forming a phospholipid "sandwich." The phosphate heads are the "bread" and align on both sides so that they are facing the water molecules. The fatty acid tails are the "cheese" in the middle, and they align so that they face each other and exclude water and polar molecules. This arrangement is known as a **phospholipid bilayer** (*bi* = "two"). Such a bilayer forms the basis of the plasma membrane in every one of our cells. When this bilayer in a cell's membrane

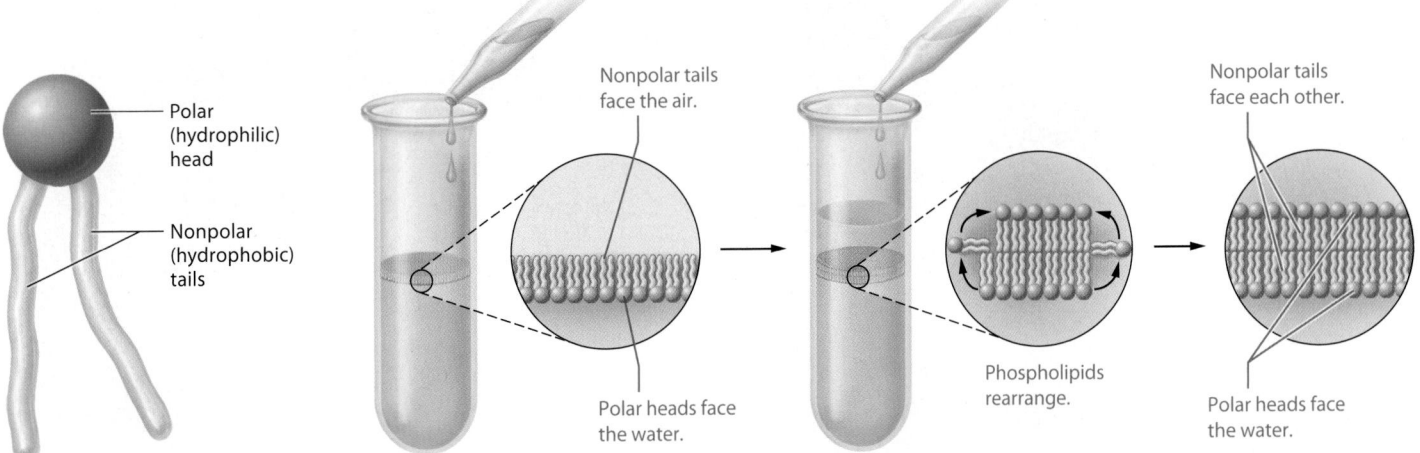

(a) **Schematic structure of a phospholipid molecule**

(b) **A monolayer forms when phospholipids are added to water.**

(c) **Phospholipids rearrange into a bilayer when another layer of water is added.**

Figure 3.3 The formation of a phospholipid bilayer.

is disrupted or torn, the phospholipids behave the same way as those in the tube—they rearrange themselves back into a bilayer, sealing the tear.

Quick Check

☐ 1. How do the phospholipids arrange themselves in the plasma membrane? Why?

The Fluid Mosaic Model of the Plasma Membrane

◀◀ **Flashback**

1. What is cholesterol? (p. 55)
2. What are glycoproteins and glycolipids? (p. 52)

To be able to perform all of the functions required by a living cell, the plasma membrane must contain many components in addition to its phospholipids. Scattered throughout the phospholipid bilayer of the plasma membrane are various proteins, other kinds of lipids, and carbohydrates. All of these components together are thought to resemble a mosaic, and accordingly, the current model of the plasma membrane is called the **fluid mosaic model** (Figure 3.4).

This model defines the plasma membrane as a structure with multiple components whose arrangement is dynamic, meaning that it changes from moment to moment. Phospholipids in the bilayer shift continually in a lateral, or sideways, direction, and do so extremely rapidly—adjacent phospholipids switch places an average of 10 million times per *second*. Other membrane components also move laterally through this constantly changing "sea" of phospholipids. The ability of phospholipids and other components to move within the plane of the membrane, a property known as *fluidity*, is critical to many plasma membrane functions, such as cell movement, cell reproduction, and transport of substances across the membrane.

Membrane Proteins

Besides phospholipids, the main components of the plasma membrane are molecules called **membrane proteins**. These proteins carry out many of the membrane's functions and give different cell types many of their unique properties. Membrane proteins are generally positioned so that their hydrophobic amino acids are in the fatty acid tail portion of the phospholipid bilayer—the middle—and their hydrophilic amino acids are facing the water-filled cytosol and/or the ECF.

There are two basic types of membrane proteins: integral and peripheral (see Figure 3.4). **Integral proteins** typically (but not always) span the entire width of the membrane; when they do reach both sides of the membrane, they are known as *transmembrane proteins*. **Peripheral proteins** are found on only one side of the membrane or the other. Some integral proteins are anchored in place by the cytoskeleton, but many float freely within the

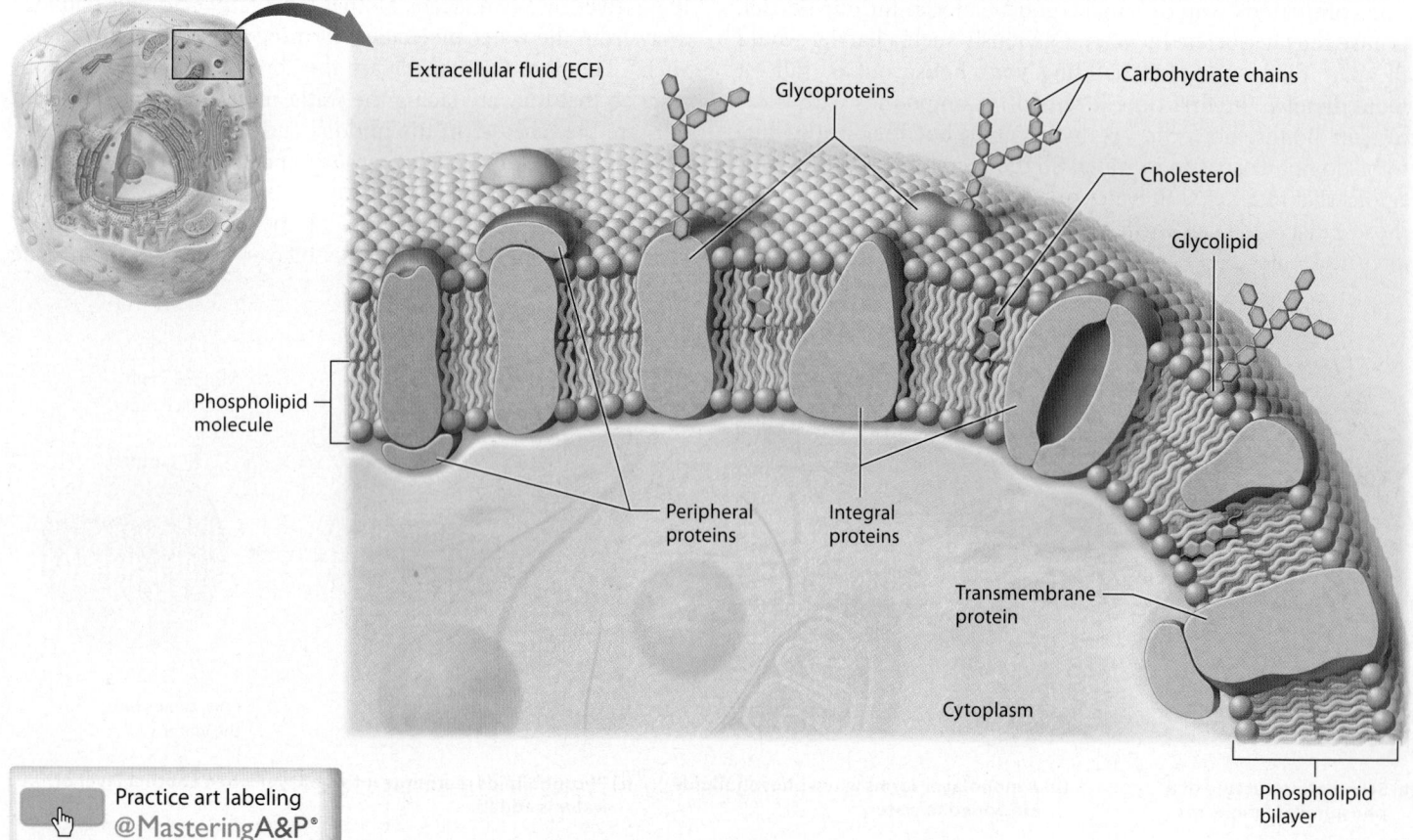

Extracellular fluid (ECF)

Glycoproteins

Carbohydrate chains

Cholesterol

Glycolipid

Phospholipid molecule

Peripheral proteins

Integral proteins

Transmembrane protein

Cytoplasm

Phospholipid bilayer

Practice art labeling
@MasteringA&P®

Figure 3.4 **The fluid mosaic model of the plasma membrane.**

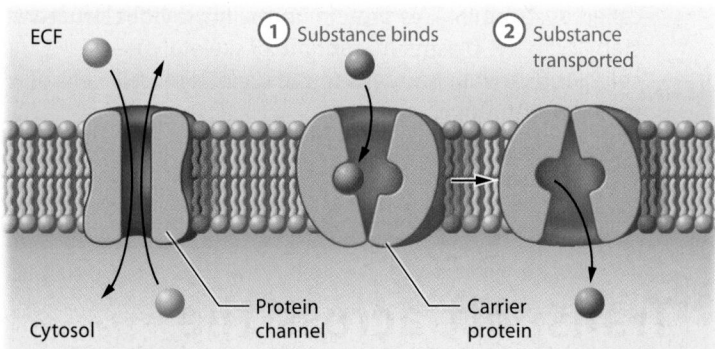

(a) Transporting substances. Some membrane proteins act as channels or carriers to transport substances into or out of the cell.

(b) Acting as receptors. Some membrane proteins act as receptors, binding to a ligand to trigger a change in the membrane protein or cell.

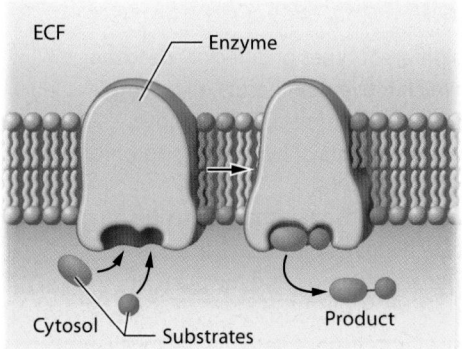

(c) Acting as enzymes. Some membrane proteins act as enzymes, catalyzing chemical reactions.

(d) Providing structural support. Some membrane proteins provide structural support by binding to other proteins in the ECF and/or cytosol.

(e) Linking adjacent cells. Some membrane proteins link adjacent cells together.

Figure 3.5 **Functions of membrane proteins.**

plane of the phospholipid bilayer. Peripheral proteins are loosely bound to the membrane surface or to integral proteins, and many can also move horizontally along the membrane surface.

The functions of membrane proteins are illustrated in **Figure 3.5**, and may include the following:

- **Transporting substances.** Many transmembrane proteins serve as **protein channels** that allow certain molecules to cross the membrane and pass into or out of the cell (Figure 3.5a). In addition, some integral proteins called **carrier proteins** bind and directly transport molecules into and out of the cell. For example, as shown on the right side of Figure 3.5a, an ion from the ECF binds to a carrier protein, which changes shape and releases the ion into the cytosol.
- **Acting as receptors.** A **receptor** is a membrane protein that binds to a chemical messenger called a **ligand** (LY-gand). When a ligand binds to its receptor, certain changes are triggered within the cell. For example, notice in Figure 3.5b that a protein channel opens when the ligand binds to its receptor. Receptor-ligand interactions (an example of the Cell-Cell Communication Core Principle, p. 27) are responsible for much of the physiology of the muscular, nervous, and endocrine systems, and we discuss them in greater

CORE PRINCIPLE
Cell-Cell Communication

detail in those chapters. See *A&P in the Real World: Drugs and Membrane Receptors* on page 74 to read about how these interactions can be targeted by medications.

- **Acting as enzymes.** As we've discussed, enzymes are proteins that speed up chemical reactions (see Chapter 2). Enzymes that are vital to plasma membrane structure and function are often lodged within the membrane, as are enzymes that catalyze various metabolic reactions. For example, some enzymes in the membrane can catalyze the formation or breakdown of bonds between molecules, as shown in Figure 3.5c.
- **Providing structural support.** Membrane proteins may be bound to components of the cytoskeleton and other protein molecules in both the ECF and the cytosol, which gives the cell shape and helps maintain its structural integrity (Figure 3.5d).
- **Linking adjacent cells.** Certain membrane proteins can attach to membrane proteins in adjacent cells, holding the cells in a tissue together (Figure 3.5e). These "linker" proteins can strengthen a tissue or allow adjacent cells in a tissue to communicate (as we will discuss in Chapter 4).

Other Membrane Components

In addition to membrane proteins, several kinds of lipids and carbohydrates form part of the plasma membrane. For example, an

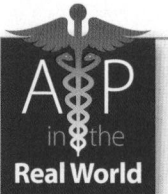

Drugs and Membrane Receptors

Many drugs are designed to have similarities to ligands that bind to membrane receptors. Certain drugs bind the receptor and mimic a ligand's actions, in which case the drug is labeled an *agonist*. For example, narcotic painkillers such as morphine bind to membrane receptors and mimic the actions of the body's own painkilling chemicals—endorphins.

Other drugs bind to the receptor but don't activate it. These drugs, called *antagonists*, inhibit the ligand's effects. The binding of an antagonist blocks the receptor and prevents the ligand from binding. An example of an antagonist is the class of drugs called antihistamines. Antihistamines bind to membrane receptors and inhibit the allergic symptoms caused by the chemical histamine.

important component of the plasma membrane shown in Figure 3.4 is the lipid molecule *cholesterol* (koh-LESS-ter-awl). Cholesterol stabilizes the structure of the plasma membrane when the temperature changes. This is critical because the plasma membrane must maintain a certain state of fluidity in order to function.

The final two components of the plasma membrane shown in Figure 3.4 are *glycolipids* (gly-koh-LIP-idz) and *glycoproteins* (gly-koh-PRO-teenz), which consist of carbohydrate chains (polysaccharides) covalently attached to either membrane lipids or proteins, respectively. These molecules, found on the outside of the plasma membrane, function in *cell recognition*, or the ability of our cells to tell one cell type from another. Cell recognition enables a sperm cell to recognize an egg, allows cells to be sorted into the proper places during development, and is the basis for the immune system's response to foreign cells. The glycoproteins and glycolipids produced by each of your cells are unique not only to that cell type but also to you as an organism, serving as a sort of identification tag.

Quick Check

☐ 2. How is the plasma membrane described according to the fluid mosaic model?
☐ 3. List five functions of membrane proteins.
☐ 4. What roles do cholesterol, glycoproteins, and glycolipids play in the plasma membrane?

Apply What You Learned

☐ 1. Detergents are similar to phospholipids in that they are amphiphilic molecules with both polar and nonpolar parts. What do you think would happen to a cell placed in a solution containing a detergent?
☐ 2. The disease Duchenne muscular dystrophy (doo-SHEN DIS-troh-fee) is caused by a defective membrane protein

called dystrophin. This protein normally provides structural stability to the plasma membrane of skeletal muscle cells. Predict what happens to the skeletal muscle cells of patients with this disease.

See answers in Appendix A.

MODULE **3.3**

Transport across the Plasma Membrane

Learning Outcomes

1. For osmosis and the different types of diffusion, describe the mechanism by which movement occurs, the energy requirements, and the types of molecules that move.
2. Compare and contrast the effects of hypertonic, isotonic, and hypotonic conditions on cells.
3. For primary and secondary active transport, describe the mechanism by which movement occurs, the energy requirements, the sources of energy, and the types of molecules moved.
4. For each type of vesicular transport, describe the mechanism by which movement occurs, the energy requirements, and the types of molecules moved.

In the previous module, we noted that the plasma membrane acts as a "fence" between the extracellular fluid and the cytosol in our cells. Like other fences, the plasma membrane has "gates" in it. One of the main advantages to having such a membrane is that these gates allow some things—but not others—to pass into or out of the cell. In other words, the membrane is *selective* about what it allows to pass through, or *permeate*, it. This property of being **selectively permeable** is critical to the cell. If the plasma membrane let nothing pass through it, our cells would have no way to exchange oxygen, nutrients, and waste products with the extracellular fluid.

Substances are continuously being exchanged between the cytosol and ECF. Both of these fluid compartments consist largely of water with dissolved ions and organic molecules of varying size and concentrations. The ECF contains substances that the cell needs, including nutrients such as glucose, amino acids, and fatty acids, as well as water, ions, oxygen, vitamins, and other macromolecules (see Chapter 2). Within the cytosol we find substances that have to leave the cell, including waste products such as carbon dioxide. We also find molecules that the cell has manufactured that have to enter the ECF to perform their functions. So both the cell and the body as a whole depend for survival on the transport of substances into and out of the cell.

Substances cross the plasma membrane using several mechanisms. Some transport mechanisms, called **passive transport processes,** require no net expenditure of energy from the cell.

Plastic divider

Higher concentration: 50% glucose

Lower concentration: 5% glucose

Glucose solution at equilibrium

① Two glucose solutions are separated by a plastic divider, creating a concentration gradient.

② The divider is removed and the glucose molecules begin to scatter, moving in general from higher concentration to lower concentration.

③ Net movement ceases when equilibrium is reached and the concentration gradient is no longer present.

Figure 3.6 Diffusion and equilibrium.

Other transport mechanisms, called **active transport processes,** do require the cell to expend energy, usually in the form of ATP. Whether a process is passive or active depends on three main variables: (1) the type of substance crossing the membrane, (2) the membrane's permeability to that substance, and (3) the concentration of that substance both inside and outside the cell. In this module we explore these processes in depth.

Passive Transport Processes

◀◀ **Flashback**

1. What is a concentration gradient? (p. 26)
2. What are kinetic and potential energy? (p. 41)
3. What are solutes and solvents? (p. 35)

The basic force that drives many types of passive transport is called a **concentration gradient.** Such a gradient involves a difference in the concentration of a substance from one area to another. For example, when you put dye in water, it looks like this:

Lower concentration

Higher concentration

Notice that more dye molecules are found on the bottom of the beaker than on the top. When such a concentration difference exists, there is a concentration gradient—an example of the Gradients Core Principle (p. 26).

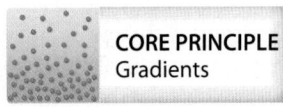

CORE PRINCIPLE
Gradients

Passive transport allows substances to move into and out of the cell using the potential energy of a concentration gradient. There are two basic types of passive transport: *diffusion* and *osmosis*.

Diffusion

Diffusion is defined as the movement of *solute* molecules (those that are dissolved) from an area of higher solute concentration to an area of lower solute concentration (**Figure 3.6**). A solute is said to move *down or "with" its concentration gradient* (from a higher to a lower concentration) during the process of diffusion. Notice in ① that we have two glucose solutions of different concentrations separated by a plastic divider. When the divider is removed in ②, the glucose molecules move from the side of the container with a higher glucose concentration to the side with a lower glucose concentration. This occurs because the concentration gradient is actually a form of potential energy. If all other forces are removed, the molecules will scatter due to their own kinetic energy, which all molecules have as long as thermal energy, or heat, is present. This movement will continue until the glucose concentration is uniform throughout the container and a balanced condition called *equilibrium* is reached, which is shown in ③. No further net movement of the molecules occurs at equilibrium because there is no more potential energy from a gradient to drive it. Note that we use the term "net movement" here, which means that no further change takes place in the solution's concentration. The molecules themselves, however, continue to bounce around even at equilibrium because of their kinetic energy.

The rate of diffusion depends on several factors, including the size and phase (e.g., liquid or gas) of the diffusing particles, the temperature, and the magnitude (size) of the concentration gradient. Generally, smaller particles, particles in the gaseous phase, higher temperatures, and steeper concentration gradients increase the rate of diffusion. All of these conditions raise the particles' kinetic energy, which causes them to move more rapidly. Two other factors to consider are the medium through which the particles diffuse and the presence of a barrier. Particles diffuse most rapidly through a gas with no barrier (such as smoke through the air). When a barrier is present, the rate of diffusion depends on the properties of the barrier, such as its thickness, surface area, and permeability to the particles.

The barrier to diffusion into and out of cells is the plasma membrane. Two basic types of diffusion occur through a membrane: simple and facilitated (**Figure 3.7**).

- **Simple diffusion** mostly involves nonpolar solutes (such as hydrocarbons and lipids, and gases such as O_2 and CO_2) that pass straight through the phospholipid bilayer without assistance from a membrane protein.
- **Facilitated diffusion** involves charged or polar solutes (such as ions and glucose) that cross the phospholipid bilayer with the help of a membrane protein.

Simple diffusion is fairly straightforward. Notice in Figure 3.7a that nonpolar molecules such as oxygen, carbon dioxide, and lipids have properties similar to those of the nonpolar fatty acid tail region of the phospholipid bilayer. This easily allows such substances to cross into or out of the cell down their concentration gradients.

Facilitated diffusion, shown in Figure 3.7b and c, is a bit more complicated. The fatty acid tails of the phospholipid bilayer repel most polar molecules and ions, making it nearly impossible for them to cross the membrane unassisted. For this reason, these molecules usually rely on two types of membrane proteins, called channels and carriers, for their transport into and out of the cell.

In protein channels, the hydrophilic amino acids in the protein are oriented toward the interior of the channel. This arrangement forms a passageway through which charged and polar molecules may pass into or out of the cell during facilitated diffusion, as shown in Figure 3.7b. Most channels are specific for an individual solute. For example, there are specific channels for different ions such as sodium, potassium, and calcium.

During facilitated diffusion using carrier proteins, shown in Figure 3.7c, the carriers bind one or two specific solutes and "carry" them into or out of the cell. As the process is currently understood, when a solute binds to its carrier protein, the protein changes shape, allowing the molecule to move across the membrane and into or out of the cell.

Keep in mind that facilitated diffusion is still a type of passive transport, even though it requires the help of membrane proteins. The cell expends no more energy in facilitated diffusion than it does in simple diffusion. Both processes rely on the potential energy of the concentration gradient to drive each solute's movement.

Osmosis

The process of *osmosis* (oz-MOH-sis) refers to the movement of *solvent*, the dissolving medium, across a membrane. In the human body, the solvent is water. Although we stated in the previous section that polar molecules cannot pass through the plasma membrane unassisted, the tiny water molecules are able to move rapidly into and out of a cell. Water can do so in two main ways: through channels or between phospholipids in the membrane. In the first mechanism, water can move through channel proteins called *aquaporins* in the plasma membrane. Most water that crosses the membrane uses this mechanism. In the second mechanism, which lets through relatively few water molecules, these molecules pass directly between the membrane phospholipids.

(a) **Simple diffusion of nonpolar molecules through the phospholipid bilayer**

(b) **Facilitated diffusion of ions through a channel**

(c) **Facilitated diffusion of polar and ionic molecules using a carrier**

Figure 3.7 Passive transport: simple and facilitated diffusion.

The process of osmosis is illustrated in **Figure 3.8**, where we again have a beaker with two unequal glucose solutions, labeled A and B. This time, however, they are separated by a selectively permeable membrane that allows water—but not glucose—to pass through. Therefore, as shown in ①, the water molecules from the less concentrated glucose solution in side B cross the membrane into the more concentrated solution in side A. This is the basic process of **osmosis**—water moves from a solution with a lower solute concentration to one with a higher solute concentration. Like all passive processes, the net movement of water molecules stops once equilibrium is reached and the concentrations of the two solutions are equal, as shown in ②.

ConceptBOOST ⟫⟫➤

Understanding Water Movement in Osmosis

Osmosis is somewhat counterintuitive, because at first glance it may look like water is moving against its concentration gradient. But remember that osmosis deals with a different kind of gradient: an osmotic pressure gradient. This pressure is created by solute molecules, and the more solutes, the higher the osmotic pressure. Remember also that osmotic pressure exerts a "pulling" force on water molecules. So the solution with the highest pressure "pulls" water the hardest, and water moves toward this concentrated solution. ■

Whereas diffusion is driven by the kinetic energy of the solute molecules themselves, osmosis is driven by a force called **osmotic pressure,** which is the "pulling" force that solutes exert on water molecules. A solution with a higher solute concentration has greater osmotic pressure than one with a lower solute concentration. This is why water moves toward the 50% glucose solution in Figure 3.8—its osmotic pressure is higher, and therefore its pulling force on water molecules is stronger.

Volume Changes in Osmosis You can also see an important consequence of osmosis in part ② of Figure 3.8: Osmosis results in a change in the volume of fluid in each side of our container. As water molecules leave the less concentrated glucose solution in side B, its volume decreases, and as water molecules move into the more concentrated glucose solution in side A, its volume increases. These volume changes have important consequences for our cells, which we discuss next.

Tonicity In the body, the plasma membrane of the cell is the selectively permeable barrier, and the two solutions are the cytosol and the extracellular fluid. Just as osmosis leads to volume changes in either side of the container in Figure 3.8, it also leads to volume changes in our cells. To understand the type of volume change that occurs, we must examine the concept of **tonicity** (toh-NISS-ih-tee), which is a way to compare the osmotic pressure gradient between two solutions (i.e., the relative solute concentration). In the case of our cells, the two solutions we are comparing are the cytosol and the ECF (**Figure 3.9**). Every cell in the body is surrounded by ECF, which is a water-based solution (see Module 3.1). Ordinarily, as shown in Figure 3.9a, the body's ECF is **isotonic** (eye-soh-TOHN-ik; *iso* = "same") to the cytosol of a cell, meaning that the ECF has about the same solute concentration as the cytosol. Under these conditions, water enters

① Water crosses the membrane, moving into the solution with a higher concentration of solute (glucose).

② When the concentration of solute reaches equilibrium, water molecules move at random but there is no net movement.

Figure 3.8 Passive transport: osmosis.

and leaves the cell at the same rate, so the cell has no net gain or loss of water over time and its volume remains the same. An example of a solution that is isotonic to human blood is a 0.9% NaCl solution, also called *normal saline*.

However, notice in Figure 3.9b and c what happens to a cell if the solute concentration of the ECF changes:

- **A hypertonic extracellular solution causes a cell to lose water.** Figure 3.9b shows a red blood cell in a **hypertonic** solution, which corresponds to ECF. In this case, the solute concentration of the ECF is *higher* than the solute concentration of the cytosol (*hyper* = "more," in this case, greater concentration of solute). We already know that water moves by osmosis toward the more concentrated solution. So a cell in a hypertonic solution loses water because it leaves the cell and enters the ECF. Such water loss causes the cell to shrivel, or *crenate* (KREE-nayt), and possibly die. This is why humans cannot drink the salt water of the ocean—it is hypertonic to our cells, and rather than causing our cells to absorb water, it actually causes them to lose water.

- **A hypotonic extracellular solution causes a cell to gain water.** In Figure 3.9c we have a red blood cell immersed in a **hypotonic** (*hypo* = "less") solution; in this case the solute concentration of the extracellular solution standing in for the ECF is *less* than that of the cytosol. Because the cytosol is the more concentrated solution, water moves into the cytosol by osmosis, which causes the cell to swell and possibly burst, or *lyse* (LYZ). Pure water is an example of a hypotonic solution. Usually we are able to drink pure water and other hypotonic solutions because our endocrine and urinary systems work together to ensure that excess water is eliminated in the urine. However, if a person drinks too much water over a short period, these systems may not be able to remove the water quickly enough, and *water poisoning* may result. Water poisoning can cause death from both ion imbalances and cellular swelling. Too little water can cause problems as well, as you can read about in *A&P in the Real World: Dehydration, Sports Drinks, and Water.*

(a) Isotonic solution

Solute concentration is equal inside and outside the cell; no net movement of water.

Increased solute concentration outside cell

Decreased solute concentration outside cell

(b) Hypertonic solution

Solute concentration is greater outside the cell; water leaves the cell, which shrivels, or crenates.

(c) Hypotonic solution

Solute concentration is greater inside the cell; water enters the cell, which swells and may burst, or lyse.

Figure 3.9 Tonicity: effects of isotonic, hypertonic, and hypotonic solutions on cell volume.

Dehydration, Sports Drinks, and Water

Dehydration, or the loss of cellular water, is a major concern for athletes or any person exercising vigorously. Strenuous exercise results in water and electrolyte loss through sweating. Because sweat contains relatively more water than electrolytes (solutes), the extracellular fluid becomes *hypertonic* to the body's cells. The hypertonic ECF draws water out of the cells by osmosis, and the concentration of solutes in the cytosol of the cells increases.

Sports drinks are a commonly used remedy for mild dehydration. Many popular sports drinks contain mixtures of water, electrolytes, and carbohydrates that are actually hypotonic to human cells. When you drink them, these solutions eventually help replenish the water and electrolytes in your ECF and make it mildly *hypotonic* to the cells. This condition causes water to move back into the cells by osmosis until the normal concentration of their cytosol is restored.

Of course, the most widely available hypotonic solution is plain old water, which generally rehydrates just as well as sports drinks. Care must be taken in cases of severe dehydration, though, because pure water can rehydrate the cells too quickly and result in cellular swelling and possibly water poisoning.

So let's summarize what we just discussed. If a solute can cross the plasma membrane, then it will move down its concentration gradient by diffusion until equilibrium is reached. If a solute cannot cross the plasma membrane, then water will move by osmosis into the more concentrated solution until equilibrium is reached.

Quick Check

Mark each of the following statements as true or false. If a statement is false, correct it to make a true statement.
☐ 1. The energy for passive processes comes from the hydrolysis of ATP.
☐ 2. Diffusion refers to the movement of a molecule down its concentration gradient.
☐ 3. Simple diffusion is a passive process, but facilitated diffusion requires ATP for the membrane's channel and carrier proteins to function.
☐ 4. Osmosis refers to the movement of solvent from a solution with a higher solute concentration to a solution with a lower solute concentration.
☐ 5. A cell placed in a hypertonic environment will lose water by osmosis and shrivel.

Active Transport via Membrane Proteins

◄◄ Flashback

1. What is ATP? (p. 59)
2. What is a hydrolysis reaction? (p. 50)

Unlike passive transport processes, active transport processes require cells to expend energy in the form of ATP. This use of energy is necessary because solutes are moved *against* their concentration gradients during active transport, from a lower concentration to a higher one. Since the concentration gradient does not provide energy to drive the molecules' movement, energy must be supplied from the hydrolysis of ATP (refer back to Figure 2.25). There are two types of active transport: *primary active transport* and *secondary active transport,* both of which require carrier proteins called **pumps** in the plasma membrane. In this section, we examine each type in greater detail.

Primary Active Transport

The process of **primary active transport** is fairly straightforward: A protein in the plasma membrane binds and transports, or "pumps," a solute against its concentration gradient using the energy from the hydrolysis of ATP. All pumps we discuss are ATPases, so in addition to transporting substances, they also function as enzymes that catalyze the reaction that breaks down ATP into ADP and an inorganic phosphate. Often, the phosphate removed by this reaction binds to the pump itself and causes a change in the shape of the pump, and this shape change transports the solute into or out of the cell.

There are three basic types of pumps: uniports, symports, and antiports. The simplest is the **uniport pump,** which drives a single substance either into or out of the cell. You've already seen one of this type of pump in Figure 3.5b. The other two types of pumps, which you'll see in the next two figures, move two or more substances at the same time. A **symport pump** drives two or more substances in the same direction across the plasma membrane. Conversely, an **antiport pump** drives two or more substances in opposite directions.

The major primary active transport pump in the body is the antiport pump known as the *sodium-potassium pump,* or **Na^+/K^+ pump** (also called the *Na^+/K^+ ATPase*). Normally the concentration of sodium ions in the ECF is about 10 times greater than that in the cytosol. The reverse is true for potassium ions—its concentration in the cytosol is about 10 times higher than its concentration in the ECF. It is absolutely critical to our homeostasis to maintain the concentration gradients of sodium and potassium ions. These gradients are required for skeletal muscles to contract, hearts to beat, nerves to send impulses, and cells to maintain their osmotic balance, among many other functions.

The Na^+/K^+ pump maintains this steep concentration gradient by transporting three sodium ions out of the cell for every two potassium ions it moves into the cell, both against their concentration gradients. The sequence of events of one cycle of the Na^+/K^+ pump, illustrated in **Figure 3.10**, is as follows:

① **The pump binds three sodium ions from the cytosol.** With the pump oriented so that its sodium ion binding sites are facing the cytosol, three sodium ions from the cytosol bind to the pump.

② **ATP is hydrolyzed, phosphate binds to the pump, and the pump changes shape.** When ATP is hydrolyzed, the

released phosphate group binds to the pump, causing it to change shape.

③ **The pump releases the three sodium ions into the ECF and binds two potassium ions.** In response to the shape change, the three sodium ions are released into the ECF. Two potassium ions from the ECF then bind to the pump at their exposed binding sites.

④ **The phosphate detaches and the pump changes back to its original shape.** After this binding, the phosphate

① The pump binds three Na^+ from the cytosol.

⑤ The pump releases the two K^+ into the cytosol.

② ATP is hydrolyzed, the phosphate binds to the pump, and the pump changes shape.

④ The phosphate detaches, and the pump changes back to its original shape.

③ The pump releases the three Na^+ into the ECF and binds two K^+.

Figure 3.10 Primary active transport by the Na^+/K^+ pump.

group dissociates from the pump, which then returns to its original shape.

⑤ **The pump releases the two potassium ions into the cytosol.** After the pump changes back to its original shape, it releases the two potassium ions into the cytosol. Three sodium ions can now bind again to their binding sites on the pump, restarting the cycle.

Due to other physiological processes, sodium and potassium ions continue to leak back across the membrane, following their concentration gradients, so Na^+/K^+ pumps must work continuously. In fact, in some cell populations, as much as 30% of the cell's ATP is used to fuel its Na^+/K^+ pumps.

Secondary Active Transport

Although primary active transport is fairly straightforward, the process of secondary active transport is somewhat more complex. As we just discussed, primary active transport uses ATP directly to fuel a transport pump, but **secondary active transport** uses ATP indirectly. In this type of transport, the cell first uses ATP to create a concentration gradient by pumping one substance across the plasma membrane. Remember that potential energy exists in every concentration gradient, so once the concentration gradient of this first substance is established, the cell then harnesses that potential energy to power the active transport of another substance against its gradient. So in secondary active transport, the active transport of one substance is coupled with the passive transport of a second substance. As an analogy, think of a random act of kindness—a person driving through a toll booth on a highway pays for both her toll and the toll of the person behind her; the second person gets a "free ride," just like the second substance being transported.

Let's look at an example that involves a concentration gradient of sodium ions outside the cell and the transport of glucose into the cell (one way cells obtain the glucose needed for cellular fuel). The process is illustrated in **Figure 3.11**:

① **A Na^+/K^+ pump creates a concentration gradient of sodium ions:**

⟨1a⟩ From the cytosol, sodium ions bind to the Na^+/K^+ primary active transport pump.

⟨1b⟩ ATP is hydrolyzed, and sodium ions are transported out of the cell, into the ECF, against their concentration gradient.

② **A carrier protein uses the potential energy of the sodium ion gradient to power the transport of glucose:**

⟨2a⟩ From the ECF, both a sodium ion and a glucose molecule bind to another carrier protein (which is a symporter).

⟨2b⟩ The carrier protein transports the sodium ion and glucose molecule into the cell—the sodium ion *with* its concentration gradient and the glucose molecule *against* its gradient.

Many secondary active transport processes involve the steep sodium ion concentration gradient created by the Na^+/K^+ pump. This gradient favors the movement of sodium ions back into the cell. In addition to glucose, other substances transported against their concentration gradients using the potential energy of this gradient include amino acids, other sugars, and ions such as chloride and bicarbonate. Secondary active transport is thus essential for absorption and secretion processes for many cells, particularly in the intestines and kidneys. This key role of the sodium ion gradient is one of the many reasons why properly functioning Na^+/K^+ pumps are so important to the body's ability to maintain homeostasis.

Quick Check

☐ 6. How does the process of primary active transport work?
☐ 7. What is the main primary active transport pump in our cells? How does it function?
☐ 8. Describe the process of secondary active transport.

① A Na^+/K^+ pump creates a concentration gradient of sodium ions.

② A carrier protein uses the potential energy of the sodium ion gradient to power the transport of glucose.

⟨1a⟩ Na^+ bind to the Na^+/K^+ pump.

⟨1b⟩ ATP hydrolysis fuels the transport of Na^+ into the ECF, against their concentration gradient.

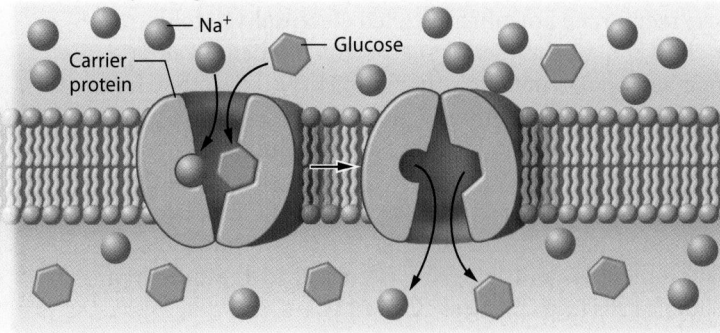

⟨2a⟩ From the ECF, a Na^+ and a glucose molecule bind to another carrier protein.

⟨2b⟩ The carrier protein transports the Na^+ and glucose molecule into the cell—the Na^+ with its concentration gradient, and the glucose against its concentration gradient.

Figure 3.11 Secondary active transport.

Consequences of Ion Transport across the Plasma Membrane: Introduction to Electrophysiology

◀◀ **Flashback**

1. What is electrical energy? (p. 41)
2. What is potential energy? (p. 41)

You have read about the different ways that ions cross the plasma membrane of a cell—through passive processes such as facilitated diffusion and active processes such as primary and secondary active transport. These mechanisms cause ions to have unequal concentrations between the cytosol and the ECF. As you can see here, this results in a separation of charges across the plasma membrane—a thin layer of positive charges is present in the ECF, and a thin layer of negative charges is found in the cytosol:

Note that this charge separation is limited to the area on either side of the plasma membrane. On the whole, the cytosol and ECF have the same overall charge.

A separation of charges like the one we see here is known as an **electrical potential.** This name refers to the fact that an electrical potential is a source of potential energy. Why is there potential energy in separated charges? The answer is simple—an electrical potential is really just an *electrical gradient,* similar to the concentration Gradients Core Principle (p. 26) you have seen many times in this chapter. Just as potential energy is found in a concentration gradient, this energy is also found in an electrical gradient. The electrical potential found across the plasma membranes of all cells is known as a **membrane potential,** and the study of these membrane potentials is called **electrophysiology.**

CORE PRINCIPLE Gradients

The value of the membrane potential when a cell is at rest, when it is not stimulated or inhibited by any other factor, is called the **resting membrane potential.** The resting membrane potential of most excitable cells, such as muscle and nerve cells, averages about –70 mV (millivolts), which means that their cytosol is about 70 mV more negative than the ECF. In later chapters you'll learn about the basic ability of cells to alter their membrane potential, and how this is critical to the functioning of excitable cells (see Chapters 10 and 11).

Active Transport via Vesicles

Carrier proteins and channels are very effective at helping polar and ionic molecules cross the plasma membrane. However, they have size limitations—polar macromolecules are simply too big to bind to a carrier or fit through a channel. Instead, larger particles are transported into or out of the cell in small sacs called **vesicles** (VESS-ih-kulz; *vesic-* = "bubble"). A vesicle is enclosed by a membrane made of a phospholipid bilayer, just like the plasma membrane that encloses the cell. In fact, vesicles form from pinched-off parts of cell membranes, and so they can fuse with the plasma membrane and also with the membranes surrounding the organelles found in the cytoplasm. Transport that uses vesicles is an active process called **vesicular transport** (vess-IK-yoo-lar). Vesicular transport requires the energy from ATP hydrolysis to fuel several steps of the process, including vesicle formation. There are two main types of vesicular transport: *endocytosis* (en'-doh-sy-TOH-sis) to move substances into the cell and *exocytosis* (eks'-oh-sy-TOH-sis) to transport them out.

Endocytosis

Extracellular fluid, large molecules, and even sometimes whole cells are taken into the cell by the process of **endocytosis** (*endo-* = "within"). There are two basic types of endocytosis: *phagocytosis* for particles and *pinocytosis* for droplets of fluid.

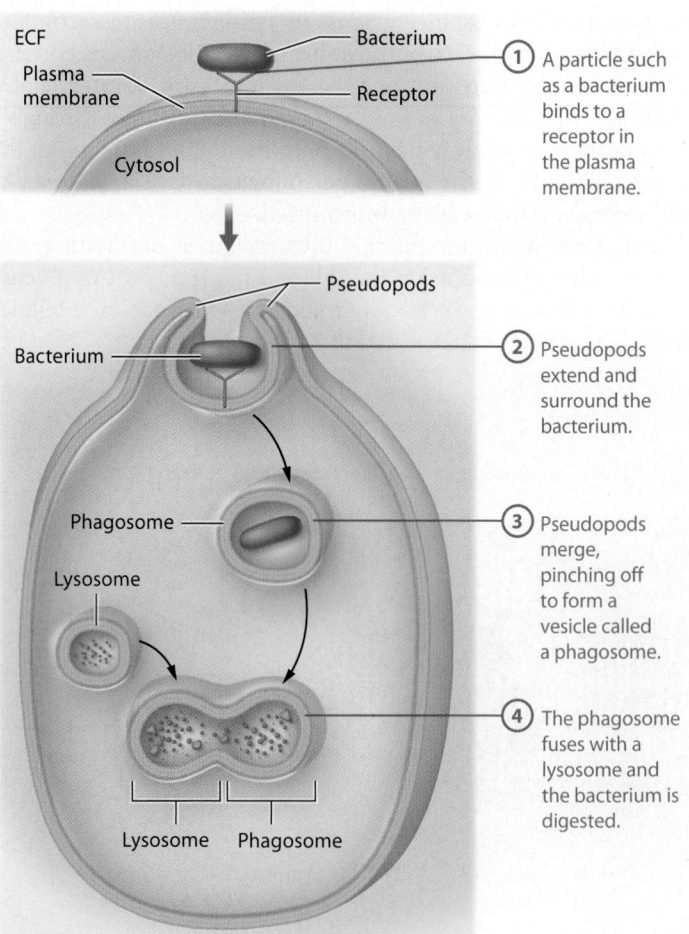

Figure 3.12 Endocytosis: phagocytosis.

Phagocytosis During **phagocytosis** (fayg′-oh-sy-TOH-sis; *phago-* = "eat," literally "cell eating"), cells ingest large particles such as bacteria, dead body cells, or parts of cells. In humans only certain cells of the immune system called *phagocytes* (FAYG-oh-sytz) carry out phagocytosis.

The process unfolds as follows, as shown in **Figure 3.12**:

1. A particle such as a bacterium binds to a receptor in the phagocyte's plasma membrane.
2. This binding triggers the cell to extend bulging "arms" called *pseudopods* (SOO-doh-podz; *pseudo-* = "false," *pod-* = "foot") that surround the bacterium.
3. The pseudopods merge to form an endocytic vesicle that pinches off from the cell surface into the cytosol, becoming a relatively large structure called a *phagosome* (FAYG-oh-sohm).
4. The phagosome eventually fuses with an organelle called a *lysosome* (LY-soh-sohm) that contains enzymes that digest the bacterium (lysosomes are discussed in Module 3.4).

Pinocytosis **Pinocytosis** (peen′-oh-sy-TOH-sis; *pino-* = "drink," literally "cell drinking"), also called *fluid-phase endocytosis,* is the cellular ingestion of fluid droplets. Pinocytosis produces smaller vesicles than phagocytosis. These vesicles transport nutrients and other substances dissolved in the ECF into the cell. Therefore, all cells carry out pinocytosis, and it occurs nearly continuously in most cells.

Pinocytosis usually takes place in regions of the plasma membrane known as coated pits, which are small indentations of the membrane that are coated with a protein (**Figure 3.13a**). Soon after these pits form, ① they begin to fold inward, eventually enclosing some extracellular fluid within a vesicle. This vesicle then ② pinches off in the cytosol, ③ loses its protein coat, and fuses with another type of vesicle called an **endosome** (EN-doh-sohm). In the endosome, the ingested substances are sorted and modified before being sent to their final destinations.

4a. Parts of the endosome may return to the plasma membrane, where they are re-inserted.

① Membrane of coated pit folds inward.

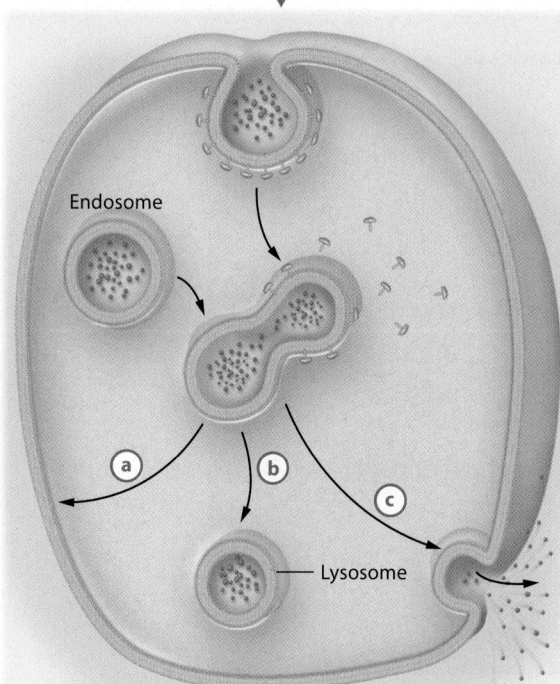

② Vesicle forms and pinches off in the cytosol.

③ Vesicle loses protein coat and fuses with endosome.

④ Once inside the endosome, the ingested products are (a) returned to the plasma membrane, (b) sent to a lysosome, or (c) may exit out of the opposite side of the cell.

(a) Pinocytosis

① Ligands bind to receptors in pit and membrane folds inward.

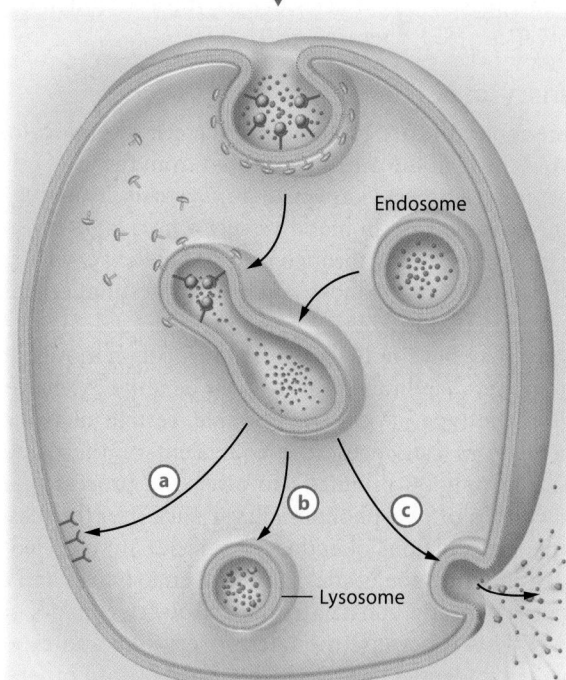

(b) Receptor-mediated endocytosis

Figure 3.13 Endocytosis: pinocytosis and receptor-mediated endocytosis.

(4b) Other parts may fuse with a lysosome, where their contents are degraded.

(4c) Finally, some parts may exit from the other side of the cell in a process called transcytosis that we'll cover soon.

Receptor-Mediated Endocytosis Pinocytosis results in vesicles containing a fairly uniform sample of the extracellular fluid. The process **receptor-mediated endocytosis,** which is basically the same as pinocytosis except for the inclusion of receptors,* forms vesicles that contain a high concentration of a specific molecule (**Figure 3.13b**). In this process, certain ligands such as cholesterol, hormones, and iron bind to receptors in a protein-coated pit in the plasma membrane. Once a certain number of these ligands are bound, the pit folds inward and pinches off to form a vesicle. Like the pinocytotic vesicle, this vesicle then fuses with an endosome and the ingested products may be sorted back to the plasma membrane, sent to lysosomes, or delivered out of the cell.

Exocytosis

Large molecules are released from the cell by the process of **exocytosis** (*exo-* = "outside," literally "out of the cell"), which is functionally the reverse of endocytosis. As illustrated in **Figure 3.14,** particles to be released are packaged into an exocytic vesicle. As you can see, (1) when the vesicle reaches the plasma membrane, (2) the vesicle binds with the membrane and (3) their phospholipid bilayers fuse. (4) This fusion releases the contents of the vesicle into the ECF. Many molecules are released, or *secreted,* from the cell in this manner, including hormones, neurotransmitters (the chemical messengers of nerve cells), and enzymes. For this reason, the vesicles in exocytosis are often called *secretory vesicles* (SEE-kreh-toh-ree).

Transcytosis

Some molecules are brought into the cell by endocytosis, transported through the cell, and released from the cell's other side by exocytosis, a process known as **transcytosis.** Transcytosis is particularly important in blood vessels, where large substances exit the blood by passing through cells lining the vessel. Transcytosis also occurs in the kidneys and in the intestinal cells during the absorption of nutrients.

You may have noticed that endocytosis removes portions of the plasma membrane; this loss can be quite extensive, depending on the type of cell. For example, certain phagocytes called *macrophages* can ingest the equivalent of their entire membrane in only 30 minutes. However, the process of exocytosis adds pieces of phospholipid bilayer back into the plasma membrane, and portions of endocytic vesicles are recycled and reinserted into the membrane. Even with endocytosis and exocytosis occurring nearly constantly, the cell never changes size appreciably. This is because the rates of recycling and exocytosis are essentially identical with that of endocytosis. In other words, the

*Bruce Alberts et al., *Molecular Biology of the Cell,* 5th ed. (New York: Garland Science, 2008), 787–92.

rate at which material is added to the plasma membrane equals the rate at which material is removed.

Table 3.1 reviews the different types of transport across the plasma membrane. You can use this table to compare these types of transport and better understand their similarities and key differences.

1. Vesicle reaches the plasma membrane.

ECF

2. Vesicle binds with the plasma membrane.

Plasma membrane

Cytosol

3. The phospholipid bilayers of the vesicle and the plasma membrane fuse.

4. Vesicle contents are released into ECF.

(a) Process of exocytosis

Product being released by exocytosis

Exocytic vesicle

TEM (400,000×)

(b) Electron micrograph of exocytosis

Figure 3.14 Exocytosis.

TABLE 3.1 PLASMA MEMBRANE TRANSPORT

Type of Transport	Definition	Example(s)
Passive		
Simple Diffusion	Movement of solute down its concentration gradient through the plasma membrane unaided by a transport protein; energy source is the solute's own kinetic energy (see Figure 3.7a).	• Oxygen • Carbon dioxide • Lipids
Facilitated Diffusion	Movement of solute down its concentration gradient with the help of a carrier or channel protein; energy source is the solute's own kinetic energy (see Figure 3.7b and c).	• Sodium ions • Potassium ions • Calcium ions • Glucose • Amino acids
Osmosis	Movement of solvent (water) from a solution of lower solute concentration to one of higher solute concentration; driven by osmotic pressure (see Figure 3.8).	Water balance
Active		
Primary Active Transport	Movement of solute against its concentration gradient using ATP (see Figure 3.10).	Na^+/K^+ pump
Secondary Active Transport	Movement of one solute with its concentration gradient used to power the transport of another solute against its concentration gradient (see Figure 3.11).	Using sodium ion gradient • Glucose • Chloride ions • Bicarbonate ions are brought into the cell.

Simple Diffusion figure labels: ECF, Solute, Plasma membrane, Cytosol

Facilitated Diffusion figure labels: Solute, Plasma membrane, Carrier/channel protein

Osmosis figure labels: Solute, H_2O, H_2O, Plasma membrane, Solute

Primary Active Transport figure labels: ATPase pump, Solute (ion), P, ATP, ADP

Secondary Active Transport figure labels: Solutes, Carrier protein, P, ATP, ADP, ATPase pump

Practice art labeling
@MasteringA&P®

(continued)

TABLE 3.1 PLASMA MEMBRANE TRANSPORT *(continued)*

Type of Transport	Definition	Example(s)
Phagocytosis 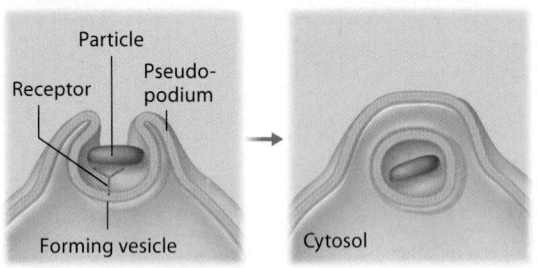	"Cell eating"; bringing large molecules or particles into the cell via a vesicle; ATP required (see Figure 3.12).	Ingestion of bacteria and cell debris by phagocytes
Pinocytosis 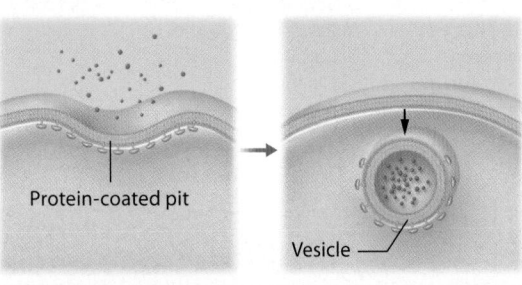	"Cell drinking"; bringing substances in the ECF into the cell via a vesicle formed from a protein-coated pit; ATP required (see Figure 3.13a).	Nutrient transport
Receptor-Mediated Endocytosis 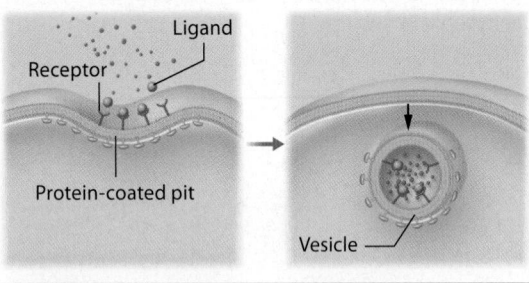	Bringing a specific substance into a vesicle using receptors on the plasma membrane; ATP required (see Figure 3.13b).	Cholesterol transport
Exocytosis 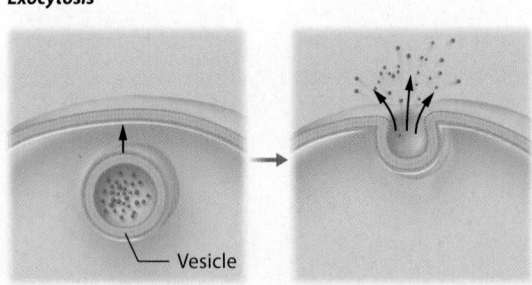	Release of a substance from the cell via a vesicle; ATP required (see Figure 3.14).	Secretion of • Hormones • Enzymes • Neurotransmitters

Practice art labeling
@MasteringA&P®

Quick Check

☐ 9. What are the three types of endocytosis, and how do they differ?

☐ 10. Explain the basic process of exocytosis.

Apply What You Learned

☐ 1. Predict what will happen to a person's red blood cells if he or she is given the following fluids intravenously:

a. Sterile water (a hypotonic solution)
b. Normal saline (an isotonic solution)
c. 10% dextrose in water (a hypertonic solution)

☐ 2. Oleander (*Nerium oleander*) is a common but poisonous ornamental shrub. All parts of the plant contain chemicals that inhibit the Na^+/K^+ pump. Explain why chemicals that have this effect could also disrupt secondary active transport.

See answers in Appendix A.

MODULE **3.4**
Cytoplasmic Organelles

Learning Outcomes

1. Describe the structure and function of each type of organelle.
2. Explain how the organelles of the endomembrane system interact.

We now move our discussion inside the cell to the tiny cellular "machines" known as cytoplasmic organelles (**Figure 3.15**). Just as each of our organs has a unique structure that enables it to

perform specific functions, so does each of a cell's organelles. Recall from earlier in the chapter that organelles allow the cell's functions to be compartmentalized, which is essential for cellular efficiency (see Module 3.1). To this end, each organelle contains a specific set of enzymes and other proteins unique to its function.

Organelles that are compartmentalized by membranes are often called *membrane-bounded organelles*. Membrane-bounded organelles include the *mitochondrion, peroxisome, endoplasmic reticulum, Golgi apparatus,* and *lysosome.* There are also organelles not enclosed by a membrane, including the *ribosome,* covered in this module, and centrosomes, covered in the next module.

Centrosome

Endocytic vesicle

Peroxisome

Nucleus

Nucleolus

Inclusions

Rough endoplasmic reticulum

Smooth endoplasmic reticulum

Cytosol

Lysosome

Plasma membrane

Exocytic vesicle

Golgi apparatus

Mitochondrion

Ribosomes (free)

Ribosomes (bound)

Practice art labeling
@MasteringA&P®

Figure 3.15 **The cell and its organelles.**

Each type of membrane-bounded organelle has a membrane that is unique in terms of its thickness and composition. In spite of these differences, an organelle's membrane is still structurally similar to the plasma membrane. This similarity allows substances to be packaged in vesicles and transported between the plasma membrane and organelles.

Because cells are so small, we need to use microscopes to get a good look at them. Microscopes can produce photographs called micrographs (*micro-* = "small," *-graph* = "something written or drawn") that you will see throughout this chapter and in the rest of this text. There are two basic types of micrographs, produced with different kinds of microscopes. The first is a *light micrograph* (*LM*), produced with a *light microscope,* which passes a beam of light through specimens to create an image. Light microscopes are used to view relatively large microscopic objects, such as whole cells and tissues. The second is an *electron micrograph,* produced with an *electron microscope,* which uses a beam of electrons instead of light to produce an image. Electron microscopes are used to view considerably smaller objects, such as single organelles. There are two main types of electron micrographs: scanning electron micrographs (SEMs), which are three-dimensional images usually of whole or broken cells, and transmission electron micrographs (TEMs), which are flat images of sectioned cells. Whenever you see a micrograph in this text, you'll see that the type of image is identified, as well as the degree of magnification used.

In this module, we explore the structure and function of many of the cell's remarkable internal components. As you read, don't lose sight of the relationship between an organelle's form and its function. Also bear in mind that no one component of the cell can function independently of the other components. Instead, all must work together to ensure that homeostasis is maintained.

Mitochondria

◀◀ **Flashback**

1. How is energy stored in chemical bonds and in ATP? (pp. 42 and 59)
2. What are oxidation and catabolic reactions? (p. 43)

The organelles known as **mitochondria** (my′-toh-KON-dree-ah; singular, *mitochondrion*) are involved in energy production. The main source of energy for cells is *chemical energy,* and the most ready supply of chemical energy for our cells is ATP (see Chapter 2). Although some ATP production occurs via glycolysis in the cytosol, most of our ATP is produced within the mitochondria, which is why they are often called the "power plants" of the cell. Most cells have hundreds to thousands of mitochondria, and the number that each cell contains is determined by its particular energy needs. For example, cells such as those in the liver, skeletal muscle, and cardiac muscle have a higher demand for ATP (compared to those in the skin, for instance), and therefore require more mitochondria.

Mitochondria are unique among human cell organelles in that each mitochondrion contains its own DNA and the enzymes required for protein synthesis within its matrix (**Figure 3.16**). In addition, each mitochondrion has its own *ribosomes,* which are small, granular organelles that participate in protein synthesis. The vast majority of mitochondrial proteins are encoded by the DNA in the cell's nucleus. The remainder, a mere 13 proteins, are encoded by mitochondrial DNA. In humans, mitochondrial DNA is passed down from the maternal side only, for reasons that we'll discuss in a later chapter (see Chapter 27).

Like most organelles, mitochondria are surrounded by a membrane made up of a phospholipid bilayer and other components.

Outer mitochondrial membrane
Intermembrane space
Inner mitochondrial membrane
Cristae
Matrix
Ribosomes
Mitochondrial DNA

SEM (60,000×)

Figure 3.16 Structure of the mitochondrion.

Notice in Figure 3.16 that there are actually *two* mitochondrial membranes: a smooth *outer mitochondrial membrane* and an *inner mitochondrial membrane* that has numerous folds called **cristae** (KRISS-tee), which significantly increase the inner surface area. This double-membrane structure creates two spaces within the mitochondrion: (1) the *intermembrane space,* which is the space between the two membranes, and (2) the *matrix,* which is the innermost space. Each membrane or space contains its own unique enzymes and other proteins, allowing each to carry out specific functions.

The outer mitochondrial membrane has large channels that permit molecules in the cell's cytosol to enter the intermembrane space (**Figure 3.17**). In contrast, the inner mitochondrial membrane is far more selective. It is essentially impermeable to most solutes except those for which it has specific transport proteins. As a result, within the matrix we find only certain organic molecules, such as DNA, mitochondrial proteins, and enzymes.

The mitochondrial matrix contains enzymes and proteins that break down organic fuels via a series of reactions known collectively as **oxidative catabolism** (see Figure 3.17). Earlier in the chapter you read about glycolytic catabolism, the ATP-generating reactions that take place in the cytosol (see Module 3.1). During oxidative catabolism, ① the products of glycolytic catabolism and other organic fuels are oxidized (made to lose electrons) by mitochondrial enzymes. ② The energy released by these oxidation reactions is ultimately ③ used to drive ATP synthesis. The reactions of oxidative catabolism can proceed only in the presence of oxygen. For this reason, these reactions are also referred to as *aerobic respiration.*

Peroxisomes

Although mitochondria consume the majority of the cell's oxygen, they are not the only organelle that uses oxygen. Another organelle, called a **peroxisome** (puh-ROKS-ih-sohm; *soma-* = "body"), also uses oxygen to carry out a variety of reactions. Peroxisomes get their name from one of their main reactions, which uses molecular oxygen to oxidize and strip hydrogen atoms off certain organic molecules to produce *hydrogen peroxide* (H_2O_2). The hydrogen peroxide, along with oxidative enzymes, then performs various functions, including the following:

- **Oxidizing toxic substances.** Hydrogen peroxide oxidizes certain chemicals that would be toxic to the cell, including ethanol (alcohol). The oxidation reactions convert such chemicals into less toxic compounds, which are then eliminated from the body. This function is particularly important in kidney and liver cells, which perform most of the detoxification reactions in the body.
- **Breaking down fatty acids.** Many of the catabolic reactions that break down fatty acids into smaller molecules occur in the peroxisomes. The products of these reactions are sent either to the cytosol to participate in anabolic reactions or to the mitochondria to be oxidized to produce ATP.
- **Synthesizing certain phospholipids.** Some of the enzymes in peroxisomes catalyze the initial reactions that form certain phospholipids. These phospholipids are critical components of the plasma membrane of specific cells in the nervous system. For this reason, diseases that affect peroxisomes invariably affect the nervous system.

Interestingly, the hydrogen peroxide formed by peroxisomes is itself toxic. Why doesn't it harm the cell? When hydrogen peroxide accumulates, an enzyme in the peroxisomes turns the excess into water and oxygen. This is another example of the benefit of compartmentalizing the cell's functions within individual membrane-bounded organelles. If the peroxisomes' reactions occurred in the cytosol, the toxic hydrogen peroxide would damage the cell.

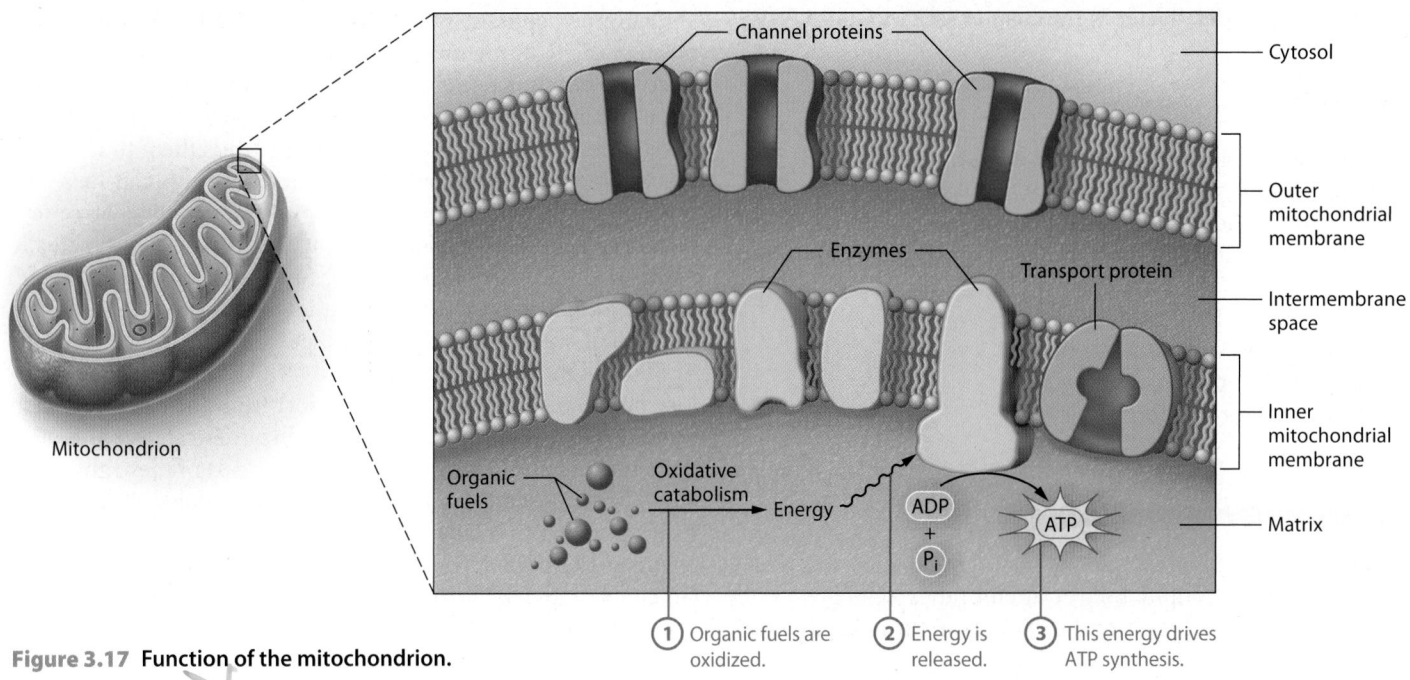

Figure 3.17 Function of the mitochondrion.

① Organic fuels are oxidized.   ② Energy is released.   ③ This energy drives ATP synthesis.

Ribosomes

The tiny, granular particles in Figure 3.15 are **ribosomes** (RY-boh-sohmz), nonmembrane-bounded organelles that are the site of protein synthesis. Each ribosome consists of two subunits, one large and one small, that fit together like the top and bottom of a hamburger bun (**Figure 3.18**). Each subunit is made of a complex assembly of ribosomal proteins and a type of RNA called **ribosomal RNA (rRNA).**

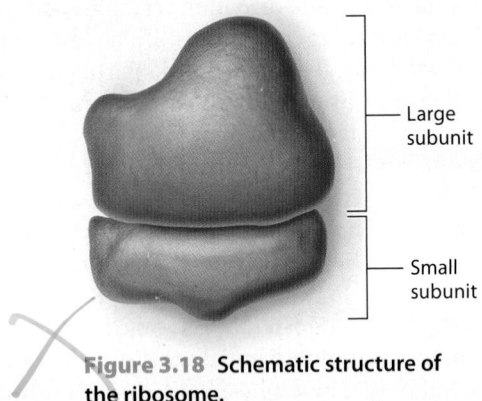

Large
subunit

Small
subunit

Figure 3.18 Schematic structure of the ribosome.

Depending on their location in the cell, ribosomes are grouped into two types: free or bound. Free ribosomes are suspended in the cytosol, where they make proteins that are mainly used in the cytosol itself. Bound ribosomes are associated with the membranes of other cellular structures. All ribosomes start out free, but some will attach themselves to the endoplasmic reticulum or the nuclear envelope. These bound ribosomes typically make proteins that will be exported from the cell, transported to certain organelles such as lysosomes, or inserted into a membrane. Free and bound ribosomes are structurally identical; they differ only in location. We discuss ribosomal structure and function in greater detail later in the chapter (see Module 3.7).

> **Quick Check**
>
> ☐ 1. Identify the following properties as belonging to
> mitochondria, peroxisomes, or ribosomes.
> a. Detoxify chemicals such as alcohol
> b. Can be bound to a membrane or suspended in the
> cytosol
> c. Metabolize fatty acids
> d. Make most of the ATP for the cell
> e. Synthesize proteins
> f. Produce certain phospholipids

The Endomembrane System

As we've discussed, most of the cell's organelles are surrounded by phospholipid bilayer membranes that are very similar to the plasma membrane. Some of these organelles form vesicles that exchange proteins and other molecules. The organelles that transfer molecules in this manner are part of a system called the **endomembrane system,** whose components together synthesize, modify, and package molecules produced by the cell. This system includes the *endoplasmic reticulum,* the *Golgi apparatus,* and *lysosomes.* Also included in the endomembrane system are the plasma membrane and the nuclear envelope, which we discuss later in the chapter (see Module 3.6).

Endoplasmic Reticulum

The large, folded membrane shown surrounding the nucleus in **Figure 3.19** is named the **endoplasmic reticulum** (en-doh-PLAZ-mik reh-TIK-yoo-lum; *reticulum* = "network"), or simply **ER.** The ER's highly folded phospholipid bilayer membrane is continuous with the nuclear envelope. Current evidence indicates that the membrane is a single, continuous structure enclosing a fluid-filled space called the *ER lumen.* This space contains enzymes that catalyze a variety of reactions. Figure 3.19 shows two distinct regions of the ER's membrane: the **rough endoplasmic reticulum,** or **RER,** where the membrane is covered in bound ribosomes, and the **smooth endoplasmic reticulum,** or **SER,** which lacks bound ribosomes. Why do you think the RER and the SER differ in structure in this way? You are on the right track if you suspect that they have very different functions, which we discuss next.

Rough Endoplasmic Reticulum Recall that proteins must be folded into their proper shape before they become functional (see Chapter 2). Yet ribosomes, the sites of protein synthesis, only assemble amino acids into a polymer called a *polypeptide* (the protein's primary structure) without a three-dimensional shape. So, ribosomes by themselves do not synthesize fully functional proteins—other processes must modify the proteins so that they can fold into their functional state.

The polypeptides synthesized on bound ribosomes pass through the RER membrane into its lumen, where enzymes catalyze the reactions that fold these polymers into their correct three-dimensional shapes. The RER recognizes proteins that have folded into an incorrect shape and sends them to the cytosol, where they are degraded. In some cases, this can be a surprisingly high percentage—for example, about 90% of certain membrane proteins misfold and for this reason never actually make it to the plasma membrane. See *A&P in the Real World: Cystic Fibrosis* on page 92 to learn about a surprising result of the disposal of one misfolded protein type.

Many of the proteins that enter the RER to be folded are secretory proteins, those destined for export from the cell. Most such proteins are glycoproteins, and RER enzymes catalyze the reactions that attach sugars to certain amino acids of these glycoproteins. When a secretory protein is fully assembled, it leaves the RER packaged in a small, membrane-bounded structure called a **transport vesicle.** Transport vesicles are generally sent to the Golgi apparatus for further processing before their contents are secreted from the cell by exocytosis.

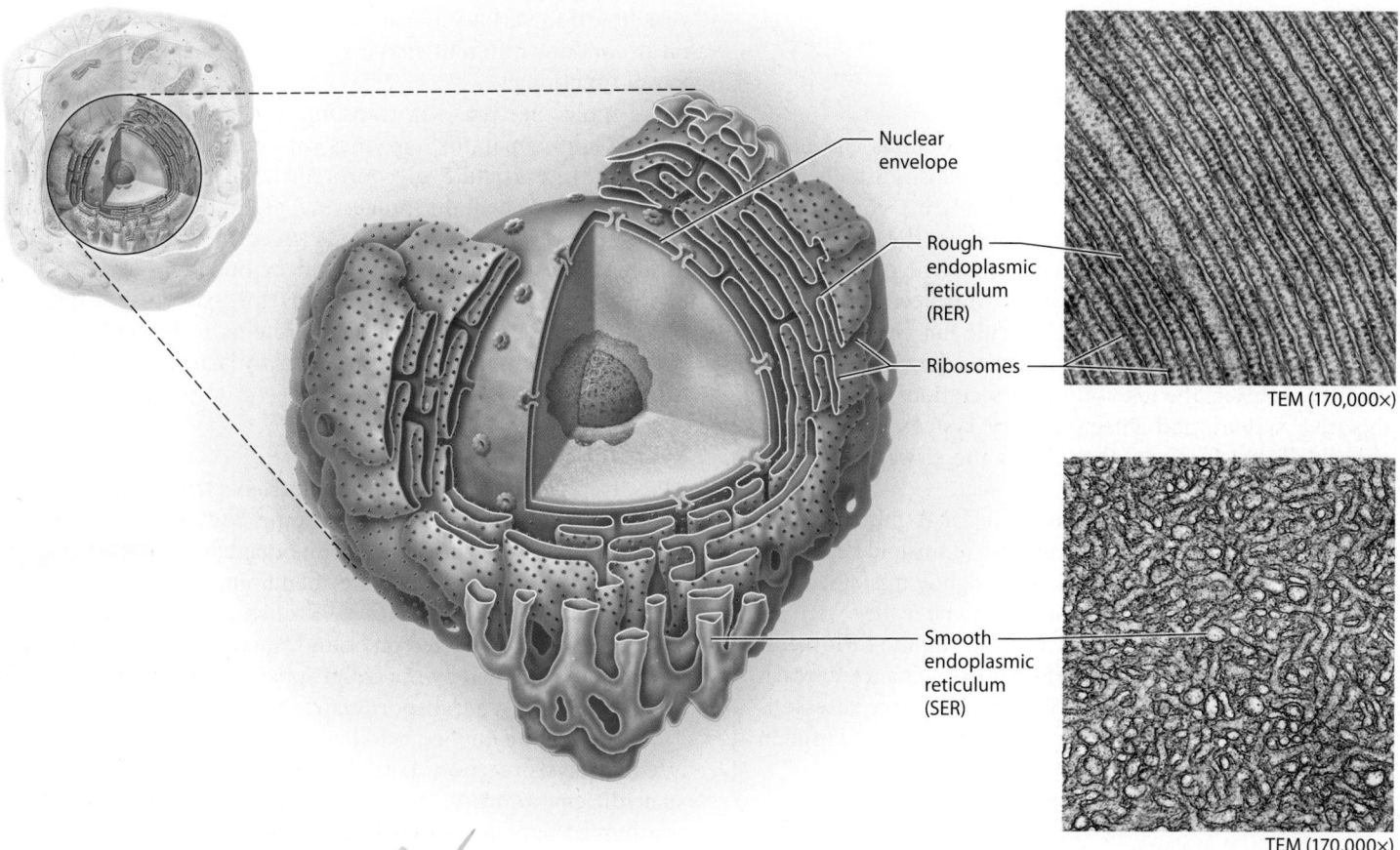

Nuclear envelope

Rough endoplasmic reticulum (RER)

Ribosomes

TEM (170,000×)

Smooth endoplasmic reticulum (SER)

TEM (170,000×)

Figure 3.19 The endoplasmic reticulum.

The RER also acts as a "membrane factory"—the components for nearly every membrane in the cell, including the plasma membrane, are made there. The ribosomes of the RER synthesize integral and peripheral membrane proteins. The RER also synthesizes some cholesterol and phospholipids and inserts all of these components into its membrane. These components are then pinched off the RER membrane and transferred to the plasma membrane or other parts of the endomembrane system via transport vesicles. The RER also makes the membranes for organelles that are not in the endomembrane system. However, the membrane components are transferred to these organelles in other ways.

Smooth Endoplasmic Reticulum Because the SER is not associated with ribosomes, it plays essentially no role in protein synthesis. The SER performs other vital functions, however, including the following:

- **Calcium ion storage.** In most cells, the SER pumps calcium ions out of the cytosol and stores them for later release. This function is particularly important for muscle cells, which rely on calcium ions for contraction.
- **Detoxification reactions.** The SER is very well developed in cells that detoxify drugs and harmful substances ingested by and produced in the body, particularly cells in the liver.

Interestingly, the SER of liver cells can nearly double in size when a person is exposed to certain drugs. When the concentrations of such drugs decrease, the cells remove the excess SER membrane and the organelle returns to its normal size. However, when a person is repeatedly exposed to the drug, the SER membrane remains enlarged and the cell increases its synthesis of the enzymes necessary to degrade the drug. This increased activity accounts in part for the *tolerance* that is seen with some drugs, including alcohol—more drug is required to produce an effect because a more extensive SER detoxifies the drug more rapidly.

- **Lipid synthesis.** The SER synthesizes the bulk of the cell's lipid membrane components, such as phospholipids and cholesterol, as well as steroid hormones and lipoproteins. Cells that perform these lipid-synthesizing functions, such as those of the liver, have an extensive SER.

The lipids produced by the SER leave its membrane in the same way that they leave the RER—via transport vesicles. Transport vesicles that depart the ER are generally headed for the next component of the endomembrane system, the Golgi apparatus.

Golgi Apparatus

The **Golgi apparatus** (GOHL-jee), located between the RER and the plasma membrane, consists of a group of flattened

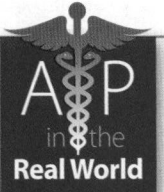

Cystic Fibrosis

The fact that the RER recognizes and helps dispose of misfolded proteins is usually beneficial, as most such proteins could cause major harm to the cell. However, sometimes the RER's selectivity can actually be detrimental. This is the case in some forms of the disease *cystic fibrosis*, in which certain cells are missing a protein that forms a chloride ion channel in the plasma membrane. This causes deficient chloride ion transport that impacts the ion and water secretions of the lungs, digestive system, and integumentary system, resulting in abnormally thick mucus that blocks the airways, digestive enzyme deficiencies, and very salty sweat.

A person with cystic fibrosis has a DNA mutation that causes the chloride ion channel protein to misfold slightly in the RER. Because the protein is only slightly misfolded, the channels would generally still be functional if they were inserted into the plasma membrane. However, the RER mediates the protein's destruction before it has a chance to reach the plasma membrane. In short, this devastating disease is caused by the RER being "overprotective" regarding protein structure.

membranous sacs filled with enzymes and other molecules. Within the Golgi apparatus, proteins and lipids produced by the ER are further modified, sorted, and packaged for export. Notice in **Figure 3.20** that unlike the ER, the sacs of the Golgi apparatus are not continuous, but rather are separated from one another by thin spaces filled with cytosol. The part of the Golgi apparatus nearest the ER receives transport vesicles from the ER, and the part farthest from the ER sends newly formed

vesicles off to various destinations. Each sac of the Golgi apparatus contains different enzymes, because each carries out different functions.

As molecules pass via transport vesicles through the sacs of the Golgi apparatus, enzymes catalyze reactions that modify them. In particular, glycoproteins made by the ER are further altered. When the processing of products in the Golgi apparatus is finished, they are packaged in vesicles bound for different destinations either within or outside the cell. Some of these secretory vesicles are ready for exocytosis from the plasma membrane, some contain proteins to be inserted into the plasma membrane, and some go to lysosomes, which we discuss next.

Lysosomes

All cells must be able to digest the nutrients they take in, as well as old, worn-out organelles and, in some cases, other cells. These digestive functions take place within organelles called **lysosomes** (LY-soh-sohmz). Lysosomes are membrane-bounded sacs that contain water and a set of enzymes called *acid hydrolases*. These hydrolases are synthesized on bound ribosomes, modified in the ER, further modified by the Golgi apparatus, and then delivered to the lysosome in a transport vesicle.

As their name implies, acid hydrolases are enzymes that catalyze hydrolysis reactions (see Chapter 2). They function best in an acidic environment, and the pH within lysosomes is kept low, at around 5.0. Note that enclosing these acidic reactions within a lysosome ensures that other contents of the cell are protected from these digestive enzymes. In addition, if the enzymes happened to leak out, they would also be relatively inactive at the cytosolic pH of 7.2. *A&P in the Real World: Lysosomal Storage Diseases* on page 95 talks about what happens when these acid hydrolases don't work correctly.

Within lysosomes, macromolecules such as proteins, carbohydrates, and fatty acids that are taken in by different types of endocytosis are degraded into smaller units. These subunits are

Vesicles from the ER enter here.

Newly formed transport vesicles leave from here for other destinations.

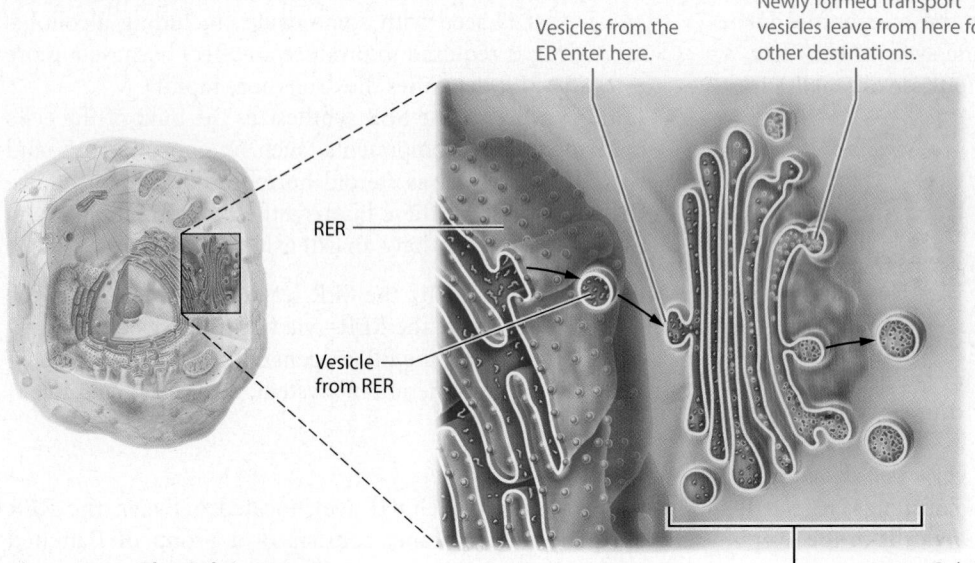

RER

Vesicle from RER

Golgi apparatus

TEM (83,000×)

Figure 3.20 The Golgi apparatus.

then released into the cytosol for disposal or for use in protein synthesis and various metabolic reactions. Recall from our earlier discussion of phagocytosis that lysosomes play a key role in this process and are critical to the proper functioning of certain cells of the immune system that digest invading bacteria. In addition, lysosomes degrade old, worn-out organelles and cell components via a process called *autophagy* (aw-TOFF-uh-jee; "self-eating"). The components of the digested organelles are then returned to the cytosol to build new cell components, an example of cellular recycling.

Summary of Endomembrane System Function

Figure 3.21 provides an overview of the function of the endomembrane system. We begin with, simultaneously, (1a) the smooth ER making lipids and (1b) bound ribosomes on the rough ER synthesizing proteins. The proteins then enter the RER, where (2) proteins are folded and modified. These products, and the products from the SER, are packed into transport vesicles and sent to the Golgi apparatus, which (3) sorts, further modifies, and packages the products into vesicles.

When the finished products are ready to exit the Golgi apparatus, the vesicles that transport them may take one of three pathways to get to their appropriate destinations:

(4a) Products may be sent to lysosomes. Certain digestive enzymes are packaged within vesicles and delivered to lysosomes to catalyze catabolic reactions.

(4b) Products may be incorporated into the plasma membrane or another membrane in the cell. Newly synthesized phospholipids and membrane proteins are packaged into transport vesicles and sent to the appropriate membrane. When the vesicles fuse with this membrane, the proteins and phospholipids are incorporated into the membrane.

(4c) Products may be secreted from the cell by exocytosis. Products that the cell secretes into the ECF are packaged into secretory vesicles. When these vesicles fuse with the plasma membrane, their contents are released outside the cell.

To review the structure and function of the cytoplasmic organelles covered in this module, see **Table 3.2**.

Figure 3.21 Function of the endomembrane system.

TABLE 3.2 CYTOPLASMIC ORGANELLES

Organelle	Structure	Function
Mitochondrion	Double membrane; inner membrane folded into cristae; has own DNA and ribosomes (see Figures 3.16 and 3.17).	• Synthesizes the majority of the cell's ATP
Peroxisome	Membrane-bounded; similar to large vesicle.	• Detoxifies certain chemicals through oxidation reactions • Metabolizes fatty acids • Synthesizes certain phospholipids
Ribosome	Two subunits made of proteins and rRNA; not membrane-bounded (see Figure 3.18).	• Synthesizes proteins
Rough Endoplasmic Reticulum (RER)	Series of saclike membranes enclosing the ER lumen; surface studded with ribosomes (see Figure 3.19).	• Modifies and folds proteins made by the ribosomes • Manufactures and assembles most of the components of the plasma membrane
Smooth Endoplasmic Reticulum (SER)	Series of tubular membranes enclosing the ER lumen; surface does not contain ribosomes (see Figure 3.19).	• Stores calcium ions and synthesizes lipids • Detoxifies certain substances
Golgi Apparatus	Stack of flattened, membrane-bounded sacs (see Figure 3.20).	• Sorts, modifies, and packages proteins and other products made by the ER
Lysosome	Membrane-bounded structure; similar to a large vesicle.	• Digests damaged organelles and products brought into the cell by endocytosis • Recycles damaged organelles

Quick Check

☐ 2. Identify the following properties as belonging to rough ER, smooth ER, the Golgi apparatus, or lysosomes.
   a. Synthesizes the majority of the cell's cholesterol
   b. Contains digestive enzymes
   c. Ensures that proteins fold into the proper shape
   d. Packages proteins into secretory vesicles
   e. Detoxifies many substances
   f. Makes the majority of membrane components
   g. Degrades molecules and organisms ingested via endocytosis
   h. Contains ribosomes on its surface membranes

☐ 3. To what destinations can products from the Golgi apparatus be sent?

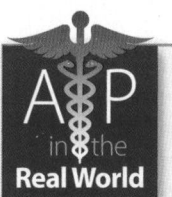

Lysosomal Storage Diseases

A group of diseases collectively called the *lysosomal storage diseases* results from a deficiency of one or more of the acid hydrolases of the lysosomes. This deficiency causes the accumulation of certain undigested molecules within the lysosomes. Symptoms of the disease depend on which specific enzyme(s) is (are) missing. Examples of lysosomal storage diseases include the following:

- *Gaucher's disease* (gow-SHAYZ) is caused by a deficiency in the enzyme that breaks down certain glycolipids, and so these lipids accumulate in cells of the blood, spleen, liver, lungs, bone, and sometimes brain. The most severe form of the disease is generally fatal in infancy or early childhood.
- *Tay-Sachs disease* (TAY saks) results from the deficiency of an enzyme that degrades another specific type of glycolipid. These glycolipids accumulate in the lysosomes of the brain, leading to progressive neural dysfunction and typically death by age four or five.
- *Hurler syndrome* is caused by a deficiency in the enzymes that digest certain large polysaccharides. These molecules accumulate in many cells of the body, including the heart, liver, and brain. Death can result in childhood from organ damage.
- *Niemann-Pick disease* is caused by a deficiency in an enzyme that is involved in the degradation of a specific lipid. The result is that lipids accumulate in the lysosomes. This disease affects the spleen, liver, brain, and lungs, as well as bone marrow. Its severity varies. As with other lysosomal storage diseases, the prognosis is very poor for patients with the severe form, who suffer damage to multiple organs and neural dysfunction.

Apply What You Learned

☐ 1. Many toxins and poisons block certain enzymes in the mitochondria. For example, many fruits, including apricots, apples, plums, peaches, and cherries, contain seeds and pits that can cause the formation and release of the poison cyanide if they are ground and ingested. Cyanide binds to an enzyme in the inner mitochondrial membrane critical to energy production. Why would this be fatal for a cell?

☐ 2. What effect do you think deficiencies in lysosomal enzymes would have on phagocytes? Explain.

☐ 3. The deadly poison ricin (RY-sin), found in castor beans, binds to and inhibits free and bound ribosomes. Explain why this effect is lethal.

See answers in Appendix A.

MODULE **3.5**

The Cytoskeleton

Learning Outcomes

1. Describe the structure and function of the three components of the cytoskeleton.
2. Describe the structure and function of centrioles, cilia, and flagella.
3. Explain the role of the cytoskeleton in cellular motion.

The components of the **cytoskeleton** function as a cell's highways. The key difference between highways and the cytoskeleton, however, is that the cytoskeleton is not a static structure—it can rapidly change its form to fit its changing functions and the needs of the cell.

The cytoskeleton is made up of several types of protein filaments organized in different ways. These filaments play a variety of critical roles within the cell, including the following:

- **Giving a cell its characteristic shape and size.** Cells in different parts of the body have different shapes and sizes in accordance with their functions. These differing shapes and sizes are largely a result of the cells' cytoskeletons, which form an internal framework for each cell.
- **Supporting the plasma and nuclear membranes as well as the organelles.** Cytoskeletal filaments help to ensure the structural integrity and strength of the cells' plasma membranes and the membrane surrounding the nucleus. The cytoskeleton also positions, organizes, and anchors various organelles within the cell.
- **Functioning in movement.** The protein filaments of the cytoskeleton are often associated with *motor proteins*. Motor proteins are like molecular propellers that use the energy derived from ATP hydrolysis to move themselves along the protein filaments. The cytoskeletal filaments and their motor proteins are responsible for moving organelles and vesicles within the cell, splitting the cell during cell division, and for certain cells, propelling substances past the cell or even propelling the cell itself.
- **Performing specialized functions in different cell types.** The cytoskeleton plays an important role in the specific physiology of certain cell types. For example, the cytoskeleton allows a phagocyte to engulf a bacterium with pseudopodia, enables our muscle cells to contract, and allows our nerve cells to form long "arms" that can reach and communicate with distant targets.

In this module, we examine each of the filaments that make up the cytoskeleton. We also explore the structure and function of extensions from the cell, called *microvilli*, which function in absorption, and *cilia* and *flagella*, which function in movement.

Types of Filaments

The cytoskeleton contains three different types of long protein filaments: *actin filaments, intermediate filaments,* and *microtubules.* Each type of protein filament is composed of smaller protein subunits, which allows the rapid disassembly and reassembly of the filament.

Actin Filaments

As you can see in the left column of **Table 3.3**, **actin filaments,** also called **microfilaments,** are the thinnest of the protein filaments in the cytoskeleton, measuring only about 5–9 nm in diameter. They are composed of two intertwined strands of protein subunits called *actin*. Actin filaments provide structural support, bear tension (stretching forces), and help to maintain the cell's shape. They are particularly common in the outer part of the cell, near the plasma membrane, where they form three-dimensional supportive networks.

Another role of actin filaments is in cellular motion. The motor protein associated with actin filaments, called *myosin*, is capable of moving the entire cell. Many different types of myosin molecules exist, but all contain a globular head that uses energy from ATP hydrolysis to propel the myosin molecule along an actin filament. Myosin-actin filament interactions are responsible for the contraction of muscle cells, for the pinching of the cell in two during cell division, and for the changes in the membrane that take place during endocytosis and exocytosis. Additionally, myosin-actin filament interactions allow phagocytes to "crawl" or migrate within our bodies in search of invaders.

Intermediate Filaments

The second type of protein filament in the cytoskeleton is the ropelike **intermediate filament** (shown in the middle column of Table 3.3). These filaments, which measure about 10 nm in diameter, are named "intermediate" because they are intermediate in size between the small actin filaments and the large microtubules. Intermediate filaments are made up of a diverse group of fibrous proteins, including keratin. The long strands are twisted together to form miniature "ropes" within the cell. Like a rope, an intermediate filament bends and twists easily but is quite difficult to break. As a result, intermediate filaments are stronger and more difficult to disassemble than other cytoskeletal filaments, and they form more permanent structures within the cell.

The strength and stability of intermediate filaments enable them to have a variety of roles within the cell, including the following:

- forming much of the framework of the cell and anchoring organelles within it;
- helping the organelles and the nucleus to maintain their shape and size; and
- helping cells and the tissues they form to withstand mechanical stresses.

Microtubules and Centrioles

The largest protein filaments in the cytoskeleton are the **microtubules,** which are hollow rods or "tubes" that measure about

Property	Actin Filaments	Intermediate Filaments	Microtubules
TABLE 3.3 CYTOSKELETAL FILAMENTS			
Location in cell	LM (1270×)	LM (1100×)	LM (1420×)
Structure	Actin subunits — 5–9 nm	10 nm	Tubulin subunits — 25 nm
Functions	• Support the plasma membrane • Form the core of microvilli • Involved in cell motion and cell division	• Form the framework of the cell • Support the shape and size of the organelles and nucleus • Provide cell strength • Help the cell to withstand mechanical stresses	• Support the cell • Maintain the position of organelles • Associate with motor proteins that move vesicles and organelles throughout the cell • Form the core of cilia and flagella

25 nm in diameter (Table 3.3, right column). Microtubules are composed of protein subunits called *tubulins* that can be rapidly added or removed from a microtubule, allowing it to change shape and size as needed by the cell.

Microtubules have a wide range of functions within the cell. For one, they function much as the steel framework of a building does, giving the cell its characteristic shape and size. Microtubules also maintain the internal architecture of the cell, positioning organelles in their proper places. In fact, if microtubules are disrupted, the endoplasmic reticulum collapses in on itself, and the sacs of the Golgi apparatus disperse in the cytosol.

In addition, microtubules are associated with motor proteins called dynein (DYN-een) and kinesin (kih-NEE-sin). These proteins move along the microtubules in the same way a train moves along a track. The motor proteins carry their "cargo" of organelles and vesicles to their appropriate destinations. Motor proteins are also involved in moving cellular extensions, which we discuss in the next section.

Notice in **Figure 3.22** that microtubules extend outward from a region of the cell called the **centrosome.** The centrosome consists of a gel matrix containing the tubulin proteins necessary for microtubule formation. When the cell is not dividing, the centrosome acts as a *microtubule-organizing center* for the cell's cytoskeleton. During this time the centrosome is located near the nucleus.

The centrosome also contains a pair of structures called **centrioles.** Centrioles are composed of a ring formed by nine groups of three modified microtubules and other proteins. The centrioles play a critical role in cell division, which we discuss in a later module

(see Module 3.8. Modified centrioles called **basal bodies** are found on the internal surface of the plasma membrane. Both cilia and flagella, which we discuss next, sprout from these basal bodies.

Quick Check

☐ 1. Determine whether the following statements apply to microtubules, intermediate filaments, or actin filaments.
   a. Thinnest cytoskeletal filaments; are associated with the motor protein myosin
   b. Consist of proteins twisted together into a ropelike form
   c. Allow certain cells to "crawl" and change the shape and size of the cell
   d. Consist of protein subunits surrounding a hollow core
   e. Help a cell to resist mechanical stress
   f. Grow from the cell's centrosome

Cellular Extensions

The cytoskeleton forms the inner framework of cell surface extensions in many cells. These extensions include microvilli, which are folds of the plasma membrane, and cilia and flagella, which are extensions that function in movement. Let's take a closer look at each.

Microvilli

Like all structures in the body, the plasma membrane obeys the rule that its form must follow its function. For this reason, we

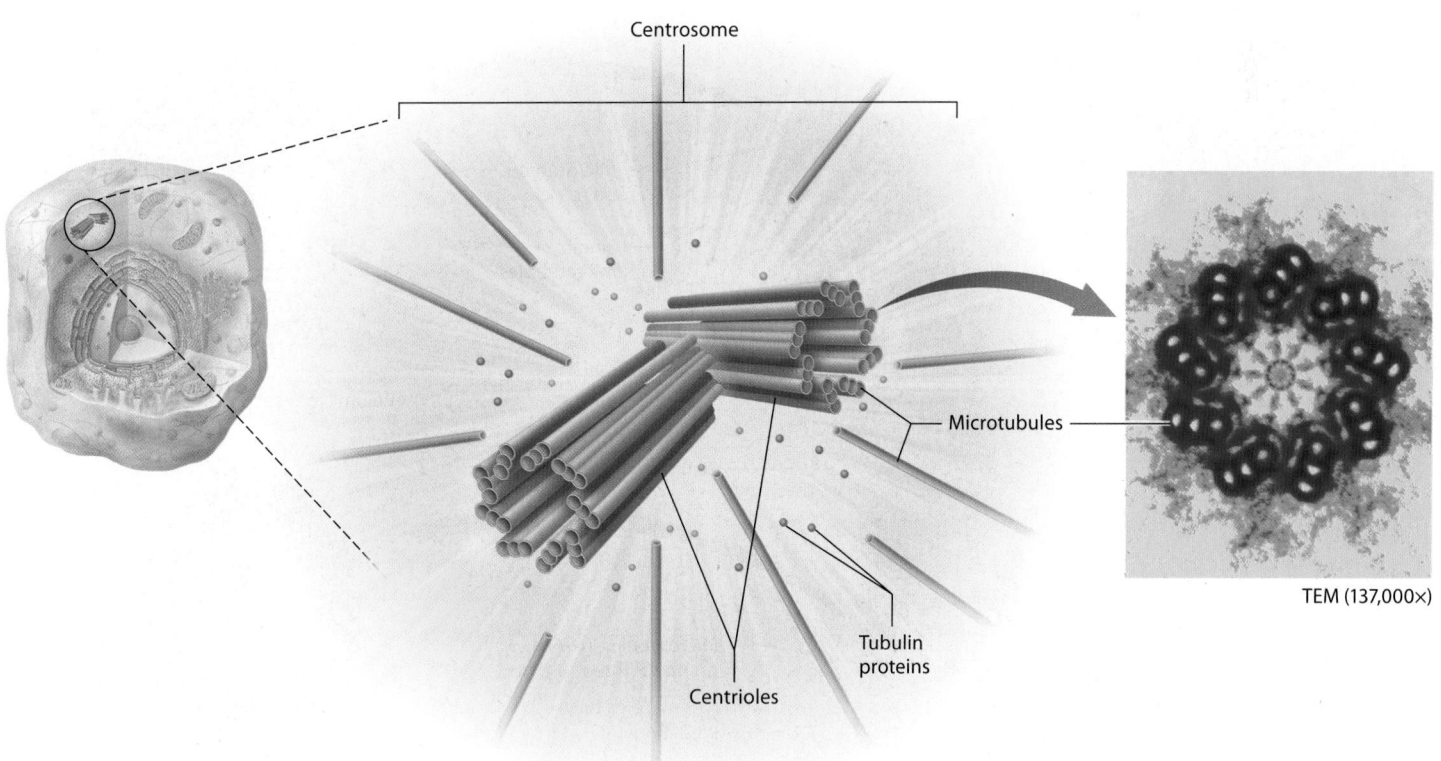

TEM (137,000×)

Figure 3.22 The centrosome with centrioles.

Microvilli

ECF

Plasma membrane

Actin filaments

Cytosol

Specialized cell with microvilli

TEM (31,700×)

Figure 3.23 Microvilli.

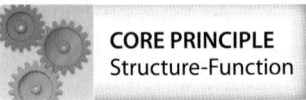

CORE PRINCIPLE
Structure-Function

Core Principle (p. 25). Actin filaments found at the core of microvilli help to maintain their shape.

Cilia and Flagella

Certain cells in the body have extensions from their surfaces called **cilia** (SILL-ee-uh; singular, *cilium*). Although cilia and microvilli may look superficially similar, they actually differ significantly in structure and function. For one, cilia are much larger than microvilli—individual cilia are visible with a light microscope, whereas microvilli appear only as a thickened border along a cell's surface. In addition, they are composed of different cytoskeletal filaments. Rather than being composed of actin filaments, like microvilli, cilia contain a core of microtubules associated with motor proteins. This reflects an important functional difference between the two structures: The microtubules and motor proteins enable cilia to be *motile*, or moveable.

Cilia share many structural similarities with another type of motile cellular extension called **flagella** (flah-JEL-ah; singular, *flagellum*). As you can see in **Figure 3.24**, cilia and flagella share the same basic structure, which is similar to that of centrioles. However, instead of the nine groups of three microtubules we saw in centrioles, cilia and flagella contain a ring of nine *pairs* of microtubules surrounding two central microtubules. Both cilia and flagella move via the action of motor proteins such as dynein that are associated with their microtubules. The motor proteins crawl along the microtubules, causing the cilium or flagellum to bend.

Although they are similar structurally, cilia and flagella differ in function. Cilia are shorter and move in a coordinated beating motion

would expect to see the plasma membrane taking on a slightly different shape in cells of the body where its function is different. This is, in fact, exactly what we see. Cells that are specialized for absorption, such as those in the kidneys and small intestine, have plasma membranes with a large surface area to perform absorption efficiently. In such cells we find the plasma membrane folded into tiny finger-like extensions known as **microvilli** (**Figure 3.23**; my-kroh-VIL-eye; *villus* = "hair"). Microvilli increase the surface area of the plasma membrane up to about 40 times (compared to a cell without microvilli), which allows the cell to absorb substances more rapidly, an example of the Structure-Function

Specialized cell with cilia

Plasma membrane

Microtubule pairs

Single microtubules

Cilium

Microtubules in groups of three

Basal body

TEM (320,000×)

Cilium cross-section

TEM (320,000×)

Basal body cross-section

Figure 3.24 Structure of cilia and flagella. Although only a cilium is shown here, the flagellum has the same structure.

TABLE 3.4 CILIA AND FLAGELLA

Property	Cilia	Flagella
Location in body	Cells lining the respiratory tract and the female reproductive tract	Sperm cell
	SEM (5000×)	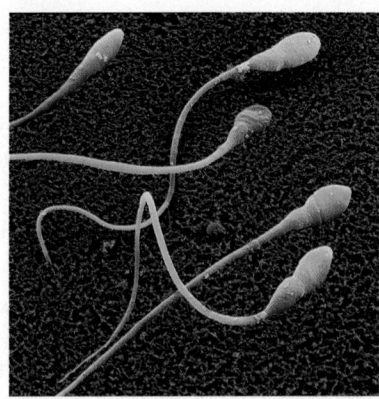 SEM (1525×)
Structure	Short, hairlike extensions from the cell; contain an internal ring of nine microtubule pairs surrounding a central microtubule core	Single, long extension from the cell; same internal structure as cilia
Function	Coordinated beating motion sweeps substances past the cell	Whiplike motion propels the cell through liquid.

(Table 3.4). They are present in very large numbers on the surfaces of certain cells, where each cilium functions like a tiny broom; these "brooms" beat in a coordinated fashion to sweep substances over the cell surface. For example, the cells that line our respiratory passages have cilia that sweep out mucus littered with dust and debris that we have inhaled. Note on the left side of Table 3.4 that the motion of each cilium resembles the movement of an oar propelling a boat—there is an initial stroke, followed by the cilium curving back to its original position. See *A&P in the Real World: Primary Ciliary Dyskinesia* to read about what happens when these cilia don't function properly. By contrast, flagella are long extensions that move in a whiplike manner to propel a whole cell (see the right side of Table 3.4). A cell generally has no more than one flagellum, and the only flagellated cell in the human body is the sperm cell in males.

Quick Check

☐ 2. What are microvilli and what are their primary functions?
☐ 3. How are cilia and flagella similar? How do they differ?

Apply What You Learned

☐ 1. Predict the cellular effects of a disorder that prevents the assembly of microtubules. (*Hint:* Why is it equally important for a microtubule to be easily assembled and disassembled?)
☐ 2. Explain how the disorder in question 1 would affect cilia and flagella.

A&P in the Real World

Primary Ciliary Dyskinesia

The importance of properly functioning cilia is evident in the condition known as **primary ciliary dyskinesia (PCD),** a rare genetic disorder that results in a defect in one or more of the protein components of cilia and flagella. This condition affects many types of cells, including those lining the respiratory passage, the middle ear, and the uterine tube in females, as well as sperm cells in males.

Ciliary dysfunction in the respiratory passages leads to the build-up of mucus in the lungs, which increases the risk of respiratory infections. Individuals with PCD commonly sustain progressive damage to the lungs due to repeated infections and mucus plugs. In addition, PCD patients may experience repeated ear infections, which can lead to hearing loss. Finally, because their sperm cells lack motility, male patients may be infertile.

☐ 3. Explain how the structure of intermediate filaments makes them better suited to provide mechanical strength than either microtubules or actin filaments.

See answers in Appendix A.

The Nucleus

Learning Outcomes

1. Describe the structure and function of the nucleus.
2. Explain the structure of chromatin and chromosomes.
3. Analyze the interrelationships among chromatin, chromosomes, and sister chromatids.
4. Describe the structure and function of the nucleolus.

As you may recall, the cell consists of multiple components that must function together optimally for the cell to survive. To ensure such function, a cell needs a structure to "govern" each of these components. The structure that serves this purpose in our cells is called the **nucleus** (Figure 3.25). We discussed earlier that the nucleus houses the cell's **deoxyribonucleic acid (DNA)**, and that DNA contains the code, or "plans," for almost every protein in the body (see Chapter 2).

But how does simply holding this code allow the nucleus to coordinate the activities of the cell? Remember that nearly all of a cell's structural components in the cytoskeleton and organelles are proteins, almost all of our enzymes are proteins, and proteins are also a vital component of the plasma membrane. The plans, called *genes,* within the DNA are used by several different kinds of **ribonucleic acid (RNA)** to build the proteins for our cells. In addition, the nucleus in large part determines the type of proteins that a cell makes and the rate at which they are made.

The nucleus consists of three main structures: (1) an enclosing membrane, (2) DNA and its associated proteins, and (3) the *nucleolus* (new-klee-OH-luss). The enclosing membrane, which is called the *nuclear envelope,* is pierced by occasional openings called *nuclear pores.* Both the DNA and the nucleolus are located within a cytosol-like gel called the *nucleoplasm.* Nucleoplasm contains many different components, including water, free-floating nucleotides, enzymes and other proteins, and different types of RNA. The most notable component of the nucleoplasm, however, is our DNA. Most of the time,

DNA is found here in a loosely organized form called *chromatin* (KROH-mah-tin). In this module, we examine the components of the nucleus so that we can address their functions in later modules.

Nuclear Envelope

In Figure 3.26 you can see that the **nuclear envelope** surrounding the nucleus is composed of a phospholipid bilayer. Like the mitochondrial membrane, the nuclear envelope is a double membrane. The outer membrane is studded with ribosomes and, as mentioned, is continuous with the endoplasmic reticulum. The inner membrane lines the interior of the nucleus and is supported by a network of intermediate filaments called the *nuclear lamina.* The two membranes are joined at large protein complexes that produce the **nuclear pores.** The nuclear pores connect the cytoplasm with the nucleoplasm of the nucleus and allow substances to move between the two areas.

Chromatin and Chromosomes

The task of storing our DNA is a daunting one, simply because of its sheer size: The human genome contains about 3 billion base pairs with about 25,000 genes. A nucleus may measure only about 6 μm, but if you removed the DNA from a single nucleus and stretched it out, it would likely be taller than you, extending about 2 meters. The feat of storing this amount of DNA is equivalent to taking 14 kilometers (about 9 miles) of very thin thread and shoving it into a golf ball! So how do our nuclei manage to squeeze so much material into such a tiny space without it becoming a tangled mess? The answer lies in proteins that bind to the DNA and allow it to twist and fold into an ordered and progressively more compact form.

Chromatin consists of one extremely long DNA molecule and its associated proteins. Figure 3.27a shows how chromatin is organized and elaborately folded to make it a compact structure. The most basic unit of organization after the DNA strand resembles beads on a string. Each "bead" is called a *nucleosome* (new-klee-oh-sohm) and consists of the strand coiled around a group

Rough ER

Nucleolus

Nucleoplasm containing chromatin (DNA)

Nuclear envelope

Nuclear pores

Ribosomes

TEM (41,000×)

Figure 3.25 The nucleus.

of proteins called *histones* (HISS-tonz). Between the beads in Figure 3.27a, you can see the "string," which is a short segment of DNA. Packaging DNA into this nucleosome structure reduces the length of the chromatin molecule by about one-third. Interactions between the amino acids of the histone proteins of neighboring nucleosomes cause the string of beads to condense and shorten even further, as they coil tightly around one another like a spring.

During the period between rounds of cell division, the chromatin threads are loose and tangled together like an unwound ball of yarn. However, during cell division, the chromatin threads coil tightly and condense into barlike structures called **chromosomes** (KROH-moh-sohmz) that are about 10,000 times more compact than the original strand of DNA. To prepare for cell division, the cell has duplicated its DNA, so at this point it has

Nucleus

Figure 3.26 The nuclear pore.

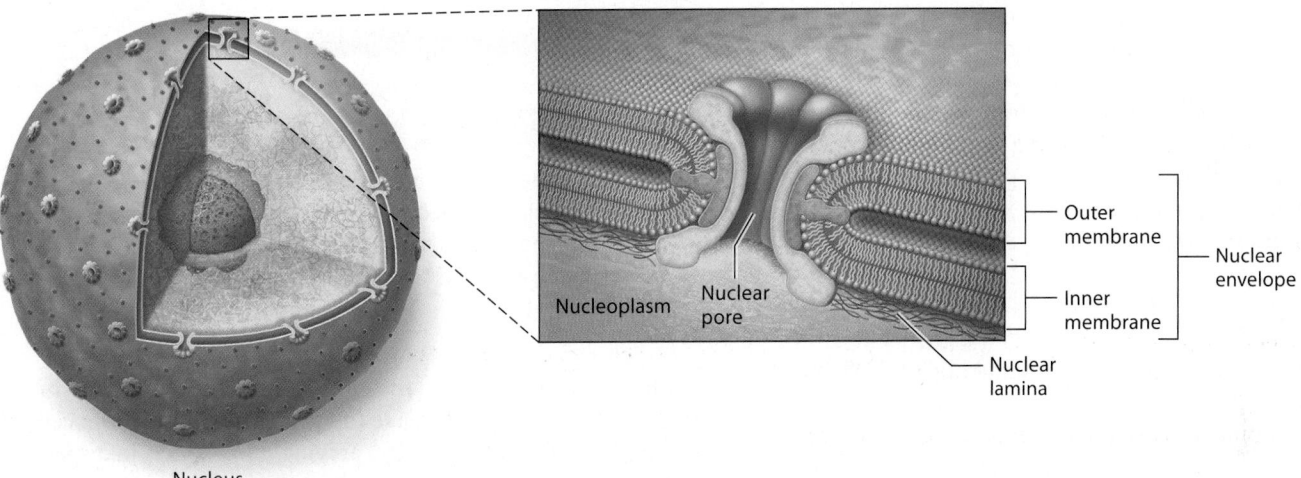

(a) **The structure of chromatin**

(b) **Replicated chromosome (in dividing cell)**

(c) **Human karyotype**

Figure 3.27 Chromatin and chromosomes.

two copies of each chromosome. The paired and identical copies of a chromosome are called *sister chromatids* and are joined at a region of the chromosome known as the **centromere.** Individual sister chromatids in a dividing cell are visible in the nucleus under a microscope; they look like what you see in **Figure 3.27b.**

Chromosomes vary in size and contain up to about 245 million base pairs each. Most human cells contain two sets of 23 chromosomes (46 total), with one set of 23 chromosomes inherited from each parent. The maternal and paternal chromosomes are called *homologous chromosomes* (or simply homologs). **Figure 3.27c** shows a *karyotype* (KAIR-ee-oh-typ), which contains the full set of 23 pairs of homologous chromosomes. These chromosomes are collected from a dividing cell, photographed, and then organized into their homologous pairs. Note that the pairs of chromosomes differ significantly in size and appearance (see Chapter 27 for more information).

Quick Check

☐ 1. What are the main components of the nucleus? What is its main function?

☐ 2. What is chromatin? How are chromatin and chromosomes related?

Nucleoli

◀◀ **Flashback**

1. What is the structure of ribonucleic acid (RNA)? (p. 61)
2. What makes up a ribosome? (p. 90)

If you look back at Figure 3.25, you'll see that among the chromatin in the nucleoplasm is a dark-staining structure called the **nucleolus** (cells sometimes have more than one of these *nucleoli*). We can think of a nucleolus as a "ribosome factory"—this is where ribosomes are assembled. It is not bounded by a membrane, but rather is simply a large aggregate of proteins, DNA, and RNA. As we discussed earlier in this chapter, much of our DNA consists of genes that specify the amino acid sequences of proteins. However, in the nucleolus we find segments of DNA containing genes that don't code for a protein, but instead just specify the nucleotide sequence of an RNA molecule. Some of these segments, known as *rRNA genes*, code for ribosomal RNA (rRNA), which is a key component of ribosomes.

Quick Check

☐ 3. What is a nucleolus and what is its function?

Apply What You Learned

☐ 1. Can a human cell without a nucleus make new ribosomes? Why or why not?

☐ 2. When a biologist is trying to isolate DNA from a cell, he or she must first add enzymes that degrade (break down) proteins. Why is this step necessary?

☐ 3. The nucleolus can change size, depending on how many ribosomes the cell needs to produce. When do you think

a cell would need more ribosomes and therefore a larger nucleolus? (*Hint:* What do ribosomes do?)

See answers in Appendix A.

MODULE **3.7**

Protein Synthesis

Learning Outcomes

1. Describe the genetic code, and explain how DNA codes for specific amino acid sequences.
2. Describe the processes of transcription and translation.
3. Explain the roles of rRNA, mRNA, and tRNA and of ribosomes in protein synthesis.

We have already discussed the molecular role of our genetic material: DNA is required to make the proteins within all the cells of our bodies. DNA does this by carrying information that specifies the amino acid sequence for each of the body's proteins and the nucleotide sequence for certain RNA molecules. The process of making proteins from DNA via RNA is called **protein synthesis.**

The human *genome*, our complete genetic makeup, contains about 25,000 genes. Over 99% of the coding DNA between individual humans is identical, which makes sense, as critical enzymes and structural proteins aren't going to vary from person to person without causing disease. The remainder of the coding DNA is unique for every individual (except identical twins), and is responsible for each person's unique physical traits.

But how, exactly, do proteins correlate with the vast array of physical traits that we humans possess? Take the simple example of the skin's tendency to either tan or freckle. A gene called the melanocortin-1 receptor gene codes for a membrane protein found in certain skin cells. When this protein is exposed to ultraviolet radiation, it triggers the production of pigment molecules. One variant of the melanocortin-1 receptor gene yields a protein that responds strongly to ultraviolet radiation, resulting in an even tan. A slightly different variation of this same gene yields a protein that responds weakly to ultraviolet radiation, resulting in freckling instead of tanning. The production of a protein from a specific gene is called *gene expression.*

In this module we examine the steps our cells take to build a protein. We begin with an exploration of the rules that govern how DNA specifies the sequence of a protein. We then discuss the two processes that actually make the proteins: transcription and translation. During *transcription,* the code specified by a gene is copied, creating a molecule of *messenger RNA* (*mRNA*). In *translation,* ribosomes read the nucleotide sequence of the mRNA molecule and synthesize a polypeptide chain containing the correct amino acid sequence. You can picture this process as follows:

$$\text{DNA} \xrightarrow{\textit{Transcription}} \text{mRNA} \xrightarrow{\textit{Translation}} \text{Protein}$$

Genes and the Genetic Code

◀◀ **Flashback**

1. What is a nucleotide? (p. 58)
2. What four nucleotides are present in DNA? (p. 61)

Recall from earlier in the chapter that DNA is composed of long chains of nucleotides containing **genes**—segments of DNA, most of which specify the amino acid sequences of proteins. Although there are only four types of nucleotides in DNA, there are 20 types of amino acids. This means that one or even two nucleotides can't code for a single amino acid, because there aren't enough possible combinations. However, there are enough combinations with groups of three nucleotides, and it is these groups (for example, AGT or CCG) that are the "words" that code for amino acids. These words are called **triplets,** and each amino acid is represented by one or more nucleotide triplets.

During transcription, each DNA triplet is transcribed into a complementary three-nucleotide sequence of mRNA called a **codon** (KOH-don). In essence, the codon is a complementary RNA "copy" of the DNA triplet. Recall from an earlier chapter how complementary base pairing works: The nucleotide bases C and G always pair with one another, and A always pairs with T (in DNA) or U (in RNA) (see Chapter 2). Then during translation, each mRNA codon is matched with an amino acid.

The **genetic code** is a list of which amino acid is specified by each possible DNA triplet. **Figure 3.28** shows this code, with each amino acid and the corresponding mRNA codons that specify it. Each three-nucleotide mRNA codon shown in the figure corresponds to the copied DNA triplet (not shown) and represents a single amino acid; often more than one codon codes for a single amino acid. This genetic code is found in all forms of life on Earth.

What happens when something alters the DNA so that its code is changed? The answer can range from having no effect to causing a serious disease such as sickle cell anemia (see Figure 19.7). Notice that the genetic code in Figure 3.28 is precise, and a difference in as little as a single nucleotide can change the corresponding amino acid entirely. For example, a change from the codon UUU to the codon UCU results in coding for the amino acid serine (Ser) instead of phenylalanine (Phe). Changes to DNA are called **mutations,** and they can be due to mistakes in the copying of DNA or can be induced by agents known as *mutagens.* Common mutagens include ultraviolet light and other forms of radiation, chemicals such as benzene, and infection with certain viruses. DNA mutations are the basis for many diseases, including cancer, which we discuss in the next module.

Now that we have a better understanding of how DNA codes for a protein, we can refine our definition of a gene: A *gene* is a series of nucleotide triplets that specifies the sequence of amino acids in a protein. But this is still not quite the whole story. The average gene is about 8000 nucleotides long, but the average protein contains only about 400 amino acids, which should require only about 1200 nucleotides. The extra parts of the gene are *noncoding*—they are chains of nucleotides that do not specify any amino acids. Much of the noncoding parts of a gene are segments called *introns;* the coding portions of a gene are *exons.* Scientists once thought that introns

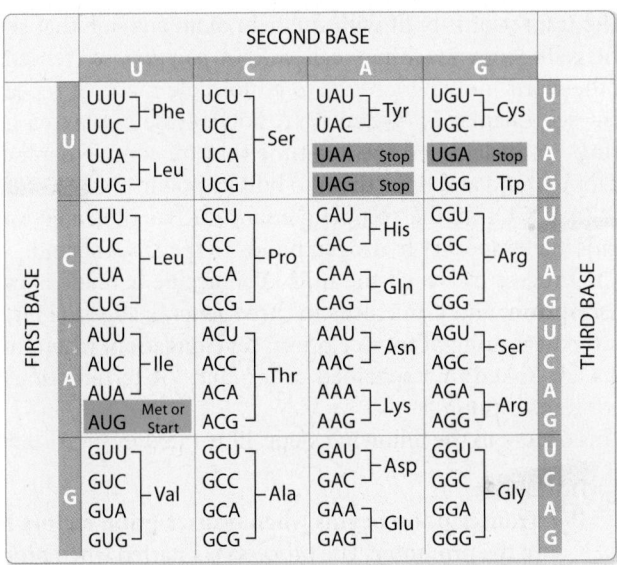

Key:

Abbreviation	Amino acid	Abbreviation	Amino acid
Ala	Alanine	Leu	Leucine
Arg	Arginine	Lys	Lysine
Asn	Asparagine	Met	Methionine
Asp	Aspartic acid	Phe	Phenylalanine
Cys	Cysteine	Pro	Proline
Glu	Glutamic acid	Ser	Serine
Gln	Glutamine	Thr	Threonine
Gly	Glycine	Trp	Tryptophan
His	Histidine	Tyr	Tyrosine
Ile	Isoleucine	Val	Valine

Figure 3.28 The genetic code. This figure shows the mRNA codons that correspond to amino acids.

were "junk DNA," but recent evidence suggests that they play an important role in the evolution of new proteins and that they help to regulate gene expression. Let's now examine how a cell reads the instructions within a gene and uses them to make a protein.

Transcription

◀◀ **Flashback**

1. What are the main differences between DNA and RNA? (p. 61)
2. Which nucleotides are complementary to one another and so are able to form hydrogen bonds in both DNA and RNA? (p. 61)

You may recall from earlier in the chapter that most of a cell's DNA resides in the nucleus, and that protein synthesis occurs on ribosomes, which are found outside the nucleus in the cytoplasm. This arrangement presents our cells with a problem: The DNA that codes for the proteins doesn't leave the nucleus, but the information must somehow leave and get into the cytosol so that it can be read there by the ribosomes. Our cells solve this problem by making a copy of the gene, called a *transcript,* which is a molecule of RNA that can exit the nucleus and enter the cytosol.

The transcript is built with the help of an enzyme that serves as the cell's "copy machine," called *RNA polymerase* (remember that the word "polymer" refers to a molecule made of repeating units—see Chapter 2). When RNA polymerase binds to a gene, it brings in nucleotides one at a time that are complementary to the DNA strand and links them to build a polymer of **messenger RNA (mRNA).** As the complementary mRNA molecule forms, it binds via temporary hydrogen bonds to the DNA strand.

The process by which the mRNA molecule is made is called **transcription,** and it proceeds in three general stages: (1) initiation, the beginning of transcription; (2) elongation, when nucleotides are linked in a specified order; and (3) termination, the end of transcription.

This occurs in the following steps, illustrated in **Figure 3.29:**

1. **Initiation:**
   1a. **Transcription begins when transcription factors bind to the promoter.** The process gets started when proteins called *transcription factors* bind to a DNA segment near the gene called the *promoter.* The promoter is found on one strand of the DNA called the *template strand.*
   1b. **RNA polymerase then binds to the promoter as well, and a segment of DNA unwinds.** Once certain transcription factors have bound to the promoter, RNA polymerase also binds. At this point, an enzyme called *helicase* (HEE-lih-kayz) catalyzes the unwinding of a segment of the DNA double helix.
2. **Elongation: RNA polymerase builds a complementary mRNA transcript with free nucleotides.** RNA polymerase brings in free nucleotides from the nucleoplasm one by one and catalyzes the formation of hydrogen bonds between the free nucleotides and the nucleotides in the template strand of DNA. Each nucleotide is bound by hydrogen bonds to its complementary base in the DNA. For example, the figure shows the mRNA codon GCU bound to the complementary DNA triplet CGA. As each new nucleotide is brought by the RNA polymerase, this enzyme covalently adds the nucleotide to the end of the growing mRNA transcript. After a section of DNA is copied, the strands re-form the double helix.
3. **Termination: Transcription ends when the end of the gene is reached, and the mRNA transcript is released.** The process continues until the RNA polymerase reaches a nucleotide sequence that signals the end of the gene. At this point enzymes catalyze the release of the mRNA transcript, and the RNA polymerase detaches from the DNA.

If the process of transcription is disrupted, the consequences for the cell, and perhaps the whole body, are severe. *A&P in the Real World: Toxicity of the "Death Cap" Mushroom* talks about one type of poison that can cause such a disruption.

After transcription, the transcript, called *pre-mRNA,* isn't yet ready to be sent out into the cytosol because it must first be modified in several ways. As part of this process, large portions of the pre-mRNA transcript are actually removed before it exits

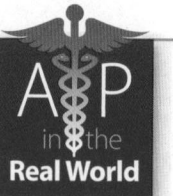

Toxicity of the "Death Cap" Mushroom

What makes some mushrooms poisonous? Consider the "death cap" mushroom (*Amanita phalloides*); it and other mushrooms of the same genus (*Amanita*) are responsible for 95% of mushroom-related fatalities worldwide. *A. phalloides* is particularly dangerous because it has a pleasant taste and resembles many species of nontoxic mushrooms. Its main toxin inhibits RNA polymerase, which prevents the formation of new strands of mRNA. This essentially stops protein synthesis and disrupts many of the cell's functions, leading to cell death. No antidote exists for this type of poisoning, although several agents have shown promise as treatments. The liver suffers the most damage from *A. phalloides* poisoning, and patients who survive generally require a liver transplant.

the nucleus. Recall that not all regions of a gene contain actual code, or exons—some of the gene instead consists of introns. The copied introns in the pre-mRNA must be removed and the exons spliced together before the final mRNA transcript is sent to the cytosol:

This step is called *RNA processing.* When processing is complete, the mRNA exits the nucleus through a nuclear pore, and enters the cytosol, ready for translation.

Quick Check

☐ 1. How is a codon related to a triplet?
☐ 2. Describe the basic steps of transcription.

1 Initiation:

1a Transcription begins when transcription factors bind to the promoter near the gene.

1b RNA polymerase binds to the promoter as well, and a segment of the DNA unwinds.

2 Elongation: RNA polymerase builds a complementary mRNA transcript with free nucleotides from the nucleoplasm.

3 Termination: Transcription ends when the end of the gene is reached, and the mRNA transcript is released.

Transcription factors

Double-stranded DNA

Promoter

Gene

RNA polymerase

mRNA transcript

Nontemplate strand of DNA

Free nucleotides

Codon

mRNA transcript

Triplet

Template strand of DNA

Gene

mRNA transcript

Play animation
@MasteringA&P®

Figure 3.29 Transcription.

Translation

◀◀ **Flashback**

1. What are polypeptides and peptide bonds? (p. 56)

At the end of transcription and RNA processing, we have an mRNA molecule that contains a copy of the triplet code in the DNA. But this information is still encoded in the language of nucleotides, and our cells need to then *translate* it into the language of amino acids. This occurs by the process that is appropriately called **translation.** Translation takes place with the help of ribosomes and involves another type of RNA molecule that acts as the molecular translator called **transfer RNA, or tRNA** (Figure 3.30). Recall from the previous module that tRNA is manufactured within the nucleus from a DNA template. It is called "transfer" RNA because its job is to pick up a specific amino acid from the cytosol and transfer it to the growing polypeptide chain at the ribosome.

Notice in Figure 3.30 that tRNA is a single strand of RNA. This strand forms hydrogen bonds with itself to create the shape of a three-leaf clover. At one end of the tRNA is a region called the **anticodon,** which contains a sequence of three nucleotides that is complementary to a specific mRNA codon. The opposite end of the tRNA molecule carries the specific amino acid that corresponds to the anticodon. At least one type of tRNA specifies each of the 20 amino acids.

Let's see how this process works using the mRNA codon UUC as an example. As you can see by looking back at Figure 3.28, the mRNA codon UUC specifies the amino acid phenylalanine (Phe). By the rules of complementary base pairing, the mRNA codon UUC binds to the tRNA anticodon AAG, which appears on a tRNA molecule with the amino acid phenylalanine attached to its other end.

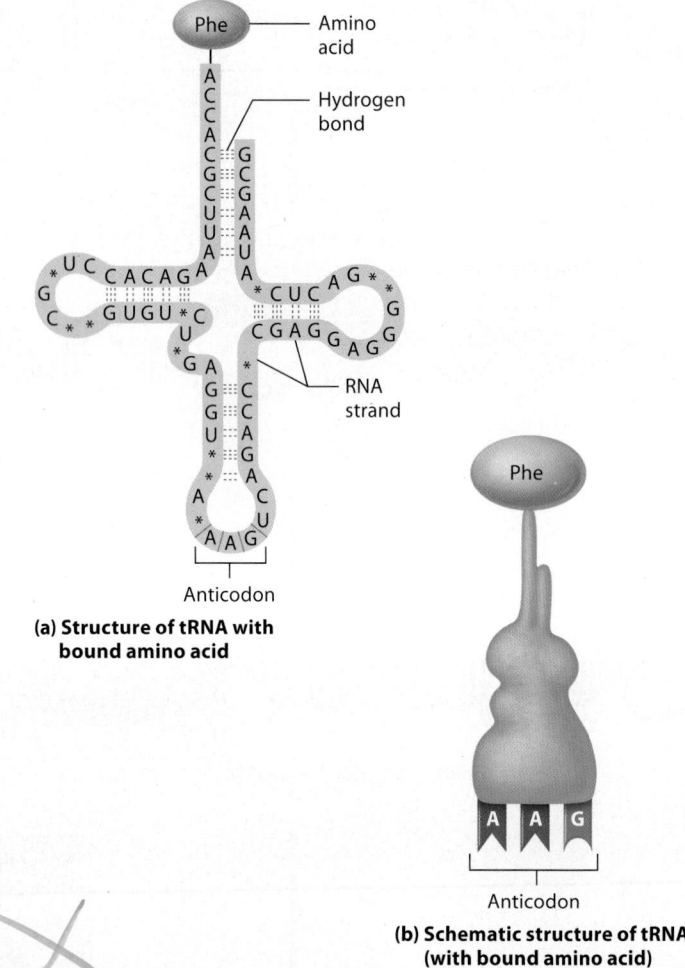

(a) **Structure of tRNA with bound amino acid**

(b) **Schematic structure of tRNA (with bound amino acid)**

Figure 3.30 Transfer RNA (tRNA).

① **Initiation**: Translation begins when an initiator tRNA binds the mRNA start codon in the ribosome.

▶ Play animation
@MasteringA&P®

② **Elongation**:

②a Another tRNA binds to the open A site.

②b The first amino acid detaches from the tRNA at the P site and is joined to the second amino acid by a peptide bond.

Figure 3.31 Translation.

ConceptBOOST ⟫⟫⟫

Connecting a DNA Triplet to a Particular Amino Acid

It's easy to get the relationships between a *triplet,* a *codon,* and an *anticodon* confused. Here are a couple of quick examples to clarify these relationships:

	Isoleucine	Alanine
Amino acid	Ile	Ala
tRNA anticodon	U A G	C G G
mRNA codon	A U C	G C C
DNA triplet	T A G	C G G

Notice in these examples that the nucleotides in the tRNA anticodon are the same as those in the DNA triplet, except that the nucleotide T in DNA is replaced by U in tRNA.

Now, using Figure 3.28 and these examples for reference, figure out the amino acid specified by each of the following DNA triplets (remember that the amino acid correlates to the mRNA codon, not the tRNA anticodon):

DNA triplet	T T C	C A A	A G G
mRNA codon			
tRNA anticodon			
Amino acid			

∎

Answers: mRNA codon: AAG, GUU, UCC; tRNA anticodon: UUC, CAA, AGG; Amino acid: Lys, Val, Ser

As we discussed earlier, protein synthesis takes place with the help of the cell's ribosomes (see Figure 3.18). Each ribosome has three binding sites for tRNA molecules, called the A, P, and E sites. The A (aminoacyl; ah-mee'-noh-AY-sil) site binds an incoming tRNA molecule carrying an amino acid. The P (peptidyl; pep-TID-'l) site is where the tRNA's amino acid is added to the growing protein. The "used" tRNA departs from the E (exit) site of the ribosome; it can pick up another amino acid in the cytosol and be reused.

During translation, the ribosome holds the mRNA transcript in place between its two subunits and facilitates its interaction with tRNAs. Like transcription, the process of translation can be broadly organized into three stages: (1) *initiation,* the beginning of translation; (2) *elongation,* when amino acids are linked together by peptide bonds; and (3) *termination,* when the completed polypeptide is released into either the cytosol or the rough ER. These stages proceed in the following steps, shown in **Figure 3.31**:

① **Initiation: Translation begins when an initiator tRNA binds the mRNA start codon in the ribosome.** The initiation stage begins when both the mRNA transcript and a tRNA called the *initiator tRNA,* which carries the amino acid methionine (Met), bind to a small ribosomal subunit. With this tRNA in place, the small ribosomal subunit moves down the mRNA transcript looking for a specific nucleotide sequence called the *start codon.* The start codon has a sequence of AUG that base pairs to the initiator tRNA's anticodon, UAC. At this point, the large ribosomal subunit binds to the small subunit, with the initiator tRNA in the P site. Initiation is now complete, and elongation can begin.

② **Elongation (continued):**

②c The ribosome moves down to the next mRNA codon, the empty tRNA exits from the E site, and a new tRNA comes into the A site.

The elongation cycle of step ② continues.

③ **Termination:** Translation ends when the ribosome reaches the stop codon and the completed polypeptide is released.

(2) **Elongation:**

(2a) **Another tRNA binds to the open A site.** The elongation stage begins with the binding of another tRNA molecule to the ribosome's open A site. Note that this tRNA has an anticodon that is complementary to the mRNA codon in the A site.

(2b) **The first amino acid is joined to the second amino acid by a peptide bond.** Elongation occurs as the amino acid (or later, growing polypeptide) is transferred from the tRNA in the P site to the amino acid on the tRNA in the A site. These two amino acids are linked covalently by a peptide bond, and both are now attached to the tRNA in the A site.

(2c) **The ribosome moves down to the next mRNA codon, the empty tRNA exits, and a new tRNA comes in.** During the next step of elongation, the ribosome moves down to the next mRNA codon. Observe that this move shifts the "empty" tRNA that was in the P site into the E site, from which it exits the ribosome, and the tRNA that was in the A site now moves into the P site. The A site is now vacant, and a new tRNA moves into it. The events from this step make up a cycle called the *elongation cycle,* with the peptide chain gaining an amino acid and the ribosome then shifting to the next mRNA codon each time the cycle is repeated.

(3) **Termination: Translation ends when the ribosome reaches the stop codon and the polypeptide is released.** The elongation cycle continues until the ribosome reaches an mRNA codon called the *stop codon* (the stop codon could be UAG, UAA, or UGA; see Figure 3.28). The stop codon does not bind a tRNA. Instead, it binds to a protein called the release factor, which releases the completed polypeptide chain from the tRNA in the P site. When this happens, translation is terminated.

At the end of translation we have finished synthesizing a polypeptide. However, such a polypeptide is not a functional protein until it folds into its proper three-dimensional configuration. This folding and other important structural changes take place during *posttranslational modification.* Polypeptides that are destined for the cytosol are synthesized on free ribosomes, and they fold either on their own or with the help of other proteins. Many other polypeptides destined for secretion or insertion into an organelle or membrane require modifications that are performed in the rough endoplasmic reticulum (RER), so they are synthesized on bound ribosomes. When translation is complete, the polypeptide remains in the RER, where it is folded and modified as necessary. The resulting protein is then sent to the Golgi apparatus for final processing, sorting, and packaging.

Quick Check

☐ 3. Explain how tRNA acts as the "translator" of the genetic code.

☐ 4. Describe the basic steps of translation.
☐ 5. Why is posttranslational modification necessary for a protein to be functional?

Figure 3.32 offers a summary of the events of protein synthesis. This process can be summarized in four basic steps: transcription, RNA processing and transit, translation, and posttranslational modification.

Regulation of Gene Expression

The human genome contains about 25,000 genes, but most cells express only a fraction of these genes and synthesize only those proteins required for their specific functions. For example, all cells in the body contain the gene for the protein hormone insulin, yet only certain cells of the pancreas actually produce insulin. Furthermore, these cells of the pancreas can produce more or less insulin, depending on the conditions in the body. How, then, do cells "know" which proteins to make, and when to make them? The answer is fairly complicated and involves a variety of mechanisms that control the production of a protein.

Many methods of regulating gene expression occur at the initiation of transcription. Specific proteins bind to parts of the gene and stimulate the initiation of transcription, whereas other proteins inhibit its initiation. Regulation of gene expression can also take place after transcription. Some examples of such mechanisms include modification of the length of time mRNA molecules remain in the cytosol before being degraded, inhibition of the start of translation, and finally, inhibition of posttranslational modification. All of these mechanisms and others work together to ensure that a cell produces only the proteins it needs at the appropriate times and in the appropriate amounts to carry out its specific functions.

Quick Check

☐ 6. Why is it important to regulate gene expression?

Apply What You Learned

☐ 1. Your friend has been using a tanning bed and, unbeknownst to him, the radiation from the tanning bed has damaged the DNA in his skin cells. In one particular gene, the triplet TGA has mutated to TGC. Will this change the structure of the protein coded for by this gene? In a separate gene, the triplet GCG has mutated to GGG. Will this change the structure of the protein coded for by this gene? Explain why or why not. (*Hint:* Refer to Figure 3.28 for the genetic code.)

☐ 2. You read earlier about the poison ricin, which inhibits the functioning of ribosomes. What part of protein synthesis would be affected by ricin? What effect would this have on a cell's ability to synthesize proteins?

See answers in Appendix A.

The Big Picture of Protein Synthesis

Figure 3.32

Play animation
@MasteringA&P®

Nucleus

RNA polymerase

DNA

Pre-mRNA transcript

Introns

Exons

mRNA

Cytoplasm

Nuclear pore

Growing polypeptide

Ribosome

tRNA

mRNA

Completed polypeptide

Modified functional protein

1 **Transcription:** RNA polymerase catalyzes the formation of the pre-mRNA transcript from a DNA template (see Figure 3.29).

2 **RNA processing and transit:** The introns are removed from the pre-mRNA, leaving exons that make up the mRNA. The mRNA then leaves the nucleus and enters the cytosol.

3 **Translation:** The mRNA binds to a ribosome and is translated by tRNAs, producing a polypeptide (see Figure 3.31).

4 **Posttranslational modification:** The polypeptide is modified and folded into its final protein form, most of which occurs in the cytosol or the rough ER.

MODULE **3.8**

The Cell Cycle

Learning Outcomes

1. Describe the events that take place during interphase and their functional significance.
2. For each stage of the cell cycle, describe the events that take place and their functional significance.
3. Distinguish between mitosis and cytokinesis.
4. Describe the process of DNA replication.

The biological principle called the *cell theory* states that cells cannot spontaneously appear, but rather, they must come from the division of cells that already exist. In other words, all forms of life, including humans, are the result of repeated rounds of cell growth and division.

Almost all cells go through a process called the **cell cycle,** which is defined as the ordered series of events from the formation of the cell to its reproduction by cell division. Cell division is vital in a fetus or child, as it is required for growth and development. However, cell division is equally important in adults for tissue repair and renewal, and millions of our cells undergo cell division each *second* throughout our lives. Our cells are not immortal, and they must be replaced as they die; otherwise, we may face major imbalances in our homeostasis. For example, cells in the bone marrow produce millions of oxygen-carrying red blood cells per day. If the cells in the bone marrow cease dividing, which happens with the condition known as *aplastic anemia,* the body cannot carry oxygen efficiently to its cells, and death may result. In this module we explore the events behind these critical processes of cell growth and division, and touch on the process of cell cycle control and what happens when it is lost.

Phases of the Cell Cycle

◀◀ **Flashback**

1. What are the functions of microtubules? (p. 96)
2. What structure forms and organizes microtubules? (p. 97)

The cell cycle, shown in **Figure 3.33**, is composed of two main phases: *interphase* and *M phase*, or cell division. ① During **interphase**—which includes three subphases known as G_1, S, and G_2—the cell grows and prepares for ② the final M phase, which is the division of the nucleus (*mitosis*; my-TOW-sis) and cytoplasm (*cytokinesis*; sy′-tow-kin-EE-sis; *cyto* = "cell", *kinesis* = "movement"). The cell spends most of its time in the G_1 phase of interphase. During interphase, the cell must duplicate its DNA precisely so that the products of cell division, called **daughter cells,** have an identical genetic makeup. The cell must also grow, synthesize proteins, and make new organelles so that the two daughter cells have adequate amounts of these components in their cytoplasm.

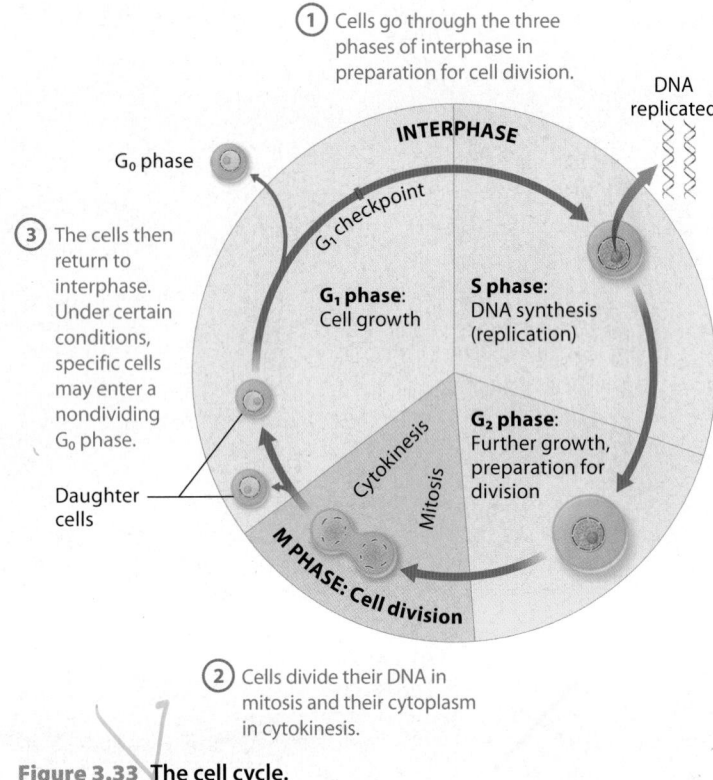

① Cells go through the three phases of interphase in preparation for cell division.

③ The cells then return to interphase. Under certain conditions, specific cells may enter a nondividing G_0 phase.

② Cells divide their DNA in mitosis and their cytoplasm in cytokinesis.

Figure 3.33 **The cell cycle.**

Let's look in more detail at the phases of the cell cycle.

- **Interphase,** the phase of growth and preparation for division, includes the following three subphases:
  - G_1 **phase.** During the G_1 **phase,** or first gap phase, the cell is performing its normal metabolic functions as well as growing and carrying out rapid protein synthesis. It uses these proteins to make new organelles and components of the cytoskeleton. (Although "G" stands for "gap," you might think of it as standing for "growth.")
  - **S phase.** As you can see in Figure 3.33, the **S phase** is when DNA synthesis takes place. (The "S" stands for "synthesis.")
  - G_2 **phase.** The G_2 **phase,** or the second gap phase, is another period of cellular growth during which the proteins necessary for cell division are rapidly produced and the centrioles are duplicated.
- **M phase,** the phase of cell division, has two overlapping parts:
  - **Mitosis** occurs when the genetic material is divided between the two daughter cells. It requires a structure called the *mitotic spindle,* which is composed of microtubules organized into *spindle fibers* by the cell's centrosomes.
  - **Cytokinesis** occurs when the cell's proteins, organelles, and cytosol are divided between the daughter cells.

Most cells in the body undergo each phase of the cell cycle, but the rate at which different cell populations progress through these phases varies drastically. Cells of the skin, hair follicles, and digestive tract are subject to a great deal of physical stress and therefore must progress through the cell cycle rapidly so the damaged cells are replaced. In addition, bone marrow cells progress rapidly through the cell cycle because they must produce millions of new blood cells each day.

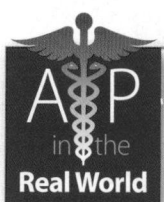

Spindle Poisons

The mitotic spindle is critical to the process of mitosis, and if its assembly or disassembly is inhibited, errors in cell division occur that could lead to the death of the cell. This is the mechanism of action of a group of chemicals known collectively as *spindle poisons*, most of which are found naturally in plants. Many of these spindle poisons have been found to be useful in the treatment of certain diseases, including the following:

- *Vinca alkaloids* are a group of several drugs that were originally derived from the periwinkle plant. They inhibit microtubule functions and fragment formed microtubules, effectively destroying the mitotic spindle. Vinca alkaloids are used in the treatment of many different cancers, including lung, breast, and testicular cancers.
- *Colchicine* is a drug derived from the autumn crocus. It inhibits the assembly of microtubules, preventing the mitotic spindle from forming. It is used primarily to treat gout, a type of arthritis.
- *Griseofulvin* is a chemical produced by certain strains of mold. Similar to colchicine, it binds to tubulin subunits of microtubules, inhibiting their function and proper assembly. It has relatively few effects on human cells, acting instead on certain species of fungi. For this reason, it is used to treat fungal infections, particularly those of the skin, hair, and nails.
- *Taxanes* are a group of drugs derived from different species of the yew tree. They are unique in that they prevent the *disassembly* of microtubules, This essentially "freezes" the mitotic spindle and prevents its functions and other normal cell functions. Like the vinca alkaloids, taxanes are used in the treatment of certain cancers, including lung, breast, and ovarian cancers.

Although these drugs are all quite effective, they can cause a number of adverse effects due to their actions on healthy cells, especially cells that undergo rapid rates of mitosis such as those lining the stomach, the skin, and in bone marrow. Common adverse effects include nausea, vomiting, hair loss, and a decrease in the production of blood cells by bone marrow.

Other cells move through the cell cycle much more slowly, and some do not move through it at all once the body's normal growth and development have ceased. Such cells are frozen in G_1 phase indefinitely, and are said to be in a nondividing state called the **G_0 phase** (step ③ in Figure 3.33). Cells in the human body in G_0 phase include muscle cells and nerve cells. Under certain very limited conditions, these cells may divide, but generally they remain in G_0 phase for life. This is why brain and spinal cord injuries and some muscle tears rarely heal completely. To learn about the effects of an inhibited mitotic spindle, see *A&P in the Real World: Spindle Poisons.* Now let's look in more detail at two phases of the cell cycle, S phase and M phase.

S Phase: DNA Synthesis or Replication

Given the large size of the human genome, it seems like a daunting task to unravel our DNA and make an exact copy. Yet this process, called **DNA synthesis** or **replication,** goes on in millions of cells in the body every second. During DNA synthesis, the chromatin unwinds from the histone proteins, and the entire set of 3.2 billion base pairs is duplicated by a process that builds new DNA strands using each "old" DNA strand as a template. Such replication is called *semiconservative replication* because one of the original strands of DNA is included in each newly formed double helix. The enzyme that catalyzes DNA synthesis is called **DNA polymerase.**

Figure 3.34 illustrates DNA synthesis, which proceeds as follows:

① **DNA strands separate as helicase unwinds them.** DNA synthesis starts as the enzyme helicase catalyzes the unwinding of the two strands of DNA.

② **Primase builds RNA primers on the existing DNA strands.** DNA polymerase can add a nucleotide only to an *existing* chain of nucleotides. This means that DNA synthesis cannot start from scratch—a few nucleotides must already be base-paired with the template strand. Our cells have a very simple solution to this problem. An enzyme called *primase* catalyzes the reactions that build a very short segment of RNA, called an *RNA primer,* on the DNA template. RNA primers give the DNA polymerase a place to begin building the new strands of DNA. As this occurs, helicase continues to unwind the DNA double helix.

③ **DNA polymerase catalyzes the addition of nucleotides to the new DNA strands.** DNA polymerase catalyzes the formation of hydrogen bonds between free nucleotides from the nucleoplasm and nucleotides in the DNA templates, determined by the rules of complementary base pairing. Covalent bonds are then formed between adjacent nucleotides, creating the new DNA strand. When DNA polymerase encounters an RNA primer, the primer is removed and replaced with DNA. Note in Figure 3.34 that DNA synthesis progresses in opposite directions along the two strands. This occurs because DNA strands have directionality to them, and DNA polymerase can add nucleotides in only one direction along the template strands.

④ **The end result is two identical double helices, each with one old strand and one newly formed strand of DNA.** This preservation of one original strand is why such replication is called **semiconservative replication.**

When DNA synthesis is complete, the histone proteins reassociate with the DNA and the cell goes into G_2 phase, in which it prepares for cell division.

M Phase: Mitosis and Cytokinesis

Recall that the process of mitosis divides the cell's replicated DNA between two daughter cells. At the beginning of mitosis, we have a cell with 92 chromatids, or 46 replicated chromosomes. At the end, we have two cells with identical sets of 46 chromosomes.

Nucleus

DNA

① DNA strands separate as the enzyme helicase unwinds them.

Helicase

② The enzyme primase builds RNA primers on the existing DNA strands to give DNA polymerase a place to begin building the new strands of DNA.

RNA primers

Primase

DNA polymerase

③ DNA polymerase catalyzes the covalent addition of free nucleotides to the growing new DNA strands. Note that the new strands of DNA are being built simultaneously in opposite directions.

Free nucleotides

Newly made DNA strands

Identical DNA double helices

④ The end result is two identical double helices, each with one old strand and one newly formed strand of DNA. For this reason, the whole process is called semiconservative replication.

Figure 3.34 DNA synthesis.

Sequence the steps @MasteringA&P®

To divide its genetic material, the cell rapidly goes through a series of complex events. Notice in **Figure 3.35a** that during interphase, the nucleus and nucleolus are clearly visible under a light microscope, and the individual chromosomes are not distinguishable. Now look at the first phase of mitosis, called *prophase,* in **Figure 3.35b**. As you can see, the cell undergoes fairly drastic changes in appearance. All of these changes ultimately make it easier for the cell to divide.

Division of a cell's genetic material occurs through four stages of the M phase, illustrated in Figure 3.35b:

① **Prophase.** During **prophase,** the chromatin becomes fully compacted so that individual chromosomes, each composed of two sister chromatids joined at the centromere, are visible. Additionally, the nucleolus breaks down, and a structure called the **mitotic spindle** forms. During this process, the previously duplicated centrosomes, each containing a pair of centrioles, move to opposite poles of the cell and organize the spindle fibers. Spindle fibers from each centriole pair then attach to each sister chromatid at the centromere. At the end of prophase, the nuclear envelope begins to break apart.

② **Metaphase.** The longest phase of mitosis is **metaphase.** During this phase, spindle fibers from opposite poles of the cell tug the sister chromatids back and forth to eventually line up on the middle or "equator" of the cell.

③ **Anaphase.** During **anaphase,** the sister chromatids part, and the individual chromosomes, now called *daughter chromosomes,* are pulled to opposite poles of the cell. This movement occurs as the spindle fibers to which they are attached get progressively shorter. By the end of anaphase, each pole of the cell has the complete set of 46 daughter chromosomes. Note that some microtubules are not attached to chromosomes, and these microtubules lengthen rather than shorten, which elongates the cell. In addition, cytokinesis may start in this phase.

④ **Telophase and Cytokinesis.** The final phase of mitosis is **telophase** (TEEL-uh-fayz), during which daughter cells separate. The nuclear envelopes reassemble, the nucleoli re-form, and the chromosomes become less visible as the DNA reassumes the looser structure of chromatin. While telophase is occurring, so is cytokinesis, which divides the cytosol and organelles equally between the two daughter cells. The cells are split at an indentation called a *cleavage furrow,* which forms along the cell equator and deepens like a belt tightening as the cell's actin filaments and myosin motor proteins pinch the cell in two. Mitosis and cytokinesis are now complete, and the result, shown in **Figure 3.35c,** is two daughter cells with identical genetic material.

Quick Check

☐ 1. What happens during each stage of the cell cycle?
☐ 2. What does "semiconservative replication" mean?

☐ 3. Describe the changes in the cell that take place during:
   a. prophase.
   b. metaphase.
   c. anaphase.
   d. telophase.

Cell Cycle Control and Cancer

Our cells have a control system that determines how rapidly each cell passes through the stages of the cell cycle. Such control is necessary to balance cell division with cell death. Monitoring occurs at specific points in the cell cycle called **checkpoints,** which act as stop/go signals. The most important checkpoint, called the G_1 *checkpoint,* occurs about three-fourths of the way through G_1, as you can see if you look back at Figure 3.33.

Researchers have worked out many of the extracellular signals that regulate progression through these checkpoints and completion of the cell cycle, but only a few of the intracellular signals. Some of the extracellular signals include the following:

- **Nutrients in the ECF.** Just as your car's engine won't start without gasoline, cells won't proceed all the way through the cell cycle without proper nutrients. If a particular nutrient is lacking in the ECF, the cell stalls at one of the checkpoints.
- **Growth factors in the ECF.** A *growth factor* is a protein that is secreted by one cell and stimulates other cells to divide.
- **Density of cells in a tissue.** In most tissues, cells stop dividing when they reach a certain number, or density. This phenomenon is due in part to the fact that if more than a certain number of cells are in a tissue, the availability of nutrients declines to the point that further cell growth and division are inhibited.
- **Anchorage of cells within the tissue.** In many tissue types, the cells are anchored to the cells around them at the extracellular material. If the cell loses its anchorage to the surroundings, its growth and division halt.

Cells that cannot pass the checkpoints and cannot be repaired undergo a process of programmed cell death called *apoptosis* (ah-puh-TOH-sis). This "cellular suicide" will also occur for a variety of other reasons to benefit the body as a whole. For example, during fetal development the hands and feet are initially webbed, but the cells in the "webs" die to separate the fingers and toes.

A cell normally maintains very precise control over the cell cycle via a balancing act between genes whose products inhibit cell growth and division, and genes whose products stimulate cell division. When either of these types of genes is damaged, the DNA may produce a defective protein that results in either insufficient or excessive cellular growth and division. If the rate of cell division in a tissue falls below the rate of cell death, the tissue may progressively deteriorate. Conversely, if the rate of cell division exceeds the rate of cell death, the tissue may increase in cell number and size. This condition, called *hyperplasia*

(a) INTERPHASE

- Nuclear envelope encloses nucleus.
- Nucleolus is visible.
- Chromosomes are indistinguishable—DNA is in form of chromatin.
- Centriole pairs are duplicated.

(b) MITOSIS

① Prophase:
- Chromatin condenses so sister chromatids are visible.
- Nucleolus disperses.
- Mitotic spindle forms, and in late prophase spindle fibers attach to sister chromatids.
- Two centriole pairs separate and begin migrating to opposite poles of the cell.
- Nuclear envelope fragments.

② Metaphase:
- Spindle fibers pull sister chromatids to align on equator of cell.

Centrosomes (with centriole pairs)

Nucleolus

Nuclei   Chromatin

LM (1430×)

Mitotic spindle   Centriole pair   Chromosomes (sister chromatids)

LM (1630×)

Sister chromatids   Centriole pairs   Spindle fibers

LM (1630×)

Sequence the steps
@MasteringA&P®

Figure 3.35 Interphase, mitosis, and cytokinesis.

(b) MITOSIS (CONTINUED)		(c) END RESULT

③ Anaphase:
- Sister chromatids separate as spindle fibers shorten.
- Daughter chromosomes are pulled to opposite poles.
- Cell elongates.
- Cytokinesis begins as organelles and cytosol are divided.

④ Telophase and Cytokinesis:
- Nuclear envelopes reassemble.
- Nucleoli re-form.
- Chromosomes are no longer distinct—DNA returns to chromatin form.
- Cytokinesis continues as the cleavage furrow forms.
- Daughter cells separate.

Two genetically identical daughter cells that enter interphase

Daughter chromosomes

Centriole pairs

LM (1580×)

Nucleolus

Cleavage furrows

Decondensing chromosomes

Chromatin

Nuclear envelopes

LM (1480×)

Daughter cells

(hy'-per-PLAY-zee-ah), can strain the surrounding tissues, robbing them of oxygen and nutrients.

When changes in the DNA of a cell cause uncontrolled cell division, the resulting cells may form a growth or mass known as a *tumor.* A benign (beh-NYN) tumor is one that is confined to its original location and does not invade surrounding tissues, but it may grow extremely large. A *cancerous,* or *malignant,* tumor is one that is made up of *cancer cells.* An example is the tumor shown in **Figure 3.36**, from a patient with renal cell carcinoma (cancer of cells of the kidney). Cancer cells are not inhibited by high cellular density or loss of their anchorage to other cells. In fact, if they have enough nutrients, such cells appear to grow and divide indefinitely. A striking example is a culture of cancerous tumor cells, known as the HeLa cells, that has been growing since the tumor was removed from a patient named Henrietta Lacks in 1951. Cancer cells from malignant tumors are able to spread into other tissues, a process called *metastasis* (meh-TASS-tuh-sis), which results in the formation of other tumors throughout the body. This can cause widespread tissue destruction that, especially if untreated, may lead to death.

Malignant tumor in kidney

Figure 3.36 Cancerous tumor of kidney cells.

Quick Check

☐ 4. What are four external factors that play a role in cell cycle control?

Apply What You Learned

☐ 1. What would happen if a cell completed several rounds of mitosis but did not undergo cytokinesis? (*Hint:* Remember that telophase and cytokinesis are separate processes.)

☐ 2. Both the periwinkle and yew plants produce toxins that interfere with microtubules by binding to them and either inhibiting their assembly (the periwinkle) or preventing their disassembly (the yew). Why would both actions ultimately *prevent mitosis?*

See answers in Appendix A.

CHAPTER SUMMARY

Take full advantage of the invaluable online practice and study tools for this chapter by going to MasteringA&P®. **There you will find:**

👆 **Practice tests, module quizzes, art-labeling and sequence-the-steps activities**

▶ **Animated figures with quizzes**

✎ *Practicing A&P* **worksheets**

🔊 **Author-narrated podcast**

🔥 **Dynamic Study Modules**

iP *Interactive Physiology*

MODULE **3.1**
Introduction to Cells 68

- All cells share the same basic functions, including **cell metabolism,** transport of substances through the cell, communication, and cell reproduction.

- Most cells contain three basic components: (Figure 3.1, p. 69)
  - Cells are surrounded by a **plasma membrane** that regulates exchange between the cell and the environment, provides structural support, and physically isolates the cell from its surroundings. The plasma membrane separates the fluid in the body into two compartments: the **extracellular fluid (ECF)** outside the cells, and the **intracellular fluid (ICF),** or **cytosol,** inside the cells.
  - The **cytoplasm** consists of the cytosol, organelles, and cytoskeleton inside the cell. The cytosol is the fluid portion of the cytoplasm. It is the site of many important cellular reactions, including those of glycolytic catabolism. **Organelles** are tiny cellular structures that compartmentalize the cell's functions. The **cytoskeleton** consists of long protein filaments that support the cell, help it maintain its shape, allow substances to be transported within the cell, and can move the cell itself.
  - The **nucleus** is the membrane-bounded organelle that houses most of the cell's *deoxyribonucleic acid (DNA)* and controls many of the cell's functions.

- Cells vary widely in size and structure to enable them to better perform specialized functions. (Figure 3.2, p. 70)

MODULE **3.2**
Structure of the Plasma Membrane 70

- The plasma membrane is composed primarily of a **phospholipid bilayer.** The phospholipids are arranged so that their hydrophilic phosphate heads face the water-containing cytosol and ECF, and their hydrophobic fatty acid tails face one another within the bilayer. (Figure 3.3, p. 71)

- According to the **fluid mosaic model,** the plasma membrane is a fluid, dynamic structure with multiple components moving laterally within the plane of the phospholipid bilayer. (Figure 3.4, p. 72)

  Practice art-labeling exercise: Figure 3.4.

- **Integral proteins** are embedded in the membrane, with transmembrane proteins spanning its width. **Peripheral proteins** are found on one side of the membrane only. Membrane proteins may function as channels, carriers, enzymes, or receptors. They may also provide structural support to cells and tissues. (Figure 3.5, p. 73)

MODULE **3.3**
Transport across the Plasma Membrane 74

- The phospholipid bilayer is **selectively permeable,** allowing certain molecules to cross it while prohibiting the passage of other molecules.

- Molecules cross the plasma membrane in different ways. **Passive transport processes** transport substances across the membrane with no ATP expenditure. **Active transport processes** transport substances across the membrane but require the use of ATP.

- Passive transport processes are driven by a **concentration gradient,** which is a form of potential energy. When two solutions are in *equilibrium*, net molecule movement stops in passive transport processes because the concentration gradient no longer exists to drive them. (Figure 3.6, p. 75)
  - **Diffusion** is a type of passive transport during which molecules move down their concentration gradient until equilibrium is reached. (Figure 3.7, p. 76)
    - **Simple diffusion** refers to the movement of molecules directly through the phospholipid bilayer.
    - **Facilitated diffusion** refers to movement of polar and ionic molecules across a membrane through a protein channel or with the help of a carrier protein.
  - **Osmosis** is a type of passive transport in which water (the solvent) moves across a selectively permeable membrane from a region of lower solute concentration to a region of higher solute concentration until equilibrium is reached. The basic force that drives osmosis is **osmotic pressure,** which is the pressure of solute molecules in a solution that exert a "pulling" force on water molecules. (Figure 3.8, p. 77)
  - The concept of **tonicity** describes the osmotic pressure of one solution relative to another, in this case the cytosol and the ECF: (Figure 3.9, p. 78)
    - An **isotonic** extracellular solution has the same osmotic pressure as the cytosol and results in no net movement of water by osmosis.
    - A **hypertonic** extracellular solution has a higher osmotic pressure than the cytosol and causes water to move out of the cell by osmosis.
    - A **hypotonic** extracellular solution has a lower osmotic pressure than the cytosol and causes water to enter the cell by osmosis.

- Active transport processes use energy in the form of ATP to move a solute against its concentration gradient from a low concentration to a high concentration.
  - During **primary active transport,** a membrane protein uses the energy from ATP to "pump" a solute against its concentration gradient. One of the body's main primary active transport **pumps** is the Na^+/K^+ pump, or ATPase, which pumps three sodium ions out of and two potassium ions into the cell. (Figure 3.10, p. 80)
  - **Secondary active transport** uses a primary active transport pump to create a concentration gradient, and then uses potential energy from that concentration gradient to pump another solute into or out of the cell against its concentration gradient. (Figure 3.11, p. 81)

- There is a separation of charges across the plasma membrane, with a thin layer of positive charges in the ECF and a thin layer of negative charges in the cytosol. A separation of charges is called an **electrical potential;** the electrical potential in our cells is called a **membrane potential.** The study of membrane potentials is **electrophysiology.**

- The **resting membrane potential** is the membrane potential of a cell at rest. It averages –70 mV.

- **Vesicular transport** is a type of active transport process that allows large molecules to enter or exit the cell packaged in membrane-bounded **vesicles.**
  - **Endocytosis** is the process by which large molecules enter the cell. There are two main types:
    - During **phagocytosis,** specialized cells called phagocytes ingest large substances such as bacteria or dead cells. (Figure 3.12, p. 82)
    - During **pinocytosis,** coated pits in the plasma membrane form vesicles containing dissolved substances from the extracellular fluid. (Figure 3.13, p. 83)
    - During **receptor-mediated endocytosis** (a process related to pinocytosis), specific molecules bind to their receptors on the plasma membrane and become concentrated in protein-coated pits.
  - **Exocytosis** is the process by which large molecules exit the cell. The molecules are packaged into a vesicle that fuses with the plasma membrane, releasing the molecules into the extracellular fluid, a process known as *secretion*. (Figure 3.14, p. 84)

- Types of plasma membrane transport are summarized in Table 3.1, pp. 85–86.

  Practice art-labeling exercise: Table 3.1.

MODULE **3.4**
Cytoplasmic Organelles 87

- **Organelles** are specialized components within the cell that carry out specific functions. Most organelles are surrounded by a membrane that is similar to the plasma membrane. (Figure 3.15, p. 87)

  Practice art-labeling exercise: Figure 3.15.

- **Mitochondria** produce the bulk of the cell's chemical energy in the form of ATP. (Figure 3.16, p. 88)
  - Mitochondria have a double membrane and two internal spaces. The *intermembrane space* is the space between the two membranes, and the *matrix* is the innermost space. Fuel oxidation occurs in the matrix, and the reactions that directly generate most ATP, called **oxidative catabolism,** occur in the inner mitochondrial membrane. (Figure 3.17, p. 89)
  - Mitochondria have their own DNA and ribosomes.
- **Peroxisomes** use oxygen to carry out oxidation reactions that include oxidation of toxic substances and fatty acids, and synthesis of certain phospholipids.
- **Ribosomes** are small, granular organelles with two subunits that synthesize proteins. They are not surrounded by a membrane and are composed of proteins and **rRNA.** Free ribosomes float in the cytosol, whereas bound ribosomes are attached to the surface of organelles of the endomembrane system. (Figure 3.18, p. 90)
- The organelles of the **endomembrane system** interact with one another by packaging and receiving substances in vesicles.
  - The **endoplasmic reticulum** (ER) is a large branching network of tubules and sacs that enclose a single space. (Figure 3.19, p. 91)
    - **Rough ER** (RER) has ribosomes attached to its surface, and it modifies and folds proteins made by those ribosomes. The rough ER also makes new proteins for secretion and for the plasma membrane and other cellular membranes.
    - **Smooth ER** (SER) has no ribosomes on its surface. It stores calcium, synthesizes many lipids, and detoxifies certain substances.
  - The **Golgi apparatus** is a stack of flattened sacs. It receives vesicles from the ER, and then modifies and packages these products to be secreted from the cell by exocytosis, to become part of the plasma membrane, or to be incorporated into lysosomes. (Figure 3.20, p. 92)
  - **Lysosomes** contain digestive enzymes that break down substances taken into the cell by endocytosis, as well as worn-out organelles.
  - The overall function of the endomembrane system is summarized in Figure 3.21, p. 93.
- The structure and function of the cytoplasmic organelles are summarized in Table 3.2, p. 94.

MODULE 3.5
The Cytoskeleton 95

- The **cytoskeleton** consists of three types of protein filaments that are made from smaller protein subunits. The functions of the cytoskeleton include: giving the cell its characteristic shape and size, supporting the plasma membrane and nucleus, moving substances within the cell with the use of associated motor proteins, dividing the cell, and moving the cell itself. (Table 3.3, p. 96)
- **Actin filaments** are small filaments composed of the protein *actin.* They provide structural support to the cell. Actin filaments associate with the motor protein *myosin.* Together they change the shape and size of the cell during endocytosis, exocytosis,

cytokinesis, and muscle contraction. They also enable certain cells to migrate within the body.
- **Intermediate filaments** are ropelike filaments made of proteins (especially keratin) that bend easily but are very strong and resist breaking. They provide the cell with mechanical strength and support for its organelles and overall structure.
- **Microtubules,** the largest filaments, are hollow tubes made of protein subunits called *tubulins.* Microtubules support the cell, hold the organelles in their proper places, move organelles within the cell, and form cilia and flagella on certain cells. They emanate from the **centrosome,** where **centrioles** help to organize their proteins and form microtubules. During cell division, microtubules also form the mitotic spindle that moves the chromosomes. (Figure 3.22, p. 97)
- Certain cells feature specialized extensions from their plasma membranes.
  - The plasma membranes of some cells are folded into extensions called **microvilli,** which increase the surface area of the membrane for functions such as absorption. Microvilli consist of a supporting core of actin filaments. (Figure 3.23, p. 98)
  - **Cilia** are hairlike projections found on the surfaces of certain cells. They beat in unison to sweep substances past the cell. (Figure 3.24, p. 98; Table 3.4, p. 99)
  - **Flagella** are single projections that propel the cell itself. Sperm are the only flagellated cells in the human body.

MODULE 3.6
The Nucleus 100

- The **nucleus** contains most of the cell's **DNA** and is the control center of the cell. (Figure 3.25, p. 100)
- The nucleus is surrounded by a double membrane called the **nuclear envelope,** which contains large **nuclear pores** and is supported internally by the *nuclear lamina.* (Figure 3.26, p. 101)
- DNA is found in the nucleoplasm in the form of **chromatin,** a material composed of one long DNA molecule with associated proteins. Chromatin is folded and condensed within the nucleus into *nucleosomes,* which consist of DNA wrapped around *histone* proteins. During cell division, chromatin condenses into **chromosomes.** Human cells contain 23 pairs of chromosomes, one maternal set and one paternal set. (Figure 3.27, p. 101)
- **Nucleoli** are regions of the nucleus that are responsible for both the synthesis of RNA molecules such as ribosomal RNA and the assembly of ribosomes.

MODULE 3.7
Protein Synthesis 102

- **Protein synthesis** is the process by which the cell uses instructions in the DNA to build proteins.
- A **gene** is a segment of DNA that specifies the amino acid sequence of a single protein. Each amino acid is specified by a three-nucleotide–long "word" called a **triplet.**
- The **genetic code** is a list of which amino acid is specified by each possible codon. (Figure 3.28, p. 103)

- Some segments of DNA within a gene do not code for proteins. These segments are called *introns.* Coding segments of the DNA are called *exons.*
- During the process of **transcription,** the cell copies the information in the gene into a **messenger RNA (mRNA)** transcript with the help of the enzyme *RNA polymerase.* The transcription process consists of three stages: initiation, when RNA polymerase binds the *promoter* and begins building the mRNA transcript; elongation, when RNA polymerase links nucleotides in a specific order; and termination, when RNA polymerase reaches the termination signal and releases the pre-mRNA transcript. (Figure 3.29, p. 105)

  ▷ Play animation on transcription: Figure 3.29.

- Before the pre-mRNA exits the nucleus into the cytosol, it undergoes *RNA processing,* which includes the removal of its introns.
- During the process of **translation,** the code in the mRNA transcript of a gene is translated into the amino acid sequence of a particular polypeptide. Translation takes place on a ribosome. The "translator" is the molecule **transfer RNA (tRNA),** which carries an amino acid on one end and has at its other end an **anticodon** that is complementary to a specific mRNA **codon.** (Figure 3.30, p. 106)
- Translation consists of three stages: (Figure 3.31, pp. 106–107)
  ○ During initiation, an initiator tRNA is brought together with the two ribosomal subunits at the start codon of the mRNA.
  ○ During elongation, specific amino acids brought into the ribosome by complementary tRNA molecules are linked together to form a polypeptide.
  ○ During termination, the ribosome encounters a *stop codon* in the mRNA and releases the newly synthesized polypeptide.

  ▷ Play animation on translation: Figure 3.31.

- Such polypeptides undergo *posttranslational modification,* being modified and then folding into functional proteins.
- Figure 3.32 on p. 109 provides a summary of the process of protein synthesis.

  ▷ Play animation on the big picture of protein synthesis: Figure 3.32.

MODULE **3.8**
The Cell Cycle 110

- The *cell theory* states that cells must come from the division of already-existing cells. The ordered series of events from cell formation to cell division is called the **cell cycle.**
- The cell cycle consists of two main phases, interphase and cell division. **Interphase** includes three subphases: (Figure 3.33, p. 110)
  ○ During **G$_1$ phase,** the cell grows and rapidly synthesizes proteins.
  ○ During **S phase,** the cell synthesizes or replicates its DNA.
  ○ During **G$_2$ phase,** the cell makes proteins necessary for cell division.
- **M phase** consists of **mitosis,** in which the genetic material divides, and **cytokinesis,** in which the cell divides into two daughter cells.
- During S phase, **DNA synthesis** or **replication,** a cell's entire set of chromosomes is duplicated. The duplicate copies of each

chromosome, called *sister chromatids,* are joined to one another at the **centromere.** DNA synthesis is an example of **semiconservative replication,** because each newly formed double helix has one old and one new DNA strand. The enzyme **DNA polymerase** synthesizes the new DNA strands. DNA strands can be built in only one direction. The sequence of events of DNA synthesis is as follows: (Figure 3.34, p. 112)
  ○ The DNA strands are unwound by helicase.
  ○ builds the RNA primers.
  ○ DNA polymerase builds the new DNA strands by adding complementary bases to the template strands.

  🖑 Practice sequencing the steps: Figure 3.34.

- **Mitosis** divides the cell's replicated DNA between the two **daughter cells.** There are four stages in mitosis: (Figure 3.35, pp. 114–115)
  ○ In **prophase,** the replicated *sister chromatids* condense, the nuclear envelope breaks apart, and the **mitotic spindle** forms and attaches to the centromere of each chromatid.
  ○ During **metaphase,** the spindle fibers pull the sister chromatids to align along the equator of the cell.
  ○ During **anaphase,** the sister chromatids separate and chromosomes are pulled to opposite poles of the cell by the spindle fibers.
  ○ In **telophase,** the two daughter cells' chromosomes reassume the loose structure of chromatin.

  🖑 Practice sequencing the steps: Figure 3.35.

- **Cytokinesis** occurs simultaneously with anaphase and telophase; it divides the cytoplasm between the daughter cells and pinches the cell in two at the *cleavage furrow.*
- The cell cycle is precisely controlled so that cell formation is balanced with cell death. During the cell cycle, **checkpoints** act as stop/go signals for the cell. The cell responds to a variety of extracellular signals. If conditions for division are not favorable, the cell may not proceed with cell division.
- If cell cycle control is lost, cancer may result.

ASSESS WHAT YOU LEARNED

See answers in Appendix A.

LEVEL 1 Check Your Recall

1. Which of the following is not a basic function shared by all cells?
   a. Cell metabolism
   b. Communication
   c. Cell reproduction
   d. Cell movement
   e. Transport of substances through the cell

2. Fill in the blanks: The three main components of a cell are the _____, the _____, and the _____.

3. What are the two fluid compartments in the body, and how are they kept separate?

4. Which of the following best describes the arrangement of the main component of the plasma membrane?

    a. A monolayer of phospholipids with the phosphate heads facing the cytosol and the fatty acid tails facing the extracellular fluid

    b. A bilayer of phospholipids with the phosphate heads facing the cytosol and extracellular fluid, and the fatty acid tails facing one another

    c. A bilayer of phospholipids with the phosphate heads of one layer facing the fatty acid tails of the other

    d. A bilayer of phospholipids with a layer of triglycerides sandwiched in the middle

5. Mark the following statements about the plasma membrane as true or false. If a statement is false, correct it to make a true statement.

    a. _____ Integral membrane proteins generally span the width of the plasma membrane, whereas peripheral proteins are found on only one side of the membrane.

    b. _____ Cholesterol provides the plasma membrane with stability in the face of changing ion concentrations.

    c. _____ Membrane cholesterol is vitally important for cell-cell recognition.

    d. _____ Membrane proteins often function as channels or carriers.

    e. _____ The overall structure of the plasma membrane is a mosaic with the components locked tightly in place.

6. What is the primary difference between active transport processes and passive transport processes?

7. Match the term with its appropriate definition.

    _____ Osmosis

    _____ Secondary active transport

    _____ Exocytosis

    _____ Phagocytosis

    _____ Simple diffusion

    _____ Primary active transport

    _____ Pinocytosis

    _____ Facilitated diffusion

    a. Type of endocytosis in which a large particle is ingested

    b. Transport across the plasma membrane against the concentration gradient via direct use of energy from ATP

    c. Passive movement of solute across the plasma membrane

    d. Movement of solvent from a solution of lower solute concentration to a solution of higher solute concentration

    e. Passive movement of solute across the plasma membrane via a channel or carrier protein

    f. Type of endocytosis in which ECF is brought into the cell in a protein-coated pit

    g. Release of large molecules from the cell through a vesicle

    h. Transport of a substance across the plasma membrane against its concentration gradient using the energy from the "downhill" movement of another substance

8. Fill in the blanks: A hypotonic solution will cause water to move _____ the cell and the cell will _____. A hypertonic solution will cause water to move _____ the cell and the cell will _____.

9. Match the following terms with the correct functions.

    _____ Peroxisome

    _____ Ribosome

    _____ Smooth endoplasmic reticulum

    _____ Mitochondrion

    _____ Golgi apparatus

    _____ Lysosome

    _____ Rough endoplasmic reticulum

    _____ Vesicle

    a. Modifies and folds proteins into the correct structure

    b. Modifies, packages, and sorts proteins

    c. Contains digestive enzymes

    d. Granular organelle that makes proteins

    e. Synthesizes most of a cell's ATP

    f. Contains enzymes that oxidize toxins and fatty acids

    g. Small, membrane-bounded structure that carries certain molecules to different organelles

    h. Detoxifies certain chemicals, makes lipids, stores calcium ions

10. Explain how the members of the endomembrane system interact.

11. Mark the following statements about the cytoskeleton as true or false. If a statement is false, correct it to make a true statement.

    a. _____ Actin filaments combine with myosin motor proteins to provide the cell with mechanical strength.

    b. _____ The cilia found on sperm cells propel the cells through a liquid medium.

    c. _____ Microtubules are hollow tubes that align organelles and shuttle them to their proper places in the cell.

    d. _____ Intermediate filaments are ropelike structures composed of tubulin proteins.

    e. _____ Centrioles are part of the centrosome from which microtubules originate.

    f. _____ Endocytosis, exocytosis, muscle cell contraction, and cellular "crawling" are all mediated by intermediate filaments.

12. Our somatic cells' DNA is distributed among 46 _____ in the nucleus.

    a. nucleosomes

    b. chromatids

    c. nucleoli

    d. chromosomes

13. Explain how and why chromatin is condensed in the nucleus.

14. Which of the following statements correctly describes the function of the nucleolus?

    a. Within the nucleolus we find genes for rRNA molecules.

    b. Nucleoli manufacture mitochondrial proteins for export to the cytosol.

c. Nucleoli assemble the nuclear envelope.

d. The nucleolus contains the genes that form the Golgi apparatus.

15. Each of the following statements about protein synthesis is false. Correct each to make a true statement.

a. In a gene, each nucleotide specifies one amino acid in a protein sequence.

b. A transcription factor must bind to the promoter region of a gene before the enzyme DNA synthetase is able to bind and begin transcription.

c. The enzyme RNA polymerase builds a strand of transfer RNA, whose codons are complementary to DNA's triplets.

d. Proteins destined for secretion from the cell enter the nucleus after translation, to be folded and modified.

e. During translation, amino acids are delivered by the messenger RNA transcript.

16. Number the following steps of protein synthesis in the order in which they occur, starting with 1 and ending with 9.

a. _____ The stop codon is reached, and the polypeptide is released.

b. _____ The small ribosomal subunit finds the start codon, and the large ribosomal subunit joins.

c. _____ The end of the gene is reached, and the pre-mRNA is released and then edited.

d. _____ The transcription factor binds the promoter.

e. _____ The protein is folded and modified to become functional.

f. _____ RNA polymerase builds the mRNA transcript.

g. _____ mRNA and initiator tRNA bind the small ribosomal subunit.

h. _____ New tRNA molecules are brought into the A site successively, and the peptide chain of the tRNA in the P site is joined to the amino acid of the tRNA in the A site.

i. _____ mRNA exits the nucleus via a nuclear pore.

17. Which of the following is not a phase of mitosis?

a. Interphase
b. Prophase
c. Anaphase
d. Metaphase

18. Why is regulation of the cell cycle necessary?

19. Mark the following statements about the cell cycle as true or false. If a statement is false, correct it to make a true statement.

a. _____ During the S phase of the cell cycle, the cell stalls until conditions for division are more favorable.

b. _____ During metaphase, the sister chromatids are pulled apart and the chromosomes move to the opposite poles of the cell.

c. _____ The main enzyme that builds the new DNA strands during DNA synthesis is RNA synthetase.

d. _____ The first three stages of the cell cycle are collectively called interphase.

20. Match the following terms with the correct definitions.

_____ G$_1$
_____ Metaphase
_____ S phase
_____ Cytokinesis
_____ Telophase
_____ M phase
_____ G$_2$
_____ Anaphase

a. Division of the cytoplasm
b. Second growth stage of the cell cycle
c. Stage of the cell cycle in which the cell divides
d. Stage of mitosis in which the sister chromatids line up on the cell's equator
e. Stage of the cell cycle in which DNA is replicated
f. Stage of mitosis in which the chromosomes move to opposite poles of the cell
g. Initial growth phase of the cell cycle
h. Stage of mitosis in which the nuclear envelopes reassemble

LEVEL 2 Check Your Understanding

1. Write a single sentence, using no more than 20 words, to summarize each of the following cellular processes:

a. Diffusion
b. Osmosis
c. Primary active transport
d. Secondary active transport
e. Transcription
f. Translation
g. DNA synthesis
h. Mitosis

2. Certain diseases are transmitted via mitochondrial DNA. Which cell types do you think would be most affected by such diseases, and why?

3. Explain how the form of each of the following structures is related to its function:

a. Cilia
b. Microvilli
c. Intermediate filaments
d. Lysosomes
e. Nuclear envelope

4. Certain types of cancerous lung tumors can secrete hormones normally made by the pancreas, adrenal gland, and hypothalamus. What prevents such secretion from happening in healthy cells?

5. Why do you think the rate of cell division is different for different tissues? Where in the body would you expect to find cells that have a rapid rate of division? Where might you find cells that have a slow rate of division? Explain.

LEVEL 3 Apply Your Knowledge

PART A: Application and Analysis

1. A patient is admitted to the hospital and given intravenous (IV) fluids. Four hours later, the patient complains that his mouth and eyes feel dry. You notice that he displays signs of dehydration, and when you check his IV, you see that he was given the wrong kind of fluids. Were these fluids likely hypotonic, isotonic, or hypertonic? Explain.

2. A popular science fiction program once had an episode that featured an "intron virus" that "turned on" the introns in the genes, causing the synthesis of abnormal proteins. The episode may have been entertaining, but its premise had a large flaw about the nature of introns. What was the flaw?

3. A hypothetical poison prevents transcription factors from binding to the gene for tubulin proteins. What impact would this have on mitosis, and why?

4. What effect would the hypothetical poison of question 3 have on other functions of the cell? Explain.

PART B: Make the Connection

5. Chemicals called *detergents* are amphiphilic molecules with a structure similar to that of phospholipids. If detergents are applied to a plasma membrane, they can break it apart. Why do you think detergents have this ability? (*Hint:* Remember that like interacts with like.) *(Connects to Chapter 2)*

4

Histology

The human body has about 200 different cell types; they often vary dramatically in the genes they express and thus in their structure and functions. Since each cell type is specialized, individual cells cannot perform their functions unless they are in the correct area of the body and surrounded by the appropriate cells. For example, a cell that supports the cells of the brain would do you no good in your big toe, where there are no brain cells to support.

To perform their functions, cells must also be surrounded by the appropriate extracellular materials. For example, a red blood cell carries oxygen through the bloodstream and delivers it to other cells in the body. However, a red blood cell that's in your skin instead of your blood would not be surrounded by high amounts of oxygen in the extracellular fluid (ECF), so it could not fulfill its role of picking up oxygen. This leads us to a basic principle of human physiology: Cells cannot carry out their functions alone—they require the presence of other cells and other materials in the ECF to maintain homeostasis. For this reason, our cells and ECF are organized into distinct groups called *tissues*. The structure and functions of tissues, in turn, directly determine the structure and functions of organs and organ systems.

Although tissues at first may seem abstract and perhaps less important than the cells or organ systems, a knowledge of tissues is critical not only for the study of anatomy and physiology but also for work in the medical field. One of the most commonly performed medical procedures is a *biopsy*, in which a tissue sample is removed from a patient, prepared, and examined under a microscope. A biopsy is performed on living tissue; however, sometimes tissue examinations are required after death during an *autopsy*. The physician who makes such tissue examinations, a *pathologist*, looks for subtle changes in the shape, size, and characteristics of the cells and components of the extracellular environment. A very slight change in the architecture of a tissue or cell can signal many things, including cancer, infection, diseases of the immune system, and the presence of certain poisons, just to name a few. A pathologist cannot determine if the tissue is abnormal unless he or she has a thorough knowledge of **histology**—the study of the normal structure of tissues. We will explore the topic of histology in this chapter.

Photo: This colored scanning electron micrograph (SEM) shows fat cells (adipocytes) surrounded by fine protein fibers.

4.1

Introduction to Tissues

Learning Outcomes

1. Define the term histology.
2. Explain where tissues fit in the levels of organization of the human body.
3. Compare and contrast the general features of the four major tissue types.
4. Describe the components of the extracellular matrix.
5. Describe the types of junctions that unite cells in a tissue.

After progressing through the first two levels of body organization, chemical and cellular, we have now reached the third level of organization—the tissues. A **tissue** is a group of structurally and functionally related cells and their external environment that together perform common functions. Each tissue type differs in structure and function, yet all tissues share the same two basic components: (1) a discrete population of cells that are related in structure and function, and (2) the surrounding material, called the *extracellular matrix (ECM)*. Note that the extracellular fluid (ECF) is part of the ECM, along with several other components. In this module, we explore these basic parts of tissues, as well as the molecular "glue" that holds them together. But first, let's look at what types of tissues are found in our bodies.

Types of Tissues

There are four primary types of tissues, each of which is distinguished by the kinds and number of cells it contains, the amount and composition of ECM present, and the function it performs. The four types are epithelial, connective, muscle, and nervous:

- **Epithelial tissues.** *Epithelial tissues,* also called *epithelia* (ep′-ih-THEE-lee-ah; *epi-* = "upon"), consist of sheets of cells that are tightly packed together with no visible ECM. Epithelia cover and line all body surfaces and cavities. Specialized epithelial cells form *glands,* groups of cells that manufacture and secrete, or release, a product such as sweat, saliva, or a type of chemical messenger called a *hormone.*
- **Connective tissues.** *Connective tissues,* as the name implies, serve as a sort of cellular Velcro® that connects all other tissues in the body to one another. In most connective tissues, the ECM is the most prominent feature, and cells are scattered throughout it. Connective tissues bind, support, protect, and allow the transport of substances through the body.
- **Muscle tissues.** *Muscle tissues* are composed of cells that can contract and generate force, with little ECM between the cells. Muscle tissues are briefly introduced in this chapter, as they are covered fully in their own chapter (see Chapter 10).
- **Nervous tissue.** Within *nervous tissue* we find a unique ECM, as well as cells that can generate, send, and receive

messages, and other cells that support this activity. Nervous tissue is covered completely in a later chapter (see Chapter 11).

Organs are made up of two or more of these tissue types working together. For example, the heart as an organ consists of muscle tissue, epithelial tissue, and connective tissue. The brain as an organ consists primarily of nervous tissue, with significant amounts of connective tissue and some epithelial tissue.

Quick Check

☐ 1. What are the four types of tissues, and what are their characteristics?

The Extracellular Matrix

◀◀ **Flashback**

1. What are glycoproteins? (p. 74)
2. What are fibrous proteins, and what are their functions? (p. 56)

The **extracellular matrix,** or **ECM,** is composed of the substances surrounding the cells in a tissue (**Figure 4.1**). The ECM has two main components: *ground substance* and *protein fibers.* The proportion of these components varies, so that in different tissues the ECM can be a fluid, a thick gel, or a solid. The ECM is made by the cells of the tissue and performs a variety of functions, some of which are only now becoming clear to scientists. These functions include the following:

- providing the tissue with the strength to resist tensile (stretching) and compressive forces;
- directing cells to their proper places within a tissue;
- regulating the development, mitotic activity, and survival of cells; and
- holding cells in their proper positions.

Let's look at the two components of the ECM in more detail.

Ground Substance

The shapeless, gel-like **ground substance** makes up most of the ECM. Ground substance contains the interstitial or extracellular fluid consisting of water, ions, nutrients, and other solutes, as well as three families of macromolecules: glycosaminoglycans, proteoglycans, and glycoproteins (see Chapter 2). **Glycosaminoglycans** (gly′-kos-AM-ih-noh-gly-kanz; GAGs) are long, straight polysaccharide chains. Examples of GAGs include the relatively small *chondroitin sulfate* (kon-DROY-tin) and the enormous *hyaluronic acid* (hy′-al-yoo-ROHN-ik). The negative charges of certain sugars in a GAG attract positively charged ions in the extracellular fluid. In an example of the Gradients Core Principle (p. 26), these ions create a concentration gradient within the ECF that draws water out of cells and blood vessels

CORE PRINCIPLE
Gradients

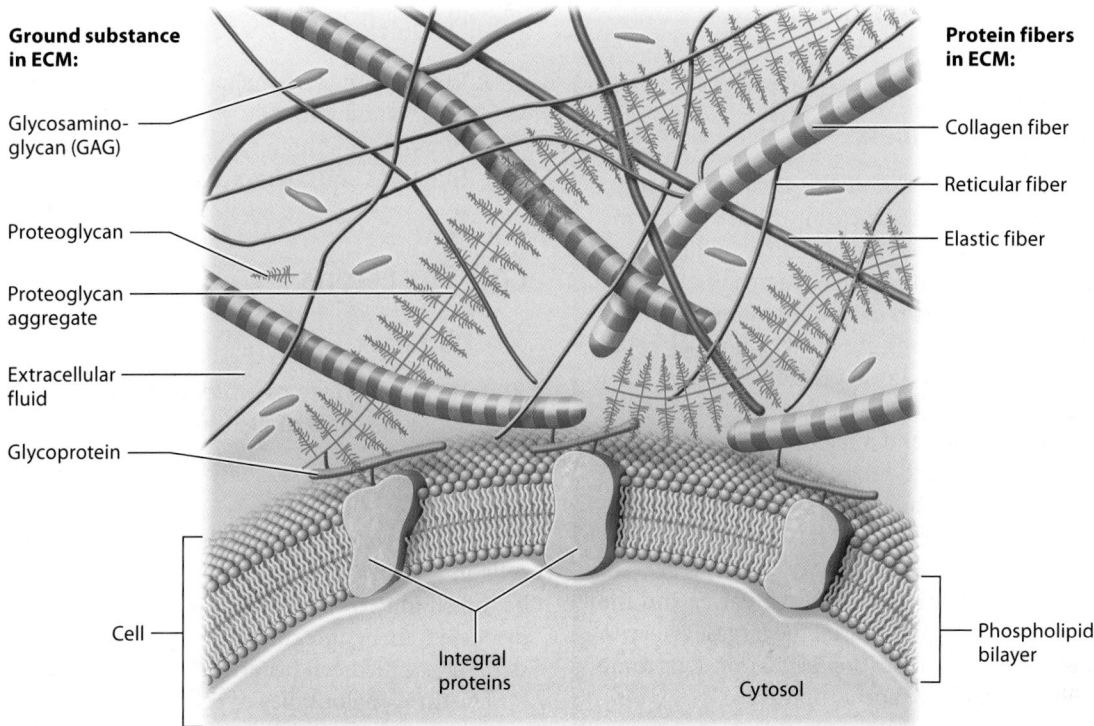

Ground substance in ECM:

Glycosamino-glycan (GAG)

Proteoglycan

Proteoglycan aggregate

Extracellular fluid

Glycoprotein

Cell

Integral proteins

Cytosol

Protein fibers in ECM:

Collagen fiber

Reticular fiber

Elastic fiber

Phospholipid bilayer

Figure 4.1 Extracellular matrix.

by osmosis and effectively "traps" it in the ECM, which helps the ECM to resist compression.

Proteoglycans consist of GAGs bound to a protein core; their structure resembles a bottle brush. Thousands of proteoglycans can bind in turn to a very long GAG such as hyaluronic acid, forming huge proteoglycan "aggregates." The size of these proteoglycan aggregates helps make the ECM firmer; tissues that contain more of these aggregates are thus more solid and resistant to compression than are tissues that contain fewer of them. Proteoglycan aggregates also act as a barrier to the diffusion of substances through the ECM, which protects the underlying tissues from invading microorganisms.

The final components of ground substance are **glycoproteins** of different types, known collectively as **cell-adhesion molecules (CAMs)**. As their name implies, they are responsible for adhering, or "gluing," the cells both to each other and into their places within the ECM. CAMs bind to cell surface proteins as well as protein fibers and proteoglycans, and help maintain the normal architecture of the tissue.

Protein Fibers

Embedded within the ground substance are numerous large, long molecules called **protein fibers.** Protein fibers are composed of multiple fibrous protein subunits that entwine to form a long, rope-like structure with a great deal of tensile strength. Three types of protein fibers are found within the ECM: *collagen, elastic,* and *reticular:*

- **Collagen fibers.** The body makes at least 20 different types of **collagen fibers** out of repeating subunits of the fibrous protein **collagen;** in fact, collagen proteins make up

20–25% of all protein in the body. Collagen fibers are composed of multiple subunits of a fibrous protein. Their structure, which resembles the entwined pieces of a steel cable, makes them very resistant to tension (pulling and stretching forces) and pressure. Groups of collagen fibers appear white in gross anatomical specimens such as tendons, which attach a muscle to a bone.

- **Elastic fibers.** Unlike collagen fibers, which resist stretching, **elastic fibers** may be stretched to one and a half times their resting length without breaking, a property called *extensibility.* When the stretching force is removed, they return to their original length, a property called *elasticity.* Elastic fibers are composed of a protein called *elastin* that is surrounded by glycoproteins. Elastin allows the fiber to stretch, whereas the glycoproteins support and organize the elastin.

- **Reticular fibers.** The thin **reticular fibers** were once thought to be a separate type of fiber; now, however, scientists recognize them as a type of collagen fiber. Since they differ in structure, functions, and distribution from the major collagen fibers, the term "reticular fiber" is still used by many histologists. Reticular fibers are thinner and shorter than regular collagen fibers, and they interweave to form a meshwork or scaffold (*reticul-* = "netlike") that supports the cells and ground substance of many tissues. They also form "webs" in certain organs, such as the spleen, that help these organs trap foreign cells.

See *A&P in the Real World: Diseases of Collagen and Elastic Fibers* on page 126 for information on what happens when these critical protein fibers malfunction.

Diseases of Collagen and Elastic Fibers

The protein fibers in the ECM are vital to maintaining the structural integrity of many of the body's tissues and organs. This becomes quite apparent when disease results in abnormal collagen and elastic fibers. One such disorder is *Ehlers-Danlos syndrome* (AY-lerz DAN-lows), which generally results from abnormal collagen fibers. Features of this condition include recurrent joint dislocations, hyperextensibility of the skin, fragility of the blood vessels, easy bruising, and various effects in organs such as the intestines, heart, and lungs.

Marfan syndrome results from abnormalities in elastic fibers. Features of this disorder include tall stature with long limbs and fingers, multiple skeletal abnormalities, recurrent joint dislocations, abnormalities of the heart valves and the lens of the eye, and dilation of the aorta, the largest artery in the body. The most lethal complication of Marfan syndrome is *aortic dissection,* in which the aorta ruptures.

Quick Check

☐ 2. What are the two major components of the ECM, and what are their functions?

☐ 3. What are the three types of protein fibers? Describe their functions and characteristics.

Cell Junctions

◀◀ **Flashback**

1. What is an integral protein? (p. 72)
2. What are some of the functions of membrane proteins? (p. 72)

Although a cell is often anchored to proteins in the ECM, the cells in a tissue are also often linked to one another. They may be linked in two ways: by cell-adhesion molecules, which we just discussed, or by integral proteins in neighboring cells that form structures called **cell junctions.**

The three major types of cell junctions are *tight junctions, desmosomes,* and *gap junctions* (**Figure 4.2**).

(a) Tight junctions help to make the spaces between cells impermeable.

(b) Desmosomes increase the resistance of the tissue to mechanical stress.

(c) Gap junctions allow small substances to move from one cell to another.

Figure 4.2 Cell junctions.

- **Tight junctions. Tight junctions,** also called *occluding junctions,* hold cells tightly together, making the spaces between them impermeable and so preventing macromolecules from passing between adjacent cells (Figure 4.2a). These junctions are composed of integral proteins in the plasma membranes of adjacent cells that interweave and lock together, forming a seal around the apical perimeter of the cell. This seals the spaces between adjacent cells, much like the zipper on a plastic freezer bag. Note that some tight junctions are "leaky" and do not form a complete seal, which allows certain molecules to pass between the cells. Tight junctions are found in multiple places in the body, such as between the cells of blood vessels, where they keep the molecules of blood in the vessel.
- **Desmosomes. Desmosomes** (DEZ-moh-sohmz; *desm-* = "band or bond") are also composed of integral proteins that link two cells; however, desmosomes act more like buttons or snaps than like zippers (Figure 4.2b). And just as rain and snow can pass through the spaces in a snapped coat, materials in the extracellular fluid may pass between cells held together by desmosomes. Desmosomes increase the strength of a tissue by holding the cells together so that mechanical stress is more evenly distributed. Their integral "linker" proteins are attached to intermediate filaments of the cytoskeleton, which reinforces them. Cells that are subject to a great deal of mechanical stress, such as those in the epithelia of the skin, have a large number of desmosomes.
- **Gap junctions.** Notice in Figure 4.2c that gap junctions also link two adjacent cells, but in an entirely different way. **Gap junctions** are small pores in adjacent plasma membranes formed by protein channels, and they allow small substances to pass freely between the cytosol of two cells. Gap junctions illustrate the Cell-Cell Communication Core Principle (p. 27)—they are prominent in cells that can communicate with electrical signals, such as the cells of cardiac muscle.

CORE PRINCIPLE
Cell-Cell Communication

Quick Check

☐ 4. What three types of junctions unite cells in a tissue, and what are their functions?

Apply What You Learned

☐ 1. What do you think would happen if the elastic fibers in tissues were extensible but not elastic?
☐ 2. Predict what would happen if the tight junctions in blood vessels were not functional.
☐ 3. Predict the consequences if the desmosomes in the outer layer of the skin were not working correctly.

See answers in Appendix A.

MODULE **4.2**
Epithelial Tissues

Learning Outcomes

1. Classify and identify the different types of epithelial tissues based on distinguishing structural characteristics.
2. Describe where in the body each type of epithelial tissue can be found.
3. Describe the function of each type of epithelial tissue and correlate that function with structure.
4. Describe and classify the structural and functional properties of exocrine and endocrine glands.

When you look in the mirror, nearly everything you see either is an epithelial tissue or is derived from epithelial tissue. Similarly, if you were able to look inside your hollow organs and body cavities, almost everything viewed there would be an epithelial tissue of some sort. This is because **epithelial tissues,** which are found on every internal and external body surface, primarily act as *barriers* between the body and the external environment, and between our organs and fluid-filled cavities.

The functions of epithelial tissues include the following:

- **Protection.** Epithelial tissues provide a continuous surface that shields the underlying tissues from mechanical and thermal injury. Many epithelial tissues produce substances that help them fulfill this role; for example, the epithelium of the skin produces the hard protein *keratin,* which makes the tissue more resistant to injury. Epithelial cells are subject to damage because of their protective and barrier functions, but their rate of mitosis is fairly rapid, allowing them to replace cells that die.
- **Immune defenses.** Epithelial tissues provide a barrier against invading microorganisms as well. For microorganisms crossing this barrier, specialized cells of the immune system that are scattered throughout epithelial tissues protect the underlying tissues.
- **Secretion.** Epithelial cells form glands that produce substances such as oil or hormones. These substances are secreted either through a small tube called a *duct* or into the bloodstream.
- **Transport into other tissues.** Epithelia are selectively permeable barriers, meaning that certain substances can cross them by passive or active transport and so enter other tissues. This function is critical in organs such as the small intestine, in which nutrients must cross the intestinal epithelium to enter the blood.
- **Sensation.** Most epithelia are richly supplied with nerves that detect changes in the internal and external environments. Additionally, specialized epithelial cells form part of the machinery for sensation, as occurs in the taste buds of the tongue.

In this module, we first examine the components that make up epithelial tissue. Then we discuss the different types of epithelia found in the body, including those that cover and line body surfaces and those that make up glands.

ConceptBOOST »»»

"But It All Looks Pink!" Part 1

Tissues are generally studied by first taking extremely thin slices of organs called *tissue sections* and mounting them on microscope slides. Each section is then treated with different stains so that components of the cells and ECM are visible. In this module and the ones to follow, we make extensive use of tissue sections to study the various tissue types.

When you're presented with a histological section in this textbook or in the laboratory, you may find yourself unsure of where to begin. The mass of dots and squiggly lines on most sections leads many students to complain, "But it all looks pink!" And indeed, sometimes it *does* all look pink, due to the chemicals that are used to stain tissue sections. The key to identifying sections is to reduce every tissue to its simplest components: cells and extracellular regions containing different types of chemicals. Remember that all tissues are made up of only cells (see Chapter 3) and chemicals of the ECM (see Chapter 2). So although they might look foreign and abstract, tissue sections simply show cells and ECM. Let's take a tour of a typical tissue section so that we can demystify it.

Shown here is a section from the esophagus, the tubular organ that transports food from the mouth to the stomach:

LM (330×)

Like most tissue sections, this one contains cells and ECM. Let's identify the easiest component first—the cells. Several different types of cells are present in this section, but they share a common, easily identifiable characteristic: They all have dark purple nuclei (due to the stain that binds to their DNA). So we can identify each structure that contains a dark purple nucleus as a cell. This leaves us with the material outside the cells—the ECM. Recall that the ECM consists of ground substance and protein fibers. Ground substance is simple to identify, as it generally either looks clear or has just a slight tinge of color. Protein fibers are also easy to distinguish, as they generally look like wavy or straight lines. In this section, you can see mostly collagen fibers that appear as pink wavy lines. This is the most common appearance of collagen fibers in sections, although sometimes the color varies due to different stains. Occasionally, collagen fibers form bundles that might resemble certain cell types. The easiest way to differentiate the two is to look for nuclei; if the bundles lack nuclei, then they are likely to be collagen fibers.

This next section is from the sublingual gland, a salivary gland (one that produces saliva) under the tongue:

LM (450×)

Go ahead and identify the cells, ground substance, and protein fibers in this section. This section, like many others, contains clusters of small, light red, round discs that lack nuclei. You may not immediately recognize these discs as cells because they lack nuclei, but they are in fact cells called *red blood cells* or *erythrocytes* (eh-RITH-roh-sytz). Erythrocytes are located in blood vessels, and you are likely to see blood vessels and possibly erythrocytes in many different tissue sections.

This procedure is the most basic way to approach a tissue section and orient yourself to what you are viewing. Once this becomes routine, you can use the cell shapes, the amount of ground substance, and the fiber characteristics to determine the exact type of tissue and the organ from which it was taken. We discuss how to do this in another Concept Boost later in the chapter. ■

Components and Classification of Epithelia

From Module 4.1, we know that epithelial tissues consist mostly of cells that are closely packed together (**Figure 4.3**). The cells are generally joined by tight junctions and desmosomes that make the sheet of cells fairly impermeable and more resistant to both stresses and mechanical injury.

Notice in Figure 4.3 that epithelial tissue lacks blood vessels—that is, it is *avascular*—so oxygen and nutrients from the blood must diffuse up to the cells in the epithelial tissue from the blood vessels in the tissues that are deep to it, which are nearly always connective tissue. This imposes a limit on the thickness of epithelial tissue. The stacks of epithelial cells can rise only so high, because the oxygen and nutrients are able to diffuse only so far within a certain time. Cells that are too far away from the deeper blood supply will lack sufficient oxygen and nutrients and so will die. The functions of epithelia are dependent on their structure; when structure changes, function can be disrupted. This is especially evident when the structure is altered by chemicals called *carcinogens*, as discussed in *A&P in the Real World: Carcinogens and Epithelial Tissues* on page 130.

Figure 4.3 also shows that essentially no ECM is present between the cells of epithelial tissue; instead, the ECM is located *beneath* the cells in a structure known as the **basement membrane.**

Figure 4.3 **Structure of epithelial tissue.**

Simple epithelium | Stratified epithelium

(a) Number of cell layers

Squamous cell | Cuboidal cell | Columnar cell

(b) Shape of cells

Figure 4.4 **Classification of epithelial cells.**

Quick Check

☐ 1. Why is living epithelial tissue limited to a certain thickness?
☐ 2. Where is the ECM of epithelial tissue located, and what is its function?
☐ 3. What are the three basic shapes of epithelial cells?

Covering and Lining Epithelia

◀◀ **Flashback**

1. What is diffusion? (p. 75)
2. Which substances cross a plasma membrane by diffusion? (p. 75)

Covering and lining epithelia are found on inner and outer body surfaces. Cells of each shape—squamous, cuboidal, or columnar—are knitted into broad, flat sheets of varying thickness. These sheets are continuous and removable as a whole, so they are often called *membranes* (a term that can include the underlying basement membrane as well). Now let's explore the different classes of epithelia.

Simple Epithelia

As simple epithelia consist of only one layer of cells, and are therefore extremely thin, they are not very resistant to mechanical stresses and thus wouldn't serve very well as outer barriers. However, their thinness makes them perfect for lining hollow organs and for lining surfaces across which substances must diffuse or be transported—a good example of the Structure-Function Core Principle (p. 25).

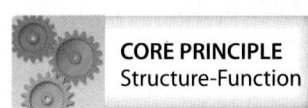

CORE PRINCIPLE
Structure-Function

Types of Simple Epithelia Simple epithelia can be grouped into four types (**Figure 4.5**):

- **Simple squamous epithelium.** As you can see in Figure 4.5a, **simple squamous epithelium** consists of a single layer of flat

---

The basement membrane has two components: the **basal lamina** (*basal-* = "bottom," *lamina-* = "layer") and its underlying **reticular lamina.** The basal lamina is the ECM of the epithelial tissue and is synthesized by the epithelial cells. It consists mostly of collagen fibers and ground substance. The reticular lamina is manufactured by the connective tissue deep to the epithelial tissue, and consists of reticular fibers and ground substance. Together, the two layers of the basement membrane "glue" the epithelial tissue to the underlying connective tissues. The membrane also anchors underlying blood vessels in place and provides a barrier between the epithelial tissue and the underlying tissue.

Note that the cells of epithelial tissue have one side in contact with the extracellular space, and another surface in contact with the basal lamina. The "free edge" of an epithelial cell or tissue is called the **apical surface** (*ap-* = "top"), and the edge attached to the basal lamina is called the **basal surface.** The sides are called the *lateral surfaces.* The properties of epithelial tissues can differ greatly on the apical and basal sides in terms of cell shape, cell functions, and plasma membrane features.

We classify epithelia based on two criteria. The first is the number of cell layers, and the second is the shape of the cells in those layers. Epithelia that consist of only a single layer of cells are **simple epithelia;** those consisting of more than one cell layer are **stratified epithelia** (**Figure 4.4a**). Epithelial cells can take on three basic shapes, shown in **Figure 4.4b:** the flattened **squamous cells** (SKWAY-muss; *squam-* = "scales"), the short **cuboidal cells,** and the tall and elongated **columnar cells.** The number of layers and the shapes of these cells have functional significance, which we discuss in the upcoming sections.

Single layer of flat cells

Simple squamous epithelium

Basement membrane

Underlying connective tissue

LM (530×)

(a) Simple squamous epithelium

Single layer of cube-shaped cells

Simple cuboidal epithelium

Basement membrane

Surrounding connective tissue

LM (680×)

(b) Simple cuboidal epithelium

Figure 4.5 **Structure of simple epithelia.**

Carcinogens and Epithelial Tissues

A P in the Real World

Epithelia cover all body surfaces and so are more subject to injuries than most other tissues. One type of injury comes from **carcinogens** (kar-SIN-oh-jenz), which are agents inducing changes in cellular DNA that can lead to cancer. Cancers of epithelial tissues are called **carcinomas** (kar-sih-NOH-mahz); some of the most common carcinomas include the following:

Clusters of tumor cells

LM (70×)

- *lung adenocarcinoma* (ah-dee′-noh-kar-sih-NOH-mah), a cancer of the lung;
- *ductal* and *papillary carcinoma* (PAP-ih-lair-ee), cancers of the breast; and *basal cell carcinoma,* a cancer of the skin, shown at right.

The basement membrane provides a barrier that helps to prevent or slow the spread of carcinomas to other tissues. Cancers that have not penetrated the basement membrane have not invaded other tissues and are called "pre-malignant" or *carcinoma in situ* (SEE-too; *in situ* = "in place"). Often cancer cells produce enzymes that degrade components of the basement membrane, which facilitates their spread into distant tissues.

Single layer of tall rectangular cells

(c) Simple columnar epithelium

Simple columnar epithelium

Basement membrane

Underlying connective tissue

LM (1900×)

Cilia

Mucus of goblet cell

Single layer of cells that appears to be stratified

Pseudostratified epithelium

Basement membrane

Underlying connective tissue

LM (1740×)

(d) Pseudostratified columnar epithelium

Figure 4.5 *(continued)*

cells. If you look at the apical surfaces of the cells, they resemble fried eggs that fit together like six-sided floor tiles. Simple squamous epithelium is extremely thin, a fact that allows substances such as oxygen, carbon dioxide, fluids, and ions to diffuse across it quickly. We find simple squamous epithelium lining the tiny air sacs of the lungs, forming the outer boundary of serous membranes such as the peritoneum (see Chapter 1), lining certain parts of the kidney tubules, and lining blood vessels.

- **Simple cuboidal epithelium.** **Simple cuboidal epithelium,** shown in Figure 4.5b, is a single layer of roughly cube-shaped cells. When viewed in a section, the cells appear square, with a large, central nucleus. Like simple squamous epithelium, simple cuboidal epithelium is relatively thin, so substances can diffuse across it rapidly. Simple cuboidal epithelium lines structures where rapid diffusion is needed, such as certain kidney tubules and respiratory passages, the ducts of many glands, and the thyroid gland.
- **Simple columnar epithelium.** The cells of **simple columnar epithelium** are tall and appear rectangular in a section, with

their nuclei generally in the basal portion of the cell (Figure 4.5c). The apical plasma membranes of these cells are often folded into *microvilli*, which increase their surface area for absorption (see Figure 3.23). Examples of simple columnar epithelia with microvilli are found in the small intestine, the gallbladder, and the kidney tubules. The apical surface of simple columnar epithelial cells also sometimes contains *cilia*, many short cellular extensions that beat in unison to propel something through a hollow organ; examples are found in the uterine tube and certain respiratory passages.
- **Pseudostratified columnar epithelium.** As its name implies, **pseudostratified columnar epithelium** (soo′-doh-STRAT-ih-fyd; *pseudo-* = "false") is simple epithelium that appears to be stratified. Figure 4.5d shows why—the nuclei of the cells are at different heights, and some of the cells are shorter than others. However, each cell rests on the basal lamina of the basement membrane, and the epithelium is only one cell layer thick. Pseudostratified columnar epithelium is found in the upper

part of the respiratory passages and the nasal cavity, where it is ciliated. Although substances can diffuse across pseudostratified columnar epithelium, it is generally too thick for this to take place efficiently. Instead, this relatively thick epithelium performs a protective function in the organs where it is located.

Transport across Simple Epithelia Transport across simple epithelial tissue may occur in two ways (**Figure 4.6**). The first is that substances can leak between the cells in an epithelial membrane, a process called *paracellular transport* (*para-* = "beside"). However, most cells in an epithelial membrane are linked by tight junctions, which make the spaces between them relatively impermeable. Thus, most substances cross an epithelial membrane by a second method, called *transcellular transport* (*trans-* = "through"). In this process, ①　a substance enters the cell through its phospholipid bilayer, ②　diffuses through the cytosol, and ③　exits through the other surface of the cell. Transcellular transport operates by the same rules we discussed earlier: Water and solutes move into and out of the cell by osmosis, simple diffusion, and facilitated diffusion when a concentration gradient is present; certain solutes move into and out of the cell by vesicular transport; and others move into and out of the cell by active transport pumps (see Chapter 3).

Stratified Epithelia

Stratified epithelia are thicker than simple epithelia due to their additional layers, and for this reason they do not line areas where transcellular transport is required. However, their thickness allows them to be very effective protective barriers, and they are often found in areas subject to a high degree of mechanical stress. As you can see in **Figure 4.7**, the cells in stratified epithelium generally do not maintain the same shape throughout the thickness of the tissue; for simplicity, these epithelia are named according to the shape of the cells in their apical layers.

The types of stratified epithelia include the following:

- **Stratified squamous epithelium.** There are two kinds of **stratified squamous epithelium:** *keratinized stratified*

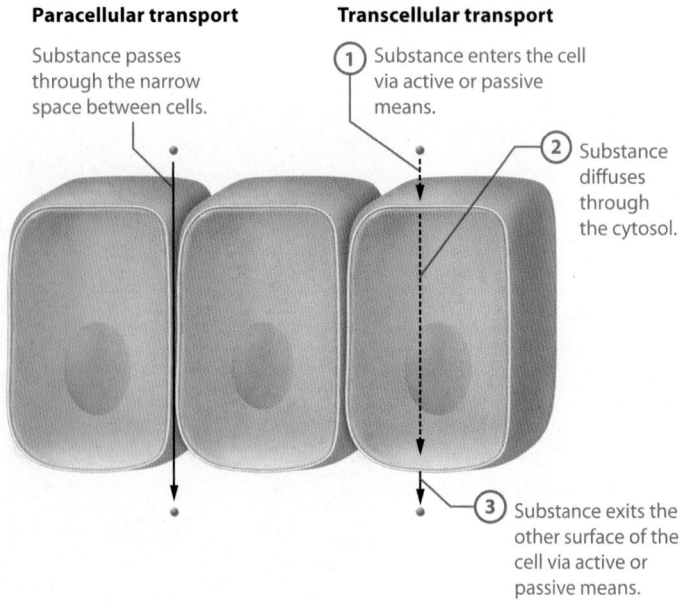

Paracellular transport

Substance passes through the narrow space between cells.

Transcellular transport

① Substance enters the cell via active or passive means.

② Substance diffuses through the cytosol.

③ Substance exits the other surface of the cell via active or passive means.

Figure 4.6　Transport across simple epithelia.

squamous epithelium and *nonkeratinized stratified squamous epithelium*. In keratinized stratified squamous epithelium, the apical layers of cells lack nuclei, and the cells are no longer living. In addition, the apical cells are filled with the protein **keratin** (*kerato-* = "hard"), which makes this type of epithelium tough and resistant to friction. Keratinized stratified squamous epithelium forms the outer layer of our skin, so it will be discussed fully in the integumentary chapter (see Chapter 5). In contrast, the apical cells of nonkeratinized stratified squamous epithelium are nucleated, the cells are living, and each cell is distinct (Figure 4.7a). Nonkeratinized stratified squamous epithelium is found in organs that require protection from mechanical abrasion but that need to retain the moist surface not found with keratinized epithelium; such epithelium lines the mouth, throat, esophagus, anus, and vagina.

- **Stratified cuboidal and columnar epithelium.** Both **stratified cuboidal epithelium** and **stratified columnar epithelium**

Vagina

Stratified squamous epithelium

Basement membrane

Underlying connective tissue

LM (350×)

(a) Nonkeratinized stratified squamous epithelium from the vagina

Figure 4.7　Structure of stratified epithelia.

(b) **Stratified cuboidal epithelium from the duct of a sweat gland**

(c) **Stratified columnar epithelium from the duct of a salivary gland**

(d) **Transitional epithelium from the urinary bladder**

Figure 4.7 *(continued)*

SIMPLE EPITHELIA: CONSIST OF ONE LAYER OF CELLS	COMPONENTS	FUNCTION	LOCATION
Simple squamous epithelium	Single layer of flat cells Flattened, central nuclei	• Provides a barrier through which gases and fluids can be exchanged • Produces serous fluid	• Air sacs of lungs • Inner lining of ventral body cavities and blood vessels • Certain parts of kidney
Simple cuboidal epithelium	Single layer of cube-shaped cells Round, central nuclei	• Absorption • Secretion of mucus and other substances	• Kidney tubules • Lower respiratory passages • Thyroid, salivary, and mammary glands
Simple columnar epithelium	Single layer of rectangular cells Ovoid, basal nuclei	• Absorption • Secretion of mucus and other substances • Propulsion of egg through uterine tube	• Digestive tract from stomach to anus • Uterine tube • Gallbladder • Certain kidney tubules
Pseudostratified columnar epithelium	Single layer of columnar cells of different heights, giving a stratified appearance Cilia	• Secretes mucus and propels it with ciliary motion	• Nasal cavity • Parts of male urethra • Upper respiratory passages

STRATIFIED EPITHELIA: CONSIST OF MORE THAN ONE LAYER OF CELLS			
Stratified squamous epithelium *Keratinized epithelium*	Multiple layers of squamous cells: Apical cells dead, flaky, and filled with keratin Cuboidal to squamous-shaped basal cells	• Protection from mechanical stresses and microorganisms • Prevents most water loss through skin	• Epidermis
Nonkeratinized epithelium	Apical cells living and very flat Cells more cuboidal toward basal layers	• Protection from mechanical stresses and microorganisms	• Mouth, pharynx, superior larynx, esophagus • Vagina • Anal canal
Stratified cuboidal epithelium	Two or more layers of cuboidal cells	• Some absorption and secretion	• Ducts of sweat glands
Stratified columnar epithelium	Two or more layers of columnar cells	• Protection • Some absorption and secretion	• Ducts of certain glands • Cornea • Parts of male urethra
Transitional epithelium	Multiple layers of cells Apical cells dome-shaped when relaxed and flattened when stretched	• Protection • Gives tissues added distensibility	• Urinary bladder • Ureter • Urethra

Practice art labeling
@MasteringA&P®

Figure 4.8 Summary of epithelial tissues.

are relatively rare in the human body (Figure 4.7b and c). Stratified cuboidal epithelium consists of two layers of cuboidal cells, and it lines the ducts of sweat glands. Stratified columnar epithelium is formed from a few layers of cells that are columnar in the apical layers and cuboidal in the basal layers. It makes up the ducts of certain glands, including salivary glands, as well as parts of the male urethra and the cornea (the clear outer covering of the eye).

- **Transitional epithelium.** The final type of stratified epithelium is **transitional epithelium,** which was named for the incorrect belief that the shape of its cells represented a "transition" between cuboidal and squamous cells (Figure 4.7d). Histologists now understand that this tissue is a distinct type of epithelium found only in the urinary system, where it lines the interior of the kidney, the ureters, the urinary bladder, and the urethra. Because of this location, it is now sometimes called *urothelium.* Notice that the cells in the basal layers of transitional epithelium are cuboidal, and the cells in the apical layers are vaguely dome-shaped when the tissue is relaxed. However, when stretched, the apical cells can become flatter, which contributes to the ability of these urinary organs to stretch.

The types of epithelial tissue are summarized in **Figure 4.8.** This figure is set up in a tabular format so you can use it to study and compare these tissues.

Quick Check

☐ 4. What are the differences between simple, stratified, and pseudostratified epithelium?
☐ 5. Why do most substances move across simple epithelia by transcellular transport?

Glandular Epithelia

◀◀ Flashback

1. How does the endomembrane system package a product for secretion from the cell? (p. 93)
2. What is the basic process of exocytosis? (p. 84)

A **gland** is a structure that forms and secretes a product. Glands arise from epithelial tissue that grew inward into the underlying connective tissue rather than remaining at the surface. Certain cells of a gland, called *secretory cells,* manufacture a product and release it. Glands can be classified according to their shapes and how they release their products.

Glands release their products by two mechanisms:

- **Exocrine glands** (EKS-oh-krin) release their products to the apical surface of the epithelium, generally to the exterior of the body or into a hollow organ that opens to the outside of the body. The product of an exocrine gland is secreted through a duct lined with epithelial cells. The products of an exocrine gland have *local* actions only—they can affect only the cells in their general vicinity.
- **Endocrine glands** (EN-doh-krin) lack ducts and secrete their products, which are usually hormones, directly into the blood. Since these hormones travel through the bloodstream, they facilitate communication between cells in distant areas of the body (Cell-Cell Communication Core Principle, p. 27). Endocrine glands are discussed fully in their own chapter (see Chapter 16); the remainder of this section discusses exocrine glands.

CORE PRINCIPLE
Cell-Cell Communication

Exocrine glands vary in complexity from those that are only a single cell to those that are large with branching ducts and many secretory units. The simplest exocrine glands are the **unicellular glands,** or those composed of a single cell (**Figure 4.9**). The most common unicellular gland in the body is the **goblet cell,** a cell found abundantly in the epithelium lining the digestive and respiratory tracts. Goblet cells secrete **mucus,** a thick, sticky liquid that protects the underlying epithelium. You can see in Figure 4.9 that goblet cells, which derive their name from their shape, are filled with mucus droplets that can be released when the cell is stimulated.

LM (950×)

Simple columnar epithelial cells

Goblet cells (unicellular glands)

Released mucus

Microvilli

Mucus droplets

Cytosol

Golgi apparatus

Rough ER

Nucleus

Figure 4.9 Unicellular exocrine glands.

Most exocrine glands are **multicellular glands,** which are made of clusters of secretory cells arranged in different ways. Multicellular glands are classified according to the structure of their ducts and the shape of their secretory cell clusters (**Figure 4.10**). **Simple glands** have ducts that don't branch, whereas the ducts of **compound glands** do have branches. The clusters of secretory cells may be arranged into three shapes: *tubular* (long and straight or coiled), *acinar* (AY-sih-nar; spherical), or *tubuloacinar* (with both tubular and acinar portions).

Exocrine glands exhibit two types of secretion: merocrine and holocrine (**Figure 4.11**). The majority of exocrine glands in the body, including salivary glands and sweat glands, use **merocrine secretion** (MAIR-oh-krin), in which they package their products into secretory vesicles for release by exocytosis. The other type of secretion is called **holocrine secretion** (HOH-loh-krin). In this type, the secretory cells accumulate their product in their cytosol, and the product is not released until the cell ruptures and dies. The dead cells are then shed and released with the product, after which they are replaced by cells at the base of the gland that undergo mitosis. Only a few examples of holocrine secretion exist in the human body; a notable one is the sebaceous (see-BAY-shuss) glands of the skin, which secrete an oily fluid called sebum.

Multicellular exocrine gland

Ductal epithelial cells

Secretory epithelial cells

Simple (not branched)

Compound (branched)

Secretory cell cluster shapes:

Tubular

Acinar

Tubuloacinar

(a) Components used to classify multicellular exocrine glands

Simple tubular

Simple acinar

Compound acinar

Compound tubuloacinar

(b) Examples of multicellular exocrine glands

Figure 4.10 Multicellular exocrine glands.

(a) **Merocrine secretion**

- Product being released by exocytosis
- Secretory vesicles
- Secretory epithelial cells

(b) **Holocrine secretion**

- Shed and ruptured epithelial cells releasing product
- Product accumulating in secretory epithelial cell
- Epithelial cell undergoing mitosis to replace shed cell

Figure 4.11 Modes of secretion in exocrine glands.

MODULE **4.3**

Connective Tissues

Learning Outcomes

1. Compare and contrast the roles of individual cell and fiber types within connective tissues.
2. Identify the different types of connective tissue, and classify them based on their distinguishing structural characteristics.
3. Describe where in the body each type of connective tissue is found.
4. Describe the functions of each type of connective tissue, and correlate function with structure for each tissue type.

Connective tissues are a diverse group of tissues with a variety of functions, including the following:

- **Connecting and binding.** As the name implies, connective tissues *connect* structures in the body. In organs, they bind other tissue layers together; they are also found between organs, where they anchor them in place and to one another.
- **Support.** Certain connective tissues such as bone and cartilage support the weight of the body.
- **Protection.** Connective tissues have many protective functions. For example, bone tissue protects certain internal organs, and cartilage and fat tissue provide shock absorption. In addition, many elements of the immune system are found within connective tissues.
- **Transport.** Blood is a fluid connective tissue that is the main transport medium in the body.

Like all tissues, connective tissue is made up of cells and ECM. In this case, the arrangement consists of loosely packed cells surrounded by protein fibers, all of which are embedded in ground substance. Although diverse in function, connective tissues do share a common feature—their ECM plays an extensive role in their functions. We discuss this expanded functional role of the ECM with the individual tissue types.

The classification scheme we'll use has two basic types of connective tissue: *connective tissue proper* and the *specialized connective tissues* (bone, cartilage, and blood). Each type differs in the main types of cells it contains and the components of its ECM. Certain cells are found only in the specialized connective tissues; however, others are more general and are present in many different connective tissues. In this module, we examine the structural and functional properties of both types of connective tissue.

Connective Tissue Proper

◀◀ Flashback

1. What are lipids? (p. 52)
2. What is a phagocyte? (p. 83)

First we explore **connective tissue proper,** which is sometimes called *general connective tissue.* Connective tissue proper is widely distributed in the body, where it connects tissues and organs to one another and forms part of the internal architecture of some organs. Let's take a brief tour of the cells that make up connective tissue proper, and then we'll focus on the different types of this tissue.

Cells of Connective Tissue Proper

Connective tissue proper houses a variety of cell types. Some of these cells are *resident* cells, meaning they inhabit the tissue permanently, whereas others are *migrant* cells and can migrate to different areas of the body in response to various needs. The resident and migrant cells include the following (**Figure 4.12**):

- **Fibroblasts.** The most common resident cell found in connective tissues is the **fibroblast** (*-blast* = "bud, sprout"; Figure 4.12a). Although in most cell types a "-blast" ending indicates an immature cell and a "-cyte" ending indicates a mature cell, the fibroblast is a mature cell with properties of an immature cell. Fibroblasts produce protein fibers as well as ground substance and other elements of the ECM. Active fibroblasts usually lie close to collagen fibers, as they continually produce collagen proteins.

- **Adipocytes.** **Adipocytes** (AD-ih-poh-sytz), also called *fat cells,* are another type of resident cell found in many different connective tissues (Figure 4.12b). The cytoplasm of most adipocytes is dominated by a single large inclusion containing lipids. The nuclei and other organelles of the adipocytes are squashed to the perimeter of the cell, and are often difficult to see in sections.

- **Mast cells. Mast cells** are the largest resident cells in connective tissues (Figure 4.12c). They are cells of the immune system that have cytosolic inclusions (or *granules*) containing chemicals called *inflammatory mediators.* An example of such a mediator is *histamine.* When stimulated, mast cells release the contents of their granules, and inflammation results. *Inflammation* is a protective response that activates the immune system and recruits immune cells to the tissue.

- **Phagocytes.** Recall that **phagocytes** (FAYG-oh-sytz) are cells of the immune system that can ingest foreign substances, microorganisms, and dead and damaged cells by phagocytosis (see Chapter 3; Figure 4.12d). Two common types of phagocytes are **macrophages** (MAK-roh-fay-jez), which can be either resident or migrant cells in connective tissues, and *neutrophils* (NOO-troh-filz), which are migrant immune cells.

- **Other immune cells.** Various other cells of the immune system migrate in and out of different connective tissues, depending on the needs of the body at the time.

Types of Connective Tissue Proper

There are four basic types of connective tissue proper, defined by their main components: (1) *loose connective tissue,* (2) *dense connective tissue,* (3) *reticular tissue,* and (4) *adipose tissue.* The next four subsections examine each type.

Loose Connective Tissue Loose **connective tissue,** also known as **areolar connective tissue** (ah-ree-OH-lahr), is composed primarily of ground substance. In addition, all three types of protein fibers, fibroblasts, and other cells such as adipocytes are suspended in the ground substance (**Figure 4.13**). (Reticular fibers require a special stain to be visible, so you can't see them in Figure 4.13.) Loose connective tissue is found deep to the epithelium of the skin, in the membranes lining the body cavities, and as layers in the walls of hollow organs. It functions in support and contains many blood vessels, whose nutrients and oxygen diffuse up to the superficial epithelial cells. It also houses numerous immune cells that help protect against microorganisms invading

(a) Fibroblasts — LM (635×)
Fibroblasts
Nucleus

(b) Adipocytes — LM (230×)
Lipid inclusion
Nucleus
Cytoplasm

(c) Mast cell — TEM (2150×)
Granules containing inflammatory mediators
Nucleus

(d) Phagocyte (macrophage) — SEM (710×)
Macrophage engulfing bacteria
Bacterium
Pseudopod

Figure 4.12 Cells of connective tissue proper.

the epithelium, which reflects the fact that it is the first tissue to come into contact with the epithelium.

Dense Connective Tissue Whereas ground substance is the primary component of loose connective tissue, **dense connective tissue** is composed primarily of protein fibers. Due to this

Superficial skin, epithelium
Deeper layer of skin, loose connective tissue

Fibroblasts

Ground substance

Elastic fibers

Collagen fibers

LM (325×)

Loose connective tissue from skin

Figure 4.13 Structure of loose connective tissue.

predominance of fibers, dense connective tissue is often called *fibrous connective tissue.*

Dense connective tissue can be grouped into three classes: irregular, regular collagenous, and regular elastic (**Figure 4.14**):

- **Dense irregular connective tissue.** The predominant fiber type in **dense irregular connective tissue** is the collagen fiber. Notice in Figure 4.14a that the bundles of collagen fibers, which appear as pink clusters due to staining, seem to be arranged haphazardly. This organization makes dense irregular connective tissue quite strong and allows it to resist tension in all three planes. It is found in organs that are subjected to such tension, such as the dermis (the deep layer of skin), as well as around organs and joints.
- **Dense regular collagenous connective tissue.** As you can see in Figure 4.14b, **dense regular collagenous connective tissue** contains thick collagen fibers arranged parallel to one another to form bundles. The fibers in these bundles are very strong because they are oriented in a single direction, but they resist tension in only one plane. This tissue

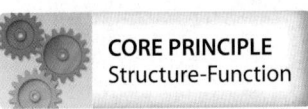

CORE PRINCIPLE
Structure-Function

is therefore found in structures subjected to tension in one direction only, such as tendons, which join muscle to bone, and ligaments, which unite two bones. Note that the arrangement of the fibers in dense regular and irregular collagenous connective tissues is another example of the Structure-Function Core Principle (p. 25).

- **Dense regular elastic connective tissue. Dense regular elastic connective tissue** (often shortened to simply "elastic tissue") consists mostly of parallel elastic fibers with randomly oriented collagen fibers (Figure 4.14c). This tissue,

which allows certain organs to stretch, is found in the lining of large blood vessels and in certain ligaments, such as those of the spine.

Reticular Tissue **Reticular tissue** is so named because it is populated by numerous reticular fibers produced by surrounding fibroblasts (often called *reticular cells;* **Figure 4.15**). The thin reticular fibers interweave to form fine networks that support small structures such as blood vessels and lymphatic vessels. Reticular tissue is also found in lymph nodes and the spleen, where reticular fibers form weblike nets that help trap old and foreign cells. It also forms part of the basement membrane that supports all epithelia and the internal structure of the liver and bone marrow.

Adipose Tissue **Adipose tissue,** also called *fat tissue,* consists of fat-storing adipocytes and their surrounding fibroblasts and ECM (**Figure 4.16**). Although fibroblasts and ECM containing collagen fibers are initially the dominant component of adipose tissue, the adipocytes can increase in size to the point that few fibroblasts and little ECM are visible. Adipose tissue has many functions, including insulation, warmth, shock absorption, and protection, and it is also the major energy reserve in the body.

The majority of adipose tissue is *white adipose tissue,* which consists of adipocytes that contain one large lipid inclusion in their cytosol. White adipose tissue is found deep to the skin (*subcutaneous fat;* sub-kyoo-TAYN-ee-us) and in characteristic places such as the abdomen, breasts, hips, buttocks, and thighs. It also surrounds certain organs such as the heart and the abdominal organs (*visceral fat;* VISS-er-al). As its name suggests, it is white.

A small percentage of adipose tissue is *brown adipose tissue,* which consists of adipocytes that contain multiple lipid

- Superficial skin, epithelium
- Deeper layer of skin, loose connective tissue
- Deepest layer of skin, dense irregular connective tissue

Irregular collagen bundles

Ground substance

Fibroblasts

(a) Dense irregular connective tissue from thick skin

LM (316×)

Tendons

Parallel collagen bundles

Fibroblasts

Ground substance

(b) Dense regular collagenous connective tissue from a tendon

LM (845×)

Aorta

Parallel elastic fibers

Ground substance

Heart

(c) Dense regular elastic connective tissue from the aorta

LM (330×)

Figure 4.14 Structure of dense connective tissue.

inclusions. Such tissue appears brown due to abundant mitochondria in the cells and an extensive blood supply. Infants and young children have brown adipose tissue around the neck and back; however, by adulthood little brown adipose tissue remains. Both types of adipose tissue are used for fuel and insulation. However, the cells of brown adipose tissue can oxidize fatty acids about 20 times as fast as those of white adipose tissue; this additional energy is used to generate heat in cold temperatures.

Obesity is the condition of having excess adipose tissue in proportion to lean body mass. See *A&P in the Real World: Adipose Tissue and Obesity* on page 142 for a discussion of the different types of this prevalent condition.

☐ 1. What are the types of cells of connective tissue proper? What are their functions?
☐ 2. What is the primary component in:
   a. loose connective tissue?
   b. dense regular collagenous connective tissue?
   c. dense regular elastic connective tissue?
   d. reticular tissue?
   e. adipose tissue?
☐ 3. What are the primary differences between brown and white adipose tissue?

Reticular tissue from a lymph node

LM (535×)

Figure 4.15 Structure of reticular tissue.

Adipose tissue from hypodermis superficial fascia

LM (445×)

Figure 4.16 Structure of adipose tissue.

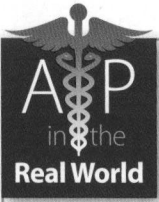

Adipose Tissue and Obesity

There are two forms of obesity. The first and most common is *hypertrophic obesity,* in which the lipid inclusions in the adipocytes accumulate excess fatty acids and increase in size. Adipocytes may grow to as much as four times their normal size in cases of severe obesity; however, the number of adipocytes remains the same.

The other form is *hypercellular obesity,* in which the actual number of adipocytes increases. Hypercellular obesity, which is generally severe, correlates with the development of obesity in infancy or early childhood. This is because the adipocytes of infants and young children can increase in number, whereas those of adults lack the ability to divide and form new cells.

Both forms of obesity increase the risk for certain health problems; however, they seem to be associated with the development of different disorders. Furthermore, the development of the disorders associated with obesity is complex and depends on the distribution of adipose tissue and on genetic factors. For a deeper discussion of this topic, go to the metabolism chapter (see Chapter 23).

Specialized Connective Tissues

The **specialized connective tissues** have more specific functions than connective tissue proper. The three types of specialized connective tissue are *cartilage, bone* or *osseous tissue* (OSS-ee-us), and *blood.* Both bone tissue and blood have chapters dedicated to their discussion (see Chapters 6 and 19, respectively), so we will describe them only briefly here.

Cartilage

Cartilage is found in joints between bones, as well as in the ears, nose, and certain respiratory passages. It is a tough but flexible tissue that absorbs shock and is resistant to tension, compression, and shearing forces. These properties largely result from the components of its ECM, which contains glycosaminoglycans, proteoglycans, collagen fibers, and elastic fibers. Cartilage contains two major cell types: immature **chondroblasts** (KON-droh-blastz; *chondro-* = "cartilage") and mature **chondrocytes** (KON-droh-sytz). As with other connective tissues, the immature chondroblasts actively divide by mitosis and make most of the ECM of cartilage. As the chondroblasts become surrounded by their own ECM, they gradually mature to relatively inactive chondrocytes. Mature chondrocytes eventually inhabit small cavities in the ECM called **lacunae** (lah-KUH-nee; *lacuna* = "cavity").

Cartilage is one of the few connective tissues that are essentially avascular, as few, if any, blood vessels course through the cartilage itself. The blood supply to this tissue is mostly limited to an outer sheath of dense irregular collagenous connective tissue called the **perichondrium** (pear'-ih-KON-dree-um; *peri-* = "around"). Thus, oxygen and nutrients must diffuse from the blood vessels in the perichondrium and through the cartilage ECM to reach the chondrocytes and chondroblasts. This zone of diffusion is fairly narrow, measuring only a few millimeters, which limits the thickness of living cartilage.

Cartilage can be divided into three classes, based on the composition of their ECM. These three classes are *hyaline cartilage, fibrocartilage,* and *elastic cartilage:*

- **Hyaline cartilage.** The most abundant type of cartilage in the body is **hyaline cartilage** (HY-ah-lin; *hyaline* = "glasslike") (**Figure 4.17a**). The ECM of hyaline cartilage is bluish-gray and consists of mostly ground substance with a fine type of collagen fiber that forms small bundles, which gives hyaline cartilage a uniform, glassy appearance. Hyaline cartilage covers the ends of bones where they form joints (joints are also referred to as *articulations,* and thus hyaline cartilage is often called *articular cartilage*). Here it forms a smooth surface so that bones may slide over one another with little to no friction and also protects the underlying bone by more equally distributing stresses. Hyaline cartilage is also found where strong yet flexible support is needed, such as at the attachment points of the sternum (breastbone) to the ribs, framing parts of the respiratory tract, and in the nose. Additionally, most of the fetal skeleton begins as hyaline cartilage, which is eventually replaced by bone. See *A&P in the Real World: Osteoarthritis and Glucosamine Supplements* on page 144 for a discussion of what happens to hyaline cartilage as we age and whether glucosamine (glew-KOH-sah-meen) helps the resulting problems.

- **Fibrocartilage.** You can see the differences between **fibrocartilage** and hyaline cartilage by comparing Figure 4.17a and **Figure 4.17b**—the ECM of hyaline cartilage is dominated by ground substance, whereas that of fibrocartilage is filled with bundles of collagen fibers, leaving hardly any room for ground substance. This is the case because fibrocartilage contains regular connective tissue fibroblasts in addition to chondroblasts and chondrocytes. These fibroblasts fill the ECM with collagen fibers and, to a lesser extent, elastic fibers as well. This composition of the ECM gives fibrocartilage great tensile strength and a fair amount of elasticity. Fibrocartilage is found in joints called *fibrous joints* and in the intervertebral discs; it also forms articular discs that improve the fit of two bones in a joint.

- **Elastic cartilage.** Elastic cartilage is limited to the external ear and parts of the framework of the larynx (LAEHR-inks; the "voice box"; **Figure 4.17c**). As its name implies, the ECM of elastic cartilage is filled with elastic fibers. This allows the tissue to vibrate and assist in the detection of sound in the air and the production of sound by the larynx.

(a) **Hyaline cartilage from the trachea**

Trachea

Chondrocytes in lacunae

ECM

LM (255×)

Intervertebral discs

Chondrocytes in lacunae

Collagen bundles

(b) **Fibrocartilage from intervertebral disc**

LM (285×)

Chondrocytes in lacunae

Elastic fibers

(c) **Elastic cartilage from epiglottis**

LM (255×)

Figure 4.17 **Structure of cartilage.**

Osteoarthritis and Glucosamine Supplements

Osteoarthritis (oss′-tee-oh-ahr-THRY-tiss) is caused by a variety of factors, including age, joint trauma, genetic disorders, and infection. Osteoarthritis develops as the hyaline cartilage lining joints degenerates. This process leads to destruction of proteoglycans and collagen fibers. Compare the smooth cartilage from a normal joint below at left with the rough, torn cartilage from an arthritic joint below at right. Osteoarthritis makes the cartilage less resistant to mechanical stresses at the joint, which further destroys the remaining collagen fibers. The cartilage may continue to degenerate until the underlying bone is exposed, leading to a painful situation in which the two bones grind on each other any time motion occurs at the joint.

A popular dietary supplement that purports to treat osteoarthritis is the chemical *glucosamine*. Glucosamine is a sugar derived from shellfish and certain fungi that is used by chondroblasts in their synthesis of proteoglycans. Hypothetically, supplementing the diet with glucosamine could increase the production of proteoglycans by chondroblasts and slow cartilage degeneration. Studies on the effectiveness of glucosamine supplementation, however, have yielded mixed results. Further study is needed to determine if glucosamine supplementation can truly stall degeneration of the joints in osteoarthritis.

Normal cartilage   SEM (65×)

Osteoarthritic cartilage   SEM (65×)

(a) Bone tissue prepared so that the calcified extracellular matrix (ECM) is visible   LM (450×)

Osteoclast

Osteoblast

Osteocytes in lacunae

Bone ECM

Calcified ECM

(b) Bone tissue with all three cell types   LM (420×)

Figure 4.18 Structure of bone.

Figure 4.19 **Components of blood.**

Bone

Bone tissue, also called **osseous tissue** (*os-, osteo-* = "bone"), serves many functions, including supporting our body; protecting our vital organs; providing a place for attachment of the muscles of voluntary movement; storing calcium; and housing the *bone marrow,* the tissue that produces our blood cells and stores fat (**Figure 4.18**). The ECM of bone tissue has two types of components, organic and inorganic. The organic portion of the ECM accounts for about 35% of the mass of bone and consists of collagen fibers and a type of ground substance called *osteoid* (OSS-tee-oyd). The inorganic portion, which represents the remaining 65% of bone mass, is composed of calcium phosphate crystals, which make bone one of the hardest substances in the body (Figure 4.18a).

Mature bone tissue contains three primary cell types: *osteoblasts, osteocytes,* and *osteoclasts* (Figure 4.18b).

- **Osteoblasts.** The **osteoblasts** (OSS-tee-oh-blasts; *-blast* = "immature cell") are the "bone builders," carrying out the process of *bone deposition.* During this process, they synthesize and secrete by exocytosis the organic portion of the ECM, and also produce chemicals required for calcium to deposit within the ECM. They are found on the outer surface of the bone, where they are closely associated with a dense irregular collagenous connective tissue covering called the **periosteum** (pear'-ee-OSS-tee-um).
- **Osteocytes.** Osteoblasts eventually become surrounded in lacunae by the ECM they secrete. At this point, the osteoblasts are called **osteocytes** (OSS-tee-oh-sytz), which are mature bone cells. Osteocytes are relatively inactive cells. By exocytosis, they produce and secrete substances that are required for bone maintenance.

- **Osteoclasts.** Whereas osteoblasts are the bone builders, the large and multinucleated **osteoclasts** (OSS-tee-oh-klasts; *-clasia* = "broken") are the bone destroyers. Osteoclasts carry out the process of *bone resorption,* during which they secrete hydrogen ions and enzymes that break down the components of the ECM.

Although bone is solid and strong, it is not an inactive tissue. *Bone remodeling,* which includes bone deposition and bone resorption, is constantly occurring in healthy bone. Furthermore, both bone deposition and bone resorption can occur simultaneously in different parts of the bone, depending on the stresses placed on the tissue. Tension on a certain part of the bone will result in increased osteoblast activity and bone deposition, whereas pressure placed on another part of the bone can result in increased osteoclast activity and bone resorption.

Blood

Blood is unique among the connective tissues in that its ECM is fluid (**Figure 4.19**). The ECM of blood is called **plasma,** and it consists largely of water, dissolved solutes, and proteins. Unlike with most connective tissues, the proteins in the ECM are not fibers; rather, they are much smaller proteins that serve a variety of functions, such as transport and blood clotting. Two main types of cells are found in blood. First, **erythrocytes** (*erythro-* = "red"), also called **red blood cells,** bind and transport oxygen through the body. Second, **leukocytes** (LOO-koh-sytz; *leuko-* = "white"), also called **white blood cells,** function in immunity. In addition, blood contains cellular fragments called **platelets,** which play a role in the blood clotting process.

CONNECTIVE TISSUE PROPER			COMPONENTS	FUNCTION	LOCATION
Loose CT			Fibroblast Ground substance Protein fibers	• Support • Protection • Houses blood vessels supplying epithelium	• Deep to epidermis • Walls of hollow organs • Membranes lining body cavities
Dense CT	Irregular		Fibroblast Collagen fibers Ground substance	• Strength • Resistance to stress in all three planes	• Deepest layer of skin • Around joints, organs
	Regular collagenous		Fibroblast Collagen fibers Ground substance	• Strength • Resistance to stress in one plane	• Tendons, ligaments
	Regular elastic		Elastic fibers Ground substance	• Allows tissue to stretch and recoil	• Large blood vessels • Certain ligaments
Reticular CT			Reticular fiber Leukocyte	• Forms internal structure of many organs • Supports smaller vessels and nerves	• Lymph nodes, spleen, bone marrow, liver • Basement membrane • Around vessels and nerves
Adipose CT			Adipocyte Protein fibers	• Warmth, insulation • Shock absorption, protection • Major energy reserve in the body	• Deep to the skin in characteristic areas • Surrounds the heart and abdominal organs

Practice art labeling @MasteringA&P®

Figure 4.20 Summary of connective tissues.

Figure 4.20 pulls together information about all the types of connective tissue so that you can study and compare them.

Quick Check

☐ 4. How do the three types of cartilage differ?
☐ 5. How do bone and blood differ from other connective tissues?

Apply What You Learned

☐ 1. Predict the effect of replacing the elastic fibers of dense regular elastic connective tissue with collagen fibers.
☐ 2. Injured hyaline cartilage is often replaced by fibrocartilage. Does fibrocartilage provide the same kind of surface for articulation as hyaline cartilage? Why, or why not?

☐ 3. What do you think would happen to bone tissue if osteoclasts were more active than osteoblasts?

See answers in Appendix A.

MODULE **4.4**

Muscle Tissues

Learning Outcomes

1. Clarify the different types of muscle tissue based on their distinguishing structural characteristics and location in the body.
2. Describe the functions of each type of muscle tissue, and correlate function with structure for each tissue type.
3. Identify the different types of muscle tissue.

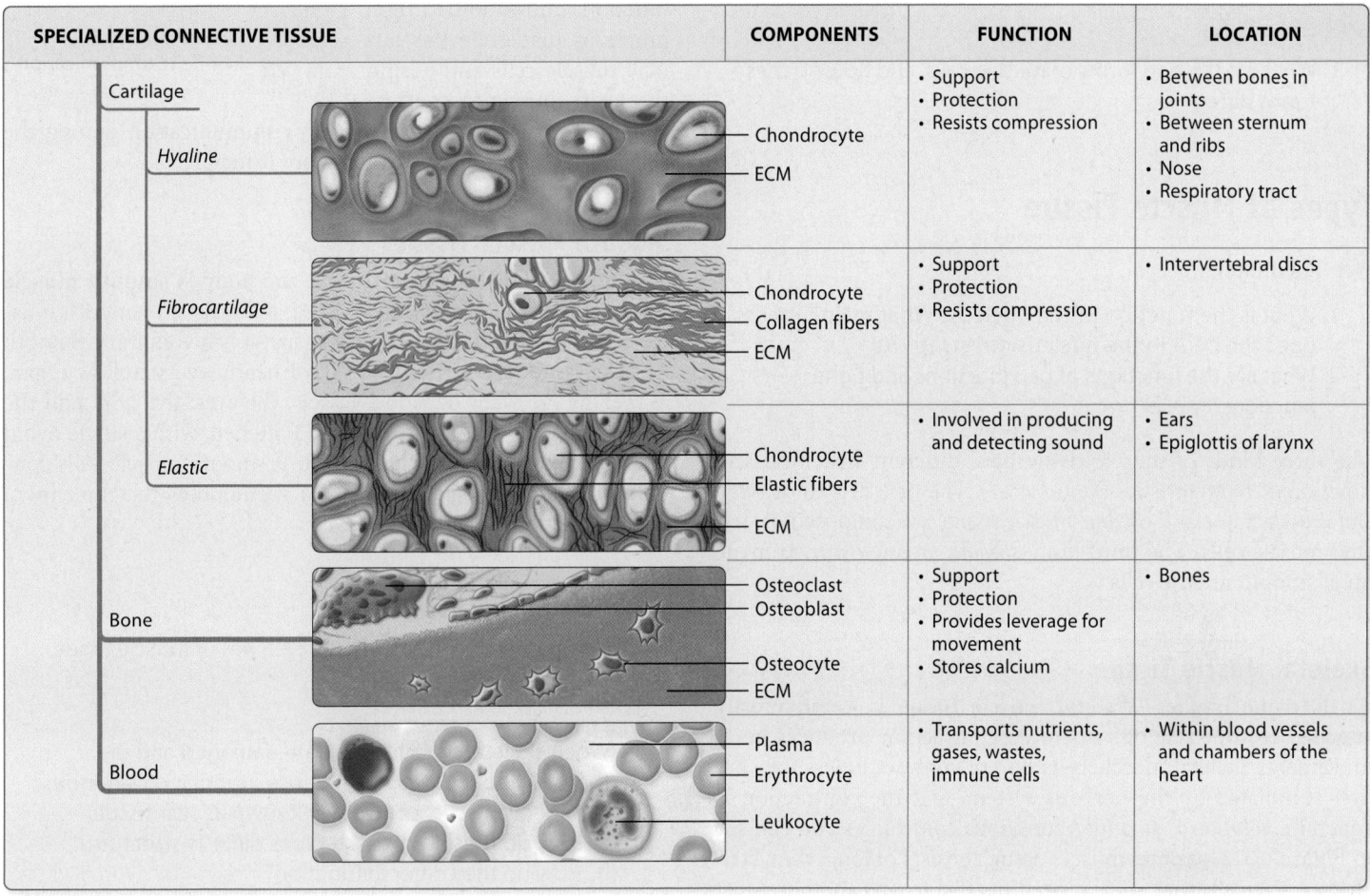

SPECIALIZED CONNECTIVE TISSUE		COMPONENTS	FUNCTION	LOCATION
Cartilage	Hyaline	• Chondrocyte • ECM	• Support • Protection • Resists compression	• Between bones in joints • Between sternum and ribs • Nose • Respiratory tract
	Fibrocartilage	• Chondrocyte • Collagen fibers • ECM	• Support • Protection • Resists compression	• Intervertebral discs
	Elastic	• Chondrocyte • Elastic fibers • ECM	• Involved in producing and detecting sound	• Ears • Epiglottis of larynx
Bone		• Osteoclast • Osteoblast • Osteocyte • ECM	• Support • Protection • Provides leverage for movement • Stores calcium	• Bones
Blood		• Plasma • Erythrocyte • Leukocyte	• Transports nutrients, gases, wastes, immune cells	• Within blood vessels and chambers of the heart

Figure 4.20 *(continued)*

Although many cells have the ability to contract, only the cells of **muscle tissues** are specialized for contraction. The three types of muscle tissues share the common ability to turn the chemical energy of ATP into the mechanical energy of movement. This movement comes in a variety of forms, including walking and breathing, the beating of the heart, and the narrowing of hollow organs. In this module, we discuss first the characteristics of the cells and ECM of muscle tissues, and then we explore the functional and structural differences of the three types of muscle tissue. These topics are only introduced here; they are fully covered in the chapters on muscle tissue and the heart (see Chapters 10 and 17, respectively).

Components of Muscle Tissue

◄◄ **Flashback**

1. What are the functions of the cytoskeleton? (p. 69)

The main component of muscle tissues is **muscle cells,** or *myocytes* (MY-oh-sytz; *myo-* = "muscle"). All muscle cells are **excitable cells,** meaning they respond to electrical or chemical stimulation. Muscle cells have a very different appearance from that of other cells we have discussed: Their cytoplasm is filled with bundles of proteins called **myofilaments.**

There are two forms of muscle cells, which are classified by the arrangement of their myofilaments:

- **Striated muscle cells.** The myofilaments of certain muscle cells are organized in such a way that there are regions where the myofilaments overlap and regions where they don't. The combination of these regions produces both dark and light areas called "bands" when such cells are viewed under a microscope. These alternating light-dark-light-dark bands are known as **striations** (stry-AY-shunz; *stria-* = "lined" or "streaked"), and muscle cells with this arrangement are thus called *striated muscle cells.*
- **Smooth muscle cells.** *Smooth muscle cells* have a very different appearance from that of striated muscle cells. Like striated muscle cells, they contain myofilaments; however, these filaments are arranged as irregular bundles scattered throughout the cytoplasm. As a result, no striations are visible; hence, these cells are called "smooth."

In muscle tissue, a small amount of ECM surrounds each muscle cell and is called the **endomysium** (en'-doh-MY-see-um) or *external lamina.* The endomysium is similar to the basal lamina of epithelial tissue and helps to hold the muscle cells together in the tissue.

☐ 1. What are the two forms of muscle cells, and how do these forms differ?

Types of Muscle Tissue

◀◀ Flashback

1. What is the function of the nucleus? What main process does the DNA in the nucleus direct? (p. 70)
2. What are the functions of gap junctions and tight junctions? (p. 127)

The three kinds of muscle tissue have different structural and functional characteristics (**Figure 4.21**). The first two kinds, *skeletal muscle tissue* and *cardiac muscle tissue,* are composed of striated muscle cells. The third kind, *smooth muscle tissue,* is made up of smooth muscle cells.

Skeletal Muscle Tissue

As its name implies, **skeletal muscle tissue** is found mostly attached to the skeleton, where its contraction produces body movement. Skeletal muscle cells do not contract unless they are first stimulated by the nervous system, and the contraction is generally *voluntary,* or under conscious control. As you can see in Figure 4.21a, skeletal muscle tissue consists of long, thin, striated muscle cells that are arranged parallel to one another. Most skeletal muscle cells are quite long, extending nearly the entire length of the whole muscle. For this reason, they are often called **muscle fibers** (not to be confused with nonliving protein fibers). Skeletal muscle fibers are formed by the fusion of embryonic cells called *myoblasts.* The nucleus of each myoblast is retained in the mature skeletal muscle fiber, causing it to be *multinucleate.* The nuclei are located near the outer edge of the fiber. Having several nuclei is advantageous for skeletal muscle fibers because their size requires nearly constant synthesis of enzymes, structural proteins, and contractile proteins.

Cardiac Muscle Tissue

Cardiac muscle tissue is found only in the heart and is composed of striated muscle cells called **cardiac muscle cells.** Cardiac muscle tissue is *involuntary,* meaning that the brain does not have conscious control over its contraction. Although both skeletal and cardiac muscle tissues contain striated muscle cells, notice in Figure 4.21b that numerous differences between these two cell types can be seen. Cardiac muscle cells are relatively short and thick, as well as branched and generally uninucleate, as opposed to the long, cylindrical multinucleated skeletal muscle fibers. The nuclei are located in the centers of the cells. Notice also that dark lines separate the cardiac muscle cells, a feature lacking in skeletal muscle tissue. These lines are structures called **intercalated discs** (in-TER-kah-lay'-t'd; *inter-* = "between"); they contain gap

junctions and modified tight junctions that unite the cardiac muscle cells and permit the heart muscle to contract

CORE PRINCIPLE
Cell-Cell Communication

as a unit. These structures enable communication among the cells (Cell-Cell Communication Core Principle, p. 27).

Smooth Muscle Tissue

The final type of muscle tissue in the body is **smooth muscle tissue** (Figure 4.21c). As mentioned, it consists of smooth muscle cells, and like cardiac muscle tissue, it is involuntary. Smooth muscle tissue is found in the wall of nearly every hollow organ, as well as the walls of blood vessels, the eyes, the skin, and the ducts of certain glands. Its cells are flattened, with a single ovoid nucleus in the center of the cell. Most smooth muscle cells contain gap junctions in their plasma membranes that link them with other smooth muscle cells.

☐ 2. Compare and contrast the three types of muscle tissue.

☐ 1. When skeletal muscle fibers are damaged and die, they may be replaced with dense irregular collagenous connective tissue, commonly known as scar tissue.
   a. How do these two tissue types differ in structure?
   b. How do they differ in function?
   c. Predict the impact of such an injury to muscle tissue on the function of a whole skeletal muscle.
☐ 2. Predict what might happen to cardiac muscle tissue in which the intercalated discs were not functional.

See answers in Appendix A.

MODULE **4.5**

Nervous Tissue

Learning Outcomes

1. Describe where in the body nervous tissue can be found and its general structural and functional characteristics.
2. Identify and describe the structure and function of neurons and neuroglial cells in nervous tissue.

Nervous tissue makes up the majority of the brain, spinal cord, and nerves. Like all tissues, nervous tissue is composed of cells and their surrounding ECM. The cells are of two main types: *neurons* (NOO-ronz), the cells that send and receive messages, and *neuroglial cells* (noo-ROG-lee-al), which perform various supportive functions. The ECM of nervous tissue is quite different from that of the other tissues we have discussed. It contains few protein

(a) Skeletal muscle tissue

Skeletal muscle fibers

Endomysium (ECM)

Multiple nuclei

Striations

LM (305×)

(b) Cardiac muscle tissue

Heart

Striations

Cardiac muscle cells

Intercalated discs

Endomysium (ECM)

Nuclei

LM (705×)

(c) Smooth muscle tissue

Stomach

Smooth muscle cells

Nuclei

LM (495×)

Practice art labeling
@MasteringA&P®

Figure 4.21 Structure of muscle tissues.

LM (395×)

Practice art labeling
@MasteringA&P®

Figure 4.22 **Structure of nervous tissue.**

fibers and instead is mostly made up of ground substance with unique proteoglycans not found in other tissues of the body. In this module, we introduce the cells of nervous tissue only briefly, as this topic is fully covered in its own chapter (see Chapter 11).

The most obvious cells of nervous tissue are the **neurons,** which generate, conduct, and receive information in the form of electrical signals called **nerve impulses.** Like muscle cells, all neurons are excitable cells. Most mature neurons are amitotic (am-ih-TOT-ik), meaning that they do not undergo mitosis.

Neurons contain three main components (**Figure 4.22**):

- **Cell body.** The large, centralized portion of the neuron is called the **cell body,** or **soma.** This is the biosynthetic center of the neuron, where the nucleus and most other organelles are housed.
- **Axon.** Extending from the cell body is a single "arm" called an **axon.** An axon is responsible for moving a nerve impulse from the cell body to a target cell, which could be another

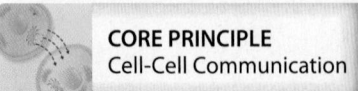
CORE PRINCIPLE
Cell-Cell Communication

neuron, a muscle cell, or a gland. Axons illustrate the Cell-Cell Communication Core Principle (p. 27).

- **Dendrites.** The other "arms" or extensions radiating from the cell body are called **dendrites,** which are typically short and highly branched. Dendrites receive messages from the axons of other neurons and bring the impulses to the cell body.

The much smaller cells surrounding neurons are supportive cells called **neuroglial cells** ("nerve glue"; see Figure 4.22). Neuroglial cells have many functions, including anchoring neurons and blood vessels in place, monitoring the composition of the extracellular fluid, speeding up the rate of nerve impulse transmission, and circulating the fluid that surrounds the brain and spinal cord. Unlike neurons, neuroglial cells can divide by mitosis.

Now, how do you put all these facts together and differentiate one tissue type from another on microscope slides? The following Concept Boost will walk you through this process.

ConceptBOOST »»»

"But It All Looks Pink!" Part 2

In Part 1 of this Concept Boost, we discussed how to orient yourself to the different components of a tissue section. Now we'll take it a step further and identify the tissue from which a section was taken. Let's try it with these examples:

Example A   Example B   Example C

This isn't as difficult as it appears, particularly if you use a methodical, step-by-step approach:

1. **Identify the cells and the ECM.** Start with the basics you were given in the first part of this Concept Boost (on p. 128) and label the cells you see in Examples A, B, and C. Now move on to the ECM. Remember, the ECM consists only of ground substance and protein fibers. The ground substance will generally stain a uniform color (or simply appear clear). The protein fibers can take on various forms in different tissues, but they will generally stain darker than the ground substance, and they will always lack nuclei. Identify the protein fibers and ground substance in Examples A, B, and C.
2. **Notice how the cells are shaped and arranged.** Are the cells packed tightly together, or are they widely spaced? Do they form a continuous sheet, as in epithelial tissue—or do they seem to be surrounded by ECM, as in connective tissue? Are the cells all identical, or are there clearly different types? Do the cells have "arms" extending from a central body? Explain how the cells are shaped and arranged in Examples A, B, and C.
3. **Notice how the ECM is arranged.** Is the ECM confined to one specific part of the tissue, or is it spaced evenly between the cells? Does ground substance predominate, or are protein fibers the main elements? What types of protein fibers can you see? Explain how the ECM is arranged in Examples A, B and C.
4. **Determine the class of tissue.** Using your analysis in the preceding steps, now you are ready to identify the class of tissue. Determine the type of tissue in Examples A, B, and C. ■

Answers: A: loose connective tissue, B: skeletal muscle tissue, C: simple columnar epithelium

The Big Picture of Tissues in Organs

FIGURE 4.23

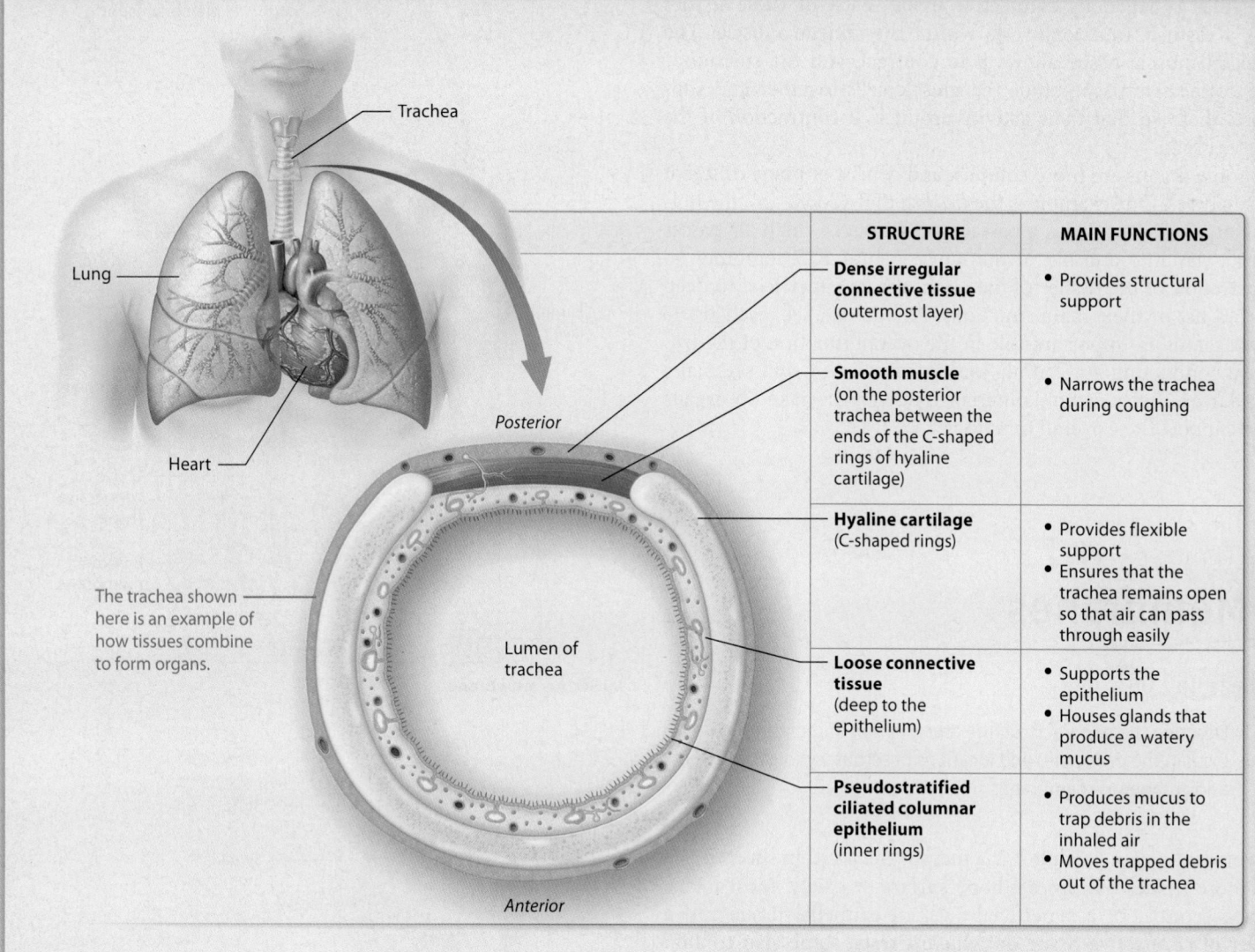

The trachea shown here is an example of how tissues combine to form organs.

Trachea

Lung

Heart

Posterior

Lumen of trachea

Anterior

STRUCTURE	MAIN FUNCTIONS
Dense irregular connective tissue (outermost layer)	• Provides structural support
Smooth muscle (on the posterior trachea between the ends of the C-shaped rings of hyaline cartilage)	• Narrows the trachea during coughing
Hyaline cartilage (C-shaped rings)	• Provides flexible support • Ensures that the trachea remains open so that air can pass through easily
Loose connective tissue (deep to the epithelium)	• Supports the epithelium • Houses glands that produce a watery mucus
Pseudostratified ciliated columnar epithelium (inner rings)	• Produces mucus to trap debris in the inhaled air • Moves trapped debris out of the trachea

Quick Check

☐ 1. What are the two main cell types in nervous tissue, and what are their roles?

☐ 2. What are the three parts of a neuron, and what are their functions?

Apply What You Learned

☐ 1. A tumor consists of cells that divide by mitosis outside the normal controls of the cell cycle. Would you expect tumors of nervous tissue to involve neurons, neuroglial cells, or both? Explain.

☐ 2. A hypothetical poison destroys the axon of a neuron. How will this affect the ability of the neuron to function?

See answers in Appendix A.

MODULE **4.6**

Putting It All Together: The Big Picture of Tissues in Organs

Learning Outcomes

1. Describe how tissues work together to form organs.

Now that we have examined each tissue in isolation, let's look at how they function together in the body. As you discovered in the introductory chapter, two or more tissues that combine structurally and functionally form an organ (see Chapter 1; **Figure 4.23**). To start with a simple example, a skeletal muscle is

151

composed of two main tissues—skeletal muscle tissue and dense irregular collagenous connective tissue. Each of these tissues has a distinct functional role within the skeletal muscle: The skeletal muscle tissue allows it to contract, and the surrounding connective tissue binds the muscle cells together and supports them so that their activity produces a contraction of the whole organ.

Some organs are more complex and consist of many different tissue types. One example is the *trachea* (TRAY-kee-ah), the hollow organ that provides a passageway through which air passes on its way into and out of the lungs. Figure 4.23 combines an illustration of the tissues of the trachea from superficial to deep with a list of their main functions. As you can see, each tissue layer serves an important role in the overall function of the trachea: conducting air. Throughout this book, you will see many similar examples of how different tissue types make up organs and support their overall functions.

MODULE 4.7

Membranes

Learning Outcomes

1. Describe the general structure and function of membranes.
2. Explain the properties and locations of serous, synovial, mucous, and cutaneous membranes.

If you recall from Module 4.2, a **membrane** is a thin sheet of one or more tissues that lines a body surface or cavity. Most membranes consist of a superficial layer of epithelial tissue and a layer of connective tissue on which it rests; some also contain a thin layer of smooth muscle. Membranes anchor organs in place, serve as barriers, function in immunity, and secrete various substances. The membranes that fit these structural and functional definitions are considered true membranes. However, other "membrane-like structures" don't necessarily fit these definitions, yet are sometimes called membranes because they perform many of the same functions.

Four general types of membranes are present in the human body: *serous, synovial, mucous,* and *cutaneous*. Of the four, the former two are true membranes and the latter two are membrane-like structures. Each membrane differs in structure, function, and products secreted.

True Membranes

◀◀ **Flashback**

1. Where are the pleural, pericardial, and peritoneal body cavities located? (p. 17)

(a) **Serous membrane**

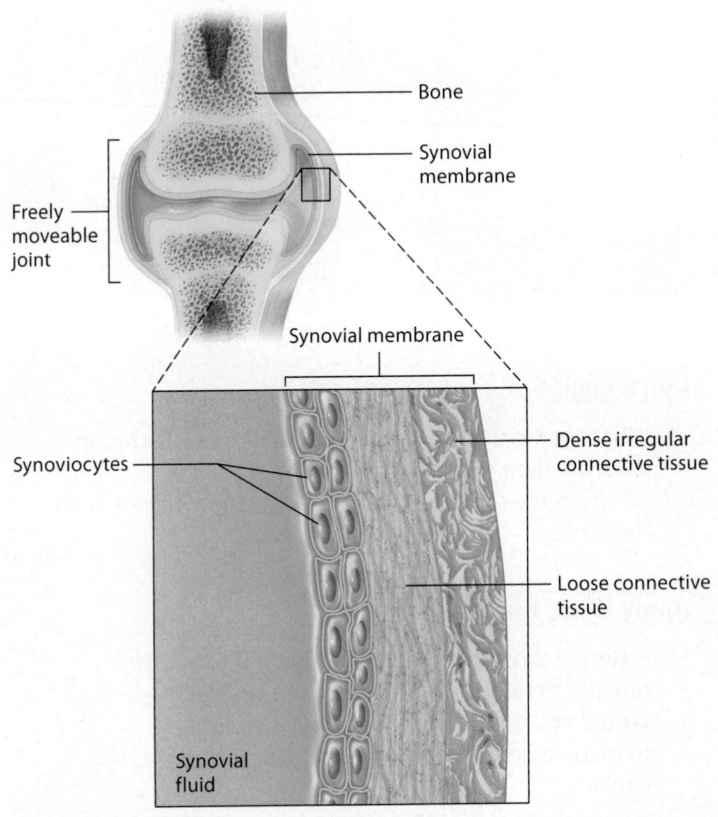

(b) **Synovial membrane**

Figure 4.24 True membranes.

Both serous and synovial membranes are true membranes that do not open to the outside of the body. We take a closer look at the structure and function of these membranes in this section.

Serous Membranes

Serous membranes (SEE-rus), also known as **serosae** (seh-ROH-see), line the pleural, pericardial, and peritoneal body cavities (**Figure 4.24a** shows an example in the pleural cavity). They consist of a layer of simple squamous epithelium called **mesothelium** (mez′-oh-THEE-lee-um), its basement membrane, and a layer of loose connective tissue. Notice in Figure 4.24a that a serous membrane is a continuous sheet that folds over on itself so that its outer part, called the *parietal layer* (pah-RY-eh-tal), is in contact with the body wall, and its inner part, called the *visceral layer,* covers the organs within a cavity. The mesothelial cells produce a thin, watery fluid called **serous fluid,** which fills the very narrow space between the parietal and visceral layers and provides lubrication so that organs can move without friction. When these layers are inflamed, however, this may not provide enough protection; see *A&P in the Real World: Friction Rubs* on page 154 for more information.

Synovial Membranes

Synovial membranes (sih-NOH-vee-al) line the cavities surrounding freely moveable joints, including the hip, knee, elbow, and shoulder (**Figure 4.24b**). Unlike mucous and serous membranes, synovial membranes do not contain a layer of epithelial cells. Instead, they are made up of two connective tissue layers:

The inner layer consists of modified fibroblasts called *synoviocytes* (sih-NOH-vee-oh′-sytz), and the external layer is generally a mixture of loose and dense irregular connective tissue. Synoviocytes secrete a fluid called *synovial fluid,* which is a watery, slippery fluid that lubricates the joint. Synovial membranes are discussed in greater detail with joints, or articulations (see Chapter 8).

Quick Check

□ 1. Where are serous and synovial membranes located, and what are their functions?

Membrane-like Structures

The body's other two types of membranes, mucous and cutaneous, are actually membrane-like structures. The cutaneous membrane is on the external surface of the body, and mucous membranes line internal body surfaces but open to the outside of the body.

Mucous Membranes

Mucous membranes, also called **mucosae** (myoo-KOH-see), line all body passages as part of the walls of hollow organs that open to the outside of the body, including the respiratory passages, the mouth, the nasal cavity, the digestive tract, and the male and female reproductive tracts (**Figure 4.25a**). Mucosae consist of a layer of epithelium, its basement membrane, a layer of loose connective tissue called the **lamina propria,** and occasionally a thin layer of smooth muscle. Not surprisingly,

Trachea
Bronchus
Goblet cell
Epithelium
Mucus
Basement membrane
Loose connective tissue

(a) Mucous membrane

Epidermis (keratinized stratified squamous epithelium)
Dermis
Dense irregular connective tissue
Loose connective tissue

(b) Cutaneous membrane

Figure 4.25 Membrane-like structures.

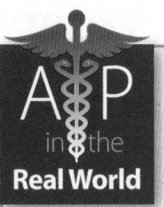

Friction Rubs

Certain viral and bacterial infections can cause inflammation of the serous membranes of the pleural and pericardial cavities. When these serosae are inflamed, the thin layer of serous fluid produced by the mesothelium is inadequate to reduce friction as the organs move within their cavities. This causes the parietal and visceral layers to rub together as the organs expand and contract, which produces a grating sound called a *friction rub* that can be heard with a stethoscope. Friction rubs are typically quite painful and cause chest pain that worsens with inhalation, body movement, and swallowing. They usually resolve with treatment of the underlying condition.

mucosae contain glands with goblet cells that secrete mucus. As we discussed earlier, mucus serves many functions, most of which are protective in nature. For example, in the respiratory tract, mucus traps inhaled debris so that the ciliated epithelial cells can sweep it out. In the urinary bladder and the stomach, mucus provides a protective layer between the acidic contents of these organs and their underlying epithelia.

Cutaneous Membrane

The term **cutaneous membrane** (kyoo-TAYN-ee-us) refers to the **skin,** which is the largest organ of the body (**Figure 4.25b**). Skin consists of an outer layer of keratinized stratified squamous epithelium called the *epidermis.* The keratinized cells of the epidermis protect the underlying tissues by providing a hard and continuous surface. Deep to the epidermis is a layer of loose connective tissue and an even deeper layer of dense irregular connective tissue; together they are known as the *dermis.* The dermis houses numerous blood vessels from which oxygen and nutrients diffuse up to the avascular epidermis. The skin (or *integument*) is covered fully in its own chapter (see Chapter 5).

Quick Check

☐ 2. Where are mucous membranes located, and what are their functions?

☐ 3. What is the cutaneous membrane? What is its function?

Apply What You Learned

☐ 1. Predict what would happen if the organs of the respiratory tract and the stomach were lined with serous membranes instead of mucous membranes.

☐ 2. Predict the effect of damaging the synovial membrane in a joint so that it could no longer secrete synovial fluid.

See answers in Appendix A.

MODULE 4.8

Tissue Repair

Learning Outcomes

1. Describe how injuries affect epithelial, connective, muscular, and nervous tissues.
2. Describe the process of regeneration.
3. Explain the process of fibrosis.

When cells are damaged and die as a result of normal wear and tear or from injury, the body must remove the dead cells and fill in the remaining gap in order to maintain homeostasis. This process of wound healing is called **tissue repair,** and it occurs differently in different tissues. Certain tissues are capable of **regeneration,** during which the damaged or dead cells are replaced with cells of the same type (**Figure 4.26a**). When regeneration is complete, the function of the tissue is in general completely restored. Other tissues are incapable of full regeneration, and a different process called **fibrosis** (fy-BROH-sis) ensues (**Figure 4.26b**). During fibrosis, fibroblasts divide by mitosis and produce collagen to fill in the defect left by the injury, and the tissue does not regain its ability to function normally. The end result of fibrosis is the formation of **scar tissue,** which is a type of dense irregular connective tissue. Let's look at the factors that affect whether a tissue goes through regeneration or fibrosis.

Capacity of Specific Tissues for Tissue Repair

The extent to which regeneration or fibrosis takes place in a particular tissue type is largely determined by the degree to which the cells in that tissue can undergo mitosis.

- **Epithelial tissues typically undergo regeneration.** Epithelial tissues are often subjected to a great degree of stress and injury; however, they are typically capable of full regeneration. Certain epithelial tissues such as those of the skin and digestive tract contain immature cells called **stem cells** that divide to continually replace dead, injured, or worn-out epithelial cells. In other epithelial tissues, such as those of the liver and blood vessels, mature cells divide to replace those that have been lost.

- **Most connective tissues heal by regeneration.** Connective tissues such as connective tissue proper, bone, and blood regenerate easily through the division of their immature cells. An exception to this is cartilage tissue, as these cells have a limited capacity to divide and thus often heal by fibrosis.

- **Smooth muscle tissue usually regenerates; cardiac and skeletal muscle tissues generally heal by fibrosis.** The cells of smooth muscle tissue largely retain their ability to undergo mitosis and thus heal by regeneration. However, mature skeletal muscle fibers and cardiac muscle cells

(a) Regeneration in stratified squamous epithelium

Injured area with damaged or dead epithelial cells

Stratified squamous epithelium

Regeneration

Cells replaced with new epithelial cells

Injured area with damaged or dead cardiac muscle cells

Cardiac muscle cells

Fibrosis

Cells replaced with collagen fibers and fibroblasts

Collagen fibers

Fibroblast

(b) Fibrosis in cardiac muscle tissue

Figure 4.26 Tissue repair by regeneration or fibrosis.

cannot undergo mitosis due to their large size and complicated cellular architecture. In skeletal muscle tissue, a population of cells called *satellite cells* can divide and become skeletal muscle fibers, allowing a limited degree of regeneration. However, cardiac muscle tissue lacks satellite cells and heals only by fibrosis.

- **Nervous tissue generally undergoes fibrosis.** Although neuroglial cells retain the ability to divide, neurons are generally unable to undergo mitosis. In general, damaged neurons in the brain and spinal cord are replaced by neuroglial cells that divide to produce a scar. However, if the cell body of a neuron is intact and only the axon is damaged, there is some chance that the axon will regenerate, depending on the location and nature of the damage. This generally applies only to axons outside the brain and spinal cord.

Quick Check

☐ 1. How do regeneration and fibrosis differ?
☐ 2. Which tissues generally heal by regeneration? Which generally heal by fibrosis?

Other Factors Affecting Tissue Repair

In addition to the ability of cells in a tissue to undergo mitosis, two other factors—nutrition and blood supply—are also important determinants of tissue repair. Because tissue repair involves production of proteins such as collagen, protein (amino acid) intake must be adequate for repair to occur. Another dietary requirement is sufficient intake of vitamin C, which is needed by the fibroblasts to produce collagen. In addition to dietary factors, the blood supply to the damaged area must be taken into consideration, as blood brings in oxygen, nutrients, and cells of the immune system that are needed for tissue repair. This is why people suffering from diseases of the arteries often have nonhealing wounds, even in tissues capable of regeneration. Details of the process of tissue repair are covered in the immune system chapter (see Chapter 20).

Quick Check

☐ 3. Which factors influence the ability of a tissue to heal?

Apply What You Learned

☐ 1. During a heart attack, the blood flow to cardiac muscle tissue is blocked, and cardiac muscle cells in the affected area generally die. Why will cardiac function rarely return to normal after this takes place?
☐ 2. Why do you think that homeless patients and alcoholics often have problems with wound healing?

See answers in Appendix A.

CHAPTER SUMMARY

Take full advantage of the invaluable online practice and study tools for this chapter by going to MasteringA&P®. There you will find:

- Practice tests, module quizzes, art-labeling and sequence-the-steps activities
- Animated figures with quizzes
- *Practicing A&P* worksheets
- Author-narrated podcast
- Dynamic Study Modules
- *Interactive Physiology*

MODULE 4.1
Introduction to Tissues 124

- **Histology** is the study of the normal structure of tissues.
- Every **tissue** is composed of cells and the surrounding **extracellular matrix (ECM).**
- There are four classes of tissues:
  - **Epithelial tissue** is composed of sheets of cells that cover or line body surfaces and form glands with little ECM.
  - **Connective tissues** have ECM as their prominent feature. They bind, support, protect, and transport substances through the body.
  - **Muscle tissues** are composed of contractile cells with little ECM.
  - **Nervous tissue** contains cells that send and receive messages as well as supporting cells and ECM.
- The ECM provides a tissue with strength, regulates cell activity, and anchors cells in place. It is composed of **ground substance** and **protein fibers.** (Figure 4.1, p. 125)
- Ground substance contains three main components: **glycosaminoglycans, proteoglycans,** and **glycoproteins.**
- Protein fibers are embedded within ground substance. There are three types of protein fibers:
  - **Collagen fibers** are very resistant to tension and pressure.
  - **Elastic fibers** are both elastic and extensible.
  - **Reticular fibers** are a very thin type of collagen fiber that form interlacing meshworks in certain tissues.
- Cells in a tissue are joined by **cell junctions.** The three types of junctions are as follows: (Figure 4.2, p. 126)
  - **Tight junctions** make the space between adjacent cells impermeable to macromolecules.
  - **Desmosomes** anchor neighboring cells together to increase the strength of a tissue with respect to mechanical stress.
  - **Gap junctions** are small pores in the plasma membranes of two adjacent cells that allow small substances to pass from the cytosol of one cell to that of another.

MODULE 4.2
Epithelial Tissues 127

- Epithelial tissues act as barriers between the body and the external environment and between our organs and fluid-filled cavities. These tissues function in protection, immune defenses, secretion, transport, and sensation.
- The cells in epithelial tissues are packed closely together and joined by tight junctions and desmosomes. There are no blood vessels in epithelia, and oxygen and nutrients must diffuse to the epithelial cells from the tissues deep to them, which limits their thickness.
- The ECM of epithelial cell or tissue is mostly confined to the **basal lamina** of the **basement membrane,** which anchors the epithelial tissue to the underlying tissues.
- The "free edge" of an epithelial cell or tissue is called the **apical surface,** and the edge attached to the basal lamina is the **basal surface.** (Figure 4.3, p. 129)
- **Simple epithelia** have only one cell layer; **stratified epithelia** have two or more cell layers. Epithelial cells may be **squamous** (flat), **cuboidal,** or **columnar** in shape. (Figure 4.4, p. 129)
- The types of simple epithelia are as follows: (Figure 4.5, p. 130)
  - **Simple squamous epithelium,** which is very thin and allows substances to cross it rapidly. It is found in the air sacs of the lungs, the lining of blood vessels, and the lining of parts of the kidney tubules.
  - **Simple cuboidal epithelium** is also thin and allows substances to cross it rapidly. It is found in certain kidney tubules, respiratory passages, and many glands.
  - **Simple columnar epithelium** also permits fairly rapid movement of substances across it. It is found in much of the digestive tract, the gallbladder, the uterine tube, and certain respiratory passages.
  - **Pseudostratified columnar epithelium** appears to be stratified but consists of only a single layer of cells. It is often ciliated, and is found in the nasal cavity and the lining of certain respiratory passages.
- Transport across simple epithelia may occur via *paracellular transport* or via *transcellular transport* (Figure 4.6, p. 132)
- Stratified epithelia provide a protective barrier. Stratified epithelia include the following: (Figure 4.7, pp. 132–133)
  - **Stratified squamous epithelium,** which may or may not contain a keratinized layer. Keratinized stratified squamous epithelium is found in the skin, and its apical cell layers are dead and filled with the protein **keratin.** Nonkeratinized stratified squamous epithelium lines the esophagus, mouth, throat, rectum, and vagina.
  - **Stratified cuboidal epithelium** lines the ducts of sweat glands, and **stratified columnar epithelium** lines the ducts of certain glands, parts of the male urethra, and the cornea (clear outer covering of the eye).
  - **Transitional epithelium** (urothelium) is found only in the urinary system. Its cells can change shape from dome-shaped to flattened when the tissue stretches.
- Figure 4.8 (p. 134) summarizes the epithelial tissues.

 Practice art-labeling exercise: Figure 4.8.

- Two types of glands are found in the body:
  - **Exocrine glands** release their product through a duct to an epithelial surface. Most of their products have local actions.
  - **Endocrine glands** release their product into the bloodstream, and their products can have actions on distant target cells in the body.
- Exocrine glands may be unicellular or multicellular. The most common unicellular gland is the mucus-secreting **goblet cell.** Multicellular glands are of varying shape and complexity. (Figure 4.9, p. 135; Figure 4.10, p. 136)
- Exocrine glands release their products in one of two ways:
  - In **merocrine secretion,** cells package their products into secretory vesicles and release them by exocytosis.
  - In **holocrine secretion,** cells accumulate their product in their cytosol and release the product once the cell ruptures and dies. (Figure 4.11, p. 137)

MODULE 4.3
Connective Tissues 137

- Functions of connective tissues include connecting and binding, support, protection, and transport.
- There are two basic types of connective tissue: **connective tissue proper** and **specialized connective tissue.**
- Cells of connective tissue proper may be resident cells or migrant cells. These cells include the following:
  - **fibroblasts,** which make ground substance and protein fibers of the ECM;
  - **adipocytes,** which contain a large lipid droplet in their cytosol;
  - **mast cells,** which secrete chemicals that cause inflammation; and
  - **phagocytes** and other immune cells. (Figure 4.12, p. 138)
- There are four types of connective tissue proper:
  - **Loose connective tissue** is also called **areolar connective tissue.** It is found deep to the skin and as part of the lining of body cavities and hollow organs. (Figure 4.13, p. 139)
  - **Dense connective tissue** features an ECM composed primarily of protein fibers. **Dense irregular connective tissue** contains bundles of collagen fibers arranged at various angles to one another, and is found in the skin, around organs, and surrounding joints. **Dense regular collagenous connective tissue** contains parallel bundles of collagen fibers, and is present in tendons and ligaments. **Dense regular elastic connective tissue** contains parallel bundles of elastic fibers, and is found in large blood vessels and certain ligaments. (Figure 4.14, p. 140)
  - **Reticular tissue** contains numerous reticular fibers in its ECM that form weblike networks in the spleen and lymph nodes. (Figure 4.15, p. 141)
  - **Adipose tissue** is filled with adipocytes. It insulates, warms, protects, and serves as the major energy reserve for the body. (Figure 4.16, p. 141)
- The specialized connective tissues are cartilage, bone, and blood.
  - **Cartilage** is a tough but flexible tissue that absorbs shock and is resistant to tension and compression. The primary cells in cartilage are **chondroblasts,** which make the ECM, and **chondrocytes,** their mature form. There are three types of cartilage: (Figure 4.17, p. 143)

- **hyaline cartilage,** which is smooth and found on the ends of bones in joints, in the nose, in certain respiratory passages, and where the ribs meet the sternum;
- **fibrocartilage,** which contains many bundles of collagen fibers in its ECM, making it very tough, and is found in certain joints and the intervertebral discs; and
- **elastic cartilage,** which contains elastic fibers in its ECM and is found in the ear and in parts of the larynx (voice box).
  - **Bone,** or **osseous tissue,** contains collagen fibers, ground substance, and calcium phosphate crystals. Three cell types are present in bone: **osteoblasts,** which make the organic component of the ECM; **osteocytes,** which are mature osteoblasts surrounded by ECM; and **osteoclasts,** which break down bone. (Figure 4.18, p. 144)
  - **Blood** is a fluid connective tissue. Its ECM is called **plasma** and consists of water, dissolved solutes, and proteins. The main cell types found in blood are **erythrocytes** and **leukocytes.** (Figure 4.19, p. 145)
- Figure 4.20 (pp. 146–147) summarizes the connective tissues.

   Practice art-labeling exercise: Figure 4.20.

MODULE 4.4
Muscle Tissues 146

- The cells of muscle tissue are specialized for contraction by turning the chemical energy of ATP into the mechanical energy of movement.
- **Muscle cells** are **excitable cells** that respond to electrical or chemical stimulation.
- The cytoplasm of muscle cells contains bundles of **myofilaments.** Based on the arrangement of myofilaments, muscle cells are either **striated** or **smooth.**
- There are three types of muscle tissue: (Figure 4.21, p. 149)
  - **Skeletal muscle tissue** is attached to the skeleton. It is voluntary and consists of long, striated muscle cells called **muscle fibers,** which are multinucleated.
  - **Cardiac muscle tissue** is involuntary and found in the heart. Its uninucleate cells are striated, short, wide, and branched, and are united by **intercalated discs.**
  - **Smooth muscle tissue** consists of uninucleate smooth muscle cells and lines most of the hollow organs, the ducts of certain glands, the eye, and the skin.

   Practice art-labeling exercise: Figure 4.21.

MODULE 4.5
Nervous Tissue 148

- Nervous tissue makes up the majority of the brain, spinal cord, and nerves.
- **Neurons** are excitable cells that send and receive messages in the form of **nerve impulses** to other neurons, muscles, and/or glands. They contain a central **cell body,** a single **axon** that carries nerve impulses away from the cell body, and one or more **dendrites** that bring impulses from other neurons into the cell body. (Figure 4.22, p. 150)

   Practice art-labeling exercise: Figure 4.22.

- **Neuroglial cells** are the supporting cells of nervous tissue. They are much smaller than neurons, and serve a variety of functions.

MODULE 4.6
Putting It All Together: The Big Picture of Tissues in Organs 151

- Organs are made up of several different types of tissues that function together. (Figure 4.23, p. 151)

MODULE 4.7
Membranes 152

- **Membranes** are thin sheets of one or more tissues that anchor organs in place, serve as barriers, function in immunity, and secrete substances. There are also membrane-like structures that have similar functions.

- **Serous membranes** and **synovial membranes** are true membranes that do not open to the outside of the body. **Mucous membranes** and the **cutaneous membrane** are membrane-like structures that open to the outside of the body or are located on the outer surface of the body.

- Serous membranes line the body wall and the organs of the pleural, pericardial, and peritoneal body cavities. They secrete **serous fluid,** which lubricates the organs in the cavity. (Figure 4.24, p. 152)

- Synovial membranes line the cavities surrounding freely moveable joints. They produce *synovial fluid,* which lubricates the joint.

- Mucous membranes line body passages and hollow organs that open to the outside of the body. They secrete **mucus,** which traps debris and serves protective functions. (Figure 4.25, p. 153)

- The cutaneous membrane is the **skin,** and it consists of a superficial layer of keratinized stratified squamous epithelium called the *epidermis* and deeper layers of loose and dense irregular connective tissue called the *dermis.*

MODULE 4.8
Tissue Repair 154

- During **regeneration,** damaged cells are replaced with cells of the same type and the tissue is able to fulfill its normal function. During **fibrosis,** fibroblasts fill in the defect with dense irregular connective tissue, forming **scar tissue.** The tissue is unable to resume its normal function after fibrosis has taken place.

- Epithelial tissues typically undergo regeneration.

- Most connective tissues heal by regeneration; however, cartilage often heals by fibrosis.

- Smooth muscle tissue is capable of regeneration. Skeletal and cardiac muscle cells heal primarily by fibrosis.

- Nervous tissue heals primarily by fibrosis, although certain axons can regenerate if the cell body of the neuron is intact.

- Nutrition and blood supply to the damaged tissue are important factors in the ability of a tissue to heal. (Figure 4.26, p. 155)

ASSESS WHAT YOU LEARNED

See answers in Appendix A.

LEVEL 1 Check Your Recall

1. Which of the following best describes the position of a tissue in the levels of organization of the human body?

   a. Tissues are the most fundamental level of organization.
   b. Tissues are between cells and organs in the levels of organization.
   c. Tissues are the most complex level of organization.
   d. Tissues are between organs and systems in the levels of organization.

2. State whether each of the following describes epithelial, connective, muscle, or nervous tissue.

   a. _____ ECM is often the primary element.
   b. _____ Consists of excitable cells that are specialized for contraction.
   c. _____ Sheets of tightly packed cells with little ECM.
   d. _____ Makes up the majority of the brain and spinal cord.
   e. _____ Cells may be smooth or striated.
   f. _____ Binds, connects, supports, and transports substances throughout the body.

3. Describe the roles of each of the following components of the ECM:

   a. Collagen fibers
   b. Glycosaminoglycans
   c. Reticular fibers
   d. Proteoglycans
   e. Glycoproteins (cell-adhesion molecules)
   f. Elastic fibers

4. Fill in the blanks: Tight junctions make the spaces between cells _____, whereas desmosomes increase the _____ of a tissue.

5. Which of the following statements regarding epithelial tissue are true? If a statement is false, correct it to make a true statement.

   a. Epithelial tissues are classified by cell shape and the number of cell layers.
   b. Epithelial tissues function in protection, immune defenses, secretion, transport, and sensation.
   c. Epithelial tissue is highly vascular.
   d. A goblet cell is a unicellular exocrine gland that secretes mucus.
   e. Pseudostratified epithelium appears to be simple epithelium but is actually stratified.
   f. Stratified epithelia are specialized to allow substances to cross their cells rapidly.

6. Match each type of epithelium with its correct location in the body.

   _____ Simple squamous   a. Skin
   _____ Pseudostratified columnar b. Urinary bladder
   _____ Keratinized stratified   c. Air sacs of the lungs,
         squamous   blood vessels

_____ Simple columnar
_____ Transitional
_____ Simple cuboidal

d. Kidney tubules, thyroid gland
e. Respiratory passages, nasal cavity
f. Digestive tract

7. Compare and contrast the following pairs of terms:

a. Endocrine gland and exocrine gland
b. Unicellular gland and multicellular gland
c. Holocrine secretion and merocrine secretion

8. Explain how connective tissues differ from epithelial tissues in structure and function.

9. Mark the following statements regarding connective tissue as true or false. If a statement is false, correct it to make a true statement.

a. Fibroblasts store lipids in a large inclusion in their cytoplasm.
b. Loose connective tissue features protein fibers as its primary component.
c. Adipose tissue functions in insulation, warmth, protection, and shock absorption, and is the major energy reserve in the body.
d. Fibrocartilage provides a smooth surface on which bones may articulate with little friction.
e. The ECM of bone tissue consists exclusively of calcium phosphate crystals.

10. Match the following types of connective tissue with their correct location(s).

_____ Hyaline cartilage
_____ Elastic cartilage
_____ Smooth muscle
_____ Dense irregular connective tissue
_____ Fibrocartilage
_____ Cardiac muscle
_____ Dense regular elastic tissue
_____ Reticular connective tissue

a. Dermis, surrounding joints, organs
b. Most hollow organs, eye, skin
c. Large arteries, certain ligaments
d. Ear and epiglottis
e. Lymph nodes, spleen, liver
f. Freely moveable joints
g. Intervertebral discs, knee joint
h. Heart

11. Which of the following statements about muscle tissue is *false*?

a. Smooth muscle tissue is involuntary and the cells are nonstriated.
b. Cardiac muscle tissue is involuntary and the cells are striated.
c. Skeletal muscle fibers are joined by intercalated discs.
d. Skeletal muscle fibers are long and cylindrical striated cells.

12. Fill in the blanks: _____ are the cells of nervous tissue that send and receive messages, and _____ are the supporting cells of nervous tissue. A(n) _____ carries a nerve impulse toward a neuron cell body, and a(n) _____ carries a nerve impulse away from a neuron cell body.

13. Each of the following statements is false. Correct each to make a true statement.

a. Mucous membranes are composed of mesothelium and the underlying loose connective tissue.
b. Mesothelial cells secrete mucus.

c. Synoviocytes are the secretory cells of mucosae.
d. The cutaneous membrane is composed of simple squamous epithelium, loose connective tissue, and dense irregular connective tissue.
e. Serous membranes line all hollow organs that open to the outside of the body.

14. Which tissues undergo regeneration? Which tissues undergo fibrosis? Why?

LEVEL 2 Check Your Understanding

1. If you were to cut through epithelial tissue without penetrating the basement membrane, would you expect bleeding to occur? Why or why not?

2. Vitamin C is required for synthesis of collagen. Predict the effect that a vitamin C deficiency (a disease called *scurvy*) would have on bone tissue, dense regular collagenous connective tissue, dense irregular connective tissue, and cartilage. What symptoms would you expect to see from this disease?

3. Predict what would happen if the mucous membranes of the body stopped secreting mucus or if they secreted excess mucus.

LEVEL 3 Apply Your Knowledge

PART A: Application and Analysis

1. The disease pemphigus vulgaris involves a patient's own immune system attacking the desmosomes between the epithelial cells of the skin. What changes would you expect to see with this disease?

2. In the disease scleroderma, excessive collagen, glycosaminoglycans, and glycoproteins are produced and deposited in tissues throughout the body. Predict the effects of this disease.

3. In the disease pulmonary fibrosis, elastic fibers of the lung are destroyed and replaced with collagen fibers. Predict the effect this would have on breathing.

4. Imagine that a disease turns the simple epithelia of the lungs, kidney tubules, and intestines into keratinized stratified squamous epithelia. What effect would this change in form have on the functions of these tissues?

PART B: Make the Connection

5. Epithelial cells of the kidneys have pumps that drive the transcellular transport of sodium ions.

a. Is this epithelium likely to be simple or stratified? Why?
b. The movement of sodium ions drives the transcellular transport of water. Explain why water follows sodium. (*Connects to Chapter 3*)

6. Explain why it would be difficult for a mature multinucleate cell such as a skeletal muscle fiber to divide by mitosis. (*Connects to Chapter 3*)

7. Predict which organelles are likely to be abundant in cells such as fibroblasts that actively produce and secrete proteins. (*Connects to Chapter 3*)

5

The Integumentary System

Perhaps more than any other system, the **integumentary system** (in-teg′-you-MEN-tuh-ree; *integu-* = "covering")—consisting of the **skin** and its accessory structures such as hair, glands, and nails, collectively called the **integument**—is associated with how we identify and express ourselves and how others perceive us. This is evidenced by our spending habits. How much money have you spent in the past year on shampoo, soap, hair and nail care, deodorant, lotions, and similar items?

In the previous chapters, we examined the three simplest levels of organization— the chemical, cellular, and tissue levels. Now we change our focus to the organ and system levels of organization. We first take an introductory tour of the integumentary system, then examine the anatomy of the skin and its accessory structures, as well as the functions of the integumentary system in maintaining homeostasis. We conclude with a look at common conditions and diseases that affect the skin.

MODULE 5.1

Overview of the Integumentary System

Learning Outcomes

1. Describe the basic structure of the skin.
2. Describe the basic functions carried out by the components of the integumentary system.

The skin doesn't get much respect as an organ, but it's actually the largest organ in the body and makes up approximately 10–15% of our total body weight. Indeed, this remarkable organ has a complex structure and performs numerous functions that are vital to homeostasis.

Skin Structure

◀◀ Flashback

1. What is the cutaneous membrane? Which tissues make up the cutaneous membrane? (p. 154)
2. What is adipose tissue? (p. 139)

Photo: Tiny sweat pores open and leave a thin film called a fingerprint when we touch most surfaces.

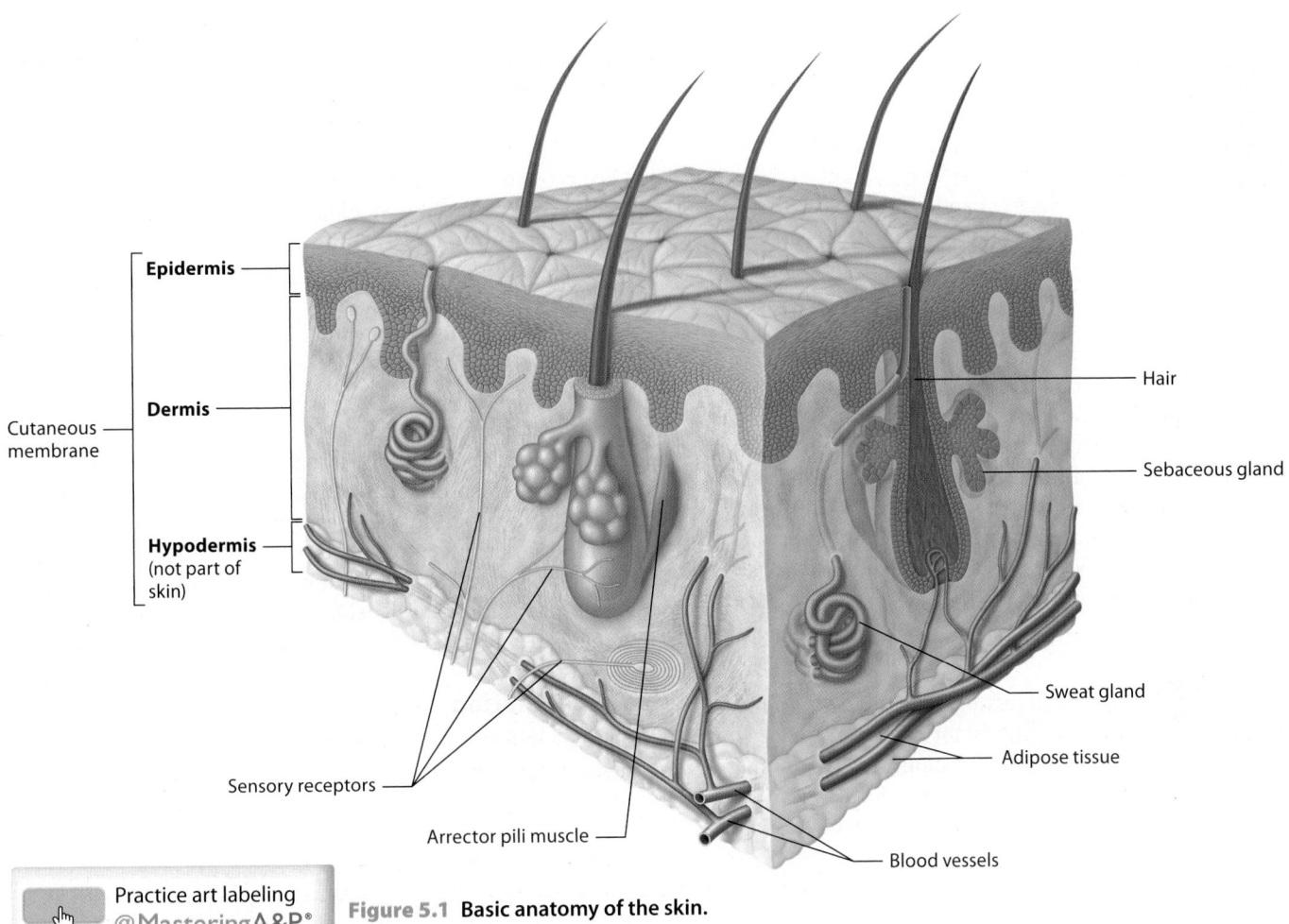

Epidermis

Dermis

Cutaneous
membrane

Hypodermis
(not part of
skin)

Hair

Sebaceous gland

Sweat gland

Adipose tissue

Sensory receptors

Arrector pili muscle

Blood vessels

Practice art labeling
@MasteringA&P®

Figure 5.1 Basic anatomy of the skin.

The skin, also known as the **cutaneous membrane,** has two main components (**Figure 5.1**):

- The superficial **epidermis** (eh-puh-DER-mis; *epi-* = on top; *derm-* = "skin") consists of keratinized stratified squamous epithelium that rests on top of a basement membrane.
- The deep **dermis** consists of loose connective tissue and dense irregular connective tissue.

Embedded within the epidermis and dermis we find the accessory structures of the skin: sweat-producing *sweat glands,* oil-producing *sebaceous glands, hair,* and *nails* (note that Figure 5.1 does not show nails). The skin also contains *sensory receptors* and small bands of smooth muscle called *arrector pili muscles* (ah-REK-tor pee-LEE) that attach to hairs in the dermis.

Like all epithelia, the epidermis is avascular, or lacks a blood supply. For this reason, oxygen and nutrients follow their concentration gradients and diffuse from blood vessels located in the dermis below. This is an example of the Gradients Core Principle (p. 26). The need for diffusion limits the thickness of the epidermis in which we can find living cells. Indeed, nearly 50% of the cells of the epidermis are too far away from the blood supply in the dermis to sustain life, so the superficial epidermis consists entirely of dead cells.

CORE PRINCIPLE
Gradients

Deep to the dermis we find the **hypodermis** (hy-poh-DER-mis; *hypo-* = "below"), which is also known as the **superficial fascia** (FASH-uh) or **subcutaneous tissue** (sub'-kyu-TAY-nee-us). The hypodermis is not actually part of the skin, but it anchors the skin to the muscle and bone deep to it. The hypodermis consists of loose connective tissue that is richly supplied with blood vessels and varying amounts of adipose tissue. The number of blood vessels makes the hypodermis a good place to administer medicines via *subcutaneous injections* using a hypodermic needle. The hypodermis can also affect the appearance of the skin, as discussed in *A&P in the Real World: Cellulite* on page 162.

Quick Check

☐ 1. What are the major structures of the skin, and what are their main properties?

Functions of the Integumentary System

◄◄ **Flashback**

1. How do polar covalent, nonpolar covalent, and ionic compounds differ? (p. 40)

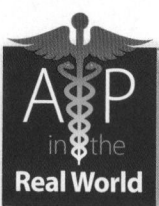

Cellulite

Cellulite is the term used to describe the dimpled or "orange peel" appearance of the skin when collagen bands form around adipose tissue in the hypodermis. Cellulite tends to develop in the thighs, hips, and gluteal area, and is influenced by many factors, including genetics, gender, the amount and distribution of adipose tissue, and age.

Once thought to be a disorder, cellulite is now believed to be the normal condition of the skin in most women and many men. Despite this fact, many individuals spend vast amounts of money to minimize its appearance. Various products claim to "cure" cellulite, but unfortunately, there is little evidence that they work. Indeed, the only proven way to minimize the appearance of cellulite is a healthy diet and regular exercise. Note, however, that even diet and exercise generally do not erase cellulite altogether.

It's easy to take our skin and its accessory structures for granted. They certainly do not *seem* as important as the brain or the heart. But in reality, the integumentary system performs numerous functions that are critical to maintaining homeostasis. They include protection, sensation, thermoregulation, excretion, and synthesis of vitamin D.

Protection

One of the most obvious functions of the integument is to protect underlying tissues from damage. Tissue damage comes from three main sources: mechanical (physical) trauma, such as stretching, pressure, or abrasion; disease-causing microorganisms known as *pathogens* (*patho-* = "disease"; *-gen* = "causing"); and environmental hazards, such as chemicals and radiation. The integument provides protection from all three sources in the following ways:

- **Protection from mechanical trauma.** Recall from the tissues chapter (Chapter 4) that the superficial aspect of the skin is composed of stratified squamous, keratinized epithelium. This structure provides a continuous, durable, yet flexible surface that protects the body from mechanical trauma.
- **Protection from pathogens.** The continuous structure of the skin also protects against invasion by pathogens. This function is evident any time the skin is compromised. The risk of infection rises dramatically when the skin is cut, which is why you often cover the laceration with a bandage. In addition, the skin contains cells of the immune system that destroy pathogens before they can invade deeper tissues. The glands of the skin also protect the body by secreting a variety of antimicrobial substances. For example, the secretions of sebaceous glands give the surface of the skin a slightly acidic pH, which deters the growth of many pathogens. This slightly acidic pH is known as the skin's *acid mantle.*
- **Protection from the environment.** The integument protects against multiple threats in the environment. For example, the

skin absorbs *ultraviolet (UV) radiation* from the sun before it can damage the underlying tissues. The skin also secretes lipid-based chemicals that have nonpolar covalent bonds. These chemicals repel molecules with polar covalent bonds, such as water, and those with ionic bonds, such as salts, which keeps them from both leaving and entering the body via the skin. This is crucial to maintaining water and electrolyte homeostasis in the body by preventing the exit or entrance of excess water and electrolytes. Be aware, however, that these lipid-based substances do allow other nonpolar covalent substances to cross the skin, which we discuss in a later module.

Sensation

The skin houses numerous *sensory receptors,* cellular structures that detect changes in the internal and/or external environment. Sensory receptors enable the nervous system to perceive such changes, a process called *sensation.* Sensation is critical to homeostasis. These receptors in our skin allow us to detect potentially harmful stimuli such as heat or cold. They allow us to perceive painful stimuli, as well, which can help avoid tissue damage (note that extreme heat and cold can also be interpreted as pain).

Thermoregulation

The body's internal temperature is determined largely by the many reactions that are part of its metabolism and muscle activity. Unless it is extreme, the external temperature of the air outside the body has little effect on internal temperature. This is due in large part to the process of **thermoregulation**—the maintenance of a stable internal body temperature through negative feedback loops— **CORE PRINCIPLE** Feedback Loops

in an example of the Feedback Loops Core Principle (p. 21). Let's look at what happens when body temperature rises above the normal range (**Figure 5.2a**):

- **Stimulus: Body temperature increases above normal range.** Body temperature may rise due to normal conditions, such as exercise and hot weather, or abnormal conditions, such as fever.
- **Receptor: Thermoreceptors detect increase in temperature.** Receptors called thermoreceptors detect temperature changes in the skin and internal body fluids.
- **Control center: Thermoregulatory center in brain receives information from receptors.** The thermoreceptors send information to a part of the brain called the hypothalamus, which acts as the body's "thermostat" or thermoregulatory center.
- **Effector/response: The control center stimulates effectors to respond with cooling mechanisms.** The hypothalamus triggers accessory organs of the skin known as sweat glands to release a watery fluid called sweat. Recall from the chemistry chapter (Chapter 2) that water takes a great deal of heat with it when it evaporates, so evaporating sweat takes heat away from the body and produces a cooling effect. The hypothalamus also adjusts blood flow through the dermis: When body temperature rises, blood vessels widen, or dilate, causing a large volume of blood to fill the dermis. This radiates heat from the blood into the environment, cooling the body.

(a) Response of the integument to rising body temperature

(b) Response of the integument to falling body temperature

Figure 5.2 Homeostatic regulation of body temperature by the integumentary system.

- **Homeostasis and negative feedback: Body temperature returns to normal range, and negative feedback stops cooling mechanisms.** Sweating and vasodilation help return body temperature to normal. The thermoreceptors detect this, and negative feedback causes the thermoregulatory center to decrease its output to blood vessels and sweat glands.

Figure 5.2b shows what happens in the opposite situation, when body temperature decreases, generally due to cold environmental conditions. In this feedback loop, the parts are mostly the same—thermoreceptors detect the lower temperature and communicate this information to the thermoregulatory control center in the hypothalamus. The main difference comes with the response that the hypothalamus triggers: It causes the blood vessels in the dermis to narrow, or constrict, a change called vasoconstriction. This decreases blood flow through the skin and redirects it to deeper tissues, which helps conserve heat. As with the previous feedback loop, thermoreceptors detect when the temperature returns to normal and the response decreases by negative feedback.

Excretion

Small amounts of waste products are eliminated from the body through the skin, a process known as *excretion*. These waste products—including chemicals such as lactic acid, which is produced during muscle metabolism, and urea, which is generated during amino acid metabolism—are eliminated in sweat. Small amounts of toxins such as metals may also be excreted in sweat. Although these waste products and toxins are removed primarily by other organs (such as the kidneys), excretion via sweat makes a small but significant contribution.

Vitamin D Synthesis

The skin also plays a vital role in synthesizing the hormone **vitamin D.** The precursor of vitamin D is a modified cholesterol molecule found in the cells of the deeper part of the epidermis. When exposed to certain wavelengths of ultraviolet radiation, this molecule is converted to *cholecalciferol,* which then enters the bloodstream. Cholecalciferol is modified first by the liver, then by the kidneys, to form *calcitriol,* the active form of vitamin D. Vitamin D is required for the small intestine to absorb calcium ions, which are important for nerve function, muscle contraction, building and maintaining bone tissue, and many other physiological processes.

Quick Check

☐ 2. Explain how the integument provides protection from potential sources of damage.

☐ 3. How does the skin participate in thermoregulation?

☐ 4. Describe the other functions of the integument. How does each help maintain homeostasis?

Apply What You Learned

☐ 1. Predict what would happen to the cells of the epidermis if blood vessels in the dermis constricted and decreased blood flow for several hours. (*Hint:* From where do the cells of the epidermis get their oxygen and nutrients?)

☐ 2. Would the skin in question feel abnormally warm, abnormally cold, or show no change? Explain.

☐ 3. Burn victims often suffer damage to large portions of their integument. Predict the complications these patients may develop. What sort of homeostatic imbalances might they experience?

See answers in Appendix A.

MODULE **5.2**
The Epidermis

Learning Outcomes

1. Explain how the cells of the epidermis are arranged into layers.
2. Describe the cells of the epidermis and the life cycle of a keratinocyte.
3. Describe how thick skin and thin skin differ.

The epidermis, the most superficial region of the skin, is composed of stratified squamous epithelial cells that rest on top of a basement membrane. Although the epidermis contains several cell types, the most numerous cells are *keratinocytes.* In this module, we first explore the organization and life cycle of keratinocytes, after which we look at the types and roles of other epidermal cells. We conclude with a way to classify skin based on the thickness of the epidermis.

Keratinocytes

◀◀ **Flashback**

1. What is keratin? (p. 132)
2. What are desmosomes, and what are their functions? (p. 127)
3. Which types of compounds are hydrophobic and which are hydrophilic? (p. 47)

Approximately 95% of cells in the epidermis are **keratinocytes** (kehr′-ah-TIN-oh-sytz). Two structural features of keratinocytes make the epidermis stronger and less susceptible to mechanical trauma, demonstrating the Structure-Function Core Principle (p. 25). For one, as their name implies, keratinocytes are cells that manufacture keratin (KEHR-uh-tin; *kerato-* = "horn," so called because it makes up the horns of other animals). Recall that **keratin** is a fibrous protein that makes a tissue tougher and more 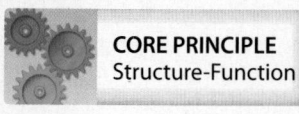 **CORE PRINCIPLE** Structure-Function resistant to mechanical stress (see Chapter 4). Second, keratinocytes are linked by numerous *desmosomes,* intercellular junctions that make the epidermis stronger.

Keratinocytes are arranged into layers, or *strata,* which we examine in the next few sections.

Layers of the Epidermis

Figure 5.3 illustrates the five structurally distinct layers of the epidermis. From deep to superficial, they are the stratum basale,

Figure 5.3 **Structure of the epidermis.**

stratum spinosum, stratum granulosum, stratum lucidum, and stratum corneum.

Stratum Basale The deepest layer of the epidermis is the thin **stratum basale** (STRA-tum bah-SA-le; *basal* = "bottom"). The stratum basale consists of a single layer of stem cells that rest on the basement membrane and appear slightly cuboidal or columnar rather than squamous. The keratinocytes in the stratum basale are closest to the blood supply in the dermis, and so are the most metabolically active keratinocytes of the epidermis. These cells actively produce the precursor to vitamin D when ultraviolet radiation from the sun interacts with specific steroid-based molecules in the cells.

The nearby blood supply also allows these keratinocytes to be mitotically active (engaged in cell division), which accounts for the alternate name of the stratum basale: *stratum germinativum* (jer′-min-ah-TYV-um; "germinating" or "generating layer"). All cells in the epidermis come from the mitotic activity of cells in the stratum basale or the layer above it, the stratum spinosum.

Stratum Spinosum Immediately superficial to the stratum basale we find the thickest stratum of the epidermis: the **stratum spinosum** (spin-OH-sum; *spino-* = "prickly"). The stratum

spinosum is named for the fact that its cells appear spiky due to bundles of cytoskeletal filaments in the periphery of the cells (this appearance is just an artifact of preparing them for a microscope slide; in living tissue these cells are not spiky). Like the cells of the stratum basale, those of the stratum spinosum are near the blood supply in the dermis and so are metabolically and mitotically active; for this reason, they are sometimes classed as part of the stratum germinativum. They also help produce vitamin D.

Stratum Granulosum The middle stratum of keratinocytes, the **stratum granulosum** (gran-yu-LOH-sum), consists of three to five rows of cells. The stratum granulosum is named for the prominent cytoplasmic granules in its cells. At least two types of granules are present in the cells: One type contains keratin bundles, and the other a lipid-based substance.

Released from the cells by exocytosis, this lipid-based substance coats keratinocytes in the stratum granulosum and the more superficial strata of the epidermis. Recall that lipids are hydrophobic, meaning they repel hydrophilic, or water-soluble, substances (see Chapter 2). So this lipid coating acts as a water barrier and prevents the passage of hydrophilic molecules into or out of the skin. As we discussed earlier, this lipid coating—along

Topical Medications

The lipid-based substance that coats the superficial layers of the epidermis is useful therapeutically. For example, some medications are toxic if swallowed, but safe if used topically (applied to the surface of the skin). Certain topical antibiotics are fairly toxic if taken by mouth, but they can be applied to the skin with minimal risk of systemic absorption because they are polar molecules that cannot pass through the epidermis to reach the blood vessels in the dermis. This allows for a local effect only.

However, hydrophobic or nonpolar substances are soluble in other nonpolar substances. For this reason, nonpolar substances cross the epidermis much more easily. This provides a convenient route of administration for certain medications such as the hormones in the birth control patch. Unfortunately, many poisons and toxins such as thallium (a heavy metal) are also nonpolar, and they cross the epidermis with the same ease. For this reason, it's always a good idea to wear gloves when dealing with chemicals.

with the multiple layers of cells—is crucial for maintaining our internal fluid and electrolyte homeostasis. However, it also isolates the cells from water and nutrients in the surrounding extracellular fluid, which promptly kills them. To find out how the lipid coating of these cells can be useful for medication delivery, see *A&P in the Real World: Topical Medications*.

Stratum Lucidum Superficial to the stratum granulosum we find a narrow, clear layer of dead keratinocytes called the **stratum lucidum** (LOO-si-dum; *lucid* = "clear"). This stratum is found only in thick skin, which we discuss shortly.

Stratum Corneum The most superficial stratum of the epidermis is the **stratum corneum** (KOR-nee-um; *cornu* = "horn"), which consists of several layers of dead, flattened keratinocytes. These cells have lost their normal organelles and are little more than bundles of keratin with thickened plasma membranes. Eventually, the most superficial cells of the stratum corneum also lose their desmosomes, which allows them to be sloughed (shed) or exfoliated mechanically. Every day you shed about a million dead keratinocytes, which amounts to roughly eight pounds over the course of a year. Where do all these dead cells go? Just look around your house for the answer—most of them are visible as dust, which is then ingested by microscopic dust mites.

⬆ STUDY BOOST

Remembering the Strata of the Epidermis
Here is a simple trick to remember the strata of the epidermis: "**B**rilliant **S**tudying **G**ives **L**oads of **C**onfidence." If you get confused as to which stratum is superficial and which is deep, think of the "B" in "basale" as standing for "bottom"; it is the bottom layer. ■

Keratinocyte Life Cycle

The location and functions of the epidermis subject it to a great deal of physical and environmental stress. As a result, the stratum corneum continually sheds dead keratinocytes. To maintain the integrity of the epidermis, these sloughed cells must be replaced, which is made possible by the mitosis of keratinocytes in the stratum basale and stratum spinosum. Mitosis occurs here rather than in more superficial strata because these keratinocytes are closest to the blood supply and therefore are the most metabolically active. Interestingly, and for reasons not fully understood, keratinocyte mitosis follows a pattern whereby the cells divide primarily at night.

As the keratinocytes in the deeper strata divide, they push the cells above them into the more superficial layers of the epidermis. In this way, a cell that begins its life in the stratum basale or stratum spinosum will eventually pass through all strata of the epidermis until it reaches the stratum corneum, where it is sloughed, a process that takes 40–50 days. This continual cycle of cells being sloughed and replaced by cells undergoing mitosis ensures that the epidermis maintains a certain thickness.

ConceptBOOST ⟫⟫⟫

Understanding Epidermal Growth
As you've just read, a keratinocyte begins in the stratum basale or spinosum, and eventually ends up in the stratum corneum. But how does the cell travel from one stratum to another? The answer is that the cell doesn't actually travel—it is pushed. The original cell in the stratum basale is a stem cell that undergoes mitosis to produce two daughter cells. One daughter cell remains in the stratum basale while the other daughter cell is pushed up through the epidermal layers. In this way the dividing stem cells push the products of mitosis up through the epidermal strata.

To see how this works, suppose for simplicity's sake that the skin has one row of cells in each epidermal stratum, as shown here:

We start with a stem cell in the stratum basale, marked cell A. This cell undergoes mitosis, and one of its two daughter cells (cell B in the diagram) is now in the stratum spinosum. The other daughter cell of cell A divides again, producing two more daughter cells, one of which pushes cell B into the stratum granulosum.

Cell B is now quite far from its blood supply and becomes coated with the lipid-based substance, which causes it to die. Stem cells in the stratum basale continue to divide, pushing cell B even farther away from the blood supply, into the stratum lucidum and then into the stratum corneum. Cell B is now a dead cell filled with keratin, and will eventually be sloughed off the skin surface. ■

☐ 1. What are the five strata of the epidermis? How does each stratum differ from the others?

☐ 2. Explain how a keratinocyte that begins its life in the deepest stratum of the epidermis eventually ends up on the epidermal surface.

Other Cells of the Epidermis

◀◀ **Flashback**

1. Define phagocytosis. What is a phagocyte? (p. 83)

In addition to keratinocytes, three other cell types are scattered throughout the epidermis (see Figure 5.3):

- **Dendritic (Langerhans) cells.** In the stratum spinosum we find **dendritic cells,** also known as **Langerhans cells.** Dendritic cells are phagocytes of the immune system that protect the skin and deeper tissues from pathogens.
- **Merkel cells.** Oval cells called **Merkel cells** are scattered throughout the stratum basale. They are sensory receptors associated with small neurons in the dermis, and together they function to detect light touch and differentiate shapes and textures. Merkel cells are especially numerous in body regions that are specialized for touch, such as the fingertips, lips, and at the base of hairs.

- **Melanocytes. Melanocytes** are cells located in the stratum basale that produce the protein **melanin** (MEL-ah-nin; *melan-* = "black"). Melanin is a skin pigment ranging from orange-red to brown-black. We discuss melanin and skin pigmentation in Module 5.4.

☐ 3. In addition to keratinocytes, what three types of cells are found in the epidermis? What are their functions?

Thick and Thin Skin

As with all structures in the body, the form of the epidermis in various parts of the body differs to match its function in agreement with the Structure-Function Core Principle (p. 25). Certain body locations are subject to a great degree of mechanical stress, including the palms of the hands, the palmar surfaces of the fingers, the soles of the feet, and the plantar surfaces of the toes. In these locations, we find **thick skin,** a type of epidermis about as thick as a paper towel. (Note that this measurement refers to the thickness of the epidermis only, not the skin as a whole). Thick skin contains all five epidermal strata and a very thick stratum corneum (**Figure 5.4a**). This type of skin lacks hair follicles but contains numerous sweat glands.

CORE PRINCIPLE
Structure-Function

Stratum corneum

Stratum lucidum

Stratum granulosum

Stratum spinosum

Stratum basale

Dermis

LM (165×)

(a) Thick skin

LM (120×)

(b) Thin skin

Figure 5.4 Thick and thin skin.

The remaining parts of the body are subject to less mechanical stress, and here we find **thin skin** (Figure 5.4b), which is about as thick as a sheet of printer paper. Thin skin lacks the stratum lucidum and so has only four epidermal strata. In addition, its stratum corneum and other strata are much thinner (have fewer layers) than those in thick skin. Thin skin has abundant hairs, sebaceous glands, and sweat glands.

In both thick and thin skin, the stratum corneum may develop additional layers when subjected to repeated pressure. This forms a **callus** (KAL-us), which is sometimes painful due to the added pressure placed on tissues deep to the callus. A callus may be treated by shaving off the excess layers of stratum corneum, but unless the source of the pressure is removed, the callus generally returns.

Quick Check

☐ 4. Compare and contrast thin and thick skin.

Apply What You Learned

☐ 1. Paper cuts often involve only the epidermis. You get a paper cut and notice that it doesn't bleed, but you can feel it. Explain why you can feel it but the cut isn't bleeding.

☐ 2. A hypothetical drug causes blood vessels to grow from the dermis into the superficial stratum granulosum of the epidermis. Would this affect the cells in the most superficial epidermal layers? Why or why not?

See answers in Appendix A.

MODULE **5.3**
The Dermis

Learning Outcomes

1. Describe the layers and basic structure and components of the dermis.

2. Explain the functions of the dermal papillae.
3. Explain how skin markings such as epidermal ridges are formed.

Deep to the epidermis lies the highly vascular dermis. The dermis serves many functions: It houses the blood supply of the epidermis, contains sensory receptors, and anchors the epidermis in place. As with the epidermis, we can divide the dermis into distinct layers. However, the two layers of the dermis are not simply layers of cells but, rather, two different types of connective tissue.

Papillary Layer

◀◀ Flashback

1. What are fibroblasts and ground substance? (pp. 124, 138)
2. Define loose connective tissue. (p. 138)

The most superficial layer of the dermis is the thin **papillary layer** (PAP-uh-lar-ee), which accounts for about 20% of the depth of the dermis (Figure 5.5). The papillary layer consists of loose connective tissue. Recall that loose connective tissue is primarily ground substance with collagen and elastic fibers (see Chapter 4). Within the papillary layer are the usual cells that we find in connective tissue, including fibroblasts and phagocytes. At the dermis-epidermis junction, special collagen fibers extend up into the epidermal basement membrane to anchor the epidermis in place. Repetitive trauma, such as rubbing from a pair of ill-fitting shoes, can disrupt these collagen fibers and separate the epidermis from the dermis. This usually results in a fluid-filled pocket called a *blister.*

The surface of the papillary layer where it contacts the epidermis is folded into projections referred to as **dermal papillae** (DER-mal pah-PILL-ee). The dermal papillae house tiny blood vessels called *capillaries* that are arranged into loops. These loops extend up into the superficial part of the dermal papillae, where oxygen and nutrients diffuse into the extracellular fluid of the dermis and up into the cells of the avascular epidermis.

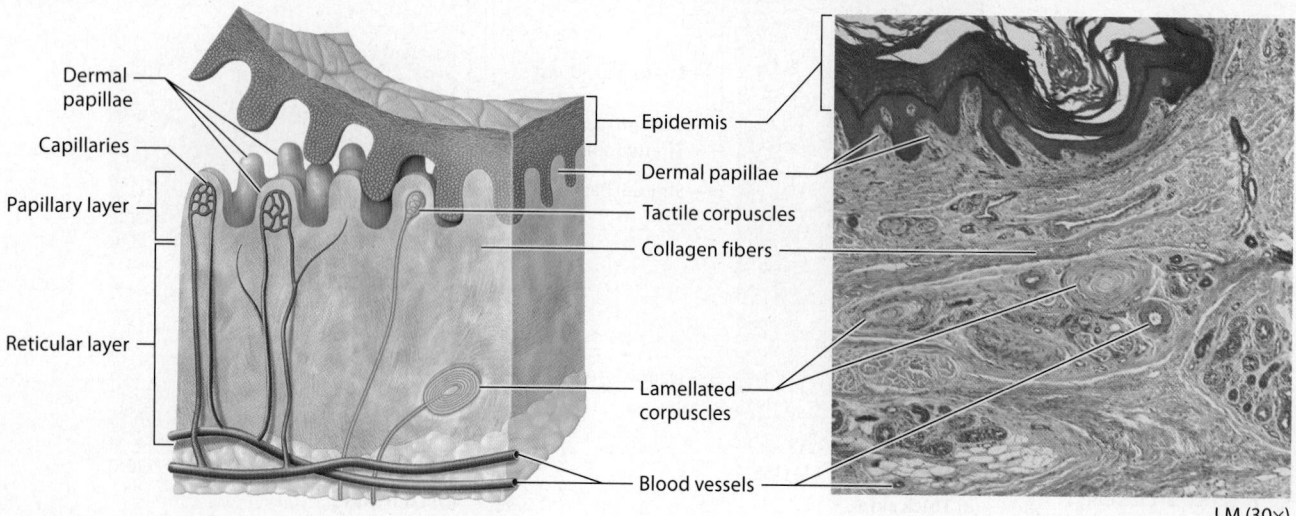

Dermal papillae

Capillaries

Papillary layer

Reticular layer

Epidermis

Dermal papillae

Tactile corpuscles

Collagen fibers

Lamellated corpuscles

Blood vessels

LM (30×)

Figure 5.5 Structure of the dermis.

The dermal papillae also house sensory receptors called **tactile (Meissner) corpuscles** (MYSS-ner KOR-pus-ulz), which respond to light touch stimuli. Tactile corpuscles are more numerous in areas of the body that have sensation as a primary function, such as the fingertips, lips, face, and external genitalia.

Quick Check

☐ 1. Which type of tissue makes up the papillary layer of the dermis?
☐ 2. What are the functions of the dermal papillae?

Reticular Layer

Flashback

1. What is dense irregular connective tissue and what are its properties? (p. 138)
2. What is smooth muscle tissue, and what are its properties? (p. 148)

The deeper and thicker **reticular layer** is composed mainly of dense irregular connective tissue. Note that the reticular layer is continuous with both the papillary layer and the hypodermis, which makes it difficult to distinguish a clear boundary between them. Recall that dense irregular connective tissue consists largely of irregularly arranged bundles of collagen fibers (see Chapter 4). Many more fibers are found in the reticular layer than in the papillary layer, and the fibers of the reticular layer are bundled together. These collagen fibers strengthen the dermis and help prevent traumatic injuries from reaching deeper tissues.

In addition to collagen fibers, we find elastic fibers, which enable the dermis to return to its original shape and size after being stretched. Note that if stretching is extreme, such as that occurring with pregnancy or significant weight gain, the elastic fibers may tear, forming *striae,* or stretch marks, that are generally permanent. The reticular layer also contains ground substance rich in proteoglycans, as well as cells such as fibroblasts. The proteoglycans draw water into the ground substance, which hydrates the skin and makes it appear firm. In addition, small pockets of adipose tissue are scattered throughout the reticular layer, although most of the adipose tissue is confined to the hypodermis.

Embedded within the reticular layer we find blood vessels and some accessory structures, including sweat glands, hairs, sebaceous glands, and a number of sensory receptors. One prominent sensory receptor is the **lamellated** or **Pacinian corpuscle** (pah-SIN-ee-an KOR-pus-ul, see Figure 5.5). Lamellated corpuscles are named for their layered appearance (*lamell-* = "layer") and resemble cut onions when viewed in cross-section. These receptors respond mainly to changes in pressure and vibration applied to the skin.

Quick Check

☐ 3. Which type of tissue makes up the reticular layer of the dermis?
☐ 4. What other structures are located in the dermis?

Skin Markings

Interactions between the dermis and the epidermis are visible as small lines in the epidermis known as *skin markings.* Some of the most obvious skin markings are seen in thick skin on the palmar surfaces of the hands and fingers and the plantar surface of the feet and toes. Here the dermal papillae are extremely prominent and thick collagen fibers arrange them into *dermal ridges* (**Figure 5.6**). This structure indents the overlying epidermis and

(a) Epidermal ridges of thick skin   LM (120×)

(b) Epidermal ridges, superficial view   SEM (9×)

Figure 5.6 Epidermal ridges and fingerprint patterns.

produces **epidermal ridges** that enhance the gripping ability of the hands and feet.

Epidermal ridges occur in characteristic patterns such as a *loop, arch,* and *whorl* that are genetically determined and unique to each individual. Notice in Figure 5.6 that tiny *sweat pores,* where sweat exits from sweat glands, open along the ridges and leave a thin film called a **fingerprint** when we touch most surfaces. (The palms, toes, and soles of the feet also leave characteristic prints.)

The reticular layer is also responsible for certain skin markings. Throughout the dermis, gaps between the bundles of collagen indent the epidermis, forming lines called **tension** or **cleavage lines.** Tension lines normally run in a circular pattern in the neck and trunk, but longitudinally in the head and upper and lower limbs. **Figure 5.7** shows why they are important clinically: If a skin wound is parallel to tension lines, its edges tend to stay closed and heal better with less risk of scarring. However, a wound that runs perpendicular to tension lines tends to gape and is more likely to form a scar. This is why surgeons plan incisions to be as parallel as possible to tension lines.

The final type of skin marking also is a product of the reticular layer. In certain locations—for example, around joints—the reticular layer is anchored tightly to deeper structures. This creates deep creases known as **flexure lines.** Flexure lines are pronounced in areas such as the palms and on the posterior surface of the fingers and toes. Another product of the reticular layer of the dermis is explored in *A&P in the Real World: Skin Wrinkles.*

Incisions made along tension lines heal faster and with less scarring.

Tension lines

Incisions made perpendicular to tension lines heal more slowly and tend to scar.

Figure 5.7 **Importance of tension lines for surgical incisions.**

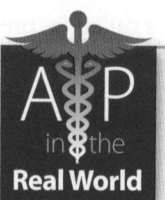

A&P in the Real World

Skin Wrinkles

A hallmark of aging is the formation of *skin wrinkles,* which many individuals consider the least desirable type of skin marking. Wrinkles are due to an age-related decrease in collagen fibers, elastic fibers, proteoglycans, and adipose tissue in the dermis. This reduces the skin's firmness, hydration, and recoil ability after stretching. Age-related wrinkles tend to be deeper in areas of repetitive muscle movement, such as in the forehead and around the eyes and mouth. Exposure to UV radiation (whether from natural sunlight or tanning beds) and cigarette smoking accelerate the formation of wrinkles.

The appearance of wrinkles can be minimized with products and procedures such as the following:

- *Botox* contains a bacterial toxin that temporarily paralyzes facial muscles for four to six months, causing the skin over them to appear smoother.
- *Fillers* contain adipose tissue, collagen, and/or proteoglycans that are injected into wrinkles to temporarily "fill" them.
- *Peels* use lasers, chemicals, or abrasion to remove varying amounts of the epidermis and superficial dermis to cause the formation of new, firmer (so the user hopes!) skin.
- *Topical creams* contain a variety of ingredients that claim to reduce the appearance of wrinkles. These creams, particularly nonprescription products, have little or no effect on wrinkles. Such creams are classified as cosmetics rather than drugs, and are therefore not required to undergo testing for safety and efficacy.

Although wrinkles are an inevitable fact of life, their appearance can be delayed or reduced by avoiding the sun, using sunscreen, staying well hydrated, and avoiding cigarette smoking.

Quick Check

☐ 5. How does the papillary layer of the dermis affect the appearance of the epidermis?
☐ 6. What causes tension lines and flexure lines? How do they differ?

Apply What You Learned

☐ 1. Your friend Lily is doing woodworking when she cuts her forearm with a saw. You examine the wound and see large quantities of adipose tissue at its base. Does her wound extend down to the epidermis, dermis, or hypodermis? How do you know?
☐ 2. Lily wants to know if the wound is likely to form a scar. The wound is located on her left anterior forearm and cuts straight across the width of her arm. What do you tell her, and why?

See answers in Appendix A.

Skin Pigmentation

Learning Outcomes

1. Explain how melanin is produced and its role in the integument.
2. Describe the other pigments that contribute to skin color.
3. Explain how skin coloration may indicate pathology.

Skin color is one of the most readily identifiable physical features in humans—so much so that it is used as a basis for individuals to form groups and for one individual or group to discriminate against others. But from a physiological standpoint, human skin color is little more than the amount of a pigment called *melanin* produced in the epidermis. In this module, we examine the cells that manufacture melanin, its functions, and how it influences skin color. We also discuss two other pigments that impact skin color in minor ways: *hemoglobin* and *carotene*. And, finally, we will discover how changes in skin color can serve as diagnostic clues to various health conditions.

Melanin

◀◀ Flashback

1. What are amino acids? (p. 55)
2. What is DNA, and what are DNA mutations? (pp. 59, 103)

The primary skin pigment is **melanin,** which as you read earlier is produced by melanocytes in the stratum basale. Melanin ranges in color from orange-red to black, and is composed of two molecules of the amino acid tyrosine joined by a series of reactions catalyzed by the enzyme *tyrosinase*. These reactions occur in specialized vesicles within the melanocyte known as **melanosomes** (muh-LAN-oh-sohmz), shown in step ① of **Figure 5.8**.

You can see that a melanocyte somewhat resembles an octopus, with numerous "arms" that reach out and contact or even pierce the plasma membranes of neighboring keratinocytes. Notice in step ② of the figure that within the melanocytes, melanosomes migrate to the tips of these arms, where they are released by exocytosis. The mechanism by which the melanosome is taken into the cytosol of the keratinocyte is not fully understood. The melanosome either enters the cytosol directly or enters the extracellular fluid next to the keratinocyte, after which the keratinocyte takes in the melanosome by endocytosis. The melanin is then transported to the region superficial to the nucleus, where it shields the DNA of the keratinocyte like an umbrella, shown in step ③. The pigment degrades after a few days, so melanocytes must continually produce new melanosomes to maintain a specific skin color.

Melanin synthesis increases upon exposure to UV radiation from the sun or sunlamps. This leads to *tanning*, or darkening of skin pigmentation. UV radiation has both immediate and delayed effects on skin pigmentation. Its immediate action is to oxidize the melanin already formed within the keratinocytes,

① Melanocyte synthesizes melanin in melanosome.

② Melanosome is released into or near a neighboring keratinocyte.

③ Melanin covers the keratinocyte's nucleus, shielding it from UV radiation.

Figure 5.8 Melanocytes and melanin function.

which causes the melanin to rapidly darken. UV radiation also damages the DNA of the melanocytes, which increases the production of additional melanin. These secondary effects become visible about 72 hours after exposure to UV but last longer than melanin oxidation. To find out if there is any such thing as a "healthy tan," see *A&P in the Real World*.

One of the functions of melanin is to protect the DNA of keratinocytes from mutations induced by UV radiation. Recall from the cell chapter (Chapter 3) that such mutations may form cancerous cells. Note, however, that melanin does not provide complete protection from DNA damage—it absorbs some UV radiation, but not all of it, especially if the radiation is intense. For this reason, people of all skin pigmentations can develop skin cancer (we discuss skin cancer in Module 5.6). Melanin also does not provide much protection from burns due to UV radiation, so individuals of all skin colors can sunburn.

A second function of melanin is to decrease the synthesis of vitamin D in response to UV radiation. This may seem counterintuitive—after all, why would the body want to produce *less* vitamin D? The answer is that vitamin D levels must be kept within a specific range, as excess amounts can lead to calcium imbalance and kidney failure.

In fact, the role of melanin in vitamin D synthesis may explain the evolution of different skin pigmentation in different parts of the world. Individuals living in regions exposed to high amounts of UV radiation, such as Africa, may have developed darker skin to prevent excess vitamin D production. Conversely, people in areas with less UV radiation, such as northern Europe, developed lighter skin so they could synthesize enough vitamin D.

As you have probably noticed, skin pigment isn't evenly distributed throughout the body. The unequal distribution of melanocytes largely accounts for this. For example, fewer melanocytes are located on the palms of the hands and the inside of the arms and thighs. Because of this distribution, these areas typically have a lighter color compared to parts of the body with more melanocytes, such as the face. Pigment is also sometimes concentrated in one spot, producing a **freckle,** which is due to localized higher melanin production, or a **mole** or **nevus** (NEE-vus), which results from a local proliferation of melanocytes.

Interestingly, the number of melanocytes is virtually identical among all individuals, irrespective of skin color. What, then, causes the wide variety of skin colors in humans? This spectrum of skin tones is due to differences in the amount of tyrosinase activity and the type of melanin produced. This is evident from the condition known as *albinism,* in which melanocytes fail to manufacture tyrosinase. Albinism results in lack of skin pigmentation and a greatly increased risk of keratinocyte DNA damage from UV radiation.

Quick Check

- [ ] 1. How is melanin produced, and how does it interact with keratinocytes?
- [ ] 2. What are the functions of melanin?

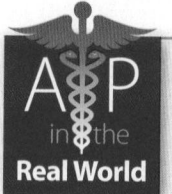

Tanning and a "Healthy Tan"

Skin tanning is big business—in fact, a $5-billion-a-year business in the United States alone. In the past decade, the number of tanning salons has soared from about 10,000 to 50,000, with most of these salons promoting the notion of a "healthy tan." But is there such a thing as a healthy tan?

Two types of UV rays reach Earth's surface: UVA and UVB rays. UVB rays are associated with sunburning, whereas UVA rays are linked with tanning. This has led tanning salons to claim that UVA rays are safe and will not damage skin. But the mechanism of increased melanin production is the same for both types of rays—oxidative damage to existing melanin and DNA damage to melanocytes that can lead to skin cancer. Not only does UVA radiation do just as much DNA damage as UVB radiation, but UVA actually ages the skin at a much faster rate, as well.

So can you get a healthy tan? The answer is no—any amount of tanning damages melanocytes and other elements of the skin, ages the skin prematurely, and increases the risk for skin cancer.

Other Pigments That Affect Skin Color: Carotene and Hemoglobin

◀◀ **Flashback**

1. What is an oxidation reaction? (p. 43)

Although melanin is the major skin pigment, two other minor contributors to skin color are *carotene* and *hemoglobin.* Normally, humans ingest the yellow-orange pigment **carotene** (KEHR-uh-teen) in their diet, from egg yolks and yellow and orange vegetables such as carrots. Carotene is a lipid-soluble molecule that tends to accumulate in the stratum corneum. This imparts a slight yellow-orange color to the skin that is particularly visible in thick skin where the stratum corneum has the most layers of cells.

Hemoglobin (HEE-muh-gloh-bin) is an iron-containing protein found within our red blood cells that binds oxygen and transports it through the body. When hemoglobin binds oxygen, its iron ion becomes oxidized. This is the same reaction that occurs when metal rusts (see Chapter 2). Just as metal changes color when it rusts, hemoglobin also changes color when it is oxidized. Oxidized hemoglobin turns a bright orange-red, which is responsible for the characteristic red color of blood.

How does a blood protein affect the color of skin? The answer is that it doesn't impact the actual color, but, rather, the color skin appears to be. Recall that the dermis is highly vascular, so a great deal of the body's blood resides in the dermis at any given time. In fair-skinned individuals, the epidermis is partly translucent, so we are able to look through the epidermis to see the blood in the vessels of the dermis. Under normal conditions, this gives the skin a faint pinkish hue. In regions where the stratum corneum is thinner, such as the lips, the pink is generally darker.

Quick Check

☐ 3. What is carotene, and what color does it give to the skin?
☐ 4. How does hemoglobin affect skin color?

Skin Color as a Diagnostic Tool

The skin color imparted by hemoglobin may change based on the amount of blood flowing through the dermis. The skin becomes brighter red when blood flow through the dermis increases, as when the body is trying to release heat during exercise, a color change called **erythema** (ehr-ah-THEE-mah; *eryth-* = "red"). We also see erythema when the body is injured by trauma (e.g., breaking a bone) or infection, as blood flow increases to the injured area.

Conversely, the pinkish hue becomes faint and nearly disappears when blood flow through the dermis decreases, as when the body is trying to conserve heat. In extreme cases, so little blood is flowing through the skin that collagen in the dermis is visible, which gives the skin a white color termed **pallor** (PAL-er). Pallor is especially visible in lighter-pigmented skin.

The color resulting from hemoglobin also depends on how much oxygen is bound to it. When high amounts are bound, its iron becomes more oxidized and assumes a brighter red color. For instance, the poison *cyanide* prevents the oxygen in hemoglobin from leaving it and entering the tissues; the cheeks of cyanide poisoning victims have a characteristic cherry color.

On the other hand, when extremely low amounts of oxygen are bound to hemoglobin, it turns dark reddish-purple. The skin then takes on a faint bluish color known as **cyanosis** (sy-ah-NO-sis; *cyan-* = "blue"). Skin may appear cyanotic when an individual has breathing difficulties, when hemoglobin and red blood cell levels are too low, or when hemoglobin is unable to bind to oxygen. Cyanosis is a sign that the person is in trouble and needs immediate assistance.

Quick Check

☐ 5. How can the oxygen content of the blood impact the color of skin?
☐ 6. What is cyanosis, and what can it tell us about a person's health? Be specific.

Apply What You Learned

☐ 1. *Vitiligo* is a condition characterized by death or dysfunction of scattered groups of melanocytes in the skin. Predict how vitiligo may cause the skin to appear.
☐ 2. Max and Breanna go to the beach, and Max tells Breanna she does not need sunscreen because she is African American and individuals with dark skin cannot get sunburns or skin cancer. What should Breanna tell him in response?

☐ 3. A hypothetical poison prevents oxygen from binding to hemoglobin. Predict how this would impact skin color in a fair-skinned person.

See answers in Appendix A.

MODULE **5.5**

Accessory Structures of the Integument: Hair, Nails, and Glands

Learning Outcomes

1. Describe the structure and function of hair and nails.
2. Explain the process by which hair and nails grow.
3. Summarize the structural properties of sweat and sebaceous glands.
4. Explain the composition and function of sweat and sebum.

Let's now turn our attention to the **accessory structures** or **appendages** of the integument: *hair, nails,* and *glands*. Each of these structures derives primarily from a single tissue type—epithelium—and assists the integument in performing its overall functions. First we will explore the structure and function of hair and nails, which are both derived from the epidermis. Then, we will focus on the two types of glands in the skin: *sweat glands* and *sebaceous glands*.

Hair

◀◀ **Flashback**

1. What type of tissue makes up the epidermis? (p. 161)

Like most mammals, humans have **hair** or **pili** (pee-LEE)—small, filamentous structures that project from the surface of the skin over the entire body except the regions with thick skin, the lips, and parts of the external genitalia. The hair of most animals assists with thermoregulation, but human hair over the majority of the body is too sparse to play any significant role in this task. However, it does help the integument perform its protective functions. For example, hairs around the eyes prevent the entrance of foreign objects, hairs inside the nose keep us from inhaling insects and harmful particles, and hairs on the head protect the body from UV radiation and mechanical trauma. Hairs also participate in sensation because they are associated with small sensory neurons that help them sense the environment.

Hair is derived from the epidermis, and so consists of stratified squamous keratinized epithelial cells. However, in its structure and the manner in which it grows, hair differs from the epidermis. In the next three sections, we look more closely at the structure of hair, the process by which it grows, and the variations in hair texture and color among individuals.

Hair Structure

A hair is composed of two main parts: the **shaft,** or the portion projecting from the skin's surface, and the **root,** the portion that is embedded in the dermis (**Figure 5.9**). The root is surrounded by a small sensory neuron and indented at its base by a projection from the dermis called a *hair papilla,* which contains blood vessels. Together, the root and the hair papilla are known as a **hair bulb.**

Both the shaft and the root are composed of columns of keratinized epithelial cells but are at different stages of development. The cells in the shaft have completed their keratinization process and are all dead. However, many cells in the root are still living and undergoing keratinization. A small population of keratinocytes called the **matrix** at the base of the root actively undergoes mitosis.

The root is embedded in the **hair follicle.** Notice in Figure 5.9 that the hair follicle is simply an infolding of the epidermis known as the **epithelial root sheath,** which extends down into the dermis or hypodermis. This epithelial root sheath has two components: The outer component anchors the follicle to the dermis, and the inner component is anchored tightly to the hair root. In fact, when a hair is pulled out, this inner root sheath is usually visible as a shiny, clear covering around the base of the hair.

On a transverse section (Figure 5.9), we can see that a strand of hair has three regions:

- The inner *medulla* is present only in thick hairs, such as those on your head, and is the soft core that consists of keratinocytes with *soft keratin,* the same type of keratin found in the epidermis.

- The middle *cortex* is a highly structured and organized region with several layers of keratinocytes that contain *hard keratin.* These keratinocytes do not flake the way keratinocytes containing soft keratin do. This is advantageous because if the cells of the cortex flaked off like the superficial layers of the epidermis, the hair strands would quickly whittle down and break.

- The outer *cuticle,* the most superficial layer, consists of a single layer of keratinocytes arranged in an overlapping manner, much like shingles on a roof. Like the cells of the cortex, those of the cuticle contain hard keratin, enabling the cuticle to compact the underlying layers and giving it mechanical strength to resist abrasion. Over time, the cuticle can become worn, particularly near the ends of a hair strand, which allows the cortex and medulla to fray, resulting in "split ends."

Surrounding the epithelial root sheath is a *dermal root sheath,* which consists of connective tissue that supports the follicle and separates it from the dermis. Tiny bands of smooth muscle called **arrector pili** muscles attach to the dermal root sheath on one end and the dermal papillary layer on the other end. When an arrector pili muscle contracts, the hair stands up, which is known as **piloerection,** and the skin where the muscle attaches to the papillary layer indents slightly. This happens when we are cold or frightened and gives the skin a dimpled appearance we call "goosebumps."

Hair Growth

Hair grows at different rates for different individuals, but averages about 1–1.5 cm per month. This growth is not continuous but occurs via a cycle with two main phases: the *growth stage* and the *resting stage.*

(a) Transverse section of a hair shaft

Cuticle
Cortex
Medulla

Hair shaft

Hair root

Hair follicle:
Epithelial root sheath
Dermal root sheath

Arrector pili muscle

Hair bulb

Hair follicle
Epithelial root sheath
Dermal root sheath
Medulla
Cuticle
Cortex
Matrix
Hair papilla
Melanocyte

LM (150×)

(b) Frontal section of a hair root and hair follicle

Figure 5.9 Hair structure.

During the growth stage, mitosis occurs within the cells of the matrix in the root. As the cells divide, those above them are pushed farther away from the blood supply, keratinize, and die, in a similar manner to what takes place in the epidermis. The growth stage lasts anywhere from about a month in the eyelashes to as long as six years in the scalp.

During the resting stage, the cells in the matrix cease dividing and start to die. The follicle shortens somewhat and the hair is pushed up near the surface of the follicle, which remains dormant for one to several months. At this point the hair may fall out on its own, or a new hair may push it out when the growth stage resumes.

The duration of the growth and resting stages determines both the rate at which a hair grows and its maximal length. For example, a scalp hair will typically be in the growth stage for about six years. If the rate of mitosis in the matrix is rapid during the growth stage, this hair could grow to be nearly a meter long during that six-year time span. Even if the rate of mitosis is slow, hair length can reach 250–500 cm. In contrast, the growth stage of an eyelash is only about 30 days. This limits the length of an upper eyelash to about 11 mm and a lower eyelash to about 7 mm, again depending on the rate of mitosis in the matrix.

Hair grows most rapidly from puberty to the fifth decade of life, after which its growth slows, a factor influenced by both environment and genetics. Gradually, hairs begin to fall out faster than they grow, which leads to thinning hair with possible *alopecia,* or baldness. Although alopecia occurs in both genders, it is more pronounced in males. Subtypes of alopecia include *male pattern baldness* and *female pattern baldness,* in which hormones, particularly testosterone, cause hair follicles to die. The remaining follicles produce thin, nonpigmented *vellus hair* (discussed in the next section).

Hair Pigment and Texture

Hair color and texture vary with the type of hair. A fetus has a thin, nonpigmented type of hair called **lanugo** (lah-NOO-go) that covers nearly the entire body. Lanugo typically falls out about the time of birth and is replaced by one of two types of hair. The first is **terminal hair,** which is thicker, coarser, and pigmented. We find

terminal hair on the scalp and around the eyes. The other type is **vellus hair** (VELL-us), which is thinner and nonpigmented, and is found over the remainder of the body. After puberty, much of the vellus hair on the body is replaced with terminal hair. This is determined by gender—terminal hair replaces about 90% of the vellus hair in males, but only about 35% in females.

Like skin color, hair color is largely determined by the pigment melanin. Melanocytes located in the matrix manufacture melanin that interacts with keratinocytes just as it does in the epidermis. The melanocytes in blonde hair produce very little melanin, whereas those in darker hair produce progressively more melanin. Red hair has a special reddish pigment that contains iron. As we age, the melanocytes produce less melanin, and the growing hair gradually loses pigment, turning gray or white. The myth of a person's hair turning white overnight from fear is technically impossible—even if all melanocytes stopped producing pigment simultaneously, the existing hair would already have color and would need to grow out before the hair appeared white.

> ### Quick Check
>
> ☐ 1. How do the hair shaft and hair root differ?
> ☐ 2. How does a hair grow in length?
> ☐ 3. What influences hair color and texture?

Nails

A feature that humans share with other primates is **nails,** hard structures located at the end of our digits. Like hair, nails derive from the epidermis and so consist of stratified squamous epithelium and contain hard keratin. The most obvious part of a nail is the **nail plate,** which sits on top of the epidermal **nail bed** (Figure 5.10). The nail plate divides into the **nail body,** which is the portion you can see, and the **nail root,** which lies under the skin. Within the root we find the **nail matrix,** which contains actively dividing cells.

Surrounding the nail are folded areas of skin that reinforce the nail. On the proximal edge covering the root we find the

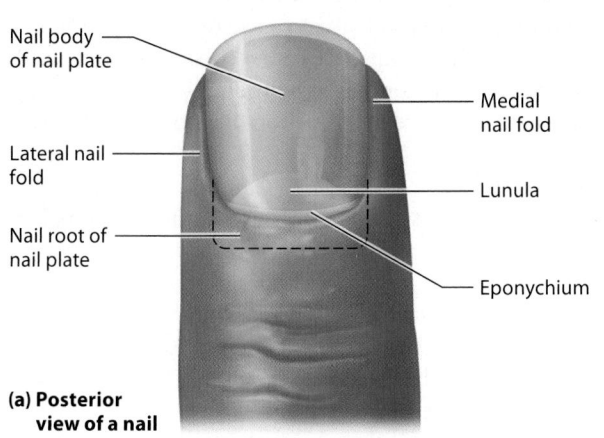

(a) Posterior view of a nail

Nail body of nail plate · Lateral nail fold · Nail root of nail plate · Medial nail fold · Lunula · Eponychium

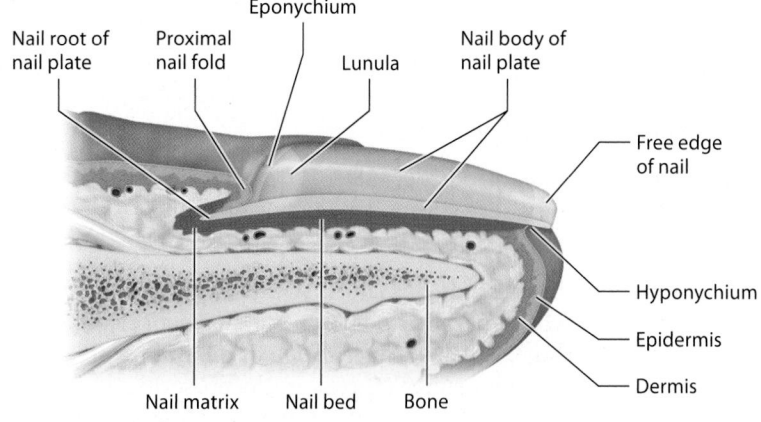

(b) Sagittal section of a fingertip

Nail root of nail plate · Proximal nail fold · Eponychium · Lunula · Nail body of nail plate · Free edge of nail · Hyponychium · Epidermis · Dermis · Nail matrix · Nail bed · Bone

Practice art labeling @MasteringA&P®

Figure 5.10 Nail structure.

proximal nail fold. The most distal part of the proximal nail fold, called the **eponychium** (ep-oh-NIK-ee-um) or *cuticle,* consists only of its stratum corneum. On the medial and lateral edges of the nail are the **medial** and **lateral nail folds,** respectively. The distal end of the nail is attached to the underlying nail bed by an accumulation of stratum corneum referred to as the **hyponychium.**

Nail growth occurs at the nail matrix, where actively dividing cells push those above them distally. As with a hair, the keratinocytes making up a nail die and keratinize as they are pushed farther from the blood supply. Fingernails grow an average of 0.5 mm per week, whereas toenails grow more slowly. Contrary to a popular urban legend, nails do not continue to grow after death. Instead, the tissue around the nails shrinks, which makes the nails appear longer.

Unlike hair, nails contain no melanocytes and so are generally translucent. Certain nails contain a slightly more opaque region near the proximal nail fold, called the **lunula.** Named for its half-moon shape, the lunula represents an area of keratin accumulation. The translucent nature of the nails makes them a good indicator of health: Nail beds are pinkish in a well-oxygenated individual, but bluish (cyanotic) in a poorly oxygenated individual.

The primary function of nails is to safeguard the underlying tissue. Their hardness protects the distal fingers and toes from trauma. Nails also act as tools, enabling more precise gripping than with fingertips alone.

One of the most common nail conditions is an *ingrown nail.* Ingrown nails most often occur in the toenails, and are caused by factors such as poorly fitting shoes that pinch the nails and cause them to penetrate the tissue lateral or medial to the nail. This results in inflammation, pain, softening of the nail, and often infection. An ingrown nail is managed by removing part or all of the nail, and occasionally the nail matrix is treated with chemicals to kill its cells and prevent regrowth of the nail.

Quick Check

☐ 4. Define the following terms: nail bed, nail plate, nail matrix, and proximal nail fold.
☐ 5. How does nail growth occur?

Glands

◀◀ **Flashback**

1. What are exocrine glands? (p. 135)
2. How do holocrine secretion and merocrine glandular secretion differ? (p. 136)

The skin has two basic kinds of glands: **sweat (sudoriferous) glands** (soo′-doh-RIF-er-us), which form *sweat,* and **sebaceous glands** (seh-BEH-shus), which produce *sebum* or oil. Both types of glands are derived from epithelial cells in the epidermis but are located in the dermis. Let's look at the properties of sweat and sebaceous glands and their products.

Sweat Glands

The body contains four types of sweat glands: eccrine, apocrine, ceruminous, and mammary. These glands have different structures and release different products. However, in all four types the secretory cells of the gland release their product by exocytosis, a type of secretion called *merocrine secretion.*

Eccrine Sweat Glands **Eccrine sweat glands** (EK-rin), the most prevalent type of sweat gland, are simple coiled tubular glands situated in the dermis (**Figure 5.11a**). Sweat exits the gland through a duct that passes from the dermis into the epidermis, and leaves the duct through a tiny **sweat pore** in the epidermis (note that these are not the same "pores" created by the infolding of hair follicles).

The sweat released by eccrine sweat glands is 99% water, with small amounts of solutes—including electrolytes such as sodium, potassium, and chloride ions—and waste products such as lactic acid. Sweat also contains antimicrobial compounds and has a slightly acidic pH, both of which deter the growth of pathogens. Although sweat functions in excretion and protection, its primary function is thermoregulation. Recall that when sweat evaporates it takes heat with it, reducing body temperature.

Apocrine Sweat Glands We find the large **apocrine glands** (AP-oh-krin) only in certain locations, including the axillae (armpits), the anal area, and the areola (the darkened area around the nipple). Note in Figure 5.11 that unlike eccrine sweat glands, apocrine glands release their product into a hair follicle rather than through a pore. This product is a thick sweat that is rich in proteins. Although sweat itself is odorless, after being secreted it is metabolized by bacteria in the skin, which produces a characteristic odor. Secretion from apocrine glands is influenced by sex hormones, and does not begin until puberty.

Ceruminous Glands The modified apocrine glands known as **ceruminous glands** (si-ROO-min-us) are located in the ear, where they secrete a thick product called **cerumen** (si-ROO-min) into hair follicles. Cerumen, better known as ear wax, lines the canal leading into the ear, where it lubricates the tympanic membrane and traps incoming particles before they reach the membrane.

Mammary Glands **Mammary glands** are highly specialized sweat glands that produce a modified sweat called **milk.** Milk contains proteins, lipids, sugars, and a variety of other substances to nourish a newborn infant. The chapter on the reproductive system (Chapter 26) discusses the structure and function of mammary glands.

Sebaceous Glands

As you can see in **Figure 5.11b**, sebaceous glands are branched glands with clusters of secretory cells called *acini* that surround small ducts. These ducts generally converge to form a central duct that empties into a hair follicle, as apocrine sweat glands do, although in some locations the products are secreted via a small pore. Most numerous in the face and scalp, sebaceous glands are also found throughout the body with the exception of the palms and soles. Their secretion is influenced by hormones, particularly the male sex hormone testosterone, and secretion from sebaceous

(a) **Eccrine sweat gland, frontal section**

(b) **Sebaceous gland, frontal section**

Figure 5.11 Sweat glands and sebaceous glands.

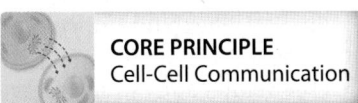

CORE PRINCIPLE
Cell-Cell Communication

glands increases after the onset of puberty, when levels of sex hormones rise dramatically. Note that this is an example of the Cell-Cell Communication Core Principle (p. 27).

Sebaceous glands produce **sebum** (SEE-bum), a waxy, oily mixture consisting largely of various lipids. Unlike sweat glands, sebaceous glands use *holocrine secretion,* during which the secretory cell accumulates its product until rupture occurs. For this reason, sebum contains cell fragments and debris in addition to lipids. Sebum coats the skin and hair, and helps keep skin moist by providing a hydrophobic barrier that prevents water from leaving it. Sebum also kills certain bacteria or deters their growth. However, certain bacteria can grow and thrive in sebum, as discussed in *A&P in the Real World: Acne.*

Quick Check

☐ 6. What are eccrine sweat glands, and what are the functions of sweat?
☐ 7. What are the other three types of sweat glands, and how do they differ from each other?
☐ 8. How do sebaceous glands and sebum differ from sweat glands and sweat?

Apply What You Learned

☐ 1. You are a researcher for a cosmetics company, and the company wants you to devise a product that makes eyelashes grow longer. How could this be achieved?
☐ 2. Certain products for hair and nails contain proteins and amino acids, which the manufacturers claim are needed by the hair shafts and nail bodies for strength. What do you make of these claims? Explain your answer.

See answers in Appendix A.

MODULE **5.6**
Pathology of the Skin

Learning Outcomes

1. Explain how to classify burns and how to estimate their severity.
2. Describe the three main types of cancerous skin tumors.

Skin diseases, or *pathologies,* are frequent reasons why people seek medical treatment. One common type of skin pathology is a

 A&P in the Real World

Acne

One of the most common skin problems to plague adolescents is *acne vulgaris* or simply *acne.* Indeed, 96% of teenagers and young adults experience some degree of acne.

It is caused by an accumulation of sebum and dead cells within sebaceous glands, which produces a *comedone,* or blackhead. Occasionally, these become infected by the bacterium *Propionibacterium acnes,* which results in inflammation and the formation of a *pustule* or pimple. Although many individuals experience only mild acne, in some people it may become severe and lead to permanent scarring.

Male sex hormones such as testosterone are the primary cause of acne. For this reason, acne tends to be more pronounced in males who are entering puberty, which is when levels of these hormones rise. (Note that females produce small amounts of these hormones as well.) Acne generally decreases and may disappear by age 20–25, but in some individuals it persists much longer.

wound, which is defined as any disruption in the skin's integrity. Wounds come in various forms, such as *lacerations* (cuts), *burns,* and *skin cancers.* They may involve the epidermis, the dermis, and occasionally deeper tissues.

Treatment of wounds depends on their type and severity. For example, lacerations may be closed with a few *sutures* or stitches, whereas severe burns may require extensive surgical repair. In this module, we first examine burns, including how to classify and treat them. We finish by discussing skin cancer, a pathology that is extremely common and potentially deadly.

Burns

A **burn** is a skin wound caused by agents such as heat, extreme cold, electricity, chemicals, and radiation. Burns are grouped into three classes according to the extent and depth of tissue damage (**Figure 5.12**):

- **First-degree burns.** The most minor burns are **first-degree burns** (or *superficial burns*), in which only the epidermis is damaged. Typically present are erythema (redness) of the skin and minor pain, but no blisters or permanent damage. First-degree burns generally require no treatment. An example is a minor sunburn.
- **Second-degree burns.** Second-degree burns (also called *partial thickness burns*) involve the epidermis and part or all of the dermis, and result in significant pain, blistering, and

possibly scarring. Second-degree burns generally require medical intervention.

- **Third-degree burns.** The most damaging burns are **third-degree burns** *(full thickness burns),* which involve the epidermis, dermis, hypodermis, and possibly deeper tissues including muscle and bone. Interestingly, third-degree burns are not generally painful at first because the nerves are destroyed too. These burns typically result in major tissue damage and significant scarring with loss of hair follicles and diminished or absent keratin production. Patients often have problems with dehydration due to massive fluid loss from the swelling that accompanies such burns, and are also at great risk for infection. Treatment of third-degree burns depends on the location and extent of the burn, but patients may require extensive *skin grafting,* in which skin is removed from another part of the body and transplanted onto the burn wound. These patients often face a prolonged and difficult recovery.

We can also describe burns by estimating the percentage of body surface area they affect. **Figure 5.13** illustrates the **rule of nines,** which divides the body into 11 areas, each representing 9% of the total body surface area (plus the genital area, which represents 1%). So a burn involving the entire right lower limb covers 18% of the body, and a burn involving the entire right upper limb and anterior trunk covers 27% of the body. Such estimates are important because they predict the complications that can arise from a burn. For example, the total body surface area affected by

(a) First-degree burn involves epidermis only.

(b) Second-degree burn involves epidermis and part or all of the dermis.

(c) Third-degree burn involves epidermis, dermis, hypodermis, and possibly deeper tissues.

Figure 5.12 The three classes of burns.

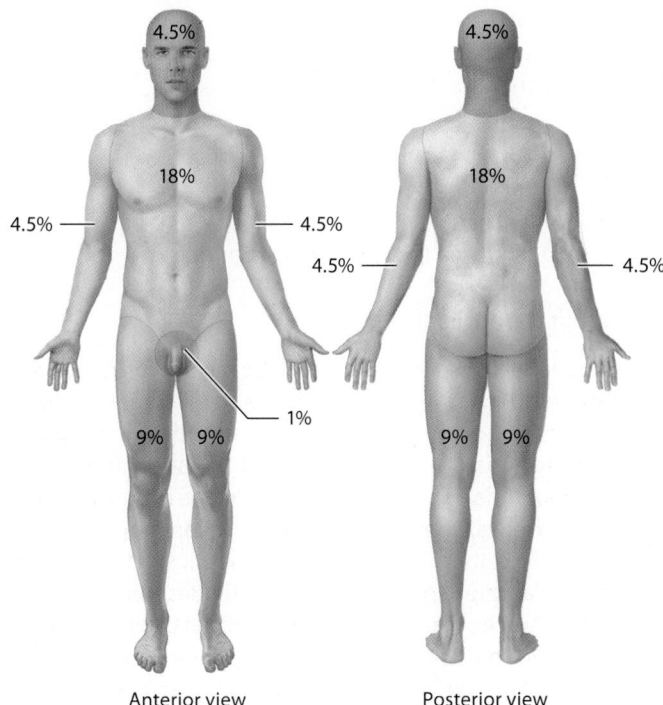

Figure 5.13 **Rule of nines: estimating the extent of a burn.**

Anterior view   Posterior view

a burn indicates the patient's level of dehydration and the amount of fluids the patient should receive. The total body surface area of the burn helps determine prognosis, although this is also influenced by factors such as age and other health problems.

Quick Check

☐ 1. What are the three classes of burns, and how do they differ?
☐ 2. Summarize the clinical importance of the rule of nines.

Skin Cancer

Cancer, one of the most common diseases in the world, claims millions of lives every year. The root cause of cancer is mutations in the DNA that induce a cell to lose control over its cell cycle. This results in the formation of a *tumor,* a cluster of undifferentiated cells. Cancerous tumors are characterized by their ability to *metastasize,* or spread to other tissues through the cardiovascular or lymphatic systems. Tumors then grow in new areas, where the tumor cells alter the structure of a tissue and prevent it from functioning normally.

Three common cancers impact the skin: basal cell carcinoma, squamous cell carcinoma, and malignant melanoma. All three types are linked to exposure to UV radiation. Other factors can play a role, too, such as exposure to other forms of radiation and to *carcinogens* (agents that induce cancer)—for example, the toxins in cigarettes.

Basal Cell Carcinoma

The most common skin cancer—indeed, the most common of all cancers—is **basal cell carcinoma** (BEH-zal sell kar-si-NOH-mah; *-oma* = "tumor") (**Figure 5.14a**). As implied by its name,

basal cell carcinoma arises from keratinocytes in the stratum basale of the epidermis. Typically found on skin that is regularly exposed to UV radiation, it generally forms a nodule with a central "crater" that ulcerates.

Although destructive to local tissue and sometimes disfiguring, basal cell carcinoma generally does not metastasize to other tissues. Surgical removal usually resolves it completely.

Squamous Cell Carcinoma

The second most common skin cancer is **squamous cell carcinoma,** a cancer of the keratinocytes of the stratum spinosum (**Figure 5.14b**).

"Cratered" center

(a) Basal cell carcinoma: cancer of keratinocytes in stratum basale; generally forms a nodule with a "cratered" center

Ulcerated area

(b) Squamous cell carcinoma: cancer of keratinocytes in stratum spinosum; forms plaques that bleed or ulcerate

Irregular borders and shape

Changing color

(c) Malignant melanoma: cancer of melanocytes; characterized by asymmetrical shape, irregular borders, blue-black color, larger diameter, and evolving nature

Figure 5.14 **The three main forms of skin cancer.**

Most common on the head and neck, it tends to form scaly plaques that bleed or ulcerate. Squamous cell carcinoma is more likely to metastasize than basal cell carcinoma but typically can still be managed with surgery.

Malignant Melanoma

The most dangerous skin cancer is **malignant melanoma** (mel-ah-NOH-mah), a cancer of melanocytes (**Figure 5.14c**). Melanomas can be distinguished from other skin cancers and from normal moles using the ABCDE rule:

- Asymmetrical shape (the two sides do not match)
- Border irregularity
- Color, usually blue-black or a variety of colors
- Diameter generally larger than 6 mm (the size of a pencil eraser)
- Evolving (changing) shape and size

Early detection of melanoma is critical because it has a much greater tendency to metastasize than basal or squamous cell carcinoma. This is because the "arms" of the cancerous melanocytes extend down into the dermis and access dermal blood vessels, enabling the cells to spread to other tissues via the bloodstream. Like basal and squamous cell carcinoma, malignant melanoma is treated with surgical excision. Treatment may also include chemotherapy and radiation therapy, although their effectiveness is limited. Prognosis depends on the size of the tumor, the depth to which it extends into the dermis, and whether it has metastasized to other tissues.

Quick Check

- ☐ 3. What is cancer?
- ☐ 4. How do the three types of skin cancer differ?

Apply What You Learned

- ☐ 1. Your friend just spent a day at the beach but forgot to use sunscreen. That evening she complains of a great deal of pain and you notice that her skin is red and beginning to form large blisters. What type of burn has she suffered? What other potential long-term damage has she done to her skin?
- ☐ 2. A hypothetical chemical is found to be carcinogenic when applied to cells in a tissue culture. Chemical analysis shows that it is a water-soluble compound (hydrophilic). What type of cancer, if any, might this chemical cause when applied to the skin? Explain your reasoning. (*Hint:* Do hydrophilic substances pass through the epidermis? Why or why not?)

See answers in Appendix A.

CHAPTER SUMMARY

Take full advantage of the invaluable online practice and study tools for this chapter by going to MasteringA&P®. There you will find:

- Practice tests, module quizzes, art-labeling and sequence-the-steps activities
- ▶ Animated figures with quizzes
- *Practicing A&P worksheets*
-))) Author-narrated podcast
- Dynamic Study Modules
- iP *Interactive Physiology*

MODULE 5.1
Overview of the Integumentary System 160

- The skin consists of the superficial **epidermis** and the deeper **dermis** (Figure 5.1, p. 161). Embedded within the dermis and epidermis we find the accessory structures of the skin, including **sweat glands, sebaceous glands, hairs,** and **nails.**

  Practice art-labeling exercise: Figure 5.1.

- The dermis rests on the **hypodermis.**

- The integument serves many functions: protection from mechanical trauma, pathogens, and the environment; sensation; thermoregulation (Figure 5.2, p. 163), excretion, and **vitamin D synthesis.**

MODULE 5.2
The Epidermis 164

- The majority of the cells in the epidermis are **keratinocytes,** which manufacture **keratin.** (Figure 5.3, p. 165)

  Practice art-labeling exercise: Figure 5.3.

- **Dendritic cells** are phagocytes in the stratum spinosum. **Merkel cells** are sensory receptors in the stratum basale.

- Thick skin has all five epidermal strata, and all five strata are thicker, particularly the stratum corneum. Thin skin lacks the stratum lucidum, and its four strata are thinner. (Figure 5.4, p. 167)

  Practice art-labeling exercise: Figure 5.4.

MODULE 5.3
The Dermis 168

- The dermis consists of two layers: the **papillary layer** and the **reticular layer** (Figure 5.5, p. 168).

- Folds of the papillary layer include **dermal papillae** and **epidermal ridges** (Figure 5.6, p. 169).

- Gaps between collagen fibers in the reticular layer cause **tension lines** (Figure 5.7, p. 170).

ASSESS WHAT YOU LEARNED

See answers in Appendix A.

LEVEL 1 Check Your Recall

1. Explain why the skin is considered an organ.

2. Which of the following correctly describes the structure of the skin?

   a. It consists of the superficial epidermis, middle dermis, and deep hypodermis.
   b. It consists of the superficial epidermis and deep dermis.
   c. It consists of the superficial dermis and deep epidermis.
   d. It consists of the superficial dermis and deep hypodermis.

3. Which of the following is *not* a function of the integument?

   a. Protection from mechanical trauma
   b. Protection from the environment
   c. Thermoregulation
   d. Protection from acid-base imbalances
   e. Vitamin D synthesis

4. Explain what happens to dermal blood vessels when heat needs to be conserved (i.e., due to a cold environment).

5. Number the strata of thick skin epidermis from deepest (1) to most superficial (5).

   _____ Stratum spinosum
   _____ Stratum corneum
   _____ Stratum basale
   _____ Stratum lucidum
   _____ Stratum granulosum

6. Keratinocytes in the superficial strata of the epidermis die because:

   a. they are too far away from the blood supply in the dermis.
   b. they are surrounded by a lipid-based substance that makes them more permeable to water.
   c. they do not die.
   d. No keratinocytes in the epidermis are alive.

7. Mark the following statements as true or false. If a statement is false, correct it to make a true statement.

   a. Melanocytes account for the bulk of the epidermis.
   b. Keratinocytes begin life in the stratum corneum and gradually are pushed into the stratum basale.
   c. Dendritic cells are phagocytes of the immune system that protect the skin and deeper tissues from invasion by pathogens.
   d. Merkel cells produce the pigment melanin.

8. Which of the following statements is false?

   a. Thin skin lacks the stratum lucidum, whereas thick skin has all five epidermal layers.
   b. Thick skin is located on the palms, the palmar surfaces of the fingers, the soles of the feet, and the plantar surface of the toes, whereas thin skin is located everywhere else.
   c. Thick skin has numerous hairs, whereas thin skin lacks hairs.
   d. Thin skin has a thin stratum corneum and the other layers are thinner than what we find in thick skin.

9. What are the functions of the dermal papillae?

10. Epidermal ridges are caused by:

    a. the epidermal papillae.
    b. mounds of papillary dermis arranged into dermal ridges.
    c. gaps between collagen bundles in the reticular layer.
    d. tight binding of the reticular layer to deeper structures.

11. Mark the following statements as true or false. If a statement is false, correct it to make a true statement.

    a. The primary skin pigment is melanin, which is derived from the amino acid tyrosine.
    b. Melanin is produced by melanocytes and covers the nuclei of neighboring dendritic cells.
    c. Carotene is a brown-black pigment that accumulates in the stratum corneum.
    d. Increased amounts of blood flowing through the dermis lead to pallor in the skin.
    e. Hemoglobin is a pigment found in red blood cells, and its color depends on the oxygen level of the blood.

12. Which of the following is not a function of body hair in humans?

    a. Protection from the environment
    b. Sensation
    c. Protection from mechanical trauma
    d. Regulation of blood flow

13. Fill in the blanks: The portion of the hair that projects from the surface of the skin is the _____, and the portion within the dermis is the _____, which is embedded in a(n) _____. The portion of the hair that contains cells that undergo mitosis during the growth stage of the hair is known as the _____.

14. Nail growth occurs when:

    a. cells in the nail plate undergo mitosis.
    b. cells in the nail matrix undergo mitosis.
    c. cells in the eponychium undergo mitosis.
    d. cells in the medial and lateral nail folds undergo mitosis.

15. Fill in the blanks: Sebaceous glands secrete by _____ secretion; sweat glands secrete by _____ secretion.

16. Match each type of gland with its correct properties.

    _____ Eccrine sweat gland
    _____ Sebaceous gland
    _____ Ceruminous gland
    _____ Mammary gland
    _____ Apocrine sweat gland

    a. Branched gland that secretes sebum into a hair follicle
    b. Gland that secretes a protein-rich sweat into a hair follicle
    c. Simple coiled tubular gland found over most of the body that secretes sweat through a sweat pore
    d. Gland that secretes ear wax
    e. Modified sweat gland that produces milk

17. How do sweat and sebum differ?

18. Which type of burn involves the epidermis and all or part of the dermis?

    a. First-degree burn
    b. Second-degree burn
    c. Third-degree burn
    d. Fourth-degree burn

19. The type of skin tumor that involves the keratinocytes of the stratum spinosum is:

    a. basal cell carcinoma.
    b. squamous cell carcinoma.
    c. malignant keratocytoma.
    d. malignant melanoma.

LEVEL 2 Check Your Understanding

1. Why don't you bleed when you cut your nails? Why don't you bleed when a hair is pulled?

2. Manufacturers of shampoos and conditioners often claim their products contain vitamins that are necessary to keep hair shafts healthy. Do hair shafts need vitamins? (*Hint:* Vitamins are

required for cells to carry out certain reactions.) Why or why not? How valid are these claims?

3. The hair and nails are sometimes called *accessory organs.* Are these structures technically organs? Why or why not?

LEVEL 3 Apply Your Knowledge

PART A: Application and Analysis

1. You are working in the emergency department when paramedics rush in with an unconscious patient. You notice that the patient is wearing athletic clothing and the skin on his face and elsewhere on his body is bright red. The paramedics tell you that he was picked up after collapsing during a bike race. What does the color of the patient's skin tell you about the probable cause of his illness? Explain.

2. After Ramon's skin came into contact with a poison ivy plant in biology lab, he developed a painful, itchy rash. However, after the skin of his colleague Cathy came into contact with snake venom in lab, she developed no skin irritation. Poison ivy plants contain lipid-soluble oils, whereas snake venoms contain mostly water-soluble peptides. Explain why Ramon developed a rash, whereas Cathy did not.

3. Nguyen comes to your clinic with a mole that has recently changed in appearance. You examine the mole and note that its borders are irregular, it has a deep blue-black color, and the color is unevenly distributed throughout the mole.

    a. What is your immediate concern? Why?
    b. Nguyen tells you that she has used a tanning booth once per week for the past several years, and that the tanning salon advertises it as safe. What do you tell her about the tanning salon's claim? How does any UV exposure affect keratinocytes and melanocytes? Explain.

PART B: Make the Connection

4. What would happen to the skin if the oil produced by sebaceous glands was instead a polar covalent compound? (*Connects to Chapter 2*)

5. Many antiaging skin creams contain collagen and hyaluronic acid. Manufacturers claim that the collagen and hyaluronic acid applied to the surface of the skin will be absorbed into the dermis, where they will be incorporated into dermal tissue.

    a. What are collagen and hyaluronic acid, and what are their functions? (*Connects to Chapter 4*)
    b. Are these molecules polar or nonpolar? (*Connects to Chapter 2*)
    c. Predict whether or not these molecules are likely to be absorbed by the epidermis.
    d. Predict the effectiveness of the creams.

6. Would a mild second-degree burn be likely to heal by regeneration or fibrosis? Would the same type of healing take place in a third-degree burn that involved muscle tissue? Why or why not? (*Connects to Chapter 4*)

6

Bones and Bone Tissue

Have you ever seen a slug creep along the ground? Well, that's how we would move without our bones. We're now starting our tour of the **skeletal system,** which includes the bones, joints, and other supporting tissues. As you discovered earlier, an *organ system* is defined as a collection of organs with related functions, and organs are composed of two or more tissues (see Chapter 1). At first glance, the main organs of the skeletal system—the bones—may seem to be only one type of tissue. But bones contain several tissue types, including *bone tissue,* which is more formally called *osseous tissue* (OSS-ee-us); dense regular and irregular collagenous connective tissues; and a tissue called *bone marrow.* These different tissues make each bone in the skeletal system an organ.

People are said to have "healthy bones" when they have properly functioning cells and adequate quantities and quality of extracellular matrix. When a bone has both of these, it is able to perform its functions much more effectively. When it is lacking in these components, bone disease nearly always results. In this chapter we will explore the gross and microscopic structure of these organs, the functions of the skeletal system, and the relationship of bones to the body's overall homeostasis. We discuss individual bones and joints in the skeletal and articulations chapters (see Chapters 7 and 8, respectively).

MODULE **6.1**

Introduction to Bones as Organs

Learning Outcomes

1. Describe the functions of the skeletal system.
2. Describe how bones are classified by shape.
3. Describe the gross structure of long, short, flat, irregular, and sesamoid bones.
4. Explain the differences between red and yellow bone marrow.

Much of this chapter is devoted to the microscopic anatomy of bone and associated tissues. However, in this module, we also take a look at the gross anatomy of bones. The upcoming sections introduce you to their different shapes and their general structure. But before we examine their structure, let's turn our attention to the overall functions of bones and the skeletal system.

Photo: Scanning electron micrograph showing osteoclasts in bone lacunae.

Functions of the Skeletal System

◄◄ **Flashback**

1. Define the terms mineral, electrolyte, and buffer. (pp. 33, 49)
2. What are adipocytes, and what do they store? (p. 138)

Some functions of the skeletal system are fairly obvious. For example, it probably comes as no surprise that the skeleton provides support for the body. But the skeleton has many other roles than simply acting as a scaffold for other body parts—a number of its functions are vital to the maintenance of homeostasis for the body as a whole. These roles, shown in **Figure 6.1**, include the following:

- **Protection.** Certain bones, including the skull, sternum (breastbone), ribs, and pelvis, protect their underlying

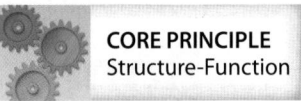

CORE PRINCIPLE
Structure-Function

organs. Bone is one of the hardest substances in the body and, in an example of the Structure-Function Core Principle (p. 25), it provides a good strong "shell" for organs such as the brain, the heart, and the lungs. Bone also protects the sense organs, such as the eyes and the organs of hearing.

- **Mineral storage and acid-base homeostasis.** Bone is the most important storehouse in the body for minerals such as calcium, phosphorus, and magnesium salts. These minerals, also found in the blood as electrolytes, acids, and bases, are vital to the maintenance of electrolyte and acid-base balance in the body.

Protection:
Skeleton protects vital organs such as the brain.

Brain

Mineral storage and acid-base homeostasis:
Bone stores minerals such as Ca^{2+} and PO_4^{3-}, which are necessary for electrolyte and acid-base balance.

Blood cell formation:
Red bone marrow is the site of blood cell formation.

Forming blood cells in red bone marrow

Fat storage:
Yellow bone marrow stores triglycerides.

Fat in yellow bone marrow

Movement:
Muscles produce body movement via their attachment to bones.

Muscle attached across joint

Support:
The skeleton supports the weight of the body.

Figure 6.1 Functions of the skeletal system.

- **Blood cell formation.** Bones house *red bone marrow,* which is a special form of connective tissue. In this tissue, the process of **hematopoiesis** (hee′-mah-toh-poy-EE-suss; *hemato-* = "blood," *poiesis-* = "to make"), or formation of blood cells, takes place. Hematopoiesis is examined in detail in the blood chapter (see Chapter 19).
- **Fat storage.** As you can see in Figure 6.1, not all bone marrow is red. Some, called *yellow bone marrow,* contains fat cells, or adipocytes, with stored triglycerides. Fatty acids from the breakdown of these triglycerides can be released and used as fuel by cells if necessary.
- **Movement.** Bones serve as sites of attachment for most skeletal muscles. When muscles contract, they pull on the bones, which generates movement around a joint.
- **Support.** The skeleton supports the weight of the body and provides its structural framework.

Quick Check

☐ 1. What are the main organs of the skeletal system?
☐ 2. What are the primary functions of the skeletal system?

Bone Structure

◀◀ **Flashback**

1. What are dense regular and dense irregular collagenous connective tissues? (p. 139)
2. What are collagen fibers? (p. 125)
3. What is hyaline cartilage? (p. 142)

Looking at all of the bones in a skeleton, you can see that they show great diversity in appearance. However, even with all this variation, we can define five structural bone classes, and each of our 206 bones fits into one of these classes. The criterion used to categorize a bone is its general shape. In this section we examine bone shapes, then discuss their structure in more detail.

Classification of Bones by Shape

The five general bone shapes, shown in **Figure 6.2**, are long, short, flat, irregular, and sesamoid. Let's take a closer look at what defines each shape.

- **Long bones. Long bones** are named for their overall shape: They are longer than they are wide (Figure 6.2a). Examples of long bones include most bones of the arms and legs (including the humerus, or upper arm bone), as well as bones of the hands, feet, fingers, and toes. Note that long bones are not named for their size—in fact, some long bones are quite short.
- **Short bones.** Like long bones, short bones are named for their shape rather than their size. **Short bones** are about as long as they are wide, or roughly cube-shaped (Figure 6.2b). Examples of short bones include the carpals (bones of the wrist) and tarsals (bones of the ankle).
- **Flat bones.** As implied by their name, **flat bones** are thin and broad (Figure 6.2c). Examples of flat bones include most bones of the skull, the clavicles (collarbones), the ribs, the sternum, and the bones of the pelvis.
- **Irregular bones.** The **irregular bones** do not fit into the other classes due to their irregular shapes (Figure 6.2d). Examples of irregular bones include the vertebrae and certain skull bones.

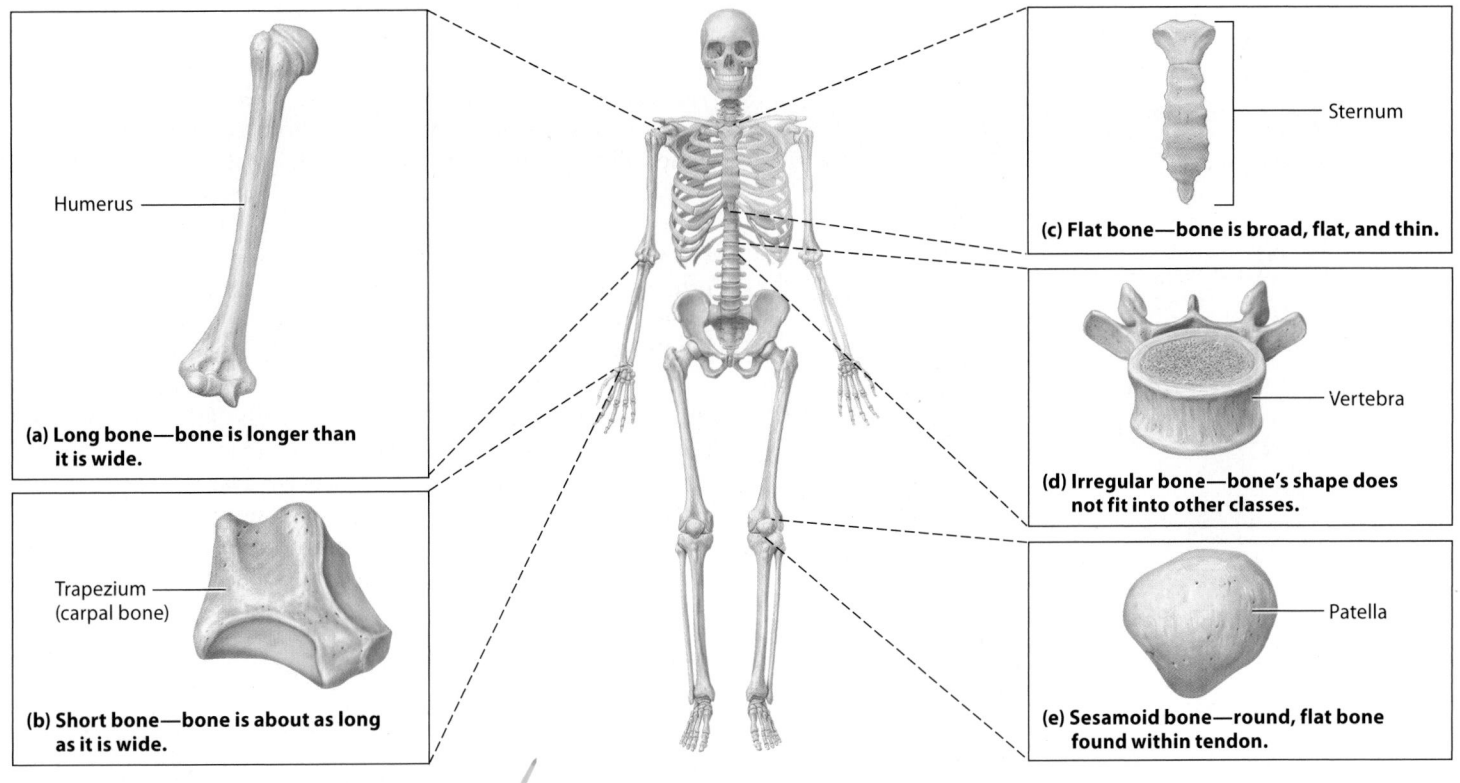

(a) Long bone—bone is longer than it is wide.

Humerus

(b) Short bone—bone is about as long as it is wide.

Trapezium (carpal bone)

(c) Flat bone—bone is broad, flat, and thin.

Sternum

(d) Irregular bone—bone's shape does not fit into other classes.

Vertebra

(e) Sesamoid bone—round, flat bone found within tendon.

Patella

Figure 6.2 Classification of bones by shape.

- **Sesamoid bones. Sesamoid bones** (SEH-sah-moyd; "sesame-shaped") are specialized bones located within tendons. They are generally small, relatively flat, and oval-shaped (Figure 6.2e). They give the tendon a mechanical advantage, providing better leverage for muscles, and also reduce wear and tear on the tendon. An example of a sesamoid bone is the patella (kneecap).

Structure of a Long Bone

When you look at the outside of a long bone, you'll see that most of its surfaces are covered with a membrane, the **periosteum** (pear'-ee-OSS-tee-um; *peri-* = "around," *oste-* = "bone") (**Figure 6.3a**). The periosteum is composed of dense irregular collagenous

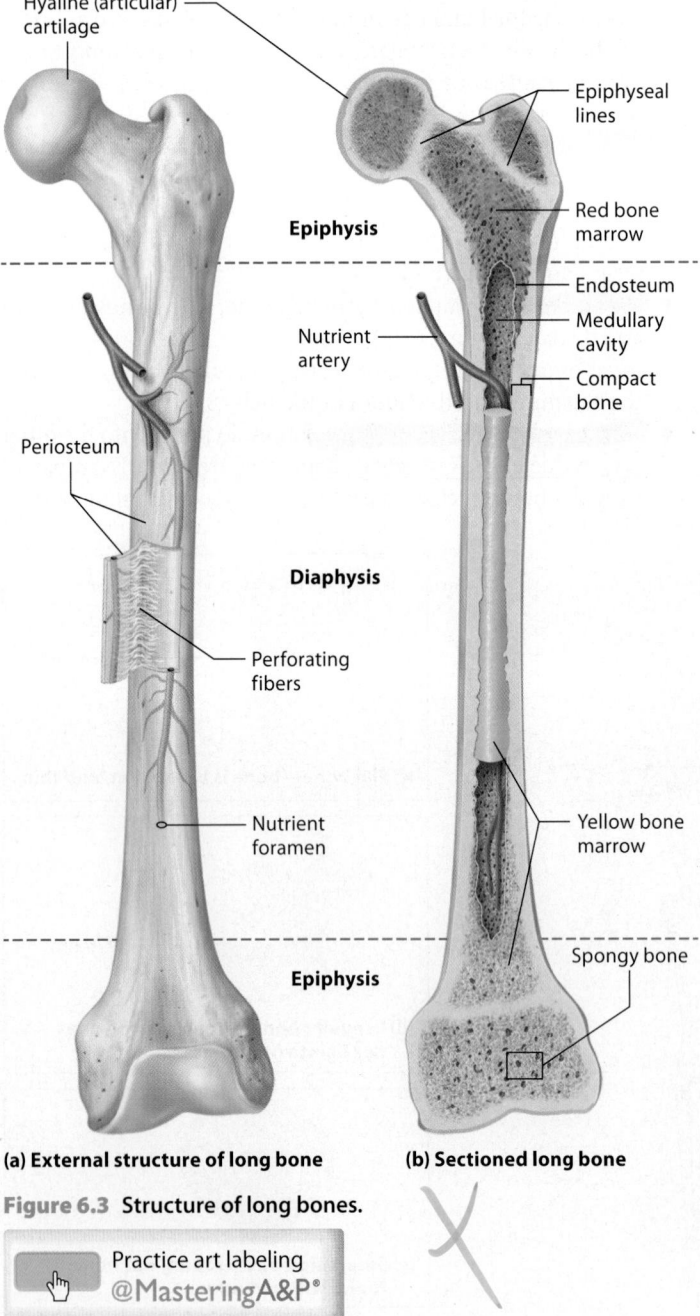

Hyaline (articular) cartilage

Epiphyseal lines

Epiphysis

Red bone marrow

Nutrient artery

Endosteum

Medullary cavity

Compact bone

Periosteum

Diaphysis

Perforating fibers

Nutrient foramen

Yellow bone marrow

Epiphysis

Spongy bone

(a) External structure of long bone   **(b) Sectioned long bone**

Figure 6.3 Structure of long bones.

Practice art labeling
@MasteringA&P®

connective tissue that is richly supplied with blood vessels and nerves. As Figure 6.3a shows, the periosteum is firmly attached to the underlying bone by collagen fibers called **perforating fibers** (or *Sharpey's fibers*). These fibers penetrate deeply into the bone matrix, securing the periosteum in place. The innermost surface of the periosteum contains different types of bone cells that function in bone homeostasis, as we'll see in the next module.

Each long bone features a long shaft and two rounded ends. The shaft of a long bone is called its **diaphysis** (dy-AF-eh-sis; *dia-* = "across"). At the ends of the bones we find the enlarged, rounded **epiphyses** (eh-PIF-eh-seez; singular, *epiphysis; epi-* = "upon"). Figure 6.3a shows that the epiphyses are covered with a thin layer of hyaline cartilage, also known as *articular cartilage,* which allows bones to rub together with reduced friction at joints. Note that diaphyses and epiphyses are unique to long bones.

Inside a long bone (**Figure 6.3b**) the first thing you might notice is that the diaphysis surrounds a hollow cavity called the medullary cavity (MED-yoo-lehr-ee), also known as the **marrow cavity** (*medullary* = "marrow"). As its name implies, this is where much of the bone marrow of a long bone is housed. The marrow can be either yellow or red, depending on the bone and the person's age.

Notice in Figure 6.3b that the inside of the bone has two obviously different textures. The hard, dense outer bone is **compact bone.** The structure of compact bone enables it to resist the majority of stresses placed on it, which are linear compression and twisting forces. The inner, honeycomb-like bone is **spongy bone,** or *cancellous bone* (KAN-sel-us). Spongy bone forms a framework of bony struts that allows it to resist forces in many directions and provides a place for the bone marrow to reside. The diaphyses of long bones consist of a very thick layer of compact bone with only scant inner spongy bone surrounding the medullary cavity. Like the diaphysis, the outer parts of the epiphyses are compact bone; however, the interior of the epiphyses consists of abundant spongy bone. Both compact and spongy bone will be discussed further in Module 6.2.

The bony struts of spongy bone, and indeed all inner surfaces of bone, are lined with a membrane called the **endosteum** (en-DOS-tee-um; *endo-* = "within"). Like the periosteum, the endosteum contains different types of bone cells that help maintain bone homeostasis. However, the endosteum is thinner and lacks the fibrous outer layer that we see with the periosteum.

One final thing to note in Figure 6.3b is the presence of lines running across the proximal and distal ends of the bone between the epiphysis and the diaphysis. These lines, called *epiphyseal lines* (eh-PIF-ih-see-uhl), are remnants of a structure termed the *epiphyseal plate.* The epiphyseal plate, also known as the *growth plate,* is a line of hyaline cartilage from which a long bone grows in length in children and adolescents. Bone growth and the role of the epiphyseal plate are discussed in Module 6.4.

Structure of Short, Flat, Irregular, and Sesamoid Bones

Short, flat, irregular, and sesamoid bones do not have diaphyses or epiphyses. As a result, they contain no medullary cavity,

Figure 6.4 Structure of short, flat, irregular, and sesamoid bones.

epiphyseal lines, or epiphyseal plates. But as you can see in Figure 6.4, they do share some similarities with long bones. Externally these bones are covered with a periosteum attached by perforating fibers and well supplied with blood vessels and nerves. In addition, they have an outer layer of compact bone that surrounds the inner spongy bone.

Notice that their internal structure is fairly simple. A section through one of these bones resembles a "sandwich" made up of two thin layers of compact bone and a middle layer of spongy bone housing bone marrow. In flat bones, the spongy bone is called the **diploë** (dip-LOW-ee; *diploë* = "fold"). The interior of some flat and irregular bones of the skull contains hollow, air-filled spaces called *sinuses* that make the bones lighter.

Blood and Nerve Supply to Bone

Bones are well supplied with blood vessels (look again at Figures 6.3 and 6.4), as evidenced by the extensive bleeding that occurs whenever a bone is injured. Bones are also supplied with many sensory fibers, which is why we feel such an injury so acutely. The blood supply to short, flat, irregular, and sesamoid bones comes largely from vessels in the periosteum that penetrate into the bone. Long bones also get a significant portion (about one-third) of their blood supply from the periosteum, which supplies primarily the compact bone. However, their greatest source of blood comes from one or two *nutrient arteries* that enter the bone via a small hole in the diaphysis called the **nutrient foramen** (for-AY-men; see Figure 6.3b). The nutrient artery passes through compact bone into the inner medullary cavity, where it supplies the internal structures of the long bone. The epiphyses receive some of their blood supply from the nutrient artery, but the majority of it comes from small blood vessels that enter and exit via numerous small holes in their compact bone.

Red and Yellow Bone Marrow

As we discussed earlier, there are two types of bone marrow with different functions. **Red bone marrow** consists of loose connective tissue supporting islands of blood-forming, or *hematopoietic*, cells. **Yellow bone marrow,** which stores triglycerides, consists mostly of blood vessels and adipocytes. In infants and young children, most bone marrow is red because their rapid rate of growth requires a constant supply of new blood cells. At about age 5, yellow bone marrow begins to replace some of the red bone marrow.

By the time adulthood is reached, most bone marrow in the body is yellow. Red marrow remains only in the bones of the pelvis, the proximal femur and humerus (thigh and arm bones), the vertebrae, the ribs, the sternum, the clavicles, and the scapulae (shoulder blades). A small amount of red marrow is also found in certain bones of the skull, but this degenerates as we age. Adults have limited red bone marrow because they are not actively growing and do not need hematopoiesis to occur as rapidly as it does in young children. However, some yellow marrow can be replaced with red marrow if increased blood cell production is needed. Bone marrow can also be transplanted in an attempt to treat some diseases of the blood—see *A&P in the Real World: Bone Marrow Transplantation.*

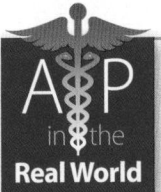

Bone Marrow Transplantation

Patients with diseases of the blood such as leukemia, sickle-cell anemia, and aplastic anemia have improperly functioning hematopoietic cells. Many such patients can therefore benefit significantly from *bone marrow transplantation,* in which hematopoietic cells from red bone marrow are removed from a matching donor and given to a recipient.

During a *bone marrow harvest,* a needle is inserted into the pelvic bone of the anesthetized donor and red bone marrow is withdrawn. This is generally repeated several times until one to two quarts of red bone marrow—about 2% of the total—has been removed.

After the recipient's bone marrow is destroyed, the removed bone marrow from the donor is then given to the recipient intravenously. The cells travel to the recipient's bone marrow cavities, and, if all goes well, in about 2–4 weeks begin producing blood cells. Recipients typically experience flu-like symptoms during this period, and they run the risk of complications, including infection and transplant rejection. However, if the transplant "takes," or is accepted, many recipients can return to a healthy life.

☐ 3. Look at Figure 6.2 and classify the following bones as long, short, flat, or irregular.
   a. Phalanges (fingers)
   b. Ribs
   c. Carpals
   d. Pelvis
   e. Humerus

☐ 4. Where are compact bone and spongy bone located, and what are their functions?

☐ 5. What are the two types of bone marrow and what are their functions?

Apply What You Learned

☐ 1. Predict how the characteristics of a person's bones would change if compact bone were located on the inside and spongy bone on the outside.

☐ 2. Sometimes a bone is injured at the epiphyseal plate. Predict the long-term effect this might have on the injured bone.

☐ 3. Sometimes the nutrient artery is damaged in a severe bone injury. Predict the potential consequences of this event. How would this affect the ability of the bone to heal?

See answers in Appendix A.

MODULE **6.2**

Microscopic Structure of Bone Tissue

Learning Outcomes

1. Describe the inorganic and organic components of the extracellular matrix of bone tissue.
2. Explain the functions of the three main cell types in bone tissue.
3. Describe the microscopic structure of compact bone and the components of the osteon.
4. Describe the microscopic structure of spongy bone.

The primary tissue type found in bones is **bone tissue,** formally known as **osseous tissue.** We introduced this tissue earlier as one of the specialized connective tissues (see Chapter 4). Bone shares many properties with the other connective tissues. For example, like most connective tissues, bone consists mainly of extracellular matrix with small numbers of cells scattered throughout. However, as we discuss in this module, the extracellular matrix and cell types in bone are unique. Let's now explore these components.

The Extracellular Matrix of Bone

◄◄ **Flashback**

1. Compare organic and inorganic substances. (pp. 45, 50)
2. What are protein fibers and ground substance? (pp. 124–125)

The extracellular matrix (ECM) of bone tissue, sometimes simply called the *bone matrix,* is not seen in any other tissue. About 65% of the total weight of bone consists of an **inorganic matrix** composed of minerals. The remaining 35% of total bone weight consists of an **organic matrix,** which contains the usual ECM "ingredients," including collagen fibers.

Although the inorganic matrix generally makes bone strong, the organic matrix contributes equally to its strength. Note in **Figure 6.5a** what happens if we destroy the organic matrix of bone and leave the inorganic matrix intact—the bone becomes brittle and shatters easily. Similarly, in **Figure 6.5b**, if we dissolve the inorganic matrix and leave the organic matrix behind, the bone becomes flexible and unable to resist compression. So, clearly, both components are required to maintain bone strength.

Inorganic Matrix

The predominant ingredient of the inorganic matrix is calcium salts (in fact, bone stores about 85% of the total calcium ions in the body), with a good amount of phosphorus as well. Most of the calcium and phosphorus salts exist as large molecules of a mineral called **hydroxyapatite crystals** $[Ca_{10}(PO_4)_6(OH)_2]$ (hy-drok′-see-AP-ah-tyt). This composition, which makes bone one of the hardest substances in the body, gives bone its strength and the ability to resist compression, allowing it to perform its functions of support and protection and demonstrating the

CORE PRINCIPLE
Structure-Function

Structure-Function Core Principle (p. 25). Other ingredients of the inorganic matrix include salts of bicarbonate (HCO_3^-), potassium, magnesium, and sodium.

Organic Matrix

The organic matrix of bone is known as **osteoid** (OSS-tee-oyd) and is not too different from the ECM of most other tissues. Osteoid consists of protein fibers, proteoglycans, glycosaminoglycans, glycoproteins, and bone-specific proteins (see Chapters 2 and 4). The predominant protein fibers in bone are collagen fibers, which form cross-links with one another and help bone to resist torsion (twisting) and tensile (pulling or stretching) forces. They also align with hydroxyapatite crystals, significantly enhancing the hardness of bone. Collagen fibers are one of the most important components of the bone ECM in terms of strength; when these fibers become inadequate or defective, bone weakens and any minor twisting force may break it.

The other components of organic matrix perform support functions. Recall the Gradients Core Principle (p. 26) and remember

CORE PRINCIPLE
Gradients

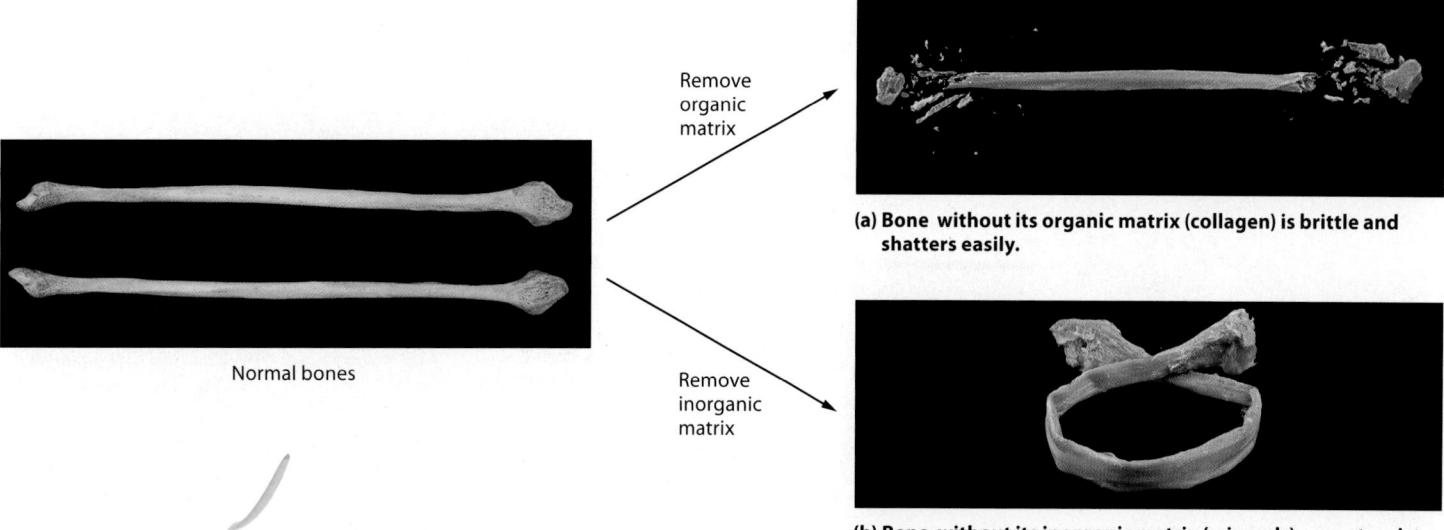

(a) **Bone without its organic matrix (collagen) is brittle and shatters easily.**

(b) **Bone without its inorganic matrix (minerals) cannot resist compression.**

Normal bones

Remove organic matrix

Remove inorganic matrix

Figure 6.5 The importance of bone matrices.

that the presence of large molecules such as glycosaminoglycans and proteoglycans creates a gradient that draws water out of the blood vessels and cells by osmosis (see Chapter 2). This process traps water within the ECM, which helps the tissue to resist compression. Recall also that most glycoproteins in the ECM act like molecular "glue" that binds the different components together. The glycoproteins in bone perform this same function by binding to both hydroxyapatite and bone cells.

Quick Check

☐ 1. What are the two components of bone ECM? How do these components differ?

Bone Cells

◀◀ **Flashback**

1. What are osteoblasts and osteocytes? (p. 145)
2. What is an osteoclast? (p. 145)

Bone is a dynamic tissue, and new bone is continually being formed as older bone is broken down. The dynamic nature of bone is due to the actions of three cell types that we discussed in the tissues chapter (see Chapter 4): osteoblasts, osteocytes, and osteoclasts (**Figure 6.6**). *Osteoblasts* build bone and mature into *osteocytes,* which help to maintain bone. In contrast, the multinucleated *osteoclasts* break down bone. Let's look in more detail at the location, primary functions, and structure of each of these cell types.

- **Osteoblasts. Osteoblasts** (OSS-tee-oh-blasts; -*blast* = "immature cell") are cuboidal to columnar cells found in the inner periosteum and endosteum. As shown in **Figure 6.7** ①, osteoblasts are derived from flattened cells called **osteogenic cells,** which differentiate into osteoblasts when stimulated by certain chemical signals. Osteoblasts are bone-building cells (remember the mnemonic "osteoBlasts Build Bone") that perform the process of *bone deposition,* during which they build up the bone ECM. In this process osteoblasts secrete the organic matrix and aid in formation of the inorganic matrix.

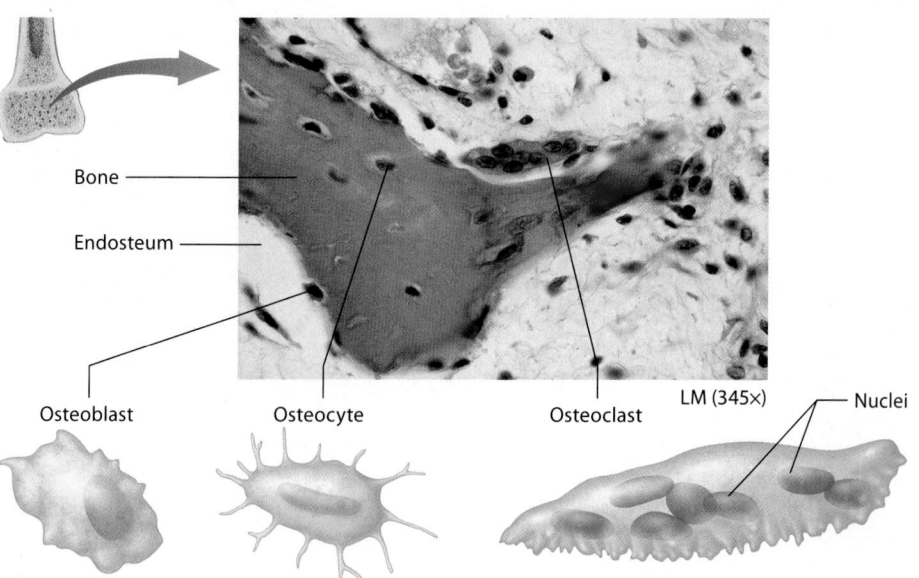

Bone

Endosteum

Osteoblast   Osteocyte   Osteoclast

LM (345×)   Nuclei

Figure 6.6 Types of bone cells.

① Osteogenic cells differentiate into osteoblasts.

② Osteoblasts deposit bone until they are trapped and become osteocytes.

③ Osteocytes maintain the bone extracellular matrix (ECM).

Figure 6.7 Functions of osteoblasts and osteocytes.

- **Osteocytes.** Osteoblasts become surrounded and eventually trapped by secreted bone matrix in a small cavity known as a *lacuna* (lah-KOO-nah; *lacun-* = "cavity"; **Figure 6.7** ②). At this point, these cells are called **osteocytes** (OSS-tee-oh-sytz). Osteocytes are mostly "retired" from their jobs and are relatively inactive. They do, however, secrete chemicals that are required for maintaining the ECM (**Figure 6.7** ③). They also appear to recruit osteoblasts to build up areas of the bone under tension.

- **Osteoclasts.** **Figure 6.8** shows that **osteoclasts** (OSS-tee-oh-klasts; *-clasia* = "broken") look quite different from osteoblasts and osteocytes, as they somewhat resemble jellyfish. Osteoclasts are large, multinucleated cells derived from the fusion of cells formed in the bone marrow. They reside in shallow depressions on the internal or external surfaces of bone. Osteoclasts are responsible for the process of *bone resorption,* during which they break down the bone ECM. They accomplish bone resorption by secreting hydrogen ions and enzymes from a region of the cell called the ruffled border. The hydrogen ions create an acidic environment that dissolves the inorganic matrix, and the enzymes break down the organic matrix. Notice that the liberated minerals, amino acids, and sugars then enter the osteoclast, after which they are eventually delivered to the blood for reuse in the body or excretion as waste. See *A&P in the Real World: Osteopetrosis* on page 192 for an explanation of what happens when osteoclasts cannot perform their functions.

The processes of bone deposition and bone resorption will be discussed further with bone remodeling in Module 6.5.

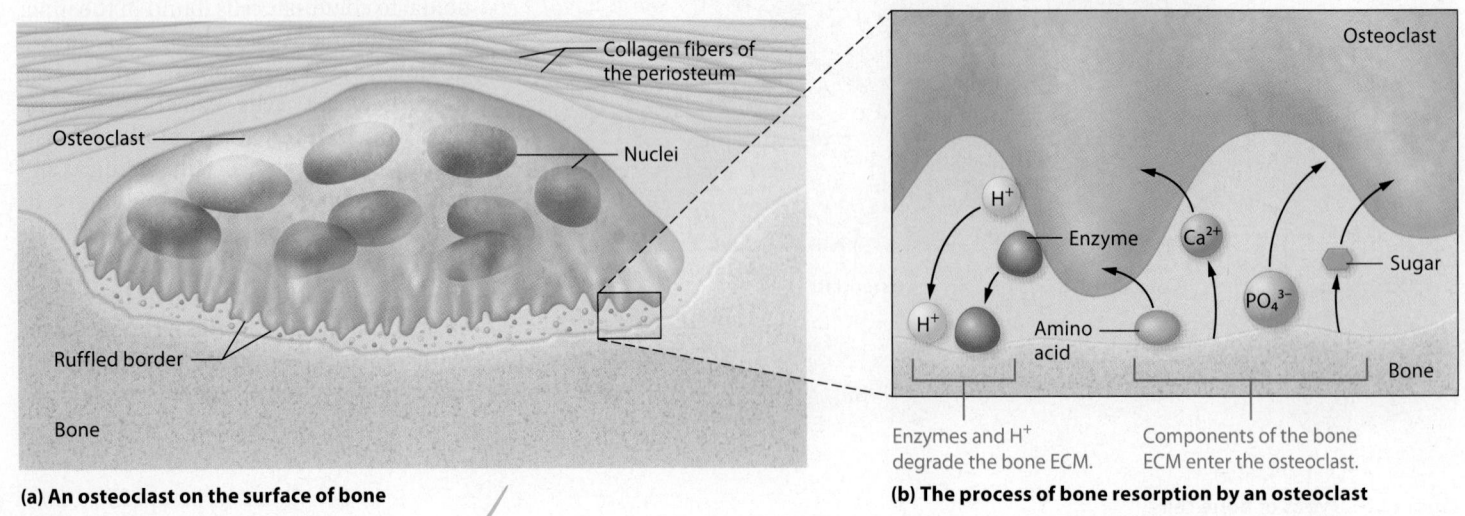

(a) An osteoclast on the surface of bone

Enzymes and H⁺ degrade the bone ECM.

Components of the bone ECM enter the osteoclast.

(b) The process of bone resorption by an osteoclast

Figure 6.8 Function of osteoclasts.

Central
canal

Nerve

Vein

Artery

Lamellae

Lacunae with
osteocytes

Canaliculi

Collagen
fibers in
lamellae

Osteon

Osteons

Circumferential
lamellae

Interstitial
lamellae

Periosteum:

Outer fibrous
layer

Inner layer
containing
osteoblasts

Blood vessels
in periosteum

Perforating
fibers

Compact
bone

Spongy
bone

Perforating canals
with blood vessels
and nerves

Central canals
with blood vessels
and nerves

Canaliculi   Central canal

Lacunae with
osteocytes

Lamellae

TEM (620×)

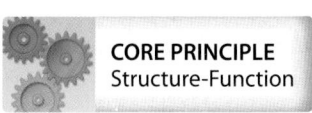
Practice art labeling
@MasteringA&P®

Figure 6.9 Structure of compact bone.

☐ 2. Compare and contrast the locations and functions of
osteoblasts, osteocytes, and osteoclasts.

Histology of Bone Tissue

As we discussed in Module 6.1, bones have two very different
types of bone tissue: the hard and dense outer compact bone,
and the porous inner spongy bone. Given the Structure-
Function Core Principle (p. 25),
since these two types of bone
tissue have different functions,
it's not surprising that they also

CORE PRINCIPLE
Structure-Function

have different structures, at both the gross and histological
levels.

Structure of Compact Bone

As an outer shell, compact bone is subject to a great amount of
stress, or forces that tend to strain or deform an object. An examination of its microscopic structure shows that compact bone is
built in a way that enables it to withstand these stresses quite
well. Notice in **Figure 6.9** that the structure of compact bone
resembles a forest of small, tightly packed trees. Each "tree" is a
unit called an **osteon** (OSS-tee-on), also known as a *Haversian
system* (hah-VER-jhun). The "rings" of our miniature trees are
very thin layers of bone termed **lamellae** (lah-MEL-ee; singular,
lamella; lamell- = "plate"). Let's look at the structure of osteons
and how they fit into the overall arrangement of compact bone.

Osteon Structure The components of an osteon include the following:

- **Lamellae.** Most osteons contain from 4 to as many as 20 lamellae, which are sometimes called *concentric lamellae.* Just as tree rings enable a tree to withstand a great deal of stress, the lamellar structure of compact bone greatly enhances its strength. Note in the blow-up of the single osteon shown in Figure 6.9 that the collagen fibers of adjacent lamellae run in opposite directions, a trait that allows the osteon to resist twisting and bending forces in more than one direction.
- **Central canal.** In the center of each osteon is a hole called the **central canal,** or *Haversian canal.* The central canal contains blood vessels and nerves that supply the cells of the osteon. This canal, like all internal canals of bone, is lined by endosteum.
- **Lacunae.** Recall that osteoblasts eventually are surrounded by the ECM that they secrete to become osteocytes. These cells reside in the remaining **lacunae,** small cavities that are filled with extracellular fluid and located between lamellae. About 20,000–30,000 osteocytes and lacunae are found in each cubic millimeter of bone.

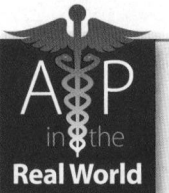

Osteopetrosis

The primary defect in the disease *osteopetrosis,* also known as "marble bone disease," is defective osteoclasts that do not properly degrade bone. This causes bone to overgrow and bone mass to increase, while paradoxically the bones become weakened and brittle. They take on a stonelike appearance on x-rays—hence the name of the disease (*petra* = "rock").

Osteopetrosis has two main forms:

- **Infantile osteopetrosis** is the predominantly inherited, more severe form of the disease. In this form, the openings of the skull fail to enlarge as an infant grows. Eventually the skull may trap the nerves for sight and hearing and so cause blindness and deafness. Additionally, the marrow cavities in all bones fail to enlarge, which decreases the amount of red bone marrow and blood cells, a condition that can prove fatal.
- **Adult osteopetrosis** is also inherited but develops during adolescence or adulthood. The symptoms may include bone pain, recurrent fractures, nerve trapping, and joint pain.

Infantile osteopetrosis requires treatment to prevent blindness, deafness, and death. Treatments include drugs to stimulate both osteoclasts and bone marrow. Currently, adult osteopetrosis is only treated symptomatically.

- **Canaliculi.** Lacunae are connected to one another by tiny canals called **canaliculi** (kan'-al-IK-yu-lee; "little canals"). Osteocytes have long, thin "arms," cytoplasmic extensions that extend through the canaliculi to contact the arms of other osteocytes. At these places of contact are gap junctions that allow small molecules to pass from cell to cell. Note in the illustration and photomicrograph of a sectioned osteon in Figure 6.9 that a lacuna and its surrounding canaliculi somewhat resemble a spider—the lacuna is the body of the spider, and the canaliculi are its legs. Canaliculi also allow oxygen and nutrients from the blood to reach every osteocyte.

Overall Compact Bone Structure Osteons are not permanent structures, as they are regularly resorbed by osteoclasts and rebuilt by osteoblasts. This process is reflected in the structure of compact bone, which is composed of multiple osteons with lamellae between them (see Figure 6.9). These lamellae, which are remnants of resorbed osteons, are called **interstitial lamellae.** In addition, outer and inner rings of lamellae are present just inside the periosteum and outside the spongy bone; these are known as **circumferential lamellae.** These lamellae add strength to the bone as a whole.

The central canals of neighboring osteons are connected by a second type of canal called a **perforating canal** (also known as a *Volkmann's canal;* FOWK-manz). Observe in Figure 6.9 that perforating canals run perpendicular to the osteons. They carry blood vessels from the periosteum that merge with vessels in the central canals.

Structure of Spongy Bone

Spongy bone is usually not the weight-bearing part of a bone, so it does not have a dense structure like that of compact bone. Instead, spongy bone resists forces from many directions and forms a protective framework for the bone marrow. These functions are performed by branching "ribs" of bone called **trabeculae** (trah-BEK-yoo-lee; *trab-* = "beam"), which project into the marrow cavity (**Figure 6.10**). Trabeculae are covered with endosteum and usually do not contain osteons. However, they do contain concentric lamellae, within which we find canaliculi and lacunae housing osteocytes. No central or perforating canals are present within trabeculae, and the cells obtain their oxygen and nutrients from the blood vessels in the bone marrow.

Read *A&P in the Real World: Osteoporosis and Healthy Bones* on page 194 for information on how both compact and spongy bone can deteriorate over time.

Quick Check

☐ 3. Match the following terms with the correct definition.

_____ Lacunae   a. Tiny canals that connect
_____ Lamellae   lacunae
_____ Central canal
_____ Canaliculi   b. Runs perpendicular to the osteons,
                          contains blood vessels

Compact bone
Spongy bone

Lacunae with osteocytes

Canaliculi

Trabecula

Trabeculae

Osteoblasts of the endosteum

Lamellae

Osteoclast

Figure 6.10 **Structure of spongy bone.**

_____ Perforating canal

c. Cavities that house osteocytes
d. Rings of bone matrix
e. Runs down the center of the osteon, contains blood vessels

Apply What You Learned

☐ 1. The disease *osteogenesis imperfecta* is characterized by defective collagen in the organic matrix of bone. Predict the effects of such a disease.
☐ 2. What would probably happen to bone with more osteoclast than osteoblast activity?
☐ 3. How does the structure of compact bone follow its function?

See answers in Appendix A.

MODULE **6.3**

Bone Formation: Ossification

Learning Outcomes

1. Explain the differences between primary and secondary bone.
2. Describe the process of intramembranous ossification.
3. Describe the process of endochondral ossification.

The process of bone formation, known as **ossification** (oss'-ih-fih-KAY-shun) or **osteogenesis** (oss'-tee-oh-JEN-eh-sis), begins during the embryonic period and for some bones continues through childhood. The majority of bones are completely ossified by age 7.

There are two types of ossification—intramembranous and endochondral. The bones formed by *intramembranous ossification* (in'-tra-MEM-brah-nus) are built on a model, or starting material, made of a membrane of embryonic connective tissue. The bones formed by *endochondral ossification* (en-doh-KON-dral; *chondr-* = "cartilage") are built on a model made of hyaline cartilage.

The first bone formed by both types of ossification is immature bone called *primary bone* or woven bone. It consists of irregularly arranged collagen bundles, abundant osteocytes, and little inorganic matrix. In most locations in the body, primary bone is resorbed by osteoclasts and replaced by mature *secondary bone,* also called lamellar bone. As its alternative name implies, secondary bone has fully formed lamellae with regularly arranged collagen bundles that are parallel to one another, which makes it much stronger than primary bone. In addition, secondary bone contains a higher percentage of inorganic matrix, which contributes to its strength.

In this module, we examine in greater depth each type of ossification and the bones in which they occur. Note, however, that bones formed by either type of ossification are histologically identical once they mature.

Intramembranous Ossification

◀◀ Flashback

1. What are membranes? (p. 129)
2. Which bones are flat bones? (p. 185)

Many flat bones, including the bones of the skull and the clavicles, form during fetal development by the process of **intramembranous ossification.** Bones formed in this way are primary bone, which is eventually resorbed and replaced with secondary bone. This type of ossification occurs within a

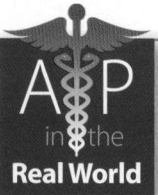

Osteoporosis and Healthy Bones

The most common bone disease in America is **osteoporosis,** in which bones become weak and brittle due to inadequate inorganic matrix in the bone extracellular matrix. You can see these changes by comparing the healthy and osteoporotic bones shown here:

SEM (15×)   SEM (15×)
Normal bone in vertebra   Osteoporitic bone in vertebra

Osteoporosis is diagnosed by measuring a person's *bone mass,* or the amount of inorganic matrix in the bone.

Osteoporosis has a variety of causes, including dietary factors (such as calcium salt and vitamin D deficiency); female gender; advanced age; lack of exercise; hormonal factors (such as lack of protective estrogen in postmenopausal women); genetic factors; and diseases of the skin, digestive system, and urinary system. Many of these causes are preventable, and a variety of methods are used in an attempt to keep osteoporosis from developing, including:

- ensuring a balanced diet and supplementing the diet with a multivitamin containing calcium if dietary deficiencies are suspected;
- engaging in weight-bearing exercise; and
- replacing estrogen in women, if appropriate.

Osteoporosis dramatically raises the risk of *fractures,* or "broken bones," and osteoporotic bone heals more slowly when it is fractured. Although a fractured bone may be little more than a painful inconvenience for a younger person, in an elderly person with osteoporosis, a fracture such as one of the hip can lead to serious complications.

Osteoporosis is treated by trying to both prevent further bone loss and increase bone mass. All of the preventative methods just described can treat osteoporosis to some extent. Other treatments include drugs that inhibit osteoclasts or stimulate osteoblasts.

mesenchymal membrane (mez-en-KY-mal). Recall that a *membrane* is simply a flat sheet of one or more tissues (see Chapter 4). In this case, the membrane is composed of a sheet of embryonic connective tissue called mesenchyme that is rich with blood vessels and mesenchymal cells.

As you consider the events of intramembranous ossification, remember the structure of flat bones—they are essentially a "sandwich" with two layers of compact bone as the "bread" and the spongy bone as the "meat." During intramembranous ossification, the inner spongy bone forms before the outer compact bone. Ossification begins at a place called the **primary ossification center.**

The events of intramembranous ossification, shown in **Figure 6.11**, are as follows:

1. **Osteoblasts develop in the primary ossification center.** At the primary ossification center, mesenchymal cells differentiate first into osteogenic cells and then into osteoblasts.

2. **Osteoblasts secrete organic matrix, which calcifies.** The newly formed osteoblasts secrete the organic matrix of bone. In a few days' time, calcium salts and other components of the inorganic matrix are deposited in the trabeculae, a process called *calcification,* and the early bone hardens. Trapped osteoblasts then become osteocytes.

3. **Early spongy bone is formed.** Osteoblasts continue to lay down new bone, forming the trabeculae of the early spongy bone. Over time, the trabeculae enlarge and merge, forming larger trabeculae. About this time, some of the mesenchyme surrounding the developing bone differentiates into the periosteum. Some of the vascular tissue in the forming spongy bone will become bone marrow.

4. **Early compact bone is formed.** The spongy bone deep to the periosteum becomes more heavily calcified, and its structure is rearranged to form immature compact bone.

Larger bones have more than one primary ossification center, and so the resulting "pieces" of bone fuse together to form one complete bone. This process is often incomplete at birth. For example, the *fontanels,* or "soft spots," in the skull of a newborn represent areas of incomplete fusion.

Quick Check

☐ 1. How do primary bone and secondary bone differ?
☐ 2. Walk through the basic steps of intramembranous ossification.

Endochondral Ossification

◀◀ Flashback

1. What are the mature and immature cells of cartilage called? (p. 142)
2. What is the perichondrium? (p. 142)

All the bones in the body below the head, except the clavicles, form by the process of **endochondral ossification.** Endochondral

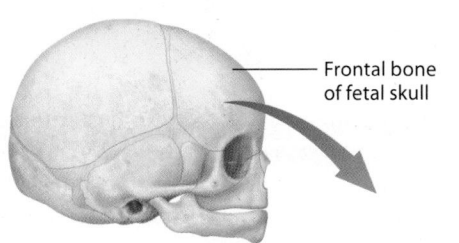
Frontal bone
of fetal skull

① Osteoblasts develop in the primary
ossification center.

Primary ossification center

Mesenchymal cells

Osteogenic cells

Osteoblasts

Collagen fibers

② Osteoblasts secrete organic matrix,
which calcifies.

Osteoblasts secreting organic matrix

Uncalcified organic matrix

Calcium salts in the calcified bone

Osteocytes

③ Early spongy bone is formed.

Osteoblasts secreting organic matrix

Trabeculae of early spongy bone

Developing periosteum

Mesenchymal cell

④ Early compact bone is formed.

Osteoblasts secreting organic matrix

Early compact bone

Early spongy bone

Blood vessel

Periosteum

**Figure 6.11 The process of
intramembranous ossification.**

Sequence the steps
@MasteringA&P®

195

ossification begins during the fetal period for most bones, although some, such as those in the wrist and ankle, ossify much later. Many bones complete this ossification process by about age 7. In this section, we'll follow this process in a long bone.

As implied by its name, endochondral ossification occurs from within a model of hyaline cartilage, which serves as a scaffold for the developing bone. The hyaline cartilage model consists of chondrocytes and cartilage ECM, and is surrounded by a connective tissue membrane called the perichondrium and immature cartilage cells known as chondroblasts. As with intramembranous ossification, endochondral ossification begins at primary ossification centers, and the bone laid down is primary bone that is eventually replaced with secondary bone. However, long bones also contain **secondary ossification centers** within their epiphyses that ossify by a very similar process.

Once the cartilage model has formed, the sequence of events of endochondral ossification, shown in **Figure 6.12**, is as follows:

① **The chondroblasts in the perichondrium differentiate into osteoblasts.** The perichondrium becomes filled with blood vessels, and chemical signals trigger its chondroblasts to differentiate into osteogenic cells and then osteoblasts. When this happens, the perichondrium is now called the periosteum.

② **The bone begins to ossify from the outside:**

②a **Osteoblasts build the bone collar on the external surface of bone.** Osteoblasts begin to secrete organic bone ECM deep to the periosteum, forming a ring of early compact bone called the *bone collar*.

②b **Simultaneously, the internal cartilage begins to calcify and the chondrocytes die.** As the bone collar calcifies, ECM surrounding the internal chondrocytes calcifies. This cuts off the blood supply to the chondrocytes, causing them to eventually die. Their death leaves cavities surrounded by calcified cartilage.

③ **In the primary ossification center, osteoblasts replace the calcified cartilage with early spongy bone; the secondary ossification centers and medullary cavity develop.** Osteoclasts etch a hole in the bone collar that allows a group of blood vessels and bone cells to enter the primary ossification center. Here, osteoblasts replace the calcified cartilage with early spongy bone while other osteoblasts continue to increase the size of the bone collar. As this happens, the cavities enlarge and combine, forming the medullary cavity. Also, secondary ossification centers develop and the epiphyses begin to ossify.

④ **As the medullary cavity enlarges, the remaining cartilage is replaced by bone; the epiphyses finish ossifying.** As ossification continues, the calcified cartilage is replaced with bone. Osteoclasts degrade much of the newly formed spongy bone, which increases the size of the medullary cavity. This space becomes filled with bone marrow. Cartilage remains in only two locations: the epiphyseal plates and the articular surface, or the part of the bone that forms a joint with another bone. This cartilage is called the **articular cartilage.** Cartilage in the epiphyseal plate will eventually be replaced with bone

Hyaline cartilage model

Perichondrium

Developing periosteum

Periosteum

Bone collar

Osteocyte

Chondrocytes

Chondroblasts

Osteogenic cells

Osteoblasts

Perichondrium

Developing periosteum

Osteoblasts secreting organic matrix

Calcium salt

Periosteum

Calcified cartilage

Dying chondrocyte

① The chondroblasts in the perichondrium differentiate into osteoblasts.

②a Osteoblasts build the bone collar on the bone's external surface as the bone begins to ossify from the outside.

②b Simultaneously, the internal cartilage begins to calcify and the chondrocytes die.

Sequence the steps @MasteringA&P®

Figure 6.12 The process of endochondral ossification.

when growth in length ceases; the articular cartilage, however, remains for life.

Although we discussed long bones in this example, a similar process occurs in all bones that form by endochondral ossification. This is evident in the following radiograph of a child's hand, which shows the hyaline cartilage persisting in both the long bones (the metacarpals [hand bones] and phalanges [finger bones]) and the short bones (the carpals):

Epiphyseal plates of long bones

Carpal bones are largely not visible because they have not completed endochondral ossification.

Hand of a young child

Observe that both the epiphyseal plates of the long bones and the unfinished carpal bones appear dark on the radiograph, which reflects that both are made of hyaline cartilage.

Certain inherited diseases cause bones formed by endochondral ossification to be shortened, resulting in dwarfism. See *A&P in the Real World: Achondroplasia* on page 199 for more information.

Quick Check

☐ 3. For each type of ossification:
   a. Does spongy bone or compact bone form first?
   b. What is the model for ossification?
   c. Which bones form by each type?
☐ 4. Which type of ossification:
   a. involves a bone collar?
   b. features secondary ossification centers?
   c. features multiple primary ossification centers?

Apply What You Learned

☐ 1. Predict what would happen if osteoclasts did not function properly during endochondral and intramembranous ossification.

Epiphyseal blood vessel
Secondary ossification centers
Calcified cartilage
Primary ossification center
Early spongy bone
Osteoblasts secreting organic matrix
Calcified cartilage

Articular cartilage
Spongy bone
Epiphyseal plate
Compact bone
Medullary cavity
Osteoclasts enlarging medullary cavity
Osteocytes
Medullary cavity

③ In the primary ossification center, osteoblasts replace the calcified cartilage with early spongy bone; the secondary ossification centers and medullary cavity develop.

④ As the medullary cavity enlarges, the remaining cartilage is replaced by bone; the epiphyses finish ossifying.

Figure 6.12 *(continued)*

□ 2. Predict what would happen if primary bone were not
   replaced by secondary bone.

See answers in Appendix A.

MODULE **6.4**

Bone Growth in Length and Width

Learning Outcomes

1. Describe how long bones grow in length.
2. Compare longitudinal and appositional bone growth.
3. Describe the hormones that play a role in bone growth.

One of the most obvious changes we see in children and young adults is that they grow taller each year. Not only do a child's bones grow in length, they also increase in width. Bone growth is, in many ways, a continuation of the ossification processes that we discussed in the previous module. In this module, we follow the process of bone growth in childhood and throughout life. We also discuss how hormones control bone growth.

Growth in Length

◀◀ **Flashback**

1. Where is the epiphysis? (p. 186)
2. What is the epiphyseal plate? (p. 186)

Long bones lengthen by a process known as **longitudinal growth.** Interestingly, longitudinal growth occurs not from the division of osteocytes or other bone cells, but from the division of chondrocytes in the **epiphyseal plate.** Recall from the previous module that the epiphyseal plate is composed of hyaline cartilage that does not ossify during endochondral ossification. Bone growth takes place from the epiphysis on the side closest to the diaphysis.

As shown in **Figure 6.13**, the epiphyseal plate contains five different zones of cells, each having a distinct appearance. The zone closest to the epiphysis is the *zone of reserve cartilage,* and it contains cells that are not directly involved in bone growth but that can be called upon to divide if needed. The region just "above" (closer to the diaphysis) the zone of reserve cartilage is the *zone*

Figure 6.13 Structure of the epiphyseal plate.

Achondroplasia

The disease *achondroplasia* (ay′-kon-droh-PLAY-jhah) is the most common cause of dwarfism. It results from a defect in a gene that is either inherited from an affected parent or is caused by a new mutation. The defective gene produces an abnormal growth factor receptor on cartilage. This interferes with the hyaline cartilage models used in endochondral ossification, as well as articular and epiphyseal plate cartilage. As a result, bones form and grow abnormally, resulting in shortened limbs, a disproportionately long trunk, and facial abnormalities.

Over the long term, individuals with achondroplasia may develop problems such as joint disorders, respiratory difficulties, and spinal cord compression. However, these symptoms may be managed with medications.

of proliferation, and it has actively dividing chondrocytes in lacunae. The next region toward the diaphysis is the *zone of hypertrophy and maturation,* and it contains mature chondrocytes. Above

that is the *zone of calcification;* it contains dead chondrocytes, some of which are calcified. And finally, the *zone of ossification* has calcified chondrocytes and osteoblasts that build bone.

The last four of these zones are actively involved in longitudinal growth. Most mitotic activity in the epiphyseal plate occurs in the zone of proliferation. As cells divide in this zone, the cells "above" them progressively become part of the next zones.

Let's look closely at the process of longitudinal growth, which proceeds by the following sequence (**Figure 6.14**):

① **Chondrocytes divide in the zone of proliferation.** The stacked chondrocytes in the zone of proliferation divide and the cells above them become part of the next zone.

② **Chondrocytes that reach the next zone enlarge and mature.** Chondrocytes that reach the zone of hypertrophy and maturation enlarge and cease dividing. Notice that the lacunae surrounding the chondrocytes are larger in this zone.

③ **Chondrocytes die and their matrix calcifies.** Chondrocytes then reach the zone of calcification. Here, chondrocytes are quite far from the blood supply, which causes them to die and their ECM to accumulate calcium salt deposits.

④ **Calcified cartilage is replaced with bone.** In the final zone, the zone of ossification, osteoblasts invade the calcified cartilage and begin to lay down bone on top of it.

Figure 6.14 Growth at the epiphyseal plate.

Sequence the steps
@MasteringA&P®

Eventually, the calcified cartilage/bone is resorbed by osteoclasts and completely replaced with bone.

Longitudinal growth continues at the epiphyseal plate as long as mitosis is happening in the zone of proliferation. However, at about 12–15 years of age, the rate of mitosis slows, but the rate of ossification in steps 3 and 4 continues unabated. This causes the epiphyseal plate to gradually shrink, as the zone of proliferation is progressively overtaken by the zones of calcification and ossification. Eventually, at about age 18–21, the zone of proliferation completely ossifies; at this point, the epiphyseal plate is said to be "closed." Recall from Module 6.1 that a closed epiphyseal plate leaves behind a calcified remnant called the **epiphyseal line.** Once just the epiphyseal line remains, the long bone can no longer grow in length.

Quick Check

☐ 1. What tissue type makes up the epiphyseal plate?
☐ 2. How does bone grow in length from the epiphyseal plate?

Growth in Width

Although only long bones undergo longitudinal growth from the epiphyseal plate, all bones grow in width. This growth occurs through a process known as **appositional growth,** during which osteoblasts between the periosteum and the bone surface lay down new bone. Appositional bone growth does not initially result in the formation of new osteons; instead, new circumferential lamellae are formed. As these new lamellae are added, the deeper circumferential lamellae are either removed or incorporated into osteons.

In actively growing bones, appositional growth is primarily responsible for thickening the compact bone of the diaphysis. As this occurs, osteoclasts in the medullary cavity digest the inner circumferential lamellae, so that as the bones increase in width, their marrow cavities enlarge as well. Bone growth in width may continue after bone growth in length ceases, depending on factors such as hormones, the forces to which the bone is subjected, and diet. These factors are discussed in Module 6.5.

Quick Check

☐ 3. How is appositional growth different from longitudinal growth?

The Role of Hormones in Bone Growth

◀◀ **Flashback**

1. What is an endocrine gland? (p. 135)

Although the processes of longitudinal and appositional bone growth are the same for everyone, clearly some of us grow more than others. This is because multiple factors play a role in determining how much cell division occurs at the epiphyseal plate and how long this process continues. One of the main factors that influence bone growth is a group of chemicals called *hormones,* which are secreted by the cells of endocrine glands into the blood.

Recall the Cell-Cell Communication Core Principle (p. 27). A primary example is *growth hormone,* produced by the anterior pituitary, an endocrine gland lying below the brain. Growth hormone is secreted throughout life but is produced in the highest amounts during infancy and childhood. It enhances

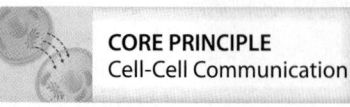

CORE PRINCIPLE
Cell-Cell Communication

protein synthesis and cell division in nearly all tissues, and it has several effects on bone tissue. Promoting both longitudinal and appositional growth, these effects include the following:

- an increase in the rate of mitosis of chondrocytes in the epiphyseal plate;
- an increase in the activity of osteogenic cells, including their activity in the zone of ossification; and
- direct stimulation of osteoblasts in the periosteum, triggering appositional growth.

Take a look at *A&P in the Real World: Gigantism and Acromegaly* for a discussion of what happens when a person has too much growth hormone.

A second hormone with a pronounced effect on bone growth is the male sex hormone *testosterone.* Testosterone increases appositional bone growth, causing bones in males to become much thicker and have greater calcium salt deposits than bones in females. Testosterone also increases the rate of mitosis at the epiphyseal plate, and so the large increases in testosterone that occur during the teenage years are accompanied by "growth spurts." Although testosterone increases longitudinal bone growth, it also accelerates the closure of the epiphyseal plates. This is why males castrated at a young age are often taller than unaltered adult males.

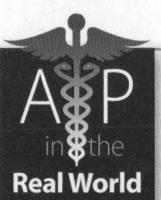

Gigantism and Acromegaly

The presence of excess growth hormone can result in one of two conditions, depending on when in life it develops. If excess growth hormone is secreted in childhood before closure of the epiphyseal plates, a condition called *gigantism* develops. People with gigantism are very tall due to excessive longitudinal and appositional bone growth.

If excessive growth hormone secretion occurs after closure of the epiphyseal plates, a different condition known as *acromegaly* (ak′-roh-MEG-ah-lee) results. Acromegaly does not cause an increase in height; however, it does cause enlargement of bone, cartilage, and soft tissue, and hence of certain parts of the body, such as the skull and the bones of the face, the tongue, and the hands and feet. In addition to significant disfigurement, acromegaly can cause heart and kidney malfunction and is associated with the development of diabetes mellitus.

Both gigantism and acromegaly are generally caused by a tumor that secretes growth hormone. Removal of the tumor is often an effective treatment for both conditions.

The female sex hormone *estrogen* also plays a role in bone growth. Like testosterone, estrogen increases the rate of longitudinal bone growth and inhibits osteoclasts. When estrogen levels increase during the teenage years, a "growth spurt" similar to that in males begins in females. Like testosterone, estrogen also accelerates closure of the epiphyseal plate, although it has a much more potent effect on epiphyseal plate closure than does testosterone. This is partly the reason why women are generally shorter in stature than men.

Quick Check

☐ 4. How does growth hormone affect bone growth?
☐ 5. How is bone growth affected by the sex hormones estrogen and testosterone?

Apply What You Learned

☐ 1. Malnutrition in children often leads to a decreased rate of mitosis in different cell populations. What effect would this have on growth at the epiphyseal plate? Explain.
☐ 2. How can excessive growth hormone increase the size of bones even after the epiphyseal plates have closed?
☐ 3. A young girl develops a tumor that secretes excess estrogen. Predict the effect of this excessive secretion on her bone growth and final height.

See answers in Appendix A.

MODULE **6.5**

Bone Remodeling and Repair

Learning Outcomes

1. Describe the processes of bone resorption and bone deposition.
2. Describe the physical, hormonal, and dietary factors that influence bone remodeling.
3. Explain the role of calcitonin, parathyroid hormone, and vitamin D in bone remodeling and calcium ion homeostasis.
4. Describe the general process of bone repair.

Even when bones finish growing in length, bone tissue is far from static. Bone undergoes a continual process of formation and loss called **bone remodeling.** During this process, new bone is formed by **bone deposition,** and old bone is destroyed by **bone resorption.** It may seem odd that our cells make all of this new bone only to eventually destroy it. But this cycle of bone formation and loss is carried out for a number of reasons, including maintenance of calcium ion homeostasis; bone repair; replacement of primary bone with secondary bone; replacement of older, brittle bone with newer bone; and bone adaptation to tension and stresses. In this module, we explore the basic process of bone remodeling, the factors that influence it, and how bone is repaired when it's damaged.

Bone Remodeling

◀◀ **Flashback**

1. Which cells are responsible for bone resorption and bone deposition? (pp. 189–190)
2. Which minerals do bones store? (p. 184)
3. Where and how is vitamin D produced? (p. 164)

In Module 6.2, we introduced the cells responsible for bone remodeling: osteoblasts and osteoclasts. In healthy bone, the two processes of formation and loss occur simultaneously, with bone continually being built by osteoblasts and broken down by osteoclasts:

Bone deposition   Bone resorption

Now we turn to how these cells carry out their functions. The next sections examine how bone remodeling takes place and the factors that influence it. We then focus on how bone remodeling impacts calcium ion homeostasis.

Process of Bone Remodeling

Let's look in more depth at the processes of bone deposition and resorption.

- **Bone deposition.** Bone deposition is carried out by osteoblasts in the periosteum and endosteum. Osteoblasts make the components of the organic matrix, as well as facilitate the formation of inorganic matrix. We know that osteoblasts secrete certain proteoglycans and glycoproteins that bind to calcium ions, and it appears that osteoblasts also secrete vesicles containing calcium ions (Ca^{2+}), ATP, and enzymes. These vesicles bind to collagen fibers, and their calcium ions eventually crystallize, which ruptures the vesicle and begins the overall process of calcification.

- **Bone resorption.** During bone resorption, osteoclasts secrete hydrogen ions (H^+) from their ruffled borders onto the bone ECM. The hydrogen ions make the pH more acidic, which breaks down the pH-sensitive hydroxyapatite crystals in the inorganic matrix. The liberated calcium ions and other minerals may then be reused in the body. To degrade organic matrix, osteoclasts secrete enzymes to catalyze reactions that degrade proteoglycans, glycosaminoglycans, and glycoproteins. The breakdown products are taken into the osteoclasts for possible reuse.

In childhood, deposition far outweighs resorption, which makes sense, since this is a period of active growth. However, once the epiphyseal plates close and longitudinal growth ceases, bone deposition approximately equals bone resorption—in healthy young adults, total bone mass remains constant. In addition, osteons are continually being rebuilt in response to many factors, including the signals that maintain calcium ion homeostasis.

Bone Remodeling in Response to Tension and Stress

One of the primary influences on bone remodeling is the stress placed on bones because bone deposition occurs in proportion to it. Stated simply, the heavier the load a bone must carry, the more bone tissue is deposited in that bone. This load doesn't necessarily refer to *weight* per se, but rather to the amount of *compression*. Compression is the act of squeezing or pressing together; this force results when the bones are, for example, pressed between the body's weight and the ground. Thus, an athlete who exercises and trains extensively with weights places intense compression forces on his or her bones, and therefore deposits more bone tissue, leading to higher bone mass. Conversely, a sedentary person who engages in few weight-bearing activities has decreased osteoblast activity but normal osteoclast activity. Overall, this results in net bone resorption, which decreases total bone mass.

Another factor that functions in bone remodeling is tension and pressure placed on the bone. Tension is a stretching force, whereas pressure is application of a continuous downward force. Where continuous pressure is placed, osteoclasts are stimulated and bone resorption occurs. However, where tension is placed, osteoblasts are stimulated and bone deposition occurs. These principles allow an orthodontist to straighten a person's teeth with braces.

Other Factors Influencing Bone Remodeling

Numerous other factors influence bone remodeling, including a variety of hormones. The male hormone testosterone promotes bone deposition, and the female hormone estrogen depresses osteoclast activity. Age also plays an important role in bone remodeling, because levels of hormones such as growth hormone decline with advancing age, which decreases protein synthesis in bone.

A number of dietary factors also influence bone remodeling. Some of these factors are as follows:

- **Calcium ion intake.** Obviously, if calcium ion intake is too low, bone deposition will not take place. This is why people are encouraged to include calcium ion–rich foods in their diets.
- **Vitamin D intake.** Recall that vitamin D is a steroid synthesized by the body in response to exposure of the skin to UV light (see Chapter 5). Vitamin D acts mainly on the intestines, where it promotes calcium ion absorption, and on the kidneys, where it prevents loss of calcium ions to the urine. The net effect of vitamin D on bone is to increase bone deposition and bone mass. However, not everyone gets enough sun exposure to produce as much as the body needs. Thus, supplementation of the diet with vitamin D is often required.

STIMULUS

Blood Ca^{2+} decreases below normal range.

Normal range

RECEPTOR

Parathyroid gland cells detect a low blood Ca^{2+} level.

Parathyroid glands

EFFECTOR/RESPONSE

Osteoclasts resorb bone.

Kidneys retain Ca^{2+}.

Intestines absorb Ca^{2+}.

Osteoclast

Ca^{2+}

Bone

Ca^{2+}

IN HOMEOSTATIC RANGE

As blood Ca^{2+} returns to normal, feedback stops effector responses.

Normal range

CONTROL CENTER

Parathyroid gland cells release PTH into the blood.

PTH

Figure 6.15 Maintaining homeostasis: response to low blood calcium ion level by a negative feedback loop.

- **Vitamin C intake.** Remember that vitamin C is required for the synthesis of collagen (see Chapter 4). If vitamin C intake is inadequate, collagen synthesis in bone decreases.
- **Vitamin K intake.** There is good evidence that vitamin K aids in the production of calcium ion–binding glycoproteins by osteoblasts. Inadequate vitamin K intake may decrease bone deposition.
- **Protein intake.** An adequate dietary intake of protein is necessary for osteoblasts to synthesize the collagen fibers needed for the organic matrix of bone.

Bone Remodeling and Calcium Ion Homeostasis

As we discussed in Module 6.1, bone stores most of the body's calcium ions. Calcium ions are required for a number of critical processes, including contraction of the heart and skeletal muscles, transmission of nerve impulses, and blood clotting. If the calcium ion concentration in the blood drops too low or rises too high, major disruptions in homeostasis, up to and including death, can occur.

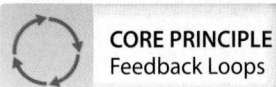
CORE PRINCIPLE Feedback Loops

One main negative feedback loop works to maintain calcium ion homeostasis (**Figure 6.15**), in an example of the Feedback Loops Core Principle (p. 21). Like all negative feedback loops, it begins with deviation from a set point—in this case, a decreased concentration of calcium ions in the blood. The response of the feedback loop involves a hormone called parathyroid hormone (PTH). The loop proceeds as follows:

- **Stimulus: Blood calcium ion level decreases below normal range.**
- **Receptor: Parathyroid gland cells detect a low blood calcium ion level.** The cells of the parathyroid gland act as receptors that detect a lower than normal concentration of calcium ions in the blood.
- **Control center: Parathyroid gland cells release parathyroid hormone into the blood.** The cells of the parathyroid gland also act as control centers. When they are stimulated, they release PTH into the bloodstream.

- **Effector/response: Parathyroid hormone stimulates effects that increase the blood calcium ion level.** PTH leads to increased osteoclast activity, which causes the breakdown of bone and the subsequent release of calcium ions into the blood. Note that this actually weakens the bones, but prevents much more serious imbalances in calcium ion homeostasis. In addition, PTH stimulates the intestines to absorb calcium ions and the kidneys to retain calcium ions. Each of these actions works to increase the blood calcium ion level to the normal range.
- **Homeostasis and negative feedback: The calcium ion concentration returns to the normal homeostatic range, and negative feedback decreases parathyroid gland cell secretion of PTH.** The cells of the parathyroid gland detect when the calcium ion level has returned to the normal range. In response, they secrete smaller amounts of parathyroid hormone via a negative feedback mechanism.

The response to an increase in the number of calcium ions in the blood triggers a different negative feedback loop. The first response is a decline in parathyroid hormone secretion. Next, secretion of the hormone **calcitonin** (kal-sih-TOW-nin) is triggered; this hormone is produced by the cells of the thyroid gland, located in the anterior neck. Calcitonin has essentially the opposite effect of PTH, as it decreases both the activity of osteoclasts and the formation of new osteoclasts. This increases bone deposition, which pulls calcium ions out of the blood and deposits them into the bone. Note that calcitonin seems to be most critical during times of active bone remodeling, as occurs during longitudinal bone growth. In adults, calcitonin doesn't appear to be as potent a regulator of blood calcium ion concentration as is PTH.

Vitamin D also plays an important role in calcium ion homeostasis. If vitamin D is absent, few calcium ions are absorbed from the intestines, regardless of how much is taken in via the diet. Inadequate vitamin D intake in children leads to the condition called *rickets*, characterized by bone pain, bony deformities, increased risk of fractures, and muscle weakness.

The factors that influence bone remodeling are summarized in **Figure 6.16**.

Figure 6.16 Factors that influence bone remodeling.

Quick Check

☐ 1. How do bone resorption and bone deposition differ?

☐ 2. Explain how compression, continuous pressure on a bone, and dietary factors influence bone remodeling.

☐ 3. What role does parathyroid hormone play in calcium ion homeostasis?

TABLE 6.1	TYPES OF FRACTURES		
Fracture Type	**Description**	**Fracture Type**	**Description**
Spiral	Fracture resulting from twisting forces applied to the bone	*Compression*	Fracture in which the bone is crushed under the weight it is meant to support; common in the elderly and those with reduced bone mass

Comminuted [kom-ih-NOOT-'d]	Fracture in which the bone is shattered into multiple fragments; difficult to repair	*Avulsion [ah-VUL-shun]*	Fracture in which a tendon or ligament pulls off a fragment of bone; often seen in ankle fractures

Greenstick	Fracture in which the bone breaks on one side but only bends on the other side, similar to the break observed when a young ("green") twig is bent; common in children, whose bones are more flexible	*Epiphyseal plate*	Fracture that involves at least part of the epiphyseal plate; occurs only in children and young adults; may interfere with growth

Bone Repair

◀◀ Flashback

1. What are the differences between primary and secondary bone? (p. 193)
2. What is a fibroblast? (p. 138)

Given the intricate and complex structure of bone, it's not surprising that many different components are damaged or destroyed when a bone is injured, including the bone ECM, bone cells, the periosteum, the endosteum, and numerous blood vessels and nerves. The most dramatic bone injury is a **fracture,** commonly referred to as a "broken bone." Fractures are grouped into many different classes, some of which are shown in **Table 6.1**. The most basic way to classify fractures is as simple (also called closed) or compound (also called open). *Simple fractures* are those in which the skin and tissue around the fracture remain intact, whereas *compound fractures* involve damage around the fracture. Beyond this basic classification, fractures are also classified according to the type of trauma inflicted on the bone. For example, a *spiral*

fracture results when twisting forces are applied to a bone, and a *compression fracture* occurs when a bone is crushed under the weight it is meant to support.

The general process of fracture healing occurs via the following steps (**Figure 6.17**):

① **A hematoma fills the gap between the bone fragments.** When a bone is damaged, its blood vessels rupture and bleed into the injured site, forming a **hematoma** (hee-mah-TOW-mah; also known as a *blood clot*), a mass of blood cells and proteins that resembles grape jelly. The hematoma cuts off the blood supply to the damaged area, and the bone cells surrounding the area die.

② **Fibroblasts and chondroblasts infiltrate the hematoma and a soft callus forms.** Fibroblasts and small blood vessels in the periosteum invade the hematoma in the first week after the fracture. The fibroblasts secrete collagen fibers and form a dense irregular collagenous connective tissue that starts to bridge the gap between the bone fragments. At the same time, some of the osteogenic cells from the endosteum become chondroblasts that secrete hyaline

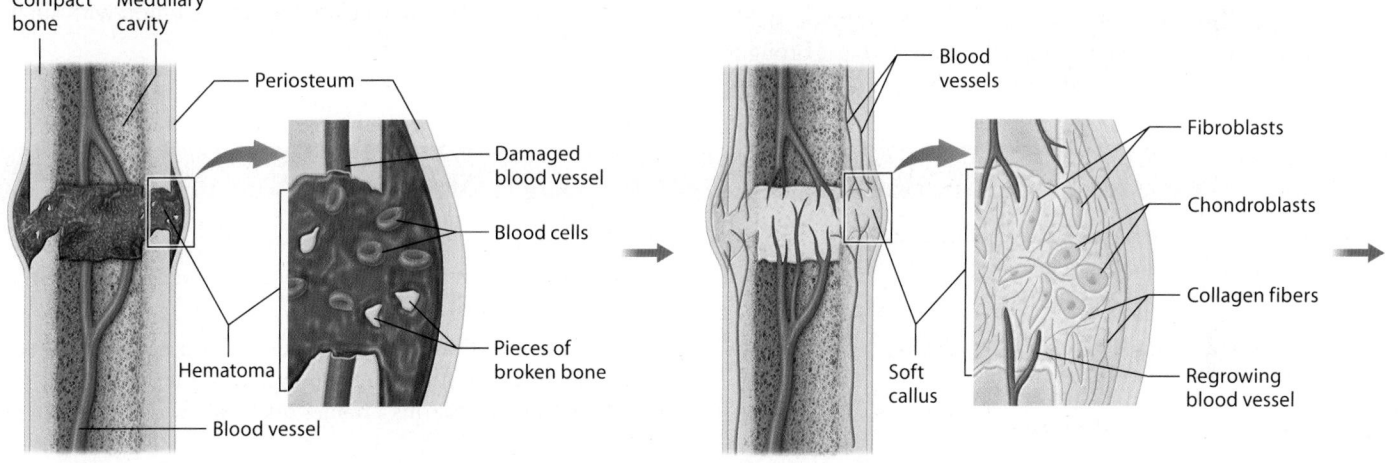

① A hematoma fills the gap between the bone fragments.

② Fibroblasts and chondroblasts infiltrate the hematoma, and a soft callus forms.

③ Osteoblasts build a bone callus.

④ The bone callus is remodeled and primary bone is replaced with secondary bone.

Figure 6.17 The process of fracture repair.

cartilage. This mixture of hyaline cartilage and collagenous connective tissue is called a **soft callus** (KAL-us; *callus* = "hard skin").

③ **Osteoblasts build a bone callus.** Osteoblasts within the periosteum begin laying down a collar of primary bone called a **bone callus** (or *hard callus*). Initially, only part of the bone callus is actually bone tissue; the remainder is the soft callus. This process continues over the next several weeks until the entire callus is made of primary bone that forms a bridge between the bone fragments. As we discussed earlier, primary bone is not as strong as secondary bone; however, notice in Figure 6.17 that the bone callus extends beyond the border of the bone, which helps to strengthen it.

④ **The bone callus is remodeled and primary bone is replaced with secondary bone.** Over the next several months, the bone callus is remodeled and the primary bone is resorbed and replaced with secondary bone. Eventually, the bone callus is fully resorbed and the bone regains its previous structure and strength. However, the bone callus often remains visible for months or longer following full healing of the injury.

The primary treatment for fractures is stabilization of the fracture, followed by immobilization. Many fractures may be stabilized through *closed reduction,* in which the bone ends are brought into contact by simply manipulating the body part. More severe fractures may require *open reduction,* in which the fracture is fixed in place surgically with, for example, plates, wires, and screws. After the fracture is stabilized, it is immobilized for a period of 6–8 weeks while the repair process occurs.

The process of fracture repair is imperfect and doesn't always proceed as we would hope. For example, if the fracture is inadequately stabilized or immobilized, or if the blood supply to the bone is poor, the ends of the bone may not fuse and grow together. In this case, the bone ends are united only by scar tissue or cartilage. This situation generally requires surgical removal of the scar tissue and refixation of the bone.

Quick Check

☐ 4. Walk through the basic steps of fracture repair.

Apply What You Learned

☐ 1. Why do you think that astronauts are often faced with decreased bone mass after returning from periods of prolonged weightlessness?

☐ 2. Predict the effect that a parathyroid hormone–secreting tumor would have on bone tissue. What would happen to the concentration of calcium ions in the blood with such a condition?

See answers in Appendix A.

CHAPTER SUMMARY

Take full advantage of the invaluable online practice and study tools for this chapter by going to MasteringA&P®. There you will find:

- Practice tests, module quizzes, art-labeling and sequence-the-steps activities
- ▶ Animated figures with quizzes
- *Practicing A&P* worksheets
- Author-narrated podcast
- Dynamic Study Modules
- iP Interactive Physiology

MODULE **6.1**
Introduction to Bones as Organs 183

- Bones are the main organs of the **skeletal system,** which also includes joints and other supporting structures.
- Functions of the skeletal system include protection, mineral storage, maintenance of acid-base homeostasis, blood cell formation, fat storage, movement, and support. (Figure 6.1, p. 184)
- There are five different shapes of bones: **long bones, short bones, flat bones, irregular bones,** and **sesamoid bones.** (Figure 6.2, p. 185)

- Bones are covered with the **periosteum,** a dense irregular collagenous connective tissue membrane that also contains bone cells and many blood vessels.
- **Compact bone** is the hard, dense outer bone whose structure can resist the stresses placed on it. **Spongy bone** is the inner, honeycomb-like bone that forms a framework within which bone marrow is located.
- A long bone has the following structural features: (Figure 6.3, p. 186)
  - The **diaphysis** is the shaft of the bone and contains a thick layer of compact bone surrounding very little spongy bone and a mostly hollow space called the **medullary cavity.**
  - The ends of the long bone are the **epiphyses.** The epiphyses are filled with spongy bone and bone marrow, and their ends are covered with hyaline cartilage.

Practice art-labeling exercise: Figure 6.3.

- Short, flat, irregular, and sesamoid bones lack a diaphysis, epiphyses, a medullary cavity, and an epiphyseal line. They consist of two thin layers of compact bone with spongy bone in the middle. (Figure 6.4, p. 187)
- Blood flows to the medullary cavity of the diaphysis through the *nutrient artery,* which enters via the **nutrient foramen.**
- **Red bone marrow** contains hematopoietic tissue. It is located in nearly all bones of infants. After about age 5, some of the red bone marrow is replaced with **yellow bone marrow,** which houses blood vessels and adipocytes. In adults, most bone marrow is yellow except in a few locations.

MODULE **6.2**
Microscopic Structure of Bone Tissue 188

- The primary tissue type in a bone is **bone tissue,** also called **osseous tissue** or simply *bone;* it is a specialized type of connective tissue.

- Bone has a unique extracellular matrix (ECM) containing an **organic matrix** and an **inorganic matrix.** (Figure 6.5, p. 189)
  ○ The inorganic matrix of bone is composed primarily of calcium and phosphorus salts existing as **hydroxyapatite crystals.** These crystals make bone one of the hardest substances in the body, allowing it to resist compression.
  ○ The organic matrix is called **osteoid** and consists of collagen fibers and other ECM components. The collagen fibers resist torsion and tension, and align with hydroxyapatite to enhance the hardness of bone.

- Three main types of cells are found in bone: (Figure 6.6, p. 189)
  ○ **Osteoblasts** are found in the inner layer of the periosteum and endosteum, and are derived from **osteogenic cells.** They are responsible for **bone deposition,** or building up the bone ECM. (Figure 6.7, p. 190)
  ○ **Osteocytes** are mature osteoblasts that have become surrounded by the ECM they have secreted. They secrete chemicals required for maintenance of the bone matrix.
  ○ **Osteoclasts** are responsible for **bone resorption,** or breaking down the bone ECM. From their ruffled border they secrete acid, which degrades inorganic matrix, and enzymes, which degrade organic matrix. (Figure 6.8, p. 190)

- Compact bone is composed primarily of subunits called **osteons** that consist of rings of bone ECM called **lamellae.** In between lamellae are **lacunae,** cavities that contain osteocytes. Running down the center of the osteon is the **central canal,** which contains blood vessels and nerves. A perforating canal connects the central canals of different osteons. (Figure 6.9, p. 191)

Practice art-labeling exercise: Figure 6.9.

- Spongy bone forms a framework for bone marrow with its branching **trabeculae.** Trabeculae do not contain osteons; however, they do contain lamellae and lacunae with osteocytes. (Figure 6.10, p. 193)

MODULE **6.3**
Bone Formation: Ossification 193

- Bones form by the process of **ossification** or **osteogenesis.**

- The first bone formed is *primary bone,* which has irregularly arranged collagen bundles with scant inorganic matrix. Primary bone is replaced by *secondary bone,* which is much stronger, with fully formed lamellae and a greater percentage of inorganic matrix.

- **Intramembranous ossification** occurs from within a membrane composed of embryonic mesenchyme. Ossification begins at the **primary ossification center,** where mesenchymal cells differentiate into osteoblasts that secrete organic matrix, which calcifies and forms the early trabeculae of spongy bone. Some of the mesenchyme differentiates into periosteum, and deep to the periosteum the spongy bone is rearranged to form compact bone. (Figure 6.11, p. 195)

Practice sequencing the steps: Figure 6.11.

- **Endochondral ossification** takes place within a hyaline cartilage model. Ossification begins as the chondroblasts in the perichondrium become osteoblasts that secrete the early compact bone in the form of a *bone collar.* Chondrocytes in the inner part of the model enlarge and die as their surrounding ECM calcifies. The calcified cartilage is replaced by osteoblasts with early spongy bone. (Figure 6.12, pp. 196–197)

Practice sequencing the steps: Figure 6.12.

MODULE **6.4**
Bone Growth in Length and Width 198

- **Longitudinal growth** occurs at the **epiphyseal plate** in long bones, which consists of five zones of cells. The *zone of reserve cartilage* is closest to the epiphysis and contains chondrocytes that may be called upon to divide, if needed. Bone growth occurs as chondrocytes in the *zone of proliferation* of the epiphyseal plate divide by mitosis and the cells above them become part of the *zone of maturation and hypertrophy,* where they enlarge and stop dividing. When the cells enter the *zone of calcification,* they die and their matrix calcifies, after which the matrix ossifies in the *zone of ossification.* (Figure 6.13, p. 198; Figure 6.14, p. 199)

Practice sequencing the steps: Figure 6.14.

- When the epiphyseal plate fully ossifies, no further longitudinal bone growth is possible.

- Short, flat, irregular, and sesamoid bones lack epiphyses and therefore do not undergo longitudinal bone growth. However, they do grow from their ossification centers by either endochondral or intramembranous ossification.

- Bones grow in width by **appositional growth,** during which osteoblasts in the periosteum secrete new circumferential lamellae.

- Three hormones exert a significant effect on bone growth:
  ○ *Growth hormone* increases the rate of mitosis in chondrocytes in the epiphyseal plate, the activity of osteogenic cells in the zone of ossification, and the activity of osteoblasts in the periosteum.
  ○ *Testosterone* increases bone formation by longitudinal and appositional growth.
  ○ *Estrogen* inhibits osteoclasts and increases the rate of bone growth but also strongly accelerates closure of the epiphyseal plate.

MODULE **6.5**
Bone Remodeling and Repair 201

- **Bone remodeling** is a combination of the continual processes of **bone deposition** and **bone resorption.**

- Bone deposition is carried out by osteoblasts, which synthesize organic matrix in the bone and also facilitate formation of inorganic matrix.

- Osteoclasts are responsible for bone resorption via the secretion of hydrogen ions that dissolve the inorganic matrix, and enzymes that degrade the organic matrix.

- Bone remodeling is influenced by several variables:
  ○ Tension placed on a bone triggers osteoblasts and bone deposition. Osteoclasts are stimulated when pressure is placed on a bone, which leads to bone resorption.

○ *Vitamin D* increases bone deposition by increasing calcium ion absorption from the intestines and preventing its loss to the urine in the kidneys.

○ The hormones testosterone and estrogen both increase overall bone deposition.

○ Dietary factors such as adequate intake of calcium ions, vitamin D, vitamin C, vitamin K, and protein all influence bone remodeling.

○ *Parathyroid hormone* is secreted by the parathyroid glands, and it stimulates osteoclasts, calcium ion absorption in the intestines, and calcium ion retention by the kidneys. These effects increase the calcium ion concentration in the blood. (Figure 6.15, p. 202)

○ Figure 6.16, p. 203, summarizes the factors that influence bone remodeling.

• Fractures can be classified as simple (or closed) or compound (or open). (Table 6.1, p. 204)

• A bone **fracture** is healed by the following process: Blood vessels break and the injured area is filled in by a **hematoma,** after which fibroblasts and capillaries invade the clot and the gap is filled with collagen fibers. Osteogenic cells then differentiate into chondroblasts that secrete cartilage, forming the **soft callus.** Osteoblasts replace the soft callus with a **bone callus** made of primary bone. The bone callus is eventually remodeled and replaced with secondary bone. (Figure 6.17, p. 205)

ASSESS WHAT YOU LEARNED

See answers in Appendix A.

LEVEL 1 Check Your Recall

1. Which of the following is *not* a function of the skeletal system?

   a. Primary storage site in the body for the minerals sodium and potassium

   b. Location of the red bone marrow, which produces red blood cells

   c. Storage of triglycerides in yellow bone marrow

   d. Support and protection of the body and vital organs

   e. Functions in movement as the site of attachment for skeletal muscles

2. Match the following terms with the correct definition.

_____ Long bone	a.	Main source of blood to the medullary cavity
_____ Epiphysis		
_____ Nutrient artery	b.	Membrane surrounding the bone
_____ Flat bone	c.	Bone that is about as wide as it is long
_____ Short bone	d.	Shaft of a long bone
_____ Periosteum	e.	Bone that is longer than it is wide
_____ Diaphysis	f.	Canal running down the center of the diaphysis
_____ Medullary cavity	g.	End of a long bone
	h.	Bone that is broad and thin

3. Explain the differences between red bone marrow and yellow bone marrow.

4. Mark the following statements about bone tissue as true or false. If a statement is false, correct it to make a true statement.

   a. The primary mineral in the inorganic matrix is hydroxyapatite.

   b. The inorganic matrix of bone is solely responsible for the strength of bone tissue.

   c. Collagen fibers are one of the predominant parts of the inorganic matrix.

   d. The collagen fibers of bone help it to resist torsion and tension.

   e. Osteoblasts are responsible for bone resorption, and osteoclasts are responsible for bone deposition.

   f. Osteocytes are mature and less active osteoblasts that have become surrounded by bone ECM.

5. Fill in the blanks: The subunit of compact bone is the _____. It consists of rings of bone matrix called _____ that surround a structure called the _____ that contains blood vessels and nerves. Other structures called _____ also contain blood vessels and nerves. Osteocytes are housed in _____ and communicate via _____.

6. The branching pieces of bone in spongy bone are called:

   a. lamellae.

   b. lacunae.

   c. osteoclasts.

   d. trabeculae.

7. Which bones form via intramembranous ossification?

   a. Irregular bones

   b. Certain flat bones

   c. Long bones

   d. Short bones

   e. More than one of the above

8. Of the following statements, identify those that are properties of intramembranous ossification, endochondral ossification, or both.

   a. The bone is formed via a hyaline cartilage model.

   b. Bone tissue forms from ossification centers.

   c. Bone forms from within a mesenchyme membrane.

   d. The early spongy bone is formed, after which the early compact bone develops.

   e. The original primary bone is replaced with secondary bone.

   f. A bone collar forms, followed by the early spongy bone.

9. What is the difference between a primary and secondary ossification center in a long bone?

10. The part of the epiphysis that does not ossify during a person's lifetime is the:

    a. articular surface.

    b. secondary ossification center.

    c. diaphyseal notch.

    d. nutrient foramen.

11. Long bones grow in length from the:

    a. diaphyseal line.

    b. epiphyseal line.

    c. epiphyseal plate.

    d. medullary cavity.

12. Correctly order the following steps of bone growth in length, by placing a 1 by the first step, a 2 by the second step, and so on.

    a. _____ Calcified cartilage is replaced with bone in the zone of ossification.

    b. _____ Chondrocytes in the zone of proliferation divide by mitosis.

    c. _____ Chondrocytes enter the zone of calcification and die as their matrix calcifies.

    d. _____ Chondrocytes enlarge and cease dividing.

13. Explain the effect that the following hormones have on growth of bone tissue:

    a. Growth hormone

    b. Testosterone

    c. Estrogen

14. Fill in the blanks. Bone deposition is carried out by _____, which secrete the _____ of the ECM by exocytosis. They facilitate the formation of the _____ of the ECM by secreting _____ filled with calcium ions, enzymes, and ATP. Bone resorption is carried out by _____, which secrete _____ and _____ from their ruffled border.

15. Mark the following statements as true or false. If a statement is false, correct it to make a true statement.

    a. Bone resorption is triggered by pressure placed on the bone.

    b. Bone deposition is triggered by tension placed on the bone.

    c. The greater the load the bone must carry, the more bone that is resorbed by osteoclasts.

    d. Calcitonin is released in response to an elevated calcium ion concentration in the blood.

    e. Parathyroid hormone increases the blood calcium ion concentration by increasing the activity of osteoblasts.

    f. Vitamin D is required for calcium ion absorption from the intestines and retention in the kidneys.

16. Which of the following influences bone remodeling?

    a. Vitamins such as vitamins D, C, and K

    b. Intake of calcium ions

    c. Hormones such as estrogen and testosterone

    d. All of the above

17. Correctly order the following steps of fracture repair, placing a 1 by the first step, a 2 by the second step, and so on.

    a. _____ Osteoblasts in the periosteum lay down a bone callus of primary bone.

    b. _____ Damaged blood vessels bleed and fill the gap between the bone fragments with a hematoma.

    c. _____ The bone callus is remodeled and replaced with secondary bone.

    d. _____ Fibroblasts, chondroblasts, and blood vessels enter the clot, and the soft callus begins to bridge the gap between bone fragments.

LEVEL 2 Check Your Understanding

1. Explain why a person who is wheelchair-bound or bed-bound often has very low bone mass, especially in his or her legs.

2. How could diseases of the kidney, skin, and/or intestines cause diseases of the bone?

3. How could a disease that affects primarily cartilage impact bone growth and bone formation?

LEVEL 3 Apply Your Knowledge

PART A: Application and Analysis

1. It used to be common practice in surgical fracture repair to strip and remove the periosteum. Now, however, surgeons take great care to avoid damaging the periosteum. Explain why, and state what would likely happen to bone healing with the periosteum missing.

2. Explain why young men who take anabolic steroids (which mimic the actions of testosterone) often display stunted growth and have an overall shorter stature.

3. Lucy Dupre is a 2-year-old girl living in northern Canada. You notice that her growth seems abnormally slow and she has exhibited signs of weakened bones, including fractures. Her parents admit that they have not supplemented her diet with vitamin D, as they feel that any supplementation is "unnatural" and "not organic." Explain why Lucy is having problems with her bones and bone growth.

PART B: Make the Connection

4. Jeff Morris is a 6-year-old boy who has been diagnosed with a rare genetic defect in which his lysosomes are unable to maintain an acidic pH. What are lysosomes? Why do lysosomes require an acidic pH? How will this impact the ability of certain bone cells to function? What consequences will this disease have on Jeff's bones? (*Connects to Chapter 3*)

The Skeletal System

In the previous chapter, we introduced bones and bone tissue. Now we look at the bones and cartilages that make up the **skeletal system,** or *skeleton*. Knowledge of the skeletal system will help build a foundation for later chapters. For example, many muscles, blood vessels, and nerves share names with the bones in their anatomical area, which can help you remember their location. In addition, the lobes of the brain are named for bones of the skull.

In progressing through the chapter, you'll realize there are a lot of bones and bone markings to remember. To help you, a number of memory devices are collected into two Study Boosts, where you can find them easily when you study. These Study Boosts are on pages 236 and 253.

MODULE **7.1**

Overview of the Skeletal System

Learning Outcomes

1. Compare and contrast the two structural divisions of the skeleton, and list the general bone structures contained within each.
2. Identify and describe bone markings, and explain their function.
3. Identify the types of cartilage tissues found in the skeletal system, and explain the functions of each.
4. Describe the major functions of the skeletal system.

On average, the human body has about 206 bones that together make up the **skeleton.** Before getting into the structure of individual bones, let's take an introductory tour of the skeleton. We start with the basic components of the skeleton and its cartilages, and then move on to the types of markings found on bones.

Structure of the Skeleton and Skeletal Cartilages

◄◄ Flashback

1. What are the properties of cartilage? (p. 142)
2. How do hyaline cartilage, elastic cartilage, and fibrocartilage differ? (p. 142)

Computer-generated image: The skeletal system consists of bones and skeletal cartilages that function together in movement, support, and protection.

Skull

Vertebral
column

Pectoral girdle:
Clavicle
Scapula

Thoracic cage:
Sternum
Ribs

Pelvic girdle:
Sacrum
Pelvic bone

Lower limb:
Femur

Patella

Tibia

Fibula

Tarsals
Metatarsals
Phalanges

Upper limb:
Humerus

Ulna

Radius

Carpals
Metacarpals
Phalanges

(a) Anterior view

(b) Posterior view

Figure 7.1 Divisions of the skeletal system. The axial skeleton is shaded orange;
the appendicular skeleton is shaded tan.

Practice art labeling
@MasteringA&P®

The skeleton consists of groups of bones that work together to
perform common functions. The basic parts of the skeleton
include (**Figure 7.1**):

- **Skull.** The *skull* is the skeleton's most complex structure. It has
  a total of 22 bones: 8 *cranial bones,* which encase the brain,
  and 14 *facial bones,* which form the framework for the face.

- **Vertebral column.** Inferior to the skull we find the **vertebral column,** which in an adult consists of approximately
  33 bones, called *vertebrae,* stacked on top of one another.
  The first 24 bones are individual vertebrae, which encase
  the spinal cord. The two inferior bones, the *sacrum* and the
  coccyx, are made up of fused vertebrae.

- **Thoracic cage.** The **thoracic cage,** also known as the *rib cage,* consists of 12 pairs of *ribs,* the *sternum,* and part of the vertebral column. Together, these bones encase and protect the lungs, heart, and other delicate structures in the thoracic cavity.
- **Pectoral girdle.** A *girdle* is a structure that surrounds and supports something. The **pectoral girdle** contains two bones: the *clavicle* and the *scapula.* These two bones support the upper limb and anchor it to the trunk.
- **Upper limb.** The **upper limb** has three parts: the **arm,** which consists of the *humerus;* the **forearm,** which consists of the *radius* and *ulna;* and the **wrist** and **hand,** which contain the *carpals, metacarpals,* and *phalanges.*
- **Pelvic girdle.** Just as the pectoral girdle supports and anchors the upper limb, the **pelvic girdle** supports the lower limb and anchors it to the trunk. It is made up of the two *pelvic bones* and the sacrum, which are together called the *pelvis.* Each pelvic bone is itself composed of three fused bones: the *ilium, ischium,* and *pubis.*
- **Lower limb.** Like the upper limb, the **lower limb** has three parts: the **thigh,** which consists of the *femur;* the **leg,** which consists of the *tibia* and the *fibula;* and the **ankle** and **foot,** which contain the *tarsals, metatarsals,* and *phalanges.*

We can group these parts into two structural divisions of the skeleton: axial and appendicular. The **axial skeleton** forms the longitudinal axis of the body. It consists of the bones of the skull, the vertebral column, and the thoracic cage (shaded orange in Figure 7.1). The bones of the axial skeleton are structured largely for protection, as they encase body cavities and protect the underlying organs. The **appendicular skeleton** consists of the bones of the upper and lower limbs and the pectoral and pelvic girdles (shaded tan in Figure 7.1). These bones are structured largely for motion and act primarily as supportive structures to which muscles attach.

Recall that the skeleton contains more than just bones—it also consists of the associated skeletal cartilages (see Chapter 6). The two main types are hyaline cartilage and fibrocartilage. We find these cartilages where two bones meet at *articulations,* or *joints.* Examples of articulations with hyaline cartilage include those between the bones of the upper and lower limbs as well as those between the ribs and sternum. We also see hyaline cartilage in the epiphyseal plates, from which long bones grow in length. Articulations containing fibrocartilage are present between the vertebrae and the two pelvic bones. We look more closely at these joints later in this chapter and in the articulations chapter.

Quick Check

- ☐ 1. Which parts of the skeleton belong to the appendicular skeleton? Which belong to the axial skeleton?
- ☐ 2. Where are skeletal cartilages located?

Bone Markings

Bones have a number of surface features—depressions, openings, and projections—known collectively as **bone markings.** Depressions allow blood vessels and nerves to travel along a bone, or provide a place where two bones can articulate (form a joint). Openings enclose delicate structures and allow them to travel through bones. Projections provide sites where bones articulate or to which ligaments and tendons attach.

Table 7.1 summarizes categories of bone markings and gives examples. Notice how each type demonstrates the Structure-Function Core Principle (p. 25).

CORE PRINCIPLE
Structure-Function

Quick Check

- ☐ 3. What are some functions of bone markings?

TABLE 7.1 BONE MARKINGS		
Bone Marking	**Description**	**Example**
Depressions: clefts of varying depth in a bone; located where a bone meets another structure, such as another bone or a blood vessel.		
Facet	Shallow convex or concave surface where two bones articulate	**Rib:** Articular facet for articulation with a transverse process
Fossa (plural, fossae)	Indentation in a bone into which another structure fits	**Humerus:** Distal portion with olecranon fossa
Fovea	Shallow pit	**Femur:** Fovea capitis

TABLE 7.1 *(continued)*

Bone Marking	Description	Example
Groove (or sulcus)	Long indentation along which a narrow structure travels	**Rib:** Costal groove

Openings: holes that allow blood vessels and nerves to travel through a bone; permit access to the middle and inner ear; encase delicate structures and protect them from trauma.

Bone Marking	Description	Example
Canal (or meatus)	Tunnel through a bone	**Temporal bone:** External acoustic meatus
Fissure	Narrow slit in a bone or between adjacent parts of bones	**Sphenoid bone:** Superior orbital fissure **Frontal bone:** Supraorbital foramen
Foramen (plural, foramina)	Hole in a bone	

Projections: bony extensions of varying shapes and sizes; some provide locations for attachment of muscles, tendons, and ligaments; some fit into depressions of other bones to stabilize joints.

Bone Marking	Description	Example
Condyle	Rounded end of a bone that articulates with another bone	**Mandible:** Mandibular condyle
Crest	Ridge or projection	**Ilium:** Iliac crest
Head	Round projection from a bone's epiphysis	**Humerus:** Head
Tubercle and tuberosity	Small, rounded bony projection; a tuberosity is a large tubercle	**Humerus:** Deltoid tuberosity
Epicondyle	Small projection usually proximal to a condyle	**Humerus:** Medial epicondyle
Process	Prominent bony projection	**Scapula:** Coracoid process
Spine	Sharp process	**Scapula:** Spine

(continued)

TABLE 7.1 BONE MARKINGS (*continued*)		
Bone Marking	**Description**	**Example**
Protuberance	Outgrowth from a bone	**Occipital bone:** External occipital protuberance
Trochanter	Large projection found only on the femur	**Femur:** Greater trochanter
Line	Long, narrow ridge	**Femur:** Linea aspera

Apply What You Learned

☐ 1. Why might a fractured bone of the axial skeleton be more damaging to overall homeostasis than a fractured bone of the appendicular skeleton?

☐ 2. Many bone abnormalities are characterized by changes in the normal structure of bone markings. Defective openings that are smaller or larger than normal can be particularly problematic. Why do you think this is true?

See answers in Appendix A.

MODULE 7.2
The Skull

Learning Outcomes

1. Describe the location, structural features, and functions of each of the cranial bones and the main sutures that unite them.
2. Describe the location, structural features, and functions of each of the facial bones and the hyoid bone.
3. Explain the structural features and functions of the orbit, nasal cavity, and paranasal sinuses.
4. Compare and contrast the skull of a fetus or infant with that of an adult.

We begin our tour of the skeletal system with its most complex bony structure, the **skull.** The skull generally has 22 bones, although in the adult all but one are fused and therefore immoveable. Note that some sources consider the six tiny bones in the ear, the *auditory ossicles,* to be skull bones (for a total of 28). For simplicity, we focus on the major skull bones in this chapter and cover the auditory ossicles in the special senses chapter (see Chapter 15). First, let's look at the overall structure of the skull.

Overview of Skull Structure

◀◀ Flashback

1. Where is the cranial cavity located, and what is housed within this cavity? (p. 16)

The skull consists of two groups of bones: (1) the **cranial bones,** which are known collectively as the **cranium** and encase the brain, and (2) the **facial bones,** which form the framework for the face (**Figure 7.2**). Both groups also contribute to the bony structures that house the special sensory organs. There are eight cranial bones: four single bones, which are the *frontal, occipital, ethmoid,* and *sphenoid bones,* and two paired bones, which are the *temporal* and *parietal bones.* The 14 facial bones include the paired *maxillary, zygomatic, nasal, lacrimal, palatine,* and *inferior nasal conchal bones,* as well as the unpaired *mandible* and *vomer.* With the exception of the mandible (the lower jaw bone), all skull bones are united in adults by immoveable joints called **sutures.**

Together, the eight cranial bones form the large **cranial cavity,** which surrounds the brain (**Figure 7.3**). Smaller cavities in the skull house the following organs:

● The *orbits* contain the eyeballs.
● The *nasal cavity* houses the sensory receptors for smell.
● The *oral cavity* surrounds the teeth and tongue.

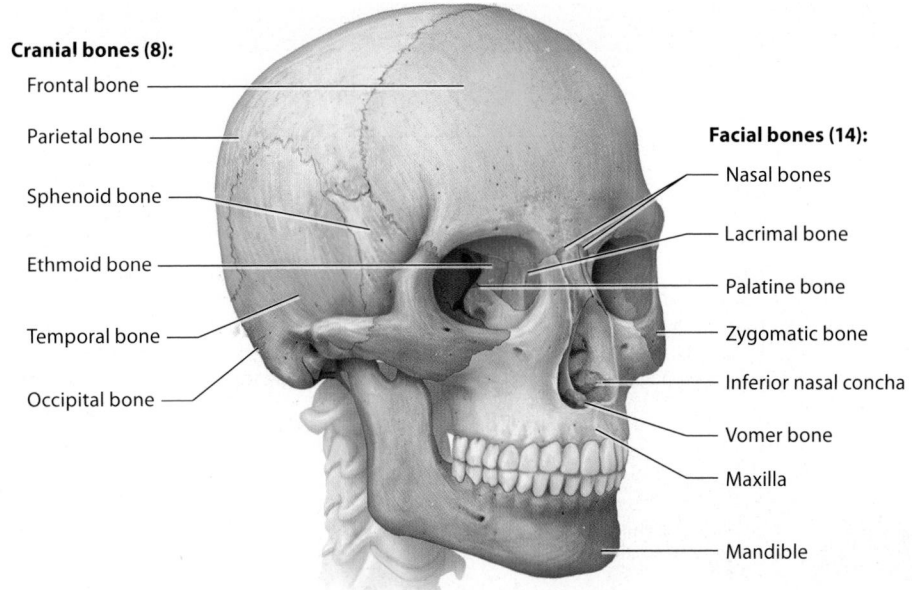

Cranial bones (8):
- Frontal bone
- Parietal bone
- Sphenoid bone
- Ethmoid bone
- Temporal bone
- Occipital bone

Facial bones (14):
- Nasal bones
- Lacrimal bone
- Palatine bone
- Zygomatic bone
- Inferior nasal concha
- Vomer bone
- Maxilla
- Mandible

Practice art labeling @MasteringA&P®

Figure 7.2 **Basic structure of the skull: anterolateral view of the cranial and facial bones.**

- Other small cavities house the sense organs for hearing and balance (not visible in Figure 7.3).

In addition, many bones contain air-filled, membrane-lined spaces known as *sinuses.* In particular, four bones around the nasal cavity have large sinuses called *paranasal sinuses.*

The cranial cavity consists of a superior portion, called the **cranial vault** or **calvarium,** and an inferior portion, the **cranial base** (Figure 7.4). Internally, the cranial base is divided into three indentations, or fossae, into which the brain fits snugly: the *anterior, middle,* and *posterior cranial fossae.*

To help with your studying, we have put the structural and functional features of the individual cranial and facial bones, as well as illustrations of the individual bones, into Table 7.2.

When you've become acquainted with each bone and its characteristics, study how these bones articulate with each other to form the skull. Figures 7.5–7.9 present several views and sections of the complete skull, and Concept Boost: Understanding How Skull Bones Relate to Each Other shows how the individual bones fit together.

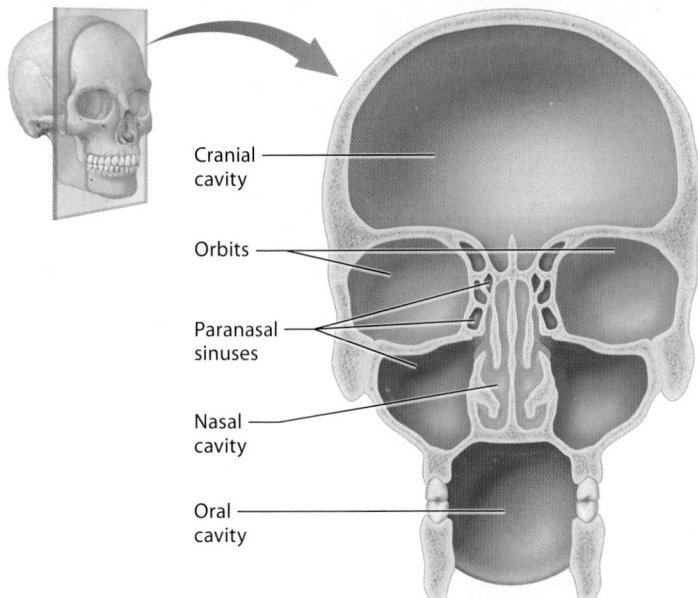

- Cranial cavity
- Orbits
- Paranasal sinuses
- Nasal cavity
- Oral cavity

Figure 7.3 **Cavities of the skull.**

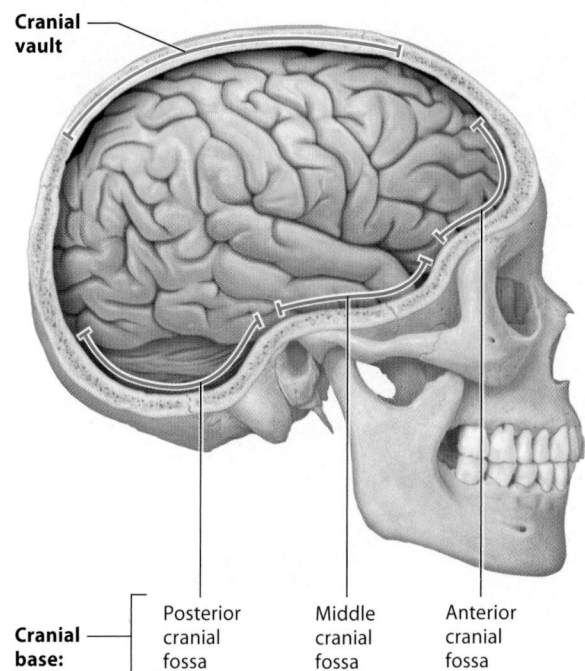

- Cranial vault
- Cranial base:
  - Posterior cranial fossa
  - Middle cranial fossa
  - Anterior cranial fossa

Figure 7.4 **Cranial vault and base.**

TABLE 7.2 BONES OF THE SKULL	
Bone	**Description**
Cranial Bones	

Frontal bone

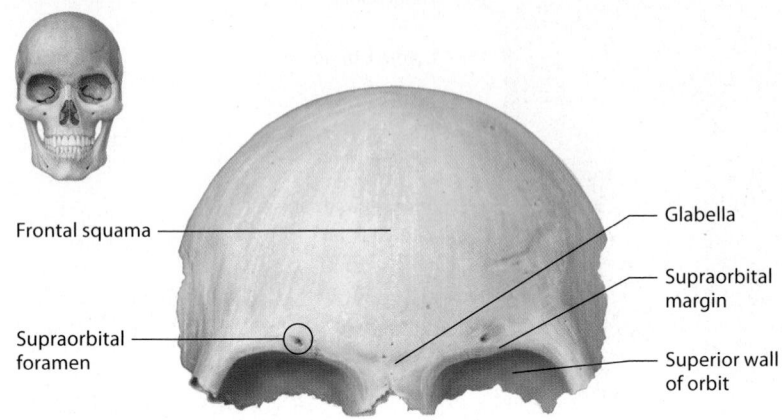

Frontal squama

Supraorbital foramen

Glabella

Supraorbital margin

Superior wall of orbit

The **frontal bone** forms the *frontal squama,* or forehead. Inside the frontal bone, we find the frontal sinuses, which are part of the paranasal sinuses (visible in Figure 7.13). The frontal bone also contains the following features:

- The **supraorbital margins** are sharp ridges that form the superior and superomedial boundaries of the orbit.
- Along the middle of the supraorbital margin, we find a tiny *supraorbital foramen,* which allows passage of blood vessels and nerves to the forehead.
- The *glabella* is the smooth region between the two supra-orbital margins.

Parietal bones

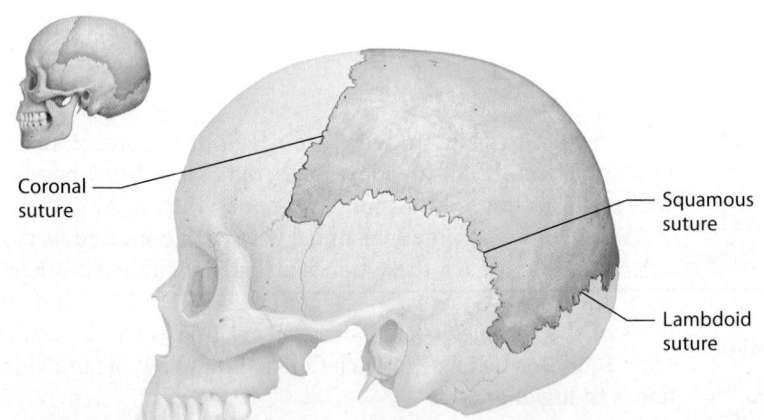

Coronal suture

Squamous suture

Lambdoid suture

The two **parietal bones** form the superior wall and part of the lateral wall of the cranial vault. They meet at the **sagittal suture** and articulate with several cranial bones at other sutures:

- together they meet the frontal bone at the **coronal suture** (*coron-* = "crown");
- each parietal bone meets a temporal bone at the **squamous suture;** and
- together they meet the occipital bone at the **lambdoid suture.**

Occipital bone

Occipital condyle

External occipital crest

External occipital protuberance

Foramen magnum

Inferior nuchal line

Superior nuchal line

The **occipital bone** forms the posterior part of the cranial cavity and the posterior cranial fossae. It features the following:

- On its anterior and inferior surface is the **foramen magnum** ("big hole"), through which the spinal cord passes to enter the vertebral cavity.
- On either side of the foramen magnum are two **occipital condyles,** which articulate with the first cervical vertebra.
- The posterior surface of the occipital bone features two parallel ridges, the **superior nuchal line** and **inferior nuchal line.** The **external occipital protuberance** runs through the superior nuchal line and continues inferiorly as the *external occipital crest.*

TABLE 7.2 *(continued)*

Bone	Description

Temporal bones

Squamous region
Mandibular fossa
Zygomatic process
External acoustic meatus
Tympanic region
Mastoid region
Styloid process
Mastoid process

Squamous region
Internal acoustic meatus
Zygomatic process
Petrous region
Styloid process
Mastoid process

Zygomatic process
Sphenoid bone
Mandibular fossa
Foramen lacerum
Styloid process
Carotid canal
Tympanic region
Jugular foramen
Mastoid process
Occipital bone

Description

The two **temporal bones** form the lateral walls of the cranium. Each temporal bone has four regions:

- The **squamous region** (*squam-* = "flat") is the bone's broad, flat surface. It features the **zygomatic process,** which forms part of the zygomatic arch, and the **mandibular fossa,** which articulates with the mandible.
- The **tympanic region** houses the **external acoustic meatus,** which is the entry into the middle ear. Inferiorly, it features the **styloid process** (*stylus* = "needle"), a needle-like spur of bone.
- The **mastoid region** contains the thick projection called the **mastoid process,** which is filled with tiny sinuses called **mastoid air cells.**
- The **petrous region** (*petr-* = "rock") is located on the internal or medial surface of the temporal bone, where it forms part of the middle cranial fossae, and contains four important features:
  - the **internal acoustic meatus,** a canal leading from the inner ear;
  - the **jugular foramen,** a posterior opening where the temporal bone meets the occipital bone, which allows the passage of nerves and the *internal jugular vein;*
  - the **carotid canal,** which allows the *internal carotid artery* to enter the cranial cavity; and
  - the **foramen lacerum,** which allows the passage of small blood vessels and nerves.

(continued)

TABLE 7.2 BONES OF THE SKULL *(continued)*

Bone	Description

Sphenoid bone

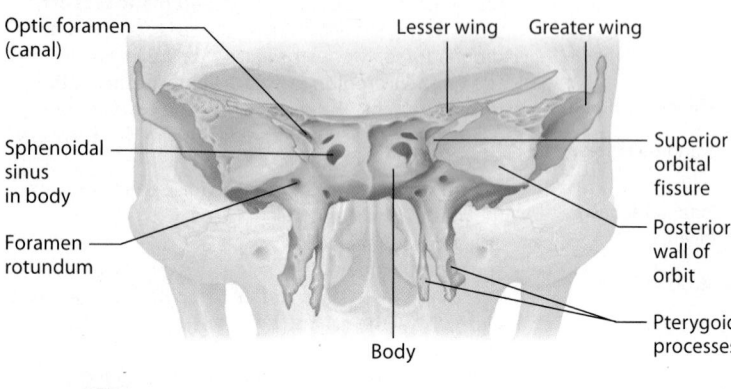

Optic foramen (canal)
Lesser wing
Greater wing
Sphenoidal sinus in body
Superior orbital fissure
Foramen rotundum
Posterior wall of orbit
Body
Pterygoid processes

Greater wing
Sella turcica
Lesser wing
Foramen rotundum
Foramen ovale
Foramen spinosum

Ethmoid bone

Cribriform plate with cribriform foramina
Crista galli
Orbital plate
Ethmoidal sinuses
Left lateral mass
Perpendicular plate
Middle nasal concha

The **sphenoid bone** is a unique deep cranial bone that articulates with every other cranial bone. It forms part of the anterior and middle cranial fossae. From an anterior view (upper figure), it resembles a bat; from a superior view (lower figure), it resembles a stingray. It has four main components:

- The central **body** is largely hollow, containing air-filled **sphenoidal sinuses.** On its superior surface, a saddle-shaped depression called the **sella turcica** (SELL-uh TER-suh-kuh; "Turk's saddle") houses the *pituitary gland.* On the anterior surface of the sphenoid's body, we find the two **optic canals,** through which the nerves that transmit vision pass.
- The **greater wings** are the wings of the "bat" that extend laterally from the body. With the temporal bone, they form part of the middle cranial fossae. With the frontal bone, they form part of the posterior walls of the orbit. They contain three foramina that allow the passage of blood vessels and nerves: the anterior **foramen rotundum,** the middle **foramen ovale,** and the posterior **foramen spinosum.**
- The **lesser wings** are the wings of the "stingray" (see the superior view), which project from the superior surface of the sphenoid body. In the anterior view, we see the **superior orbital fissure,** a slit between the lesser and greater wings. The nerves that control eye movement pass through this fissure.
- The **pterygoid processes** (TEHR-uh-goyd; *ptery-* = "fin" or "wing") are the inferior projections of the sphenoid bone. These processes form part of the posterior wall of the oral and nasal cavities, and are the site of attachment for muscles of chewing.

The **ethmoid bone** is located deep in the anterior cranium between the eyes and behind the nasal bones. Its superior surface, the **cribriform plate,** forms the roof of the nasal cavity. Tiny nerves that detect smell pass through the *cribriform (olfactory) foramina* from the nasal cavity to the brain.

The cribriform plate has a superior projection called the **crista galli** (KRIS-tuh GAL-ee; "rooster's comb"), to which membranes that surround the brain attach. Extending inferiorly from the crista galli is the **perpendicular plate,** which forms part of the *nasal septum* that divides the nasal cavity into two sides.

Flanking the perpendicular plate are the ethmoid bone's two *lateral masses,* which are filled with small hollow spaces collectively known as the right and left **ethmoidal sinuses.** The lateral side of each lateral mass contributes to the orbit. The medial side of each lateral mass contains the **superior** and **middle nasal conchae** (KAHN-kee; singular, *concha; concha* = "shell"), which form part of the lateral walls of the nasal cavity.

TABLE 7.2 *(continued)*

Bone	Description

Facial Bones

Nasal, lacrimal, and zygomatic bones

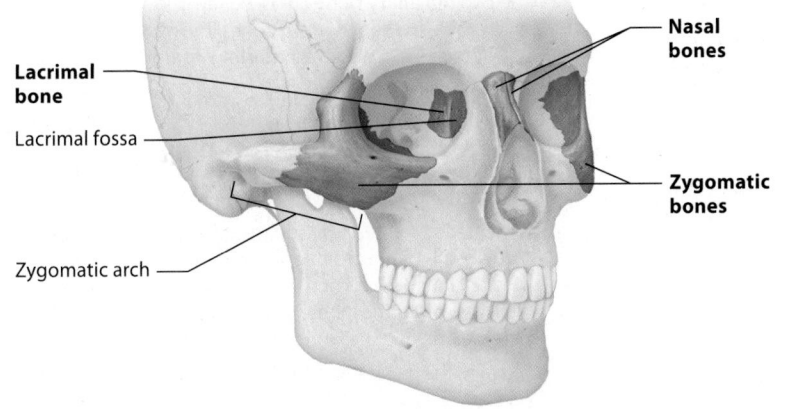

The two **nasal bones** form the bridge of the nose. They articulate with the hyaline cartilages that form most of the framework of the nose.

The two **lacrimal bones** are the smallest and most delicate facial bones. They are found in the medial wall of the orbit, and contain a depression called the lacrimal fossa through which tears drain.

The two **zygomatic bones,** along with the zygomatic processes of the temporal bones and maxillae, form the bulk of the cheekbone, or *zygomatic arch*. They also form the lateral wall and part of the inferior border of the orbit.

Palatine bones

The two L-shaped **palatine bones** are located in the posterior nasal cavity between the maxillae and the pterygoid processes of the sphenoid. Each bone is composed of two *plates*. The smaller *horizontal plates* form the posterior part of the hard palate. The larger *perpendicular plates* are vertical projections that form part of the lateral walls of the nasal cavity and a tiny piece of the orbit.

Mandible

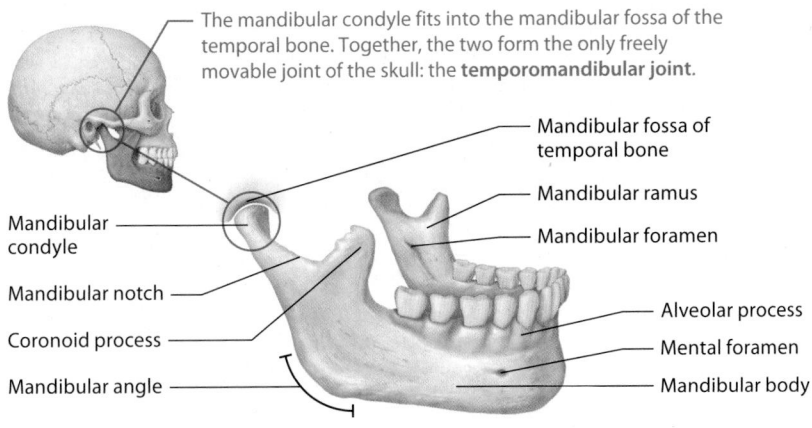

The mandibular condyle fits into the mandibular fossa of the temporal bone. Together, the two form the only freely movable joint of the skull: the **temporomandibular joint**.

The **mandible** is the inferior jaw bone and the only moveable bone of the adult skull. It consists of three parts:

- the central **mandibular body,** and
- the right and left **mandibular rami** (man-DIB-yoo-lar RAY-mee; *ramus* = "branch").

The body meets the two rami at the right and left **mandibular angles.** The mandibular body forms the chin and houses the inferior teeth in deep sockets called *dental alveoli* within a ridge called the *alveolar process* (al-VEE-oh-lahr; *alve-* = "hollow"). In each side of the lateral body, we find the **mental foramen** (*menta* = "chin").

Each mandibular ramus consists of a plate of bone topped by the U-shaped **mandibular notch,** inferior to which we find the **mandibular foramen.** Each ramus is topped with two processes:

- on the anterior is the **coronoid process,** which serves as attachment site for a major muscle of chewing; and
- on the posterior is the **condylar process,** at the top of which is the **mandibular condyle.**

(continued)

TABLE 7.2 BONES OF THE SKULL (continued)

Bone	Description
Maxillae Note: The intermaxillary suture, inferior orbital fissure, and maxillary sinuses are not visible in these views.	The two **maxillary bones** or **maxillae** (mak-SILL-ee; *maxil-* = "jaw") are the superior jaw bones. Their fusion creates a bony projection called the **anterior nasal spine.** The superior teeth reside in the dental alveoli of the alveolar process. Medial to the alveolar processes, the maxillae form two horizontal **palatine processes,** which meet at the *intermaxillary suture.* Together, these two processes, form the anterior part of the hard palate (roof of the mouth). The anterior and superior surfaces form part of the infero-medial walls of the orbit. Just below the inferior margin of the orbit, arteries and nerves pass through the inferior orbital fissure and the infraorbital foramen to supply the face. Here we also find the large hollow **maxillary sinuses.** Laterally, the maxillae articulate with the zygomatic bones via their *zygomatic processes.*
Inferior nasal conchae 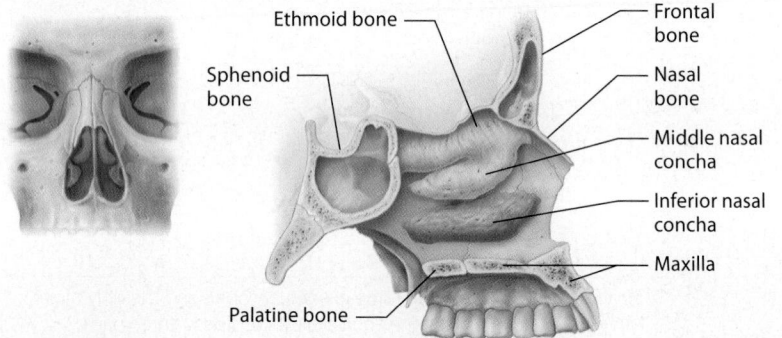	The **inferior nasal conchae** are located in the lateral walls of the nasal cavity. They are situated inferior to the middle nasal conchae of the ethmoid bone.
Vomer bone	The **vomer** is the deepest facial bone. It forms the inferior and posterior parts of the *nasal septum.*

The first view of the whole skull is the anterior view. While all facial bones are visible here, some are visible only in the orbit (a tiny piece of palatine bone) or the nasal cavity (the vomer and inferior nasal conchae). Certain cranial bones are visible here as well, including the frontal bone and portions of the sphenoid and ethmoid bones.

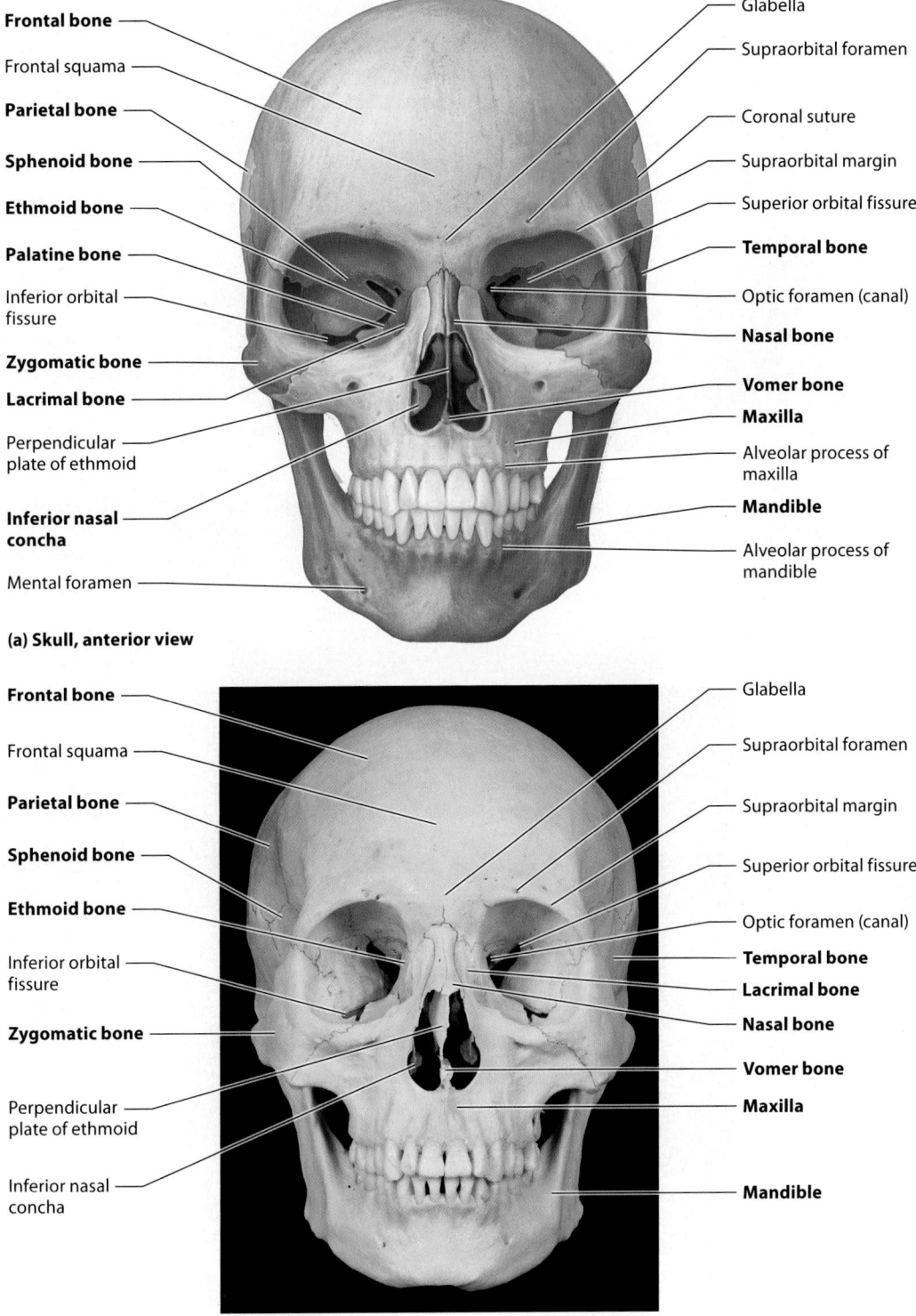

Frontal bone

Frontal squama

Parietal bone

Sphenoid bone

Ethmoid bone

Palatine bone

Inferior orbital fissure

Zygomatic bone

Lacrimal bone

Perpendicular plate of ethmoid

Inferior nasal concha

Mental foramen

Glabella

Supraorbital foramen

Coronal suture

Supraorbital margin

Superior orbital fissure

Temporal bone

Optic foramen (canal)

Nasal bone

Vomer bone

Maxilla

Alveolar process of maxilla

Mandible

Alveolar process of mandible

(a) Skull, anterior view

Frontal bone

Frontal squama

Parietal bone

Sphenoid bone

Ethmoid bone

Inferior orbital fissure

Zygomatic bone

Perpendicular plate of ethmoid

Inferior nasal concha

Glabella

Supraorbital foramen

Supraorbital margin

Superior orbital fissure

Optic foramen (canal)

Temporal bone

Lacrimal bone

Nasal bone

Vomer bone

Maxilla

Mandible

(b) Photo of skull, anterior view

Figure 7.5 Anterior view of the skull.

Notice that in the lateral view, we can see all of the cranial bones and many of the facial bones. This is also the best view to see the bones that contribute to the zygomatic arch and how the temporal bone and mandible articulate.

The sphenoid bone articulates with the temporal, parietal, and frontal bones, forming a "weak spot" called the temple. This is a dangerous location for a head injury because bones fracture more easily along sutures.

Coronal suture
Parietal bone
Temporal bone
Squamous suture
Zygomatic arch
Lambdoid suture
External acoustic meatus
Occipital bone
Mastoid process
Styloid process
Mandibular condyle
Mandibular notch
Mandibular ramus
Coronoid process
Mandibular angle

Frontal bone
Sphenoid bone (greater wing)
Ethmoid bone
Lacrimal bone
Lacrimal fossa
Nasal bone
Zygomatic bone
Anterior nasal spine
Maxilla
Alveolar processes
Mental foramen
Mandible

(a) External anatomy of skull, lateral view

Coronal suture
Parietal bone
Temporal bone
Squamous suture
Zygomatic arch
Lambdoid suture
Occipital bone
External acoustic meatus
Mastoid process
Styloid process
Mandibular condyle
Mandibular notch
Mandibular ramus
Coronoid process
Mandibular angle

Temple

Frontal bone
Sphenoid bone (greater wing)
Ethmoid bone
Lacrimal bone
Nasal bone
Maxilla
Anterior nasal spine
Alveolar processes
Mandible
Mental foramen

(b) Photo of skull, lateral view

Figure 7.6 Lateral view of the skull.

Here we can see the occipital, parietal, and parts of the temporal bones, as well as the posterior sides of the mandible and maxillae. This is a good view in which to see the surface markings of the occipital bone and the sagittal, squamous, and lambdoid sutures.

In this view, we can also see several sutures, particularly where the parietal bones meet the frontal bone at the coronal suture.

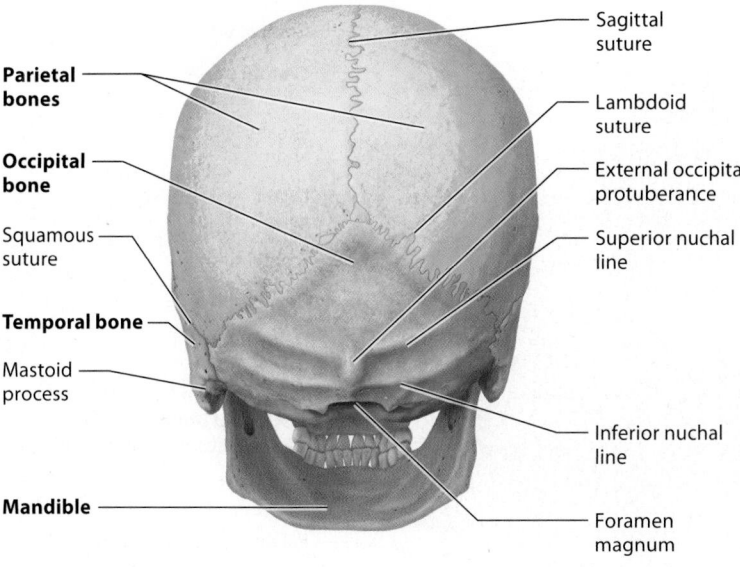

(a) Posterior view

Parietal bones
Occipital bone
Squamous suture
Temporal bone
Mastoid process
Mandible

Sagittal suture
Lambdoid suture
External occipital protuberance
Superior nuchal line
Inferior nuchal line
Foramen magnum

Anterior
Frontal bone
Coronal suture
Sagittal suture
Parietal bones
Lambdoid suture
Occipital bone
Posterior

(b) Superior view

For this inferior view, imagine holding the skull (without its mandible) and flipping it upside down so you are looking at its base. This is a good view to appreciate the structure of the hard palate. The two maxillae meet at the intermaxillary suture and articulate with the palatine bones posteriorly. We can also see the sphenoid bone's pterygoid processes and laterally its greater wings.

View

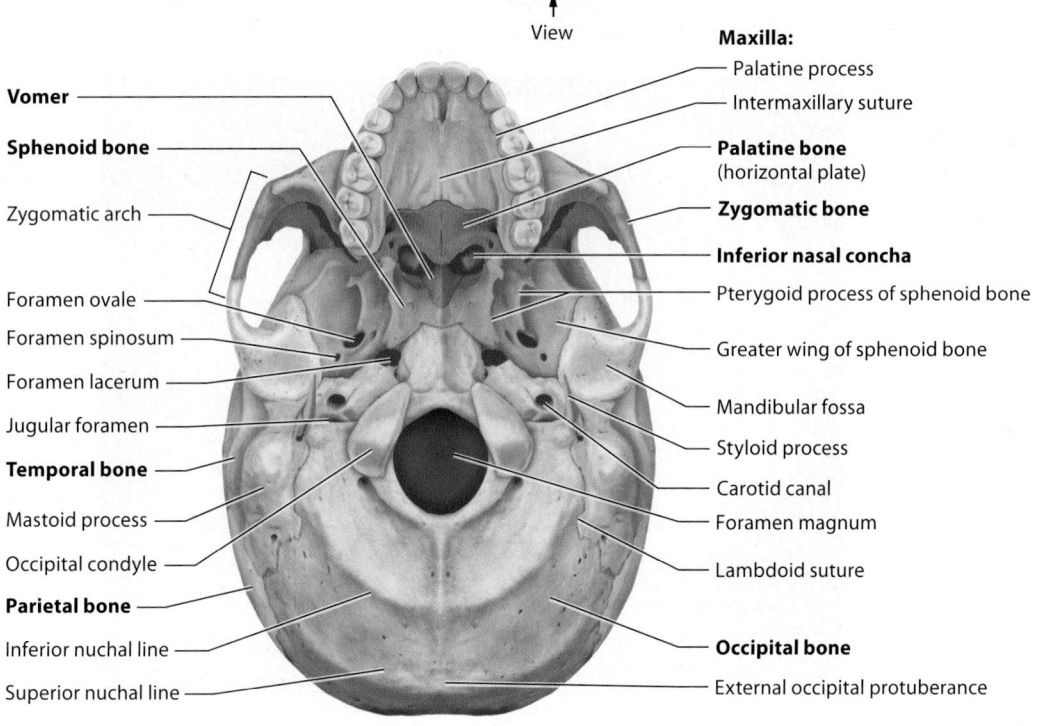

Vomer
Sphenoid bone
Zygomatic arch
Foramen ovale
Foramen spinosum
Foramen lacerum
Jugular foramen
Temporal bone
Mastoid process
Occipital condyle
Parietal bone
Inferior nuchal line
Superior nuchal line

Maxilla:
Palatine process
Intermaxillary suture
Palatine bone (horizontal plate)
Zygomatic bone
Inferior nasal concha
Pterygoid process of sphenoid bone
Greater wing of sphenoid bone
Mandibular fossa
Styloid process
Carotid canal
Foramen magnum
Lambdoid suture
Occipital bone
External occipital protuberance

(c) Inferior view

Figure 7.7 Posterior, superior, and inferior views of the skull.

This midsagittal section is the best view to appreciate the structure of the nasal septum. Its superior portion is composed of the perpendicular plate of the ethmoid bone and the inferior portion of the vomer bone. We can also see two paranasal sinuses—the frontal sinus and sphenoid sinus.

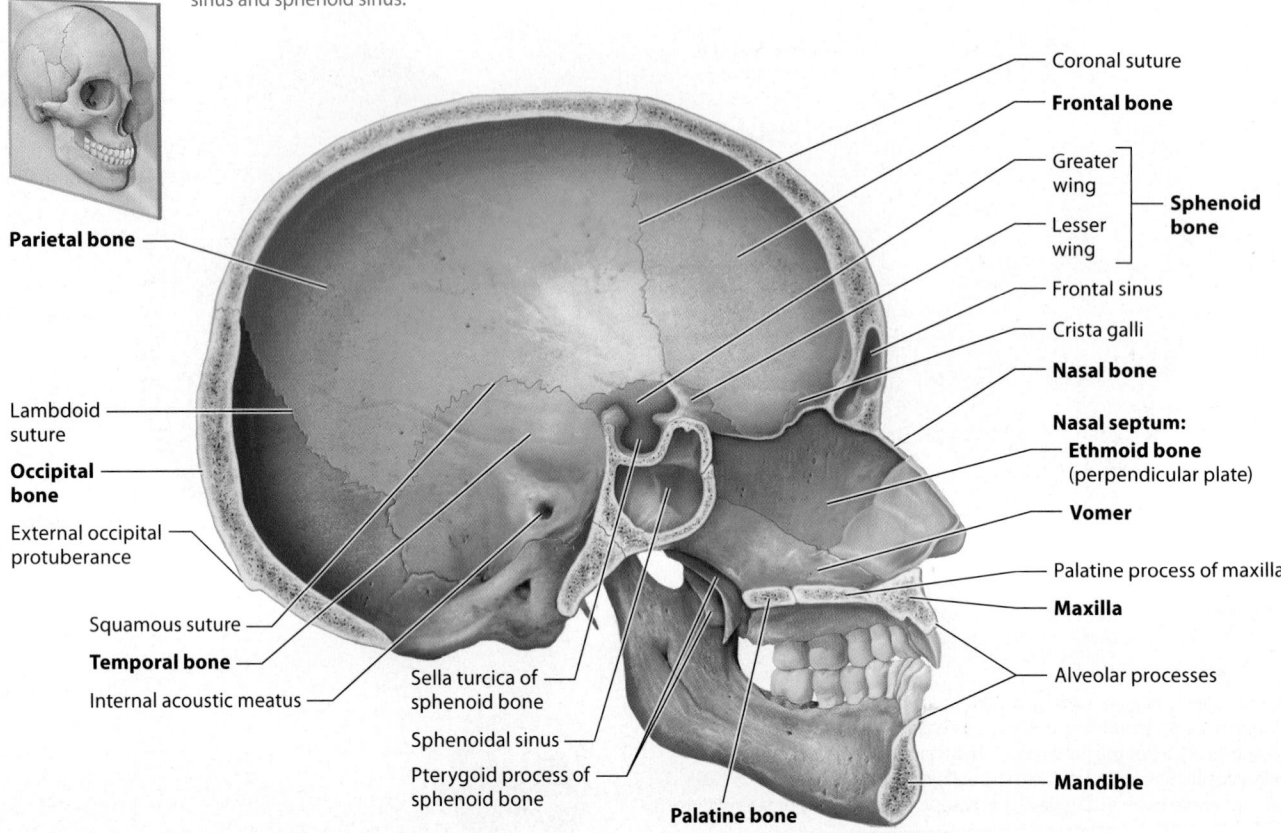

Coronal suture
Frontal bone
Greater wing
Lesser wing
Sphenoid bone
Frontal sinus
Crista galli
Nasal bone
Nasal septum:
  Ethmoid bone (perpendicular plate)
  Vomer
Palatine process of maxilla
Maxilla
Alveolar processes
Mandible

Parietal bone
Lambdoid suture
Occipital bone
External occipital protuberance
Squamous suture
Temporal bone
Internal acoustic meatus
Sella turcica of sphenoid bone
Sphenoidal sinus
Pterygoid process of sphenoid bone
Palatine bone

(a) Internal anatomy of the left half of skull, midsagittal section

Frontal bone
Frontal sinus
Sphenoid sinus
Nasal bone
Ethmoid bone (perpendicular plate)
Vomer
Maxilla
Palatine process of maxilla
Mandible

Parietal bone
Occipital bone
Temporal bone
Internal acoustic meatus
Pterygoid process of sphenoid bone
Palatine bone

(b) Photo of the skull, midsagittal section

Figure 7.8 Medial view of midsagittal section of the skull.

Practice art labeling
@MasteringA&P®

Imagine you are holding a skull, looking down at its superior surface. Now imagine removing the top of the skull and the brain and looking inside. This is the best view to appreciate the way the cranial fossae cradle the brain. Also note the cribriform plate of the ethmoid bone, the lesser and greater wings and sella turcica of the sphenoid bone, and the petrous part of the temporal bone.

Ethmoid bone:
Cribriform plate
Crista galli

Sphenoid bone:
Lesser wing
Greater wing

Middle cranial fossa

Sella turcica

Temporal bone
(petrous region)

Jugular foramen

Frontal bone

Anterior cranial fossa

Optic canal

Foramen rotundum

Foramen ovale

Foramen spinosum

Foramen lacerum

Foramen magnum

Posterior cranial fossa

Parietal bone

Occipital bone

Ethmoid bone:
Cribriform plate
Crista galli

Sphenoid bone:
Lesser wing
Greater wing

Middle cranial fossa

Sella turcica

Temporal bone
(petrous region)

Jugular foramen

(a) Superior, interior view showing internal structures of skull

(b) Photo of superior, interior view of skull

Figure 7.9 Internal view of the skull.

ConceptBOOST ❯❯❯

Understanding How Skull Bones Relate to Each Other

It can be hard to appreciate the skull bones in terms of their structure and how they relate to each other. For this reason, we present **Figure 7.10**, which shows a disarticulated, or "exploded," skull. The bones are separated from one another so that you can see each individual bone clearly, as well as how they fit together. Think of the skull as a three-dimensional jigsaw puzzle with the pieces taken apart and separated. This is the best way to appreciate the structure of complex bones like the ethmoid and sphenoid, and observe how they relate to other skull bones. ■

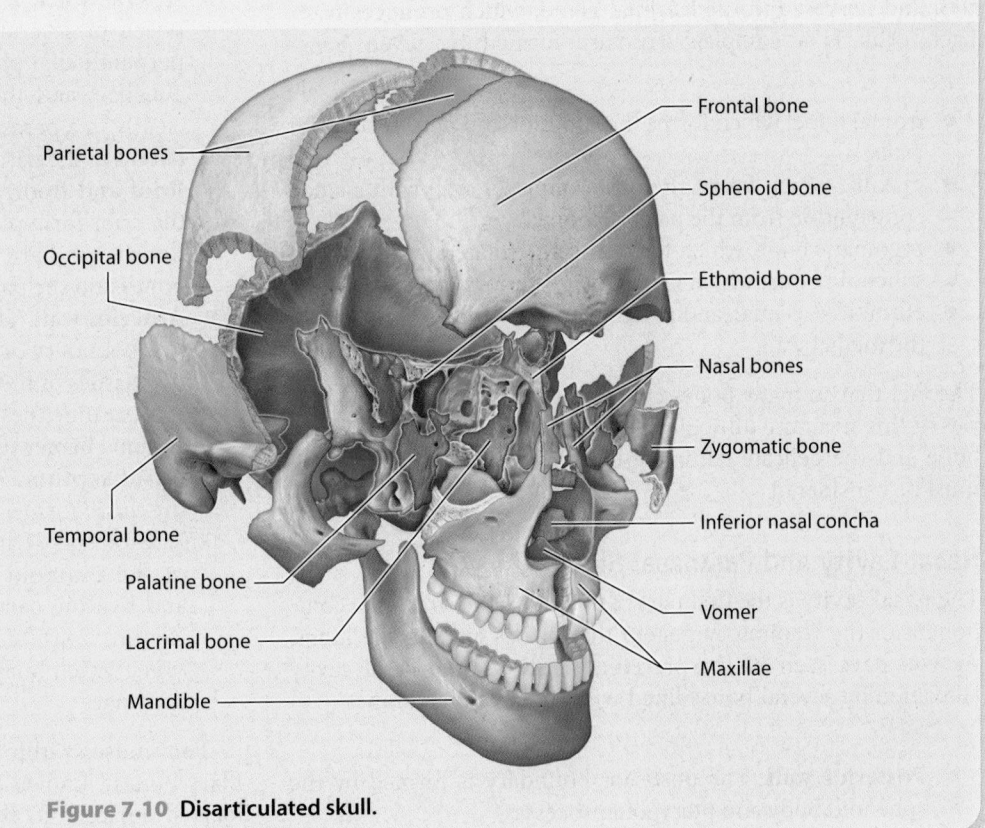

Parietal bones

Occipital bone

Temporal bone

Palatine bone

Lacrimal bone

Mandible

Frontal bone

Sphenoid bone

Ethmoid bone

Nasal bones

Zygomatic bone

Inferior nasal concha

Vomer

Maxillae

Figure 7.10 Disarticulated skull.

☐ 1. Match each bone with the correct description from the right column.

_____ Ethmoid bone	a. Forms the inferior part of the nasal septum
_____ Vomer	b. Composed of a body, greater and lesser wings, and pterygoid processes
_____ Mandible	
_____ Occipital bone	c. Posterior cranial bone; contains the foramen magnum
_____ Sphenoid bone	
_____ Zygomatic bones	d. Features the crista galli and superior and middle nasal conchae
	e. Form the anterior parts of the cheeks
	f. Lower jaw bone

Cavities of the Skull

The skull is full of cavities. We've already discussed the largest, the cranial cavity, and turn now to the structure and function of several smaller ones: the orbit, nasal cavity and paranasal sinuses, and oral cavity.

Orbit

The **orbit** houses the eyeball; its associated blood vessels, muscles, and nerves; and the lacrimal gland, which produces tears. Each orbit is a complex structure formed by seven bones (**Figure 7.11**):

- frontal bone, which forms the superior and posterosuperior walls;
- maxilla, which forms the posteroinferior wall with a small contribution from the palatine bone;
- zygomatic bone, which forms the anterolateral wall;
- sphenoid bone, which forms the posterior wall; and
- ethmoid, lacrimal, and palatine bones, which together form the medial wall.

The fact that so many bones contribute to the orbit makes injuries of this structure difficult to treat, as the contribution of each bone and the delicate nature of the structures within the orbit must be considered.

Nasal Cavity and Paranasal Sinuses

The **nasal cavity** is the first part of the *respiratory tract,* the component of the respiratory system that brings air into the lungs. As you have seen in the preceding sections, the nasal cavity is formed by several bones lined with mucous membranes (**Figure 7.12**):

- **Posterior wall.** The posterior boundary is formed by the sphenoid body and pterygoid processes.

Sphenoid bone   Frontal bone   Ethmoid bone   Lacrimal bone

Zygomatic bone

Maxilla   Palatine bone

Figure 7.11   The orbit.

- **Lateral walls.** The lateral walls of the nasal cavity are composed of several bones, including the ethmoid bone, the perpendicular plate of the palatine bones, the inferior nasal conchae, and the maxilla. You can only see these bones clearly in a parasagittal section of the skull (a section that is lateral to the nasal septum; Figure 7.12a).
- **Roof and floor.** The roof of the nasal cavity is formed by the cribriform plate of the ethmoid bone. The nasal cavity's floor is the hard palate, which is composed of the palatine bones and the palatine processes of the maxillae.
- **Anterior wall.** The nasal bones and maxillae form the anterior boundary of the nasal cavity. The cartilage and connective tissues of the nose attach to the margins of these bones. We examine these cartilages more closely in the respiratory system chapter (Chapter 21).
- **Nasal septum.** The two sides of the nasal cavity, called the *nasal fossae,* are divided by the **nasal septum** (Figure 7.12b). This structure consists of the perpendicular plate of the ethmoid bone and the vomer posteroinferiorly and hyaline cartilage anteriorly. Occasionally, the nasal septum is shifted to one side, resulting in a *deviated septum,* which may make it difficult to breathe through the nose.

The sinuses within the frontal, sphenoid, ethmoid, and maxillary bones—known collectively as the **paranasal sinuses**—are located adjacent to the nasal cavity and communicate with it

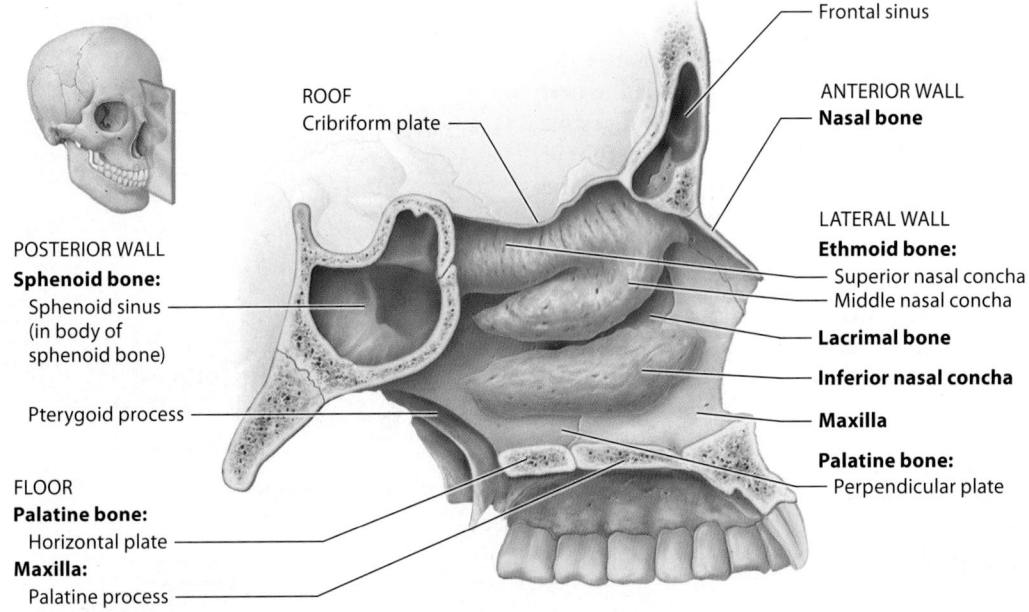

(a) Parasagittal section through nasal cavity (nasal septum removed)

(b) Anterior view of nasal cavity

Figure 7.12 Nasal cavity.

through small bony openings (**Figure 7.13**). Like the nasal cavity, they are lined with mucous membranes. Air flowing through the nasal cavity passes through the openings into the sinuses, where the mucous membranes filter, warm, and humidify the air. The paranasal sinuses also lighten the skull considerably and enhance voice resonance.

The connection of the sinuses to the nasal cavity has the unfortunate consequence of making the transmission of bacteria and viruses between the two structures quite common. When infectious agents enter the paranasal sinuses, a condition called *sinusitis* results. The swollen mucous membranes produce excess mucus and block the drainage of mucus into the nasal cavity. Excess mucus leads to the common symptoms of pressure and a sinus headache. You can use this to your advantage, though, to help you remember the location of the sinuses: Sinus headaches occur above the eyes (frontal sinuses), between the eyes (ethmoidal sinuses), behind the eyes (sphenoidal sinuses), and below the eyes (maxillary sinuses).

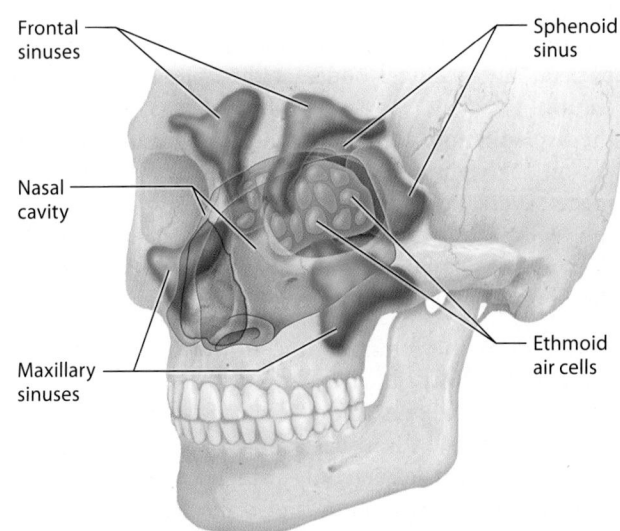

Figure 7.13 Paranasal sinuses.

Oral Cavity

The final cavity we discuss is the **oral cavity,** which houses the teeth, the tongue, and most of the salivary glands. It is the first part of the *gastrointestinal tract,* the component of the digestive system that takes in food and digests and absorbs it. The roof of the oral cavity is formed by the same structure as the floor of the nasal cavity (the hard palate), and the anterior and lateral walls of the oral cavity are formed by the maxillae and the mandible (refer to Figure 7.8). Unlike the other cavities, the oral cavity has no bony floor or posterior wall. Instead, the floor and wall are formed by soft tissues, including muscle and connective tissue.

Quick Check

☐ 2. Which bones form the orbit?
☐ 3. What are the paranasal sinuses, and how are they related to the nasal cavity?
☐ 4. How are the oral and nasal cavities related structurally?

Fetal Skull

◀◀ **Flashback**

1. What is intramembranous ossification? (p. 193)
2. Which bones form by intramembranous ossification? (p. 193)

You may have heard that infants have "soft spots" in their skulls. These "soft spots"—actually, membranous areas called **fontanels**—result from the ossification of cranial bones. In the previous chapter, you learned about *intramembranous ossification,* in which bone develops from a membrane of fibrous connective tissue. Until the cranial bones complete this process, when a child is 18–24 months old, several fontanels are present, including (**Figure 7.14**):

- the **anterior fontanel,** which is located between the developing frontal and parietal bones where the coronal and sagittal sutures meet;
- the **posterior fontanel,** which is between the developing parietal and occipital bones at the apex of the lambdoid suture;
- the two **sphenoid fontanels,** which are located in the *temple* on the right and left sides where the sphenoid bone meets several other cranial bones; and
- the two **mastoid fontanels**, which are located at the junction of the lambdoid and squamous sutures where the developing parietal, temporal, and occipital bones meet.

In addition to the fontanels, the sutures between the skull bones are not yet fused, and certain individual bones have sutures within them. For example, a suture called the *metopic (or frontal) suture* exists within the frontal bone. This suture fuses completely after birth, leaving only a single frontal bone.

Fontanels and unfused sutures give the fetal skull a significant amount of flexibility. This flexibility ensures that an infant's

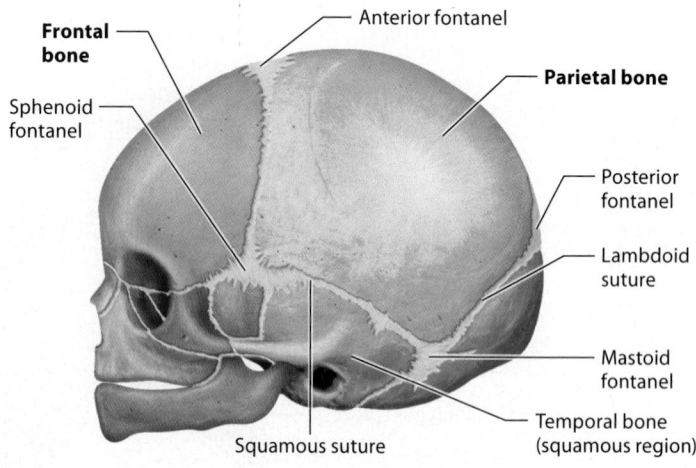

(a) Fetal skull, lateral view

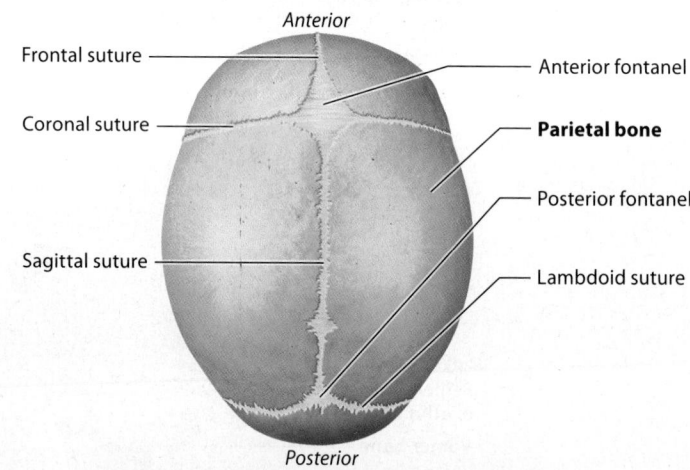

(b) Fetal skull, superior view

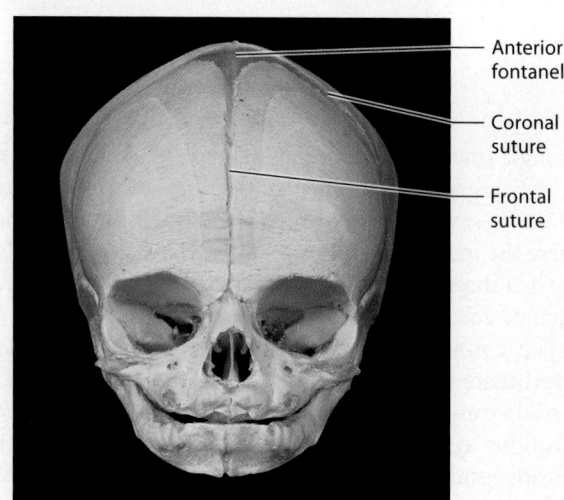

(c) Photo of fetal skull, anterior view

Figure 7.14 Fetal skull.

skull is able to fit through the mother's vaginal canal during birth. A common consequence is that the heads of many

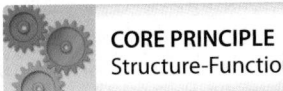

CORE PRINCIPLE
Structure-Function

infants are misshapen after a vaginal birth. Fortunately, their heads generally return to a normal shape in a few days. The fontanels also allow the skull to enlarge as an infant experiences normal brain growth, another example of the Structure-Function Core Principle (p. 25).

In a forensic setting, the degree to which skull sutures have fused can give important information about an individual's age. *A&P in the Real World: Forensic Skull Anatomy* examines sutures and other identifying features of the skull.

Quick Check

☐ 5. What are fontanels, and why are they important in the fetal skull?
☐ 6. Where are the six main fontanels located?

Hyoid Bone

The **hyoid bone** is a small, C-shaped bone in the superior neck (**Figure 7.15**). It is actually not a skull bone, but we discuss it with the skull bones because of its close proximity to them. However, it does not articulate with any skull bones or, indeed, with any bones at all. Instead, the hyoid bone is suspended in the superior neck by muscles and ligaments that attach it to the styloid processes of the temporal bone and to the larynx (voice box). It

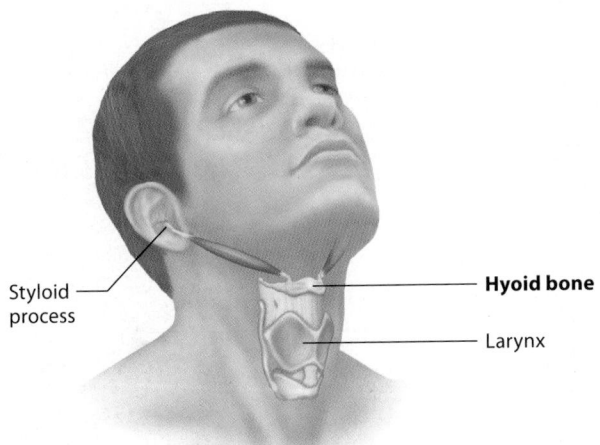

Styloid process
Hyoid bone
Larynx

(a) Position of the hyoid bone

Greater horn
Lesser horn
Body

(b) Hyoid bone, anterior view

Figure 7.15 Structure of the hyoid bone.

serves as the attachment point for numerous muscles, including those involved in swallowing and speech. It is also an important bone in forensics, as a broken hyoid bone can indicate that a crime victim was strangled.

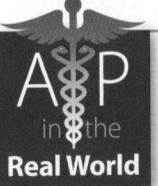

Forensic Skull Anatomy

Forensic investigators often must identify human remains with little to go on except bones. Fortunately, bones, particularly the skull, can give investigators many clues. One of the most basic traits that can be identified from a skull is the individual's gender. The two skulls shown here are from a male (left) and a female (right):

Male skull   Female skull

There are four obvious differences between the male and female skulls:

1. The forehead of the male skull is sloped and the female's is straight.
2. The male's supraorbital ridge is more prominent.
3. The mandibular angle of the male skull is closer to 90 degrees, and that of the female skull is generally greater than 90 degrees.
4. The mastoid process of the male skull is larger and more prominent.

These features and several others are factored into gender identification.

In addition, the size of the skull, the appearance of the sutures, and the teeth can help approximate an individual's age. Other skull features provide clues to the individual's ethnic heritage. Once these factors have been analyzed, a forensic artist can develop a sketch of what the person looked like in life.

☐ 7. What is unique about the hyoid bone?

Apply What You Learned

☐ 1. Ms. Mayer has been brought to the emergency department after sustaining a head injury during a cycling collision.
   a. During the physical exam, you note that a clear fluid is leaking from her nose; you recognize this as cerebrospinal fluid, the fluid surrounding the brain. Which bone has likely been fractured? Why would this cause cerebrospinal fluid to leak from the nasal cavity? (*Hint:* Which bone forms part of the cranial and nasal cavities?)
   b. Ms. Mayer is reporting vision problems. An x-ray of her skull shows a fracture of the lesser wing of the sphenoid bone. Why could this interfere with her vision?

☐ 2. Ms. Midna presents to your clinic with a complaint of pain in her face. She points to two areas, between her eyes and inferior to her eyes over her cheeks, as the locations of greatest pain. She states that she has had a cold for the past few days and has been unable to breathe through her nose. What is causing her pain? How and why is it related to her cold?

See answers in Appendix A.

MODULE **7.3**

The Vertebral Column and Thoracic Cage

Learning Outcomes

1. Describe the curvatures of the vertebral column.
2. Compare and contrast the three classes of vertebrae, the sacrum, and the coccyx.
3. Explain the structure and function of intervertebral discs and vertebral ligaments.
4. Describe the location, structural features, and functions of the bones of the thoracic cage.

The previous module focused on the uppermost part of the axial skeleton, the skull. Now, our attention shifts to the rest of the axial skeleton: the bones and skeletal cartilages of the vertebral column and thoracic cage.

Overview of the Vertebral Column

The **vertebral column**, or **spine**, consists of, on average, 33 bones called **vertebrae** (Figure 7.16a). We classify vertebrae on the basis of their structure and location:

- 7 **cervical vertebrae** located in the neck;
- 12 **thoracic vertebrae** that articulate with the ribs;

- 5 **lumbar vertebrae** in the lower back;
- 5 fused **sacral vertebrae** (collectively called the *sacrum*) that articulate with the pelvic bones; and
- 3–5 fused **coccygeal vertebrae** (collectively called the *coccyx*) at the most inferior end of the vertebral column.

We refer to each individual vertebra with a letter abbreviation, which signifies its type, and a subscript number that indicates its position, from superior to inferior, in the vertebral column. For example, C_4 is the 4th cervical vertebra and T_{11} is the 11th thoracic vertebra.

The lateral view (**Figure 7.16b**) reveals a space between most of the moveable (unfused) vertebrae; this is called the *intervertebral foramen*. The spinal cord is visible through these foramina, where spinal nerve roots emerge.

Spinal Curvatures

In a newborn infant, the vertebral column is C-shaped, reflecting its position within the uterus as the fetus develops. As an infant grows, the vertebral column changes shape to develop a series of *curvatures* that result in a vaguely S-shaped vertebral column

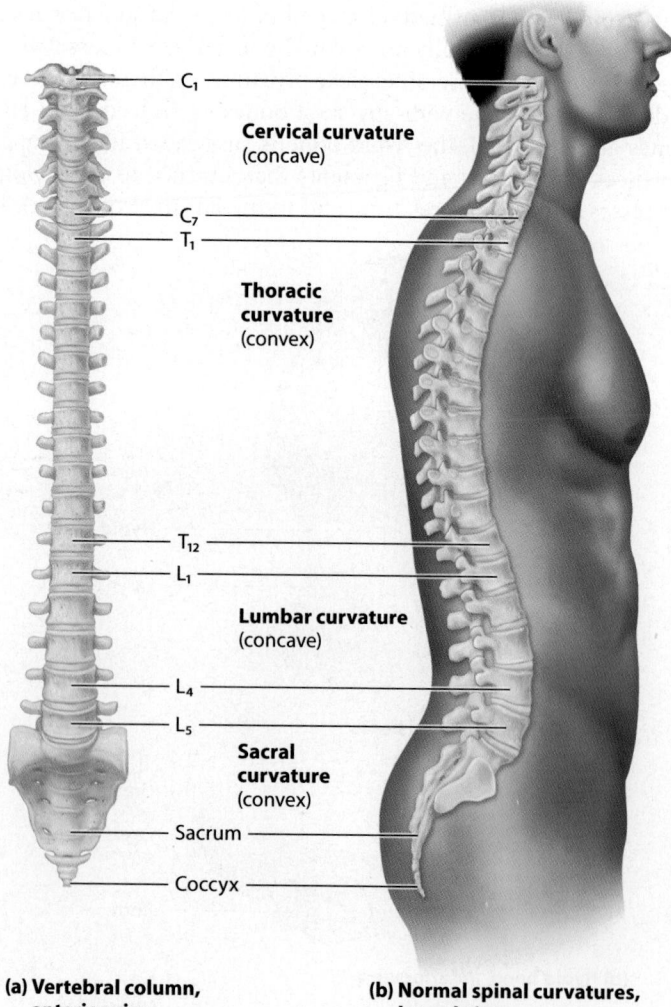

C_1

Cervical curvature (concave)

C_7
T_1

Thoracic curvature (convex)

T_{12}
L_1

Lumbar curvature (concave)

L_4
L_5

Sacral curvature (convex)

Sacrum

Coccyx

(a) Vertebral column, anterior view

(b) Normal spinal curvatures, lateral view

Figure 7.16 The vertebral column and normal spinal curvatures.

(Figure 7.16b). The concave **cervical curvature,** which extends from about C_2 to T_2, develops as an infant begins to lift the head and crawl. As the child ages and begins to walk, the **lumbar curvature** develops, which is a concave curvature extending from about T_{12} to L_5.

The remaining two curvatures are the **thoracic curvature,** from T_2 to T_{12}, and the **sacral curvature,** from the lumbosacral junction to the coccyx. Both are convex curves that are present in fetal life as part of the fetus's original C-shaped vertebral column. For this reason, the thoracic and sacral curvatures are known as *primary curvatures.* The cervical and lumbar curvatures, in contrast, are called *secondary curvatures* because they develop after the fetal period.

The development of these secondary curvatures is critical to our species' ability to walk upright. The cervical curvature allows

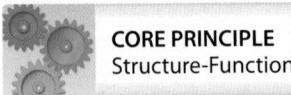

CORE PRINCIPLE
Structure-Function

us to hold our heads up, and the lumbar curvature shifts the weight of the body onto the sacrum, which lends the balance and support needed to walk on two legs. Here we have another example of the Structure-Function Core Principle (p. 25).

Abnormal Spinal Curvatures

The spinal curvatures discussed so far are normal anatomical features of an adult vertebral column. However, abnormal curvatures may develop. Three conditions are marked by such curvatures:

- **Scoliosis.** Scoliosis (skoh-lee-OH-sis; *scolios* = "bending") is characterized by lateral curvatures in the vertebral column that give it a C or S shape when viewed from the posterior or anterior side (**Figure 7.17a**). Scoliosis may be *congenital* (caused by deformities present at birth), *neuromuscular* (caused by abnormalities of or trauma to the nerves and muscles around the vertebral column), or *idiopathic* (of unknown cause). Idiopathic cases are the most common. Mild scoliosis may cause no symptoms and require no treatment. Severe cases, however, can put pressure on the heart and lungs, and require treatment, which may include back braces, physical therapy, or surgery.
- **Lordosis.** Lordosis (lor-DOH-sis; *lordos* = "bent forward"), commonly known as "swayback," is characterized by exaggerated cervical and lumbar curvatures (**Figure 7.17b**). Some degree of lordosis, particularly of the lumbar curvature, is normal in young children but generally diminishes with age. Abnormal

(a) **Scoliosis, posterior view, radiograph**   (b) **Lordosis, lateral view, CT scan**   (c) **Kyphosis, lateral view, radiograph**

Figure 7.17 Three examples of abnormal spinal curvatures.

(a) Thoracic vertebra, superior view

Practice art labeling
@MasteringA&P®

(b) Posterolateral view of three articulated thoracic vertebrae

Figure 7.18 Basic structure of vertebrae.

lordosis, sometimes called *hyperlordosis,* is commonly seen in adults bearing extra abdominal weight, such as pregnant women or overweight people. It may also have skeletal and neuromuscular causes. For example, dancers often develop lordosis due to their training, which can lead to muscle imbalances. Irrespective of the cause, lordosis puts extra stress on the lumbar vertebrae and can lead to lower back pain. It is generally treated with weight loss (if the person is overweight) and physical therapy.

- **Kyphosis.** In **kyphosis** (ky-FOH-sis; *kyphos* = "hump"), the thoracic curvature is exaggerated, giving a "hunchback" appearance (**Figure 7.17c**). It is caused by joint conditions such as arthritis, bone conditions such as osteoporosis and vertebral fractures, and developmental abnormalities of the skeleton. Mild kyphosis generally requires no treatment, but severe cases can be debilitating, leading to heart and lung dysfunction, nerve compression, and significant pain. Such severe cases generally require surgical correction.

Quick Check

☐ 1. How many cervical, thoracic, lumbar, sacral, and coccygeal vertebrae are normally present in the vertebral column?
☐ 2. What are the normal spinal curvatures?
☐ 3. Compare scoliosis, lordosis, and kyphosis.

Structure of the Vertebrae

◀◀ **Flashback**

1. Where is the vertebral cavity? (p. 16)
2. Which structure resides in the vertebral cavity? (p. 16)

All vertebrae have certain features in common, reflecting their shared function of protecting the spinal cord and supporting the head and trunk. **Figure 7.18a** shows a typical thoracic vertebra:

- **Body.** The most anterior structure (the part of the vertebra facing the thoracic cavity) is the blocklike **body,** also called the **centrum.** The body is the vertebra's primary weight-bearing structure. Vertebral discs cushion adjacent vertebral bodies and help absorb the shock of running and jumping. Fractures of the vertebral body can be painful and debilitating (see *A&P in the Real World: Vertebral Compression Fractures* on page 235).
- **Vertebral foramen.** Just posterior to the body is the large **vertebral foramen,** through which the spinal cord and its associated tissues pass. The vertebral foramina of the entire vertebral column form the *vertebral cavity,* or *vertebral canal* (which you read about in Chapter 1).
- **Pedicles and laminae.** The vertebral foramen is bordered anteriorly by the body, laterally by two **pedicles** that project posteriorly from the body, and most posteriorly by the laminae. Together, the pedicles and laminae form the **vertebral arch.** The inferior side of the pedicles curves up, creating a deep **inferior vertebral notch,** and the superior side curves down slightly, creating a **superior vertebral notch.** The inferior and superior notches of successive vertebrae stacked on top of one another form the intervertebral foramina that you saw in the previous section. The pedicles merge into the **laminae,** which enclose the posterior part of the vertebral foramen.
- **Superior and inferior articular processes.** At the junctions of the pedicles and laminae we find two small processes, the **superior** and **inferior articular processes,** on the superior and inferior sides, respectively. The surfaces of both

processes contain smooth regions known as *facets* that are covered with hyaline cartilage. The superior articular facets of one vertebra form joints with the inferior articular facets of the vertebra above it.

- **Transverse processes.** Projecting from the lateral sides of the vertebral arch are two **transverse processes.** These processes serve as attachment sites for muscles.
- **Spinous process.** The posterior projection from the vertebral arch is the **spinous process.** Like the transverse processes, the spinous processes are a site for muscle attachment.

These structures of the vertebrae fit together (**Figure 7.18b**). The vertebrae articulate with one another at their bodies and at the superior and inferior articular facets. Observe how the spinous processes help protect the posterior spinal cord, which would otherwise be fairly vulnerable. Although most vertebrae share these general features, their specific shape and size differ among cervical, thoracic, and lumbar vertebrae. In addition, cervical and thoracic vertebrae contain structures unique to those regions. The upcoming subsections focus on these differences.

Cervical Vertebrae

The seven cervical vertebrae are the smallest ones (**Figure 7.19**). Their most easily identifiable features are holes in their transverse processes called **transverse foramina.** The transverse foramina allow the passage of blood vessels called *vertebral arteries* and *vertebral veins*.

Vertebrae C_3-C_7 generally have oval bodies, large triangular vertebral foramina, and short straight spinous processes that are forked like the tongue of a snake (Figure 7.19a). However, C_7 has a longer spinous process that is not forked. This long process is easily palpable and is used clinically as a landmark, which has led to the common name for C_7: *vertebra prominens,* or "prominent vertebra." The first and second cervical vertebrae are notably different from C_3-C_7; let's examine these two vertebrae now.

C_1: Atlas The first cervical vertebra (C_1) is the **atlas** (Figure 7.19b). Its name stems from the fact that it holds the head, just as Atlas, the Titan from Greek mythology, supported the world. The atlas is immediately recognizable by its large, teardrop-shaped vertebral foramen and its lack of a vertebral body and spinous

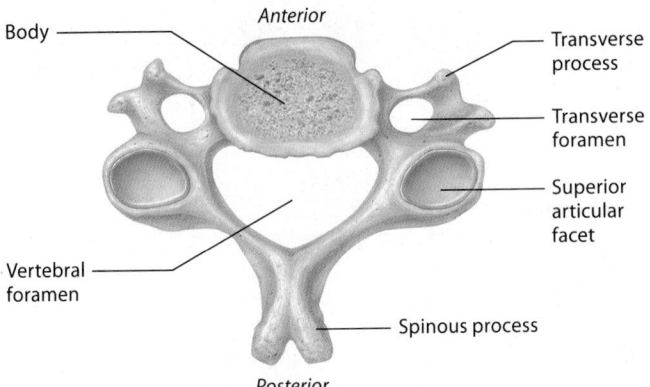

(a) Typical cervical vertebra (C₅)

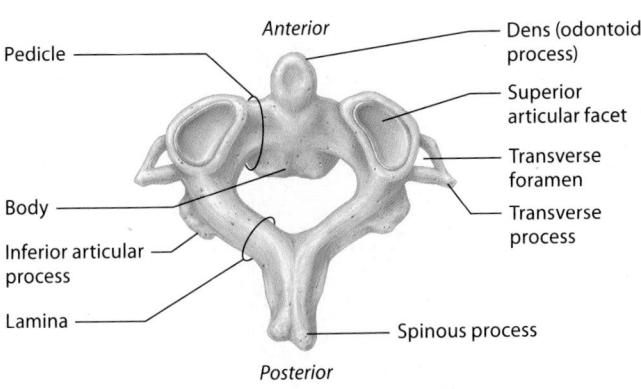

(c) Axis (C₂), superior view

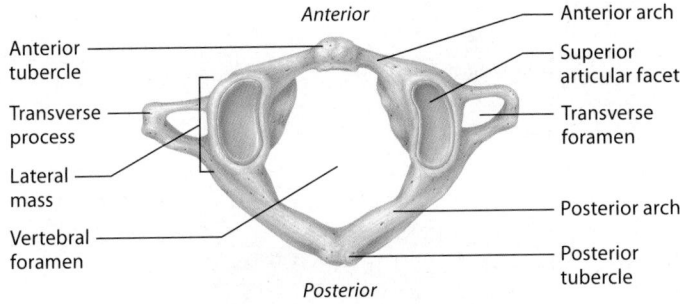

(b) Atlas (C₁), superior view

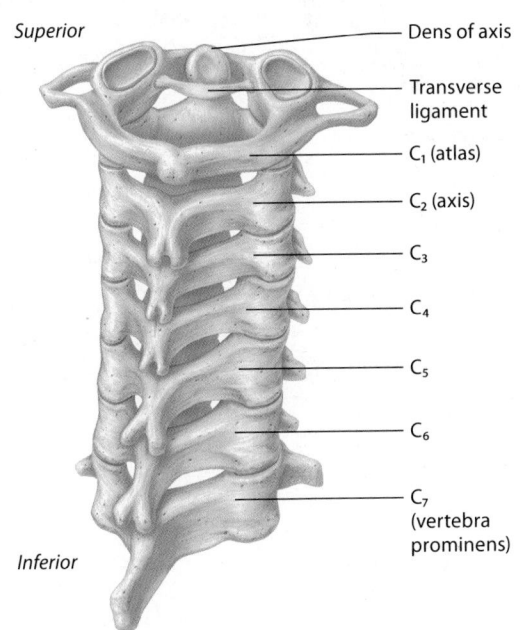

(d) Posterior view of articulated cervical vertebrae

Figure 7.19 Cervical vertebrae.

process. Instead of having the typical vertebral features, it is composed of two *anterior arches* that meet at the *anterior tubercle,* and two *posterior arches* that meet at the *posterior tubercle.* Where the anterior and posterior arches meet, we find the *lateral masses* that contain the superior and inferior articular facets.

Its superior articular facets articulate with the occipital condyles, and the inferior facets articulate with the second cervical vertebra.

C₂: Axis The second cervical vertebra (C₂) is the **axis** (Figure 7.19c). Its most notable feature is a superior projection from the body known as the **dens** or **odontoid process.** Both names derive from the fact that the process is shaped like a tooth (*den/dont* = "tooth"). Figure 7.19d shows that the dens fits up into the vertebral foramen of the atlas where the body of the atlas

(a) Superior view

(b) Lateral view

Figure 7.20 Thoracic vertebrae.

(a) Superior view

(b) Right lateral view

(c) Posterolateral view

Figure 7.21 Lumbar vertebrae.

(c) Two articulated thoracic vertebrae, superior view

would be located, and is held snugly in place by the *transverse ligament*. Together, dens and atlas form the *atlantoaxial joint*.

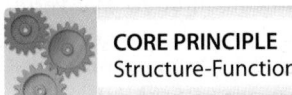
CORE PRINCIPLE
Structure-Function

The way the dens fits into the atlas allows this joint to perform its function of rotational motion of the neck, such as shaking your head to indicate "no." Here we see another example of the Structure-Function Core Principle (p. 25).

Thoracic Vertebrae

The 12 thoracic vertebrae are larger than the cervical vertebrae and grow progressively larger as we move down the vertebral column. Notice in **Figure 7.20** that they have vaguely heart-shaped bodies, circular vertebral foramina, and long spinous processes that point inferiorly. On the lateral sides of the bodies of thoracic vertebrae, we find two small facets: the **superior costal facet** and **inferior costal facet** (*cost* = "rib"; note that the bodies of $T_{10}-T_{12}$ have only a single costal facet on each side). As implied by their name, the costal facets provide the points of articulation for the ribs. Two additional costal facets are located on the transverse processes of T_1-T_{10}, the **transverse costal facets,** which articulate with another portion of the ribs, to be discussed shortly.

Lumbar Vertebrae

The five lumbar vertebrae are the largest and heaviest vertebrae, which reflects their primary function: bearing the weight of the torso (**Figure 7.21**). Their large kidney-shaped bodies, thick spinous processes that point posteriorly, and vertebral foramina shaped like a flattened triangle make them easy to recognize. In addition, lumbar pedicles and laminae are thicker and shorter than in other regions of the vertebral column.

Table 7.3 summarizes and compares the structure of cervical, thoracic, and lumbar vertebrae.

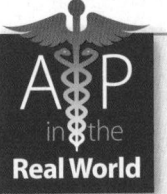
Real World

Vertebral Compression Fractures

A *vertebral compression fracture* is defined as a fracture of the vertebral body that reduces the height of the vertebra by more than 20%. Most often these fractures result from trauma or from diseases that destroy bone, such as osteoporosis and bone tumors. Compression fractures are quite common.

There are two types of compression fractures: wedge and burst. *Wedge fractures* involve only the anterior portion of the vertebra. These are most commonly seen in elderly women with kyphosis and osteoporosis, and can be induced by very minor trauma, such as sneezing or rolling over in bed. *Burst fractures* involve the entire vertebral body and are generally due to severe trauma. The symptoms of a wedge fracture vary but ordinarily involve pain that may radiate to other locations. The symptoms of a burst fracture are usually more severe, as bone fragments can damage the spinal cord and nerves.

Treatment depends on many factors, including the type and severity of the fracture and the overall health of the patient. Many patients respond well to conservative treatment, such as bed rest, bracing, and physical therapy, whereas others require surgery.

TABLE 7.3 COMPARISON OF CERVICAL, THORACIC, AND LUMBAR VERTEBRAE

Characteristic	Cervical Vertebrae	Thoracic Vertebrae	Lumbar Vertebrae
Body shape and size	Small and oval; C_1 lacks a body; C_2 has the dens on the superior surface of its body	Larger and heart-shaped; contain costal facets	Largest and kidney-shaped
Vertebral foramen shape	Triangular	Circular	Flattened triangular
Transverse processes	Contain transverse foramina	Long; contain articular facets for ribs	Short with no facets or foramina
Spinous processes	Most are fork-shaped; C_1 lacks a spinous process	Long; point inferiorly	Thick; point posteriorly
Appearance (superior view)			

Sacrum and Coccyx

The five sacral vertebrae (S_1-S_5) form the posterior wall of the pelvic cavity. They begin as individual vertebrae but generally fuse by ages 20–25 to form the thick, triangular sacrum (**Figure 7.22**). Recall that the sacrum is curved as part of the concave sacral curvature. Its superior surface, the flattened **base,** articulates with the fifth lumbar vertebra at its superior articular processes. The inferior surface, which articulates with the coccyx, is called the **apex.** The sacrum's two lateral surfaces, known as the **auricular surfaces,** articulate with the two pelvic bones.

The anterior sacrum (Figure 7.22a) shows the fused bodies of the sacral vertebrae, with *transverse lines* demarcating where the vertebrae fused. At the anterior margin of the base there is a bony projection called the **sacral promontory.** Lateral to the sacral promontory are two relatively smooth regions, the **alae** (EH-lee; singular, *ala*).

On the posterior surface of the sacrum (Figure 7.22b) is the opening to the **sacral canal,** a continuation of the vertebral canal, which contains nerve roots from the spinal cord as well as surrounding connective tissue membranes. The posterior boundary of the sacral canal is formed by the remnants of the sacral spinous processes, a ridge of bone called the **median sacral crest.** Flanking either side of the median sacral crest are four pairs of holes known as **sacral foramina,** through which pass nerves exiting the sacral canal. Lateral to the sacral foramina are the **lateral sacral crests;** these are remnants of the transverse processes of the sacral vertebrae. Near the apex of the sacrum is the **sacral hiatus,** which is the termination of the sacral canal.

The final region of the vertebral column, the coccyx, is generally composed of four vertebrae that begin to fuse at about age 25, although it can contain anywhere from three to five vertebrae. Like the sacrum, the coccyx has transverse ridges where the vertebrae fused. At its superior end, it has two hornlike projections, each called a **coccygeal cornu.** In many adults, some mobility exists between the coccyx and sacrum, although this generally diminishes with age.

↟ STUDY BOOST

Remembering Skull Bones and Vertebrae

Are you feeling overwhelmed by the names of all these bones? Here are some tricks to help you remember the skull bones and vertebrae.

- **PEST OF 6.** This identifies the six cranial bones: **P**arietal, **E**thmoid, **S**phenoid, **T**emporal, **O**ccipital, **F**rontal.
- **V**irgil **I**s **N**ow **M**aking **M**y **P**et **Z**ebra **L**augh. This silly sentence will help you remember the facial bones: **V**omer, **I**nferior nasal conchae, **N**asal, **M**andible, **M**axillae, **P**alatine, **Z**ygomatic, **L**acrimal.
- **E**very **S**tudent **F**ancies **L**earning **Z**illions **M**ore **P**arts. This optimistic statement will help you with the bones of the orbit: **Et**hmoid, **S**phenoid, **F**rontal, **L**acrimal, **Z**ygomatic, **M**axilla, **P**alatine.
- **F**or **E**asier **S**inus **M**emorization. As you can tell, this trick is for remembering the paranasal sinuses: **F**rontal, **E**thmoidal, **S**phenoidal, **M**axillary.
- Breakfast at **7,** lunch at **12,** dinner at **5.** This will help you remember the number of each type of vertebrae: 7 cervical, 12 thoracic, and 5 lumbar.

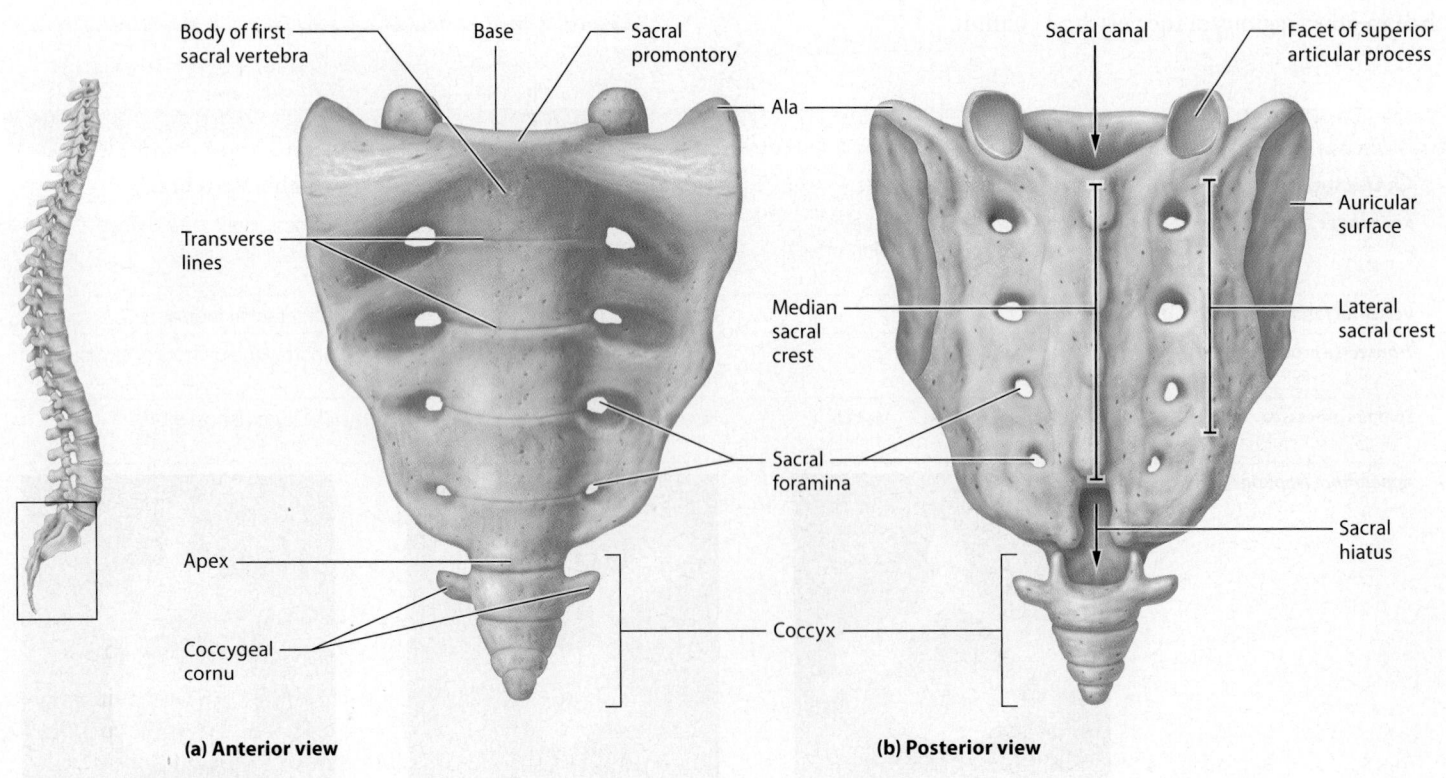

Body of first sacral vertebra — Base — Sacral promontory — Ala — Transverse lines — Apex — Coccygeal cornu

Sacral canal — Facet of superior articular process — Auricular surface — Median sacral crest — Lateral sacral crest — Sacral foramina — Sacral hiatus — Coccyx

(a) Anterior view   **(b) Posterior view**

Figure 7.22 The sacrum and coccyx.

To remember the different bones, you may also use visual analogies that compare an anatomical structure to a more familiar object. Here are a few examples:

- The sphenoid bone looks like a bat when viewed from the anterior side (it's the bone with "wings").
- The ethmoid bone is positioned somewhat like an iceberg in the skull, with only its tip (the crista galli) visible in the anterior view (refer to Table 7.2).
- A thoracic vertebra looks like the head of a giraffe when viewed from the posterior side (think "thoRAcic giRAffe"):

Thoracic giraffe

- A lumbar vertebra, in contrast, resembles the head of a "lumbering" moose:

Lumbering moose

You aren't necessarily going to use everything on this list. Choose whatever tricks work for you and use them. If you find that a memory aid doesn't stick after repeating it a couple of times, move on. Search for a new one (the Internet has many), make up your own, or simply don't use one at all. Be careful not to spend your energy memorizing mnemonics that don't work for you—these aids should make studying easier, not harder. ∎

Quick Check

☐ 4. How do the atlas and axis differ from other vertebrae?
☐ 5. Identify each of the following characteristics as belonging to cervical, thoracic, or lumbar vertebrae; the sacrum; or the coccyx.
  _____ Have thick, blocklike bodies with thick spinous processes
  _____ Makes up the inferiormost portion of the vertebral column
  _____ Contain transverse foramina
  _____ Consists of five fused vertebrae
  _____ Have downward-pointing spinous processes and costal facets

Intervertebral Discs

◀◀ **Flashback**

1. What are the properties of fibrocartilage? (p. 142)
2. What are the properties of dense regular collagenous connective tissue? (p. 139)

An **intervertebral disc** is a fibrocartilage pad between two vertebrae that absorbs shock, binds the vertebral column together, and helps support the weight of the body. There are 23 intervertebral discs; the first one is between C_2 and C_3, and the final one is between L_5 and S_1 (the sacrum).

A disc has two main components (**Figure 7.23**). A soft, inner, jelly-like substance called the **nucleus pulposus** (NOO-klee-us pull-POH-sis) is a resilient shock absorber. An outer ring of fibrocartilage, the **anulus fibrosus** (AN-yoo-lus fy-BRO-sus; *anulus* = "ring"), keeps the nucleus pulposus in place and joins the vertebrae. *A&P in the Real World: Herniated Disc* on page 238 discusses what happens when the anulus fibrosus has a tear.

In addition to the intervertebral discs, a number of ligaments and muscles support the vertebral column (see the muscular system chapter).

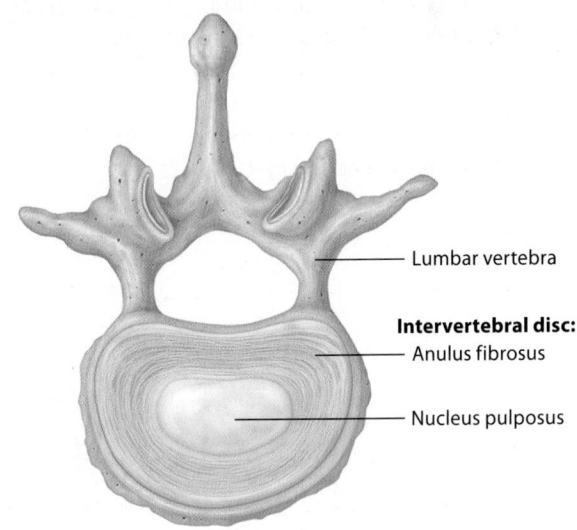

Lumbar vertebra

Intervertebral disc:
Anulus fibrosus

Nucleus pulposus

Figure 7.23 Structure of an intervertebral disc, superior view.

Quick Check

☐ 6. Describe the structure of an intervertebral disc.

The Thoracic Cage

◀◀ **Flashback**

1. Where is the thoracic cavity? (p. 17)
2. Which organs does the thoracic cavity enclose? (p. 17)

The **thoracic cage** consists of the *sternum*, the 12 pairs of *ribs*, and the thoracic vertebrae. These bones form the peripheral boundary of the thoracic cavity, and they shield the heart, lungs, and great blood vessels.

Sternum

The flattened **sternum,** known generally as the *breastbone,* forms the anterior, median part of the thoracic cage. It consists of three portions (**Figure 7.24**):

- **Manubrium.** The top portion of the sternum is the **manubrium** (muh-NOO-bree-um). On its superior surface, it has a notch called the *suprasternal notch.* Just lateral to the suprasternal notch on both sides are the two *clavicular notches,* which articulate with the clavicles of the pectoral girdles. Inferior to the clavicular notches we find the sites of attachment for the cartilages of the first ribs.
- **Body.** The middle **body** is the largest part of the sternum. The cartilages of the second ribs attach where the manubrium and the body meet, a location called the **sternal angle.** The lateral edges of the body have several notches where the cartilages of the third through seventh ribs attach.
- **Xiphoid process.** The most inferior portion of the sternum is the pointed **xiphoid process** (ZIH-foyd; *xiph-* = "sword"), which is the site of attachment for certain abdominal muscles. Composed of hyaline cartilage initially, it is generally fully ossified by about age 40.

To discover an important clinical implication of sternal anatomy, see *A&P in the Real World: The Sternum and CPR.*

Ribs

In both men and women, the **rib cage** consists of 12 pairs of ribs and the anterior **costal cartilages** (*cost-* = "rib"). The spaces

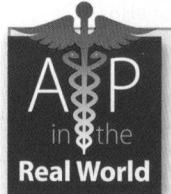

A&P in the Real World

Herniated Disc

A tear in the anulus fibrosus can allow the nucleus pulposus to protrude, a condition known as a **herniated disc** (commonly called a *slipped disc*). Tears generally result from trauma, injuries secondary to improper lifting, or longer-term damage due to repeated motions such as lifting and squatting. The anulus fibrosus generally tears on its posterolateral side because of the arrangement of ligaments around the vertebral column. The bulging nucleus pulposus compresses nerve roots as they emerge from the spinal cord.

Nerve root compression can cause significant pain that may radiate to other parts of the body, as well as numbness, tingling, and muscle weakness. Treatments include injection of anti-inflammatory steroid medications, physical therapy, and surgery to repair the tear and relieve nerve compression.

between successive ribs are the **intercostal spaces** (Figure 7.24). Recall from earlier in the chapter that the costal cartilages are composed of hyaline cartilage, which gives the rib cage some flexibility during movement and breathing. Each rib has a posterior attachment to a thoracic vertebra, from which it curves around anteriorly in a C shape.

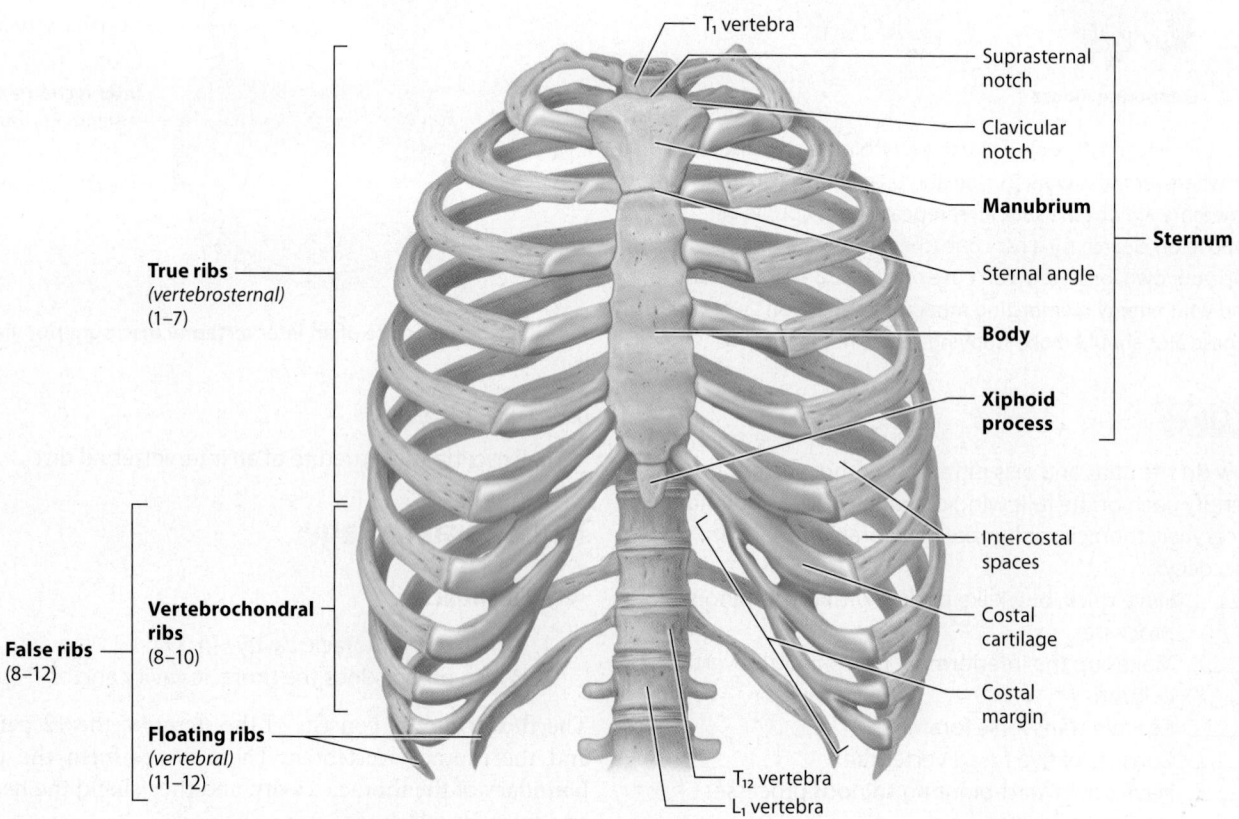

Figure 7.24 The thoracic cage, anterior view.

We group ribs into two classes based on their anterior attachment:

- Ribs 1–7 are called **true ribs** or *vertebrosternal ribs* because they attach to the sternum via their own costal cartilage.
- Ribs 8–12 are called **false ribs** because they do not attach to the sternum directly. Instead, the costal cartilages of ribs 8–10 (the *vertebrochondral ribs)* attach to the cartilage of the seventh rib. The costal cartilage of ribs 7–10 is called the **costal margin.** Ribs 11 and 12 are referred to as **floating ribs** or *vertebral ribs* because they lack any attachment to the sternum.

Figure 7.25 shows the structure of a typical rib. It attaches to the vertebra at its rounded **head.** The heads of most ribs have two articular facets: an *inferior articular facet* that articulates with the body of the thoracic vertebra of the same number as the rib, and a *superior articular facet* that articulates with the body of the vertebra superior to it. For example, the third rib articulates with T_3 at its inferior articular facet and T_2 at its superior articular facet.

Lateral to the rib's head is its **neck,** which runs along the transverse process of the vertebra, followed by a projection of bone called the **tubercle.** The tubercle articulates with the articular facet of the tip of the transverse process of the vertebra. The rib then curves anteriorly at its **angle** to become the **shaft,** or **body.** The concave internal surface of the shaft contains a groove called the *costal groove,* along which blood vessels and nerves travel. The ends of the shafts for ribs 1–10 are square and attach to their costal cartilage; the ends of the floating ribs are capped with a small layer of hyaline cartilage, but lack an attachment to the sternum.

Ribs 2–9 contain all of the features just discussed, but ribs 1 and 10–12 have slight variations in their structure. For instance, each of these ribs attaches to the body of only one thoracic vertebra, and so their heads lack separate superior and inferior articular facets. In addition, the floating ribs have no articulation with the transverse processes of the vertebrae and so lack tubercles.

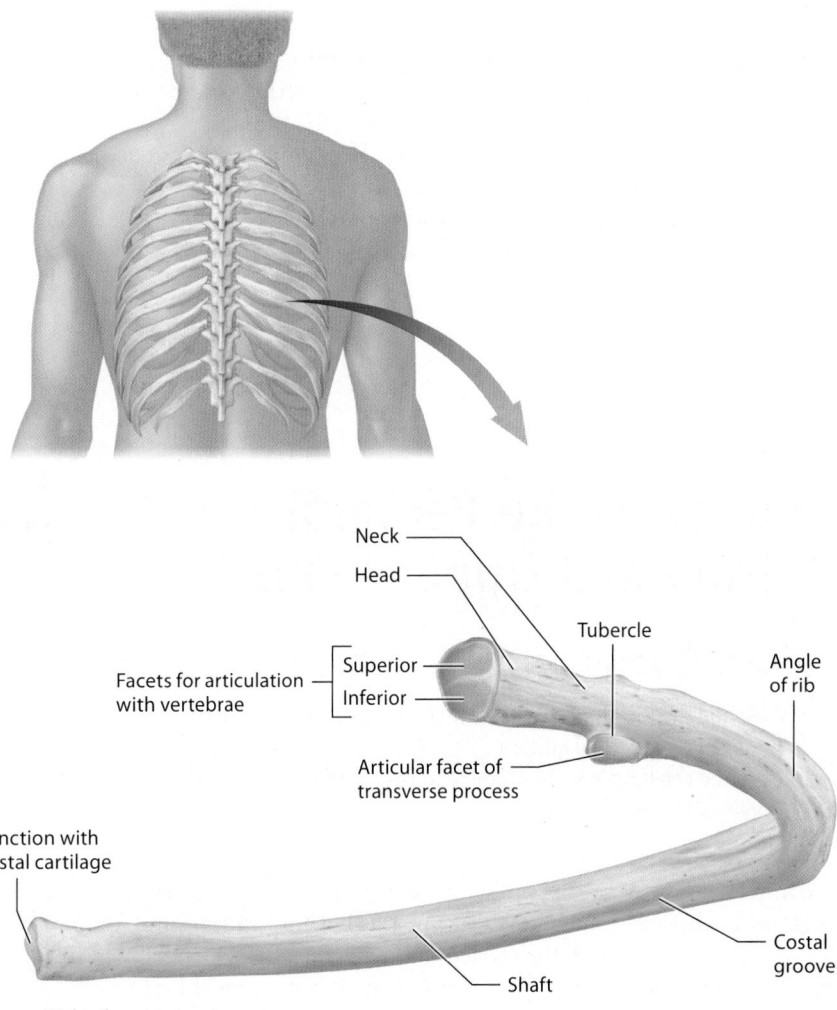

Figure 7.25 **Structure of a typical rib.**

Right rib, posterior view

Quick Check

☐ 7. Name and describe the three components of the sternum.
☐ 8. How do true, false, and floating ribs differ?

Apply What You Learned

☐ 1. Ms. Cho has fractured the transverse process of her fourth thoracic vertebra. She reports back pain, as well as pain with movement such as breathing. What else was likely fractured with her vertebra (be specific)? Explain. Why does this cause pain when she breathes?

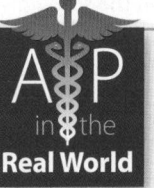

The Sternum and CPR

Cardiopulmonary resuscitation (CPR) is a lifesaving technique administered when an individual's heart has stopped producing functional contractions and/or breathing has stopped. CPR may restore circulation to the body by the application of repeated compressions to the chest over the sternum. Correct placement of the hands on the sternum is critical—they must be positioned in the center of the sternum between the person's two nipples. This way, compressions are delivered over the sternal body instead of the xiphoid process. Indeed, compressions applied directly to the xiphoid process can cause it to break off and damage underlying organs, particularly the liver. Note though that even when performed correctly, CPR can break the sternal body and ribs, especially in elderly individuals with low bone mass.

☐ 2. Tran is a 12-year-old patient who is experiencing back pain. Upon examination, you notice that his vertebral column has a pronounced lateral curvature, along with weakness in the muscles on one side, particularly in the thoracic region. An MRI scan of his back shows two herniated discs between T_8 and T_9 and T_9 and T_{10}.
a. What abnormal spinal curvature is present?
b. Are the muscle weakness and herniated discs related? Explain.

See answers in Appendix A.

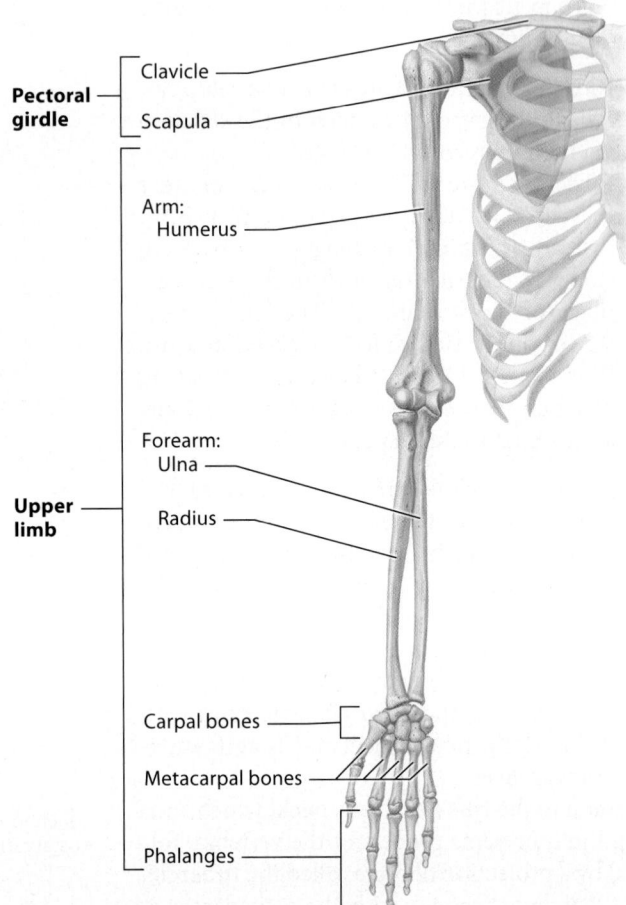

Figure 7.26 Overview of the bones of the pectoral girdle and upper limb.

MODULE **7.4**

Bones of the Pectoral Girdle and Upper Limb

Learning Outcomes

1. Describe the location, structural features, and functions of the bones of the pectoral girdle.
2. Describe the location, structural features, and functions of the bones of the upper limb.

We move on now to the appendicular skeleton, starting with the pectoral girdle and upper limb (**Figure 7.26**). The pectoral girdle consists of two bones, the *clavicle* and *scapula,* which support the upper limb. The upper limb consists of 30 bones, including the *humerus, ulna, radius, carpal bones, metacarpal bones,* and *phalanges.*

The Pectoral Girdle

The two bones of the pectoral girdle—the *clavicle* and the *scapula*—reside along the superior and posterior thoracic cage (**Figure 7.27**). See how they articulate with the humerus, the bone of the arm, as well as with the manubrium and one another. Together they support the upper limb, particularly the humerus, and are the sites of attachment of numerous muscles.

Clavicle

The **clavicle** (often called the *collarbone*) has two distinct ends that are easily palpable through the skin over the shoulders. From an anterior view it appears straight (Figure 7.27a), but from a superior or inferior view it is actually S-shaped (Figure 7.27b and c). Its medial **sternal end** articulates with the manubrium at the sternum, forming the *sternoclavicular joint.* Note that the clavicle is the only bone of the pectoral girdle that directly articulates with the axial skeleton. The lateral **acromial end** articulates with a process of the scapula called the *acromion* to form the *acromioclavicular joint* (discussed in the next section). Near the acromial end we find the **conoid tubercle,** which is the site of ligament attachment.

The structure of the clavicle enables it to function like a brace in a building that supports two opposing beams (an example of the Structure-Function Core Principle, p. 25). It sits between the shoulder and thoracic cage and braces the upper limb so that it rests laterally to the trunk. Its role in supporting the upper limb is evident in clavicular fractures, in which the arm falls anteriorly and medially. Most clavicular fractures result from direct trauma or from falling onto an outstretched arm.

CORE PRINCIPLE Structure-Function

Scapula

The triangular **scapula** (plural, *scapulae*) sits on the posterosuperior rib cage, extending from approximately ribs 2 to 7 (**Figure 7.28**). Its largest portion is the *body,* which has three borders: the **medial, lateral, and superior borders.** The apices of the scapular triangle are called *angles:* the *superior, inferior,* and *lateral angles.*

The anterior surface of the scapula (Figure 7.28a) features a hook-shaped projection called the **coracoid process** (KOR-uh-koyd; *corac-* = "crow"; be careful not to confuse this with the cor*onoid* process of the mandible and ulna). Inferior to the cor*a*coid process is a broad indentation, the *subscapular fossa,* to which a muscle called the *subscapularis* attaches.

(a) Pectoral girdle, anterior view

(b) Right clavicle, superior view

(c) Right clavicle, inferior view

Figure 7.27 **The pectoral girdle.**

(a) Right scapula, anterior view

(b) Right scapula, lateral view

(c) Right scapula, posterior view

Figure 7.28 **Structure of the scapula.**

From a lateral view (Figure 7.28b), you can see that the scapula's lateral angle features a shallow indentation called the **glenoid cavity** (GLEN-oyd KA-vuh-tee; *glen-* = "pit" or "cavity"). The glenoid cavity, along with the humerus, forms the shoulder joint.

On the posterior side of the scapula, we find a ridge of bone, the **spine,** that crosses from medial to lateral along its superior border (Figure 7.28c). The scapular spine is the "blade" of the shoulder that you can feel on your posterior thorax. The area superior to the spine is the **supraspinous fossa;** the area inferior to it is the **infraspinous fossa.** The spine terminates in the enlarged **acromion** (ah-KROH-mee-ahn; *acr-* = "point"), which articulates with the clavicle to form the **acromioclavicular (AC) joint.** Injuries to the AC joint are common and result in a condition commonly referred to as a *separated shoulder.*

Quick Check

☐ 1. With which structures does the clavicle articulate?
☐ 2. Describe the glenoid cavity, acromion, and coracoid process.

The Humerus

◀◀ **Flashback**

1. What is a long bone? (p. 185)
2. What are the epiphyses and diaphysis? (p. 186)

The largest and strongest bone of the upper limb, and the only bone of the *brachium,* or arm, is the **humerus** (**Figure 7.29**). Like all long bones, it consists of two *epiphyses* that articulate with other bones flanking a long *diaphysis.* The proximal epiphysis of the humerus features a ball-shaped humeral **head** on its medial side that articulates with the glenoid cavity to form the shoulder joint. Surrounding the head is a groove called the *anatomical neck* of the humerus. Just lateral to the neck we find a projection known as the **greater tubercle.** It is separated from the smaller and more medial **lesser tubercle** by a groove called the **intertubercular sulcus** or **bicipital groove.** The second name stems from the fact that the tendon of the biceps brachii muscle passes through this bony groove.

The proximal epiphysis is attached to the diaphysis at a region called the **surgical neck,** so named because it is a frequent site

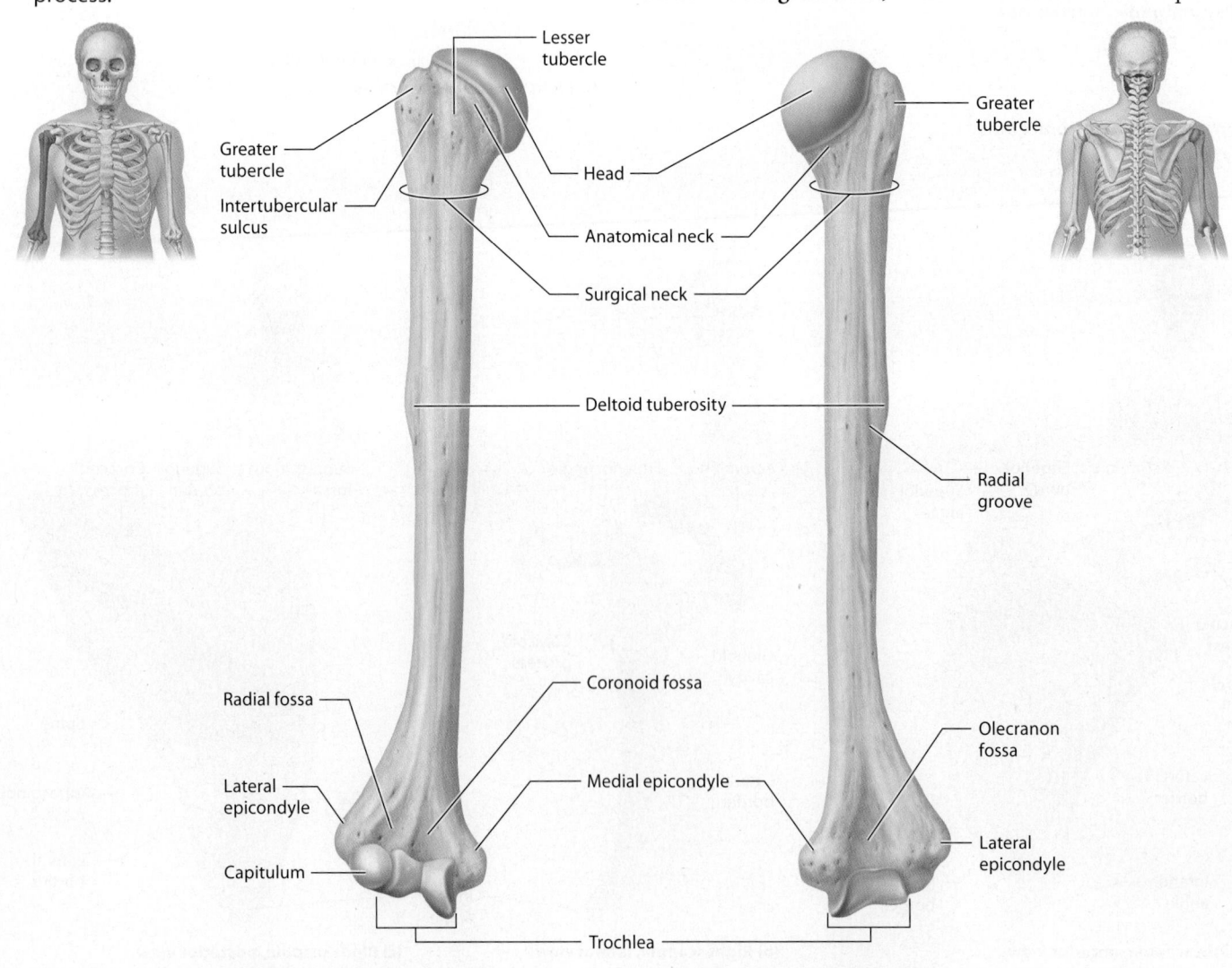

(a) Right humerus, anterior view   **(b) Right humerus, posterior view**

Figure 7.29 The humerus.

of fractures (which may require surgical repair). The humeral diaphysis is fairly featureless compared to the two epiphyses. It contains only one major projection in about the middle of the bone on the lateral side, which is known as the **deltoid tuberosity.** As its name implies, this provides the site of attachment for the *deltoid muscle*. The only other feature of note on the diaphysis is the **radial groove,** located on the posterior side of the bone, along which travels a nerve for which the groove was named.

The distal epiphysis has two flared ends, the **medial epicondyle** and **lateral epicondyle,** which serve as sites of muscle attachment. The remaining features are involved in the humeral articulation with the ulna and radius at the elbow joint. On the anterior side, we find two rounded knobs: the lateral **capitulum** (*capit-* = "head"), named for its spherical shape and resemblance to a head; and the medial **trochlea** (*troch-* = "wheel"), also named for its shape, which resembles a wheel or spool of thread. Just proximal to the capitulum and trochlea are two small indentations: the lateral **radial fossa** and the medial **coronoid**

fossa. On the posterior side of the distal epiphysis, we see the continuation of the trochlea and a deep indentation called the **olecranon fossa.**

Quick Check

☐ 3. With which structures does the humerus articulate at its proximal and distal epiphyses?

☐ 4. Describe the structure and location of the following: the humeral head, capitulum, trochlea, deltoid tuberosity, and greater tubercle.

Bones of the Forearm: The Radius and Ulna

The forearm, or *antebrachium*, consists of two bones: the lateral **radius** and the medial **ulna,** held together by a fibrous structure called the **interosseous membrane** (**Figure 7.30**). The

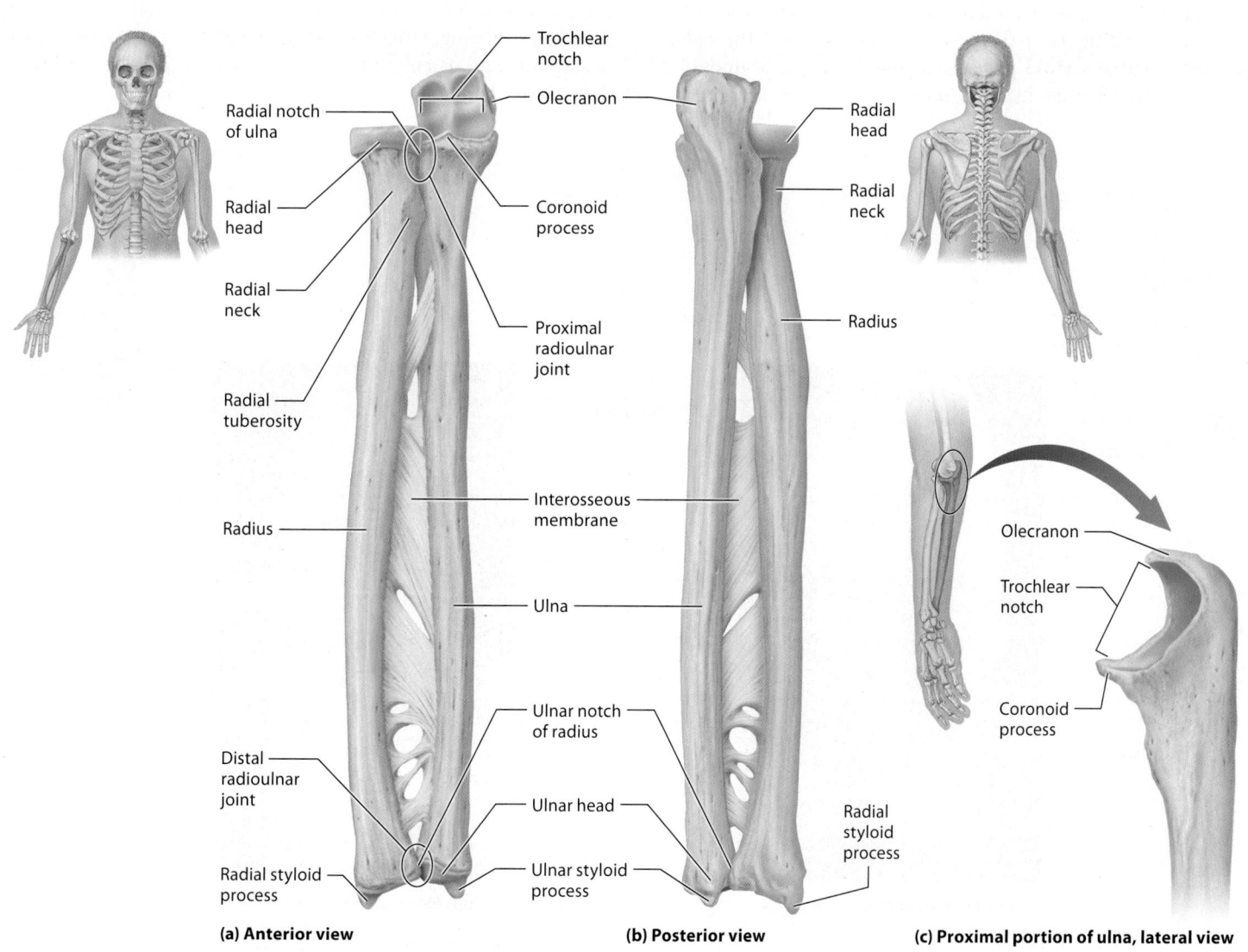

(a) **Anterior view** (b) **Posterior view** (c) **Proximal portion of ulna, lateral view**

Figure 7.30 The bones of the forearm: the radius and ulna.

interosseous membrane distributes the force borne by the bones more equally, reducing the load on each bone. These two bones articulate with the humerus at their proximal ends and the carpal bones at their distal ends. In addition, they also articulate with one another at two joints called the *radioulnar joints.* The next two subsections examine the structural and functional features of these bones.

Radius

Notice in Figure 7.30a and b that the radius is narrow proximally and becomes progressively broader as we move distally. Its proximal epiphysis is called the **radial head,** which is a round, flattened structure. It articulates with the capitulum of the humerus to form part of the elbow joint and with the ulna to form the proximal radioulnar joint. Just distal to the radial head is the *radial neck,* which ends in the **radial tuberosity.** This process sits on the medial side of the bone and is the site of attachment of the biceps brachii muscle.

At the widened distal epiphysis of the radius, we find an indentation called the **ulnar notch,** which is the site of the distal radioulnar joint. The flattened end of this epiphysis articulates with the carpal bones at the wrist. The lateral tip of the radius, called the **radial styloid process,** forms the lateral boundary of the wrist and helps stabilize the joint.

Ulna

Unlike the radius, the ulna is wide proximally and narrow distally. Its proximal epiphysis has several prominent features. The first is a U-shaped notch called the **trochlear notch,** into which the trochlea of the humerus fits (Figure 7.30c). On the posterior side of the trochlear notch is a bump, the **olecranon** (oh-LEK-ruh-nahn), which is the knob of the elbow that you can feel through the skin. The anterior lip of the trochlear notch is a projection known as the **coronoid process,** which fits into the coronoid fossa of the humerus. Just lateral to the coronoid process is a smooth area called the **radial notch of the ulna.** This area articulates with the radial head. Note that these notches are named for the bone with which they articulate, not the bone on which they are located.

The distal epiphysis of the ulna contains the **ulnar head.** Note that this is another difference between the radius and ulna: The ulnar head is at the distal epiphysis, and the radial head is at the proximal epiphysis. The medial side of the ulnar head has a small **styloid process.** The styloid processes of both the radius and the ulna are palpable through the skin.

The proximal radius and ulna fit together with the distal humerus to form the elbow joint. **Figure 7.31a** shows how the structures fit snugly together, which gives this joint a great deal of stability. The articulations chapter discusses the elbow in more detail (see Chapter 8).

(a) Right elbow, anterior view

(b) Right elbow, posterior view

Figure 7.31 **The elbow joint.**

☐ 5. How do the radius and ulna differ in their shape and features?
☐ 6. Which parts of the radius and ulna articulate with the humerus?
☐ 7. With what other bones do the radius and ulna articulate?

Bones of the Wrist: Carpals

◀◀ **Flashback**

1. What is a short bone? (p. 185)
2. How does a short bone differ from a long bone? (pp. 185–186)

The wrist, or *carpus,* consists of eight short bones—known collectively as the **carpals**—arranged in two rows containing four bones each (**Figure 7.32**). The four proximal carpal bones are the following, from lateral to medial (see the Study Boost on p. 253 for help with remembering their names):

- the boat-shaped **scaphoid** (SKA-foyd; *scaph-* = "boat"), which articulates with the radius;
- the slightly moon-shaped **lunate,** which articulates primarily with the radius but also contacts the ulna;
- the triangular **triquetrum** (try-KWEE-trum), which articulates with the ulna; and
- the small, round **pisiform** ("pea-shaped"), which articulates with the anterior surface of the triquetrum (note that it is visible in the anterior view only).

The distal carpal bones, from lateral to medial, are as follows:

- the **trapezium,** which articulates proximally with the scaphoid;
- the **trapezoid,** which also articulates primarily with the scaphoid proximally;
- the rounded **capitate,** which articulates proximally with the scaphoid and lunate; and
- the **hamate,** which articulates proximally with the triquetrum and is named for its anterior hooklike projection (*ham-* = "hook").

The carpal bones, along with the radius and ulna, are often injured during falls. To learn more, see *A&P in the Real World: Wrist Fractures* on page 247.

☐ 8. List the proximal and distal carpal bones from lateral to medial.

Bones of the Hand and Fingers: Metacarpals and Phalanges

The hand, or *manus,* consists of five long bones called **metacarpals** (numbered I–V from lateral to medial) that articulate with the distal carpal bones proximally and the bones of the fingers distally (see Figure 7.32). Each metacarpal consists of three parts: The proximal epiphysis is the **base,** the diaphysis is the **body,** and the distal epiphysis is the **head.** You can see the heads of your

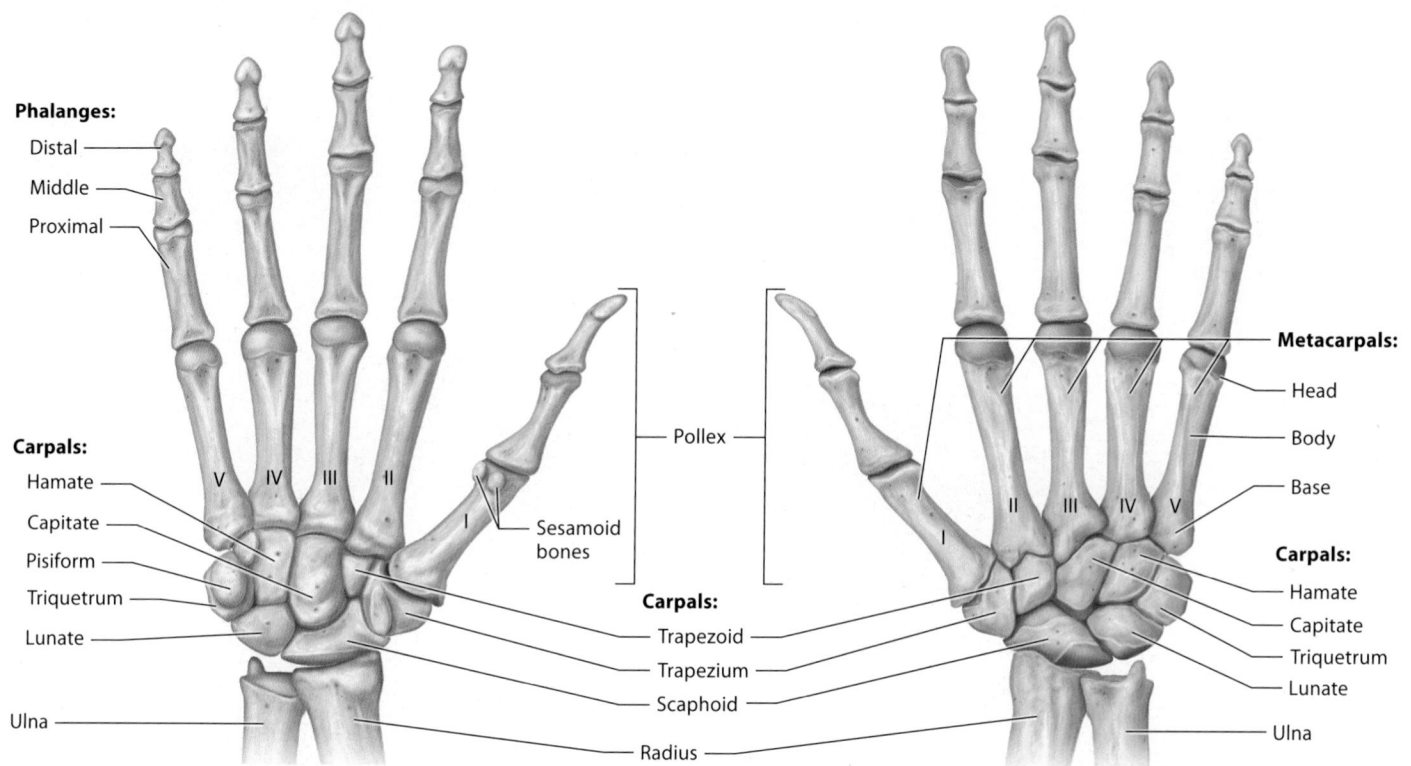

Phalanges:
Distal
Middle
Proximal

Carpals:
Hamate
Capitate
Pisiform
Triquetrum
Lunate

Ulna

V IV III II

I

Sesamoid bones

Pollex

Carpals:
Trapezoid
Trapezium
Scaphoid

Radius

(a) Right wrist and hand, anterior view

Metacarpals:
Head
Body
Base

II III IV V

I

Carpals:
Hamate
Capitate
Triquetrum
Lunate

Ulna

(b) Right wrist and hand, posterior view

Figure 7.32 The hand and wrist.

metacarpals when you make a fist, as they form your "knuckles." Fractured metacarpal heads are called *boxer's fractures.*

The bones of the fingers consist of 14 total **phalanges** (fuh-LAN-jeez; singular, *phalanx*). Each finger consists of three bones: a **proximal, middle** (or intermediate), and **distal phalanx** (FAY-langks). The exception is the thumb, or **pollex,** which has only a proximal and a distal phalanx. Like the metacarpals, the phalanges have a base, body, and head.

☐ 9. How many metacarpals and phalanges are in the hand and fingers?
☐ 10. What are the three parts of a metacarpal and a phalanx?

Apply What You Learned

☐ 1. Predict what might happen to the stability of the elbow joint if the trochlear notch were abnormally shallow. Explain your answer.
☐ 2. Mr. Heller presents to the emergency department following a bicycle crash during which he landed on his shoulder. During the physical exam, you notice that his upper limb is hanging in a more medial position than normal and he can hold it in a lateral position only with assistance. What has likely happened? Explain.

See answers in Appendix A.

MODULE **7.5**

Bones of the Pelvic Girdle and Lower Limb

Learning Outcomes

1. Describe the location, structural features, and functions of the bones of the pelvic girdle.
2. Compare and contrast the adult male and female pelvic bones.
3. Describe the location, structural features, and functions of the bones of the thigh, the leg, the ankle, and the foot.

This module explores the lower part of the appendicular skeleton: the pelvic girdle and the lower limb. The **pelvic girdle** consists of the **hip bones** or **pelvic bones** (also called the *coxal bones*), which articulate posteriorly with the sacrum. The lower limb has three parts: the thigh, which has only one bone, the *femur*; the leg, which is made up of the *tibia* and *fibula*; and the foot and ankle, which consist of the *tarsals, metatarsals,* and *phalanges.* As you read this module, note the similarities and differences between these bones and those of the pectoral girdle and upper limb.

The Pelvis and Bones of the Pelvic Girdle

◀◀ **Flashback**

1. Where is the pelvic cavity located? (p. 17)
2. Which organs are located within the pelvic cavity? (p. 18)

(a) Anterolateral view of pelvis

Sacrum
Pelvic bone
Femur
Greater pelvis (blue)
Lesser pelvis (green)
Pelvic inlet
Pelvic brim

(b) Midsagittal section of pelvis

Sacrum
Coccyx
Pelvic outlet

Figure 7.33 The pelvis. The greater (false) pelvis is shaded blue; the lesser (true) pelvis is shaded green.

The two pelvic bones and the sacrum together form the bowl-shaped **pelvis** (Figure 7.33), which establishes the boundaries for the pelvic cavity that houses organs of the digestive, urinary, and reproductive systems. The pelvic girdle and sacrum form an oval opening called the **pelvic inlet.** The bony ridge surrounding the inlet is the **pelvic brim.** As you can see in Figure 7.33a, the pelvic brim defines the boundaries between the **greater** or *false* **pelvis,** which is the area superior to the pelvic brim, and the **lesser** or *true* **pelvis,** which is inferior to the pelvic brim. At the inferior boundary of the lesser pelvis, we find the **pelvic outlet** (Figure 7.33b).

Each pelvic bone is composed of three bones that fuse during childhood: the *ilium, ischium,* and *pubis* (**Figure 7.34**). All three bones contribute to the deep socket on the lateral side of the coxal bone, referred to as the **acetabulum** (ah'-suh-TAB-yoo-lum), which, along with the femur, forms the hip joint. In addition, two bones, the ischium and pubis, contribute to the large opening in the anterior pelvic bone, called the **obturator foramen** (AHB-too-ray-tor for-AY-mun), through which nerves and blood vessels pass. Let's now look at the individual bones and their features.

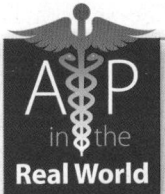

Wrist Fractures

The wrist is the most frequently injured region of the upper limb, usually from direct trauma or from an attempt to stop a fall with an outstretched arm. Fractures commonly involve the distal radius and ulna, and may also involve the carpal bones, particularly the scaphoid. The fracture shown at right is a *Colles fracture* of the distal radius.

Wrist fractures occur most commonly in children and in adults aged 60–69. The incidence of wrist fractures declines with advancing age (and the incidence of other fractures, such as hip fractures, increases). This change in incidence likely occurs because people fail to catch themselves and land on their hip instead.

Like most fractures, wrist fractures generally present with pain and swelling over the affected area. A common misconception is that an individual with a fractured bone will be unable to move the affected part. In fact, this is only the case with severe fractures in which a bone is dislocated. Generally, a person with a wrist fracture will have slightly less mobility due to pain and swelling, but it is possible for motion to be totally normal (a fact to which the author can attest personally).

Treatment of wrist fractures varies according to the severity of the injury and which bone is injured. Some cases are treated with simple casting, whereas others require surgical intervention. Many fractures could be successfully prevented by teaching individuals how to fall in a way that distributes the force across their trunk rather than concentrating it on an outstretched arm.

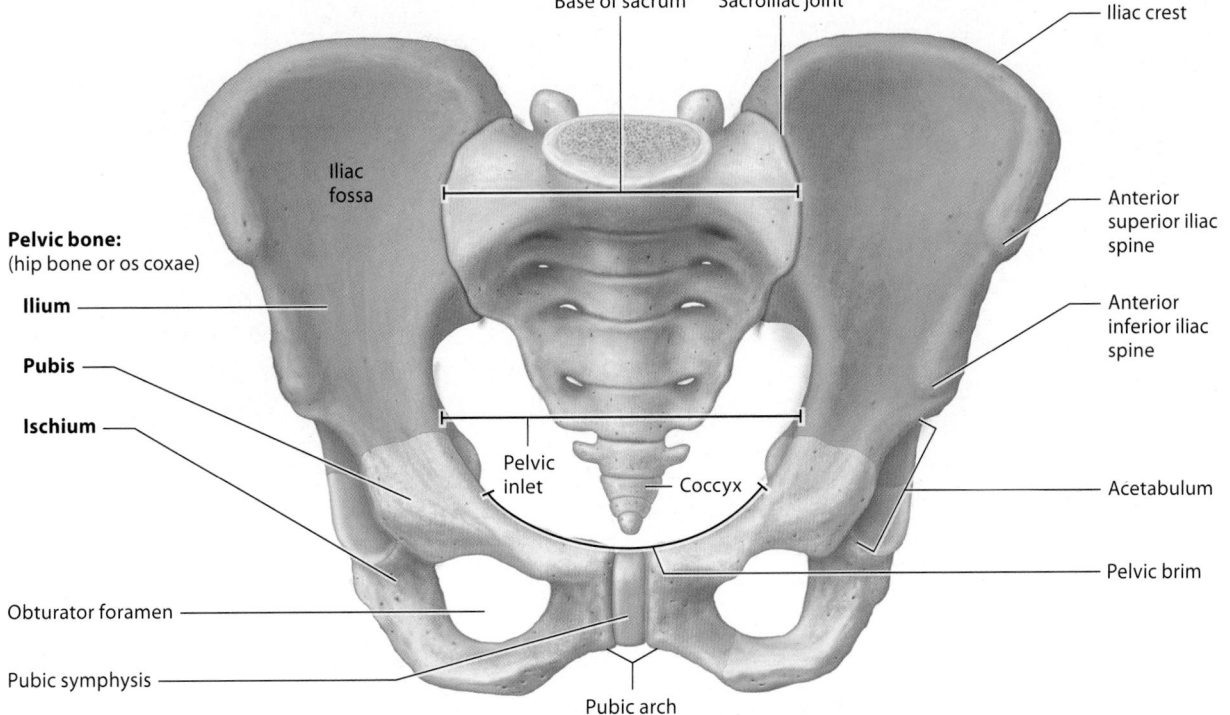

Figure 7.34 The pelvis and pelvic bones, anterior view. Each pelvic bone (also called the hip bone) consists of three fused bones: the ilium, ischium, and pubis.

Ilium

The **ilium** (ILL-ee-um; plural, *ilia*) forms the superior portion of the coxal bone. Its superior portion is the flared **ala** (AY-luh), which is named for its resemblance to a wing; its inferior portion is the **body of the ilium** (**Figure 7.35**). The superior boundary of the ala is the **iliac crest,** which you feel every time you put your hands on your hips. At its anterior end, it terminates in a bony projection known as the **anterior superior iliac spine** (ASIS), which can be palpated in some people on the anterior pelvis. Just inferior to the ASIS is another, smaller ridge called the *anterior inferior iliac spine.*

In a medial view of the coxal bone, at the posterior end of the iliac crest we find the **posterior superior iliac spine,** and inferior to it, the *posterior inferior iliac spine* (Figure 7.35a). The ilium indents sharply at the posterior inferior iliac spine to form a deep groove called the **greater sciatic notch,** through which the sciatic nerve passes. The anterior portion of the medial surface, which is smooth and concave, is the **iliac fossa.** The surface of the posterior side, known as the *auricular surface,* is rough. The **arcuate line,** which marks where the ilium forms part of the pelvic brim, runs from the auricular surface along the body of the ilium.

The smooth lateral surface of the ilium (Figure 7.35b) is known as its *gluteal surface* (GLOO-tee-uhl). It is marked by three lines: the **posterior, anterior,** and **inferior gluteal lines,** to which the gluteal muscles attach.

Ischium

The posteroinferior portion of the coxal bone is formed by the **ischium** (ISS-kee-um) (Figure 7.35). It consists of two components that together approximate a C shape: the posterior **ischial body,** which, along with the ilium and pubis, forms part of the acetabulum, and the anterior **ischial ramus,** which, along with the pubis, forms part of the obturator foramen.

Projecting posteriorly and medially from the ischial body is the **ischial spine,** to which a ligament from the sacrum attaches. Just inferior to the ischial spine is the **lesser sciatic notch.** Like the greater sciatic notch, the lesser notch provides a passageway for blood vessels and nerves.

The most prominent feature of the ischium is the thick **ischial tuberosity,** located on the its posteroinferior side. Together, the ischial tuberosities from our two coxal bones bear all of our weight when we sit. When someone is said to have a "bony butt," it simply means that the individual's ischial tuberosities are prominent.

(a) Right pelvic bone, medial view

(b) Right pelvic bone, lateral view

Figure 7.35 The pelvic bone.

Practice art labeling
@MasteringA&P®

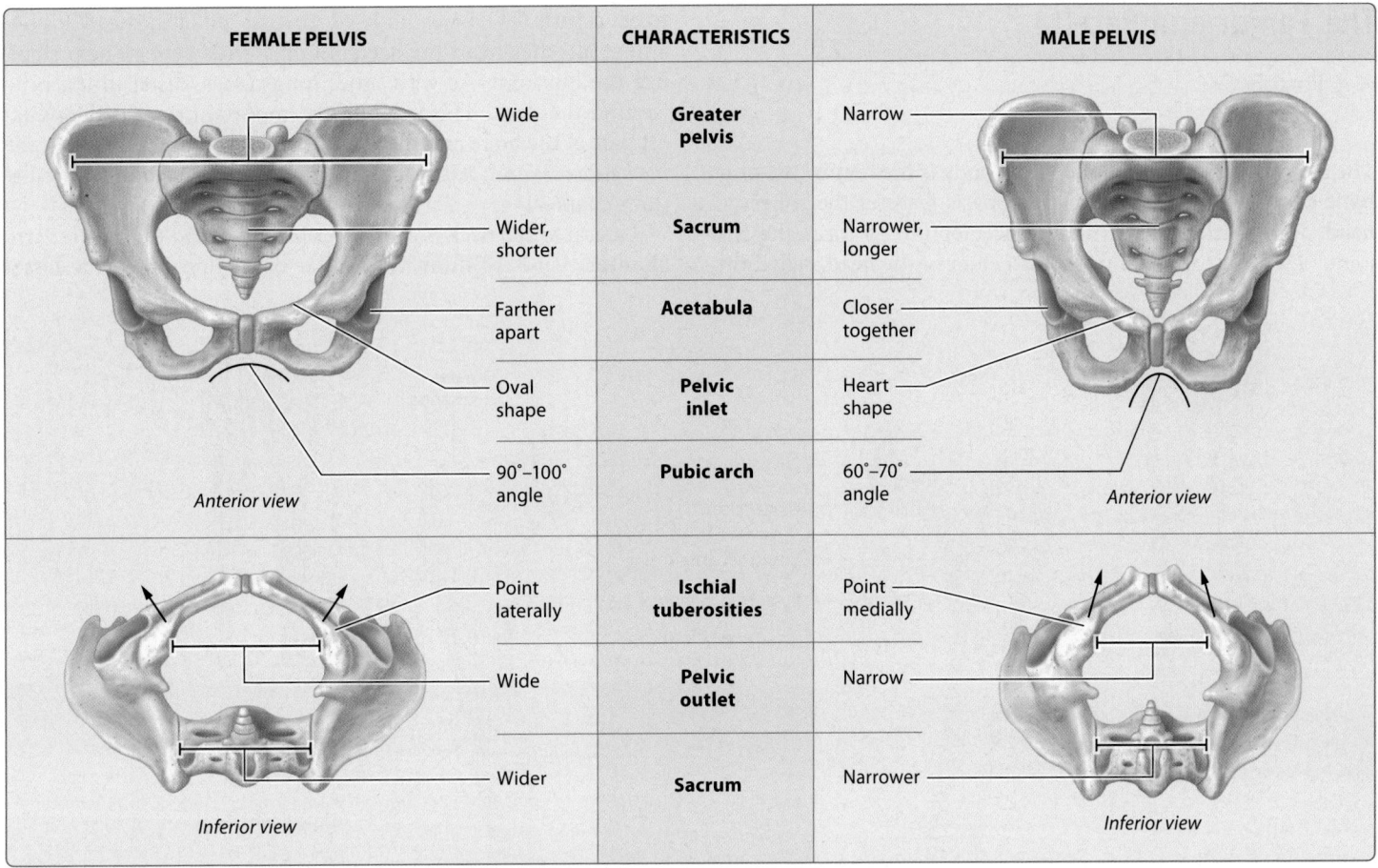

FEMALE PELVIS	CHARACTERISTICS	MALE PELVIS
Wide	Greater pelvis	Narrow
Wider, shorter	Sacrum	Narrower, longer
Farther apart	Acetabula	Closer together
Oval shape	Pelvic inlet	Heart shape
90°–100° angle	Pubic arch	60°–70° angle
Anterior view		*Anterior view*
Point laterally	Ischial tuberosities	Point medially
Wide	Pelvic outlet	Narrow
Wider	Sacrum	Narrower
Inferior view		*Inferior view*

Figure 7.36 **Differences between the female and male pelves.**

Pubis

The smallest component of the coxal bone is the anterior **pubis** (PYOO-biss), also called the *pubic bone* (Figure 7.35). It consists of three parts that together approximate a C shape: the **superior pubic ramus,** the **pubic body,** and the **inferior pubic ramus.** The superior pubic ramus contributes to the acetabulum, and both rami form parts of the boundary of the obturator foramen. The two pubic bodies meet at a joint called the **pubic symphysis,** where they are united by a pad of fibrocartilage. The angle where the two bodies meet is the *pubic arch;* this pelvic feature is important in determining gender, as discussed next.

Female and Male Pelves

Apart from the skull, the other skeletal structure that can be reliably used to determine gender is the pelvis (plural, *pelves;* Figure 7.36). The female pelvis is structured for pregnancy and childbirth and, as a result, is generally wider and shallower than the male pelvis. We can therefore examine a number of features to determine gender, including the following:

- **Shape of greater pelvis.** The greater pelvis is wider in females, with the anterior superior iliac spines farther apart, and with flared iliac crests.

- **Coccyx and sacrum.** The female sacrum tends to be wider and shorter than the male sacrum. In addition, the female coccyx is generally situated more posteriorly and is more moveable than the male coccyx. A woman with a coccyx situated anteriorly could suffer coccygeal fractures during childbirth.
- **Pelvic inlet and outlet.** The female pelvic inlet is usually wider and oval, whereas the male pelvic inlet is narrow and has a vague heart shape. The female pelvic outlet is also typically wider than that of the male.
- **Acetabula.** Female acetabula are generally farther apart and pointed more anteriorly than those of males. This affects gait patterns, giving many women a "swaying" walk.
- **Pubic arch.** The pubic arch in females tends to have an angle between 90° and 100°, whereas the male pubic arch has a narrower angle of 60°–70°. If an investigator has only a single coxal bone, the angle of the pubic arch can still be estimated by holding the bone against a mirror and measuring the angle in the reflection.
- **Ischial tuberosities.** The female ischial tuberosities tend to point laterally, whereas those of the male point medially. This gives the female buttocks a more rounded appearance.

Also, the female pelvis is generally lighter and less robust than the male pelvis due to the male's greater muscle mass and weight.

The Femur and Patella

◀◀ **Flashback**

1. What is a sesamoid bone? (p. 186)

The largest and strongest bone in the body is the **femur,** the only bone of the thigh. Its proximal epiphysis features the spherical **head,** which articulates with the acetabulum to form the hip joint. **Figure 7.37** shows a pit in the center of the head, called the

fovea capitis (FOH-vee-uh KAP-uh-tiss; "pit of the head"); a ligament attaches from the acetabulum to this spot to help stabilize the hip joint. As with other long bones, distal to the head we find the **neck.** This is clinically important, as it is the weakest part of the bone and the area most likely to fracture (see *A&P in the Real World: Hip Joint Replacement Surgery* in the articulations chapter).

Lateral to the neck is a large projection called the **greater trochanter** (tro-KAN-tuhr). A similar projection called the **lesser**

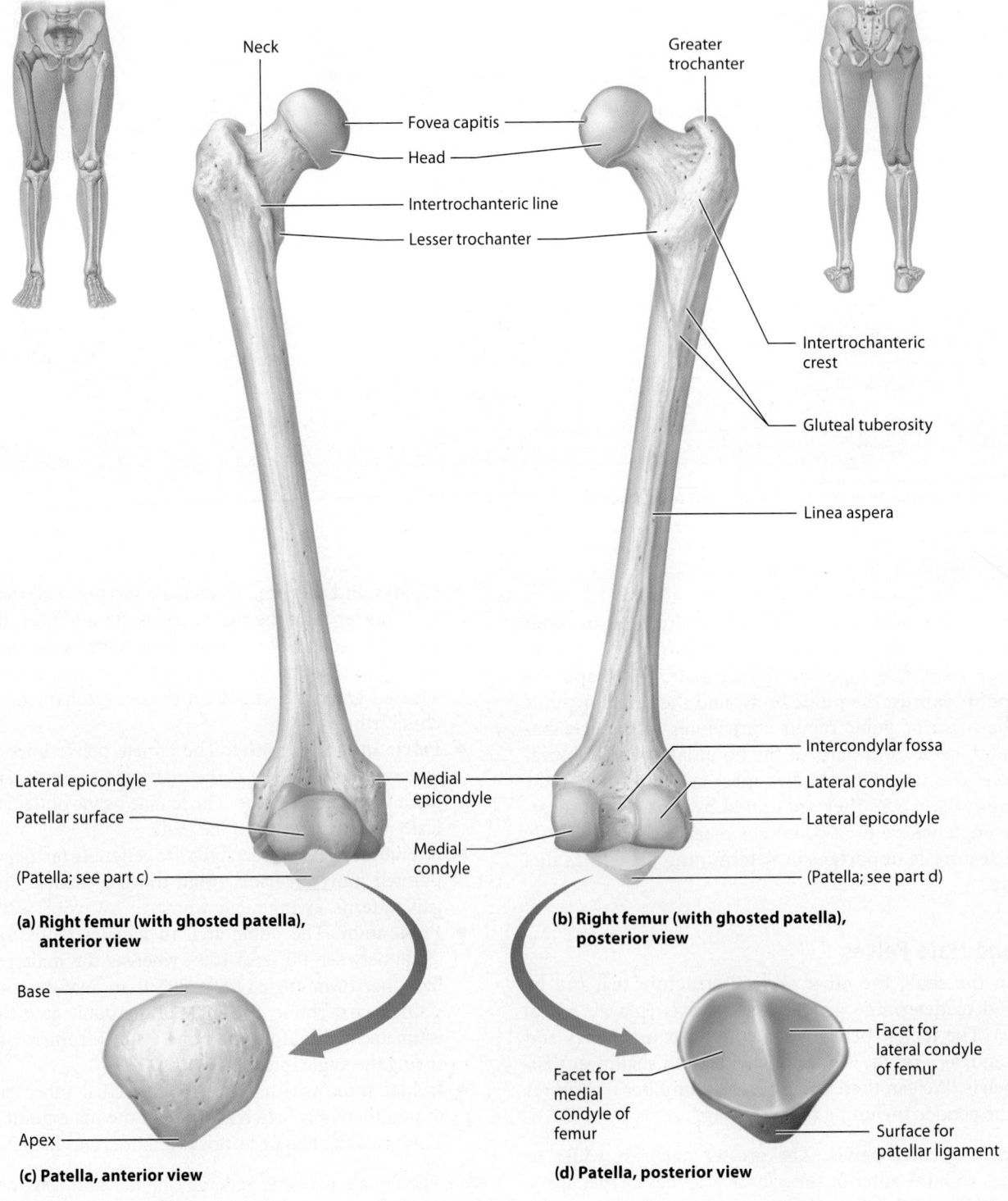

(a) Right femur (with ghosted patella), anterior view

(b) Right femur (with ghosted patella), posterior view

(c) Patella, anterior view

(d) Patella, posterior view

Figure 7.37 The femur and patella.

trochanter is found medial and distal to this one. The two are connected by a ridge of bone on the anterior side known as the **intertrochanteric line,** which continues on the posterior surface of the bone as the **intertrochanteric crest.**

The femoral diaphysis, also known as the *shaft,* is fairly smooth and featureless on the anterior side. On the posterior side, a prominent line called the **linea aspera** (LIN-ee-uh AS-per-uh; "rough line") runs down the shaft. Near the distal epiphysis, the line splits into two lines that lead to two projections, the **medial epicondyle** and **lateral epicondyle,** which are the widest points of the femur. From the epicondyles, the femur tapers into the **medial condyle** and **lateral condyle,** which articulate with the tibia to form the knee joint. On the posterior side (Figure 7.37b) is the **intercondylar fossa,** which is an indentation between the two condyles. Anteriorly, the space between the condyles is smooth; this area is known as the *patellar surface.*

The bone that articulates with the femur's patellar surface is the triangular **patella,** or "kneecap" (Figure 7.37c). The patella is a sesamoid bone located within the tendon of the anterior thigh muscle, the *quadriceps femoris.* The *patellar ligament* is a continuation of this tendon that inserts into the tibia and secures the patella over the anterior knee (see the next section). The proximal end of the patella is the *base,* and its pointed distal end is the *apex.* Even though the patella is not technically a thigh bone, we include it here because it articulates with the femur.

Quick Check

☐ 1. With which bones does the femur articulate? Be specific.
☐ 2. Which parts of the femur form these articulations?
☐ 3. What type of bone is the patella, and where is it located?

Bones of the Leg: The Tibia and Fibula

Two bones make up the leg: the medial **tibia** (TIB-ee-uh) and the lateral **fibula** (FIB-yoo-luh; **Figure 7.38**). (Be careful with these names; many a student has missed a test question by writing

(a) Right tibia and fibula, anterior view   **(b) Right tibia and fibula, posterior view**

Figure 7.38 The tibia and fibula.

"tibula" and "fibia"!) Like the radius and ulna, the tibia and fibula are united by an interosseous membrane. They articulate with one another at both the distal and the proximal *tibiofibular joints*. However, unlike the radioulnar joints, the tibiofibular joints allow very little movement, which makes the lower limb more stable.

Tibia

The larger, medial tibia is the second strongest bone in the body, which reflects its primary function as the main weight-bearing bone of the leg. At its proximal end, it features two concave depressions called the **medial condyle** and **lateral condyle,** which articulate with the femoral condyles to form the knee joint. The two condyles are separated by a ridge, the **intercondylar eminence.** As the next chapter will point out, the intercondylar eminence is an important site for ligament attachment in the knee joint. Just distal to the tibial condyles is a rough projection from the tibia's anterior surface, known as the **tibial tuberosity,** which is where the patellar ligament attaches to the tibia.

The tibial diaphysis features a sharp ridge on its anterior side called the **anterior crest,** commonly known as the *shin.* Near the distal epiphysis, the anterior crest curves to the medial side, where it terminates in the **medial malleolus** (mal-lee-OH-lus; plural, *malleoli*), which you can easily palpate through your skin as your medial *ankle bone.* Lateral to this is a flat articular surface where the tibia articulates with a tarsal bone called the *talus* to form the ankle joint.

Fibula

The fibula is the much thinner, lateral bone of the leg (Figure 7.38). Often incorrectly reported to bear none of the body's weight, it does, in fact, bear about one-sixth of our total weight. Proximally, the **head of the fibula** articulates with the lateral tibia at the *proximal tibiofibular joint*. Distally, it articulates with the tibia again at the *distal tibiofibular joint,* after which it expands to form the **lateral malleolus** (lateral ankle bone). This structure articulates with the talus and helps stabilize the ankle joint. The fibula's thin structure causes it to be the most commonly injured bone in *ankle fractures.*

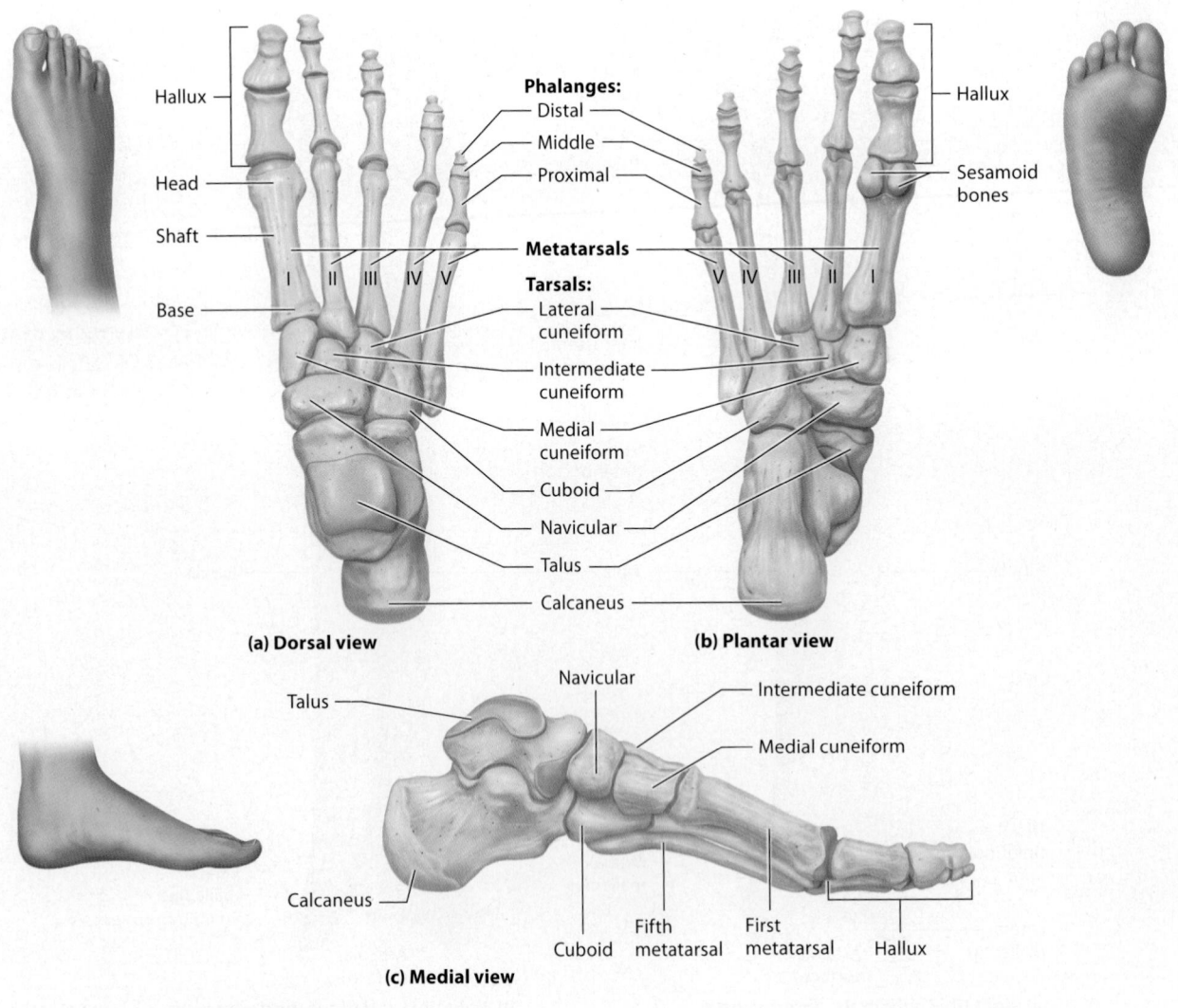

Figure 7.39 The ankle and foot.

- [ ] 4. With which bones does the tibia articulate? With which bones does the fibula articulate?
- [ ] 5. What are the bony projections of the medial and lateral ankle called, respectively?

Bones of the Ankle and Foot: The Tarsals, Metatarsals, and Phalanges

◀◀ **Flashback**

1. Where is the plantar surface of the foot located? (p. 14)

The structure of the ankle parallels that of the wrist in many ways. The ankle consists of seven short bones, collectively called **tarsals,** that connect the leg with the foot (**Figure 7.39**). However, unlike the arrangement of the wrist, only a single tarsal bone articulates with the leg: the dome-shaped **talus** (TAYL-us), which forms the ankle joint with the tibia and fibula. The talus rests on top of the largest tarsal bone, the **calcaneus** (kal-KEHN-ee-us), or *heel bone.* The posterior side of the calcaneus attaches to a large tendon called the *calcaneal* (*Achilles*) *tendon.* Distally, the talus articulates with a bone called the **navicular,** named for its resemblance to the shape of a boat. Together, the talus, calcaneus, and navicular are the proximal tarsal bones.

The next four tarsal bones are the distal tarsals. From medial to lateral, they include the **medial cuneiform, intermediate cuneiform, lateral cuneiform,** and **cuboid.** Notice in Figure 7.39 that the cuboid articulates posteriorly with the calcaneus, whereas the cuneiforms all articulate with the navicular.

The structure of the bones of the foot and toes largely parallels that of the hand and finger bones. The five foot bones are called **metatarsals** (numbered I–V from medial to lateral), and

like the metacarpals, they have a proximal *base,* a middle *shaft,* and a distal *head.* On the plantar surface of the first metatarsal we usually find two small sesamoid bones (often just called the *sesamoids*). These bones, and the tendon in which they are located, can become inflamed with repetitive activity, resulting in *sesamoiditis.* This condition is particularly common in athletes playing on artificial turf, giving it the common name "turf toe."

As we find in the fingers, the bones of the toes consist of 14 **phalanges.** The second through fifth toes generally have three bones each: a proximal, middle, and distal phalanx. The great toe, or **hallux,** usually has only two, a proximal and distal phalanx (sometimes the fifth toe has only two, as well).

The bones of the foot generally do not rest flat on the ground during standing or walking. This is due to the presence of three *arches* that are supported by ligaments and muscles. The most prominent is the **medial longitudinal arch,** which runs along the medial side of the foot from the calcaneus to metatarsals I–III (**Figure 7.40**). The less prominent **lateral longitudinal arch** extends from the lateral side of the calcaneus to metatarsals IV and V. Finally, the **transverse arch** runs along the middle of the foot, and involves the distal tarsals and the bases of all five metatarsals. The plantar arches help support the weight of the body and distribute this weight evenly during walking.

⬆ **STUDY BOOST**

Remembering Bones of the Arm and Leg
The following mnemonics may help you remember the bones of the arm and leg:

- **Stop Letting The People Touch The Cadaver's Hand.** This sentence helps you remember the carpal bones: **S**caphoid, **L**unate, **T**riquetrum, **P**isiform, **T**rapezium, **T**rapezoid, **C**apitate, **H**amate. Since it mentions the "hand," remember that it describes the carpals, not the tarsals. If you find that you are confusing the trapezium and trapezoid, remember that the trapezi**UM** is by the th**UM**b.
- **TIB**ia and **Fibu**LA. Use the following trick to tell the two leg bones apart: **TIB**ia = **T**hick, **I**nner **B**one; **Fibu**LA = **LA**teral bone.

(a) **Plantar view**   (b) **X-ray of medial right foot**

Figure 7.40 The three arches of the foot.

- **This College Needs Me In Lab Classes.** This statement will help you remember the names of the tarsals from proximal to distal and medial to lateral: **T**alus, **C**alcaneus, **N**avicular, **M**edial cuneiform, **I**ntermediate cuneiform, **L**ateral cuneiform, **C**uboid.

Here are a few other tricks that might be helpful:

- If you turn the ulna on its side, the trochlear notch is the shape of the letter U for "ulna."
- To remember on which side we find the radius, take your radial pulse—the radius is on the same side as the radial pulse (this obviously works only if you know how to take your radial pulse).
- The word *patella* in Latin means "little plate"; the patella is the plate-shaped bone of the lower limb.
- Remember the navicular bone by its shape, which resembles a boat. Think of the word "navy" to connect "navicular" and "boat." ■

CHAPTER SUMMARY

Take full advantage of the invaluable online practice and study tools for this chapter by going to MasteringA&P®. There you will find:

- Practice tests, module quizzes, art-labeling and sequence-the-steps activities
- Animated figures with quizzes
- *Practicing A&P* worksheets
- Author-narrated podcast
- Dynamic Study Modules
- *Interactive Physiology*

MODULE 7.1
Overview of the Skeletal System 210

- The skeletal system consists of the **skull, vertebral column, thoracic cage, pectoral girdle, upper limb, pelvic girdle,** and **lower limb,** as well as the skeletal cartilages. (Figure 7.1, p. 211)

 Practice art-labeling exercise: Figure 7.1.

- The **axial skeleton** consists of the bones of the skull, vertebral column, and thoracic cage. The **appendicular skeleton** consists of the bones of the pectoral girdle, upper limb, pelvic girdle, and lower limb.
- The three basic types of **bone markings** are depressions, openings, and projections. (Table 7.1, pp. 212–214)

MODULE 7.2
The Skull 214

- The skull consists of the **cranial bones,** which encase the brain, and the **facial bones,** which form the framework of the face. All skull bones except the mandible are joined by immoveable **sutures.** (Figures 7.2 and 7.3, p. 215)

 Practice art-labeling exercise: Figure 7.2.

- The **cranium** contains two portions: the **cranial vault** and the **cranial base** (Figure 7.4, p. 215). They are formed by eight cranial bones: the frontal bone, two parietal bones, two temporal bones, occipital bone, sphenoid bone, and ethmoid bone.
- The 14 facial bones include the mandible and vomer and the maxillary, zygomatic, nasal, lacrimal, palatine, and inferior nasal conchal bones. Features of the cranial and facial bones are covered in Table 7.2 (pp. 216–220) and Figures 7.5–7.10 (pp. 221–225).

 Practice art-labeling exercise: Figure 7.8.

- The **orbit** is formed from contributions by the frontal, maxillary, zygomatic, sphenoid, ethmoid, lacrimal, and palatine bones. (Figure 7.11, p. 226)
- The **nasal cavity** is formed by the ethmoid, vomer, nasal, palatine, sphenoid, and maxillary bones. (Figure 7.12, p. 227)
- The **paranasal sinuses** are air-filled cavities in the frontal, maxillary, sphenoid, and ethmoid bones that warm, filter, and humidify inspired air. (Figure 7.13, p. 227)
- The **oral cavity** is formed by the maxillary, palatine, and mandibular bones.
- The fetal skull largely forms by intramembranous ossification, leading to the presence of **fontanels.** (Figure 7.14, p. 228)
- The **hyoid bone,** which is not a skull bone, is suspended in the neck by muscles. (Figure 7.15, p. 229)

MODULE 7.3
The Vertebral Column and Thoracic Cage 230

- The **vertebral column** consists of 33 **vertebrae** and the ligaments and cartilages that join them. There are 7 **cervical vertebrae,** 12 **thoracic vertebrae,** 5 **lumbar vertebrae,** 5 fused **sacral vertebrae,** and 4 fused **coccygeal vertebrae.** (Figure 7.16a, p. 230)
- There are four normal spinal curvatures: the concave **cervical** and **lumbar curvatures** and the convex **thoracic** and **sacral curvatures.** (Figure 7.16b, p. 230)

- Vertebrae share common features, including a central **body,** a **vertebral foramen** through which the spinal cord passes, lateral **transverse processes,** and a posterior **spinous process.** (Figure 7.18, p. 232)

  Practice art-labeling exercise: **Figure 7.18.**

- Cervical vertebrae are located in the neck. They contain **transverse foramina** in their transverse processes and have forked spinous processes. (Figure 7.19, p. 233)
  - C_1, or the **atlas,** has a large vertebral foramen and no body or spinous process. It articulates with the occipital bone at its superior articular facets.
  - C_2, or the **axis,** has a vertical projection called the **dens** that fits inside the vertebral foramen of C_1.
- Thoracic vertebrae have inferiorly pointing spinous processes and superior and inferior costal facets where they articulate with the ribs. (Figure 7.20, p. 234)
- Lumbar vertebrae are the thickest and heaviest vertebrae; they have thick spinous processes that point posteriorly. (Figure 7.21, p. 234)
- Table 7.3 (p. 235) summarizes the three classes of vertebrae.
- The fused vertebrae of the sacrum feature **sacral foramina** through which nerves pass. (Figure 7.22, p. 236)
- An **intervertebral disc** is a fibrocartilage pad between two vertebrae. (Figure 7.23, p. 237)
- The thoracic cage consists of the **sternum,** 12 pairs of **ribs,** and the thoracic vertebrae. (Figure 7.24, p. 238; Figure 7.25, p. 239)

MODULE **7.4**
Bones of the Pectoral Girdle and Upper Limb 240

- The pectoral girdle consists of the **clavicle** and the **scapula.** (Figures 7.26–7.28, pp. 240, 241)
  - The S-shaped clavicle articulates with the manubrium of the sternum and the acromion of the scapula.
  - The scapula rests on the posterosuperior rib cage. On its posterior side, it has a ridge called the **spine** that terminates in the **acromion.** Its lateral border features the **glenoid cavity,** which articulates with the humerus.
- The only bone of the arm is the **humerus.** Its proximal epiphysis is the **head,** which articulates with the glenoid cavity. Distally it articulates with the radius and ulna at the **capitulum** and **trochlea,** respectively. (Figure 7.29, p. 242)
- The bones of the forearm are the medial **ulna** and the lateral **radius.** (Figure 7.30, p. 243; Figure 7.31, p. 244)
  - The radius articulates with the capitulum of the humerus.
  - The ulna articulates proximally with the trochlea of the humerus. Its posterior process is the **olecranon.**
- The proximal **carpal bones** are the **scaphoid, lunate, triquetrum,** and **pisiform.** The distal carpals are the **trapezium, trapezoid, capitate,** and **hamate.** (Figure 7.32, p. 245)
- The hand consists of 5 **metacarpals,** and the fingers consist of 14 **phalanges.**

MODULE **7.5**
Bones of the Pelvic Girdle and Lower Limb 246

- The **pelvic girdle** consists of the two **pelvic bones,** or **hip bones,** that articulate posteriorly with the sacrum. Each pelvic bone consists of the **ilium,** the **ischium,** and the **pubis,** which together form the **acetabulum.** (Figures 7.33–7.35, pp. 246, 247, 248)

  Practice art-labeling exercise: **Figure 7.35.**

  - The ilium is the superior portion of the pelvic bone. Its superior boundary is the **iliac crest,** which terminates at the **anterior** and **posterior superior iliac spines.**
  - The ischium is the posteroinferior portion of the pelvic bone. Posteriorly it has projections called **ischial tuberosities.**
  - The anterior portion of the pelvic bone is formed by the pubis. The two pubic bones meet at the **pubic symphysis.**
- The male and female **pelves** have notable differences. (Figure 7.36, p. 249)
- The only bone of the thigh is the **femur.** Its proximal epiphysis is the **head,** which articulates with the acetabulum. The distal epiphysis features the **medial** and **lateral condyles,** which articulate with the tibia. (Figure 7.37a and b, p. 250)
- The **patella** is a sesamoid bone that resides within the tendon of the anterior thigh muscle. (Figure 7.37c, p. 250).
- The **tibia** is the medial leg bone. The **fibula** is the lateral leg bone. (Figure 7.38, p. 251)
- The bones of the foot and ankle are composed of 7 **tarsals,** 5 **metatarsals,** and 14 **phalanges.** (Figure 7.39, p. 252) The bones of the foot form three arches supported by ligaments and muscles: the **medial** and **lateral longitudinal arches** and the **transverse arch.** (Figure 7.40, p. 253)

ASSESS WHAT YOU LEARNED

See answers in Appendix A.

LEVEL 1 Check Your Recall

1. Which of the following are considered parts of the axial skeleton? (Circle all that apply.)
   a. Pectoral girdle   d. Vertebral column
   b. Lower limb   e. Pelvic girdle
   c. Skull   f. Thoracic cage

2. _____ is the anatomical name for a hole in a bone.
   a. Fossa   b. Foramen   c. Condyle   d. Tubercle

3. Fill in the blanks: The two parietal bones are united at the _____ suture; they meet the frontal bone at the _____ suture, the temporal bones at the _____ sutures, and the occipital bone at the _____ suture.

4. Mark the following statements as true or false. If a statement is false, correct it to make a true statement.
   a. The four paranasal sinuses are the frontal, parietal, sphenoidal, and mandibular sinuses.
   b. The cribriform plate is a component of the ethmoid bone.
   c. The sella turcica of the sphenoid bone houses the pituitary gland.
   d. The styloid process of the temporal bone is a thick, posterior projection.
   e. The most conspicuous feature of the temporal bone is the foramen magnum.

5. The only moveable bone in the adult skull is the:

   a. maxilla.   c. mandible.
   b. lacrimal bone.   d. frontal bone.

6. The structure(s) that divide the nasal cavity into right and left sides is/are the:

   a. nasal bones.
   b. perpendicular plate of the ethmoid bone.
   c. vomer.
   d. Both a and b are correct.
   e. Both b and c are correct.

7. The "soft spots" in an infant's skull are known as:

   a. fontanels.   c. metopic joints.
   b. sutures.   d. hyaline cartilage.

8. Mark the following statements as true or false. If a statement is false, correct it to make a true statement.

   a. The thoracic and sacral curvatures are the vertebral column's convex curvatures.
   b. A vertebral disc is composed of an inner anulus fibrosus and an outer nucleus pulposus.
   c. The posterior projection from a vertebra is the spinous process.
   d. The sacral, coccygeal, and cervical vertebrae are fused in an adult.

9. Transverse foramina are a characteristic of which kind of vertebra?

   a. Thoracic   c. Sacral
   b. Lumbar   d. Cervical

10. Fill in the blanks: The inferior portion of the sternum is the _____. The superior portion of the sternum is the _____, and it articulates with the _____ and the first rib.

11. How do true, false, and floating ribs differ from one another?

12. Which of the following portions of the scapula articulates with the clavicle?

   a. Coracoid process   c. Spine
   b. Acromion   d. Glenoid cavity

13. Fill in the blanks: The only bone of the arm is the _____. The forearm consists of the medial _____ and the lateral _____.

14. The "elbow bone" is called the:

   a. trochlea.   c. olecranon.
   b. capitulum.   d. deltoid tuberosity.

15. Which of the following is not a proximal carpal bone?

   a. Hamate   b. Pisiform   c. Scaphoid   d. Lunate

16. Mark the following statements as true or false. If a statement is false, correct it to make a true statement.

   a. The obturator foramen articulates with the head of the femur at the hip joint.
   b. The superior border of the coxal bone is the iliac crest.
   c. The weight of the body in the sitting position is borne by the ischial tuberosities.

   d. The two pubic bones articulate at the acetabulum.
   e. The pelvic brim is the boundary between the greater and lesser pelvis.

17. The most lateral projection of the proximal epiphysis of the femur is the:

   a. lesser trochanter.   c. greater trochanter.
   b. gluteal tuberosity.   d. femoral neck

18. Fill in the blanks: The bones of the leg are the medial _____ and the lateral _____. The sesamoid bone that articulates with the distal femur is the _____.

19. The heel bone is more properly known as the:

   a. talus.   b. navicular.   c. cuboid.   d. calcaneus.

20. The arch(es) of the foot are the:

   a. transverse arch.
   b. medial longitudinal arch.
   c. lateral longitudinal arch.
   d. Both a and b are correct.
   e. All of the above are correct.

LEVEL 2 Check Your Understanding

1. How do the atlas (C_1) and the axis (C_2) differ from other cervical vertebrae? How does this difference enable them to perform their functions?

2. Explain how abnormal bone structure could affect muscle function.

3. What structures form the knee and elbow joints? Of the two joints, which do you think would be more stable? Why?

LEVEL 3 Apply Your Knowledge

PART A: Application & Analysis

1. A *deviated septum* results when the nasal septum is shifted to one side or the other. What bones might be involved in this condition? Why might this make breathing difficult?

2. Mrs. Dent presents to the clinic with back pain. During the exam, you notice that she has severe kyphosis, and you suspect a vertebral fracture. What specific part of her vertebra is likely to be fractured, considering her deformity? Explain.

3. You arrive on the scene where a person without a pulse was found. Someone on the scene performed CPR, but the individual unfortunately could not be revived. On postmortem examination, it is discovered that several ribs and the xiphoid process were fractured. What likely caused these fractures?

PART B: Make the Connection

4. Predict where each of the following structures is located, based on your knowledge of skeletal anatomy and anatomical terms from Chapter 1 (your answers should be along the lines of "lateral crural region" or "posterior cervical region").

   a. Frontal lobe of brain   d. Tibial nerve
   b. Suprahyoid muscle   e. Intercostal muscle
   c. Ulnar artery   f. External iliac artery

8

Articulations

The human skeleton contains over 200 bones having a variety of functions. Almost all of these bones contact other bones at **articulations** or **joints.** By connecting two bones, joints perform three important functions:

- **Joints enable movement.** The link between bones can allow movement when surrounding muscles and tendons exert the necessary amount of force across the joint. Most joints in the human body can accommodate specific motions used in the activities of daily life.
- **Joints provide stability.** Some joints do not allow movement. This lack of motion makes the joint stable and enables the bone to protect underlying structures. The connective tissues linking some joints are stable enough to provide a foundation to which muscle tissue, tendons, and ligaments can adhere.
- **Joints allow long bones to lengthen.** Recall that the *epiphyseal plate* is the location from which long bones grow in length during skeletal development (see Chapter 6). These epiphyseal plates are actually temporary joints, which we discuss in a later module.

In this chapter we examine the structural and functional classes of joints as well as the different types of motion they allow. As you read, note that one of the core themes you have seen throughout this textbook—the structure-function relationship—applies to joints just as it applies to other body parts.

MODULE **8.1**

Classification of Joints

Learning Outcomes

1. Describe the functional classification of joints based on degree of movement allowed—synarthrosis, amphiarthrosis, and diarthrosis.
2. Describe the structural classification of joints based on their anatomical features.
3. Explain how the functional and structural classifications are related.

Joints can be classified in two ways: (1) according to the amount of motion they allow, which is their functional classification; or (2) according to their anatomical features, which is their structural classification.

Photo: Colored x-ray of a prosthetic knee (dark red) in a patient with osteoarthritis. The knee replacement attaches to the femur and tibia and has a flexible hinge joint.

Functional Classification

◀◀ **Flashback**

1. What is dense regular collagenous connective tissue? (p. 139)
2. What is cartilage? (p. 142)

Joints can be classified on the basis of how much motion they allow, giving us the following functional categories:

- A **synarthrosis** (sin-ar-THROH-sis) does not allow any movement between articulating bones.
- An **amphiarthrosis** (am'-fee-ahr-THROH-sis) allows only a small amount of movement between articulating bones.
- A **diarthrosis** (die-ahr-THROH-sis) is freely moveable, allowing a wide variety of specific movements.

Note that synarthroses and amphiarthroses provide little, if any, motion, but a great deal of stability. Diarthroses, in contrast, provide a great deal of motion, but not much stability. This tradeoff of motion at the expense of stability is a theme that we examine repeatedly throughout this chapter.

Structural Classification

Joints can also be grouped on the basis of structure. Two anatomical features are considered: the type of connective tissue that links bones, and the presence or absence of a space between the articulating bones. These features give us the following structural classes:

- **Fibrous joints** are fastened together by dense regular collagenous connective tissue. The bones are united by collagen fibers, so no joint space is present. Functionally, fibrous joints are synarthroses or amphiarthroses.
- **Cartilaginous joints** (kar'-ti-LAJ-i-nuhs) have cartilage between the articulating bones. Like fibrous joints, no joint space is present. Functionally, cartilaginous joints are synarthroses or amphiarthroses.
- **Synovial joints** (si-NO-vee-uhl) have a layer of hyaline cartilage on the surface of each articulating bone. They are the only class to have a joint space, or *cavity*, filled with fluid between articulating bones. Other supportive structures surround the bones and enclose the joint cavity. Synovial joints are diarthroses and provide the most motion of any joint type.

Although structural classifications will be the primary focus of this chapter, keep the structure-function relationship in mind. We cannot really separate the structural and functional aspects of a joint.

Quick Check

☐ 1. How are joints classified functionally?
☐ 2. How are joints classified structurally?

Apply What You Learned

☐ 1. Do you think the cranial bones are joined by synarthroses, diarthroses, or amphiarthroses? Explain.
☐ 2. A surgical procedure called a *fusion* takes two bones and joins them, making them functionally one bone.

What functional class of joint does this resemble? What properties would the fused joint have?

See answers in Appendix A.

MODULE **8.2**

Structural Classification: Fibrous Joints

Learning Outcomes

1. Discuss the common features of fibrous joints.
2. Compare and contrast the three subclasses of fibrous joints.
3. Give examples of fibrous joints, and describe how they function.

In the tissues chapter, you learned about dense regular collagenous connective tissue. Recall that it contains fibers of the tough protein collagen (see Chapter 4). Collagen fibers found in fibrous joints lend stability to these joints but permit little, if any, motion. Fibrous joints fall into three categories: *sutures, gomphoses,* and *syndesmoses,* which are discussed in the next subsections.

Sutures

◀◀ **Flashback**

1. Which bones in the body are joined by sutures? (p. 214)
2. Name the major sutures. (p. 223)

Recall that a **suture** (SOO-cher) is a joint between the edges of bones that make up the cranium (**Figure 8.1a**). These bone edges have finger-like projections that articulate somewhat like a closed zipper. They are reinforced by dense regular collagenous connective tissue. During skeletal development, sutures allow a small amount of movement but fuse with maturity to become synarthroses as the connective tissue is replaced by bone. Fused sutures are very stable and therefore well suited for reinforcing the

 CORE PRINCIPLE Structure-Function

bones that protect the delicate tissues of the brain, an example of the Structure-Function Core Principle (p. 25).

Gomphoses

A **gomphosis** (gahm-FOH-sis) is an immoveable fibrous joint between a tooth and its corresponding osseous pocket or alveolus in the mandible or maxilla (**Figure 8.1b**). A gomphosis is a synarthrosis, whose primary function is to provide stability by restricting motion. Each individual tooth is firmly attached to the jaw by a strong fibrous membrane called the *periodontal ligament*. The articulations between the teeth and the maxilla or mandible are the only gomphoses in the human body.

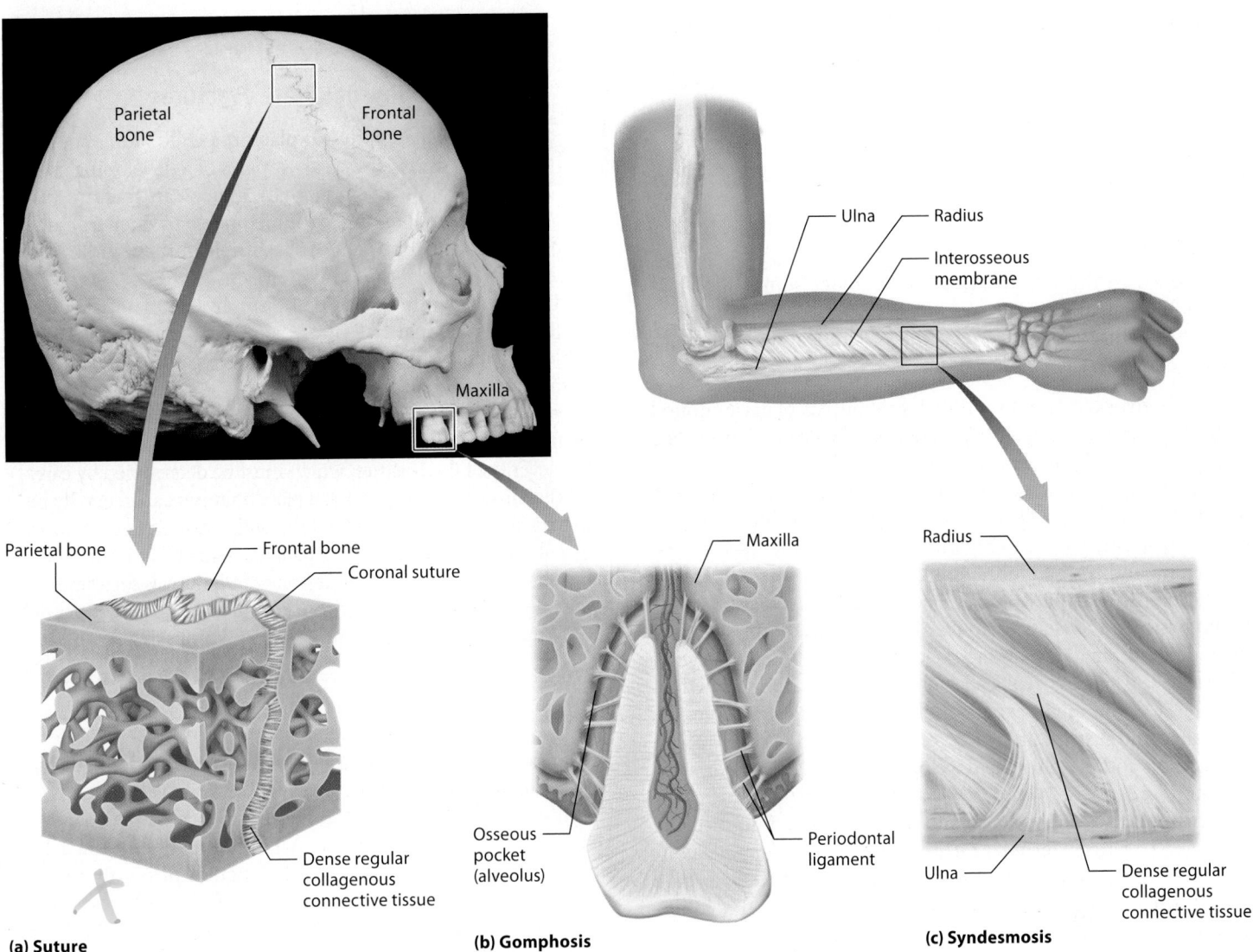

(a) Suture

(b) Gomphosis

(c) Syndesmosis

Figure 8.1 The three types of fibrous joints.

Syndesmoses

The third type of fibrous joint is the **syndesmosis** (sin-dez-MOH-sis), in which the articulating bones are joined by a long membrane composed of dense regular collagenous connective tissue. This membrane, the *interosseous membrane* or *ligament*, allows a limited amount of movement in that the articulating bones are able to pivot around one another; we therefore classify syndesmoses as amphiarthroses. Examples of syndesmoses are found between the radius and ulna in the forearm (**Figure 8.1c**) and between the fibula and tibia in the leg.

Quick Check

☐ 1. What are the structural and functional properties of a suture?

☐ 2. What is a gomphosis? Where are gomphoses found in the human body?

☐ 3. What is a syndesmosis? Where are syndesmoses located?

Apply What You Learned

☐ 1. Certain diseases cause the production of collagen that is weaker than normal. Predict how such diseases could affect the structure and function of gomphoses.

☐ 2. What might happen if sutures were fused at birth?

See answers in Appendix A.

MODULE **8.3**

Structural Classification: Cartilaginous Joints

Learning Outcomes

1. Discuss the common features of cartilaginous joints.
2. Compare and contrast the two subclasses of cartilaginous joints.
3. Give examples of cartilaginous joints, and describe their function.

As the name implies, in a cartilaginous joint the articulating bones are held together by cartilage. Cartilaginous joints lack a joint cavity and allow for little, if any, motion. There are two types of cartilaginous joints: *synchondroses* and *symphyses*.

Synchondroses

◄◄ **Flashback**

1. How does the growth of long bones differ from that of flat and irregular bones? (p. 193)
2. How would you characterize the structure and function of the epiphyseal plate? (p. 198)

A **synchondrosis** (sin-kahn-DROH-sis) consists of bones united by hyaline cartilage. Recall from the tissues chapter that *hyaline cartilage* is a tough but flexible connective tissue often associated with joints. Synchondroses permit essentially no motion and are therefore functionally synarthroses. Examples include the epiphyseal plates in growing long bones, as well as the first sternocostal joint and costochondral joint.

The **epiphyseal plate** (ep′-uh-FIZZ-ee-uhl) (**Figure 8.2a**) is found between the shaft (diaphysis) and the head (epiphysis) of the long bones in a developing skeleton. The hyaline cartilage of an epiphyseal plate is replaced with bone when bone growth is complete. As you might expect, motion occurring at an epiphyseal plate could disrupt its structure, which would in turn disturb its function and possibly the development of the skeletal system (see *A&P in the Real World: Epiphyseal Plate Fractures*).

Synchondroses that persist into adulthood are found in the rib cage. The first sternocostal joint, where the first rib connects to the manubrium of the sternum, stabilizes the rib cage (**Figure 8.2b**). Additional synchondroses, the costochondral joints, are found where the ribs attach to their costal cartilages.

Quick Check

☐ 1. What are the features of a synchondrosis?
☐ 2. Which joints in the rib cage are synchondroses? Which joints belong to other joint classes?

Symphyses

◄◄ **Flashback**

1. What are the basic properties of fibrocartilage? (p. 142)
2. Where is the pubic symphysis located? (p. 247)

A **symphysis** (SIM-fuh-sis) is functionally an amphiarthrosis in which the bones are united by a tough fibrocartilage plug or pad. In an

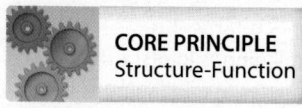

CORE PRINCIPLE
Structure-Function

example of the Structure-Function Core Principle (p. 25), a symphysis is best suited for regions of the skeleton that must resist compression

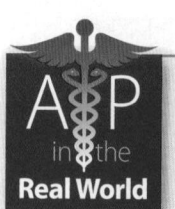

Epiphyseal Plate Fractures

The epiphyseal plate in a child's long bone is one of the weakest parts of a developing skeleton. Even a minor injury can fracture this delicate structure, possibly with lifelong consequences, such as differences in limb length, limb deformities, and early-onset arthritis (joint inflammation).

The most common causes of epiphyseal plate fractures include recreational activities, accidents, and competitive athletics. Injuries may affect any joint but occur most often at the epiphyseal plates of the forearm.

Symptoms of epiphyseal plate fractures include swelling, pain, and redness over the injured joint. Treatment depends on the severity of the fracture, which must be determined by careful diagnosis. Minor epiphyseal plate fractures can generally be managed by immobilizing the joint with a cast. Severe fractures usually require surgery. Many patients benefit from rehabilitation exercises to strengthen the bones and muscles surrounding the joint and regain full function. Fortunately, most fractures do not impair bone development if managed properly.

and tension while still allowing a small amount of motion. The intervertebral joints and the pubic symphysis are examples.

An intervertebral joint (**Figure 8.2c**) is located between adjacent vertebral bodies of the spinal column. These joints have fibrocartilage pads, or *discs,* which function primarily for protection and shock absorption, especially from compression that could damage the vertebrae. Although each individual intervertebral joint allows only a small degree of motion, the spinal column as a whole is quite flexible.

The pubic symphysis (**Figure 8.2d**) is located between the two pubic bones of the pelvic girdle. This region of the skeleton is under enormous tension, stress, and compression, especially when the body is moving. Normally, this joint limits motion, but during childbirth it allows increased motion, which creates a larger opening through the pelvic outlet where the head of the newborn emerges.

Quick Check

☐ 3. Describe the general features of a symphysis.
☐ 4. What are two examples of symphyses?

Apply What You Learned

☐ 1. Explain how increased motion at a synchondrosis would impair the function of the joint.
☐ 2. How would absolute lack of motion at a symphysis impair the function of the joint?

See answers in Appendix A.

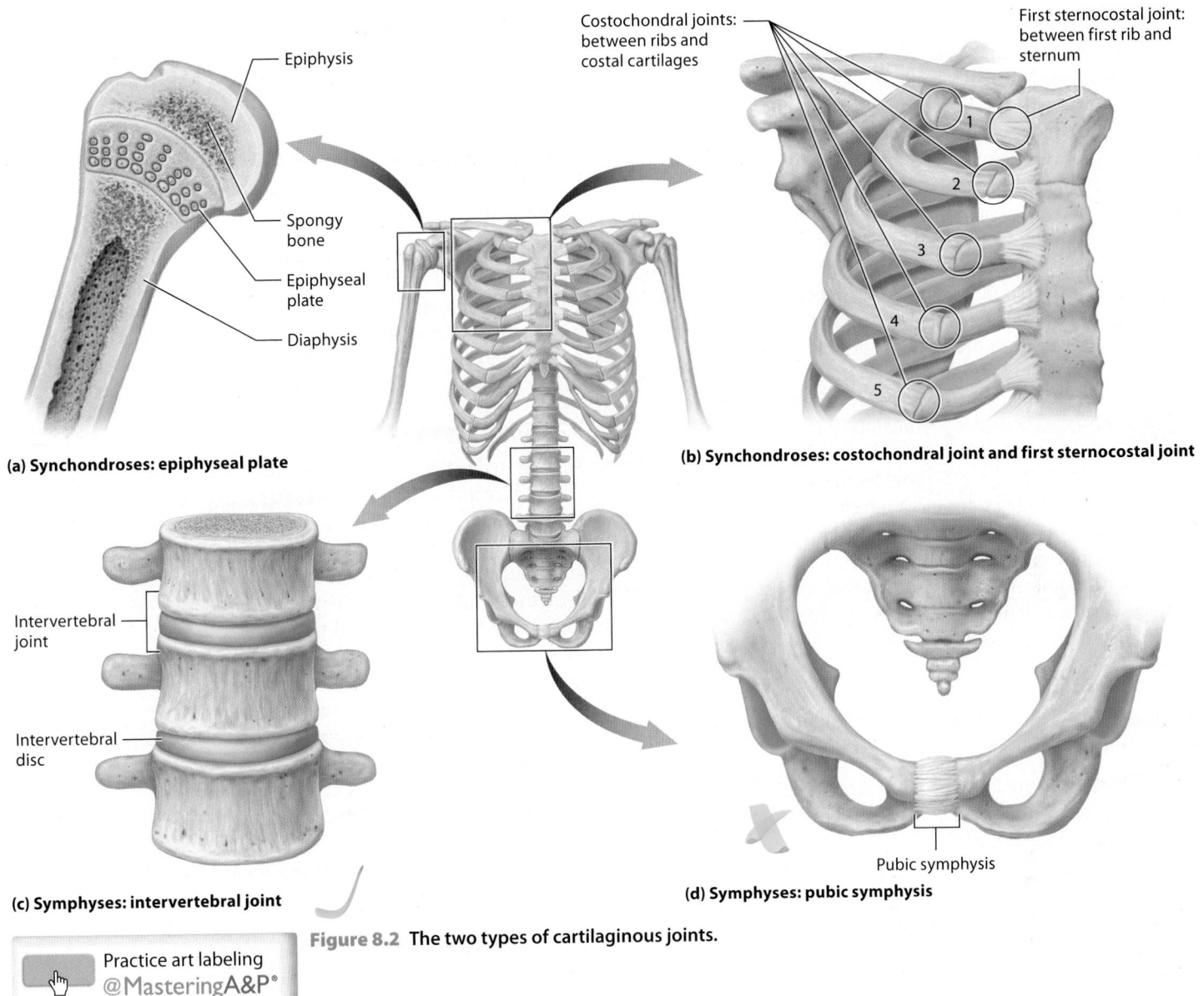

Epiphysis

Spongy bone

Epiphyseal plate

Diaphysis

(a) Synchondroses: epiphyseal plate

Costochondral joints: between ribs and costal cartilages

First sternocostal joint: between first rib and sternum

(b) Synchondroses: costochondral joint and first sternocostal joint

Intervertebral joint

Intervertebral disc

(c) Symphyses: intervertebral joint

Pubic symphysis

(d) Symphyses: pubic symphysis

Figure 8.2 **The two types of cartilaginous joints.**

Practice art labeling
@MasteringA&P®

MODULE 8.4

Structural Classification: Synovial Joints

Learning Outcomes

1. Identify the structural components of a synovial joint, including accessory structures such as ligaments, tendons, bursae, and tendon sheaths.
2. Compare and contrast synovial joints with fibrous and cartilaginous joints.

Most articulations in the human body are synovial joints. Recall that these are functionally diarthroses (freely moveable joints)

and that they are structurally different from fibrous and cartilaginous joints.

Structural Elements

◀◀ Flashback

1. What are the properties of hyaline cartilage? (p. 142)
2. How do hyaline cartilage and fibrocartilage differ? (p. 142)

One of the key differences between synovial joints and other types of joints is the presence of a space—the **joint cavity,** or **synovial cavity**—between the two articulating bones. The joint cavity is characterized by three unique features: an *articular capsule, synovial fluid,* and an *articular cartilage.*

Articular Capsule

Observe in **Figure 8.3** that where two bones join, they are enclosed by a double-layered structure composed of connective tissue, called the **articular capsule** (ar-TIK-you-luhr). The outer fibrous layer is dense irregular collagenous connective tissue that keeps the articulating bones from being pulled apart. This outer layer also isolates the internal environment of the joint from the blood supply in the connective tissues outside the joint. If blood vessels were inside the joint cavity, they would be damaged by the forces that generate movement in the joint. Isolating the joint cavity from the blood supply protects the blood vessels and also safeguards the joint from toxins and disease-causing organisms sometimes found in blood.

However, cells within the joint still need to obtain nutrients and oxygen, as well as eliminate metabolic wastes, to maintain homeosta-

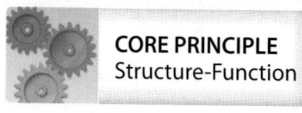

sis. In an example of the Structure-Function Core Principle (p. 25), the inner layer of the articular capsule provides the joint with a means for obtaining life-sustaining substances and eliminating waste products.

The delicate inner layer of the articular capsule is called the **synovial membrane.** This membrane is made up of loose connective tissue that lines the entire joint cavity except where hyaline cartilage covers articulating surfaces of each bone. Cells in the membrane secrete *synovial fluid,* which we discuss next.

Synovial Fluid

Synovial fluid, a thick, colorless, oily liquid, serves three main functions in the joint:

1. **Lubrication.** Synovial fluid lubricates the joint cavity and articulating surfaces to reduce friction. This protects the articulating ends of the bones.

Figure 8.3 Structure of a typical synovial joint.

Practice art labeling
@MasteringA&P®

2. **Metabolic functions.** Synovial fluid supplies nutrients such as glucose. In addition, it removes metabolic waste products formed by the cells in the joint cavity. This is critical because, as you just discovered, the joint is not directly connected to the blood vessels.

3. **Shock absorption.** Synovial fluid helps to evenly distribute the force and stress exerted on the articular surfaces of the bones during movement.

Cells of the synovial membrane continuously circulate and replenish a small amount of synovial fluid. Compression generated by joint motion also circulates synovial fluid around the joint. It is important that synovial fluid be removed as quickly as it is secreted because too much synovial fluid can cause pain and impair joint mobility. However, too little synovial fluid can result in joint damage.

Articular Cartilage

All exposed articulating bone surfaces within the joint cavity are covered by a thin layer of **articular cartilage** composed of hyaline cartilage (Figure 8.3). Recall from the tissues chapter that hyaline cartilage appears smooth and glassy. In an illustration of the Structure-Function Core Principle (p. 25), this layer provides a smooth

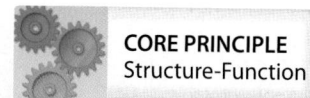

surface on which bones articulate. It also protects the joint by reducing friction and absorbing shock.

Articular cartilage is *avascular,* meaning it does not have direct access to a blood supply within its capsule. This isolation from the surrounding blood supply can cause permanent damage if the articular cartilage is injured, and it can contribute to the development of *arthritis,* which we discuss later. The cells in cartilage depend on the circulation of synovial fluid, which permeates the spaces between them, providing oxygen and nutrients and removing metabolic byproducts.

Other Components of a Synovial Joint

Other components of a synovial joint include adipose tissue, nerves, and blood vessels. Adipose tissue (fat) is packed into the empty spaces surrounding the joint, which provides protective padding. These fat pads are pliable and can change shape as a synovial joint moves.

Synovial joints are surrounded by a copious supply of blood vessels and nerves found in the supportive connective tissues. Blood vessels deliver nutrients to and remove waste products from the articular capsule, with the aid of the synovial membrane cells and synovial fluid. Nerves send messages to the brain about painful stimuli that could damage the joint. They also transmit information about the position of the joint, which helps with body movement and postural adjustments.

Quick Check

☐ 1. What are the layers of the articular capsule, and what are their functions?

☐ 2. What functions are served by synovial fluid?

☐ 3. Which type of cartilage makes up articular cartilage, and what is its function?

Stabilizing and Supportive Structures

Synovial joints allow more mobility than other types of joints, but they are also the least stable. These joints are continuously stressed as part of their normal function. Although the articular capsule provides some structural support, synovial joints must also rely on several structures outside the joint itself for stability and support. These include *ligaments, tendons, bursae,* and *tendon sheaths* (Figure 8.4).

Ligaments

A **ligament** is a strand of dense regular collagenous tissue that connects one bone to another to strengthen and reinforce the articulating bones. Two types of ligaments are present around synovial joints. *Intrinsic ligaments* (not visible in Figure 8.4) are thickened regions of the articular capsule; they are called "intrinsic" because they are found within the articular capsule. *Extrinsic ligaments* are not part of the articular capsule. They may be inside or outside the joint cavity. For instance, the coracoacromial ligament (Figure 8.4) is an extrinsic ligament that stabilizes the shoulder joint.

Tendons

A **tendon**—a structural component of skeletal muscle composed of dense regular collagenous connective tissue—connects the muscle to a bone or another structure. Tendons typically cross over or around a joint so that when a muscle contracts, the force generated is transmitted across the joint. This force results in motion as the tendon pulls on the bone to which it attaches. Tendons often stabilize joints, particularly joints that have multiple tendons crossing them. For instance, the tendon of the long head of the biceps brachii muscle crosses the shoulder joint, stabilizing the head of the humerus in the joint cavity (Figure 8.4).

The muscles to which the tendons attach also help stabilize joints. Muscles constantly generate small contractions, collectively known as *muscle tone,* that pull the tendons taut around or across the joint. Muscle tone keeps the ends of the articulating bones near one another.

Bursae and Tendon Sheaths

A **bursa** (BER-suh) is a synovial fluid–filled structure resembling a limp water balloon. This fibrous structure is lined with a synovial membrane on its inner surface, much like the articular capsule. Bursae may be attached to the articular capsule or completely separate. Generally, bursae are found in regions of high stress where bones, tendons, muscles, and skin interact in a small space. Bursae, such as the subacromial bursa of the shoulder joint, minimize friction between all the moving parts of a synovial joint (Figure 8.4). What happens to bursae when excessive stress is placed on them? See *A&P in the Real World: Bursitis* on page 264.

Tendon sheaths are long bursae that surround some tendons in high-stress regions of the human body. Tendon sheaths protect long tendons as they course over and around synovial joints. The stabilizing force that the biceps brachii tendon exerts on the shoulder joint is further enhanced by its surrounding tendon sheath (Figure 8.4).

Quick Check

☐ 4. Why do synovial joints require stabilization?
☐ 5. What are the basic structural and functional properties of ligaments, tendons, and bursae?

Arthritis

Arthritis is defined as the inflammation of one or more joints, resulting in pain and decreased range of motion. The articular cartilage breaks down and the underlying bone is damaged. Three common forms of arthritis are osteoarthritis, rheumatoid arthritis, and gouty arthritis. Each type results in the same basic symptoms, but the mechanism by which the cartilage is damaged differs.

Osteoarthritis

The most common type, **osteoarthritis** (os´-tee-oh-ar-THRY-tis), is the arthritis that we generally associate with wear and tear. Although younger people may develop the condition due to accidents, competitive athletics, and recreational injuries, most individuals with osteoarthritis are over 65. Obesity, female gender,

Acromion of scapula
Coracoacromial ligament (extrinsic)
Coracoid process of scapula
Articular cartilage
Joint cavity
Scapula
Synovial membrane
Fibrous layer of articular capsule

Subacromial bursa
Humerus
Tendon sheath

Tendon of the long head of the biceps brachii muscle

Figure 8.4 Supportive structures of a synovial joint.

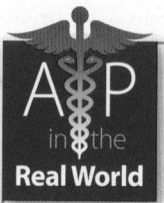

Bursitis

A&P in the Real World

Bursitis refers to inflammation of a bursa. Bursitis can result from a single traumatic event such as a fall, repetitive movements like pitching a baseball, or an inflammatory disease such as rheumatoid arthritis. The most common sites of bursitis are the shoulder, elbow, hip, and knee.

Clinical features of bursitis include pain both at rest and with motion of the affected joint. The joint may feel tender, swollen, and warm.

Treatment is aimed primarily at reducing pain and swelling. Rest, ice, compression of the injured area, and medications are beneficial in the early stages of the injury. Anti-inflammatory steroid medications may be injected directly into the bursa itself. Fluid can also be removed from the bursa to relieve swelling. If left untreated, bursitis can become chronically painful and increasingly difficult to cure.

and advancing age increase the risk. Although gender and age are not modifiable risk factors, overweight people can lower their risk by losing weight. Any joint can be affected, but the back, hands, knees, and shoulders are the most common sites.

Common features of osteoarthritis include pain and stiffness, as well as decreased joint mobility. Surrounding muscles, tendons, tendon sheaths, and bursae can also become painful and swollen.

Although osteoarthritis is not reversible, it can be managed. The first goal is to minimize pain, primarily through pain-relieving and anti-inflammatory medications. Another important goal, which might come as a surprise, is to keep the joint mobile. Movement helps replace damaged articular cartilage with scar tissue, which slows the progression of the disease. Mild to moderate low-impact exercise is therefore recommended for most patients.

Rheumatoid Arthritis

A less common type of arthritis is **rheumatoid arthritis** (ROO-mah-toyd). Unlike osteoarthritis, rheumatoid arthritis generally develops between the ages of 30 and 50. The individual's immune system attacks the connective tissues surrounding synovial joints, damaging the articular cartilage. Like osteoarthritis, rheumatoid arthritis is not curable but it can be managed. Medications can reduce inflammation and pain, as well as suppress the immune system to prevent further damage to the cartilage.

Gouty Arthritis

In gout, excess uric acid, a waste product, may crystallize and form deposits in the connective tissue surrounding a joint, most often in the big toe. The inflammatory reaction to the crystals damages the articular cartilage, leading to gouty arthritis. As with other forms of arthritis, medications that reduce inflammation

and pain can help. In addition, patients may take medications to lower the blood level of uric acid.

Quick Check

☐ 6. What is osteoarthritis, and in what age group might you expect the disease to develop? How does osteoarthritis differ from rheumatoid and gouty arthritis?

Apply What You Learned

☐ 1. Certain conditions cause *ligamentous laxity,* which means that the ligaments are very loose. Predict how this would affect synovial joints. Would it have the same effect on cartilaginous and fibrous joints? Why or why not?

☐ 2. Explain why joint movement is often painful when a tendon—which is outside the joint—is injured.

See answers in Appendix A.

MODULE **8.5**
Function of Synovial Joints

Learning Outcomes

1. Describe and demonstrate the generalized movements of synovial joints.

Now that we have discussed the structural features of synovial joints, we can start to appreciate how those features translate into specific movements.

Functional Classes of Synovial Joints

Bones in a synovial joint travel through a plane around an imaginary line called an *axis*. Each plane has an axis that is perpendicular to its flat surface. Motion of a particular joint is described by these imaginary flat surfaces and the perpendicular lines that run through them. We can group synovial joints into four functional classes based on the number of axes around which a bone can move:

● **Nonaxial joints** (nahn-AX-ee-uhl) allow motion to occur in one or more planes but do not move around an axis.
● **Uniaxial joints** (you'-nee-AK-see-uhl) allow motion around only one axis.
● **Biaxial joints** (by-AK-see-uhl) allow motion around two axes.
● **Multiaxial** or **triaxial joints** (muhl'-tee-AK-see-uhl, try-AK-see-uhl) allow motion around three axes.

For more help understanding the axes in which synovial joints can move, see Concept Boost: Understanding Axes of Motion.

ConceptBOOST »»»

Understanding Axes of Motion

Axes of motion might seem complex at first, so let's break down their functions with simple examples. The elbow joint has only one axis (axis 1 in the figure at right) that acts like a hinge and allows motion in one plane perpendicular to the axis. This uniaxial joint only allows the forearm and hand to move upward toward the shoulder or to make the opposite movement away from the shoulder.

Now let's look at an example of a biaxial joint, which is capable of movement around two separate axes. The metacarpophalangeal joints are a group of biaxial joints found between the proximal phalanges and the metacarpals in the hand, as shown in the figures below. The metacarpophalangeal joints can move around axis 1, allowing the proximal phalanges to move toward and away from the palm of the hand in the same way that the elbow joint moves. The metacarpophalangeal joints can also move around axis 2, allowing the fingers to be squeezed together or fanned out as you can see in the lower part of the figure.

Finally, a third axis allows an additional motion that our uniaxial and biaxial joints are not able to perform. The shoulder is an example of a multiaxial joint. As the figure below shows, your humerus can move forward and backward within the shoulder joint around axis 1 (as when you swing your arms back and forth while walking). The humerus can also move away from and toward your body around axis 2 (as when you do jumping jacks). Finally, it can rotate, or move in a circular fashion, around axis 3 (as when you throw a Frisbee). ■

Uniaxial: elbow joint

Biaxial: metacarpophalangeal joint

Multiaxial: shoulder joint

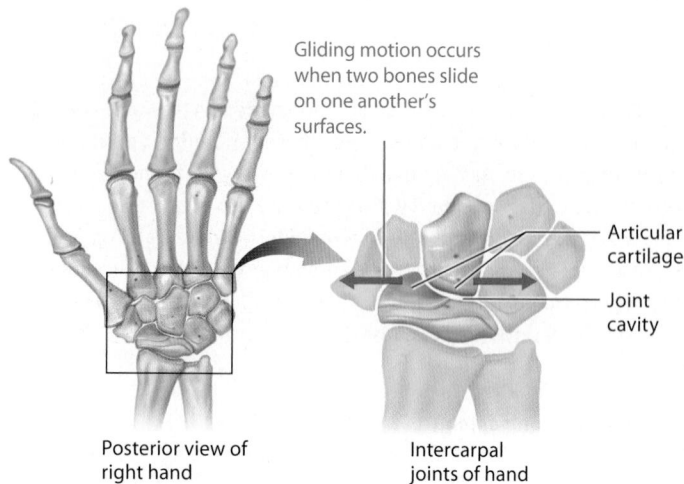

Gliding motion occurs when two bones slide on one another's surfaces.

Articular cartilage

Joint cavity

Posterior view of right hand

Intercarpal joints of hand

Figure 8.5 Gliding movements of synovial joints.

Movements at Synovial Joints

Four general types of movement can take place at synovial joints. In *gliding* movements, the joint does not move around an axis; these gliding joints are known as nonaxial joints. The other three types of movement—*angular, rotation,* and *special*—take place around one or more axes of motion. Each synovial joint can engage in one or more of these movements. We examine these classes in the upcoming subsections.

Gliding Movements

Gliding is a sliding motion between the articulating surfaces of the bones in a joint. These movements are considered nonaxial

because the bones slide past each other and move in several possible directions without the constraint of an axis or axes. The direction of movement depends on the shape of the articulating surfaces at the joint as well as the limitations of the surrounding supportive connective tissues. For instance, we find gliding movements between the intertarsal joints of the ankle and foot as well as the intercarpal joints of the wrist and hand (**Figure 8.5**).

Angular Movements

Angular movements increase or decrease the angle between the articulating bones. This motion includes several types of movements that are discussed next. As you read, note that many of these movements are paired and oppose each other.

Flexion and Extension The first pair of angular movements consists of *flexion* and *extension*. **Flexion** (FLEK-shuhn) decreases the angle between articulating bones by bringing the two bones closer together. An example is the action of the elbow joint in which the forearm is pulled toward the upper arm (**Figure 8.6a**). As the elbow flexes, the angle between the arm and forearm decreases. Flexion also occurs at the shoulder and the hip (**Figure 8.6b and c**). Flexion of these joints moves the arm and thigh anteriorly, in front of the body.

Lateral flexion is sideways movement, such as when you tilt your head to one side or the other (**Figure 8.6d**). Bending the trunk sideways at the hips in either direction is also lateral flexion (**Figure 8.6e**). In each of these examples, the angle decreases between the head and neck or the trunk and hips.

Extension (eks-TEN-shuhn), the opposite of flexion, increases the angle between articulating bones. For instance, in Figure 8.6a,

Flexion

Extension

(a) Elbow flexion and extension

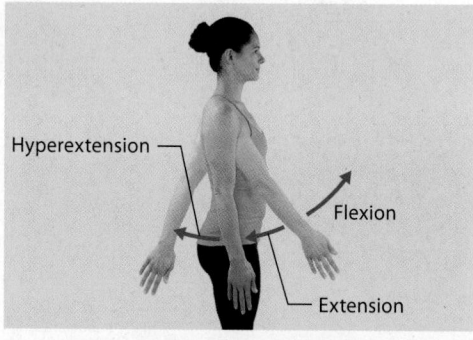

Hyperextension

Flexion

Extension

(b) Shoulder flexion, extension, and hyperextension

Lateral flexion

(e) Lateral flexion of trunk at the hips

Hyperextension

Flexion

Extension

(c) Hip flexion, extension, and hyperextension

Lateral flexion

Lateral flexion

(d) Lateral flexion of head and neck

Figure 8.6 Angular movements: flexion and extension of synovial joints.

the elbow is flexed with the forearm close to the upper arm. If we extend the elbow, the arm and forearm move away from one another, increasing the angle. Extension of the shoulder and hip is motion in a posterior direction, moving the hand or foot behind the rest of the body (Figure 8.6b and c).

A motion related to extension is **hyperextension** (hy′-per-eks-TEN-shuhn), or extension beyond the anatomical position of the joint (Figure 8.6b and c). For instance, you hyperextend your shoulder when you prepare to pitch a softball as your arm moves behind the trunk of your body beyond anatomical position of the arm at the shoulder joint. You hyperextend your hip as it moves behind your trunk and beyond anatomical position while you are running.

Abduction and Adduction Another pair of angular movements is abduction and adduction. **Abduction** (ab-DUHK-shuhn) is the motion of a body part away from the midline of the body or another reference point (such as the midline of the hand or foot). The first movement of a jumping jack exercise—spreading the legs at the hips and raising the arms at the shoulders—is an example of abduction (Figure 8.7a). Note that the arms and legs are moving away from the midline of the body.

Another example of abduction is fanning out the fingers and toes (Figure 8.7b). Note that the reference point has changed from the midline of the body to the midline of the hand or foot, respectively. Again, the fingers and toes are moving away from the midline of the reference point.

Adduction (uh-DUHK-shuhn), the opposite of abduction, is the motion of a body part toward the midline of the body or some other reference point. Completing the second movement of the jumping jack adducts the arms and legs by bringing them back toward the body (Figure 8.7c). Similarly, if we bring our fingers and toes back toward the reference point, we are adducting these joints (Figure 8.7d).

Be careful to remember anatomical position, especially with abduction and adduction of the wrist. Recall that in anatomical position the palm of the hand faces anteriorly. If you move the wrist away from the trunk of the body or the midline of the hand, by definition this is abduction. The opposite motion, movement of the wrist toward the midline of the body, is adduction.

Circumduction Circumduction is the only angular movement not paired with an opposite movement. In **circumduction** (SUR-kuhm-duhk-shuhn), a freely moveable distal bone moves around a stationary proximal bone in a cone-shaped motion. This is a complex movement that is the sum total of flexion-extension and abduction-adduction, best seen at the hip and shoulder joints. If you draw a circle on the wall by moving your shoulder only, that is circumduction (Figure 8.7e). Note that the proximal shoulder (the scapula) is fixed while the distal bone (the humerus) moves.

Rotation

Rotation is a nonangular, pivoting motion, in which one bone rotates or twists on an imaginary line running down its middle, known as its *longitudinal axis*. Shaking the head to indicate "no" is a rotational movement of the atlantoaxial joint in the neck (Figure 8.8a).

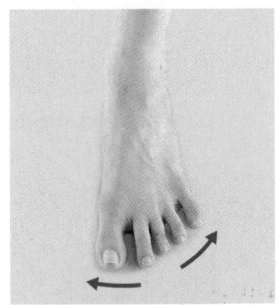

(a) Abduction of shoulders and hips   **(b) Abduction of fingers and toes**

(c) Adduction of shoulders and hips   **(d) Adduction of fingers and toes**

(e) Circumduction of shoulder

Figure 8.7 Angular movements: abduction, adduction, and circumduction of synovial joints.

(a) Rotation of atlantoaxial joint in neck

Internal rotation

External rotation

(b) Internal and external rotation of hip

External rotation

Internal rotation

(c) Internal and external rotation of shoulder

Figure 8.8 Rotational movements: internal and external rotation of synovial joints.

Multiaxial joints such as the hip and shoulder are also able to rotate. *Internal (medial) rotation* of the hip is best appreciated by using the foot as a reference point. When the hip is internally rotated, the toes point medially toward the midline of the body (Figure 8.8b). In the opposite movement, *external (lateral) rotation,* the toes point away from the midline of the body. In both

motions, the femur rotates along its longitudinal axis, while the head of the femur articulates with the pelvic bone in the hip joint.

Internal and external rotation of the shoulder is best appreciated with the elbow in a flexed position as if you are holding a coffee mug. If you swing the mug away from your body, the shoulder rotates externally (Figure 8.8c). Swinging the mug inward as you raise it to your mouth would be internal rotation. Similar to the femur, the humerus rotates along its longitudinal axis while the head of the humerus articulates with the scapula in the shoulder joint.

Be careful not to confuse rotation with circumduction. Although both involve circular movements, they are entirely separate classes of motion. In rotation, the bone turns around itself. In circumduction, the bone moves in a circular or cone-shaped manner:

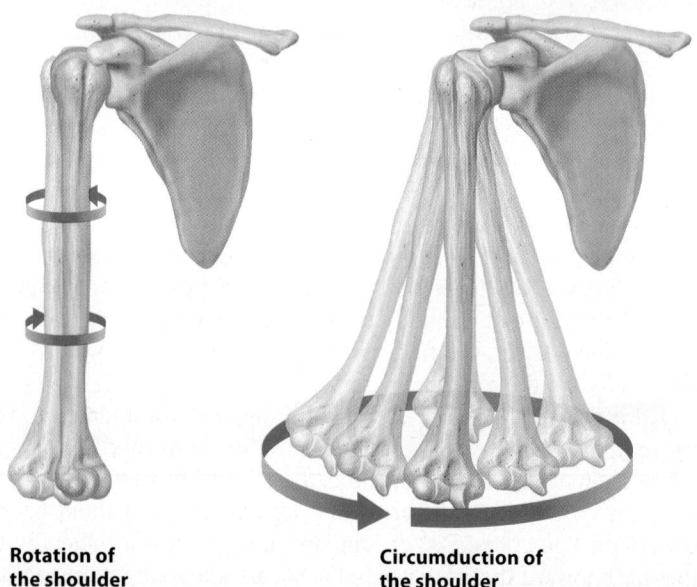

Rotation of the shoulder

Circumduction of the shoulder

Special Movements

Special movements include those that are not well described by the previous categories or that pertain to only one or a few synovial joints. The following subsections explain the different kinds of special movements.

Opposition and Reposition **Opposition** is a special movement that occurs only at the thumb or first carpometacarpal joint (Figure 8.9a). This motion allows the thumb to travel across the palmar surface of the hand (which is why we say humans have *opposable thumbs*). You can perform this motion by using your thumb to touch the tips of your other four fingers. Opposition is important for grasping objects. **Reposition** is the return of the thumb to its anatomical position (Figure 8.9b).

Depression and Elevation **Depression** is the movement of a body part in an inferior direction. For instance, opening the mouth depresses the mandible, which moves inferiorly relative to the maxilla (Figure 8.9c). **Elevation** is the opposite of depression in which, for example, the mandible is pulled up toward the maxilla in a superior direction (Figure 8.9d). These two movements are combined in chewing.

Protraction and Retraction **Protraction** (proh-TRAK-shuhn) moves a body part in the anterior direction. Moving the mandible forward so the bottom row of teeth stick out is an example of protraction (**Figure 8.9e**). The mandible moves anteriorly to the maxilla during this motion.

Retraction (ree-TRAK-shuhn) is the opposite motion, in which the body part moves posteriorly. The mandible, once protracted, must be retracted to return the mandible posteriorly to its anatomical place (**Figure 8.9f**). Protraction and retraction are also important in chewing.

Inversion and Eversion **Inversion** (in-VUR-zhuhn) is a rotational motion of the foot in which the plantar surface (the sole of the foot) rotates medially toward the midline of the body (**Figure 8.9g**). **Eversion** (ee-VER-zhuhn) is the opposite motion, in which the plantar surface of the foot rotates laterally away from the midline of the body (**Figure 8.9h**). This combination of motions is important for walking.

Dorsiflexion and Plantarflexion of the Foot and Ankle Dorsiflexion and plantarflexion are a combination of movements that involve the foot and ankle. In **dorsiflexion** (dawr-suh-FLEK-shuhn), the angle between the foot and the tibia decreases (**Figure 8.9i**). In other words, the toes are pulled up toward the head.

In **plantarflexion** (plant-uhr-FLEK-shuhn), the opposite motion, the angle between the foot and the tibia increases—the toes point toward the ground (**Figure 8.9j**). As with inversion and eversion, this pair of opposing movements is instrumental in walking.

Supination and Pronation **Supination** (soo-puh-NAY-shuhn) and **pronation** (proh-NAY-shuhn) are rotational movements that occur at the proximal radioulnar joint of the forearm and at a combination of the ankle and tarsal joints. Your forearm is supinated when your palm faces anteriorly with the thumb pointing laterally, as in anatomical position (**Figure 8.10a**). Starting with the hand in a supinated position, you pronate the forearm by turning the palmar surface medially until it faces posteriorly with the thumb pointing toward the body (**Figure 8.10b**).

(a) Opposition of thumb

(b) Reposition of thumb

(c) Depression of mandible

(d) Elevation of mandible

(e) Protraction of mandible

(f) Retraction of mandible

(g) Inversion of foot

(h) Eversion of foot

(i) Dorsiflexion of foot

(j) Plantarflexion of foot

Figure 8.9 Special movements of synovial joints.

(a) Supination of forearm

(b) Pronation of forearm

Figure 8.10 Special movements: supination and pronation.

The ulna and the radius of the forearm articulate with each other at the proximal radioulnar joint, which is near the elbow joint. This joint, with assistance from the interosseous membrane between the radius and ulna, can perform a pivoting movement that allows the radius to cross over the ulna. In anatomical position, in which the hand is supinated, the ulna and radius are parallel to each other. When the hand is pronated, the radius crosses over the ulna. An example is the movement of twisting a doorknob back and forth.

↟ STUDY BOOST

Keeping Synovial Joint Movements Straight

Remember the difference between supination and pronation with this tip: You hold a cup of *soup* when your hand is *supinated*, and you *pour* it out when your hand *pronates*.

If you find abduction and adduction confusing, remember this tip: With *abduc*tion, you *abduct* (take away) the part from the body. With *adduc*tion, you *add* the part back to the body. ■

Quick Check

☐ 1. Compare and contrast: gliding and angular motions, circumduction and rotation, abduction and adduction.

Range of Motion

The **range of motion** of a joint is the amount of movement it is capable of under normal circumstances. For example, when you move your knee joint from a relaxed state to full flexion, and then return the joint to its fully extended state, that is the range of motion of the knee. Uniaxial joints (such as the knee) tend to have the smallest range of motion, whereas multiaxial joints (such as the shoulder) tend to have the greatest.

Apply What You Learned

☐ 1. In the condition known as *drop foot,* the foot and ankle are unable to dorsiflex. Predict the consequences of this condition for walking.

☐ 2. Marfan syndrome can affect connective tissues around joints and make ligaments loose. How would this affect the elbow's range of motion?

See answers in Appendix A.

MODULE **8.6**

Types of Synovial Joints

Learning Outcomes

1. Describe the anatomical features of each structural type of synovial joint.
2. Describe where each structural type can be found.

3. Predict the kinds of movements that each structural type of synovial joint will allow.
4. Compare and contrast the structural features of the knee and elbow and of the shoulder and hip.

The function of a synovial joint depends on its structure. Although all synovial joints are diarthroses, some joints allow more movement than others. The amount of motion a joint allows is largely based on the shape of the surfaces of its two articulating bones. In the upcoming subsections, we look at the six structural classes of synovial joints, from those that allow the least motion to those that allow the most.

Plane Joint

A **plane joint,** the simplest and least mobile synovial joint, features two bones whose flat surfaces sit next to each other (**Figure 8.11a**). It is also known as a gliding joint, which describes its motion. Plane joints are nonaxial, as they allow only side to side "gliding" movements. The intercarpal joints of the wrist and intertarsal joints of the ankle are examples of these simple, stable joints.

Hinge Joint

At a **hinge joint** (**Figure 8.11b**), a convex surface of one bone fits into a concave depression of another bone. As the name implies, the motion of a hinge joint is restricted to movement in only one plane (uniaxial motion), much like the hinge of a door. Examples include the elbow, knee, and interphalangeal joints of the fingers and toes.

Pivot Joint

A **pivot joint** (**Figure 8.11c**) is a uniaxial joint in which the rounded surface of one bone fits into a groove on the surface of another bone. Joint stability is enhanced by a ringlike ligament that surrounds the rounded bone and holds it in the groove of the second bone. This allows the joint to rotate or pivot on its longitudinal axis within this ring.

The atlantoaxial joint between the first (atlas) and second (axis) cervical vertebrae is an example of a pivot joint. The rounded dens process of the axis articulates with the atlas, and is surrounded and reinforced by the transverse ligament. When muscles move this joint, it rotates, resulting in the side to side movement of the head, such as when motioning "no."

Condylar Joint

A **condylar** (KAHN-duh-luhr), or **ellipsoid, joint** is a biaxial joint in which the oval, convex surface of one bone fits into the shallow concave surface of the other bone (**Figure 8.11d**). This alignment allows both flexion and extension, and abduction and adduction. Examples of condylar joints include the metacarpophalangeal joints (the "knuckles") of the fingers. Notice that you

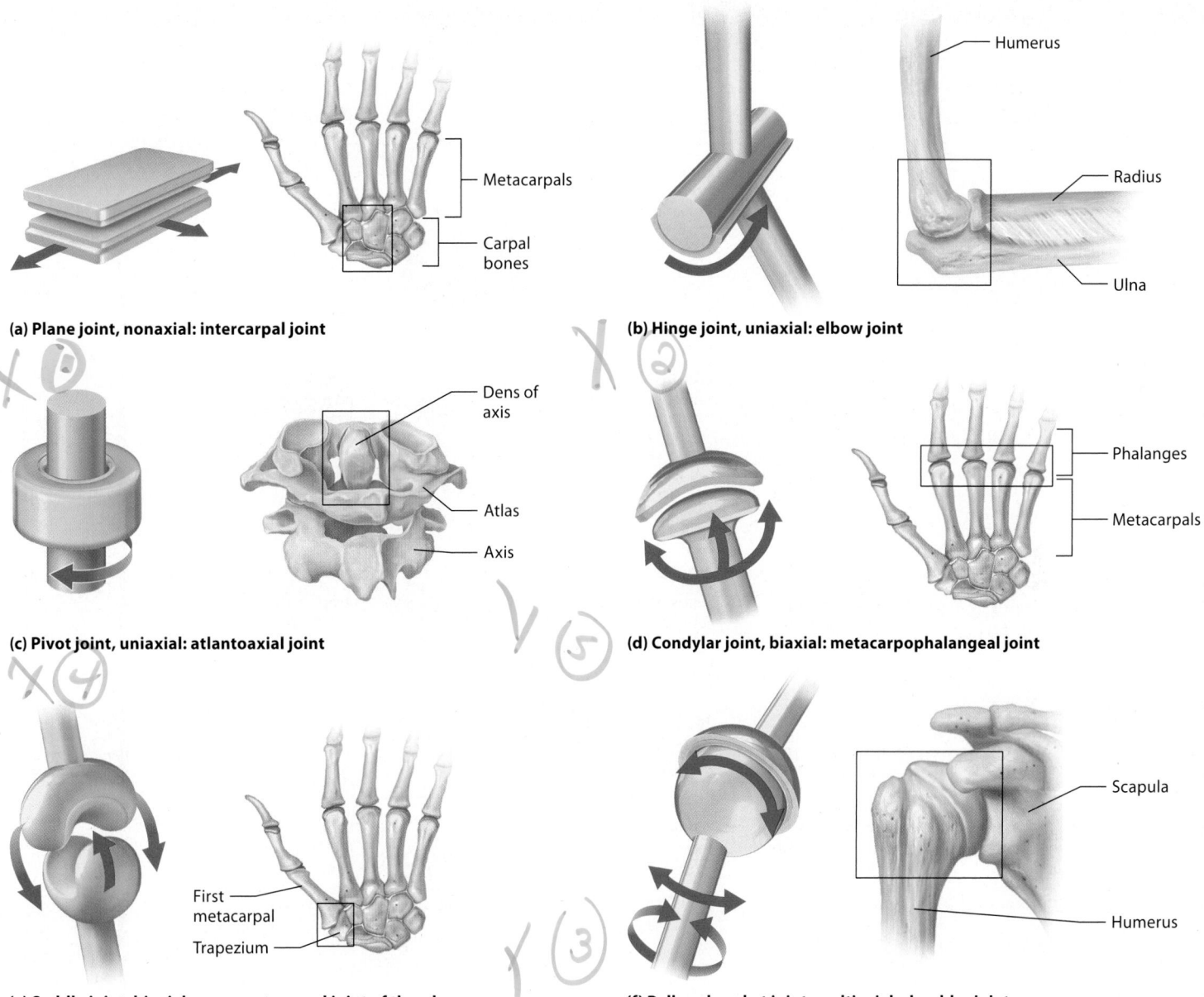

(a) Plane joint, nonaxial: intercarpal joint

Metacarpals

Carpal bones

(b) Hinge joint, uniaxial: elbow joint

Humerus

Radius

Ulna

(c) Pivot joint, uniaxial: atlantoaxial joint

Dens of axis

Atlas

Axis

(d) Condylar joint, biaxial: metacarpophalangeal joint

Phalanges

Metacarpals

(e) Saddle joint, biaxial: carpometacarpal joint of thumb

First metacarpal

Trapezium

(f) Ball-and-socket joint, multiaxial: shoulder joint

Scapula

Humerus

Figure 8.11 The six types of synovial joints and the motion allowed at each.

can move these joints to bunch your fingers into a fist or spread them out like a fan.

Saddle Joint

The articulating surfaces of a **saddle joint** resemble a horseback riding saddle (**Figure 8.11e**). The surface of each articulating bone has both convex and concave regions that complement one another. This allows for a good fit, much like putting one saddle on the ground and then stacking a second one on top, while also permitting a large amount of motion.

A saddle joint is biaxial but allows greater motion than condylar joints. The carpometacarpal joint between the first metacarpal of the thumb and the trapezium is an example. Wiggle your fingers and thumb and note how much more motion is available to the thumb compared to the fingers. This arrangement is crucial in allowing fingers and thumb to work together for opposition and reposition.

Ball-and-Socket Joint

A **ball-and-socket joint** is a multiaxial joint in which the articulating surface of one bone is ball-shaped or spherical and fits into a cup or socket formed by the articulating surface of the other bone (**Figure 8.11f**). Both the shoulder and hip joints are ball-and-socket joints and will be discussed later in this module.

The Big Picture

of Joint Classifications and Stability versus Mobility

Figure 8.12

STRUCTURAL CLASSIFICATION AND CHARACTERISTICS	STRUCTURAL CATEGORY	EXAMPLE	FUNCTIONAL CLASSIFICATION (MOBILITY)	STABILITY/ MOBILITY CONTINUUM
Fibrous Bone edges held together by dense regular collagenous connective tissue; no joint cavity; no articular capsule (see Figure 8.1)	Suture: skull bones held together by short fibrous connective tissue fibers	Coronal suture between the frontal and parietal bones	Synarthrosis: no movement allowed	*MOST STABLE, LEAST MOBILE*
	Gomphosis: tooth within bony cavity held by periodontal ligament	Tooth in the mandible or maxilla	Synarthrosis: no movement allowed	
	Syndesmosis: dense regular collagenous connective tissue between two long bones	Interosseous membrane between radius and ulna and between tibia and fibula	Amphiarthrosis: some movement allowed	
Cartilaginous Wedge of cartilage located between articulating bones; no joint cavity; no articular capsule (see Figure 8.2)	Synchondrosis: hyaline cartilage plate between bones	Epiphyseal plates; sternocostal joint	Synarthrosis (epiphyseal plate) or amphiarthrosis (sternocostal joint): no or little movement allowed, respectively	*LESS STABLE, SOMEWHAT MOBILE*
	Symphysis: fibrocartilage pad between the hyaline cartilage on the two articulating bones	Pubic symphysis between two pubic bones	Amphiarthrosis: some movement allowed	

Ball-and-socket joints allow movement in all three axes of motion. This joint configuration allows the most freedom of movement. Remember, though, that this mobility comes with a stability tradeoff. Ball-and-socket joints are rather unstable, especially the shoulder joint, which is very susceptible to dislocation.

Putting It All Together: The Big Picture of Joint Classifications and Stability versus Mobility

Throughout this chapter, we have discussed two important functions of joints: enabling movement and providing stability. As

STRUCTURAL CLASSIFICATION AND CHARACTERISTICS	STRUCTURAL CATEGORY	EXAMPLE	FUNCTIONAL CLASSIFICATION (MOBILITY)	STABILITY/ MOBILITY CONTINUUM
Synovial Layer of hyaline cartilage covers surface of each articulating bone; separated by a fluid-filled joint cavity; articular capsule present (see Figure 8.3)	Plane: nonaxial; flat bone surfaces glide across one another (see Figures 8.5 and 8.11a)	Intercarpal and intertarsal joints	Diarthrosis: freely moveable	*LESS STABLE, MORE MOBILE*
	Hinge: uniaxial; convex surface of one bone fits into a concave surface of another bone to allow angular motion along axis (see Figure 8.11b)	Elbow and knee joints	Diarthrosis: freely moveable	
	Pivot: uniaxial; projection of one bone fits into another bone; the second bone rotates around the first (see Figure 8.11c)	Atlantoaxial joint	Diarthrosis: freely moveable	
	Condylar: biaxial; convex oval surface of one bone articulates with concave oval surface of another bone (see Figure 8.11d)	Second through fifth metacarpophalangeal joints	Diarthrosis: freely moveable	
	Saddle: biaxial; saddle-shaped surface of one bone fits into a depression of another bone (see Figure 8.11e)	Articulation between the first metacarpal and the trapezium (thumb joint)	Diarthrosis: freely moveable	
	Ball-and-socket: multiaxial; round head of one bone fits into a cup-shaped surface of another bone (see Figure 8.11f)	Shoulder and hip joints	Diarthrosis: freely moveable	*LEAST STABLE, MOST MOBILE*

you have seen, a joint can be mobile or stable, but not both at the same time. As joints allow an increasing amount of movement, they become increasingly unstable. The most mobile joints in the body are also the most easily injured. **Figure 8.12** summarizes the different classes of joints and where they fall on this continuum of joint stability versus mobility.

Quick Check

☐ 1. Compare the six structural classes of synovial joints.

☐ 2. Which type of synovial joint has the greatest range of motion? Which type has the least?

Specific Hinge Joints: The Elbow and the Knee

◀◀ Flashback

1. What are the functions of the trochlea and trochlear notch of the humerus and ulna, respectively? (pp. 243, 244)
2. Where are the femoral and tibial condyles located? (p. 251)

Now that we have discussed the general properties of synovial joints, let's focus on specific examples. First, we examine the elbow and the knee, and highlight their key similarities and differences. Both are hinge joints whose primary motions are flexion and extension. We concentrate on these two hinge joints because they are commonly injured due to their structural complexity and relative instability.

The Elbow

The **elbow** joint is actually composed of two separate articulations (**Figure 8.13a**). The larger articulation is the *humeroulnar joint* (hyoo′-muhr-oh-UHL-nuhr) between the trochlea of the humerus and the trochlear notch of the ulna. The smaller articulation is the *humeroradial joint* (hyoo′-muhr-oh-RAY-dee-uhl) between the capitulum of the humerus and the head of the radius.

(a) Anterior view

(b) Sagittal section

Figure 8.13 Anatomical structure of the elbow joint.

Three strong extrinsic ligaments reinforce and support the articular capsule, which strengthens the elbow joint (**Figure 8.13b**). The **radial collateral ligament** (or lateral collateral ligament) supports the lateral side of the joint, and the **ulnar collateral ligament** (or medial collateral ligament) supports the medial side. An accessory ligament, the **anular ligament** (AN-yoo-luhr), binds the head of the radius to the neck of the ulna to stabilize the radial head in the elbow joint.

The elbow is more stable than the knee: Its durable articular capsule encloses both of the articulations between the humerus, ulna, and radius (Figure 8.13). Notice how the articulating surfaces of the humeroulnar joint fit together snugly, further stabilizing the joint.

The Knee

The **knee** joint is the largest diarthrosis in the human body. This complex joint is functionally a hinge joint that allows flexion and extension, but when the knee is flexed, it also allows some degree of rotation and lateral gliding. The articular capsule encloses all but the anterior surface of the knee joint, where it is covered by the patella. The patella is surrounded by connective tissues of the tendon of the anterior thigh muscles, collectively known as the *quadriceps femoris* (KWAHD-ruh-seps FEM-uh-ris) (**Figure 8.14a and b**). Distal to the patella, this structure is considered a ligament because it now connects bone to bone. The patellar ligament continues distally to its destination on the anterior surface of the tibia. This structural arrangement leaves the knee joint without a continuous articular capsule.

Like the elbow, the knee is made up of two distinct articulations (**Figure 8.14c**): the *tibiofemoral joint* (TIB-ee-oh-fem′-uhr-uhl) between the femoral and tibial condyles, and the *patellofemoral joint* (puh-TELL-oh-fem′-uhr-uhl) between the patella and the patellar surface of the femur. For this reason, the knee joint lacks a single, continuous joint cavity.

The knee joint is inherently unstable due to its structure and the many forces placed on it. However, several anatomical structures enhance its stability. For instance, a pair of C-shaped fibrocartilage pads—the **medial meniscus** and **lateral meniscus** (men-ISS-kuhs; plural, *menisci*)—sit on the tibial condyles (Figure 8.14a and b). The menisci provide shock absorption and cushioning between the articular cartilages of the femoral and tibial condyles. They also improve the fit between the two bones, which further stabilizes the joint.

Two extrinsic *collateral ligaments* on either side of the knee joint tighten when the knee is extended to provide additional stabilization during weight bearing and walking. The **tibial collateral ligament** (or medial collateral ligament) connects the femur with the tibia and attaches to the medial meniscus. This ligament prevents the tibia from moving too far laterally on the femur (hyperabduction of the knee). The **fibular collateral ligament** (or lateral collateral ligament) links the femur to the fibula and prevents the tibia from moving too far medially on the femur (hyperadduction). Observe in Figure 8.14a and b that the fibular collateral ligament does not attach to the lateral meniscus.

Femur

Posterior cruciate ligament

Lateral condyle

Medial condyle

Fibular collateral ligament

Anterior cruciate ligament

Lateral meniscus

Medial meniscus

Tibial collateral ligament

Tibia

Patellar ligament

Fibula

Posterior patella

Tendon of the quadriceps femoris muscle (cut)

Femur

Lateral condyle

Fibular collateral ligament

Lateral meniscus

Tibia

Fibula

(a) Anterior view, right knee

(b) Cadaver photo, anterior view, left knee

Tendon of quadriceps femoris muscle

Femur

Patellofemoral joint

Tibiofemoral joint

Patella

Articular cartilage

Joint cavity

Tibia

Patellar ligament

(c) Sagittal section of knee

Figure 8.14 Anatomical structure of the knee joint.

Practice art labeling
@MasteringA&P®

We find two other stability-promoting extrinsic ligaments within the joint cavity of the knee: the two *cruciate ligaments* (KROO-shee-iht; *cruci* = "X-shaped"). These ligaments, which limit motion of the knee joint in the anterior and posterior directions, are named for their tibial attachments. The **anterior cruciate ligament** (ACL) runs from an anterior insertion site on the tibia to the posterior aspect of the femur. When the knee joint is extended, the ACL tightens to prevent hyperextension and to prevent the tibia from moving too far anteriorly on the femur. The **posterior cruciate ligament** (PCL) travels from a posterior position on the tibia to the anterior femur. When the PCL tightens during knee flexion, it prevents the tibia from displacing itself posteriorly (i.e., sliding backward) from the femur.

An important feature of this synovial joint is its ability to "lock out." When an individual is weight-bearing, the knee joints can be locked in extension for long periods of time without causing undue muscle fatigue. When fully extended, the tibia rotates laterally, which tightens the ACL and squeezes the menisci between the articular surfaces of the tibial and femoral condyles. All of this locks the knees while the individual is in a standing position.

Knee ligaments and menisci are easily injured (see *A&P in the Real World: Knee Injuries and the Unhappy Triad* on page 276).

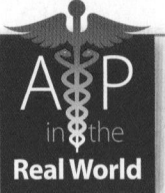

Knee Injuries and the Unhappy Triad

Despite its supportive structures, the knee joint is still susceptible to injury. Any activity that involves quick changes in direction can injure the knee. Athletes who participate in contact sports such as football or soccer are also at risk, especially if the knee is struck from the side or from behind.

A lateral blow—such as when a football player is illegally blocked below the knees—often ruptures the tibial collateral ligament. The medial meniscus and the anterior cruciate ligament may tear as well, forming the so-called "unhappy triad" (see the illustration). The fibular collateral and posterior collateral ligaments can also be damaged, but this is less common.

Treatment of knee injuries depends on their severity. Initial interventions include rest, ice, compression, and anti-inflammatory medications to reduce swelling and minimize pain. More severe injuries may require surgical repair of the damaged ligaments. Physical therapy and rehabilitation to strengthen the surrounding muscles are also helpful.

Anterior MRI of torn tibial collateral ligament

Sagittal MRI of torn anterior cruciate ligament

Quick Check

☐ 3. What two joints form the elbow joint?
☐ 4. Describe four ligaments that stabilize the knee joint.
☐ 5. Where are the menisci, and what are their functions?

Specific Ball-and-Socket Joints: The Shoulder and the Hip

◄◄ Flashback

1. Where is the glenoid cavity, and which bone articulates with it? (p. 242)
2. Where is the acetabulum, and which bone articulates with it? (p. 247)

We end the chapter by discussing the shoulder and hip. Recall that these are ball-and-socket joints enabling several types of motion. Both joints are multiaxial and allow movement in all three axes.

The Shoulder

The **glenohumeral joint** (glen′-oh-HYOO-muhr-uhl), more commonly known as the **shoulder,** is a member of the group of articulations that constitute the pectoral girdle and upper extremities (**Figure 8.15**). The glenohumeral joint is made up of the ball-shaped proximal end of the humerus and a cavity or socket on the lateral aspect of the scapula called the *glenoid cavity.*

The shoulder is the most freely moving joint in the human body. However, as we have discussed, greater mobility comes at the price of less stability and increased risk of injury. The glenoid cavity is rather shallow, so the head of the humerus does not fit snugly inside. This requires support from surrounding structures, such as the **glenoid labrum** (LAY-bruhm), a fibrocartilaginous ring that sits on the rim of the glenoid cavity (Figure 8.15c).

The tendon of the long head of the biceps brachii muscle, found in the anterior arm, passes through the articular capsule of the shoulder joint on its way to where it attaches to the scapula. This tendon and the muscle tone of the upper arm provide a stabilizing force that keeps the humeral head firmly within the glenoid cavity.

Four additional muscles—the supraspinatus, infraspinatus, teres minor, and subscapularis—and their corresponding tendons are collectively called the **rotator cuff** (Figure 8.15b; note that the subscapularis is not visible in this view). The rotator cuff surrounds the shoulder joint and provides most of its strength and stability. Rotator cuff tears are common injuries, as we discuss in the muscular system chapter (Chapter 9).

Several extrinsic ligaments provide some limited support and stability for the shoulder joint. One is the *coracohumeral ligament* (KOHR-uh-koh-hyoo′-muhr-uhl), which is a thickening of the superior aspect of the articular capsule between the humeral head and the coracoid process. In addition, three *glenohumeral ligaments,* which may or may not be present, are sometimes found supporting the anterior aspect of the articular capsule.

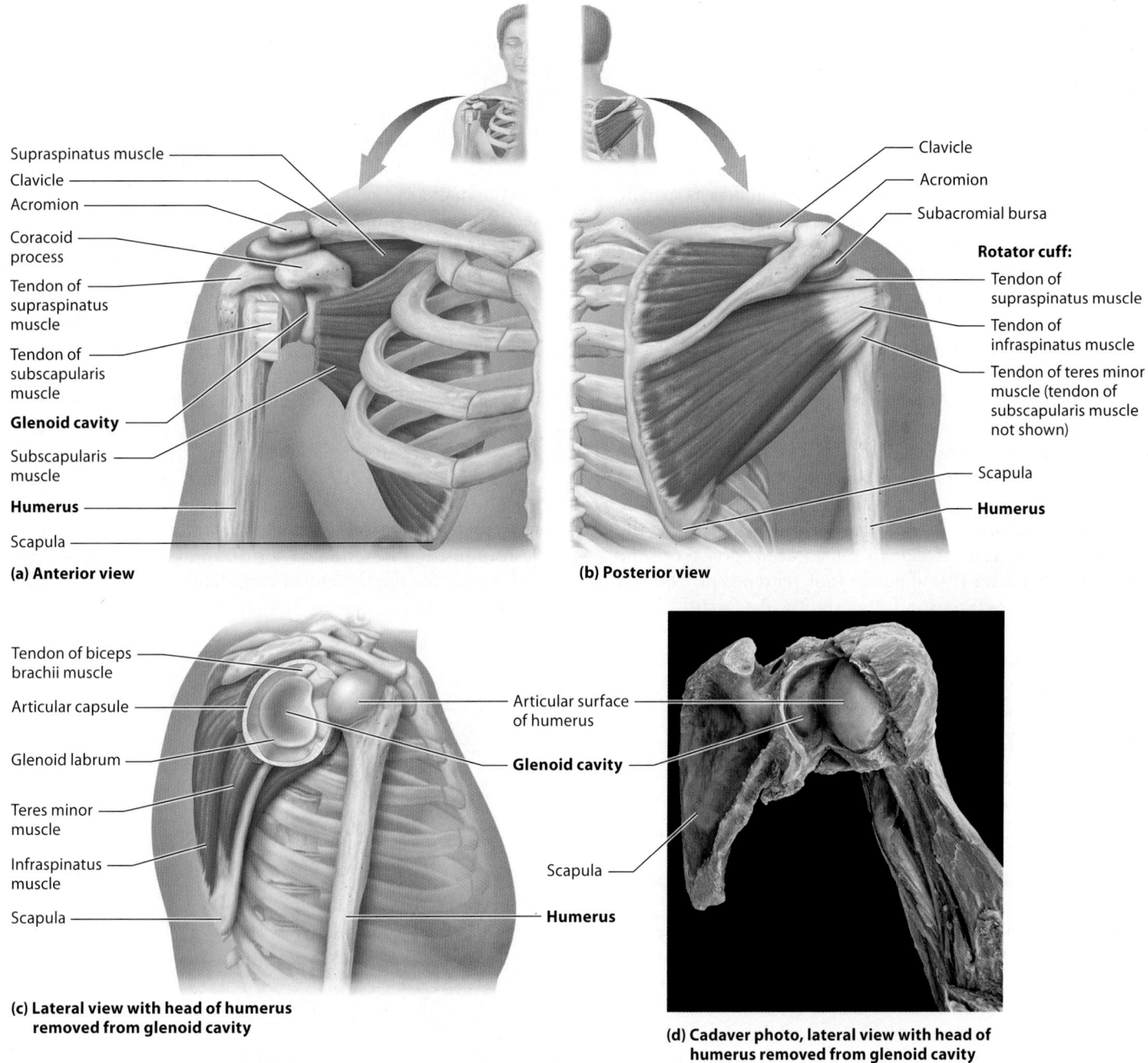

Supraspinatus muscle
Clavicle
Acromion
Coracoid process
Tendon of supraspinatus muscle
Tendon of subscapularis muscle
Glenoid cavity
Subscapularis muscle
Humerus
Scapula

(a) Anterior view

Clavicle
Acromion
Subacromial bursa
Rotator cuff:
Tendon of supraspinatus muscle
Tendon of infraspinatus muscle
Tendon of teres minor muscle (tendon of subscapularis muscle not shown)
Scapula
Humerus

(b) Posterior view

Tendon of biceps brachii muscle
Articular capsule
Glenoid labrum
Teres minor muscle
Infraspinatus muscle
Scapula

Articular surface of humerus
Glenoid cavity
Humerus

(c) Lateral view with head of humerus removed from glenoid cavity

Scapula
Humerus

(d) Cadaver photo, lateral view with head of humerus removed from glenoid cavity

Figure 8.15 Anatomical structure of the shoulder joint.

Practice art labeling
@MasteringA&P®

In spite of the many structural support mechanisms of the shoulder joint, it can still become *dislocated* (see *A&P in the Real World: Shoulder Dislocations* on page 278).

Several bursae surround the shoulder joint. With so many muscles and tendons in such a tight space, friction could damage the joint. Recall that bursae minimize frictional stress on a joint by cushioning its moving structures.

The Hip

The **hip joint**, also known as the **coxal joint** (KAHK-suhl), is the articulation between the acetabulum and the ball-shaped head of the femur. Recall from the skeletal system chapter (Chapter 7) that the femur fits snugly into the acetabulum. A fibrocartilaginous ring called the *acetabular labrum* further strengthens the fit between the bones (**Figure 8.16a**).

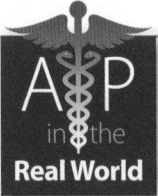

Shoulder Dislocations

The mobility of the shoulder joint comes at the expense of stability. Shoulder injuries are very common, accounting for more than half of all joint dislocations. A *dislocated shoulder* is specific to the glenohumeral joint, where the head of the humerus is traumatically displaced from the glenoid cavity. Another common injury, a *separated shoulder*, is specific to the acromioclavicular joint, which is not actually a component of the shoulder.

Contact sport athletes are especially susceptible to shoulder injuries, but anyone can suffer them under the right circumstances. Any fall in which the hand and forearm are outstretched can result in a dislocation injury. The impact may force the head of the humerus through the inferior wall of the articular capsule, where it is weakest relative to the rest of the capsule. Chest wall muscles pull the dislocated humeral head superiorly and medially so the head comes to rest inferior to the coracoid process of the scapula. This makes the shoulder look flattened or "squared off." The injured person often holds the wrist of the affected shoulder against the abdomen, as this is the least painful position.

Some minor shoulder dislocations can "pop" back into place with limited effort, but more severe injuries may need to be surgically repaired.

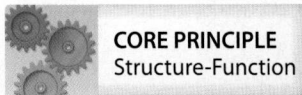

CORE PRINCIPLE
Structure-Function

Despite its similarities to the ball-and-socket configuration of the shoulder joint, the hip is much more stable because of its role in weight bearing and locomotion. This is an example of the Structure-Function Core Principle (p. 25). This increased stability comes with some reduction in mobility, especially compared to that of the shoulder. Obviously, the shoulder joint is neither weight-bearing nor involved in walking, so it does not need to be as stable as the hip.

The hip joint derives its support from a strong articular capsule, the many large and powerful muscle groups that surround it, and a number of reinforcing ligaments. This arrangement greatly reduces the risk of joint dislocation. Intracapsular ligamentous structures known as *retinacular fibers* surround the neck of the femur and reinforce the articular capsule.

In addition to a strong articular capsule and the retinacular fibers, the hip is also supported by several extrinsic ligaments (**Figure 8.16b–d**):

- The **iliofemoral ligament** (IL-ee-oh-fem′-uh-ruhl) is a Y-shaped structure that reinforces the anterior aspect of the hip joint.

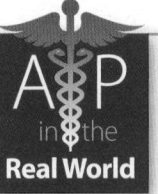

Hip Joint Replacement Surgery

Hip replacement is a surgical procedure that replaces a painful damaged joint with an artificial prosthetic device. An individual may elect to have a hip replaced due to debilitating pain and subsequent loss of joint function. Severe arthritis, trauma, fractures, and bone tumors can all progress to the point where hip joint replacement is an option.

A total hip replacement removes and replaces the head of the femur and reconstructs the acetabulum (see the illustration). A partial replacement removes only the head of the femur and replaces it with a prosthetic device, while leaving the acetabulum intact. The choice of replacement depends on many factors, including the type of injury and the patient's age and general health.

Surgical complications are rare and are further minimized with good postprocedure follow-up care. A rigorous rehabilitation program usually follows the surgery to restore normal function as soon as possible. It may take 3–6 weeks for the patient to completely return to normal daily activities.

Hip bone

Plastic socket in acetabulum (frontal section)

Cement to hold socket in place

Prosthetic replacement (ball)

Femur

Prosthetic device fitted inside the femur

Cement to hold device in place

Fused hip bones

Articular cartilage

Acetabular labrum

Joint cavity

Articular capsule

(a) Frontal section, anterior view

Ligament of head of femur

Ilium

Iliofemoral ligament

Greater trochanter

Pubofemoral ligament

Pubic bone

Femur

(b) Anterior view

Acetabulum

Acetabular labrum

Articular cartilage

Ligament of head of femur

Head of femur

Ischial tuberosity

Greater trochanter

Femur

(c) Lateral view with head of femur removed from acetabulum

(d) Cadaver photo, lateral view with head of femur removed from acetabulum

Practice art labeling
@MasteringA&P®

Figure 8.16 Anatomical structure of the hip joint.

- The **ischiofemoral ligament** (IS-kee-oh-fem′-uh-ruhl) is a spiral-shaped structure that supports the posterior aspect of the hip joint (not visible in the figure).
- The **pubofemoral ligament** (PYOO-boh-fem′-uh-ruhl) is a triangular thickening of the articular capsule that supports the inferior aspect.
- The **ligament of the head of the femur** is a small ligament that links the center of the head of the femur with the acetabulum. While it offers very little structural support to the hip joint, it does provide a pathway for a small artery that supplies blood to the head of the femur.

The hip joint is most stable when the joint is extended, as these ligaments are all pulled taut. The range of motion is greater when the hip is flexed than when it is extended. A damaged hip joint can be replaced with an artificial prosthesis (see *A&P in the Real World: Hip Joint Replacement Surgery* on page 278).

Quick Check

☐ 6. What features increase the stability of the shoulder joint?
☐ 7. Which bones articulate to form the hip joint?
☐ 8. Why is the shoulder joint less stable than the hip joint?

Apply What You Learned

☐ 1. Predict the consequences of an injury to each of the following knee ligaments for the function and range of motion of the knee.
  a. Anterior cruciate ligament
  b. Tibial collateral ligament
  c. Posterior cruciate ligament
☐ 2. Predict how the functions of the human hand would be different if the carpometacarpal joint of the thumb were a uniaxial hinge joint instead of a saddle joint.

See answers in Appendix A.

CHAPTER SUMMARY

Take full advantage of the invaluable online practice and study tools for this chapter by going to MasteringA&P®. There you will find:

- Practice tests, module quizzes, art-labeling and sequence-the-steps activities
- Animated figures with quizzes
- *Practicing A&P* worksheets
- Author-narrated podcast
- Dynamic Study Modules
- *Interactive Physiology*

MODULE 8.1
Classification of Joints 257

- Joints can be classified based on their functional ability:
  ○ A **synarthrosis** allows no motion between articulating bones.
  ○ An **amphiarthrosis** provides a small amount of motion between articulating bones.
  ○ A **diarthrosis** provides a wide range of motion between articulating bones.
- Joints can be classified based on their structure:
  ○ **Fibrous joints** are held together by dense regular collagenous connective tissue. They do not have a joint space or an articular capsule.
  ○ **Cartilaginous joints** have cartilage between articulating bones. They do not have a joint space or an articular capsule.
  ○ **Synovial joints** have a fluid-filled joint space between articulating bones and are enclosed within an articular capsule.

MODULE 8.2
Structural Classification: Fibrous Joints 258

- There are three types of fibrous joints: sutures, syndesmoses, and gomphoses. (Figure 8.1, p. 259)
- **Sutures** are synarthroses between the bones of the skull.
- **Gomphoses** are synarthroses between each tooth and the bony socket in either the mandible or the maxilla.
- **Syndesmoses** are amphiarthroses in which the articulating bones are joined by an interosseous membrane or ligament.

MODULE 8.3
Structural Classification: Cartilaginous Joints 259

- There are two types of cartilaginous joints: synchrondroses and symphyses. (Figure 8.2, p. 261)

Practice art-labeling exercise: Figure 8.2.

- **Synchondroses** are synarthroses in which the articulating bones are held together by tough but flexible *hyaline cartilage*.
- **Symphyses** are amphiarthroses connected by a tough fibrocartilaginous pad.

MODULE 8.4
Structural Classification: Synovial Joints 261

- The unique structural elements of a synovial joint include a **joint cavity,** or **synovial cavity.** (Figure 8.3, p. 262)

Practice art-labeling exercise: Figure 8.3.

- The **articular capsule** is a double-layered structure whose outer layer is a dense irregular collagenous connective tissue that holds articulating bones together.

- The inner layer of the articular capsule is the **synovial membrane,** which secretes synovial fluid into the joint space.
- The **articular cartilage** is a thin layer of hyaline cartilage that covers the ends of the bones in the joint that are not covered by the articular capsule.
- Ligaments, tendons, bursae, and **tendon sheaths** all provide the synovial joint with additional support. (Figure 8.4, p. 263)

MODULE 8.5
Function of Synovial Joints 264

- The following is the functional classification for motion around the body planes and their perpendicular axes:
  - **Nonaxial joints** do not move around an axis or axes.
  - **Uniaxial joints** allow motion around only one axis.
  - **Biaxial joints** allow motion around two axes.
  - **Multiaxial joints** allow motion around all three axes.
- **Gliding** is simple back-and-forth or up-and-down motion between articulating surfaces. (Figure 8.5, p. 266)
- The following are **angular movements** in which the angle between articulating bones changes. (Figures 8.6, p. 266; Figure 8.7, p. 267)
  - **Flexion:** the angle between articulating bones decreases.
  - **Extension:** the angle between articulating bones increases.
  - **Abduction:** motion away from the midline of the body or another point of reference.
  - **Adduction:** motion toward the midline of the body or another point of reference.
  - **Circumduction:** cone-shaped motion that combines flexion and extension with abduction and adduction.
- **Rotation** is a nonangular motion in which a bone pivots or twists along its long axis. (Figure 8.8, p. 268)
- *Special movements* include **opposition** and **reposition, depression** and **elevation, protraction** and **retraction, inversion** and **eversion, dorsiflexion** and **plantarflexion,** and **supination** and **pronation.** (Figures 8.9 and 8.10, p. 269)

MODULE 8.6
Types of Synovial Joints 270

- Synovial joints are grouped into the following classes (Figure 8.11, p. 271):
  - **Plane joint:** two bones whose flat surfaces glide along one another.
  - **Hinge joint:** the convex end of one bone fits into the concave end of another bone.
  - **Pivot joint:** the rounded surface of one bone fits into a groove on another bone.
  - **Condylar joint:** one bone has an oval, convex surface that fits into the concave surface of another bone.
  - **Saddle joint:** two bones fit together like two saddles, one upside down on top of the other.
  - **Ball-and-socket joint:** the rounded end of one bone fits into a depression on the surface of another bone.
- Figure 8.12 (pp. 272–273) summarizes the structure and function of fibrous, cartilaginous, and synovial joints.
- The **elbow** is a hinge joint made up of two separate joints enclosed in a single articular capsule. (Figure 8.13, p. 274)

- The **knee** is a hinge joint made up of two articulations and is the largest diarthrosis in the human body. (Figure 8.14, p. 275)

 Practice art-labeling exercise: Figure 8.14.

- The **shoulder joint (glenohumeral joint)** is the most mobile and least stable joint in the human body. (Figure 8.15, p. 277)

 Practice art-labeling exercise: Figure 8.15.

- The **hip joint (coxal joint)** is a ball-and-socket articulation between the head of the femur and the acetabulum. (Figure 8.16, p. 279)

 Practice art-labeling exercise: Figure 8.16.

ASSESS WHAT YOU LEARNED

See answers in Appendix A.

LEVEL 1 Check Your Recall

1. Which of the following is *not* a function of articulations?
   a. Movement
   b. Blood cell formation
   c. Stability
   d. Providing growth in length for long bones

2. Which functional joint class includes freely moveable joints?
   a. Amphiarthroses
   b. Synarthroses
   c. Diarthroses
   d. All functional joint classes are freely moveable.

3. Identify each of the following joints as synovial, fibrous, or cartilaginous.
   a. Pubic symphysis _____
   b. Elbow joint _____
   c. Epiphyseal plate _____
   d. Frontal suture _____
   e. Gomphosis _____

4. In general, when mobility of a joint _____, its stability _____.
   a. increases; increases
   b. decreases; decreases
   c. increases; remains unchanged
   d. increases; decreases

5. Mark the following statements as true or false. If a statement is false, correct it to make a true statement.
   a. Fibrous joints are united by collagen fibers.
   b. A syndesmosis is a type of cartilaginous joint.
   c. Cartilaginous joints are synarthroses.
   d. The joint between the two pubic bones and the intervertebral joints are examples of symphyses.

6. Fill in the blanks: The articulating ends of bones of synovial joints are covered in _____. The remaining internal surfaces of the joint are lined by the _____, which produces synovial fluid. The entire joint is encased by the _____, which is composed of dense irregular collagenous connective tissue.

7. What is/are the function(s) of synovial fluid?

8. Which of the following correctly describes the function of a ligament?

   a. It connects a muscle to its attachment point, such as a bone.
   b. It is a fluid-filled structure that minimizes friction during movement.
   c. It connects two bones to one another in a joint.
   d. It surrounds a tendon and protects it in high-stress areas.

9. Bone movement at a joint is described around an invisible line known as a/an:

   a. synarthrosis.   c. sagittal plane.
   b. axis.   d. amphiarthrosis.

10. Match the following terms with the correct definition from the right column.

   _____ Plane joint
   _____ Saddle joint
   _____ Ball-and-socket joint
   _____ Condylar joint
   _____ Pivot joint
   _____ Hinge joint

   a. Uniaxial joint in which the rounded articular surface of one bone fits into a groove of another bone and is held in place by a ligamentous ring
   b. Multiaxial joint in which the spherical articular surface of one bone fits into a cup or socket of another bone
   c. Nonaxial joint in which two flat surfaces glide over one another
   d. Uniaxial joint in which the convex articular surface of one bone fits into a concave articular depression
   e. Biaxial joint in which the oval, convex articulating surface of one bone fits into the shallow depression of another bone
   f. Biaxial joint in which each articular surface has both convex and concave regions

11. Define each of the following movements or movement pairs.

   a. Flexion and extension
   b. Adduction and abduction
   c. Rotation
   d. Circumduction
   e. Dorsiflexion and plantarflexion
   f. Elevation and depression

12. Mark the following statements as true or false. If a statement is false, correct it to make a true statement.

   a. The knee and the elbow are multiaxial joints.
   b. The elbow joint consists of two separate articulations.
   c. The patella is encased within the quadriceps femoris tendon.
   d. The shoulder joint is stabilized by the medial and lateral menisci.
   e. The hip joint is less stable than the shoulder joint, but it allows more motion.

13. Which of the following best describes the function of the anterior cruciate ligament?

   a. It prevents posterior displacement of the tibia on the femur.
   b. It prevents the tibia from moving too far laterally on the femur.
   c. It prevents anterior displacement of the tibia on the femur and hyperextension.
   d. It improves the fit between the femur and the tibia.

14. The structure that stabilizes the shoulder joint is known as the:

   a. rotator cuff.   c. posterior cruciate ligament.
   b. radial collateral ligament. d. fibular collateral ligament.

15. Which factors contribute to the stability of the hip joint?

LEVEL 2 Check Your Understanding

1. Explain how the structure of each of the following joint types follows its function.

   a. Fibrous joint
   b. Cartilaginous joint
   c. Synovial joint

2. The primary action of the biceps brachii muscle of the anterior arm is to flex the forearm at the elbow. However, when this muscle is inflamed, pain is felt in the shoulder. Why do you think this is so?

3. Some individuals have an abnormally small and shallow glenoid labrum. How would this impact the stability of the shoulder joint?

LEVEL 3 Apply Your Knowledge

PART A: Application and Analysis

1. Some health practitioners claim that the cranial bones are moveable and that they are able to move them to treat a variety of conditions. Is this likely to be true in an adult? Why or why not?

2. Lauren has hurt her knee playing soccer. She explains that during the match, someone tackled her, hitting her on the lateral side of her knee. You notice during your examination that she has excessive range of motion in her knee; specifically, you are able to hyperextend her knee and anteriorly displace her tibia on her femur. Her tibia also displaces laterally on her femur. What has likely happened to her knee? Explain.

PART B: Make the Connection

3. When articular cartilage is damaged, often fibrocartilage forms instead of new hyaline cartilage. Does fibrocartilage have the same properties as hyaline cartilage? Is it likely to provide the same type of surface as hyaline cartilage? Explain. (*Connects to Chapter 4*)

9

The Muscular System

Eighty-eight keys on a piano, struck a few at a time in the right combinations, can produce beautifully complex music. Over 600 muscles in the human body, stimulated a few at a time in the right combinations, can produce amazingly coordinated movements. Even apparently simple actions require a highly coordinated series of muscle contractions. Speaking, walking, tying your shoes, and holding and reading this book—all demand the coordinated teamwork of skeletal muscles.

The shape, size, placement, and arrangement of fibers in a skeletal muscle contribute to the ways in which individual muscles do their jobs. Muscles must be structured for specialization yet still be able to work in groups to accomplish their remarkable functions. Some muscles provide the force for a movement, whereas others modify the movement by guiding, stabilizing, or even slowing it as needed. In this chapter, we will explore the relationships between muscles, movement, and other functions of the muscular system.

MODULE **9.1**

Overview of Skeletal Muscles

Learning Outcomes

1. Explain how the name of a muscle can help identify its action, appearance, or location.
2. Summarize the major functions of skeletal muscles.
3. Define the terms agonist, antagonist, synergist, and fixator.
4. For a given movement, differentiate specific muscles that function as agonist, antagonist, synergist, or fixator.
5. Differentiate among the three classes of levers in terms of the relative position of fulcrum, force, and load, as well as the relative power and range of motion.
6. Give examples in the human body of muscles and their associated joints that illustrate each type of lever system.

Skeletal muscles are not made of muscle cells alone. Nerves, blood vessels, and connective tissues are also essential to the job of a skeletal muscle. A brief overview of muscle structure will prove useful prior to our detailed discussion of how muscles do their work.

Computer-generated image: Most movements at joints require teamwork by several muscles, each playing a unique role.

Structure of a Skeletal Muscle

◀◀ **Flashback**

1. Define the terms flexion, extension, abduction, and adduction. (p. 266)

Each individual muscle cell, or fiber, is surrounded by a thin layer of extracellular matrix called **endomysium** (en-doh-MY-see-um) (**Figure 9.1**). Several (between 10 and 100) muscle cells are bundled together into a **fascicle** by the connective tissue of the **perimysium**, and all the fascicles that make up a muscle are, in turn, enclosed in an outer fibrous connective tissue wrapping, the **epimysium**. These interconnected fibrous connective tissues

CORE PRINCIPLE
Structure-Function

taper down and connect to **tendons** or other connective tissues that attach the muscle to a bone or other structure to be moved. In an example of the Structure-Function Core Principle (p. 25), this arrangement works a bit like a sled dog harness: making sure the muscle pulls as a unit even if some muscle cells (like some dogs) are not pulling with the same strength as others.

Additionally, a skeletal muscle contains blood vessels that supply the muscle cells with oxygen and glucose and remove wastes. It also contains many nerves that coordinate muscle contraction. As you will see in the muscle tissue chapter, there is a close relationship between neurons (nerve cells) and skeletal muscle cells (see Chapter 10). Recall from the histology chapter that skeletal muscle is *voluntary*. This means that it is under conscious control, and so much be stimulated by the nervous system in order to contract. For this reason, skeletal muscle is richly supplied with nerves,

CORE PRINCIPLE
Cell-Cell Communication

which is another example of the Cell-Cell Communication Core Principle (p. 27).

Fascicles and Muscle Shapes

The pattern in which fascicles are arranged greatly affects the appearance and function of a skeletal muscle (**Figure 9.2**). Fascicle patterns include parallel, convergent, pennate, circular, spiral, and fusiform.

- A **parallel** muscle has evenly spaced fascicles attaching to a tendon that is about the same width as the muscle (Figure 9.2a). This orientation often produces a straplike muscle, like the sartorius in the thigh.
- A **convergent** muscle is a broad muscle that uniformly tapers to a single tendon (Figure 9.2b). Triangular muscles, such as the pectoralis major in the chest, usually have a convergent fascicle arrangement.
- A **pennate** muscle has fibers and fascicles that attach to the tendon at an angle in such a way that the muscle resembles a feather (Figure 9.2c). (To remember the connection between the terms pennate and feather, recall that writing pens of long ago were originally quill pens made from a goose feather.) The pennate pattern has three basic variations. A **unipennate** muscle has fascicles that feather out at an angle from only one side of its tendon (an example is the flexor pollicis longus of the forearm). The fascicles of a **bipennate** muscle angle out from both sides of its tendon (for instance, the rectus femoris of the thigh). A **multipennate** muscle is usually composed of several areas where fascicles angle away from connective tissue separators, looking

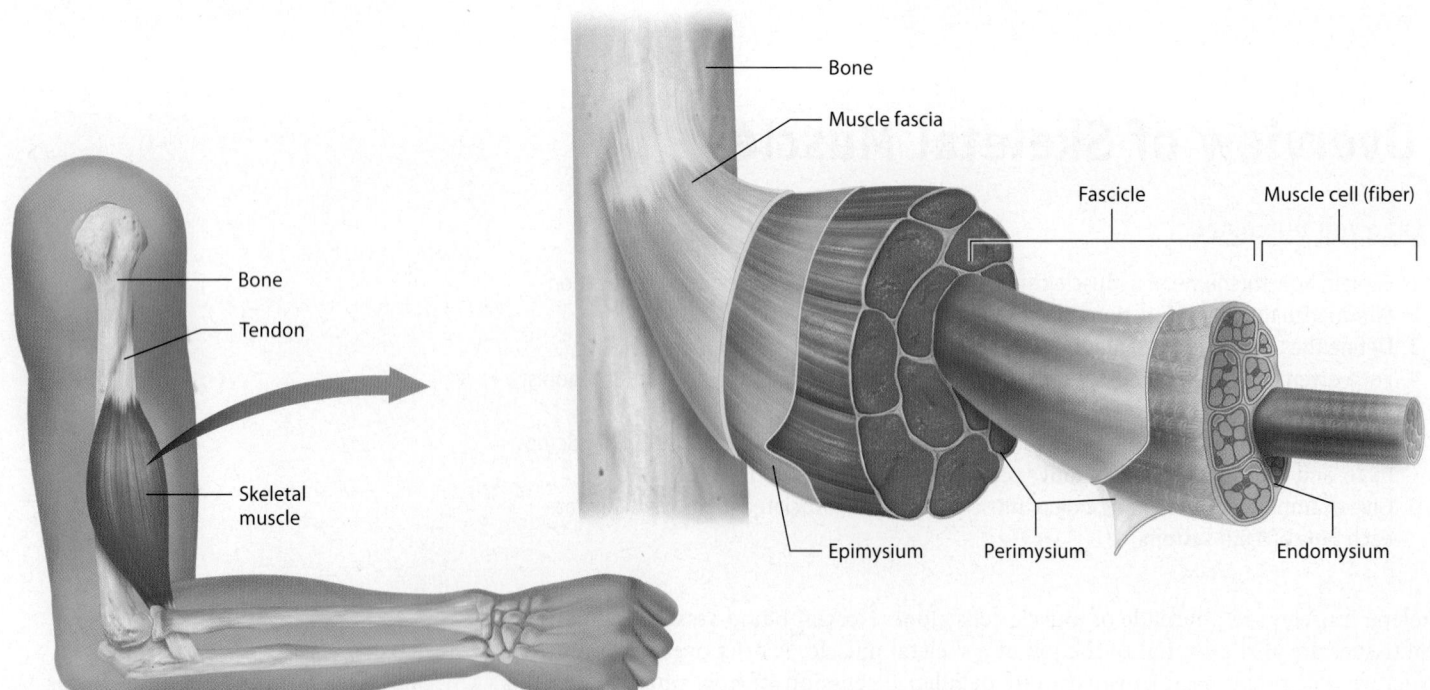

(a) Position of a representative skeletal muscle (brachialis)

(b) Structure of a skeletal muscle

Figure 9.1 Position and structure of a skeletal muscle.

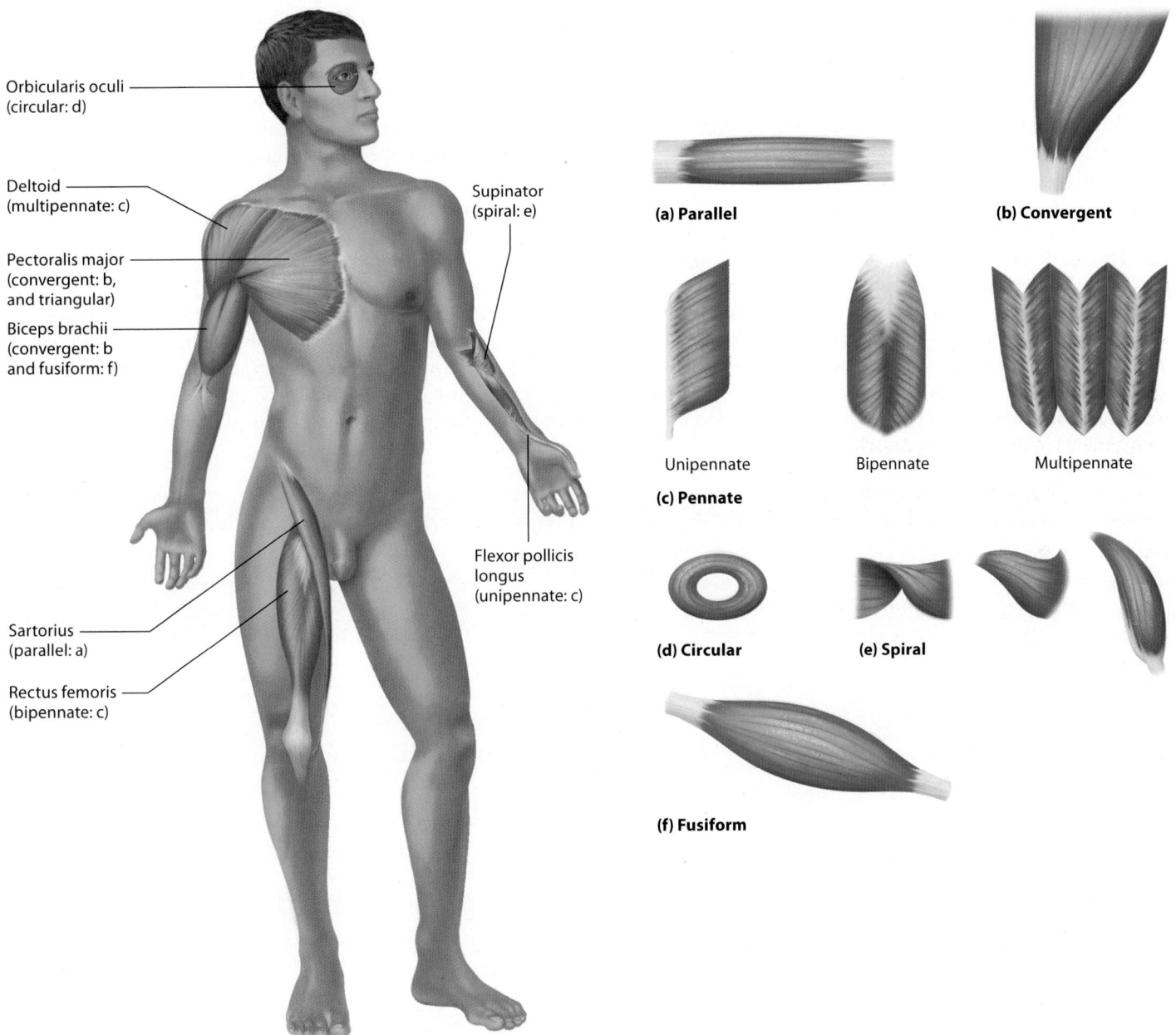

Figure 9.2 Fascicle pattern and muscle shape.

like several feathers joined together (for example, the deltoid of the shoulder).

Less common are muscles that encircle or spiral around a structure or bone:

- A **circular** muscle encircles a structure, such as the opening of the eye, to close it when it contracts (for example, the orbicularis oculi [or-BIC-u-lar'-iss OCK-u-lie], Figure 9.2d). Circular muscles are often referred to as **sphincters.**
- A **spiral** muscle may wrap around a bone or have the twisted appearance of a towel wrung out to dry (for example, the supinator muscle in the forearm, Figure 9.2e).

Fascicle arrangement affects the overall shape and size of a muscle, as well as its function and name. Some terms, such as *triangular* and *straplike,* are straightforward. Others require some

explanation. **Fusiform** muscles tend to be thicker in their **belly,** or middle region, and tapered at their ends (for instance, the biceps brachii [BUY-seps BRAKE-ee-i], Figure 9.2f).

Naming Muscles

Many muscles are named for their shape or appearance, and others for their size or position. Still others are named for structural considerations, such as the number of separate sections or heads with individual tendons providing multiple points of attachment. For size, self-explanatory terms such as major, minor, longus (long), and brevis (short) are commonly used.

Muscle names based on location use many of the same terms you ran across in earlier chapters. Directional words like superior and inferior, or medial and lateral, are used, often slightly modified based on their Greek or Latin roots (such as "lateralis" for lateral). For example, the vastus lateralis (VAST-us lah-teh-RAL-iss) and vastus medialis (meed-ee-AL-iss) muscles of the thigh are located on the lateral and medial sides of the femur, respectively. Also, regional anatomy you read about in earlier chapters will be helpful in understanding muscle names—for instance, pectoral for the chest area and oculi for the eyes.

Some muscles have names that reveal the structures they attach to, which makes remembering these muscles easier. A great example is the sternocleidomastoid (ster-no-CLYDE-o-mass'-toid) muscle. This neck muscle is attached to the sternum (sterno), clavicle (cleido), and mastoid process of the temporal bone.

Other muscles are named for what they do (their actions). The terms *flexion* and *extension* from Chapter 8 apply to naming muscles as **flexors** and **extensors.** Similarly, a **levator** muscle elevates, or raises, a body part. An **adductor** pulls a part of the body toward the midline. Therefore, you may find it helpful to review joint movements before tackling muscle actions.

A number of muscle names have historical origins. The soleus (SO-lee-us) and gastrocnemius (gas-trock-NEE-me-us) that make up the "calf" of the leg come from Latin terms for "sandal" and "stomach" of the leg, respectively. The tendon they share to attach to the calcaneus, the calcaneal (cal-CANE-nee-uhl) (Achilles) tendon, is named for the one vulnerable spot on the legendary Greek warrior's body.

Finally, some muscles and their locations relate to common vocabulary. For example, the flexor digitorum profundus (FLEX-or dih-jih-TOH-rum pro-FUND-us) in the forearm is located at a deeper plane than the flexor digitorum superficialis (soo'-per-fish-ee-AL-iss). *Profundus* means "deep." This term makes sense when you remember that a profound statement has deep meaning. And, when you consider where you wear a baseball *cap,* is it any surprise that *capitis* (CAP-it-iss) refers to "head"? Table 9.1 summarizes these and other helpful terms for understanding muscle names.

TABLE 9.1 COMMON TERMS IN MUSCLE ANATOMY	
Term and Meaning	**Example**
Body Region	
Abdominis—abdominal area	Rectus abdominis
Brachii—arm area	Biceps brachii
Capitis—head area	Semispinalis capitis
Carpi—wrist area	Extensor carpi ulnaris
Cervicis—neck area	Splenius cervicis
Digitorum/digiti—related to fingers/toes	Extensor digitorum
Femoris—femur or thigh	Biceps femoris
Gluteal—buttocks	Gluteus maximus
Hallucis—great toe	Abductor hallucis
Oculi—eye area	Orbicularis oculi
Oris—mouth area	Orbicularis oris
Pectoralis—chest area	Pectoralis minor
Pollicis—thumb	Flexor pollicis brevis
Muscle Action	
Abductor—pulls away from the midline	Abductor pollicis longus
Adductor—pulls toward the midline	Adductor magnus
Depressor—pulls down	Depressor labii inferioris
Erector—holds erect or straight	Erector spinae
Extensor—increases the angle between bones	Extensor digitorum longus
Flexor—decreases the angle between bones	Flexor digitorum longus
Levator—raises a body part	Levator scapulae
Pronator—turns palm posteriorly	Pronator teres
Supinator—turns palm anteriorly	Supinator
Muscle Fiber Orientation	
Oblique—at an angle	External oblique
Orbicular—circular	Orbicularis oculi
Rectus—straight	Rectus femoris
Transversus—across/transverse	Transversus abdominis

Quick Check

☐ 1. Describe how the terms pennate, parallel, and circular relate to muscle fiber orientation.

☐ 2. A muscle has the words "adductor" and "longus" in its name. What two facts do you know about this muscle?

Functions of Skeletal Muscles

◀◀ **Flashback**

1. What are some of the functions of bone markings? (p. 212)
2. What are tendons and ligaments? (p. 263)

The ability to contract allows muscles to produce movement and to perform many other jobs related to the workings of organs and structures. Smooth muscle does much of this other work, but skeletal muscle also plays a role.

Movement and Other Functions

The movements we think of as the main function of skeletal muscles often occur at joints and cause bones to move. Those movements are indeed the main focus of this chapter, but we can point to many notable exceptions. Here are just a few:

● Contraction of the diaphragm muscle produces volume and pressure changes in the thoracic cavity that help us breathe. Although other breathing-related muscles move bones, the diaphragm changes the dimensions of the thoracic cavity.

TABLE 9.1 *(continued)*

Term and Meaning	Example
Muscle Location	
Anterior—toward the front	Tibialis anterior
External—toward the outside	External intercostals
Infra—below	Infraspinatus
Intercostal—between the ribs	Internal intercostals
Internal—toward the inside	Internal oblique
Posterior—toward the back	Tibialis posterior
Profundus—deep	Flexor digitorum profundus
Superficialis—nearer the surface	Flexor digitorum superficialis
Supra—above	Supraspinatus
Muscle Heads of Origin	
Biceps—two heads	Biceps brachii
Quadriceps—four heads	Quadriceps femoris
Triceps—three heads	Triceps brachii
Muscle Size	
Brevis—short	Fibularis brevis
Longus—long	Adductor longus
Vastus—wide/large	Vastus lateralis
Muscle Shape	
Deltoid—triangular	Deltoid
Lata/Lati—wide or broad	Latissimus dorsi
Maximus—largest	Gluteus maximus
Minimus/Minimi—smallest	Gluteus minimus
Minor—small	Pectoralis minor
Quadratus—shaped like a rectangle	Pronator quadratus
Rhomboid—shaped like a rhombus (slanted parallelogram)	Rhomboid major
Serratus—serrated or jagged	Serratus anterior
Trapezius—shaped like a trapezoid	Trapezius

- Muscles attached to the skin of the face enable nonverbal communication by facial expression.
- Skeletal muscles in the throat initiate swallowing.
- The smallest muscle of the body, the *stapedius* (stuh-PEED-ee-iss) of the middle ear, modifies the level of sound we perceive as we hear. Interestingly, that tiny stapedius (about 1 mm long) dampens excessive vibration and prevents us from being overwhelmed by the sound of our own voice as we speak. It is also one reason why our voice sounds different to us when we speak, as opposed to when it is played back from a recording device.
- Although most circular muscles (sphincters) surrounding tubular structures are smooth muscle, several are made of skeletal muscle. Earlier we mentioned the orbicularis oculi in the face. Other skeletal muscle sphincters in the body give us voluntary control over defecation and urination, an example of the Structure-Function Core Principle (p. 25).

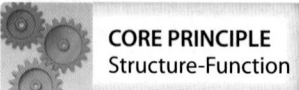
CORE PRINCIPLE
Structure-Function

- Humans owe most of their ability to generate heat to skeletal muscles. Muscle tissue produces heat as a waste product, just as an automobile engine does when idling. Shivering occurs when the body tries to generate extra heat to compensate for dropping body temperature with extra muscle stimulation.

Sometimes muscle fascicles contract without relaxing, causing painful muscle knots (see *A&P in the Real World: Muscle Knots*).

- Fixator (supraspinatus)
- Synergist (biceps brachii)
- Agonist (brachialis)
- Antagonist (triceps brachii)

Figure 9.3 Functional groups of muscles.

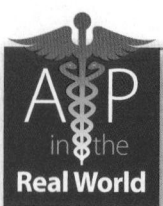
A&P in the Real World

Muscle Knots

Muscle knots, technically known as myofascial trigger points, can occur when groups of muscle fibers, or fascicles, get "stuck" in a painful contraction that will not relax.

The entire muscle does not necessarily go into sustained contraction, so the cause of the pain is not always obvious. Palpating a muscle until the painful area is discovered is often necessary to identify the trigger point. And even though "just" a group of muscle fibers is involved, as opposed to the entire muscle, the pain and stress on joints and connective tissues can be severe.

Causes of muscle knots are not completely understood. They happen in response to diverse activities, and in athletes and nonathletes alike. Repetitive exercises that stress muscles, such as running, can cause microscopic tears and local damage (and subsequent repair cycles) in muscle fascicles that set up trigger points. At the other end of the spectrum, sudden muscle use by people with poor posture or sedentary lifestyles can also trigger muscle knots. Diet and ion imbalance have been suggested as contributing factors as well.

As an interesting historical note, former President John F. Kennedy was a famous sufferer from knots in his back muscles. His personal physician, Dr. Janet Travell, treated him successfully and became a pioneer in the research and treatment of myofascial pain.

As with muscle spasms, the best treatment is to relax the muscle fibers in the trigger points. Massage therapy, anti-inflammatory medications, and muscle relaxants are all helpful. Regular, gentle stretching activities such as yoga may be useful as well. And, of course, especially as we age, moderate exercise and a healthy diet provide an ounce of prevention for many ailments, including muscle knots.

Functional Groups of Muscles

In the chapter opener, we mentioned that teamwork by muscles is needed for most movements, or **actions** as they are called in muscle physiology. A car needs different components like motors, brakes, and steering mechanisms to move in a controlled manner. So it isn't surprising that most movements at joints involve several muscles, each with a particular job to do. These muscles are the agonists, antagonists, synergists, and fixators.

The **agonist** (**Figure 9.3**) provides most of the force required for a given movement—thus, its alternative name, *prime mover*. An agonist is typically easy to identify as one of the larger muscles spanning the joint to be moved. Although it may seem as though this muscle could do all the work of the particular movement, many movements would be jerky and imprecise without the help of other team members (see below). For instance, when you lift a glass of water toward your lips, the agonist is the brachialis (brake-ee-AL-iss).

The **antagonist** usually lies on the opposite side of a joint from its agonist partner, and tends to oppose and slow the action. Although this may seem wasteful (why not just stimulate the agonist less?), many of our slower, more graceful movements result from highly controlled action by antagonists. Slow ballet moves, yoga, and gymnastics all involve carefully balancing the force of agonists and antagonists.

Synergists, as their name implies, work together with the agonist. But synergists are more than just "secondary movers." Otherwise, you might ask, why not just make a bigger agonist? Especially at more freely moveable joints, it is desirable to have additional muscles supplement the force of the agonist and help guide the movement to minimize unwanted deviations. In Figure 9.3, the biceps brachii acts as the synergist to help guide the glass to the mouth. Synergists sometimes make a movement more efficient by stabilizing a joint, overlapping with the role of the next group.

Fixators hold a bone in place, an anchoring function that makes movement more efficient and reduces the risk of injury. The important role of a fixator can be easily visualized if you imagine trying to push a door open while wearing roller skates. If you're not careful, you

are just as likely to push yourself backward as push the door forward. But if you lock the wheels of the skates in place, all of your force goes forward into opening the door. Several muscles attached to the scapula, such as the supraspinatus, provide this stabilization function for movement at the shoulder joint. Think about moving a glass of water in your hand. Without the ability to hold the shoulder joint steady, the water in that glass might end up splashing your cheek or chin.

Muscle Origin and Insertion

A skeletal muscle is usually attached to a different bone at each end, often via a tendon. The **origin** is the (usually) less moveable, anchoring point on a bone, and the **insertion** end of the muscle is attached to the bone or structure that will be moved when the muscle contracts. **Figure 9.4** illustrates these attachments for the biceps brachii before contraction (Figure 9.4a) and after contraction (Figure 9.4b). At one end, this muscle has two heads ("biceps"), which attach to different parts of the scapula. The other end attaches to the radial tuberosity of the radius (a bone in the forearm). The scapula is relatively stationary when the biceps brachii contracts, and therefore is the origin. The radius is the bone that moves and therefore is the insertion.

This rule does have a few exceptions. For muscles that act on more than one joint, the origin and insertion may seem to switch jobs, depending on which fixators are in play and what the

(a) Biceps brachii before contraction

(b) Biceps brachii after contraction

Figure 9.4 Muscle origin and insertion.

ultimate action is. For example, the iliopsoas (ill-ee-oh-SO-az) is a muscle group that attaches to the pelvis and lower vertebral column on one end and to the femur (thigh bone) at the other end. The femur is considered the insertion of the iliopsoas, since it is the more commonly moved body part. However, when we flex the trunk (torso) at the hip or bend forward, clearly the femur is fixed and the trunk moves. Yet, those bones of the trunk (pelvis and vertebrae) are still considered the origin of the muscle.

Lever Systems

If you have ever played on a seesaw, you are already familiar with levers. In any lever system, the *load* (or resistance) is the weight or object you are trying to move. The *force* is the effort applied to the lever, and the **fulcrum** is the pivot, or hinge point.

In moving the body, bones act as levers. The load is the weight of the body part being moved. The force is the effort generated by muscle contractions, and the fulcrum is the joint at which the movement occurs.

A seesaw is a **first-class lever** in which the fulcrum is *between* the point of applied force and the load to be moved. The direction in which you apply force is opposite to the movement produced (you push down on the seesaw, and the other end goes up). In the case of the triceps brachii (TRY-ceps BRAY-kee-eye) moving the forearm, the elbow joint is the fulcrum between the force being applied and the load being moved. Dribbling a basketball is an example of a movement using a first-class lever (**Figure 9.5a**).

In both second- and third-class levers, the fulcrum is near one end, and you apply force in the same direction as the movement produced. What makes them different from each other is the location of the fulcrum relative to where you apply the force. A **second-class**

lever is one in which the fulcrum is located farther from the applied force, with the load moved in between (**Figure 9.5b**). Rising up on your toes is an example of a movement using a second-class lever.

A **third-class lever** applies force much closer to the fulcrum, with the load farther away from the fulcrum. Pulling up on a fishing rod is a movement using a third-class lever (**Figure 9.5c**). In this example, the biceps brachii applies the force in front of the elbow—the fulcrum. The force occurs on the same side of the forearm where the load (the fishing rod) is being moved.

The concepts of mechanical advantage and mechanical disadvantage are important in understanding how levers work and can be applied to human bones and muscles that serve as lever systems. In a lever system that works at a mechanical advantage, the fulcrum is located farther from the applied force. This allows a relatively small force to move a large load over a short distance. Standing on tiptoe is an example of a second-class lever system that demonstrates mechanical advantage. Here, the ball of the foot represents the fulcrum, contraction of the calf muscles is the applied force, and the weight of the body is the load moved.

In a lever system that works at a mechanical disadvantage, the fulcrum is located nearer to the applied force and farther from the load moved. For instance, third-class levers work with a mechanical disadvantage. Most levers in the human body are third-class levers. In our example of pulling up on a fishing rod, the elbow joint is the fulcrum. The biceps brachii muscle provides the applied force just distal to the joint, and the load moved (the fishing rod) is located farther away. Although third-class levers have a disadvantage in terms of force, they exchange it for advantages in speed of movement and distance through which the load is moved (see Concept Boost: Understanding Lever Systems and Mechanical Advantage).

ConceptBOOST »»»

Understanding Lever Systems and Mechanical Advantage

Your childhood playground taught you valuable lessons about levers. Good playground seesaws often have adjustable fulcrums—three or four notches so that the position of the fulcrum can be moved off center to compensate for weight differences in the seesaw partners.

The lighter person puts the fulcrum position farther away from her to overcome the heavier weight at her partner's end of the seesaw. Her longer end of the seesaw moves in a greater arc distance but, as a payoff, moves the heavier partner (thus conferring a mechanical advantage). Conversely, the heavier partner moves a shorter distance on his end and produces greater (and faster!) movement at the lighter partner's end (though at a mechanical disadvantage).

This helps explain why so many levers with off-center fulcrums are found in the body. Fulcrum placement in the body's lever systems

- Smaller arc
- Greater movement (faster)
- Mechanical disadvantage

- Greater arc
- Moves the heavier weight
- Mechanical advantage

Off-center fulcrum

allows us to use muscle contraction for an astounding variety of movements. And, as you can see from the seesaw example, placing the fulcrum closer to or farther from the applied force has a lot to do with whether the lever system is built for strength or for speed. What other examples of lever systems in the body can you think of? ■

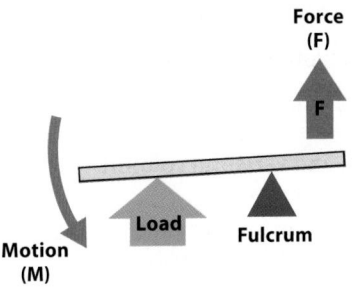

- Fulcrum is located between applied force and load to be moved.
- Force applied and load moved are in opposite directions.
- Lever works at a mechanical advantage or disadvantage, depending on location of fulcrum.

(a) First-class lever: dribbling a basketball

- Fulcrum is located farther from applied force.
- Load moved is between fulcrum and force applied, and moves in same direction as the force.
- Lever works at a mechanical advantage to move large loads a short distance with little effort.

(b) Second-class lever: rising on tiptoe

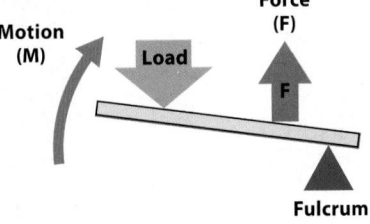

- Fulcrum is located closer to applied force.
- Force applied is between fulcrum and load moved, and moves in same direction as the force.
- Lever works at a mechanical disadvantage to move small loads a greater distance with greater speed.

(c) Third-class lever: pulling up on a fishing rod

Figure 9.5 Lever systems.

Studying Muscles

In the sections that follow, we examine the muscles of the human body, categorized as:

- Muscles of the head, neck, and vertebral column
- Muscles of the trunk and pelvic floor
- Muscles of the pectoral girdle and upper limb
- Muscles of the hip and lower limb

You may find it useful to review the anterior and posterior superficial muscles of the human body in **Figures 9.6** and **9.7**. That way, some of the muscles in each section will be familiar to you.

As you go through the remainder of this chapter, note the strategy for presenting information on the anatomy and function of these muscles: The narrative description "talks you through" the muscles of a section and cites examples as appropriate. The figures show the location of the major muscles. The tables contain detailed information on each muscle. Therefore, the

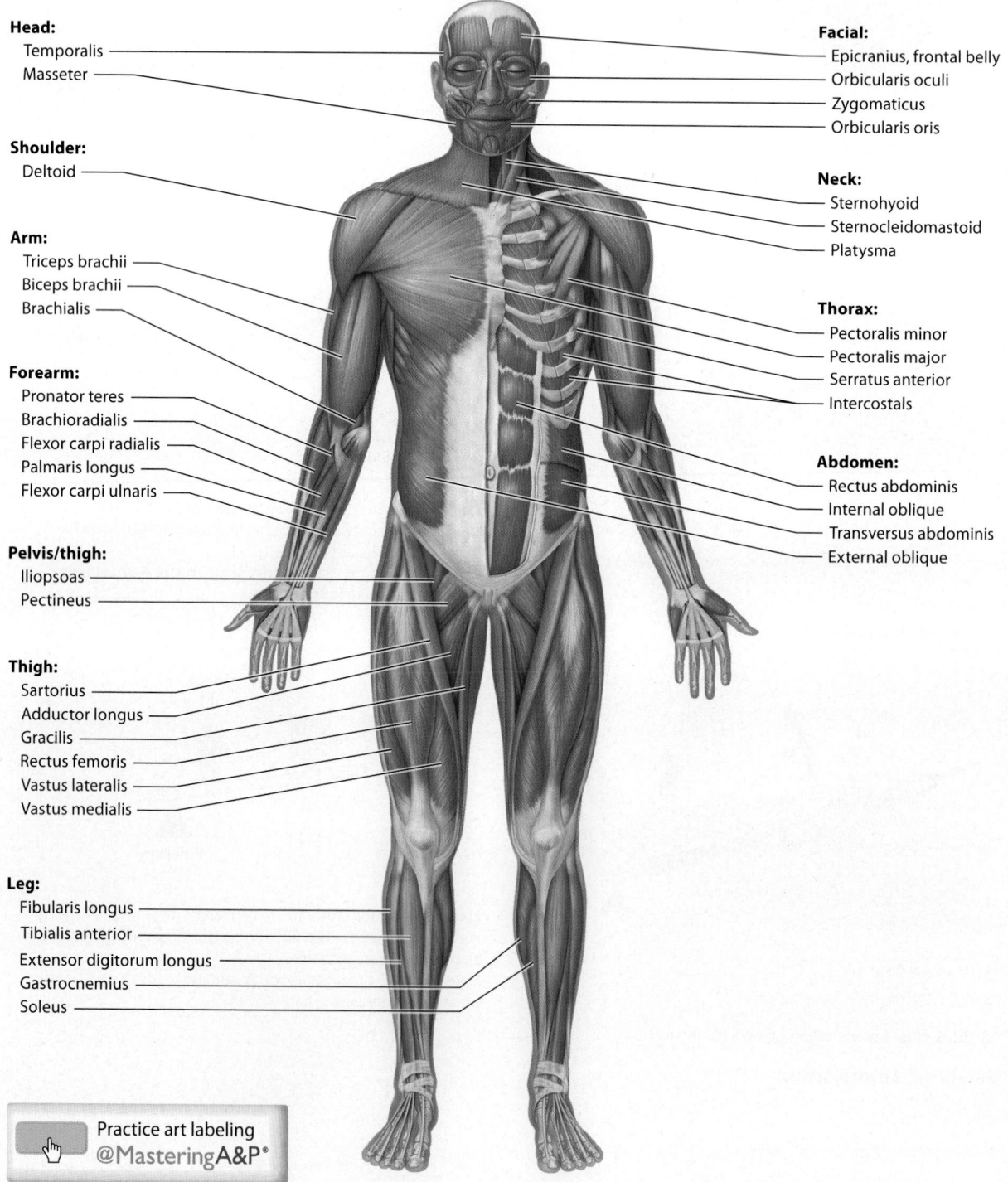

Head:
Temporalis
Masseter

Shoulder:
Deltoid

Arm:
Triceps brachii
Biceps brachii
Brachialis

Forearm:
Pronator teres
Brachioradialis
Flexor carpi radialis
Palmaris longus
Flexor carpi ulnaris

Pelvis/thigh:
Iliopsoas
Pectineus

Thigh:
Sartorius
Adductor longus
Gracilis
Rectus femoris
Vastus lateralis
Vastus medialis

Leg:
Fibularis longus
Tibialis anterior
Extensor digitorum longus
Gastrocnemius
Soleus

Facial:
Epicranius, frontal belly
Orbicularis oculi
Zygomaticus
Orbicularis oris

Neck:
Sternohyoid
Sternocleidomastoid
Platysma

Thorax:
Pectoralis minor
Pectoralis major
Serratus anterior
Intercostals

Abdomen:
Rectus abdominis
Internal oblique
Transversus abdominis
External oblique

Practice art labeling
@MasteringA&P®

Figure 9.6 Superficial muscles: anterior view. The left abdominal and thoracic surfaces have been partially dissected to show somewhat deeper muscles.

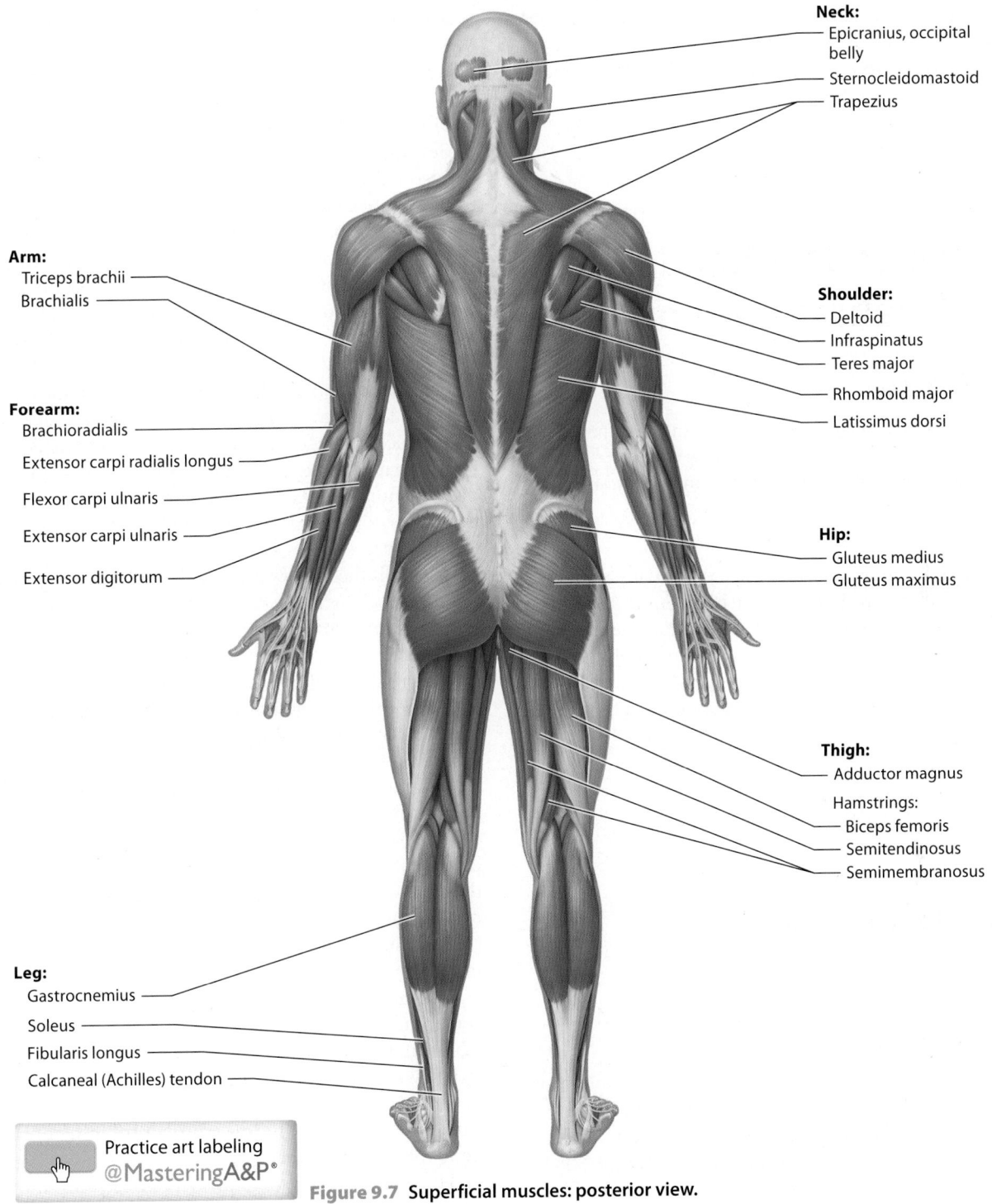

Neck:
Epicranius, occipital belly
Sternocleidomastoid
Trapezius

Arm:
Triceps brachii
Brachialis

Forearm:
Brachioradialis
Extensor carpi radialis longus
Flexor carpi ulnaris
Extensor carpi ulnaris
Extensor digitorum

Shoulder:
Deltoid
Infraspinatus
Teres major
Rhomboid major
Latissimus dorsi

Hip:
Gluteus medius
Gluteus maximus

Thigh:
Adductor magnus
Hamstrings:
Biceps femoris
Semitendinosus
Semimembranosus

Leg:
Gastrocnemius
Soleus
Fibularis longus
Calcaneal (Achilles) tendon

Practice art labeling
@MasteringA&P®

Figure 9.7 Superficial muscles: posterior view.

best approach might be to read the narrative description first, view the figures second, and then study the tables to become acquainted with the muscles assigned by your instructor.

Quick Check

☐ 3. Define the roles of agonists, antagonists, synergists, and fixators in muscle actions.

☐ 4. Describe the design of a third-class lever, and explain why third-class levers are used in many muscle actions.

Apply What You Learned

☐ 1. Predict the name that might be assigned to a muscle that attaches to the sternum and the hyoid bone.

☐ 2. The temporalis muscle is involved in chewing. It is attached to the temporal bone and the mandible. Which bone serves as the origin, and which is the insertion? Explain your answer.

☐ 3. You are attempting to pry a large rock out of the soil. Your pry bar has its fulcrum (another rock underneath it) close to the large rock you want to move. Your arms and torso move the pry bar down 3 feet, and the rock moves up 3 inches and rolls free. What kind of lever system did you use to dislodge the rock, and was your work done at a mechanical advantage or mechanical disadvantage? Explain your answer.

See answers in Appendix A.

MODULE **9.2**

Muscles of the Head, Neck, and Vertebral Column

Learning Outcomes

1. Name, describe, and identify the muscles of the head, neck, and vertebral column.
2. Identify the origin, insertion, and action of these muscles and demonstrate their actions.

Muscles of the head, neck, and vertebral column are among the most diverse and interesting in the human body. Some perform the "traditional" task of moving body parts to do work, whereas others play the more subtle role of enabling us to communicate. Some function to maintain posture, and others help us chew and swallow our food. We now explore these diverse groups of skeletal muscles:

- Muscles of facial expression
- Extrinsic eye muscles
- Muscles of the head and neck
- Muscles of the vertebral column

Muscles of Facial Expression

◀◀ **Flashback**

1. Which part of the skin is connective tissue? (p. 161)
2. Which cranial bones are facial bones? (p. 214)

The muscles involved in producing facial expressions differ from most of the others covered in this chapter because they do not cause movement at a joint. They commonly insert into skin or connective tissue rather than on bone, producing subtle yet important shape changes to the skin and structures of the face.

The **frontalis** (front-TAL-iss) and **occipitalis** (ok'-sip-i-TAL-iss) muscles are parts or bellies of a larger muscle, the **epicranius** (eh'-pih-KRAY-nee-us), which covers most of the skull (**Figure 9.8**). The greater portion of the epicranius is not muscle tissue, but a sheet of connective tissue called an **aponeurosis** (ap'-uh-noo-ROW-sis). The aponeurosis that separates the two muscular parts of the epicranius is the **epicranial aponeurosis.** The most important actions of this muscle are to raise the eyebrows and skin of the forehead into horizontal wrinkles, as we do when looking surprised.

Two important muscles move the skin and tissue around the eyes: the orbicularis oculi and the corrugator supercilii. The **orbicularis oculi** is a circular muscle surrounding the orbit (eye socket) and within the eyelids of each eye. Like pulling a drawstring purse closed, the contraction of this muscle closes the eye. Its antagonist, the **levator palpebrae superioris** (lee-VAY-tohr pal-PEE-bray sue-peer-ee-OR-iss), is located at the back of the orbit and attaches to the connective tissue and skin of the upper eyelid to open it. Contraction of the part of the orbicularis oculi located within the eyelids produces blinking, which moistens the eye with tears. More powerful contraction of the part that encircles the orbit produces winking. The **corrugator supercilii** (core-u-GAY-tor soo'-per-SIL-ee-eye) pulls the eyebrows down and medially to produce vertical wrinkles, as in frowning. Squinting is a combined action that involves both the orbicularis oculi and the corrugator supercilii.

Several muscles interact with the mouth and surrounding area to produce facial expressions. The **zygomaticus** (zy-go-MAT-i-cuss) **major** and **zygomaticus minor** assist with smiling, since their terminal points of attachment enter the upper lip and raise the corners of the mouth. The **levator labii superioris** (LEE-vay-tore LAY-bee-eye sue-peer-ee-OR-iss), located above the upper lip as well, pulls the lip upward in folds, to produce a grimace or sneer.

Also attaching to the corner of the mouth are the **risorius** (rise-OR-ee-us) muscles. They course nearly straight to the corner of the mouth and so are more involved in "widening the smile" and closed-mouth grinning. Of note, the risorius is a variable muscle, meaning it develops differently in many individuals. Part of the reason why we have distinctive smiles is that the size and placement of the risorius differ slightly as it forms during gestation and childhood.

Although it is important for facial expressions, the **orbicularis oris** (or-BIC-u-lar'-iss OR-iss) is much more than that. The orbicularis oris controls the fine movements of the lips that are critical for eating, drinking, whistling, and proper speech. Originally thought to be a simple circular muscle, the orbicularis oris, upon closer examination, appeared more like four curved quadrants (two per lip). This makes sense, considering that speech requires movements of the lips beyond simply opening and closing them.

Several muscles help us convey negative emotions. The **depressor anguli oris** (dee-PRESS-er ANG-u-lee or'-iss), **depressor labii inferioris** (in-fear-ee-OR-iss), and **mentalis** (men-TAL-iss) all work in slightly different ways to contribute to the "down in the mouth" look. The latter two muscles also protract (protrude or jut forward) the lower lip, contributing to expressions of doubt or disappointment.

The final two muscles in this section have less impact on facial expression. The **buccinator** (BUCK-sin-nate-tor) is the muscle found under most of the skin of the cheek. Compressing the cheeks to manipulate food, whistle, or blow a wind instrument like a trumpet involves tensing and contracting the buccinator. The **platysma** (pla-TIZ-mah) is a broad, flat, sheetlike muscle found in the neck region. It helps produce that open-mouthed, "jaw-dropping" look of horror and tightens the skin of the neck.

Table 9.2 summarizes the details of these muscles, including their origin, insertion, action, the nerves that innervate them (stimulate their activity), and a concept figure to demonstrate their actions.

Epicranial aponeurosis

Frontalis

Corrugator supercilii

Orbicularis oculi

Levator labii superioris

Zygomaticus minor

Zygomaticus major

Levator anguli oris

Buccinator

Risorius

Orbicularis oris

Depressor labii inferioris

Depressor anguli oris

Mentalis

Platysma

(a) Anterior view

Epicranius:
Epicranial aponeurosis
Occipitalis
Frontalis

Corrugator supercilii

Orbicularis oculi

Levator labii superioris

Zygomaticus minor

Zygomaticus major

Buccinator

Risorius

Orbicularis oris

Depressor labii inferioris

Mentalis

Depressor anguli oris

Platysma

(b) Lateral view

Figure 9.8 Muscles of facial expression.

Practice art labeling
@MasteringA&P®

TABLE 9.2 MUSCLES OF FACIAL EXPRESSION

Muscle	Action(s)	Origin/Insertion/Nerve(s)	Concept Figures
Frontalis	**Raises** eyebrows; wrinkles skin of forehead horizontally	O: Epicranial aponeurosis I: Skin of eyebrows N: Facial nerve	
Occipitalis	Pulls scalp **posteriorly**	O: Occipital bone I: Epicranial aponeurosis N: Facial nerve	
Corrugator supercilii	Pulls eyebrows **inferiorly** and **medially** (as in squinting)	O: Medial supraorbital margin of frontal bone I: Skin of medial eyebrows N: Facial nerve	
Orbicularis oculi	**Closes** eye; pulls skin around the eyes (as in blinking and winking)	O: Orbital portions of the frontal bone and maxilla I: Skin of the orbital area and eyelids N: Facial nerve	
Levator labii superioris	**Elevates** the upper lip; everts and furrows upper lip (as in sneering)	O: Zygomatic and upper maxilla near orbit I: Skin and muscle of the upper lip N: Facial nerve	
Zygomaticus minor	**Raises** lateral portion of the upper lip to expose upper teeth (as in smiling)	O: Zygomatic I: Skin and muscle of the lateral upper lip N: Facial nerve	
Zygomaticus major	Pulls the angle of the mouth **superiorly** and **laterally** (as in smiling or laughing)	O: Zygomatic I: Lateral muscle fibers of corner/angle of mouth N: Facial nerve	
Risorius	Pulls the angle of the mouth **laterally** to make a closed-mouth smile	O: Connective tissue anterior to the ear I: Modiolus* N: Facial nerve	
Orbicularis oris	**Closes** and **protrudes** lips (as in puckering the lips for a kiss)	O: Maxilla and mandible I: Skin and connective tissue of the lips N: Facial nerve	
Depressor anguli oris	Draws corners of the mouth **inferiorly** (unhappy face)	O: Lower body of mandible I: Modiolus* N: Facial nerve	
Depressor labii inferioris	**Protrudes** lower lip (sad or pouting expressions)	O: Medial mandible near mental foramen I: Skin and connective tissue of lower lip N: Facial nerve	
Mentalis	**Protrudes** the lower lip and chin for drinking and "doubtful" expression	O: Anterior mandible I: Skin of the chin near lower lip N: Facial nerve	
Buccinator	Helps manipulate food during chewing and **expels** air through pursed lips (as in blowing a trumpet)	O: Molar regions of maxilla and mandible I: Orbicularis oris and connective tissue of cheek/lips N: Facial nerve	
Platysma	Lowers lower lip and opens mouth by **depressing** the mandible; tenses skin of the neck	O: Connective tissue of deltoid and pectoralis major I: Mandible; skin and connective tissue below mouth N: Facial nerve	

Frontalis — Occipitalis — Corrugator supercilii — Orbicularis oculi — Levator labii superioris — Zygomaticus minor — Zygomaticus major — Risorius — Orbicularis oris — Depressor anguli oris — Depressor labii inferioris — Mentalis — Buccinator — Platysma

Note: Colors of actions and/or directions of action in Action(s) column match colors of directional arrow(s) in Concept Figures.

*Mix of muscle and connective tissue at the corners of the mouth.

☐ 1. Name and describe the muscle involved in opening the eye and the muscle involved in closing the eye.

☐ 2. Name and describe three muscles involved in producing happy facial expressions.

Extrinsic Eye Muscles

◀◀ **Flashback**

1. What do the directional terms superior, inferior, medial, and lateral mean? (p. 12)

2. What do the terms oblique and trochlea mean? (pp. 243, 286)

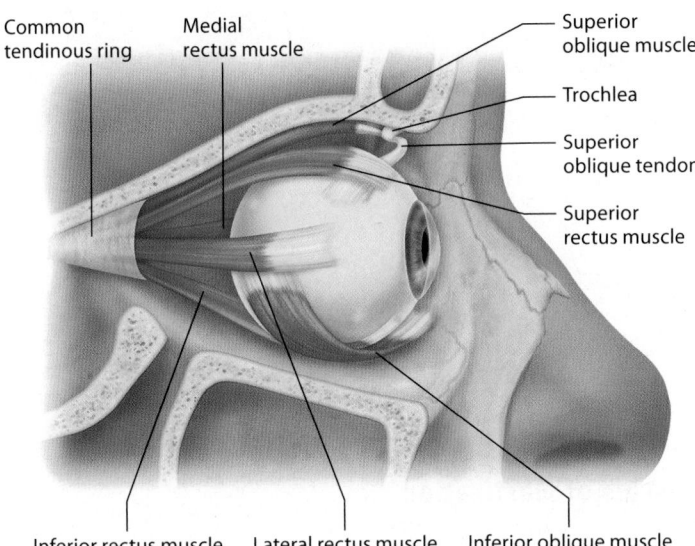

Common tendinous ring — Medial rectus muscle — Superior oblique muscle — Trochlea — Superior oblique tendon — Superior rectus muscle

Inferior rectus muscle — Lateral rectus muscle — Inferior oblique muscle

(a) Lateral view of the right eye

Superior rectus muscle — Superior oblique tendon — Trochlea — Superior oblique muscle — Medial rectus muscle

Lateral rectus muscle — Inferior oblique muscle — Inferior rectus muscle

(b) Anterior view of the right eye

Figure 9.9 **Extrinsic eye muscles.**

For such small, lightweight structures, our eyes are controlled by a relatively large number of muscles. Six skeletal muscles move each eye. Called **extrinsic eye muscles** because they attach to the outer surface of the eye, they work together to quickly and accurately point our eyes toward the object we need to see. This makes sense when we consider the survival value of being able to rapidly focus on dangerous objects we must avoid and useful ones we must obtain.

Four of our six extrinsic eye muscles are *rectus* muscles, named for their straight-line placement from the back of the orbit (eye socket) to the portion of the eyeball on which they insert (**Figure 9.9**). As a group, these muscles are arranged in a conelike shape around the eyeball. At the narrow end of this cone, they attach in common to a circular ring of dense connective tissue called the common tendinous ring at the posterior orbit. From here the **superior rectus** (RECK-tuss), **inferior rectus, medial rectus,** and **lateral rectus** diverge to insert on the corresponding surface of the eyeball for which they are named.

These muscles are relatively easy to add to your vocabulary because their names not only tell their placement but also imply their principal actions. For example, the medial rectus moves the eye medially, and the lateral rectus moves it laterally. However, the actions of the superior and inferior recti are more complex: they can produce more than one action. The superior rectus moves the eye *superiorly,* but it also turns it *medially.* That is why, when your doctor tests the movements of your extrinsic eye muscles, she asks you to move the eye in different positions so she can assess the action of each muscle.

The two other extrinsic eye muscles, the superior and inferior oblique muscles, are different in their placement and action. The term *oblique* means that the tendons of these muscles attach to the eyeball at an angle. The **superior oblique** attaches near the common tendinous ring at the back of the orbit. As the muscle courses forward in the orbit, its tendon hooks around a fibrocartilaginous structure called the trochlea (TROH-klee-ah; "pulley") near the front of the orbit and is redirected at an angle over the top of the eyeball. So, unlike the superior rectus, which pulls the eye upward, the trochlea causes the superior oblique to pull the eye downward and also somewhat laterally (since it attaches at an angle).

Whereas the other five extrinsic eye muscles are anchored to the back of the orbit, the **inferior oblique** is anchored to the *anterior,* or front, portion of the orbit. So, by pulling the bottom of the eyeball toward the front and medial portion where it originates, the inferior oblique forces the eye up and laterally.

To help you visualize the way your inferior oblique muscle works, you can think of holding a bowling ball just before delivering it. The holes for your fingers are facing down as you step forward and are about to release the ball. As you release the ball, the holes rotate toward you and up—the same direction as the front of the eye.

TABLE 9.3 EXTRINSIC EYE AND ORBIT MUSCLES

Muscle(s)	Action(s)	Origin/Insertion/Nerve(s)	Concept Figures
Superior rectus	Elevates (rotates upward) front of eye, as in looking up, and turns eye medially	O: Common tendinous ring attached to posterior orbit I: Connective tissue (sclera) of superior surface of eye N: Oculomotor nerve	
Inferior rectus	Depresses (rotates downward) front of eye, as in looking down, and turns eye medially	O: Common tendinous ring attached to posterior orbit I: Connective tissue (sclera) of inferior surface of eye N: Oculomotor nerve	
Medial rectus	Pulls front of eye medially, as when left eye looks right toward the nose	O: Common tendinous ring attached to posterior orbit I: Connective tissue (sclera) of medial surface of eye N: Oculomotor nerve	
Lateral rectus	Pulls front of eye laterally, as when left eye looks left toward the side	O: Common tendinous ring attached to posterior orbit I: Connective tissue (sclera) of lateral surface of eye N: Abducens nerve	
Superior oblique	Pulls (and helps rotate) front of eye downward and laterally, as when left eye rolls to left and down	O: The posterior orbit I: Sclera of superior/lateral surface of eye after passing through the trochlea N: Trochlear nerve	
Inferior oblique	Pulls (and helps rotate) front of eye upward and laterally, as when left eye rolls to left and up	O: Maxilla just inside the anterior portion of the orbit I: Connective tissue (sclera) of inferior/lateral surface of eye N: Oculomotor nerve	
Levator palpebrae superioris	Elevates (opens) upper eyelid	O: Posterior orbit I: Connective tissue and skin of upper eyelid N: Oculomotor nerve	

See **Table 9.3** for more information about the extrinsic eye muscles.

Quick Check

☐ 3. Name and describe four eye muscles whose actions are basically the same as their name and placement.
☐ 4. Name and describe two eye muscles whose actions are largely the opposite of their name and placement.

Muscles of the Head and Neck

◀◀ Flashback

1. Where are the pterygoid processes of the sphenoid bone located? (p. 218)
2. What is the hyoid bone? (p. 229)

Some of the muscles in this section may be familiar, at least in terms of function if not always by name. You have been moving your head, chewing, and maintaining your posture as long as you can remember. First, let's examine the muscles involved in chewing, also known as mastication.

Muscles of Mastication

Have you heard that in alligator wrestling, you can hold those powerful jaws closed if you just get a good grip? It's true, at least with a small gator. They have far more muscles associated with clamping their jaws shut than opening them. Similarly, most of the powerful muscles attached to the human mandible close our jaws—or *elevate the mandible,* as described in anatomy and physiology terms—rather than open them.

If you watch someone chew food, or *masticate,* notice the bulging and relaxing of the muscle above and in front of the ear. That movement is caused by the two major muscles of jaw elevation: the masseter and temporalis. The **masseter** (MASS-ah-terr) is a thick, bandlike muscle attached to the zygomatic process and the angle/ramus of the mandible and covering much of the ramus of the mandible (**Figure 9.10a**). The **temporalis** (tem-poor-AL-iss) is a fan-shaped, convergent muscle that spreads over the flat portion of the temporal bone and attaches to the coronoid process of the mandible. Together, these two muscles provide much of the force needed for elevating the mandible to masticate.

Two additional sets of mastication muscles attach to the mandible (**Figure 9.10b**). The **medial pterygoid** (TARE-ee-goid) is anchored to the back of the mouth and side of the throat by attachment to the pterygoid process of the sphenoid and parts of the palatine and maxilla bones. Its fibers are directed posteriorly and laterally, and the muscle attaches to the inner surface of the angle of the mandible. As such, it is a synergist to the actions of the masseter and temporalis. In addition, it helps pull the mandible forward, an action known as protraction (or protrusion) of the jaw. The **lateral pterygoid** muscles pull the mandibular condyle from a different angle and are the main protruder of the mandible. They also depress (lower) the mandible. As a result of their attachments, alternating coordinated contractions of the medial and lateral pterygoids produce the grinding motion needed to masticate. See **Table 9.4** for more details on the muscles of mastication.

(a) Lateral view

(b) Posterior view

Figure 9.10 Muscles of mastication. Labels in **bold** indicate muscles involved in this action.

Muscle	Action(s)	Origin/Insertion/Nerve(s)	Concept Figures
Masseter	**Elevates** the mandible	O: Zygomatic arch I: Angle and lateral surface of ramus of mandible N: Mandibular nerve (branch of the trigeminal nerve)	
Temporalis	**Elevates** and **retracts** the mandible	O: Lateral surface of skull I: Coronoid process of the mandible N: Mandibular nerve	
Medial pterygoid	**Elevates** and **protracts** the mandible; assists in lateral movements to grind food	O: Pterygoid plate of the sphenoid and a small portion of the maxilla I: Medial angle and ramus of the mandible N: Mandibular nerve	
Lateral pterygoid	**Protracts** and **depresses** the mandible; lateral movements to grind food	O: Pterygoid plate and greater wing of the sphenoid I: Condyle and neck of mandible N: Mandibular nerve	

TABLE 9.4 MUSCLES OF MASTICATION

Muscles of Swallowing

The digestive system chapter covers the details of swallowing. However, since swallowing involves skeletal muscles, we summarize the process here by describing it in two stages of muscular action: (1) muscles that push food to the back of the oral cavity, and (2) muscles that initiate swallowing in the throat, or **pharynx** (FAIR-inks), and push food into the *esophagus* (eh-SOF-ah-gus), the tube that connects the pharynx to the stomach.

Pushing Food to the Back of the Oral Cavity Just before we begin to swallow, muscles of chewing close the jaws, and the **digastric** (die-GAS-trick), **stylohyoid** (STY-low-hi-oid), **mylohyoid** (MY-low-hi-oid), and **geniohyoid** (JEAN-ee-oh-hi-oid) muscles elevate the hyoid bone, raising the tongue and floor of the mouth. This helps nudge the chewed food posteriorly toward the pharynx (**Figure 9.11a and b**).

The tongue plays an important role in preparing food for swallowing and initiating the swallow reflex. During chewing it continually moves food within the mouth so the teeth can break it down to a consistency that is safe for swallowing. The tongue is made up of several muscles. The **genioglossus** (jee-nee-oh-GLOSS-us), the largest tongue muscle, protrudes the tongue (think of licking an ice cream cone). The **hyoglossus** (HIGH-o-gloss-us) depresses the tongue, and the **styloglossus** (STY-low-gloss-us) elevates and retracts the tongue. Together the tongue muscles manipulate food within the oral cavity and move food posteriorly toward the pharynx. Table 9.5 summarizes the muscles of the tongue.

Pushing Food into the Esophagus Table 9.6 summarizes the muscles of swallowing. As the food enters the pharynx, the throat is briefly elevated to receive it. You can check this yourself by holding your hyoid bone as you swallow. After food enters the pharynx, the **sternohyoid** (ster-no-HIGH-oid), **sternothyroid** (ster-no-THY-roid), **omohyoid** (oh-mow-HIGH-oid),

Digastric:
- Anterior belly
- Posterior belly
- Hyoid bone
- Thyrohyoid
- Thyroid cartilage of larynx
- Thyroid gland
- Sternothyroid

- **Mylohyoid**
- **Stylohyoid**
- Sternohyoid
- Omohyoid, superior belly
- Omohyoid, inferior belly
- Sternocleidomastoid

(a) Muscles that push food to the back of the oral cavity

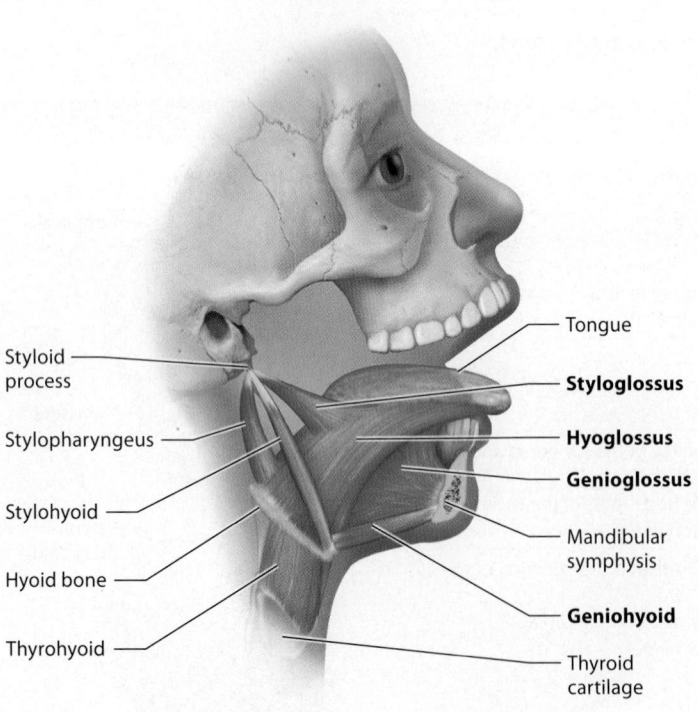

- Styloid process
- Stylopharyngeus
- Stylohyoid
- Hyoid bone
- Thyrohyoid

- Tongue
- **Styloglossus**
- **Hyoglossus**
- **Genioglossus**
- Mandibular symphysis
- **Geniohyoid**
- Thyroid cartilage

(b) Muscles that change the shape of the oral cavity and manipulate food

- Buccinator
- Styloid process
- **Superior pharyngeal constrictor**
- **Stylopharyngeus**
- **Middle pharyngeal constrictor**
- Hyoid bone
- Thyrohyoid membrane
- **Inferior pharyngeal constrictor**
- **Esophagus**

- Mandible
- Mylohyoid
- Geniohyoid
- Hyoglossus
- Thyroid cartilage of larynx
- Trachea

(c) Muscles that are involved in swallowing

Figure 9.11 Muscles of swallowing. Labels in **bold** indicate muscles involved in the phases of swallowing.

TABLE 9.5 MUSCLES OF THE TONGUE

Muscle(s)	Action(s)	Origin/Insertion/Nerve(s)	Concept Figure
Genioglossus	**Protrudes** tongue	O: Internal/anterior mandible (near symphysis) I: Majority of muscle blends into the tongue; body of hyoid bone N: Hypoglossal nerve	
Hyoglossus	**Depresses** tongue	O: Hyoid bone (greater horn) I: Posterior lateral surface of the tongue N: Hypoglossal nerve	
Styloglossus	**Retracts** and **elevates** tongue	O: Styloid process of temporal bone I: Lateral/posterior border of tongue N: Hypoglossal nerve	

TABLE 9.6 MUSCLES OF SWALLOWING

Muscle(s)	Action(s)	Origin/Insertion/Nerve(s)	Concept Figures
Stylohyoid	**Elevates** and **retracts** the hyoid and floor of the mouth during swallowing	O: Styloid process of temporal bone I: Hyoid bone N: Facial nerve	
Mylohyoid	**Elevates** hyoid and floor of mouth, assisting tongue to push food toward the pharynx	O: Medial portion of the mandible I: Hyoid bone and midline connective tissue raphe (seam) N: Mylohyoid nerve	
Geniohyoid	**Elevates** and **protracts** hyoid	O: Inner surface of mandibular symphysis I: Hyoid bone N: Branch of the cervical plexus*	
Sternohyoid	**Depresses** hyoid bone and larynx	O: Posterior manubrium of sternum I: Lower portion of hyoid bone N: Branch of the cervical plexus	
Sternothyroid	**Depresses** larynx and hyoid bone	O: Posterior manubrium of sternum I: Thyroid cartilage of the larynx N: Branch of the cervical plexus	
Omohyoid	**Depresses** and **retracts** hyoid bone	O: Superior border of scapula I: Lower portion of hyoid bone N: Branch of the cervical plexus	
Thyrohyoid	**Depresses** hyoid bone; may elevate larynx	O: Thyroid cartilage of the larynx I: Hyoid bone N: Branch of the cervical plexus*	
Superior, middle, and inferior pharyngeal constrictors	Stepwise **contraction** squeezes food from top to bottom of pharynx into esophagus	O: Superior—sphenoid Middle—hyoid Inferior—larynx I: Posterior portion of pharynx N: Branches of vagus nerve	
Digastric	**Depresses** the mandible; fixator of the hyoid during swallowing	O: Anterior belly: anterior inner margin of the mandible; posterior belly: into mastoid notch, medial to the mastoid process I: Connective tissue of the hyoid bone N: Mylohyoid nerve (anterior belly) and facial nerve (posterior belly)	

*Cervical nerve fibers are distributed to this muscle via the hypoglossal nerve.

and **thyrohyoid** thy-roh-HY-oyd muscles depress the hyoid and pharynx. All of these movements, timed properly as part of the swallowing reflex, push food toward the middle and lower parts of the pharynx while closing off entry to the respiratory tract. (We will say more about this part of the process in the respiratory system chapter.) Finally, the **pharyngeal constrictor muscles** (fah-RIN-jee-al) rapidly contract in sequence from the top to the bottom of the pharynx to push the food into the esophagus (**Figure 9.11c**).

Muscles That Move the Head and Neck

Although we have discussed many muscles *found* in the head and neck region, few of them play any role in actually moving the head or neck. Everyday movements we take for granted, like nodding "yes" or shaking our head "no," are mostly controlled by coordinated movements of another half-dozen muscles.

As mentioned earlier, the **sternocleidomastoid** is the neck muscle attached to the sternum, clavicle, and mastoid process of the temporal bone (**Figure 9.12**). Each sternocleidomastoid muscle, because of its diagonal pathway across the neck, individually rotates (slightly twists) the head toward the opposite shoulder; to flex the head (as in bowing it), both the left and right sternocleidomastoids contract together.

The anterior, middle, and posterior **scalenes** (SKAY-leans)—a muscle group on each side of the neck, attached to the upper two ribs and the cervical vertebrae—move the head and neck laterally when each group contracts individually. This is another example of a muscle or muscle group for which the origin and insertion are relative, depending on which end is permitted to move by various fixators. When fixators hold the head and neck

in place, the ribs are elevated, which is an important part of deep breathing in humans.

The *trapezius* is a large diamond-shaped muscle (a smaller triangle atop a larger triangle) that has several functions. The upper triangular section acts on the head, causing extension (raising the head from a "bowed" position). As we will soon see, the trapezius also plays a major role in moving the scapula and arm.

The **splenius capitis** (SPLEN-ee-us CAP-it-iss) and **splenius cervicis** (SERVE-iss-us) insert on the lower skull and cervical vertebrae, respectively, and originate lower along the vertebral column. They extend the head and neck. They also rotate the head and neck, especially when their muscle fibers contract on just one side.

The *semispinalis capitis* is part of a large muscle group that we will discuss in the next section with other muscles of the vertebral column. But it is important to mention here that this is another muscle that extends the head and neck, and rotates the head to the other side. See **Table 9.7** for more details about these muscles.

Quick Check

☐ 5. Name and describe the four muscles of mastication.
☐ 6. Describe the role of the pharyngeal constrictor muscles in swallowing.
☐ 7. Name and describe the main muscle of head flexion.

Muscles of the Vertebral Column

◀◀ Flashback

1. Where are spinous & transverse processes of vertebrae located? (p. 233)

(a) Anterior view

(b) Posterior view

Figure 9.12 Muscles that move the head and neck. Labels in **bold** indicate muscles involved in these actions.

TABLE 9.7 MUSCLES THAT MOVE THE HEAD AND NECK

Muscle(s)	Action(s)	Origin/Insertion/Nerve(s)	Concept Figures
Sternocleidomastoid	Together: **flex** head (chin moves toward the chest); individually: flex and **rotate** the head toward opposite side; accessory muscles of inspiration	O: Manubrium of sternum, medial portion of clavicle I: Mastoid process of temporal bone N: Accessory nerve	
Scalenes	A mix of actions, depending on the fixators in play; move head **laterally** when contracted individually; **elevate** rib cage with vertebral column fixed; accessory muscles of inspiration	O: Tubercles of transverse processes of C_2–C_7* I: Laterally on the first two ribs N: Spinal nerve branches of C_3–C_8	
Trapezius (superior section)†	**Extends** the head (raises head from "bowed" position); other actions covered in the section on muscles of the arm	O: External occipital protuberance and cervical vertebrae I: Lateral portion of clavicle N: Accessory nerve	
Splenius capitis	**Extends** the head; with other muscles, **rotates** the head to the same side as the muscle that is contracting	O: Spinous processes of cervical vertebrae and connective tissue of the posterior neck (the ligamentum nuchae) I: Mastoid process and occipital bone N: Branches of C_2–C_3 dorsal rami (posterior branches of spinal nerves)	
Splenius cervicis	**Extends** the head and neck; **rotates** the upper cervical vertebrae to the same side as the muscle that is contracting	O: Spinous processes of T_3–T_6 I: Transverse processes of C_1–C_4 N: Branches of the lower cervical dorsal rami	
Semispinalis capitis	**Extends** the head; with other muscles, **rotates** the head to the opposite side from the muscle that is contracting	O: Transverse and articular processes of C_4–T_6 I: Occipital bone N: Variable dorsal rami of cervical and thoracic nerves	

*Some sources disagree about which is the origin or insertion, since both ends can move, depending on the fixators in play.

†The trapezius is covered in more detail with the muscles of the shoulder and arm; however, it is important to note the substantial role of the superior portion of the trapezius in moving the head here.

The mostly straplike muscles of the vertebral column are critical to functions as diverse as actively lifting objects to passively maintaining posture. Often we do not think about these muscles unless we injure them.

The **erector spinae** (eh-REK-tor SPINE-ee) is a large, important muscle group of the vertebral column (**Figure 9.13a and c**). It is made up of nine muscles, which are divided into three groups of three and arranged in vertical columns along the sides of the vertebral column and posterior portions of the ribs. The three groups are named in order of medial-lateral placement relative to the axis of the vertebral column:

- The **spinalis** (spy-NAL-iss) group is located closest to the center of the vertebral column, just to the sides of the spinous processes of the vertebrae.
- The **longissimus** (lon-JISS-ih-mus) group is slightly lateral to the spinalis group.
- The **iliocostalis** (ill-lee-o-cost-AL-iss) group is the most lateral.

We see that these three groups are also placed a little more inferiorly as we look farther from the center (see Figure 9.13a). That placement accounts for the different names of the three muscles in each group. The superiorly positioned spinalis and longissimus groups each have a **capitis** muscle (which inserts on the skull), **cervicis** muscle (which inserts on neck vertebrae), and **thoracis** (thor-AH-sis) muscle (which inserts on thoracic vertebrae or ribs).

The iliocostalis group starts a little lower and does *not* have a capitis muscle. Instead, the three muscles of the iliocostalis group are the cervicis, thoracis, and **lumborum** (lum-BORE-um; a muscle unique to the iliocostalis).

Most of the erector spinae muscles are powerful extensors of the vertebral column. Some also help rotate or laterally bend the vertebral column when contracted on one side only. Which of the nine muscles are stimulated depends on which areas of the vertebral column need to be extended, turned, or held in place (e.g., to maintain posture). However, the capitis and, to some

Spinalis capitis

Longissimus
capitis

Spinalis cervicis

Longissimus
cervicis

Spinalis
thoracis

Iliocostalis
thoracis

Iliocostalis
lumborum

Iliocostalis
cervicis

Longissimus
thoracis

(a) The erector spinae group, posterior deep view

Mastoid process of
temporal bone

Ligamentum nuchae

Semispinalis
capitis

Semispinalis
cervicis

Semispinalis
thoracis

Multifidus

Quadratus
lumborum

(b) The transversospinalis group and quadratus lumborum
(note that the quadratus lumborum is not a deep back muscle)

Longissimus cervicis

Iliocostalis cervicis

Spinalis thoracis

Longissimus thoracis

Iliocostalis lumborum

(c) Cadaver photo

Figure 9.13 Muscles of the vertebral column.

extent, the cervicis muscles are very small and probably play minor roles in extending (or rotating) the head and neck.

↟ STUDY BOOST

Sorting Out the Erector Spinae

Nine muscles in the same area with overlapping functions—now, that's a lot to keep straight. This is a situation in which a schematic diagram and labeling practice will help. Using the sketch below, practice the following:

- Use a red pencil or marker to make three dashed lines on both sides of the vertebral column to represent the three muscles of the spinalis group.
- Use a green pencil or marker to make three lines just to the left and right (and a little lower) for the longissimus group.
- Finally, use a blue pencil or marker to make three lines lateral to the green marks but lower for the iliocostalis group.

Label the three muscle groups when you're finished. Now you've practiced seeing how these muscles sit in relation to each other.

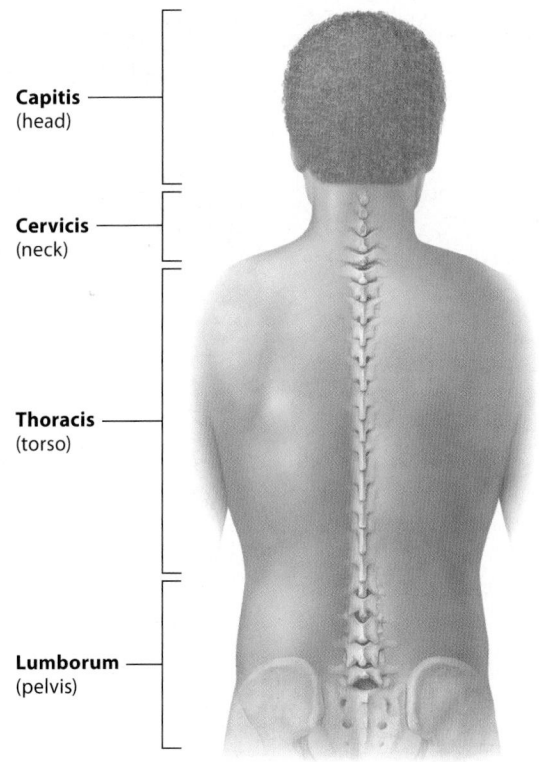

Capitis (head)

Cervicis (neck)

Thoracis (torso)

Lumborum (pelvis)

Another similar but deeper muscle group is the transversospinal group. These muscles largely "fill in" the grooves along the vertebral column between the transverse and spinous processes of vertebrae, hence, the source of the group's name. This group is composed of the **semispinalis, multifidus** (mull-TIFF-id-us), and **rotatores** (row-tah-TORE-eez) muscles. Collectively, their actions are similar to those of the erector spinae muscles: They extend the head, neck, and vertebral column (**Figure 9.13b**).

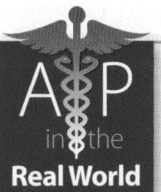

Back Pain

One of the most common medical problems in today's society is back pain, which affects about 8 out of 10 people. Back pain is not a diagnosis, but rather a symptom of some underlying problem. Many of the causes of this type of pain have their basis in the complex anatomy of the spine. Because of this complex anatomy, identifying the actual source of the pain becomes a diagnostic challenge for the physician who must sort through a patient's history and conduct one or more tests. The pain may be due to degeneration of an intervertebral disc or disc herniation ("slipped disc") that is compressing a spinal nerve, to a muscle spasm, or to the stretching or tearing of ligaments as the result of trauma or improper lifting practices.

Advancing age, obesity, poor level of core body fitness, genetics, and occupation are common risk factors for developing back pain. The best approach for preventing this type of pain is to recognize the risk factors and adopt healthy practices to protect the back. Fortunately, most back pain goes away on its own and can be treated with readily available over-the-counter pain medications. Persistent or severe pain, however, requires professional medical attention.

Another muscle of the vertebral column that helps extend and laterally bend the vertebral column is the **quadratus lumborum** (quad-RATE-us lum-BORE-um).

See **Table 9.8** for details of these muscles as well as the others described in this section.

Quick Check

- ☐ 8. Name and describe the nine muscles of the erector spinae group.
- ☐ 9. Describe the roles played by the muscles of the vertebral column *in addition to* extension.

Apply What You Learned

- ☐ 1. Bell's palsy is a disorder of the facial nerve involving temporary or permanent paralysis of the muscles of facial expression. Which muscles of the head and face would be *unaffected* by Bell's palsy?
- ☐ 2. Which would be a bigger problem for a person's ability to properly move the eyes: damage to the trochlear nerve or damage to the oculomotor nerve? Explain.
- ☐ 3. Predict how severe injury to the erector spinae muscles on the left side of the vertebral column would affect childhood growth and posture.

See answers in Appendix A.

TABLE 9.8 MUSCLES OF THE VERTEBRAL COLUMN

Muscle(s)	Action(s)	Origin/Insertion/Nerve(s)	Concept Figures
Erector Spinae Group			
Spinalis column Capitis Cervicis Thoracis	**Extend** the vertebral column (particularly the thoracis); capitis and cervicis may play a minor role in **extending** the head	O: Spinous processes of upper lumbar, thoracic, and lower cervical vertebrae I: Spinous processes of thoracic and cervical vertebrae and skull N: Posterior rami of cervical, thoracic, and lumbar spinal nerves	
Longissimus column Capitis Cervicis Thoracis (largest and most powerful component of the erector spinae)	**Extend** the vertebral column (particularly the thoracis) and maintain posture; capitis and cervicis may play a minor role in **extending** the head and rotating it to the same side; **laterally bend** the vertebral column when contracted on one side only	O: Transverse processes of vertebrae—usually several vertebrae lower than insertions I: Mastoid process (capitis); transverse processes of vertebrae; medial posterior portions of ribs in the thoracic area N: Posterior rami of cervical, thoracic, and lumbar spinal nerves	
Iliocostalis column Cervicis Thoracis Lumborum	**Extend** the vertebral column (particularly the thoracis and lumborum); maintain **posture**; **laterally bend** the vertebral column when contracted on one side only	O: Posterior surfaces of ribs; portions of sacrum and ilium I: Transverse processes of lower cervical vertebrae; posterior surfaces of ribs N: Posterior rami of cervical, thoracic, and lumbar spinal nerves	
Transversospinal Group			
Semispinalis capitis (see Table 9.7), cervicis, and thoracis	**Extend** the vertebral column; capitis and cervicis **extend** the head	O: Transverse and articular processes of C_4–T_6 I: Spinous processes (several vertebrae higher than origins for cervicis and thoracis) and skull N: Posterior rami of cervical and thoracic nerves	
Multifidus	Synergist of other muscles of vertebral column movement and assists in maintaining **posture**	O: In its inferior half, sacrum and transverse processes of lumbar vertebrae; in its superior half, transverse processes of thoracic vertebrae and articular processes of lower cervical vertebrae I: Spinous processes of vertebrae from L_5 to C_2 N: Posterior rami of cervical and thoracic nerves	
Rotatores	**Extend** the vertebral column	O: Transverse process of one vertebra I: Spinous process of next one or two vertebrae above N: Posterior rami of cervical, thoracic, and lumbar nerves	
Other Spinal Extensors			
Quadratus lumborum	**Extend** the vertebral column; maintain **posture**; **laterally bend** the vertebral column when contracted on one side only	O: Iliac crest and connective tissue of lumbar region I: Rib 12 and transverse processes of lumbar vertebrae N: Anterior rami of T_{12} and lumbar spinal nerves	

MODULE **9.3**
Muscles of the Trunk and Pelvic Floor

We have already discussed some muscles of the trunk, such as those of the vertebral column. Others will be discussed later, as a number of muscles that move the arm are also anatomically part of the trunk. This section explores muscles that are limited to the trunk by both structure and function.

Learning Outcomes

1. Identify and describe the muscles of the trunk.
2. Identify the origin, insertion, and action of these muscles, and demonstrate their actions.

Muscles of the Trunk

◀◀ **Flashback**

1. What defines the boundaries of the thoracic and abdominopelvic body cavities? (p. 17)
2. What are the three bones that make up the pelvis? (pp. 248–249)

Muscles of Respiration (Breathing)

The **diaphragm** (DY-ah-fram; **Figure 9.14a**) is an unusual skeletal muscle that does not cause movement at a joint. This sheet-like muscle is dome-shaped when relaxed. If you have seen a dome-style camping tent, you know its approximate shape. When the diaphragm contracts, it pulls downward and flattens. This increases the internal dimensions and volume of the thoracic cavity and reduces the pressure inside the pleural cavities, so air rushes into them. This activity causes us to breathe in, or inspire (inhale).

The intercostal muscles are located between the ribs and come in two sets: **external intercostals** and **internal intercostals** (**Figure 9.14b**). Some sources state that several fibers partially

Inferior view

Diaphragm

Xiphoid process of sternum

Central tendon of diaphragm

Left phrenic nerve

Foramen for inferior vena cava

Esophageal hiatus

Abdominal aorta

Aortic hiatus

Quadratus lumborum

12th rib

External intercostal muscles

Internal intercostal muscles

(a) Inferior surface of diaphragm as viewed from the abdominal cavity

(b) The external and internal intercostal muscles

Figure 9.14 Muscles of respiration (breathing). Notice that part (a) shows an inferior view of this transverse section of the trunk.

intertwined with the internal intercostals are actually separate muscles, the *innermost intercostals*. However, for simplicity we will group these fibers with the internal intercostals.

With other parts of the rib cage fixed, the external intercostals raise and spread the ribs. Imagine this working like an accordion placed on its side, anchored to the floor at its bottom end, and raised at its top end. Like the diaphragm, the external intercostals increase the volume inside the thoracic cavity, decrease its internal pressure, and create the suction needed to inspire.

During expiration, relaxation of the diaphram causes it to move superiorly, back into its dome shape. This decreases the lungs' volume and increases their internal pressure, moving air out of the lungs. When expiration is forced, the internal intercostals depress the rib cage, further decreasing the lungs' volume and increasing their pressure.

Two muscles discussed in a previous section, the scalenes and sternocleidomastoid, also raise the rib cage. They are considered secondary muscles of respiration because they function during extreme exertion when we need to maximize the dimensions of the thoracic cavity. However, the diaphragm is the agonist of inspiration, particularly at rest.

Table 9.9 summarizes the muscles of respiration.

Abdominal Muscles

Have you seen photos of celebrities with "six-pack abs"? Those "abs" are part of the **rectus abdominis** (ab-DOM-ih-nis) muscle, the superficial muscle of the anterior abdominal area (**Figure 9.15**). This muscle is enclosed by dense connective tissue called the **rectus sheath.** A line of connective tissue along the midline, the **linea alba** (LINN-ee-ah AHL-bah), separates the right and left rectus abdominis muscles. Each muscle is partially separated into sections by three or more transverse **tendinous intersections.** Six of those sections are indeed prominent in a well-conditioned person with a low percentage of body fat.

The rectus abdominis is the agonist of trunk flexion, as in sitting up from a reclining position or taking a bow while standing. Not surprisingly, this muscle also acts as the agonist during a sit-up exercise. When you recline or are in a low-gravity environment (not just in space, but in water, too, such as sitting in a hot tub), the rectus abdominis can also serve as the antagonist to the scalenes and depress the rib cage during expiration.

The **external oblique,** so named because its fibers run at an angle, is lateral to the rectus abdominis (Figure 9.15). As its placement suggests, it is responsible for both lateral trunk flexion and trunk rotation. The same is true of the **internal oblique**

TABLE 9.9 MUSCLES OF RESPIRATION (BREATHING)*			
Muscle(s)	**Action**	**Origin/Insertion/Nerve(s)**	**Concept Figures**
Diaphragm	**Flattens** to lower the floor of the thoracic cavity to increase volume and decrease pressure, thereby causing inspiration (inhalation)	O: Xiphoid process of sternum, lower ribs and costal cartilages, lumbar vertebrae I: Central tendon of diaphragm itself N: Phrenic nerve	 Diaphragm
External intercostal muscles	**Elevates** the rib cage, spreading the ribs, assisting inspiration	O: Lower edge of rib superior to its insertion I: Upper edge of rib inferior to its origin N: Intercostal nerves	 External intercostal muscles
Internal intercostal muscles	**Depresses** the rib cage, pulling ribs closer together, assisting expiration	O: Upper edge of rib inferior to its insertion I: Lower edge of rib superior to its origin, deep to the insertion of the external intercostals N: Intercostal nerves	 Internal intercostal muscles

*See Table 9.7 for scalenes and sternocleidomastoid, which also assist with forced inspiration.

deep to it, which has fibers that run opposite and nearly perpendicular to those of the external oblique.

The **transversus abdominis** (tranz-VER-sus), which has horizontal fibers, is the deepest abdominal muscle. As a result of its placement, the transversus abdominis does not move the trunk at all. Its main action is to compress the abdominal cavity, which is the secondary action of all of these abdominal muscles.

Compressing the abdominal cavity may seem a minor action, but it is quite important and overlaps with the functions of many other organ systems. Abdominal compression increases the pressure of the abdominal cavity, which pushes the diaphragm upward for forced expiration. The extra pressure can be used to facilitate many other purposes as well, from childbirth (push!) to assisting urination and defecation.

Rectus sheath

Rectus abdominis

Linea alba

Internal oblique

Transversus abdominis

External oblique

Aponeurosis of external oblique

Inguinal ligament (formed by free inferior border of aponeurosis of external oblique)

(a) Anterior view

(b) Cadaver photo

External oblique

Iliac crest

Inguinal ligament

Superficial

Rectus abdominis

Internal oblique

Lumbar fascia

Inguinal ligament

Pubic tubercle

Intermediate

Transversus abdominis

Lumbar fascia

Inguinal ligament

Deep

(c) Lateral view at varying depths of dissection

Practice art labeling
@MasteringA&P®

Figure 9.15 Abdominal muscles. Labels in **bold** indicate muscles involved in abdominal movement.

TABLE 9.10 ABDOMINAL MUSCLES

Muscle(s)	Action(s)	Origin/Insertion/Nerve(s)	Concept Figures
Rectus abdominis	**Flexes** the trunk; with the vertebral column fixed, **compresses** abdominal cavity	O: Superior aspect of pubic bones I: Costal cartilages of lower ribs N: Branches of anterior rami of the lower six to seven thoracic nerves	
External oblique	**Flexes** and **laterally bends** the trunk; with the vertebral column fixed, **compresses** abdominal cavity	O: Lower eight ribs I: Iliac crest, pubic tubercle, and linea alba N: Branches of the lower six thoracic nerves	
Internal oblique	**Flexes** and **laterally bends** the trunk; **compresses** abdominal cavity	O: Iliac crest and inguinal ligament I: Lower three to four ribs, linea alba, and pubic bone N: Branches of the lower six thoracic nerves and first lumbar nerve	
Transversus abdominis	**Compresses** abdominal cavity	O: Lower six costal cartilages, thoracolumbar fascia, crest and inguinal ligament I: Linea alba and pubic bone N: Branches of the lower six thoracic nerves and first lumbar nerve	

Table 9.10 provides more information about abdominal muscles.

Quick Check

☐ 1. What are the main muscles of inspiration?
☐ 2. Which muscles are involved in trunk flexion?

Muscles of the Pelvic Floor, Urogenital Diaphragm, and Perineum

◀◀ Flashback

1. What are the organs of the pelvic body cavity? (p. 17)
2. What structures form the pelvic brim? (p. 247)

The pelvic floor, also known as the **pelvic diaphragm,** is composed of the bilateral **levator ani** groups, consisting of the **pubococcygeus** (pyoo'-boh-kok-SIJ-ee-us) and **iliococcygeus** (il'ee-oh-kok-SIJ-ee-us), and the **coccygeus** muscles (**Figure 9.16a**). These muscles once again demonstrate that long names with Latin or Greek roots are actually helpful when you recall skeletal anatomy. These muscles are paired and attach to the coccyx and anococcygeal (ay'-noh-kok-SIJ-ee-al) ligament at one end and the pubis, ilium, and ischium (respectively) at the other end. There is a gap between the anterior and medial portions of the pubococcygeus muscles through which pass the male and female urethrae and rectum, and the vagina in the female. This gap is closed off below the pelvic diaphragm and in front of the rectum by the **urogenital diaphragm,** a muscular sheet attaching to both right and left pubic bones. The urogenital diaphragm transmits the urethra in the male and female, and the vagina in the female, from the pelvic cavity to the perineum.

Despite its small size, the urogenital diaphragm is quite complex (**Figure 9.16b**). It contains: (1) the **external urethral sphincter,** a small circular muscle that surrounds the urethra and allows voluntary control of urination; (2) a circular muscle that surrounds the vagina in the female; and (3) the **deep transverse perineal muscle,** which mainly supports the genitourinary organs. The **superficial transverse perineal** muscles attach to the underside of the urogenital diaphragm and stabilize the central tendon of the perineum or perineal body, a large area of reinforcing connective tissue anterior to the anus. The terminal part of the anal canal, the anus, is surrounded by the **external anal sphincter,** which attaches to the perineal body and the tip of the coccyx, and provides voluntary control of defecation (Figure 9.16b). As their names

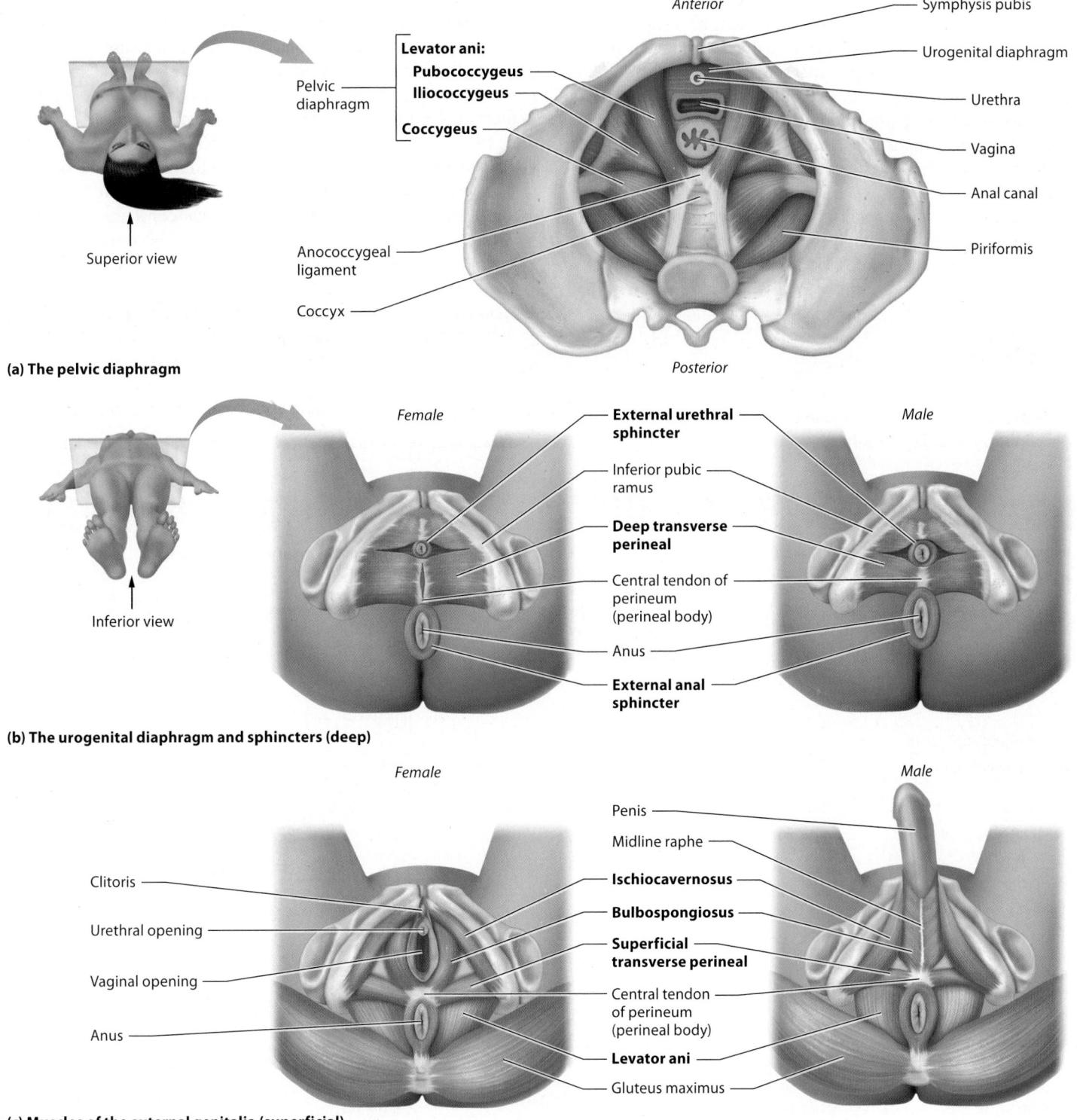

(a) The pelvic diaphragm

(b) The urogenital diaphragm and sphincters (deep)

(c) Muscles of the external genitalia (superficial)

Figure 9.16 Muscles of the pelvic floor and perineum. Labels in **bold** indicate muscles that act in each region of the pelvic floor and peritoneum.

suggest, the external urethral sphincter and external anal sphincter are the outermost superficial muscles that surround those tubes and provide voluntary control over passage of contents through them.

The perineum is the diamond-shaped area between the upper thighs that contains the external genitalia (the penis in the male and the clitoris in the female) and the opening of the anus (**Figure 9.16c**). Two perineal muscles, the **bulbospongiosus** (bulb-oh-spun'-jee-OH-suss) and **ischiocavernosus,** are associated with the external genitalia. These thin, delicate muscles help make these structures erect during sexual excitement. In males, the bulbospongiosus also helps expel semen from the urethra.

See **Table 9.11** for more details about the muscles of the pelvic floor, urogenital diaphragm, and perineum.

As you can see, there is a close relationship between pelvic floor muscles and the lower portions of the urinary, reproductive, and digestive tracts. So, it logically follows that problems with these muscles could result in problems with controlling the movement of materials through these tubes. This is particularly true of the urinary tract, where lack of control is referred to as urinary incontinence (see *A&P in the Real World: Kegel Exercises*).

Quick Check

☐ 3. Name and describe the muscles of the levator ani group.
☐ 4. Which muscles are involved in conscious control of urination and defecation?

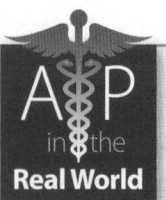

Kegel Exercises

Not all urinary incontinence is due to weak pelvic floor muscles. However, weak muscles are a likely suspect, particularly for persistent urinary incontinence in women who have delivered babies vaginally. Stretching and trauma to those muscles during childbirth may cause problems later.

Childbirth is just one of many risk factors. Others are age, weight gain, changing hormone levels after menopause, and other conditions that affect structures in the pelvic cavity. Men may also develop urinary incontinence due to age-related changes in the prostate gland and other reproductive system structures.

Pelvic floor exercises, also called Kegel exercises (for Dr. Arnold Kegel), help strengthen several muscles of the pelvic floor, particularly the pubococcygeus. The goal is to become aware of these muscles through various contraction movements within the pelvic cavity and then become adept at slowly contracting and relaxing them.

Not only does strengthening these muscles reinforce the sphincter, it also guards against pelvic organ prolapse (in which weak muscles allow women's pelvic organs to slip out of place). Several thin muscles are the sole support of certain pelvic organs. Evidence suggests that men, too, can benefit from Kegel exercises to enhance both bladder control and sexual performance.

TABLE 9.11 MUSCLES OF THE PELVIC FLOOR, UROGENITAL DIAPHRAGM, AND PERINEUM

Muscle(s)	Action(s)	Origin/Insertion/Nerve(s)
Pelvic Floor (or Pelvic Diaphragm)		
Levator ani Pubococcygeus Iliococcygeus	Supports pelvic floor, anal canal, and genitourinary organs	O: Interior of pelvic bones I: Coccyx and anococcygeal ligament N: Branches of S_3 and S_4
Coccygeus	Supports pelvic floor	O: Ischial spine I: Side of coccyx and lower sacrum N: Branches of S_3 and S_4
Urogenital Diaphragm		
Deep transverse perineal	Supports pelvic organs; fixes central tendon	O: Ischiopubic rami I: Perineal body N: Pudendal nerve
External urethral sphincter	Compresses urethra (encircles urethra and, in females, vagina)	N: Pudendal nerve
Perineum		
External anal sphincter	Closes anal opening	O: Coccyx and perineal body I: Connective tissues around the anal canal N: Pudendal nerve
Bulbospongiosus	Expels semen and urine in males; assists erection of penis and clitoris; constricts opening of vagina	O: Perineal body I: Corpus spongiosum of the penis; clitoris of females N: Pudendal nerve
Ischiocavernosus	Maintains erection of the penis or clitoris	O: Ischiopubic rami I: Corpus cavernosum of the penis; clitoris of females N: Pudendal nerve
Superficial transverse perineal	Assists with stabilizing the perineal body	O: Ischial tuberosity I: Perineal body N: Pudendal nerve

☐ 1. What are the muscles of inspiration? What do they do to the thoracic cavity and lungs?
☐ 2. Identify and describe the exercises that can be useful in strengthening the abdominal muscles.

See answers in Appendix A.

MODULE **9.4**

Muscles of the Pectoral Girdle and Upper Limb

Learning Outcomes

1. Identify and describe the muscles that move the pectoral girdle and upper limb.
2. Identify the origin, insertion, and action of these muscles, and demonstrate their actions.

Anatomically speaking, the "arm" refers to the upper limb between the shoulder and elbow. It contains the humerus, which attaches to the **pectoral girdle,** consisting of the scapula and the clavicle. Muscles in this area largely interact with the scapula, humerus, and clavicle. The shoulder joint, or glenohumeral joint, a ball-and-socket joint between the head of the humerus and the glenoid cavity of the scapula, has a diverse and interesting range of movements. The term "forearm" describes the region of the upper limb from the elbow joint distally to the wrist. Distal to the wrist is the hand.

Muscles of the Shoulder and Arm

◀◀ **Flashback**

1. Define elevation and depression. (p. 268)
2. Describe protraction and retraction. (p. 268)

Whether you are hugging a friend or throwing a fastball, your shoulder and arm muscles work together in complex ways to move the scapula, stabilize the shoulder joint, and provide force for these actions. Some muscle pairs work together for certain movements and work as antagonists for other movements. The muscles in this region also show a tremendous diversity of size, from the many small muscles of the scapula to the massive latissimus dorsi of the back.

Muscles That Move the Scapula

The muscles that move the scapula do much more than simply shrug your shoulders. The subtle movements of the scapula maneuver the pectoral girdle and stabilize it for coordinated, powerful

movements like rowing or throwing a punch. The movements of the scapula include elevation, depression, protraction, and retraction, as well as more subtle actions like upward and downward rotation.

The **serratus** (sir-RATE-us) **anterior** muscle attaches to the upper nine ribs. Its origins of many pointed tips attaching to the ribs give it a "scalloped" appearance that looks a little like the edge of a serrated knife—hence, its name (**Figure 9.17a**).

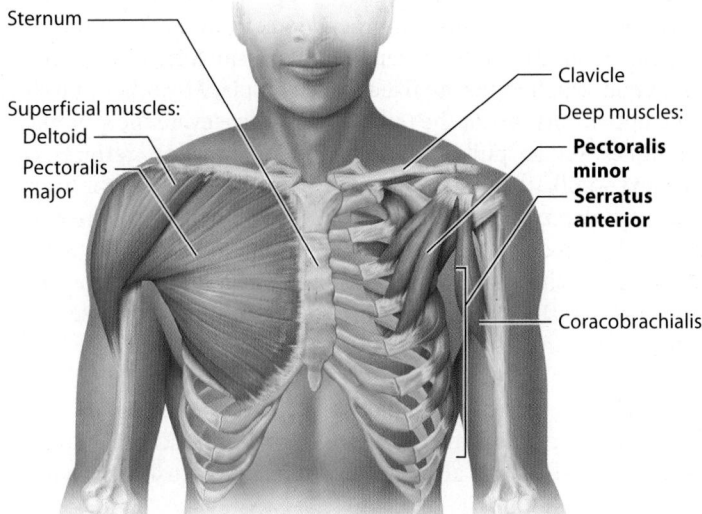

Sternum
Clavicle
Superficial muscles:
Deltoid
Pectoralis major
Deep muscles:
Pectoralis minor
Serratus anterior
Coracobrachialis

(a) Anterior view: superficial muscles shown on left; deep muscles shown on right, with superficial removed

Superficial muscles:
Trapezius
Deltoid
Latissimus dorsi
Deep muscles:
Levator scapulae
Rhomboid minor
Rhomboid major
Scapula
Humerus

(b) Posterior view: superficial muscles shown on left; deep muscles shown on right, with superficial removed

Figure 9.17 Muscles that move the scapula. Labels in **bold** indicate muscles involved in these actions.

The muscle wraps around the contour of the rib cage to insert along the medial border of the scapula. The serratus anterior is involved in several movements of the scapula, including protraction, as in reaching forward, and upward rotation of the scapula. Indeed, it is nicknamed the "boxer's muscle" because that extra inch of thrust forward can make the difference in a successfully executed jab in boxing. The serratus anterior is easily visible in a well-conditioned athlete. The smaller **pectoralis** (peck-tore-AL-iss) **minor** muscle, located nearby and also having rib and scapular attachments, assists with this forward action as well.

The **trapezius** (trap-EEZ-ee-us) (Figure 9.17b) is a diamond-shaped muscle we already mentioned in the section on muscles that move the head. This is another good example of a muscle that can have multiple actions, depending on which fibers are stimulated and which joints are fixed or held in place by other muscles. The upper portions of the trapezius mainly elevate the scapula—in other words, pull it up. The middle portion retracts the scapula—pulls it medially—and inferior portions of the trapezius can depress the scapula, pulling it down. The upper and lower portions can work together to rotate the scapula upward, as when reaching for something on a shelf. The serratus anterior also assists this action. Since some origins of the serratus anterior are on ribs *lower* than the scapula and insert near its inferior angle, contraction of this portion of the muscle pulls the inferior angle of the scapula outward, thereby assisting the upper and lower portions of the trapezius to rotate the scapula upward. Think of raising your arm above your head to brush your hair.

The **levator scapulae** (leh-VAY-tohr SCAP-yoo-lee) muscle works with the superior portion of the trapezius. The origin of the levator scapulae is located above the scapula, and so it makes sense that this muscle elevates the scapula. The **rhomboid** (RAHM-boid) **major** and **rhomboid minor** muscles work with the middle portion of the trapezius. The rhomboids, which have their origin on the vertebral column, mainly retract the scapula. Downward rotation of the scapula, such as when you reach down to pick up a backpack, occurs by contracting the lower fibers of the rhomboids to rotate the inferior angle of the scapula medially and upward, with the assistance of gravity.

For more information on muscles that move the scapula, see Table 9.12.

ConceptBOOST »»»

Demystifying Muscle Actions
The large number of muscle actions presented in this chapter can seem daunting. So, how can you possibly memorize all of them? Well, you don't have to–instead you can use logic to understand their actions.

Remember that the origin is usually the anchoring point, and the insertion generally moves toward the origin. Therefore, when a

TABLE 9.12 MUSCLES THAT MOVE THE SCAPULA

Muscle(s)	Action(s)	Origin/Insertion/Nerve(s)	Concept Figures
Levator scapulae	Elevates scapula	O: Transverse processes of C_1–C_4 I: Superior angle of scapula N: Dorsal scapular nerve	
Rhomboid major and rhomboid minor	Retracts scapula	O: Spinous processes of: T_2–T_5 (major); C_7–T_1 (minor) I: Medial border of scapula N: Dorsal scapular nerve	
Trapezius (superior section)	Elevates the scapula; rotates scapula upward in collaboration with inferior portion	O: External occipital protuberance and cervical vertebrae I: Lateral clavicle N: Accessory nerve	
Trapezius (middle and inferior section)	Retracts (via middle portion) and depresses (via inferior section) the scapula; rotates scapula upward in collaboration with upper fibers	O: Spinous processes of thoracic vertebrae I: Spine and acromion of scapula N: Accessory nerve	
Pectoralis minor	Protracts and depresses scapula	O: Anterior part of ribs 3–5 I: Coracoid process N: Medial pectoral nerve	
Serratus anterior	Prime mover of scapula in protraction; assists in upward rotation of scapula	O: First nine ribs I: Medial border of scapula N: Long thoracic nerve	

Note: Some of these actions vary depending on which fibers are stimulated.

muscle contracts, it usually pulls its insertion toward its origin. So look to the origin to determine where the muscle moves another bone or body part attached to its other end. Visualize the origin of the muscle "reeling in" the insertion, the way a fishing pole reels in a fish.

Imagine pulling a bucket on a rope out of a well—you have to be *above* it to pull it upward, right? So, muscles with an origin above the scapula (such as the levator scapulae) elevate the scapula. Similarly, to pull a book down from the top shelf at the library, you have to be below it. Muscles with an origin below the scapula (like the inferior portions of the trapezius) depress the scapula.

Although this method of visualization is particularly useful for the scapula, it works for most of the muscles you will encounter in your study. Ask yourself where the anchoring origin is in relation to the insertion, and visualize the resulting movement. This suggests the action of the muscle. You will *understand* the action rather than just memorize it. ■

Quick Check

☐ 1. Name and describe two muscles that elevate the scapula.
☐ 2. Name and describe two muscles that protract the scapula.

Muscles That Move the Humerus

The muscles that move the humerus at the shoulder joint allow it to flex, extend, abduct, adduct, and rotate. Some of these muscles are quite large, such as the pectoralis major, and others are much smaller, such as the coracobrachialis (core′-ack-oh-brake-ee-AL-iss).

The **pectoralis major** is the large muscle making up the flesh of the chest area (**Figure 9.18a**). It is a powerful agonist of arm flexion: the action your humerus executes when you do a push-up or lift this book. The origin of the pectoralis major is medial to its insertion on the humerus, which enables it to assist in adduction (pulling the arm toward the midline) and medial or internal rotation.

The **latissimus dorsi** (latt-TISS-ih-muss DOOR-see), a large triangular muscle found in the middle and lower back, is a

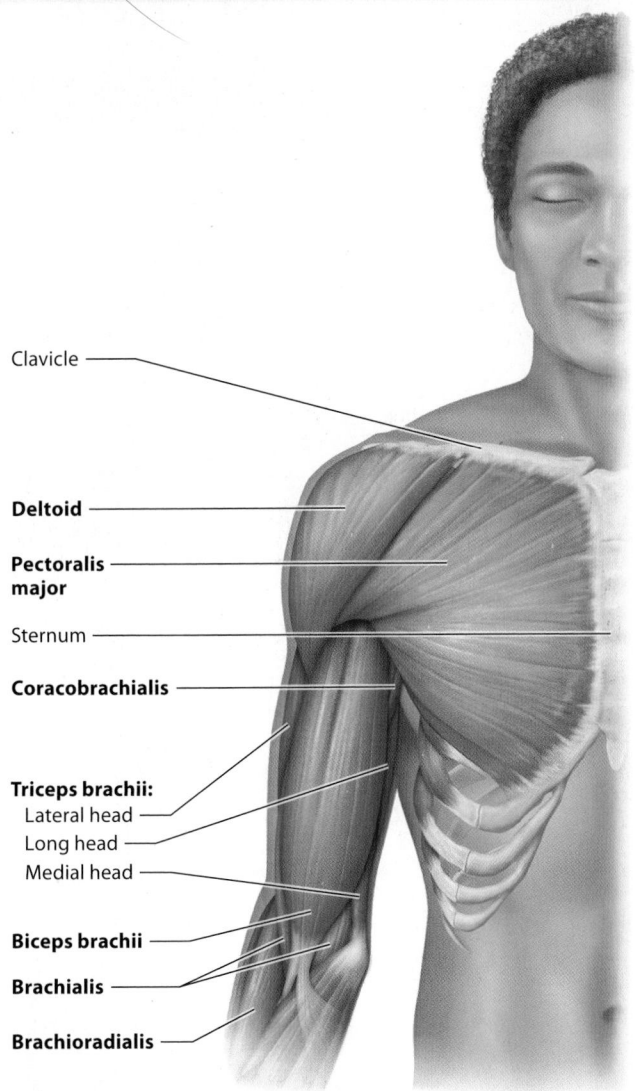

Clavicle
Deltoid
Pectoralis major
Sternum
Coracobrachialis
Triceps brachii:
Lateral head
Long head
Medial head
Biceps brachii
Brachialis
Brachioradialis

(a) Anterior view

Spine of scapula
Supraspinatus
Infraspinatus
Deltoid (cut)
Greater tubercle of humerus
Teres minor
Teres major
Triceps brachii:
Lateral head
Long head
Latissimus dorsi
Humerus
Olecranon process of ulna
Anconeus

(b) Posterior view

Figure 9.18 Muscles that move the arm and forearm. Labels in **bold** indicate muscles involved in these actions.

powerful agonist of extension and adduction (**Figure 9.18b**). The origin of the muscle is well below and medial to its insertion on the humerus, giving it an ideal placement to pull the arm downward (extension) and toward the body (adduction). The pectoralis major and latissimus dorsi have an interesting relationship in that they are antagonists for arm extension and flexion, respectively, but are partners in arm adduction and internal rotation.

The **deltoid** is the thick, rounded muscle of the shoulder and has many potential actions. Its central portion, which attaches to the acromion of the scapula, is the agonist of arm abduction, like the upward motion of the arm when you're doing jumping

jacks. Its anterior fibers, which attach to the lateral clavicle, assist with arm flexion, whereas the posterior fibers, attached to the spine of the scapula, help with arm extension. These two actions are best observed during the alternating flexion and extension of the arms while you're walking. The deltoid is also a popular site for intramuscular ("into a muscle") injections because of its bulk and easy, safe access.

The **teres major** (TAYR-eez) and **teres minor** attach to the posterior and lateral border of the scapula. The teres major attaches to the humerus near the attachment site of the latissimus dorsi and shares similar actions with it (adduction, extension, and medial

TABLE 9.13 MUSCLES THAT MOVE THE ARM

Muscle(s)	Action(s)	Origin/Insertion/Nerve(s)	Concept Figures
Pectoralis major	**Flexes** and **adducts** the arm; **rotates** arm medially	O: Medial clavicle, sternum, costal cartilages 1–7 I: Greater tubercle and lateral lip of intertubercular sulcus of humerus N: Medial and lateral pectoral nerves	
Latissimus dorsi	**Adducts** and **extends** the arm; **rotates** arm medially	O: Iliac crest, spinous processes of lower thoracic and all lumbar vertebrae I: Floor of intertubercular sulcus of the humerus N: Thoracodorsal nerve	
Deltoid	**Abducts** the arm; secondarily **flexes** and **extends** arm	O: Acromion and spine of scapula; lateral clavicle I: Deltoid tuberosity of humerus N: Axillary nerve	
Teres major	**Adducts**, **extends**, and **rotates** arm medially	O: Posterior, inferior portion of scapula I: Medial lip of intertubercular sulcus of the humerus N: Lower subscapular nerve	
Coracobrachialis	**Flexes** and **adducts** the arm	O: Coracoid process I: Medial diaphysis of humerus N: Musculocutaneous nerve	
*Supraspinatus**	Assists **abduction**; holds the humerus and stabilizes shoulder joint	O: Supraspinous fossa of scapula I: Greater tubercle of humerus N: Suprascapular nerve	
*Infraspinatus**	Laterally **rotates** the humerus; stabilizes shoulder joint	O: Infraspinous fossa of scapula I: Greater tubercle of humerus N: Suprascapular nerve	
*Teres minor**	Laterally **rotates** the humerus; stabilizes shoulder joint	O: Posterior, lateral border of scapula I: Greater tubercle of humerus N: Axillary nerve	
*Subscapularis**	**Adducts** and **rotates** the humerus medially; stabilizes shoulder joint	O: Subscapular fossa of scapula I: Lesser tubercle of humerus N: Subscapular nerves	

*Rotator cuff muscles.

rotation of the humerus). In contrast, the teres minor inserts into the greater tubercle of the humerus and laterally rotates the humerus when it contracts. Another adductor of the humerus, the **coracobrachialis**, is a small cylindrical muscle that originates on the coracoid process of the scapula and inserts midway down the diaphysis of the humerus. Its origin and insertion allow it to flex the arm as well.

The **supraspinatus** (sue′-pra-spy-NAY-tuss) is located above the spine of the scapula, and the **infraspinatus** (in′-fra-spy-NAY-tuss) is located below the spine. These muscles mainly stabilize the shoulder joint, but the supraspinatus also assists the deltoid with arm abduction by initiating the first 15 degrees of that abduction. The **subscapularis** (sub′-scap-u-LARE-iss), which covers much of the anterior surface of the scapula (the subscapular fossa), originates medially enough to assist with adduction and medial rotation.

See **Table 9.13** for more information on muscles that move the arm. Notice that Table 9.13 footnotes four muscles as belonging to the **rotator cuff** group: the supraspinatus, infraspinatus, teres minor, and subscapularis. The tendons of the rotator cuff muscles attach firmly to the shoulder joint capsule and strengthen it. In addition, they help fix and hold the head of the humerus in place in the glenoid cavity so that other muscles can move the arm. As such, they serve as dynamic stabilizers of the shoulder joint during forceful movement at the joint. Injuries to these muscles are a common and painful problem, as described in *A&P in the Real World: Rotator Cuff Injuries* on page 322.

Quick Check

☐ 3. Name and describe two muscles that abduct the arm.
☐ 4. Name and describe two muscles that adduct the arm.

Muscles of the Arm, Forearm, and Hand

◀◀ **Flashback**

1. Define supination and pronation. (p. 269)
2. Describe the major bone markings of the radius and ulna. (p. 244)

This group includes muscles that move the forearm at the elbow joint and muscles that move the wrist and hand.

Muscles That Move the Forearm at the Elbow Joint

On the anterior side of the arm, the **biceps brachii** is the "showy" muscle that bulges when arm wrestling (Figure 9.18a). It does the combined work of flexing the forearm when you are supinating the wrist/hand at the same time (e.g., picking up a bowling ball).

However, the smaller, deeper **brachialis** is the agonist of elbow flexion. Its attachment to the ulna (which makes up the hinge of the elbow joint) and position give it better leverage than the biceps brachii, allowing the brachialis to focus all its force on flexing the elbow. A muscle positioned much lower on the lateral forearm, the **brachioradialis** (brake′-ee-oh-raid-ee-AL-iss), is a synergist of elbow flexion.

On the posterior side of the arm (Figure 9.18b), the **triceps brachii** is the large three-headed muscle that is the agonist of forearm extension. One of its heads (long head) also attaches to the scapula, so it plays a supporting role in adducting the arm and stabilizing the shoulder as well. The *anconeus* is a small triangular muscle that assists with extending the forearm and stabilizes the elbow joint.

See **Table 9.14** for more information.

Muscle(s)	Action(s)	Origin/Insertion/Nerve(s)	Concept Figure
Biceps brachii	Flexes the elbow; supinates the forearm	O: Supraglenoid tubercle (long head) and coracoid process (short head) I: Radial tuberosity N: Musculocutaneous nerve	
Brachialis	Flexes the elbow (prime mover)	O: Distal half of the diaphysis of the anterior humerus I: Coronoid process of ulna N: Musculocutaneous nerve	
Brachioradialis	Flexes the elbow (synergist)	O: Ridge above the lateral condyle of the distal humerus I: Distal radius above styloid process N: Radial nerve	
Triceps brachii	Extends the elbow	O: Infraglenoid tubercle (long head); posterior, proximal diaphysis (lateral head); most of posterior diaphysis (medial head) I: Olecranon process N: Radial nerve	
Anconeus	Assists extension of the elbow	O: Lateral epicondyle of humerus I: Olecranon process N: Radial nerve	

TABLE 9.14 MUSCLES THAT MOVE THE FOREARM AT THE ELBOW JOINT

Muscles That Move the Wrist, Hand, and Fingers

There are *a lot* of muscles that move the wrist and hand (**Figure 9.19** and **Table 9.15**). Fortunately, most of their names are based on either their location or their action. To review some terms from Table 9.1 that will help you with these muscles, remember that: *carpi* = wrist, *superficialis* = nearer the surface, *profundus* = deep, *pollicis* = thumb, *digiti* = fingers and toes, and *minimi* = smallest.

(a) Anterior view with muscles removed as necessary to reveal deeper muscles

(b) Posterior view with muscles removed as necessary to reveal deeper muscles

Figure 9.19 **Muscles that move the wrist and hand.** Labels in **bold** indicate muscles and tendons involved in these actions.

TABLE 9.15 MUSCLES THAT MOVE THE WRIST AND HAND

Muscle(s)	Action(s)	Origin/Insertion/Nerve(s)
Pronator teres	Pronates forearm (palm posterior)	O: Medial epicondyle of humerus, coronoid process of ulna I: Lateral surface of radius N: Median nerve
Flexor carpi radialis	Flexes wrist (bends wrist toward anterior forearm); abducts hand	O: Medial epicondyle of humerus I: Second and third metacarpals N: Median nerve
Palmaris longus	Flexes wrist; tense dense connective tissue in palm of hand	O: Medial epicondyle of humerus I: Flexor retinaculum and surrounding connective tissue in palm of hand N: Median nerve
Flexor carpi ulnaris	Flexes wrist; adducts hand	O: Medial epicondyle of humerus, medial olecranon process of ulna I: Pisiform, hamate, fifth metacarpal N: Ulnar nerve
Flexor digitorum superficialis	Flexes fingers; flexes wrist	O: Medial epicondyle of humerus, coronoid process of ulna, anterior proximal radius I: Middle phalanges of fingers 2–5 N: Median nerve
Flexor pollicis longus	Flexes thumb	O: Anterior diaphysis of radius I: Distal phalanx of thumb N: Median nerve
Flexor digitorum profundus	Flexes fingers; flexes wrist	O: Medial/anterior ulna and coronoid process I: Distal phalanges of fingers 2–5 N: Median and ulnar nerve branches
Pronator quadratus	Pronates forearm	O: Anterior distal ulna I: Anterior distal radius N: Median nerve
Extensor carpi radialis longus	Extends wrist; abducts hand	O: Lateral lower diaphysis of humerus I: Second metacarpal N: Radial nerve
Extensor carpi radialis brevis	Extends wrist; abducts hand	O: Lateral epicondyle of humerus I: Third metacarpal N: Deep radial nerve
Extensor digitorum	Extends fingers; extends wrist	O: Lateral epicondyle of humerus I: Posterior phalanges of fingers 2–5 N: Posterior interosseous nerve
Extensor digiti minimi	Extends fifth finger	O: Lateral epicondyle of humerus I: Phalanges of finger 5 N: Posterior interosseous nerve
Extensor carpi ulnaris	Extends wrist; adducts hand	O: Lateral epicondyle of humerus, lateral proximal ulna I: Fifth metacarpal N: Posterior interosseous nerve
Supinator	Supinates forearm (palm anterior)	O: Lateral epicondyle of humerus, area of the radial notch of ulna I: Anterior/lateral area of proximal third of radius, below tuberosity N: Deep radial nerve
Abductor pollicis longus	Abducts and extends thumb	O: Posterior diaphysis of radius and ulna I: First metacarpal N: Posterior interosseous nerve
Extensor pollicis longus	Extends thumb ("thumbs up" motion)	O: Posterior diaphysis of ulna I: Distal phalanx of thumb N: Posterior interosseous nerve
Extensor pollicis brevis	Extends thumb	O: Lower posterior diaphysis of radius I: Proximal phalanx of thumb N: Posterior interosseous nerve
Extensor indicis	Extends index finger	O: Lower posterior diaphysis of ulna I: Phalanges of finger 2 N: Posterior interosseous nerve

So, for example, the flexor digiti minimi brevis (as intimidating as that sounds at first!) simply translates as a short muscle that flexes the smallest finger (the fifth digit). Therefore, it will be useful to refer to Table 9.1 as you learn these muscles.

You can also make the study of these muscles easier on yourself by not trying to memorize each of their actions. Remember what we discussed in the Concept Boost on page 314–use logic to understand their actions rather than rote memorization.

(a) Superficial layer

Tendons of:
Flexor digitorum profundus
Flexor digitorum superficialis
Third lumbrical
Fourth lumbrical
Opponens digiti minimi
Flexor digiti minimi brevis
Abductor digiti minimi
Pisiform bone
Flexor carpi ulnaris tendon
Flexor digitorum superficialis tendons

Fibrous sheath
Second lumbrical
Dorsal interossei
First lumbrical
Adductor pollicis
Flexor pollicis brevis
Abductor pollicis brevis
Flexor retinaculum
Opponens pollicis
Abductor pollicis longus
Tendons of:
Palmaris longus
Flexor carpi radialis
Flexor pollicis longus

(b) Second layer

Palmar interossei
Dorsal interossei
Opponens digiti minimi
Flexor digiti minimi brevis (cut)
Abductor digiti minimi (cut)

Flexor digitorum profundus tendon
Flexor digitorum superficialis tendon
Adductor pollicis
Flexor pollicis brevis
Abductor pollicis brevis
Opponens pollicis
Flexor pollicis longus tendon

(c) Lumbricals

Lumbricals

(d) Palmar interossei

Palmar interossei

(e) Dorsal interossei

Dorsal interossei

Figure 9.20 Right hand, palmar view: muscles of the hand that move the fingers. Labels in **bold** indicate muscles and tendons involved in these actions.

Many of the muscles that move the hand and wrist share a common tendon of origin that attaches to the medial epicondyle of the humerus. These muscles include the **flexor carpi radialis** (FLEX-or CAR-pea raid-ee-AL-iss), **palmaris longus** (palm-AIR-iss LONG-us), and **flexor carpi ulnaris** (ALL-nare-iss), which flex the wrist (Figure 9.19a). Located slightly deeper are muscles with tendons that attach to the fingers and thumb to flex them. These include the **flexor digitorum superficialis, flexor digitorum profundus,** and **flexor pollicis longus** (POL-uh-sis). There are also muscles that pronate the forearm to turn the hand and palm down, such as the **pronator teres** and **pronator quadratus,** and antagonists that turn the palm up, as in supination. The **supinator** (soo-pin-NAY-tohr) is a forearm muscle that supinates the forearm. The most powerful supinator of the forearm, however, is the biceps brachii, which is a synergist in supination.

The extensor muscles of the wrist and fingers are located on the posterior and lateral side of the forearm. These muscles can be felt through the skin by flexing your elbow to 90 degrees and hyperextending your wrist. Many of these muscles share a common tendon of origin that attaches to the lateral epicondyle of the humerus. Like the flexor muscles, they can be organized into groups that extend the wrist or extend the fingers and thumb. The muscles that extend the wrist include the **extensor carpi radialis longus, extensor carpi radialis brevis,** and **extensor carpi ulnaris** (Figure 9.19b). Muscles that extend the fingers and thumb are the **extensor digitorum, extensor indicis** (IN-dih-kiss), **extensor digiti minimi, extensor pollicis longus,** and **extensor pollicis brevis.**

The muscles shown in Figure 9.20 and listed in Table 9.16 are located in the hand and move the fingers and thumb. As in Table 9.15, the groupings and names suggest their function and location. For example, the **abductor pollicis brevis** is a short muscle that abducts the thumb, and the **flexor pollicis brevis** is a short muscle that flexes the thumb (Figure 9.20a and b). The **lumbricals** are named for their "wormlike" appearance running from the palm to the fingers (Figure 9.20c). Interestingly, they flex the fingers where they meet the palm, but more distally, they *extend* the fingertips.

The last two muscles we will describe in this group, the **dorsal interossei** (inter-AH-see-eye), and **palmar interossei,** insert between the fingers (Figure 9.20d and e). The dorsal interossei abduct the fingers (spread them), and the palmar interossei adduct the fingers (pull them together side by side). The Vulcan hand greeting from *Star Trek,* "live long and prosper," uses these muscles—the palmar interossei pull the fingers together, while the dorsal interossei separate the third and fourth digits.

TABLE 9.16 MUSCLES OF THE HAND THAT MOVE THE FINGERS

Muscle(s)	Action(s)	Origin/Insertion/Nerve(s)
Abductor pollicis brevis	Abducts thumb	O: Flexor retinaculum, scaphoid and trapezium I: Lateral side of proximal phalanx of thumb N: Median nerve
Flexor pollicis brevis	Flexes thumb	O: Flexor retinaculum, trapezium I: Lateral proximal phalanx of thumb N: Median nerve
Opponens pollicis	Opposes thumb	O: Flexor retinaculum, trapezium I: First metacarpal N: Median nerve
Adductor pollicis	Adducts thumb	O: Capitate and metacarpals 2–3 I: Proximal phalanx of thumb N: Ulnar nerve
Abductor digiti minimi	Abducts fifth digit	O: Pisiform I: Proximal phalanx of finger 5 N: Ulnar nerve
Flexor digiti minimi brevis	Flexes fifth digit	O: Flexor retinaculum, hamate I: Proximal phalanx of finger 5 N: Ulnar nerve
Opponens digiti minimi	Opposes fifth digit to thumb	O: Flexor retinaculum, hamate I: Fifth metacarpal N: Ulnar nerve
Lumbricals	Flex metacarpophalangeal joints; extend interphalangeal joints	O: Tendons of the flexor digitorum profundus (in the palm) I: Proximal phalanx of fingers 2–5 N: Median nerve and ulnar nerve
Palmar interossei	Adduct fingers (pull extended fingers together)	O: Adductor side of metacarpals 2, 4, and 5 I: Proximal phalanx of fingers 2, 4, and 5 N: Ulnar nerve
Dorsal interossei	Abduct fingers (spread them, as when signaling the number 5)	O: Interior sides of metacarpals I: First phalanx of fingers 2, 3, and 4 N: Ulnar nerve

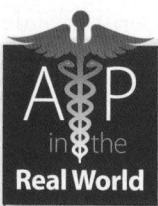

Rotator Cuff Injuries

A&P in the Real World

The rotator cuff group consists of small muscles with relatively minor actions. Yet, together they perform one tough job: holding the shoulder joint together during forceful movements that could damage the joint and even dislocate the humerus.

The rotator cuff muscles do the most work when executing forceful overhead arm movements like throwing a fastball in baseball. Therefore, it is not surprising that activities involving repetitive, strong, overhead motions can damage them. These injuries can afflict nonathletes, too, such as construction workers who spend a lot of time working with their hands above their heads. Anyone who falls on a shoulder or suffers other trauma to the area can damage the same muscles and tendons. Age is a factor, as well: Individuals over 40, especially those who exercise intensely, are more likely to suffer this injury.

Rotator cuff injuries vary widely in severity. Some involve localized inflammation that reverses itself with rest and anti-inflammatory medication. At the other extreme are acute muscle and tendon tears that require surgical intervention.

Someone experiencing an acute tear usually knows it right away, as typically immediate pain and loss of movement occur. Symptoms of less serious conditions include weakness, discomfort, and limited range of motion in the shoulder joint (particularly with overhead movements). Some people are not even aware of the injury until a simple motion like raising the arm to put on a shirt triggers acute damage or inflammation.

Targeted exercises can help prevent rotator cuff injuries. And if the cuff muscles do suffer damage, gentle exercise can serve as physical therapy afterward. Strengthening the muscles, tendons, and insertions without pushing them to the point of damage can help prevent future injuries.

Quick Check

☐ 5. Name and describe two muscles that flex the arm at the elbow joint.

☐ 6. Name and describe two muscles that extend the wrist.

Apply What You Learned

☐ 1. Explain and describe how the placement of the pectoralis major and latissimus dorsi can allow them to be antagonists in arm flexion and extension but to work together in adduction.

☐ 2. Explain why the brachialis is the agonist of elbow flexion, despite being much smaller than its synergist, the biceps brachii.

☐ 3. Name and describe three muscles a musician would use to pluck the strings of a guitar.

See answers in Appendix A.

MODULE **9.5**

Muscles of the Hip and Lower Limb

Learning Outcomes

1. Identify and describe the muscles of the hip area that move the hip and lower limb.
2. Identify the origin, insertion, and action of these muscles, and demonstrate their actions.

The bones and muscles of the lower limb differ structurally and functionally from those of the arm and shoulder. The pelvis (unlike the scapula) is firmly affixed to the axial skeleton at the sacrum, partially because of the need to support our weight—an example of the Structure-Function Core Principle (p. 25). This "frees up" muscles that would otherwise be required to fix and stabilize the pelvis, as many do for the scapula. Therefore, nearly all of the muscles in the pelvis and thigh are also involved in moving the thigh and leg.

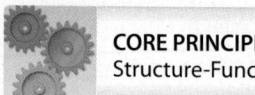

CORE PRINCIPLE
Structure-Function

Muscles That Move the Thigh and Knee

◀◀ **Flashback**

1. Describe the markings of the femur, tibia, and fibula. (pp. 250–252)
2. Identify and describe the obturator foramen and sacrum. (p. 247)

Let's start with the anterior and medial muscles of the thigh and knee, which are shown in **Figure 9.21**. The **iliopsoas** and **sartorius** (sar-TORE-ee-us) are anterior muscles of the thigh that flex the thigh (think of bringing your knees up while marching in place) (Figure 9.21a). The iliopsoas is formed by the union of two muscles, the **iliacus** (il-ee-AK-us) and **psoas major,** that insert in common to the lesser trochanter of the femur. The iliacus originates in the iliac fossa of the ilium, and the psoas major muscle arises from the lateral sides of the lumbar vertebrae. The psoas major can also flex the vertebral column laterally (bending to the side) when the hip joint is fixed. It is an example of a muscle for which the origin and insertion switch roles depending on which joints are fixed.

(a) Superficial muscles

12th rib

Iliac crest

Anterior superior iliac spine

Iliopsoas:
Iliacus
Psoas major

Sartorius

Quadriceps femoris:
Rectus femoris
Vastus lateralis
Vastus medialis
Vastus intermedius (not shown)

12th thoracic vertebra

5th lumbar vertebra

Pectineus

Adductor longus

Gracilis

Adductor magnus

Tendon of quadriceps femoris

Patella

Patellar ligament

(b) Deeper muscles under the rectus femoris

Vastus lateralis

Vastus intermedius

Vastus medialis

Rectus femoris tendon (cut)

Patella

Patellar ligament

(c) Adductor muscles

Adductor brevis

Adductor longus

Femur

Pectineus (cut)

Adductor magnus

(d) Cadaver photo

Iliopsoas

Sartorius

Rectus femoris

Vastus lateralis

Gracilis

Adductor longus

Vastus medialis

Figure 9.21 Anterior and medial muscles that move the thigh and knee. Labels in **bold** indicate muscles involved in these actions.

Three muscles of the inner thigh area *adduct* the thigh, as indicated by their names: the **adductor magnus, adductor longus,** and **adductor brevis** (Figure 9.21c). The **pectineus** (peck-tin-NEE-us) and **gracilis** (grass-ILL-uss) also adduct the thigh, in addition to assisting with flexion of the hip and knee, respectively (Figure 9.21a and d). Their angle of attachment

enables all five muscles to help rotate the thigh medially. See Table 9.17 for more information on anterior and medial muscles, and Figure 9.22 on posterior muscles.

The **quadriceps femoris** (KWAH-drih-seps FEM-oh-ris) muscle group of the anterior thigh extends or straightens the leg at the knee and is an antagonist of the hamstrings in extending

TABLE 9.17 MUSCLES THAT MOVE THE THIGH AND KNEE: ANTERIOR AND MEDIAL MUSCLES

Muscle(s)	Action(s)	Origin/Insertion/Nerve(s)	Concept Figures
Iliacus*	Flexes the thigh	O: Iliac fossa I: Lesser trochanter of the femur N: Femoral nerve	
Psoas major*	Flexes the thigh; **laterally** flexes the vertebral column	O: Lateral sides and transverse processes of T_{12} and lumbar vertebrae I: Lesser trochanter of the femur N: L_1–L_3	
Sartorius	Flexes the thigh and flexes the leg at the knee; abducts and laterally rotates the thigh; actions produce a "tailor's" seated posture	O: Anterior superior iliac spine I: Proximal portion of the medial condyle of the tibia N: Femoral nerve	
Adductor magnus	Adducts the thigh; rotates the thigh medially *(Lower part is a synergist of hamstring action, see Table 9.18.)*	O: Ischial tuberosity, ischial ramus, inferior ramus of pubis I: Linea aspera and adductor tubercle of femur N: Obturator and sciatic nerves	
Adductor longus	Adducts the thigh; flexes the thigh; rotates the thigh medially	O: Body of pubis I: Linea aspera N: Obturator nerve	
Adductor brevis	Adducts the thigh; flexes the thigh; rotates the thigh medially	O: Inferior ramus of the pubis I: Superior linea aspera N: Obturator nerve	
Pectineus	Adducts the thigh; flexes the thigh; rotates the thigh medially	O: Superior ramus of pubis I: Posterior femur between the lesser trochanter and linea aspera N: Femoral nerve	
Gracilis	Adducts the thigh; flexes the leg; rotates the thigh medially	O: Body and inferior ramus of pubis; ischial ramus I: Medial surface of proximal tibia N: Obturator nerve	
Rectus femoris†	Extends the knee; flexes the thigh	O: Anterior inferior iliac spine, superior margin of acetabulum I: Patella and tibial tuberosity N: Femoral nerve	
Vastus lateralis†	Extends the knee; stabilizes patella	O: Greater trochanter, intertrochanteric line, proximal half of the linea aspera I: Patella and tibial tuberosity N: Femoral nerve	
Vastus medialis†	Extends the knee; stabilizes patella	O: Linea aspera, intertrochanteric line I: Patella and tibial tuberosity N: Femoral nerve	
Vastus intermedius†	Extends the knee	O: Anterior and lateral portions of proximal two-thirds of the diaphysis of femur I: Patella and tibial tuberosity N: Femoral nerve	

*Part of the group of iliopsoas muscles that share a common insertion tendon.

†Part of the group of quadriceps femoris muscles that share a common insertion tendon.

(a) Superficial muscles

- Gluteus medius
- Gluteus maximus
- Adductor magnus
- Iliotibial tract
- Gracilis

Hamstrings:
- **Biceps femoris:**
  - Long head
  - Short head
- **Semitendinosus**
- **Semimembranosus**

(b) Cadaver photo, posterior view

(c) Muscles under the gluteus maximus and gluteus medius

- Gluteus medius
- Gluteus minimus
- **Piriformis**
- **Obturator internus**
- **Gemelli** (superior and inferior)
- **Obturator externus**
- **Quadratus femoris**
- Gluteus maximus

(d) The obturator externus muscle, anterior view

- **Obturator externus**

Figure 9.22 Posterior muscles that move the thigh and knee. Labels in **bold** indicate muscles involved in these actions.

the leg at the knee joint. The centrally located, superficial **rectus femoris** also assists with flexing the thigh because it crosses the hip joint. The names of the other three muscles of the group—the **vastus lateralis, vastus medialis** and **vastus intermedius** (in-ter-MEED-ee-iss)—suggest their relative placement (lateral, medial, and "in between," respectively) along the shaft of the femur (Figure 9.21a and b). All four components of this muscle unite at the knee joint to form the quadriceps tendon, which contains the patella, or kneecap. This tendon continues from the patella to the tibial tuberosity and is commonly referred to as the patellar ligament.

The **gluteal muscle group** is located on the posterior aspect of the pelvis and consists of muscles that extend, abduct, and medially and laterally rotate the thigh. The thrusting of the thighs in extension when we climb stairs demonstrates the role of the **gluteus maximus** (GLUE-tee-us MAX-ih-mus) (Figure 9.22a and b), whereas walking mainly uses the **gluteus medius** and **gluteus minimus** muscles (Figure 9.22c). During the swing phase of walking, when the unsupported limb swings forward before the next step, these two muscles play an important role in abducting and medially rotating the thigh of the unweighted limb. They are also important during the stance phase (when weight is placed on one limb) to keep the pelvis level. Several small muscles—the **piriformis** (peer-ah-FORM-iss), **obturator internus, gemelli** (gah-MELL-ee), **quadratus femoris,** and **obturator externus** (Figure 9.22c and d)—also help laterally rotate the thigh.

The muscles of the **hamstring group,** or "hamstrings"—the **biceps femoris, semitendinosus** (sem'-ee-ten-dih-NOH-sus), and **semimembranosus** (sem'-ee-mem-brah-NOH-sus)—flex the leg at the knee, a movement important in walking (Figure 9.22a). However, their origin on the ischial tuberosity means the group also spans the hip joint and therefore works with the gluteus maximus to extend the thigh. When a runner is running fast to beat a throw in baseball, he or she is demonstrating that combined action.

Remembering the Difference between Semitendinosus and Semimembranosus

If the names semitendinosus and semimembranosus sound too similar, here is a tip to keep them straight. Recall that tendons are often attached *centrally,* especially in broad, symmetrical muscles. Thus, the semitendinosus is the muscle located in the *center* of the posterior thigh between the biceps femoris and semimembranosus. As for the semiMeMbranosus, remember that it is the Most Medial hamstring. ∎

In this chapter we have discussed many muscles and their tendons of attachment to bones. Almost any tendon can be damaged when subjected to a force that is substantially greater than what it is "accustomed to" (see *A&P in the Real World: Calcaneal Tendon Injuries*).

See **Table 9.18** for more details regarding posterior muscles that move the thigh and knee.

Quick Check

☐ 1. Name and describe the major muscles of leg extension.
☐ 2. Name and describe the muscles of leg flexion.

Muscles That Move the Ankle and Foot

◀◀ **Flashback**

1. What are the bones of the leg? With which tarsal bone do they articulate? (p. 251)
2. What occurs during the motions dorsiflexion, plantarflexion, inversion, and eversion? (p. 269)

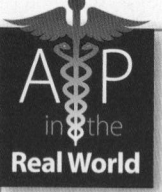

Calcaneal Tendon Injuries

Injuries to the calcaneal (Achilles) tendon of the posterior leg are relatively common in sports requiring powerful running and changes in direction (such as football). These injuries also occur in weekend athletes and nonathletes who subject the tendon to undue stress, such as stepping in a hole or excessive stair climbing. The tendon may become inflamed, tear, or even rupture (break).

Why is the calcaneal tendon so vulnerable? Running and walking depend on a combination of hip, knee, and foot movements, with the muscles that move the hip and knee providing the most force. The lever system of the foot involved in plantar flexion and pushing off is a second-class lever where the fulcrum is far to one side. Therefore, a great deal of force is exerted on the calcaneus and calcaneal tendon in order to produce small movements at the "ball" of the foot (recall the Concept Boost on levers

and mechanical advantage/disadvantage). So, the little extra push needed to break that tackle or climb multiple steps actually exerts a great deal of force on the calcaneal tendon. Treatment for a ruptured tendon may involve nonsurgical options, such as wearing a walking boot while the tendon heals, or surgery to reattach the torn tissue.

As with many musculoskeletal injuries, a little prevention can forestall problems:

- Stretch the calcaneal tendons and calf muscles gently before exercising.
- Vary activities so you do not stress the tendon in the same way all the time.
- Increase exercise intensity gradually, over several weeks. Gentle increases in exercise or activities that stress the calcaneal tendon strengthen it without damaging the tissue.

TABLE 9.18 MUSCLES THAT MOVE THE THIGH AND KNEE: POSTERIOR MUSCLES

Muscle(s)	Action(s)	Origin/Insertion/Nerve(s)	Concept Figures
Gluteus maximus	**Extends** the thigh (especially when the thigh is in flexed position); laterally rotates the thigh; **abducts** the thigh	O: Posterior and lateral portions of the ilium, sacrum, and coccyx I: Gluteal tuberosity of femur N: Inferior gluteal nerve	
Gluteus medius	**Abducts** the thigh; medially rotates the thigh; stabilizes pelvis while walking	O: Between posterior and anterior gluteal lines on the outer surface of ilium I: Greater trochanter of the femur N: Superior gluteal nerve	
Gluteus minimus	**Abducts** the thigh; medially rotates the thigh; stabilizes pelvis while walking	O: Between the anterior and inferior gluteal lines on the outer surface of ilium I: Greater trochanter of the femur N: Superior gluteal nerve	
Piriformis	**Abducts** the thigh; laterally rotates the thigh	O: Anterior/lateral sacrum I: Greater trochanter of femur N: L_5, S_1–S_2	
Obturator externus	Laterally rotates the thigh	O: Edges of and outer surface of obturator foramen and membrane I: Greater trochanter of femur N: Obturator nerve	
Obturator internus	Laterally rotates the thigh	O: Edges of and inner surface of obturator foramen and membrane I: Greater trochanter of femur N: L_5, S_1	
Gemelli (superior gemellus and inferior gemellus)	Laterally rotate the thigh	O: Ischial spine and tuberosity I: Greater trochanter of femur N: L_5, S_1	
Quadratus femoris	Laterally rotates the thigh	O: Ischial tuberosity I: Intertrochanteric crest of femur N: L_5, S_1	
Biceps femoris*	**Extends** the thigh; **flexes** the knee	O: Ischial tuberosity and lower half of posterior femur I: Head of fibula; lateral condyle of tibia N: Sciatic nerve	
Semitendinosus*	**Extends** the thigh; **flexes** the knee	O: Ischial tuberosity I: Upper medial surface of tibia N: Sciatic nerve	
Semimembranosus*	**Extends** the thigh; **flexes** the knee	O: Ischial tuberosity I: Posterior surface of medial condyle of tibia N: Sciatic nerve	

Gluteus muscle group

Piriformis

Obturator and gemelli muscles

Quadratus femoris

Biceps femoris

Semitendinosus

Semimembranosus

*Part of the hamstring muscle group.

Muscles That Move the Foot and Toes

The movements of this group of muscles are subtle, yet important in helping move the foot correctly for walking, running, dancing, or standing on tiptoe. The anterior and lateral muscles of this group are shown in **Figure 9.23**. The **tibialis anterior** and **extensor digitorum longus** are the main muscles of dorsiflexion (movement of the dorsal or upper surface of the foot toward the leg, as when walking on the heels). The extensor digitorum longus also extends toes 2–5 (digits), as its name implies. The **extensor hallucis longus** assists dorsiflexion by pulling up on the hallux (great toe) and also extends the hallux, as its name suggests.

Far more muscles are involved in plantarflexion, or pushing the front of the foot down (think "planting" the foot) for actions such as standing on your toes and pushing off when running. The *triceps surae* group, consisting of the **soleus** (SO-lee-us) and the two heads of the **gastrocnemius** (gas′-trock-NEE-me-us), may be the most familiar muscles producing this action; they are

shown in **Figure 9.24** with the posterior muscles. Another tip for remembering this action is that in a car you use the gastrocnemius to step on the gas. These muscles form the fleshy "calf" of the leg and taper down to the calcaneal (Achilles) tendon.

See **Table 9.19** for more information on muscles that move the foot and toes. Most of the muscles in the lower three-fourths of Table 9.19 share plantar flexion as their action with the gastrocnemius and soleus.

Some of the muscles in this group also assist with moving the sole of the foot outward (eversion) or inward (inversion). The **fibularis longus** and **fibularis brevis** muscles are the main muscles of foot eversion. The **tibialis posterior** and **tibialis anterior** are the main muscles of foot inversion. The **flexor hallucis longus** flexes the hallux, and the **flexor digitorum longus** flexes toes 2–5. You use these muscles when you curl your toes.

Having already explored the muscles that move the fingers, you will find that many of the small muscles that move the toes

(a) Anterior view

(b) Lateral view

Figure 9.23 Anterior and lateral muscles that move the ankle and toes. Labels in **bold** indicate muscles involved in these actions.

Gastrocnemius

Tendon of
gastrocnemius

Soleus

Medial
malleolus

Lateral
malleolus

Calcaneal
(Achilles)
tendon

Calcaneus

(a) Superficial muscles

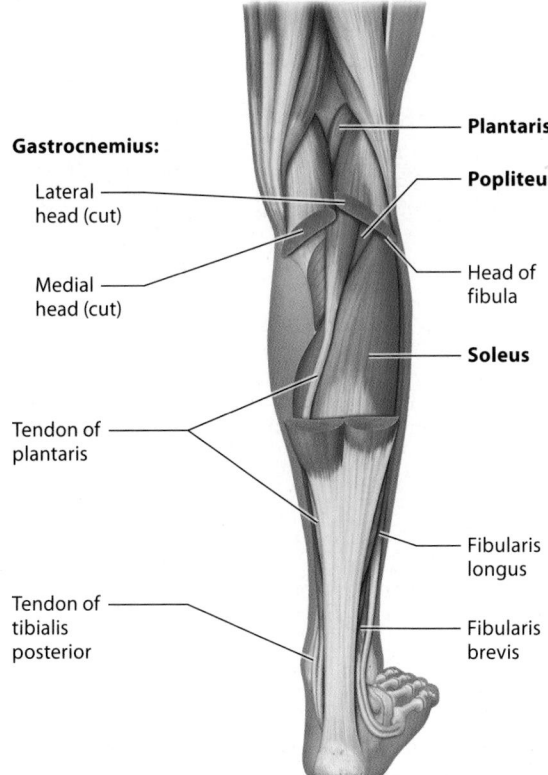

Gastrocnemius:

Lateral
head (cut)

Medial
head (cut)

Plantaris

Popliteus

Head of
fibula

Soleus

Tendon of
plantaris

Tendon of
tibialis
posterior

Fibularis
longus

Fibularis
brevis

(b) Muscles under the gastrocnemius

Plantaris
(cut)

Gastrocnemius
(cut)

Popliteus

Soleus
(cut)

Fibula

**Tibialis
posterior**

**Flexor
hallucis
longus**

**Flexor
digitorum
longus**

**Fibularis
longus**

Tendon of
tibialis
posterior

**Fibularis
brevis**

Medial
malleolus

Calcaneal
tendon (cut)

Calcaneus

(c) Deeper muscles under the gastrocnemius and soleus

Figure 9.24 Right leg: posterior muscles that move the ankle and toes. Labels in **bold** indicate muscles involved in these actions.

TABLE 9.19 MUSCLES THAT MOVE THE FOOT AND TOES

Muscle(s)	Action(s)	Origin/Insertion/Nerve(s)	Concept Figures
Tibialis anterior	**Dorsiflexes** the foot; **inverts** the foot	O: Lateral condyle and proximal diaphysis of the tibia I: Medial cuneiform and first metatarsal N: Deep fibular nerve	
Extensor digitorum longus	**Extends** the toes; **dorsiflexes** the foot	O: Lateral condyle of the tibia and upper portion of the fibula I: Phalanges and connective tissues of toes 2–5 N: Deep fibular nerve	
Extensor hallucis longus	**Extends** the hallux (great toe); **dorsiflexes** the foot	O: Diaphysis of the fibula and interosseous membrane I: Distal phalanx of hallux N: Deep fibular nerve	
Fibularis (peroneus) longus	**Everts** the foot; **plantarflexes** the foot	O: Upper lateral fibula I: Medial cuneiform and first metatarsal N: Superficial fibular nerve	
Fibularis brevis	**Everts** the foot; **plantarflexes** the foot	O: Lower fibula I: Fifth metatarsal N: Superficial fibular nerve	
*Gastrocnemius**	**plantarflexes** the foot; assists with flexing knee	O: Medial and lateral condyles of the femur I: Posterior surface of calcaneus N: Tibial nerve	
*Soleus**	**plantarflexes** the foot	O: Head of the fibula, proximal tibia, and interosseous membrane I: Posterior surface of calcaneus N: Tibial nerve	
Plantaris	**plantarflexes** the foot; assists with flexing knee (very weak synergist)	O: Posterior femur near lateral condyle I: Calcaneus and calcaneal tendon N: Tibial nerve	
Tibialis posterior	**Inverts** the foot; **plantarflexes** the foot; stabilizes foot	O: Upper tibia, fibula, and interosseous membrane I: Metatarsals 2–4 N: Tibial nerve	
Flexor digitorum longus	**Flexes** toes; stabilizes foot	O: Posterior tibia I: Distal phalanges of toes 2–5 N: Tibial nerve	
Flexor hallucis longus	**Flexes** the hallux	O: Posterior fibula and interosseous membrane I: Distal phalanx of hallux N: Tibial nerve	
Popliteus	**Flexes** leg at the knee; rotates leg and other minor movements, depending on which fixators are in use	O: Lateral condyle of the femur I: Posterior surface of proximal tibia N: Tibial nerve	

*Part of the group of triceps surae muscles that share a common insertion tendon, the calcaneal (Achilles) tendon.

have similar names (**Figure 9.25**). The muscles of the foot help support the arches of the foot and allow it to adapt to changing terrain when we walk. Although we often think of hand muscles as more adept with precise movements, the muscles of the feet can also be trained to perform skilled movements, such as drawing or knitting, as shown by individuals with upper limb disabilities. Most of the muscles in this group are named for their actions. For example, the **flexor digitorum brevis** is a small muscle that flexes toes 2–4.

Other muscles in this region are named for their appearance or their placement. The **lumbricals** are, again, the "wormlike" muscles that flex the toes and extend the tips of the toes, similar to their actions in the hand. The **plantar interossei** and **dorsal interossei** attach "between the bones," as do their counterparts in the hand, except that the plantar interossei arise from underneath the metatarsals.

See **Table 9.20** for more details about muscles that move the toes.

Quick Check

☐ 3. Name and describe the major muscles of dorsiflexion.
☐ 4. Name and describe the muscles of plantarflexion.

Apply What You Learned

☐ 1. Explain how the rectus femoris can be part of the quadriceps femoris group that mainly extends the knee but also flexes the thigh.
☐ 2. Consider the movements of plantarflexion and dorsiflexion. Propose a reason why there are far more (and stronger) muscles involved in plantarflexion.

Dorsum
(top) of foot

Superior and
inferior extensor
retinacula

**Extensor
hallucis
brevis**

**Extensor
digitorum
brevis**

Plantar aspect
(sole) of foot

Flexor digiti
minimi brevis

**Abductor
digiti minimi**

Quadratus
plantae

Calcaneus

Tendon of flexor
hallucis longus

Lumbricals

Flexor hallucis
brevis

**Flexor digitorum
brevis**

Abductor hallucis

Plantar aponeurosis
(cut)

(a) Superficial muscles of the dorsum and the plantar aspect

Plantar aspect

Lumbricals

**Flexor hallucis
brevis**

Flexor hallucis
longus tendon

Flexor digitorum
longus (tendon)

**Flexor digiti
minimi brevis**

**Quadratus
plantae**

Adductor hallucis:

Oblique head

Transverse head

(b) Deeper muscles of the plantar aspect

Dorsal aspect

**Dorsal
interossei**

Plantar aspect

**Plantar
interossei**

(c) Muscles between the toes

Figure 9.25 Muscles that move the toes. Labels in **bold** indicate muscles involved in these actions.

TABLE 9.20 MUSCLES THAT MOVE THE TOES		
Muscle(s)	**Action(s)**	**Origin/Insertion/Nerve(s)**
Extensor digitorum brevis	Extends ("uncurls") the toes	O: Superior lateral surface of calcaneus I: Dorsal surfaces of toes 2–4 N: Deep fibular nerve
Extensor hallucis brevis	Extends hallux	O: Superior lateral surface of calcaneus I: Dorsal surface of proximal phalanx of hallux N: Deep fibular nerve
Flexor digitorum brevis	Flexes ("curls") the toes	O: Calcaneal tuberosity I: Middle phalanx of toes 2–5 N: Medial plantar nerve
Abductor digiti minimi	Flexes and abducts fifth toe	O: Calcaneal tuberosity I: Lateral side of the base of proximal phalanx of toe 5 N: Lateral plantar nerve
Abductor hallucis	Abducts the hallux	O: Calcaneal tuberosity and medial connective tissue I: Medial side of the base of proximal phalanx of hallux N: Medial plantar nerve
Quadratus plantae	Flexes the toes (assists the flexor digitorum longus)	O: Sides of the calcaneus and medial connective tissue I: Tendon of flexor digitorum longus N: Lateral plantar nerve
Lumbricals	Flex toes close to ball of foot; extend toes at phalanges ("curl" and straighten toes)	O: Tendons of the flexor digitorum longus I: Connective tissues of toes 2–5 N: Lateral and medial plantar nerves
Adductor hallucis	Adducts the hallux; stabilizes the metatarsals and arch of the foot	O: Metatarsals and surrounding connective tissues, including the tendon of the fibularis longus I: Base of proximal phalanx of hallux (lateral side) N: Lateral plantar nerve
Flexor hallucis brevis	Flexes the hallux	O: Cuboid, lateral cuneiform, and tendon branches of other muscles (e.g., the tibialis posterior) I: Base of hallux and tendons of other muscles (e.g., the adductor hallucis) N: Medial plantar nerve
Flexor digiti minimi brevis	Flexes the fifth toe	O: Fifth metatarsal and tendon of fibularis longus I: Base of proximal phalanx of toe 5 N: Lateral plantar nerve
Plantar interossei	Adduct and flex toes 3–5	O: Metatarsals 3–5 I: Medial base of proximal phalanges of toes 3–5 N: Lateral plantar nerve
Dorsal interossei	Abduct and flex toes 2–4	O: Metatarsals 2–5 I: Base of proximal phalanges of toes 2–4 N: Lateral plantar nerve

☐ 3. Compare and contrast the location and function of the plantar and dorsal interossei muscles of the foot and hand.

See answers in Appendix A.

MODULE **9.6**

Putting It All Together: The Big Picture of Muscle Movement

Learning Outcomes

1. Identify the locations of the major skeletal muscles, and demonstrate their actions.

Throughout this chapter, we have demonstrated that muscles must work together in a coordinated fashion to accomplish even the simplest movements. We will finish by presenting a common activity—climbing stairs—and how we use our muscles to accomplish it. The next time you climb steps, notice these muscles in action.

Figure 9.26 breaks this process down. (For simplicity, the figure focuses on basic components. The complete process is more complicated and involves more muscles than can be shown here.) We assume you walk up the right side of the staircase and take the first step using your left leg. You use your right arm to grasp the handrail. In essence, as you climb a flight of stairs, the movements of your legs result in your "catching up" to your flexed arm. You repeat this sequence of muscle actions again and again until you reach your destination.

Challenge yourself and your classmates to think of other activities performed every day. Try to identify the actions involved and the muscles responsible for them.

The Big Picture of Muscle Movement

Figure 9.26

Pectoralis major and **coracobrachialis** flex your right arm.

Brachioradialis flexes your right forearm.

Hamstrings flex your left leg at the knee.

Iliopsoas, **rectus femoris**, and **sartorius** flex your left thigh at the hip.

Tibialis anterior and **extensor digitorum longus** dorsiflex your left foot at the ankle.

Quadriceps femoris extends your left leg at the knee.

Gluteus maximus and **hamstrings** extend your left thigh at the hip.

Latissimus dorsi and **teres major** extend your right arm at the shoulder.

Triceps brachii extends your right forearm.

Gastrocnemius and **soleus** plantarflex your right foot at the ankle.

① Bring your left foot to the first step and flex your right arm to reach for the handrail.

② Stand upright on your left leg and extend your right arm and forearm on the handrail.

Latissimus dorsi, teres major, and **triceps brachii** hyperextend your right arm at the shoulder.

Iliopsoas, rectus femoris, and **sartorius** flex your right thigh at the hip.

Gluteus maximus and **hamstrings** extend your right thigh at the hip.

Hamstrings flex your right leg at the knee.

Hamstrings flex your left leg at the knee.

Quadriceps femoris extends your right leg at the knee.

Tibialis anterior and **extensor digitorum longus** dorsiflex your right foot at the ankle.

③ Bring your right leg to the next step and extend it. Take the weight off your left leg and hyperextend your right arm to swing it forward again, reaching for the handrail.

CHAPTER SUMMARY

Take full advantage of the invaluable online practice and study tools for this chapter by going to MasteringA&P®. There you will find:

- Practice tests, module quizzes, art-labeling and sequence-the-steps activities
- Animated figures with quizzes
- *Practicing A&P* worksheets
- Author-narrated podcast
- Dynamic Study Modules
- *Interactive Physiology*

MODULE 9.1
Overview of Skeletal Muscles 283

- Skeletal muscles vary in size and shape, but have a similar structure. (Figure 9.1, p. 284)
  - A muscle cell is also called a muscle fiber, and each is surrounded by an extracellular matrix called **endomysium**.
  - A **fascicle** is a bundle of many muscle fibers enclosed by connective tissue called the **perimysium**.
  - A muscle is composed of many fascicles all enclosed in a connective wrapping known as the **epimysium**.
- The pattern of fascicle arrangement greatly contributes to the appearance and function of a skeletal muscle. (Figure 9.2, p. 285)
- Evenly spaced fascicles attaching to a tendon that is about the same width as the muscle is a **parallel** orientation.
- A broad muscle that uniformly tapers down to a single, ropelike tendon is referred to as a **convergent** orientation.
- **Pennate** orientation describes fibers and fascicles that "feather" outward at an angle from an extension of the tendon partway into the "meat" of the muscle.
- **Circular** muscles (e.g., **sphincters**) simply encircle a structure such as the eye or urinary tract opening.
- **Spiral** muscles may wrap around a bone or have a twisted appearance.
- Some muscles are named for their shape, appearance, size, position, or other structural considerations, such as number of heads (e.g., triceps). (Table 9.1, pp. 286–287)
- Some muscles are named for their actions, such as **flexors** and **extensors.**
- Skeletal muscles commonly move at joints and move the bones to which they are attached.
- Other functions of skeletal muscle include facial expression, communication, breathing, and generating heat to regulate body temperature.
- Muscles often do work in coordinated groups, with each muscle having a specific role. (Figure 9.3, p. 288)

- **Agonists,** or *prime movers,* usually provide most of the force required for the movement.
- **Antagonists** oppose or slow the action.
- **Synergists** work with the agonist, providing additional force and guidance.
- **Fixators** hold a bone in place, anchoring it for more efficient movement.
- Points of attachment of a muscle are its more stationary **origin** and more mobile **insertion.** (Figure 9.4, p. 289)
- When muscles produce movement at joints, they function in accordance with the physics of lever systems. (Figure 9.5, p. 291)
- A **first-class lever** places the **fulcrum,** or supporting hinge point, *between* the point of applied force and the load to be moved.
- A **second-class lever** applies force near the end *opposite* the fulcrum with the load between the applied force and the fulcrum.
- A **third-class lever** applies force much closer to the fulcrum with the load farther away from the fulcrum.
- The concepts of mechanical advantage and disadvantage relate to arrangement of the lever system such that a force can be translated into a great deal of movement (usually of something lighter) or less movement (usually of something heavier).
- The anterior and posterior superficial muscles of the human body are summarized in Figures 9.6 and 9.7 (pp. 292 and 293).

   Practice art-labeling exercises: Figures 9.6 and 9.7.

MODULE 9.2
Muscles of the Head, Neck, and Vertebral Column 294

- The muscles of facial expression often insert on skin or connective tissue rather than bone, producing subtle yet important shape changes to the skin and structures of the face. (Figure 9.8, p. 295; Table 9.2, p. 296)
  - The **frontalis, occipitalis** (parts of the **epicranius),** and **corrugator supercilii** contribute to movements of the forehead and eyebrows.
  - The circular **orbicularis oculi** closes the eye and contributes to blinking.
  - The **zygomaticus major, zygomaticus minor, levator labii superioris,** and **risorius** muscles interact with the mouth and surrounding area, producing expressions of smiling, grimacing, grinning, and sneering.
  - The **orbicularis oris** contributes to the expressions of smiling, grimacing, grinning, and sneering, as well as actions involved in eating, drinking, and speech.
  - The **depressor anguli oris, depressor labii inferioris,** and **mentalis** contribute to sad, "pouting," and "doubtful" expressions.
  - The **buccinator** is under most of the skin of the cheek, and moves the cheeks to manipulate food, to blow, and to whistle.

   Practice art-labeling exercise: Figure 9.8.

- The six extrinsic eye muscles move the eyeball, with the four **rectus** muscles positioned and producing movement according to their names (**superior, inferior, medial,** and **lateral rectus**). The two **oblique** muscles, **superior** and **inferior,** angle across the eyeball and produce movements mostly opposite from their names. (Figure 9.9, p. 297; Table 9.3, p. 298)

- Muscles of the head and neck are important for mastication, swallowing, and maintaining posture. (Figure 9.10 and Table 9.4, p. 299)
  - The **masseter** and **temporalis** provide much of the force for elevating the mandible to masticate food, with the **medial pterygoid** acting as a synergist.
  - The **lateral pterygoid** protrudes the mandible and along with the **digastric** muscles depresses the mandible.

- The muscles of swallowing work by pushing food to the back of the oral cavity, and then into the throat, or **pharynx,** and the esophagus. (Figure 9.11, p. 300; Tables 9.5 and 9.6, p. 301)
  - The **genioglossus, hyoglossus,** and **styloglossus** move the tongue to manipulate food while chewing and push the food into the throat.
  - The **stylohyoid, mylohyoid, geniohyoid,** and digastric muscles elevate the hyoid bone, which raises the floor of the mouth and helps nudge the food posteriorly.
  - Then the **sternohyoid, sternothyroid, omohyoid,** and **thyrohyoid** depress the hyoid, larynx, and pharynx during this phase of swallowing as the food enters the pharynx.
  - The **pharyngeal constrictor muscles** rapidly contract, in order from top to bottom of the pharynx, to push the food into the esophagus.

- Muscles that move the head include the **sternocleidomastoid** and **scalenes,** which rotate and flex the head. (Figure 9.12, p. 302; Table 9.7, p. 303)

- The *trapezius,* **splenius capitis, splenius cervicis,** and **semispinalis capitis** extend the head and neck, with the latter three also involved in rotation.

- Numerous beltlike muscles parallel the vertebral column and are involved in extension of the vertebral column. The most superior of these also extends and rotates the head. (Figure 9.13, p. 304; Table 9.8, p. 306)
  - The **erector spinae** group is made of three groups of muscles placed from close to the vertebrae to more lateral (in order): the **spinalis, longissimus,** and **iliocostalis** groups, each with three muscles.
  - Also running parallel to the vertebrae and playing supporting roles in extension of the vertebral column and posture are the **transversospinal** group of muscles: **semispinalis, multifidus,** and **rotatores.**
  - The **quadratus lumborum** also is involved in extension of the vertebral column.

MODULE **9.3**
Muscles of the Trunk and Pelvic Floor 307

- Muscles of respiration include the **diaphragm, external intercostals,** and **internal intercostals.** The first two increase the volume of the thoracic cavity to promote inspiration, and the last decreases the volume of the thoracic cavity to promote expiration. (Figure 9.14, p. 307; Table 9.9, p. 308)

- Abdominal muscles include the **rectus abdominis, external oblique, internal oblique** (which flex the trunk and compress the abdominal cavity), and **transversus abdominis** (which only compresses the abdominal cavity). (Figure 9.15, p. 309; Table 9.10, p. 310)

    Practice art-labeling exercise: **Figure 9.15.**

- Muscles of the pelvic floor provide support for pelvic cavity organs and regulate movement of materials out of the body. (Figure 9.16, p. 311; Table 9.11, p. 312)

- The three muscles of the pelvic diaphragm are the **pubococcygeus** and the **iliococcygeus** of the **levator ani** group and the **coccygeus.** They form the floor of the pelvis and help to work like sphincters around the vagina and anal canal.

- The **external urethral sphincter** and **external anal sphincter** are circular muscles that allow voluntary control of urination and defecation.

- The **deep transverse perineal** and **superficial transverse perineal** muscles support the pelvic organs and stabilize the perineal body or central tendon of the perineum.

- The **bulbospongiosus** and **ischiocavernosus** help with erection of the penis and clitoris.

MODULE **9.4**
Muscles of the Pectoral Girdle and Upper Limb 313

- A number of muscles that move the scapula elevate, retract, protract, depress, and rotate it. (Figure 9.17, p. 313; Table 9.12, p. 314)
  - Parts of the **trapezius** elevate, retract, and depress the scapula, and rotate the glenoid cavity of the scapula upward, depending on which portions of the muscle are stimulated.
  - The **levator scapulae** elevates the scapula. The **rhomboid major** and **rhomboid minor** mainly retract the scapula and assist in rotating the glenoid of the scapula downward.
  - The **serratus anterior** protracts the scapula and rotates it upward. The **pectoralis minor** also assists with protraction and depression of the scapula.

- The muscles that move the humerus flex, extend, abduct, adduct, and rotate the humerus. (Figure 9.18, p. 315; Table 9.13, p. 316)
  - The **pectoralis major** is a powerful agonist of arm flexion and assists in adduction and medial rotation of the arm.
  - The **latissimus dorsi** is a powerful agonist of adduction and medial rotation of the arm, as is the **teres major.** Both muscles extend the arm and serve as antagonists to the pectoralis major.
  - The **deltoid** is the agonist of arm abduction after it has been initiated by the **supraspinatus.**
  - The **coracobrachialis** adducts and flexes the humerus.
  - The **teres minor** and **infraspinatus** are principal lateral rotators of the arm. They are part of the **rotator cuff** group of four muscles (supraspinatus, infraspinatus, teres minor, and subscapularis) that stabilize the shoulder joint.
  - The **subscapularis** medially rotates the arm and assists the other rotator cuff muscles in stabilizing the shoulder joint.

- Muscles that flex the arm at the elbow joint include the **biceps brachii** and the deeper **brachialis** that is actually the agonist of elbow flexion. The **brachioradialis** is a synergist of elbow flexion. (Figure 9.18, p. 315; Table 9.14, p. 317)

- Muscles that extend the arm at the elbow joint include the **triceps brachii** and the *anconeus* (a small muscle that also assists with extension of the elbow joint).

- Numerous small muscles move the wrist and hand, and are usually named for their action and placement. (Figure 9.19, p. 318; Table 9.15, p. 319)

  - The **pronator teres** and **pronator quadratus** turn the palm of the hand and forearm posteriorly, whereas the **supinator** and **biceps brachii** turn them anteriorly.

  - Muscles of this group with the term flexor in their names, such as the **flexor pollicis longus,** flex the thumb or other parts of the hand at the wrist.

- Numerous small muscles move the fingers, and are usually named for their action and placement. (Figure 9.20, p. 320; Table 9.16, p. 321)

  - Muscles of this group with the term extensor in their names, such as the **extensor indicis** (which extends the index finger), extend the fingers.

  - The **palmar interossei** and **dorsal interossei** are muscles that insert between the fingers and adduct and abduct the fingers, respectively.

MODULE 9.5
Muscles of the Hip and Lower Limb 322

- The **iliopsoas,** formed of the **iliacus** and **psoas major** muscles, and the **sartorius** flex the thigh. (Figure 9.21, p. 323; Table 9.17, p. 324)

- The **adductor magnus, adductor longus,** and **adductor brevis** adduct the thigh and assist with medial rotation of the thigh.

- The **pectineus** and **gracilis** also adduct the thigh, and assist with flexion and medial rotation of the thigh.

- The **quadriceps femoris** muscle group (**rectus femoris, vastus lateralis, vastus medialis,** and **vastus intermedius**) of the anterior thigh extends the leg at the knee and is an antagonist of the hamstrings.

- The **gluteal** muscle group (**gluteus maximus, gluteus medius,** and **gluteus minimus**) is responsible for extension, abduction, and rotation of the thigh. The gluteus medius and minimus are also important stabilizers of the pelvis during walking and running. (Figure 9.22, p. 325; Table 9.18, p. 327)

- The **piriformis, obturator externus, obturator internus, gemellus,** and **quadratus femoris** assist with abduction and lateral rotation of the thigh.

- The muscles of the **hamstring group** (the **biceps femoris, semitendinosus,** and **semimembranosus**) flex the leg at the knee.

- The main muscles that move the ankle in dorsiflexion are the **tibialis anterior** and **extensor digitorum longus.** To a lesser extent, the **extensor hallucis longus** also assists dorsiflexion. (Figure 9.23, p. 328; Table 9.19, p. 330)

- The main muscles that move the ankle in plantarflexion are the **gastrocnemius** and **soleus.** (Figure 9.24, p. 329)

- Muscles that move the ankle in eversion and inversion are the **fibularis longus** and **fibularis brevis** (eversion) and the **tibialis anterior** and **tibialis posterior** (inversion).

- Numerous small muscles move the toes, and are usually named for their action and placement (e.g., the **flexor digitorum brevis** is a small muscle that flexes the toes; the **extensor digitorum brevis** is a small muscle that extends the toes). (Figure 9.25, p. 331; Table 9.20, p. 332)

- Other muscles that move the toes have similarly named (and functioning) counterparts in the hand, such as the **plantar interossei** and **dorsal interossei** muscles.

MODULE 9.6
Putting It All Together: The Big Picture of Muscle Movement 332

- Stair climbing requires a combination of lower limb and upper limb actions. (Figure 9.26, p. 333)

ASSESS WHAT YOU LEARNED

See answers in Appendix A.

LEVEL 1 Check Your Recall

1. Which type of muscle fascicle pattern has an appearance similar to a feather?

   a. Circular

   b. Fusiform

   c. Triangular

   d. Pennate

   e. Parallel

2. If "adductor" is part of the name of a muscle, what does that tell you about the muscle?

   a. It is a wide muscle.

   b. It is a long muscle.

   c. It raises a body part.

   d. It is found in the head region.

   e. It pulls a body part toward the midline.

3. If "profundus" is part of the name of a muscle, what does that tell you about the muscle?

   a. It is near or nearer the surface.

   b. It is deeper or farther from the surface.

   c. It is very wide.

   d. It is very long.

   e. It lowers a body part.

4. Where would you expect to find a muscle with "pollicis" in its name?

   a. Attached to the thumb

   b. Attached to the great toe

   c. Around the eye

   d. In the pelvis

   e. In the arm

5. Match the term with its description:

_____	Brevis	a.	Related to the fingers/toes
_____	Digitorum	b.	Straight
_____	Hallucis	c.	Short
_____	Rectus	d.	Decreases the angle between bones
_____	Flexor	e.	Turns palm down
_____	Pronator	f.	Related to the hallux (great toe)

6. Of which kind of lever system is the action of the triceps brachii on the hinge joint of the elbow an example?

a. First class
b. Second class
c. Third class
d. Fourth class

7. Which function is being fulfilled by a muscle that does not contribute to the force of a movement but holds a bone steady?

a. Antagonist
b. Synergist
c. Supinator
d. Fixator
e. Pronator

8. Which muscle of facial expression is involved in "sneering" or raising the upper lip?

a. Zygomaticus major
b. Risorius
c. Levator labii superioris
d. Platysma
e. Levator palpebrae superioris

9. Which eye muscle passes through the trochlea and turns the eye inferiorly and laterally?

a. Superior rectus
b. Inferior rectus
c. Inferior oblique
d. Superior oblique
e. Lateral rectus

10. The thick, bandlike muscle of mastication that covers much of the ramus of the mandible is the:

a. masseter.
b. temporalis.
c. lateral pterygoid.
d. risorius.
e. buccinator.

11. Which of the following muscles depresses the mandible (as in opening the jaw)?

a. Masseter
b. Platysma
c. Digastric
d. Zygomaticus major
e. Zygomaticus minor

12. Which is a muscle that elevates and retracts the tongue?

a. Buccinator
b. Stylohyoid
c. Hyoglossus
d. Styloglossus
e. Platysma

13. Which of the erector spinae muscles are most lateral to the vertebral column?

a. Spinalis
b. Longissimus
c. Iliocostalis
d. Semispinalis
e. Multifidus

14. Which muscles raise and spread the ribs?

a. External intercostals
b. Internal intercostals
c. External obliques
d. Internal obliques
e. Digastrics

15. With which muscle is the linea alba most closely associated?

a. Epicranius
b. Hamstring group
c. Rectus abdominis
d. Rectus femoris
e. Quadratus lumborum

16. Which muscle fits this description: a thick, tapering muscle running from the back of the scapula to the proximal anterior portion of the humerus?

a. Serratus anterior
b. Brachialis
c. Latissimus dorsi
d. Teres major
e. Triceps brachii

17. What is the main action of the fibularis longus?

a. Dorsiflexion of the ankle
b. Plantarflexion of the ankle
c. Eversion of the foot
d. Inversion of the foot
e. Extension of the knee

18. True or false: The lumbricals are "wormlike" muscles found in the hands and feet.

LEVEL 2 Check Your Understanding

1. If a patient has a damaged lateral rectus muscle, which eye movement would be impaired?

2. For which two muscles would being damaged affect both movement of the head/neck and breathing? Explain your answer.

3. A patient has been told that the back pain he is experiencing is due to an inflamed longissimus thoracis muscle. He asks you what that means. How would you explain the diagnosis to him?

4. Most of the muscles attached to the scapula are fixators, but a few do contribute to arm movement. Compare and contrast two fixators of the scapula with two muscles attached to the scapula that contribute to arm movement.

5. The supraspinatus stabilizes the shoulder joint but also assists arm abduction. Based on this information alone (and without checking any tables), which bones serve as the origin and insertion for this muscle?

6. You ask a child how old she is, and she holds up three fingers. Which muscles are necessary to move the fingers into that position? Explain your answer.

7. In addition to the muscles of the thigh with "adductor" in their name, what other muscles assist with leg adduction?

8. Why do you suppose there are more and larger muscles involved in flexion of the fingers and hand than extension?

9. Why do you suppose there are more and larger muscles involved in plantar flexion of the foot than dorsiflexion?

10. Name and describe the muscles of the levator ani group.

11. Name and describe the muscles that extend the forearm at the elbow joint.

LEVEL 3 Apply Your Knowledge

PART A: Application and Analysis

1. A child has one eye that seems to move medially on its own at rest, and she is unable to move the eye laterally. You are told this is a strabismus due to malfunction of one of the extrinsic eye muscles. Which muscle is affected? Explain your answer.

2. In many respects the muscles of the upper limb and lower limb are similar in form and function. However, there are some key differences, particularly with regard to muscles attached to the scapula versus those attached to the pelvis. Analyze and explain the reasons for these differences.

3. A 28-year-old man woke up one morning and noticed that the right side of his face "felt heavy" and he couldn't close his right eye. When he tried to drink a glass of water, it spilled out of the right corner of his mouth. What do you think is the problem, and why does it cause these symptoms?

4. A 17-year-old skateboarder fell while riding down a railing and suffered a fracture at the middle of his humerus. After surgery his surgeon told him nerve damage from the fall would prevent him from being able to extend his wrist and fingers, but he would still be able to extend his elbow. Explain why he can no longer extend his wrist and fingers. Which nerve was injured? Which muscle was unaffected that will enable him to still extend his elbow?

PART B: Make the Connection

5. Ms. Nolan suffered a severe hip fracture that required hip replacement surgery. After an extended recovery period, she is undergoing physical therapy to regain strength and mobility. Which bone is involved in a hip fracture, and what part of the bone is likely to be fractured? Which muscles were likely affected by the hip replacement surgery, and to which parts of the bone do they attach? Which actions could Ms. Nolan perform to strengthen these muscles? (*Connects to Chapter 7*)

10
Muscle Tissue and Physiology

How many times did you move in the last minute? Did you sit down, maybe take out your earbuds, open your book, and follow these words with your eyes? Every one of your movements is the result of a process that happens over and over again in millions of your muscle cells. These cells make up the tissues and organs of the **muscular system**—the skeletal muscles (see Chapter 9). To understand how movements happen, we need to focus on skeletal muscles at the cellular and tissue levels of organization.

Although this chapter deals mostly with skeletal muscle tissue, we also discuss smooth muscle tissue and provide an overview of cardiac muscle tissue in the last module. (You will find complete coverage of cardiac muscle in the heart chapter—see Chapter 17.) Let's begin with a look at what all types of muscle tissue have in common.

MODULE **10.1**
Overview of Muscle Tissue

Learning Outcomes

1. Describe the major functions of muscle tissue.
2. List and describe the structural elements and properties common to all types of muscle cells.
3. Compare and contrast the characteristics of skeletal, cardiac, and smooth muscle tissue.

This module introduces you to the basic structure and properties common to all types of muscle tissue. As we know, there are three types of muscle tissue: skeletal, cardiac, and smooth (see Chapter 4). Although each type has unique features, they all have the same basic function: generating a force called *muscle tension*. Recall that at the organ level, this force often creates movement, but it also maintains posture, stabilizes joints, generates heat, and regulates the flow of materials through hollow organs (see Chapter 9).

Muscle tissue consists of **muscle cells,** which are sometimes called *myocytes* (MY-oh-sytz; *myo-* = "muscle"), and the surrounding extracellular matrix, referred to as the *endomysium.* (The endomysium was originally thought to be connective tissue, but it's now known to be extracellular matrix.) The endomysium holds the muscle cells together

Computer-generated image: Myosin heads are shown binding to actin molecules in a key step of skeletal muscle contraction.

within muscle tissue. As a result of their common function, muscle cells share many structural properties. However, the three types of muscle tissue also have notable differences. In this module, we examine both the similarities of and differences between the types of muscle tissue.

Types of Muscle Tissue

There are two basic forms of muscle cells: *striated,* which feature alternating light and dark bands known as **striations** (stry-AY-shunz); and *smooth,* which lack striations. The three types of muscle tissue are **skeletal muscle** and **cardiac muscle,** which consist of striated muscle cells, and **smooth muscle,** which consists of nonstriated smooth muscle cells (**Figure 10.1**). As you can see in Figure 10.1a, skeletal muscle tissue is made up of long, multinucleated (containing more than one nucleus) muscle cells that are arranged parallel to one another. Some skeletal muscle cells are quite long, extending nearly the entire length of the muscle. In keeping with their shape, these cells are often called **muscle fibers** (the terms cell and fiber are often used interchangeably). Skeletal muscle fibers are mostly found attached by connective tissue to the skeleton, where their contraction can produce the movement of a body part. However, this contraction occurs only when muscle fibers are stimulated by the nervous system. Although we're not always conscious of the contractions of skeletal muscle fibers, they can be controlled *voluntarily,* or by conscious thought.

Although cardiac muscle tissue, like skeletal muscle tissue, consists of striated muscle cells, Figure 10.1b reveals numerous differences between these two types of tissue. For example, cardiac muscle cells are shorter and wider, are branched, and generally have only one or two nuclei. Notice also that dark, wavy lines join the cardiac muscle cells. These lines are called *intercalated discs* (in-TER-kuh-lay′-t'd; *inter-* = "between"), and they contain gap junctions and modified tight junctions that unite the muscle cells and permit them to coordinate contraction. Cardiac muscle tissue is found only in the heart and is *involuntary,* meaning that the brain does not have conscious control over its contraction.

As its name implies, smooth muscle tissue consists of smooth muscle cells, which are long and flattened with two pointed ends (often described as "spindle-shaped") and with a single, centrally located, oval nucleus (Figure 10.1c). These cells line nearly every hollow organ, and are found as well in the eyes, the skin, and the ducts of certain glands. Many smooth muscle cells are linked to one another by gap junctions in their plasma membranes. Like cardiac muscle tissue, smooth muscle tissue is involuntary.

Quick Check

☐ 1. What are the two types of striated muscle?
☐ 2. Which two types of muscle are involuntary?

Properties of Muscle Cells

What properties allow all types of muscle cells to contract and produce muscle tension? Imagine muscle cells as mechanical devices somewhat like your car—in the same way your car converts the chemical energy of gasoline into the mechanical energy of movement, a muscle cell converts the chemical energy of ATP into the mechanical energy of muscle tension. All muscle cells share a set of common properties that enable them to generate tension:

- **Contractility.** *Contractility* is, of course, the ability of cells to contract. You might think that "contracting" means "shortening," but the term contraction actually refers to the ability of proteins within muscle cells to draw together. As you will discover in later modules, a muscle cell does not necessarily shorten when it contracts.
- **Excitability.** Muscle cells are **excitable,** or responsive, in the presence of various stimuli; these might include chemical signals from the nervous or endocrine systems, mechanical stretch signals, and local electrical signals. Such stimuli generate electrical changes across the plasma membrane of the muscle cell.
- **Conductivity.** When a muscle cell is excited, the electrical changes across the plasma membrane do not stay in one place. Instead, they are rapidly *conducted* along the entire length of the plasma membrane, similar to how an electrical impulse is conducted through a copper wire.

(a) **Skeletal muscle tissue**

Multiple nuclei · Striations · Skeletal muscle cell

(b) **Cardiac muscle tissue**

Single nucleus · Striations · Intercalated disc · Cardiac muscle cell

(c) **Smooth muscle tissue**

Smooth muscle cell · Single nucleus

Figure 10.1 Three types of muscle tissue.

- **Extensibility.** Most cells will rupture when stretched, but muscle cells are **extensible**—they can be stretched up to three times their resting length without damage.
- **Elasticity.** Muscle cells can also return to their original shape after being stretched, a property called **elasticity.** Often, elasticity is mistaken for stretch, but extensibility, not elasticity, refers to stretch.

Quick Check

☐ 3. What is the basic function of all types of muscle cells?
☐ 4. What five properties are common to all muscle cells?

Structure of Muscle Cells

◀◀ **Flashback**

1. What is the basic structure of the plasma membrane? (p. 70)
2. What are the smooth endoplasmic reticulum and cytoskeleton, and what are their functions? (pp. 90, 95)

The right side of **Figure 10.2** illustrates a generic muscle cell. Note that this cell has most of the same organelles as the generalized cell pictured to the left of it, including nuclei and mitochondria. However, muscle cells also exhibit multiple structural differences, so we use different terminology when describing them. For example, the term *cytoplasm* is replaced with **sarcoplasm** (SAR-koh-plazm; *sarco-* = "flesh"), and *plasma membrane* is replaced with **sarcolemma** (sar-koh-LEM-ah; *-lemma* = "husk or shell"). Sarcoplasm, like cytoplasm, contains cytosol and all of the organelles in the muscle cell. The sarcolemma is composed of a phospholipid bilayer with multiple specialized integral and peripheral proteins.

Muscle cells also have unique structures that allow them to serve their functions. You can see in Figure 10.2 that the sarcoplasm of muscle cells contains cylindrical organelles called *myofibrils* (my-oh-FY-brilz). All three types of muscle cells contain myofibrils, although they are arranged differently in smooth muscle cells than in skeletal and cardiac cells. Myofibrils are essentially bundles of specialized proteins, including those involved in muscle contraction. A myofibril measures about 1 μm in diameter, or about $\frac{1}{100}$ the thickness of a human hair. Each muscle cell has hundreds to thousands of myofibrils—they make up about 50–80% of its volume—and other organelles such as mitochondria are found packed between them.

Another unique structure common to all muscle cells is an organelle called the *sarcoplasmic reticulum* (sar-koh-PLAZ-mik reh-TIK-yoo-lum), or simply *SR*. The SR is a modified smooth endoplasmic reticulum that forms a weblike network surrounding each myofibril. The structure of the SR varies in the three types of muscle tissue, as we'll discuss in later modules.

Quick Check

☐ 5. What basic structures are unique to muscle cells? What are their functions?

Apply What You Learned

☐ 1. Suppose that a new type of muscle tissue is discovered and, like the other types, it generates tension. Would you expect this new tissue type to have the same basic structure as skeletal, smooth, and cardiac muscle tissues? Why or why not?
☐ 2. Predict what might happen to a muscle cell that is extensible but not elastic. What if the reverse were true—a muscle cell is elastic but not extensible?

See answers in Appendix A.

Figure 10.2 A generalized cell (left) compared with a generic muscle cell (right).

MODULE **10.2**

Structure and Function of Skeletal Muscle Fibers

Learning Outcomes

1. Describe the structural properties and components of a skeletal muscle fiber.
2. Explain the organization of a myofibril.
3. Describe the structure and components of thick, thin, and elastic filaments.
4. Name and describe the function of each of the contractile, regulatory, and structural protein components of a sarcomere.
5. Explain the sliding-filament mechanism of muscle contraction.

In this module, we explore the structure of skeletal muscle fibers and the proteins found within them. Our discussion also covers how the arrangement of these proteins within a myofibril correlates with the mechanism of muscle contraction.

Structure of the Skeletal Muscle Fiber

Skeletal muscle tissue consists of many skeletal muscle fibers and the endomysium surrounding each of them. These fibers are quite different from most of the cells we have studied up to this point. Some differences include the following:

- **Cell shape and size.** Figure 10.3 shows that skeletal muscle fibers are shaped like long, thin cylinders and are much longer than most cells. The longest fibers, found in the thigh muscles, are about 100 μm in diameter and 30 cm (almost a foot) long. (Compare this with a red blood cell, which is only about 7.5 μm in diameter.)
- **Striations.** Notice also in Figure 10.3 that skeletal muscle fibers appear striated, or striped, a characteristic we've mentioned before.
- **Multiple nuclei.** Each skeletal muscle fiber contains multiple sausage-shaped nuclei on the inner surface of its sarcolemma. As you may recall, this feature is a result of the embryonic origin of the fiber, as each muscle fiber arises from the fusion of multiple embryonic cells called *myoblasts* (see Chapter 4).

In addition to these differences, skeletal muscle fibers have many unique features that enable them to perform their functions. As you know, a skeletal muscle fiber, like all types of muscle fibers, is surrounded by its sarcolemma (**Figure 10.4**). Within the boundary of the sarcolemma, we find the cylindrical **myofibrils,** which are made up primarily of proteins involved in contraction. Myofibrils are the most abundant organelle in the sarcoplasm. Surrounding the myofibrils like a web is the **sarcoplasmic reticulum,** or **SR,** which is a modified smooth endoplasmic reticulum. Its primary function is the storage and release of calcium ions, activities vital to muscle contraction and relaxation.

The plasma membrane of the muscle fiber is called by a different name because its structure is different. Notice in Figure 10.4 that the sarcolemma isn't confined to the exterior of the cell—it forms inward extensions called **transverse tubules** (or **T-tubules**) that dive deeply into the muscle fiber and surround each myofibril. The T-tubules form a tunnel-like network within the muscle fiber. These tunnels are continuous with the exterior of the cell and so are filled with extracellular fluid. Flanking each side of a T-tubule are enlarged portions of the SR called **terminal cisternae** (SIS-ter-nee; "end sacs"). The combination of a T-tubule and the two terminal cisternae on either side is known as a **triad** (see the blow-up in Figure 10.4). This association is important in muscle contraction, as we'll discuss in Module 10.4.

Quick Check

☐ 1. How does the structure of a skeletal muscle cell differ from the structure of other cells?
☐ 2. How are the terminal cisternae related to the sarcoplasmic reticulum?
☐ 3. What is a triad?

Structure of the Myofibril

◀◀ **Flashback**

1. What are actin filaments? (p. 96)
2. What are motor proteins? (p. 95)

As you can see in Figure 10.4, a myofibril is composed of hundreds to thousands of protein bundles called **myofilaments.** These consist of one or more of the following types of proteins:

- **contractile proteins,** which produce tension;
- **regulatory proteins,** which control when the muscle fiber can contract; and
- **structural proteins,** which hold the myofilaments in their proper places and ensure the structural stability of the myofibril and the muscle fiber.

See *A&P in the Real World: Duchenne Muscular Dystrophy* to read about what happens when the connections between these components of the muscle fiber break down.

Size of red blood cell for comparison   Striations   Nuclei

Skeletal muscle fiber

Figure 10.3 Size and shape of a skeletal muscle fiber.

Muscle fiber

Endomysium   Sarcoplasmic   Transverse   Terminal   Sarcolemma
                reticulum   (T-) tubule   cisternae
                                                          Myofilaments

Openings
of T-tubules

                                                          Myofibrils

                                                          Cytosol

Nucleus   Mitochondrion

Triad

Terminal
cisternae

Sarcolemma   Opening of
                                              T-tubule

Practice art labeling
@MasteringA&P®

Figure 10.4 Structure of a skeletal muscle fiber.

A&P in the Real World

Duchenne Muscular Dystrophy

Duchenne muscular dystrophy (doo-SHEN), or **DMD,** is a degenerative muscular disease occurring almost exclusively in boys. It is caused by a defective gene for the protein **dystrophin** (diss-TROH-fin) that is carried on the X chromosome (see Chapter 27 for a discussion of X-linked disorders). Dystrophin is a structural protein found in skeletal and cardiac muscle fibers that anchors the sarcolemma to the surrounding connective tissue and to the myofibrils; this is the arrangement found in normal muscle tissue, as shown in the light micrograph on the left. In the absence of functional dystrophin, the sarcolemma breaks down and the muscle fiber is destroyed. As muscle fibers die, they are replaced with fatty and fibrous connective tissue, as shown in the light micrograph on the right. This process causes the whole muscle to enlarge while it's actually losing function.

Symptoms of DMD typically arise between the ages of 2 and 6 years, and include weakness of the proximal limb muscles, a waddling gait pattern, and difficulty in performing certain activities such as climbing stairs. Generally by 12 years of age, patients with DMD are wheelchair-bound, and death from cardiac and respiratory failure occurs about age 20. Researchers have made a lot of progress toward understanding the mechanism of this disease; although it's still incurable, current research that involves modifying the mutated gene so that the muscle cells can produce functional dystrophin appears promising.

Normal muscle fibers   LM (260×)

Muscle replaced by fat in DMD patient   LM (260×)

There are three types of myofilaments, each with characteristic contractile, regulatory, and/or structural proteins: *thick filaments, thin filaments,* and *elastic filaments* (Figure 10.5). The thick filaments, as their name implies, have the largest diameter. Each **thick filament** is composed of many molecules of the contractile protein **myosin** (MY-oh-sin). Note in Figure 10.5a that myosin looks somewhat like two golf clubs twisted together, with two globular "heads" and two intertwining polypeptide chains making up a "tail." The heads protrude from the myosin tail on a "neck." The neck of each myosin molecule is flexible where it meets the tail at a point called the *hinge.* Each myosin head includes a site that binds to a thin filament, among other functional components we'll discuss later. The myosin molecules are arranged within the thick filament in such a way that clusters of myosin heads are found at each end, with only myosin tails found in the middle.

A **thin filament** is made up of both contractile and regulatory proteins (Figure 10.5b), as follows:

- **Actin.** A thin filament consists of many molecules of the contractile protein **actin** (AK-tin). This bead-shaped protein has an area, called the **active site,** that can bind to a myosin head. Multiple actin subunits string together like beads on a necklace to form the largest part of the thin filament.

This actin "string" appears as two intertwining strands in the functional thin filament.
- **Tropomyosin.** The long, ropelike regulatory protein called **tropomyosin** (trohp´-oh-MY-oh-sin) spirals around the two actin strands so that, at rest, it covers the active sites on actin.
- **Troponin.** A second regulatory protein is the smaller, globular **troponin** (TROH-poh-nin) that holds the tropomyosin in place.

The two regulatory proteins tropomyosin and troponin help to switch on and off the process of muscle contraction.

Elastic filaments, the thinnest type, have only recently been recognized as a distinct type of myofilament. They are composed of a single massive structural protein called **titin** (TY-tin). Figure 10.5c shows that titin is shaped like a spring that can uncoil when stretched and recoil to its original shape when the stretching force is removed. Titin runs through the core of the thick filament, which helps to stabilize the myofibril structurally (you can't see titin through the thick filaments in the figures because it is inside them). Elastic filaments serve several purposes in addition to holding the thick filaments in place—these include resisting excessive stretching and providing elasticity to the muscle fiber, that is, helping it to "spring" back to its original length after it is stretched.

Putting It All Together: The Big Picture of Skeletal Muscle Structure

Figure 10.6 provides a summary of the organizational levels of a skeletal muscle. Remember that multiple skeletal muscle fibers together with the surrounding endomysium form a fascicle. Each fascicle is surrounded by a layer of connective tissue called the perimysium, and bundles of fascicles together make up a skeletal muscle. The entire muscle is covered by a layer of connective tissue referred to as the epimysium. Near the end of the muscle, where it inserts into bone or other tissues, the perimysium and

Figure 10.5 Structure of myofilaments.

Muscle

Fascicle

Tendon

Bone

Epimysium

Perimysium

Muscle fiber

Endomysium

Sarcolemma

Myofibril

Myofilaments

Thick filament

Thin filament

Elastic filament

epimysium usually extend beyond the skeletal muscle fibers to form a tendon that binds the muscle to its attaching structure. Skeletal muscles are enclosed by a layer of thick connective tissue called fascia, which anchors them to the surrounding tissues and holds groups of muscles together (see Chapter 9 for additional review of these structures). The blow-ups in Figure 10.6 show how these structures relate to the structure of a muscle fiber, made up of myofibrils composed of myofilaments.

Myofilament Arrangement and the Sarcomere

In **Figure 10.7** observe how the arrangement of thin and thick myofilaments forms a pattern within the myofibril—there are regions where we find only thin filaments, regions where the thin and thick filaments overlap, and regions that have only thick filaments. Now, recall that skeletal muscle fibers are striated. These alternating light and dark bands are produced by the overlapping, repeating arrangement of thin and thick filaments. As explained in Module 10.6, this

region with overlapping filaments, referred to as the *zone of overlap,* is where tension is generated during a muscle contraction.

The dark and light areas of the striations are named as follows (see Figure 10.7):

- **I bands.** The **I band** is the light region of a striation. It appears lighter because it contains only thin filaments, which allow more light to pass through them.
- **A bands.** The **A band** is the dark region of a striation, which contains thick filaments. Thick filaments block more light than thin filaments, making the A band appear darker in micrographs. In Figure 10.7 we find both thick and thin filaments in the outer portion of the A band, and only thick filaments in the middle of the A band. This middle area is called the **H zone.**

You can also see that both the A and I bands in Figure 10.7 have a dark line running down their middle. The dark line in the A bands is called the **M line.** The M line consists of structural proteins that hold the thick filaments in place and serve as an anchoring point for the elastic filaments. The dark line in the I bands is known as a **Z-disc.** The Z-disc is composed of structural proteins that have many functions: They anchor the thin filaments in place and to one another, they are an attachment point for elastic filaments, and they attach myofibrils to one another across the whole diameter of the muscle fiber.

All this may sound at first like alphabet soup, but these details are vital: It is this arrangement of

Figure 10.7 Structure and bands of the sarcomere.

myofilaments within the myofibrils that allows muscle to contract. The functional unit of contraction—where muscle tension is produced—is the **sarcomere** (SAR-koh-meer; "muscle part"). A sarcomere is the section of a myofibril that extends from one Z-disc to the next Z-disc (Z comes from the German word *zwischen,* which means "between"). Each sarcomere, then, includes a full A band and half of two I bands.

Figure 10.8 shows the three-dimensional arrangement of the sarcomere in a myofibril. Notice in the cross-sections through the different regions of the sarcomere that the A band consists of both thin and thick filaments, and each thick filament is surrounded by six thin filaments. However, the I band contains only thin filaments, and the H zone, only thick filaments.

↟ STUDY BOOST

Remembering the Bands of the Sarcomere

The following are some simple descriptions and mnemonics to help you remember the basic organization of the sarcomere.

Structure	Description	Mnemonic
A band	The dark band that contains both thick and thin filaments.	A is the d**A**rk band.
H zone	The middle portion of the A band in which only thick filaments are found.	"Ha!," for H is in the A band.
M line	Middle line of the A band; contains structural proteins that hold the thick filaments in place.	Think of the "M" as standing for "middle." Also remember that the "Myosin tails point toward the M line."
I band	The light band that contains only thin filaments.	I is the l**I**ght band.
Z-disc	Line bisecting the I band. Both the thin and elastic filaments anchor to the Z-discs, which also attach myofibrils to one another.	Identify the Z-disc by its "Z" shape (see Figure 10.7).

■

Quick Check

☐ 6. How does the arrangement of myofilaments produce the characteristic striations of skeletal muscle fibers?

☐ 7. Describe the structure of a sarcomere. What is its function?

The Sliding-Filament Mechanism of Contraction

At this point you know that muscle contraction occurs at the level of the sarcomere, but how, precisely, does contraction occur? Physiologists previously believed that the myofilaments themselves shortened. However, after detailed study of skeletal muscle, the currently accepted model is that the thin filaments instead

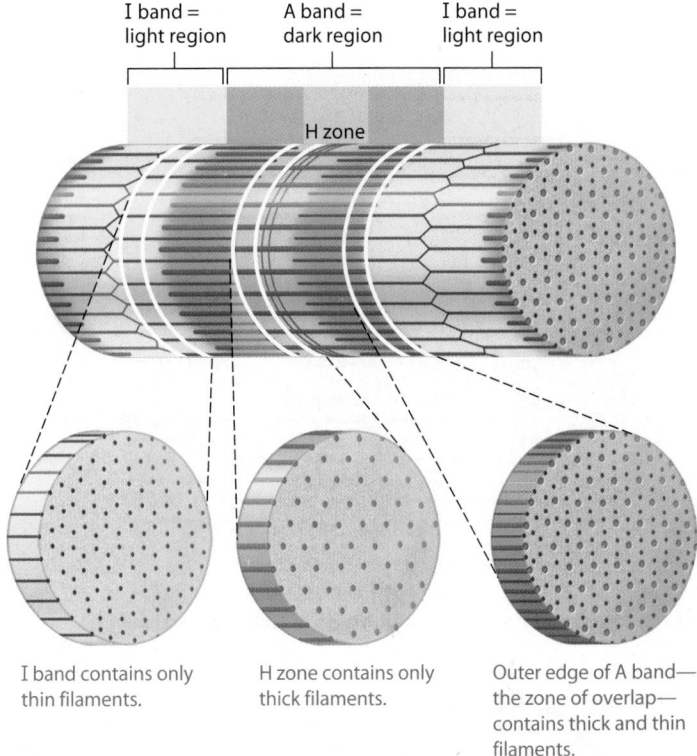

I band contains only thin filaments.

H zone contains only thick filaments.

Outer edge of A band—the zone of overlap—contains thick and thin filaments.

Figure 10.8 Three-dimensional structure of the sarcomere.

slide past the thick filaments, generating tension throughout the whole sarcomere. This interaction of the thin and thick filaments during muscle contraction and relaxation is known as the **sliding-filament mechanism.**

Although this mechanism operates at the microscopic level, it's easy to picture how it works on a gross level. Take your hands and turn them as you see in the first illustration here:

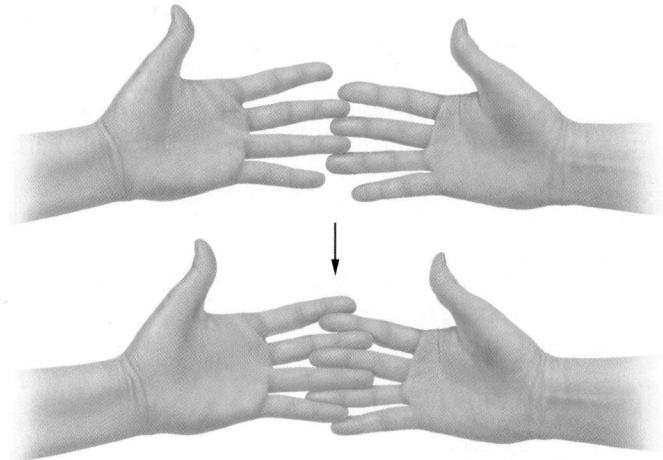

Your hands represent a single, large sarcomere, and where your fingers overlap represents the zone of overlap of the thick and thin filaments; the inner edges of your palms are the Z-discs. Now move your fingers slowly together. As you can see, your "sarcomere" (your hands) grows progressively

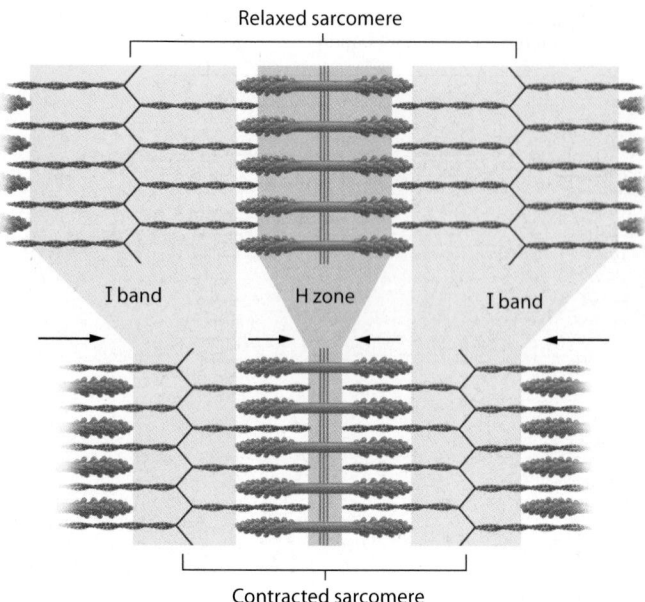

Relaxed sarcomere

I band   H zone   I band

Contracted sarcomere

(a) The I band and H zone narrow during a muscle contraction.

Relaxed sarcomere

A band

Contracted sarcomere

(b) The A band remains unchanged during a muscle contraction.

Figure 10.9   The sliding-filament mechanism: changes in the bands of the sarcomere during a muscle contraction.

shorter with each contraction, ending up as in the second illustration.

In **Figure 10.9** notice how the sarcomeres change during a contraction: Both the I bands and the H zone narrow (part a), but the A band remains unchanged (part b). This happens because the myosin heads of the thick filaments "grab" the thin filaments and pull them toward the M line. This pulling action brings the Z-discs closer together and causes the sarcomere as a whole to shorten. Remember, though, that none of the filaments themselves actually shorten—the thin filaments simply move toward the M line. We discuss details of the sliding-filament mechanism in Module 10.4.

Sarcomeres are arranged end to end within each myofibril. As many sarcomeres simultaneously contract, the whole muscle fiber contracts.

Quick Check

☐ 8. What is the basic mechanism of contraction via the sliding-filament mechanism?

Apply What You Learned

☐ 1. A hypothetical poison denatures the protein titin, making it straight instead of spring-shaped. Predict the effect of this change on muscle tissue.

☐ 2. Predict what might happen if the thin and thick filaments were unable to move relative to one another within a sarcomere.

See answers in Appendix A.

MODULE   **10.3**
Skeletal Muscle Fibers as Electrically Excitable Cells

Learning Outcomes

1. Contrast the relative concentrations of sodium and potassium ions inside and outside a cell.
2. Differentiate between a concentration gradient and an electrical potential.
3. Describe the role of the Na^+/K^+ ATPase (or Na^+/K^+ pump) in maintaining the resting membrane potential.
4. Describe the sequence of events of a skeletal muscle fiber action potential.

Although skeletal muscle contraction takes place at the level of the sarcomere, the events leading to contraction actually begin with electrical changes that occur across the sarcolemma. For this reason, we need to discuss the electrical properties of the sarcolemma before we get into the mechanism of contraction. The branch of physiology that studies electrical changes across plasma membranes, and the accompanying physiological processes, is called **electrophysiology.** Many of our body's processes, including skeletal muscle contraction, involve electrophysiology. This module presents the basic principles of electrophysiology and how they apply to muscle fibers. To begin, we'll review a concept from the cell chapter: *membrane potential* (see Chapter 3).

Membrane Potentials in Our Cells

◀◀ **Flashback**

1. What are potential energy and kinetic energy? (p. 41)
2. What is the resting membrane potential? How are charges separated across the plasma membranes of cells? (p. 82)

Earlier we discussed a characteristic common to all cells, including electrically excitable cells: an unequal distribution of electrically charged ions near the plasma membranes of the cells. You can see this feature in the skeletal muscle fiber shown here—near the sarcolemma are a thin layer of negative ions in the cytosol and a thin layer of positive ions in the extracellular fluid:

Away from the sarcolemma, positive and negative ions are present in equal numbers. This means that the cytosol and extracellular fluid are always electrically neutral.

The separation of electrical charges across the plasma membrane really just repeats the Gradients Core Principle (p. 26). Specifically, the separation of charges is called an *electrical gradient*, where we have unequal num-

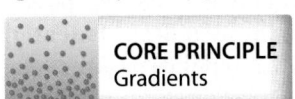

CORE PRINCIPLE
Gradients

bers of positive and negative ions separated by a barrier. As with all gradients, an electrical gradient represents a source of potential energy. When the barrier separating the ions is removed, they follow their gradients, creating a flow of electrical charges, and the potential energy becomes kinetic energy. For this reason, we can call the electrical gradient an **electrical potential,** because we are referring to its potential energy.

A difference in electrical potential between two points is called a **voltage;** the size of the voltage, measured in *volts,* indicates the size of the potential. For example, the potential difference present between an electrical outlet in your wall and an appliance such as a toaster is relatively large, measuring about 110 volts. In comparison, the potential difference across the two sides of a muscle fiber sarcolemma is small, and is measured in a unit called a *millivolt (mV),* or 1/1000 of a volt.

The **membrane potential** of a cell is the electrical potential across its plasma membrane (see Chapter 3). The electrical potential across the sarcolemma in a muscle fiber at rest is called the **resting membrane potential** (note that a fiber "at rest" is one that is not being stimulated). This potential measures about −85 mV, which means that the cytosol is about 85 mV more negative than the extracellular fluid. The membrane is considered to be **polarized** in its resting state because two sides, or

poles, of opposite charge are present, just like the positive and negative poles of a magnet. We discuss more about the resting membrane potential, including how it is generated, in the nervous tissue chapter (see Chapter 11).

Quick Check

☐ 1. What is an electrical potential?
☐ 2. What is the resting membrane potential, and what is its normal value in a skeletal muscle fiber?

The Na⁺/K⁺ ATPase Pump and the Sodium and Potassium Ion Concentration Gradients

◀◀ **Flashback**

1. What is the function of the Na⁺/K⁺ATPase pump? (p. 79)

The resting membrane potential can be changed when ions cross the plasma membrane. Recall that the phospholipid bilayer of the plasma membrane is impermeable to charged particles, and so ions such as sodium and potassium are able to cross the membrane only through protein channels or carriers (see Chapter 3). However, ions can pass through these channels only when a concentration gradient is present to drive their passive diffusion or when ATP is expended to drive their active transport. The electrophysiology of muscle fibers involves both passive diffusion and active transport pumps that use ATP.

The most important active transport pump for sodium and potassium ions is the **Na⁺/K⁺ ATPase pump.** Recall that this pump maintains concentration gradients of sodium and potassium ions across the sarcolemma, using the energy from ATP hydrolysis, or splitting (**Figure 10.10**). Specifically, the pump moves three sodium ions out of the cell, making the concentration of sodium ions low in the cytosol and higher in the extracellular fluid. The pump also moves two potassium ions into the cell, making the concentration of potassium ions higher in the cytosol and lower in the extracellular fluid. ATP hydrolysis is necessary because this pump moves the ions against their concentration gradients.

The sarcolemma of a skeletal muscle fiber has millions of these pumps, which work together constantly to maintain steep concentration gradients of sodium and potassium ions. These concentration gradients are critical because they provide much of the driving force for the passive diffusion of sodium and potassium ions through membrane channels. The next section focuses on how the movement of ions through such channels changes the membrane potential of a cell.

Quick Check

☐ 3. Where do we find higher concentrations of sodium and potassium ions (in the cytosol or extracellular fluid)?
☐ 4. What role does the Na⁺/K⁺ pump play in this ion distribution?

Muscle fiber   Sarcolemma

Extracellular fluid (ECF)

Low K+, high Na+ concentrations

Na+

K+

Na+/K+ Pump

P

ATP

ADP

High K+, low Na+ concentrations

Cytosol

P_i

Na+

K+

① The Na+/K+ pump releases 3 Na+ into the ECF as it binds 2 K+ from the ECF.

② The Na+/K+ pump releases 2 K+ into the cytosol as it binds 3 Na+ from the cytosol.

Figure 10.10 Ion gradients maintained by the Na+/K+ pump.

Action Potentials

A key property of all cells, including excitable cells such as muscle fibers, is that their resting membrane potential can be altered to become more negative or more positive. One such change is the **action potential,** which is a quick, temporary change in the membrane potential in a single region of the plasma membrane. During an action potential, the membrane potential goes from its negative resting value to a more positive value and then back to its negative resting potential again. This change can be used to signal electrically, to communicate with and sometimes stimulate a response from other cells. All of this happens extremely rapidly, within a few milliseconds—much faster than the muscle contraction itself.

Action potentials are generated by the opening and closing of protein channels in the plasma membrane that control the movement of sodium and potassium ions across the membrane. These channels are called **gated channels** because they contain "gates" that are normally closed and open only in response to some sort of stimulus. Two types of gated channels are present in the plasma membrane: ligand and voltage. **Ligand-gated channels** (also called *chemically gated channels*) open in response to the presence of certain chemicals called **ligands** (LY-gandz). **Voltage-gated channels** open and close in response to changes in the membrane potential across the

plasma membrane. As a consequence of these channel openings (and closings), the membrane experiences rapid and dramatic changes in the movement of both sodium and potassium ions. This activity in turn alters the membrane potential, producing the action potential.

Using the example of a muscle cell, let's look at a single region of the sarcolemma to see how an action potential is generated. We begin with the membrane at the resting membrane potential and the voltage-gated channels closed (**Figure 10.11**).

From this resting stage, a stimulus triggers specific gated channels to open, which leads to an action potential. As you can see, the action potential consists of two basic stages, *depolarization* and *repolarization*:

① **Depolarization stage: In response to a stimulus, voltage-gated sodium ion channels open and sodium ions enter the cell, making the membrane potential less negative.** When a stimulus reaches the sarcolemma, the **depolarization stage** begins, during which voltage-gated sodium ion channels open. This opening allows sodium ions to flow rapidly through these channels down their concentration gradient into the cell. The entry of these positively charged sodium ions makes the membrane potential become less negative, or less polarized (hence the term *de*polarization). The membrane potential rises toward zero and rapidly reaches a peak of about $+30$ mV. Notice that the membrane potential is reversed now—the outside of the cell is negative, and the inside is positive.

② **Repolarization stage: Sodium ion channels close while voltage-gated potassium ion channels open and potassium ions leave the cell, making the membrane potential more negative again.** During the **repolarization stage,** the sarcolemma returns to its negative resting membrane potential, which is a polarized state (hence the term *re*polarization). Repolarization is caused by the simultaneous closing of voltage-gated sodium ion channels and opening of voltage-gated potassium ion channels. As potassium ion channels open, potassium ions flow rapidly through these channels down their concentration gradient out of the cell. The loss of positively charged potassium ions makes the membrane potential more negative, returning the sarcolemma to the resting potential of -85 mV and causing potassium ion channels to close. This marks the end of the action potential.

Note that the total intracellular and extracellular concentrations of sodium and potassium ions do not alter significantly during an action potential. The reason is that just a few ions have to cross the sarcolemma to cause large changes in its membrane potential. Any minor shifts in ion concentration are corrected by the action of the Na+/K+ pump.

After this detailed discussion of how action potentials occur, let's step back and consider the big picture of *why* action potentials happen: long-distance signaling. Up to this point, we have been discussing action potentials as if they were single events in one area of the sarcolemma. However, recall that a property

RESTING STAGE

Before a stimulus arrives, the membrane is at the resting membrane potential (RMP), and voltage-gated Na⁺ and K⁺ channels are closed.

ACTION POTENTIAL

① **Depolarization stage:**
In response to a stimulus, voltage-gated Na⁺ channels open and Na⁺ enter the cell, making the membrane potential less negative.

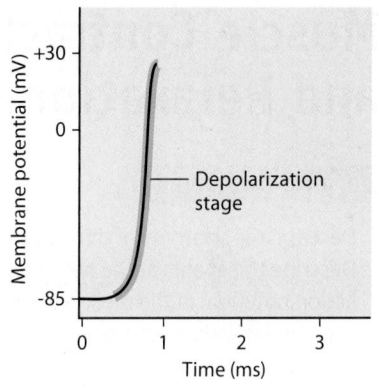

② **Repolarization stage:**
Na⁺ channels close while voltage-gated K⁺ channels open and K⁺ leave the cell, making the membrane potential more negative again.

Sequence the steps
@MasteringA&P®

Figure 10.11 Stages of an action potential.

of muscle fibers is conductivity, which means that electrical changes across the sarcolemma are not isolated events. Action potentials don't stay in one place; rather, they are conducted, or **propagated,** throughout the entire sarcolemma like ripples in a pond. The process happens incredibly fast and results in depolarization of the entire sarcolemma, including the T-tubules. This rapid propagation enables a single stimulus in one region of the sarcolemma to have nearly instant, far-reaching impact. In fact, the arrival of the action potential at the T-tubules is what initiates a muscle contraction. As we discuss in the next module, this is needed so that the muscle fiber can contract as a unit. (We'll find out more about action potentials during our study of nervous tissue—see Chapter 11.)

Quick Check

☐ 5. What is an action potential?
☐ 6. What are the two phases of an action potential?
☐ 7. Why is the movement of ions across the plasma membrane important, and in which direction do ions move during each phase of the action potential?

Apply What You Learned

☐ 1. The poison ouabain (wah-BAY-in), or *arrow poison*, blocks the Na⁺/K⁺ pump. What effect would this poison have on the sodium and potassium ion gradients and so on the action potentials of a muscle fiber?

☐ 2. The poison tetrodotoxin from the pufferfish blocks voltage-gated sodium ion channels in the sarcolemma. Predict the effect of this poison on the action potentials of the muscle fiber.

☐ 3. What would happen if tetrodotoxin instead blocked voltage-gated potassium ion channels in the sarcolemma?

See answers in Appendix A.

MODULE **10.4**

The Process of Skeletal Muscle Contraction and Relaxation

Learning Outcomes

1. Describe the anatomy of the neuromuscular junction.
2. Describe the events at the neuromuscular junction that elicit an action potential in the muscle fiber.
3. Explain excitation-contraction coupling.
4. Describe the sequence of events involved in the contraction cycle of a skeletal muscle fiber.
5. Explain the process of skeletal muscle fiber relaxation.

Muscles contract when skeletal muscle fibers are stimulated by the nervous system, causing the contractile proteins in myofilaments to slide past one another. This module explores the sliding-filament mechanism of contraction in detail, along with the events that lead to muscle relaxation. First, however, we need to understand the relationship between the muscular system and the nervous system.

The Neuromuscular Junction

◀◀ Flashback

1. What is a vesicle? (p. 82)
2. What is a neuron and what are its components? (p. 150)

All skeletal muscle fibers are **innervated,** which means they are connected to a neuron. A single neuron called a **motor neuron** communicates with many muscle fibers; each connection is referred to as a **synapse** (SIN-aps; *syn-* = "clasp together"). The synapse of a motor neuron with a muscle fiber is known as the **neuromuscular junction,** or **NMJ.** The function of the neuromuscular junction is to transmit a signal, called a *nerve impulse* (which is actually an action potential of the neuron), from the neuron to the sarcolemma of the muscle fiber. In this example of the Cell-Cell Communication Core

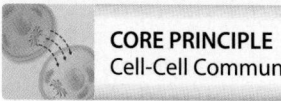

CORE PRINCIPLE
Cell-Cell Communication

Principle (p. 27), neurons communicate with muscle fibers at the neuromuscular junction.

Each NMJ consists of three parts: the *axon terminal, synaptic cleft,* and *motor end plate* (**Figure 10.12**). Let's look at each of these in more detail:

● **Axon terminal.** The motor neuron extends a long "arm" or cytoplasmic extension, called an **axon,** to the muscle fiber. The end of the axon swells to form an **axon terminal,** or *synaptic bulb,* which contains **synaptic vesicles.** These vesicles have **neurotransmitters,** chemicals that, when released, trigger changes in the target cells of the neuron. The neurotransmitter in the axon terminals of motor neurons that connect to skeletal muscle fibers is **acetylcholine** (ah-SEE-til-koh´-leen; **ACh**).

● **Synaptic cleft.** The **synaptic cleft** is the narrow space between the axon terminal and the muscle fiber into which

Figure 10.12 Structures of the neuromuscular junction.

Practice art labeling
@MasteringA&P®

ACh is released. It is filled with collagen fibers and an extracellular gel that anchors the neuron in place. It also contains enzymes that break down ACh.

- **Motor end plate.** The **motor end plate** is a specialized region of the sarcolemma, whose folded surface contains many receptors for ACh. Recall that a receptor is a protein within the plasma membrane that binds to a specific ligand. These receptors are actually ligand-gated ion channels; ACh is the ligand.

Notice that as a neuron stimulates a muscle cell, an electrical signal (the neuronal action potential) is converted to a chemical signal (ACh) and back to an electrical signal (muscle fiber action potential). This is a pattern we see repeatedly throughout anatomy and physiology.

Quick Check

- ☐ 1. What are the three components of the neuromuscular junction?
- ☐ 2. What is the function of each part of the neuromuscular junction?

Skeletal Muscle Contraction

It's now time to talk about how ion channels, action potentials, and the neuromuscular junction work together to create a muscle contraction. We know that skeletal muscle fibers must be excited in order to contract, but what this means more specifically is that the sarcolemma of a muscle fiber will not have an action potential without stimulation by ACh from a motor neuron. Once the muscle fiber is excited by such

stimulation, a process called *excitation-contraction coupling* conveys this excitation to the parts of the fiber that produce the contraction, the myofilaments. Then one type of myofilament is able to slide past the other. The sarcomere contracts, and contractions in many sarcomeres produce contraction of the whole muscle. So the process of muscle contraction can be broken down into three parts: the excitation phase, excitation-contraction coupling, and the contraction phase.

Excitation Phase

The **excitation phase** involves transmission of a signal from the motor neuron to the sarcolemma of a muscle fiber. This phase, which occurs at the neuromuscular junction, proceeds with the following steps (**Figure 10.13**):

1. **An action potential arrives at the axon terminal.** An action potential from the brain or spinal cord reaches the axon terminal of the motor neuron, triggering exocytosis of the synaptic vesicles.
2. **Synaptic vesicles release acetylcholine into the synaptic cleft.** These vesicles release the neurotransmitter ACh into the synaptic cleft between the axon terminal and the motor end plate of the muscle fiber.
3. **Acetylcholine binds to ligand-gated ion channels in the motor end plate.** ACh diffuses across the cleft until

Figure 10.13 **Excitation phase: events at the neuromuscular junction.**

Play animation @MasteringA&P®

it reaches and binds to ACh receptors on the motor end plate. These receptors are ligand-gated ion channels in the sarcolemma.

④ **Ion channels open and sodium ions enter the muscle fiber.** When ACh binds, the ligand-gated ion channels of the ACh receptors open and allow sodium ions to enter the muscle fiber.

⑤ **Entry of sodium ions depolarizes the sarcolemma locally, producing an end-plate potential.** As sodium ions enter the fiber, a small area of the sarcolemma depolarizes, producing an effect called an **end-plate potential.** This potential is simply a local depolarization in the area of the motor end plate.

Multiple end-plate potentials must be generated to produce a functional contraction of a muscle fiber. However, the ACh released from the synaptic vesicles is degraded and inactivated almost immediately by an enzyme in the synaptic cleft called **acetylcholinesterase** (ah-SEET-'l-kohl-in-ess'-teh-rayz; AChE). For this reason, the neuron must continue to fire action

potentials and release new ACh in order to stimulate repeated end-plate potentials in the muscle fiber.

What happens to the muscle contraction when excitation is blocked? See *A&P in the Real World: Botulism and Botox* on p. 358 to find out about a toxin that does this by blocking the release of acetylcholine at the motor end plate.

Excitation-Contraction Coupling

Up to this point, excitation of the sarcolemma and contraction of the myofilaments might seem like two separate processes. However, you've read that a muscle fiber will not contract without first being excited, so these two processes must be connected, or "coupled" together. This connection occurs by a series of events known collectively as **excitation-contraction coupling** (Figure 10.14). In this phase, the end-plate potential leads to an action potential in the sarcolemma, which in turn triggers events that result in contraction.

The steps of excitation-contraction coupling are as follows:

① **The end-plate potential stimulates an action potential.** The end-plate potential opens enough voltage-gated sodium ion channels in areas of the sarcolemma adjacent to the motor end plate, stimulating an action potential.

Figure 10.14 Excitation-contraction coupling: events at the sarcolemma and sarcoplasmic reticulum.

② **The action potential is propagated down the T-tubules.** The action potential propagates like a wave along the sarcolemma, as depolarization of one area of the membrane triggers the next few voltage-gated sodium ion channels to open. This process continues like a chain reaction down the muscle fiber. Ultimately, the wave of depolarization dives deeply into the muscle fiber via the T-tubules.

③ **T-tubule depolarization leads to the opening of calcium ion channels in the SR, and calcium ions enter the cytosol.** Electrical changes in the T-tubules in turn trigger changes in proteins in the terminal cisternae of the SR, which flank either side of the T-tubule in a triad. The T-tubules are directly linked to the terminal cisternae by voltage-gated proteins. When the T-tubules depolarize, these proteins "twist" open the calcium ion channels in the terminal cisternae. In essence, these proteins open the "floodgates" of the SR, releasing many calcium ions into the cytosol.

As calcium ions flood the cytosol, the sarcolemma is repolarizing.

The events described here involve only a single action potential. However, a skeletal muscle contraction in the body requires this process to occur repeatedly. So new action potentials are generated over and over, and the concentration of calcium ions in the cytosol increases with each one. Now let's see how large numbers of calcium ions in the cytosol lead to the contraction phase.

Contraction Phase: The Crossbridge Cycle of the Sliding-Filament Mechanism

We've said that myofilaments slide past one another during the **contraction phase.** Recall that the myosin heads of the thick filaments attach to the actin of the thin filaments at their active sites and pull them toward the middle of the sarcomere. But before the contractile proteins can do their work, the regulatory proteins must let them.

As you can see in **Figure 10.15a,** in a resting muscle fiber the regulatory protein tropomyosin curls around actin, blocking its active sites. Calcium ions initiate the mechanism that moves the attached troponin and tropomyosin away from the active sites. At rest, the concentration of calcium ions in the cytosol is very low. However, as we just saw, once excitation-contraction coupling has occurred, the concentration of calcium ions in the cytosol increases dramatically. In preparation for muscle contraction (**Figure 10.15b**):

① **Calcium ions bind to troponin.** The regulatory protein troponin has three subunits: One binds calcium ions,

---

Terminal cisterna of SR
Ca²⁺ channel (closed)
Ca²⁺
T-tubule
Myofibril

Thin filament
Actin active sites (covered)
Tropomyosin
Troponin
Myosin head
Thick filament

(a) At rest, tropomyosin blocks actin's active sites.

Ca²⁺ channel (open)
Depolarization
Ca²⁺

Ca²⁺
Actin active sites (exposed)

① Ca²⁺ bind to troponin.
② Tropomyosin moves, and the active sites of actin are exposed.

(b) After stimulation, Ca²⁺ release causes the active sites of actin to be exposed.

Figure 10.15 Preparation for contraction: regulatory events at the myofibril.

Play animation
@MasteringA&P®

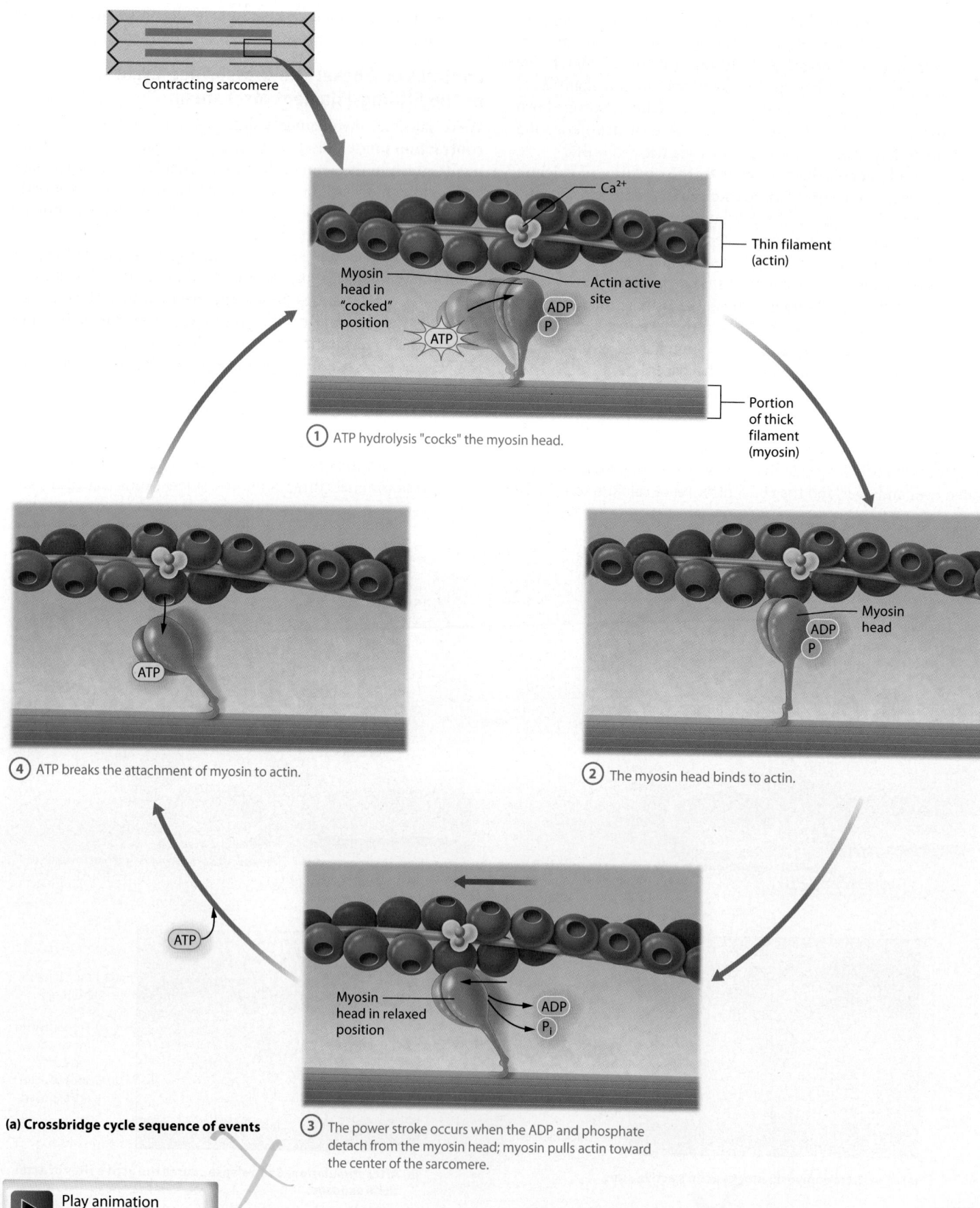

Contracting sarcomere

Ca²⁺

Thin filament (actin)

Myosin head in "cocked" position

Actin active site

ADP
P

ATP

Portion of thick filament (myosin)

① ATP hydrolysis "cocks" the myosin head.

Myosin head

ADP
P

② The myosin head binds to actin.

ATP

④ ATP breaks the attachment of myosin to actin.

ATP

Myosin head in relaxed position

ADP
Pᵢ

③ The power stroke occurs when the ADP and phosphate detach from the myosin head; myosin pulls actin toward the center of the sarcomere.

(a) Crossbridge cycle sequence of events

Play animation
@MasteringA&P®

Figure 10.16 Contraction phase: the crossbridge cycle of the sliding-filament mechanism.

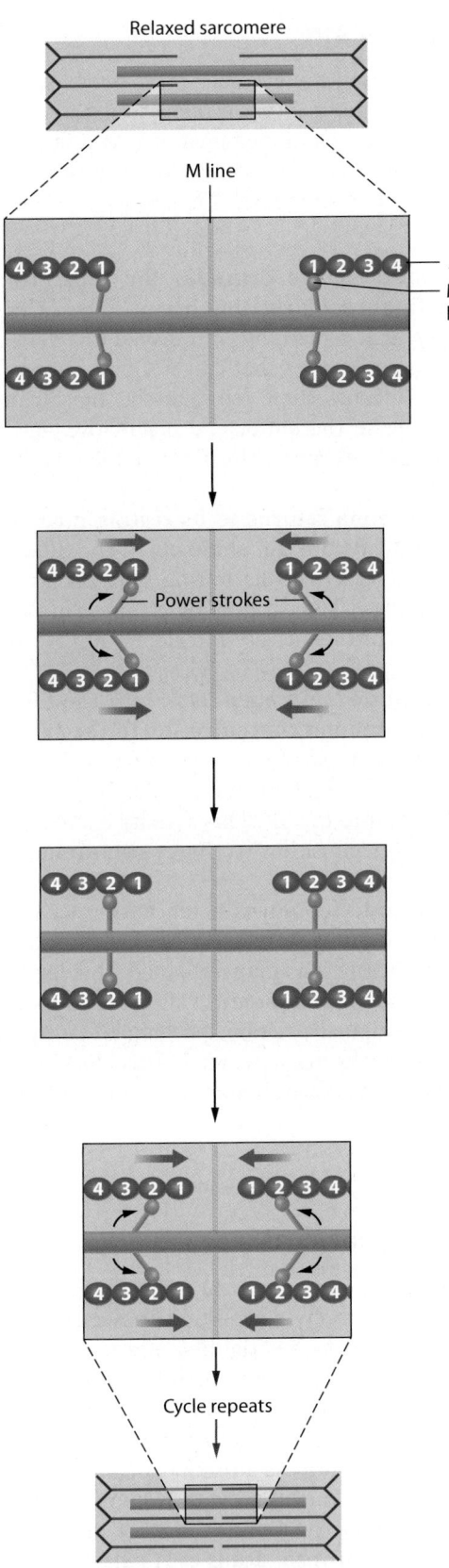

Relaxed sarcomere

M line

Actin
Myosin head

Power strokes

Cycle repeats

Contracted sarcomere

(b) Crossbridge cycle repeats, pulling actin toward the center of the sarcomere.

Figure 10.16 *(continued)*

one binds actin, and the other tropomyosin. In this step, calcium ions bind to the appropriate troponin subunit.

② **Tropomyosin moves, and the active sites of actin are exposed.** The binding of calcium ions causes troponin to shift its position, allowing tropomyosin to move away from the active sites.

When the active sites of actin are exposed, the myosin heads are able to bind tightly to them. A myosin head bound to an actin molecule is known as a **crossbridge.** When this binding occurs, a **crossbridge cycle** is initiated that leads to the sliding of myofilaments. A muscle **contraction** is simply a succession of crossbridge cycles and the resulting production of force. During these cycles, the myosin head grabs onto a series of actin molecules in the thin filament, pulling the filament progressively closer to the M line of the sarcomere.

The sequence of events of the crossbridge cycle, illustrated in **Figure 10.16a,** is as follows:

① **ATP hydrolysis "cocks" the myosin head.** Muscle contraction requires energy provided by the hydrolysis of ATP. Each myosin head contains both a site that binds to ATP and an ATPase enzyme that breaks down ATP. When ATP binds to the head of the myosin molecule, the ATPase in the myosin rapidly hydrolyzes the ATP to ADP and a phosphate group (P). The energy liberated from this reaction "cocks" the myosin head into its high-energy position, ready to work. The ADP and phosphate remain attached to the myosin head when it is cocked.

② **The myosin head binds to actin.** With the myosin head in its cocked position, it may now bind to the active site of actin. Note that the resulting crossbridge is at about a 90° angle relative to the thick filament.

③ **The power stroke occurs when the ADP and phosphate detach from the myosin head; myosin pulls actin toward the center of the sarcomere.** The ADP and phosphate (inorganic phosphate, represented as P_i) detach from the myosin head as it pivots on its hinge and moves from its cocked high-energy position to its relaxed (or low-energy) position. As the myosin pivots, it pulls the actin toward the center of the sarcomere. This action is known as the **power stroke.** Notice that the myosin crossbridge is now at about a 45° angle relative to the thick filament.

④ **ATP breaks the attachment of myosin to actin.** Another ATP binds to the myosin head, which breaks its attachment to actin.

The contraction cycle is then repeated, as shown in **Figure 10.16b.** ATP is hydrolyzed, the myosin head is recocked, it binds to the next actin molecule, and the power stroke repeats. You can see in the figure how myosin first attaches to the first actin, then the second, and so on. For an average contraction, this process will repeat about 20–40 times for each myosin head in each sarcomere of the muscle fiber.

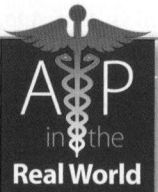

Botulism and Botox

The bacterium *Clostridium botulinum* (kloh-STRID-ee-um bot-yoo-LIN-um) produces the most lethal known biological poison—as little as one gram of crystalline toxin is enough to kill about one million adults. Humans are generally exposed to these bacteria and their toxin, called *botulinum toxin,* via contaminated food, especially food that is canned without proper sterilization. Exposure to the toxin causes the disease *botulism* (BOT-yoo-lizm), in which botulinum toxin binds to motor neurons of the NMJ and blocks the release of acetylcholine from synaptic vesicles. This activity renders affected muscles unable to contract, and without proper treatment, death from respiratory failure will follow.

Given the deadly nature of this toxin, it seems an unlikely candidate for a drug with therapeutic benefits. However, when injected into certain muscles in miniscule amounts, it can provide temporary relief from painful muscle spasms and migraine headaches. The toxin is also used for cosmetic purposes under the name Botox. When Botox is injected into the muscles under the skin of the face, it relaxes them, temporarily reducing the appearance of fine lines and wrinkles.

At step 4 of the crossbridge cycle shown in Figure 10.16a, you might think that the detachment of myosin from actin would make actin slide backward. However, this doesn't happen because at any given time during a muscle contraction, some of the myosin crossbridges will still be attached to actin molecules.

To picture this process, let's use the analogy of sailors pulling on a rope to bring up an anchor from the sea floor. The rope represents the thin filament, the anchor the Z-disc, and the hands of the sailors the myosin heads. At any given time, the hands of some sailors are holding the rope and pulling, like myosin heads attached to actin and pulling it with power strokes. In addition, at any given time, some sailors are repositioning their hands farther up the rope and also allowing their arms to rest for a second. This is also true of the myosin heads—while some are pulling, others are detached and being reactivated ("recocked") by ATP. So the sailors work together to pull on the rope and move the anchor toward the ship (M line). This represents what happens in a sarcomere during a contraction.

Quick Check

☐ 3. How does excitation from a neuron trigger muscle fiber contraction?

☐ 4. Describe the steps of excitation-contraction coupling.

☐ 5. Describe the steps of the crossbridge cycle.

Muscle Relaxation

A properly functioning muscle fiber must be able not only to contract but also to relax. **Muscle relaxation** has two components: ACh release stops and the remaining ACh in the synaptic cleft is broken down, and the calcium ion concentration in the cytosol returns to its resting level. The sequence of events is as follows (**Figure 10.17**):

① **Acetylcholinesterase degrades the remaining ACh, and the final repolarization occurs.** The AChE breaks down the ACh remaining in the synaptic cleft, degrading it into substances that can no longer stimulate the muscle. Without such stimulation, the ligand-gated ion channels in the motor end plate close, and the sarcolemma goes through the final repolarization of the contraction.

② **The sarcolemma returns to its resting membrane potential, and calcium ion channels in the SR close.** The sarcolemma returns to its resting state once repolarization is complete. The calcium ion channels in the SR then close, so no further release of calcium ions from the SR takes place.

③ **Calcium ions are pumped back into the SR, returning the calcium ion concentration in the cytosol to its resting level.** Active transport pumps in the SR membrane consume ATP to pump calcium ions from the cytosol back into the SR. This activity decreases the calcium ion concentration of the cytosol, returning it to resting level.

④ **Troponin and tropomyosin block the active sites of actin, and the muscle relaxes.** As the cytosolic concentration of calcium ions returns to its resting level, calcium ions dissociate from troponin. This causes troponin to return to its original position, pushing tropomyosin back to where it blocks the active sites. This blocking prohibits the myosin heads from binding to actin, and the muscle

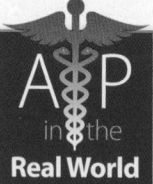

Rigor Mortis

After death, muscle spasms occur over the entire body, a condition known as *rigor mortis* (*rigor-* = "rigidity," *mortis-* = "death"). Rigor mortis is the progressive stiffening of skeletal muscles that begins about 3–4 hours after death, as the pumps that drive calcium ions back into the SR no longer have ATP to fuel their activity. As a result, calcium ions remain in the cytosol, where they bind with troponin and initiate a muscle contraction. The muscle fibers are unable to relax without ATP, so the myosin heads can't detach from actin. Thus, the muscles remain in spasm until the proteins in the myofilaments begin to degenerate, about 48–72 hours after death.

① Acetylcholinesterase degrades the remaining ACh, and the final repolarization begins.

② The sarcolemma returns to its resting membrane potential, and the Ca²⁺ channels in the SR close.

③ Ca²⁺ are pumped back into the SR, returning the Ca²⁺ concentration in the cytosol to its resting level.

④ Troponin and tropomyosin block the active sites of actin, and the muscle relaxes.

Axon terminal

Synaptic cleft

Repolarization

Voltage-gated K⁺ channel (open)

Degraded ACh

ACh

Acetylcholinesterase

Motor end plate

Ligand-gated ion channel (closed)

K⁺   K⁺   K⁺   K⁺

Sarcolemma

T-tubule

Sarcoplasmic reticulum

Cytosol

Ca²⁺ channel (closed)

Ca²⁺ pump

Ca²⁺

Troponin   Tropomyosin

Play animation
@MasteringA&P®

Figure 10.17 Relaxation phase: the process of muscle relaxation.

contraction is over. The myofilaments then slide back into their original positions, with support from titin and other structural proteins.

The calcium ion pumps and AChE work all the time to allow relaxation of a muscle fiber. This makes it possible for the fiber to start a new contraction when a new stimulus comes along.

A muscle that is unable to relax is said to be in **spasm.** Muscle spasms are very common and may be due to factors such as dehydration, electrolyte imbalances, muscle injury, or muscle overload. These spasms vary in degree, ranging from a "tight" muscle (also known as a muscle "knot") to a full spasm that impairs function and is typically very painful. See *A&P in the Real World: Rigor Mortis* to discover why muscles can't relax for a period of time after death.

Quick Check

☐ 6. Describe the steps of skeletal muscle fiber relaxation.

The Big Picture of Skeletal Muscle Contraction

Figure 10.18

Play animation
@MasteringA&P®

Axon terminal

① **Excitation:** ACh triggers an end-plate potential in the motor end-plate (see Figure 10.13).

Na⁺

End-plate potential of motor end-plate

② **Excitation-contraction coupling:** The resulting action potential in the sarcolemma travels down the T-tubules and triggers Ca²⁺ release from the SR into the cytosol (see Figure 10.14).

Troponin

Tropomyosin

Ca²⁺

③ **Preparation for contraction:** Ca²⁺ bind to troponin, which moves tropomyosin away from the active the sites of actin (see Figure 10.15).

ATP

④ **Contraction:** Actin and myosin bind, and myosin undergoes a power stroke. ATP detaches actin and myosin, and the cycle repeats, leading to contraction of the muscle fiber (see Figure 10.16).

Troponin

Ca²⁺

Tropomyosin

⑤ **Relaxation:** The neuron stops releasing ACh, and the AChE degrades the ACh in the synaptic cleft. The cytosolic concentration of Ca²⁺ returns to the resting level, the active sites of actin are blocked, and the muscle fiber relaxes (see Figure 10.17).

Putting It All Together: The Big Picture of Skeletal Muscle Contraction

Let's now bring together all of this information and look at the big picture of skeletal muscle contraction. **Figure 10.18** illustrates the main steps of this process.

Apply What You Learned

☐ 1. The disease myasthenia gravis (my'-uss-THEE-nee-ah GRAH-viss) results in the destruction of acetylcholine receptors on the motor end plate. Predict the symptoms and effects of this disease.
☐ 2. Predict the effect of improperly functioning troponin that isn't able to bind to tropomyosin.
☐ 3. Researchers discover a genetic mutation that leads to a lack of T-tubules in a person's skeletal muscle fibers. Predict the possible effects of such a mutation. (*Hint:* How would it impact the structure and function of a triad?)

See answers in Appendix A.

MODULE **10.5**

Energy Sources for Skeletal Muscle

Learning Outcomes

1. Describe the immediate energy sources available to muscle fibers.
2. Describe the glycolytic and oxidative mechanisms that muscle fibers use to obtain ATP for muscle contraction.
3. Explain the duration of activity that each ATP source can fuel.

Muscle fibers perform many functions that require energy. From Module 10.4, we know that energy in the form of ATP is required for multiple aspects of skeletal muscle contraction. For example, cells need ATP to run the Na^+/K^+ pumps that maintain the ion gradients involved in the action potential of skeletal muscle fibers. ATP is also necessary for both contraction *and* relaxation of the muscle fiber.

With so much ATP required for muscle function, you might expect muscle fibers to contain a great deal of stored ATP. Surprisingly, they store only a few seconds' worth of ATP in their cytosol. Muscle fibers, then, must have a way to quickly regenerate ATP. Regeneration may be accomplished through three processes: (1) reactions in the cytosol that immediately add a phosphate group to ADP, (2) *glycolytic catabolism* in the cytosol, and (3) *oxidative catabolism* in the mitochondria. All three processes may occur simultaneously in muscle fibers during

contractions, but they are used in different proportions, depending on the resources and needs of the cells.

Immediate Sources of Energy for Muscle Contraction

When contraction begins, the main immediate energy source of the muscle fiber is stored as ATP. This ATP is rapidly consumed, but is regenerated almost immediately by a reaction using a molecule called **creatine phosphate** (KREH-ah-tin FOS-fayt; **CP**; Figure 10.19a). Creatine phosphate, found primarily in muscle fibers, is about 5–6 times more abundant than ATP in the cytosol. During the *creatine phosphate reaction*, creatine phosphate, with the help of the enzyme **creatine kinase** (KY-nayz; **CK**), donates a phosphate group to ADP, producing ATP:

$$ADP + \text{Creatine phosphate} \xrightleftharpoons{CK} ATP + \text{Creatine}$$

ATP produced by this reaction provides the muscles with enough energy for about an additional 10 seconds of maximal muscle activity. Though this may not sound like much, it's enough for a short burst of activity, such as a 100-meter sprint. Does more creatine make for a better athletic performance? To find out, take a look at *A&P in the Real World: Creatine Supplementation* on p. 363.

Quick Check

☐ 1. What are the two immediate energy sources for muscle contraction?
☐ 2. How long can these immediate energy sources fuel muscle activity?

Glycolytic Energy Sources

◄◄ Flashback

1. What is a catabolic reaction? (p. 42)
2. What is glycolysis? (p. 69)

When immediate energy sources are depleted, muscle fibers turn to glycolysis, also known as **glycolytic** or **anaerobic catabolism,** to make ATP. As you may recall, *glycolysis* is a series of reactions that takes place in the cytosol of all cells, including muscle fibers (see Chapter 3). During this process, glucose is broken down to produce two ATP per molecule of glucose (left side of Figure 10.19b).

A muscle fiber has two potential sources of glucose for glycolysis: glucose that enters the fiber from the bloodstream and a storage form of glucose called *glycogen* (GLY-koh-jen). Glycogen granules are found in the cytosol of both muscle fibers

(a) Immediate energy sources

(b) Glycolytic and oxidative energy sources

Figure 10.19 Sources of energy for muscle fibers.

and liver cells. Muscle fibers that depend primarily on glycolysis as their means of ATP production have large quantities of glycogen in their cytosol. Regardless of the amount of stored glycogen, though, glycolysis by itself can provide adequate ATP for only about 30–40 seconds of sustained muscle contraction.

Glycolysis requires no oxygen directly, which is why it is sometimes called *anaerobic catabolism*. However, the fate of the product of glycolysis depends on the availability of oxygen to the muscle fiber. The product is always two molecules of a compound known as *pyruvate*. If oxygen is abundant, this compound then enters the mitochondria for oxidative catabolism, which at that point will be occurring simultaneously with glycolysis as long as glucose is available. If oxygen is not

abundant, the pyruvate is converted into two molecules of the compound **lactic acid** (also called *lactate*). About 20% of this lactic acid diffuses out of the muscle fiber into the bloodstream, where much of it is converted into glucose by the liver. The remaining lactic acid was once thought to be a dead-end product, but current evidence indicates that a great deal of it enters the mitochondria and is used for oxidative catabolism, our next subject.

Quick Check

☐ 3. What is glycolysis?
☐ 4. How long can glycolytic catabolism fuel muscle activity?

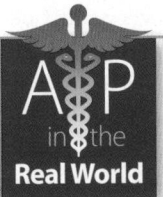

Creatine Supplementation

Dietary supplement manufacturers market creatine supplements as a way to improve muscle strength and performance. The evidence to support these claims, however, is mixed. Research has demonstrated that supplementation with creatine does mildly improve performance for activities that require short bursts of muscle activity. However, the effects of creatine on endurance-type activities are minimal to nonexistent. And creatine may actually be detrimental to athletes in certain sports, because supplementation causes weight gain from water retention.

One of the most pressing questions on creatine supplementation isn't about its effectiveness, but rather its safety. Most of the recent data suggest that supplementation within the recommended dose range is safe in the short term. But long-term studies remain to be done. In addition, massive doses of creatine can have negative health implications such as kidney damage. Beyond the possible health issues, huge doses are a waste of money. Skeletal muscle fibers have a maximal storage capacity for creatine, and any amount ingested in excess of this limit is simply excreted in the urine.

Oxidative Energy Sources

◀◀ Flashback

1. What is an oxidation-reduction reaction? (p. 43)
2. Where does oxidative catabolism take place, and what does it yield? (p. 89)

Combined, immediate and anaerobic glycolytic energy sources are adequate for short bursts of activity, such as running a 400-meter race. However, longer-lasting muscle activity requires a muscle fiber to use mostly **oxidative** or **aerobic catabolism** to generate ATP (right side of Figure 10.19b). Recall that the reactions of oxidative catabolism occur in the mitochondria (see Chapter 3). During this process, electrons are removed from carbon-based molecules, and the energy liberated is then used to fuel the synthesis of ATP.

Oxidative catabolism produces more ATP than does glycolysis; the amount of ATP depends on the type of fuel used by the muscle fiber. Muscle fibers can use multiple fuels for oxidative catabolism, including the products of glycolysis, as well as fatty acids and amino acids. Glucose is generally the preferred fuel for muscle fibers—the one they use first. As glucose becomes less available, muscle fibers will catabolize fatty acids and amino acids if necessary.

The final step in this process is the transfer of electrons to a molecule of oxygen, which is why this type of metabolism is also called *aerobic catabolism*. In fact, oxidative catabolism, which takes place in nearly all cells, is the main reason we need oxygen

for survival. To meet this requirement, muscle fibers must have a plentiful supply of oxygen during activity. Much of this oxygen diffuses into the cytosol from the bloodstream. The rest is bound to the oxygen-binding protein **myoglobin** (MY-oh-glohb-in), which is similar to hemoglobin, the oxygen-carrying protein in the blood. Myoglobin, which is found in the cytosol, binds to oxygen that has diffused into the muscle fiber from the extracellular fluid, and releases it as the available oxygen is depleted by mitochondria performing oxidative catabolism.

Oxidative catabolism can provide ATP for hours, as long as oxygen and fuels are available. After only 1 minute of skeletal muscle activity, oxidative catabolism is the predominant source of ATP for almost all muscle fibers. After several minutes, nearly 100% of the ATP is produced by this aerobic process. This is why exercise that lasts continuously for at least several minutes is often called "aerobics."

(This module provides only an overview of the three main sources of ATP for skeletal muscle fibers. You'll find full coverage of this topic in the metabolism chapter—see Chapter 23.)

Quick Check

☐ 5. How do oxidative catabolism and glycolytic catabolism differ?
☐ 6. How long can oxidative catabolism fuel muscle activity?
☐ 7. Why is oxidative catabolism referred to as "aerobic"?

Apply What You Learned

☐ 1. Would extra creatine phosphate likely enhance performance of activities lasting several minutes to an hour? Explain.
☐ 2. Explain why engaging in 30–60 minutes of "aerobics" is promoted as an activity that "burns fat." Would you get this same "fat burning" from several 20-second sprints? Explain your answer in biochemical terms.

See answers in Appendix A.

MODULE **10.6**

Muscle Tension at the Fiber Level

Learning Outcomes

1. Describe the stages of a twitch contraction and explain how a twitch is affected by the frequency of stimulation.
2. Relate tension production to the length of a sarcomere.
3. Compare and contrast the anatomical and metabolic characteristics of type I and type II muscle fibers.

As we discussed earlier, a basic principle of muscle physiology is that muscle tissue contracts to produce tension. The smallest muscle contraction, known as a **muscle twitch,** is the response of a muscle fiber to a single action potential in a motor neuron. Twitches are laboratory phenomena—they don't occur in whole muscles in the body. In spite of this fact, twitches are useful in experiments to study different types of contractions.

Every twitch produces some degree of tension, which translates to some degree of force. Twitches can create vastly different amounts of tension, both within an individual muscle fiber and within an entire muscle. In this module, we explore the specifics of the production and strength of a twitch contraction, as well as the relationship between the length of a sarcomere and the amount of tension that can be produced. We then examine the different classes of muscle fibers and the relative speeds with which they proceed through a twitch contraction. Finally, we conclude by considering how tension is transmitted from the muscle fibers to a whole muscle.

Twitch Contraction

A muscle twitch is produced in a laboratory by direct electrical stimulation of a muscle fiber. An instrument records the twitch and generates a recording known as a **myogram** (Figure 10.20). Notice in this myogram that the muscle twitch has three distinct phases—the latent, contraction, and relaxation periods:

① **Latent period.** The latent period is the 1- to 2-ms (millisecond) time that it takes for the action potential to spread

through the sarcolemma. At the end of this very short period, the action potential spreads past the T-tubules and triggers the release of calcium ions from the terminal cisternae of the SR. These ions then bind to troponin, and tropomyosin is allowed to move away from the active sites of actin. The myofibril is now ready to enter a crossbridge cycle. Note that the sarcolemma completes the repolarization phase of the action potential at the end of this period.

② **Contraction period.** This period is marked by a rapid increase in tension as crossbridge cycles occur repeatedly.

③ **Relaxation period.** During the relaxation period, tension decreases due to the decreasing calcium ion concentration in the cytosol. It takes the pumps in the SR from 10 to 100 ms to pump calcium ions from the cytosol back into the SR. Then, as tropomyosin once again blocks the active sites of actin, the muscle fiber relaxes.

Between the start of the latent period and the start of the contraction period, there is an interval of about 5 ms during which the muscle cannot respond to another stimulus. This period, called the *refractory period,* is a property of all excitable cells. Cardiac muscle and smooth muscle have refractory periods as long as their contractions, so the cells must fully relax before they can contract a second time. However, skeletal muscle fibers have a much shorter refractory period, allowing them to maintain a sustained contraction phase (more on this shortly).

Figure 10.20 shows an average duration for contraction and relaxation periods. However, these times vary greatly among skeletal muscle fibers, and each may last from about 7 to 100 ms, depending largely on the type of muscle fiber involved and the level of activity of the ATPase on the myosin molecules during a crossbridge cycle. The tension produced during a twitch also varies considerably with several factors, including the timing and frequency of stimulation, the length of the fiber at rest, and the type of muscle fiber. These factors are examined in the next sections.

Quick Check

☐ 1. What is a twitch contraction?
☐ 2. What are the phases of a twitch contraction?

Tension Production and the Timing and Frequency of Stimulation

In general, repeated stimulation of a muscle fiber by a motor neuron results in twitches with progressively greater tension. This happens because the pumps in the SR membranes have inadequate time to pump all of the released calcium ions back into the SR before the fiber is restimulated, and so the concentration of calcium ions in the cytosol increases with each stimulus.

This phenomenon is known as **wave summation,** so named because the waves of contraction add together. The amount

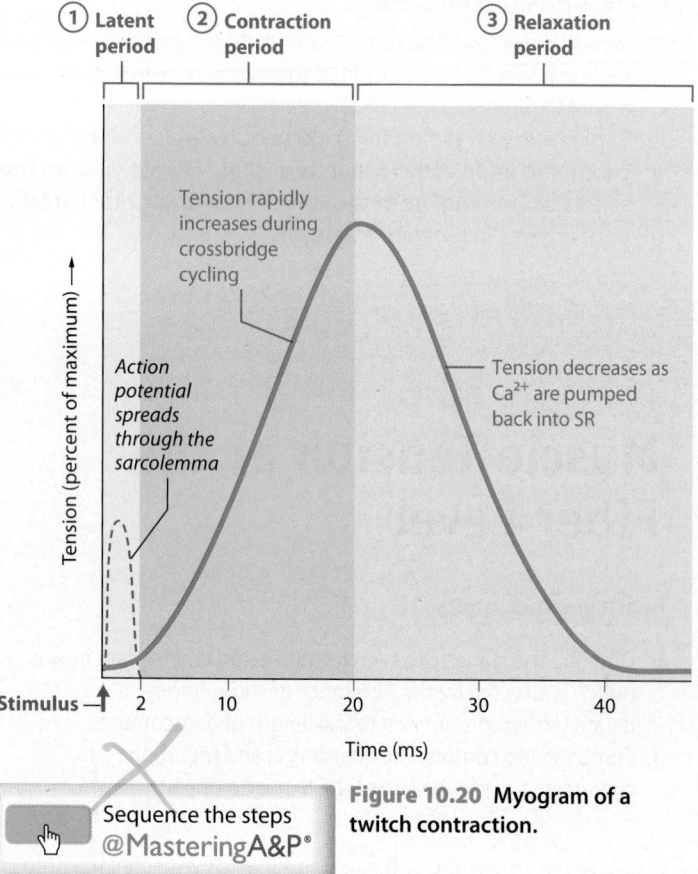

Figure 10.20 Myogram of a twitch contraction.

Sequence the steps
@MasteringA&P®

(a) Wave summation: unfused tetanus

Maximum tension in unfused tetanus

Maximum tension of single twitch

Muscle fiber is not allowed to relax completely between stimuli; fiber stimulated about 50 times per second.

(b) Wave summation: fused tetanus

Maximum tension in fused tetanus

Muscle fiber is not allowed to relax between stimuli; fiber stimulated 80–100 times per second, which generates sustained contraction and maximal tension.

Figure 10.21 Wave summation: unfused and fused tetanus.

of tension produced during wave summation depends on the frequency of stimulation by the motor neuron. Depending on the number of times per second the muscle fiber is stimulated, we end up with one of two states, unfused or fused *tetanus* (**Figure 10.21**):

- **Unfused tetanus.** If the fiber is stimulated about 50 times per second, it can only partially relax between contractions. A sustained contraction called **unfused** (or **incomplete**) **tetanus** results. Notice in Figure 10.21a that during unfused tetanus, the tension pulsates, decreasing slightly and then increasing a bit more with each successive twitch until a level of maximal tension is reached.
- **Fused tetanus.** If the fiber is stimulated at a higher rate of 80–100 times per second, the muscle fiber does not have time to relax between stimuli because the calcium ion concentration in the cytosol remains high. The availability of calcium ions allows more and more crossbridges to form, contributing to the increase in tension. This increase results in a condition called **fused** (or **complete**) **tetanus,** in which the tension remains constant at a maximal level (Figure 10.21b). Note that fused tetanus is possible only because of the extremely short refractory period of the skeletal muscle fiber. This period is so short that the fiber does not have to relax between stimuli, and so can respond to a new stimulus before the last twitch is over, producing a sustained contraction.

Both unfused and fused tetanus generate more tension than a single twitch contraction. Tension is highest during fused tetanus, as the muscle fiber produces a constant tension several times greater than that created during a single twitch.

Take care not to confuse the states of fused and unfused tetanus with the disease tetanus. The disease is caused by the bacterium *Clostridium tetani* (teh-TAN-ee), which releases a toxin that enters the nervous system and produces excessive stimulation of skeletal muscles. The result is simultaneous contraction of muscle groups, known as a *tetanic spasm*. The disease tetanus is almost completely preventable with vaccination, but it remains a serious health threat for unvaccinated individuals, particularly in countries where inhabitants have little access to health care.

Quick Check

- ☐ 3. How does the timing of a stimulus impact the tension produced by the resulting muscle contraction?
- ☐ 4. How do fused and unfused tetanus differ?

The Length-Tension Relationship

The second factor that determines the amount of tension produced by a twitch contraction is the number of crossbridges that can form within each sarcomere of the muscle fiber. The number of crossbridges depends on the length of the sarcomere prior to contraction, a principle known as the **length-tension relationship.** The *optimal length* of the sarcomere is the length of the muscle fiber at which the most crossbridges can form, allowing

Figure 10.22 **The length-tension relationship.**

the fiber to generate almost 100% of the tension that is possible to produce (**Figure 10.22**).

A myofibril's elastic filaments enable it to be stretched to varying degrees, which results in different possible lengths for a myofibril and its sarcomeres. In Figure 10.22, sarcomeres A, B, and C are resting at different lengths, as shown by the degree of overlap of their thin and thick filaments, or the **zone of overlap.**

The size of the zone of overlap is largely due to the position of the muscle before contraction. Figure 10.22 shows that the muscle containing sarcomere A is shortened by contraction at the wrist, and therefore all of the sarcomeres are shorter, with large zones of overlap. The muscle that sarcomere B is part of, however, is nearer its optimal length, as the wrist is in a neutral, extended position. As a result, all of the sarcomeres are longer and have smaller zones of overlap. Finally, the muscle containing sarcomere C is stretched due to hyperextension at the wrist, leading to long sarcomeres with very small zones of overlap.

Let's apply this concept to muscle tension: For tension to be generated, the myosin heads must be able to grab the actin molecules to pull them toward the center of the sarcomere. With this in mind, look at sarcomere A. Although sarcomere A has a great deal of overlap, this is actually an obstacle to producing muscle tension, because there is little room for the thin filaments to be pulled together before the thick filaments would run into the Z-discs. As you can see from the graph line in Figure 10.22, this results in decreased tension production. To demonstrate this yourself, flex your wrist and try to grip your pen tightly in your fist. Notice how difficult

it is to maintain a tight grip—this is due to the limited amount of tension that the shortened sarcomeres in your forearm can generate.

Now consider sarcomere C. You may think that sarcomere C could produce more tension than sarcomere A because it has a small zone of overlap, but actually the two sarcomeres produce about the same amount of tension. Why? Because the zone of overlap in sarcomere C is so short that the myosin heads are unable to get much of a "grip" on the actin molecules, so less tension results from this contraction. Try the experiment with your pen again, but this time hyperextend your wrist. The results are about the same—your forearm muscles are still unable to generate enough tension to grip the pen very tightly.

From our discussion so far, it follows that less tension will be produced in a sarcomere that is either overly shortened or overly stretched at rest. Maximal tension production, therefore, occurs in a narrow range of resting sarcomere lengths, near the optimal length of the sarcomere. We find this situation with sarcomere B—it has enough room to shorten and plenty of actin molecules for the myosin heads to grab. The graph line in Figure 10.22 shows that sarcomere B produces close to 100% of the maximal tension of the muscle fiber, whereas sarcomeres A and C are able to produce only about 85% and 90% of this tension, respectively.

Quick Check

☐ 5. At what length will a sarcomere be able to generate the greatest amount of tension? Why?

ConceptBOOST »»»

Understanding How Events at the Myofilaments Produce Tension of a Whole Muscle

The next module will explore muscle tension at the organ level. For right now, let's take a look at how all of the elements of a skeletal muscle work together. The movement produced by each crossbridge cycle in a sarcomere is extremely small, so sometimes it's difficult to see how these actions cause an entire muscle to contract. The trick is to realize just how many sarcomeres a single muscle fiber contains; each sarcomere is simply shortening simultaneously with the millions of other sarcomeres in the fiber.

Figure 10.23 shows how the tension produced in the sarcomeres of a myofibril is transmitted to a whole muscle. Remember that myofibrils are connected to the sarcolemma of the muscle fiber, so any tension in the myofibrils is transmitted to the muscle fiber as a whole. The muscle fiber then relays that tension to the collagen fibers in the endomysium, which causes contraction of the fascicle as a whole. As the fascicles contract, the tension is in turn conducted from the surrounding perimysium to the epimysium and the tendons of the entire muscle. The muscle then contracts, pulling on the attached bones and causing movement. ■

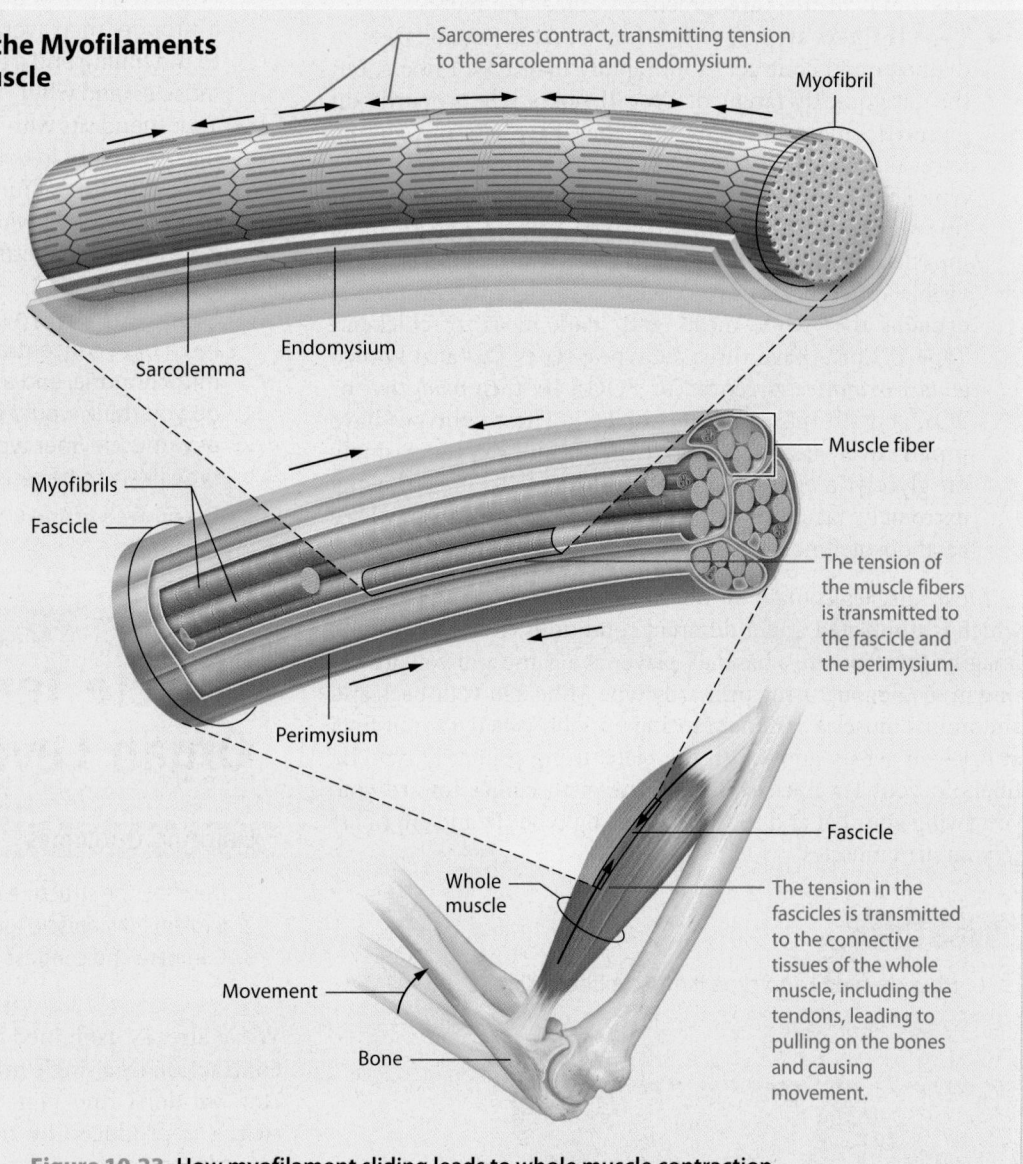

Figure 10.23 How myofilament sliding leads to whole muscle contraction.

Classes of Skeletal Muscle Fibers

Up to this point in the chapter, we've been describing the twitch contraction of a generic skeletal muscle fiber. However, skeletal muscle fibers are neither structurally nor functionally identical, and so some proceed through a twitch contraction more rapidly than others. How fast this process happens depends mostly on the level of myosin ATPase activity in the fiber, which may be "fast" or "slow." Recall that myosin ATPase hydrolyzes ATP so that a power stroke can occur. Muscle fibers with high myosin ATPase activity are called **fast-twitch fibers** because they proceed rapidly through contraction cycles. Fast-twitch fibers are found in muscles that must move body parts rapidly, such as those that move the eyeballs. Fibers with low myosin ATPase activity are called **slow-twitch**

fibers because they proceed through contraction cycles more slowly. Slow-twitch fibers are found in muscles that require slow, sustained contractions, such as the postural muscles of the back.

The speed of a twitch contraction is combined with its predominant energy source (oxidative versus glycolytic catabolism) to give us two main classes of skeletal muscle fibers, types I and II:

- **Type I fibers** are slow-twitch fibers that are small in diameter. Type I fibers contract more slowly and less forcefully than other fibers, but they can maintain extended periods of contraction. This ability requires the continual oxidative generation of large quantities of ATP, so their metabolism is primarily oxidative and they have large quantities of myoglobin, many mitochondria, and a well-developed blood supply.

They are often called *slow oxidative* fibers. The high myoglobin content of type I muscle fibers makes them red, and so type I muscle fibers are sometimes called "red muscle."

- **Type II fibers** are fast-twitch fibers that are often larger in diameter and contract more rapidly than type I fibers, but they are quickly fatigued. Type II fibers rely primarily on glycolytic energy sources, and they have less myoglobin, fewer mitochondria, and a less extensive blood supply than type I fibers. Due to their low myoglobin content, type II fibers are lighter in color than type I fibers, and so these muscle fibers are sometimes called "white muscle." (This visible difference between white muscle and red muscle explains the "white meat" and "dark meat" of chicken.) Type II fibers have three subtypes: types IIa (also known as *fast oxidative glycolytic*, or FOG), IIx (*fast oxidative*, or FO), and IIb (*fast glycolytic*, or FG). These subtypes have progressively faster, stronger twitches and rely increasingly on glycolytic energy sources, with type IIb fibers having extremely fast, powerful twitches and using glycolytic catabolism almost exclusively.

Most muscles contain all fiber classes (**Figure 10.24**), each of which is stimulated under different conditions. To tie it together, imagine this: You are a baseball player at a game and you are sitting in the dugout, using primarily type I fibers in your back and abdominal muscles to remain sitting upright. When it's your turn at bat, you get up and jog to the plate, using primarily type IIa fibers in your leg muscles. When the pitch comes toward you, you swing your bat and hit the ball, using type IIx and IIb fibers in your arm muscles.

Quick Check

☐ 6. How do type I and type II muscle fibers differ?

TEM (850×)

Figure 10.24 Comparison of type I and type II muscle fibers.

Type I fibers

Type II fibers

☐ 1. When you lift a heavy box, your muscles need to generate a stronger than average contraction. What will happen to the timing and frequency of nerve stimulation of your muscles, and why?

☐ 2. Your friend sits with her arm flexed 45° at the elbow. You place a weight in her hand, and she notes that she has difficulty lifting it further. However, when she attempts to lift the weight with her elbow in a relaxed or neutral position, she has little trouble. Explain these differing results.

☐ 3. Suppose a new type of muscle fiber is discovered that has a high concentration of myoglobin, large numbers of mitochondria, and a well-developed blood supply. What do you think would be the primary energy source for this new muscle fiber type? Explain your answer. Is this new type likely to be resistant to fatigue? Why or why not?

See answers in Appendix A.

MODULE **10.7**

Muscle Tension at the Organ Level

Learning Outcomes

1. Describe the structure and function of a motor unit.
2. Explain how muscle tone is produced.
3. Compare and contrast the three types of contractions.

We've already examined the production of tension via a twitch contraction by a single muscle fiber. However, in a living organism, we don't find twitches of single muscle fibers. Rather, a twitch is produced by multiple muscle fibers that are innervated by the same motor neuron, a structure known as a *motor unit*. In this module, we explore the structure of a motor unit and how motor unit activation controls tension of the entire muscle. We then look at three types of skeletal muscle contractions and the amount of tension that each contraction can generate.

Motor Units

As a motor neuron approaches a muscle, its axon branches out and innervates multiple muscle fibers. A single motor neuron along with the muscle fibers it innervates is called a **motor unit** (**Figure 10.25**). When the motor neuron fires an action potential, all of the muscle fibers within its motor unit respond and produce about the same amount of tension. Note that this applies only to a motor unit, not to an entire muscle.

An average motor unit consists of about 150 muscle fibers, but this number can vary widely with the degree of motor

Figure 10.25 **The motor unit. Two motor units (colored red and blue) are shown here.**

control needed for the muscle—another example of the Structure-Function Core Principle (p. 25). Muscles that require fine con-

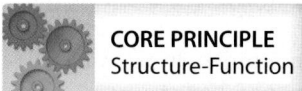
CORE PRINCIPLE
Structure-Function

trol, such as those around the larynx (voice box) or in the fingers, will have multiple small motor units, often containing as few as 10 muscle fibers per motor unit. This gives the nervous system more precise control over the amount and rate of tension produced. Conversely, large, powerful muscles, such as the postural muscles of the back and the gastrocnemius in the calf, can have 2000–3000 muscle fibers in each motor unit.

Motor units contain one class of muscle fiber, whether type I or type II. Those with type I fibers are called **slow motor units,** and those with type II fibers are called **fast motor units.**

Now let's see how motor units function in the body.

Recruitment

We know that an increased frequency in the stimulation of a muscle fiber affects muscle tension. However, when thinking about the tension generated by whole muscles, we need to consider another factor: the number of motor units that are stimulated. Initiation of a contraction activates a small number of motor units. As greater force is needed, more motor units are activated, a phenomenon called **recruitment.** Slow motor units are typically activated first, followed by fast motor units if additional tension is needed. For example, if you were lifting this

textbook off your desk, you would likely use your slow motor units for the most part. However, if you were to lift a stack of textbooks from the floor, your nervous system would first activate the slow motor units and then the fast motor units.

Muscle Tone

If you reach down right now and squeeze your calf muscle, you will probably notice that it feels somewhat taut despite being relaxed. Even when a muscle is at rest, it still has some degree of tension. This small amount of tension produces what is known as **muscle tone.** This tone is due to the involuntary activation of motor units by the brain and spinal cord. The nervous system alternates which motor units it activates, so that some can rest while others contract. Although the contractions that produce muscle tone are not strong enough to cause movement, they do serve important functions, including maintaining erect posture, stabilizing joints, generating heat, and ensuring that the muscle is ready to respond if movement is initiated.

If the tone in a skeletal muscle is abnormally low, generally because of a nervous system disorder, a condition known as **hypotonia** results. A hypotonic muscle looks flattened and feels soft and loose rather than firm. Often, the joints around the affected muscles are hyperextended, as the muscles are no longer able to generate enough tension to stabilize them. Alternately, a muscle may have abnormally high tone, a condition called **hypertonia.** A hypertonic muscle often feels rock hard, and it may shorten so much that it causes painful joint contractures (see Chapter 8). Normal hypertonia occurs during *shivering,* which takes place in response to a drop in body temperature, as more motor units are activated to generate additional heat.

Quick Check

☐ 1. What is a motor unit?
☐ 2. Explain the process of recruitment.
☐ 3. How is muscle tone produced?

Types of Muscle Contractions

Besides the number of motor units that are activated, another factor influencing the amount of tension a muscle generates involves how the length of the muscle changes during contraction. Although you may equate muscle contraction with muscle shortening, a muscle shortens only in specific types of contractions. Some muscle contractions result in no change in fiber length, and others actually occur during muscle fiber *lengthening.*

On the basis of muscle length during a contraction, we can classify contractions into three types: *isotonic concentric, isotonic eccentric,* and *isometric contractions.* Note that an isotonic contraction (*iso-* = "same," *tono-* = "tension") is one that produces enough tension to initially move a load, such as a weight, and then maintains that same level of tension throughout the contraction. An isometric contraction (*-metr* = "measure"), however, results when the external load is equal to the force generated by the muscle. In other words, isotonic contractions are the same tension but changing length, whereas isometric contractions are the same length but changing tension.

Let's look more closely at each type of contraction (**Figure 10.26**):

- **Isotonic concentric contractions.** You are probably familiar with the type of muscle contraction in which the muscle shortens; this is known as an **isotonic concentric contraction** or a *miometric contraction* (my-oh-MET-rik; *mio-* = "shorter"). During an isotonic concentric contraction, the force generated by the muscle is greater than that of the external load. Let's consider the example of lifting a 10-lb weight with your hand (Figure 10.26a). As you begin to pick up the weight (which is the external load), the motor units in your arm flexors

Muscle shortens

(a) Isotonic concentric contraction

Muscle lengthens

(b) Isotonic eccentric contraction

Muscle stays the same length

(c) Isometric contraction

Figure 10.26 The three types of muscle contractions.

generate force to flex your forearm at the elbow. When they generate force of more than 10 lb, the muscle shortens, and an isotonic concentric contraction results.

- **Isotonic eccentric contractions.** The second type of contraction, called an **isotonic eccentric contraction** or a *pliometric contraction* (ply-oh-MET-rik; *plio-* = "longer"), results in muscle lengthening. It occurs when the force generated by the muscle is less than that of the external load. As a result, the muscle lengthens as tension is produced. A muscle is able to lengthen while it's contracting because the elastic filaments in its myofibrils allow it to stretch considerably. Consider the example of the 10-lb weight that you have just lifted. As you slowly set the weight down (Figure 10.26b), the force produced by your muscle becomes less than the load of the weight. However, your motor units are still generating tension, even though the sarcomeres are stretching and lengthening.

- **Isometric contractions.** The third type of contraction, in which the length of the muscle doesn't change, is the **isometric contraction** (eye-soh-MET-rik). This time, lift the 10-lb weight by flexing your shoulder rather than your elbow, and hold it out to your side (Figure 10.26c). The initial lifting involved an isotonic concentric contraction; however, now that you are holding it out to the side, you are neither flexing nor extending your shoulder. So, the muscle is neither shortening nor lengthening, even though it is producing 10 lb of force.

A&P in the Real World

Delayed-Onset Muscle Soreness

Whether we've tried a new workout at the gym or had an overly ambitious day of snowboarding, all of us have probably experienced muscle soreness at one time or another. For years physiologists thought that this phenomenon, called **delayed-onset muscle soreness** (or DOMS), was due to the lactic acid produced during glycolysis. However, research studies within the past 10 years have shown that lactic acid accumulation is not to blame. Instead, studies have suggested another culprit—minor structural damage, in particular that caused by isotonic eccentric muscle contractions. The force produced by these contractions can cause minor damage to the myofilaments, particularly in muscles not trained for specific activities. Recent data indicate that the resulting pain may also be due to remodeling of the myofibrils, which is an adaptive response to the stress placed on the muscle.

Oddly, the most consistently effective treatment for DOMS is more exercise. Other treatment modalities, including massage, topical therapies, acupuncture, and oral medications, have shown little benefit. The effectiveness of exercise in treating DOMS, however, is temporary. Once the exercise has ceased, the pain returns until the muscle is sufficiently conditioned through training.

These types of contractions can be induced in the laboratory in individual muscle fibers. But as we discussed earlier, in the body these contractions occur at the organ level through activation of motor units.

Of the three types of contractions, isotonic eccentric contractions require the greatest amount of tension, and therefore produce the greatest amount of force. As a result, activities that require a large number of isotonic eccentric contractions (e.g., running downhill) have a greater tendency to produce exercise-induced injuries than do those using more isotonic concentric or isometric contractions (e.g., running on a flat surface). To understand a common type of muscle injury, see *A&P in the Real World: Delayed-Onset Muscle Soreness.*

Quick Check

☐ 4. How do isotonic concentric, isotonic eccentric, and isometric contractions differ?

Apply What You Learned

☐ 1. Using biochemical terms, explain why lifting a stack of textbooks over and over again would likely make you feel fatigued more quickly than repeatedly lifting a single textbook. (*Hint:* What types of motor units are activated when more tension is needed?)

☐ 2. A fitness trainer advises her client to lower a weight slowly and in a controlled manner to reduce the risk of injury. Which type of muscle contraction is being used when the weight is lowered? Explain why the trainer advises her client to train in this manner.

See answers in Appendix A.

MODULE **10.8**
Skeletal Muscle Performance

Learning Outcomes

1. Describe the effects of physical conditioning on skeletal muscle tissue, and compare endurance and resistance training.
2. Explain the factors that contribute to muscular fatigue.
3. Summarize the events that occur during the recovery period.

Exercise has both short-term and long-term effects on the structure and performance of skeletal muscle tissue. For example, in the short term the body's metabolic rate increases during exercise as the skeletal muscle fibers try to keep up with the demand for ATP. In the process, waste products such as heat and carbon dioxide are generated, which the body must deal with in order to recover. Over the long term, consistent exercise yields far more than just waste products—it results in changes to the architecture of the muscles themselves. In this module we examine how the body responds to exercise, factors that contribute to muscular fatigue, and how the body recovers from exercise.

Changes Caused by Physical Training

Physical training refers to the repetitive use of skeletal muscles that leads to changes in the structure and biochemistry of the muscle fibers. The changes that occur in the muscle fibers are governed by the **principle of myoplasticity,** which is just another way to say that a muscle will alter its structure to follow its function—an example of the Structure-Function Core Principle (p. 25). For example, when a person engages in little physical activity 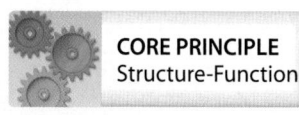 and is "out of shape," the muscle fibers will change structurally and biochemically to meet this lesser demand. Conversely, when someone physically trains to "get in shape," his or her muscle fibers change again to meet this increased demand.

CORE PRINCIPLE
Structure-Function

Note that the great majority of mature skeletal muscle fiber nuclei are *amitotic,* meaning that they generally do not undergo mitosis. Therefore, these changes are within the muscle fibers themselves, not in the number of muscle fibers. (A small population of unspecialized *satellite cells,* however, does retain the ability to divide. These cells can help repair injured skeletal muscle.) The precise structural and biochemical changes depend on the type of training chosen—endurance training and resistance training are the two basic types.

Endurance Training

Endurance training is defined as training with a large increase in the frequency of motor unit activation and a moderate increase in force production—in other words, more repetitions with lighter weight. Endurance training involves activities such as cycling, jogging, cross-country skiing, and distance swimming. **Figure 10.27a** shows that after endurance training the muscle fiber has increased amounts of oxidative enzymes, more mitochondria and mitochondrial proteins, and a greater number of blood vessels. These adaptations enhance the oxidative capabilities of the muscle dramatically, resulting in a better ability to use fatty acids and other fuels to make ATP, and so an increased resistance to fatigue. These changes occur even in type II fibers, but are more dramatic in type I fibers.

Resistance Training

Resistance training, also called *strength training,* features a moderate increase in the frequency of motor unit activation and a large increase in force production—in other words, fewer repetitions with heavier weight. Resistance training typically involves the use of either free weights or a resistance-exercise machine. Whereas many of the changes due to endurance training are primarily biochemical, several changes that result from resistance training are in large part anatomical. In **Figure 10.27b** you can see that both the number of myofibrils and the diameter of the muscle fibers increase, a change called **hypertrophy** (hy-PER-troh-fee). With hypertrophy comes a *decreased* proportion of

Muscle fiber with normal use

Blood vessel   Myofibril

Mitochondrion

Results from levels of activity

(a) **Endurance training**

Increased oxidative enzymes

Increased number of mitochondria and mitochondrial proteins

Increased number of blood vessels

(b) **Resistance training**

Increased number of myofibrils

Increased diameter of muscle fiber and myofibrils

(c) **Disuse**

Decreased oxidative enzymes

Decreased number of myofibrils and decreased diameter of muscle fiber

Figure 10.27 Adaptive changes of muscle fibers due to training and disuse.

mitochondrial proteins and blood supply to the muscle. However, this decrease is a function of the fiber enlarging rather than actually losing mitochondria or blood vessels. As you might expect, resistance training can decrease the capacity for endurance. For this reason, a balanced program combining both resistance and endurance training is recommended for most people.

Disuse

Regardless of the type of training, the old adage "Use it or lose it" holds true. In response to physical inactivity, the diameter of the muscle fiber decreases due to loss of myofibrils (**Figure 10.27c**), a condition called **atrophy** (AH-troh-fee). In addition, the amount of oxidative enzymes decreases, and the fiber has a lower capacity for oxidative catabolism. The result is a decline in both strength and endurance. Muscle atrophy is a particular problem for people who are bedridden or have lost the use of their limbs.

Quick Check

☐ 1. What changes in the muscle fiber accompany endurance training, resistance training, and disuse?

Muscular Fatigue

If you were to pick up a 10-lb weight and hold it out in front of you in an isometric contraction, your muscles would probably begin to feel weak after about a minute (or sooner, for some people). This weakness is called **muscular fatigue,** defined as the inability to maintain a given level of intensity of a particular exercise. Though the idea of fatigue seems straightforward, it actually results from a combination of several different factors, which may include the following:

- **Depletion of key metabolites.** Certain metabolites, such as creatine phosphate, glycogen, and blood glucose, are depleted by intense activity, rendering the muscle fiber less able to replenish the ATP consumed by the activity.

- **Decreased availability of oxygen to muscle fibers.** Exercise increases the oxygen requirement of muscle fibers because of the need for more ATP. The amount of oxygen bound to myoglobin may also be depleted by intense exercise, and the amount of oxygen taken in by the lungs may be inadequate to replace it. The muscle fiber must then rely more heavily on the less efficient process of glycolysis to generate ATP. Keep in mind that we are referring only to the amount of oxygen available to muscle fibers relative to what they need; the actual amount of oxygen in the body does not decrease significantly.

- **Accumulation of certain chemicals.** Many chemicals that may contribute to fatigue are produced by muscle fibers. For example, calcium ions accumulate in the mitochondria, where they interfere with the ability of those organelles to carry out oxidative catabolism. Also, as ATP is split, phosphate ions and ADP accumulate in the cytosol and interfere with excitation-contraction coupling.

- **Environmental conditions.** Severe environmental conditions, particularly extreme heat, disrupt the body's homeostasis, leading to more rapid muscular fatigue. Sweating in response to excessive heat may cause electrolyte disturbances that can interfere with the production of action potentials in a muscle fiber.

Many other causes of muscular fatigue exist, some of which don't involve the muscular system directly. For example, psychological factors play a role in fatigue. The mechanism behind psychological input is poorly understood, but some athletes seem able to minimize distressing stimuli such as pain or breathlessness and push their physical performance to the limits, whereas others are sidelined by these stimuli.

As you can see, the many potential causes of muscular fatigue make a precise cause difficult to pin down. Further complicating the scenario is the fact that all of the body's physiological functions are interrelated (as you've seen many times in this textbook already). Thus, a failure of something as simple as one enzyme system may unleash a cascade of events that hampers the ability of the entire organism to maintain maximal output during exercise.

Quick Check

☐ 2. How is muscular fatigue defined?
☐ 3. What factors influence muscular fatigue?

Excess Postexercise Oxygen Consumption and the Recovery Period

During physical activity, whether or not you reach the fatigued state, you've likely noticed that your rate and depth of breathing increase. You've also probably observed that these changes persist for a time, even after you've finished exercising. Have you ever wondered why? The fact is that your body needs a period of time to return to the pre-exercise state. This time is called the **recovery period,** and the persisting increased rate of breathing is referred to as **excess postexercise oxygen consumption (EPOC).** EPOC was formerly called the *oxygen debt* or *oxygen deficit,* but these terms are misnomers, as there is no debt or deficit of oxygen that must be "repaid."

For many years, physiologists thought that EPOC was due to an accumulation of lactic acid in skeletal muscle fibers and in the bloodstream. However, as discussed earlier, current evidence indicates that lactic acid is oxidized by the mitochondria to generate ATP. Thus, physiologists now think that EPOC results from the responses necessary to correct the disturbances to homeostasis that were brought on by exercise. To return to homeostasis, the body must accomplish several goals, which include dissipating heat, restoring ion concentrations, and correcting blood pH. Let's see why each of these is important:

- **Heat dissipation.** Energy is lost as heat during both glycolytic and oxidative catabolism. Exercise increases the rate of catabolic processes, so it also raises the amount of heat generated. The body must dissipate this heat and return to pre-exercise temperature levels to maintain temperature homeostasis.
- **Restoration of intracellular and extracellular ion concentrations.** During muscle contractions, the ATP-consuming pumps in the sarcolemma and SR must work harder to maintain the normal concentrations of calcium and sodium ions in the cytosol and of potassium ions in the extracellular fluid.

- **Correction of blood pH.** Certain products of metabolism, including lactic acid and carbon dioxide, can cause the pH of the blood to decrease (which means acidity increases) from its normal value of 7.35–7.45. The body must return the pH of the blood to pre-exercise levels in order to maintain homeostasis.

So, how does increasing the rate and depth of breathing after exercise help to accomplish these goals? Most obvious is that a relationship exists between breathing rate and the level of carbon dioxide in the blood: As the rate and depth of breathing increase, more carbon dioxide is exhaled, helping to return the pH of the body's fluids to normal. However, the relationship may be less obvious for temperature and electrolyte homeostasis. To make it easier to understand, consider facts you already know—the Na^+/K^+ pump restores ion gradients, pumps move calcium ions back into the sarcoplasmic reticulum, and sweating produces cooling. These three mechanisms have one thing in common: They all consume ATP. Since oxidative catabolism—the most efficient way to generate ATP—requires oxygen, it follows that the body's cells will need "extra" oxygen after exercise.

Quick Check

☐ 4. What conditions does excess postexercise oxygen consumption help to correct?

Apply What You Learned

☐ 1. Predict the appearance of the muscle fibers of an athlete who trains to run marathons. What type of muscle fiber would likely predominate in this person's muscles?
☐ 2. A new wonder drug doubles the concentration of myoglobin in muscle fibers. In what sort of sport would athletes benefit the most from taking this drug?
☐ 3. You have finished exercising with a friend, who says that he is breathing rapidly because he must have "used up all of his oxygen." What do you tell him?

See answers in Appendix A.

MODULE **10.9**

Smooth and Cardiac Muscle

Learning Outcomes

1. Describe the structure, location in the body, and functions of smooth and cardiac muscle tissue.
2. Describe the contraction process of smooth muscle fibers, and contrast it with skeletal muscle fiber contraction.

Although smooth and cardiac muscle tissues share many structural and functional properties with skeletal muscle tissue, they also have notable differences. One such difference is that both smooth and cardiac muscle tissues are involuntary. In this module we examine these differences in greater detail. (We emphasize smooth muscle here, as the structure and function of cardiac muscle tissue are covered in detail in the heart chapter—see Chapter 17.)

Smooth Muscle

◀◀ Flashback

1. What are intermediate filaments? (p. 96)
2. What are the primary structural and functional differences between smooth muscle tissue and skeletal muscle tissue? (p. 148)
3. What is the function of a gap junction? (p. 127)

Smooth muscle tissue is widely distributed in the body. Typically lining hollow organs, it is also present elsewhere, such as in the arrector pili muscles in the dermis and the muscles in the iris of the eye. Some functions of smooth muscle include the following:

- **Peristalsis.** In many organs, such as the small intestine, the smooth muscle tissue is arranged into two differently oriented layers of fibers, the circular and longitudinal layers (**Figure 10.28a**). These two layers alternately contract and relax, producing rhythmic waves called **peristalsis** (pear-ih-STAL-sis; "constricting around"). Peristalsis propels materials through hollow organs of the digestive, urinary, and reproductive systems.
- **Formation of sphincters.** In addition, in both the digestive and urinary systems, smooth muscle forms rings called **sphincters** (SFINGK-terz). Sphincters are usually contracted but relax periodically to allow substances to pass through them.

- **Regulation of flow.** Smooth muscle controls the flow of materials through certain hollow organs by changing the diameter of the tubes. For example, the smooth muscle found in the walls of all but the smallest blood vessels controls both blood pressure and blood flow to organs and tissues (see Chapter 18). In addition, the smooth muscle in the walls of the respiratory tract regulates air flow to the smaller airway passages (see Chapter 21).

The next section discusses the structure and contractile mechanism of smooth muscle cells. As you read, notice the significant overlap between smooth and skeletal muscle physiology.

Structure of Smooth Muscle Cells

Smooth muscle cells share the same contractile proteins found in skeletal muscle fibers: myosin and actin. However, the arrangement of these contractile proteins differs greatly from that in skeletal muscle. To begin with, smooth muscle cells lack striations and sarcomeres, which gives them their characteristic "smooth" appearance (**Figure 10.28b**). In addition, the actin filaments are arranged obliquely in the sarcoplasm and are anchored to proteins called **dense bodies.** Some dense bodies are associated with the sarcolemma, where they attach to the dense bodies of other smooth muscle cells and transmit tension from one cell to the next. Other dense bodies are found in the sarcoplasm, where they are bound to scaffold-like intermediate filaments that connect the dense bodies to one another. From each dense body radiate several thin filaments surrounding one thick filament. The ratio of thin to thick filaments is higher in smooth muscle than in skeletal muscle, with more thin filaments for every thick filament.

Differences are also found in the structure and composition of thin and thick filaments in smooth muscle cells. Both

(a) Circular and longitudinal layers of smooth muscle lining the small intestine

Small intestine

Longitudinal layer

Circular layer

Smooth muscle cells from longitudinal layer (cross-sectional view)

Smooth muscle cells from circular layer

(b) Smooth muscle cell

Myofilaments   Nucleus   Dense bodies

Intermediate filaments

Figure 10.28 Structure of smooth muscle tissue and cells.

types of filaments are longer than those in skeletal muscle fibers, and the thin filaments lack troponin. You can see in **Figure 10.29a** that another difference is found in the myosin heads, which have opposite-facing hinges—the heads on one side hinge in one direction, whereas those on the other side hinge in the opposite direction. Observe also that a thick filament contains myosin heads along its entire length.

In addition to differences in contractile proteins, a smooth muscle cell is structurally different from a skeletal muscle fiber in three main ways:

- It lacks motor end plates.
- The sarcoplasmic reticulum is much less extensive.
- T-tubules are absent.

The importance of these differences will become apparent in the next section.

Smooth Muscle Contraction and Relaxation

Smooth muscle and skeletal muscle share some structural features and also have some of the same functional features. For example, the thick and thin filaments slide past one another, calcium ions are required for a contraction to occur, and ATP is consumed in the process. However, just as significant structural variations exist between smooth and skeletal muscle, so, too, do significant functional variations. Structural differences in smooth muscle allow it to carry out specialized functions.

Earlier in the chapter, you read that only stimulation by a motor neuron can excite a skeletal muscle fiber. However, many different stimuli may elicit a smooth muscle cell contraction, such as mechanical (stretch), hormonal, and nervous stimulation. In addition, some smooth muscle cells are able to depolarize and contract spontaneously. These cells, known as **pacemaker cells,** have unstable membrane potentials that cause them to spontaneously depolarize in a rhythmic fashion. Their depolarizations cause the firing of an action potential that spreads through the surrounding muscle cells and initiates a wave of contraction. Pacemaker cells of this type are responsible for the waves of contraction that move through the stomach and intestines.

The triggering event for a smooth muscle contraction is the same as that for skeletal muscle: calcium ions flooding the cytosol. However, stimulation of the smooth muscle cell leads to release of calcium ions from the SR, and calcium ions also enter from the extracellular fluid. In addition, because smooth muscle cells lack troponin, the role of calcium ions in smooth muscle contraction can't be the same as in skeletal muscle. In smooth muscle, calcium ions initiate a different cascade of events:

1. Calcium ions bind to a protein in the sarcoplasm called **calmodulin** (kal-MOD-yoo-lin; **Cam**).
2. The calcium ion–Cam complex activates an enzyme associated with myosin, called **myosin light-chain kinase (MLCK).**
3. MLCK causes the activation of myosin ATPase.
4. Crossbridge cycles then ensue.

As the crossbridge cycles repeat, the crossbridges pull actin along the length of the myosin, and the muscle cell changes shape from long and thin to fat and globular, which you can see

in **Figure 10.29b.** Smooth muscle cells can contract up to 80% of their resting length (in contrast, skeletal muscle fibers can contract a maximum of 30–40% of their resting length).

Although smooth muscle is often functionally required to maintain sustained contractions, it consumes surprisingly little ATP. Most smooth muscle cell contractions consume as little as $\frac{1}{100}$ of the ATP used by skeletal muscle fibers, a fact that physiologists think is due to low myosin ATPase activity. In spite of this low energy consumption, smooth muscle cells are still able to generate significant force, because the percentage of time that the crossbridges are attached is much greater than in skeletal muscle fibers. Anyone who has gone through childbirth has felt how strong and sustained smooth muscle contractions can be.

Relaxation of smooth muscle begins with removal of calcium ions from the cytosol. As this concentration declines, calcium ions dissociate from calmodulin, and MLCK is inactivated. At this point, the muscle cell either relaxes or enters the *latch state*. During the latch state, the muscle cell maintains tension while consuming very little ATP. The latch state is very important in terms of the ability of smooth muscle to maintain a state of energy-efficient sustained contraction, as is found in sphincters, which must stay contracted to remain closed.

Types of Smooth Muscle

Smooth muscle tissue varies in structure and organization with each organ of the body in which it is found. However, from all this variation, two general types emerge:

- **Single-unit** (or **unitary**) **smooth muscle,** also called **visceral smooth muscle** (VISS-er-ul), is the predominant type of smooth muscle in the body and is found in nearly all hollow organs. Single-unit smooth muscle consists of hundreds to thousands of muscle cells whose plasma membranes are linked electrically via gap junctions. Action potentials spread rapidly through the cells via the gap junctions, causing the muscle cells to contract in a coordinated wave as a single unit.

(a) Myofilaments in relaxed smooth muscle cell

(b) Myofilaments in contracted smooth muscle cell

Figure 10.29 Contraction of smooth muscle cells.

- **Multi-unit smooth muscle** is the second, and rarer, of the two types. It is found in such locations as the uterus, the muscles in the eye, and the arrector pili muscles in the dermis. Multi-unit smooth muscle consists of individual muscle cells whose plasma membranes are not joined by gap junctions. This characteristic allows each cell to contract independently of the others, permitting precise control of contractions. As with skeletal muscle, the amount of tension produced by multi-unit smooth muscle varies with the number of muscle cells activated.

Whereas multi-unit smooth muscle responds primarily to nerve stimulation, single-unit smooth muscle is able to respond to multiple types of stimulation. Keep in mind, however, that not all single-unit smooth muscle is the same. Though it shares the property of contracting as a single unit, its structure and function vary in different organs of the body.

Quick Check

☐ 1. What are some functions of smooth muscle tissue?
☐ 2. How does contraction differ in smooth muscle cells and skeletal muscle fibers?
☐ 3. How are single-unit and multi-unit smooth muscle different?

Cardiac Muscle

Cardiac muscle cells more closely resemble skeletal muscle fibers than smooth muscle cells, as we saw in Figure 10.1. Like skeletal muscle fibers, cardiac muscle cells are striated and consist of sarcomeres. They also have both T-tubules and extensive networks of sarcoplasmic reticulum. However, notable structural differences between cardiac muscle cells and skeletal muscle fibers do exist. Cardiac muscle cells are typically shorter, branched cells

with one or two nuclei and abundant myoglobin. Mitochondria account for about 30% of their cytoplasmic volume. In addition, cardiac muscle cells are connected by **intercalated discs** that contain desmosomes and gap junctions (see Chapter 4). These discs join cardiac muscle cells to one another physically and electrically, which permits the heart to contract as a unit.

Unlike skeletal muscle fibers, cardiac muscle cells do not require stimulation from the nervous system to generate action potentials, because their electrical activity is coordinated by a small population of pacemaker cells. Like those in smooth muscle, these cells rhythmically and spontaneously generate action potentials that trigger the remaining cardiac muscle cells to have action potentials as well. This quality allows cardiac muscle tissue to be **autorhythmic**—it sets its own rhythm. In this way, cardiac muscle tissue is similar to single-unit smooth muscle in that all of the cells depolarize and contract with every contraction cycle.

Quick Check

☐ 4. What are the basic structural features of cardiac muscle cells?

Apply What You Learned

☐ 1. Would you expect to find motor units and recruitment in single-unit smooth muscle tissue and cardiac muscle tissue? Why or why not?
☐ 2. Predict the effect of excess calcium ions in the extracellular fluid on a smooth muscle contraction.
☐ 3. One consequence of oxygen deprivation is decreased production of ATP. Would you expect this to have a greater effect on smooth or cardiac muscle tissue? Explain.

See answers in Appendix A.

CHAPTER SUMMARY

Take full advantage of the invaluable online practice and study tools for this chapter by going to MasteringA&P®. There you will find:

 Practice tests, module quizzes, art-labeling and sequence-the-steps activities

 Animated figures with quizzes

 Practicing A&P worksheets

 Author-narrated podcast

 Dynamic Study Modules

 Interactive Physiology

MODULE **10.1**
Overview of Muscle Tissue 339

- All muscle cells turn chemical energy into mechanical energy by contracting to generate *muscle tension*.

- All types of muscle tissue consist of **muscle cells** and the surrounding extracellular matrix, called the endomysium.

- All types of muscle tissue share common properties, including *contractility*, **excitability, conductivity, extensibility,** and **elasticity.**

- The plasma membrane of a muscle cell is called the **sarcolemma,** and the cytoplasm is called the **sarcoplasm.** The sarcoplasm contains protein-filled **myofibrils** and the Ca^{2+}-storing **sarcoplasmic reticulum (SR).** (Figure 10.2, p. 341)

MODULE 10.2
Structure and Function of Skeletal Muscle Fibers 342

- Skeletal muscle fibers are very long cells that consist of striated, multinucleated voluntary muscle cells. (Figure 10.3, p. 342)

- The sarcolemma has inward extensions called **T-tubules** that surround myofibrils. The SR swells where it meets T-tubules to form **terminal cisternae.** (Figure 10.4, p. 343)

    Practice art-labeling exercise: Figure 10.4.

- Myofibrils are composed of protein units called **myofilaments.** Myofilaments are in turn composed of one or more **contractile proteins, regulatory proteins,** and/or **structural proteins.**

- There are three types of myofilaments: (Figure 10.5, p. 344)
    - **Thick filaments** are composed of **myosin,** a contractile protein.
    - **Thin filaments** are composed of bead-shaped **actin** proteins and the smaller regulatory proteins **troponin** and **tropomyosin.**
    - **Elastic filaments** are composed of the massive structural protein **titin.**
- Figure 10.6 (p. 345) provides a summary of the levels of organization of a skeletal muscle.

    Play Interactive Physiology tutorial on Muscular System: Anatomy Review: Skeletal Muscle Tissue.

- The characteristic striations of skeletal muscle tissue are due to the arrangement of myofilaments within the myofibrils:
    - The **I bands** are the light regions of striations where only thin filaments are found. In the middle of each I band is a line composed of structural proteins called the **Z-disc.**
    - The **A bands** are the dark regions of striations where the thick and thin filaments overlap. The central region of the A band contains only thick filaments; this lighter region is called the **H zone.** It is bisected by the **M line.** (Figure 10.7, p. 346; Figure 10.8, p. 347)

    Practice art-labeling exercise: Figure 10.7.

- The **sarcomere** is the functional unit of contraction. It is defined as the area from one Z-disc to the next.

- The **sliding-filament mechanism** is the currently accepted model of muscle contraction. In this process, the thick and thin filaments slide past one another. (Figure 10.9, p. 348)

    Play Interactive Physiology tutorial on Muscular System: Sliding-Filament Theory.

MODULE 10.3
Skeletal Muscle Fibers as Electrically Excitable Cells 348

- The inner face of the sarcolemma has a thin layer of negative charges, and the outer face has a thin layer of positive charges, which is an electrical gradient or **electrical potential.**

- Electrical potentials across plasma membranes are known as **membrane potentials.** The **resting membrane potential** of a muscle fiber is -85 mV.

- The **Na$^+$/K$^+$ ATPase,** or **Na$^+$/K$^+$ ATPase pump,** maintains a high K$^+$ concentration in the cytosol and a high Na$^+$ concentration in the extracellular fluid. (Figure 10.10, p. 350)

- An **action potential** is a temporary, quick reversal in the membrane potential. It consists of two phases: (1) **depolarization,** in which Na$^+$ enter the fiber, causing the membrane potential to become more positive; and (2) **repolarization,** in which K$^+$ exit the fiber, returning it to its resting (polarized) state. (Figure 10.11, p. 351)

    Practice sequencing the steps: Figure 10.11.

MODULE 10.4
The Process of Skeletal Muscle Contraction and Relaxation 352

- All skeletal muscle fibers are innervated by **motor neurons.** The neuron meets the fiber at the **neuromuscular junction,** or **NMJ.** The NMJ consists of the **axon terminal,** the **synaptic cleft,** and the **motor end plate.** (Figure 10.12, p. 352)

    Practice art-labeling exercise: Figure 10.12.

    Play Interactive Physiology tutorial on Muscular System: The Neuromuscular Junction.

- Skeletal muscle contraction can be divided into three parts: excitation, excitation-contraction coupling, and contraction.

- The **excitation phase** relays an action potential from the neuron to the sarcolemma. It begins as the axon terminal releases ACh into the synaptic cleft, which binds ACh receptors on the motor end plate. The result is an **end-plate potential.** (Figure 10.13, p. 353)

    Play animation on the excitation phase: Figure 10.13.

- In **excitation-contraction coupling,** the end-plate potential triggers an action potential in the sarcolemma, which propagates down the T-tubules. This triggers the opening of Ca^{2+} channels in the SR, and Ca^{2+} flood the cytosol. (Figure 10.14, p. 354)

    Play animation on excitation-contraction coupling: Figure 10.14.

- In preparation for contraction, the Ca^{2+} in the cytosol bind troponin, which allows tropomyosin to move away from the active sites of actin. (Figure 10.15, p. 355)

    Play animation on preparation for contraction: Figure 10.15.

- The **contraction phase** then begins, and the **crossbridge cycle** occurs as ATP hydrolysis "cocks" the myosin head and it binds to actin. When ADP and phosphate detach from the myosin head, it pivots and undergoes **a power stroke** that pulls actin toward the center of the sarcomere. (Figure 10.16, pp. 356–357)

    Play animation on the crossbridge cycle: Figure 10.16.

- **Muscle relaxation** has two components: (1) The ACh in the synaptic cleft is broken down, and (2) the Ca^{2+} concentration in the cytosol returns to the resting level. (Figure 10.17, p. 359)

    Play animation on the relaxation phase: Figure 10.17.

- The big picture of skeletal muscle contraction is shown in Figure 10.18, p. 360.

    Play animation on the big picture of skeletal muscle contraction: Figure 10.18.

    Play Interactive Physiology tutorial on Muscular System: Contraction of a Whole Muscle.

MODULE **10.5**
Energy Sources for Skeletal Muscle 361

- Immediate energy sources for muscle fibers include stored ATP within the cytosol and **creatine phosphate.** (Figure 10.19, p. 362)
- **Glycolytic catabolism** is an anaerobic process during which glucose in the cytosol is split and ATP is produced. It produces the end product pyruvate, which may be converted to **lactic acid.**
- **Oxidative catabolism** is an aerobic process that takes place in the mitochondria, during which the products of glycolysis, fatty acids, and amino acids are oxidized and the energy from their removed electrons is used to generate ATP.

iP Play Interactive Physiology tutorial on Muscular System: Muscle Metabolism.

MODULE **10.6**
Muscle Tension at the Fiber Level 363

- A single contraction-relaxation cycle of a muscle fiber is called a **twitch.** It consists of the **latent period,** the **contraction period,** and the **relaxation period.** Depending on the duration of the twitch, muscle fibers may be classified as **fast-twitch** or **slow-twitch** fibers. (Figure 10.20, p. 364)

  Practice sequencing the steps: Figure 10.20.

- The amount of tension generated by the contraction of a muscle fiber depends on the frequency of stimulation by the motor neuron and the resulting concentration of Ca^{2+} (Figure 10.21, p. 365)
  - If a muscle fiber is stimulated before the relaxation period is completed, a condition called **unfused tetanus** results.
  - If a muscle fiber is stimulated 80–100 times per second before the relaxation period begins, a condition called **fused tetanus** results.
- A contraction can produce a maximal amount of tension at the optimal length of a sarcomere. This relationship is called the **length-tension relationship.** (Figure 10.22, p. 366)
- Figure 10.23 (p. 367) summarizes how myofilament sliding leads to muscle contraction.
- There are two general types of skeletal muscle fibers, classified according to the speed of their contraction: **type I,** which are slow-twitch fibers that use primarily oxidative catabolism; and **type II,** which are fast-twitch fibers that use primarily glycolytic catabolism. (Figure 10.24, p. 368)

MODULE **10.7**
Muscle Tension at the Organ Level 368

- A **motor unit** consists of a single motor neuron and the muscle fibers that it innervates. (Figure 10.25, p. 369)
  - When more force is needed in a muscle contraction, more motor units are activated, which is known as **recruitment.**
  - **Muscle tone** is produced by small, involuntary contractions of alternating motor units.
- There are three types of tension-generating contractions, each defined by the change that results in the length of the muscle fibers: an **isotonic concentric** (shortening) **contraction,** an **isotonic eccentric** (lengthening) **contraction,** and an **isometric** (unchanging length) **contraction.** (Figure 10.26, p. 370)

iP Play Interactive Physiology tutorial on Muscular System: Contraction of Motor Units.

MODULE **10.8**
Skeletal Muscle Performance 371

- **Endurance** and **resistance** training result in changes in the structure and biochemistry of skeletal muscle fibers. (Figure 10.27, p. 372)
- Disuse leads to **atrophy** of the muscle fibers and a decrease in strength and endurance.
- **Muscular fatigue** is caused by depletion of key metabolites, inadequacy of oxygen delivery to muscle fibers, accumulation of certain metabolites, and environmental conditions.
- The rate and depth of breathing continue to be elevated after exercise, which is known as **excess postexercise oxygen consumption (EPOC).** EPOC is due to the body's efforts to restore homeostasis after exercise.

MODULE **10.9**
Smooth and Cardiac Muscle 373

- The functions of **smooth muscle** tissue include peristalsis, forming sphincters, and regulating the flow of material through hollow organs.
- Smooth muscle cells are uninucleate; lack T-tubules, striations, and sarcomeres; and have a less extensive SR than do skeletal muscle fibers. (Figure 10.28, p. 374)
- Smooth muscle contraction is stimulated by the autonomic nervous system, stretch, hormones, and the activity of pacemaker cells.
- Smooth muscle cell contraction begins with the entry of Ca^{2+} into the cytosol, which bind to **calmodulin.** This activates **myosin light-chain kinase (MLCK),** which in turn initiates crossbridge cycles. (Figure 10.29, p. 375)
- Smooth muscle cell relaxation involves removal of Ca^{2+} from the cytosol.
- There are two types of smooth muscle tissue:
  - **Single-unit smooth muscle** consists of muscle cells that are linked by gap junctions and that contract together as a single unit.
  - **Multi-unit smooth muscle** consists of individual muscle cells that are not electrically coupled, and can contract independently of one another.
- Cells of **cardiac muscle** are striated and are shorter, branching cells with one or two nuclei. They are joined physically and electrically by **intercalated discs.** The electrical activity of cardiac muscle cells is coordinated by **pacemaker cells.**

ASSESS WHAT YOU LEARNED

See answers in Appendix A.

LEVEL 1 Check Your Recall

1. Mark the following statements as true or false. If a statement is false, correct it to make a true statement.

   a. A property of all muscle cells is elasticity, which means that the tissue is able to stretch.

b. The common function of all types of muscle tissue is to generate tension.

c. The plasma membrane of a muscle cell is called the sarcoplasmic reticulum.

d. Muscle cells contain organelles called myofibrils.

e. Muscle cells are contractile, conductive, extensible cells.

2. How does a skeletal muscle fiber differ structurally from typical cells?

3. Which of the following statements about myofilaments is *false?*

a. Elastic filaments are composed of the protein titin.

b. Both thick and thin filaments contain the contractile protein myosin.

c. Thin filaments of skeletal muscle contain the regulatory proteins troponin and tropomyosin.

d. Thick filaments contain the contractile protein myosin.

4. Match the following terms with the correct definition.

_____ Z-disc
_____ Sarcomere
_____ A band
_____ H zone
_____ I band
_____ M line

a. The dark band containing the entire length of the thick filament
b. The band of proteins in the middle of the H zone
c. The boundary between sarcomeres
d. The functional unit of contraction
e. The middle region of the A band containing only thick filaments
f. The light band containing only thin filaments

5. What is the basic mechanism of contraction at the level of myofilaments?

6. Which of the following statements regarding the phases of an action potential is *true?*

a. The membrane is depolarized at rest.

b. The resting membrane potential of the fiber is positive—the inside of the fiber is positive with respect to the outside of the fiber.

c. Potassium ions rush into the fiber during repolarization, and the membrane potential becomes more negative.

d. Sodium ions rush into the fiber during depolarization, and the membrane potential becomes more positive.

7. Describe the three components of the neuromuscular junction.

8. Order the following events of excitation and excitation-contraction coupling. Put 1 by the first event, 2 by the second, and so on.

_____ The motor end plate generates an end-plate potential.

_____ The action potential spreads along the T-tubules, SR calcium ion channels are pulled open, and calcium ions flood the cytosol.

_____ Acetylcholine binds to receptors on the motor end plate, and ligand-gated ion channels open.

_____ Calcium ions bind troponin, which allows tropomyosin to move away from the actin active site, initiating a contraction cycle.

_____ The action potential propagates through the sarcolemma and dives deeply into the cell along the T-tubules.

9. Which of the following statements accurately describes the role of ATP in a muscle contraction?

a. ATP is directly responsible for the power stroke.

b. ATP moves troponin and tropomyosin away from actin.

c. ATP breaks the actin/myosin attachment and "cocks" the myosin head.

d. ATP causes the myofilaments to shorten.

10. A muscle fiber relaxes when:

a. the concentration of calcium ions in the cytosol returns to resting levels.

b. the supply of ATP is exhausted.

c. calcium ions flood the cytosol.

d. acetylcholine is released from the axon terminal and the sarcolemma depolarizes.

11. Mark the following statements as true or false. If a statement is false, correct it to make a true statement.

a. Stored ATP and creatine phosphate are immediate energy sources for muscle fibers.

b. Glycolytic catabolism occurs in the mitochondria and requires oxygen.

c. Glycolytic catabolism provides enough ATP to fuel from several minutes to a few hours of activity.

d. Oxidative catabolism may oxidize the products of glycolysis, fatty acids, and amino acids.

12. Mark the following statements as true or false. If a statement is false, correct it to make a true statement.

a. Muscle fibers generate more tension if the starting length of their sarcomeres is very short.

b. Stimulation by a motor neuron before a muscle fiber has fully relaxed results in a condition called wave summation.

c. Muscles that require a great deal of precise control will have large motor units.

d. A muscle fiber changes length during isotonic concentric and isotonic eccentric contractions.

13. Which of the following types of muscle fibers has low myosin ATPase activity and is classified as slow twitch?

a. Type I fibers   c. Type IIx fibers
b. Type IIa fibers   d. Type IIb fibers

14. Muscle tone is:

a. the result of voluntary shortening of the muscle.

b. the result of a small amount of involuntary activation of motor units by the nervous system.

c. abnormal—a person's muscles should be relaxed normally.

d. present only where there is damage to the nervous system.

15. Fill in the blanks: Resistance-type activities will likely rely on _____ energy sources, whereas endurance activities will probably rely on _____ energy sources.

16. Which of the following is *not* likely to result from endurance training alone?

a. Increase in oxidative enzymes in the muscle fiber

b. A transition from type IIb to type IIa muscle fibers

c. Hypertrophy of the muscle fibers

d. Increase in blood supply to the muscle fibers

17. Which of the following factors is/are responsible for muscular fatigue? (Circle all that apply.)

 a. Accumulation of chemicals, including calcium and phosphate ions

 b. Depletion of key metabolic fuels, such as creatine phosphate

 c. Decreased availability of oxygen

 d. Psychological and environmental factors

 e. Increased blood flow to the muscle

18. What is thought to cause excess postexercise oxygen consumption?

19. List some of the functions of smooth muscle tissue.

20. Which of the following best describes single-unit smooth muscle tissue?

 a. The fibers function individually.

 b. It is found in organs that require precise control of contraction.

 c. It contains gap junctions that couple the fibers electrically.

 d. The amount of tension produced varies with the number of muscle cells recruited.

21. Mark the following statements as true for smooth muscle tissue, cardiac muscle tissue, and/or skeletal muscle tissue.

 a. _____ Actin attaches to dense bodies.

 b. _____ Cells are joined by intercalated discs.

 c. _____ The thick and thin filaments are arranged into sarcomeres.

 d. _____ The thick filaments contain myosin heads along their entire length.

 e. _____ The cells depolarize and contract as a unit.

 f. _____ Calcium ions binding to troponin is the initiating event of contraction.

 g. _____ Calcium ions binding to calmodulin is the initiating event of contraction.

 h. _____ The sarcolemma has a distinct motor end plate.

LEVEL 2 Check Your Understanding

1. Would you expect to find larger motor units in the postural muscles of the back or the muscles of the hand? Explain your answer.

2. The condition *hyponatremia* (hy'-poh-nah-TREE-mee-ah) is characterized by low levels of sodium ions in the blood and extracellular fluid. What do you think would happen to the action potential in the sarcolemma if there were insufficient sodium ions in the extracellular fluid? (*Hint:* Think about what will happen to the normal gradient of sodium ions in hyponatremia.)

3. The condition *hyperkalemia* (hy'-per-kah-LEE-mee-ah) is characterized by high levels of potassium ions in the blood and extracellular fluid. What do you think would happen to the action potential in the sarcolemma if there were excess potassium ions in the extracellular fluid?

4. Explain why cardiac muscle cells and some smooth muscle cells will continue to contract even when their nerve supply has been removed.

LEVEL 3 Apply Your Knowledge

PART A: Application and Analysis

1. The poison *curare* (kyoo-RAH-ray) blocks the binding of acetylcholine to its receptors at the neuromuscular junction. What effects would you predict from such a poison? Can you think of any useful applications for it? Why might an overdose of it be lethal?

2. The drug *neostigmine* blocks the activity of the enzyme acetylcholinesterase in the synaptic cleft. Predict the effects of this drug.

3. Some athletes will consume only protein for several days before a competition, which reduces the amount of glycogen in both the muscle fibers and the liver. What impact would this have on their ability to perform activities that require short, powerful bursts of activity? How would it affect their ability to perform endurance activities?

4. Ms. Sanchez was in a motorcycle accident in which she sustained significant damage to the nerves that innervate the muscles in her upper limbs. As a result of this trauma, she no longer has use of these muscles. However, when an electrode is inserted into her muscles, they are able to contract. Explain specifically why nerve damage has impacted her ability to use her muscles. Why can her muscles still respond to stimulation from an electrode?

5. Mr. White has cerebral palsy and suffers severe muscle spasms as a result of his condition. He is prescribed the drug *dantrolene,* which prevents the release of calcium ions from the SR. Explain how this will treat his muscle spasms.

PART B: Make the Connection

6. Jesse is a 2-year-old boy who presents with difficulty in walking and poor motor control of movements. When the doctor examines Jesse, she notices that when his muscles contract, they are very slow to relax and remain contracted well after the movement has been performed. She sends a sample of his tissue for genetic analysis, and the lab reports a defect in the gene that codes for the protein that pumps calcium ions back into the SR during relaxation. This defect causes the pumps to operate much more slowly than normal. How does a defect in DNA lead to a malfunctioning protein? How does this finding explain Jesse's symptoms? (*Connects to Chapter 3*)

7. Paola is a 3-year-old girl with a disease that reduces the ability of her mitochondria to generate ATP. Explain the specific effects of this disease on the ability of Paola's muscles to function properly. What other tissues and organs are likely to be affected by her disease, and why? (*Connects to Chapter 3*)

11

Introduction to the Nervous System and Nervous Tissue

You can't turn on the television or radio, much less go online, without seeing something to remind you of the nervous system. From advertisements for medications to treat depression and other psychiatric conditions to stories about celebrities and their battles with illegal drugs, information about the nervous system is everywhere in our popular culture. And there is good reason for this—the nervous system controls our perception and experience of the world. In addition, it directs voluntary movement, and is the seat of our consciousness, personality, and learning and memory. Along with the endocrine system, the **nervous system** regulates many aspects of homeostasis, including respiratory rate, blood pressure, body temperature, the sleep/wake cycle, and blood pH.

In this chapter we introduce the multitasking nervous system and its basic functions and divisions. We then examine the structure and physiology of the main tissue of the nervous system: nervous tissue. As you read, notice that many of the same principles you discovered in the muscle tissue chapter (see Chapter 10) apply here as well.

MODULE **11.1**

Overview of the Nervous System

Learning Outcomes

1. Describe the major functions of the nervous system.
2. Describe the structures and basic functions of each organ of the central and peripheral nervous systems.
3. Explain the major differences between the two functional divisions of the peripheral nervous system.

In this module we introduce the organs of the nervous system and how they fit within anatomical and functional divisions. These organs and their classifications are covered in more detail in later chapters (see Chapters 12, 13, and 14).

Anatomical Divisions of the Nervous System

◀◀ Flashback

1. Define neuron, neuroglial cell, and axon. (p. 150)
2. Where is the foramen magnum located, and what is the main nervous system structure that passes through it? (p. 216)
3. What are vertebral foramina? (p. 232)

Computer-generated image: A synapse between two nerve cells is shown.

Figure 11.1 Structure of the nervous system.

The nervous system can be divided anatomically into the **central nervous system (CNS)** and the **peripheral nervous system (PNS).** The CNS is made up of the *brain* and *spinal cord,* whereas *nerves* make up the PNS (**Figure 11.1**). Let's look at each of these divisions more closely.

The Central Nervous System

The organ of the central nervous system that is likely most familiar to you, yet still holds the greatest mysteries for physiologists, is the **brain.** Enclosed completely by the skull, the brain is composed primarily of nervous tissue. This remarkable organ consists of about 100 billion cells called *neurons* (NOOR-onz), or *nerve cells,* that enable everything from the regulation of breathing and the processing of algebra to performing in the creative arts. The cells that make up nervous tissue are discussed in Module 11.2.

At the foramen magnum, the brain merges with the next organ of the central nervous system: the **spinal cord.** The spinal cord passes through the vertebral foramen of the first cervical vertebra and continues inferiorly to the first or second lumbar vertebra (see Chapter 7). It contains fewer cells than the brain, with only about 100 million neurons. The spinal cord enables the brain to communicate with most parts of the body below the head and neck; it is also able to carry out certain functions on its own (which are discussed in later chapters).

The Peripheral Nervous System

The peripheral nervous system is made up of the most numerous organs of the nervous system, the **nerves,** which carry

signals to and from the central nervous system. A nerve consists of a bundle of long neuron "arms" known as *axons* that are packaged together with blood vessels and surrounded by connective tissue sheaths. Nerves are classified according to their origin or destination: Those originating from or traveling to the brain are called *cranial nerves,* and those originating from or traveling to the spinal cord are called *spinal nerves* (see Figure 11.1). There are 12 pairs of cranial nerves and 31 pairs of spinal nerves. The PNS has separate functional divisions, which we discuss next.

Quick Check

☐ 1. What are the organs of the CNS?
☐ 2. What are the organs of the PNS?

Functional Divisions of the Nervous System

As the nervous system performs its many tasks, millions of processes may be occurring simultaneously. However, all of these tasks or functions generally belong to one of three types: sensory, integrative, or motor. **Sensory functions** involve gathering information about the internal and external environments of the body. **Integrative functions** analyze and interpret incoming sensory information and determine an appropriate response. **Motor functions** are the actions performed in response to integration. An example of these functions is illustrated in **Figure 11.2**, which shows a woman ① seeing a soccer ball moving toward her, ② integrating this input to interpret the position of the ball, and then ③ kicking the ball.

Sensory input is gathered by the **sensory,** or **afferent, division** (AF-er-ent; "carrying toward") of the PNS. Integration is performed entirely by the CNS, mostly by the brain. Motor output is performed by the **motor,** or **efferent, division** (EE-fer-ent; "carrying away") of the PNS. Let's look more at these three functional divisions:

● **PNS sensory division.** Sensory information is first detected by structures of the PNS called **sensory receptors.** The structure of these receptors is diverse—they range from small tips of neurons found in the skin that sense temperature to complex receptors within muscles that sense muscle stretch. Depending on the location of the sensory receptors, the PNS sensory division may be further classified as follows:

○ The **somatic sensory division** (*soma-* = "body") consists of neurons that carry signals from skeletal muscles, bones, joints, and skin. This division also includes special sensory neurons that transmit signals from the organs of vision, hearing, taste, smell, and balance (see Chapter 15). Sometimes the neurons of this division are referred to as the *special sensory division.*

○ The **visceral sensory division** consists of neurons that transmit signals from viscera (organs) such as the heart, lungs, stomach, intestines, kidneys, and urinary bladder.

① **Sensory input:** A woman sees a soccer ball moving toward her; sensory signals are sent to her brain.

Brain

② **Integration:** Her brain receives and integrates the sensory input, and sends signals for an appropriate motor response.

Spinal cord

③ **Motor output:** When the muscles of her thigh and leg receive the motor signals, they contract and kick the soccer ball.

Figure 11.2 Functions of the nervous system.

information, it responds by disregarding about 99% of such integrated data, a process that happens subconsciously. For example, you are likely unaware of the watch on your wrist or the hum of the air conditioner, because this information is filtered out as unimportant. However, that small percentage of sensory stimuli to which the CNS does respond generally leads to a motor response.

- **PNS motor division.** The PNS motor division consists of motor neurons that carry out the motor functions of the nervous system. Motor output traveling from the brain and spinal cord via cranial and spinal nerves of the PNS may be used to control the contraction of muscle or secretion from a gland. Organs that carry out the effects of the nervous system are often called **effectors.** Like the sensory division, the motor division may be further classified based on the organs that the neurons contact:
  - ○ The **somatic motor division** consists of neurons that transmit signals to skeletal muscles. Because skeletal muscle tissue is under conscious control, this division is sometimes referred to as the *voluntary motor division.*
  - ○ The *visceral motor division,* better known as the **autonomic nervous system** (aw-toh-NOM-ik; **ANS**), consists of neurons that carry signals primarily to thoracic and abdominal viscera. The ANS regulates secretion from certain glands, the contraction of smooth muscle, and the contraction of cardiac muscle in the heart. Because these functions are not generally under voluntary control, the ANS is sometimes called the *involuntary motor division.* The ANS, which is very important for maintaining homeostasis of the internal environment, is discussed in its own chapter (see Chapter 14).

Although the divisions of the nervous system are classified separately, both functionally and anatomically, remember that all functions of the nervous system rely on these divisions working together smoothly—no division operates independently. **Figure 11.3** summarizes the divisions, organs, and functions of the nervous system.

Sensory input from both divisions is carried from sensory receptors to the spinal cord and/or the brain by cranial and spinal nerves of the PNS.

- **CNS.** The neurons of the CNS put together the many different types of sensory input, or integrate them, to form a more complete picture that can then elicit response if necessary. Interestingly, once the CNS integrates sensory

CENTRAL NERVOUS SYSTEM (CNS)
Brain and spinal cord integrate information

CNS: Brain and spinal cord

PERIPHERAL NERVOUS SYSTEM (PNS)
Cranial and spinal nerves link CNS and rest of body; perform motor and sensory functions

PNS: Cranial nerves and spinal nerves

Sensory (afferent) division

Motor (efferent) division

Somatic sensory division
Carries general sensory signals from muscles, bones, joints, and the skin, as well as special sensory signals

Visceral sensory division
Carries signals from organs

Somatic motor division
Carries signals to skeletal muscles

Autonomic nervous system (ANS)
Carries signals to smooth muscle, cardiac muscle, and glands

Figure 11.3 Summary of the structural and functional divisions of the nervous system.

MODULE **11.2**

Nervous Tissue

Learning Outcomes

1. Describe the structure and function of each component of the neuron.
2. Describe the structure and function of each type of neuron.
3. Describe how the structure of each type of neuron supports its function.
4. Describe the structure and function of the four types of CNS neuroglial cells and the two types of PNS neuroglial cells.
5. Explain how the structure of each neuroglial cell supports its function.

The majority of tissue that makes up nervous system organs is nervous tissue, although connective and epithelial tissues are also present. Recall that all tissues consist of two components: cells and extracellular matrix (ECM) (see Chapter 4). Some tissues, such as epithelial tissue, are primarily cellular with very little ECM. Others, such as many connective tissues, have few cells and are mostly ECM.

Like epithelial tissue, nervous tissue is highly cellular; about 80% of nervous tissue volume consists of cells (**Figure 11.4**). When you look at such a micrograph of nervous tissue, the most obvious type of cell is the *neuron*, which is the excitable cell type responsible for sending and receiving signals. The other cell type in nervous tissue is the smaller and more prevalent *neuroglial cell* (noor-oh-GLEE-ahl; "nerve glue"), or *neuroglia*, which generally does not transmit signals but rather serves a variety of supportive functions. This module examines each of these cell types in greater detail, as well as the covering—the *myelin sheath*—that insulates and protects certain neurons. We also discuss how some of these cells can be regenerated if they are damaged.

Neurons — Nuclei of neuroglia — Extracellular matrix

Dendrites — Axon — LM (220×)

Figure 11.4 Nervous tissue.

Neurons

◀◀ **Flashback**

1. What are the functions of nucleoli, ribosomes, rough ER, the Golgi apparatus, intermediate filaments, and microtubules? (pp. 94, 96, 102)
2. What are the three components of a neuron? (p. 150)
3. Do neurons undergo mitosis? (p. 150)

The billions of neurons in nervous tissue are directly responsible for its sensory, integrative, and motor functions. **Neurons** are the excitable cell type responsible for sending and receiving signals in the form of *action potentials*. Recall that most neurons are *amitotic*, meaning that at a certain point in development, they lose their centrioles and after that lack the ability to undergo mitosis (see Chapter 4). Luckily, neurons are very long-lived cells, and some can easily survive the entire lifespan of an organism if given adequate nutrition and oxygen in a supportive environment.

Neurons vary greatly in size. Some tiny neurons in the CNS are only 1 mm long, whereas some PNS neurons may be up to 1 m or longer. As **Figure 11.5** shows, most neurons consist of three parts: the central *cell body*, where the majority of the biosynthetic processes of the cell occur; one or more *dendrites*, which carry electrical signals to the cell body; and one *axon*, the long "arm" that generally carries electrical signals away from the cell body. Let's examine each of these parts in greater detail.

The Cell Body

The most conspicuous part of a neuron is its large **cell body,** or *soma*, which ranges from 5 to 100 μm in diameter. The cell body is the most metabolically active part of the neuron, because it is responsible for maintaining the sometimes huge cytoplasmic

Practice art labeling
@MasteringA&P®

Figure 11.5 Neuron structure.

volume of the neuron and also for manufacturing all of the proteins the neuron needs. This high level of biosynthetic activity is reflected in the composition of the organelles within its cytoplasm:

- Free ribosomes and rough endoplasmic reticulum (RER) are found in abundance, reflecting the commitment of the cell body to protein synthesis. Note that the association of ribosomes and RER forms what appears under a microscope as dark-staining clusters called *Nissl bodies;* these are represented in Figure 11.5.
- Other organelles involved in protein synthesis, including the Golgi apparatus and one or more prominent nucleoli, are present.
- Mitochondria are found in large numbers, indicating the high metabolic demands of the neuron.

Additionally, the cytoplasm of the cell body contains lysosomes, smooth ER, and other organelles found in most cells.

The characteristic shape of the cell body is maintained by another component of the cytoplasm—the neuronal cytoskeleton, which is composed of intermediate filaments. These filaments bundle together to form larger structures called **neurofibrils,**

which provide structural support that extends out into the dendrites and axon of the neuron as well (see Figure 11.5). The cytoskeleton also contains microtubules that provide structural support and a means for transporting chemicals between the cell body and the axon.

Processes: Dendrites and Axons

Extending from all neuron cell bodies are long "arms," cytoplasmic extensions that are called *processes*. These processes allow the neuron to communicate with other cells. Most neurons have two types of processes, including one or more dendrites and one axon.

Dendrites Dendrites (DEN-drytz; *dendr-* = "branch or tree") are typically short, highly forked processes that resemble the branches of a tree limb. They receive input from other neurons, which they transmit in the form of electrical impulses toward the cell body. Note, however, that dendrites usually do not generate or conduct action potentials. Their cytoplasm contains most of the same organelles as the cell body, including mitochondria, ribosomes, and smooth endoplasmic reticulum. The extensively forked "dendritic trees" of most neurons give them a huge receptive surface area. Interestingly, the branches of the dendritic tree change throughout an individual's lifetime: They grow and are "pruned" as a person grows and develops and as functional demands on the nervous system change.

Axon Although a neuron may have multiple dendrites, each neuron has only a single **axon,** sometimes called a *nerve fiber.* Traditionally, an axon was defined as a process that carried a signal away from the cell body. However, the axons of certain neurons can carry a signal both toward and away from the cell body. For this reason, new criteria have been developed to define an axon: They are considered processes that can generate and conduct action potentials.

Notice in Figure 11.5 that each axon arises from an area of the cell body called the **axon hillock,** and then tapers to form the slender axon, which is often wrapped in the insulating myelin sheath. Depending on the type of neuron, the axon may range in length from short to very long; in some neurons the axon accounts for most of the length of the neuron. For example, the axons of motor neurons going to the foot must extend from the lumbar portion of the spinal cord all the way down the lower limb and to the foot.

Extending from some axons are branches that typically arise at right angles to the axon, called **axon collaterals.** Both the axon and its collaterals split near their ends to produce multiple fine branches known as **telodendria** (tee'-loh-DEN-dree-ah). The telodendria terminate in **axon terminals,** or **synaptic knobs,** that communicate with a target cell. Each axon generally splits into 1000 or more axon terminals.

The plasma membrane that envelops the axon is called the **axolemma** (aks-oh-LEM-ah), and its cytoplasm is known as **axoplasm.** Although dendrites have most of the same organelles as the cell body, axons do not. Axons contain mitochondria, abundant intermediate filaments, vesicles, and lysosomes; however, they do not contain protein-making organelles such as ribosomes or Golgi apparatus. The composition of the axoplasm is dynamic, as substances move both toward and away from the cell body along the axon's length.

Substances may travel through the axoplasm using one of two types of transport, which are together termed *axonal transport* or *flow:*

- **Slow axonal transport.** Substances within the axoplasm, such as cytoskeletal proteins and other types of proteins, move by slow axonal transport. These substances move only away from the cell body and do so at a rate of about 1–3 mm/day.
- **Fast axonal transport.** Vesicles and membrane-bounded organelles use fast axonal transport to travel much more rapidly through the axon. This type of transport relies on motor proteins in the axoplasm that consume ATP to move substances along microtubules either toward the cell body (called *retrograde* transport), at a maximum rate of about 200 mm/day, or away from the cell body (called *anterograde* transport), at a maximum rate of about 400 mm/day. See *A&P in the Real World: Poliovirus and Retrograde Axonal Transport* to see how microorganisms use this method of transport to cause disease.

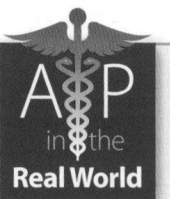

Poliovirus and Retrograde Axonal Transport

Poliomyelitis (poh'-lee-oh-my-eh-LY-tus), the disease caused by *poliovirus,* is an infection that can impact the central nervous system (CNS), particularly the spinal cord, and can result in deformity and paralysis. No cure for polio is available, but it can be easily prevented by vaccination.

The virus is thought to gain access to the CNS by entering muscle fibers and then passing into motor neurons at the neuromuscular junction. The virus then travels the length of the axon, using retrograde axonal transport, until it eventually enters the spinal cord. Other toxins and viruses such as herpes simplex virus, rabies virus, and tetanus toxin also have the ability to invade the nervous system via retrograde axonal transport.

Functional Regions of Neurons

Now let's briefly discuss how these various components of the typical neuron function together. As you see here, the neuron has three main functional parts:

The receptive region of the neuron consists of the dendrites and cell body. The dendrites may receive signals from other neurons, or may monitor the external and internal environments via sensory receptors. The received signals are collected in the cell body, which then may send a signal down the axon, the conducting region of the neuron. When the signal reaches the axon terminals, they secrete chemicals that trigger changes in their target cells.

Classification of Neurons

As with many topics that we've covered, neurons can be classified according to both their structure and their function. These classification schemes overlap—certain functional groups of neurons often have the same structural features, another example of how "form follows function" in the body.

Structural Classification Neurons vary widely in shape, with the greatest structural variation seen in the number and form of the processes extending from the cell body. On this basis, neurons are classed structurally into three groups:

- **Multipolar neurons.** Over 99% of neurons in the human body fall into the group known as **multipolar neurons.** These neurons have a single axon and typically multiple highly branched dendrites. This group of neurons has the widest variability in terms of shape and size.
- **Bipolar neurons.** A **bipolar neuron** has only two processes: one axon and one dendrite. In humans the majority of bipolar neurons are sensory neurons, located in places such as the retina of the eye and the olfactory epithelium of the nasal cavity.
- **Pseudounipolar neurons.** Pseudounipolar neurons (soo'-doh-yoo-nih-POH-lar; formerly referred to as *unipolar neurons*) begin developmentally as bipolar neurons, but their two processes fuse to give rise to a single axon. As the axon extends from the cell body, it splits into two processes: one that brings information from sensory receptors to the cell body, called the *peripheral process* or *axon,* and one that travels to the spinal cord away from the cell body, called the *central process* or *axon.* The pseudounipolar neurons are sensory neurons that sense information such as touch, pressure, and pain.

Functional Classification Functionally, neurons are grouped into three classes based on the direction in which they carry information. The three classes are as follows, in order of information flow:

1. **Sensory,** or **afferent, neurons** carry information *toward* the central nervous system. These neurons receive information from a sensory receptor and transmit this information to their cell body in the PNS, then down their axon to the brain or spinal cord. Because sensory neurons receive information from one area, they are generally pseudounipolar or bipolar in structure. Sensory neurons detect the internal and external environments (such as from the skin and viscera) and facilitate motor coordination (such as in joints and muscles).
2. **Interneurons,** also called *association neurons,* relay messages *within* the CNS, primarily between sensory and motor neurons, and are the location of most information processing. The vast majority of neurons are interneurons. Multipolar in structure, interneurons generally communicate with many other neurons (for example, one Purkinje cell [per-KIN-jee] of the cerebellum can receive as many as 150,000 contacts from other neurons).
3. **Motor,** or **efferent, neurons** carry information *away from* their cell bodies in the CNS to muscles and glands. As motor tasks are generally complicated and require input from many other neurons, most motor neurons are multipolar.

The classification systems of neurons are summarized in **Table 11.1**. Note that Table 11.1 includes three different examples of multipolar neurons—one from the spinal cord (spinal motor neuron), one from the hippocampus of the brain (pyramidal cell), and another from the cerebellum of the brain (Purkinje cell).

Structural Groups of Neuron Components

In the CNS and PNS, specific neuron components group together. For example, cell bodies of neurons are typically found within clusters, most of which are in the CNS, where they are called **nuclei.** Within the PNS, clusters of cell bodies are called **ganglia** (GANG-glee-ah; singular, *ganglion; gangli-* = "knot"). In addition, axons tend to be bundled together in the CNS and the PNS. In the CNS, these bundles are referred to as *tracts,* and in the PNS, as *nerves.*

TABLE 11.1 NEURON CLASSIFICATION				
Structural Class	**Multipolar Neurons**		**Bipolar Neurons**	**Pseudounipolar Neurons**
Structural Features	One axon with two or more dendrites; typically have highly branched dendritic tree		One axon and one dendrite	Single short process that splits into two axons (no dendrites)
	Spinal motor neuron — Pyramidal cell — Purkinje cell		Special sensory neuron	General sensory neuron
Typical Functional Class	Motor (efferent) neurons, interneurons		Sensory (afferent) neurons	Sensory (afferent) neurons
Location	Most neurons in the CNS, motor neurons in the PNS		Special sense organs in the PNS, such as the retina and olfactory epithelium	Sensory neurons in the PNS associated with touch, pain, and vibration sensations

Quick Check

☐ 1. What are the functions of the cell body, dendrites, and axon?

☐ 2. What are the structural differences between multipolar, bipolar, and pseudounipolar neurons?

☐ 3. What are the functional differences between sensory neurons, interneurons, and motor neurons?

Neuroglia

◀◀ **Flashback**

1. Why do nonpolar, lipid-based substances diffuse easily across cell membranes, but polar compounds do not? (p. 76)
2. What are tight junctions and gap junctions? How does their form follow their function? (p. 127)

Neuroglia (noo-ROG-lee-ah), or **neuroglial cells,** were named for the early scientific idea that these cells "glued together" the neurons, as the word root *glia* means "glue." However, we now recognize that neuroglia also serve many more functions. Some of their roles include maintaining the environment around neurons, protecting them, and assisting in their proper functioning. Unlike the mostly amitotic neurons, neuroglia retain their ability to divide, and they fill in gaps left when neurons die.

Six different types of neuroglia can be found in the nervous system, four in the CNS and two in the PNS. Like all cells we've covered, the form of each type of neuroglial cell is specialized for its function, another example of the Structure-Function Core Principle (p. 25). Keep this in mind as we examine each of the six types of cells.

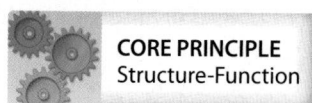

CORE PRINCIPLE
Structure-Function

Neuroglia in the CNS

Neuroglia are about 10 times more abundant in the CNS than neurons, and they make up about half the mass of the brain. Within the CNS we find four types of neuroglia: *astrocytes, oligodendrocytes, microglia,* and *ependymal cells* (**Figure 11.6**).

Astrocytes The star-shaped **astrocytes** (ASS-troh-sytz; "star cells") are the most numerous and the largest of the neuroglia in the CNS. Note in Figure 11.6 that each astrocyte has a central portion and numerous processes, all of which terminate in structures called *end-feet*. This anatomical feature equips astrocytes to perform multiple functions, including the following:

● **Anchoring neurons and blood vessels in place.** Astrocytes help form the three-dimensional structure of the brain by using their end-feet to anchor neurons and blood vessels in place. In addition, astrocytes may facilitate the transport of nutrients and gases from the blood vessels to neurons.

NEUROGLIAL CELL TYPE	ASTROCYTE	OLIGODENDROCYTE	MICROGLIAL CELL	EPENDYMAL CELL
FUNCTION	• Anchor neurons and blood vessels • Regulate the extracellular environment • Facilitate the formation of the blood-brain barrier • Repair damaged tissue	• Myelinate certain axons in the CNS	• Act as phagocytes	• Line cavities • Cilia circulate fluid around brain and spinal cord • Some secrete this fluid

Labels on figure: Neuron, Processes, End-feet, Capillary, Particle being ingested, Cilia, Fluid being secreted

Figure 11.6 Neuroglial cells of the CNS.

Practice art labeling
@MasteringA&P®

- **Regulating the extracellular environment of the brain.** Astrocytes are connected by gap junctions that allow them to communicate with one another about the local extracellular environment within the brain. Via this communication they can act as a "clean-up crew," removing excess extracellular potassium ions as well as chemicals known as *neurotransmitters*. Although neurons use neurotransmitters to send signals, their extracellular accumulation can lead to toxicity.
- **Assisting in the formation of the blood-brain barrier.** Astrocytes facilitate the formation of a protective structure called the *blood-brain barrier* by ensheathing capillaries and inducing their cells to form tight junctions. These tight junctions render the capillaries virtually impermeable to most proteins and polar compounds, and the only substances that can cross these capillaries easily are those that are nonpolar and lipid-soluble and/or those for which special transporters exist. The double barrier separates the blood from the brain ECF, which ensures selective transport of substances between the two fluids. The blood-brain barrier is discussed fully in the CNS chapter (see Chapter 12).
- **Repairing damaged brain tissue.** When brain injury occurs, astrocytes are triggered to divide rapidly. Although this growth stabilizes the damaged tissue, it may also impede complete healing. Recent research has demonstrated that excess astrocyte activity actually inhibits the regrowth of neurons, leading to more permanent defects.

Astrocytes are critical to normal functioning of the nervous system, so when they undergo rapid, uncontrolled cell division, the results can be devastating. Find out more about this in *A&P in the Real World: Gliomas and Astrocytomas* on page 391.

Oligodendrocytes Like astrocytes, **oligodendrocytes** (oh-lig′-oh-DEN-droh-sytz; *oligo-* = "few") also have radiating processes, but they are fewer in number and smaller than those of astrocytes. The flattened ends of some of these processes wrap around part of the axons of certain neurons. These wrapped processes form concentric layers of plasma membrane that are collectively called **myelin** (MY-eh-lin). Repeating segments of myelin along the length of an axon form the myelin sheath. Observe in Figure 11.6 that each oligodendrocyte has several of these processes that wrap around multiple axons. We will consider the formation of the myelin sheath and its functional significance later in this module.

Microglia The least numerous neuroglial cells are the small and branching **microglia** (my-KROG-lee-ah). Although many functions of microglia are still under investigation, we do know that they are activated by injury within the brain and become wandering phagocytes—cells that "clean up" the environment in the brain. When activated, microglia ingest disease-causing organisms, dead neurons, and other cellular debris.

Ependymal Cells Within the brain and spinal cord are fluid-filled cavities lined with ciliated neuroglia known as **ependymal cells** (eh-PEN-dih-mal). These ciliated cells have a variety of functions, including circulating *cerebrospinal fluid*, which is the fluid in the cavities of the brain and spinal cord. Certain ependymal

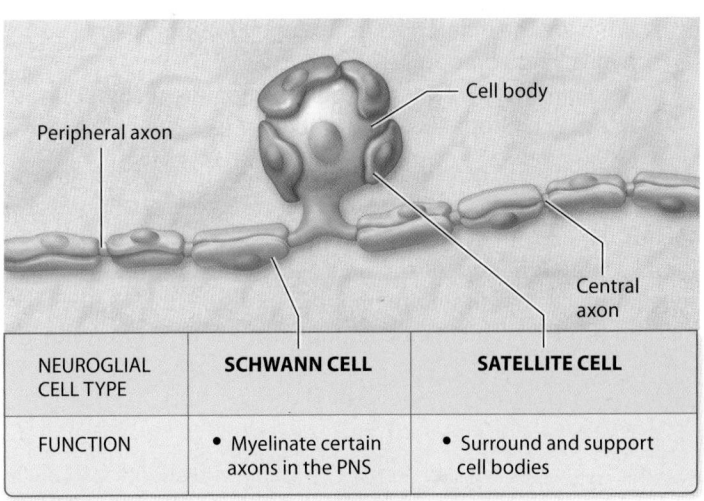

NEUROGLIAL CELL TYPE	**SCHWANN CELL**	**SATELLITE CELL**
FUNCTION	• Myelinate certain axons in the PNS	• Surround and support cell bodies

Practice art labeling @MasteringA&P®

Figure 11.7 Neuroglial cells of the PNS.

cells also play a role in the formation of this fluid, and others are thought to monitor its composition.

Neuroglia in the PNS

In the PNS the two types of neuroglia are *Schwann cells* and *satellite cells* (**Figure 11.7**). Like those in the CNS, the neuroglia of the PNS serve supportive and protective functions, with, once again, their form specialized for their function.

Schwann Cells Larger axons of the PNS are also covered with a myelin sheath that has structural and functional properties nearly identical to those of the myelin sheath in the CNS. However, the myelin sheath of the PNS is created by a different type of neuroglial cell: the sausage-shaped **Schwann cells** (see Figure 11.7, left). As we will see, unmyelinated axons are also encased in Schwann cells. Additionally, Schwann cells play a vital role in repair of damaged axons in the PNS.

Satellite Cells **Satellite cells** are flat cells that surround the cell bodies of neurons in the PNS (see Figure 11.7, right). The most poorly understood of the neuroglia, they appear to enclose and support the cell bodies, and have intertwined processes that link them with other parts of the neuron, other satellite cells, and also neighboring Schwann cells.

Quick Check

☐ 4. What are the functions of astrocytes?
☐ 5. What are the functions of microglia?
☐ 6. Which neuroglial cell forms and circulates the fluid surrounding the brain and spinal cord?

The Myelin Sheath

◄◄ Flashback

1. What is the difference between polar and nonpolar covalent bonds? (p. 39)

2. What are the differences between hydrophobic and hydrophilic compounds? (p. 47)
3. Are lipids polar covalent or nonpolar covalent compounds? Are they hydrophilic or hydrophobic? (p. 52)

As we discussed, certain neuroglia wrap themselves around the axons of neurons to create a structure known as the **myelin sheath** (**Figure 11.8**). Myelin is composed of repeating layers of the plasma membrane of the neuroglial cell, so it has the same substances as any plasma membrane: phospholipids, other lipids, and proteins. The main components (70–80%) of myelin are

various lipids, including cholesterol, phospholipids, and other unique lipids.

In the fluids of the body, *electric current* is the movement of ions. Ions do not easily pass through the phospholipid bilayer of the plasma membrane, and so the high lipid content of myelin makes it an excellent insulator of electrical current (akin to rubber tubing around a copper wire). The overall effect of this insulation is to increase the speed of conduction of action potentials: *Myelinated axons* conduct action potentials about 15–150 times faster than *unmyelinated axons*. This is a good example of the Structure-Function Core Principle (p. 25).

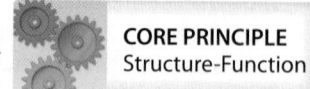

CORE PRINCIPLE
Structure-Function

Recall that myelin is formed by Schwann cells in the PNS and by oligodendrocytes in the CNS. The formation of the myelin sheath is known as **myelination** (my′-eh-lin-AY-shun).

(a) **The myelin sheath and myelination in the PNS**

Cross-section of a myelin sheath   SEM (30,700×)

(b) **The myelin sheath in the CNS**

Practice art labeling
@MasteringA&P®

Figure 11.8 **The myelin sheath in the PNS and CNS.**

During this process in the PNS, a Schwann cell wraps itself outward away from the axon in successively tighter bands, forming a myelin sheath up to 100 layers thick (see Figure 11.8a). The basic process is similar for an oligodendrocyte in the CNS. However, in the CNS the arms of an oligodendrocyte wrap inward toward the axon—the opposite direction from the Schwann cells (see Figure 11.8b).

Many other differences can be found between myelination in the PNS and CNS, including the following:

- **Presence or absence of a neurolemma.** Note in Figure 11.8a that on the outer surface of a myelinated axon in the PNS we find the nucleus and the bulk of the cytoplasm and organelles of the Schwann cell, known as the **neurolemma** (noor-uh-LEM-ah). Because the nucleus and cytoplasm of the oligodendrocyte remain in a centralized location, no outer neurolemma is found in the CNS (Figure 11.8b).
- **Number of axons myelinated by a single glial cell.** Also see that each oligodendrocyte may send out multiple processes to envelop parts of several axons. However, Schwann cells can encircle only a portion of a single axon.
- **Timing of myelination.** The timing of myelination is also different within the CNS and the PNS. In the PNS myelination begins during the early fetal period, whereas myelination in the CNS, particularly in the brain, begins much later. Very little myelin is present in the brain of the newborn (which is why babies and toddlers need adequate fat in their diets).

In both the CNS and the PNS, axons are generally much longer than a single oligodendrocyte or Schwann cell, so more than one cell is needed to myelinate the entire axon. The segments of an axon that are covered by neuroglia are called **internodes,** and they range from 0.15 to 1.5 mm in length. Between each internode is a gap about 1 μm wide called a **node of Ranvier** (RAHN-vee-ay), or *myelin sheath gap,* where no myelin is found. Also unmyelinated is a short region from the axon hillock to the first neuroglial cell; this is known as the *initial segment.*

Short axons in both the CNS and the PNS are nearly always unmyelinated. However, in the PNS, even axons that lack a myelin sheath associate with Schwann cells (**Figure 11.9**). Take note, though, that the Schwann cells do not wrap themselves around these axons. Instead, they enclose them much like a hot dog in a bun. A single Schwann cell can envelop multiple axons in this manner.

Figure 11.9 Unmyelinated peripheral axons and Schwann cells.

Schwann cell
Unmyelinated axons
Schwann cell nucleus

In the CNS you can actually see which regions of the brain and spinal cord contain myelinated axons and which do not. In sections of both the spinal cord and the brain, regions of darker- and lighter-colored tissue (see Figure 12.2) can be noted. This color difference reflects the distribution of the myelin sheath. The lighter-colored areas, or **white matter,** are composed of myelinated axons. The darker-colored areas, or **gray matter,** are made up primarily of cell bodies and dendrites, which are never myelinated, as well as small unmyelinated axons.

Quick Check

☐ 7. What is the function of the myelin sheath?
☐ 8. How does the myelin sheath differ in the CNS and the PNS?

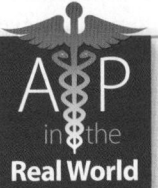

Real World

Gliomas and Astrocytomas

Tumors that originate within the brain are called *primary brain tumors,* and most such tumors are *gliomas* (glee-OH-mahz). A glioma is caused by an abnormally high rate of cellular division in glial cells. Some conditions are known to predispose patients to gliomas, such as exposure to ionizing radiation and certain diseases, but the cause of the tumor usually goes undiscovered. The most common glial cell affected by glioma is the astrocyte, and the resulting tumor is called an *astrocytoma* (ass'-troh-sy-TOH-mah), shown here:

Astrocytoma

Astrocytomas range in severity from relatively mild tumors with a good prognosis to highly aggressive tumors with a very poor prognosis. Treatment varies with the type of tumor and the age and health of the patient, but generally involves surgical removal of the mass, with the use of chemotherapeutic drugs and perhaps radiation therapy.

Regeneration of Nervous Tissue

◀◀ **Flashback**

1. What is the difference between regeneration and fibrosis? Which tissues are generally able to regenerate? (p. 154)
2. What is a basal lamina? (p. 129)

Human nervous tissue has a fairly limited capacity for **regeneration,** or replacement of damaged tissue with new tissue. Damaged axons and dendrites in the CNS almost never regenerate, a phenomenon apparently due to several factors. For example, oligodendrocytes may inhibit the process of neuronal growth, and chemicals called *growth factors* that trigger mitosis are largely absent in the CNS. In addition, the growth of astrocytes creates space-filling scar tissue that also prohibits regeneration. For these reasons, injuries to the brain or spinal cord have largely permanent effects. However, in some circumstances lost function may be regained through retraining of the remaining neurons.

In contrast, neural tissue in the PNS is capable of regeneration to some extent. Within the PNS, a neuron will regenerate only if the cell body remains intact. When a peripheral axon is damaged, the following sequence of events repairs the damaged neuron (**Figure 11.10**):

① **The axon and myelin sheath distal to the injury degenerate.** The damaged axon is cut off from the cell body, and so from all of the protein-synthesis machinery located there. Thus, the axon and myelin sheath distal to the injury begin to degenerate via a process called *Wallerian degeneration* (vah-LAIR-ee-an), in which phagocytes digest the cellular debris.

② **Growth processes form from the proximal end of the axon.** As Wallerian degeneration occurs, protein synthesis within the cell body increases, and several small *growth processes* sprout from the proximal end of the axon.

③ **Schwann cells and the basal lamina form a regeneration tube.** Schwann cells begin to proliferate along the length of the surrounding basal lamina near the site of the injury, forming a cylinder called the **regeneration tube.**

④ **A single growth process grows into the regeneration tube.** Note in step ② of Figure 11.10 that several growth processes form; however, only one will make it into the regeneration tube. In the tube, Schwann cells secrete growth factors that stimulate regrowth of the axon. The regeneration tube then guides the axon to grow toward its target cell at a rate of about 1.5–3 mm/day.

⑤ **The axon is reconnected with the target cell.** If the axon continues to grow, it most likely will meet up with its target cell and re-establish its synaptic contacts. Over time, the Schwann cells re-form the myelin sheath.

This process occurs only under ideal conditions. Even with the cell body intact, the process often stalls after axon degeneration, and the neuron dies. And if regeneration occurs, the

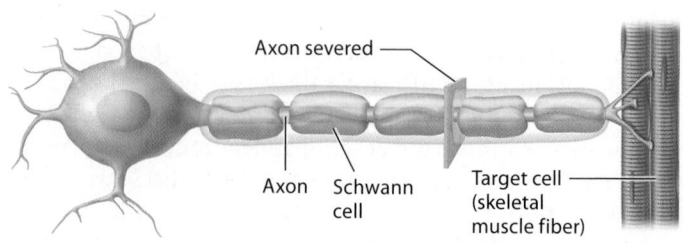

Axon severed

Axon | Schwann cell | Target cell (skeletal muscle fiber)

① Axon and myelin sheath distal to the injury degenerate (Wallerian degeneration).

Phagocytes digesting debris

② Growth processes form from the proximal end of the axon.

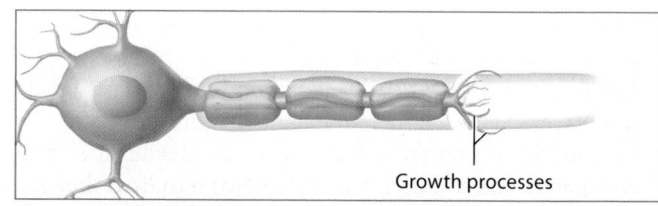

Growth processes

③ Schwann cells and the basal lamina form a regeneration tube.

Schwann cells

Regeneration tube

④ A single growth process grows into the regeneration tube.

Growth process in regeneration tube

⑤ The axon is reconnected with the target cell.

Reconnected synaptic contacts

Figure 11.10 Repair of axon damage in the PNS.

results are often imperfect. Occasionally, the axon will contact the wrong target cell, or contact between the cells will not be re-established.

Quick Check

☐ 9. Are neurons more likely to regenerate in the CNS or in the PNS? Why?

☐ 10. What must be intact for a neuron to regenerate?

Apply What You Learned

☐ 1. When a pathologist performs an autopsy on a person who died of a brain injury, explain why he or she typically finds large numbers of microglia in the brain.

☐ 2. Guillain-Barré (GEE-yan bar-RAY) syndrome is caused by the patient's own immune system attacking the myelin sheath of PNS neurons. Predict the symptoms and effects of such a disease.

☐ 3. Ms. Karabekian suffers a vertebral fracture that damages a large number of ganglia, then loses feeling in much of her right leg. Is she likely to recover the function of these damaged neurons? Why or why not? (*Hint:* Are ganglia part of the CNS or PNS? What do ganglia contain?)

See answers in Appendix A.

MODULE 11.3
Electrophysiology of Neurons

Learning Outcomes

1. Explain how ion channels cause development of the resting membrane potential in neurons.
2. Describe the voltage-gated ion channels that are essential for the development of the action potential.
3. Interpret a graph showing the voltage-versus-time relationship of an action potential, and relate the terms depolarize, repolarize, and hyperpolarize to the events of an action potential.
4. Explain the physiological basis of the absolute and relative refractory periods.
5. Compare and contrast continuous and saltatory conduction.
6. Explain how axon diameter and myelination affect conduction velocity.

Neurons share two key properties with skeletal muscle fibers (see Chapter 10). For one, all neurons are *excitable* (responsive) in the presence of various stimuli, including chemical signals, local electrical signals, and mechanical deformation. These stimuli generate electrical changes across the plasma membrane of the neuron. Another property is *conductivity,* which means that electrical changes across the plasma membrane don't stay in one place. Instead, they are rapidly conducted along the entire length of the membrane, similar to how an electrical impulse is conducted through a copper wire.

The electrical changes across a neuron's plasma membrane come in two forms: (1) *local potentials,* which travel only short distances, and (2) *action potentials,* which travel the entire length of an axon. Both types of potentials rely on the same principles of electrophysiology that we discussed with muscle tissue (see Chapter 10). In this module we re-examine these principles in terms of the electrophysiology of neurons, and you will see that these two types of potentials allow the nervous system to perform virtually all of its functions.

Principles of Electrophysiology

◀◀ **Flashback**

1. Is the concentration of sodium ions greater in the cytosol or in the extracellular fluid? How about the concentration of potassium ions? What maintains these two gradients? (p. 79)
2. What is the resting membrane potential? (p. 82)
3. What are the two classes of ion channels? (p. 349)

In the muscle tissue chapter you were introduced to some of the concepts of *electrophysiology*—the branch of physiology that studies electrical changes across the plasma membrane and the accompanying physiological processes (see Chapter 10). Although discussion in that chapter revolved around the electrophysiology of the muscle fiber, the same basic principles apply to the electrophysiology of neurons. Like muscle fibers, electrical changes across the plasma membrane of neurons rely on the presence of *ion channels* in the membrane and a *resting membrane potential.* So, before we move on, let's review these important concepts.

Ion Channels and Gradients

Ions cannot pass through the hydrophobic portion of the phospholipid bilayer of the plasma membrane because they are charged particles. For this reason, their movement across the plasma membrane is dependent on specific protein channels. There are two main classes of channels:

- **Leak channels** are always open and continually allow ions to follow their concentration gradient into or out of the cell.
- **Gated channels** are closed at rest, and open only in response to certain stimuli (see Chapter 10). Some gated channels, called **ligand-gated channels,** open in response to a certain chemical binding to the channel (or to an associated receptor). Other channels, called **voltage-gated channels,** open or close in response to changes in voltage across the membrane. A third type of gated channel is the **mechanically gated channel,** which opens or closes in response to mechanical stimulation such as stretch, pressure, and vibration.

TABLE 11.2 TYPES OF ION CHANNELS IN NEURONS AND OTHER ELECTRICALLY EXCITABLE CELLS

Type of Channel	Structure	Stimulus for Opening/Closing
Leak Channel	ECF — / Plasma membrane / Cytosol —	None, always open
Ligand-Gated Channel	Closed → Open (Ligand)	Binding of a ligand to a receptor associated with the channel
Voltage-Gated Channel	Closed → Open	Voltage changes across the plasma membrane
Mechanically Gated Channel	Closed → Open	Mechanical deformations of the channel (by stretch, pressure, etc.)

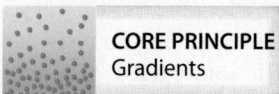

Table 11.2 reviews the different types of channels involved in generating and transmitting action potentials.

You've also been introduced to the vitally important concentration gradients of sodium and potassium ions that exist across

CORE PRINCIPLE
Gradients

the plasma membrane, an example of the Gradients Core Principle (p. 26). In a neuron, as in a muscle fiber, the concentration of sodium ions is higher in the extracellular fluid than in the cytosol, and the concentration of potassium ions is higher in the cytosol than in the extracellular fluid.

These gradients are maintained (and to some degree even created) by the ATP-consuming Na^+/K^+ pump, which brings two potassium ions into the cytosol as it moves three sodium ions into the extracellular fluid.

The Resting Membrane Potential

We've also discussed the separation of charges across the plasma membrane—there is a thin layer of negative charges in the cytosol lining the inside of the membrane and a thin layer of positive charges in the extracellular fluid lining the outside of the membrane (see Chapter 10). Recall that this separation of charges, called a *voltage*, is simply a type of gradient referred to as an *electrical gradient*. The electrical gradient across the cell membrane is known as a **membrane potential,** named for the fact that, like any gradient, an electrical gradient is a source of potential energy for the cell.

The voltage across the membrane may be measured with a voltmeter, as shown in **Figure 11.11.** Notice that as you measure from outside to inside the cell with a voltmeter, the voltage

Figure 11.11 **Measurement of the voltage across a plasma membrane.**

becomes more negative. This negative voltage is present when the cell is at rest (not being stimulated), and for this reason it is called the **resting membrane potential.** The cell in this state is said to be **polarized,** which simply means that the voltage difference across the plasma membrane of the cell is not at 0 mV, but rather measures to either the positive or the negative side (or *pole*) of zero. All of these factors apply to neurons as well as muscle fibers. A typical neuron has a resting membrane potential of about −70 mV, and as we discuss shortly, changes in this potential are responsible for the electrical events of a neuron.

Generation of the Resting Membrane Potential

Now that we know what the resting membrane potential is, let's talk about how it's generated. Imagine a neuron that isn't polarized— its membrane potential has been temporarily changed to 0 mV. What happens as the membrane returns to its resting state of −70 mV? Two factors work together:

- Ion concentration gradients favor diffusion of potassium ions out of the cell and sodium ions into the cell.
- Potassium ions diffuse through leak channels more easily than do sodium ions.

The first factor, the concentration gradients across the membrane, is due to the activity of the Na$^+$/K$^+$ pumps. The effects of these gradients are that potassium ions tend to diffuse out of the cell, and sodium ions tend to diffuse into the cell. Any difference in these relative rates of diffusion causes the membrane potential to change.

To get to the resting state, the membrane potential must become more negative, which means *more* potassium ions must leave the cell than sodium ions enter. So why does this happen? It occurs because of the second factor, the ease with which potassium ions can cross the membrane through leak channels. Basically, you can think of the membrane as being "leakier" for potassium ions than for sodium ions, and for this reason, more potassium ions exit the cell than sodium ions enter.

As these two factors work together, the cytosol loses more positive charges than it gains. This net loss causes the membrane potential to become more negative, until the value of the resting membrane potential is reached.

ConceptBOOST »»»

How Do Positive Ions Create a Negative Resting Membrane Potential?

Much of the *negative* resting membrane potential is caused by the movement of *positive* ions. But how can positive ions create a negative potential? To understand how this works, let's start with a neuron that has no membrane potential, which means that the charges are distributed equally across the plasma membrane. In our diagram here, five positive charges and five negative charges are found on each side of the membrane:

Now, imagine that a potassium ion diffuses out of the cytosol down its concentration gradient through a leak channel:

We now find six positive charges outside the membrane and four positive charges inside. This makes the overall charge inside the cytosol −1 and in the extracellular fluid +1—a membrane potential has been created. Next imagine that thousands or more potassium ions exit through leak channels, which causes the membrane potential to become progressively more negative. ∎

The two main factors that lead to generation of the resting membrane potential are illustrated in **Figure 11.12.**

Quick Check

☐ 1. What is the resting membrane potential?
☐ 2. How are sodium and potassium ions distributed across the plasma membrane? What creates this distribution?
☐ 3. What two factors generate the resting membrane potential?

Electrochemical Gradients

The resting membrane potential in our cells has important implications for ion transport. Several times in this book, you

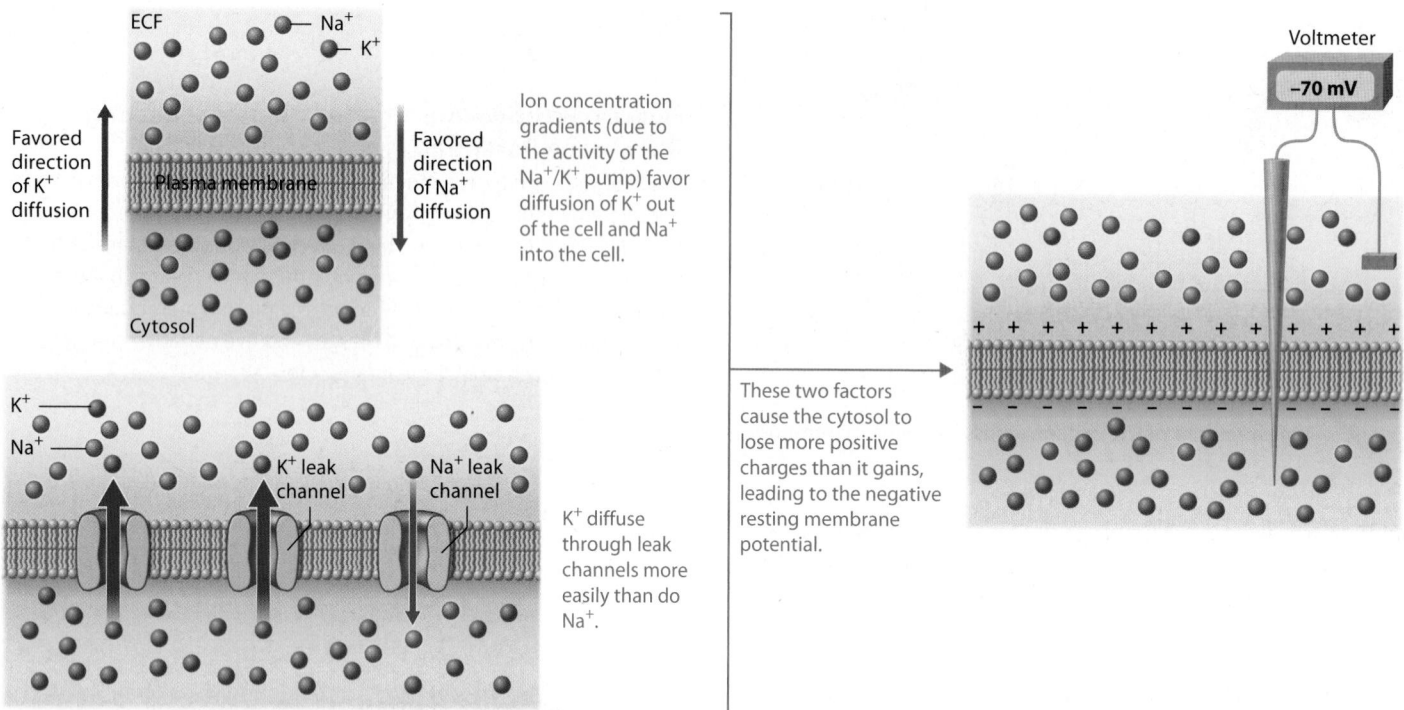

Figure 11.12 Generation of the resting membrane potential.

have seen how solutes move across membranes by diffusion according to their concentration gradient. In fact, the concentration gradient is the main factor that determines the movement of uncharged solutes such as carbon dioxide, glucose, and oxygen. But the story for ions is more complicated because they are also affected by electrical gradients. For this reason, diffusion of an ion across the plasma membrane is determined by both its concentration gradient and its electrical gradient. These two combined forces are called the **electrochemical gradient.**

As an example, consider a potassium ion in the cytosol of a neuron (**Figure 11.13**). You have already seen that ① the concentration gradient for potassium ions favors their diffusion into the extracellular fluid. But now let's add the force of the electrical gradient. The −70 mV resting potential means that the cytosol is negatively charged relative to the extracellular fluid. As you know, opposite

charges attract, so the positively charged potassium ion is attracted to the negatively charged cytosol. ② This electrical gradient then favors the movement of potassium ions in the opposite direction, into the cytosol. The overall electrochemical gradient is the sum of these two forces—one drawing potassium ions into the cytosol and one drawing them into the extracellular fluid. If these two forces were equal, no net movement of potassium ions would occur.

However, ③ the concentration gradient for potassium ions is stronger than the electrical gradient by a small amount. For this reason, the net electrochemical gradient is a small force that draws potassium ions into the extracellular fluid. The small size of the electrochemical gradient for potassium ions in a neuron at rest helps to ensure that the cell doesn't lose too many potassium ions to the extracellular fluid through leak channels.

When we look at sodium ions, however, a different picture emerges. You already know that the concentration gradient favors the movement of sodium ions into the cytosol. The electrical gradient also favors their movement into the cytosol, as the positively charged sodium ions are attracted to its negative charges. This creates a strong electrochemical gradient for sodium ions that draws them into the cytosol.

Quick Check

☐ 4. How is an electrochemical gradient different from a concentration gradient?
☐ 5. How do the electrochemical gradients for potassium ions and sodium ions differ?

Changes in the Membrane Potential: Ion Movements

Now let's connect the two concepts we have been discussing: ion channels and gradients plus the resting membrane potential.

① The concentration gradient favors the movement of K⁺ to the ECF.

② The electrical gradient favors the movement of K⁺ to the cytosol.

③ The concentration gradient is slightly stronger, so a small force favors the movement of K⁺ into the ECF.

Figure 11.13 The electrochemical gradient for potassium ions.

Because the resting membrane potential results from the unequal distribution of ions and their different abilities for crossing the membrane, if we change the ability of the ions to cross the membrane, the membrane potential will change as well. This happens by opening gated channels in the plasma membrane. As shown in Figure 11.14a, if gated sodium ion channels open, sodium ions follow their electrochemical gradient and rush into the cell, and the cell gains positive charges. The influx of positive charges makes the membrane potential more positive, a change called **depolarization.** By this process, the cell becomes less polarized as its membrane potential approaches 0 mV. When a cell returns to its resting membrane potential, **repolarization** has occurred.

If we instead open gated potassium ion channels, potassium ions follow their electrochemical gradient out of the cell, and the cell loses positive charges. This causes the membrane potential to become more negative than it is at rest, a change termed **hyperpolarization** (Figure 11.14b). Note that hyperpolarization may also result from the opening of channels for anions, such as chloride ions, which would allow these negatively charged ions to flow into the cell. (This additional change in membrane potential doesn't occur in muscle fibers, which is why we didn't discuss it in the muscle tissue chapter.)

Both types of changes in membrane potential are seen in neurons. In the upcoming sections, we see how this applies to nervous system physiology and the ability of the neuron to send signals.

Quick Check

☐ 6. In and around the axon, where is the higher concentration of sodium ions? Where is the higher concentration of potassium ions? What maintains this gradient?
☐ 7. What is the resting membrane potential, and what is responsible for generating it?
☐ 8. Define depolarization, repolarization, and hyperpolarization.

Local Potentials

You read in the muscle tissue chapter that each stimulus from a motor neuron leads to a quick, temporary reversal in the membrane potential of a muscle fiber, called an *action potential* (see Chapter 10). However, when a neuron is stimulated just once, a full action potential rarely results. Instead, a small, local change in the membrane potential of the neuron, called a **local potential,** is produced (see Figure 11.14).

A local potential may have one of two effects:

● It may cause a depolarization in which positive charges enter the cytosol and make the membrane potential less negative (e.g., a change from −70 to −60 mV).
● Alternatively, it may cause a hyperpolarization in which either positive charges exit or negative charges enter the cytosol, which makes the membrane potential more negative (e.g., a change from −70 to −80 mV).

Local potentials are sometimes called *graded potentials* because they vary greatly in size—some produce a larger change in membrane potential than others. The degree of change in the membrane potential during a local potential depends on multiple factors, including length of stimulation, number of ion channels that open, and type(s) of ion channels that open. Another feature of local potentials is that they are reversible; on cessation of the stimulus that caused the ion channels to open, the neuron quickly returns to its resting potential. Local potentials are also *decremental* in nature: The changes in membrane potential they produce are small, and the current generated is lost across the membrane over the distance of a few millimeters. Consequently, local potentials cannot send signals over great distances, but are useful for short-distance signaling only (which is why they're called *local* potentials). However, even though they occur only over short distances, we will see in the next section that local potentials are vital triggers for action potentials, our long-distance signals.

(a) Depolarization: Gain of positive charges makes the inside of the cell less negative, causing depolarization.

(b) Hyperpolarization: Loss of positive charges (or gain of negative charges) makes the inside of the cell more negative, causing hyperpolarization.

Figure 11.14 Ion movements leading to changes in the membrane potential. The changes shown here are local potentials.

Play animation
@MasteringA&P®

☐ 9. Define local potential. Why is it also called a graded potential?
☐ 10. Why are local potentials useful only for short-distance signaling?

Action Potentials

◀◀ **Flashback**

1. What are negative and positive feedback loops? (p. 22)
2. What takes place during an action potential? (p. 350)

An **action potential** is a uniform, rapid depolarization and repolarization of the membrane potential of a cell (see Chapter 10). This change in the membrane potential causes a response of some sort. For a muscle fiber, the change initiates events that lead to muscle fiber contraction. Within the nervous system, signals are sent through an axon to another neuron, a muscle fiber, or a gland.

Recall that only axons generate action potentials; dendrites and cell bodies generate local potentials only. Action potentials are generated in a region called the *trigger zone,* which includes the axon hillock and the initial segment of the axon.

In this section we look at what happens during an action potential. First, however, we need to delve deeper into the function of the voltage-gated channels that allow ions to move and change the membrane potential of the neuron.

States of Voltage-Gated Channels

Two types of voltage-gated channels function in the depolarization and repolarization of the action potential—one for sodium ions and one for potassium ions. Voltage-gated channels are found most abundantly in the axolemma of the neuron, which is why only axons have action potentials.

The structures of voltage-gated potassium and sodium ion channels are depicted in **Figure 11.15**. Notice in Figure 11.15a that the voltage-gated potassium ion channel has two possible states: resting and activated. In the resting state, the channel is closed. In the activated state, the channel is open and allows potassium ions to cross the axolemma.

The voltage-gated sodium ion channel shown in Figure 11.15b is more complicated. It has two gates: an *activation gate* and an *inactivation gate.* This means a sodium ion channel has three potential "states":

- **Resting state: Inactivation gate opened, activation gate closed.** During the resting state the neuron is not being stimulated, and the activation gate is closed and the inactivation gate is open. No sodium ions cross the membrane when the channel is in the resting state.
- **Activated state: Both activation and inactivation gates opened.** When an action potential is initiated, the voltage change opens the activation gates and the channel is in its activated state. The channel in the activated state allows sodium ions to cross the axolemma.
- **Inactivated state: Inactivation gate closed, activation gate opened.** When the inactivation gate closes, the channel is in its inactivated state. The channel in this state no longer allows sodium ions to pass through. Observe that during this state, the activation gate remains open. When the action potential is finished, the channel returns to the resting state.

Events of an Action Potential

Let's examine the sequence of events of an action potential in a section of axon, illustrated in **Figure 11.16**. The entire sequence takes just a few milliseconds. Neuronal action potentials have

Resting state

ECF

Activation gate closed

Cytosol

K⁺

Activation gate open

Activated state

(a) The two states of a voltage-gated K⁺ channel

Resting state

Inactivation gate open — Activation gate closed

Inactivation gate closed — Activation gate open

Inactivated state

Na⁺

Activation gate open — Inactivation gate open

Activated state

(b) The three states of a voltage-gated Na⁺ channel

Figure 11.15 States of voltage-gated channels.

Trigger zone

① A local potential depolarizes the axolemma of the trigger zone to threshold.

ECF

Axolemma

Voltage-gated Na⁺ channel in resting state

Voltage-gated K⁺ channel in resting state

Local potential

Axoplasm

② Voltage-gated Na⁺ channels activate, Na⁺ enter, and the axon section depolarizes.

Na⁺

Membrane potential reverses.

③ Na⁺ channels inactivate and voltage-gated K⁺ channels activate, so Na⁺ stop entering and K⁺ exit the axon— repolarization begins.

K⁺

Na⁺ inactivation gate closed

Membrane potential returns to negative.

④ Na⁺ channels return to the resting state and repolarization continues.

Voltage-gated Na⁺ channel in resting state

K⁺

⑤ The axolemma may hyperpolarize before K⁺ channels return to the resting state; after this, the axolemma returns to the resting membrane potential.

Figure 11.16 Events of an action potential.

Play animation
@MasteringA&P®

three general phases: the depolarization phase, the repolarization phase, and the hyperpolarization phase. During the **depolarization phase,** the membrane potential rises toward zero and then becomes briefly positive. The membrane potential returns to a negative value during the **repolarization phase,** and then becomes temporarily more negative than resting during the **hyperpolarization phase.** Each phase occurs because of the selective opening and closing of specific ion channels. Note that before the action potential, when the membrane is at rest, the gates for both the sodium and the potassium ion channels are closed.

The action potential proceeds as follows:

① **A local potential depolarizes the axolemma of the trigger zone to threshold.** The action potential begins when the voltage-gated sodium ion channels in the axolemma of the trigger zone enter the activated (open) state (see Figure 11.15b). However, these voltage-gated channels will become activated only if the axon is depolarized. The source of this depolarization in the trigger zone is generally a local potential that arrives from the cell body. Note that the local potential must be strong enough to depolarize the axon to a level known as **threshold,** usually -55 mV.

② **Voltage-gated sodium ion channels activate, sodium ions enter, and the axon section depolarizes.** When threshold is reached, the sodium ion channels in the trigger zone are activated (open) and sodium ions rush into the neuron with their electrochemical gradient. As the membrane potential becomes more positive, more voltage-gated sodium ion channels are activated. This cycle continues, and the more the axon depolarizes, the more voltage-gated sodium ion channels are activated. This influx of positive charges causes rapid depolarization to about

CORE PRINCIPLE
Feedback Loops

$+30$ mV. You may recognize this as an example of a *positive* feedback loop—the initial input (activation of sodium ion channels and depolarization) amplifies the output (more sodium ion channels are activated and the axolemma depolarizes further), an example of the Feedback Loops Core Principle (p. 21).

③ **Sodium ion channels inactivate and voltage-gated potassium ion channels activate, so sodium ions stop entering and potassium ions exit the axon—repolarization begins.** When the axolemma is fully depolarized (about $+30$ mV), the inactivation gates of the voltage-gated sodium ion channels close, and sodium ions stop entering the axon. As this occurs, voltage-gated potassium ion channels slowly open and potassium ions flow out of the axon along their electrochemical gradient, causing the axolemma of the trigger zone to lose positive charges and so to begin repolarization.

④ **Sodium ion channels return to the resting state and repolarization continues.** As potassium ions exit the axon and repolarization continues, the activation gates of the sodium ion channels close and the inactivation gates open, returning the sodium ion channels to their resting state.

⑤ **The axolemma may hyperpolarize before potassium ion channels return to the resting state; after this, the axolemma returns to the resting membrane potential.** In many axons, the outflow of potassium ions continues until the membrane potential of the axolemma hyperpolarizes, possibly becoming as negative as -90 mV. The axolemma hyperpolarizes because the gates of the potassium ion channels are slow to close, allowing additional potassium ions to leak out of the cell. Hyperpolarization finishes as the voltage-gated potassium ion channels return to their resting state. After the action potential, the potassium leak channels and Na^+/K^+ pumps re-establish the resting membrane potential.

Throughout the preceding sequence of events of a single action potential, very little change occurs in the intracellular or extracellular concentration of sodium or potassium ions, and therefore the gradient isn't too disturbed. However, with repetitive action potentials, the gradient will eventually deplete, and the neuron relies on the Na^+/K^+ pumps in the axolemma to restore it. Read *A&P in the Real World: Local Anesthetic Drugs* to find out what happens when sodium ion channels are blocked on purpose.

Quick Check

☐ 11. What takes place during the depolarization phase of an action potential? How is it an example of a positive feedback loop?
☐ 12. What must be reached in order for voltage-gated sodium ion channels to open?
☐ 13. What takes place during the repolarization and hyperpolarization phases of an action potential?

The Refractory Period

Neurons are limited in how often they can fire action potentials. For a brief time after a neuron has produced an action potential, the membrane cannot be stimulated to fire another one. This time is called the **refractory period** (Figure 11.17). The refractory period may be divided into two phases: the *absolute refractory period* and the *relative refractory period*.

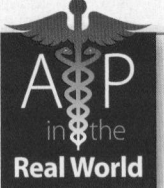
Local Anesthetic Drugs

Local anesthetics such as *lidocaine* are commonly administered agents that produce temporary numbness in an area, usually for a surgical or dental procedure. These drugs block the voltage-gated sodium ion channels of the neurons in the region where they are injected. Blocking these channels prohibits depolarization, and so action potentials relaying the impulses for pain are not transmitted to the CNS. Because agents such as lidocaine block sodium ion channels nonselectively, they will also affect the sodium ion channels of muscles in the area. This causes temporary weakness or paralysis of the affected muscle. As a result, you may be prone to drooling for a few hours (because of not being able to close your mouth completely) after leaving the dentist's office.

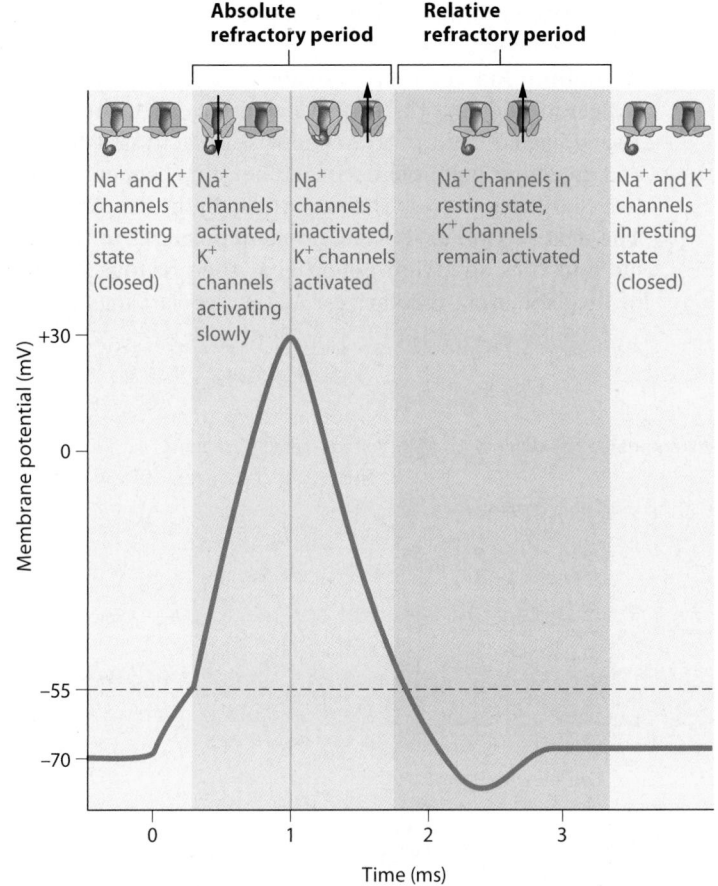

Figure 11.17 Refractory periods of an action potential.

Local and Action Potentials Compared

Now that we have discussed both local potentials and action potentials, let's highlight their differences. You discovered earlier that local potentials are graded, and so produce changes in membrane potential of varying degree; however, each action potential will cause a maximum depolarization of the same amount, to about $+30$ mV. This is due to a phenomenon called the **all-or-none principle.** Simply put, this principle refers to an event, in this case an action potential, that either happens completely or doesn't happen at all. If a neuron does not depolarize to threshold, an action potential does not occur. If the neuron does depolarize to threshold, the result is an action potential of a characteristic strength. The size of the action potential is not determined by the strength, frequency, or length of the stimulus, and therefore is not graded like a local potential.

The all-or-none principle leads us to a second difference between local potentials and action potentials: their reversibility. Recall that a local potential is reversible; once the stimulus stops, the ion channels close and the resting membrane potential is restored. However, a key feature of an action potential is that when one occurs, it is irreversible—once threshold is reached, it cannot be stopped and will proceed to completion.

Finally, a third important difference between local potentials and action potentials is the distance over which the signal travels. Whereas local potentials are decremental and decrease over short distances, action potentials are *nondecremental;* that is, their strength does not diminish. Without this property, signals could not be sent over long distances in the nervous system.

Quick Check

☐ 15. How do local potentials and action potentials differ?
☐ 16. Which is useful for long-distance signaling, and why?

Propagation of Action Potentials

A single action potential in one spot can't perform its main function, which is to act as a method of long-distance signaling. To do this, it has to be conducted, or **propagated,** down the length of the axon. This movement creates a flow of charged particles, a current. Action potentials are *self-propagating,* meaning that each action potential triggers another one in a neighboring section of the axon. You can imagine this process like a string of dominoes—when the first one is tipped over, the next one falls, which triggers the next to fall, and the process continues until the end of the line is reached. Only the first domino needs the "push," and once they start to fall, the process sustains itself until the end.

Action potential transmission occurs largely in one direction—from the trigger zone to the axon terminals—and at a constant speed. Propagation takes place in a single direction because the membrane in the previous section (behind the action potential) is still in the refractory period. Recall that the sodium channels in refractory parts of the membrane are in their inactivated state, which means that the wave of depolarization cannot trigger them to open.

During the **absolute refractory period,** no additional stimulus, no matter how strong, is able to produce an additional action potential. Notice in Figure 11.17 that this period coincides with the voltage-gated sodium ion channels being in their activated and inactivated states; sodium ion channels may not be activated until they return to their resting states with their activation gates closed and their inactivation gates open.

Immediately following the absolute refractory period is the **relative refractory period,** during which only a strong stimulus will produce an action potential. The relative refractory period is marked by a return of voltage-gated sodium ion channels to their resting state while some potassium ion channels remain activated. Because the potassium ion channels are activated and the membrane is repolarizing or even hyperpolarizing, it's difficult to depolarize the membrane to threshold and trigger an action potential. However, if a greater than normal stimulus is applied, the membrane may depolarize to threshold, and the axon may fire off another action potential.

The absolute and relative refractory periods limit the frequency of action potential production. In addition, the relative refractory period ensures that stronger stimuli trigger more frequent action potentials.

Quick Check

☐ 14. What are the absolute and relative refractory periods?

Propagation of action potentials down an axon forms the nerve impulse. In this section we look at how propagation occurs and the factors that influence the speed of action potential conduction.

Events of Propagation

The action potential is propagated along the axon by the following sequence of events, shown in **Figure 11.18**:

① **The axolemma depolarizes to threshold due to local potentials.** The axolemma of the trigger zone is depolarized to threshold by local potentials from the dendrites, cell body, or axon.

② **As sodium ion channels activate, an action potential is triggered and spreads positive charges down the axon.** Voltage-gated sodium ion channels are activated (open) and an action potential occurs. When this happens, positive charges flow down the axon through the axoplasm.

③ **The next section of the axolemma depolarizes to threshold and fires an action potential as the previous section of the axolemma repolarizes.** As the depolarizing current

① The axolemma depolarizes to threshold due to local potentials.

② As Na⁺ channels activate, an action potential is triggered and spreads down the axon.

③ The next section of the axolemma depolarizes to threshold and fires an action potential as the previous section of the axolemma repolarizes.

④ The current continues to move down the axon, and the process repeats.

Play animation
@MasteringA&P®

Figure 11.18 Propagation of an action potential.

reaches the next section of the axolemma, it depolarizes that section to threshold. The voltage-gated sodium ion channels in that part of the axolemma are activated, and that section then generates an action potential. The current flows down to the next section of the axon. Note that the section of the axolemma that had an action potential in step ② is repolarizing (its potassium channels are activated and potassium ions are exiting the axoplasm) and is in its refractory period, so any current that flows backward can't trigger an action potential.

④ **The current continues to move down the axon, and the process repeats.** The current flowing into the next section of the axolemma causes it to depolarize to threshold, activating voltage-gated sodium ion channels and producing an action potential there.

The process then repeats down the length of the axon until it reaches the axon terminals.

Conduction Speed

The rate at which propagation occurs is called **conduction speed,** and it determines how rapidly signaling can occur within the nervous system. Conduction speed is influenced by two main factors: the diameter of the axon and the presence or absence of a myelin sheath. The diameter of the axon affects the conduction of current through the axon because larger axons have lower resistance to conduction, and therefore current flows through them more easily.

The second determinant of conduction speed is the presence or absence of a myelin sheath. Two types of conduction can take place in an axon: **saltatory conduction,** in which the myelin sheath is present, and **continuous conduction,** in which it is absent (**Figure 11.19**). Recall from our discussion in Module 11.2 that myelin is an excellent insulator of electrical charge. Thus, the flow of current is far more efficient in a myelinated axon, which causes saltatory conduction to be significantly faster than continuous conduction.

Figure 11.19 compares conduction along a myelinated axon and an unmyelinated axon, and depicts the effect of the myelin sheath on conduction speed. You can see that on the myelinated axon, only the nodes of Ranvier must be depolarized to threshold. When the node, rich in voltage-gated sodium ion channels, is depolarized to threshold, an action potential is triggered. This action potential generates a current that flows passively and efficiently with little loss of charge through the next myelinated segment, or internode (see Figure 11.8 for a review of anatomical terminology). When the current reaches the next node of Ranvier, another action potential is generated. This cycle is repeated down the length of the axon, and the current "jumps" from one node to the next (in fact, "saltatory" comes from the Latin word *saltare,* which means "leaping"). See *A&P in the Real World: Multiple Sclerosis* on page 406 for information on how loss of the myelin sheath affects conduction speed.

Compare these "leaping" action potentials with the much slower and more gradual continuous conduction seen in the unmyelinated axon. Figure 11.19 shows that in continuous conduction, absence of the myelin sheath means that each section of the axolemma must be depolarized to threshold. This is why conduction of this sort is called *continuous*—action potentials must be generated in a continuous sequence along the entire axolemma for the current to spread down the length of the axon. The flow of current in a myelinated axon is much faster than the process of triggering action potentials in each part of an unmyelinated axon.

Saltatory conduction in myelinated axon

Depolarizing membrane

Resting membrane

Myelinated axon

① First action potential   ② Second action potential   ③ Third action potential

Continuous conduction in unmyelinated axon

Depolarizing membrane

Resting membrane

Unmyelinated axon

① First action potential   ② Second action potential   ③ Third action potential

Figure 11.19 Comparison of saltatory and continuous conduction.

ConceptBOOST »»»

How Does Myelin Insulate an Axon and Increase Its Speed of Propagation?

We just discussed how myelin increases the speed of action potential propagation by "insulating" the charge and enabling saltatory conduction. But how, exactly, is charge insulated, and why does this make propagation occur more quickly? To understand this, we have to go back to some of the basic principles of electricity. Let's first think of the electrical current flowing down a bare copper wire that is going to a light bulb. Ideally, the current flows directly down the wire and illuminates the light bulb, as shown here:

Electric current

Wire

But, if you touch the wire with a metal probe, as depicted in the next illustration, most of the current might instead flow down the probe, a situation known as a *short circuit*:

Probe

Short circuit

However, if the wire is encased in a material that is a poor conductor of electricity, as shown in this final illustration, the current is *insulated* and unable to move from the copper wire to the probe. This prevents a short circuit.

Probe

Insulation

Now let's apply this to axons. An unmyelinated axon most closely resembles the wire in the middle illustration. The axolemma is very leaky with respect to current, and so the current flows easily from the axoplasm to the extracellular fluid, just as current flowed easily from the copper wire to the metal probe. Remember from our earlier discussion that the strength of an electrical signal decreases as the ions that make up the current leak across the membrane. So in an unmyelinated axon, the current generated by the action potential dissipates over a short distance, which could cause the action potential to fail. For this reason, the action potential must constantly be regenerated along the length of the axolemma. This requires the opening of voltage-gated ion channels, which takes time, so propagation is much slower.

A myelinated axon more closely resembles the wire in the final illustration. Myelin is a very good insulator because it is a poor conductor of electricity, and so it prevents current from leaking out through the axolemma. This means that the signal decreases very little in strength as it travels through an internode. The action potential can then propagate through the internode without having to be regenerated. It is not until the current reaches the unmyelinated node of Ranvier that it starts to dissipate and the action potential must be regenerated. This is why the action potentials of saltatory conduction can "leap" from node to node and why this type of propagation is so much faster than continuous conduction. ■

Classification of Axons by Conduction Speed

Axons are often classified according to conduction speed. The two primary defining features in this classification are the diameter of the axon and the presence or absence of the myelin sheath. The three main classes are as follows:

- **Type A fibers** are the largest-diameter axons (5–20 μm), all of which are myelinated. These characteristics give them the fastest conduction speed, with a maximum velocity of about 120 m/sec (about 250 mi/h). Type A fibers are found in parts of the body with which the CNS must communicate extremely rapidly, such as certain sensory axons from joints and muscle fibers, as well as motor axons to skeletal muscles.
- **Type B fibers** are intermediate in diameter (2–3 μm) and most are myelinated. They typically have a maximum conduction speed of about 15 m/sec (about 32 mi/h), and include certain efferent fibers of the autonomic nervous system and certain sensory axons coming from organs.
- **Type C fibers** are the smallest fibers (0.5–1.5 μm) and are unmyelinated. Their conduction velocity is the slowest, conducting action potentials only at about 0.5–2 m/sec (about 1–5 mi/h). Type C fibers include other efferent fibers of the autonomic nervous system and certain sensory axons that transmit pain, temperature, and certain pressure sensations.

Quick Check

☐ 17. How is an action potential propagated down an axon?
☐ 18. How do saltatory conduction and continuous conduction differ? Which is faster, and why?

Putting It All Together: The Big Picture of Action Potentials

After all of this talk of action potentials, you may still find some of these concepts to be rather abstract. Before moving on to the next module, make sure you think about the big picture of

① **Local potentials:** Local potentials usually generated in the neuron's dendrites accumulate and reach the trigger zone of the axon (see Figure 11.14).

Dendrite

Local potentials

Trigger zone

Na^+ AP

Direction of AP

② **Action potential:** The trigger zone depolarizes to threshold and generates an action potential (AP) (see Figure 11.16).

Axon terminals

③ **Action potential propagation:** The action potential is propagated down the axon to the axon terminals (see Figure 11.18).

AP   AP   AP   AP

Voltage-gated Na^+ channel

Axon

Na^+

---

how and why the action potential occurs. The *how* of the action potential is shown in **Figure 11.20**. The *why* of the action potential is long-distance signaling. As you will see in the upcoming module, the arrival of the action potential at the axon terminal is what allows the neuron to communicate with its target cells.

Apply What You Learned

☐ 1. Predict the effect of the poison ouabain (way-BAH-in), which blocks Na^+/K^+ pumps, on the neuronal action potential. (*Hint:* What would happen to the sodium and potassium ion gradients?)

☐ 2. What do you think would happen to the neuronal action potential if the concentration of sodium ions in the extracellular fluid decreased significantly, to the point of reversing the gradient?

☐ 3. Sometimes when you pull your dinner out of the microwave you have to hold your fingertips to the food for a second or two before you can tell if it is hot or cold. Explain why this happens. (*Hint:* What type of fiber carries temperature information?)

See answers in Appendix A.

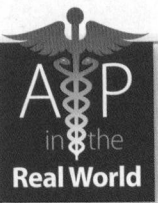

Multiple Sclerosis

Multiple sclerosis (MS) is a disease in which certain cells of the immune system attack the myelin sheath around axons of the CNS. This disease is therefore a type of autoimmune disorder (one in which the patient's own immune system attacks a certain part of the body). Why these attacks happen is unknown, but in most cases they cause a progressive loss of the myelin sheath, which in turn produces loss of current from the neurons. Symptoms of the disease result from progressive slowing of action potential propagation as saltatory conduction becomes less efficient.

The exact symptoms of MS depend on the regions of the CNS that are affected. Over time most patients exhibit changes in sensation (e.g., numbness), alterations in behavior and cognitive abilities, and motor dysfunction including paralysis, all of which may cause significant disability.

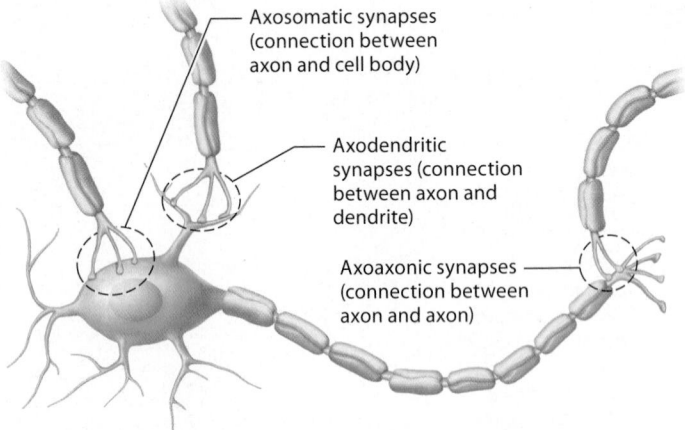

Figure 11.21 Structural types of synapses.

Labels in figure:
- Axosomatic synapses (connection between axon and cell body)
- Axodendritic synapses (connection between axon and dendrite)
- Axoaxonic synapses (connection between axon and axon)

MODULE **11.4**

Neuronal Synapses

Learning Outcomes

1. Compare and contrast electrical and chemical synapses.
2. List the structures that make up a chemical synapse.
3. Discuss the relationship between a neurotransmitter and its receptor.
4. Describe the events of chemical synaptic transmission in chronological order.
5. Define excitatory postsynaptic potential (EPSP) and inhibitory postsynaptic potential (IPSP), and interpret graphs showing the voltage-versus-time relationship of an EPSP and an IPSP.
6. Explain temporal and spatial summation of synaptic potentials.

Up to this point, we have discussed how signals are generated and propagated *within* a neuron. Remember, though, that neurons must communicate with other cells, including other neurons, in order to carry out their functions—an example of the Cell-Cell Communication Core Principle (p. 27). Therefore, in this module, we will discuss how signals are transmitted *between* neurons at locations called *synapses*. Recall that a **synapse** (SIN-aps; *syn-* = "to clasp or join") is where a neuron meets its target cell (see Chapter 10). The discussion in that chapter revolved around a specific type of synapse—the neuromuscular junction. Here we explore the synapses that occur between two neurons, or *neuronal synapses*. These synapses may be of

CORE PRINCIPLE
Cell-Cell Communication

two types, *electrical* and *chemical*. We examine both types in this module, although we focus on chemical synapses. The module concludes with a look at *neural integration,* or the way in which the many synapses of a neuron impact its integrative processes.

Overview of Neuronal Synapses

Neuronal synapses generally occur between an axon and another part of a neuron; they may occur between an axon and a dendrite, an axon and a cell body, and an axon and another axon. These types are called *axodendritic, axosomatic,* and *axoaxonic synapses,* respectively (**Figure 11.21**).

Regardless of the type of synapse, we use certain terms to describe the neurons sending and receiving the message:

- **Presynaptic neuron.** The presynaptic neuron is the neuron that is sending the message from its axon terminal.
- **Postsynaptic neuron.** The postsynaptic neuron is the neuron that is receiving the message from its dendrite, cell body, or axon.

The transfer of chemical or electrical signals between neurons at a synapse is called **synaptic transmission,** and it is the fundamental process for most functions of the nervous system. Synaptic transmission allows voluntary movement, cognition, sensation, and emotion, as well as countless other processes. Each neuron has an enormous number of synapses. Recall from Module 11.2 that each axon generally splits into 1000 or more axon terminals, and each terminal meets up with another axon, dendrite, or cell body. So an average presynaptic neuron, then, generally forms synapses with about 1000 postsynaptic neurons. A postsynaptic neuron can receive input from even more synapses—an average neuron can have as many as 10,000 synaptic connections from different presynaptic neurons.

Quick Check

☐ 1. What are the three most common locations where presynaptic axons connect with a postsynaptic neuron?
☐ 2. Define synaptic transmission.

Electrical Synapses

An **electrical synapse** (Figure 11.22a) occurs between cells that are electrically coupled via gap junctions. Observe that in these synapses the axolemmas of the two neurons are nearly touching (they are separated by only about 3.5 nm) and that the gap junctions contain precisely aligned channels that form pores through which ions and other small substances may travel. This allows the electrical current to flow directly from the axoplasm of one neuron to that of the next.

This arrangement creates two unique features of electrical synapses:

- **Synaptic transmission is bidirectional.** In an electrical synapse, transmission is usually bidirectional, which means that either neuron may act as the presynaptic or the postsynaptic neuron and that current may flow in either direction between the two cells.
- **Synaptic transmission is nearly instantaneous.** The delay between depolarization of the presynaptic neuron and change in potential of the postsynaptic neuron is less than 0.1 ms (millisecond), which is extraordinarily fast (we will see that transmission at most chemical synapses requires from one to a few milliseconds).

These features of electrical synapses allow the activity of a group of cells to be synchronized—when stimulated, the cells will produce action potentials in unison. Electrical synapses are found primarily in areas of the brain that are responsible for programmed, automatic behaviors such as breathing. They are also present in developing nervous tissue in the embryo and fetus and are thought to assist in the development of the brain. In addition, electrical synapses are found outside the nervous system in locations such as cardiac and visceral smooth muscle, where they allow those tissues to engage in coordinated muscle activity.

Quick Check

☐ 3. What are the two main features of an electrical synapse?

Chemical Synapses

◄◄ Flashback

1. What is a synaptic vesicle? (p. 352)
2. What is a neurotransmitter? (p. 352)

The vast majority of synapses in the nervous system are **chemical synapses.** These synapses are more common because they are more efficient—the current in electrical synapses eventually becomes weaker as it dissipates into the extracellular fluid. A chemical synapse, in contrast, converts an electrical signal into a controlled chemical signal, so there is no loss of strength. The chemical signal is reconverted into an electrical signal in the postsynaptic neuron. In the upcoming sections, we explore how this takes place. But first let's look a little more closely at the differences between chemical and electrical synapses.

(a) Electrical synapse

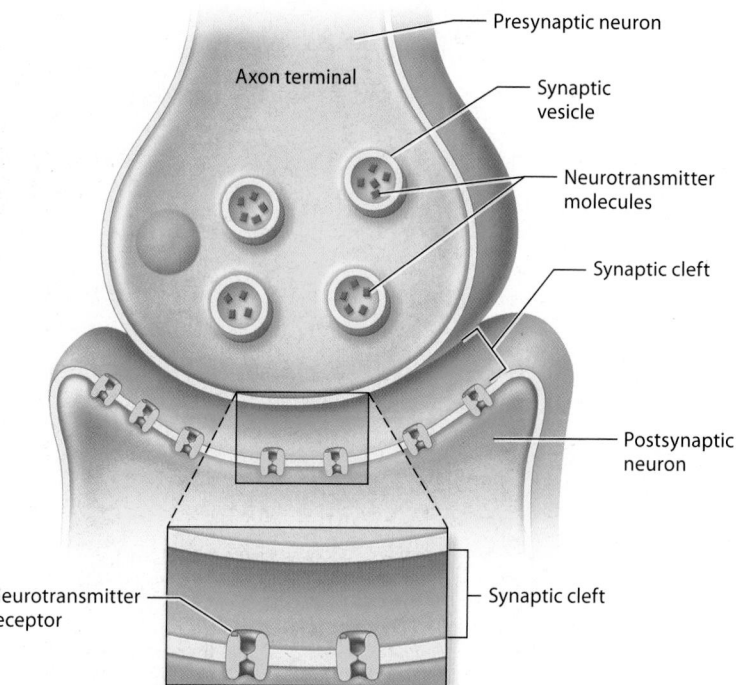

(b) Chemical synapse

Figure 11.22 The structures of electrical and chemical synapses.

Practice art labeling
@MasteringA&P®

Electrical and Chemical Synapses Compared

Figure 11.22b shows the structure of a chemical synapse. We can point out three important structural differences between a chemical and an electrical synapse:

- **Synaptic vesicles.** The axon terminal of the presynaptic neuron of every chemical synapse houses **synaptic vesicles.** These vesicles contain chemical messengers called **neurotransmitters** that transmit signals from the presynaptic to the postsynaptic neuron.
- **Synaptic cleft.** Whereas the cells of an electrical synapse are electrically connected by gap junctions, the cells of a chemical synapse are separated by a larger but still microscopic space called the **synaptic cleft.** The synaptic cleft measures 20–50 nm and is filled with extracellular fluid.
- **Neurotransmitter receptors.** In chemical synapses the postsynaptic neuron must have **receptors** for the neurotransmitters that the presynaptic neuron releases or it cannot respond to the signal being transmitted. Receptors are generally linked either directly or indirectly to ion channels.

These three features of chemical synapses cause them to transmit signals more slowly than do electrical synapses. In fact, there is about a 0.5-ms gap between the arrival of the action potential at the axon terminal and the effects on the postsynaptic membrane, known as **synaptic delay.** Also, chemical synapses are unidirectional—the message can be sent only by the presynaptic neuron. However, these three structural differences also allow something not permitted by the structure of the electrical synapse: The signal can vary in size. If more neurotransmitters are released, then the presynaptic neuron has a greater effect on the postsynaptic neuron. The signal in an electrical synapse, by contrast, will always be the same size. In addition, the effect that the presynaptic neuron triggers can vary with different neurotransmitters and receptors.

Events at a Chemical Synapse

The neuromuscular junction is a type of chemical synapse, and although some of the terms are different, the events occurring at a neuronal chemical synapse are similar to those occurring at the neuromuscular junction. Under most conditions, the action potential reaches the axon terminal and triggers the release of neurotransmitters, which bind to receptors and cause changes in the membrane potential of the postsynaptic neuron.

However, neuronal synapses are more complicated than neuromuscular synapses for several reasons. For one, a muscle fiber receives input from only a single axon, whereas neurons may receive input from hundreds or even thousands of axons. Another difference is the relationships of the neurotransmitters

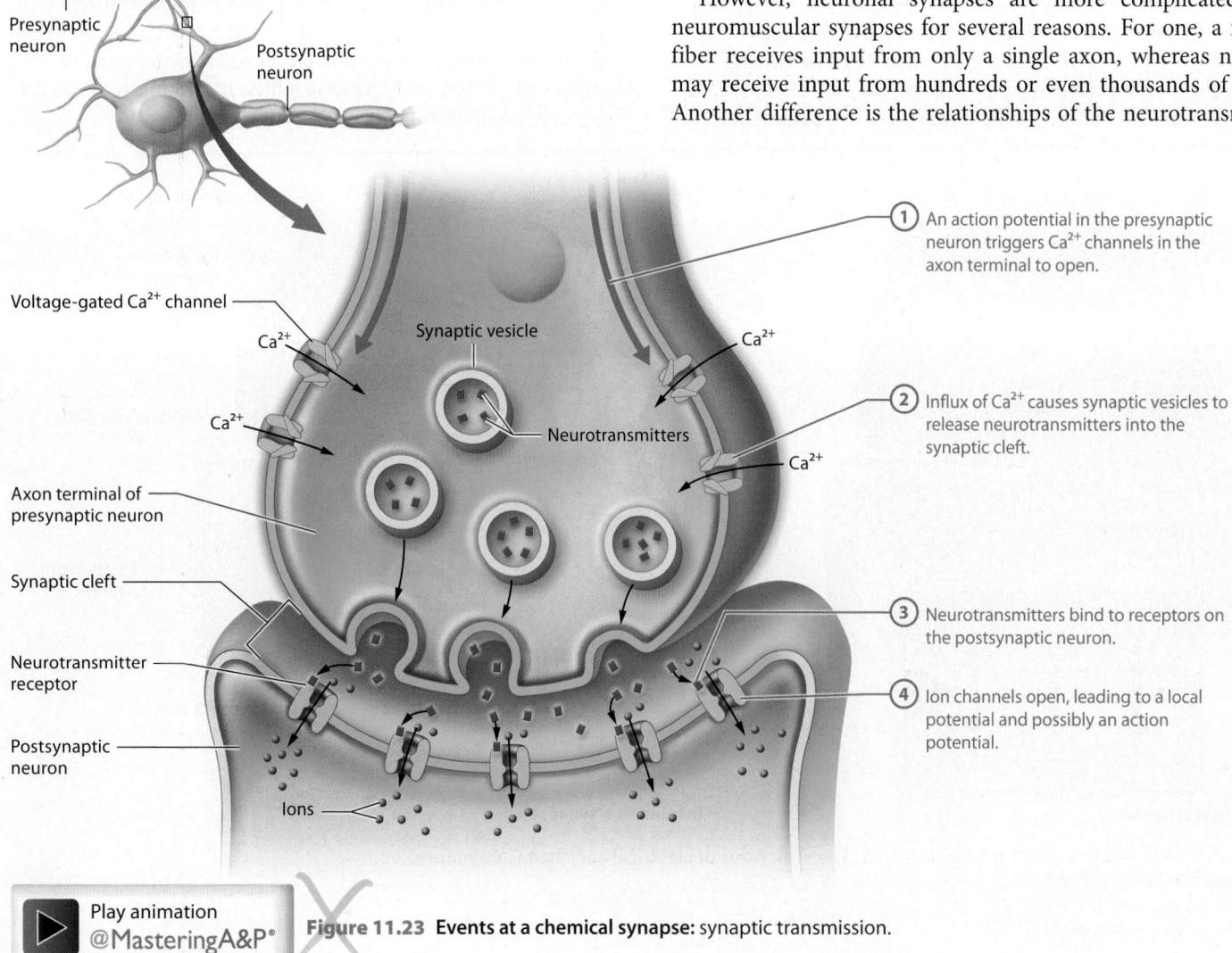

Presynaptic neuron

Postsynaptic neuron

① An action potential in the presynaptic neuron triggers Ca^{2+} channels in the axon terminal to open.

② Influx of Ca^{2+} causes synaptic vesicles to release neurotransmitters into the synaptic cleft.

③ Neurotransmitters bind to receptors on the postsynaptic neuron.

④ Ion channels open, leading to a local potential and possibly an action potential.

Voltage-gated Ca^{2+} channel

Ca^{2+}

Synaptic vesicle

Ca^{2+}

Ca^{2+}

Neurotransmitters

Ca^{2+}

Axon terminal of presynaptic neuron

Synaptic cleft

Neurotransmitter receptor

Postsynaptic neuron

Ions

Play animation @MasteringA&P®

Figure 11.23 Events at a chemical synapse: synaptic transmission.

to their target cells. At a neuromuscular synapse, the only neurotransmitter used is *acetylcholine,* and acetylcholine serves only to excite the muscle fiber. However, in neuronal synapses, a variety of neurotransmitters may be released to excite or inhibit the postsynaptic neuron. Finally, at the neuromuscular junction, each action potential in the neuron triggers an action potential in the muscle fiber. A postsynaptic neuron, however, may require the input of 100 or more presynaptic neurons to trigger an action potential.

Let's now dig into what actually takes place at a neuronal synapse. **Figure 11.23** depicts the following events:

① **An action potential in the presynaptic neuron triggers calcium ion channels in the axon terminal to open.** An action potential reaches the axon terminal of the presynaptic neuron, which triggers the opening of voltage-gated calcium ion channels in its axolemma.

② **Influx of calcium ions causes synaptic vesicles to release neurotransmitters into the synaptic cleft.** Calcium ions enter the axon terminal, causing synaptic vesicles in the area to fuse with the presynaptic membrane. This releases neurotransmitters into the synaptic cleft via exocytosis.

③ **Neurotransmitters bind to receptors on the postsynaptic neuron.** The neurotransmitters diffuse across the synaptic cleft, where they bind to neurotransmitter receptors on the membrane of the postsynaptic neuron.

④ **Ion channels open, leading to a local potential and possibly an action potential.** The binding of neurotransmitters to receptors generally either opens or closes ligand-gated ion channels in the postsynaptic membrane, resulting in a local potential. Such local potentials may or may not lead to an action potential in the postsynaptic neuron.

Postsynaptic Potentials

Local potentials in the membrane of the postsynaptic neuron, which are called **postsynaptic potentials,** can move the membrane either closer to or farther away from threshold. Therefore, depending on which channels are opened, one of two events may occur (**Figure 11.24**):

- **The membrane potential of the postsynaptic neuron moves closer to threshold.** A small, local depolarization called an **excitatory postsynaptic potential (EPSP)** occurs, which brings the membrane of the postsynaptic neuron closer to threshold (Figure 11.24a). If the membrane potential reaches threshold, an action potential is triggered.
- **The membrane potential of the postsynaptic neuron moves away from threshold.** A small, local hyperpolarization known as an **inhibitory postsynaptic potential (IPSP)** occurs, moving the membrane of the postsynaptic neuron farther away from threshold, and so tending to inhibit an action potential from firing (Figure 11.24b).

EPSPs typically result from the opening of ion channels such as those for sodium or calcium ions and the entrance of positive charges into the postsynaptic neuron (see Figure 11.24a). A single EPSP produces only a very small, local potential. However, each successive EPSP makes the membrane more depolarized and so more likely to reach threshold and fire an action potential.

Note in Figure 11.24b that an IPSP typically results from either the opening of potassium ion channels, which causes a loss of positive charges, or the opening of chloride ion channels. Chloride ions are more abundant in the extracellular fluid than in the neuron, so when chloride ion channels open, these anions enter the neuron and make the membrane potential more negative. Opening of either type of channel will yield the same result:

Ion channels open and positive charges enter the postsynaptic neuron, causing an excitatory postsynaptic potential (EPSP).

Neurotransmitter — Ligand-gated Na⁺ channel

Postsynaptic neuron — Ligand-gated Ca²⁺ channel

Ligand-gated K⁺ channel

OR

Ligand-gated Cl⁻ channel

Ion channels open and the postsynaptic neuron gains negative charges or loses positive charges, causing an inhibitory postsynaptic potential (IPSP).

Postsynaptic neuron

Membrane potential (mV)

Small depolarization (EPSP)

Threshold

Time (ms)

(a) The membrane potential of the postsynaptic neuron moves closer to threshold.

Membrane potential (mV)

Small hyperpolarization (IPSP)

Threshold

Time (ms)

(b) The membrane potential of the postsynaptic neuron moves away from threshold.

▶ Play animation @MasteringA&P®

Figure 11.24 Postsynaptic potentials.

The membrane potential moves farther away from threshold, and an action potential becomes less likely.

Termination of Synaptic Transmission

We move now to the final step of synaptic transmission—termination. But why terminate synaptic transmission? Once presynaptic neurons have triggered EPSPs and/or IPSPs and so generated a specific response in the postsynaptic neuron, the response cannot be initiated again until the postsynaptic neuron stops being stimulated. The neurons involved in breathing provide a simple example. When we need to inhale, specific neurons are stimulated to trigger our respiratory muscles to contract. Once we have taken a breath, our nervous system needs to stop stimulating these neurons or we will continue to inhale. This is accomplished by stopping synaptic transmission.

The messenger of synaptic transmission is the neurotransmitter released by the presynaptic neuron. Therefore, synaptic transmission may be terminated by ending the effects of the neurotransmitter. In general, this happens in three ways (**Figure 11.25**):

- **Diffusion and absorption.** Some neurotransmitters simply diffuse away from the synaptic cleft through the extracellular fluid, where they diffuse through the plasma membrane of a neuron or astrocyte and are then returned to the presynaptic neuron.
- **Degradation in the synaptic cleft.** Certain neurotransmitters are broken down by enzymes that reside in the synaptic cleft. The components of the destroyed neurotransmitter are often then taken back up by the presynaptic neuron and resynthesized into the original neurotransmitter.
- **Reuptake into the presynaptic neuron.** Some neurotransmitters are removed by a process called *reuptake,* in which

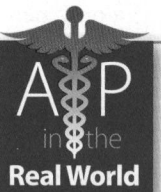

Arthropod Venom

Within the United States live several species of venomous arthropods, including spiders and scorpions. Many of their venoms affect neuronal synapses and are therefore called *neurotoxins.*

One of the most notorious venomous spiders is the female black widow spider (*Latrodectus mactans*), which produces a neurotoxic venom. This toxin causes massive release of neurotransmitters from presynaptic neurons, leading to repetitive stimulation of postsynaptic neurons.

The United States is also home to about 40 species of scorpions, the most lethal of which is the bark scorpion. The venom of the bark scorpion prohibits sodium ion channels in the postsynaptic neuron from closing, causing the neuron to remain depolarized and so to continue to fire action potentials.

Although the toxins operate by different means, they produce similar effects because both lead to overstimulation of postsynaptic neurons. Common symptoms include muscle hyperexcitability, sweating, nausea and vomiting, and difficulty in breathing. Treatment and prognosis for both types of bites depend on the amount of venom received and the availability of medical care. Severe cases generally require an *antivenin* to block the effects of the toxin.

proteins in the axolemma of the presynaptic neuron transport them back into the presynaptic neuron. Depending on their type, these neurotransmitters may be repackaged into synaptic vesicles or degraded by enzymes.

Once the neurotransmitter has been removed from the synaptic cleft, synaptic transmission is complete. See *A&P in the Real World: Arthropod Venom* to learn what happens when continued synaptic transmission causes postsynaptic neurons to be overstimulated.

Quick Check

- ☐ 4. What is the role of calcium ions in a chemical synapse?
- ☐ 5. How do the two types of postsynaptic potentials differ?
- ☐ 6. How is synaptic transmission terminated?

Putting It All Together: The Big Picture of Chemical Synaptic Transmission

At this point we've discussed the particulars of synaptic transmission at a chemical synapse: how the action potential triggers the release of neurotransmitters from the presynaptic neuron, how the neurotransmitters induce an EPSP or IPSP in the postsynaptic neuron, and how transmission is ended. Now we can summarize the whole process, as shown in **Figure 11.26**.

Diffusion and Absorption
Neurotransmitters diffuse away from the synaptic cleft and are returned to the presynaptic neuron.

Presynaptic neuron

Synaptic cleft

Postsynaptic neuron

Degradation
Neurotransmitters are degraded by enzymatic reactions in the synaptic cleft.

Reuptake
Neurotransmitters are taken back into the presynaptic neuron.

Play animation
@MasteringA&P®

Figure 11.25 Methods of termination of synaptic transmission.

Play animation
@MasteringA&P®

① **Action potential:** An action potential reaches the axon terminal of the presynaptic neuron.

Action potential

Axon terminal of presynaptic neuron

Ca²⁺

Neurotransmitters

Synaptic cleft

Ca²⁺

② **Synaptic transmission:** Ca²⁺ channels open in the presynaptic neuron; neurotransmitters are released from synaptic vesicles and bind to receptors on the postsynaptic neuron (see Figure 11.23).

Open receptor/ ion channel

Postsynaptic neuron

OR

EPSP
Threshold

IPSP
Threshold

③ **Postsynaptic potentials:** Neurotransmitters trigger an EPSP or IPSP, moving the membrane potential of the postsynaptic neuron either closer to or farther from threshold (see Figure 11.24).

Diffusion

Reuptake

④ **Termination of synaptic transmission:** Neurotransmitter concentration in the synaptic cleft decreases and synaptic transmission is terminated (see Figure 11.25).

Degradation

Closed receptor/ ion channel

Neural Integration: Summation of Stimuli

In Module 11.1 you read that the integrative functions of the nervous system occur in the neurons of the CNS. Put simply, integration refers to the process of putting together all of the stimuli that impact a neuron and either excite or inhibit the firing of an action potential. A neuron very rarely receives input from a single source; rather, it receives input from multiple other neurons, each of which influences whether or not it generates an action potential. To complicate matters, synaptic transmission in the CNS occurs continuously for most neurons—they are constantly bombarded by synaptic inputs from hundreds to thousands of presynaptic neurons. Additionally, the input from each presynaptic neuron may be different; the input may be excitatory or inhibitory, and the strength and location of each input may vary.

The postsynaptic neuron integrates all of this information into a single effect by a process known as **neural integration.**

As we discussed, a single EPSP produces only a small, local potential and overall has very little effect on the ability of the axolemma to depolarize to threshold and fire an action potential. Recall that most synapses connect to dendrites (axodendritic) or the cell body (axosomatic) of the postsynaptic neuron; however, the trigger zone of the axon is where an action potential is generated. Therefore, many EPSPs are required to generate a large enough change in membrane potential to impact the neuron all the way from its dendrites to its axonal trigger zone. This phenomenon of adding the input from several postsynaptic potentials to affect the membrane potential at the trigger zone is known as **summation.**

ConceptBOOST »»»

How Summation Connects Local Potentials and Action Potentials

By now, you know that local potentials are initiated when neurotransmitters bind to ligand-gated cation channels in the postsynaptic membrane. You also know that an action potential is initiated when the trigger zone is depolarized to threshold and voltage-gated sodium ion channels open in the axolemma. Now we can connect the dots between these two events via summation (**Figure 11.27**). As you can see, the link between local potentials and action potentials is summation—as excitatory local potentials summate, they depolarize the trigger zone to threshold and initiate an action potential. ■

1. Action potentials reach the axon terminals of presynaptic neurons. The released neurotransmitters open ligand-gated cation channels in the postsynaptic membrane.

2. Open ligand-gated cation channels allow cations to flow into the postsynaptic neuron. This results in an EPSP—a local potential.

3. The depolarization of the EPSP spreads away from the point of stimulation toward the axon hillock.

4. If enough local potentials summate at the axon hillock to reach threshold, voltage-gated Na^+ channels open and an action potential results.

Figure 11.27 Local potentials summating and leading to an action potential.

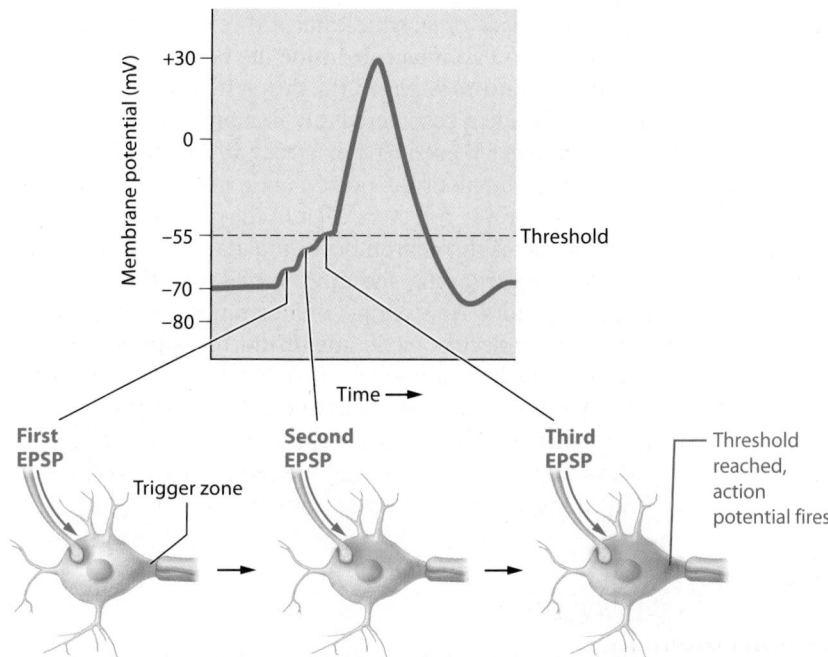

(a) **Temporal summation:** Successive EPSPs bring the trigger zone to threshold.

(b) **Spatial summation:** Simultaneous EPSPs from multiple neurons bring the trigger zone to threshold.

Figure 11.28 **Temporal and spatial summation of excitatory postsynaptic potentials (EPSPs).**

There are two types of summation, temporal and spatial, which differ in the timing of neurotransmitter release and number of presynaptic neurons, respectively. The first, **temporal summation,** occurs when neurotransmitter is released repeatedly from the axon terminal of a single presynaptic neuron (**Figure 11.28a**). The resulting EPSPs propagate toward the trigger zone. Each EPSP is very quick, lasting no more than about 15 ms. For this reason, the EPSPs must occur rapidly in succession in order for temporal summation to occur, or the trigger zone will not reach threshold.

The second type of summation, **spatial summation,** involves the simultaneous release of neurotransmitters from the axon terminals of multiple presynaptic neurons (**Figure 11.28b**). In Figure 11.28 the graph of temporal summation shows a staircase-like rise in membrane potential, whereas the membrane potential in the graph of spatial summation shows a smooth rise as large quantities of neurotransmitters are released at once.

Spatial summation can combine with temporal summation. When several presynaptic neurons fire together and trigger EPSPs in the postsynaptic neuron, the membrane at the trigger zone depolarizes, and the potential approaches threshold. The closer the membrane potential gets to threshold due to spatial summation of those EPSPs, the more likely it becomes that the next EPSP will trigger an action potential due to temporal summation, even if the stimulus is smaller.

Although we have discussed summation of EPSPs, IPSPs can summate both temporally and spatially as well. With summation of IPSPs, the postsynaptic neuron becomes less and less likely to fire an action potential. Additionally, IPSPs and EPSPs can summate. The overall result of this will depend on the individual strength of the IPSP and EPSP—if the IPSP is stronger, the membrane potential will hyperpolarize slightly, and if the EPSP is stronger, the membrane potential will depolarize slightly.

Quick Check

☐ 7. How do temporal summation and spatial summation differ?

Apply What You Learned

☐ 1. Predict how a poison that blocks calcium ion channels in the axon terminal would affect synaptic transmission.

☐ 2. A new drug opens calcium ion channels in the membrane of the postsynaptic neuron. Would this produce an EPSP or an IPSP? Would this make an action potential more or less likely to occur?

☐ 3. Explain how you could increase the likelihood that a certain neuron will reach threshold and have an action potential. (*Hint:* Think about the different types of summation.)

See answers in Appendix A.

MODULE **11.5**

Neurotransmitters

Learning Outcomes

1. Explain how a single neurotransmitter may be excitatory at one synapse and inhibitory at another.
2. Describe the structural and functional properties of the major classes of neurotransmitters.
3. Describe the most common excitatory and inhibitory neurotransmitters in the CNS.

The search for new neurotransmitters is still going on in laboratories around the world every day. The precise number of neurotransmitters operating in the human nervous system is not yet known but is well above 100, and they are diverse in structure and function. Despite their diversity, however, they share similar features. For one, they nearly all undergo a similar pattern of use: They are made in either the cell body or the axon terminal and packaged into synaptic vesicles, they are released from the presynaptic neuron, they bind to their receptors on the postsynaptic membrane, and finally their effects are often rapidly terminated through removal and/or degradation. In this module we will explore the properties and effects of the major neurotransmitters and their receptors.

Neurotransmitter Receptors

Nearly all neurotransmitters induce postsynaptic potentials by binding to their receptors in the postsynaptic membrane. The type of receptor to which a neurotransmitter binds determines the postsynaptic response. Two types of neurotransmitter receptors have been identified: ionotropic and metabotropic (**Figure 11.29**):

- *Ionotropic receptors* (eye-ON-oh-troh′-pik) are simply receptors that are part of ligand-gated ion channels. They are called ionotropic because they directly control the movement of ions into or out of the neuron when bound by a neurotransmitter. Neurotransmitters that bind ionotropic receptors have very rapid but short-lived effects on the membrane potential of the postsynaptic neuron.
- *Metabotropic receptors* (meh-TAB-oh-troh′-pik) are receptors within the plasma membrane that are connected to a separate ion channel in some fashion. They are called metabotropic because they are directly connected to metabolic processes that begin when they are bound by neurotransmitters. Most are connected through a group of intracellular enzymes called **G-proteins.** When the neurotransmitter molecule binds to the receptor, it activates one or more G-proteins and

begins a cascade of enzyme-catalyzed reactions that ends in the formation of a molecule inside the postsynaptic neuron, called a **second messenger** (in this system the neurotransmitter molecule is considered the "first messenger"). The second messenger then opens or closes an ion channel in the plasma membrane of the postsynaptic neuron. The changes that metabotropic receptors elicit in the membrane potential of the postsynaptic neuron occur much more slowly, but are typically longer-lasting and more varied than those of ionotropic receptors. An example of a common second messenger is the molecule **cyclic adenosine monophosphate** (or **cAMP**), which is derived from ATP. In the neuron, cAMP has multiple functions, including binding a group of enzymes that add phosphate groups to ion channels, triggering them to open or close. Second messengers are covered more fully in the endocrine chapter (see Chapter 16).

Quick Check

☐ 1. What are the two classes of neurotransmitter receptors?
☐ 2. What are G-proteins and second messengers?

Major Neurotransmitters

◀◀ **Flashback**

1. What are amino acids and peptides? (pp. 55–56)
2. What are ATP and adenosine? (p. 59)
3. What effect does acetylcholine have on the motor end plate of a muscle fiber? (p. 353)

Regardless of the type of receptor that a neurotransmitter binds, that binding leads to either EPSPs or IPSPs. Neurotransmitters that induce EPSPs in the postsynaptic neuron are said to have **excitatory** effects; those that induce IPSPs have **inhibitory** effects. Most neurotransmitters can have both excitatory and

(a) Ionotropic receptor

(b) Metabotropic receptor

Figure 11.29 Types of neurotransmitter receptors.

inhibitory effects, depending on which postsynaptic neuron receptors they bind. In fact, a single neurotransmitter can have several receptor types. This makes a purely functional classification of neurotransmitters difficult. For this reason the major neurotransmitters operating within the nervous system are usually classified into four groups by their chemical structures, which we will now explore.

For quick reference, **Table 11.3** summarizes the location, function, and effects of selected neurotransmitters.

Acetylcholine

The best-studied, and one of the most widely used neurotransmitters by the nervous system overall, is the small-molecule neurotransmitter **acetylcholine (ACh)** (ah-seet'l-KOH-leen). Synapses that use ACh, called **cholinergic synapses,** are located at the neuromuscular junction, within the brain and spinal cord, and within the autonomic nervous system (ANS). Its effects are largely excitatory; however, it does exhibit inhibitory effects at some PNS synapses.

ACh is synthesized from the precursors choline and acetyl-CoA (an acetic acid molecule bound to coenzyme A) and then packaged into synaptic vesicles. Once ACh is released from the synaptic vesicles, its activity is rapidly terminated by an enzyme in the synaptic cleft known as **acetylcholinesterase (AChE;** ah-seet'l'-koh-leh-NESS-ter-ayz). AChE degrades ACh back into acetic acid and choline. The presynaptic neuron then takes the choline back up, to be used in the synthesis of new ACh molecules.

The Biogenic Amines

The **biogenic amines,** also called the *monoamines,* are a class of five neurotransmitters synthesized from amino acids. Most biogenic amines are widely used by the CNS and the PNS, and have diverse functions including regulation of homeostasis and cognition (thinking). The biogenic amines are implicated in a wide variety of psychiatric disorders and are often the targets of drug therapy for these disorders.

Three of the biogenic amines form a subgroup called the **catecholamines** (kat'-eh-KOHL-ah-meenz), all of which are synthesized from the amino acid tyrosine and share a similar chemical structure. Though many of their synapses are excitatory, like most neurotransmitters, catecholamines can cause inhibition as well. The three catecholamines are as follows:

- **Norepinephrine.** *Norepinephrine* (nor'-ep-ih-NEF-rin; also called *noradrenalin*) is widely used by the ANS, where it influences functions such as heart rate, blood pressure, and digestion. Neurons that secrete norepinephrine in the CNS are largely confined to the brainstem, where they work to regulate the sleep/wake cycle, attention, and feeding behaviors.

TABLE 11.3 MAJOR NEUROTRANSMITTERS

Neurotransmitter	Precursor Molecule(s)	Predominant Postsynaptic Effect	Location(s)	Type of Receptor(s)
Acetylcholine	Acetyl-CoA and choline	Excitatory	CNS: brain and spinal cord PNS: neuromuscular junction and ANS	Ionotropic and metabotropic
Biogenic Amines				
Catecholamines (norepinephrine, epinephrine, dopamine)	Tyrosine	Excitatory	CNS: brain and spinal cord PNS: ANS (sympathetic division)	Metabotropic
Serotonin	Tryptophan	Excitatory	CNS: brain	Metabotropic
Histamine	Histidine	Excitatory	CNS: brain	Metabotropic
Amino Acids				
Glutamate	Glutamine	Excitatory	CNS: brain (major neurotransmitter of the brain)	Ionotropic and metabotropic
GABA (γ-aminobutyric acid)	Glutamate	Inhibitory	CNS: brain and spinal cord	Ionotropic and metabotropic
Glycine	Serine	Inhibitory	CNS: brain and spinal cord (most common inhibitory neurotransmitter in the spinal cord)	Ionotropic
Neuropeptides				
Substance P	Amino acids	Excitatory and inhibitory	CNS: brain and spinal cord (major neurotransmitter for pain perception) PNS: enteric nervous system (neurons in the digestive tract)	Metabotropic
Opioids (enkephalin, α-endorphin, dynorphin-A)	Amino acids	Excitatory and inhibitory	CNS: brain and spinal cord (major neurotransmitters for pain control)	Metabotropic
Neuropeptide Y	—	Excitatory and inhibitory	CNS: brain PNS: ANS	Metabotropic

- **Epinephrine.** *Epinephrine* (also called *adrenalin*) is also used by the ANS, where it has the same effects as norepinephrine. However, it is more widely used as a hormone by the endocrine system (see Chapter 16 for details).

- **Dopamine.** *Dopamine,* used extensively in the CNS, has a variety of functions. It helps to coordinate movement, and is also involved in emotion and motivation. The receptor for dopamine in the brain is a target for certain illegal drugs, such as cocaine and amphetamine, and is likely responsible for the behavioral changes seen with addiction to these drugs.

Another biogenic amine is *serotonin* (sair-oh-TOH-nin), which is synthesized from the amino acid tryptophan. Most neurons that use serotonin are found in the brainstem, and their axons project to multiple places in the brain. Serotonin is thought to be one of the major neurotransmitters involved in mood regulation (possibly along with norepinephrine), and it is a common target in the treatment of depression. Additionally, serotonin acts to affect emotions, attention and other cognitive functions, motor behaviors, feeding behaviors, and daily rhythms.

The final biogenic amine we'll discuss is *histamine* (HISS-tah-meen), which is synthesized from the amino acid histidine. Histamine is involved in a large number of processes in the CNS, including regulation of arousal and attention. In addition, outside the nervous system, histamine is an important mediator of allergic responses. Drugs called *antihistamines* block histamine receptors outside the nervous system to alleviate allergy symptoms, but most also block histamine receptors in the CNS. As histamine plays a part in arousal, blocking its actions often leads to the common side effect of drowsiness seen with these drugs.

Amino Acid Neurotransmitters

There are three major **amino acid neurotransmitters:** glutamate; glycine; and γ-aminobutyric acid, or GABA. *Glutamate* is the most important excitatory neurotransmitter in the CNS—it is estimated that over half of all synapses in the CNS release glutamate. When it binds to its ionotropic postsynaptic receptors, glutamate triggers the opening of a type of channel that can pass both sodium and calcium ions. This elicits an EPSP in the postsynaptic neuron.

Glycine and *GABA* are the two major inhibitory neurotransmitters of the nervous system. Both induce IPSPs in the postsynaptic neurons primarily by opening chloride ion channels and hyperpolarizing the axolemma. GABA use is widespread in the CNS; as many as one-third of neurons in the brain use it as their major inhibitory neurotransmitter. Glycine is found in about half of the inhibitory synapses in the spinal cord; the remainder of the synapses use GABA.

Neuropeptides

The **neuropeptides** are a group of neurotransmitters that have a wide variety of effects within the nervous system. Because they are peptides rather than modified amino acids, they must be synthesized in the cell body, as axons lack the organelles for protein synthesis. Multiple neuropeptides have been identified, and a few are described next.

A&P in the Real World

Psychiatric Disorders and Treatments

Psychiatric disorders, those that affect the thought processes of the brain, are generally treated by modifying synaptic transmission in order to change how neurons communicate with one another. Much of the science of psychopharmacology (the study of drugs that affect higher brain functions) targets either action potential generation or some aspect of neurotransmitter physiology. Some examples of disorders and drug actions include the following:

- **Schizophrenia.** The disease schizophrenia (skit'-zoh-FREEN-ee-ah) is characterized by repetitive *psychotic episodes*—periods during which the patient is unable to appropriately test his or her beliefs and perceptions against reality. Schizophrenia is thought to result from excessive release of dopamine; therefore, pharmacologic management of the disorder primarily involves blocking postsynaptic dopamine receptors.

- **Depressive disorders.** The depressive disorders are marked by disturbances in mood and are thought to be caused by a deficiency in synaptic transmission of serotonin, norepinephrine, and/or dopamine. Pharmacologic treatment makes use of drugs that prolong the lifespan of these neurotransmitters, particularly serotonin, in the synaptic cleft. The most widely used antidepressants are the *selective serotonin reuptake inhibitors* (SSRIs), which block only the serotonin transporter, preventing the reuptake of serotonin by the presynaptic neuron.

- **Anxiety disorders.** The hallmark of anxiety disorders is an exaggerated and inappropriate fear response, believed to stem from abnormalities in norepinephrine, serotonin, and GABA transmission. Drugs used to treat anxiety disorders may include the antidepressants already discussed, drugs that enhance GABA activity, and other drugs that modulate norepinephrine transmission.

- **Bipolar disorder.** Bipolar disorder is a group of disorders characterized by episodes of abnormally elevated mood (called *mania*) followed by episodes of depression. Many treatments for bipolar disorder involve decreasing the ease with which axons generate action potentials, generally by blocking sodium ion channels in the axolemma.

- **Substance P.** Substance P was the first identified neuropeptide (its name comes from the fact that it was extracted from brain and gut *powder*). It is released from type C sensory afferent fibers that carry information about pain and temperature (leading many students to use the mnemonic that the "P" stands for "pain"). It is also released by neurons in the brain, spinal cord, and gut.

- **Opioids.** The opioids (OH-pee-oydz) make up a family of more than 20 neuropeptides that includes three classes: the *endorphins,* the *enkephalins,* and the *dynorphins.* All share the same property of eliciting pain relief (called *analgesia* [an′-al-JEE-zee-ah]), and all are nervous system depressants. They also appear to be involved in sexual attraction and aggressive or submissive behaviors.
- **Neuropeptide Y.** Neuropeptide Y is a large neuropeptide with 36 amino acids. It appears to function in feeding behaviors, and may mediate hunger or feeling "full."

Read about how medications can affect synaptic transmission in *A&P in the Real World: Psychiatric Disorders and Treatments.*

Quick Check

☐ 3. How do neurotransmitters excite a postsynaptic neuron? How do they inhibit a postsynaptic neuron?
☐ 4. Which neurotransmitters have largely excitatory effects?
☐ 5. Which neurotransmitters have largely inhibitory effects?

Apply What You Learned

☐ 1. Toxins from the cone snail block glutamate receptors in the postsynaptic membrane. What specifically will this action inhibit? (*Hint:* What kind of receptor binds glutamate?)
☐ 2. Predict the effects of the poison strychnine, which blocks glycine receptors on postsynaptic neurons of the CNS.
☐ 3. What would happen to synaptic transmission if you blocked the degradation and/or reuptake of excitatory neurotransmitters in the synaptic cleft? What if the neurotransmitters were inhibitory?

See answers in Appendix A.

MODULE **11.6**
Functional Groups of Neurons

Learning Outcomes

1. Define a neuronal pool, and explain its purpose.
2. Compare and contrast the two main types of neural circuits in the central nervous system.

So far, we have mostly discussed the behavior of individual neurons—how action potentials are generated and conducted, how an action potential leads to synaptic transmission, and how one neuron sends a message to another neuron. Now we explore the behavior of *groups* of neurons. Neurons don't typically operate as discrete entities but instead form networks called *neuronal pools* that perform a common function. Neuronal pools are organized into functional groups called *neural circuits* (or *neural networks*). In this module, we examine neuronal pools and neural circuits

and discuss how groups of neurons work together to carry out the many activities of the nervous system.

Neuronal Pools

Neuronal pools are groups of interneurons within the CNS. These pools typically are a tangled mat of neuroglial cells, dendrites, and axons in the brain, while their cell bodies may lie in other parts of the CNS. The type of information that can be processed by a pool is defined by the synaptic connections of that pool. The connections between pools allow for complex mental activity such as planned movement, cognition, and personality.

Each neuronal pool begins with one or more neurons called *input neurons* that initiate the series of signals. The input neuron branches repeatedly to serve multiple neurons in the pool; however, it may have different effects on different neurons. For some neurons, it may generate EPSPs that trigger an action potential, and for others, it may simply bring the neuron closer to threshold. This difference is determined by the number of contacts the input neuron makes with the postsynaptic neuron.

A small neuronal pool with one input neuron and its postsynaptic neurons is illustrated in **Figure 11.30**. You can see that the postsynaptic neurons in the center (surrounded by green) have the highest number of synaptic contacts with the input neuron. Because of these connections, spatial summation is possible and the firing of the input neuron is likely to generate adequate EPSPs to trigger an action potential.

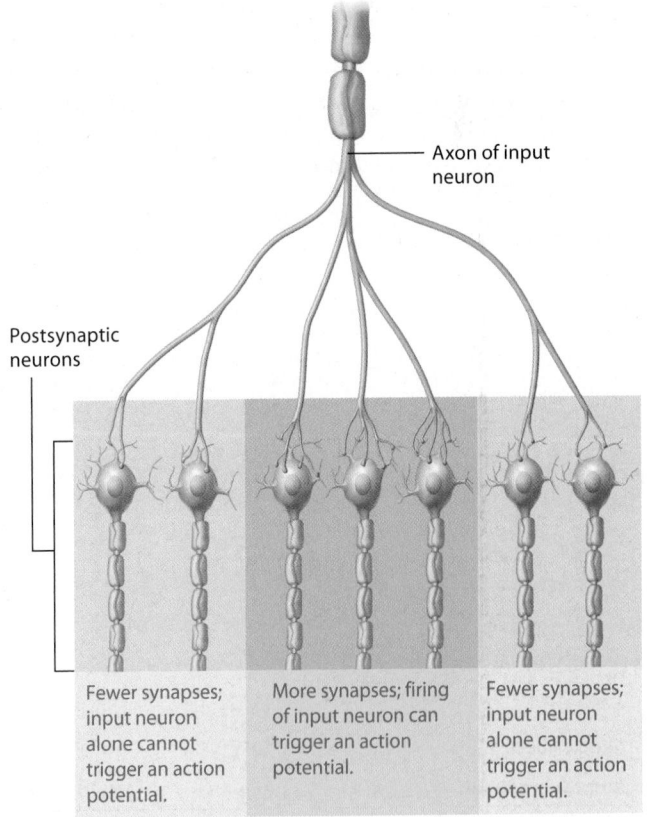

Figure 11.30 A neuronal pool.

Notice, however, that the neurons in the yellow area on either side have fewer synaptic contacts with the input neuron. As a result, the input neuron acting alone will not be able to bring these neurons to threshold and elicit an action potential. However, the input neuron can help another input neuron bring the postsynaptic neurons to threshold.

Until now we've been discussing only excitatory input; however, remember that inhibitory synapses occur as well. The degrees of inhibition also correlate strongly with the number of synaptic contacts. Action potentials are effectively prevented in the postsynaptic neurons that receive the greatest number of IPSPs from the input neuron.

Neural Circuits

As you know, form follows function, so the functional characteristics of a neuronal pool are determined largely by its pattern of structural organization. The patterns of synaptic connection between neuronal pools are called **neural circuits.** Each neuronal pool in a circuit receives input from other pools, and then produces output that travels to additional pools. How the pools are connected determines the function of the circuits.

There are two basic types of neural circuits, diverging and converging:

- **Diverging circuit.** As shown in **Figure 11.31a**, a **diverging circuit** begins with one axon of an input neuron that branches to make contacts with multiple postsynaptic neurons. The axons of these postsynaptic neurons then branch to contact more neurons, which in turn make contact with yet more neurons, and so on. Thus, when a signal is transmitted down the pathway of the circuit, an increasing number of neurons are excited. Diverging circuits are critical because they allow a single neuron to communicate with multiple parts of the brain and/or body. Observe in

Figure 11.31a, left, that some diverging circuits are *amplifying circuits,* in which the signal passes through a progressively greater number of neurons. We start with one neuron, which branches to excite two, the two then excite four, and so on. Some diverging circuits (Figure 11.31a, right) split into multiple tracts, each of which goes in a different direction. This type of circuit is characteristic of those transmitting incoming sensory information, which is sent from neurons in the spinal cord to different neuronal pools in the brain for processing.

- **Converging circuit.** A **converging circuit** (**Figure 11.31b**) is essentially the opposite of a diverging circuit. In converging circuits, axon terminals from multiple input neurons converge onto a single postsynaptic neuron, allowing for spatial summation of synapses. Converging circuits are important for control of skeletal muscle movement—the interneurons in the spinal cord receive input from neurons in different regions of the brain, which then converges to synapse on the motor efferent neurons that stimulate skeletal muscle contraction. Converging circuits also allow the nervous system to respond to the sensory information that it collects and processes; an example is shown in Figure 11.31b.

Given that every part of the brain contacts virtually every other part of the brain via some neural circuit, disorganized electrical activity could be disastrous (see *A&P in the Real World: Epileptic Seizures*). To prevent electrical activity in the brain from becoming chaotic and overly excitatory, the CNS has two basic mechanisms to stabilize neural circuits. The first mechanism is simply inhibitory circuits. Most neural circuits have an intrinsic negative feedback mechanism that inhibits either the input neurons or the postsynaptic neurons of their pools if they become overly excited. Additionally, some neuronal pools consist largely of inhibitory neurons that control the activity of other neural circuits.

Input neuron in the brain sends signals to multiple muscle fibers.

Brain

Spinal cord

Muscle fibers

Amplifying circuit

Input neuron in the spinal cord sends signals to different parts of the brain.

Multiple tracts

(a) Diverging circuits

Multiple input neurons from sensory receptors converge onto a neuron in the brain.

Sensory receptors

(b) Converging circuits

Figure 11.31 Types of neural circuits.

The second mechanism for stabilizing neural circuits is a property of synapses called *synaptic fatigue*. Fatigue refers to the fact that synaptic transmission becomes progressively weaker with prolonged and intense excitation. This is in part due to a gradual decrease in the sensitivity of the postsynaptic neurons in the circuit to neurotransmitters when they are overused. Further, over the long term, the number of neurotransmitter receptors in the plasma membrane of the postsynaptic neuron actually decreases when high levels of neurotransmitters are present for extended periods of time. This "downregulation" of postsynaptic receptors is thought to be why people develop a tolerance to certain medications that modulate neurotransmitter release.

Quick Check

☐ 1. Why are neurons organized into neuronal pools?
☐ 2. How do diverging and converging circuits differ?
☐ 3. What mechanisms stabilize neural circuits?

Apply What You Learned

☐ 1. Sometimes diverging circuits split into excitatory and inhibitory paths. When might such a circuit be required? (*Hint:* Think about muscle contraction.)
☐ 2. Cocaine blocks the reuptake of dopamine, causing high levels of dopamine to stimulate the postsynaptic receptors for an extended period. Explain why people who abuse cocaine eventually need more of the drug to reach an equivalent "high."

See answers in Appendix A.

Epileptic Seizures

Epilepsy is characterized by recurrent episodes of abnormal, disorganized electrical activity in the brain called *seizures*. A seizure results from a sudden burst of excitatory electrical activity within a neuronal pool, which may be triggered by instability in the membrane potential of a single neuron. The excess excitation overwhelms the inhibitory circuits that would normally prevent overexcitation. Once the inhibitory mechanisms are lost, a continuous wave of excitation spreads over part of the brain (a *partial seizure*) or the entire brain (a *generalized seizure*). When the brain is inundated with reverberating signals of this nature, no meaningful signals can be transmitted. This can lead to a wide variety of symptoms, ranging from mild sensory disturbances to loss of consciousness to characteristic jerking movements from uncontrolled muscle activity.

Although a number of medications will stop seizures, they generally end on their own due to synaptic fatigue. Thus, most therapy for epilepsy is aimed at preventing seizures and allowing the inhibitory circuits to function optimally.

CHAPTER SUMMARY

Take full advantage of the invaluable online practice and study tools for this chapter by going to MasteringA&P®. There you will find:

- 🖱 Practice tests, module quizzes, art-labeling and sequence-the-steps activities
- ▶ Animated figures with quizzes
- ✎ *Practicing A&P* worksheets
-))) Author-narrated podcast
- ♨ Dynamic Study Modules
- iP *Interactive Physiology*

MODULE **11.1**
Overview of the Nervous System 381

- The nervous system is anatomically divided into the **central nervous system** (CNS) and the **peripheral nervous system** (PNS). The **brain** and **spinal cord** make up the CNS. The PNS includes the *cranial* and *spinal nerves*. (Figure 11.1, p. 382)
- The nervous system can be functionally classified into three divisions:

  ○ The **PNS sensory,** or **afferent, division** uses **sensory receptors** to detect changes in the internal and external environment. This division has two branches: the **somatic sensory division** and the **visceral sensory division.**
  ○ The CNS receives the sensory input and processes it; this is called **integration.**
  ○ The **PNS motor,** or **efferent,** division carries out the directions of the CNS by performing **motor functions.** The two branches of the motor division are the **somatic motor division** and the **autonomic nervous system (ANS).** (Figures 11.2 and 11.3, p. 383)

MODULE **11.2**
Nervous Tissue 384

- Nervous tissue consists of **neurons** and **neuroglial cells.** (Figure 11.4, p. 384)
- Neurons are excitable cells that send, propagate, and receive **action potentials.** They consist of three parts:
  ○ The **cell body** contains the nucleus and the bulk of the organelles.
  ○ Most neurons have one or more **dendrites,** which are the receptive ends of the neuron.

○ Each neuron has only a single **axon,** which generally carries information away from the cell body to another neuron, muscle fiber, or gland. Materials are transported through an axon via axonal transport. (Figure 11.5, p. 385)

👆 **Practice art-labeling exercise: Figure 11.5.**

- Neurons are classified both structurally and functionally.
  ○ Structural classes include **multipolar neurons, bipolar neurons,** and **pseudounipolar neurons,**
  ○ Functional classes include **sensory,** or **afferent, neurons; interneurons;** and **motor,** or **efferent, neurons.** (Table 11.1, p. 387)

- The **neuroglia** in the CNS include the following:
  ○ **Astrocytes** anchor neurons and blood vessels in place, assist in the formation of the blood-brain barrier, regulate the extracellular environment of the brain, and participate in repair of damaged brain tissue.
  ○ **Oligodendrocytes** wrap around the axons of neurons and form the **myelin sheath.**
  ○ **Microglia** are phagocytes that "clean up" the extracellular environment of the brain.
  ○ **Ependymal cells** are ciliated cells that line the cavities of the brain and spinal cord and produce and circulate cerebrospinal fluid. (Figure 11.6, p. 388)

👆 **Practice art-labeling exercise: Figure 11.6.**

- The neuroglia of the PNS include the following:
  ○ **Schwann cells** form the myelin sheath in the PNS.
  ○ **Satellite cells** surround cell bodies of neurons in the PNS. (Figure 11.7, p. 389)

👆 **Practice art-labeling exercise: Figure 11.7.**

- Oligodendrocytes in the CNS and Schwann cells in the PNS wrap around the axon up to 100 times to form the myelin sheath. This covering significantly speeds up conduction of an action potential through the axon.

- The segment of an axon that is myelinated by one glial cell is an **internode,** and the small gap between internodes is called a **node of Ranvier.** (Figure 11.8, p. 390)

👆 **Practice art-labeling exercise: Figure 11.8.**

- Unmyelinated axons in the PNS are embedded in Schwann cells. (Figure 11.9, p. 391)

- Axons of the PNS may be regenerated if the cell body remains intact. When **regeneration** occurs, an axonal *growth process* is guided toward its target cell by a **regeneration tube** made of Schwann cells and the basal lamina. (Figure 11.10, p. 392)

📱 **Play Interactive Physiology tutorial on Nervous System I: Anatomy Review.**

MODULE **11.3**

Electrophysiology of Neurons 393

- Ions move across the axolemma via channels. Two types of channels are **leak** (always open) and **gated channels.** (Table 11.2, p. 394)

- Two important ion gradients are those of Na^+ and K^+: The concentration of Na^+ is higher in the extracellular fluid, and the concentration of K^+ is higher in the cytosol.

- A separation of charges occurs across the membrane of neurons. An unstimulated neuron shows a decrease in voltage across the membrane, called the **resting membrane potential.** Neurons at rest are **polarized** with a resting membrane potential of about -70 mV. The negative resting membrane potential is due to the loss of K^+ through leak channels and the actions of the Na^+/K^+ pumps. (Figure 11.11, p. 395; Figure 11.12, p. 396)

- Ion movement is driven by the forces of the concentration gradient and the electrical gradient. The sum of these two forces is the **electrochemical gradient.** (Figure 11.13, p. 396)

📱 **Play Interactive Physiology tutorial on Nervous System I: Ion Channels.**

- A **local potential** is a small, local change in the membrane potential of a neuron. (Figure 11.14, p. 397)
  ○ A local potential may either **depolarize** the neuron, making it less negative, or **hyperpolarize** the neuron, making it more negative.
  ○ Local potentials are *graded,* reversible, *decremental* with distance, and useful for short-distance signaling only.

▶ **Play animation on local potentials: Figure 11.14.**

- Voltage-gated K^+ channels have two states, resting and activated, whereas voltage-gated Na^+ channels have three: resting, activated, and inactivated. (Figure 11.15, p. 398)

📱 **Play Interactive Physiology tutorial on Nervous System I: Resting Membrane Potential.**

- An **action potential** is a rapid depolarization and repolarization of the membrane potential of the cell. (Figure 11.16, p. 399)
  ○ During the **depolarization phase** Na^+ flood the axon, and the membrane potential rises toward a positive value.
  ○ During the **repolarization phase** K^+ flow out of the axon, returning the axon to its negative resting membrane potential. For many neurons, K^+ flow out even after the axolemma has returned to resting, causing hyperpolarization.

▶ **Play animation on action potentials: Figure 11.16.**

- Action potentials are *nondecremental,* they obey the **all-or-none principle,** they are irreversible, and they are long-distance signals.

- The **refractory period** is the span of time during which it is difficult or impossible to elicit another action potential. (Figure 11.17, p. 401)

- Action potentials are **propagated** along an axon via the flow of current. (Figure 11.18, p. 402)

▶ **Play animation on action potential propagation: Figure 11.18.**

- The speed of action potential propagation depends on the diameter of the axon (larger axons conduct more rapidly) and on the presence or absence of a myelin sheath. Conduction may occur in two ways: (Figure 11.19, p. 403)
  ○ **Saltatory conduction** occurs rapidly because the current is insulated as it flows through each internode and action potentials are generated only at nodes of Ranvier.
  ○ **Continuous conduction** occurs much more slowly, as each consecutive region of the membrane must be depolarized to threshold and generate an action potential.

- The big picture of action potentials is shown in Figure 11.20 (p. 405).

▶ **Play animation on the big picture of action potentials: Figure 11.20.**

iP2 **Play Interactive Physiology tutorial on Generation of an Action Potential.**

MODULE **11.4**
Neuronal Synapses 406

- A **synapse** is the location where a neuron meets its target cell. The transfer of information between neurons at a synapse is called **synaptic transmission.** (Figure 11.21, p. 406)

- **Electrical synapses** occur between neurons whose axolemmas are electrically joined via gap junctions. Information transfer at an electrical synapse is nearly instantaneous and bidirectional. (Figure 11.22a, p. 407)

- The majority of synapses in the nervous system are **chemical synapses,** which rely on **neurotransmitters** to send signals. Chemical synapses are slower than electrical synapses and are unidirectional. (Figure 11.22b, p. 407)

🖑 **Practice art-labeling exercise: Figure 11.22.**

iP **Play Interactive Physiology tutorials on Nervous System II: Anatomy Review, Ion Channels, Synaptic Transmission.**

- The events at a chemical synapse start with an action potential reaching the axon terminal of the **presynaptic neuron,** which opens calcium ion channels. The influx of calcium ions triggers exocytosis of neurotransmitters stored in synaptic vesicles. The neurotransmitters diffuse through the **synaptic cleft** and bind to receptors on the membrane of the **postsynaptic neuron.** When the transmitter binds, a local **postsynaptic potential** results. (Figure 11.23, p. 408)

▶ **Play animation on events at a chemical synapse: Figure 11.23.**

- One of two things may happen during a postsynaptic potential: (Figure 11.24, p. 409)
  ○ The postsynaptic neuron may be depolarized by an **excitatory postsynaptic potential** (EPSP).
  ○ The postsynaptic neuron may be hyperpolarized by an **inhibitory postsynaptic potential** (IPSP).

▶ **Play animation on postsynaptic potentials: Figure 11.24.**

- The effects of synaptic transmission are terminated by removal of the neurotransmitters from the synaptic cleft. (Figure 11.25, p. 410)

▶ **Play animation on methods of termination of synaptic transmission: Figure 11.25.**

- The big picture of synaptic transmission is shown in Figure 11.26 (p. 411).

▶ **Play animation on the big picture of chemical synaptic transmission: Figure 11.26.**

- **Neural integration** is the process of combining the factors that influence whether or not a neuron fires an action potential.
  ○ **Summation** is the phenomenon that combines local postsynaptic potentials. (Figure 11.27, p. 412)

- In **temporal summation,** a single presynaptic neuron fires at a rapid pace to influence the postsynaptic neuron. In **spatial summation,** multiple presynaptic neurons fire simultaneously to influence the postsynaptic neuron. (Figure 11.28, p. 413)

iP **Play Interactive Physiology tutorial on Nervous System II: Synaptic Potentials and Cellular Integration.**

MODULE **11.5**
Neurotransmitters 413

- Neurotransmitters produce their effects by influencing the opening or closing of ion channels in the axolemma of the postsynaptic neuron.

- There are two types of neurotransmitter receptors: (1) *ionotropic receptors,* and (2) *metabotropic receptors.* (Figure 11.29, p. 414)

- The effects of a neurotransmitter are described as **excitatory** if they generally induce EPSPs and **inhibitory** if they generally induce IPSPs. Many neurotransmitters are capable of generating both EPSPs and IPSPs.

- The major neurotransmitters include the following:
  ○ **Acetylcholine** is mostly excitatory, and is degraded by **acetylcholinesterase.**
  ○ The **biogenic amines** include the **catecholamines** (*norepinephrine, dopamine,* and *epinephrine*), *serotonin,* and *histamine.*
  ○ The **amino acid neurotransmitters** include *glutamate, glycine,* and *γ-aminobutyric acid* (*GABA*). Glutamate is the major excitatory neurotransmitter. Both GABA and glycine are major inhibitory neurotransmitters.
  ○ Other neurotransmitters are summarized in Table 11.3 (p. 415).

iP **Play Interactive Physiology tutorial on Nervous System II: Synaptic Transmission.**

MODULE **11.6**
Functional Groups of Neurons 417

- Interneurons are organized into **neuronal pools** that enable specialization within the CNS and so higher mental activity.

- An *input neuron* is the presynaptic neuron that initiates the series of signals in a neuronal pool.
  ○ If the input neuron has sufficient synaptic connections with the postsynaptic neuron, it may trigger it to depolarize to threshold on its own.
  ○ With sufficient synaptic connections, the IPSPs of an input neuron can inhibit an action potential in postsynaptic neurons. (Figure 11.30, p. 417)

- The pattern of connection between neuronal pools is called a **neural circuit.** There are two main types of neural circuits:
  ○ A **diverging circuit** begins with one or more input neurons that contact an increasing number of postsynaptic neurons
  ○ A **converging circuit** is one in which the signals from multiple neurons converge onto one or more final postsynaptic neurons. (Figure 11.31, p. 418)

ASSESS WHAT YOU LEARNED

See answers in Appendix A.

LEVEL 1 Check Your Recall

1. Which of the following statements about the general functions of the nervous system is *false*?

   a. The three primary functions of the nervous system include sensory, integrative, and motor functions.

   b. The integrative functions of the nervous system are its processing functions.

   c. Sensory information is transmitted on sensory efferent fibers to a sensory receptor.

   d. Motor functions are carried out by fibers that carry signals to an effector.

2. Regulation of heart rate, blood pressure, and digestive functions is carried out by the:

   a. somatic motor division of the peripheral nervous system.

   b. central nervous system.

   c. visceral sensory division of the peripheral nervous system.

   d. autonomic nervous system.

3. Match each type of neuroglial cell with its correct function.

   _____ Schwann cells
   _____ Ependymal cells
   _____ Microglial cells
   _____ Oligodendrocytes
   _____ Satellite cells
   _____ Astrocytes

   a. Phagocytic cells of the CNS
   b. Surround the cell bodies of neurons in the PNS
   c. Create the myelin sheath in the PNS
   d. Anchor neurons and blood vessels, maintain extracellular environment around neurons, assist in repair of damaged brain tissue
   e. Create the myelin sheath in the CNS
   f. Ciliated cells in the CNS that produce and circulate the fluid around the brain and spinal cord

4. With respect to the cell body of a neuron, which of the following statements is *false*?

   a. Aggregates of Golgi apparatus and lysosomes form dark-staining Nissl bodies within it.

   b. Reflecting its function of protein synthesis, the cell body contains a high density of ribosomes, rough endoplasmic reticulum, and Golgi apparatus.

   c. Within its cytoplasm are bundles of intermediate filaments that come together to form neurofibrils.

   d. The cell body has high metabolic demands, and thus has large numbers of mitochondria.

5. An axon is *best* defined as a process that:

   a. carries information only toward the cell body.

   b. can generate action potentials.

   c. carries information only away from the cell body.

   d. cannot generate action potentials.

6. Fill in the blanks: The segment of an axon that is covered by a glial cell is called a/an _____; the gaps between glial cells where the axolemma is exposed are called _____.

7. Fill in the blanks: The _____ is the period of time during which it is impossible to stimulate a neuron to have an action potential, whereas the _____ is the period of time during which a larger-than-normal stimulus is required to elicit an action potential.

8. With respect to the conduction of action potentials, which of the following statements is *false*?

   a. Every region of the membrane must be depolarized and triggered to generate an action potential via saltatory conduction.

   b. In an unmyelinated axon, every region of the membrane must be depolarized and generate action potentials.

   c. Type A fibers are the largest, have the most myelin, and conduct action potentials the fastest.

   d. The myelin sheath allows saltatory conduction, in which only the nodes of Ranvier must generate action potentials.

9. Identify the following as properties of electrical synapses (ES), chemical synapses (CS), or both (B).

   a. _____ The plasma membranes of presynaptic and postsynaptic neurons are joined by gap junctions.

   b. _____ Transmission is unidirectional and delayed.

   c. _____ A presynaptic neuron and a postsynaptic neuron are involved.

   d. _____ The use of neurotransmitters packaged into synaptic vesicles is required.

   e. _____ Transmission is nearly instantaneous and bidirectional.

10. The trigger for exocytosis of synaptic vesicles from the presynaptic neuron is:

    a. arrival of an action potential at the axon terminal and influx of calcium.

    b. summation of IPSPs at the presynaptic neuron.

    c. binding of neurotransmitters to the axon hillock.

    d. influx of Na^+ into the postsynaptic neuron.

11. Match the following neurotransmitters with their correct description.

    _____ GABA
    _____ Dopamine
    _____ Substance P
    _____ Acetylcholine
    _____ Glutamate
    _____ Endorphins
    _____ Norepinephrine

    a. Neuropeptide involved in transmission of pain
    b. Neurotransmitter released at the neuromuscular junction
    c. Major excitatory neurotransmitter in the brain
    d. Major inhibitory neurotransmitter in the brain
    e. Neuropeptide involved in relief of pain
    f. Catecholamine involved in the autonomic nervous system, the sleep/wake cycle, attention, and feeding behaviors
    g. Catecholamine involved in movement and behavior

12. Which of the following is *not* a method by which the effects of neurotransmitters are terminated?

    a. Reuptake into the presynaptic neuron
    b. Diffusion away from the synaptic cleft and uptake by glial cells
    c. Movement back to the cell body by retrograde axonal transport
    d. Degradation by enzymes in the synaptic cleft

13. A _____ is characterized by multiple input neurons synapsing on one postsynaptic neuron.

    a. diverging circuit   c. facilitation zone
    b. discharge zone   d. converging circuit

14. Mark the following statements as true or false. If a statement is false, correct it to make a true statement.

    a. The myelin sheath is formed by oligodendrocytes in the CNS and Schwann cells in the PNS.
    b. Dendrites are highly branched receptive processes that generate and conduct action potentials.
    c. The myelin sheath insulates the axon and increases the rate at which action potentials are conducted through the axon.
    d. Multipolar neurons have multiple axons and a single dendrite.
    e. Areas of unmyelinated processes and cell bodies appear white and are called white matter.
    f. Interneurons are generally multipolar neurons that perform integrative functions.

15. Mark the following statements as true or false. If a statement is false, correct it to make a true statement.

    a. The resting membrane potential refers to the voltage difference across the membranes of excitable cells at rest.
    b. The concentration of Na^+ is highest in the cytosol, and the concentration of K^+ is highest in the extracellular fluid.
    c. The Na^+/K^+ pumps and gated channels maintain the Na^+ and K^+ gradients necessary for action potentials to occur.
    d. A depolarization is a change in membrane potential that makes the potential less negative.
    e. A local potential is a change in membrane potential that conducts the long-distance signals of the nervous system.

16. Sequence the following list of events of a neuronal action potential by placing 1 next to the first event, 2 next to the second event, and so on.

    a. _____ The activation gates of voltage-gated Na^+ channels open, Na^+ flood the cytoplasm, and depolarization occurs.
    b. _____ K^+ continue to flow out of the axon until the membrane is hyperpolarized.
    c. _____ Local potentials cause the membrane to depolarize to threshold.
    d. _____ The inactivation gates of voltage-gated Na^+ channels close as voltage-gated K^+ channels open, K^+ begin to exit the axon, and repolarization begins.
    e. _____ Repolarization continues and Na^+ channels return to resting.

17. Mark the following statements as true or false. If a statement if false, correct it to make a true statement.

    a. An excitatory postsynaptic potential is caused by K^+ or Cl^- channels opening in the membrane of the postsynaptic neuron.
    b. Postsynaptic potentials may summate by spatial summation in which multiple neurons fire onto a single postsynaptic neuron.
    c. An inhibitory postsynaptic potential causes the membrane potential of the postsynaptic neuron to approach threshold.
    d. Spatial summation can combine two EPSPs, two IPSPs, or an EPSP and an IPSP.

LEVEL 2 Check Your Understanding

1. A drug that blocks Na^+ channels in neurons does so not only in the axon but also in the dendrites and cell body. What overall effect would this have on action potential generation?

2. What would happen if the drug blocked K^+ channels instead?

3. What conditions must be met for an axon to regenerate? How does regeneration occur?

4. Explain how an action potential is propagated down an axon in continuous conduction. Why is saltatory conduction faster than continuous conduction?

LEVEL 3 Apply Your Knowledge

PART A: Application and Analysis

1. The drug neostigmine blocks the actions of acetylcholinesterase in the synaptic cleft. What effect would this have on synaptic transmission? What effects might you expect to see as a result of this drug?

2. During a surgical procedure, an anesthesiologist administers to the patient an inhaled anesthetic agent that opens Cl^- channels in the postsynaptic membranes of neurons of the brain. Explain why this would put the patient "to sleep" for the duration of the surgical procedure.

3. Albert accidentally ingests the poison *tetrodotoxin* from the pufferfish, which you know blocks voltage-gated Na^+ channels. Predict the symptoms Albert will experience from this poisoning.

4. Albert, the patient in question 3, takes the drug lithium, which reduces the permeability of the neuronal axolemma to Na^+. Predict the effect this would normally have on his neuronal action potentials. Do you think this drug would be beneficial or harmful, considering his condition?

PART B: Make the Connection

5. Predict the effect that tetrodotoxin would have on Albert's muscle fiber action potentials (see question 3). Would it affect end-plate potentials at the motor end plate? Why or why not? (*Connects to Chapter 10*)

6. Explain what would happen if depolarization of the axon hillock triggered a negative feedback loop instead of a positive one. (*Connects to Chapter 1*)

Photo: Colored three-dimensional magnetic resonance imaging (MRI) scan showing the white matter pathways of the brain, lateral view.

We throw around the term "the mind" a lot in our day-to-day language—we have things on our minds, we give someone a piece of our minds, we change our minds, and sometimes we even lose our minds. But what *is* the mind, exactly, and where is it located? Scientists, philosophers, and theologians have spent centuries trying to answer these questions. The current working definition for "the mind" is the collection of thoughts, emotions, intellect, and imagination that is unique to each individual. The idea that the mind resides within and can be explained by the processes of the body has met with much resistance; in fact, the mind was believed to be completely separate from the body for millennia, and this belief is still held by some. However, abundant scientific evidence shows that the mind does indeed reside in the body and that normal physiological processes can account for it. The most dramatic evidence comes from disease processes or injuries to regions of the brain that lead to changes in the functions of the mind.

In this chapter, we explore the structure and functions of the remarkable organs of the **central nervous system (CNS):** the *brain* and the *spinal cord.* After an overview, the next three modules focus on the anatomy of these structures, and the subsequent modules describe the functions and processes of the CNS, including movement, interpreting sensory information, maintaining homeostasis, and functions relating to the mind. Note, however, that some of the body's critical processes—sensation, movement, and maintenance of homeostasis—rely on components in both the CNS and the peripheral nervous system (PNS). For this reason, it's difficult to talk about the CNS without occasionally discussing parts of the PNS, as well. In this chapter we cover primarily the CNS contribution to sensation, homeostasis, and movement, and in the next two chapters we discuss how the PNS contributes to each process (see Chapters 13 and 14). You will notice in all three chapters, though, that we can't ever really discuss the functions of the CNS or the PNS exclusively, as the two are able to function only as an integrated whole.

MODULE **12.1**

Overview of the Central Nervous System

Learning Outcomes

1. Describe the structure and function of each major area of the brain.
2. Describe the five developmental regions of the brain, and identify the major areas of the adult brain that arise from each region.

Before we get into the specifics of the CNS, an introductory tour of its components is in order. This module first examines the basic functions carried out by the CNS. Next it presents the general structure of the brain and spinal cord. It concludes with an overview of the embryonic development of the CNS.

Overview of CNS Functions

◄◄ Flashback

1. How do motor, sensory, and integrative functions differ? (p. 382)

The functions of the nervous system can be broken down into three categories: *motor functions,* or stimulation of a muscle cell contraction or a gland secretion; *sensory functions,* or detection of sensations within and outside the body; and *integrative functions,* or "decision-making" processes (see Chapter 11). Both motor and sensory functions are performed by the PNS, whereas integrative functions are carried out exclusively by the CNS. Integrative functions encompass a wide variety of processes, including interpretation of sensory information, planning and monitoring movement, maintenance of homeostasis, and higher mental functions such as use of language and learning.

As you saw in Figure 11.2 (see p. 383), the brain receives and processes input from a number of sources, including sensory neurons of the PNS, interneurons from other parts of the brain, and interneurons from the spinal cord. Once the information has been processed, the brain makes a "decision" on how to respond to the input and sends this output to other interneurons of the brain or spinal cord or to a muscle or gland via motor neurons of the PNS.

Quick Check

☐ 1. What types of functions are performed by the CNS?

Basic Structure of the Brain and Spinal Cord

◄◄ Flashback

1. Where are the cranial and vertebral cavities located? (p. 16)
2. Which types of cells make up nervous tissue? (p. 148)

3. What are the differences between gray matter and white matter? What name is given to structural groups of gray and white matter in the CNS? (p. 391)

The brain and spinal cord are anatomically a single, continuous structure. The **brain** is a soft, whitish-gray organ that resides in the cranial cavity and directly or indirectly controls most of the body's functions. It is composed primarily of nervous tissue, although it contains epithelial and connective tissue, as well (see Chapter 11). It has internal cavities called *ventricles* (VEN-trih-kulz) that are filled with a protective *cerebrospinal fluid*. An adult brain weighs between 1250 and 1450 g and tends to be slightly larger in males than in females. However, the size of the brain appears to have little to do with intelligence, and the ratio of brain weight to body weight of females tends to be equal to or slightly greater than that of males.

The brain is richly supplied with blood vessels. Indeed, during rest about 20% of total blood flow in the body goes to the brain. This reflects its need for large amounts of oxygen, glucose, and other nutrients. For a view of blood supply to the brain, please refer to the blood vessels chapter (see Figure 18.17).

The brain consists of four divisions: the cerebrum, diencephalon, cerebellum, and brainstem (**Figure 12.1**). Each division is distinct in terms of the type of input it receives and where it sends its output. Let's look at each division in more detail.

- **Cerebrum.** The *cerebrum* (seh-REE-brum; Latin *cerebr-* = "brain") is the enlarged superior portion of the brain. It is composed of two halves known as the right and left **cerebral hemispheres.** Each cerebral hemisphere has five lobes (rounded subdivisions of an organ) that contain groups of neurons performing specific tasks. The cerebrum is responsible for our higher mental functions, including learning, memory, personality, cognition (thinking), language, and conscience, and it also plays major roles in sensation and movement.

- **Diencephalon.** Buried beneath the cerebral hemispheres is the central core of the brain: the *diencephalon* (dy'-en-SEF-ah-lon; Greek *encephal-* = "brain"). The diencephalon consists of four distinct structural parts and is responsible for processing, integrating, and relaying information to different parts of the brain; homeostatic functions; regulation of movement; and biological rhythms.

- **Cerebellum.** The posterior and inferior portion of the brain is the *cerebellum* (sair-ah-BEL-um; "little brain"). Like the cerebrum, the cerebellum is composed of right and left hemispheres. The cerebellum figures importantly in the planning and coordination of movement, particularly for complex activities such as playing an instrument or a sport.

- **Brainstem.** The oldest part of the brain from an evolutionary standpoint, the *brainstem* connects the brain and spinal cord. Its functions include basic involuntary homeostatic functions, control of certain reflexes, monitoring movement, and integrating and relaying information to other parts of the nervous system.

The **spinal cord** is a long, tubular organ encased within and protected by the vertebral cavity. It begins at the foramen magnum of the skull, where it blends with the inferior part of the brainstem, and ends approximately between the first and second

Anterior Posterior

BRAIN DIVISION	FUNCTION
Cerebrum	• Performs higher mental functions • Interprets sensory stimuli • Plans and initiates movement
Diencephalon	• Processes, integrates, and relays information • Maintains homeostasis • Regulates biological rhythms
Cerebellum	• Monitors and coordinates movement
Brainstem	• Maintains homeostasis • Controls certain reflexes • Monitors movement • Integrates and relays information

Spinal cord

Figure 12.1 Divisions of the brain (lateral view).

lumbar vertebrae. It measures about 43–46 cm (17–18 inches) in length and is quite thin in diameter, ranging from about 0.65–1.25 cm (0.25–0.5 inch). Like the brain, the spinal cord has an internal cavity filled with cerebrospinal fluid; its cavity, the *central canal,* is continuous with the ventricles of the brain.

Both the brain and the spinal cord consist of **white matter,** which contains myelinated axons, and **gray matter,** which is made up of neuron cell bodies, dendrites, and unmyelinated axons (**Figure 12.2**). Within the brain, the gray matter is located in the cerebrum's outer few millimeters and is also scattered throughout its deeper portions; the remainder of the cerebrum is white matter (Figure 12.2a). Each lobe of the cerebrum contains

bundles of white matter, called **tracts,** that receive input from and send output to the different clusters of cell bodies and dendrites in the cerebral gray matter, called **nuclei.** This communication connects the different parts of the brain and spinal cord. The myelin sheath on the axons that form these tracts enables nearly instantaneous communication between parts of the CNS.

The spinal cord consists internally of gray matter containing nuclei that process information, as well as tracts of white matter that shuttle this information to and from the brain. As you can see in Figure 12.2b, the spinal gray matter is surrounded by white matter and in cross section has the shape of a butterfly. Note that the organization of gray and white matter in the brain and in the spinal cord are reversed: The spinal white matter is superficial, whereas the cerebral white matter is deep.

Cerebral gray matter   Cerebral white matter

Brain
Spinal cord

Spinal gray matter   Spinal white matter

LM (3×)

(a) Brain, frontal section

(b) Spinal cord, transverse section

Figure 12.2 White and gray matter in the CNS.

☐ 2. What are the divisions of the brain?
☐ 3. Where is the spinal cord located? Where does it begin and end?

Overview of CNS Development

Although CNS structure is complicated, it can be useful to think of it as just a tube with an enlarged end. We can see how the brain and spinal cord end up with this structure by looking at how they develop in the embryo and fetus (**Figure 12.3**). Nervous tissue develops from a hollow structure called the **neural tube,** which is completely formed by the fourth week of development. Just after the neural tube forms, its caudal end goes on to become the spinal cord, and the cranial end rapidly enlarges and forms three saclike structures called **primary brain vesicles:** the *forebrain, midbrain,* and *hindbrain.* These primary brain vesicles continue to enlarge and give rise to five **secondary brain vesicles** by about the fifth week of development.

These secondary vesicles eventually become the four divisions of the mature brain.

● The forebrain expands into two secondary brain vesicles: the *telencephalon* and the *diencephalon.* The two lobes of the telencephalon enlarge to become the cerebral hemispheres,

and the diencephalon, as you might predict, becomes the diencephalon of the mature brain.

● The midbrain expands into a secondary brain vesicle known as the *mesencephalon.* This term means "midbrain," and indeed the mesencephalon goes on to become a portion of the brainstem called the *midbrain.*

● The hindbrain develops into two secondary brain vesicles: the *metencephalon* and *myelencephalon.* Both vesicles develop into the remainder of the brainstem, and part of the metencephalon becomes the cerebellum.

Note that the cavity in the hollow neural tube becomes the ventricles in the brain and the central canal in the spinal cord.

☐ 4. What is the neural tube?
☐ 5. What does each of the secondary brain vesicles become in the mature brain?

☐ 1. In terms of survival, to which of the four divisions of the brain would an injury be most damaging? Explain.

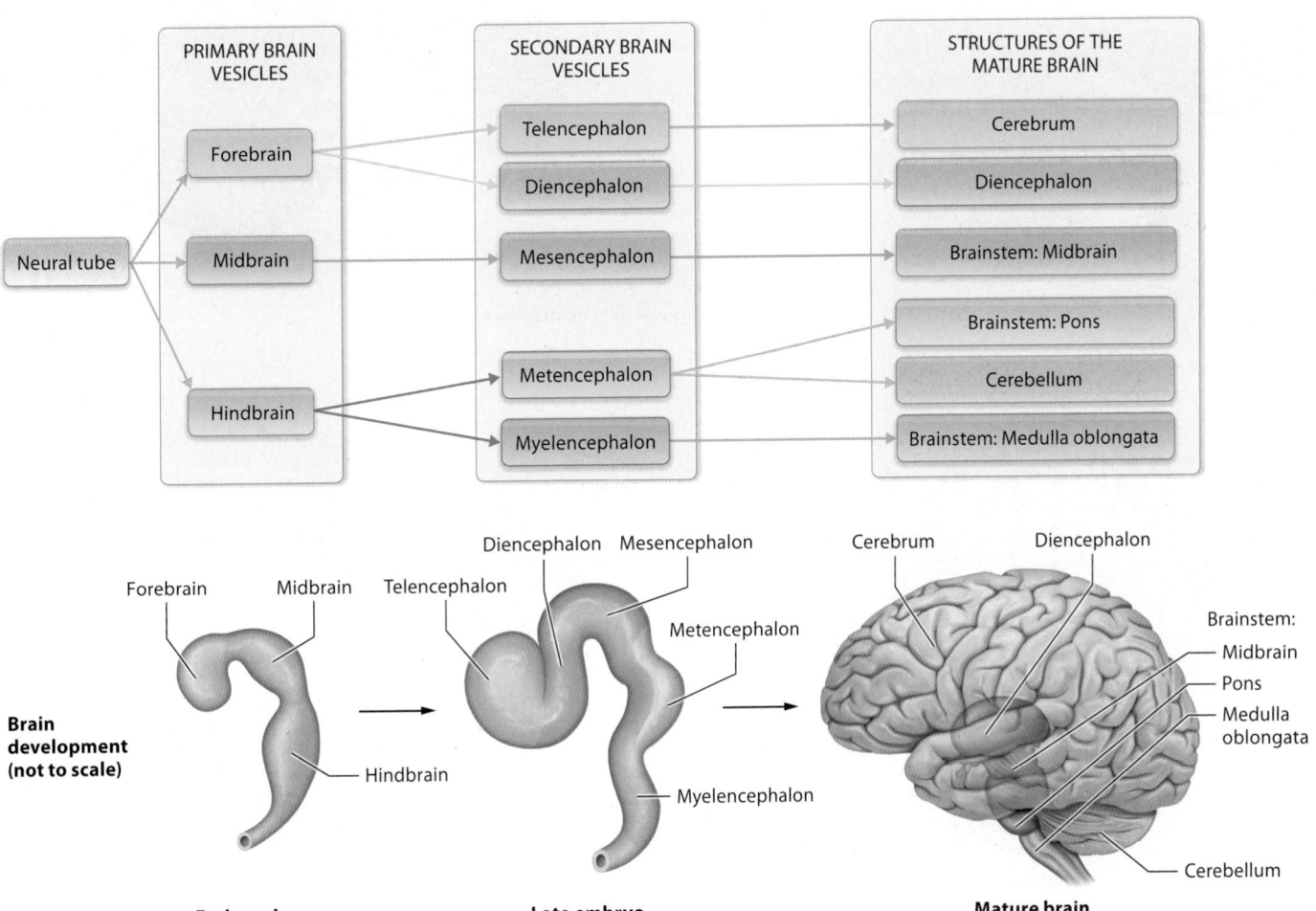

Figure 12.3 Development of the brain.

☐ 2. In which of the four divisions of the brain might an injury cause changes in personality? Explain.

See answers in Appendix A.

See answers in Appendix A.

MODULE **12.2**

The Brain

Learning Outcomes

1. Describe and identify the five lobes of the cerebral cortex, and explain how motor and sensory functions are distributed among the lobes.
2. Describe the structure, components, and general functions of the regions of the diencephalon, cerebellum, and brainstem.
3. Describe the location and functions of the limbic system and the reticular formation.

We only use 10% of our brains, right? Wrong—this is a myth; we use 100% of our brains. This error has been widely disseminated by the media, by psychics looking to explain their purported abilities, and by those wishing to sell you books and courses that claim to teach you how to "tap into the potential" of the 90% of your brain you don't use. Although it is widely believed, this myth doesn't even really make much sense when you think about it: Why would the body invest so much of its blood flow, nutrients, and energy in something that was 90% nonfunctional? Although we don't precisely know the function of every region of the brain, it is fairly safe to say that each part is doing *something* important.

Now we explore the four divisions of the brain that were introduced in the previous module. We touch briefly on the function of each component of the divisions, with full discussions coming in subsequent modules.

The Cerebrum

As the previous module indicated, the **cerebrum** is that part of the brain responsible for our higher mental functions, which

(a) Lateral view (left hemisphere)

(b) Superior view

Figure 12.4 Structure of the cerebrum.

(c) Lateral view (frontal, parietal, and temporal lobes pulled back)

is why people who think a lot are sometimes called "cerebral." Its large, paired cerebral hemispheres somewhat resemble a shelled walnut (**Figure 12.4**). The superficial parts of the cerebrum are separated by both shallow and deep grooves. The shallow grooves are called **sulci** (SUL-kee; singular, *sulcus*), and the deep grooves are called **fissures**. The two hemispheres are separated from one another by a long fissure called the **longitudinal**

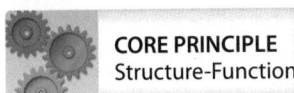

CORE PRINCIPLE
Structure-Function

fissure. Each hemisphere contains multiple sulci, and between the sulci are elevated ridges known as **gyri** (JY-ree; singular, *gyrus*). In an example of the Structure-Function Core Principle (p. 25), the gyri and sulci increase the surface area of the brain and make it a much more compact structure; if our brains were smooth, we would require much larger skulls to hold them. Although the gyri and sulci can vary in location and size from one brain to the next, certain gyri and sulci are found in almost all brains and thus are named.

The cerebral hemispheres house two ventricles, the right and left *lateral ventricles.* They also contain a number of structures that carry out specialized CNS functions. The upcoming sections provide a closer look at these structures and their roles.

Lobes of the Cerebrum

The five lobes of each cerebral hemisphere are as follows:

- **Frontal lobes.** The anteriormost lobe of each cerebral hemisphere is the **frontal lobe.** Note in Figure 12.4a and b that its posterior boundary is called the **central sulcus.** Just anterior to the central sulcus is the **precentral gyrus.** The neurons of the frontal lobe are responsible for planning and executing movement and complex mental functions such as behavior, conscience, and personality.
- **Parietal lobes.** Posterior to the frontal lobes are the paired **parietal lobes** (pah-RY-eh-tal). The major gyrus of each of these lobes is the **postcentral gyrus,** which sits just posterior to the central sulcus. Neurons of the parietal lobe are responsible for processing and integrating sensory information and function in attention.
- **Temporal lobes.** The paired **temporal lobes** are on the lateral surfaces of the hemisphere, separated from the frontal and parietal lobes by the **lateral fissure.** The neurons of the temporal lobes perform functions related to hearing, language, memory, and emotions.
- **Occipital lobes.** The posterior lobe of each cerebral hemisphere is the **occipital lobe.** It is separated from the parietal lobe by the *parieto-occipital sulcus* (look ahead to Figure 12.12). The neurons of the occipital lobe process all information relating to vision.
- **Insulas.** As you can see in Figure 12.4c, the paired **insulas** (IN-syoo-lahz) are visible only when you pry the frontal, parietal, and temporal lobes apart at the lateral fissure (a common mnemonic is that the "insula is *insula*ted by the other lobes"). The neurons of the insulas are currently thought to have functions relating to taste and to our viscera (internal organs).

Gray Matter: Cerebral Cortex

Functionally, the most complex part of the brain is the region of gray matter known as the **cerebral cortex** (**Figure 12.5**). Its location is reflected in its name; "cortex" means "bark," and the cerebral cortex covers the cerebral hemispheres in the same way that bark covers a tree. The majority of the cerebral cortex is **neocortex,** or "new cortex," so named because it is the most recent part of our brains to have evolved. The neocortex is composed of six layers of neurons and neuroglia and varies in thickness from about 1.5 to 4.5 mm. The many convolutions of the gyri give the neocortex quite a large surface area—about 0.25 square meter (2.5 square feet).

The functions of the neocortex revolve around *conscious* processes (those in which we are aware of ourselves and our environment), such as planning movement, interpreting incoming sensory information, and complex higher functions like performing long division. In other words, it allows us to become aware of—and respond to—our surroundings. Other components of the cerebrum and the remainder of the brain function largely in *subconscious* processes, that is, those that are carried out below the level of conscious thought. Such processes are often called "subcortical" because they do not directly involve the cerebral cortex. Many of these subconscious processes, such as control of blood pressure and body temperature, are vital, arguably even more so than our ability to perform long division.

As you can see in Figure 12.5, the neocortex has three generally recognized areas: the *primary motor cortex,* which plans and executes movement; the *primary sensory cortices* (the plural of cortex), which are the first to receive and process sensory input; and the **association areas,** which integrate different types of information. Notice that most of the neocortex is devoted to association areas and that there are two types of these areas: unimodal and multimodal. Some association areas are *unimodal association areas* that integrate one specific type of information only, which generally relates to perception, such as identifying a color as orange or an odor as smoke. Unimodal association areas are typically near one of the primary motor or sensory areas. Others are *multimodal association areas,* which integrate information from multiple different sources and appear to carry out many of our higher mental functions. The multimodal association areas, for instance, would integrate the stimuli of the color orange and a smoky odor and identify both as indicating the presence of a fire.

The following sections will briefly introduce the areas of the neocortex responsible for motor, sensory, and higher-level functions. (Be sure not to confuse sensory and motor areas with sensory and motor neurons—all neurons in the cortex are interneurons.) Note that although we discuss each area as a discrete region, it is a vast oversimplification to do so. Realistically, boundaries for the neocortical areas are at best fuzzy and their functions often overlap those of other areas. Each area is discussed more fully later in the chapter and in the special senses chapter (see Chapter 15).

Motor Cortices and Upper Motor Neurons Most of the motor areas are located in the frontal lobe; these areas contain **upper motor neurons,** which are actually interneurons because they connect to other neurons instead of to skeletal muscles. Conscious

Primary motor cortex

Primary somatosensory cortices

Association areas:
- Unimodal association areas
- Multimodal association areas

Figure 12.5 **Structure of the cerebral cortex (left hemisphere, lateral view).**

planning of movement is carried out by upper motor neurons of the **primary motor cortex,** located in the precentral gyrus of the frontal lobe. The upper motor neurons of each cerebral hemisphere control the motor activity of the *opposite* side of the body via neurons of the PNS called *lower motor neurons* (which are responsible for executing the movements).

Movement requires input from several motor association areas, including the large **premotor cortex,** located anterior to the primary motor cortex in the frontal lobe. Motor association areas are unimodal areas involved in the planning, guidance, coordination, and execution of movement. Another motor association area consists of the paired **frontal eye fields,** one on each side of the brain anterior to the premotor cortex. These fields are required for back-and-forth eye movements, such as those relating to reading. The frontal eye fields appear to overlap functionally with certain multimodal association areas and may in fact be multimodal association areas themselves.

Sensory Cortices Each of the five classic senses—touch, vision, hearing, smell, and taste—is represented by both a primary and a unimodal association area in the cerebral cortex. In addition, the sense of *equilibrium,* which gives us information about balance, also has primary and association areas. These areas, most of which are shown in Figure 12.5, are found in all lobes except the frontal lobe (although some overlap does occur) and include the following:

- **Somatosensory areas.** Two main areas of the cerebral cortex deal with the *somatic* senses (*soma-* = "body"), which are those pertaining to temperature or touch (mechanical deformation of a tissue), including vibration, pressure, stretch, and

joint position. The first is the **primary somatosensory cortex** (also called **S1**), which is located in the postcentral gyrus of the parietal lobe. The second is the **somatosensory association cortex** (also called **S2**), which is located posterior to S1.
- **Visual areas.** Humans have a large amount of cortical area dedicated to processing visual input, including all of the occipital lobes and part of the temporal lobes. This reflects the importance of vision in humans and other primates. The **primary visual cortex,** which lies at the posterior end of the occipital lobe, is the first area to receive visual input. It feeds this information into surrounding **visual association areas,** which process color, object movement, and depth.
- **Auditory areas.** The **primary auditory cortex,** located in the superior temporal lobe, is the first to receive auditory information. From here, information travels to the adjacent **auditory association cortex** as well as other multimodal association areas.
- **Other sensory areas.** Taste information is processed by the poorly localized **gustatory cortex,** which appears to be in the insula and part of the parietal lobe. Information pertaining to equilibrium and positional sense is processed by **vestibular areas,** which are located in the parietal and temporal lobes. The sense of smell is processed by the **olfactory cortex,** which consists of several areas in the limbic lobes and the medial temporal lobes (see Chapter 15). (These areas are not shown in Figure 12.5.)

Multimodal Association Areas The multimodal association areas are some of the most fascinating parts of the brain, as these areas allow us to perform complex mental functions. Some of

TABLE 12.1 MOTOR, SENSORY, AND ASSOCIATION AREAS OF THE CEREBRAL CORTEX

Area Name	Type of Cortex	Location	Function
Primary motor cortex	Motor	Precentral gyrus of frontal lobe	Plans and executes movement
Premotor cortex	Unimodal association	Widespread throughout lateral and medial frontal lobe	Plans and executes complex movement
Frontal eye fields	Unimodal association; possibly multimodal	Anterior to the premotor cortex	Back-and-forth eye movements
Primary somatosensory cortex	Sensory	Postcentral gyrus of parietal lobe	Interprets incoming somatic sensory information
Somatosensory association cortex	Unimodal association	Posterior to the primary somatosensory cortex in the parietal lobe	Integrates somatic sensory information
Primary visual cortex	Sensory	Posterior occipital lobe	Interprets and processes visual stimuli
Primary auditory cortex	Sensory	Superior temporal lobe	Processes auditory stimuli
Gustatory cortex	Sensory	Insula, parietal lobe	Processes taste stimuli
Vestibular areas	Sensory	Several in both parietal and temporal lobes	Processes stimuli relating to equilibrium and balance
Olfactory cortex	Sensory	Limbic lobe, medial temporal lobe	Processes smell stimuli
Broca's area	Multimodal association	Superolateral frontal lobe	Language production
Wernicke's area	Multimodal association	Superolateral temporal lobe	Language comprehension
Prefrontal cortex	Multimodal association	Anterior frontal lobe	Planning, personality, higher cognitive functions
Parietal association cortex	Multimodal association	Widespread in the parietal lobe	Spatial awareness and attention
Temporal association cortex	Multimodal association	Widespread in the temporal lobe	Recognition and associations

the best-characterized multimodal association areas include the following:

- **Language areas.** Two areas of the cortex are directly involved in language: **Broca's area,** a premotor area for speech sounds, located in the anterolateral frontal lobe, and **Wernicke's area,** or the integrative speech area, located in the temporal and parietal lobes. Broca's area is responsible for the ability to *produce* language, whereas Wernicke's area is responsible for the ability to *understand* language.
- **Prefrontal cortex.** As you can see in Figure 12.5, the **prefrontal cortex** is located in and occupies most of the frontal lobe. It communicates with many areas of the brain, including parts of the diencephalon, other cerebral gray matter, and the association cortices of other cerebral lobes. The prefrontal cortex has a number of functions, among which are modulating behavior, personality, learning, memory, and an individual's psychological state.
- **Parietal and temporal association cortices.** The *parietal* and *temporal association cortices* occupy much of their respective lobes. These areas perform a variety of tasks, including integration of sensory information, language, maintaining attention, recognition, and spatial awareness.

The major motor, sensory, and association areas of the cerebral cortex are summarized in Table 12.1.

Gray Matter: Basal Nuclei

Buried within the cerebral hemispheres on either side of the diencephalon lie clusters of neuron cell bodies collectively called the

basal nuclei, which are all involved in movement (Figure 12.6). (Note that classically these nuclei have been referred to as *basal ganglia;* however, this is a misnomer, as ganglia are collections of cell bodies in the PNS.) The basal nuclei are separated from the gray matter of the diencephalon by an area of white matter known as the *internal capsule.*

Cerebral hemispheres

Globus pallidus

Striatum:
Caudate nucleus
Putamen

Figure 12.6 Structure of the basal nuclei (anterolateral view).

The basal nuclei include the following:

- **Caudate nucleus.** As you can see in Figure 12.6, the **caudate nuclei** (KOW-dayt) are C-shaped rings of gray matter that sit lateral to the lateral ventricles in each hemisphere. The caudate nucleus is named for its long, slender "tail" (*cauda-* = "tail") that curls around and points anteriorly.
- **Putamen.** The **putamen** (poo-TAY-men; "pod") lies posterior and inferior to the caudate nucleus and is connected to it via small bridges of gray matter. The caudate nucleus and putamen are similar structurally and functionally, and are thus sometimes referred to as a single structure called the *corpus striatum.*
- **Globus pallidus.** The **globus pallidus** ("pale globe"), which sits medial to the putamen, is named for the fact that it contains more myelinated fibers than the other basal nuclei and is thus paler.

Although the basal nuclei are primarily associated with movement, disorders involving the basal nuclei also cause problems with behavior, cognition, and perception. This demonstrates that these nuclei have multiple functional roles in the brain.

Cerebral White Matter

The white matter in the cerebrum can be classified as one of three types—commissural, projection, or association fibers—depending on its function (**Figure 12.7**):

- **Commissural fibers.** The **commissural fibers** (kom-ih-SHUR-al; *commis-* = "united") connect the right and left cerebral hemispheres. Of the four groups of commissural fibers, the largest is the **corpus callosum,** which sits in the middle of the brain at the base of the longitudinal fissure.
- **Projection fibers.** The **projection fibers** connect the cerebral cortex of one hemisphere with other areas of the same

hemisphere, as well as with other parts of the brain and the spinal cord. Note in Figure 12.7 that the cerebral projection fibers in each hemisphere form a radiating pattern called the *corona radiata;* these fibers condense around the diencephalon on the right and left sides to form two V-shaped bands known as the **internal capsules.**
- **Association fibers.** The **association fibers** are also restricted to a single hemisphere and connect the gray matter of cortical gyri with one another.

A possible pathway for information transfer by the conduction of an action potential from one region of the brain to another is shown in **Figure 12.8**. Note that ① an action potential originates in gray matter and ② is then sent to another area of gray matter by projection fibers. This area of gray matter in turn ③ sends a new action potential to neighboring gray matter by association fibers, and ④ another action potential is generated and sent to the other cerebral hemisphere by commissural fibers. Some nuclei send impulses to only a single nucleus; however, many project to multiple places, and their tracts of white matter resemble dizzying three-dimensional highway systems.

Structures of the Limbic System

Groups of gray matter that share the same function and the white matter that connects them are sometimes referred to as *functional brain systems.* An important functional brain system is the group of structures known collectively as the **limbic system.** This system includes a region of the medial cerebrum sometimes referred to as the *limbic lobe,* the *hippocampus* (hip-poh-KAM-pus), a collection of gray matter called the *amygdala* (ah-MIG-duh-lah), and the pathways that connect these regions of gray matter with each other and with other parts of the brain (**Figure 12.9**). The limbic system is found only in mammals and contains some of

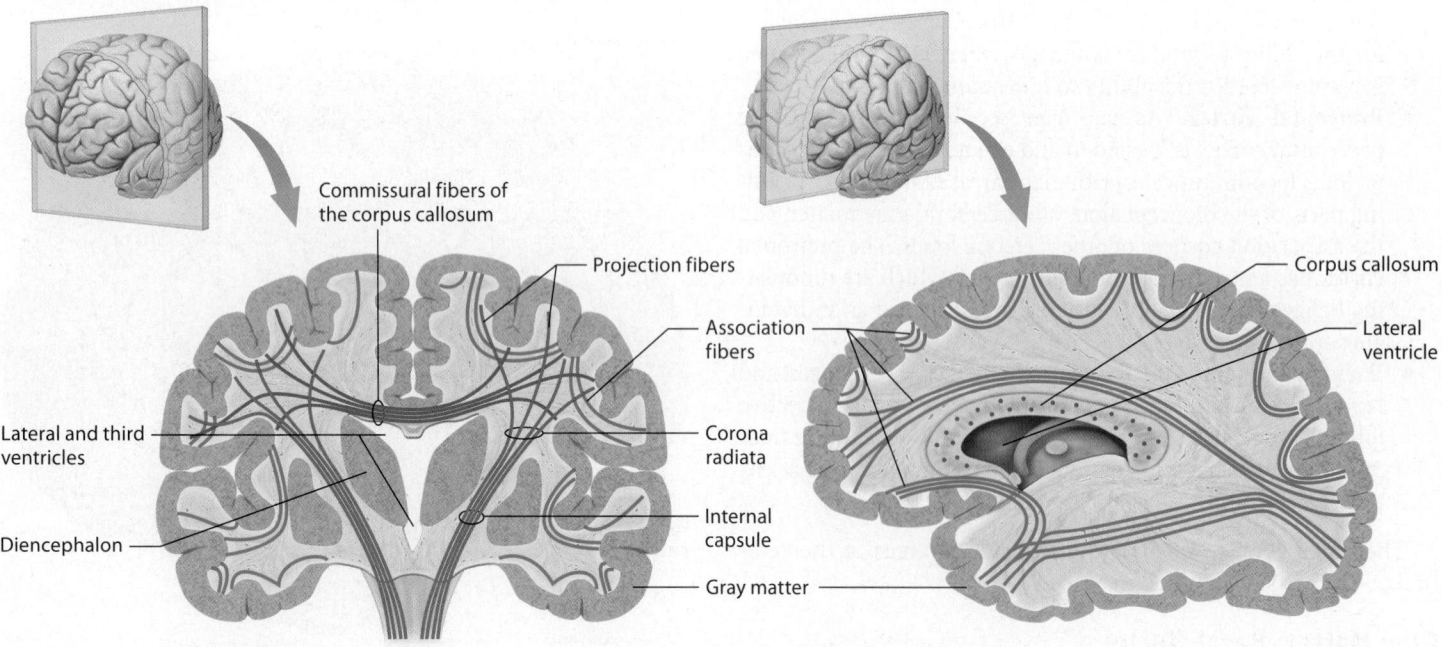

(a) Frontal section

Commissural fibers of the corpus callosum
Projection fibers
Association fibers
Lateral and third ventricles
Corona radiata
Internal capsule
Diencephalon
Gray matter

(b) Parasagittal section

Corpus callosum
Lateral ventricle

Figure 12.7 Structure of cerebral white matter.

① An action potential originates in one of the gray matter regions of the brain.

② The action potential travels to another area of gray matter by projection fibers.

③ A new action potential is generated and sent to neighboring gray matter by association fibers.

④ Another action potential is generated and sent to the other cerebral hemisphere by commissural fibers.

Gray matter

White matter

Figure 12.8 A possible pathway for conduction of an action potential in the brain.

the most ancient parts of the mammalian brain from an evolutionary perspective. It participates in a number of the more fascinating aspects of brain functions, including memory, learning, emotion, and behavior. It has been called the "visceral brain" because it mediates the responses of the viscera to certain situations. When you are nervous and have "butterflies in your stomach" or are scared and have a racing heart rate, that reaction is partially a product of your limbic system.

Note in Figure 12.9 that the limbic lobe and its associated structures form a ring visible on the medial side of the cerebral hemisphere. This lobe has two main gyri: the **cingulate gyrus** (SING-gyoo-layt) and the **parahippocampal gyrus.** Adjacent to the parahippocampal gyrus in the temporal lobe is the **hippocampus,** a structure whose name reflects its shape, as it is

the Greek term for "sea horse." Note that the hippocampus is connected to a prominent, C-shaped ring of white matter called the **fornix,** which is its main output tract. The hippocampus has a well-studied role in memory and learning, which we find out more about in Module 12.8.

Anterior to the hippocampus is the **amygdala,** a structure whose shape is also reflected in its name, which is the Greek word for "almond." The amygdala functions in the behavioral expression of emotion, particularly fear.

Quick Check

☐ 1. Match the following terms with the correct definition/description:

_____ Insula
_____ Occipital lobe
_____ Cerebral cortex
_____ Cerebral white matter
_____ Parietal lobe
_____ Basal nuclei

a. Outer few millimeters of gray matter of the cerebral hemispheres
b. Cerebral lobe deep to the frontal, parietal, and temporal lobes
c. Clusters of gray matter within the cerebrum that control movement
d. Posterior cerebral lobe
e. Groups of myelinated axons that connect regions of the cerebrum and other parts of the brain
f. Middle and superior cerebral lobe

☐ 2. Explain the differences between primary motor, primary sensory, and unimodal and multimodal association areas of the cerebral cortex.

The Diencephalon

◀◀ Flashback

1. What are endocrine glands, and what do they secrete? (p. 135)
2. What is the autonomic nervous system? (p. 383)

Limbic lobe:
Cingulate gyri
Parahippocampal gyrus
Fornix
Diencephalon
Amygdala
Hippocampus

Figure 12.9 Structures of the limbic system (anterolateral view).

The **diencephalon** is found at the physical center of the brain and so is almost completely hidden from external view by the cerebral hemispheres. Structurally and functionally the diencephalon has four components: the large, central *thalamus* (THAL-ah-muss; "inner chamber"); the posterior and superior *epithalamus* (*epi-* = "above"); the inferior *hypothalamus;* and the *subthalamus* (**Figure 12.10**). Each component consists of various nuclei that receive specific input and send output to other parts of the brain.

Thalamus

As you can see in Figure 12.10a and b, the **thalamus** consists of two large, egg-shaped masses of gray matter that together make up about 80% of the diencephalon. Between these two masses lies a cavity known as the *third ventricle*. Internally, each half of the thalamus has three main groups of nuclei separated by two thin layers of white matter. Thalamic nuclei receive input from multiple sources, including the cerebral cortex, the cerebellum, the basal nuclei, structures of the limbic system, and the sensory system (except for the sense of smell), and their main output travels to the cerebral cortex. The thalamus is literally the "main entrance" into the cerebral cortex—nearly all information destined for the cerebral cortex must first pass through the thalamus. This allows the thalamus to control which information reaches the cerebral cortex and where the information is sent, which means that ultimately the thalamus regulates cortical activity.

The thalamic nuclei are classified functionally into one of three groups—specific relay, association, or nonspecific.

- **Specific relay nuclei.** The *specific relay nuclei* function as relay stations—they receive incoming information, process and integrate it, and then send it to specific motor or sensory areas of the cerebral cortex. Some specific relay nuclei receive only sensory stimuli; others receive motor information.
- **Association nuclei.** The *association nuclei* receive input from a variety of sources; however, the information arrives indirectly via other thalamic nuclei or directly from other parts of the brain. Generally, these nuclei process information related to emotions, memory, and integration of sensory information from multiple sources. After processing information, they send it mainly to the association areas of the cerebral cortex.
- **Nonspecific nuclei.** The *nonspecific nuclei* receive information from the basal nuclei, cerebellum, and motor cortex, and send information to broad areas of the cortex and the rest of the brain. Current evidence indicates that they function in controlling arousal, consciousness, and the level of responsiveness and excitability of the cerebral cortex.

Hypothalamus

The **hypothalamus** is a collection of nuclei that sits anterior and inferior to the thalamus. It is much smaller than the thalamus, weighing only about 4 g and making up less than 1% of the mass of the brain. Don't let its small size fool you, though, as the hypothalamic nuclei perform a number of functions that are vital to our survival. Some of these include regulating much of

(a) Diencephalon, anterior view

(b) Diencephalon, superior view

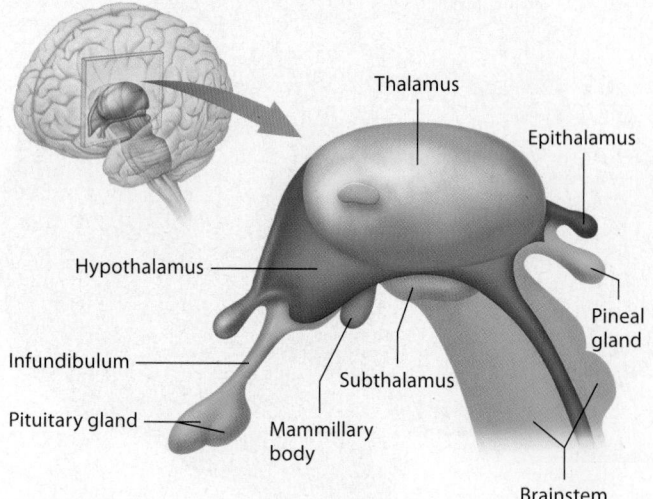

(c) Diencephalon, midsagittal section

Figure 12.10 **Structure of the diencephalon.**

the autonomic nervous system, the sleep/wake cycle, thirst and hunger, and body temperature.

In addition to these roles, the hypothalamus has a close anatomical and functional connection to an endocrine organ called the **pituitary gland** (pih-TOO-ih-tair′-ee). Note in Figure 12.10c that the pituitary gland is attached to the inferior portion of the hypothalamus by an extension referred to as the **infundibulum,** and hypothalamic tissue makes up the posterior portion of this gland. The hypothalamus secretes a variety of hormones that affect the pituitary gland—hypothalamic releasing or inhibiting hormones increase or decrease the secretion of other hormones from the pituitary gland, respectively. The pituitary gland, in turn, secretes hormones that influence secretion from other endocrine glands throughout the body. The hypothalamus also produces two additional hormones that do not act on the pituitary gland—*antidiuretic hormone* (*ADH*), which influences water balance in the body, and *oxytocin* (ok-sih-TOW-sin), which in women stimulates contraction of the uterus during childbirth and in both sexes may promote emotional bonding.

The hypothalamus receives input from a variety of sources, including the cerebral cortex and basal nuclei. It also connects with the limbic system via two small projections called the **mammillary bodies.** These structures receive input from the hippocampus, with which they play a role in regulating memory and behavior. Input also arrives at the hypothalamic nuclei from sources outside the nervous system, including the endocrine system, receptors that detect changes in body temperature, and receptors that detect the blood's osmotic concentration.

Epithalamus and Subthalamus

As implied by their names, the epithalamus lies superior to the thalamus and the subthalamus lies inferior. The major part of the epithalamus, located at the posterosuperior portion of the diencephalon, is an endocrine organ called the **pineal gland** (PIN-ee-al). Named for its resemblance to a pine cone, the pineal gland secretes the hormone **melatonin** (mel-ah-TOW-nin), which helps to regulate the sleep/wake cycle. The largest structure of the subthalamus is functionally connected to the basal nuclei and works with them to control movement.

Quick Check

☐ 3. Which component of the diencephalon performs each of the following functions?
  a. Controls numerous homeostatic functions
  b. Integrates emotion, memory, and sensory information and sends it to association areas of the cerebral cortex
  c. Produces the hormone melatonin
  d. Works with the basal nuclei to monitor and control movement
  e. Controls the ANS and parts of the endocrine system
  f. Determines which information reaches the cerebral cortex

The Cerebellum

The **cerebellum** makes up the posterior and inferior portion of the brain. It functions with the cerebral cortex, basal nuclei, brainstem, and spinal cord in the coordination of movement. It is composed of two **cerebellar hemispheres** connected by a structure called the **vermis** (VER-miss; "worm") (Figure 12.11a). The cerebellar surface contains ridges known as **folia** ("leaves"), which are separated by shallow sulci; this arrangement increases the cerebellar surface area in the same way that gyri and sulci increase the surface area of the cerebral cortex. The cerebellum is divided into three different lobes: the **anterior lobe,** the **posterior lobe,** and the small **flocculonodular lobe** (flok′-yoo-loh-NOD-yoo-lar; Figure 12.11b).

As you can see in Figure 12.11c and d, the cerebellar interior is arranged in the same way as the cerebrum, with an outer layer of gray matter—the *cerebellar cortex*—and inner white matter in which clusters of gray matter called the *deep cerebellar nuclei* are embedded. Since the cerebellar cortex is so extensively folded, the branching white matter resembles tree branches and is thus called the **arbor vitae** (AHR-bor VEE-tay; "tree of life"). The cerebellar white matter converges into three large tracts of white matter known as **cerebellar peduncles** (PEH-dung-k′lz) that connect the cerebellum to the brainstem and so form the only route by which information flows into and out of the cerebellum.

The role of the cerebellum in movement (as well as its interactions with the cerebral cortex, basal nuclei, brainstem, and spinal cord) is discussed fully in Module 12.6.

Quick Check

☐ 4. Describe the basic anatomical arrangement of the cerebellum.
☐ 5. What is the primary function of the cerebellum?

The Brainstem

One of the most evolutionarily ancient components of the brain is the **brainstem.** In terms of our immediate survival, it is perhaps the most important part of the brain, as its nuclei control many basic homeostatic functions, including maintenance of heart rate and breathing rhythm. This is why head injuries involving brainstem damage have poor prognoses (prospects of recovery)—even relatively small lesions of the brainstem can lead to death. The brainstem also controls numerous *reflexes,* which are programmed, automatic responses to stimuli (see Chapter 13), and functions in movement, sensation, and maintaining alertness.

In the midsagittal section of the brain shown in Figure 12.12 note that the brainstem is located inferior to the diencephalon; it is also anterior to the cerebellum and superior to the spinal cord. It extends to the level of the foramen magnum of the occipital bone, after which it becomes the spinal cord. It contains another ventricle, the *fourth ventricle,* which is continuous with the central canal of the spinal cord.

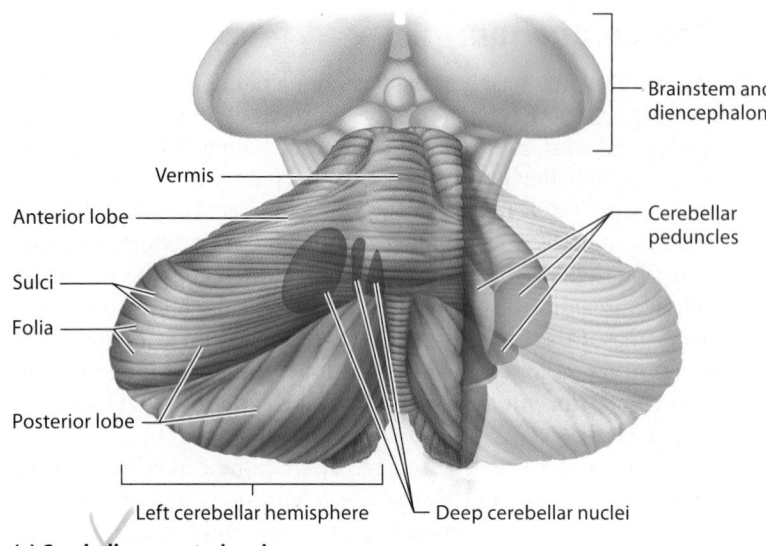

(a) Cerebellum, posterior view

(b) Cerebellum, anterior view

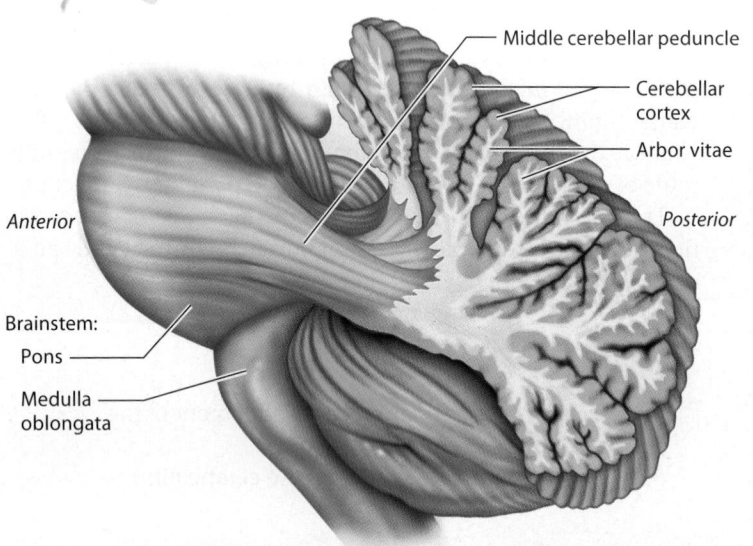

(c) Cerebellum, parasagittal section

(d) Photo of cerebellum, midsagittal section

Figure 12.11 **Structure of the cerebellum.**

The brainstem has three subdivisions: the superior *midbrain,* the middle *pons,* and the inferior *medulla oblongata* (**Figure 12.13**). Within these subdivisions we find the following structures:

- **Fibers of the cerebellum and related nuclei.** Cerebellar fibers travel through parts of the brainstem; some synapse on brainstem nuclei, whereas others are destined for the cerebral cortex or the spinal cord.
- **Nuclei of the reticular formation.** Throughout the brainstem is found a group of connected nuclei called the *reticular formation.* The reticular formation has a multitude of functions, including regulation of respiration, blood pressure, the sleep/wake cycle, pain perception, and consciousness, which we cover in more detail later.

- **Tracts of white matter between the spinal cord and the brain.** Almost all pathways to and from the brain and spinal cord pass through the brainstem. One of these is the *corticospinal tract,* which originates in the primary motor cortex and descends through the cerebral hemispheres, the brainstem, and the spinal cord. As with other tracts, right and left corticospinal tracts lie on each side of the brain and spinal cord.
- **Cranial nerve nuclei.** Numerous cranial nerves originate from the brainstem; their nuclei have multiple sensory, motor, and autonomic responsibilities (we will discuss the individual cranial nerves in the peripheral nervous system chapter [see Chapter 13]). Sometimes the functions of the reticular formation and cranial nerve nuclei overlap.

(a)

(b)

Figure 12.12 Midsagittal section of the brain showing the brainstem. Part (a) is an illustration, and part (b) is a photo showing the same view.

(a) Brainstem, anterior view

(b) Brainstem, posterior view

Figure 12.13 External anatomy of the brainstem.

Now we turn our attention to the basic anatomy of the three brainstem subdivisions (**Figure 12.14**). The following subsections introduce some major functions of each subdivision.

Midbrain

If you've ever been startled by a loud noise or an unexpected sight, your midbrain can take the blame for it. The **midbrain** is found just inferior to the diencephalon. It surrounds a channel, the *cerebral aqueduct,* that connects the third and fourth ventricles (Figure 12.14b). The midbrain is the superiormost and shortest component of the brainstem, and is also known as the *mesencephalon,* reflecting its embryonic origin.

Structures of the midbrain include the following:

- **Superior and inferior colliculi.** Protruding from its posterior surface are two pairs of rounded projections called the **superior** and **inferior colliculi** (koh-LIH-kyoo-lye; singular, *colliculus*), which form part of the **tectum,** or "roof," of the midbrain. These structures are involved with visual and auditory functions, respectively, and project to the thalamus.

- **Descending tracts.** Tracts of white matter originating from the cerebrum are located in the anteriormost portion of the midbrain in an area referred to as the *crus cerebri* (KROOS SER-eh-bry).

- **Substantia nigra.** The **substantia nigra** (sub-STAN-chah NIH-grah; "black substance") is located posterior to the crus cerebri. As its name implies, this nucleus has a dark color due to a pigment similar to melanin found in its neurons. The neurons of the substantia nigra work closely with the basal nuclei to control movement.

- **Red nucleus.** Just posterior to the substantia nigra is an area called the **red nucleus,** so named because its cells have an iron-containing pigment that lends it a pinkish appearance. The red nucleus communicates with the cerebellum and the spinal cord as well as other structures regulating movement. It appears to be active in movement in other animals, but its role in humans is not yet understood.

- **Cranial nerve nuclei.** Several cranial nerve nuclei are found in the midbrain, some of which are labeled in Figure 12.14.

- **Midbrain tegmentum. Tegmentum** is a generic term for the area of the midbrain between the cerebral aqueduct and

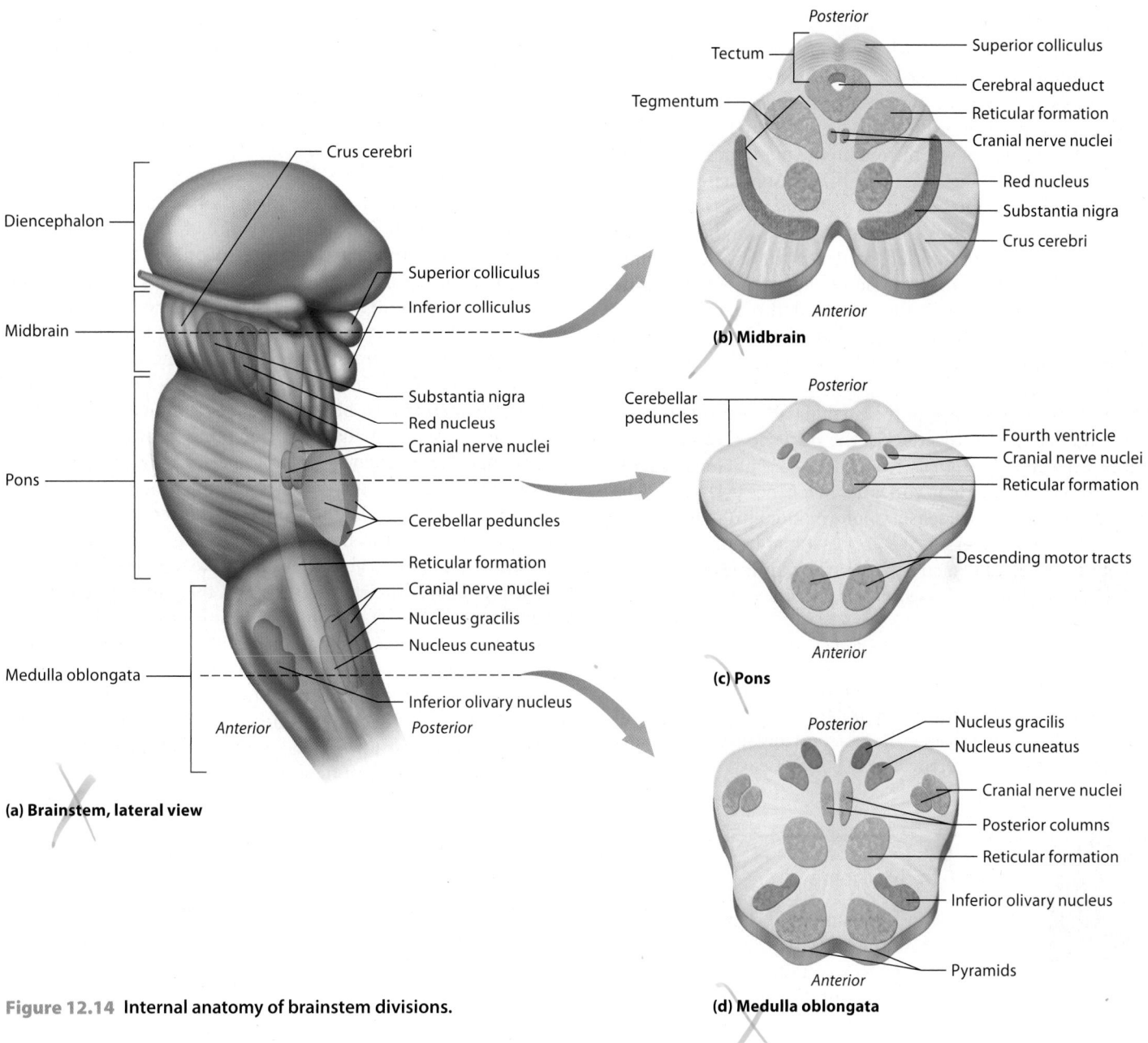

Figure 12.14 Internal anatomy of brainstem divisions.

the substantia nigra. It contains numerous nuclei, many of which are part of the reticular formation, as well as ascending and descending white matter tracts.

The nuclei of the midbrain have diverse functions, including roles in movement, sensation, and certain reflexes, such as the previously mentioned startle reflex. These functions will be discussed in later modules and in the peripheral nervous system chapter, along with the cranial nerves (see Chapter 13).

Pons

Inferior to the midbrain is the **pons,** whose anterior surface is rounded and prominent (Figure 12.14c). The anterior pons contains the descending motor tracts from the crus cerebri, some of which pass through the pons on their way to the spinal cord; others enter the cerebellum via the middle cerebellar peduncle. (See *A&P in the Real World: Locked-In Syndrome* on page 440 to discover what can happen if these motor tracts are damaged.) Posterior to these tracts is an area that contains reticular formation nuclei and cranial nerve nuclei. Surrounding the pontine tegmentum are the middle cerebellar peduncles.

The pontine nuclei have diverse roles. Certain nuclei help regulate movements, including those required for breathing. Others, particularly the cranial nerve nuclei, participate in reflexes, and some are involved in complex functions such as sleep and arousal.

Medulla Oblongata

The inferiormost portion of the brainstem is the **medulla oblongata** (or simply *medulla*), which blends with the spinal cord after it passes through the foramen magnum (Figure 12.14c). On the anterior surface of the medulla we find elevated ridges known as the **pyramids,** which contain the upper motor neuron fibers of the corticospinal tract (for this reason the corticospinal tract is also sometimes called the *pyramidal tract*) that come from the motor areas of the cerebral cortex. Within the pyramids most fibers of the right and left corticospinal tracts do something interesting—they cross over and switch sides, or **decussate** (DEE-kus-sayt), after which they descend through the spinal cord. Thus, the motor fibers originating from the right side of the cerebral cortex descend through the left side of the spinal cord, and vice versa.

On the posterior surface of the medulla are two other tracts of white matter known together as the **posterior columns** and the nuclei that receive signals from them (the **nucleus gracilis** [GRAS-ih-liss] and **nucleus cuneatus** [kyoo-nee-AH-tus]). The posterior columns carry sensory information from the spinal cord, and like most fibers of the corticospinal tracts, some also decussate in the medulla, so sensory fibers from the right side of the spinal cord are processed by the left side of the brain, and vice versa.

Other features of the medulla include the following:

- **Olive.** Lateral to each pyramid is a protuberance called the **olive,** which contains the *inferior olivary nucleus.* This nucleus receives sensory fibers from the spinal cord and passes them to the cerebellum.

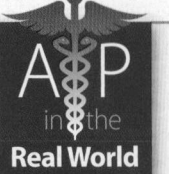

Locked-In Syndrome

Patients who have sustained damage to the motor tracts of the pons may develop *locked-in syndrome.* In this condition, the cerebral cortex is unable to communicate with the spinal cord, and no purposeful movement is possible. However, the sensory pathways between the brain and spinal cord remain intact, as do other cortical functions. Such patients are therefore literally "locked in" their bodies and are fully aware of everything going on around them.

Some patients are able to communicate using certain eye movements controlled by the midbrain. In fact, the French editor Jean-Dominique Bauby composed an entire book using only eye movements—he wrote *Le Scaphandre et le Papillon* (*The Diving Bell and the Butterfly*) by communicating one letter at a time to a transcriptionist.

Although a small number of patients with locked-in syndrome recover some motor functions, most do not. Overall, the prognosis of patients with this condition is poor, and most succumb to infections or other complications of paralysis.

- **Cranial nerve nuclei.** As with the midbrain and pons, the medulla also contains several cranial nerve nuclei, discussed in Chapter 13.
- **Reticular formation nuclei.** Several nuclei of the reticular formation, our next topic, reside in the medulla.

Reticular Formation

Figure 12.15 shows the **reticular formation,** a collection of more than 100 nuclei found in the central core of the three brainstem subdivisions. The reticular formation is one of the most complex regions of the brain. Observe in the figure that it receives input from several sources, including the cerebral cortex, the limbic system, and sensory stimuli. It sends output to essentially the entire brain as well as just about every part of the spinal cord. Like many structures of the limbic system, it is one of the more ancient parts of the mammalian brain from an evolutionary perspective. Accordingly, it shares many of the functions of the limbic system.

Although the reticular formation nuclei are diverse and many of them blend, we can generalize about the functions of several zones of these nuclei. In the center of the reticular formation are nuclei that function in sleep, pain transmission, and mood. Surrounding the central nuclei are those operative in motor functions, of both skeletal muscles and the autonomic nervous system. Several nuclei are intimately involved in the maintenance of homeostasis, such as the regulation of breathing and blood pressure. The lateral nuclei of the reticular formation play a part in sensation and in the level of alertness and activity of the cerebral cortex. Because of the vast number of connections between the reticular formation and the rest of the CNS, we will revisit these nuclei several times in this chapter.

Quick Check

- ☐ 6. List the functions of the brainstem.
- ☐ 7. What four types of structures are found within the brainstem?
- ☐ 8. What general functions does the reticular formation have? Where are the reticular formation nuclei located?

Sources of input:
Cerebral cortex
Limbic system
Sensory stimuli

Reticular formation

Figure 12.15 The reticular formation.

The Big Picture of Brain Anatomy

Figure 12.16

Cerebral hemispheres

Cerebral cortex

Basal nuclei:
Caudate nucleus
Putamen and globus pallidus

Lateral ventricles

Diencephalon:
Thalamus
Hypothalamus

Limbic system:
Hippocampus
Amygdala

Brainstem:
Midbrain
Pons
Medulla oblongata

Cerebellum

Putting It All Together: The Big Picture of Major Brain Structures and Their Functions

A big picture overview of the brain and its many structures is shown in **Figure 12.16** for your review. As in previous figures, the cerebrum is presented in a "see-through" fashion so that you can appreciate the location of the brain's deeper structures. When you know the structures fairly well, take a look at **Figure 12.17**, which summarizes the relationships between those structures and connects them with their functions.

Apply What You Learned

☐ 1. A person who is in a vegetative state has no activity in his or her cerebrum. Why can a person in this condition continue living for a certain period of time? Can such a person feel pain or initiate voluntary movements? Explain.

☐ 2. Predict the effects of damage to the basal nuclei or the cerebellum.

☐ 3. The condition *lissencephaly* (liss'-en-SEF-ah-lee; "smooth brain") is characterized by a lack of gyri and sulci in the cerebral cortex, which gives the brain hemispheres a smooth appearance. Predict the effects of such a condition.

☐ 4. Ms. Voschin sustained major head trauma in an automobile accident. She is unresponsive to sensory stimuli, and scans of her cerebral cortex demonstrate no cortical activity. During a surgical procedure to relieve pressure on her brain, it is discovered that she has sustained major damage to her thalamus. How does the damage to her thalamus explain her symptoms?

See answers in Appendix A.

The Big Picture of Major Brain Structures and Their Functions

Figure 12.17

DIVISION	STRUCTURES	SUBSTRUCTURES	MAIN FUNCTIONS
Cerebrum	Cerebral cortex	Primary motor cortex	• Plans and executes movement
		Primary somatosensory cortices	• Receive and process different types of sensory input
		Multimodal association areas	• Integrate sensory and motor information from a variety of different primary cortices
	Basal nuclei	Caudate nuclei Putamen Globus pallidus	• Regulate movement
	Limbic system	Hippocampus	• Plays a role in memory and learning
		Amygdala	• Plays a role in behavioral expression and emotion

Anterolateral view

DIVISION	STRUCTURES	MAIN FUNCTIONS
Diencephalon	Thalamus	• Controls information entry into the cerebral cortex • Edits, sorts, and routes stimuli
	Hypothalamus	• Regulates the autonomic nervous system • Regulates the sleep/wake cycle • Regulates thirst and hunger • Regulates body temperature • Produces hormones • Controls secretion from the pituitary gland

Midsagittal section

DIVISION	MAIN FUNCTION
Cerebellum	• Coordinates voluntary movement

Lateral view

DIVISION	STRUCTURES	MAIN FUNCTIONS
Brainstem	Midbrain	• Processes and routes visual and auditory stimuli to the thalamus • Carries motor fibers from the cerebral cortex • Monitors movement with the basal nuclei
	Pons	• Regulates breathing • Regulates reflexes • Regulates the sleep/wake cycle
	Medulla oblongata	• Contains tracts of white matter involved in movement and sensation • Regulates many homeostatic functions
	Reticular formation	• Involved in sleep and arousal • Involved in pain transmission • Plays a role in mood regulation • Involved in many homeostatic functions

Lateral view

12.3
Protection of the Brain

Learning Outcomes

1. Describe the functions of cerebrospinal fluid as well as the details of its production, its circulation within the CNS, and its ultimate reabsorption into the bloodstream.
2. Describe the structural basis for and the importance of the blood-brain barrier.
3. Identify and describe the cranial meninges, and explain their functional relationship to the brain.

The brain isn't a very durable organ—it is rather gelatinous and soft, and its neurons can be damaged by even slight pressure. Given the importance of the structures of the brain and their fragility, it should come as no surprise the body offers substantial protection for the brain. As you already know, it is housed within the cranial cavity, and the thick skull provides strong protection from external forces such as trauma, movement, and temperature changes.

Inside the skull, three features further shelter the brain from external and internal forces:

- the *cranial meninges,* which are a set of three membranes that surround the brain;
- the *cerebrospinal fluid* (*CSF*), a protective fluid that bathes the brain and fills its cavities; and
- the *blood-brain barrier,* which prevents many substances in the blood from gaining access to the cells of the brain.

This module explains the structure and function of each of these features.

The Cranial Meninges

◄◄ **Flashback**

1. What is dense irregular connective tissue, and what are its properties? (p. 139)
2. What is the periosteum? (p. 145)

The three **cranial meninges** (meh-NIN-jeez; singular, *meninx*) are protective membranes made primarily of dense irregular collagenous connective tissue (**Figure 12.18**). The outermost meninx, known as the **dura mater** (DOO-rah MAH-ter; "tough womb"), or simply *dura,* is the thickest and toughest of these membranes. Between the dura and the cranial bones is a potential space called the *epidural space.* (This is referred to as "potential" space because normally the dura is bound tightly to the bone and nothing is found between them except blood vessels.) Deep to the dura is another very narrow space, called the **subdural space,** that houses a thin layer of serous fluid and certain veins that drain the

brain. The middle meninx is the thinner **arachnoid mater** (ah-RAK-noyd), or simply *arachnoid,* and the thinnest and innermost meninx is the **pia mater** (PEE-ah; "gentle mother"), or *pia.* The arachnoid and pia are separated by another narrow, fluid-filled space called the **subarachnoid space.** The subarachnoid space contains CSF and the major blood vessels of the brain. *A&P in the Real World: Infectious Meningitis* on page 448 discusses what can happen when viruses or bacteria invade this space.

Each of the potential spaces around and between the meninges may become filled with blood; such conditions are called *epidural, subdural,* and *subarachnoid hematomas* (hee-mah-TOW-mahz), respectively, and tend to result from trauma or from the rupture of a weakened blood vessel. These serious conditions are potentially life-threatening, as they increase intracranial pressure, or pressure on the brain. Treatment of a hematoma in a meningeal space generally consists of drilling a hole through the skull and implanting a device that allows the blood to drain from the space, thus relieving pressure on the brain.

Now let's examine each of the three cranial meninges.

Dura Mater

As implied by its name, the dura mater is a tough, leathery membrane with abundant collagen fibers and very few elastic fibers. As shown in Figure 12.18a, it has two layers (from superficial to deep):

1. **Periosteal dura.** The outer *periosteal dura* is attached to the inner surface of the bones of the cranial cavity; it functions as the periosteum of those bones and has an extensive blood supply that resides in the epidural space.
2. **Meningeal dura.** The *meningeal dura* is the inner, avascular layer that lies superficial to the arachnoid mater.

Throughout most of the dura these two layers are fused, giving rise to a single, thick, inelastic membrane. However, in certain locations the two layers separate and form cavities called **dural sinuses.** The dural sinuses are venous channels that drain CSF and deoxygenated blood from the brain's many veins. In Figure 12.18 the dural sinuses are blue, reflecting the fact that they are filled with deoxygenated blood.

Dural sinuses may also be found where the meningeal dura folds over on itself and dives between certain structures of the brain (Figure 12.18b). These locations include the following:

- **Falx cerebri.** The **falx cerebri** (seh-REE-bry) is the partition between the right and left cerebral hemispheres, and lies in the longitudinal cerebral fissure. A large dural sinus called the **superior sagittal sinus** is situated superior to the falx cerebri.
- **Tentorium cerebelli.** The **tentorium cerebelli** (ten-TOW-ree-um seh-reh-BEL-ay) is the partition between the cerebellum and the occipital lobe of the cerebrum. It is so named because it resembles a "tent" covering the cerebellum.
- **Falx cerebelli.** The small **falx cerebelli** (not visible in Figure 12.18b) separates the right and left cerebellar hemispheres.

(a) Cranial meninges

(b) Dural folds and dural sinuses

(c) Cranial meninges, frontal section

Figure 12.18 Structure of the cranial meninges and dural sinuses.

Practice art labeling
@MasteringA&P®

Arachnoid Mater

The middle meninx, the arachnoid mater, is named for its resemblance to a spider web ("arachnoid" means "spider-like"). Like the dura, the arachnoid is composed of dense irregular collagenous connective tissue; however, it is thinner and somewhat more elastic. If you look at Figure 12.18c, observe that the arachnoid has inward extensions called *arachnoid trabeculae*. These trabeculae, made up of bundles of collagen fibers and fibroblasts, anchor the arachnoid mater to the deeper pia mater. Take note also that small bundles of the arachnoid project up through the meningeal dura

into the dural sinuses. These *arachnoid granulations* (also known as *arachnoid villi*) play an important role in returning CSF to the bloodstream.

Pia Mater

The innermost and very thin pia mater is the only meninx that physically touches the brain, and it closely follows each contour of the brain and dives into its sulci and fissures (see Figure 12.18). The pia mater is permeable to substances in the brain extracellular fluid and CSF. This allows substances to move between the two fluids and helps to balance the concentration of different solutes in them.

The Ventricles and Cerebrospinal Fluid

◀◀ **Flashback**

1. What properties of water make it good for protecting structures of the body? (p. 46)
2. Which type of neuroglial cell lines the cavities of the brain and spinal cord? (p. 388)

As we discussed earlier, the brain has cavities called **ventricles** (**Figure 12.19**). The ventricles are continuous with one another and with the cavity in the spinal cord, called the *central canal.* Both the ventricles and the central canal are lined with ependymal cells and filled with cushioning cerebrospinal fluid.

Recall from the previous module that four ventricles are found within the brain, as follows:

● **Right and left lateral ventricles.** The cavities within the right and left cerebral hemispheres are called the right and left **lateral ventricles;** these are considered the first and second ventricles. In Figure 12.19, from the anterior view,

the lateral ventricles resemble ram's horns; from the lateral view, they are horseshoe-shaped and have three regions—the *anterior horn,* the *inferior horn,* and the *posterior horn.*

● **Third ventricle.** The narrow **third ventricle** is housed between the two lobes of the diencephalon. It is continuous with the lateral ventricles via an opening called the **interventricular foramen.**

● **Fourth ventricle.** The third ventricle drains into the **fourth ventricle** via the **cerebral aqueduct,** a small canal that passes through the midbrain of the brainstem. The fourth ventricle is situated between the pons and the cerebellum and is continuous inferiorly with the central canal of the spinal cord. Figure 12.19a shows that this ventricle contains several posterior openings that allow CSF in the ventricles to flow into the subarachnoid space.

Cerebrospinal fluid (CSF) is a clear, colorless liquid similar in composition to plasma (the fluid portion of blood). CSF protects the brain by cushioning it and maintaining a constant temperature within the cranial cavity, removing wastes, and increasing the buoyancy of the brain. This buoyancy is particularly important, as without adequate CSF, the brain can literally crush itself under its own weight.

Most CSF is formed within each of the four ventricles by structures called **choroid plexuses** (KOR-oyd). The choroid plexuses are located where blood capillaries come into direct contact with the ependymal cells (which also produce some CSF) lining the ventricles. The capillaries within the choroid plexuses are *fenestrated*—their cells have small gaps between them and within their membranes (*fenestra* = "window") that allow fluid and electrolytes to escape the blood and enter the extracellular fluid.

Anterior horns of lateral ventricles

Lateral ventricles

Interventricular foramen

Posterior horns of lateral ventricles

Third ventricle

Cerebral aqueduct

Fourth ventricle

Inferior horns of lateral ventricles

Cerebellum (midsagittal section)

Brainstem (midsagittal section)

Central canal of the spinal cord

Brainstem

Diencephalon

(a) Lateral view

(b) Anterior view

Figure 12.19 Ventricles of the brain.

Figure 12.20 **Formation and flow of cerebrospinal fluid (CSF).**

The general pathway for the formation, circulation, and reabsorption of CSF is shown in **Figure 12.20**, and consists of the following steps:

① **Fluid and electrolytes leak out of the capillaries of the choroid plexuses.** Note that the fluid and electrolytes are now in the extracellular fluid of the ventricles, *not* that of the brain.

② **Ependymal cells secrete CSF into the ventricles.** The fluid and electrolytes that have escaped the capillaries enter the ependymal cells. The ependymal cells then secrete CSF into the ventricles by exocytosis.

③ **CSF circulates through and around the brain and spinal cord in the subarachnoid space.** The CSF circulates through the ventricles with the help of the ciliated ependymal cells. Some of this CSF then flows into the subarachnoid space around the spinal cord.

④ **Some of the CSF is reabsorbed into the blood in the dural sinuses via the arachnoid granulations.** As the CSF flows through the subarachnoid space, some of it is reabsorbed through the arachnoid granulations. The reabsorbed CSF ends up in the dural sinuses, where it mixes with venous blood.

At any given time about 150 ml (roughly two-thirds of a cup) of CSF circulates through the brain and spinal cord; approximately 50 ml is in the ventricles, the subarachnoid space, and around the spinal cord. However, about 750–800 ml of CSF is produced per day, which means that the CNS must drain old CSF as the choroid plexuses secrete new CSF. This occurs constantly, and in fact roughly 50% of the total CSF is completely replaced every 5–6 hours.

Hydrocephalus (hy′-droh-SEF-ah-luss; "water on the head") results if too much CSF is formed by the choroid plexuses, if the flow of CSF is blocked, or if not enough CSF is reabsorbed by the arachnoid granulations. If left untreated, hydrocephalus can increase intracranial pressure and lead to brain damage.

a

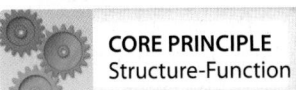

The Blood-Brain Barrier

◀◀ **Flashback**

1. To which molecules is the plasma membrane generally permeable? If the plasma membrane is not permeable to a substance, how does the cell transport the substance across the membrane? (p. 74)
2. What are tight junctions, simple squamous epithelial cells, and the basal lamina? (pp. 127, 129)
3. What are astrocytes, and what are their functions? (p. 388)

The brain is protected not only from threats outside the body but also from many within. Most chemicals and disease-causing organisms such as bacteria and viruses are denied access to the cells of the brain by the **blood-brain barrier,** which keeps the CSF and brain extracellular fluid separate from the blood.

The blood-brain barrier consists mainly of the simple squamous epithelial cells (referred to as *endothelial cells*) of the blood capillaries in the brain and their basal laminae, as well as astrocytes. These endothelial cells are unique in two ways:

● **They have more tight junctions than cells of most capillaries.** Recall that *tight junctions* are cell-cell junctions that prevent fluids and other molecules from passing through the spaces between cells (see Chapter 4). The cells of most capillaries are joined by tight junctions; however, the endothelial cells of the brain capillaries are joined by a very high number of these structures, a fact due largely to the influence of astrocytes on the developing brain. Here we see an example of the Structure-Function Core Principle (p. 25).

CORE PRINCIPLE
Structure-Function

● **They limit endocytosis and exocytosis.** Because of the tight junctions, molecules cannot pass between endothelial cells, so instead they must go through the phospholipid bilayers of the cells' plasma membranes. However, large, polar molecules such as proteins generally don't cross the plasma membrane freely and so rely on *endocytosis* and *exocytosis* to get into and out of the cell, respectively (see Chapter 3). The endothelial cells of the blood-brain barrier, compared to those of capillaries elsewhere in the body, have a very limited ability to perform these processes, which means that many molecules in the blood simply cannot enter the extracellular fluid of the brain. The converse is true as well: Many substances in the extracellular fluid of the brain cannot enter the blood.

Certain substances can cross the blood-brain barrier relatively easily, including those to which the plasma membrane is generally freely permeable, such as water, oxygen, carbon dioxide, and nonpolar, lipid-based molecules. Other molecules that cross easily include those for which protein channels or carriers exist, such as glucose, amino acids, and ions. However, most large, polar molecules are effectively prevented from crossing the blood-brain barrier in any significant amounts. This can be both a blessing and a curse—although the blood-brain barrier prevents many toxins and organisms from entering the brain, it also prevents the entry of many therapeutic drugs. For example, only a handful of antibiotics penetrate the blood-brain barrier in significant enough concentrations to treat bacterial infections in the brain, making such infections difficult to treat.

ConceptBOOST ⟫⟫⟫

Where Exactly Is the Blood-Brain Barrier?

You might be tempted to visualize the blood-brain barrier as a sort of raincoat that surrounds the brain. But notice that we never labeled a single structure as the "blood-brain barrier" on any figure. This is because the blood-brain barrier isn't around the brain; it's *within* the brain. Furthermore, it is not located in one distinct place but is found throughout the entire brain. To understand this, we need to first uncover more about the body's tiniest blood vessels: the *capillaries*. Capillaries are the vessels that deliver oxygen and nutrients to the body's cells and remove any wastes produced by the cells.

Now let's look at how this applies to the blood-brain barrier. Capillaries found in most organs and tissues are fairly leaky and allow a wide variety of substances to move from the blood to the extracellular fluid (and vice versa). But, as you just read, the capillaries of the brain are specialized to allow only selected substances to enter its extracellular fluid, and so they effectively act as a "barrier" that prevents other substances from doing so (**Figure 12.21**). The blood-brain barrier, therefore, is actually a property of the capillaries found throughout the brain rather than a distinct physical barrier. ■

A typical capillary allows water and small solutes to move from the blood to the ECF.

Astrocytes and tight junctions in brain capillaries limit the solutes that enter the brain ECF.

Figure 12.21 The blood-brain barrier.

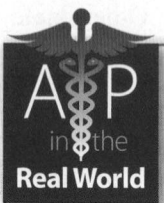

Infectious Meningitis

Infectious meningitis (meh-nin-JY-tis) is a potentially life-threatening infection of the meninges in the subarachnoid space. Inflammation of the meninges occurs, causing the classic signs of headache, lethargy, stiff neck, and fever. Infectious meningitis is diagnosed by examining a sample of CSF for infectious agents and white blood cells (cells of the immune system).

Certain viruses and bacteria are the most common agents of infectious meningitis. Differentiating between the two types of pathogens to ensure a correct diagnosis and proper treatment is critical—viral meningitis is generally mild and resolves in 1–2 weeks, whereas bacterial meningitis can rapidly progress to brain involvement and death, so aggressive antibiotic treatment is necessary. Fortunately, some of the most common forms of bacterial meningitis are largely preventable with safe and effective vaccines.

Quick Check

☐ 5. What two factors create the blood-brain barrier?
☐ 6. Which substances can easily cross the blood-brain barrier?

Apply What You Learned

☐ 1. A tear in the meninges may result in the leakage of CSF. Considering the functions of CSF, predict the symptoms you would see with a CSF leak.
☐ 2. When patients are given a drug that must enter the ECF of the brain, they are sometimes infused with the chemical mannitol. Mannitol temporarily disrupts tight junctions. How would this help deliver the drug?

See answers in Appendix A.

MODULE **12.4**
The Spinal Cord

Learning Outcomes

1. Describe the gross anatomy and location of the spinal cord.
2. Identify and describe the anatomical features seen in a cross-sectional view of the spinal cord.
3. Identify and describe the spinal meninges and the spaces between and around them.
4. Describe the differences between ascending and descending tracts in the spinal cord.

Although the spinal cord is less complex structurally than the brain, there is no denying its importance; the dramatic consequences of spinal cord injuries attest to this fact. The **spinal cord** is composed primarily of nervous tissue and is responsible for both relaying and processing information. The spinal cord functions in these two primary roles as follows:

- **Relay station.** The spinal cord acts as a *relay station* and an intermediate point between the body and the brain; in fact, it is the only means by which the brain can communicate with the body below the head and neck. It receives outgoing information from the brain and sends it to the rest of the body; it also receives incoming information from the body and sends it to the brain.
- **Processing station.** Although the spinal cord cannot perform the higher functions characteristic of the brain, it does do some integration and processing. In fact, certain activities called *spinal reflexes* can be carried out by the spinal cord alone, without influence from the brain (see Chapter 13).

This module tells us about the structure, organization, and functions of the nervous tissue in the spinal cord. First, however, we see how the spinal cord is protected.

Protection of the Spinal Cord

The meninges of the brain pass through the foramen magnum and cover the spinal cord, as well. These sheaths actually continue even after termination of the spinal cord and envelop the nerves at its base. As with the brain, three **spinal meninges** are present: the *dura,* the *arachnoid,* and the *pia mater* (**Figure 12.22a**). The spinal meninges are similar in structure to those of the brain except for the dura mater, which lacks a periosteal layer and consists only of a meningeal layer. Also, the spinal pia plays a role that the cranial pia does not—it helps to anchor the spinal cord in the vertebral cavity. Thin pieces of the spinal pia called **denticulate ligaments** extend outward through the arachnoid and attach to the spinal dura.

As with the cranial meninges, actual or potential spaces are found between the spinal meninges (**Figure 12.22b**), including (from superficial to deep):

1. **Epidural space.** As noted, the spinal cord lacks a periosteal dura, so an **epidural space** exists between the meningeal dura and the walls of the vertebral foramina. This space is filled with veins and adipose tissue, which helps to cushion and protect the spinal cord.
2. **Subdural space.** Much like the epidural space around the brain, the subdural space in the spinal cord is only a *potential* space; normally, the dura and arachnoid adhere to one another.
3. **Subarachnoid space.** Like the subarachnoid space around the brain, the spinal subarachnoid space lies between the arachnoid and pia and is filled with a very thin layer of CSF. The area inferior to the base of the spinal cord contains a larger volume of CSF, making it a useful place from which to sample CSF if needed.

See *A&P in the Real World: Epidural Anesthesia and Lumbar Punctures* for information on how these spaces are used clinically.

Spinal cord

Pia mater

Subarachnoid space

Arachnoid mater

Subdural space

Dura mater

(a) Spinal meninges and spinal cord, anterior view

Posterior

Dura mater

Arachnoid mater

Pia mater

Epidural space

Subdural space

Subarachnoid space

Denticulate ligament

Body of vertebra

Anterior

(b) Spinal meninges and spinal cord, transverse section

Figure 12.22 Structure of the spinal meninges.

Practice art labeling
@MasteringA&P®

Quick Check

☐ 1. What are the two primary roles of the spinal cord?
☐ 2. List and describe the three spinal meninges.
☐ 3. What are the three meningeal spaces around the spinal cord? Which are actual spaces, and which are potential spaces?

A&P in the Real World

Epidural Anesthesia and Lumbar Punctures

The anatomy of the meningeal spaces in the spinal cord has proved quite useful for certain medical procedures. One such procedure is the administration of *epidural* or *spinal anesthesia*, during which a local anesthetic medication is injected into the epidural space through an inserted needle. This procedure causes "numbing" of the nerves extending off the spinal cord below the level of the injection, which means they are unable to transmit motor or sensory impulses. This type of anesthesia is commonly given to women during childbirth, as well as to both sexes during certain surgical procedures.

Another procedure that involves the spinal meningeal spaces is a **lumbar puncture**, also known as a *spinal tap*. During a lumbar puncture, a needle is inserted into the subarachnoid space and CSF is withdrawn for analysis, as shown here:

Arachnoid mater

CSF in subarachnoid space

Nerve roots

Fifth lumbar vertebra (midsagittal-section)

The needle is generally inserted between the fourth and fifth lumbar vertebrae to avoid any possibility of injuring the spinal cord. Lumbar punctures are performed to assess for a number of conditions, including meningitis, encephalitis, and multiple sclerosis.

External Spinal Cord Anatomy

As the first module told us, the spinal cord extends from the foramen magnum and terminates between the first and second lumbar vertebrae (**Figure 12.23**). On the posterior side of the spinal cord is the narrow **posterior median sulcus**, and on its anterior side is a wider slit called the **anterior median fissure**, both of which are present along its entire length. The end of the spinal cord forms the cone-shaped **conus medullaris** (KOHN-us med-yoo-LAIR-us). At the level of the first and second lumbar vertebrae, the spinal pia gathers into a very thin structure known as the **filum terminale** (FY-lum ter-mee-NAL-ay), which continues through the vertebral cavity and anchors into the first coccygeal vertebra.

Note in Figure 12.23a that the spinal cord has two "bulges"—one in the cervical region and another in the lumbar region. These bulges are known as the **cervical enlargement** and **lumbar enlargement,** respectively. The **nerve roots** that fuse to form the spinal nerves serving the upper and lower limbs attach to these enlargements. **Spinal nerves,** which are part of the PNS, carry sensory and motor impulses to and from the spinal cord. Projecting off either side of the spinal cord between the vertebrae are *posterior* and *anterior nerve roots*. Roots of spinal nerves extend inferiorly from the conus medullaris and fill the remainder of the vertebral cavity. This bundle of spinal nerve roots is collectively called the **cauda equina** (KOW-dah eh-KWY-nah), so named because it resembles a horse's tail (*equin-* = "horse").

Quick Check

☐ 4. Where are the two bulges of the spinal cord?
☐ 5. What is the cauda equina?

(a) **Spinal cord, posterior view**

(b) **Section of spinal cord and associated structures, posterior view**

Figure 12.23 External structure of the spinal cord.

Internal Spinal Cord Anatomy

As you read in Module 12.1, the butterfly-shaped spinal gray matter is surrounded by tracts of white matter (see Figures 12.2b and 12.23b). A **central canal** filled with CSF runs down the middle of the cord. The two "butterfly wings" are connected by a thin strip of gray matter called the **gray commissure** (KAHM-ih-shoor), which surrounds the central canal of the spinal cord. You can tell the anterior and posterior sides of the spinal cord apart by the different shapes of the wings: Those on the anterior side are broader, whereas those on the posterior side are thinner and extend nearly to the outer surface of the spinal cord.

Spinal Gray Matter

Three distinct regions of gray matter are found within the spinal cord, each of which houses neurons with specific related functions (**Figure 12.24**). The three regions include the following:

- **Anterior horn.** The **anterior horn** (or *ventral horn*) is the area of gray matter that makes up the anterior wing of the butterfly. The neurons of the anterior horn are concerned with somatic motor functions (those of the skeletal muscles).
- **Posterior horn.** The longer and thinner **posterior horn** (or *dorsal horn*) makes up the posterior wing of the butterfly. It contains the cell bodies of neurons that are involved in processing both somatic and visceral incoming sensory information.
- **Lateral horn.** The **lateral horn** is present only from the first thoracic vertebra to the lumbar portion of the cord. It contains the cell bodies of neurons responsible for motor

control of the viscera via the autonomic nervous system (discussed in Chapter 14).

As you can see in Figure 12.24, the horns give rise to anterior (motor) and posterior (sensory) nerve roots. Notice that the posterior root has a swollen area just after it emerges—this is the *posterior* or *dorsal root ganglion*, where the cell bodies of these neurons are housed. These different roots form spinal nerves that contain both motor and sensory axons. The nerves carry motor and sensory information to and from, respectively, different parts of the body, such as the muscles and skin.

Spinal White Matter: Ascending and Descending Tracts

Recall that one of the main functions of the spinal cord is to act as a relay station. This function is carried out by its white matter, which contains the axons of neurons that travel to and from the brain. The spinal white matter, like the gray matter, is organized into regions. Each general region of spinal white matter is called a **funiculus** (foo-NIK-yoo-luss; "long rope"), and three funiculi lie on each side of the spinal cord: the **posterior funiculus**, the **lateral funiculus**, and the **anterior funiculus**. The white matter within each funiculus is further organized into tracts, also known as *columns*. The ascending and descending tracts bring information to and from a specific part of the brain, respectively. Like the rest of the spinal cord, the tracts are bilaterally symmetrical, meaning that the right and left sides of the cord contain the same ascending and descending tracts that serve the right and left sides of the body. Sensory pathways travel in posterior and lateral funiculi, whereas motor pathways tend to travel in anterior and lateral funiculi.

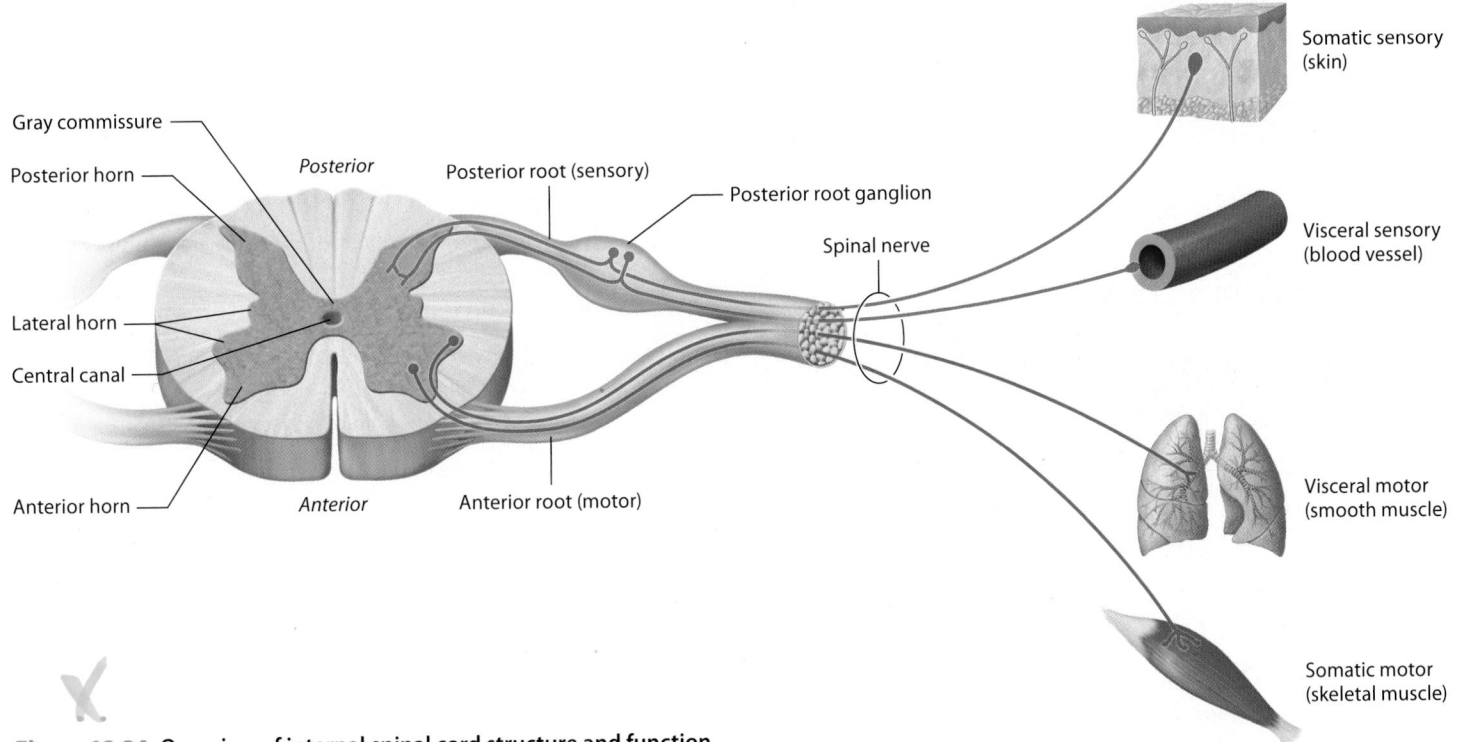

Figure 12.24 Overview of internal spinal cord structure and function.

Ascending Tracts (Sensory) Several ascending tracts carry various kinds of sensory information (**Figure 12.25a**). The major tracts include the following:

- **Posterior columns.** Within the posterior funiculus we find two tracts that together are called the **posterior columns.** The two tracts of the posterior columns are the medial *fasciculus gracilis* (fah-SIK-yoo-las) and the lateral *fasciculus cuneatus.* These tracts carry somatosensory information such as touch and *proprioception* (joint position) to the brain. The nucleus gracilis transmits such signals from the lower limbs and lower trunk, whereas the nucleus cuneatus transmits stimuli from the upper trunk and upper limbs.
- **Spinocerebellar tracts.** Several tracts in the lateral funiculi carry information about joint position and muscle stretch from the entire body to the cerebellum, including the anterior and posterior *spinocerebellar tracts.*

- **Anterolateral system.** Another set of ascending tracts is the *anterolateral system,* which includes the *spinothalamic tracts.* These tracts travel in the anterior and lateral funiculi; they send pain and temperature information from the entire body to the brain.

Descending Tracts (Motor) The main descending spinal tracts are shown in **Figure 12.25b**. The largest are the **corticospinal tracts,** which help to control the skeletal muscles below the head and neck. Recall that the corticospinal tracts originate primarily from the motor areas of the cerebral cortex. The fibers descend first as part of the internal capsule, then through the brainstem, where the majority of them decussate. As they travel through the lateral funiculi of the spinal cord, they bring motor information to the appropriate places in the anterior horn. Other, smaller descending tracts are also shown in Figure 12.25b.

ASCENDING TRACT	FUNCTION
Posterior columns—fasciculus gracilis	• Carry somatosensory information including fine touch, vibration, and proprioception from the lower limbs
Posterior columns—fasciculus cuneatus	• Carry somatosensory information including fine touch, vibration, and proprioception from the trunk, neck, and upper limbs
Spinocerebellar tracts	• Carry proprioceptive information to the cerebellum
Anterolateral system—spinothalamic tracts	• Carry information about pain, temperature, and certain types of touch

(a) Ascending tracts (sensory)

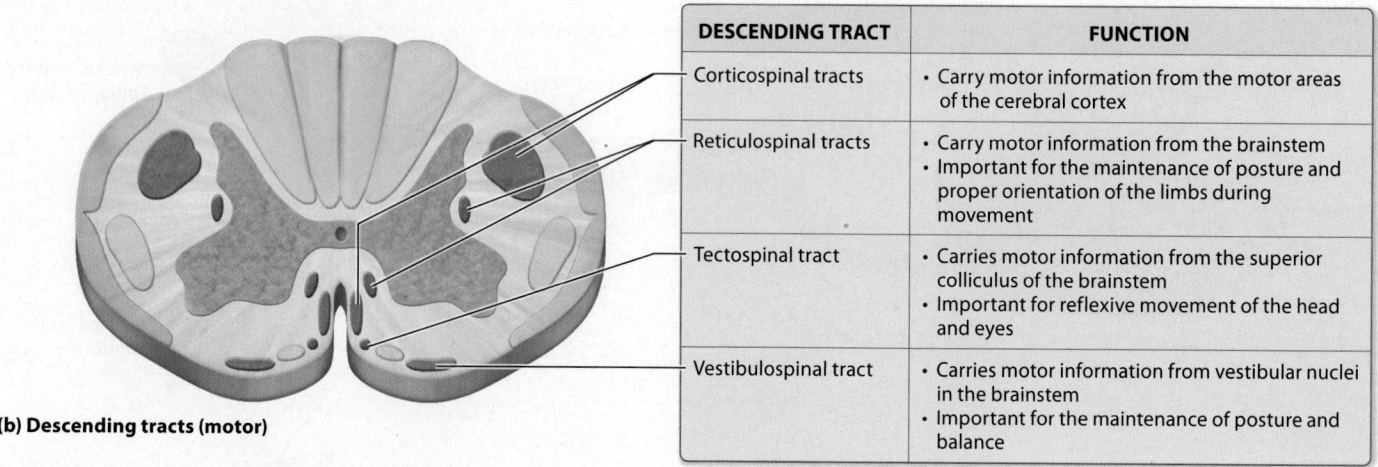

DESCENDING TRACT	FUNCTION
Corticospinal tracts	• Carry motor information from the motor areas of the cerebral cortex
Reticulospinal tracts	• Carry motor information from the brainstem • Important for the maintenance of posture and proper orientation of the limbs during movement
Tectospinal tract	• Carries motor information from the superior colliculus of the brainstem • Important for reflexive movement of the head and eyes
Vestibulospinal tract	• Carries motor information from vestibular nuclei in the brainstem • Important for the maintenance of posture and balance

(b) Descending tracts (motor)

Practice art labeling
@MasteringA&P®

Figure 12.25 Ascending and descending tracts of the spinal cord.

Quick Check

☐ 6. What are the functions of the neurons in the three horns of spinal gray matter?

☐ 7. What are the major ascending tracts of spinal white matter? What are their functions?

☐ 8. What are the major descending tracts of spinal white matter? What are their functions?

Apply What You Learned

☐ 1. Why are no dural sinuses present in the spinal cord?

☐ 2. Explain why a patient with damage to one of the tracts in the posterior columns still has sensation that is mostly intact.

See answers in Appendix A.

MODULE **12.5**

Sensation Part I: Role of the CNS in Sensation

Learning Outcomes

1. Describe the roles of the central and peripheral nervous systems in processing sensory stimuli.
2. Describe the locations and functions of first-, second-, and third-order neurons in a sensory pathway.
3. Explain the ways in which special sensory stimuli are processed by the CNS.

We humans have an amazing ability not only to perceive different **sensory stimuli** (things that cause our senses to respond) but also to assemble multiple sensory stimuli into a single mental picture or idea. For example, the sound of a snake moving through leaves is processed in one part of the brain, the sight of the moving snake is processed by another, the color of the snake by another, the feel of the snake's scales by another, and the smell of the snake's musk by yet another. Each of these disparate stimuli reaches the brain by a two-part process:

1. **The stimulus is detected by neurons in the PNS and sent as sensory input to the CNS.** These neurons initiate action potentials that travel along their axons to eventually reach the spinal cord or the brainstem.
2. **In the CNS, the sensory input is sent to the cerebral cortex to be interpreted.** The spinal cord or brainstem carries a signal generated in response to a stimulus along its ascending tracts until the signal finally reaches the cerebral cortex, where it is interpreted.

When the CNS has received all of the different sensory inputs, it integrates them into a single **perception** (a conscious awareness of the sensation). In this case, we very quickly perceive a moving snake so that we can decide if we need to move too.

We can group sensations into two basic types: the *special senses,* which are detected by special sense organs and include vision, hearing, equilibrium, smell, and taste, and the *general senses,* which are detected by sensory neurons in the skin, muscles, or walls of organs. The general senses may be further subdivided into the *general somatic senses,* those involving the skin, muscles, and joints, and the *general visceral senses,* those involving the internal organs.

The initial detection of stimuli is discussed with the peripheral nervous system (see Chapter 13 for Sensation Part II). This module explores the delivery and interpretation of stimuli, including the central pathways of sensation and their components in the brain and spinal cord. Note that most general visceral and certain general somatic sensory inputs are delivered to parts of the brain such as the cerebellum and the brainstem; these pathways are discussed in a later chapter (see Chapter 14). The following discussion covers only those sensory inputs that are consciously interpreted by the cerebral cortex to produce sensations.

General Somatic Senses

◄◄ Flashback

1. What are the components of a pseudounipolar neuron, and what are their functions? (p. 387)

Recall that the *general somatic senses* pertain to touch, stretch, joint position, pain, and temperature. Two types of touch stimuli are delivered by different pathways to the appropriate part of the cerebral cortex:

- **Tactile senses** pertain to fine or *discriminative* touch, and include vibration, two-point discrimination (see Chapter 5), and light touch. The tactile senses allow you to discriminate between different shapes and textures without visual input; your tactile senses are at work when you can tell with just your fingers whether the coin in your pocket is a dime or a nickel.
- **Nondiscriminative touch,** also called *crude touch,* lacks the fine spatial resolution of the tactile senses. Often when a body part is numbed by local anesthesia, the sensation of crude touch is preserved and so pressure can still be detected.

Most of the general somatic senses are considered *mechanical senses*—the neurons that detect them are responsive to mechanical deformations in the skin, a joint, and/or an organ (temperature sensation is the exception). Let's first explore the pathways that carry this sensory information, and then how the brain interprets touch and pain stimuli.

Sensory Pathways through the Spinal Cord to the Brain

Two major ascending tracts in the spinal cord carry somatic sensory information to the brain: the *posterior columns/medial lemniscal system* and the *anterolateral system.* The basic pathway consists of an initial **first-order neuron,** which is the sensory neuron that detects the initial stimulus in the PNS. The central process of the pseudounipolar neuron then synapses on a **second-order neuron,** an interneuron located in the posterior horn of the spinal cord or the brainstem. The axons of second-order neurons generally synapse on **third-order neurons,** interneurons in the thalamus. Then, in turn, the axons of third-order neurons deliver impulses to the cerebral cortex. Below is the basic pathway, as well as its relationship to the motor functions of the spinal cord:

These sensory pathways warrant a closer look.

Posterior Columns/Medial Lemniscal System The axons of neurons that transmit tactile sensory information about discriminative touch travel with axons that convey information regarding proprioception, or joint position, along tracts known as the **posterior columns/medial lemniscal system.** This system, shown in **Figure 12.26,** begins with a stimulus from the first-order neuron that enters the spinal cord on its posterior side. Axons from these neurons ascend through the spinal cord in the posterior columns. As mentioned before, two tracts are located within the posterior columns: (1) the medial **fasciculus gracilis,** which carries information from the lower limbs, and (2) the lateral **fasciculus cuneatus,** which carries information from the trunk, neck, and the upper limbs. Figure 12.26 shows that both tracts of the posterior columns ascend on the same side of the body as the original stimulus.

When the axons of the fasciculus gracilis and fasciculus cuneatus neurons enter the medulla, they synapse on the second-order neurons of the *nucleus gracilis* and *nucleus cuneatus,* respectively. The axons of these neurons then decussate and form tracts called the **medial lemniscus.** The fibers of the medial lemniscus ascend through the pons and midbrain until they reach the thalamus. In the thalamus, the neurons synapse on third-order neurons, which ascend to the cerebral cortex.

Figure 12.26 Ascending (sensory) pathways: the posterior columns/ medial lemniscal system in the right and left sides of the body.

Anterolateral System As with the tactile senses and proprioception, the fibers that carry the stimuli of pain, temperature, and nondiscriminative touch enter the posterior spinal cord. However, these stimuli ascend via pathways in the anterolateral spinal cord, referred to as the **anterolateral system.** In **Figure 12.27** you can see a difference between the posterior columns and the anterolateral system: The first-order neurons synapse on the second-order neurons in the posterior horn, after which they decussate.

Several tracts lie within the anterolateral system, the largest of which are the right and left **spinothalamic tracts.** As the name implies, they send signals through the spinal cord to the sensory

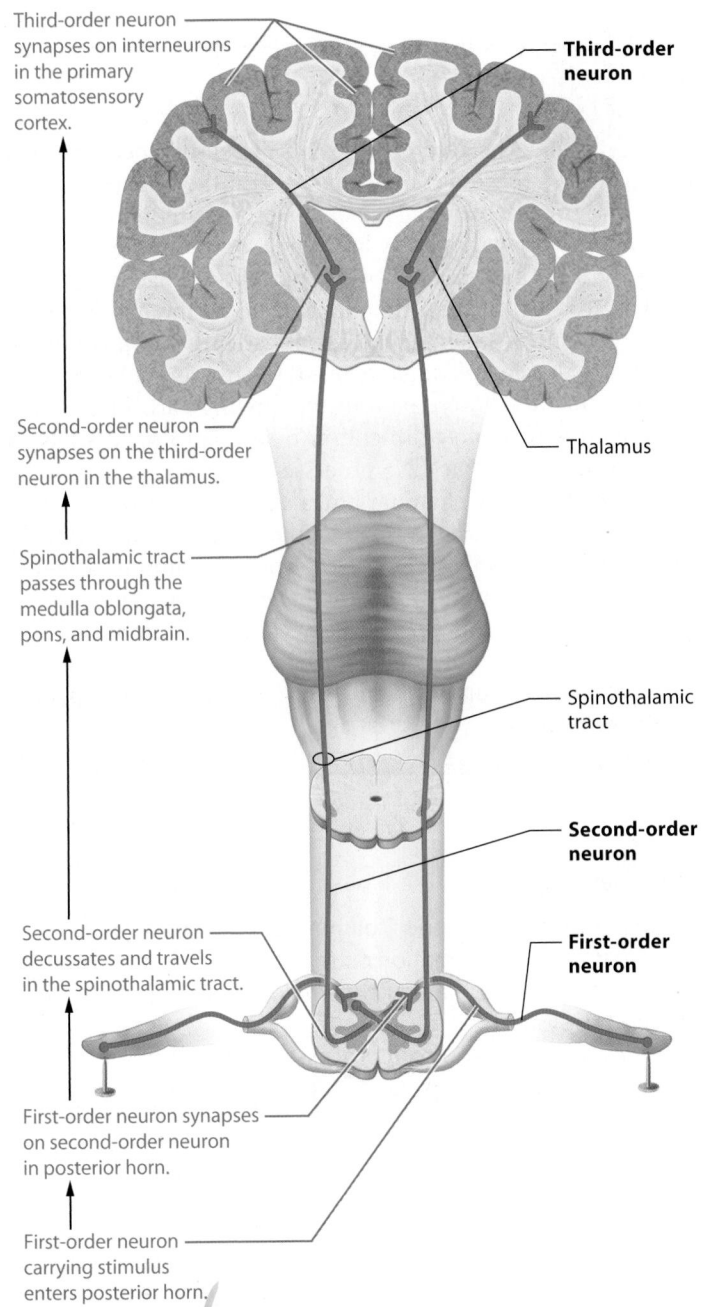

Third-order neuron synapses on interneurons in the primary somatosensory cortex.

Third-order neuron

Second-order neuron synapses on the third-order neuron in the thalamus.

Thalamus

Spinothalamic tract passes through the medulla oblongata, pons, and midbrain.

Spinothalamic tract

Second-order neuron

Second-order neuron decussates and travels in the spinothalamic tract.

First-order neuron

First-order neuron synapses on second-order neuron in posterior horn.

First-order neuron carrying stimulus enters posterior horn.

Figure 12.27 Ascending (sensory) pathways: the right and left spinothalamic tracts (part of the anterolateral system).

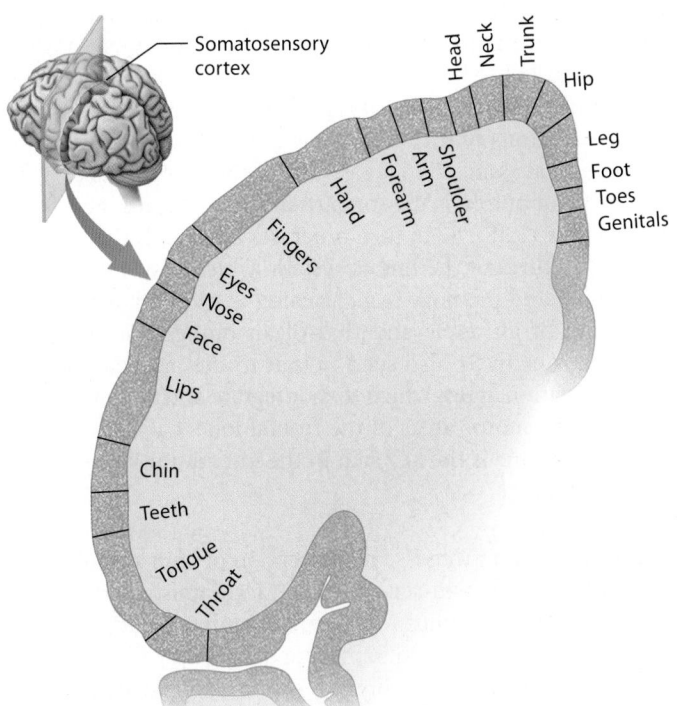

Somatosensory cortex

Head Neck Trunk Hip Leg Foot Toes Genitals Shoulder Arm Forearm Hand Fingers Eyes Nose Face Lips Chin Teeth Tongue Throat

(a) Map of the primary somatosensory cortex (S1)

(b) The sensory homunculus

Figure 12.28 Representations of the primary somatosensory cortex.

relay nuclei of the thalamus. Third-order neurons from the thalamus then carry the information to the cerebral cortex.

Role of the Cerebral Cortex in Sensation

The thalamus relays most incoming information to the primary somatosensory cortex, or S1, in the postcentral gyrus (see Figure 12.5). Each part of the body is represented by a specific region of S1, a type of organization called **somatotopy** (*-topos* = "place"). Note that the map of S1 in **Figure 12.28a** illustrates that different parts of the body are represented unequally—a disproportionate amount of space is dedicated to the hands and the face, for example, which reflects the importance of manual dexterity, facial expression, and speech to humans. This unequal representation of body parts in S1 is exemplified by the *sensory homunculus* (literally "little man"), which shows what we would look like if the body were proportionate to the areas of S1 (**Figure 12.28b**). We now turn our attention to how the primary somatosensory cortex processes two different types of sensory input, those for touch and pain.

Processing of Touch Stimuli The thalamic nuclei relay touch information from the spinothalamic tracts and posterior columns primarily to S1 for conscious perception. Once the sensory information has reached S1, it is processed (i.e., the touch is perceived) and axons of the S1 neurons then transmit the information to cortical association areas. One of the main areas to which S1 axons send output is the somatosensory association cortex, or S2 (see Figure 12.5). S2 in turn processes and sends the information to structures of the limbic system, which play a role in tactile learning and memory (e.g., learning to differentiate between the texture of an apple and that of an orange without visual cues). Neurons in S1 also send output to the parietal and temporal association areas, which then integrate and relay this information to the motor areas of the frontal lobe. Current evidence indicates that this is the key step in the integration of motor and sensory functions.

Processing of Pain Stimuli The perception of pain stimuli, called **nociception** (NOH-sih-sep-shun), is accomplished differently. For example, the thalamus relays pain stimuli to multiple parts of the brain, including S1 and S2, where the sensory discriminative (location, intensity, and quality) aspects of the pain are perceived and analyzed. Such stimuli are also sent to the basal nuclei, structures of the limbic system, the hypothalamus, and the prefrontal cortex, where the emotional, behavioral, and other aspects of pain are processed.

The processing of pain input differs from that for the other somatic senses in additional ways. For instance, the cortex seems to have a great deal of influence over how pain is perceived. This is evident in a phenomenon called the *placebo effect,* in which a patient is given a *placebo* (plah-SEE-bow)—a "dummy treatment" with no actual therapeutic value. The use of a placebo as an *analgesic* (an-ahl-JEE-zik; pain reliever) can be surprisingly effective; in most studies, up to 30% of patients report relief from pain when a placebo is administered, but this number has been reported to be as high as 75%. This does not mean that the patient is "crazy" or that the pain is imagined; instead, it reflects the very real ability of the brain to modulate the perception of pain.

The explanation for the placebo effect appears to involve a descending pathway that originates largely in S1, the amygdala, and a region of the midbrain known as the *periaqueductal gray matter* (so named because it surrounds the cerebral aqueduct). The neurons of the periaqueductal gray matter release *endorphins,* neurotransmitters that render the neurons of the posterior horn in the spinal cord less sensitive to pain input, an example of

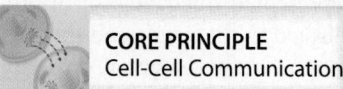
CORE PRINCIPLE
Cell-Cell Communication

the Cell-Cell Communication Core Principle (p. 27). Thus, the stimulus causing the pain (e.g., a broken bone) is still present and its intensity has not decreased, but the CNS neurons *perceive* it as being less intense or even absent. Painkillers such as morphine work by binding to the same receptors as endorphins. Researchers don't yet know why some people can modulate pain perception in this way more effectively than others. How pain is processed can have other unusual consequences, one of which we explore in *A&P in the Real World: Phantom Limb Pain.*

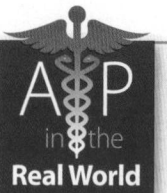

Phantom Limb Pain

The role of the CNS in sensory perception becomes quite apparent in the phenomenon known as *phantom limb.* A phantom limb occurs after a patient has had a limb, digit, or even a breast amputated; it is characterized by the perception that the body part is still present and functional even in the absence of sensory input. A small percentage of patients with phantom limbs also develop *phantom pain*—the sensation of burning, tingling, or even severe pain in the missing part. Phantom limb pain is extremely difficult to treat due to the complicated way in which the CNS processes pain.

The existence of phantom limbs and phantom pain support the idea that S1 has a "map" of the body that exists independently of the PNS. Thus, even when the limb is absent and no stimuli from the limb are therefore being sent to the brain, S1 continues to perceive the limb as present. Over time, the map generally will rearrange itself so that the body is represented accurately, and the phantom limb and pain sensations decrease.

Quick Check

☐ 1. Where are the posterior columns and their two tracts? What type of information do they carry?

☐ 2. What is the anterolateral system? Which stimuli are carried via this system?

☐ 3. How are touch and pain processed by the cerebral cortex?

Introduction to the Special Senses

The *special senses* include vision, hearing (audition), taste (gustation), smell (olfaction), and balance (vestibular sensation). Like the general somatic senses, each special sense involves neurons that detect a stimulus and bring it to the CNS for processing and integration. The details of the special senses are discussed in a later chapter (see Chapter 15); in this chapter, we introduce their basic pathways within the CNS.

We can note many similarities in the ways general somatic and special sensory stimuli are processed. As with the somatic senses, the cerebral cortex interprets the majority of special sensory information and the thalamus is the "gateway" for the entry of most special sensory stimuli into the cerebral cortex (olfaction is an exception). The usual pathway for processing each type of sensory stimulus is as follows:

- **Visual stimuli.** Most visual stimuli are delivered directly to the thalamus, which relays them to the primary visual cortex in the occipital lobe. The primary visual cortex processes these stimuli and perceives an object's depth and color, and also detects rapidly changing stimuli (which occurs when you are watching a TV program, for example). It also sends information to association areas in the temporal and parietal lobes, which are crucial for object recognition and spatial awareness, respectively.

- **Auditory stimuli.** Stimuli from sound waves are delivered first to nuclei in the brainstem. Some of these stimuli are processed in those nuclei; the remainder are routed to the thalamus and then to the primary auditory cortex in the superior temporal lobe. The neurons in this part of the cerebral cortex are specialized for processing sounds with complex tempos such as speech and music. Information from the primary auditory cortex is relayed to many association areas, including Wernicke's area for language comprehension.

- **Gustatory stimuli.** Stimuli pertaining to taste are delivered to a nucleus in the medulla and relayed to the thalamus. From the thalamus, gustatory stimuli are sent to gustatory cortices in both the insula and the parietal lobes that analyze the different components of the taste. These cortices also communicate with the hypothalamus and limbic system, which presumably influence taste preferences and food-seeking behaviors.

- **Olfactory stimuli.** As stated earlier, olfactory stimuli actually enter the cerebral cortex of the limbic system for initial processing, without going through the thalamus. From the limbic cortex, these stimuli are relayed to a number of places, including the thalamus (which then transmits them to the prefrontal cortex), the hypothalamus, and other components of the limbic system. This feature allows olfactory stimuli to influence a number of behaviors, including those related to feeding, emotion, and cognition.

- **Vestibular stimuli.** The processing of vestibular stimuli involves multiple brainstem nuclei, the cerebellum, descending pathways through the spinal cord, and pathways through the thalamus to the cerebral cortex. For this reason, we will save the discussion of all of these pathways for the special senses chapter (see Chapter 15).

Quick Check

☐ 4. How is the processing of olfactory stimuli different from that of other special sensory stimuli?

Apply What You Learned

☐ 1. Would you expect the fingertips or the back to be represented by a greater area of the primary somatosensory cortex? Explain.

☐ 2. A person with a spinal cord injury is experiencing loss of pain sensation in his left leg and loss of tactile sensation in his right leg. On which side of the spinal cord is the injury located? Explain.

☐ 3. The term "painkiller" is used for drugs that bind to receptors for endorphins in the CNS. How do these drugs relieve pain? Do they actually treat the cause of the pain? (*Hint:* Where do they bind?)

See answers in Appendix A.

MODULE **12.6**

Movement Part I: Role of the CNS in Voluntary Movement

Learning Outcomes

1. Describe the locations and functions of the upper and lower motor neurons in a motor pathway.
2. Explain the roles of the cerebral cortex, basal nuclei, and cerebellum in movement.
3. Describe the overall pathway from the decision to move to the execution and monitoring of a motor program.
4. Explain how decussation occurs in sensory and motor pathways, and predict how brain and spinal cord injuries affect these pathways.

We don't often stop to consider the complex neurological processes behind even the simplest movements, or the rapidity with which these occur. For example, we don't have to think about which muscles to activate in standing up or how to remain balanced as we walk across the room, because our nervous system does this automatically once a motor skill is learned. Planning and coordination of voluntary movement are carried out within the CNS, and involve the motor areas of the cerebral cortex, the basal nuclei, the cerebellum, and the spinal cord. Each of these components is required for motion to occur smoothly and rapidly, and if one is disrupted, the consequences can be dramatic.

Three types of neurons are directly involved in eliciting a muscle contraction. The first are the upper motor neurons, described in Module 12.2, whose cell bodies reside in the motor areas of the cerebral cortex (some upper motor neurons are located in the brainstem, as well). Their axons descend through the cerebral white matter, and as they enter the brainstem and spinal cord, they interact with local interneurons. These interneurons pass on the messages of the upper motor neurons to their neighbors, the lower motor neurons. The cell bodies of these lower motor neurons lie in the anterior horn of the spinal gray matter, and their axons exit the spinal cord and innervate the skeletal muscles. Lower motor neurons are part of the PNS.

This module discusses the pathway from the upper motor neurons to the lower motor neurons leaving the spinal cord. Then we explore how movement is achieved, including the roles of the upper motor neurons and how the cerebellum and basal nuclei modulate their activity. In the PNS chapter, we trace the pathway from the lower motor neurons to the muscles (see Chapter 13).

Motor Pathways from the Brain through the Spinal Cord

Recall from Module 12.4 that the axons from the cortical motor areas unite to form several tracts of white matter. Among the largest tracts are the right and left **corticospinal tracts,** which, as noted, control the muscles below the head and neck via the lower motor neurons of spinal nerves, and the **corticonuclear tracts,** which control the muscles of the head and neck via the lower motor neurons of the cranial nerves. Let's trace the pathway taken by motor information in each of these tracts.

Corticospinal Tracts

As you can see in **Figure 12.29**, the axons that form the corticospinal tracts originate from the cell bodies of upper motor neurons of the primary motor cortex and premotor cortex. These axons combine and descend first through the corona radiata and internal capsule until they reach the brainstem, where they go through the midbrain and pons. Recall from Module 12.4 that at the level of the medullary pyramids most of the fibers of the corticospinal tracts decussate. Thus, the neurons on the right side of the brain send fibers to the left side of the body, and vice versa. This is why patients with damage to certain parts of the brain show loss of function of muscles on the *opposite* side of the body from their injury.

The fibers that decussate travel in the lateral funiculi of the spinal cord and are known as the right and left lateral corticospinal tracts. As shown in Figure 12.29, when the fibers of the lateral corticospinal tracts reach the anterior horn of spinal gray matter, they synapse on local interneurons. Note that over half of the corticospinal fibers terminate at the cervical levels of the spinal cord to control motor functions of the upper limbs. This location reflects the importance of precise motor control of our arms and hands.

About 10–15% of the motor fibers from the cerebrum, however, do not decussate and so remain on the same side. These fibers travel through the anterior funiculi of the spinal white matter and are therefore called the right and left *anterior corticospinal tracts.*

Corticonuclear Tracts

The pathway for the corticonuclear tracts (formerly called the *corticobulbar tracts*) is similar to that for the corticospinal tracts. Like the corticospinal tracts, they originate from the cell bodies of upper motor neurons. The corticospinal and corticonuclear tracts travel together through the corona radiata and internal capsule to the brainstem. However, the fibers of the corticonuclear tracts do not enter the spinal cord. Rather, they synapse on interneurons that communicate with cranial nerve nuclei at various levels of the brainstem, whose lower motor neurons innervate muscles of the head and neck.

These fibers do not decussate, but interestingly most of the cranial nerve nuclei communicate with upper motor neurons from both cerebral hemispheres. For this reason, damage to upper motor neurons on one side of the cerebrum does not lead to noticeable deficits for many of the cranial nerves.

Figure 12.29 Descending (motor) pathways: the right and left lateral corticospinal tracts.

Labels in figure:
Upper motor neurons originate in the primary motor cortex (and premotor cortex).

Corona radiata

Internal capsule

Corticospinal tract passes through the midbrain and pons.

Most fibers decussate at medullary pyramids in medulla oblongata.

Lateral corticospinal tract

Upper motor neuron synapses on the interneurons of the anterior horn.

Lower motor neuron

Interneuron synapses on lower motor neuron.

Quick Check

☐ 1. What is the main difference between the corticospinal and corticonuclear tracts?
☐ 2. Where do the fibers of the corticospinal tracts decussate?

Role of the Brain in Voluntary Movement

◄◄ **Flashback**

1. What are multipolar neurons? What are the functions of their dendrites? (p. 387)
2. What happens at excitatory and inhibitory synapses? (p. 409)

The pathways just discussed tell only part of the story of how we achieve movement. Realistically, even a motion as simple as pushing open a door requires the simultaneous firing of a dizzying number of neurons as part of a selected group of actions called a **motor program.** The motor program for pushing open a door includes the muscles you will need to open it, the amount of force that will be required, the muscles that will be used to maintain your balance as you push it open, and more. Execution of this motor program requires the firing of neurons in motor association areas; the firing of upper motor neurons; and input from the basal nuclei, the cerebellum, the spinal cord, multi-modal association areas such as the prefrontal cortex, and various sensory areas. In addition, the firing of lower motor neurons in the PNS is necessary. If simply pushing open a door demands this much activity, imagine what is required to execute the motor program for playing an instrument, painting a portrait, or catching a football!

Role of the Cerebral Cortex

As discussed, the upper motor neurons stimulate the lower motor neurons to move a certain part of the body. The majority of our upper motor neurons that control complex movements reside in the primary motor cortex and the premotor and motor association areas. These neurons plan and initiate voluntary movement by selecting an appropriate motor program and coordinating the sequence of skilled movements, such as those used to ride a bicycle or send a text message. Upper motor neurons are also located in certain nuclei of the brainstem, where they work to maintain posture, balance, and body position, particularly during locomotion. These brainstem upper motor neurons also produce motor responses to sensory stimuli, such as turning your head and eyes toward the source of a loud noise.

In **Figure 12.30** you can see that the map of the upper motor neurons in the primary motor cortex and the *motor homunculus* very much resemble the somatosensory map and sensory homunculus discussed earlier (see Figure 12.28). Like the primary somatosensory cortex, the primary motor cortex is organized somatotopically, and certain areas have disproportionately more cortical area devoted to them, especially the lips, tongue, and hands. This indicates the importance of vocalization and manual dexterity to human survival.

The upper motor neurons in the cortical motor areas don't act alone to deliver the commands for movement to the lower motor neurons. In fact, movement initiated by the cerebral cortex alone is jerky, uncoordinated, and difficult to control. Smooth, fluid motion requires input from the basal nuclei and the cerebellum.

Role of the Basal Nuclei

Module 12.2 told us that three collections of cell bodies make up the basal nuclei: the caudate nucleus, the globus pallidus, and the putamen. The substantia nigra of the midbrain works very closely with the basal nuclei and is often grouped with them because of their shared functions. Unlike the upper motor neurons that modify the activity of interneurons, and so of lower motor neurons, the basal nuclei modify the activity of upper motor neurons to produce voluntary movements and inhibit involuntary, unintentional ones.

(a) Map of the primary motor cortex

(b) The motor homunculus

Figure 12.30 Representations of the primary motor cortex.

Complicated interconnections form a circuit between the basal nuclei and other structures of the brain. Let's consider how this circuit functions in the absence of voluntary movement. As shown in **Figure 12.31a**, the neurons of the globus pallidus fire continuously to inhibit the motor nuclei of the thalamus. This, in turn, inhibits the upper motor neurons of the cortical motor areas from firing at inappropriate times. The importance of this function is obvious: Imagine you are standing in line at the grocery store. You certainly don't want the muscles of your arms or legs to suddenly contract and your limbs to strike the people in front of or behind you, nor do you want to spontaneously begin barking like a dog. This part of the basal nuclei circuit helps prevent such occurrences; thus, a critical function of the basal nuclei is to *inhibit inappropriate movements.*

Now consider how voluntary movement is begun by this same circuit (**Figure 12.31b**). Notice that the neurons of the caudate nucleus and the putamen receive input from the cerebral cortex, with which they form excitatory synapses. These neurons then send action potentials to the globus pallidus that inhibit its neurons from firing. This inhibitory effect is enhanced by the substantia nigra, which increases the output of the caudate nucleus and putamen. The overall effect is that input from these two nuclei inhibits the inhibitor (i.e., the globus pallidus). This allows the motor nuclei of the thalamus to fire and stimulate the upper motor neurons of the cerebral cortex, which enables motion to begin. Thus, a second critical function of the basal nuclei is to *initiate voluntary motion.*

The final player in the basal nuclei drama is the substantia nigra. The neurons of the substantia nigra are called *dopaminergic neurons* because they release the neurotransmitter *dopamine.* These dopaminergic neurons project to the caudate nucleus and putamen, and when dopamine binds their receptors,

their inhibitory output to the globus pallidus is enhanced. The overall effect is an increase in the eventual activation of the upper motor neurons that will stimulate the desired motor activity. The input of the substantia nigra is thus vital for initiating movement.

Damage to any component of the basal nuclei system results in a **movement disorder.** There are a vast number of movement disorders, but they come in two main forms. The first causes an inability to initiate voluntary movement, making simple activities such as walking or talking difficult. The second causes an inability to inhibit inappropriate, involuntary movements. Some such movements are mild and may take the form of throat clearing or blinking; however, others may be severe enough to cause disability. One of the more common movement disorders, Parkinson's disease, is discussed in *A&P in the Real World.*

Role of the Cerebellum

The cerebellum monitors *ongoing* movement and integrates information about the contraction and relaxation of the muscles, the position of the joints, and the direction, force, and type of movement that is going to occur next. Once this information has been integrated, the cerebellum then determines the *motor error*—the difference between the intended movement and the actual movement that is taking place. It then influences different parts of the brain to minimize this error. Interestingly, the correction of motor error can occur over both the short and the long term by a process called **motor learning.** In essence, the corrections for motor error are added to the motor program over time. The more you perform a certain action, the more corrections for motor error that are added to the program, and the more fluid and error-free the motion becomes. In this case, practice really does make perfect, or at least mostly error-free.

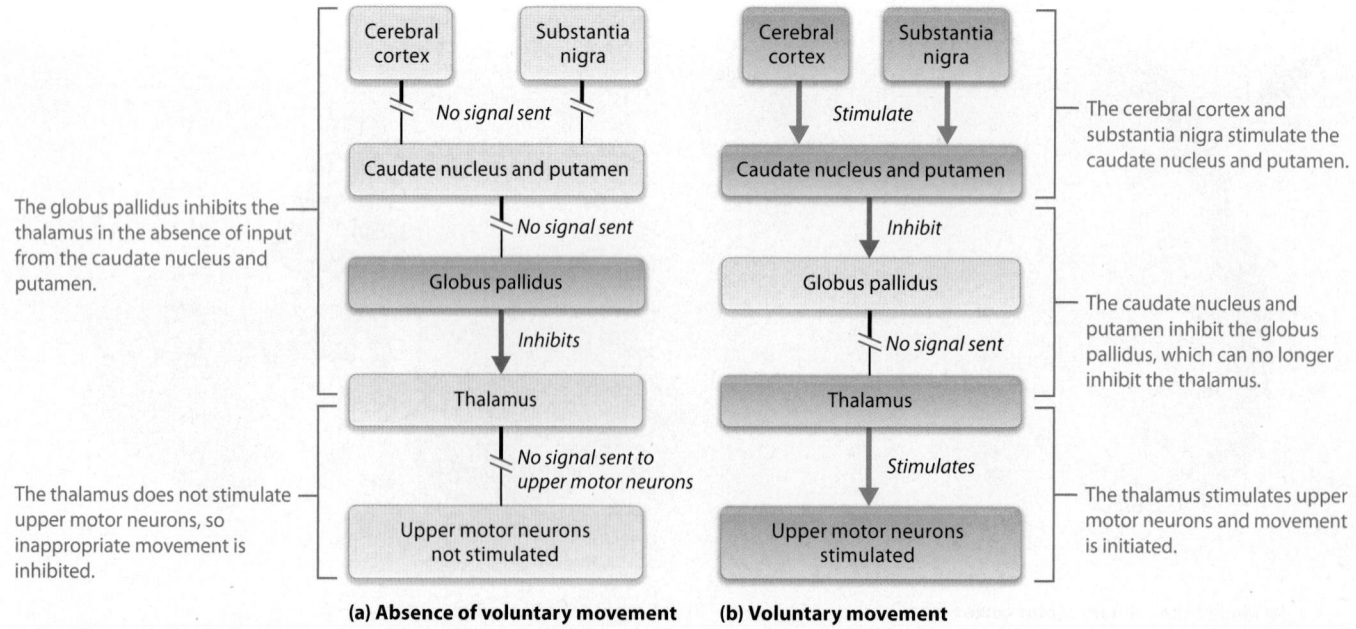

The globus pallidus inhibits the thalamus in the absence of input from the caudate nucleus and putamen.

The thalamus does not stimulate upper motor neurons, so inappropriate movement is inhibited.

(a) Absence of voluntary movement

The cerebral cortex and substantia nigra stimulate the caudate nucleus and putamen.

The caudate nucleus and putamen inhibit the globus pallidus, which can no longer inhibit the thalamus.

The thalamus stimulates upper motor neurons and movement is initiated.

(b) Voluntary movement

Figure 12.31 Role of the basal nuclei in voluntary movement.

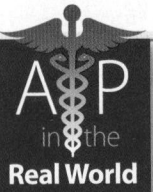

Parkinson's Disease

One of the most common movement disorders is *Parkinson's disease*. This disease is called a *hypokinetic* ("low movement") movement disorder, which means that movement is difficult to initiate and, once started, difficult to terminate. This leads to characteristic symptoms such as minimal facial expression, a shuffling gait with no arm swing, a resting tremor (uncontrollable shaking movements when at rest), difficulty in swallowing, and rigid movements of the arms and legs.

Degeneration of dopamine-secreting neurons of the substantia nigra causes Parkinson's disease. The degeneration tends to progress slowly over a 10- to 20-year period. Its underlying cause is unknown, although genetic mechanisms are suspected in approximately 10% of cases. Parkinson's disease is generally treated by medications that increase the level of dopamine in the brain to compensate for the decreased level.

The motor areas provide information on intended movement, and the vestibular nuclei and ascending tracts supply data on the actual movement performed. Note that in step ②, cerebellar neurons process and integrate the input from these sources. Then, in step ③, cerebellar neurons send output to correct motor error, primarily to the upper motor neurons via the premotor cortex and the primary motor cortex. Like the basal nuclei, the cerebellum affects movement by modifying the activity of upper motor neurons; the cerebellum does not have direct connections with lower motor neurons.

As you would expect, damage to the cerebellum makes fluid, well-coordinated movement nearly impossible—instead, movements are jerky and inaccurate. This condition, called **cerebellar ataxia** (ah-TAKS-ee-ah; *a-* = "without," *taxis* = "order"), typically features a wide, staggering gait; difficulty with rapid, alternating motions; overreaching or underreaching for an object; and/or an *intention tremor* (shaking that occurs when attempting to move a body part with accuracy or precision, such as reaching for a glass). Keep in mind that intention tremors may also be present in individuals without a history of cerebellar injury. In such cases, it is referred to as *essential tremor,* and it appears to result from degeneration of certain cerebellar neurons. Although the precise cause of essential tremor is unknown, a genetic link appears to be present.

As step ① in **Figure 12.32** shows, the cerebellum receives input from three sources simultaneously:

- motor areas of the cerebral cortex via upper motor neurons,
- vestibular nuclei of the pons, and
- ascending sensory tracts from the spinal cord.

Quick Check

☐ 3. Where do upper motor neurons reside, and what are their functions?
☐ 4. What are the two parts of the basal nuclei circuit?
☐ 5. What is the overall function of the cerebellum?

Primary motor cortex

③ The cerebellar neurons send output to the upper motor neurons via the motor cortices to correct motor error.

Vestibular nuclei in pons

① The cerebellum receives input from motor areas via upper motor neurons, vestibular nuclei in the pons, and ascending sensory tracts from the spinal cord.

② The cerebellar neurons process and integrate the input.

Figure 12.32 Role of the cerebellum in voluntary movement.

The Big Picture of CNS Control of Voluntary Movement

Figure 12.33

Sequence the steps @MasteringA&P®

① Motor association areas select a motor program.

② Motor association areas and the substantia nigra stimulate the caudate nucleus and putamen (see Figure 12.31).

③ The caudate nucleus and the putamen inhibit the globus pallidus. The globus pallidus stops inhibiting the thalamus.

④ Motor nuclei of the thalamus stimulate upper motor neurons.

⑤ Upper motor neurons execute the motor program by sending signals to the lower motor neurons in the spinal cord and PNS.

Thalamus

Globus pallidus
Caudate nucleus — Basal nuclei
Putamen

Substantia nigra in midbrain

Pons

⑥ The cerebellum monitors movement and corrects for motor error (see Figure 12.32).

Cerebellum

Medulla oblongata

Putting It All Together: The Big Picture of CNS Control of Voluntary Movement

As you've seen, multiple parts of the CNS work together to execute and monitor a motor program. The big picture view of this role and all of the different components is shown in **Figure 12.33**.

Quick Check

☐ 6. Trace the overall voluntary movement pathway from deciding to move to monitoring of the ongoing movement.

Apply What You Learned

☐ 1. Predict the effects of degeneration of the globus pallidus.
☐ 2. How would degeneration of the globus pallidus differ from degeneration of the caudate nucleus, putamen, and substantia nigra? Explain.
☐ 3. A potential effect of chronic alcohol abuse is the development of cerebellar damage. Predict the symptoms you would likely see in a person with such a condition.
☐ 4. A woman presents to the emergency department with loss of muscle function on the right side of her body. The physician suspects she had suffered a stroke (a blood clot in the brain). On which side of the patient's brain did the stroke likely occur? Explain.

See answers in Appendix A.

12.7

Homeostasis Part I: Role of the CNS in Maintenance of Homeostasis

Learning Outcomes

1. Describe the differences between the endocrine system and the nervous system in terms of how they maintain homeostasis.
2. Provide specific examples demonstrating how the nervous system responds to maintain homeostasis in the body.

If you recall, *homeostasis* is defined as the maintenance of a relatively stable internal environment in the face of ever-changing conditions (see Chapter 1). Homeostatic functions include maintaining fluid, electrolyte, and acid-base balance; blood pressure; blood glucose and oxygen concentrations; biological rhythms; and body temperature. The two main systems in the body dedicated to maintaining homeostasis are the nervous and endocrine systems. Although they work together to accomplish this function, they do so in different ways. The endocrine system secretes hormones into the blood, which regulate the functions of other cells, whereas the nervous system sends action potentials that excite or inhibit target cells. The actions of the endocrine system are typically slow and take several hours to several days to have an effect, whereas those of the nervous system are generally immediate.

Although one could argue that the entire brain and spinal cord are required for the maintenance of homeostasis in some respects, two structures of the CNS are concerned directly with it: certain nuclei of the brainstem reticular formation and the hypothalamus. In Module 12.2, we found out that the reticular formation controls the functions of many internal organs as well as aspects of behavior. The hypothalamus is closely associated both anatomically and functionally with the pituitary gland, which reflects the close relationship of the nervous and endocrine systems. The reticular formation and hypothalamus have numerous connections with one another that enable them to coordinate many homeostatic functions.

In this module, we focus on the role of the CNS in maintaining homeostasis in the body. Our discussion of homeostasis continues in the chapters on the autonomic nervous system and the endocrine system (see Chapters 14 and 16, respectively).

Homeostasis of Vital Functions

The homeostasis of vital functions such as heart pumping, blood pressure, and digestion is largely controlled by the **autonomic nervous system** (**ANS**). Recall from the nervous system chapter that the ANS is a branch of the PNS that regulates the function of the body's viscera (see Chapter 11). Although the ANS is classified as a division of the PNS, it is controlled by structures of the CNS.

The main "boss" of the ANS is the hypothalamus, which receives sensory input from the viscera, the components of the limbic system, and the cerebral cortex. This allows the hypothalamus to respond to both normal physiological changes and emotional changes and to adjust ANS output accordingly to preserve homeostasis. The hypothalamus accomplishes this task largely by relaying instructions to nuclei in the reticular formation of the medulla, including the following:

- **Vasopressor center.** The neurons of the **vasopressor center** (VAYZ-oh-press-er; *vaso-* = "vessel") are located in the anterolateral medulla. When stimulated by the hypothalamus, this center increases the rate and force of cardiac contractions and also causes blood vessels to narrow. All three of these effects raise blood pressure.
- **Vasodepressor center.** The **vasodepressor center** is located inferior and medial to the vasopressor center. Its effects on the heart and blood vessels are the opposite of those of the vasopressor center—it decreases the rate and force of contractions of the heart and opens the blood vessels. All three of these effects decrease blood pressure.
- **Other centers.** Many nuclei in the reticular formation participate in regulation of digestive processes and control of urination. Much of this activity is discussed in the ANS chapter (see Chapter 14).

Respiration is one of the few vital functions not under direct control of the ANS. Instead, the rate and depth of breathing are regulated by a group of neurons in the anterior medullary reticular formation. Several factors influence the rate at which these neurons fire, including input from the cerebral cortex, the limbic system, the hypothalamus, certain sensory receptors, and nuclei in the pons. We discuss control of breathing more completely in the respiratory system chapter (see Chapter 21).

Quick Check

☐ 1. Which two body systems coordinate the maintenance of homeostasis? How do they differ in the ways they accomplish this task?
☐ 2. Which branch of the PNS controls most of the body's vital functions? What structure is the "boss" of this branch?

Body Temperature and Feeding

The hypothalamus regulates both body temperature and feeding behaviors. It acts as the body's thermostat and creates a set point, or normal body temperature (about 37° C, or 98.6° F). It receives input from temperature-sensitive neurons located in several places, including the skin and areas deeper in the body, and from neurons in the hypothalamus itself. When the body temperature increases above the normal range, a negative feedback loop is initiated whereby certain hypothalamic nuclei induce changes that cool the body, such as sweating, an example of the Feedback Loops Core Principle (p. 21). Conversely, when body temperature decreases below the normal range, another negative feedback loop sets in motion mechanisms to conserve body heat, such as shivering. When the set point for body temperature is temporarily set higher than normal, a fever ensues; see *A&P in the Real World: Fever* on page 466 for more information on this topic.

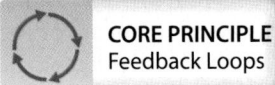

CORE PRINCIPLE
Feedback Loops

The hypothalamus also regulates feeding, which in turn helps maintain levels of glucose and other nutrients in the bloodstream. Stimulation of certain hypothalamic nuclei induces hunger and feeding behaviors, which indirectly works to preserve homeostasis of glucose levels in the blood. These effects are believed to be due, in part, to neurotransmitters called *orexins* (*orexis* = "appetite"), which are secreted in high amounts by hypothalamic neurons during periods of fasting and appear to induce eating. Stimulation of different hypothalamic nuclei seems to inhibit feeding; however, the mechanisms controlling food intake are complex and involve hormones and several nuclei in the hypothalamus as well as other parts of the brain. Interestingly, the same area of the hypothalamus that mediates hunger also mediates aggression and rage, which perhaps factors into the grumpiness some people feel when hungry. The orexins have also been found to play an important role in regulating the sleep/wake cycle, which is our next topic.

Quick Check

☐ 3. How does the hypothalamus regulate hunger?
☐ 4. What role does the hypothalamus play in temperature homeostasis?

Sleep and Wakefulness

Sleep is one of the most fundamental homeostatic processes carried out by humans and most other animals, yet it is also one of our most mysterious processes. **Sleep** is defined as a reversible and normal suspension of consciousness. We have uncovered a great deal about how we sleep, how sleep is regulated, and brain activity during sleep, but we are still unsure *why* most animals sleep. Physiologically, sleep seems to serve an energy restoration function, allowing the brain to replenish its glycogen supply; however, if there are psychological reasons why we sleep, they remain elusive. Nonetheless, it is clearly required for our survival; sleep deprivation may cause imbalances in temperature homeostasis, weight loss, a decrease in cognitive abilities, hallucinations, and even death.

To feel rested, most adults require 7–8 hours of sleep per night. The requirement for infants is much higher, at about 17 hours per day, but this requirement decreases with increasing age. As we grow older, we tend to sleep for shorter periods and less deeply for reasons that are at present unclear. The following discussion will examine what is known about this mysterious state and some of the possible reasons we spend so much time in it.

Circadian Rhythms and the "Biological Clock"

Among the most basic questions about sleep are those that revolve around *when* and *how* we sleep. Human sleep generally follows a **circadian rhythm** (sir-KAY-dee-an; *circa-* = "around," *dia-* = "day"), meaning that we spend a period of the 24-hour cycle awake and another period of the cycle asleep. This rhythm is controlled by our own hypothalamic "biological clock," which causes changes in the level of wakefulness in response to day and night cycles. As shown in **Figure 12.34**, the clock brings about sleep in the following fashion:

① **Nerves from the eye signal the suprachiasmatic nucleus (SCN) that the light level is decreasing.** Receptors in the eye that detect the decreasing light level carry this information to a region of the hypothalamus called the *suprachiasmatic nucleus* (soop'-rah-ky-az-MAT-ik; *SCN*). The SCN is often called the body's "master clock."

Ventrolateral preoptic nucleus

Thalamus

Suprachiasmatic nucleus (SCN)

① Nerves from the eye signal the suprachiasmatic nucleus (SCN) that the light level is decreasing.

② The SCN stimulates the ventrolateral preoptic nucleus.

③ The ventrolateral preoptic nucleus decreases the activity of the reticular formation.

Reticular formation

④ The decreased activity of the reticular formation "disconnects" the thalamus from the cerebral cortex, in turn decreasing the level of consciousness.

Figure 12.34 The process of falling asleep.

② **The SCN stimulates the ventrolateral preoptic nucleus.** The SCN sends this stimulus to another hypothalamic nucleus, the *ventrolateral preoptic nucleus.* The SCN also sends signals to the pineal gland that stimulate the secretion of melatonin.

③ **The ventrolateral preoptic nucleus decreases the activity of the reticular formation.** When excited, the ventrolateral preoptic nucleus sends signals that decrease the activity of several nuclei of the midbrain and pontine reticular formation. These nuclei stimulate the thalamus during wakefulness. Melatonin also depresses the activity of the reticular formation.

④ **The decreased activity of the reticular formation "disconnects" the thalamus from the cerebral cortex, in turn decreasing the level of consciousness.** A decrease in the activity of the reticular formation lowers the rate at which the thalamus transmits external stimuli to the cerebral cortex. This in effect "disconnects" the cortex from the outside world (represented by red X's in Figure 12.34) and reduces the level of consciousness.

Arousal from sleep is mediated by a different group of hypothalamic neurons (not those regulating satiety) that secrete the neurotransmitter orexin, in an example of the Cell-Cell Communication Core Principle (p. 27).

CORE PRINCIPLE
Cell-Cell Communication

Orexin release increases before waking and helps maintain a state of wakefulness throughout the day. This effect is thought to be due to stimulation of the reticular formation by these orexin-secreting neurons. When the reticular formation neurons are active, they cause the thalamus to begin transmitting external sensory stimuli to the cerebral cortex. Histamine is an important excitatory neurotransmitter in this system, which is why *antihistamines,* drugs that block histamine receptors, often produce drowsiness as an adverse effect. Interestingly,

an orexin deficiency is believed to be the culprit behind the disease *narcolepsy,* in which an individual sleeps poorly at night, leading to very strong urges to fall asleep during the day.

Brain Waves and Stages of Sleep

Research has demonstrated that in one night we progress through several different cycles of sleep, each having many stages. These stages can be monitored by an **electroencephalogram,** or **EEG,** a test that measures the electrical activity of the brain via electrodes attached to the skin. The tracings of electrical activity, called **brain waves,** show characteristic patterns during wakefulness and the different stages of sleep. Waves differ in their height, also known as *amplitude,* and in the speed at which they are generated, or their *frequency.* As you can see in **Figure 12.35,** when we are awake and engaged in mental activity, the brain waves are low amplitude and high frequency, a pattern called **beta waves.** When we fall asleep and progress through its stages, the amplitude of the waves progressively increases and their frequency progressively decreases.

The stages of sleep are as follows:

- **Stages I–III sleep.** The first three sleep stages are characterized by drowsiness that progresses to moderately deep sleep and a slowing of the normal waking beta wave rhythm into what is known as **theta waves.** The amplitude of the theta waves increases and their frequency decreases as we move from stage I to stage III sleep.

- **Stage IV sleep.** The deepest sleep occurs in stage IV. During this stage, the brain waves, called **delta waves,** become lower frequency and higher amplitude. Delta waves are also seen during general anesthesia and in awake adults with significant brain damage. The entire sequence from stage I to stage IV sleep takes less than an hour to complete.

- **REM sleep.** As a person emerges from stage IV sleep, he or she enters a puzzling portion of the sleep cycle called **rapid**

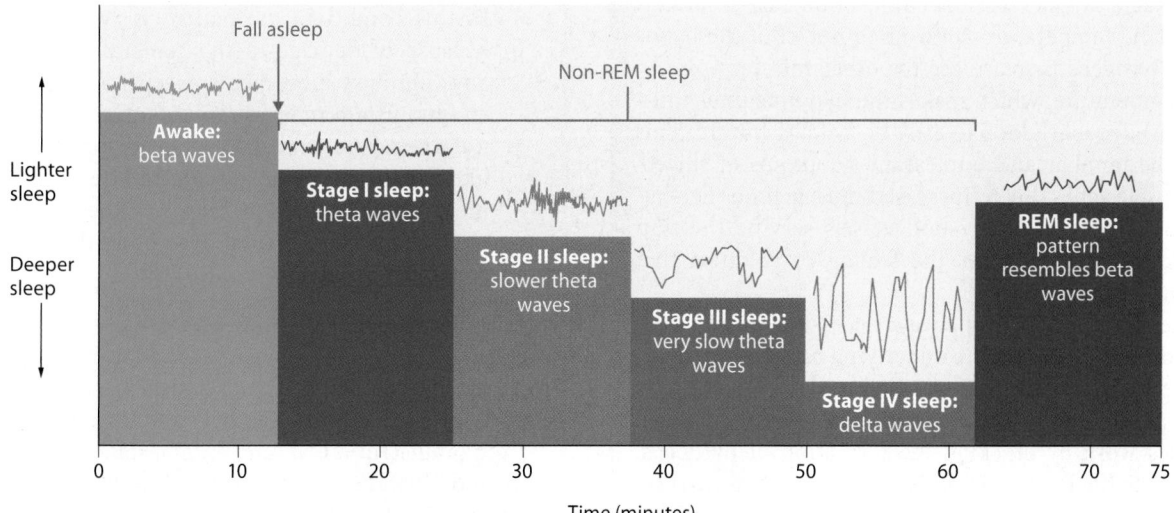

Figure 12.35 Stages of wakefulness and sleep as shown by EEG patterns.

eye movement (**REM**) **sleep** (the preceding four stages are referred to as **non–rapid eye movement [non-REM] sleep**). REM sleep is named for the rapid, back-and-forth eye movements that occur during this stage of sleep but are absent from other stages. REM sleep is when the majority of *dreams* occur—curious hallucination-like episodes that often have a highly emotional content but make little sense. The brain wave patterns during REM sleep resemble the beta waves of an awake adult, although clearly the two brain states are not equivalent. REM sleep generally lasts only about 10–15 minutes before cycling back into stage I sleep, although the REM sleep phase generally gets longer with each sleep cycle.

Keep in mind that an individual won't necessarily progress through each stage in order. For example, one may transition from stage IV sleep into REM, stage I, or stage II sleep.

The contrast between the activity of the brain and the body during non-REM and REM sleep extends beyond the eye movements. During stages I–IV of the sleep cycle, the body is able to move ("tossing and turning" occurs during these stages), and blood pressure, heart rate, respiratory rate, body temperature, and metabolic rate decrease. During REM sleep, however, blood pressure, heart rate, and metabolism increase, and most of the major muscle groups are paralyzed. The muscle paralysis that occurs during REM sleep presumably prevents us from acting out our dreams.

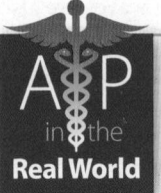

Fever

At some point, everyone has had a *fever*—an elevation of normal body temperature that can accompany a variety of infectious and noninfectious conditions. Fevers are caused by chemicals called *pyrogens* secreted by cells of the immune system and by certain bacteria that cross the blood-brain barrier and interact with the temperature-controlling nuclei of the hypothalamus. Pyrogens increase the hypothalamic set point to a higher temperature, which makes the hypothalamus initiate a negative feedback loop to raise body temperature. This produces the familiar and unpleasant symptoms of shivering and muscle aches due to increased muscle tone. The hypothalamus also causes the blood vessels serving the skin to constrict, which produces the feeling of coldness that accompanies a fever.

Treatment of a fever is not always necessary, as it is often more important to address the underlying cause of the fever. When a fever is treated, agents called *antipyretics,* such as acetaminophen and aspirin, are generally administered. Antipyretics work by blocking the formation of pyrogens, which permits the hypothalamus to return to its normal set point.

The functions of REM sleep and dreaming are unknown. One hypothesis is that dreaming and sleep in general are required for consolidation of memories and recently learned tasks. But another hypothesis actually posits the opposite—that dreams are necessary to dispose of "unwanted" memories that have accumulated during the waking period. Studies of brain activity during dreaming have indicated that the structures of the limbic system, which play a role in emotion, are highly active

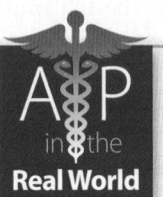

States of Altered Consciousness Mimicking Sleep

States of altered consciousness can indicate serious problems with brain function. Some of these states include the following:

- **Stupor.** A person with *stupor* has a diminished level of cortical activity, although he or she is often arousable to strong or painful stimuli. Common causes of stupor include infections, mental illnesses, and conditions of the brain such as brain tumors.
- **Coma.** *Coma* is defined as unarousable unconsciousness in which there are no purposeful responses to any stimuli, even pain. The underlying defect in the coma state is damage to the reticular activating system or a related component that prohibits normal arousal of the cerebral cortex. EEGs demonstrate a slow, fixed pattern with no observable sleep/wake cycles. Brainstem reflexes, however, remain intact, and the patient may exhibit certain involuntary movements. Coma may result from head trauma, various neurological conditions, certain drugs, and disturbances in acid-base homeostasis.
- **Persistent vegetative state.** Sometimes patients with head trauma or neurological injury will move from the coma state into a perplexing *persistent vegetative state.* In this condition, the patient is awake but unaware, because of damage to the cerebral cortex, and lacks any voluntary movements. Sleep/wake cycles do occur and brainstem reflexes remain intact, leading to involuntary movements such as head turning and grunt-like vocalization that caregivers often misinterpret as meaningful interaction. However, these actions are not mediated by the cerebral cortex and thus do not imply conscious awareness.

People in these states may occasionally regain consciousness, depending on the cause of the state. However, the most extreme altered state of consciousness is **brain death,** in which the EEG shows no activity and the reflexes controlled by the brainstem are absent. In this state, cerebral metabolism and blood flow are reduced to essentially zero and consciousness will not be regained.

during dreams, but the prefrontal cortex, which is involved in logic and reason, is barely active at all. This perhaps explains the highly emotional and often nonsensical content of dreams but sheds little light on their actual purpose. Questions about the purposes of REM sleep and dreams therefore remain, for the moment, unanswered.

Although sleep is a normal altered state of consciousness, a key feature is that it is a *reversible* state. Other altered states of consciousness have variable levels of reversibility, as described in *A&P in the Real World: States of Altered Consciousness Mimicking Sleep.*

Quick Check

☐ 5. How is sleep defined? What are its functions?
☐ 6. What type of rhythm does human sleep follow? Explain the general neural pathway by which sleep is initiated.
☐ 7. What is an electroencephalogram? What is the normal brain wave pattern for an awake, alert adult?

Apply What You Learned

☐ 1. You are about to take your anatomy and physiology exam and you notice that your heart is beating rapidly and your rate of breathing has increased. What parts of the brain are responsible for these changes, and how do they accomplish them?
☐ 2. Predict the effects of a tumor that secretes excess orexins.
☐ 3. Explain why sleepwalking and sleeptalking can occur only during non-REM sleep.

See answers in Appendix A.

MODULE **12.8**
Higher Mental Functions

Learning Outcomes

1. Describe the areas of the cortex responsible for cognition and language.
2. Discuss the concept of cerebral hemispheric specialization.
3. Describe the parts of the brain involved in storage of long-term memory, and discuss possible mechanisms of memory consolidation.

Our current understanding of biology, evolution, and physiology leaves little doubt that higher mental functions, such as morality, language, and learning, are a product of normal electrochemical processes occurring in the human brain. In this module,

we explore these complex mental functions and the remarkable physiological processes behind them.

Cognition and Language

◀◀ **Flashback**

1. What are multimodal association areas? (p. 429)
2. Where are Wernicke's area and Broca's area located? What are their functions? (pp. 430-431)

In Modules 12.5 and 12.6, we discussed the areas of the cerebral cortex dedicated to processing sensory and motor functions. As you saw in Figure 12.5, these areas account for only about 20% of the total cerebral cortex. The remainder of the cerebral cortex consists of association areas that perform the diverse group of tasks collectively known as **cognition.** Our cognitive functions include processing and responding to complex external stimuli, recognizing related stimuli, processing internal stimuli, and planning appropriate responses to stimuli. Our cognitive processes are responsible for social and moral behavior, "intelligence," thoughts, problem-solving skills, language, and personality. The upcoming sections focus on those parts of the brain responsible for cognitive functions and how language is produced.

Localization of Cognitive Function

Aspects of cognition have been localized to different multimodal association areas in the frontal, parietal, and temporal lobes. These areas and their functions include:

- **Parietal association cortex.** This area of cortex is responsible for spatial awareness and attention. It allows us to focus on distinct aspects of a specific object, such as the direction in which it is moving, its color, or its shape. This part of the brain also has visual-spatial functions and enables us to recognize the position of an object in space. A curious consequence of damage to this area is called *neglect;* in this condition, patients may not acknowledge the existence of an entire half of their visual field or even an entire half of their body in spite of intact sensory pathways.
- **Temporal association cortex.** This area of cortex is primarily responsible for recognizing stimuli, especially complex stimuli such as faces. Patients with damage to this part of the brain suffer from *agnosias* (ag-NOH-see-ahz)—the awareness of a stimulus but the inability to identify it.
- **Prefrontal cortex.** The prefrontal cortex is the largest and most complex of the association cortices. It is responsible for the majority of cognitive functions that make up a person's "character" or "personality." It receives information from the temporal and parietal association cortices as well as from multiple other sensory and motor cortices. This allows for the integration of complex sensory and motor information that creates an awareness of "self" in the world, which leads to planning and execution of behaviors appropriate for given circumstances. For example, our prefrontal

cortex enables us to recognize that it is appropriate to laugh aloud when having lunch with friends, but inappropriate to do the same during a funeral. The prefrontal cortex also allows us to select "morally appropriate" behaviors, even those not necessarily to our immediate advantage. Additionally, it plays a role in governance of emotions, which are discussed later in this module. Damage to the prefrontal cortex can be devastating and tends to produce dramatic changes in personality, impulsive behavior with little restraint, difficulty in making plans, and disordered thought processes. Brain disorders such as *dementia* (dee-MEN-shah) can also cause loss of cognitive and intellectual functions. The consequences of different types of dementias for mental faculties are discussed in *A&P in the Real World: Dementia*.

Cerebral Lateralization

Many of the cognitive functions we just discussed are unequally represented in the right and left hemispheres, a phenomenon called **cerebral lateralization**. The idea that certain cognitive functions are lateralized has led to the unfortunate misconception by the public that one hemisphere is "dominant" over the other and that people can be "right-brained" or "left-brained." What cerebral lateralization actually represents is a division of labor of sorts between the hemispheres; it is how our brains make the most use out of a limited amount of space.

The functions that appear to be lateralized include the following (although lateralization is not absolute):

- *emotional functions,* with the left frontal cortex mostly being responsible for "positive" emotions such as happiness and the right frontal cortex for "negative" emotions such as anger and fear;
- *attention,* which is lateralized to the right parietal cortex;
- *facial recognition,* which tends to be performed by the right temporal cortex; and
- *language-related recognition,* including the ability to identify an object with its proper name, which is lateralized to the left temporal cortex.

Other language functions, which we discuss next, appear to be strongly lateralized.

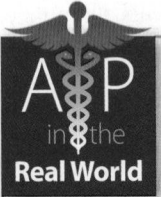

Dementia

Some of the most devastating diseases of the brain are those collectively referred to as dementias. Dementia has many causes, but all patients with dementia exhibit a progressive loss of recent memory, degeneration of cognitive functions, and changes in personality. In effect, a person with dementia literally "loses his or her mind." Common forms of dementia, listed in descending order from the most frequently occurring, include the following:

- **Alzheimer's disease.** *Alzheimer's disease* (ALZ-hy-merz; AD), the most common form of dementia, affects about 30 million people worldwide. Some of the characteristic findings in brain tissue of those with AD are neurofibrillary tangles (tangled aggregates of proteins in the neurons), senile plaques (extracellular deposits of a specific type of protein around neurons), and degeneration of cortical neurons and synaptic connections. These changes are found throughout the brain, but are especially numerous in the cortical association areas and the hippocampus. The earliest signs of the disease typically involve recent memory loss and forgetfulness. As the disease progresses, impairment of attention, language skills, critical thinking, and visual-spatial abilities, as well as changes in personality, becomes apparent. Eventually, motor skills, sensory perception, and long-term memory are also affected.
- **Vascular dementia.** Vascular dementia is a group of disorders that share the common feature of a disruption in blood flow to parts of the brain, particularly the cerebral cortex. The signs and symptoms of vascular dementia are highly variable because of the range of brain regions affected. There is significant overlap of vascular dementia with AD, and in fact often patients have both forms, known as *mixed dementia.*
- **Lewy body dementia.** Lewy body dementia is characterized by cytoplasmic inclusions called *Lewy bodies* in neurons in certain parts of the brain. Unlike other forms of dementia, Lewy body dementia features fluctuations in cognitive functions, a less profound impact on memory formation, hallucinations and delusions, and sleep disorders. In addition, patients with Lewy body dementia often have the motor features of Parkinson's disease.
- **Pick's disease.** In Pick's disease, the cerebral cortex of the frontal and temporal lobes progressively degenerates, resulting in severe atrophy due to loss of neurons, as well as excess neuroglial cells. Pick's disease is unusual in that it strikes a younger population, with onset typically around age 55–65. Patients tend to deteriorate rapidly, and because the degeneration affects the prefrontal cortex, dramatic personality changes and behavioral problems are common.

No proven methods exist for preventing the many forms of dementia, nor are any of the forms curable. Some drugs may slow the progression of AD in certain patients; however, they do not reverse the changes that have already taken place. These drugs have not shown effectiveness for any other forms of dementia. Dementias have profound and devastating effects on a person's quality of life; research into the causes, prevention, and treatment of these disorders is therefore intensive and ongoing.

Language

Language is arguably one of the most important cognitive functions of the brain in terms of progress made by the human species. The ability to assign a word as a symbolic representation of an object or concept allows us to communicate with one another through speaking, writing, and/or signing. This ability has made possible many of our most important cultural, technological, and social achievements. Indeed, it's difficult for most of us to even imagine life without language, as it is so integral to our day-to-day lives.

In the context of neurology, **language** refers to the ability to comprehend and produce words through speaking, writing, and/or signing, and to assign and recognize the symbolic meaning of a word correctly. But language is far more complex than simple words; it involves syntax (the arrangement of words and phrases to create sentences), grammar, and context, as well. In addition, correct understanding of a statement also relies on cues such as tone, volume, and the speaker's body language. The phrase "oh, great" could have a number of meanings based on the way it's said and in what context, including excitement, dread, apathy, sarcasm, or joy.

Multiple parts of the brain are required for a human to communicate successfully through speech, writing, and/or sign language (**Figure 12.36**). However, most of these areas are concerned with the motor and sensory aspects of communication rather than with language itself. For example, a blind or deaf person does not lose the capacity for language, nor does someone with an inability to speak. Instead, language appears to depend on two multimodal association areas. The first is Broca's area in the frontal lobe, which is responsible for the *production* of language, including the planning and ordering of words with proper grammar and syntax. The second is Wernicke's area in the temporal lobe, which is responsible for *understanding* language and linking a word with its correct symbolic meaning. The functions of these two areas become readily apparent when one area is damaged, which results in a language deficit called an **aphasia** (ah-FAY-zhee-ah). You can find out more about these deficits in *A&P in the Real World: Aphasias.*

As we noted in the previous section, language functions are heavily lateralized, with the majority of language functions residing in the left cerebral hemisphere in about 97% of people. However, the right hemisphere does have some important language functions: those relating to the emotional nature of language. Damage to language areas of the right hemisphere results in the loss of emotional and tonal components of language (imagine not being able to detect whether someone intends sarcasm or humor); thus, both hemispheres are required for normal language function.

Quick Check

☐ 1. What is cognition? Which part of the brain is responsible for cognitive processes?

☐ 2. What is cerebral lateralization? Which functions are known to be lateralized?

☐ 3. Define language in the context of neurology. What are the functions of Broca's and Wernicke's areas?

Cerebral cortex

Ventricle

Level of activity of neurons during speech:

■ High

■ Medium

■ Low

Figure 12.36 Functional neuroimaging (functional magnetic resonance imaging, fMRI) of the brain during speech.

Learning and Memory

When we *learn*, our nervous systems acquire new information that is observable as some sort of behavioral change, be it providing the correct answer on an examination or riding a bicycle. What we learn generally is put into our **memory**, where it is encoded and stored in our neural circuitry and is retrievable at will. There are two basic types of memory: **declarative (fact) memory,**

Aphasias

The two major types of aphasia (language deficits) can be described this way:

- **Broca's aphasia** is characterized by the inability to produce speech because the patient is unable to correctly plan and order the grammar and syntax of a statement. Speech in a patient with Broca's aphasia is not fluent; words and phrases tend to be repeated; and grammar, syntax, and the structure of individual words are disordered. Broca's aphasia tends to be frustrating for patients because language comprehension remains intact—they understand what is being said to them but are unable to organize a reply.

- **Wernicke's aphasia** is characterized by the inability to understand language. Patients with this form of aphasia can speak fluently with adequate grammar and syntax, but the words make little to no sense, as words and meanings are not correctly linked. In Wernicke's aphasia, language comprehension is not intact, so patients cannot fully understand what is being said to them and cannot realize that what they're saying makes no sense.

which is defined as the memory of things that are readily available to consciousness that could in principle be expressed aloud (hence the term "declarative"), and **nondeclarative memory** (also called *procedural* or *skills memory*), which tends to include skills and associations that are largely unconscious. Examples of declarative memory are a phone number, a quote, or the pathway of the corticospinal tracts. Nondeclarative memory can be exemplified by how to enter the phone number on a phone, how to move your mouth to speak the quote, and how to read this chapter.

Both declarative memory and nondeclarative memory can be classified by the length of time for which they are stored. These classes include the following:

- **Immediate memory.** Your immediate memory is stored for only a few seconds. This may not seem significant, but it is crucial for carrying out normal conversation, reading, and daily tasks, such as remembering the areas of the floor you just swept.
- **Short-term (working) memory.** Your short-term memory is stored for several minutes and allows you to remember and manipulate information with a general behavioral goal in mind. Short-term memory comes into play when you have lost your keys and are searching for them—without such memory, you wouldn't be able to remember where you had already looked, making an effective search impossible.
- **Long-term memory.** Your long-term memory is a more permanent form of storage for days, weeks, or even for your entire lifetime.

Some of the information in immediate and working memory is transferred into long-term memory, a process called **consolidation** (**Figure 12.37**). However, much of the information is simply forgotten. Humans have a very limited ability to consolidate relatively meaningless data into long-term memory; however, if these data are presented in a form that is associated with strong emotion, another stimulus, or something one has already learned, consolidation occurs much more easily. Consider, for example, the method

by which children are taught the alphabet—by song. By setting the string of 26 letters to music, one can associate a letter with a musical note. Try right now to recite the alphabet backwards. Although you know the sequence of letters, you likely had a hard time doing this, as you have no notes with which to associate it.

The process of consolidation is notoriously flawed. This same use of association to form memories can lead individuals to make false associations that become encoded in memory. Additionally, strong emotions and previous perceptions can color the memory significantly. For instance, people who believe they have seen a "flying saucer" will often dramatically exaggerate the speed and size of the object and their relative peril during the incident. Such exaggeration doesn't make these people liars—they genuinely remember the event occurring that way. However, even those memories about which people feel most confident can be false.

The mechanism by which the physical embodiment of a memory, called an *engram,* is formed differs for declarative and nondeclarative memory. The following sections discuss both mechanisms.

Formation and Storage of Declarative Memory

Formation of new declarative memories appears to require involvement of the hippocampus, which, as mentioned, is a component of the limbic system. Brain imaging studies have shown dramatically increased activity in the hippocampal neurons when one is studying information with the intent of learning, indicating that immediate and short-term memories are likely stored here. The mechanism by which hippocampal neurons encode long-term declarative memories seems to involve an increase in synaptic activity between associated neurons, called **long-term potentiation (LTP).** In LTP, an example of the Cell-Cell Communication Core Principle (p. 27), the strength of synaptic connection between two neurons increases when the two neurons are stimulated simultaneously. The resulting increase in synaptic activity produces growth in dendrites, greater release of neurotransmitters between the two neurons, and induction of protein synthesis in the neurons that maintain this high degree of synaptic activity.

CORE PRINCIPLE
Cell-Cell Communication

Current evidence indicates that LTP in the hippocampus is required for new declarative memories to form, but it is clear from patients with damage to the hippocampus that long-term declarative memories aren't *stored* here. Instead, memories appear to be stored in different areas of the cerebral cortex that correlate with their function. So, for example, the image of a cat is stored in the visual association area, the sound of the cat in the auditory association area, and the feel of a cat's fur in the somatosensory association area. Retrieval of these memories seems to be mediated by pathways that involve the hippocampus and the prefrontal cortex, although researchers are still working to understand these pathways.

Formation and Storage of Nondeclarative Memory

Damage to the hippocampus and associated structures does not seem to impact the ability to form new nondeclarative memories. Instead, it appears from studies of patients with certain brain diseases that the motor cortices, the cerebellum, and the basal nuclei are required for the formation and storage of nondeclarative memories.

Figure 12.37 Pathways for consolidation of memories.

Immediate memory
(several seconds)

Short-term memory
(several minutes)

Consolidation enhanced by:
• *Association with other stimuli*
• *Emotion*
• *Association with information already in long-term memory*

Long-term memory (days to lifetime)

Forgotten

The exact locations and the mechanism involved are not clear; however, many researchers think a form of LTP comes into play, similar to what happens in the formation of declarative memories.

Quick Check

☐ 4. Explain the differences between declarative memory and nondeclarative memory.
☐ 5. How do immediate, short-term, and long-term memory differ?

Emotion

What we commonly think of as *emotion* is actually a complex combination of three separate phenomena:

- visceral motor responses, such as blushing when one is embarrassed or a racing heart when one is afraid;
- somatic motor responses, such as smiling, laughing, frowning, and crying; and
- highly subjective "feelings" that respond to and are integrated with sensory and/or cognitive stimuli, such as feeling sad when remembering a lost pet or feeling tense when watching a suspenseful movie.

As an example, let's discuss the different parts of the brain responsible for the visceral and somatic motor responses that accompany a feeling of excitement. The hypothalamus plays a role in both the visceral and somatic motor components. It sends instructions to nuclei of the reticular formation, which in turn activate autonomic neurons that innervate the viscera; this causes the heart rate and blood pressure to increase. Both the hypothalamus and the limbic cortex influence the somatic motor response by sending instructions to other nuclei in the reticular formation. These nuclei relay the instructions to lower motor neurons, which causes us to open our eyes widely and smile.

The third component of emotion—the highly subjective "feelings" that we experience—is the most complex. A key player in this aspect of emotion is the amygdala. The amygdala receives sensory information from a number of sources, including the brainstem, the thalamus, the cerebral cortex, and the basal nuclei. These connections allow the amygdala to analyze the emotional significance of sensory stimuli and create associations between different sensory stimuli, a type of "emotional learning." For example, a young child learns to associate a caregiver's cheerful tone of voice with happiness and a loud, harsh tone of voice with anger and fear. The amygdala also has connections with basal nuclei circuits, which appear to select behaviors that maximize rewards and minimize punishments.

Still, none of these circuits or connections explains why we "feel" sad or happy or angry. Neither do they explain why the same stimulus can make one person happy and another person sad. This enigmatic aspect of emotion is not understood on a cellular level, but it is assumed to involve the amygdala's projections to the prefrontal cortex. Given the importance of emotions in the experiences of all human beings, this is likely to be an area of intensive research for years to come.

Quick Check

☐ 6. What three phenomena constitute emotion, and which components of the nervous system mediate them?

Apply What You Learned

☐ 1. Mr. Jacobs suffers widespread damage to his right cerebral hemisphere. His wife believes he will be fine because his left cerebral hemisphere is undamaged. Is his wife correct? Explain. What deficits, if any, can Mr. Jacobs expect to face?
☐ 2. Ms. Mah undergoes treatment for a seizure disorder that involves removal of the right and left hippocampus. Will this affect her declarative memory, nondeclarative memory, or both? Will this remove her old memories, impair the formation of new memories, or both?
☐ 3. The patient in question 2, Ms. Mah, also has her right and left amygdala removed. What effect is this likely to have?

See answers in Appendix A.

CHAPTER SUMMARY

Take full advantage of the invaluable online practice and study tools for this chapter by going to MasteringA&P®**. There you will find:**

- Practice tests, module quizzes, art-labeling and sequence-the-steps activities
- ▶ Animated figures with quizzes
- *Practicing A&P* worksheets
- Author-narrated podcast
- Dynamic Study Modules
- **iP** *Interactive Physiology*

MODULE **12.1**
Overview of the Central Nervous System 425

- The central nervous system performs the integrative nervous system functions.
- The **brain** is an organ that consists mostly of nervous tissue. It is housed within the cranial cavity.
- The four divisions of the brain are the **cerebrum,** the **diencephalon,** the **cerebellum,** and the **brainstem.** (Figure 12.1, p. 426)
- The **spinal cord** is nervous tissue that forms an extension of the inferior brainstem.
- The CNS contains both gray matter (nuclei) and white matter (myelinated axons). (Figure 12.2, p. 426)

- The CNS develops from tissue that folds into the **neural tube** by the fourth week of development. The cranial end of the neural tube enlarges to form the **primary brain vesicles,** which enlarge further to form the **secondary brain vesicles.** The secondary brain vesicles develop into the four divisions of the brain. (Figure 12.3, p. 427)

MODULE 12.2
The Brain 428

- Each cerebral hemisphere has five lobes: the anterior **frontal lobe,** the middle **parietal lobe,** the posterior **occipital lobe,** the lateral **temporal lobe,** and the deep **insula.** (Figure 12.4, p. 428)

- The **cerebral cortex** is the cerebrum's outer covering of gray matter. The majority of the cerebral cortex is **neocortex,** which functions in conscious processes such as planning of movement and interpretation of incoming sensory information.

- The three types of neocortex are the **primary motor cortex,** the *primary sensory cortices,* and **association areas.** Association areas may be *unimodal* and integrate only one type of information, or *multimodal,* and integrate several types of information. (Figure 12.5, p. 430; Table 12.1, p. 431)

- The **basal nuclei** are a group of nuclei in the cerebrum that function in movement. The three basal nuclei are the **caudate nucleus,** the **putamen,** and the **globus pallidus.** (Figure 12.6, p. 431)

- Within the CNS, **tracts** of white matter carry information to bundles of cell bodies called **nuclei,** which send their outgoing information along other tracts of white matter. The cerebral white matter may be of three types: **Commissural fibers** connect areas of the right and left hemispheres, **projection fibers** connect the outer cerebrum with other parts of the CNS, and **association fibers** connect gyri. (Figure 12.7, p. 432; Figure 12.8, p. 433)

- Structures of the **limbic system** include the gyri of the *limbic lobe,* the **hippocampus,** the **amygdala,** and the tracts that connect these structures with one another and with other parts of the brain. The limbic system functions in memory, learning, emotion, and behavior. (Figure 12.9, p. 433)

- The diencephalon is the central core of the brain and is composed of four divisions, including the **subthalamus** and: (Figure 12.10, p. 434)
  ○ The **thalamus** is the main "gateway" into the cerebral cortex, controlling the flow of information both to and from the cerebral cortex.
  ○ The **hypothalamus** controls functions relating to the maintenance of homeostasis; regulates the sleep/wake cycle; regulates the autonomic nervous system; functions in the endocrine system; and plays a role in emotion, behavior, and memory.
  ○ The *epithalamus* contains the **pineal gland,** which secretes the hormone **melatonin.**

- The cerebellum plays a role in the coordination of movement. It is located inferior to the occipital lobe and consists of three lobes and two hemispheres, which are connected by the **vermis.** It has an outer layer of gray matter, the *cerebellar cortex,* and inner white matter, the **arbor vitae.** (Figure 12.11, p. 436)

- The brainstem controls many basic homeostatic functions as well as numerous reflexes, and participates in movement,

sensation, and complex functions such as sleep. It houses many important nuclei and tracts of white matter. (Figure 12.12, p. 437; Figure 12.13, p. 438; Figure 12.14, p. 439)
  ○ The **midbrain** is the superior portion of the brainstem. It functions in movement, sensation, and reflexes such as the startle reflex.
  ○ The **pons** is the middle portion of the brainstem. The pontine nuclei play a role in movement, breathing, reflexes, and complex functions.
  ○ The inferiormost portion of the brainstem is the **medulla oblongata.** It contains the **pyramids,** where the corticospinal tracts decussate. It is also the site of decussation of the **posterior columns.**
  ○ The **reticular formation** is a group of more than 100 nuclei that are found throughout all three divisions of the brainstem. Its nuclei are involved in sleep, pain transmission, motor functions, the autonomic nervous system, mood, and sensation. (Figure 12.15, p. 440)

- Figure 12.16 (p. 441) provides a summary of brain anatomy, and Figure 12.17 (p. 442) offers a big picture view of major brain structures and their functions.

Practice art-labeling exercise: Figure 12.16.

MODULE 12.3
Protection of the Brain 443

- The three cranial meninges include the thick outer **dura mater,** the middle **arachnoid mater,** and the thin inner **pia mater.** Between the dura and the arachnoid mater is the **subdural space;** between the arachnoid and the pia is the **subarachnoid space.** (Figure 12.18, p. 444)

Practice art-labeling exercise: Figure 12.18.

- The cavities within the brain are the **ventricles.** The ventricles contain **cerebrospinal fluid (CSF)** and are continuous with the central canal of the spinal cord. (Figure 12.19, p. 445)

- Cerebrospinal fluid increases the buoyancy of the brain, cushions it, and maintains a constant temperature. It is formed in the ventricles by a cluster of capillaries and ependymal cells called the **choroid plexus,** and is reabsorbed into the blood through the *arachnoid granulations.* (Figure 12.20, p. 446)

- The **blood-brain barrier** is a barrier between the blood in the vessels of the brain and the extracellular fluid of the brain cells. It consists of the cells of the capillaries and their basal laminae. These cells have many tight junctions and limited vesicular transport, so only a limited number of substances can cross them. (Figure 12.21, p. 447)

MODULE 12.4
The Spinal Cord 448

- The spinal cord has two main roles: It acts as a relay station to connect the brain with the rest of the body, and it carries out some processing and integration.

- The spinal cord is protected by its **spinal meninges,** which are similar to the cranial meninges. However, the spinal dura lacks a periosteal layer, and there is a fat-filled **epidural space** between the dura and the vertebral periosteum. (Figure 12.22, p. 449)

 Practice art-labeling exercise: Figure 12.22.

- The spinal cord begins at the foramen magnum and terminates as the **conus medullaris** between the first and second lumbar vertebrae. It is narrow except at the **cervical** and **lumbar enlargements,** where the spinal nerves for the upper and lower limbs originate, respectively. (Figure 12.23, p. 450)

- **Spinal nerves** project off each side of the spinal cord. The terminal bundle of spinal nerves is the **cauda equina,** and it fills the remainder of the vertebral cavity.

- Spinal gray matter is surrounded by an outer sheath of spinal white matter. The spinal gray matter is divided into three regions: (Figure 12.24, p. 451)
  - The **anterior horn** houses the cell bodies of lower motor neurons that control the motor functions of skeletal muscles.
  - The **posterior horn** contains the cell bodies of neurons that process and transmit sensory information.
  - The **lateral horn** is present only from the first thoracic to the second or third lumbar vertebrae; it contains cell bodies of the autonomic nervous system.

- There are three regions of white matter called funiculi: the **posterior funiculus,** the **lateral funiculus,** and the **anterior funiculus.** Each funiculus contains ascending and/or descending tracts that carry sensory and motor information, respectively. (Figure 12.25, p. 452)

 Practice art-labeling exercise: Figure 12.25.

MODULE **12.5**

Sensation Part I: Role of the CNS in Sensation 453

- Stimuli are processed in two steps: (1) The stimulus is detected by neurons of the PNS, and (2) the stimulus is interpreted by the CNS.

- The general somatic senses include the tactile senses, nondiscriminative touch, pain, stretch, proprioception (joint position), and temperature. Information relating to these senses is delivered to the brain via two major pathways:
  - The **posterior columns/medial lemniscal pathway** carries information concerning the tactile senses and proprioception. Sensory information is relayed by the thalamus to the primary somatosensory cortex and cortical association areas for interpretation. (Figure 12.26, p. 454)
  - The **anterolateral system** carries pain, temperature, and nondiscriminative touch sensations. The stimuli ascend through a number of tracts, including the **spinothalamic tracts,** with the thalamus as the ultimate destination. The thalamus relays temperature and nondiscriminative touch stimuli to the primary somatosensory cortex, as well as pain stimuli to many areas of the brain. (Figure 12.27, p. 455)

- Each part of the body is represented by a specific region of the primary somatosensory cortex. (Figure 12.28, p. 455)

- Most visual stimuli are delivered directly to the thalamus, which relays them to the primary visual cortex in the occipital lobe.

- Auditory stimuli are delivered to brainstem nuclei and then ascend to the thalamus. The thalamus sends them to the primary auditory cortex.

- Gustatory stimuli are delivered to a nucleus in the medulla and then ascend to the thalamus, which transmits them to gustatory cortices in the insula and parietal lobes.

- Olfactory stimuli are first processed by the cortex of the limbic system, after which they are relayed to the thalamus, hypothalamus, and other components of the limbic system.

- Vestibular stimuli are processed by multiple brainstem nuclei, the cerebellum, and areas within the cerebral cortex.

MODULE **12.6**

Movement Part I: Role of the CNS in Voluntary Movement 457

- **Upper motor neurons** reside mostly in the motor areas of the cerebral cortex. They stimulate local interneurons in the brainstem or spinal cord. The interneurons pass the message to *lower motor neurons* of the PNS, which innervate muscle fibers.

- Axons from the cortical motor areas unite to form the **corticospinal tracts,** which control the muscles below the head and neck, and the **corticonuclear tracts,** which control muscles of the head and neck.
  - The corticospinal tracts descend through the cerebral white matter into the brainstem, where the tracts decussate in the medullary pyramids. The majority of the fibers then travel through the lateral funiculus of the spinal cord as the lateral corticospinal tracts. (Figure 12.29, p. 458)
  - The corticonuclear tracts are similar to the corticospinal tracts except their axons terminate in the brainstem.

- Every motion requires the selection of a **motor program** by the upper motor neurons of the primary motor cortex in the cerebral cortex, which plan and execute voluntary motion.

- Each part of the body is represented by a specific region of the primary motor cortex. (Figure 12.30, p. 459)

- The basal nuclei and substantia nigra inhibit inappropriate movements and initiate motion. (Figure 12.31, p. 460)

- The cerebellum regulates ongoing movement and modifies the activity of the upper motor neurons to reduce motor error. The correction of motor error can occur over the long term as the process of **motor learning.** (Figure 12.32, p. 461)

- Figure 12.33 (p. 462) shows the big picture of CNS control of voluntary movement.

 Practice sequencing the steps: Figure 12.33.

MODULE **12.7**

Homeostasis Part I: Role of the CNS in Maintenance of Homeostasis 463

- The two systems in the body dedicated to maintaining homeostasis are the endocrine system and the nervous system.

- The two structures of the nervous system most concerned with homeostasis are the hypothalamus and the reticular formation.

- The homeostasis of vital functions is largely controlled by the **autonomic nervous system (ANS).** The hypothalamus is the main "boss" of the ANS and adjusts its output by relaying instructions to the nuclei of the reticular formation.

- Respiration is regulated by neurons in the medullary reticular formation.

- The hypothalamus triggers behavioral and physiological changes to maintain body temperature near a specific set point.

- Certain hypothalamic nuclei release neurotransmitters called orexins that stimulate feeding behaviors.

- *Sleep* is the normal and reversible suspension of consciousness. It is necessary for our survival and appears to serve a restorative function.
  - Human sleep follows a **circadian rhythm** set by the hypothalamus. When light levels decrease, the *suprachiasmatic nucleus (SCN)* of the hypothalamus decreases the activity of the reticular formation, which "disconnects" the thalamus from the cerebral cortex and triggers sleep. (Figure 12.34, p. 464)
  - Arousal from sleep is mediated by orexins that stimulate the neurons of the reticular formation.
  - Sleep progresses through five stages, discernible by their patterns of brain activity measured on an **electroencephalogram (EEG).** When a person is awake, the EEG shows a pattern of high-frequency, low-amplitude waves called **beta waves.** (Figure 12.35, p. 465)
    - Stages I, II, and III are light to moderately deep sleep stages and are characterized by lower-frequency and higher-amplitude **theta waves.**
    - Stage IV is deep sleep and is characterized by very low-frequency and very high-amplitude **delta waves.**
    - **Rapid eye movement,** or **REM, sleep** is characterized by rapid movements of the eyes and produces an EEG pattern resembling that in a waking adult.

MODULE **12.8**
Higher Mental Functions 467

- **Cognition** consists of processes that include social and moral behavior, intelligence, thoughts, language, problem-solving skills, and personality. These functions are carried out by the multimodal association areas, and many are lateralized to one hemisphere.

- Most aspects of **language** are lateralized to the left cerebral hemisphere; however, the emotional aspects of language are controlled by the right hemisphere. **Broca's area** and **Wernicke's area** control the ability to produce and understand language, respectively.

- **Declarative memory** is memory of things readily available to consciousness; **nondeclarative memory** includes skills and associations that are largely unconscious.
  - Immediate memory is stored for only seconds, short-term (working) memory is stored for several minutes, and long-term memory is stored for weeks or longer.
  - Moving the information in immediate and working memory into long-term memory is called **consolidation.** (Figure 12.37, p. 470)

  - Declarative memories may be formed by **long-term potentiation (LTP)** that occurs within the hippocampus.
  - Nondeclarative memories are formed by systems that do not rely on the hippocampus. They seem to be stored in the cerebral cortex, cerebellum, and the basal nuclei.

- Emotion involves responses mediated by the hypothalamus and the limbic system, particularly the amygdala.

ASSESS WHAT YOU LEARNED

See answers in Appendix A.

LEVEL 1 Check Your Recall

1. The central nervous system is responsible for:
   a. integrative functions.
   b. sensory functions.
   c. motor functions.
   d. Both b and c are correct.

2. Mark the following statements about the brain as true or false. If a statement is false, correct it to make a true statement.
   a. Humans use only 10% of their brains.
   b. The four main components of the brain are the cerebrum, the diencephalon, the cerebellum, and the brainstem.
   c. The right and left lateral ventricles are the largest of the ventricles in the brain and are located in the diencephalon.
   d. The cerebrum is responsible for our basic, involuntary homeostatic functions and reflexes.

3. The central sulcus separates the:
   a. parietal and temporal lobes.
   b. parietal and occipital lobes.
   c. frontal and temporal lobes.
   d. frontal and parietal lobes.

4. Which of the following statements about cerebral white matter is *false*?
   a. Commissural fibers connect the right and left cerebral hemispheres.
   b. Projection fibers connect the cerebral cortex of one hemisphere with structures in the other hemisphere.
   c. The corpus callosum is the largest bundle of white matter in the brain.
   d. Association fibers connect the gyri of the cerebral cortex with one another.

5. Mark the following statements about the cerebral cortex as true or false. If a statement is false, correct it to make a true statement.
   a. The neocortex is the "newest" component of the cerebral cortex from an evolutionary perspective.
   b. Unimodal association areas integrate information from many different sources.
   c. The primary visual cortex is located in the occipital lobe.
   d. The prefrontal cortex is located in the frontal lobe and is concerned with movement.

6. Which of the following is *not* one of the basal nuclei?

   a. Caudate nucleus
   b. Globus pallidus
   c. Cerebellum
   d. Putamen

7. Match the term on the left with its correct description in the column on the right.

   _____ Amygdala
   _____ Thalamus
   _____ Hippocampus
   _____ Midbrain
   _____ Medulla oblongata
   _____ Hypothalamus
   _____ Cerebellum
   _____ Pons

   a. Serves as the main "entryway" into the cerebral cortex
   b. Contains the pyramids where the corticospinal tracts decussate
   c. Portion of the limbic system involved in emotion
   d. Regulates homeostasis, the autonomic nervous system, the endocrine system, and the sleep/wake cycle
   e. Consists of three lobes and coordinates movement
   f. Portion of the limbic system; involved in memory
   g. Middle portion of brainstem; plays a role in movement, sleep, and arousal
   h. First component of the brainstem; participates in reflexes, sensation, and movement

8. Which of the following statements about the cranial meninges is *true*?

   a. The subdural space is between the dura and the arachnoid mater.
   b. The arachnoid mater closely follows the contours of the cerebral gyri and sulci.
   c. The pia mater is the tough outer meninx.
   d. The subarachnoid space is between the dura and the arachnoid mater.

9. What structure forms cerebrospinal fluid? How does this occur?

10. Explain the importance of the blood-brain barrier, and list the structures that form it.

11. Mark the following statements about the spinal cord as true or false. If a statement is false, correct it to make a true statement.

    a. The spinal cord functions as a relay and processing station.
    b. The spinal cord extends from the foramen magnum to the level between the first and second lumbar vertebrae.
    c. The epidural space around the spinal cord is only a potential space.
    d. The posterior horn of spinal gray matter contains the cell bodies of motor neurons.
    e. The corticospinal tracts are the main sensory tracts in the spinal cord.

12. Which of the following somatic sensations is *not* carried by the posterior columns?

    a. Tactile senses
    b. Temperature
    c. Proprioception
    d. Stretch

13. Fill in the blanks: The tracts of the posterior columns decussate in the _____, whereas the tracts of the anterolateral system decussate in the _____.

14. Which parts of the body have the greatest amount of space dedicated to them in the primary somatosensory cortex? Why?

15. Which of the following statements is *false*?

    a. The spinothalamic tracts are part of the anterolateral system.
    b. Pain, temperature, and crude touch stimuli are carried by the anterolateral system.
    c. Descending pathways from the brain and spinal cord can make the spinal cord less receptive to pain stimuli.
    d. The thalamus serves as the "gateway" for entry of all special sensory stimuli into the cerebral cortex, with the exception of audition (hearing).

16. Fill in the blanks: The cell bodies of upper motor neurons reside in the _____ and function to _____, whereas the cell bodies of lower motor neurons reside in the _____ and function to _____.

17. Label the following components of the corticospinal tracts with numbers 1 through 6, with 1 being the origin of the tracts and 6 their destination.

    a. _____ Medullary pyramids where most fibers decussate
    b. _____ Anterior horn of the spinal gray matter
    c. _____ Midbrain and pons
    d. _____ Upper motor neurons in the primary motor and premotor cortices
    e. _____ Corona radiata and internal capsule
    f. _____ Lateral funiculus of the spinal cord

18. Mark the following statements on the role of the brain in movement as true or false. If a statement is false, correct it to make a true statement.

    a. The dopaminergic neurons of the substantia nigra enhance the actions of the caudate nucleus and putamen.
    b. The cerebellum monitors the initiation of movement but does not monitor ongoing movements.
    c. The basal nuclei inhibit inappropriate movements and are required for the initiation of movement.
    d. The correction of motor error by the cerebellum can occur over the long term by motor learning.

19. Fill in the blanks: The two structures of the CNS that are responsible for maintenance of homeostasis are the _____ and the _____.

20. Which of the following statements is *false*?

    a. The suprachiasmatic nucleus is often called the body's "master clock."
    b. Melatonin and the ventrolateral preoptic nucleus increase the activity of the reticular formation.
    c. A decrease in the activity of the reticular formation disconnects the thalamus from the cerebral cortex and decreases the level of consciousness.
    d. Orexins are neurotransmitters that stimulate the reticular formation.

21. Match the term on the left with its correct description from the column on the right.

    _____ REM sleep   a. The EEG pattern seen in the deepest
    _____ Stage I sleep   stage of sleep
    _____ Delta waves   b. The deepest stage of sleep
    _____ Beta waves   c. The EEG pattern seen in an alert,
    _____ Theta waves   awake adult
    _____ Stage IV sleep   d. Drowsiness or the lightest stage of sleep
                                          e. The EEG pattern seen in lighter
                                          stages of sleep
                                          f. The stage of sleep during which
                                          most dreaming takes place

22. The part of the brain responsible for the production of language is known as:

a. Wernicke's area.
b. the prefrontal cortex.
c. the auditory association area.
d. Broca's area.

23. Fill in the blanks: Declarative memories are formed by the process of _____ that takes place in the _____, and are stored in the _____.

24. Which of the following is *not* a component of emotion?

a. Visceral motor responses mediated by the hypothalamus
b. Somatic motor responses mediated by the limbic system and the hypothalamus
c. Somatic motor responses mediated by the upper motor neurons of the corticospinal tracts
d. Subjective feelings mediated by the amygdala and the cerebral cortex

LEVEL 2 Check Your Understanding

1. Huntington's disease is characterized by a loss of normal inhibition mediated by the basal nuclei. Predict the symptoms you are likely to see with this disease.

2. How could you tell the difference between an injury that damaged the cerebellum and one that damaged the basal nuclei?

3. Why do injuries to the hippocampus interfere with the formation of new declarative memories but do not impact memories already encoded in long-term memory? Why do such injuries not affect the formation of nondeclarative memories?

LEVEL 3 Apply Your Knowledge

PART A: Application and Analysis

1. Ms. Norris is brought to the emergency department with injuries to the posterior of her head resulting from a motor vehicle accident. Which lobes of the cerebrum likely sustained injuries? Predict the main signs and symptoms you will see resulting from the injury to these cerebral lobes.

2. On further examination, Ms. Norris is shown to be in a state of unarousable unconsciousness, or coma. Which area or system of the brain is likely damaged? Explain.

3. In a diving accident, Arlene damages the left side of her upper thoracic spinal cord. On which side of the body will she lose: muscle control, tactile sensation, and pain sensation? Explain.

4. A new diet wonder drug is designed to block the release of orexins. How might this cause weight loss? Predict the potential adverse effects that might come from blocking orexin release.

PART B: Make the Connection

5. An athlete engaging in strenuous activity sometimes experiences a "runner's high," which is characterized by release of endorphins from the midbrain periaqueductal gray matter. What effect is this likely to have on the athlete's perception of pain and ability to continue with the strenuous activity? Will the release of endorphins stop the transmission of pain impulses? (*Connects to Chapter 11*)

13

The Peripheral Nervous System

as anyone ever snuck up right behind you and yelled your name? You jumped a foot in the air, right? Well, that was an automatic response that depended on your nervous system receiving input and responding to it, both functions that rely on your peripheral nervous system. In the previous chapter, we explored the central nervous system (CNS) and examined its role in movement, sensory perception, and the maintenance of homeostasis (see Chapter 12). Now we turn to the second part of the nervous system story: the *peripheral nervous system,* or *PNS.* Although we discuss the PNS and CNS in three separate chapters (continuing in the autonomic nervous system chapter—see Chapter 14), remember that the central and peripheral systems are continuous with one another anatomically and functionally. In fact, the cell bodies of many neurons in the PNS actually reside in the CNS, and the two systems cannot function independently of one another.

In this chapter, we examine the anatomy of the PNS, as well as its role in sensation and movement. A "big picture" view then ties together the roles of the CNS and PNS, showing how the nervous system as a whole controls these critical life functions. Finally, we wrap up the chapter with a discussion of the processes that cause the automatic movements called *reflexes.*

MODULE **13.1**

Overview of the Peripheral Nervous System

Learning Outcomes

1. Explain the differences between the sensory and motor divisions of the peripheral nervous system.
2. Differentiate between the somatic motor and visceral motor (autonomic) divisions of the nervous system.
3. Describe the structure of a peripheral nerve, and explain the differences between spinal nerves and cranial nerves.

The **peripheral nervous system (PNS)** links the CNS to the body and to the external environment. The PNS does this by first detecting sensory stimuli and delivering

Computer-generated image: Thirty-one pairs of spinal nerves carry sensory and motor signals throughout the body.

them to the CNS as sensory input. The CNS processes the input and then transmits the impulses through the PNS to *effectors*—muscle cells and glands—for motor output. This module explains the basics of the PNS, including its functional divisions and the structure of its main organs, the *peripheral nerves*. It concludes with an overview of how the PNS functions and how those activities are integrated with the CNS.

Divisions of the PNS

◀◀ Flashback

1. How do a motor (efferent) neuron and a sensory (afferent) neuron differ in structure and function? (p. 382)
2. What is a ganglion? (p. 387)
3. Define the terms somatic and visceral. (p. 382)

Recall that the PNS is classified functionally into two main divisions: the *sensory* (or *afferent*) *division* and the *motor* (or *efferent*) *division* (see Chapter 11). Each division is further categorized anatomically—by the types of structures that are innervated—into *somatic* (body) and *visceral* (internal organ) branches (**Figure 13.1**).

Sensory Division

As its name implies, the **sensory division** of the PNS consists of sensory (or afferent) neurons that detect various sensory stimuli and bring them to the CNS (left side of Figure 13.1). The sensory division contains two anatomical subdivisions:

● The **somatic sensory division** detects sensory information from the skin and structures of the musculoskeletal system. It detects stimuli of the *general senses* that arise external to the body, such as touch, temperature, and pain, as well as stimuli from within the body, such as muscle stretch and

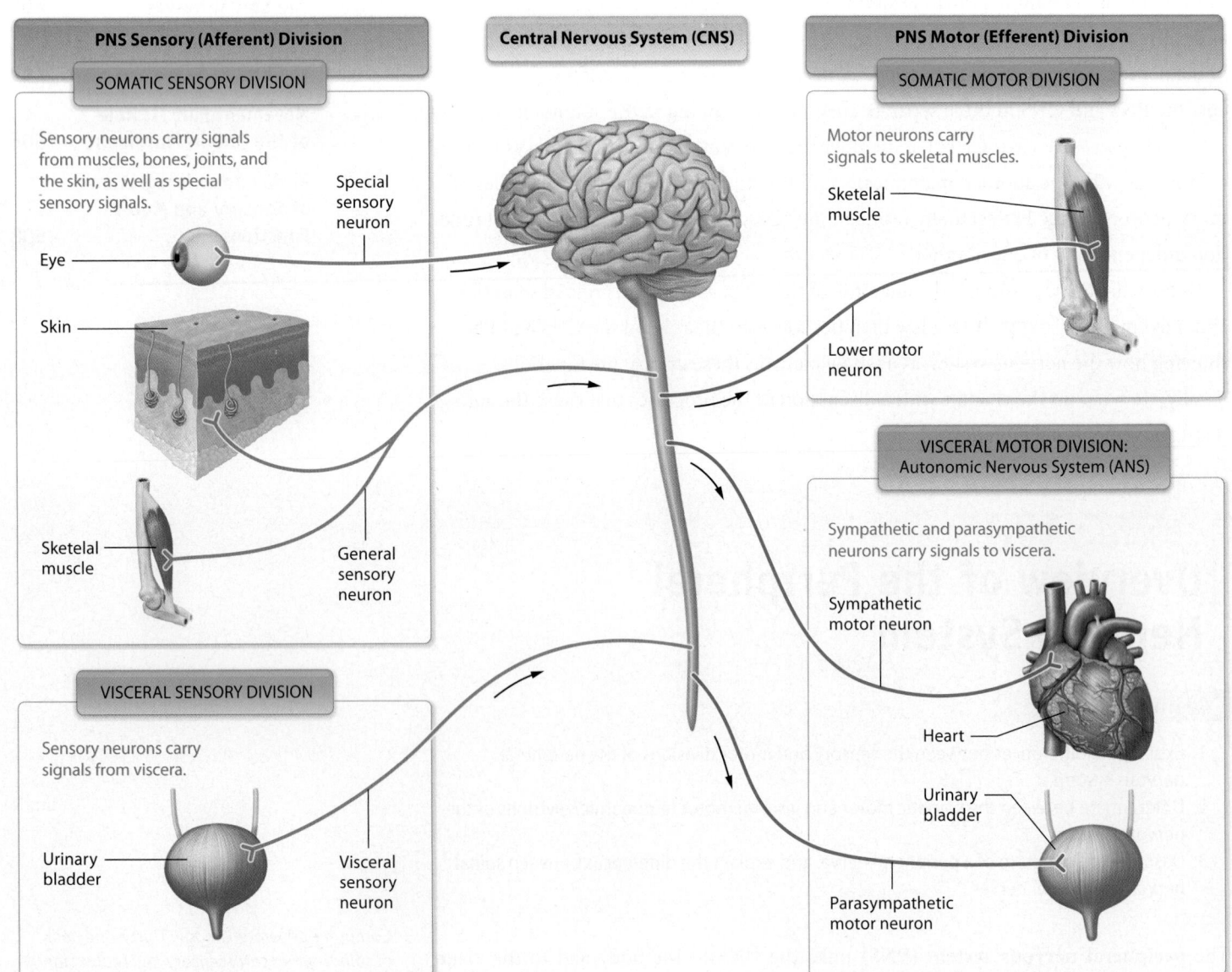

Figure 13.1 The organization of the peripheral nervous system.

the concentration of different chemicals in the body's fluids. The somatic sensory division also contains special sensory neurons that are responsible for detecting stimuli of the *special senses* (vision, hearing, equilibrium, taste, and smell).

- The **visceral sensory division** relays sensory information from the organs of the abdominopelvic and thoracic cavities. It detects internal stimuli such as blood pressure and the degree to which organs like the urinary bladder are stretched.

Modules 13.2 and 13.4 focus on the somatic sensory division. The special somatic sensory neurons are discussed in the special senses chapter (see Chapter 15).

Motor Division

The **motor division** of the PNS consists of motor neurons that carry out the motor functions of the nervous system (right side of Figure 13.1). Like the sensory division, it may be further classified—based on the organs that the neurons contact—into the *somatic motor division* and the *visceral motor division,* or *autonomic nervous system.*

Somatic Motor Division The **somatic motor division** is responsible for the body's *voluntary* motor functions. It is made up of *lower motor neurons* (or *somatic motor neurons*) that directly contact skeletal muscle fibers and trigger a contraction when stimulated by *upper motor neurons* in the CNS. We'll see in Module 13.6 that the activities of the somatic sensory and motor divisions are closely linked.

Visceral Motor Division: The Autonomic Nervous System As explained in the chapters on nervous tissue and the CNS (see Chapters 11 and 12, respectively), the **visceral motor division,** or **autonomic nervous system (ANS),** is responsible for maintaining many aspects of homeostasis through its control of the body's *involuntary* motor functions. Its neurons innervate cardiac muscle cells, smooth muscle cells, and the secretory cells of glands. The ANS has two divisions: the **sympathetic nervous system** and the **parasympathetic nervous system.** The sympathetic nervous system is usually described as the "fight or flight" division of the ANS—this name reflects its role in preparing the body for emergency situations that would involve fighting off an attacker or fleeing from danger. Its role is broader than this, however, as it maintains homeostasis when the body is engaged in any type of physical work and mediates the body's visceral responses to emotion. The parasympathetic nervous system is often described as the "rest and digest" division of the ANS, which reflects its role in digestion and in maintaining the body's homeostasis at rest. The ANS is discussed at greater length in its own chapter (see Chapter 14).

Quick Check

☐ 1. What two subclasses make up the sensory division of the PNS, and how do they differ?
☐ 2. What is a lower motor neuron? How are upper and lower motor neurons different?
☐ 3. In what ways do the somatic and visceral motor divisions of the PNS differ?

Overview of Peripheral Nerves and Associated Ganglia

◀◀ Flashback

1. What are the anterior and posterior horns of the spinal cord, and what types of neurons do they contain? (p. 451)
2. Where are the spinal nerves and cranial nerves located? (p. 382)

The main organs of the PNS are its many **peripheral nerves** (this term is generally shortened to just *nerves*), which consist of the axons of many neurons bound together by a common connective tissue sheath (**Figure 13.2**). The many nerves of the PNS contact, or **innervate** (IN-er-vayt), the majority of structures in the body. Most nerves are **mixed nerves**, meaning they contain both sensory and motor neurons. This is why a damaged nerve affects both sensation and movement to some degree. There are, however, **sensory nerves** that contain only sensory neurons and **motor nerves** that contain mostly motor neurons. All motor nerves contain a small population of sensory neurons that carry information pertaining to muscle stretch and tension, so no nerve is a pure motor nerve.

As we discussed in the nervous tissue chapter (see Chapter 11), the two types of nerves are classified according to their location:

- **Spinal nerves.** The **spinal nerves** originate from the spinal cord and innervate structures below the head and neck. As you can see in Figure 13.2a, two collections of axons connect the PNS with the spinal cord's gray matter—an **anterior root** containing motor neurons from the anterior horn, and a **posterior root** containing sensory neurons that enter the posterior horn. Note that the posterior root features a swollen area that houses the cell bodies of sensory neurons. Recall from the nervous tissue chapter that a collection of cell bodies in the PNS is called a *ganglion,* and accordingly, this swelling is the *posterior root ganglion* (or dorsal root ganglion). Just lateral to the posterior root ganglion, the two roots fuse to form a spinal nerve. Because each spinal nerve contains both sensory and motor neurons, all are mixed nerves. There are 31 pairs of spinal nerves, which are discussed fully in Module 13.3.
- **Cranial nerves.** The **cranial nerves** attach to the brain and mainly innervate structures of the head and neck. Unlike spinal nerves, cranial nerves are not formed from the fusion of motor and sensory nerve roots, and so some cranial nerves are purely sensory, others are mixed, and others are predominantly motor. Module 13.2 covers the 12 pairs of cranial nerves.

Note in Figure 13.2c that the motor and sensory axons making up a nerve are held together by a connective tissue sheath called the **epineurium** (ep′-ih-NOOR-ee-um). Within the nerve, axons are bundled into smaller groups known as **fascicles** (FASS-ih-kulz). Fascicles are in turn bound by another connective tissue sheath, the **perineurium** (pehr′-ih-NOOR-ee-um).

(a) **Anterior and posterior roots and spinal nerve**

(b) **Photomicrograph of dissected spinal nerve**

(SEM, 20×)

(c) **Detailed structure of spinal nerve**

Figure 13.2 **The structure of roots and spinal nerves.**

Nestled between the fascicles are blood vessels that supply the axons with oxygen and nutrients. Each axon within a fascicle is surrounded by its own connective tissue sheath, the **endoneurium** (en′-doh-NOOR-ee-um) (note that although the figure shows a spinal nerve, the arrangement within spinal and cranial nerves is the same). Notice how the arrangement of a nerve closely resembles that within a skeletal muscle, which we discussed in the muscle tissue chapter (see Chapter 10).

Quick Check

☐ 4. What structures are found in a peripheral nerve?
☐ 5. Which structures unite to form a spinal nerve?
☐ 6. How do cranial and spinal nerves differ?

Functional Overview of the PNS

Let's look at how the PNS functions and how those functions are integrated with the CNS. We start with the sensory arm of the PNS. Sensory neurons detect stimuli at structures known as *sensory receptors*. The detected stimuli are transmitted along the

sensory neurons via spinal or cranial nerves to the cerebral cortex, where the information is interpreted and integrated. Depending on the nature of the sensory information, an appropriate motor response is initiated. For example, if you hear the microwave beep, your CNS interprets this input to mean that your dinner is ready, and initiates a motor response so you can walk to the microwave and remove, and presumably eat, your dinner.

The motor response is initiated by commands from the motor areas of the brain to the upper motor neurons. These impulses travel to the spinal cord, where they synapse on lower motor neurons of the PNS. The lower motor neurons then carry the impulse to the appropriate muscle fibers via specific cranial or spinal nerves, where they trigger their contraction. Later modules will delve into the specifics of these processes.

Quick Check

☐ 7. How are sensations detected in the PNS and delivered to the CNS?
☐ 8. How are motor impulses transmitted from the CNS to the PNS?

Apply What You Learned

☐ 1. How do a nerve and a neuron differ? Which would most affect motor or sensory function: damage to a neuron or damage to a nerve? Explain.

☐ 2. Explain how the arrangement of axons within a nerve resembles the arrangement of muscle fibers within a skeletal muscle.

☐ 3. To which division of the PNS would damage be most life-threatening? Explain.

See answers in Appendix A.

MODULE **13.2**

The Cranial Nerves

Learning Outcomes

1. Identify the cranial nerves by name and number.
2. Describe the specific functions of each pair of cranial nerves, and classify each pair as sensory, motor, or mixed nerves.
3. Describe the locations of selected cranial nerve nuclei and the ganglia associated with the cranial nerves.

As we discussed in Module 13.1, the cranial nerves transmit information to and from the brain. Each pair of cranial nerves is identified in two ways: First, a Roman numeral reflects its attachment site in the brain, and second, a name reflects its function. The order and names of the 12 pairs of cranial nerves are shown in **Figure 13.3**.

Recall that certain cranial nerves contain only sensory axons, others contain both motor and sensory axons, and the remainder contain mostly motor axons. Generally speaking, the olfactory (I), optic (II), and vestibulocochlear (VIII) nerves are sensory nerves; the oculomotor (III), trochlear (IV), abducens (VI), accessory (XI), and hypoglossal (XII) nerves are motor nerves; and the trigeminal (V), facial (VII), glossopharyngeal (IX), and vagus (X) nerves are mixed nerves.

In this module, we discuss the structure, location, and functions of all 12 pairs of cranial nerves. Although it is traditional to cover the cranial nerves sequentially in ascending order, here we categorize them functionally as sensory, motor, or mixed nerves.

⬆ **STUDY BOOST**

Remembering the Cranial Nerves

It's easy to feel overwhelmed when you're learning the cranial nerves. You have to remember not only what all 12 pairs do and where they are located, but also two names for each nerve! A way to make this easier, however, is to use a mnemonic to recall the order and names of the cranial nerves. A popular mnemonic is the following:

Oh	(I, **O**lfactory)
Once	(II, **O**ptic)
One	(III, **O**culomotor)
Takes	(IV, **T**rochlear)
The	(V, **T**rigeminal)

Anatomy	(VI, **A**bducens)
Final	(VII, **F**acial)
Very	(VIII, **V**estibulocochlear)
Good	(IX, **G**lossopharyngeal)
Vacations	(X, **V**agus)
Are	(XI, **A**ccessory)
Happening	(XII, **H**ypoglossal)

If you have trouble remembering which Roman numeral is assigned to the optic nerve and which to the olfactory nerve, just remember that you have *one* nose (I, olfactory) and *two* eyes (II, optic).

A second mnemonic can help you recall the cranial nerves by their main function—sensory (S), motor (M), or both (B):

Some	(I, Olfactory—**S**ensory)
Say	(II, Optic—**S**ensory)
Money	(III, Oculomotor—**M**otor)
Matters	(IV, Trochlear—**M**otor)
But	(V, Trigeminal—**B**oth)
My	(VI, Abducens—**M**otor)
Brother	(VII, Facial—**B**oth)
Says	(VIII, Vestibulocochlear—**S**ensory)
Big	(IX, Glossopharyngeal—**B**oth)
Brains	(X, Vagus—**B**oth)
Matter	(XI, Accessory—**M**otor)
More	(XII, Hypoglossal—**M**otor)

A final hint that can help you remember the functions of the cranial nerves is to look closely at their names and connect them with word roots (see the back of the book). For example, cranial nerve III is the oculomotor nerve. If you break this word into its two components, you get *oculo-*, which means "eye," and *-motor*, which means "movement." So you can deduce just from this nerve's name one of its main functions, which is to move the eye. ∎

The Sensory Cranial Nerves

◀◀ **Flashback**

1. What is the limbic system, and what are its functions? (p. 432)
2. Define the term decussate. (p. 440)

There are three cranial nerves that contain the axons of only sensory neurons: the olfactory (I), optic (II), and vestibulocochlear (VIII) nerves. To make these nerves easier to study, we have arranged the illustrations and text in **Table 13.1**. Each of these nerves is involved in one of the special senses and will be discussed in more detail in that chapter (see Chapter 15). However, we explore the basics of these nerves next; please be sure to study Table 13.1 before going on.

Quick Check

☐ 1. List the Roman numeral, origins, destinations, and function(s) of the following nerves:
   a. Olfactory nerve
   b. Optic nerve
   c. Vestibulocochlear nerve

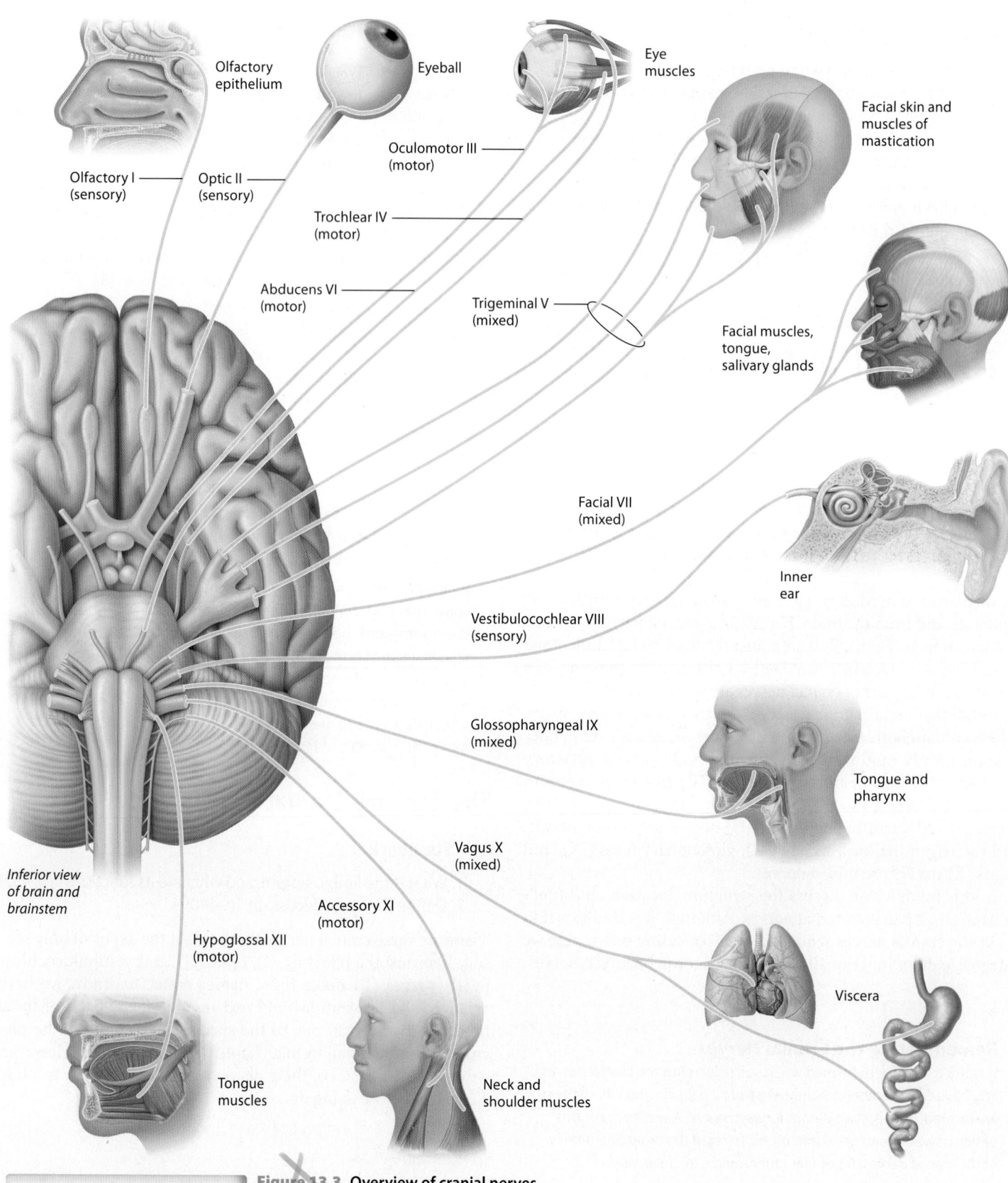

Olfactory epithelium

Eyeball

Eye muscles

Facial skin and muscles of mastication

Oculomotor III (motor)

Olfactory I (sensory)

Optic II (sensory)

Trochlear IV (motor)

Abducens VI (motor)

Trigeminal V (mixed)

Facial muscles, tongue, salivary glands

Facial VII (mixed)

Inner ear

Vestibulocochlear VIII (sensory)

Glossopharyngeal IX (mixed)

Tongue and pharynx

Vagus X (mixed)

Inferior view of brain and brainstem

Accessory XI (motor)

Hypoglossal XII (motor)

Viscera

Tongue muscles

Neck and shoulder muscles

Figure 13.3 **Overview of cranial nerves.**

Practice art labeling
@MasteringA&P®

TABLE 13.1 THE SENSORY CRANIAL NERVES

Nerve	Origin, Course, and Destination	Function
Olfactory nerve (I) 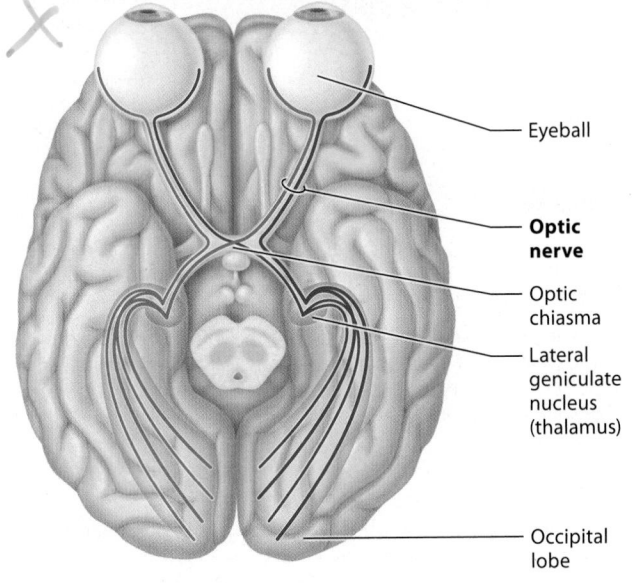	*Origin:* Originates from the unmyelinated axons of neurons whose cell bodies are located in the olfactory epithelium in the roof of the nasal cavity. *Course:* The axons form bundles that penetrate the holes in the cribriform plate of the ethmoid bone, and end in a bulbous structure known as the *olfactory bulb* (shown to the left). This bulb sits on the inferior surface of the brain's frontal lobe and continues posteriorly as the *olfactory tract*. *Destination:* Terminates in the medial temporal lobe and structures of the limbic system, including the limbic cortex, the hippocampus, and the amygdala.	The nerve for *olfaction* (*olfact-* = "smell"), or the sense of smell. Its nerve endings contain chemoreceptors that depolarize in response to chemicals in the air that we breathe. These stimuli are interpreted in the brain by the primary olfactory cortex.
Optic nerve (II)	*Origin:* Originates from myelinated axons of neurons in the posterior eye. *Course:* As shown to the left, the two optic nerves meet and form an "X" called the **optic chiasma** (ky-AZ-mah), where some of the axons decussate (switch sides). *Destination:* Axons from the optic chiasma are destined for structures such as the lateral geniculate nucleus of the thalamus, the midbrain, and ultimately the primary visual cortex in the occipital lobe.	The nerve for vision (*opt-* = "vision"). It transmits visual stimuli in the form of action potentials triggered when light hits the eye's photoreceptors. These stimuli are processed in the brain by the primary visual cortex.
Vestibulocochlear nerve (VIII) 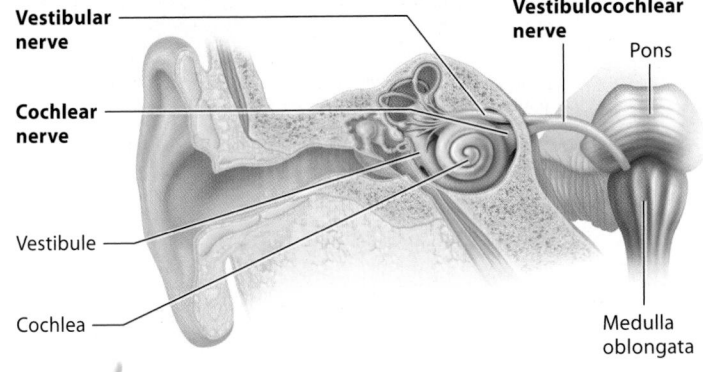	*Origin:* Actually two separate nerves, the vestibular and cochlear, which originate in the inner ear and share a common epineurium. *Courses:* Fibers from the vestibular and cochlear nerves fuse shortly after they leave the temporal bone to become the vestibulocochlear nerve. *Destination of vestibular nerve:* Axons terminate in the cerebellum and nuclei in the medulla oblongata. *Destination of cochlear nerve:* Axons travel to the medulla oblongata and terminate in auditory cochlear areas.	*Vestibular nerve:* Its neurons depolarize in response to body position and so are concerned with balance and equilibrium. *Cochlear nerve:* Its neurons depolarize in response to sound waves and are responsible for *audition* (aw-DIH-shun; *aud-* = "hear"), or the sense of hearing.

The Olfactory nerve (I) figure includes the labels: Corpus callosum, Frontal lobe, Olfactory bulb, Cribriform plate of ethmoid bone, Olfactory epithelium, **Olfactory nerves**, Amygdala, Olfactory tract.

The Optic nerve (II) figure includes the labels: Eyeball, **Optic nerve**, Optic chiasma, Lateral geniculate nucleus (thalamus), Occipital lobe.

The Vestibulocochlear nerve (VIII) figure includes the labels: **Vestibular nerve**, **Cochlear nerve**, Vestibule, Cochlea, **Vestibulocochlear nerve**, Pons, Medulla oblongata.

The Motor Cranial Nerves

Certain cranial nerves—namely, the oculomotor (III), trochlear (IV), abducens (VI), accessory (XI), and hypoglossal (XII) nerves—are considered motor nerves, although technically they contain axons of proprioceptive sensory neurons. In spite of these sensory axons, they are still viewed as motor nerves because the main function of their sensory neurons is to allow the brain to monitor the contraction of the muscles that they innervate. The locations and functions of the motor cranial nerves are covered in Table 13.2; make a point of studying it before proceeding.

TABLE 13.2 THE MOTOR CRANIAL NERVES

Nerve	Origin and Destination	Function
Oculomotor nerve (III) (ok'-yoo-low-MOW-ter)	*Origin:* Arises from the superior and lateral portion of the midbrain. *Destination:* Extraocular muscles and smooth muscles of the eye (*oculo-* = "eye")	Contains axons of both somatic motor neurons and visceral motor neurons of the parasympathetic nervous system, and has the following four primary functions: • **Moving the eyeball.** As shown above, the somatic motor axons innervate four of the six extraocular muscles that move the eyeball: the medial rectus, superior rectus, inferior rectus, and inferior oblique muscles. It moves our eyes medially, superiorly, inferiorly, and inferolaterally. • **Opening the eye.** Other somatic motor axons innervate the levator palpebrae superioris muscle, which opens the eyelid. • **Constricting the pupil.** The parasympathetic axons innervate a muscle surrounding the pupil called the *sphincter pupillae muscle.* These axons stimulate this muscle to contract, which constricts the pupil and limits the light entering the eye. (Pupil dilation is mediated by a different nerve.) • **Changing the lens shape.** The thickness of the lens (which focuses light on the photoreceptors in the posterior eyeball) is controlled by the ciliary muscle. This muscle is innervated by the parasympathetic axons, which stimulate the ciliary muscle to contract, making the lens rounder for near vision.
Trochlear nerve (IV) (TROK-lee-ur)	*Origin:* Originates from the inferior portion of the midbrain. Named for the cartilaginous trochlea through which the tendon of the superior oblique muscle passes. *Destination:* Superior oblique muscle of the eye	As shown above, the somatic motor neurons innervate the superior oblique muscle (a common mnemonic is "SO$_4$," as it is cranial nerve IV), which moves the eye medially and inferiorly.
Abducens nerve (VI) (ab-DOO-senz)	*Origin:* The cell bodies of the abducens nerve are located in the pons. *Destination:* Lateral rectus muscle of the eye	The somatic motor neurons innervate the lateral rectus muscle. Its name comes from the fact that this muscle abducts the gaze when it turns the eye laterally.

TABLE 13.2 *(continued)*

Nerve	Origin and Destination	Function
Accessory nerve (XI)	*Origin:* Unique in that its origin has both a cranial component from the medulla and a spinal component from the cervical spinal cord. The somatic motor neurons of the spinal component travel superiorly and enter the cranial cavity through the foramen magnum, after which they merge with the somatic motor neurons of the cranial component. As shown above, the two travel together for only a short distance, after which they diverge. *Cranial Destination:* The cranial component accompanies the vagus nerve (covered in Table 13.3) and innervates certain muscles of the larynx. *Spinal Destination:* The spinal component turns inferiorly and exits the cranial cavity to innervate the trapezius and sternocleidomastoid muscles of the neck and shoulders.	The cranial component of the accessory nerve innervates certain muscles of speech, whereas the spinal component innervates the head and shoulder.
Hypoglossal nerve (XII)	*Origin:* Arises from the inferiormost part of the medulla (see above). *Destination:* As its name suggests, (*hypo-* = "below," *glosso-* = "tongue"), its destination lies inferior to the tongue. It innervates most of the intrinsic and extrinsic muscles of the tongue.	This is a motor nerve that innervates the muscles of the tongue—it plays no role in taste sensation.

Quick Check

☐ 2. List the Roman numeral, origins, destinations, and main function(s) of the following nerves:
   a. Oculomotor nerve
   b. Trochlear nerve
   c. Abducens nerve
   d. Accessory nerve
   e. Hypoglossal nerve

The Mixed Cranial Nerves

As you may expect, the mixed cranial nerves—the trigeminal (V), facial (VII), glossopharyngeal (IX), and vagus (X) nerves—are generally large and have a fairly wide distribution. All but one (the trigeminal nerve) contain somatic sensory, somatic motor, and parasympathetic neurons wrapped up in the same nerve. Because of their size and the different types of neurons they house, the mixed cranial nerves have multiple and diverse functions. Their structure and functions are described in **Table 13.3**.

TABLE 13.3 THE MIXED CRANIAL NERVES

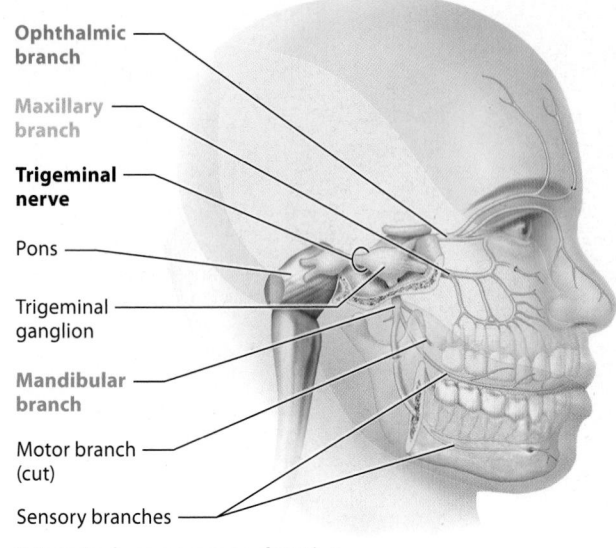

Ophthalmic branch
Maxillary branch
Trigeminal nerve
Pons
Trigeminal ganglion
Mandibular branch
Motor branch (cut)
Sensory branches

Trigeminal nerve: sensory function

Motor branch of mandibular nerve
Temporalis muscle
Masseter muscle

Mandibular nerve: motor function

Nerve	Origin, Course, and Destination	Sensory Function	Motor Function
Trigeminal (V) (try-JEM-ih-nul)	*Origin:* The motor portion originates at the midbrain and pons junction, and the sensory portion from sensory receptors around the face. *Course:* A short distance from its origin is the large, bulbous **trigeminal ganglion,** which houses the cell bodies of its sensory neurons. Anterior to this ganglion, it splits into three branches (*tri-* = "three"): the *ophthalmic* (off-THAL-mik), *maxillary,* and *mandibular* nerves. *Destinations:* All three branches have sensory fibers that terminate in the primary somatosensory cortex. The mandibular nerve has motor fibers that terminate at the muscles of mastication.	The sensory root detects facial sensation, including stimuli from the oral and nasal cavities. ***Ophthalmic nerve:*** Its somatic sensory axons supply the area shaded blue, shown above, which includes the skin over much of the scalp, the forehead, around the eyes, and over the anterior nose; they also supply the nasal mucosa and structures of the eye. ***Maxillary nerve:*** As shown above, its somatic sensory axons supply the area shaded orange, which includes the skin over the middle of the face. ***Mandibular nerve:*** The sensory axons of the inferior mandibular nerve supply the area shaded green in the illustration above, which includes the skin of the chin and the lateral part of the face.	***Ophthalmic nerve:*** No motor function. ***Maxillary nerve:*** No motor function. ***Mandibular nerve:*** Shown above, its motor axons supply the masseter and temporalis muscles, which elevate the mandible (close the jaw) during mastication (chewing) and swallowing.
Facial nerve (VII)	*Origin:* As shown in the illustrations on the next page, the motor portion (or *motor root*) originates in nuclei in the pons and medulla, and the sensory portion (or *sensory root*) from the tongue, external ear, palate, and nasal cavity. *Course:* Several ganglia house cell bodies of the facial nerve's sensory root; the largest is called the **geniculate ganglion** (jen-IK-yoo-lat).	The sensory root provides the following: • taste sensation from chemoreceptors in specialized receptor cells in the mucosa of the anterior two-thirds of the tongue; and • somatic sensation from the external ear, palate, and nasal cavity.	The somatic motor neurons of the five branches of the motor root supply the muscles of facial expression and other facial muscles.

TABLE 13.3 *(continued)*

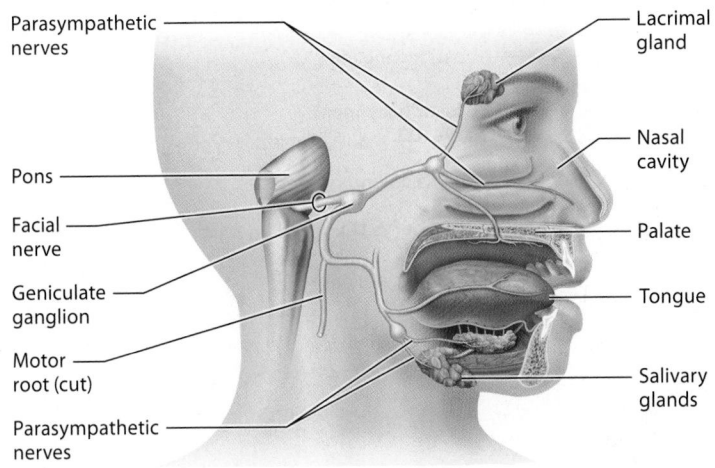

Facial nerve: mixed sensory and visceral motor roots

Facial nerve: motor function

Nerve	Origin, Course, and Destination	Sensory Function	Motor Function
Facial nerve (VII) *(continued)*	*Sensory Destination:* Various somatosensory areas of the cerebral cortex. *Motor Destination:* As shown above, these neurons innervate the muscles of facial expression as well as several other facial and neck muscles. The motor root splits into five branches: the *temporal, zygomatic, buccal, mandibular,* and *cervical nerves* (a common mnemonic is "**T**o **Z**anzibar **B**y **M**otor **C**ar").		In addition, the parasympathetic neurons trigger secretion from certain salivary glands in the mouth, the lacrimal (tear) glands over the eye, and the nasal mucous glands.
Glossopharyngeal nerve (IX) (gloss'-oh-fah-RIN-jee-ul)	*Origin:* The motor neurons originate in nuclei in the medulla; the sensory neurons originate in the tongue, pharynx, around the ear, and in blood vessels of the neck. *Course:* The cell bodies of its sensory neurons are located in two ganglia: the *superior ganglion* and the *inferior ganglion.* *Destination:* Its location and the structures it innervates are reflected in its name—recall that *glosso-* means "tongue" and *pharynx* means "throat."	The sensory portion of this nerve detects sensation as follows: • The chemoreceptors of the posterior one-third of the tongue are associated with special sensory axons of this nerve. The cell bodies of these neurons are in the inferior ganglion. • A small branch of this nerve contains somatic sensory neurons that innervate the external ear alongside the facial nerve. It also contains visceral sensory neurons that provide sensation for the posterior pharynx and surrounding structures. In addition, certain visceral neurons detect changes in blood pressure via receptors in the carotid artery of the neck.	The motor branches of this nerve supply a muscle around the pharynx that is responsible for swallowing movements. Also, parasympathetic neurons trigger salivation from the salivary gland called the *parotid gland.* You can feel this stimulation and subsequent salivation in your cheeks when you eat very salty or acidic foods such as pickles or lemons.

(continued)

TABLE 13.3 THE MIXED CRANIAL NERVES *(continued)*

Illustration is not to scale; orange is used simply to distinguish the glossopharyngeal nerve.

Nerve	Origin, Course, and Destination	Sensory Function	Motor Function
Vagus nerve (X)	*Origin:* The motor neurons originate from the brain at the medulla. The sensory neurons originate around the tongue, pharynx, skin of the ear, and certain blood vessels of the neck. *Course:* As shown above, cell bodies of its sensory neurons are housed in the *superior ganglion* and the *inferior ganglion.* *Destination:* It is the main parasympathetic nerve, and the most widely distributed nerve in the body (see above). It innervates both the throat and anterior neck, and nearly all of the thoracic and abdominal viscera.	The somatic sensory neurons serve the skin around the ear. The special sensory neurons convey taste sensation from the pharynx, and the visceral sensory neurons detect sensation in the mucous membranes of the pharynx. This nerve also contains visceral sensory neurons whose chemoreceptors detect the blood CO_2 concentration.	The somatic motor fibers supply the muscles surrounding the pharynx and larynx (the "voice box") that contract during speaking and swallowing. We continue our discussion of the visceral innervation of this nerve in the ANS chapter (see Chapter 14).

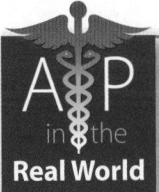

Trigeminal Neuralgia

Real World

Trigeminal neuralgia (noo-RAL-jah), also known as *tic douloureux* (TIK doo-loh-ROO), is a chronic pain syndrome that involves one or more branches of the trigeminal nerve. Patients with trigeminal neuralgia suffer brief attacks of intense pain that last from a few seconds to 2 minutes. The pain is typically unilateral and may occur several times per day. Certain stimuli are known to trigger attacks, such as chewing and light touch or vibratory stimuli to the face (even a light breeze has been said to trigger an attack for certain patients).

In spite of the recurrent painful episodes, neurological examinations of patients with trigeminal neuralgia are normal, and the cause of the disease is unknown. Pain medications are typically ineffective, and treatment is instead aimed at reducing the aberrant transmission through the nerve, often by severing it.

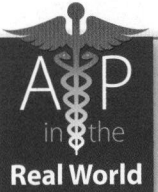

Bell's Palsy

Real World

A common problem associated with the facial nerve is *Bell's palsy*, in which the facial nerve's motor root is impaired by a virus, tumor, trauma, or an unknown cause. Patients with Bell's palsy have weakness or complete paralysis of the muscles of facial expression on the affected side. The unaffected side, however, appears normal. The paralysis leads to problems with blinking, closing the eye, and making general facial expressions such as smiling. In addition, other structures innervated by the facial nerve may be affected, including the lacrimal gland, salivary glands, and taste sensation to the anterior two-thirds of the tongue.

Typically, the individual experiences the onset of symptoms rapidly, within about 72 hours. Treatment may include anti-inflammatory medication, antiviral medication, physical therapy, and surgery. However, even without treatment, many individuals recover function of the paralyzed muscles in about 3–8 weeks.

Having multiple functions means that these mixed nerves are subject to many problems when they are damaged. See *A&P in the Real World: Trigeminal Neuralgia* and *A&P in the Real World: Bell's Palsy* to read about two such conditions.

Quick Check

☐ 3. List the Roman numeral, origins, destinations, and main function(s) of the following nerves:
   a. Trigeminal nerve
   b. Facial nerve
   c. Glossopharyngeal nerve
   d. Vagus nerve
☐ 4. List the 12 pairs of cranial nerves in ascending order by both their name and number.
☐ 5. Which cranial nerves are sensory, which are motor, and which are mixed nerves?

Apply What You Learned

☐ 1. Often when the brain is damaged, the cranial nerves are damaged as well. It is possible to test clinically for the function of each cranial nerve with a simple physical exam. Damage to which cranial nerve or nerves would lead to the following findings? Explain your answer.
   a. Complete loss of taste sensation
   b. Inability to move the tongue
   c. Inability to move the eyes in any direction
   d. Loss of balance and equilibrium
   e. Inability to close the jaw

☐ 2. Drugs that stimulate the parasympathetic nervous system often lead to the adverse effect of excessive drooling. Which cranial nerve(s) is/are involved in this effect? What would happen if the activity of these nerves was instead inhibited?

See answers in Appendix A.

MODULE **13.3**
The Spinal Nerves

Learning Outcomes

1. Discuss the relationships between structures of the spinal nerves: root, nerve, ramus, plexus, tract, and ganglion.

2. Describe the gross anatomy of the spinal nerves.
3. Identify and describe the four spinal nerve plexuses, and give examples of nerves that emerge from each.
4. Describe the functions of the major spinal nerves.

Now let's turn our attention to the structure and function of the 31 pairs of spinal nerves. First, we focus on the structure

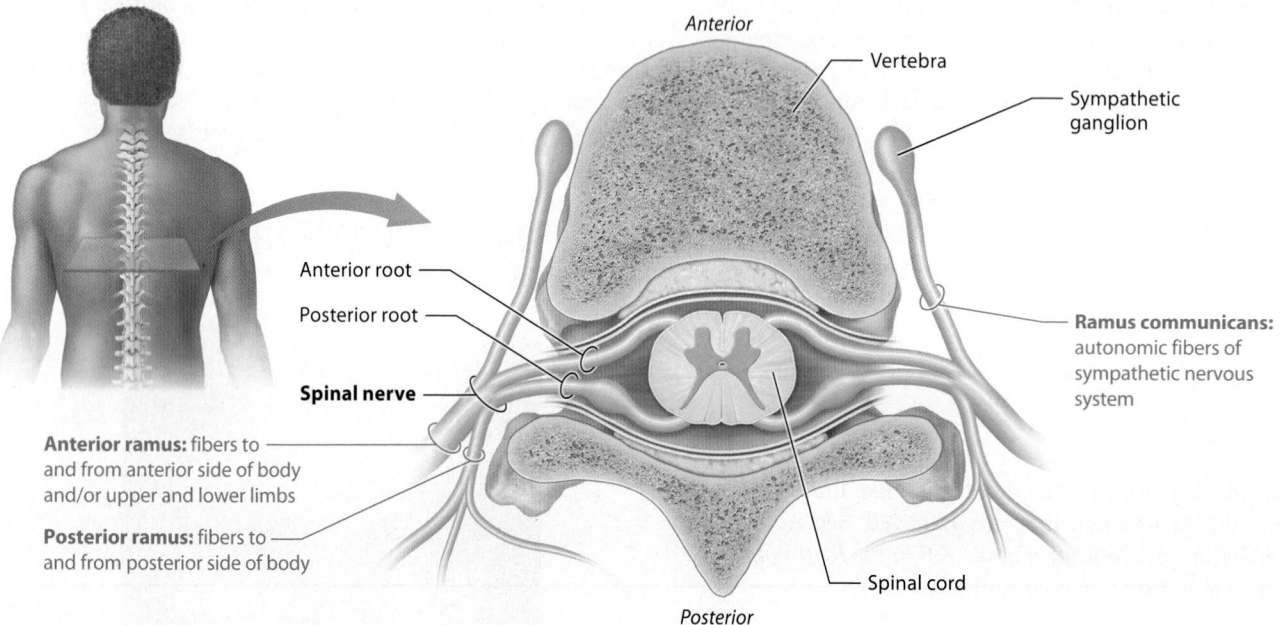

(a) Structure of anterior and posterior rami of spinal nerves

Labels in (a):
Anterior
Vertebra
Sympathetic ganglion
Anterior root
Posterior root
Spinal nerve
Ramus communicans: autonomic fibers of sympathetic nervous system
Anterior ramus: fibers to and from anterior side of body and/or upper and lower limbs
Posterior ramus: fibers to and from posterior side of body
Spinal cord
Posterior

(b) Function of roots, spinal nerve, and rami

Labels in (b):
Anterior
Ramus communicans: carries visceral motor signals from the ANS to organs of the ventral body cavity.
Anterior root: carries somatic and visceral motor signals from the CNS to skeletal muscles and gland cells.
Visceral motor neuron
Somatic motor neuron
Spinal nerve: carries visceral motor, somatic motor, and sensory signals to and from the spinal cord.
Posterior root: carries sensory signals from the PNS to the spinal cord.
Anterior and posterior rami: carry somatic motor signals from the CNS to skeletal muscles, and sensory signals from sensory receptors in the PNS to the spinal cord.
Sensory neuron
Posterior

Figure 13.4 Structure and function of roots, spinal nerves, and rami.

of spinal nerves and how many of them cluster to form networks, or *plexuses*. Then the discussion examines each plexus and the main nerves that branch from it.

Structure of Spinal Nerves and Spinal Nerve Plexuses

As we have discussed, all spinal nerves are formed from the fusion of an anterior root containing motor neurons from the spinal cord's anterior horn and a posterior root containing sensory neurons that enter the posterior horn (look back at Figure 13.2). Because the posterior root contains only sensory axons, it can carry only sensory information from the PNS to the CNS. Its partner, the anterior root, carries only somatic and visceral motor signals from the CNS to the PNS. When these roots come together in a spinal nerve, however, the nerve transmits signals in both directions, as it contains axons of neurons for each type: sensory from sensory receptors and motor to skeletal muscles.

As you can see in **Figure 13.4a**, the spinal nerve itself is actually quite short, because about 1–2 cm after it forms and leaves the vertebral cavity, it splits into two nerves: (1) a **posterior ramus** (RAY-muss; *ramus* = "branch"; plural, *rami*), which travels to the posterior side of the body; and (2) an **anterior ramus,** which travels to the anterior side of the body and/or the upper and lower limbs. Each ramus typically branches multiple times along its course through the body and ultimately supplies different skeletal muscles and regions of the skin. Both the anterior and posterior rami are mixed nerves, as they carry sensory and somatic motor information (**Figure 13.4b**).

Note that another small branch stems from the anterior ramus. These small branches are called *rami communicantes* (singular, *ramus communicans*), and they contain visceral motor or autonomic neurons of the sympathetic nervous system (see Figure 13.4). Unlike the other branches of the anterior and posterior rami, the rami communicantes contain visceral motor axons only and so are not mixed nerves.

The 31 pairs of spinal nerves consist of 8 pairs of cervical spinal nerves (C_1-C_8), 12 pairs of thoracic spinal nerves (T_1-T_{12}), 5 pairs each of lumbar and sacral spinal nerves (L_1-L_5 and S_1-S_5), and 1 tiny pair of coccygeal nerves (Co_1) (**Figure 13.5**). The anterior rami of the cervical, lumbar, and sacral spinal nerves come together and merge to form complicated-looking networks of nerves called **nerve plexuses** (PLEK-suss-ez). The axons of each spinal nerve cross over one another to enter different plexus branches. For this reason, the muscles supplied by a single branch of a nerve plexus are often served by two or more different spinal nerves. This works to our advantage, as it means that injury to one spinal nerve does not completely cut off motor or sensory innervation to that body part.

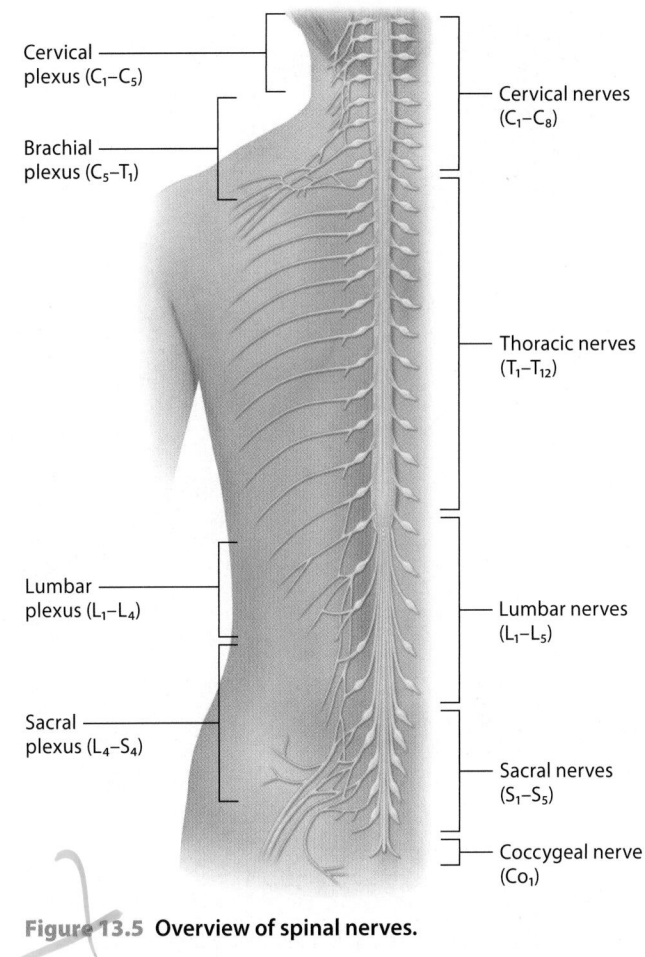

Cervical plexus (C_1-C_5)

Brachial plexus (C_5-T_1)

Lumbar plexus (L_1-L_4)

Sacral plexus (L_4-S_4)

Cervical nerves (C_1-C_8)

Thoracic nerves (T_1-T_{12})

Lumbar nerves (L_1-L_5)

Sacral nerves (S_1-S_5)

Coccygeal nerve (Co_1)

Figure 13.5 Overview of spinal nerves.

Next, we explore the anatomy of the nerve plexuses and the main structures supplied by the larger nerves. We also examine the nerves that arise from the anterior rami of the thoracic nerves, which do not form plexuses.

Quick Check

☐ 1. Which structures form a spinal nerve? Why are all spinal nerves mixed?

☐ 2. What are the anterior and posterior rami, and what do they supply?

Cervical Plexuses

The right and left **cervical plexuses** are located deep in the neck lateral to the first through the fourth cervical vertebrae. These plexuses consist largely of the anterior rami of C_1-C_4, although they also receive small contributions from C_5 and the hypoglossal nerve (cranial nerve XII). As you can see in

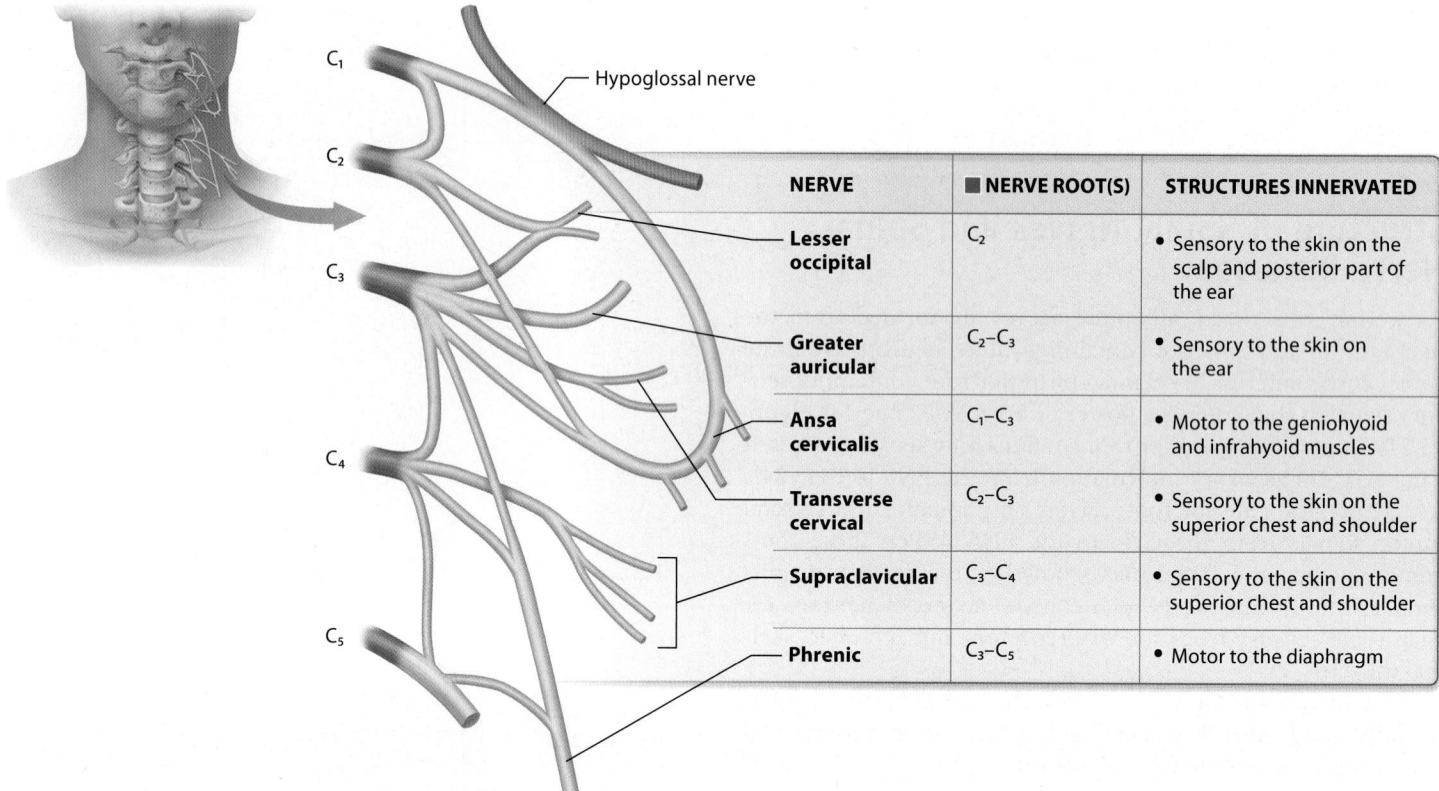

Figure 13.6 The cervical plexus.

Figure 13.6, the nerves of each cervical plexus contain cutaneous branches that supply the skin of the neck and portions of the head, chest, and shoulders. They also contain motor branches that supply muscles of the neck, including the infrahyoid muscles and the scalenes. The major named motor branch is the **phrenic nerve** (FREN-ik), which contains axons from C_3 to C_5, although its main contributor is C_4. The phrenic nerve supplies the diaphragm and is the main nerve that drives breathing (a good mnemonic to remember the roots of the phrenic nerve is "3, 4, 5 to stay alive"). An interesting application of phrenic nerve anatomy is discussed in *A&P in the Real World: A Hiccups Cure That Really Works* on page 496.

Brachial Plexuses

◀◀ Flashback

1. What is the flexor retinaculum, and where is it located? (p. 318)

The complicated-looking **brachial plexuses** have been the cause of headaches for many anatomy students. As implied by its name, the brachial plexus provides motor and sensory innervation to the upper limb. You can see in the following lateral view that it originates from the anterior rami of spinal nerves C_5–T_1; the first structures formed in the brachial plexus are its large **trunks.** Typically, C_5 and C_6 unite to form the large *superior trunk*, C_7

forms the *middle trunk*, and C_8 and T_1 combine to form the *inferior trunk.*

This next step is where things get tricky: Each trunk splits into an *anterior division* and a *posterior division* that become the **cords** of the plexus. The anterior division of the inferior trunk forms the *medial cord* that descends in the medial arm. The anterior divisions of the superior and middle trunks combine to form the *lateral cord,* which descends in the lateral arm. The posterior divisions of each trunk unite to form the *posterior cord,* which lies in the posterior arm.

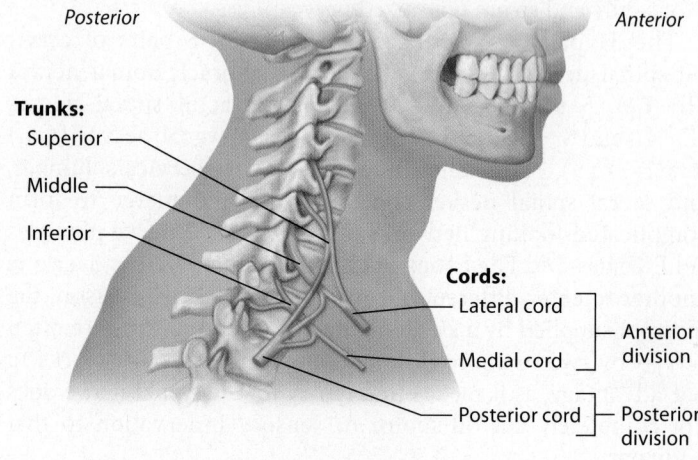

Figure 13.7a shows the entire brachial plexus in an anterior view. From different points along the trunks and cords of the brachial

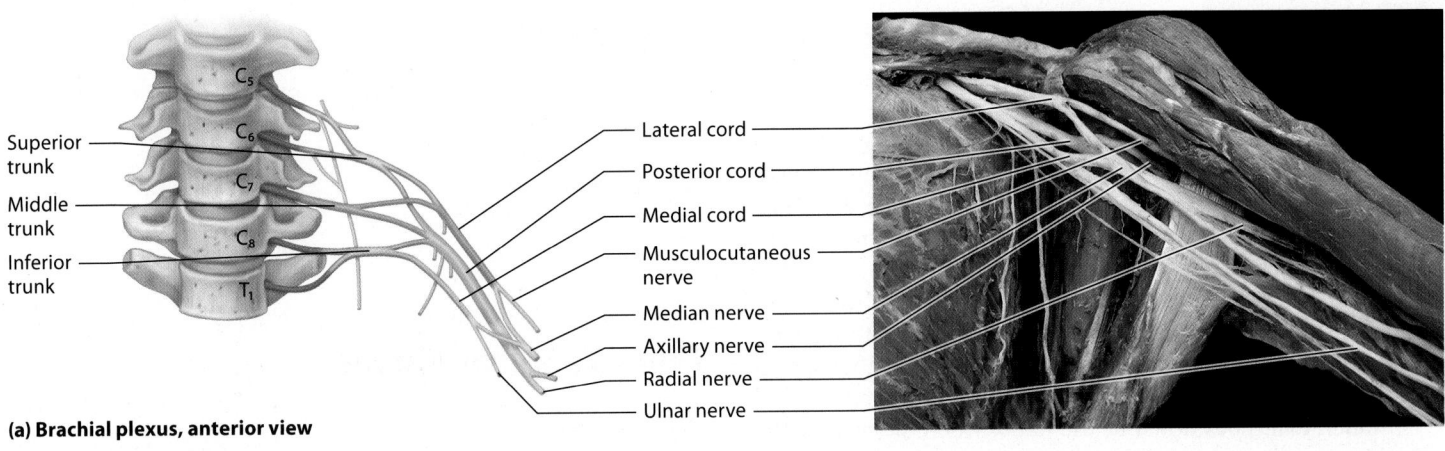

Superior trunk

Middle trunk

Inferior trunk

C_5
C_6
C_7
C_8
T_1

Lateral cord
Posterior cord
Medial cord
Musculocutaneous nerve
Median nerve
Axillary nerve
Radial nerve
Ulnar nerve

(a) Brachial plexus, anterior view

Clavicle

Lateral cord

Posterior cord

Medial cord

Scapula

Humerus

Ulna

Radius

NERVE	■ NERVE ROOTS	STRUCTURES INNERVATED
Axillary nerve	C_5–C_6, branch of the posterior cord	• Motor to the deltoid and teres minor muscles • Sensory to the skin over the deltoid
Radial nerve	C_5–T_1, continuation of the posterior cord	• Motor to the triceps brachii, brachioradialis, and the extensor muscles of the forearm and digits • Sensory to the posterior arm and forearm, posterior thumb, and posterior second, third, and lateral half of the fourth digits
Musculo-cutaneous nerve	C_5–C_7, continuation of the lateral cord	• Motor to the biceps brachii, coracobrachialis, and brachialis muscles • Sensory to the lateral forearm
Median nerve	C_5–T_1, from the fusion of the medial and lateral cords	• Motor to most of the muscles that flex the wrist and digits and certain muscles of the hand • Sensory to the anterior thumb and the anterior second, third, and lateral side of the fourth digits
Ulnar nerve	C_8–T_1, continuation of the medial cord	• Motor to certain forearm flexors and most of the intrinsic hand muscles • Sensory to the fifth digit and the medial half of the fourth digit

(b) Nerves of brachial plexus, anterior view

Figure 13.7 The brachial plexus.

Practice art labeling
@MasteringA&P®

plexus, various nerves originate and travel to different muscles and skin regions in the arm, forearm, and hand. The five major nerves are as follows (**Figure 13.7b**):

- **Axillary nerve.** The **axillary nerve** (AKS-ih-lair-ee) is a branch of the posterior cord of the brachial plexus. True to its name, the axillary nerve serves structures near the axilla, including the deltoid and teres minor muscles and the skin over the deltoid region.
- **Radial nerve.** As the posterior cord descends in the posterior arm, it becomes the **radial nerve.** In the posterior arm, this nerve supplies the triceps brachii muscle, after which it supplies most of the extensors in the forearm. A small branch continues down to the hand, where it supplies the skin over the posterior thumb, the second and third digits, and the lateral half of the fourth digit.
- **Musculocutaneous nerve.** The continuation of the lateral cord of the brachial plexus is the **musculocutaneous nerve** (muss'-kyoo-low-kyoo-TAYN-ee-us). This nerve supplies most of the anterior arm muscles, including the biceps brachii, and the skin over the lateral forearm.
- **Median nerve.** If you look back at Figure 13.7a, you can see that the lateral and medial cords unite to form the **median nerve,** so named because it runs approximately down the middle of the arm and forearm. In the forearm this nerve innervates the wrist and digital flexor muscles, and in the hand it supplies some of the intrinsic hand muscles and the skin over the anterior thumb, the second and third digits, and the lateral half of the fourth digit. As the median nerve enters the hand, it passes under a band of connective tissue that you encountered in the muscular system chapter, called the *flexor retinaculum* (see Figure 9.19). This anatomical arrangement unfortunately sometimes causes the median nerve to become inflamed and "trapped" under the retinaculum, resulting in *carpal tunnel syndrome.* This painful condition is generally treated by making a small incision in the retinaculum and releasing the pressure on the nerve.
- **Ulnar nerve.** The **ulnar nerve** is the continuation of the medial cord of the brachial plexus. It begins in the posterior arm, then passes over the medial epicondyle of the humerus to enter the forearm. At this point, the ulnar nerve is very superficial, a fact that you may become acutely aware of if you bang your elbow on a hard surface. The painful, tingling, electrical sensation that results from hitting your "funny bone" (although it is decidedly unfunny) is the effect of the slight contusion this nerve receives with such an injury. Once the ulnar nerve enters the forearm, it travels along the ulna and supplies the flexor muscles in the forearm not supplied by the median nerve, most of the intrinsic hand muscles, and the skin of the fifth digit and the medial side of the fourth digit.

Smaller nerves of the brachial plexus serve structures outside the arm, including the pectoralis major and minor, the serratus anterior, the latissimus dorsi muscles, and some of the rotator cuff muscles.

Quick Check

☐ 3. What are the key structures supplied by each of the following nerves?
    a. Phrenic nerve
    b. Ulnar nerve
    c. Musculocutaneous nerve
    d. Median nerve

Thoracic Spinal Nerves

◀◀ Flashback

1. Which vertebrae do the ribs articulate with? (p. 238)
2. Where is an intercostal space located? (p. 238)

Except for T_1, the nerves coming from the thoracic spinal cord do not form plexuses. Instead, each posterior ramus serves the deep back muscles, and each anterior ramus travels between two ribs as an **intercostal nerve.** The pattern of innervation for the intercostal nerves is as follows:

- The intercostal branch of the anterior ramus of T_1 travels in the first intercostal space, where it innervates the intercostal muscles and the skin of the axilla.
- The anterior rami of T_2-T_6 travel in their respective spaces and serve the intercostal muscles and the skin of the chest wall.
- The anterior rami of T_7-T_{12} also travel in their respective spaces and serve the intercostal muscles and overlying skin. However, they continue from the intercostal spaces into the anterior abdominal wall, where they supply the abdominal muscles and the overlying skin.

Quick Check

☐ 4. What are the intercostal nerves, and what do they innervate?

Lumbar Plexuses

◀◀ Flashback

1. Where is the obturator foramen located? (p. 247)
2. Where is the inguinal ligament located? (p. 309)

The left and right **lumbar plexuses** arise from the anterior rami of L_1-L_4 (**Figure 13.8a**). These plexuses lie anterior to the vertebrae, embedded within the posterior part of the psoas muscle. Their many branches primarily serve structures of the pelvis and the lower extremity.

As with the brachial plexus, most nerve roots of the lumbar plexus also separate into anterior and posterior divisions; **Figure 13.8b** shows the nerves of the anterior division. The largest

(a) Lumbar plexus, anterior view

Iliohypogastric nerve
Ilioinguinal nerve
Genitofemoral nerve
Lateral femoral cutaneous nerve
Femoral nerve
Obturator nerve

NERVE	■ NERVE ROOTS	STRUCTURES INNERVATED
Iliohypogastric nerve	L₁	• Motor to the transversus abdominus and internal oblique muscles • Sensory to the skin over the lateral gluteal region and the suprapubic region
Ilioinguinal nerve	L₁	• Motor to the transversus abdominus and internal oblique muscles • Sensory to the medial thigh, scrotum (males), and labia majora (females)
Genitofemoral nerve	L₁–L₂	• Motor to the cremaster muscle (males) • Sensory to the anteromedial thigh, scrotum (males), and labia majora (females)
Femoral nerve	L₂–L₄, posterior division	• Motor to the quadriceps femoris, iliopsoas, and sartorius muscles • Sensory to the anterior thigh, medial thigh, medial leg, and foot
Lateral femoral cutaneous nerve	L₂–L₃	• Sensory to the anterolateral thigh
Obturator nerve	L₂–L₄, anterior division	• Motor to the thigh adductors and gracilis muscle • Sensory to the superomedial thigh

(b) Nerves of lumbar plexus, anterior view

Figure 13.8 The lumbar plexus.

nerve of this division is the **obturator nerve** (OB-too-ray-ter). This nerve travels through the pelvis and enters the thigh by passing through the obturator foramen. Its branches serve the adductor muscles in the thigh, the hip joint, and the skin over the medial part of the thigh.

The largest nerve from the posterior division, and of the lumbar plexus as a whole, is the **femoral nerve.** It passes from the psoas muscle through the pelvis and under the inguinal ligament to enter the thigh. The branches of the femoral nerve supply the anterior thigh muscles, including the quadriceps femoris, the skin over the anterior and medial thigh and leg, and the knee joint.

Quick Check

☐ 5. What are the major nerves of the lumbar plexus, and what structures do they innervate?

Sacral Plexuses

Just inferior to the lumbar plexuses against the posterior pelvic wall we find the right and left **sacral plexuses,** which form from the anterior rami of spinal nerves $L_4 - S_4$ (**Figure 13.9a**). The nerves from the sacral plexuses innervate structures of the pelvis and gluteal region and much of the lower extremity. Like the nerves of the brachial and lumbar plexuses, those of the sacral plexus separate into anterior and posterior divisions.

The largest nerve of the sacral plexus—indeed, the largest and longest nerve in the body—is the **sciatic nerve** (sy-AEH-tik; *sciatic* = "of the hip"). Unlike the nerves in other plexuses, the sciatic nerve actually contains axons from both the anterior and posterior divisions that travel together and share an epineurium. The sciatic nerve travels from the pelvis through the greater sciatic notch and enters the thigh by passing between the greater trochanter and the ischial tuberosity. Within the posterior thigh, it serves the hip joint before it splits into its two main branches, typically near the lower third of the thigh: the **tibial nerve** and the **common fibular nerve.**

As you can see in **Figure 13.9b**, the tibial nerve is the larger branch—it contains the axons from the anterior division of the sacral plexus. In the thigh its branches supply almost all of the hamstring muscles (note that sometimes the sciatic nerve also supplies these muscles before it splits into the tibial and common fibular nerves). As implied by its name, it then descends through the leg alongside the tibia, where it innervates part of the knee joint, the ankle joint, and the plantarflexor muscles of the ankle, such as the gastrocnemius muscle. One of the tibial nerve's early branches, about the level of the gastrocnemius, is the *sural nerve* (SOO-rul; *sural* = "calf"), which supplies the posterior and lateral skin of the lower leg and part of the foot. At the level of the medial malleolus, the tibial nerve divides into its terminal branches, which supply the muscles and skin of the foot.

The smaller common fibular nerve, also known as the *common peroneal nerve* (pear-uh-NEE-ul; *peron-* = "fibula"),

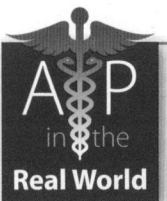

A Hiccups Cure That Really Works

The common annoyances known as *hiccups* are actually spasms of the diaphragm that cause a forceful inhalation of air. Numerous remedies are purported to cure the hiccups, from holding your breath to having someone sneak up on you and scare you, but none have ever been shown to work reliably. However, there is one way to call a halt to many cases of hiccups, and it involves the phrenic nerve. Here's how it works: Find the approximate area of the third through fifth cervical vertebrae, which is roughly in the middle of the neck. Place your fingers about 1 centimeter lateral to the vertebral column on both sides of the neck, illustrated here:

Apply firm pressure to the muscles of neck that overlie the phrenic nerve until the hiccups stop, which should happen in about 5–10 seconds. This pressure interrupts the aberrant impulses that are causing the diaphragm to contract inappropriately (although the pressure is not adequate to stop the nerve from firing completely or interfere with breathing). Try this trick on your friends and family—they will think you're a genius when you cure their hiccups.

contains axons from the posterior division of the sacral plexus. This nerve descends along the lateral leg, where it supplies part of the knee joint and the skin of the anterior and distal leg. It terminates by dividing into superficial and deep branches. The superficial branch supplies the skin on the dorsum of the foot and the lateral leg muscles that evert the foot, whereas the deep branch supplies the ankle dorsiflexors and two muscles on the dorsum of the foot.

(a) Sacral plexus, posterior view

NERVE	■ NERVE ROOTS	STRUCTURES INNERVATED
Superior gluteal nerve	L_4–S_1	• Motor to the gluteus medius, gluteus minimus, and tensor fasciae latae muscles
Inferior gluteal nerve	L_5–S_2	• Motor to the gluteus maximus muscle
Pudendal nerve	L_4–S_3	• Sensory to the hip joint
Posterior femoral cutaneous nerve	S_2–S_4	• Motor to the muscles of the pelvic floor, the external anal sphincter, and the external urethral sphincter • Sensory to the skin of the external genitalia
Sciatic nerve	S_1–S_3	• Sensory to the skin of the posterior thigh
Common fibular nerve	L_4–S_2, terminal branch of sciatic nerve	• Motor to the lateral leg muscles (superficial branch), the anterior leg muscles, and two foot muscles (deep branch) • Sensory to the knee joint, the skin of the anteroinferior leg, and the dorsum of foot
Tibial nerve	L_4–S_3, terminal branch of sciatic nerve	• Motor to the hamstring muscles, posterior leg muscles, and plantar foot muscles • Sensory to the knee joint, ankle joint, skin of the posterior and lateral leg (via the sural nerve), and skin of the plantar surface of the foot

(b) Nerves of sacral plexus, posterior view

Figure 13.9 The sacral plexus.

Practice art labeling
@MasteringA&P®

☐ 6. What are the major nerves of the sacral plexus, and what structures do they innervate?

Summary of the Distribution of Spinal Nerve Branches

After looking at all of the plexuses, let's take a moment to think about how they function together. **Figure 13.10a** summarizes the cutaneous distribution of the spinal plexuses, indicating the areas of the skin from which these nerves carry sensory information. **Figure 13.10b** illustrates the motor distribution of these plexuses, showing to which groups of muscles these nerves carry motor signals.

☐ 1. Explain why damage to the spinal cord around the level of C_4 or higher generally leaves a patient unable to breathe without the assistance of mechanical ventilation.

☐ 2. Predict the symptoms (both sensory and motor) you might experience if you suffered an injury to each of the following nerves:
   a. Median nerve
   b. Common fibular nerve
   c. Femoral nerve
   d. Sciatic nerve
   e. Ulnar nerve

See answers in Appendix A.

Anterior   Posterior

Cervical plexus

Brachial plexus:

Axillary nerve

Radial nerve

Musculocutaneous nerve

Median nerve

Ulnar nerve

Lumbar plexus:

Lateral femoral cutaneous nerve

Obturator nerve

Femoral nerve

Sacral plexus:

Posterior femoral cutaneous nerve

Common fibular nerve

Tibial nerve

(a) Cutaneous distribution of spinal nerve plexuses

Anterior   Posterior

Cervical plexus

Brachial plexus:

Axillary nerve

Musculocutaneous nerve

Radial nerve

Median nerve

Ulnar nerve

Lumbar plexus:

Femoral nerve

Obturator nerve

Sacral plexus:

Tibial nerve

Common fibular nerve

Sacral plexus:

Inferior gluteal nerve

Superior gluteal nerve

Tibial nerve

Common fibular nerve

(b) Motor distribution of spinal nerve plexuses

Figure 13.10 The distribution of spinal nerve branches.

Practice art labeling
@MasteringA&P®

Sensation Part II: Role of the PNS in Sensation

Learning Outcomes

1. Describe the structure and function of a sensory neuron.
2. Define and describe exteroceptors, interoceptors, and proprioceptors in terms of the general location of each in the body and the origin of the stimuli that each receives.
3. Describe each of the following types of receptors, indicating what sensation it detects and giving an example of where it can be found in the body: nociceptors, thermoreceptors, mechanoreceptors, chemoreceptors, and photoreceptors.
4. Explain how sensory transduction takes place at a sensory receptor.
5. Compare and contrast receptors for the general senses with receptors for the special senses.
6. Describe the pathway that a sensation takes from its detection in the PNS to its delivery to the CNS.

In the CNS chapter, we discussed the role of the CNS in sensation, including the pathways through which sensory stimuli travel in the spinal cord and brain and how these stimuli are processed and interpreted (see Chapter 12). Now we focus on the role of the PNS in sensation: the initial detection and transmission of sensory stimuli by sensory neurons. These neurons are located throughout the body but are especially numerous in the skin, skeletal muscles, and special sense organs.

This module examines the structures that detect sensations: sensory neurons. First we look at the portion of a sensory neuron that actually detects stimuli, known as *sensory receptors;* from there the focus shifts to the structure and function of entire sensory neurons. The module concludes with an explanation of how stimuli are delivered to the spinal cord and brain, and the overall pathway that a stimulus takes from detection in the PNS to interpretation by the CNS. The detection and transmission of special sensory stimuli are covered in the special senses chapter (see Chapter 15).

From PNS to CNS: Sensory Reception

In the process of sensation, stimuli are initially detected by sensory neurons, which then transmit impulses to the CNS. The basic pathway of information flow is as follows:

1. Stimulus is detected by sensory receptors of the PNS →
2. transmitted by PNS sensory neurons to the CNS →
3. integrated and interpreted by the CNS neurons.

Sensory Receptors

◄◄ **Flashback**

1. What are mechanically gated and voltage-gated ion channels? (p. 393)
2. What is depolarization? What is an action potential? (pp. 350, 397)
3. Where are tactile corpuscles, lamellated corpuscles, and Merkel cell fibers located? (pp. 167, 169)

The initial portion of the peripheral process belonging to a somatic sensory neuron features multiple small *nerve endings.* It is here that a stimulus is detected and the process of **sensory transduction**—the conversion of a stimulus into an electrical signal—takes place. Sensory transduction begins at a region of the nerve ending called a **sensory receptor,** which can be of several types. Some sensory receptors are surrounded by specialized supportive cells; these receptors are known as **encapsulated nerve endings.** Others are **free nerve endings** that lack specialized supportive cells and are "naked" (these were formerly called *free dendritic endings,* but they are part of an axon, not a dendrite, so this name is misleading).

Sensory Transduction

Figure 13.11 illustrates the basic mechanism behind sensory transduction. The axolemma of a nerve ending contains many gated ion channels that respond to various stimuli (such as the mechanically gated ion channels shown in this figure), changing the permeability of the membrane to one or more ions. The basic steps in sensory transduction are as follows: ① Before any stimulus arrives, the ion channels in the axolemma of the somatic sensory neuron are closed. Then the process of transduction begins with, for example, pressure on a finger from a probe. ② When a stimulus such as pressure is applied, mechanically gated sodium ion channels open. Sodium ions enter the axoplasm, generating a temporary depolarization referred to as a **receptor potential.** ③ If enough sodium ions enter that the membrane potential reaches threshold, voltage-gated sodium ion channels open. This triggers an action potential, which will be propagated along the axon to the spinal cord.

Different types of sensory receptors respond to stimuli with different speed, intensity, and duration. Some receptors respond rapidly and with high intensity but stop sending the stimuli after a certain period of time, a phenomenon known as *adaptation.* These **rapidly adapting receptors** are important for detecting the initiation of stimuli, but they ignore ongoing stimuli. Rapidly adapting receptors are the reason why you can walk around your home in search of your sunglasses only to find out that they were on top of your head the whole time. Other receptors, called **slowly adapting receptors,** respond to stimuli with constant action potentials that do not diminish with time. The dull, throbbing pain you feel for a week after spraining your ankle is the work of slowly adapting receptors.

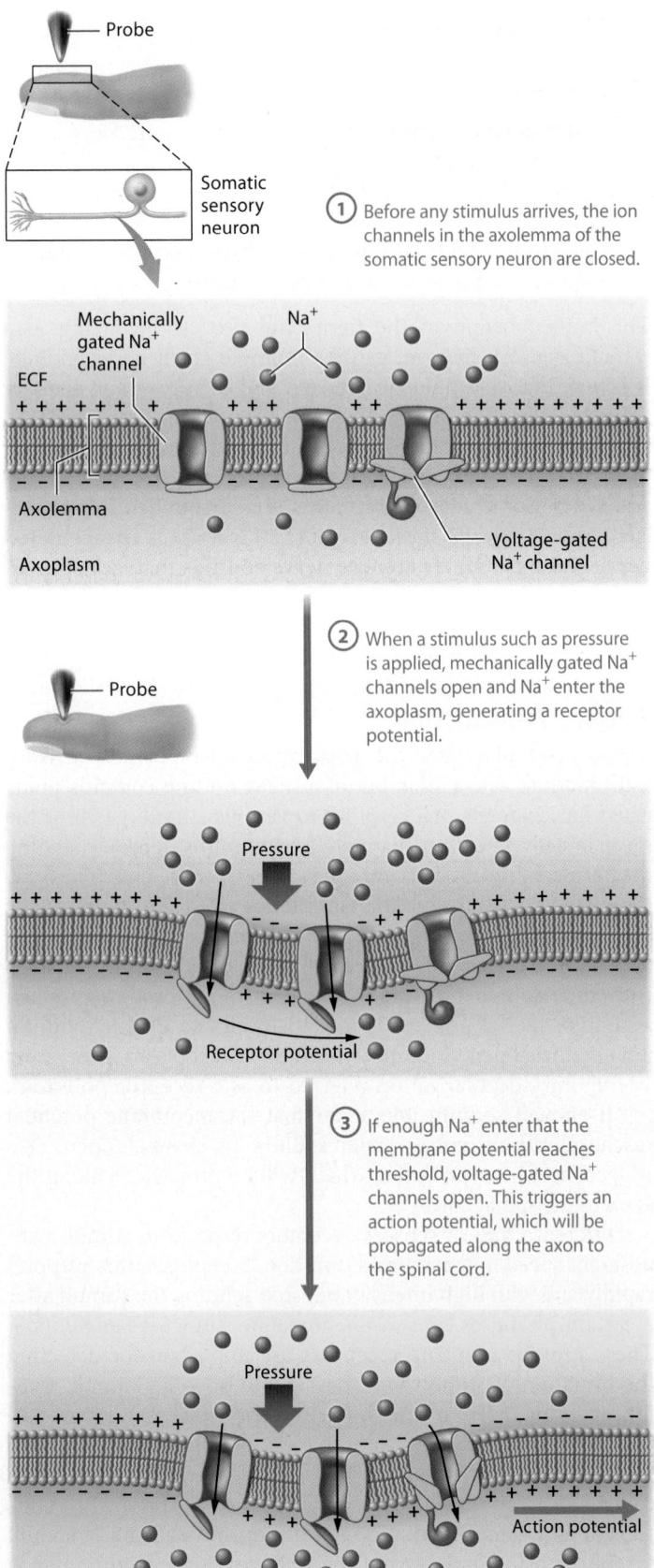

① Before any stimulus arrives, the ion channels in the axolemma of the somatic sensory neuron are closed.

② When a stimulus such as pressure is applied, mechanically gated Na⁺ channels open and Na⁺ enter the axoplasm, generating a receptor potential.

③ If enough Na⁺ enter that the membrane potential reaches threshold, voltage-gated Na⁺ channels open. This triggers an action potential, which will be propagated along the axon to the spinal cord.

Figure 13.11 Sensory transduction.

Classification of Sensory Receptors

Sensory receptors can be broadly classified according to the location of the stimuli they detect. **Exteroceptors** (EK-ster-oh-sep′-terz) are typically close to the surface of the body. They detect stimuli originating outside the body, including an object's texture, temperature, and color; chemical odors in the air; and the level of light. **Interoceptors** (IN-ter-oh-sep′-terz) lie generally within the body's interior, where they detect stimuli originating inside the body, including blood pressure, the stretch of an organ such as a skeletal muscle or the urinary bladder, the concentration of certain chemicals in body fluids, and body temperature.

A more detailed way to classify sensory receptors is by the type of stimuli that causes them to depolarize and generate a receptor potential. This classification system gives us five types of receptors, which are the following:

- **Mechanoreceptor. Mechanoreceptors** (mek′-ah-noh-ree-SEP-terz) are encapsulated exteroceptors or interoceptors found in the skin, the musculoskeletal system, and many different organs. They depolarize in response to anything that *mechanically* deforms the tissue, including external stimuli such as light touch and vibration and internal stimuli such as stretch and pressure. The mechanism behind their sensory transduction is found within their specialized ion channels, called *mechanically gated ion channels,* which are shown in Figure 13.11.
- **Thermoreceptor.** As you can probably guess by their name, **thermoreceptors** are exteroceptors that respond to *thermal* stimuli, depolarizing in response to temperature changes. Most thermoreceptors are slowly adapting receptors. Separate receptors detect hot and cold stimuli.
- **Chemoreceptor.** A **chemoreceptor** (KEE-moh-ree-sep′-ter) is an exteroceptor or interoceptor that depolarizes in response to certain chemicals in body fluids or in the air. Chemicals that are specific for the receptor bind and trigger ion channels to open, which generates a receptor potential and perhaps an action potential. Internal chemoreceptors detect the hydrogen ion concentration, the level of carbon dioxide, and the level of oxygen in the body's fluids. External chemoreceptors are responsible for the special senses of smell and taste.
- **Photoreceptor. Photoreceptors,** found only in the eye, are special sensory exteroceptors that depolarize in response to light.
- **Nociceptor.** As you learned in the CNS chapter, *nociception* is the detection of noxious stimuli, and *pain* is the perception and interpretation of these stimuli (see Chapter 12). The receptors that depolarize in response to these stimuli are accordingly called **nociceptors** (NOH-sih-sep-terz) and are generally exteroceptors. Like thermoreceptors, nociceptors are slowly adapting receptors.

Although nociception is not a primary function of mechanoreceptors, thermoreceptors, chemoreceptors, and photoreceptors, each type of receptor can transmit stimuli that the brain can perceive as painful. Common examples include a bright light shining in your eyes and toxic chemicals with a pungent odor—neither stimulus will cause depolarization of nociceptors, but they are nonetheless perceived as painful. See *A&P in the Real World: Capsaicin* to find out why hot peppers feel like they burn your mouth.

This chapter primarily deals with mechanoreceptors, so we discuss these in more detail next. We also briefly address the types of thermoreceptors.

Types of Mechanoreceptors There are six classes of mechanoreceptors in the body. These receptors, ordered from superficial to deep, are as follows (see **Figure 13.12**):

- **Merkel cell fibers.** Recall from the integumentary chapter that **Merkel cell fibers,** which consist of a nerve ending surrounded by a capsule of Merkel cells, are located in the floors of the epidermal ridges (see Chapter 5). They are found primarily in the skin of the hands, where they are especially numerous in the fingertips. Action potentials appear to stem from mechanically gated ion channels in the nerve ending. These slowly adapting receptors have the finest spatial resolution of any of the skin mechanoreceptors. For this reason, they primarily detect discriminative touch stimuli such as form and texture.
- **Tactile corpuscles. Tactile corpuscles,** also known as *Meissner corpuscles* (MYS-ner KOR-pus-ulz), are found in the dermal papillae, projections of the dermis into the epidermis (see Chapter 5). These rapidly adapting receptors are more numerous than Merkel cell fibers. Like Merkel cell fibers, tactile corpuscles transmit discriminative touch stimuli, although their resolution is not as fine.
- **Ruffini endings.** The spindle-shaped **Ruffini endings,** also referred to as *bulbous corpuscles,* are located in the dermis and the hypodermis, as well as in ligaments. They are slowly adapting receptors that respond to stretch and movement; they do not transmit discriminative touch stimuli.
- **Lamellated corpuscles. Lamellated corpuscles,** formerly called *Pacinian corpuscles* (pah-SIN-ee-an) are named for

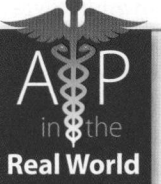

Capsaicin

Anyone who has ever eaten hot peppers knows they can make your mouth feel like it's on fire. This happens because of a chemical in peppers called *capsaicin* (kap-SAY-sin). Capsaicin opens specific ligand-gated ion channels in nociceptors and triggers action potentials, which causes the CNS to perceive the chemical as painful. Mice that lack this receptor due to genetic engineering can ingest capsaicin as if it were water.

Interestingly, repeated application of capsaicin to nociceptors seems to desensitize them. This makes the nociceptors less likely to generate receptor potentials in response to painful stimuli. For this reason, capsaicin may be applied as a topical cream to relieve the pain of peripheral nerve disorders called *neuropathies;* the viral infection shingles, caused by the chickenpox virus; and other conditions. Remember that capsaicin doesn't do anything to treat the actual cause of pain; it simply makes the nociceptors less able to relay the painful stimuli to the CNS.

their layered, onion-like appearance and are located deep within the dermis. Having the structure of a layered capsule enables these receptors to perform their function—the layers act somewhat like a filter that allows only high-frequency vibratory stimuli and deep pressure to activate them. This is an example of the Structure-Function Core Principle (p. 25). Lamellated corpuscles are rapidly adapting and, like Ruffini endings, do not transmit discriminative touch stimuli.

 CORE PRINCIPLE Structure-Function

- **Hair follicle receptors.** Recall that **hair follicle receptors** are free nerve endings wrapped around the base of a hair follicle in the dermis or hypodermis (see Chapter 5). These receptors

Figure 13.12 Mechanoreceptors in the skin.

MECHANORECEPTOR	FUNCTION
Merkel cell fiber	Discriminative touch with fine spatial resolution
Tactile corpuscle	Discriminative touch with less spatial resolution
Ruffini ending (bulbous corpuscle)	Stretch and movement
Lamellated corpuscle	Vibration and deep pressure

are not found in thick skin, the type of skin shown in Figure 13.12. Hair follicle receptors respond to stimuli that cause the hair to bend, such as an insect landing on your arm.

- **Proprioceptors. Proprioceptors** are located in the musculoskeletal system, where they detect the movement and position of a joint or body part (they are not found in the skin, so they are not included in Figure 13.12). These receptors are integral to the body's ability to sense its position in space and to monitor ongoing movement. In other words, they are critical to the integration of sensory and motor functions.

Types of Thermoreceptors Thermoreceptors are generally small knobs on the ends of free nerve endings in the skin. There are two types of thermoreceptors: so-called "cold" receptors that respond to temperatures between 10 °C and 40 °C (50–104 °F) and "hot" receptors that respond to warmer temperatures between 32 °C and 48 °C (about 90–118 °F). Cold receptors are located in the superficial dermis, whereas hot receptors lie in the deep portion of the dermis. Temperatures outside the ranges of hot and cold receptors are detected by nociceptors, which is why very hot or very cold objects feel painful to the touch.

Quick Check

☐ 1. What is sensory transduction?
☐ 2. What role do sensory receptors play in sensory transduction?

Sensory Neurons

◀◀ Flashback

1. What is a pseudounipolar neuron? (p. 387)
2. What is an axon? (p. 386)
3. What two factors determine the speed of action potential conduction in an axon? (p. 403)

This section examines the structure and classification of somatic sensory neurons (those involved in the general senses). It also discusses the area of skin served by a somatic sensory neuron, known as the *receptive field*.

Structure of Sensory Neurons

A typical somatic sensory neuron is shown in **Figure 13.13**. As you can see, somatic sensory neurons are *pseudounipolar neurons* with three main components—a cell body and two axons:

- The cell body of the neuron is located in a **posterior root ganglion** (or *dorsal root ganglion*) found lateral to the spinal cord. The cell bodies of cranial nerves are situated in **cranial nerve ganglia** that are present in the head and neck.
- The *peripheral process* of the neuron is a long axon. At one end, it splits into nerve endings; associated with each nerve ending is a sensory receptor. At the other end, the peripheral process terminates near the neuron's cell body. ① The peripheral process transmits an action potential from the sensory receptor to the neuron's other axon, the central process.
- The *central process* exits the cell body and travels through the posterior root of the spinal cord to enter the posterior horn (or into the brainstem for cranial nerves). ② The central process transmits an action potential from the peripheral process to the posterior horn, eventually synapsing on a neuron in the spinal cord or brainstem.

An action potential propagated down the peripheral process does not generally reach the cell body. In fact, the stimulus is usually transmitted to the central process in the area where the peripheral and central processes come into contact near the cell body. However, the cell body still has the important function of manufacturing the organelles and proteins needed by the peripheral and central processes to do their jobs.

① The peripheral process transmits an action potential from the sensory receptor to the neuron's other axon, the central process.

② The central process transmits an action potential from the peripheral process to the posterior horn, eventually synapsing on a neuron in the spinal cord or brainstem.

Nerve endings with sensory receptors

Action potential

Peripheral process of somatic sensory neuron

Cell body of somatic sensory neuron

Posterior root ganglion

Central process of somatic sensory neuron

Posterior root

Posterior horn

Posterior

Anterior

Spinal cord

Figure 13.13 Somatic sensory neuron structure and function.

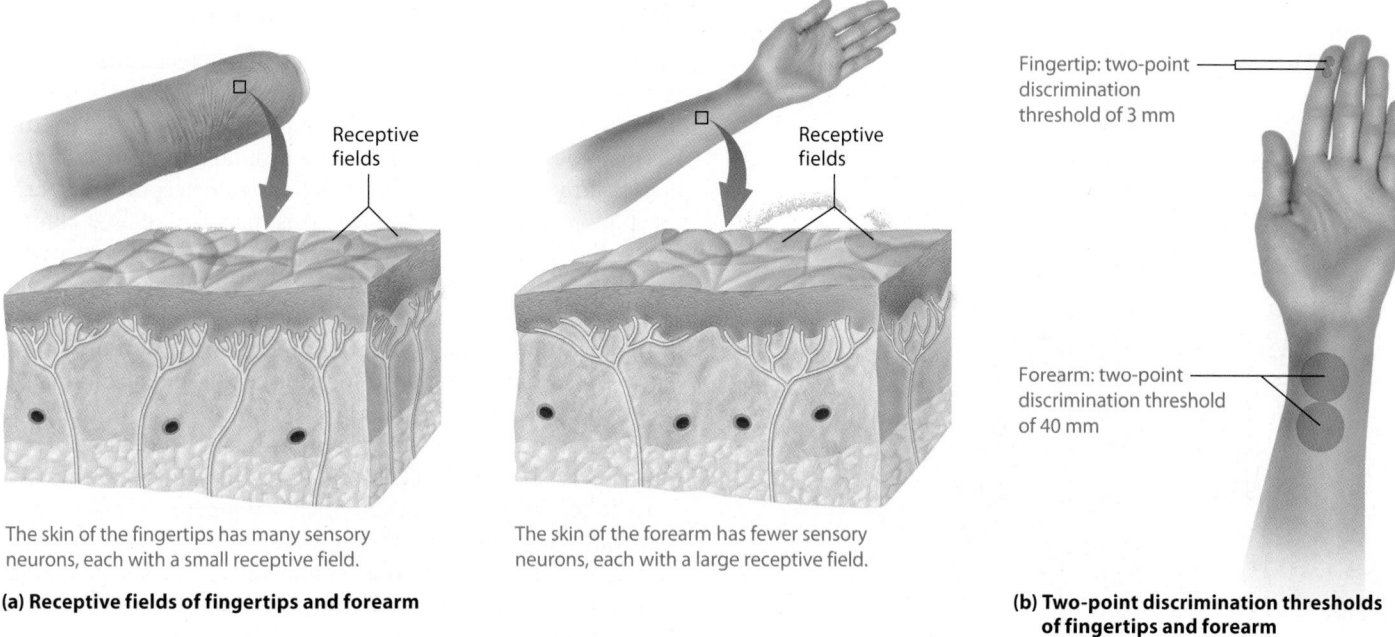

The skin of the fingertips has many sensory neurons, each with a small receptive field.

The skin of the forearm has fewer sensory neurons, each with a large receptive field.

(a) Receptive fields of fingertips and forearm

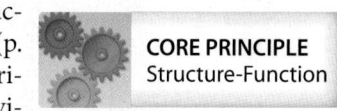

Fingertip: two-point discrimination threshold of 3 mm

Forearm: two-point discrimination threshold of 40 mm

(b) Two-point discrimination thresholds of fingertips and forearm

Figure 13.14 Receptive fields and two-point discrimination.

Classification of Sensory Neurons

Sensory neurons are generally classified according to the speed with which their peripheral axons conduct action potentials. Recall from the nervous tissue chapter that two factors determine the speed at which action potentials are conducted along an axon—the diameter of the axon and the thickness of its myelin sheath (see Chapter 11). The axons with the largest diameter and the thickest myelin sheaths conduct impulses the fastest. Such axons include those that convey proprioceptive information to the CNS. It's critical for this information to travel quickly because the CNS needs it to make adjustments during movement to maintain posture and balance. The axons that convey both discriminative and nondiscriminative touch information are also fairly large and heavily myelinated, and so conduct action potentials rapidly.

The smallest-diameter axons with the least amount of myelin transmit action potentials the slowest. Both pain and temperature stimuli are conveyed to the CNS by such axons. Indeed, conduction through these axons is so slow that a noticeable delay occurs between the initiation of the stimulus and its interpretation by the CNS. This explains why, after you touch an object, it may take a moment to determine whether it is hot or cold. It also accounts for the "throbbing" nature of pain—each "wave" of pain that you feel is actually one continuous stimulus. The delay between waves is just the time it takes for the painful stimulus to travel up the sensory neurons and be delivered to the CNS.

Receptive Fields

An important property of sensory neurons, particularly those that serve the skin, is the *receptive field,* or the area served by that neuron (**Figure 13.14a**). Each somatic sensory neuron has an extensive network of branches in the skin, and the more branching that

exists, the larger the neuron's receptive field. How much a neuron branches is generally determined by the number of neurons that serve a particular region of the body. The skin of the fingertips is richly innervated with sensory neurons, and each of these neurons has a relatively small receptive field. However, the skin of the forearm is innervated by a much smaller number of neurons, so each neuron has a somewhat larger receptive field. Here is another example of the Structure-Function Core Principle (p. 25): Body regions that have a primary function of sensing the environment (such as fingertips) contain more neurons and smaller receptive fields, which enables them to perform this job.

CORE PRINCIPLE Structure-Function

Different regions of the body contain neurons with receptive fields of varying size. These differences account for the fact that you can accurately determine an object's size and texture with your fingertips, but not with your forearm or back. The relative sizes of receptive fields can be measured by the **two-point discrimination threshold.** In this test, two stimuli (usually a pair of calipers or applicator sticks) are placed closely together on the skin (**Figure 13.14b**). The stimuli are then moved apart until the subject can feel two distinct points. The two-point discrimination threshold for the fingertips is quite small (about 3 mm) compared to that of the forearm (about 40 mm).

Quick Check

☐ 3. What are the three components of a typical somatic sensory neuron?

☐ 4. What is a sensory neuron's receptive field?

☐ 5. What is the two-point discrimination threshold, and what does it measure?

Anterior   Posterior

Trigeminal (V) nerve

(a) Dermatome map

Dermatomes and Referred Pain

The skin can be divided into different segments called **dermatomes** (DER-mah-tohmz; *derma* = "skin," *-tome* = "section"), based on the spinal nerve that supplies the region with somatic sensation. Dermatomes can be combined and assembled into a *dermatome map* that represents all of the spinal nerves (except the first cervical spinal nerve) (**Figure 13.15a**). This map can be used clinically to test the integrity of the sensory pathways to different parts of the body. For example, a patient with a cervical spinal cord injury may be able to feel his lateral hand, his thumb, and his second and third digits, but be unable to feel his medial hand or fourth or fifth digits. This would tell you the location of the spinal cord injury—about the level of C_8.

Many spinal nerves carry both somatic and visceral neurons, so visceral sensations travel along the same pathways as do somatic sensations. This anatomical arrangement has a curious consequence known as **referred pain,** whereby pain that originates in an organ is perceived as cutaneous pain. The pain is generally located along the dermatome for that nerve. For example, pain caused by a heart attack is often perceived as pain in the anterior chest wall, the left arm, and the neck and lower jaw. This is because the nerves that carry the pain signals from the heart serve these particular dermatomes. Other examples of referred pain are shown in **Figure 13.15b**.

Dermatomes are also clinically relevant in *shingles,* also known as *herpes zoster.* Shingles develops due to the chickenpox virus (note that this is not the same virus that causes either form of herpes, but the viruses do belong to the same family). After the initial infection, the virus usually persists in the cell bodies of somatic sensory neurons in posterior root ganglia, where it remains inactive, often for decades, until it is reactivated for unknown reasons. In patients with compromised immune systems, the virus proliferates and migrates down the axon of the

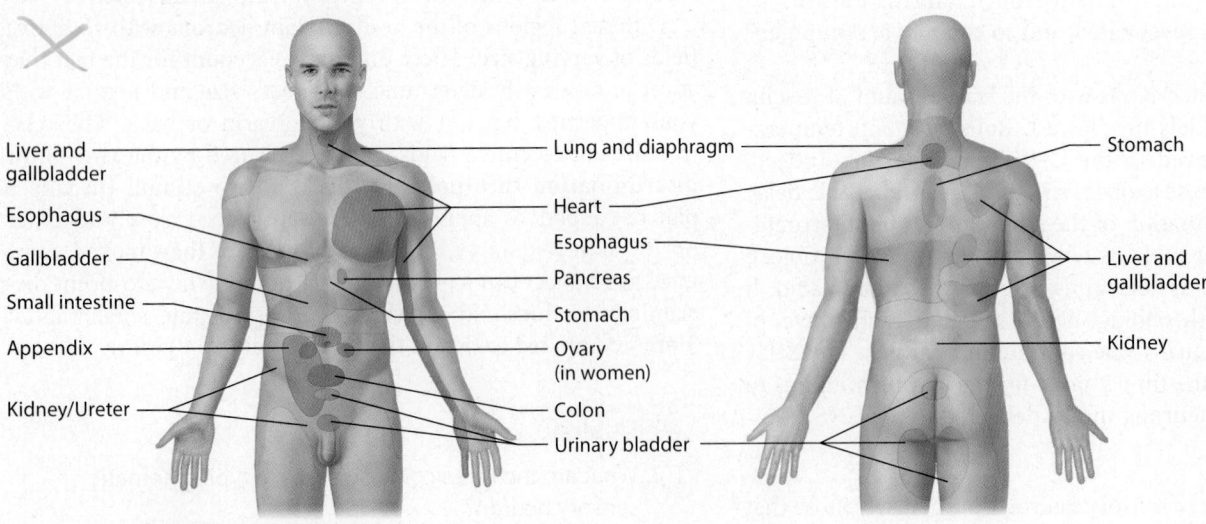

Liver and gallbladder

Esophagus

Gallbladder

Small intestine

Appendix

Kidney/Ureter

Lung and diaphragm

Heart

Esophagus

Pancreas

Stomach

Ovary (in women)

Colon

Urinary bladder

Stomach

Liver and gallbladder

Kidney

(b) Common locations of referred visceral pain

Figure 13.15 The dermatome map and referred pain.

① **PNS**: The stimulus is detected by receptors, which triggers an action potential (see Figure 13.11).

Na+

Pressure

Action potential

Somatosensory cortex

Cell body

Thalamus

CNS sensory neurons

Action potential

Action potential

Peripheral process

Central process

Action potential

④ **CNS**: The action potential travels to the brain and is interpreted there (see Figure 12.26).

② **PNS**: The action potential is conducted along the peripheral process of the sensory neuron and then transferred to the central process (see Figure 13.13).

③ **PNS to CNS**: The central process transmits the action potential to the CNS (spinal cord and brain, see Figure 13.13).

Nerve endings of a PNS somatic sensory neuron

---

peripheral process to the skin, causing a painful rash with blisters. The virus may also migrate to neighboring ganglia and infect one or more entire dermatomes. This leads to the characteristic presentation of a painful unilateral rash (on one side of the body) that follows a dermatomal distribution. No known treatments cure shingles, but it may be prevented effectively with a vaccine.

Quick Check

☐ 6. What is a dermatome?
☐ 7. Why is visceral pain often perceived as cutaneous pain?

Putting It All Together: The Big Picture of the Detection and Perception of Somatic Sensation by the Nervous System

The overall pathway for the detection and perception of somatic sensory stimuli is shown in **Figure 13.16**. This figure uses the example of a pressure or nondiscriminative touch stimulus. The stimulus is conducted to the sensory neuron's central process, which enters the brainstem or the spinal cord and transmits the action potential to a neuron in the CNS. As you may recall, the pathway that a stimulus takes once it enters the CNS depends on the type of stimulus (see Chapter 12). In the case of a pressure stimulus, it will pass through the thalamus to be processed and interpreted by the primary and secondary somatosensory cortices. Be aware that many proprioceptive stimuli are routed to the cerebellum instead for initial processing.

Apply What You Learned

☐ 1. Predict what might happen if a disease caused the axons that transmit proprioceptive stimuli to lose their myelin sheaths.

☐ 2. Which types of mechanoreceptors allow one to read Braille (raised dots that enable a blind person to read text)? (*Hint:* What receptors allow for fine, discriminative touch?)

☐ 3. Why is it advantageous to our survival that our thermoreceptors and nociceptors are slowly adapting?

See answers in Appendix A.

MODULE **13.5**

Movement Part II: Role of the PNS in Movement

Learning Outcomes

1. Describe the differences between upper motor neurons and lower motor neurons.
2. Describe the overall "big picture" view of how movement occurs.

Let's now concentrate on the role of the PNS in movement. This module begins by exploring the structure and function of lower motor neurons. It ends by examining the "big picture" view of how the nervous system initiates and controls movement.

From CNS to PNS: Motor Output

As we discussed in the muscle tissue chapter, there is a close relationship between the muscular and nervous systems (see Chapter 10). Skeletal muscle fibers are voluntary and contract only when stimulated to do so by a somatic motor neuron. In the CNS chapter, you explored how the CNS initiates movement—**upper motor neurons** in the primary motor cortex of the cerebrum make the "decision" to move and initiate movement (with the help of the basal nuclei; see Chapter 12). However, the upper motor neurons do not contact skeletal muscle fibers, and so by themselves cannot stimulate a muscle contraction. Instead, the messages from upper motor neurons are relayed to **lower motor neurons,** which release acetylcholine onto the muscle fiber and initiate a muscle contraction.

The Role of Lower Motor Neurons

Lower motor neurons are multipolar neurons whose cell bodies are located within the CNS (in either the anterior horn of the spinal cord or the brainstem) and whose axons are located in the PNS. As we covered in the muscle tissue chapter, each lower motor neuron innervates skeletal muscle fibers within a single skeletal muscle (see Chapter 10). Groups of lower motor neurons that innervate the same muscle, called *motor neuron pools,* are located together in the anterior horn in rod-shaped clusters.

Most lower motor neurons within a motor neuron pool are the large **α-motor neurons.** α-Motor neurons stimulate skeletal muscle fibers to contract by the excitation-contraction mechanism (see Chapter 10). Also present within a motor neuron pool are smaller **γ-motor neurons.** These lower motor neurons innervate muscle fibers called *intrafusal fibers* that are part of specialized stretch receptors, discussed in Module 13.6.

Quick Check

☐ 1. What are the main differences between an upper motor neuron and a lower motor neuron?

☐ 2. What is a motor neuron pool?

Putting It All Together: The Big Picture of Control of Movement by the Nervous System

◀◀ Flashback

1. What is the function of the cerebellum? (p. 460)
2. What are the functions of the caudate nucleus, putamen, and globus pallidus? (p. 432)
3. What is a motor program? (p. 459)

Now that we've examined somatic motor function at the CNS and the PNS levels, we can put together a "big picture" view of movement. The basic steps in how we get from deciding to move to actually making the movement are shown in **Figure 13.17.**

Quick Check

☐ 3. What is the general sequence of events for movement, and which parts of the CNS and PNS are involved?

Apply What You Learned

☐ 1. Predict the effects of damage to each of the following motor neurons:
  a. Upper motor neurons of the premotor cortex
  b. Upper motor neurons of the primary motor cortex
  c. Lower motor neurons

☐ 2. Predict the effect of damage to proprioceptors and sensory neurons that communicate with the cerebellum.

See answers in Appendix A.

1. **CNS:** Upper motor neurons in the premotor cortex select a motor program.

2. **CNS:** The basal nuclei enable the thalamus to stimulate upper motor neurons of the primary motor cortex (see Figure 12.31).

Thalamus

Globus pallidus

Caudate nucleus — Basal nuclei

Putamen

Pons

Cerebellum

Upper motor neuron

Spinal cord

5. **PNS to CNS:** Sensory information is relayed back to the cerebellum in the CNS. The cerebellum then sends instructions to upper motor neurons to modify movement as needed (see Figure 12.32).

3. **CNS to PNS:** Upper motor neurons stimulate lower motor neurons.

Sensory neuron

Contracting skeletal muscle

4. **PNS:** Lower motor neurons stimulate a skeletal muscle to contract.

Lower motor neuron

507

MODULE **13.6**

Reflex Arcs: Integration of Sensory and Motor Function

Learning Outcomes

1. Describe reflex responses in terms of the major structural and functional components of a reflex arc.
2. Distinguish between each of the following pairs of reflexes: somatic versus visceral reflexes and monosynaptic versus polysynaptic reflexes.
3. Describe a simple stretch reflex, a flexion reflex, and a crossed-extension reflex.
4. Describe the role of stretch receptors in skeletal muscles.

The proper functioning of the nervous system, and indeed of the body as a whole, requires more than just sensory and motor impulses in isolation. It also requires the integration of motor and sensory information. Consider this situation: You see flames, smell smoke, and feel heat. Your sensory neurons deliver all of this sensory input to the CNS, which interprets it to mean fire. Simply understanding that a fire is present does you no good without an appropriate motor response, likely an attempt to extinguish the fire or get away from it. Certain types of sensory-motor integration, such as the example just described, involve cooperation between multiple parts of the CNS and PNS. Other types of integration, however, are much simpler and result in programmed, automatic responses to selected sensory input. These types of integrative responses are known as *reflexes*. Both types of integration involve rapid communication between neurons and exemplify the importance of the Cell-Cell Communication Core Principle (p. 27).

CORE PRINCIPLE
Cell-Cell Communication

This module begins by explaining the basic principles of a reflex arc, then explores the sensory neurons and receptors that detect stimuli within skeletal muscles and tendons. After discussing how these receptors lead to different reflex responses, the module concludes with an overview of some disorders that can affect sensory and motor neurons.

Reflex Arcs

You jump when someone surprises you, as we mentioned at the beginning of the chapter; you also jerk your hand away when you touch a hot pot and blink when something touches your eye. These are all programmed, automatic responses to stimuli called **reflexes.** Most reflexes are protective, preventing tissue damage in some way.

Notice that each of these reflexes begins with a sensory stimulus and finishes with a rapid motor response. Between the sensory stimulus and the motor response neural integration takes place within the CNS, typically within the brainstem or the spinal cord. So the overall sequence of a reflex is this:

As you can see, steps ① and ③ are carried out within the PNS and ② is carried out within the CNS. This three-step sequence of events is known as a **reflex arc.** Most reflex arcs are negative feedback loops—the sensory receptors stop firing when the original conditions are restored.

Quick Check

☐ 1. What is a reflex? What is the overall sequence of a reflex?

The Role of Stretch Receptors in Skeletal Muscles

◄◄ **Flashback**

1. What is muscle tension? (p. 339)
2. What is a tendon? (p. 263)

Some of the most common reflexes occur without our realizing it. These reflexes are part of normal movement and allow the CNS to correct motor error and prevent muscle damage. The sensory component of these reflexes is detected by mechanoreceptors within muscles and tendons called *muscle spindles* and *Golgi tendon organs.* These mechanoreceptors monitor muscle length and the force of contraction and communicate this information to the spinal cord, cerebellum, and cerebral cortex.

Muscle Spindles

Muscle spindles are tapered structures that are found embedded among the regular contractile muscle fibers; these fibers are also known as **extrafusal muscle fibers** (ek-strah-FYOO-zul; *extrafusal* = "outside the spindle") (**Figure 13.18a**). Within each spindle are 2–12 specialized muscle fibers, the **intrafusal muscle fibers** ("within the spindle"). As you can see in Figure 13.18a, intrafusal fibers have contractile filaments composed of actin

(a) Muscle spindle—stretch, length, and position

(b) Golgi tendon organ—force

Figure 13.18 Muscle spindles and Golgi tendon organs.

Practice art labeling
@MasteringA&P®

and myosin at their poles and a central area where contractile filaments are absent. These contractile poles are innervated by γ-motor neurons.

Two structural and functional classes of sensory neurons innervate the intrafusal fibers. Both types of neurons contain mechanically gated ion channels that open when the intrafusal fiber is stretched. The first type of neuron, known as a *primary afferent,* responds to stretch when it is first initiated. The second type, known as a *secondary afferent,* responds to both the static length of a muscle and the position of a limb.

Muscle groups differ in their number of muscle spindles. Those groups that produce fine movements, such as those of the hand and the extraocular muscles, have large numbers of muscle spindles. This allows for precise control of muscle contractions. Conversely, muscle groups that produce coarse movements, such as the postural muscles of the back, have relatively few muscle spindles.

Golgi Tendon Organs

Golgi tendon organs are mechanoreceptors located within tendons near the muscle-tendon junction; they monitor the tension generated by a muscle contraction. Note in **Figure 13.18b** that a

Golgi tendon organ consists of an encapsulated bundle of collagen fibers attached to about 20 extrafusal muscle fibers. Each Golgi tendon organ contains a single somatic sensory axon whose endings are wrapped around its enclosed collagen fibers. The rate at which these neurons fire depends on the amount of muscle tension generated with each contraction—the greater the tension, the more rapidly they fire. This feedback provides the CNS with information about the force being generated by each muscle contraction.

Quick Check

☐ 2. How do intrafusal and extrafusal muscle fibers differ?
☐ 3. How do α-motor neurons and γ-motor neurons differ?
☐ 4. How do Golgi tendon organs and muscle spindles differ?

Types of Reflexes

Reflexes can be classified by at least two criteria. For the first criterion, we determine the number of synapses that occur between the neurons involved in the arc. The simplest reflex arcs involve only a single synapse within the spinal cord between the sensory and motor neurons; this is called a **monosynaptic reflex.** A more complicated type of reflex arc, the **polysynaptic reflex,** involves multiple synapses.

The second way to classify reflex arcs is according to the type of organ in which the reflex takes place. *Visceral reflexes* involve neurons that innervate our internal organs and are largely connected with the autonomic nervous system. We discuss some of these reflexes in the ANS chapter (see Chapter 14), and others are discussed in specific chapters throughout the book. The reflexes covered in the upcoming subsections are primarily *somatic reflexes,* which involve somatic sensory and motor neurons. Somatic reflexes can involve either cranial or spinal nerves. We'll look closely at four types of somatic reflexes: the simple stretch, Golgi tendon, flexion and crossed-extension, and cranial nerve reflexes.

Simple Stretch Reflexes

The upper motor neurons in the CNS have a set "idea" of the optimal length for each skeletal muscle. This optimal length applies to both contracting and resting muscles. At rest, a few of the muscle's motor units remain activated to maintain this optimal length, which produces *muscle tone* (see Chapter 10). When a muscle is stretched, it deviates from this optimal length. The body's reflexive response is to shorten the muscle so that it returns to its optimal length via a monosynaptic reflex known as a **simple stretch reflex.**

Using an example of movement required for getting a glass of juice, **Figure 13.19** shows the steps of a simple stretch reflex in a spinal nerve:

1. **An external force stretches the muscle.** In Figure 13.19, the muscle is stretched by the external force of the extra weight that results from adding the liquid to the glass.
2. **Muscle spindles detect the stretch, and primary and secondary afferents transmit an action potential to the spinal cord.** The intrafusal fibers are stretched, which opens the mechanically gated ion channels in their primary and secondary afferents. If the stretch is strong enough to depolarize the neuronal membranes to threshold, an action potential is initiated and these neurons carry it to the spinal cord (or the brainstem for cranial nerves).
3. **In the spinal cord, sensory afferents synapse on α-motor neurons and trigger an action potential.** The action potential is transmitted to α-motor neurons; note that upper motor neurons are not directly involved in this process.
4. **The α-motor neurons stimulate the muscle to contract, and it returns to its optimal length.** In Figure 13.19, the biceps brachii is stimulated to contract, and the forearm is returned to its original position.

Despite being classified as a monosynaptic reflex, this reflex arc has more going on than the stimulation of a single synapse. The sensory afferents also synapse with interneurons in the spinal cord that inhibit the antagonist muscles (in Figure 13.19, the triceps brachii is the antagonist muscle that would be inhibited). This permits the agonist muscle to contract with no interference from the antagonist muscle groups.

The simple stretch reflex is easily elicited by tapping on certain tendons. Perhaps the best-known example is the **patellar,** or

① An external force stretches the muscle.

② Muscle spindles detect the stretch, and primary and secondary afferents transmit an action potential to the spinal cord.

③ In the spinal cord, sensory afferents synapse on α-motor neurons and trigger an action potential.

④ The α-motor neurons stimulate the muscle to contract, and it returns to its optimal length.

Figure 13.19 A simple stretch reflex.

knee-jerk, reflex. The patellar reflex is elicited by using a blunt object such as a reflex hammer or the medial surface of your hand to tap the patellar tendon. This stretches the quadriceps femoris muscle of the thigh, producing a quick "knee jerk" with extension of the knee. Although you likely associate this reflex with trips to the doctor's office, it is critical in keeping you upright when you're standing. When you're standing upright, any slight buckling of the knee will also stretch the quadriceps femoris muscle. The muscle spindles detect this stretch and initiate the reflex, which results in firing of the quadriceps femoris muscle to straighten the leg and prevent the knee from buckling further.

Similar reflex responses are seen when tapping the triceps brachii tendon and the Achilles tendon. Simple stretch reflexes may also be elicited by simply standing up and bending over to touch your toes. When doing so, you can feel your posterior leg muscles tighten—this tightening is because they are contracting in response to the stretch.

A cranial nerve reflex that is controlled by a single cranial nerve is the **jaw-jerk reflex,** which is mediated by the trigeminal nerve. The jaw-jerk reflex is a simple stretch reflex elicited by tapping the chin, which leads to a reflexive contraction of the masseter and temporalis muscles.

Golgi Tendon Reflex

A polysynaptic reflex that also protects muscles and tendons from damage is the **Golgi tendon reflex.** However, the effect of this reflex is opposite that of the simple stretch reflex—it causes muscle *relaxation.* As its name implies, this reflex involves Golgi tendon organs. When tension in the muscle and tendon increases dramatically, the Golgi tendon organs signal the spinal cord and cerebellum. This leads to inhibition of the motor neurons supplying the contracting muscles and simultaneous activation of antagonist muscles. As a result, the muscle relaxes and the antagonist muscles contract, protecting the muscle and tendon from potentially damaging forces.

The Golgi tendon reflex is part of the reason why you should perform certain weight-lifting exercises with a spotter—if the weight is too heavy, the Golgi tendon reflex might cause you to drop the weight. Although this would prevent muscle and tendon damage, you could drop the weight on yourself and be seriously injured.

Flexion (Withdrawal) and Crossed-Extension Spinal Reflexes

When you touch a very hot object or step on a tack, you almost immediately pull your hand or foot away from the source of the pain. These are examples of the **flexion,** or **withdrawal, reflex.** As you can see in Figure 13.20a, the flexion reflex involves rapidly conducting nociceptive afferents and multiple synapses in the spinal cord, which therefore makes it a polysynaptic reflex. ① When stimulated, the nociceptive afferents transmit the stimulus to interneurons in the spinal cord. These interneurons then pass it on to α-motor neurons. ② The α-motor neurons stimulate the hip flexors of the limb receiving the painful stimulus (in our example the right limb) to contract. This withdraws the limb from the stimulus.

A second reflex that occurs simultaneously with the flexion reflex is the **crossed-extension reflex.** Note in Figure 13.20b that some of the spinal interneurons synapse on motor neurons that supply the extensor muscles of the opposite limb. When the flexion reflex is stimulated, ③ the hip extensors of the opposite limb are triggered to contract, in our example causing the left hip to extend while the affected right hip flexes. Together, these two motions allow for withdrawal from the painful stimulus (the broken glass) while providing balance and postural support.

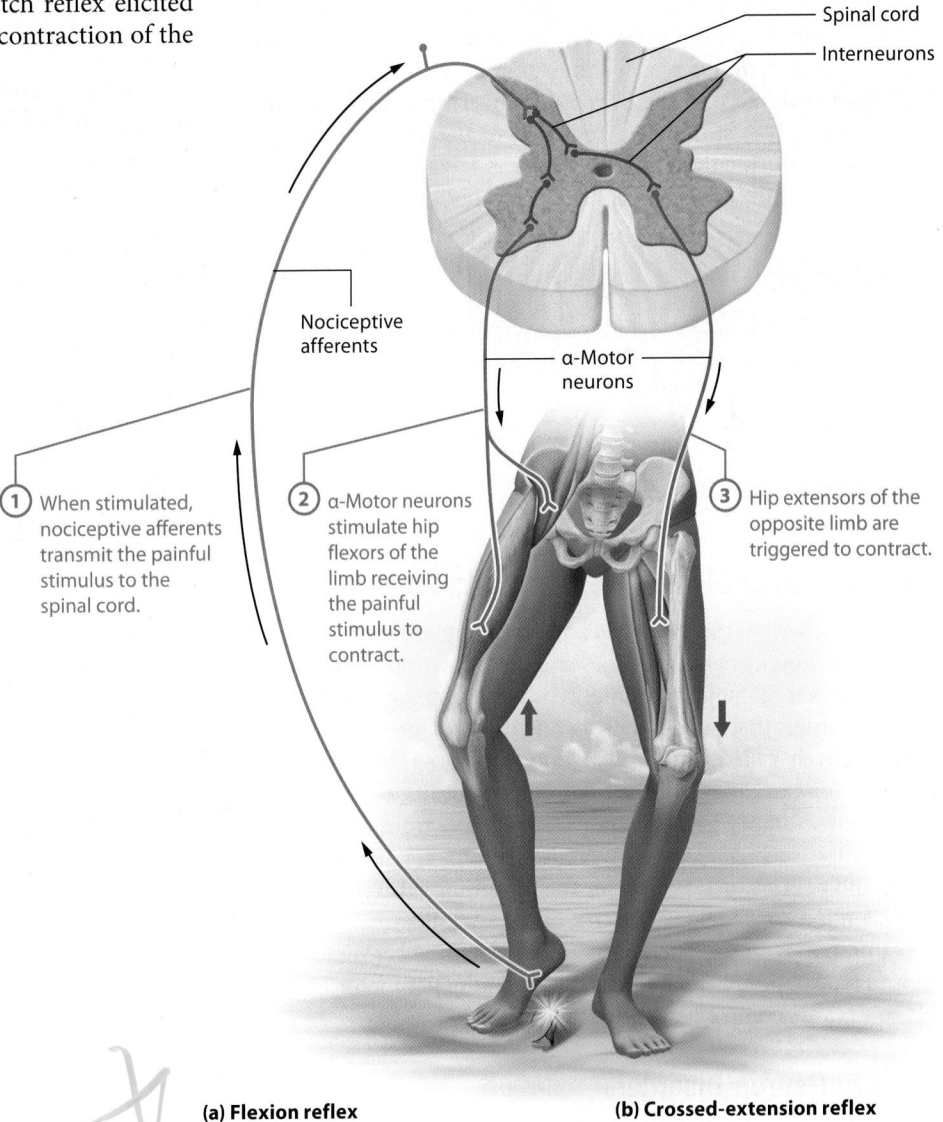

Figure 13.20 The flexion and crossed-extension reflexes.

(a) **Flexion reflex**

(b) **Crossed-extension reflex**

Without the crossed-extension reflex occurring with the flexion reflex, we would likely lose our balance and fall when withdrawing our limb from the painful stimulus.

Cranial Nerve Reflexes

Several polysynaptic reflex arcs involve the cranial nerves. Recall that some structures of the head and the neck are supplied with motor and sensory neurons from different nerves. For this reason, these cranial nerve reflexes involve two separate nerves—one afferent and one efferent. Among the more important of these cranial nerve reflexes, and ones you have probably experienced, are the gag and corneal blink reflexes.

The **gag reflex** is triggered when the visceral sensory nerve endings of the glossopharyngeal nerve in the posterior throat are stimulated unilaterally. This stimulus is brought back to a medullary nucleus shared by the glossopharyngeal and vagus nerves. When integration is complete, somatic motor neurons of the vagus nerve trigger contractions of the muscles of the pharynx, producing the familiar "gagging" sound and action.

When something comes in contact with your eye, generally you blink, thanks to the **corneal blink reflex.** This reflex is triggered when a stimulus reaches the somatic sensory receptors of the trigeminal nerve in the thin outer covering of the eye called the *cornea.* These stimuli are returned to the pons for integration, and the orbicularis oculi muscle is triggered to contract, producing a blink, via the facial nerve's somatic motor neurons.

Quick Check

☐ 5. How do polysynaptic and monosynaptic reflex arcs differ?
☐ 6. Describe the reflex arcs of the following reflexes:
    a. Simple stretch reflex
    b. Flexion reflex
    c. Gag reflex
    d. Corneal blink reflex

Sensory and Motor Neuron Disorders

As you have seen in the preceding sections, homeostasis requires the proper functioning of both sensory and motor neurons. For this reason, the effects of damage to either type of neuron can lead to problems. Disorders that impact sensory and motor neurons of the PNS are collectively called *peripheral neuropathies* (noor-AH-pah-theez; *patho-* = "disease"). One of the most common causes of peripheral neuropathy is injury to a spinal or cranial nerve from trauma, systemic illness such as diabetes mellitus (see Chapter 16), or certain drugs and poisons such as arsenic. The upcoming subsections direct our attention to damaged sensory and motor neurons and their effect on homeostasis. The effects of damage to lower motor neurons (a type of peripheral neuropathy) and to upper motor neurons of the CNS are compared, as well.

Sensory Neuron Disorders

The symptoms and severity of sensory peripheral neuropathy depend on which spinal or cranial nerve is injured. Damage to visceral sensory neurons can produce a variety of symptoms, such as urinary and/or fecal incontinence. Injury to somatic sensory neurons often decreases or eliminates sensation, including pain, in the affected part of the body. This can be dangerous, as the person with such an injury doesn't recognize and respond to painful stimuli, which can lead to tissue damage. Occasionally, the opposite phenomenon results, and the individual experiences burning and "shooting" pain. This is generally due to inflammation of the neurons and inappropriate firing of nociceptors and other sensory receptors.

Another type of sensory peripheral neuropathy involves injury to proprioceptive neurons. This condition is generally due to damaged ligaments and tendons in which proprioceptors are housed. It results in difficulty with monitoring and controlling movement, as the axons do not effectively relay information on joint position to the cerebellum and other parts of the CNS.

Lower Motor Neuron Disorders

Motor peripheral neuropathies are also known as *lower motor neuron disorders.* They most often result from injury to a spinal or cranial nerve or to lower motor neuron cell bodies in the spinal cord. Such injuries prevent an α-motor neuron from stimulating a skeletal muscle fiber to contract. For this reason, lower motor neuron disorders may result in *paralysis* (inability to move a given muscle) or *paresis* (weakness of a given muscle). The damaged α-motor neuron is also unable to respond to feedback from muscle spindles. This causes a reduction or absence of stretch reflexes and loss of muscle tone.

Upper Motor Neuron Disorders

Lower motor neuron disorders produce very different symptoms from those of *upper motor neuron disorders* (note that upper motor neuron disorders are not considered peripheral neuropathies, as they impact neurons of the CNS). These conditions can result from damage or disease anywhere along the pathways from the motor cortices to the spinal cord. The body's initial response to upper motor neuron damage is *spinal shock,* characterized by paralysis. Spinal shock is believed to result from the "shock" experienced by spinal cord circuits when input from the upper motor neurons is removed.

For reasons that are not well understood, after a few days spinal shock wears off and in its place *spasticity* often develops. Spasticity is characterized by an increase in stretch reflexes, an increase in muscle tone, and a phenomenon called *clonus,* or the alternating contraction and relaxation of a stretched muscle. Spasticity is likely due to a loss of normal inhibition mediated by upper motor neurons. In addition to spasticity, the *Babinski sign* (bah-BIN-skee) develops. This sign is elicited by stroking the bottom of the foot, which in a healthy adult will cause flexion of the toes, a response known as the *plantar reflex* (**Figure 13.21a**). However, a patient with an upper motor neuron disorder will extend the hallux (big toe) and splay out the other toes (**Figure 13.21b**). A positive Babinski sign is often present in infants up to 18 months old and does not signify pathology, but the same response in an adult is always considered abnormal.

Toes flex.

(a) Plantar reflex—normal response

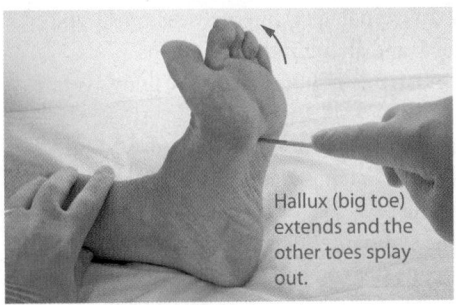
Hallux (big toe) extends and the other toes splay out.

(b) Positive Babinski sign—present in adults with upper motor neuron disorders

Figure 13.21 The Babinski sign.

A&P in the Real World

Amyotrophic Lateral Sclerosis

The devastating disease *amyotrophic lateral sclerosis* (ay′-my-oh-TROH-fik; ALS), also known as *Lou Gehrig's disease,* involves degeneration of the cell bodies of α-motor neurons in the anterior horn of the spinal cord as well as upper motor neurons in the cerebral cortex. The cause of the degeneration is unknown at present, and many factors likely play a role.

The most common early feature of the disease is muscle weakness, particularly in the distal muscles of the limbs and the muscles of the hands. Over time the muscle weakness spreads to other muscle groups, and upper motor neuron symptoms develop as well. Death from ALS typically results within 5 years of the disease's onset. In most forms of ALS, cognitive functions are spared, so the patient is fully aware of the effects and complications of the disease. Although intensive research efforts are ongoing, at this time there is no cure or treatment that prevents the disease's progression.

Some diseases affect both the upper and lower motor neurons. For one example, see *A&P in the Real World: Amyotrophic Lateral Sclerosis.*

Quick Check

☐ 7. What are some potential effects of sensory peripheral neuropathy?

☐ 8. How do upper and lower motor neuron disorders differ?

Apply What You Learned

☐ 1. An experimental drug blocks the body's muscle spindles from detecting stretch. Predict the effects of such a drug on movement, posture, and balance.

☐ 2. Mr. Pratchett has damage to his facial nerve, but not his trigeminal nerve. What would happen to his blink reflex? Explain.

☐ 3. Explain why a lower motor neuron disorder leads to absent or weakened simple stretch reflexes but an upper motor neuron disorder does not.

See answers in Appendix A.

CHAPTER SUMMARY

Take full advantage of the invaluable online practice and study tools for this chapter by going to MasteringA&P®. **There you will find:**

- Practice tests, module quizzes, art-labeling and sequence-the-steps activities
- ▶ Animated figures with quizzes
- ✐ *Practicing A&P* worksheets
- ◗)) Author-narrated podcast
- ♨ Dynamic Study Modules
- iP *Interactive Physiology*

MODULE 13.1
Overview of the Peripheral Nervous System 477

- The **peripheral nervous system (PNS)** performs the sensory and motor functions of the nervous system and links the CNS with the rest of the body and its external environment.

- The PNS is classified functionally into the **sensory (afferent) division** and the **motor (efferent) division.** (Figure 13.1, p. 478)
  - The sensory division detects sensory stimuli and delivers them to the CNS. The **somatic sensory division** detects general sensory stimuli, including external stimuli from the skin and internal stimuli from the musculoskeletal system. The somatic sensory division also contains special sensory neurons that are responsible for detecting stimuli of the *special senses.*

- The **visceral sensory division** detects internal sensory stimuli from within the thoracic and abdominopelvic organs.
- The **somatic motor division** is responsible for the body's voluntary motor functions.
- The **visceral motor division**, or **autonomic nervous system (ANS)**, is responsible for maintaining many aspects of homeostasis via its control of the body's involuntary motor functions. It has two divisions: the **sympathetic nervous system** and the **parasympathetic nervous system.**

- The **peripheral nerves** are the main organs of the PNS. A nerve consists of the axons of many neurons, multiple connective tissue sheaths, and blood vessels. Nerves may contain only sensory axons (**sensory nerves**), both motor and sensory axons (**mixed nerves**), or primarily motor axons (**motor nerves**). (Figure 13.2, p. 480)

- The two types of nerves are based on their location:
  - The 12 pairs of **cranial nerves** originate from the brain and innervate structures of the head and neck. Cranial nerves may be sensory, motor, or mixed nerves.
  - The 31 pairs of **spinal nerves** originate from the spinal cord and innervate structures below the head and neck. All spinal nerves are mixed nerves.

MODULE **13.2**
The Cranial Nerves 481

- Each of the 12 pairs of cranial nerves is given a specific descriptive name and a Roman numeral that reflects the pair's position on the brain. (Figure 13.3, p. 482)

  Practice art-labeling exercise: Figure 13.3.

- The sensory cranial nerves are the following: (Table 13.1, p. 483)
  - **Cranial nerve I**, the **olfactory nerve**, is the nerve for the sense of smell. It originates from the neurons of the olfactory epithelium in the roof of the nasal cavity.
  - **Cranial nerve II**, or the **optic nerve**, is the nerve for vision. It arises from neurons in the posterior eye.
  - **Cranial nerve VIII**, or the **vestibulocochlear nerve**, is composed of two separate nerves. The *vestibular nerve* detects the position of the body for the sense of balance and equilibrium; the *cochlear nerve* detects sound waves for the sense of audition, or hearing.

- The motor cranial nerves are the following: (Table 13.2, pp. 484–485)
  - **Cranial nerve III**, or the **oculomotor nerve**, innervates four of six extraocular muscles and the levator palpebrae superioris muscle, constricts the pupil, and changes the shape of the lens for better near vision.
  - **Cranial nerve IV**, or the **trochlear nerve**, innervates one of six extraocular muscles (the superior oblique muscle).
  - **Cranial nerve VI**, or the **abducens nerve**, innervates one of the six extraocular muscles (the lateral rectus muscle).
  - **Cranial nerve XI**, or the **accessory nerve**, has both a cranial component that innervates certain muscles of the larynx and a spinal component that innervates the trapezius and sternocleidomastoid muscles.
  - **Cranial nerve XII**, or the **hypoglossal nerve**, innervates the intrinsic and extrinsic muscles of the tongue.

- The mixed cranial nerves are the following: (Table 13.3, pp. 486–488)
  - **Cranial nerve V**, the **trigeminal nerve**, originates from the pons and supplies the skin of the superior face and scalp, the skin of the middle part of the face, the skin over the inferior part of the face, and the muscles of mastication.
  - The neurons of **cranial nerve VII**, the **facial nerve**, innervate the muscles of facial expression and other muscles of the face and neck; trigger secretion from the lacrimal, salivary, and nasal mucous glands; provide taste sensation to the anterior two-thirds of the tongue; and provide somatic sensation to the external ear, the palate, and the nasal cavity.
  - **Cranial nerve IX**, or the **glossopharyngeal nerve**, supplies taste sensation to the posterior one-third of the tongue, provides sensation to the external ear and posterior pharynx, detects changes in blood pressure, supplies muscles involved in swallowing, and triggers secretion from certain salivary glands.
  - **Cranial nerve X**, or the **vagus nerve**, provides somatic sensation to the skin around the ear, taste sensation to the pharynx, and visceral sensation to the mucosa in the pharynx and larynx; detects carbon dioxide concentrations in the blood; and stimulates movement of the muscles of swallowing and speaking. Its parasympathetic neurons supply parasympathetic innervation to most of the thoracic and abdominal viscera.

MODULE **13.3**
The Spinal Nerves 490

- **Spinal nerves** form from the fusion of a *motor root* stemming from the anterior horn of the spinal cord and a *sensory root* stemming from the posterior horn of the spinal cord.

- About 1–2 cm after a spinal nerve forms, it splits into a **posterior ramus** and an **anterior ramus**. (Figure 13.4a, p. 490)

- Anterior roots of the spinal nerves carry only motor signals, and the posterior roots only sensory signals. Sensory and motor signals are both carried in spinal nerves and rami. (Figure 13.4b, p. 490)

- There are 31 pairs of spinal nerves, divided into cervical, thoracic, lumbar, sacral, and coccygeal nerves. The anterior rami of the cervical, lumbar, and sacral spinal nerves merge to form networks called **nerve plexuses**. (Figure 13.5, p. 491)

- The **cervical plexuses** consist of the anterior rami of C_1-C_4; they innervate the skin of the neck and portions of the head, chest, and shoulders. Their motor branches innervate certain muscles of the neck, and the **phrenic nerve** innervates the diaphragm. (Figure 13.6, p. 492)

- The **brachial plexuses** stem from C_5-T_1 and innervate the skin and muscles of the upper limb. The nerves of the brachial plexus include the **axillary nerve**, the **radial nerve**, the **musculocutaneous nerve**, the **median nerve**, and the **ulnar nerve**. (Figure 13.7, p. 493)

  Practice art-labeling exercise: Figure 13.7.

- The anterior rami of the thoracic spinal nerves form the **intercostal nerves**. Each intercostal nerve travels in an intercostal space and innervates the intercostal muscles and the overlying skin. The intercostal nerves from T_7-T_{12} also innervate the abdominal muscles and skin.

- The **lumbar plexuses** stem from the anterior rami of L_1–L_4 and innervate structures of the pelvis and the lower limb. Their nerves include the **obturator nerve** and the **femoral nerve.** (Figure 13.8, p. 495)

- The **sacral plexuses** stem from the anterior rami of spinal nerves L_4–S_4 and innervate structures of the pelvis, the gluteal region, and much of the lower limb. The largest nerve in the body is the **sciatic nerve,** which splits into the **tibial nerve** and the **common fibular nerve.** (Figure 13.9, p. 497)

Practice art-labeling exercise: Figure 13.9.

- The spinal nerves carry sensory input from the skin and motor output to skeletal muscles throughout the body. (Figure 13.10, p. 498)

Practice art-labeling exercise: Figure 13.10.

MODULE 13.4
Sensation Part II: Role of the PNS in Sensation 499

- **Sensory transduction,** the conversion of a stimulus into an electrical signal, takes place at a **sensory receptor.** During sensory transduction, ion channels open and a **receptor potential** is generated. If the receptor potential is large enough for the membrane to reach threshold, it triggers an action potential. (Figure 13.11, p. 500)

- Sensory receptors may be classified according to their location. **Interoceptors** detect stimuli originating within the body. **Exteroceptors** detect stimuli originating outside the body.

- Sensory receptors may also be classified according to the type of stimuli they detect: **Mechanoreceptors** detect stimuli that mechanically deform the tissue through the opening of mechanically gated ion channels; **thermoreceptors** detect thermal stimuli; **chemoreceptors** detect the presence or concentration of certain chemicals in the body's fluids or in the air; **photoreceptors** detect light; and **nociceptors** detect noxious or painful stimuli. (Figure 13.12, p. 501)

- Sensory, or afferent, neurons detect and transmit sensory stimuli.

- Somatic sensory neurons are pseudounipolar neurons with three components: (1) a cell body located in a posterior root ganglion or a cranial nerve ganglion, (2) a *peripheral process* that transmits an action potential from a sensory receptor to the central process, and (3) a *central process* that carries the impulse from the peripheral process to the brainstem or spinal cord. Both the central and peripheral processes are axons. (Figure 13.13, p. 502)

- Large-diameter and heavily myelinated sensory neurons transmit impulses the fastest.

- The *receptive field* of a neuron is the area served by that neuron. The more the peripheral process of a neuron branches once it is in the skin, the larger its receptive field. The size of a receptive field may be measured by assessing the **two-point discrimination threshold.** (Figure 13.14, p. 503)

- The skin can be divided into regions known as **dermatomes** that are innervated by a single cranial or spinal nerve. Visceral pain is often perceived as cutaneous pain along the dermatome for the

nerve that supplies the organ. This is called **referred pain.** (Figure 13.15, p. 504)

- The overall pathway for the detection and perception of somatic sensation is as follows: (1) The stimulus triggers an action potential; (2) the action potential is conducted along the neuron's peripheral process and transferred to its central process; (3) the central process transmits the action potential to the CNS, (4) where it is interpreted. (Figure 13.16, p. 505)

MODULE 13.5
Movement Part II: Role of the PNS in Movement 506

- The CNS initiates movement when **upper motor neurons** in the motor cortices make the "decision" to move. Messages from the upper motor neurons are relayed to interneurons in the brainstem and spinal cord, then to **lower motor neurons** that contact the muscle fiber and initiate a muscle contraction.

- Lower motor neurons are multipolar neurons whose cell bodies are located in the anterior horn of the spinal cord or in the brainstem. Groups of lower motor neurons that innervate the same muscle are located together in the spinal cord in *motor neuron pools.*

- The overall process of movement proceeds by the following steps: (1) The upper motor neurons in the premotor cortex select a motor program; (2) the basal nuclei allow the thalamus to stimulate upper motor neurons of the primary motor cortex; (3) upper motor neurons stimulate lower motor neurons; (4) lower motor neurons trigger a skeletal muscle to contract; and (5) sensory information from sensory receptors in the muscles, tendons, and joints is relayed back to the CNS. (Figure 13.17, p. 507)

MODULE 13.6
Reflex Arcs: Integration of Sensory and Motor Functions 508

- A **reflex** is a programmed, automatic motor response to a specific sensory input. The series of events that culminates in a reflex response is a **reflex arc.** Reflexes may be *visceral* or *somatic,* and **monosynaptic** or **polysynaptic.**

- In reflexes that monitor and control movement, the sensory component is detected by **muscle spindles** and **Golgi tendon organs.** (Figure 13.18, p. 509)

Practice art-labeling exercise: Figure 13.18.

  ○ Muscle spindles contain **intrafusal muscle fibers,** which contain contractile filaments only at their poles. They are innervated by γ-motor neurons. The central area of the intrafusal muscle fiber is noncontractile and is innervated by two types of sensory neurons that detect stretch.

  ○ Golgi tendon organs are located within tendons near the muscle-tendon junction. They detect the tension generated by a muscle contraction.

- A **simple stretch reflex** is a monosynaptic reflex that returns a muscle to its optimal length after being stretched. When intrafusal fibers are stretched, their primary and secondary afferents deliver this information to the spinal cord or brainstem, where they synapse directly on the α-motor neurons. This causes the muscle to contract and so return to its optimal length. (Figure 13.19, p. 510)

- The Golgi tendon reflex causes the muscle to lengthen in response to increased muscle tension.

- The **flexion,** or **withdrawal reflex,** is polysynaptic and involves nociceptive afferents that stimulate contraction of flexor muscles to withdraw the affected limb from a painful stimulus. The **crossed-extension reflex** occurs simultaneously with the flexion reflex and causes contraction of the extensor muscles of the opposite limb. (Figure 13.20, p. 511)

- The cranial nerves are responsible for several polysynaptic reflexes, including the following:
  - The **gag reflex** is triggered by unilateral stimulation of the sensory receptors of the glossopharyngeal nerve at the posterior throat. The motor response is mediated by the somatic motor neurons of the vagus nerve.
  - The **corneal blink reflex** is triggered when the trigeminal nerve's somatic sensory receptors on the outer eye are stimulated. The motor response is mediated by the somatic motor neurons of the facial nerve.

- Disorders of nerves of the PNS are *peripheral neuropathies.* Peripheral neuropathies of sensory neurons may result in loss of sensation in the affected part. Peripheral neuropathies of motor neurons result in lower motor neuron disorders, which lead to weakness or paralysis of the affected muscles and a loss of reflexes. One clinical indicator of problems with upper motor neuron function is the *Babinski sign.* (Figure 13.21, p. 513)

ASSESS WHAT YOU LEARNED

See answers in Appendix A.

PART 1 Check Your Recall

1. Mark the following statements as true or false. If a statement is false, correct it to make a true statement
   a. The somatic sensory division of the PNS detects sensory information from the organs in the thoracic and abdominopelvic cavities.
   b. The somatic motor division of the PNS consists of lower motor neurons that directly innervate skeletal muscle fibers.
   c. The visceral motor division is also known as the autonomic nervous system and maintains many aspects of the body's homeostasis.
   d. The term nerve is the equivalent of the term neuron.
   e. There are 31 pairs of spinal nerves and 12 pairs of cranial nerves.

2. Fill in the blanks: A spinal nerve divides into a(n) _____ that serves the anterior side of the body and the limbs and a(n) _____ that serves the posterior side of the body.

3. Define each of the following terms in your own words, using 20 or fewer words.
   a. Peripheral nerve
   b. Nerve plexus
   c. Posterior root ganglion

4. First, write the Roman numeral that corresponds to each named cranial nerve (after the abbreviation CN). Second, match the cranial nerve with its correct function from the column on the right.

CN _____
_____ Vestibulocochlear nerve

CN _____
_____ Trigeminal nerve

CN _____
_____ Hypoglossal nerve

CN _____
_____ Abducens nerve

CN _____
_____ Vagus nerve

CN _____
_____ Olfactory nerve

CN _____
_____ Accessory nerve

CN _____
_____ Oculomotor nerve

CN _____
_____ Facial nerve

CN _____
_____ Optic nerve

CN _____
_____ Glossopharyngeal nerve

CN _____
_____ Trochlear nerve

a. Motor to the lateral rectus muscle
b. Motor to the muscles of facial expression; lacrimation; salivation; taste to the anterior two-thirds of the tongue
c. Sense of smell
d. Motor to the muscles for swallowing; salivation; taste to the posterior one-third of the tongue; somatic sensation from the throat
e. Senses of hearing and equilibrium
f. Motor to the superior oblique muscle
g. Motor to the tongue
h. Motor to the sternocleidomastoid and trapezius muscles
i. Sense of vision
j. Motor to muscles of swallowing and speaking; parasympathetic innervation to thoracic and abdominal viscera; sense of taste from the throat
k. Sensory to the face; motor to the muscles of mastication
l. Motor to four of six extraocular muscles; constricts the pupil; changes the shape of the lens; opens the eyelid

5. Which cranial nerves are sensory only, primarily motor, and mixed?

6. Match the following nerves with the structures they innervate.

_____ Phrenic nerve
_____ Median nerve
_____ Femoral nerve
_____ Tibial nerve
_____ Radial nerve
_____ Intercostal nerves
_____ Common fibular nerve
_____ Musculocutaneous nerve

a. Motor to the triceps brachii muscle and extensors of the forearm; sensory from the posterior hand
b. Motor to the anterior arm muscles; sensory from skin over the lateral forearm
c. Motor to the lateral and anterior leg muscles; sensory from the skin of the anteroinferior leg
d. Motor to the diaphragm
e. Motor to the anterior thigh muscles; sensory from the skin over the anterior thigh and leg
f. Motor to the hamstring muscles, flexor muscles of the leg, muscles of the foot; sensory from the skin over the posterior and lateral leg and foot

g. Motor to the muscles between the ribs and the abdominal muscles; sensory from the skin over the abdomen

h. Motor to the flexors in the forearm, certain intrinsic hand muscles; sensory from the skin of the anterior hand

7. Somatic sensory neurons are _____ neurons whose cell bodies are located in the _____.

    a. multipolar, posterior horn
    b. pseudounipolar, posterior root ganglion
    c. bipolar, anterior horn
    d. pseudounipolar, posterior horn

8. A receptor potential:

    a. always leads to an action potential.
    b. never leads to an action potential.
    c. causes hyperpolarization of the neuron.
    d. leads to an action potential if the stimulus is strong enough.

9. Why is visceral pain often perceived as cutaneous pain?

10. Merkel cell fibers, tactile corpuscles, Ruffini endings, and lamellated corpuscles are all types of:

    a. nociceptors.
    b. mechanoreceptors.
    c. photoreceptors.
    d. chemoreceptors.

11. Place the following sequence of events for the detection of somatic sensation in the proper order. Place a 1 by the first event, a 2 by the second event, and so on.

    a. _____ The central process transmits the action potential to the CNS.
    b. _____ The action potential is transferred to the central process in the posterior root ganglion.
    c. _____ The stimulus triggers an action potential.
    d. _____ The signal is interpreted by some part of the brain.
    e. _____ The action potential is conducted along the peripheral process of the neuron.

12. How do upper and lower motor neurons differ?

13. List and describe the basic steps involved in producing movement, beginning with the upper motor neurons in the cerebral cortex.

14. The lower motor neurons that innervate contractile skeletal muscle fibers are called:

    a. α-motor neurons.
    b. β-motor neurons.
    c. upper motor neurons.
    d. γ-motor neurons.

15. Fill in the blanks: _____ detect the degree to which a muscle is stretched, whereas _____ detect the force of a muscle contraction.

16. Which of the following is the correct order of events of a reflex arc?

    a. Stimulus detection and delivery → motor response → integration in the CNS

    b. Motor response → stimulus detection and delivery → integration in the CNS
    c. Stimulus detection and delivery → integration in the CNS → delivery of motor response
    d. Integration in the CNS → motor response → stimulus detection and delivery

17. Mark the following statements as true or false. If a statement is false, correct it to make a true statement.

    a. The simple stretch reflex is a monosynaptic reflex with only a single synapse in the spinal cord between the sensory and motor neurons.
    b. The Golgi tendon organs detect stretch in a simple stretch reflex.
    c. A flexion reflex is a monosynaptic reflex with only one synapse in the spinal cord.
    d. The crossed-extension reflex occurs simultaneously with the simple stretch reflex.

18. Which two cranial nerves mediate the gag reflex?

PART 2 Check Your Understanding

1. Devise a series of physical examination tests that assess the function of each of the cranial nerves.

2. Predict what might happen if nociceptors were rapidly adapting instead of slowly adapting receptors. Would this change be beneficial or potentially harmful?

3. Explain why you lose both motor and sensory function of a part of your body when a spinal nerve is numbed with anesthetic agents.

PART 3 Apply Your Knowledge

PART A: Application and Analysis

1. Complaining of muscle weakness, Ms. McKenna goes to the doctor. Devise a series of physical examination tests that would help to determine if her problem involves upper or lower motor neurons.

2. Jason is a patient who presents for evaluation after he suffered a severe shoulder injury during which his entire brachial plexus suffered damage. What effects would you expect Jason to suffer from this injury?

3. When Mr. Williams goes to the emergency department with pain in the area along the midline of the diaphragm, he worries he is having a heart attack. Could the pain be related to his heart? Could it be related to any other organ(s)? Explain.

4. Maria is a 3-year-old who has been diagnosed with the condition CIPA, or congenital insensitivity to pain with anhidrosis. This disease results from a genetic mutation that causes essentially all general sensory neurons to not function properly. What types of sensations will Maria be unable to detect (be specific)? Predict what problems she might face from her condition.

PART B: Make the Connection

5. Another feature of CIPA is anhidrosis, or the inability to sweat. Explain why Maria's inability to produce sweat could potentially be life-threatening. (*Connects to Chapter 5*)

14

The Autonomic Nervous System and Homeostasis

Have you ever had to slam on the brakes of your car unexpectedly to avoid hitting another car? Do you remember how shaky you felt right after that, maybe also sweaty and out of breath? Well, that feeling, and your ability to hit the brakes so fast, was courtesy of the neurotransmitter norepinephrine, produced by neurons of the largely involuntary arm of the peripheral nervous system (PNS)—the *visceral motor division,* or **autonomic nervous system (ANS).**

We've already discussed the roles of the CNS in maintaining homeostasis (see Chapter 12). Recall that homeostasis is mostly maintained by the hypothalamus and the reticular formation of the brainstem. Much of this is mediated by the neurons of the ANS. Recall that there are two branches of the ANS, the sympathetic and parasympathetic nervous systems; look back at Figure 13.1 to see again how the PNS divisions are related.

As in any other part of the nervous system, the neurons of the ANS communicate with one another at synapses, and the neurons' effects are mediated by neurotransmitters that bind to receptors on target cells. The sympathetic and parasympathetic nervous systems are covered in separate modules, but bear in mind that the two systems work together constantly to maintain the delicate balance that is homeostasis. We conclude the chapter by examining how the ANS and the nervous system as a whole maintain homeostasis.

MODULE **14.1**

Overview of the Autonomic Nervous System

Learning Outcomes

1. Describe the structural and functional details of the sensory and motor (autonomic) components of visceral reflex arcs.
2. Distinguish between the target cells of the somatic and autonomic nervous systems.
3. Contrast the cellular components of the somatic and autonomic motor pathways.
4. Discuss the physiological roles of each division of the autonomic nervous system.

Computer-generated image: The autonomic nervous system is comprised of the sympathetic and parasympathetic nervous systems, which generally have opposite effects on a body function or organ.

Before delving into the intricacies of the ANS, let's first take an introductory tour of its properties. This module begins with an overview of ANS functions and then compares its structures and roles with those of the somatic motor division of the PNS. The discussion then moves on to differences between the two divisions of the ANS, the sympathetic and parasympathetic nervous systems.

Functions of the ANS and Visceral Reflex Arcs

◀◀ **Flashback**

1. What is a reflex arc? (p. 508)
2. What is a ganglion? (p. 387)

The autonomic nervous system oversees such vital functions as heart rate, blood pressure, and digestive and urinary processes. The ANS is named for the fact that it can operate autonomously, without conscious control. It performs these functions via a series of **visceral reflex arcs.** As you have learned, a reflex arc is a series of events in which a sensory stimulus leads to a predictable motor response (see Chapter 13).

Notice in **Figure 14.1** that in the visceral reflex arcs of the ANS, ① sensory signals from the viscera and skin are sent by afferent sensory neurons to the brain or spinal cord. ② The stimuli are then integrated by the CNS. ③ Next, motor impulses from the CNS are sent out via efferent neurons in cranial and spinal nerves. These nerves usually lead to ganglia in the PNS,

 CORE PRINCIPLE Cell-Cell Communication

called **autonomic ganglia.** ④ Finally, the autonomic ganglia send the impulses via other efferent neurons to various target organs, where they trigger a motor response in the target cells. Note that visceral reflex arcs provide another example of the Cell-Cell Communication Core Principle (p. 27).

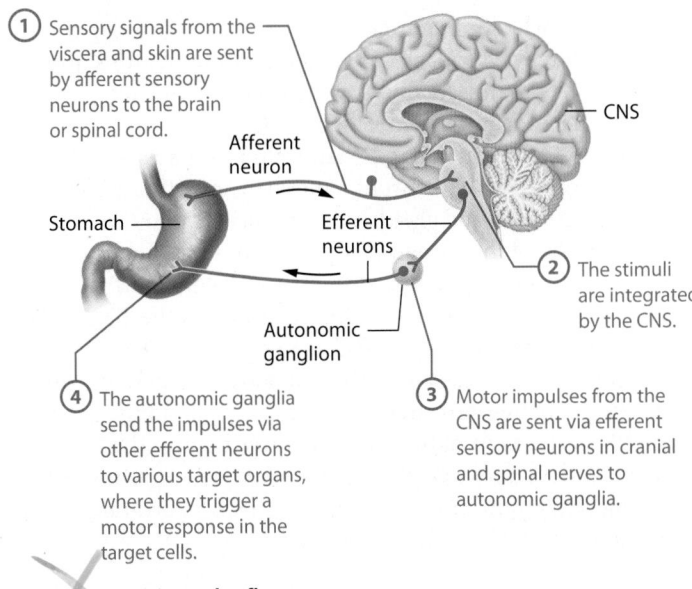

Figure 14.1 **Visceral reflex arcs.**

Quick Check

☐ 1. What are the basic functions of the ANS?
☐ 2. What are autonomic ganglia?

Comparison of Somatic and Autonomic Nervous Systems

◀◀ **Flashback**

1. What is the somatic motor division of the PNS? (p. 479)
2. What is a somatic motor neuron? (p. 479)

Previously (see Chapter 13), you read that the somatic motor division consists of somatic motor neurons that innervate skeletal muscle fibers (**Figure 14.2a**). These neurons largely control

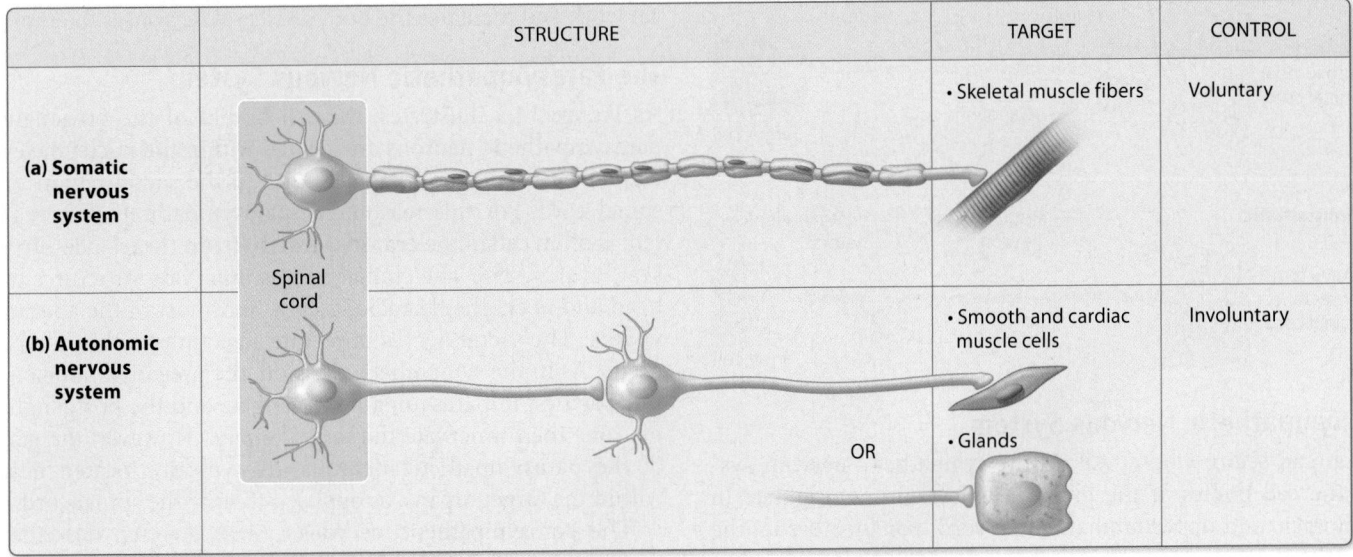

	STRUCTURE	TARGET	CONTROL
(a) Somatic nervous system	Spinal cord	• Skeletal muscle fibers	Voluntary
(b) Autonomic nervous system	OR	• Smooth and cardiac muscle cells • Glands	Involuntary

Figure 14.2 **Comparison of the somatic and autonomic nervous systems.**

voluntary muscle contractions, or those that are initiated consciously. You may also recall that somatic motor neurons directly innervate skeletal muscle fibers, which they stimulate by releasing the neurotransmitter acetylcholine (ACh).

In **Figure 14.2b** you can see that, unlike somatic motor neurons, autonomic motor neurons innervate smooth muscle cells, cardiac muscle cells, and glands, and produce involuntary actions. In addition, these motor neurons do not directly innervate their target cells. Instead, two neurons are involved. The first is the initial efferent neuron whose cell body resides in the CNS, called the **preganglionic neuron** (PREE-gang-glee-on´-ik). The preganglionic neuron synapses on the cell body of a second neuron, called the **postganglionic neuron,** within an autonomic ganglion in the PNS. The axon of the postganglionic neuron then triggers a change in the target cell by releasing various neurotransmitters, including ACh and norepinephrine. These neurotransmitters may stimulate or inhibit their target cells.

Quick Check

☐ 3. How do somatic motor neurons and autonomic neurons differ?

☐ 4. What are preganglionic and postganglionic neurons?

Divisions of the ANS

The structures and functions of the two divisions of the ANS, the sympathetic and parasympathetic nervous systems, differ considerably, as we touch on here. These divisions are discussed in more detail in Modules 14.2 and 14.3, respectively.

To start, we can note a major difference between the two systems in the arrangement of their neurons: In the sympathetic nervous system the preganglionic axons are often short and the postganglionic axons are generally long. The opposite is true for the axons of the parasympathetic nervous system:

The Sympathetic Nervous System

As you can see in **Figure 14.3**, in the sympathetic nervous system, the cell bodies of the preganglionic neurons originate in the thoracic and upper lumbar spinal cord. For this reason, the sympathetic nervous system is often called the **thoracolumbar division** (thor´-ah-koh-LUM-bar) of the ANS. These neurons

Figure 14.3 Overview of the structure of ANS divisions.

synapse first in sympathetic ganglia, which are generally located near the spinal cord (with a few exceptions). The postganglionic neurons innervate their target cells.

The sympathetic nervous system is usually described as the "fight or flight" division of the ANS—this reflects its role in preparing the body for emergency situations in which one would need to fight off an attacker or flee from danger. The sympathetic nervous system's role is broader than this, however, as it maintains homeostasis when the body is engaged in any type of physical work and mediates the body's visceral responses to emotion.

The Parasympathetic Nervous System

As Figure 14.3 illustrates, the cell bodies of the preganglionic parasympathetic neurons are located within the nuclei of several cranial nerves in the brainstem and in the sacral region of the spinal cord. For this reason, the parasympathetic nervous system is often called the **craniosacral division** (kray´-nee-oh-SAY-kral) of the ANS. The cranial nerves innervate structures of the head and neck, the thoracic viscera, and most of the abdominal viscera. The sacral nerves innervate structures within the pelvic cavity. As in the sympathetic division, the preganglionic neurons synapse first in parasympathetic ganglia, and the postganglionic neurons then innervate the target organs. However, the ganglia of the parasympathetic neurons are typically located near or within the target organs, as opposed to near the spinal cord.

The parasympathetic nervous system is often described as the "rest and digest" division of the ANS. This reflects its role in digestion and in maintaining the body's homeostasis when at rest.

The Balance between the Parasympathetic and Sympathetic Nervous Systems

The actions of the parasympathetic nervous system directly antagonize those of the sympathetic nervous system. For example, parasympathetic neurons decrease the rate of the heart's contractions, whereas sympathetic neurons increase their rate. Together, these two divisions maintain a delicate balance to ensure that homeostasis is maintained at all times.

Quick Check

☐ 5. What are the two divisions of the ANS?
☐ 6. How are the two divisions of the ANS alike, and how do they differ?

Apply What You Learned

☐ 1. Certain types of peripheral neuropathy (diseases of PNS neurons) affect somatic motor neurons, whereas others affect autonomic motor neurons. Predict the effects of autonomic neuropathy. How would autonomic neuropathy differ from somatic motor neuron neuropathy?
☐ 2. Mrs. Williams is taking a medication for high blood pressure that blocks the effects of the sympathetic nervous system. What is likely to happen if she is faced with an emergency situation? Explain.

See answers in Appendix A.

MODULE **14.2**

The Sympathetic Nervous System

Learning Outcomes

1. Explain how the sympathetic nervous system maintains homeostasis.
2. Describe the anatomy of the sympathetic nervous system.
3. Describe the neurotransmitters and neurotransmitter receptors of the sympathetic nervous system.
4. Explain the effects of the sympathetic nervous system on the cells of its target organs.

As you know, the **sympathetic nervous system** is the division of the ANS that prepares the body for emergency situations. But the "fight or flight" characterization grossly oversimplifies this system, as it is also active, for example, when you experience emotions (particularly strong ones, such as fear, excitement, embarrassment, and rage) and even during minor physical activities. In fact, even the act of standing up requires the sympathetic nervous system; if its neurons don't temporarily initiate changes in your heart rate and blood vessel diameter, your blood pressure could drop quite

low due to the effect of gravity on your blood (a condition called *orthostatic hypotension*), causing you to fall.

This module explores the structure and function of the neurons and ganglia of the sympathetic nervous system and explains how this system, when dominant, maintains homeostasis. The interactions between this system and the parasympathetic nervous system are discussed in Module 14.4.

Gross and Microscopic Anatomy of the Sympathetic Nervous System

◀◀ **Flashback**

1. What are the three structural components of a chemical synapse? (p. 407)

The organization of the sympathetic nervous system and its preganglionic and postganglionic neurons is shown in **Figure 14.4.** Notice that most of the cell bodies of the postganglionic neurons are found in a series along the vertebral column. Their chain-like appearance has given them the name **sympathetic chain ganglia.** The sympathetic chain ganglia extend beyond the thoracic and lumbar spinal cord, from the *superior cervical ganglion* down to the *inferior sacral ganglion.*

Preganglionic neurons originate in the lateral horns of the thoracic and lumbar spinal cord. The axons of preganglionic neurons exit the spinal cord with the axons of lower motor neurons via the anterior root. They then travel with the spinal nerve and the anterior ramus for a short distance before branching off to form small nerves called **white rami communicantes** (kuh-myoo'-nih-KAN-teez; singular, *communicans;* called *white* because they are myelinated), as shown here:

The axons in the white rami communicantes then enter the sympathetic chain ganglia that house the cell bodies of the postganglionic sympathetic neurons.

Some preganglionic neurons pass through the sympathetic chain ganglia without synapsing. Instead, they may synapse on cell bodies in different chain ganglia, or on **collateral ganglia,** which are located near the target organ. Many collateral ganglia are located near the aorta, and for this reason are called *preaortic ganglia.* Other collateral ganglia are located near the organs of the abdominopelvic cavity. The preganglionic axons that synapse here are part of nerves called the **splanchnic nerves** (SPLANK-nik; *splanchn-* = "viscera") that synapse on ganglia including the

Sympathetic preganglionic neurons supplying the head, neck, and thoracic viscera synapse on sympathetic chain ganglia.

Axons of postganglionic neurons

To structures of the head and neck

Superior cervical ganglion

Sympathetic chain ganglia

Spinal cord T_1

To thoracic viscera

Axons of postganglionic neurons

Collateral ganglia:
Celiac ganglion
Superior mesenteric ganglion

To abdominal viscera

Inferior mesenteric ganglion

Cell bodies of preganglionic neurons

Axons of preganglionic neurons

Axons of splanchnic nerves

Aorta

Sympathetic preganglionic neurons supplying the abdominal viscera extend beyond chain ganglia, forming splanchnic nerves that synapse on collateral ganglia on the aorta.

Practice art labeling
@MasteringA&P®

Figure 14.4 Organization of the sympathetic nervous system.

celiac ganglion, the *superior mesenteric ganglion,* and the *inferior mesenteric ganglion.*

To summarize, a preganglionic neuron may synapse with a postganglionic neuron in one of three ways (**Figure 14.5**):

- The axon synapses with a postganglionic neuron in the sympathetic chain ganglion;
- the axon descends or ascends and synapses with a postganglionic neuron in a different chain ganglion; or
- the axon passes through the chain ganglion and synapses with a postganglionic neuron in a collateral ganglion.

The postganglionic neuron then innervates its target cells. Figure 14.5 shows that some of these postganglionic axons exit the ganglia as small **gray rami communicantes** (called *gray* because they are unmyelinated) that rejoin and travel with

spinal nerves to reach their target cells. Postganglionic axons that take the first two pathways exit through the gray rami communicantes, whereas those that take the third pathway travel to their target cells directly. The largest population of postganglionic axons that take this third path are the splanchnic nerves.

Quick Check

☐ 1. Where are the cell bodies of sympathetic preganglionic neurons located? Where are their axons located, and what do the axons contact?

☐ 2. Where are the cell bodies of sympathetic postganglionic neurons located? Where are their axons located, and what do the axons contact?

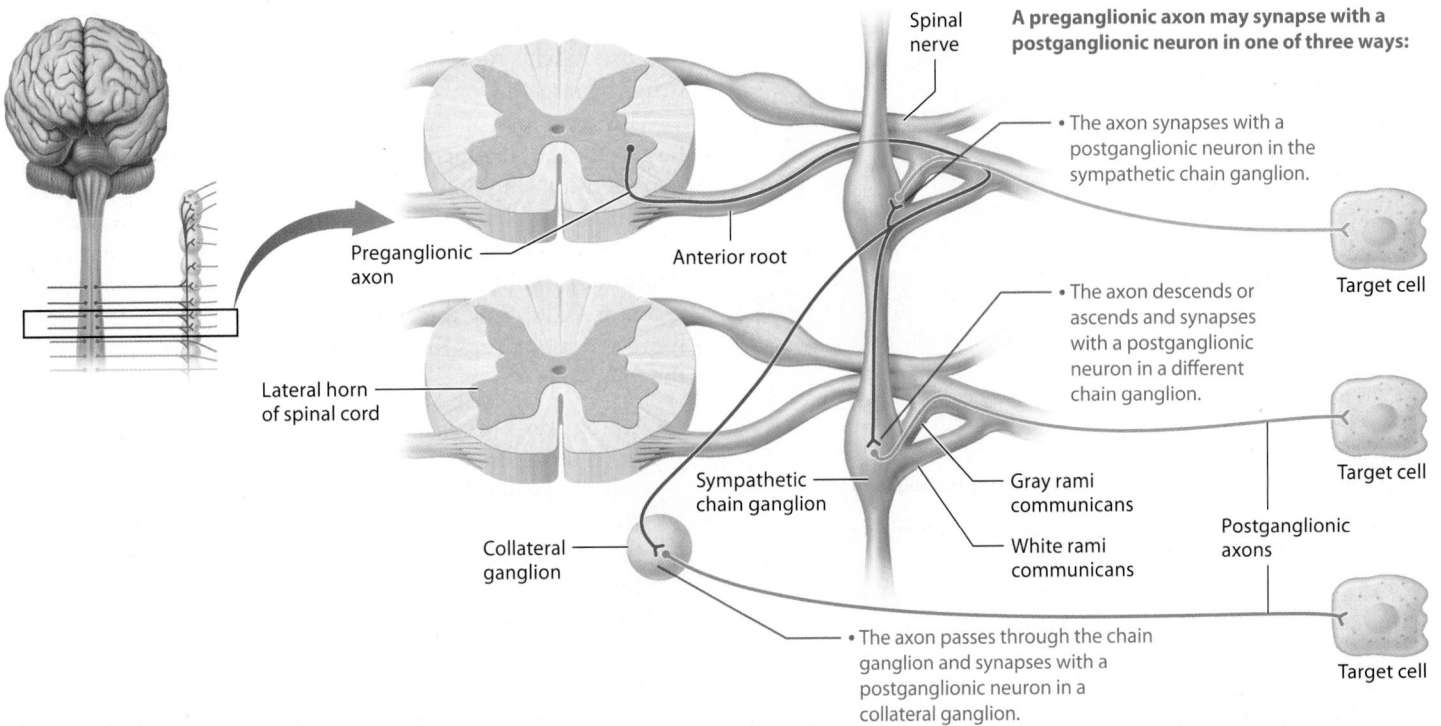

Figure 14.5 Three possible pathways of sympathetic preganglionic and postganglionic neurons.

Sympathetic Neurotransmitters and Receptors

◀◀ **Flashback**

1. What are the functions of the neurotransmitters norepinephrine, epinephrine, and acetylcholine? (p. 415)
2. What is a neurotransmitter receptor? What happens when a neurotransmitter binds to a receptor? (p. 414)

Recall that neurons communicate with other neurons and target cells through the release of chemicals called *neurotransmitters* (see Chapter 11). A neurotransmitter interacts with a target cell by binding to a protein-based receptor for that neurotransmitter embedded in the plasma membrane of the target cell. The following subsections discuss the neurotransmitters released by sympathetic neurons and the receptors to which these neurotransmitters bind on their target cells.

Classes of Sympathetic Neurotransmitters

The sympathetic preganglionic axon communicates with the postganglionic neuron by way of an excitatory synapse using the neurotransmitter **acetylcholine (ACh).** This is the same neurotransmitter released by somatic motor neurons, as we discussed in the muscle tissue chapter (see Chapter 10). The unmyelinated axons of the postganglionic neurons then carry the message in the form of an action potential to their target cells.

At the synapses with their target cells, postganglionic axons release one of three neurotransmitters: **norepinephrine,**

epinephrine, or ACh. Norepinephrine and epinephrine are also known as *noradrenalin* and *adrenalin,* respectively; these names derive from the *adrenal gland,* where these chemicals were first discovered. Approximately 80% of postganglionic sympathetic neurons release norepinephrine, and the remainder release epinephrine or ACh.

Classes of Sympathetic Receptors

The neurotransmitter receptors that bind to norepinephrine or epinephrine are called **adrenergic receptors** (ad-ren-ER-jik; the name comes from their ability to bind adrenalin). Those that bind to ACh are **cholinergic receptors** (kohl-in-ER-jik).

Adrenergic Receptors There are two major types of adrenergic receptors: **alpha (α) receptors** and **beta (β) receptors.** These receptors are further classified according to the structure of their proteins. The subtypes include the following:

● **Alpha-1 (α_1) receptors** are located in the plasma membranes of smooth muscle cells of many different organs, including the blood vessels that supply the skin, organs of the gastrointestinal system, and the kidneys. These receptors are also found on arrector pili muscles in the dermis, the uterus (during pregnancy), and certain organs of the genitourinary system.
● **Alpha-2 (α_2) receptors** are different from the other adrenergic receptors—we find most of them in the membrane of the preganglionic sympathetic neuron rather than in a peripheral target cell. Recall that normally an action potential in a preganglionic neuron leads to ACh release, which stimulates the

(a) α₂ receptor with no NE

When α₂ receptors are not bound to NE, the preganglionic neuron releases ACh, exciting the postganglionic neuron.

(b) α₂ receptor bound to NE

When α₂ receptors bind NE, the preganglionic neuron hyperpolarizes, decreasing ACh release and postganglionic neuron activity.

Figure 14.6 The effect of α₂ receptors on preganglionic neurons.

postganglionic neuron (**Figure 14.6a**). However, notice in **Figure 14.6b** that when norepinephrine binds to α₂ receptors, the axon terminal is hyperpolarized, slowing or even canceling the action potential. As a result, the preganglionic neuron stops stimulating the postganglionic neuron, which dampens or even shuts off the sympathetic response. This is part of a negative feedback loop that prevents excessive sympathetic activity, reminding us of the Feedback Loops Core Principle (p. 21). α₂ receptors are also found in the plasma membranes of certain sympathetic target cells, including cells in the pancreas and adipose tissue.

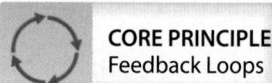

CORE PRINCIPLE
Feedback Loops

- **Beta-1 (β₁) receptors** are located in the plasma membranes of cardiac muscle cells as well as certain cells of the kidney and adipose tissue.
- **Beta-2 (β₂) receptors** are found in the plasma membranes of smooth muscle cells lining the airway passages in the lungs (the *bronchioles*). β₂ receptors are also found on skeletal muscle fibers, smooth muscle cells of the urinary bladder, smooth muscle cells lining the blood vessels serving skeletal muscles, cells of the liver and pancreas, and cells of the salivary glands.
- **Beta-3 (β₃) receptors** are located predominantly on the cells of adipose tissue and smooth muscle cells in the wall of the digestive tract.

Cholinergic Receptors There are two types of cholinergic receptors: the **muscarinic receptor** (muss-kah-RIN-ik; so named because it binds the poison muscarine) and the **nicotinic receptor** (nik-oh-TIN-ik; so named because it binds the poison found in tobacco, nicotine). Muscarinic receptors are found on sweat glands in the skin. Nicotinic receptors are located in the membranes of all postganglionic neurons within sympathetic ganglia and the adrenal medulla.

Pharmacology and Sympathetic Nervous System Receptors

The existence of different subtypes of sympathetic nervous system receptors has allowed researchers to design drugs that are

fairly specific for one type of receptor. This is advantageous because it helps to minimize the potential side effects from the drugs by targeting only certain organs.

The receptors of the sympathetic nervous system provide targets for the treatment of many medical conditions, including hypertension (high blood pressure) and the respiratory disease asthma. Most of these drugs work in one of two ways: Either they are *antagonists* that block the receptor and prevent norepinephrine from binding to it, or they are *agonists* that bind the receptor and mimic the effects of norepinephrine. Some of the more common drugs that bind to sympathetic receptors include the following:

- α₁ *blockers* (*antagonists*) bind to α₁ receptors, particularly those on the smooth muscle cells lining blood vessels. They block the action of norepinephrine and prevent the blood vessels from constricting, an effect that lowers blood pressure and is useful in treating hypertension. Certain α₁ blockers also cause relaxation of the smooth muscle in the prostate gland; these are used to treat benign prostatic hyperplasia (see Chapter 26).
- α₂ *agonists* bind to the presynaptic α₂ receptors and activate them, which decreases the output of both preganglionic and postganglionic sympathetic neurons. These drugs may be used in the treatment of hypertension and other conditions such as opiate withdrawal.
- β-*blockers* are antagonists that bind to β₁ receptors on the heart and decrease its rate and force of contraction. These drugs are widely used in the treatment of hypertension and other diseases of the cardiovascular system.
- β₂ *agonists* bind to β₂ receptors on the smooth muscle of the airways and cause bronchodilation. These drugs are commonly used to treat asthma.

Quick Check

- ☐ 3. Which neurotransmitter(s) are released by sympathetic preganglionic neurons? Which are released by sympathetic postganglionic neurons?
- ☐ 4. What types of receptors do these neurotransmitters bind?
- ☐ 5. What are agonists and antagonists?

Effects of the Sympathetic Nervous System on Target Cells

◀◀ Flashback

1. What are the differences between smooth, skeletal, and cardiac muscle tissue? (p. 148)
2. What are endocrine and exocrine glands? (p. 135)

The effects of the sympathetic nervous system on its target cells are shown in **Figure 14.7**. These effects are easy to understand if you consider the system's most basic function: to ensure

TARGET	NT	RECEPTOR	MAIN EFFECTS
Smooth muscle cells around pupil	NE	α_1	Dilation of pupil
Cells of salivary glands	NE	β_1 and β_2	Increase in secretion in certain cells
Cardiac muscle cells	NE	β_1	Increase in heart rate and force of contraction
Smooth muscle cells of bronchus	NE	β_2	Dilation of bronchioles (bronchodilation)
Cells of sweat glands	ACh	Muscarinic	Increase in secretion
Smooth muscle cells of blood vessels to skin	NE	α_1	Constriction of blood vessels (vasoconstriction)
Smooth muscle cells of blood vessels to skeletal muscles	NE	β_2	Dilation of blood vessels (vasodilation)
Smooth muscle cells of blood vessels to digestive and urinary organs	NE	α_1	Vasoconstriction
Smooth muscle cells of digestive and urinary tracts and sphincters	NE	α_1 (sphincters) β_3 (digestive) β_2 (urinary)	Relaxation of digestive and urinary tracts, contraction of sphincters
Cells of digestive glands	NE	β_2	Decrease in secretion
Cells of pancreas and liver	NE	β_2	Increase in release of glucose
Cells of adrenal medulla	ACh	Nicotinic	Release of epinephrine and norepinephrine

Labels on diagram: Eye, Salivary glands, Heart, Lung, Bronchus, Sweat gland, Collateral ganglion, Blood vessels, Liver, Sympathetic chain ganglia, Stomach, Pancreas, Large intestine, Adrenal gland, Kidney, Small intestine, Rectum, Urinary bladder

Figure 14.7 The main effects of the sympathetic nervous system on target cells. Note that NT stands for neurotransmitter, NE for norepinephrine, and ACh for acetylcholine.

Practice art labeling @MasteringA&P®

survival and maintain homeostasis during times of physical or emotional stress. Think about what happens to your body when you're running a race—your heart beats faster and harder, you begin to sweat, and your blood pressure increases, among many other changes. Much the same happens, although to a lesser degree, when you're nervous about a big exam or excited about an upcoming vacation. Each of these situations triggers a release of norepinephrine, epinephrine, and ACh from pre- and postganglionic sympathetic neurons, which in turn triggers changes in the cells of the target organs and tissues.

Bear in mind that the changes mediated by the sympathetic nervous system are intended to promote survival in an emergency situation. If you have trouble remembering this, try the following mnemonic: You are walking through the woods when suddenly a swarm of bees starts chasing you. Your nervous system is *sympathetic* to your situation and so is going to initiate changes in your body, via your sympathetic nervous system, that will make it easier to outrun the bees. In the upcoming subsections, we'll examine specific effects of the sympathetic nervous system on the body's cells. Since most adrenergic synapses use norepinephrine, we will refer primarily to this neurotransmitter, although a small percentage of postganglionic neurons release epinephrine.

Effects on Cardiac Muscle Cells

As you already know, when you are nervous or exercising your heart rate and blood pressure rise. This is due to the effect of norepinephrine binding to β_1 receptors on cardiac muscle cells, which opens ion channels that raise both the rate and force of contractions. This heightens not only the amount of blood delivered to the tissues, but also blood pressure, which maintains homeostasis during periods of increased activity.

Effects on Smooth Muscle Cells

The smooth muscle cells of many organs also have receptors for norepinephrine. When norepinephrine binds, it mediates the following changes:

- **Constriction of blood vessels serving the digestive, urinary, and integumentary systems.** Norepinephrine binds to α_1 receptors on the smooth muscle cells of blood vessels serving the organs of the digestive, urinary, and integumentary systems and causes them to contract. This narrows the blood vessels, an action called *vasoconstriction* (vay'-zoh-kon-STRIK-shun; *vaso-* = "vessel"), which decreases the blood flow to those organs in order to divert blood flow to temporarily "more important" tissues, such as skeletal and cardiac muscle. This wide-scale vasoconstriction also increases the overall blood pressure. Decreasing blood flow to the urinary and digestive organs slows the rate of urine formation and digestion, respectively (if you are running from a bear, digesting the lunch you just ate can wait!). Reduced blood flow to the skin explains why people with light-colored skin "turn white as a sheet" when they are scared—as the blood vessels serving the skin constrict, the pinkish color of the blood under the skin is less visible and the skin appears paler.

- **Dilation of the bronchioles.** During times of stress more oxygen needs to be delivered to the body's cells, especially the skeletal muscle cells, in order to keep up with their metabolic demands. The sympathetic nervous system accomplishes this through norepinephrine's effect on the smooth muscle cells lining the smaller airway passages (the *bronchioles*). When norepinephrine binds to the β_2 receptors on these cells, they relax. This action, *bronchodilation* (brong'-koh-dy-LAY-shun), increases the diameter of the airway passages and increases the amount of oxygen that can be inhaled with each breath. Although the sympathetic nervous system does increase oxygen intake, note that this system has no effect on the *rate* of ventilation. The rate does tend to increase during exercise and emergency situations, but this is due to a mechanism independent of the sympathetic nervous system (see Chapter 21).

- **Dilation of blood vessels serving the skeletal and cardiac muscle.** Although norepinephrine constricts most of the body's blood vessels, when it binds to β_2 receptors on the smooth muscle cells of blood vessels serving skeletal and cardiac muscle, it causes them to relax. This opens the blood vessels, an action called *vasodilation* (vay'-zoh-dy-LAY-shun), which increases the blood flow and the delivery of oxygen and nutrients to the skeletal muscles. If this vasodilation occurred alone, the blood pressure would decrease. However, because it occurs simultaneously with the vasoconstriction of nearly every other blood vessel in the body, the effect of skeletal muscle blood vessel vasodilation on blood pressure is minimal.

- **Contraction of urinary and digestive sphincters.** The binding of norepinephrine to the smooth muscle cells of the sphincters of the urinary and digestive systems causes them to contract. This makes emptying the bowel and bladder more difficult during times of exercise, stress, or emergency.

- **Relaxation of the smooth muscle of the digestive tract.** Although stimulation by the sympathetic nervous system causes the smooth muscle of the digestive tract sphincters to contract, when norepinephrine binds to β_3 receptors it causes relaxation of the remainder of the smooth muscle in the digestive tract. This decreases the movement of food through the digestive system and slows digestive processes overall.

- **Dilation of the pupils.** The dilator pupillae muscles are innervated by sympathetic nervous system neurons. When norepinephrine binds to the α_1 receptors on these smooth muscle cells, it causes them to contract, which dilates the pupils and allows more light to enter the eyes.

- **Constriction of blood vessels serving most exocrine glands.** The sympathetic nervous system nearly universally decreases secretion from exocrine glands—apart from sweat glands—via norepinephrine binding to β_2 receptors. This binding affects the exocrine cells of the digestive system, including the salivary glands around your mouth and the cells that secrete digestive products. This effect on salivary glands is why your mouth is often dry when you're nervous. The decreased secretion is generally a result of reduced blood flow to these glands rather than a direct effect on the gland itself.

Effects on Cellular Metabolism

During times of sympathetic nervous system activation, nearly all cells, and in particular skeletal muscle cells, consume dramatically more ATP as their metabolic rates increase. To ensure that skeletal muscle cells have a steady supply of fuels with which to make ATP, norepinephrine brings about the following three main effects: (1) It binds to β_3 receptors on adipocytes and triggers the breakdown of lipids, which releases free fatty acids into the bloodstream; (2) it binds to β_2 receptors on cells of the liver and triggers the release of glucose from glycogen and also the synthesis of glucose from other precursors; and (3) it binds to β_2 receptors on cells of the pancreas and triggers the release of a hormone called *glucagon* that increases the blood glucose level. Given all the effects of the sympathetic nervous system on nutrient metabolism, it's not surprising that sympathetic receptors are purported targets for weight loss supplements. See *A&P in the Real World: The Sympathetic Nervous System and Weight Loss Supplements* for more information.

Effects on Secretion from Sweat Glands

When you're nervous or exercising, you sweat. This useful (but often annoying) attempt to help the body maintain temperature

The Sympathetic Nervous System and Weight Loss Supplements

As you just read, one effect of the sympathetic nervous system is to increase the metabolic rate, or increase the rate at which ATP is produced and consumed. This has led dietary supplement manufacturers to create products intended to result in weight loss by capitalizing on this effect of the sympathetic nervous system. These products are generally sold as topical creams that the user rubs over "problem areas" with excess adipose tissue.

One chemical found in many of these creams is *yohimbine* (yoh-HIM-been), which is the active ingredient derived from a plant called *yohimbe* (yoh-HIM-bay). Manufacturers of the creams claim that yohimbine binds to β_3 receptors on adipocytes and triggers the breakdown of lipids. This is, at best, a misleading statement. Yohimbine actually blocks α_1 receptors in blood vessels and α_2 receptors in the spinal cord. This has a dual effect: It causes vasodilation while also increasing the activity of sympathetic neurons. The higher sympathetic activity can, briefly, increase the metabolic rate. However, it can also dangerously elevate the heart rate; cause seizures, high blood pressure, and kidney failure; and lead to insomnia and panic attacks.

Fortunately, yohimbine is not actually absorbed through the epidermis in any significant amount, so it never reaches the blood vessels in the dermis and hypodermis. This limits the amount of harm it can do to your body, although the supplement can still do significant harm to your wallet.

homeostasis is due to postganglionic sympathetic neurons that release ACh onto the cells of sweat glands in the skin. ACh binds to these cells' muscarinic receptors and increases their secretions. This occurs as part of a negative feedback loop that controls body temperature, which reminds us of the Feedback Loops Core Principle (p. 21).

CORE PRINCIPLE
Feedback Loops

Effects on Cells of the Adrenal Medulla

The **adrenal medulla** (ah-DREE-nul meh-DOOL-ah) is the internal part of the adrenal glands, which sit atop each kidney. It's unique in that it is contacted directly by preganglionic sympathetic neurons. This is because the adrenal medulla is not like the remainder of the adrenal gland—instead of glandular epithelium, it actually consists of modified sympathetic postganglionic neurons. In other words, each adrenal medulla is functionally a ganglion.

As shown in **Figure 14.8**, (1) when the preganglionic neuron releases ACh, (2) it binds to the nicotinic receptors of adrenal medulla cells. (3) ACh then stimulates the medullary cells to release additional epinephrine and norepinephrine into the bloodstream. Although norepinephrine and epinephrine are usually considered to be neurotransmitters, they can both also function as *hormones,* long-distance chemical messengers. This means that the adrenal medulla effectively acts as an interface between the sympathetic nervous system and the endocrine system (see Chapter 16).

The ratio of chemicals the adrenal medulla produces is generally opposite that of the nervous system, with about 80% epinephrine and 20% norepinephrine. This additional epinephrine and norepinephrine is useful in many ways. For one, it prolongs the duration of the sympathetic nervous system's effects. The effects

Figure 14.8 Sympathetic nervous system stimulation of the adrenal medulla.

of neurotransmitters released by postganglionic neurons last only a few seconds, whereas in the blood the effects of these chemicals continue for several minutes. A second benefit is that the epinephrine and norepinephrine in the blood can reach cells that are not innervated by sympathetic nervous system neurons. This allows the sympathetic nervous system to affect the entire body without innervating every body cell. Finally, the adrenal medulla acts as a "backup" for the sympathetic nervous system—even if the pathways connecting this system with many of its target organs are disrupted, it can still affect these organs indirectly via the adrenal medulla.

Effects on Other Cells

The sympathetic nervous system influences many of the body's other processes via its effects on target cells. For instance, this system enhances mental alertness by increasing neuron activity in the association areas of the cerebral cortex. Stimulation by this system also increases the blood's tendency to clot, which can be useful if a person is injured during a "fight" or a "flight" situation. In addition, the sympathetic nervous system can temporarily increase the tension generated by skeletal muscle cells during a muscle contraction, which is why people have been known to perform unusual feats of strength under the influence of an "adrenaline (epinephrine) rush." Furthermore, the postganglionic sympathetic neurons trigger contraction of the arrector pili muscles, which produces "goose bumps." Finally, these neurons also cause ejaculation of semen via effects on smooth muscle cells of the male reproductive ducts.

Quick Check

☐ 6. Describe the effects of the sympathetic nervous system on each of the following tissues:
   a. Cardiac muscle
   b. Blood vessels serving skeletal muscles
   c. Sweat glands
   d. Blood vessels serving digestive and urinary organs
   e. Smooth muscle lining the airways

Apply What You Learned

☐ 1. A nurse injects a patient with a drug that selectively binds to and activates the α_2 adrenergic receptors. What will happen to the sympathetic nervous system response? Why? What useful applications might such a drug have?

☐ 2. If there were a drug that selectively blocked nicotinic receptors on sympathetic postganglionic neurons, the effects observed in a patient would be similar to those of the drug administered in question 1. Explain why.

☐ 3. Mr. Nguyen is in an accident that destroys the sympathetic ganglia housing the neurons that innervate his heart. However, you notice that his heart rate and blood pressure increase in response to exercise. What is causing this increase?

See answers in Appendix A.

The Parasympathetic Nervous System

Learning Outcomes

1. Identify the role of the parasympathetic nervous system, and explain how it maintains homeostasis.
2. Describe the anatomy of the parasympathetic nervous system, including central nervous system origin locations, ganglia locations, and relative lengths of pre- and postganglionic neurons.
3. Describe the neurotransmitters and neurotransmitter receptors of the parasympathetic nervous system.
4. Describe the effects of the parasympathetic nervous system on its target cells.

As we discussed, the **parasympathetic nervous system** is often characterized as the "rest and digest" division of the ANS. This relates to its role in the body's maintenance functions, such as digestion and urine formation, that are typically carried out during periods of rest. Like the sympathetic nervous system, this system is rarely completely silent, although it is in general considered subordinate to the sympathetic nervous system. This module explores the anatomy and physiology of the parasympathetic nervous system.

Gross and Microscopic Anatomy of the Parasympathetic Nervous System

In the parasympathetic nervous system, or craniosacral division, the cranial nerves involved are the oculomotor (III), facial (VII), glossopharyngeal (IX), and vagus (X) nerves. The sacral nerves (S_2-S_4) make up the pelvic nerves. **Figure 14.9** shows the organization of this division.

As with preganglionic sympathetic neurons, the axons of preganglionic parasympathetic neurons synapse with postganglionic neurons located within ganglia. The ganglia, known as *terminal ganglia*, are typically found near the postganglionic neurons' target cells, and the postganglionic axons are in general fairly short.

Parasympathetic Cranial Nerves

As we discussed, four cranial nerves carry parasympathetic fibers (see Chapter 13). The main parasympathetic nerves are the two vagus nerves (CN X), which innervate most thoracic and abdominal viscera and together provide about 90% of the parasympathetic innervation to the body. Branches of the vagus nerve contribute to several groups of nerves or *plexuses* that innervate specific organs. For example, the *cardiac plexus* innervates the heart; the *pulmonary plexus,* the lungs and airway passages; and the *esophageal plexus,* the esophagus. These preganglionic fibers typically synapse on terminal ganglia in the walls of the organ being innervated.

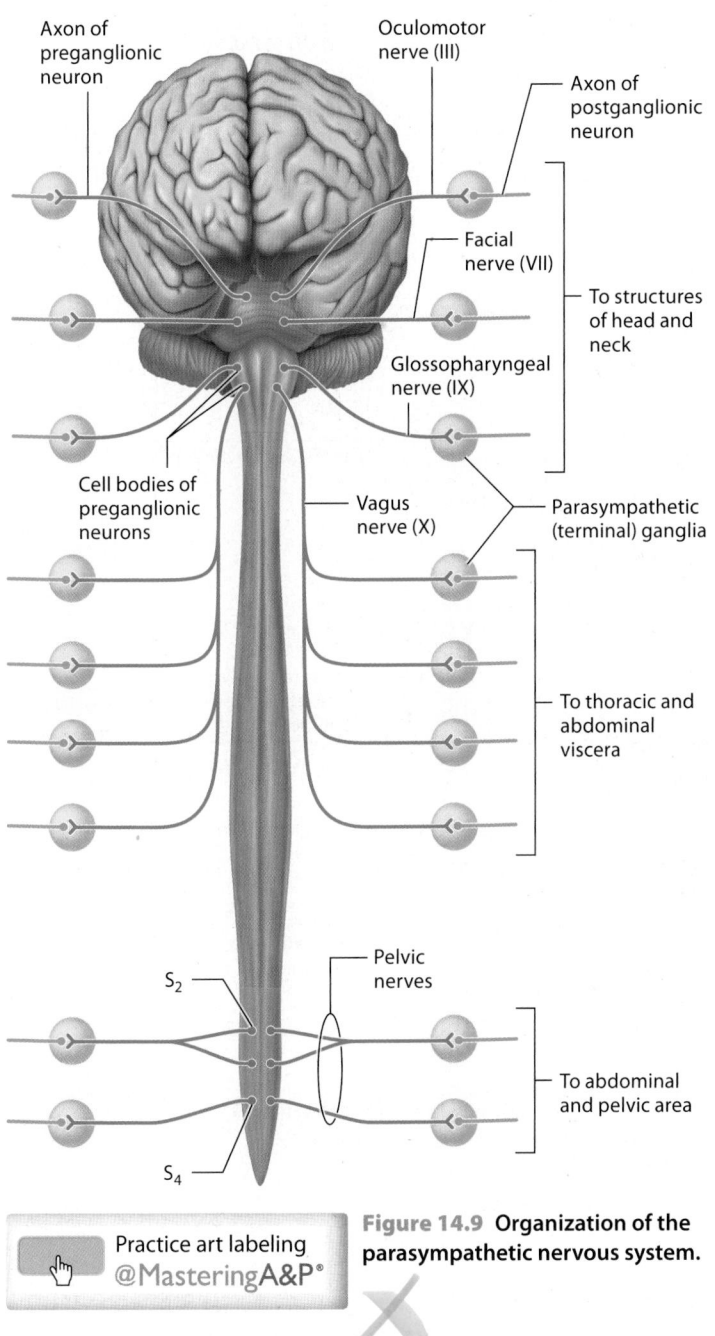

Figure 14.9 **Organization of the parasympathetic nervous system.**

Practice art labeling
@MasteringA&P®

Parasympathetic Sacral Nerves

The parasympathetic cranial nerves supply most of the viscera of the thoracic and abdominopelvic cavities. The remaining organs, including the last segment of the large intestine, the urinary bladder, and the reproductive organs, are supplied by parasympathetic sacral nerves. Branches from the sacral spinal cord form the **pelvic sphlanchnic nerves,** which in turn form plexuses in the pelvic floor. Some preganglionic neurons synapse in terminal ganglia in these plexuses, but most synapse in terminal ganglia in the walls of the organs.

Quick Check

☐ 1. What is the basic function of the parasympathetic nervous system?
☐ 2. Where are the preganglionic parasympathetic cell bodies located? Where are the postganglionic parasympathetic cell bodies located?
☐ 3. How does the arrangement of neurons in the sympathetic nervous system differ from that in the parasympathetic nervous system?

Parasympathetic Neurotransmitters and Receptors

Both the preganglionic and postganglionic parasympathetic neurons release ACh at their synapses, and the effect is generally excitatory. As we saw in the sympathetic nervous system, there are two types of cholinergic receptors: nicotinic and muscarinic. Nicotinic receptors are located in the membranes of all postganglionic parasympathetic neurons, whereas muscarinic receptors are found in the membranes of all parasympathetic target cells. Next we examine what happens when ACh binds to these muscarinic receptors.

Quick Check

☐ 4. Which neurotransmitter is released by all parasympathetic neurons?
☐ 5. Which cells contain nicotinic receptors? Which contain muscarinic receptors?

Effects of the Parasympathetic Nervous System on Target Cells

As with the sympathetic nervous system, the effects of the parasympathetic nervous system on its target cells are easily understood if you keep in mind its most basic function: to maintain homeostasis when the body is at rest. Consider what happens after you eat dinner and sit down to read your anatomy and physiology textbook—your heart rate slows, your blood pressure decreases, you digest your food, and your eyes adjust for near vision in order to read the book. Each effect results from a decrease in sympathetic nervous system activity and the release

Other cranial nerves also carry parasympathetic fibers. The parasympathetic preganglionic neurons of the oculomotor nerves (CN III) synapse on terminal ganglia called the **ciliary ganglia** (SIL-ee-air-ee). Several terminal ganglia, including the **submandibular ganglia** and the **pterygopalatine ganglia** (ter′-ee-goh-PAL-a-teen), house the cell bodies of sensory neurons and are the sites where preganglionic parasympathetic neurons of the facial nerves (CN VII) synapse on postganglionic parasympathetic neurons. The preganglionic parasympathetic neurons of the glossopharyngeal nerves (CN IX) synapse on postganglionic parasympathetic neurons in small terminal ganglia called the *otic ganglia* (OH-tik).

	TARGET	MAIN EFFECTS
Eye	Smooth muscle cells around pupil and lens	Constriction of pupil, adjustment of lens for near vision
Lacrimal gland / Salivary glands	Cells of lacrimal and salivary glands	Increase in secretions
Heart	Cardiac muscle cells	Decrease in heart rate
Lung / Bronchiole	Smooth muscle cells of bronchioles	Bronchoconstriction
Liver / Stomach / Gallbladder / Pancreas / Large intestine	Smooth muscle cells of digestive tract and sphincters	Contraction of digestive tract smooth muscle, relaxation of sphincters
Small intestine	Cells of digestive glands	Increase in secretions
Kidney	Smooth muscle cells of urinary sphincters	Relaxation of sphincters
Urinary bladder		

Oculomotor nerve (III), Facial nerve (VII), From brainstem, Glossopharyngeal nerve (IX), Vagus nerve (X), Pelvic nerves, From sacral spinal cord

Figure 14.10 The main effects of the parasympathetic nervous system on target cells. All of these effects are stimulated by ACh at nicotinic receptors on postganglionic neurons and muscarinic receptors in target organs.

Practice art labeling @MasteringA&P®

of ACh from pre- and postganglionic parasympathetic neurons. The main effects of the parasympathetic nervous system are shown in **Figure 14.10**.

Effects on Cardiac Muscle Cells

Since the sympathetic nervous system increases heart rate and blood pressure, it follows that the parasympathetic nervous system decreases them. Preganglionic parasympathetic neurons travel to the heart via the vagus nerve (CN X) and stimulate

postganglionic neurons to reduce the rate of the heart's contraction, which lowers blood pressure.

Effects on Smooth Muscle Cells

Postganglionic parasympathetic neurons innervate the smooth muscle cells in many different organs, and trigger the following effects:

● **Constriction of the pupil.** Postganglionic parasympathetic fibers of the oculomotor nerve in the ciliary ganglion innervate the sphincter pupillae muscle, which constricts the pupil and so allows less light into the eye.

- **Accommodation of the lens for near vision.** Postganglionic parasympathetic fibers of the oculomotor nerve also innervate the smooth muscle cells of the ciliary muscle. When this muscle contracts, the lens becomes rounder, allowing accommodation for near vision.
- **Constriction of the bronchioles.** When vagal parasympathetic neurons stimulate contraction of the smooth muscle cells lining the airways, the airways narrow, a response called *bronchoconstriction* (brong′-koh-kon-STRIK-shun). This effect is mild, except with very strong stimulation.
- **Contraction of the smooth muscle lining the digestive tract.** Parasympathetic neurons of the vagus nerve also trigger contraction of the smooth muscle cells lining the digestive tract. This produces rhythmic contractions known as *peristalsis* that help to propel ingested food from one part of the digestive tract to the next.
- **Relaxation of digestive and urinary sphincters.** Parasympathetic neurons from the vagus and sacral nerves stimulate relaxation of the smooth muscle cells of the urinary and digestive sphincters. These actions promote urination, as well as the movement of digesting food from one area of the digestive tract to the next, ultimately resulting in defecation.
- **Engorgement of the penis or clitoris.** In men, the blood vessels in the penis receive parasympathetic innervation, as do the clitoral blood vessels in women. When stimulated by the preganglionic parasympathetic neurons of the sacral division, postganglionic neurons trigger the smooth muscle cells in these vessels to relax, which leads to vasodilation. This allows the penis or clitoris to engorge with blood.

Unlike the sympathetic nervous system, the parasympathetic nervous system innervates virtually no blood vessels except in specific areas of the body such as the penis. Many blood vessels do dilate when the parasympathetic nervous system is active, particularly those in the digestive and urinary systems. However, this dilation is due to a decrease in sympathetic nervous system activity and removal of epinephrine from the bloodstream rather than any direct action by parasympathetic neurons.

Effects on Glandular Epithelial Cells

Whereas the sympathetic nervous system promotes sweating while decreasing secretion from other exocrine glands, the parasympathetic nervous system has little to no effect on sweat glands; instead, it triggers copious secretion from other glands. For example, parasympathetic neurons of the facial nerve stimulate tear production from the lacrimal (tear) glands and mucus production from glands in the nasal mucosa. Fibers from the both the facial and glossopharyngeal nerves stimulate secretion of saliva from the salivary glands. Finally, fibers from the vagus nerve stimulate secretion of enzymes and other products from the cells of the digestive tract.

Effects on Other Cells

Unlike the sympathetic nervous system, the parasympathetic nervous system has no direct effect on cells that mediate metabolic rate, mental alertness, the force generated by skeletal muscle contractions,

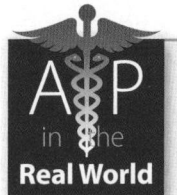

Side Effects of Anticholinergic Drugs

Many drugs block either ACh release or ACh receptors as an unintended side effect. Such drugs are prescribed for many different conditions, such as allergies, respiratory diseases, and gastrointestinal conditions. These *anticholinergic drugs* have a wide range of potential effects. Some of the more common unwanted effects are the following:

- *Urinary retention.* Anticholinergics block the relaxing effect that the parasympathetic nervous system has on the urinary sphincters, and makes passing urine more difficult.
- *Constipation.* Anticholinergics also block the parasympathetic nervous system's effect on the smooth muscle of the digestive tract. This causes digested food to move more slowly through the tract and can lead to constipation.
- *Dry mouth.* One of the most common anticholinergic effects is a dry mouth, caused by a decrease in the parasympathetic nervous system's ability to stimulate the secretion of saliva.

blood clotting, adipocytes, or most endocrine secretions. Instead, each of these factors returns to a normal, or "resting," state during times of parasympathetic activity simply because the sympathetic nervous system is no longer dominant. This return to the resting state is an important time for the body to store glucose and other fuels in preparation for the next round of sympathetic nervous system activity.

Some drugs have the side effect of blocking cholinergic synapses, and therefore many effects of the parasympathetic nervous system. For more information, see *A&P in the Real World: Side Effects of Anticholinergic Drugs.*

Quick Check

☐ 6. How does the parasympathetic nervous system affect blood pressure?
☐ 7. Summarize the effects of the parasympathetic nervous system on the smooth muscle cells of the following structures:
   a. Pupil
   b. Lens
   c. Blood vessels serving the penis
   d. Digestive tract
   e. Urinary and digestive sphincters

Apply What You Learned

☐ 1. The drug *atropine,* derived from the poisonous plant deadly nightshade, blocks muscarinic receptors. Predict the effects of administering this drug.
☐ 2. Drugs that stimulate ACh release often have adverse effects, such as drooling, poor distance vision, and diarrhea. Explain these effects, considering the functions of the parasympathetic nervous system.

See answers in Appendix A.

14.4

Homeostasis Part II: PNS Maintenance of Homeostasis

Learning Outcomes

1. Define sympathetic and parasympathetic tone.
2. Explain how the nervous system as a whole regulates homeostasis.

The sympathetic and parasympathetic nervous systems do not function in isolation. Instead, they act together to keep many of the body's regulated variables within their normal homeostatic ranges. In this module, we focus on the "working relationship" between the two systems. We conclude with a view of how the nervous system as a whole maintains homeostasis.

Interactions of Autonomic Divisions

As you have discovered, the sympathetic and parasympathetic nervous systems generally work antagonistically, each having the opposite effect on a particular body function or organ. Together the two maintain a balance that ensures the body's needs are met appropriately at all times. They can accomplish this because most organs are innervated by neurons from both systems, referred to as *dual innervation*. This allows the sympathetic nervous system to become dominant and trigger effects that maintain homeostasis during exercise or emergency, and the parasympathetic nervous system to regulate these same organs and preserve homeostasis when the emergency or exercise has finished. See **Figure 14.11** for a summary and comparison of the structural and functional features of the sympathetic and parasympathetic nervous systems.

SYMPATHETIC NERVOUS SYSTEM		PARASYMPATHETIC NERVOUS SYSTEM
"Fight or flight" — maintains homeostasis during exercise, emotion, and emergency	MAIN ROLE	"Rest and digest" — homeostasis of maintenance functions such as digestion and formation of urine
Thoracolumbar region of spinal cord	LOCATION OF PREGANGLIONIC CELL BODIES	Craniosacral region of spinal cord
Mostly in ganglia near the vertebral column	LOCATION OF POSTGANGLIONIC CELL BODIES	In ganglia near the target cell
Acetylcholine (ACh)	PREGANGLIONIC NEUROTRANSMITTERS	Acetylcholine (ACh)
Nicotinic	POSTGANGLIONIC RECEPTORS	Nicotinic
Norepinephrine (NE) Epinephrine ACh	POSTGANGLIONIC NEUROTRANSMITTERS	ACh
Adrenergic Muscarinic	TARGET CELL RECEPTORS	Muscarinic

Figure 14.11 Comparison of structures and effects of sympathetic and parasympathetic nervous system.

☐ 1. What is dual innervation?
☐ 2. How do the sympathetic and parasympathetic nervous systems interact?

Autonomic Tone

As we discussed in the previous module, neither the sympathetic nor the parasympathetic nervous system is ever completely silent, as both are active to some degree most of the time. This constant amount of activity from each system is known as **autonomic tone**; it can be divided into *sympathetic tone* and *parasympathetic tone*. Interestingly, different degrees of sympathetic and parasympathetic tone exist in different organs. For example, the sympathetic nervous system is normally dominant in blood vessels and keeps them partially constricted at all times, which is important for maintaining blood pressure at rest.

The parasympathetic nervous system is normally dominant in the heart and keeps the heart rate at an average of 72 beats per minute. Parasympathetic tone in the heart is even stronger in athletes, who have been conditioned to recover more quickly from the bursts of sympathetic nervous system activity that accompany exercise. This causes the resting heart rate of an athlete to be lower than that of someone who is not physically conditioned. The parasympathetic nervous system also dominates in the digestive and urinary systems, and keeps the activity of these systems

at normal levels. This is why drugs with anticholinergic actions (discussed in *A&P in the Real World: Side Effects of Anticholinergic Drugs*) can have such profound effects on these systems.

When normal autonomic tone is disrupted, this can have dramatic effects on homeostasis. One example of such a condition is discussed in *A&P in the Real World: Postural Orthostatic Tachycardia Syndrome* on page 534.

☐ 3. What is autonomic tone?

Summary of Nervous System Control of Homeostasis

◄◄ **Flashback**

1. Where is the hypothalamus located? What aspects of homeostasis does it control? (p. 434)
2. Where is the reticular formation located? What aspects of homeostasis does it control? (p. 440)
3. What are the primary functions of the cerebral cortex and the amygdala? (pp. 429, 433)

Maintenance of homeostasis is one of the body's most essential functions, and the two divisions of the ANS figure importantly in how the nervous system overall fulfills this role (**Figure 14.12**).

Figure 14.12 Summary of nervous system control of homeostasis.

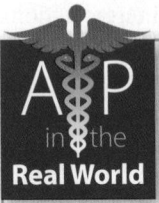

Postural Orthostatic Tachycardia Syndrome

Postural orthostatic tachycardia syndrome (POTS) is characterized by an abnormally large increase in heart rate (known as *tachycardia;* tak′-ih-KAR-dee-ah) when an individual moves from lying or sitting down to standing up. This inappropriate rise of heart rate is accompanied by the vasodilation of most blood vessels, which causes blood pressure to drop due to gravity. Symptoms, which generally result from the low blood pressure, include dizziness and lightheadedness, fatigue, and thirst. Low blood pressure also reduces blood flow to organs, leading to shortness of breath, chest pain, cold extremities, and muscle weakness.

The fundamental cause of POTS has yet to be determined, but the symptoms appear to be secondary to excessive sympathetic activity or excessive sensitivity of sympathetic receptors to epinephrine and norepinephrine. Normally, when a person stands up, the sympathetic nervous system temporarily increases blood pressure to ensure that blood flow remains constant against the force of gravity. The sympathetic response generally tapers off, and parasympathetic tone resumes control of the heart rate. With POTS, however, the response to sympathetic stimulation is exaggerated, leading to the characteristic symptoms. Treatment for POTS generally consists of dietary modifications, such as increasing water and salt intake; an exercise regimen; and medications to block sympathetic receptors. Most patients will see some improvement in the condition gradually, although some will struggle with it for the remainder of their lives.

Now that we know more about the ANS, we can revisit what you learned in the CNS section on homeostasis of vital functions (see Chapter 12). Recall that homeostatic functions are largely controlled centrally by the hypothalamus and the brainstem reticular formation. Many of the actions of these structures are mediated via the sympathetic and parasympathetic nervous systems.

Note in Figure 14.12 that many signals sent from the hypothalamus are directed toward areas of the reticular formation called **autonomic centers.** These centers contain neurons that control the activity of preganglionic sympathetic and parasympathetic neurons. Although the hypothalamus does exert some degree of control over these centers, the reticular formation can function even if the circuits connecting it with the hypothalamus are severed. This indicates that the reticular formation is capable of controlling many of our most critical autonomic functions, such as heart rate and blood pressure, on its own.

Figure 14.12 also shows that both the hypothalamus and reticular formation receive input from higher centers in the brain, including the amygdala and multiple areas of the cerebral cortex. This in part explains why emotion has such a profound effect on our visceral functions. A state of excitement, mediated by the cerebral cortex and amygdala, will be passed on to the hypothalamus, reticular formation, and finally to the ANS, after which we notice an increased heart rate, elevated blood pressure, dilated pupils, and other signs of sympathetic nervous system activity.

This is not the entire homeostasis story, however. We explore the other part of it when we discuss the endocrine system (in Chapter 16).

Quick Check

☐ 4. How does the CNS mediate control over the ANS?

Apply What You Learned

☐ 1. Explain why the nervous systems of individuals with brainstem injuries generally have difficulty controlling their autonomic functions.

☐ 2. Predict what would happen if parasympathetic tone were dominant in the airways.

See answers in Appendix A.

CHAPTER SUMMARY

Take full advantage of the invaluable online practice and study tools for this chapter by going to MasteringA&P®. There you will find:

 Practice tests, module quizzes, art-labeling and sequence-the-steps activities

 Animated figures with quizzes

Practicing A&P worksheets

Author-narrated podcast

🔥 Dynamic Study Modules

iP *Interactive Physiology*

 MODULE **14.1**

Overview of the Autonomic Nervous System 518

- The autonomic nervous system (ANS) performs its functions via a series of **visceral reflex arcs.** (Figure 14.1, p. 519)

- Autonomic motor neurons consist of **preganglionic neurons** that synapse on **autonomic ganglia,** and then **postganglionic neurons** that contact the target cells. (Figure 14.2, p. 519)

- The autonomic nervous system is made up of the sympathetic and parasympathetic nervous systems. The two systems are antagonistic. (Figure 14.3, p. 520)

MODULE 14.2

The Sympathetic Nervous System 521

- The **sympathetic nervous system** is often called the "fight or flight" division of the ANS because it prepares the body for emergency situations. It is the dominant arm of the ANS during exercise and times of emotion as well.

- The sympathetic nervous system is called the **thoracolumbar division** because its preganglionic cell bodies lie in the lateral horn of the thoracic and lumbar spinal cord. The axons of these neurons leave the spinal cord and travel as **white rami communicantes,** after which they synapse on the postganglionic cell bodies located in sympathetic ganglia. Most ganglia are found along the vertebral column as part of the sympathetic chain ganglia. (Figure 14.4, p. 522; Figure 14.5, p. 523)

   Practice art-labeling exercise: Figure 14.4.

- The preganglionic sympathetic axon communicates with the postganglionic neuron by releasing acetylcholine (ACh) at the synapse. Postganglionic axons communicate with their target cells by releasing norepinephrine, epinephrine, or ACh. (Figure 14.6, p. 524)

- **Adrenergic receptors** bind to epinephrine and norepinephrine. There are five classes of adrenergic receptors: α_1, α_2, β_1, β_2, and β_3.
  - α_1 receptors are located in the plasma membranes of smooth muscle cells of blood vessels that supply the skin, digestive system, and kidney.
  - α_2 receptors are located in the membranes of preganglionic neurons. Norepinephrine inhibits these neurons from firing when it binds to these receptors, and thus dampens the sympathetic response.
  - β_1 receptors are located in the plasma membranes of cardiac muscle cells.
  - β_2 receptors are located in the plasma membranes of the smooth muscle cells lining the airways (the bronchioles).
  - β_3 receptors are located in the plasma membranes of adipocytes and smooth muscle cells of the digestive tract.

- **Cholinergic receptors** bind ACh. The two types of cholinergic receptors are **muscarinic receptors** located on sweat glands in the skin and **nicotinic receptors** found in the membranes of all postganglionic neurons and the adrenal medulla.

- Many different pharmacological effects can be caused by binding and activating a sympathetic receptor with an *agonist* or by blocking it with an *antagonist*.

- The sympathetic nervous system increases blood pressure by raising the rate and force of contraction of the heart and also by causing *vasoconstriction* (Figure 14.7, p. 525).

   Practice art-labeling exercise: Figure 14.7.

- Sympathetic neurons cause *bronchodilation,* which increases the amount of air that can be exchanged with each breath.

- Norepinephrine released by sympathetic neurons causes *vasodilation* of the blood vessels serving skeletal muscles.

- The adrenal medulla is made up of modified sympathetic postganglionic neurons. When stimulated, the adrenal medulla releases epinephrine and norepinephrine, making it an interface between the sympathetic nervous system and the endocrine system. (Figure 14.8, p. 527)

- The sympathetic nervous system also decreases urinary and digestive functions and the secretion from most glands, dilates the pupil, increases the levels of metabolic fuels in the blood, and increases the secretion from sweat glands.

MODULE 14.3

The Parasympathetic Nervous System 528

- The **parasympathetic nervous system** is the "rest and digest" division of the ANS that performs the body's maintenance functions such as digestion and urine formation.

- The parasympathetic nervous system is called the **craniosacral division** because its preganglionic cell bodies are located in the nuclei of cranial nerves III, VII, IX, and X and in the sacral spinal cord. Preganglionic axons synapse with the cell bodies of postganglionic neurons in ganglia near the target cells. (Figure 14.9, p. 529)

   Practice art-labeling exercise: Figure 14.9.

- Both preganglionic and postganglionic parasympathetic neurons release ACh at their synapses. Nicotinic receptors are located in the membranes of all postganglionic parasympathetic neurons. Muscarinic receptors are found in the membranes of all parasympathetic target cells.

- The parasympathetic nervous system lowers blood pressure by decreasing the rate of contraction of the heart. The parasympathetic nervous system innervate blood vessels. (Figure 14.10, p. 530)

   Practice art-labeling exercise: Figure 14.10.

- Other effects of the parasympathetic nervous system include the following: constriction of the pupil, accommodation of the lens for near vision, *bronchoconstriction,* contraction of the smooth muscle lining the digestive tract, relaxation of digestive and urinary sphincters, engorgement of the penis or clitoris, and an increase in secretion from most glands except the sweat glands.

MODULE 14.4

Homeostasis Part II: PNS Maintenance of Homeostasis 532

- The sympathetic and parasympathetic nervous systems generally work antagonistically, which is possible because of dual innervation of many organs. (Figure 14.11, p. 532)

- **Autonomic tone** refers to the constant amount of activity present in each system most of the time. Sympathetic tone is dominant in the blood vessels, and they are partially constricted most of the time. Parasympathetic tone is dominant in the digestive and urinary organs and in the heart.

- Homeostasis is controlled centrally by the hypothalamus and brainstem reticular formation. Many of the signals sent by the hypothalamus are directed toward **autonomic centers** in the reticular formation. The hypothalamus and reticular formation receive input from higher centers in the brain, such as the amygdala and the cerebral cortex. (Figure 14.12, p. 533)

ASSESS WHAT YOU LEARNED

See answers in Appendix A.

LEVEL 1 Check Your Recall

1. Which of the following best describes the basic function of the autonomic nervous system?

    a. Controls the somatic nervous system
    b. Controls automatic functions to maintain homeostasis
    c. Detects somatic sensory information
    d. All of the above

2. Fill in the blanks: The sympathetic nervous system is also known as the _____ division because the cell bodies of its preganglionic neurons are located in the _____.

3. Sympathetic preganglionic neurons synapse:

    a. directly on target cells.
    b. on skeletal muscle fibers.
    c. on sympathetic chain or collateral ganglia.
    d. on vagal ganglia.

4. Differentiate between adrenergic and cholinergic receptors.

5. Which neurotransmitter(s) is/are used by sympathetic post-ganglionic neurons?

    a. Epinephrine   d. Both a and b
    b. Norepinephrine   e. All of the above
    c. Acetylcholine

6. Explain what happens when sympathetic neurons stimulate the adrenal medulla. What purpose does this serve?

7. Which of the following actions would you expect when sympathetic neurons release neurotransmitter onto β_1 receptors?

    a. Decreased sweat production
    b. Constriction of blood vessels serving the digestive, urinary, and integumentary systems
    c. Adjustment of the shape of the lens
    d. Increase in the rate and force of contraction of the heart.

8. Mark each of the following as an effect of the sympathetic nervous system or the parasympathetic nervous system.

    a. _____ Bronchodilation
    b. _____ Bronchoconstriction
    c. _____ Constriction of pupil
    d. _____ Vasoconstriction of blood vessels serving organs of digestive, urinary, and integumentary systems
    e. _____ Increased metabolic rate
    f. _____ Dilation of pupil
    g. _____ Contraction of smooth muscle of digestive tract

9. Mark the following statements as true or false. If a statement is false, correct it to make a true statement.

    a. The parasympathetic nervous system generally decreases the secretion from digestive glands.
    b. Sympathetic stimulation increases sweat secretion.

    c. The parasympathetic nervous system releases acetylcholine onto all of its synapses.
    d. Sympathetic tone controls the resting rate of the heart.

10. Fill in the blanks: _____ receptors are located on parasympathetic postganglionic neurons, and _____ receptors are located on parasympathetic target cells.

11. Parasympathetic ganglia are typically:

    a. located along the spinal cord.
    b. located near their target cells.
    c. located within the central nervous system.
    d. Parasympathetic neurons do not synapse in ganglia.

12. Central nervous system control over the ANS is mediated by:

    a. the reticular formation.   c. the thalamus.
    b. the hypothalamus.   d. both a and b.

LEVEL 2 Check Your Understanding

1. Using 20 or fewer words, define each of the following terms in your own words.

    a. Sympathetic nervous system
    b. Parasympathetic nervous system

2. You are running a race to the top of a mountain. Explain all the changes your sympathetic nervous system will initiate to maintain homeostasis as you run the race.

3. Describe all the changes initiated by the parasympathetic nervous system that will take place when you finish the race. How will these changes maintain homeostasis?

LEVEL 3 Apply Your Knowledge

PART A: Application and Analysis

1. Which cranial nerves would be affected by a drug that stimulates the parasympathetic nervous system? Predict potential adverse effects that one might experience.

2. A patient, Dr. Young, has both asthma and high blood pressure. Her physician prescribed the drug propranolol to treat her hypertension; this drug blocks all types of beta-adrenergic receptors. She also takes the drug albuterol for asthma, which activates β_2 receptors on bronchial smooth muscle. Will the pairing of these two drugs cause problems for Dr. Young? Explain.

3. Mr. Chevalier has been diagnosed with Horner syndrome, which is caused by dysfunction of the sympathetic neurons in the superior cervical ganglion that innervate structures of the head, face, and neck. What symptoms is Mr. Chevalier likely to face due to his disease? Explain.

PART B: Make the Connection

4. Many chemical warfare agents, such as the poisonous gas *sarin*, block the enzyme acetylcholinesterase in the synaptic cleft. What effects will this poison have on muscle contraction? (*Connects to Chapter 10*) What effects will it have on the autonomic nervous system? What symptoms would you expect to see from this poison?

15

The Special Senses

I f a tree falls in the forest when no one is around, does it still make a sound? Do freshly baked chocolate chip cookies still have that delicious aroma even if you have a cold that blocks your sense of smell? In studying the special senses, we can distinguish between physical stimuli—the sound waves produced by the falling tree and the chemical odor of the cookies—and the receptors that are specialized to detect them.

You can probably list four of the special senses quite easily—smell (olfaction), taste (gustation), vision, and hearing (audition). However, can you name the fifth special sense? It is vestibular sensation, the detection of head movement and position that helps maintain equilibrium, or balance. This chapter examines the role of the special senses in detecting stimuli (such as sounds, aromas, and light), *transducing* (converting) them into neural signals, and interpreting them. But first, let's see how the special senses are similar to, and different from, the general senses.

MODULE **15.1**

Comparison of General and Special Senses

Learning Outcomes

1. Describe the similarities between the general senses and special senses.
2. Contrast the general senses with the special senses.

Recall from the chapter on the central nervous system (CNS) that the general senses detect such stimuli as touch, pain, and temperature (see Chapter 12). "General" refers to the fact that these receptors are relatively simple and located throughout the body in both the skin and internal organs. The special senses, in contrast, are so named because they convey a specific type of information from specialized sensory organs in discrete locations of the head.

In the peripheral nervous system chapter, you read that **transduction** occurs when a physical or chemical stimulus is converted into a neural signal that can be interpreted by the brain (see Chapter 13). Neurons involved in the general senses have specialized receptive endings to detect touch, temperature, or chemicals that produce pain. The presence of these stimuli alters the resting membrane potential of the sensory neurons to produce an action potential that reaches the CNS.

Photo: Vision, one of the five special senses, is the perception of light reflected by various objects.

537

The special senses also detect stimuli—such as light, chemicals, or sounds—in the environment and transduce them into neural signals. This information also travels through the axons of peripheral nerves to the CNS.

In both general and special sensation, the stimuli are first processed by sensory nuclei and transmitted to the thalamus and primary cortex areas for awareness and identification. Sensory signals are then passed to association areas for further interpretation and integration.

ConceptBOOST ⟫⟫⟫

Understanding Transduction

To understand how the special senses transduce physical stimuli into neural impulses, you first need to pinpoint where transduction occurs in each pathway. Here's a hint: Look for the cells that are the first to either increase or decrease the amount of neurotransmitter they release onto a neuron. These cells might come first in the pathway (as in olfaction) or somewhere in the middle (as in hearing).

Once you find these receptor cells, look for the stimulus that causes them to alter their release of neurotransmitters. Is it a chemical? Sound? Light, as shown here?

① A stimulus (such as light) directly or indirectly opens ion channels, altering the membrane potential of the receptor cell.

② The change in membrane potential causes an increase or decrease in neurotransmitter release, triggering a change in the activity of the neuron.

Regardless of its type, the stimulus will alter membrane potential by opening or closing ion channels, which will then alter the release of neurotransmitter by receptor cells. The details of transduction might look different for each of the special senses, but the fundamental process is similar. ∎

Despite the similarities between the general and special senses, they differ in two significant ways (**Figure 15.1**):

- In general sensation, the receptors are the receptive endings of sensory neurons. However, most special senses rely on receptors that are not neurons but instead are specialized cells that detect specific types of stimuli, such as light, sound waves, head movements, or chemicals. The sense organs transduce these stimuli into electrical signals that

are passed on to neurons via neurotransmitters. Olfaction is the only special sense in which the receptor cells are neurons, in this case specialized to detect different odors.
- General sensation is carried into the CNS by the axons of both spinal and cranial nerves. However, because the special sense organs are all housed in the head, only the axons of cranial nerves convey information about special senses.

Quick Check

☐ 1. Compare and contrast the general and special senses.
☐ 2. Which structural type of nerve (spinal, cranial, or both) carries information about the special senses into the CNS?

Apply What You Learned

☐ 1. Stimuli such as light or sound are not detected directly by neurons. Explain why this is true. (*Hint:* Focus on the differences between the general and special senses.)

See answer in Appendix A.

MODULE **15.2**

Anatomy and Physiology of Smell

Learning Outcomes

1. Describe and identify the location of olfactory receptors.
2. Explain how odorants activate olfactory receptors.
3. Describe the path of action potentials from the olfactory receptors to various parts of the brain.

Think about a whiff of your favorite food, the smell of exhaust as a bus pulls away, the perfume your mother always wears. Our olfactory system allows us to detect the presence of **odorants** (OH-dor-entz), or chemicals, in the air and transduces them into signals our brain can interpret. In addition to identifying odors, the nervous system allows us to experience the emotional responses that accompany them. In this section, we first examine the structures and physiological processes that enable us to detect odors and then trace the olfactory pathway through the CNS.

Structures of Olfaction

◄◄ Flashback

1. What are the basic structure and function of epithelial tissue? (p. 127)
2. What are bipolar neurons? (p. 387)

Olfaction (ohl-FAK-shun) begins in the **olfactory epithelium,** a small patch of specialized epithelium located in the superior

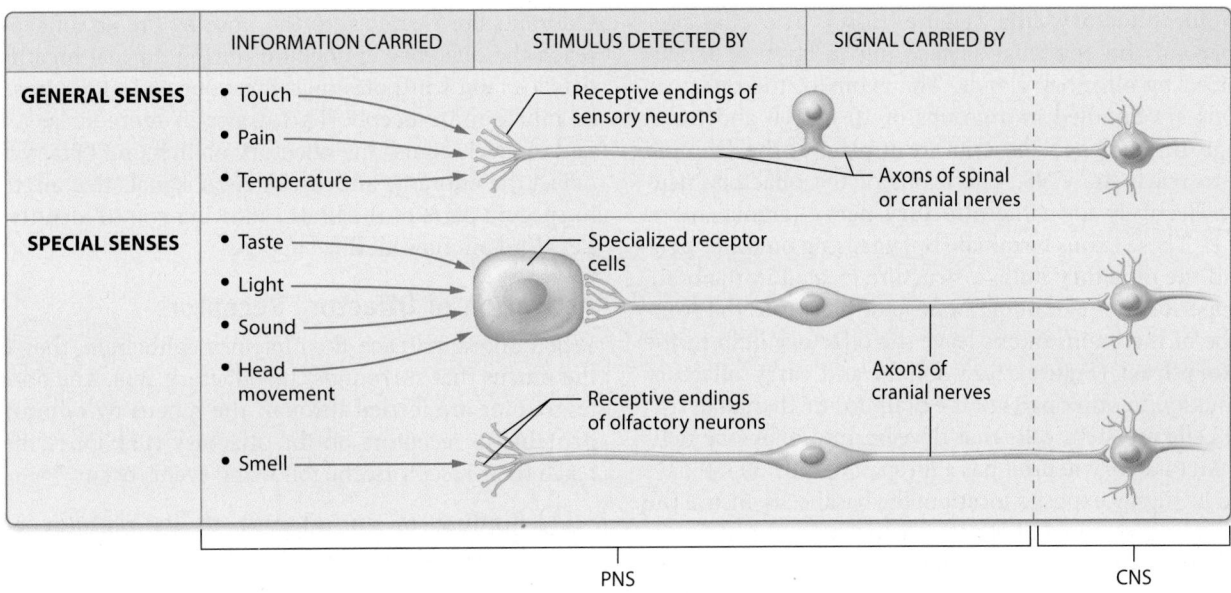

	INFORMATION CARRIED	STIMULUS DETECTED BY	SIGNAL CARRIED BY
GENERAL SENSES	• Touch • Pain • Temperature	Receptive endings of sensory neurons	Axons of spinal or cranial nerves
SPECIAL SENSES	• Taste • Light • Sound • Head movement	Specialized receptor cells	Axons of cranial nerves
	• Smell	Receptive endings of olfactory neurons	

PNS | CNS

Figure 15.1 Comparison of general and special senses.

region of each nasal cavity (**Figure 15.2a**). The olfactory epithelium is yellowish-brown and about the size of your thumbnail. It contains three types of cells (**Figure 15.2b and c**):

- **Olfactory neurons** are modified bipolar neurons that detect the chemical substances we perceive as odors. Olfactory

neurons are known as **chemoreceptors,** since they respond to the presence of certain chemicals in the air rather than to neurotransmitters. Humans have approximately 10 million olfactory neurons, but dogs and other organisms with a keen sense of smell may have several billion. The modified dendrite of each neuron has an expanded tip containing

(a) Location of olfactory epithelium in the nasal cavity

Olfactory bulb
Cribriform plate of ethmoid bone
Middle and inferior nasal conchae
Inhaled air containing odorants
Olfactory bulb
Olfactory tract
Olfactory epithelium

(b) Structure of olfactory epithelium

Olfactory bulb
Olfactory tract
Cribriform plate
Axon of olfactory neuron
Olfactory gland
Olfactory neuron
Supporting cell
Basal cell
Dendrite
Olfactory cilia
Olfactory epithelium

(c) Olfactory cilia

Dendrite of olfactory epithelium

SEM (710×)

Figure 15.2 Olfactory epithelium and olfactory neurons.

nonmotile **olfactory cilia** (Figure 15.2c). The cilia project beyond the epithelial surface into a layer of mucus produced by olfactory glands. The axons of the olfactory neurons are bundled into groups of 10 to 100 and travel through tiny holes in the cribriform plate of the ethmoid bone to reach the CNS. The axons of the olfactory neurons collectively form the **olfactory nerve** (cranial nerve [CN] I). These axons terminate by synapsing on other neurons in the **olfactory bulb,** a structure located in the brain just superior to the ethmoid bone and inferior to the frontal lobe of the brain. Axons leave the olfactory bulb in the **olfactory tract** (Figure 15.2a and b) and carry olfactory information to other parts of the brain for further analysis.

- **Basal cells** are stem cells that develop into olfactory neurons. An olfactory neuron has a lifespan of only 30–60 days due to its highly exposed location; the basal cells ensure the continual replacement of olfactory neurons.
- **Supporting cells** are columnar cells that support and surround the olfactory neurons. They contain a pigment that gives the olfactory epithelium its color.

Quick Check

☐ 1. Where is the olfactory epithelium located?
☐ 2. What types of cells make up the olfactory epithelium?
☐ 3. Describe the structure of an olfactory neuron.

Physiology of Olfaction

◀◀ **Flashback**

1. What happens to the membrane potential of a cell when it depolarizes? (p. 350)
2. What is a local potential? How does it differ from an action potential? (p. 397)

Odorants are carried into the nose by the air, but few of them reach the olfactory epithelium during normal breathing. If you detect a faint whiff of something, your typical response is to sniff or inhale more deeply. This draws in more air at a faster rate and forces it across the olfactory epithelium. Olfactory neurons detect the odorants and send neural signals that are transmitted to specific parts of the brain, allowing you to identify the scent. Let's find out how all this happens.

Activation of Olfactory Receptors

When odorants reach the olfactory epithelium, they dissolve in the mucus that surrounds the olfactory cilia. The odorants then migrate or are ferried through the mucus by **odorant-binding proteins** to receptors on the olfactory cilia. Once the odorants reach their receptors, the following events occur (**Figure 15.3**):

① **Binding of an odorant to its receptor activates a G-protein.** The activated G-protein then detaches from the receptor.
② **The activated G-protein triggers the enzyme adenylate cyclase to convert ATP into cyclic AMP (cAMP).**
③ **cAMP opens ion channels that allow sodium and calcium ions to enter the cell.** These ions depolarize the membrane of the olfactory neuron.

As you have seen before, a depolarization that reaches threshold at the axon hillock triggers an action potential. When an action potential is triggered in an olfactory neuron, the chemical odorant has been transduced into a neural signal.

Even though our sense of smell is less sensitive than that of many other vertebrates, humans are capable of detecting up to 10,000 different odors. However, that doesn't mean we have 10,000 different olfactory receptors! We have only about 400 types of receptors, but each can bind more than one type of odorant, and we identify a particular smell by the combination of receptors that are stimulated. Just as you can play thousands of different chords using the 88 keys on a piano, the olfactory

① Binding of an odorant to its receptor activates a G-protein.

② The activated G-protein triggers adenylate cyclase to convert ATP into cyclic AMP (cAMP).

③ cAMP opens ion channels that allow sodium and calcium ions to enter the cell.

Figure 15.3 Transduction of olfaction in an olfactory neuron.

neurons can code for thousands of odors using combinations of a few hundred receptors.

Not all chemical odorants bind to olfactory neurons. Some substances, such as ammonia or chlorine, bind to branches of the trigeminal nerve in the epithelium lining the nasal cavity and produce a burning sensation. If you have ever gotten a large whiff of ammoniated window cleaner, you know how greatly these substances irritate your nasal cavity.

The Olfactory Pathway

For us to detect and identify an odor, the neural signals carrying olfactory information must enter the CNS and travel to various regions of the brain (**Figure 15.4**):

① **The axons of olfactory neurons (making up the olfactory nerve) carry olfactory stimuli to the olfactory bulb in the CNS.** The axons of the olfactory neurons terminate in the olfactory bulb just inferior to the frontal lobe. The axon terminals synapse with the dendrites of neurons called **mitral cells** (MY-tral) in the olfactory bulb; many olfactory neurons with the same receptor type end on the same mitral cell.

② **An olfactory stimulus travels from the olfactory bulb to the primary olfactory cortex in the temporal lobe.** The axons of the mitral cells form the bulk of the olfactory tract and travel to the **primary olfactory cortex,** which is located in the inferomedial temporal lobe near the optic chiasma. This is the only sensory pathway that has no synapse in the thalamus. The primary olfactory cortex is responsible for awareness and identification of an odor. Interestingly, it is also a common location for seizures,

which are often preceded by the sensation of an unpleasant odor.

Disruptions of this pathway can be quite significant, as explained in *A&P in the Real World: Anosmia* on page 542.

Neurons from the olfactory cortex then contact other neurons in the amygdala, hippocampus, hypothalamus, and other parts of the limbic system. This evokes emotional responses to odors as well as visceral reactions such as salivation, sexual arousal, and nausea. Connections also exist between the primary olfactory cortex and the cortex in the base of the frontal lobe, which are thought to play a role in integrating olfaction and taste.

Quick Check

☐ 4. What happens in an olfactory neuron when an odorant binds to its receptor?
☐ 5. How can humans detect such a variety of odors with a limited number of olfactory receptors?
☐ 6. Where are olfactory stimuli sent for conscious perception and interpretation?

Apply What You Learned

☐ 1. How might nasal congestion (a "stuffy" nose) interfere with olfaction?
☐ 2. Which structure(s) in the olfactory pathway would be most affected by a large tumor in the inferior frontal lobe? How would this affect olfaction?

See answers in Appendix A.

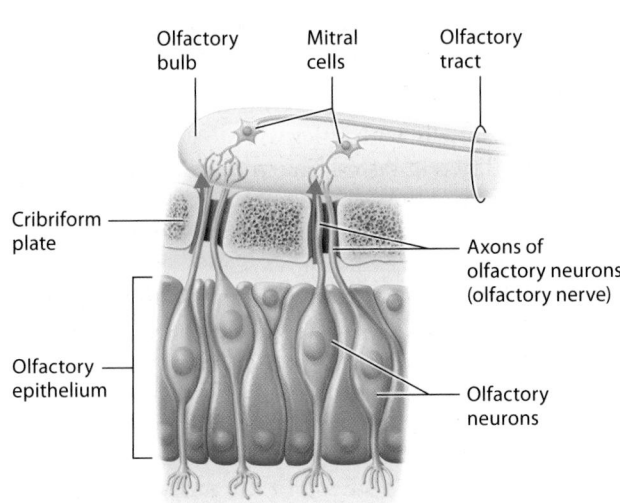

Olfactory bulb   Mitral cells   Olfactory tract

Cribriform plate

Axons of olfactory neurons (olfactory nerve)

Olfactory epithelium

Olfactory neurons

① The axons of olfactory neurons (making up the olfactory nerve) carry olfactory stimuli to the olfactory bulb in the CNS.

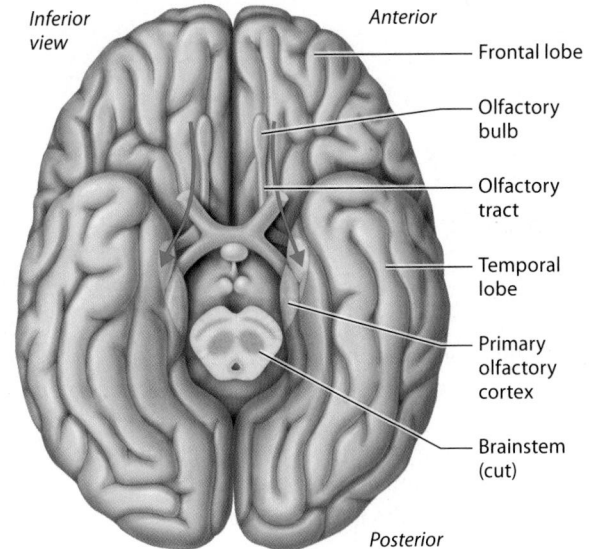

Inferior view

Anterior

Frontal lobe

Olfactory bulb

Olfactory tract

Temporal lobe

Primary olfactory cortex

Brainstem (cut)

Posterior

② An olfactory stimulus travels from the olfactory bulb to the primary olfactory cortex in the temporal lobe.

Figure 15.4 The olfactory pathway.

Practice art labeling @MasteringA&P®

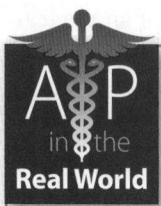

Anosmia

Although the thought of losing your sense of smell might not be as frightening as that of losing your vision or hearing, **anosmia** (an-OZ-mee-ah; the lack of olfaction) can be serious. People with anosmia may not be able to detect smoke or the odor of spoiled food. They may also suffer from malnutrition—olfaction contributes significantly to the sense of taste, so food loses much of its appeal.

Many conditions result in anosmia or **hyposmia** (hi-POZ-mee-ah; reduced olfactory sensitivity). Anything that blocks air from reaching the olfactory epithelium will interfere with olfaction; examples include nasal polyps, a deviated or fractured nasal septum, and a swollen respiratory epithelium from a respiratory infection or allergic reaction. In these cases, anosmia is often temporary, and olfaction returns once air flow is restored.

However, anosmia can be permanent. Head injuries that damage the neural pathways at any point can disrupt olfaction; the olfactory nerve is particularly vulnerable to fracture of the cribriform plate. The neural degeneration caused by Alzheimer's disease and Parkinson's disease often impairs the sense of smell. In fact, research suggests that decreased olfaction may be an early sign of Alzheimer's disease.

Hyposmia also develops naturally with aging, as fewer olfactory neurons are replaced by basal cells in the olfactory epithelium. The process occurs gradually but becomes especially noticeable after age 60.

MODULE **15.3**

Anatomy and Physiology of Taste

Learning Outcomes

1. Describe the location and structure of taste buds.
2. Explain how chemicals dissolved in saliva activate gustatory receptors.
3. Trace the path of nerve impulses from the gustatory receptors to various parts of the brain.
4. Summarize the five primary taste sensations.

The flavor of a food encompasses many types of sensation. Surprisingly, much of what we call taste is actually the smell of a food, which reaches the olfactory epithelium via the nasal cavities and throat. This is why food lacks flavor if the nasal cavities are blocked, such as when you have a cold. Other sensations, such as

food temperature and texture, are detected by general sensory neurons. Some just "taste" better when they are warm, and the texture of a food may not be appealing in spite of a pleasant taste. Spicy foods actually stimulate receptors for pain and heat in the trigeminal nerve, and we interpret the sensation as "spicy" or "hot."

However, a large part of our experience in eating a food does depend on the sense of **gustation** (gus-TAY-shun), or *taste*. Like olfaction, gustation involves chemoreceptors that respond to chemicals. Taste begins with stimulation of specialized receptor cells in **taste buds,** which are small clusters of receptor cells and supporting cells scattered across the tongue and parts of the oral cavity. Each taste bud is contacted by sensory neurons that carry information to the central nervous system. This module explores taste sensation, from the detection of chemical molecules on the tongue to the processing of neuronal signals in various regions of the brain.

Structures of Gustation: Taste Buds

Notice that your tongue is covered with rounded projections called **papillae** (pah-PILL-ee; singular, *papilla*). Papillae are classified into four groups based on shape (**Figure 15.5a and b**):

- **Vallate (or circumvallate) papillae** (VAL-ate) are the largest and are dome-shaped; each contains hundreds of taste buds.
- **Fungiform papillae** (FUN-jeh-form) are mushroom-shaped and contain only a few taste buds.
- **Foliate papillae** (FOH-lee-ate) are ridges on the sides of the tongue and contain taste buds only in childhood.
- **Filiform papillae** (FILL-uh-form) are long, thin cylinders scattered across the tongue. They do not contain any taste buds but have sensory nerve endings that detect the texture and temperature of food.

Taste buds are typically located on the lateral surfaces of the papillae and contain three types of cells (**Figure 15.5c**):

- **Gustatory (taste) cells** (GUS-tah-tore-ee) are specialized epithelial cells with microvilli that contain receptors that detect different tastes. The microvilli are found on the apical surface of the cell and project into a small pocket on the surface of the papilla, called the **taste pore.** The basal end of each gustatory cell forms a synapse with the receptive endings of a sensory neuron that carries taste sensation into the CNS via the facial (CN VII), glossopharyngeal (CN IX), or vagus (CN X) nerve. A typical taste bud contains approximately 100 gustatory cells.
- **Basal cells** are stem cells that differentiate into new gustatory cells, which have a lifespan of only 10–14 days.
- **Supporting cells** surround the gustatory cells. They provide physical support but have no role in taste sensation.

Taste sensation tends to decline as people age because the number of basal cells decreases and so the gustatory cells are not replaced when they degenerate. We are typically born with approximately 10,000 taste buds but lose half of them by adulthood, and the rate of decline is greatest after age 50.

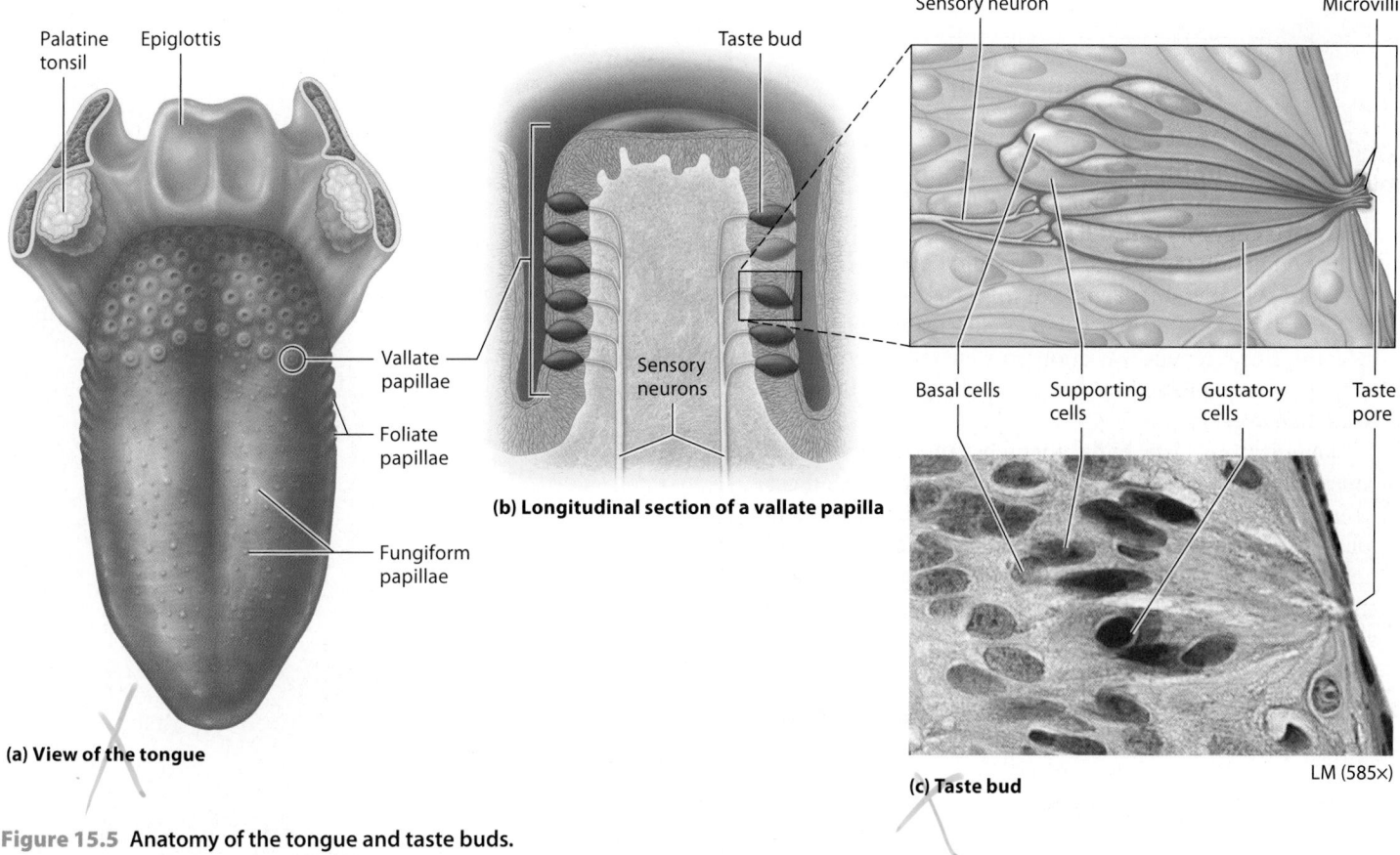

Figure 15.5 Anatomy of the tongue and taste buds.

(a) View of the tongue

(b) Longitudinal section of a vallate papilla

(c) Taste bud   LM (585×)

Quick Check

☐ 1. Where are taste buds located?
☐ 2. How does a gustatory cell stimulate a sensory neuron?
☐ 3. What are the roles of basal and supporting cells?

Physiology of Gustation

◀◀ **Flashback**

1. What happens to the membrane potential of a cell when potassium ion channels open? (p. 350)

Like olfaction, the sense of taste involves transducing chemicals—in this case, in the food we eat—into neural signals. This transduction from chemical to neural signal takes place in the gustatory cells, and the neural signal is then transmitted to the CNS by one of three cranial nerves. To understand this process, we first need to look at the chemicals we perceive as tastes.

Taste Sensations

The sense of taste relies on detecting various chemicals, which are typically grouped into five classes:

- **Sweet** tastes are due to simple sugars such as glucose and fructose. Ethylene glycol (found in antifreeze) and lead (in

lead paint) can also taste sweet, which is why young children may ingest these substances, leading to accidental poisonings.
- **Sour** tastes are produced by hydrogen ions, as found in the citric acid of lemon juice.
- **Salty** foods taste salty due to the presence of metal ions such as sodium and potassium ions.
- **Bitter** flavors are imparted by many nitrogen-containing compounds and are often found in rancid or poisonous substances.
- **Umami** (oo-MOM-ee; the Japanese word for "savory" or "delicious"), a taste often associated with meat or broth, is produced by glutamate or other amino acids.

We detect taste stimuli when they bind to receptors on the microvilli of gustatory cells. The taste receptors are classified by the substance to which they respond, and only one type of receptor is typically found on a gustatory cell. Contrary to popular belief, different tastes are not localized to separate regions of the tongue.

Obviously, the foods we eat contain more flavors than the five mentioned above. The unique flavor of a food lies in its particular combination of taste molecules and the unique combination of taste receptors it activates (as well as other sensory components, especially olfactory). Thus, a pretzel tastes different from a potato chip, even though both are salty. Flavor is also affected by the number of taste buds an individual possesses, as described in *A&P in the Real World: Are You a Supertaster?* on page 544.

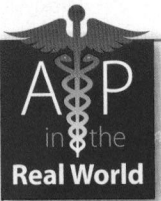

Are You a Supertaster?

Most of our taste preferences are determined by the foods we ate while growing up, but biological factors also play a role. When asked to taste a compound called PTU (6-*n*-propylthiouracil), 25% of the population taste nothing, 50% find it mildly bitter, and 25% find it intolerable. People in this last group are known as "supertasters," and their heightened sensitivity is thought to come from genetic variation and an increased number of fungiform papillae. In other words, they have more taste buds than average.

Having more taste buds might be an advantage, since supertasters often avoid sugary foods, such as frosted cake, or intensely flavored drinks like coffee and beer; one study found that women who are supertasters are thinner than non-supertasters. However, supertasters may also avoid strong-tasting fruits and vegetables like grapefruit and brussels sprouts, which may increase their risk of colon cancer and other health problems.

Activation of Taste Receptors

To reach the taste buds in the crevices between the papillae, a substance must first dissolve in saliva. To grasp the importance of saliva to taste, see what happens when you completely dry part of your tongue and then touch it with a piece of pretzel or candy.

Once the taste stimulus reaches a gustatory cell, it must be transduced into a neural signal. This occurs through several steps, beginning with the depolarization of the gustatory cell (**Figure 15.6**):

① **Changes in ion movements depolarize the gustatory cell's plasma membrane.** The specific ion movements vary depending on the particular taste. For instance, the sodium ions in salty foods travel directly through sodium ion channel receptors to depolarize the cell. On the other hand, receptors for bitter, sweet, and umami compounds activate a G-protein that indirectly closes potassium ion channels involved in the leak current. The decreased movement of potassium ions out of the cell produces depolarization (not shown in Figure 15.6). These are examples of the Gradients Core Principle (p. 26), since the ions move down their concentration gradients.

CORE PRINCIPLE Gradients

② **Depolarization of the membrane opens voltage-gated calcium ion channels, and calcium ions enter the cell.** Depolarization results in a local (graded) potential, just as occurs in a postsynaptic neuron when neurotransmitters bind to their receptors. Unlike a neuron, however, a gustatory cell usually does not generate an action potential. Gustatory cells are small enough that the local potential can quickly spread across their membrane, making an action potential unnecessary.

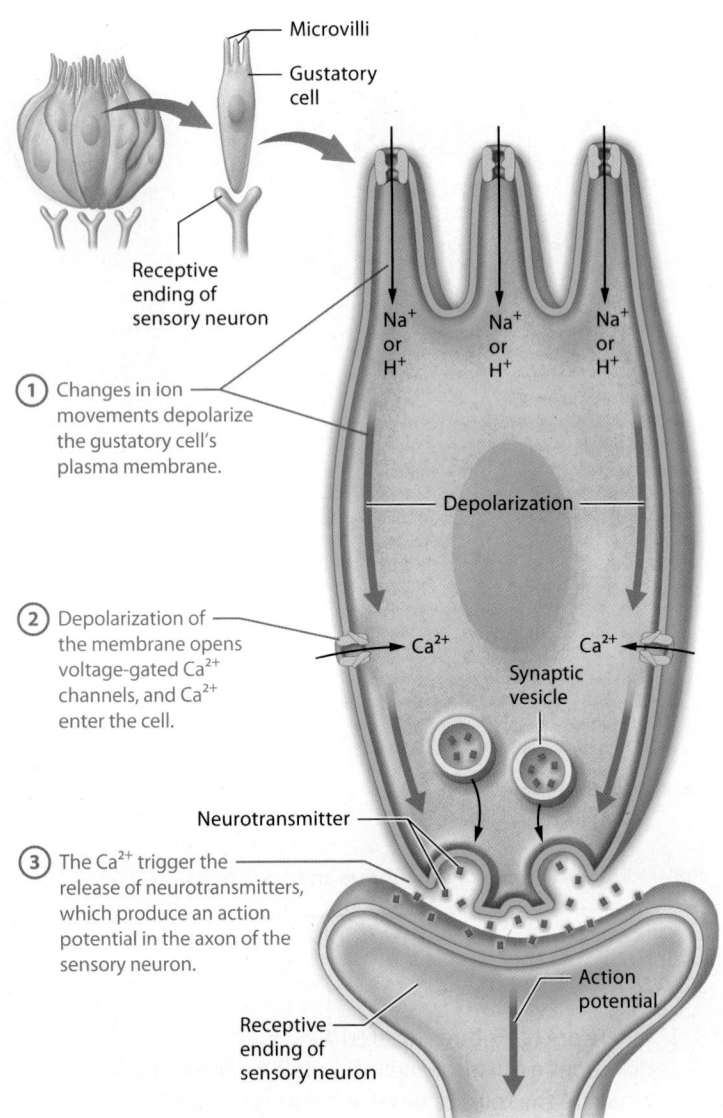

Figure 15.6 Transduction of taste in a gustatory cell.

③ **The calcium ions trigger the release of neurotransmitters, which produce an action potential in the axon of the sensory neuron.** Gustatory cells use a variety of neurotransmitters, including serotonin, glutamate, acetylcholine, and GABA. In the next section, we find out where these sensory neurons take the taste stimuli in the CNS.

Quick Check

☐ 4. What types of chemicals produce each of the five taste sensations?

☐ 5. How does a gustatory cell transduce a chemical taste into a nerve signal?

The Gustatory Pathway

Taste stimuli travel to the CNS, where they are sent to various parts of the brain for processing (**Figure 15.7**):

Figure 15.7 **The gustatory pathway.**

① Axons of the facial, glossopharyngeal, and vagus nerves carry taste stimuli from the tongue into the CNS. The sensory neurons that receive signals from the taste buds travel as part of one of these three cranial nerves. The facial nerve carries taste sensations from the anterior two-thirds of the tongue, and the glossopharyngeal nerve carries taste sensations from the posterior third of the tongue. The vagus nerve plays a minor role in taste, innervating taste buds that are scattered on the base of the tongue and the posterior roof of the mouth.

② Axons of these three nerves terminate in the solitary nucleus in the medulla oblongata by synapsing on central sensory neurons. Several neurons in the *solitary nucleus* connect with motor neurons of the vagus nerve to control reflexive production of saliva.

③ Axons from the solitary nucleus synapse on neurons in the thalamus, which then send the taste signals to the primary gustatory cortex in the parietal lobe. The primary gustatory cortex produces awareness and identification of a particular taste.

From the primary gustatory cortex, taste signals are sent to two different areas. In the inferior part of the frontal lobe, this taste information is integrated with visual and olfactory information to form a complete "picture" of what you are eating. Taste information also travels to the limbic system, which provides emotional reactions to taste (both pleasant and unpleasant).

Quick Check

☐ 6. Which cranial nerves carry taste sensation to the CNS?
☐ 7. Which part of the brain is responsible for identifying a particular taste?

Apply What You Learned

☐ 1. The salivary glands produce saliva. Explain why a tumor in the salivary glands that destroys the saliva-producing cells can interfere with taste sensation.
☐ 2. A patient has had a stroke that damaged the trigeminal nerve but not the facial, glossopharyngeal, or vagus nerve. Would this individual still be able to taste the difference between hot peppers and French fries? Explain your answer.
☐ 3. You ordered coleslaw at a new restaurant but unfortunately suffered repeated bouts of vomiting for the next 12 hours. Three years later, your stomach turns at the very thought of eating coleslaw. Explain this reaction.

See answers in Appendix A.

MODULE 15.4

Anatomy of the Eye

Learning Outcomes

1. Discuss the structure and functions of the accessory structures of the eye.
2. Describe the innervation and actions of the extrinsic eye muscles.
3. Identify and describe the three layers of the eyeball.
4. Describe the structure of the retina.

The eyeballs are located in the orbits lateral to the nasal cavity. Each eyeball occupies only the anterior one-third of the orbit; the remainder of the orbit is filled with adipose tissue that cushions the eyeball and holds it in position. The orbits also contain accessory structures.

Accessory Structures of the Eye

◄◄ **Flashback**

1. What is the function of sebaceous glands? (p. 176)
2. What is a motor unit? (p. 368)

The accessory structures of the eye include the *eyelids, eyebrows* and *eyelashes, conjunctiva, lacrimal apparatus,* and *extrinsic eye muscles.* These structures have important roles in moving the eyeball and protecting it from injury.

Eyelids

The **eyelids,** or **palpebrae** (pal-PEE-bree), are two thin folds that cover the anterior part of the orbit (**Figure 15.8a**). They prevent foreign objects from entering the eye and distribute tears across the surface of the eye during blinking. Each eyelid is stiffened by a **tarsal plate,** a thin piece of dense regular collagenous connective tissue (**Figure 15.8b**). Modified sebaceous glands called **tarsal glands** are located within the tarsal plate and secrete oil to prevent the eyelids from sticking together. The upper and lower eyelids meet at the edges of the orbit at the medial and lateral **commissures** (KOM-uh-sure; also known as the medial and lateral canthi). The **lacrimal caruncle** (KAR-ung-kl) is a fleshy structure found at the medial commissure; it contains sebaceous glands that secrete a whitish lubricating substance.

Two skeletal muscles move the eyelids. Both the upper and lower eyelids contain portions of the orbicularis oculi muscle, which allows you to tightly close the eyelids. The levator palpebrae superioris muscle inserts in the upper lid and elevates it.

Eyebrows and Eyelashes

The **eyebrows** and **eyelashes** serve a protective function (Figure 15.8). Hairs along the ridge of the brow form the eyebrows, which prevent perspiration from running into the eyes. They

Lacrimal caruncle

Medial commissure

Lateral commissure

(a) Anterior view

Eyebrow

Levator palpebrae superioris muscle

Orbicularis oculi muscle

Tarsal plate

Tarsal gland

Superior eyelid

Eyelashes

Inferior eyelid

Tarsal gland

Tarsal plate

Orbicularis oculi muscle

Palpebral conjunctiva

Bulbar (ocular) conjunctiva

(b) Lateral view

Figure 15.8 Accessory structures of the eye.

Practice art labeling
@MasteringA&P®

also reduce glare in bright light and are important in forming facial expressions. Eyelashes are stiff hairs located in the edges of the upper and lower eyelids. They are richly innervated by nerve endings so that the slightest touch or puff of air triggers the eyelids to blink, which reduces the risk of injury to the eye.

Conjunctiva

The **conjunctiva** (kon-junk-TY-vah) is a thin epithelial membrane that lines the posterior surfaces of the eyelids (the **palpebral conjunctiva**) and then turns back on itself as the **bulbar (or ocular) conjunctiva** to cover the anterior surface of the white part of the eye (Figure 15.8b). The conjunctiva is translucent, which permits you to see tiny blood vessels beneath it. Inflammation of the conjunctiva is called conjunctivitis; when caused by a viral or bacterial infection, conjunctivitis is known as "pink eye," a highly contagious condition.

Lacrimal Apparatus

The **lacrimal apparatus** (LACK-rah-mal) produces tears and drains them from the eye. The **lacrimal gland** (Figure 15.9a) is located in the superolateral region of the orbit, just posterior to the conjunctiva. When stimulated by autonomic neurons, tiny ducts release tears and mucus into the conjunctival sac to lubricate and wash away dust and debris.

Blinking sweeps the tears medially and inferiorly across the surface of the eye, where they drain into a series of passages that lead to the nasal cavity (Figure 15.9a). Tears first enter the **lacrimal puncta** (singular, *punctum*), which are tiny holes in the medial edge of each eyelid. The puncta are continuous with the **lacrimal canaliculi** (can'-ah-LICK-you-lie; singular, *canaliculus*), small ducts that empty into the **lacrimal sac.** The sac

is located in a small depression in the lacrimal bone and drains into the **nasolacrimal duct,** which travels through the lacrimal and maxillary bones to reach the inferior nasal meatus (just inferior to the inferior concha). A special type of x-ray taken after the injection of dye into these structures shows how narrow they are (Figure 15.9b).

If tear production increases in response to debris, allergens, or emotions, you end up with a runny nose, and excess tears may spill over onto your cheek. The connection between the nasal cavity and the eye can be a two-way street—pathogens can travel from the nose to the eye, resulting in conjunctivitis or a painful infection of the nasolacrimal duct. It also explains, from an anatomical perspective, why people can compete in contests to see how far they're able to shoot milk out of their eye.

Extrinsic Eye Muscles

Our vision would be more limited if our eyes were fixed in a single position. We'd have to move our heads every time we wanted to focus on something new or follow a moving object (imagine watching a tennis match!). Fortunately, we can move our eyes, thanks to the extrinsic eye muscles, six skeletal muscles that have their origins on the walls of the orbit and insert into the outer layer of each eyeball. The **superior, inferior, lateral,** and **medial rectus** muscles extend from a common tendinous ring on the posterior wall of the orbit to their respective insertions on the eyeball (Figure 15.10). The **superior oblique** muscle travels from the posterior orbit along the medial wall and through a fibrous loop called the **trochlea** (TROKE-lee-ah) before inserting on the superolateral part of the eyeball. The **inferior oblique** muscle sweeps from the medial floor of the orbit to insert on the inferolateral part of the eyeball.

Lacrimal punctum

Lacrimal sac

Lacrimal canaliculi

Nasolacrimal duct

Inferior concha

Inferior nasal meatus

Lacrimal gland

(a) Anterior view

(b) X-ray of lacrimal ductwork with dye injected

Figure 15.9 The lacrimal apparatus.

(a) Anterior view of right eye

(b) Lateral view of left eye

Figure 15.10 Extrinsic eye muscles.

The actions of the four rectus muscles seem straightforward, given their names: The superior rectus moves the eye superiorly, for example. However, the origin of these muscles lies slightly medial to the eyeball; thus, the superior and inferior rectus muscles tend to draw the eye medially as well as superiorly and inferiorly. The oblique muscles can counteract this medial movement. The superior oblique depresses the eye and draws it laterally; acting together with the inferior rectus, it moves the eye directly inferiorly. Likewise, the inferior oblique elevates the eye and moves it laterally. Looking straight up at the ceiling thus requires coordination of the inferior oblique and the superior rectus. (See Table 9.3 for a summary of extrinsic eye muscle movements.)

The extrinsic eye muscles must produce very small, precise movements, so they are among the most highly innervated skeletal muscles in the body. Some of the motor units consist of only two or three muscle fibers (compared to a thousand in muscles of the leg). Three of the 12 cranial nerves—the oculomotor

Figure 15.11 Sagittal section of internal structures of the eye.

Practice art labeling
@MasteringA&P®

(CN III), trochlear (CN IV), and abducens (CN VI)—innervate the extrinsic eye muscles.

Injury to any of these cranial nerves can weaken or paralyze the eye muscles. If the eyes cannot move together, they send slightly different images to the brain, which is interpreted as blurred or double vision, a condition known as *diplopia* (dih-PLOH-pee-ah). If diplopia is present at birth, the child is said to have "lazy eye," or **strabismus** (strah-BIZ-mus). Strabismus can sometimes be corrected by covering the good eye with a patch, thereby forcing the use of the affected eye to strengthen the weak muscles. If not corrected in time, the muscles will atrophy and diplopia will become permanent.

Quick Check

☐ 1. What is the function of the eyelashes?
☐ 2. Trace the path of tears as they travel from the eye to the nasal cavity.
☐ 3. Which two extrinsic eye muscles move the eye superiorly?

The Eyeball

◄◄ **Flashback**

1. What is the function of collagen fibers? (p. 125)
2. What happens when a sphincter muscle contracts? (p. 285)

The eyeball is a hollow sphere approximately 2.5 cm (1 inch) in diameter that rests in the anterior portion of the orbit. The eyeball consists of an outer wall that encloses several inner chambers and the **lens,** which focuses light as it enters the eye.

Layers of the Eyeball

Like many organs we have discussed, the wall of the eyeball consists of three tissue layers or tunics: the *fibrous, vascular,* and *neural* layers. These layers perform diverse functions, including maintaining the shape of the eyeball and transducing light into neural signals. Let's look more closely at each.

Fibrous Layer The **fibrous layer,** the outermost layer of the eyeball, has two parts: the **sclera** (SKLER-ah) and the **cornea** (KOR-nee-ah) (**Figure 15.11**). The sclera, the "white" part of the eye,

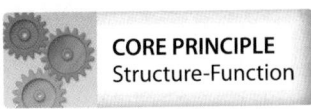

CORE PRINCIPLE
Structure-Function

covers approximately five-sixths of its surface. In an example of the Structure-Function Core Principle (p. 25), the sclera's numerous collagen fibers allow it to resist deformation from external and internal forces and maintain its shape. These fibers are arranged irregularly, which is why the sclera is opaque.

The sclera is continuous with the cornea in the anterior eye. Unlike the sclera, the cornea is translucent, which allows it to play an important role in admitting light into the eye and focusing it. The cornea is translucent due to its orderly parallel arrangement of collagen fibers, relative lack of water in its tissue, and absence of blood vessels.

Since the cornea is avascular, it must obtain oxygen from the fluid behind it as well as from the air. Contact lenses sit directly on the cornea and are designed to allow oxygen to reach it. Extended-wear lenses can be worn overnight because they allow more oxygen through, which reduces corneal swelling and the risk of infection. The cornea also contains many nerve endings that convey pain, but relatively few touch receptors. This is why we can tolerate smooth contact lenses, but not jagged particles of dust.

Vascular Layer If you peel away the cornea and sclera, you will find the **vascular layer** directly underneath. This middle layer of the eyeball has several components (Figure 15.11):

1. **Choroid.** The most extensive component of the vascular layer, the **choroid** (KORE-oyd), contains many tiny blood vessels called capillaries as well as a pigment to reduce the scattering of light. The reflection of light from a camera's flash off the choroid is responsible for producing the "red-eye" effect that has ruined many a family photo.
2. **Ciliary body.** The choroid is continuous anteriorly with the **ciliary body** (SILL-ee-air-ee), which contains a ring of smooth muscle that surrounds the lens. Fine threads called **suspensory ligaments** connect the ciliary body to the lens; contraction and relaxation of the smooth muscle changes the shape of the lens to focus light on the retina.
3. **Iris.** The **iris,** or colored portion of the eye, is the extension of the vascular layer just anterior to the ciliary body. The color of the iris is determined by the amount of the brown pigment melanin that is present—brown eyes result from a large amount of melanin, and blue or green eyes result when there is very little pigment. Infants typically have blue or gray eyes at birth because melanin does not develop immediately.

The iris surrounds the circular **pupil,** an opening through which light enters the eye. The iris contains two muscles that protect the eye and improve vision by controlling how much light passes through the pupil (**Figure 15.12**). The **pupillary sphincter** (PEW-pih-lair-ee) muscle contracts following parasympathetic stimulation and constricts the pupil to reduce the amount of light entering the eye. The **pupillary dilator** muscle contracts

Figure 15.12 Constriction and dilation of the pupil.

Optic disc   Fovea centralis   Macula lutea

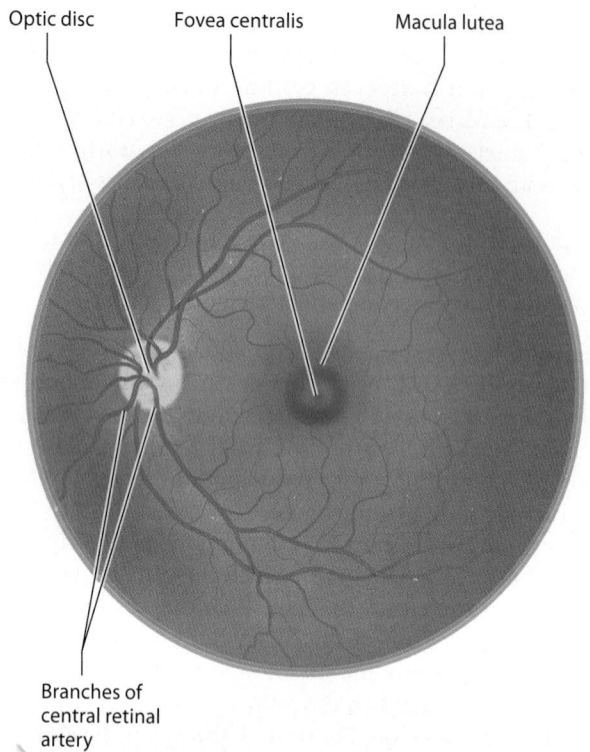

Branches of
central retinal
artery

Figure 15.13 **Photo of interior of the eye.**

following sympathetic stimulation; its fibers are arranged in a radial pattern, which causes it to enlarge the pupil to permit the entrance of more light into the eye. If you've ever had your "eyes dilated" during an eye exam, you have experienced the effect of blocking parasympathetic activity, which allows sympathetic activity to dilate the pupils without any opposition.

Neural Layer (Retina) The neural layer of the eyeball, commonly known as the **retina** (RET-ih-nah), is the innermost layer of the eyeball. It is not a complete layer and is only found deep to the choroid. It contains **photoreceptor cells,** which are specialized to detect the presence of light and produce a neural signal, as explained in the next module.

The retina consists of two layers. The superficial layer (adjacent to the choroid) is a thin, pigmented epithelium that reduces scattering of light and nourishes the photoreceptors. Deep to this is a layer that contains photoreceptors and the cells that give rise to the optic nerve. The boundary between anterior retina and the posterior edge of the ciliary body is visible as a serrated line known as the *ora serrata* (see Figure 15.11).

The **fovea centralis** (FOE-vee-ah sen-TRA-liss) is an area in the retina that contains a high density of photoreceptors relative to the rest of the retina. For this reason, it allows for extremely detailed vision, enabling us to focus on a particular object (see **Figure 15.13**). The fovea is located in the center of a yellowish region called the **macula lutea** (MACK-you-lah LOO-tee-ah), which also contains a large number of photoreceptors.

The axons of the optic nerve gather at the **optic disc,** a portion of the retina that does not contain any photoreceptors (see Figure 15.13). For this reason, this area does not capture any visual image and is called the *blind spot.* We don't typically notice our blind spot because the brain is adept at "filling in" the missing information based on the images next to it. However, you can find your blind spot by using **Figure 15.14**.

The retina's epithelium and photoreceptors are nourished by blood vessels in the choroid. The deeper neural structures of the retina are supplied by branches of the central retinal artery, which enters the eye through the optic nerve. These blood vessels are visible through the pupil during an eye exam (see Figure 15.13).

The two layers of the retina are usually in contact with each other. However, certain conditions—such as trauma, diabetes, or an abnormally shaped eyeball—can pull the inner layer away from the pigmented epithelium, a condition commonly referred to as a *detached retina.* Since this isolates the photoreceptors from their blood supply, permanent loss of vision can occur if it is not corrected within several days.

Lens

The lens is a slightly flattened sphere located posterior to the iris and pupil that focuses light on the retina from objects near the eye. As you can see in **Figure 15.15**, it is surrounded by the ciliary

(a) Find your "blind spot."

① To test your left eye, cover your right eye and focus on the X. Slowly move your book toward (or away from) you until the dot disappears.

② To test your right eye, cover your left eye and focus on the dot. Move the book until the X disappears.

(b) Fill in your "blind spot."

① To test your left eye, cover your right eye and focus on the X. Slowly move your book toward (or away from) you until the pencil appears intact.

② When the gap in the pencil hits your blind spot, areas in your brain fill in the missing information so the pencil appears intact.

③ To test your right eye, turn your book upside down, cover your left eye, and repeat the process.

Figure 15.14 Demonstration of the blind spot and "filling in."

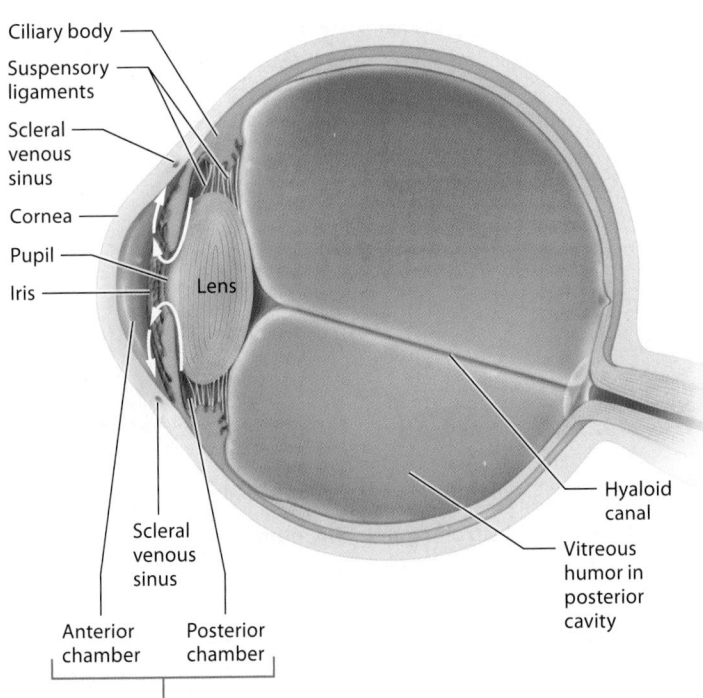

Ciliary body
Suspensory ligaments
Scleral venous sinus
Cornea
Pupil
Iris
Lens
Scleral venous sinus
Anterior chamber
Posterior chamber
Hyaloid canal
Vitreous humor in posterior cavity

Anterior and posterior chambers form the anterior cavity and are filled with aqueous humor. The arrows show the flow of aqueous humor before it drains into the scleral venous sinus.

Figure 15.15 Cavities and chambers of the eye.

body and connected to it by the suspensory ligaments. The lens contains cells known as lens fibers that lack nuclei and are tightly packed to make them transparent. It is typically maintained in its roundest shape by elastic fibers found in its capsule unless external forces flatten it. To discover how a clouded lens impairs vision, see *A&P in the Real World: Cataracts*.

Cavities and Chambers of the Eye

Notice in Figure 15.15 that the eyeball is divided into anterior and posterior cavities by the lens and ciliary body. The large **posterior cavity,** located posterior to the lens, is filled with a gelatinous material called the **vitreous humor,** or vitreous body (VIT-ree-us). Vitreous humor presses the retina against the choroid and helps maintain the shape of the eyeball. It consists mostly of water and collagen fibers, with very few cells. Some people experience "floaters," which are shadows cast by cellular debris trapped in the vitreous humor. The number of floaters increases with age, as the vitreous humor tends to shrink but is not replaced. Also note in the figure that the small *hyaloid canal* passes through the vitreous humor from the lens to the optic nerve. After birth, it contains fluid that assists in changes in the shape of the lens.

The **anterior cavity** is the area anterior to the lens and ciliary body. It itself consists of two smaller cavities: the **posterior chamber,** between the lens and the iris, and the **anterior chamber,** between the iris and the cornea. These two chambers are both filled with watery **aqueous humor** (AY-kwee-us), secreted by the ciliary body. Aqueous humor flows from the posterior chamber through the pupil into the anterior chamber. To prevent buildup, the fluid must then drain at the same rate from the anterior chamber into a

Cataracts

A transparent lens is essential for vision; if light cannot pass through the lens, vision is impaired even if there are functional photoreceptors in the retina. Indeed, a clouded lens, or **cataract** (KAT-ah-rakt), is one of the most frequent causes of blindness. Trauma, exposure to UV radiation, and diseases such as diabetes can promote cataracts. However, the most common cause is simply aging, as the lens fibers tend to progressively darken and become less organized, turning the lens milky white. A cataract cannot be reversed, so the usual treatment is to replace the entire lens surgically with a synthetic lens. This restores vision, although glasses or contact lenses may be required for minor adjustments.

blood vessel that circles the anterior edge of the iris, called the **scleral venous sinus** (canal of Schlemm). What happens when the aqueous humor doesn't drain properly? For the answer, see *A&P in the Real World: Glaucoma*.

Quick Check

☐ 4. In which layer of the eyeball are the photoreceptors found?
☐ 5. What are the functions of each component of the vascular layer of the eyeball?

Glaucoma

In the condition **glaucoma** (glaw-KOH-mah), the aqueous humor cannot drain and fluid builds up in the anterior and posterior chambers. This excess fluid raises the pressure inside the eyeball, or *intraocular pressure*. This elevated pressure pushes on the vitreous body, which in turn compresses and damages the retina and the optic nerve as it leaves the eye. Glaucoma is the second leading cause of blindness and can result from eye infections, certain medications, or congenital defects in the scleral venous sinus. However, in most cases the cause is unknown.

Most people who have glaucoma do not have any symptoms other than gradual loss of vision, which can occur so slowly that it may not be detected until the disease is advanced. Unfortunately, lost vision cannot be restored, but progression can be stopped or slowed with medications that either improve drainage or reduce the amount of aqueous humor produced. If medications fail to control intraocular pressure, surgical procedures may improve drainage.

□ 6. What are the two fluids found in the eyeball? How do they differ in location, composition, and function?

Apply What You Learned

□ 1. Explain why it is not uncommon to develop pink eye after having a cold.
□ 2. When a patient is asked to look to the left, her right eye moves to the left, but her left eye does not. Which extrinsic eye muscle is not functioning properly?
□ 3. While a student is examining the anterior half of the eye in cadaver lab, the cadaver lens falls out. Which structure(s) is/are no longer intact?

See answers in Appendix A.

MODULE **15.5**
Physiology of Vision

Learning Outcomes

1. Describe how light activates photoreceptors.
2. Explain how the optical system of the eye creates an image on the retina.
3. Compare the functions of rods and cones in vision.
4. Trace the path of light as it passes through the eye to the retina and the path of nerve impulses from the retina to various parts of the brain.
5. Explain the processes of light and dark adaptation.

The structures described in the previous module work together to produce **vision,** which is the perception of light reflected by various objects. However, vision is more than simply detecting light. Using our eyes and the visual pathways in the CNS, we can determine the size, shape, and color of an object (such as a round, purple Frisbee), as well as the rate and direction of movement (it's rapidly flying toward your head), and the distance between us and the object (it's close enough that you'd better duck). This module shows how the structures of the visual system provide these capabilities.

Principles of Light

To understand how our special sense of vision works, it's helpful to first summarize a few basic principles of light. Some energy takes the form of electromagnetic radiation, which has different wavelengths, measured in nanometers (nm). These range from short wavelengths, such as those of gamma rays and x-rays, to the much longer wavelengths of microwaves and radio waves (**Figure 15.16**). **Visible light** is the range of wavelengths that we can detect with our eyes, and we perceive segments of that range as particular colors. For example, we see shorter wavelengths of visible light (about 420 nm) as blue and violet, and longer

Figure 15.16 The electromagnetic spectrum.

wavelengths (660 nm) as red. Regardless of wavelength, the basic unit of light is the **photon** (FOH-tawn), which stimulates the photoreceptors in the retina.

Rays of light bend, or **refract,** when they pass through a translucent object. The **refractive index** of an object measures the amount of refraction it provides. Air has a refractive index of 1.0, meaning that light passing through air is not significantly bent (**Figure 15.17a**). If a medium has a refractive index greater than that of air, light is refracted (**Figure 15.17b**). You have seen refraction if you have ever put a spoon into a glass of water. The portion of the spoon sticking out of the water appears straight, but the portion under water appears bent. This is due to the water having a higher refractive index than air, and so the light rays entering the water refract.

The degree to which light refracts also depends on the angle at which it strikes the surface: The greater the angle, the greater the refraction. When light strikes a curved surface, the rays at the edges of the surface strike at a greater angle and bend more. A convex lens has a surface that bulges outward and causes rays of light to **converge,** or come together, as they pass through. When the rays converge on one point, known as the *focal point,* they are said to be *focused* (**Figure 15.17c**). A concave lens has a surface that "caves in" and causes the rays to **diverge,** or spread out, so they do not focus (**Figure 15.17d**).

We now turn to how the structures of the eye focus light on the retina.

Quick Check

□ 1. What is refraction?
□ 2. What is the difference between convergence and divergence of light?

Focusing Light on the Retina

To provide clear vision, rays of light must focus on the retina, not behind or in front of it. Unfortunately, when light strikes

Air:
refractive index = 1.0

(a) Density similar to air: Light does not refract.

Water:
refractive index > 1.0

(b) Density greater than air: Light refracts.

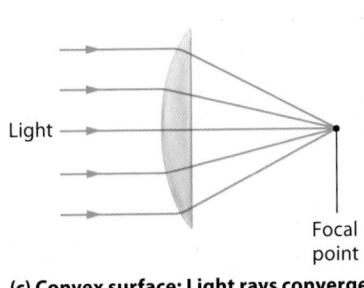

Light →

Focal point

(c) Convex surface: Light rays converge to focus on a focal point.

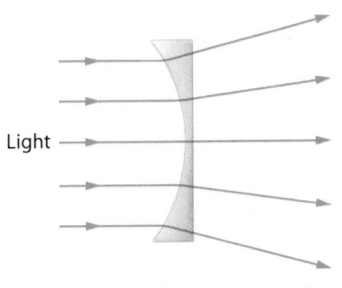

Light →

(d) Concave surface: Light rays diverge and do not focus.

Figure 15.17 Refraction of light.

a nontransparent object (such as an apple), the light rays are reflected and scatter away from the object in a diverging pattern. If the object is more than 20 feet away, the reflected rays are nearly parallel when they reach your eyes. If the object is closer than 20 feet away, the rays are still spreading out, which is shown in **Figure 15.18a**. In either case, the rays must be refracted to converge and focus on your retina. The cornea and lens provide this refraction.

The greatest degree of refraction—about two-thirds of the eye's refractive power—occurs as light passes from air through the cornea, which has a refractive index similar to that of water. (This explains why your vision underwater is blurry—the light passing from the water to the cornea does not refract much.) Some additional refraction occurs as light passes through the lens, but the refractive index of the lens is similar to that of the aqueous humor, so the lens cannot provide as much refraction as the cornea. For this reason, the lens typically provides for fine refractive adjustments.

Since rays of light from distant objects are nearly parallel when they reach the eye, they need little refraction to be focused on the retina. The cornea provides most of the necessary refraction, and the lens in its normal flattened shape provides the rest (**Figure 15.18b**). When the eye is relaxed and focusing on a distant object, this is known as the **emmetropic** (em-eh-TRO-pick) state of the eye.

However, diverging rays from nearby objects require a greater degree of refraction, far beyond what the cornea can provide. The lens increases total refraction by changing its shape from flattened to round; this ability of the lens to "round up" is called **accommodation.** When the lens is rounded, its curved surface refracts light to a greater extent and can focus it on the retina (**Figure 15.18c**).

How does the lens change its shape? The answer lies in the sphincter-like smooth muscle found in the ciliary body that surrounds the lens. Here we see another example of the Structure-Function Core Principle (p. 25). When viewing distant objects, the smooth

CORE PRINCIPLE
Structure-Function

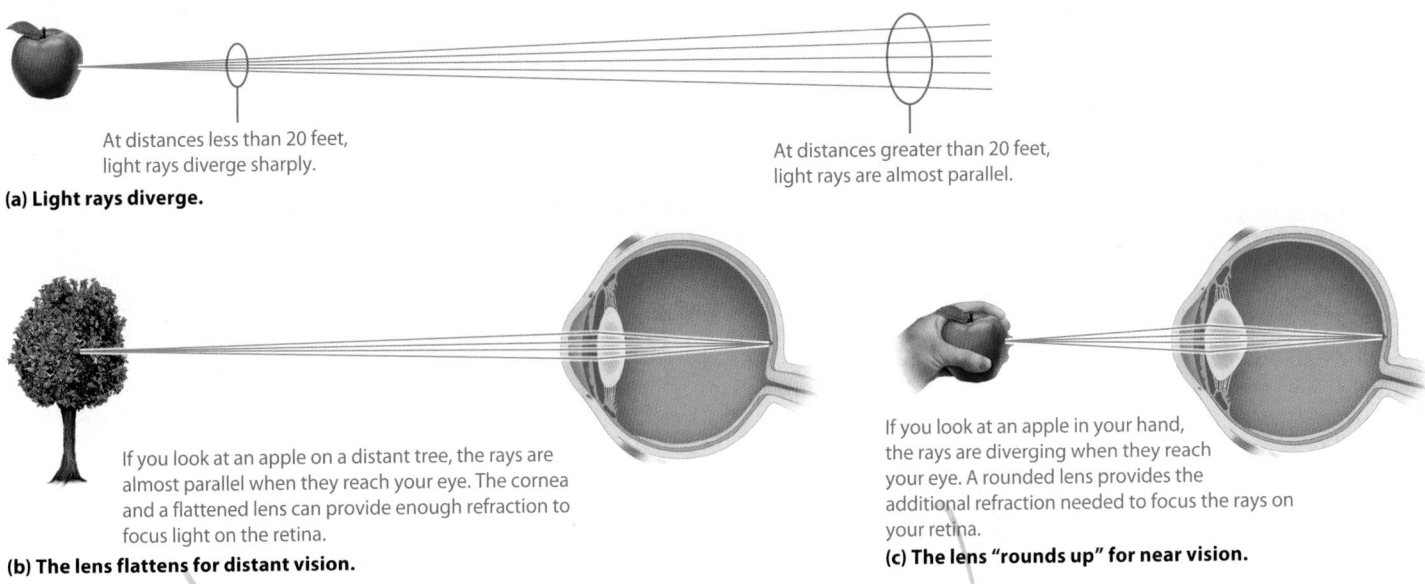

At distances less than 20 feet, light rays diverge sharply.

At distances greater than 20 feet, light rays are almost parallel.

(a) Light rays diverge.

If you look at an apple on a distant tree, the rays are almost parallel when they reach your eye. The cornea and a flattened lens can provide enough refraction to focus light on the retina.

(b) The lens flattens for distant vision.

If you look at an apple in your hand, the rays are diverging when they reach your eye. A rounded lens provides the additional refraction needed to focus the rays on your retina.

(c) The lens "rounds up" for near vision.

Figure 15.18 Refraction by the lens of the eye.

Posterior view of lens

Suspensory ligaments

Lens

Lateral view

Smooth muscle of ciliary body relaxes.

The relaxed ciliary body moves away from the lens, putting tension on suspensory ligaments and flattening the lens.

(a) Emmetropic vision: Ciliary body relaxes and flattens the lens for distant vision.

Posterior view of lens

Suspensory ligaments

Lens

Lateral view

Smooth muscle of ciliary body contracts.

The contracted ciliary body moves closer to the lens, removing tension on suspensory ligaments and allowing the lens to round up.

(b) Accommodation: Ciliary body contracts, and the lens "rounds up" for near vision.

Figure 15.19 Action of the ciliary body in accommodation.

muscle of the ciliary body relaxes, moving the ciliary body away from the lens. This movement increases tension in the suspensory ligaments, flattens the lens, and reduces refraction (**Figure 15.19a**).

When accommodation occurs in the viewing of nearby objects, the ciliary muscle contracts and the ciliary body moves closer to the lens, which you can see in **Figure 15.19b**. This decreases tension on the suspensory ligaments, which reduces their pull on the lens. Without this tension, the elastic fibers of the lens make it "round up" into a nearly spherical shape. This change in shape increases refraction and focuses light from nearby objects on the retina.

However, accommodation is not the only event necessary to focus on near objects. Two other processes must also occur:

1. **Pupillary constriction.** Light rays that enter the edge of the lens cannot be refracted enough and therefore produce blurry images on the retina. To prevent this, the pupils constrict slightly to cover the edge of the lens. Dilation of the pupils during an eye exam prevents this constriction, and you cannot focus well on close objects because of the light entering the edge of the lens.
2. **Convergence.** To focus near objects on the fovea of each eye, the eyeballs must move medially. This movement is called **convergence**.

You can easily observe these processes by watching a friend's eyes as he or she shifts focus from a far object to a near object.

Errors of Refraction: Presbyopia, Hyperopia, and Myopia

Accommodation serves the normal, or emmetropic, eye well under many circumstances, but it has its limits. If an object is too close to the eye, the lens cannot round up enough to provide the necessary refraction. The closest point at which we can focus on an object is called the **near point of accommodation,** and it increases with age. Infants have very flexible lenses and can focus on things only a few inches from their eyes. As we age, the lens gradually stiffens and cannot accommodate as well. By the time we reach our mid-40s, the near point is approximately 10–20 inches, and most people begin to have difficulty with reading or activities requiring close vision. This condition, **presbyopia** (prez'-bee-OH-pee-ah), is usually corrected with reading glasses or bifocals (glasses with one set of lenses for distance vision and another set of lenses for close vision).

Accommodation by the lens may also be insufficient if the shape of the eye or the curvature of the cornea is abnormal. In **emmetropia** (em'-eh-TRO-pee-ah), the length of the eyeball is normal in the anterior-posterior direction and the lens can focus light on the retina (**Figure 15.20a**). In **hyperopia** (hy'-per-OH-pee-ah; farsightedness), the eyeball is too short (**Figure 15.20b**)

(a) Emmetropia: normal eyeball shape.

 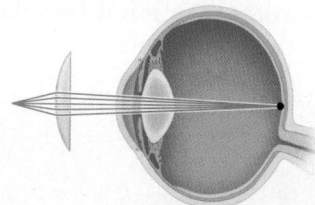

Uncorrected, the lens cannot round up enough and light rays focus behind the retina when viewing near objects.

A convex lens converges the light rays before they reach the eye, and the lens can provide the rest of the needed refraction.

(b) Hyperopia (farsightedness): Eyeball is too short.

Uncorrected, the lens cannot flatten enough and light rays focus in front of the retina when viewing distant objects.

A concave lens diverges the light rays before they reach the eye, so the minimum refraction of the lens is enough to focus the rays on the retina.

(c) Myopia (nearsightedness): Eyeball is too long.

Figure 15.20 Errors of refraction.

or the cornea is too flat. The lens cannot round up enough, and the image focuses behind the retina, causing blurry images when viewing objects up close. Refraction is increased by using a lens with convex surfaces that cause the light to converge on the retina. The opposite occurs in **myopia** (my-OH-pee-ah; nearsightedness): The distance between the cornea and the retina is too long (**Figure 15.20c**), or the cornea curves too much. The lens cannot flatten enough, which focuses the light in front of the retina and blurs the image when viewing distant objects. Myopia is corrected with a concave lens that diverges light before it strikes the lens of the eye. The lens of the eye can then provide enough refraction to focus light on the retina.

Astigmatism (ah-STIG-mah-tizm) occurs when the curvature of the lens or the cornea is irregular and the rays of light are not evenly refracted. Corrective lenses can treat astigmatism, too.

Quick Check

- ☐ 3. What is accommodation?
- ☐ 4. How do the ciliary body and lens focus light from a near object on the retina?
- ☐ 5. What is the near point of accommodation, and how does it change as we age?

Photoreceptors and the Retina

◀◀ **Flashback**

1. How do depolarization and hyperpolarization differ? (p. 397)
2. What is an inhibitory postsynaptic potential (IPSP)? (p. 409)

As described earlier, the retina consists of two layers, an outer pigmented epithelium and an inner layer that contains photoreceptors and other cells. This section focuses on the inner layer, as it contains the photoreceptors and neurons responsible for detecting light, although we will see that the pigmented layer also has an essential role.

The inner layer of the retina contains several types of cells arranged in layers. Adjacent to the pigmented layer we find the two types of photoreceptors, *cones* and *rods*. These cells synapse with neurons called **bipolar cells,** which communicate with **retinal ganglion cells,** the most anterior cells in the retina (**Figure 15.21**). The axons of the retinal ganglion cells course across the retinal surface and form the optic nerve (CN II). Two other cell types, **horizontal cells** and **amacrine cells** (AM-ah-kreen), are also involved in image processing. The horizontal cells form connections between photoreceptors, and the amacrine cells are located among the dendrites of the retinal ganglion cells.

As noted, the two types of photoreceptors are cones and rods. **Cones** contain pigments that allow us to perceive color and they

Light

| Choroid |
| Pigmented layer of retina |
| Rods |
| Cones |
| Horizontal cell |
| Bipolar cells |
| Amacrine cell |
| Retinal ganglion cells |

Light

(a) Cellular structure of retina

(b) Micrograph of retina

LM (450×)

Figure 15.21 Layers of the retina.

function best in bright light. **Rods** cannot detect color, but they are very sensitive and capable of responding in less intense light. Indeed, our ability to see at night is almost entirely due to rods. There are 5 million cones, which are highly concentrated in the fovea centralis. As you move outward away from the fovea, the concentration of cones decreases, until, at the very edge of the retina, no cones are found at all. Conversely, there are 100 million rods, but none occur in the fovea; the concentration of rods increases as we move away from the fovea and tapers off near the edges of the retina. The location of cones in the fovea allows us to focus on an object. On the other hand, the location of rods makes them important for peripheral vision, or detecting images at the edge of our vision.

This arrangement of cones and rods explains why, at night, you can see an object if you look just to the side of it, but it disappears when you look directly at it. Looking directly at the object focuses it on the fovea, where there are only cones, but cones do not function in dim light. Looking to the side of the object focuses it on the retina adjacent to the fovea, where rods are present.

Now let's discuss how cones and rods transduce light into a neural signal that can be interpreted by the brain.

Structure of Photoreceptors

Cones and rods are located adjacent to the retina's pigmented layer. They consist of cell bodies connected to portions called *inner* and *outer segments* that are on the side closer to the pigmented layer (**Figure 15.22a**). The outer segments house stacks of flattened discs that absorb light and are formed from segments of the plasma membrane. They have a lifespan of only 10–12 days, and then they are shed by the photoreceptor and phagocytosed by the cells of the pigmented layer. The inner segments contain mitochondria and the typical cellular organelles; they also generate new discs for the outer segments.

Rods have an outer segment that is shaped like a cylinder and contains as many as 1000 discs that fold and bud off from the plasma membrane and float in the cytosol. Each disc contains the pigment **rhodopsin** (roh-DOP-sin) in its membrane (**Figure 15.22b**). Rhodopsin has two components: the protein **opsin** (OP-sin) and the pigment **retinal** (RET-ih-nal), which is derived from vitamin A. In the dark, the retinal is in a bent configuration known as *cis*-retinal. All rods contain this pigment and do not distinguish between different wavelengths (colors) of light.

Like rods, cones are named because of the shape of their outer segment (Figure 15.22a and **Figure 15.22c**). The discs in the cones are infoldings of the plasma membrane and contain the pigment **iodopsin**. Iodopsin is made of two components: retinal and the protein **photopsin** (foe-TOP-sin). Photopsin is similar to opsin but has a slightly altered structure that allows it to absorb different wavelengths of light.

There are three forms of photopsin that respond to wavelengths we perceive as either blue, green, or red. The color pigments react to slightly overlapping ranges of wavelengths, so that our perception of each color is generated by stimulation of varying combinations of cone populations (**Figure 15.22d**). For example, red light stimulates primarily red cones, but orange

light stimulates red cones and some green cones as well. Read about what happens when one or more types of cones do not function in *A&P in the Real World: Color Blindness* on page 560.

Transduction in Photoreceptors

Transduction of light into a neural signal begins when a photon strikes a disc in the outer segment of a rod or cone. Much more is known regarding the events in rods, so we'll focus on the process there, although it is likely similar in cones.

It is important to understand a fundamental difference in transduction in the retina compared to other special senses. Recall that olfactory neurons and gustatory cells detect smell and taste stimuli and respond by depolarizing and signaling another cell; in other words, the olfactory neurons and gustatory cells are "off" until a stimulus turns them "on."

In the visual system, the process is reversed. In the dark (the absence of a stimulus), the photoreceptor cells are depolarized ("on") and continually release neurotransmitters onto adjacent neurons. The presence of light hyperpolarizes the photoreceptor and reduces the release of neurotransmitters ("off"), which alters the activity of other cells in the retina and sends that information to the brain. The end result is the same: The nervous system detects a change in the environment and transduces it into a neural signal.

What happens when light strikes a photoreceptor cell? To answer this, we first need to look at what is occurring in the photoreceptor cell in the dark (**Figure 15.23a**):

① **Opsin and *cis*-retinal combine to form rhodopsin in the disc membrane.** The G-protein complex transducin and the enzyme phosphodiesterase (PDE) are inactive.

② **Sodium ions enter the outer segment of the photoreceptor and depolarize it.** Sodium ion channels in the plasma membrane of the outer segment are opened by the second messenger **cyclic guanosine monophosphate (cGMP)**, which is bound to them. In an example of the Gradients Core Principle (p. 26), sodium ions flow down their 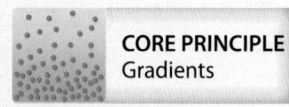 concentration gradient into the cell and depolarize the cell from its resting membrane potential of –70 mV to a membrane potential of –40 mV.

CORE PRINCIPLE Gradients

However, something very different occurs when light strikes the photoreceptor (**Figure 15.23b**):

① **Light causes retinal to separate from opsin.** When a photon hits the outer segment of the rod, it is absorbed by the rhodopsin molecules embedded in the disc membrane. This causes the retinal to change from a bent shape (*cis*-retinal) to a straightened form (*trans*-retinal). The *trans*-retinal doesn't fit into opsin, which causes the two to dissociate.

② **Transducin and phosphodiesterase (PDE) are activated.** The unbound opsin changes shape, which triggers a series of chemical reactions, beginning with the activation of transducin, which then activates PDE.

Pigmented layer

Outer
segment

Discs

Light

Inner
segment

Mitochondria

Opsin

Rhodopsin

cis–Retinal

Cell
body

Synaptic
terminals

Cone   Rod

(a) Structure of a cone and a rod

(b) Rhodopsin in the disc membrane of a rod

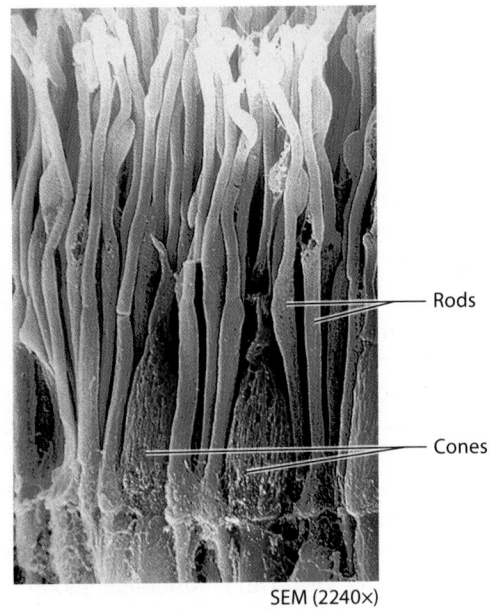

SEM (2240×)

**(c) Micrograph of rods and cones in a
human retina**

Rods

Cones

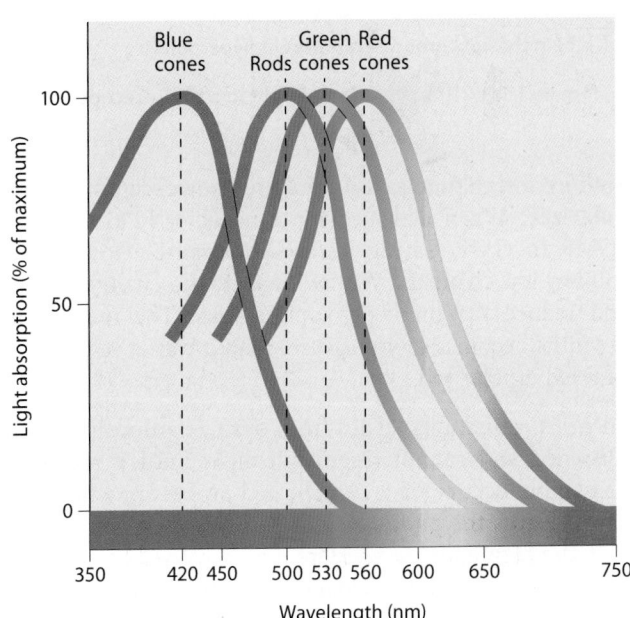

Blue
cones

Green Red
Rods cones cones

100

50

0

Light absorption (% of maximum)

350   420 450   500 530 560   600   650   750

Wavelength (nm)

(d) Absorption of various wavelengths of light by cones and rods

Figure 15.22 Cones and rods.

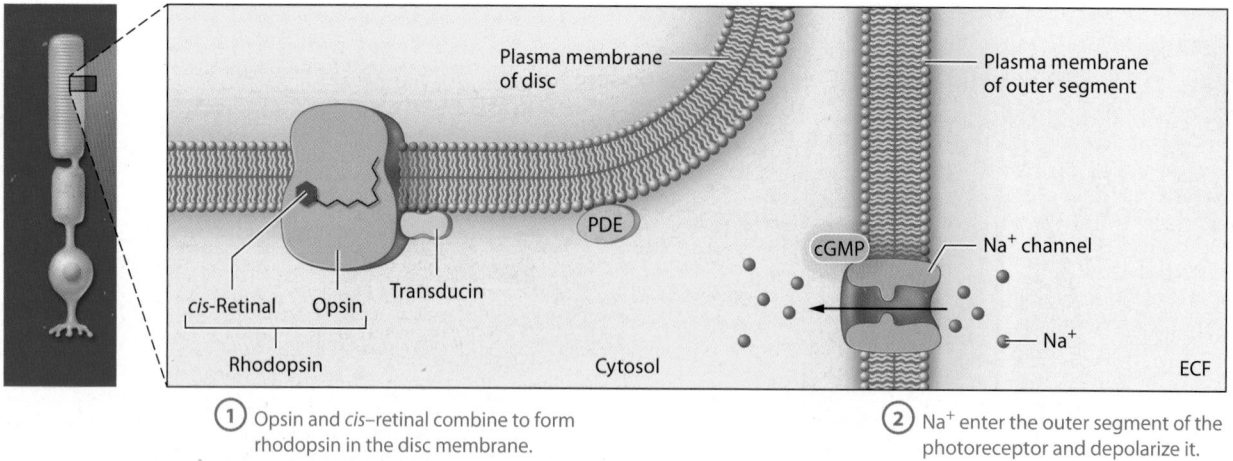

① Opsin and *cis*–retinal combine to form rhodopsin in the disc membrane.

② Na⁺ enter the outer segment of the photoreceptor and depolarize it.

(a) In the dark, photoreceptor cells depolarize.

① Light causes retinal to separate from opsin.

② Transducin and phosphodiesterase (PDE) are activated.

③ Na⁺ channels close, and the photoreceptor hyperpolarizes.

(b) In the light, photoreceptor cells hyperpolarize.

Figure 15.23 Transduction of light in a photoreceptor cell.

③ **Sodium ion channels close, and the photoreceptor hyperpolarizes.** When activated by transducin, PDE converts cGMP to GMP, leaving less cGMP available to bind to sodium ion channels. Without cGMP, the channels close and reduce the influx of sodium ions. The membrane potential becomes more negative than it was at rest, and is now hyperpolarized.

When *trans*-retinal dissociates from opsin, rhodopsin is said to be **bleached** and cannot respond to light until it reassembles. Reassembly occurs via a complicated process in which the *trans*-retinal leaves the photoreceptor and travels to the epithelial cells of the pigmented layer. Here it is converted back to *cis*-retinal in a process that uses ATP and, after transport back to the photoreceptor, combines with opsin to make rhodopsin. Thus, the pigmented layer plays an essential role in the function of photoreceptors.

Dark and Light Adaptation

Suppose you leave a sunny parking lot to enter a dark theater, or you are relaxing in your darkened dorm room when your roommate suddenly turns on the light. In both cases, although you may feel blinded at first, your eyes gradually adjust in processes known as **dark adaptation** and **light adaptation,** respectively. These adjustments to different light intensities partially depend on pupil size. The pupils constrict in bright light and dilate in dim light, but this occurs almost immediately and does not explain why it can take several minutes for your eyes to adjust to bright light or up to 40 minutes to completely adjust to darkness. Instead, it is explained by changes that occur in the photoreceptor cells.

The adjustment to darkness depends on the rods, as they are the photoreceptors that function in dim light. But in bright light, all of the rhodopsin in the rods is bleached and the rods are unable to function, and vision therefore depends on the cones. When light is suddenly reduced, the cones can no longer function and rods are slow to regenerate enough rhodopsin to function. It can take as long as 40 minutes for dark adaptation to occur and for rods to be completely functional.

Bleaching also plays an important role in light adaptation. In the dark, rods are fully functional and the retina is extremely sensitive to light. When bright light hits the retina, it immediately

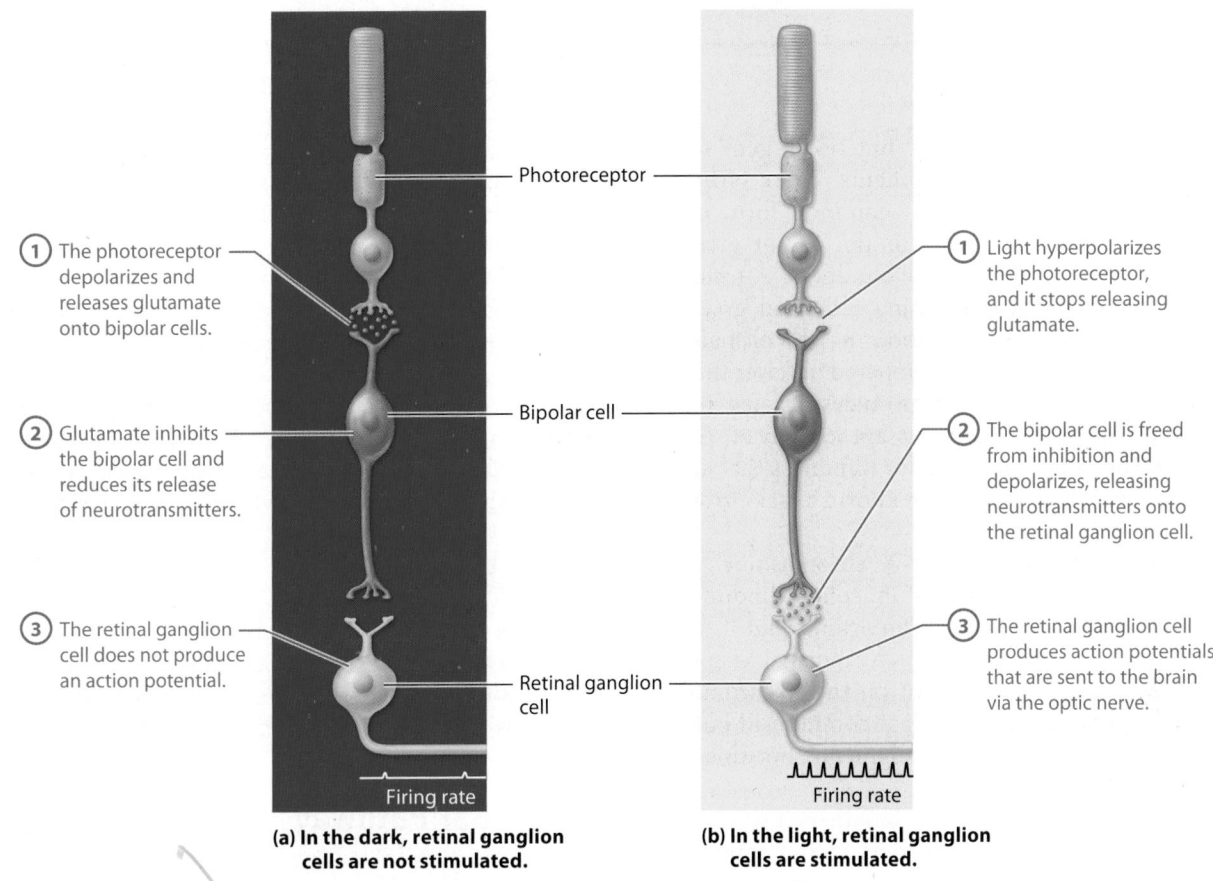

(a) In the dark, retinal ganglion cells are not stimulated.

(b) In the light, retinal ganglion cells are stimulated.

Photoreceptor

Bipolar cell

Retinal ganglion cell

① The photoreceptor depolarizes and releases glutamate onto bipolar cells.

② Glutamate inhibits the bipolar cell and reduces its release of neurotransmitters.

③ The retinal ganglion cell does not produce an action potential.

Firing rate

① Light hyperpolarizes the photoreceptor, and it stops releasing glutamate.

② The bipolar cell is freed from inhibition and depolarizes, releasing neurotransmitters onto the retinal ganglion cell.

③ The retinal ganglion cell produces action potentials that are sent to the brain via the optic nerve.

Firing rate

Figure 15.24 Image processing in the retina.

bleaches the rods and cones, resulting in a blinding glare. The rods become nonfunctional, since the rhodopsin is bleached as fast as it is regenerated. The cones, however, can regenerate functional pigments faster and are able to respond within a few minutes as their sensitivity decreases.

Image Processing by the Retina

We've seen how rods and cones move from a depolarized state in the dark to a hyperpolarized state in the light. We now look at how this change affects neighboring bipolar cells and retinal ganglion cells.

In the dark, the following steps occur (**Figure 15.24a**):

① **The photoreceptor depolarizes and releases glutamate onto bipolar cells.** Recall that the cell depolarizes, but an action potential does not occur. An action potential is not necessary because the cell is short enough for a graded potential to spread over the entire cell without dying out. As the depolarization spreads, the neurotransmitter glutamate is released onto bipolar cells.

② **Glutamate inhibits the bipolar cell and reduces its release of neurotransmitters.** In most bipolar cells, glutamate produces hyperpolarization via an inhibitory postsynaptic potential (IPSP). Some bipolar cells depolarize in response to glutamate, but they will not be discussed here.

③ **The retinal ganglion cell does not produce an action potential.** No signal is sent through the optic nerve to the brain.

When light strikes the photoreceptors, here's what happens (**Figure 15.24b**):

① **Light hyperpolarizes the photoreceptor, and it stops releasing glutamate.**

② **The bipolar cell is freed from inhibition and depolarizes, releasing neurotransmitters onto the retinal ganglion cell.** Bipolar cells are small enough that no action potential is needed. A graded potential is enough to trigger the release an excitatory neurotransmitter (the specific neurotransmitter is not yet known).

③ **The retinal ganglion cell produces action potentials that are sent to the brain via the optic nerve.**

The horizontal and amacrine cells are not direct participants in this pathway. However, they are important in altering the sensitivity of the pathway and aiding the detection of contrast and edges of objects.

The pathways in Figure 15.24 are simplified to show what happens to each type of cell. The retina has more than 105 million photoreceptors, but only 1 million retinal ganglion cells,

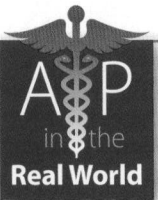

A&P in the Real World

Color Blindness

When a person lacks a functional gene for one or more cone pigments, **color blindness** results. The most common form involves a missing or defective gene for the red or green pigment, resulting in red-green color blindness. Affected people have difficulty in distinguishing red from green; both colors appear grayish-brown. About 8–10% of males have some form of color blindness, compared to fewer than 1% of females. It is more common in males because the genes for the red and green pigments are located on the X chromosome. Disorders such as color blindness that are transmitted via the X chromosome are known as *sex-linked disorders*.

Males have only one copy of the X chromosome, so a missing or defective gene results in color blindness. Females have two copies of the X chromosome, and if the second copy is normal, females will have normal vision. The gene for the blue pigment is not on the X chromosome, and a mutation in it is much less common in both sexes. It is possible to lack all three pigments, but this is extremely rare.

The most common test for color blindness uses *Ishihara plates*, which are collections of dots with a number embedded using dots of a different color. In the plate below, a person with red-green color blindness will not be able to detect the number 74 and will see a 21 instead. Note that reliable color blindness tests require looking at original Ishihara plates.

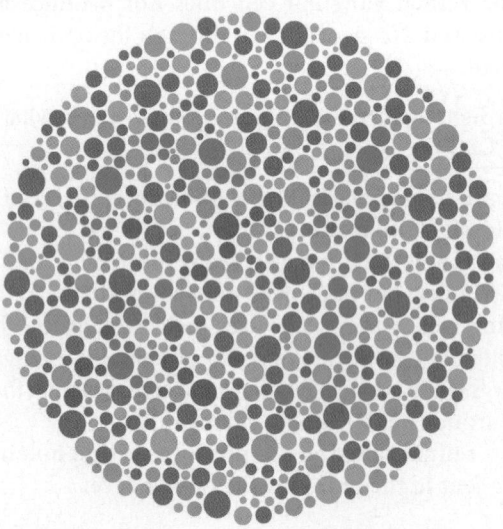

Color blindness has no cure or treatment. However, most people learn to compensate by looking for clues, such as color intensity or location—for instance, the red light is the top light of a traffic signal.

so most retinal ganglion cells receive input from multiple photoreceptors. However, vision is sharpest when a ganglion cell receives input from only a single photoreceptor. This is the case in the fovea, where one cone may synapse with only two bipolar cells, which in turn contact only one retinal ganglion cell. For this reason, sharp vision is possible here, because when a specific ganglion cell is stimulated, the brain knows exactly which tiny part of the retina (and the visual field) has been stimulated.

Toward the edges of the retina, however, hundreds or thousands of photoreceptors contact a small number of bipolar cells, which then synapse with one retinal ganglion cell. The retinal ganglion cell cannot determine exactly which one of the photoreceptors has been stimulated, so the brain does not receive precise visual information. So, peripheral vision is less detailed than central vision, which explains why you can clearly read the menu at a restaurant but not immediately recognize a friend who walks up beside you.

Quick Check

☐ 6. How are rods different from cones?
☐ 7. Why are photoreceptors depolarized in the dark?
☐ 8. How do rods hyperpolarize when light strikes them? How does this affect the firing rate of a ganglion cell?

The Visual Pathway

◀◀ **Flashback**

1. Where is the occipital lobe? (p. 428)
2. Where is the optic chiasma? (p. 483)

Detection of light and image formation in the retina is only part of the story of how we see. The information from the retina must be transferred to the visual areas of the brain for conscious awareness and interpretation. This involves many structures in the brain, which we now examine.

Structures of the Visual Pathway

To understand how visual information travels through the brain, you must understand the concept of a **visual field.** The visual field of each eye is what you see with that eye alone. However, the term also refers to what you see with both eyes open, where it is divided into right and left visual fields with the focal point (what your gaze is focused on) in the center.

In **Figure 15.25**, the right visual field is shaded orange and the left visual field is shaded purple. Notice that the center of the visual field is viewed by both eyes (see the dark orange and dark purple sections), but the edges of the visual field are seen only by the eye on the same side. The visual pathway is as follows:

① **The retina of each eye detects visual stimuli from portions of the right and left visual fields.** Images from the left visual field (purple in Figure 15.25) fall on the *nasal retina* (nearest the nose) of the left eye and the *temporal retina* (nearest the temple) of the right eye. Images from the right visual field (orange in Figure 15.25) fall on the

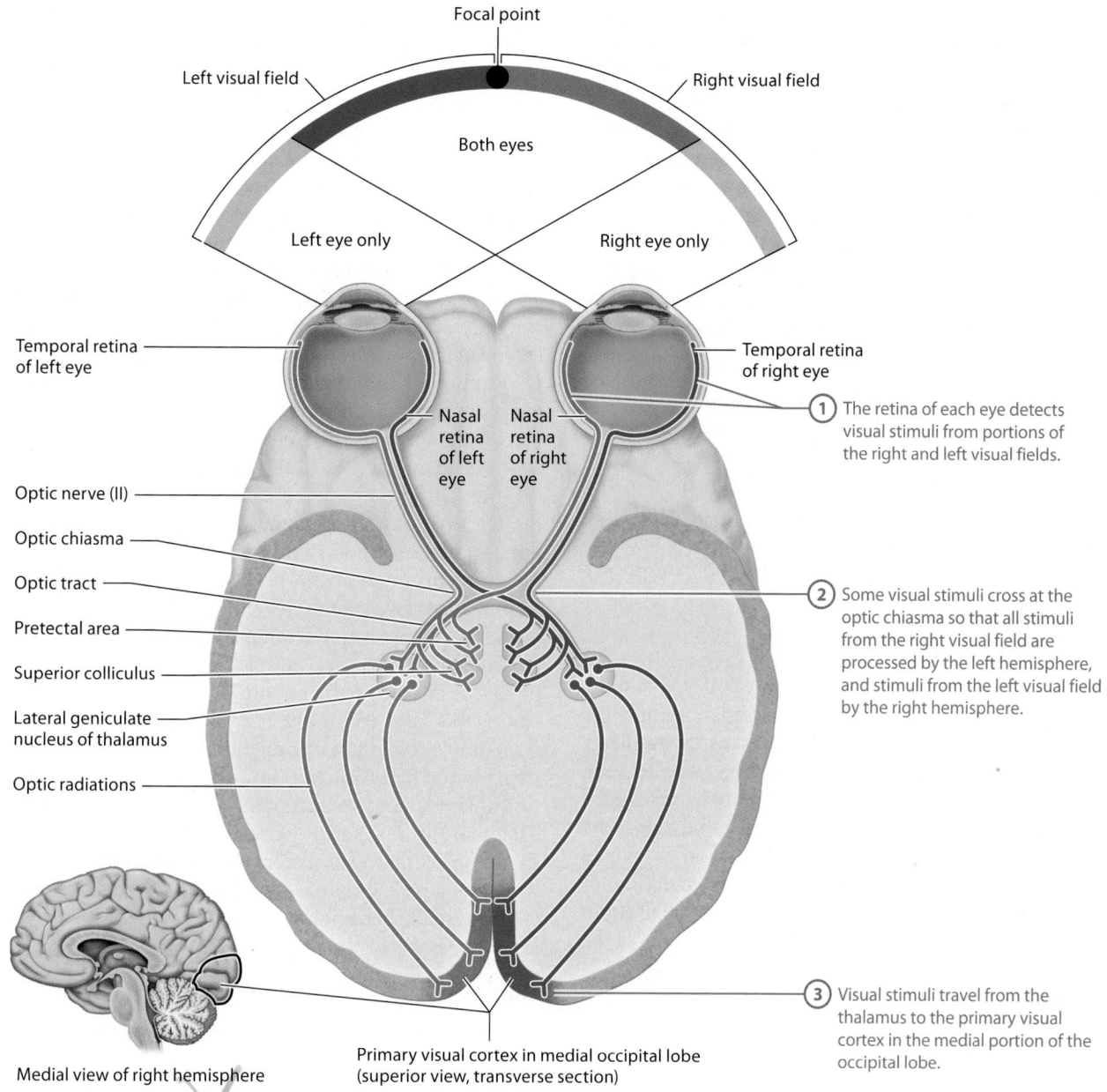

Figure 15.25 The visual pathway.

nasal retina of the right eye and the temporal retina of the left eye.

② **Some visual stimuli cross at the optic chiasma so that all stimuli from the right visual field are processed by the left hemisphere, and stimuli from the left visual field by the right hemisphere.** All axons of the retinal ganglion cells gather at the optic disc and leave the posterior wall of the eye as the optic nerve. They exit the orbit and enter the skull through a foramen in the sphenoid bone, then meet at the midline in the X-shaped **optic chiasma** (KYE-azm) inferior to the hypothalamus.

The optic chiasma is an important structure in the visual pathway—this is where stimuli from each half of the visual field are segregated so that they reach the opposite hemisphere of the brain (much like general sensory and motor areas of the brain are concerned with the opposite side of the body). In the optic chiasma, axons from retinal ganglion cells in each nasal retina cross to the opposite side, but axons from each temporal retina do not cross. As the axons leave the optic chiasma, they form the **optic tracts.** All of the stimuli from the left visual field are now in the right optic tract, and all of the stimuli from the right visual field are now in the left optic tract. Axons in the optic tracts terminate in several different regions, but most of the axons end in the thalamus in the **lateral geniculate nucleus** (jeh-NICK-you-late).

③ **Visual stimuli travel from the thalamus to the primary visual cortex in the medial portion of the occipital lobe.**

The axons of the lateral geniculate neurons are grouped as bundles called *optic radiations* that spread out deep to the temporal and parietal lobes. Note that when the optic radiations reach the **primary visual cortex** of the occipital lobe, each half of the visual field is now represented entirely in the opposite half of the cerebral hemisphere. At this point, we become consciously aware of visual stimuli and basic analysis of objects begins in terms of shape, color, movement, and location. From the primary visual cortex, information is relayed to other parts of the occipital lobe and to portions of the temporal and parietal lobes for more interpretation and integration with other senses.

Some axons in the optic tracts terminate in the superior colliculus, located on the posterior surface of the midbrain. Neurons in the superior colliculus coordinate eye movements when you shift your gaze from one object to another. Each eye must move precisely the same distance in the same direction at the same time in order to focus the new object on the fovea of each eye. The superior colliculus also communicates with neurons that control head and neck movement so that you can quickly turn toward an object of interest.

Consensual Pupillary Response

Have you ever watched what happens to one eye if you shine a bright light into the other eye? Both pupils constrict, not just the one exposed to the bright light. This is the **consensual pupillary response,** and it occurs because of the neural connections between the eyes. Instead of continuing to the thalamus, some axons of each optic tract terminate in both sides of the midbrain in the *pretectal area,* which you can see in Figure 15.25. Pretectal neurons contact other neurons that send their axons back through the oculomotor nerve to the pupillary muscles in both eyes. Since the signal about bright light hitting the retina in one eye travels to both sides of the brain, the reflexive response constricts the pupil in both eyes.

The consensual pupillary response is clinically important. If it does not occur, this indicates damage to the retina, optic nerve, or brainstem.

Stereoscopic Vision (Depth Perception)

Since our eyes face anteriorly, much of the visual field for one eye overlaps with that for the other eye, which is known as **binocular vision.** Each eye sees many of the same objects, but from a slightly different viewpoint. The brain compares these images and uses the differences to generate **stereoscopic vision** (or **depth perception**)—in other words, the distance of an object from the eye. The loss of vision in one eye destroys the ability to compare images, and stereoscopic vision is lost.

Image Processing by the Brain

The act of seeing involves much more than simple detection and conscious awareness of visual stimuli, and as just described, areas of the brain involved in processing visual information are extensive. Retinal axons project to the lateral geniculate nucleus and the primary visual cortex so that the image from the retina is "mapped" onto these areas. However, the image is distorted: Parts of it falling on the fovea of the retina take up much more area than parts from peripheral areas. This occurs because the increased numbers of photoreceptors in the fovea provide enhanced sensitivity, much like general sensation from your fingertips occupies a disproportionately large area of cortex compared to sensations from your calf.

In addition to conscious awareness of visual stimuli, the primary visual cortex also provides initial processing of location, movement, shape, and color, and combines images from each eye for binocular vision. From the primary cortex, visual information is then sent into two general pathways: (1) the dorsal (posterior) pathway that involves the parietal lobe, and (2) the ventral (anterior) pathway that ends in the inferior portion of the temporal lobe. Each pathway is specialized for processing particular aspects of vision. The dorsal pathway appears to be involved in interpreting motion; a person with damage in this region may have difficulty with pouring coffee into a cup or crossing a street because he or she is unable to detect the movement of the coffee or the cars.

The ventral pathway is specialized for processing color and the forms of objects. For example, some neurons in the inferior temporal lobe respond more vigorously to faces as opposed to furniture or skyscrapers. People suffering from *prosopagnosia* cannot recognize people (even close family members) by their faces; they must rely on other cues such as the sound of their voices. Damage to the ventral pathway can also result in complete loss of color perception despite the presence of functional cones; this rare condition is called *achromatopsia* and can be more unpleasant than you might expect. Would you find it appetizing to put black ketchup on your gray French fries?

The dorsal and ventral pathways are not the end of the story. The neurons here form myriad connections with other parts of the brain so that we can form visual memories, read aloud the words on a page or name an object, direct the extrinsic eye muscles, and make behavioral decisions based on what we see.

Quick Check

☐ 9. What happens at the optic chiasm? What is the consequence of this?

☐ 10. What type of visual processing occurs in the dorsal and ventral pathways?

Putting It All Together: The Big Picture of Vision

Now let's take a look at all the processes involved in vision. Imagine you are sitting on a beach, watching a friend surf the waves and tumble off his surfboard. **Figure 15.26** summarizes the process by which you detect, transduce, and analyze these visual stimuli.

Apply What You Learned

☐ 1. If your vision is normal and you put on your hyperopic friend's glasses, is the focal point in front of your retina or behind it? (*Hint:* How does a lens that corrects hyperopia affect rays of light?)

Left visual field   Right visual field

Light rays

Lens
Retina

Optic nerve (II)

Optic chiasma

Optic tract

Lateral geniculate nucleus of thalamus

Primary visual cortex

Right visual field image

Left visual field image

Cornea
Lens

Action Potential

Light

① The cornea and flattened lens provide enough refraction to focus light rays on the retina because the subject is more than 20 feet away (see Figures 15.17 and 15.18).

② Your photoreceptors hyperpolarize, and retinal ganglion cells send action potentials through the optic nerve (see Figures 15.23 and 15.24).

③ Selected axons cross at the optic chiasma so that stimuli from each half of the visual field are sent to the opposite hemisphere (see Figure 15.25).

④ The primary visual cortex provides conscious awareness and initial processing of shape, color, and movement. Each hemisphere processes stimuli from the opposite eye's visual field.

2. Explain what would happen in a rod if the cGMP were replaced with a form that was insensitive to phosphodiesterase (PDE).
3. In what part of the visual field would blindness occur if a tumor disrupted the left optic tract?

See answers in Appendix A.

MODULE **15.6**
Anatomy of the Ear

Learning Outcomes

1. Describe the structure and function of the outer and middle ear.
2. Discuss the role of the pharyngotympanic tube in draining and equalizing pressure in the middle ear.
3. Summarize the functions of the ceruminous glands.
4. Describe the structure and function of the cochlea, vestibule, and semicircular canals.

The ear is divided into three regions: outer (external) ear, middle ear, and inner (internal) ear (**Figure 15.27**). All three regions participate in hearing, but only part of the inner ear provides vestibular sensation, which is important for equilibrium. This module examines the structure and basic functions of each region.

Outer Ear

◄◄ Flashback

1. Where is the temporal bone? (p. 428)

The outer ear consists of the auricle and the external auditory canal (Figure 15.27). The **auricle** (AW-rih-kul), or *pinna*, consists of elastic cartilage covered with skin, except for the fleshy *lobule*—the part of the ear most commonly pierced for earrings. The auricle funnels sound waves into the **external auditory canal (external acoustic meatus)**. The canal is a slightly curved tunnel through the temporal bone that ends at the *tympanic membrane* (tim-PAN-ick), commonly known as the *eardrum*. The external auditory canal is lined with modified sweat glands called **ceruminous glands** (seh-ROO-meh-nus) that secrete yellowish-brown to gray **cerumen** (seh-ROO-men), or *ear wax*. Ear wax lubricates and waterproofs the external auditory canal and the tympanic membrane; it also traps debris before it reaches the tympanic membrane and sweeps it out of the auditory canal. In

Figure 15.27 Regions of the ear.

Practice art labeling @MasteringA&P®

most people, ear wax dries out and exits the canal, but in others it can build up and interfere with hearing.

The **tympanic membrane** is a thin sheet of epithelium and connective tissue that separates the outer ear from the middle ear (Figure 15.27). It is cone-shaped, with its tip or apex pointing into the middle ear. This membrane plays an important role in enabling the energy of sound waves to reach the inner ear, as we describe shortly.

Quick Check

☐ 1. What is the auricle?
☐ 2. What is the purpose of cerumen?
☐ 3. Where is the tympanic membrane located?

Middle Ear

The middle ear is a hollow, air-filled chamber in the temporal bone lined with a mucous membrane (Figure 15.27). It is bounded by the tympanic membrane laterally and the wall of the inner ear medially. It is connected to the nasopharynx (the part of the throat posterior to the nasal cavity) by the **pharyngotympanic tube** (fah-ring′-go-tim-PAN-ick), or *auditory tube* (formerly called the Eustachian tube).

The pharyngotympanic tube serves to equalize air pressure on both sides of the tympanic membrane. Normally closed and flattened, it opens briefly when we yawn or swallow. You no doubt have had firsthand experience with the function of the tube when your ears "pop" during an airplane takeoff or when riding in a speedy elevator. As your elevation changes, air pressure in the external auditory canal changes relative to that in the middle ear. Any pressure difference causes the tympanic membrane to stretch and bulge inward or outward. Excess tension reduces the ability of the membrane to vibrate, interfering with hearing and resulting in a feeling of pressure in the ear. Yawning or swallowing opens the pharyngotympanic tube and equalizes pressure in the middle ear and outer ear. The pharyngotympanic tube can also allow pathogens in the respiratory tract to enter the middle ear, which is examined in *A&P in the Real World: Otitis Media.*

The middle ear contains the three **auditory ossicles** (OSS-ih-kulz), which are the smallest bones in the body (all three would fit on your fingernail). They are named for their shape: The **malleus** (MAL-ee-us) is shaped like a hammer, the **incus** (ING-kus) like an anvil, and the **stapes** (STAY-peez) like a stirrup. The ossicles are connected by synovial joints and together form a bridge that extends across the middle ear. Notice in Figure 15.27 how the malleus is connected to the tip of the

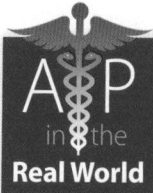

Otitis Media

Otitis media (oh-TIE-tis) is inflammation of the middle ear cavity. One of the most common infections in children, it is almost always preceded by an upper respiratory infection. Common symptoms include ear pain, fever, and impaired hearing, as well as irritability and, in some cases, vomiting.

In *acute otitis media* (AOM), the pharyngotympanic tube acts as a conduit for pathogens in the nasopharynx to reach the middle ear. The resulting inflammation causes a buildup of fluid or pus, and the tympanic membrane bulges painfully into the external auditory canal and interferes with hearing. The photos below (left and middle) compare a healthy tympanic membrane with a tympanic membrane affected by AOM. If the pressure is great enough, the tympanic membrane may rupture. Children are more susceptible to AOM because their pharyngotympanic tube is much shorter and more horizontal than that of adults. They also suffer more frequent upper respiratory infections because their immune systems are less developed.

AOM is commonly treated with antibiotics, although some experts recommend waiting to see if the infection clears on its own. However, even with treatment, AOM can progress to *otitis media with effusion* (OME), in which the infection has cleared but fluid remains and interferes with movement of the tympanic membrane for normal hearing. Persistent OME can be treated in several ways. One option is a **myringotomy** (mir′-in-GOT-oh-mee), a procedure in which a small hole is pierced in the tympanic membrane to drain the middle ear. A small tube may be inserted into the hole to prevent infection from recurring (see the photo on the right below). As the tympanic membrane heals, it naturally pushes the tube out within a year or two.

Healthy tympanic membrane viewed through the external auditory canal

Tympanic membrane in acute otitis media (AOM)

Tube inserted into tympanic membrane via a myringotomy

cone-shaped tympanic membrane, and the stapes is attached to an oval membrane in the medial wall called the **oval window.** The oval window is the boundary between the air-filled middle ear and the fluid-filled inner ear. The ossicles play an important role in amplifying and converting sound waves in the air into fluid movement. We'll soon see how this leads to neural signals.

Two very small skeletal muscles attach to the ossicles. The **tensor tympani** (TEN-sohr TIM-pah-nee) has its origin on the walls of the pharyngotympanic tube and inserts on the malleus to pull the tympanic membrane medially and tense it. The **stapedius** (stah-PEE-dee-us) muscle (not shown in Figure 15.27) arises from the posterior wall of the middle ear and inserts on the stapes; it reduces movement of the ossicles as a unit. These muscles are part of a reflex that prevents damage from loud sounds at low frequencies by reducing the movements of the tympanic membrane and ossicles. Since sound travels faster than the muscles can contract, they will not prevent damage from sudden short sounds like gunshots, but they can be protective for longer-lasting sounds, such as a nearby train. The muscles also contract just before we speak and while we chew so that these sounds are not uncomfortably loud.

Quick Check

☐ 4. What are the three auditory ossicles, and what are their functions?
☐ 5. What is the role of the pharyngotympanic tube?
☐ 6. What is the function of the tensor tympani and stapedius muscles?

Inner Ear

◀◀ Flashback

1. What are the functions of the vestibulocochlear nerve? (p. 483)

The inner ear is a cavity within the temporal bone that is divided into three regions (**Figure 15.28**). The *cochlea* (COKE-lee-ah) houses structures involved in hearing, and the *vestibule* and *semicircular canals* contain structures that detect head movement and position (vestibular sensation) (see Module 15.8).

The Bony and Membranous Labyrinths

A labyrinth (LAB-uh-rinth) is a network of winding passages. The inner ear is sometimes called the **bony labyrinth** because of its mazelike series of tunnels (Figure 15.28). A membrane lines the inner walls of the bony labyrinth and is referred to as the **membranous labyrinth.** Within the membranous labyrinth, we find **endolymph,** a fluid with a higher concentration of potassium ions than sodium ions, similar to intracellular fluid. Between the walls of the bony labyrinth and the membranous labyrinth is **perilymph,** a fluid that has a higher concentration of sodium ions than potassium ions and so resembles extracellular fluid. These fluids play an important role in transducing sound waves and head movements into neural signals.

Vestibule

The **vestibule** is located medial to the middle ear, and its wall contains the membranous oval window to which the stapes is connected (Figure 15.28). The vestibule houses two portions of

Figure 15.28 The membranous labyrinth. Structures containing perilymph are shaded orange, and structures containing endolymph are shaded blue. This color coding is also used in Figures 15.29 and 15.31.

the membranous labyrinth filled with endolymph, the **utricle** (YOU-treh-kl) and the **saccule** (SACK-yule). These chambers contain receptor cells that convey stimuli about head tilting and linear movement. The receptor cells transmit the stimuli to neurons that form the vestibular portion of the *vestibulocochlear nerve* (CN VIII; ves-tib′-yu-low-KOHK-lee-ur).

Semicircular Canals

The utricle is continuous with three tubes of the membranous labyrinth called **semicircular ducts,** which are enclosed in the bony **semicircular canals.** Note in Figure 15.28 that the anterior, posterior, and lateral semicircular ducts are oriented at right angles to each other so they can detect rotational movement of the head in any plane. At the base of each duct is a swollen bulb, called the **ampulla** (am-PEWL-ah), which contains receptor cells in contact with neurons. As in the vestibule, the receptor cells are innervated by the vestibular portion of the vestibulocochlear nerve.

Cochlea

The bony **cochlea** features a membrane-covered opening called the **round window,** which separates the middle and inner ear. Within the cochlea, we find a spiral-shaped portion of the membranous labyrinth known as the **cochlear duct,** or **scala media** (SKAY-lah), which is connected to the saccule and filled with endolymph. As you can see in **Figure 15.29a,** the cochlear duct spirals about two and a half times around a screw-shaped core of bone called the **modiolus** (moh-DIE-oh-lus) and then ends at the tip of the cochlea.

Two other chambers spiral through the cochlea, both of which are filled with perilymph. Superior to the cochlear duct is the **scala vestibuli** (ves-TIB-you-lie) (**Figure 15.29b**), which is continuous with the vestibule near the oval window. The inferior chamber is the **scala tympani,** which ends in the vestibule near the round window. Together, the scala tympani and scala vestibuli form a "sandwich" around the cochlear duct, and are continuous with each other at the tip of the cochlea.

The boundary between the cochlear duct and the scala vestibuli is formed by a sheet of epithelium called the **vestibular membrane.** This membrane separates the endolymph and perilymph and serves as a diffusion barrier between the two in order to maintain the concentrations of ions in the fluids.

The cochlear duct and scala tympani are separated by the **basilar membrane** (BAZ-ih-lar), a structure composed of collagenous connective tissue that is connected to a small shelf of bone projecting from the wall of the bony labyrinth. The basilar membrane is narrow and stiff near the base of the cochlea, but as it nears the tip, it widens and becomes more flexible. This difference in flexibility is important in detecting variations in sound, as we'll see.

The basilar membrane deserves special attention because it supports the **spiral organ (organ of Corti),** which contains receptor cells for hearing. The spiral organ is located within the cochlear duct and is where energy from sound waves is transduced into neural impulses. The impulses are carried into the

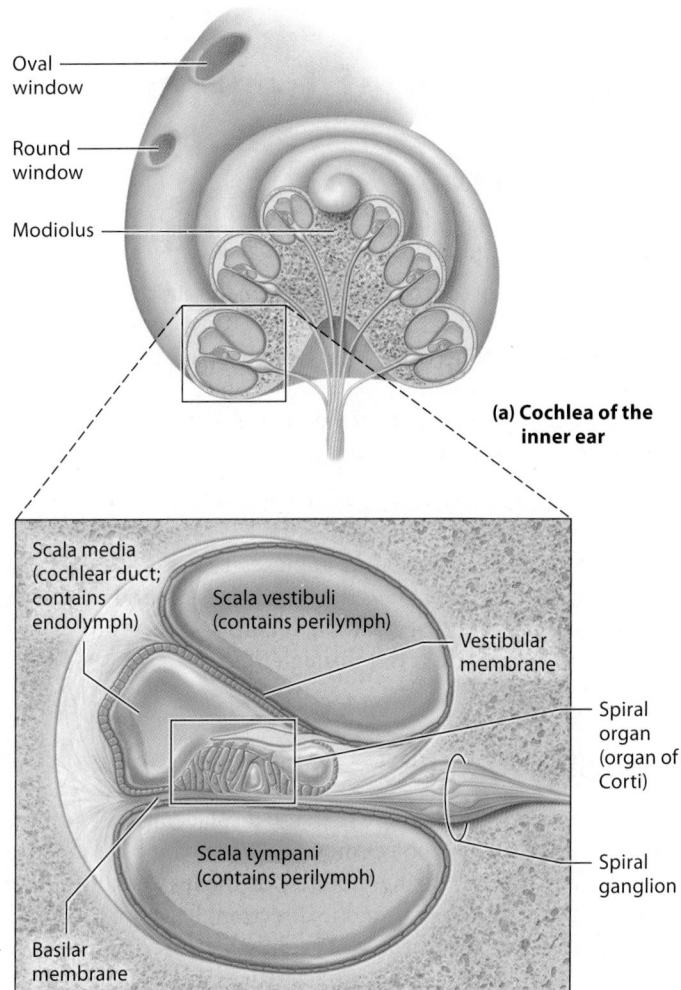

(a) Cochlea of the inner ear

(b) Spiral organ and ducts of the cochlea

Figure 15.29 Structure of the cochlea.

CNS by axons in the cochlear portion of the vestibulocochlear nerve, also known as the *cochlear nerve;* the cell bodies of these neurons form the *spiral ganglion* at the base of the cochlea.

Quick Check

☐ 7. What are the functions of the utricle, saccule, and semicircular ducts?

☐ 8. How are the scala tympani, scala media, and scala vestibuli arranged with respect to each other?

☐ 9. What is the spiral organ, and where is it located?

Apply What You Learned

☐ 1. Suppose someone's ceruminous glands did not produce any cerumen. What symptoms would this person experience?

☐ 2. How would your hearing be affected if your tensor tympani muscle suddenly contracted and did not relax?

See answers in Appendix A.

MODULE **15.7**

Physiology of Hearing

Learning Outcomes

1. Describe how the structures of the outer, middle, and inner ear function in hearing.
2. Trace the path of sound conduction from the auricle to the fluids of the inner ear.
3. Trace the path of nerve impulses from the spiral organ to various parts of the brain.
4. Explain how the structures of the ear enable differentiation of the pitch and loudness of sounds.

Our ears are able to detect sound waves. These sound waves must be transduced into neural impulses for us to be aware of and interpret the sound. As a starting point for understanding this process of **hearing,** or *audition,* let's first look at how sound waves are produced and travel through the air.

Principles of Sound

Sound waves are nothing more than the displacement of air molecules in response to an object moving. When you pluck a string on an instrument, the string vibrates, or moves back and forth (**Figure 15.30a**). As it moves in one direction, it compresses molecules of air and creates an area of slightly increased air pressure. When the string moves in the opposite direction, it causes the air pressure to drop slightly.

Although sound waves are often drawn as if the air molecules radiate away from the source, this is not the case. The molecules barely move from their original position; instead they bump into adjacent molecules and transfer their energy, which causes those molecules to collide with other molecules (much like the chain reaction in a row of dominos). The energy is transmitted in waves that radiate away from the vibrating object like ripples spreading across a pond, except that sound waves spread in three dimensions.

We often describe sound in terms of pitch and loudness. Every sound has a certain *pitch:* We hear the low-pitched rumble of thunder and the high-pitched yipping of a poodle frightened by the storm. Pitch, or *frequency,* is determined by how many times the object vibrates back and forth during a certain period (**Figure 15.30b**). Pitch is measured in hertz (Hz), which is the number of cycles or vibrations in one second. Our ears can typically detect sounds between 20 Hz and 20,000 Hz (20 kHz). The hearing range for dogs is much higher and explains why they respond to a seemingly "silent" dog whistle. We can feel the vibrations of sounds between 4 and 16 Hz, even though we can't hear them. Elephants and some whales communicate by producing sounds at frequencies well below 20 Hz.

Loudness, or the *amplitude* of a sound wave, is determined by the difference between areas of high and low air pressure (**Figure 15.30c**). If you gently pluck a string, it moves a limited distance as it vibrates, and the difference between the waves of high and low air pressure is also limited. Plucking the string with more force produces a louder sound because the string vibrates over a greater distance and displaces more air as it moves. Loudness is measured in decibels (dB); the rustling of leaves is barely audible at 10 dB, whereas the roar of a jet engine at 120 dB is painful and damaging to the ears.

(a) Plucking a string makes it vibrate, transferring energy that radiates through air molecules.

Vibrating string

(b) Pitch (frequency) equals the number of cycles in one second, measured in hertz (Hz).

Low frequency   High frequency

Amplitude

(c) Loudness (amplitude) equals the difference between the areas of high and low pressure, measured in decibels (dB).

Low amplitude (soft)   High amplitude (loud)

Amplitude

Figure 15.30 Sound waves.

Quick Check

☐ 1. What physical properties determine the pitch and loudness of a sound?
☐ 2. What is the range (in hertz) of human hearing?

Transmission of Sound to the Inner Ear

What happens when sound waves reach the ear? The auricle funnels sound waves into the auditory canal, a task at which it excels because of its shape, providing a good example of the Structure-Function Core Principle (p. 25). As you can see in step ① in **Figure 15.31a**, the sound waves strike the tympanic membrane and cause it to vibrate at the same frequency. This in turn causes the auditory ossicles in the middle ear to vibrate, and these vibrations are transmitted to the oval window.

CORE PRINCIPLE
Structure-Function

The oval window vibrates at the same frequency as the tympanic membrane. However, when the ossicles transmit the force from the larger tympanic membrane to the smaller oval window, the force is concentrated in a smaller area and for this reason it increases significantly. The ossicles also act as a lever system to amplify the vibration of the oval window.

Why is this important? The answer lies in what is found on the other side of the oval window in the inner ear: fluid, in particular the perilymph in the vestibule and the scala vestibuli. ② Vibration of the oval window produces pressure waves in the fluid that are just like those in the air. However, more force is needed to generate motion in fluid than in air (think about the effort required to move your cupped hand through water as opposed to air). So, increased force on the oval window is necessary to move the fluid. Now let's look at how movement of fluid leads to neural stimulation in the inner ear.

Quick Check

☐ 3. Why is the force exerted on the oval window greater than that exerted on the tympanic membrane during sound transmission?

Processing of Sound in the Inner Ear

The pressure waves generated at the oval window travel through the perilymph of the vestibule to the scala vestibuli of the cochlea.

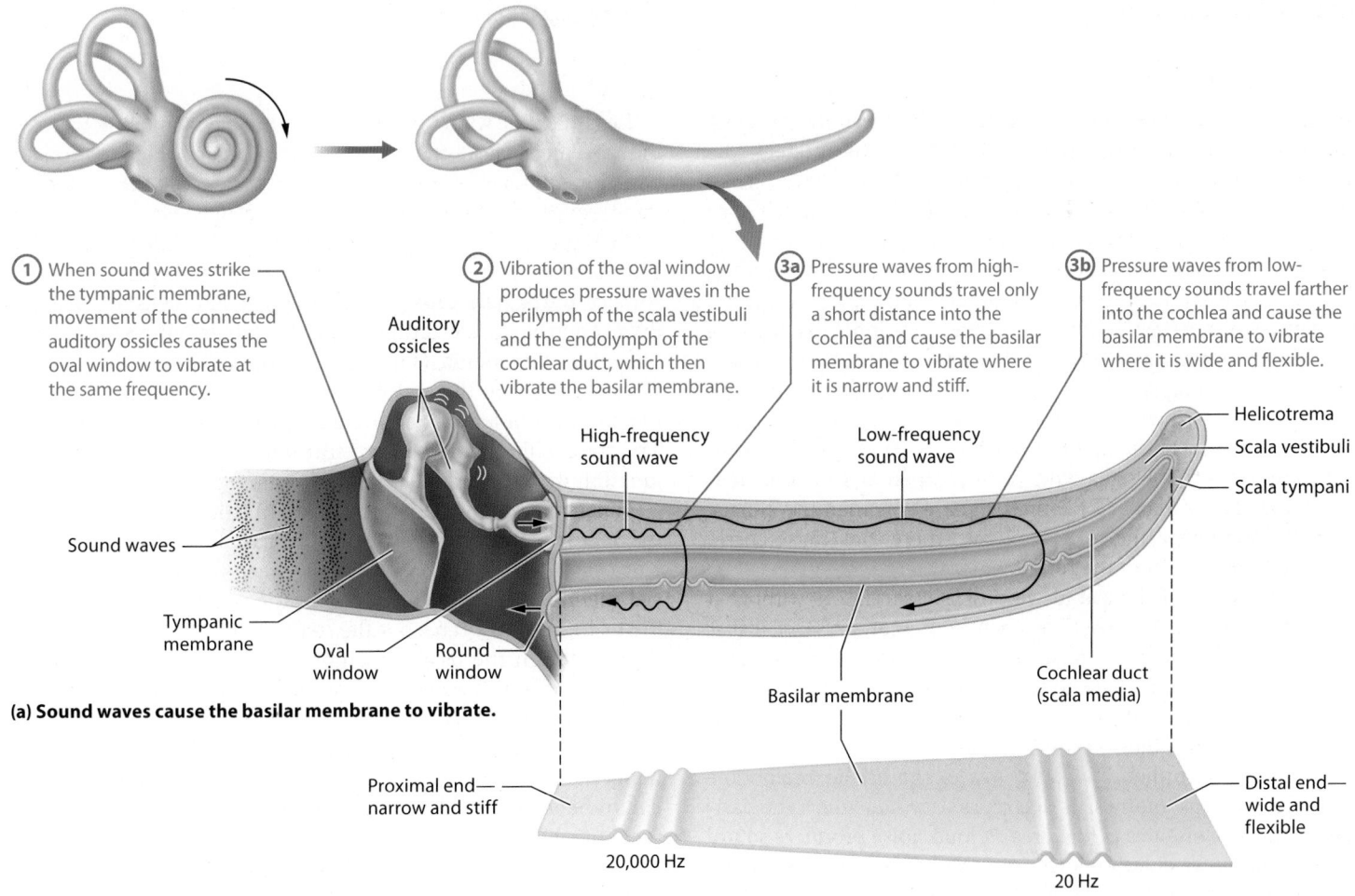

① When sound waves strike the tympanic membrane, movement of the connected auditory ossicles causes the oval window to vibrate at the same frequency.

② Vibration of the oval window produces pressure waves in the perilymph of the scala vestibuli and the endolymph of the cochlear duct, which then vibrate the basilar membrane.

③a Pressure waves from high-frequency sounds travel only a short distance into the cochlea and cause the basilar membrane to vibrate where it is narrow and stiff.

③b Pressure waves from low-frequency sounds travel farther into the cochlea and cause the basilar membrane to vibrate where it is wide and flexible.

Auditory ossicles

High-frequency sound wave

Low-frequency sound wave

Helicotrema
Scala vestibuli
Scala tympani

Sound waves

Tympanic membrane

Oval window

Round window

Basilar membrane

Cochlear duct (scala media)

(a) Sound waves cause the basilar membrane to vibrate.

Proximal end—narrow and stiff

Distal end—wide and flexible

20,000 Hz

20 Hz

(b) The shape of the basilar membrane varies.

Figure 15.31 **Vibration of the basilar membrane.**

(a) Spiral organ and hair cells

(b) Micrograph of spiral organ   LM (420×)

Figure 15.32 Structure of the spiral organ.

High-frequency vibrations (such as those produced by squealing brakes on a train) take a shortcut through the cochlear duct (scala media) to reach the scala tympani, which is shown in step ③a of Figure 15.31a. As they pass through the cochlear duct to the scala tympani, they cause the endolymph and basilar membrane to vibrate. These vibrations occur close to the base of the cochlea, where the basilar membrane is narrow and stiff (**Figure 15.31b**).

Notice in step ③b that low-frequency vibrations (like those of the rumble of the train) travel farther into the cochlea and make the basilar membrane vibrate where it is wider and more flexible. For very low-frequency sounds that we cannot hear, the waves travel all the way to the tip of the cochlea, where the scala vestibuli connects with the scala tympani at an opening called the **helicotrema** (hel′-ih-koh-TREE-mah). The waves then travel back to the vestibule via the scala tympani, where they cause corresponding movement of the round window.

The presence of the round window is essential, as without it the fluid would not move. Fluid cannot be compressed; if it is pushed from one area, it must move into another area (squeezing a water balloon at one end forces it to bulge at the other end). In the same way, as the oval window is pushed into the inner ear, the round window bulges outward, and vice versa.

Pitch is determined by which *area* of the basilar membrane vibrates, whereas loudness is determined by *how much* the basilar membrane vibrates at that area. A loud noise produces sound waves with more energy, which in turn cause greater movement of the tympanic membrane, middle ear ossicles, and the perilymph and endolymph. Increased movement of these fluids produces greater vibration of the basilar membrane and stronger stimulation of nearby receptor cells.

Regardless of which portion of the basilar membrane vibrates, the movement causes changes in the spiral organ, which rests on the basilar membrane. To understand these changes, we need to look at the structure of the spiral organ in more detail (**Figure 15.32**).

The spiral organ contains receptor cells called **hair cells.** They are grouped into a single row of **inner hair cells** and three rows of **outer hair cells.** The 3500 inner hair cells are primarily responsible for detecting sound. Each hair cell has microvilli, known as **stereocilia** (stare′-ee-oh-SILL-ee-ah), which project from the cell into the endolymph of the cochlear duct. The stereocilia vary in length and are arranged from tallest to shortest. The stereocilia themselves are stiff but "bend" by flexing at the junction with the cell body (like a joystick).

What makes the stereocilia bend? A stiff membrane, called the **tectorial membrane** (teck-TORE-ee-al), extends over the hair cells and "sandwiches" them between the basilar and tectorial membranes. The stereocilia on the outer hair cells contact the tectorial membrane, but the stereocilia in the inner hair cells do not. When the basilar membrane vibrates up and down in response to sound waves, the hair cells move with it toward and away from the tectorial membrane.

Each stereocilium is connected to its neighbor by an elastic filament called a *tip link* (**Figure 15.33a**). When the tallest stereocilium bends in one direction, the other stereocilia are pulled in the same direction.

The sequence of events that takes place when sound is transmitted to the inner ear is shown in **Figure 15.33b** and proceeds as follows:

① **The basilar membrane moves up toward the tectorial membrane, bending the stereocilia toward the tallest**

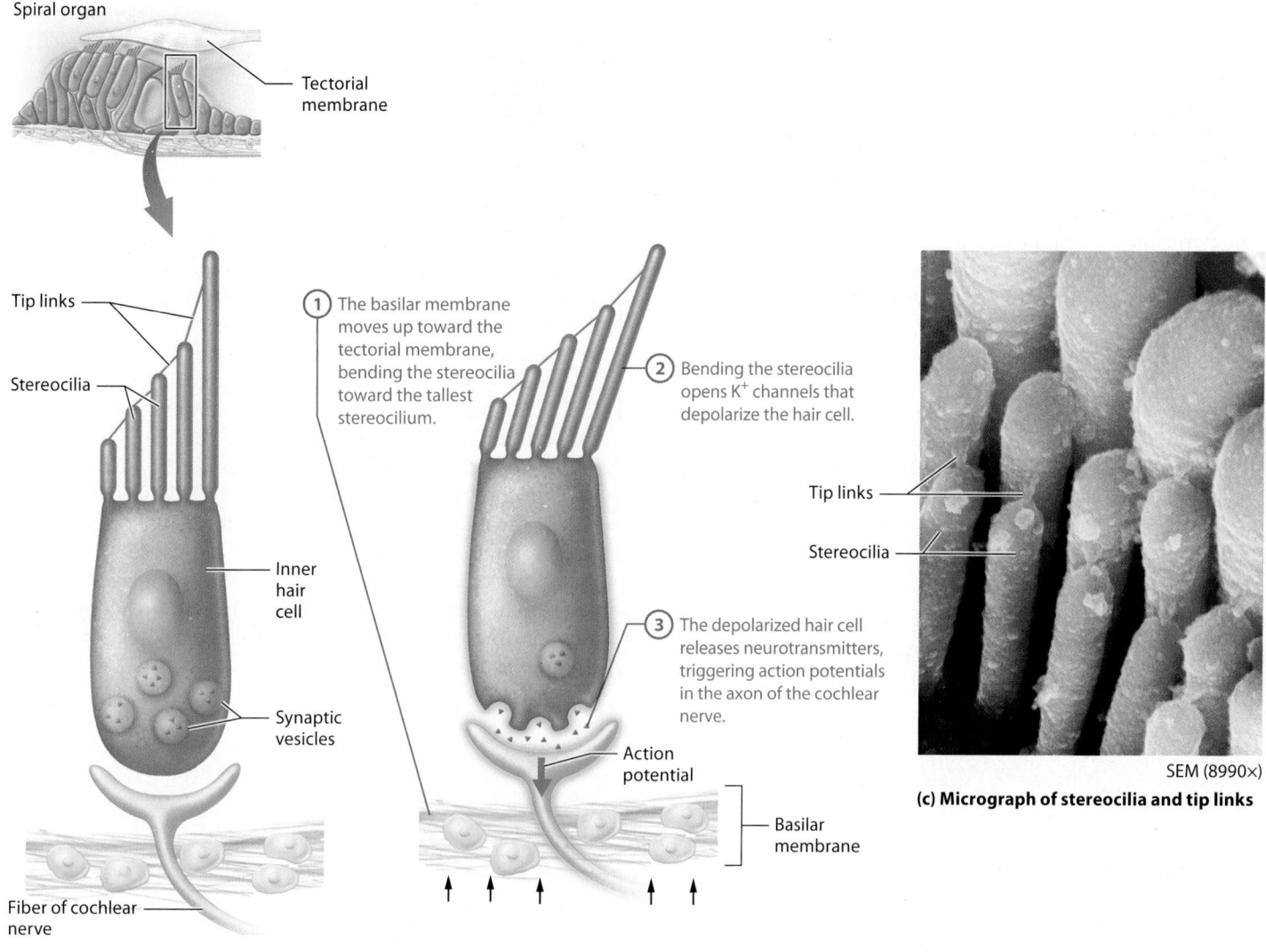

Spiral organ

Tectorial membrane

Tip links

Stereocilia

① The basilar membrane moves up toward the tectorial membrane, bending the stereocilia toward the tallest stereocilium.

② Bending the stereocilia opens K⁺ channels that depolarize the hair cell.

Inner hair cell

③ The depolarized hair cell releases neurotransmitters, triggering action potentials in the axon of the cochlear nerve.

Synaptic vesicles

Action potential

Basilar membrane

Fiber of cochlear nerve

Tip links

Stereocilia

SEM (8990×)

(c) Micrograph of stereocilia and tip links

(a) Hair cell when there is no sound.

(b) Sound occurs.

Figure 15.33 Transduction of sound in hair cells.

stereocilium. As the basilar membrane moves toward the tectorial membrane, the stereocilia on the outer hair cells bend against it. Stereocilia of inner hair cells also bend due to movement of the endolymph between the two membranes. When the basilar membrane moves away from the tectorial membrane, the stereocilia bend in the opposite direction.

② **Bending the stereocilia opens potassium ion channels that depolarize the hair cell.** As the stereocilia bend toward the tallest one, the tip links stretch and pull open a potassium ion channel in the tip of each stereocilium. In typical neurons, opening potassium ion channels would hyperpolarize the cell, but the endolymph contains such a high concentration of potassium ions that they flow into the hair cells, causing depolarization and a local potential. This is an example of the Gradients Core Principle (p. 26).

CORE PRINCIPLE
Gradients

③ **The depolarized hair cell releases neurotransmitters, triggering action potentials in the axon of the cochlear nerve.** The neurotransmitter is likely glutamate, and it triggers an action potential in the neuron of the cochlear nerve innervating the hair cell.

When the basilar membrane moves back, away from the tectorial membrane, the stereocilia bend in the opposite direction. The tip links are no longer stretched and the potassium ion channels close. The hair cells hyperpolarize, which stops the release of neurotransmitters.

Each inner hair cell is contacted by multiple neurons; approximately 90% of auditory neurons contact the inner hair cells. In contrast, only one neuron contacts multiple outer hair cells. The inner hair cells perform most of the work of detecting sound, but destruction of the outer hair cells by certain drugs—such as the antibiotic kanamycin—can result in nearly complete deafness. This is because the outer hair cells play an essential role in increasing the sensitivity of the inner hair cells and amplifying the vibration of the basilar membrane. The outer hair cells are capable of rapidly changing their length in an oscillating fashion (in other words, they jiggle), which enhances the movement of the basilar membrane as much as a hundred-fold. The movement of the outer hair cells can send vibrations backward through the inner ear and middle ear to the tympanic membrane and cause it to move. A very sensitive microphone placed next to the tympanic membrane can actually detect sounds produced by the ear, called *otoacoustic emissions.* The lack of such sounds indicates damage in the auditory pathway. To see how damage to hair cells or other structures of the ear can interfere with hearing, read *A&P in the Real World: Tinnitus.*

Quick Check

☐ 4. How does the cochlea distinguish different frequencies?
☐ 5. What is the function of tip links?
☐ 6. What is the role of the tectorial membrane in sound detection?

The Auditory Pathway

After the hair cells transduce the energy from sound waves into neural signals, these signals travel into the CNS along the pathway shown in **Figure 15.34**.

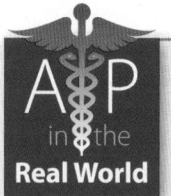

Tinnitus

Tinnitus (tin-EYE-tus) is the sensation of hearing noise in the absence of actual sound. People who suffer from tinnitus describe the noise as ringing, buzzing, whistling, or hissing. It is not a disease itself, but often a symptom of an underlying problem.

Almost one in five adults suffers from tinnitus; the most common cause is hearing loss due to damaged stereocilia of the cochlear hair cells. Damaged stereocilia are often bent or misshapen, so they depolarize the hair cells enough to release some neurotransmitters onto the cochlear nerve endings. Tinnitus may also result from fusion of the middle ear ossicles, inflammation or irritation of the tympanic membrane, certain medications, or damage to the neural pathways in the CNS. However, in many individuals, no cause is ever identified. In some cases, tinnitus is only temporary (for example, a large dose of aspirin can cause it), but for many people it is permanent. Antidepressant medications can help in some cases, and many people find relief by using a fan or other source of "white noise" to block the sounds of tinnitus, especially while sleeping.

Figure 15.34 The auditory pathway.

① **Auditory signals travel through axons of the cochlear portion of the vestibulocochlear nerve to the cochlear nuclei at the medulla-pons junction.** Axons of the cochlear nerve synapse in the cochlear nuclei, which are located at the junction of the pons and medulla.

② **Axons from the cochlear nuclei contact the superior olivary nucleus in the pons.** Neurons in the **superior olivary nucleus** (ALL-ih-vair-ee) compare information from both ears to determine the location of the sound.

③ **Auditory stimuli are then sent to the inferior colliculus of the midbrain.** The stimuli are coordinated with head and trunk movement to produce the *startle reflex*. This reflex, a response to unexpected sounds, causes you to turn your head and body toward the sound to investigate it. If you have ever "jumped out of your skin" in response to a loud clap of thunder, you are familiar with the startle reflex.

④ **The auditory stimuli are relayed to the medial geniculate nucleus of the thalamus.** The role of the thalamus in human hearing is not clear, but it may be important for processing combinations of sounds according to their frequency and time interval.

⑤ **The thalamus sends signals to the primary auditory cortex in the superior portion of the temporal lobe.** As with other sensory cortices, the **primary auditory cortex** is where our first conscious awareness of sound begins, as well as analysis of its location, pitch, and loudness.

The primary auditory cortex has connections with other parts of the temporal lobe that are specialized for language, and with the limbic system for emotions and memory.

Impaired hearing in just one ear indicates damage to the cochlear nuclei, vestibulocochlear nerve, or structures in the ear itself. Axons from the cochlear nuclei travel to both sides of the brain, so any damage to the auditory pathway after this point will affect certain aspects of hearing but will not cause hearing loss. For example, a person with damage to the primary auditory cortex in one hemisphere may have trouble determining the location of a sound, but still be able to hear it because stimuli from both ears still reach the other hemisphere.

Quick Check

☐ 7. Where is the primary auditory cortex?
☐ 8. What is the role of the superior olivary nucleus in processing sound information?

Hearing Loss

Hearing loss or deafness can be classified into two broad categories: conduction and sensorineural. *Conduction hearing loss* is due to a problem in the outer or middle ear that prevents sound waves from reaching the inner ear. Examples include excessive buildup of ear wax, a middle ear infection, a perforated tympanic membrane, or fusion of the auditory ossicles. Vibrations can reach the fluid of the inner ear through the temporal bone (bone

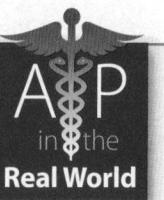

Cochlear Implants

A cochlear implant involves surgically implanting electrodes that bypass damaged hair cells and directly stimulate the cochlear nerve. Although implants do not fully restore hearing, they enable a person to perceive enough sound to understand speech or hear warnings such as smoke alarms.

An implant consists of several components. First, an external microphone, usually worn just behind the ear, sends sound waves to a processor, which converts the waves to electronic signals. Then, an external transmitter sends the signals to an internal receiver that is implanted beneath the skin. Finally, the receiver relays the signals to a long, thin bundle of electrodes that are threaded through the cochlea, where they stimulate the nerve endings of the vestibulocochlear nerve.

The success of a cochlear implant varies. For children who are born deaf or suffer hearing loss shortly after birth, implants are usually inserted before age 3, during the period of maximal language and speech acquisition. These children often learn to speak normally. However, if a deaf child receives an implant after age 9, it is unlikely that normal speech will develop. Children and adults who lose their hearing after acquiring speech and language often do well with cochlear implants and no longer have to solely use sign language or lip reading. Regardless of age, extensive training is required to use the implant successfully.

conduction), but with much less accuracy than conduction through air. Many forms of conduction hearing loss are temporary and can be corrected, for example, by removing excess ear wax or repairing a perforated tympanic membrane.

Sensorineural hearing loss refers to a deficit in the cochlea or any of the neural pathways in the cochlear nerve or CNS. There are actually two types of sensorineural hearing loss, and they differ significantly in terms of cause and treatment:

1. *Sensory hearing loss* occurs when action potentials cannot be generated in the cochlea, usually due to dysfunction of the hair cells. Exposure to loud sounds or certain medications may damage hair cells. We also lose stereocilia as we age, especially those in the proximal part of the cochlea that are responsible for detecting high-frequency sounds. Hearing aids can help mild to moderate cases of sensory hearing loss. A cochlear implant may also be an option (see *A&P in the Real World: Cochlear Implants*).

2. *Neural hearing loss* occurs when the signals fail to travel through the cochlear branch of the vestibulocochlear nerve or CNS pathways. Typical causes include strokes or tumors such as an acoustic neuroma, a benign tumor outside the brainstem at the junction of the medulla and pons. Cochlear implants are not effective for neural hearing loss.

The Big Picture of Hearing

Figure 15.35

2 Sound waves cause vibration of the tympanic membrane, auditory ossicles, and oval window (see Figure 15.31).

3 Vibration of the oval window moves the perilymph and endolymph in the cochlea, which in turn causes the basilar membrane to vibrate (see Figure 15.31).

Tympanic membrane

Stapes and oval window

1 The ocean waves and your friend's laugh as he falls produce sound waves that enter the external auditory canal.

Basilar membrane

Cochlea (unwound)

Tectorial membrane

Inner hair cell

Primary auditory cortex

Medial geniculate nucleus

Cochlear neurons in vestibulocochlear nerve

Basilar membrane

Inferior colliculus

Superior olivary nucleus

Cochlear nuclei

4 Transduction occurs when the vibration of the basilar membrane and endolymph depolarizes the hair cells, triggering action potentials in the cochlear neurons (see Figure 15.33).

5 Auditory signals travel through the axons of the vestibulocochlear nerve to the brainstem and eventually reach the primary auditory cortex in the temporal lobe. There conscious awareness and analysis of the sound begin (see Figure 15.34).

Putting It All Together: The Big Picture of Hearing

Let's now look at the overall, "big picture" process of hearing, shown in **Figure 15.35**. We return to our earlier example: You are at the beach and your friend is surfing. Your ears detect many auditory stimuli as you listen to ocean waves on the sand, people talking, and your friend laughing as he falls off his surfboard. All of these sounds travel through the air as waves that enter your auditory canals, after which they are transduced into neural signals that travel to the CNS for interpretation.

Apply What You Learned

☐ 1. A sound produces substantial vibration in the basilar membrane at the base of the cochlea. Describe the sound in terms of pitch and loudness.

See answer in Appendix A.

MODULE | **15.8**

Anatomy and Physiology of Vestibular Sensation

Learning Outcomes

1. Distinguish between static and dynamic equilibrium.
2. Describe the structure of the maculae, and explain their function in static equilibrium.
3. Describe the structure of the crista ampullaris, and explain its function in dynamic equilibrium.

While walking down the sidewalk, you suddenly trip over your own feet and flail your arms wildly to keep from falling. Your first thought on regaining your balance is likely to be wondering if anyone saw you nearly hit the ground. But you probably wouldn't consider the rapid responses of your inner ear, nervous system, and skeletal muscles that helped prevent you from falling. This is a classic example of how we think about our sense of equilibrium—that is, we don't think about it much at all.

The sense of equilibrium depends on input from three sources: (1) the visual system; (2) the proprioceptors in muscles and joints that provide feedback about body position; and (3) the **vestibular system** (vestibule and semicircular canals) in the inner ear, which provides information about head position and movement, known as **vestibular sensation.** If one of these systems is impaired, equilibrium becomes more difficult to maintain. To demonstrate this, have a friend observe your body movement as you stand quietly with your eyes open and then closed. You probably sway more with your eyes closed, since without visual input, you have to rely more on your inner ear and proprioceptors.

The inner ear contains organs that detect two types of equilibrium: static and dynamic equilibrium. The meaning of each term can be gleaned from its name: *static* means stationary or unmoving, and *dynamic* means changing or moving. **Static equilibrium** therefore refers to maintaining balance when the head and body are not moving but the head is tilted. **Dynamic equilibrium,** on the other hand, refers to maintaining balance when the head or body is undergoing rotational or angular motions (such as spinning in a rotating chair) or linear acceleration (such as riding in a car). This module examines how the structures of the inner ear detect head position and movement and communicate with muscles to adjust posture in order to maintain static and dynamic equilibrium.

The Utricle and Saccule

◄◄ **Flashback**

1. What is proprioception? (p. 452)

Static equilibrium is monitored by the utricle and the saccule, the portions of the membranous labyrinth located in the vestibule. Each chamber has a structure known as a **macula** (plural, *maculae;* **Figure 15.36a**), located in the epithelium lining its wall,

(a) **Location of the utricle and saccule**

(b) **Relationship of hair cells to the otolithic membrane in a macula**

Figure 15.36 **Maculae of the utricle and saccule.**

which contains the receptor cells for head position and movement. The receptor cells are hair cells similar to those in the cochlea, except that each cell contains stereocilia and one true cilium, called a **kinocilium** (ky′-noh-SIL-ee-um), which is taller than the stereocilia (**Figure 15.36b**).

The stereocilia and kinocilium are embedded in a gelatinous mass referred to as the **otolithic membrane** (oh-toe-LITH-ick), which is suspended in the endolymph of the chamber. The otolithic membrane gets its name because it contains **otoliths** ("ear stones"), which are crystals of calcium carbonate that increase its density. The hair cells release the neurotransmitter glutamate continuously onto dendrites of the bipolar neurons that form the vestibular portion of the vestibulocochlear nerve. As you can guess, bending the stereocilia alters the release of glutamate, which changes the activity of the vestibular neurons.

When you tilt your head, how do the hair cells respond? In the utricle, the stereocilia point vertically when your head is upright and stationary, and they bend when the head tilts to either side or to the front or back. In the saccule, the stereocilia are oriented horizontally when your head is upright and stationary, and bend in response to up-and-down movements (such as when you sit up straight from a slouched position). When the head tilts, gravity pulls on the otolithic membrane, which bends the stereocilia, which you can see in **Figure 15.37**. When the stereocilia bend toward the kinocilium, the hair cells experience a local potential and depolarize. They then release more glutamate, increasing the number of action potentials produced in the axons of the vestibular nerve. Bending of the stereocilia away from the kinocilium hyperpolarizes the hair cells and they release less glutamate, reducing the number of action potentials in the axons of the vestibular nerve.

The maculae are paired so that tilting your head in one direction depolarizes hair cells in one ear and hyperpolarizes hair cells in the other ear. This increases signals in one vestibular nerve and decreases signals in the other. The brain interprets the combination of signals as the tilting of your head in a particular direction.

The utricle and saccule are also important in detecting linear acceleration, a form of dynamic equilibrium, such as riding in a car or elevator. As movement begins, the hair cells move with the rest of the body, but the endolymph and otolithic membrane do not move immediately because of inertia—they "lag behind" the hair cells. This lag bends the stereocilia, which either depolarizes or hyperpolarizes the hair cells and changes the activity in the vestibular nerve. When the movement stops, the hair cells stop with the body, but the endolymph and the otolithic membrane continue moving. This bends the stereocilia in the opposite direction, and the hair cells respond by altering glutamate release.

Take note that the maculae are sensitive only to changes in the *rate* of movement, not movement itself. If acceleration stops, the endolymph and otolithic membrane eventually "catch up" to the hair cells and the stereocilia no longer bend. Activity in the vestibular nerve returns to its baseline level, which your brain perceives as no movement. You can easily observe this the next time you are a passenger in a car. When you close your eyes, you can sense the acceleration of the car as the driver pulls away from a stop sign. However, once the driver stops accelerating and maintains a steady speed, it can be difficult to tell if the car is moving unless you open your eyes. To see what can happen when your CNS receives contradictory signals about movement, see *A&P in the Real World: Motion Sickness.*

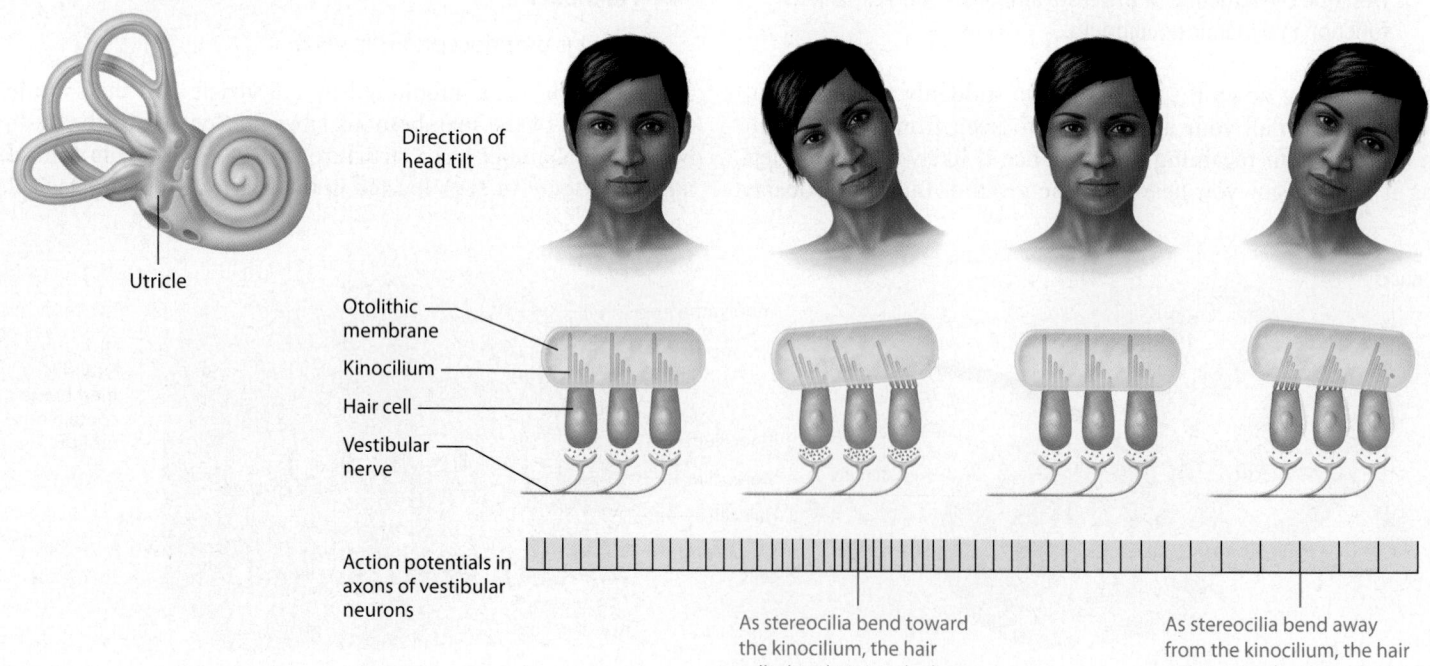

Direction of head tilt

Otolithic membrane

Kinocilium

Hair cell

Vestibular nerve

Utricle

Action potentials in axons of vestibular neurons

As stereocilia bend toward the kinocilium, the hair cells depolarize and release more glutamate.

As stereocilia bend away from the kinocilium, the hair cells hyperpolarize and release less glutamate.

Figure 15.37 Head tilting and activity of the maculae.

ConceptBOOST »»»

How Inertia Influences Movement of the Otolithic Membrane and Endolymph

If you are having difficulty grasping how inertia influences the movement of the otolithic membrane and endolymph, here's an analogy that may help. Imagine you are stopped at a red light while riding in the front passenger seat of a friend's car with a cup of coffee (without a lid) in your hand. The light turns green and your friend stomps on the gas pedal. Why are you now wearing your coffee on the front of your shirt? The answer is inertia. The car, your body, and the cup all move together as one unit. However, the coffee (a fluid) moves independently. It also has inertia, so it does not move immediately with the cup. Instead, it "lags behind" the cup and ends up all over your shirt.

The coffee is equivalent to the endolymph and otolithic membrane in the vestibule. Their movement lags behind that of the hair cells in the wall of the utricle and saccule (the "cup"), which bends the stereocilia. Depending on the direction in which they bend, neuron activity either increases or decreases, which the brain interprets as the body speeding up, or accelerating.

Now, let's turn back to the coffee cup. Imagine that after driving at a constant speed, your friend slams on the brakes to avoid hitting another car. Where is your coffee now? It's on the floor of the car, which is also explained by inertia. While traveling at a constant speed, the coffee "catches up" and moves with the cup. However, when the car, your body, and the cup suddenly stop (decelerate), inertia causes the coffee to keep moving forward, out of the cup.

In the same way, when your hair cells and the rest of your body come to a stop, your endolymph and otolithic membrane continue to move forward. This bends the hair cells in the direction opposite to that when the car accelerated, and the activity of the vestibular neurons changes accordingly. In other words, if nerve activity increased when you first started moving, it will now decrease. The brain interprets this as slowing down, or decelerating. ■

Quick Check

☐ 1. How do static and dynamic equilibrium differ?
☐ 2. What role do the utricle and saccule play in static equilibrium?
☐ 3. What role do they play in dynamic equilibrium?

The Semicircular Ducts

In addition to linear acceleration, dynamic equilibrium includes angular or rotational movements, such as turning your head or spinning in a chair. This type of movement is detected by hair cells located in the anterior, posterior, and lateral semicircular ducts of the membranous labyrinth. The three ducts, which contain endolymph, are oriented to detect rotation in all three planes. The semicircular ducts are connected to the utricle; near this connection, each duct has a swelling, the **ampulla** (plural, *ampullae;*

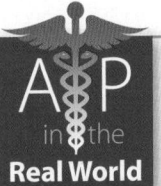

Motion Sickness

Motion sickness refers to symptoms such as nausea, dizziness, and sometimes vomiting, which can occur during particular types of movement, such as riding in a car, boat, or roller coaster. Motion sickness occurs when the brain receives mismatched sensory information from the eyes and the vestibular system. For instance, when you are on a boat and it rocks on the water, your vestibular system reports that you are moving. However, your visual system indicates that you are not moving relative to the boat. The brain cannot integrate the contradictory information, which, through pathways not well understood, causes the familiar symptoms of motion sickness.

Certain antihistamines or a scopolamine patch can help deal with motion sickness by relieving nausea. It may also help to sit in the front seat of a car or look at the horizon while on a boat, both of which reduce the conflicting information about movement reported by the visual system.

Figure 15.38a). Within each ampulla we find a **crista ampullaris** (KRIS-tah am-pew-LAIR-is), a cluster of hair cells and supporting cells. The hair cells have stereocilia and one kinocilium that are embedded in a gelatinous mass called the **cupula** (KUH-pew-lah).

Since the structure of the crista ampullaris is similar to that of the maculae, it's not surprising that it functions very much like the maculae—another example of the Structure-Function Core Principle (p. 25). For example, when you turn your head, the endolymph lags behind and pushes on the cupula, as shown in **Figure 15.38b.** This

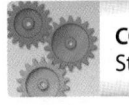

CORE PRINCIPLE
Structure-Function

bends the stereocilia, which either increases or decreases glutamate release from the hair cells and causes a corresponding change in the activity of neurons in the vestibular nerve. Your brain interprets the change in nerve activity as the start of head rotation (acceleration). The endolymph "catches up" quickly during continued rotation. When you stop turning your head, the endolymph continues to move and bends the cupula in the direction opposite to that when you started turning your head. This has the opposite effect on glutamate release and the activity of the vestibular nerve, and your brain interprets the change to mean that head rotation has stopped.

The Vestibular Sensation Pathway

Detecting head movement and position is only part of the battle to maintain equilibrium; the brain must receive and analyze the stimuli and then coordinate a response. The responses can be grouped into three categories: (1) muscle movement to produce changes in posture (for example, flailing your arms to avoid falling when you trip), (2) cognitive awareness of head position and/or movement, and (3) compensatory eye movements for some types of head movement.

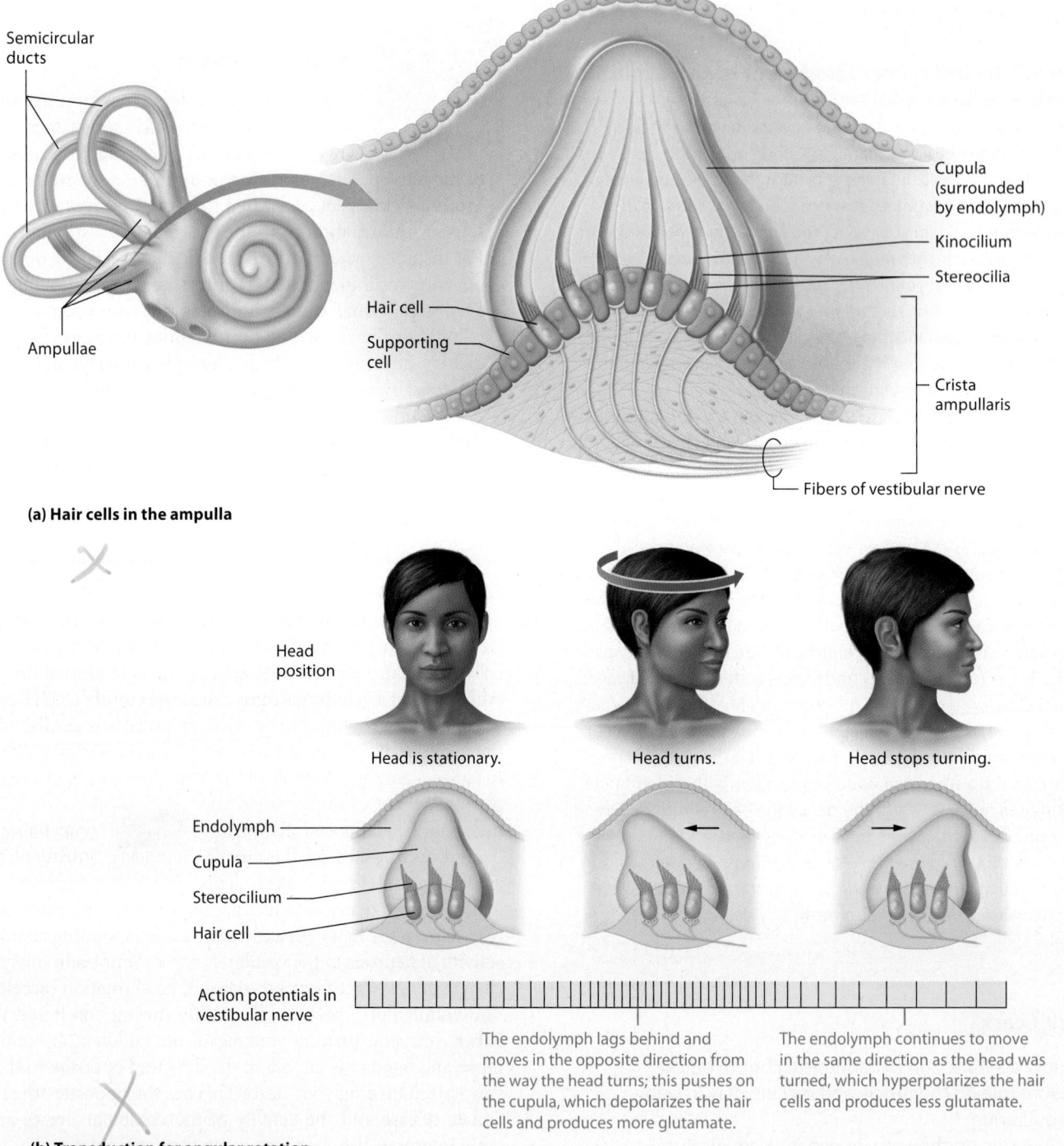

(a) Hair cells in the ampulla

Head position

Head is stationary. Head turns. Head stops turning.

Endolymph
Cupula
Stereocilium
Hair cell

Action potentials in vestibular nerve

The endolymph lags behind and moves in the opposite direction from the way the head turns; this pushes on the cupula, which depolarizes the hair cells and produces more glutamate.

The endolymph continues to move in the same direction as the head was turned, which hyperpolarizes the hair cells and produces less glutamate.

(b) Transduction for angular rotation

Figure 15.38 Angular movement and the semicircular canals.

To produce these responses, the vestibular signals need to reach various regions of the central nervous system, as shown in **Figure 15.39**:

(1) **Vestibular signals travel to the vestibular nuclei located at the medulla-pons junction.**

(2) **The vestibular nuclei forward the signals simultaneously to the following areas of the CNS:**

- **The thalamus and then the inferior parietal lobe, for conscious awareness of head position and movement.**

The inferior parietal lobe integrates the signals so you are aware of the direction in which your head is moving (for example, knowing that your head is turning to the right and not the left). Several regions of the parietal lobe receive vestibular signals, but their locations have not been determined with the same accuracy as in the visual and auditory pathways.

- **The cranial nerve nuclei, to coordinate eye movement in response to head movement.** Axons from the vestibular nuclei terminate in the oculomotor, trochlear, and

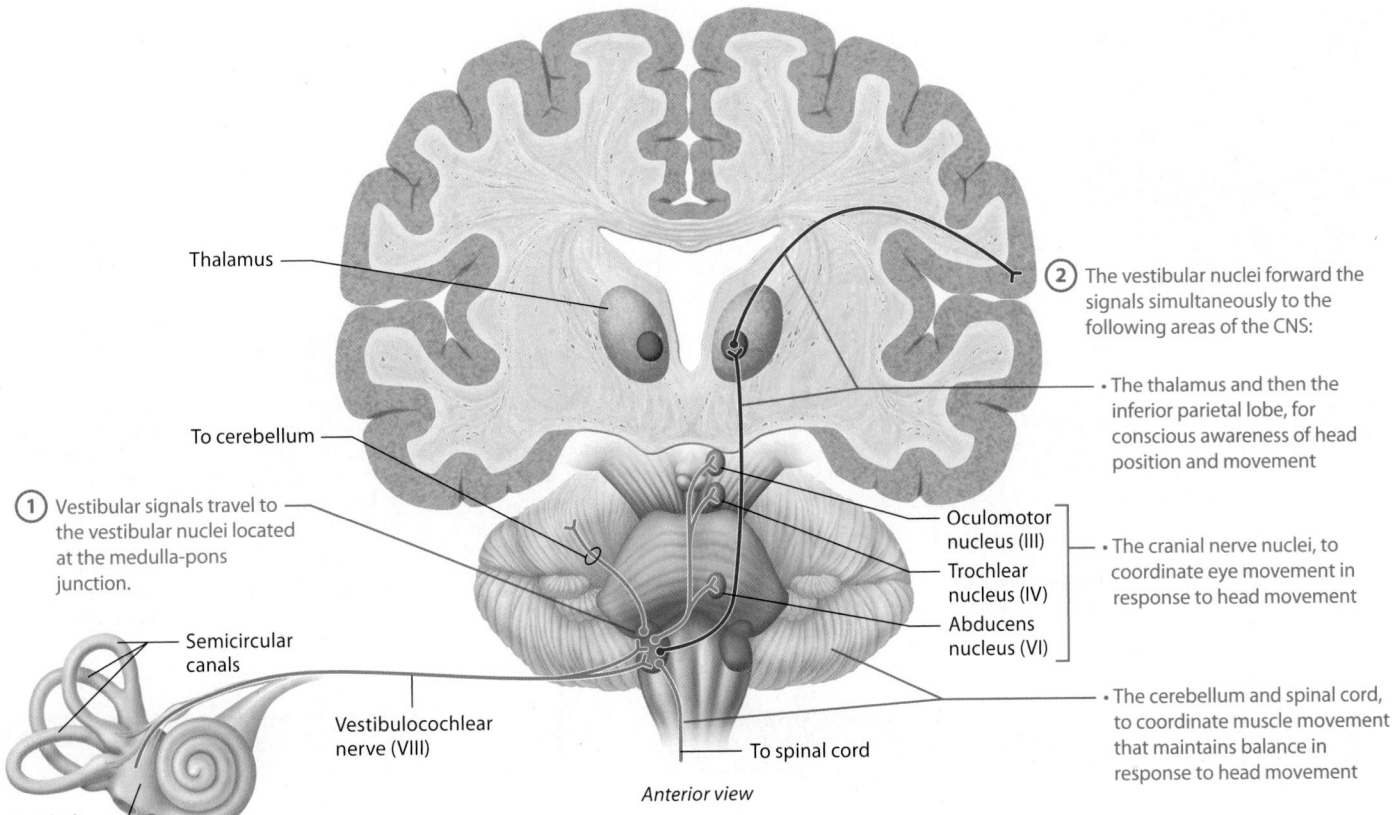

Thalamus

② The vestibular nuclei forward the signals simultaneously to the following areas of the CNS:

- The thalamus and then the inferior parietal lobe, for conscious awareness of head position and movement

To cerebellum

① Vestibular signals travel to the vestibular nuclei located at the medulla-pons junction.

Oculomotor nucleus (III)

Trochlear nucleus (IV)

Abducens nucleus (VI)

- The cranial nerve nuclei, to coordinate eye movement in response to head movement

Semicircular canals

Vestibulocochlear nerve (VIII)

To spinal cord

- The cerebellum and spinal cord, to coordinate muscle movement that maintains balance in response to head movement

Vestibule

Anterior view

Figure 15.39 The vestibular sensation pathway.

abducens nuclei which give rise to the cranial nerves that control eye movement. One example of coordinated eye movement is the *vestibulo-ocular reflex*, which enables us to keep our eyes fixed on an object while our head is moving. This happens when we walk or run toward an object; as our head moves up, our eyes move inferiorly and vice versa. We can choose to override the reflex if we want to turn our heads without fixing our gaze.

- **The cerebellum and spinal cord, to coordinate muscle movement that maintains balance in response to head movement.** For example, tilting your head back to look at the sky requires contracting muscles in your torso and lower limbs so you don't fall over backward.

Quick Check

☐ 4. How do the hair cells in the crista ampullaris detect rotation of the head?

☐ 5. What are the three primary areas of the central nervous system that receive input from the vestibular nuclei?

Apply What You Learned

☐ 1. You feel dizzy and lose your balance whenever you tilt your head back to look at the sky. In which part of the inner ear is dysfunction likely causing your symptoms?

☐ 2. You fix your gaze on your book as you turn your head to the right. In which direction does the vestibulo-ocular reflex cause your eyes to move? Which cranial nerve nuclei

receive input from the vestibular nuclei to produce the vestibulo-ocular reflex?

☐ 3. Your grandmother suffers from dizziness and loses her balance easily. She asks you why her doctor recommended that she visit an ear, nose, and throat specialist. What would you tell her?

See answers in Appendix A.

MODULE **15.9**

How the Special Senses Work Together

Learning Outcomes

1. Summarize the pathways for each of the special senses.
2. Describe how the frontal lobe and limbic system integrate the signals from the special senses into a meaningful picture of a situation.

In this chapter, you have learned how each of the special senses detects a stimulus and transduces it into a neural signal that your brain can interpret. However, the special senses do not function in isolation. They collect and report information simultaneously, so that other areas of the brain can integrate the information into

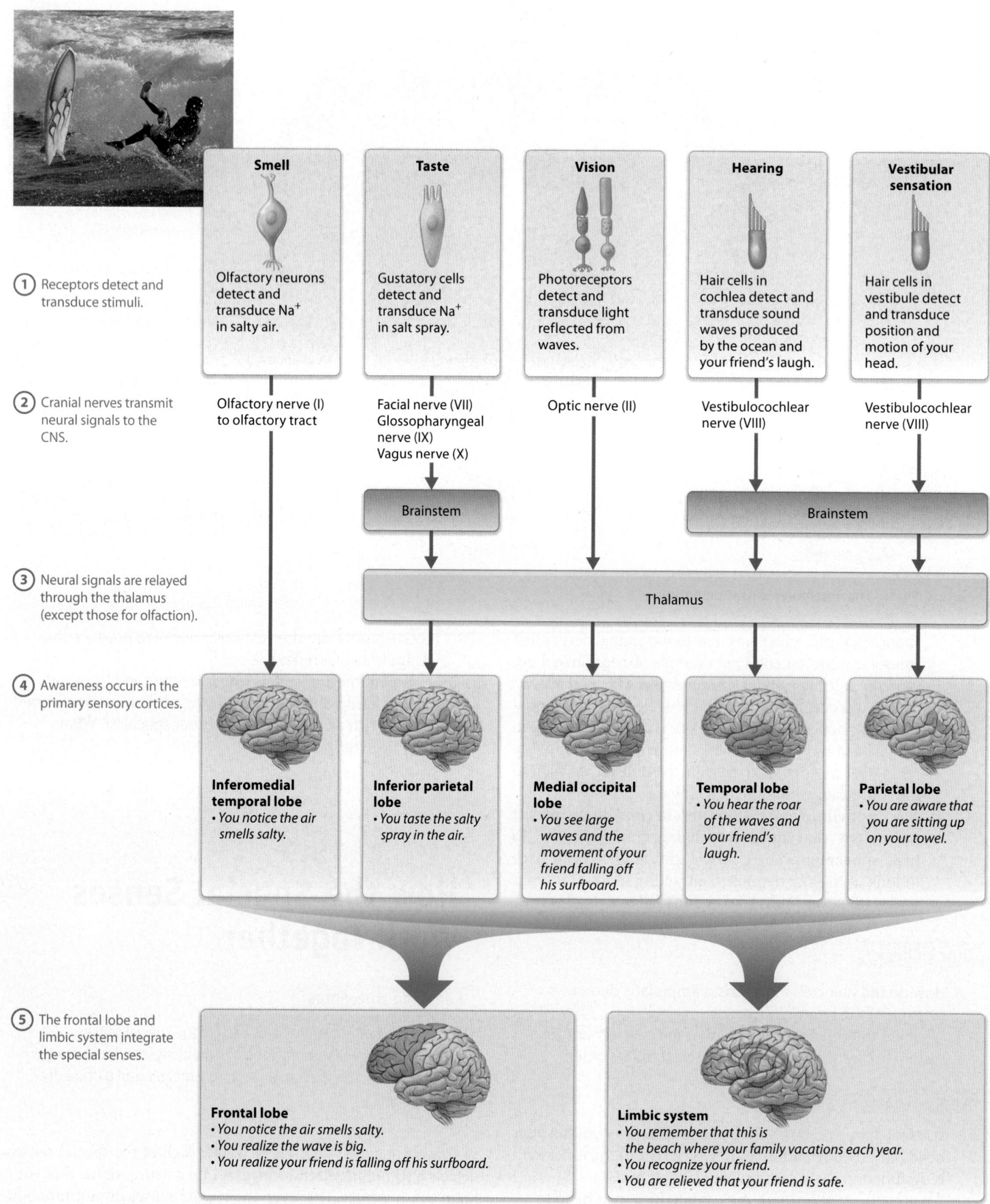

① Receptors detect and transduce stimuli.

Smell

Olfactory neurons detect and transduce Na⁺ in salty air.

Taste

Gustatory cells detect and transduce Na⁺ in salt spray.

Vision

Photoreceptors detect and transduce light reflected from waves.

Hearing

Hair cells in cochlea detect and transduce sound waves produced by the ocean and your friend's laugh.

Vestibular sensation

Hair cells in vestibule detect and transduce position and motion of your head.

② Cranial nerves transmit neural signals to the CNS.

Olfactory nerve (I) to olfactory tract

Facial nerve (VII) Glossopharyngeal nerve (IX) Vagus nerve (X)

Optic nerve (II)

Vestibulocochlear nerve (VIII)

Vestibulocochlear nerve (VIII)

Brainstem

Brainstem

③ Neural signals are relayed through the thalamus (except those for olfaction).

Thalamus

④ Awareness occurs in the primary sensory cortices.

Inferomedial temporal lobe
· You notice the air smells salty.

Inferior parietal lobe
· You taste the salty spray in the air.

Medial occipital lobe
· You see large waves and the movement of your friend falling off his surfboard.

Temporal lobe
· You hear the roar of the waves and your friend's laugh.

Parietal lobe
· You are aware that you are sitting up on your towel.

⑤ The frontal lobe and limbic system integrate the special senses.

Frontal lobe
· You notice the air smells salty.
· You realize the wave is big.
· You realize your friend is falling off his surfboard.

Limbic system
· You remember that this is the beach where your family vacations each year.
· You recognize your friend.
· You are relieved that your friend is safe.

Figure 15.40 How the special senses work together.

a coherent picture of the situation. Let's see how this happens by continuing with our example of watching your friend fall off his surfboard (**Figure 15.40**):

① **Receptors detect and transduce stimuli.** Your olfactory neurons and gustatory cells detect the salt in the air and salt spray. The olfactory neurons respond by generating an action potential, and the gustatory cells generate local potentials, both of which result in release of neurotransmitters. Photoreceptors (rods and cones) in the retina detect light reflected from the waves and hyperpolarize in response; the corresponding decrease in neurotransmitter release stimulates bipolar cells. Hair cells in the cochlea respond to the energy of sound waves and release neurotransmitters. Hair cells in the vestibule detect the position and movement of your head and also release neurotransmitters.

② **Cranial nerves transmit neural signals to the CNS.** Olfactory stimuli are transmitted to the olfactory bulb, which leads to the olfactory tract. The brainstem receives taste, auditory, and vestibular stimuli from the facial, glossopharyngeal, and vagus nerves (taste) and the vestibulocochlear nerve (auditory and vestibular). Visual stimuli are transmitted to the thalamus by the optic nerve.

③ **Neural signals are relayed through the thalamus (except those for olfaction).** With the exception of olfaction, all of the special senses pathways travel to the thalamus for processing and analysis at a subconscious level. Neurons in the thalamus then project to the primary sensory cortex for each sensation. Olfaction travels through the olfactory tract directly to the primary olfactory cortex.

④ **Awareness occurs in the primary sensory cortices.** The primary olfactory cortex is in the inferomedial temporal lobe, and the primary gustatory cortex is in the inferior portion of the parietal lobe; these two areas allow you to notice the aroma of salty air and the taste of ocean spray on your lips. The primary visual cortex in the medial occipital lobe enables you to see the large wave behind your friend and his movement as he falls off his surfboard. The primary auditory cortex in the temporal lobe detects and analyzes the sounds of the waves and your friend's laughter. Several regions of the parietal lobe provide awareness that you are sitting upright on your beach towel.

⑤ **The frontal lobe and limbic system integrate the special senses.** Each primary cortical area has significant connections with the frontal lobe and limbic system that allow you to convert all these pieces of information into a meaningful picture. The frontal lobe integrates the taste and smell of the salty air with the sight and sound of the waves, which enables you to know you are at the beach. It also analyzes the height of the wave and the fact that it threw your friend off his surfboard. At the same time, the limbic system integrates your knowledge with memories of past experiences and appropriate emotions. You recognize the beach where your family has vacationed every year and that it is your friend, not a stranger, who just went into the water. You also recall that your friend is an experienced surfer and realize that his laughter means he is safe.

From this simplified example, you can see the complexity involved in detecting and processing each type of stimulus and integrating it into a meaningful whole. And yet, this seems to happen almost instantly—another example of how our nervous system allows us to interact with our environment.

Apply What You Learned

☐ 1. Your sister sits beside you as you watch your friend surf, but she cannot see him because she is blind. What anatomical structures may not be functioning correctly and thus causing her blindness?

☐ 2. A tumor has destroyed the areas of a patient's thalamus that are part of the special senses pathways. Which special sense will *not* be affected, and why? Explain why the patient has no conscious awareness of the remaining special senses.

See answers in Appendix A.

CHAPTER SUMMARY

Take full advantage of the invaluable online practice and study tools for this chapter by going to MasteringA&P®. There you will find:

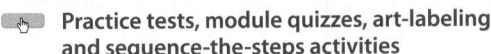 Practice tests, module quizzes, art-labeling and sequence-the-steps activities

 Animated figures with quizzes

Practicing A&P worksheets

 Author-narrated podcast

Dynamic Study Modules

 Interactive Physiology

MODULE **15.1**
Comparison of General and Special Senses 537

● There are five special senses: **olfaction** (smell), **gustation** (taste), **vision, hearing,** and **vestibular sensation.** Their receptors are housed in discrete organs in the head.

● Most special senses have receptors that are specialized cells instead of neurons. They transmit stimuli to the CNS via cranial nerves only instead of cranial and spinal nerves. (Figure 15.1, p. 539)

MODULE **15.2**
Anatomy and Physiology of Smell 538

● The **olfactory epithelium** is located in the roof of the nasal cavity and contains the **olfactory neurons.** Each olfactory neuron is a bipolar neuron with cilia on the expanded dendrite and an axon;

together the axons form the **olfactory nerve** (I). Basal cells in the olfactory epithelium give rise to new olfactory neurons. (Figure 15.2, p. 539; Figure 15.3, p. 540)

- Receptors on the cilia of an olfactory neuron bind **odorants,** which causes the olfactory neuron to depolarize and produce an action potential. Odorants bind to more than one receptor type and are identified by the unique combination of receptors activated.

- The axons of the olfactory neurons travel through the cribriform plate and terminate in the **olfactory bulb. Mitral cells** in the olfactory bulb send their axons out of the bulb as the **olfactory tract.** This terminates in the **primary olfactory cortex** for conscious awareness; signals are also sent to the limbic system for emotional and visceral responses. (Figure 15.4, p. 541)

  Practice art-labeling exercise: Figure 15.4.

MODULE **15.3**
Anatomy and Physiology of Taste 542

- **Taste buds** are clusters of cells located on the **vallate, fungiform,** and **foliate papillae** of the tongue. (Figure 15.5, p. 543)

- Taste buds contain three types of cells:
  - **Gustatory cells** detect taste molecules.
  - **Basal cells** develop into new gustatory cells.
  - **Supporting cells** provide physical support to the taste buds.

- There are five taste sensations: **sweet, salty, sour, bitter,** and **umami** (savory). Each gustatory cell responds to only one taste, and every area of the tongue can respond to each taste.

- Taste chemicals bind to receptors on microvilli of the gustatory cells and depolarize the cell in one of two ways:
  - The taste chemical (such as sodium or hydrogen ions) enters the cell through an ion channel.
  - The taste chemical may activate a G-protein, which closes leak potassium ion channels. (Figure 15.6, p. 544)

- The depolarized gustatory cell releases neurotransmitters onto the sensory endings of neurons in the facial, glossopharyngeal, and vagus nerves. (Figure 15.7, p. 545)

- Axons in these three cranial nerves terminate in the medulla oblongata in the brainstem. From there, taste stimuli are then relayed to the parietal lobe for conscious awareness and to the limbic system for emotional and visceral responses.

MODULE **15.4**
Anatomy of the Eye 546

- Accessory structures of the eye are not directly involved in vision. They include the **eyelids, eyebrows** and **eyelashes, conjunctiva, lacrimal apparatus,** and extrinsic eye muscles. (Figure 15.8, p. 546; Figure 15.9, p. 547; Figure 15.10, p. 548)

  Practice art-labeling exercise: Figure 15.8.

- The eyeball itself consists of a three-layered wall, two cavities, and the lens. (Figure 15.11, p. 548)

  Practice art-labeling exercise: Figure 15.11.

  - The wall of the eyeball has three layers:
    - The **fibrous layer** is the outermost layer. The **sclera** is the white portion that covers about five-sixths of the surface. It is continuous with the transparent **cornea** on the anterior surface.
    - The **vascular layer** consists of the **choroid, ciliary body,** and **iris.**
    - The neural layer, or **retina,** has a pigmented epithelial layer and a layer containing the photoreceptor cells, which transduce light into neural signals.
  - The **lens** is a transparent sphere surrounded by the ciliary body and connected to it by **suspensory ligaments.** It alters the refraction of light so we can focus on near and far objects.
  - The eye has two cavities:
    - The posterior cavity lies posterior to the lens and contains **vitreous humor.**
    - The anterior cavity lies anterior to the lens and is divided into the **anterior** and **posterior chambers** that both contain **aqueous humor.** (Figure 15.15, p. 551)

MODULE **15.5**
Physiology of Vision 552

- Visible light is the range of electromagnetic wavelengths that are detected by the eye. **Rods** and the three types of **cones** respond to particular ranges of wavelengths. (Figure 15.16, p. 552)

- Rays of light refract as they pass through materials of different density and/or a curved surface. (Figure 15.17, p. 553)

- Rays of light from a close object diverge as they approach the eye, and the cornea and lens must converge the rays to focus them on the retina. **Accommodation** is the ability of the lens to assume a rounder shape to focus these rays on the retina. (Figure 15.18, p. 553; Figure 15.19, p. 554)

- Several conditions can interfere with rays focusing on the retina: **presbyopia, myopia, hyperopia,** and **astigmatism.** (Figure 15.20, p. 554)

- The retina contains rods and cones, **bipolar cells, retinal ganglion cells, amacrine cells,** and **horizontal cells.** (Figure 15.21, p. 555; Figure 15.22, p. 557)
  - Rods have a cylindrical outer segment, and cones have a cone-shaped outer segment. Each outer segment contains discs with a photosensitive pigment in the membrane. Each photoreceptor cell is contacted by the dendrites of bipolar cells.
  - When light strikes a photoreceptor, the photon is absorbed by rhodopsin in rods and iodopsin in cones, which triggers a series of reactions that ultimately reduce the levels of **cGMP.** This closes sodium ion channels and hyperpolarizes the cell. (Figure 15.23, p. 558)

- Hyperpolarization of the rods and cones removes the inhibition of the bipolar cells, which increase release of excitatory neurotransmitters onto the retinal ganglion cells. The axons of the retinal ganglion cells form the optic nerve and carry visual stimuli into the brain. (Figure 15.24, p. 559)

- The optic nerves meet at the **optic chiasma,** where axons from the nasal retinas cross to the opposite side. All axons from the opposite visual field then travel in the **optic tract** to the **lateral** geniculate nucleus. Some axons form synapses in the midbrain to control the pupillary reflex.

- Axons from the lateral geniculate nucleus form optic radiations and terminate in the **primary visual cortex.** Conscious awareness and initial analysis begin here; color and shape analysis occurs in the temporal lobe, and analysis of movement occurs in parts of the parietal lobe. (Figure 15.25, p. 561; Figure 15.26, p. 563)

MODULE 15.6
Anatomy of the Ear 564

- The ear is divided into three regions: the outer, middle, and inner ear. (Figure 15.27, p. 564)

  Practice art-labeling exercise: Figure 15.27.

- The outer ear consists of the **auricle,** or pinna, made of elastic cartilage, which funnels sound into the **external auditory canal.** The auditory canal ends at the **tympanic membrane,** or eardrum.

- The middle ear is a cavity in the temporal bone. The tympanic membrane is on its lateral wall, and the **oval** and **round windows** are on the medial wall. The **pharyngotympanic tube** connects the middle ear and superior pharynx. The three auditory ossicles (the **malleus, incus,** and **stapes**) connect the tympanic membrane to the oval window.

- The inner ear is a series of tunnels in the temporal bone. It is lined with the **membranous labyrinth,** which contains **endolymph;** the **perilymph** separates the membranous labyrinth from the walls of the inner ear. The inner ear is divided into three regions that are associated with hearing and head movement and position. (Figure 15.28, p. 566)
  - The **vestibule** contains the **utricle** and the **saccule,** which detect head position and linear acceleration.
  - The **semicircular ducts** are in the **semicircular canals** and detect angular (rotational) movement of the head.
  - The **cochlea** is the spiral-shaped portion of the inner ear and contains structures for hearing. It has three chambers: the **cochlear duct** is filled with endolymph, and the **scala vestibuli** and the **scala tympani** are filled with perilymph. (Figure 15.29, p. 567)

MODULE 15.7
Physiology of Hearing 568

- Sound waves are alternating areas of compressed and less dense air molecules that are generated by the vibration of an object. (Figure 15.30, p. 568)

- In order to hear a sound, the sound waves must reach the inner ear and be transduced into neural impulses that are sent through the vestibulocochlear nerve to the CNS.
  - Sound waves move through the auditory canal and trigger vibration of the tympanic membrane, auditory ossicles, and oval window.
  - The movement of the oval window produces pressure waves in the fluid of the scala vestibuli, which pass through the cochlear duct and the scala tympani to the round window.
  - As vibrations pass through the scala media, they vibrate the **basilar membrane** on which the **spiral organ** sits. The **hair**

cells move with the basilar membrane against the stiffer **tectorial membrane,** which bends the stereocilia. Bending of the stereocilia toward the tallest stereocilium depolarizes the hair cell; bending in the opposite direction hyperpolarizes it. (Figure 15.31, p. 569; Figure 15.32, p. 570)

- The hair cells release neurotransmitters onto neurons of the cochlear nerve, triggering action potentials that travel to the cochlear nuclei in the brainstem.
- After relays in the brainstem, and the medial geniculate nucleus of the thalamus, the auditory stimuli reach the **primary auditory cortex** in the temporal lobe for conscious perception. From there, the signals travel to other parts of the temporal and parietal lobes, as well as the limbic system. (Figure 15.33, p. 571; Figure 15.34, p. 572; Figure 15.35, p. 574)

MODULE 15.8
Anatomy and Physiology of Vestibular Sensation 575

- The utricle and saccule are important in detecting head tilting and linear acceleration. (Figure 15.36, p. 575)
  - They each have a **macula,** which contains a cluster of hair cells with stereocilia and a single **kinocilium** embedded in an **otolithic membrane.**
  - Gravity and linear acceleration displace the otolithic membrane, which bends the stereocilia and kinocilium. Bending toward the kinocilium depolarizes the hair cell, and bending in the opposite direction hyperpolarizes it. This alters the rate of action potentials produced in axons of the vestibular nerve. (Figure 15.37, p. 576)

- The semicircular ducts detect rotational head movements.
  - An **ampulla** at the base of each semicircular duct houses the **crista ampullaris,** where the hair cells are located. The stereocilia and kinocilium are embedded in the gelatinous **cupula.**
  - Rotation of the head pushes the endolymph against the cupula, which bends the stereocilia and kinocilium. This depolarizes or hyperpolarizes the hair cells, depending on the direction of rotation, which changes the rate of action potential generation in the axons of the vestibular nerve. (Figure 15.38, p. 578)

- Vestibular axons terminate in the vestibular nuclei at the junction of the medulla and pons. (Figure 15.39, p. 579)

- The vestibular nuclei project to the following regions of the CNS:
  - cerebellum and spinal cord to coordinate muscle movement that maintains balance;
  - inferior parietal lobe for conscious awareness of head position and movement; and
  - cranial nerve nuclei for eye movement in response to head movement.

MODULE 15.9
How the Special Senses Work Together 579

- Specialized receptor cells—olfactory neurons, gustatory cells, photoreceptors, and hair cells—detect the stimuli for each special sense. (Figure 15.40, p. 580)

- The receptor cells stimulate one or more cranial nerves that carry the signal into the CNS.

- Except for olfaction, pathways for the special senses travel through the brainstem to the thalamus.
- Neurons from the thalamus (or the olfactory tracts in the case of olfaction) terminate in primary sensory areas, where conscious awareness and identification of the stimulus occur.
  - The frontal lobe integrates the information from the special senses into a coherent picture.
  - The limbic system provides appropriate emotional responses and associated memories.

ASSESS WHAT YOU LEARNED

See answers in Appendix A.

LEVEL 1 Check Your Recall

1. Match the cell type with the correct stimulus.

   _____ Rod or cone   a. Head movement
   _____ Hair cell in cochlea b. Odorant
   _____ Gustatory cell   c. Photon
   _____ Olfactory neuron   d. Taste substance
   _____ Hair cell in vestibule e. Sound wave

2. The axons of the olfactory nerve terminate in the:

   a. olfactory epithelium.
   b. olfactory bulb.
   c. olfactory tract.
   d. primary olfactory cortex.

3. Fill in the blanks: In an olfactory neuron, the binding of a(n) _____ to its membrane receptor triggers a(n) _____ potential in the axons of the _____ nerve.

4. The primary olfactory cortex is located in the:

   a. frontal lobe.
   b. occipital lobe.
   c. parietal lobe.
   d. temporal lobe.

5. Which of the following statements is *true* regarding gustatory cells?

   a. They have microvilli that project into the taste pore.
   b. There are only 10–20 gustatory cells in a typical taste bud.
   c. They have a lifespan of approximately 6 months.
   d. Some form synapses with neurons that give rise to the trigeminal nerve.

6. Match the taste with the chemical substance that produces it.

   _____ Sweet   a. Nitrogen-containing compounds
   _____ Sour   b. Sucrose
   _____ Salty   c. Glutamate
   _____ Bitter d. Sodium ions
   _____ Umami   e. Hydrogen ions

7. Which of the following cranial nerves is *not* involved in the gustatory sense?

   a. Vagus nerve (X)
   b. Hypoglossal nerve (XII)

   c. Facial nerve (VII)
   d. Glossopharyngeal nerve (IX)

8. Tears normally flow from the lacrimal sac into the:

   a. lacrimal gland.
   b. lacrimal canaliculi.
   c. lacrimal puncta.
   d. nasolacrimal duct.

9. Which cells in the retina are depolarized in darkness?

   a. Rods and cones
   b. Bipolar cells
   c. Retinal ganglion cells
   d. All of the above
   e. Only rods

10. Why can you see an object better in dim light by looking to the side of it instead of directly at it?

11. Each of the following statements is false. Correct each statement to make it true.

    a. Photons are absorbed by rhodopsin in retinal ganglion cells.
    b. Rods and cones are found in the fovea centralis.
    c. In the dark, rods and cones produce action potentials.

12. The axons from the nasal retina in the left eye terminate in the:

    a. right lateral geniculate nucleus.
    b. left lateral geniculate nucleus.
    c. right medial occipital lobe.
    d. left medial occipital lobe.

13. Mark the following statements as true or false. If a statement is false, correct it to make a true statement.

    a. The incus is connected to the tympanic membrane.
    b. The stapes is attached to the oval window.
    c. The auditory canal is separated from the middle ear by the round window.
    d. The cochlear duct is filled with perilymph.
    e. The semicircular ducts are connected to the utricle, and the cochlear duct is continuous with the saccule.
    f. The spiral organ is located in the scala tympani.

14. Explain how sounds of different frequencies are detected in the cochlea.

15. How do the hair cells of the crista ampullaris detect rotation of the head?

16. The macula in the utricle detects:

    a. rotation of the head to the right.
    b. very low frequency sound waves that we can feel but not hear.
    c. tilting of the head to one side.
    d. linear acceleration of the head in a vertical plane.

17. Fill in the blanks: The spiral organ sits on the _____ membrane, and the stereocilia of the outer hair cells contact the _____ membrane.

18. True or false: Hair cells in the spiral organ will depolarize or hyperpolarize, depending on the direction in which the stereocilia are bent.

19. Which of the following is *not* part of the auditory pathway in the CNS?

   a. Superior temporal lobe
   b. Superior colliculus
   c. Medial geniculate nucleus
   d. Nuclei in the pons
   e. Cochlear nuclei

20. Fill in the blanks: In the ampulla of a semicircular duct, the hair cells are located in the _____ _____ and their stereocilia and kinocilium are embedded in the _____.

21. Stimuli from the inner ear regarding head movement and position are sent to all of the following except the:

   a. vestibular nuclei.
   b. oculomotor nucleus.
   c. parietal lobe.
   d. cerebellum.
   e. trigeminal nucleus.

LEVEL 2 Check Your Understanding

1. Explain what would happen to your sense of smell if there were no basal cells present in the olfactory epithelium.

2. In which direction would you be unable to move your right eye if your right abducens nerve were damaged?

3. If a patient suffers visual impairment only in one eye, why must the damage be located in the visual pathway prior to the optic chiasma?

4. Following a stroke, a patient lost vision in the left visual field. Name several structures in the brain that could have been damaged by the stroke. Be specific as to right or left side.

5. When standing with your eyes closed, why do you sway more than with your eyes open but don't lose your balance and fall over?

6. Why do you have the sensation that you are still spinning immediately after stopping?

LEVEL 3 Apply Your Knowledge

PART A: Application and Analysis

1. Mr. Spencer suffers loss of taste sensation. Tests reveal that his taste buds and salivary glands are functional. What other structures might be impaired that would explain his loss of taste?

2. If severe congestion from a cold prevented your pharyngotympanic tube from opening, what could happen to your tympanic membrane? Explain your answer.

3. Suppose the round window in your middle ear became very stiff. What effect would this have on hearing, and why?

PART B: Make the Connection

4. Mrs. Flores is a 45-year-old female who suffered a fracture of the ethmoid bone as the result of a car accident. Several days later, she complained that she couldn't taste her food very well. How is her injury related to the loss of taste sensation? *(Connects to Chapter 7)*

5. Your 60-year-old patient, Mr. Guster, has Bell's palsy, which causes dysfunction of those portions of the facial nerve that control muscles of facial expression. He is surprised when you tell him to tape the eyelids of his right eye closed at night to prevent dryness. Why might Bell's palsy cause dryness of the eye on the affected side? *(Connects to Chapter 9)*

The Endocrine System

Throughout this book, we have seen that a cell's activities can be altered in the face of changing conditions to maintain homeostasis. For example, when the concentration of calcium ions in the blood decreases, osteoclasts become more active and resorb more bone to release calcium ions. But how does a cell "know" how and when to alter its activities? The answer lies in the body's two major *regulatory systems*, meaning those that regulate the functions of other cells. The two major regulatory systems in the body are the nervous system, which we explored in previous chapters, and the endocrine system, which we turn to in this chapter. Organs of the **endocrine system** work by synthesizing and secreting chemical messengers called hormones into the blood. **Hormones,** in turn, interact with their target cells and influence the cells' functions in order to maintain fluid, electrolyte, and acid-base homeostasis; promote growth; regulate metabolic reactions; and more.

In this chapter, we explore the organs and cells of the endocrine system, the hormones they produce, and how these hormones maintain various aspects of homeostasis. Along the way, we examine the effects of endocrine system disorders and the often dramatic impact they can have on the body's functions. The chapter concludes with three examples of endocrine control of physiological variables, providing a "big picture" look at how the endocrine organs and hormones work together and thereby maintain overall homeostasis.

MODULE **16.1**

Overview of the Endocrine System

Learning Outcomes

1. Compare and contrast how the endocrine and nervous systems control body functions.
2. Describe the major structures and functions of the endocrine system.
3. Explain the different types of chemical signaling used by the body.
4. Define the terms hormone, endocrine gland (organ), endocrine tissue, and target cell.
5. Describe the major chemical classes of hormones found in the human body, and compare and contrast the types of receptors to which each class of hormone binds.
6. Describe several types of stimuli that control production and secretion of hormones, including the roles of negative and positive feedback.

Computer-generated image: Many hormones function by binding to a membrane receptor and triggering a change inside the cell.

Let's start our exploration of the endocrine system with an overview of its basic properties. In this module, we first examine the differences between the ways in which the nervous and endocrine systems maintain homeostasis. The discussion continues with a look at the different types of chemical signals used in the body, the anatomy of the endocrine organs, and finally the main properties of the endocrine system's chemical messengers: hormones.

Comparison of the Endocrine and Nervous Systems

◀◀ Flashback

1. How does a neuron communicate with its target cell? (p. 406)

Together, the endocrine and nervous systems complement one another to maintain nearly every aspect of homeostasis. The cells of both systems use chemicals to communicate with other cells, in an example of the Cell-Cell Communication Core Principle

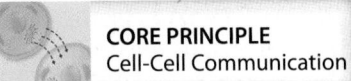 **CORE PRINCIPLE** Cell-Cell Communication

(p. 27). However, the two systems differ significantly in how they regulate body functions. As you discovered in

previous chapters, the nervous system operates through a series of neurons that directly affect their target cells through the release of neurotransmitters (see Chapter 11). The effects of the nervous system are almost immediate, but are short-lasting unless the stimulation is repetitive. The cells of the endocrine system, however, do

not come into close contact with their target cells. Instead, they secrete hormones, which then travel in the bloodstream to reach the receptors on their target cells. Hormones require seconds to several hours or days to elicit their effects, but these effects are generally longer-lasting than those of the nervous system. Note also that the endocrine and nervous systems can work together to increase each other's effects.

Quick Check

☐ 1. How do the endocrine and nervous systems differ in the ways they maintain homeostasis? How are they similar?

Types of Chemical Signals

Hormones and neurotransmitters are not the only chemical signals used by the body. As you can see in **Figure 16.1**, there are three basic types of chemical signals: endocrine, *paracrine* (PAR-ah-krin), and *autocrine* (AW-toh-krin). In endocrine signals, as we just discussed, the hormone is secreted into the blood and travels to affect distant tissues (Figure 16.1a). **Paracrine** signals consist of a chemical secreted by cells into the extracellular fluid to affect nearby but different types of cells (Figure 16.1b). In **autocrine** signals, a chemical secreted by cells into the interstitial fluid elicits effects from the same cell or cell type (Figure 16.1c). Classically, hormones have been described as exclusively endocrine signals, but physiologists now understand that some hormone molecules can function as paracrine and autocrine signals as well.

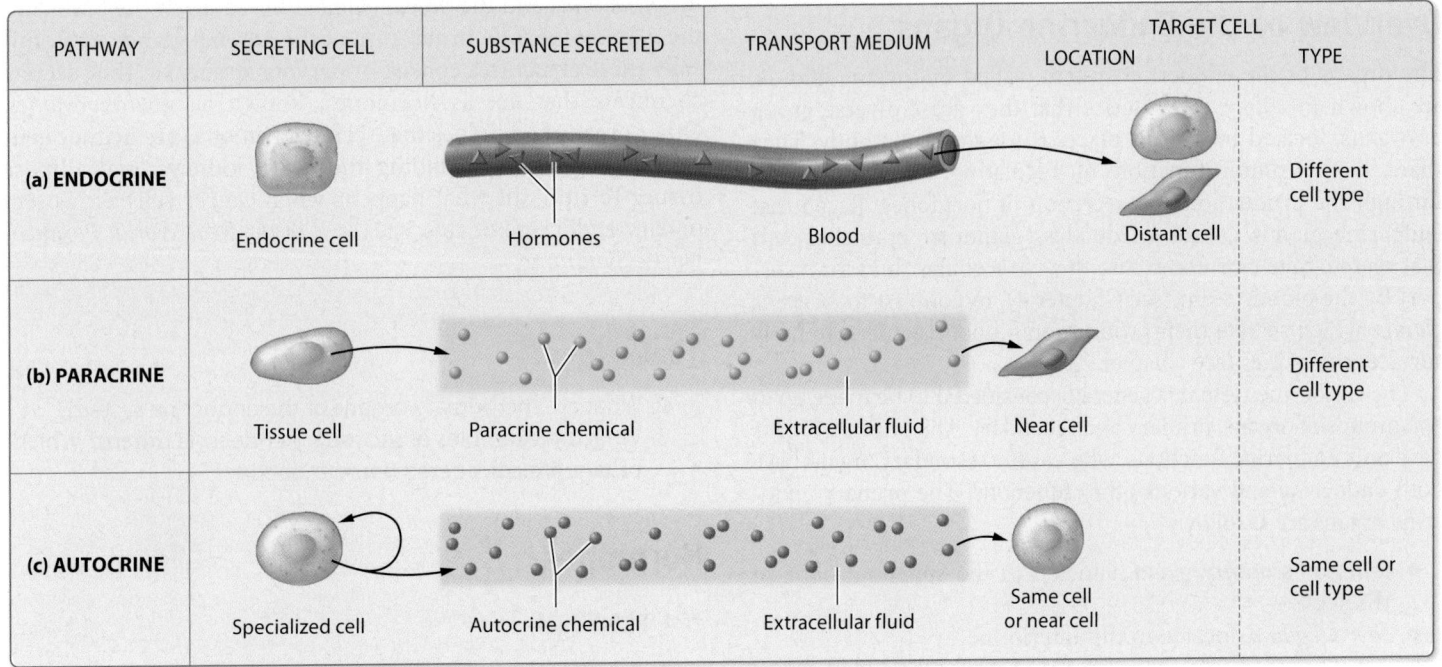

Figure 16.1 Three basic signaling pathways.

Endocrine organs:

Neuroendocrine organs:

- Anterior pituitary gland
- Thyroid gland
- Parathyroid glands
- Thymus gland
- Adrenal cortex
- Pancreas
- Ovaries (female)
- Testes (male)

- Hypothalamus
- Pineal gland
- Posterior pituitary gland

Anterior view

Posterior view

- Adrenal medulla

Figure 16.2 Overview of the endocrine organs.

Practice art labeling
@MasteringA&P®

Overview of the Endocrine Organs

The organs of the endocrine system, called **endocrine glands,** are shown in **Figure 16.2.** Notice that they are a diverse group of organs, located in various places throughout the body. They share the common function of regulating other cell types through the production and secretion of hormones. Recall that endocrine glands consist of ductless glandular epithelial cells that secrete their hormones into the extracellular fluid for transport by the bloodstream (see Chapter 4), in contrast to *exocrine glands,* which secrete their products into ducts that lead to body surfaces or cavities (see Chapter 22).

The endocrine system is generally considered to be made up of two groups of organs, primary and secondary. The primary organs have only endocrine functions, whereas the secondary organs have both endocrine and various other functions. The primary endocrine organs are as follows:

- *anterior pituitary gland,* situated in the sphenoid bone of the skull;
- *thyroid gland,* located in the anterior neck;
- *parathyroid glands,* found on the posterior thyroid gland;
- *adrenal cortices,* located on the superior side of the kidneys;

- *endocrine pancreas,* located in the left side of the abdominal cavity mostly posterior to the stomach; and
- *thymus,* located in the superior mediastinum.

A number of organs can be considered secondary endocrine organs, from the heart to the kidneys and small intestines. Perhaps the most well-known of these are the *testes* and *ovaries,* the former suspended below the pelvic cavity in men and the latter in the pelvic cavity in women. Many of these secondary endocrine organs are **neuroendocrine organs,** including the *hypothalamus* and the *pineal gland* (PIN-ee-al) in the brain and the *adrenal medulla* in the core of the adrenal gland. Anatomically these structures consist of nervous tissue, yet they secrete chemicals that act as hormones, known as *neurohormones.* Other organs and tissues that secrete hormones are neither neural nor endocrine, including the heart, kidneys, and adipose tissue. To find out what happens when cancer cells act as secondary endocrine organs, see *A&P in the Real World: Paraneoplastic Syndrome.*

Quick Check

☐ 2. What are the primary organs of the endocrine system?
☐ 3. What are examples of secondary endocrine organs? Which organs are considered neuroendocrine?

Hormones

◀◀ **Flashback**

1. Are peptides and steroids generally hydrophilic or hydrophobic? (p. 55)

2. What is gene expression? (p. 102)

3. What is a negative feedback loop? (p. 22)

Hormones are chemicals that regulate some functions of other cells. In a sense, they represent the "middle managers" of the body—they "tell" other cells what to do by stimulating some sort of change in those cells. Hormones can have widely varying effects by reaching cells all over the body. Some of them act quickly, triggering changes in cell function within seconds to minutes, whereas others require hours or more to cause an effect in a target cell.

Hormones may travel in the blood either freely or bound to plasma protein carriers. Their mode of transport depends largely on the hormone's chemical structure. Hormones composed of hydrophobic molecules travel bound to proteins because they cannot associate with the water molecules in blood plasma. However, protein hormones consisting of largely hydrophilic molecules do associate with water, and so are able to travel freely through the blood.

Hormones continue to circulate in the blood until they are either taken up by a target cell or broken down and deactivated. The amount of a particular hormone in the blood at any given time depends on both the amount of hormone secreted from the endocrine gland and the rate at which the hormone is removed from the blood. Some hormones are inactivated very rapidly in the cells they affect by reactions catalyzed by enzymes, but most are removed more slowly from the circulation by the kidneys (where they are eliminated in the urine) or liver (where they are broken down by enzymes). Hydrophilic hormones are generally removed from the circulation by these processes more rapidly than are hydrophobic hormones, which are bound to protein carriers.

The upcoming subsections focus on the types of target cells with which hormones interact and how these cells help us classify different types of hormones. Next, we see how hormones interact with cells and produce certain effects. Finally, we look at how hormone secretion is regulated and how hormones interact with one another.

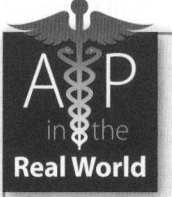

Paraneoplastic Syndrome

The heart and kidneys aren't the only secondary endocrine tissues able to produce hormones—many types of cancer cells can, too. The signs and symptoms that accompany hormone secretion from cancer cells are collectively termed *paraneoplastic syndrome*. Many different types of cancer cells can cause paraneoplastic syndrome, but the most common are cells of lung and gastrointestinal cancers.

The effects of paraneoplastic syndrome depend on the type of hormone secreted by the cancer cells; common findings include imbalances in fluid, calcium ion, and sodium ion homeostasis. The symptoms of paraneoplastic syndrome often precede other symptoms of the cancer, and investigating them may lead to an earlier diagnosis.

Target Cells and Receptors

Hormones are able to affect only particular cells, called their **target cells;** these cells contain specific proteins, referred to as **receptors,** to which hormones can bind. Receptors have three-dimensional shapes that are highly specific for the hormone molecules to which they bind, and can bind to hormones present at extremely low concentrations, which is an example of the Structure-Function Core Principle (p. 25). Hormones may have multiple types of target cells in different tissues. In some cases, the receptors for hormones will be the same among different types of target cells, whereas in other situations, the receptors vary slightly in structure and function.

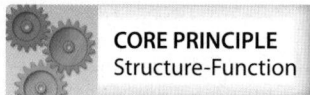

CORE PRINCIPLE
Structure-Function

Hormone receptors may either be embedded in the plasma membrane or reside within the cytosol or nucleus of the target cell. The location of the receptor that a hormone binds depends on the hormone's structure and therefore whether it is hydrophilic or hydrophobic. Hydrophilic hormones cannot readily cross the phospholipid bilayer of the plasma membrane to enter target cells because they are repelled by the fatty acid tails of the phospholipid bilayer. In contrast, as shown here, hydrophobic hormones are able to cross membranes due to the attraction between the hormone and the membrane's fatty acid tails:

As a result of these differences, hydrophilic hormones interact with receptors embedded in the target cell's plasma membrane, with the binding site exposed on the cell surface. The receptor is often associated with another membrane protein that produces a response inside the cell. These other proteins can be ion channels, enzymes, or peripheral proteins. Hydrophobic hormones, however, are able to bind to receptors found within the plasma membrane, in the cytosol, or in the nucleus.

The number of receptors in the target cell varies with the body's needs. When the level of a particular hormone rises in the blood, certain target cells will produce more receptors for that hormone, a process called **up-regulation.** However, prolonged exposure to a high level of a given hormone causes the opposite effect—target cells decrease the number of receptors specific for that hormone, which is known as **down-regulation.** These mechanisms allow cells to have tight control over how they interact with hormones.

Classes of Hormones

There are two basic types of hormones, classified according to their chemical structure: *amino acid–based* and *steroid*. **Amino acid–based hormones** consist of one or more amino acids (see Figure 2.21). They vary in size from single amino acids (amine

(a) Hydrophilic hormone and second-messenger system

(b) Hydrophobic hormone and intracellular receptor mechanism

Figure 16.3 Mechanisms of action of hydrophilic and hydrophobic hormones. This figure presents examples of how these hormones can work.

hormones) to several amino acids (peptide hormones) to complete proteins (protein hormones). **Steroid hormones** are cholesterol derivatives, with a core of hydrocarbon rings. Whereas amino acid–based hormones are generally hydrophilic and therefore bind to plasma membrane receptors, all steroid hormones are hydrophobic and can interact with either intracellular or plasma membrane receptors. Thyroid hormones are an exception, as they are amino acid–based but hydrophobic.

Mechanisms of Hormone Action

How hormones bring about their effects depends on whether they are hydrophilic or hydrophobic. The next two subsections examine these different mechanisms of action.

Hydrophilic Hormones Several types of mechanisms can be set in motion when a hydrophilic hormone binds with its cell-surface receptor on a target cell. One such mechanism is a *second-messenger system*. The typical steps of this system are illustrated in **Figure 16.3a**, which shows an example using a receptor linked to a peripheral protein called a **G-protein.**

① **The hydrophilic hormone (first messenger) binds to its receptor in the plasma membrane.** In our example, the target cell's receptor is linked to a G-protein.

② **The receptor activates a peripheral protein.** When the hormone binds to its receptor, it induces a change in the G-protein, which dissociates into two subunits.

③ **The peripheral protein activates an enzyme.** One subunit of the G-protein then leads to the activation or inhibition of another protein associated with the plasma membrane, such as an enzyme.

④ **The enzyme catalyzes the formation of a second messenger.** If the enzyme is activated, it will catalyze the formation of another molecule known as a **second messenger,** and the amount of second messenger in the cytosol will increase. If the enzyme is inhibited, the cytosolic level of second messenger will decrease.

⑤ **The second messenger initiates a series of events in the cell that leads to changes in its activity.** Examples of second messenger effects include triggering smooth muscle cells to relax or contract, opening or closing an ion channel, and activating or deactivating enzyme systems.

In this type of mechanism, the second messenger, not the first, is the molecule actually initiating the cellular change. One reason second-messenger systems are effective is that they allow *signal amplification.* This refers to the fact that a single hormone molecule can bind a receptor and lead to the formation of hundreds of second-messenger molecules, which in turn can activate or inhibit thousands of molecules that then produce events in the cell.

A well-studied example of a G-protein second-messenger system is the *adenylate cyclase–cAMP system,* in which the G-protein activates an enzyme called **adenylate cyclase.** This enzyme then catalyzes the formation of the second messenger **cyclic**

adenosine monophosphate (cAMP) from ATP. Cyclic AMP then initiates changes within the cell. Many of cAMP's effects are due to the activation of enzymes called **protein kinases.** These enzymes catalyze the transfer of a phosphate group from ATP to another molecule, a process called *phosphorylation*. Phosphorylation changes the activity or shape of a specific protein and so can have a wide range of effects within a cell.

Note that the adenylate cyclase–cAMP system is just one of many second-messenger systems that involve G-proteins and protein kinases. Other such systems function in a similar manner. Examples of other second messengers include *inositol trisphosphate (IP3)* and *diacylglycerol (DAG)*.

G-protein–mediated second-messenger systems are only one possible way that hydrophilic hormones interact with their target cells. Other hydrophilic hormones bind to receptors that also function as ion channels. Binding of the hormone to such receptors may cause the channel to open or close. In addition, some hydrophilic hormones can open or close ion channels indirectly by binding to receptors linked to G-proteins.

Hydrophobic Hormones As discussed earlier, hydrophobic hormone molecules are able to diffuse across the plasma membrane and enter the cytosol of target cells. You can see in **Figure 16.3b** that this enables them to interact with intracellular receptors that may be located in the cytosol or the nucleus. When a hydrophobic hormone binds to an intracellular receptor, it forms a hormone-receptor complex. This complex then generally interacts with the DNA inside the cell's nucleus, changing the rate of synthesis of one or more specific proteins. It was long thought that interaction with DNA was the only mechanism of action of hydrophobic hormones; however, current evidence indicates they can exert many other effects not involving DNA. These effects are carried out via binding plasma membrane receptors and/or certain cytosolic receptors and producing the responses we described for hydrophilic hormones.

Effects of Hormone Actions

Activation of either cell-surface or intracellular receptors triggers different mechanisms inside the cell that lead to cellular effects. The many effects of hormone action include:

- stimulating secretion from an endocrine or exocrine cell;
- activating or inhibiting enzymes;
- stimulating or inhibiting mitosis and/or meiosis;
- opening or closing ion channels in the cell's plasma membrane and/or altering its membrane potential; and
- activating or inhibiting transcription of genes that code for RNA or proteins (also known as gene expression).

Regulation of Hormone Secretion

Hormone secretion can be initiated by hormonal, humoral, or neural stimuli (**Figure 16.4**).

(a) Hormonal stimuli

Hormonal stimulation: Growth hormone–releasing hormone (GHRH) stimulates secretion of growth hormone (GH) from an anterior pituitary cell.

Hormonal inhibition: Somatostatin inhibits secretion of growth hormone from an anterior pituitary cell.

GHRH · Blood vessel · GH · Anterior pituitary · Hypothalamus · Anterior pituitary cell · Somatostatin · Anterior pituitary cell · GH

(b) Humoral stimulus

Pancreas · Glucose uptake by a pancreatic cell triggers insulin secretion. · Glucose · Insulin · Pancreatic cell

(c) Neural stimulus

Spinal cord · Adrenal medulla · Sympathetic nerve · Sympathetic neurons stimulate secretion of epinephrine and norepinephrine from an adrenal medulla cell. · Epinephrine · Norepinephrine · Axon terminal of sympathetic neuron · Adrenal medulla cell

Figure 16.4 Types of stimuli for hormone secretion.

Let's look at each type of stimulus in more detail:

- **Hormonal stimuli.** Some endocrine cells increase or decrease their secretion in response to other hormones. For example, the hypothalamus regulates secretion from the anterior pituitary gland through secretion of *releasing* and *inhibiting hormones.* In Figure 16.4a, we use the example of two hypothalamic hormones: growth hormone–releasing hormone and somatostatin, which stimulate and inhibit secretion of growth hormone from the anterior pituitary gland, respectively.
- **Humoral stimuli.** Many endocrine cells respond to the concentration of a certain ion or molecule in the blood or extracellular fluid, such as glucose or calcium ions. These stimuli are known as *humoral stimuli,* a term reflecting the historical name for body fluids, *humors.* The example in Figure 16.4b shows a cell of the pancreas releasing the hormone insulin in response to an elevated level of glucose in the blood.
- **Neural stimuli.** Finally, some cells respond to signals from the nervous system, as shown in Figure 16.4c. The best example of this is the adrenal medulla, which is stimulated by sympathetic neurons to release the neurohormones epinephrine and norepinephrine.

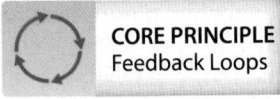

Secretion is generally regulated as part of a *negative feedback loop*—recall the Feedback Loops Core Principle (p. 21). Negative feedback loops of the endocrine system proceed by the following general steps (**Figure 16.5**; note that we are using an example in which the endocrine cell acts as the receptor and control center).

- **Stimulus: A regulated physiological variable deviates from its normal range.** The change can be an increase or decrease from the normal range; in this figure, we are showing a decrease.
- **Receptor: Receptors on target cells detect the deviation of the variable.** The receptors of the endocrine gland cells continuously monitor regulated variables in the blood. Such variables include glucose and calcium ion concentrations and body temperature.
- **Control center: The stimulated control center (often the endocrine cell) increases or decreases its secretion of a particular hormone.** In the endocrine system, the endocrine cell that detects a change in the variable from the normal range is generally also the control center. When the control center (the endocrine cell) is stimulated, it changes its secretion of the appropriate hormone.

STIMULUS

A regulated physiological variable deviates from its normal range.

Normal range

IN HOMEOSTATIC RANGE

As the variable returns to normal, feedback to the control center decreases the effector response.

Normal range

RECEPTOR

Receptors on endocrine cells detect the deviation of the variable.

Endocrine cells

Regulated variable   Blood vessel

EFFECTOR/RESPONSE

The hormone triggers a response in its target cells that moves conditions toward the normal range.

Target cells

CONTROL CENTER

The stimulated control center (often the endocrine cell) increases or decreases its secretion of a particular hormone.

Hormone molecule

Regulated variable

Figure 16.5 Maintaining homeostasis: regulation of hormone secretion by negative feedback loops.

- **Effector/response: The hormone triggers a response in its target cells that moves conditions toward the normal range.** The hormone enters the blood and reaches its target tissue or tissues, where it triggers changes that return the variable to its normal range.
- **Homeostatic range: As the variable returns to its normal range, feedback to the control center decreases the effector response.** When homeostatic conditions have been restored, the control center returns hormone secretion to the normal range by either decreasing or increasing it. This in turn stops stimulation to the target cells, decreasing or ending the response.

As the upcoming modules will discuss, many negative feedback loops in the endocrine system are more complex than this simple example, and many involve several levels of control from multiple endocrine glands. There are also rare instances of hormones regulated by positive feedback mechanisms.

Hormone Interactions

The maintenance of homeostasis requires not just individual hormones, but the interaction of multiple hormones. Many hormones have *complementary* actions, meaning that they each affect different target cells in order to accomplish a common goal. For example, hormones of the adrenal gland and pancreas all work together to maintain homeostasis during exercise. Other hormones can act on the same target cell to exert the same effect. Such hormones are known as **synergists** (SIN-er-jists), and they have more pronounced effects when acting together than when acting alone. Sometimes, however, hormones that act on the same target cells have opposite effects. These hormones are said to be **antagonists.**

Quick Check

☐ 4. What are the two major classes of hormones, and how do they interact with their target cells?
☐ 5. What are the three types of stimuli that initiate hormone secretion? How do they perform this function?
☐ 6. How is hormone secretion generally regulated?
☐ 7. How do synergistic and antagonistic hormones differ?

Apply What You Learned

☐ 1. A newly discovered hormone X is determined to be structurally similar to cholesterol. Predict how this new hormone interacts with its target cells.
☐ 2. Hormone X appears to regulate the blood concentration of chemical Y via a negative feedback loop.
   a. Predict what happens to the secretion of hormone X when the blood concentration of chemical Y *decreases*.
   b. What effect will this likely have on the blood concentration of chemical Y?
   c. Predict what happens to the secretion of hormone X when the blood concentration of chemical Y *increases*.

See answers in Appendix A.

The Hypothalamus and the Pituitary Gland

Learning Outcomes

1. Describe the locations of, as well as the anatomical and functional relationships between, the hypothalamus and the anterior and posterior pituitary glands.
2. Explain how releasing hormones, inhibiting hormones, and tropic hormones affect the secretion of other hormones.
3. Describe the stimulus for release, the target tissue, and the functional effect of each hormone released from the posterior pituitary.
4. Describe the stimulus for release, the target tissue, and the functional effect of each anterior pituitary hormone.
5. Explain the negative feedback loops that regulate the production and release of anterior pituitary hormones.

Recall that the hypothalamus of the brain is a small collection of nuclei with many important homeostatic functions, including regulation of hunger, thirst, fluid balance, body temperature, the sleep/wake cycle, and certain reproductive functions (see Chapter 12). The hypothalamus works together with the pituitary gland, with which it has a close anatomical and physiological relationship. Several hypothalamic and anterior pituitary hormones directly control certain body functions, whereas others are *tropic hormones,* meaning they control hormone secretion from other endocrine glands. Be careful not to confuse a tropic hormone with the similar-sounding *trophic hormone* (TROH-fik; *troph-* = "change"), which induces growth in its target cell. Note that some hormones have both tropic and trophic effects on their target cells.

Some of the primary functions of the hypothalamus are performed directly by hypothalamic neurons that end in the posterior pituitary gland, whereas others are accomplished by hormones targeted to the anterior pituitary gland. In this module we examine this relationship in discussing the cells, hormones, and target tissues of the hypothalamus and the anterior and posterior pituitary glands. We will revisit these structures in later modules to more closely explore their roles in regulating other endocrine organs.

Structure of the Hypothalamus and Pituitary Gland

◀◀ **Flashback**

1. What is the sella turcica, and where is it located? (p. 218)

As you can see in **Figure 16.6a,** the **hypothalamus** is the small anteroinferior portion of the diencephalon of the brain. It is connected to the **pituitary gland** (also called the *hypophysis;* hy-POF-eh-sis) by a stalk called the **infundibulum** (in'-fun-DIB-yoo-lum). The pituitary gland is a small organ, about the size of a bean, that sits in the sella turcica of the sphenoid bone.

(a) Locations of hypothalamus and pituitary gland in the brain

Pituitary gland

Hypothalamus

Sella turcica of sphenoid bone

Optic chiasma

Hypothalamus

Anterior pituitary (adenohypophysis)

Infundibulum

Posterior pituitary (neurohypophysis)

(b) Structure of hypothalamus, and anterior and posterior pituitary glands

Figure 16.6 Structure of the hypothalamus and pituitary gland.

As **Figure 16.6b** shows, the pituitary gland is composed of two structurally and functionally distinct components: the **anterior pituitary gland,** or **adenohypophysis** (ad'-eh-noh-hy-POF-eh-sis), and the **posterior pituitary gland,** or **neurohypophysis** (noo'-roh-hy-POF-eh-sis). The differences in the two components are inherent in their names: The *adeno-* in adenohypophysis means "gland," which reflects the fact that the anterior pituitary is a true gland composed of hormone-secreting glandular epithelium. The *neuro-* in neurohypophysis refers to the fact that the posterior pituitary is actually made up of nervous tissue.

The hypothalamus and pituitary gland have a specialized blood supply that allows them to deliver their hormones to their target cells. The tiny blood vessels called *capillaries* merge in the hypothalamus to form larger blood vessels, the *portal veins,* that travel through the infundibulum. These veins lead to a second group of capillaries in the anterior pituitary gland. Systems like this, in which capillaries are drained by veins that lead to another set of capillaries, are called *portal systems.* Accordingly, this group of vessels is called the **hypothalamic-hypophyseal portal system** (hy'-poh-FIZ-ee-ul; look ahead to Figure 16.7).

Quick Check

☐ 1. How do the anterior and posterior pituitary differ structurally and functionally?
☐ 2. What is the hypothalamic-hypophyseal portal system?

Hormones of the Hypothalamus and Posterior Pituitary

◀◀ **Flashback**

1. What are groups of neuronal cell bodies in the central nervous system called? (p. 431)
2. What are the primary functions of the hypothalamus? (p. 434)

The posterior pituitary makes no hormones of its own. Instead, it stores and releases two neurohormones produced by the hypothalamus. These hormones are *antidiuretic hormone* (an'-ty-dy-yoo-RET-ik) and *oxytocin* (ok-sih-TOW-sin). Let's look more closely at these hormones and their roles.

Antidiuretic Hormone: Water Retention

The hypothalamic hormone that controls water balance is known as **antidiuretic hormone (ADH),** or **vasopressin.** As you can see in **Figure 16.7a**, ① the cell bodies of hypothalamic neurons continually make ADH in low amounts. ② The ADH travels through the hypothalamic axons in the infundibulum ③ to the axon terminals in the posterior pituitary, where it is stored in synaptic vesicles. Here, the axon terminal does not form a synapse with another cell. Instead, ④ ADH is secreted into the blood vessels in the posterior pituitary when the hypothalamic neurons fire action potentials.

The primary function of ADH is to increase the amount of water retained by the kidneys. This function is apparent from its name: A *diuretic* (dy-yoo-RET-ik) is a chemical that increases urine production by the kidneys, which decreases the amount of water in the body. Therefore, an *anti*diuretic chemical has the opposite effect—it decreases urine production by increasing the amount of water retained by the body.

So that you can better understand the effect of ADH on the kidneys, let's take a brief look at the kidneys' microscopic structure. The fluid that eventually becomes urine is located within small tubes in the kidney, called *tubules.* The relationship between the tubules, interstitial fluid, and blood is shown here:

Tubule

Interstitial fluid

Capillary

Fluid in tubule

Kidney

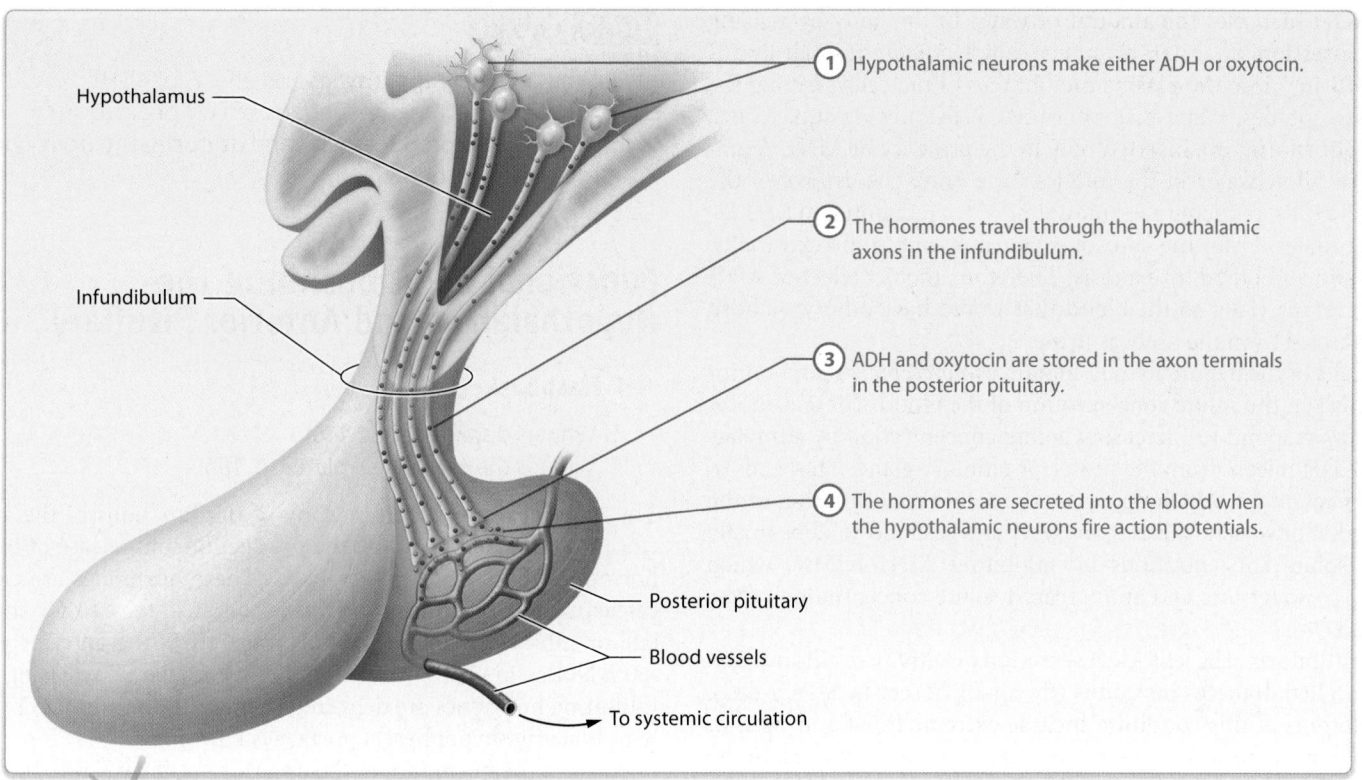

1. Hypothalamic neurons make either ADH or oxytocin.

2. The hormones travel through the hypothalamic axons in the infundibulum.

3. ADH and oxytocin are stored in the axon terminals in the posterior pituitary.

4. The hormones are secreted into the blood when the hypothalamic neurons fire action potentials.

Hypothalamus

Infundibulum

Posterior pituitary

Blood vessels

To systemic circulation

(a) **Relationship between the hypothalamus and posterior pituitary**

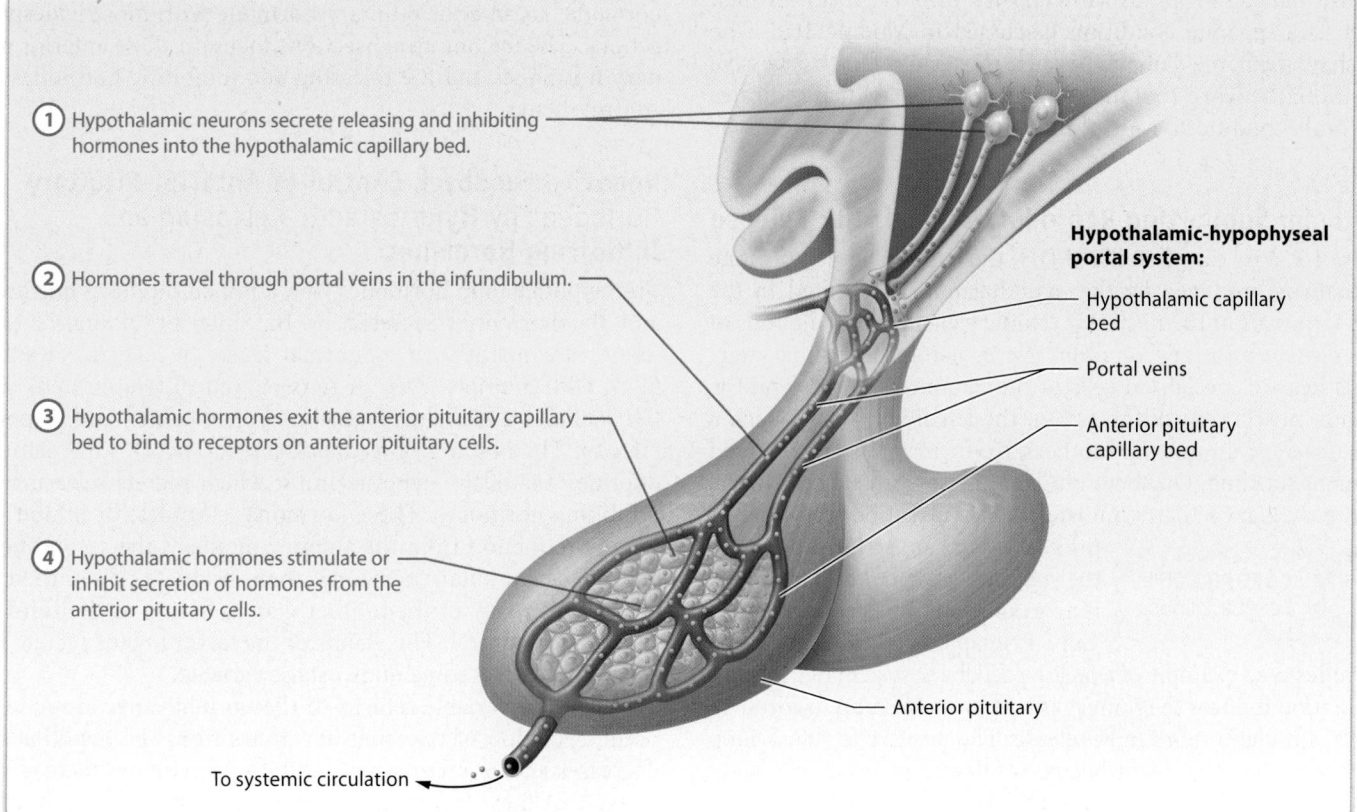

1. Hypothalamic neurons secrete releasing and inhibiting hormones into the hypothalamic capillary bed.

2. Hormones travel through portal veins in the infundibulum.

3. Hypothalamic hormones exit the anterior pituitary capillary bed to bind to receptors on anterior pituitary cells.

4. Hypothalamic hormones stimulate or inhibit secretion of hormones from the anterior pituitary cells.

Hypothalamic-hypophyseal portal system:

Hypothalamic capillary bed

Portal veins

Anterior pituitary capillary bed

Anterior pituitary

To systemic circulation

(b) **Relationship between the hypothalamus and anterior pituitary**

Figure 16.7 Functional relationships between the hypothalamus and pituitary gland.

ADH increases the amount of water in the body by causing the insertion of water channels called *aquaporins* (ah-kwah-POHR-inz) into the plasma membranes of the cells forming the kidney tubules. Water can enter these cells only via aquaporins, and aquaporins are inserted only in the presence of ADH. Aquaporins allow water in the tubules to re-enter the cytosol of the kidney cells, reducing the amount of water remaining in the kidney tubules. From the cytosol, water moves into the extracellular fluid and blood by osmosis. Therefore, the net effect of ADH is to return water to the blood that would have otherwise been eliminated from the body as urine.

Cells of the hypothalamus contain *osmoreceptors* that monitor changes in the solute concentration of the blood. These osmoreceptors respond to increasing solute concentration by stimulating ADH release from the posterior pituitary gland. This leads to water retention, which decreases the solute concentration of the blood. Conversely, these same receptors respond to abnormally low solute concentrations by inhibiting ADH release, which leads to water loss and an increased solute concentration of the blood.

An abnormal lack of ADH secretion or activity results in a disease called **diabetes insipidus** (dy-ah-BEET-eez in-SIP-ih-dus). Symptoms of this condition include extreme thirst and signs of dehydration because the body is unable to conserve most of the water that is consumed. Most forms of diabetes insipidus can be treated by administration of synthetic ADH. By the way, don't confuse diabetes insipidus with *diabetes mellitus* (MELL-ih-tus), which is a separate condition discussed in Module 16.5. The two share the name diabetes because the term refers to excessive urination; the word root *diab-* means "siphon", which indicates that both conditions were thought to siphon water from the body.

Oxytocin: Supporting Reproduction and Milk Release

Figure 16.7a reveals that, like ADH, the hypothalamic hormone **oxytocin** is produced by the hypothalamus and stored in the axon terminals of the posterior pituitary gland. The functions of oxytocin are primarily reproductive in nature. Its primary target tissues are specialized cells of the mammary glands and the smooth muscle cells of the *uterus,* the female organ in which a fetus develops. In nursing mothers, oxytocin release is triggered by infant suckling. Oxytocin binds to the specialized cells in the mammary glands that stimulate their contraction, resulting in

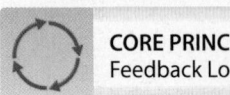 CORE PRINCIPLE
Feedback Loops

milk ejection, an action known as the *milk let-down reflex.* This reflex is an example of the Feedback Loops Core Principle (p. 21). Specifically, this reflex is an example of a positive feedback system in the body, as lactation induces the hungry infant to suckle more vigorously, which stimulates more milk release. The positive feedback loop is broken when the infant becomes full and so no longer suckles. Oxytocin may have effects on the brain as well. For example, the possible role of oxytocin in promoting emotional bonding in both sexes is currently under investigation. These topics are covered more fully in the development and heredity chapter (Chapter 27).

Quick Check

☐ 3. What are the target tissues and effects of ADH?
☐ 4. What are the target tissues and effects of oxytocin?
☐ 5. Which gland produces ADH and oxytocin, and from where are they secreted?

Functional Relationship of the Hypothalamus and Anterior Pituitary

◀◀ **Flashback**

1. What is dopamine? (p. 416)
2. What is the epiphyseal plate? (p. 186)

The hypothalamus controls many of the functions of the anterior pituitary gland through the production and release of tropic hormones that affect its secretions. These hormones are called **releasing** and **inhibiting hormones** because they either stimulate or inhibit the release of hormones from the anterior pituitary. Notice in **Figure 16.7b** that the hypothalamic releasing and inhibiting hormones are delivered to the anterior pituitary by the hypothalamic-hypophyseal portal system.

Most of the hormones of the anterior pituitary gland, in turn, are also primarily tropic and control the secretion of hormones by other endocrine glands or tissues. One exception is *growth hormone,* an anterior pituitary hormone with more widespread actions. The upcoming subsections focus on these anterior pituitary hormones and the releasing and inhibiting hormones that control them.

Negative Feedback Control of Anterior Pituitary Hormones by Hypothalamic Releasing and Inhibiting Hormones

The hypothalamic hormones, the anterior pituitary hormones, and the hormones secreted by the anterior pituitary's target tissues are maintained at normal levels by negative feedback loops with multiple levels or tiers of control (**Figure 16.8**). Each tier includes a stimulus, receptor, control center, and response/effector. The first tier of feedback control involves the neuroendocrine cells of the hypothalamus, which secrete releasing and inhibiting hormones. These hormones stimulate or inhibit hormone production from the anterior pituitary, the second tier of control. The pituitary hormones then stimulate cells of their target organs, many of them other glands, acting as the third tier of feedback control. The glands of the target organs secrete hormones that affect some homeostatic variable.

When the variable returns to the normal range around a set point, secretion of releasing hormones from the hypothalamus decreases and/or secretion of inhibiting hormones increases. As a result of these changes, secretion from the anterior pituitary gland will decrease, and secretion from the endocrine target cells will in turn decrease. These multiple levels of control allow the endocrine system to tightly regulate conditions within the body, which is critical to the maintenance of homeostasis.

Anterior Pituitary Hormones That Affect Other Glands

The tropic hormones of the anterior pituitary include the following:

- **Thyroid-stimulating hormone.** As implied by its name, **thyroid-stimulating hormone (TSH)**, or *thyrotropin*, stimulates development of the thyroid gland and its secretion of *thyroid hormones*. TSH is produced and secreted by cells called *thyrotrophs* (THY-roh-trofs). Its secretion is triggered by a hypothalamic releasing hormone known as **thyrotropin-releasing hormone (TRH)**. Secretion of TSH is inhibited by the hypothalamic hormone **somatostatin** (soh′-mah-toh-STAH-tin; *-statin* = "inhibiting or blocking").

- **Adrenocorticotropic hormone.** **Adrenocorticotropic hormone (ACTH)**, or *corticotropin*, is produced by cells called *corticotrophs*. It stimulates the development of the adrenal glands and their synthesis of various steroid hormones. ACTH secretion is stimulated by a hypothalamic releasing hormone known as **corticotropin-releasing hormone (CRH)**.

- **Prolactin.** The hormone **prolactin (PRL;** *lact-* = "milk") is produced by cells called *lactotrophs*. Its primary target tissue is mammary gland cells, where it stimulates growth of the mammary gland, the initiation of milk production after childbirth, and the maintenance of milk production for the duration of breastfeeding. The main stimulus for prolactin release is *prolactin-releasing hormone* (*PRH*, which may be the same chemical as TRH) from the hypothalamus, which itself is stimulated by suckling of the nipple by an infant. Normally, prolactin secretion is inhibited by the hypothalamic inhibiting hormone **prolactin-inhibiting factor,** now known to be dopamine. Suckling by the infant decreases the release of dopamine, which facilitates prolactin release.

- **Luteinizing hormone.** **Luteinizing hormone (LH;** LOO-tey-in-ay′-zing) is one of two hormones known as **gonadotropins** (goh-nad′-uh-TROHP-inz) secreted by cells called *gonadotrophs.* The gonadotropins are so named because they stimulate the male and female *gonads,* the testes and ovaries, respectively. LH secretion is stimulated by the hypothalamic releasing hormone **gonadotropin-releasing hormone (GnRH).** In males, LH stimulates production of the male hormone *testosterone* by cells of the testes. In females, it stimulates production of female hormones called *estrogens* and *progesterone.* It also triggers the release of an oocyte (OO-oh-syt; an immature egg cell) in the process of ovulation.

- **Follicle-stimulating hormone.** The other gonadotropin produced by gonadotrophs is **follicle-stimulating hormone (FSH).** Like LH, FSH secretion is stimulated by GnRH from the hypothalamus. In males, FSH stimulates cells of the testes to produce chemicals that bind and concentrate testosterone. In females, FSH works with LH to

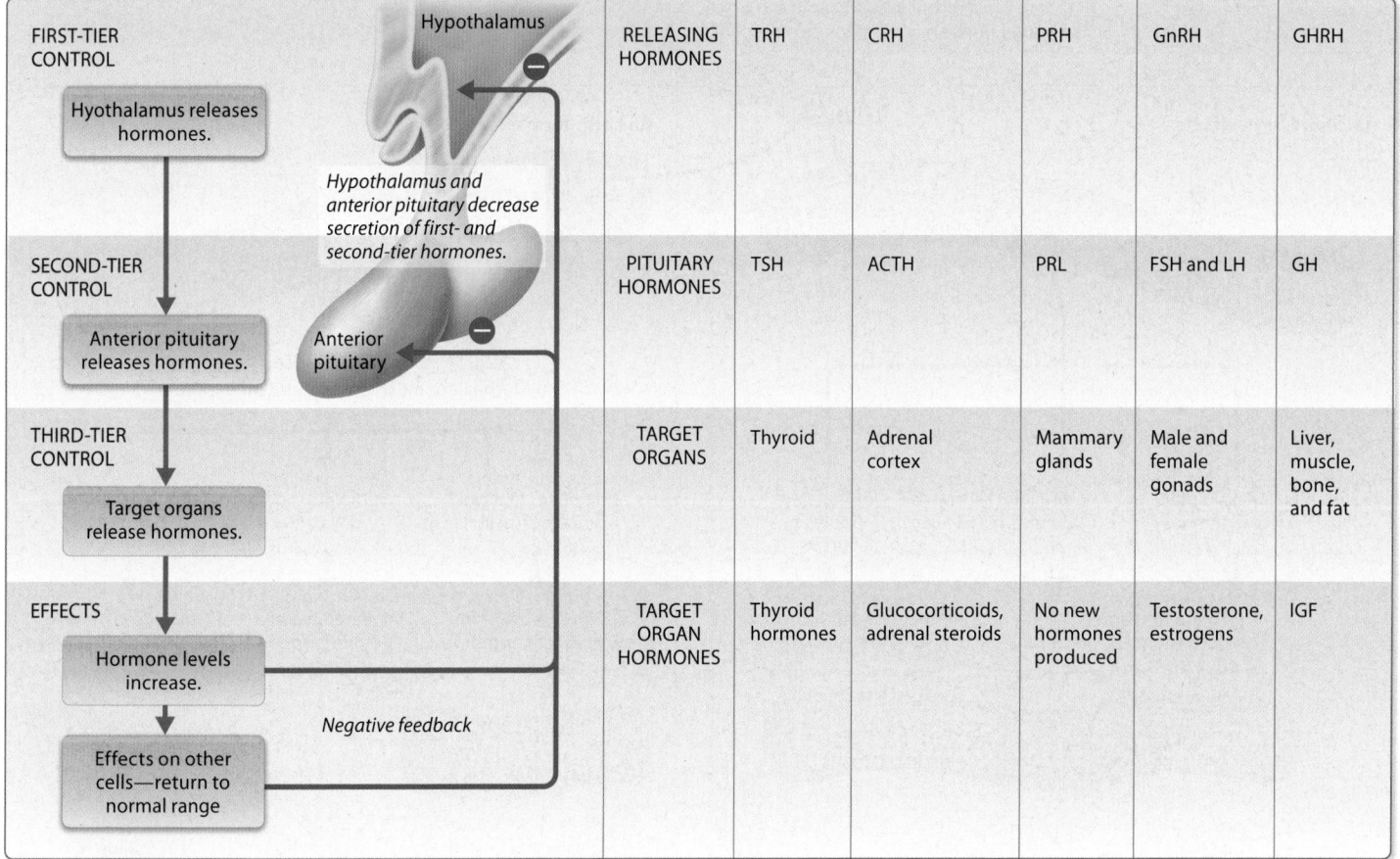

Figure 16.8 Multi-tiered negative feedback control of hormones of the hypothalamus, anterior pituitary gland, and target organs.

trigger the production of estrogens. Its name derives from the fact that it also triggers the maturation of *ovarian follicles,* which house the developing oocytes.

We discuss TSH and ACTH in more detail in Modules 16.3 and 16.4, respectively. LH and FSH are explored further in Module 16.6, and when we discuss the reproductive system and development (see Chapters 26 and 27).

Anterior Pituitary Hormone with Widespread Effects: Growth Hormone

Another important hormone produced by the anterior pituitary gland is **growth hormone (GH).** GH is produced and secreted by cells called *somatotrophs* (*somato-* = "body"; GH itself is sometimes called *somatotropin*). These cells release GH periodically throughout the day, with peak secretion occurring during sleep.

Effects and Regulation of Growth Hormone The primary function of GH is inherent in its name—it regulates and controls growth. The effects of GH on its target tissues, which include

skeletal and cardiac muscle, adipose, liver, cartilage, and bone, can be short-term or long-term. As you can see in **Figure 16.9a,** the short-term effects of GH are primarily metabolic in nature, and include promotion of fat breakdown (lipolysis), production of new glucose by the liver (*gluconeogenesis*), and inhibition of glucose uptake by muscle fibers. These effects increase the concentrations of glucose and fatty acids in the blood, allowing cells to use them as fuel and raw materials for growth.

The long-term effects of GH are not all mediated by GH directly (**Figure 16.9b**). Instead, GH acts on the liver and other target tissues to promote the production of a hormone called **insulin-like growth factor (IGF);** its name reflects the fact that its actions are similar to those of the hormone *insulin,* which is discussed later. IGF affects nearly every cell type in the body, and triggers rapid protein synthesis and cell division, leading to increased longitudinal bone growth and muscle development in children. It also decreases the blood glucose concentration by stimulating glucose uptake by cells, which is the opposite action of acute GH release. Both hormones continue to play important roles in adults even after longitudinal bone growth has ceased,

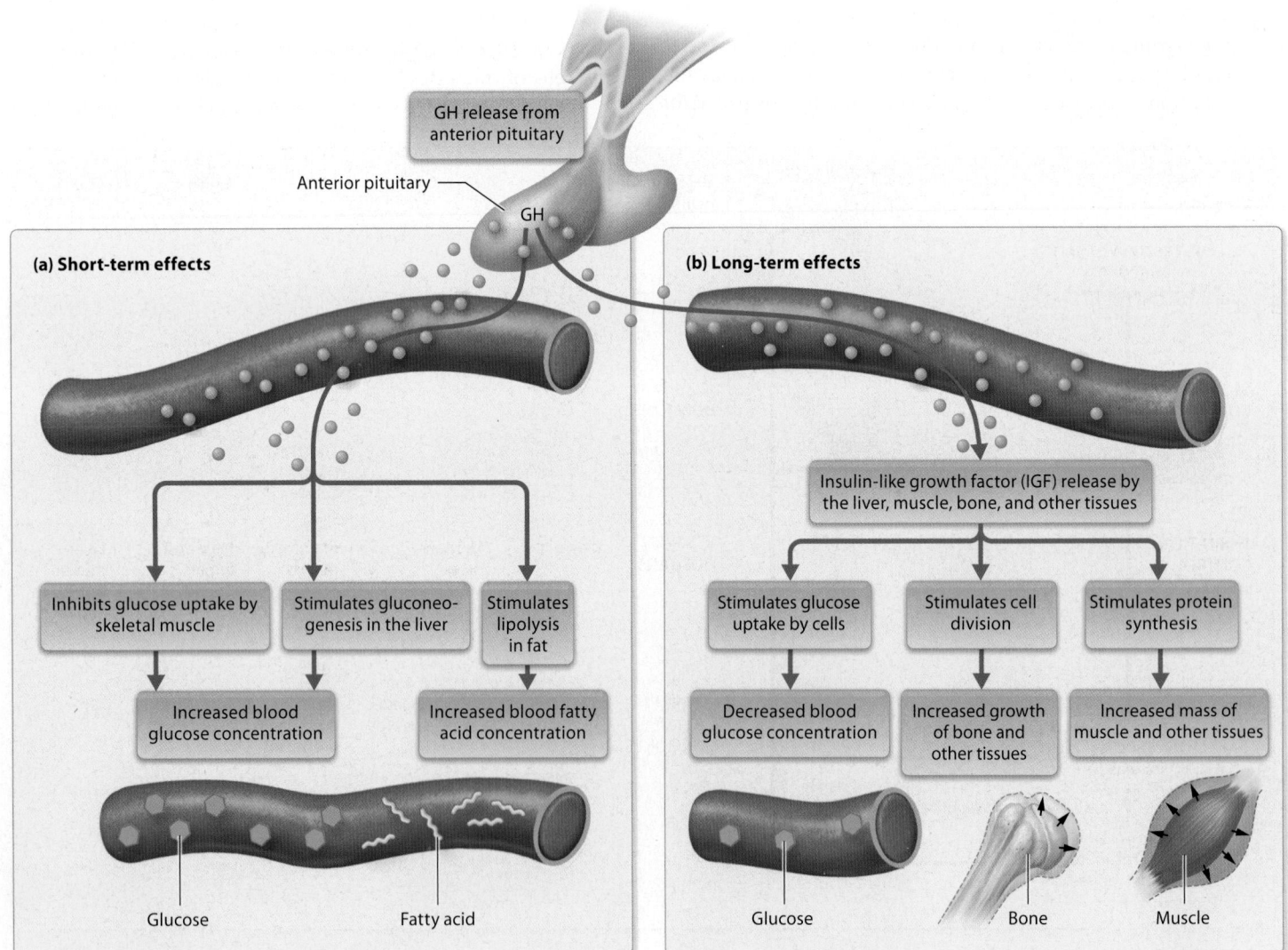

Figure 16.9 Effects of growth hormone (GH).

Human Growth Hormone and the "Fountain of Youth"

You may have received spam in your e-mail inbox advertising the wonders of synthetic human growth hormone, or hGH. Proponents claim that hGH promotes fat loss and muscle growth, and is literally a "fountain of youth." It has become a popular supplement with body builders, those seeking weight loss, and individuals looking to stave off the inevitable effects of aging.

As with most such claims, there is a grain of truth to these assertions. GH does indeed promote fat breakdown and protein synthesis. In addition, adult patients with GH deficiency who are treated with hGH do lose fat and gain muscle. However, serious problems can also arise with hGH supplementation. Excess GH has many potentially harmful effects, including elevated blood glucose level (hyperglycemia) and growth of tissues such as the tongue and the bones of the hands, face, and feet. Additionally, most hGH supplements are sold as oral preparations to be ingested. However, because GH is a protein hormone, it is destroyed by acid and enzymatic reactions in the stomach. The only way to effectively administer the drug is via injection, so the oral preparations are ineffective and a complete waste of money. (But at least these preparations offer minimal risk of adverse effects!)

Figure 16.10 The regulation of growth hormone (GH) release.

promoting muscle development as well as regulating body mass. These effects have led to the unscrupulous promotion of GH as a "health" supplement, which is explored in *A&P in the Real World: Human Growth Hormone and the "Fountain of Youth."*

GH release is regulated by two hypothalamic hormones, as shown in **Figure 16.10**. **Growth hormone–releasing hormone (GHRH)** stimulates the release of GH, and its secretion increases during exercise, fasting, and stress, and after the ingestion of a protein-rich meal. GH release, like that of TSH, is inhibited by the hypothalamic hormone somatostatin.

Growth Hormone Disorders Imbalances in GH secretion can result in various disruptions in growth. Hypersecretion of GH, which is generally due to a GH-secreting pituitary tumor, produces different effects depending on the age of the individual. If GH hypersecretion occurs *before* closure of the epiphyseal plates, the result is **gigantism.** Individuals afflicted with gigantism are unusually tall, generally more than 7 feet. Not only does the skeleton increase in size, but all other tissues, including the heart and other organs, enlarge as well. If the condition is detected early enough, it can be managed by removal of the tumor. However, if it is left untreated, affected individuals generally succumb to heart failure at an early age.

If GH hypersecretion occurs *after* closure of the epiphyseal plates, the result is a different condition, **acromegaly** (aeh´-kroh-MEG-ah-lee; "large extremities"). Longitudinal growth

can no longer take place, but tissues do increase in girth. The tissues most affected seem to be those of the head, face, hands, and feet, as well as organs such as the heart and liver. This condition results in progressively distorted facial features, thickening of the tongue and skin, and enlargement of the hands and feet. Like gigantism, acromegaly can, if left untreated, lead to disruption of the functions of various organ systems, up to and including heart failure.

Hyposecretion of GH prior to closure of the epiphyseal plates results in **pituitary dwarfism.** Those with this condition are generally short in stature, but their limbs and trunk are proportional. A similar form of dwarfism results when IGF secretion is abnormally low, although in these patients GH secretion is normal to elevated.

Table 16.1 on page 600 gives a summary of the hypothalamic hormones stored in and released from the posterior pituitary. It also provides details on the regulatory hormones of the hypothalamus and anterior pituitary.

Quick Check

☐ 6. How does the hypothalamus control the secretion of hormones from the anterior pituitary gland?

☐ 7. What are the tropic hormones of the anterior pituitary gland? What glands do they affect?

☐ 8. Describe the target tissues and effects of growth hormone and IGF.

Apply What You Learned

☐ 1. A hypothetical poison destroys the hypothalamic cells that produce TRH. Predict the effects of this poison on hormone secretion by the anterior pituitary and thyroid glands.

TABLE 16.1 HORMONES OF THE HYPOTHALAMUS AND PITUITARY GLAND

Hormone	Stimulus for Release	Inhibitor(s) of Release	Target Tissue(s)	Effects
Hypothalamic Hormones (stored in the posterior pituitary)				
Antidiuretic hormone (ADH)	Increased solute concentration of the blood	Decreased solute concentration of the blood	Kidneys Brain	• Water reabsorption from the kidney tubules • Increases blood volume
Oxytocin	• Stretching of the uterus • Infant suckling at the nipple	Lack of appropriate stimuli	Uterus Mammary gland	• Uterine contractions • Milk let-down reflex
Anterior Pituitary Hormones				
Thyroid-stimulating hormone (TSH)	• Thyrotropin-releasing hormone (TRH) from the hypothalamus • Exposure to cold • Stress	• Increased levels of thyroid hormones • Somatostatin from the hypothalamus	Thyroid gland	• Growth and development of the thyroid gland • Synthesis of thyroid hormones
Adrenocorticotropic hormone (ACTH)	• Corticotropin-releasing hormone (CRH) from the hypothalamus • Stress	• Increased level of cortisol • Increased level of aldosterone	Adrenal cortex	• Growth and development of adrenal cortices • Release of adrenal steroids and catecholamines
Prolactin	• Prolactin-releasing hormone (PRH) from the hypothalamus • Infant suckling at the nipple	Prolactin-inhibiting factor (dopamine) from the hypothalamus	Mammary gland	• Development of mammary glands • Milk production
Luteinizing hormone (LH)	Gonadotropin-releasing hormone (GnRH) from the hypothalamus	Increased levels of testosterone (males) and estrogens and progesterone (females)	Male gonads Female gonads	Male gonads: • Development of gonads • Testosterone production Female gonads: • Development of gonads • Production of estrogens and progesterone • Ovulation
Follicle-stimulating hormone (FSH)	Gonadotropin-releasing hormone (GnRH) from the hypothalamus	Increased levels of testosterone (males) and estrogens (female)	Male gonads Female gonads	• Production of factors that bind and concentrate testosterone • Production of estrogens; maturation of ovarian follicles
Growth hormone (GH)	• Growth hormone–releasing hormone (GHRH) from the hypothalamus • Stress/exercise • Ingestion of protein • Fasting	Somatostatin from the hypothalamus	Liver Adipose tissue Muscle tissue Bone and cartilage	• Gluconeogenesis • Fat breakdown (lipolysis) • Protein breakdown • Production of insulin-like growth factor (IGF), which stimulates cell division and protein synthesis

2. The condition called *syndrome of inappropriate ADH secretion,* or *SIADH,* is characterized by an abnormally increased secretion of ADH from the hypothalamus. Predict the effects of this condition.

3. A tumor increases the secretion of GHRH. Predict the effects of this tumor on secretion of GH and IGF. How would these effects differ in a child versus an adult?

See answers in Appendix A.

MODULE 16.3
The Thyroid and Parathyroid Glands

Learning Outcomes

1. Describe the gross and microscopic anatomy of the thyroid and parathyroid glands.
2. Identify and describe the types of cells within the thyroid gland that produce thyroid hormone and calcitonin.
3. Describe the stimulus for release, the target tissue, and the effects of thyroid hormone.
4. Explain how negative feedback loops regulate the production of thyroid hormones.
5. Describe the stimulus for release, the target tissue, and the effects of parathyroid hormone.

As you can see in **Figure 16.11a**, the *thyroid gland* is located in the anterior neck, just superficial to the larynx. The thyroid gland secretes two types of hormones: *thyroid hormones,* which regulate growth and metabolism, and *calcitonin* (kal-sih-TOW-nin), which helps to regulate calcium ion homeostasis. Embedded in the posterior surface of the thyroid gland we find several small glands, the *parathyroid glands* (*para-* = "next to"; refer back to Figure 16.2). These glands secrete a hormone called *parathyroid hormone,* which is a major player in maintaining the level of calcium ions in the ECF within the normal range.

This module begins by describing the structure of the thyroid gland and parathyroid glands. Then we explore the thyroid hormones and their many physiological roles. We then turn our attention to parathyroid hormone and calcitonin, and the module concludes by discussing hormonal maintenance of calcium ion homeostasis.

Structure of the Thyroid and Parathyroid Glands

The butterfly-shaped **thyroid gland** consists of right and left lobes connected by a small band called the **isthmus** (ISS-muss). Microscopically, the thyroid gland is composed of multiple spheres known as **thyroid follicles.** Note in **Figure 16.11b** that follicles are bounded by a layer of simple cuboidal epithelial cells referred to as *follicle cells,* which produce thyroid hormones. The interior of a follicle is filled with a protein-rich, gelatinous material called **colloid** (KOL-oyd), which is where the precursor of thyroid hormones is stored. Colloid also contains a high

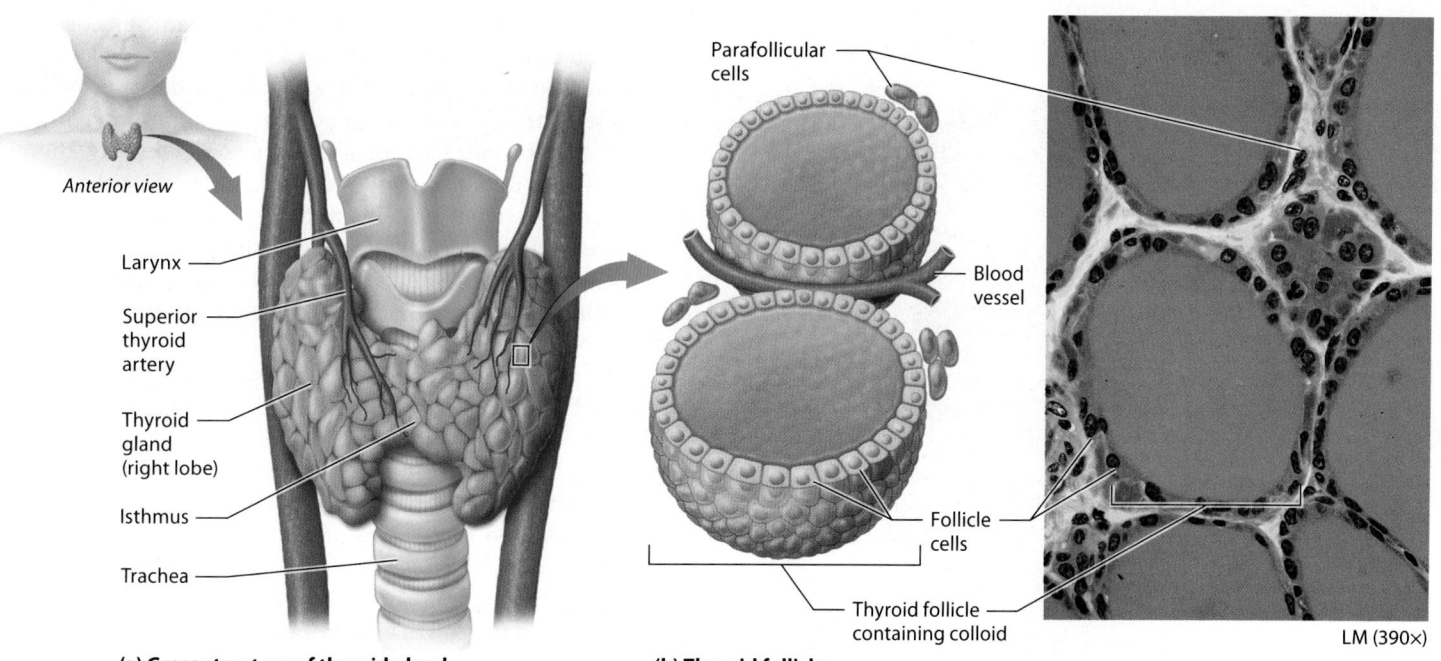

(a) Gross structure of thyroid gland

(b) Thyroid follicles

Figure 16.11 Anatomy and histology of the thyroid gland.

Practice art labeling @MasteringA&P®

concentration of iodine atoms, which are required for thyroid hormone synthesis.

Between the thyroid follicles lie clusters of **parafollicular cells.** These cells are larger than follicle cells, and they play no role in thyroid hormone production. Instead, they produce the hormone calcitonin, which we discuss later in this module.

The **parathyroid glands** are small glands embedded in the posterior surface of the thyroid gland (**Figure 16.12**). Typically, there are three to five parathyroid glands, with four being most common. The hormone-secreting cells of the parathyroid gland, which produce parathyroid hormone, are called **chief cells.**

Quick Check

☐ 1. What are thyroid follicles, and how are they structured?
☐ 2. What are parafollicular cells?

Thyroid Hormones: Metabolic Regulators

◀◀ **Flashback**

1. Why is heat a byproduct of metabolism? (p. 42)
2. What are the effects of the sympathetic nervous system on heart rate and blood pressure? (p. 526)

Thyroid hormones consist of an amino acid core bound to iodine atoms. There are two forms of thyroid hormone, which are named according to the number of iodine atoms they contain: **triiodothyronine** (try′-eye-oh-doh-THY-roh-neen; T_3), which contains three iodine atoms, and **thyroxine** (thy-ROKS-een; T_4),

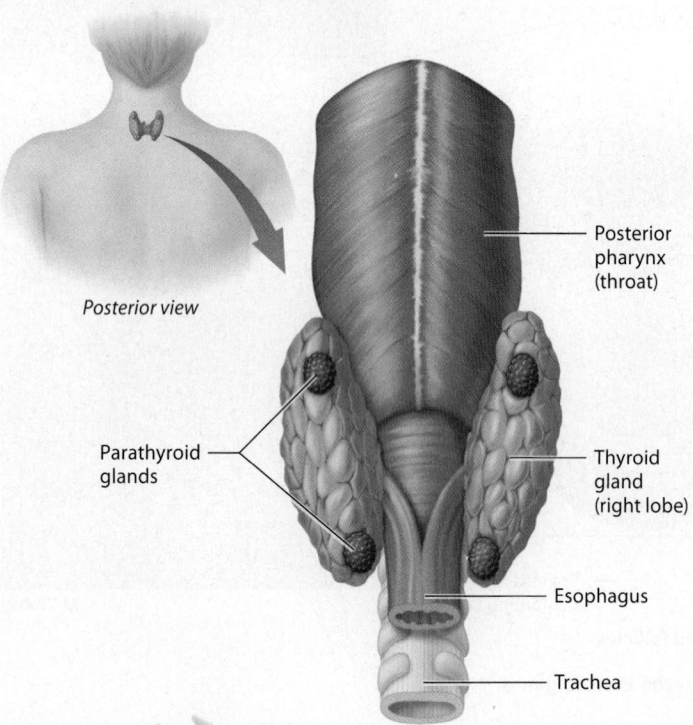

Posterior view

Parathyroid glands

Posterior pharynx (throat)

Thyroid gland (right lobe)

Esophagus

Trachea

Figure 16.12 Anatomy of the parathyroid glands.

which contains four iodine atoms. Both forms of the hormone are active, but T_3 has much greater physiological activity than T_4. In fact, as we discuss shortly, T_4 is usually converted into T_3 in the target tissues.

Although T_3 and T_4 are amino acid–based hormones, their amino acids are nonpolar and hydrophobic, so they do not bind to cell-surface receptor proteins. Instead, they diffuse into target cells and then bind to intracellular receptors in the nucleus. Here, like the steroid hormones, they either activate or inhibit the transcription of certain genes.

This section first covers the functions of thyroid hormones. Then we see how they are produced, how their production is regulated, and what happens when this production is imbalanced.

Effects of Thyroid Hormone

Nearly every cell in the body contains receptors for thyroid hormones, so their effects are widespread. We can broadly divide these effects into three main categories: (1) regulation of the metabolic rate and thermoregulation, (2) promotion of growth and development, and (3) synergism with the sympathetic nervous system.

● **Regulation of the metabolic rate and thermoregulation.** Thyroid hormones set the *basal metabolic rate*—the amount of energy required by the body to carry out all of its reactions when at rest. The hormones accomplish this in several ways. For one, they increase the rate at which the ATP-requiring Na^+/K^+ pumps function, which raises the amount of ATP the cell consumes. Thyroid hormones also act on a number of different cell types to trigger other energy-requiring reactions. For example, they act on cells of the liver and trigger reactions that generate glucose from amino acids and fats, a process known as gluconeogenesis. They also stimulate reactions that break down proteins in skeletal muscle and fats in adipose tissue in order to provide these molecules for gluconeogenesis. Finally, thyroid hormones initiate energy-requiring reactions in these same tissues that *build* proteins and fats. So, thyroid hormones increase the rate at which cells carry out both catabolic (breaking down) and anabolic (building up) reactions, leading to essentially no net change in the cell's overall composition. Why would cells expend energy in a way that leads to no net change? The answer lies in the byproduct of these reactions: heat. Recall that the energy transfer in our cellular reactions is inefficient, and much of this energy is lost as heat (see Chapter 2). The heat from these reactions helps to maintain our core body temperature, a process called *thermoregulation*. So although these actions of thyroid hormones may seem "wasteful," they are actually vital to maintaining our body temperature.

● **Promotion of growth and development.** Thyroid hormones are required for normal bone growth, muscle growth, and nervous system development. The exact roles played by thyroid hormones are not fully understood, but as we discuss shortly, deficiencies in thyroid hormones result in stunted growth and problems with nervous system development.

- **Synergism with the sympathetic nervous system.** Increases in thyroid hormones act on target cells of the sympathetic nervous system and increase the production of (up-regulate) receptors for sympathetic neurotransmitters. In this way, thyroid hormones help regulate blood pressure, heart rate, and other actions of the sympathetic nervous system without directly causing these effects.

Thyroid Hormone Production

The synthesis of thyroid hormones is a multistep process that occurs both in the colloid and in the follicle cells. The process is illustrated in **Figure 16.13** and proceeds as follows:

① **Iodide ions and thyroglobulin are secreted into the colloid.** The synthesis of thyroid hormones begins with the accumulation of iodide ions (I^-) in the colloid. Iodide ions are ingested in the diet, after which they are absorbed into the blood. In the capillaries of the thyroid gland, the iodide ions diffuse into the interstitial fluid and are actively pumped into the follicle cells. Finally, the ions enter the colloid via transport proteins in the plasma membranes of the follicle cells. In addition, follicle cells secrete the large precursor protein **thyroglobulin** (thy'-roh-GLOB-yoo-lin) into the colloid by exocytosis.

② **Iodide ions are converted to iodine atoms that attach to thyroglobulin.** An enzymatic reaction converts iodide ions to iodine atoms (I^0; the superscript zero indicates that the atom has no charge). Iodine atoms then move into the colloid and attach to specific amino acids in the thyroglobulin, forming *iodinated thyroglobulin.*

③ **Iodinated thyroglobulin enters the follicle cell by endocytosis and is converted to T_3 and T_4 by lysosomal enzymes.** The iodinated thyroglobulin is taken up by the follicle cells via endocytosis. The endocytic vesicles fuse with lysosomes, where enzymatic reactions convert thyroglobulin to T_3 and T_4. About 90% of the hormone produced is T_4, and the remainder is T_3.

④ **T_3 and T_4 are released into the blood.** T_3 and T_4 are released into the blood, where most of them bind to plasma proteins. The unbound hormones are known as *free T_3 and T_4.* Only free T_3 and T_4 have biological activity.

The thyroid gland produces T_4 in much greater amounts than T_3, yet T_3 is the far more active hormone. Why would the thyroid gland produce more of the less active hormone? The answer is that many peripheral tissues can convert T_4 to T_3, so the presence of large amounts of T_4 in the blood provides a ready supply of potential T_3. This is important because T_3 has a short *half-life* (the time it takes for half of the hormone to be removed or deactivated) of about 2.5 days, whereas T_4 has a much longer half-life of about 6.5 days.

① Iodide ions (I^-) and thyroglobulin are secreted into the colloid.

② Iodide ions are converted to iodine atoms (I^0) that attach to thyroglobulin.

③ Iodinated thyroglobulin enters the follicle cell by endocytosis and is converted to T_3 and T_4 by lysosomal enzymes.

④ T_3 and T_4 are released into the blood.

Iodide ion (I^-)

Blood vessel

Follicle cell

Thyroglobulin in a vesicle

T_3 T_4

Lysosomal enzymes

Lysosome

I^0

Colloid

Figure 16.13 Production of thyroid hormones.

Regulation of Thyroid Hormone Production

As with most hormones, production of T_3 and T_4 is regulated by a negative feedback loop. However, note in **Figure 16.14** that this negative feedback loop has multiple tiers of control. It involves both thyrotropin-releasing hormone (TRH) from the hypothalamus as the first tier of control and thyroid-stimulating hormone (TSH) from the anterior pituitary as the second tier. Hypothalamic neurons secrete TRH at a fairly constant rate; TRH then travels to the anterior pituitary via the hypothalamic-hypophyseal portal system and stimulates production and release of TSH. TSH stimulates multiple responses from the thyroid gland, including:

- production of T_3 and T_4 by follicle cells,
- secretion of T_3 and T_4 from follicle cells into the blood, and
- growth and development of the thyroid gland.

Production and secretion of TRH and TSH increase when levels of free T_3 and T_4 fall or when the body is exposed to cold temperatures. Secretion of both TRH and TSH is inhibited by rising levels of free T_3 and T_4. As mentioned earlier, the hypothalamic hormone somatostatin also inhibits TSH secretion.

Thyroid Disorders

Due to the systemic effects of thyroid hormone, serious imbalances in homeostasis can accompany its over- or underproduction. Overproduction of thyroid hormone is known as **hyperthyroidism;** underproduction is called **hypothyroidism.**

The most common cause of hyperthyroidism is **Graves' disease,** which results from the immune system producing abnormal proteins that mimic the actions of TSH on the thyroid gland. The abnormal proteins stimulate the thyroid, and the levels of both T_3 and T_4 in the blood increase. The common signs and

Figure 16.14 **Maintaining homeostasis: regulation of thyroid hormone production by a negative feedback loop.**

symptoms of Graves' disease result from excessive T₃ and T₄ activity, and include:

- weight loss due to an elevated metabolic rate,
- heat intolerance due to excessive heat production, and
- disruptions in heart rhythm and increase in blood pressure due to the synergism of thyroid hormones with the sympathetic nervous system.

Figure 16.15 Disorder of thyroid hormone secretion: goiter.

Another sign of Graves' disease is *exophthalmos* (eks-off-THAL-mohs). Exophthalmos is characterized by bulging of the eyeballs due to weakening of the extraocular muscles and water retention in the tissues around the orbits. A final sign of hyperthyroidism is enlargement of the thyroid gland, known as a **goiter** (GOY-t'r), shown in **Figure 16.15**. Thyroid enlargement results from the abnormal TSH-like proteins secreted in Graves' disease, which stimulate growth of the thyroid gland.

Hyperthyroidism may be treated with drugs that decrease the production of T₃ and T₄, but such drugs must be taken for the rest of a patient's life and are not always a good permanent solution. Permanent treatments involve the surgical removal or nonsurgical destruction of the thyroid gland. The gland is destroyed nonsurgically with radioactive iodine, which is selectively taken up by the follicle cells. The radiation produced as the iodine decays kills the follicle cells.

The two most common causes of hypothyroidism are iodine deficiency and destruction of the thyroid gland by the immune system. The latter condition is known as *Hashimoto's thyroiditis*. Iodine-deficiency hypothyroidism is relatively uncommon in the developed world, because most salt is supplemented with iodine. Also, people who regularly consume food from the ocean may receive their iodine naturally. However, in some developing countries, iodine deficiency remains relatively common. Both causes of hypothyroidism result in decreased T₃ and T₄ production and concurrent increases in TRH and TSH production. This can be explained by the negative feedback loops that attempt to increase T₃ and T₄ production from the thyroid gland.

Most signs and symptoms of hypothyroidism are the opposites of those of hyperthyroidism and include:

- weight gain due to a decreased metabolic rate,
- cold intolerance due to decreased heat production, and
- slow heart rate and low blood pressure due to decreased synergism with the sympathetic nervous system.

Surprisingly, many patients with hypothyroidism present with one of the same signs found in hyperthyroidism—a goiter. A goiter develops in Hashimoto's thyroiditis because the increased

level of TSH promotes growth of the thyroid gland. In iodine-deficiency hypothyroidism, the elevated TSH level contributes to the goiter, but the goiter also results from the accumulation of thyroglobulin in the colloid of thyroid follicles. For this reason, goiters seen with iodine-deficiency hypothyroidism can become quite large.

The treatment for hypothyroidism depends on the cause. Iodine deficiency is treated with iodine supplementation, after which the condition generally resolves. Hashimoto's thyroiditis has no cure, but can be easily treated with thyroid hormone supplementation for the remainder of the patient's lifetime.

A special variant of hypothyroidism, **congenital hypothyroidism** (formerly known as *cretinism*), develops when an infant is born with inadequate thyroid function. If untreated, it results in delayed physical and nervous system development and eventual mental retardation. Congenital hypothyroidism is completely avoidable with early diagnosis and treatment. For this reason, infants are screened for hypothyroidism at birth.

ConceptBOOST ⟫⟫⟩

Understanding the Relationship between Negative Feedback Loops and Thyroid Function

As we've discussed, TSH secretion *increases* with hypothyroidism and *decreases* with hyperthyroidism. This may seem counterintuitive, so let's look at why this happens. You've read that the production of T₃ and T₄ is regulated by a negative feedback loop. This means that when T₃ and T₄ levels in the blood drop, the receptors in the cells of the hypothalamus detect the change and secrete more TRH, which stimulates the anterior pituitary to secrete more TSH. Under normal conditions, the elevated TSH will stimulate the thyroid gland to produce more T₃ and T₄. When T₃ and T₄ levels increase, receptors in both the hypothalamus and anterior pituitary gland detect this change, and TRH and TSH levels decrease through negative feedback.

However, in hypothyroidism, the thyroid gland is unable to produce more T₃ and T₄. The hypothalamus and anterior pituitary don't "know" that the thyroid gland is not functioning; all their receptors sense is that T₃ and T₄ levels are low, so the cells continue to secrete more TRH and TSH in an attempt to stimulate the thyroid. This leads to the characteristic elevated TSH and decreased T₃ and T₄ levels seen with hypothyroidism.

This situation works in reverse with hyperthyroidism. Under normal conditions, the cells of the hypothalamus and anterior pituitary gland decrease production of TRH and TSH when levels of T₃ and T₄ rise. In the case of hyperthyroidism, levels of T₃ and T₄ are extremely elevated, which is detected by the receptors of the hypothalamus. For this reason, production and secretion of TRH and TSH fall to nearly zero. This leads to the characteristic decreased TSH with elevated T₃ and T₄ levels seen in hyperthyroidism. ■

Quick Check

☐ 3. What are the main functions of thyroid hormones?
☐ 4. How are thyroid hormones produced? How is this production regulated?
☐ 5. What homeostatic imbalances may accompany hyperthyroidism and hypothyroidism?

Parathyroid Hormone and Calcitonin: Calcium Ion Homeostasis

◀◀ **Flashback**

1. What are osteoblasts and osteoclasts, and what are their functions? (p. 189)
2. What is the function of vitamin D? (p. 202)

As we discussed in the introduction to this module, calcitonin and parathyroid hormone are produced by the thyroid and parathyroid glands, respectively. We now take a closer look at the functions and target tissues of these hormones, as well as their roles in maintaining calcium ion concentration in the blood within a specific range.

Parathyroid Hormone

The chief cells of the parathyroid glands secrete the peptide hormone **parathyroid hormone (PTH),** which is a major factor in maintaining the blood calcium ion concentration in the normal range. PTH is secreted in response to a declining blood calcium ion concentration, a condition known as *hypocalcemia*. Its actions raise the calcium ion level of the blood by triggering the following effects:

- **Increasing release of calcium ions from bone by stimulating osteoclasts.** PTH induces changes in osteoblasts that enhance osteoclast activity. Recall that osteoclasts are "bone-destroying" cells—they release calcium into the blood by breaking down the extracellular matrix of bone (see Chapter 6). As the osteoclasts degrade bone, calcium ions are released into the blood.
- **Increasing absorption of dietary calcium ions by the small intestine.** The effects of PTH on the small intestine are indirect and involve the hormone vitamin D, which is produced by the skin in response to certain wavelengths of ultraviolet light (see Chapter 5). However, the skin produces an *inactive* form of vitamin D. PTH acts on the kidneys to stimulate the conversion of vitamin D to its active form, vitamin D3, or **calcitriol** (kal-sih-TRY-ohl). In this

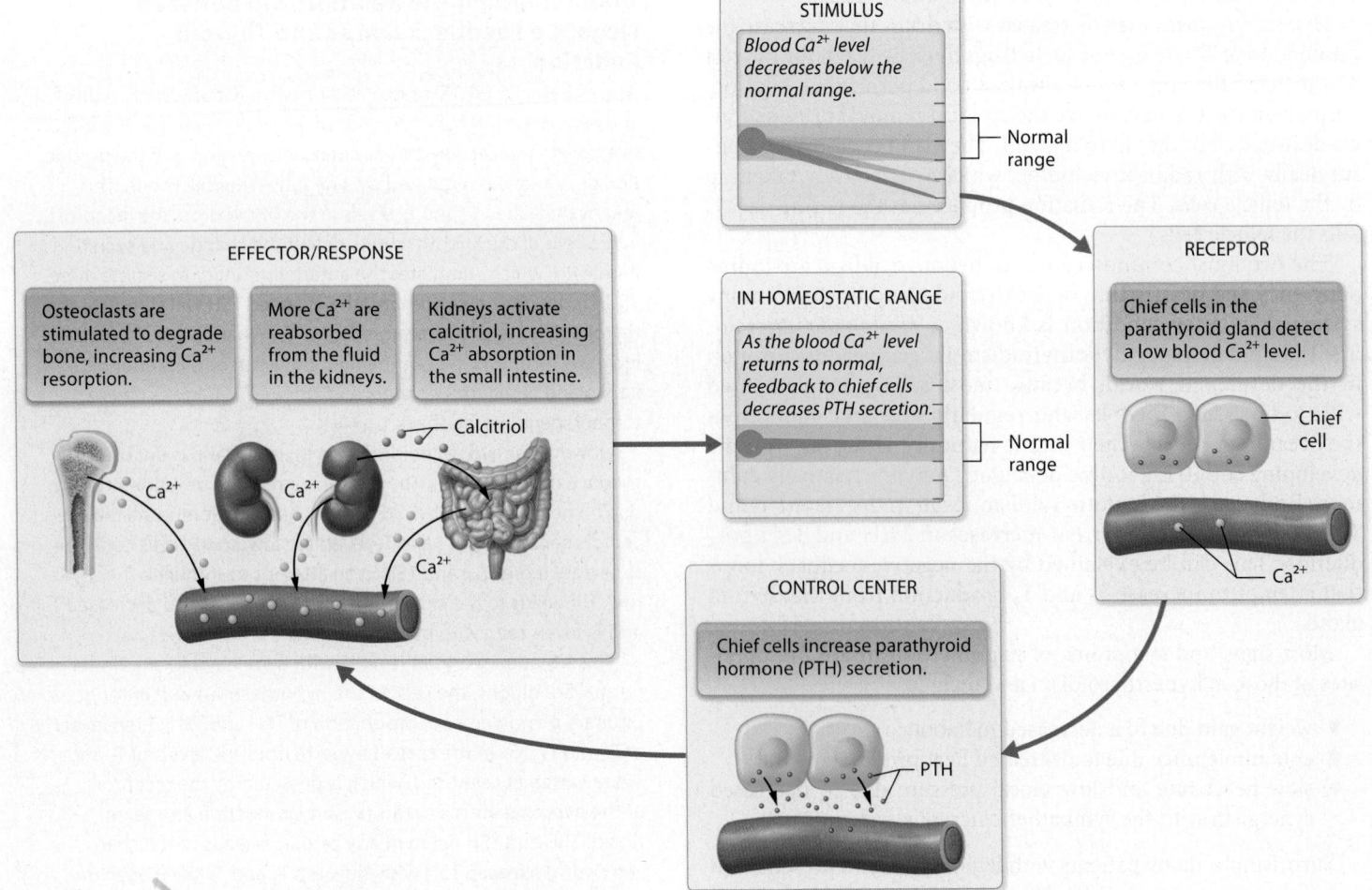

Figure 16.16 Maintaining homeostasis: regulation of blood calcium ion concentration by a negative feedback loop.

form, vitamin D3 increases the number of calcium ions absorbed from the small intestine.

- **Increasing reabsorption of calcium ions from the fluid in the kidneys.** In the process of filtering blood, the kidneys remove large amounts of water and solutes from the blood, including calcium ions. The cells of the kidney tubules then reclaim most of the water and solutes through a process called *tubular reabsorption*. PTH acts on the kidney tubule cells to increase the number of calcium ions reclaimed from the fluid in the tubules before it becomes urine and is lost to the body.

Regulation of Calcium Ion Homeostasis

The negative feedback loop involving PTH that maintains the blood calcium ion concentration in its normal range proceeds as follows (**Figure 16.16**).

- **Stimulus: The blood calcium ion level decreases below the normal range.**
- **Receptor: Chief cells in the parathyroid gland detect a low blood calcium ion level.** A low calcium ion concentration in the blood leads to a decreased calcium ion concentration in the ECF around the chief cells. This is then detected by their cell-surface receptors.
- **Control center: The chief cells increase PTH secretion.** The chief cells, which also act as the control center, increase their production and secretion of PTH into the blood.
- **Effector/response: The effects of PTH on target cells increase the blood calcium ion concentration.** PTH affects the functions of its target cells in bone (osteoclasts), the kidneys, and the small intestine. In addition, it stimulates the conversion of inactive vitamin D to the active form, calcitriol. All of these effects increase the concentration of calcium ions in the blood.

- **Homeostatic range and negative feedback: As the blood calcium ion level returns to the normal range, negative feedback to chief cells decreases PTH secretion.** The calcium ion concentration in the blood returns to the normal range. Receptors on the chief cells detect this change, and negative feedback decreases their production and secretion of PTH. This in turn decreases stimulation to the target cells, and the response is reduced.

Calcitonin

Hypercalcemia occurs when the calcium ion level in the blood increases above normal. In response to hypercalcemia, the hormone **calcitonin** is released. Calcitonin, which is produced by the thyroid gland's parafollicular cells, decreases the blood calcium ion concentration. Calcitonin's primary target cells are the osteoclasts in bone. This hormone inhibits the activity of these cells, which leaves the effects of the osteoblasts, the "bone-building" cells, unopposed. Osteoblasts deposit calcium ions into the bone, which reduces the blood concentration of calcium ions.

Calcitonin's effects have been difficult to characterize because they are fairly transient—the effects of a high level of calcitonin appear to last only a few hours. Its activity seems to be most important during times of active bone turnover, such as during bone growth. At present, calcitonin is not believed to be a major regulator of calcium ion homeostasis, and might instead play a more important role in bone homeostasis.

Both calcitonin and PTH can be used in the treatment of the bone disease *osteoporosis* (os'-tee-oh-por-OH-sis). See *A&P in the Real World: Calcitonin, Parathyroid Hormone, and Osteoporosis* on page 608 for more information.

The hormones of the thyroid and parathyroid glands are summarized in **Table 16.2**.

TABLE 16.2 HORMONES OF THE THYROID AND PARATHYROID GLANDS

Cell Type	Hormone(s)	Stimulus for Secretion	Inhibitor(s) of Secretion	Target Tissue(s)	Effects
Thyroid Gland					
Follicle cells	Triiodothyronine (T_3) and thyroxine (T_4)	TSH from the anterior pituitary	Increased levels of T_3 and T_4 inhibit TRH and TSH.	Nearly every cell in the body	• Set the basal metabolic rate • Thermoregulation • Growth and development • Synergism with the SNS
Parafollicular cells	Calcitonin	Increased concentration of calcium ions in the blood	Decreased concentration of calcium ions in the blood	Osteoclasts	Inhibits osteoclast activity under certain conditions, which transiently decreases the blood calcium ion concentration
Parathyroid Glands					
Chief cells	Parathyroid hormone (PTH)	Decreased concentration of calcium ions in the blood	Increased concentration of calcium ions in the blood	Bone, kidneys, intestines	• Increases calcium ion reabsorption from the fluid in the kidneys • Increases calcium ion absorption from the contents of the small intestine • Indirectly increases osteoclast activity • All effects lead to an increased blood calcium ion concentration.

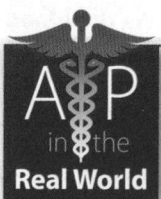

Calcitonin, Parathyroid Hormone, and Osteoporosis

You may recall that osteoporosis is characterized by a decrease in the mineral density of bones (see Chapter 6). It results in weak bones and significantly increases the risk for bone fractures. One commonly used treatment for osteoporosis is calcitonin administered via a nasal spray in much higher amounts than what is produced by the thyroid gland. The drug increases bone mass by slowing bone breakdown.

Another common treatment, surprisingly, is a drug that mimics PTH. This may seem counterintuitive, as PTH stimulates bone breakdown. But it appears that PTH has this effect on osteoclasts only if the calcium ion concentration in the blood falls below the normal level. In the face of a normal calcium ion concentration, PTH increases the concentration through its effects on the gut and the kidneys, with minimal effects on osteoclasts.

Quick Check

☐ 6. What are the target tissues and effects of calcitonin and parathyroid hormone?
☐ 7. How does parathyroid hormone maintain calcium homeostasis?

Apply What You Learned

☐ 1. Certain dietary supplements marketed for weight loss contain thyroid hormone. How would these supplements cause weight loss? Would these products generally be safe?
☐ 2. Mr. Scully, a patient with hypothyroidism, has had blood tests done to check his thyroid function. The lab technician reports that his TSH level is abnormally high, but forgets to record the levels of T_3 and T_4. Based on the TSH level, predict whether Mr. Scully's T_3 and T_4 levels will be elevated, normal, or decreased. Explain your answer.
☐ 3. Tumors of the parathyroid gland often secrete excessive parathyroid hormone. Predict the effects of such a tumor.

See answers in Appendix A.

MODULE **16.4**
The Adrenal Glands

Learning Outcomes

1. Describe the gross and microscopic anatomy of the cortex and medulla of the adrenal gland.
2. Describe the stimulus for release, the target tissue, and the effects of the mineralocorticoids secreted by the adrenal cortex.
3. Describe the stimulus for release, the target tissue, and the effects of the glucocorticoids secreted by the adrenal cortex.
4. Explain the relationship of the adrenal medulla to the sympathetic nervous system.
5. Describe the stimulus for release, the target tissue, and the effect of catecholamines.

As you discovered in Module 16.1, the roughly pyramid-shaped **adrenal glands** are located on the superior end of the right and left kidneys. The adrenal glands produce two types of hormones: steroid hormones and catecholamines. These two types of hormones have a wide variety of roles in the body. In this module we examine these roles and the regulation of these hormones. First, however, let's delve into the structure of the adrenal glands.

Structure of the Adrenal Glands

A look inside the adrenal glands reveals that they have two distinct regions: an outer **adrenal cortex** and an inner **adrenal medulla** (Figure 16.17a). Whereas the adrenal cortex is a typical endocrine gland that functions in much the same way as the other such organs we've discussed, the adrenal medulla is a neuroendocrine organ that secretes neurohormones.

As you can see in Figure 16.17b, the adrenal cortex has three distinct zones: the outer **zona glomerulosa** (glom-air'-yoo-LOW-sah), the middle **zona fasciculata** (fah-sik'-yoo-LAH-tah), and the inner **zona reticularis** (reh-tik'-yoo-LAIR-us). The zona glomerulosa consists of densely packed cells, whereas the middle zona fasciculata consists of cells stacked on top of one another in columns. The cells of the thin zona reticularis are arranged more loosely in clusters.

Hormones of the Adrenal Cortex

◄◄ Flashback

1. Is the concentration of sodium ions higher in the cytosol of our cells or the extracellular fluid? What about potassium ions? (p. 79)

Each zone of the adrenal cortex produces steroid hormones derived from cholesterol. The zona glomerulosa cells produce hormones called *mineralocorticoids* (min'-er-ah-loh-KOR-tih-koydz). The cells of the zona fasciculata and zona reticularis produce and secrete two types of steroids: *glucocorticoids* (glew'-koh-KOR-tih-koydz) and *androgenic steroids*.

Production of these steroids by the adrenal cortex is partially regulated by a multi-tiered negative feedback system referred to as the *hypothalamic-pituitary-adrenocortical*, or *HPA, axis*. Let's examine the primary hormones of the zones of the adrenal cortex and how their production and secretion are regulated by the HPA axis and other factors.

(a) Gross structure of the adrenal gland

LM (160×)

(b) Histology of the adrenal gland: illustration (left) and light micrograph (right)

Practice art labeling
@MasteringA&P®

Figure 16.17 Anatomy and histology of the adrenal gland.

Mineralocorticoids: Fluid and Electrolyte Homeostasis

Mineralocorticoids regulate the concentration of certain *minerals* in the body (such as sodium and potassium ions). The main mineralocorticoid is **aldosterone** (al-DOS-ter-own). Let's look at the functions of aldosterone and how it is regulated.

Effects of Aldosterone Notice in **Figure 16.18** that aldosterone has several functions:

- **Maintaining the concentrations of extracellular sodium and potassium ions within their normal ranges.** As we discussed in earlier chapters, the gradients of sodium and potassium ions across plasma membranes are critical to the functioning of muscle cells and neurons (see Chapters 10 and 11). Specifically, such excitable cells require high extracellular and low intracellular concentrations of sodium ions, and low extracellular and high intracellular concentrations

of potassium ions. These gradients are maintained largely by the actions of the sodium-potassium ion pumps in the cells' plasma membranes. The concentrations of these ions in the ECF are also partially maintained by aldosterone acting on the cells of the kidney tubules. Here, it stimulates transcription of more sodium/potassium ion pumps, sodium ion channels, and sodium/chloride/potassium ion cotransporters. These actions have two net effects: (1) Sodium and chloride ions are transported from the fluid in the kidney tubules to the extracellular fluid for return to the blood, and (2) potassium ions are transported from the extracellular fluid into the fluid in the kidney tubules, where they will be excreted in the urine.

- **Regulating extracellular fluid volume.** Aldosterone's effects on sodium and chloride ions in the tubule cells create a concentration gradient that favors the movement of water by osmosis from the fluid in those tubules to the extracellular

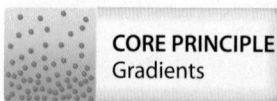

Figure 16.18 Effects of aldosterone.

CORE PRINCIPLE
Gradients

fluid and blood. This is an example of the Gradients Core Principle (p. 26) Note that this is an indirect effect—aldosterone has no direct actions on water molecules or water channels in the tubule cells.

- **Maintaining blood pressure.** Blood volume is one of the key factors that determine blood pressure. When aldosterone indirectly promotes water reabsorption, this maintains the extracellular fluid and blood volume, which in turn maintains blood pressure. Aldosterone's effects on blood pressure are a key part of a larger system involving the kidneys, lungs, and liver, called the *renin-angiotensin-aldosterone system* (*RAAS*). The overall physiological functions of this system are to increase blood pressure, and to preserve blood flow to the heart, brain, and kidneys. We further discuss this system with the kidneys as other endocrine organs in Module 16.6, and fully with the urinary system (see Chapter 24).
- **Maintaining acid-base homeostasis.** The pH of the blood is slightly alkaline, at about 7.35–7.45. Aldosterone helps to maintain this normal range by activating hydrogen ion pumps in certain cells of the kidney tubules. These pumps transport hydrogen ions from the extracellular fluid into the fluid of the tubules, after which they are excreted in the urine. This lowers the hydrogen ion concentration of the blood, preserving its slightly alkaline pH.

Regulation of Aldosterone Synthesis Multiple factors regulate aldosterone synthesis and secretion. The major factors that stimulate aldosterone release include an elevated blood potassium ion concentration, a decrease in blood pH, and a hormone called *angiotensin-II* (which we discuss more fully in Chapter 24). In addition, in the HPA axis, corticosteroid-releasing hormone (CRH) from the hypothalamus triggers release of ACTH from the anterior pituitary, and ACTH stimulates aldosterone production and release.

Aldosterone Disorders Hypersecretion of aldosterone, or *hyperaldosteronism*, can lead to hypokalemia (abnormally low extracellular concentration of potassium ions), hypernatremia (abnormally high extracellular concentration of sodium ions), and high blood pressure. It can also disrupt acid-base balance and make the pH of the blood overly alkaline, which disrupts many cellular processes. Hyperaldosteronism is generally *idiopathic* (id'-ee-oh-PATH-ik), meaning of unknown cause, but is occasionally due to an aldosterone-secreting tumor. Hyposecretion of aldosterone is discussed later in this module.

Glucocorticoids: Metabolic Homeostasis

Glucocorticoids are produced in the zona fasciculata and zona reticularis, and their main role is to help mediate the body's response to stress. As implied by their name, part of the glucocorticoids' response to stress involves regulating the concentration of blood glucose. The most potent glucocorticoid in the body is the hormone **cortisol** (KOR-tih-zohl), also known as **hydrocortisone.**

Effects of Cortisol Cortisol's primary target tissues are liver, muscle, and adipose, where it stimulates the following processes (**Figure 16.19**):

- **Gluconeogenesis in the liver.** Cortisol binds to genes in liver cells (hepatocytes) and stimulates them to synthesize enzymes that convert amino acids and fats into glucose by gluconeogenesis. This increases the blood glucose level.
- **Release of amino acids from muscle tissue.** Cortisol acts on skeletal muscle to induce the breakdown of muscle proteins into amino acids. These amino acids enter the bloodstream, where they are then available to the liver for use in gluconeogenesis.
- **Release of fatty acids from adipose tissue.** The fatty acids stored in adipose tissue can be used by cells as an alternative fuel to glucose, and by the liver in gluconeogenesis. For this reason, cortisol acts on adipocytes and triggers lipolysis, or fat breakdown, releasing fatty acids into the bloodstream.

Cortisol's role in the body extends beyond its influence on blood glucose concentration. In fact, most tissues in the body have receptors for cortisol, so its effects are wide-ranging. One of its most pronounced effects is on the immune system, where cortisol acts as a potent anti-inflammatory agent by decreasing the levels of certain leukocytes. The strength of cortisol as an anti-inflammatory agent led to the development of synthetic cortisol derivatives called *corticosteroids,* which are used to treat a variety of conditions in which inflammation is a key component. Cortisol is also used to suppress the immune response; for example, it is given to patients with *autoimmune disorders,* in which a patient's

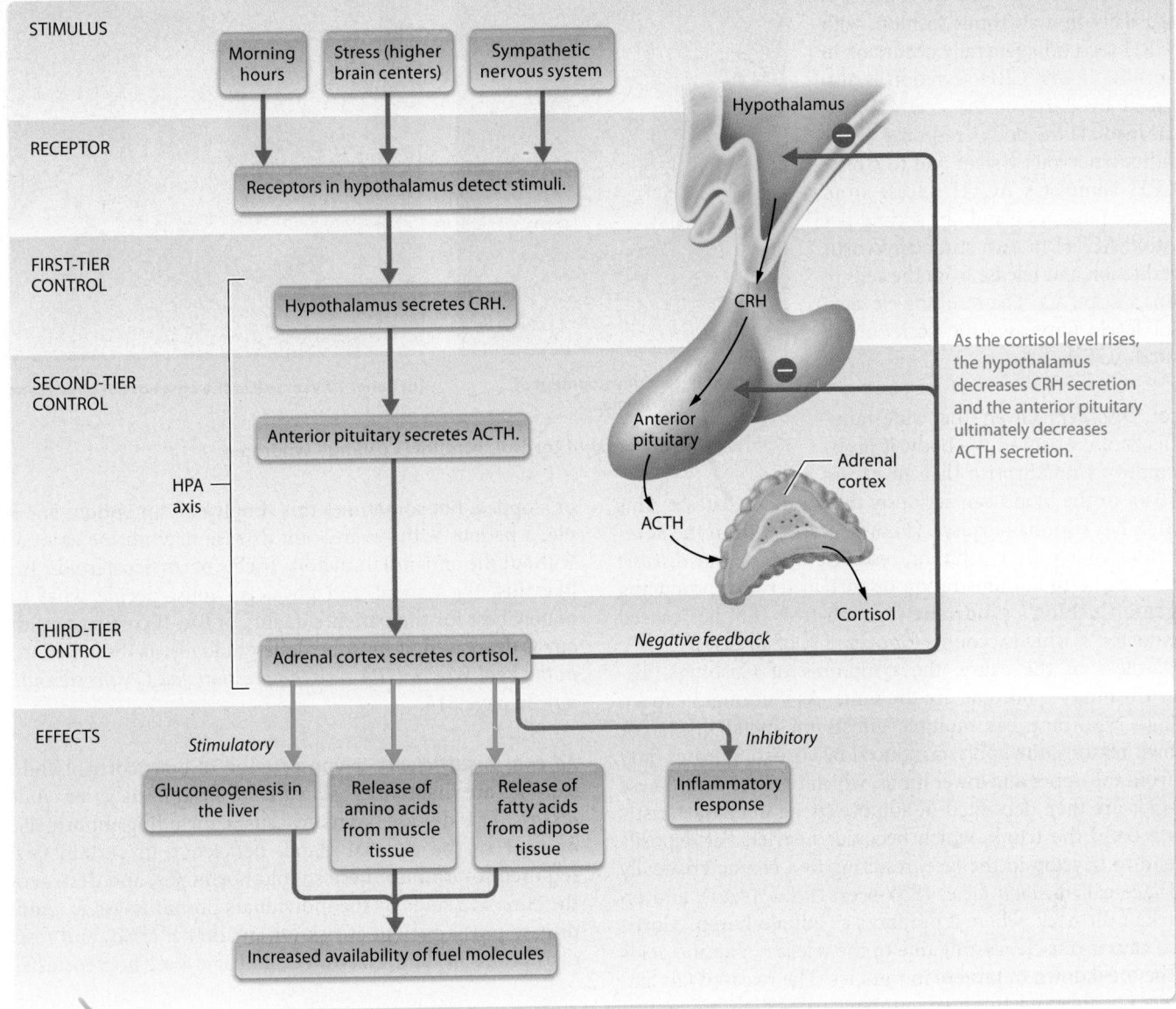

Figure 16.19 Regulation of cortisol production by a negative feedback loop.

own immune system attacks normal body cells, and to organ transplant recipients to help prevent rejection of the new organ.

Cortisol's actions make it an important component of the **stress response,** which is the series of changes that maintains homeostasis when the body is faced with a *stressor* (which can be anything from exercise or emotional excitement to being chased by an alligator). Indeed, cortisol is sometimes called the "stress hormone," which reflects this role. To understand this relationship, step back and look at what would happen if you were subjected to the physical and emotional stress of being chased by an alligator:

- You are running from the alligator, which means your skeletal muscle cells will need additional glucose to fuel their activity.
- Your muscle activity will quickly use up the supply of glycogen, so these cells will need other fuel sources such as fatty acids.

- If you fall and break your toe while running, you will be unable to stop and nurse your injury. Instead, you need to continue running until you are safe.

How does cortisol help with each of these points? First, cortisol triggers breakdown of fats and proteins for gluconeogenesis, which increases the glucose level in the blood. Second, the fatty acids released from adipose tissue provide additional fuels for continued muscle activity. Finally, the anti-inflammatory actions of cortisol prevent inflammation from the injury to your toe, which enables you to continue running. These effects make cortisol an extremely important hormone in the stress response, which we examine more fully in Module 16.7.

Regulation of Cortisol Cortisol synthesis is primarily under control of the HPA axis (see Figure 16.19). CRH from the hypothalamus, the first tier of control, is secreted daily in a rhythmic fashion, with peak CRH secretion generally occurring in the morning hours. CRH secretion is also influenced by parts of the cerebral cortex that mediate the body's response to the sympathetic nervous system and to stressors. CRH stimulates ACTH release from the anterior pituitary gland, the second tier of control. ACTH in turn stimulates cortisol production and release from the cells of the zona fasciculata. The resulting elevated cortisol level suppresses the secretion of CRH and ACTH.

(a) Patient before development of Cushing's syndrome

(b) Patient 3 years after the onset of Cushing's syndrome

Figure 16.20 Disorder of cortisol secretion: Cushing's syndrome.

Cortisol Disorders Given the wide-ranging effects of cortisol in the body, it probably comes as no surprise that an excess of cortisol in the blood can seriously disrupt homeostasis. This condition has two main causes: (1) oversecretion from the adrenal cortex, usually from a tumor, resulting in *Cushing's disease;* and (2) long-term administration of corticosteroids, producing **iatrogenic Cushing's syndrome** (eye′-ah-troh-JEN-ik; "caused by the doctor"), which accounts for over 90% of all cases.

Regardless of the cause, the symptoms of Cushing's disease or Cushing's syndrome are the same. As you might expect, Cushing's syndrome has multiple effects on metabolism. For unknown reasons, the lipolysis induced by cortisol releases fatty acids from the upper and lower limbs, which become slim. These fatty acids are then deposited in adipose tissue in characteristic places around the trunk, which becomes heavier. Fat deposits also tend to develop in the face, resulting in a characteristically round face called *moon facies* (FAY-seez; **Figure 16.20**), and on the back of the neck, where it produces a "buffalo hump." Cortisol also causes muscle wasting due to the release of amino acids from the breakdown of protein in muscles. The released fats and amino acids are used to make an abnormally high amount of glucose, which raises the blood glucose level, a condition known as *hyperglycemia* (hy′-per-gly-SEE-mee-ah). This and the other metabolic effects of cortisol tend to increase appetite, which leads to weight gain.

Excess cortisol has effects on other body systems, as well. For example, it has slight mineralocorticoid effects on the kidneys, resulting in greater renal retention of sodium ions, and therefore water. Over time, this excess water increases the volume of blood in the circulation, which increases blood pressure. In addition, cortisol's effects on leukocytes lead to overall suppression of the immune response. As a result, patients with Cushing's syndrome have a higher risk for developing infections. Finally, the excess cortisol decreases osteoblast activity and interferes with the absorption of dietary calcium ions by the small intestine. Both effects lead to bone loss, which can eventually cause osteoporosis.

Cushing's disease is generally treated with removal of the ACTH- or cortisol-secreting tumor. The treatment for iatrogenic Cushing's syndrome is somewhat more complex. Many symptoms will lessen or reverse if corticosteroid therapy is decreased

or stopped, but sometimes this simply isn't an option. For example, a patient with severe joint disease may not be able to walk without the anti-inflammatory effects of corticosteroids. In cases like this, the patient and physician must decide what is ultimately best for the patient's quality of life. If corticosteroid therapy is decreased, it must be done gradually, as discussed in *A&P in the Real World: HPA Axis Suppression and Corticosteroid Therapy* on page 614.

Adrenal Insufficiency Hyposecretion of both cortisol and aldosterone are hallmarks of *adrenal insufficiency,* or *Addison's disease.* This disease has many causes, including abnormal development of the adrenal gland, deficiency in certain enzymes required to produce these steroid hormones, and destruction of the adrenal glands by the individual's immune system. Addison's disease puts a patient at risk for an adrenal crisis, and results in disruptions in fluid, electrolyte, and acid-base homeostasis.

Adrenal Production of Androgenic Steroids: Sex Hormones

The **androgenic steroids** (*andro-* = "male") are sex hormones that affect the reproductive organs, or gonads, and multiple other tissues. The majority of the androgenic steroids are produced by the gonads, but they are also synthesized in small amounts in both males and females by the adrenal cortex, largely as byproducts of the pathway that yields cortisol. Androgenic steroids made by the adrenal cortex have the same general effects as those made by the gonads, and can be converted in the circulation to the androgen testosterone or the female hormone estrogen. We discuss the target tissues and effects of androgens in Module 16.6.

Quick Check

☐ 1. What are the three zones of the adrenal cortex, and which hormones are produced by each zone?
☐ 2. What are the target tissues and effects of aldosterone?
☐ 3. What are the target tissues and effects of cortisol?

Hormones of the Adrenal Medulla: Messengers of the Sympathetic Nervous System

◀◀ **Flashback**

1. What are pre- and postganglionic neurons? (p. 520)
2. Which neurotransmitters are used by the neurons of the sympathetic nervous system? (p. 523)

The adrenal medulla consists of neuroendocrine cells called **chromaffin cells** (KROH-maf-in), which are derived from nervous tissue rather than glandular tissue. They are stimulated by acetylcholine released by preganglionic neurons of the sympathetic nervous system. Recall that a sympathetic preganglionic neuron exits the spinal cord and triggers a postganglionic neuron to release neurotransmitters onto a target cell (see Chapter 14). Chromaffin cells are very similar to these postganglionic neurons, except they secrete their products into the bloodstream (which is why these products are considered neurohormones).

As with the sympathetic postganglionic neurons, the products released by chromaffin cells are chemicals called *catecholamines* (kat′-eh-KOHL-ah-meenz), chiefly **epinephrine** and **norepinephrine.** Chromaffin cells produce and secrete primarily epinephrine (the reverse is true for sympathetic neurons, for which the primary neurotransmitter is norepinephrine). Hormonal epinephrine and norepinephrine have almost the same effects on target cells as the neurotransmitters of the sympathetic nervous system. Specifically, they mediate the body's immediate response to a stressor by:

- increasing the rate and force of heart contraction;
- dilating the bronchioles (the airway passages in the lungs);
- constricting the blood vessels supplying the skin, digestive organs, and urinary organs (which increases blood pressure);
- dilating the blood vessels supplying skeletal muscles;
- dilating the pupils; and
- decreasing digestive and urinary functions.

As we discussed previously, the additional epinephrine and norepinephrine secreted by the adrenal medulla prolong the duration of the sympathetic response (see Chapter 14). Additionally, the epinephrine and norepinephrine from the adrenal medulla can reach cells that are not innervated by the sympathetic nervous system. In this way, the sympathetic nervous system can affect the entire body without innervating every individual cell.

Secretion from the adrenal medulla is regulated not only by the sympathetic nervous system but also by the hormones of the HPA axis. ACTH can directly stimulate epinephrine synthesis by chromaffin cells, and cortisol increases synthesis of enzymes required to produce epinephrine. The overall result is a coordinated, integrated response between the endocrine and nervous systems that gives us our best chance of survival when faced with a stressor.

The hormones of the adrenal cortex and adrenal medulla are summarized in **Table 16.3.**

TABLE 16.3 HORMONES OF THE ADRENAL GLAND

Hormone	Stimulus for Release	Inhibitor(s) of Release	Target Tissue(s)	Effects
Adrenocortical Hormones				
Aldosterone (mineralocorticoids)	• Angiotensin-II • Elevated potassium ion concentration • ACTH • Elevated hydrogen ion concentration	• Increased level of aldosterone • Decreased potassium ion concentration • Decreased hydrogen ion concentration • Increased blood pressure	Tubules of the kidneys	• Increases sodium ion retention directly and water retention indirectly • Increases potassium ion loss in the urine • Increases hydrogen ion loss in the urine • Regulates blood pressure
Cortisol (glucocorticoids)	ACTH	Increased level of cortisol	• Liver • Muscle • Adipose • White blood cells	• Increases gluconeogenesis in the liver • Increases protein breakdown in muscle • Increases lipolysis in adipose tissue • Inhibits the inflammatory response
Androgens (androgenic steroids)	ACTH	Poorly understood	Multiple target tissues, including: • Organs of the reproductive tract • Brain • Bone • Skeletal muscle	• Can be converted to testosterone in the circulation • Likely responsible for development of female pubic hair and libido
Adrenal Medullary Hormones				
Catecholamines (epinephrine and norepinephrine)	• Stimulation from preganglionic sympathetic neurons • ACTH	Lack of sympathetic nervous system stimulation	Nearly every cell in the body	• Increase rate and force of heart contractions • Dilate bronchioles • Constrict blood vessels to the digestive and urinary organs and the skin • Increase the metabolic rate; dilate the pupils

Quick Check

☐ 4. What two hormones are produced by the adrenal medulla? What are their target tissues and their effects?

☐ 5. What is the relationship between the adrenal medulla and the sympathetic nervous system?

Apply What You Learned

☐ 1. A patient has a tumor of the anterior pituitary gland that is found to secrete ACTH. Predict the effects this tumor will have on CRH and cortisol levels in the patient's blood. Explain your answers.

☐ 2. Certain drugs used to treat high blood pressure work by blocking the effects of aldosterone on the kidneys. Predict the effects these drugs would have on the:
   a. sodium and potassium ion concentrations in the blood.
   b. pH of the blood.
   c. solute concentration of the blood.

☐ 3. A *pheochromocytoma* (fee′-oh-kroh-moh-sy-TOHM-ah) is a tumor of the adrenal medulla that secretes large amounts of epinephrine and norepinephrine. Predict the effects of this kind of tumor.

See answers in Appendix A.

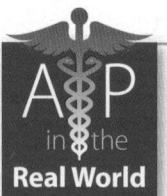

HPA Axis Suppression and Corticosteroid Therapy

When discontinuing corticosteroid therapy, the dose of corticosteroids must be tapered off gradually over a period of weeks or months. This is necessary because of the HPA axis feedback loops that regulate cortisol production and secretion. When a patient receives high doses of corticosteroids for an extended period of time, the negative feedback loops essentially halt production of CRH and ACTH via negative feedback, a phenomenon known as *HPA axis suppression*. This makes a patient's body unable to produce adequate amounts of cortisol for several weeks or even months after discontinuation of corticosteroid therapy.

The most serious potential consequence of HPA axis suppression is an *adrenal crisis*. Cortisol is absolutely vital to survival when a patient is subjected to stress such as surgery, trauma, or illness. If a patient is unable to produce an adequate amount of cortisol in the face of such a stressor, symptoms such as low blood glucose level, low blood pressure, fainting, vomiting, and convulsions may develop. Untreated, an adrenal crisis can easily be fatal. However, it can be prevented by slowly reducing the dose of corticosteroids to give a patient's HPA axis adequate time to resume normal levels of CRH, ACTH, and cortisol production.

MODULE **16.5**

The Endocrine Pancreas

Learning Outcomes

1. Describe the structure of the endocrine pancreas and its hormone-secreting cells.
2. Describe the stimulus for release, the target tissue, and the effect of glucagon.
3. Describe the stimulus for release, the target tissue, and the effect of insulin.
4. Explain how insulin and glucagon work together to maintain the blood glucose level within the normal range.
5. Describe the causes, symptoms, and treatments for the two types of diabetes mellitus.

We've already discussed three hormones that have an impact on blood glucose: thyroid hormone, growth hormone, and cortisol. Although their effects are important, they are not the primary hormones that maintain the blood glucose level within its normal range. Instead, this role falls to two hormones produced by the endocrine pancreas: *insulin* and *glucagon*. This module examines these hormones and their roles in the maintenance of blood glucose homeostasis.

Structure of the Pancreas

The **pancreas** is a club-shaped organ located in the abdominal cavity, mostly posterior to the stomach. It consists of three major parts: the rounded *head,* the middle *body,* and the thin *tail* (**Figure 16.21a**). The pancreas is unusual in that it is both an endocrine and an exocrine gland. Notice in **Figure 16.21b** that there are two groups of cells within the pancreas: (1) small, round "islands" of endocrine cells called **pancreatic islets,** or **islets of Langerhans** (LANG-ger-hanz), and (2) exocrine cells called *acinar cells* (AY-sih-nahr). Clusters of acinar cells are located around small ducts, into which they secrete enzymes and other products that are delivered into the digestive tract (see Chapter 22).

Pancreatic islets secrete hormones into the bloodstream. These islets contain three main cell types:

• **alpha (α) cells,** which secrete the peptide hormone **glucagon;**
• **beta (β) cells,** which secrete the protein hormone **insulin;** and
• **delta (δ) cells,** which secrete the peptide hormone somatostatin (note that this is identical to the somatostatin produced by the hypothalamus).

This module focuses on the alpha and beta cells.

(a) Gross structure of the pancreas

Body of pancreas

Stomach

Tail of pancreas

Pancreas

Pancreatic artery

Head of pancreas

Duodenum

(a) Gross structure of the pancreas

Acinar cells

α Cell

β Cell

δ Cell

Pancreatic islet

Pancreatic blood vessel

LM (285×)

(b) Histology of pancreatic islet and acinar cells: illustration (left) and light micrograph (right)

Figure 16.21 **Anatomy and histology of the pancreas.**

Practice art labeling
@MasteringA&P®

Hormones of the Endocrine Pancreas: Glucose Homeostasis

◀◀ **Flashback**

1. What is glycogen and where is it stored? (p. 52)

As we noted in the introduction to this module, glucagon and insulin regulate the concentration of glucose in the blood. Let's now take a look at their specific roles. For a summary of this information, look ahead to Table 16.4.

Glucagon

Glucagon's major targets are the cells of the liver, muscle tissue, and adipose tissue. In these tissues, it promotes reactions that increase the levels of glucose and other metabolic fuels in the blood. These reactions include the following:

- the breakdown of glycogen into glucose, a process called *glycogenolysis;*
- the formation of new glucose by gluconeogenesis in the liver;

- protein breakdown in muscle tissue to release amino acids for gluconeogenesis;
- the release of fats from adipose tissue for gluconeogenesis and for an additional source of fuel; and
- the formation in the liver of fuel molecules called *ketone bodies* (KEE-tohn) from fatty acids.

Ketone bodies are formed when the products of fatty acid metabolism are combined to yield four-carbon ketone molecules. The liver cannot itself catabolize ketone bodies in any appreciable amount, and so they are released into the bloodstream and delivered to other cells. Some tissues, such as those of the heart and skeletal muscle, are able to take up and then oxidize ketone bodies. Other tissues, such as the brain, are not as readily able to oxidize ketone bodies, although under starvation conditions, the brain can adapt and generate some of its ATP from ketone bodies.

During extreme calorie restriction or starvation, glucagon stimulates rapid ketone body production. However, the cells that can use ketone bodies as fuel are able to use only so many of them, and as a result, they accumulate in the blood. If too many ketone bodies accumulate, they can lower the pH of the blood, producing the dangerous condition **ketoacidosis** (kee′-toh-aeh-sih-DOH-sis).

Glucagon secretion is triggered by several factors, including a decrease in blood glucose concentration, stimulation from the sympathetic nervous system, and circulating catecholamines from the adrenal medulla. Interestingly, glucagon is also secreted in response to ingested proteins. This secretion is part of an integrated hormonal response ensuring that the blood glucose level remains stable during feeding. Glucagon secretion is inhibited by both an elevated blood glucose level and somatostatin.

Insulin

Insulin is the primary antagonist of glucagon; its main effect is to promote the uptake and storage of ingested nutrients into its target cells, which lowers the level of glucose in the blood. Insulin's primary target tissues are those of the liver, cardiac muscle, skeletal muscle, and certain parts of the brain. In these cells, insulin stimulates:

- uptake of lipids, amino acids, and glucose;
- synthesis of glycogen in the liver;
- synthesis of fat from lipids and carbohydrates; and
- promotion of *satiety* (sah-TY-eh-tee; the feeling of fullness).

Through these effects, insulin replenishes the fuel sources that were depleted during fasting, and helps to maintain the blood glucose concentration within a tight range. Insulin is absolutely required for its target cells to take up glucose (note, however, that the liver is an exception). In the absence of insulin, these cells can be bathed in glucose yet essentially starve.

Insulin secretion decreases during fasting and increases during feeding. Its secretion from the β cells is stimulated directly by rising blood glucose concentrations, and is inhibited by somatostatin and the sympathetic nervous system.

When the insulin level rises too high, the blood glucose level drops, a condition called **hypoglycemia** (hy′-poh-gly-SEE-mee-ah). Symptoms of hypoglycemia include weakness, dizziness, rapid breathing, nausea, and sweating. Severe hypoglycemia can lead to confusion, hallucinations, seizures, coma, and even death. These symptoms generally arise because the brain has inadequate glucose to fuel its metabolic processes. Hypoglycemia is most often caused by overadministration of insulin or other agents that lower the blood glucose level, but it may also result from prolonged starvation, insulin-secreting tumors, alcohol, poisons, infection, and genetic metabolic conditions.

Conversely, insufficient insulin secretion or decreased insulin sensitivity cause the blood glucose level to become elevated, a condition called **hyperglycemia.** The two most common causes of chronic hyperglycemia are diabetes mellitus types 1 and 2.

Type 1 Diabetes Mellitus
The disease **type 1 diabetes mellitus,** or *insulin-dependent diabetes mellitus,* affects about 5–10% of people with diabetes mellitus in the United States. It is caused by the destruction of the insulin-producing β cells of the pancreas by the immune system. The lack of insulin means that no matter how much glucose is in the blood, it cannot enter most of its target cells. This starves the cells for glucose, and patients with this condition often feel continually hungry.

The disease is characterized by hyperglycemia due to two factors: (1) Certain target cells are unable to take in the circulating glucose, and (2) glucose is overproduced in the liver because of the unopposed actions of glucagon. Glucagon also elevates the level of ketone bodies in the blood. The increased blood concentrations of glucose and ketone bodies lead to the presence of more glucose and ketone bodies in the fluid in the kidneys. These excess molecules are lost to the urine, resulting in *glucosuria* (glew′-koh-SUR-ee-ah) and *ketonuria* (kee′-tohn-YUR-ee-ah), respectively. The glucose and ketone bodies in the kidney fluid create a gradient that draws water from the extracellular fluid by osmosis into the kidney tubules, where it is then lost to the urine, another example of the Gradients Core Principle (p. 26). This activity causes two other common symptoms of type 1 diabetes: frequent urination, known as *polyuria* (pol′-ee-YUR-ee-ah); and dehydration causing excessive thirst, called *polydipsia* (pol′-ee-DIP-see-ah).

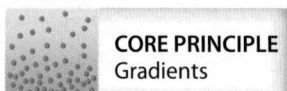

CORE PRINCIPLE
Gradients

The only treatment at present for type 1 diabetes is insulin administered by injection. The disease requires strict monitoring of diet and blood glucose level to ensure the level doesn't rise too high or fall too low. Failure to closely monitor the disease may result in the accumulation of ketones in the blood and a decrease in the pH of the blood, a condition referred to as *diabetic keto-acidosis.* This dangerous condition can lead to coma and death from nervous system shutdown.

The chronic hyperglycemia that occurs in poorly controlled type 1 diabetes has wide-ranging effects on the body. Hyperglycemia damages blood vessels, particularly those in the heart and lower limbs. This results in decreased circulation to these tissues, which increases the risk of heart attack, nonhealing wounds, and amputation. Hyperglycemia also damages peripheral nerves, again particularly in the lower limbs. This leads to *peripheral neuropathy,* which produces numbness, tingling, and burning pain in the affected areas. Other tissues affected by hyperglycemia include the lens of the eye and the capillaries of the retina and the kidneys, possibly resulting in blindness and kidney failure.

Type 2 Diabetes Mellitus
Type 2 diabetes mellitus, also known as *non–insulin-dependent diabetes mellitus,* is the much more common form of diabetes, affecting up to 95% of people with diabetes mellitus in the United States. This form of diabetes generally develops in adults, and so it was formerly referred to as *adult-onset diabetes mellitus.* Patients with type II diabetes generally have functional β cells capable of producing insulin. However, the β cells do not respond to normal increases in blood glucose, and other target cells are less responsive to insulin, an effect called *insulin resistance.* This results in hyperglycemia and the accompanying characteristic signs and symptoms, such as glucosuria, polyuria, and polydipsia. However, unlike individuals with type 1 diabetes, those with type 2 diabetes generally produce enough insulin to keep ketoacidosis from developing. The development of type 2 diabetes is strongly associated with heredity and obesity.

The complications of poorly controlled type 2 diabetes are the same as those of type 1 diabetes—damage to the blood vessels, peripheral nerves, retina and lens, and kidneys. Patients with type 2 diabetes seem to be at higher risk for these complications, probably because the signs and symptoms of hyperglycemia in these patients are often vague, leading to a delay in diagnosis and treatment. By contrast, the signs and symptoms of type 1 diabetes are more immediately evident because these individuals produce no insulin, and so diagnosis and treatment are likely to be earlier.

Nearly all patients with type 2 diabetes are advised to make lifestyle changes, including weight loss, healthy diet, and regular exercise. In fact, some patients can manage the disease with lifestyle modifications alone. However, many require additional treatment. One of the more common treatments is a class of drugs, called *oral hypoglycemics*, that lower the circulating blood glucose level. Occasionally, patients with type 2 diabetes require insulin therapy to maintain a stable blood glucose concentration, but this is not a mainstay of treatment for most patients.

Quick Check

☐ 1. What are the main target tissues of glucagon? What are the stimuli for secretion of glucagon?
☐ 2. What are the main target tissues of insulin?
☐ 3. What are the signs and symptoms of the two types of diabetes mellitus? How do the two types differ?

Blood Glucose Regulation

As you have read, insulin and glucagon act as antagonists in a negative feedback loop that regulates the blood glucose concentration. But this is somewhat more complex than most feedback loops, because both insulin and glucagon are secreted during feeding to maintain blood glucose homeostasis. The regulation of blood glucose level is illustrated in **Figure 16.22**. When the blood glucose level increases, the following feedback loop is initiated (Figure 16.22a):

- **Stimulus: Blood glucose level increases above its normal range.** The main reason for an increase in blood glucose concentration is feeding, although it can also rise in response to hormones such as cortisol.
- **Receptor: Beta cells of the pancreas detect the increased blood glucose concentration.** Like α cells, β cells act as sensors that detect the increase in blood glucose level.
- **Control center: Beta cells increase insulin secretion.** The β cells also act as the control center, controlling the secretion of insulin. The same stimulus reduces glucagon secretion from α cells as well.
- **Effector/response: Insulin decreases the blood glucose level by increasing glucose uptake by cells and storage of glucose, amino acids, and fats.** Insulin triggers the uptake of glucose by muscle and fat cells as well as the storage of glucose, amino acids, and fatty acids. These effects are balanced by a slight increase in glucagon secretion in response

to proteins in food. This prevents hypoglycemia if the food ingested contains limited carbohydrates.
- **Homeostatic range and negative feedback: As the blood glucose level returns to its normal range, negative feedback to β cells decreases insulin secretion.** The falling blood glucose concentration is detected by β cells. These cells then slow insulin secretion, decreasing the responses through negative feedback so that the blood glucose level never falls very far below the normal range.

If the blood glucose level decreases below the normal range, as during fasting, a different loop is initiated (Figure 16.22b):

- **Stimulus: Blood glucose level decreases below its normal range.** The blood glucose concentration can decrease below the normal range due to fasting or an increased use of injected insulin in diabetics.
- **Receptor: Alpha cells of the pancreas detect the decreased blood glucose concentration.** The α cells act as the sensors and detect the falling blood glucose level. (Note that α cells also detect the presence of ingested protein, which we discuss in Module 16.7.)
- **Control center: Alpha cells increase glucagon secretion.** The α cells also act as the control center, increasing glucagon secretion. At the same time, insulin secretion by β cells falls.
- **Effector/response: Glucagon increases the blood glucose level by triggering breakdown of glycogen into glucose and formation of new glucose.** Glucagon triggers glycogenolysis and gluconeogenesis, raising the blood glucose level. In addition, it increases the availability of other fuels such as ketone bodies, free amino acids, and fatty acids.
- **Homeostatic range and negative feedback: As the blood glucose level returns to its normal range, negative feedback to α cells decreases glucagon secretion.** When the blood glucose concentration returns to its normal range, it is detected by the α cells. They, in turn, decrease glucagon secretion, which decreases the responses through negative feedback.

Quick Check

☐ 4. How do glucagon and insulin work together to regulate the level of blood glucose?

Apply What You Learned

☐ 1. Predict the effects of a pancreatic tumor that secretes glucagon. What other disease would this condition resemble? Explain.
☐ 2. Predict the effects of a pancreatic tumor that secretes insulin. What signs and symptoms would you expect to see from such a condition?
☐ 3. Explain why drugs that activate or suppress the sympathetic nervous system have effects on the blood glucose concentration.

See answers in Appendix A.

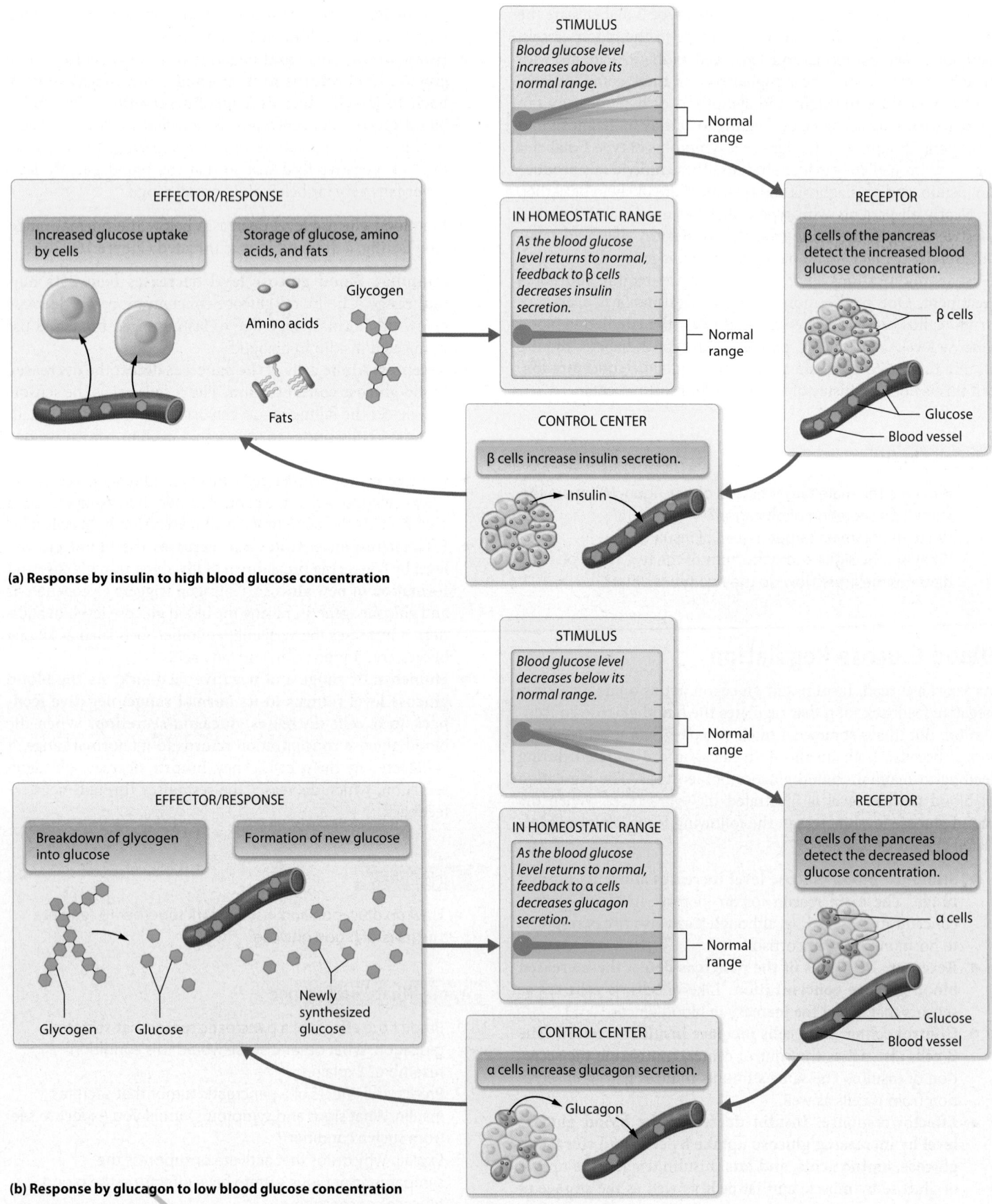

STIMULUS

Blood glucose level increases above its normal range.

Normal range

RECEPTOR

β cells of the pancreas detect the increased blood glucose concentration.

β cells

Glucose

Blood vessel

IN HOMEOSTATIC RANGE

As the blood glucose level returns to normal, feedback to β cells decreases insulin secretion.

Normal range

EFFECTOR/RESPONSE

Increased glucose uptake by cells

Storage of glucose, amino acids, and fats

Glycogen

Amino acids

Fats

CONTROL CENTER

β cells increase insulin secretion.

Insulin

(a) Response by insulin to high blood glucose concentration

STIMULUS

Blood glucose level decreases below its normal range.

Normal range

RECEPTOR

α cells of the pancreas detect the decreased blood glucose concentration.

α cells

Glucose

Blood vessel

IN HOMEOSTATIC RANGE

As the blood glucose level returns to normal, feedback to α cells decreases glucagon secretion.

Normal range

EFFECTOR/RESPONSE

Breakdown of glycogen into glucose

Formation of new glucose

Glycogen   Glucose   Newly synthesized glucose

CONTROL CENTER

α cells increase glucagon secretion.

Glucagon

(b) Response by glucagon to low blood glucose concentration

Figure 16.22 Maintaining homeostasis: regulation of blood glucose concentration by negative feedback loops.

Other Endocrine Glands and Hormone-Secreting Tissues

Learning Outcomes

1. Describe the stimulus for release, the target tissue, and the effects of the hormones produced by the pineal and thymus glands.
2. Describe the stimulus for release, the target tissue, and the effects of the hormones produced by the gonads, adipose tissue, the heart, and the kidneys.

There are two remaining primary endocrine organs: the pineal gland and the thymus gland. In addition, as we discussed in Module 16.1, many secondary organs and tissues secrete hormones. These organs and tissues consist of a variety of cell types, including those that form glandular epithelium, muscle tissue, connective tissue, and nervous tissue. This module covers the pineal and thymus glands as well as secondary endocrine organs and tissues, their hormones, and their effects.

The Pineal Gland: Melatonin

◀◀ **Flashback**

1. What is the epithalamus? (p. 435)
2. What is the reticular formation? (p. 436)

The **pineal gland** is part of the *epithalamus,* which is the posterior portion of the diencephalon in the brain. Recall that the pineal gland secretes the neurohormone **melatonin** (mel-ah-TOH-nin; see Chapter 12). The trigger for melatonin secretion appears to be related to light and dark cycles. The pineal gland begins secreting melatonin when ambient light decreases in the evening, and it reaches its peak secretion over the nighttime hours. Melatonin's main target tissues are the sleep-regulating centers in the reticular formation of the brainstem, where it appears to adjust the sleep phase of the sleep/wake cycle in some people. Synthetic melatonin is often sold as a dietary supplement, to be used to promote sleep and prevent jet lag. However, its effectiveness for these uses is questionable.

The Thymus: Thymosin and Thymopoietin

The **thymus** is an irregularly shaped organ found in the mediastinum (look back at Figure 16.2). It is the site of maturation of a class of white blood cells (leukocytes) called *T lymphocytes* (LIM-foh-sytz) that are involved in the immune response. It also secretes the hormones **thymosin** and **thymopoietin** (thy′-moh-POY-eh-tin), which function mainly as paracrine signals that assist in T lymphocyte maturation. The thymus gland is larger and far more active in infants and children, who have developing immune systems (see Figure 20.9). In adults, the thymus gland shrinks (or *atrophies*) and eventually is replaced mostly with adipose and other connective tissues.

The Gonads: Sex Hormones

The **testes** and **ovaries** are the primary male and female reproductive organs, or **gonads,** respectively. The gonads are responsible for the production of *gametes* (GAM-eetz)—sperm in males and ova, or eggs, in females. In addition to these functions, the gonads produce the sex steroid hormones that are responsible for gamete production and have multiple other functions throughout the body. Let's take a closer look at the hormones produced by these organs.

The Testes: Testosterone

The testes produce the steroid hormone **testosterone** (tes-TOS-ter-ohn). Although testosterone is considered the "male hormone," females do produce small amounts of it in the adrenal cortex and ovaries. Testosterone production and secretion in the male is regulated by a multi-tiered negative feedback loop involving the hypothalamus and anterior pituitary. Recall that GnRH from the hypothalamus triggers secretion of both LH and FSH from the anterior pituitary. LH stimulates testosterone synthesis, and FSH stimulates production of a protein that binds testosterone and concentrates it in the testes. Via negative feedback, a high blood level of testosterone decreases GnRH secretion and therefore decreases LH and FSH secretion.

Like all steroid hormones, testosterone is hydrophobic and so easily crosses the plasma membranes of its target cells and binds to intracellular receptors that influence the expression of various genes. Testosterone has two basic types of effects on its target cells: (1) *anabolic effects,* which include those that stimulate bone growth and increase muscle mass; and (2) *androgenic effects* ("male generating"), which encompass those involving the development of male secondary sex characteristics, such as a deeper voice, more prominent facial hair, and increased bone and muscle mass. (These characteristics are called "secondary" because the affected organs are not part of the reproductive system.)

The Ovaries: Estrogens and Progesterone

Cells of the ovary produce the female steroid hormones known as **estrogens.** In parallel with testosterone, estrogens are considered "female" hormones but are produced in smaller amounts in males. As with testosterone, the secretion of estrogens is regulated by negative feedback loops involving GnRH from the hypothalamus and LH and FSH from the anterior pituitary. Estrogens stimulate the development of female secondary sex characteristics such as the breasts; regulate the menstrual cycle; and have multiple effects on other tissues and organs, such as bone, blood, kidney, and adipose.

Another steroid hormone produced by the ovaries is **progesterone** (proh-JES-ter-ohn). Progesterone production peaks after ovulation and during pregnancy (*pro-* = "supporting," *ges-* = "a carrying; pregnancy"). Progesterone has multiple physiological effects, many of which help to prepare the body for pregnancy and support fetal development during pregnancy. It also affects smooth muscle tissue, body temperature, blood clotting, bone tissue, and metabolism.

We discuss the physiological effects of estrogens, progesterone, and testosterone more completely in the reproductive system chapter (see Chapter 26).

Quick Check

☐ 1. What are the target tissues and effects of melatonin?
☐ 2. What are the target tissues and main effects of testosterone and estrogens?

Adipose Tissue: Leptin

Adipocytes produce a protein hormone known as **leptin.** Although it is a protein, leptin manages to cross the blood-brain barrier to interact with its main target cells: neurons in the hypothalamus that control feeding. Leptin acts on these neurons to induce satiety, a feeling of fullness, which prevents overfeeding. Interestingly, the amount of leptin secreted appears to be related to the amount of adipose tissue present. In theory, this should prevent humans with excess adipose tissue from overeating. However, the mechanisms that regulate feeding are quite complex, involving multiple neural and hormonal factors. See *A&P in the Real World: Leptin and Obesity* for one of the limits to this hormone's effects.

The Heart: Atrial Natriuretic Peptide

Certain cardiac muscle cells contain stretch-sensitive ion channels. These channels open more widely when the blood volume inside the heart increases, and this stimulates the cardiac muscle cells to secrete the hormone **atrial natriuretic peptide,** or **ANP** (nay'-tree-yoo-RET-ik; *natri-* = "sodium"). ANP has two main target tissues: smooth muscle cells lining blood vessels, and the tubules of the kidneys. ANP triggers relaxation of smooth muscle cells in blood vessels, an effect that increases the vessel diameter, which is known as *vasodilation*. In the kidneys, it enhances excretion of sodium ions, an effect called *natriuresis* (nay'-tree-yoo-REE-sis). The process of natriuresis also enhances water excretion because it creates a concen-

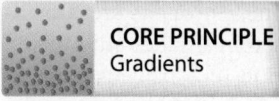
CORE PRINCIPLE
Gradients

tration gradient that water follows into the kidney fluid by osmosis, another example of the Gradients Core Principle (p. 26). Both vasodilation and the decreased blood volume that accompanies natriuresis decrease blood pressure.

Quick Check

☐ 3. What are the hormones of the thymus gland, adipose tissue, and cardiac muscle cells? What are the target tissues and effects of each of these hormones?

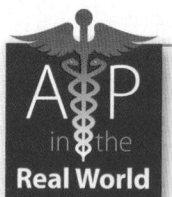

A&P in the Real World

Leptin and Obesity

Leptin has been heavily researched for its possible pharmaceutical application in weight loss. The hypothesis was that leptin supplements would suppress the appetite, which would lead to weight loss. Unfortunately, this research demonstrated that administered leptin was not effective at reducing appetite in humans. However, this hasn't stopped many manufacturers of over-the-counter dietary supplements from marketing their leptin-containing supplements as "miracle weight loss" solutions.

The Kidneys: Erythropoietin

◄◄ **Flashback**

1. What are erythrocytes, and where are they formed? (p. 145)

The kidneys play several endocrine roles, including the following:

- **Erythropoietin production.** The hormone **erythropoietin** (eh-rith'-roh-POY-eh-tin), or **EPO,** is secreted mainly by certain cells of the kidney in response to a decreased level of oxygen in the blood. EPO acts on the red bone marrow, where it stimulates the development of erythrocytes, which increases the oxygen-carrying capacity of the blood. We discuss the role of EPO in erythrocyte production more fully in the chapter on blood (see Chapter 19).
- **Renin secretion.** The kidney cells also secrete **renin** (REE-nin), an enzyme in blood that converts the plasma protein angiotensinogen to angiotensin-I. It is an important part of the renin-angiotensin-aldosterone system, which maintains blood pressure (see Chapters 18 and 24).
- **Conversion of vitamin D to its active form.** As mentioned in Module 16.3, it is not until vitamin D arrives in the kidneys that it is converted to its active form, calcitriol. Recall that this change is stimulated by parathyroid hormone.

The hormones, target tissues, and effects of the pancreas, the endocrine glands covered in this module, and the secondary endocrine organs and tissues are summarized in **Table 16.4**.

Quick Check

☐ 4. What are the endocrine roles of the kidneys?

Apply What You Learned

☐ 1. Individuals who live at extreme northern and southern latitudes experience winter with little to no periods of daylight and summer with little to no periods of darkness. Predict the effect these seasonal conditions will have on melatonin secretion and the sleep/wake cycle.

TABLE 16.4 HORMONES PRODUCED BY OTHER ENDOCRINE ORGANS AND HORMONE-SECRETING TISSUES

Organ/Tissue	Hormone(s)	Target Tissue(s)	Effects
Other Endocrine Organs			
Pancreas	• Insulin • Glucagon	Insulin: most cells, including those of the liver, skeletal and cardiac muscle, and adipocytes Glucagon: liver, muscle, adipose	Insulin: stimulates glucose uptake by most target cells, glycogenesis, and lipogenesis Glucagon: stimulates gluconeogenesis, glycogen breakdown, ketogenesis, and fat and protein breakdown
Pineal gland	Melatonin	Reticular formation of the brainstem	• Regulates the sleep/wake cycle • Promotes sleep
Thymus	• Thymosin • Thymopoietin	T lymphocytes	Promote T lymphocyte maturation
Hormone-Secreting Tissues			
Testes	Testosterone	• Male reproductive organs • Multiple other target tissues	• Androgenic effects • Anabolic effects
Ovaries	• Estrogens	Female reproductive organs	• Development of female secondary sex characteristics • Regulate the menstrual cycle • Multiple other effects
	• Progesterone	Multiple other target tissues	• Prepare the body for pregnancy • Support fetal development • Multiple other effects
Adipose tissue	Leptin	Brain	Promotes satiety
Heart	Atrial natriuretic peptide (ANP)	• Smooth muscle cells in blood vessel walls • Kidney tubules	• Relaxes smooth muscle cells in blood vessels • Promotes natriuresis and fluid loss in the kidneys • Lowers blood pressure as a result of above effects
Kidneys	• Erythropoietin	Red bone marrow	Increases rate at which erythrocytes are formed
	• Renin	Enzyme that activates angiotensinogen	Part of the renin-angiotensin-aldosterone system
	• Vitamin D (calcitriol)	Small intestines	Vitamin D activation

☐ 2. A genetic mutation in mice leads to a dysfunctional leptin receptor in the hypothalamus that does not bind leptin. Predict the consequences of this mutation.

☐ 3. Renal failure is generally characterized by inability of the kidney cells to perform their many physiological functions. Predict those consequences of renal failure that extend beyond the urinary system.

See answers in Appendix A.

MODULE **16.7**

Three Examples of Endocrine Control of Physiological Variables

Learning Outcomes

1. Provide specific examples to demonstrate how the endocrine organs respond to maintain homeostasis in the body.
2. Describe how hormones work together with the nervous system to mediate the stress response.

As we saw in previous modules, many homeostatic variables are controlled by the endocrine system. It can be challenging to form a "big picture" view of how the endocrine system regulates these variables, because they are generally controlled by multiple hormones from multiple endocrine organs. In this module, we examine some important examples of hormone "teamwork" by looking at fluid balance, metabolism, and the stress response. This discussion will help you put together all of the pieces of the endocrine system to build a greater understanding of how hormones work together to maintain homeostasis.

Hormonal Control of Fluid Homeostasis

Fluid homeostasis refers to maintenance of the necessary volume of water in the extracellular fluid. Although several hormones indirectly contribute to fluid homeostasis, the three main water-regulating hormones are ADH, aldosterone, and ANP. Let's summarize their interactions.

- **Normal conditions: Low levels of ADH and aldosterone are secreted continually and lead to average urine output (Figure 16.23a).** The hypothalamus and adrenal cortex continually secrete small amounts of ADH and aldosterone, respectively. This is necessary because huge volumes of blood are filtered through the kidneys every minute. In fact, the entire volume of blood plasma flows through the kidneys about 60 times per day. Therefore, ADH and aldosterone must be continually secreted to ensure that most of the plasma volume is not lost to the urine. Aldosterone's important effects on electrolyte and acid-base homeostasis provide another reason why it must be continually secreted.

- **Decreased plasma volume and increased plasma solute concentration: ADH and aldosterone secretions rise, leading to decreased urine output (Figure 16.23b).** Two conditions signal the cells of the endocrine system that there is inadequate water in the body: a decreased plasma volume, which is sensed by specialized cells in the kidneys, and an increased solute concentration of the plasma, which is sensed by receptors in the hypothalamus. These conditions trigger increased secretions of aldosterone and ADH, respectively, which lead to greater water retention and lower urine output.

- **Increased plasma volume and decreased plasma solute concentration: ADH and aldosterone secretions decline and ANP secretion increases, leading to increased urine output (Figure 16.23c).** Through negative feedback loops, the secretion of ADH and aldosterone falls when the plasma volume rises and/or the plasma solute concentration decreases. If plasma volume continues to rise, ANP secretion from the heart is triggered. Together, these changes in hormone secretion result in water loss from the blood and an increased urine output.

Quick Check

☐ 1. Which hormones primarily control fluid homeostasis?
☐ 2. What is the role of each of these hormones with respect to fluid homeostasis?

Hormonal Control of Metabolic Homeostasis

The control of metabolic homeostasis by the endocrine system is perhaps our most complex example because it involves several factors: controlling the metabolic rate of different cell populations, ensuring fuels are available during times of activity, storing metabolic fuels during times of rest, regulating growth, and feeding behaviors. Let's simplify things by looking at the hormones that maintain metabolic homeostasis during three periods: at rest while fasting, at rest while feeding, and during exercise.

- **At rest, while fasting: Thyroid hormones determine the basal metabolic rate.** As you read in Module 16.3, thyroid hormones set the basal metabolic rate by triggering processes that cause the body cells to consume ATP. Thyroid hormones also stimulate gluconeogenesis, which leads to an increase in the blood glucose concentration. During fasting, this increase helps to maintain a stable blood glucose level. Fasting also triggers the release of glucagon and an elevated level of growth hormone, which further raise the blood glucose level.

- **At rest, while feeding: Insulin secretion increases.** When feeding begins, the increase in blood glucose concentration causes insulin release. Insulin triggers glucose uptake by its target cells and the storage of excess glucose as glycogen and fat in adipocytes. The presence of protein in a meal leads to GH and IGF-1 release, which stimulates protein synthesis with the ingested amino acids. Protein in a meal also stimulates glucagon release to prevent hypoglycemia in the event of a carbohydrate-poor meal. Feeding ends with

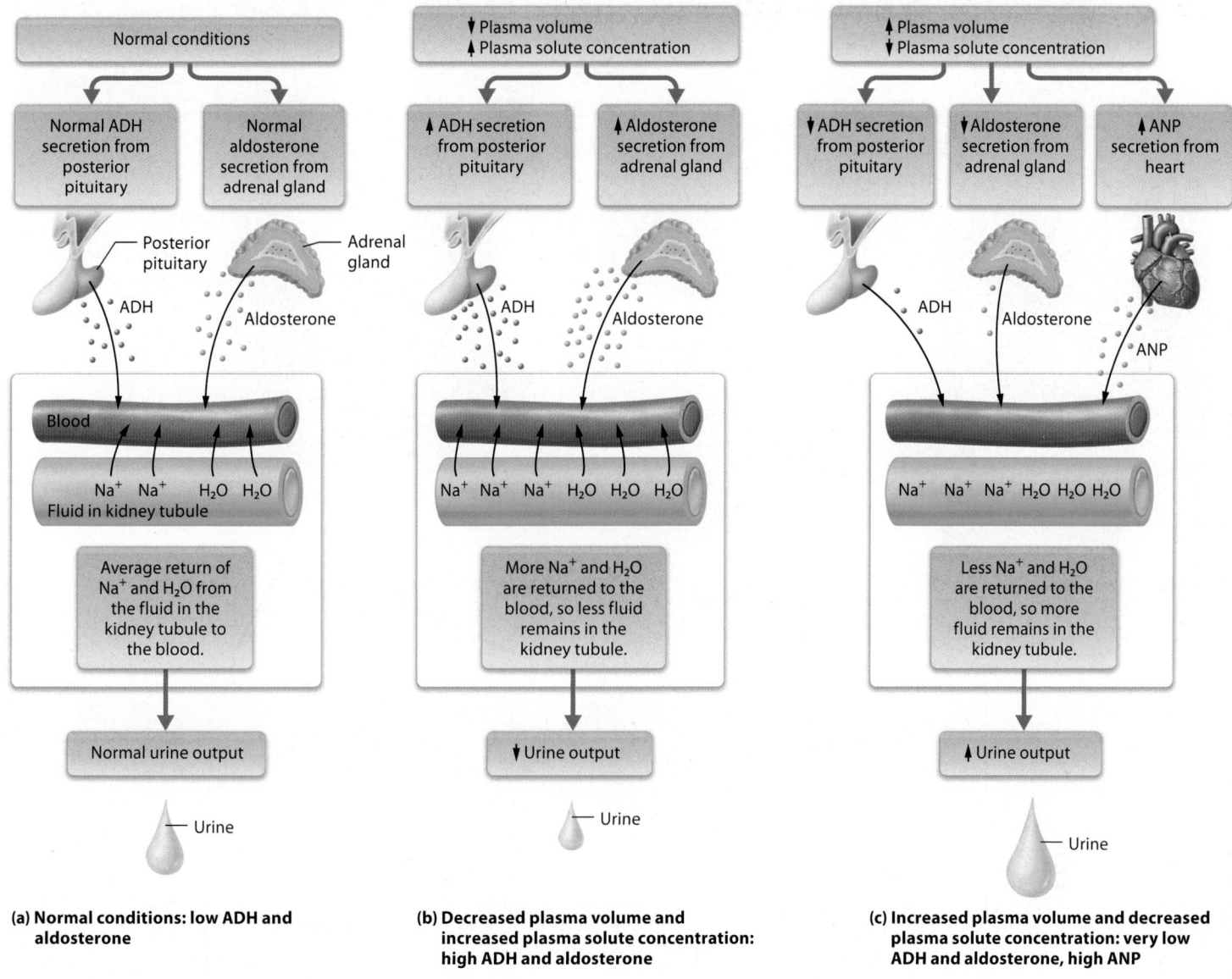

Figure 16.23 Summary of endocrine control of fluid homeostasis.

(a) **Normal conditions: low ADH and aldosterone**

(b) **Decreased plasma volume and increased plasma solute concentration: high ADH and aldosterone**

(c) **Increased plasma volume and decreased plasma solute concentration: very low ADH and aldosterone, high ANP**

the help of both insulin and leptin, which promote satiety in the hypothalamus.

- **During exercise: Catecholamines control the metabolic rate and glucagon secretion increases.** When exercise begins, the sympathetic nervous system increases catechol-amine secretion from the adrenal medulla. Catecholamines in the blood cause the metabolic rate of all cells to increase to a level higher than the basal rate set by thyroid hormones. This increased metabolic rate necessitates more metabolic fuels, and for this reason the α cells of the pancreas increase glucagon secretion. These actions are further discussed in our upcoming examination of the stress response.

Quick Check

☐ 3. Which hormones control metabolic homeostasis?

☐ 4. What is the role of each of these hormones with respect to metabolic homeostasis?

Putting It All Together: The Big Picture of the Hormonal Response to Stress

◀◀ Flashback

1. What is the role of the hypothalamus in the sympathetic response? (p. 532)
2. What is the role of the parasympathetic nervous system? (p. 528)

We first discussed regulation of the stress response with the sympathetic nervous system (see Chapter 14). There we pointed out that although it's common to think of a stressor as being

The Big Picture of the Hormonal Response to Stress

Figure 16.24

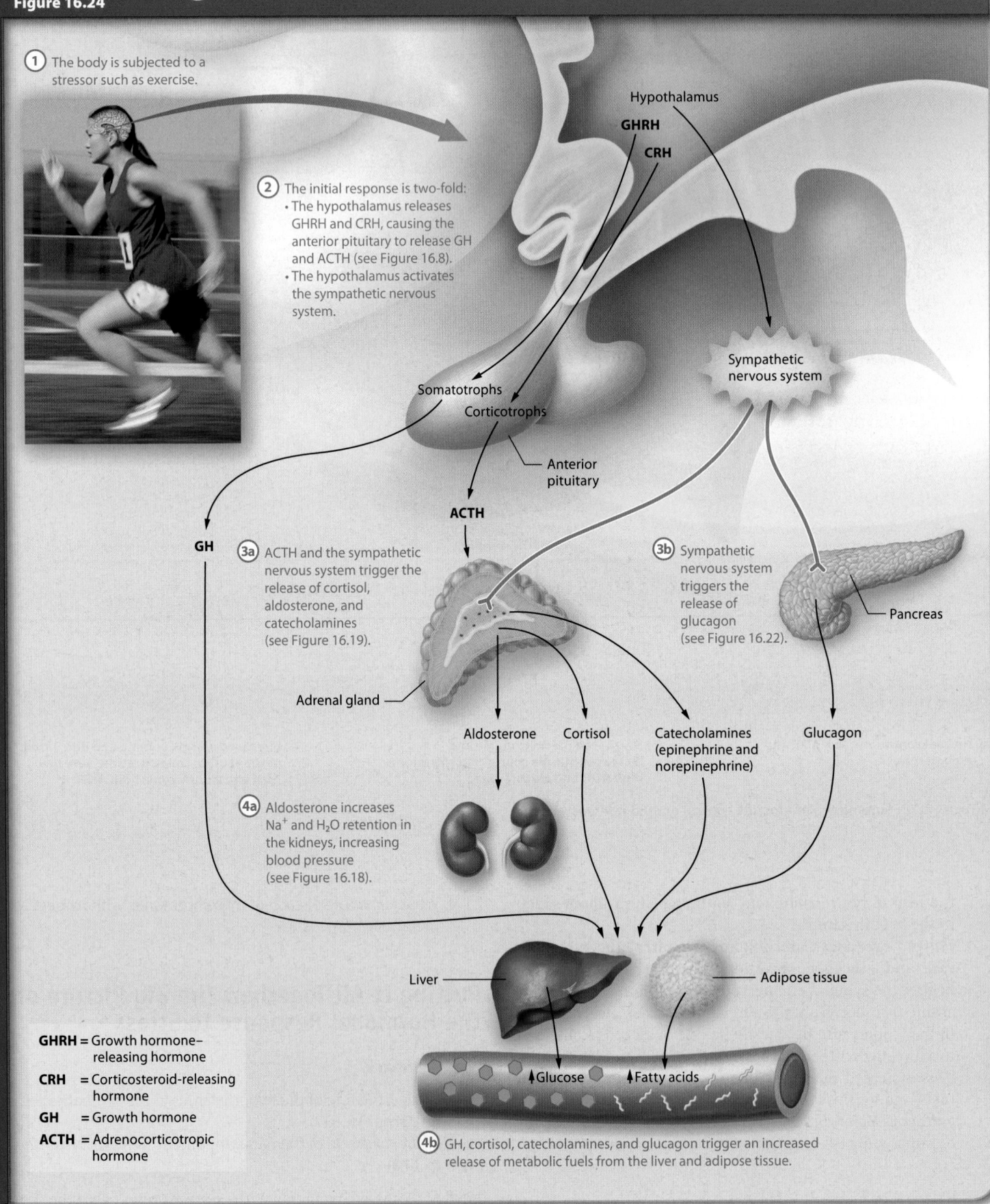

1. The body is subjected to a stressor such as exercise.

2. The initial response is two-fold:
   - The hypothalamus releases GHRH and CRH, causing the anterior pituitary to release GH and ACTH (see Figure 16.8).
   - The hypothalamus activates the sympathetic nervous system.

Hypothalamus

GHRH

CRH

Sympathetic nervous system

Somatotrophs

Corticotrophs

Anterior pituitary

ACTH

GH

3a. ACTH and the sympathetic nervous system trigger the release of cortisol, aldosterone, and catecholamines (see Figure 16.19).

3b. Sympathetic nervous system triggers the release of glucagon (see Figure 16.22).

Pancreas

Adrenal gland

Aldosterone   Cortisol   Catecholamines (epinephrine and norepinephrine)   Glucagon

4a. Aldosterone increases Na⁺ and H₂O retention in the kidneys, increasing blood pressure (see Figure 16.18).

Liver

Adipose tissue

↑Glucose   ↑Fatty acids

GHRH = Growth hormone–releasing hormone

CRH = Corticosteroid-releasing hormone

GH = Growth hormone

ACTH = Adrenocorticotropic hormone

4b. GH, cortisol, catecholamines, and glucagon trigger an increased release of metabolic fuels from the liver and adipose tissue.

something inherently *stressful,* a stressor can be anything from excitement to mild exercise. In this chapter, we have discussed endocrine regulation of the stress response with multiple hormones, including CRH, ACTH, and adrenal steroid hormones from the HPA axis, GH from the anterior pituitary, catecholamines from the adrenal medulla, and glucagon from the pancreas. Now let's put all of the pieces of the nervous system and endocrine system together and examine how they work together to achieve the physiological changes witnessed in the body during the stress response.

The steps in the stress response are outlined in **Figure 16.24.** When the stressor is resolved, the hypothalamus decreases its stimulation of the sympathetic neurons, CRH secretion, and GHRH secretion. The decline in sympathetic activity and CRH secretion, along with an increase in parasympathetic nervous system activity, lowers ACTH, GH, and adrenal hormone levels and restores resting conditions.

☐ 5. Which hormones are involved in the stress response?
☐ 6. What is the role of each hormone in the stress response? How do these hormonal effects tie in with the response of the sympathetic nervous system?

Apply What You Learned

☐ 1. Mr. Dent is discovered to have an ADH-secreting tumor. Predict what will happen to the amount of aldosterone secreted by his adrenal cortex. What do you think will happen to the amount of ANP secreted by Mr. Dent's heart? Explain.
☐ 2. Ms. Prefect has a tumor of the hypothalamus that reduces its secretion of CRH. Will her body continue to secrete adrenal steroids and catecholamines as part of the stress response? Explain.
☐ 3. Your friend is on a no-carbohydrate, all-fat diet, leading to essentially no intake of dietary glucose. Predict which hormones might be secreted in excess to compensate for the decreased level of blood glucose. How will these hormones compensate?

See answers in Appendix A.

CHAPTER SUMMARY

Take full advantage of the invaluable online practice and study tools for this chapter by going to MasteringA&P®. There you will find:

- Practice tests, module quizzes, art-labeling and sequence-the-steps activities
- Animated figures with quizzes
- *Practicing A&P* worksheets
- Author-narrated podcast
- Dynamic Study Modules
- *Interactive Physiology*

MODULE **16.1**
Overview of the Endocrine System 586

- Cells of the **endocrine system** secrete **hormones** into the bloodstream that affect distant targets, in contrast to the neurons of the nervous system, which come close to their target cells and work through the release of neurotransmitters into synapses.
- The three types of chemical signaling are the endocrine, **paracrine,** and **autocrine** signals. (Figure 16.1, p. 587)

- The primary endocrine organs include the anterior pituitary, thyroid, and parathyroid glands, the adrenal cortices, the thymus, and the endocrine pancreas. **Neuroendocrine organs** secrete *neurohormones,* including those of the hypothalamus, the adrenal medulla, and the pineal gland. (Figure 16.2, p. 588)

- Practice art-labeling exercise: Figure 16.2.
- **iP** Play Interactive Physiology tutorial on Endocrine System: Orientation.
- **iP** Play Interactive Physiology tutorial on Endocrine System: Endocrine System Review.

- Hormones exert their effects by binding to **receptors** on their **target cells.** Hormones with a hydrophilic structure bind to receptors in the plasma membrane; hormones with a hydrophobic structure bind to either intracellular receptors or receptors in the plasma membrane. (Figure 16.3, p. 590)
- The two types of hormones are **amino acid–based hormones** and **steroid hormones.**

- **iP** Play Interactive Physiology tutorial on Endocrine System: Biochemistry, Secretion, and Transport of Hormones.

- Hydrophilic hormones interact with their target cells by binding to receptors in the target cells' plasma membranes and triggering the formation of a **second messenger.** The second messenger then initiates changes within the cell.

- Hydrophobic hormones can bind to intracellular receptors. The resulting receptor-steroid complex binds to and interacts with the cell's DNA to trigger a change within the cell.

- Hormones may affect gene transcription, enzymes, cell division, and the permeability of the plasma membrane, and may cause secretion from other glands.

- Hormone secretion can be stimulated by hormonal, humoral, and neural stimuli. (Figure 16.4, p. 591)

- The production and secretion of most hormones are regulated through negative feedback loops. (Figure 16.5, p. 592)

iP Play Interactive Physiology tutorial on Endocrine System: The Actions of Hormones on Target Cells.

MODULE 16.2
The Hypothalamus and the Pituitary Gland 593

- The **hypothalamus** is the anteroinferior portion of the diencephalon. (Figure 16.6, p. 594)

- The **pituitary gland** consists of two components: the **anterior pituitary,** which is an endocrine gland that secretes hormones, and the **posterior pituitary,** which is composed of axons and axon terminals that store certain neurohormones produced by the hypothalamus. The posterior pituitary produces no hormones of its own. (Figure 16.7, p. 595)

- The hypothalamus secretes the following hormones that are stored in the posterior pituitary:
  - **Antidiuretic hormone (ADH)** is released when osmoreceptors in the hypothalamus detect an increase in the blood solute concentration. It acts on kidney cells to promote water retention and causes vasoconstriction of peripheral blood vessels. These effects increase the amount of water in the body.
  - **Oxytocin's** target tissues are specialized cells of the mammary gland, where it triggers milk ejection, and the smooth muscle cells of the uterus, where it triggers contractions during childbirth.

- The hypothalamus controls the anterior pituitary through **releasing** and **inhibiting hormones** that are secreted into the blood vessels of the **hypothalamic-hypophyseal portal system.** (Figure 16.8, p. 597)

- The anterior pituitary hormones that affect other glands do so by controlling secretion from other glands and tissues. These hormones include the following:
  - **Thyroid-stimulating hormone,** or **TSH,** stimulates growth of and secretion from the thyroid gland. Its secretion is triggered by **thyrotropin-releasing hormone** from the hypothalamus.
  - **Adrenocorticotropic hormone,** or **ACTH,** stimulates secretion from the adrenal cortex and the adrenal medulla. Its secretion is triggered by **corticotropin-releasing hormone** from the hypothalamus.
  - **Prolactin** stimulates milk production in the mammary glands. Its secretion is triggered by prolactin-releasing hormone

(possibly TRH) and inhibited by **prolactin-inhibiting factor** (dopamine) from the hypothalamus.
  - **Luteinizing hormone (LH)** and **follicle-stimulating hormone (FSH)** are **gonadotropins** secreted in response to **gonadotropin-releasing hormone** from the hypothalamus. In the male, these hormones trigger the production of both testosterone and chemicals that bind and concentrate testosterone. In the female, these hormones trigger the production of estrogens and progesterone, the release of oocytes via ovulation, and the maturation of oocytes.

- **Growth hormone,** or **GH,** is produced in response to **growth hormone–releasing hormone (GHRH)** from the hypothalamus. Its release is inhibited by **somatostatin** from the hypothalamus.

- GH's acute effects include fat breakdown and the production of new glucose by the liver. Its long-term effects are mediated by **insulin-like growth factor (IGF),** which is secreted by liver cells. IGF promotes growth and body mass regulation through stimulation of rapid protein synthesis and cell division. (Figure 16.9, p. 598)

- Growth hormone disorders include **pituitary dwarfism** and **gigantism.**

- Hormones of the hypothalamus and pituitary gland are summarized in Table 16.1 (p. 600).

MODULE 16.3
The Thyroid and Parathyroid Glands 601

- The **thyroid gland** is located in the anterior neck. It consists of hollow, spherical **thyroid follicles** that are filled with **colloid** and lined by cuboidal *follicle cells.* Follicle cells produce and secrete thyroid hormones. Between follicles are **parafollicular cells** that secrete **calcitonin.** (Figure 16.11, p. 601)

Practice art-labeling exercise: Figure 16.11.

- Three to five **parathyroid glands** are located on the posterior thyroid gland. They consist of **chief cells** that secrete **parathyroid hormone,** or **PTH.** (Figure 16.12, p. 602)

- Thyroid hormones consist of amino acids attached to iodine atoms. **Triiodothyronine,** or T_3, has three iodine atoms, and **thyroxine,** or T_4, has four iodine atoms. About 90% of the hormone produced by the follicle cells is T_4, but T_3 is more active.
  - Both T_3 and T_4 are hydrophobic hormones. Their actions include regulation of the metabolic rate, thermoregulation, promotion of growth and development, and synergistic effects with the sympathetic nervous system.
  - Thyroid hormone synthesis begins with the attachment of iodine molecules to **thyroglobulin** in the colloid. Iodinated thyroglobulin is then taken into the follicle cells and converted to T_3 and T_4, after which they are released into the circulation. T_4 is converted to T_3 in target tissues. (Figure 16.13, p. 603)
  - The production of thyroid hormones is regulated by a multi-tiered feedback loop involving TRH from the hypothalamus and TSH from the anterior pituitary. (Figure 16.14, p. 604)
  - **Hyperthyroidism** is characterized by elevated levels of T_3 and T_4, weight loss, heat intolerance, disruptions in blood pressure

and heart rhythm, and development of a goiter and exophthalmos. (Figure 16.15, p. 605)

- **Hypothyroidism** is caused by iodine deficiency or immune system–mediated destruction of the thyroid gland. Signs and symptoms include weight gain, cold intolerance, slow heart rate and low blood pressure, and goiter.

- PTH increases the blood calcium ion concentration by stimulating bone breakdown by osteoclasts, calcium ion absorption from the small intestine, and calcium ion retention from the fluid in the kidney tubules. (Figure 16.16, p. 606)

- Calcitonin inhibits osteoclasts and transiently decreases the blood calcium ion concentration.

- Table 16.2 (p. 607) summarizes the hormones of the thyroid and parathyroid glands.

MODULE **16.4**
The Adrenal Glands 608

- Each **adrenal gland** consists of an outer **adrenal cortex** and an inner **adrenal medulla.** The adrenal cortex consists of three zones of cells that secrete steroid hormones. (Figure 16.17, p. 609)

Practice art-labeling exercise: Figure 16.17.

- **Mineralocorticoids** regulate fluid, electrolyte, and acid-base homeostasis. They are secreted by the **zona glomerulosa,** the outer zone of the adrenal cortex. The main mineralocorticoid is **aldosterone,** which triggers sodium ion retention and potassium ion and hydrogen ion loss to the urine. This indirectly causes water retention. (Figure 16.18, p. 610)

- Aldosterone secretion is stimulated by the hormone *angiotensin-II* and elevated blood potassium ion and hydrogen ion concentrations.

- **Glucocorticoids** are secreted by the middle and inner zones of the adrenal cortex, the **zona fasciculata** and **zona reticularis,** respectively. The main glucocorticoid is **cortisol.** Cortisol is an important part of the **stress response,** and has effects on multiple tissues, which include inducing gluconeogenesis, triggering protein and fat breakdown, and inhibiting the inflammatory response. It is regulated by the *hypothalamic-pituitary-adrenal (HPA) axis,* which involves CRH from the hypothalamus and ACTH from the anterior pituitary gland. One important disorder involving cortisol is Cushing's syndrome. (Figure 16.19, p. 611; Figure 16.20, p. 612)

iP Play Interactive Physiology tutorial on Endocrine System: The Hypothalamic-Pituitary Axis.

- **Androgenic steroids** are also secreted by the cells of the zona fasciculata and zona reticularis. They have effects on reproductive organs and many other tissues.

- The adrenal medulla consists of **chromaffin cells** that secrete *catecholamines,* particularly **epinephrine,** into the circulation when stimulated by the sympathetic nervous system. The catecholamines secreted by the adrenal medulla have the same effects as those produced by the sympathetic nervous system, and function to prolong the response to a stressor.

- Table 16.3 (p. 613) summarizes the hormones of the adrenal glands.

MODULE **16.5**
The Endocrine Pancreas 614

- The pancreas consists of two groups of cells: *acinar cells,* which secrete exocrine products into the digestive tract; and **pancreatic islets,** which secrete hormones. (Figure 16.21, p. 615)

Practice art-labeling exercise: Figure 16.21.

- The hormone **glucagon** is produced by **alpha (α) cells** of the pancreatic islets. Its main target tissues include liver, muscle, and adipose tissue, where it stimulates glycogen breakdown, gluconeogenesis, protein and fat breakdown, and the formation of *ketone bodies.* Glucagon secretion is triggered by decreased blood glucose level, ingestion of protein, stimulation from the sympathetic nervous system, and circulating catecholamines.

- The hormone **insulin** is produced by the **beta (β) cells** of the pancreatic islets. Insulin is the primary antagonist for glucagon, and it stimulates the uptake of glucose, fats, and amino acids by cells, and the formation of glycogen and fat. It also promotes satiety. Its main stimulus for release is an increase in blood glucose level. It is inhibited by somatostatin and stimulation from the sympathetic nervous system. (Figure 16.22, p. 618)

- Abnormally high insulin levels result in **hypoglycemia,** which may lead to confusion, weakness, dizziness, rapid breathing, coma, and eventually death.

- When insulin secretion is inadequate or the released insulin is ineffective, **hyperglycemia** results. The two most common causes of hyperglycemia are **type 1 diabetes mellitus,** in which the β cells are destroyed and no insulin is produced, and **type 2 diabetes mellitus,** in which the tissues are insensitive to insulin and the β cells do not have a normal response to increases in blood glucose concentration.

MODULE **16.6**
Other Endocrine Glands and Hormone-Secreting Tissues 619

- The **pineal gland** consists of neuroendocrine cells that secrete the hormone **melatonin,** which regulates the sleep/wake cycle.

- The **thymus** in the superior mediastinum produces **thymosin** and **thymopoietin,** which stimulate T lymphocyte maturation.

- The **gonads** include the **testes** in the male and the **ovaries** in the female. Both gonads produce steroid hormones.
  - The testes produce and activate the hormone **testosterone** in response to stimulation from LH and FSH. Testosterone has both androgenic and anabolic effects.
  - The ovaries produce **estrogens** and **progesterone.** Estrogen production is stimulated by LH and FSH, and estrogens trigger the development of female sex characteristics, regulate the menstrual cycle, and affect multiple other organs and tissues. Progesterone production is stimulated after ovulation, and it prepares the body for pregnancy and supports fetal development.

- Adipose tissue produces the hormone **leptin,** which induces satiety in the hypothalamus.

- Cardiac muscle cells in the atria produce **atrial natriuretic peptide,** which promotes vasodilation and natriuresis and water loss in the kidneys. These effects lower blood pressure.

- The kidneys produce the hormone **erythropoietin,** which stimulates erythrocyte development, and the enzyme **renin,** which is part of a system that activates aldosterone. They are also where the hormone vitamin D is converted to its active form, calcitriol.

- Table 16.4 (p. 621) summarizes the hormones of the other endocrine organs and hormone-secreting tissues.

MODULE | **16.7**
Three Examples of Endocrine Control of Physiological Variables 622

- Fluid balance is maintained by ADH, aldosterone, and ANP. Secretion of ADH and aldosterone rises when the plasma solute concentration increases and the plasma volume decreases. Both effects increase the amount of water retained by the kidneys. ADH and aldosterone secretion decreases when the plasma solute concentration decreases and plasma volume increases. ANP secretion rises when the plasma volume rises. (Figure 16.23, p. 623)

- Metabolic homeostasis is maintained by several hormones, including thyroid hormones, GH, glucagon, insulin, and catecholamines. When a person is fasting, thyroid hormones determine the metabolic rate, and glucagon and GH maintain blood glucose concentrations. At rest, during feeding, thyroid hormones control the metabolic rate and insulin secretion rises. During exercise, catecholamines are secreted, increasing the metabolic rate and glucagon secretion.

- The stress response involves the hormones of the HPA axis, GH, catecholamines, angiotensin-II, and glucagon. When the body is subjected to a stressor, secretion of CRH, GHRH, ACTH, GH, and glucagon rises. Both ACTH and the sympathetic nervous system stimulate secretion from the adrenal cortex and adrenal medulla. This increases the circulating levels of catecholamines, cortisol, and aldosterone, which help the body adapt to the stressor. (Figure 16.24, p. 624)

iP Play Interactive Physiology tutorial on Endocrine System: Response to Stress.

ASSESS WHAT YOU LEARNED

See answers in Appendix A.

LEVEL 1 Check Your Recall

1. Mark the following statements as true or false. If a statement is false, correct it to make a true statement.

   a. The cells of the nervous system communicate via action potentials, whereas the cells of the endocrine system communicate via hormones.

   b. Autocrine signals affect the same cells that secrete them.

   c. The pancreas, thyroid gland, and parathyroid glands secrete neurohormones.

   d. Steroid hormones are hydrophilic molecules that bind to plasma membrane proteins as part of a second-messenger system.

   e. The secretion of most hormones is regulated by a negative feedback system.

2. Which of the following is *not* a potential effect that a hormone could have on its target cell?

   a. Activating genes in the DNA

   b. Stimulating cellular division

   c. Altering the permeability of the plasma membrane

   d. All of the above are potential effects of a hormone on its target cell.

3. Which of the following hormones is/are produced by the posterior pituitary?

   a. Antidiuretic hormone

   b. Oxytocin

   c. Both a and b

   d. Neither a nor b

4. How does ADH affect the amount of water in the body, and how does it accomplish this? How does this affect the osmolarity of the blood?

5. Fill in the blanks: Hypothalamic releasing and inhibiting hormones are released into the _____ system and affect secretion from the _____ gland.

6. List the target tissues and effects of the following anterior pituitary gland hormones.

   a. Thyroid-stimulating hormone

   b. Adrenocorticotropic hormone

   c. Prolactin

   d. Gonadotropins

   e. Growth hormone

7. The thyroid gland consists of:

   a. follicle cells that secrete calcitonin.

   b. spherical thyroid follicles that contain iodine-containing colloid.

   c. parafollicular cells that produce thyroid hormones.

   d. spherical thyroid follicles that surround parathyroid hormone–secreting cells.

8. Which of the following is *not* an effect of thyroid hormones?

   a. Regulation of the metabolic rate

   b. Promotion of growth and development

   c. Thermoregulation

   d. Synergism with the parasympathetic nervous system

9. Mark the following statements as true or false. If a statement is false, correct it to make a true statement.

   a. About 90% of the thyroid hormone produced is triiodothyronine (T_3).

   b. Thyroxine (T_4) is the more active of the two thyroid hormones.

   c. Thyroid hormones are produced by follicle cells.

   d. Iodine molecules are a key component of parathyroid hormone.

   e. Excess secretion of thyroid hormones produces weight gain, cold intolerance, and slow heart rate.

10. Fill in the blanks: A rise in free T_3 and T_4 would be expected to produce a(n) _____ in TRH and TSH secretion. A decrease in free T_3 and T_4 would be expected to produce a(n) _____ in TRH and TSH secretion.

11. Which of the following statements correctly describes the role of parathyroid hormone?

    a. Parathyroid hormone regulates the metabolic rate.
    b. Parathyroid hormone increases the blood calcium ion concentration.
    c. Parathyroid hormone decreases the blood sodium ion concentration.
    d. Parathyroid hormone causes water retention by the kidneys.

12. Fill in the blanks: The outer part of the adrenal gland is the _____, which secretes _____. The inner part of the adrenal gland is the _____, which secretes _____.

13. Which of the following is *not* an effect of aldosterone?

    a. Increased excretion of hydrogen ions from the fluid in the kidneys
    b. Increased retention of sodium ions from the fluid in the kidneys
    c. Increased retention of potassium ions from the fluid in the kidneys
    d. Increased retention of water from the fluid in the kidneys

14. Cortisol is:

    a. a potent inhibitor of the inflammatory response.
    b. an important part of the stress response.
    c. Both a and b are correct.
    d. Neither a nor b is correct.

15. Describe the components of the hypothalamic-pituitary-adrenocortical axis, and explain how it regulates secretion from the adrenal gland.

16. Which of the following statements about the adrenal medulla is *false*?

    a. Secretion from the adrenal medulla is triggered by ACTH and the sympathetic nervous system.
    b. The adrenal medulla consists of glandular epithelial cells.
    c. The adrenal medulla is a modified postsynaptic sympathetic ganglion.
    d. The products of the adrenal medulla prolong the effects of the sympathetic response.

17. Explain how insulin and glucagon are antagonists.

18. Type 1 diabetes mellitus is characterized by _____, and type 2 diabetes mellitus is characterized by _____.

    a. hypoglycemia; destruction of the pancreatic β cells
    b. destruction of the pancreatic β cells; destruction of the pancreatic α cells
    c. insulin resistance; destruction of the pancreatic β cells
    d. destruction of the pancreatic β cells; insulin resistance

19. Which of the following hormones is *not* an integral part of the stress response?

    a. ACTH   d. Insulin
    b. Epinephrine   e. Glucagon
    c. Cortisol

20. Match the following hormones with their correct descriptions.

    _____ Leptin
    _____ Atrial natriuretic peptide
    _____ Melatonin
    _____ Estrogens
    _____ Erythropoietin
    _____ Testosterone

    a. Produced by pineal gland; regulates the sleep/wake cycle
    b. Produced by the kidneys; regulates red blood cell production
    c. Produced by the heart; promotes sodium ion loss in the kidneys and vasodilation
    d. Produced by the testes; promotes androgenic and anabolic actions
    e. Produced by adipose tissue; promotes satiety
    f. Produced by the ovaries; regulate the menstrual cycle and the development of secondary sex characteristics

21. Mark the following statements as true or false. If a statement is false, correct it to make a true statement.

    a. Both ADH and aldosterone increase the amount of water in the body and decrease the solute concentration of the blood.
    b. Thyroid hormones and insulin maintain blood glucose concentration during fasting.
    c. During exercise, the parasympathetic nervous system increases the metabolic rate.
    d. Insulin secretion rises during feeding; growth hormone and glucagon are secreted during feeding if protein is present in the meal.

LEVEL 2 Check Your Understanding

1. Predict the effects of a pancreatic tumor that secretes insulin. How would the effects change if the tumor secreted glucagon instead?

2. Females with hormone-secreting tumors of the adrenal cortex occasionally develop male secondary sex characteristics. Explain why this may happen.

3. A patient has a brain tumor that necessitates removal of his pituitary gland. Will its removal affect production of ADH and oxytocin? Explain.

LEVEL 3 Apply Your Knowledge

PART A: Application and Analysis

1. Ms. Reczkiewicz has her thyroid gland removed to treat hyperthyroidism. Her condition is stable after surgery, but one day later she develops symptoms of severe hypocalcemia. What has happened? How would you correct this problem?

2. A new diet guru claims hypersecretion of cortisol is the reason why so many people are obese. He is marketing a dietary supplement that is supposedly able to block cortisol secretion and lead to weight loss. How would cortisol hypersecretion cause weight gain? Would blocking normal cortisol secretion lead to weight loss? Why or why not?

3. Let's say that the dietary supplement in question 2 actually works and successfully blocks cortisol secretion. Could this drug potentially be harmful? Predict its effects on overall homeostasis.

4. Mr. Montez is a patient with type I diabetes mellitus. He presents with dizziness, rapid breathing, confusion, and weakness. You find out that he forgot to inject his normal dose of insulin this morning. Will his blood glucose concentration be normal? Explain. Your colleague suggests that Mr. Montez needs to ingest some sugar. Is this going to help him? Why or why not?

PART B: Make the Connection

5. What has likely happened to the pH of Mr. Montez's blood? What does this mean about the hydrogen ion concentration in his blood? How will his buffer systems respond to this change in pH? *(Connects to Chapter 2)*

6. You have read that aldosterone causes sodium ion retention from the kidneys. How would blocking aldosterone secretion decrease the amount of water retained from the fluid in the kidneys? *(Connects to Chapter 3)*

17

The Cardiovascular System I: The Heart

Humans have long had a fascination with the heart, and we still equate it with our most treasured emotion: love. In this chapter, we begin our tour of the **cardiovascular system,** which consists of the heart, blood vessels, and blood. The *heart* pumps **blood,** the liquid carrying oxygen and nutrients, into the **blood vessels,** a system of tubes that distributes it throughout the cardiovascular system. We'll discuss blood vessels and blood in subsequent chapters (see Chapters 18 and 19, respectively); for now let's explore this system's most prominent organ—the heart.

MODULE **17.1**

Overview of the Heart

Learning Outcomes

1. Describe the position of the heart in the thoracic cavity.
2. Describe the basic surface anatomy of the chambers of the heart.
3. Explain how the heart functions as a double pump and why this is significant.

As you know, the muscular organ called the **heart** is a pump that contracts rhythmically to deliver blood to the body. But the heart actually consists of two pumps that propel blood to two different destinations. In this module, we introduce the basic structure and location of the heart, the two locations to which the heart delivers blood, and other functions of this remarkable organ.

Location and Basic Structure of the Heart

The heart doesn't really resemble the shape we generally think of as a heart—instead, it is somewhat cone-shaped. As you can see in **Figure 17.1a**, it is situated slightly to the left side in the thoracic cavity, posterior to the sternum in the mediastinum, where it rests on the diaphragm. Its **apex,** which is the point of the cone, points toward the left hip, and its flattened **base** is its posterior side (not the inferior) facing the posterior rib cage (**Figure 17.1b**). In spite of the heart's incredible power and endurance, it is a relatively small organ, only about the size of your fist, and it generally weighs from 250 to 350 grams (slightly less than 1 pound).

The heart consists of four hollow chambers: the superior right and left **atria** (AY-tree-ah; singular, *atrium*) and the inferior right and left **ventricles** (VEN-trih-kulz)

Computer-generated image: The heart is a muscular organ that pumps blood through the body.

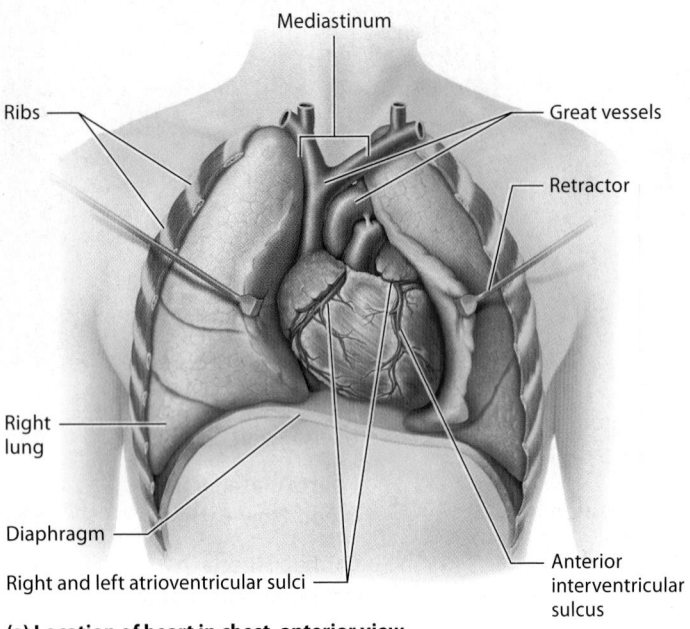

(a) Location of heart in chest, anterior view

Labels: Mediastinum; Ribs; Great vessels; Retractor; Right lung; Diaphragm; Right and left atrioventricular sulci; Anterior interventricular sulcus

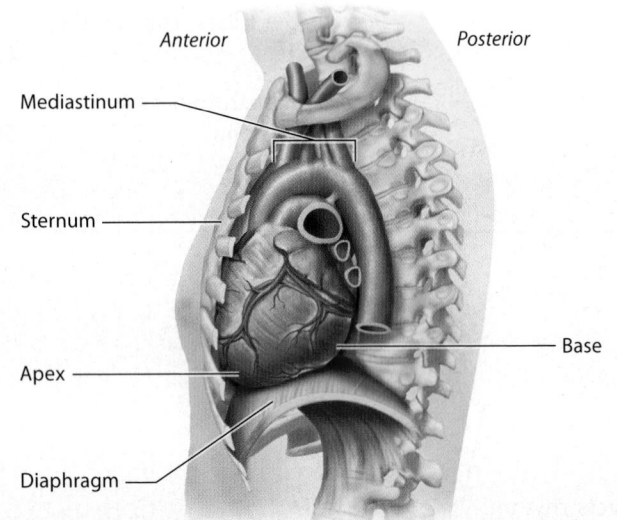

(b) Location of heart in chest, left lateral view

Labels: Anterior; Posterior; Mediastinum; Sternum; Apex; Diaphragm; Base

(c) Heart chambers, anterior view, frontal section

Labels: Chambers of the heart: Left atrium; Right atrium; Left ventricle; Right ventricle

Figure 17.1 **Location and basic anatomy of the heart in the thoracic cavity.**

(**Figure 17.1c**). Externally, an indentation known as the **atrioventricular sulcus** (ay′-tree-oh-ven-TRIK-yoo-lar SUL-kuss); *sulcus* = "groove") is found at the boundary between the atria and ventricles. Another depression, the **interventricular sulcus** (or *interventricular groove*), is located between the right and left ventricles.

Both the right and left atria receive blood from **veins,** which are blood vessels that bring blood to the heart. Blood drains from the atria to the ventricles. The ventricles pump blood into blood vessels called **arteries,** which carry blood away from the heart. The main veins and arteries that bring blood to and from the heart are known as the *great vessels*. Module 17.2 discusses the anatomy of cardiac structures in detail.

Quick Check

☐ 1. Where is the heart located, and how large is it?
☐ 2. What are the heart's upper and lower chambers called?

Functions of the Heart

The primary function of the heart—to pump blood—seems readily apparent. But of course this isn't as straightforward as it sounds. As you read earlier in this module, the heart pumps blood through two different sets of vessels, or *circuits*. We take a closer look at these circuits and some of the heart's other functions in the following subsections.

Circulation of Blood through the Pulmonary and Systemic Circuits

The heart is divided functionally into right and left sides. The right side of the heart pumps blood to the lungs, whereas the left side of the heart pumps blood to the rest of the body. The right side of the heart, shaded blue in **Figure 17.2a**, is sometimes called the *pulmonary pump* (PULL-mon-air-ee; *pulmon-* = "lungs") because it pumps blood into a series of blood vessels leading to and within the lungs, collectively called the **pulmonary circuit**. The pulmonary arteries of the pulmonary circuit deliver oxygen-poor and carbon dioxide–rich, or *deoxygenated* (dee-OK-suh-jeh-nay′-t'd), blood to the lungs.

A process called *gas exchange* takes place between the tiny air sacs in the lung, called *alveoli* (ahl-VEE-oh-lye; see Chapter 21), and the smallest vessels of the pulmonary circuit, called the pulmonary **capillaries** (KAP-uh-lair-eez). During gas exchange, oxygen diffuses from the air in the alveoli into the blood in the pulmonary capillaries, and carbon dioxide diffuses from the blood in the pulmonary capillaries to the air in the alveoli, to be expired. The veins of the pulmonary circuit then deliver this oxygen-rich, or *oxygenated*, blood to the left side of the heart. Vessels and organs that transport oxygenated blood are color-coded red in this text, and those that carry deoxygenated blood are blue.

The left side of the heart, often called the *systemic pump* and shaded red in **Figure 17.2b**, receives oxygenated blood from the pulmonary veins and pumps it into the blood vessels that serve

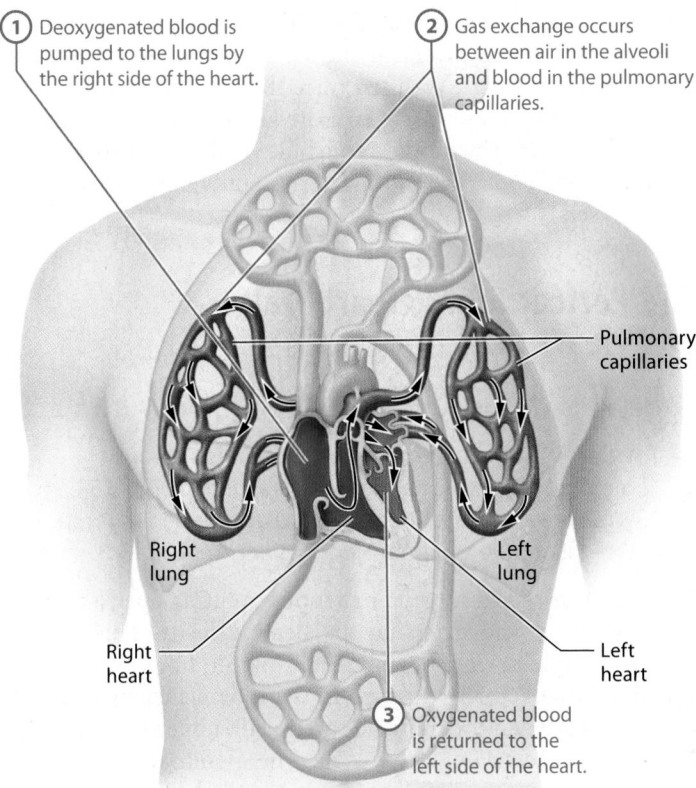

① Deoxygenated blood is pumped to the lungs by the right side of the heart.

② Gas exchange occurs between air in the alveoli and blood in the pulmonary capillaries.

Pulmonary capillaries

Right lung

Left lung

Right heart

Left heart

③ Oxygenated blood is returned to the left side of the heart.

(a) The pulmonary circuit

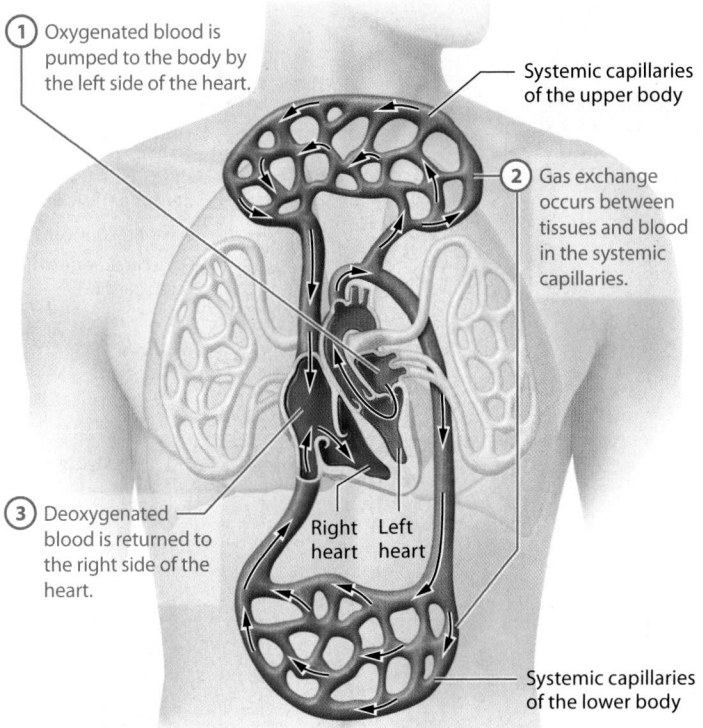

① Oxygenated blood is pumped to the body by the left side of the heart.

Systemic capillaries of the upper body

② Gas exchange occurs between tissues and blood in the systemic capillaries.

③ Deoxygenated blood is returned to the right side of the heart.

Right heart   Left heart

Systemic capillaries of the lower body

(b) The systemic circuit

Figure 17.2 The pulmonary and systemic circuits.

the rest of the body, collectively called the **systemic circuit.** In the systemic circuit, arteries deliver oxygenated blood to the smallest blood vessels, the systemic capillaries. Here gas exchange occurs again, except in reverse: Oxygen diffuses from the blood into the tissues, and carbon dioxide diffuses from the tissues into the blood. In addition to delivering oxygen, blood also delivers nutrients, picks up wastes to be excreted, and distributes hormones to their target cells throughout the body. As a result of gas exchange in the tissues, the blood is deoxygenated and the veins of the systemic circuit then deliver it back to the right side of the heart, to be pumped into the pulmonary circuit.

The pulmonary circuit is a low-pressure circuit because it delivers blood to the lungs, whereas the systemic circuit is a high-pressure circuit because it has to deliver blood to the entire rest of the body. Keep in mind that both sides of the heart function at the same time to keep blood moving through these circuits. We'll see that this has implications for the structure of the cardiac chambers, which have to push blood through these circuits.

Other Functions of the Heart

The heart has other functions besides pumping blood. One of the most important is to help maintain the homeostasis of the pressure that blood exerts on the blood vessels, also known as *blood pressure.* As we discuss in the next chapter, the rate and force of the heart's contraction are major factors that influence blood pressure and the blood flow to organs. Recall that the heart (specifically the atria) also acts as an endocrine organ and produces a hormone called *atrial natriuretic peptide* (nay′-tree-yoor-ET-ik), or *ANP* (see Chapter 16). ANP lowers blood pressure by decreasing sodium ion retention in the kidneys, therefore reducing osmotic water reabsorption and the volume and pressure of blood in the blood vessels.

Quick Check

☐ 3. From what sources does blood flow into the upper and lower chambers of the heart? What are the destinations to which these chambers pump blood?

☐ 4. Which side of the heart is considered the pulmonary pump? What is the destination of blood from the pulmonary pump? To which side of the heart does it return?

☐ 5. Which side of the heart is considered the systemic pump? What is the destination of blood from the systemic pump? To which side of the heart does it return?

Apply What You Learned

☐ 1. Predict the potential effects of right-sided heart failure (failure of the right ventricle to pump adequately).

☐ 2. How would these effects differ from those of left-sided heart failure (failure of the left ventricle to pump adequately)?

See answers in Appendix A.

MODULE 17.2

Heart Anatomy and Blood Flow Pathway

Learning Outcomes

1. Describe the layers of the pericardium.
2. Describe the location and function of the right and left coronary arteries and their branches, the cardiac veins, and the coronary sinus.
3. Describe the structure and function of the major arteries and veins entering and leaving the heart.
4. Describe the structure and function of the chambers, septa, atrioventricular valves, semilunar valves, papillary muscles, and chordae tendineae of the heart.
5. Trace the pathway of blood flow through the heart, and describe the major factors that ensure one-way flow.

As you read about in the previous module, the heart is a muscular organ that pumps blood through the body. The blood is contained within the heart's hollow spaces, its *lumen*. Like all organs, the heart is composed of different tissue layers that work together to ensure its proper functioning.

In this module we focus on the body cavity in which the heart is found, the tissue layers of the heart wall, and the detailed structure of the heart. We then trace the pathway of blood flow through the heart and discover more about the pulmonary and systemic circuits. The module concludes with a discussion of the *coronary circulation,* the set of arteries and veins that serve the heart muscle.

The Pericardium, Heart Wall, and Heart Skeleton

◄◄ Flashback

1. What is a serous membrane? What does a serous membrane produce and what is its function? (p. 153)
2. What are the components of cardiac muscle tissue? (p. 148)
3. What is simple squamous epithelium? (p. 129)

You may have seen a surgeon in a movie splitting open a patient's thoracic cavity and exposing the beating heart. This is great for movie-style drama, but it's somewhat of an embellishment. The heart isn't actually exposed in the thoracic cavity—instead, what you see when the thoracic cavity is first opened looks more like what is shown in **Figure 17.3a**. The membranous structure that you see surrounding the heart is called the **pericardium**

(a) The heart in the pericardium

Heart inside the pericardium

Right lung

Left lung

Diaphragm

Fibrous pericardium fused to the diaphragm

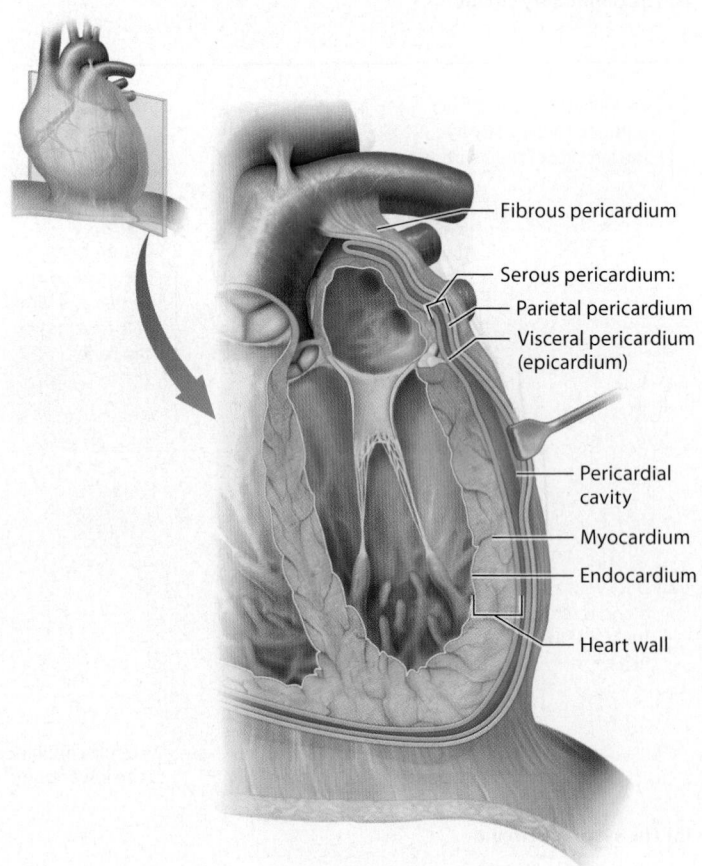

(b) The layers of the percardium and the heart wall

Fibrous pericardium
Serous pericardium:
Parietal pericardium
Visceral pericardium (epicardium)
Pericardial cavity
Myocardium
Endocardium
Heart wall

Figure 17.3 **The pericardium and the layers of the heart wall.**

(pehr'-ih-KAR-dee-um; *peri-* = "around," *cardi-* = "heart"). The pericardium is a sac with two components: the **fibrous pericardium,** a tough outer layer that attaches the heart to surrounding structures, and the **serous pericardium,** a thin inner serous membrane that produces serous fluid **(Figure 17.3b)**.

The fibrous pericardium is composed of collagen bundles that make it tough and enable it to anchor the heart to structures such as the diaphragm and the great vessels. The collagen bundles also give the fibrous pericardium low distensibility—it doesn't change shape or size considerably when stretching forces are applied. This helps to prevent the chambers of the heart from overfilling with blood. (However, see *A&P in the Real World: Cardiac Tamponade* for a discussion of when this property becomes a disadvantage.)

Let's turn now to the serous pericardium in Figure 17.3b. Notice that the outer layer of the serous pericardium, the **parietal pericardium,** is fused to the inner surface of the fibrous pericardium. The parietal pericardium encases the heart like a sac, but when it reaches the great vessels, it folds under itself and forms another layer that adheres directly to the heart. This inner layer is called the **visceral pericardium,** also known as the **epicardium** (ep'-ih-KAR-dee-um; *epi-* = "on top"). The visceral pericardium is considered the most superficial layer of the heart wall. Although it might seem as if the serous pericardium is two separate structures—the parietal and visceral pericardia—remember that they are actually one continuous structure, as shown here:

Heart — Serous fluid (inside serous pericardium)

Between the parietal and visceral pericardia we find the **pericardial cavity,** which is filled with a very thin layer of serous fluid, called **pericardial fluid.** This fluid acts as a lubricant, decreasing friction as the heart moves.

The visceral pericardium rests on top of a thin layer of areolar connective tissue that contains large fat deposits. These deposits tend to collect in the atrioventricular and interventricular sulci (which you could see in Figure 17.1a). Deep to this connective tissue we find the second and thickest layer of the heart wall, the **myocardium** (my'-oh-KAR-dee-um; *myo-* = "muscle"). The myocardium has two components: cardiac muscle tissue and a fibrous skeleton. Cardiac muscle tissue consists of **cardiac muscle cells,** or **myocytes** (MY-oh-sytz), and their surrounding extracellular matrix. Cardiac muscle cells are attached to and woven through the **fibrous skeleton,** which is composed

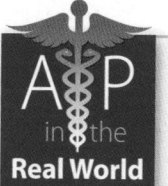

Cardiac Tamponade

If the pericardial cavity becomes filled with excess fluid, **cardiac tamponade** (tam-poh-NAHD) may result. This condition has many potential causes, including trauma, certain cancers, kidney failure, recent thoracic surgery, and HIV. Regardless of the cause, the result is the same—because the fibrous pericardium is strong but not very flexible, the excess fluid in the pericardial cavity expands toward the heart instead, applying pressure to the outside of the heart. This pressure squeezes the heart and reduces the capacity of the ventricles to fill with blood, compromising the amount of blood pumped with each beat. The treatment for cardiac tamponade may include a procedure in which the excess fluid is removed via a needle inserted into the pericardial cavity.

of dense irregular collagenous connective tissue. The fibrous skeleton has multiple functions, which include giving the cardiac muscle cells something on which to pull when they contract, providing structural support, and acting as an insulator for the heart's electrical activity. We explore the structure of cardiac muscle cells in Module 17.3.

The lumen of the heart is lined by the third and deepest layer of the heart wall, the **endocardium** (en'-doh-KAR-dee-um; *endo-* = "within"). The endocardium is composed of a special type of simple squamous epithelium called **endothelium** (en'-doh-THEE-lee-um) and several layers of connective tissue with elastic and collagen fibers. The endothelial cells of the endocardium are continuous with the endothelial cells that line blood vessels and so share many of the same functions. For example, the endothelial cells that make up the heart endothelia play a role similar to those found in the blood-brain barrier—they form a sort of "blood-heart barrier" that helps to regulate the concentration of electrolytes and other chemicals in the extracellular fluid of the myocardium.

Quick Check

☐ 1. How do the fibrous pericardium and the serous pericardium differ?

☐ 2. Where is the pericardial cavity located?

☐ 3. What are the three layers of the heart wall, and of what types of tissue are they composed?

The Coronary Circulation

Although the heart's chambers are filled with blood, the myocardium is too thick for oxygen and nutrients to diffuse from inside the chambers to all of the organ's cells. For this reason, the heart is supplied by a set of blood vessels collectively called the

(a) Distribution of the coronary arteries

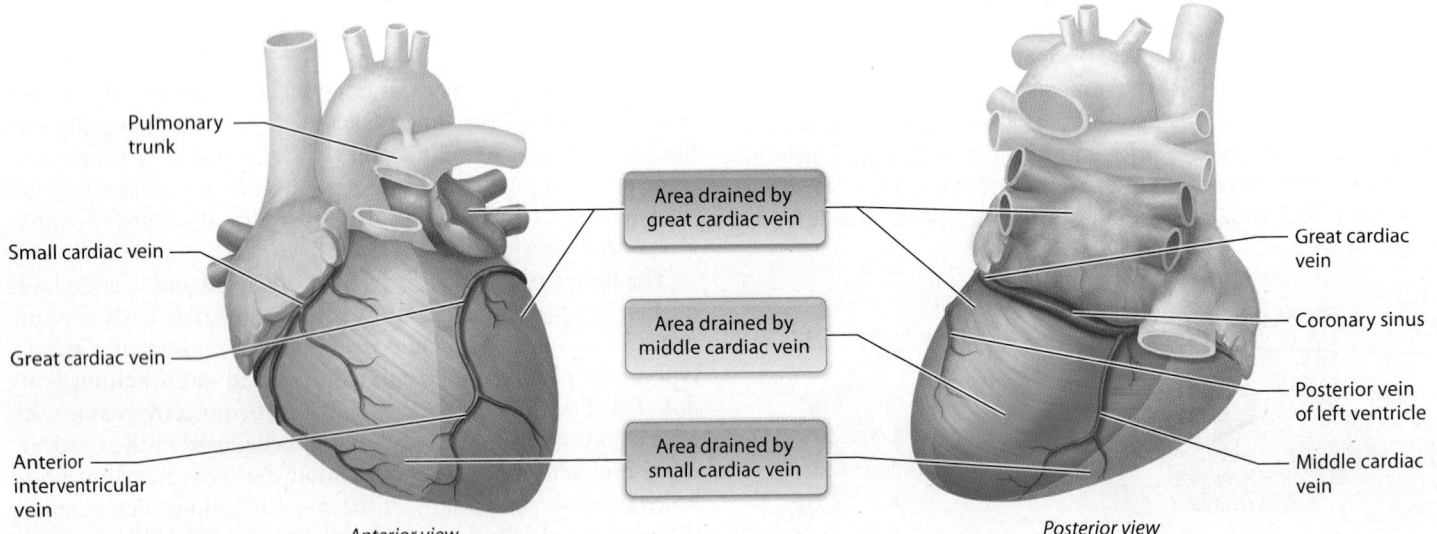

(b) Distribution of the coronary veins

Figure 17.4 The coronary circulation.

coronary circulation (**Figure 17.4**). Let's look at the anatomy of this circuit, which is part of the systemic circulation.

The Coronary Vessels

The *coronary arteries* deliver oxygenated blood to the coronary capillary beds, where gas and nutrient exchange takes place within the myocardium. Then the deoxygenated blood drains from capillaries into a series of *coronary veins.*

The Coronary Arteries The main systemic artery into which the left ventricle pumps blood is the *ascending aorta* (ay-OR-tah). Immediately after the ascending aorta emerges from the left ventricle, two branches arise: the **right and left coronary arteries.** ...e in Figure 17.4a that both arteries are posterior to another ...vessel, the *pulmonary trunk,* so the trunk had to be removed ...w us to see them. The right and left coronary arteries travel ...ght and left atrioventricular sulci, respectively.

The right coronary artery continues to travel inferiorly and laterally along the right atrioventricular sulcus, where it gives off several branches that supply the right atrium and ventricle (orange area in Figure 17.4a):

- **Marginal artery.** The largest branch is the **marginal artery,** so named because it typically arises near the inferior margin, or border, of the heart.
- **Posterior interventricular artery.** After the marginal artery branches off, the right coronary artery curls around to the posterior heart and travels in the posterior interventricular sulcus as the appropriately named **posterior interventricular artery.**

Shortly after the left coronary artery emerges from the ascending aorta, it generally branches into two vessels:

- **Anterior interventricular artery.** As you can see in Figure 17.4a, the **anterior interventricular artery** (also called the *left anterior descending artery,* or *LAD*) gets its name from the fact that it travels along the anterior interventricular sulcus. At the apex of the heart, it generally curls around and travels a short distance along the posterior interventricular sulcus. Along its course it gives off branches that supply the *interventricular septum* (the muscular wall between the ventricles), most of the anterior left ventricle, and some of the posterior left ventricle (area shaded yellow).

- **Circumflex artery.** The **circumflex artery** (SIR-kum-fleks; *circum-* = "around") also travels in a path consistent with its name, as it curves along the left atrioventricular sulcus and *flexes around* the heart. It supplies the left atrium and parts of the left ventricle (area shaded pink) and, in some people, replaces the right coronary artery in supplying the branch that becomes the posterior interventricular artery.

These arteries can be imaged with a procedure called *percutaneous coronary angiography.* In this procedure, a small tube is fed through an artery in the systemic circuit, into the ascending aorta, and finally into the right and left coronary arteries. A special dye is then injected into the arteries, and their condition is examined by x-ray, producing an image like this one:

Left coronary artery

Circumflex artery (on posterior side of heart)

Anterior interventricular artery

Anterior view

Although the arterial supply of the heart is shown fairly simply in Figure 17.4a, it's not always quite this simple. The distribution of these blood vessels is inconsistent—some people have certain arterial branches that others lack. Coronary arterial supply is further complicated by the formation of **anastomoses** (an'-ass-toh-MOH-seez; "coming together"), which are systems of channels formed between blood vessels. The coronary arteries may form anastomoses with one another, with branches from the pericardium, or even with arteries from outside the coronary circulation entirely. When blood flow to the myocardium is insufficient, occasionally new anastomoses will form to provide alternate routes of blood flow, or **collateral circulation,** to the myocardium. Collaterals formed in this manner help to protect the muscle cells from damage that could result from blocked vessels.

The Coronary Veins The most common pattern of venous flow in the coronary veins (also called the *cardiac veins*) is illustrated in Figure 17.4b. Generally, the majority of the heart's

veins empty into a large venous structure on the posterior heart, called the **coronary sinus,** which drains into the posterior right atrium. The coronary sinus receives blood from three major veins:

- **Great cardiac vein.** The large **great cardiac vein** ascends along the anterior interventricular sulcus and then travels to the posterior side of the heart along the left atrioventricular sulcus. It drains the left atrium and much of both ventricles (blue area).
- **Small cardiac vein.** The appropriately named **small cardiac vein** travels along the right atrioventricular sulcus, where it drains the right atrium and parts of the right ventricle (light purple area).
- **Middle cardiac vein.** The **middle cardiac vein** travels along the posterior interventricular groove, where it drains mostly the posterior left ventricle (blue-green area).

The anatomy of the coronary veins is even more inconsistent than that of the coronary arteries, so a great deal of variation exists from one person to another. Some of the veins that are present with some consistency, including the anterior interventricular vein and posterior vein of the left ventricle, are also labeled in Figure 17.4. In general, the coronary veins are also connected by multiple anastomoses.

Coronary Artery Disease and Myocardial Infarction

A buildup of fatty material called *plaques* (PLAKS) in the coronary arteries results in **coronary artery disease,** or **CAD,** which is the leading cause of death worldwide. CAD decreases blood flow to the myocardium, which results in inadequate oxygenation of the myocardium, a condition known as **myocardial ischemia** (iss-KEE-mee-ah). Unfortunately, CAD often causes no symptoms, so a person with this disease may be unaware that he or she has it. When symptoms do arise, they generally come in the form of chest pain, also referred to as **angina pectoris** (an-JY-nah pek-TOH-riss).

The most dangerous potential consequence of CAD is a **myocardial infarction (MI),** or *heart attack.* Generally, an MI occurs when plaques in the coronary arteries rupture and a clot forms that obstructs blood flow to the myocardium. This causes the myocardial tissue supplied by that artery to *infarct,* or die. Symptoms of an MI include chest pain that radiates along the dermatomes to the left arm or left side of the neck (an example of referred pain—see Figure 13.15), shortness of breath, sweating, anxiety, and nausea and/or vomiting. Note that women may not present with chest pain and may suffer back, jaw, or arm pain instead. Survival after an MI depends on the extent and location of the damage. Cardiac muscle cells generally do not undergo mitosis, and so after an MI the dead cells are replaced with fibrous, noncontractile scar tissue. Death of part of the myocardium increases the workload of the remaining heart muscle.

The risk factors for CAD and MI include smoking, high blood pressure, poorly controlled diabetes, high levels of certain lipids in the blood, obesity, age over 40 for males and over 50 for females, genetics, and male gender. CAD is definitively diagnosed via angiography.

Superior vena cava

Ascending aorta

Right pulmonary artery

Pulmonary trunk

Right pulmonary veins

Auricle of right atrium

Right coronary artery (in atrioventricular sulcus)

Right ventricle

Inferior vena cava

Aortic arch

Left pulmonary artery

Left pulmonary veins

Auricle of left atrium

Circumflex artery

Left coronary artery (in atrioventricular sulcus)

Great cardiac vein

Left ventricle

Anterior interventricular artery (in anterior interventricular sulcus)

Apex

(a) Anterior view of the external heart

Figure 17.5 The external anatomy of the heart.

Practice art labeling
@MasteringA&P®

Treatments for CAD include lifestyle modifications and appropriate medications. If these two approaches fail, then invasive treatments are considered. A commonly performed invasive procedure is **coronary angioplasty** (AN-jee-oh-plas'-tee), during which a balloon is inflated in the blocked artery and a piece of wire-mesh tubing called a *stent* may be inserted into the artery to keep it open. A more invasive treatment is *coronary artery bypass grafting,* during which other vessels are grafted onto the diseased coronary artery to bypass the blockage and provide an alternate route for blood flow.

Quick Check

☐ 4. Where is the right coronary artery located? What are its main branches, and what do they supply?
☐ 5. Where is the left coronary artery located? What are its main branches, and what do they supply?
☐ 6. What is the largest vein that drains the coronary circulation called, and where is it located?

The Great Vessels, Chambers, and Valves of the Heart

As we discussed in Module 17.1, the heart consists of four chambers: two atria and two ventricles. The atria receive blood from veins, and pump blood into the ventricles through structures called **valves.** These valves have flaps that close when the ventricles contract, keeping the blood from moving backward. The contracting ventricles then eject blood into arteries, which carry the blood through either the systemic or pulmonary circuit. We now turn to the structure of these chambers, the valves, and the vessels that enter and exit the heart.

The Great Vessels

The vessels that bring blood to and away from the heart, the **great vessels,** are the largest ones in the body. There are four main great vessels (**Figure 17.5**):

- **Major systemic veins: superior vena cava and inferior vena cava.** The two veins that drain the majority of the systemic circuit are the **superior and inferior venae cavae** (VEE-nee KAY-vee; singular, *vena cava*). The superior vena cava (SVC) drains deoxygenated blood from veins superior to the diaphragm, and the inferior vena cava (IVC) drains deoxygenated blood from veins inferior to the diaphragm. Note in Figure 17.5 that both veins have large openings into the posterior aspect of the right atrium, the chamber into which they drain blood.
- **Pulmonary trunk.** The large **pulmonary trunk** receives deoxygenated blood pumped from the right ventricle. The pulmonary trunk is the largest vessel of the pulmonary circuit. It is located on the anterior aspect of the heart, nearly along the midline of the heart, where the trunk originates from the right ventricle. After a short distance it splits into

Aortic arch

Left pulmonary artery

Ascending aorta

Right pulmonary artery

Auricle of left atrium

Pulmonary trunk

Circumflex artery

Auricle of right atrium

Left coronary artery (in atrioventricuar sulcus)

Right coronary artery (in atrioventricular sulcus)

Left ventricle

Right ventricle

Anterior interventricular artery (in anterior interventricular sulcus)

Apex

(b) Cadaver photo, anterior view of the external heart

Ascending aorta

Aortic arch

Superior vena cava

Left pulmonary artery

Right pulmonary artery

Left pulmonary veins

Right pulmonary veins

Auricle of left atrium

Left atrium

Right atrium

Inferior vena cava

Circumflex artery (in atrioventricular sulcus)

Auricle of right atrium

Right coronary artery (in atrioventricular sulcus)

Coronary sinus

Right ventricle

Left ventricle

Posterior interventricular artery (in posterior interventricular sulcus)

(c) Posterior view of the external heart

Figure 17.5 *(continued)*

the *right* and *left pulmonary arteries,* which bring deoxygenated blood to the right and left lungs, respectively. The pulmonary arteries branch extensively inside the lungs to become the tiny pulmonary capillaries where gases are exchanged.

- **Pulmonary veins.** Once the blood is oxygenated in the pulmonary capillaries, it returns to the heart via a set of **pulmonary veins.** Most people have four pulmonary veins—two from each lung—that drain oxygenated blood into the posterior part of the left atrium.
- **Aorta.** The **aorta** supplies the entire systemic circuit with oxygenated blood. The aorta is the largest and thickest artery in the systemic circuit—and, in fact, in the entire body. The aorta stems from the left ventricle as the *ascending aorta,* after which it curves to the left and makes a U-turn as the *aortic arch.* (Notice that the pulmonary trunk splits into the two pulmonary arteries just underneath the aortic arch.)

The Chambers of the Heart

We now turn to the heart's largest structures: its four chambers. **Figure 17.6** shows the internal anatomy of the heart. We can see that the ventricles are larger than the atria and have much thicker walls. This makes the ventricles much stronger pumps. The greater strength is needed to generate the pressure that pumps blood around the pulmonary and systemic circuits.

The Atria As you can see in Figure 17.6, the atria are not symmetrical in size, shape, or location. The **right atrium** is larger, thinner-walled, and more anterior than the **left atrium,** which is thicker-walled, somewhat smaller, and located mostly on the posterior side of the heart, where it makes up much of the heart's base and posterior surface. Externally, each atrium has a muscular pouch called the **auricle** (OR-ih-kul). The auricles, named for their resemblance to the external ear, expand to give the atria more space in which to hold blood. The auricle of the right atrium is much larger than that of the left atrium.

The internal surface of the right atrium has muscular ridges on its anterior side, called **pectinate muscles** (PEK-tin-et; *pectinate* = "combed"). On the smooth posterior surface of the atrium we find the openings of the superior and inferior venae cavae, and the coronary sinus. The left atrium is mostly formed by the pulmonary veins, and internally its walls are smooth.

The two atria are separated by a thin wall, the **interatrial septum.** Notice in Figure 17.6 that there is a small indentation in the interatrial septum, called the **fossa ovalis.** This indentation is the remnant of a hole known as the **foramen ovale** (for-AY-men oh-VAH-lay), which is present in the interatrial septum of the fetal

Ascending aorta
Pulmonary valve
Opening of superior vena cava
Right atrium
Fossa ovalis
Tricuspid valve
Right ventricle
Trabeculae carneae

Pulmonary trunk
Openings of left pulmonary veins
Left atrium
Aortic valve
Mitral (bicuspid) valve
Chordae tendineae
Papillary muscle
Left ventricle
Interventricular septum

Practice art labeling @MasteringA&P®

Figure 17.6 The internal anatomy of the heart, anterior dissection.

heart. The foramen ovale directs blood from the right atrium directly to the left atrium (bypassing the right ventricle and the pulmonary circulation, because fetal lungs are not yet functional) and normally closes shortly before or after birth (the foramen ovale is discussed in the development chapter—see Chapter 27).

The Ventricles Like the atria, the right and left ventricles are asymmetrical, as you can see here:

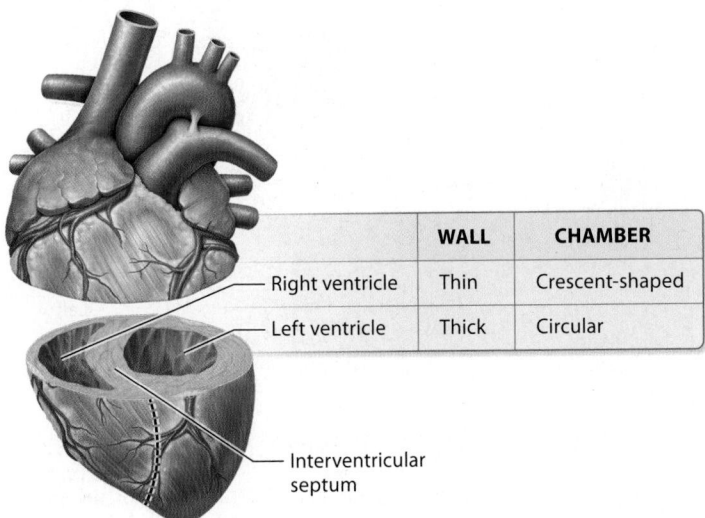

	WALL	CHAMBER
Right ventricle	Thin	Crescent-shaped
Left ventricle	Thick	Circular

Observe that the **right ventricle** is wider and has thinner walls than the **left ventricle**. This is an example of the Structure-Function Core Principle (p. 25)—because of the pressure differences in the pulmonary and systemic circuits, the right ventricle has little resistance against which to pump, whereas the left ventricle pumps against much greater resistance. This means that the left ventricle has to work harder than the right ventricle and, as a result, has greater muscle mass. In fact, the walls of the left ventricle are about three times thicker than those of the right ventricle. Notice also that the left ventricle is longer and circular in cross-section, whereas the right ventricle is shorter and crescent-shaped in cross-section.

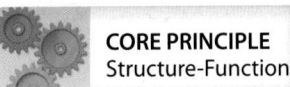
CORE PRINCIPLE
Structure-Function

Internally, both ventricles have a ridged surface created by irregular protrusions of cardiac muscle tissue (see Figure 17.6). These ridges are collectively referred to as **trabeculae carneae** (trah-BEK-yoo-lee kar-NEE-ee; "beams of flesh"). Each ventricle also contains finger-like projections of muscle, the **papillary muscles** (PAP-ih-lair-ee; *papill-* = "nipple"), which attach by tendon-like cords called *chordae tendineae* (KOR-dee ten-din-EE-ee) to the valves located between the atria and the ventricles. The trabeculae carneae and the papillary muscles ensure that the valves work properly, discussed in Module 17.4. A thick, muscular wall, the **interventricular septum,** separates the right and left ventricles. This septum contracts with the rest of the ventricular muscle and helps to expel blood into the pulmonary trunk and aorta.

The Valves of the Heart

Blood flow through the heart must occur in only one direction so that deoxygenated blood goes to the pulmonary circuit and oxygenated blood goes to the systemic circuit. Valves ensure this

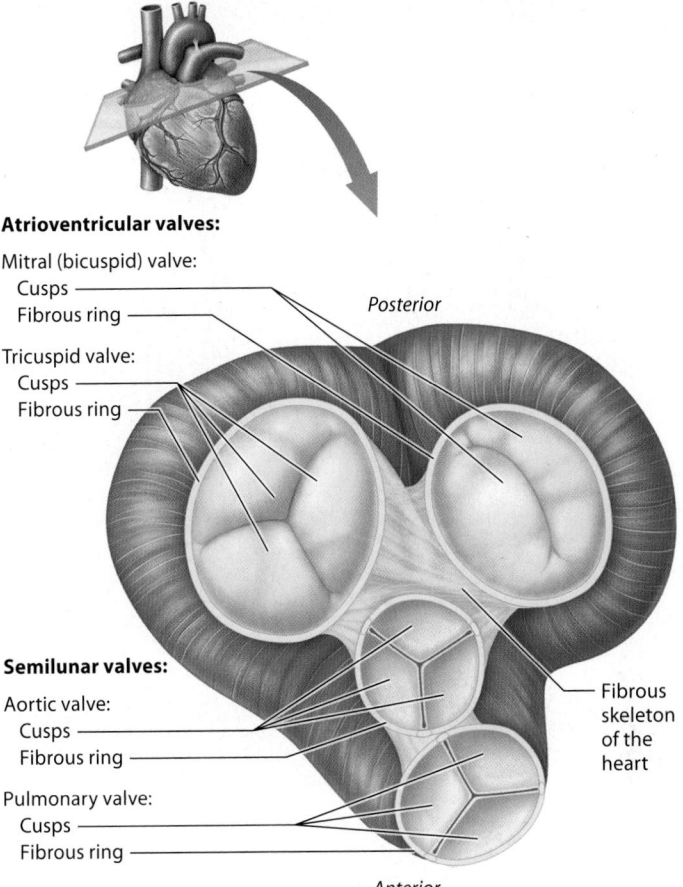

Atrioventricular valves:

Mitral (bicuspid) valve:
Cusps
Fibrous ring

Tricuspid valve:
Cusps
Fibrous ring

Posterior

Semilunar valves:

Aortic valve:
Cusps
Fibrous ring

Pulmonary valve:
Cusps
Fibrous ring

Fibrous skeleton of the heart

Anterior

(a) The atrioventricular and semilunar valves

(b) The pulmonary valve, closed (left) and open (right)

Figure 17.7 Anatomy of the atrioventricular and semilunar valves.

by preventing blood from flowing backward. There are two types of valves, which we examine next.

The Atrioventricular Valves Backflow of blood generally doesn't occur in the veins draining into the atria, as the atria are under very low pressure and blood mostly flows into the atria with the help of gravity and the pressure in the veins. Therefore, there are no valves present between the atria and the veins that drain blood into them. However, backflow of blood from the ventricles into the atria could occur because their forceful contractions could drive blood backward into the atria. This backward flow is prevented by valves between the atria and the ventricles, called the right and left **atrioventricular (AV) valves** (Figure 17.7a).

The Big Picture of Blood Flow through the Heart

Figure 17.8

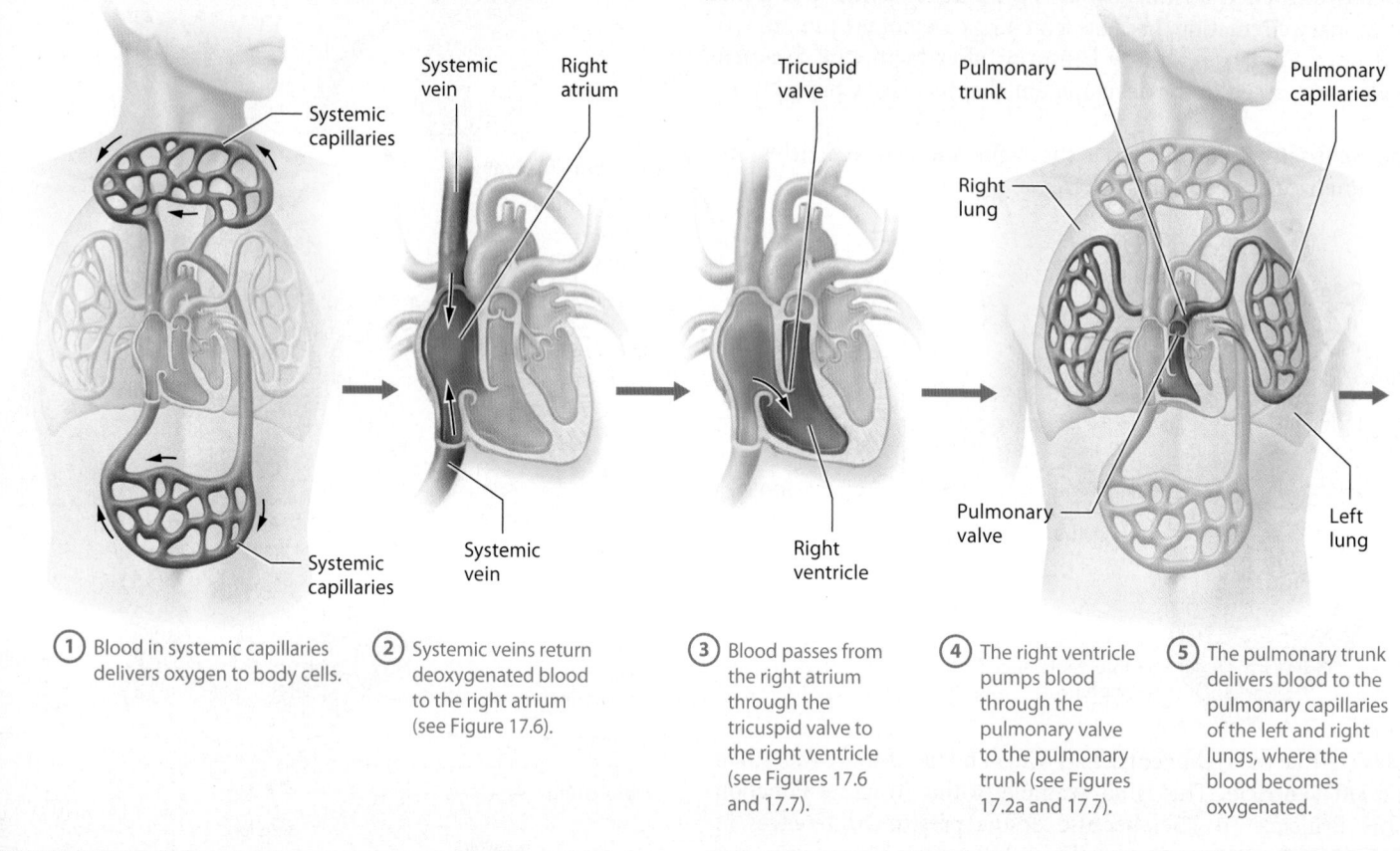

Systemic capillaries / **Systemic vein** / **Right atrium** / **Tricuspid valve** / **Pulmonary trunk** / **Right lung** / **Pulmonary capillaries** / **Systemic capillaries** / **Systemic vein** / **Right ventricle** / **Pulmonary valve** / **Left lung**

1. Blood in systemic capillaries delivers oxygen to body cells.

2. Systemic veins return deoxygenated blood to the right atrium (see Figure 17.6).

3. Blood passes from the right atrium through the tricuspid valve to the right ventricle (see Figures 17.6 and 17.7).

4. The right ventricle pumps blood through the pulmonary valve to the pulmonary trunk (see Figures 17.2a and 17.7).

5. The pulmonary trunk delivers blood to the pulmonary capillaries of the left and right lungs, where the blood becomes oxygenated.

You can think of these valves as "inflow" valves, as blood must flow through them to enter the ventricles.

The two AV valves consist of flaps, called *cusps,* that are composed of endocardium overlying a core of collagenous connective tissue. Each valve is named for the number of cusps it contains: The **tricuspid valve** between the right atrium and right ventricle contains three cusps, and the **bicuspid valve** between the left atrium and left ventricle has two cusps. Note that the bicuspid valve is more commonly called by its clinical name, **mitral valve** (MY-tral), and this is the term used in this book. Figure 17.7a illustrates how the fibrous skeleton of the heart supports these atrioventricular valves—two rings extend from a central core, and each ring surrounds the opening of a valve.

Attached to the inferior end of each cusp we find the **chordae tendineae** ("tendon strings"). The thin chordae tendineae attach to papillary muscles that contract just before the ventricles begin ejecting blood. This creates tension on the chordae tendineae, which keeps the valves closed. These actions prevent the cusps from *everting,* or ballooning back into the atria, when the pressure in the ventricles increases.

The Semilunar Valves Backflow of blood can also become a problem in the pulmonary artery and the aorta, because blood begins to flow backward when the ventricles relax as a result of the higher pressure in the arteries and gravity. The blood is stopped from completely flowing back into the ventricles by two valves called the **semilunar (SL) valves** (seh-mee-LOO-nur). You can think of these valves as "outflow" valves, since blood passes through them as it flows out of the heart.

The word semilunar refers to the half-moon shape of the valves' three cusps, which are also composed of endocardium and a central collagenous core. The two SL valves are named according to the artery in which they reside: The **pulmonary valve** is located between the right ventricle and pulmonary trunk, and the **aortic valve** is posterior to it between the left ventricle and the aorta. The heart's fibrous skeleton supports the aortic and pulmonary valves in much the same way it supports the AV valves. You can see what the pulmonary valve looks like closed and open in **Figure 17.7b**. No chordae tendineae or papillary muscles attach to the SL valves.

As you can imagine, dysfunctional valves can lead to many problems. *A&P in the Real World: Valvular Heart Disease* describes what happens when valves are not working correctly. Often, dysfunctional valves must be replaced surgically with either a mechanical valve or a real valve from a human cadaver or a pig.

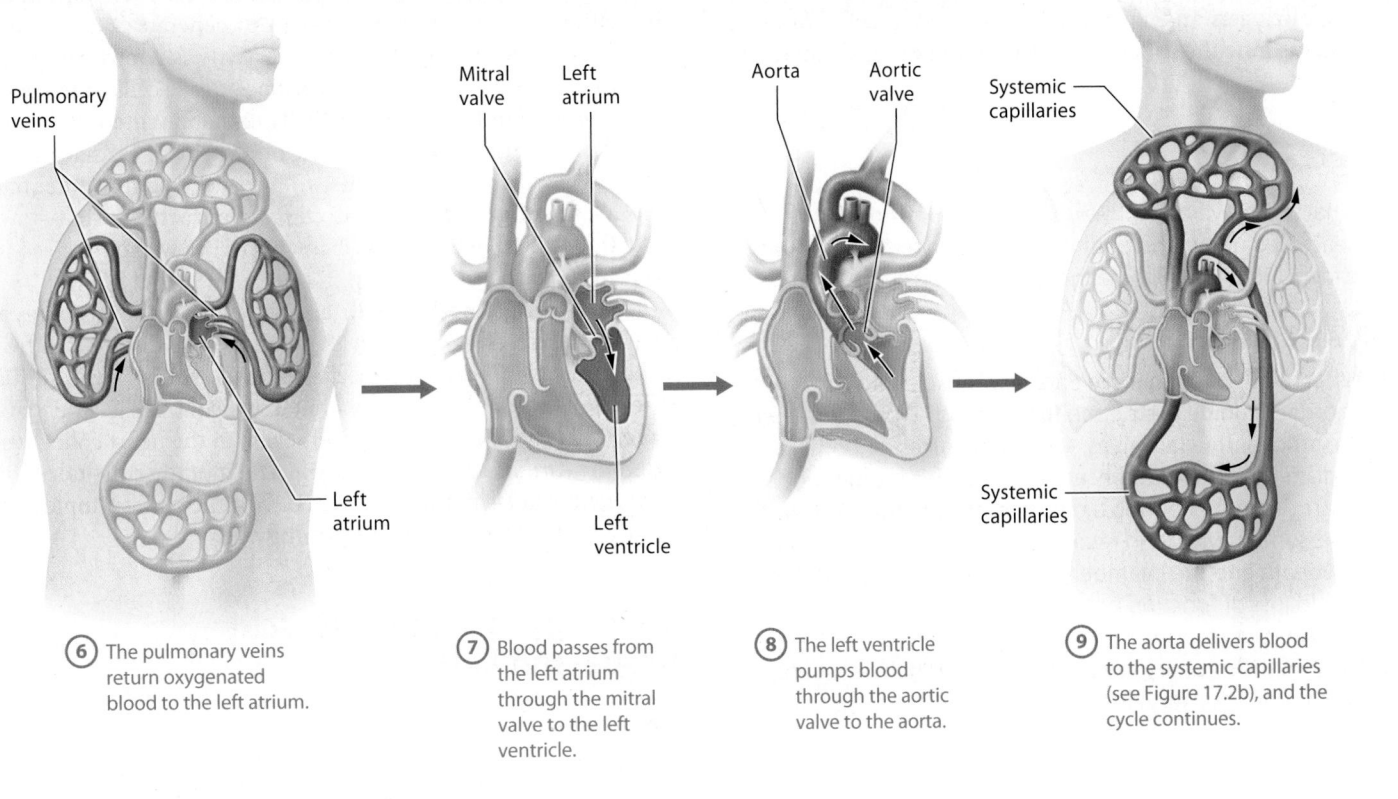

(6) The pulmonary veins return oxygenated blood to the left atrium.

(7) Blood passes from the left atrium through the mitral valve to the left ventricle.

(8) The left ventricle pumps blood through the aortic valve to the aorta.

(9) The aorta delivers blood to the systemic capillaries (see Figure 17.2b), and the cycle continues.

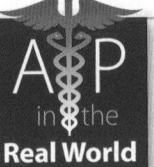

Real World

Valvular Heart Diseases

Valvular heart diseases impair the function of one or more of the valves. Such diseases may be congenital (present at birth) or acquired from a disease process such as infection, cancer, or disorders of the immune system. There are generally two major types of valvular defects: **insufficiency** and **stenosis.** An insufficient valve fails to close fully and so allows blood to leak backward. A stenotic valve is one in which calcium deposits have built up in the cusps, making them hard and inflexible. Blood flows through a stenotic valve with difficulty, and often the heart has to pump harder to eject blood through it.

Both types of valvular heart diseases may cause an audible "swooshing" of blood when the heart beats, called a *heart murmur.* Other signs and symptoms vary with the type and severity of the disease but may include enlargement of the heart, fatigue, dizziness, and heart palpitations. The mitral and aortic valves are the ones most commonly affected by valvular heart disease.

Quick Check

☐ 7. What are the four main great vessels? From which structure(s) does each vessel receive blood? To which structure(s) does each vessel deliver blood?

☐ 8. How do the right and left ventricles differ in size and shape?

☐ 9. Why do you think it is important for heart valves to ensure that blood flows in only one direction in the heart?

Putting It All Together: The Big Picture of Blood Flow through the Heart

Now that we've discussed the heart's anatomy, let's draw a "big picture" view of the pathway of blood flow through the heart, the pulmonary circuit, and the systemic circuit, shown in **Figure 17.8.** We begin our pathway in the systemic capillaries, where blood becomes deoxygenated, and trace it through the major structures of the heart and circulation. This cycle then repeats continuously. If you find yourself forgetting parts of this pathway, just remember the whole point of blood flowing through the heart: The right side of the heart must pump blood into the lungs so it can become oxygenated, and the left side of the heart must pump blood into the rest of the body so it can give oxygen and nutrients to the body's cells. Module 17.4 explains how the functions of the chambers and valves are coordinated.

Quick Check

☐ 10. What is the overall pathway of blood flow through the heart, beginning and ending with the systemic capillaries, and including the valves and great vessels?

Apply What You Learned

☐ 1. What might happen if the papillary muscles and/or chordae tendineae stopped functioning? Would this affect the atrioventricular valves, semilunar valves, or both? Explain the potential consequences of this problem.

☐ 2. Heart tissue dies during a myocardial infarction (MI), and a person's survival and recovery depend on the extent of cell death and the chamber(s) involved. In which chamber would loss of function be most damaging to survival? Explain your response.

☐ 3. How might coronary artery anastomoses help a patient with coronary artery disease?

See answers in Appendix A.

MODULE **17.3**

Cardiac Muscle Tissue Anatomy and Electrophysiology

Learning Outcomes

1. Differentiate between a cardiac pacemaker and contractile cell.
2. Describe the histology of cardiac muscle tissue, and differentiate it from that of skeletal muscle.
3. Describe the phases of the cardiac muscle action potential, including the ion movements that occur in each phase, and explain the importance of the plateau phase.
4. Contrast the way action potentials are generated in cardiac pacemaker cells, cardiac contractile cells, and skeletal muscle cells.
5. Describe the parts of the cardiac conduction system, and explain how the system functions.
6. Identify the waveforms in a normal electrocardiogram (ECG), and relate the ECG waveforms to atrial and ventricular depolarization and repolarization and to the activity of the conduction system.

As you are no doubt aware, your heart contracts without any conscious effort on your part. Like all muscle cells, cardiac muscle cells contract in response to electrical excitation in the form of

action potentials. However, unlike skeletal muscle fibers and many smooth muscle cells, cardiac muscle cells do not require stimulation from the nervous system to generate action potentials. This is because their electrical activity is coordinated by a very small, unique population of cardiac muscle cells called **pacemaker cells.** These cells rhythmically and spontaneously generate action potentials that trigger the other type of cardiac muscle cell, known as **contractile cells,** to also have action potentials. This gives cardiac muscle the property of **autorhythmicity** (aw′-toh-rith-MISS-ih-tee)—it sets its own rhythm without a need for input from the nervous system.

We begin this module with an overview of the histology of cardiac muscle cells and their similarities and differences from skeletal muscle fibers. The discussion then shifts to the cardiac action potentials that are generated by contractile and pacemaker cells and the group of pacemakers in the heart collectively called the *cardiac conduction system*. The module concludes with an overview of *electrocardiograms*, one of the clinical topics pertaining to the heart's **electrophysiology.**

Histology of Cardiac Muscle Tissue and Cells

◄◄ **Flashback**

1. What are gap junctions and desmosomes? (p. 127)
2. What is an electrical synapse? (p. 407)
3. What is excitation-contraction coupling? (p. 354)

Many components of the cardiac muscle cells illustrated in **Figure 17.9** should look familiar to you. Compare the diagram of cardiac muscle cells in this figure to the diagram of skeletal muscle fibers in Figure 10.4. Cardiac muscle cells, like skeletal muscle fibers, have *striations*, which means that the cells have alternating light and dark bands when viewed under the microscope. As with skeletal muscle fibers, these striations are due to the arrangement of the contractile proteins within the cardiac muscle cell. Both types of muscle tissue also have transverse tubules and a sarcoplasmic reticulum. In an example of the Structure-Function Core Principle (p. 25), these structural similarities reflect physiological similarities—both skeletal and cardiac muscle tissues have the same general function, to generate tension, and both generate tension through the sliding-filament mechanism of contraction (see Module 10.2 for a review of this mechanism).

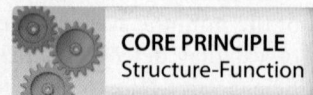

CORE PRINCIPLE
Structure-Function

However, in spite of these similarities, there are notable differences between cardiac muscle cells and skeletal muscle fibers. Cardiac muscle cells are typically branched cells with a single nucleus, and they are shorter and wider than skeletal muscle fibers. In addition, cardiac muscle cells contain abundant myoglobin, a protein that carries oxygen, and nearly half of their cytoplasmic volume is composed of mitochondria. Both of these reflect their high energy demands.

Figure 17.9 reveals another difference between cardiac and skeletal muscle cells: Structures called **intercalated discs**

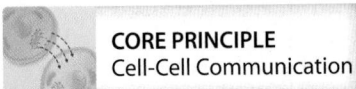

Figure 17.9 Cardiac muscle cells.

(in-TER-kah-lay′-t'd; "discs inserted between") join adjacent cardiac muscle cells. Intercalated discs join pacemaker cells to contractile cells, and contractile cells to one another. Intercalated discs contain *desmosomes* that hold the cardiac muscle cells together and *gap junctions* that allow ions to rapidly pass from one cell to another, permitting communication among cardiac muscle cells (an example of the Cell-Cell Communication Core Principle, p. 27). These gap junctions are examples of the electrical synapses that we discussed in the nervous system chapter (look back at Figure 11.22).

> **CORE PRINCIPLE**
> Cell-Cell Communication

Like skeletal muscle fibers and other excitable cells, a cardiac muscle cell contains selective gated ion channels in its sarcolemma, including the following:

- **Voltage-gated sodium ion channels.** The voltage-gated sodium ion channels in the cardiac muscle cell sarcolemma open in response to voltage changes across the membrane. These channels are found in all cardiac muscle cells except certain pacemaker cells.
- **Calcium ion channels.** Calcium ion channels are present in all cardiac muscle cells. These channels demonstrate voltage-gated opening but time-gated closing, meaning that they close after a certain period regardless of the voltage.

- **Potassium ion channels.** All types of cardiac muscle cells have one or more types of potassium ion channels. Some of these channels are ligand-gated, whereas others are voltage-gated.

The opening and closing action of these ion channels is responsible for both the pacemaker and contractile cardiac action potentials. Before moving on to these topics, let's review some electrophysiology terms.

⬆ STUDY BOOST

Revisiting Electrophysiology

Yes, it's time to go back again to everyone's favorite topic in A&P—electrophysiology. Before you delve into cardiac electrophysiology, first bear in mind that the majority of this material isn't new to you. In fact, this is the third time that we have explored this topic in this book. Second, remember that although cardiac muscle cells may look exotic, they are simply excitable cells that behave in a similar way to the other excitable cells you have studied, including skeletal muscle fibers, smooth muscle cells, and neurons. Following are some simplified "refresher" definitions that you may find helpful.

- **Voltage:** A difference in electrical potential between two points.
- **Membrane potential:** The voltage or charge difference that exists across the membranes of all cells, including excitable cells.

- **Resting membrane potential:** The membrane potential of an excitable cell at rest (that is not being stimulated), which averages between –60 and –90 mV.
- **Ion gradient:** A difference in the concentration of ions across a plasma membrane.
- **Current:** The flow of ions or electrons with a chemical or electrical gradient.
- **Depolarization:** A change in resting membrane potential to a *less negative* value. This change occurs when positive charges (generally, sodium and/or calcium ions) rush into the cell.

- **Repolarization:** The return of a cell to its negative resting membrane potential. This occurs when positive charges (potassium ions) leave the cell.

Last but not least, let's review those all-important ion gradients and the Gradients Core Principle (p. 26) in a bit more depth. Remember that the concentration of sodium and calcium ions in the extracellular fluid (ECF) is *higher* than that in the cytosol, and the concentration

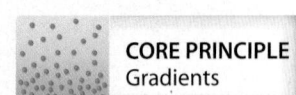

CORE PRINCIPLE
Gradients

① **Rapid depolarization phase:**
Voltage-gated Na⁺ channels activate and Na⁺ enter, rapidly depolarizing the membrane.

② **Initial repolarization phase:**
Na⁺ channels are inactivated and some K⁺ channels open; K⁺ leak out, causing a small initial repolarization.

③ **Plateau phase:**
Ca²⁺ channels open and Ca²⁺ enter as K⁺ exit, prolonging the depolarization.

④ **Repolarization phase:**
Na⁺ and Ca²⁺ channels close as K⁺ continue to exit, causing repolarization.

Sequence the steps
@MasteringA&P®

Figure 17.10 A contractile cell action potential.

of potassium ions in the ECF is *lower* than that in the cytosol. For this reason, sodium and calcium ions tend to follow their concentration gradients to enter a cell when their channels open, whereas potassium ions follow their concentration gradient and leave the cell when their channels open. The sodium and potassium ion gradients are maintained by the Na^+/K^+ pump and the calcium ion gradient by a separate active transport pump. ■

Quick Check

☐ 1. How do pacemaker and contractile cells differ? What is autorhythmicity?

☐ 2. What are intercalated discs? What is their function?

Electrophysiology of Cardiac Muscle Tissue

◀◀ **Flashback**

1. How does an action potential occur in a skeletal muscle fiber? (p. 352)
2. What are tetanus and the refractory period? (p. 364)
3. What is the sliding-filament mechanism of contraction, and what role do calcium ions play in muscle contraction? (p. 355)

As we discussed earlier, pacemaker cells undergo rhythmic, spontaneous depolarizations that lead to action potentials. These pacemaker action potentials are spread quickly through the heart by a group of interconnected pacemaker cells collectively called the *cardiac conduction system*. The action potentials are transmitted from the pacemaker cells to contractile cells through the intercalated discs that unite them. The gap junctions in these discs allow the electrical activity generated by the pacemaker cells to rapidly spread to all cardiac muscle cells via electrical synapses. This permits the heart to contract as a unit and produce a coordinated *heartbeat*. For this reason, the cells of the heart are sometimes referred to as a *functional syncytium* (sin-SISH-um), the term for a large, multinucleated cell. Let's look at the electrophysiology and contraction of both pacemaker and contractile cardiac muscle cells.

Activity of Contractile Cells

Contractile cells make up the great majority (99%) of cardiac muscle cells. Our discussion of cardiac electrophysiology begins with contractile cell action potentials because they are more similar to the action potentials you learned about in the muscle and nervous tissue chapters. Then we go on to see how excitation-contraction coupling works in contractile cells.

Contractile Cell Action Potentials As with skeletal muscle fibers, an action potential in a contractile cardiac muscle cell results from a reversal in membrane potential—the inside of the plasma membrane swings from negative (about -85 mV) to momentarily positive (ranging from 0 to $+20$ mV). These changes happen because of voltage-gated ion channels in the sarcolemma, and because of unequal concentrations of sodium and potassium

ions on either side of the membrane that drive those ions in or out of the cell through the channels.

The sequence of events of a contractile cell action potential is diagrammed in **Figure 17.10**, and proceeds as follows:

① **Rapid depolarization phase.** In response to pacemaker cell action potentials that cause voltage changes in adjacent cells, voltage-gated sodium ion channels in the sarcolemma are activated. This causes a rapid and massive influx of sodium ions, which leads to rapid membrane depolarization.

② **Initial repolarization phase.** Note in Figure 17.10 that there is a small, initial repolarization immediately after the depolarization spike. This small repolarization is due to the abrupt inactivation of the sodium ion channels and to a very small outflow of potassium ions through selected potassium ion channels that are open only briefly.

③ **Plateau phase.** If repolarization were to continue to completion at step ②, the action potential would look nearly identical to that of a skeletal muscle fiber. However, as you can see in Figure 17.10, the depolarization is sustained at about 0 mV, which is known as the **plateau phase.** This critically important phase is mostly due to the slow opening of calcium ion channels and the resulting influx of calcium ions. As the calcium ions enter, the potassium ions from step ② continue to leave, so there is very little net change in membrane potential. The calcium ion channels are also slow to close, which allows the plateau phase to last much longer than the initial depolarization phase. Note that certain sodium ion channels remain open during this phase, which contributes to the plateau.

④ **Repolarization phase.** The final phase of the action potential occurs when both the sodium and calcium ion channels return to their resting states and most of the potassium ion channels open. This allows positively charged potassium ions to exit the cardiac muscle cell, and the membrane potential returns to its resting value of about -85 mV.

The sequence of events of a contractile cell action potential resembles that of a skeletal muscle fiber action potential (see Module 10.3 for a review) with one important exception: the plateau phase. The importance of this phase cannot be overemphasized. Compare the timing of action potentials in skeletal muscle fibers and cardiac contractile cells:

Action potential of skeletal muscle fiber

Action potential of cardiac contractile cell

If cardiac action potentials lasted only about 1–5 msec, like skeletal muscle fiber action potentials, this would lead to a resting heart rate that was about 15 times faster than it should be. The plateau phase lengthens the cardiac action potential to about 200–300 msec, which slows the heart rate, providing the time required for the heart to fill with blood. It also increases the strength of the heart's contraction. The prolonged action potential makes the muscle twitch last a long time, so it can develop more force, and it allows more calcium ions to enter the cell, which are needed for the cell's contraction via the sliding-filament mechanism.

The plateau phase also effectively prevents *tetany* (sustained contraction) in the heart by lengthening the **refractory period.** Recall that the refractory period is the time during which an excitable cell cannot be stimulated to contract again (see Chapter 10). Skeletal muscle fibers have such short refractory periods that tetanus can occur with repeated stimulation. The refractory period in cardiac muscle cells is so long, however, that the cells cannot maintain a sustained contraction. This allows the heart to relax and the ventricles to refill with blood before the cardiac muscle cells are stimulated to contract again.

Mechanism of Contractile Cell Contraction: Excitation-Contraction Coupling

The mechanism for cardiac muscle cell contraction is very similar to that of skeletal muscle fiber contraction, and it occurs by the sliding-filament mechanism. In a stimulated cardiac muscle cell, the depolarization propagates through the sarcolemma and dives into the cell along the T-tubules, which causes the sarcoplasmic reticulum to release calcium ions. These ions bind to troponin, which allows actin and myosin to bind and a crossbridge cycle to begin.

However, there is an important difference between the contraction of the two cell types. Skeletal muscle fibers release nearly all of the calcium ions required for contraction from their sarcoplasmic reticulum. For this reason, extracellular calcium ion concentration has little effect on the strength of contraction of a skeletal muscle fiber. However, the sarcoplasmic reticulum of cardiac muscle cells is much less extensive and does not release enough calcium ions during the action potential to produce a reliably strong contraction. The rest of the calcium ions needed for contraction diffuse into the cell through calcium ion channels from the extracellular fluid in the T-tubules. For this reason, the concentration of calcium ions in the cardiac extracellular fluid plays a significant role in determining the strength of the contraction, illustrating one reason why calcium ion homeostasis is so critical.

Pacemaker Cells and the Cardiac Conduction System

Pacemaker cells make up only about 1% of the total number of cardiac muscle cells. There are three populations of these cells in the heart that are capable of spontaneously generating action potentials, thereby setting the pace of the heart. These three cell populations are collectively called the **cardiac conduction system.** After looking at how pacemaker cells generate action potentials, we'll cover the components of the cardiac conduction system.

Pacemaker Cell Action Potentials

Figure 17.11 shows that the electrical tracing produced by an action potential of a pacemaker cell, called a **pacemaker potential,** is much different in appearance from that of a contractile cell. For one, the depolarization in a pacemaker cell occurs much more slowly. This is due in part to the lack of voltage-gated sodium ion channels in the pacemaker sarcolemma. In addition, pacemaker cell action potentials lack a plateau phase. Finally, notice that the pacemaker cell membrane potential oscillates—that is, it never remains at a resting level and instead occurs in a cycle, with the last event triggering the first. This is possible because of nonspecific cation channels that are unique to pacemaker cells. These channels open when the membrane reaches a minimum potential after repolarization, and their opening starts a slow depolarization. Figure 17.11 shows how this works:

① **Slow initial depolarization phase.** We start a pacemaker potential with the plasma membrane in a *hyperpolarized* state—it is at its minimum membrane potential. This state opens nonspecific cation channels in the membrane. These channels allow more sodium ions to leak into the cell than potassium ions to leak out, which results in an overall slow depolarization to threshold.

② **Full depolarization phase.** When the membrane reaches threshold, voltage-gated calcium ion channels open, allowing calcium ions to enter the cell and thereby causing the membrane to fully depolarize. This causes the slow upstroke of the tracing that you see in Figure 17.11. Notice that this upstroke is not as rapid as that in contractile cell depolarization. This is because calcium ions enter more slowly than sodium ions as a result of differences in channel structure.

③ **Repolarization phase.** Recall that calcium ion channels are time-gated for closing, so after a certain time (about 100–150 msec), they close. At the same time, voltage-gated potassium ion channels begin to open. This allows potassium ions to exit the cell, and the membrane begins to repolarize.

④ **Minimum potential phase.** Potassium ion channels remain open until the membrane reaches its minimum potential. When this happens, the membrane is hyperpolarized, which opens the nonspecific cation channels, and the cycle begins again.

The events of a pacemaker cell action potential are such that the end of step ④ is what causes step ① to begin. It is this self-perpetuating, cyclic nature that allows pacemaker cells to pace the heart.

Anatomy of the Cardiac Conduction System

The cardiac conduction system includes the following three populations of pacemaker cells (**Figure 17.12**):

- **Sinoatrial node.** The **sinoatrial node** (sy′-noh-AY-tree-ul), or **SA node,** is located in the upper right atrium slightly inferior and lateral to the opening of the superior vena cava. Under normal conditions, the SA node has the fastest intrinsic rate of depolarization—about 60 or more times per minute, a rate that is subject to influence from the sympathetic and parasympathetic nervous systems.

- **Atrioventricular node.** The **atrioventricular node,** or **AV node,** is a cluster of pacemaker cells located posterior and medial to the tricuspid valve. It is slower than the SA node, with an intrinsic rate of only about 40 action potentials per minute.

1 **Slow initial depolarization phase:** More cations leak in than leak out through nonspecific cation channels in the plasma membrane, causing the membrane to slowly depolarize to threshold.

2 **Full depolarization phase:** At threshold, voltage-gated Ca^{2+} channels open, and Ca^{2+} enter the cell, causing the membrane to fully depolarize.

3 **Repolarization phase:** Ca^{2+} channels close and voltage-gated K^+ channels open, causing K^+ outflow and membrane repolarization.

4 **Minimum potential phase:** K^+ channels remain open, and the membrane hyperpolarizes. This opens nonspecific cation channels, and the cycle repeats.

Sequence the steps
@MasteringA&P®

Figure 17.11 A pacemaker cell action potential.

- **Purkinje fiber system.** The slowest group of pacemaker cells is collectively called the **Purkinje fiber system** (per-KIN-jee); its cells depolarize only about 20 times per minute. The cells of this system are sometimes called *atypical pacemakers*, because their action potentials rely on different ion channels and they function in a slightly different way. Note in Figure 17.12 that there are three components to the Purkinje fiber system: the AV bundle, the right and left bundle branches, and the terminal branches of the Purkinje fibers. The **atrioventricular bundle (AV bundle)** penetrates the heart's fibrous skeleton in the inferior interatrial septum and the superior interventricular septum. The **right and left bundle branches** course along the right and left sides of the interventricular septum, respectively. Then the **terminal branches** penetrate the ventricles and finally come into contact with the contractile cardiac muscle cells.

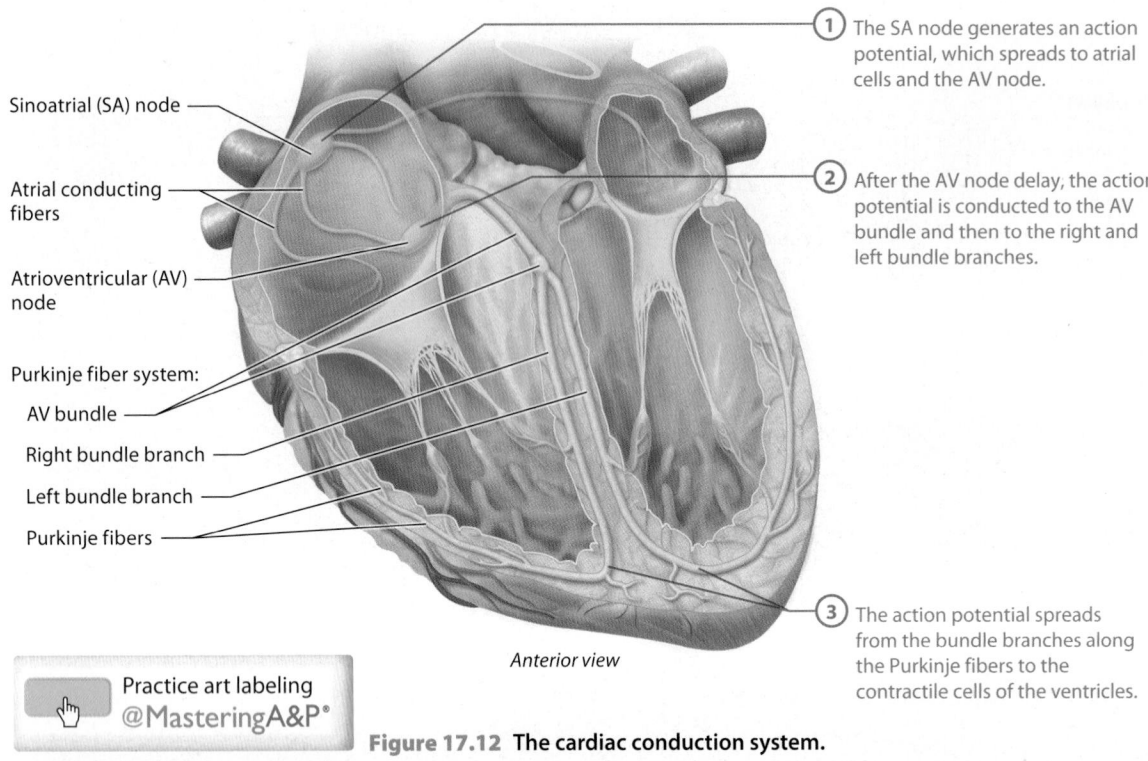

Sinoatrial (SA) node

Atrial conducting fibers

Atrioventricular (AV) node

Purkinje fiber system:

AV bundle

Right bundle branch

Left bundle branch

Purkinje fibers

① The SA node generates an action potential, which spreads to atrial cells and the AV node.

② After the AV node delay, the action potential is conducted to the AV bundle and then to the right and left bundle branches.

③ The action potential spreads from the bundle branches along the Purkinje fibers to the contractile cells of the ventricles.

Anterior view

Practice art labeling
@MasteringA&P®

Figure 17.12 The cardiac conduction system.

Pacing the Heart: Sinus Rhythm Each of these populations of pacemaker cells can potentially pace the heart, or make it beat at a certain rate, but the one that depolarizes fastest is the one that "wins" and sets the heart rate. The other pacemakers of the cardiac conduction system will pace the heart only if the fastest pacemaker ceases to function.

The SA node is the normal pacemaker of the entire heart; electrical rhythms generated and maintained by the SA node are known as **sinus rhythms**. The AV node and Purkinje fiber system normally only conduct action potentials generated by the SA node. However, if the SA node ceases to function, the AV node can successfully pace the heart, albeit somewhat slowly. Note that the AV bundle of the Purkinje system is the only connection between the AV node and the ventricles. If this system is blocked, the SA node cannot pace the ventricles even if it is functioning normally. The Purkinje fiber system is capable of pacing the heart, but its slow rate of depolarization is not adequate to sustain life beyond a short period of time. Occasionally, a group of regular contractile cells or pacemaker cells other than the SA node will attempt to pace the heart at the same time as the SA node. This "extra" pacemaker is called an **ectopic pacemaker,** and can result in irregular heart rhythms.

Conduction Pathway through the Heart Now that we know the components, let's look at the sequence of electrical events in the heart for one heartbeat (see Figure 17.12). Under normal conditions, ① the SA node generates an action potential, which spreads rapidly via gap junctions to the surrounding atrial cells. The impulses are then conducted by specialized *atrial conducting fibers* to the AV node, a process that altogether requires about 0.03 second. ② Once the impulse reaches the AV node, conduction slows considerably as a result of two factors: first, the low number of gap junctions between AV nodal cells, and second, the presence of the nonconducting

fibrous skeleton that surrounds the AV node. This slow conduction, known as the *AV node delay,* generally lasts about 0.13 second. The time it takes for the action potential to spread from the SA node to the AV bundle allows the atria to depolarize (and contract) *before* the ventricles, giving the ventricles time to fill with blood. It also helps to prevent current from flowing backward into the atria.

The action potential is then conducted from the AV bundle to the right and left bundle branches. ③ At this point, depolarization spreads along the Purkinje fibers to the contractile cardiac muscle cells of the ventricles. In spite of their slow rate of depolarization as pacemaker cells, the Purkinje fibers conduct action potentials faster than any cell in the myocardium (as much as 150 times as fast as the cells of the AV node). This speed is due to differences in their ion channels and the large numbers of gap junctions in their membranes. In only about 0.06 second, the entire mass of the right and left ventricles is depolarized, for a total action potential duration of about 0.22 second.

Quick Check

☐ 3. What is the sequence of events of a contractile cell action potential? What is the significance of the plateau phase of this action potential?

☐ 4. How does the refractory period of cardiac muscle cells differ from that in skeletal muscle fibers? What is its purpose?

☐ 5. What are the three populations of cells in the heart that can act as pacemakers? How do they differ?

☐ 6. What is the importance of the AV node delay?

The Electrocardiogram

One of the most important clinical tools for examining the health of the heart is the **electrocardiogram,** or **ECG** (also called an *EKG* from the German term *Elektrokardiogramm*), which is a graphic depiction of the electrical activity occurring in all cardiac muscle cells over a period of time. An ECG is recorded by placing *electrodes* on the surface of a patient's skin: six on the chest and two on each extremity. These electrodes record the changes in electrical activity from unique positions in the heart. These electrical changes are shown on the ECG as deflections, or **waves.** Keep in mind that a wave shows *changes* in electrical activity—if there is no *net* difference, there is no deflection shown on the ECG. However, even when the line on the ECG is flat between waves, the cells of the heart are in some phase of an action potential. Also note that the ECG can show only electrical changes that occur in contractile cells, because there are simply not enough pacemaker cells to measure with a standard ECG. For this reason, the ECG appears flat when pacemaker cells have their action potentials.

As you can see in **Figure 17.13**, an ECG recording generally consists of five waves, each of which represents the active depolarization or repolarization of different parts of the heart. Note that some of the waves point upward and others downward. Be careful not to equate these upward and downward deflections with depolarization and repolarization. Remember: The ECG tells you only that there is a net change in electrical activity. It does *not* tell you if that change is negative or positive across the membrane.

The waves of an ECG include the following:

- **P wave.** The small, initial **P wave** represents the depolarization of all cells within the atria except the SA node. The P wave nearly always registers as an upward deflection on the ECG. Figure 17.13 shows a flat segment immediately preceding the P wave that represents the time period during which the SA node depolarizes.
- **QRS complex.** The large **QRS complex,** which represents ventricular depolarization, is actually three separate waves. The Q wave is the first downward deflection, the R is the large upward deflection, and the S is the second downward deflection. The QRS is usually much larger in magnitude than the P wave because of the size differences between the atria and ventricles. Note that atrial repolarization occurs as the ventricles depolarize. This repolarization does produce a small electrical signal, but it is generally obscured by the much larger QRS complex and therefore not visible on the tracing.
- **T wave.** The small **T wave** occurs after the S wave of the QRS complex and represents ventricular repolarization. The T wave is an upward deflection under normal conditions. However, certain pathological states, such as myocardial ischemia, may cause the T waves to become inverted as a result of functional changes in the cells' electrical activity when the oxygen level is inadequate.

The periods between the waves represent important phases of the action potentials and of the spread of electrical activity through the heart. *Intervals* include a component of at least one wave, whereas *segments* do not include any wave components. Three intervals are of note. The first is the **R-R interval,** the time between two successive R waves. This interval represents the entire duration of the generation and spread of an action potential through the heart, and can be measured to determine the heart rate. The second is the **P-R interval,** or the period from the beginning of the P wave to the beginning of the QRS complex. The P-R interval represents the time it takes for the depolarization from the SA node to spread through the atria to the ventricles; it includes the AV node delay. The third is the **Q-T interval,** the time from the beginning of the QRS complex to the end of the T wave. During the Q-T interval, the ventricular cells are undergoing action potentials.

Between the end of the S wave and the beginning of the T wave is the **S-T segment.** The S-T segment is flat because it is recorded during the plateau phase of the ventricles, and no net

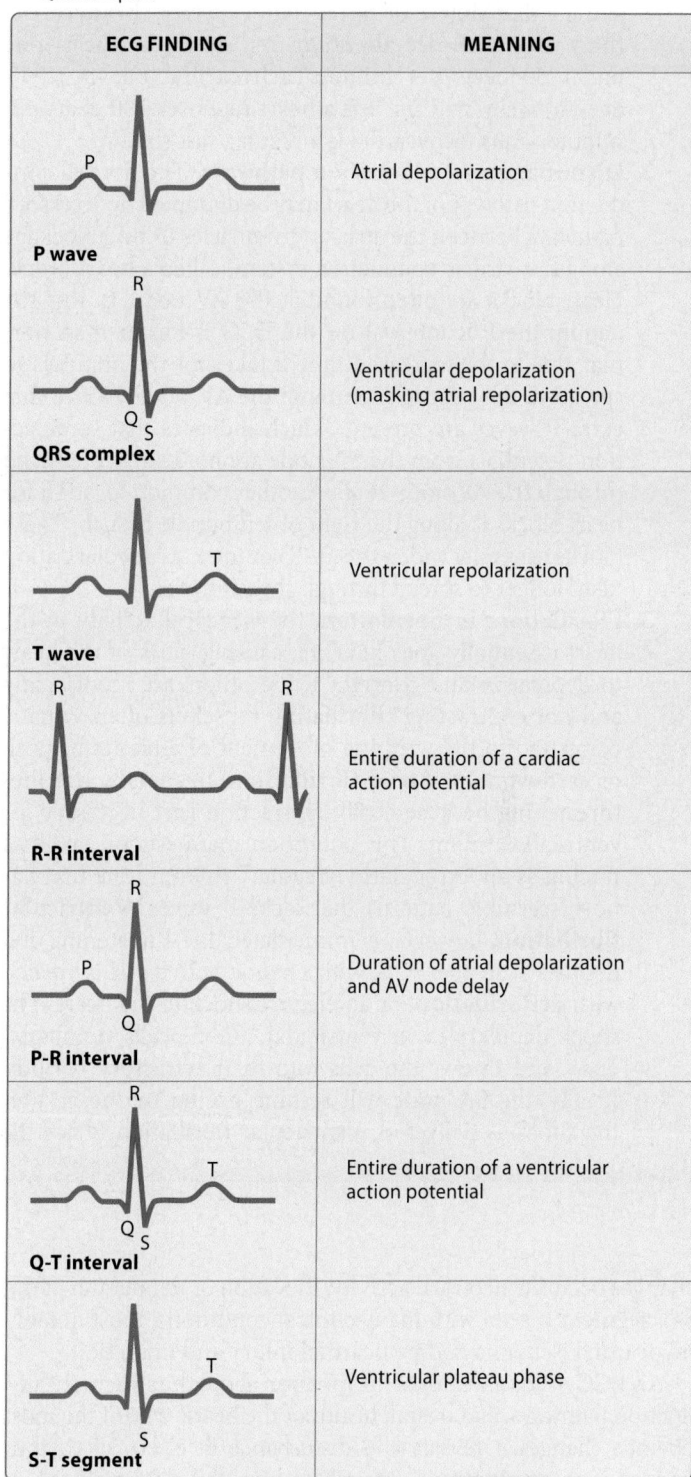

ECG FINDING	MEANING
P wave	Atrial depolarization
QRS complex	Ventricular depolarization (masking atrial repolarization)
T wave	Ventricular repolarization
R-R interval	Entire duration of a cardiac action potential
P-R interval	Duration of atrial depolarization and AV node delay
Q-T interval	Entire duration of a ventricular action potential
S-T segment	Ventricular plateau phase

Figure 17.13 A normal electrocardiogram (ECG) tracing.

A&P in the Real World

Dysrhythmias

Cardiac dysrhythmias have three basic patterns:

1. **Disturbances in heart rate.** Disturbance in resting heart rate can be one of two types: going either slower or faster than expected. **Bradycardia** (bray'-dee-KAR-dee-ah; *brady-* = "slow") is a heart rate under 60 beats per minute. **Tachycardia** (tak'-ih-KAR-dee-ah; *tachy-* = "fast") is a heart rate over 100 beats per minute; sinus tachycardia is a regular, fast rhythm.

2. **Disturbances in conduction pathways.** The normal conduction pathway of the heart may be disrupted by accessory pathways between the atria and ventricles or by a blockage along the cardiac conduction system, called a **heart block.** Heart blocks are often found at the AV node. In this situation, the P-R interval on the ECG is longer than normal, due to the increased time it takes for the impulses to spread to the ventricles through the AV node. Notice that extra P waves are present, which indicates that some action potentials from the SA node are not being conducted through the AV node at all. Another common location for heart blocks is along the right or left bundle branch. These blocks generally widen the QRS complex, as depolarization takes longer to spread through the ventricles.

3. **Fibrillation.** In **fibrillation,** the electrical activity in the heart essentially goes haywire, causing parts of the heart to depolarize and contract while others are repolarizing and not contracting (fibrillating muscle is often visually compared to the writhing movement of a plastic bag full of earthworms). **Atrial fibrillation** is generally not life-threatening because atrial contraction isn't necessary for ventricular filling. This condition manifests on an ECG tracing as an "irregularly irregular" rhythm (one that has no discernible pattern) that lacks P waves. **Ventricular fibrillation,** however, is immediately life-threatening and manifests on the ECG with chaotic activity. It is treated with **defibrillation,** or an electric shock to the heart. The shock depolarizes all ventricular muscle cells simultaneously and throws the cells into their refractory periods. Ideally, the SA node will resume pacing the heart after the shock is delivered. Ventricular fibrillation is not the same as "flat-lining," which is a condition called **asystole** (ay-SIS-toh-lee). Defibrillation is not used for asystole because the heart is not fibrillating and there is no electrical activity to reset. Instead, asystole is treated with CPR and pharmacological agents that stimulate the heart such as atropine and epinephrine.

Sinus tachycardia

Heart block at AV node

Atrial fibrillation

Ventricular fibrillation

changes occur in electrical activity. Elevation or depression of the S-T segment is seen with many clinical conditions, most notably myocardial ischemia and myocardial injury and infarction.

An ECG yields a wealth of information about the structure, conduction pathways, and overall health of the heart. One of the most obvious changes it reveals is a disturbance in electrical rhythm known as a **dysrhythmia** (dis-RITH-mee-ah; *dys-* = "bad" or "wrong") or **arrhythmia** (however, the term arrhythmia is a misnomer, as it technically means there is *no* heart rhythm). Some common dysrhythmias are discussed in *A&P in the Real World*.

Quick Check

☐ 7. What does an ECG record?
☐ 8. What are the five waves in an ECG, and what do they represent?

Apply What You Learned

☐ 1. As their name suggests, calcium channel blockers block calcium ion channels in the sarcolemma of cardiac muscle cells and slow calcium ion entry into the cell during an action potential. Predict the effects of these drugs on the action potential of a contractile cell and on the length and strength of contraction.

☐ 2. Certain drugs used to treat dysrhythmias, called *local anesthetics,* work by blocking sodium ion channels and slowing the entry of sodium ions into the cell during an action potential. Predict the effect of these drugs on the action potential of contractile cells. Would local anesthetic drugs have a significant effect on the cells of the SA or AV node? Why or why not?

See answers in Appendix A.

MODULE **17.4**

Mechanical Physiology of the Heart: The Cardiac Cycle

Learning Outcomes

1. Describe the phases of the cardiac cycle.
2. Relate the opening and closing of specific heart valves in each phase of the cardiac cycle to pressure changes in the heart chambers.
3. Relate the heart sounds to the events of the cardiac cycle.
4. Relate the ECG waveforms to the normal mechanical events of the cardiac cycle.
5. Compare and contrast pressure and volume changes of the left and right ventricles and the aorta during one cardiac cycle.

So far in this chapter we've focused on the heart's electrophysiology. Now we turn to the heart's **mechanical physiology,** which refers to the actual processes by which blood fills the cardiac chambers and is pumped out of them. As we discussed earlier, cardiac muscle cells contract as a unit to produce one coordinated contraction, called a **heartbeat.** The muscle cells are arranged in a spiral pattern, producing a "wringing" action in the heart when it contracts. Pressure changes caused by contractions drive blood flow through the heart, with valves preventing backflow. The sequence of events that take place within the heart from one heartbeat to the next is known as the **cardiac cycle.**

This module explains the events of the cardiac cycle and the sounds the heart produces as it beats. The module concludes with a big picture view that correlates cardiac electrophysiology with mechanical physiology.

The Relationship between Pressure Changes, Blood Flow, and Valve Function

You have read that the heart's contraction drives blood flow through its chambers and into the pulmonary trunk and aorta. However, its contraction does not open the valves through which the blood must pass. Instead, blood flows in response to pressure gradients. As the ventricles contract and relax, the pressure in the

chambers changes, causing blood to push on the valves and either open or close them (an example of the Gradients Core Principle, p. 26).

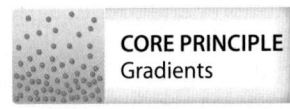
CORE PRINCIPLE
Gradients

Let's examine how this happens, using the left ventricle as our example. When the left ventricle contracts, its pressure rises above the pressures in the left atrium and the aorta, as shown in **Figure 17.14a**. This causes blood to flow from the ventricle to

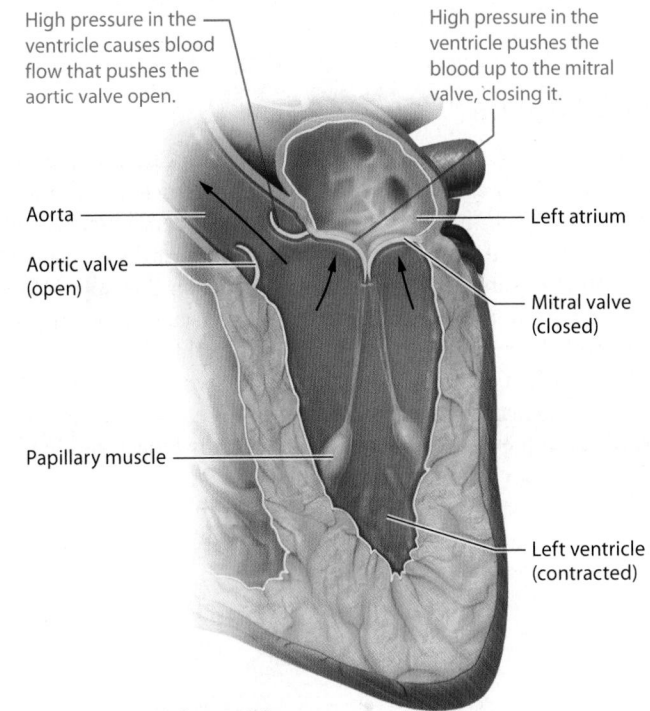

High pressure in the ventricle causes blood flow that pushes the aortic valve open.

High pressure in the ventricle pushes the blood up to the mitral valve, closing it.

Aorta

Aortic valve (open)

Left atrium

Mitral valve (closed)

Papillary muscle

Left ventricle (contracted)

(a) Contracted left ventricle

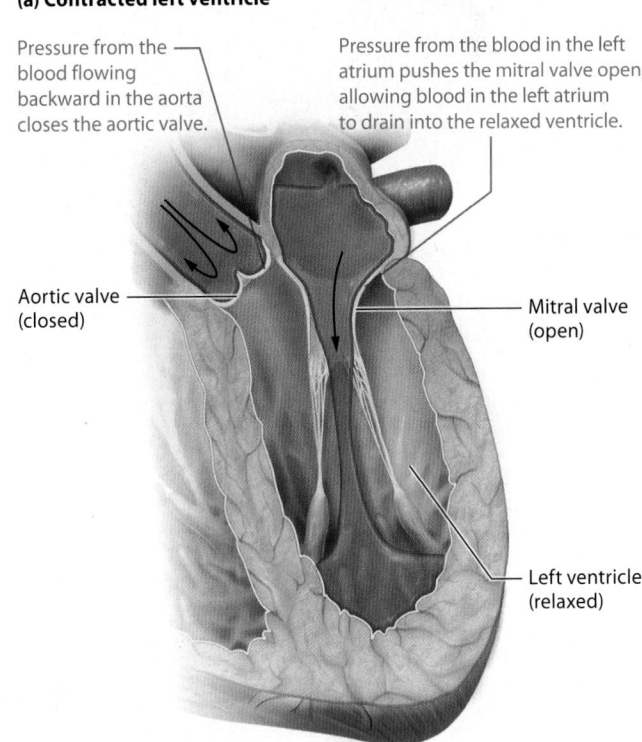

Pressure from the blood flowing backward in the aorta closes the aortic valve.

Pressure from the blood in the left atrium pushes the mitral valve open, allowing blood in the left atrium to drain into the relaxed ventricle.

Aortic valve (closed)

Mitral valve (open)

Left ventricle (relaxed)

(b) Relaxed left ventricle

Figure 17.14 Pressure changes, blood flow, and valve function.

the aorta and produces two changes in the valves, which occur at the same time: (1) The mitral valve is forced shut by blood pushing up on it; and (2) the aortic valve is forced open by the outgoing blood. In **Figure 17.14b** you can see that when the ventricle relaxes, the opposite occurs: The pressures in the ventricle falls below those in the left atrium and in the aorta. The higher pressure in the left atrium forces the mitral valve open, allowing blood to drain into the relaxed ventricle. At the same time, the higher pressure in the aorta pushes the cusps of the aortic valve closed.

Heart Sounds

One of the most basic tools available to a health care provider is a *stethoscope,* which can be used to listen to, or *auscultate* (OS-kul-tayt), the rhythmic "lub dub" **heart sounds.** Under normal conditions, blood flow through the open AV and SL valves is relatively quiet, and the examiner hears sounds only when the valves close. These sounds are not due to the actual valve "slamming shut" but instead probably result from vibrations of the ventricular and blood vessel walls when the valves shut. There are two heart sounds: **S1,** or the "lub," is heard when the AV valves close, and **S2,** the "dub," is heard when the SL valves close. S1 is typically longer and louder than S2, although it's lower in frequency.

Figure 17.15 shows where the sound produced by each valve is heard clearest on the chest with a stethoscope. As you can see,

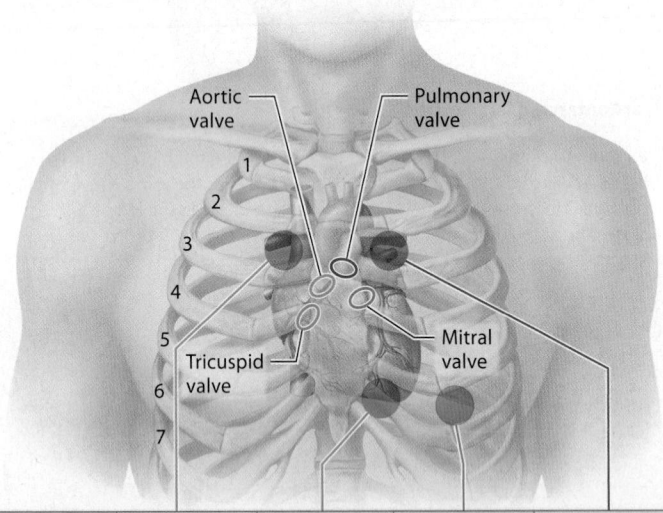

AREA	Aortic	Tricuspid	Mitral	Pulmonary
LOCATION OF SOUND	Second intercostal space, right sternal border	Fifth intercostal space, left sternal border	Fifth intercostal space, mid-clavicular line	Second intercostal space, left sternal border
TIMING OF SOUND	Aortic valve is heard here during S2.	Tricuspid valve is heard here during S1.	Mitral valve is heard here during S1.	Pulmonary valve is heard here during S2.

Figure 17.15 Heart sounds.

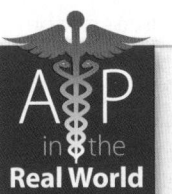

Heart Murmurs and Extra Heart Sounds

One of the more common findings on chest auscultation is an audible sound called a **heart murmur,** which occurs when blood flow through the heart is turbulent. Heart murmurs are generally caused by defective valves, although they may also result from defective chordae tendineae or holes in the interatrial or interventricular septum. Children, however, often have heart murmurs that do not represent defects.

Chest auscultation may also reveal extra heart sounds. The first, called **S3,** can occur just as blood begins to flow into the ventricles, right after S2. It results from the recoil of the ventricular walls as they are stretched and filled. Another heart sound is **S4,** which is heard when most of the blood has finished draining from the atria to the ventricles, just before S1. It typically results from blood being forced into a stiff or enlarged ventricle. Both S3 and S4 may represent pathology, but they can also occasionally be heard in a healthy heart.

the sounds are best heard a slight distance away from the actual valves. This is because the sound waves are carried away somewhat from the valves by the blood moving through the ventricles or vessels. *A&P in the Real World: Heart Murmurs and Extra Heart Sounds* discusses some of the problems that may be found by listening to heart sounds.

Quick Check

☐ 1. What causes the heart sounds S1 and S2?

Events of the Cardiac Cycle

Each cardiac cycle consists of one period of relaxation called **diastole** (dy-ASS-toh-lee) and one period of contraction called **systole** (SIS-toh-lee) for each chamber of the heart. The atrial and ventricular diastoles and systoles occur at different times as a result of the AV node delay that we discussed earlier. However, both sides of the heart are working to pump blood into their respective circuits simultaneously.

We can divide the cardiac cycle into four main phases that are defined by the actions of the ventricles and the positions of the valves: filling, contraction, ejection, and relaxation. These four phases, illustrated in **Figure 17.16,** are as follows:

① **Ventricular filling phase.** The **ventricular filling phase** is the period during which blood drains from the atria into the ventricles. During ventricular filling, pressures in the left and right ventricles are lower than those in the atria,

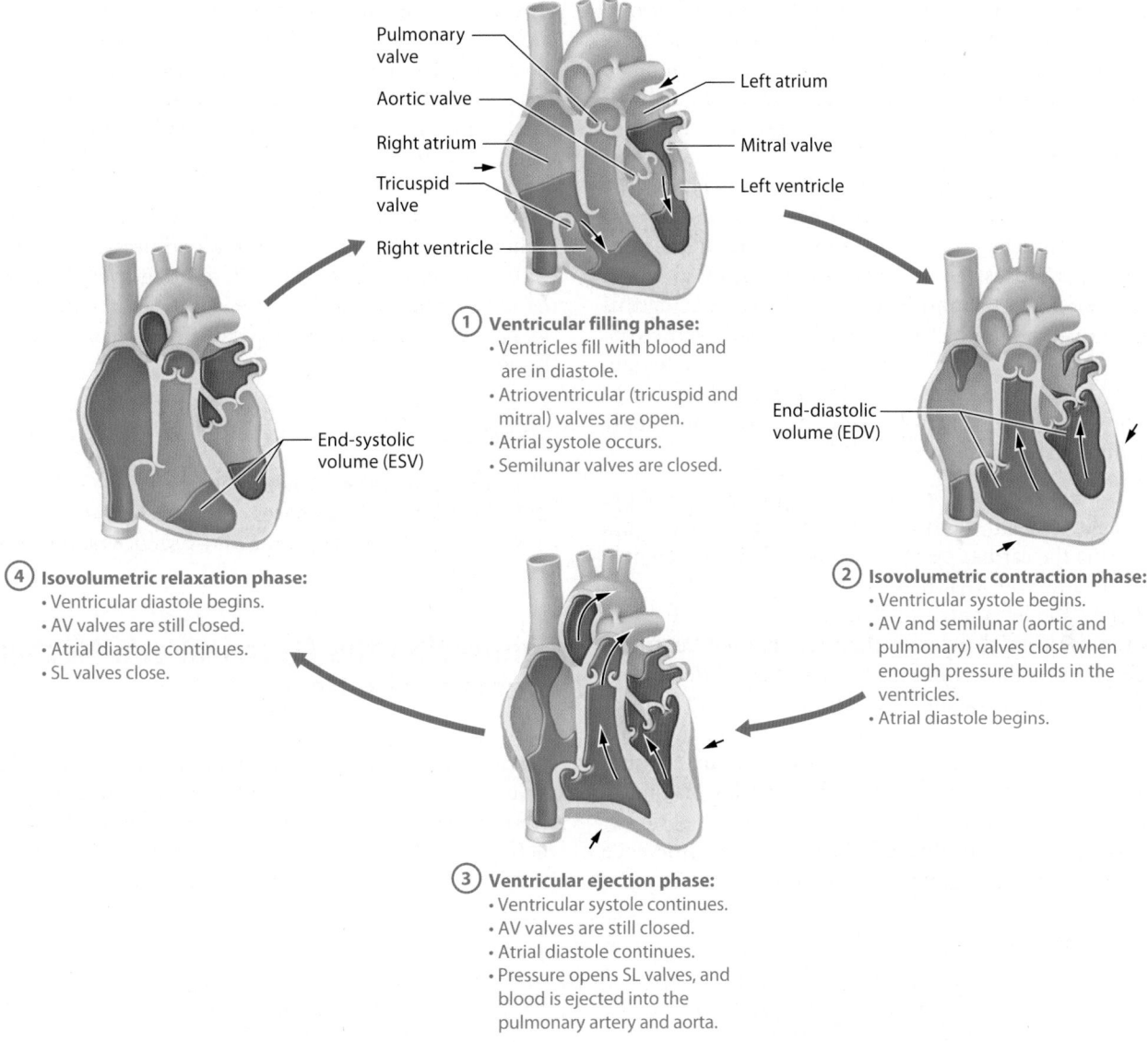

Figure 17.16 Events of the cardiac cycle.

the pulmonary trunk, and the aorta. The higher pressures in the pulmonary trunk and aorta cause the semilunar valves to be closed, which prevents the flow of blood from the pulmonary trunk and aorta back into the ventricles. The atrioventricular valves, however, are open because of the higher atrial pressure, and blood flows down its pressure gradient from the atria into the ventricles. Nearly 80% of the total blood volume of the atria drains passively in this manner into the ventricles. Initially, the atria are in diastole, but as blood continues to drain into the ventricles, the pressure gradient becomes smaller and filling slows. At this point, atrial systole takes place and the contracting atria eject a variable volume of blood into the ventricles—as much as the remaining 20% of blood volume or as little as just a few percent. At the end of atrial systole, each ventricle contains about 120 ml of blood, a volume known as the **end-diastolic volume (EDV)** because it is the ventricular volume at the end of ventricular diastole.

(2) **Isovolumetric contraction phase.** The beginning of ventricular systole occurs during the shortest phase of the cardiac cycle, called the **isovolumetric contraction phase** (eye′-soh-vol-yoo-MET-rik; *iso-* = "same"). During this phase, the pressure in the ventricles rises rapidly as the ventricles begin to contract. This high pressure closes the AV valves and causes the S1 heart sound to be heard. However, ventricular pressure is not yet high enough to push open the semilunar valves, so both sets of valves are closed and the ventricular volume does not change (hence the name *iso*volumetric).

(3) **Ventricular ejection phase.** At the beginning of the **ventricular ejection phase,** the pressure in the ventricles rises to a level higher than that in the pulmonary trunk and aorta, and this pushes the SL valves open. The beginning of the ejection phase is marked by rapid outflow of blood from the ventricles. However, as this phase continues, the pressure in the pulmonary trunk and aorta approaches

that in the ventricles. At this point, ejection of blood into the vessels decreases considerably. The ventricular ejection phase sees approximately 70 ml of blood pumped from each ventricle. This means that about 50 ml of blood remains in each ventricle, a volume known as the **end-systolic volume (ESV).**

④ **Isovolumetric relaxation phase.** The final phase, the short **isovolumetric relaxation phase,** occurs as ventricular diastole begins and pressure declines in the ventricles. This causes the SL valves to snap shut, at which point the S2 heart sound is heard. However, during this phase the pressure in the ventricles is still somewhat higher than that in the atria, so the AV valves remain closed. Thus, blood is neither being ejected from nor entering into the ventricles and their volume briefly remains constant.

Figure 17.17 illustrates the pressure changes that take place within the right and left ventricles and the pulmonary trunk and aorta during the cardiac cycle. Notice first that there are two graphs, which show you what happens during the cardiac cycle in both the right and left ventricles. The green line in each graph represents the pressure changes in the right and left ventricles during each phase of the cardiac cycle. Notice the difference in pressure generated in the isovolumetric contraction and ventricular ejection phases by the right ventricle versus that generated by the left ventricle—the right ventricle's maximum pressure is only about 28 mm Hg, whereas the left ventricle's maximum pressure is about 118 mm Hg.

You can see the reason for this ventricular pressure difference if you compare the other two lines in the graphs, which show the pressures in the pulmonary trunk (blue line; for the right ventricle) and the aorta (purple line; for the left ventricle). Notice in the graphs that the pressure curves for the vessels and the ventricles eventually cross. As you can see, the pressure in the pulmonary trunk varies between about 10 and 28 mm Hg. This means that the maximum pressure the right ventricle has to generate in order to push open the pulmonary valve and eject blood is about 28 mm Hg. However, the pressure in the aorta is considerably higher due to the higher overall blood pressure in the systemic circuit, and ranges from about 80 to 118 mm Hg. For this reason, the left ventricle has to squeeze significantly harder to generate the pressure needed to open the aortic valve and eject blood into the aorta.

Quick Check

☐ 2. How is the cardiac cycle defined?
☐ 3. What are the four phases of the cardiac cycle?
☐ 4. Is the end-diastolic or the end-systolic volume of blood larger? Why?

Connecting the Electrical and Mechanical Events in the Heart

Let's now put the mechanical and electrical events together and see how their timing correlates during a normal cardiac cycle. **Figure 17.18** presents what is known as a *Wigger's diagram,* which shows you what is happening in the heart during each stage of the cardiac cycle. Four variables are presented: the heart's electrical events, visible on the ECG; the heart sounds;

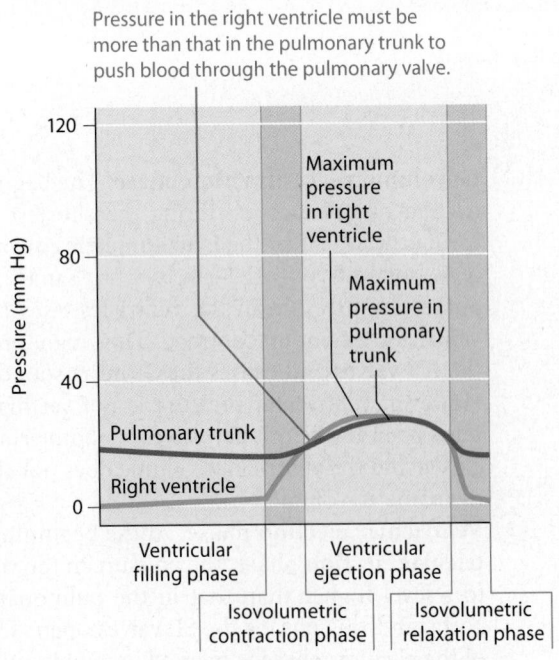

(a) Pressures in right ventricle and pulmonary trunk

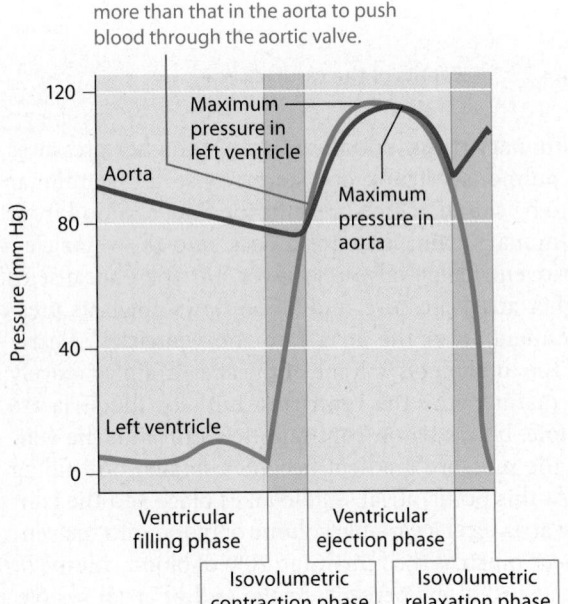

(b) Pressures in left ventricle and aorta

Figure 17.17 Comparison of pressure changes in left and right ventricles during the cardiac cycle.

Figure 17.18 Wigger's diagram showing an overview of electrical and mechanical events in the heart during the cardiac cycle.

the pressure changes in the left side of the heart, including the aortic, left ventricular, and left atrial pressures; and the changes in the volume of blood found in the left ventricle. The pressure and volume graphs show only the left ventricle, for simplicity; these graphs look much the same for the right ventricle, except that the pressures are lower.

It is important to realize that electrical and mechanical events in the heart cannot be positively correlated. That is, there may be electrical activity with no heartbeat. However, under normal conditions, action potentials in the heart should lead to a physical contraction. The following Concept Boost will give you a clearer understanding of the cardiac cycle.

ConceptBOOST »»»

Deconstructing the Wigger's Diagram

At first glance, the Wigger's diagram may look like a jumbled collection of lines. So Figure 17.19 breaks it down and clarifies what is happening in the heart during each phase of the cardiac cycle. ■

Figure 17.19 Electrical and mechanical events in the left side of the heart during each phase of the cardiac cycle.

Quick Check

☐ 5. Walk through the mechanical events of the cardiac cycle.
☐ 6. How do the ECG waves correlate with each part of the cardiac cycle?
☐ 7. How does the left ventricular pressure correlate with the opening and closing of the mitral and aortic valves?

Apply What You Learned

☐ 1. Sometimes health care providers will elect not to treat conditions such as atrial fibrillation in which there is no functional atrial contraction. Explain the logic behind this decision.
☐ 2. Predict what would happen to the end-systolic volume if the ventricles were diseased and failing to pump

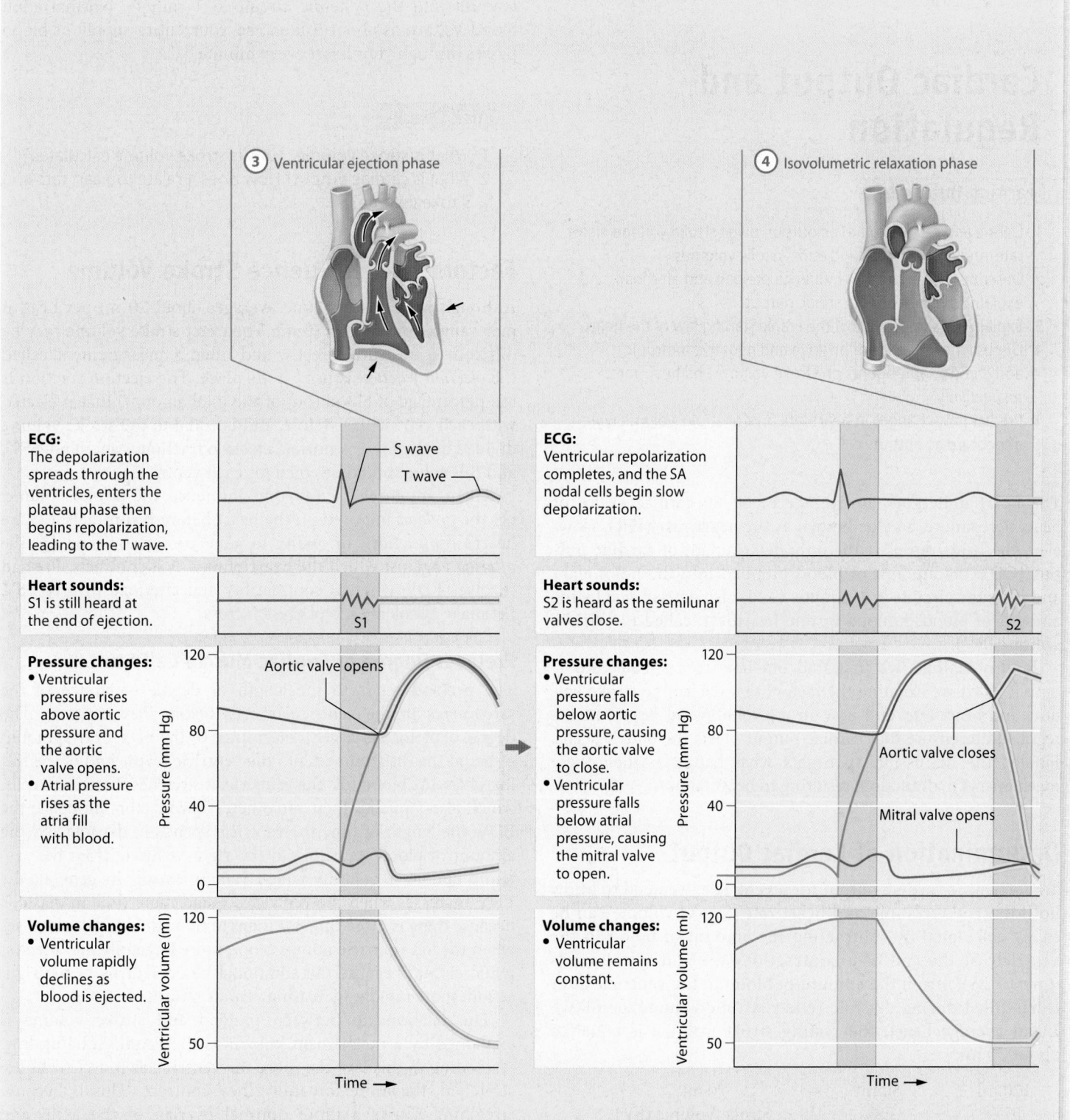

③ Ventricular ejection phase

ECG:
The depolarization spreads through the ventricles, enters the plateau phase then begins repolarization, leading to the T wave.

S wave
T wave

Heart sounds:
S1 is still heard at the end of ejection.

S1

Pressure changes:
• Ventricular pressure rises above aortic pressure and the aortic valve opens.
• Atrial pressure rises as the atria fill with blood.

Aortic valve opens

Volume changes:
• Ventricular volume rapidly declines as blood is ejected.

Time →

④ Isovolumetric relaxation phase

ECG:
Ventricular repolarization completes, and the SA nodal cells begin slow depolarization.

Heart sounds:
S2 is heard as the semilunar valves close.

S2

Pressure changes:
• Ventricular pressure falls below aortic pressure, causing the aortic valve to close.
• Ventricular pressure falls below atrial pressure, causing the mitral valve to open.

Aortic valve closes
Mitral valve opens

Volume changes:
• Ventricular volume remains constant.

Time →

adequately. Would this cause the end-diastolic volume to increase, decrease, or stay the same? Explain.

☐ 3. What happens in the heart electrically if the Q-T interval is prolonged? Which part of the cardiac cycle would this affect? What would this do to the heart rate?

See answers in Appendix A.

MODULE **17.5**

Cardiac Output and Regulation

Learning Outcomes

1. Define and calculate cardiac output, given stroke volume, heart rate, and end-diastolic and end-systolic volumes.
2. Describe the factors that influence preload and afterload, and explain how they affect cardiac output.
3. Explain the significance of the Frank-Starling law of the heart.
4. Discuss the influence of positive and negative inotropic and chronotropic agents on stroke volume and heart rate, respectively.
5. Predict how changes in heart rate and/or stroke volume will affect cardiac output.

The heart undergoes an average of 60–80 cardiac cycles or beats per minute, a value known as the **heart rate (HR).** However, the heart rate isn't the only determinant of **cardiac output (CO),** the amount of blood pumped into the pulmonary and systemic circuits in 1 minute. CO is also determined by the amount of blood pumped in one heartbeat, called the **stroke volume (SV).**

In this module, we first find out how to calculate cardiac output. Next we examine the factors that determine stroke volume and heart rate, and how stroke volume and heart rate are regulated to ensure that cardiac output meets the body's needs. Finally, the discussion spotlights what happens when these mechanisms malfunction, resulting in heart failure.

Determination of Cardiac Output

To determine cardiac output for a ventricle, we need to know both its stroke volume and heart rate. Stroke volume can be easily calculated by subtracting the amount of blood in the ventricle at the end of a contraction (the end-systolic volume, or ESV) from the amount of blood in the ventricle after it has filled during diastole (end-diastolic volume, or EDV). In an average heart, the resting stroke volume is equal to about 70 ml:

$$\begin{array}{ccccc} 120 \text{ ml} & - & 50 \text{ ml} & = & 70 \text{ ml} \\ \text{EDV} & - & \text{ESV} & = & \text{Stroke Volume (SV)} \end{array}$$

To find the cardiac output, you simply multiply the heart rate by the stroke volume, as shown here:

$$\begin{array}{ccccc} 72 \text{ beats/min} & \times & 70 \text{ ml/beat} & = & 5040 \text{ ml/min, or } {\sim}5 \text{ liters/min} \\ \text{HR} & \times & \text{SV} & = & \text{Cardiac Output (CO)} \end{array}$$

So as you can see, resting cardiac output averages about 5 liters/min, which means that the right ventricle pumps about 5 liters into the pulmonary circuit and the left ventricle pumps the same amount into the systemic circuit in 1 minute. Normal adult blood volume is about 5 liters, so your entire supply of blood passes through your heart every minute.

Quick Check

☐ 1. What is stroke volume? How is stroke volume calculated?
☐ 2. What is cardiac output? How does it relate to heart rate and stroke volume?

Factors That Influence Stroke Volume

Although the stroke volume averages about 70 ml per beat, it may range from 50 to 120 ml. The exact stroke volume may be difficult to measure directly, and often a measurement called the *ejection fraction* is used in its place. The ejection fraction is the percentage of blood (out of the total amount) that is ejected with each ventricular systole, and is equal to the stroke volume divided by the EDV. A normal ejection fraction is about 50–65%, and this value should be equal for each ventricle.

There are three factors that influence the stroke volume: (1) the *preload* imposed on the heart before it contracts; (2) the heart's *contractility*, or ability to generate tension; and (3) the *afterload* against which the heart pumps as it contracts. Preload involves EDV, whereas contractility and afterload involve ESV. Let's take a look at each of these factors.

Preload: Degree of Cardiac Muscle Cell Stretch

The **preload** refers to the length or degree of stretch of the sarcomeres in the ventricular cells before they contract. The degree of preload is largely determined by the EDV, or the amount of blood that has drained into the ventricle by the end of the filling phase. As blood fills the ventricle, it stretches the muscle cells, which also stretches their sarcomeres. Two factors influence the EDV: the length of time the ventricle spends in diastole and the amount of blood returning to the right ventricle from the systemic circuit, a quantity called *venous return*. In general, the EDV increases when the ventricles spend more time in diastole, because there is more time for them to fill with blood. It also rises when the left ventricle pumps blood more forcefully into the systemic circuit, because the additional blood returns to the right atrium more rapidly, increasing venous return.

The relationship between preload and stroke volume is explained by a mechanism known as the **Frank-Starling law.** According to this law, the more the ventricular muscle cells are stretched, the more forcefully they contract. This is because stretching causes a more optimal overlap of the actin and myosin filaments in the muscle cells, which enables a stronger contraction and a higher stroke volume. This relationship ensures that the volume of blood discharged from the heart is equal to the volume that enters it, which is particularly important during exercise, when cardiac output must increase to meet the body's needs.

Contractility

The heart's **contractility** is its intrinsic pumping ability, or ability to generate tension. Contractility is difficult to measure directly; however, it can be estimated clinically by examining the velocity of blood being ejected from the ventricles. The relationship between contractility and stroke volume is pretty straightforward: Increasing contractility will increase the stroke volume and therefore decrease the ESV (because more blood is ejected during each cardiac cycle). Decreasing contractility will do the opposite: decreasing stroke volume and increasing the ESV (assuming that preload and afterload remain constant). Agents that affect contractility are known as *inotropic agents* (EYE-noh-troh-pik; *ino-* = "a muscle fiber," *trop-* = "change").

Factors that increase heart rate, such as sympathetic nervous stimulation, often also affect contractility and so increase the force of contraction. This explains why stroke volume and heart rate generally increase together. Note, though, that this is true only up to a certain point—when the heart rate is too high, contractility decreases, as does preload. This happens because the heart is beating too rapidly to develop significant tension during each contraction. The result is a decrease in both stroke volume and cardiac output.

Afterload: Amount of Back Pressure

Afterload refers to the force that the right and left ventricles must overcome in order to eject blood into their respective arteries. It is largely determined by the blood pressure in the arteries of both the pulmonary and systemic circuits. As afterload increases, ventricular pressure must be greater to exceed the pressure in arterial pulmonary and systemic vessels and open the semilunar valves. An increase in afterload therefore generally causes a decrease in stroke volume and so a rise in the ESV of the ventricles. Conversely, a decrease in afterload generally corresponds to a higher stroke volume and a lower ESV. Read *A&P in the Real World: Ventricular Hypertrophy* on page 662 to discover what happens when preload and afterload are increased over time.

ConceptBOOST ⟫⟫⟫

Understanding How Changes in Preload, Contractility, and Afterload Affect Stroke Volume

You've just read about the factors that determine stroke volume—preload, contractility, and afterload—which the following diagram illustrates, using only the left ventricle for simplicity.

① The **preload** of the left heart is determined by the volume of blood coming in from the pulmonary circuit.

② The **contractility** is determined by the heart's intrinsic pumping ability.

③ The **afterload** of the left heart is determined by the pressure in the systemic circuit.

Let's put these three factors together to see how they interact to determine stroke volume. In our first scenario, we have a high preload and contractility and a low afterload, and all three together raise stroke volume:

The high preload leads to a high EDV, and this volume of blood stretches the cardiac muscle cells. This, combined with increased contractility, leads to a more forceful contraction. Since the afterload is low, this forceful contraction isn't pumping against a great deal of resistance. A forceful contraction against low resistance leads to a high stroke volume and a low ESV.

In this second scenario, we have a low preload and contractility and a high afterload, which all decrease stroke volume:

Notice here that the low preload leads to a low EDV, so now the cardiac muscle cells are much less stretched. When this is combined with a diminished contractility, the heart contracts weakly. Since the afterload is now high, this weak contraction is pumping against a great deal of resistance. A weak contraction against high resistance leads to a low stroke volume, and more blood left in the ventricle after the contraction (a high ESV). ∎

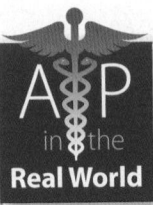

Ventricular Hypertrophy

Real World

Long-standing increases in afterload are associated with enlargement of the ventricles, or *ventricular hypertrophy* (hy-PER-troh-fee). The cardiac muscle cells of the ventricles need to generate more tension to continue pumping blood against the higher afterload. These cells respond the same way as skeletal muscle fibers when they have to generate more tension—they make more myofibrils and more organelles, and as a result they get bigger.

Right ventricular hypertrophy most often results from respiratory disease or high blood pressure in the pulmonary circuit, and left ventricular hypertrophy generally results from high blood pressure in the systemic circuit. Ventricular hypertrophy can increase the effectiveness of the heart's pumping up to a certain point. However, because this condition decreases the heart lumen and so filling space, it increases the risk for many other cardiac conditions, including heart failure, which is discussed at the end of this module.

Quick Check

☐ 3. What three factors determine stroke volume? How does the stroke volume differ with each factor?

☐ 4. What is the Frank-Starling law, and how does it relate to preload?

Factors That Influence Heart Rate

The other determinant of cardiac output is the heart rate. Under normal conditions, the rate at which the SA node generates action potentials determines the heart rate. Factors that influence the rate at which the SA node depolarizes are known as *chronotropic agents* (KRON-oh-troh-pik; *chrono-* = "time"). Anything that increases the rate at which this node fires is called a *positive chronotropic agent;* one with the opposite effect is known as a *negative chronotropic agent.* Positive chronotropic agents include the sympathetic nervous system, certain hormones, and elevated body temperature; negative chronotropic agents include the parasympathetic nervous system and decreased body temperature. We discuss many chronotropic agents in the next section.

Quick Check

☐ 5. What is a chronotropic agent?

Regulation of Cardiac Output

◀◀ Flashback

1. What effects does the sympathetic nervous system have on the adrenal gland and the heart? (pp. 526–527)

2. What effects does the parasympathetic nervous system have on the heart? (p. 530)

Although the heart is autorhythmic, it still requires regulation to ensure that cardiac output meets the body's needs at all times. Cardiac output is regulated primarily by the nervous and endocrine systems, which influence both heart rate and stroke volume. Next we look at the effects of the nervous system, endocrine system, and other factors on the heart.

Cardiac Innervation and Regulation by the Nervous System

Recall that the two branches of the autonomic nervous system, or ANS, regulate our automatic functions (see Chapter 14). The *sympathetic nervous system* innervates the heart via a set of **sympathetic nerves** that stem from ganglia located along the spinal cord (**Figure 17.20**). The neurons of these nerves release the neurotransmitter norepinephrine, which increases cardiac output with both positive chronotropic and inotropic effects.

First, they increase the heart rate by raising the rate at which the SA node fires, up to 180–200 or more times per minute. Second, they increase the entry of calcium ions into the cardiac muscle cell. The higher calcium ion concentration increases contractility of the cells, which in turn raises stroke volume. Together, these two effects can dramatically increase cardiac output.

Note the major difference in the way the nervous system controls the strength of cardiac and skeletal muscle contractions.

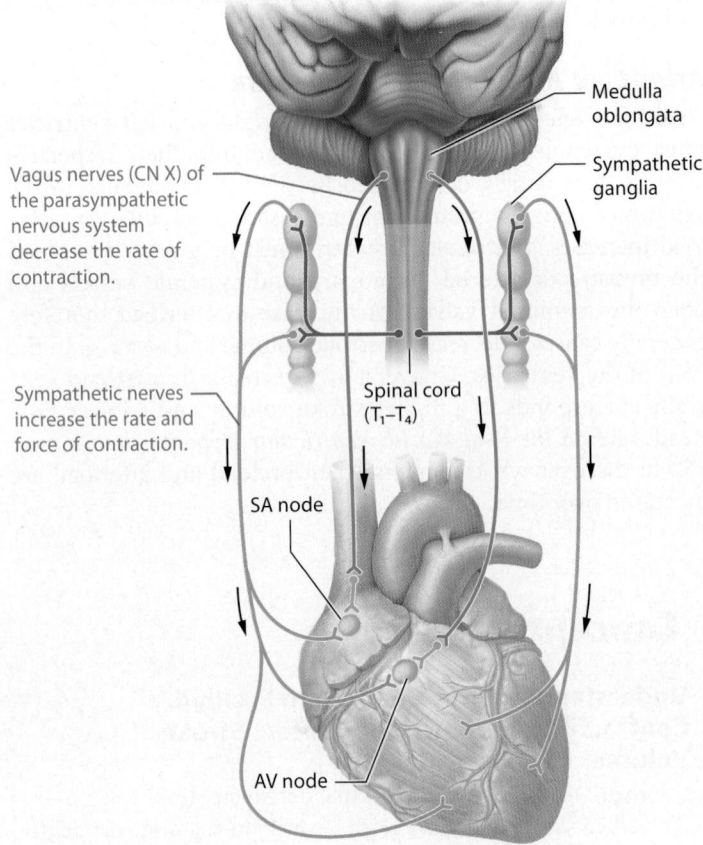

Figure 17.20 Innervation and nervous regulation of the heart.

The somatic nervous system simply stimulates more motor units, and so more skeletal muscle fibers, to elicit a stronger skeletal muscle contraction, a phenomenon called *recruitment*. However, all cardiac muscle cells work together for each contraction of the heart, regardless of the strength of the contraction. So, the sympathetic nervous system cannot increase the number of cardiac muscle cells that are activated, and recruitment is not possible. For this reason, the sympathetic nervous system must control the strength of contraction instead by the number of calcium ions that enter the cardiac muscle cell.

The parasympathetic nervous system exerts essentially the opposite effects on the heart. As you can see in Figure 17.20, it innervates the heart by the left and right *vagus nerves* (CN X). Both of these nerves release acetylcholine when stimulated. Acetylcholine primarily affects the SA node, decreasing its rate of action potential generation. This negative chronotropic effect slows the heart rate and can even stop the heart temporarily if the parasympathetic stimulation is strong enough. The vagus nerves primarily innervate atrial muscle, so they have less effect on ventricular contractility than on heart rate and therefore have only weak negative inotropic effects.

Cardiac Regulation by the Endocrine System

Hormonal regulation of cardiac output occurs in various forms. For one, recall that the adrenal medulla is activated by the sympathetic nervous system, and in response it secretes the hormones epinephrine and norepinephrine into the bloodstream (see Chapter 14). These hormones have the same effects as the sympathetic nervous system neurotransmitters—they are positive inotropic and chronotropic agents—but their effects are longer-lasting. Other hormones that have these same effects include thyroid hormone and glucagon produced by the pancreas.

Some hormones regulate cardiac output through their control of water balance. The amount of water in the blood (the blood volume) plays a significant role in determining the heart's preload and therefore its strength of contraction. Hormones such as aldosterone and antidiuretic hormone increase blood volume and preload, and so raise cardiac output. Others, such as atrial natriuretic peptide, decrease blood volume and preload, and therefore reduce cardiac output.

Other Factors That Influence Cardiac Output

Although the nervous and endocrine systems are the primary regulators of cardiac output, other factors influence cardiac output, as well. For one, the concentration of certain electrolytes in the extracellular fluid plays a large role in determining the length and magnitude of an action potential and cardiac output. Another factor that influences cardiac output is body temperature—the SA node fires more rapidly at higher body temperatures and more slowly at lower body temperatures. In addition, factors such as age and physical fitness influence heart rate and cardiac output; younger children and the elderly often have a higher resting heart rate, whereas trained athletes often have a much lower resting heart rate. This is because exercise increases the stroke volume, so for the body to maintain a constant cardiac output, the heart rate must decrease. Factors that affect cardiac output are summarized in **Figure 17.21**.

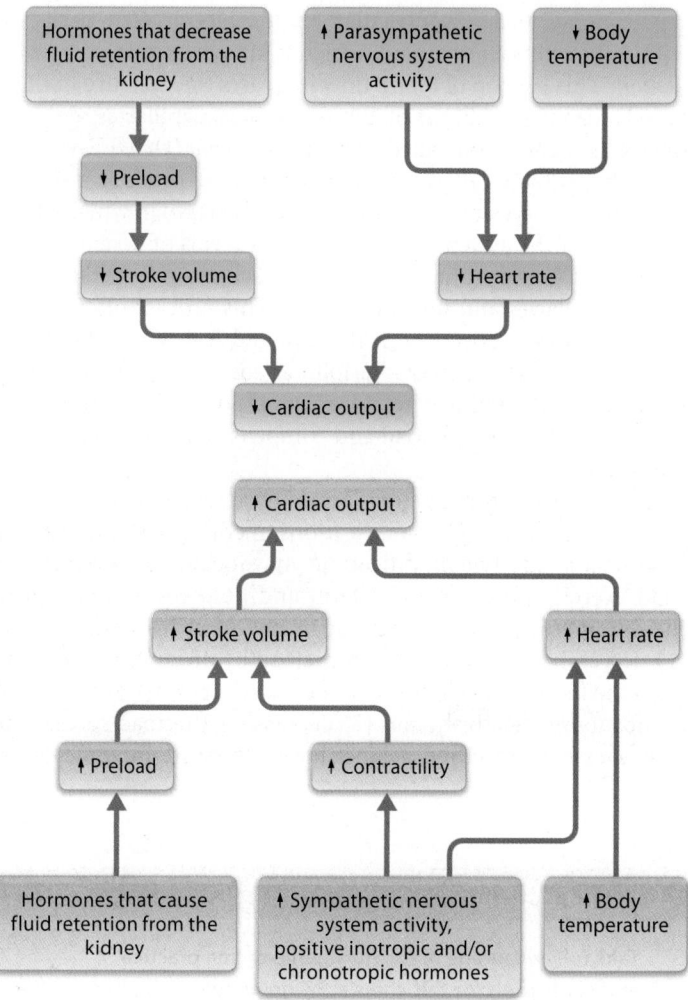

Figure 17.21 Regulation of cardiac output.

Quick Check

☐ 6. What effects does the sympathetic nervous system have on cardiac output?

☐ 7. What effects does the parasympathetic nervous system have on cardiac output?

☐ 8. How would a hormone that decreases the amount of water in the body affect cardiac output?

Heart Failure

Heart failure, formerly known as *congestive heart failure,* is defined as any condition that reduces the heart's ability to function effectively as a pump. Causes of heart failure include reduced contractility due to myocardial ischemia and/or myocardial infarction, valvular heart diseases, any disease of the heart muscle itself (known as *cardiomyopathy*), and electrolyte imbalances. Heart failure generally results in decreased stroke volume, which in turn reduces cardiac output.

The signs and symptoms of heart failure generally depend on the type of heart failure and the side of the heart that is affected.

For example, in left ventricular failure, blood often backs up within the pulmonary circuit, which is known as *pulmonary congestion.* This backup of blood flow increases the pressure in these vessels, driving fluid out of the pulmonary capillaries and into the lungs, a condition called *pulmonary edema* (eh-DEE-mah).

Both right and left ventricular failure may produce a similar finding in the systemic circuit: *peripheral edema,* in which blood backs up in the systemic capillaries as a result of systemic congestion. As in the pulmonary circuit, this backup forces fluid out of the capillaries and into the tissues. This process often causes visible swelling, especially in the legs and feet, where fluid collects as a result of gravity. Peripheral edema is exacerbated by the fact that the kidneys retain excess fluid during heart failure (in order to increase preload and compensate for a lower cardiac output).

Treatment of heart failure is generally aimed at increasing cardiac output via lifestyle modifications, medications, and surgery if necessary. Lifestyle modifications may include weight loss and mild exercise plus dietary sodium and fluid restrictions. Drug therapy increases cardiac output in one of at least three ways: (1) decreasing abnormally high preload by promoting fluid loss from the kidneys, (2) increasing the heart's contractility so that it pumps more effectively, and (3) decreasing afterload so that the ventricles have to pump against lower pressure. In some cases, a heart transplant and/or a surgically implanted pacemaker that electrically stimulates and paces the heart may be necessary.

Quick Check

☐ 9. How is heart failure defined?

Apply What You Learned

☐ 1. The drug digoxin increases the concentration of intracellular calcium ions in cardiac muscle cells. Predict the specific effect this will have on stroke volume and cardiac output.

☐ 2. Predict what would happen to cardiac output if sympathetic nervous system stimulation of the heart had positive chronotropic effects only.

☐ 3. A patient has a heart rate of 70 beats per minute, an EDV of 110 ml, and an ESV of 70 ml.
   a. What is this patient's stroke volume?
   b. What is her ejection fraction?
   c. What is her cardiac output?
   d. Are the values that you calculated within normal ranges? Explain.

See answers in Appendix A.

CHAPTER SUMMARY

Take full advantage of the invaluable online practice and study tools for this chapter by going to 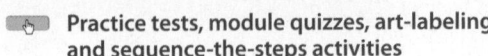. There you will find:

🖱 Practice tests, module quizzes, art-labeling and sequence-the-steps activities

▶ Animated figures with quizzes

✎ *Practicing A&P* worksheets

🔊 Author-narrated podcast

🔥 Dynamic Study Modules

iP *Interactive Physiology*

MODULE 17.1
Overview of the Heart 631

● The **heart** is a two-sided pump that drives blood into blood vessels. It is about the size of a fist and sits in the left side of the mediastinum. (Figure 17.1, p. 632)

● The heart's **apex** is its tip, and the **base** is the flattened posterior side.

● The four chambers of the heart include the two superior **atria** and the two inferior **ventricles.** The atria receive blood from **veins** and pump blood into the ventricles. The ventricles pump blood into **arteries.**

● The right side of the heart is the *pulmonary pump,* which pumps deoxygenated blood through the vessels of the **pulmonary circuit.** The left side of the heart is the *systemic pump,* which drives oxygenated blood through the **systemic circuit.** (Figure 17.2, p. 633)

● The heart also plays a role in blood pressure regulation and helps to regulate water balance by secreting the hormone *atrial natriuretic peptide.*

MODULE 17.2
Heart Anatomy and Blood Flow Pathway 634

● The heart wall has three layers: the outer **pericardium,** the middle **myocardium,** and the inner **endocardium.** The pericardium is composed largely of connective tissue and has two layers: the tough outer **fibrous pericardium** and the double-layered **serous pericardium.** (Figure 17.3, p. 634)
  ○ The serous pericardium itself has two layers: the outer **parietal pericardium** and the inner **visceral pericardium.** Between the parietal and visceral pericardial layers is the **pericardial cavity,** which is filled with lubricating serous fluid.
  ○ The myocardium is composed of cardiac muscle tissue and the heart's **fibrous skeleton.** The endocardium is composed of **endothelium.**

● The **coronary circulation** consists of the *coronary arteries,* which supply the myocardium with blood, and the *coronary veins,* which drain the myocardium. (Figure 17.4, p. 636)
  ○ The two main coronary arteries are the **right coronary artery** and the **left coronary artery.** The left coronary artery has two

main branches: the **anterior interventricular artery** and the **circumflex artery.**
- ○ The **coronary sinus** is the large vessel that receives blood from the coronary veins and empties it into the right atrium. It receives blood from the **great cardiac vein,** the **small cardiac vein,** and the **middle cardiac vein.**

- The four main great vessels are the **venae cavae (superior and inferior),** the **pulmonary trunk,** the **pulmonary veins,** and the **aorta.** (Figure 17.5, pp. 638–639)

👆 Practice art-labeling exercise: Figure 17.5.

- Each atrium has an external **auricle,** which is lined internally with **pectinate muscles.** The two atria are separated by the **interatrial septum.** (Figure 17.6, p. 640)

👆 Practice art-labeling exercise: Figure 17.6.

- The two ventricles are separated by the **interventricular septum.** The interior surfaces of both ventricles contain **trabeculae carneae** and finger-like **papillary muscles.** The left ventricle has a thicker wall than the right because it pumps against higher pressure.

- Blood flowing into the ventricles passes through the **atrioventricular valves,** including the **tricuspid valve** between the right atrium and right ventricle, and the **bicuspid** or **mitral valve** between the left atrium and left ventricle. The **cusps** of each valve are attached to papillary muscles by **chordae tendineae;** these cords prevent the cusps from everting when the ventricles contract. (Figure 17.7, p. 641)

- Blood flowing out of the ventricles passes through the **semilunar valves,** including the **pulmonary valve** between the right ventricle and the pulmonary trunk, and the **aortic valve** between the left ventricle and the aorta.

- Figure 17.8 (pp. 642–643) provides a big picture of blood flow through the heart.

iP Play Interactive Physiology tutorial on Cardiovascular System: Anatomy Review: The Heart.

MODULE **17.3**
Cardiac Muscle Tissue Anatomy and Electrophysiology 644

- **Cardiac muscle cells** contract in response to **action potentials** via the sliding-filament mechanism. **Pacemaker cells** are noncontractile and nonstriated cardiac muscle cells that are **autorhythmic** and spontaneously generate action potentials. These action potentials trigger cardiac **contractile cells** to in turn have action potentials.

- Cardiac muscle cells are short, branched, striated cells joined by **intercalated discs,** which allow the heart to act as a functional syncytium. (Figure 17.9, p. 645)

- A contractile cell action potential results from a reversal in membrane potential from about −85 mV to about 0–20 mV. The stages of the action potential are as follows: (Figure 17.10, p. 646)
  - ○ rapid depolarization due to an influx of sodium ions;
  - ○ a brief initial repolarization due to the closure of sodium ion channels and a slight outflow of potassium ions;
  - ○ a **plateau phase** due to an influx of calcium ions as potassium ions exit; and

- ○ repolarization due to potassium ion outflow through open potassium ion channels.

👆 Practice sequencing the steps: Figure 17.10.

- The plateau phase lengthens the action potential, lengthens and strengthens the heart's contraction, and lengthens the refractory period.

- Pacemaker cells depolarize rhythmically. Their action potentials consist of the following phases: (Figure 17.11, p. 649)
  - ○ a slow depolarization to threshold due to the inflow of cations through nonselective cation channels;
  - ○ depolarization due to the inflow of calcium ions once threshold is reached;
  - ○ repolarization due to the outflow of potassium ions; and
  - ○ hyperpolarization due to the continued outflow of potassium ions, which triggers the nonselective cation channels to open, and the cycle begins again.

👆 Practice sequencing the steps: Figure 17.11.

- Three populations of pacemaker cells exist in the heart, collectively called the **cardiac conduction system.** These pacemakers are as follows: (Figure 17.12, p. 650)
  - ○ the **sinoatrial (SA) node** in the upper right atrium, which paces at 60 or more beats per minute and is the normal pacemaker of the heart;
  - ○ the **atrioventricular (AV) node** posterior and medial to the tricuspid valve, which paces the heart at 40 beats per minute; and
  - ○ the **Purkinje system,** which consists of the **atrioventricular (AV) bundle,** the right and left **bundle branches,** and the terminal **Purkinje fibers.** The Purkinje system conducts action potentials faster than any other cells in the heart, but is the slowest pacemaker at 20 beats per minute.

👆 Practice art-labeling exercise: Figure 17.12.

- The action potential in a normal heart initiates in the SA node, spreads through the atria, then encounters a delay at the AV node. This delay permits the atria and ventricles to depolarize and contract separately. The action potential is then conducted along the AV bundle to the right and left bundle branches, then to the Purkinje fibers, and finally into the contractile cells of the ventricles.

- The **electrocardiogram** measures electrical changes occurring in the heart. The electrical changes appear as **waves** on the recording. (Figure 17.13, p. 651)

iP2 Play Interactive Physiology tutorial on Cardiovascular System: Intrinsic Conduction System.

MODULE **17.4**
Mechanical Physiology of the Heart: The Cardiac Cycle 653

- **Mechanical physiology** describes the physiology of cardiac pumping.

- Pressure changes generated by ventricular contraction and relaxation cause the blood to flow between chambers and the valves to open and close. (Figure 17.14, p. 653)

- Two heart sounds are normally heard when ascultating the heart: **S1,** which is caused by the closing of the AV valves at the beginning of isovolumetric contraction, and **S2,** which is caused by the closing of the semilunar valves at the beginning of isovolumetric relaxation. (Figure 17.15, p. 654)

- The sequence of events that takes place within the heart from one heartbeat to the next is the **cardiac cycle.** During the cardiac cycle, each chamber has one period of relaxation called **diastole** and one period of contraction called **systole.**
- There are four stages of the cardiac cycle: (Figure 17.16, p. 655)
  - During the **ventricular filling phase,** blood drains from the atria to the relaxed ventricles through the open atrioventricular valves. The volume of blood in the ventricles after atrial systole is the **end-diastolic volume (EDV).**
  - Ventricular systole begins and initiates **isovolumetric contraction,** during which both sets of valves are shut and the ventricular volume remains constant.
  - When the pressure in the ventricles rises above the pressure in the pulmonary trunk and aorta, the **ventricular ejection phase** begins, the semilunar valves open, and blood is ejected from the ventricles. The blood remaining in the ventricles when this stage is complete is the **end-systolic volume (ESV).**
  - **Isovolumetric relaxation** begins when the ventricles re-enter diastole. Both sets of valves are closed during this phase, and the volume in the ventricles remains constant.
- Blood is pushed out of the ventricles when the pressure in the ventricles is higher than the pressure in the vessels into which they pump the blood. The left ventricle has to squeeze significantly harder than the right ventricle to overcome the pressure in the aorta. (Figure 17.17, p. 656)
- The mechanical and electrical events of the cardiac cycle are shown in Figure 17.18 (p. 657) and explained more in Figure 17.19 (pp. 658–659).

iP2 Play Interactive Physiology tutorial on Cardiovascular System: Cardiac Cycle.

MODULE **17.5**
Cardiac Output and Regulation 660

- **Cardiac output** is the amount of blood pumped out by each ventricle in 1 minute. It is equal to the **heart rate** multiplied by the **stroke volume.**
- Stroke volume is the amount of blood ejected from the right or left ventricle with each beat, and is equal to the EDV minus the ESV.
- Three factors influence stroke volume:
  - **Preload** is the degree of stretch imposed on the cardiac muscle cells by the blood filling the ventricles before they contract. According to the **Frank-Starling law,** increases in preload raise the stroke volume.
  - **Contractility** is the effectiveness with which the heart pumps. Increases in contractility raise the stroke volume. Agents that regulate contractility are inotropic agents.
  - **Afterload** is the force against which the right and left ventricles must pump. Increases in afterload decrease stroke volume.

- Heart rate is influenced by chronotropic agents such as the ANS and endocrine system.
- Cardiac output is regulated primarily by the endocrine and nervous systems.
  - The sympathetic nervous system innervates the heart and releases epinephrine and norepinephrine, which are positive chronotropic and inotropic agents. (Figure 17.20, p. 662)
  - The parasympathetic nervous system innervates the heart via the vagus nerves. They release acetylcholine, which is primarily a negative chronotropic agent.
  - The endocrine system releases a variety of chronotropic and inotropic agents, including epinephrine, norepinephrine, thyroid hormone, and glucagon. Other hormones regulate water balance and preload.
- Figure 17.21 (p. 663) summarizes the factors that regulate cardiac output.

iP2 Play Interactive Physiology tutorial on Cardiovascular System: Cardiac Output.

ASSESS WHAT YOU LEARNED

See answers in Appendix A.

LEVEL 1: Check Your Recall

1. Mark the following statements as true or false. If a statement is false, correct it to make a true statement.

   a. The heart is located in the mediastinum slightly to the left of the midline.

   b. The heart consists of two superior ventricles and two inferior atria.

   c. Arteries always carry oxygenated blood away from the heart, and veins always carry deoxygenated blood toward the heart.

   d. The pulmonary circuit delivers blood from the right side of the heart to the lungs to become oxygenated.

   e. The heart plays a role in the regulation of blood pressure and secretes the hormone atrial natriuretic peptide.

2. The pericardial cavity is located between:

   a. the parietal pericardium and the fibrous pericardium.

   b. the fibrous pericardium and the myocardium.

   c. the parietal pericardium and the visceral pericardium.

   d. the epicardium and the endocardium.

3. Which of the following statements is *true*?

   a. The tricuspid valve is located between the right atrium and the right ventricle.

   b. The mitral valve is located between the pulmonary veins and the left atrium.

   c. The pulmonary valve is located between the pulmonary artery and the pulmonary veins.

   d. The aortic valve is located between the right ventricle and the aorta.

4. Match the following terms with the correct definition.

_____ Auricle
_____ Aorta
_____ Coronary sinus
_____ Papillary muscle
_____ Fossa ovalis
_____ Pectinate muscle
_____ Venae cavae
_____ Pulmonary trunk
_____ Chordae tendineae
_____ Pulmonary veins

a. Drainage point for the coronary veins
b. Extensions that attach papillary muscles to valves
c. Remnant of a hole present in the fetal interatrial septum
d. Two largest veins of the systemic circuit
e. Flaplike extension from the right or left atrium
f. Finger-like projections of ventricular muscle
g. Main artery of the pulmonary circuit
h. Veins that drain the pulmonary circuit
i. Largest artery of the systemic circuit
j. Ridges of muscle in the atria

5. Fill in the blanks: The coronary arteries are the first branches off the _____. The right coronary artery becomes the _____ on the posterior side of the heart. The left coronary artery branches into the _____ and the _____.

6. How do pacemaker cardiac muscle cells differ from contractile cardiac muscle cells? What is autorhythmicity?

7. Cardiac muscle cells are joined by structures called:

a. T-tubules.
b. tight junctions.
c. sarcoplasmic reticulum.
d. intercalated discs.

8. Mark the following statements as true or false. If a statement is false, correct it to make a true statement.

a. The rapid depolarization phase of the contractile cell action potential is due to the opening of voltage-gated potassium ion channels.
b. Pacemaker cells lack a distinct plateau phase.
c. The plateau phase in contractile cells is due to the influx of calcium ions through calcium ion channels.
d. The repolarization phase of the contractile cell is due to the potassium ions rushing into the cell through potassium ion channels.
e. Open sodium ion channels cause hyperpolarization in pacemaker cells, which triggers nonselective cation channels to open and begins a new action potential.

9. What are the effects of the plateau phase of the contractile cell action potential?

10. The _____ is the primary pacemaker of the heart.

a. atrioventricular node
b. sinoatrial node
c. Purkinje fiber system
d. atrioventricular bundle

11. The AV node delay:

a. allows the atria and ventricles to depolarize and contract as a unit.
b. allows the two ventricles to depolarize and contract separately.
c. allows the atria and ventricles to depolarize and contract separately.
d. speeds up the impulse transmission from the atria to the ventricles.

12. Explain what each of the following terms represents on an electrocardiogram (ECG).

a. P wave
b. QRS complex
c. T wave
d. P-R interval
e. S-T segment

13. Mark the following statements as true or false. If a statement is false, correct it to make a true statement.

a. Systole is the contraction portion of the cardiac cycle and diastole is the relaxation portion.
b. Atrial systole is responsible for ejecting most of the blood into the ventricles during the ventricular filling phase of the cardiac cycle.
c. The amount of blood in the ventricles at the end of the ventricular filling phase is the end-systolic volume.
d. The ventricular ejection phase generally correlates with the S-T segment and the T wave on the ECG.

14. Which chamber generates the highest pressure during systole?

a. Right atrium
b. Right ventricle
c. Left atrium
d. Left ventricle

15. Fill in the blanks: The first heart sound is called _____ and it is caused by the closing of the _____ valves. It occurs at the beginning of the _____ phase of the cardiac cycle. The second heart sound is called _____ and it is caused by the closing of the _____ valves. It occurs at the beginning of the _____ phase of the cardiac cycle.

16. Cardiac output is equal to:

a. end-diastolic volume minus end-systolic volume.
b. heart rate multiplied by stroke volume.
c. stroke volume divided by end-diastolic volume.
d. heart rate multiplied by preload.

17. Fill in the blanks: An increase in preload causes a/an _____ in stroke volume in accordance with the _____ law. An increase in afterload causes a/an _____ in stroke volume. An increase in contractility causes a/an _____ in stroke volume.

18. Which of the following statements is *false*?

a. The sympathetic nervous system releases epinephrine and norepinephrine, which are positive chronotropic and inotropic agents.
b. The endocrine system regulates cardiac output through chronotropic and inotropic hormones and through hormones that regulate water balance.

c. The parasympathetic nervous system releases acetylcholine, which is a strongly negative inotropic agent.

d. Factors such as electrolyte concentrations, body temperature, and age all affect cardiac output.

LEVEL 2: Check Your Understanding

1. A birth defect called *transposition of great vessels* results in the pulmonary trunk emanating from the left ventricle and the aorta stemming from the right ventricle.

   a. Which ventricle is thicker-walled, and why?

   b. Considering your answer to part (a), predict the potential effects of this birth defect.

2. Predict which would be more damaging to long-term survival: a blood clot lodged in the right coronary artery or one in the left coronary artery. Explain.

3. When the SA node doesn't function properly, the AV node takes over pacing the heart and produces what is known as a junctional rhythm. Explain why we don't see P waves on the ECG of an individual with such a rhythm.

4. Common findings in heart failure are fluid retention from the kidneys and stimulation of the heart by the sympathetic nervous system. How would both of these findings help the body to compensate for the failing heart?

LEVEL 3: Apply Your Knowledge

PART A: Application and Analysis

1. You are an athletic trainer who is working with someone planning to run a marathon. Your trainee tells you to give him a workout that will make his heart "beat faster than ever before." What do you tell him about the effects of too rapid a heart rate?

2. You are a nursing student in a hospital, and a fellow nursing student tells you that he has prepared two syringes: one with potassium chloride and one with sodium chloride. He forgot to label the syringes and isn't sure which syringe is which, but he figures that potassium and sodium are "pretty much the same anyway." What do you tell him to do? Explain.

3. Mr. Watson has been diagnosed with mitral insufficiency, or a malfunctioning mitral valve, which causes the valve to not close properly. Predict the signs and symptoms you might expect from a disease of this valve. What would happen to the patient's stroke volume and cardiac output? Explain. What might help improve his cardiac output?

PART B: Make the Connection

4. An experimental toxin makes the refractory period of cardiac muscle cells equal in length to that of skeletal muscle fibers. Predict the consequences of this toxin. *(Connects to Chapter 10)*

18

The Cardiovascular System II: The Blood Vessels

We now turn our attention to the second portion of the cardiovascular system: the blood vessels, collectively called the **vasculature** (VASS-kyoo-lah-chur; *vascul-* = "vessel"). The vasculature of one individual consists of billions of blood vessels that transport blood to the tissues, where gases, nutrients, and wastes are exchanged, and then transport it back to the heart. If these blood vessels were situated end to end, they would measure over 60,000 miles long! But blood vessels are not merely the "pipes" of the cardiovascular system. They also regulate blood flow to tissues, control blood pressure, and secrete a variety of chemicals.

In the heart chapter, we discussed the two circuits that carry blood through the body: the *pulmonary circuit,* which transports blood between the heart and the lungs; and the *systemic circuit,* which transports blood between the heart and the rest of the body (see Chapter 17). Here we cover primarily the systemic circuit, as the pulmonary circuit is discussed with the respiratory system (see Chapter 21). The first part of this chapter (Modules 18.1 through 18.5) explores the basic structure and function of the different types of blood vessels, the physiology of blood pressure and circulation, and the physiology of gas, nutrient, and waste exchange. In the second part (Modules 18.6 through 18.8), we examine the anatomy of the vessels that make up the systemic circuit.

MODULE **18.1**

Overview of Arteries and Veins

Learning Outcomes

1. Compare and contrast the structures of arteries and veins, and of arterioles and venules.
2. Describe the structure and function of the types of arteries and veins.
3. Define vascular anastomosis, and explain the significance of anastomoses.

As you read about in the heart chapter, the pulmonary and systemic circuits are composed of three kinds of blood vessels—arteries, capillaries, and veins (see Figure 17.8):

- **Arteries** are the *distribution system* of the vasculature. As they travel away from the heart, they branch into vessels of progressively smaller diameter that supply most tissues in the body with blood. Recall that arteries in the pulmonary circuit carry deoxygenated blood, whereas those in the systemic circuit carry oxygenated blood.

Computer-generated image: Blood vessels are tubular organs that distribute blood throughout the body.

669

- **Capillaries** are the *exchange system* of the vasculature. They are very-small-diameter vessels that form branching networks called **capillary beds.** Gases, nutrients, wastes, and other molecules are quickly exchanged between tissue cells and the blood through the capillary walls, many of which are only a single cell thick.
- **Veins** function as the *collection system* of the vasculature. Veins drain blood from capillary beds and return it to the heart; they follow the opposite pattern of arteries—small veins merge with other veins to become progressively larger vessels as they progress toward the heart. In the pulmonary circuit, veins transport oxygenated blood, whereas in the systemic circuit, they transport deoxygenated blood.

In this module, we introduce the basic structure of the blood vessel wall and the general features of arteries and veins. The basic structure and function of capillaries and capillary beds are discussed in Module 18.4.

Structure and Function of Arteries and Veins

◀◀ **Flashback**

1. What is smooth muscle tissue, and what are its basic properties? (p. 148)
2. What is the inner lining of the heart called? Of what tissue type is it composed? (p. 635)

Although blood vessels differ in both diameter and the thickness of their walls, they still share the same basic pattern of organization. All blood vessels are tubular organs that contain a central space—the **lumen**—surrounded by several tissue layers, or **tunics.**

The three tunics of the blood vessel wall are the *tunica intima, tunica media,* and *tunica externa* (**Figure 18.1**):

- **Tunica intima.** The innermost **tunica intima** is composed of **endothelium,** which consists of a sheet of simple squamous epithelium and its basal lamina. Recall that this endothelium is continuous with the inner lining of the heart, the *endocardium,* which is also composed of endothelium (see Chapter 17). Endothelial cells provide a smooth surface over which blood can flow with a minimum of friction and turbulence. In addition, endothelial cells produce many chemicals and proteins, such as *nitric oxide,* collagen, and clotting factors, and they contain enzymes within their plasma membrane. Deep to the endothelium we find a thin layer of subendothelial connective tissue and a layer of elastic fibers called the *internal elastic lamina.* The elastic fibers give the vessel the properties of *distensibility,* or the ability to stretch when subjected to increased pressure, and *elasticity,* or the ability to recoil back to the original size when the stretching force is removed.
- **Tunica media.** The middle layer of the blood vessel wall is the **tunica media** (*medi-* = "middle"). As you can see in Figure 18.1, it has two components: a layer of smooth muscle cells arranged in a circular manner around the lumen, and

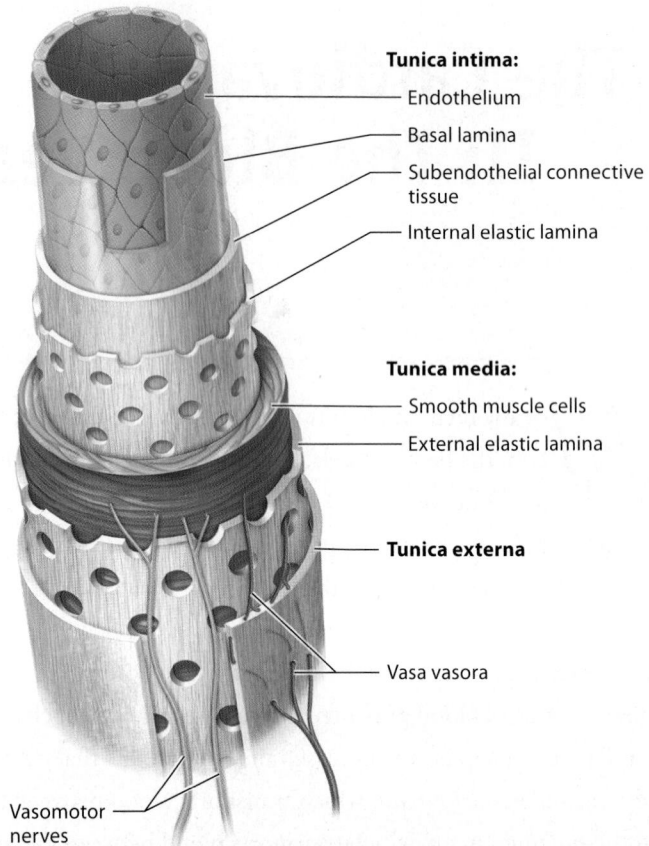

Figure 18.1 The tunics (layers) of the blood vessel wall.

Tunica intima:
- Endothelium
- Basal lamina
- Subendothelial connective tissue
- Internal elastic lamina

Tunica media:
- Smooth muscle cells
- External elastic lamina

Tunica externa

Vasa vasora

Vasomotor nerves

another layer of elastic fibers called the *external elastic lamina.* The smooth muscle cells of the tunica media control the diameter of the blood vessel and so the amount of blood that flows to organs. Notice that these muscle cells are innervated by nerves from the sympathetic nervous system called *vasomotor nerves.* These nerves stimulate the smooth muscle cells of the tunica media to contract, an action known as **vasoconstriction,** which narrows the diameter of the vessel. When sympathetic stimulation of the smooth muscle cells decreases, these cells relax and the vessel's diameter increases, a change called **vasodilation.** As discussed later, vasoconstriction and vasodilation are important determinants of blood pressure and the blood flow to organs and tissues.

- **Tunica externa.** The outermost **tunica externa** (also known as the *tunica adventitia*) is composed of dense irregular collagenous connective tissue that supports the blood vessel and prevents it from overstretching.

In Figure 18.1, you can see tiny vessels supplying the tunica media and tunica externa. These vessels are the **vasa vasora** (VAY-zah vay-ZOH-rah), or "vessels to the vessels." The vasa vasora supply oxygen and nutrients to the outer layers of the larger blood vessels, whose cells are too far away from the lumen to receive oxygen and nutrients by diffusion alone.

Note that the structure of the blood vessel wall—an inner epithelial tissue lining, a layer of connective tissue, a layer of smooth muscle tissue, and an additional outer layer of tough connective tissue—is a pattern you will see repeated in other hollow organs. In an example of the Structure-Function Core Principle (p.

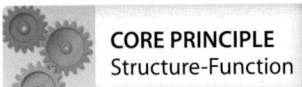
CORE PRINCIPLE
Structure-Function

25), this pattern of organization ensures that the structure of hollow organs best follows their function: to control the flow of substances that pass through them.

Quick Check

☐ 1. Define the three types of blood vessels in the cardio-vascular system.

☐ 2. Of which tissue types are the following layers of the blood vessel wall composed?
   a. Tunica intima
   b. Tunica media
   c. Tunica externa

☐ 3. What are vasoconstriction and vasodilation?

Arteries

As you can see in **Figure 18.2**, there are notable differences between the walls of a typical artery and those of a typical vein. Most arteries have much thicker tunicae mediae than do veins, which reflects the arteries' role in controlling blood pressure and blood flow to organs. Additionally, the internal and external elastic laminae are much more extensive in arteries than in veins, which reflects the fact that arteries are under much higher pressure than are veins.

The thickness of each component of the arterial wall depends on the artery's size and function. We can generally classify arteries based on similarity in structure and function into three classes—*elastic arteries, muscular arteries,* and *arterioles.*

- **Elastic arteries.** The largest-diameter arteries are the **elastic arteries,** also known as the **conducting arteries.** As their name implies, elastic arteries have very extensive elastic laminae. The tunica media of an elastic artery contains 40–70 sheets of elastic fibers arranged between thin layers of smooth muscle cells. These arteries, including the aorta and its immediate branches, are nearest the heart and therefore under the highest pressure of any vessels in the cardiovascular system. For this reason, they need to be extremely distensible, as they are greatly stretched with each ventricular systole. The relatively small amount of smooth muscle tissue in these vessels means that their diameter does not change significantly with stimulation from the vasomotor nerves.

- **Muscular arteries. Muscular arteries,** also called **distributing arteries,** are generally intermediate in diameter. They contain a well-developed tunica media composed primarily of smooth muscle cells. Most of the smaller branches off the aorta are considered muscular arteries, including most of the named arteries that supply organs, discussed in Module 18.6. Due to the predominance of smooth muscle cells, the diameter of muscular arteries does change significantly with vasoconstriction and vasodilation. This allows the nervous and endocrine systems to adjust local blood flow to different organs by changing the vessel diameter. These vessels are the ones most likely to become blocked, as you can discover in *A&P in the Real World: Atherosclerosis.*

- **Arterioles.** The smallest arteries are the **arterioles** (ar-TEER-ee-olz), which range in size from 0.3 mm (the size of the smallest pencil lead) to less than the diameter of a human hair (about 70–120 μm). Arterioles contain each of the three layers of the blood vessel wall, but the layers are extremely thin, and the tunica media contains only one to

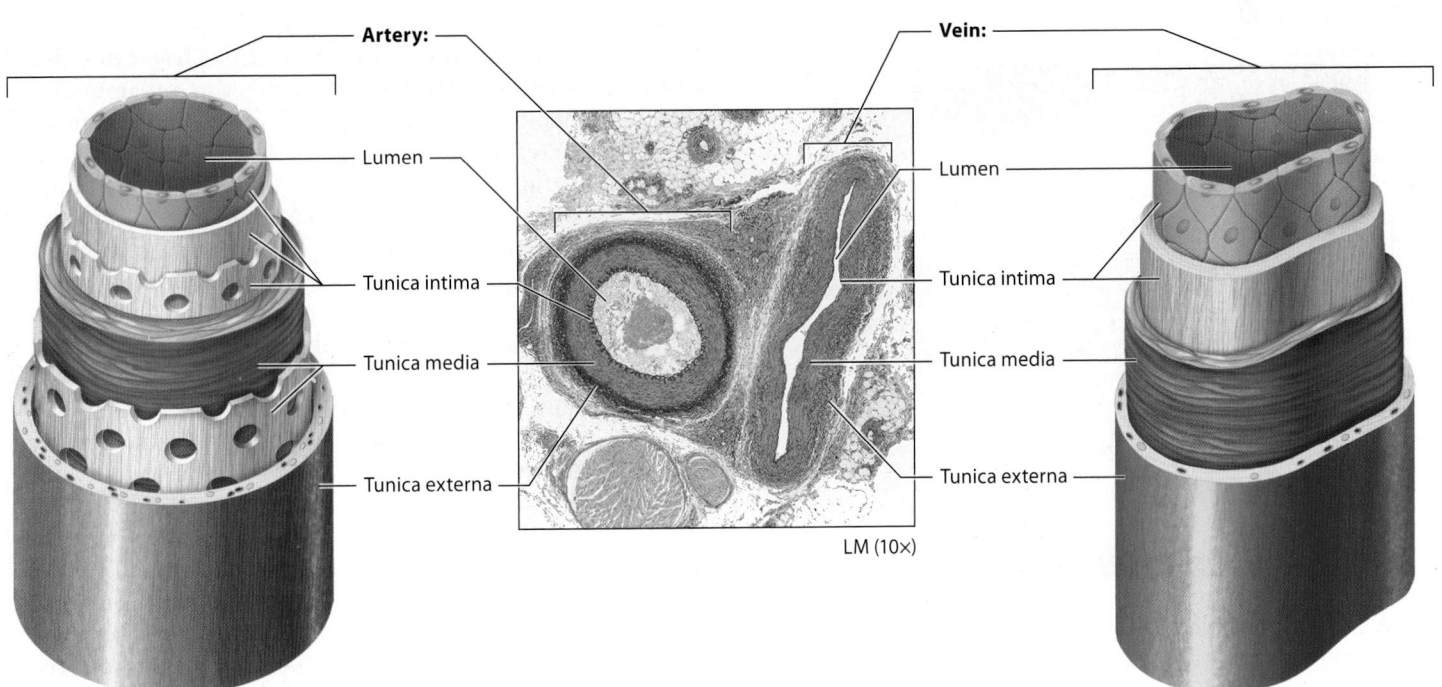

Figure 18.2 A comparison of the walls of arteries and veins. The photomicrograph shows an artery and a vein in cross section.

three layers of smooth muscle cells. The smallest arterioles, called **metarterioles** (*meta-* = "boundary"), directly feed capillary beds in most tissues. The smooth muscle cells of metarterioles are confined mostly to a circular **precapillary sphincter** that encircles the metarteriole-capillary junction. Both vasomotor nerves and hormones in the blood can affect the diameter of arterioles; precapillary sphincters are mostly responsive to hormones and local tissue conditions. Vasoconstriction and vasodilation of these structures have a profound impact on the blood flow to individual tissues; in fact, arteriolar and/or precapillary sphincter constriction can completely cut off blood flow to tissues. We revisit the structure of arterioles and metarterioles in Module 18.4, where we discuss capillary beds in detail.

In addition to their roles as conduits for blood and in controlling blood flow to organs and tissues, certain arteries also play a role in monitoring blood pressure and detecting the concentration of certain chemicals in the blood. Pressure receptors, or **baroreceptors,** are found in the aorta, as well as in the *common carotid artery,* located in the neck. Also in the carotid artery and the aorta are groups of *chemoreceptors* that detect blood oxygen,

carbon dioxide, and hydrogen ion concentrations. These pH receptors, discussed in Module 18.3, send signals to the CNS.

Quick Check

☐ 4. Describe the basic properties of elastic arteries, muscular arteries, and arterioles.

☐ 5. Which type of artery controls blood flow to organs? Which controls blood flow to tissues?

Veins

Veins typically outnumber arteries and their lumens have a larger average diameter. For these reasons, up to 70% of the total blood in the body is located in the veins at any given moment (this includes both systemic and pulmonary veins). This feature allows veins to function as **blood reservoirs;** Figure 18.3 shows how much of the blood is typically found in veins and how much in other parts of the cardiovascular system. When necessary, blood can be diverted from the veins to other parts of the body. In the histological section in Figure 18.2, the vein appears collapsed, whereas the artery is circular—this is because

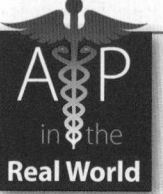

Atherosclerosis

The leading cause of death in the developed world is **atherosclerosis** (ah-ther′-oh-skleh-ROW-sis), an arterial disease that affects large- and medium-sized muscular arteries. The disease is characterized by the formation of atherosclerotic **plaques** (PLAKS), which are buildups of lipids, cholesterol, calcium salts, and cellular debris within the arterial tunica intima, as shown here:

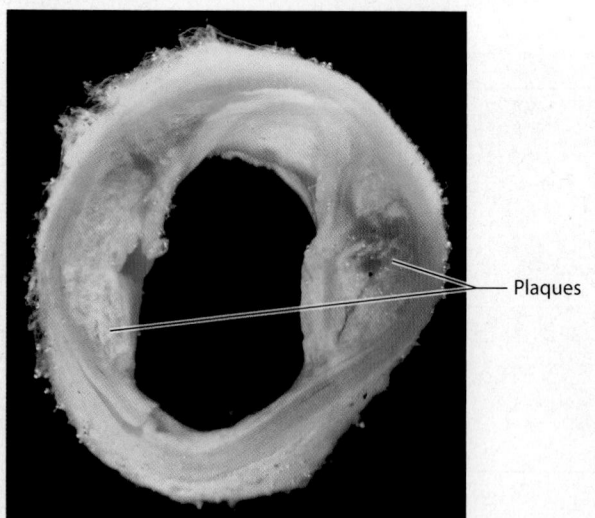

Plaques

Atherosclerotic plaques tend to form in regions where blood undergoes sudden changes in velocity and direction of flow, such as branching points or where the vessels curve.

Current evidence indicates that these plaques are generated in response to some sort of injury to the endothelium. The endothelium may be injured by high blood pressure, certain types of cholesterol, previous infections, the toxins in cigarette smoke, a high and prolonged blood glucose level, and a high blood concentration of a chemical called *homocysteine* (hoh′-moh-SIS-tee-een). When the endothelium is injured, the vessel wall becomes inflamed, which attracts phagocytes that attempt to "clean up" the area. However, this inflammation actually results in damage to the blood vessel, and eventually the damaged area turns into a plaque. Although the plaque is located in the tunica intima, changes also occur in the tunica media. Smooth muscle cells proliferate and secrete an extracellular matrix with protein fibers and ground substance that surrounds the plaque. If the contents of the plaque are exposed to blood, a clot may form in the vessel and obstruct blood flow entirely, as occurs in a myocardial infarction (heart attack) or stroke.

An estimated 10% of the world population has atherosclerosis, but its true prevalence is difficult to determine because most patients don't display symptoms in the early stages. Much of the treatment of atherosclerosis focuses on reducing the factors that injure the endothelium. This includes dietary modification, physical activity, agents to lower cholesterol, control of blood glucose level, smoking cessation, and management of high blood pressure. In severe disease, surgery or other invasive procedures may be necessary to open or bypass occluded vessels.

veins typically have much thinner walls, fewer elastic fibers, less smooth muscle, and larger lumens than arteries.

The smallest veins are the **venules,** which drain blood from capillary beds. The tiny **postcapillary venules** consist of little more than endothelium and some surrounding connective tissue. This structure enables them to exchange material with the surrounding interstitial fluid in much the same way that materials are exchanged in capillaries. The three tunics become more distinct as venules merge to become larger venules and then veins. Most veins have a thin tunica media with few smooth muscle cells, and their diameter changes only slightly with vasodilation and vasoconstriction.

Many veins contain **venous valves.** These valves are extensions of the tunica intima that overlap and prevent blood from flowing backward in the venous circuit. They are especially numerous in the veins of the legs, where blood flow toward the heart is strongly opposed by gravity.

Table 18.1 summarizes information on the types of arteries and veins.

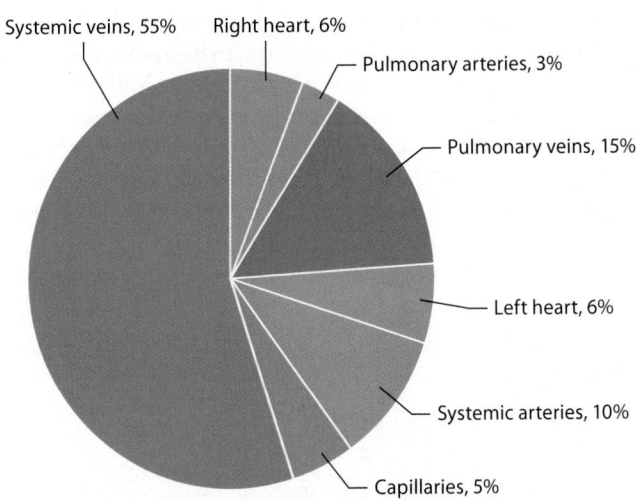

Figure 18.3 Blood distribution in the cardiovascular system.

Quick Check

☐ 6. How do veins differ structurally and functionally from arteries?
☐ 7. What are venules, and how do they differ from veins?
☐ 8. What are venous valves, and what are their functions?

Vascular Anastomoses

◀◀ Flashback

1. What is an anastomosis? What purpose does this structure serve in the coronary circulation? (p. 637)

In the heart chapter you read about the **vascular anastomoses** (an-ass'-toh-MOH-seez) that are present within the coronary circulation (see Chapter 17). Recall that anastomoses are locations where vessels connect via pathways called **collateral vessels.** *Arterial anastomoses* exist in many organs such as the heart and the brain, as well as around joints. Interestingly, new arterial anastomoses can be formed when blood flow through an artery is insufficient to meet the tissue's metabolic needs. Tissues deprived of oxygen appear to secrete chemicals that trigger a process called *angiogenesis,* or the formation of new blood vessels. As these new arteries grow, they often connect parts of the circulation via collaterals and increase blood flow to the tissue.

The most common type of anastomosis is a *venous anastomosis,* in which neighboring veins are connected by small

	Elastic arteries	Muscular arteries	Arterioles	Venules	Veins
Diameter	2.5–1.0 cm	1.0 cm–0.3 mm	0.3 mm–10 µm	8.0–100 µm	100 µm–1.5 cm
Structure	Large arteries with well-developed elastic laminae	Thick-walled arteries with well-developed tunica media	Thin walls with all three tunics	Small venules have only a tunica intima; larger venules have all three tunics	Thin-walled vessels with large lumens, little smooth muscle, and valves
Function(s)	• Conduct blood under high pressure to organs	• Control blood flow to organs • Regulate blood pressure	• Control blood flow to tissues • Feed capillary beds	• Drain capillary beds	• Return blood to the heart

TABLE 18.1 TYPES OF ARTERIES AND VEINS

collaterals. Smaller veins are often so interconnected by collaterals that they form complex, weblike patterns that are sometimes visible beneath the skin, particularly in individuals with pale skin. A third type of anastomosis is the *arteriovenous anastomosis,* in which an artery empties directly into a vein without passing through a capillary bed. Examples of arteriovenous anastomoses are found in the skin and in the fetal circulation, where blood needs to be shunted between the arterial and venous systems to bypass certain organs.

Quick Check

☐ 9. What is an anastomosis?

Apply What You Learned

☐ 1. Predict the possible effects of defective elastic fibers in arteries.

☐ 2. Would you expect the same effects if venous elastic fibers were defective? Why or why not?

☐ 3. How does atherosclerosis change the structure of an artery? How does this impair the artery's ability to perform its functions?

See answers in Appendix A.

MODULE 18.2
Physiology of Blood Flow

Learning Outcomes

1. Define blood flow, blood pressure, and peripheral resistance.
2. Explain the relationships between vessel diameter, cross-sectional area, blood pressure, and blood velocity.
3. Explain how blood pressure varies in different parts of the systemic and pulmonary circuits.
4. Describe how blood pressure changes in the arteries, capillaries, and veins.
5. Explain how mean arterial pressure is calculated.
6. Describe the mechanisms that assist in the return of venous blood to the heart.

Public awareness of blood pressure has increased dramatically over the past two decades. Mountains of research have shown the dramatic effects that high blood pressure, or *hypertension,* has on heart disease, vascular disease, kidney disease, stroke risk, and more. But what exactly is blood pressure and why does it affect the health of so many body systems? To answer these questions, we now turn our attention to *hemodynamics* (hee'-moh-dy-NAM-iks; "blood power"), or the physiology of

blood flow in the cardiovascular system. In this module we first explore the basic concepts of circulation and define terms such as blood pressure and blood flow. The discussion then shifts to those factors that determine blood pressure, to the blood pressure in different circuits, and to types of vessels in the cardiovascular system.

Introduction to Hemodynamics

◄◄ **Flashback**

1. What is a gradient? (p. 26)
2. What is cardiac output? What is the average value for cardiac output? (p. 660)

Before we venture into the physiology of circulation, we need to address some basic concepts related to it, such as gradients. Recall that gradients—whether concentration gradients, pressure gradients, or electrical gradients—drive most of the processes in the body. The physiology of the cardiovascular system is no exception. As you learned in the last chapter, the pumping heart provides the force that drives blood through the vasculature. In an example of the Gradients Core Principle (p. 26), the heart drives blood through these vessels by creating a *pressure gradient.* The pressure is highest near the heart and decreases as we move 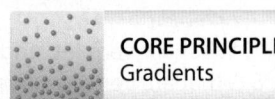 away from it. Blood flows down this pressure gradient from the area of higher pressure near the heart to the area of lower pressure in the peripheral vasculature.

CORE PRINCIPLE Gradients

But what exactly is "pressure" with respect to blood? Defined simply, **blood pressure** is the outward force that the blood exerts on the walls of the blood vessels. Blood pressure is expressed in the units *millimeters of mercury,* or *mm Hg,* which is the force exerted by a column of mercury one millimeter in height. So, if the blood pressure is 40 mm Hg, the pressure in the blood vessel is equal to that generated by a column of mercury 40 mm in height. The blood pressure varies dramatically in different parts of the vasculature—it is highest in the large systemic arteries and lowest in the large systemic veins.

The magnitude of the blood pressure gradient is one main factor that determines the **blood flow,** or the volume of blood that flows per minute. Generally, blood flow matches cardiac output, which averages about 5–6 liters/min. Blood flow is *directly proportional* to the pressure gradient, meaning that blood flow increases when the pressure gradient increases and vice versa. The other factor that determines blood flow is **resistance,** which is defined as any impedance to blood flow. Blood flow is *inversely proportional* to resistance; that is, as resistance increases, blood flow decreases.

The velocity with which blood flows is largely determined by the cross-sectional area of the blood vessel. As the arterial system branches into more, progressively smaller vessels, the total cross-sectional area increases. This increase in area causes the velocity of blood flow to decrease. For this reason, the velocity of blood flow is fastest in the aorta and slowest in the capillaries.

ConceptBOOST ⟫⟫⟫

A Closer Look at Cross-Sectional Area and Velocity

The idea of cross-sectional area of a vessel might sound complicated, but it's actually a combination of two fairly simple things. First, it refers to something you may recall from the first chapter of this book—a cross-section, or transverse section, through a blood vessel. The "area" is simply the mathematical calculation of the area of the circle made by the cross-section (A), which is equal to pi (π) times the radius of the circle (r) squared (or $A = \pi r^2$).

The radius of an arteriole or capillary is quite small, so the cross-sectional area of such a vessel will also be very small. However, these small vessels are far more numerous than large vessels with a large cross-sectional area. For this reason, the *total* cross-sectional area of the small vessels is much greater than that of the large vessels. You can see this in the very simple example below, in which our one large vessel has a cross-sectional area of 5 cm², and each of our five smaller vessels has an area of 2 cm²:

As you can see, the total cross-sectional area of the small vessels (10 cm²) is twice that of our large vessel. So it's easy to see why cross-sectional area increases as arteries branch into millions of smaller vessels.

Let's apply this to velocity, which decreases as cross-sectional area increases.* This is because the same volume of blood is now filling the equivalent of a much larger container:

*Note that we are assuming a flow rate of 10 ml/s in determining velocity.

You can see that the blood flows fastest in the small container and slowest in the large one. This happens because blood flowing through a larger container is spread out over a wider area. So when cross-sectional area increases as vessels branch, the blood is more spread out, causing it to flow more slowly. This is important because a slower velocity of blood flow allows efficient gas and nutrient exchange in capillaries, which we discuss shortly. ■

Quick Check

☐ 1. What is hemodynamics?
☐ 2. How are blood flow and the pressure gradient related? How are blood flow and resistance related?
☐ 3. How does cross-sectional area influence the velocity of blood flow?

Factors That Determine Blood Pressure

◀◀ Flashback

1. What are stroke volume and heart rate? (p. 660)
2. What is contractility, and how do changes in contractility affect cardiac output? (p. 661)

Multiple factors interact to determine the blood pressure in various parts of the vasculature. The three main factors that influence blood pressure are *resistance, cardiac output,* and *blood volume.* As we discuss in the next module, the body maintains blood pressure at a relatively constant set point by adjusting one or more of these three factors. Each factor is explored next; **Figure 18.4** provides an overview.

Peripheral Resistance

Anything that hinders blood flow through the vasculature contributes to the overall resistance of the circuit. The vessels near the heart are generally large, elastic vessels that contribute little to overall resistance. Instead, most of the resistance is encountered away from the heart, in the body's *periphery,* which has given rise to the term **peripheral resistance.** Peripheral resistance and blood pressure are directly related: *As peripheral resistance increases, blood pressure increases.*

Resistance is mainly determined by three variables: blood vessel radius, blood viscosity, and blood vessel length. Of these variables, vessel radius is the quickest to change, whereas vessel length is, of course, the slowest. Let's examine each of these more closely.

- **Radius of the vessels.** Vessel radius dramatically affects resistance. Resistance varies *inversely* with the vessel's radius (remember from basic math that radius is just one-half of the diameter of a circle, in this case a transverse cross-section of the vessel). So, as a vessel's radius increases (i.e., the vessel dilates), the resistance to blood flow decreases, and vice versa.

- **Blood viscosity. Viscosity** is often described as the "thickness" of a liquid, but its technical definition is the inherent resistance that all liquids have to flow. The more viscous a liquid is, the more its molecules resist being put into motion and staying in motion. Blood has a relatively high viscosity due to the number of proteins and cells it contains. Generally, blood viscosity remains relatively constant, but it can be altered by states that change either the number of cells or proteins in the blood or the amount of water in blood, such as dehydration. Peripheral resistance is increased by

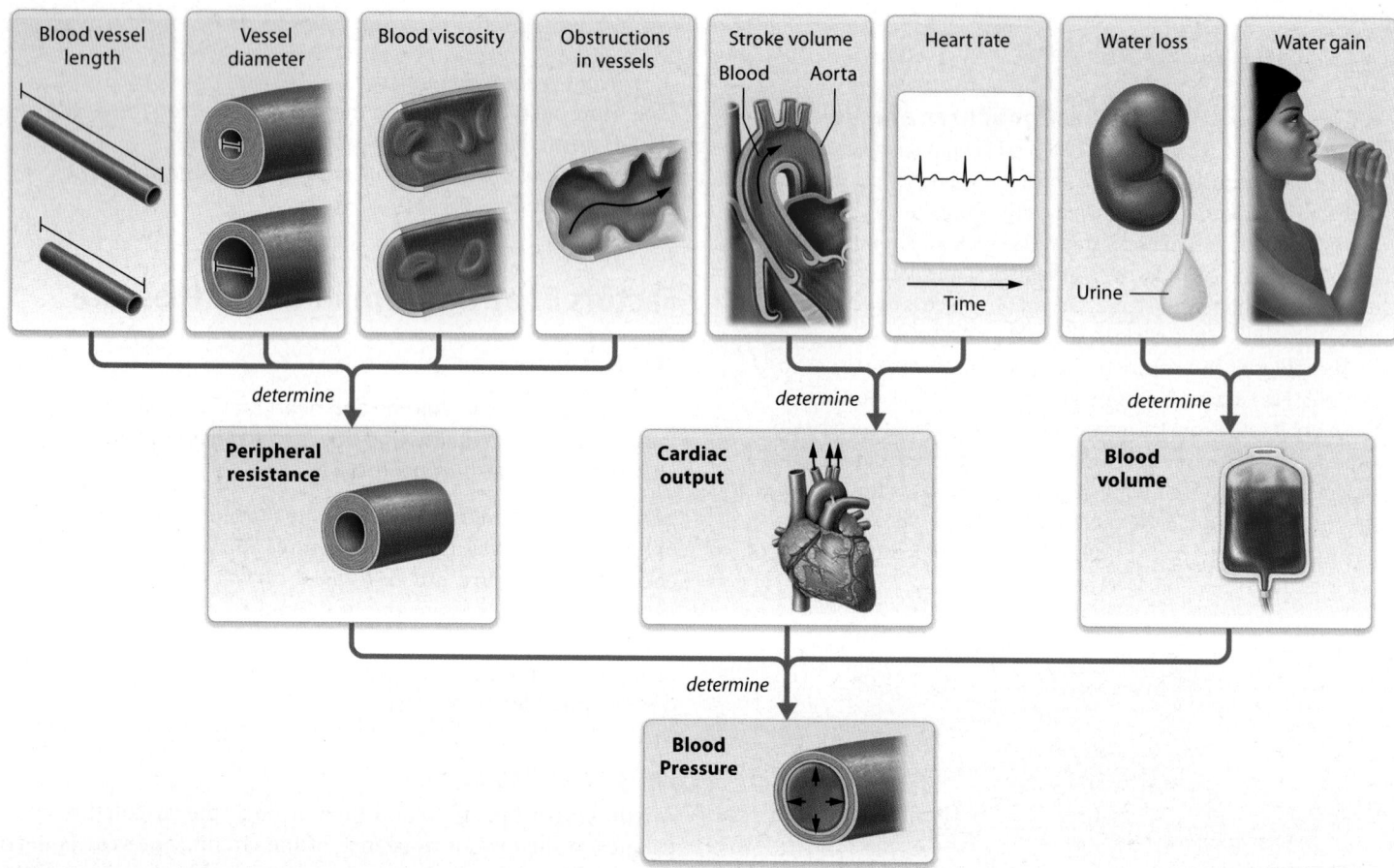

Figure 18.4 Factors that determine blood pressure.

conditions that increase blood viscosity and is decreased by conditions that decrease blood viscosity. This makes sense logically—the flow in a pipe filled with maple syrup (higher viscosity) is going to be much slower than the flow in one of the same diameter filled with water (lower viscosity).

- **Blood vessel length.** The length of a blood vessel also influences resistance. Stated simply, the longer the blood vessel, the greater the resistance. Just as more pressure is required to propel water through a long hose than a short one, more pressure is needed to propel blood through a long vessel than a short one. This is one reason why resistance in the pulmonary circuit is so much lower than in the systemic circuit: The vessels in the pulmonary circuit are simply shorter and therefore offer less resistance to blood flow. This also explains in part why blood pressure rises with obesity—vessel length increases as body size increases.

A fourth factor that influences peripheral resistance is the presence of obstructions within the blood vessels. Obstructions in the vessel, such as tumors, the fatty plaques seen with atherosclerosis, or blood clots, affect the way in which blood flows through the vessel. Ideally, blood flow through a vessel should be *laminar* (*laminar* = "layered"), in which the layer of blood nearest the blood vessel wall adheres to the wall due to friction (resistance) but the blood in the center of the vessel flows more freely. Flow becomes *turbulent* when there are obstructions within the vessel. In turbulent flow, blood doesn't move forward readily and

requires more force to move through the vessels, which increases resistance. Although laminar blood flow produces no audible sound, turbulent blood flow often does as the blood bounces along the vessel walls. These sounds are heard in the heart as murmurs and in the vessels as *bruits* (BROO-eez; *bruit* = "noise").

Cardiac Output

As we've explained, cardiac output (CO) is the product of *stroke volume* (the amount of blood pumped with each beat) times *heart rate* (the number of beats per minute) (look at Figure 18.4 again). Cardiac output and peripheral resistance are the two factors that determine the pressure gradient that drives circulation. The relationship is expressed by the simple mathematical equation:

$$\Delta P \quad = \quad CO \quad \times \quad PR$$

$$\text{Pressure} \qquad \text{Cardiac} \qquad \text{Peripheral}$$
$$\text{gradient} \qquad \text{output} \qquad \text{resistance}$$

Note here that the symbol Δ (delta) indicates change, meaning that a pressure change is caused by altering cardiac output and/or peripheral resistance. This relationship indicates that anything changing cardiac output is also going to alter the pressure gradient that drives blood flow and therefore the blood pressure. So, generally speaking, *when cardiac output increases, blood pressure increases*. Such an increase occurs, for example, in response to sympathetic nervous system stimulation and the intake of drugs such as caffeine. The reverse is also true; decreases in cardiac output lower blood pressure. Factors that decrease cardiac output include parasympathetic nervous system stimulation and drugs that block the sympathetic response.

Blood Volume and Vessel Compliance

A final factor that determines overall blood pressure is the volume of blood in the circulation. Note that the total volume of blood is directly linked to the amount of water in the blood—when the blood contains more water, blood volume increases, and vice versa. The relationship between blood volume and blood pressure is pretty straightforward: *As blood volume increases, blood pressure increases.* This relationship makes sense because when the blood volume increases, the vessels are "fuller" and the pressure on their walls is greater. Conversely, when the blood volume decreases, the vessels are "less full" and the pressure on their walls is lower. A simple way to understand this is to imagine the stress on the wall of a balloon when you fill it with air. When you add more air to the balloon so that the volume increases, the pressure exerted on its walls rises. The opposite is true as well—when you let air out of the balloon so that the volume decreases, the pressure exerted on its walls falls.

Small increases in blood volume are offset by the ability of the vessels to stretch, a property known as *compliance.* Veins are the most compliant vessels, and they stretch to accommodate the added fluid when blood volume increases, with only a small rise in pressure. However, when the veins cannot stretch further to accommodate additional increases in blood volume, the extra blood shifts to the arteries. Arteries are much less compliant vessels, and when blood volume increases in arteries, overall blood pressure rises. Anything that decreases the compliance of arteries or veins, such as normal changes that occur with aging, makes the vasculature less able to adapt to increases in blood volume. So, when compliance decreases, even small increases in blood volume can raise the blood pressure.

Quick Check

☐ 4. What three main factors determine blood pressure? How does each factor influence it?

☐ 5. How does a change in vessel diameter affect peripheral resistance?

☐ 6. What is vessel compliance, and which vessels are the most compliant?

Blood Pressure in Different Portions of the Circulation

◀◀ **Flashback**

1. Which ventricle is stronger and thicker? Why? (p. 635)
2. Define systole and diastole. (p. 654)

TABLE 18.2 PRESSURES IN PULMONARY AND SYSTEMIC CIRCUITS	
Circuit	**Pressure**
Pulmonary Circuit	
Pulmonary arteries	15 mm Hg
Pulmonary veins	5 mm Hg
Systemic Circuit	
Arteries	120 mm Hg (systolic), 80 mm Hg (diastolic)
Arterioles	80–35 mm Hg
Capillaries	35–15 mm Hg
Venules	15–5 mm Hg
Veins	5–0 mm Hg

The term blood pressure is fairly generic and can refer to the pressure anywhere within the systemic or pulmonary circuit. Recall that the pressure in the pulmonary circuit is much lower than in the systemic circuit, which is why the right ventricle has thinner walls than the left (see Chapter 17). The cardiac outputs of the right and left ventricles are equal under normal conditions, but the resistance in the systemic circuit is far greater than that in the pulmonary circuit. For this reason, the pressures in the two circuits are dramatically different, averaging about 15 mm Hg in the pulmonary circuit and about 95 mm Hg in the systemic circuit.

As you can see in Table 18.2, the pressure doesn't change much in the pulmonary circuit—it remains fairly low from the pulmonary artery to the pulmonary veins. However, the pressure does change significantly as blood travels through the systemic circuit (Figure 18.5). We examine the pressure changes in the vessels of the systemic circuit more closely in the next subsections.

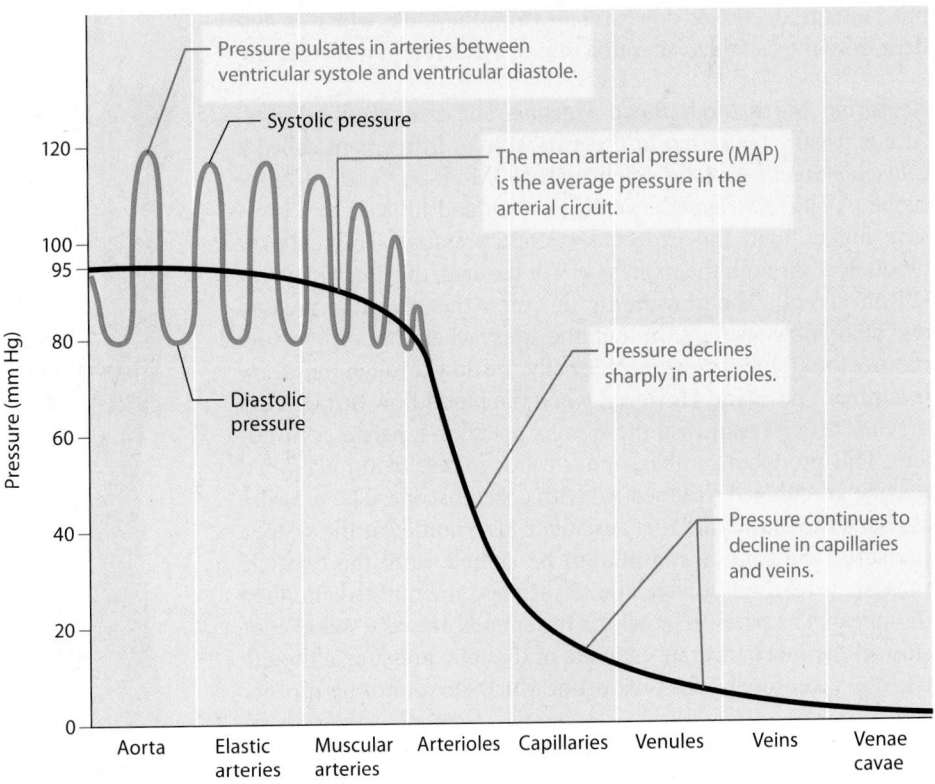

Figure 18.5 Pressure profile of the systemic circuit.

Systemic Arterial Pressure

Generally, when someone reports their "blood pressure," he or she is actually reporting the pressure in the systemic arterial circuit. Figure 18.5 shows the pressure profile of the entire systemic circuit. Notice that it is highest in the aorta and the elastic arteries and that it declines slightly as it spreads throughout the muscular arteries. The black line in the graph shows the **mean arterial pressure,** or **MAP,** which is the average pressure in the systemic arteries during an entire cardiac cycle. The MAP generally measures about 95 mm Hg.

However, because the heart has both contraction and relaxation periods, the pressure gradient generated by the heart *pulsates*: It rises during ventricular systole and declines during ventricular diastole. This leads to two separate pressures in the arteries: a **systolic pressure,** which averages about 120 mm Hg, and a **diastolic pressure,** which averages about 80 mm Hg when the person is at rest. These pulsations are represented by the orange wavy line in Figure 18.5 that overlies the MAP line. The difference between the systolic and diastolic pressures—about 40 mm Hg—is known as the **pulse pressure.**

Pressure in the systemic circuit declines most sharply in the arterioles. As you can see in Figure 18.5, MAP decreases from about 80 mm Hg in the large arterioles to about 30 mm Hg in the smallest arterioles. Pulsations are generally measurable in the large arterioles but decline in smaller arterioles until they are imperceptible. This sharp decrease in pressure is due to the sharp increase in peripheral resistance in the arterioles. This may sound counterintuitive because you read earlier that increases in peripheral resistance increase blood pressure. But think about what happens when you kink a garden hose, which increases its resistance to flow. Pressure increases upstream from the kink (closer to the spigot), but pressure decreases downstream from the kink, which is evident from the fact that water flow slows or even stops downstream.

Measuring the Arterial Blood Pressure The arterial blood pressure is usually measured in the arm with an instrument called a *sphygmomanometer* (sfig'-moh-mah-NOM-eh-ter) and a stethoscope. A cuff is wrapped around the arm and inflated to a pressure higher than the expected systolic pressure. This cuts off blood flow through the main artery in the arm, the *brachial artery* (BRAY-kee-ul). The pressure in the cuff is then slowly decreased and blood flow resumes through the artery when the cuff pressure reaches the systolic pressure. Normally, the flow through the artery is laminar and silent. However, when the blood flow through the brachial artery resumes at the systolic pressure, it becomes turbulent. This produces audible sounds called *sounds of Korotkoff* that may be auscultated (listened to) with a stethoscope. The pressure at which the sounds are first auscultated is recorded as the systolic pressure. The sounds continue to be audible until the diastolic pressure is reached, at which point they become muffled and then disappear. The pressure at which the sounds are very soft is considered the most accurate estimate of diastolic pressure, although it is often recorded as the pressure at which the sounds disappear.

Calculating the Mean Arterial Pressure MAP is difficult to measure directly, but we can use the measured systolic and diastolic pressures to estimate the MAP. You might expect the MAP to simply equal the average of the diastolic and systolic pressures. However, the heart spends more time in diastole than in systole, and this affects the MAP. We therefore calculate MAP by using the following equation:

$$MAP = \text{diastolic pressure} + \tfrac{1}{3}(\text{systolic pressure} - \text{diastolic pressure})$$

So we can say that the MAP is approximately equal to the diastolic pressure plus one-third of the pulse pressure. For a person with a blood pressure of 120/80, this gives us the following calculation:

$$MAP = 80 \text{ mm Hg} + \tfrac{1}{3}(120 \text{ mm Hg} - 80 \text{ mm Hg}) = 93 \text{ mm Hg}$$

Notice that this is close to the average value mentioned earlier for MAP of about 95 mm Hg.

Systemic Capillary Pressure

Pressure continues to decline throughout the remainder of the systemic circuit. Figure 18.5 shows that the pressure at the arteriolar end of a capillary bed is approximately the same as that in the small arterioles—about 35 mm Hg. However, at the venular end of the capillary bed, the pressure has decreased to about 15 mm Hg. This pressure decrease is largely due to the reduction in blood volume that takes place in capillaries (see Module 18.5).

Systemic Venous Pressure and Mechanisms of Venous Return

Pressure declines even further in venules and veins, dropping to only about 4 mm Hg in the inferior vena cava and as low as 0 mm Hg in the right atrium. The low pressure is largely due to the high compliance of veins and the declining resistance as these vessels merge and become larger.

When the muscle contracts, blood is squeezed upward, which pushes the valve open.

When the muscle relaxes, blood flows backward and the valves close.

Figure 18.6 The skeletal muscle pump.

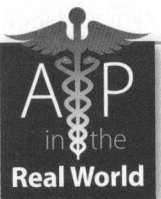

Varicose Veins

The common condition of **varicose veins** is characterized by dilated, bulging, and often hardened veins. Varicose veins are often located in the superficial veins of the lower limb. Conditions that decrease the rate of venous return, such as pregnancy, standing upright for prolonged periods, and abdominal obesity, cause blood to pool in the veins of the lower limb. There also appears to be a genetic predisposition to the development of varicose veins. The lower limb's superficial veins are not supported by skeletal muscle pumps, so the extra blood volume stretches them, causing enlarged regions that are visible beneath the skin.

Another common location for varicose veins is around the anus, where they are called **hemorrhoids** (HEM-ah-roydz). High pressure in the abdominopelvic cavity during defecation or childbirth decreases the return of venous blood from the anal veins. These veins are also superficial and not well supported by surrounding tissues, and thus may weaken and dilate because of the high pressure.

Venous blood must be returned to the heart at the same rate that it is pumped into the arteries. But the venous circuit is under such low pressure that there isn't much of a driving force to propel venous blood back to the heart. Veins have a higher cross-sectional area than arteries, which makes the flow of venous blood slower than that of arterial blood. Indeed, in some locations blood could even potentially flow backward, particularly in those veins that flow against gravity. See *A&P in the Real World: Varicose Veins* for a common problem related to decreased venous return.

You have already read about two of the mechanisms that help return venous blood to the heart: (1) venous valves that prevent backward flow in the veins, and (2) smooth muscle in the walls of veins that may contract under sympathetic nervous system stimulation to increase the rate of venous return. Another mechanism that assists venous return is called the *skeletal muscle pump* (Figure 18.6). Skeletal muscles surrounding the deeper veins of the upper and lower limbs squeeze the blood in the veins and propel it upward (toward the heart) as they contract and relax.

A mechanism similar to the muscle pump exists for veins of the thoracic and abdominopelvic cavities that are not surrounded by skeletal muscle. This mechanism, known as the *respiratory pump*, is driven by the rhythmic changes in pressure in the thoracic and abdominopelvic cavities that occur with ventilation. During inspiration, high pressure in the abdominopelvic cavity creates a pressure gradient that pushes blood in the abdominal veins upward, and low pressure in the thoracic cavity allows thoracic veins to expand. The reverse occurs during expiration—abdominal veins expand and fill with blood while thoracic veins are squeezed.

Quick Check

☐ 7. How does mean arterial pressure differ from systolic pressure and diastolic pressure?

☐ 8. How does the pressure change in arterioles, capillaries, and veins? Why?

☐ 9. What are the mechanisms that assist in returning venous blood to the heart?

Apply What You Learned

☐ 1. What effect would nicotine, which is a vasoconstrictor, have on peripheral resistance? Considering this effect, explain why cigarette smoking is a risk factor for developing high blood pressure (hypertension).

☐ 2. Diseases such as *arteriosclerosis* that harden the arteries cause a loss of arterial compliance. Predict the effect this would have on blood pressure.

☐ 3. Predict what would happen to a person's blood pressure in the case of severe blood loss.

See answers in Appendix A.

MODULE **18.3**

Maintenance of Blood Pressure

Learning Outcomes

1. Describe the role of arterioles in regulating tissue blood flow and systemic arterial blood pressure.
2. Describe the local, hormonal, and neural factors that affect and regulate blood pressure.
3. Explain the main effects and importance of the baroreceptor reflex.
4. Explain how the respiratory and cardiovascular systems maintain blood flow to tissues via the chemoreceptor reflex.
5. Describe common causes of and treatments for hypertension.

A sufficient pressure gradient must be present at all times for blood flow to meet the body's needs. The mean arterial pressure must be constantly maintained at about 95 mm Hg, and any deviation from this set point triggers nervous, endocrine, and renal mechanisms that act to restore blood pressure to this level. We discuss these mechanisms in this module.

Short-Term Maintenance of Blood Pressure

◀◀ Flashback

1. What effects does the sympathetic nervous system have on blood pressure? (p. 521)
2. How does antidiuretic hormone affect blood volume? (p. 594)

Short-term control of blood pressure, such as that needed when you stand up from a sitting position, is primarily accomplished by the nervous system and certain hormones of the endocrine system. Such short-term effects are generally achieved by adjustment of resistance and cardiac output.

Nervous System Maintenance of Blood Pressure

Recall that the main branch of the nervous system in control of homeostasis is the autonomic nervous system (ANS) via its two divisions: the sympathetic and parasympathetic nervous systems (see Chapter 14). Both divisions have immediate effects on blood pressure. The upcoming subsections discuss these effects, as well as a reflex mediated by the ANS called the *baroreceptor reflex* and the effects mediated by vessel chemoreceptors.

Sympathetic and Parasympathetic Effects on Blood Pressure As we discussed in the ANS chapter, the sympathetic nervous system is the "fight or flight" division of the ANS that prepares the body for stressors such as exercise, even mild exercise like position changes (see Chapter 14). In **Figure 18.7a**, axons release the neurotransmitters **norepinephrine** and **epinephrine** onto cardiac muscle cells and the smooth muscle cells of blood vessels, to produce two immediate changes: (1) an increase in heart rate and contractility, which increases cardiac output; and (2) vasoconstriction of all types of vessels, but especially arterioles, which increases peripheral resistance. Both changes increase blood pressure. Note that primarily blood vessels serving and draining the digestive and urinary organs and the skin are constricted. The vessels serving the active skeletal muscles and heart actually dilate. This is critical during exercise, as it allows increased delivery of oxygen and nutrients to these tissues.

The effects of the parasympathetic nervous system on blood pressure are opposite those of the sympathetic nervous system (**Figure 18.7b**). The axons of the parasympathetic system, via the vagus nerve, release **acetylcholine** primarily onto certain cardiac pacemaker cells (particularly the sinoatrial [SA] and atrioventricular [AV] nodes) and atrial cardiac muscle cells. This slows the heart rate, which decreases cardiac output and so blood pressure. Most axons of the parasympathetic nervous system do not innervate the smooth muscle cells of blood vessels. However, an increase in parasympathetic activity does allow vasodilation and a decrease in peripheral resistance. This effect occurs because increased firing of parasympathetic neurons inhibits the firing of sympathetic neurons in the medulla of the brainstem. When the sympathetic neurons are inhibited, the vessels dilate.

The Baroreceptor Reflex Blood pressure is monitored by the ANS with specialized mechanoreceptors in the vessels, called *baroreceptors.* Earlier in the chapter we described the baroreceptors located in areas of the common carotid artery and the aortic arch, the **carotid sinus** and **aortic sinus,** respectively. These receptors depolarize and fire rapid action potentials in response to pressure exerted on the arterial wall; this triggers the baroreceptor reflex.

The **baroreceptor reflex** arc, another example of the Feedback Loop Core Principle (p. 21), is a negative feedback loop that responds to an increase in blood pressure. The reflex arc, illustrated in **Figure 18.8a**, is described next.

CORE PRINCIPLE
Feedback Loops

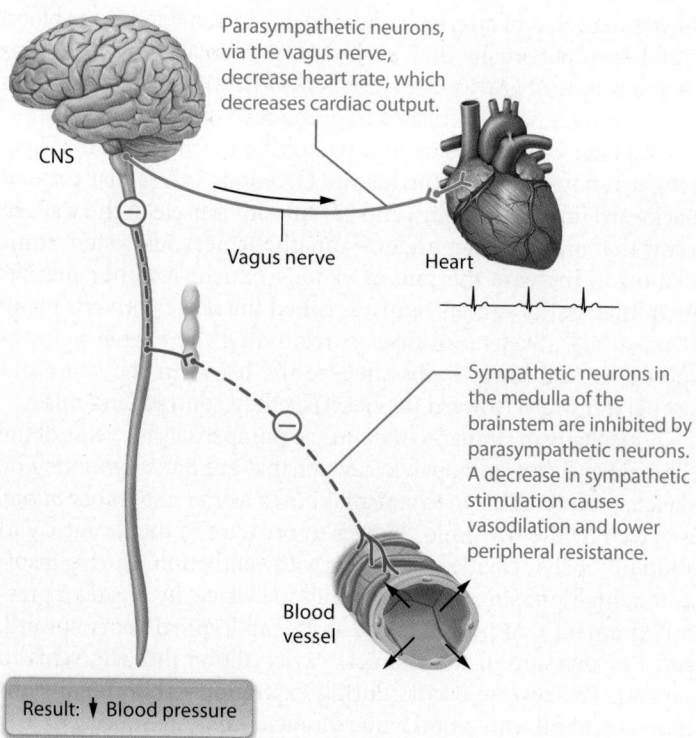

(a) Short-term mechanisms by which the sympathetic nervous system increases blood pressure

(b) Short-term mechanisms by which the parasympathetic nervous system decreases blood pressure

Figure 18.7 Effects of the autonomic nervous system on blood pressure.

(a) Response to rising blood pressure

(b) Response to decreasing blood pressure

Figure 18.8 **Maintaining homeostasis: regulation of blood pressure by the baroreceptor reflex.**

- **Stimulus: Blood pressure increases above normal range.**
- **Receptor: Baroreceptors in the carotid sinus (and aortic sinus) detect the increased pressure and fire action potentials at a faster rate.** An increase in blood pressure stretches the arteries. This stretch is detected by the baroreceptors, and they depolarize at an increased rate. This produces increasingly frequent action potentials.
- **Control center: The impulses travel to the medulla of the brainstem for integration.** The carotid sinus is innervated by cranial nerve IX, the glossopharyngeal nerve. Impulses travel via its parasympathetic afferent neurons to the control center for this negative feedback loop: the cardiovascular center in the medulla oblongata, where the impulses are integrated.
- **Effector/response: Parasympathetic neurons in the medulla inhibit sympathetic activity, inducing vasodilation and decreased heart rate, lowering cardiac output.** Parasympathetic neurons in the medulla inhibit sympathetic neurons from firing. This decrease in sympathetic activity causes vasodilation (note that the parasympathetic neurons have no direct effect on most blood vessels). In addition, they stimulate vagal parasympathetic neurons that decrease the heart rate and cardiac output.
- **Homeostatic range and negative feedback: Blood pressure decreases, and negative feedback decreases response from the medulla.** These effects decrease both cardiac output and peripheral resistance, so blood pressure decreases to the normal homeostatic range. When the baroreceptors are strongly stimulated, the decrease in cardiac output and blood pressure can be dramatic—the heart can nearly come to a complete stop for several seconds. As blood pressure falls to a normal level, the rate of firing of the baroreceptors decreases, which in turn decreases the output of the medullary parasympathetic neurons by negative feedback.

The opposite feedback loop will occur if the systemic blood pressure decreases (**Figure 18.8b**). In this case, the receptors detect less stretch, and the rate of depolarization of the baroreceptors decreases. In response, the medulla stimulates the sympathetic nervous system to increase blood pressure. This may be tested clinically via the **Valsalva maneuver,** during which a subject bears down and tries to expire against a closed glottis (the airway in the larynx), as occurs during coughing, sneezing, defecation, and heavy lifting. The Valsalva maneuver raises the pressure in the thoracic cavity and reduces the return of venous blood to the heart, which causes a drop in blood pressure. This should trigger the baroreceptor reflex and generate an increase in heart rate.

The overall purpose of the baroreceptor reflex is to protect the body from sudden increases or decreases in blood pressure from moment to moment. It adjusts blood pressure with changes in position, such as when you get up from bed in the morning. This reflex can also be useful clinically, which is discussed in *A&P in the Real World: Carotid Sinus Massage* on page 684. However, the baroreceptor reflex is ineffective at reducing chronic elevations in blood pressure because these receptors adapt with repeated stimulation.

Effects of Chemoreceptor Stimulation As you read about in the PNS chapter, a *chemoreceptor* is a type of receptor that detects the presence of specific chemicals (see Chapter 13). There are two groups of chemoreceptors known as the *peripheral chemoreceptors* located near the baroreceptors in the aortic arch and the carotid artery. The peripheral chemoreceptors primarily play a role in the regulation of breathing, but they also affect blood pressure. These receptors respond mostly to the level of oxygen in the blood. A significant decrease in the blood oxygen concentration triggers a series of feedback loops that indirectly stimulate an increase in heart rate and cause vasoconstriction. Another group of chemoreceptors, the *central chemoreceptors,* are located in the medulla of the brainstem. The central chemoreceptors respond to the pH of the interstitial fluid of the brain. When the pH of this fluid decreases, another feedback loop is stimulated that indirectly increases the activity of sympathetic neurons, which results in vasoconstriction and a rise in blood pressure.

Endocrine System Maintenance of Blood Pressure

As we discussed, the endocrine system works with the nervous system to maintain homeostasis, including blood pressure (see Chapter 16). The endocrine system is somewhat slower than the nervous system in responding to changes in blood pressure, requiring from several minutes to several hours for its effects to be demonstrated. Different hormones affect all three factors that determine blood pressure, and some affect more than one factor. However, only the hormonal effects on resistance and cardiac output are involved in short-term blood pressure maintenance.

Let's look in more detail at these effects:

- **Hormones that control cardiac output.** Hormones that increase cardiac output include epinephrine, norepinephrine, and *thyroid hormone.* The effects of epinephrine and norepinephrine produced by the adrenal medulla are identical to the effects of these chemicals produced by the sympathetic nervous system—they increase both heart rate and contractility. The effects of thyroid hormone, which is the body's primary regulator of the metabolic rate, are less direct. It causes the cardiac muscle cells to produce more receptors for epinephrine and norepinephrine, which allows these two chemicals to have a greater impact on cardiac output.
- **Hormones that control resistance.** Hormones affect peripheral resistance mainly through their control of blood vessel diameter. As you have read, epinephrine and norepinephrine from the adrenal medulla cause vasoconstriction and increase peripheral resistance, elevating blood pressure. Another hormone that affects resistance is *angiotensin-II,* which is produced as part of the renin-angiotensin-aldosterone system (RAAS). Angiotensin-II is a very powerful vasoconstrictor that sharply increases peripheral resistance and blood pressure. The hormone *atrial natriuretic peptide,* or *ANP,* is produced by the atria in response to increased blood volume. ANP causes vasodilation, especially of the vessels supplying the kidney. This causes a mild decrease in peripheral resistance and blood pressure.

As you will discover shortly, hormones are a common target for drug therapy of high blood pressure, particularly aldosterone and angiotensin-II.

☐ 1. What effects do the sympathetic and parasympathetic nervous systems have on blood pressure?
☐ 2. What are the steps of the baroreceptor reflex arc?
☐ 3. What effects do epinephrine and norepinephrine have on peripheral resistance and blood pressure?
☐ 4. How does angiotensin-II affect blood volume, peripheral resistance, and blood pressure?

Long-Term Maintenance of Blood Pressure by the Endocrine and Urinary Systems

The long-term maintenance of blood pressure falls to the urinary system and certain hormones of the endocrine system that affect the kidneys. These systems control blood pressure by increasing or decreasing the amount of body water lost as urine, which affects blood volume.

The endocrine system regulates blood volume through the release of hormones, including ANP and angiotensin-II, as well as *antidiuretic hormone* (*ADH*) and *aldosterone*. When blood pressure increases, atrial cells secrete ANP, which causes the kidneys to excrete more water and sodium ions to decrease blood volume, and so blood pressure. Conversely, when blood pressure decreases, ADH secretion from the posterior pituitary increases. ADH triggers thirst and increases the amount of water retained by the kidneys. These effects raise blood volume and blood pressure. Another effect triggered when blood pressure decreases is secretion of the enzyme *renin* (REE-nin) from the kidneys as part of the previously mentioned RAAS. Renin begins the process that activates angiotensin-II, which induces thirst, causes sodium ion retention, and as a result increases blood volume. Angiotensin-II also triggers the secretion of aldosterone from the adrenal gland, which causes retention of sodium ions and water from the kidneys, increasing blood volume.

The urinary system's control over blood volume is fairly simple: If blood pressure increases, more water flows through the tiny filtering tubes of the kidneys, called tubules, than these cells can return to the blood. This water is then lost from the body as urine, and blood volume and blood pressure decrease. Conversely, if blood pressure decreases, less water flows through the kidneys' tubules and these cells have more time to reclaim the water and return it to the blood. This results in decreased urine production and a slight increase in blood volume and blood pressure.

☐ 5. Which hormones affect blood volume, and how do they affect it?
☐ 6. How do the kidneys influence blood pressure maintenance?

Summary of Blood Pressure Maintenance

The maintenance of blood pressure by the nervous, endocrine, and urinary systems is summarized in **Figure 18.9**. Each panel examines the effects the systems have on one of the factors that

MECHANISMS AFFECTING PERIPHERAL RESISTANCE	EFFECT
Endocrine Epinephrine Norepinephrine Angiotensin-II Renin secretion **Neural** Increased sympathetic activity	**Increased peripheral resistance**
Endocrine Decreased renin secretion **Neural** Increased parasympathetic activity Decreased sympathetic activity	**Decreased peripheral resistance**

MECHANISMS AFFECTING CARDIAC OUTPUT	EFFECT
Endocrine Thyroid hormone Epinephrine Norepinephrine **Neural** Increased sympathetic activity Decreased parasympathetic activity	**Increased rate and force**
Endocrine Atrial natriuretic peptide **Neural** Increased parasympathetic activity Decreased sympathetic activity	**Decreased rate**

MECHANISMS AFFECTING BLOOD VOLUME	EFFECT
Endocrine Angiotensin-II ADH Aldosterone **Renal** Increased H_2O retention	**Increased blood volume**
Endocrine Atrial natriuretic peptide **Renal** Decreased H_2O retention	**Decreased blood volume**

Figure 18.9 Blood pressure maintenance.

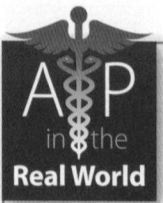

Carotid Sinus Massage

The baroreceptor reflex may be initiated artificially in an emergency setting to reduce dangerously high blood pressure. This maneuver, called *carotid sinus massage,* is performed by "massaging" the skin over both carotid sinuses simultaneously. The pressure applied directly to the carotid sinus triggers the baroreceptor reflex and leads to a drop in blood pressure.

In the anatomy and physiology lab and in the clinic, you should never try to feel the pulse in both carotid arteries at the same time because you could accidentally perform a carotid sinus massage. This could cause a sudden, dangerous drop in your lab partner's or patient's blood pressure and he or she could temporarily lose consciousness (and you could lose points in your lab grade).

influences blood pressure—peripheral resistance, cardiac output, and blood volume.

Disorders of Blood Pressure: Hypertension and Hypotension

As we have discussed, blood pressure must be maintained within a certain normal range. If the blood pressure rises too high (a condition known as *hypertension*) or falls too low (a condition known as *hypotension*), severe disturbances in homeostasis may result.

Hypertension

An estimated 20% of the world's population has **hypertension,** or high blood pressure. Untreated or undertreated hypertension has a high cost to society and is associated with coronary artery disease (the leading cause of death in North America), stroke (the third leading cause of death), heart failure, certain types of dementia, kidney disease, and vascular disease. Since hypertension is the number one modifiable risk factor for each of these potentially life-threatening conditions, the health care field has focused intensive efforts on educating the public about this "silent killer."

Hypertension is diagnosed based on the average of two or more readings recorded on two or more successive visits to a health care provider. It is classified according to the following system:

- Normal (or *normotensive*): systolic less than 120 mm Hg and diastolic less than 80 mm Hg
- Pre-hypertension: systolic 120–139 mm Hg, diastolic 80–99 mm Hg
- Stage 1 hypertension: systolic 140–159 mm Hg, diastolic 90–99 mm Hg
- Stage 2 hypertension: systolic greater than or equal to 160 mm Hg, diastolic greater than or equal to 100 mm Hg

Note that pre-hypertension is a relatively new category that is intended to determine those individuals at risk of developing hypertension.

There are two types of hypertension: **essential (or primary) hypertension,** for which no cause is identifiable, and **secondary hypertension,** for which a cause is identifiable, such as narrowing of the arteries serving the kidneys (*renal artery stenosis*). Essential hypertension is far more common than secondary hypertension, accounting for about 95% of cases. The pathogenesis of essential hypertension is complex, and much about it remains uncertain. Current evidence indicates its development is based on multiple factors, including genetics, ethnic heritage (it is over twice as common in African Americans than in Americans of European descent), advancing age, dietary factors such as excessive salt intake, excess sympathetic tone, and abnormal activity in the RAAS.

Treatment of hypertension depends on the stage at which it is diagnosed. Two areas of key treatment in all stages are (1) lifestyle modifications, including smoking cessation, weight loss if needed, limited alcohol intake, and increased physical activity; and (2) dietary modifications, including decreased salt, cholesterol, and saturated fat intake. Drug therapy is aimed at modifying the three factors that influence blood pressure: cardiac output, blood volume, and peripheral resistance. Common targets for drug therapies include the RAAS, the tubules in the kidneys, calcium ion channels in vascular smooth muscle, and sympathetic receptors on the heart.

Hypotension

Hypotension is considered to be any abnormally low blood pressure. Technically, hypotension is defined by a systolic pressure lower than 90 mm Hg and/or a diastolic pressure lower than 60 mm Hg, but is generally diagnosed as such only if the individual shows symptoms. When symptoms do arise, they vary greatly with the severity of the hypotension. Mild hypotension may cause mild dizziness and lightheadedness. Severe hypotension, a condition known as **circulatory shock,** results in much more dramatic symptoms, including loss of consciousness and organ failure. In this condition, there is insufficient blood pressure to deliver oxygen and nutrients to the cells, which can be rapidly fatal.

There are many potential causes of hypotension, all of which relate to one or more of the three factors that determine blood pressure. Let's look at these more closely.

- **Reduced blood volume.** The most common cause of hypotension is reduced blood volume, a condition known as *hypovolemia.* This condition could occur as a result of blood loss; fluid losses from diarrhea, vomiting, or overuse of diuretics; or insufficient fluid intake. Severe blood loss can lead to *hypovolemic shock,* which is fatal unless blood volume is restored.
- **Decreased cardiac output.** A decrease in either stroke volume or heart rate can result in hypotension. Decreases in heart rate are generally due to medications prescribed to treat hypertension. A decrease in stroke volume is most commonly caused by heart failure, or the inability of the heart to function efficiently as a pump. Severe heart failure can produce dramatic drops in cardiac output, and may result in a condition called *cardiogenic shock.*

- **Vasodilation.** Excessive vasodilation can produce profound hypotension. Many conditions can cause excessive vasodilation, including hypertension medications, ANS abnormalities, and a low blood pH. Another potential cause of excessive vasodilation is the release of histamine into the blood during a severe allergic reaction, a condition known as *anaphylactic shock.* Similar excessive vasodilation occurs with certain bacterial infections of the blood, called *septic shock.*

The treatment of hypotension depends largely on its severity and cause. Mild hypotension may be managed simply by increasing fluid intake or changing a person's medication. Severe hypotension and circulatory shock, however, require aggressive management aimed at the underlying cause of the shock. A patient may be given fluids and/or blood to increase fluid volume, medications to raise cardiac output, and systemic vasoconstrictors.

Quick Check

☐ 7. What is hypertension?
☐ 8. What is circulatory shock, and why is it dangerous?

Apply What You Learned

☐ 1. Explain how each of the following treatments for hypertension lowers blood pressure:
   a. Diuretics (drugs that increase the production of urine)
   b. Angiotensin-receptor blockers (drugs that block angiotensin receptors on smooth muscle cells)
   c. Beta-blockers (drugs that block the effects of the sympathetic nervous system on the heart)
☐ 2. Urine production essentially stops in cases of extremely low blood pressure. Explain why this happens. (*Hint:* Consider that urine production affects blood volume.)
☐ 3. Why is a low-salt diet recommended for patients with hypertension?

See answers in Appendix A.

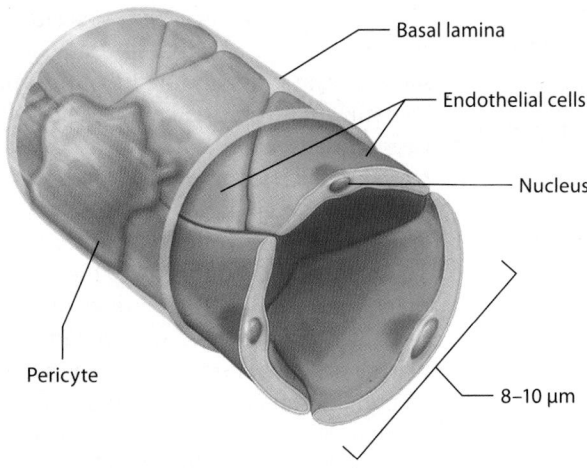

(a) A generalized capillary

Figure 18.10 The structure of a generalized capillary.

MODULE **18.4**

Capillaries and Tissue Perfusion

Learning Outcomes

1. Describe the different types of capillaries, and explain how their structure relates to their function.
2. Explain the roles of diffusion, filtration, and osmosis in capillary exchange.
3. Describe how autoregulation controls blood flow to tissues.

We now turn our attention to the anatomy and physiology of the smallest blood vessels: the capillaries. As you learned in the first module, capillaries generally are found in clusters called *capillary beds* that wind their way through the cells of most tissues in the body—exceptions include cartilage, the sclera and cornea of the eye, and epithelial tissue. The blood flow to a tissue through a capillary bed is known as **tissue perfusion.** As you might expect, tissue perfusion is tightly regulated to ensure that the metabolic needs of all tissues are met at all times.

In this module, we explore the structure and function of the three types of capillaries and capillary beds. We then examine tissue perfusion and the factors that control this process. The module concludes with a look at the perfusion in "special" circulations such as those of the heart and the brain.

Capillary Structure and Function

◄◄ Flashback

1. What are lipid-soluble (nonpolar) substances? What are water-soluble (ionic and polar) substances? (p. 39)
2. What is transcytosis? (p. 84)

As you can see in **Figure 18.10**, capillaries are extremely thin vessels, with walls that are only about 0.2 μm in thickness. Each

(b) Photomicrograph of a capillary

LM (400×)

capillary consists of only an endothelium rolled into a tube and a small amount of basal lamina secreted by the endothelial cells (Figure 18.10a). Around some capillaries are cells called **pericytes** (PEAR-ih-sytz; *peri-* = "around") that have contractile filaments and appear to control blood flow through the capillary.

Capillaries measure, on average, about 50 μm in length and about 8–10 μm in diameter. The walls of most capillaries consist of one to three endothelial cells joined by tight junctions that curl around the capillary's entire circumference. In such capillaries, blood cells have to pass through in single file (see Figure 18.10b).

Capillary Exchange

The reason for the thinness of the capillary wall can be traced back to capillaries' primary function: Nutrients, gases, ions, and wastes must be able to cross the wall and travel between the blood in the capillary and the tissue cells. This movement of materials is known as **capillary exchange.** Materials are generally exchanged via three mechanisms that you have already read about (**Figure 18.11**):

- **Diffusion and osmosis through gaps and fenestrations.** The tight junctions between many endothelial cells are incomplete and leave small gaps between endothelial cells. In addition, some capillaries have small pores within their endothelial cells, called **fenestrations** (fen-eh-STRAY-shunz; *fenestr-* = "window"). Water is able to move freely through these pores when a gradient exists, as well as small substances such as monosaccharides and amino acids that are dissolved in the water.

- **Diffusion through the membranes of endothelial cells.** Lipid-soluble substances such as oxygen, carbon dioxide, and certain lipids can generally enter and exit the capillary by diffusing across the membrane of one side of the endothelial cell and out the membrane on the other side. These substances then enter either the interstitial fluid or the blood.

- **Transcytosis.** Larger molecules must cross the endothelial cells by transcytosis. Recall that during transcytosis, molecules are taken into the cell by endocytosis and then are moved out the other side of the cell by exocytosis (see Chapter 3).

Both diffusion and transcytosis involve a substance crossing the entire width of the endothelial cell. If capillaries contained thicker cells or multiple tissue layers, diffusion and transcytosis would occur far too slowly to meet the body's needs. This is another example of the Structure-Function Core Principle (p. 25).

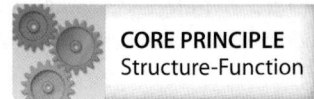

CORE PRINCIPLE
Structure-Function

Types of Capillaries

Capillaries in different parts of the body have slightly different functions and, accordingly, they have slightly different structures. The three types of capillaries are *continuous, fenestrated,* and *sinusoidal* (**Table 18.3**); their characteristics are as follows:

- **Continuous capillaries.** The majority of capillaries in the body are **continuous capillaries,** which are located in the muscles, skin, and most nervous and connective tissues. Continuous capillaries are the least "leaky"—they permit the fewest substances to enter or exit the blood by the paracellular route because their endothelial cells are joined together by tight junctions. They contain small gaps between endothelial cells through which some small molecules diffuse, but the majority of substances move into or out of the capillaries by diffusion through the endothelial cells or by transcytosis. The capillaries in the brain that form the blood-brain barrier are a modified type of continuous capillary, with specialized tight junctions that prohibit many substances from crossing into or out of the interstitial fluid around brain cells.

- **Fenestrated capillaries.** The **fenestrated capillaries** are much leakier than continuous capillaries because their endothelial cells contain fenestrations. These pores allow diffusion to take place much more quickly than it does in continuous capillaries, and fenestrated capillaries are therefore located in places where substances must rapidly enter or exit the blood. Examples include the capillaries of endocrine glands, the small intestine, and the kidney.

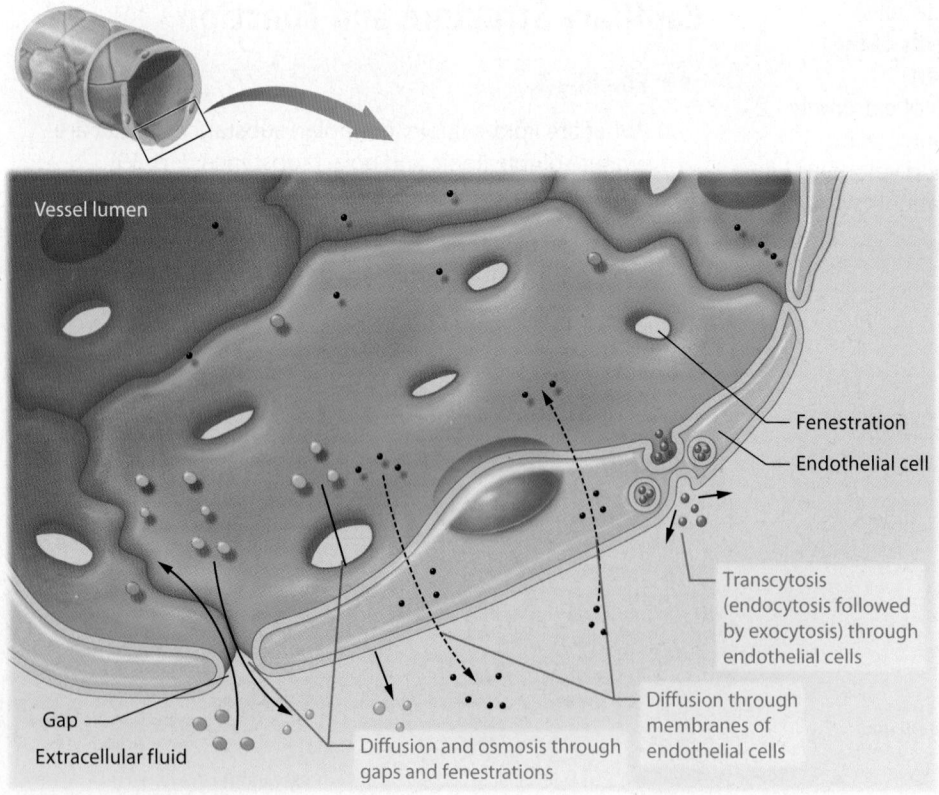

Figure 18.11 Capillary exchange mechanisms.

TABLE 18.3 TYPES OF CAPILLARIES

	Continuous capillaries	Fenestrated capillaries	Sinusoidal capillaries
Structure	Endothelial cells joined by tight junctions	Contain fenestrations in the endothelial cells	Discontinuous sheet of endothelium, irregular basal lamina, very large pores
Location	• Skin • Most nervous and connective tissue • Muscle tissue	• Kidneys • Endocrine glands • Small intestine	• Liver • Lymphoid organs • Bone marrow • Spleen
Function	Least "leaky"—permit a narrow range of substances to cross the capillary walls	Moderately leaky—allow large volumes of fluid and larger molecules to cross capillary walls	Leakiest—allow large substances such as cells to cross the capillary walls

• **Sinusoidal capillaries.** The leakiest capillaries are the **sinusoidal capillaries,** often referred to simply as **sinusoids** (SY-nuh-soydz; *sinu-* = "fold" or "hollow"). As you can see in Table 18.3, sinusoids have a discontinuous sheet of endothelium, an irregular basal lamina, and very large pores in their endothelial cells. Sinusoids are typically three to four times larger in diameter than are other capillaries, and often have an irregular shape because their boundaries are determined by the organ in which they reside. They are located in organs and tissues such as the liver, spleen, lymphoid organs, and bone marrow. Their size, shape, and sluggish blood flow allow them to facilitate the transfer of large substances such as blood cells and large proteins between the interstitial fluid and the blood.

Quick Check

☐ 1. Describe the structure and size of a typical capillary.
☐ 2. List three ways in which substances may cross a capillary to enter or exit the blood.
☐ 3. Describe the properties of the three types of capillaries, and explain how they differ.

Blood Flow through Capillary Beds

The flow of blood that takes place within the body's capillary beds is collectively called the **microcirculation.** The microcirculation occurs in two types of vessels: (1) the true capillaries, where materials are exchanged; and (2) a small, central vessel. As you can see in **Figure 18.12a,** true capillaries form interweaving

networks with multiple anastomoses. These true capillaries are fed by the proximal end of the central vessel, which is formed by either a small terminal arteriole or a metarteriole. You read in Module 18.1 that each capillary-metarteriole junction contains a precapillary sphincter that controls the amount of blood flowing into the capillaries. At rest, only about 25% of the body's capillary beds are fully open.

True capillaries are drained at the distal end of the capillary bed by a portion of the central vessel called a **thoroughfare channel.** The thoroughfare channel lacks precapillary sphincters, which reflects the fact that this is the drainage end of the capillary bed. When the precapillary sphincters close, as in **Figure 18.12b,** blood flows straight from the metarteriole to the thoroughfare channel and into a venule without passing through the true capillaries. Blood does not flow backward into the true capillaries from the thoroughfare channels because the pressure gradient favors the movement of blood in one direction only.

Local Regulation of Tissue Perfusion

Tissue perfusion is critical for the survival of our cells. As you might expect, tissue perfusion is largely a function of arterial blood pressure—if the arterial pressure is too low, blood will not be adequately delivered to the tissues and they will be poorly perfused. In addition, tissue perfusion is also regulated by local factors within each individual tissue, a phenomenon known as **autoregulation,** or self-regulation. Autoregulation ensures that the correct amount of blood is delivered to match a tissue's level of activity. There are two main types of local autoregulatory controls: the *myogenic mechanism* and *metabolic controls*.

The Myogenic Mechanism

The **myogenic mechanism** relies on properties inherent in the vascular smooth muscle cells in the arterioles supplying capillary beds. Increases in arteriolar pressure open stretch-sensitive channels in the arteriolar smooth muscle cells. This initiates a

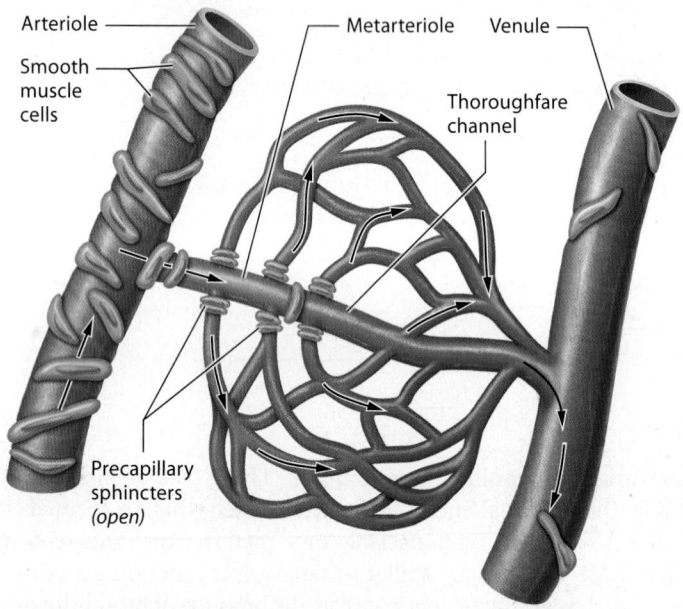

(a) Blood flow through a capillary bed when precapillary sphincters are open

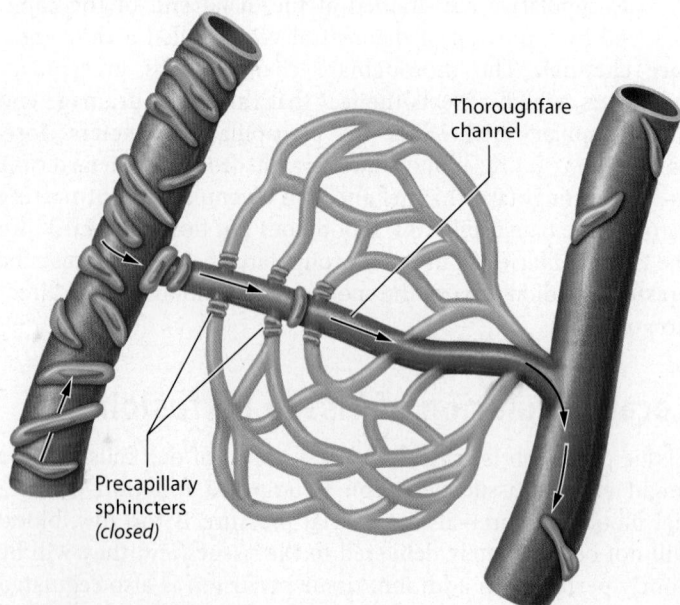

(b) Blood flow through a capillary bed when precapillary sphincters are closed

Figure 18.12 Structure of and blood flow through a capillary bed.

depolarization in the membranes of these cells that allows them to contract without nervous system stimulation. Thus, increases in arteriolar pressure lead to arteriolar vasoconstriction. The reverse response also occurs—arteriolar smooth muscle relaxes when arteriolar pressure decreases.

To understand how the myogenic response affects tissue perfusion, think about the relationships between the velocity of blood flow, the pressure gradient, and resistance. Recall that the velocity of blood flow is proportional to the pressure gradient. Normally, blood is only present in capillaries for less than 1 second, which means that all gases, nutrients, ions, and wastes must be exchanged between the blood and the tissues in this short period of time. However, if the velocity of blood flow increases, less capillary exchange takes place. Conversely, if velocity decreases, flow is inadequate and blood that has already undergone capillary exchange remains in the capillary.

The myogenic mechanism counters a change in blood flow by altering arteriolar resistance. The velocity of blood flow is related inversely to resistance, so if resistance increases, velocity decreases, and vice versa. So the myogenic mechanism slows blood flow by increasing resistance when arteriolar pressure rises. Conversely, it speeds up blood flow by decreasing resistance when arteriolar pressure lowers. Both changes maintain local tissue perfusion at a constant level and ensure that gases and other substances continue to be exchanged in adequate amounts even in the face of fluctuating blood pressure.

Metabolic Controls

A second type of autoregulatory mechanism is mediated by chemicals present in the interstitial fluid surrounding capillaries. These chemicals are the result of cellular metabolic (ATP-generating) activities, and so the mechanisms are called *metabolic controls*. Recall that cells consume oxygen to turn ADP and inorganic phosphate into ATP while generating carbon dioxide as a waste product (see Chapter 3). The carbon dioxide diffuses into the interstitial fluid and reacts with water to yield an acid called *carbonic acid*. So the interstitial fluid of an actively metabolizing cell, such as a contracting skeletal muscle fiber, will contain a low concentration of oxygen and high concentrations of carbon dioxide and hydrogen ions. All three of these conditions cause the smooth muscle cells of local arterioles to relax, dilating the arterioles. This increases perfusion and ensures adequate oxygen and nutrient delivery to the actively metabolizing cells.

Metabolic controls also work in the opposite direction. Tissues whose cells are producing ATP slowly will have a high concentration of oxygen and low concentrations of carbon dioxide and hydrogen ions. These conditions cause constriction of local arterioles and a decrease in tissue perfusion.

Quick Check

☐ 4. What is tissue perfusion?
☐ 5. Explain what happens to arterioles and tissue perfusion via the myogenic mechanism when blood pressure increases or decreases.

☐ 6. Which chemicals produce changes in tissue perfusion? How do they affect tissue perfusion?

Tissue Perfusion in Special Circuits

◀◀ **Flashback**

1. What are the two layers of the skin? Where is the blood supply of the skin located? (p. 161)
2. What are neurotransmitters? (p. 414)
3. What is the coronary circulation? What are the two main vessels of the coronary circuit? (p. 634)

Each organ has its own unique requirements in terms of blood flow and necessary nutrients. Although the autoregulatory mechanisms that we just discussed regulate blood flow, many organs also have their own specific regulatory mechanisms. In this section we address the regulatory mechanisms of the heart, brain, skeletal muscles, and skin. The special circulations of other organs are discussed in their respective chapters.

Heart

The heart is a fairly small organ—it is only about the size of a fist and weighs under 1 pound, which is less than 0.5% of total body mass (see Chapter 17). In spite of its small size, however, the heart receives about 5% of the total cardiac output via the coronary circulation. Interestingly, the perfusion pattern through the coronary circulation is the opposite of the rest of the systemic circuit, which generally increases during systole and decreases during diastole. In the heart this is reversed because it actually squeezes its own arteries during ventricular systole, so heart tissue perfusion *decreases* during systole. During diastole, the ventricles relax, blood resumes flowing through the coronary arteries, and tissue perfusion to the cardiac muscle cells increases. This is one reason why extreme increases in heart rate and/or force of contraction can be dangerous: The heart does not spend enough time in diastole and the myocardium is not adequately perfused. For this reason, people are advised to see a physician before starting an exercise program. Too intense a workout by an individual who is not adequately fit could lead to dangerous elevations in heart rate.

The main local autoregulatory mechanism of cardiac muscle tissue appears to be metabolic controls, in particular the concentration of oxygen in the cardiac interstitial fluid. Under normal resting conditions, blood flow to the heart is about 60–70 ml/min per 100 g of tissue. However, during strenuous exercise or other sympathetic activity, blood flow can increase to more than 250 ml/min per 100 g of tissue. This appears to happen because a low interstitial fluid oxygen level triggers the production of chemicals called vasodilators; these chemicals directly dilate the arterioles serving the myocardium and so greatly increase perfusion.

Brain

The brain is extremely intolerant of ischemia, more so than any other tissue in the body. A sudden decrease in tissue perfusion to the brain will result in loss of consciousness within seconds. The brain accounts for only about 2% of the total body mass, but receives about 15% of the total cardiac output. Given the high, constant metabolic demands of the brain, it will probably come as no surprise that its blood flow is tightly regulated. Autoregulatory mechanisms, including myogenic and metabolic controls, maintain cerebral blood flow at a nearly constant rate of about 750 ml/min.

Despite this constant blood supply, blood flow within the brain to different areas varies considerably with neuronal activity. This is in part due to standard metabolic controls; areas of the brain that have higher activity consume more oxygen and produce more carbon dioxide. However, it also appears to be caused by neurotransmitters produced by active neurons. Certain neurotransmitters are direct vasodilators, whereas others act on astrocytes (neuroglial cells) that are in contact with cerebral arterioles. When stimulated, the astrocytes induce the production of vasodilators that open arterioles and so increase perfusion to active regions of the brain.

Skeletal Muscle

Blood flow changes dramatically in skeletal muscle tissue during exercise, increasing as much as 50-fold, which is known as *hyperemia* (hy'-per-EE-mee-ah; "excessive blood"). The mechanism behind skeletal muscle hyperemia lies in the structure of the arteries supplying skeletal muscle tissue. An artery that enters a skeletal muscle is known as a *feed artery*. Feed arteries branch into multiple arterioles and end in **terminal arterioles** that supply a capillary bed. During resting conditions, resistance in the feed arteries is relatively high and many of the terminal arterioles are constricted, which limits tissue perfusion. When exercise begins, metabolic conditions around the skeletal muscle fiber dilate the terminal arterioles. This produces a small increase in tissue perfusion as more capillary beds open, but not enough of an increase to meet the demands of vigorously working skeletal muscle.

However, as exercise continues, the vessels proximal to the terminal arterioles also dilate. First the arterioles dilate, which significantly boosts tissue perfusion. Second, continuing exercise causes the feed arteries to dilate, which dramatically increases tissue perfusion, as shown here:

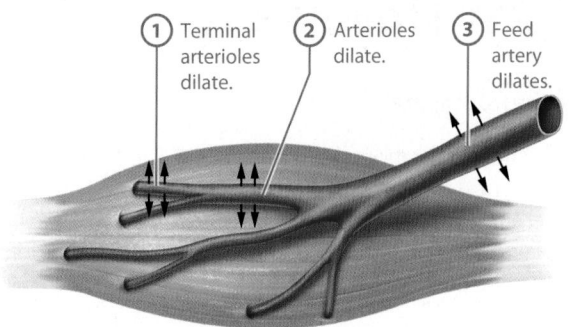

① Terminal arterioles dilate. ② Arterioles dilate. ③ Feed artery dilates.

This progressive, stepwise vasodilation allows the skeletal muscles to precisely match perfusion to the degree of exercise.

Skin

As we've explained, the skin is the largest organ in the body (see Chapter 5). Its blood supply is located in the dermis, and oxygen and nutrients must diffuse from capillaries in the dermis to the epidermis. Local autoregulation of the skin's blood flow does take place, especially in response to temperature—placing something warm directly on the skin will cause vasodilation, whereas something cold will produce vasoconstriction. However, the most important control over the skin's blood flow is neural via the sympathetic nervous system. This is because sympathetic neurons regulate blood flow in the skin as part of its temperature-regulation physiology. The sympathetic nervous system alters perfusion through the skin via vasoconstriction to conserve heat or vasodilation to release heat.

Quick Check

☐ 7. How is the pattern of blood flow in the coronary circulation different from that in the rest of the systemic circuit?
☐ 8. How is perfusion in the brain increased for areas with increased activity?
☐ 9. What is hyperemia? How is skeletal muscle tissue able to develop hyperemia?

Apply What You Learned

☐ 1. Predict what might happen if capillaries were composed of thicker epithelium.
☐ 2. A hypothetical poison destroys the tight junctions of the continuous capillaries of the brain. Predict the effects of such a poison.
☐ 3. In the initial stages of ventricular fibrillation (during which the heart produces no functional contraction), tissue perfusion in the heart actually *increases*. Explain.

See answers in Appendix A.

MODULE **18.5**

Capillary Pressures and Water Movement

Learning Outcomes

1. Describe hydrostatic pressure and colloid osmotic pressure.
2. Explain how net filtration pressure across the capillary wall determines movement of fluid across that wall.
3. Explain how changes in hydrostatic and colloid osmotic pressure may cause edema.

So far in this chapter we have presented the gases, electrolytes, nutrients, and wastes that are exchanged within the capillaries by diffusion or transcytosis. However, we have not yet addressed the other substance that is exchanged in large volumes between the capillaries and the interstitial fluid: water. Water also moves into and out of the blood by passing through either the small pores of the capillaries or the fenestrations of the endothelial cells. Water doesn't cross a capillary wall by diffusion or transcytosis like other molecules. Instead, the movement of water across a capillary is driven by a process called *filtration,* which is the movement of a fluid by a force such as pressure or gravity. We examine this process in this module and determine the overall amount of water lost or gained from a capillary.

Pressures at Work in a Capillary

◀◀ **Flashback**

1. During osmosis, do water molecules move toward or away from a solution with a higher solute concentration? (p. 77)

There are two basic pressures that drive water movement at work within a capillary: *hydrostatic pressure* and *osmotic pressure.* These forces promote movement in opposite directions—hydrostatic pressure drives water out of the capillary, whereas osmotic pressure generally draws fluid into the capillary. Let's examine each in greater detail.

Hydrostatic Pressure

Hydrostatic pressure, or **HP,** is simply the force that a fluid exerts on the wall of its container. This definition should sound familiar, as we gave essentially the same one for blood pressure earlier. This is because blood is the fluid in a vessel that is creating the hydrostatic pressure. For this reason, blood pressure is equal to hydrostatic pressure.

You can see what happens here when we have a hydrostatic pressure *gradient* (i.e., the pressure is higher in one compartment than in another compartment):

Hydrostatic pressure gradient

Higher hydrostatic pressure

Lower hydrostatic pressure

Water flows from inside the beaker (higher hydrostatic pressure) to outside the beaker (lower hydrostatic pressure).

Notice that fluid flows through the holes from the area of higher hydrostatic pressure to the area of lower hydrostatic pressure until the gradient is extinguished. This fluid flow is a passive

(a) Hydrostatic pressure (HP)

Interstitial fluid
Arteriole — Venule
Capillary
H_2O
HP = 35 mm Hg   HP = 15 mm Hg
H_2O
HP = 0 mm Hg
HP pushes water out of the capillary.

(b) Osmotic pressure (OP)

Interstitial fluid
Arteriole — Venule
Solutes (albumin) — Capillary
H_2O
OP = 25 mm Hg   OP = 25 mm Hg
H_2O
OP = 3 mm Hg
OP draws water into the capillary.

(c) How hydrostatic and osmotic pressures work together

If HP is higher than OP, water leaves the capillary by filtration.
Arteriole — Tissue cells — Venule
Capillary
H_2O
HP = 15 mm Hg
OP = 25 mm Hg
HP = 35 mm Hg
OP = 25 mm Hg
H_2O
If OP is higher than HP, water enters the capillary by osmosis.
HP = 0 mm Hg
OP = 3 mm Hg
Interstitial fluid

Figure 18.13 Hydrostatic and osmotic pressures in capillary blood and interstitial fluid.

process known as **filtration,** which is something you witness any time you make a pot of coffee with a coffee filter.

How this process applies to water movement across a capillary is illustrated in **Figure 18.13a.** Notice first that the hydrostatic pressure in a capillary changes as we move from its arteriolar end, where it measures about 35 mm Hg, to its venular end, where it measures about 15 mm Hg. Second, note that the hydrostatic pressure of the interstitial fluid is so low that it is functionally 0 mm Hg. Thus, we have steep hydrostatic pressure gradients on both ends of the capillary that drive water *out* of the capillary and into the interstitial fluid.

Osmotic Pressure

In the cell chapter you read about one of the most essential passive transport processes in the body: osmosis (see Chapter 3). Recall that osmosis involves the movement of water from a

solution with a lower solute concentration to one with a higher solute concentration, in this way:

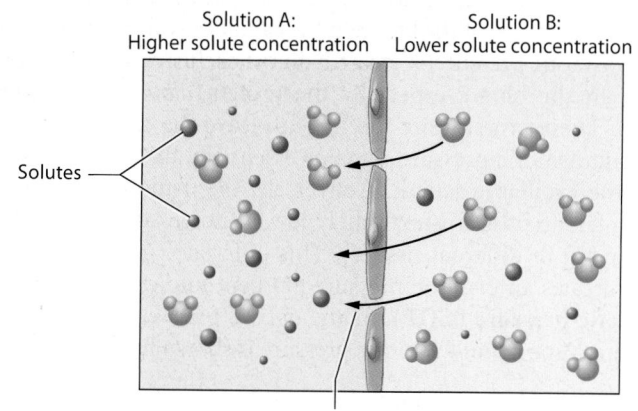

Solution A: Higher solute concentration   Solution B: Lower solute concentration

Solutes

Water moves from the solution with the low solute concentration to the one with the higher solute concentration.

See what's happening in this figure: Water is moving from solution B, which has fewer solute molecules, to solution A, which has more solute molecules. The number of solute particles in a solution determines its *osmolarity*, also known as its *osmotic concentration*. In our example, solution B has a lower osmolarity than solution A.

Another way to explain osmosis is to say that the solute particles in a solution exert a force, or "pull," on water molecules. This pull is called **osmotic pressure,** or **OP.** Solution A has a higher osmolarity; it therefore has a higher osmotic pressure and so a greater pull on water molecules. Take note of something about both solutions in our example: Both contain several types of particles of differing sizes. What effect do different particle types and sizes have on osmotic pressure? The answer is that they basically have no effect at all—osmotic pressure is determined almost exclusively by the *number* of particles, not their size. So solution A has 13 particles and therefore has a higher osmolarity that exerts a higher osmotic pressure than solution B with its mere six particles.

ConceptBOOST ⟩⟩⟩⟩

Understanding the Pulling Force of Osmotic Pressure

The terminology for osmotic pressure may sound confusing at first, because we tend to think of a "pressure" as something that pushes. But as we've mentioned, osmotic pressure doesn't push—it pulls. To understand why it's called a pressure, first know that a pressure is simply a force applied to a molecule or group of molecules. Second, understand that a force is defined as a push *or* a pull. Think about it this way: You have a wagon at the top of a hill and you apply a force and *push* it down the hill. To get the wagon back up to the top of the hill, you must apply another force and *pull* it up the hill. In both cases, you are applying a force that moves the wagon. In this analogy, we can think of hydrostatic pressure as the force that pushed the wagon to the bottom of the hill (i.e., the water out of a vessel) and osmotic pressure as the force that pulled it back to the top (i.e., the water into the vessel). ∎

Now turn your attention to **Figure 18.13b,** and notice that the osmotic pressure of the blood in the capillary is about 25 mm Hg. This osmotic pressure is created almost exclusively by large proteins in the blood, especially the protein *albumin* (AL-byoo-min). These proteins are too large to leave the capillary, and as a result osmotic pressure remains essentially the same throughout the capillary's length. Conversely, the osmotic pressure of the interstitial fluid is extremely low (it ranges from about 0 to 3 mm Hg in different tissues). This difference in osmotic pressure creates an osmotic pressure gradient known as the **colloid osmotic pressure (COP),** or the *oncotic pressure.* The equation to determine colloid osmotic pressure is shown here:

25 mm Hg	−	3 mm Hg	=	22 mm Hg
Capillary OP		*Interstitial OP*		*Colloid osmotic pressure*

So the colloid osmotic pressure for the blood in most tissues is about 22 mm Hg.

An osmotic pressure gradient has an effect opposite to that of a hydrostatic pressure gradient (you can see the two pressures combined in **Figure 18.13c**). Instead of pushing water out of the capillary, colloid osmotic pressure draws water *into* the capillary by osmosis. The proteins in the blood pull water into the capillary in the same way that the particles in solution A pulled water molecules away from solution B. Notice that we have revisited the Gradients Core Principle (p. 26).

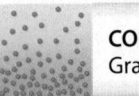
CORE PRINCIPLE
Gradients

Quick Check

☐ 1. What is hydrostatic pressure? How does hydrostatic pressure differ at the two ends of a capillary?

☐ 2. In which direction does the hydrostatic pressure gradient drive the movement of water in a capillary?

☐ 3. What is colloid osmotic pressure? In which direction does colloid osmotic pressure drive the movement of water in a capillary?

Capillary Net Filtration Pressure

Since colloid osmotic pressure and the hydrostatic pressure gradient drive water in opposite directions, this leads to a question: Which one "wins" the push-pull battle? Well, in a sense, they both do, but in different locations—look at **Figure 18.14** to see why. Let's first compare the two opposing forces at the capillary's arteriolar end, where we see that the hydrostatic pressure gradient is about 35 mm Hg and colloid osmotic pressure is about 22 mm Hg:

35 mm Hg	−	22 mm Hg	=	13 mm Hg
Hydrostatic pressure pushing water out		*Colloid osmotic pressure pulling water in*		*Overall force driving water out*

So, at the capillary's arteriolar end, the hydrostatic pressure gradient is larger and therefore "wins." The overall pressure, the **net filtration pressure (NFP),** is 13 mm Hg; this force drives water out of the capillary by filtration.

However, look what happens as we move toward the capillary's venular end—the hydrostatic pressure gradient decreases to about 15 mm Hg, whereas colloid osmotic pressure stays the same:

15 mm Hg	−	22 mm Hg	=	−7 mm Hg
Hydrostatic pressure pushing water out		*Colloid osmotic pressure pulling water in*		*Net filtration pressure inward*

Here the net filtration pressure is a *negative* number, which means that water flows *into* the capillary, a process called *absorption.* So, at the venular end, colloid osmotic pressure "wins" and water is absorbed into the capillary by osmosis.

Now we can put all of this together and examine the entire capillary to determine whether there is net water loss or gain in capillaries.

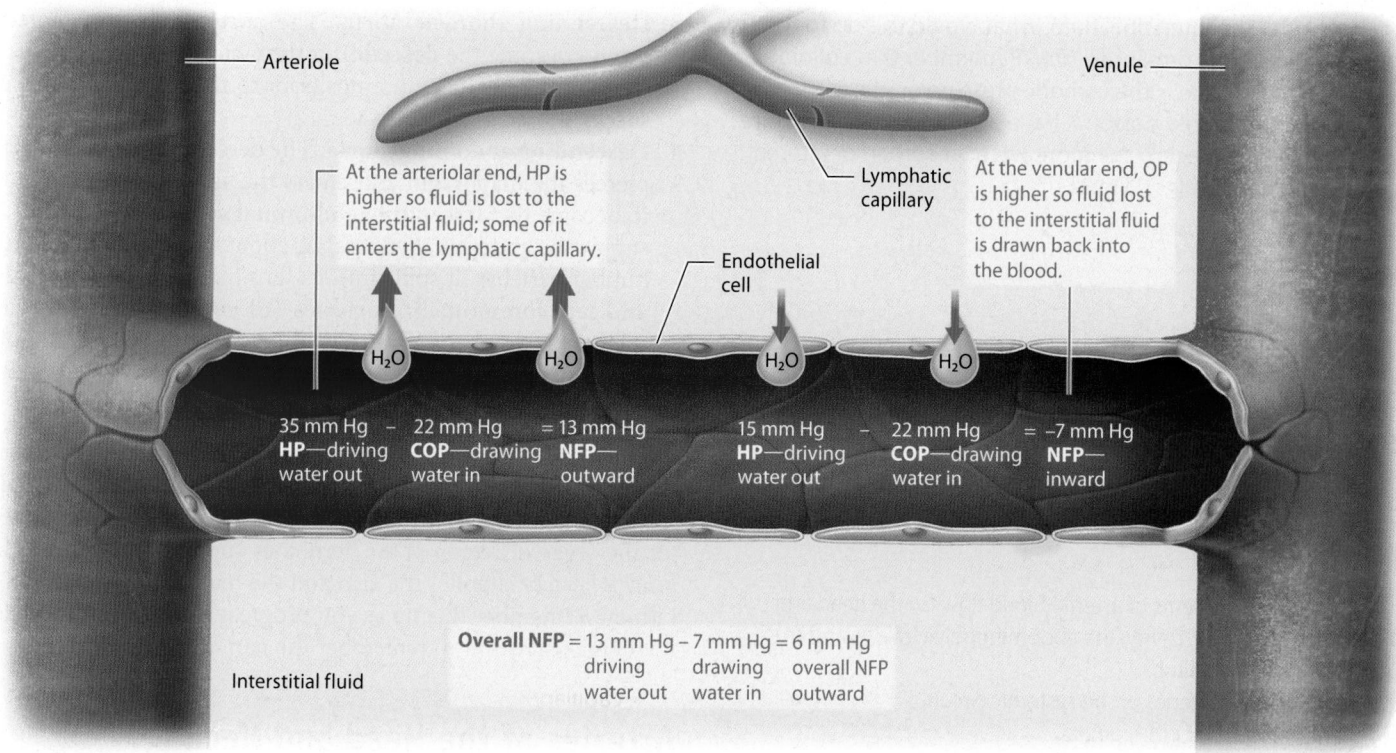

Figure 18.14 **Net filtration pressure (NFP) in capillaries.**

13 mm Hg	−	7 mm Hg	=	6 mm Hg
Hydrostatic pressure pushing water out		*Colloid osmotic pressure pulling water in*		*Overall net filtration driving water out*

Therefore, an overall net filtration pressure of about 6 mm Hg drives water *out* of the capillary by filtration. This doesn't sound like much, but it translates to about 2–4 liters of water lost from the blood to the interstitial fluid per day. Considering that the entire blood volume averages only 5 liters, that is a significant amount of water. This lost volume of water is eventually returned to the blood by vessels of the *lymphatic system* via a mechanism discussed in the lymphatic and immune systems chapter (see Chapter 20).

Edema

The condition known as **edema** (eh-DEE-mah; "swelling") is characterized by an excessive amount of water in the interstitial fluid. Common causes of edema include the following: (1) an increase in the capillary hydrostatic pressure gradient and (2) a decrease in the colloid osmotic pressure. Several factors may increase the capillary hydrostatic pressure gradient, the most common of which is hypertension. Patients with hypertension have a greater capillary hydrostatic pressure at both the arteriolar and the venular ends of capillaries, and this raises the overall net filtration pressure, which causes more water to be lost from the blood. This effect is especially pronounced in the legs and feet, a condition known as *peripheral edema*, where the hydrostatic pressure gradient is already slightly higher due to the effects of gravity.

Even more profound edema may result from decreases in colloid osmotic pressure. This occurs in several conditions; one of the most common is liver failure due to chronic alcohol abuse, viral infection, or cancer. The liver makes the majority of the plasma proteins (such as albumin) that create osmotic pressure. When the liver fails, production of these proteins decreases significantly and colloid osmotic pressure drops. The excess water tends to accumulate in the interstitial fluid of the abdomen, a condition known as **ascites** (ah-SY-teez). Ascites may also result from severe protein malnutrition, in which case it is called *kwashiorkor* (kwah-shee-OR-kor).

Quick Check

- ☐ 4. What is the net filtration pressure?
- ☐ 5. Where in the capillary does net filtration take place? Why? Where in the capillary does net absorption take place? Why?
- ☐ 6. What is the overall net filtration pressure in a capillary? Is there net filtration or net absorption by the capillary?

Apply What You Learned

- ☐ 1. What effect would hypotension have on the capillary hydrostatic pressure gradient and net filtration pressure?
- ☐ 2. Poison X destroys the tight junctions between endothelial cells and allows blood proteins to leak out of the capillary

and enter the interstitial fluid. What would this do to the colloid osmotic pressure? (*Hint:* Remember that colloid osmotic pressure is the osmotic pressure gradient.)

☐ 3. What effect would poison X have on net filtration pressure? Would you expect to see more water absorbed or filtered from the capillaries? Why?

See answers in Appendix A.

MODULE 18.6

Anatomy of the Systemic Arteries

Learning Outcomes

1. Describe the patterns of arterial blood flow for the head and neck, the thoracic cavity, the abdominopelvic cavity, and the upper and lower limbs.
2. Identify major arteries of the systemic circuit.
3. Identify the major pulse points.

We now turn our attention to the anatomy of the systemic circuit arteries—the anatomy of the pulmonary circuit is covered in the chapters that cover the heart and respiratory systems (see Chapters 17 and 21, respectively). We mentioned previously that the largest artery in the body is the **aorta,** which begins at the left ventricle. Notice in **Figure 18.15** that the aorta has four divisions (listed under "Arteries of the trunk"):

1. **Ascending aorta.** The initial segment of the aorta is the **ascending aorta,** which travels superiorly. Recall that the right and left coronary arteries that supply the myocardium branch from the ascending aorta (see Chapter 17).
2. **Aortic arch.** The ascending aorta curves to the left to become the **aortic arch.** As you can see in Figure 18.15, the aortic arch gives off three large elastic arteries (listed under "Branches of the aortic arch"):
   - **Brachiocephalic artery.** The first branch off the aortic arch is the short **brachiocephalic artery** (bray'-kee-oh-seh-FAL-ik; *brachi-* = arm, *cephal-* = head), also known as the **brachiocephalic trunk.** This short artery branches into the **right common carotid artery** that supplies the right side of the head and neck and the **right subclavian artery** (sub-KLAY-vee-an; "below the clavicle") that supplies the right upper limb and thorax.
   - **Left common carotid artery.** The middle branch of the aortic arch is the **left common carotid artery,** which supplies the left side of the head and neck.
   - **Left subclavian artery.** The third branch is the **left subclavian artery** that supplies the left upper limb and thorax.

   Notice that the branches of the aortic arch are asymmetrical, which reflects the asymmetrical location and shape of the heart in the thoracic cavity.

3. **Descending thoracic aorta.** The aortic arch turns inferiorly to become the **descending thoracic aorta,** which travels through the mediastinum posterior to the heart. Its branches supply thoracic structures.
4. **Descending abdominal aorta.** The descending thoracic aorta pierces the diaphragm and enters the abdominopelvic cavity to become the **descending abdominal aorta,** whose branches supply the abdominal viscera. At about the level of the fourth lumbar vertebra, it splits into its terminal branches: the **right and left common iliac arteries** (EE-lee-ak). These arteries then split into the **internal iliac arteries,** which supply pelvic structures, and the **external iliac arteries** (listed under "Arteries of the lower limb"), which supply the lower limbs.

Now let's explore the anatomy and distribution of the major named branches of the aorta's divisions. As you read and study the accompanying figures, notice that many arteries are named for the organ or region of the body they supply. For example, the *brachial artery* supplies the arm and the *renal artery* supplies the kidney. Remember this tip as you progress through this module; it will make it easier to remember the names of these arteries.

Arteries of the Head and Neck

◄◄ **Flashback**

1. Where are the transverse foramina located? What purpose do they serve? (p. 644)
2. Where is the carotid canal located? What structure passes through it? (p. 217)

The majority of the arterial supply to the head and neck comes from the right and left common carotid arteries, although the subclavian arteries provide contributions, as well. Notice in **Figure 18.16** that the common carotid arteries split into two branches at about the level of the fourth cervical vertebra: the **external carotid artery,** which supplies the superficial structures of the head and face, and the **internal carotid artery,** which supplies the brain. At this split we find the *carotid sinus,* which is located in the common carotid artery and the initial segment of the internal carotid artery.

Blood Supply of the Neck, Face, and Superficial Head

The external carotid artery divides into several branches that serve the neck, face, and superficial structures of the head. These branches, shown in Figure 18.16, are named for the regions they supply, as follows:

- **superior thyroid artery,** which supplies the thyroid gland and anterior neck structures;
- **lingual artery,** which supplies the tongue;
- **facial artery,** which supplies the face; and
- **occipital artery,** which supplies the posterior scalp.

After giving off the occipital artery, the external carotid artery splits into its two terminal branches: the **maxillary artery** and

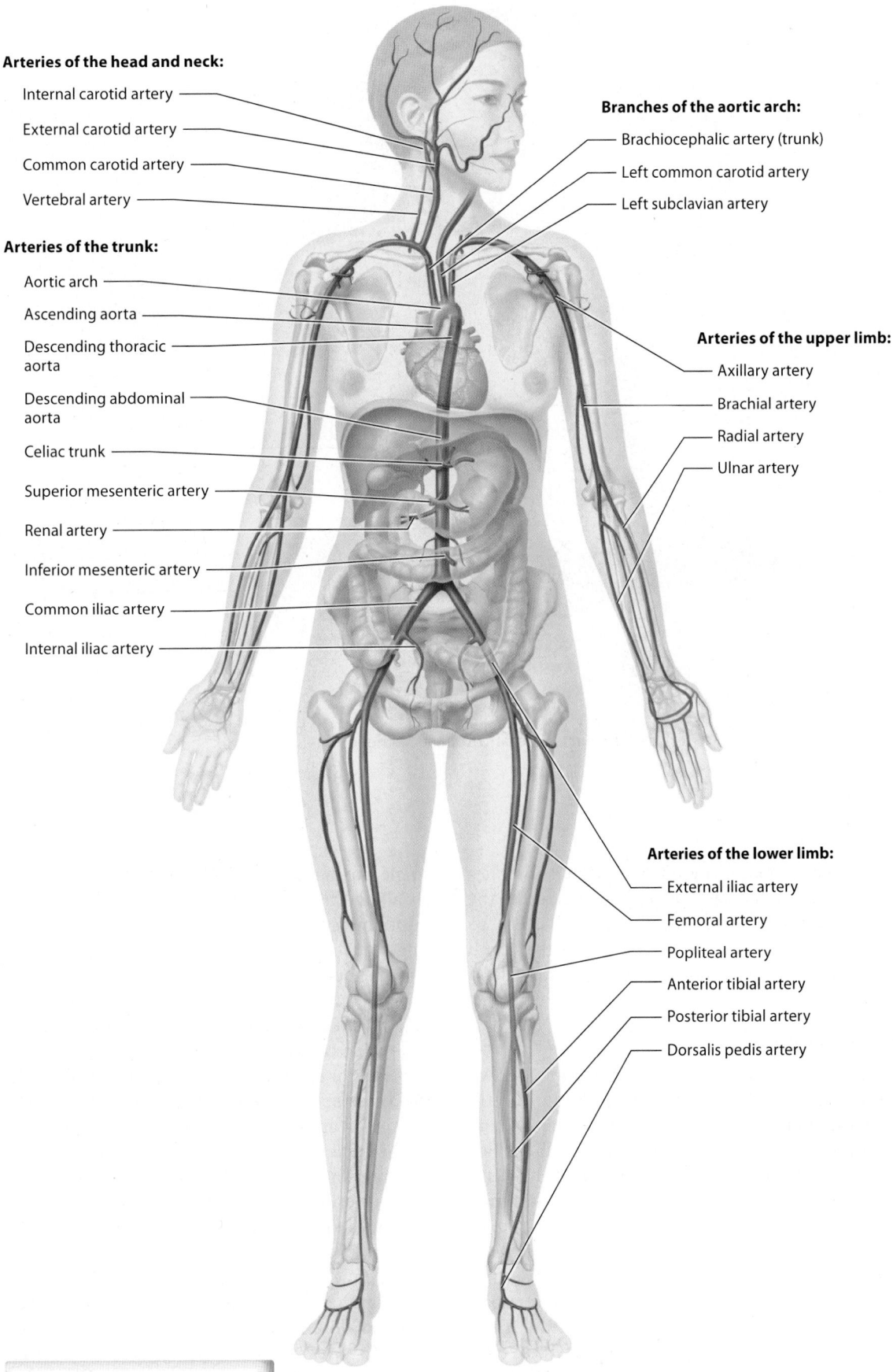

Arteries of the head and neck:

Internal carotid artery

External carotid artery

Common carotid artery

Vertebral artery

Arteries of the trunk:

Aortic arch

Ascending aorta

Descending thoracic aorta

Descending abdominal aorta

Celiac trunk

Superior mesenteric artery

Renal artery

Inferior mesenteric artery

Common iliac artery

Internal iliac artery

Branches of the aortic arch:

Brachiocephalic artery (trunk)

Left common carotid artery

Left subclavian artery

Arteries of the upper limb:

Axillary artery

Brachial artery

Radial artery

Ulnar artery

Arteries of the lower limb:

External iliac artery

Femoral artery

Popliteal artery

Anterior tibial artery

Posterior tibial artery

Dorsalis pedis artery

Practice art labeling
@MasteringA&P®

Figure 18.15 The major systemic arteries.

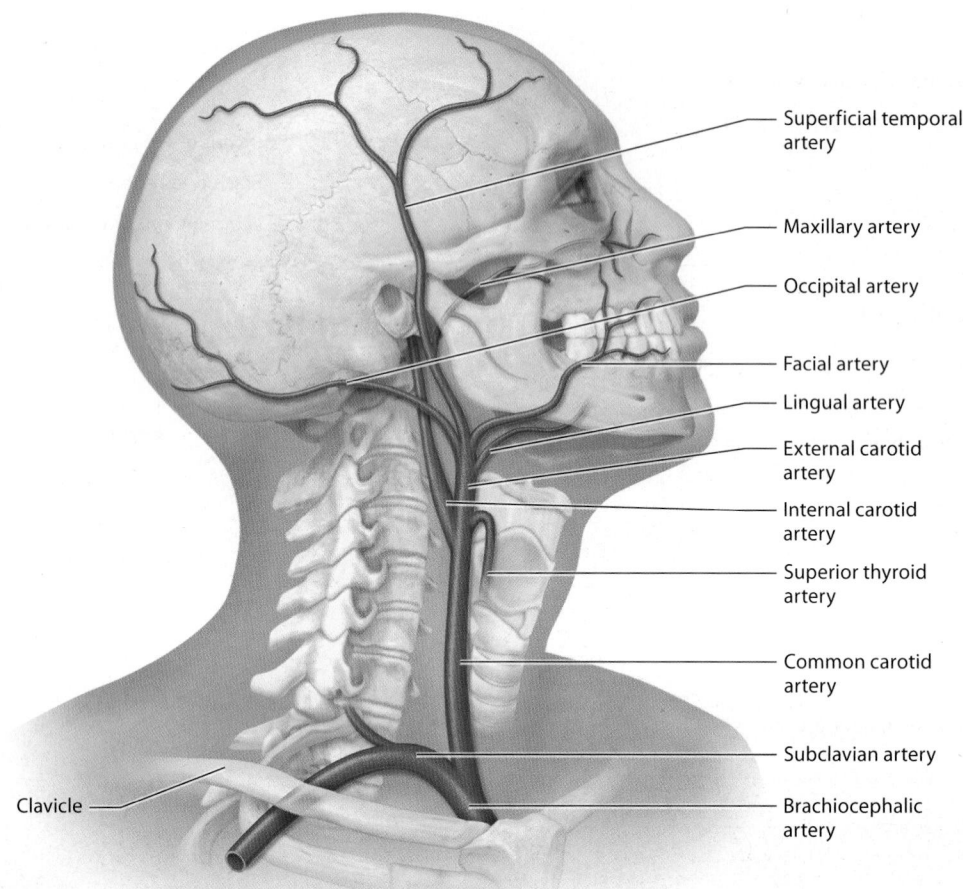

Superficial temporal
artery

Maxillary artery

Occipital artery

Facial artery

Lingual artery

External carotid
artery

Internal carotid
artery

Superior thyroid
artery

Common carotid
artery

Subclavian artery

Brachiocephalic
artery

Clavicle

Figure 18.16 Arteries of the neck, face, and superficial head.

the **superficial temporal artery.** The maxillary artery supplies many deeper structures of the face, including the teeth, the gums, and the nasal cavity. The superficial temporal artery, so named because it crosses the temporal bone, supplies the lateral side of the head, most of the scalp, and the parotid salivary gland in the lateral cheek.

Blood Supply of the Brain

The majority of the brain's blood supply comes from the two internal carotid arteries, which enter the skull via the carotid canals in the temporal bone (**Figure 18.17a and b**). The internal carotid artery gives off several branches inside the skull, including the **anterior cerebral artery** and **middle cerebral artery,** which supply the lobes of the cerebrum, and the **ophthalmic arteries,** which supply the eyes.

The remainder of the blood supply to the brain comes from two arteries called the **vertebral arteries,** which are branches of the right and left subclavian arteries. Recall that the vertebral arteries ascend through the neck by passing through the *transverse foramina* in the transverse processes of the cervical vertebrae (see Chapter 7). The two arteries enter the skull by passing through the foramen magnum, after which they fuse to form the

single **basilar artery** (BAY-zih-lar), so named because it travels along the base of the brainstem. The basilar artery gives off numerous small branches that supply the brainstem before splitting into its terminal divisions: the **posterior cerebral arteries,** which supply the posterior cerebrum.

Notice in **Figure 18.17c** that the posterior cerebral arteries give off small branches called **posterior communicating arteries,** which connect with the internal carotid artery. In addition, the anterior cerebral arteries are connected by a small collateral, the **anterior communicating artery.** These arteries are part of a circular anastomosis known as the **cerebral arterial circle,** or the *circle of Willis.* The cerebral arterial circle is made up of the anterior and posterior communicating arteries, the anterior and posterior cerebral arteries, and the internal carotid artery. This anastomosis serves a dual purpose. First, it helps to equalize pressure in the arteries of the brain and minimize changes in systemic arterial pressure. Second, it provides collateral circulation that allows blood to continue flowing to the brain even if blood flow through one of the brain's major arteries is disrupted. To learn about what happens when the cerebral blood flow is disrupted, see *A&P in the Real World: Cerebrovascular Accident.*

(a) **Arteries of the brain, anterior view**

(b) **Arteries of the brain, lateral view**

(c) **Detail of cerebral arterial circle, anterior view**

Figure 18.17 Arteries of the brain.

Quick Check

☐ 1. List the three branches of the aortic arch.
☐ 2. What are the two branches of the common carotid artery, and what do they supply?
☐ 3. What is the cerebral arterial circle, and what is its purpose? Which arteries contribute to it?

Arteries of the Thorax

The blood supply to the thorax and the thoracic organs is derived from branches of the subclavian arteries and the descending thoracic aorta (Figure 18.18a). The majority of the anterior thoracic wall is supplied by two arteries that branch off the subclavian artery, called the **internal thoracic arteries** (also referred to as

A&P in the Real World

Cerebrovascular Accident

A **cerebrovascular accident (CVA)**, or **stroke**, is damage to the brain caused by a disruption to its blood flow. A CVA is the fourth most common cause of death in the United States. There are two main causes of CVA: (1) blockage of one of the brain's arteries due to a clot; and (2) loss of blood, or *hemorrhage*, due to a ruptured cerebral artery.

Common symptoms of CVA are sudden-onset paralysis or paresis (weakness), loss of vision, difficulty speaking or understanding speech, and headache. Symptoms often affect only one side of the body; however, they can occasionally occur on both sides of the body when multiple clots break off and block several vessels downstream.

Risk factors for CVA include hypertension, atherosclerosis (particularly in the carotid arteries), diabetes mellitus, cigarette smoking, hypercholesterolemia, and a cardiac dysrhythmia called atrial fibrillation. In addition, the risk of CVA increases with age, and women have a higher lifetime risk of CVA than men do.

The treatment for a CVA due to ischemia generally includes medications to dissolve the clot and thin the blood. A hemorrhagic stroke usually requires surgery to repair the damaged vessel. Regardless of the cause of a CVA, prompt treatment is necessary, as any delay in treatment can result in permanent neurological damage or even death.

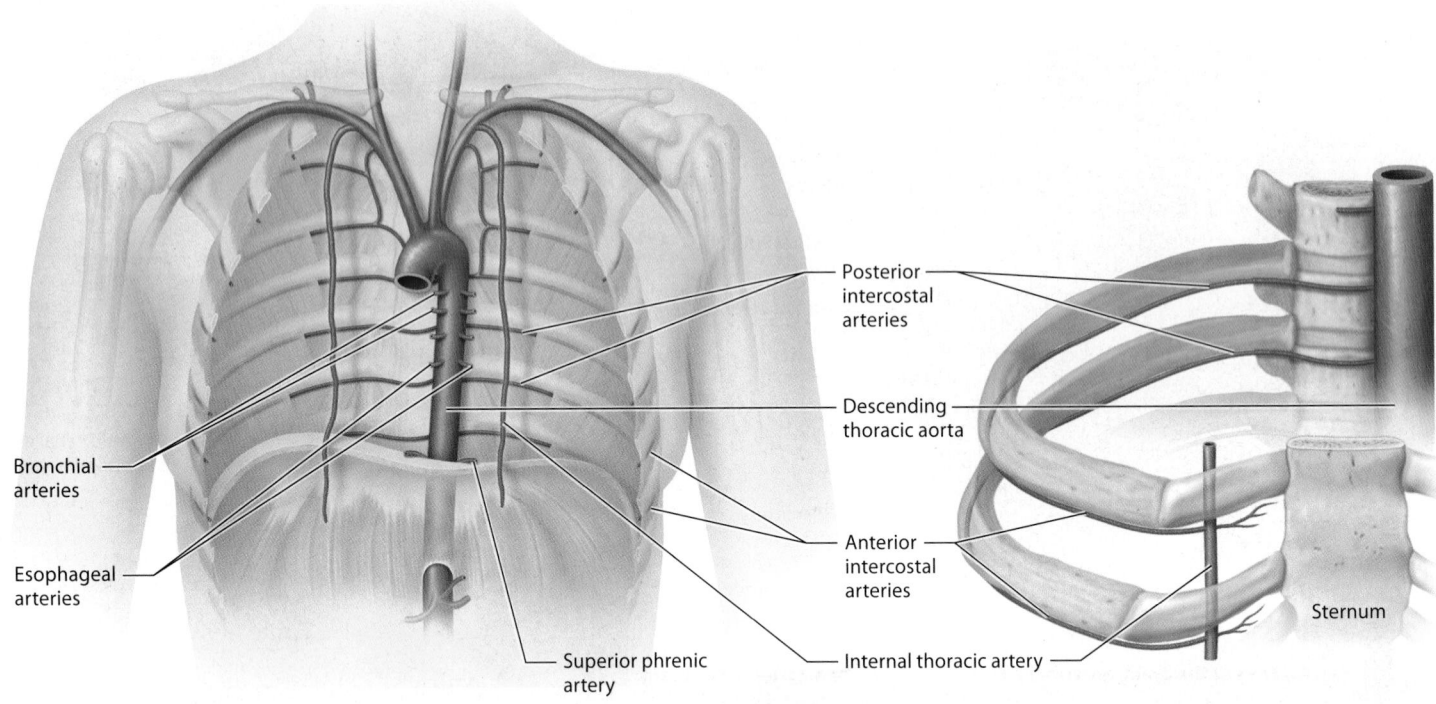

(a) Arterial supply of the thorax

(b) Anastomosis of the anterior and posterior intercostal arteries

Figure 18.18 Arteries of the thorax.

the *internal mammary arteries*). The left internal thoracic artery is important clinically, as it's often used to bypass a blocked coronary artery and restore blood flow to the myocardium. The internal thoracic arteries give off branches, the **anterior intercostal arteries,** which travel in the anterior intercostal spaces and supply structures located therein. The posterior intercostal spaces are supplied by the **posterior intercostal arteries,** most of which are branches of the descending thoracic aorta. You can see in **Figure 18.18b** that the anterior and posterior intercostal arteries meet and form anastomoses.

The organs of the thoracic cavity are supplied by arteries that primarily branch from the descending thoracic aorta. These arteries include the following:

- **bronchial arteries,** which supply the lungs (don't confuse these with the vessels of the pulmonary circuit—the bronchial arteries are separate and part of the systemic circuit);
- **esophageal arteries,** which supply the esophagus; and
- **superior phrenic arteries,** which supply portions of the diaphragm.

Certain thoracic arteries also supply structures of the abdomen (just as some abdominal arteries supply structures of the thorax), which we discuss next.

Arteries of the Abdominal Organs

The superior portion of the anterior abdominal wall is supplied by the termination of the internal thoracic artery, which becomes the **superior epigastric artery** after it crosses into the abdomen. The superior epigastric artery forms extensive anastomoses with the **inferior epigastric artery,** which is a branch of the external iliac artery that supplies the inferior abdominal wall. The posterior abdominal wall is supplied by five pairs of **lumbar arteries,** which branch from the descending abdominal aorta.

The main blood supply to the organs of the abdominopelvic cavity comes from side branches or terminal branches of the descending abdominal aorta. The major branches, listed from superior to inferior, are as follows (**Figure 18.19a**):

1. **Inferior phrenic arteries.** With the superior phrenic arteries, the small **inferior phrenic arteries** supply blood to the diaphragm.
2. **Celiac trunk.** The short, unpaired **celiac trunk** (SEE-lee-ak; *celi-* = "abdominal cavity") branches inferior to the inferior phrenic arteries (**Figure 18.19b**). It splits immediately into three separate arteries:
   - The **common hepatic artery** travels off to the right side of the body in the direction of the liver. It gives off two branches, the **right gastric artery,** which supplies the stomach, and the **gastroduodenal artery,** which supplies the stomach, pancreas, and duodenum (doo-AW-den-um; the first part of the small intestine). After it gives off the gastroduodenal artery, it becomes the **hepatic artery proper,** which supplies the liver and gallbladder.
   - The **splenic artery** travels toward the spleen and branches to supply the stomach and pancreas. Its ultimate destination is, of course, the spleen.

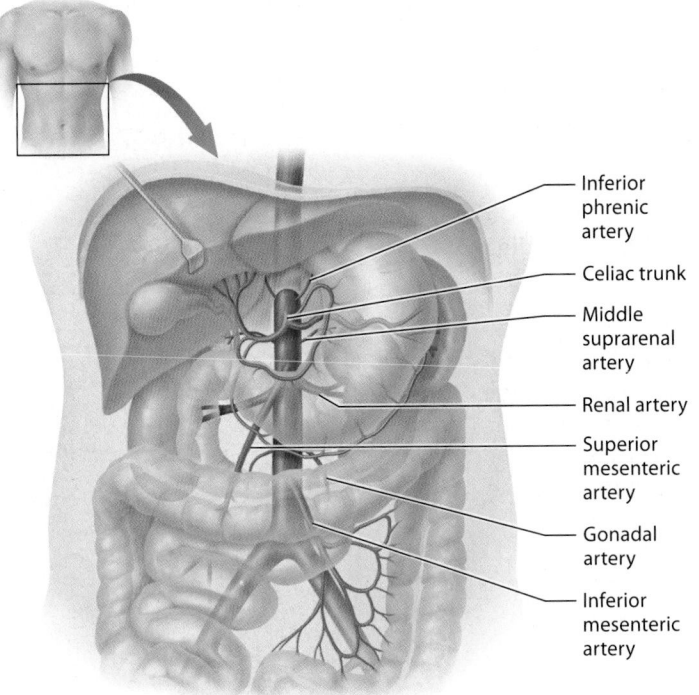

(a) Overview of the abdominal arteries

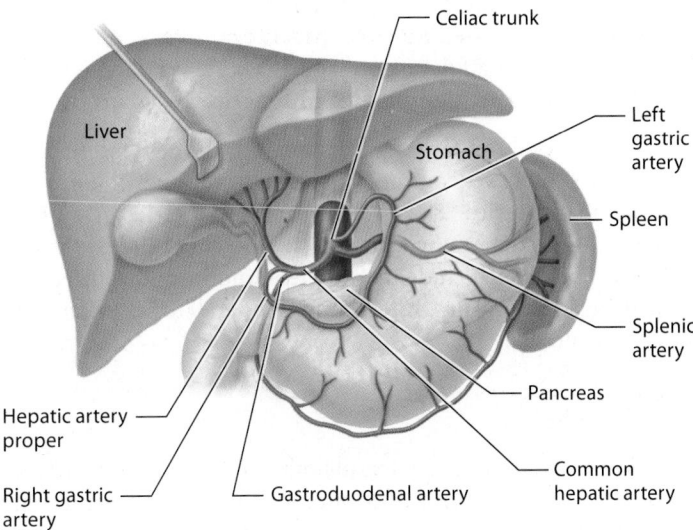

(b) Distribution of the celiac trunk

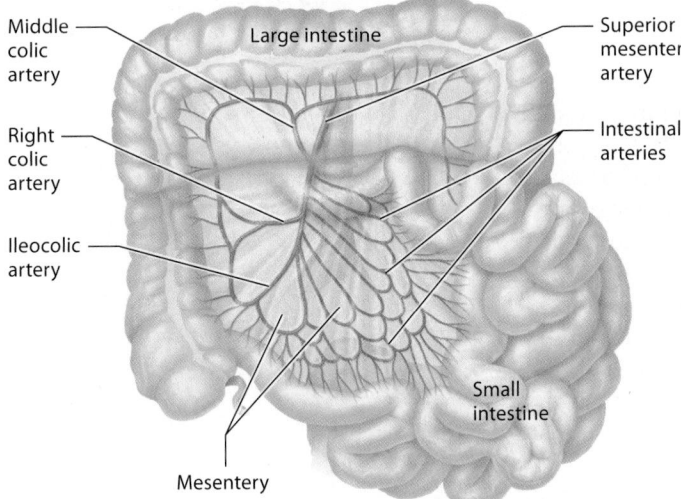

(c) Distribution of the superior mesenteric artery

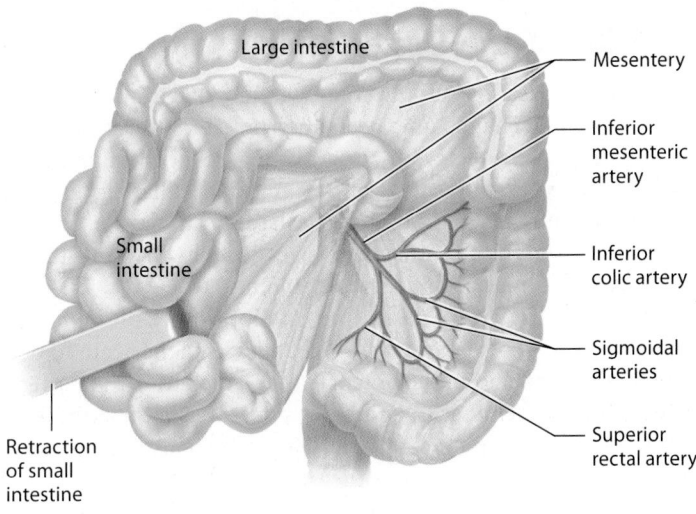

(d) Distribution of the inferior mesenteric artery

Figure 18.19 Arteries of the abdomen.

- The **left gastric artery** is the smallest branch of the celiac trunk. It supplies the stomach and forms an anastomosis with the right gastric artery.

3. **Middle suprarenal arteries.** The small, paired **middle suprarenal arteries** supply the adrenal glands that rest atop the kidneys (see Figure 18.19a).

4. **Superior mesenteric artery.** The small intestine is the site of the majority of nutrient absorption, and it therefore requires an extensive blood supply. This blood supply comes from the large, single **superior mesenteric artery** (**Figure 18.19c**). Its branches to the small intestine consist of numerous anastomosing **intestinal arteries** that travel through the *mesentery,* a double fold of serous membrane surrounding the small intestine. Occlusion of the branches of the superior mesenteric artery can result in **acute mesenteric ischemia,** which may result in the death of intestinal tissue, or a **bowel infarction.** The superior mesenteric artery also supplies much of the large intestine via its branches, the **ileocolic** (ee′-lee-oh-KOL-ik), **middle colic,** and **right colic arteries.**

5. **Renal arteries.** The paired **renal arteries** branch from the descending abdominal aorta at about the level of the first and second lumbar vertebrae. As their name implies, they deliver blood to the kidneys.

6. **Gonadal arteries.** The small **gonadal arteries** deliver blood to the male and female gonads, or reproductive organs. They are called the **testicular arteries** in males and the **ovarian arteries** in females.
7. **Inferior mesenteric artery.** The final major branch off the descending abdominal aorta is the unpaired **inferior mesenteric artery.** As you can see in **Figure 18.19d**, it supplies the remainder of the large intestine via its anastomosing branches, including the **inferior colic, sigmoidal,** and **superior rectal arteries.**

As you discovered earlier in this module, the descending abdominal aorta terminates by splitting into two common iliac arteries, which themselves split into internal and external iliac arteries. The two internal iliac arteries give off branches that supply the majority of the remaining structures in the abdominopelvic cavity and the pelvic musculature. Examples include the **superior and inferior gluteal arteries,** which supply the gluteal muscles, and the **internal pudendal arteries,** which supply the perineum and external genitalia.

Quick Check

☐ 4. Which arteries supply the majority of the thoracic wall?
☐ 5. Which arteries supply the anterior and posterior abdominal wall?
☐ 6. List the main branches off the descending abdominal aorta and the structures they supply.

Arteries of the Upper Limb

The main arteries that supply the upper and lower limbs are like the streets in a city—their names can change unexpectedly. The arterial supply of the upper limb is derived entirely from the subclavian artery, which changes its name in the axillary region to become the **axillary artery** (**Figure 18.20**). The axillary artery gives off several branches that supply muscles and other tissues around the axilla, including the **anterior and posterior humeral circumflex arteries,** which surround the head of the humerus and supply the bone and the shoulder joint.

As the axillary artery passes the inferior border of the teres major muscle (the inferior portion of the "armpit"), it changes its name again to become the **brachial artery.** The branches of the brachial artery, including the posterior **deep brachial artery,** supply the muscles and other structures of the arm. Just proximal to the elbow joint, the deep brachial artery forms an anastomosis with other branches of the brachial artery to supply the elbow joint. As you can see in Figure 18.20, the brachial artery becomes very superficial in the antecubital fossa, after which it splits into its two terminal branches: the lateral **radial artery** and the medial **ulnar artery.** The radial and ulnar arteries travel alongside the bones for which they are named and supply structures of the forearm. They terminate in the palm, in arterial anastomoses called the **superficial and deep palmar arches.**

Quick Check

☐ 7. Which artery supplies the upper limb?
☐ 8. Trace the arterial supply of the upper limb from the axilla to the hand.

Arteries of the Lower Limb

The pattern of blood flow through the lower limb resembles the pattern of blood flow through the upper limb. We start in each with a main vessel that changes its name in different regions and bifurcates into terminal branches. In the lower limb, the main vessel is the **femoral artery.** As you can see in **Figure 18.21**, the femoral artery is the continuation

Subclavian artery

Axillary artery

Deep brachial artery

Brachial artery

Ulnar artery

Radial artery

Deep palmar arch

Superficial palmar arch

Figure 18.20 Arteries of the upper limb.

Inguinal ligament

Abdominal aorta

Common iliac artery

Internal iliac artery

External iliac artery

Femoral artery

Deep femoral artery

Popliteal artery

Anterior tibial artery

Posterior tibial artery

Dorsalis pedis artery

(a) Anterior view

(b) Lateral view

Figure 18.21 Arteries of the lower limb.

of the external iliac artery, which becomes the femoral artery after it passes under the inguinal ligament. At this point, the femoral artery becomes superficial and gives off a branch called the **deep femoral artery,** which supplies the structures of the hip joint, the femur, and many thigh muscles. The femoral artery then dives under the thigh muscles and passes into the posterior thigh through an opening in the adductor magnus muscle.

When the femoral artery emerges in the posterior thigh, it takes a new name, to become the **popliteal artery** (pop-lih-TEE-al). This vessel is named for the fact that it resides in the *popliteal fossa* (the posterior knee). The popliteal artery supplies the knee joint and splits into two terminal branches in the proximal leg: the **anterior tibial artery** and the **posterior tibial artery.** As implied by its name, the anterior tibial artery travels along the anterior tibia, where it supplies the structures of the anterior leg. The anterior tibial artery is renamed the **dorsalis pedis artery** (door-SAL-iss PEE-diss) as it crosses the ankle joint and enters the foot.

The branches of the posterior tibial artery supply the posterior and lateral structures of the leg. The posterior tibial artery becomes superficial just distal to the medial malleolus, after which it travels to the plantar surface of the foot. Here it splits into the **medial and lateral plantar arteries.** A branch of the lateral plantar artery forms an anastomosis with the dorsalis pedis artery to form the **plantar arch** of the foot. (You can't see these plantar structures well in Figure 18.21, as they are on the plantar surface, or sole, of the foot.)

The major arteries of the body are summarized in **Table 18.4.**

Quick Check

☐ 9. Which artery supplies the lower limb?
☐ 10. Trace the arterial supply of the lower limb from the femur to the foot.

Pulse Points

◀◀ Flashback

1. Why does the blood pressure pulsate in arteries but not in veins? (p. 678)

We can take advantage of the superficial locations of many arteries to measure the heart rate and assess the blood flow to different parts of the body. As you know, the blood pressure in systemic arteries is *pulsatile*—it increases during ventricular systole and decreases during ventricular diastole. These pressure changes cause the arteries to expand and recoil with each heartbeat, a change known as a **pulse.** Since the pulse occurs each time the heart beats, the number of arterial pulsations per minute is equal to the heart rate.

The pulse can be felt, or **palpated,** through the skin in superficial arteries at locations known as **pulse points.** The common pulse points are illustrated in **Figure 18.22,** and include the carotid, radial, brachial, femoral, dorsalis pedis, and posterior tibial arteries. Pulses may also be palpated in the ulnar, superficial temporal, and popliteal arteries; however, these are less common sites in the clinical setting.

Quick Check

☐ 11. What is a pulse? What are the common pulse points?

Apply What You Learned

☐ 1. Would a nearly complete blockage of the right internal carotid artery cut off blood flow to the brain? Why or why not?
☐ 2. If the ulnar artery were severed, would the blood supply to the medial side of the hand and digits be disrupted? Why or why not?
☐ 3. Explain why irreparable damage to the femoral artery may result in loss of the lower limb.

See answers in Appendix A.

Figure 18.22 Common pulse points.

TABLE 18.4 MAJOR SYSTEMIC ARTERIES

Artery	Description	Location	Structure(s) Supplied
Aorta	Largest artery in the body with four divisions: ascending aorta, aortic arch, descending thoracic aorta, descending abdominal aorta	Mediastinum and posterior abdominal wall	• Branches supply all structures in the body.
Brachiocephalic artery	First branch off the aortic arch	Right side of the mediastinum	• Branches supply the right side of the upper body.
Common carotid arteries	Right is a branch of the brachiocephalic artery; left is the second branch off the aortic arch.	Lateral neck	• Head • Neck • Brain
Subclavian arteries	Right is a branch of the brachiocephalic artery; left is the third branch off the aortic arch.	Deep to the clavicle	• Thorax • Upper limb
External carotid artery	Superficial branch of the common carotid artery	Lateral neck	• Neck • Face • Superficial head
Internal carotid artery	Deep branch of the common carotid artery	Deep neck, passes through the carotid canals into the brain	• Brain • Orbit
Vertebral arteries	Branches of the subclavian that fuse in the brain to form the basilar artery	Pass through transverse foramina of cervical vertebrae and enter the foramen magnum	• Brain
Celiac trunk	Large unpaired artery, second branch off the descending abdominal aorta	Abdominal cavity, just inferior to the diaphragm	• Structures of the abdominal cavity
Common hepatic artery	Branch off the celiac trunk; splits into the right gastric artery, the gastroduodenal artery, and the hepatic artery proper	Right side of the celiac trunk	• Liver • Stomach • Duodenum • Pancreas
Left gastric artery	Branch off the celiac trunk	Along the medial side of the stomach	• Stomach
Splenic artery	Third branch of the celiac trunk	Posterior to the stomach	• Spleen • Stomach • Pancreas
Superior mesenteric artery	Large unpaired artery; fourth branch off the descending abdominal aorta	Branches fan throughout small and large intestines.	• Small intestine • Most of the large intestine
Renal arteries	Approximately the fifth and sixth branches off the descending abdominal aorta	Travel to the kidneys	• Kidneys
Inferior mesenteric artery	Final branch off the descending abdominal aorta	Branches fan throughout terminal part of the large intestine.	• Remainder of the large intestine
Axillary artery	Continuation of the subclavian artery	Axilla	• Structures around the axilla
Brachial artery	Continuation of the axillary artery	Arm	• Arm
Radial artery	Terminal branch of the brachial artery	Lateral forearm	• Forearm and hand
Ulnar artery	Terminal branch of the brachial artery	Medial forearm	• Forearm and hand
Common iliac arteries	Terminal branches of the descending abdominal aorta	Posterior pelvic wall	• Structures of the pelvis and lower limb
Internal iliac artery	Terminal branch of the common iliac artery	Posterior pelvis	• Pelvic structures
External iliac artery	Terminal branch of the common iliac artery	Anterior pelvis	• Anterior abdominal wall
Femoral artery	Continuation of the external iliac artery	Anterior thigh	• Thigh
Popliteal artery	Posterior continuation of the femoral artery	Popliteal fossa	• Knee joint • Posterior leg • Posterior thigh
Anterior tibial artery	Terminal branch of the popliteal artery	Anterior tibia	• Anterior leg • Anterior foot
Dorsalis pedis artery	Continuation of the anterior tibial artery	Ankle and foot	• Ankle joint • Foot
Posterior tibial artery	Terminal branch of the popliteal artery	Posterior leg, becomes superficial around the medial malleolus	• Posterior leg • Posterior foot

Veins of the head and neck:

Dural sinuses

Vertebral vein

External jugular vein

Internal jugular vein

Veins of the upper limb:

Axillary vein

Brachial vein

Cephalic vein

Basilic vein

Median cubital vein

Median antebrachial vein

Radial vein

Ulnar vein

Veins of the trunk:

Brachiocephalic vein

Subclavian vein

Superior vena cava

Accessory hemiazygos vein

Azygos vein

Hemiazygos vein

Inferior vena cava

Hepatic veins

Gastric veins

Splenic vein

Hepatic portal vein

Renal vein

Inferior mesenteric vein

Superior mesenteric vein

Common iliac vein

Internal iliac vein

External iliac vein

Veins of the lower limb:

Femoral vein

Great saphenous vein

Popliteal vein

Anterior tibial vein

Posterior tibial vein

Small saphenous vein

Practice art labeling
@MasteringA&P®

Figure 18.23 The major systemic veins.

MODULE 18.7

Anatomy of the Systemic Veins

Learning Outcomes

1. Describe the patterns of venous blood drainage for the head and neck, the thoracic cavity, the abdominopelvic cavity, and the upper and lower limbs.
2. Identify major veins of the systemic circuit.
3. Describe the structure and function of the hepatic portal system.

Deoxygenated blood that has traveled through capillary beds drains into systemic veins to be returned to the heart. Most veins superior to the diaphragm, including those of the head, neck, thorax, and upper limbs, drain into the two **brachiocephalic veins** (listed under "Veins of the trunk"; **Figure 18.23**). Notice that two brachiocephalic veins are present, whereas only a single brachiocephalic artery is present (on the right side of the body). The two brachiocephalic veins in turn merge to form the **superior vena cava,** which empties into the right atrium.

Venous blood from the lower limbs and pelvis drains into the **external and internal iliac veins,** respectively. These two vessels merge to form the **common iliac veins,** which themselves merge to form the **inferior vena cava.** Most veins inferior to the diaphragm drain into the inferior vena cava as it ascends through the abdominal cavity. The inferior vena cava pierces the diaphragm and empties directly into the right atrium.

In this module we discuss the pathways taken by venous blood as it returns to the heart through the systemic circuit. You will notice that venous return in many ways parallels the arterial circuit, and many veins share the same names as the corresponding arteries, which makes their names easier to remember.

Veins of the Head and Neck

◀◀ Flashback

1. What is the dura mater? What is the falx cerebri? (p. 443)
2. What are the dural sinuses, and where are they located? (p. 443)
3. What are arachnoid granulations, and what is their main function? (p. 444)

We begin with the veins that drain the structures of the head and neck. As with several other body regions, the venous drainage of these structures is more complicated than their arterial supply. This section examines the anatomy of those veins that drain the superficial structures of the head and neck and the venous system that drains the brain.

Venous Drainage of the Neck, Face, and Scalp

As you can see in **Figure 18.24**, three main veins drain the head and neck: the deep **internal jugular vein** and **vertebral vein** and the superficial **external jugular vein** (listed under "Veins of the neck"). Note that there is no common jugular vein. The external jugular and vertebral veins empty into the **subclavian vein,** which merges with the internal jugular vein to become the brachiocephalic vein. The vertebral vein primarily drains structures of the posterior neck and travels with the vertebral artery in the transverse foramina of cervical vertebrae. The veins that drain the face share the same names with many of the arteries that supply this region. The **superficial temporal, facial,** and **maxillary veins** (listed under "Veins of the head and face") drain into both the internal and external jugular veins.

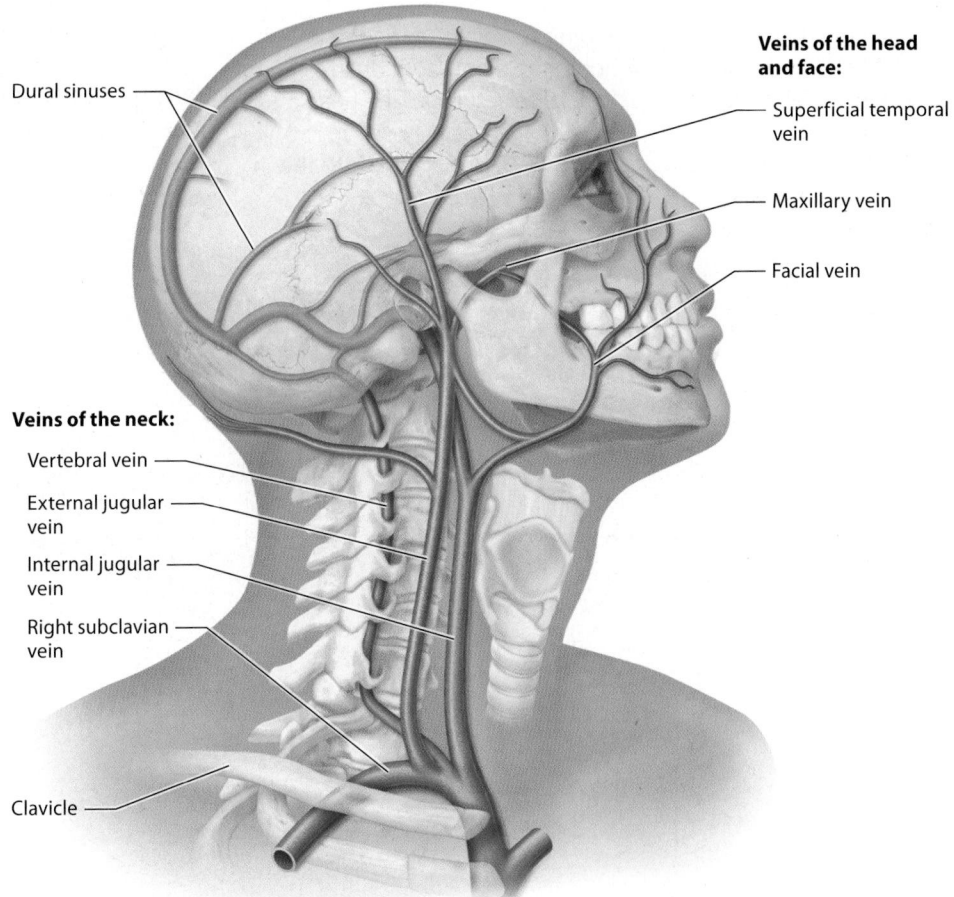

Figure 18.24 Veins of the neck and superficial head.

Dural sinuses

Veins of the head and face:
- Superficial temporal vein
- Maxillary vein
- Facial vein

Veins of the neck:
- Vertebral vein
- External jugular vein
- Internal jugular vein
- Right subclavian vein

Clavicle

Venous Drainage of the Brain

Recall that the brain is surrounded by a tough connective tissue membrane called the dura mater (see Chapter 12). The dura mater consists of two layers of tissue that are fused over much of the brain. However, these layers separate in certain locations, and the spaces between these two layers form a system of venous channels known as the **dural sinuses** (Figure 18.25). Blood from the brain capillaries drains into **cerebral veins,** which then drain into the dural sinuses. In addition, the dural sinuses receive cerebrospinal fluid from the arachnoid granulations that project into these venous channels, which allows this fluid to return to the circulation.

The dural sinus system begins with the **superior sagittal sinus** in the longitudinal fissure of the brain and the **inferior sagittal sinus** in the inferior falx cerebri. The inferior sagittal sinus drains posteriorly into the **straight sinus,** which merges with the superior sagittal sinus to form the two **transverse sinuses** that sit on the floor of the occipital bone. The transverse sinuses travel anteriorly and become the S-shaped **sigmoid sinuses,** which drain into the internal jugular vein. The sigmoid sinuses also receive blood from the anastomosing **cavernous sinuses,** which lie in the anterior and inferior cranial cavity and drain the ophthalmic veins.

Quick Check

☐ 1. Where do most veins superior to the diaphragm drain? Where do most veins inferior to the diaphragm drain?
☐ 2. Which veins drain the superficial structures of the head and face?
☐ 3. Where are the dural sinuses located? What drains into the dural sinuses?

Veins of the Thorax and Abdomen

The venous drainage of the thorax and abdomen is more complicated than their arterial supply, although many vessels do share the same names. This section covers these vessels in examining the drainage of the thoracic wall, that of the abdomen and pelvis, and a special circulatory route called the *hepatic portal system.*

Venous Drainage of the Thoracic and Abdominal Walls

The drainage of the anterior thorax and abdomen is fairly straightforward—the anterior thorax is drained primarily by the **internal thoracic veins,** which empty into the brachiocephalic veins, and the anteroinferior abdomen is drained by the **inferior epigastric veins,** which empty into the external iliac veins. However, notice in Figure 18.26 that the posterior thoracic and abdominal walls are drained by an entirely different system collectively called the **azygos system** (AZ-eye-gohs; *azygos* = "unpaired"). This system consists of three vessels: the **azygos vein** on the right and the **hemiazygos** (hem′-ee-AH-zih-gohs) and **accessory hemiazygos veins** on the left. The left **lumbar veins** (not shown in Figure 18.26) and most of the left **posterior intercostal veins** drain into the hemiazygos or accessory hemiazygos veins, which themselves drain into the azygos vein. The same veins on the right, along with the **bronchial** and **esophageal veins,** drain into the azygos vein. The azygos vein merges directly with the superior vena cava.

Venous Drainage of the Pelvic and Abdominal Organs

The pelvic organs are drained by veins that merge into the internal iliac vein, which then empties into the common iliac vein. The common iliac veins, as well as several veins of the

Labels on figure:
Venous blood
Dura mater
Falx cerebri
Internal jugular vein

Dural sinuses:
Superior sagittal sinus
Inferior sagittal sinus
Cerebral veins
Straight sinus
Cavernous sinus
Transverse sinus
Sigmoid sinus

Figure 18.25 Dural sinuses and other veins of the brain.

Figure 18.26 Veins of the posterior thorax.

abdominopelvic cavity, drain directly into the inferior vena cava. These vessels are illustrated in Figure 18.27a and include the following:

- **right gonadal vein,** which drains the testes in males and the ovaries in females (the left gonadal vein typically drains into the left renal vein);
- **renal veins,** which drain the kidneys; and
- **hepatic veins,** which drain the liver.

Other veins of the abdominal cavity include the **splenic vein,** which drains the spleen, and the **gastric veins,** which drain the stomach. In addition, the **superior mesenteric vein** drains the small intestine and much of the large intestine, and the **inferior mesenteric vein** drains the remainder of the large intestine. These veins are part of a special circulation that is discussed next.

The Hepatic Portal System

Take note of something interesting about the splenic, gastric, and superior and inferior mesenteric veins in Figure 18.27b— they do not drain into the inferior vena cava. Instead, ① they merge with and drain into a large vein that enters the liver, called the **hepatic portal vein.** The hepatic portal vein branches extensively in the liver to form another set of capillary beds. This special type of circuit in which veins feed a capillary bed is known as a **portal system,** and this one is called the **hepatic portal system.**

The word portal means "gate," and in essence the hepatic portal system is the "gatekeeper" of the abdominal circulation. To understand this role, we must first address some of the varied functions of the liver. Many of the most basic metabolic reactions, such as the storage of glucose, the production of proteins from dietary amino acids, and the formation of new glucose, occur in the cells of the liver. In addition, the cells of the liver, called *hepatocytes* (heh-PAT-oh-sytz), contain an extensive network of smooth endoplasmic reticulum that allows them to carry out *detoxification reactions:* The enzymes interact with chemicals we ingest, catalyzing reactions that alter them to render them less harmful to the body. Finally, hepatocytes also contain enzymes that catalyze reactions to break down and process the products of red blood cell destruction.

Venous blood draining from the stomach and the small and large intestines contains both absorbed nutrients and ingested chemicals such as drugs, preservatives, alcohol, and more, and venous blood from the spleen contains products of red blood cell breakdown. The hepatic portal system in essence allows the liver to have "first dibs" on this venous blood and the nutrients and chemicals it contains. (See *A&P in the Real World: Drugs and the Hepatic Portal System* on page 711 to read about how this fact influences the administration of medications.) ② This blood is delivered to the liver by the hepatic portal vein, which branches extensively to form a network of capillary beds that consist of

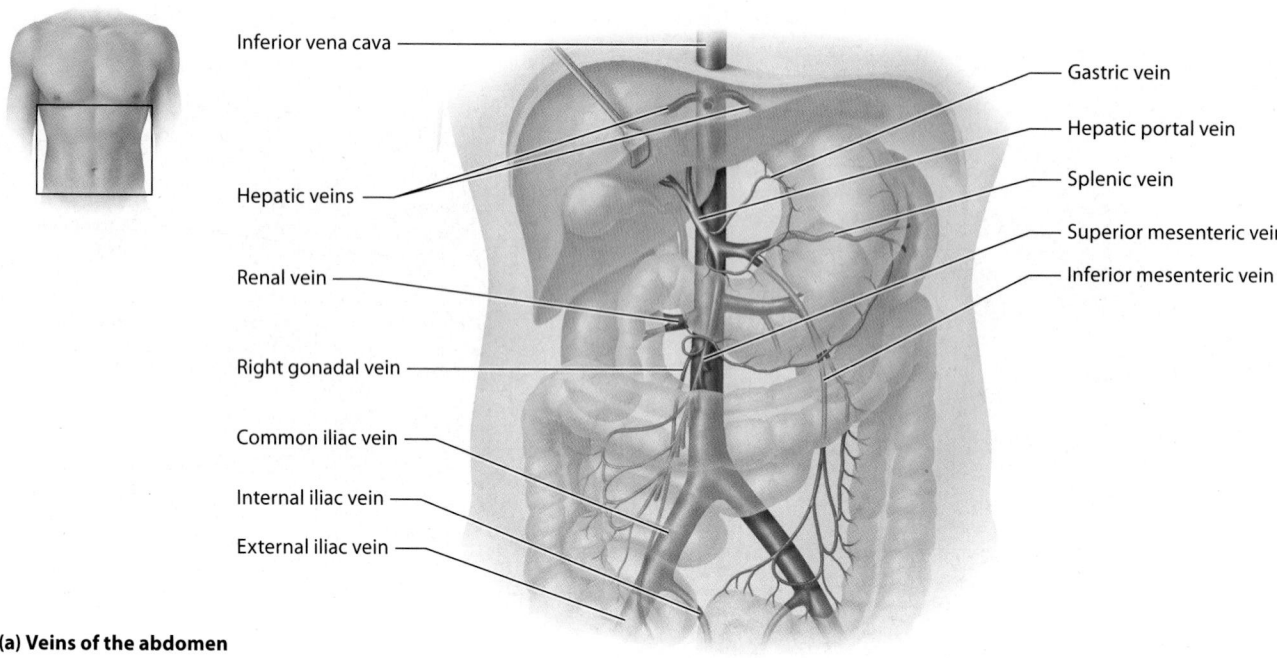

Inferior vena cava

Hepatic veins

Renal vein

Right gonadal vein

Common iliac vein

Internal iliac vein

External iliac vein

Gastric vein

Hepatic portal vein

Splenic vein

Superior mesenteric vein

Inferior mesenteric vein

(a) Veins of the abdomen

(2) Nutrient-rich blood from the digestive organs and the spleen is delivered to hepatocytes by capillaries of the hepatic portal system.

Capillary beds of hepatic portal system

(1) Most of the blood from the abdominal organs goes into the hepatic portal vein instead of the inferior vena cava.

Hepatic portal vein

Gastric veins

Superior mesenteric vein

Splenic vein

Inferior mesenteric vein

Hepatic veins

Inferior vena cava

Liver

(3) After processing, the blood is delivered to the inferior vena cava via the hepatic veins.

Stomach

Spleen

Pancreas

Large intestine

Small and large intestine

(b) Organization of the hepatic portal system

Figure 18.27 Veins of the abdomen.

extremely leaky sinusoidal capillaries. Here, the nutrients, toxins, and products of red blood cell degradation exit the sinusoids to be processed by hepatocytes. In this way, the liver can process and detoxify this blood before it ever reaches the rest of the systemic circulation. ③ The processed blood then exits the liver via hepatic veins and joins the rest of the blood in the systemic circuit in the inferior vena cava.

Quick Check

☐ 4. How does the drainage of the posterior body wall differ on the right and left sides?

☐ 5. Which abdominal vessels drain straight into the inferior vena cava?

☐ 6. How does blood circulate through the hepatic portal system? What is the function of this system?

Veins of the Upper Limb

The venous drainage of the upper limb consists of a superficial system and a deep system. The superficial venous system begins in the hand with the **dorsal venous arch,** which drains into a set of highly anastomosing veins in the hand and wrist whose anatomy often varies from person to person (**Figure 18.28a**). Then the most common venous pattern in the forearm includes the lateral **cephalic vein,** the middle **median antebrachial vein,** and the medial **basilic vein** (listed under "Superficial veins" in **Figure 18.28b**). In the antecubital fossa, a small oblique vein called the **median cubital vein** often connects the cephalic and basilic veins. The median cubital vein is a common site for drawing blood.

Deep veins:
Axillary vein
Brachial vein
Radial vein
Ulnar vein
Palmar venous arches

Superficial veins:
Cephalic vein
Basilic vein
Median cubital vein
Median antebrachial vein

(b) Anterior view of the upper limb

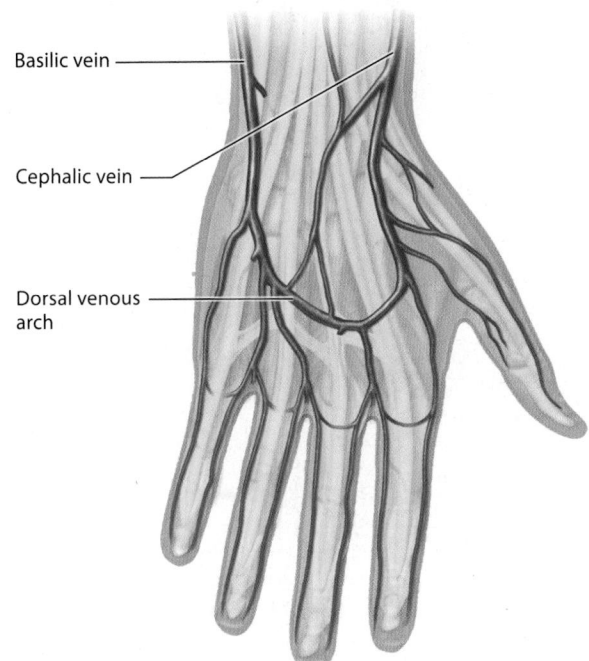

Basilic vein
Cephalic vein
Dorsal venous arch

(a) Posterior aspect of the hand

Figure 18.28 Veins of the upper limb.

Common iliac vein

Internal iliac vein

External iliac vein

Femoral vein

Deep femoral vein

Great saphenous vein

Popliteal vein

Small saphenous vein

Anterior tibial vein

Posterior tibial vein

Medial plantar vein

Lateral plantar vein

Dorsal venous arch

(a) Anterior view

(b) Lateral view

Figure 18.29 Veins of the lower limb.

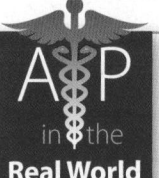

Drugs and the Hepatic Portal System

The hepatic portal system serves numerous vital functions, including protection from some of the nastier toxins we ingest. However, the liver doesn't have a way to discern "good chemicals" from "bad chemicals," and treats most medications exactly as it would most lethal toxins—its enzymes catalyze reactions that destroy them. This poses a problem because generally drugs need to reach the systemic capillary beds to have an effect, and they can't get there unless they can pass the gatekeeper that is the hepatic portal system.

The exact extent of hepatic metabolism varies for each drug, and this serves as a basis for drug dosage calculation. For example, if 100 mg of a drug is needed to produce a therapeutic effect, but 90% of the dose is destroyed by hepatic metabolism, the drug is administered in a 1000-mg dose. However, certain drugs are so thoroughly destroyed in the hepatic portal system that effectively none of the drug reaches the systemic circulation. These drugs must therefore be administered by routes such as injection that allow them access to the systemic capillaries without first having to pass through the hepatic gatekeeper.

The deep venous system also begins in the hand with the **palmar venous arches** (listed under "Deep veins" in Figure 18.28b). These anastomosing vessels drain into the **radial** and **ulnar veins** in the wrist and forearm, which largely parallel the radius and ulna. The radial and ulnar veins merge around the antecubital fossa to form the **brachial vein**, which unites with the basilic vein to form the **axillary vein.** The cephalic vein empties into the axillary vein in the proximal arm. The axillary vein becomes the subclavian vein after it passes deep to the clavicle.

Quick Check

☐ 7. What are the superficial veins of the upper limb?
☐ 8. What are the deep veins of the upper limb?

Veins of the Lower Limb

Like the veins of the upper limb, those of the lower limb are divided into a superficial system and a deep system (**Figure 18.29**). The superficial system begins with the **dorsal venous arch** in the foot; this drains into the anastomosing **great and small saphenous veins** (sah-FEE-nus) in the ankle and leg. The great saphenous vein is located in the superficial medial leg and thigh and is a frequent location for varicose veins. This vein is also commonly harvested for bypass grafts, as described in *A&P in the Real World: Vein Grafting*. It empties into the **femoral vein** in the proximal thigh. The small saphenous vein is slightly lateral to the great saphenous vein and empties into the **popliteal**

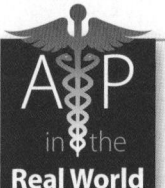

Vein Grafting

When the blood flow through an artery is blocked as a result of atherosclerosis or another disease, it is often necessary to bypass the artery to restore blood flow to a body part. This is done via a procedure called a *bypass graft*, in which a vein is removed and grafted onto an artery in two places to give blood an alternate route of blood flow. One of the most common veins used for this procedure is the great saphenous vein, because it is superficial and thus readily accessible. In addition, blood in the superficial structures that it drains can usually take alternate venous paths.

The vein is removed by a surgical procedure in which two small incisions are made in the medial leg. The vein is then isolated and tied off, and removed with a tug. The surgeon sews the harvested vein into place on the affected artery, allowing the blood to bypass the blockage.

vein in the popliteal fossa. Note that the superficial veins have collateral branches that connect them with the deep veins.

The system of deep veins of the lower limb mirrors the lower limb's arterial supply. The deep veins of the foot drain into the **medial and lateral plantar veins,** which drain into the **anterior and posterior tibial veins** in the ankle and leg. Near the popliteal fossa, these veins merge to form the popliteal vein. The popliteal vein crosses into the anterior thigh through the adductor magnus muscle, at which point it becomes the femoral vein. After the femoral vein passes deep to the inguinal ligament, it becomes the external iliac vein.

The major veins of the body are summarized in **Table 18.5**.

Quick Check

☐ 9. What are the two major superficial veins of the lower limb?
☐ 10. What is the main deep vein of the lower limb?

Apply What You Learned

☐ 1. Most veins in the systemic circuit have high compliance and can stretch to accommodate increases in blood volume. Do you think this is the case for the dural sinuses? Why or why not?
☐ 2. The antibiotic medication ticarcillin is not effective if given by mouth, because nearly the entire dose of the drug is destroyed by the hepatic portal system. However, this drug is effective when injected into the bloodstream. Explain.
☐ 3. The great saphenous vein is commonly harvested and used in coronary artery bypass grafts. Do you think removal of this vein would significantly affect venous drainage of the leg? Why or why not?

See answers in Appendix A.

TABLE 18.5 MAJOR SYSTEMIC VEINS

Vein	Description	Location	Structure(s) Drained
Brachiocephalic veins	Merge to form the superior vena cava	Mediastinum	• Most structures superior to the diaphragm
Internal iliac veins	Merge with the external iliac veins to form the common iliac vein	Posterior pelvis	• Pelvis
External iliac veins	Merge with the internal iliac veins to form the common iliac vein	Anterior pelvis	• Lower limb
Common iliac veins	Merge to form the inferior vena cava	Posterior pelvic wall	• Pelvis • Lower limb
Dural sinuses	Venous channels in the brain that receive blood and cerebrospinal fluid	Between the two layers of the dura in the brain	• Brain
Internal jugular veins	Merge with the subclavian veins to form the brachiocephalic veins	Deep neck, alongside the common carotid artery	• Brain • Superficial head • Neck
External jugular veins	Drain into the subclavian veins	Superficial neck	• Superficial face • Neck
Vertebral veins	Drain into the subclavian veins	Pass through the vertebral foramina of cervical vertebrae	• Posterior neck
Subclavian veins	Merge with the internal jugular veins to form the brachiocephalic veins	Deep to the clavicle	• Upper limb • Upper thorax
Azygos vein	Large unpaired vein	Right side of the posterior thoracic wall	• Posterior thoracic and abdominal wall on the right side • Esophagus • Lungs
Hemiazygos and accessory hemiazygos veins	Drain into the azygos vein	Posterior body wall on the left side	• Posterior body wall
Splenic vein	Drains into the hepatic portal vein	Posterior to the stomach	• Spleen
Gastric veins	Drain into the hepatic portal vein	Around the stomach	• Stomach
Superior mesenteric vein	Drains into the hepatic portal vein	Multiple branches around the small and large intestines	• Small intestine • Much of the large intestine
Inferior mesenteric vein	Drains into the hepatic portal vein	Multiple branches around the large intestine	• Remainder of the large intestine
Hepatic portal vein	Receives blood from the digestive organs and the spleen	Large vein that enters the liver	• Processes blood from the spleen and digestive organs
Hepatic veins	Drain into the inferior vena cava	Posterior liver	• Arterial blood from the liver • Venous blood from the hepatic portal system
Renal veins	Drain into the inferior vena cava	Posterior abdomen	• Kidneys
Cephalic veins	Superficial forearm vein, drains into the axillary vein	Lateral arm and forearm	• Superficial structures of the upper limb
Basilic veins	Superficial forearm vein, merges with the brachial vein to form the axillary vein	Medial arm and forearm	• Superficial structures of the upper limb
Radial and ulnar veins	Deep forearm veins, merge to form the brachial vein	Lateral forearm (radial), medial forearm (ulnar)	• Forearm
Brachial veins	Forms from the union of the radial and ulnar veins	Anterior arm	• Arm
Axillary veins	Continuation of the brachial vein	Axilla	• Arm • Axilla
Great saphenous veins	Superficial vein of the lower limb	Medial lower limb	• Superficial structures of the lower limb
Anterior and posterior tibial veins	Merge to form the popliteal vein	Anterior and posterior leg	• Leg • Foot
Popliteal veins	Form from the union of the anterior and posterior tibial veins	Popliteal fossa	• Knee joint • Leg • Thigh
Femoral veins	Continuation of the popliteal veins	Anterior thigh	• Thigh

18.8

Putting It All Together: The Big Picture of Blood Vessel Anatomy

Learning Outcomes

1. Describe the general pathway of blood flow through the body.
2. Identify the arteries and veins of the body.

Up to this point in the chapter you've seen illustrations of the vessels in isolation. However, humans obviously don't have just arteries or just veins—we have both, which is what you will see in a laboratory or clinical setting. Therefore, we have included a big picture view of the blood vessels so you can better appreciate the relationships these vessels have to one another and to the other organs of the body. In these figures, we continue to use the convention of showing arteries in red and veins in blue.

The blood vessels of the head and neck are illustrated in **Figure 18.30a and b.** In part b, the dissection on the left is superficial, and that on the right is deep. Surrounding structures have been faded back to more clearly show the vessels.

Figure 18.31 depicts the blood vessels of the abdomen. The superficial structures of the abdomen and parts of the abdominal organs have been removed to more clearly show the vessels. For the same reason, the remaining structures have also been faded back.

The blood vessels of the upper and lower limbs are shown in **Figure 18.32.** As with other figures, the musculature of both the upper and lower limbs has been faded back to more clearly show the vessels deep to them.

Finally, **Figure 18.33** shows a schematic view of the overall pattern of systemic blood flow in the body. Blood flows through the systemic arteries and through

Temporal artery and vein

Maxillary artery and vein

Facial artery and vein

External jugular vein

Internal carotid artery

Internal jugular vein

Common carotid artery

Subclavian artery and vein

Sternocleidomastoid muscle (reflected)

Brachiocephalic artery and vein

(a) Vessels of the head

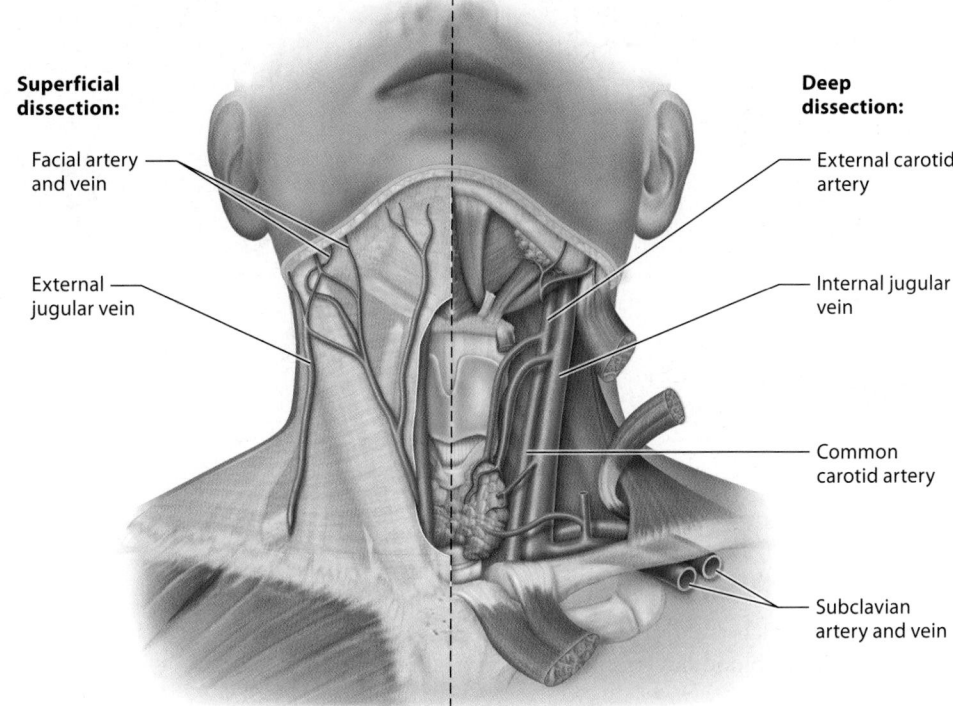

Superficial dissection:

Deep dissection:

Facial artery and vein

External carotid artery

External jugular vein

Internal jugular vein

Common carotid artery

Subclavian artery and vein

(b) Vessels of the neck

Practice art labeling
@MasteringA&P®

Figure 18.30 Blood vessels of the head and neck.

Hepatic veins

Gallbladder

Inferior vena cava

Hepatic portal vein

Inferior mesenteric vein

Pancreas

Superior mesenteric vein

Inferior vena cava

Right common iliac artery and vein

Right external iliac artery and vein

Thoracic aorta

Stomach (sectioned)

Celiac trunk

Splenic artery and vein

Spleen

Superior mesenteric artery

Renal artery and vein

Kidney

Inferior mesenteric artery

Left common iliac artery and vein

Left external iliac artery and vein

Figure 18.31 Blood vessels of the abdomen.

Practice art labeling @MasteringA&P®

capillary beds, which are drained by systemic veins. Note from the figure that this blood flow occurs only in one direction, from arteries to veins.

Apply What You Learned

☐ 1. As Figure 18.33 shows, blood can flow only one way through the blood vessels—it cannot go backward from

veins to arteries. Explain why this is true. (*Hint:* Consider what drives blood flow.)

☐ 2. During procedures on the carotid artery, one must be careful not to enter or injure the internal jugular vein. Explain why this is so, considering their anatomical arrangement. Why is there minimal risk of hitting the external jugular vein?

See answers in Appendix A.

Subclavian vein

Subclavian artery

Axillary artery

Axillary vein

Brachial vein

Cephalic vein

Brachial artery

Radial artery

Radial vein

Palmar venous arches

Median cubital vein

Median antebrachial vein

Basilic vein

Ulnar artery

Deep palmar arch

Superficial palmar arch

(a) Arteries and veins of the upper limb

Common iliac vein

Common iliac artery

Inguinal ligament

Deep femoral artery

Small saphenous vein

Venous arches

Abdominal aorta

Inferior vena cava

Internal iliac vein

Internal iliac artery

External iliac vein

External iliac artery

Femoral vein

Femoral artery

Great saphenous vein

Popliteal vein

Popliteal artery

Posterior tibial vein

Posterior tibial artery

Anterior tibial vein

Anterior tibial artery

Dorsalis pedis artery

(b) Arteries and veins of the lower limb

Practice art labeling
@MasteringA&P®

Figure 18.32 Blood vessels of the upper and lower limbs.

The Big Picture of Systemic Blood Flow in the Body

Figure 18.33

SYSTEMIC FLOW:

Artery → Arteriole → Capillary bed → Venule → Vein

Note that throughout most of the systemic circuit, blood flows through the arteries and then through capillary beds that are drained by veins.

Capillary beds of the head, neck and face

Common carotid artery

Internal jugular vein

Subclavian artery

Subclavian vein

Axillary artery

Axillary vein

Capillary beds of the upper limb

Superior vena cava

Inferior vena cava

Hepatic vein

Capillary beds of the liver and hepatic portal system

Hepatic portal system flow: Blood in the arteries of most abdominal organs flows through a capillary bed that is drained by veins that deliver the blood to a second set of capillaries in the hepatic portal system.

Hepatic portal vein

Gastric and splenic veins

Superior and inferior mesenteric veins

Inferior vena cava

Renal vein

External iliac artery and vein

Capillary beds of the lower limb

Descending aorta

Celiac trunk and branches

Capillary beds of the stomach, spleen, and intestines

Superior and inferior mesenteric arteries

Renal artery

Capillary beds of the kidneys

CHAPTER SUMMARY

Take full advantage of the invaluable online practice and study tools for this chapter by going to

MasteringA&P®. There you will find:

- Practice tests, module quizzes, art-labeling and sequence-the-steps activities

- ▶ Animated figures with quizzes

- Practicing A&P worksheets

-))) Author-narrated podcast

- Dynamic Study Modules

- iP Interactive Physiology

MODULE 18.1
Overview of Blood Vessels 669

- The pulmonary and systemic circuits consist of three types of vessels:
  - ○ **Arteries,** the vasculature's distribution system, carry blood away from the heart.
  - ○ **Capillaries,** the vasculature's exchange system, exist in **capillary beds.**
  - ○ **Veins,** the vasculature's collection system, drain blood from capillaries and return it to the heart.

- Blood vessels are organs with three tissue layers that surround a hollow **lumen.** (Figure 18.1, p. 670)
  - ○ The innermost **tunica intima** is composed of endothelium and its basal lamina, a thin layer of connective tissue, and the internal elastic lamina.
  - ○ The middle **tunica media** is composed of smooth muscle cells and the external elastic lamina. The muscle cells are innervated by vasomotor nerves that cause **vasoconstriction** and **vasodilation.**
  - ○ The outermost **tunica externa** is composed of supportive connective tissue.

- Arteries typically have a thicker wall, a thicker tunica media, and more extensive elastic laminae than do veins. (Figure 18.2, p. 671; Table 18.1, p. 673)

- There are three types of arteries:
  - ○ **Elastic arteries** are the largest arteries, are closest to the heart, and have extensive elastic laminae.
  - ○ **Muscular arteries** are intermediate in diameter and deliver blood to organs. They have a well-developed tunica media.
  - ○ **Arterioles** are the smallest-diameter arteries that control blood flow to tissues.

- Certain arteries contain **baroreceptors** that monitor blood pressure and *chemoreceptors* that detect the oxygen, carbon dioxide, and hydrogen ion concentrations of the blood.

- **Venules** are the smallest veins that drain capillary beds. They merge to form larger venules and eventually veins. Veins, which are both more numerous and have a larger lumen than arteries, function as **blood reservoirs** (Figure 18.3, p. 673). Many veins contain **venous valves** that prevent blood from flowing backward.

- Many vessels in the body are connected by **anastomoses,** which are of three types: arterial, venous, and arteriovenous.

- iP Play Interactive Physiology tutorial on Cardiovascular System: Anatomy Review: Blood Vessel Structure and Function.

MODULE 18.2
Physiology of Blood Flow 674

- *Hemodynamics* is the study of the physiology of blood flow in the cardiovascular system.

- The heart drives blood through the vessels by creating a *pressure gradient*—pressure is highest near the heart and decreases going away from the heart.

- **Blood pressure** is the force that blood exerts on the walls of the blood vessels.

- **Blood flow** is determined by the size of the pressure gradient created by the heart and generally matches cardiac output. It is directly proportional to the pressure gradient but inversely proportional to **resistance,** which is defined as any impedance to blood flow.

- The velocity of blood flow is determined by the volume of blood and the cross-sectional area of the blood vessel. Velocity decreases as cross-sectional area increases through the branching arterial system.

- Three factors determine blood pressure: *resistance, cardiac output,* and *blood volume.* (Figure 18.4, p. 676)
  - ○ As vessel radius decreases, resistance increases, and as vessel radius increases, resistance decreases. As blood viscosity and blood vessel length increase, resistance increases. Obstructions to flow within the vessel also affect resistance. Flow within an unobstructed vessel is *laminar.* Flow within an obstructed vessel is *turbulent* and has higher resistance.
  - ○ As cardiac output increases, blood pressure also generally increases.
  - ○ Increases in blood volume generally lead to increases in blood pressure. Small increases in blood volume are offset by the vessels' *compliance,* or their ability to stretch to accommodate added blood volume.

- iP2 Play Interactive Physiology tutorial on Cardiovascular System: Factors Affecting Blood Pressure.

- The pressure in the pulmonary circuit is significantly lower than that in the systemic circuit, averaging about 15 mm Hg. (Table 18.2, p. 677)

- The **mean arterial pressure** (MAP) in the systemic circuit is about 95 mm Hg. The pressure in the arteries pulsates; it rises during ventricular systole to a **systolic pressure** of about 120 mm Hg and falls during ventricular diastole to a **diastolic pressure** of about 80 mm Hg. The difference between the systolic and diastolic pressures is the **pulse pressure.** The pressure in the arterial system decreases dramatically in the arterioles to about 35 mm Hg. (Figure 18.5, p. 677)

- Arterial blood pressure is measured with a *sphygmomanometer* via compression of an artery to a value above the systolic pressure and slow release of the compression until blood flow resumes.

- The MAP is calculated by adding one-third of the pulse pressure to the diastolic pressure.

- Blood pressure declines further in the capillaries, from about 35 mm Hg at the arteriole end to about 15 mm Hg at the venule end. Pressure declines in the venous system to a low value of about 4 mm Hg in the inferior vena cava.

iP Play Interactive Physiology tutorial on Cardiovascular System: Measuring Blood Pressure.

- The mechanisms that assist in the return of venous blood to the heart include venous valves, sympathetic nerves that constrict the veins' smooth muscle, skeletal *muscle pumps,* and the *respiratory pump.* (Figure 18.6, p. 678)

MODULE **18.3**
Maintenance of Blood Pressure 679

- Mean arterial pressure must be maintained at a constant level of about 95 mm Hg. Deviations from this set point trigger feedback mechanisms in the nervous, endocrine, and urinary systems.

- Short-term control of blood pressure is mediated by the nervous system and certain hormones of the endocrine system.
  - The axons of the sympathetic nervous system increase blood pressure through release of **epinephrine** and **norepinephrine,** which causes vasoconstriction of vessels and an increase in cardiac output. Similar axons of the parasympathetic nervous system decrease blood pressure through release of **acetylcholine,** which lowers cardiac output and indirectly triggers vasodilation through inhibition of sympathetic activity. (Figure 18.7, p. 680)
  - These changes are mediated by the **baroreceptor reflex** arc, which consists of the following parts: (Figure 18.8, p. 681)
    - Blood pressure changes are detected by baroreceptors in the carotid sinus and in the aortic arch.
    - Impulses are transmitted to the medulla via afferent parasympathetic neurons of the glossopharyngeal nerve.
    - The medulla triggers an increase in either sympathetic or parasympathetic activity.
    - Blood pressure increases if the sympathetic nervous system is activated, and decreases if the parasympathetic nervous system is activated.
  - Peripheral chemoreceptors detect the levels of oxygen in the blood. When the oxygen level decreases sharply, peripheral chemoreceptors trigger a rise in blood pressure. Central chemoreceptors detect the pH of the brain interstitial fluid. When the pH decreases, the chemoreceptors trigger a rise in heart rate and blood pressure.
  - Hormones that increase cardiac output include *thyroid hormone,* epinephrine, and norepinephrine.
  - Hormones that increase peripheral resistance include epinephrine, norepinephrine, and angiotensin-II. Atrial natriuretic peptide has vasodilatory effects and produces a mild decrease in resistance.

- Long-term control of blood pressure is mediated by the urinary system and certain hormones of the endocrine system that regulate blood volume.
  - Hormones that increase blood volume through promoting water retention include *aldosterone* and *antidiuretic hormone. Angiotensin-II* promotes thirst and also increases blood volume. *Atrial natriuretic peptide* causes the kidneys to excrete more water, which decreases blood volume.
  - The urinary system directly controls blood pressure by increasing or decreasing the amount of water lost from the body as urine.

- Figure 18.9 (p. 683) provides an overall look at both short- and long-term mechanisms of blood pressure maintenance.

- **Hypertension,** or high blood pressure, is defined as a systolic pressure over 140 mmHg and/or a diastolic pressure over 90 mmHg. The two types of hypertension are **essential hypertension,** which does not have an identifiable cause, and **secondary hypertension,** which does have an identifiable cause.

- **Hypotension** is a systolic pressure lower than 90 mmHg and/or a diastolic pressure lower than 60 mmHg.

iP Play Interactive Physiology tutorial on Cardiovascular System: Blood Pressure Regulation.

MODULE **18.4**
Capillaries and Tissue Perfusion 685

- **Tissue perfusion** is the blood flow through a capillary bed to a tissue.

- A capillary consists of only a layer of endothelium with a small amount of basal lamina. Capillary walls are extremely thin to allow **capillary exchange.** Substances may be exchanged by diffusion through the endothelial cells, diffusion through small pores, and transcytosis. (Figure 18.10, p. 685; Figure 18.11, p. 686)

- There are three types of capillaries: (Table 18.3, p. 687)
  - **Continuous capillaries** are located in muscles, the skin, and most nervous and connective tissue. Their cells are joined by tight junctions, and they are therefore not very leaky.
  - **Fenestrated capillaries** are located in the kidneys, small intestine, and endocrine glands. They contain fenestrations that make them "leakier" than continuous capillaries.
  - **Sinusoidal capillaries** have very large pores in their endothelial cells that make them extremely leaky. They are located in the liver, spleen, lymphoid organs, and bone marrow.

- The **microcirculation** is the total blood flow that takes place within the capillary beds.

- Capillaries branch off a central vessel that is fed by a metarteriole and drained by a **thoroughfare channel.** (Figure 18.12, p. 688)

- Tissue perfusion in most tissues is largely **autoregulated** to ensure that blood flow meets the cells' needs. There are two mechanisms of local control:
  - The **myogenic mechanism** proceeds as follows: When blood pressure increases, the smooth muscle cells are stretched and reflexively vasoconstrict. The opposite response occurs when blood pressure decreases.

○ *Metabolic controls* proceed as follows: In actively metabolizing tissues, the concentration of O_2 decreases and the concentrations of CO_2 and H^+ increase, which triggers vasodilation of the arterioles serving that tissue. The opposite response occurs in tissues that are not actively metabolizing.

- Tissue perfusion in the heart decreases during ventricular systole and increases during ventricular diastole because the heart squeezes its own coronary arteries during systole.

- Autoregulatory mechanisms maintain blood flow to the brain as a whole at about 750 ml/min.

- Blood flow increases dramatically in skeletal muscles during exercise, which is known as *hyperemia.*

- Perfusion to the skin is largely controlled by the sympathetic nervous system as part of the body's temperature-regulating mechanism.

iP Play Interactive Physiology tutorial on Cardiovascular System: Autoregulation and Capillary Dynamics.

MODULE **18.5**
Capillary Pressures and Water Movement 690

- Water crosses a capillary by **filtration,** or the movement of a fluid by a force such as pressure or gravity.

- **Hydrostatic pressure** is the force that the blood exerts on the wall of the vessel. A hydrostatic pressure gradient exists across the capillary wall and forces water to leave the capillary, where hydrostatic pressure is higher, and enter the interstitial fluid, where hydrostatic pressure is lower. (Figure 18.13, p. 691)

- Solute particles in a solution exert a force on water molecules, called **osmotic pressure.** Osmotic pressure is higher in blood than in interstitial fluid, which creates an osmotic pressure gradient known as the **colloid osmotic pressure.** This gradient pulls water into the capillary.

- The **net filtration pressure** in a capillary is the difference between the opposing forces of the hydrostatic pressure gradient and colloid osmotic pressure. The overall net filtration pressure in a capillary bed is about 6 mm Hg in favor of filtration. (Figure 18.14, p. 693)

- **Edema** is an excessive amount of water in the interstitial fluid. It may be caused by an increase in the hydrostatic pressure gradient or a decrease in the colloid osmotic pressure.

iP Play Interactive Physiology tutorial on Cardiovascular System: Autoregulation and Capillary Dynamics.

MODULE **18.6**
Anatomy of the Systemic Arteries 694

- The aorta has four divisions: the **ascending aorta,** the **aortic arch,** the **descending thoracic aorta,** and the **descending abdominal aorta.** The aortic arch has three branches: the **brachiocephalic artery,** the **left common carotid artery,** and the **left subclavian artery.** The descending abdominal aorta terminates by splitting into the two **common iliac arteries.** The common iliac arteries divide into the **internal iliac arteries,** which supply the pelvis, and the **external iliac arteries,** which supply the lower limb. (Figure 18.15, p. 695)

🖑 Practice art-labeling exercise: **Figure 18.15.**

- The majority of the blood supply to the superficial head and face comes from branches of the **external carotid arteries.** (Figure 18.16, p. 696)

- The blood supply to the brain stems from the **internal carotid arteries** and the two **vertebral arteries.** The two arterial systems communicate via an anastomosis called the **cerebral arterial circle,** which equalizes pressure in the brain and provides alternate routes of blood flow. (Figure 18.17, p. 697)

- Braches of the descending thoracic aorta supply the lungs, esophagus, and diaphragm. (Figure 18.18, p. 698)

- The descending abdominal aorta has the following major branches: the **celiac trunk,** which supplies the liver, pancreas, duodenum, stomach, and spleen; the **superior mesenteric artery,** which supplies the small intestine and much of the large intestine; the **renal arteries,** which supply the kidneys; and the **inferior mesenteric artery,** which supplies the remainder of the large intestine. (Figure 18.19, p. 699)

- The upper limb is supplied by the subclavian artery, which becomes the **axillary artery** in the axilla and the **brachial artery** in the arm. The brachial artery splits into the **radial** and **ulnar arteries.** (Figure 18.20, p. 700)

- The external iliac artery becomes the **femoral artery** as it passes deep to the inguinal ligament. The femoral artery crosses to the posterior thigh, where it becomes the **popliteal artery.** The popliteal artery splits into the **anterior tibial artery,** which becomes the **dorsalis pedis artery** in the ankle, and the **posterior tibial artery.** (Figure 18.21, p. 701)

- The pulsations produced by changes in systolic and diastolic pressure may be palpated through the skin in **pulse points,** where the arteries are superficial. (Figure 18.22, p. 702)

- Table 18.4 (p. 703) provides a summary of the major systemic arteries.

MODULE **18.7**
Anatomy of the Systemic Veins 705

- Most veins superior to the diaphragm drain into the **brachiocephalic veins,** which merge to form the **superior vena cava.** Venous blood from the lower limbs and pelvis drain into the **external and internal iliac veins,** which merge to form the **common iliac veins.** The two common iliac veins merge to form the **inferior vena cava,** into which most blood from veins inferior to the diaphragm drains. (Figure 18.23, p. 704)

🖑 Practice art-labeling exercise: **Figure 18.23.**

- Blood from the head, face, and neck drains into the **vertebral, external jugular,** and **internal jugular veins.** The vertebral and external jugular veins drain into the **subclavian vein,** which merges with the internal jugular vein to become the brachiocephalic vein. (Figure 18.24, p. 705)

- Blood from the brain drains into the **dural sinuses,** which also receive cerebrospinal fluid from the arachnoid granulations. Blood from the dural sinuses drains into the internal jugular vein. (Figure 18.25, p. 706)

- The anterior thoracic and abdominal walls drain into the **internal thoracic veins** and the external iliac veins. The posterior thoracic and abdominal walls drain into the **azygos system.** (Figure 18.26, p. 707)

- Abdominal veins that drain directly into the inferior vena cava include the **gonadal, renal,** and **hepatic veins.** (Figure 18.27a, p. 708)

- The **splenic, gastric, superior mesenteric,** and **inferior mesenteric veins** drain into the **hepatic portal vein,** which delivers blood to the capillaries of the **hepatic portal system.** Here hepatocytes process nutrients, toxins, and the products of red blood cell destruction. Blood drains from the hepatic portal system via the hepatic veins into the inferior vena cava. (Figure 18.27b, p. 708)

- The upper limb is drained by superficial and deep veins. The superficial veins include the lateral **cephalic vein,** the middle **median antebrachial vein,** and the medial **basilic vein.** The deep veins include the **radial** and **ulnar veins,** which merge to form the **brachial vein,** which becomes the **axillary** and finally the **subclavian vein.** (Figure 18.28, p. 709)

- The superficial veins of the lower limb drain into the **great and small saphenous veins.** The deep veins include the **anterior and posterior tibial veins,** which merge to form the **popliteal vein,** becoming the **femoral vein** in the anterior thigh. (Figure 18.29, p. 710)

- Table 18.5 (p. 712) summarizes the major systemic veins.

MODULE | **18.8**
Putting It All Together: The Big Picture of Blood Vessel Anatomy 713

- Figures 18.30–18.33 (pp. 713, 714, 715, and 716) summarize the big picture of the circulation and of the systemic arteries and veins.

  👆 **Practice art-labeling exercise: Figure 18.30.**

  👆 **Practice art-labeling exercise: Figure 18.31.**

  👆 **Practice art-labeling exercise: Figure 18.32.**

ASSESS WHAT YOU LEARNED

See answers in Appendix A.

LEVEL 1 Check Your Recall

1. Mark the following statements as true or false. If a statement is false, correct it to make a true statement.

   a. Arteries are the exchange vessels of the cardiovascular system.
   b. The tunica intima is composed of a sheet of endothelium, connective tissue, and the internal elastic lamina.
   c. The tunica media consists of smooth muscle cells innervated by vasomotor nerves.
   d. Muscular arteries control blood flow at the tissue level.
   e. Veins have smaller lumens, more elastic fibers, and more smooth muscle than arteries.

2. Locations where vessels connect via collateral vessels are known as:
   a. thoroughfare channels.   c. anastomoses.
   b. metarterioles.   d. venules.

3. Define each term:
   a. Pressure gradient   c. Blood flow
   b. Blood pressure   d. Resistance

4. Which of the following factors would increase peripheral resistance?
   a. Increased blood viscosity
   b. Shorter vessel
   c. Vasodilation
   d. An increase in vessel radius

5. Which of the following would produce a decrease in blood pressure?
   a. Increased cardiac output
   b. Vasodilation
   c. Vasoconstriction
   d. Increased blood volume

6. Fill in the blanks: The two pressures within the systemic arterial circuit are the _____ pressure and the _____ pressure. The difference between these two pressures is the _____ pressure.

7. The lowest pressure in the systemic circuit occurs in the:
   a. arteries.   d. venules.
   b. arterioles.   e. veins.
   c. capillaries.

8. Explain the mechanisms that assist in the return of venous blood to the heart.

9. Mark the following statements as true or false. If a statement is false, correct it to make a true statement.

   a. The sympathetic nervous system increases blood pressure in the short term by increasing cardiac output and peripheral resistance.
   b. A sudden increase or decrease in blood pressure triggers the baroreceptor reflex, which is mediated by the medulla and the autonomic nervous system.
   c. Hormones that decrease blood volume and blood pressure include antidiuretic hormone and aldosterone.
   d. Angiotensin-II is a chemical that produces profound vasoconstriction.
   e. The kidneys control blood pressure directly through adjustment of blood volume.

10. The carotid sinus contains:
    a. baroreceptors.   c. metabolic controls.
    b. chemoreceptors.   d. smooth muscle cells.

11. Capillaries consist of:
    a. three thin tunics.
    b. only a thin tunica intima with a well-developed internal elastic lamina.
    c. a thin sheet of endothelium with a basal lamina.
    d. stratified epithelium.

12. List three ways in which substances can cross the capillary wall.

13. Which of the following structures is the "leakiest"?
    a. Continuous capillary
    b. Sinusoidal capillary
    c. Precapillary sphincter
    d. Fenestrated capillary

14. Mark the following statements as true or false. If a statement is false, correct it to make a true statement.
    a. Arterioles reflexively dilate when blood pressure increases.
    b. Increased concentrations of carbon dioxide and hydrogen ions and a decreased concentration of oxygen in the interstitial fluid cause local arteriolar constriction.
    c. Tissue perfusion to the heart decreases during systole and increases during diastole.
    d. The increase in tissue perfusion in skeletal muscle that occurs during exercise is known as hyperemia.
    e. The sympathetic nervous system causes vasodilation in the skin when body temperature decreases.

15. The hydrostatic pressure gradient drives water _____ capillaries, and the colloid osmotic pressure drives water _____ capillaries.
    a. out of; out of
    b. out of; into
    c. into; out of
    d. into; into

16. Net filtration pressure in most capillary beds favors:
    a. absorption.
    b. osmosis.
    c. filtration.
    d. secretion.

17. Match the following arteries with the correct description.

    _____ Radial artery
    _____ Celiac trunk
    _____ Basilar artery
    _____ Superior mesenteric artery
    _____ Dorsalis pedis artery
    _____ Femoral artery
    _____ Internal iliac artery
    _____ Renal artery
    _____ Internal carotid artery
    _____ Subclavian artery

    a. Supplies the small intestine and most of the large intestine
    b. Supplies the pelvis
    c. Supplies the upper limb
    d. Located in the lateral forearm
    e. Provides the blood supply to the lower limb
    f. Supplies the brain via the anterior and middle cerebral arteries
    g. Large single branch off the aorta that supplies the liver, stomach, duodenum, and spleen
    h. Supplies the foot and ankle
    i. Supplies the kidney
    j. Forms from the fusion of the two vertebral arteries

18. Which of the following is not a common pulse point?
    a. Femoral artery
    b. Subclavian artery
    c. Common carotid artery
    d. Brachial artery

19. Which of the following vessels does not drain into the hepatic portal vein?
    a. Splenic vein
    b. Inferior mesenteric vein
    c. Gastric vein
    d. Renal vein
    e. Superior mesenteric vein

20. Match the following veins with the correct description.

    _____ Cephalic vein
    _____ Great saphenous vein
    _____ Dural sinus
    _____ Azygos vein
    _____ Hepatic portal vein
    _____ Splenic vein
    _____ Internal jugular vein
    _____ Brachiocephalic veins
    _____ Brachial vein
    _____ Inferior mesenteric vein

    a. Drains the posterior abdominal and thoracic walls
    b. Two veins merge to form the superior vena cava
    c. Receives blood from the spleen and digestive organs
    d. Superficial vein in the medial leg
    e. Drains the brain and face
    f. Drains part of the large intestine
    g. Superficial vein in the lateral upper limb
    h. Deep vein of the arm
    i. Drains the spleen
    j. Drain the brain capillaries and cerebrospinal fluid

LEVEL 2 Check Your Understanding

1. Explain why a severed artery spurts blood, whereas a severed vein merely leaks blood.

2. Explain why a person who is 7 feet tall is likely to have higher blood pressure than a person whose height is 4 feet.

3. Collagen vascular diseases affect the collagen in the tunica externa of blood vessels. Predict the effects of such a disease.

LEVEL 3 Apply Your Knowledge

PART A: Application and Analysis

1. You are babysitting two children who are having a contest to see who can hang upside-down for the longest period of time. At the end of the contest, both children feel dizzy and are worried they are sick. Explain to them why they feel this way and why the effect is only temporary. (*Hint:* Consider how the baroreceptor reflex responds when a person hangs upside-down.)

2. Mr. Wilms has been diagnosed with *nephrotic syndrome,* which is characterized by the loss of plasma proteins such as albumin into the urine. What impact will this have on his colloid osmotic pressure? How will this affect the net filtration pressure in his capillary beds? What effect will this likely cause?

3. Predict the effects of each of the following on systemic arterial blood pressure:

   a. The practice of "blood doping," which increases the number of red blood cells in the blood
   b. Caffeine consumption, which increases heart rate and causes vasoconstriction
   c. Blood loss from a bleeding ulcer

PART B: Make the Connection

4. Explain why the skin and lips of fair-skinned individuals may "turn blue" in very cold weather. *(Connects to Chapter 5)*

5. Ms. Rodgers has been diagnosed with *secretion of inappropriate ADH syndrome* (SIADH). What effect will SIADH have on the amount of water in her body? What symptoms would you expect from this condition? How would it affect her blood pressure and net filtration pressure? *(Connects to Chapter 16)*

19

Blood

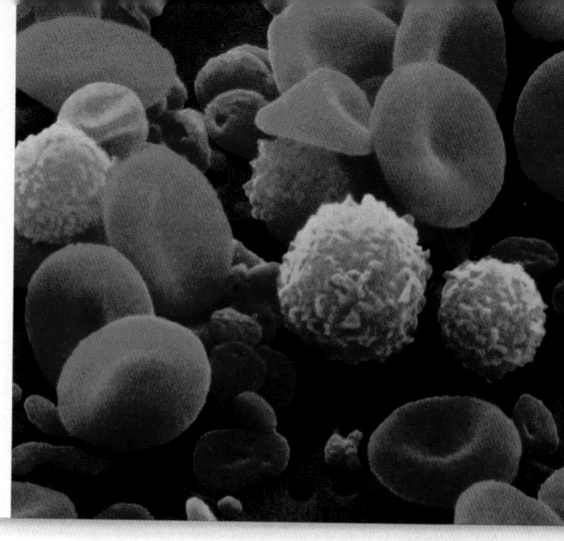

Blood has fascinated humans for millennia. The practice of "bloodletting"—removing large quantities of blood—was the most commonly performed medical procedure from ancient Egypt to 19th-century America. Although bloodletting has fallen out of favor for treating most conditions, we still draw blood to diagnose illness and monitor health. Indeed, blood figures prominently in the medical field—just think how many times you and the people around you have had "blood tests" of some sort.

For the past several chapters we have concentrated on the organ and system levels of organization. Now, we shift our focus to the tissue level as we explore the structure and function of blood. Many topics covered here will tie in closely with our next chapter, the immune system.

MODULE **19.1**

Overview of Blood

Learning Outcomes

1. Describe the major components of blood.
2. Describe the basic functions of blood.
3. Describe the overall composition of plasma, including the major types of plasma proteins, their functions, and where in the body they are produced.

About 5 liters of **blood,** a fluid connective tissue, circulate through the body's blood vessels at all times. Blood makes up about 8% of total body weight. When a sample of blood is removed and examined under a microscope, its major components are visible: **plasma,** the liquid extracellular matrix; and **formed elements,** the cells and cell fragments that are suspended in plasma. Blood contains three types of formed elements: (1) *erythrocytes,* or *red blood cells (RBCs);* (2) *leukocytes,* or *white blood cells (WBCs);* and (3) tiny cellular fragments called *platelets.*

When a blood sample is placed in a centrifuge and spun rapidly, these components separate into three distinct layers. The top layer is plasma, which constitutes about 55% of the total blood volume. The middle layer, the **buffy coat,** makes up only about 1% of the total blood volume. It consists of both leukocytes and platelets. The bottom layer,

Photo: Colored scanning electron micrograph (SEM) showing red and white blood cells and platelets.

(a) Blood sample (non-centrifuged)

(b) Blood sample (centrifuged)

Practice art labeling @MasteringA&P®

Figure 19.1 The three visible layers of blood.

which accounts for the remaining 44% of blood, is a dark red precipitate that consists of erythrocytes. The percentage of blood composed of erythrocytes is called the **hematocrit** (heh-MAH-toh-krit; *hemat-* = "blood") **(Figure 19.1)**.

Overview of Blood Functions

◀◀ **Flashback**

1. What is a buffer? (p. 49)
2. What are metabolic reactions? (p. 68)

Blood performs multiple functions, many of which revolve around its ability to access nearly every tissue in the body. These functions include the following:

- **Exchanging gases.** Oxygen is transported from the lungs to the tissues primarily by erythrocytes. Similarly, both erythrocytes and plasma transport carbon dioxide away from the tissues to the lungs.
- **Distributing solutes.** Plasma transports many solutes, including nutrients, hormones, and wastes. Additionally, blood transports ions and plays a role in regulating ion concentrations in the tissues.
- **Performing immune functions.** Both the cells (leukocytes) and proteins of the immune system use blood as a transport vehicle to reach almost any tissue in the body.
- **Maintaining body temperature.** Heat is a byproduct of many chemical reactions in the body, particularly metabolic reactions. Blood carries away the heat produced by an actively metabolizing tissue, which helps to maintain a constant temperature in that tissue.
- **Functioning in blood clotting.** When a blood vessel is broken, platelets and certain proteins form a *blood clot* that seals the damaged vessel, preventing excessive blood loss.
- **Preserving acid-base homeostasis.** The pH of blood is generally maintained within the narrow range of 7.35–7.45. The pH remains fairly constant because blood composition

controls many of the body's most important buffer systems. Also, many plasma proteins act as buffers.
- **Stabilizing blood pressure.** One of the primary factors that determine blood pressure is the volume of blood in the circulation (see Chapter 18). Blood volume is vital to maintaining blood pressure at a constant level.

Our discussion, as it unfolds, will cover several of these functions, including gas transport, immune functions, and blood clotting. However, many specifics involving these roles of blood will be addressed in later chapters.

Quick Check

☐ 1. Which substances does blood transport through the body?
☐ 2. What role does blood play in the immune system and in blood clotting?
☐ 3. How does blood regulate temperature and acid-base homeostasis?

Plasma

◀◀ **Flashback**

1. Which molecules are hydrophilic? Which are hydrophobic? (p. 47)
2. What is colloid osmotic pressure? (p. 692)

Plasma is a pale yellow liquid consisting primarily of water, which accounts for about 90% of its volume. The amount of water in plasma is one factor that determines the thickness, or *viscosity,* of blood. If plasma contains less water, its viscosity increases and blood flow becomes sluggish.

Plasma proteins make up about 9% of plasma volume. These proteins—most made by the liver—are too large to fully dissolve in the water portion of plasma and therefore form a colloid. Important plasma proteins include:

- **Albumin.** As you read in Chapter 18, **albumin** (al-BYOO-mun) is a relatively large protein produced by the liver that is responsible for blood's colloid osmotic pressure, or the

  CORE PRINCIPLE Gradients

  pressure that draws water into the blood via osmosis, an example of the Gradients Core Principle (p. 26). To discover what happens when the liver cannot produce enough albumin, see *A&P in the Real World: Cirrhosis.*

- **Immune proteins.** The γ-globulins, also known as **antibodies,** are plasma proteins of the immune system. Unlike most plasma proteins, antibodies (see later in this chapter) are produced by leukocytes. Chapter 20, on immunity, will discuss antibodies in detail.

- **Transport proteins.** Recall that lipid-based molecules such as fats and steroids are hydrophobic (see Chapter 2). This feature makes their transport through blood problematic, because hydrophobic molecules tend to associate with one another and form clumps rather than associate with water molecules. These molecules may be transported safely through the blood by binding to **transport proteins** that are hydrophilic and can associate with water molecules. Examples of such transport proteins include globular proteins called **α- and β-globulins** and **lipoproteins.**

- **Clotting proteins.** A blood clot, which is a collection of platelets and *clotting proteins,* stops bleeding from an injured blood vessel. Module 19.5 gives details on clotting proteins.

The remaining 1% of plasma volume consists of a variety of small molecules that generally dissolve in the water portion of plasma and form a solution. The solutes include nutrients such as glucose and amino acids, nitrogenous wastes, ions, and small amounts of dissolved gases such as oxygen and carbon dioxide. These small molecules are readily exchanged between the blood and the interstitial fluid in most capillary beds.

Table 19.1 summarizes the components of plasma.

TABLE 19.1 COMPONENTS OF PLASMA

Plasma Component	Function
Water	90% of plasma volume; solvent that dissolves and transports many solutes through the body
Plasma Proteins	9% of plasma volume; multiple functions (see below)
Albumin	Maintains osmotic pressure
Immune proteins	Produced by leukocytes; function in immunity
Transport proteins	Bind and transport hydrophobic molecules through the blood
Clotting proteins	Function in blood clotting
Other Solutes	1% of plasma volume; multiple functions (see below)
Glucose, amino acids	Nutrition; building blocks for protein synthesis
Ions	Electrolyte and acid-base homeostasis
Dissolved gases (small amounts of oxygen and carbon dioxide)	Oxygen delivered to the tissues; carbon dioxide delivered to the lungs to be exhaled
Wastes	Delivered to the appropriate organ for excretion

Quick Check

☐ 4. List the four main categories of plasma proteins.
☐ 5. What are the other components of blood plasma?

Apply What You Learned

☐ 1. Predict the effect of decreased albumin levels in the plasma.
☐ 2. Explain why proteins can readily travel through plasma, but lipids cannot.
☐ 3. Predict the homeostatic consequences of extensive blood loss from an injury. (*Hint:* How does blood normally help maintain homeostasis?)

See answers in Appendix A.

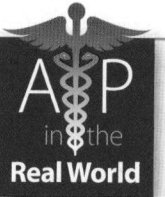

Cirrhosis

Liver disease, or *cirrhosis*, has many causes, including cancer, alcoholism, and viral hepatitis. A common disease in the United States, cirrhosis is the 10th leading cause of death for men and the 12th for women. One consequence of cirrhosis is a progressive decrease in the production of plasma proteins, particularly albumin and clotting proteins. The decrease in albumin leads to a reduction in osmotic pressure, which causes the blood to lose water to the extracellular spaces. This produces severe edema (swelling) in the abdominal area, a condition known as *ascites* (ah-SY-teez). The progressive decline in clotting proteins also causes easy bruising and delays clotting, which can be fatal.

MODULE **19.2**

Erythrocytes and Oxygen Transport

Learning Outcomes

1. Describe the structure and functions of erythrocytes.
2. Discuss the structure and function of hemoglobin, as well as its breakdown products.
3. Explain the basic process of erythropoiesis and its regulation through erythropoietin.
4. Describe the causes and symptoms of anemia.

As you learned in the first module, erythrocytes average about 44% of the total blood volume, a value known as the hematocrit. The hematocrit is generally higher in males (40–50%) than in females (36–44%) due to males' typically larger body size and greater muscle and bone mass. The primary function of erythrocytes is oxygen and carbon dioxide transport, at which they are extremely effective.

The structure of erythrocytes enables them to carry out their vital tasks of gas transport. Even so, certain conditions can impair the oxygen-carrying capacity of blood, a topic covered later in this module.

Erythrocyte Structure

◀◀ **Flashback**

1. What is a polypeptide? (p. 56)
2. What is an oxidation reaction? (p. 43)

A typical **erythrocyte** (eh-RITH-roh-syt; *erythr-* = "red"), or **red blood cell (RBC),** is a *biconcave disc:* a flattened, donut-shaped cell that is concave on both sides (**Figure 19.2**). This shape gives erythrocytes a large surface area, which is critical to their role in gas exchange. Notice that mature erythrocytes are anucleate (have no nucleus) and also lack most other organelles. This means that mature erythrocytes are not capable of carrying out oxidative catabolism or protein synthesis. In fact, an erythrocyte consists of little more than a plasma membrane surrounding cytosol filled with enzymes and about one billion molecules of the oxygen-binding protein **hemoglobin,** or **Hb** (HEE-muh-glow-bin). The shape and composition of an erythrocyte facilitate its transport of oxygen through the blood, an example of the Structure-Function Core Principle (p. 25).

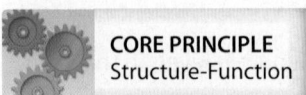

CORE PRINCIPLE
Structure-Function

As you can see in **Figure 19.3**, hemoglobin is a large protein that consists of four polypeptide subunits: two alpha (α) chains and two beta (β) chains. Each polypeptide is bound to an iron-containing compound called a **heme group.** The iron ion in each heme binds to oxygen in parts of the body where oxygen concentrations are high (such as the lungs), forming a molecule called **oxyhemoglobin (HbO_2).** Where oxygen concentrations are low, as in the tissues surrounding systemic capillaries, hemoglobin releases oxygen to become **deoxyhemoglobin.**

Note in Figure 19.3b that the iron ion loses an electron in this reaction—its charge changes from 2+ to 3+ because the negative charge is lost. The iron ion is therefore *oxidized,* which is the same reaction that takes place when metal rusts. Just as metal turns orange-red when it oxidizes, oxidized hemoglobin turns red, giving erythrocytes and blood their characteristic red color. In vessels with a high oxygen concentration, such as the systemic arteries, nearly 100% of hemoglobin molecules are oxyhemoglobin, making arterial blood bright red. Conversely, in vessels with a low oxygen concentration, fewer oxyhemoglobin molecules are present and the blood takes on a dark red color.

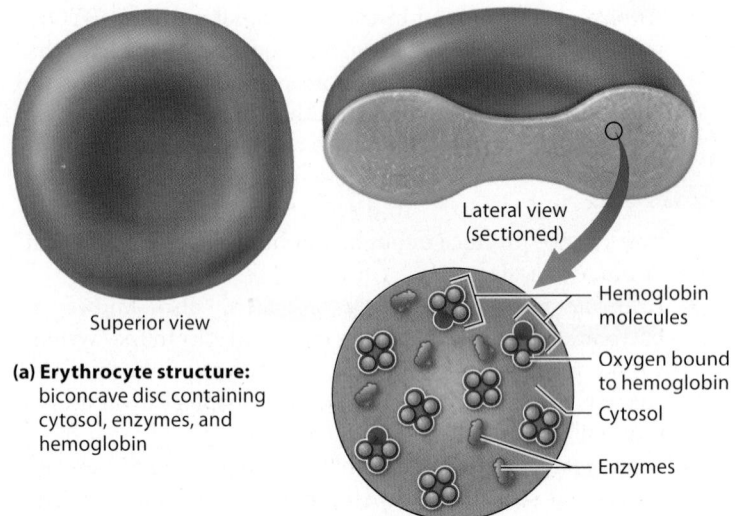

(a) Erythrocyte structure: biconcave disc containing cytosol, enzymes, and hemoglobin

Superior view

Lateral view (sectioned)

Hemoglobin molecules

Oxygen bound to hemoglobin

Cytosol

Enzymes

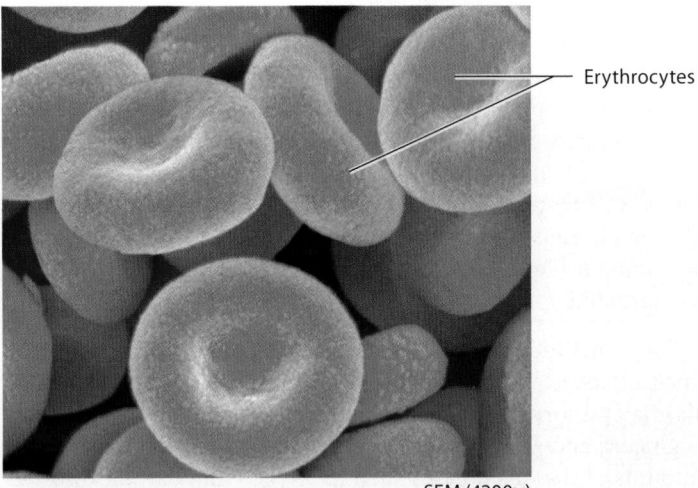

SEM (4200×)

Erythrocytes

(b) Erythrocytes

Figure 19.2 Erythrocyte structure.

A low oxygen concentration occurs in systemic veins, which accounts for the slightly different colors we see in venous and arterial blood. (Contrary to popular perception, venous blood is not blue. Rather, veins simply *look* blue because of an optical illusion related to light refraction.)

In tissue spaces where oxygen levels are low, hemoglobin binds to carbon dioxide, to form **carbaminohemoglobin.** Approximately 23% of the carbon dioxide in blood is transported in this manner. Hemoglobin also binds to carbon monoxide (CO), which forms **carboxyhemoglobin.** Unfortunately, carbon monoxide binds to hemoglobin more strongly than does oxygen and has an affinity for hemoglobin about 200 times greater than that of oxygen. Furthermore, CO changes the shape of hemoglobin so that the oxygen already bound to hemoglobin cannot be released into the tissues. For this reason, exposure to high levels of carbon monoxide is lethal—it severely decreases the oxygenation of body tissues.

β chain   Heme groups   β chain

α chain   α chain

(a) Hemoglobin molecule

Heme in oxyhemoglobin

Heme in deoxyhemoglobin

(b) Heme groups

Figure 19.3 **Hemoglobin structure.** When the iron ion in each heme binds oxygen, hemoglobin becomes oxyhemoglobin. When the hemes release oxygen, hemoglobin becomes deoxyhemoglobin.

Quick Check

☐ 1. Describe the structure of a typical erythrocyte.
☐ 2. Differentiate between oxyhemoglobin, deoxyhemoglobin, carbaminohemoglobin, and carboxyhemoglobin.

Lifespan of an Erythrocyte

◀◀ **Flashback**

1. What is erythropoietin? (p. 620)
2. What is a sinusoid? Where are sinusoids located? (p. 687)

Erythrocytes lead rough lives. They tumble and squeeze through more than 100,000 blood vessels under high pressure and enormous shear forces, which leads to damage that accumulates over time. Erythrocytes lack the cellular machinery to repair this damage, so their average lifespan is only 100–120 days. The body must continuously make new erythrocytes.

Erythrocyte Formation: Erythropoiesis

Recall that **hematopoiesis** (he′-mah-toe-poh-EE-sis; *poiesis* = "make, produce") is the process that produces the formed elements in blood (see Chapter 6). Hematopoiesis occurs in **red bone marrow,** which houses the cells from which all formed elements arise: the **hematopoietic stem cells,** or **HSCs.**

Erythropoiesis (ee-rith′-row-poh-EE-sis), the formation of erythrocytes, is part of the larger process of hematopoiesis (**Figure 19.4**). Erythropoiesis begins when HSCs differentiate into progenitor cells called *erythrocyte colony-forming units (CFUs).* At this point they are *committed,* meaning they are able to become only a single cell type. Erythrocytes form at the incredible rate of about 250 *billion* cells per day.

Erythrocyte CFUs differentiate into cells called **proerythroblasts** (-*blast* = "immature cell"), a process requiring the presence of the hormone **erythropoietin** (ee-rith′-row-poh-EE-tin), or **EPO,** secreted by the kidneys. Proerythroblasts develop into **erythroblasts,** which rapidly synthesize hemoglobin and other proteins. Notice that as erythroblasts mature, their nuclei shrink and are eventually ejected from the cells, at which point they are called **reticulocytes** (reh-TI-kyu-loh-sytz). After reticulocytes eject their remaining organelles by exocytosis, they enter the bloodstream by squeezing through the large pores in the sinusoidal capillaries of the bone marrow. The entire process takes 5–7 days.

Keep in mind that reticulocytes are normally found in the blood in small numbers, but they are able to bind and transport oxygen on their hemoglobin molecules. For this reason, reticulocytes are sometimes released into the circulation if the number of circulating erythrocytes is inadequate. Indeed, an elevated reticulocyte count is a sign of slow blood loss, as from a stomach ulcer.

Regulation of Erythropoiesis

Erythropoiesis is regulated by a negative feedback loop to maintain the hematocrit within normal limits (**Figure 19.5**). This is critical, because if the hematocrit drops too low, there will not be enough erythrocytes available to transport oxygen (O_2) to the tissues. Conversely, if the hematocrit climbs too high, the blood becomes thicker (its viscosity increases), making blood flow sluggish.

This negative feedback loop primarily involves the hormone erythropoietin, with the concentration of oxygen in the blood influencing the production of this hormone. The synthesis of erythropoietin demonstrates the Feedback Loops Core Principle (p. 21).

CORE PRINCIPLE
Feedback Loops

Red bone marrow

Hematopoietic stem cell: Cell may become any type of formed element.

Erythrocyte-CFU: Cell is now committed to becoming an erythrocyte.

Proerythroblast: Step requires the hormone erythropoietin.

Sequence the steps @MasteringA&P®

Figure 19.4 Erythropoiesis: formation of erythrocytes.

The negative feedback loop that regulates erythropoiesis proceeds as follows:

- **Stimulus: Blood levels of oxygen fall below normal.** A decrease in the oxygen content of blood is usually due to respiratory problems, but it may also result from heart conditions or reduced availability of oxygen (e.g., at high altitudes).
- **Receptor: Kidney cells detect falling oxygen levels.** Specialized cells in the kidneys act as chemoreceptors and monitor the blood's oxygen content. They are stimulated when the oxygen content falls below a certain level.
- **Control center: Kidneys produce more erythropoietin and release it into the blood.**
- **Effector/response: Production of erythrocytes increases.** Erythropoietin has many effects. It speeds up the rate of erythropoiesis and reduces the amount of time needed for new erythrocytes to mature.

Erythropoietin can also trigger the replacement of yellow bone marrow with red bone marrow. The kidneys produce erythropoietin, which communicates with hematopoietic stem cells in the bone marrow. Here we see an example of the Cell-Cell Communication Core Principle (p. 27).

CORE PRINCIPLE Cell-Cell Communication

- **Homeostasis: Blood levels of oxygen rise to normal.** As more erythrocytes enter the blood, the hematocrit and oxygen-carrying capacity of the blood increase. Under normal circumstances, the oxygen content of blood will rise as a result. Chemoreceptors in the kidneys detect the return to normal oxygen levels, and erythropoietin production and release decline to normal.

In addition to erythropoietin, chemicals known as *growth factors* can increase the production of

Figure 19.5 Regulation of erythropoiesis.

Early erythroblast: Hemoglobin is synthesized rapidly.

Late erythroblast: Nucleus shrinks and is ejected with other organelles.

Reticulocyte: Remaining organelles are ejected; cell enters bloodstream.

Erythrocyte: Cell is now mature.

Figure 19.4 *(continued)*

erythrocytes. Growth factors are produced by cell types such as endothelial cells in blood vessels and fibroblasts in connective tissue. They aid in the transition from yellow bone marrow to red bone marrow, and also trigger mitosis of hematopoietic stem cells.

Erythrocyte Death

As erythrocytes age, their plasma membranes become less flexible, making their passage through tiny capillaries more difficult.

This is particularly the case in the sinusoids of the **spleen,** an organ located in the upper left abdominal cavity. Erythrocyte destruction proceeds by the following steps (**Figure 19.6**):

① **Erythrocytes become trapped in the sinusoids of the spleen.** Older erythrocytes are not flexible enough to exit the tortuous sinusoids.

② **Spleen macrophages digest erythrocytes.** In the sinusoids, erythrocytes encounter leukocytes called *macrophages*

① Erythrocytes become trapped in the sinusoids of the spleen.

② Spleen macrophages digest erythrocytes.

Spleen

Liver

Macrophage

③ Hemoglobin is broken down into amino acids, iron ions, and bilirubin.

Erythrocytes

Hb

Amino acids

Heme

Fe^{3+}

Bilirubin

Spleen sinusoid

Red bone marrow

Amino acids

Fe^{3+}-Transferrin

Bilirubin

Liver

④a Iron ions and amino acids are recycled and used to make new hemoglobin in red bone marrow.

④b Bilirubin is sent to the liver for excretion.

Figure 19.6 Erythrocyte death.

Sequence the steps
@MasteringA&P®

(discussed in Module 19.3). Macrophages are phagocytes that ingest and destroy older erythrocytes and other cells.

(3) **Hemoglobin is broken down into amino acids, iron ions, and bilirubin.** The macrophages break down the hemoglobin of erythrocytes. The polypeptide chains of hemoglobin are broken down into amino acids, the iron ions are removed from the heme groups, and the remainder of the heme is converted first into the waste product **biliverdin,** a greenish pigment (*verd* = "green"). Generally, biliverdin is then converted into a second waste product, the yellowish pigment **bilirubin** (bi-li-ROO-bin). Interestingly, biliverdin is responsible for the greenish color seen in bruises. As the biliverdin is converted to bilirubin, the bruise becomes yellow.

(4a) **Iron ions and amino acids are recycled and used to make new hemoglobin in red bone marrow.** Most of the iron ions and amino acids are transported to the red bone marrow to be incorporated into new hemoglobin molecules, a form of molecular recycling. You can see in Figure 19.6 that the iron ions are transferred to a protein called **transferrin** (trans-FER-in) that carries them through the blood.

(4b) **Bilirubin is sent to the liver for excretion.** At the same time that the events in (4a) are occurring, bilirubin enters the blood and is transported to the liver, where it is modified and excreted in feces and urine.

Occasionally, trauma or disease requires the removal of the spleen. If the spleen is no longer present, older erythrocytes are destroyed by macrophages in the liver and bone marrow sinusoids.

Quick Check

☐ 3. Walk through the basic steps of erythropoiesis.
☐ 4. What is the lifespan of an erythrocyte?
☐ 5. How are erythrocytes and hemoglobin recycled?

Anemia

The common condition **anemia** (ah-NEE-mee-ah; *an-* = "without"; *-emia* = "blood") is defined as decreased oxygen-carrying capacity of the blood. Although anemia has a variety of causes, the symptoms are similar and mostly reflect decreased oxygen delivery to tissues. They include pallor (pale skin), fatigue, weakness, and shortness of breath. In addition, many types of anemia cause elevated numbers of reticulocytes in the blood as the body boosts erythrocyte production. Severe anemia may cause rapid heart rate, which is the body's attempt to compensate for impaired oxygenation of the tissues by increasing cardiac output. If untreated, such severe anemia can be fatal.

Anemia has three primary causes: decreased amount of hemoglobin, decreased hematocrit, and abnormal hemoglobin. A widely used test to detect anemia (and many other conditions) is the complete blood count (see *A&P in the Real World* on page 732).

Decreased Hemoglobin

The most common type of anemia is **iron-deficiency anemia,** which is due to inadequate dietary iron, reduced intestinal absorption of dietary iron, or slow blood loss (including menstruation). Without functional iron-containing heme groups, erythroblasts cannot make hemoglobin. Erythrocytes in an individual with iron-deficiency anemia are generally small and pale. Another common form is *anemia of chronic disease,* in which an underlying condition such as cancer produces chemicals that interfere with the transport of iron from the liver to the red bone marrow. Decreased hemoglobin may also be due to malnutrition, vitamin B6 deficiency, certain drugs, poisoning with heavy metals such as lead, or even pregnancy.

Decreased Hematocrit

Many factors can reduce the number of erythrocytes in the blood, thus lowering the hematocrit. One is blood loss, which may be acute, as in an injury, or chronic, as in a bleeding stomach ulcer.

SEM (2180×)

(a) Normal and sickled erythrocytes

(b) Sickled erythrocytes block small blood vessels.

Normal erythrocyte
Sickled erythrocyte
Blood vessel
Normal erythrocyte
Sickled erythrocytes
Hemoglobin S molecules line up and for chains, forcing the erythrocyte to sickle.

Figure 19.7 Erythrocytes in sickle-cell disease.

Pernicious anemia results from vitamin B12 deficiency, which interferes with DNA synthesis of rapidly dividing cells, including hematopoietic cells in bone marrow. Erythrocyte destruction can lead to **hemolytic anemia.** Causes of hemolytic anemia include bacterial infections, diseases of the immune system or liver, and lead poisoning. Finally, the red bone marrow may stop producing erythrocytes, which leads to **aplastic anemia** (*a-* = "without"; *-plasis* = "formation"). Certain medications or exposure to ionizing radiation may induce aplastic anemia, but the cause in many cases is unknown.

Abnormal Hemoglobin

The most common cause of abnormal hemoglobin is the inherited condition **sickle-cell disease.** Individuals with a single copy of a defective gene have *sickle-cell trait* and are generally asymptomatic. Such individuals are resistant to the mosquito-borne disease *malaria,* which is endemic to Africa and other tropical locations. Individuals with two copies of the defective gene have *sickle-cell disease,* in which they produce an abnormal hemoglobin called **hemoglobin S, or HbS.** When oxygen levels are low, HbS molecules line up in a row, forcing the erythrocytes into a curved "sickle" shape (**Figure 19.7**). The sickle-shaped cells get stuck in capillary beds, which leads to ischemia and tissue damage. The sickled cells are eventually destroyed; this lowers the number of circulating erythrocytes and causes anemia.

Quick Check

☐ 6. What are the common symptoms of anemia?
☐ 7. Summarize the three primary causes of anemia.

Apply What You Learned

☐ 1. Explain how an erythrocyte's structure relates to its functions.
☐ 2. Athletes have been known to abuse erythropoietin. Explain what potential benefit an athlete might hope to derive from erythropoietin. What are the risks of abusing erythropoietin?

See answers in Appendix A.

MODULE **19.3**

Leukocytes and Immune Function

Learning Outcomes

1. Compare and contrast the relative prevalence and morphological features of the five types of leukocytes.
2. Describe functions for each of the five major types of leukocytes.
3. Discuss the difference in leukopoiesis of granulocytes and agranulocytes.

Leukocytes (LEWK-oh-sytz; *leuk-* = "white"), or **white blood cells (WBCs),** are larger than erythrocytes and have prominent nuclei. Also unlike erythrocytes, leukocytes don't generally function within the blood, but instead use it as a transport vehicle to reach nearly all tissues in the body. When they reach their destination, they adhere to the walls of capillaries or venules and exit the blood by squeezing between the endothelial cells of the vessels.

Leukocytes come in two basic varieties (**Figure 19.8**): (1) **granulocytes** (GRAN-yoo-low-sytz), which contain cytoplasmic

GRANULOCYTES	CHARACTERISTICS
Neutrophil LM (2700×)	Nucleus has 3–5 lobes. Cytoplasmic granules are not visible.
Eosinophil LM (2700×)	Nucleus is bilobed. Cytoplasmic granules stain red.
Basophil LM (2700×)	Nucleus is S-shaped. Cytoplasmic granules stain dark purple.

AGRANULOCYTES	CHARACTERISTICS
Lymphocyte LM (2700×)	Nucleus is spherical. Thin rim of light blue cytoplasm becomes visible when cell is stained.
Monocyte LM (2700×)	Nucleus is U-shaped. Light blue to purple cytoplasm becomes visible when cell is stained.

Practice art labeling
@MasteringA&P®

Figure 19.8 Classes of leukocytes.

granules that the cells release when activated, and (2) **agranulo-cytes** (ay-GRAN-yoo-low-sytz; *a-* = "without"), which lack visible cytoplasmic granules. The following sections examine the properties of both granulocytes and agranulocytes, as well as the process of leukocyte formation. We'll focus on leukocyte functions in detail in the next chapter, The Lymphatic System and Immunity.

Granulocytes

◀◀ Flashback

1. What is phagocytosis? (p. 83)
2. What are the main functions of lysosomes and lysosomal enzymes? (p. 92)

Granulocytes are easy to distinguish from other cell types by their unusual nuclei (Figure 19.8). Although many granulocytes may appear to have multiple nuclei, each cell actually has a single nucleus with multiple lobes that are connected by thin bands of nuclear material. All granulocytes contain general lysosomal granules as well as specific granules unique to each type of granulocyte.

Cells must be treated with specific dyes or stains so we can see their features under a microscope (see Chapter 4). Cells in a blood smear are generally treated with two dyes: the red, acidic dye *eosin* and the dark purple, basic (alkaline) dye *methylene blue*. We classify granulocytes into three categories based on the color of their granules after being stained: *neutrophils, eosinophils,* and *basophils.*

Neutrophils

Neutrophils (noo-TROH-fills; "neutral liking") are the most common type of leukocyte, making up 60–70% of total leukocytes in the blood. The cytoplasmic granules of neutrophils take up both the acidic and basic dyes (hence the name *neutro*phil), staining their cytoplasm a light lilac color (see Figure 19.8). Their nuclei typically have three to five variably shaped lobes, which is the reason for their other name: **polymorphonucleocytes** (pah′-lee-mor-foh-NOO-klee-oh-sytz; *poly-* = "many"; *morpho-* = "shape"), sometimes shortened to **PMNs or polys.**

Neutrophils are attracted to injured cells by chemicals released by the damaged cells, a process called **chemotaxis** (kee-moh-TAX-is). This process occurs with any type of cellular injury, from trauma caused by stepping on a nail to infection by bacteria. When neutrophils reach the damaged tissue, they exit the blood and release the contents of their granules. These contents directly kill bacterial cells, enhance inflammation, and attract more neutrophils and leukocytes to the area. Neutrophils are also very active phagocytes that ingest and destroy bacterial cells.

Eosinophils

As you can infer from their name, the granules of **eosinophils** (ee′-uh-SIN-uh-fills; "eosin liking") take up the dye eosin and therefore appear red. Eosinophils are relatively rare leukocytes that account for less than 4% of total leukocytes in the blood. Their nuclei are bilobed, resembling a barbell, with two circular lobes connected by a thin strand of nuclear material.

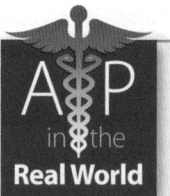

Complete Blood Count

An important test for anemia and many other conditions is a **complete blood count,** or **CBC.** Blood is drawn and examined under a microscope or by an automated analyzer to evaluate the number and characteristics of blood cells. The factors examined include:

- erythrocyte count, the number of erythrocytes reported as number of cells per milliliter, which is used to calculate the hematocrit;
- concentration of hemoglobin in the sample;
- erythrocyte characteristics such as size, volume, and concentration of hemoglobin in their cytosol; and
- number and volume of platelets.

Most CBCs also examine the numbers and types of leukocytes. Chapter 20 provides a closer look at this portion of the CBC.

Eosinophils are involved in the body's response to infection with parasitic worms and in allergic reactions. Their granules contain substances that assist in these functions, including enzymes and toxins that are specific to parasites, as well as chemicals that mediate (bring about) inflammation. In addition, eosinophils are phagocytes and ingest foreign molecules that have been bound by proteins of the immune system.

Basophils

Basophils (BEH-zuh-fills; "basic liking"), the least common leukocyte, make up less than 1% of total leukocytes in the blood. As implied by their name, basophils' granules take up the basic dye methylene blue, which stains their granules dark purple-blue. Note that the granules in these cells almost completely obscure their typically S-shaped nuclei.

Basophils release chemicals from their granules that mediate inflammation. We discuss inflammation in the chapter on immunity (see Chapter 20).

Quick Check

☐ 1. How do granulocytes and agranulocytes differ?
☐ 2. Describe the basic structure and function of the three types of granulocytes.

Agranulocytes

Lymphocytes (LIM-fo-syts) and monocytes (MAH-no-syts) are the two kinds of agranulocytes. Both cell types lack visible cytoplasmic granules, although like granulocytes, they do contain lysosomes.

Lymphocytes

The second most numerous type of leukocyte, **lymphocytes,** make up 20–25% of total leukocytes in the blood. They contain large, spherical nuclei and generally a thin rim of light blue cytoplasm that is visible when stained (Figure 19.8).

There are two basic types of lymphocytes: **B lymphocytes** (or **B cells**) and **T lymphocytes (or T cells)**. Both B and T lymphocytes are activated by cellular markers called **antigens** (AN-ti-jenz). Antigens, which are generally glycoproteins, are present on all cells and most biological molecules.

Although B and T lymphocytes are structurally similar, they differ in their functions. Activated B lymphocytes produce proteins called **antibodies** that bind to antigens and remove them from tissues. Each population of B lymphocytes secretes antibodies with a specific structure that allows them to bind to only one unique antigen. Populations of T lymphocytes also show specificity for individual antigens. However, activated T lymphocytes do not produce antibodies. Instead, populations of T lymphocytes have specific receptors for individual antigens. Once

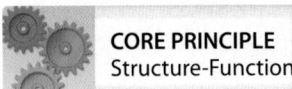

CORE PRINCIPLE
Structure-Function

the receptors are bound, T lymphocytes activate other components of the immune system and directly destroy abnormal body cells, such as cancer cells and those that are virally infected. Antibodies and T lymphocyte receptors, then, are structured in such a way that they bind to only one unique antigen, an example of the Structure-Function Core Principle (p. 25).

Monocytes

The largest leukocytes are the **monocytes,** which account for 3–8% of the total leukocyte population. Monocytes are easily distinguished from other leukocytes by their large U-shaped nuclei and light blue or purple cytoplasm that becomes visible when the cells are stained.

Monocytes remain in the blood for only a few days before they exit the capillaries and enter the tissues, where some mature into very active phagocytes called **macrophages** (MAK-roh-fehj-uhz; "big eater"). Macrophages ingest dead and dying cells (such as old erythrocytes or those damaged from trauma), bacteria, antigens, and other cellular debris. In addition, they activate other parts of the immune system by displaying phagocytosed antigens to other leukocytes.

Quick Check

☐ 3. Compare the two types of agranulocytes.
☐ 4. How do B and T lymphocytes differ?
☐ 5. What does a monocyte become in the tissues? What is its primary function?

Leukocyte Formation: Leukopoiesis

Leukocytes form in the bone marrow by the process of **leuko-poiesis** (loo′-koh-poh-EE-sis). Like erythrocytes, all leukocytes

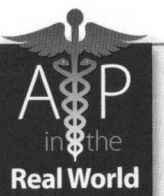

A&P in the Real World

Leukemias

Leukemia refers to several cancers of the blood cells or bone marrow. Leukemias are classified as *acute* or *chronic* based on the speed at which the disease progresses. Acute leukemias are characterized by a rapid increase in immature, nonfunctional, or poorly functional blood cells, whereas chronic leukemias show a slow accumulation of abnormal mature leukocytes.

Leukemias are also classified according to the cell line from which the abnormal cells derive. *Lymphocytic leukemias* derive from the lymphoid cell line and generally involve abnormal B lymphocytes. *Myelogenous leukemias* come from the myeloid cell line and can involve any of the myeloid formed elements.

The abnormal cells in leukemias crowd the bone marrow and reduce its ability to manufacture healthy formed elements. The cancerous cells enter the bloodstream and spread, or *metastasize*, to other tissues.

Acute lymphocytic leukemia is the most common type of leukemia in children, whereas the other leukemias are more prevalent in adults. Treatment and prognosis vary with the type of leukemia, although generally the acute forms metastasize earlier and require more aggressive management.

arise from hematopoietic stem cells (HSCs). The HSCs divide and split into two cell lines:

- the **myeloid cell line,** which produces most of the formed elements, including erythrocytes and platelets; and
- the **lymphoid cell line,** which forms lymphocytes.

The myeloid cell line differentiates early into blast cells that are committed to becoming monocytes or granulocytes. Monocytes develop from committed cells called **monoblasts** and precursor cells called **promonocytes.** All granulocytes derive from **myeloblasts** that differentiate into precursor cells known as **promyelocytes.** In the final precursor stage, the cells are referred to as **band cells** or **stab cells,** which develop into mature granulocytes that enter the bloodstream. If someone has an active infection, it is common to find band cells in the blood as the immune system attempts to increase the numbers of circulating leukocytes.

The lymphoid cell line differentiates first into **lymphoblasts,** at which point the cells are committed to becoming lymphocytes. Lymphoblasts develop into precursor cells called *prolymphocytes,* which then become mature B and T lymphocytes. These cells mature in different locations: B lymphocytes remain in the bone marrow, whereas T lymphocytes migrate to the thymus gland in the mediastinum to complete their maturation (remember "B" for "bone marrow" and "T" for "thymus").

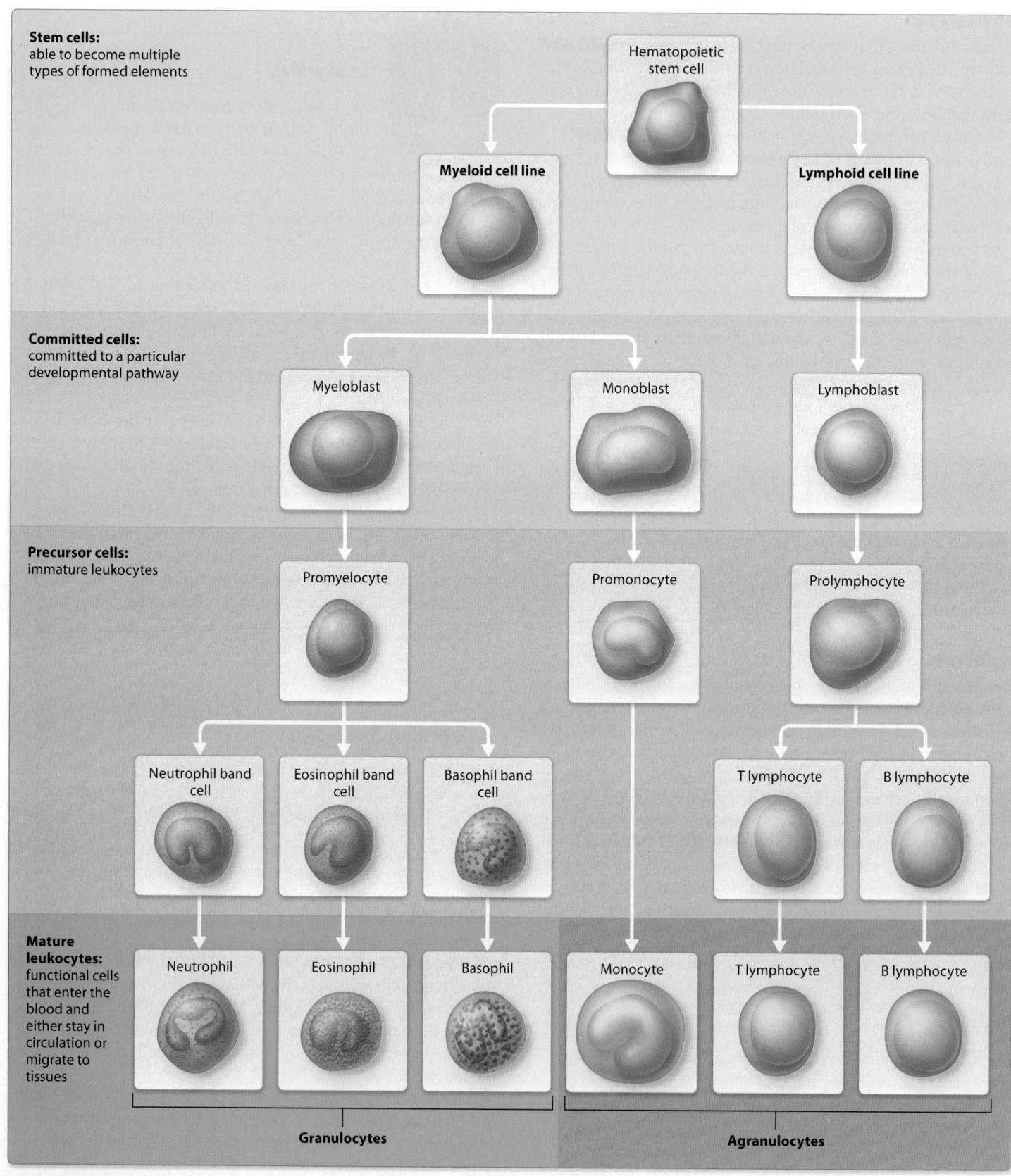

Stem cells:
able to become multiple
types of formed elements

Hematopoietic
stem cell

Myeloid cell line

Lymphoid cell line

Committed cells:
committed to a particular
developmental pathway

Myeloblast

Monoblast

Lymphoblast

Precursor cells:
immature leukocytes

Promyelocyte

Promonocyte

Prolymphocyte

Neutrophil band
cell

Eosinophil band
cell

Basophil band
cell

T lymphocyte

B lymphocyte

**Mature
leukocytes:**
functional cells
that enter the
blood and
either stay in
circulation or
migrate to
tissues

Neutrophil

Eosinophil

Basophil

Monocyte

T lymphocyte

B lymphocyte

Granulocytes

Agranulocytes

Figure 19.9 Leukopoiesis: formation of leukocytes.

Practice art labeling
@MasteringA&P®

Figure 19.9 summarizes the process of leukopoiesis. Sometimes myeloid or lymphoid cells undergo abnormal changes, resulting in the uncontrolled cell division of cancer (see *A&P in the Real World: Leukemias* on page 733).

Quick Check

☐ 6. From which cell type do all leukocytes derive?
☐ 7. How do the lymphoid and myeloid cell lines differ?
☐ 8. Where do T and B lymphocytes mature?

Apply What You Learned

☐ 1. Predict the type of leukocyte that would respond to the following conditions.
   a. Bacterial infection
   b. Viral infection
   c. Parasitic infection
   d. Tissue injury due to trauma

   See answers in Appendix A.

MODULE **19.4**

Platelets

Learning Outcomes

1. Explain how platelets differ structurally from the other formed elements of blood.
2. Discuss the role of the megakaryocyte in the formation of platelets.

Platelets, the smallest of the formed elements, are involved in *hemostasis,* a process that stops blood loss from an injured blood vessel. We now examine the characteristics of platelets and how they form in bone marrow. Module 19.5 explores their role in hemostasis.

Platelet Characteristics

◀◀ Flashback

1. What normally happens during telophase of mitosis? (p. 113)
2. What are the functions of actin and myosin proteins? (p. 344)

Platelets (PLAYT-letz) are not true cells, but instead small fragments of cells surrounded by a plasma membrane. As you can see in **Figure 19.10a**, platelets lack nuclei and most other organelles. However, they do contain multiple types of granules that house clotting factors and enzymes, small numbers of mitochondria, and glycogen deposits that enable them to carry out

(a) Platelet structure

(b) Formation of platelets

Figure 19.10 Structure and formation of platelets.

oxidative catabolism. These tiny structures also contain cytoskel-etal elements, including a system of microtubules that are associated with actin and myosin filaments.

Platelet Formation

Like all formed elements, platelets begin as hematopoietic stem cells (**Figure 19.10b**). Their precursor cells, called **megakaryoblasts** (meg'-uh-KAR-ee-yo-blasts; "immature big nucleus cells"), derive from the myeloid cell line. During development, megakaryo-blasts become **megakaryocytes,** which undergo repeated rounds of mitosis. However, unlike other cells, megakaryocytes do not undergo cytokinesis during anaphase and telophase of mito-sis. The end result is a massive cell with multiple copies of DNA located in a single nucleus (observe the size difference between a megakaryocyte and an erythrocyte in Figure 19.10b).

Under the influence of certain hormones such as *thrombo-poietin,* mature megakaryocytes extend ribbonlike "arms" filled with cytosol, granules, and several organelles. As you can see in Figure 19.10b, these arms extend through the clefts in the bone marrow sinusoids and into the bloodstream. The force of the blood coursing past the arms lops off small pieces that become platelets. Each arm can give rise to thousands of platelets. Most remain in the general circulation, where they have a lifespan of only 7–10 days. Once platelets have reached old age, the liver and spleen remove them from circulation.

Quick Check

☐ 1. Describe the structure of a platelet.
☐ 2. How are platelets formed?

Apply What You Learned

☐ 1. Many toxins and poisons interfere with a cell's ability to undergo protein synthesis. Will this affect platelets that are already in the bloodstream? Why or why not? (*Hint:* Which cellular structure is required to undergo protein synthesis? What is the structure of a platelet?)
☐ 2. Certain chemotherapy drugs for cancer inhibit cell division. How will this affect the size and structure of a megakaryocyte? Predict how this will influence platelet formation.

See answers in Appendix A.

MODULE **19.5**

Hemostasis

Learning Outcomes

1. Distinguish between the terms hemostasis and coagulation.
2. Describe the process of hemostasis, including the vascular phase, the formation of the platelet plug, and the formation of fibrin.
3. Explain the differences between the intrinsic and extrinsic clotting cascades.
4. Describe the role of calcium ions and vitamin K in blood clotting.
5. Explain how the positive feedback loops in the platelet and coagulation phases promote hemostasis.
6. Discuss the process of thrombolysis.

Significant blood loss threatens many homeostatic body func-tions, including blood pressure maintenance, tissue oxygenation, and electrolyte balance. Fortunately, the body has mechanisms that minimize the amount of blood lost from an injured blood vessel. These mechanisms are part of a process called **hemostasis** (he-mo-STAY-sis; "blood stoppage").

Hemostasis involves a series of events that form a gelatinous **blood clot** to "plug" the broken vessel. We can summarize hemo-stasis in five steps:

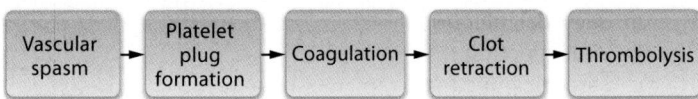

In examining each of these steps, we also discover the factors that regulate clot formation and see what can happen when hemosta-sis does not function properly.

Hemostasis Part 1: Vascular Spasm

◄◄ Flashback

1. What is vasoconstriction, and how does it affect blood flow? (p. 670)

When a blood vessel is injured ① and blood leaks into extra-cellular fluid ②, two immediate responses occur (**Figure 19.11**): vasoconstriction and increased tissue pressure ③. Both responses decrease the blood vessel diameter. Recall that such a decrease leads to a local reduction in blood pressure and blood flow, which minimizes blood loss from the injured vessel (see Chapter 17). This activity has the same effect as applying compression or a tourniquet to a bleeding wound. Surgeons also put this principle into practice by clamping off blood vessels during surgery: The external pressure collapses the vessel and reduces blood loss.

Quick Check

☐ 1. What purpose does vascular spasm serve in the process of hemostasis?

Hemostasis Part 2: Platelet Plug Formation

◄◄ Flashback

1. What are the layers of the blood vessel wall? Where are collagen fibers located in the vessel wall? (p. 670)

Although vasoconstriction and increased tissue pressure can minimize blood loss from an injured vessel, they cannot stop bleeding entirely. For this, we require a "plug" that acts like a

Vascular spasm → Platelet plug formation → Coagulation → Clot retraction → Thrombolysis

① Vessel is injured.

② Blood leaks into extracellular fluid.

③ Vessel spasms and leaked blood increases tissue pressure, both of which decrease blood flow and blood loss.

Figure 19.11 Hemostasis, part 1: vascular spasm.

cork. This plug consists primarily of platelets, and is therefore known as the **platelet plug.**

Normally, platelets aren't "sticky"—a good thing, because otherwise they would adhere to tissues and block blood flow. However, Figure 19.12 shows that platelets adhere to the injury site. Why? When a vessel is injured, collagen fibers and chemicals in the vessel's tunica adventitia are exposed ①. The injured endothelial cells release a glycoprotein called *von Willebrand factor* (fahn VIL-uh-brant), or vWF, which binds to receptors on the surface of the platelets' plasma membranes ②. Together, the exposed collagen and vWF make the platelets sticky so they adhere to one another and the vessel wall. Part of this process demonstrates the Cell-Cell Communication Core Principle (p. 27): Endothelial cells produce vWF, which communicates with platelets.

CORE PRINCIPLE
Cell-Cell Communication

The binding of vWF and collagen to platelets triggers a series of events within platelets known collectively as **platelet activation.** Activated platelets release the contents of their granules—ATP, ADP, serotonin, calcium, clotting factors, and other chemicals—by exocytosis. Many of these factors attract and activate nearby platelets and cause them to clump together, or **aggregate** ③. Platelet aggregation forms the platelet plug. This plug seals the injured vessel temporarily, but the repair is short-lived without a molecular "glue" that can hold the platelets together.

Quick Check

☐ 2. What is the role of platelets in hemostasis?
☐ 3. How are platelets triggered to aggregate?

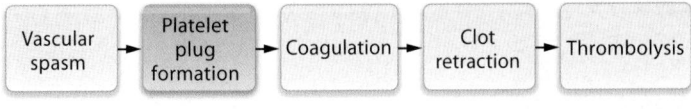

Vascular spasm → Platelet plug formation → Coagulation → Clot retraction → Thrombolysis

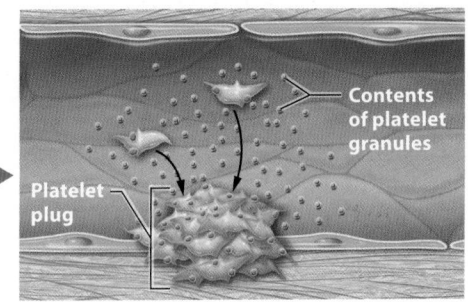

① Vessel injury exposes collagen fibers and causes endothelial cells to produce von Willebrand factor (vWF).

② vWF binds platelets, which activates them and causes them to stick to the exposed collagen and to one another. Activated platelets release the contents of their granules.

③ Contents of platelet granules activate and attract more platelets, which aggregate to form the platelet plug.

Figure 19.12 Hemostasis part 2: platelet plug formation.

Figure 19.13 Hemostasis part 3: coagulation cascade.

Hemostasis Part 3: Coagulation

The glue that binds platelets, endothelial cells, and other formed elements together is a sticky, threadlike protein called **fibrin** (FY-brin). Fibrin converts the soft, liquid platelet plug into a more solid mass by the process of **coagulation** (koh-ah-gyoo-LAY-shun; *coag-* = "bring together"). We normally find fibrin in plasma and in platelets in its inactive form, **fibrinogen** (fy-BRIN-oh-jen). Fibrinogen is converted into fibrin by the **coagulation cascade,** a series of reactions that occur at the surface of the platelets and/or damaged endothelial cells (**Figure 19.13**).

The coagulation cascade relies on **clotting factors.** Most clotting factors are enzymes produced by the liver that circulate in the blood in their inactive forms. Each factor is named with a Roman numeral (factors I–XIII), although many factors have a descriptive name as well (**Table 19.2**). Unfortunately, the factors were assigned the Roman numeral in the order of their discovery, not in the order in which they operate in the coagulation cascade. This is why the cascade does not proceed in numerical sequence.

Keep in mind that the synthesis of four clotting factors—factors II, VII, IX, and X—depends on the presence of vitamin K. These vitamin K–dependent clotting factors are important targets of drug therapy intended to block coagulation.

The coagulation cascade proceeds by two pathways: the **intrinsic pathway** and the **extrinsic pathway.** Both pathways converge to a common one that ends by forming fibrin.

Intrinsic Pathway

The intrinsic pathway is so named because all of the factors required for it to proceed are located in ("intrinsic to") the blood. The intrinsic pathway is activated when the inactive protein clotting factor XII comes into contact with exposed collagen fibers. The pathway goes forward as follows (Figure 19.13, left side):

1. **Exposed collagen fibers activate factor XII.** The negative charges on the surface of the exposed collagen molecules attract and bind a clotting protein in the plasma called **factor XII.** Activated factor XII is known as **factor XIIa** ("a" stands for "activated").

2. **Factors XI and IX become activated.** Factor XIIa is an enzyme that activates another clotting protein in the plasma referred to as **factor XI.** Factor XIa, in turn, enzymatically activates another clotting protein, **factor IX.**

3. **Factors IXa and VIIIa along with calcium ions form an enzyme complex that activates factor X.** The intrinsic pathway ends by forming a large enzyme complex that consists of factor IXa, another protein called **factor VIIIa,** and calcium ions. This enzyme complex activates **factor X,** which becomes factor Xa. Note that calcium ions are required for the factors to associate with platelets in the platelet plug.

The intrinsic pathway isn't just activated by exposed collagen fibers—activation occurs any time blood contacts a surface with negative charges, including glass test tubes. For this reason, blood in a test tube will coagulate unless treated with a chemical to prevent clotting.

Extrinsic Pathway

The extrinsic pathway, which occurs simultaneously with the intrinsic pathway, has this name because it is initiated by a factor outside ("extrinsic to") the blood. The extrinsic pathway proceeds as shown in Figure 19.13 (right side):

1. **Subendothelial cells display tissue factor.** Damaged *subendothelial cells,* located deep to the endothelium, display a protein called **tissue factor** in their plasma membranes.

2. **Tissue factor activates factor VII.** The clotting protein **factor VII** circulates in the blood in its inactive form. When it comes into contact with tissue factor, it changes in structure and becomes the active form, **factor VIIa.**

3. **Factor VIIa, tissue factor, and calcium ions form an enzyme complex that activates factor X.** Factor VIIa forms a complex with tissue factor and calcium ions that enzymatically cleaves factor X into factor Xa. The calcium ions play the same role in the extrinsic pathway as in the intrinsic pathway: They are required for the interaction of factors with platelets in the platelet plug.

The end result of both the intrinsic and extrinsic pathways is the same: activation of factor X to the active enzyme factor Xa. Although the extrinsic pathway is thought to be the more important path to factor Xa production, both pathways interact and rely heavily on one another to move to completion. A deficiency in any clotting protein can disrupt the entire coagulation cascade.

At this point, the two pathways converge to the common pathway.

Common Pathway

In the common pathway, fibrin is produced (Figure 19.13):

4. **Factors Xa and Va along with calcium ions form prothrombin activator, which converts prothrombin into thrombin.** The common pathway begins with the formation of *prothrombin activator,* a large enzyme complex that consists of factor Xa, another protein called **factor Va,** and calcium ions. Prothrombin activator converts the inactive protein **prothrombin** to the active enzyme thrombin.

5. **Thrombin turns fibrinogen into fibrin, which "glues" the platelet plug together.** In the final step of the coagulation cascade, thrombin acts on the inactive protein fibrinogen, which is both present in plasma and released by platelets. Thrombin converts fibrinogen into fibrin, its active form. The fibrin threads form a mesh that glues together the platelet plug, other formed elements, and damaged endothelial cells, sealing the damaged blood vessel.

TABLE 19.2 CLOTTING FACTORS

Factor	Alternate Name	Description	Function
I	Fibrinogen	Plasma protein made by the liver and platelets	• Becomes fibrin, sticky protein threads that "glue" the platelet plug together in the common pathway
II	Prothrombin	Plasma protein made by the liver; synthesis requires vitamin K	• Becomes thrombin, which converts fibrinogen to fibrin
III	Tissue factor	Glycoprotein in the plasma membranes of cells external to the endothelium	• Binds factor VIIa • Activates the extrinsic clotting pathway
IV	Calcium ion	Ion in the plasma	• Multiple roles, including activation of factor X and prothrombin
V	Labile factor	Plasma protein made by the liver; stored in platelets	• Part of the complex that activates prothrombin in the common pathway
VII*	Stable factor	Plasma protein made by the liver; synthesis requires vitamin K	• Part of the complex that activates factor X in the extrinsic pathway
VIII	Antihemophilic factor A	Plasma protein made by the liver	• Part of the complex that activates factor X in the intrinsic pathway
IX	Christmas factor	Plasma protein made by the liver; synthesis requires vitamin K	• Part of the complex that activates factor X in the intrinsic pathway
X	Stuart factor	Plasma glycoprotein made by the liver; synthesis requires vitamin K	• Part of the complex that activates prothrombin in the common pathway
XI	Plasma thromboplastin antecedent	Plasma protein made by the liver and stored in platelets	• Activates factor IX in the intrinsic pathway
XII	Hageman factor	Plasma glycoprotein made by the liver	• Activates factor XI in the intrinsic pathway
XIII	Fibrin stabilizing factor	Plasma protein made by the liver and stored in platelets	• Activated by thrombin • Stabilizes the fibrin mesh

*Note that there is no factor VI.

Here's an interesting fact about thrombin: It catalyzes the activation of factors V and VIII. However, thrombin is produced *after* these factors are activated. How is this possible?

The answer is that the enzymes have a small amount of activity before thrombin is produced. As this activity leads to thrombin production, thrombin acts on the factors to convert them to their active forms, which accelerates thrombin production. So, as more thrombin is produced, the response is progressively amplified, which is an example of a *positive feedback loop.* The positive feedback loop is self-limiting for two reasons. First, clotting factors are rapidly inactivated by enzymes in plasma, so their action is fairly brief and limited to the site of injury. Second, fibrin inhibits the production of thrombin via a negative feedback loop, so thrombin activity decreases as fibrin levels increase.

In an example of the Feedback Loops Core Principle (p. 21), 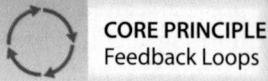 **CORE PRINCIPLE** Feedback Loops coagulation proceeds by a positive feedback loop. However, it is ultimately regulated by negative feedback.

ConceptBOOST »»»

Making Sense of the Coagulation Cascade
What's the best way to approach the coagulation cascade? Remember that the entire process has three simple goals:

- **Goal 1: Produce factor Xa.** The goal of both the intrinsic and extrinsic pathways is to produce the active enzyme factor Xa. Although the reactions in each pathway are different, the end result is the same.
- **Goal 2: Produce thrombin.** The product of both intrinsic and extrinsic pathways, factor Xa, accomplishes the second goal of coagulation by producing the enzyme thrombin. This happens in the common pathway.
- **Goal 3: Produce fibrin.** Thrombin, in turn, accomplishes the third goal of coagulation: producing fibrin to hold the platelet plug together and seal the wound. This is the final result of the common pathway.

Develop a global understanding of these three goals, and the details will be much easier to remember. ■

Quick Check

☐ 4. What is the coagulation cascade?
☐ 5. How do the intrinsic and extrinsic pathways differ?
☐ 6. What are the overall goals of the intrinsic, extrinsic, and common pathways?

Hemostasis Part 4: Clot Retraction

As the coagulation cascade completes, the actin and myosin fibers in platelets contract via a mechanism similar to that of the actin and myosin fibers in a skeletal muscle fiber. This action, known as **clot retraction,** brings the edges of the wounded vessel closer together, much as sutures (or "stitches") do with the edges of a skin wound (**Figure 19.14**). Clot retraction also squeezes **serum** (SEAR-um)—a fluid consisting of plasma minus the clotting proteins—out of the clot. (Think of it as squeezing water out of a wet rag.)

Quick Check

☐ 7. What is the purpose of clot retraction?

Hemostasis Part 5: Thrombolysis

After a wound has healed, the blood clot is no longer necessary and it dissolves in a process called **thrombolysis** (thrahm-BAH-luh-sis; -*lysis* = "split"). Thrombolysis begins with **fibrinolysis** (fy-brin-AH-luh-sis)—the breakdown of the fibrin glue that was produced in the coagulation cascade. Compared to the formation of fibrin, its dissolution is relatively straightforward (**Figure 19.15**):

① **Endothelial cells release tissue plasminogen activator (tPA).** Thrombolysis begins when healed endothelial cells produce and release the enzyme **tissue plasminogen**

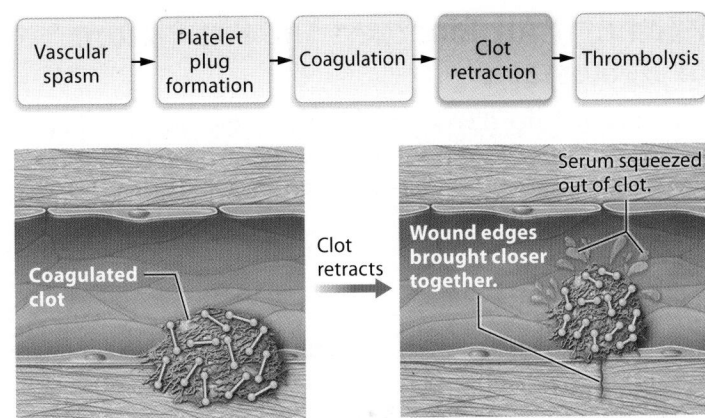

Figure 19.14 **Hemostasis part 4: clot retraction.**

(plaz-MIN-no-jen) **activator,** or **tPA.** Thrombolysis can also be initiated by a similar enzyme called **urokinase** (yoo-roh-KI-naz), which is produced by cells of the kidney and found in the plasma and interstitial fluid.

② **tPA activates plasminogen.** The inactive enzyme **plasminogen,** which is normally found in plasma, has a high affinity for fibrin molecules and binds them as they are incorporated into the blood clot. As a result, every blood clot contains a significant amount of plasminogen. When tPA contacts plasminogen, it catalyzes the reaction that converts it to the active enzyme **plasmin.**

③ **Plasmin degrades fibrin, and the clot dissolves.** Plasmin catalyzes the reaction that degrades both fibrin and fibrinogen. This causes the remaining components of the clot to dissociate from the endothelium.

Quick Check

☐ 8. What is thrombolysis?
☐ 9. Explain the roles of tPA and plasmin in thrombolysis.

① Endothelial cells release tissue plasminogen activator (tPA).

② tPA activates plasminogen.

③ Plasmin degrades fibrin, and the clot dissolves.

Figure 19.15 **Hemostasis part 5: thrombolysis.**

Putting it All Together: The Big Picture of Hemostasis

Let's now put the pieces together for a "big picture" view of hemostasis. **Figure 19.16** summarizes the five key steps of the process: vascular spasm, platelet plug formation, coagulation, clot retraction, and thrombolysis.

Quick Check

☐10. Walk through the steps of clot formation and dissolution.

Regulation of Clotting

Blood clotting is tightly regulated to keep its positive feedback cycle in check. Endothelial cells produce chemicals that regulate the first and second stages of clot formation. Two of these chemicals are **prostacyclin** (PRAHS-tuh-sy-klin), which inhibits platelet aggregation, and nitric oxide, which causes vasodilation.

Prostacyclin is a prostaglandin, a category of chemicals with diverse functions, including triggering inflammation. Anti-inflammatory medications such as ibuprofen inhibit prostaglandin formation. Some of them also inhibit prostacyclin formation, a fact that accounts for the increased risk of blood clots with these drugs.

Endothelial cells and hepatocytes produce molecules called **anticoagulants** (an'-tee-ko-AG-yoo-lunts) that inhibit coagulation. Anticoagulant molecules include the following:

- **Antithrombin-III.** One of the most important anticoagulants, the protein **antithrombin-III (AT-III)**, binds and inhibits the activity of both factor Xa and thrombin. It inhibits thrombin that has already formed and prevents the formation of new thrombin.
- **Heparan sulfate. Heparan sulfate** is a polysaccharide that enhances the activity of AT-III. Heparan sulfate is similar to another anticoagulant called *heparin,* which is often used

clinically. We discuss its clinical applications, along with related drugs, in *A&P in the Real World: Anticlot Medications* on page 744.

- **Protein C.** Active **protein C** catalyzes the reactions that degrade factors Va and VIIIa. To become active, protein C requires another protein, called *protein S.*

Table 19.3 summarizes thrombolytic (clot-destroying) agents and factors that regulate clotting.

Quick Check

☐11. How does prostacyclin regulate hemostasis?
☐12. What are the functions of antithrombin-III, heparan sulfate, and protein C? How do they affect hemostasis?

Disorders of Clotting

A clotting disorder—a condition in which clotting is not regulated properly—can have drastic consequences for homeostasis. There are two types of clotting disorders: (1) *bleeding disorders,* when the blood is unable to clot, and (2) *hypercoagulable conditions,* when clots form at improper times and tissue locations.

Bleeding disorders can increase blood loss from even a minor injury. They often result from clotting protein deficiencies; for example, **hemophilia A** is caused by a shortage of factor VIII, and **hemophilia B** by inadequate factor IX. Many bleeding disorders can be treated by replacement of the missing clotting factor or protein with periodic infusions.

The opposite problem, a hypercoagulable condition, can result in the formation of an inappropriate clot, a condition called **thrombosis.** The clot, or **thrombus** (plural, *thrombi*), is dangerous because it can obstruct blood flow through a vessel. In addition, a piece of the thrombus, called a **thromboembolus** (plural, *thromboemboli*), may break off and lodge in smaller vessels downstream from the thrombus.

TABLE 19.3 THROMBOLYTIC AGENTS AND REGULATORS OF CLOT FORMATION

Factor	Alternate Name	Description	Function
Thrombolytic Agents			
Tissue plasminogen activator	tPA	Protein made by endothelial cells	• Activates plasminogen
Urokinase		Protein made by cells of the kidney	• Activates plasminogen
Plasmin		Plasma protein made by the liver	• Degrades fibrin and causes clot dissolution
Regulators of Clot Formation			
Prostacyclin	PGI$_2$	Prostaglandin made by endothelial cells	• Inhibits platelet aggregation • Promotes vasodilation
Antithrombin-III	AT-III	Plasma protein made by the liver	• Binds and inhibits thrombin and factor Xa • Activity enhanced by heparin and heparan sulfate
Heparan sulfate		Polysaccharide in the extracellular fluid produced by many different cell types	• Enhances activity of antithrombin-III
Protein C		Plasma protein produced by the liver; synthesis requires vitamin K	• Degrades factors Va and VIIIa

Intact blood vessel

Injured blood vessel

① **Vascular spasm:** Blood flow through the injured vessel decreases (see Figure 19.11).

Platelet plug

② **Platelet plug formation:** Platelet plug forms (see Figure 19.12).

Intrinsic

Extrinsic

③a **Coagulation:** The intrinsic and extrinsic pathways produce Factor Xa (see Figure 19.13).

Thrombin

Prothrombin

Enzyme complex

③b **Coagulation:** The common pathway produces thrombin (see Figure 19.13).

Fibrinogen

Fibrin

③c **Coagulation:** Thrombin converts fibrinogen to fibrin, and the platelet plug is "glued" together (see Figure 19.13).

④ **Clot retraction:** The clot retracts (see Figure 19.14).

Platelets

Fibrin

PL

⑤ **Thrombolysis:** Plasmin degrades fibrin, and the clot dissolves (see Figure 19.15).

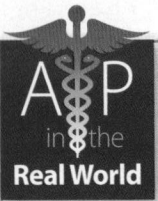

Anticlot Medications

Real World

Patients with thrombi or emboli are treated with drugs that prevent one step in the blood clotting process. The anticoagulants constitute one widely used group of medications; these include heparin (and heparin derivatives) and **warfarin** (trade name, Coumadin). Warfarin inhibits the production of the vitamin K–dependent clotting factors by the liver (factors II, VII, IX, and XII). However, warfarin does not affect clotting factors already in the plasma, so the drug requires 2–3 days to show noticeable effect.

Heparin is generally used in hospitals to manage and prevent emboli because its onset of action is nearly immediate. However, it is injected; thus, warfarin, which can be given by mouth, is preferred in outpatient settings. With both drugs, careful monitoring is required to ensure that a patient is neither over- nor under-anticoagulated.

Other anticlotting medications act directly on platelets. The most commonly used antiplatelet drug is simple aspirin, which inhibits enzymes in platelets that contribute to platelet aggregation. Another medication is *clopidogrel*, which blocks platelet receptors and inhibits platelet activation. The *glycoprotein IIb/IIIa inhibitors*, used commonly in heart attack patients, work by a similar mechanism.

Occasionally, it is necessary to attempt to dissolve the clot with a thrombolytic agent. The enzyme tPA or urokinase is used when blood flow must be restored rapidly to prevent tissue damage, as, for example, when thrombi or emboli have caused a stroke or heart attack.

Thrombi form most often in deep veins of the legs, leading to **deep vein thrombosis,** or **DVT.** The most dangerous complication of DVT is **pulmonary embolism,** in which emboli break off and lodge in the vessels of the lungs. Risk factors for DVT include prolonged immobility, which causes blood to pool and potentially clot, and other conditions that increase the tendency to clot, such as surgery, trauma, abnormal clotting factors, anticoagulant deficiencies, and pregnancy. In addition to DVT, thrombi and emboli can block the vessels of the brain, resulting in a *stroke,* or the heart, causing a *myocardial infarction,* or heart attack.

Quick Check

☐ 13. What are common causes of bleeding disorders?
☐ 14. What is thrombosis? What conditions could lead to thrombosis?

Apply What You Learned

☐ 1. The venom of certain snakes contains molecules that activate factors V, IX, and X. Predict the effects these molecules would have on human blood.

☐ 2. Snake venom can also contain molecules that activate protein C and directly inhibit thrombin. Predict the effects these molecules would have on human blood. How do these effects differ from those in question 1?

☐ 3. It is generally recommended that patients who have received cardiopulmonary resuscitation (CPR) not be given thrombolytic agents such as tPA for several days. Explain. (*Hint:* CPR is a fairly traumatic procedure, often resulting in broken bones and tissue damage.)

See answers in Appendix A.

MODULE **19.6**

Blood Typing and Matching

Learning Outcomes

1. Explain the role of surface antigens on erythrocytes in determining blood groups.
2. List the type of antigen and the type of antibodies present in each ABO and Rh blood type.
3. Explain the differences between the development of anti-Rh antibodies and the development of anti-A and anti-B antibodies.
4. Predict which blood types are compatible, and explain what happens when the incorrect ABO or Rh blood type is transfused.
5. Explain why blood type O− is the universal donor and type AB+ is the universal recipient.

The practice of **blood transfusion,** in which blood is removed from a **donor** and given to a **recipient,** first saw wide use in the 1800s. Tragically, many early recipients died. Not until 1901 was the reason for these deaths discovered—the presence of surface markers, or antigens, on erythrocytes. As Module 19.3 noted, all cells and most biological molecules have unique antigens. The immune system recognizes foreign antigens and responds to remove them. This response is what killed so many early transfusion patients.

The antigens on erythrocytes, which give rise to different **blood groups,** are carbohydrate chains that are genetically determined. More than 30 different types of antigens are found on erythrocytes, but two antigen groups have particular importance in the clinical setting: the *ABO blood group* and the *Rh blood group.* This module explores how blood is typed based on these antigens and how blood types apply to blood transfusions.

Blood Typing

The **ABO blood group** features two antigens: the **A antigen** and the **B antigen.** The presence or absence of the A and B antigens gives us four possible ABO types (see **Table 19.4**):

- **type A**, in which only the A antigen is present on the erythrocytes;

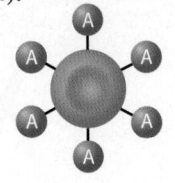

Blood Type	Prevalence in U.S. Population	Antigens Present on Erythrocyte Surface	Antibodies Present in Plasma	May Receive From	May Donate To
AB+	3%	A, B, Rh	None	Universal recipient	AB+
AB−	1%	A, B	Anti-Rh	AB−, A−, B−, O−	AB+, AB−
A+	34%	A, Rh	Anti-B	A+, A−, O+, O−	AB+, A+
A−	6%	A	Anti-B, anti-Rh	A−, O−	AB+, AB−, A+, A−
B+	9%	B, Rh	Anti-A	B+, B−, O+, O−	AB+, B+
B−	2%	B	Anti-A, anti-Rh	B−, O−	AB+, AB−, B+, B−
O+	38%	Rh	Anti-A, anti-B	O+, O−	AB+, A+, B+, O+
O−	7%	None	Anti-A, anti-B, anti-Rh	O−	Universal donor

TABLE 19.4 THE EIGHT MAJOR BLOOD TYPES

- **type B,** in which only the B antigen is present;

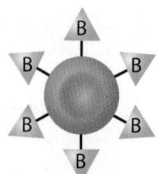

- **type AB,** in which both the A and B antigens are present; and

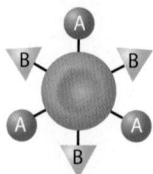

- **type O,** in which neither the A nor B antigens are present.

Note that there is no "O" antigen. The "O" simply means that both the A and B antigens are absent.

The **Rh blood group** (so named because it was first discovered in the rhesus monkey) features the **Rh antigen,** also known as the *D antigen.* Individuals with an Rh antigen on their erythrocytes are **Rh-positive** (abbreviated "Rh+"), and those lacking an Rh antigen are **Rh-negative** (abbreviated "Rh−").

The ABO and Rh blood groups combine to give the eight common blood types. For example, the erythrocytes in a type AB− individual feature the A and B antigens, but no Rh antigen, whereas the erythrocytes in a type O+ individual feature only the Rh antigen, without A or B:

Type O+ is the most common blood type, present in about 38% of the U.S. population. AB− is the least common, found in only about 1% of the U.S. population.

We can determine blood type in the laboratory using antibodies that recognize and bind to individual antigens on erythrocytes. Antibodies are produced by cells derived from activated B lymphocytes, and they bind to foreign antigens. Each unique antibody has a unique antigen to which it binds. As you can see in **Figure 19.17,** antibodies cause bound antigens to clump together, or **agglutinate.** For this reason, antibodies are sometimes called **agglutinins** (uh-GLOO-tuh-nins). Ultimately, agglutination promotes destruction of the erythrocytes, a reaction known as **hemolysis** (hee-MAH-luh-sis; *lysis* = "to separate or split").

To determine blood type, a blood sample is treated with three antibodies:

- **Anti-A antibodies** bind and agglutinate A antigens.
- **Anti-B antibodies** bind and agglutinate B antigens.
- **Anti-Rh antibodies** bind and agglutinate Rh antigens.

As you can see in **Figure 19.18,** if agglutination occurs in response to an antibody, that antigen is present on the erythrocyte. If agglutination does not occur, that antigen is absent.

Quick Check

☐ 1. What determines the eight common blood types?
☐ 2. A sample reacts with anti-A and anti-B antibodies, but does not react with anti-Rh antibodies. What is the blood type?
See answers in Appendix A.

Blood Transfusions

Your immune system recognizes the antigens on your erythrocytes as "self" antigens, meaning they belong to you. Under normal circumstances, your immune system does not produce antibodies to self antigens, because if it did, your antibodies would bind your own antigens. However, your immune system does produce antibodies to *foreign* antigens. This means that antibodies are present in your plasma only if the antigens are normally *absent* on your erythrocytes.

Note that anti-A and anti-B antibodies are *pre-formed,* meaning that they are present in the plasma even if the individual has never been exposed to those antigens. Anti-Rh antibodies, on the other hand, are produced only if a person has been exposed to blood containing Rh antigens. Therefore, an Rh− individual generally has no anti-Rh antibodies unless he or she has been exposed (sensitized) to Rh+ erythrocytes. This has particular relevance for pregnant Rh− women, which is discussed in *A&P in the Real World: Hemolytic Disease of the Newborn, or Erythroblastosis Fetalis.*

Here's an example: Sue has type B− blood, which means that her erythrocytes have B antigens only. As you can see here, her immune system cannot make anti-B antibodies, because if it does, she would agglutinate her own erythrocytes. However, Sue's erythrocytes lack the A and Rh antigens, so she can safely produce anti-A and anti-Rh antibodies.

Anti-B antibodies would agglutinate Sue's erythrocytes. / Anti-A and anti-Rh antibodies do not cause agglutination.

Let's apply this to blood transfusions: A patient cannot receive any blood containing antigens that his or her immune system would recognize as foreign. So, in our example, Sue cannot receive blood with A or Rh antigens, which includes types A+, A−, B+, AB+, AB−, and O+. Sue can safely receive blood only from the blood types that do not have A and Rh antigens: B− and O−. These two blood types are a **match** for Sue's blood type.

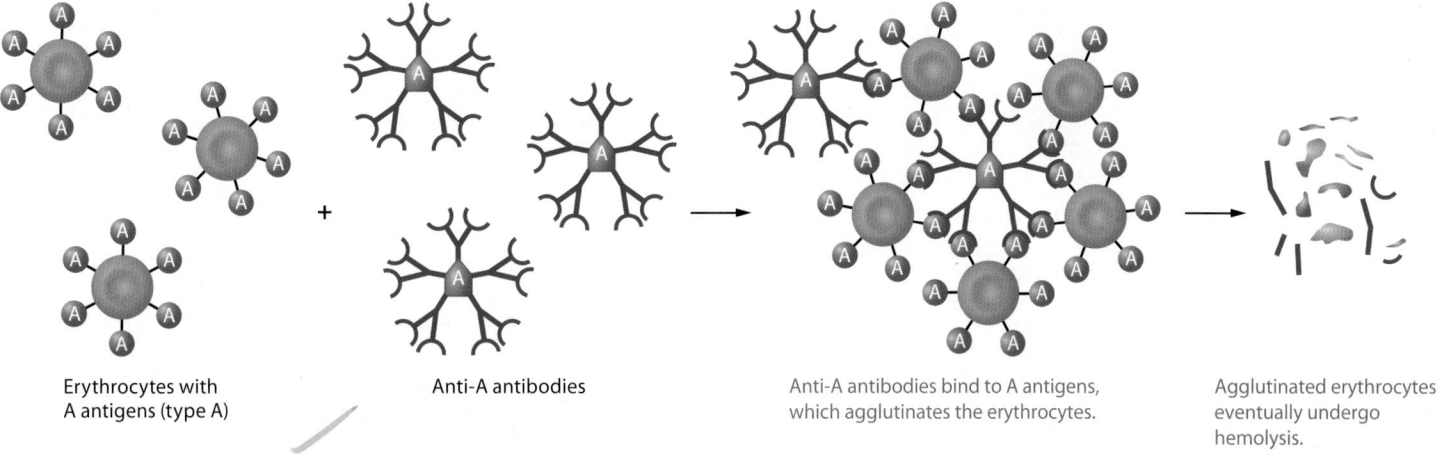

Erythrocytes with
A antigens (type A)

Anti-A antibodies

Anti-A antibodies bind to A antigens,
which agglutinates the erythrocytes.

Agglutinated erythrocytes
eventually undergo
hemolysis.

Figure 19.17 How antibodies agglutinate erythrocytes.

What if Sue wants to donate blood? We have already established that she has B antigens on her erythrocytes. This means her blood can be given only to recipients who lack the anti-B antibody. Refer to Table 19.4, and you'll see that the blood types that lack anti-B antibodies, and are therefore a match for Sue's blood, are types B+, B−, AB+, and AB−. All of the other blood types have anti-B antibodies that will agglutinate her donated erythrocytes. This agglutination, known as a **transfusion reaction,** destroys the donor erythrocytes, possibly leading to kidney failure and death.

Figure 19.19a shows another example:

- Donor Ed has O− blood, which has none of the antigens of the ABO or Rh groups.
- Recipient Tom has AB+ blood, which has none of the anti-A, anti-B, or anti-Rh antibodies.

BLOOD TYPE	**Anti-A** antibodies agglutinate A antigens.	**Anti-B** antibodies agglutinate B antigens.
A		
B		
AB		
O		

Figure 19.18 Blood type testing. Blood samples from four patients are combined with antibodies. Agglutination indicates that a specific antigen is present on that patient's erythrocytes.

To determine if we have a match, we simply check to see if Tom makes any antibodies that might bind to Ed's antigens. Obviously, since Ed has no antigens of the ABO or Rh groups and Tom has no anti-A, anti-B, or anti-Rh antibodies, nothing could possibly interact. In fact, since Ed's O− erythrocytes lack these antigens, he can safely donate to a recipient with any blood type. For this reason, blood type O− is considered the **universal donor.** Similarly, since Tom's AB+ blood lacks the major antibodies, he can safely receive any blood type. For this reason, blood type AB+ is considered the **universal recipient.**

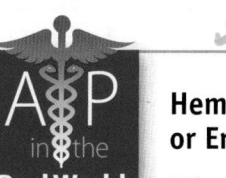

Hemolytic Disease of the Newborn, or Erythroblastosis Fetalis

When an Rh-negative mother carries and gives birth to an Rh-positive fetus, **hemolytic disease of the newborn,** also known as **erythroblastosis fetalis,** results. During birth, fetal erythrocytes enter the mother's blood, which stimulates her immune system to produce anti-Rh antibodies. This carries little risk for her first such pregnancy. However, in subsequent pregnancies, maternal anti-Rh antibodies can cross the placenta into the blood of the fetus, hemolyzing (rupturing) fetal erythrocytes if they are Rh+ and potentially killing the fetus. A characteristic finding is the presence of immature erythroblasts in fetal blood (giving the condition its other name, erythroblastosis fetalis).

Fortunately, this disease can be effectively prevented with blood type screening. If a pregnant woman is Rh-negative, she is given an injection of $Rh_0(D)$ *immune globulin,* which contains anti-Rh antibodies. These antibodies bind any fetal cells that have entered the maternal circulation, which prevents maternal leukocytes from producing and secreting their own anti-Rh antibodies.

(a) **Blood match: type O– donor and type AB+ recipient**

(b) **Blood mismatch: type AB+ donor and type O– recipient**

Figure 19.19 **Matching blood types for blood transfusions.**

Let's flip the donors and recipients around now (**Figure 19.19b**):

- Donor Tom has AB+ blood, which has A, B, and Rh antigens.
- Recipient Ed has O– blood, which has anti-A, anti-B, and anti-Rh antibodies.

Ed has three types of antibodies that would bind Tom's antigens, so infusing Tom's erythrocytes into Ed would produce a transfusion reaction. Indeed, donor erythrocytes with any antigens at all will produce a transfusion reaction in Ed. So, individuals with blood type O– may receive blood only from other type O– individuals. Additionally, because Tom's erythrocytes have all three types of antigens, he can donate blood only to recipients with no anti-A, anti-B, and/or anti-Rh antibodies in their blood—in other words, only type AB+ individuals.

ConceptBOOST »»»

What about the Donor's Antibodies?

At this point you may be wondering, "But can't the donor's antibodies bind and destroy the recipient's antigens?" The answer is yes, donor antibodies can bind to a recipient's antigens, and unless the blood types are exactly matched, some donor antibodies might destroy a few of the recipient's erythrocytes. The key word is *few*—only a *few* of the recipient's erythrocytes are likely to be harmed.

Take a closer look at our example in Figure 19.19a. Donor Ed's erythrocytes have no antigens of the ABO or Rh groups, but his blood does have anti-A, anti-B, and anti-Rh antibodies. For the sake of simplicity, let's say the sample of Ed's donated blood contains 100 antibodies. Ed's blood is given to recipient Tom, who has type AB+ blood. Now, in Tom's bloodstream we find:

- donated O– erythrocytes from Ed;
- 100 anti-A, anti-B, and anti-Rh antibodies from Ed's blood; and
- Tom's own AB+ erythrocytes.

Ed's 100 antibodies might destroy 100 of Tom's erythrocytes. But, Tom has received millions of new erythrocytes from Ed, so he won't really miss the 100 that he lost.

Now reverse the donors and recipients (Figure 19.19b): AB+ Tom becomes the donor and O– Ed, the recipient. Ed receives a sample of Tom's blood, and in Ed's bloodstream we find:

- donated AB+ erythrocytes from Tom;
- no major erythrocyte antibodies from Tom;
- Ed's own O– erythrocytes; and
- thousands to millions of Ed's own anti-A, anti-B, and anti-Rh antibodies.

That's quite a difference in the number of antibodies. And what's more, as soon as Ed's leukocytes come into contact with Tom's erythrocytes, Ed's immune system makes millions more of these antibodies. And the result? A transfusion reaction in which Ed's antibodies destroy every one of Tom's donated erythrocytes.

See the difference? The number of antibodies in the donor's blood is relatively low, which limits the amount of damage they do to the recipient's erythrocytes. But, the recipient's immune system can make millions more antibodies, more than enough to destroy all of the donated erythrocytes. So remember—we worry about the donor's *antigens* and the recipient's *antibodies*. ■

Quick Check

☐ 3. How are blood types matched for blood transfusions?
☐ 4. Which blood type is the universal donor? Why?
☐ 5. Which blood type is the universal recipient? Why?

Apply What You Learned

☐ 1. Mr. Jones, whose blood type is A−, has been injured and requires a blood transfusion. His friend, Mr. Ramirez, wants to donate blood to Mr. Jones. Mr. Ramirez's blood type is AB+. Can he donate to Mr. Jones? Explain.

☐ 2. Ms. Wright also wants to donate blood to Mr. Jones. Her blood type is O+. Can she donate blood to Mr. Jones? Explain.

☐ 3. Explain why a person with blood type O− can donate to someone with any blood type, but can receive blood only from other O− donors.

See answers in Appendix A.

CHAPTER SUMMARY

Take full advantage of the invaluable online practice and study tools for this chapter by going to MasteringA&P®. There you will find:

 Practice tests, module quizzes, art-labeling and sequence-the-steps activities

▷ Animated figures with quizzes

🖊 *Practicing A&P* worksheets

🔊 Author-narrated podcast

🔥 Dynamic Study Modules

iP Interactive Physiology

 19.1
Overview of Blood 723

- Three types of formed elements are found in blood:
  ○ **erythrocytes,** or **red blood cells,** which make up 44% of the volume of blood and transport oxygen;
  ○ **leukocytes,** or **white blood cells,** which make up less than 1% of the volume of blood and perform immune functions; and
  ○ **platelets,** which make up less than 1% of blood volume and function in blood clotting. (Figure 19.1, p. 724; Figure 19.2, p. 726)

  👆 Practice art-labeling exercise: Figure 19.1.

- Plasma makes up 55% of the total blood volume. It consists of water, **plasma proteins,** and dissolved solutes. (Table 19.1, p. 725)

- Functions of blood include exchanging gases; distributing solutes; performing immune functions; maintaining body temperature; functioning in blood clotting; maintaining acid-base homeostasis; and stabilizing blood pressure.

MODULE 19.2
Erythrocytes and Oxygen Transport 725

- Erythrocytes consist mainly of a plasma membrane around cytosol with enzymes and the protein **hemoglobin.**

- Hemoglobin is a protein with four polypeptide subunits, each associated with an iron-containing **heme group.** The iron ions bind to oxygen to form **oxyhemoglobin.** (Figure 19.3, p. 727)

- **Erythropoiesis** occurs in red bone marrow. The order of development of an erythrocyte is: **hematopoietic stem cell** → *cell in a colony-forming unit* → **proerythroblast** → **erythroblast** → **reticulocyte** → erythrocyte. The production of proerythroblasts requires the hormone **erythropoietin.** (Figure 19.4, pp. 728–729; Figure 19.5, p. 728)

  👆 Practice sequencing the steps: Figure 19.4.

- Old and damaged erythrocytes are destroyed in the **spleen.** The iron and polypeptide chains of hemoglobin are recycled, and the heme group is converted into **bilirubin.** (Figure 19.6, p. 729; Figure 19.7, p. 730)

  👆 Practice sequencing the steps: Figure 19.6.

- **Anemia** is the decreased oxygen-carrying capacity of the blood.

MODULE 19.3
Leukocytes and Immune Function 731

- **Granulocytes** are leukocytes with visible cytoplasmic granules and lobed nuclei.
  ○ **Neutrophils** are the most common leukocyte. They enhance inflammation, attract other leukocytes to an area of injury, and directly kill bacteria by phagocytosis and by releasing their granules.
  ○ **Eosinophils** respond to parasitic worm infections and are involved in allergic responses.
  ○ The granules of **basophils** mediate inflammation.

- **Agranulocytes** lack visible cytoplasmic granules.
  ○ **Lymphocytes** are the second most numerous leukocyte. The two types of lymphocytes are **B lymphocytes** and **T lymphocytes.** Activated B lymphocytes secrete **antibodies** that bind to individual **antigens.** Activated T lymphocytes directly destroy cancer or virally infected cells and activate other parts of the immune system.
  ○ **Monocytes** mature into **macrophages.** (Figure 19.8, p. 731)

  👆 Practice art-labeling exercise: Figure 19.8.

- **Leukopoiesis** is the process of leukocyte formation. All leukocytes are derived from hematopoietic stem cells, which divide into the **lymphoid cell line** and the **myeloid cell line.** B and T lymphocytes derive from the lymphoid cell line; all other formed elements derive from the myeloid cell line. (Figure 19.9, p. 734)

  👆 Practice art-labeling exercise: Figure 19.9.

MODULE 19.4
Platelets 735

- Platelets are anucleate fragments of massive **megakaryocytes,** which are derived from the myeloid cell line. (Figure 19.10, p. 735)

MODULE 19.5
Hemostasis 736

- Hemostasis—stopping blood loss from an injured vessel—is carried out by platelets and **clotting factors.** A **blood clot** is a collection of platelets, clotting proteins, and other formed elements. Hemostasis has five phases: (1) vascular spasm, (2) formation of a **platelet plug,** (3) **coagulation,** (4) clot retraction, and (5) **thrombolysis.** (Figures 19.11 and 19.12, p. 737; Figure 19.13, p. 738; Table 19.2, p. 740; Figures 19.14 and 19.15, p. 741; Figure 19.16, p. 743)
- Clotting is regulated by **anticoagulants,** including **antithrombin-III** and **protein C.** (Table 19.3, p. 742)

MODULE 19.6
Blood Typing and Matching 744

- All cells and biological molecules feature unique surface **antigens.** Two major groups of erythrocyte antigens are the **ABO blood group** and the **Rh blood group.** (Table 19.4, p. 745)
- Blood type is based on the presence or absence of antigens from the ABO and Rh groups.
- B lymphocytes recognize and respond to foreign antigens by producing **antibodies** that bind and **agglutinate** antigens. Three antibodies may be present in blood: **anti-A antibodies, anti-B antibodies,** and **anti-Rh antibodies.** (Figure 19.17, p. 747)
- Blood type is determined by treating erythrocytes with anti-A, anti-B, and anti-Rh antibodies and watching for agglutination reactions. (Figure 19.18, p. 747)
- Antigens and antibodies are the basis for blood matching in **blood transfusions,** in which blood is transferred from a compatible **donor** to a **recipient.** Type O− is the **universal donor** because erythrocytes from individuals with type O− lack antigens of the ABO and Rh groups. Type AB+ is the **universal recipient** because the plasma in these individuals lacks anti-A, anti-B, and anti-Rh antibodies. (Figure 19.19, p. 748)

ASSESS WHAT YOU LEARNED

See answers in Appendix A.

LEVEL 1 Check Your Recall

1. Which of the following is *not* a formed element of blood?
   a. Erythrocyte   c. Mast cell
   b. Leukocyte   d. Platelet

2. Which of the following plasma proteins is responsible for osmotic pressure?
   a. γ-Globulins   c. α-Globulins
   b. Albumin   d. Clotting proteins

3. List the seven major functions of blood.

4. Mark the following statements as true or false. If a statement is false, correct it to make a true statement.
   a. Erythrocytes are biconcave discs with prominent nuclei.
   b. The main function of erythrocytes is to transport oxygen on the molecule hemoglobin.
   c. The iron ions of hemoglobin are oxidized when they bind to oxygen.
   d. Hemoglobin consists of two polypeptide chains bound to a heme group.

5. Erythropoiesis requires stimulation from the hormone:
   a. thrombin.   c. thymosin.
   b. thrombopoietin.   d. erythropoietin.

6. Fill in the blanks: Erythrocytes are derived from stem cells called _____, circulate in the blood for approximately _____ days, and are destroyed by an organ called the _____.

7. Anemia is defined as:
   a. decreased oxygen-carrying capacity of the blood.
   b. decreased iron content of the blood.
   c. decreased bone marrow function.
   d. abnormalities in hemoglobin.

8. Leukocytes are:
   a. nucleated cells that function in blood clotting.
   b. nucleated cells that function in immunity.
   c. anucleate cells that function in blood clotting.
   d. anucleate cells that function in immunity.

9. Match the following leukocytes with the correct definition.
   _____ Basophil
   _____ B lymphocyte
   _____ Neutrophil
   _____ Monocyte
   _____ T lymphocyte
   _____ Eosinophil

   a. Destroy bacteria; directly phagocytose bacteria
   b. Responds to parasitic worm infection and mediates the allergic response
   c. Activates all parts of the immune response; directly kill cancer or virally infected cells
   d. Secrete inflammatory mediators
   e. Agranulocyte that matures into macrophage
   f. Agranulocyte that secretes antibodies

10. Fill in the blanks: Lymphocytes are derived from the _____ cell line, whereas the other leukocytes are derived from the _____ cell line.

11. Platelets are derived from cells called:
    a. thromboblasts.   c. megakaryocytes.
    b. leukoblasts.   d. thrombokaryocytes.

12. Number the steps of hemostasis in order, putting 1 by the first event, 2 by the second, and so on.
    _____ The intrinsic and extrinsic pathways produce factor Xa.
    _____ The clot retracts.
    _____ Thrombin converts fibrinogen to fibrin, and fibrin glues the plug together.
    _____ Platelets are activated, and the platelet plug forms.

_____ Vasoconstriction and increased tissue pressure decrease blood flow through the vessel.

_____ Tissue plasminogen activator activates plasmin, which degrades fibrin.

_____ The common pathway produces thrombin.

13. How do the intrinsic and extrinsic coagulation pathways differ? How are they similar?

14. What are the overall goals of the common pathway of coagulation?

15. Which of the following is *not* an anticlotting agent produced by endothelial cells?

a. Prostacyclin   c. Antithrombin-III
b. Protein C   d. Warfarin

16. Tissue plasminogen activator, urokinase, and plasmin are important components of:

a. coagulation.   c. platelet plug formation.
b. fibrinolysis.   d. hemostasis.

17. Fill in the blanks: The two most clinically important groups of antigens on erythrocytes are the _____ and _____ blood groups.

18. Which antigens does a person with blood type A− have on the surface of his or her erythrocytes?

a. A antigens   d. Both a and c are correct.
b. B antigens   e. All of the above
c. Rh antigens

19. Which of the following antibodies does a person with type O+ blood have in his or her plasma?

a. Anti-A antibodies   d. Both a and b are correct.
b. Anti-B antibodies   e. All of the above
c. Anti-Rh antibodies

20. Mr. Reczkiewicz has blood type AB−. Which of the following blood types could be safely donated to Mr. Reczkiewicz, assuming he has had prior exposure to Rh+ blood? (Circle all that apply.)

a. Type O+   d. Type AB+
b. Type A−   e. Type O−
c. Type B−

LEVEL 2 Check Your Understanding

1. Explain how blood, being a liquid, enables all of its components (formed elements and plasma) to perform their functions.

2. Predict how abnormal hemoglobin molecules that contain only two iron ions, instead of four, would affect homeostasis.

3. The anticoagulant drug warfarin primarily disrupts the extrinsic coagulation pathway. Explain why disrupting only the extrinsic pathway disrupts the entire coagulation cascade.

4. Cirrhosis of the liver often reduces production of many types of plasma proteins, including albumin and clotting factors. Predict the effects on the body of decreased numbers of these proteins in the plasma. Would this also affect the number of γ-globulins in the plasma? Why or why not?

LEVEL 3 Apply Your Knowledge

PART A: Application and Analysis

1. A blood sample from your patient shows that she has decreased numbers of neutrophils. Predict the effects of this condition. How would it differ if numbers of T lymphocytes were decreased instead?

2. Mr. Jackson presents to the emergency room with a minor wound that has bled for several days. An examination of his medical history reveals that Mr. Jackson has hemophilia A, which is caused by a deficiency of factor VIII. Your co-worker suggests that you give Mr. Jackson some platelets to stop the bleeding. Will this help your patient? Why or why not?

3. Ms. Wu, whose blood type is O−, requires a blood transfusion. Her family members volunteer to donate blood. Their blood types are as follows: her son, type B−; her husband, type B+; her daughter, type O+. Which family members could safely donate blood to Ms. Wu? Who could not? Explain.

PART B: Make the Connection

4. Elise is a 36-year-old woman who has volunteered to donate red bone marrow to a patient in need of a bone marrow transplant for aplastic anemia. The physician performing the bone marrow extraction is an intern, and it is her first time doing the procedure. She asks that you prepare a site on Elise's tibia for the procedure. Is this a good place for red bone marrow to be extracted, considering Elise's age? Why or why not? Can you suggest an alternative location? Explain. (*Connects to Chapter 6*)

20

The Lymphatic System and Immunity

We come into contact with literally trillions of microbes every day, so why aren't we constantly sick? The short answer is that our bodies protect us from infection by many of these microbes. How do they do this? Well, that requires a longer answer. Remember in our exploration of the structure and function of blood, we introduced formed elements known as *leukocytes,* or *white blood cells* (WBCs), and immune proteins located in the plasma (see Chapter 19). Now we focus our attention on these cells and proteins, which make up part of a larger system collectively called the **immune system,** which works to defend the body against internal and external threats. Note that the immune system is not technically a system according to the definition you have learned. This system contains no organs or tissues of its own, and instead consists of cells and proteins located in the blood and the tissues of other systems, including the lymphatic system (lim-FAH-tik). As we will see, the **lymphatic system** is a group of organs and tissues that not only works with the immune system but also participates in a number of functions such as fluid homeostasis.

The immune and lymphatic systems function together in what is broadly called **immunity:** the set of diverse processes that protect the body from both cellular injury and disease-causing cells and molecules known as *pathogens* (PATH-uh-jenz; *patho-* = "disease"). These functions are crucial to our homeostasis, which is evidenced by the devastating consequences of diseases such as *AIDS* (*acquired immunodeficiency syndrome*) that result from the loss of certain components of the immune system. Although the lymphatic system is technically its own system and has functions independent of immunity, the two systems have such close structural and functional relationships that it makes sense to cover them together in a single chapter.

We begin by introducing the structures and functions of the lymphatic system. Next, we tie the two systems together as we examine the cells and proteins of the immune system. The chapter concludes with a focus on how immune disorders affect the body's homeostasis.

Scanning electron micrograph: Cytotoxic T cells surround and attack a cancer cell.

MODULE **20.1**

Structure and Function of the Lymphatic System

Learning Outcomes

1. Describe the major functions of the lymphatic system.
2. Compare and contrast lymphatic vessels and blood vessels in terms of structure and function.
3. Describe the mechanisms of lymph formation, and trace the pathway of lymph circulation through the body.
4. Describe the basic structure and cellular composition of lymphatic tissue, and relate them to the overall functions of the lymphatic system.
5. Describe the structures and functions of the lymphoid organs.

The lymphatic system consists of two main components, both of which are illustrated in **Figure 20.1**: (1) a system of blind-ended tubes called **lymphatic vessels,** and (2) **lymphatic tissue** and **lymphoid organs**, including clusters of *lymphoid follicles* such as the *tonsils,* as well as the *lymph nodes, spleen,* and *thymus.* In this module, we introduce the basic functions of these vessels and organs and then turn to their structure. Much of what we discuss here will be revisited several times throughout this chapter, which reflects the close level of integration between the lymphatic and immune systems.

Functions of the Lymphatic System

◀◀ **Flashback**

1. What is the net filtration pressure in blood capillaries? (p. 690)
2. How much water is lost from the blood vessels to the interstitial fluid every day? (p. 691)

The lymphatic system has three basic functions that are carried out by its vessels and organs. These functions are as follows:

- **Regulation of interstitial fluid volume.** As you read about in the blood vessels chapter, the net filtration pressure in blood capillaries favors filtration, meaning that water is lost from the plasma in the blood to the interstitial fluid (see Chapter 18). The amount of fluid that is lost from the plasma is significant, measuring about 2–4 liters per day. This fluid must be returned to the circulation, or both the blood volume and blood pressure will drop too low to maintain homeostasis. The vessels that perform this function are the lymphatic vessels, which pick up excess fluid in the extracellular space, transport it through the body, and deliver it back to the cardiovascular system. When the fluid exits the extracellular space and enters the lymphatic vessels, it is known as **lymph** (LIMF). This means that interstitial fluid and lymph are very similar in composition.
- **Absorption of dietary fats.** The breakdown products of fats in the diet are too large to pass through the tiny spaces between the endothelial cells of blood capillaries. However, they are able to enter small lymphatic vessels in the small intestine. The dietary fats travel through the lymphatic vessels and are delivered to the blood with lymph.
- **Immune functions.** The lymphatic system plays an important role in the immune system. As we discuss shortly, the lymphoid organs filter pathogens from the lymph and blood. They also house several types of leukocytes, and play a role in their maturation.

Quick Check

☐ 1. List the main functions of the lymphatic system.

Lymphatic Vessels and Lymph Circulation

◀◀ **Flashback**

1. What are endothelial cells? (p. 685)
2. How are continuous capillaries structured? (p. 686)
3. What are the mechanisms that assist in the return of venous blood to the heart? (p. 674)

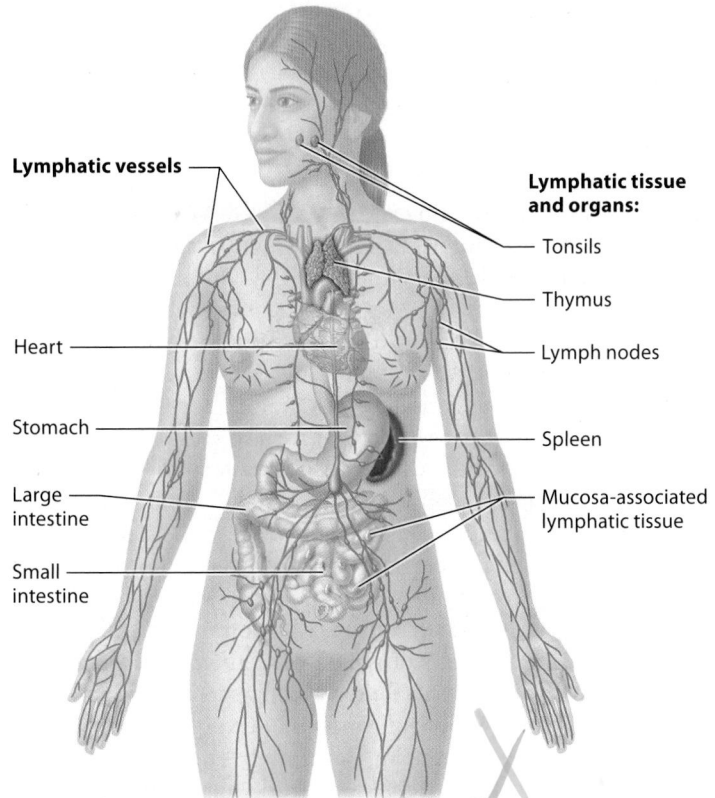

Lymphatic vessels

Heart

Stomach

Large intestine

Small intestine

Lymphatic tissue and organs:

Tonsils

Thymus

Lymph nodes

Spleen

Mucosa-associated lymphatic tissue

Figure 20.1 Overview of the lymphatic system.

Practice art labeling
@MasteringA&P®

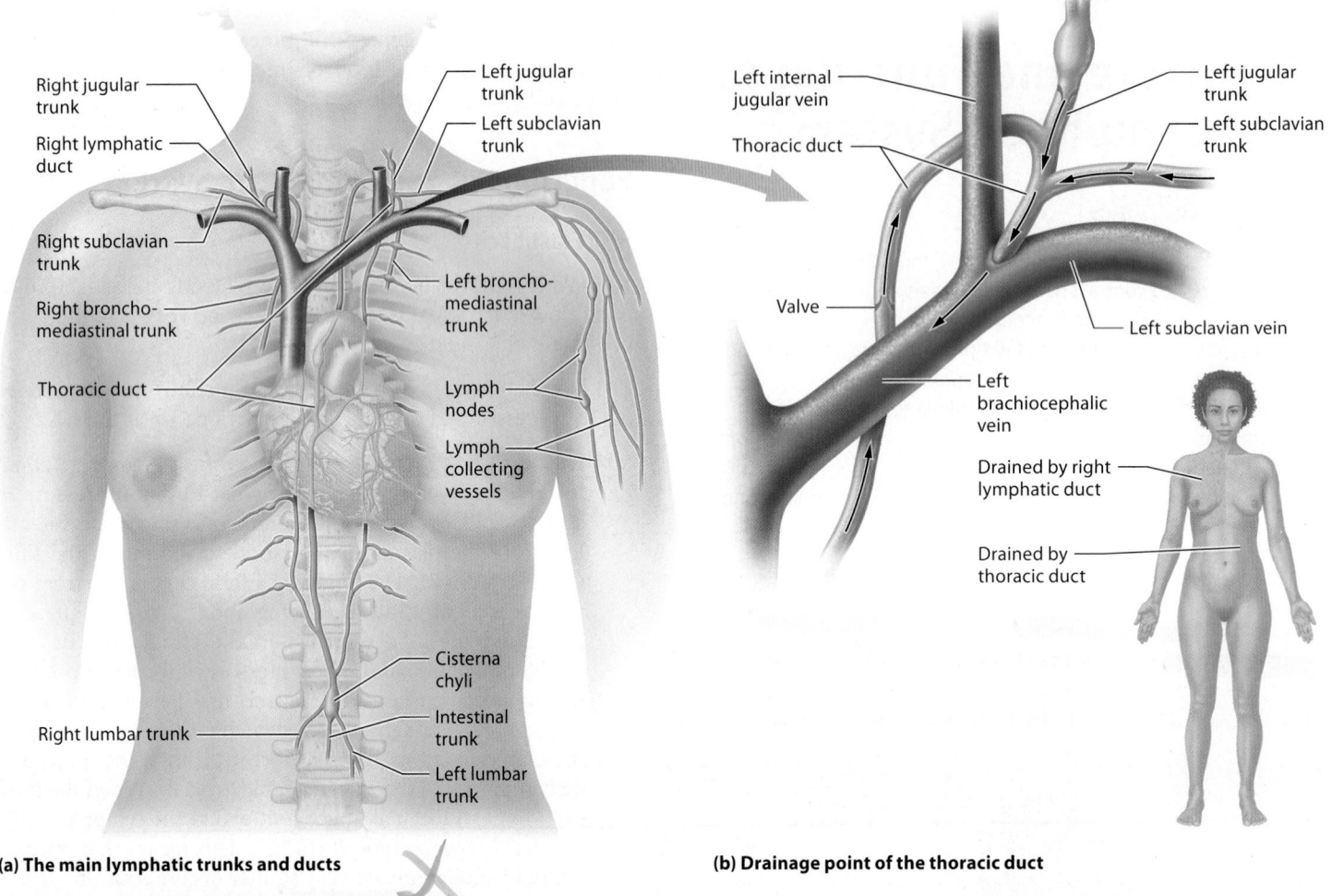

(a) The main lymphatic trunks and ducts

(b) Drainage point of the thoracic duct

Figure 20.2 Main lymphatic trunks and ducts.

Lymph is collected in vessels appropriately called **lymph-collecting vessels,** which merge to form larger vessels, the **lymph trunks.** There are nine lymphatic trunks that drain the lymph from body regions, illustrated in **Figure 20.2a**:

- *lumbar trunks,* which receive lymph from the lower limbs and pelvic area;
- *the intestinal trunk,* which receives fat-containing lymph from the small lymphatic vessels in the small intestine;
- *jugular trunks,* which receive lymph from the head and neck;
- *bronchomediastinal trunks* (brong′-koh-meh-dee-ah-STY-nul), which receive lymph from the thoracic cavity; and
- *subclavian trunks,* which receive lymph from the upper limbs.

Note in Figure 20.2a that the intestinal and lumbar trunks all drain into a large, swollen-looking vessel called the **cisterna chyli** (sis-TER-nah KY-lee; "reservoir of digestive juice"). The cisterna chyli and the other lymphatic trunks drain into one of two **lymphatic ducts.** The cisterna chyli and the trunks from the left side of the body drain into the **thoracic duct,** the largest lymphatic duct, which runs along the anterior vertebral column and drains into the junction of the left internal jugular and left subclavian veins (**Figure 20.2b**). In fact, the thoracic duct drains all of the lower body as well as the left side of the upper body. The remaining trunks from the right upper side of

the body drain into the tiny **right lymphatic duct,** which drains into the junction of the right internal jugular and right subclavian veins. At these venous junctions, the lymph drains into the blood of the low-pressure venous circuit at the subclavian veins.

Lymphatic vessels make up a low-pressure circuit because there is no main pump to drive lymph through the vessels, and most of them are transporting lymph against gravity. For this reason, we see features in lymphatic vessels that we also see in the low-pressure venous circuit of the cardiovascular system. First, lymphatic vessels have *lymphatic valves* that prevent lymph from flowing backward. In addition, lymphatic vessels, like deeper veins, are often found lodged between muscles, where the contracting muscles massage lymph up toward the heart. Finally, lymph flow through the vessels is driven in part by contractions of smooth muscle found in the walls of lymph-collecting vessels. See *A&P in the Real World: Lymphedema* for what happens if lymph vessels become blocked or the flow of lymph is interrupted.

Lymphatic vessels begin in the tissues with tiny **lymphatic capillaries,** which form weblike networks that surround blood capillary beds. In addition, specialized lymphatic capillaries known as **lacteals** (lak-TEE-ulz) collect fat in the small intestine. As you can see in **Figure 20.3a**, lymphatic capillaries differ both structurally and functionally from blood capillaries. For

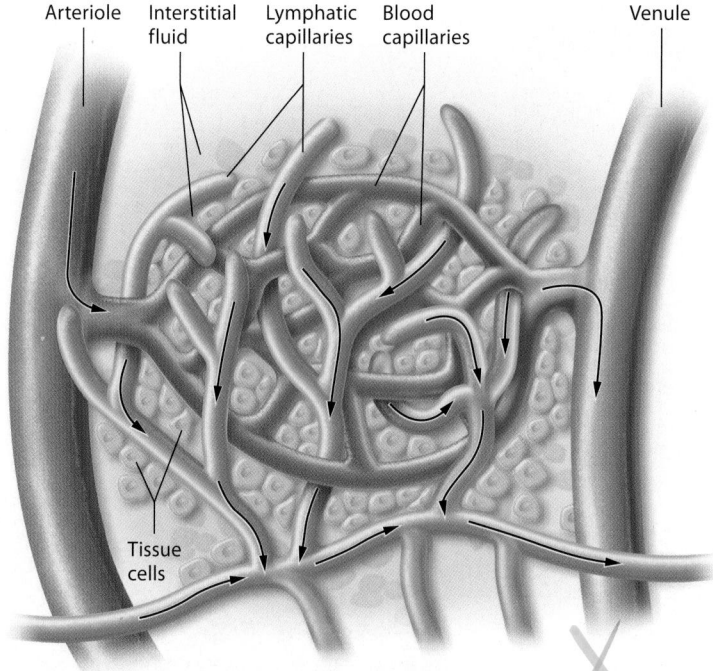

(a) Lymphatic capillaries surrounding a capillary bed

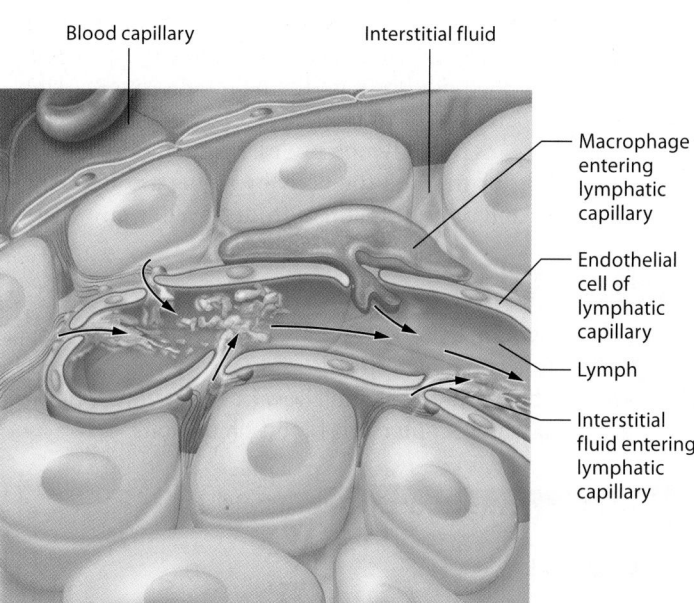

(b) Lymphatic capillary structure and function

Figure 20.3 Structure and function of lymphatic capillaries.

one, lymphatic capillaries are blind-ended, which makes the lymphatic vasculature a one-way system that only moves lymph away from the tissues. Blood capillaries, of course, form a two-way system that moves blood both toward and away from the tissues.

Notice in **Figure 20.3b** that unlike the endothelial cells that form the walls of blood capillaries, the cells of lymphatic capillary walls are not tightly joined, and instead are able to flap open and closed. Fluid that leaks from the blood capillaries increases the pressure in the interstitial fluid, which forces the lymphatic endothelial cells apart and allows large volumes of fluid to enter the lymphatic capillaries. When the pressure in the interstitial fluid decreases, the endothelial cells flap shut. In this way, the lymphatic system plays the crucial role of precisely controlling the amount of fluid between our cells. Note in Figure 20.3b that these vessels are also leaky enough to allow cells such as *macrophages* and other immune cells to enter the lymph.

Lymphedema

The role of the lymphatic system in regulating the volume of interstitial fluid becomes readily apparent in **lymphedema** (limf-eh-DEE-mah). **Edema,** commonly known as *swelling,* is defined as an accumulation of excess interstitial fluid. Many conditions can cause mild to moderate edema, including trauma, vascular disease, and heart failure. However, the edema seen with lymphedema is typically severe and can be disfiguring.

Lymphedema is generally due to removal of lymphatic vessels during surgery or blockage of the vessels from pathogens such as parasites. Both conditions prevent the lymphatic vessels from transporting excess interstitial fluid back to the cardiovascular system. The fluid therefore accumulates in the tissues of the affected body part, causing it to enlarge. The photo shows a case of lymphedema in the arm of a breast cancer patient resulting from the surgical removal of lymph nodes.

Because of the extreme permeability of the lymphatic capillaries, pathogens such as bacteria and cancer cells in the interstitial fluid have an easier time entering the lymphatic capillaries than the blood capillaries. These pathogens therefore can spread to other parts of the body through the lymphatic system. However, along the pathway of lymphatic vessels, we find clusters of the lymphoid organs called lymph nodes. Lymph nodes limit the spread of pathogens through the body by acting as filters, trapping pathogens and preventing them from traveling elsewhere.

Quick Check

☐ 2. How do lymphatic capillaries differ from blood capillaries in structure and function?

☐ 3. What is the pathway of lymph flow from the right foot to the point where it enters the blood circulation?

Lymphoid Tissues and Organs

◀◀ Flashback

1. What is reticular tissue? (p. 139)
2. What are splenic sinusoids and where are they located? (p. 687)
3. Where do T lymphocytes migrate to mature? (p. 733)

The predominant tissue type of the lymphatic system is a type of loose connective tissue called *reticular tissue* that contains specialized cells and thin *reticular fibers* (look back at Figure 4.15).

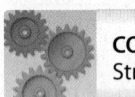
CORE PRINCIPLE
Structure-Function

These reticular fibers interweave to form "nets" that trap disease-causing pathogens, which is an example of the Structure-Function Core Principle (p. 25). In the lymphatic system, reticular tissue is often simply called **lymphoid tissue.** Lymphoid tissue is found in lymphoid organs and also in independent clusters that have varying degrees of organization.

Most lymphoid tissues and organs consist of several main cell types, shown in **Figure 20.4:**

- **Leukocytes.** Lymphoid organs house leukocytes, including macrophages, *B lymphocytes,* and *T lymphocytes* (more commonly referred to as *B cells* and *T cells,* respectively). As we discussed previously, **macrophages** (MAK-roh-fay-jez; "big eaters") are mature monocytes that are very active phagocytes, and **B lymphocytes** and **T lymphocytes** are agranulocytes with diverse immune functions (see Chapter 19).
- **Dendritic cells. Dendritic cells** are immune cells with spiny processes that resemble the dendrites of neurons. There are several types of dendritic cells, some derived from the bone marrow and others from connective tissue.
- **Reticular cells.** Reticular fibers, composed of a specialized, thin type of collagen protein, are produced by **reticular cells.** Reticular cells are particularly abundant in organs such as the spleen and lymph nodes.

In this section, we take a look at the lymphoid organs and tissues and see how these cells function within them.

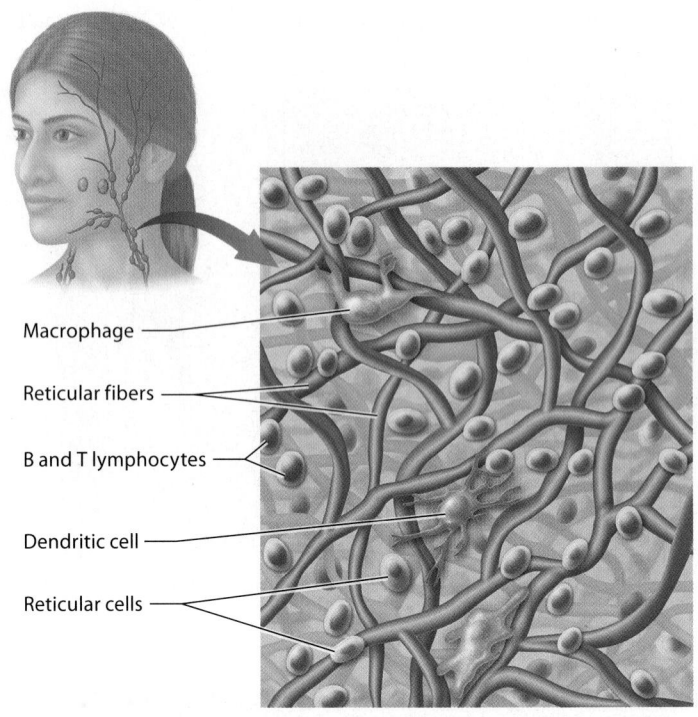

Macrophage

Reticular fibers

B and T lymphocytes

Dendritic cell

Reticular cells

Figure 20.4 Microscopic structure of lymphoid organs.

Mucosa-Associated Lymphatic Tissue (MALT)

The mucous membranes, such as those of the oral and nasal cavities, are all vulnerable parts of the body because they are exposed to large numbers of pathogens. These areas are protected by loosely organized clusters of lymphoid tissue collectively called **mucosa-associated lymphatic tissue (MALT).** MALT is found scattered throughout the gastrointestinal tract, the respiratory passages, and, to a limited extent, the genitourinary tract.

Much of the MALT in the body consists of loosely organized clusters of B and T cells. Most MALT lacks any kind of connective tissue capsule. However, certain types of specialized MALT are partially encapsulated by connective tissue.

Specialized MALT consists of roughly spherical clusters called **lymphoid follicles,** or **lymphoid nodules.** Lymphoid follicles that have not been exposed to pathogens consist primarily of B cells. However, most lymphoid follicles have been exposed to pathogens and so contain light-staining regions, the **germinal centers,** which consist of dividing B cells, a special kind of dendritic cell, and macrophages. In the gastrointestinal tract, we find these specialized clusters of MALT in three locations: the **tonsils,** located around the oral and nasal cavities; **Peyer's patches** (PY-erz), or *aggregated lymphoid nodules,* located in the last portion of the small intestine (called the *ileum*); and the **appendix,** located off the large intestine.

The three main tonsils are illustrated in **Figure 20.5.** The first is the **pharyngeal tonsil** (also known as the *adenoid* [AD-eh-noyd]), which is located in the posterior nasal cavity (the nasopharynx). The **palatine tonsils** are found in the posterolateral oral cavity, whereas the **lingual tonsil** is found at the base of the tongue. The epithelium lining the tonsils indents deeply in several locations, forming *tonsillar crypts* that trap bacteria and

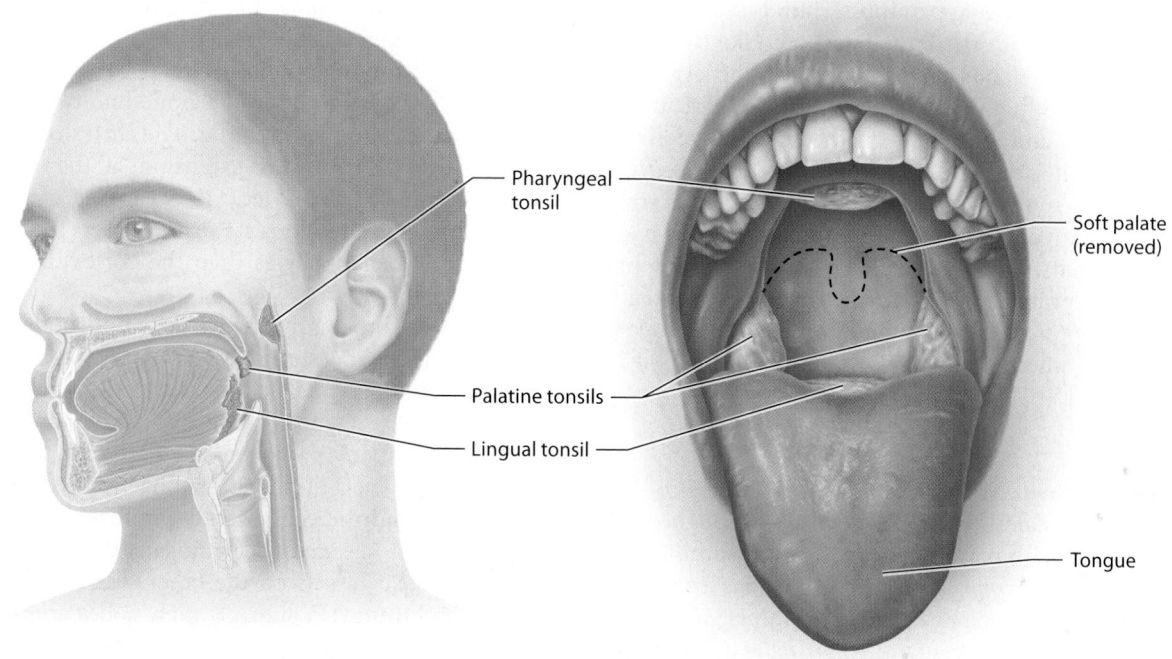

(a) **Location of tonsils in the oral and nasal cavities**

Pharyngeal tonsil

Palatine tonsils

Lingual tonsil

(b) **Anterior view of the oral cavity**

Soft palate (removed)

Tongue

Figure 20.5 Location of the tonsils.

debris. The tonsils' location puts them into contact with a large number of potential pathogens. As a result, they commonly become inflamed, a condition known as *tonsillitis*.

Peyer's patches are also exposed to a large number of potential pathogens. The last portion of the small intestine joins to the large intestine, which houses a tremendous number of bacteria. These bacteria generally are not pathogenic in the large intestine, but if they gain entry to the small intestine, they could cause disease. The position of the Peyer's patches allows them to defend against any bacteria that have escaped from the large intestine (**Figure 20.6**).

(a) **Location of MALT in intestines**

Large intestine

MALT

Small intestine

Peyer's patches

MALT in appendix

Blood vessels

Connective tissue

Peyer's patches

Epithelium

LM (20×)

(b) **Micrograph of small intestine showing Peyer's patches**

Figure 20.6 MALT of the intestines.

Finally, we turn to the appendix—a blind-ended, worm-shaped tube that juts from the large intestine. Like the Peyer's patches, the appendix defends against bacteria in the large intestine, specifically those that could be pathogenic. Unfortunately, it is easy for bacteria and fecal matter to become trapped in the appendix. The result is a condition known as *appendicitis,* in which the appendix swells much like lymph nodes do (as you've likely noticed) when you are sick and they are filled with bacteria or other pathogens. The blind-ended structure of the appendix gives it little room to expand, which creates the potential for rupture. Such a rupture spills bacteria into the abdominopelvic cavity and leads to internal bleeding, which may be fatal if not treated promptly.

Lymph Nodes

Lymph nodes are small, vaguely bean-shaped clusters of lymphatic tissue located along lymphatic vessels throughout the body. We find clusters of lymph nodes located in specific areas, including the *axillary lymph nodes* in the axillae (under the arms), the *cervical lymph nodes* in the neck, the *inguinal lymph nodes* in the groin, and the *mesenteric lymph nodes* in the abdominal cavity around the abdominal organs (**Figure 20.7a**). **Figure 20.7b** shows an inguinal lymph node in a cadaver.

Note in **Figure 20.7c** that lymph nodes feature an external connective tissue capsule that surrounds a network of reticular fibers filled with macrophages, lymphocytes, and dendritic cells. We can divide the interior of a node into two main regions: the outer **cortex** and inner **medulla.** The cortex consists of lymphoid follicles divided by inward extensions of the capsule, called **trabeculae** (trah-BEK-yoo-lee). Between the cortex and medulla we find a zone that consists primarily of T cells. The medulla contains fewer leukocytes than the cortex, but it does contain mature B cells.

The basic process by which a lymph node filters lymph is shown by the arrows in Figure 20.7c. Lymph flows into the node through multiple small lymphatic vessels called **afferent lymphatic vessels.** Lymph percolates through the reticular network, where many pathogens in the lymph become trapped in the reticular "net." Trapped pathogens then encounter leukocytes and dendritic cells, which eliminate these threats. Lymph that has been "cleaned" of pathogens drains out through **efferent lymphatic vessels** on the other side of the node at an area called the **hilum** (HY-lum). Lymph nodes are remarkably effective at their job and trap approximately 90% of the pathogens in lymph. This prevents these pathogens from being delivered to the blood, where they could easily spread to other tissues and organs.

(a) Locations of lymph node clusters

Cervical lymph nodes

Axillary lymph nodes

Mesenteric lymph nodes

Inguinal lymph nodes

(b) Photo of a lymph node in a human cadaver

Lymph node

(c) Frontal section through a lymph node

Valve

Afferent lymphatic vessels

Efferent lymphatic vessel

Cortex

Medulla

Germinal center

Lymphoid follicle

Trabeculae

Capsule

Figure 20.7 Location and structure of lymph nodes.

Practice art labeling @MasteringA&P®

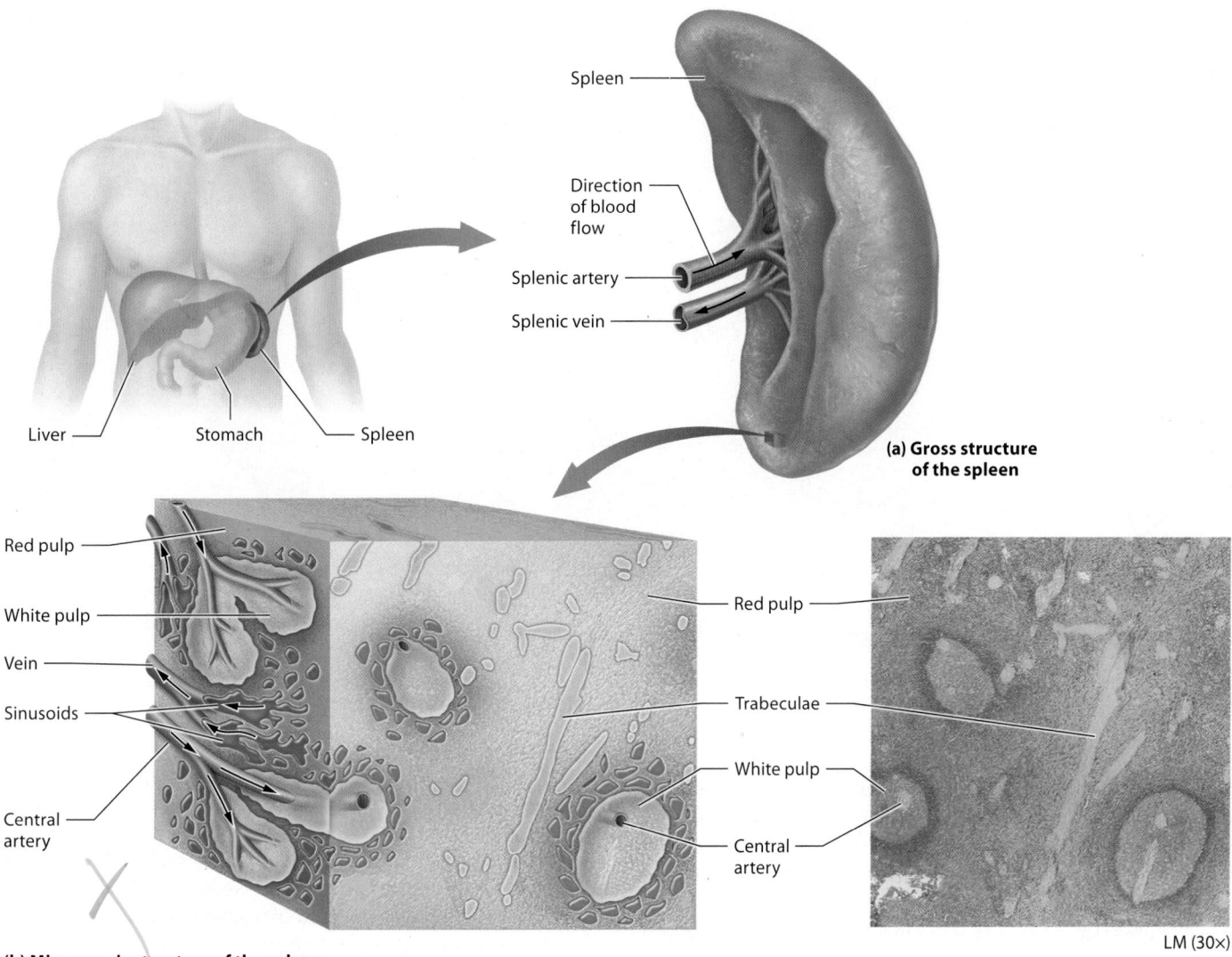

Figure 20.8 Structure of the spleen.

Occasionally, large numbers of pathogens become trapped in a lymph node. This triggers a response that causes the lymph node to enlarge or swell, a condition commonly known as "swollen glands." Note, however, that this term is incorrect—lymph nodes are not glands and they secrete no products. Certain conditions such as bacterial and viral infections cause the swollen lymph nodes to become painful. Other conditions, such as allergies or the spread of cancerous cells through lymphatic vessels, generally cause the nodes to enlarge with little pain.

The Spleen

Having covered the organs and tissues that protect us from pathogens on our mucous membranes and in the lymph, we now turn to the organ that protects us from pathogens in the blood: the spleen. The **spleen,** which is the largest lymphoid organ in the body, is located on the lateral side of the left upper quadrant of the abdominopelvic cavity. It is a purplish-brown organ about the size of a large bar of soap (**Figure 20.8a**). Recall that the

spleen's internal structure consists of an interweaving network of reticular fibers made by reticular cells. Within this reticular tissue framework we find two distinct histological regions: **red pulp,** which contains macrophages that destroy old erythrocytes; and **white pulp,** which filters pathogens from the blood and contains leukocytes and dendritic cells (**Figure 20.8b**).

Let's take a closer look at the white pulp. Figure 20.8b shows that the white pulp surrounds branches of the splenic artery called the *central arteries*. Immediately surrounding the arteries in the center of the white pulp we find a zone that contains mostly T cells. Just outside this zone are lymphoid follicles composed of a central core of B cells, and on the outer rim of the white pulp are macrophages, dendritic cells, and some B and T cells.

When blood enters the spleen, it flows through a series of arterioles until it comes to the sinusoids—the spleen's large, exceptionally leaky capillaries. Here pathogens encounter macrophages and dendritic cells, which process portions of the pathogens.

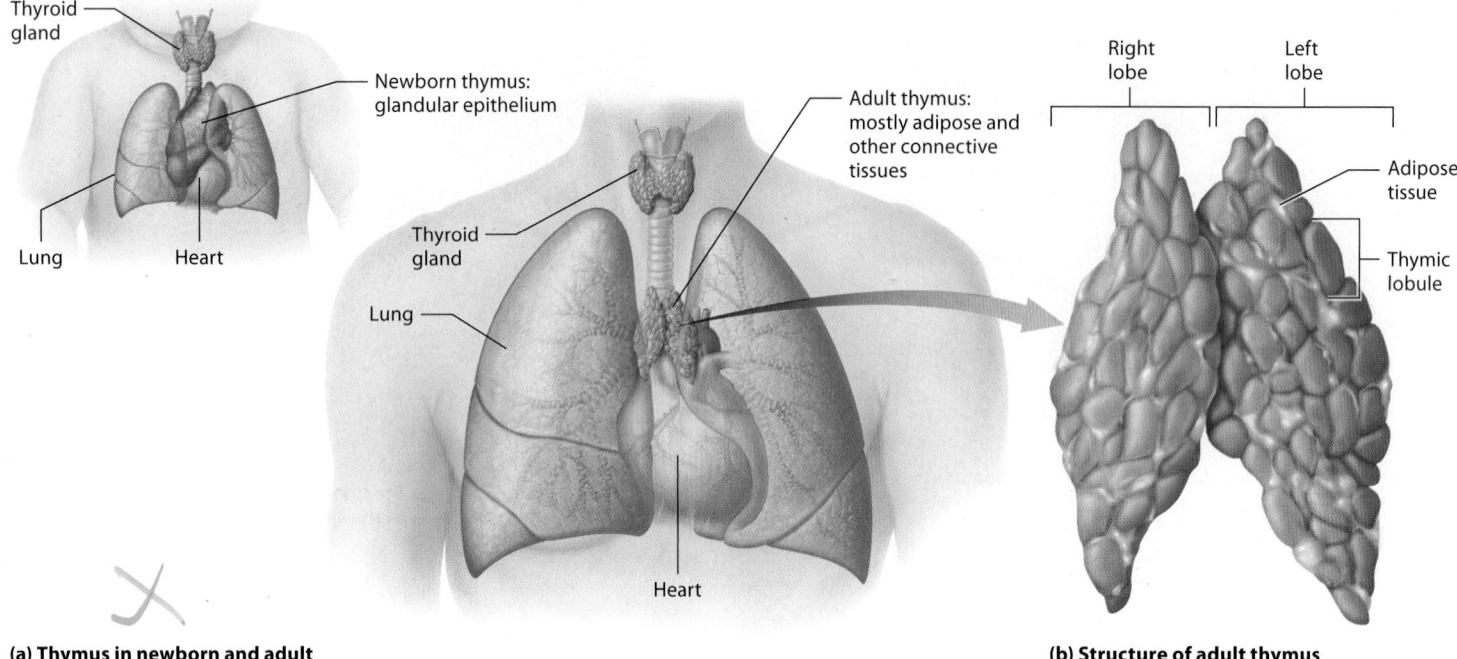

(a) Thymus in newborn and adult

(b) Structure of adult thymus

Figure 20.9 Comparison of the thymus in a newborn and an adult.

The Thymus

In the endocrine chapter, you read about the **thymus,** a small, encapsulated organ in the superior mediastinum that consists of two lobes (see Chapter 16). Recall that the thymus secretes hormones that enable it to carry out its primary function: generating a population of functional T cells capable of protecting the body from pathogens. Unlike the spleen and lymph nodes, the thymus doesn't trap pathogens.

The thymus is large and very active in infants and children, as you can see in **Figure 20.9a**, and it reaches its maximum size in individuals about 12–14 years old. After this point, it begins to atrophy, and the thymic tissue is gradually replaced with fat. As the thymus atrophies, its production of T cells decreases as well. Indeed, by age 65, the rate of T cell production falls to about 2% of the rate at which an infant produces T cells. This drop in production does not normally lead to a decline in immune function, however, because of the enormous number of T cells produced during infancy and childhood.

The adult thymus, shown in **Figure 20.9b**, consists of subunits called *thymic lobules* (or thymic corpuscles), which are visible externally as the "lumps" on the surface of the thymus. Each lobule contains two regions: an outer cortex and an inner medulla. The cortex contains densely packed T cells. The medulla contains fewer of these cells, and is thought instead to be mostly the site of destruction of certain populations of T cells that could react to the body's own cells (see Module 20.4). Notice that there are no lymphoid follicles in the thymus because it lacks B cells.

Quick Check

☐ 4. What main cell types are located in lymphoid organs?
☐ 5. List the main functions of MALT, lymph nodes, the spleen, and the thymus.

Apply What You Learned

☐ 1. A patient has undergone surgery for breast cancer in which the axillary lymph nodes and associated lymphatic vessels were removed. Predict the effects of removing lymph nodes and vessels.
☐ 2. Mr. Jackson steps on a nail with his left foot and chooses not to seek medical treatment. Two weeks later, he presents to his physician with a complaint of painful masses in his groin on the left side. His physician notes that Mr. Jackson's inguinal lymph nodes are swollen. What has likely happened?
☐ 3. Predict the potential consequences of removing each of the following lymphoid organs:
    a. Spleen
    b. Thymus (in an adult)
    c. Thymus (in a newborn)
    d. Tonsils
See answers in Appendix A.

MODULE **20.2**

Overview of the Immune System

Learning Outcomes

1. Identify the differences between innate and adaptive immunity, and explain how the two types of immunity work together.
2. Describe the basic differences between antibody-mediated and cell-mediated immunity.

3. Describe the roles that surface barriers play in immunity.
4. Describe the cells and proteins that make up the immune system.
5. Explain how the immune and lymphatic systems are connected structurally and functionally.

As we have discussed, the immune system consists primarily of a group of many different types of leukocytes and proteins that defend the body from anything that causes cellular injury. This includes injury resulting from trauma and pathogens. We commonly think of pathogens as microorganisms that cause *infection*, meaning they invade a tissue, increase in number, and cause cellular damage. Such microorganisms include bacterial cells, viruses, fungi, and parasites. However, many other cells and molecules are pathogenic, including cancer cells, bacterial toxins, and toxins found in animal venoms.

Together, the components of the immune system offer three lines of defense against pathogens:

- The *first line of defense* involves cutaneous and mucous membranes that act as surface barriers to block the entry of pathogens into the body.
- The *second line of defense* consists of the responses of the cells and proteins that make up innate immunity.
- The *third line of defense* consists of the responses of the cells and proteins of adaptive immunity.

Before we consider the inner workings of the immune system, let's survey its parts and processes. In this module, we examine the two basic types of immunity, the surface barriers that protect us, the components of the immune system, and the connections between the lymphatic and immune systems. The following modules explore the functional aspects and components of the immune system more closely.

Types of Immunity

Immunity is classified according to the way it responds to different pathogens or forms of cellular injury. As implied by its name, **innate,** or **nonspecific, immunity** responds to all pathogens or classes of pathogen in the same way. The innate immune system consists of antimicrobial proteins and certain cells. Note that innate immunity responds quickly and is the dominant response to pathogens for the first 12 hours after exposure. This occurs because the cells and proteins exist in the bloodstream, even in the absence of a stimulus (hence the term "innate").

The other type of immunity is **adaptive,** or **specific, immunity,** the components of which respond individually to unique glycoprotein markers called **antigens** (AN-tih-jenz). Antigens are present on all cells and most biological molecules, including our own cells, and they identify a cell or molecule as belonging to a specific group. There are two "arms" of the adaptive immune system. The first, **cell-mediated immunity,** is brought about by two types of T cells. The second, **antibody-mediated immunity** (also known as *humoral immunity*), is carried out by B cells and proteins they produce, called *antibodies*. Adaptive immunity

responds more slowly than innate immunity because one must be exposed to a specific antigen for the response to be initiated. It is therefore also known as **acquired immunity.** Adaptive immunity takes 3–5 days to mount a response, but after this point, it is the dominant response.

One key difference between the two types of immunity is that adaptive immunity has the capacity for *immunological memory,* in which exposure to an antigen is "remembered" by specific lymphocytes and antibodies. This allows a more rapid and efficient response on subsequent exposures. Innate immunity lacks the capacity for immunological memory and will respond in the same way with repeat exposure to a pathogen.

Before we move on, keep in mind that adaptive immunity and innate immunity are not independently functioning arms of the immune system. Each type of immunity relies on the other, and the response to a pathogen involves a highly integrated series of events within both parts of the immune system. Indeed, dysfunction within a component of either adaptive or innate immunity may lead to catastrophic failure of the entire system (see Module 20.7).

Quick Check

☐ 1. How do innate immunity and adaptive immunity differ?
☐ 2. What are the three lines of defense?

Surface Barriers

◄◄ Flashback

1. Where are mucous membranes located? What do they secrete? (p. 153)
2. What substances does the skin secrete? What are the functions of these secretions? (p. 162)

The first line of defense against any potential threat to the body is the coverings that line body surfaces: the skin and the mucous membranes, and certain products they secrete. Both the skin and mucosae are **surface barriers,** meaning that they provide a continuous physical barrier to block the entry of potential pathogens into the body.

We've previously discussed the skin's numerous protective functions (see Chapter 5). Besides being continuous and covering nearly every external surface of the body, skin is also relatively resistant to mechanical stresses, thanks to its several layers of epithelial cells filled with the hard protein *keratin.* In addition, sebaceous glands in the skin secrete sebum or oil, which has a slightly acidic pH that deters the growth of most pathogenic organisms.

The body's other type of surface barrier is the mucous membrane. Recall that mucous membranes line all passageways in the body that open to the outside, including the respiratory, gastrointestinal, and genitourinary tracts (see Chapter 4). Mucous membrane epithelia lack keratin and are generally thinner than skin epithelia, so they are less resistant to mechanical abrasion. Mucous membranes also secrete products that discourage pathogen invasion, namely, the sticky substance *mucus*. Mucus traps pathogens and other debris and also protects the underlying cells from chemical and mechanical trauma. In addition, the mucous membranes of the stomach secrete acid, which kills ingested pathogens.

The importance of surface barriers is evident any time the surface barrier is compromised. Think about what happens when you receive a wound to your skin—pathogens likely entered the body when the wound occurred, so you must wash it thoroughly to remove them. You must then cover the wound with a bandage to prevent future invasion from pathogens. In this case, the bandage acts as the new surface barrier until the skin is healed and again able to prevent pathogen invasion.

Of course, surface barriers can't protect us from every pathogen we encounter, in part because the pathogens have evolved mechanisms to circumvent them. To read about some of these mechanisms, see *A&P in the Real World: How Pathogens Can Evade Surface Barriers.*

Quick Check

☐ 3. What are the body's main surface barriers, and how do they function in its defense?

Overview of Cells and Proteins of the Innate and Adaptive Immune Systems

◀◀ **Flashback**

1. Where are most leukocytes produced? (p. 731)
2. What are phagocytes? (p. 732)

The cells and proteins of the innate and adaptive immune systems produce the responses of the second and third lines of defense. The main cells of the immune system are the different types of leukocytes, which we've already discussed: the *agranulocytes* (B and T lymphocytes and monocytes), which lack cytoplasmic granules; and the *granulocytes* (neutrophils, eosinophils, and basophils), which contain cytoplasmic granules (see Figure 19.8). Many cells of innate immunity can function as **phagocytes**—cells that "eat" foreign or damaged cells—although neutrophils and macrophages are the most active phagocytes. As you read earlier in this module, lymphocytes are the primary cells of adaptive immunity.

Other types of leukocytes also function in immunity. One type is the **natural killer (NK) cell,** which acts primarily in innate immunity. NK cells are generally located in the blood and spleen, and are formed in the bone marrow from the same cell lines as lymphocytes. Another cell type is one you encountered in the previous module: *dendritic cells,* immune cells that are located in many lymphoid organs. Dendritic cells are part of the innate immune response, but their main role is to activate the T cells of adaptive immunity.

The other main components of the immune system are groups of different types of proteins. You have already been

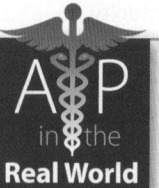

How Pathogens Can Evade Surface Barriers

Pathogens have evolved multiple mechanisms to evade the defenses provided by surface barriers and facilitate their invasion of our bodies. The following are three examples of such pathogens:

- *Collagenase-producing bacteria.* Certain bacterial species, notably *Clostridium perfringens* (klos-TRID-ee-um per-FRIN-genz), produce enzymes called *collagenases* (kah-LAJ-eh-nay′-zez), which catalyze reactions that degrade the protein collagen. When a wound is infected with *C. perfringens,* the bacteria begin to produce collagenases that destroy collagen in the basement membrane and dermis. This allows them access to deeper tissues such as skeletal muscle, eventually resulting in massive tissue death. As the bacteria destroy tissues, they produce large quantities of gas, leading to the common name of *gas gangrene* (gang-GREEN). To make matters worse, *C. perfringens* also produces enzymes that catalyze the destruction of neutrophils, making the immune system much less able to respond to the invasion.

- *Phagocytosis-resistant fungi.* Pathogenic organisms that come into contact with mucosae are generally ingested by local macrophages, minimizing the risk of infection. However, some fungi are quite resistant to phagocytosis. For example, the fungal yeast cells that cause the disease *blastomycosis* (blas′-toh-my-KOH-sis) have very thick walls that are difficult for phagocytes to ingest. Still other fungal species, such as those producing the respiratory diseases *cryptococcosis* (krip′-toh-kok-KOH-sis) and *histoplasmosis* (hiss′-toh-plaz-MOH-sis), actually survive inside macrophages. This makes the fungi invisible to the immune system, and the macrophages spread the disease as they travel through the blood and lymph to different tissues.

- *Acid-tolerant pathogens.* Secretions produced by surface barriers are generally slightly to very acidic, which deters or kills most pathogens. But certain pathogens tolerate or even prefer an acidic pH. For example, the poliovirus is able to survive the highly acidic environment of the stomach when it is ingested, allowing it to replicate in the digestive tract and potentially invade the rest of the body. In addition, many species of fungus and the bacterium *Helicobacter pylori* (hel′-ih-koh-BAK-ter py-LOH-rye) prefer an acidic pH, allowing them to infect the skin and stomach, respectively.

introduced to one group of proteins—antibodies, the proteins produced by B lymphocytes that function in adaptive immunity. Another group of proteins, collectively called the *complement system*, functions in innate immunity. The final type is the **cytokines** (SY-toh-kynz), a diverse group of proteins secreted by cells of both innate and adaptive immunity. Cytokines have a variety of effects, including regulating the development and activity of immune cells; we will discuss their many other effects in later modules.

The upcoming modules examine the roles of each of these cells and proteins in more detail. Then we look into how they interact to produce a functional immune response.

Quick Check

☐ 4. List the main types of cells and proteins of the immune system.

How the Lymphatic and Immune Systems Work Together

As we mentioned in the previous module, the lymphatic and immune systems are closely connected both structurally and functionally. Some of the ways in which they are connected include the following:

- **Lymphoid organs and tissues provide a residence for cells of the immune system.** B cells, T cells, and macrophages frequently take up residence in lymphoid organs such as the lymph nodes, MALT, and the spleen. In addition, they provide a place for lymphocyte development.
- **Lymphoid organs and tissues trap pathogens for the immune system.** The fine networks of reticular fibers in lymphoid tissues form "nets" that trap pathogens so that leukocytes may interact with them more easily.
- **Lymphoid organs activate cells of the immune system.** Lymphoid organs house cells such as dendritic cells, which play a crucial role in activating B and T cells (we discuss this role shortly). In addition, the thymus is required for the selection of a functional population of T cells.

The lymphatic system plays a greater role in adaptive immunity than in innate immunity, although lymphoid organs do house cells of innate immunity, such as macrophages. In the upcoming modules, we explore more closely how these two systems interact.

Quick Check

☐ 5. How are the immune system and lymphatic system connected structurally and functionally?

Apply What You Learned

☐ 1. *Interstitial cystitis* is characterized by erosion of the mucous membrane lining the urinary bladder. Predict the potential consequences of this condition.

☐ 2. Your friend argues that the third line of defense of the immune system is the most important and that the first and second are relatively unimportant. What do you tell him?

See answers in Appendix A.

MODULE **20.3**
Innate Immunity: Internal Defenses

Learning Outcomes

1. Describe the roles that phagocytic and nonphagocytic cells and plasma proteins such as complement and interferon play in innate immunity.
2. Walk through the stages of the inflammatory response, and describe its purpose.
3. Describe the process by which fever is generated, and explain its purpose.

As you learned in the previous module, innate immunity is present in the body without needing exposure to a specific antigen. For this reason, innate immunity is typically quick to respond to a pathogen, generally within seconds to a few hours. Innate immunity consists of two main components: (1) a group of antimicrobial molecules, including the antimicrobial complement proteins and a variety of cytokines; and (2) several types of cells, including neutrophils, macrophages, and NK cells. In this module, we explore these components and finish with an examination of how they interact to produce the *inflammatory response*.

Cells of Innate Immunity

◄◄ Flashback

1. How do granulocytes and agranulocytes differ? (p. 732)
2. What are the three types of granulocytes, and what are their properties? (p. 732)
3. What is phagocytosis? Walk through its basic steps. (p. 83)

Pathogens that have moved past our surface barriers next meet our bodies' second line of defense: the cells and proteins of innate immunity, which were introduced in the previous module. We can divide the cells of innate immunity into two types, phagocytic cells and nonphagocytic cells. The phagocytic cells include macrophages, *neutrophils* (NOO-troh-filz), and *eosinophils* (ee'-uh-SIN-uh-filz). You have read about the process by which cells ingest particles and other cells, called **phagocytosis** (fay'-goh-sy-TOH-sis)—you may wish to review this process by looking at Figure 3.12 before continuing. The nonphagocytic cells include NK cells, basophils, and dendritic cells.

Phagocytes

Certain granulocytes and agranulocytes can function as phagocytes. For example, the agranulocytes known as monocytes exit the bloodstream and take up residence in various tissues. After they do so, they enlarge and undergo a variety of other changes to become macrophages, which are very active phagocytes. Some macrophages remain fixed in certain tissues, whereas others roam freely in the body.

Macrophages are activated by a variety of stimuli, including certain molecules present on pathogens, chemicals secreted by damaged cells, and signals from the cells of adaptive immunity. Activated local macrophages are generally the first cells to respond to a cellular injury, where they ingest other cells and cellular debris, as shown here:

SEM (2700×)

They kill pathogens they have ingested with chemicals, including hydrogen peroxide and hypochlorous acid (the active component in bleach). Macrophages also have *cytotoxic* effects, meaning that they can secrete these substances onto pathogens that are too large to ingest. In addition, they function as **antigen-presenting cells,** cells that display portions of the pathogens (antigens) they ingest on their plasma membranes. T cells respond to these antigens by becoming activated. Activated T cells in turn secrete substances that increase the activity of the macrophages, in an example of a positive feedback loop. This is one of many instances of innate and adaptive immunity working together.

The most predominant granulocyte is the **neutrophil.** Like macrophages, neutrophils are highly effective phagocytes that kill their ingested pathogens with chemicals such as hydrogen peroxide, hypochlorous acid, and lysozyme. Neutrophils can ingest many types of cells, but are particularly effective at destroying bacterial pathogens. When pathogens are too large for them to ingest, neutrophils release the cytotoxic contents of their granules onto the pathogens to damage their plasma membranes. Neutrophils generally reside in the blood and must be recruited to damaged tissues by chemical signals.

The granulocytes known as **eosinophils** are phagocytes that can migrate from the blood to the tissues where they are needed. However, their phagocytic functions appear to be less critical than

those of neutrophils, and they are primarily involved in responses to parasitic pathogens. When a person is infected with a parasite, eosinophils cover it and release the contents of their granules. The chemicals from the granules damage the parasite and either destroy it or make it easier for other immune cells to destroy.

Other Cells of Innate Immunity

The first group of nonphagocytic cells we'll cover is the natural killer (NK) cells, introduced in the previous module. NK cells have the remarkable ability to recognize cancerous cells and cells infected with certain viruses in spite of the fact they cannot recognize antigens. They appear to do this by scanning the cells for irregularities and by reacting to cells that have bound to antibodies (another example of innate and adaptive immunity working together). NK cells are also cytotoxic, releasing substances that destroy their target cells. In addition, they secrete an antimicrobial cytokine that activates macrophages and enhances phagocytosis.

The next group of nonphagocytic cells is the dendritic cells. Like macrophages, these cells function as antigen-presenting cells. The substances they ingest are presented to T cells (and, to a lesser extent, B cells), which are then activated. We'll give more attention to these functions in the sections on adaptive immunity.

A relatively rare nonphagocytic cell is the **basophil.** Basophils are granulocytes whose granules contain chemicals that mediate, or convey, inflammation (*inflammatory mediators*). Basophils are located primarily in the blood, although a related type of cell called the **mast cell** is located in mucous membranes. Like regular basophils, mast cells contain granules with chemicals that trigger inflammation, particularly that involved in allergic responses.

Quick Check

☐ 1. Match the following cell types with their correct definition.
_____ Macrophage
_____ NK cell
_____ Eosinophil
_____ Dendritic cell
_____ Neutrophil
_____ Basophil

a. Granulocyte that attacks parasites
b. Granulocyte that secretes inflammatory mediators
c. Highly active phagocytic agranulocyte that acts as an antigen-presenting cell
d. Highly active phagocytic granulocyte
e. Cell that is cytotoxic to cancerous cells and to cells infected with certain viruses
f. Cell found in lymphoid organs that has long processes and acts as an antigen-presenting cell

☐ 2. Which of the cells in question 1 do not directly destroy a pathogen or abnormal cell?

Antimicrobial Proteins

In addition to cells, the innate immune response is mediated by a variety of plasma **antimicrobial proteins,** including complement proteins and several types of cytokines. In this section we examine the roles of these proteins and how they are produced

CLASSICAL PATHWAY

Complement proteins bind to antibody bound to antigen.

Complement protein

Antibody

Antigen

ALTERNATIVE PATHWAY

Complement protein C3 spontaneously cleaves to C3a and C3b.

C3a

C3

C3b

Enzymatic cascade

Activated **C3b**

leads to

Activation of **C5b**

leads to

Cell lysis

Complement proteins

Membrane attack complex (MAC)

Bacterium

Lysed cell

Enhanced inflammation

Inflammatory mediators

Basophils

Complement proteins

Neutralized viruses

C3b

Neutralized viruses

Enhanced phagocytosis due to opsonization

C3b coating a bacterium

Macrophage

Clearance of immune complexes

Immune complex

C3b

Macrophage

Figure 20.10 Pathways for activation of the complement system.

and activated. As with most of the topics we have discussed in this module, note that complement and cytokines have roles in adaptive immunity, as well, and are revisited later.

Complement

The group of molecules collectively known as the **complement system** consists of 20 or more plasma proteins that are produced primarily by the liver. Note that complement proteins are designated with a "C" and a number.

Complement proteins circulate primarily in their inactive forms and must be activated by a complex cascade of events mediated by enzymes. Note in **Figure 20.10** that there are two main series of enzymatic reactions, or pathways, by which complement proteins may be activated: the classical and alternative pathways. The *classical pathway* begins when inactive complement proteins bind to antibodies bound to antigen. The *alternative pathway* can begin at the cleavage of an inactive complement protein called C3 into its active form C3b. This can also occur when inactive complement proteins encounter foreign cells such as bacteria. The two pathways converge when C3b is activated, which in turn cleaves the inactive protein C5 into its active component C5b.

Activated complement proteins lead to the following main effects, illustrated in Figure 20.10:

- **Cell lysis.** Some complement proteins are able to *lyse,* or "pop," the plasma membranes of pathogens, leading to their destruction. This is mediated by C5b, which binds to the surface of a pathogen and provides a docking site for several other activated complement proteins. Together these complement proteins form a structure collectively known as the **membrane attack complex, or MAC,** which inserts itself into the plasma membrane of the target cell, creating a pore that causes it to lyse.

- **Enhanced inflammation.** The *inflammatory response* is a nonspecific response to cellular injury. Several complement proteins enhance this response by triggering basophils and mast cells to release chemicals that mediate inflammation. We discuss how this works in the next section.

- **Neutralized viruses.** Another function of C3b and the components of the membrane attack complex is binding to certain viruses and *neutralizing* them, or blocking them from infecting host cells.

- **Enhanced phagocytosis.** The complement protein C3b acts as an *opsonin,* binding both pathogens and phagocytes. This action, known as **opsonization,** makes phagocytes bind more strongly to the pathogen and enhances phagocytosis. This function is evident from the name, as opsonin means "prepare for eating."

- **Clearance of immune complexes.** C3b also binds to *immune complexes*—clusters of antigens bound to antibodies—and triggers their phagocytosis. This clears these complexes from the circulation, which is critical to preventing them from lodging in different tissues around the body.

Cytokines

Recall that cytokines are proteins produced by several types of immune cells that enhance the immune response in some way. Interestingly, many cytokines induce "flu-like" symptoms, including fever, chills, and aches (the aches are due to stimulation of inflammation). Although one tends to blame the pathogen for these symptoms, usually it is your own immune response that actually causes them. There are several cytokines that are involved in innate immunity, including the following:

- **Tumor necrosis factor.** The cytokine known as *tumor necrosis factor,* or *TNF,* is named for the fact that it was originally discovered through its ability to induce necrosis (death) of tumor cells. TNF is secreted primarily by activated macrophages in response to certain bacteria and other pathogens. Effects of TNF include inducing the flu-like symptoms we just described, attracting phagocytes to the area of infection, increasing the activity of phagocytes, and stimulating phagocytes to release additional cytokines. In severe infections, the amount of TNF secreted can increase dramatically. This results in the possibly fatal condition of *septic shock,* in which a marked drop in blood pressure, failure of organs such as the kidneys and liver, inappropriate

blood clotting, and a severe drop in blood glucose concentration seriously disrupt homeostasis.

- **Interferons.** The cytokines known as **interferons** (in-ter-FEER-onz), abbreviated **INF,** are produced by a variety of cells, including macrophages, dendritic cells, NK cells, and cells of adaptive immunity. Interferons are generally produced in response to infection with intracellular agents such as viruses or intracellular bacteria, and are named for their ability to "interfere" with the ability of the pathogens to infect other cells. One of their primary actions is to inhibit viral replication inside host cells. In addition, they activate various components of both innate and adaptive immunity, such as stimulating nearby cells to produce antimicrobial proteins. Interferons are actually partly responsible for producing some of the symptoms associated with flu-like illnesses, including muscle aches and fever.

- **Interleukins.** *Interleukins* (in-ter-LOO-kinz) constitute a class of 29 cytokines produced mainly by different types of leukocytes. Examples of the effects of interleukins include stimulating production of neutrophils by the bone marrow, stimulating NK cells, triggering the production of certain types of interferons, and activating T cells.

Cytokines have been studied for their therapeutic uses in conditions such as persistent viral infections and certain cancers. For example, a specific type of interferon is used to treat infections like hepatitis B and C (viral infections of the liver) and cancers like melanoma (cancer of the melanocytes in the skin). The adverse effects of cytokine therapy revolve around their induction of flu-like symptoms. Patients being treated with cytokines often feel as though they have a flu-like illness for the duration of therapy, which may last several months or more. The many potential therapeutic uses of cytokines have made them an active area of research, and our understanding of them is growing and changing rapidly.

Quick Check

☐ 3. What is the complement system, and what are its roles?
☐ 4. In what two ways is the complement system activated?
☐ 5. What are the main cytokines of innate immunity, and what are their roles?

Inflammatory Response

Like all innate responses, the series of events known collectively as the **inflammatory response** occurs in reaction to any cellular injury. Inflammation begins when a cell is damaged by anything, including trauma, bacterial or viral invasion, toxins, heat, or chemicals. This triggers a series of events that collectively act to wall off the damaged area from the rest of the body, clean up damaged cells and cell components, and bring cells and proteins to the area to repair the damaged tissue. There are two basic stages to the inflammatory response: (1) Damaged cells release inflammatory mediators that cause local changes in the damaged tissue; and (2) phagocytes arrive at the area and clean up the damaged tissue.

Part 1: Release of Inflammatory Mediators and the Cardinal Signs of Inflammation

As shown in **Figure 20.11**, after tissue damage occurs (such as that caused by the splinter in step ①), the inflammatory response begins when the damaged cells and nearby mast cells release inflammatory mediators (step ②). These mediators can include histamine, serotonin, cytokines, a peptide called *bradykinin* (brad-ee-KY-nin), and a group of related molecules known as *prostaglandins* (pros-tah-GLAN-dinz) and *leukotrienes* (loo-koh-TRY-eenz). Prostaglandins and leukotrienes are important targets for medications that block the inflammatory response, discussed in *A&P in the Real World: Anti-inflammatory Medications* on page 768. In addition, certain activated complement proteins trigger the release of inflammatory mediators from cells such as basophils and mast cells, and act to a limited degree as inflammatory mediators themselves.

Inflammatory mediators trigger several effects, which you have almost certainly seen first-hand on occasion (shown in step ③): The injured area becomes red and swollen, feels warm to the touch, and hurts. These four signs—redness, heat, swelling (or *edema;* eh-DEE-mah), and pain—are known as the *cardinal signs of inflammation,* and they are due to the following actions of inflammatory mediators (see Figure 20.11):

- **Vasodilation of arterioles.** Inflammatory mediators such as histamine and bradykinin are potent *vasodilators*—they relax the smooth muscle in arterioles supplying the damaged tissue, causing the vessels to "open up" and blood to flow through them more freely. This results in increased blood flow to the damaged tissue, and the area becomes congested with blood, a condition known as *hyperemia* (hyper- = "above" or "elevated," -emia = "blood"). Hyperemia accounts for the redness and heat that accompany inflammation (because blood is warmer than the body's surface).

Sequence the steps @MasteringA&P®

Figure 20.11 The inflammatory response, part 1: effects of inflammatory mediators.

- **Increased capillary permeability.** The structure of most capillaries prevents them from leaking all of their contents to the interstitial fluid. Inflammatory mediators increase the permeability, or "leakiness," of local capillary beds. This allows protein-rich fluid to leak from the blood vessels into the tissue spaces, and leads to the cardinal sign of swelling. The proteins in the fluid include clotting proteins such as fibrinogen, complement proteins, and proteins needed for tissue repair.
- **Occurrence of pain.** Many of the inflammatory mediators, particularly bradykinin and prostaglandins, trigger action potentials in the peripheral processes of sensory neurons. This leads to the extremely important and underappreciated effect of *pain,* which serves numerous protective functions. Pain lets us know when our tissues are being damaged so that we can try to avoid further damage. In some cases, it also causes a temporary loss of function—you're unlikely to run on your broken ankle when it's throbbing—which allows the body to repair the damage.
- **Recruitment of other cells: chemotaxis.** The final effect of inflammatory mediators is the recruitment of leukocytes to

the damaged area through a process known as **chemotaxis** (kee-moh-TAK-sis; *-taxis* = "movement"). Inflammatory mediators attract and activate a number of leukocytes, particularly macrophages and neutrophils. Note that as this is occurring, complement proteins are also being activated.

Part 2: Phagocyte Response

The second component of the inflammatory response deals with the phagocytes that ingest pathogens and cellular debris. We can divide the arrival and activation of phagocytes into stages, based on which phagocytes enter the area and the processes occurring there. The stages, shown in **Figure 20.12**, proceed as follows:

1. **Local macrophages are activated.** Within minutes of a cellular injury, macrophages already present in the area enlarge and begin to phagocytize pathogens and damaged cells. Though few in number, these "first responders," the only phagocytes present within the first hour or so of the inflammatory response, perform the critical function of containing invading pathogens.

2. **Neutrophils migrate by chemotaxis to the damaged tissue and phagocytize bacteria and cellular debris.** After about an hour, neutrophils from the blood migrate to the damaged tissue. Inflammatory mediators and activated complement proteins attract neutrophils and enable them to leave the blood and enter the tissue. Notice in Figure 20.12 how this works: The inflammatory mediators make the capillary endothelium in the damaged area "sticky," and the neutrophils adhere to the capillary wall, a process called *margination.* The inflammatory mediators also increase capillary permeability, as we discussed earlier, which provides enough spaces between endothelial cells for neutrophils to squeeze through into the damaged tissue. This is a process known as *diapedesis* (dy′-ah-peh-DEE-sis). Once the neutrophils are in the tissue, they then begin to destroy bacteria and other cellular debris. Note that during this time, neutrophils stored in the bone marrow are released into the blood, which leads to a rapid, acute rise in the level of circulating neutrophils.

3. **Monocytes migrate to the tissue by chemotaxis and become macrophages, which phagocytize pathogens and cellular debris.** The next group of cells to be attracted to the area by chemotaxis is the circulating monocytes, which exit the blood in the same way as neutrophils. In the damaged tissue, monocytes enlarge and mature into macrophages, which ingest pathogens and damaged cells with the neutrophils. This response is much slower than that of neutrophils, because it takes several hours for monocytes to mature and there are far fewer monocytes than neutrophils stored in the bone marrow and the circulation. However, monocytes live longer and are more aggressive than neutrophils.

4. **The bone marrow increases production of leukocytes, leading to leukocytosis.** The final response occurs in the bone marrow, where leukocytes are produced. Cytokines produced by the activated phagocytes act on cells in the bone marrow to increase the production of neutrophils and monocytes over the next 3–4 days. This leads to an

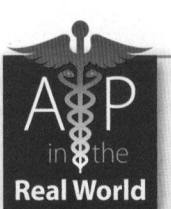

Anti-inflammatory Medications

Pain is one of the more unpleasant effects of the inflammatory response. Fortunately, many medications have been developed to reduce inflammation and the accompanying pain by blocking the production of prostaglandins. Prostaglandins and the related leukotrienes are derived from a fatty acid called arachidonic acid (ah-rak′-ih-DOHN-ik) that is present in nearly all cell membranes. When the cell is damaged or triggered in some other way, the enzyme *phospholipase A2* (fos-foh-LY-payz) catalyzes a reaction that cleaves arachidonic acid. The products of this reaction may be acted on by two broad classes of enzymes: *cyclooxygenases* (sy-kloh-OK-seh-jen-ay′-zez), which produce prostaglandins, and *lipooxygenases* (ly-poh-OK-seh-jen-ay′-zez), which produce leukotrienes.

The first, and largest, group of anti-inflammatory medications consists of the *nonsteroidal anti-inflammatories,* or *NSAIDs.* NSAIDs work by inhibiting the cyclooxygenase enzyme and preventing the formation of prostaglandins. An example of a common anti-inflammatory medication available over the counter is *ibuprofen* (eye-byoo-PROH-fen).

The *corticosteroids* make up a second group of medications. Corticosteroids mimic the actions of the body's own hormone cortisol, which inhibits the formation of both prostaglandins and leukotrienes. This leads to a wider-ranging and more potent anti-inflammatory effect. Corticosteroids are generally used for conditions with more severe inflammation or inflammation due primarily to leukotrienes (such as allergy-related inflammation). Examples of common corticosteroids include *cortisone* and *prednisone.*

(1) Local macrophages are activated.

Neutrophil phagocytosing pathogen

Neutrophil undergoing diapedesis

Neutrophil undergoing margination

(2) Neutrophils migrate by chemotaxis to the damaged tissue and phagocytize bacteria and cellular debris.

Macrophage

Neutrophils

Monocytes

(3) Monocytes migrate to the tissue by chemotaxis and become macrophages, which phagocytize pathogens and cellular debris.

Bone marrow sinusoid

New leukocyte entering circulation

Leukocytes undergoing mitosis

(4) The bone marrow increases production of leukocytes, leading to leukocytosis.

Figure 20.12 The inflammatory response, part 2: phagocyte response. This response involves the processes of margination, diapedesis, chemotaxis, and phagocytosis.

Sequence the steps
@MasteringA&P®

elevated number of circulating leukocytes, a condition called *leukocytosis* (loo′-koh-sy-TOH-sis), commonly referred to as a "high white cell count." The elevated numbers of leukocytes allow the damaged area to be cleared and any pathogens removed so that cells such as fibroblasts can begin the process of healing.

The neutrophils and macrophages in the damaged tissue go to work digesting pathogens and damaged cells. Eventually, all of the neutrophils and many of the macrophages die themselves. The accumulation of dead leukocytes, dead tissue cells, and fluid leads to a whitish mixture known as **pus** (a wound filled with pus is said to be *purulent* [PYOOR-uh-lent]). It is common to associate pus with infection; however, pus indicates only that an inflammatory response took place. However, much more pus is typically found with inflammation due to certain bacterial infections than would be produced by other types of cellular injury. The pus is generally absorbed into the surrounding tissues and lymphatic vessels when healing has completed.

Quick Check

☐ 6. What are the four cardinal signs of inflammation?
☐ 7. What are inflammatory mediators, and what do they do?
☐ 8. Explain the role of phagocytes in the inflammatory response.

Fever

◀◀ **Flashback**

1. Which part of the brain is responsible for temperature homeostasis? (p. 463)
2. What effects are triggered when this part of the brain senses that body temperature is too low or too high? (p. 463)

One of the most recognizable, and unpleasant, signs of illness is fever. **Fever** is defined simply as a body temperature above the normal range, which is generally between 36 and 38° C (or 97–99° F);

an individual with a fever is referred to as *febrile* (FEB-rihl). Many conditions can cause fever, but in this section we primarily focus on those related to the immune system.

Like inflammation, fever is an innate response to cellular injury. Fever is initiated when chemicals called **pyrogens** (PY-roh-jenz; *pyro-* = "fire," *-gen* = "causing") are released from damaged cells or certain bacteria. Pyrogens act on the hypothalamus, the small component of the diencephalon (the central core of the brain) that regulates many aspects of homeostasis (refer to Figure 12.17). The hypothalamus normally functions as the body's thermostat, and maintains body temperature within the normal range through a series of negative feedback loops, an example of the Feedback Loops Core Principle (p. 21). Pyro-

CORE PRINCIPLE
Feedback Loops

gens cause the hypothalamic thermostat to reset to a higher range. At this higher range, the hypothalamus interprets *normal* body temperature as being too low, which triggers the negative feedback loop. This is why you feel cold, or have "chills," when you have a fever—the hypothalamus is triggering responses to elevate body temperature to the new, higher range. This includes the familiar sign of shivering, increased muscle activity that generates heat to bring body temperature to the new set point.

What does a fever accomplish? Surprisingly, this is still not fully understood. There is a common misconception that the elevated temperature of a fever kills bacterial cells. However, any temperature high enough to kill bacterial cells would almost certainly kill all of our cells as well. In addition, fevers can occur in the absence of bacterial infection (it is an innate response, after all), so this idea just doesn't hold water. What good, then, does a fever do us? The best current hypothesis is that our phagocytes function more efficiently at a slightly higher body temperature, which helps to increase the rate of recovery from whatever caused the inflammation.

Fevers may be alleviated, or made to "break," either spontaneously or through the action of certain medications. Medications that reduce an inflammation-mediated fever are known as *antipyretics*. Most antipyretics work in the same manner as anti-inflammatories, by inhibiting the formation of prostaglandins. Indeed, many anti-inflammatory medications such as aspirin and ibuprofen are also antipyretics. Regardless of what makes the fever break, it results from the hypothalamus being reset back to the normal temperature range. This causes the hypothalamus to sense the febrile temperature as being too high, and it triggers negative feedback mechanisms to lower the body's temperature. These include sweating and dilation of the blood vessels serving the skin, which makes the skin appear red or flushed. Both conditions make you feel hot, although your body temperature is actually decreasing.

From a clinical perspective, fever is a critical warning sign of inflammatory processes occurring somewhere in the body. Although we tend to associate fevers with infection, remember that fever is a nonspecific response and can tell you only that widespread inflammation is present. This is evidenced by the fact that fevers may be seen in the absence of any infectious pathogens following severe trauma, burns, or even childbirth. It is also important to remember that occasionally fevers are not related to inflammation. Conditions such as heatstroke, certain brain diseases, and reactions to medications can cause fevers, and these fevers will not reduce with administration of anti-inflammatory medications. We discuss these and other causes of elevated body temperature in the metabolism chapter (see Chapter 23).

Quick Check

☐ 9. What causes the elevated temperature and common signs of fever?
☐ 10. What benefits result from the elevated temperature of fever?

Apply What You Learned

☐ 1. *Neutropenia* (noo'-troh-PEE-nee-ah) is the condition of having insufficient levels of neutrophils in the blood. Predict the effect this condition will have on the body's ability to fight certain pathogens.
☐ 2. A patient presents with symptoms of a heart attack, and it is noted that she also has a fever. Her husband suggests that she must have an infection. What would you tell him?
☐ 3. You are working in a nursing home, and Mrs. Jackson develops a fever. You prepare to call her doctor about the fever. Your co-worker suggests that it's not necessary to call the doctor, and that you should just administer aspirin to Mrs. Jackson to make her fever go away. What do you tell your co-worker?

See answers in Appendix A.

MODULE **20.4**

Adaptive Immunity: Cell-Mediated Immunity

Learning Outcomes

1. Explain the differences between antigens, haptens, antigenic determinants, and self antigens.
2. Describe the processes of T lymphocyte activation, differentiation, and proliferation, including the roles of antigen-presenting cells.
3. Compare and contrast the classes of T lymphocytes.
4. Describe the two types of major histocompatibility complex antigens, and explain their functions.
5. Describe the purpose of immunological memory, and explain how it develops.
6. Describe the process by which a transplanted organ or tissue is rejected, and explain how this may be prevented.

The first arm of the adaptive immune system is cell-mediated immunity. As we discussed, cell-mediated immunity involves the different classes of T cells, including the **helper T (T$_H$) cells,** also known as **CD4 cells,** and the **cytotoxic T (T$_C$) cells,** also known as **CD8 cells.** ("CD" stands for "cluster of differentiation," which refers to specific molecules within the plasma membrane that differentiate one cell type from another.) These cells respond primarily to cells infected with intracellular pathogens (viruses and intracellular bacteria), cancer cells, and foreign cells such as those from a transplanted organ.

In this module, we trace the cell-mediated response, starting with T cell activation and finishing with the effects of T$_H$ and T$_C$ cells. As you read, note not only the differences between the different types of immunity but also the extensive interactions between the cell-mediated, antibody-mediated, and innate immune responses. The module concludes with an examination of how cell-mediated immunity is involved in rejection of transplanted tissues and organs, and what can be done to prevent this rejection.

T Cell Response to Antigen Exposure

◀◀ Flashback

1. What is the function of a lysosome? (p. 92)
2. What are the functions of the rough endoplasmic reticulum? (p. 90)

T cells are formed in the bone marrow, but they leave the bone marrow and migrate to the thymus to mature (**Figure 20.13**). As T cells mature, they undergo gene rearrangements that lead to a huge variety of genetically distinct T cells. Each population of T cells that can respond to a specific antigen is known as a **clone.** There are millions of different clones in the immune system, but only a few cells of each clone exist in the body at any given time.

Some T cell clones are capable of recognizing and responding to pathogens, whereas others are not. The thymus "screens" these cells

and mediates the destruction of those clones that cannot recognize antigens. This ensures that an individual is **immunocompetent,** or able to mount a normal response to foreign antigens. Other T cell clones, known as *self-reactive T cells,* recognize your own cells as foreign and would attack your cells if released into the circulation. These self-reactive T cells are also destroyed, ensuring **self tolerance** and preventing our T cells from attacking our own cells. Over 95% of all T cells die in the thymus in this manner. T cells that do survive this screening are released into the circulation when they mature. These cells, which are known as *naïve* T cells because they have not yet encountered their specific antigens, eventually reside in the blood, lymphoid organs, or other lymphatic tissue.

Only certain antigens are capable of generating a response from the immune system. Those with this capability are called *immunogens* (im-MYOO-noh-jenz; "immune generating"). Antigens present on your own cells, called **self antigens,** are not immunogens in your body. Some very small antigens, called *haptens,* are immunogenic only if they are attached to a protein carrier. An important example of a hapten is urushiol (oo-ROO-she-ohl), the toxin in poison ivy. When urushiol enters the skin, it is oxidized and the product of this reaction binds to skin proteins, making it immunogenic.

Every T cell has a receptor on its surface, called (appropriately enough) the **T cell receptor.** This receptor must bind a specific antigen before the cell can become activated. The unique portion of the antigen to which the receptor binds is known as its *antigenic determinant.* In this section, we see how T cells are activated and discuss the molecules required for this activation: major histocompatibility complex molecules.

MHC Molecules and Antigen Presentation

T cells cannot directly interact with an antigen. Instead, they can interact only with pieces of antigen bound to glycoproteins, called **major histocompatibility complex (MHC) molecules** (hiss'-toh-kom-pat-ih-BIL-ih-tee). The name comes from the

① Cells of the lymphoid line divide in the bone marrow.

② Immature T cells exit the bone marrow and migrate to the thymus.

③ The thymus stimulates T cell maturation and destruction of self-reactive T cells.

④ Mature naïve T cells migrate to lymphoid organs.

Figure 20.13 T cell maturation.

fact that MHC molecules are major determinants of compatibility among tissue and organ donors and recipients (remember that *histo-* means "tissue"). MHC molecules are found on nearly all nucleated cells. Note, however, that they are not present on the surface of erythrocytes, which is why blood is considerably easier to donate than other organs and tissues.

MHC molecules in essence serve as "docking sites" for specific components of antigens that are then displayed to T cells. There are two types of MHC molecules: **Class I MHC molecules** are found on the surface of the plasma membrane on nearly all nucleated cells, whereas **class II MHC molecules** are found only on the surfaces of antigen-presenting cells. Cytotoxic T (T$_C$) cells generally interact only with class I MHC molecules, whereas helper T (T$_H$) cells generally interact with class II MHC molecules. The two classes of molecules also differ in the types of antigens they display. Class I MHC molecules present *endogenous antigens,* those synthesized inside the cell, whereas class II MHC molecules present *exogenous antigens,* those the cell takes in by phagocytosis.

It's easy to think of exogenous as meaning "foreign" and endogenous as meaning "self," but that would be incorrect. The names actually relate to the way that antigens get into the cell. Exogenous antigens originate outside the cell and must be taken into the cell by phagocytosis. Endogenous antigens, however, are those that originate inside the cell. An endogenous antigen is one of two things: (1) a foreign antigen present on a pathogen that lives inside your cell, such as an intracellular bacterium, or (2) a foreign or self antigen encoded by your DNA. The second scenario includes normal self antigens, foreign cancer antigens made by mutated DNA, and foreign viral antigens (when a virus infects your cell, it inserts its genetic material into your DNA and triggers your cell to synthesize millions of copies of its proteins, including its antigens). For this reason, endogenous antigens can be foreign or self, but exogenous antigens are nearly always foreign.

ConceptBOOST »»»»

Why Do We Need Both Class I and Class II MHC Molecules?

At this point you may be wondering why we need two types of MHC molecules. The answer lies in the differing functions of T$_H$ and T$_C$ cells. As you've discovered, T$_H$ cells are activated by antigen-presenting cells (APCs) bearing portions of antigen on their class II MHC molecules, as shown here:

T$_H$ cell
T$_H$ cell receptor
Exogenous antigen
Class II MHC molecule
Antigen-presenting cell (APC)

The APCs aren't diseased themselves; rather, they function as a warning system that alerts the T$_H$ cells to a problem somewhere in the body. The T$_H$ cells then stimulate other parts of the innate and adaptive defenses to combat the threat.

Compare this to what happens with a T$_C$ cell. A T$_C$ cell is activated by a *diseased* cell bearing an antigen on a class I MHC molecule, as shown here:

T$_C$ cell
T$_C$ cell receptor
Endogenous antigen
Class I MHC molecule
Diseased body cell

When this occurs, the activated T$_C$ cell lyses and so kills the diseased cell, which is its target. If T$_C$ cells were also able to bind and recognize class II MHC molecules, they would end up killing APCs, not diseased cells, which would be detrimental to immune function and the whole body.

The basic steps by which a class I MHC molecule processes and displays an endogenous antigen are as follows (**Figure 20.14a**):

1. The cell synthesizes either a self antigen (the possibility that is illustrated) or a foreign antigen. Note that some foreign endogenous antigens, such as those from intracellular bacteria, are not synthesized by the cell.
2. The antigen is broken down by enzymes in the cytosol.
3. An antigen fragment containing the antigenic determinant is transported into the rough endoplasmic reticulum (RER) and is coupled with a class I MHC molecule in the RER membrane.
4. The MHC-antigen complex leaves the RER by a vesicle and is inserted into the cell's plasma membrane.

Now look at **Figure 20.14b** to contrast how exogenous antigens are displayed by class II MHC molecules. The basic steps are as follows:

1. The cell ingests a pathogen by phagocytosis.
2. The phagocytic vesicle fuses with a lysosome; the pathogen is degraded and its antigens are fragmented.
3. The lysosome fuses with a vesicle from the RER that contains class II MHC molecules, and an antigen fragment binds to the MHC molecule.
4. The MHC-antigen complex is inserted into the cell's plasma membrane.

Note that both processes end up with the same result: portions of antigens displayed on the plasma membrane attached to MHC molecules. These MHC-antigen complexes then interact with and activate T cells.

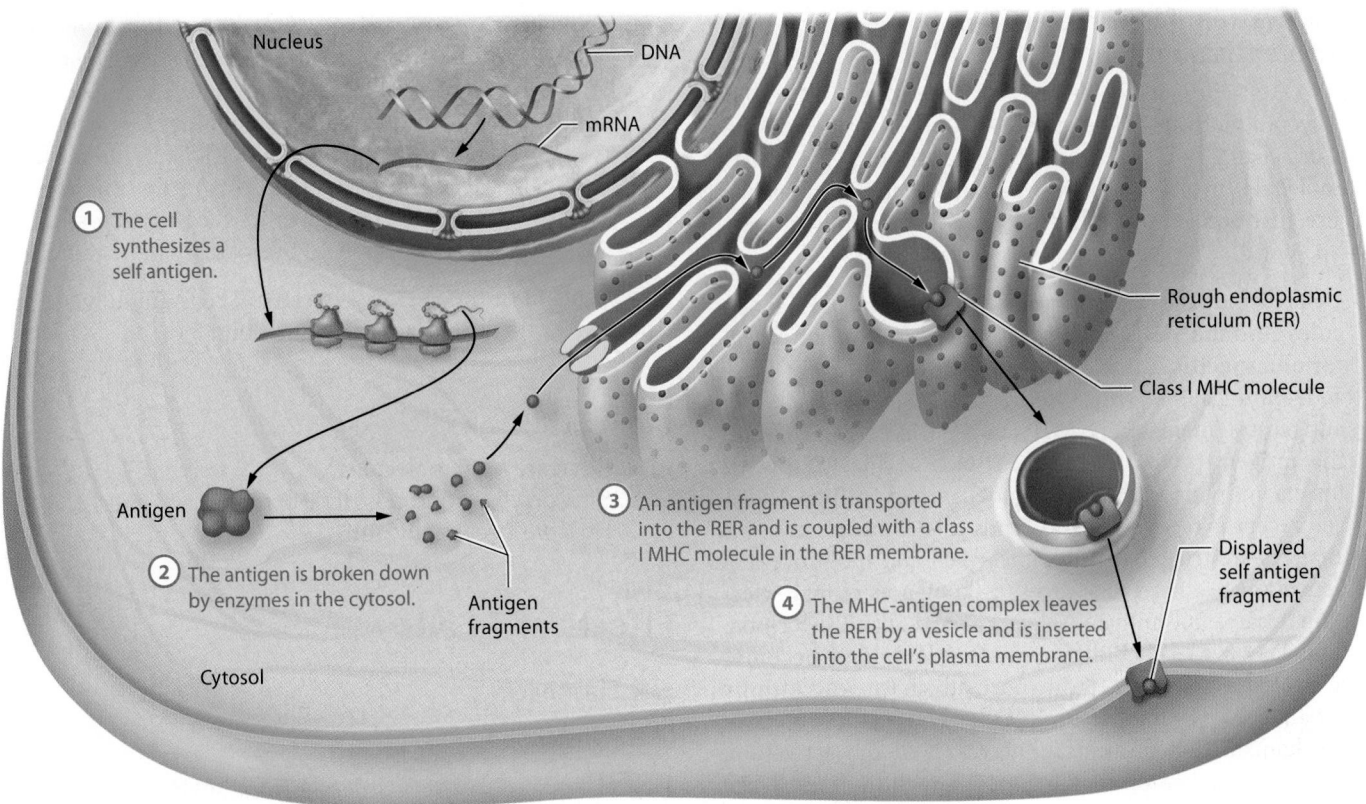

(a) Class I MHC molecules process and display endogenous antigens.

(b) Class II MHC molecules process and display exogenous antigens.

Figure 20.14 Antigen processing and display by MHC molecules.

T Cell Activation, Clonal Selection, and Differentiation

Now let's take a look at how T cells interact with MHC molecules to become activated. The process of T cell activation is very similar for T_H and T_C cells, the main difference being the type of MHC molecule involved. For this reason, we present a generic view here that applies to both cell types. The overall process, shown in **Figure 20.15**, consists of the following steps:

1. **Cells display the antigen fragments on their MHC molecules, and the MHC-antigen complex binds to the receptor of a specific T_H or T_C cell clone.** The process of T cell activation begins with a cell processing and displaying antigen fragments on its MHC molecules. It then presents the antigen to a particular T cell clone, which is specific for its individual MHC-antigen complex. When the T cell receptor recognizes and binds this complex, multiple changes are triggered inside the T cell and the process of activation begins. This process is known as **clonal selection** because the antigen "selects" a particular T cell clone.
2. **The T_H or T_C cell binds a co-stimulator and becomes activated.** T cell receptors normally have low affinity for their MHC-antigen complexes, which is a protective mechanism that prevents unnecessary T cell activation. Full T cell activation requires the interaction of the T cell with other molecules on the cells called *co-stimulators*.
3. **The activated T_H or T_C cell clone proliferates and differentiates into effector cells and memory T cells.** T_H or T_C

cells increase in number when activated and differentiate into *effector cells,* or those that cause the immediate effects, and *memory T cells.* **Memory T cells** are responsible for cell-mediated **immunological memory,** in which these cells respond more quickly and efficiently to subsequent exposures to an antigen. One reason for this efficiency is that memory T cells have no need of a co-stimulator, which makes the response considerably more rapid.

At the end of this process, effector T cells are fully activated. The next section examines their functions.

Quick Check

☐ 1. What are MHC molecules?
☐ 2. How do class I and class II MHC molecules differ?
☐ 3. How are T cells activated?

Effects of T Cells

◀◀ **Flashback**

1. What is apoptosis? (p. 113)

T_H and T_C cells have very different roles, although they do interact and depend on one another to function properly. We now examine how they interact with each other and with elements of antibody-mediated adaptive immunity.

Lymph node

Co-stimulator

MHC molecule

T cell receptor

T cell

Dividing cell

Effector T cells

Memory T cells

1. Cells display the antigen fragments on their MHC molecules, and the MHC-antigen complex binds to the receptor of a specific T_H or T_C cell clone.

2. The T_H or T_C cell binds a co-stimulator and becomes activated.

3. The activated T_H or T_C cell clone proliferates and differentiates into effector cells and memory T cells.

Figure 20.15 T cell activation, clonal selection, and differentiation. Note that T_H and T_C cells interact with different MHC molecules.

Activated helper (T_H) cell

Directly stimulates B cell proliferation

Secretes cytokines such as

| Interleukin-3 | Interleukin-2 | Various interleukins |

Stimulation of macrophages, enhancing phagocytosis

Triggering macrophages to produce interleukin–12, which stimulates T_H cells

Interleukin-12

Macrophage

T_H cell

Activation and proliferation of cytotoxic (T_C) cells

T_C cell

Stimulation of B cell proliferation and antibody production

Antibodies

Innate immunity

Adaptive cell-mediated immunity

Adaptive antibody-mediated immunity

Figure 20.16 Effects of T_H cells.

Role of T_H Cells

Unlike most other immune cells, helper T cells have no phagocytic or cytotoxic abilities. Instead, T_H cells primarily exert their effects through the secretion of cytokines that then activate and enhance various components of the immune response. In this way, they "help" the immune response, hence their common name. Some of the main functions of T_H cells include the following, illustrated in **Figure 20.16**:

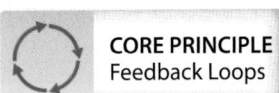

- **Innate immunity: stimulation of macrophages.** T_H cells secrete the cytokine *interleukin-3* which stimulates macrophages to become more efficient phagocytes. It also causes macrophages to produce an interleukin-12, which stimulates T_H cells to generate more interleukin-3 in a positive example of the Feedback Loops Core Principle (p. 21).
- **Adaptive cell-mediated immunity: activation of T_C cells.** T_H cells secrete the cytokine *interleukin-2* (IL-2), which is required to activate T_C cells. Indeed, in the absence of T_H cells and IL-2, most T_C cells fail to activate and become unresponsive to the antigen. IL-2 also stimulates the proliferation of T_C cells.

- **Adaptive antibody-mediated immunity: stimulation of B cells.** T_H cells directly bind to B cells and stimulate them to proliferate and differentiate. They also secrete various interleukins that stimulate B cell proliferation and increase antibody production.

T_H cells are required for normal function of all components of the immune system, including innate, antibody-mediated, and cell-mediated immunity. For this reason, failure of T_H cells to function, which we see with conditions such as acquired immunodeficiency syndrome (AIDS), can lead to failure of the entire immune response. We discuss AIDS and the virus that causes it in Module 20.7.

Role of T_C Cells

The primary function of cytotoxic T cells is evident from their name—they kill other cells, specifically those with foreign antigens bound to class I MHC molecules. Their ability to interact with class I MHC molecules means they can detect abnormalities in any cell type with a nucleus, which is critical for the detection of cancer cells, foreign cells, and cells infected with intracellular pathogens such as viruses and bacteria. T_C cells are activated in the same way as T_H cells, with the addition that they require IL-2

① The T$_C$ cell binds its target cell.

② The T$_C$ cell releases perforin, which forms pores in the target cell's plasma membrane.

③ The T$_C$ cell then releases enzymes that can now enter the target cell and fragment its DNA.

④ When the target cell begins to degrade, the T$_C$ cell detaches and searches for a new target cell.

Figure 20.17 Function of T$_C$ cells.

from T$_H$ cells in order to activate fully. This protective mechanism prevents abnormal T$_C$ cell activation.

As you can see in **Figure 20.17,** ① an activated T$_C$ cell binds its target cell, after which ② it releases a protein called **perforin.** Perforin forms pores in, or perforates, the target cell's plasma membrane. ③ The T$_C$ cell then releases enzymes that can now enter the target cell's cytosol. These enzymes catalyze reactions that degrade target cell proteins and eventually lead to fragmentation of the target cell's DNA and its death. T$_C$ cells also bind to proteins on the plasma membrane of target cells that induce the process of *apoptosis,* or programmed cell death (see Chapter 3). ④ When the target cell begins to degrade, the T$_C$ cell detaches and searches for a new target cell.

Although T$_C$ cells perform critical functions in terms of maintaining an organism's health, they can also cause problems. Certain types of allergies and the rejection of transplanted organs are mediated by T$_C$ cells. We discuss allergies in Module 20.7, and organ transplantation next.

Quick Check

☐ 4. How do T$_H$ cells differ from other immune cells? What are their main functions?
☐ 5. What are the main functions of T$_C$ cells?

Organ and Tissue Transplantation and Rejection

The ability to transplant organs and tissues between nonidentical individuals was one of the great medical advances of the 20th century. Heart, lung, liver, kidney, pancreas, and bone marrow transplants save thousands of lives every year, with ever-increasing success rates.

The four basic kinds of tissue and organ transplants, also known as *grafts,* are as follows:

• *Autografts* involve tissue transplanted from one site to another in the same individual. Examples of autografts include skin grafts, in which skin is removed from one part of the body and placed on another part to repair skin damaged by trauma such as burns.
• *Isografts* involve organs and tissues transplanted between two genetically identical individuals. Because of the small number of identical twins, these grafts are relatively uncommon.
• *Allografts,* which are the most common type of grafts, involve organs and tissues transplanted between two nonidentical individuals of the same species.
• *Xenografts* are relatively rare procedures that involve organs and tissues transplanted between two individuals of different species, such as a pig and a human.

Autografts and isografts result in no response from T$_C$ cells because the antigens bound to class I MHC molecules are not recognized as foreign. However, both allografts and xenografts do contain antigens that the organ recipient's immune system recognizes as foreign. This leads to a reaction from the immune system that, if left untreated, results in *rejection* of the organ or tissue. A rejected organ or tissue first fails to function properly, and then its cells die as T$_C$ cells destroy them, a condition known as *necrosis.* Rejection can lead to the death of the transplant recipient from organ failure and from blood clots and other complications of necrotic tissue.

Clearly, graft rejection is something that must be prevented, and this is done in two ways. The first way involves ensuring that allograft antigens are as similar as possible to the recipient's antigens. This is determined through a process that screens the antigens most likely to cause rejection—those associated with MHC molecules, which are encoded on genes called *human leukocyte antigen (HLA)* genes. A donor individual who has a high percentage of HLA similarity to the recipient is said to be a *match.* Exact HLA matches are rare and most likely to be found within families. For example, siblings have a 25% chance of being exact HLA matches. Fortunately, exact matches aren't necessary, and grafts can do well even with one or two HLA mismatches out of four.

The other means of preventing graft rejection is by suppressing the immune response with medications, which is known as *immunosuppressive therapy.* Of course, immunosuppressive therapy is not without serious risks, and dramatically increases a patient's risk of infection, even with agents that are normally nonpathogenic.

One of the most commonly performed transplant procedures is a bone marrow transplant, also referred to as a *hematopoietic stem cell transplant,* or *HSCT.* The procedure removes healthy hematopoietic stem cells from the blood or bone marrow of a donor and infuses them into a recipient. Typically, recipients are treated with immunosuppressive drugs and radiation to kill the diseased bone marrow prior to transplantation; this also minimizes the risk of rejection. HSCTs are used to treat a variety of conditions, including cancers such as leukemia and lymphoma, bone marrow failure, and disorders in which the immune system is deficient.

Quick Check

☐ 6. Which cells are involved in transplant rejection?
☐ 7. How are donor grafts matched to a recipient?

Apply What You Learned

☐ 1. A researcher is investigating a drug to prevent organ rejection; this drug selectively inhibits T_H cells. How would this help to prevent rejection of a transplanted organ?
☐ 2. What other effects would the drug from question 1 have on the immune system? Explain.
☐ 3. A hypothetical poison acts on the cells of the kidney and induces a change in the genes for the self antigens displayed on their class I MHC molecules. Predict what might happen if this poison were administered to someone.

See answers in Appendix A.

MODULE **20.5**

Adaptive Immunity: Antibody-Mediated Immunity

Learning Outcomes

1. Describe the process of B cell activation and proliferation.
2. Describe the five major classes of antibodies, and explain their structure and functions.

3. Compare and contrast the primary and secondary immune responses.
4. Explain how vaccinations induce immunity.
5. Compare and contrast active immunity and passive immunity.

We now turn our attention to the second arm of adaptive immunity: antibody-mediated immunity, also called *humoral immunity.* (Humoral immunity refers to the fact that its components were first identified in the body's fluids, which were known historically as *humors.*) As we discussed earlier, antibody-mediated immunity involves B cells and proteins secreted by B cells, called **antibodies.** Just as T cells have T cell receptors, B cells have **B cell receptors** that bind to specific antigens, and a group of B cells that bind to a specific antigen is known as a clone. The antibodies secreted by a B cell clone bind to the same antigen as the B cell receptor. In fact, we'll see that the structure of a B cell receptor is almost identical to that of an antibody.

The antibody-mediated immune response has three basic phases. The first phase involves a B cell clone recognizing its specific antigen, which triggers it to undergo changes and start secreting antibodies. The second phase begins when the antibody level in the blood rises dramatically. Antibodies, also known as **immunoglobulins** or **gamma globulins**, are the "action" component of antibody-mediated immunity, in that they are directly responsible for its actions or effects in mediating the destruction of antigens to which they bind. The third and final phase of antibody-mediated immunity is the persistence of a population of B cells called *memory B cells* that react much more rapidly and efficiently if the antigen is encountered again.

This module examines the three phases of the antibody-mediated response. It also considers how we can take advantage of the third phase of the response, the persistence of memory cells, to *vaccinate,* or *immunize,* ourselves against a variety of pathogens.

Phase 1: B Cell Activation, Clonal Selection, and Differentiation

B cells develop in the bone marrow from the lymphoid cell line, where billions of B cells are produced each day (**Figure 20.18**). They remain in the bone marrow to mature, but only about 10% of these cells finish their maturation process. This is largely because B cell clones that recognize self antigens (self-reactive B cells) are destroyed. This prevents the development of *autoimmunity,* in which B cells recognize self antigens as foreign and produce *autoantibodies* that bind self antigens. (We discuss autoimmune disorders in Module 20.7.)

Naïve B cell clones that complete maturation enter the circulation and eventually take up residence in lymphoid organs such as the spleen and lymph nodes. When antigens enter the body, they are captured in these lymphoid organs. However, if the B cells are not exposed to their specific antigens within a few days to a few weeks, the cells die.

Bone marrow

Maturing B cell

Bone marrow sinusoid

Lymph node

Cells undergoing mitosis

Destroyed self-reactive B cells

Mature B cell

(1) Cells of the lymphoid line divide in the bone marrow.

(2) B cells mature in the bone marrow, where self-reactive B cells are destroyed.

(3) Mature naïve B cells exit the bone marrow and take up residence in lymphoid organs.

Figure 20.18 B cell maturation.

Naïve B cells that do encounter their antigens become activated by the following process, illustrated in **Figure 20.19**:

(1) **The B cell clone binds its antigen and is activated.** The antigen binds to a B cell receptor on the surface of a specific B cell clone. As with T cell activation, this process is called clonal selection. A B cell clone bound to an antigen is said to be *sensitized*.

(2) **The sensitized B cell processes the antigen and presents it on its class II MHC molecules. It then binds to a T_H cell, which secretes cytokines that activate the B cell.** The B cell processes the antigen and displays it on its class II MHC molecules. A T_H cell then binds the class II MHC molecules on the sensitized B cell. This binding starts a series of events inside the cell that triggers transcription of antibody genes and multiple other changes that activate the B cell. The two cells

CORE PRINCIPLE
Cell-Cell Communication

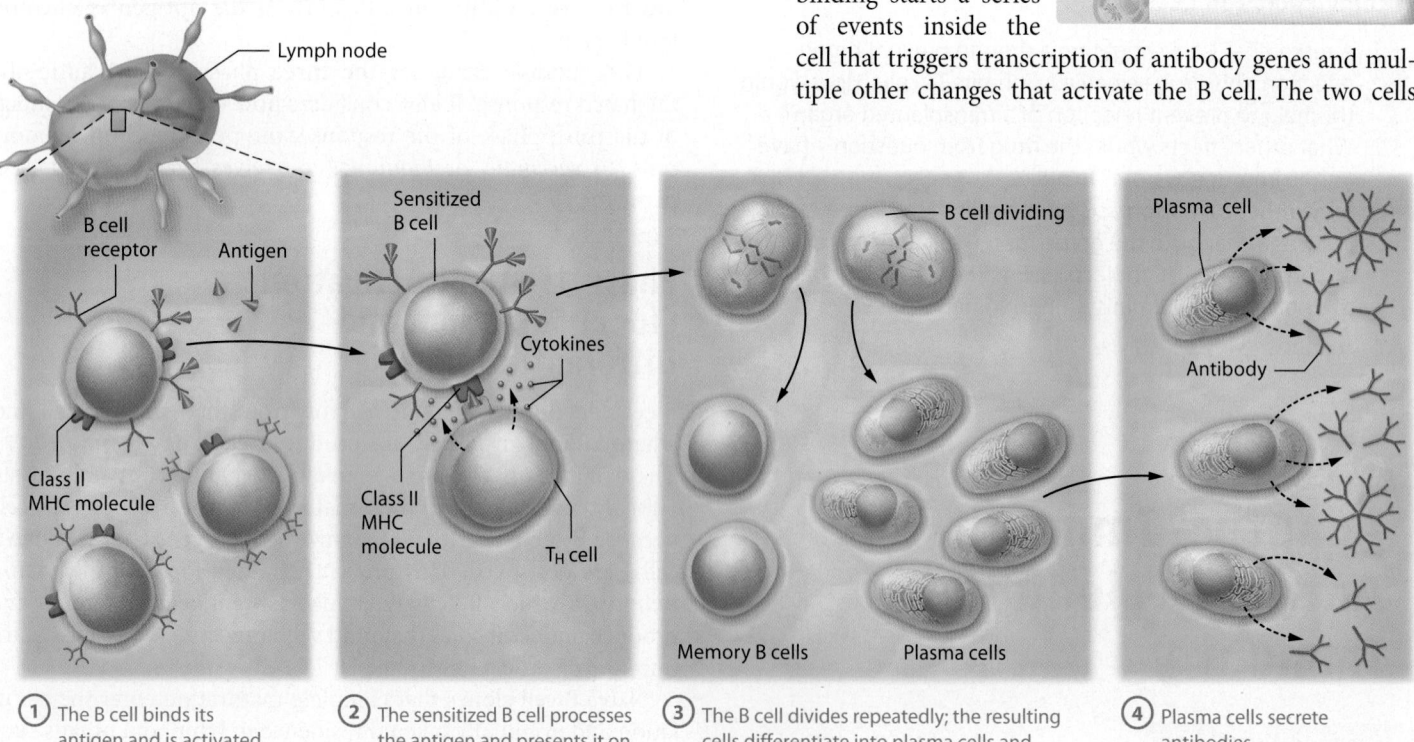

Lymph node

B cell receptor

Antigen

Class II MHC molecule

Sensitized B cell

Cytokines

Class II MHC molecule

T_H cell

B cell dividing

Memory B cells

Plasma cells

Plasma cell

Antibody

(1) The B cell binds its antigen and is activated.

(2) The sensitized B cell processes the antigen and presents it on its class II MHC molecules. It then binds to a T_H cell, which secretes cytokines that activate the B cell.

(3) The B cell divides repeatedly; the resulting cells differentiate into plasma cells and memory B cells.

(4) Plasma cells secrete antibodies.

Figure 20.19 B cell activation, clonal selection, and differentiation.

interacting and influencing one another is an example of the Cell-Cell Communication Core Principle (p. 27).

③ **The B cell divides repeatedly; the resulting cells differentiate into plasma cells and memory B cells.** Among the changes triggered in the B cells when they are activated is rapid cell division, which results in an explosion in the population of these cells. In addition to rapid proliferation, the resulting B cells differentiate into two populations of cells: (1) **plasma cells,** which secrete antibodies; and (2) **memory B cells,** which are long-lived cells that do not secrete antibodies but will respond to antigens upon a second exposure. Dendritic cells and other antigen-presenting cells continue to expose the plasma cells and other B cells to the antigen, and in response they keep proliferating and differentiating.

④ **Plasma cells secrete antibodies.** Plasma cells continue to divide and increase their numbers as they actively secrete antibodies. Note in Figure 20.19 that the rough endoplasmic reticulum is prominent in plasma cells, which reflects their active protein synthesis.

Quick Check

☐ 1. How is a B cell activated?
☐ 2. What happens after a B cell is activated?

Phase 2: Antibodies and Their Effects

We now move to the second phase of the antibody-mediated immune response: the antibodies and their effects. The study of antibodies, also known as *serology* (see-ROL-uh-jee), examines the structure and functions of antibodies as well as their applications to medicine and research. The name serology derives from the antibody-containing fluid called *serum,* which is the part of blood that is left over when you remove the cells, platelets, and clotting proteins from the rest of the plasma (note that a serum in which significant quantities of antibodies are found is known as an *antiserum*). Let's now survey serology and examine antibodies' basic structure and functions.

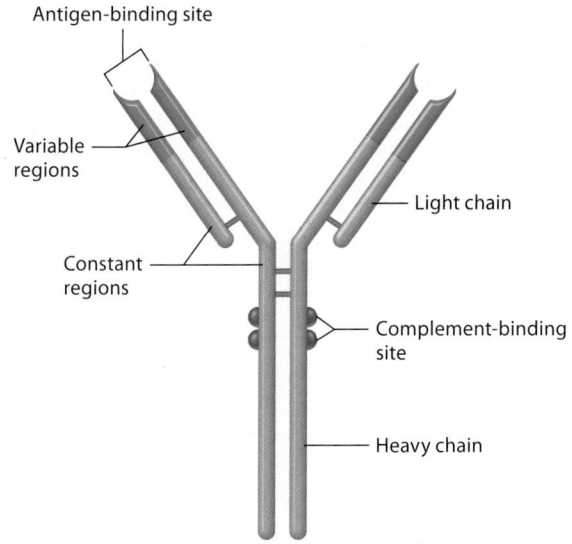

Figure 20.20 The basic structure of an antibody monomer.

Antibody Structure and Classes

The basic subunit of an antibody, shown in **Figure 20.20**, is a Y-shaped molecule formed from four peptide chains, two *heavy* (*H*) and two *light* (*L*) *chains.* Each of these chains has two types of regions: (1) **constant (C) regions,** which are relatively similar among antibodies and are responsible for many of the antibodies' effects; and (2) **variable (V) regions,** which are unique sequences of amino acids responsible for antigen recognition and binding. An antibody has V regions at the tips of the two arms of the molecule, so the basic subunit of an antibody has two **antigen-binding sites,** one on each arm.

As you can see here, the single antibody subunit, called a *monomer,* can be combined with other subunits to form a larger structure such as a *dimer,* which consists of two subunits, or a *pentamer,* which consists of five subunits:

Dimer

Pentamer

There are five basic classes of antibody, which are grouped according to the structure of their C regions. Each antibody is named with the two-letter abbreviation "Ig," which stands for "immunoglobulin," followed by a letter that designates its class. This gives us the types IgG, IgA, IgM, IgE, and IgD, which you can remember with the mnemonic GAMED. Let's take a closer look at these classes:

● **IgG.** The most prevalent antibody in the body is **IgG,** which accounts for about 75–80% of the antibodies in serum. It consists of a single subunit. IgG is the only antibody able to cross from the blood of a pregnant woman to her developing fetus through a structure called the *placenta.*

● **IgA. IgA** is usually a dimer, consisting of two Y-shaped subunits, which gives this antibody four antigen-binding sites. IgA is present in secretions from the skin, mucous membranes, and exocrine glands (i.e., tears, saliva, sweat, and breast milk).

● **IgM.** The largest antibody is the pentamer **IgM** (think "M" for "massive"), which consists of five subunits in a starlike arrangement, for a total of 10 antigen-binding sites. IgM is generally the first antibody secreted by plasma cells when the body is invaded by a pathogen. Note that IgM also exists as a single subunit embedded in the B cell plasma membrane, where it functions as a B cell receptor.

● **IgE.** The single-subunit antibody **IgE** is generally present in very low amounts in the body's fluids. IgE binds to two types of antigen: (1) antigens associated with parasitic pathogens such as tapeworms, and (2) environmental antigens known as *allergens,* which are linked with inflammatory reactions called *allergies.* IgE molecules bind to mast cells in mucous membranes, and when they come into contact with their specific antigens, they trigger mast cells to release the contents of their granules, a process called *degranulation.* Mast cell granules contain inflammatory mediators such as histamine that initiate a localized inflammatory response. The ensuing inflammation is responsible for common allergy symptoms such as a runny nose and watery eyes. We discuss allergies in more detail in Module 20.7.

- **IgD.** The antibody **IgD** is unique because it is the only antibody not secreted by B cells in significant amounts. Rather, its single subunit is located on the surface of B cells, where it acts as an antigen receptor that helps activate B cells in a similar manner to IgM.

Note that although all antibodies in a particular class feature similar C regions, each antibody has a unique V region that determines its ability to recognize and bind a certain antigen. There is remarkable variability in antibodies' V regions, and there are an estimated 1 million or more different antibodies in the body. This means that your immune system is capable of recognizing at least 1 million unique antigens, which is certainly a feat. This is made possible by a genetic mechanism found only in lymphocytes whereby the genes that code for antibodies undergo random recombination, producing a huge number of potential genes and antibody products.

Functions of Secreted Antibodies

The actions of antibodies are based on their ability to bind antigens, which leads to multiple effects on pathogens. Many of these effects represent an integration of adaptive and innate immunity—the antibodies of adaptive immunity often activate phagocytes and complement proteins of innate immunity. The basic effects of secreted antibodies are shown in **Figure 20.21** and include the following:

- **Agglutination and precipitation.** Antibodies can bind to antigens on more than one cell. This creates a clump of cells that are cross-linked by their attachment to antibodies. The clumping of cells is known as **agglutination.** Similar to agglutination is the process called **precipitation,** which involves soluble antigens (proteins and other biological molecules) instead of whole cells. Both agglutination and precipitation allow these antigen-antibody complexes to precipitate out of body fluids and therefore make the complexes easier for phagocytes to ingest. IgM is the most potent agglutinating and precipitating antibody as a result of its 10 antigen-binding sites.
- **Opsonization.** As you read about, opsonization involves molecules such as complement proteins coating the pathogen and activating phagocytes. IgG antibodies are also opsonins able to coat pathogens and bind and activate phagocytes, which greatly enhances phagocytosis.
- **Neutralization.** Bacterial toxins, viral proteins, and animal venoms are molecules with specific components that are harmful. Antibodies bind to these components, as well as certain viruses and bacteria, and prevent them from interacting with our cells. This renders the toxin inactive, which is known as **neutralization.** Most neutralizing antibodies are of either the IgG or the IgA class.

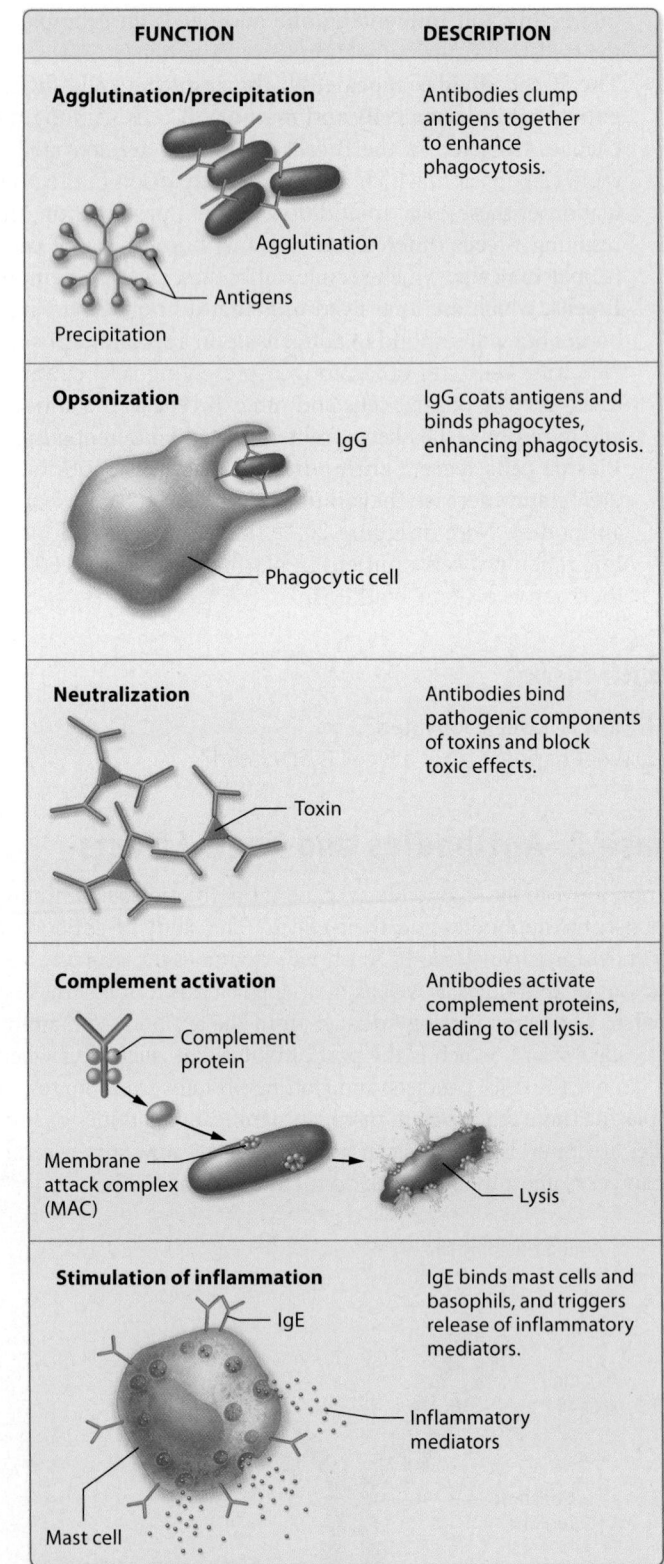

FUNCTION	DESCRIPTION
Agglutination/precipitation — Agglutination, Antigens, Precipitation	Antibodies clump antigens together to enhance phagocytosis.
Opsonization — IgG, Phagocytic cell	IgG coats antigens and binds phagocytes, enhancing phagocytosis.
Neutralization — Toxin	Antibodies bind pathogenic components of toxins and block toxic effects.
Complement activation — Complement protein, Membrane attack complex (MAC), Lysis	Antibodies activate complement proteins, leading to cell lysis.
Stimulation of inflammation — IgE, Inflammatory mediators, Mast cell	IgE binds mast cells and basophils, and triggers release of inflammatory mediators.

Practice art labeling
@MasteringA&P®

Figure 20.21 Functions of antibodies.

- **Complement activation.** Several antibodies, particularly IgM and IgG, bind and activate the complement proteins of innate immunity (see the classical pathway in Figure 20.10). When several antibodies bind a single cell, their complement-binding sites are exposed. This allows complement to activate and lyse the foreign cell with its membrane attack complex. This effect is particularly important in defense against cellular pathogens such as bacteria and is partly responsible for our reaction to foreign cells like donated erythrocytes.
- **Stimulation of inflammation.** The antibody IgE directly triggers inflammation by initiating the release of inflammatory mediators from mast cells and basophils. Antibodies also trigger inflammation indirectly through their activation of complement.

Keep in mind that most antibodies don't directly "kill" pathogens. Instead, they either render pathogens less harmful or facilitate their destruction by phagocytes or complement. Only rarely do antibodies cause lysis of pathogens on their own.

The structures and functions of antibodies are summarized in Figure 20.22.

Quick Check

☐ 3. What are the five classes of antibody, and how do they differ?
☐ 4. List the five main functions of antibodies.

Phase 3: Immunological Memory

As we discussed, activated B cells divide into two populations: plasma cells and memory B cells. The memory B cells are responsible for antibody-mediated immunological memory. Like memory T cells, the memory B cells respond more efficiently when the antigen is encountered a second time. In this section we explore how the immune response differs after initial exposure to an antigen and how we apply the principles of immunological memory to prevent disease with vaccinations.

The Primary and Secondary Immune Responses

The first time you are exposed to an antigen, the response proceeds the way that was described in an earlier section: A B cell specific for that antigen recognizes it, the activated B cell proliferates and differentiates into plasma and memory B cells, and plasma cells begin to secrete antibodies. This response is called the **primary immune response.** The primary immune response is effective, but it's slow—there is an initial 4–5 day *lag phase* as the B cells proliferate, differentiate into plasma cells and memory cells, and begin to secrete antibodies. Antibody levels peak about 7–14 days after the antigen is encountered. It is during the lag phase that you generally feel "sick."

The next time you are exposed to the same antigen, the response is much different. The memory B cells formed during the primary immune response are long-lived, persisting in the body from years to an entire lifetime. When these memory B cells

CLASS	FUNCTION
IgG	• Makes up the majority of antibodies in serum • The only antibody that can cross the placenta from mother to fetus • Functions in opsonization, neutralization, and complement fixation
IgA	• Found in secretions such as breast milk and saliva • Functions in agglutination and neutralization
IgM	• The first antibody secreted on exposure to an antigen • Potent agglutinating and precipitating agent • Functions in complement fixation
IgE	• Binds mast cells and basophils and triggers their degranulation, facilitating inflammation, particularly in the allergic response
IgD	• Antibody found exclusively on the surface of B cells • Has a role in B cell sensitization and activation

Figure 20.22 Antibody classes.

encounter the antigen for which they are specific, the **secondary immune response** begins. There are several key differences between the two responses (summarized in Table 20.1, on page 782). First, the secondary immune response has a shorter lag phase (about 1–3 days), and its antibody levels peak more rapidly (3–5 days) and reach a peak 100–1000 times larger. Also, the primary antibody involved in the secondary response is IgG, whereas it's IgM in the primary response. In addition, the secondary response lasts longer than the primary response. Finally, the antibodies secreted in the secondary immune response are more effective—they bind more tightly and have a higher affinity for their antigens. For this reason, you may not feel "sick" when exposed to the same antigen or may experience only mild symptoms. We take advantage of this secondary immune response with vaccinations, our next topic.

TABLE 20.1 COMPARISON OF THE PRIMARY AND SECONDARY IMMUNE RESPONSES

Characteristic	Primary Immune Response	Secondary Immune Response
Lag phase	4–5 days	1–3 days
Time until antibody peak	7–14 days	3–5 days
Primary antibody	IgM	IgG
Duration of response	14–21 days	28 days and beyond

Vaccinations

A **vaccination,** also known as an **immunization,** involves exposing an individual to an antigen to elicit a primary immune response and generate memory cells. Then if the individual is exposed to the antigen a second time, a secondary immune response will occur and symptoms will be minimal.

A vaccine may be any of several types, depending on its components, which can include the following:

- **Live, attenuated vaccines.** Certain pathogens can be *attenuated* (ah-TEN-yoo-ay'-tid), meaning that their ability to cause disease has been greatly reduced. When these pathogens are administered as a vaccine, they can still divide to a limited extent in the body and so induce a primary immune response. The ability of the pathogens to divide generates a strong primary immune response due to their extended presence in the body. Live, attenuated vaccines are completely safe for most individuals, although there is an extremely small risk that the organisms could revert to a pathogenic form. Therefore, these vaccines are often avoided in patients with compromised immune systems.
- **"Killed" vaccines.** Vaccines may also be prepared with inactivated, or "killed," pathogens. Killed pathogens are not capable of dividing in the body, so they generate a weaker primary immune response and generally require several repeat vaccinations, known as "boosters."
- **Subunit vaccines.** With certain pathogens, only the portion of the pathogen that causes disease is required to develop immunity. For example, the bacteria responsible for tetanus and diphtheria secrete disease-causing toxins. Vaccinations called *toxoids* contain inactivated toxins from these

bacteria, which induce the immune system to produce antibodies to the toxins. Other subunits that may be used in a vaccine are the antigenic polysaccharides found in the outer component of certain bacterial walls. Like killed vaccines, subunit vaccines tend to induce a weaker primary immune response and typically require boosters to maintain immunity.

One of the biggest challenges to effective vaccination is posed by pathogens that have a high rate of *mutation*—random changes in their genetic makeup. Mutations often change the pathogens' antigens, which alters the B cells and antibodies needed to respond to those antigens. So the initial exposure confers immunity only to the original strain, not the mutated strains. This has been a major issue in the development of effective vaccines for *human immunodeficiency virus* (*HIV*) and tuberculosis (a vaccine for this disease exists, but it is unreliable). The pathogens mutate so rapidly that a vaccine cannot reliably protect an individual from infection.

The influenza viruses constitute another group of pathogens with a high mutation rate. Their rapid mutation rate means that a new vaccine must be prepared every year. The vaccine typically contains the dominant viral strain from the previous year, the dominant strain from the current year, and another strain that could become prevalent. This method of preparing the annual "flu shot" isn't perfect—sometimes the viruses mutate into new strains, or an undetected strain becomes the dominant one. Still, the vaccine does provide significant protection from influenza most years, and health care providers encourage their patients to be vaccinated every year. It should be noted that one cannot "get the flu" from an influenza vaccination, as it contains killed virus (note, though, that the nasal FluMist vaccine contains an attenuated strain). If an individual gets sick shortly after receiving the vaccine, the illness is likely due to a different strain of influenza, an inadequate amount of time for the immune system to mount a primary response, or another virus entirely.

Vaccinations are undeniably among the most important medical discoveries of the past 200 years. Before the discovery of vaccinations, the only way to develop memory cells for a pathogen was to be exposed to it, generally via infection. Unfortunately, many did not survive this initial exposure or suffered serious health consequences as a result. Vaccines have eliminated the virus that causes smallpox, once a major threat to human survival, and have nearly eliminated the virus that causes the paralytic disease polio. The only places where polio persists are those countries that have stopped vaccination programs or those in which members of the population have no access to vaccinations. Vaccination is, unfortunately, a practice about which many misconceptions exist; for a discussion of one of the most common misconceptions, see *A&P in the Real World: The Myth of Vaccines and Autism* on page 784.

Active and Passive Antibody-Mediated Immunity

There are two types of antibody-mediated immunity, active and passive (**Figure 20.23**). The type of immunity we have been discussing is **active immunity,** so named because the body's cells actively respond to an antigen. Active immunity may be received naturally through exposure to an antigen via infection or artificially via a vaccination. It results in the production of memory cells and large numbers of antibodies and is, therefore, relatively long-lasting, ranging from years to a lifetime. The length of time during which active immunity lasts depends on several

factors, particularly the extent of exposure. For example, children exposed to the chickenpox virus through moderate infection tend to develop lifelong immunity. However, those children exposed to the virus via a very mild infection may develop immunity for only a few years and become infected again in later childhood or in adulthood. (A vaccination for the chickenpox virus is now available, so this scenario mostly applies to unvaccinated children and adults.)

Passive immunity is found when preformed antibodies are passed from one organism to another. Passive immunity may be naturally acquired, for example, when the IgG of a pregnant woman crosses from her blood into that of her fetus, or artificially acquired, from an injection with preformed antibodies. This form of immunity lasts only the amount of time that the antibodies stay in the bloodstream, which is about 3 months on average. Even though it is temporary, passive immunity is critical and life-saving in many instances. For example, passive immunity is the sole source of antibodies that an infant has at birth. In addition, preformed antibodies are administered when a person is exposed to animal venom (as an *antivenin*), certain toxin-producing bacteria (as an *antitoxin* or *toxoid*) to neutralize the toxins, and certain viruses such as hepatitis A.

Quick Check

☐ 5. How do the primary and secondary immune responses differ?

☐ 6. What is the purpose of a vaccination, and what are the three types of vaccines?

☐ 7. How do active immunity and passive immunity differ?

Figure 20.23 Active and passive antibody-mediated immunity.

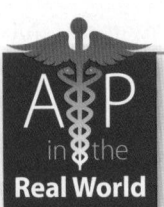

The Myth of Vaccines and Autism

If you do an Internet search for "vaccines and autism" you will receive hundreds of thousands of hits, many of which will claim that vaccines cause the developmental disorder autism. The concern about a possible link between vaccinations and autism arose when it was noticed that many children first started showing symptoms of autism around the time of their childhood vaccinations. The thought was that perhaps *thimerosal*, a mercury-containing preservative found in most vaccines, was the culprit. Certain forms of mercury, after all, are known to cause brain damage.

However, there was one major problem with this hypothesis: The form of mercury that causes brain damage is methylmercury, but thimerosal contains *ethylmercury*. Nonetheless, independent researchers across the globe have examined thimerosal, but no link between thimerosal and autism has ever been found. Indeed, thimerosal was removed from vaccines in Denmark in 1992, and the autism rate actually *increased* by a small amount after this was done. Although no evidence supports the idea that thimerosal is harmful, the United States removed thimerosal from most vaccines in 2001 as a result of public concerns.

Researchers have also explored other potential links between vaccinations and autism, but the results have been the same: No link can be established. Unfortunately, many people continue to staunchly advocate that vaccines are the root cause of autism, even with an absolute lack of evidence. This has generated concern among well-meaning parents and led to a decline in the number of individuals who receive vaccinations. The consequence of decreased vaccination rates is an increase in the number of cases of potentially lethal diseases, such as measles, whooping cough, and meningitis, which raises serious public health concerns.

Apply What You Learned

☐ 1. A patient has a hypothetical disorder that prevents the development of memory cells. Predict the effects this disorder would have on the primary and secondary immune responses.

☐ 2. The disorder *hypogammaglobulinemia* is characterized by a decrease in the secretion of one or more classes of antibody. Predict the effect of decreased secretion of:
   a. IgA.
   b. IgG.
   c. IgG in a pregnant female.
   d. IgM.

☐ 3. You are injured in a farming accident, and in the emergency room you are administered injections of both *tetanus antitoxin*, which contains antibodies to the tetanus toxin, and tetanus antigens. Which injection provided active immunity, and which passive? Explain.

See answers in Appendix A.

MODULE **20.6**
Putting It All Together: The Big Picture of the Immune Response

Learning Outcomes

1. Describe how the immune and lymphatic systems work together to respond to internal and external threats.
2. Explain how the immune response differs for different types of threats.
3. Describe the immune response to cancerous cells.
4. Explain how certain pathogens can evade the immune response.

The immune response involves so many different components working simultaneously that it's often difficult to see the big picture of how the body responds to a threat. This module helps you see the "forest for the trees," as we walk through the basic immune response in three different scenarios: infection with the common cold virus, infection with a strain of toxin-producing bacteria, and a cancerous tumor. Along the way, we discuss how the parts of the immune system work with the lymphatic system to form an integrated response that protects the body from a variety of threats. We conclude the module with a look at how certain pathogens are able to evade the immune response and cause serious homeostatic disturbances.

Scenario 1: The Common Cold

Let's start with one of the most frequent infectious conditions in the world: the *common cold*. The common cold is due to a variety of viruses that typically have a high mutation rate, which is why researchers have yet to develop a vaccine for the pathogens. Although it might seem that you frequently get sick with a cold, in truth most pathogens, viruses included, are deterred by the body's surface barriers and so never even gain entry into the body. When a virus does gain entry, the response proceeds by the following steps, which are illustrated in **Figure 20.24**:

① **Infected cells trigger an inflammatory response and secrete interferons.** The virus enters through the respiratory tract and infects local cells. This triggers the infected cells to secrete certain interferons and other inflammatory mediators that initiate the inflammatory response.

The Big Picture of the Immune Response to the Common Cold

Figure 20.24

Epithelial cell

Virus

Inflammatory mediators

Blood capillary

Interferon

Infected cell undergoing lysis

NK cell

Interferons

① Infected cells trigger an inflammatory response and secrete interferons (see Figure 20.11).

② Interferons prevent infection in neighboring cells and activate natural killer (NK) cells, which lyse infected cells.

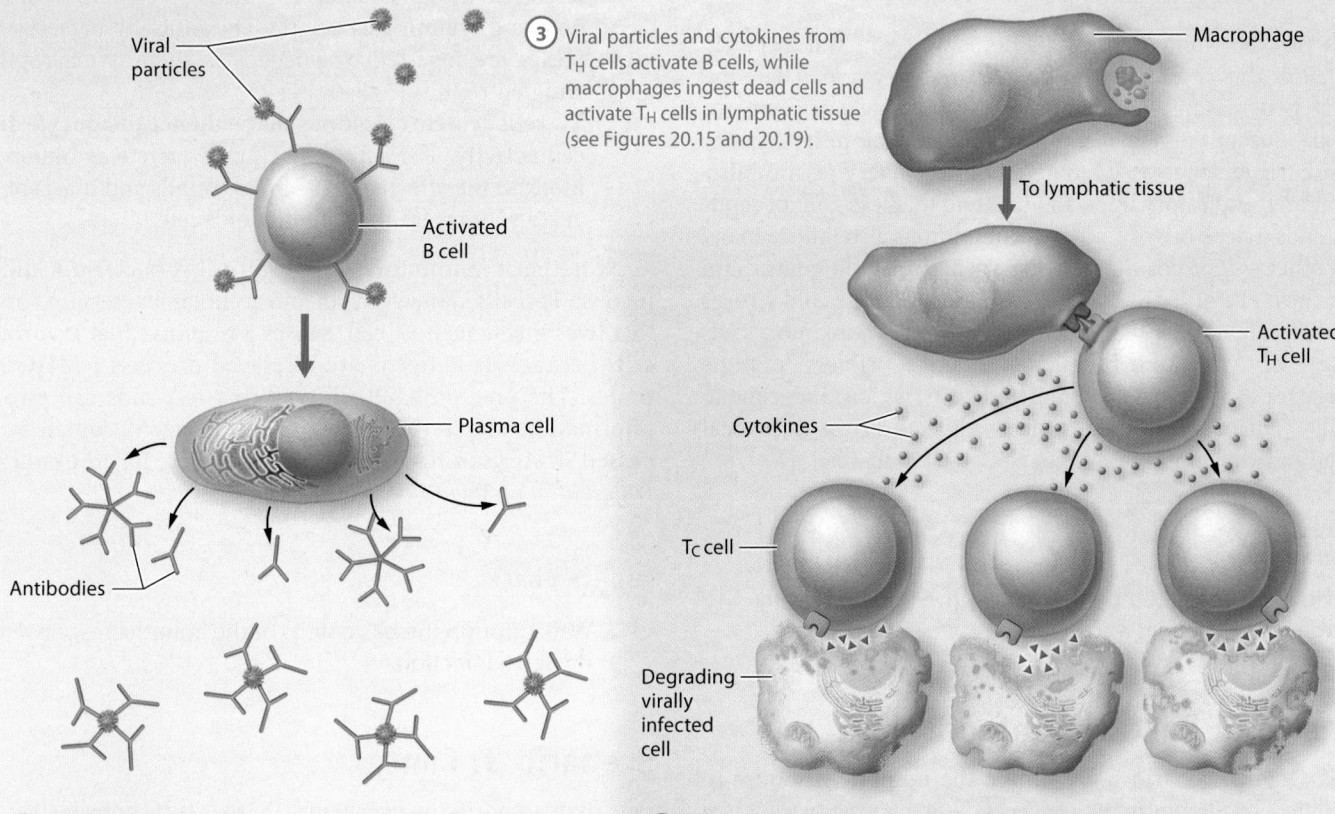

Viral particles

Activated B cell

Plasma cell

Antibodies

③ Viral particles and cytokines from T_H cells activate B cells, while macrophages ingest dead cells and activate T_H cells in lymphatic tissue (see Figures 20.15 and 20.19).

Macrophage

To lymphatic tissue

Activated T_H cell

Cytokines

T_C cell

Degrading virally infected cell

④ B cells differentiate into plasma cells, which secrete antibodies that bind and neutralize the virus (see Figure 20.21).

⑤ T_H cells secrete cytokines that amplify the immune response, and T_C cells destroy virally infected cells (see Figures 20.16 and 20.17).

② **Interferons prevent infection in neighboring cells and activate natural killer (NK) cells, which lyse infected cells.** The interferons secreted by the infected cells travel to nearby cells and prevent viral infection. Interferons also bind and activate NK cells, which begin destroying infected cells.

③ **Viral particles and cytokines from T_H cells activate B cells, while macrophages ingest dead cells and activate T_H cells in lymphatic tissue.** Local macrophages and those attracted by the inflammatory mediators ingest dead and dying cells. These macrophages and viral particles enter lymphatic tissue via the lymph. Here, the macrophages present antigens to T_H cells and activate them, and viral particles bind and activate B cells. The activated T_H cells further amplify the B cell response.

④ **B cells differentiate into plasma cells, which secrete antibodies that bind and neutralize the virus.** The activated B cells differentiate into plasma cells and memory B cells. The plasma cells secrete antibodies that bind to the viral particles and neutralize them, preventing them from interacting with the human host cells.

⑤ **T_H cells secrete cytokines that amplify the immune response, and T_C cells destroy virally infected cells.** The activated T_H cells secrete cytokines that activate all elements of the immune response, including innate and antibody-mediated immunity. They also activate T_C cells, which recognize and lyse infected cells.

You may be wondering at this point why you still feel sick, when all of this is going on in your immune system. There are actually two reasons. For one, many unpleasant symptoms of colds are due to your own immune system. The inflammatory response triggers increased blood flow and capillary permeability in the upper respiratory tract, and the fluid that leaks out of capillaries tends to leak out of your nose. Interferons also induce many of the other symptoms that you experience, including fever and body aches. The other reason you feel sick is that the cold viruses mutate rapidly, and new strains for which you lack memory cells emerge every year. Your body must mount a primary immune response for each new type of mutated virus, which takes time to complete. During this time, you feel sick as the virus does local damage and your immune system works to combat it.

Quick Check

☐ 1. How do innate immunity and adaptive immunity work together to defeat a cold virus?

Scenario 2: Bacterial Infection

In this second scenario, we examine the immune response to the bacterium *Streptococcus pyogenes*. This bacterium causes a variety of infections, the most common of which is *streptococcal pharyngitis* (faer-en-JY-tis; commonly called "strep throat"—recall that the *pharynx* is the throat). The process proceeds by the following steps, illustrated in **Figure 20.25:**

① **Bacteria damage cells and induce an inflammatory response, as local macrophages ingest the bacteria.** The bacteria damage cells in and around the pharynx, and the damaged cells release inflammatory mediators that induce an inflammatory response. This inflammation is responsible for one of the first signs of streptococcal pharyngitis: a sore throat. Local macrophages become activated and begin to phagocytize the bacteria.

② **Neutrophils enter the area in large numbers and phagocytize bacteria.** The inflammatory mediators attract circulating neutrophils to the area, which phagocytize bacteria. They also trigger the bone marrow to release its store of neutrophils and increase their production. In many cases, this is sufficient to contain and resolve the infection, and adaptive immunity is either only weakly activated or not activated at all.

③ **Macrophages and bacteria enter lymphatic tissue and activate B cells and T_H cells.** If the infection is not resolved by the neutrophils and macrophages alone, macrophages and bacteria eventually enter the lymph and migrate to local lymphoid organs such as the tonsils. Here they activate B and T_H cells.

④ **B cells differentiate into plasma cells that secrete antibodies to agglutinate bacteria.** The activated B cells differentiate into plasma cells, which secrete antibodies that bind and agglutinate bacteria. The antibody-bacteria complexes are ingested by phagocytes such as macrophages and neutrophils.

⑤ **T_H cells secrete cytokines that enhance phagocyte and B cell activity.** The activated T_H cells secrete cytokines that increase the effectiveness of neutrophils and macrophages and also increase the activities of B cells.

Note that the immune response to this bacterium did not involve T_C cells. However, an intracellular bacterium, or one that lives inside its host cell, causes a response that involves T_C cells because its antigens are displayed on class I MHC molecules. The predominant cell type that responds can give you information about the likely infectious agent, which is discussed in *A&P in the Real World: Complete Blood Count with Differential* on page 788.

Quick Check

☐ 2. Walk through the basic steps of the immune response to a bacterial infection.

Scenario 3: Cancer

Our final scenario involves a disease that is responsible for over 8 million deaths per year worldwide: *cancer*. Cancer cells are formerly normal body cells that have undergone mutations causing them to de-differentiate (become less specialized), lose control of their cell cycles, and lose their attachments to the surrounding

Bacteria
Inflammatory mediators
Macrophage
Protein-rich fluid

Bacteria
Neutrophil

(1) Bacteria damage cells and induce an inflammatory response, as local macrophages ingest the bacteria (see Figures 20.11 and 20.12).

(2) Neutrophils enter the area in large numbers and phagocytize bacteria (see Figure 20.12).

Bacteria
To tonsils
Macrophage
To tonsils
Activated B cell

(3) Macrophages and bacteria enter lymphatic tissue and activate B cells and T$_H$ cells (see Figures 20.15 and 20.19).

Activated T$_H$ cell
Cytokines
Antibody
Plasma cell
Antibody
Antibody
Plasma cell
Agglutinated bacteria
Macrophage

(4) B cells differentiate into plasma cells that secrete antibodies to agglutinate bacteria (see Figures 20.19 and 20.21).

(5) T$_H$ cells secrete cytokines that enhance phagocyte and B cell activity (see Figure 20.16).

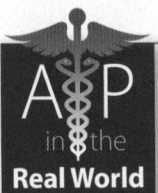

Complete Blood Count with Differential

One of the first laboratory tests ordered when a patient is admitted to the hospital is a *complete blood count* (*CBC*). A component of the CBC is the leukocyte count, which can indicate if inflammation is present somewhere in the body. Unfortunately, inflammation is nonspecific and an elevated leukocyte count tells you little about the cause of inflammation. For this reason, a *differential*, which measures the relative prevalence of the different types of leukocytes in the blood, is ordered along with the CBC.

A differential can give you a great deal of information. First, if the infection is caused by bacteria, neutrophils tend to be selectively elevated because of their extensive role in the eradication of bacteria. Second, viral pathogens tend to induce primarily a lymphocyte-dominant response, so the numbers of lymphocytes in the blood would likely be elevated with a viral infection. Finally, inflammation due to parasites or allergies usually causes an elevated level of eosinophils in the blood.

cells and extracellular matrix (see Chapter 3). These mutations result in a mass of unspecialized cells called a *malignant tumor,* whose cells are capable of indefinite growth and can *metastasize* (meh-TAS-tah-syz), or spread, through the lymphatic system or blood to other parts of the body. The cancer cells invade and destroy healthy tissues as they crowd out normal cells and compete for nutrients. (Note that *benign tumors* [beh-NYN] lack the ability to metastasize and do not divide indefinitely. Benign tumors are not considered cancerous.)

The immune system is quite effective at eliminating cancers before they ever get the chance to damage other tissues. This effectiveness is largely due to certain cells of the immune system, especially T cells and NK cells, that "scan" the cells in the body for tumor antigens, a function known as **immune surveillance.** Immune surveillance is absolutely critical to the prevention of tumor development.

When tumor cells are discovered via immune surveillance, they are eliminated by the following process, illustrated in **Figure 20.26:**

① **Cancer cells damage surrounding cells and induce an inflammatory response.** Cancer cells invade surrounding, healthy cells, which secrete inflammatory mediators that induce an inflammatory response.

② **NK cells migrate to the area and begin to destroy the cancer cells and secrete interferons.** Inflammatory mediators attract NK cells to the area, and they begin to destroy the cancer cells and secrete certain interferons, which also induce cancer cell death. In addition, macrophages are

activated, and they secrete tumor necrosis factor, which also induces cell death.

③ **Dendritic cells ingest cellular debris and migrate to lymph nodes, where they activate T_H cells.** Cancer cell debris produced as a result of tumor death is then ingested by dendritic cells, which then migrate to the lymph and lymph nodes. In the lymph nodes, dendritic cells present antigens to T_H cells.

④ **Activated T_H cells secrete cytokines that stimulate T_C cells to kill cancer cells.** The activated T_H cells secrete interleukin-2, which fully activates T_C cells. These cells can then destroy any cancer cells they encounter.

As you are likely aware, this process isn't always 100% effective, and cancer cells are often able to escape the immune response. We address how this happens in the next section.

Quick Check

☐ 3. How does the immune system survey the body for cancer cells?

☐ 4. Walk through the steps of the immune response to cancerous cells.

Pathogens That Evade the Immune Response

Although the immune response works most of the time in preventing serious homeostatic disruptions, certain pathogens have evolved ways to evade elements of the innate and/or adaptive immune responses. For example, many viruses block or inhibit cytokines such as interferons. Other viruses are able to prevent display of viral antigens on class I MHC molecules, which prohibits activation of T_C cells. Bacteria also have ways of evading the immune response, including adaptations that allow them to attach more strongly to host cells, secretion of substances that destroy antibodies, and inhibition of phagocytosis. Such microorganisms are especially pathogenic and difficult for the immune system to eradicate.

Cancer cells are also able to evade the immune response. Sometimes the cells simply grow too quickly and overwhelm the immune system. Other cancer cells have mutated so that they express low numbers of class I MHC molecules, which prevents the expression of adequate numbers of tumor antigens to fully activate T_C cells. In addition, tumor cells provide poor co-stimulatory signals, which prevents T_H cell activation and subsequent T_C cell activation. Finally, tumor cells secrete cytokines that suppress T cell activity. All of these factors combine to make many cancers able not only to evade the immune response but to cause the death of the patient as well.

Quick Check

☐ 5. Explain how microorganisms evade the immune response.

☐ 6. How do cancer cells escape the immune response and spread to other tissues?

The Big Picture of the Immune Response to Cancer Cells

Figure 20.26

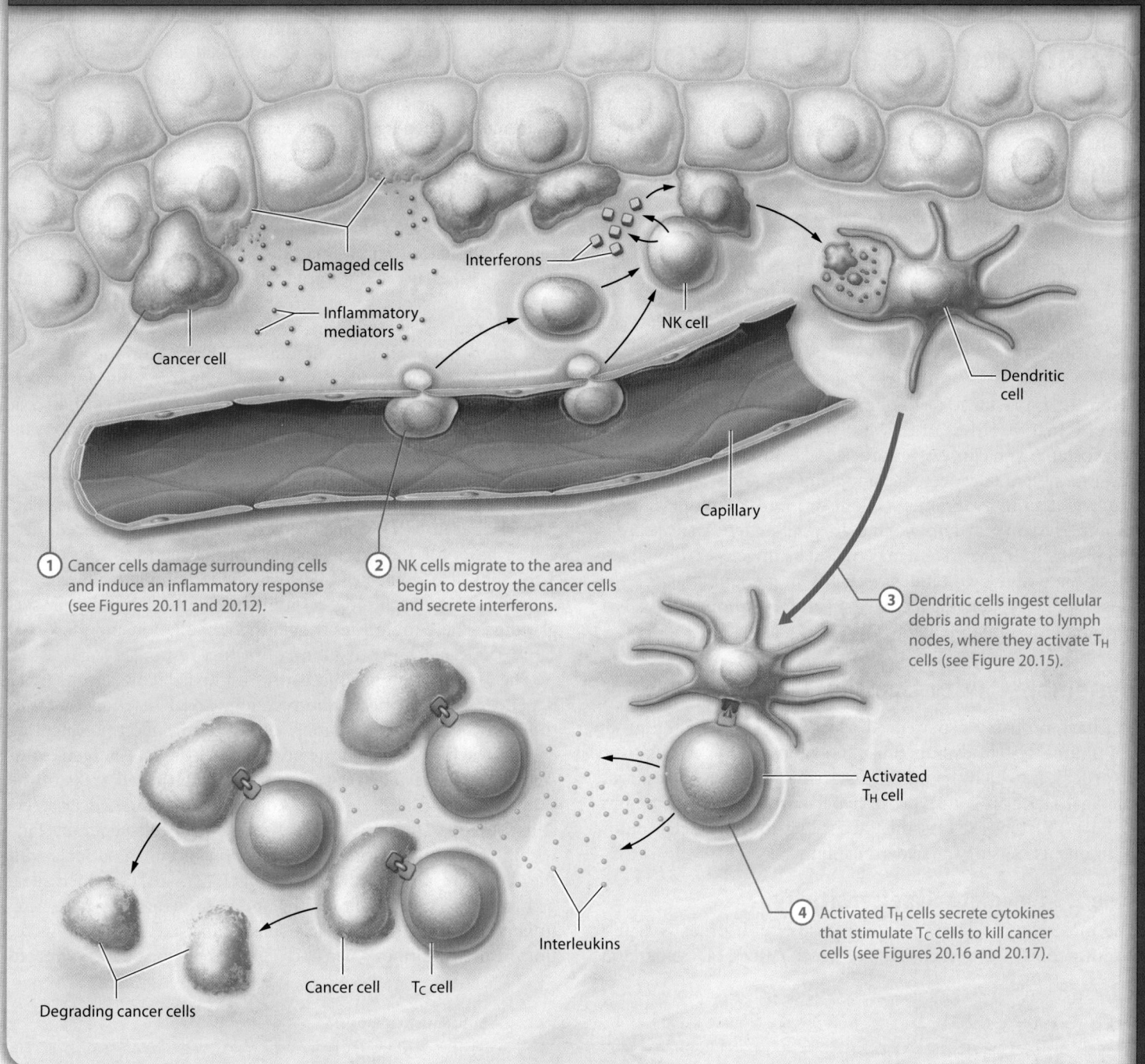

Damaged cells

Interferons

Inflammatory mediators

NK cell

Cancer cell

Dendritic cell

Capillary

1. Cancer cells damage surrounding cells and induce an inflammatory response (see Figures 20.11 and 20.12).

2. NK cells migrate to the area and begin to destroy the cancer cells and secrete interferons.

3. Dendritic cells ingest cellular debris and migrate to lymph nodes, where they activate T_H cells (see Figure 20.15).

Activated T_H cell

4. Activated T_H cells secrete cytokines that stimulate T_C cells to kill cancer cells (see Figures 20.16 and 20.17).

Interleukins

Cancer cell   **T_C cell**

Degrading cancer cells

Apply What You Learned

☐ 1. A virus develops a mutation that prevents the activation of T_C cells. Predict the consequences of this mutation. What other components of the immune response would need to compensate for the lack of activated T_C cells?

☐ 2. What would happen if cancer cells developed a mutation similar to that in question 1?

☐ 3. A patient presents to the emergency room with a fever. What does the fever indicate?

☐ 4. Blood work on the patient in question 3 shows an elevated level of neutrophils in her blood. What does this tell you about the possible cause of the fever? Explain.

See answers in Appendix A.

Disorders of the Immune System

Learning Outcomes

1. Describe the characteristics of the types of hypersensitivity disorders.
2. Describe the common immunodeficiency disorders.
3. Explain why HIV targets certain cell types, and describe the effects this virus has on the immune system.
4. Describe the characteristics of common autoimmune disorders.

Given the importance of the immune system, it is probably no surprise that disorders of this system can result in catastrophic homeostatic imbalances. Disorders take three forms: (1) The immune system may overreact and damage tissues, which results in a *hypersensitivity disorder;* (2) one or more components of the immune system may fail, which is an *immunodeficiency disorder;* and (3) the immune system may treat self antigens as foreign and attack the body's own tissues, which is an *autoimmune disorder.* This module explains each type of disorder, including the causes, symptoms, and treatments.

Hypersensitivity Disorders

You have probably experienced at least one type of the group of immune system dysfunctions known as the **hypersensitivity disorders,** in which the immune system's response causes tissue damage. There are four types of hypersensitivity disorder (numbered I–IV) that are classified according to the exact immune components causing the hypersensitivity.

Type I: Immediate Hypersensitivity

The most common type of hypersensitivity disorder is **type I,** or **immediate, hypersensitivity,** which affects a tremendous number of people—about 20% of the U.S. population. This type of hypersensitivity is more commonly known as **allergies,** and the disorders that accompany it are called *allergic disorders.* We briefly discussed this form of hypersensitivity earlier with our discussion of IgE (the "allergy antibody") and mast cells.

Immediate hypersensitivity reactions occur when an individual reacts to a foreign antigen, an allergen. Common allergens include pollen, dust mites, pet dander, peanuts, shellfish, and bee venom. As shown in **Figure 20.27,** in the first exposure of an individual to an allergen, ① an allergen binds a B cell, and ② the B cell differentiates into plasma cells that begin to secrete antibodies, just as they normally would when exposed to a foreign antigen. But there is one key difference between this response and a normal response: The plasma cells secrete IgE instead of IgG or IgM. This first exposure generates a primary immune response, which occurs slowly and produces few symptoms, but does result in the formation of large numbers of IgE molecules that coat mast cells and basophils. Such mast cells and basophils are said to be *sensitized.*

Subsequent exposures to the allergen in a sensitized person result in rapid responses that occur within a few minutes. This happens because ③ when the allergen binds the sensitized mast cell or basophil, the IgE molecules on the cell form cross-links. These links cause the cell to release the inflammatory mediators in its granules, including histamine, leukotrienes, and prostaglandins. These molecules trigger the vasodilation and increased capillary permeability of an inflammatory response.

The symptoms produced by immediate hypersensitivity depend on the scope of the reaction. Local reactions such as those in the nasal cavity upon exposure to pollen produce the common symptoms of runny nose and itchy eyes. Allergen exposure may also result in a skin rash in which small areas of the skin appear red and elevated, known as *hives* or *urticaria* (er′-tih-KAER-ee-ah). Urticaria are often itchy but generally fade within 30 minutes. More potent reactions occur in patients with the allergic respiratory disease *asthma* (AZ-mah), in which allergen exposure severely limits the ability to breathe by triggering inflammation, smooth muscle spasm, and excess mucus secretion in the respiratory passages. These reactions can be treated

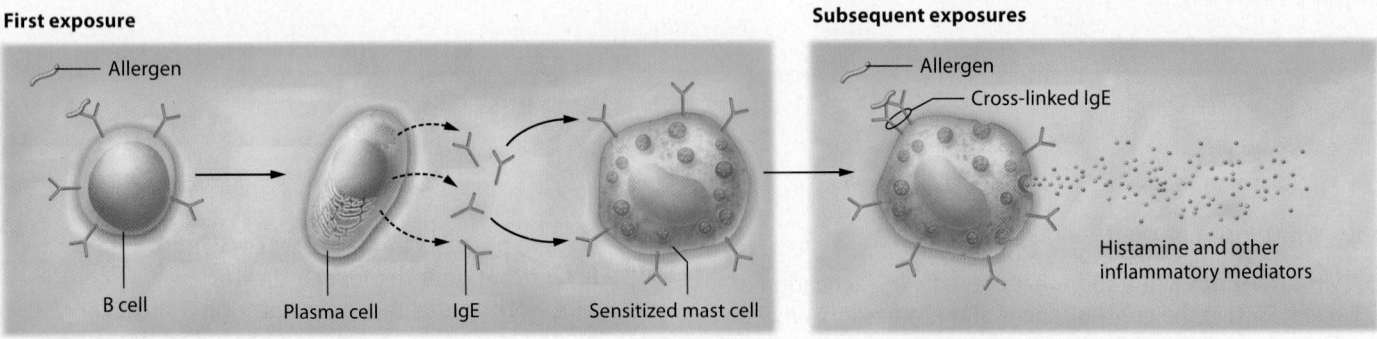

First exposure

Allergen

B cell   Plasma cell   IgE   Sensitized mast cell

Subsequent exposures

Allergen
Cross-linked IgE

Histamine and other inflammatory mediators

① An allergen binds a B cell.   ② The B cell differentiates into plasma cells that secrete IgE antibodies, which bind to a mast cell, sensitizing it.

③ The allergen binds the sensitized mast cell, and IgE molecules on the cell form cross-links that cause the cell to release inflammatory mediators from its granules, triggering an inflammatory response.

Figure 20.27 Type I hypersensitivity response.

by a variety of medications, as discussed in *A&P in the Real World: Treatments for Allergies.*

The most dramatic immediate hypersensitivity reaction, **anaphylactic shock** (ae′-nah-fah-LAK-tik), involves a systemic release of histamine and other inflammatory mediators. Anaphylactic shock consists of several life-threatening events, including (1) severe spasm of the smooth muscle of the respiratory tract, particularly the smooth muscle in the larynx (the "voice box"); (2) systemic vasodilation, which causes blood pressure to drop and decreases blood flow to all organs, including the brain; and (3) increased capillary permeability in all of the body's capillaries, which further lowers blood pressure and causes body-wide swelling as there is a massive loss of fluid to the tissue spaces and lungs. Anaphylactic shock is easily fatal if not treated immediately. Treatment generally consists of an injection of epinephrine, which causes relaxation of the smooth muscle of the airways and systemic vasoconstriction. Patients with a known risk for anaphylactic shock generally carry an epinephrine "pen" with them that allows self-administration of epinephrine in case of exposure to the allergen.

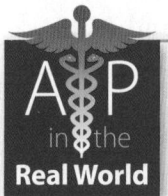

Treatments for Allergies

Given the enormous number of people who suffer from allergic disorders, the market for medications that treat allergies is huge. Most medications available over the counter are *antihistamines,* or medications that block cells' receptors for histamine. This prevents the pro-inflammatory effects of histamine molecules and limits the symptoms experienced by allergy sufferers. Unfortunately, histamine is also involved in a variety of other processes, both as a hormone and as a neurotransmitter. This leads to the most common side effect of antihistamines: drowsiness. Indeed, some antihistamines are so sedating that they are used more as sleep aids than as allergy treatments. Newer antihistamines have been modified so they are less able to cross the blood-brain barrier, and are therefore somewhat less sedating.

Other classes of anti-allergy medications and treatments include the following:

- *Antileukotriene agents* block the enzyme that produces leukotrienes, which inhibits many aspects of the allergic inflammatory response.
- *Corticosteroids* block the synthesis of leukotrienes and prostaglandins and are potent inhibitors of allergic inflammation. They are commonly used for related allergic disorders such as asthma.
- *Allergen immunotherapy,* commonly known as "allergy shots," involves administration of the allergen in increasing doses, with the aim of inducing tolerance to the allergen and a diminished IgE response.

Type II: Antibody-Mediated Hypersensitivity

In **type II hypersensitivity,** also known as *antibody-mediated hypersensitivity,* the antibodies produced by the immune response to foreign antigens also bind to self antigens. This reaction occurs in three situations: (1) Foreign antigens bind to normal self antigens; (2) donor erythrocytes infused into an individual are mismatched in the ABO/Rh antigen groups; and (3) self-reactive B cells are not destroyed in the bone marrow, which leads to autoimmunity (discussed shortly).

An example of the first situation is the reaction to antibiotic medications such as penicillin. These drugs are able to bind to erythrocytes in certain patients, which alters their antigens and causes them to be recognized as foreign. Activated B cells secrete antibodies that lead to complement activation and complement-mediated lysis of erythrocytes.

Type III: Immune Complex–Mediated Hypersensitivity

The reactions of **type III hypersensitivity** are mediated by *immune complexes,* or clusters of soluble antigens (those not attached to the cell surface) bound to antibodies. Immune complexes are generally cleared by phagocytes, but certain complexes are difficult for macrophages to ingest. These complexes deposit in various places in the body, including the capillary beds in the kidneys, blood vessel walls, the synovial membrane of joints, and the choroid plexus in the brain. When they deposit in these organs and tissues, they initiate an inflammatory reaction that attracts neutrophils and causes damage to the affected areas.

Type IV: Delayed-Type Hypersensitivity

The final type of hypersensitivity is **type IV,** or **delayed-type, hypersensitivity (DTH).** DTH is unique in that it is mediated by T cells rather than antibodies. In these cases, T_H cells recognize antigens bound to MHC molecules as foreign and mediate their destruction by activating and recruiting macrophages and in some cases T_C cells. This reaction generally takes 2–3 days to manifest, hence its name. As with type I hypersensitivity, T_H cells must be sensitized by an initial exposure, and the reaction occurs with subsequent exposures.

One of the most common types of DTH reactions is *contact dermatitis,* in which skin comes into contact with an allergen such as the oils in poison ivy or poison oak, certain metals, or other chemicals that can form complexes with skin proteins. The allergens are taken in and displayed by dendritic cells in the skin, which activate T_H cells. The T_H cells then secrete cytokines that recruit macrophages. The macrophages arrive in large numbers and destroy the self cells with the foreign antigens through phagocytosis and the release of enzymes and other chemicals that also damage surrounding cells. This results in a rash that is itchy and occasionally painful.

Other forms of DTH are caused by intracellular pathogens that are not easily cleared by the immune response. One such pathogen is the bacterium *Mycobacterium tuberculosis,* which causes the respiratory infection tuberculosis. In this case, the DTH reaction begins as a normal immune response

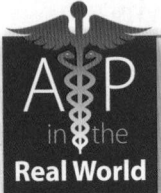

The Tuberculin Skin Test

We can take advantage of the DTH reaction to determine whether a person has been exposed to the bacterium that causes tuberculosis. The **Mantoux test** (man-too), also known as the **PPD test** (for purified protein derivative), involves injection of a small amount of a protein derived from the cell wall of the *Mycobacterium tuberculosis* bacterium just underneath the skin. If the individual has been exposed to tuberculosis, he or she will have sensitized T cells and will mount a DTH response. This leads to an area of *induration*—a hard, swollen area caused by the infiltration of macrophages and subsequent cellular destruction. A person who has not been previously exposed to the tuberculosis antigens will have no area of induration because he or she will lack sensitized T cells.

to a pathogen. However, this pathogen is difficult to eradicate, and the response often persists, with macrophages continually secreting enzymes and chemicals that nonspecifically damage surrounding cells. See *A&P in the Real World: The Tuberculin Skin Test* to find out how the DTH reaction can be used to identify tubercular infection.

Quick Check

☐ 1. Define the four types of hypersensitivity reactions.
☐ 2. Explain how type IV hypersensitivity differs from the other three types.

Immunodeficiency Disorders

A decrease in the function of one or more components of the immune system results in an **immunodeficiency disorder.** In the preceding modules, you have seen how the various parts of the immune response are interconnected. For this reason, when one component of the immune response fails, it generally leads to failure of other components, and the results can be catastrophic. There are two basic types of immunodeficiency disorders: *primary immunodeficiencies,* which are genetic or developmental in nature, and *secondary immunodeficiencies,* which are acquired through infection, trauma, cancer, or certain medications. In the following sections, we focus on some of the more common primary and secondary immunodeficiency disorders, and examine their effects on the immune response as a whole.

Primary Immunodeficiency Disorders

Primary immunodeficiency disorders may impair either innate or adaptive immunity. The most common dysfunctions of innate immunity involve deficient complement proteins or abnormalities in phagocytes (which may be present in too low a number, demonstrate deficient bacterial killing, or be unable to properly adhere to pathogens). Patients with defective phagocytes and complement proteins are at much higher risk for bacterial infections and parasitic infections.

Common dysfunctions of adaptive immunity include the *hypogammaglobulinemias* (hy′-poh-gah′-mah-glob-yoo′-lin-EE-mee-ahz), which are characterized by a decrease in one or more types of antibodies. Another common form of primary immunodeficiency involving adaptive immunity is a cluster of disorders referred to as **severe combined immunodeficiency,** or **SCID.** The different forms of SCID are caused by failures of lymphoid cell lines in the bone marrow, which affect B cells, NK cells, and T cells to varying degrees. Regardless of the underlying cellular involvement, the forms of SCID share common features, including low circulating levels of lymphocytes, failure of the thymus to develop, and failure of cell-mediated immunity. This results in recurrent infections that typically cause death early in life unless the patient lives in a sterile environment (a "bubble"). However, it is extraordinarily difficult to maintain a completely sterile environment, and such measures only temporarily prolong a patient's life.

Secondary Immunodeficiency Disorders and AIDS

Secondary immunodeficiency disorders have multiple forms. Many of these disorders are induced artificially to combat cancers originating in the bone marrow or to prevent transplant rejection. Others are a result of cancers of immune cells and lymphoid organs that depress the immune response in some way. However, the most common cause of secondary immunodeficiency by far is the virally induced disease **acquired immunodeficiency syndrome,** or **AIDS.**

AIDS is caused by **human immunodeficiency virus 1 (HIV-1),** which is spread through contact with infected blood, semen, vaginal fluid, or breast milk. HIV-1 preferentially binds and interacts with cells displaying CD4 molecules. The affinity of the virus for CD4 molecules is due to a glycoprotein on its surface that fits into a CD4 molecule the way a key fits into a lock. This makes cells that display large numbers of CD4 molecules, particularly T_H cells, specifically at risk for infection. Keep in mind that certain strains of HIV-1 can also interact with CD4 molecules on other immune cells, such as monocytes, macrophages, and dendritic cells.

After an HIV-1 virion (VEE-ree-on), or viral particle, has bound to a CD4 molecule, it interacts with other cell surface molecules that allow it to gain entry into the host cell. The events that occur inside the cell differ from many other viruses because HIV-1 is a *retrovirus,* which means that it contains an RNA genome and reproduces with the help of an enzyme called *reverse transcriptase.* As implied by its name, reverse transcriptase catalyzes the reverse process of normal transcription—instead of transcribing DNA into RNA, this enzyme catalyzes the transcription of viral RNA into DNA. The DNA copy of the viral genome is then inserted into the host DNA, where it triggers the production of viral RNA and proteins. The infected cell eventually lyses, and the new virions are released to infect new cells.

The progression of HIV-1 infection can be divided into three phases, although the course may vary from one individual to another. The first phase, the *acute phase,* lasts about 3 months and is characterized by a sharp decline in T_H cells and a sharp rise in HIV-1 virions. Patients may exhibit flu-like symptoms during this phase, although the condition often goes unnoticed. The second phase, the *chronic phase,* begins with the production of antibodies to HIV-1 virions, an initial slight recovery in the number of T_H cells, and a decline in the number of HIV-1 virions. This is due to the actions of B cells, T_C cells, various cytokines, and the remaining T_H cells. The chronic phase may last 8 or more years in untreated individuals, during which many patients show few to no signs of HIV-1 infection. It is during the final phase when an individual is said to have AIDS, which is characterized by progressively declining numbers of T_H cells and progressively increasing numbers of HIV-1 virions. Without treatment, the duration of the final phase is generally no more than 3 years.

The signs and symptoms of AIDS are due largely to the destruction of T_H cells. As you have read, T_H cells are required for almost all parts of the innate and adaptive immune responses to function properly. For this reason, the loss of T_H cells causes the entire adaptive immune response to fail (some innate responses remain). This failure leads to recurrent infections, particularly with agents that are *opportunistic,* or not generally pathogenic in immunocompetent patients. For example, respiratory infection with a species of the fungus *Pneumocystis* (noo-moh-SIS-tus) is extremely common in AIDS patients but very rare in individuals with a healthy immune system. Other consequences of AIDS include cancers such as *Kaposi's sarcoma* (kap-OH-seez), which affects the blood vessels and leads to purple-red skin lesions, shown in **Figure 20.28**; damage to the central and peripheral nervous systems; and severe muscle wasting. Patients generally succumb to complications from infection or cancer.

AIDS was discovered in 1981 primarily in homosexual males and intravenous drug abusers. Since that time, it has spread across the globe to now-epidemic proportions. Over 1 million individuals in the United States are infected with HIV-1, and although the disease remains more prevalent in homosexual males, it is being found with increasing frequency in heterosexual males and females. Worldwide, over 35 million were living with HIV-1 as of the end of 2012. Note that a staggering percentage of these individuals with HIV-1 live in sub-Saharan Africa—about 90%—and the majority of these patients are females.

Clearly, the development of treatments for HIV and AIDS is a global priority. Currently, there are a number of drug therapies for HIV-1. There are three main mechanisms by which most of these drugs work: (1) inhibiting reverse transcriptase, (2) inhibiting viral enzymes needed to synthesize mature virions, and (3) blocking the entry of HIV-1 into its target cells. The drugs are typically administered in combination as a "cocktail" to inhibit as many aspects of the viral replication cycle as possible.

Although drug cocktails do prolong patients' lives, none actually cure the disease. In addition, the newer drug cocktails are expensive, require following a strict regimen, and cause unpleasant adverse effects. Furthermore, due to the error-prone nature of reverse transcriptase, HIV-1 is prone to mutations, some of which may render certain drugs ineffective. For these reasons, the best way to effectively curb the spread of HIV-1 and put a halt to the epidemic is to develop a vaccine for the virus. Unfortunately, multiple issues make this a considerably more difficult task than with other pathogens. For one, the virus's high mutation rate selects for mutants that are resistant to immunity. An additional factor is that most vaccines stimulate the development of antibodies, but antibodies to HIV-1 in most individuals provide little protection from the disease because the virus has a high mutation rate. Research into how to best combat these and other issues is ongoing, but progress has unfortunately been slow. Indeed, the eradication of this epidemic is proving to be one of the greatest public health challenges of modern times.

Figure 20.28 **The skin of an AIDS patient with Kaposi's sarcoma.**

Quick Check

☐ 3. How do primary and secondary immunodeficiency disorders differ?

☐ 4. Why is AIDS so devastating to the immune system?

Autoimmune Disorders

Normally, several processes ensure immune self tolerance—that the immune system does not react to self antigens and damage the body's own tissues and cells. However, sometimes these processes fail, and the result is populations of self-reactive T cells or of B cells that secrete antibodies that bind to self antigens, called *autoantibodies.* This produces an **autoimmune disorder.** Autoimmune disorders may be localized and affect only one organ or tissue, or they may be systemic and affect multiple parts of the body. Autoimmunity results in a type II, III, or IV hypersensitivity reaction, which is what produces the organ and tissue damage characteristic of the disease.

Many situations could lead to the development of autoimmunity, including the following:

- **Release of self antigens not previously encountered by T cells.** T cells are generally exposed to self antigens during development, and those self-reactive T cells are destroyed. However, some antigens are *sequestered,* meaning that they are not exposed to the developing T cells. For example, a protein in myelin, the substance surrounding the axons of many neurons, is normally sequestered by the blood-brain barrier and T cells are not exposed to it. However, infection or trauma might release this protein and its antigens into the circulation, activating T cells specific for this antigen. This is believed to be the mechanism behind the development of the autoimmune disease *multiple sclerosis,* in which the myelin sheath is attacked.

- **Foreign antigens mimic self antigens.** Certain viral and bacterial antigens closely resemble normal self antigens. Normally, T cells specific for these antigens do not attack self cells because of the lack of co-stimulatory signals. However, if an individual comes into contact with these pathogens, co-stimulatory signals activate these T cells, which then attack self cells. This occurs with the disease *rheumatic fever* (roo-MAT-ik), which occurs after infection with a specific species of streptococcal bacterium whose antigens resemble those on the heart muscle.

- **Cells may inappropriately express class II MHC molecules.** As we discussed earlier, only B cells and antigen-presenting cells normally display class II MHC molecules. For poorly understood reasons, other cell types may also inappropriately express class II MHC molecules. This activates T cells and triggers an immune response to these normal self antigens. This appears to be the case with type 1

diabetes mellitus, in which the immune system destroys the insulin-producing cells of the pancreas.

- **Certain pathogens nonspecifically activate B cells.** Many pathogens can induce the production of cytokines that nonspecifically activate B cells, resulting in the production of autoantibodies. This is the proposed mechanism behind the disease *systemic lupus erythematosus* (eh′-rih-them-ah-TOH-sis), in which infection with a certain virus is followed by production of antibodies to proteins in DNA, erythrocytes, platelets, and leukocytes.

Each of these situations also relies on multiple genetic factors that increase an individual's tendency to develop autoimmunity.

Quick Check

☐ 5. What is an autoimmune disorder? What produces the tissue damage characteristic of these disorders?

☐ 6. What are the potential causes of autoimmune disorders?

Apply What You Learned

☐ 1. Certain drugs have been designed to inhibit the release of the contents of mast cell granules. Predict the effect of these drugs. Which type of hypersensitivity would they most affect?

☐ 2. A hypothetical treatment for HIV-1 infection increases antibody production by B cells. Is this likely to help treat this infection? Why or why not?

☐ 3. Autoimmune disorders are often treated with agents that suppress the immune system. How would this alleviate some of the symptoms of these disorders? Predict a potential adverse effect of these agents.

See answers in Appendix A.

CHAPTER SUMMARY

Take full advantage of the invaluable online practice and study tools for this chapter by going to

MasteringA&P®. **There you will find:**

- 🖱 Practice tests, module quizzes, art-labeling and sequence-the-steps activities

- ▶ Animated figures with quizzes

- 🖊 *Practicing A&P* worksheets

- 🔊 Author-narrated podcast

- 🔥 Dynamic Study Modules

- iP *Interactive Physiology*

MODULE 20.1
Structure and Function of the Lymphatic System 753

- The lymphatic system consists of **lymphatic vessels** and **lymphoid organs** that regulate the interstitial fluid volume, absorb dietary fats, and assist in immune functions. (Figure 20.1, p. 753)

  🖱 Practice art-labeling exercise: Figure 20.1.

- Lymphatic vessels are a series of blind-ended vessels that begin as highly leaky **lymphatic capillaries** that unite to form **lymph-collecting vessels,** which themselves unite to form **lymph trunks.** The lymph trunks from the right arm and right side of the head and neck drain into the **right lymphatic duct,** and the remaining trunks drain into the **thoracic duct.** (Figure 20.2, p. 754; Figure 20.3, p. 755)

- Lymphoid organs contain leukocytes, including **macrophages,** lymphocytes, and granulocytes, as well as **dendritic cells** and **reticular cells.** (Figure 20.4, p. 756)

- **Mucosa-associated lymphatic tissue,** or **MALT,** consists of clusters of unencapsulated lymphoid tissue located along the mucous membranes. Specialized MALT, which may be partially encapsulated, includes the **tonsils, Peyer's patches,** and **appendix.** (Figure 20.5, p. 757; Figure 20.6, p. 757)

- **Lymph nodes** are encapsulated clusters of lymphatic tissue located along lymphatic vessels. Lymph nodes filter lymph as it travels back to the cardiovascular system and trap pathogens. (Figure 20.7, p. 758)

   Practice art-labeling exercise: Figure 20.7.

- The **spleen** filters the blood. The **white pulp** removes pathogens, and the **red pulp** destroys old erythrocytes. (Figure 20.8, p. 759)

- The **thymus** is the site of maturation of T cells. The thymus is active in infants and children, and begins to atrophy after the early teenage years. (Figure 20.9, p. 760)

iP Play Interactive Physiology tutorial on Immune System: Immune System Overview and Anatomy Review.

MODULE **20.2**
Overview of the Immune System 760

- **Innate immunity** does not require exposure to specific cellular markers called **antigens** to produce a response, and it responds to pathogens or classes of pathogens in the same way. It consists of certain plasma proteins and cells.

- The components of the immune system offer three lines of defense: The first line of defense involves surface barriers; the second line of defense involves the cells and proteins of innate immunity; and the third line of defense involves the cells and proteins of adaptive immunity.

- **Adaptive immunity** requires exposure to a specific antigen to mount a response. The two arms of adaptive immunity are **antibody-mediated immunity** and **cell-mediated immunity.** Adaptive immunity is capable of **immunological memory.**

- **Surface barriers** consist of the skin that lines the body's external surfaces and the mucous membranes that line all passageways that open to the outside. Surface barriers protect the underlying tissues from mechanical stress and from pathogen invasion.

- The cells of innate immunity include **phagocytes** and **natural killer (NK) cells,** as well as dendritic cells. The cells of adaptive immunity include **B** and **T lymphocytes.**

- Proteins of innate immunity include the group of proteins known as **complement** and many types of **cytokines.** Proteins of adaptive immunity include many types of cytokines and antibodies.

- Lymphoid organs and tissues house cells of the immune system, trap pathogens for the immune system, and activate cells of the immune system.

iP Play Interactive Physiology tutorial on Immune System: Innate Host Defenses.

MODULE **20.3**
Innate Immunity: Internal Defenses 763

- Phagocytes of innate immunity include the following:
  - Macrophages function as **antigen-presenting cells,** which activate T cells of adaptive immunity.
  - **Neutrophils** ingest damaged cells and bacteria.
  - **Eosinophils** are active in defense against parasites and are also involved in allergies.

- Nonphagocytic cells of innate immunity include the following:
  - NK cells are related to lymphocytes; they recognize and destroy cancer cells and cells infected with certain viruses.
  - **Basophils** mediate inflammation. Tissue-bound basophils called **mast cells** are located in mucous membranes.
  - Dendritic cells are located in many lymphoid organs and function as antigen-presenting cells.

- The **complement system** is a group of 20 or more plasma proteins produced primarily by the liver that are involved in cell lysis via a **membrane attack complex,** enhancing phagocytosis through **opsonization,** enhancing inflammation, and neutralizing certain viruses. The system may be activated by either the *classical* or the *alternative pathway.* (Figure 20.10, p. 765)

- The cytokines involved in innate immunity include *tumor necrosis factor* (*TNF*), **interferons,** and *interleukins.*

- The **inflammatory response** occurs in response to any cellular injury.
  - Injured cells secrete *inflammatory mediators* that trigger vasodilation, leading to *hyperemia,* which causes redness and heat; increased capillary permeability, which causes *edema* and pain; and **chemotaxis,** which recruits leukocytes to the damaged area. (Figure 20.11, p. 767)

   Practice sequencing the steps: Figure 20.11.

  - In the second part of the inflammatory response, local macrophages are activated, and blood-borne neutrophils and monocytes are attracted to the area. Finally, cytokines trigger increased leukocyte production in the bone marrow, leading to *leukocytosis.* (Figure 20.12, p. 769)

   Practice sequencing the steps: Figure 20.12.

- **Fever** is a body temperature above the normal range of 36–38° C. It is an innate response to cellular injury that increases the effectiveness of phagocytes. Fever is initiated when **pyrogens** released from damaged cells or from bacteria reset the hypothalamic temperature from normal to a higher range.

iP Play Interactive Physiology tutorial on Immune System: Innate Host Defenses.

MODULE **20.4**
Adaptive Immunity: Cell-Mediated Immunity 770

- T cells form in the bone marrow but migrate to the thymus to mature. (Figure 20.13, p. 771)

- Cell-mediated immunity involves T_H, or **CD4, cells** and T_C, or **CD8, cells.** These cells respond to cancer cells, cells infected with intracellular pathogens, and foreign cells.

- T cells interact with pieces of antigen bound to **major histocompatibility,** or **MHC, molecules. Class I MHC molecules,** which are located on all nucleated cells, display endogenous antigens and interact with T_C cells. **Class II MHC molecules,** which are located on antigen-presenting cells and other immune cells, display exogenous antigens and interact with T_H cells. (Figure 20.14, p. 773)

- T cells are activated when the T cell receptor binds to an MHC molecule-antigen complex in the presence of a *co-stimulator.* Activated T cells proliferate and differentiate into *effector cells* and **memory T cells.** (Figure 20.15, p. 774)

- The functions of T_H cells include the following: activating T_C cells, stimulating macrophages and B cells, and enhancing the cells and proteins of innate immunity. (Figure 20.16, p. 775)

- T_C cells kill other cells by releasing **perforin** that perforates the plasma membrane so that enzymes can enter the cell and fragment its DNA. They also induce *apoptosis.* (Figure 20.17, p. 776)

- The four basic kinds of transplants are *autografts, isografts, allografts,* and *xenografts.* The antigens on allografts and xenografts are foreign, and the cells can be destroyed by T cells, a process known as *rejection.*

iP Play Interactive Physiology tutorial on Immune System: Cellular Immunity and Common Characteristics of B and T Lymphocytes.

MODULE 20.5
Adaptive Immunity: Antibody-Mediated Immunity 777

- B cells develop and mature in the bone marrow. (Figure 20.18, p. 778)

- Antibody-mediated immunity is mediated by B cells that secrete **antibodies,** or **immunoglobulins,** which bind to specific antigens. Normal antibodies bind only foreign antigens; they do not bind to self antigens.

- When a naïve B cell encounters its antigen, the antigen binds and triggers the B cell to proliferate and differentiate into antibody-secreting **plasma cells** and long-lived **memory B cells.** (Figure 20.19, p. 778)

- Antibodies are Y-shaped molecules with antigen-binding sites on the tips of their arms. There are five classes of antibody: (Figure 20.20, p. 779; Figure 20.22, p. 781)
  ○ **IgG** is a monomer and the most prevalent antibody. It is the only antibody that crosses the placenta.
  ○ **IgA** is a dimer present in secretions from the skin, mucous membranes, and certain exocrine glands.
  ○ **IgM** is a pentamer and is the first antibody secreted by B cells in response to a pathogen.
  ○ **IgE** is a monomer that binds antigens associated with parasitic worms and *allergens.*
  ○ **IgD** is a monomer present on the surface of B cells that acts as an antigen receptor.

- The functions of antibodies include **agglutination, precipitation,** and opsonization, all of which enhance phagocytosis;

neutralization; activation of complement; and stimulation of inflammation. (Figure 20.21, p. 780)

Practice art-labeling exercise: Figure 20.21.

- Memory B cells are responsible for immunological memory. The first exposure to an antigen generates the slower **primary immune response.** Subsequent exposures to antigens generate the faster, more efficient **secondary immune response** (Table 20.1, p. 782).

- **Vaccinations** expose an individual to an antigen to generate memory cells so that subsequent antigen exposures will trigger a secondary immune response.

- **Active immunity** is generated by exposure to an antigen and results in the production of memory cells. **Passive immunity** is generated by receiving preformed antibodies from another source. (Figure 20.23, p. 783)

iP Play Interactive Physiology tutorial on Immune System: Humoral Immunity and Common Characteristics of B and T Lymphocytes.

MODULE 20.6
Putting It All Together: The Big Picture of the Immune Response 784

- The immune response to a viral infection involves a series of events that culminates in the destruction of virally infected cells by T_C cells. (Figure 20.24, p. 785)

- The immune response to a bacterial infection involves phagocytosis of bacteria by neutrophils and macrophages, and the activation of antibody-secreting plasma cells. (Figure 20.25, p. 787)

- The immune response to cancer culminates in the activation of T_C cells, which kill tumor cells. (Figure 20.26, p. 789)

- Pathogens evade the immune response through evading phagocytosis, mutation, and blocking secretion of certain cytokines.

MODULE 20.7
Disorders of the Immune System 790

- A **hypersensitivity disorder** occurs when the immune system's response causes tissue damage. There are four classes: **type I,** or **immediate hypersensitivity,** also known as **allergies; type II,** or *antibody-mediated,* **hypersensitivity; type III,** or *immune complex–mediated,* **hypersensitivity;** and **type IV,** or **delayed-type, hypersensitivity (DTH).** (Figure 20.27, p. 790)

- A decrease in the function of one or more components of the immune system results in an **immunodeficiency disorder.** *Primary immunodeficiencies* are genetic or developmental. *Secondary immunodeficiencies* are acquired from another source. A common example is **acquired immunodeficiency syndrome,** or **AIDS,** which is caused by the destruction of T_H cells by the **human immunodeficiency virus (HIV-1).** (Figure 20.28, p. 793)

- **Autoimmune disorders** occur when the immune system recognizes normal cells as foreign due to self-reactive T and B cells and the production of *autoantibodies.*

See answers in Appendix A.

LEVEL 1 Check Your Recall

1. Which of the following is not a function of the lymphatic system?

   a. Absorption of dietary fats
   b. Secretion of digestive enzymes
   c. Regulation of interstitial fluid volume
   d. Housing and maturation of leukocytes

2. Mark the following statements as true or false. If a statement is false, correct it to make a true statement.

   a. Lymphatic vessels constitute a one-way system that delivers fluid from the blood vessels to the extracellular space.
   b. Lymph from the lower limbs drains into the right lymphatic duct.
   c. Fat-containing lymph from the intestines drains into the cisterna chyli.
   d. The thymus is the site of maturation of B cells.
   e. Clusters of MALT located around the oral and nasal cavities are known as Peyer's patches.

3. Fill in the blanks: The lymphoid organ that filters the blood is the _____, and the lymphoid organ that filters the lymph is the _____.

4. Fill in the blanks: Nonspecific immunity is also known as _____. Specific immunity is also known as _____. Specific immunity has the capacity for immunological _____.

5. Which of the following make up the body's first line of defense?

   a. Surface barriers
   b. Cells and proteins of adaptive immunity
   c. Cells and proteins of innate immunity
   d. All of the above

6. Which of the following does *not* describe the relationship between the lymphatic and immune systems?

   a. Lymphoid organs and tissues house cells of the immune system.
   b. Lymphoid organs and tissues trap pathogens for the immune system.
   c. Lymphoid organs activate cells of the immune system.
   d. Lymphoid organs and tissues create the surface barriers of the immune system.

7. Mark the following statements as true or false. If a statement is false, correct it to make a true statement.

   a. Surface barriers contain substances in their secretions that kill pathogens and deter their growth.
   b. Phagocytic cells of innate immunity include NK cells and basophils.
   c. NK cells are cytotoxic cells that lyse and kill cancer cells and cells infected with certain viruses.
   d. Interleukin-1 is a cytokine that prevents viral replication in infected cells.
   e. Fever is generated by pyrogens that reset the temperature set point of the hypothalamus to a lower value.

8. Which of the following functions are performed by complement proteins?

   a. Cell lysis
   b. Opsonization
   c. Enhancing inflammation
   d. Only a and b
   e. All of the above

9. Fill in the blanks: Injured tissue releases chemicals called _____ that mediate the four cardinal signs of inflammation, which are _____, _____, _____, and _____.

10. When B cells are activated, they differentiate into:

    a. plasma cells.
    b. memory cells.
    c. both a and b.
    d. none of the above.

11. Match the following antibodies with the correct definition.

    _____ IgD   a. Antibody found in secretions
    _____ IgM   b. Most common antibody; crosses the placenta
    _____ IgG   c. Antibody involved in allergies and parasitic
    _____ IgA   infections
    _____ IgE   d. Pentamer and potent agglutinating agent
                  e. Antibody bound to the B cell plasma membrane

12. Mark the following statements as true or false. If a statement is false, correct is to make a true statement.

    a. In the primary immune response, the lag phase lasts about 5 days, during which time B cells proliferate and differentiate.
    b. The secondary immune response is mediated by plasma cells.
    c. Subunit vaccines consist of pathogens that are alive but unable to cause disease.
    d. Vaccinations are given to induce the production of the inflammatory response.

13. Explain how active immunity and passive immunity differ.

14. On which type(s) of cells are class I MHC molecules located?

    a. Antigen-presenting cells only
    b. B cells only
    c. All nucleated body cells
    d. Antigen-presenting cells and B cells

15. Fill in the blanks: Class I MHC molecules display _____ antigens and activate _____ cells. Class II MHC molecules display _____ antigens and activate _____ cells.

16. Which of the following is *not* a role of T_H cells?

    a. Stimulation of macrophages
    b. Activation of T_C cells
    c. Stimulation of B cells
    d. Stimulation of clonal selection

17. Mark the following statements as true or false. If a statement is false, correct it to make a true statement.

    a. The cells involved in organ and transplant rejection are primarily B cells.
    b. The immune response to a viral infection involves NK cells and different kinds of lymphocytes.
    c. Neutrophils are a critical component of the response to a parasitic infection.
    d. Cancer cells are destroyed by NK cells, T_C cells, and macrophages.

18. Type I hypersensitivity is due to release of inflammatory mediators from _____, and type IV hypersensitivity is due to the actions of _____.

    a. T_H cells; macrophages
    b. neutrophils; T_H cells
    c. B cells; T_C cells
    d. mast cells; T_H cells

19. Explain why a patient with AIDS due to HIV is at an increased risk of infection and of developing certain cancers.

20. Which of the following conditions may lead to the development of autoimmunity?

    a. Self antigens not previously encountered by T cells are released into the circulation.
    b. Foreign antigens mimic self antigens.
    c. Cells may inappropriately display class II MHC molecules.
    d. Certain pathogens nonspecifically activate B cells.
    e. All of the above are correct.

LEVEL 2 Check Your Understanding

1. Eileen presents with flu-like symptoms, including fever, chills, and body aches.

    a. What part of the immune response is likely producing these symptoms?
    b. You order blood work and find that the number of neutrophils in the blood is greatly elevated. What does this tell you about the likely type of pathogen involved? Explain.
    c. What might it mean if lymphocytes instead of neutrophils were elevated? Explain your reasoning.

2. Complement proteins are crucial for stimulating phagocytes to clear immune complexes. Predict what type of hypersensitivity disorder might result from complement deficiencies. Explain what other consequences might arise from a complement deficiency.

3. Terrence has severe asthma and allergies, and is placed on a medication that blocks the functioning of IgE. How would this medication alleviate his symptoms?

LEVEL 3 Apply Your Knowledge

PART A: Application and Analysis

1. Your friend tests her snake-wrangling skills on an eastern diamondback rattlesnake, and is bitten and envenomated (injected with the snake's venom). In the emergency department, doctors administer the antivenin *CroFab,* which contains antibodies to the toxins in snake venom. Is this an example of active or passive immunity? Will this confer any lasting protection to your friend?

2. Carla presents to her physician with a complaint of swollen axillary lymph nodes. Her physician performs a breast exam and finds a tumor on the same side as the swollen nodes.

    a. Explain what process has likely caused her nodes to swell.
    b. Carla undergoes surgery to have the tumor and affected lymph nodes and surrounding lymphatic vessels removed. Predict any complications Carla might have after the surgery that could arise from removal of the lymph nodes and vessels.

3. Mr. White has been diagnosed with *febrile neutropenia,* a condition characterized by the presence of a fever with an extremely low number of neutrophils in the blood. Blood work demonstrates the presence of bacteria in his blood. Explain why neutropenia might lead to a bacterial infection. Why is a bacterial infection particularly dangerous for Mr. White?

PART B: Make the Connection

4. Mr. White developed neutropenia as a consequence of cancer chemotherapy, which destroyed much of his bone marrow. What other components of the immune system would be harmed by bone marrow destruction? Would you expect his hematocrit to be elevated or decreased? What effects would you expect to see from this change in hematocrit? *(Connects to Chapter 19)*

5. The chemotherapeutic drugs Mr. White is taking affect all cells that undergo rapid mitosis, such as those of the skin. What effect would this have on the functions of the skin? How could this affect his immunity? *(Connects to Chapter 5)*

21

The Respiratory System

As the American Lung Association says, "You can go three weeks without food, and three days without water, but only three minutes without air." Under most conditions, cells can survive only a few minutes if a person's respiratory functions cease. The organs of the **respiratory system** work ceaselessly to bring our cells the oxygen needed for survival. The respiratory system coordinates other functions that contribute to the maintenance of homeostasis, including the body's acid-base balance and the removal of carbon dioxide. In this chapter we explore the structures of the respiratory system, and link these structures to how breathing and gas exchange occur, how homeostasis of the respiratory system is maintained, and what happens when this homeostasis is disrupted.

MODULE **21.1**

Overview of the Respiratory System

Learning Outcomes

1. Describe and distinguish between the upper and lower respiratory tracts.
2. Describe and distinguish between the conducting and respiratory zones of the respiratory tract.
3. Describe the major functions of the respiratory system.
4. Define and describe the four respiratory processes—pulmonary ventilation, pulmonary gas exchange, gas transport, and tissue gas exchange.

In this module, we introduce some basic structures and functions of the respiratory system; we will revisit many of them in greater depth later in the chapter. Let's begin with a brief introductory tour of the structures that support the respiratory system's many functions.

Anatomy of the Respiratory System: An Overview

◄◄ Flashback

1. What is elastic connective tissue, and what is its function? (p. 139)
2. What is a serous membrane? What does it secrete? (p. 18)

Computer-generated image: The human respiratory tract, highlighting the trachea and bronchial tree.

Nose —
Nasal cavity —
Pharynx —
Larynx —

— Upper respiratory tract

Left lung —
Right lung —
Thoracic cage (cut) —
Thoracic cavity —
Diaphragm —

— Trachea
— Bronchi

— Lower respiratory tract

— Bronchiole
— Alveoli

Practice art labeling
@MasteringA&P®

Figure 21.1 Organs of the respiratory system.

The organs of the respiratory system, shown in **Figure 21.1**, reside in the head, the neck, and the thoracic cavity. Notice that these organs include the blood vessels of the pulmonary circuit; the structures of the thoracic (rib) cage, including the respiratory muscles; the paired lungs; and the **respiratory tract,** which consists of the hollow passages that collectively transport gases. Each component of the respiratory tract has a unique gross and histological structure. These components include the following:

- the *nose* and the *nasal cavity,* which is encased by cranial and facial bones;
- the *pharynx*, also called the *throat;*
- the *larynx,* or "voice box," in the anterior neck;
- the *trachea,* or "windpipe," in the mediastinum; and
- the *bronchial tree,* branching tubes that begin as *bronchi* (BRONG-kye; singular, *bronchus*) and divide successively until they become tiny *bronchioles.*

The passages of the respiratory tract are classified anatomically into upper and lower tracts. The **upper respiratory tract** includes the passageways from the nasal cavity to the larynx. The **lower respiratory tract** includes the passageways from the trachea to the respiratory tract's terminal structures, the alveoli. **Alveoli** (al-vee-OHL-eye; singular, *alveolus*) are tiny air sacs, arranged in grapelike clusters, through which gases are exchanged. The most conspicuous structures of the respiratory system are the *lungs,* paired spongy organs in the thoracic cavity that are enclosed by the diaphragm and thoracic cage. Each lung is actually just a collection of about 150 million alveoli with the surrounding blood vessels and elastic connective tissue in which the alveoli and branches of the respiratory tract are embedded. We explore in detail the structure of the respiratory tract components and lungs in Module 21.2.

Quick Check

- ☐ 1. What are the main structures of the respiratory system?
- ☐ 2. Is the larynx part of the upper or lower respiratory tract?
- ☐ 3. Where are alveoli? What is their basic function?

Basic Functions of the Respiratory System

◀◀ Flashback

1. What is the function of a buffer? (p. 49)

In addition to being classified anatomically, the organs of the respiratory system are also classified functionally into the conducting and respiratory zones (**Figure 21.2**). The tubes of the **conducting zone** are the conduits through which air travels on its way in and out of the body as it is *inspired,* or inhaled, and *expired,* or exhaled. Air is filtered, warmed, and moistened as it travels through the many passages of the conducting zone. The **respiratory zone** is where gases are exchanged. The conducting zone includes structures from the nose and nasal cavity through the bronchioles, and the respiratory zone includes structures that contain alveoli. The functional classification of the respiratory tract is more precise and useful than the anatomical one, and this is the system we use in this chapter.

Respiration is the process that provides the body's cells with oxygen and removes the waste product carbon dioxide. Respiration actually includes four separate processes: (1) pulmonary ventilation, (2) pulmonary gas exchange, (3) gas transport in the blood, and (4) tissue gas exchange. *Pulmonary ventilation* (PULL-mon-air-ee vent-ih-LAY-shun; *pulmon-* = "lung"), often shortened to *ventilation,* is the movement of air in and out of the lungs. The remaining processes refer to the movement of gases through the body: *Pulmonary gas exchange* is the movement of gases between the lungs and the blood, *gas transport* is the movement of gases through the blood, and *tissue gas exchange* is the movement of gases between the blood and the tissues.

It's no surprise that respiration is a primary function of the respiratory system. However, the respiratory system also has several other vital roles. The mechanism for speech and sound production is found in the respiratory system, as we will discuss in Module 21.2. In addition, recall that the neurons for the sense of smell are located in the olfactory epithelium, a patch of modified epithelial tissue in the roof of the nasal cavity (see Chapter 15).

Other functions of the respiratory system have to do with pressure changes in the thoracic cavity. When you hold your breath and bear down, the pressure increases in the abdominopelvic cavity. This pressure helps to expel the cavity's contents during urination, defecation, and childbirth. The pressure changes generated also help to propel lymph and venous blood in the thoracic and abdominopelvic cavities, which are under very low pressure.

However, you may be less familiar with another important function of the respiratory system that has to do with maintaining the body's homeostasis. Proper functioning of the respiratory system is critical in maintaining acid-base balance in the extracellular fluid, as this system controls one of the primary buffer systems in the body. In addition, it synthesizes an enzyme involved in the production of *angiotensin-II.* Recall that angiotensin-II is involved in maintaining acid-base homeostasis and is also critical in maintaining blood pressure and fluid homeostasis (see Chapter 16). We discuss angiotensin-II again in the urinary chapter (see Chapter 24).

To recap, the functions of the respiratory system include the following:

- Respiration
- Producing speech and other vocalizations
- Detecting odors
- Helping to expel contents of the abdominopelvic cavity
- Assisting in the flow of venous blood and lymph in the thoracic and abdominopelvic cavities
- Maintaining acid-base homeostasis
- Assisting in the production of angiotensin-II for maintenance of blood pressure and fluid homeostasis

Given all of these functions, it makes sense that when the respiratory system's homeostasis is disrupted, not only is tissue oxygenation disrupted, but other homeostatic functions are disrupted as well. Throughout this chapter, we look at how the body maintains respiratory homeostasis and the various effects of homeostatic imbalances of the respiratory system.

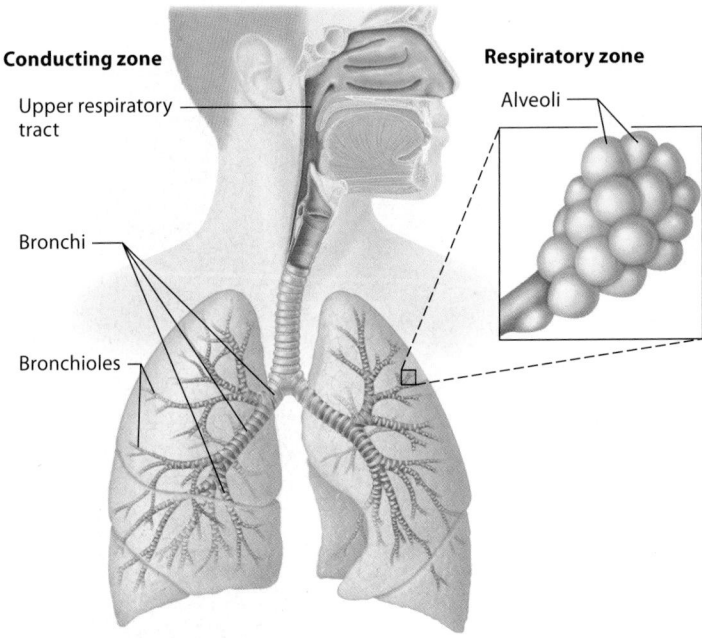

Conducting zone

Upper respiratory tract

Bronchi

Bronchioles

Respiratory zone

Alveoli

Figure 21.2 The conducting and respiratory zones of the respiratory system.

Quick Check

☐ 4. List and define the four processes that make up respiration.

☐ 5. How does the respiratory system contribute to the maintenance of homeostasis?

☐ 6. List and describe four functions of the respiratory system besides respiration.

Apply What You Learned

☐ 1. How would blocking the structures of the conducting zone of the respiratory tract impair ventilation? How would the situation change if the structures of the respiratory zone were blocked? Explain.

☐ 2. What homeostatic disturbances would you expect to see with diseases of the respiratory system? Explain.

See answers in Appendix A.

MODULE 21.2

Anatomy of the Respiratory System

Learning Outcomes

1. Trace the pathway through which air passes during inspiration.
2. Describe the gross anatomical features and function of each region of the respiratory tract, the pleural and thoracic cavities, and the pulmonary blood vessels and nerves.
3. Describe the histology of the different regions of the respiratory tract, the types of cells present in alveoli, and the structure of the respiratory membrane.
4. Explain how the changes in epithelial and connective tissue in air passageways relate to their function.
5. Describe the structure of the lungs and pleural cavities.

In this module, we explore the structure of each component of the conducting and respiratory zones of the respiratory tract. As you read, note how the subtle differences in structure at the gross and microscopic levels enable each component of the respiratory tract to carry out its function.

The Nose and Nasal Cavity

◄◄ Flashback

1. What are goblet cells, and what are their functions? (p. 135)
2. Which bones form the anterior, superior, and lateral walls of the nasal cavity? (p. 214)
3. In which bones are the paranasal sinuses located? What are the functions of the paranasal sinuses? (p. 215)

The respiratory tract begins with the **nose** and the **nasal cavity,** which is the cavity posterior to the nose. (The mouth and oral cavity are part of the gastrointestinal tract—see Chapter 22.) The nose and nasal cavity perform the following important functions:

- warm and humidify the inhaled air;
- filter out debris from inhaled air and secrete antibacterial substances;
- house olfactory receptors; and
- enhance the resonance of the voice (which is why your voice sounds funny when you have a "stuffed-up" nose).

Let's take a closer look at the external structure of the nose and the internal structure of the nasal cavity.

External Nasal Anatomy

The nose is covered with skin and supported internally by muscle, bone, and cartilage. Both bone and hyaline cartilage form the framework of the nose, including the paired **nasal bones** superiorly, followed by the **lateral cartilages** and the **alar cartilages** (AY-lar) inferiorly (**Figure 21.3**). This framework gives rise to several surface features, including the *root* and *bridge* of the nose (the areas between the eyebrows and eyes, respectively), the

(a) Internal structures

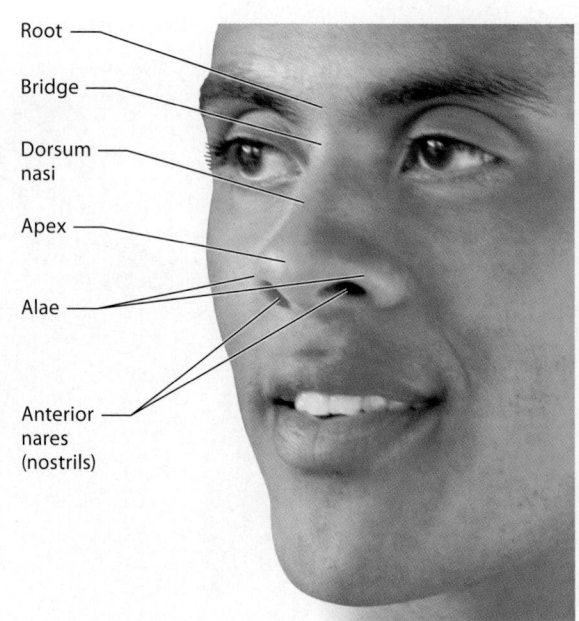

(b) External structures

Figure 21.3 Anatomy of the nose.

dorsum nasi (NAYZ-eye; the anterior margin of the nose), and the *apex*, or tip, of the nose. Lateral to the apex are the flared **alae** (AIL-ee; "wings"), which surround the paired openings called the **nostrils**, or **anterior nares** (NAIR-eez), anteriorly and laterally. Individual variations in nose size and shape are typically due to differences in the structure of its cartilage framework.

Internal Nasal Anatomy and Histology

When you inhale, air goes through the nostrils and enters the nasal cavity, a hollow space framed by bone and cartilage. Notice in **Figure 21.4** that the nasal cavity is larger than you might have thought. It extends from its two anterior openings, the nostrils or anterior nares, to its two posterior openings, called the **posterior nares** (singular, *naris*). The nasal cavity is divided into right and left portions by the *nasal septum,* which, like the framework of the nose, is composed of both bone and hyaline cartilage (review Figure 21.3). The anteriormost part of the nasal cavity, just inside the nostril, is called the **vestibule.** The vestibule contains bristle-like hairs that prevent objects such as insects and airborne particles from entering the nasal cavity.

Although the nasal cavity is a hollow structure, it contains very little empty space. Rather, it is filled with bony projections known as **nasal conchae** (KON-kee; singular, *concha*) (refer back to Figure 7.12). There are three sets of nasal conchae: the *superior* and *middle nasal conchae,* both part of the ethmoid bone; and the *inferior nasal conchae,* which are independent bones (see Chapter 7). The nasal conchae curl around three narrow passages called the **superior, middle,** and **inferior nasal meatuses** (mee-AY-tuss-ez). This arrangement causes air flow through the meatuses to be turbulent. The turbulent air flow extracts dust and other debris from the air, much in the same way that the

turbulent air flow of a clothes dryer helps to remove lint from your clothes. The foreign debris sticks to the mucus lining the nasal conchae, so it does not reach the deeper structures of the respiratory tract. The turbulence also slightly delays the movement of the air, during which time it picks up moisture from the mucous membranes.

Figure 21.4 also shows hollow cavities called the **paranasal sinuses,** or simply *sinuses,* which are connected to the nasal cavity via small passageways located between the nasal conchae. The paranasal sinuses are located within the frontal, ethmoid, sphenoid, and maxillary bones, and so are named the **frontal, ethmoid, sphenoid,** and **maxillary sinuses** (see Chapter 7). Like the nasal cavity, they warm, humidify, and filter the air. They also lighten the skull and enhance voice resonance. Their epithelium is continuous with that of the nasal cavity, which allows the movement of air and mucus between the two areas. Unfortunately, this arrangement also allows infections to spread from the nasal cavity into the paranasal sinuses.

Now let's look a little closer at how the type of cells in each of these areas allows them to perform their functions. The vestibule is lined with stratified squamous epithelium, which enables it to be more resistant to mechanical stresses such as abrasion or scratching. Posterior to the vestibule, the epithelium changes to become two types of mucous membrane: the **olfactory mucosa** and the **respiratory mucosa** (also called *respiratory epithelium*). The olfactory mucosa is located on the roof of the nasal cavity, where it houses smell receptors (see Figure 21.4a). These bipolar neurons project through the holes in the cribriform plate to contact the overlying olfactory bulbs in the brain (see Chapter 15).

The remainder of the nasal cavity is lined by the respiratory mucosa, which is composed of pseudostratified ciliated columnar

(a) Sagittal section of nasal cavity

Nasal meatuses:
Superior
Middle
Inferior
Posterior naris
Olfactory mucosa
Vestibule
Anterior naris

Sinuses:
Sphenoid
Frontal
Ethmoid
Maxillary

Nasal conchae:
Superior
Middle
Inferior

(b) Frontal section of nasal cavity (CT scan)

Orbits

Figure 21.4 **Anatomy of the nasal cavity.**

epithelium (see Figure 4.5). Interspersed throughout the ciliated columnar cells are specialized unicellular glands called **goblet cells,** which secrete mucus. As one of the main functions of the nasal cavity is air filtration, this ciliated, goblet cell–rich epithelium is an example of how form follows function in the nasal cavity. Foreign particles get trapped in the mucus made by the goblet cells, and the cilia beat in unison to propel the debris and mucus toward the posterior nasal cavity, where the mixture travels down the pharynx and is eventually swallowed or expelled by sneezing or blowing your nose. (Almost all microorganisms that make it to the stomach are destroyed by the acid found there.)

Deep to the respiratory mucosa is a lamina propria that is richly supplied with blood vessels. Interestingly, the lamina propria contains a number of large venous sinusoids (see Chapter 18) that can become engorged with blood, an anatomical feature similar to the erectile tissue of the penis. The venous sinusoids fill with blood on one side of the nasal cavity, partially obstructing the flow of air through that side, which helps to prevent the mucosa from drying out. About every 30 minutes, the side filled with blood switches, and air flows through the previously obstructed side. This partial blockage is not usually noticeable, but it may become apparent when the nasal mucosa is inflamed due to allergy or a viral infection such as a cold.

The Pharynx

After inspired air has passed through the posterior nares, it enters the next segment of the respiratory tract, the **pharynx** (FAIR-inks), commonly called the throat. Anatomically, the pharynx has three divisions: the *nasopharynx,* the *oropharynx,* and the *laryngopharynx* (**Figure 21.5**).

The **nasopharynx** (nayz-oh-FAIR-inks) lies posterior to the nasal cavity. Like this cavity, it is lined with pseudostratified ciliated columnar epithelium; the two structures therefore perform the same functions of warming, humidifying, and filtering the inspired air. The nasopharynx extends from the posterior nares to a part of the soft palate called the *uvula* (YOOV-yuh-lah). The uvula and the soft palate move posteriorly during swallowing to prevent food from entering the nasopharynx and nasal cavity. However, sometimes this mechanism fails, especially when you laugh, and the unfortunate result is that food may come out of your nose.

The next segment of the pharynx is the **oropharynx** (OR-oh-fair-inks), located posterior to the oral cavity. Because the oropharynx serves as a passageway for both food and air, it is lined with a protective nonkeratinized stratified squamous epithelium rather than pseudostratified ciliated columnar epithelium, which is too thin to be protective. It extends from the uvula to the tip of the larynx, called the *epiglottis.*

The last portion of the pharynx, the **laryngopharynx** (lah-ring'-goh-FAIR-inks), extends from the hyoid bone to the *esophagus,* the tube that conveys food to the stomach. Anteriorly, it opens into the larynx (the "voice box"), and posteriorly it opens into the esophagus. Like the oropharynx, it is a passageway for both food and air, and so is lined with a nonkeratinized stratified squamous epithelium to prevent abrasion from food.

Quick Check

☐ 1. Match the following terms with the correct definition or description.

_____ Respiratory mucosa

_____ Nasopharynx

_____ Laryngopharynx

_____ Oropharynx

_____ Paranasal sinuses

_____ Goblet cell

a. Posterior to the oral cavity, lined with stratified squamous epithelium
b. Extends from the posterior nares to the uvula
c. Unicellular gland that secretes mucus
d. Extends from the hyoid bone to the esophagus
e. Pseudostratified ciliated columnar epithelium
f. Hollow cavities that are connected to the nasal cavity

The Larynx

◀◀ **Flashback**

1. What are the basic properties of hyaline cartilage? (p. 142)
2. What are the basic properties of elastic cartilage? (p. 142)

Inspired air moves from the laryngopharynx into the next part of the respiratory tract: the **larynx** (LAIR-inks), or voice box (**Figure 21.6**). This short tube lies anterior to the esophagus and extends from about the third to the sixth cervical vertebra (see Figure 21.5). The larynx keeps food and liquids out of the rest of the respiratory tract and, as its common name implies, it houses the vocal cords, which are involved in sound production.

Superior to the vocal cords, the larynx is lined with a stratified squamous nonkeratinized epithelium that is continuous with that of the laryngopharynx. This prevents the larynx surface from abrasion due to contact with food. Inferior to the vocal cords, the epithelium changes to a pseudostratified ciliated columnar epithelium, which we saw in the nasal cavity and the nasopharynx. The cilia in the larynx propel mucus and debris

Figure 21.5 Anatomy of the pharynx.

Nasopharynx

Oropharynx

Laryngopharynx

Esophagus

Nasal cavity

Oral cavity

Uvula

Tongue

Epiglottis

Larynx

Vocal cord

Hyoid bone

Thyroid
cartilage

Cricothyroid
ligament

Cricoid
cartilage

Cartilage
of trachea

(a) Anterolateral view

Epiglottis

Corniculate
cartilage

Arytenoid
cartilage

Esophagus

(b) Posterior view

Hyoid bone

Adipose tissue

Thyroid
cartilage

Vestibular
fold

Vocal fold

Cricoid
cartilage

Cartilage
of trachea

(c) Midsagittal section

Figure 21.6 Anatomy of the larynx.

Practice art labeling
@MasteringA&P®

upward and out; when we "clear our throats," we are expelling this mucus from the larynx.

Cartilage Framework of the Larynx

The flexible framework for the larynx is made of nine pieces of cartilage (see Figure 21.6). This cartilage framework is supported by muscles that attach the larynx to other structures of the neck and by muscles within the larynx itself (see Chapter 9 for a review). The three largest cartilages are unpaired, whereas the other six occur as three pairs of smaller cartilages.

The three unpaired cartilages are the thyroid cartilage, the epiglottis, and the cricoid cartilage. The shieldlike **thyroid cartilage** is the largest piece of the larynx's framework, forming its anterior and superior walls (see Figure 21.6). It is attached superiorly to the hyoid bone and inferiorly to the cricoid cartilage by fibrous membranes. The thyroid cartilage is prominent anteriorly, and is known as the "Adam's apple." This common name refers to the fact that the larynx is typically more prominent in males than in females, due to the influence of male sex hormones on its shape and size during puberty.

Posterior to the thyroid cartilage is the flap of elastic cartilage called the **epiglottis**. Notice in Figure 21.6b that the base of the epiglottis is attached to the posterior side of the thyroid cartilage, and its superior edge is free standing. Normally this superior edge stands upright to allow air to enter the larynx at its opening, called the **glottis** (refer to Figure 21.7). During swallowing, however, the larynx is elevated by its surrounding muscles so that the epiglottis

covers the glottis, preventing food and liquids from entering the larynx. The tongue also pushes the epiglottis down during swallowing, helping to keep the larynx sealed off. Occasionally this mechanism fails, usually when you are laughing or talking while eating or drinking, and food or liquid "goes down the wrong pipe" (or is aspirated) into the larynx instead of the esophagus. This triggers a cough reflex that removes the offending bit of food or liquid.

The third piece of unpaired cartilage is the **cricoid cartilage** (KRY-koyd) (see Figure 21.6). The cricoid cartilage is inferior to the thyroid cartilage, and is attached to it by a thin membrane called the *cricothyroid ligament*. This ligament is the site for a procedure sometimes performed on a choking victim, called a *cricothyroidotomy*. During this procedure the cricothyroid ligament is cut and a sterile tube is inserted into the larynx, which restores the airway.

The remaining six laryngeal cartilages occur in three pairs, and make up part of the posterior and lateral walls of the larynx. These small cartilages include the following:

- **Arytenoid cartilages.** The arytenoid (ar-ih-TEN-oyd) cartilages are triangular pieces of cartilage that are involved in sound production (see Figure 21.6b and c). They attach to the *vocal folds* and the intrinsic muscles of the larynx, and we discuss them in the next section.
- **Corniculate cartilages.** Notice in Figure 21.6b and c that the arytenoid cartilages are capped by small pieces of cartilage—these are the tiny corniculate (kor-NIK-yu-layt) cartilages. Like the arytenoids, the corniculate cartilages function in sound production.

• **Cuneiform cartilages.** The cuneiform (kyu-NEE-ih-form) cartilages are found in the lateral wall of the larynx, where they help to support the epiglottis.

The thyroid, cricoid, and the bulk of the arytenoid cartilages are hyaline cartilage, whereas the rest are composed of elastic fibrocartilage. Within this cartilaginous framework of the larynx, we find the elastic structures that are involved in sound production, which we address next.

Mucosal Folds and Sound Production

As shown in **Figure 21.7**, the inner surface of the larynx isn't smooth. Rather, folds of its mucosa project into its lumen. The superior pair of folds extends from the arytenoid cartilages to the thyroid cartilage. These are the **vestibular folds,** or the *false vocal cords.* As their common name implies, the vestibular folds play no role in sound production. However, they do have the important function of closing off the glottis during swallowing.

Inferior to the vestibular folds we find another set of mucosal folds called the **vocal folds,** or the *true vocal cords.* Like the vestibular folds, the vocal cords have a posterior attachment to the arytenoid cartilages and an anterior attachment to the thyroid cartilage. At the core of the vocal cords are elastic bands called the *vocal ligaments.* Notice in Figure 21.7 that the vocal cords appear whitish. They have this appearance because the mucosa that overlies the vocal ligaments is thin, making the white vocal ligaments visible.

The muscles of the larynx control the length and tension of the vocal cords. When these muscles contract, they pull on the arytenoid and corniculate cartilages and cause them to pivot. Figure 21.7 shows the effect of this pivoting motion—when the cartilages

View

Structures in the living throat

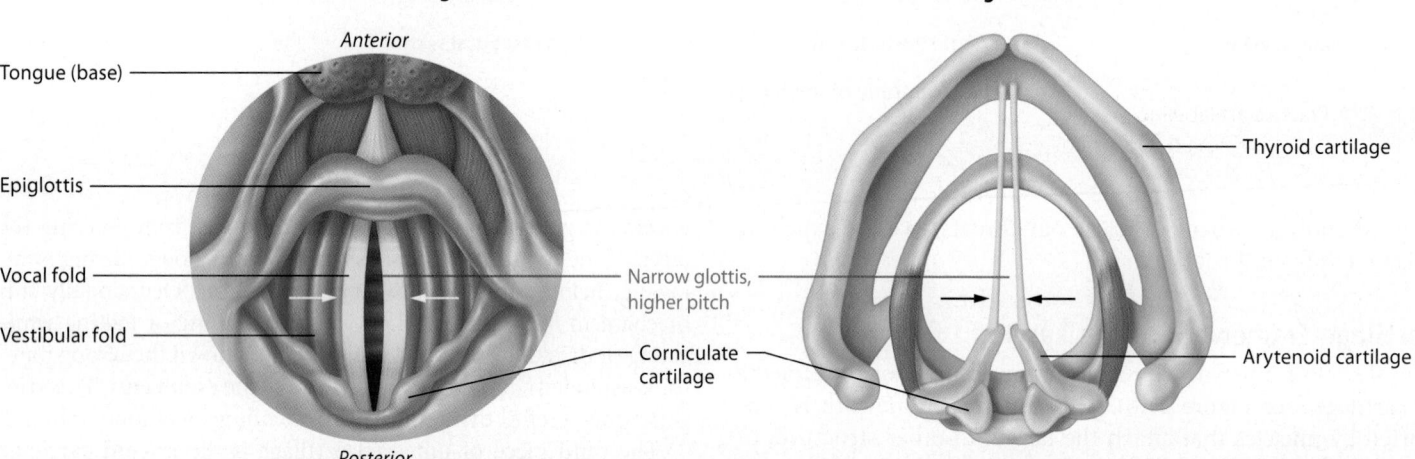

Anterior

Tongue (base)

Epiglottis

Vocal fold

Vestibular fold

Posterior

Narrow glottis, higher pitch

Corniculate cartilage

Cartilage structure

Thyroid cartilage

Arytenoid cartilage

(a) Adduction of vocal ligaments (corniculate and arytenoid cartilages rotated inward)

Anterior

Tongue (base)

Epiglottis

Vocal fold

Vestibular fold

Posterior

Open glottis, lower pitch

Corniculate cartilage

Thyroid cartilage

Arytenoid cartilage

(b) Abduction of vocal ligaments (corniculate and arytenoid cartilages rotated outward)

Figure 21.7 Changes in the vocal ligaments during speech.

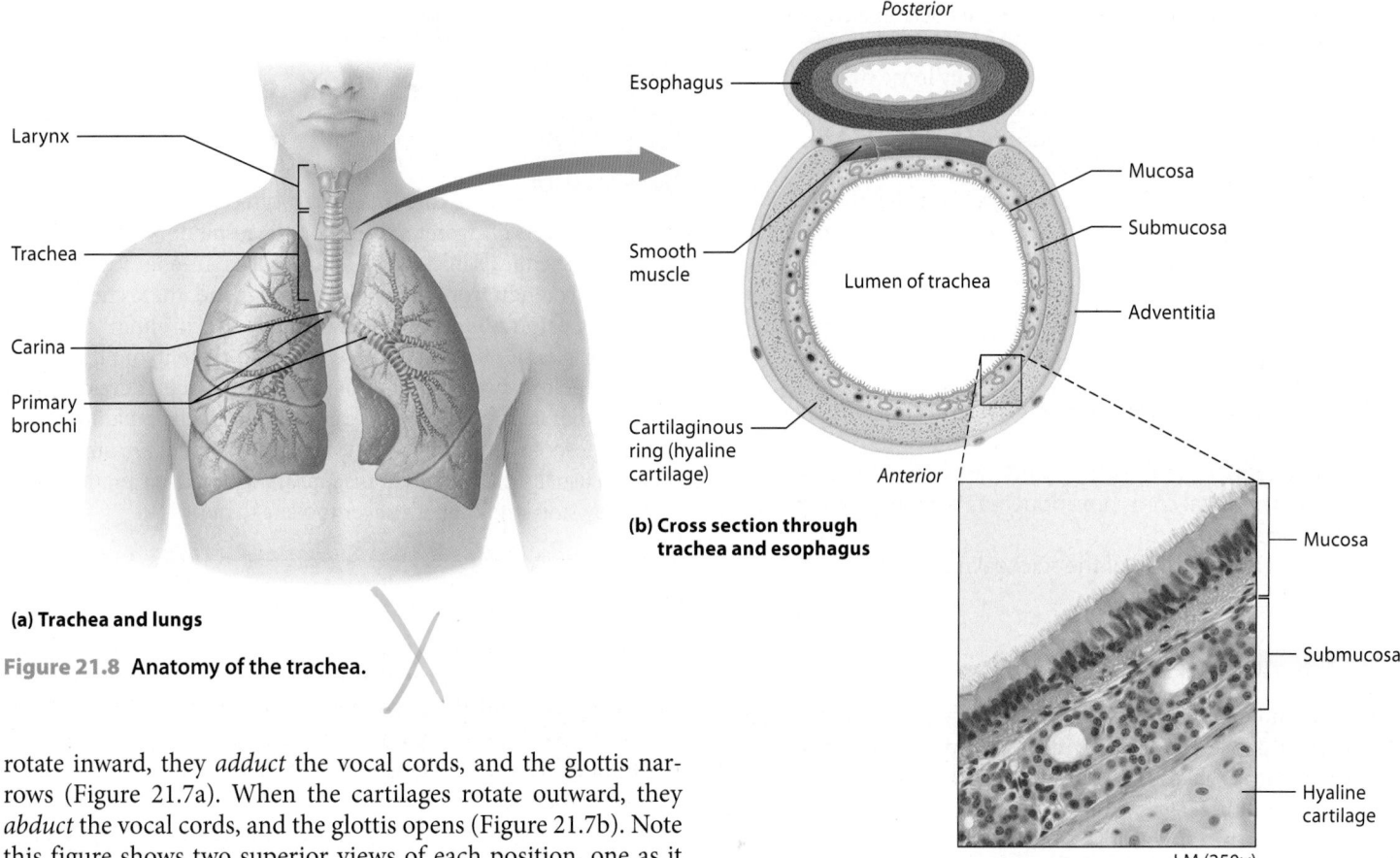

Figure 21.8 Anatomy of the trachea.

(a) Trachea and lungs

Larynx
Trachea
Carina
Primary bronchi

Posterior
Esophagus
Smooth muscle
Lumen of trachea
Cartilaginous ring (hyaline cartilage)
Anterior
Mucosa
Submucosa
Adventitia

(b) Cross section through trachea and esophagus

Mucosa
Submucosa
Hyaline cartilage
LM (250×)

(c) Light micrograph of tracheal tissue layers

rotate inward, they *adduct* the vocal cords, and the glottis narrows (Figure 21.7a). When the cartilages rotate outward, they *abduct* the vocal cords, and the glottis opens (Figure 21.7b). Note this figure shows two superior views of each position, one as it would look in a living throat and one of the cartilage structure.

Sound is produced as expired air passes over the vocal ligaments. The loudness of the sound is determined by the force of the airstream—the more forceful the expiration, the louder the sound that is produced. The sound's pitch is largely determined by the tension of the vocal ligaments and the speed of their vibration, just like plucking the strings of a guitar. A higher pitch is produced when the vocal ligaments are tightly adducted, because they are tense and vibrate more rapidly (see Figure 21.7a). A lower pitch results when the vocal ligaments are more abducted, because they are looser and vibrate more slowly (see Figure 21.7b). Adult males typically have a deeper (lower-pitched) voice than females because their vocal ligaments are longer and thicker as a result of their wider larynx, and thus vibrate more slowly.

Note that the movement of air across the vocal cords produces only a buzzing sound, not speech. Speech requires the coordinated efforts of the structures superior to the glottis, including the muscles of the pharynx, the soft palate, the tongue, and the lips.

The Trachea

The inspired air passes from the larynx into the **trachea** (TRAY-kee-uh), often called the "windpipe" because it delivers this air to the lower structures of the respiratory tract (**Figure 21.8a**). The trachea is a hollow tube about 2 cm in diameter and about 10–12 cm long. It begins in the inferior part of the neck and extends to the mediastinum.

As you can see in **Figure 21.8b**, rings of hyaline cartilage cover the anterior and lateral surfaces of the trachea, but not

its posterior surface, giving these rings a C shape. The cartilage rings are rigid enough to support the trachea and keep it open, or *patent* (PAY-tent), but they are also flexible enough to allow the trachea to change diameter during pulmonary ventilation.

On its posterior surface, where the cartilage is absent, the trachea is covered with elastic connective tissue and a band of smooth muscle. For this reason, the posterior surface of the trachea is soft, an arrangement that allows the esophagus to expand during swallowing. When you swallow a very large mouthful of food without adequately chewing it first, the esophagus expands into the trachea a significant amount, causing a temporary feeling of discomfort until the food has passed through the esophagus.

The last tracheal cartilage ring is called the **carina** (kar-EYE-nah). The carina is different in appearance from the other rings, forming a "hook" of cartilage that curves down and back (see Figure 21.8a). From the anterior view it forms partial rings that surround the first branches of the bronchial tree. The mucosa of the carina contains sensory receptors that trigger a violent cough reflex if any foreign material comes into contact with them.

Note in **Figure 21.8b and c** that the trachea follows the same histological pattern we have seen for most hollow organs in the body, with an inner *mucosa*, a middle *submucosa*, and an outer *adventitia* (sometimes called the *serosa*). The **mucosa** (myew-KOH-suh) of the trachea is continuous with that of the inferior larynx and consists of respiratory epithelium (pseudostratified ciliated columnar) with numerous goblet cells that secrete mucus. The mucus lines the trachea and traps foreign debris; the cilia beat in a coordinated fashion that propels the mucus up to the pharynx to be either swallowed or expelled. The *submucosa* is primarily loose connective tissue, but it also houses the trachea's cartilage rings. The outermost *adventitia* (ad-ven-TISH-ah) is dense irregular connective tissue that anchors the trachea to the surrounding structures.

Cigarette smoking alters the structure, and also the function, of the epithelial lining of the respiratory tract. These effects are discussed in *A&P in the Real World: Smoker's Cough.*

Quick Check

☐ 2. Describe the external and internal structure of the larynx.
☐ 3. What happens to the glottis and the pitch of the sound when the vocal cords are adducted? What happens when they are abducted?
☐ 4. What is the function of the tracheal mucosa?

The Bronchial Tree

Inhaled air now continues on its way through the conducting zone. At the level of the carina, the trachea divides into right and left *primary bronchi,* which enter the right and left lungs, respectively, through the hilum (**Figure 21.9a**). Once in the lung, each bronchus branches to form what is collectively called the **bronchial tree** (BRONG-kee-uhl)—a series of progressively smaller tubes that terminate in tiny alveoli, the structures for gas exchange.

The Bronchi

The first structures of the bronchial tree are the **bronchi.** As we just mentioned, the first bronchi are named the right and left **primary bronchi,** which branch from the distal portion of the

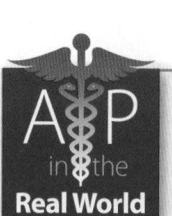

Smoker's Cough

The deep, rattling sound of the "smoker's cough" is fairly easy for most of us to recognize. This cough is linked directly to the numerous adverse effects of cigarette smoke on the respiratory system. For one, the chemicals in cigarette smoke act as irritants, increasing mucus secretion from the goblet cells. Second, these chemicals partially paralyze, and eventually destroy, the cilia lining the respiratory tract. As a result, more mucus is present, but the cilia are less able to sweep it out of the airways. A cough develops, as this is the only way the body can prevent mucus buildup. The good news for smokers is that cilia reappear, and so smoker's cough usually diminishes or disappears within a few months after they stop smoking. It's never too late to quit!

trachea. Figure 21.9a shows that the two primary bronchi differ in appearance: The right primary bronchus is wider, shorter, and straighter, and the left is narrower, longer, and more horizontal. These differences are due primarily to the position of the left lung in relation to the heart. The right primary bronchus is the more common place for an inhaled object (such as a peanut) to become lodged because the tube is more of a "straight shot" from the trachea.

Inside the lungs, the primary bronchi branch into **secondary bronchi.** Notice that there are three secondary bronchi in the right lung, and only two in the left lung due to the space needed for the heart. The secondary bronchi then branch into smaller **tertiary bronchi.** Generally 10 tertiary bronchi can be found in each lung, although the left lung may have one or two fewer than the right lung. The tertiary and subsequent bronchi continue

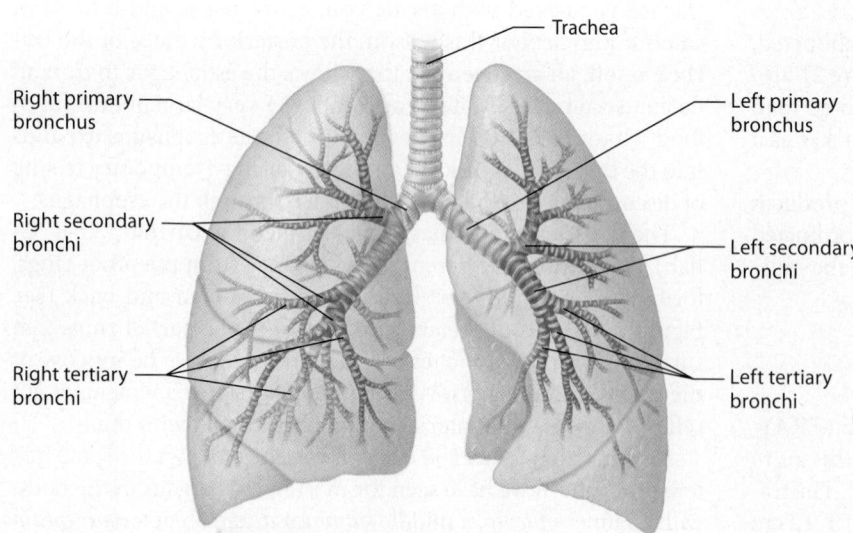

(a) Conducting zone passages and bronchial tree

(b) Cast of bronchial tree

Figure 21.9 Branching pattern of the bronchial tree.

to branch into smaller and smaller bronchi. The photo in **Figure 21.9b** shows a cast of the bronchial tree that demonstrates how extensively this branching occurs. In fact, the bronchial tree branches about 26 times in total.

As the airways divide and get smaller, their histology changes significantly. The primary bronchi are nearly identical to the trachea, but three changes are evident as bronchi become smaller:

- The cartilage changes from C-shaped to complete rings, and then to irregular plates that are progressively fewer in number.
- The epithelium gradually changes from respiratory epithelium in the larger bronchi to columnar cells that become progressively shorter in smaller bronchi.
- The amount of smooth muscle gradually increases.

These histological changes reflect the different functions of larger and smaller bronchi. The larger bronchi need to serve as fairly rigid conduits to allow the passage of large volumes of air, which is why they contain more hyaline cartilage and less smooth muscle. However, the smaller airways must be able to change size to control air flow into the bronchioles and alveoli. For this reason, they contain less hyaline cartilage and more smooth muscle.

Bronchioles

The smallest airways of the bronchial tree are its tiny **bronchioles** (BRONG-kee-olz) (**Figure 21.10**). By definition, bronchioles are less than 1 mm in diameter and lack cartilage. Bronchioles also feature a thicker ring of smooth muscle and simple cuboidal epithelium with few cilia and few, if any, goblet cells. This is because most of the dust and debris have been removed by the time the air reaches the bronchioles. Bronchioles continue to branch until they become tiny *terminal bronchioles,* which are the final part of the conducting airways. To give you a sense of perspective, your lungs contain about 65,000 terminal bronchioles.

Each terminal bronchiole gives rise to two or more smaller **respiratory bronchioles,** which are surrounded by very thin bands of smooth muscle. These bronchioles have alveoli budding off their walls and so are the beginning of the respiratory tract's respiratory zone. Each respiratory bronchiole then branches into two or more smaller **alveolar ducts,** short tubes that also contain alveoli along their walls. The alveolar ducts finally terminate in **alveolar sacs,** which are grapelike clusters of alveoli.

Terminal bronchiole

Pulmonary venule

Respiratory bronchiole

Alveolar duct

Pulmonary capillaries

Elastic fibers

Alveolar sac

Pulmonary arteriole

Smooth muscle

Alveoli

Conducting zone | *Respiratory zone*

(a) Structures of the respiratory zone

(b) Electron micrograph of alveoli   SEM (95×)

Practice art labeling @MasteringA&P®

Figure 21.10 **Anatomy of the respiratory zone.**

At the alveoli, we have finally arrived at the site for gas exchange. As a review, to get to the alveoli, our inhaled air had to pass through the following:

Nares → Nasal cavity → Nasopharynx → Oropharynx → Laryngopharynx → Larynx → Trachea → Primary bronchi → Secondary bronchi → Tertiary bronchi → Multiple branches of bronchi → Bronchioles → Terminal bronchioles → Respiratory bronchioles → Alveolar ducts → Alveolar sacs

Alveoli and the Respiratory Membrane

◄◄ Flashback

1. What is the function of simple squamous epithelium? (p. 129)
2. What is a phagocyte? What is its function? (p. 83)

When the inhaled air finally reaches the alveoli, the terminal structures of the respiratory tract, the gases in the air are available to diffuse into the blood. Let's look at the structures that accomplish this gas exchange.

The majority of alveoli are located in groups called alveolar sacs, but some do protrude from the walls of respiratory bronchioles and alveolar ducts. Notice in Figure 21.10a that externally alveoli are surrounded by a network of elastic fibers, very thin bands of smooth muscle and pulmonary capillaries that arise from the pulmonary artery.

Figure 21.11a shows the structure of a single alveolus. Each round, thin-walled alveolus has three cell types:

- **Type I alveolar cells** are squamous cells that make up about 90% of the cells in the alveolar wall. They are exceedingly thin, which permits rapid diffusion of gases across their cell membranes.
- **Type II alveolar cells** are small cuboidal cells that make up about 10% of the cells in the alveolar wall. Within their cytoplasm are the precursors to a chemical called *surfactant*, which helps to reduce the surface tension on the alveoli. (We discuss surfactant in detail in Module 21.3.)

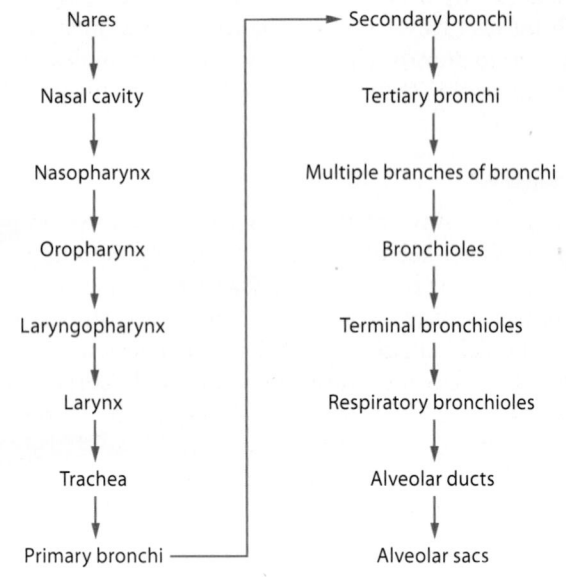

Type I alveolar cells

Type II alveolar cells

Alveolar macrophage

Pulmonary capillaries

(a) Alveoli and pulmonary capillaries

Respiratory membrane

Erythrocyte

Alveolus

Capillary

Fluid containing surfactant

Basal lamina

Capillary endothelial cell

Type I alveolar cell

(b) Respiratory membrane

Figure 21.11 Structures of the alveoli and the respiratory membrane.

Practice art labeling
@MasteringA&P®

- **Alveolar macrophages** are phagocytes derived from cells formed in the bone marrow. They roam the inner surface of the alveoli, cleaning up and digesting any stray debris that was not filtered out in the bronchial tree (hence their alternate name, "dust cells"). Most alveolar macrophages migrate up to the bronchioles, where they are swept up to the pharynx and eventually swallowed. Alveolar macrophages are directly involved with an agent that causes a potentially serious respiratory infection; see *A&P in the Real World: Tuberculosis* on page 812.

The very thin type I alveolar cells help make up what is known as the **respiratory membrane,** which is the barrier through which gases must diffuse. Note in **Figure 21.11b** that the respiratory membrane has three major parts: (1) the type I alveolar cells; (2) the basal lamina of the type I alveolar cells, which is fused with the capillary basal lamina; and (3) the capillary endothelial cells. Essentially, the respiratory membrane is formed from the joining of the simple squamous epithelium of the alveolus with the endothelium of the pulmonary capillary. These three parts together are extremely thin, measuring between 0.2 and 0.6 μm (about 1/100 the thickness of a human hair). The combined surface area of the respiratory membranes is huge—for both lungs it is about 80–100 m^2 (about 1000 square feet—the size of a small house). In an example of the Structure-Function Core Principle (p. 25), this large surface area (along with the thinness of the respiratory membrane) makes pulmonary gas exchange a very efficient process.

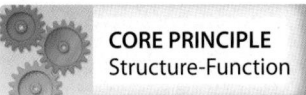

CORE PRINCIPLE
Structure-Function

The Lungs and Pleurae

◀◀ **Flashback**

1. Where are the pleural cavities located? Where is the mediastinum? (p. 17)

The left and right **lungs** are separated by the heart and the mediastinum. The lung's flat **base** rests on the diaphragm, and its superior **apex** sits just above the clavicle (**Figure 21.12a**). The anterior, posterior, and lateral surfaces of the lung come into contact with the rib cage, and for this reason are called the *costal surfaces* of the lung (*cost-* = "rib"). The lung's medial surface comes into contact with the structures of the mediastinum, and so is named the *mediastinal surface* (mee′-dee-ass-TY-nal) (**Figure 21.12b**). The mediastinal surface of both lungs contains a triangular depression called the **hilum** (HY-lum), where the primary bronchi, blood vessels, lymphatic vessels, and nerves enter and exit the lung. Additionally, the mediastinal surface of the left lung comes into contact with the heart, resulting in a groove called the **cardiac notch.**

As you can see in Figure 21.12a, each lung is divided into **lobes.** The left lung has two lobes because of the space necessary for the heart: the left superior lobe and the left inferior lobe, separated from one another by an *oblique fissure.* The right lung has three lobes: the right superior, the right middle, and the right inferior lobes, separated from one another by the *horizontal* and *oblique fissures.* Each secondary bronchus supplies one lobe of a

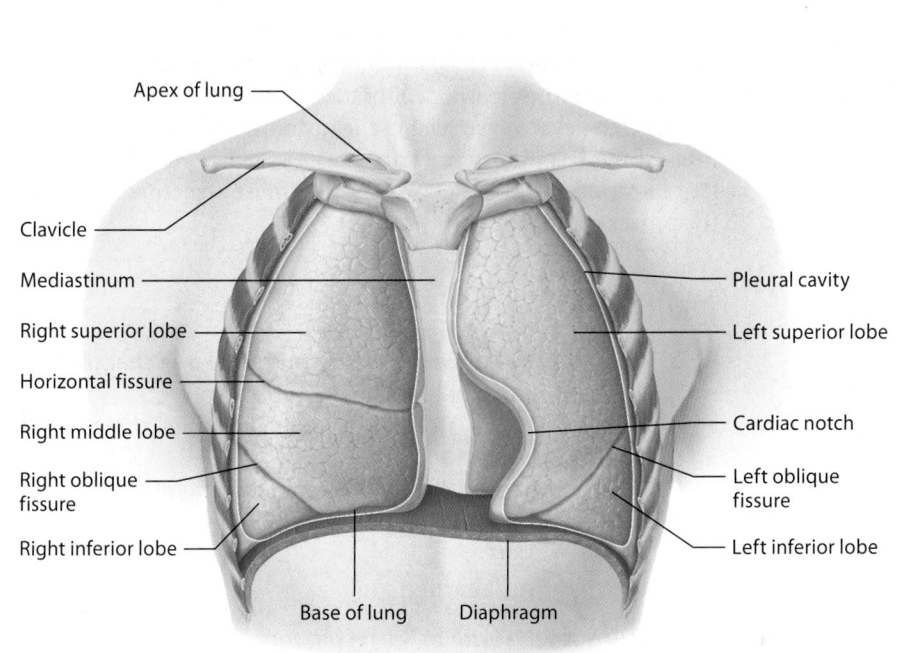

Apex of lung
Clavicle
Mediastinum
Right superior lobe
Horizontal fissure
Right middle lobe
Right oblique fissure
Right inferior lobe
Base of lung   Diaphragm
Pleural cavity
Left superior lobe
Cardiac notch
Left oblique fissure
Left inferior lobe

(a) Anterior view of right and left lungs

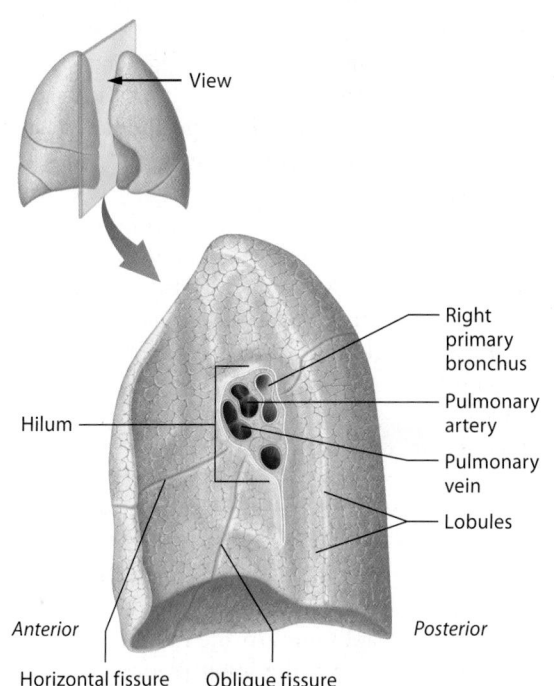

View
Hilum
Right primary bronchus
Pulmonary artery
Pulmonary vein
Lobules
Anterior
Posterior
Horizontal fissure   Oblique fissure

(b) Mediastinal surface of right lung

Figure 21.12 Anatomy of the lungs and associated structures.

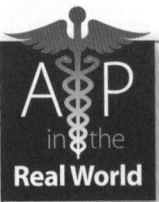

Tuberculosis

The respiratory infection **tuberculosis,** or **TB,** is caused by the bacterium *Mycobacterium tuberculosis.* The bacterium is spread easily via the air, generally when an infected individual coughs or sneezes. About 90% of tuberculosis infections are *latent,* or asymptomatic. The remaining 10% of infections are *active,* meaning that the infected individual shows symptoms of the disease. Symptoms include a persistent cough with blood-tinged sputum, fever, night sweats, and weight loss. Tuberculosis may also involve organs outside the respiratory system, such as the bones or the central nervous system, resulting in additional, organ-specific symptoms.

Active tuberculosis infection begins when the bacteria reach the alveoli. Here they enter alveolar macrophages and trigger an inflammatory response that attracts lymphocytes and other immune cells. Eventually, the lymphocytes attempt to surround and "wall off" the infected area, which results in the formation of a structure known as a *granuloma.* The granuloma can protect neighboring tissues and, in some cases, lead to the infection's becoming latent. However, it can also lead to death of the normal cells within the granuloma, a condition called *necrosis.*

Diagnosis of active tuberculosis is based on the finding of granulomas on chest x-ray; a positive tuberculin skin test, which indicates the presence of antibodies to the bacterium; and the presence of the bacteria in the sputum. Tuberculosis is notoriously difficult to treat, as the structure of *M. tuberculosis* renders many common antibiotics ineffective. In addition, most strains of the bacterium are resistant to multiple drugs that were previously effective, further limiting the choice of antibiotic therapy. Treatment of active infections usually consists of a cocktail of several antibiotics taken for a period of 6 months. Individuals with latent infections are treated with a single antibiotic to prevent the infection from becoming active. Untreated active infections have about a 50% mortality rate, and untreated latent infections have a greater chance of becoming active.

---

lung, which explains why the secondary bronchi are often called *lobar bronchi.*

The lobes are divided by thin walls of connective tissue into **bronchopulmonary segments** (brong′-koh-PULL-mon-air-ee). Each tertiary bronchus serves one bronchopulmonary segment (for this reason they are often called *segmental bronchi*). As with tertiary bronchi, there are usually 10 bronchopulmonary segments. Each bronchopulmonary segment is further subdivided into **lobules,** hexagonal structures about the size of a dime.

Each lung is served by a large **pulmonary artery,** which brings deoxygenated blood from the right ventricle of the heart to the lungs (Figure 21.12b; see Chapter 17). Oxygenated blood leaves the lungs and returns to the left atrium of the heart by smaller **pulmonary veins.** However, neither the pulmonary artery nor the pulmonary veins supply the tissues of the lungs with oxygenated blood and nutrients. Instead, this task is accomplished by the *bronchial arteries,* which are part of the systemic circuit, not the pulmonary circuit.

Each lung is enclosed by a subcavity of the thoracic cavity called the **pleural cavity** (PLOOR-al; Figure 21.13; see Chapter 1). The pleural cavities are found between two layers of a serous membrane; this arrangement is very similar to that of the pericardial cavity around the heart (refer back to Figure 17.3). The outer pleural membrane, the **parietal pleura** (pah-RY-eh-tal), is fused to the structures surrounding the lungs, including the rib cage and the diaphragm. At the hilum, the parietal pleura turns over on itself to form the inner pleural membrane, the **visceral pleura** (VISS-er-al). The visceral pleura is continuous with the surface of the lobes of the lungs and dives into each of the lung's fissures.

These pleural membranes secrete a thin layer of serous fluid, often called **pleural fluid,** that fills the pleural cavities. Just as serous fluid in the pericardial cavities lubricates the beating heart, pleural fluid lubricates the lungs as they expand and contract, reducing friction. As you will see in Module 21.3, the pressure within the pleural cavity also helps to prevent the lungs from collapsing when you take a breath. The pleural fluid helps the visceral pleura adhere to the parietal pleura. See *A&P in the Real World: Pleuritis and Pleural Friction Rub* on page 816 to learn about what happens when the pleural membranes are inflamed.

The changes we have seen in the histology of the different portions of the respiratory tract are summarized in Table 21.1. Now let's move on to how the air gets into these structures through the process of breathing.

Quick Check

☐ 5. How does the epithelium of the bronchial tree change as it divides into smaller branches?

☐ 6. Trace the pathway from the primary bronchi to the alveolar sacs.

☐ 7. What structures make up the respiratory membrane?

☐ 8. Explain the structure of the pleural cavities.

Apply What You Learned

☐ 1. How does the form of the respiratory epithelium follow its function?

☐ 2. Explain why a person with laryngitis (inflammation of the larynx) "loses" his or her voice.

Figure 21.13 The pleurae and pleural cavities. Notice that you are looking at an inferior view of this transverse section of the trunk.

Practice art labeling @MasteringA&P®

TABLE 21.1 FUNCTION AND COMPOSITION OF REGIONS OF THE CONDUCTING AND RESPIRATORY ZONES

Structure	Function	Type of Epithelia	Other Histological Features
Nasal cavity	• Detect odors • Warm, moisten, filter air • Sound resonance	Olfactory, pseudostratified ciliated columnar	Richly supplied with blood vessels for maximal warming and humidification
Nasopharynx	Warm, moisten, filter air	Pseudostratified ciliated columnar	—
Oropharynx and laryngopharynx	Conduit for food and air	Stratified squamous nonkeratinized	—
Larynx	• Proper routing of food and air • Sound production • Protection of the airways • Warming and moistening of air	Superior to the vocal cords: stratified squamous; inferior to the vocal cords: pseudostratified ciliated columnar	Framework is hyaline and elastic cartilage; contains the vestibular and vocal folds
Trachea	Warm, moisten, filter air	Pseudostratified ciliated columnar	Has C-shaped rings of cartilage for flexible support
Bronchi	Warm, moisten, filter air	Pseudostratified columnar to simple columnar	Large bronchi contain plates of hyaline cartilage; amount of cartilage decreases as the diameter of the bronchi decreases
Bronchioles	Smooth muscle controls air flow, some filtering of air	Simple columnar to simple cuboidal	Lack cartilage; contain large amounts of smooth muscle
Respiratory bronchioles and alveolar ducts	Gas transport, gas exchange	Simple cuboidal to simple squamous	Very little smooth muscle; have alveoli budding from their walls
Alveoli	Gas exchange	Extremely thin simple squamous (type I alveolar cells)	Also contain cuboidal type II alveolar cells and alveolar macrophages

3. Predict what would happen if the tracheal hyaline cartilage rings were absent. What might happen if they were made of bone rather than cartilage?

4. How does form follow function in type I alveolar cells?

See answers in Appendix A.

MODULE 21.3
Pulmonary Ventilation

Learning Outcomes

1. Describe Boyle's law, and relate it to the specific sequence of events causing inspiration and expiration.
2. Explain how the muscles of inspiration, forced inspiration, and forced expiration change the volume of the thoracic cavity, and identify the nerves responsible for stimulating those muscles.
3. Explain how the values for atmospheric pressure, intrapulmonary pressure, and intrapleural pressure change with inspiration and expiration.
4. Explain how each of the following factors affects pulmonary ventilation: airway resistance, pulmonary compliance, and alveolar surface tension.
5. Describe and identify the values for the respiratory volumes and the respiratory capacities.

The first process of respiration is breathing, or **pulmonary ventilation.** It consists of two phases: **inspiration,** or inhaling, which brings air into the lungs, and **expiration,** or exhaling, which moves air out of the lungs. In this module, we first examine the relationship between pressure and volume and then explore how this relationship provides the driving force for pulmonary ventilation. We also consider physical factors that affect the ability of the lungs to do their job, such as the diameter of the airways. Finally, we describe the volumes of air that can be measured when a person breathes and explain the use of these measured volumes to determine how well the lungs are functioning.

The Pressure-Volume Relationship

What causes air to move into and out of the body when we breathe? Remember that any time molecules are moved in the body without the use of additional energy, a *gradient* of some sort is involved. Air is a mixture of gas molecules, and the movement of gas molecules

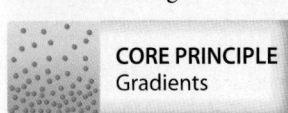

CORE PRINCIPLE
Gradients

depends on **pressure gradients,** that is, differences in pressure between one area and another. Gas molecules move from areas of higher pressure to areas of lower pressure. In an example of the Gradients Core Principle (p. 26), these pressure gradients are what drive the movement of air molecules during inspiration and expiration.

One of the simplest ways to change the pressure in a container is to change its volume. This relationship between pressure and volume is described by **Boyle's law,** which states that at a constant temperature and constant number of gas molecules, the pressure and volume of a gas are inversely related—as the volume of a container increases, the pressure the gas exerts on the container decreases. The opposite is also true—as the volume of a container decreases, the pressure the gas exerts on the container increases. These changes are easy to visualize with a syringe, like the ones you've seen in a doctor's office. When you depress the plunger, you decrease the syringe's volume and increase the pressure. When you pull the plunger out, you increase the syringe's volume and decrease the pressure:

Capped syringe

Lifting the plunger in a capped syringe increases the volume and decreases the pressure inside the syringe.

Pushing on the plunger decreases the volume and increases the pressure inside the syringe.

The example you just saw used a capped syringe, so gas molecules were trapped in the chamber. But when the syringe is opened, you can see how the gas molecules move in relation to the pressure gradient. Before the syringe's plunger is moved, the pressures inside and outside the chamber are equal and the gas molecules are at equilibrium. However, when we lift or push on the plunger, the volume and pressure of the chamber change and gas molecules flow along the pressure gradients:

In an open syringe, gas pressure and volume are at equilibrium.

Lifting the plunger decreases the pressure in the syringe chamber, creating a pressure gradient. Gas molecules move to the area of lower pressure inside the syringe.

Pushing on the plunger increases the pressure in the syringe chamber, reversing the gradient. Gas molecules again move to the area of lower pressure outside the syringe.

Let's now see how pressure and volume changes in the lungs lead to pulmonary ventilation.

The Process of Pulmonary Ventilation

◄◄ Flashback

1. What are the origin, insertion, and action of the diaphragm? (p. 287)
2. What are the origin, insertion, and action(s) of the internal and external intercostal muscles? (p. 308)

Remember that ventilation consists of inspiration (air moving into the lungs) and expiration (air flowing out of the lungs). In our syringe example, when you depress the plunger, you decrease the volume and increase the pressure. What happens as a result? Air flows *out* of the syringe. And what happens when you pull out the plunger? You increase the volume and decrease the pressure, so air flows back *into* the syringe. In a similar way, pulmonary ventilation involves volume changes in the thoracic cavity and lungs that lead to the creation of a pressure gradient. This gradient, in turn, causes air to move into or out of the lungs.

Let's take a look at each of these processes and then see how they combine to drive ventilation.

Volume Changes during Ventilation

Since the lungs themselves have no skeletal muscle tissue, they cannot change their volume on their own. Instead, they rely on skeletal muscles of the thoracic cavity, called **inspiratory muscles,** to increase their volume (**Figure 21.14**). The inspiratory muscles don't increase the volume of the lungs directly. Rather, they increase the size of the thoracic cavity, which indirectly increases the volume of the lungs. This functional relationship is due to the pleural membranes. Recall that the parietal pleura is attached to the inner surface of the thoracic cavity, so when the thoracic cavity increases in height and diameter, the parietal pleura is pulled with it. The parietal pleura, in turn, pulls on the visceral pleura, which pulls the lungs outward, increasing their volume.

The main inspiratory muscle is the **diaphragm** (DY-ah-fram), which is located between the thoracic and abdominopelvic cavities. The other inspiratory muscles are the **external intercostal muscles,** which are located between the ribs. As shown in Figure 21.14 ①, when relaxed between breaths, the diaphragm is dome-shaped and bulges up into the thoracic cavity. This keeps the

① **Between breaths**　　② **Inspiration:** Diaphragm and external intercostals contract.　　③ **Expiration:** Diaphragm and external intercostals relax.

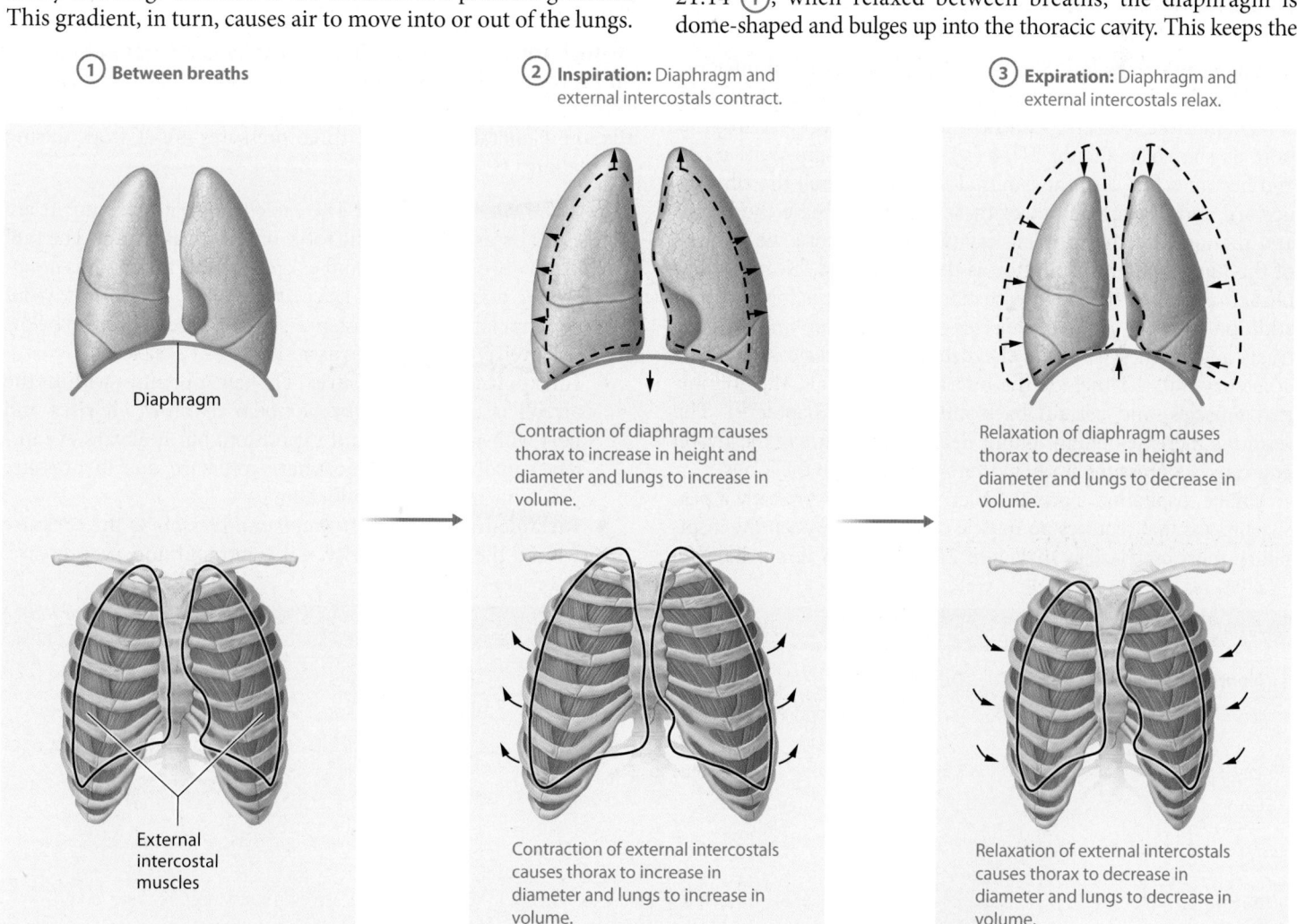

Diaphragm

External intercostal muscles

Contraction of diaphragm causes thorax to increase in height and diameter and lungs to increase in volume.

Contraction of external intercostals causes thorax to increase in diameter and lungs to increase in volume.

Relaxation of diaphragm causes thorax to decrease in height and diameter and lungs to decrease in volume.

Relaxation of external intercostals causes thorax to decrease in diameter and lungs to decrease in volume.

Figure 21.14 Volume changes in pulmonary ventilation: structure and function of the inspiratory muscles in quiet breathing.

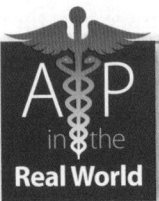

Pleuritis and Pleural Friction Rub

Many medical conditions, ranging from heart failure to pneumonia, can cause inflammation of the visceral and parietal pleura, a condition called *pleuritis*. One of the most common symptoms of this condition is *pleuritic pain*—chest pain when we inhale and often when we exhale as well. The pain results from the rubbing of two inflamed pleurae against one another as the lungs expand and contract. Sometimes this rubbing, called a *pleural friction rub*, can actually be heard with a stethoscope—its sound resembles two pieces of sandpaper grating against each other. The treatment for pleuritis usually involves addressing the underlying medical condition. However, the pleuritic pain may persist for months, even after the medical condition has resolved, due to lasting inflammation of the pleurae.

volume of the lungs low. Similarly, the relaxed external intercostals allow the rib cage, and therefore the lungs, to reduce their size.

When these muscles contract for normal (quiet) inspiration, as shown in Figure 21.14 ②, the diaphragm pulls down and becomes flat, and the external intercostals pull the rib cage upward and outward. Both of these actions increase the height and diameter of the thoracic cavity, and therefore the volume of the lungs. When inspiration is deep or forced, both the diaphragm and the external intercostals contract more forcefully. In addition, another group of muscles called *accessory muscles* contracts and further increases the size of the thoracic cavity. The accessory muscles include the internal intercostals, the abdominal muscles, and certain back muscles (see Chapter 9). The resulting increase in lung volume decreases pressure in the alveoli and causes a pressure gradient that moves air into the lungs.

Unlike inspiration, normal (quiet) expiration is typically a passive process that requires no muscle contractions. When the inspiratory muscles relax, as in Figure 21.14 ③, two things happen:

(1) the diaphragm resumes its original dome shape, which pushes up on the lungs; and (2) the elastic tissue in the lungs **recoils,** and the lungs snap back to a smaller size. These two actions decrease lung volume and raise intrapulmonary pressure above atmospheric pressure, so air flows out of the lungs.

When expiration is forced, the accessory muscles of expiration come into play. These muscles forcefully decrease the size of the thoracic cavity. This is why your abdominal and back muscles are often sore after you have had a cough for a few days. Expiration may also be forced by other means, such as slapping a person on the back or delivering abdominal thrusts that push up on the diaphragm (commonly called the *Heimlich maneuver;* HYM-lik). Both of these actions are used on people who are choking, in the hope that a forceful expiration will dislodge the obstruction and restore breathing.

Not all inspiratory and expiratory movements are for breathing, however. *Nonrespiratory movements,* which include sighs, yawns, and coughs, help to keep the alveoli inflated and prevent obstruction of the airways. The nonrespiratory movements are defined in **Table 21.2.**

Pressure Changes during Ventilation

As we learned from the inverse pressure-volume relationship, the air pressure in the lungs will change in response to their changing volume. This leads to the movement of air during pulmonary ventilation. Generally speaking, three pressures are at work during ventilation:

- **Atmospheric pressure.** The molecules that make up air are every bit as subject to the force of gravity as we are. The pull of gravity on the air around us creates what is known as *atmospheric pressure*. At sea level, atmospheric pressure is about 760 mm Hg. Atmospheric pressure increases as you go below sea level, and it decreases as you rise above sea level.
- **Intrapulmonary pressure.** The air pressure within the alveoli is called the *intrapulmonary pressure*. It rises and falls with inspiration and expiration, but it always eventually equalizes with atmospheric pressure due to pressure gradients reaching equilibrium.
- **Intrapleural pressure.** *Intrapleural pressure* is the pressure within the pleural cavity. Like intrapulmonary pressure,

TABLE 21.2 NONRESPIRATORY MOVEMENTS		
Nonrespiratory Movement	**Definition**	**Function**
Sigh	A slow and deep inspiration that is held and followed by a slow expiration	Reopens local groups of collapsed alveoli and stimulates release of surfactant
Yawn	Large sigh that takes the lung volume to its inspiratory capacity	• Opens collapsed alveoli • Yawning when tired occurs to minimize alveolar collapse during sleep • Yawning after sleep opens alveoli that have collapsed during sleep
Sneeze	Deep inspiration followed by a large, forceful expiration through the nose at a velocity of about 100 miles per hour	Clears foreign or irritating substances from the nasal cavity
Cough	Similar to a sneeze except the initial inspiration is small or absent; velocity can approach 500 miles per hour	Clears the larynx, trachea, or lower airways

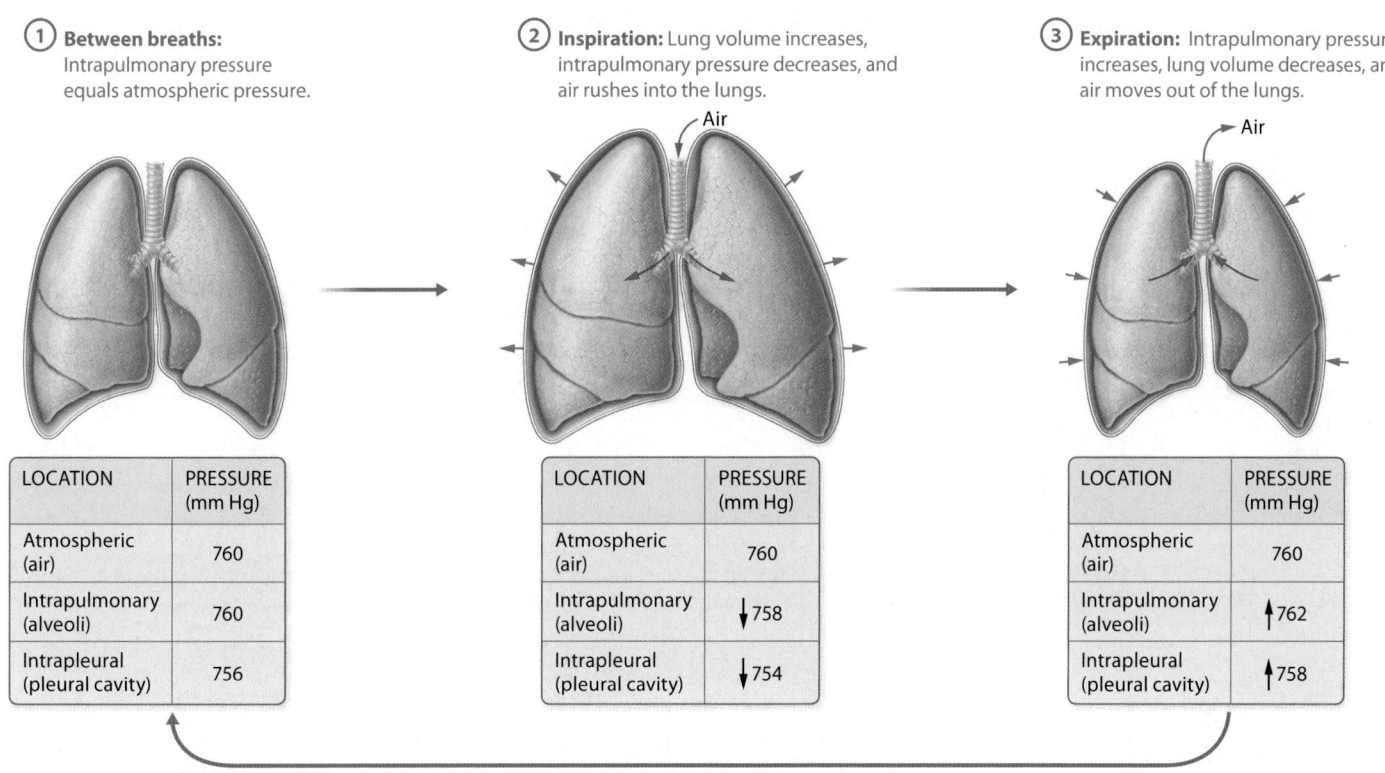

① **Between breaths:** Intrapulmonary pressure equals atmospheric pressure.

② **Inspiration:** Lung volume increases, intrapulmonary pressure decreases, and air rushes into the lungs.

③ **Expiration:** Intrapulmonary pressure increases, lung volume decreases, and air moves out of the lungs.

LOCATION	PRESSURE (mm Hg)
Atmospheric (air)	760
Intrapulmonary (alveoli)	760
Intrapleural (pleural cavity)	756

LOCATION	PRESSURE (mm Hg)
Atmospheric (air)	760
Intrapulmonary (alveoli)	↓758
Intrapleural (pleural cavity)	↓754

LOCATION	PRESSURE (mm Hg)
Atmospheric (air)	760
Intrapulmonary (alveoli)	↑762
Intrapleural (pleural cavity)	↑758

Figure 21.15 Pressure changes in pulmonary ventilation.

intrapleural pressure also rises and falls with inspiration and expiration. However, intrapleural pressure does not equalize with atmospheric pressure. Instead, it is normally about 4 mm Hg *less* than intrapulmonary pressure. This difference is due to the slight suction effect created by the tendency of the lung to collapse and the chest wall to expand. The thin layer of pleural fluid allows the two pleural membranes to move over each other, just as a thin layer of water will let two glass slides or two sheets of paper move over one another. Additionally, the pleural fluid is constantly pumped out of the pleural cavity and into the lymphatic vessels, which creates a slight vacuum that in effect "glues" the two membranes together.

Figure 21.15 shows how these pressures change during quiet ventilation. In step ① of this figure, the lungs are at rest between breaths. The intrapulmonary pressure in the alveoli is the same as the atmospheric pressure at 760 mm Hg, so no air flows in or out of the lungs. The intrapleural pressure is 4 mm Hg below atmospheric, at 756 mm Hg.

For inspiration to occur, the intrapulmonary pressure must be brought below the atmospheric pressure by an increase in the lung volume (Figure 21.15, step ②). During normal inspiration, the intrapleural pressure remains *lower* than the intrapulmonary pressure. When the volume of the lungs increases, the intrapulmonary pressure decreases to about 758 mm Hg, a level below atmospheric pressure, and air moves into the lungs.

During expiration, this process is reversed—the intrapulmonary pressure increases to about 762 mm Hg, slightly above atmospheric pressure, and air flows out of the lungs (Figure

21.15, step ③). The lungs do not fully deflate during expiration because of the slight suction effect from the intrapleural pressure and the outward recoil of the chest wall. Both forces oppose the lungs' elastic recoil, keeping the lungs inflated and most of the alveoli open at all times. Expiration stops when intrapulmonary pressure has increased to the same level as the atmospheric pressure. Throughout these changes, the intrapleural pressure stays about 4 mm Hg less than the intrapulmonary pressure.

What happens if the intrapleural pressure increases to a level at or above atmospheric pressure? Under these conditions, the intrapleural pressure no longer exerts a suction effect that prevents the lungs from collapsing. Indeed, the added pressure actually enhances the lungs' elastic recoil, and the lungs immediately collapse, as shown here:

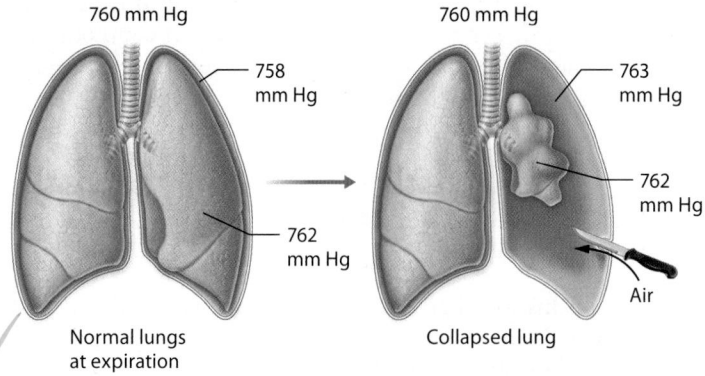

Normal lungs at expiration: 760 mm Hg, 758 mm Hg, 762 mm Hg.
Collapsed lung: 760 mm Hg, 763 mm Hg, 762 mm Hg, Air.

When the lung is fully collapsed, intrapulmonary pressure is so high that the body is unable to bring it below atmospheric

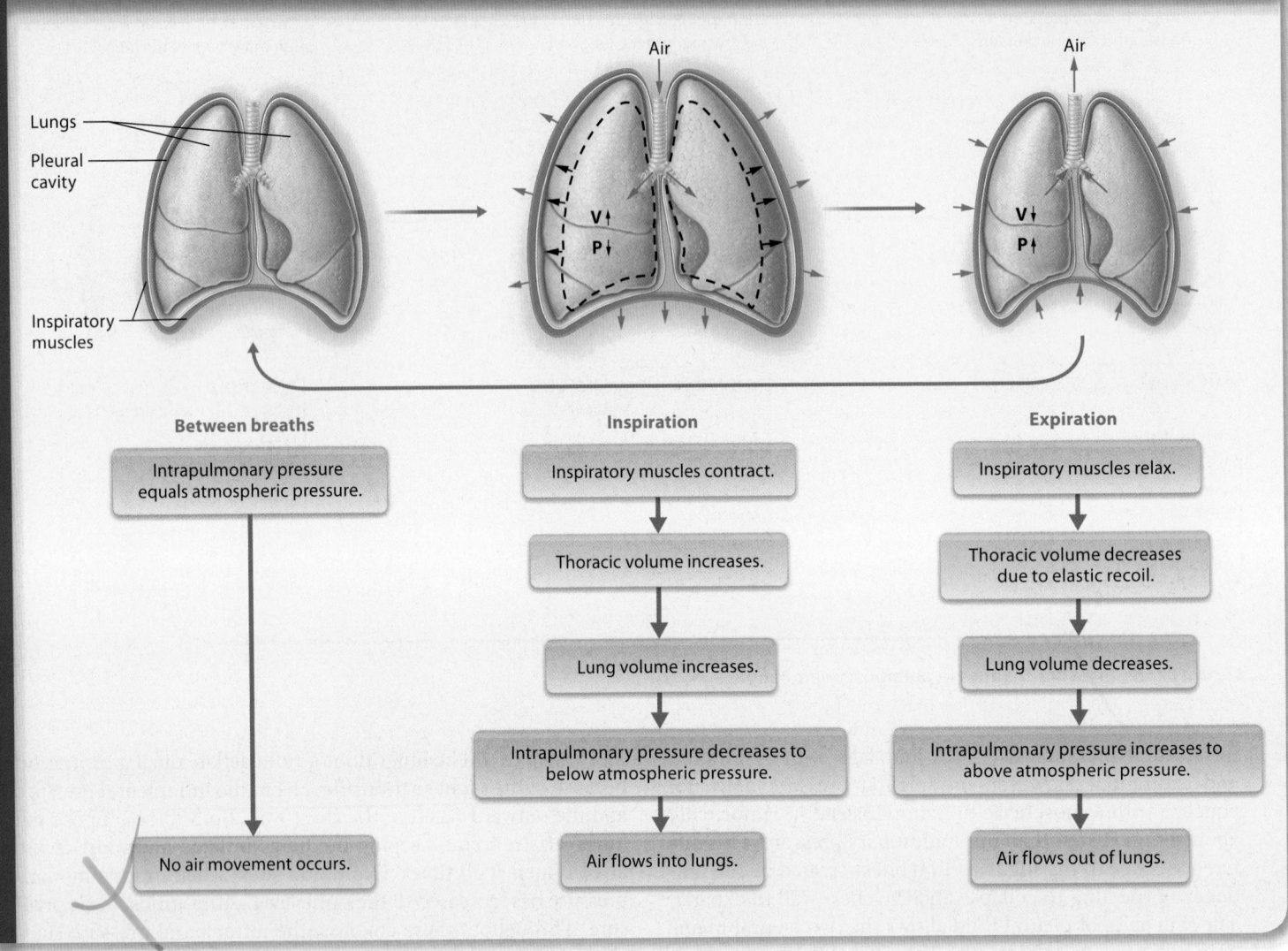

Lungs

Pleural cavity

Inspiratory muscles

Air

V↑
P↓

Air

V↓
P↑

Between breaths

Intrapulmonary pressure equals atmospheric pressure.

No air movement occurs.

Inspiration

Inspiratory muscles contract.

Thoracic volume increases.

Lung volume increases.

Intrapulmonary pressure decreases to below atmospheric pressure.

Air flows into lungs.

Expiration

Inspiratory muscles relax.

Thoracic volume decreases due to elastic recoil.

Lung volume decreases.

Intrapulmonary pressure increases to above atmospheric pressure.

Air flows out of lungs.

pressure, and inspiration cannot occur. Many things can increase intrapleural pressure above atmospheric, including excess pleural fluid (a *pleural effusion*), air in the pleural cavity (a *pneumothorax;* new-moh-THOR-aks), or blood in the pleural cavity (a *hemothorax*). The preceding illustration shows a pneumothorax caused by the trauma of a knife wound, but note that trauma is only one potential cause of lung collapse.

Putting It All Together: The Big Picture of Pulmonary Ventilation

Let's put all this together now as we review the process of pulmonary ventilation (**Figure 21.16**). For air to move into the lungs during inspiration, the body must decrease intrapulmonary pressure below atmospheric pressure. This change requires an increase in the volume of the lungs, which is caused by the

contraction of the inspiratory muscles (the diaphragm and the external intercostal muscles). When the inspiratory muscles hold their contraction just before expiration, intrapulmonary pressure is equal to atmospheric pressure because air has moved in until the pressure gradient was at equilibrium. For expiration to occur, intrapulmonary pressure must now increase above atmospheric pressure. As before, this increase in pressure is accomplished by changing the volume of the lungs—except in this case, the lungs' volume decreases because of the elastic recoil of the lungs.

Quick Check

☐ 1. What drives the movement of gases during breathing?

☐ 2. What does the intrapleural pressure prevent under normal circumstances?

☐ 3. How are inspiration and expiration achieved?

Physical Factors Influencing Pulmonary Ventilation

◀◀ **Flashback**

1. What are hydrogen bonds? How do molecules of water interact to produce surface tension? (pp. 39–40)
2. What are the effects of the sympathetic and parasympathetic nervous systems on bronchial smooth muscle? (p. 530)
3. How is resistance defined? (p. 674)

Our ability to move air in and out of our lungs via pulmonary ventilation depends on more than just the intrapulmonary pressure. Certain physical properties of the respiratory tract and the lungs determine the overall effectiveness of pulmonary ventilation (or how well gases are transported to and from the alveoli). The three primary physical factors are *airway resistance, alveolar surface tension,* and *pulmonary compliance.* Many diseases of the respiratory system can be traced back to problems with one or more of these factors. We examine each in the following sections.

Airway Resistance

The first physical factor, **airway resistance,** is defined as anything that impedes air flow through the respiratory tract; it is similar to blood vessel resistance (see Chapter 18). As in blood vessels, the degree of resistance in the airways is largely determined by the diameter of the passageway (**Figure 21.17**). Resistance normally varies during pulmonary ventilation due to changes in intrapulmonary pressure. During inspiration, resistance decreases slightly because the airways are pulled open as the lungs expand, and during expiration, resistance increases slightly as the lungs recoil and the airways narrow. These normal changes in resistance do not adversely affect pulmonary ventilation due to the large diameter of the bronchi and supporting cartilage, as well as the huge total cross-sectional area of the bronchioles.

The diameter of the airways is also controlled by the contraction or relaxation of the smooth muscles of the bronchioles.

Relaxation of the bronchial smooth muscle, or **bronchodilation,** increases the diameter of the bronchioles. Figure 21.17 shows that this increases the size of the bronchiole's lumen, which decreases airway resistance. Neurons of the sympathetic nervous system release norepinephrine during times of exercise, stress, or emergency, which triggers bronchodilation to increase the efficiency of pulmonary ventilation (see Chapter 14).

Conversely, when the smooth muscle contracts, a process called **bronchoconstriction,** the diameter of the bronchioles can decrease dramatically. As shown in Figure 21.17, this narrows the lumen of the bronchiole, which dramatically increases resistance. Bronchoconstriction may be triggered by inhaled irritants that activate the parasympathetic neurons serving the bronchial smooth muscle or by inflammatory mediators (see Chapter 20).

Certain disease states may also increase airway resistance by inflammation or an obstruction such as a tumor or an abnormal amount of mucus. Many bacterial and viral diseases of the respiratory system, such as the viral disease *influenza,* cause symptoms because they increase airway resistance in this fashion.

Alveolar Surface Tension

We said that **alveolar surface tension** was the second physical factor affecting pulmonary ventilation, but what exactly is surface tension? Recall that water is a *polar covalent* molecule, meaning that its atoms share electrons unequally (see Chapter 2). The oxygen atom is more electronegative than the hydrogen atoms, so it is the "greedier" atom, pulling on the electrons so they spend more time around it than around the hydrogen atoms. This unequal sharing gives the oxygen atom a partially negative charge and the hydrogen atoms partially positive charges. When two or more water molecules are close together, the partial charges result in the weak attraction called a hydrogen bond. Wherever a gas-water boundary exists, the water molecules are more attracted to one another than to the nonpolar gas molecules. This attraction causes the water molecules to cluster together and form hydrogen bonds, and a state of surface tension is created at the gas-water boundary:

Hydrogen bonds

Hydrogen bonds pull the water molecules close together, creating a "tense" surface.

Air

Water

The cells of the alveoli, like all cells in the body, are coated in a liquid film composed mainly of water. These water molecules attract one another, and their attraction is greatest when the alveoli are smallest during expiration. Left unopposed, this attraction would create a high amount of surface tension that would collapse the alveoli during expiration (a condition known

↓ Airway resistance as smooth muscle relaxes

↑ Airway resistance as smooth muscle contracts

Smooth muscle

↑ Diameter of airway

↓ Diameter of airway

Figure 21.17 Relationship between airway resistance and airway diameter.

as **atelectasis;** at'-eh-LEK-tah-sis), and make it very difficult to re-inflate them during inspiration (**Figure 21.18a**).

Fortunately, the attraction isn't left unopposed. The liquid film coating the alveoli isn't pure water, but rather it contains molecules of a chemical called pulmonary **surfactant.** The function of this surfactant, which is produced by type II alveolar cells, is to reduce alveolar surface tension. Surfactant is similar in structure to a detergent—the molecule has one end that is polar and another end that is nonpolar. When surfactant is added to water, its polar ends interact with water molecules while its nonpolar ends repel water molecules. The net effect of this action is that the hydrogen bonds between water molecules are disrupted, which reduces alveolar surface tension. The reduced surface tension allows the alveoli to remain partially open, even during expiration (**Figure 21.18b**).

Pulmonary Compliance

The third physical factor that influences the effectiveness of gas exchange is called pulmonary compliance. **Pulmonary compliance** refers to the ability of the lungs and the chest wall to stretch. Note that compliance does not refer to *elasticity,* which is the ability of a tissue to return to its original state when stretched. Rather, it is the ability of the tissue to stretch in the first place. Pulmonary compliance is primarily determined by three factors:

- **Degree of alveolar surface tension.** As you just learned, surface tension on the alveoli resists their inflation, but surfactant counters this tension.
- **Distensibility of elastic tissue in the lungs.** The elastic tissue in the lungs gives them not only their ability to recoil but also the property of *distensibility,* or the ability to stretch during inflation.
- **Ability of the chest wall to move.** The chest wall must also stretch during inspiration, as the lungs and chest wall move together.

Factors that decrease pulmonary compliance make pulmonary ventilation less efficient and increase the work of breathing. For example, diseases that decrease surfactant production can result in alveolar surface tension that is too high, which makes the alveoli difficult to inflate during inspiration. (See *A&P in the*

Collapsed alveolus

Surfactant molecules

Hydrogen bond

Water molecules

Inflated alveolus

Without surfactant, hydrogen bonds pull the water molecules together and the alveolus collapses.

With surfactant, the hydrogen bonds are disrupted and the alveolus remains inflated.

(a) Without surfactant

(b) With surfactant

Figure 21.18 Effect of surfactant on alveolar surface tension.

Real World: Infant Respiratory Distress Syndrome to learn about how this problem can affect premature infants.) Anything that damages the lung, including infections such as tuberculosis or the accumulation of foreign particles (such as coal dust), can destroy the elastic tissue and/or cause the formation of collagen fibers in the lungs. In this situation the elastic tissue is replaced with dense irregular collagenous connective tissue, which stiffens the lungs, decreasing their compliance, and produces a state called *fibrosis* (see Chapter 4). With the elastic fibers damaged, the lungs become difficult to inflate. Also, any factors that inhibit the movement of the chest wall, such as trauma to the chest wall and/or bony deformities of the wall, can decrease compliance.

Quick Check

☐ 4. What is airway resistance? What is the main factor that determines the amount of airway resistance?
☐ 5. How does surfactant decrease surface tension?
☐ 6. Define pulmonary compliance. What three factors influence pulmonary compliance?

Pulmonary Volumes and Capacities

Measuring the volumes of air that a person exchanges with each breath is a useful way to assess the state of his or her pulmonary function. Such a measurement can be obtained with an instrument called a **spirometer** (spy-ROM-eh-ter; *spiro-* = "breathing"). The graph produced by a spirometer shows a record of a

A&P in the Real World

Infant Respiratory Distress Syndrome

When surfactant is inadequate, it takes tremendous effort to keep the alveoli inflated between breaths because alveolar surface tension increases and lung compliance decreases. Many premature newborns face this difficulty because surfactant is not produced in significant amounts until the last 10–12 weeks of gestation (as it's not needed during the fetal period). These infants are said to have **infant respiratory distress syndrome (RDS).** Risk factors for RDS include prematurity, male gender, maternal history of diabetes, family history of RDS, and caesarean delivery.

Treatment is aimed at delivering natural or synthetic surfactant by inhalation. Another treatment is *continuous positive airway pressure (CPAP),* which delivers slightly pressurized air that prevents the alveoli from collapsing during expiration. In cases of severe RDS, CPAP may be inadequate, necessitating mechanical ventilation. However, the main goal of the medical community is to prevent RDS from happening at all, by recognizing at-risk patients and preventing premature births.

single person's breathing as he or she is asked to inhale and exhale either normally or forcefully. Three volumes can be measured—the tidal volume and the inspiratory and expiratory reserve volumes (see green arrows in **Figure 21.19**):

- **Tidal volume (TV)** is the amount of air inspired or expired during normal, quiet ventilation. In the average healthy adult, the TV is about 500 ml of air. If we multiply the TV by the number of breaths per minute, we get the **minute volume,** the total volume of air that moves in and out of the lungs each minute. For the average adult, who takes about 12 breaths per minute, the minute volume is about 6 liters per minute. Note, however, that not all of the air in a tidal inspiration actually makes it into the respiratory zone. The air that remains in the conducting zone airways is said to be in the **anatomical dead space.** Out of a 500-ml tidal inspiration, only about 350 ml is involved in gas exchange—the remaining 150 ml fills the anatomical dead space. For this reason, some health professionals prefer to think in terms of the **alveolar ventilation rate (AVR),** as opposed to the minute volume. The AVR is the volume of air that reaches the alveoli multiplied by the breaths per minute, and it averages about 4.2 liters per minute.
- **Inspiratory reserve volume (IRV)** refers to the volume of air that can be forcibly inspired after a normal tidal inspiration. The IRV averages 2100–3300 ml of air, depending on a person's gender and size.
- **Expiratory reserve volume (ERV)** is essentially the IRV in reverse. It refers to the amount of air that can be forcibly expired after a normal tidal expiration. The ERV averages about 700–1200 ml of air.

You may have noticed that the IRV is larger than the ERV by about 1400–2100 ml of air. This difference exists because even after the most forceful expiration, some air remains in the lungs, a volume known as the **residual volume (RV).** The RV remains in the lungs due mostly to the intrapleural pressure and outward recoil of the chest wall, which keeps the lungs slightly inflated. Unlike the TV, IRV, and ERV, the RV is measured not by spirometry but by other techniques.

To get a better picture of a person's pulmonary function, two or more of the pulmonary volumes can be combined to create what are called **pulmonary capacities** (see blue arrows in Figure 21.19). The following four pulmonary capacities are commonly considered:

- **Inspiratory capacity** is the total amount of air that a person can inspire after a tidal expiration. It is equal to the TV plus the IRV.
- **Functional residual capacity** is the amount of air that is normally left in the lungs after a tidal expiration. It is the sum of the ERV and the RV.
- **Vital capacity** represents the total amount of exchangeable air, or the total amount of air that can move in and out of the lungs. It is equal to the sum of the TV, IRV, and the ERV. The vital capacity is measured in the laboratory with a spirometer as the **forced vital capacity (FVC),** in which the subject follows a maximal inspiration with a maximal exhalation. This measurement can give important information about the presence and type of pulmonary disease, which we discuss in Module 21.8.
- **Total lung capacity** represents the total amount of exchangeable and nonexchangeable air in the lungs. It is the total of all four pulmonary volumes: IRV, TV, ERV, and RV.

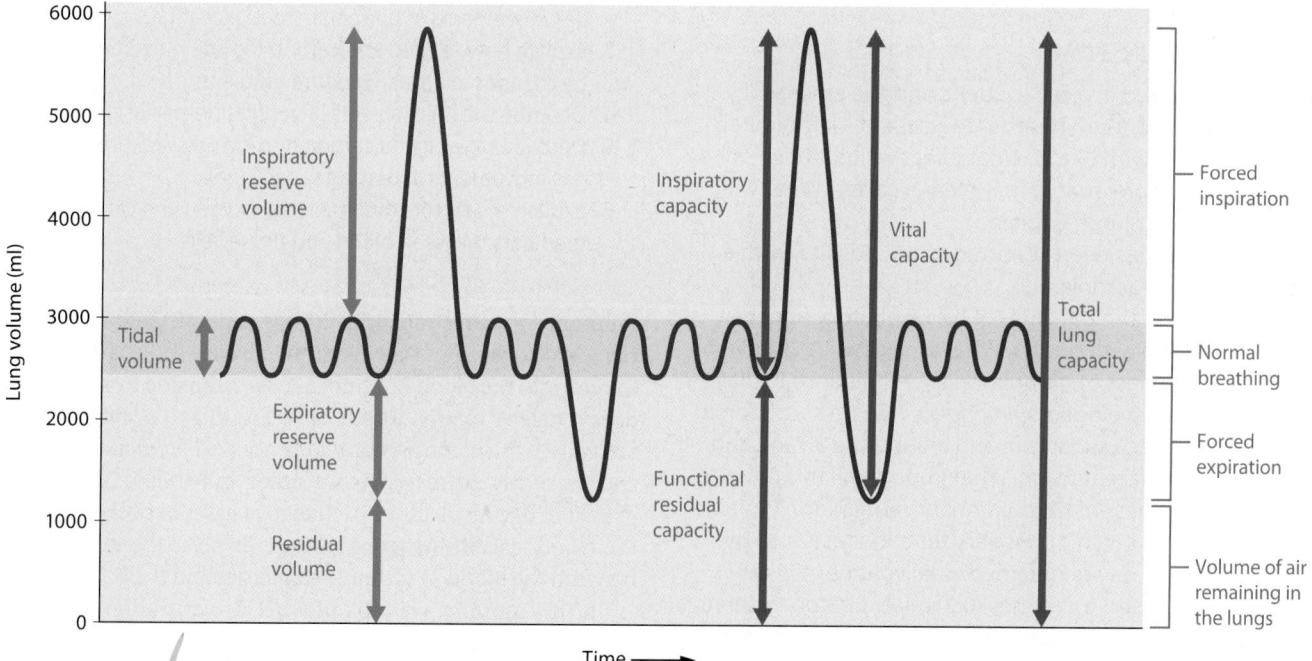

Figure 21.19 Graph of pulmonary volumes and capacities.

TABLE 21.3 PULMONARY VOLUMES AND CAPACITIES

Measurement	Average Value (Female, Male)	Definition
Pulmonary Volumes		
Tidal volume (TV)	500 ml, 500 ml	Volume of air exchanged with normal, quiet breathing
Inspiratory reserve volume (IRV)	1900 ml, 3100 ml	Maximum volume of air that can be forcibly inspired after a tidal inspiration
Expiratory reserve volume (ERV)	700 ml, 1200 ml	Maximum volume of air that can be forcibly expired after a tidal expiration
Residual volume (RV)	1100 ml, 1200 ml	Volume of air that remains in the lungs after a forced expiration
Pulmonary Capacities		
Inspiratory capacity	2400 ml, 3600 ml	Total amount of air that can be inspired; equal to tidal volume plus inspiratory reserve volume
Functional residual capacity	1800 ml, 2400 ml	Total amount of air that normally remains in the lungs after a tidal expiration; equal to the residual volume plus expiratory reserve volume
Vital capacity	3100 ml, 4800 ml	Total amount of exchangeable air; equal to sum of the tidal volume, expiratory reserve volume, and inspiratory reserve volume
Total lung capacity	4200 ml, 6000 ml	Total amount of exchangeable and nonexchangeable air; equal to the sum of all of the pulmonary volumes

Table 21.3 summarizes the pulmonary volumes and capacities and gives average values for each in adults. Note that except for tidal volume, the values for all volumes and capacities are lower in females than in males. These values also vary with a person's height, age, and health.

Quick Check

☐ 7. List and define the common pulmonary volumes.
☐ 8. List and define the common pulmonary capacities.

Apply What You Learned

☐ 1. A person is injured in a car accident, and the emergency room physician determines that the patient's left pleural cavity has filled with blood. What effect will this have on the intrapleural pressure in that pleural cavity? What will this do to the left lung? Explain.
☐ 2. Melvin Gartner has been admitted to the hospital with a diagnosis of pneumonia.
   a. You note that the passages of his respiratory tract can become filled with mucus. What impact will this have on the effectiveness of his pulmonary ventilation? Explain.
   b. He has also developed *pulmonary edema,* a condition in which fluid accumulates in and around the alveoli and increases surface tension. What impact will this have on the effectiveness of his pulmonary ventilation? Explain.
   c. Mr. Gartner is given pulmonary function tests and the following values are obtained: tidal volume, 400 ml; inspiratory reserve volume, 3050 ml; expiratory reserve volume, 650 ml. Calculate his inspiratory capacity and vital capacity. Are his values normal?

See answers in Appendix A.

MODULE **21.4**

Gas Exchange

Learning Outcomes

1. State Dalton's law and Henry's law, and relate both laws to the events of internal and pulmonary gas exchange and to the amounts of oxygen and carbon dioxide dissolved in plasma.
2. Describe oxygen and carbon dioxide pressure gradients and net gas movements in pulmonary gas exchange.
3. Explain how oxygen and carbon dioxide movements are affected by changes in partial pressure gradients.
4. Describe the mechanisms of ventilation-perfusion matching.
5. Describe oxygen and carbon dioxide pressure gradients and net gas movements in tissue gas exchange.
6. Explain the factors that maintain oxygen and carbon dioxide gradients between blood and tissue cells.

Pulmonary ventilation only brings new air into the alveoli and removes oxygen-poor air from the alveoli. It doesn't actually bring oxygen into the blood or remove carbon dioxide from the blood, nor does it deliver oxygen to our cells. For this to occur, we rely on *gas exchange,* which consists of two processes: *pulmonary gas exchange* and *tissue gas exchange.* As we noted in Module 21.1, **pulmonary gas exchange** involves the exchange of gases between the alveoli and the blood, and **tissue gas exchange** involves the exchange of gases between the blood in systemic capillaries and the body's cells.

In this module, we look at what drives pulmonary and tissue gas exchange, including the pressure that a gas exerts and its ability to dissolve in water. We also consider various characteristics of the body that make these processes more or less efficient, such as the surface area and blood supply available for gas exchange. Let's begin with two gas laws that apply to gas exchange.

The Behavior of Gases

The first gas law that applies to gas exchange is called **Dalton's law of partial pressures.** Dalton's law states that each gas in a mixture exerts its own pressure, called its *partial pressure,* so the total pressure of a gas mixture is the sum of the partial pressures of its gases. The partial pressure of a gas is symbolized by P_{gas} and is equal to its percent of the gases times the total pressure of all gases.

Let's look at atmospheric pressure first. We know that air is a mixture of gases, made up of about 78% nitrogen and 21% oxygen, with the remaining 1% composed of carbon dioxide, argon, and other gases. The total atmospheric pressure is equal to the partial pressures of all of these gases:

$$P_{N_2} + P_{O_2} + P_{Ar} + P_{others} = \text{Atmospheric pressure}$$
$$\text{(760 mm Hg at sea level)}$$

$$P_{N_2} = 0.78 \times 760 = 593 \text{ mm Hg}$$

$$P_{O_2} = 0.21 \times 760 = 160 \text{ mm Hg}$$

$$P_{Ar} = [0.0093] \times 760 = 7 \text{ mm Hg}$$

The partial pressures of these gases determine where a gas moves by diffusion. As we discussed previously, a gas follows a pressure gradient and moves from an area of higher pressure to an area of lower pressure. The greater the pressure difference, the more rapidly the gas diffuses. This is the rationale for giving 100% oxygen to a person in respiratory distress—it increases the partial pressure of oxygen in the lungs, causing oxygen to diffuse into the person's blood at a faster rate.

A related gas law that applies to gas exchange deals with the ability of a gas to dissolve in water, termed its *solubility* in water. Just as a pinch of salt dissolves in water, some gases also dissolve to a certain extent when mixed with water. **Henry's law** states that the degree to which a gas dissolves in a liquid is proportional to both its partial pressure and its solubility in the liquid. The simplest example of this is the carbon dioxide dissolved in carbonated drinks. The carbon dioxide is added to the solution under high pressure, which causes the molecules to go into solution. When you open the can or bottle containing the drink, the pressure immediately decreases as the carbon dioxide molecules leave the solution and move from the area of higher pressure (the drink) to the area of lower pressure (the surrounding atmosphere).

Henry's law also explains the behavior of gases we find in air when they come in contact with water in the body:

- Although nitrogen gas has a high partial pressure in air (making up about 78% of its volume), very little nitrogen is present in the plasma. The reason is its low solubility in water, too low for any of it to dissolve in appreciable amounts.
- Even though oxygen gas has a lower partial pressure than nitrogen (as we said, oxygen makes up about 21% of the gas in air), a small amount of oxygen is found dissolved in plasma. Why? Its solubility in water, although fairly low, is still higher than that of nitrogen.
- Carbon dioxide has a very low partial pressure (making up less than 1% of the volume of air), yet it is found dissolved in plasma. Why? It has a relatively high solubility in water. In fact, it is about 20 times more soluble in water than is oxygen.

Henry's law forms the basis for a treatment called *hyperbaric oxygen therapy.* Please see *A&P in the Real World: Hyperbaric Oxygen Therapy* for more information.

Quick Check

☐ 1. How does the pressure gradient between two gas mixtures influence the diffusion of those gases?

☐ 2. Why do we find more carbon dioxide dissolved in the plasma than oxygen?

Pulmonary Gas Exchange

As we have already discussed, *pulmonary gas exchange* (also called *external respiration;* not to be confused with cellular respiration) is the diffusion of gases between the alveoli and the blood. During pulmonary gas exchange, oxygen diffuses from the air in the alveoli into the blood in the pulmonary capillaries, and carbon dioxide flows in the opposite direction. This exchange converts the oxygen-poor, carbon dioxide–rich blood delivered by the pulmonary artery into oxygen-rich blood that contains less carbon dioxide. This oxygenated blood is then delivered via the pulmonary veins to the left side of the heart to be distributed to all the tissues of the body (see Chapter 17).

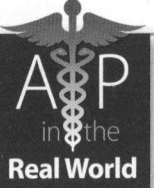

Hyperbaric Oxygen Therapy

In hyperbaric oxygen therapy, a person is placed in a chamber and exposed to a higher-than-normal partial pressure of oxygen. This higher partial pressure increases the amount of oxygen that is dissolved in the plasma, which in turn increases overall oxygen delivery to the tissues.

Hyperbaric oxygen is useful in treating conditions that benefit from having a greater percentage of oxygen dissolved in the plasma, such as severe blood loss, crush injuries, anemia (a decreased oxygen-carrying capacity of the blood), chronic wounds, certain infections, and burns. Hyperbaric oxygen therapy is also used for decompression sickness (also known as "the bends"), a condition generally seen in divers who have ascended from a dive too rapidly. Decompression sickness is caused by dissolved gases in the blood coming out of solution and forming gas bubbles in the bloodstream. Hyperbaric oxygen forces the gases back into solution, eliminating the bubbles.

Like the movement of all gases, pulmonary gas exchange is driven by pressure gradients (an

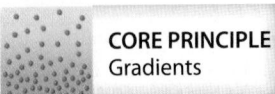

CORE PRINCIPLE
Gradients

example of the Gradients Core Principle, p. 26); in this case we are dealing with the differences in partial pressure between the oxygen and carbon dioxide in the alveoli and in the blood.

As you can see in **Figure 21.20a**, the deoxygenated blood entering the pulmonary capillaries has a very low partial pressure of oxygen (P_{O_2}) of about 40 mm Hg, compared with the air in the alveoli, with a P_{O_2} of about 104 mm Hg. This steep gradient drives the movement of oxygen from the alveoli to the blood very efficiently. The movement of oxygen is so efficient that it takes only about one-quarter of a second for equilibrium to occur (i.e., to achieve the same P_{O_2} in both the pulmonary capillary blood and the alveoli).

Notice, however, that the pressure gradient for carbon dioxide isn't as steep as the gradient for oxygen. The partial pressure of carbon dioxide (P_{CO_2}) in the blood of pulmonary capillaries is about 45 mm Hg, and in the alveoli it is about 40 mm Hg. Yet even with this smaller gradient, approximately equal amounts of carbon dioxide and oxygen are exchanged. This is due again

to Henry's law. Remember that it is not just partial pressure that determines gas movement, but also solubility. As you learned, carbon dioxide is about 20 times more soluble than oxygen in water, so it moves extremely rapidly across the respiratory membrane and doesn't require a big pressure gradient.

Factors Affecting Efficiency of Pulmonary Gas Exchange

Aside from the partial pressures of oxygen and carbon dioxide, three additional factors impact the efficiency of pulmonary gas exchange: (1) the surface area of the respiratory membrane, (2) the thickness of the respiratory membrane, and (3) the degree to which the amount of air flow matches the amount of blood flow. Let's examine each of these factors.

Surface Area of the Respiratory Membrane

We have already discussed the extremely large surface area of the respiratory membrane in Module 21.2. Recall that for the average person, it is about 80–100 m^2 for both lungs (roughly 1000 square feet). Consider for a moment that only about 75–100 ml of blood

Pulmonary capillaries

Pulmonary gas exchange

Erythrocyte

O_2

CO_2

Alveolus

(a) Pulmonary circuit

Alveolus

Pulmonary capillary

P_{O_2}

104 mm Hg

40 mm Hg

Alveolus   Pulmonary capillary

P_{CO_2}

40 mm Hg   45 mm Hg

Alveolus   Pulmonary capillary

Tissue gas exchange

O_2

CO_2

(b) Systemic circuit

Systemic capillaries   Tissue cells

Systemic capillary

Tissue cell

P_{O_2}

100 mm Hg

40 mm Hg

Systemic capillary   Tissue cell

P_{CO_2}

40 mm Hg   45 mm Hg

Systemic capillary   Tissue cell

Figure 21.20 Pulmonary and tissue gas exchange.

is in the pulmonary capillaries at any given time—about the volume of a small coffee mug. The huge surface area of the respiratory membrane is enough for essentially every erythrocyte in this volume of blood to be in contact with the respiratory surface.

Any factor that decreases the surface area of the respiratory membrane decreases the efficiency of pulmonary gas exchange. For example, the disease *emphysema* (em-fih-SEE-mah) causes the destruction of the alveolar walls, which reduces the surface area of the respiratory membrane to as little as one-fifth of normal. This reduction results in severely impaired pulmonary gas exchange, leading to the conditions **hypoxemia** (hy′-pok-SEE-mee-uh; low levels of oxygen in the blood) and **hypercapnia** (hy′-pur-KAP-nee-uh; high levels of carbon dioxide in the blood).

Thickness of the Respiratory Membrane

A second factor that determines the efficiency of pulmonary gas exchange is the thickness of the respiratory membrane, which is the distance over which the gas must diffuse. In general, the respiratory membrane is extremely thin, measuring as little as 0.2 μm in some places, with the majority having a width of about 0.6 μm. Anything that increases the thickness of the respiratory membrane, such as inflammation, increases the amount of time it takes for oxygen and carbon dioxide to be exchanged and so decreases the efficiency of gas exchange.

Ventilation-Perfusion Matching

The third factor that affects pulmonary gas exchange is the degree of match between the amount of air reaching the alveoli, or *ventilation*, and the amount of blood flow, called *perfusion*, in the pulmonary capillaries. This relationship is known as **ventilation-perfusion matching** (also called *ventilation-perfusion coupling*, because the two processes are coupled together).

Two phenomena keep ventilation and perfusion closely matched:

- **Changes in alveolar ventilation lead to changes in perfusion, so blood flow is directed to areas with the most oxygen.** Changes in ventilation can be measured by changes in the alveolar partial pressure of oxygen, P_{O_2}. As shown in the left side of **Figure 21.21a**, when an alveolus is insufficiently ventilated, the P_{O_2} of the air within that alveolus drops, and the pulmonary arteriole serving that alveolus constricts. The opposite occurs when an alveolus is adequately ventilated, as we see in the right side of Figure 21.21a. The P_{O_2} rises and the pulmonary arteriole dilates in response. Therefore, low alveolar P_{O_2} causes pulmonary arterioles to constrict, and high alveolar P_{O_2} causes them to dilate. These responses by the arterioles redirect blood flow to the alveoli with the greatest P_{O_2}, maximizing the efficiency of pulmonary gas exchange.

- **Changes in the efficiency of perfusion lead to changes in the amount of ventilation, so air flow is directed to areas with the most blood flow.** Changes in the efficiency of perfusion can be measured by the partial pressure of carbon dioxide (P_{CO_2}) in alveolar arterioles. When the pulmonary capillaries bring blood with less carbon dioxide to the alveoli, lowering P_{CO_2}, as shown in the left side of **Figure 21.21b**, the bronchiole constricts. Conversely, notice in the right side of Figure 21.21b that when pulmonary capillaries bring more carbon dioxide to the alveoli, raising P_{CO_2}, the bronchiole dilates. Therefore, high arteriolar P_{CO_2} causes bronchioles to dilate, and low arteriolar P_{CO_2} causes them to constrict. These responses by the bronchioles allow the lungs to ventilate the areas that have the greatest blood flow, so that gas exchange takes place efficiently.

These two processes combine to ensure that pulmonary ventilation and pulmonary capillary perfusion match one another as closely as possible. A value called the **ventilation/perfusion (V/Q) ratio** describes this match. Hypothetically, the V/Q ratio should be about 1.0, but in the human body it averages around 0.8–0.9. The V/Q ratio is a bit lower than expected because the

Normal P_{O_2} and P_{CO_2} in alveolus and arteriole
(ventilation and perfusion matched)

Bronchiole

Pulmonary arteriole

Alveolus

Pulmonary capillaries

Pulmonary venule

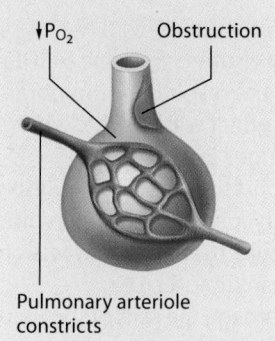

Low P_{O_2} in alveolus

↓P_{O_2} Obstruction

Pulmonary arteriole constricts

High P_{O_2} in alveolus

↑P_{O_2}

Pulmonary arteriole dilates

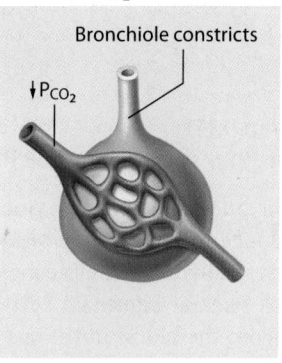

Low P_{CO_2} in arteriole

Bronchiole constricts

↓P_{CO_2}

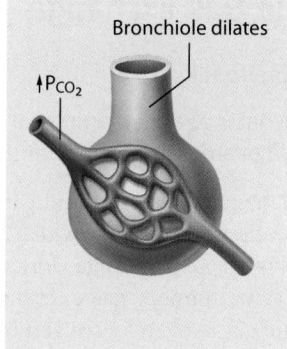

High P_{CO_2} in arteriole

Bronchiole dilates

↑P_{CO_2}

(a) Changes in ventilation (alveolar P_{O_2}) lead to changes in perfusion.

(b) Changes in perfusion (arteriolar P_{CO_2}) lead to changes in ventilation.

Figure 21.21 Ventilation-perfusion coupling.

bases of the lungs have slightly higher perfusion than ventilation due to the effects of gravity on the blood in the pulmonary arteries. Read *A&P in the Real World: V/Q Mismatch* for information on what happens when these processes are out of sync.

Quick Check

☐ 3. What takes place during pulmonary gas exchange? Which pressure gradients drive this process?

☐ 4. How do the surface area and thickness of the respiratory membrane influence the efficiency of pulmonary gas exchange?

☐ 5. What is the ventilation/perfusion ratio, and what two mechanisms keep the ratio close to 1?

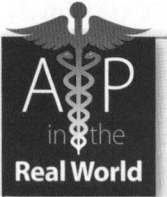

V/Q Mismatch

When ventilation does not match perfusion to an area of the lung, the result is called a *V/Q mismatch*. Common causes of a V/Q mismatch include pneumonia, asthma, pulmonary edema, or any other condition that results in less air getting to the alveoli. In cases like these, a right-to-left *shunt* is said to exist, meaning that blood is "shunted" from the right side of the heart to the left side without receiving adequate oxygenation in the pulmonary circulation.

Another potential cause of a V/Q mismatch is a reduction in blood flow to the lungs. The most common scenario in this case is a blockage in a pulmonary artery or arteriole, called a *pulmonary embolus*. Many substances can act as pulmonary emboli, including pieces of a blood clot, gas bubbles, fat globules, and amniotic fluid (the fluid surrounding a fetus in a pregnant woman). Such a condition creates what is known as *alveolar dead space*—the alveoli in a segment of the lung are ventilated but are not perfused, so no pulmonary gas exchange takes place.

Tissue Gas Exchange

◀◀ **Flashback**

1. What is stratified epithelial tissue? (p. 132)
2. What are capillary beds, and how are they structured? (p. 685)

The exchange of oxygen and carbon dioxide between the blood and the tissues is called *tissue gas exchange* (or *internal respiration*). Tissue gas exchange works via essentially the same mechanisms as pulmonary gas exchange (look back at Figure 21.20b). In tissue gas exchange, oxygen moves from the blood in the systemic capillaries into the tissues.

As with pulmonary gas exchange, the main driving force for tissue gas exchange is a pressure gradient—the difference in the

P_{O_2} between blood and tissues. The cells are constantly using oxygen for cellular respiration (see Chapter 3), and for this reason the P_{O_2} in the tissues is low, averaging about 40 mm Hg. In contrast, the P_{O_2} in the systemic capillaries is high, measuring about 100 mm Hg. This steep pressure gradient facilitates rapid exchange of oxygen between the capillaries and the tissues.

The pressure gradient functions in the reverse direction for carbon dioxide. The tissues produce large quantities of carbon dioxide as a waste product of cellular respiration, so the P_{CO_2} in the tissues is relatively high (about 45 mm Hg). The P_{CO_2} in the blood is lower (about 40 mm Hg), and this pressure gradient, combined with carbon dioxide's high solubility in water, facilitates its rapid transport into the systemic capillaries.

Factors Affecting Efficiency of Tissue Gas Exchange

As with pulmonary gas exchange, physical factors play a role in the efficiency of tissue gas exchange. These factors include the following:

- **The surface area available for gas exchange.** Most capillaries exist in branching, weblike capillary beds (see Chapter 18). This branching increases the surface area available for tissue gas exchange and makes it a more efficient process.
- **The distance over which diffusion must occur.** Stratified epithelial tissues can only have so many layers of living cells because oxygen can only diffuse so far (see Chapter 4). Cells that are too distant from capillaries do not receive adequate oxygen and die.
- **The perfusion of the tissue.** Tissue gas exchange is most efficient when tissues are adequately supplied with blood via the capillaries. Inadequate perfusion explains why diseases that affect the small blood vessels can lead to damage and death of a tissue. If a tissue is not adequately perfused, oxygen is not delivered to it and carbon dioxide is not removed.

Quick Check

☐ 6. What are three factors that influence the efficiency of tissue gas exchange?

Apply What You Learned

☐ 1. The condition *acute respiratory distress syndrome* (*ARDS*) is characterized by damage to the respiratory membrane that causes it to thicken. What effect will this thickening have on pulmonary gas exchange? Explain.

☐ 2. What effect will low-oxygen conditions have on pulmonary arterioles? What effect will such conditions have on blood pressure in the pulmonary circuit? (*Hint:* You may wish to review blood vessel resistance in Chapter 18.)

☐ 3. The toxins in cigarette smoke damage the cells of the tissue capillaries and also cause arterioles supplying capillary beds to constrict. What effect will this have on tissue gas exchange?

See answers in Appendix A.

MODULE 21.5
Gas Transport through the Blood

Learning Outcomes

1. Describe the ways in which oxygen is transported in blood, including the reversible reaction for oxygen binding to hemoglobin.
2. Interpret the oxygen-hemoglobin dissociation curve, and describe the factors that affect the curve.
3. Describe the ways in which carbon dioxide is transported in blood, including the reversible reaction that converts carbon dioxide and water to carbonic acid.
4. Predict how changing the partial pressure of carbon dioxide will affect the pH of plasma.
5. Describe the conditions hyperventilation and hypoventilation.

We discussed pulmonary and tissue gas exchange together because of their similarities, but in doing so we skipped an important step: **gas transport** in the blood. Remember that oxygen and carbon dioxide have limited solubility in water. Therefore, in order for pulmonary and tissue gas exchange to occur, both of these gases must undergo chemical reactions that enable their safe transport through the bloodstream and exchange in the lungs and tissues. We explore these reactions in this module. First, we look at the importance of the *hemoglobin* molecule in transporting oxygen, and then at three ways of transporting carbon dioxide in the blood, including the role of a substance known as *bicarbonate*. We also describe how ventilation and carbon dioxide can affect the body's acid-base homeostasis.

Oxygen Transport

◄◄ **Flashback**

1. What is hemoglobin's main function? (p. 726)
2. Describe the structure of hemoglobin. (p. 726)

As you may recall from our discussion of Henry's law, oxygen does not dissolve in great amounts in the plasma because its solubility in water is rather low. In fact, only about 1.5% of the total oxygen transported by the blood travels unbound. Most of the oxygen is transported on the protein **hemoglobin (Hb).** In this section, we examine how oxygen is carried in this manner.

Transport of O_2 on Hemoglobin

Hemoglobin is a protein found in erythrocytes that consists of four polypeptide chains, each including a **heme group** (see Figure 19.3 for review). Each of these heme groups contains one atom of iron that can bind one molecule of oxygen, so each Hb protein can transport up to four molecules of oxygen. About 280 million Hb molecules are present in each erythrocyte, giving each one the ability to carry over a billion oxygen molecules.

Hb binds and releases O_2 via two reactions known as *loading* and *unloading reactions*, respectively (**Figure 21.22**). These reactions can be summarized as follows:

$$\underset{Deoxyhemoglobin}{HHb} + O_2 \rightleftharpoons \underset{Oxyhemoglobin}{HbO_2}$$

① **During loading, oxygen from alveoli binds to hemoglobin in the pulmonary capillaries.** This loading reaction converts hemoglobin from **deoxyhemoglobin (HHb)** to **oxyhemoglobin (HbO$_2$).** An Hb with one to three oxygen

① During loading, oxygen from alveoli binds to hemoglobin (Hb) in the pulmonary capillaries, converting it to oxyhemoglobin (HbO$_2$).

② Oxygen-rich blood travels to the heart, which pumps it to the systemic circulation.

③ During unloading, Hb in the systemic capillaries releases oxygen to the tissue cells.

④ Oxygen-poor blood then returns to the heart, which pumps it back to the pulmonary circulation.

Figure 21.22 Transport of oxygen: loading and unloading of oxygen on hemoglobin in erythrocytes.

molecules attached is said to be *partially saturated*. When an Hb is fully loaded with four oxygen molecules, it is said to be *fully saturated*.

② Once the Hb is fully saturated, **the oxygen-rich blood travels to the left side of the heart, which pumps it to the systemic circulation.**

③ **During unloading, Hb in the systemic capillaries releases oxygen to the tissue cells.** Without the unloading reaction, Hb would never give any of its oxygen to the tissues that need it.

④ **Oxygen-poor blood then returns to the right side of the heart, which pumps it back to the pulmonary circulation.**

Let's look at the unloading reaction more closely. The amount of oxygen unloaded from Hb must strictly match the tissues' metabolic needs. Cells that are actively carrying out metabolic reactions require large amounts of oxygen, and if Hb does not unload adequate oxygen, the cells will not be able to produce adequate ATP. At the same time, cells with a slower metabolic rate do not require large amounts of oxygen. If Hb were to give these cells too much oxygen, toxicity and cellular damage could result.

Hb's ability to load or unload oxygen, and therefore its saturation, depends on two factors:

- the P_{O_2} in the lungs or the tissues, and
- the tightness with which Hb binds oxygen, also called the *affinity*, or the *bond strength*, of Hb.

Let's examine how each of these factors ensures appropriate tissue oxygenation.

Effect of P_{O_2} on Hemoglobin Saturation

The percentage of Hb bound to oxygen is called the **percent saturation of Hb.** One of the main determinants of the percent saturation of Hb is the P_{O_2} of the blood and tissues. The higher the P_{O_2} of the blood, the more the loading reaction is favored, because more O_2 molecules are available to bind to Hb. The opposite is also true.

The relationship between P_{O_2} and percent saturation of Hb is shown in the **oxygen-hemoglobin dissociation curve,** which is illustrated in **Figure 21.23.** Let's see what this graph tells us about the relationship between these two factors:

- **Percent O_2 saturation of Hb.** These values are on the left-hand side of the graph. Notice that the Hb in arterial blood at point A normally has a percent saturation of about 97–100%. This blood has passed through the lungs and heart but has not yet reached the tissues to unload. After unloading O_2 in the tissues, the Hb in venous blood at point B has a percent saturation of about 75%. This means that the venous blood returning to the heart and lungs of a person at rest still has about 75% of the oxygen it can hold bound to its Hb. But note how this situation can change when oxygen demands by the tissues change. Point C represents the percent saturation of Hb in venous blood during

vigorous exercise, when the tissues require much more oxygen. In this situation, the percent saturation of Hb can drop significantly, to 25% or even less.

- **Blood P_{O_2}.** Along the bottom of the graph is the partial pressure of oxygen in blood, the P_{O_2}. Notice that the P_{O_2} in arterial blood is close to 100 mm Hg (at point A), and the P_{O_2} in venous blood of a person at rest is about 40 mm Hg (at point B). Then notice at point C that during vigorous exercise the P_{O_2} of venous blood drops significantly as the tissues consume more oxygen.

You can see that this curve has an S shape, which results from Hb's changing affinity for oxygen as the P_{O_2} changes. What's happening? When Hb is fully saturated, it hangs on to the oxygen molecules tightly, which matches the red-shaded part of the curve in Figure 21.23. In this region, even a big change in P_{O_2} doesn't alter the percent saturation of Hb significantly. However, when Hb loses its first oxygen molecule, it changes shape, and it becomes easier for the second and third oxygen molecules to unload from Hb. The purple-shaded curve on the graph shows the resulting sharp decrease in the percent saturation of Hb. Finally, the fourth and last oxygen molecule is bound least tightly to Hb. Notice in the blue-shaded part of the curve that it takes a drastically low P_{O_2} for Hb to unload the fourth oxygen molecule.

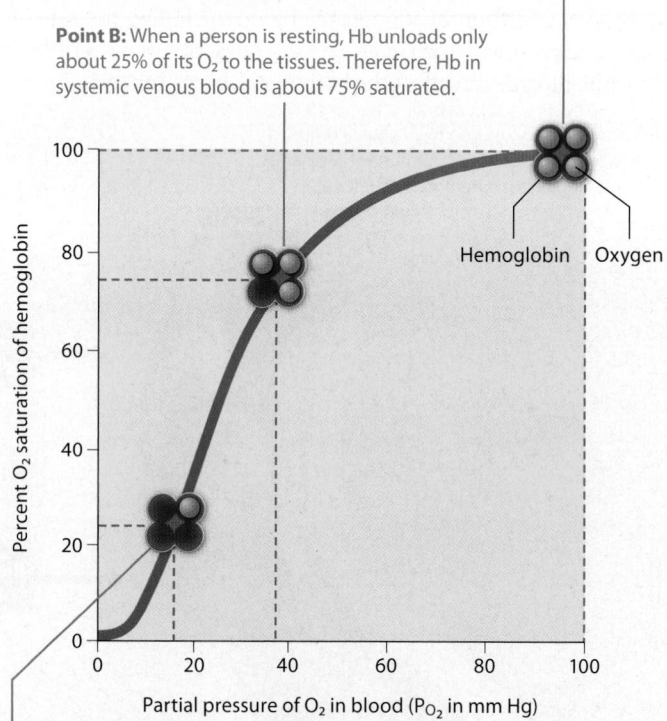

Point A: Hemoglobin (Hb) is almost 100% saturated in systemic arterial blood, as no O_2 has been unloaded to the tissues.

Point B: When a person is resting, Hb unloads only about 25% of its O_2 to the tissues. Therefore, Hb in systemic venous blood is about 75% saturated.

Hemoglobin Oxygen

Percent O_2 saturation of hemoglobin (y-axis)

Partial pressure of O_2 in blood (P_{O_2} in mm Hg) (x-axis)

Point C: When a person exercises vigorously, Hb unloads most of its oxygen in the tissues. In such cases, the Hb entering venous blood is only about 25% saturated.

Figure 21.23 The oxygen-hemoglobin dissociation curve.

ConceptBOOST »»»

Making Sense of the Oxygen-Hemoglobin Dissociation Curve

The graph of the oxygen-hemoglobin dissociation curve may at first look confusing, but remember that all physiological graphs show something being *measured*. In this case, the graph is showing two measurements:

- the amount of oxygen bound to hemoglobin, and
- the partial pressure of oxygen (P_{O_2}) in the blood and tissues.

The shape of the oxygen-hemoglobin dissociation curve in the graph tells us how these two measurements change with respect to one another. So what do these changes tell us? They tell us two important facts about hemoglobin:

- *The percent saturation of Hb remains nearly constant at relatively high P_{O_2} levels.*

Notice that the percent saturation of Hb changes very little when the P_{O_2} drops from 100 mm Hg to 80 mm Hg. This means that even when oxygen levels in arterial blood are lower than normal, hemoglobin holds on to most of its oxygen molecules. This is important because arterial P_{O_2} never gets above about 80–85 mm Hg in some individuals, such as those at high altitudes or those with pulmonary disease. So this property of Hb is crucial because it allows Hb to hang on to its oxygen molecules until it reaches the much lower P_{O_2} levels found in the capillaries.

- *The percent saturation of Hb drops rapidly when P_{O_2} levels are low.*

Notice that when the P_{O_2} drops to about 40 mm Hg, which is generally the P_{O_2} in the capillary beds of a person at rest, the percent saturation of Hb drops sharply and rapidly. Remember that blood is present in the capillaries for an extremely short

time—only about 0.75 second. For this reason, Hb must be able to unload its oxygen molecules rapidly. This property of Hb is therefore important because it allows the Hb not only to unload its oxygen where it is needed in the capillary beds, but also to do so at a fast enough pace to ensure that the cells receive sufficient oxygen.

Take note of one final aspect of this graph: From the shape of this curve you can see why arterial P_{O_2} levels of 40–50 mm Hg, which occur when one stops breathing, are not compatible with life. At such a low P_{O_2}, Hb unloads all of its oxygen in the arterial blood instead of the capillaries. Because free oxygen molecules in arterial blood cannot diffuse in appreciable amounts through the arterial wall to reach the cells, the cells are starved of oxygen and, as a result, they die. ■

Effect of Affinity on Hemoglobin Saturation

Four main factors can change the affinity of Hb for oxygen by altering Hb's shape. The first three factors are (1) the temperature, (2) the pH of the blood, and (3) the P_{CO_2} (**Figure 21.24**) (we'll get to the fourth shortly). In general, increasing temperature, hydrogen ion concentration, and P_{CO_2} all decrease Hb's affinity for oxygen, which results in the unloading of more oxygen into the tissues. This scenario makes sense if you remember that a cell actively producing ATP needs more oxygen and also produces heat, hydrogen ions, and CO_2 as byproducts. The opposite is also true—as the temperature, acidity, and P_{CO_2} decrease, as occurs in tissues with a slower metabolic rate, Hb binds oxygen more strongly, and less oxygen is unloaded. This phenomenon is known as the **Bohr effect.**

The fourth factor that influences Hb's affinity for oxygen is the level of a chemical called **BPG (2,3-bisphosphoglycerate).** Erythrocytes produce BPG during a side reaction of glycolysis (look ahead to Chapter 23). BPG binds with Hb, reducing its affinity for oxygen, which increases the unloading of oxygen into the tissues. Erythrocytes produce greater amounts of BPG when Hb is less saturated with oxygen. (See *A&P in the Real World: Carbon Monoxide Poisoning* on page 831 to read about the effects of increasing Hb's affinity for oxygen.)

BPG is a key factor that allows people to live at higher altitudes or with poor pulmonary function, as both situations result in lower P_{O_2} levels and less saturation of Hb with oxygen. In an example of the Cell-Cell Communication Core Principle (p. 27), the BPG levels also increase in response to certain hormones, such as norepinephrine, epinephrine, thyroxine, testosterone, and human growth hormone. Each of these hormones promotes cellular metabolism and for this reason increases the cellular demand for oxygen.

CORE PRINCIPLE
Cell-Cell Communication

(a) **Increased temperature causes more O₂ unloading.**

(b) **Increased H⁺ causes more O₂ unloading.**

(c) **Increased P_{CO_2} causes more O₂ unloading.**

Figure 21.24 **Effect of temperature, hydrogen ion concentration, and P_{CO_2} on oxygen unloading from hemoglobin.**

Quick Check

☐ 1. How is the majority of oxygen transported through the bloodstream?

☐ 2. How do temperature, pH, P_{CO_2}, and BPG affect Hb's affinity for oxygen?

☐ 3. Why is the S shape of the oxygen-hemoglobin dissociation curve important?

Carbon Dioxide Transport

◀◀ **Flashback**

1. What is an enzyme? What is its main function? (p. 44)
2. Define a buffer. (p. 49)
3. What is the normal pH range for the blood? (p. 48)

The carbon dioxide produced by our cells' metabolic reactions is transported through the blood from tissue cells to the alveoli. Carbon dioxide is transported in the blood in three ways: dissolved in plasma, bound to Hb, and as bicarbonate ions. Let's take a look at each of these methods.

- **Dissolved in the plasma.** Recall from our earlier discussion that carbon dioxide has a higher solubility in water than that of oxygen. For this reason, more carbon dioxide than oxygen dissolves in plasma, and about 7–10% of the total carbon dioxide is transported in this manner.

- **Bound to Hb within erythrocytes.** About 20% of the carbon dioxide is transported by Hb. Unlike oxygen, carbon dioxide does not bind to the heme group of Hb. Instead, carbon dioxide binds to Hb's polypeptide chains. Carbon dioxide binds easily and reversibly with Hb, forming **carbaminohemoglobin** (not to be confused with carboxyhemoglobin, which is the combination of carbon monoxide with Hb).

- **As bicarbonate ions in blood.** The majority of carbon dioxide in the blood (about 70%) travels as a totally different compound, **bicarbonate ion** (HCO_3^-).

Because the latter is the major method of transport for carbon dioxide, the upcoming sections discuss how carbon dioxide is converted to bicarbonate in erythrocytes. We then examine how ventilation controls the amount of carbon dioxide and bicarbonate in the blood and why this is critical to the maintenance of blood pH homeostasis.

Transport of CO₂ as Bicarbonate

Most of the carbon dioxide entering the plasma diffuses into erythrocytes, where it encounters the enzyme **carbonic anhydrase (CA)**. CA rapidly catalyzes the reaction of water with carbon dioxide to yield carbonic acid (H_2CO_3) via the following reversible reaction:

$$CO_2 + H_2O \rightleftarrows H_2CO_3 \rightleftarrows HCO_3^- + H^+$$

Carbon dioxide   *Water*   *Carbonic acid*   *Bicarbonate ion*   *Hydrogen ion*

Note that carbonic acid doesn't linger for long, but instead dissociates to become bicarbonate and hydrogen ions. Most of the bicarbonate ions then diffuse into the plasma, and many of the hydrogen ions bind with Hb, which acts as a buffer to resist pH change (see Chapter 2). Since bicarbonate ions have a negative charge, this influx of negative charges into the plasma could be problematic. Fortunately, the outward movement of bicarbonate ions is balanced by the movement of chloride ions into the erythrocyte, a phenomenon known as the **chloride shift.**

In **Figure 21.25a**, the forward reaction of carbon dioxide and water to carbonic acid and then to bicarbonate and hydrogen ions is shown in an erythrocyte in a systemic capillary. The process is as follows:

1. **Carbon dioxide diffuses from the cells into the erythrocyte.** Carbon dioxide is formed in tissue cells as a byproduct of the cell's catabolic reactions, after which it diffuses into the systemic capillary and erythrocyte.
2. **Carbon dioxide is converted to carbonic acid.** The enzyme carbonic anhydrase catalyzes the reaction of carbon dioxide and water to form carbonic acid (H_2CO_3).
3. **Carbonic acid dissociates into bicarbonate (HCO_3^-) and hydrogen ions.** At the pH in the cytosol, much of the carbonic acid dissociates into bicarbonate and hydrogen ions.

4. **Hydrogen ions bind to hemoglobin.** The newly formed hydrogen ions bind to hemoglobin, which serves as a buffer and helps to minimize pH changes from the hydrogen ions.
5. **Bicarbonate ions enter the plasma (as chloride ions enter the erythrocyte).** Bicarbonate ions leave the erythrocyte and enter the plasma. The outward flow of bicarbonate ions is balanced by the inward flow of chloride ions in the chloride shift.

(a) **Bicarbonate formation in an erythrocyte in a systemic capillary**

(b) **Carbon dioxide formation in an erythrocyte in a pulmonary capillary**

Figure 21.25 Transport of carbon dioxide: the conversion of carbon dioxide and water into carbonic acid in erythrocytes.

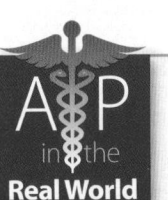

Carbon Monoxide Poisoning

Carbon monoxide (CO) poisoning is familiar to most people from a common scene in television and movies: a suicide or homicide that occurs when a car sits running in a closed garage. This is indeed one way to get CO poisoning, but it is certainly not the only way. Carbon monoxide is produced from burning organic compounds. The colorless, odorless, and tasteless gas is found in smoke from fires and cigarettes, and in exhaust fumes from portable heaters, stoves, and engines.

CO produces its toxic effects by binding reversibly with Hb, producing *carboxyhemoglobin*. Not only does this binding cause CO to occupy some of Hb's oxygen-carrying spots, it also causes a change in Hb's shape that increases its affinity for oxygen. This change means that Hb binds oxygen much more strongly and will not adequately release it to the body's tissues.

CO has an affinity for Hb about 200–230 times that of oxygen, so even small concentrations of this gas can cause serious problems. Symptoms of CO poisoning include confusion, dizziness, and nausea. In severe cases, it can cause seizures, coma, and death. The main treatment is 100% oxygen, delivered at either atmospheric or hyperbaric pressure.

The reverse reaction, which occurs in an erythrocyte in a pulmonary capillary, is illustrated in **Figure 21.25b** and proceeds as follows:

① **Carbonic acid is re-formed in the erythrocyte.** Hydrogen ions dissociate from hemoglobin, and bicarbonate ions re-enter the erythrocyte. They combine to form carbonic acid.

② **Carbonic acid breaks down into water and carbon dioxide.** Carbonic anhydrase also catalyzes the reverse reaction, or the breakdown of carbonic acid into water and carbon dioxide.

③ **Carbon dioxide diffuses into the alveolus.** Carbon dioxide leaves the pulmonary capillary and diffuses into the alveolus along its pressure gradient.

Effect of P_{CO_2} on Blood pH

One of the primary buffer systems in the body is the **carbonic acid–bicarbonate buffer system** (see Chapter 2). Under normal conditions, the pH of the blood changes very little when hydrogen ions are released into the blood. The change is small because the hydrogen ions bind with buffers such as bicarbonate ions to form carbonic acid, which is a weak acid that has very little effect on blood pH at physiological levels. The opposite reaction also takes place—when the blood pH increases and the level of hydrogen ions drops too low, hydrogen ions are released from carbonic acid to yield bicarbonate ions, which act as a weak base.

The major determinant of the amount of carbonic acid in the blood is the P_{CO_2} of the blood. In turn, the P_{CO_2} of the blood is primarily determined by two factors: (1) the rate and depth of ventilation, and (2) the rate at which carbon dioxide is generated by cellular metabolic reactions (we discuss this second factor in Chapter 23). For this reason, a change in the normal pattern of ventilation can have a dramatic impact on the blood's concentration of hydrogen ions. Two main changes in the pattern of ventilation can have profound effects on the blood pH—hyperventilation and hypoventilation (**Figure 21.26**):

● **Hyperventilation.** When a person *hyperventilates*, the rate or depth of his or her breathing, or both, increases. Hyperventilation generally increases the amount of carbon dioxide that is expired and decreases the blood's P_{CO_2}. This means that less carbonic acid is formed, fewer hydrogen ions are released, and the pH of the blood *increases* (becomes more basic). Hyperventilation may also affect the P_{O_2} in the blood—it raises the amount of oxygen in the lungs, which causes more oxygen to dissolve in the plasma.

● **Hypoventilation.** Hypoventilation involves a decrease in a person's rate or depth of breathing, or both. It generally causes retention of carbon dioxide and an increase in the blood's P_{CO_2}. As a result, more carbonic acid is formed and more hydrogen ions are released. When the excess hydrogen ions exceed the body's buffer systems, the pH of the blood *decreases* (becomes more acidic). Hypoventilation usually also significantly decreases the arterial P_{O_2}, causing hypoxemia.

Figure 21.26 Effect of pattern of ventilation on blood pH.

As you can see, a change in the rate and depth of ventilation is one of the body's quick ways to compensate for sudden or dramatic changes in blood pH. However, if changes in the pattern of ventilation are chronic rather than temporary, serious disturbances in blood pH may result. When a person chronically hyperventilates, the relative lack of carbon dioxide results in a condition called *hypocapnia*. This increases the pH of the blood, bringing about **respiratory alkalosis.** Respiratory alkalosis has varied causes, including fever, anxiety, head trauma, aspirin overdose, hyperthyroidism, and certain pulmonary diseases. If a person chronically hypoventilates, excess carbon dioxide (*hypercapnia*) causes the blood pH to decrease, resulting in a condition called **respiratory acidosis.** Common causes of respiratory acidosis include pulmonary disease and neurological disease or injury.

ConceptBOOST ⟫⟫⟫

Relating Ventilation and Blood pH

Students often find the relationship between rate and depth of ventilation and the pH of the blood confusing and difficult to remember. An easy way to remember the relationship is as follows:

- The prefix *hyper-* from the word "hyperventilation" means *above,* and hyperventilation leads to a pH increase *above* normal values.

Hyperventilation → pH increase

- The prefix *hypo-* from the word "hypoventilation" means *below,* and hypoventilation leads to a pH decrease *below* normal values.

Hypoventilation → pH decrease

This relationship is easy enough to remember, but don't forget what each change in pH means for the hydrogen ion concentration. Remember the pH scale:

The lower the number, the higher the hydrogen ion concentration; the higher the number, the lower the hydrogen ion concentration. If you get confused, draw yourself a pH scale like the one shown here for reference. ■

Quick Check

☐ 4. What are the three ways in which the body transports carbon dioxide through the bloodstream? How is the majority of the carbon dioxide carried?

☐ 5. What happens to the amount of carbon dioxide in the blood when a person hyperventilates? What does this do to the pH of the blood?

☐ 6. What happens to the amount of carbon dioxide in the blood when a person hypoventilates? What does this do to the pH of the blood?

Apply What You Learned

☐ 1. Predict the effects on tissue oxygenation of a drug that significantly increases hemoglobin's affinity for oxygen.

☐ 2. Predict the effect that each of the following situations would have on the unloading of oxygen from hemoglobin:
   a. Fever (elevated systemic temperature)
   b. Alkalosis (inadequate hydrogen ions in the blood)
   c. Hypercapnia (excess carbon dioxide in the blood)

☐ 3. Why is it important for carbonic anhydrase to catalyze both the conversion of carbon dioxide and water into carbonic acid and the reverse reaction?

See answers in Appendix A.

MODULE **21.6**

Putting It All Together: The Big Picture of Respiration

Learning Outcomes

1. Describe the overall big picture of the processes involved in respiration.

Let's now put the different steps of respiration together and view the big picture of how it occurs, as illustrated in **Figure 21.27.** The major processes involved are ① pulmonary ventilation, ② pulmonary gas exchange, ③ gas transport, and ④ tissue gas exchange. Note that gas transport happens both before and after tissue gas exchange, with oxygen and carbon dioxide, respectively.

Quick Check

☐ 1. Which steps of respiration rely on partial pressure gradients to drive them? How do the other steps work?

1 **Pulmonary ventilation:** Oxygen and carbon dioxide are moved between the air and the alveoli in the lungs (see Figure 21.16).

Alveolus

Pulmonary capillary

2 **Pulmonary gas exchange:** Oxygen enters and carbon dioxide exits the blood via the alveoli (see Figure 21.20a).

Oxygen is transported throughout the blood, primarily on hemoglobin.

After tissue gas exchange, most carbon dioxide in the blood is converted to bicarbonate ions and is transported to the lungs to be expired.

3 **Gas transport:** Oxygen and carbon dioxide are transported by the blood (see Figures 21.22 and 21.25).

Systemic capillary

4 **Tissue gas exchange:** Oxygen and carbon dioxide are exchanged between the blood and the tissue cells (see Figure 21.20b).

Tissue cells

Neural Control of Ventilation

Learning Outcomes

1. Describe the locations and functions of the brainstem respiratory centers.
2. List and describe the major chemical and neural stimuli to the respiratory centers.
3. Compare and contrast the central and peripheral chemoreceptors.

Most of us rarely think about our breathing, as it takes place without the need for conscious control. However, people with pulmonary disease or some other homeostatic imbalance of the respiratory system are often acutely aware of each breath they take. **Dyspnea** (DISP-nee-uh; "bad breathing"), or feeling short of breath, is one of the most unpleasant sensations a human can experience, as we know cessation of breathing leads to death. So, although we tend to take it for granted, control of our ventilation to maintain **eupnea** (YOOP-nee-uh; "good breathing") is one of the most vital functions that the body carries out.

Breathing is controlled largely by groups of neurons in the brainstem. Although breathing in and out might seem to be a simple process, we will see in this module that many factors contribute to our respiratory rhythm. Key among them are specialized cells that monitor levels of carbon dioxide, hydrogen ions, and oxygen in the body. Negative feedback loops and stretch receptors in the lungs are also involved in ensuring that our oxygen intake and carbon dioxide elimination match our metabolic needs.

Control of the Basic Pattern of Ventilation

◄◄ Flashback

1. What is a nucleus of the brain? (p. 431)
2. Which nuclei of the brainstem contribute to the control of ventilation? (p. 435)
3. Which nerve innervates the diaphragm? Which nerves innervate the external intercostal muscles? (p. 519)

Ventilation is largely under the control of the brainstem, particularly the medulla oblongata (see Chapter 12). Physiologists thought for a long time that many of the neurons that maintained eupnea were located in the pons in groups called the *pneumotaxic center* and *apneustic center*. However, we now understand that although these neurons in the pons can influence the respiratory rhythm, they are not responsible for maintaining eupnea.

The collection of neurons that generates the basic rhythm for breathing, called the **respiratory rhythm generator (RRG)**, is

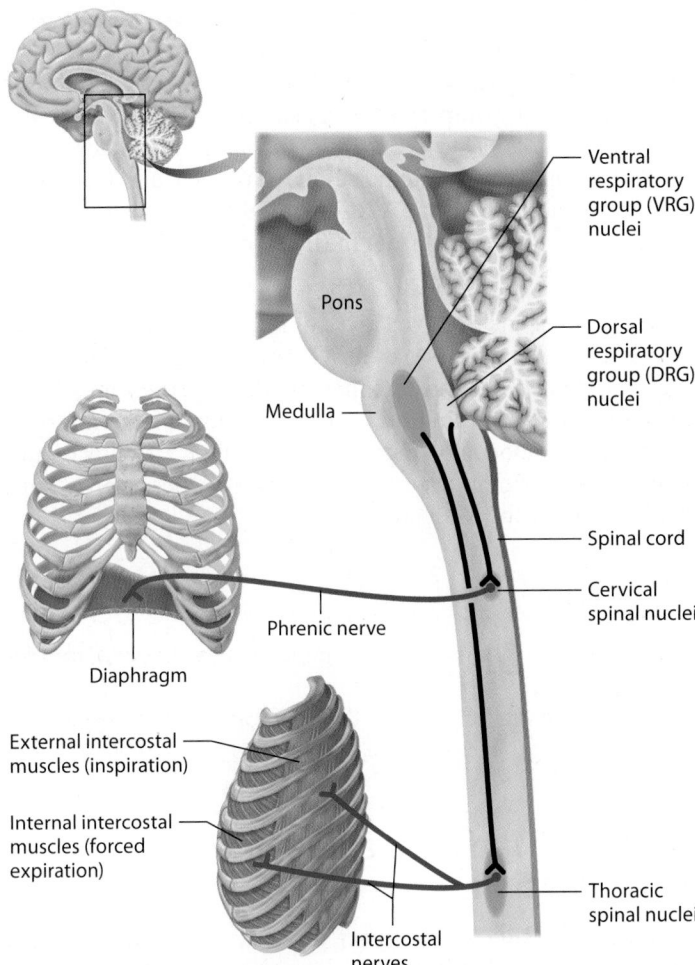

Figure 21.28 Neural control of the basic pattern of ventilation.

believed to lie within a structure called the ventral respiratory column. A current hypothesis states that there are actually two RRGs, one that generates inspiration and one that generates expiration. Further research is needed to confirm if this is indeed the case.

Helping the RRG are groups of neurons in the medullary reticular formation, known as the ventral and dorsal respiratory groups (**Figure 21.28**):

- **Ventral respiratory group.** In the anterior and lateral portions of the medulla we find the **ventral respiratory group (VRG)** nuclei, which contain both inspiratory and expiratory neurons. In an example of the Cell-Cell Communication Core Principle (p. 27), the VRG's inspiratory neurons send impulses to spinal cord relay centers that eventually trigger action potentials in the **phrenic nerve** supplying the diaphragm and in the **intercostal nerves** supplying the external intercostal muscles. These

CORE PRINCIPLE
Cell-Cell Communication

nerves also supply certain accessory muscles of inspiration and expiration.

- **Dorsal respiratory group.** The neurons of the **dorsal respiratory group (DRG)** nuclei are found in the posterior medulla. They seem to be primarily inspiratory neurons that likely send impulses to the phrenic and intercostal nerves along with the VRG. The DRG may also integrate sensory information from the blood and the lungs and relay it to the VRG.

Many other neurons related to respiratory functions are located in the cranial nerve nuclei surrounding the DRG and VRG. These neurons mostly innervate the accessory muscles of inspiration and/or expiration, and some bring sensory information from the blood and lungs back to the DRG. So many redundant and overlapping respiratory centers exist because we must continue to breathe while we sleep.

Quick Check

☐ 1. Which collection of neurons generates the basic rhythm for breathing?

☐ 2. What are the functions of the dorsal and ventral respiratory groups?

Control of the Rate and Depth of Ventilation

Although the RRG seems to set the basic rate and pattern of ventilation, several other control mechanisms play a role in

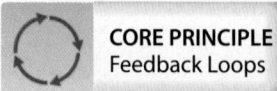

CORE PRINCIPLE
Feedback Loops

determining the rate and depth of ventilation. These mechanisms are generally negative feedback loops that ensure the ventilatory needs of the body are met at all times (an example of the Feedback Loops Core Principle, p. 21).

These feedback loops rely on receptors to gather chemical information about the current respiratory state of the body. Recall that a *chemoreceptor* is a specialized cell that responds to a change in the concentration of a certain chemical (see Chapter 15). Two types of chemoreceptors are involved in the control of ventilation: central and peripheral. Let's examine what each of these chemoreceptors does.

Although we discuss the control of ventilation as if it were purely involuntary, we are able to exert some voluntary control over ventilation. This voluntary control is mediated by the cerebral cortex, and it bypasses the respiratory centers. Signals travel instead through the spinal cord to the appropriate spinal nerves.

Function of Central Chemoreceptors

We discussed with Figure 21.26 how a person's pattern of ventilation can change his or her blood pH. Other mechanisms can also affect blood pH, such as an increased metabolic rate. But how does the body deal with these changes?

The answer is a negative feedback cycle started by the **central chemoreceptors,** which are neurons in the medullary reticular formation. These chemoreceptors detect changes in the concentrations of carbon dioxide and hydrogen ions in the blood by monitoring the hydrogen ion concentration of cerebrospinal fluid (CSF). Current evidence indicates that the central chemoreceptors communicate this information to the DRG, which then relays it to the VRG. The VRG, in turn, changes the rate and depth of ventilation as appropriate by stimulating or inhibiting expiratory or inspiratory muscles. These actions cause the loss or retention of carbon dioxide and a return to the normal blood pH.

Alterations in arterial P_{CO_2} are the most powerful stimuli that induce changes in ventilation. In fact, the urge to breathe after holding your breath comes largely from the slight accumulation of carbon dioxide during the period of breath holding. Carbon dioxide in the blood diffuses through the blood-brain barrier into the CSF and the extracellular fluid of the brain, where it combines with water to form bicarbonate and hydrogen ions. These hydrogen ions interact directly with the central chemoreceptors. (Note, however, that the blood-brain barrier is relatively impermeable to hydrogen ions, so changes in *arterial* hydrogen ion concentration provide a relatively weak stimulus for the central chemoreceptors to induce ventilatory changes.)

When subjected to changes in P_{CO_2} and hydrogen ion concentration, the central chemoreceptors trigger one of the following responses via classic negative feedback loops:

- **High P_{CO_2} or high hydrogen ion concentration, or both, triggers hyperventilation.** As you can see in **Figure 21.29a**, in the presence of high arterial P_{CO_2} and/or high hydrogen ion concentration, the central chemoreceptors stimulate the VRG to trigger an increase in the rate of ventilation. Recall that during hyperventilation, carbon dioxide leaves the blood at a rapid pace. Its departure causes a decrease in blood P_{CO_2} and hydrogen ions—and therefore CSF P_{CO_2} and hydrogen ion concentration—and restores homeostasis.

- **Low P_{CO_2} or low hydrogen ion concentration, or both, triggers hypoventilation.** In the presence of low arterial P_{CO_2} and/or low hydrogen ion concentration, as shown in **Figure 21.29b**, the central chemoreceptors trigger the opposite response—they inhibit the VRG and slow the rate of ventilation so that the body retains carbon dioxide. This response causes an increase in blood P_{CO_2} and hydrogen ions—and therefore CSF P_{CO_2} and hydrogen ion concentration—and restores homeostasis.

A sudden change in arterial P_{CO_2} can alter ventilation within minutes. This response is initially strong but diminishes over the next day or two because the central chemoreceptors become less sensitive over time. In addition, bicarbonate ions move from the blood (through the blood-brain barrier) into the cerebrospinal fluid and brain extracellular fluid to counteract the pH imbalance. This is partly why some patients with pulmonary disease and chronically elevated arterial P_{CO_2} have a nearly normal ventilatory rate. Certain drugs such as opiates and alcohol also blunt

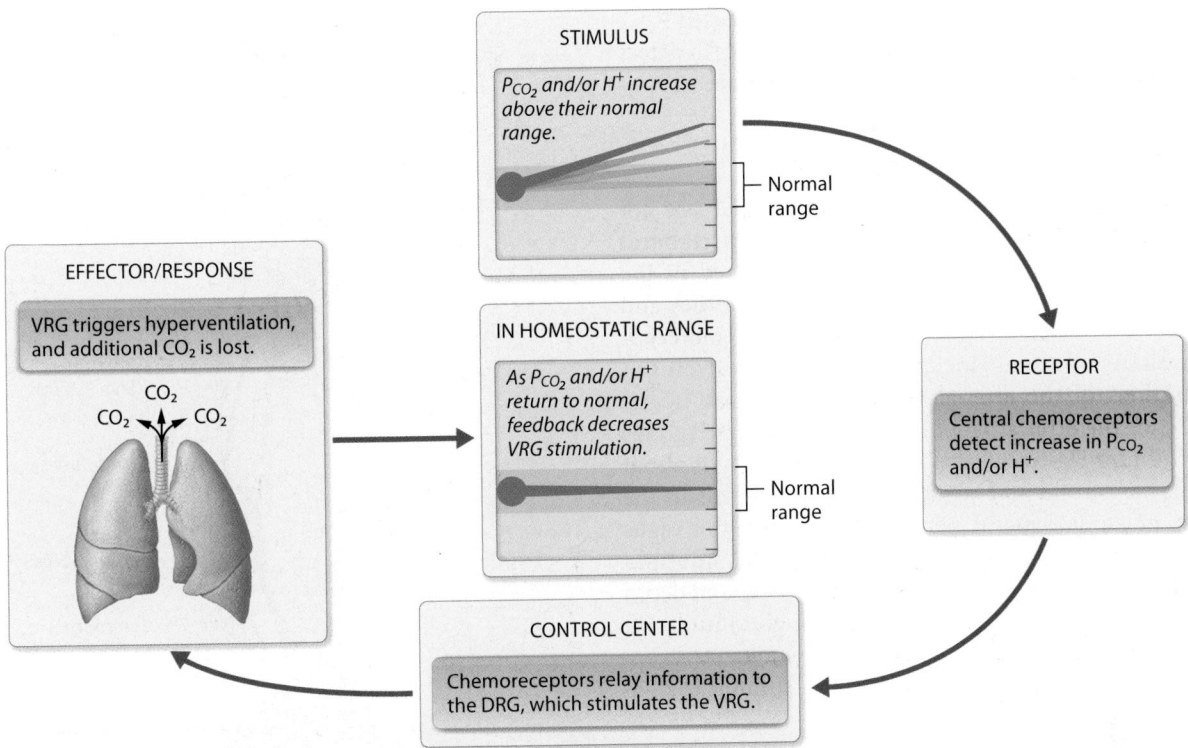

(a) Response to increased arterial P$_{CO_2}$ and/or H$^+$ concentration by a negative feedback loop

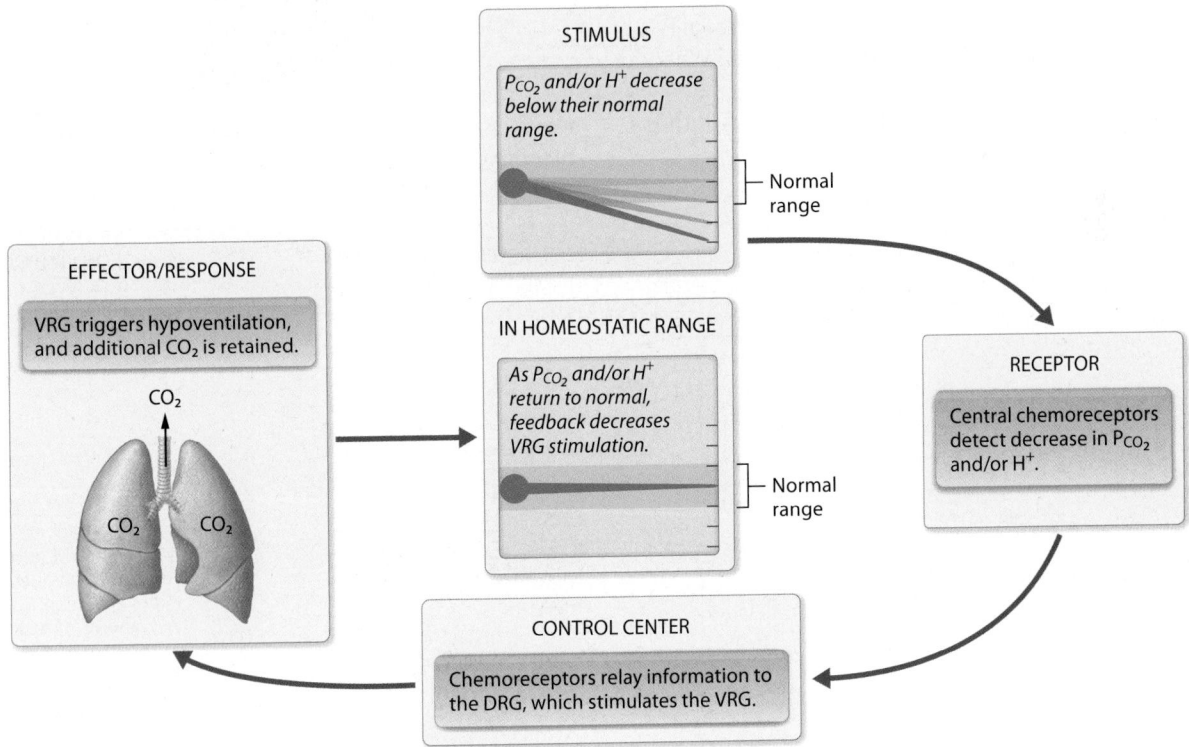

(b) Response to decreased arterial P$_{CO_2}$ and/or H$^+$ concentration by a negative feedback loop

Figure 21.29 Role of the central chemoreceptors in regulation of blood pH via breathing rate.

the response of the central chemoreceptors to an increased P_{CO_2}, inhibiting one of the main feedback loops the body uses to detect and rectify impaired ventilation. Such drugs can produce *respiratory depression*, especially in overdoses.

Influence of Peripheral Chemoreceptors

The P_{CO_2} and hydrogen ion concentrations of the CSF are not the only factors that influence the medulla. The **peripheral chemoreceptors** are specialized groups of cells located in the carotid arteries and in the aorta, called the **carotid bodies** and **aortic bodies,** respectively (**Figure 21.30**). These chemoreceptors detect a variety of stimuli, including the P_{CO_2} and hydrogen ion concentrations of the arterial blood, but are most sensitive to the P_{O_2} of the arterial blood. When the arterial P_{O_2} falls below about 70 mm Hg (recall that normal arterial P_{O_2} is about 100 mm Hg), the carotid and aortic bodies send signals to the DRG via the glossopharyngeal nerve (CN IX) and the vagus nerve (CN X), respectively. The DRG then relays this information to the VRG, which triggers an increase in the rate and depth of ventilation. See how this process works at high altitudes in *A&P in the Real World: High-Altitude Acclimatization.*

The control mechanisms of ventilation are summarized in **Figure 21.31**.

Quick Check

☐ 3. Where are the central chemoreceptors located? What do they detect?

☐ 4. What do the central chemoreceptors trigger if arterial P_{CO_2} or hydrogen ion concentration, or both, increases? What do they do if either or both decreases?

☐ 5. What is the role of the peripheral chemoreceptors in the control of ventilation?

Figure 21.30 Role of the peripheral chemoreceptors in regulation of breathing rate.

STIMULI	CONTROL MECHANISM	EFFECT ON RESPIRATORY CENTERS	EFFECT ON VENTILATION
Cerebral cortex inputs (e.g., emotion)	Voluntary control	+/–	Varied
Changes in arterial P_{CO_2}, H^+ concentrations	Central chemoreceptors	+/–	Hyperventilation when P_{CO_2} and/or H^+ concentrations increase; hypoventilation when P_{CO_2} and/or H^+ decrease
Changes in arterial P_{O_2}	Peripheral chemoreceptors	+	Hyperventilation when arterial P_{O_2} decreases

Figure 21.31 Control mechanisms of ventilation.

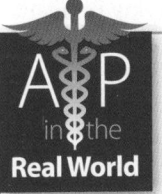

High-Altitude Acclimatization

Experienced mountain climbers know that when they gradually increase their elevation over a period of days rather than hours, they are able to tolerate much lower atmospheric oxygen levels. This phenomenon, called *high-altitude acclimatization,* allows the peripheral chemoreceptors to stimulate an increase in ventilation, which permits the body to maintain an acceptable blood P_{O_2}.

The reason that a slow ascent is necessary lies in the sensitivity of the peripheral chemoreceptors. The longer they are exposed to a low P_{O_2}, the *more* they stimulate an increase in ventilation. A gradual ascent up a mountain allows more time for the peripheral chemoreceptors to be exposed to a low P_{O_2}, and therefore a greater increase in ventilation. This practice permits experienced mountain climbers to reach great elevations without supplemental oxygen.

Apply What You Learned

□ 1. Ms. Green was involved in a motorcycle accident and has sustained damage to her medulla in the area of the VRG. What is her likely prognosis? Explain.

□ 2. Mr. Jackson has chronic pulmonary disease, and laboratory analysis of his arterial blood demonstrates that he has a high arterial P_{CO_2} and a low arterial P_{O_2}. However, his respiratory rate is normal. Explain this finding.

See answers in Appendix A.

MODULE **21.8**

Diseases of the Respiratory System

Learning Outcomes

1. Explain the difference between restrictive and obstructive disease patterns.
2. Describe the basic pathophysiology for certain pulmonary diseases.

Lung diseases are the third leading cause of death in the United States. Every year, they take the lives of nearly 350,000 people and cost the American economy more than 150 billion dollars. A history of smoking is implicated in many of these diseases. Two generally recognized patterns of lung disease can be identified—*restrictive* and *obstructive*—although many diseases have characteristics of both. In this module we describe these patterns and the reasons for them.

Restrictive Lung Diseases

Restrictive lung diseases decrease pulmonary compliance and reduce the effectiveness of *inspiration*. The most common causes of decreased pulmonary compliance are increased alveolar surface tension and destruction of the elastic fibers of the lung. Restrictive lung diseases decrease the inspiratory capacity, vital capacity, and total lung capacity, and make effective pulmonary ventilation difficult.

Individually, the restrictive lung diseases are not as prevalent as some of the more common obstructive diseases, but collectively their number is large. Some of the more common restrictive diseases are the following:

- **Idiopathic pulmonary fibrosis.** The primary problem in idiopathic pulmonary fibrosis is inflammation of the lung tissue, followed by destruction of the lungs' elastic tissue and its replacement with thick bundles of collagen fibers. The cause of the disease is unknown, but about 66–75% of patients have a history of heavy smoking.
- **Pneumoconiosis.** Pneumoconiosis (new'-moh-koh-nee-OH-sis) refers to a collection of diseases that arise from the inhalation of inorganic dust particles, such as coal dust, silicon dioxide, asbestos, fiberglass, and various heavy metals. The particles settle permanently in the tissue of the lungs, where they cause inflammation followed by fibrosis.
- **Neuromuscular diseases and chest wall deformities.** These conditions are not purely lung diseases, but one of their potential consequences is pulmonary dysfunction in a restrictive pattern due to weakness of the inspiratory muscles or a stiff chest wall.

Obstructive Lung Diseases

Obstructive lung diseases increase airway resistance. It may seem surprising, but these diseases actually decrease the efficiency of *expiration*. Normally during expiration, the elastic recoil of the lungs decreases the diameter of the airways slightly, but they remain open to air flow. However, when airway resistance is abnormally high before expiration, the elastic recoil of the lungs after expiration can actually collapse the airways, trapping oxygen-poor, carbon dioxide–rich air in the alveoli:

A patient with an obstructive lung disease will often purse his or her lips and breathe out slowly to minimize the compression of the airways and prevent air trapping. Patients also often hyperinflate their lungs to decrease air trapping by enlarging their airways. Typically, with an obstructive disease pattern, the residual volume increases, which decreases the vital capacity. Let's look at some common obstructive lung diseases, including emphysema, asthma (AZ-mah), and lung cancer.

Chronic Obstructive Pulmonary Disease

Over 16 million Americans are affected by **chronic obstructive pulmonary disease,** or **COPD,** and the disorder is the sixth leading cause of death worldwide. COPD is defined as persistent airway obstruction that is not fully reversible. Nearly all patients with COPD have a history of cigarette smoking, although genetic and environmental factors may be involved as well. There are three subtypes of COPD: emphysema, small airway disease, and chronic bronchitis.

- **Emphysema.** As we mentioned, **emphysema** is characterized by the destruction of the structures of the respiratory zone and a loss of alveolar surface area. Most cases of emphysema are due to cigarette smoke, as the resins in the smoke destroy the lungs' elastic fibers. About 1–2% of cases are due to a genetic mutation causing deficiency of an enzyme that prevents the destruction of lung tissue by other enzymes. When this protective enzyme is deficient, the elastic tissue of the lungs is destroyed at a rapid pace.
- **Small airway disease.** Some degree of disease of the bronchioles usually accompanies emphysema. In small airway disease, the bronchioles narrow and are typically plugged with mucus.
- **Chronic bronchitis.** Patients with **chronic bronchitis** have excessive mucus in the airways that must be cleared by coughing. This condition is almost exclusively due to cigarette smoke, which increases the number and size of mucous glands and goblet cells and paralyzes the cilia of the respiratory epithelium.

Most patients with COPD eventually exhibit a mixture of all three subtypes. The symptoms of COPD generally develop over a period of years. They include shortness of breath, cough, and excessive mucus production. Patients typically exhibit a prolonged expiratory phase, an elevated arterial P_{CO_2}, and a decreased arterial P_{O_2}, especially as the disease worsens. With severe disease, patients may exhibit **cyanosis** (a bluish color to the skin because of less oxygen in the arterial blood) and may develop a "barrel chest" as a result of lung hyperinflation.

One of the first and most important steps in the treatment of COPD is smoking cessation. Other treatments are aimed at increasing oxygenation through supplemental oxygen therapy and reducing airway resistance. In severe cases, lung transplant surgery may be an option.

Asthma

Asthma is a condition that affects about 7–10% of the worldwide population. It is an obstructive disease in which the airways are hyperresponsive to a trigger, which is most often dust mites, mold, pollen, or animal dander (see Chapter 20). Other triggers include exercise, certain drugs such as aspirin, infections, and air pollutants. When people with asthma are exposed to a trigger, three responses occur in their airways: (1) bronchoconstriction; (2) inflammation of the airways; and (3) increased production of excessively thick mucus. Each of these three responses causes a significant increase in airway resistance. Even when not exposed to a trigger, the airways of many asthma patients are mildly inflamed, a state that contributes to their hyperresponsiveness.

Asthma manifests episodically in *attacks* of varying frequency, during which patients suffer shortness of breath, wheezing (whistling during inspiration or expiration), and sometimes coughing. Most attacks are short-lived, and patients typically recover their normal pulmonary function after the attack has ended. However, in some patients, a persistent state of airway obstruction—called *status asthmaticus*—may develop and last for days or weeks.

Treatment of acute asthma attacks is aimed at reducing airway resistance through bronchodilators that bind beta-2 receptors on bronchiolar smooth muscle cells and trigger their relaxation. Longer-term treatments are aimed at preventing or decreasing the frequency of attacks by reducing the inflammation of the airways with anti-inflammatory steroids and decreasing their responsiveness to allergens.

Lung Cancer

What is the number one cancer killer of women? Most people would answer "breast cancer," but in fact it is lung cancer. This disease is the number one cancer killer of both women and men. The term **lung cancer** refers to tumors arising from the epithelium lining the bronchi, bronchioles, and alveoli (**Figure 21.32**). There are many types of lung tumors, each with a different predominant cell type, clinical course, and rate of spread, or *metastasis* (see Chapter 3).

The number one risk factor for lung cancer is cigarette smoking. Smoking raises the risk 13-fold, and heavy smoking as much as 60- or 70-fold. Even passive, or "second-hand," smoke increases the risk of developing lung cancer. Research estimates this risk to be about one and a half times that of a nonsmoker. Other factors that increase the risk for lung cancer include genetics and exposure to certain chemicals or radiation.

The symptoms of lung cancer are fairly nonspecific. They include cough, blood-streaked sputum, chest pain, and recurrent infections. Initially, the symptoms have an obstructive pattern as the airways narrow due to tumor growth. However, some classes of lung cancer can result in the accumulation of fluid in the pleural cavity, which may lead to collapse of the alveoli and

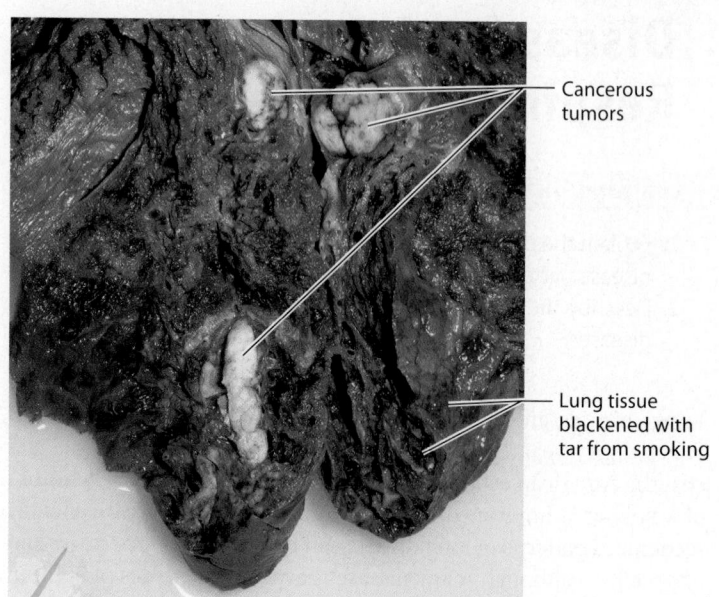

Cancerous tumors

Lung tissue blackened with tar from smoking

Figure 21.32 Cancerous tumor of the lung.

a decrease in lung compliance. This, in turn, causes a restrictive pattern of symptoms. Additionally, as lung tumors enlarge, the disease may take on a restrictive pattern as the normal lung tissue is infiltrated by the expanding tumor.

Lung cancer is difficult to detect early because of the nonspecific symptoms. Also, since most people with lung cancer have a history of smoking, the condition may be confused with COPD. Treatment varies depending on the class of tumor, but it may include surgery to remove the tumor and affected lymph nodes, radiation therapy, and chemotherapy.

Quick Check

☐ 1. What are the differences between obstructive and restrictive lung diseases?

☐ 2. What are the three subtypes of COPD? What is the main risk factor for developing COPD or lung cancer?

☐ 3. What are the three airway responses to asthma?

Apply What You Learned

☐ 1. Obstructive diseases increase the residual volume. Explain how this change decreases the vital capacity.

☐ 2. Cigarette smoke destroys the cilia of the respiratory epithelium. Why would this create the cough seen with chronic bronchitis?

☐ 3. Why does a person with asthma hyperventilate during an asthma attack? (*Hint:* Think about the level of carbon dioxide in the patient's blood during an attack.)

See answers in Appendix A.

CHAPTER SUMMARY

Take full advantage of the invaluable online practice and study tools for this chapter by going to MasteringA&P®**. There you will find:**

🖑 **Practice tests, module quizzes, art-labeling and sequence-the-steps activities**

▷ **Animated figures with quizzes**

✎ *Practicing A&P* **worksheets**

))) **Author-narrated podcast**

🔥 **Dynamic Study Modules**

iP *Interactive Physiology*

MODULE **21.1**
Overview of the Respiratory System 799

- The **respiratory system** includes the pulmonary arteries and veins, the thoracic cage, the lungs, and the respiratory tract. (Figure 21.1, p. 800)

  🖑 **Practice art-labeling exercise: Figure 21.1.**

- The **respiratory tract** includes the nose and nasal cavity, the pharynx, the larynx, the trachea, and the bronchial tree.

- The terminal structures of the respiratory tract are **alveoli,** tiny air sacs arranged in grapelike clusters through which gases are exchanged.

- **Lungs** are paired spongy organs in the thoracic cavity that consist of millions of alveoli.

- The organs of the respiratory tract may be classified anatomically and functionally:
  ○ Anatomically, they are classified as either the **upper respiratory tract** (the nasal cavity through the larynx) or the **lower respiratory tract** (the trachea through the alveoli).
  ○ Functionally, they are classified as part of the **conducting zone** (the nasal cavity through the bronchioles), which transports, warms, moistens, and filters air, or the **respiratory zone**

(respiratory bronchioles through the alveoli), which functions in gas exchange. (Figure 21.2, p. 801)

- **Respiration** consists of four separate processes:
  ○ *pulmonary ventilation,* the movement of air in and out of the lungs;
  ○ *pulmonary gas exchange,* the movement of gases between the lungs and the blood;
  ○ *gas transport* through the blood; and
  ○ *tissue gas exchange,* the movement of gases between the blood and the tissues.

- In addition to respiration, the respiratory system also helps to maintain acid-base, fluid, and blood pressure homeostasis; assists in the flow of venous blood and lymph; produces speech; detects odors; and helps expel the abdominal contents.

MODULE **21.2**
Anatomy of the Respiratory System 802

- The **nose** and **nasal cavity** moisten, warm, and filter the inspired air; house olfactory receptors; and enhance voice resonance.
  ○ The nose has a framework of hyaline cartilage and bone. (Figure 21.3, p. 802)
  ○ The nasal cavity extends from the **anterior nares** to the **posterior nares.** It is divided into right and left portions by the *nasal septum.* (Figure 21.4, p. 803)
  ○ The nasal cavity is lined mostly with the **respiratory mucosa,** which is pseudostratified ciliated columnar epithelium with **goblet cells.** A portion of the roof of the nasal cavity is lined with **olfactory mucosa.**

- The **pharynx** has three divisions: the **nasopharynx,** the **oropharynx,** and the **laryngopharynx.** (Figure 21.5, p. 804)

- The **larynx** protects the airway and produces sound. (Figure 21.6, p. 805)

  🖑 **Practice art-labeling exercise: Figure 21.6.**

  ○ It is framed by nine pieces of cartilage: the **epiglottis,** the **thyroid cartilage,** the **cricoid cartilage,** and the paired **arytenoid, corniculate,** and **cuneiform cartilages.**

○ The epiglottis seals off the larynx during swallowing to prevent food or liquid from entering the respiratory tract.

○ The larynx contains two sets of mucosal folds: the **vestibular folds,** also called the false vocal cords, and the **vocal folds,** also called the **true vocal cords.** At the core of the vocal cords are the elastic *vocal ligaments,* which vibrate to produce sound when air is passed over them. (Figure 21.7, p. 806)

● The **trachea** extends from the inferior neck to the mediastinum. (Figure 21.8, p. 807)

○ Its wall consists of a **mucosa** composed of respiratory epithelium. It contains C-shaped rings of hyaline cartilage to keep the airway open.

● The **bronchial tree** consists of the airways that lead from the trachea to the structures of gas exchange. (Figure 21.9, p. 808)

○ The bronchial tree begins with the right and left **primary bronchi.**

○ The right primary bronchus splits into three **secondary bronchi;** the left primary bronchus splits into two secondary bronchi.

○ The smallest passageways of the bronchial tree are the **bronchioles;** the *terminal bronchioles* are the last branches of the conducting zone. Terminal bronchioles lead to **respiratory bronchioles,** which begin the respiratory zone and lead to **alveolar ducts.** Both structures have alveoli in their walls. Alveolar ducts lead to **alveolar sacs,** which are composed of clusters of alveoli. (Figure 21.10, p. 809)

　　Practice art-labeling exercise: Figure 21.10.

● Alveoli are where gas exchange takes place. (Figure 21.11, p. 810)

○ Alveoli have three cell types: **type I alveolar cells,** which are squamous epithelial cells; **type II alveolar cells,** which secrete surfactant; and **alveolar macrophages,** which clean up debris in the alveoli.

○ The **respiratory membrane** is the barrier through which gases diffuse. It consists of the type I pneumocytes, a shared basal lamina, and the endothelial cells of the pulmonary capillaries.

　　Practice art-labeling exercise: Figure 21.11.

● The **lungs** consist of the alveoli and the surrounding elastic connective tissue.

○ Each lung is divided into **lobes:** The right lung has three lobes; the left lung has two. The lobes are divided into **bronchopulmonary segments,** which are further subdivided into lobules. (Figure 21.12, p. 811)

○ Each lung is enclosed by its own **pleural cavity,** which is the space between the **parietal** and **visceral pleurae.** This cavity is filled with serous fluid, which reduces friction during ventilation. (Figure 21.13, p. 813)

　　Practice art-labeling exercise: Figure 21.13.

● The function and composition of regions of the respiratory system are summarized in Table 21.1 (p. 813).

iP　Play Interactive Physiology tutorial on Respiratory System: Anatomy Review.

MODULE 21.3
Pulmonary Ventilation 814

● **Pulmonary ventilation** depends on pressure gradients that drive **inspiration** and **expiration.**

● **Boyle's law** states that for a gas under constant temperature, the pressure and volume are inversely related. This means that as the volume of a container increases, its pressure decreases. As the volume of a container decreases, its pressure increases.

● Inspiration is brought about by the **inspiratory muscles,** which include the **diaphragm** and **external intercostal muscles.** The inspiratory muscles increase the volume of the lungs. (Figure 21.14, p. 815)

● Expiration during normal quiet breathing is passive due to the elastic **recoil** of the lungs. As the lungs recoil, their volume decreases.

● The respiratory system can perform several nonrespiratory movements. (Table 21.2, p. 816)

● Three pressures are at work during the process of ventilation: (Figure 21.15, p. 817)

○ **Atmospheric pressure** is the pressure exerted by the air around us.

○ **Intrapulmonary pressure** is the pressure within the alveoli. It varies during inspiration and expiration, but it always eventually equilibrates with atmospheric pressure.

○ **Intrapleural pressure** is the pressure within the pleural cavity. It also varies with inspiration and expiration, but it is normally less than the intrapulmonary pressure, a state that prevents the lungs from collapsing.

● Pulmonary ventilation is driven by changes in volume and pressure gradients. During inspiration, the volume of the lungs increases, decreasing intrapulmonary pressure below atmospheric pressure, which causes air to move into the lungs. During expiration, the volume of the lungs decreases, increasing intrapulmonary pressure above atmospheric pressure, which causes air to flow out of the lungs. (Figure 21.16, p. 818)

● Three physical factors determine the effectiveness of pulmonary ventilation:

○ **Airway resistance** is defined as the impedance to air flow, and any increase in airway resistance decreases the effectiveness of pulmonary ventilation. (Figure 21.17, p. 819)

○ **Alveolar surface tension** is defined as the attraction created by hydrogen bonding between water molecules that tends to collapse the alveoli. **Surfactant** reduces alveolar surface tension by interrupting the hydrogen bonds between water molecules. (Figure 21.18, p. 820)

○ **Pulmonary compliance** refers to the ability of the lungs to stretch. If pulmonary compliance decreases, the lungs are less able to expand, and the effectiveness of pulmonary ventilation decreases.

● Lung function may be assessed using measurements taken with a **spirometer.** The pulmonary volumes measured include the **tidal volume (TV),** the **inspiratory reserve volume (IRV),** and the **expiratory reserve volume (ERV).** A fourth pulmonary volume, which cannot be assessed by spirometry, is the **residual volume (RV).** (Figure 21.19, p. 821)

● Pulmonary volumes may be combined to yield **pulmonary capacities,** which give a more complete picture of respiratory function. Spirometry can measure the **inspiratory capacity** and the **vital capacity.** Two additional capacities that cannot be measured by spirometry are the **functional residual capacity** and the **total lung capacity.** (Table 21.3, p. 822) iP Play Interactive Physiology tutorial on Respiratory System: Pulmonary Ventilation.

MODULE **21.4**
Gas Exchange 822

- **Dalton's law of partial pressures** states that the total pressure of a gas mixture is equal to the sum of the partial pressures of the individual gases.

- **Henry's law** states that the degree to which a gas dissolves in a liquid is proportional to its partial pressure and its solubility in that liquid.

- **Pulmonary gas exchange** is the exchange of gases in which oxygen diffuses from the alveoli to the blood, and carbon dioxide diffuses from the blood to the alveoli. The effectiveness of pulmonary gas exchange depends on the surface area of the respiratory membrane, the thickness of the respiratory membrane, and the match between ventilation and perfusion. (Figure 21.20, p. 824; Figure 21.21, p. 825)

- **Tissue gas exchange** is the exchange of gases in which oxygen moves from the blood to the tissues, and carbon dioxide moves from the tissues into the blood.

iP Play Interactive Physiology tutorial on Respiratory System: Gas Exchange.

MODULE **21.5**
Gas Transport through the Blood 827

- Only about 1.5% of inspired oxygen dissolves in plasma. The remainder is transported on **hemoglobin (Hb)**. Hemoglobin can carry four oxygen molecules, one on each of its iron-containing heme groups.

- The ability of Hb to load or unload oxygen depends on the P_{O_2} in the lungs or the tissues and the tightness with which Hb binds oxygen (the *affinity* of Hb for oxygen). (Figure 21.22, p. 827)

- The **oxygen-hemoglobin dissociation curve** shows the relationship between the percent saturation of Hb and the P_{O_2}. This curve shows that even when the P_{O_2} drops fairly low (from 100 mm Hg to 80 mm Hg, for example), the percent saturation of Hb only changes slightly, which helps to load O_2 in the lungs. However, when the P_{O_2} drops very low (for example, to 40 mm Hg), Hb rapidly unloads its oxygen molecules and the percent saturation decreases sharply. This provides more oxygen to tissues that need it. (Figure 21.23, p. 828)

- Increases in temperature, hydrogen ion concentration (acidity), P_{CO_2}, and **BPG** facilitate the unloading of oxygen from Hb by decreasing its affinity for oxygen. (Figure 21.24, p. 830)

- Carbon dioxide is transported in blood in three ways: dissolved in plasma (7–10%), bound to Hb (20%), and as **bicarbonate** ions (70%). In erythrocytes, carbon dioxide is converted to bicarbonate by being combined with water in a reaction catalyzed by the enzyme **carbonic anhydrase**. (Figure 21.25, p. 831)

- One of the primary buffer systems in the body is the **carbonic acid–bicarbonate buffer system**. Increasing P_{CO_2}, as occurs with **hypoventilation**, raises carbonic acid and lowers blood pH. Decreasing P_{CO_2}, as occurs with **hyperventilation**, lowers carbonic acid and raises blood pH. (Figure 21.26, p. 832)

iP Play Interactive Physiology tutorial on Respiratory System: Gas Transport.

MODULE **21.6**
Putting It All Together: The Big Picture of Respiration 833

- Respiration consists of (1) pulmonary ventilation, (2) pulmonary gas exchange, (3) gas transport, and (4) tissue gas exchange. (Figure 21.27, p. 834)

MODULE **21.7**
Neural Control of Ventilation 835

- Groups of neurons in the medulla of the brainstem maintain **eupnea:** (Figure 21.28, p. 835)
  - The **respiratory rhythm generator** sets the basic pattern of ventilation.
  - The **ventral respiratory group** (VRG) consists of both inspiratory neurons that stimulate the inspiratory muscles and expiratory neurons that stimulate the accessory muscles of expiration.
  - The **dorsal respiratory group** (DRG) consists mostly of inspiratory neurons that stimulate the inspiratory muscles.

- **Central chemoreceptors** of the medulla respond to changes in hydrogen ion concentration and P_{CO_2} of the CSF. If the P_{CO_2} and hydrogen ion concentration increase, they trigger hyperventilation. If the P_{CO_2} and hydrogen ion concentration decrease, they trigger hypoventilation. (Figure 21.29, p. 837)

- Other mechanisms that influence pulmonary ventilation include **peripheral chemoreceptors** in the **aortic** and **carotid bodies** that respond to the P_{O_2} of arterial blood (Figure 21.30, p. 838) and voluntary control mediated by the cerebral cortex.

- Control mechanisms of ventilation are summarized in Figure 21.31 (p. 838).

iP Play Interactive Physiology tutorial on Respiratory System: Control of Respiration.

MODULE **21.8**
Diseases of the Respiratory System 839

- **Restrictive lung diseases** decrease pulmonary compliance (ability to stretch). This problem decreases the effectiveness of inspiration, inspiratory capacity, vital capacity, and total lung capacity.

- **Obstructive lung diseases** increase airway resistance. These disorders can decrease the effectiveness of expiration, leading to air trapping in the alveoli.
  - **Chronic obstructive pulmonary disease** (COPD) is defined as persistent airway obstruction that is not fully reversible. It is nearly always due to cigarette smoking.
  - **Asthma** is an obstructive disease in which the airways are hyperresponsive to a trigger of some sort. Bronchoconstriction,

airway inflammation, and increased secretion of excessively thick mucus occur in response to the trigger.

○ **Lung cancer** refers to tumors that arise from the epithelium of the lung tissue. It is the number one cancer killer of both men and women. (Figure 21.32, p. 840)

ASSESS WHAT YOU LEARNED

See answers in Appendix A.

LEVEL 1 Check Your Recall

1. Which of the following are functions of the respiratory system? Circle all that apply.

   a. Providing for speech
   b. Regulating the autonomic nervous system
   c. Maintaining the acid-base balance of the blood
   d. Temperature homeostasis
   e. Raising the pressure in the abdominopelvic cavity
   f. Assisting in blood-pressure regulation

2. Fill in the blanks: Air enters the lungs through the hollow passages known collectively as the _____, which terminate in grapelike clusters called _____. The lungs are encased in the _____ membranes.

3. Mark the following statements as true or false. If a statement is false, correct it to make a true statement.

   a. The trachea contains O-shaped rings of hyaline cartilage.
   b. Goblet cells secrete serous fluid into the respiratory tract.
   c. The framework of the larynx is formed by nine pieces of cartilage, the largest of which is the thyroid cartilage.
   d. The pseudostratified ciliated columnar epithelium of the respiratory tract warms, filters, and humidifies the inspired air.
   e. The epithelium of the oropharynx changes from stratified columnar epithelium to simple squamous epithelium to enable it to resist abrasion from food.

4. The function of the epiglottis is to:

   a. contract muscularly to cover the laryngopharynx.
   b. vibrate to produce sound.
   c. trigger a cough reflex.
   d. cover the glottis during swallowing.

5. Fill in the blanks: The structures that vibrate to produce sound are called the_____. A higher-pitched sound is produced when they are _____; a lower-pitched sound is produced when they are _____.

6. Which of the following structural changes does *not* take place as we progress down the bronchial tree?

   a. More smooth muscle tissue
   b. More goblet cells
   c. Less cartilage
   d. Decrease in the height of the epithelium

7. All of the following statements about the alveoli and respiratory membrane are false. Correct each to make a true statement.

   a. Type I alveolar cells secrete a chemical called surfactant.
   b. Alveolar macrophages are squamous epithelial cells that constitute 90% of the total cells in the alveoli.
   c. The respiratory membrane consists of the type II alveolar cells, the pulmonary capillaries' endothelial cells, and their shared basal lamina.
   d. The respiratory membrane must be thick to function effectively in gas exchange.

8. Which of the following statements about pulmonary ventilation is *false*?

   a. Normal expiration requires the use of the expiratory muscles to decrease lung volume.
   b. The inspiratory muscles increase lung volume, which decreases intrapulmonary pressure.
   c. For inspiration to occur, intrapulmonary pressure must decrease below atmospheric pressure.
   d. The intrapleural pressure is less than the intrapulmonary pressure; this prevents the lungs from collapsing during expiration.

9. Match each term with the correct definition.

   ____ Airway resistance
   ____ Surface tension
   ____ Surfactant
   ____ Pulmonary compliance
   ____ V/Q ratio

   a. A detergent-like chemical secreted by bronchial smooth muscle that reduces surface tension
   b. The matching of ventilation to perfusion
   c. Largely determined by the diameter of the airways
   d. Caused by the formation of hydrogen bonds between water molecules
   e. Determined by the surface tension of the alveoli, the elastic tissue of the lungs, and the condition of the chest wall

10. Mark the following statements as true or false. If a statement is false, correct it to make a true statement.

    a. The functional residual capacity is the volume of air normally left in the lungs after a tidal expiration.
    b. The tidal volume is the amount of air left in the lungs after maximal expiration.
    c. The inspiratory capacity is equal to the vital capacity plus the tidal volume.
    d. The vital capacity is the total amount of exchangeable air.

11. Which of the following does *not* affect the efficiency of pulmonary gas exchange?

    a. The surface area of the respiratory membrane
    b. The degree of match of ventilation to perfusion
    c. The percent saturation of hemoglobin
    d. The thickness of the respiratory membrane

12. Henry's law states that the degree to which a gas dissolves in a liquid is determined by its:

    a. partial pressure.   d. Both a and b are correct.
    b. solubility.   e. All of the above are correct.
    c. surface tension.

13. Fill in the blanks: When the alveolar P_{O_2} decreases, the pulmonary arterioles _____. When the arteriolar P_{CO_2} increases, the bronchioles _____.

14. Mark the following statements as true or false. If a statement is false, correct it to make a true statement.

    a. Oxygen is transported primarily by dissolving in the plasma.
    b. Each Hb molecule can carry up to four oxygen molecules.
    c. Hb in arterial blood is normally nearly fully saturated.
    d. When the P_{O_2} of arterial blood drops slightly, the percent saturation of Hb drops dramatically.
    e. Increased P_{CO_2}, temperature, and hydrogen ion concentration all increase Hb's affinity for oxygen.

15. Match the following terms with the correct description.

    _____ Bicarbonate
    _____ Hemoglobin
    _____ BPG
    _____ Carbonic anhydrase
    _____ Chloride shift
    _____ Carbamino-hemoglobin

    a. The movement of anions into the erythrocyte to balance the outward movement of bicarbonate
    b. Hemoglobin bound to carbon dioxide
    c. Enzyme that catalyzes the conversion of carbon dioxide and water into carbonic acid
    d. The main way carbon dioxide is transported in the blood
    e. Protein that transports oxygen in the blood
    f. Substance that decreases hemoglobin's affinity for oxygen

16. Fill in the blanks: Hyperventilation causes a/an _____ in P_{CO_2}, which causes a/an _____ in blood pH. Hypoventilation causes a/an _____ in P_{CO_2}, which causes a/an _____ in blood pH.

17. The basic rhythm for breathing is maintained by the:

    a. respiratory pattern generator of the pons.
    b. dorsal respiratory group of the medulla.
    c. ventral respiratory group of the pons.
    d. respiratory rhythm generator of the VRG.

18. Mark the following statements as true or false. If a statement is false, correct it to make a true statement.

    a. The dorsal respiratory group maintains the respiratory rhythm.
    b. The ventral respiratory group triggers action potentials in the phrenic nerve and intercostal nerves.
    c. The dorsal respiratory group stimulates inspiratory muscles with the ventral respiratory group.
    d. The respiratory groups in the pons maintain eupnea.

19. Fill in the blanks: The main chemical factor(s) to which the central chemoreceptors of the medulla respond is/are _____, whereas the main chemical factor to which the peripheral chemoreceptors respond is _____.

20. Which of the following statements is *false*?

    a. COPD has three subclasses: emphysema, small airway disease, and chronic bronchitis.
    b. An obstructive disease pattern is characterized by increased airway resistance, which makes expiration difficult.
    c. Asthma is characterized by bronchoconstriction, airway inflammation, and excessive mucus secretion.
    d. Research has demonstrated that cigarette smoking only marginally increases the risk of lung cancer.

LEVEL 2 Check Your Understanding

1. Explain what would happen to the pressure inside a cylinder if you decreased its volume, according to Boyle's law.

2. If you swallow a large bite of food without properly chewing it first, you will feel discomfort during ventilation. Explain this, considering the arrangement of the trachea and the esophagus.

3. Predict what would happen to the tidal volume and inspiratory reserve volume if the phrenic nerves were severed. Which muscles would fire to try to compensate for this?

4. Inflammation of the epiglottis, called *epiglottitis,* is considered an emergency. Explain why this situation is so serious.

LEVEL 3 Apply Your Knowledge

PART A: Application and Analysis

1. When a person hyperventilates, what happens to his or her blood pH? A person who is hyperventilating is often told to breathe into a paper bag. Why might this help to correct the pH imbalance?

2. Some athletes train at higher altitudes. Over time, this results in higher levels of erythrocytes and the chemical BPG. What would be the advantages to these changes?

3. You and a friend are having a contest to see who can hold his or her breath the longest. Your friend hyperventilates before holding his breath, and subsequently wins the contest. Why did hyperventilation give him an advantage?

4. Prolonged vomiting can result in significant loss of hydrogen ions from the blood, and a subsequent increase in the blood's pH. Would you expect a person suffering from this condition to hypoventilate or hyperventilate? Explain.

PART B: Make the Connection

5. Mrs. Jordan is brought to the emergency room by paramedics, who report that she lost consciousness while at the grocery store. You notice that her respiratory rate is elevated and her breath has a "fruity" odor. You suspect that she is suffering from diabetic ketoacidosis.

    a. What is diabetic ketoacidosis? *(Connects to Chapter 16)*
    b. Does Mrs. Jordan likely have type I or type II diabetes mellitus? Explain. *(Connects to Chapter 16)*
    c. What has likely happened to the pH of her blood, and why? *(Connects to Chapters 16 and 19)*
    d. Explain why she is hyperventilating.

6. What happens to the metabolic rate of skeletal muscle tissue during exercise? What waste products are produced from metabolic reactions? How and why does this affect your rate of ventilation during exercise? *(Connects to Chapter 10)*

22

The Digestive System

We all must eat to survive, but most of us don't think about eating except when we're hungry. We generally understand *why* we eat, as we know that our bodies require nutrients—carbohydrates, lipids, and proteins—to maintain homeostasis. But do you ever think about what happens *when* we eat?

Nutrients are available to us in the form of food, but this comes with a small problem: The food isn't in a form that we can immediately use. If you have a peanut butter and jelly sandwich, you can't just put that sandwich straight into your bloodstream. Our bodies therefore require an organ system to take the sandwich into the body and break it down so that its nutrients can be delivered to cells via the blood. The set of organs that accomplishes this task is the **digestive system.**

This chapter explores the many structures and functions of the digestive system. We look at the anatomy of its organs, and how they break down food and deliver nutrients to the blood. We also examine how hormones and the autonomic nervous system regulate these organs and their processes. The chapter concludes with a "big picture" look at the processes carried out by the digestive system. But first, let's start with an overview of its basic structures and functions.

MODULE 22.1

Overview of the Digestive System

Learning Outcomes

1. Describe the major functions of the digestive system.
2. Explain the basic anatomy, organization, and regulation of the digestive system.
3. Describe the histological structure and function of each of the four layers of the alimentary canal wall.

The organs of the digestive system are located from the head to the abdominopelvic cavity (**Figure 22.1**). The digestive system consists of two types of organs: the organs of the alimentary canal, also known as the **gastrointestinal (GI) tract** or **digestive tract;** and the accessory organs. The **alimentary canal** is a continuous tube through which food passes directly. It consists of the **oral cavity** (mouth), *pharynx, esophagus, stomach, small intestine,* and *large intestine*. The **accessory organs** are not part of the alimentary canal but assist in digestion in some way. They are located around the alimentary canal

Photo: SEM showing the extensive blood supply of the small intestine, which allows it to rapidly absorb digested nutrients.

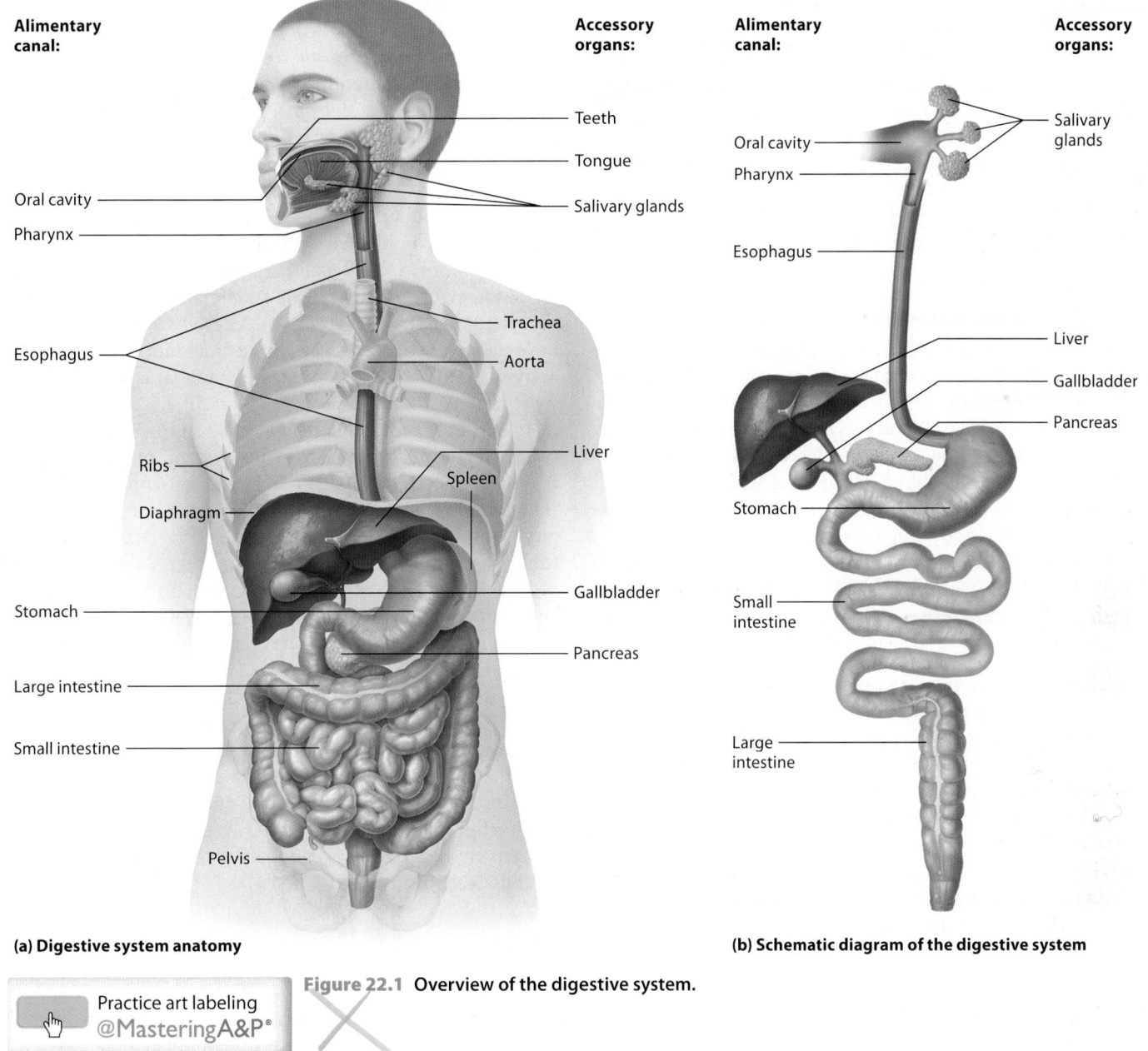

Alimentary canal:
- Oral cavity
- Pharynx
- Esophagus

Accessory organs:
- Teeth
- Tongue
- Salivary glands

Trachea
Aorta
Liver
Ribs
Spleen
Diaphragm
Gallbladder
Stomach
Pancreas
Large intestine
Small intestine
Pelvis

(a) Digestive system anatomy

Alimentary canal:
- Oral cavity
- Pharynx
- Esophagus

Accessory organs:
- Salivary glands

Liver
Gallbladder
Pancreas
Stomach
Small intestine
Large intestine

(b) Schematic diagram of the digestive system

Figure 22.1 Overview of the digestive system.

Practice art labeling
@MasteringA&P®

and include the *teeth, tongue, salivary glands, liver, gallbladder,* and *pancreas*. Food generally does not come into direct contact with the accessory organs, although the teeth and tongue are exceptions.

Let's survey all these organs to see just what part they play in digestion and to understand their basic structure and organization. We examine each in greater detail in subsequent modules.

Basic Digestive Functions and Processes

The most fundamental function of the digestive system is to take food into the body and break it down into its component nutrients so that they can be used by body cells. But the digestive system also plays a critical role in fluid, electrolyte, and acid-base homeostasis by taking in water and electrolytes via the diet and delivering them to the blood. Other functions of the digestive system include ingesting vitamins and minerals, producing hormones, and excreting metabolic wastes.

The digestive system must perform the following six basic processes to carry out these functions:

- **Ingestion.** Food and water are brought into the digestive system by **ingestion,** which occurs via the mouth under normal conditions.
- **Secretion.** Digestive organs contain both endocrine and exocrine glands that **secrete** a variety of substances—such as mucus, enzymes, acid, and hormones—to aid other digestive processes.

- **Propulsion.** Ingested food and liquids pass from one digestive organ to the next by the process of **propulsion.** Propulsion is accomplished largely by rhythmic contractions of the smooth muscle of the alimentary canal called **peristalsis** (pehr-uh-STAL-sis) and is aided by mucus secreted by multiple organs.

- **Digestion.** Food breakdown occurs by the process of **digestion.** There are two kinds of digestion. In **mechanical digestion,** digestive organs physically break food down into smaller pieces via processes such as chewing and mixing food by movements by the muscles of the alimentary canal. In **chemical digestion,** enzymes secreted by digestive organs catalyze reactions that break the chemical bonds within food particles until only small molecules remain.

- **Absorption.** Once food particles are mechanically and chemically digested, nutrients move through the wall of the alimentary canal into blood or lymphatic vessels by a process called **absorption.** Water, electrolytes, and vitamins are also absorbed into the blood in the same manner.

- **Defecation.** Certain ingested materials are not digestible or usable by the body. Such materials continue their transit through the alimentary canal until they exit the body as *feces* through **defecation.** Defecation also provides the body with a way to eliminate certain metabolic wastes. Note that defecation is simply a specialized form of propulsion.

The organs of the alimentary canal carry out all of these basic processes. Food is ingested at the mouth and propelled through the canal from one organ to another as it is digested mechanically and chemically. The nutrients are absorbed into the bloodstream, and the indigestible substances are removed from the body by defecation. Substances secreted by various digestive organs aid each of these processes. Most of the accessory organs do not come into direct contact with food, and so they are involved in secretion, mechanical digestion, and chemical digestion. The digestion and absorption of each nutrient are covered together in a later module after we have discussed the anatomy of the digestive system.

Quick Check

☐ 1. What are the functions of the digestive system?
☐ 2. What are the six basic processes of the digestive system?

Regulation of Motility by the Nervous and Endocrine Systems

◀◀ **Flashback**

1. What are the two branches of the autonomic nervous system, and how do they differ with respect to regulating digestion? (p. 520)
2. What are hormones and paracrines? (pp. 587–588)

As you will see throughout this chapter, movement, or **motility,** of the alimentary canal is a key process in every region of the

canal. In the oral cavity, the pharynx, the superior portion of the esophagus, and the last portion of the large intestine, motility is due to skeletal muscle. In the rest of the alimentary canal, motility is the work of smooth muscle. Motility takes several forms, including *swallowing, churning, peristalsis,* and *defecation.* Some of these movements propel food through the canal, whereas others mix it as part of mechanical digestion.

Each type of motility is regulated by the nervous system and/or the endocrine system. Nervous system regulation of motility is accomplished by the nerves of the *autonomic nervous system,* or ANS. Recall that there are two branches of the ANS: the *sympathetic nervous system* and the *parasympathetic nervous system* (see Chapter 14). These two branches generally have opposite effects on gastrointestinal motility—sympathetic activity inhibits digestive processes, and parasympathetic activity stimulates them.

In addition, motility is regulated by a group of nerves known collectively as the **enteric nervous system,** or **ENS** (en-TEHR-ik; *enter-* = "intestine"). The ENS is a self-contained branch of the autonomic nervous system that extends from the esophagus to the anus (the terminal portion of the large intestine). The functions stimulated by the ENS are often called *short reflexes* because the reflex pathways are confined to local neurons. In contrast, functions stimulated by the ANS are known as *long reflexes* because they must travel outside the local digestive neurons to the CNS to function.

The endocrine system regulates digestive processes by secreting *hormones.* Many of these hormones are actually paracrines that are produced by the digestive organs themselves, particularly the stomach and small intestine. Each hormone binds to a receptor on or within its target cell and regulates some aspect of its functioning. Hormones generally either stimulate or inhibit motility of the alimentary canal, often together with one or more neural mechanisms.

Quick Check

☐ 3. How do the effects of the sympathetic and parasympathetic nervous systems on digestion differ?
☐ 4. How does the endocrine system regulate digestive processes?

Histology of the Alimentary Canal

◀◀ **Flashback**

1. What is a mucous membrane, and what are goblet cells? (pp. 135, 153)
2. What are the properties of smooth muscle? (p. 148)

The organs of the alimentary canal follow the same general tissue pattern of other hollow organs we have studied: concentric layers of tissue surround a space called the **lumen** (LOO-men). Like other hollow organs, those of the alimentary canal contain an inner epithelium, a layer of connective tissue, a layer of smooth muscle, and an outer layer of connective tissue. Most regions of

Figure 22.2 **The basic tissue organization of most of the alimentary canal.**

the alimentary canal have four named tissue layers (**Figure 22.2**): *mucosa, submucosa, muscularis externa,* and either the *serosa* or *adventitia.*

- **Mucosa.** The innermost **mucosa** consists of three components: First is a layer of epithelium facing the lumen, followed by a thin layer of loose connective tissue called the **lamina propria,** and finally two thin layers of smooth muscle together known as the **muscularis mucosae.** The epithelium from the stomach to the end of the large intestine is simple columnar epithelium with copious goblet cells that secrete an alkaline mucus. This mucus coats the epithelium and protects it and the underlying tissues from ingested food and chemicals secreted by digestive organs. The mucosa also houses *regenerative epithelial cells* that have a high rate of mitosis. These cells allow the mucosa to replace epithelial cells as they are damaged or sloughed off in the alimentary canal. The lamina propria houses blood and lymphatic vessels, glands, and mucosa-associated lymphatic tissue (MALT; see Chapter 20). The two layers of the muscularis mucosae are arranged in different directions— the inner layer is circular and the outer layer is longitudinal. As discussed later, this arrangement allows motility as the two layers contract alternately.
- **Submucosa.** The **submucosa** is composed of dense irregular connective tissue with blood and lymphatic vessels and submucosal glands. A cluster of nerves is called a *plexus,* and here we find nerve clusters of the enteric nervous system referred to as the **submucosal plexus,** or *Meissner's*

plexus. Each plexus regulates secretion from and blood flow to its area of the alimentary canal.
- **Muscularis externa.** The **muscularis externa** is a thick muscular layer composed of smooth muscle in most of the alimentary canal. We generally find two layers of smooth muscle that are arranged in the same manner as in the muscularis mucosae, with inner circular and outer longitudinal layers. The motility of the muscularis externa is regulated by groups of nerves of the enteric nervous system called the **myenteric plexus** (*my-* = "muscle," *enter-* = "intestine"), or *Auerbach's plexus.*
- **Serosa or adventitia.** The outer connective tissue layer is the **serosa** in the organs within the peritoneal cavity and the **adventitia** in organs outside the cavity. The serosa, also called the *visceral peritoneum,* is composed of simple squamous epithelial tissue and loose connective tissue, whereas the adventitia is composed of dense irregular connective tissue. Both structures support digestive organs and anchor them to surrounding structures.

Each organ of the alimentary canal has a slightly different variation of this basic plan. These variations ensure that the form of each organ best follows its function.

Quick Check

- ☐ 5. What are the four main tissue layers of the alimentary canal?
- ☐ 6. Which type of epithelium makes up the mucosa from the stomach to the end of the large intestine?

Organization of the Abdominopelvic Digestive Organs

◀◀ **Flashback**

1. What is a serous membrane, and what is its function? (p. 18)
2. Which blood vessels supply and drain the digestive organs? (p. 707)

Most of the digestive organs—with the exceptions of the oral cavity, pharynx, and esophagus—reside in the abdominopelvic cavity. Here they share a common set of serous membranes, blood vessels, and nerves, which the following sections examine.

The Peritoneal Membranes

The abdominopelvic cavity houses the largest serous membrane in the body, the **peritoneal membrane** (pehr'-uh-toh-NEE-uhl), or **peritoneum** (Figure 22.3a). Like the serous membranes that surround the heart and lungs, the peritoneal membrane consists of two layers. The outer layer is the **parietal peritoneum** (puh-RY-uh-tal), which lines the inner surface of the body wall. Where the parietal peritoneum meets the abdominal organs, it folds in on itself to become the inner **visceral peritoneum,** or serosa, which forms the outer tissue layer of such organs. Between these two peritoneal layers we find the **peritoneal cavity,** which contains serous fluid. This fluid lubricates organs as they slide past one another. Note that the peritoneal cavity is much larger than the pleural or pericardial cavity. Abdominal trauma or infection can inflame the peritoneum, as you can read about in *A&P in the Real World: Peritonitis.*

Organs located entirely within the peritoneal cavity are **intraperitoneal.** Some abdominal organs are partly or completely outside the peritoneal cavity; such organs are said to be **retroperitoneal** (reh'-troh-pehr-uh-toh-NEE-uhl; *retro-* = "behind"). Some organs, though, are partly intraperitoneal and partly retroperitoneal. For example, in Figure 22.3a, the pancreas is shown as retroperitoneal, but if we took a more lateral section, we would find that part of it is actually intraperitoneal.

Notice in Figure 22.3a how the visceral peritoneum folds over on itself around certain organs, particularly the small intestine, to form structures called **mesenteries** (mez-en-TEHR-eez). The mesenteries support and bind these organs together and keep the small intestine in a particular shape that fits within the abdominopelvic cavity. This function is key, as we don't want loops of intestine wandering around our abdominopelvic cavities. The mesenteries also house blood vessels, nerves, and lymphatic vessels, anchoring them in place (Figure 22.3b). The mesentery attached to much of the large intestine is often called the *mesocolon.*

Two mesenteries—the greater omentum and lesser omentum—are especially prominent. The **greater omentum** (oh-MEN-tum; *omentum* = "apron"), which is unique among the mesenteries in that it consists of four layers of folded visceral peritoneum, is named for the fact that it covers the abdominal organs like an apron (Figure 22.3c). The greater omentum, which is the first

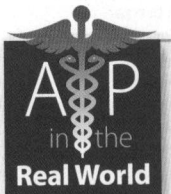

Peritonitis

Inflammation of the peritoneum leads to **peritonitis** (pehr'-ih-tuh-NY-tis; *-itis* = "inflammation of"). Peritonitis results when substances such as blood or the contents of an abdominal organ leak into the peritoneal cavity. Usually this is due to abdominal trauma that ruptures a blood vessel or abdominal organ, and often involves a bacterial infection.

Abdominal pain is the most universal symptom of peritonitis. Most patients experience *rebound tenderness,* in which they feel little pain when a hand is placed on their abdomen, but significant pain when the hand is removed. This is caused by the inflamed peritoneal membranes snapping back into place when the hand leaves the abdomen.

The treatment for peritonitis may involve antibiotic therapy and surgery to correct its underlying cause. Most patients recover with appropriate treatment, but it can prove fatal if left untreated.

structure visible when the abdominal cavity is opened, extends from the base of the stomach down into the pelvis. The **lesser omentum** is a smaller mesentery that extends from the medial surface of the stomach to the liver. Both the greater and lesser omenta generally accumulate adipose tissue between their folds.

Blood Supply to the Abdominal Digestive Organs

You read previously about the blood vessels that supply and drain the abdominal digestive organs, which are collectively called the **splanchnic circulation** (SPLANK-nik; *splan-* = "organ"; see Chapter 18). The arterial supply of the digestive organs consists of branches from the abdominal aorta, including the *celiac trunk, superior mesenteric artery, inferior mesenteric artery,* and branches from each of these arteries. The digestive organs are drained by a set of veins that drain into the *hepatic portal vein.* The hepatic portal vein then delivers the blood to the liver for processing. Blood drains from the liver by a set of *hepatic veins,* which in turn deliver blood to the inferior vena cava.

Nerve Supply to the Digestive Organs

The organs of the digestive system are extensively supplied with nerves, most of which are branches from the autonomic nervous system. The nerves of the sympathetic and parasympathetic divisions that serve the digestive organs are located in three main clusters: the *celiac plexus,* the *superior mesenteric plexus,* and the *inferior mesenteric plexus.* Notice that these clusters have the same names as the arteries that supply the digestive organs. In addition, the organs of the alimentary canal receive innervation from nerve plexuses of the ENS, including the submucosal and myenteric plexuses we introduced earlier in this module. These plexuses work together to control secretion from and motility of the alimentary canal.

Diaphragm

Parietal peritoneum
Visceral peritoneum

Parietal peritoneum

Peritoneal cavity

Liver

Liver

Parietal peritoneum

Lesser omentum

Visceral peritoneum

Peritoneal cavity

Greater omentum

Mesenteries

Pancreas (retroperitoneal)

Duodenum (part of the small intestine, retroperitoneal)

Large intestine

Small intestine

Rectum (part of the large intestine, retroperitoneal)

Anterior

Posterior

(a) Sagittal section of the abdominopelvic cavity showing the peritoneal cavity

Greater omentum (reflected)

Large intestine

Mesenteries

Blood vessels

Small intestine

(b) Mesenteries with greater omentum reflected and small intestine pulled aside, anterior view

Liver

Gallbladder

Lesser omentum

Stomach

Greater omentum

Large intestine

Small intestine

(c) Anterior view of the abdominal organs and omenta, illustration (left), and cadaver photo (right)

Practice art labeling
@MasteringA&P®

Figure 22.3 The abdominopelvic cavity houses the peritoneum, the largest serous membrane in the body.

The neurons of the sympathetic and parasympathetic nervous systems, the ENS, and even the CNS are involved in multiple digestive reflex arcs. The sensory component of the reflex arcs is provided by receptors located in the walls of the alimentary canal organs. These receptors respond to stimuli such as stretch, the concentration and pH of the fluids the organs contain, and the presence of certain products of digestion. When these receptors are stimulated, their axons contact neurons in the ANS or ENS, which then initiate the appropriate response such as triggering peristalsis or secretion from certain glands.

Quick Check

☐ 7. Where is the peritoneal cavity located?
☐ 8. Where does blood from the abdominal digestive organs drain?
☐ 9. Which branches of the nervous system supply the digestive organs?

Apply What You Learned

☐ 1. Explain why peritonitis (inflammation of the peritoneal membranes) can affect multiple digestive organs.
☐ 2. Disorders of the digestive system often disrupt many aspects of homeostasis. Explain why.
☐ 3. Many forms of chemotherapy for cancer treatment kill rapidly dividing cells. Predict the effect these drugs will have on the epithelium of the alimentary canal.

See answers in Appendix A.

MODULE **22.2**

The Oral Cavity, Pharynx, and Esophagus

Learning Outcomes

1. Discuss the structure and basic functions of the oral cavity, the different types of teeth, and the tongue.
2. Describe the structure and function of the salivary glands, their respective ducts, and the products secreted by their cells.
3. Describe and classify the regions of the pharynx with respect to passage of food and/or air through them.
4. Describe the structure and function of the esophagus, including the locations of skeletal and smooth muscle within its wall.
5. Explain the process of deglutition, including the changes in position of the glottis and larynx.

The digestive system begins at the *oral cavity, or mouth.* Four digestive processes take place here: ingestion, secretion, chemical and mechanical digestion, and propulsion. Although the oral cavity is technically part of the alimentary canal, it houses two accessory organs: the *teeth* and the *tongue.* In addition, three pairs of accessory organs, the *salivary glands,* are located in and

around the oral cavity. Together, these organs turn ingested food into a moist, chewed mass called a **bolus** (BOW-lus; "ball"). The posterior oral cavity, pharynx, and esophagus then deliver the bolus to the stomach through the process of *swallowing.* In this module, we look at the structure and function of the oral cavity and its associated accessory organs, the pharynx, and the esophagus. Then we examine the process of swallowing.

Structure of the Oral Cavity

◀◀ **Flashback**

1. Which bones form the hard palate? (p. 226)
2. What is the main function of the soft palate and uvula? (p. 804)

The oral cavity, illustrated in **Figure 22.4**, is the area posterior to the teeth and bounded by the palate and tongue. The lateral walls of the oral cavity are formed by the *cheeks,* which are composed largely of the buccinator muscles and lined internally by stratified squamous nonkeratinized epithelium. The cheeks terminate anteriorly as the *lips,* which are formed by the orbicularis oris muscle and covered with stratified squamous keratinized epithelium. The integument covers most of the lips, but the portion near the mouth, called the *red margin,* contains less keratin than the surrounding skin. As a result, the red margin is fairly translucent and the blood in the vessels of the dermis is more visible, giving the lips a slight reddish tint. The mouth's inferior wall is composed of muscles of the tongue and muscles that attach to the hyoid bone.

Posterior to the lips and cheeks we find the **gums,** or *gingivae* (JIN-jih-vay). The gums are covered with stratified squamous nonkeratinized epithelium overlying connective tissue and the maxilla and mandible, where the teeth are housed. A narrow band of mucosa called the **labial frenulum** (FREN-yoolum) attaches the internal surfaces of the upper and lower lips to the gums on the midline. The narrow space between the teeth and gums and the internal surfaces of the lips and cheeks is the **vestibule.** The space posterior to the teeth and gums is the **oral cavity proper.**

The superior boundary, or "roof," of the mouth is the **palate** (PAL-it), which consists of two portions: the anterior two-thirds is the **hard palate,** and the posterior one-third is the **soft palate.** The hard palate consists of stratified squamous epithelium and connective tissue covering the palatine processes of the maxillary bones and the palatine bones. The surface of the hard palate is slightly rough, which assists in mechanical digestion. The arch-shaped soft palate consists of stratified squamous epithelium overlying skeletal muscle. Extending inferiorly from the soft palate is a projection called the **uvula** (YOO-vyoo-luh). When we swallow, the soft palate and uvula move posteriorly to prevent food from entering the nasal cavity.

Quick Check

☐ 1. Which structures form the lateral, anterior, and superior walls of the oral cavity?

(a) Anterior view of the oral cavity

Upper lip
Gums (gingivae)
Lower lip
Superior labial frenulum
Hard palate
Soft palate
Uvula
Palatine tonsil
Oropharynx
Teeth
Tongue
Inferior labial frenulum

(b) Oral cavity and pharynx, sagittal section

Vestibule
Tongue
Epiglottis
Larynx
Hard palate
Soft palate
Oral cavity
Uvula
Palatine tonsil
Oropharynx
Laryngopharynx
Esophagus

Figure 22.4 Oral cavity and pharynx.

Practice art labeling
@MasteringA&P®

The Teeth and Mastication

◀◀ **Flashback**

1. In which two bones are the teeth located? (pp. 219–220)
2. Which muscles are involved in mastication? (p. 298)
3. What are the components of bone tissue? (p. 188)

The **teeth** are key organs of mechanical digestion. They are located in their bony sockets, called *alveoli*, within the mandible and maxilla and are held in place by bands of connective tissue known collectively as the **periodontal ligament** (pehr′-ee-oh-DAHN-tuhl; *peri-* = "around," *odont-* = "tooth"). With the assistance of the masseter and temporalis muscles, the teeth **masticate**, or chew, ingested food, grinding it into smaller pieces. Mastication aids digestion by increasing the overall surface area of the food, giving digestive enzymes more places to catalyze the reactions of chemical digestion.

There are three types of teeth, which are classified according to their shape (**Figure 22.5a**):

- **Incisors.** The **incisors** are the central teeth that are broad and flat with a narrow crown. They are specialized for cutting off pieces of food. The middle two incisors are the *central incisors,* and those to either side are the *lateral incisors.*
- **Canines.** The **canines** (KAY-nynz), also known as **cuspids,** are on either side of the incisors. Their pointed crowns are specialized for ripping and tearing.

- **Molars.** The teeth posterior and lateral to the canines are the **premolars** and the **molars.** Both types of molars have broad crowns with rounded projections called **cusps** that are specialized for grinding.

Humans develop two sets of teeth: one set of "baby teeth" and one set of "permanent teeth." In the next subsections, we examine the differences between these two sets of teeth and the general structure of a tooth.

Primary and Secondary Dentition

The "baby teeth" are the **primary dentition,** or **deciduous teeth** (dih-SIJ-oo-us; *decid-* = "falling off"). There are 20 deciduous teeth, with 4 incisors, 2 canines, and 4 molars in both the mandible and maxilla (**Figure 22.5b**). The first deciduous teeth to erupt are generally the lower central incisors at about 6 months of age. Deciduous teeth continue to erupt at a rate of about one pair of teeth per month until the age of 24 months, at which time all 20 teeth are usually present.

The **secondary dentition** is the set of **permanent teeth,** which are situated above the primary dentition in the maxilla and below it in the mandible. When the child is about 6 years of age, these teeth enlarge and begin to press on the deciduous teeth. This causes the root to gradually dissolve and the deciduous tooth falls out of the bone. The permanent tooth then erupts and takes its place. There are 32 permanent teeth, with 4 incisors, 2 canines, 4 premolars, and 6 molars in both the mandible and

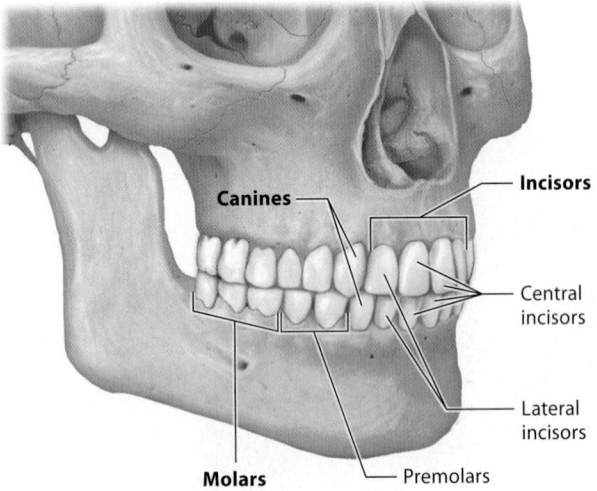

(a) The three main classes of teeth

	INCISORS		CANINES	MOLARS			
	Central incisors	Lateral incisors		Premolars	First	Second	Third
UPPER JAW							
LOWER JAW							

Primary dentition:
- Central incisors (7.5 mo)
- Lateral incisors (9 mo)
- Canines (18 mo)
- Molars (14–24 mo)

Maxillary teeth

Mandibular teeth
- Molars (12–20 mo)
- Canines (16 mo)
- Lateral incisors (7 mo)
- Central incisors (6 mo)

(b) Primary and secondary dentition, with approximate age of eruption

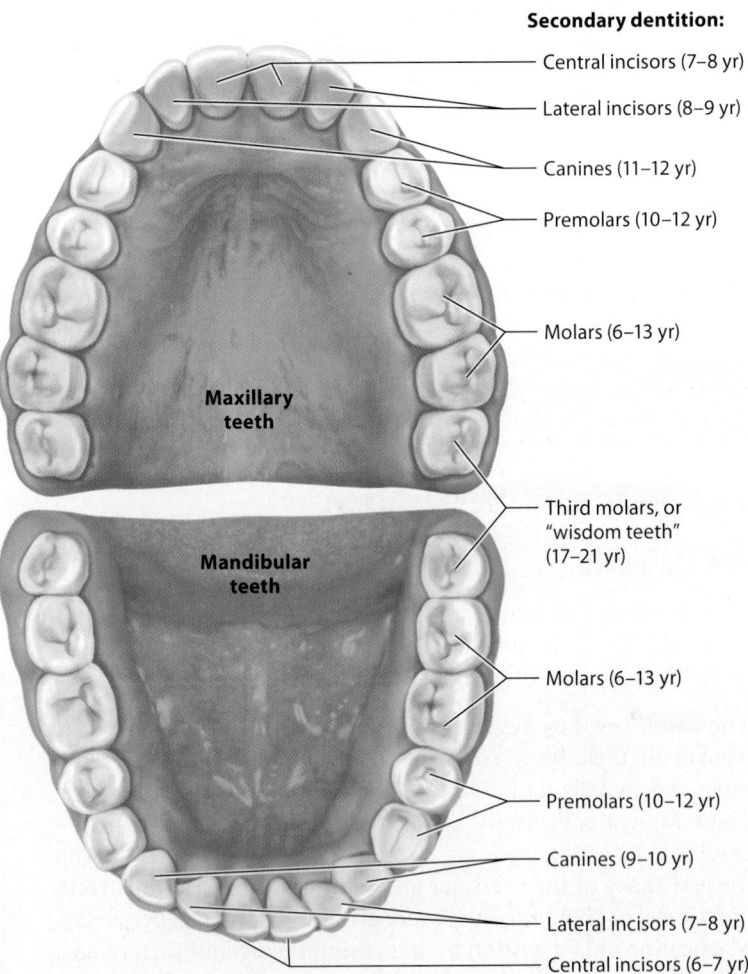

Secondary dentition:
- Central incisors (7–8 yr)
- Lateral incisors (8–9 yr)
- Canines (11–12 yr)
- Premolars (10–12 yr)
- Molars (6–13 yr)

Maxillary teeth

Mandibular teeth
- Third molars, or "wisdom teeth" (17–21 yr)
- Molars (6–13 yr)
- Premolars (10–12 yr)
- Canines (9–10 yr)
- Lateral incisors (7–8 yr)
- Central incisors (6–7 yr)

Figure 22.5 Types of teeth and the primary and secondary dentition.

maxilla. Generally by age 12 all of the deciduous teeth have fallen out and all but the third set of secondary molars have erupted.

The third set of molars, known as the *wisdom teeth,* erupt somewhat later, between ages 17 and 21. Sometimes wisdom teeth remain embedded in the bone, a condition called *impaction.* Any tooth has the potential to become impacted, but wisdom teeth become impacted more often than any other teeth because of their position at the back of the jaw. Impacted teeth can be removed surgically.

Tooth Structure

Figure 22.6 illustrates the structure of a tooth. A tooth consists of two components: the **crown,** which is the visible portion above the gum line, and the **root,** which is embedded within the alveolus. Notice that the tooth consists of outer layers of hard mineralized substances—the **enamel**—covering a soft, inner gelatinous substance called the **pulp.** Enamel is composed almost fully of secreted calcium hydroxyapatite crystals with only a small

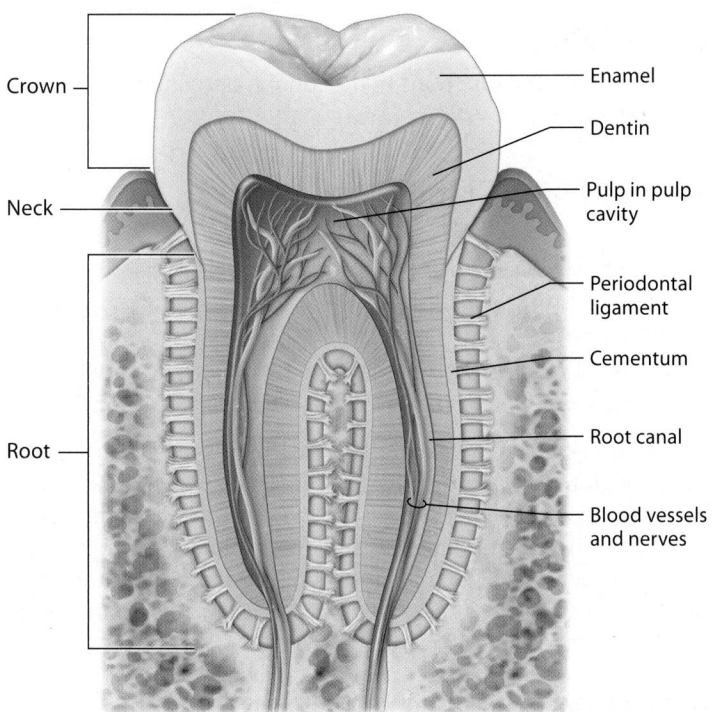

Crown — Enamel
— Dentin
Neck — Pulp in pulp cavity
— Periodontal ligament
— Cementum
Root — Root canal
— Blood vessels and nerves

Figure 22.6 Structure of a tooth.

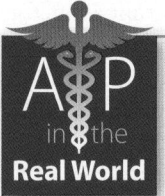

Dental Caries

Bacteria tend to accumulate in slight defects in tooth enamel. There they metabolize nutrients on the tooth surface, particularly simple sugars. Their metabolism produces acid byproducts that dissolve the enamel and result in *dental cavities,* or *dental caries* (KEHR-eez). The bacteria multiply and produce more acids, which expand the cavity into the underlying dentin.

Cavities may be treated by removing the damaged area of the tooth (with the dreaded dental drill) and filling it with a hard restorative material. However, it is better to prevent cavities from developing in the first place. One key weapon is fluoride, an ion that increases the hardness of enamel, particularly in children. The addition of fluoride to the public water supply in many countries and to toothpastes and mouthwashes has greatly decreased the incidence of dental caries. Other preventative measures include brushing and flossing regularly, applying dental sealants that fill defects in the enamel, taking calcium supplements, and reducing sugar intake.

amount of organic material. This makes enamel the hardest substance in the body, allowing it to endure the forces that accompany chewing. The cells that secrete enamel deteriorate after the tooth erupts, so the body cannot repair damaged enamel.

The oral cavity houses hundreds of different bacterial species, collectively called the *normal flora.* In individuals with a healthy immune system and good oral hygiene, the normal flora generally cause few problems, and indeed may protect us from more pathogenic bacteria. However, damaged enamel sets the stage for tooth decay, which is explored in *A&P in the Real World: Dental Caries.*

On the outer portion of the root of the tooth, we find a different kind of mineralized bonelike tissue known as **cementum.** Cementum is composed of about half calcium hydroxyapatite crystals and half organic molecules such as collagen fibers. This is approximately the same composition that bone has, and for this reason cementum is about as hard as bone. The periodontal ligament extends collagen fibers into the cementum, which helps this substance "cement" the tooth in place.

The inner layer of mineralized tissue in both the root and crown is **dentin,** which is composed of about 70% calcium hydroxyapatite crystals. This makes dentin the second hardest material in the body. However, dentin also has some degree of elasticity, which is critical for preventing the overlying enamel from fracturing when chewing hard substances. Unlike enamel, dentin is formed throughout life by cells called *odontoblasts* that line its inner surface.

The final component of a tooth is the inner pulp, which is composed of loose connective tissue and housed within the central **pulp cavity.** Pulp contains blood vessels and nerves that supply the other tissues of the tooth with nutrients and innervation. The pulp cavity extends into the root via the thin **root canal.** Occasionally, the pulp becomes infected, which results in inflammation and generally a great deal of pain. When the infection risks the health of the tooth, a *root canal procedure* may be performed to remove the pulp from the pulp cavity and root canal, and fill the newly hollow space with an inert material.

Quick Check

☐ 2. What are the three types of teeth?
☐ 3. How do the primary dentition and secondary dentition differ?
☐ 4. What are the mineral and nonmineral components of a tooth?

The Tongue

◀◀ Flashback

1. What are the four types of tongue papillae, and what are their functions? (p. 542)
2. Which muscles make up the tongue? (p. 301)

The **tongue** consists of skeletal muscle covered with a layer of stratified squamous epithelium. It is attached to the floor of the oral cavity by a thin band called the **lingual frenulum** (LING-gwuhl; *lingual* = "tongue"), which prevents the tongue from moving too far posteriorly. You have likely heard the urban legend that you should put something in the mouth of a person having a seizure to prevent the person from "swallowing the tongue" and choking, but the lingual frenulum renders this quite impossible. (Note, however, that a person having a seizure can easily choke on objects placed in the mouth, so you should never attempt this.)

Recall that the epithelium of the tongue is arranged into small projections called **papillae** (puh-PIL-ee; refer to Figure 15.5). There are four kinds of papillae: *filiform, fungiform, circumvallate,* and *foliate papillae.* All papillae except filiform contain epithelium with sensory receptors called *taste buds,* which detect chemicals associated with different taste sensations. Filiform papillae play no role in taste, and are instead covered with stratified squamous keratinized epithelium. The keratinized cells make the surface of the tongue rough, which assists in mechanical digestion. Human tongues have a limited number of keratinized cells, so our tongues do not feel rough to the touch. Cats' tongues, however, have a large number of keratinized cells, which are responsible for their sandpaper-like texture.

Two groups of skeletal muscles control tongue movement: extrinsic and intrinsic. The **extrinsic muscles** control the position of the tongue. Extrinsic muscles, involved during the ingestion phase of digestion, move the tongue during chewing and help turn the food into a bolus. The **intrinsic muscles** control the shape and size of the tongue. Intrinsic muscles push the food against the hard palate during chewing, which assists in mechanical digestion, and also push the bolus posteriorly during swallowing.

Quick Check

☐ 5. What are the functions of the tongue pertaining to digestion?

The Salivary Glands

◀◀ **Flashback**

1. Which antibody class is found in salivary gland secretions? (p. 779)
2. How do the parasympathetic and sympathetic nervous systems affect secretion from salivary glands? (p. 531)

The other accessory organs associated with the mouth are the **salivary glands** (SAL-uh-vehr-ee). There are three pairs of salivary glands, all of which secrete **saliva** (suh-LY-vah), a fluid containing water, enzymes, mucus, and other solutes, through a duct into the oral cavity. The upcoming subsections explore the anatomy and histology of these glands, as well as the components and functions of saliva.

Anatomy and Histology of the Salivary Glands

Figure 22.7 illustrates the three pairs of salivary glands:

- **Parotid glands.** The **parotid glands** (puh-ROT-id; *para-* = "beside," *ot-* = "ear") are large glands located over the masseter muscle just anterior to the ear. These glands secrete

Buccinator muscle

Parotid duct

Parotid gland (25–30%)

Masseter muscle

Submandibular duct

Tongue

Sublingual ducts

Sublingual gland (5%)

Mandible (cut)

Submandibular gland (65–70%)

Figure 22.7 Anatomy of the salivary glands. The percentages indicate the portion of total salivary production for each type of gland.

saliva through the **parotid duct**, which passes over the masseter muscle and pierces the buccinator muscle to open into the oral cavity near the second molar. The parotid glands secrete 25–30% of total saliva.

- **Submandibular glands.** The smaller **submandibular glands** are located just medial to the inferior portion of the body of the mandible. They secrete saliva through the **submandibular duct,** which empties into the floor of the oral cavity. In spite of their smaller size, the submandibular glands are very active, and secrete 65–70% of total saliva.
- **Sublingual glands.** As implied by their name, the **sublingual glands** (sub-LING-gwuhl) are situated inferior to the tongue. The sublingual glands secrete saliva through several small **sublingual ducts** that empty into the oral cavity just under the tongue. These are the smallest salivary glands and secrete only about 5% of total saliva.

The basic secretory cell of the salivary glands is the **acinar cell** (ASS-uh-nahr). There are two main types of acinar cells in salivary glands: **serous cells,** which secrete a water-based fluid with enzymes and other solutes; and **mucous cells,** which secrete mucus. The secretions from serous cells are involved in digestive processes, and are generally released just before or during eating. However, secretions from mucous cells are primarily involved in keeping the oral mucosa moist, and so are released continually.

The three main types of salivary glands differ in the proportion of mucous and serous cells they contain. The parotid glands have only serous cells, the submandibular glands have mostly serous cells and a small number of mucous cells, and the sublingual glands contain mostly mucous cells. For this reason, the parotid glands secrete mainly water and enzymes, the submandibular glands secrete enzymes mixed with some mucus, and the sublingual glands secrete mainly mucus with a small amount of enzymes.

Saliva

Saliva consists primarily of water; electrolytes such as sodium, chloride, and potassium ions; and variable amounts of mucus, depending on the type of salivary gland. It also contains the following components:

- **Salivary amylase.** The digestive enzyme **salivary amylase** (AM-uh-layz) is the first digestive enzyme that ingested food encounters. It catalyzes the beginning of carbohydrate digestion, breaking down large polysaccharides into smaller polysaccharides.
- **Lysozyme.** The enzyme **lysozyme** (LY-soh-zy'm) catalyzes the perforation of bacterial plasma membranes. This allows bacteria-killing substances in the saliva to enter and kill the bacteria.
- **Secretory IgA.** Recall that the antibody immunoglobulin A, or IgA, is found in the body's secretions, including saliva (see Chapter 20). IgA binds specific antigens on pathogens and mediates their destruction.
- **Bicarbonate ions.** During eating, when flow rates of saliva are high, basic bicarbonate ions (HCO_3^-) are added to saliva. Their primary function is to neutralize any acid from the stomach that regurgitates into the esophagus.

Saliva performs several functions in the oral cavity, including moistening, lubricating, and cleansing the oral mucosa. In addition, its lysozyme and IgA deter the growth of pathogenic bacteria in the oral cavity. Saliva also functions in (1) mechanical digestion, by moistening and helping to mix ingested food into a bolus so it can be swallowed, and (2) chemical digestion, by the actions of salivary amylase. Finally, many food molecules dissolve in the water of saliva, and these molecules then stimulate taste receptors on the tongue.

Salivation

Salivation refers to the process of secretion from the three pairs of salivary glands. Salivation is primarily controlled by the parasympathetic nervous system in a reflex arc. The arc begins with sensory stimuli such as the smell or taste of food, which communicate this information to the *salivatory nucleus* in the brainstem. Parasympathetic fibers from the brainstem exit via the facial nerve (cranial nerve VII) to innervate the submandibular and sublingual glands, and via the glossopharyngeal nerve (cranial nerve IX) to innervate the parotid glands. Neurons from these nerves release acetylcholine (ACh) onto the acinar cells, which triggers the acinar cells to secrete saliva. Even the mere smell of food can trigger this reflex, which is why you find yourself drooling when you smell your favorite food cooking.

The importance of ACh in salivation is evident any time ACh receptors are blocked. A number of medications such as antihistamines for allergies have *anticholinergic* properties that block the effects of ACh around the body. Anticholinergic drugs bind to the ACh receptors on the salivary glands and decrease saliva secretion, leading to the common side effect of a dry mouth.

The sympathetic nervous system also innervates the salivary glands. Interestingly, this is one of the only instances in the body in which the sympathetic and parasympathetic divisions have similar effects on an organ. Sympathetic nerves innervate the salivary glands and their ducts, triggering increased saliva production and facilitating saliva transport through ducts. The primary difference is that sympathetic nerves stimulate mostly mucous cells, whereas parasympathetic nerves stimulate mostly serous cells.

Quick Check

☐ 6. What are the three types of paired salivary glands, and where are they located?
☐ 7. What are the components of saliva, and what are its functions?
☐ 8. Walk through the steps of the parasympathetic reflex arc that stimulates salivation.

The Pharynx

◀◀ Flashback

1. What are the three divisions of the pharynx? (p. 804)
2. What are pharyngeal constrictor muscles? (p. 302)

Muscularis externa

Mucosa:
Stratified squamous epithelium

Submucosa

Skeletal muscle cells

Smooth muscle cells

Lumen

LM (120×)

(a) Micrograph showing section through the middle third of the esophagus

Stratified squamous epithelium of esophagus

Lumen

Esophagus-stomach junction

Simple columnar epithelium of stomach

LM (85×)

(b) Micrograph showing section through the esophagus-stomach junction

Figure 22.8 Histology of the esophagus.

As noted previously, the **pharynx** (FEHR-inks), or throat, sits posterior to the nasal and oral cavities (see Chapter 21). It consists of three divisions: the *nasopharynx, oropharynx,* and *laryngopharynx.* Of these three divisions, only the oropharynx and laryngopharynx are part of the alimentary canal. Like the oral cavity, both are lined with stratified squamous epithelium to protect them from abrasion by food. The oropharynx houses two sets of tonsils—the *palatine tonsils* and the *lingual tonsils.* The tonsils perform defensive functions and help protect the remainder of the alimentary canal from any pathogens that enter the body via the oral and nasal cavities.

The primary function of the pharynx is propulsion in the form of swallowing, during which the bolus passes through the pharynx and into the esophagus. Recall that the pharynx is surrounded by three pairs of skeletal muscles: the *upper, middle,* and *lower pharyngeal constrictor muscles* (see Chapter 9). These muscles contract sequentially during swallowing and propel the bolus inferiorly.

Quick Check

☐ 9. What is the primary digestive function of the pharynx? Which muscles of the pharynx carry out this function?

The Esophagus

The **esophagus** (eh-SOF-ah-gus) is a muscular tube about 25 cm (10 in.) long found posterior to the trachea. It transmits the bolus from the pharynx to the stomach. Like the pharynx, the esophageal

mucosa is lined with stratified squamous nonkeratinized epithelium (**Figure 22.8**). The mucosa and submucosa contain *esophageal glands* (eh-sah′-fah-JEE-uhl) that secrete mucus to lubricate the bolus as it passes through the esophagus.

The muscularis externa of the esophagus consists of two layers of muscle, but it differs from the remainder of the alimentary canal. Rather than being only smooth muscle, its superior third is composed of skeletal muscle, its middle third is a mixture of skeletal and smooth muscle (see Figure 22.8a), and its inferior third is composed of smooth muscle. At the junction of the pharynx and the esophagus, the muscularis externa is modified into a sphincter called the *upper esophageal sphincter,* which controls the passage of the bolus into the esophagus. At the esophagus' inferior end is another sphincter, the **gastroesophageal sphincter** (gas′-troh-ee-sah-fah-JEE-al), also known as the *lower esophageal sphincter,* that regulates the passage of the bolus into the stomach. This sphincter also prevents the contents of the stomach from re-entering the esophagus.

The primary functions of the esophagus are propulsion and a small amount of secretion of mostly mucus. During swallowing, the skeletal muscle and smooth muscle of the muscularis undergo peristalsis, which pushes the bolus inferiorly. Although the thick epithelium of the esophagus prevents any significant absorption from taking place, it does protect the esophagus from abrasion by food. Note that this is another example of the Structure-Function Core Principle (p. 25).

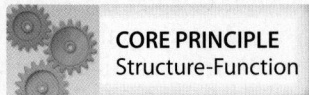

CORE PRINCIPLE
Structure-Function

Quick Check

☐ 10. Where is the esophagus located? What are its major functions?

☐ 11. How do the muscularis externa and mucosa in the esophagus differ from these tissues in most of the alimentary canal?

Swallowing

The process of **swallowing,** or **deglutition** (dee-gloo-TISH-un; *degluto-* = "to swallow"), is a specialized type of propulsion that pushes a bolus of food from the oral cavity through the pharynx and esophagus to the stomach. Swallowing relies on the coordinated action of the upper alimentary canal, including the soft palate, pharynx, and esophagus (**Figure 22.9**). The tongue also plays a role in this process, and is the only accessory organ to directly participate in motility. Swallowing consists of three phases: voluntary, pharyngeal, and esophageal.

① **Voluntary phase: The tongue pushes the bolus posteriorly toward the oropharynx.** During the **voluntary phase,** the tongue pushes the bolus superiorly against the hard palate and posteriorly toward the oropharynx. The voluntary phase is the only stage of swallowing under conscious control.

② **Pharyngeal phase: The bolus enters the oropharynx; the soft palate and epiglottis seal off the nasopharynx and larynx, respectively.** The bolus passes from the pharynx to the esophagus during the **pharyngeal phase.** This phase involves the involuntary contraction of skeletal muscles such as the pharyngeal constrictor muscles, and is controlled by the *swallowing reflex,* a reflex initiated by the medulla oblongata. The reflex arc is initiated when the bolus contacts sensory receptors in the oropharynx. This triggers the elevation of the uvula and soft palate, which prevents the bolus from going into the nasopharynx. It also triggers the elevation of the larynx. Notice in Figure 22.9 that the bolus pushes the epiglottis down. This, together with the elevation of the larynx, prevents *aspiration,* or the entry of food into the larynx.

③ **Esophageal phase: Peristaltic waves move the bolus down the esophagus to the stomach.** During the final phase of swallowing, the **esophageal phase,** the bolus passes through the esophagus to the stomach. This phase begins as the upper esophageal sphincter relaxes. Neurons of the enteric nervous system then stimulate the inner circular and outer longitudinal layers of the muscularis externa to undergo peristalsis and "massage" the bolus inferiorly toward the stomach.

Notice that control of swallowing is almost entirely neural. The voluntary phase is under control of the cerebral cortex, and the remaining two phases are regulated by the medulla and the enteric nervous system. The ANS does not directly control any phase of swallowing, but it does influence the esophageal phase—the parasympathetic division stimulates peristalsis, and the sympathetic division inhibits it.

Quick Check

☐ 12. Walk through the steps of deglutition.
☐ 13. What controls the process?

Apply What You Learned

☐ 1. Inadequate production of saliva is associated with dental caries. Why do you think saliva and dental health are linked?

① **Voluntary phase:** The tongue pushes the bolus posteriorly toward the oropharynx.

② **Pharyngeal phase:** The bolus enters the oropharynx; the soft palate and epiglottis seal off the nasopharynx and larynx, respectively.

③ **Esophageal phase:** Peristaltic waves move the bolus down the esophagus to the stomach.

Figure 22.9 The process of swallowing (deglutition).

☐ 2. Chemical warfare agents called *nerve gases* block the enzyme acetylcholinesterase, which increases the amount of acetylcholine released at cholinergic synapses. Explain why drooling (excessive salivation) is an effect of these agents.

☐ 3. Mr. Kaku suffers from Parkinson's disease, which diminishes his swallowing reflex and decreases the activity of the muscularis externa of his esophagus. Predict what complications may result.

See answers in Appendix A.

MODULE **22.3**

The Stomach

Learning Outcomes

1. Describe the structure and function of the different regions of the stomach.
2. Describe the structure of the gastric glands and the functions of the types of cells they contain.
3. Explain how hormones, nervous system stimulation, and the volume, chemical composition, and osmolarity of the chyme affect motility in both the stomach and the duodenum.
4. Discuss the function, production, and regulation of the secretion of hydrochloric acid.
5. Explain the effects of the cephalic phase, gastric phase, and intestinal phase on the functions of the stomach and small intestine, and give examples for each phase.

The esophagus pierces the diaphragm and passes through the *esophageal hiatus* (hy-AY-tus) to empty into the J-shaped organ known as the **stomach.** In the sections that follow, we examine the gross and microscopic anatomy of the stomach. The discussion then turns to secretion from and motility of the stomach, and how the endocrine and nervous systems regulate these activities.

Gross Anatomy of the Stomach

As you saw in Figure 22.1, the stomach sits primarily in the left upper quadrant just inferior to the diaphragm. The stomach's convex left side is known as its *greater curvature,* and its concave right side is its *lesser curvature.* **Figure 22.10** shows that the stomach has five anatomical regions:

- **Cardia.** The region where the esophagus empties into the stomach is the **cardia** (KAR-dee-ah). The cardia receives the bolus when the gastroesophageal sphincter relaxes.
- **Fundus.** The dome-shaped top of the stomach is its **fundus.**
- **Body.** The largest portion of the stomach is its **body.**
- **Pyloric antrum.** The inferior portion of the stomach is the **pyloric antrum** (py-LOR-ik AN-trum).

- **Pylorus.** The terminal portion of the stomach is the **pylorus** (*pylor-* = "gatekeeper"), which abuts the first portion of the small intestine, the *duodenum.* Like the cardia, the pylorus contains a sphincter that controls the flow of ingested food. In the pylorus, a thick ring of smooth muscle called the **pyloric sphincter** regulates the flow of materials between the stomach and the small intestine.

Notice that the interior of the stomach contains folds called **rugae** (ROO-ghee). Rugae allow the stomach to expand considerably.

Histology of the Stomach

◀◀ **Flashback**

1. How do endocrine glands and exocrine glands differ? (p. 135)
2. What is an acid? How does an acid affect the pH of a solution? (p. 48)

The stomach has the same four tissue layers as the rest of the alimentary canal, with a mucosa, submucosa, muscularis externa, and serosa. However, the stomach's muscularis externa and mucosa are modified to better suit its functions. Let's take a look at these modifications.

Stomach Muscularis Externa

Recall that the muscularis externa of the alimentary canal usually has two layers of smooth muscle: an inner circular layer and an outer longitudinal layer. Figure 22.10 shows you how this differs in the stomach—there is an additional inner layer of smooth muscle in the stomach's body with its fibers oriented obliquely. This oblique layer of smooth muscle allows the stomach to perform **churning,** a motion that pummels the food into a liquid called **chyme** (KYME).

Stomach Mucosa: Gastric Glands

Figure 22.11a shows that the mucosa of the stomach is heavily indented to form deep structures called **gastric pits.** Between the gastric pits we find columnar cells and numerous goblet cells, both of which secrete a thick mucus that lines and protects the cells of the stomach from its own secretions. Conditions that decrease the amount of mucus secreted by these cells can lead to a *gastric ulcer* (also called a *peptic ulcer*), in which acid and enzymes eat away at the mucosa and expose the underlying tissues. Ulcers are often associated with the presence of certain bacteria and/or excessive production of acid.

At the base of the gastric pits are multiple branched glands known as **gastric glands.** Gastric glands are unusual in that they contain both endocrine cells that secrete hormones into the bloodstream and exocrine cells that secrete an acidic, enzyme-containing fluid called **gastric juice** into the lumen of the stomach. We find four main types of cells in or near gastric glands, each of which secretes a different product. From superficial to deep, these cell types are as shown in **Figure 22.11b and c.**

Esophagus

Diaphragm

Gastroesophageal sphincter

Fundus

Cardia

Body

Muscularis externa:

Outer longitudinal layer

Middle circular layer

Inner oblique layer

Greater curvature

Greater omentum

Lesser omentum

Lesser curvature

Pyloric sphincter

Duodenum

Pylorus

Rugae

Pyloric antrum

(a) Regions and tissue layers of the stomach

Gastroesophageal sphincter

Cardia

Fundus

Body

Rugae

Pylorus

Pyloric sphincter

Pyloric antrum

(b) Stomach and surrounding organs, cadaver photo

Figure 22.10 Gross anatomy of the stomach.

Practice art labeling
@MasteringA&P®

Stomach

Mucus

Mucosa

Gastric pit

Gastric gland

Submucosa

Muscularis externa

Serosa

(a) Layers of the stomach wall

Mucus

Goblet cell

Simple columnar epithelial cells

Gastric pit

Mucous neck cells

Gastric gland

Parietal cell

Chief cell

Enteroendocrine cell

(b) Section through a gastric pit and gastric gland

K⁺

ATP

ADP + Pᵢ

Pepsin

H⁺

Cl⁻

HCl

Pepsinogen

Parietal cells secrete HCl: The H⁺/K⁺ pump consumes ATP while secreting H⁺, and taking in K⁺. Cl⁻ follows H⁺.

Chief cells secrete pepsinogen, which is activated by the acidic pH and becomes the enzyme pepsin.

Enteroendocrine cells called G cells secrete the hormone gastrin, stimulating secretions from parietal cells.

Gastrin

Other hormones

Capillary

(c) Functions of gastric gland cells

Figure 22.11 Structure and functions of the stomach mucosa and gastric glands.

- **Mucous neck cells.** The cells located near the top, or "neck," of the gland are called **mucous neck cells.** As their name implies, these cells secrete mucus much like goblet cells. However, goblet cells secrete alkaline mucus, whereas mucous neck cells secrete acidic mucus. This prevents their mucus from neutralizing the acid produced by other cells known as parietal cells.
- **Parietal cells.** The next cells are the **parietal cells,** which secrete the hydrochloric acid (HCl) that is responsible for the acidic pH of gastric juice. Acid is an important component of gastric juice because it is required to activate a precursor enzyme called **pepsinogen** (pep-SIN-oh-jen) and also because it destroys most disease-causing organisms we ingest. In addition, parietal cells produce the chemical **intrinsic factor,** which is required for intestinal absorption of vitamin B12, found in various foods.
- **Chief cells.** Next down are the **chief cells,** which secrete the inactive precursor enzyme pepsinogen. When pepsinogen encounters an acidic pH, it becomes the active enzyme **pepsin,** which begins protein digestion in the stomach.
- **Enteroendocrine cells.** Located at the very bottom of the gland are the **enteroendocrine cells,** which secrete several hormones that influence digestion. For example, enteroendocrine cells called **G cells** secrete the hormone **gastrin,** which stimulates acid secretions from the parietal cells. As you can see in Figure 22.11c, enteroendocrine cells are closest to the blood vessels in the underlying submucosa, so their hormones have ready access to the blood.

Quick Check

☐ 1. Why does the stomach have an extra layer of smooth muscle in its muscularis externa?

☐ 2. What are the primary cell types of the gastric glands, and what does each type of cell produce?

Functions of the Stomach

The stomach performs three primary functions: secretion, propulsion, and digestion. Note that absorption of food molecules is not a main role of the stomach, but small amounts of chemicals such as alcohol may be absorbed here. This is why drinking alcohol without eating food first can lead to rapid intoxication and vomiting.

As you have read, the gastric glands of the stomach secrete many chemicals, including hormones, acid, mucus, and pepsinogen, which becomes the digestive enzyme pepsin. Pepsin catalyzes the beginning of the chemical digestion of proteins, and the stomach's churning actions begin mechanical digestion. When the stomach has completed churning, it propels the food, now chyme, through the pyloric sphincter into the duodenum. Secretion from the stomach, particularly of acid, and its motility are tightly regulated by the endocrine and nervous systems. We examine this regulation in the next two subsections.

Acid Secretion from the Stomach

Recall that the gastric glands of the stomach secrete multiple exocrine products, including hydrochloric acid (HCl) released by parietal cells. The hydrogen ions are secreted by an ATP-consuming pump in the plasma membrane, called the H^+/K^+ pump, or proton pump (Figure 22.11c). The H^+/K^+ pump drives a hydrogen ion out of the cell and into the lumen of the gland while taking a potassium ion into the cytosol. The chloride ion follows the hydrogen ion via passive diffusion through a membrane channel. Acid secretion occurs in this manner continuously throughout the day between meals, at what is known as the *basal rate.*

During eating, gastric acid secretion changes from the basal rate under the influence of the parasympathetic nervous system and multiple hormones. Secretion can be divided into three phases based on the primary source of regulation: the *cephalic phase, gastric phase,* and *intestinal phase* (**Figure 22.12**).

Cephalic Phase The initial **cephalic phase** is mediated by the sight, smell, taste, or even thought of food. This phase, which is directed by the CNS, prepares the stomach to receive food by increasing the release of hydrogen ions into it. These stimuli trigger output from the *vagus nerve,* the main nerve of the parasympathetic nervous system, resulting in four physiological effects:

- **Direct stimulation of hydrogen ion release.** The vagus nerves release ACh onto parietal cells, which directly stimulates them to release hydrogen ions.
- **Stimulation of gastrin secretion.** Vagal nerve stimulation triggers the release of a peptide that stimulates G cells to release the hormone gastrin. Gastrin triggers hydrogen ion secretion.
- **Stimulation of histamine secretion.** The vagus nerve stimulates a specific type of enteroendocrine cell to release the hormone **histamine.** Like gastrin, histamine triggers hydrogen ion secretion.
- **Inhibition of somatostatin secretion.** The hormone **somatostatin** (soh-mah'-toh-STAT-in) is produced by a type of enteroendocrine cell located in the antrum of the stomach. Somatostatin inhibits acid secretion. Vagal nerve stimulation inhibits somatostatin release, which has the effect of increasing hydrogen ion secretion.

Together, the four changes during the cephalic phase are responsible for about 30–40% of total hydrogen ion secretion from parietal cells.

Gastric Phase The **gastric phase** begins when food enters the stomach, and continues the stimulation provided during the cephalic phase. Two stimuli trigger acid secretion during the gastric phase. The first is the simple presence of food in the stomach. Distention of the stomach wall stimulates neurons of the ENS and sensory receptors involved with vagus nerve reflexes. Both the vagus nerve and ENS neurons release ACh, which has the same four effects we discussed with the cephalic phase. The net effect is to increase acid secretion.

The second stimulus is the presence of partially digested proteins in gastric juice. These protein fragments stimulate

CEPHALIC PHASE

Sight, smell, taste, and thoughts of food → Vagus nerve stimulation → Gastric enteroendocrine cells →

↑ Gastrin →

↑ Histamine →

↓ Somatostatin →

Parietal cell

H^+

H^+

↑ H^+ release

GASTRIC PHASE

Food entering stomach → Vagus nerve and enteric nervous system stimulation

Partially digested proteins in gastric juice → Gastric enteroendocrine cells

Positive feedback

↑ Gastrin →

↑ Histamine →

↓ Somatostatin →

Stomach

H^+ H^+

H^+

↑ H^+ release

H^+

H^+

INTESTINAL PHASE

Duodenum

Partially digested proteins entering duodenum → Vagus nerve stimulation

Brief stimulating effect

→ Intestinal enteroendocrine cells → ↑ Intestinal gastrin →

H^+

H^+

↑ H^+ release

H^+

Declining pH and presence of lipids in duodenum → Enterogastric reflex → Vagus nerve activity and H^+ secretion

→ ↑ Secretin →

→ ↑ Gastric inhibitory peptide (GIP) →

↓ H^+ release

H^+

→ Stimulates

→ Inhibits

Figure 22.12 The three phases of acid secretion from the stomach.

enteroendocrine cells to produce and release gastrin, which in turn stimulates acid secretion. The acid activates pepsin, which catalyzes protein digestion. Digested proteins stimulate more gastrin release, which triggers further acid secretion and more protein digestion. Note that this is a positive feedback loop and we have again visited the Feedback Loops Core Principle (p. 21). Other stimuli trigger

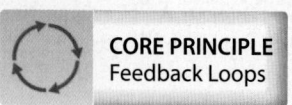

CORE PRINCIPLE
Feedback Loops

gastrin secretion as well, including caffeine and alcohol. Together, gastrin and the neural stimulation of parietal cells during the gastric phase account for about 50–60% of total acid secretion.

The positive feedback loop in the stomach does not continue indefinitely. Once the acid concentration reaches a certain level—a pH of about 2.0—a negative feedback mechanism is initiated. Extremely low pH levels trigger somatostatin release, which in turn decreases acid secretion.

Intestinal Phase The final phase of gastric acid secretion, the **intestinal phase,** is responsible for the remaining approximately 10% of acid secretion, after which further acid secretion is inhibited. The intestinal phase is triggered by the presence of partially digested proteins in the fluid entering the duodenum. As in the stomach, these partially digested proteins trigger enteroendocrine cells to release *intestinal gastrin.* This hormone has the same effect as gastrin produced by the stomach, and stimulates hydrogen ion secretion from parietal cells.

The stimulatory effect of the intestinal phase is brief. As chyme enters the duodenum, the declining pH and presence of lipids trigger the **enterogastric reflex,** which decreases vagal activity and acid secretion. The low pH in the duodenum also triggers the production of hormones by the cells of the duodenal mucosa, including **secretin** and **gastric inhibitory peptide (GIP).** Both hormones reduce acid secretion.

A&P in the Real World

Gastroesophageal Reflux Disease (GERD)

The gastroesophageal sphincter normally remains closed except during swallowing. When this mechanism fails, acidic chyme from the stomach *regurgitates* into the esophagus. If this occurs on a chronic basis, it is called **gastroesophageal reflux disease,** or **GERD,** and may lead to pain, difficulty swallowing, vocal cord damage, respiratory problems, and even esophageal cancer.

Multiple factors contribute to the development of GERD, including increased acid secretion, obesity, caffeine consumption, a *hiatal hernia* (abnormal enlargement of the esophageal hiatus in the diaphragm), and medications such as anti-inflammatory drugs. Another factor that contributes to a significant proportion of cases is infection with the bacterium *Helicobacter pylori.* This bacterium creates a state of chronic inflammation, which stimulates acid production.

GERD is treated in a variety of ways. One of the simplest treatments is dietary modifications such as consuming less caffeine and alcohol. Most patients with GERD, however, also require drug therapy, including one or more of the following:

- *Antacids.* The most common drugs for GERD are *antacids,* which contain alkaline substances that neutralize the acid in the stomach.
- *H_2 blockers.* H_2 blockers block histamine receptors on parietal cells. This lowers the release not only of histamine-mediated acid but also of gastrin- and ACh-mediated acid.
- *Proton pump inhibitors.* Proton pump inhibitors are powerful inhibitors of the H^+/K^+ pump. These drugs generally result in the most drastic decrease in acid secretion.

Note that if *H. pylori* infection is present, antibiotics must be given in addition to acid-reducing medications to eradicate the bacteria.

Normally, gastric acid secretions do not back up into the esophagus. When they do, a painful condition results, discussed in *A&P in the Real World: Gastroesophageal Reflux Disease (GERD).*

Motility of the Stomach

The stomach's motility enables it to perform three actions: (1) receive food from the esophagus, (2) churn the incoming bolus into chyme, and (3) control the rate at which chyme empties into the small intestine.

Receptive Function In its resting state, the stomach has a volume of only about 50 ml (0.2 cup), but it can expand to about 1500 ml (6.3 cups) when filled with food and liquid. When food or liquid is swallowed, the gastroesophageal sphincter and smooth muscle of the fundus and body of the stomach relax to allow the stomach to fill. This is known as **receptive relaxation,** and is mediated by both the medulla (as part of the swallowing reflex) and the vagus nerve. Another factor that allows the stomach to fill is the inherent ability of smooth muscle to relax when it is stretched. These two factors, along with the rugae, allow the stomach to receive food and fluid without raising its internal pressure. This is important, as increases in gastric pressure can trigger *vomiting,* expelling of the stomach contents through the mouth, as discussed in *A&P in the Real World: Vomiting.*

A&P in the Real World

Vomiting

Normally, the stomach empties its contents into the duodenum. But occasionally stomach contents move backward, an unpleasant process known as **vomiting,** or **emesis** (EM-uh-sis). Vomiting involves a complex motor response during which the smooth muscle of the stomach and small intestine relaxes while the abdominal skeletal muscles and diaphragm contract to increase intra-abdominal pressure. In addition, the upper and lower esophageal sphincters relax, the soft palate closes off the nasopharynx, and the larynx elevates so that the epiglottis covers the glottis.

Vomiting can be a response to a variety of stimuli, including emotional stress, an overfilled stomach, the presence of irritants such as alcohol and certain drugs, or viral or bacterial infections. These stimuli activate sensory components of the vagus nerve, which communicates with the medulla oblongata and mediates the motor responses of vomiting. Drugs that treat vomiting are known as *anti-emetics.* Many act by blocking neurotransmitter receptors in the medulla to prevent these neurons from initiating vomiting.

Although vomiting is unpleasant, it can sometimes be beneficial. Many stomach irritants that stimulate vomiting could be harmful if they were absorbed into the bloodstream, and expelling them from the body prevents this from happening.

① The stomach's smooth muscle contracts and propels a small amount of chyme through the pyloric sphincter. The rest is propelled backward.

② The stomach's peristaltic waves churn the remaining chyme.

③ The remaining chyme is again propelled toward the pyloric sphincter. More chyme is forced through as the process repeats.

Figure 22.13 The process of churning in the stomach.

Churning Function After a meal, the smooth muscle layers of the stomach begin to produce waves of peristalsis for churning (**Figure 22.13**). In this process, ① the stomach's smooth muscle produces peristaltic contractions that propel a small amount of chyme through the pyloric sphincter into the duodenum. The rest of the chyme is pushed backward into the stomach. ② The peristaltic waves then churn and mix the remaining chyme, and ③ the process repeats.

The peristaltic waves are initiated and controlled by a group of specialized cells collectively called the *gastric pacemaker*. The pacemaker controls the rate of the waves, which remains relatively constant at about three per minute. Their strength, however, is subject to regulation by several factors, including stimulation from the vagus nerve and secretion of certain hormones. For example, the enteroendocrine cells of the stomach produce the hormone **serotonin,** which stimulates gastric motility. Another hormone with the same effect is **intestinal gastrin,** which is produced by the duodenum.

Emptying Function The final function of gastric motility is to control the movement of chyme into the duodenum. Different materials pass through the pyloric sphincter at different rates. Liquids such as water and saline move rapidly from the stomach to the duodenum with essentially no delay. Solids must be converted to a nearly liquid state before they are able to enter the small intestine. This occurs relatively easily with carbohydrates and proteins, which mix with gastric juice with few problems. It may take much longer with lipids, because they require extensive churning due to their nonpolar structure.

A key factor that determines the rate of gastric emptying is the amount and composition of chyme in the duodenum. When sensory receptors in the wall of the duodenum detect a high degree of stretch, a low pH, a high lipid composition, and/or a high solute concentration in the chyme, they trigger a negative feedback loop

that delays gastric emptying. This feedback loop is mediated by both the vagus nerve and hormones secreted by the duodenum, including *secretin, cholecystokinin,* and *gastric inhibitory peptide.* Each of these hormones is also involved in secretion from the stomach and other digestive organs, and will be discussed shortly.

Control of gastric emptying is critical because the duodenum must mix the incoming chyme thoroughly before it moves to the rest of the small intestine. There are two reasons for this: (1) Chyme is acidic, and the duodenum must mix it with bicarbonate ions to avoid damaging the intestinal mucosa; and (2) chyme is generally very concentrated and must be diluted with water from pancreatic juice to prevent the chyme from drawing water into the intestinal lumen by osmosis. The duodenum can process incoming chyme only so quickly, and therefore must receive small amounts at a time.

Quick Check

☐ 3. What are the three phases of acid secretion in the stomach? Which hormones and neural stimuli trigger each phase?

☐ 4. What are the basic functions of gastric motility?

☐ 5. Why is control of gastric emptying into the duodenum so critical?

Apply What You Learned

☐ 1. *Pyloric stenosis* (a hardened pyloric sphincter) causes the stomach contents to be emptied slowly and erratically. What symptoms might accompany this disorder? (*Hint:* What effect would pyloric stenosis have on gastric pressure?)

☐ 2. A procedure called a *vagotomy,* in which the vagus nerve is severed, is occasionally performed to decrease acid secretion by the stomach. How would such a procedure lower acid secretion? What other digestive processes would it affect and in what way?

See answers in Appendix A.

22.4

The Small Intestine

Learning Outcomes

1. Describe the structure and functions of the duodenum, jejunum, and ileum.
2. Discuss the histology and functions of the circular folds, villi, and microvilli of the small intestine.
3. Describe the functions and regulation of motility in the small intestine.

The 6-meter-long (almost 20-feet-long) **small intestine,** or *small bowel,* is the longest portion of the alimentary canal (it is considerably shorter in a living person—about 3 meters or 10 feet—than in a cadaver because of smooth muscle tone). Four main processes occur in the small intestine: secretion, digestion, absorption, and propulsion.

Cells of the small intestine, known as *enterocytes,* produce multiple digestive enzymes, hormones, and mucus. These enzymes, along with those released by the pancreas, are responsible for the bulk of chemical digestion. Indeed, no further nutrient digestion takes place by human enzymes once food leaves the small intestine (bacteria in the large intestine do digest certain nutrients). After the nutrients are digested chemically, they are absorbed across the enterocytes into either the blood or the lymph, along with water, vitamins, and other substances. Like the stomach, the small intestine also mixes and propels its contents along its length and into the large intestine. Let's look at the structure and functions of the small intestine.

Divisions of the Small Intestine

The small intestine consists of three divisions—the *duodenum, jejunum,* and *ileum* (**Figure 22.14**). These divisions are distinguishable by the following slight histological differences:

- **Duodenum.** The **duodenum** (doo-AH-den-um), the initial segment of the small intestine, begins at the pylorus. The shortest of the three divisions, it is only about 25 cm (10 in.) long. The duodenum arches into a "C" shape as it curves around the pancreas. Only the proximal portion of the duodenum is within the peritoneum; the remainder sits posterior to the peritoneal cavity and so is retroperitoneal. Internally, the duodenum houses the **major duodenal papilla,** which is where secretions from the gallbladder and pancreas enter the small intestine. The duodenal submucosa contains specialized glands called *duodenal (Brunner's) glands,* which produce an alkaline mucus to protect the duodenum from the acidic chyme.

- **Jejunum.** The **jejunum** (jeh-JOO-num), the middle portion of the small intestine, measures about 2.5 meters (7.5 feet) in length. It begins at the *duodenojejunal flexure* and sits within the peritoneal cavity. It is the most active site for chemical digestion and absorption.

- **Ileum.** The **ileum** (ILL-ee-um), the small intestine's final segment, is also intraperitoneal. About 3.6 meters (10.8 feet) in length, it terminates at the portion of the large intestine called the *cecum* (SEE-kum). A sphincter known as the **ileocecal valve** (ill'-ee-oh-SEE-kuhl) controls the movement of materials from the ileum into the cecum. This sphincter also prevents materials in the large intestine from flowing backward into the ileum. This function is key because, as we discuss shortly, the large intestine houses a great number of bacteria that could cause illness if they entered the ileum.

The small intestine has the same four tissue layers as the rest of the alimentary canal. But, as always, the form of the small intestine has specializations that better enable it to perform its functions, as we'll see next.

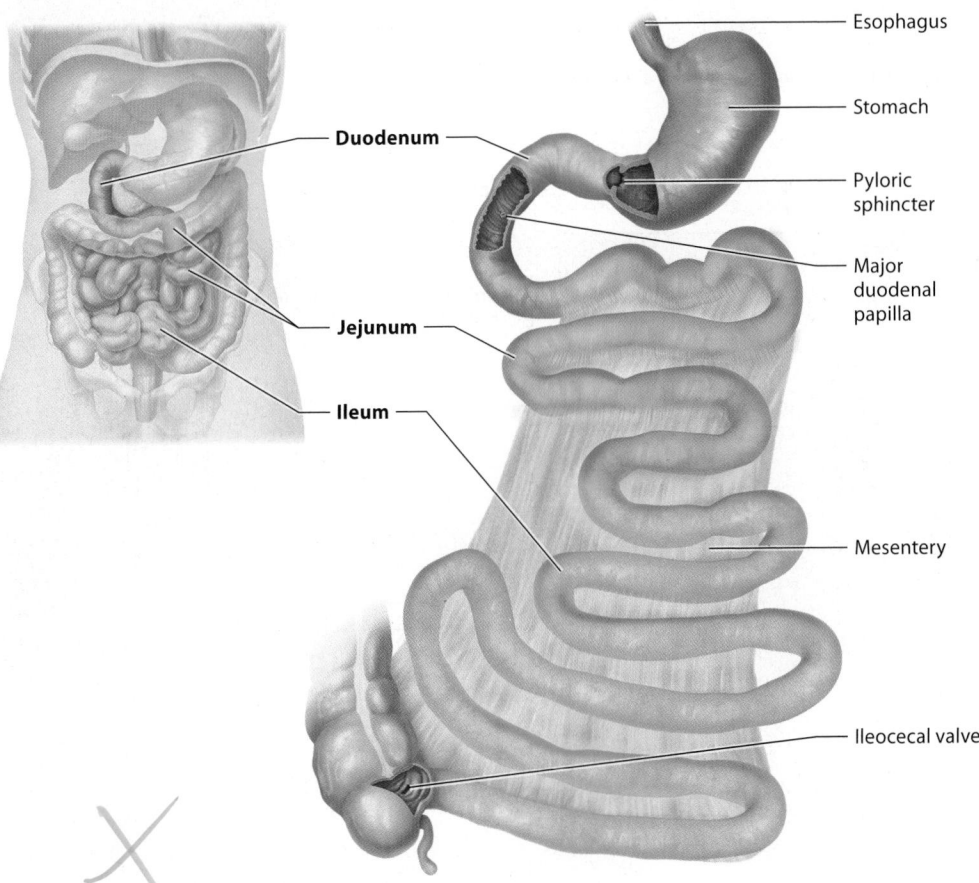

Figure 22.14 Gross anatomy of the small intestine.

☐ 1. What are the three divisions of the small intestine?

Structure and Functions of the Small Intestine

◀◀ **Flashback**

1. What are microvilli? What is the function of microvilli? (p. 97)

The internal surface of the small intestine contains three progressively smaller types of folds. Most of the absorption that occurs within the alimentary canal takes place in the small intestine, and these folds increase the surface area available for absorption about 400 to 600 times—another example of the Structure-Function Core Principle (p. 25).

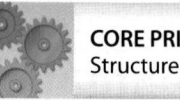

CORE PRINCIPLE
Structure-Function

The largest type of folds, which are visible with the naked eye as ridges in the wall, are called **circular folds** (or *plicae circulares;* Figure 22.15a). Circular folds involve both the mucosa and submucosa of the small intestine. These folds not only increase surface area but also slow down the transit of chyme through the small intestine, which gives the nutrients more time to be digested, and the small intestine cells, called **enterocytes** (EN-tehr-oh-sytz), more time to absorb nutrients.

The smaller two types of folds are not visible to the naked eye. The mucosa folds into projections called **villi** (VILL-ahy; singular, *villus;* Figure 22.15b). Notice that each villus consists of a layer of

Muscularis externa

Circular folds

Jejunum

(a) Section of the jejunum showing circular folds

Villi

Simple columnar epithelium

Goblet cells

Lacteal

Blood capillaries

Intestinal crypts

Enteroendocrine cell

LM (140×)

(b) Intestinal villi, photomicrograph (left) and illustration (right)

Microvilli

Brush border

Enterocyte

(c) Enterocyte with microvilli

Figure 22.15 Folds of the small intestine.

enterocytes and occasional mucus-secreting goblet cells surrounding a central core of blood capillaries and a lymphatic vessel called a *lacteal*. Between villi, the mucosa indents to form **intestinal crypts,** which house glands that also contain enteroendocrine cells (which secrete a variety of hormones).

The smallest folds, the **microvilli,** are found in the plasma membrane of the enterocytes (**Figure 22.15c**). Each enterocyte has as many as 3000 microvilli, which gives the cell the appearance of a bristle brush, or **brush border,** on microscopic examination. Associated with the brush border are numerous digestive enzymes produced and secreted by enterocytes, such as sucrase, maltase, and lactase, which catalyze the reactions that break down disaccharides, and peptidases, which catalyze reactions that break down peptides.

Quick Check

☐ 2. Why is the internal surface of the small intestine arranged into progressively smaller folds? What are these folds called?

Motility of the Small Intestine

Small intestinal motility differs during fasting and eating. During fasting, the small intestine exhibits slow, rhythmic contractions along its length in a pattern called the **migrating motor complex.** These contractions clear any remaining material from the small intestine, including leftover food and secretions. The migrating motor complex requires about 2 hours to push digesting food from the duodenum to the ileocecal valve. This movement is controlled by both the ENS and a hormone called *motilin,* which is produced by cells in the duodenal mucosa.

The small intestine undergoes two types of movement during eating: peristalsis and segmentation (**Figure 22.16**). As with other regions of the alimentary canal, peristalsis in the small intestine is accomplished by alternating contractions of the longitudinal and circular layers of smooth muscle in the muscularis externa (see Figure 22.16a).

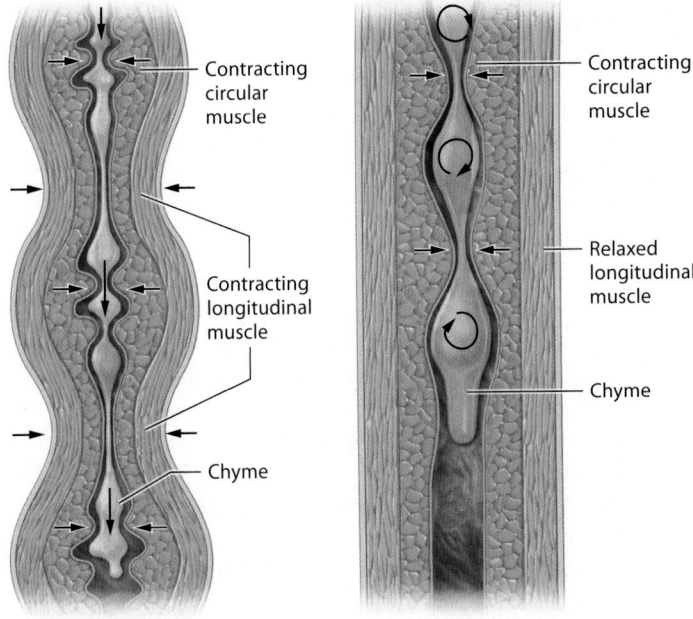

(a) Peristalsis: propulsive, alternating contractions of longitudinal and circular smooth muscle

(b) Segmentation: mixing contractions of circular smooth muscle only

Figure 22.16 Peristalsis and segmentation in the small intestine.

Its primary function is to propel chyme toward the ileum and ultimately through the ileocecal valve to the cecum. However, **segmentation,** also known as intestinal churning, involves contractions of only the circular layer of smooth muscle, which produces a squeezing motion (see Figure 22.16b). The primary functions of segmentation are mechanical digestion and mixing the chyme with intestinal and pancreatic enzymes as well as bile. The vagus nerve appears to regulate both peristalsis and segmentation.

The hormones and paracrines that control secretion and motility of the digestive organs are summarized in **Table 22.1**.

TABLE 22.1 HORMONES AND PARACRINES INVOLVED IN SECRETION AND MOTILITY OF THE DIGESTIVE ORGANS		
Hormone	**Stimulus for Production**	**Effects**
Stomach Hormones		
Gastrin	Partially digested proteins; stimulation from the vagus nerve	Increases acid secretion by parietal cells
Histamine	Stimulation from the vagus nerve	Increases acid secretion by parietal cells
Serotonin	Distention of the stomach	Stimulates gastric motility
Somatostatin	Decreasing stomach pH	Decreases acid secretion by parietal cells
Intestinal Hormones		
Cholecystokinin (CCK)	Partially digested proteins and lipids in chyme entering the duodenum	Causes gallbladder to contract and release bile; stimulates secretion of pancreatic enzymes from acinar cells; relaxes hepatopancreatic sphincter
Gastric inhibitory peptide	Chyme entering the small intestine	Inhibits acid secretion from parietal cells
Intestinal gastrin	Chyme entering the small intestine	Stimulates acid secretion from parietal cells
Motilin	Released regularly during fasting	Stimulates the migrating motor complex of the small intestine
Secretin	Partially digested proteins in the duodenum	Inhibits gastric motility and acid secretion; stimulates bicarbonate release from pancreatic duct cells; increases bile production by the liver
Vasoactive intestinal peptide	Partially digested proteins in the duodenum	Inhibits acid secretion by parietal cells; stimulates pancreatic secretion; increases intestinal blood flow

Quick Check

☐ 3. How does motility of the small intestine differ during fasting and during eating?

☐ 4. What are segmentation and peristalsis, and how do they differ?

Apply What You Learned

☐ 1. The condition known as *celiac disease* (SEE-lee-ak), or *sprue,* is caused by an allergy to gluten, a protein found naturally in certain grains. Celiac disease leads to the destruction of the microvilli and even the villi of the small intestine. Predict the effects of this disease.

☐ 2. Predict what would happen to the motility of the stomach and the small intestine if the vagus nerve were cut. Are there processes that would be unaffected by a severed vagus nerve? Explain.

See answers in Appendix A.

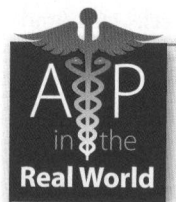

Appendicitis

The small size of the appendix and the fact that it is blind-ended cause it to occasionally become blocked, generally by fecal matter. Bacteria within the feces multiply in the appendix and cause infection. This results in inflammation, a condition known as *appendicitis* (ah-pen'-dih-SY-tus). Signs and symptoms of appendicitis include abdominal pain, particularly over the right lower quadrant; rebound tenderness similar to that experienced with peritonitis; nausea; and vomiting.

Appendicitis requires immediate treatment, which is typically surgical removal of the appendix. Left untreated, the appendix continues to swell and may rupture. This not only causes internal bleeding but also spills bacteria-filled fecal material into the peritoneal cavity, leading to peritonitis. These two conditions can be fatal if not properly managed.

22.5

The Large Intestine

Learning Outcomes

1. Describe the gross and microscopic anatomy of the divisions of the large intestine.
2. Describe the defecation reflex and the functions of the internal and external anal sphincters.
3. Discuss conscious control of the defecation reflex.

The **large intestine**, or *large bowel*, runs along the border of the abdominal cavity, surrounding the small intestine and other abdominal organs like a frame (**Figure 22.17a**). About 1.5 meters (5 feet) long, it is so named because it has a larger diameter than the small intestine. The large intestine receives material from the small intestine that was not digested or absorbed. You might think this means the small intestine does the "important" work and the large intestine is simply a passageway for the leftovers, known as **feces,** or *fecal matter.* However, the large intestine is very active in absorbing water and electrolytes. Absorption in the large intestine is critical for maintaining fluid, electrolyte, and acid-base homeostasis. When this absorption is disrupted, as with bacterial or viral gut infections, dehydration as well as electrolyte and acid-base disturbances may result.

The tasks of the large intestine also include secretion (primarily in the form of mucus), propulsion, and defecation. In addition, it houses numerous bacteria that perform important functions such as synthesizing vitamins. The upcoming sections examine the gross and microscopic anatomy of the large intestine, the bacteria that inhabit it, and its basic functions.

Gross Anatomy of the Large Intestine

The large intestine is made up of three segments: the cecum, the colon, and the rectum. As you read in the last module, the first portion of the large intestine is the **cecum,** a blind pouch (*cecum-* = "blind") that is intraperitoneal and located in the right lower quadrant of the abdomen. The cecum features a smaller blind-ended pouch extended from its posteroinferior end, called the **vermiform appendix** (VER-muh-form uh-PEN-diks; "wormlike appendage"), which is generally shortened to simply *appendix* (see Figure 22.17a). The appendix was long considered a vestigial organ, or one left over from evolution that served no function. However, it is now understood that it houses multiple lymphatic nodules and plays a role in the immune system. Despite its small size, an inflamed appendix can lead to big problems, as discussed in *A&P in the Real World: Appendicitis.*

The next and longest portion of the large intestine is the **colon** (KOH-lun), which is divided into four portions:

● **Ascending colon.** The retroperitoneal **ascending colon** travels superiorly along the right side of the abdomen from the right lower quadrant to the right upper quadrant. When it reaches the liver, it makes a sharp left-hand turn at a junction called the **hepatic flexure,** also known as the *right colic flexure.*

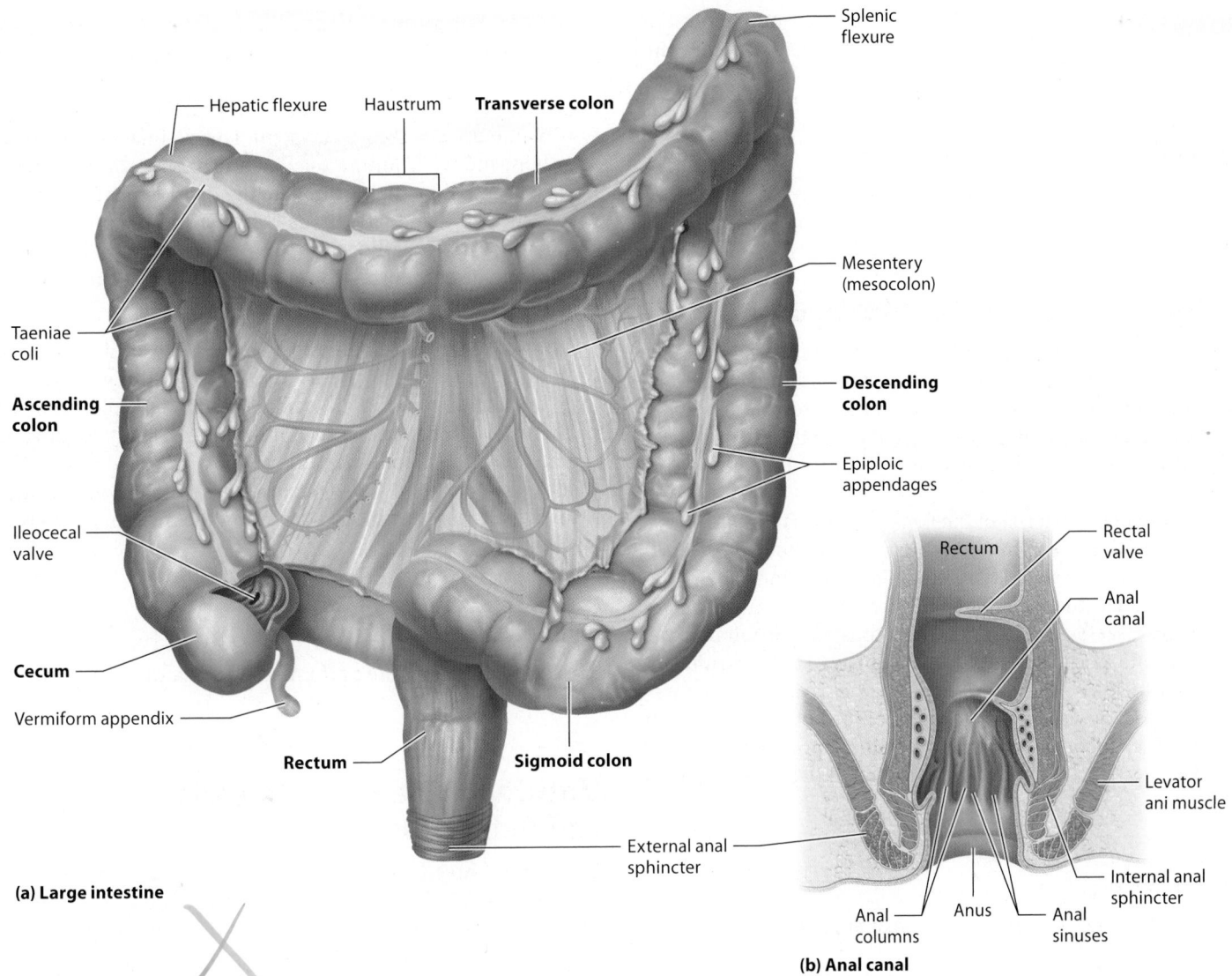

(a) **Large intestine**

(b) **Anal canal**

Figure 22.17 Gross anatomy of the large intestine.

- **Transverse colon.** At the hepatic flexure, the ascending colon becomes the intraperitoneal **transverse colon,** so named because it passes transversely across the superior abdominal cavity. At the spleen, it takes a sharp turn inferiorly at a junction called the **splenic flexure,** also known as the *left colic flexure.*
- **Descending colon.** The splenic flexure gives rise to the retroperitoneal **descending colon,** which passes along the left side of the abdominal cavity.
- **Sigmoid colon.** In the left lower quadrant, the descending colon becomes the S-shaped **sigmoid colon** (*sigmoid* = "S-shaped"), which is intraperitoneal and passes toward the sacrum.

After the sigmoid colon, the large intestine continues as the **rectum** (REK-tum). This portion of the large intestine runs anterior to the sacrum and is retroperitoneal. The walls of the rectum feature horizontal folds called **rectal valves,** which allow the passage of *flatus* (gas) without risking the simultaneous passage of feces.

The rectum ends at the **anal canal,** the last portion of the large intestine (**Figure 22.17b**). Recall from the muscular system chapter that the anal canal passes through the levator ani muscle in the floor of the pelvic cavity (see Chapter 9). The walls of the anal canal feature longitudinal grooves called **anal columns.** Between the anal columns are **anal sinuses,** which secrete mucus when feces pass through the anal canal during defecation. The terminal portion of the anal canal has two sphincters. The first is the involuntary **internal anal sphincter,** which is simply the thickened circular layer of the muscularis externa. The internal anal sphincter is supplied by parasympathetic motor neurons. The second is the voluntary **external**

anal sphincter, which is composed of skeletal muscle. As this sphincter is voluntary, it is innervated by somatic motor neurons controlled by the cerebral cortex.

Quick Check

☐ 1. What are the regions of the large intestine?
☐ 2. How do the internal and external anal sphincters differ?

Histology of the Large Intestine

The mucosa of the large intestine lacks villi and its cells lack microvilli. These structural adaptations reflect the fact that nutrient absorption is not the large intestine's primary function. Like most of the alimentary canal, its mucosa is rich with goblet cells that secrete protective and lubricating mucus.

The muscularis externa of the large intestine is unique in that its longitudinal layer is not continuous throughout most of its length. Instead, this layer is gathered into three bands or ribbons of muscle called **taeniae coli** (TEE-nee-ee KOHL-ahy; *taeniae* = "ribbons"; see Figure 22.17a). Their constant tension bunches the colon into pockets referred to as **haustra** (HAW-struh; singular, *haustrum*).

The serosa, or visceral peritoneum, of the large intestine contains fat-filled pouches known as **epiploic appendages** (ep-ih-PLOH-ik; *epiploic* = "membrane-covered"). Their function is unknown.

Quick Check

☐ 3. What are the taeniae coli, and what are their functions?

Bacteria in the Large Intestine

The large intestine hosts a staggering number of bacteria; indeed, there are about 10 times more bacteria in the large intestine than cells in the human body, and bacteria make up as much as 60% of the dry mass of feces. These bacteria, the **normal flora** or **gut flora,** consist of about 500 different bacterial species that coexist with humans in a *symbiotic* (mutually beneficial) relationship. Humans provide the bacteria with the environment they need to survive, and the bacteria perform a number of useful functions for humans, including the following:

- **Produce vitamins.** Bacteria produce vitamins such as vitamin K, which is necessary for blood clotting.
- **Metabolize undigested materials.** Bacteria metabolize carbohydrates such as soluble fibers that the small intestine is unable to digest, converting them into fatty acids and other molecules the body can absorb and use. This also aids in the absorption of certain vitamins and electrolytes. A somewhat unfortunate byproduct of this metabolism is the production of gas within the intestine that is released as flatus.

- **Deter the growth of harmful bacteria.** The normal flora prevent the growth of pathogenic, or disease-causing, microorganisms by competing for nutrients and producing chemicals that kill certain harmful bacterial species.
- **Stimulate the immune system.** During infancy, the normal flora induce immune tolerance to their own antigens. At the same time, they stimulate the development of mucosa-associated lymphatic tissue (MALT) and the production of antibodies to pathogens. This creates a favorable environment for the normal flora while also protecting the host from pathogenic bacteria.

Long-term antibiotic therapy, particularly antibiotics taken by mouth, can disrupt these functions by killing large numbers of normal flora. This can result in multiple problems, particularly infection with antibiotic-resistant pathogenic organisms such as the bacterium *Clostridium difficile*. Note that this situation may also occur in individuals who unintentionally consume antibiotics by eating the meat of animals fed antibiotics.

Quick Check

☐ 4. What functions are performed by the large intestine's normal flora?

Motility of the Large Intestine and Defecation

The large intestine has two functional segments: proximal and distal. The *proximal large intestine* consists of the ascending and transverse colon, and the *distal large intestine* consists of the descending and sigmoid colon, rectum, and anal canal.

The proximal large intestine is the primary site of water and electrolyte absorption and bacterial activity, and exhibits two main types of motility. The first is a type of segmentation, or churning, similar to what we saw in the small intestine, in which the circular muscle of each haustrum contracts repeatedly. This swirls the material around in the haustrum, which aids in water and electrolyte absorption. These contractions are controlled primarily by local neurons of the ENS and are triggered by stretch.

The other type of motility in the proximal large intestine is a propulsive motion known as a **mass movement,** or **mass peristalsis.** During a mass movement, multiple haustra undergo peristalsis, which propels their contents toward the distal large intestine. Mass movements occur three to four times per day, and appear to be triggered by food consumption, which initiates reflexes controlled by the ENS.

The distal large intestine performs a small amount of absorption, primarily of water, but its main role is to store fecal material until it is ready to be expelled during defecation. For this reason, the distal large intestine is much less motile than the proximal large intestine. When mass

Involuntary reflex:

Voluntary control:

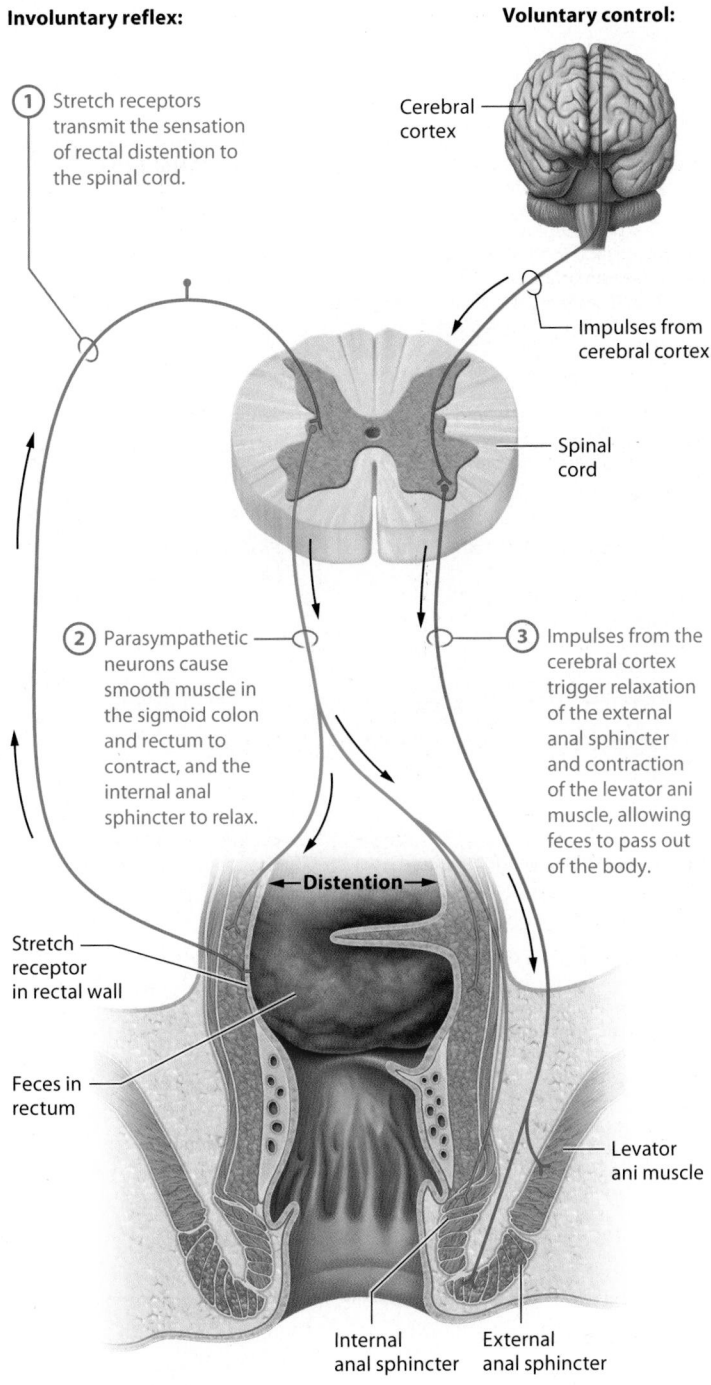

① Stretch receptors transmit the sensation of rectal distention to the spinal cord.

Cerebral cortex

Impulses from cerebral cortex

Spinal cord

② Parasympathetic neurons cause smooth muscle in the sigmoid colon and rectum to contract, and the internal anal sphincter to relax.

③ Impulses from the cerebral cortex trigger relaxation of the external anal sphincter and contraction of the levator ani muscle, allowing feces to pass out of the body.

←Distention→

Stretch receptor in rectal wall

Feces in rectum

Levator ani muscle

Internal anal sphincter

External anal sphincter

Figure 22.18 Defecation reflex.

movements force fecal material into the normally empty rectum, it initiates the parasympathetic-mediated **defecation reflex** (**Figure 22.18**):

① **Stretch receptors transmit the sensation of rectal distention to the spinal cord.** The walls of the rectum house stretch receptors that are activated by the presence of fecal material stretching the wall. Sensory neurons transmit these impulses to the spinal cord.

② **Parasympathetic neurons cause smooth muscle in the sigmoid colon and rectum to contract, and the internal anal sphincter to relax.** In an infant or young child, relaxation of the internal anal sphincter results in release of fecal matter from the anal canal. In older children and adults, the defecation reflex ends here and there is conscious contraction of the external anal sphincter if it is not an appropriate time to defecate. The reflex is initiated again with the next mass movement, and steps 1 and 2 repeat. However, if the time is appropriate for defecation to occur, then we proceed to the next step:

③ **Impulses from the cerebral cortex trigger relaxation of the external anal sphincter and contraction of the levator ani muscle, allowing feces to pass out of the body.** The cerebral cortex stimulates voluntary relaxation of the external anal sphincter. The levator ani muscle also contracts, which elevates the anus. These three movements combined allow the feces to pass. Note that this activity is assisted by contraction of the abdominal muscles against a closed glottis, a procedure called the *Valsalva maneuver*, which increases intra-abdominal pressure.

Anything that increases or decreases the motility of the large intestine affects the amount of water present in feces. When motility increases, the large intestine does not have enough time to absorb water from fecal material. This produces watery feces, a condition known as *diarrhea* (dy-ah-REE-ah). Factors that may increase motility include irritation of the colon due to bacterial or viral infections or drugs that stimulate the parasympathetic nervous system. Conversely, when motility decreases, the large intestine absorbs too much water and the fecal material becomes hard, a condition called *constipation*. The large intestine's motility may be slowed by drugs such as opiate narcotics or those that block the effects of acetylcholine (ACh).

Figure 22.19 summarizes the structural and functional properties of the large intestine and other organs of the alimentary canal.

Quick Check

☐ 5. Why is control of motility in the large intestine important to fluid balance?

☐ 6. Which types of movements occur in the proximal large intestine?

☐ 7. Summarize the steps of the defecation reflex.

Apply What You Learned

☐ 1. Patients taking oral antibiotics for a bacterial infection are often advised to take supplements containing bacteria present in the normal flora. Why are patients counseled to do this?

☐ 2. What might happen to step 3 of the defecation reflex (see Figure 22.18) if the spinal cord were cut? Explain.

See answers in Appendix A.

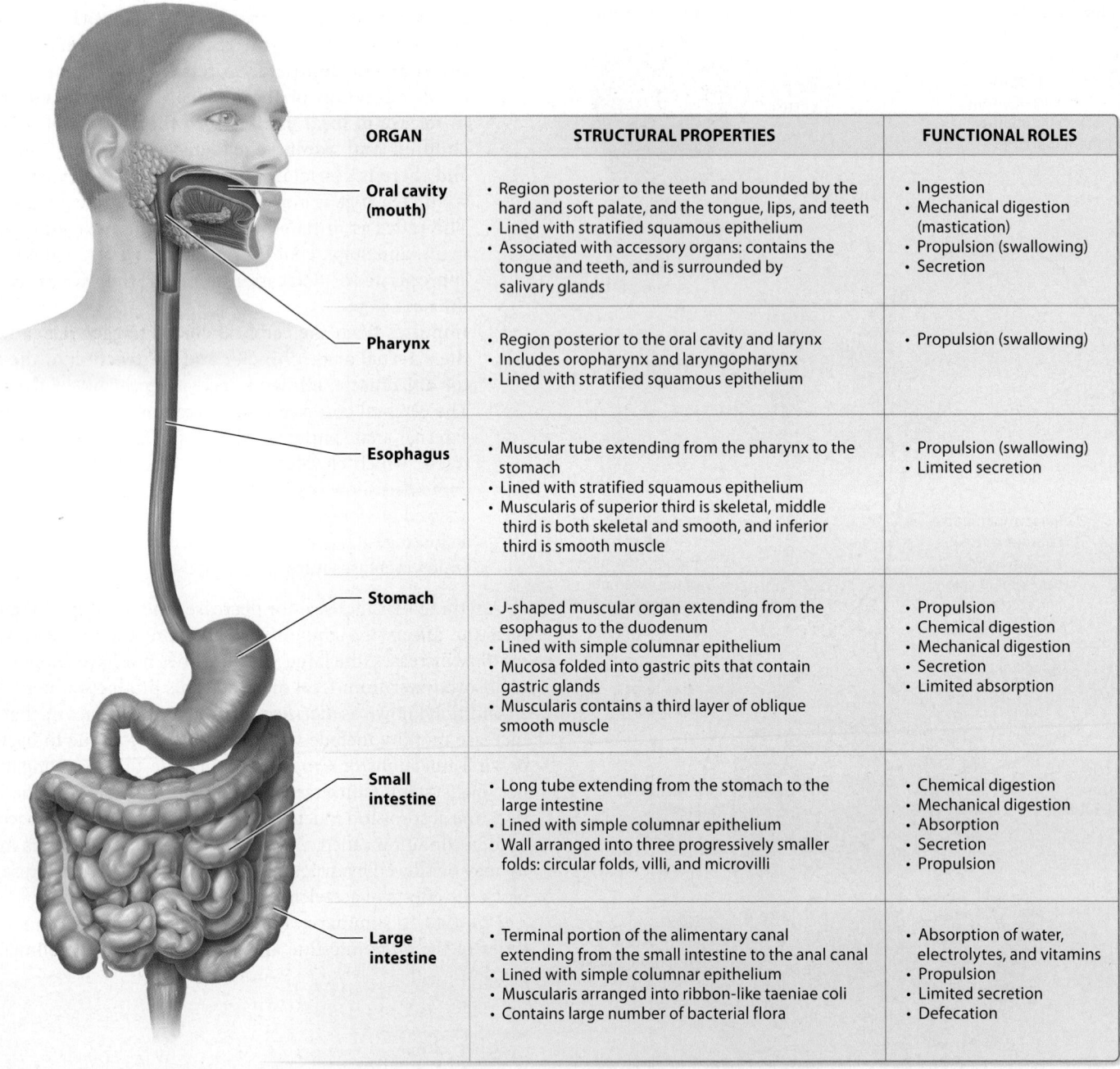

ORGAN	STRUCTURAL PROPERTIES	FUNCTIONAL ROLES
Oral cavity (mouth)	• Region posterior to the teeth and bounded by the hard and soft palate, and the tongue, lips, and teeth • Lined with stratified squamous epithelium • Associated with accessory organs: contains the tongue and teeth, and is surrounded by salivary glands	• Ingestion • Mechanical digestion (mastication) • Propulsion (swallowing) • Secretion
Pharynx	• Region posterior to the oral cavity and larynx • Includes oropharynx and laryngopharynx • Lined with stratified squamous epithelium	• Propulsion (swallowing)
Esophagus	• Muscular tube extending from the pharynx to the stomach • Lined with stratified squamous epithelium • Muscularis of superior third is skeletal, middle third is both skeletal and smooth, and inferior third is smooth muscle	• Propulsion (swallowing) • Limited secretion
Stomach	• J-shaped muscular organ extending from the esophagus to the duodenum • Lined with simple columnar epithelium • Mucosa folded into gastric pits that contain gastric glands • Muscularis contains a third layer of oblique smooth muscle	• Propulsion • Chemical digestion • Mechanical digestion • Secretion • Limited absorption
Small intestine	• Long tube extending from the stomach to the large intestine • Lined with simple columnar epithelium • Wall arranged into three progressively smaller folds: circular folds, villi, and microvilli	• Chemical digestion • Mechanical digestion • Absorption • Secretion • Propulsion
Large intestine	• Terminal portion of the alimentary canal extending from the small intestine to the anal canal • Lined with simple columnar epithelium • Muscularis arranged into ribbon-like taeniae coli • Contains large number of bacterial flora	• Absorption of water, electrolytes, and vitamins • Propulsion • Limited secretion • Defecation

Figure 22.19 Summary of the structure and function of the organs of the alimentary canal.

MODULE **22.6**

The Pancreas, Liver, and Gallbladder

Learning Outcomes

1. Describe the gross and microscopic structure of the pancreas and its digestive functions.
2. Describe the gross and microscopic structures of the liver and gallbladder.
3. Describe the functions of the liver pertaining to digestion.
4. Explain the structural and functional relationship between the liver and the gallbladder.
5. Explain how pancreatic and biliary secretions are regulated.

We turn now to the structure and function of the final three organs of the digestive system: the pancreas, liver, and gallbladder. Like the salivary glands, these accessory organs are *exocrine glands* that secrete a product through a duct to the outside of the body (see Chapter 4). You might find this strange, because the accessory organs secrete their products into the alimentary canal, and the alimentary canal seems to be inside the body.

However, the lining of the alimentary canal is actually considered an external body surface because it is open to the outside on both ends. Think of it like a paper towel roll—its outer surface is akin to the body's skin, and its inner surface is akin to the alimentary canal. Both its surfaces are in contact with the outside environment and thus are considered "external."

This module explores the structures and functions of the pancreas, liver, and gallbladder. The neural and hormonal signals that regulate their secretion are also discussed. Module 22.7 examines the secretory products of each organ and their roles.

The Pancreas

◀◀ Flashback

1. What are the endocrine functions of the pancreas? (p. 614)

Recall from the endocrine system chapter that the **pancreas** is a gland with both endocrine and exocrine functions (see Chapter 16). Its endocrine secretions, the hormones *insulin* and *glucagon*, are released from pancreatic islets into the blood and affect most cells in the body. Its exocrine secretions are enzymes secreted primarily by clusters of acinar cells. These enzymes are released into ducts of the pancreas that empty into the alimentary canal, where they help with digestion. We examine the structure of the pancreas and the acinar cells, as well as their secretions, in the next subsections.

Anatomy and Histology of the Pancreas

The pancreas is mostly located in the left upper quadrant of the abdomen, where it extends from the duodenum to the spleen; most of it is retroperitoneal. It has three portions (**Figure 22.20a**): (1) a wide **head** that contacts the duodenum, (2) a middle **body,** and (3) a thinner **tail** that tapers off toward the spleen. Running down the middle of the pancreas is the **main pancreatic duct,** which receives secretions from acinar cells. Near the duodenum, the main pancreatic duct merges with the duct from the liver and gallbladder, after which it empties into the duodenum.

Acinar cells are modified simple cuboidal epithelial cells. They are found in clusters known as **acini** (ASS-uh-nye; singular, *acinus;* **Figure 22.20b**). Each acinus surrounds a small duct into which its cells secrete their products. The duct cells also secrete products, discussed shortly. Most of these small ducts merge and drain into the main pancreatic duct, although some secrete into the smaller *accessory pancreatic duct.*

Pancreatic Juice

The collective secretions of the pancreatic acinar and duct cells are known as **pancreatic juice.** It consists of water and multiple digestive enzymes and other proteins. In addition, the duct cells secrete bicarbonate ions, a base, which make pancreatic juice alkaline. This helps to neutralize the acidic chyme that enters the duodenum from the stomach and protects the duodenum from damage by the acid. The digestive enzymes, secreted by acinar cells, are crucial in chemical digestion and catalyze the

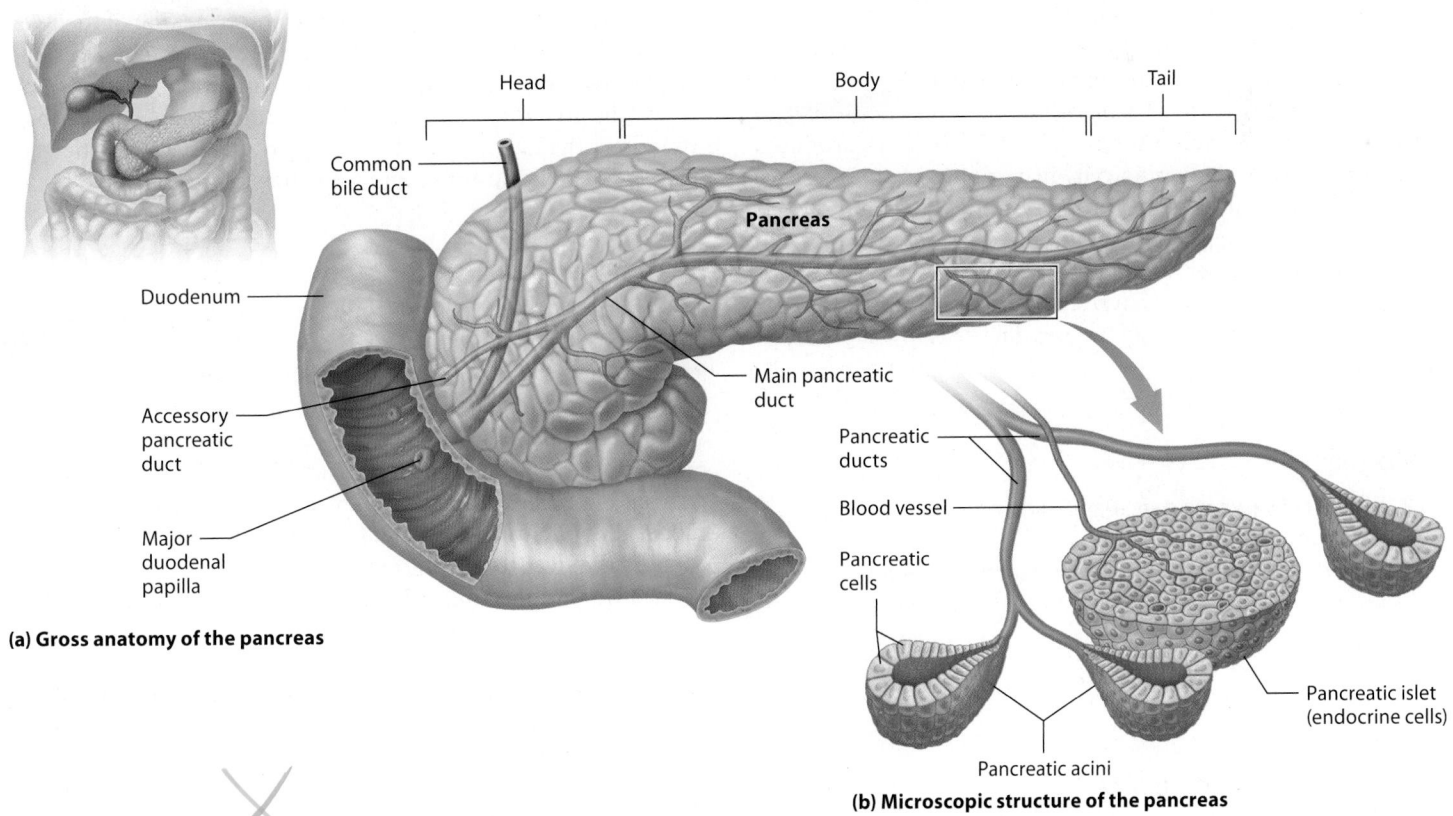

Head   Body   Tail

Common bile duct

Pancreas

Duodenum

Accessory pancreatic duct

Major duodenal papilla

Main pancreatic duct

Pancreatic ducts

Blood vessel

Pancreatic cells

Pancreatic islet (endocrine cells)

Pancreatic acini

(a) Gross anatomy of the pancreas

(b) Microscopic structure of the pancreas

Figure 22.20 Anatomy and histology of the pancreas.

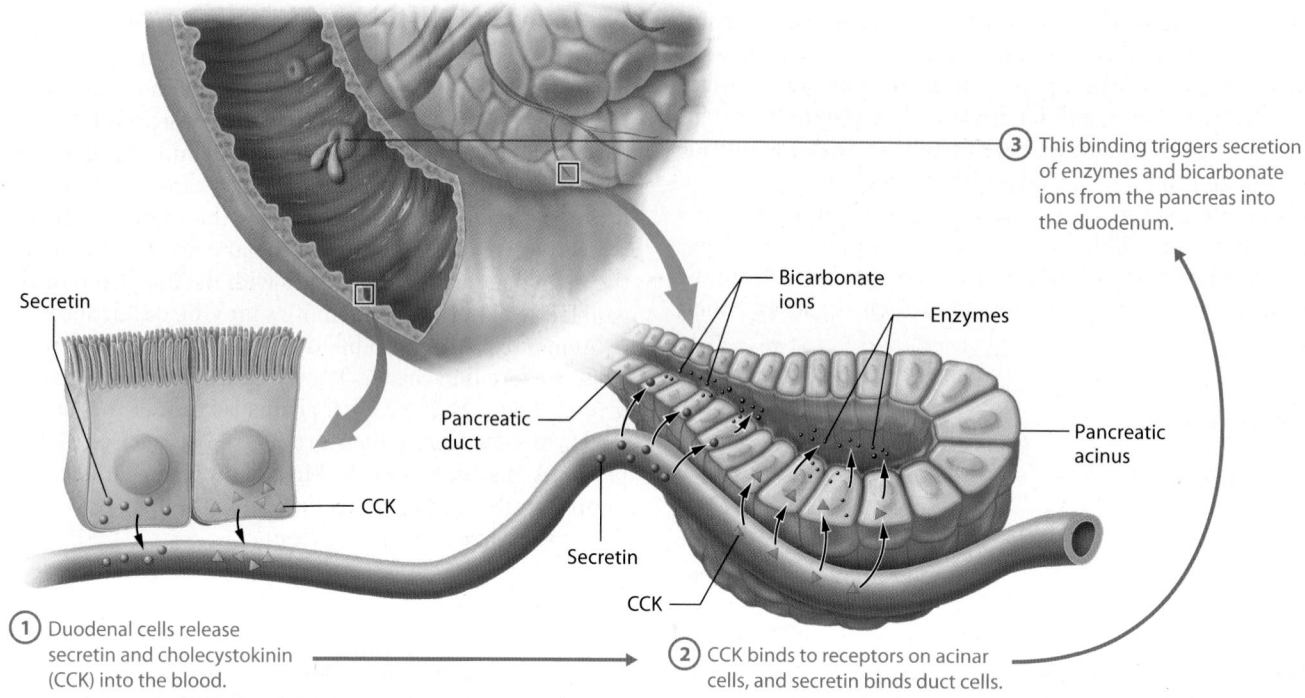

③ This binding triggers secretion of enzymes and bicarbonate ions from the pancreas into the duodenum.

Secretin

Bicarbonate ions

Enzymes

Pancreatic duct

Pancreatic acinus

CCK

Secretin

CCK

① Duodenal cells release secretin and cholecystokinin (CCK) into the blood.

② CCK binds to receptors on acinar cells, and secretin binds duct cells.

Figure 22.21 Secretion of pancreatic juice.

reactions that digest carbohydrates, proteins, lipids, and nucleic acids. Module 22.7 discusses the individual enzymes.

Like gastric secretion, pancreatic secretion occurs at a basal rate between meals. During eating, pancreatic secretion rises due to parasympathetic and hormonal stimulation (**Figure 22.21**). One hormonal mediator of pancreatic secretion is **cholecystokinin** (kohl′-eh-sih-stoh-KY-nin), or **CCK.** CCK is produced by duodenal enteroendocrine cells in response to the presence of lipids and partially digested proteins in the duodenum. It acts on acinar cells to trigger the secretion of digestive enzymes and other proteins. Another stimulatory hormone is *secretin,* which is released by duodenal cells in response to acid and lipids in the duodenum. Secretin primarily triggers duct cells to secrete bicarbonate ions.

Quick Check

☐ 1. Which cells secrete pancreatic juice, and what are its components?
☐ 2. How do CCK and secretin affect pancreatic secretion?

The Liver and Gallbladder

◀◀ **Flashback**

1. What are sinusoids? (p. 687)
2. Which veins drain into the hepatic portal vein? (p. 707)
3. What is the hepatic portal system? (p. 707)

The pyramid-shaped **liver,** one of the largest organs in the body, is located in the right upper quadrant nestled against the inferior surface of the diaphragm. On the posterior side of the liver, we find the small sac known as the **gallbladder**. The liver and gallbladder have close anatomical and functional relationships, so we discuss them together. In the upcoming subsections, we look at the anatomy and histology of the liver and the gallbladder, their basic functions, and their main digestive secretion, *bile.*

Anatomy of the Liver

The liver is wrapped in a thin connective tissue capsule, and most of it is covered by the visceral peritoneum. It is composed of four lobes: the large **right lobe** and **left lobe** (**Figure 22.22a**)—located on the right and left sides of the liver, respectively—and the smaller **caudate lobe** (KAW-dayt) and **quadrate lobe,** located on the posterior side of the liver's right lobe (**Figure 22.22b**). The right and left lobes are separated by a fold of visceral peritoneum called the **falciform ligament** (FALL-suh-form), which also anchors the liver to the anterior abdominal wall. On its inferior surface we find the **round ligament,** a remnant of the *umbilical vein* that was present in the fetus.

On the liver's posterior side we find an indentation called the **porta hepatis** (POR-tuh heh-PAH-tis; *port-* = "gate," *hepat-* = "liver"). Numerous blood vessels enter and exit the liver at the porta hepatis, including the *hepatic artery,* which brings oxygen-rich blood to the liver; and the *hepatic portal vein,* which brings nutrient-rich, deoxygenated blood to the liver from multiple abdominal organs. In addition to blood vessels, we find nerves,

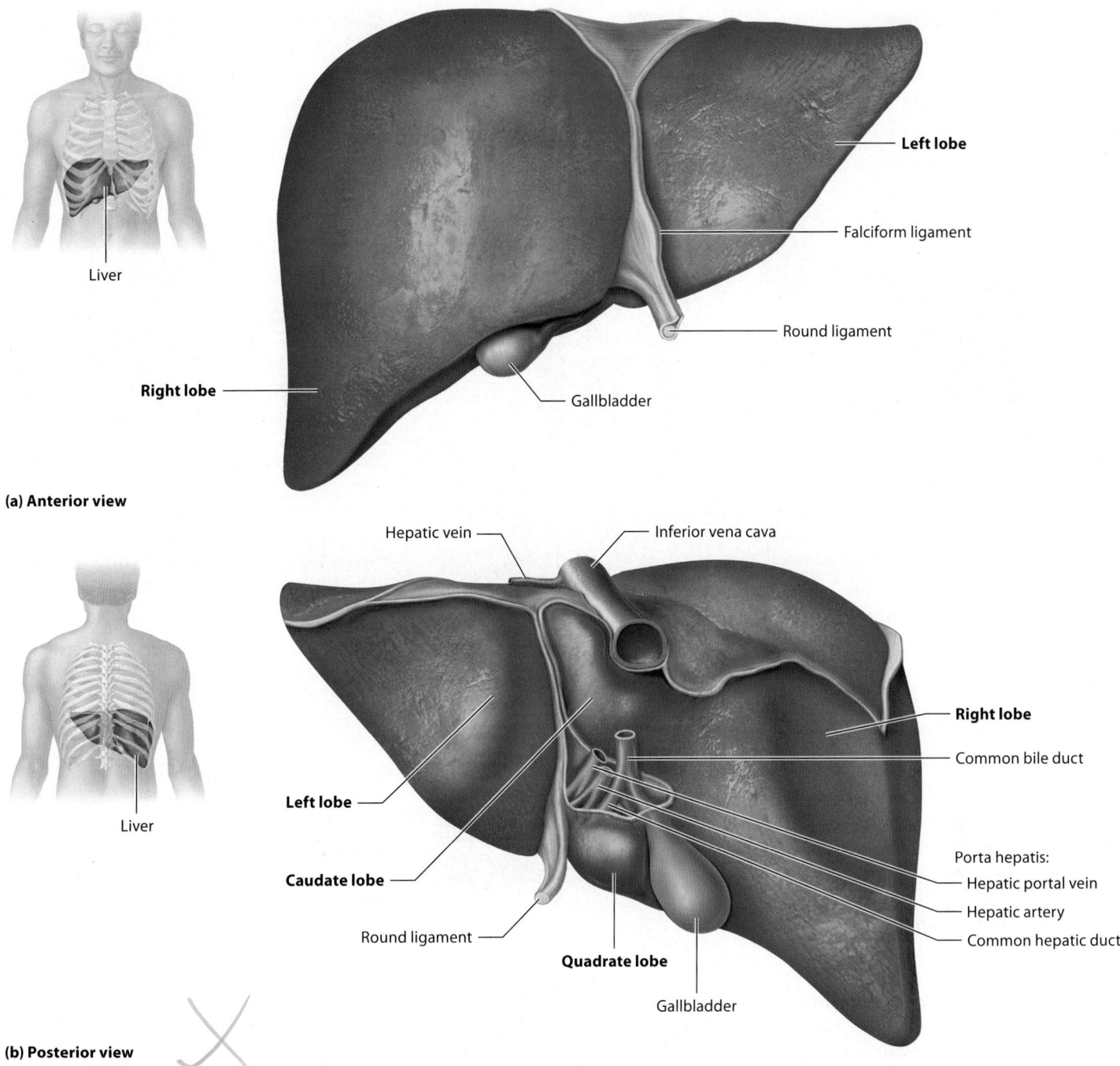

(a) Anterior view

Liver

Left lobe

Falciform ligament

Round ligament

Gallbladder

Right lobe

Hepatic vein

Inferior vena cava

Right lobe

Common bile duct

Left lobe

Caudate lobe

Round ligament

Quadrate lobe

Gallbladder

Porta hepatis:
Hepatic portal vein
Hepatic artery
Common hepatic duct

Liver

(b) Posterior view

Figure 22.22 Gross anatomy of the liver.

lymphatic vessels, and the *common hepatic duct* entering and exiting the porta hepatis. Note that the hepatic veins, which drain blood from the sinusoids, empty into the inferior vena cava on the liver's superior surface.

Histology of the Liver

The basic unit of the liver is the **liver lobule** (LOB-yool; **Figure 22.23**). Liver lobules are separated from one another by septa that branch in from the connective tissue capsule of the liver. A liver lobule is composed of flattened plates of cells, called **hepatocytes** (heh-PAH-toh-sytz), arranged in the shape of a hexagon and stacked on

one another. In the center of the lobule is a small **central vein.** At each of the six corners of the lobule, we find three structures collectively referred to as **a portal triad:** (1) a branch of the hepatic artery called a *hepatic arteriole,* (2) a branch of the portal vein called a *portal venule,* and (3) a small **bile duct** that carries bile.

The hepatic arteriole and portal venule both drain into large, leaky capillaries, the **hepatic sinusoids,** that pass between rows of hepatocytes. Recall from the blood vessels chapter that sinusoids have walls that are not continuous, which allows large substances to enter and exit (see Chapter 18). Blood flows slowly through the sinusoids as materials are exchanged between the

Bile canaliculus

Hepatic sinusoids

Liver lobule

Central vein

Portal triad:
Bile duct
Portal venule
Hepatic arteriole

Blood from the hepatic arterioles and portal venules flows through hepatic sinusoids and then enters the central vein.

Hepatocytes
Hepatic sinusoids

Bile canaliculus

(b) Photomicrograph of a liver lobule from a pig

LM (20×)

Bile formed by hepatocytes drains into a bile canaliculus, which empties into a bile duct.

Bile duct
Portal venule
Hepatic arteriole

(a) Liver lobule

Figure 22.23 The structure of a liver lobule.

blood and hepatocytes, eventually draining into the central vein. The central veins merge and drain into the hepatic veins, which in turn feed into the inferior vena cava. Notice that bile flows through the liver lobule in the opposite direction, from the hepatocytes into tiny ducts called *bile canaliculi* (kah′-nah-LIK-yoo-lye; singular, *canaliculus*), which eventually drain into a bile duct.

Functions of the Liver

The liver has one of the most diverse sets of functions of any organ in the body. Like the pancreas, it releases both endocrine and exocrine secretions, and it converts harmful chemicals into nontoxic substances that the body can eliminate. We take a closer look at the functions of the liver in the next two subsections.

Bile Production The liver's main digestive function is to produce **bile,** a liquid that contains multiple components, including water, electrolytes, and organic compounds. Bile serves two

critical functions: (1) It is required for the digestion and absorption of lipids; and (2) it is the mechanism by which the liver excretes wastes and other substances that the kidneys cannot excrete. One of its main organic compounds is **bile salts,** which are derived from cholesterol. Bile salts are *amphiphilic,* meaning they have both polar and nonpolar parts. This allows them to interact with both lipids and the watery environment of the small intestine. When bile is released into the duodenum, the bile salts coat the lipids and physically break them apart into smaller pieces, a process known as *emulsification* (ee-mul′-sih-fih-KAY-shun). Although emulsification is mechanical digestion, it is necessary for the chemical digestion and absorption of lipids, which we further explore in Module 22.7.

Bile also contains varying amounts of materials that the liver excretes, including cholesterol, waste products, and harmful chemicals (toxins) such as heavy metals. Many of these products are not reabsorbed by the small or large intestines and so pass into the feces. An important example of the liver's excretory function involves *bilirubin* (bih-LIH-roo-bin), a waste product that results from the breakdown of hemoglobin by the spleen. Hepatocytes secrete bilirubin into bile, and the normal flora in the large intestine convert it to *urobilinogen* (yoor'-oh-bih-LIN-oh-jen). Bacteria convert most of the urobilinogen to *stercobilin,* which is brown and is responsible for the brown color of feces. A small portion of the urobilinogen is reabsorbed by the large intestine and ends up back in the hepatic portal system. The majority of this urobilinogen is simply re-secreted into the bile, although a small portion remains in the blood and is excreted by the kidneys (note that yellow urobilinogen is largely responsible for the color of urine).

Other Functions of the Liver The liver performs a host of other functions, including the following (note that some of these functions are only peripherally related to the digestive system, so they are not discussed in detail here):

- **Nutrient metabolism.** The liver processes nutrients obtained from the diet. Carbohydrates and proteins absorbed from the alimentary canal are delivered to the liver by the hepatic portal vein, and lipids by the hepatic arteries. Within the liver, some of these nutrients are stored for later use, modified into another form, or used to synthesize other molecules. For example, some of the glucose taken in by the liver is stored in the form of glycogen, and many of the dietary amino acids are used to synthesize plasma proteins such as albumin and clotting proteins. We discuss the liver's role in multiple metabolic processes in the metabolism chapter (see Chapter 23).

- **Detoxification.** The liver detoxifies substances produced by the body, such as the previously mentioned bilirubin. In addition, the liver processes substances that we eat or drink, some of which are toxins harmful to the body (such as alcohol). These substances are generally delivered to the liver first via the hepatic portal vein, where the liver converts them into less harmful materials that can be excreted in bile or in urine. The liver also metabolizes many drugs, such as antibiotics. Individuals with impaired liver function metabolize drugs more slowly than those with normal liver function, and the drugs remain in their systems much longer. Does the liver ever need help with detoxification? For more information, see *A&P in the Real World: Do We Really Need to "Detox"?* on page 880.

- **Excretion.** Recall that the liver directly excretes bilirubin in bile. Several other substances the liver processes are excreted in bile, particularly drugs such as certain antibiotics. The liver also modifies substances so that they can be excreted by the kidneys.

The Gallbladder and Its Relationship to the Liver

The gallbladder, a small sac that sits on the posterior liver, receives most of the bile from the common hepatic duct (**Figure 22.24**). The gallbladder stores bile, concentrates it (removing water), and releases it when stimulated. Bile release is stimulated by the hormone CCK, which triggers contraction of the smooth muscle in the wall of the gallbladder. This causes the gallbladder to release bile into the **cystic duct** (SIS-tik).

As you can see in Figure 22.24, the cystic duct joins the common hepatic duct to form the **common bile duct.** The common bile duct joins the main pancreatic duct near the duodenum to form the **hepatopancreatic ampulla** (heh-PAH-toh-payn'-kree-at-ik am-POOL-uh). The ampulla is surrounded by a ring of smooth muscle called the **hepatopancreatic sphincter,** which controls the emptying of bile and pancreatic fluids into the duodenum. Recall that the contents of the hepatopancreatic ampulla then empty into the duodenum at the major duodenal papilla.

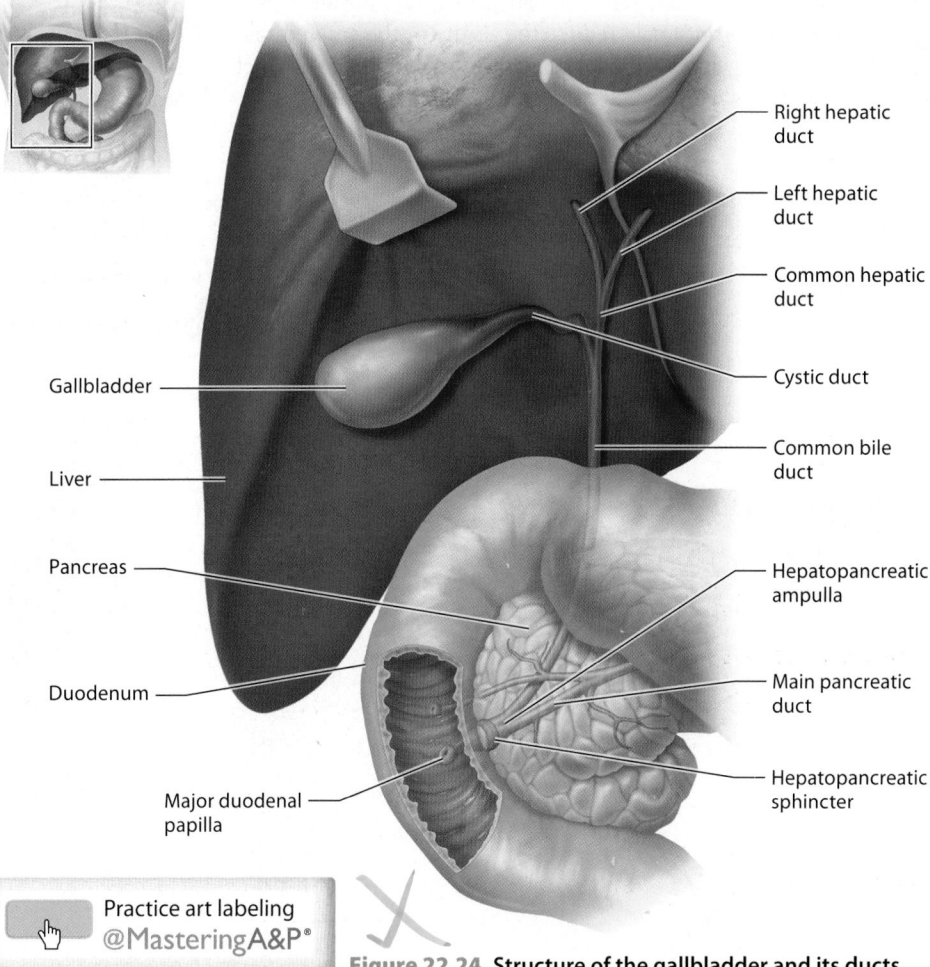

Practice art labeling
@MasteringA&P®

Figure 22.24 Structure of the gallbladder and its ducts.

Right hepatic duct

Left hepatic duct

Common hepatic duct

Cystic duct

Common bile duct

Gallbladder

Liver

Pancreas

Duodenum

Hepatopancreatic ampulla

Main pancreatic duct

Hepatopancreatic sphincter

Major duodenal papilla

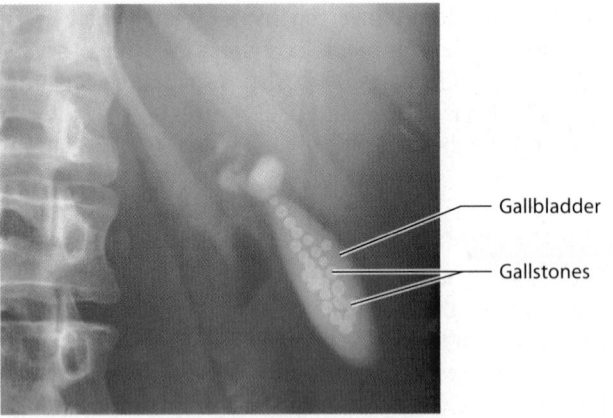

Figure 22.25 Radiograph of a gallbladder showing gallstones.

Bile contains calcium salts and cholesterol, and both can precipitate and form hard lumps called **gallstones** (Figure 22.25). Gallstones are generally asymptomatic, but occasionally they become lodged in the cystic duct or the common bile duct and block the outflow of bile. This results in abdominal pain, particularly when eating a fatty meal, and the presence of undigested fats in the feces. The feces also take on a clay color due to the lack of bilirubin excretion. Painful gallstones are generally treated by surgically removing the gallbladder, a procedure known as a *cholecystectomy* (koh'-leh-sis-TEK-toh-mee). The absence of a gallbladder may temporarily reduce the ability of the small intestine to digest fats. However, over time the common hepatic duct enlarges and takes over some of the gallbladder's functions.

A&P in the Real World

Do We Really Need to "Detox"?

The manufacturers of dietary supplements and alternative health products tell us that our bodies are toxic wastelands filled with disease-causing toxins that only their products can remove (for a price, of course). These "detoxification" products are big business, accounting for a large portion of the profits for the multibillion dollar alternative health industry. Many people ask if these products work, but a better question is whether we actually need them in the first place. So, are our bodies swimming with toxins? Do we all need to "detox"?

The simple answer is no, and we have our livers to thank for it. The liver is a magnificent organ that works tirelessly to convert harmful chemicals, or toxins, into nonharmful substances that can be excreted in bile or urine. There is simply no herbal supplement, foot pad, or magic bracelet that can even come close to doing the job our livers do every day. And what's more, the liver doesn't charge us for its services.

Bile Secretion

The liver produces bile continually but generally does not secrete it at a basal rate, the way pancreatic and gastric juices are released. Instead, bile secretion does not occur until the gallbladder contracts and the sphincter at the hepatopancreatic ampulla relaxes. This activity is mostly accomplished by cholecystokinin (CCK) and to a small extent by the vagus nerve (Figure 22.26). Other factors influence bile secretion, including the hormone secretin, which stimulates bile production and release by hepatocytes.

However, the most potent stimulus for bile production and release is bile itself, specifically bile salts. Recall from our earlier discussion of the liver that bile salts are reabsorbed in the last section of the ileum and transported back to the liver through the hepatic portal vein. As bile salts re-enter the liver, bile secretion rises dramatically in a positive feedback loop. Bile secretion continues into the duodenum until the duodenum empties, at which point CCK and secretin levels decline.

Figure 22.27 summarizes the structural and functional properties of the accessory digestive organs.

① Duodenal cells secrete cholecystokinin (CCK) and secretin into the blood.

② CCK triggers contraction of the gallbladder, which releases bile.

③ Bile salts and secretin trigger the release of more bile, an example of positive feedback.

Secretin

Liver
Gallbladder
Duodenal cells

Positive feedback

Bile
Duodenum

Bile salts

CCK

Secretin

Ileum

Figure 22.26 Secretion of bile.

ORGAN	STRUCTURAL PROPERTIES	FUNCTIONAL ROLES
Teeth	• Two sets: primary and secondary dentition • Consist of a crown above the gum and a root embedded in bone • Composed of inner pulp cavity surrounded by dentin, which in turn is surrounded by enamel or cementum	• Mechanical digestion (mastication)
Tongue	• Consists of skeletal muscle with overlaying stratified squamous epithelium • Surface contains papillae	• Mechanical digestion • Propulsion (swallowing) • Sense of taste
Salivary glands	• Three sets: parotid glands, submandibular glands, and sublingual glands • Consist of mucous cells and serous cells	• Secrete saliva, which assists in chemical digestion, deters the growth of harmful microorganisms, and moistens food to assist in swallowing and mechanical digestion • Chemical digestion of carbohydrates
Pancreas	• Consists of pancreatic acini, composed of acinar cells surrounding a duct	• Secretes enzymes that catalyze chemical digestion of lipids, carbohydrates, proteins, and nucleic acids • Secretes bicarbonate ions to neutralize acidic chyme
Liver	• Consists of hexagonal liver lobules surrounding a central vein • Liver lobules contain plates of hepatocytes	• Mechanical digestion (via bile production) • Excretion (excretes wastes in bile)
Gallbladder	• Muscular sac on the posteroinferior liver	• Mechanical digestion (stores, concentrates, and releases bile)

Figure 22.27 Summary of the structure and function of the accessory digestive organs.

Quick Check

☐ 3. What is a portal triad?
☐ 4. What are the main functions of the liver and gallbladder?
☐ 5. What are the primary triggers for bile release?

Apply What You Learned

☐ 1. The treatment for pancreatic cancer may require removing the pancreas. Predict the consequences of this procedure for the body's digestive functions.
☐ 2. A patient needs a medication that is normally given at a dose of 500 mg by mouth every 8 hours. However, the patient has liver disease, and the pharmacist recommends giving the drug at a dose of 250 mg every 12 hours instead. Explain why the pharmacist made this recommendation.

See answers in Appendix A.

MODULE **22.7**

Nutrient Digestion and Absorption

Learning Outcomes

1. Explain the process of enzymatic hydrolysis reactions of nutrient molecules.
2. Describe the enzymes involved in chemical digestion, including their activation, substrates, and end products.
3. Describe the process of emulsification, explain its importance, and discuss how bile salts are recycled.
4. Explain the processes involved in absorption of each type of nutrient, fat-soluble and water-soluble vitamins, and vitamin B12.

After covering the anatomy and basic functions of the digestive organs, we can now look more closely at two of the digestive processes: digestion and absorption. This module explores the digestion and absorption of dietary carbohydrates, proteins, lipids, and nucleic acids. The discussion then shifts to the absorption of water, electrolytes, and vitamins, which occurs in both the small and large intestines. But first, let's introduce some basic principles of digestion and absorption.

Overview of Digestion and Absorption

◀◀ **Flashback**

1. What is a hydrolysis reaction? (p. 50)

The process of digestion breaks food down into smaller and smaller pieces until only individual nutrient molecules remain. Recall that there are two types of digestion: mechanical digestion, in which food is physically broken into smaller parts, and chemical digestion, in which the chemical bonds between food molecules are broken. All nutrients go through the same mechanisms of mechanical digestion—mastication in the mouth, churning in the stomach, and segmentation in the small intestine. However, chemical digestion varies for each nutrient because digestive enzymes are specific for a single type of bond in a single type of nutrient.

Most digestive enzymes catalyze *hydrolysis* reactions—those that use a water molecule to break a bond between two molecules, as illustrated here (this example shows a peptide bond between two amino acids):

This is one reason why so much water is secreted with fluids like gastric and pancreatic juices—the water molecules are key components of the reactions that chemically break down food molecules. These hydrolysis reactions would occur without digestive enzymes, but they would occur far too slowly and we would be unable to extract nutrients from the food we eat. Our digestive enzymes speed up the reactions, a process known as *enzymatic hydrolysis*.

Once nutrients are digested to their component molecules, they must enter the body. Remember that the mucosa of the alimentary canal is actually an external membrane, so even though it might seem that the food is "inside" the body, it technically is not until it is absorbed. Nutrients and other molecules must be absorbed across the epithelial lining of the alimentary canal and enter the bloodstream before they can be delivered to body cells.

As with digestion, the process by which this occurs varies slightly for each nutrient, as explored in the upcoming sections.

ConceptBOOST »»»

Understanding Absorption in the Alimentary Canal

To travel from the lumen of the alimentary canal to the blood, a molecule must cross barriers. In **Figure 22.28**, you can see the path a typical molecule takes through an enterocyte of the small intestine. Note that sometimes small molecules such as water can cross between enterocytes instead of through them, but most molecules we discuss take the route in Figure 22.28.

The apical side of the plasma membrane—step 1 in Figure 22.28—is the most significant barrier to the absorption of ingested molecules. Most substances require a transport protein or channel to cross the membrane, and if no such protein exists, the substance will remain in the lumen and eventually be excreted. This can be beneficial and help to prevent absorption of harmful chemicals. One example is elemental mercury (the liquid mercury found in certain thermometers). In this form, mercury has great difficulty crossing a plasma membrane by diffusion, and no transporter or channel protein exists for it in the plasma membrane. So if elemental mercury is accidentally ingested, most of it passes straight through the alimentary canal unabsorbed. ■

Figure 22.28 Path taken by a molecule being absorbed into the blood through an enterocyte.

Digestion and Absorption of Carbohydrates

◀◀ Flashback

1. How do polysaccharides, disaccharides, and monosaccharides differ? (pp. 61–62)
2. How does secondary active transport allow molecules to cross the plasma membrane? (p. 81)
3. What is facilitated diffusion? (p. 76)

Carbohydrate digestion begins in the mouth with the help of salivary amylase from the salivary glands, which catalyzes the reactions that break long polysaccharides into shorter *oligosaccharides*. Salivary amylase generally accomplishes little actual chemical digestion because the food simply is not in the mouth long enough for it to have much effect. Some minor digestion by salivary amylase continues in the stomach, but stomach acid generally inactivates this enzyme before it can complete the task.

Chemical digestion of carbohydrates resumes in the small intestine (**Figure 22.29**). Here polysaccharides and oligosaccharides encounter a pancreatic enzyme called **pancreatic amylase**. This enzyme is very similar in structure and function to salivary amylase, and ① catalyzes the reactions that break the remaining polysaccharides into oligosaccharides.

Oligosaccharide digestion is completed by reactions catalyzed by enzymes that are products of enterocytes on the brush border of the small intestine (*brush border enzymes*), including the enzymes **lactase, maltase,** and **sucrase.** Lactase catalyzes only one reaction: the digestion of the sugar lactose (found in milk and milk-based products) into glucose and galactose (see *A&P in the Real World: Lactose Intolerance* on page 884). ② Both maltase and sucrase catalyze reactions that break oligosaccharides into disaccharides and monosaccharides. The end result of all this activity is the production of the monosaccharides glucose, fructose, and galactose.

After carbohydrates have been digested into monosaccharides, they are ready to be absorbed. Notice in Figure 22.29 that both glucose and galactose are transported across the enterocyte's apical membrane by the same mechanism. Note also that the concentrations of glucose and galactose are higher in the enterocyte's cytosol than they are in the lumen of the small intestine. This means that absorption of glucose and galactose requires these molecules to move *against* their concentration gradients.

In the cell chapter, we discussed the fact that any time we move a molecule against its concentration gradient, energy must be expended (see Chapter 3). The energy in this case comes via a secondary active transport mechanism known as the **Na⁺/ glucose cotransporter.** Recall that secondary active transport is a form of membrane transport in which ATP is used indirectly to move a molecule across the membrane. In this case, the Na⁺/K⁺

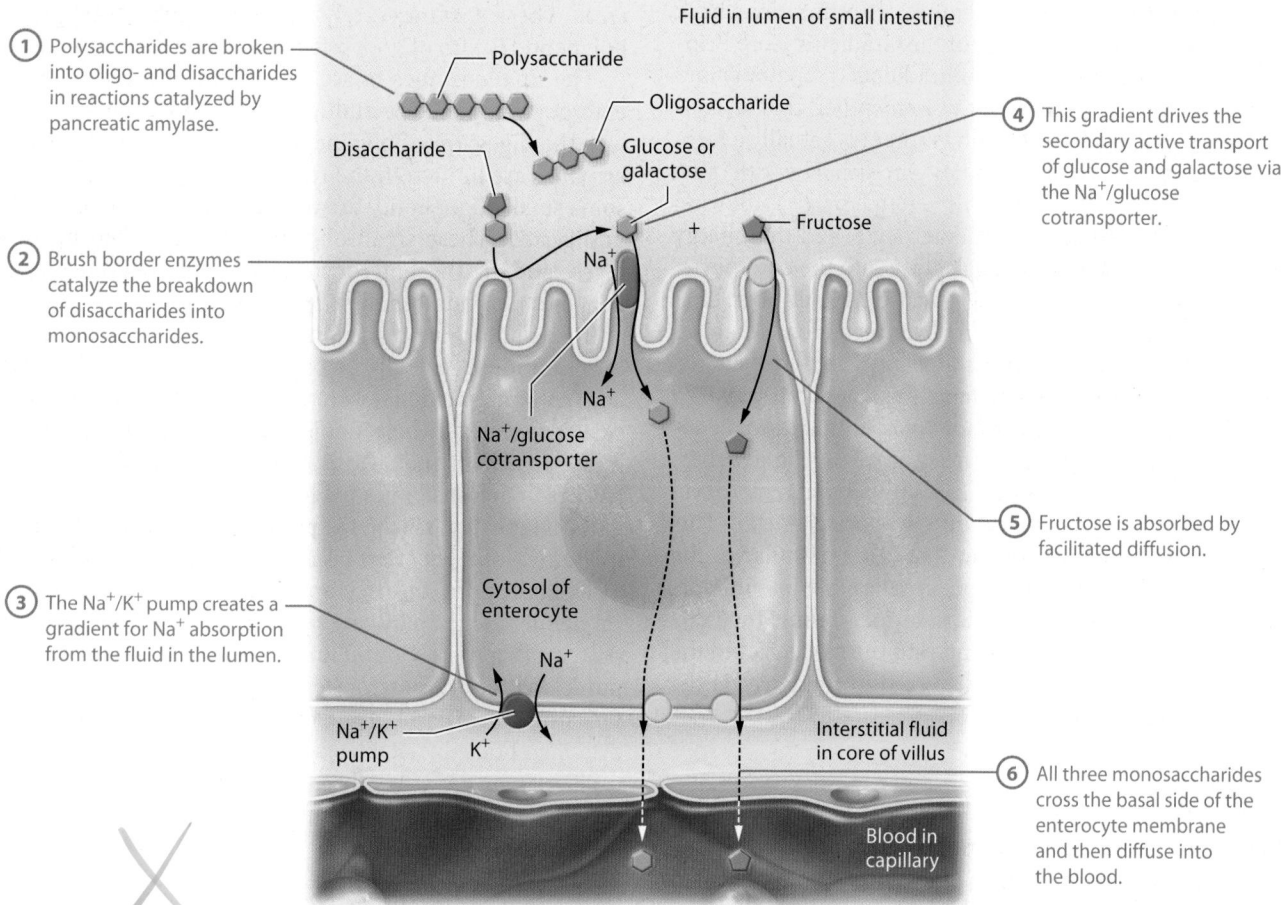

① Polysaccharides are broken into oligo- and disaccharides in reactions catalyzed by pancreatic amylase.

② Brush border enzymes catalyze the breakdown of disaccharides into monosaccharides.

③ The Na⁺/K⁺ pump creates a gradient for Na⁺ absorption from the fluid in the lumen.

Fluid in lumen of small intestine

Polysaccharide

Oligosaccharide

Disaccharide

Glucose or galactose

Fructose

Na⁺

Na⁺

Na⁺/glucose cotransporter

Cytosol of enterocyte

Na⁺

Na⁺/K⁺ pump

K⁺

Interstitial fluid in core of villus

Blood in capillary

④ This gradient drives the secondary active transport of glucose and galactose via the Na⁺/glucose cotransporter.

⑤ Fructose is absorbed by facilitated diffusion.

⑥ All three monosaccharides cross the basal side of the enterocyte membrane and then diffuse into the blood.

Figure 22.29 Carbohydrate digestion and absorption in the small intestine.

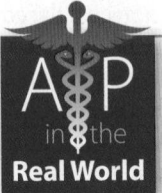

Lactose Intolerance

Many adults lack the enzyme lactase and as a result cannot digest the milk sugar lactose, a condition known as *lactose intolerance*. Most people produce lactase as infants, but production of the enzyme declines as we age, with up to 75% of adults showing decreased lactase activity. The percentage of lactose intolerance varies with ethnic heritage—it is found in only about 5% of the population of northern Europe, but about 90% of the population in certain regions of Asia.

Without lactase, disaccharides such as lactose cannot be absorbed into the enterocytes of the small intestine. The undigested lactose remains in the small intestine and is eventually delivered to the large intestine. Lactose in the large intestine acts as an osmotic particle that creates a gradient to draw water into the feces. This can lead to abdominal cramping and diarrhea when lactose is consumed. Lactose intolerance can be managed by avoiding lactose-containing foods or by taking lactase supplements.

pump consumes ATP to drive sodium ions out of the enterocyte and into the lumen of the small intestine. This creates a concentration gradient that favors the movement of sodium ions back into the enterocyte, another example of the Gradients Core Principle (p. 26) and of the importance of gradients in anatomy and physiology. Remember that a concentration gradient is actually a form of potential energy. The cell harnesses this energy by ④ allowing sodium ions to diffuse back into the enterocyte, and this energy drives the transport of glucose and galactose into the enterocyte.

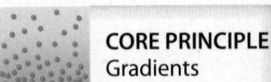

CORE PRINCIPLE
Gradients

Fructose is unable to bind the Na⁺/glucose cotransporter, so it crosses the apical enterocyte membrane by a separate mechanism (see Figure 22.29). ⑤ Fructose binds a channel that mediates its facilitated diffusion across the membrane. Facilitated diffusion is a passive process and requires no net input of energy, and so a concentration gradient must be present or fructose will not be absorbed.

⑥ All three monosaccharides cross the basal enterocyte membrane by the same facilitated diffusion mechanism. This process involves a membrane protein that is very similar to the one that helps fructose cross the apical enterocyte membrane. After the monosaccharides cross the basal membrane, they diffuse through the extracellular fluid and into the capillaries in the villus. Once in the blood, they are delivered to the liver via the hepatic portal vein for processing.

Quick Check

☐ 1. Where does carbohydrate digestion take place?
☐ 2. Which enzymes are involved in digesting carbohydrates?
☐ 3. How do glucose, galactose, and fructose cross the apical enterocyte membrane?

Digestion and Absorption of Proteins

◀◀ **Flashback**

1. What is endocytosis? (p. 82)
2. What are Peyer's patches? (p. 756)

Chemical digestion of proteins does not begin until they reach the stomach, where they encounter the enzyme pepsin. Recall that the chief cells of the gastric glands produce the inactive precursor pepsinogen. Pepsinogen requires a pH of about 2 to become pepsin, and pepsin is inactivated completely at a pH of 7. Activated pepsin catalyzes reactions that digest proteins into smaller polypeptides, oligopeptides, and some free amino acids. However, the reactions catalyzed by pepsin digest only about 10–15% of the proteins in ingested food. Individuals who do not produce pepsin do not appear to have any less ability to digest proteins.

The remainder of protein digestion takes place in the small intestine with the help of pancreatic enzymes and brush border enzymes (**Figure 22.30**). There are five pancreatic enzymes that digest proteins, all of which are released as inactive precursors. This protects the pancreas from *autodigestion,* or digestion of its cells by its own enzymes.

The first pancreatic enzyme to become activated is the precursor *trypsinogen* (trip-SIN-oh-jen), which becomes the active enzyme **trypsin** (TRIP-sin) when it encounters enzymes on the intestinal brush border. Trypsin in turn catalyzes the reactions that convert the other pancreatic enzymes to their active forms, and activates additional trypsinogen as it is secreted by the pancreas. These enzymes catalyze reactions that digest proteins and polypeptides into oligopeptides and some free amino acids.

The final enzymes to act on proteins are associated with the enterocytes. There are multiple brush border enzymes that catalyze the digestion of oligopeptides into free amino acids. These enzymes are limited in the reactions they are able to catalyze, so some small oligopeptides (two to three amino acids long) remain undigested. These small oligopeptides are taken up by enterocytes, and in the enterocyte cytoplasm they encounter more enzymes that catalyze their breakdown into free amino acids.

To be absorbed in the small intestine, proteins consumed for nutrition must generally be broken down into small oligopeptides and free amino acids. However, humans do have the ability to absorb small amounts of whole, undigested proteins involved in immunity by the process of endocytosis. Specialized cells overlying the Peyer's patches in the small intestine ingest these proteins and deliver them to the lymphatic tissue in the patches. This is particularly important in the development of the immune system in newborn infants.

Oligopeptides and free amino acids cross the enterocyte apical membrane primarily by secondary active transport membrane proteins that use a sodium ion gradient established by the Na⁺/K⁺ pump. Within the enterocyte, oligopeptides are broken down into free amino acids. The free amino acids then exit the basal enterocyte membrane by facilitated diffusion, after which they enter the capillaries in the villus. Like carbohydrates, the amino acids are then delivered to the liver for processing via the hepatic portal vein.

Quick Check

☐ 4. Where does protein digestion begin, and with which enzyme?

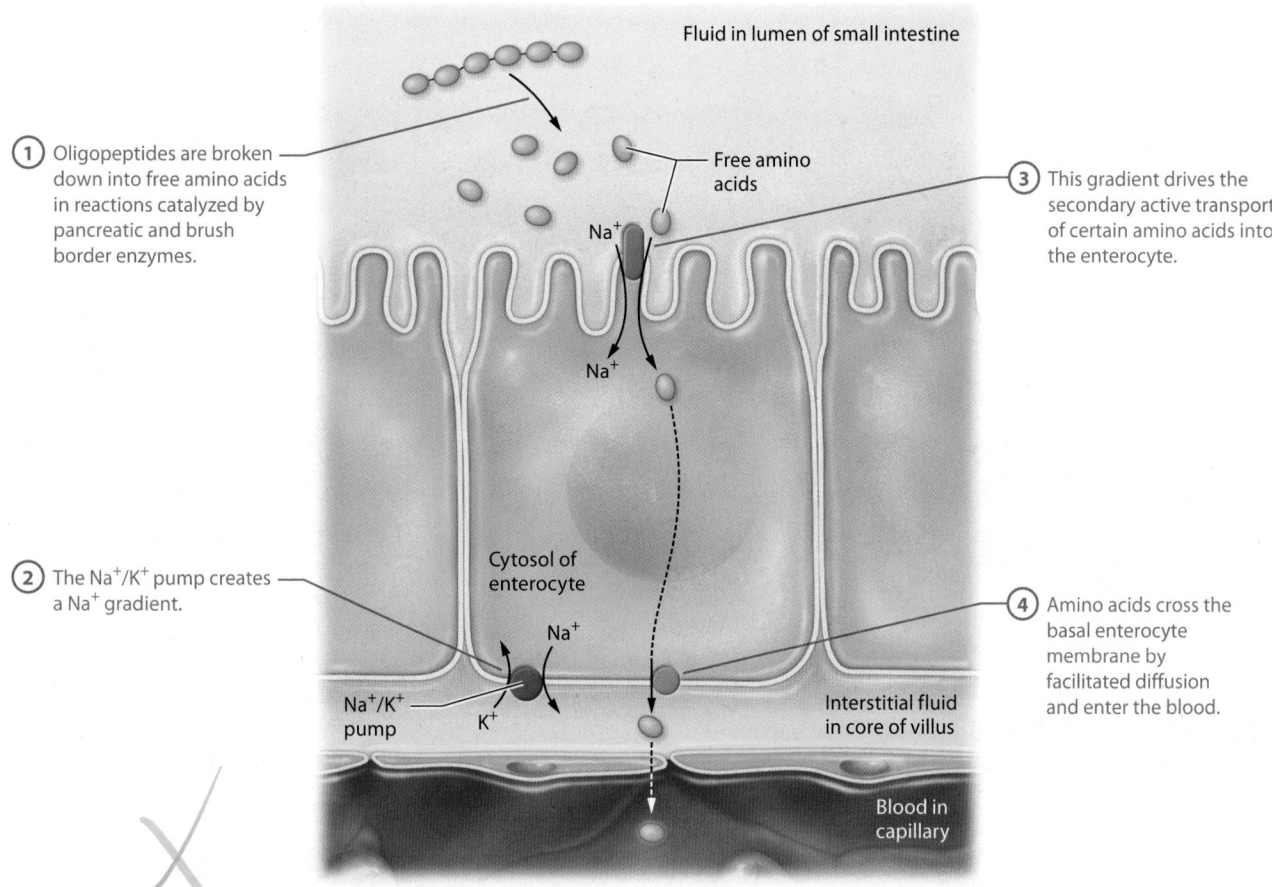

Figure 22.30 Amino acid digestion and absorption in the small intestine.

5. How does protein digestion occur in the small intestine?
6. How are amino acids absorbed across the apical and basal enterocyte membranes?

Digestion and Absorption of Lipids

◄◄ **Flashback**

1. What are triglycerides? (p. 52)
2. How do polar and nonpolar covalent bonds differ? (p. 39)
3. Which types of bonds are hydrophilic? Which bonds are hydrophobic? (p. 47)

The class of organic molecules known as lipids includes a variety of fats and oils. However, the majority (90%) of lipids taken in by the diet are *triglycerides*, which is what we discuss in this section. Recall from the chemistry chapter that a triglyceride consists of three fatty acid molecules bound to a core molecule called glycerol (see Chapter 2). We take a look at the digestion of lipids first, followed by their absorption.

Lipid Digestion

The process of lipid digestion is more complex than either carbohydrate or protein digestion because lipids are nonpolar molecules. For this reason, lipids tend to stick together, forming large globules rather than distributing evenly in a water-based

solution. If this happened in the alimentary canal, digestive enzymes would have very little surface area on which to work to catalyze reactions that break the bonds in the lipids.

Lipid globules can be broken up into smaller particles in two ways. The first is by mechanical digestion—mastication in the mouth, churning in the stomach, and segmentation in the small intestine. These movements scatter lipids and prevent them from coming together. The second way involves bile salts in the small intestine. As you discovered earlier, bile salts are amphiphilic molecules with polar and nonpolar parts. We use the following schematic structure to represent a bile salt:

Bile salt structural formula

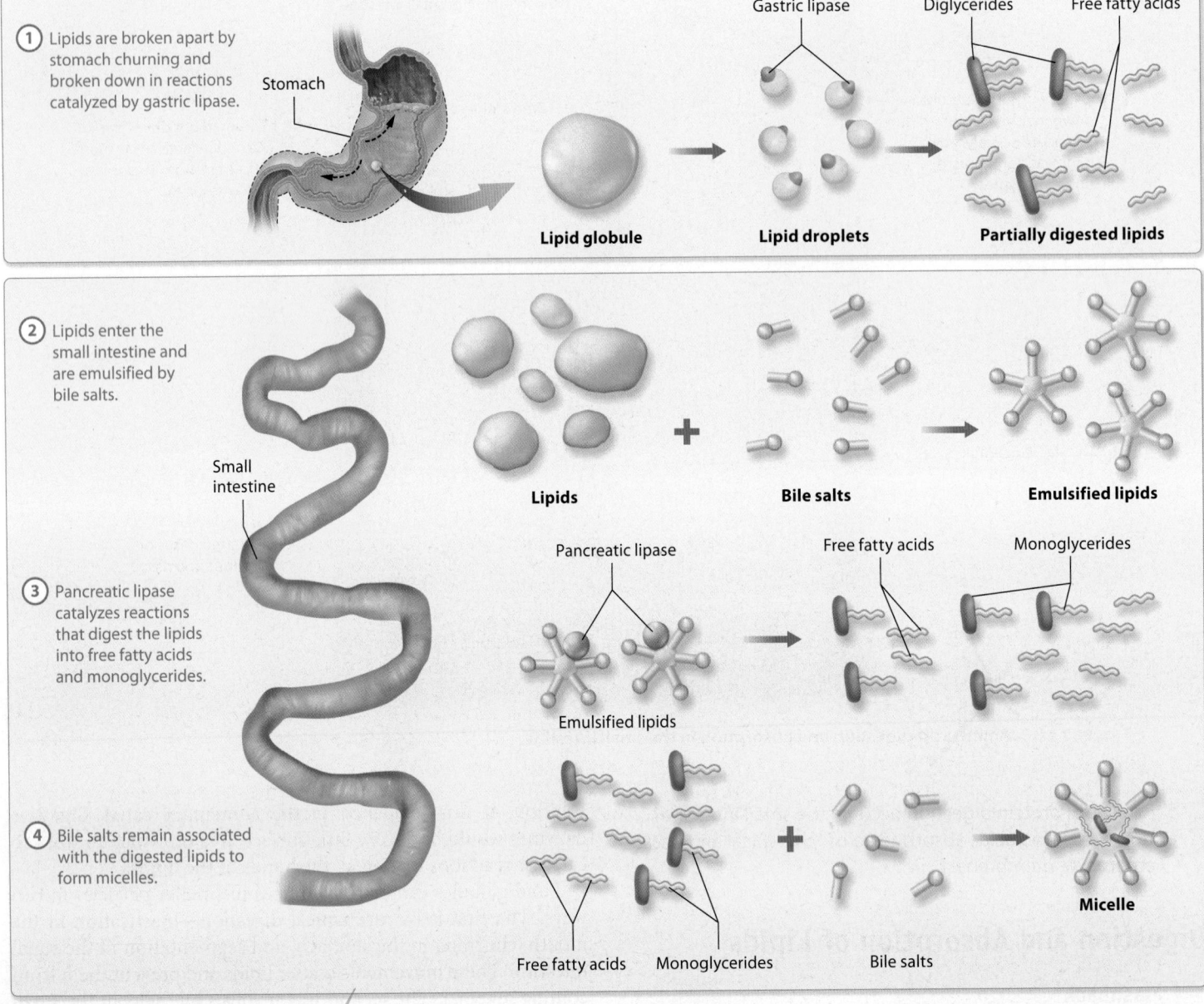

① Lipids are broken apart by stomach churning and broken down in reactions catalyzed by gastric lipase.

Stomach

Gastric lipase   Diglycerides   Free fatty acids

Lipid globule   **Lipid droplets**   **Partially digested lipids**

② Lipids enter the small intestine and are emulsified by bile salts.

Small intestine

Lipids   **Bile salts**   **Emulsified lipids**

Pancreatic lipase

③ Pancreatic lipase catalyzes reactions that digest the lipids into free fatty acids and monoglycerides.

Free fatty acids   Monoglycerides

Emulsified lipids

④ Bile salts remain associated with the digested lipids to form micelles.

Micelle

Free fatty acids   Monoglycerides   Bile salts

Figure 22.31 Digestion of lipids.

When bile salts mix with lipids, their nonpolar parts interact with the lipids, while their polar parts interact with the surrounding watery fluid. This breaks up the lipid globules into smaller pieces by the process of **emulsification**. The end result is multiple tiny lipid droplets each coated with bile salts, a mixture called an **emulsion** (ee-MUL-shun). This gives digestive enzymes enough surface area on which to work to efficiently digest the lipids.

The process of lipid digestion proceeds as follows (**Figure 22.31**):

① **Lipids are broken apart by stomach churning and broken down in reactions catalyzed by gastric lipase.** Lipid digestion begins in the stomach with the help of the enzyme **gastric lipase.** Gastric lipase catalyzes the

reactions that remove one fatty acid from triglycerides, leaving some free fatty acids and diglycerides. About 15% of total dietary fats are digested this way.

② **Lipids enter the small intestine and are emulsified by bile salts.** The undigested and partially digested triglycerides enter the small intestine, where they interact with bile salts. Bile salts coat the lipids and physically break them apart into smaller pieces. This is purely a physical process—no bonds are broken during emulsification.

③ **Pancreatic lipase catalyzes reactions that digest the lipids into free fatty acids and monoglycerides.** The pancreas releases an enzyme, **pancreatic lipase,** that catalyzes lipid breakdown. In this process, triglycerides are digested into monoglycerides and free fatty acids.

Figure 22.32 Summary of the digestion of carbohydrates, proteins, and lipids.

(4) **Bile salts remain associated with the digested lipids to form micelles.** After chemical digestion by lipase is complete, the bile salts and digested lipids stay together in structures known as **micelles** (my-SELZ). If the digested lipids did not remain associated with bile salts, they would re-form into large globules and absorption would not be possible.

Figure 22.32 summarizes the processes of digestion for carbohydrates, proteins, and lipids.

Lipid Absorption

Just as lipid digestion is more complicated than that of carbohydrates and proteins, lipid absorption is also more complex. Again this is due to lipids' nonpolar nature. Lipids face several water-based barriers that deter them from passing into the cytosol of the enterocytes, including the mucus lining of the small intestine and the polar phosphate heads of the enterocytes' plasma membranes. For this reason, lipids require assistance to approach the enterocyte membranes and eventually cross them.

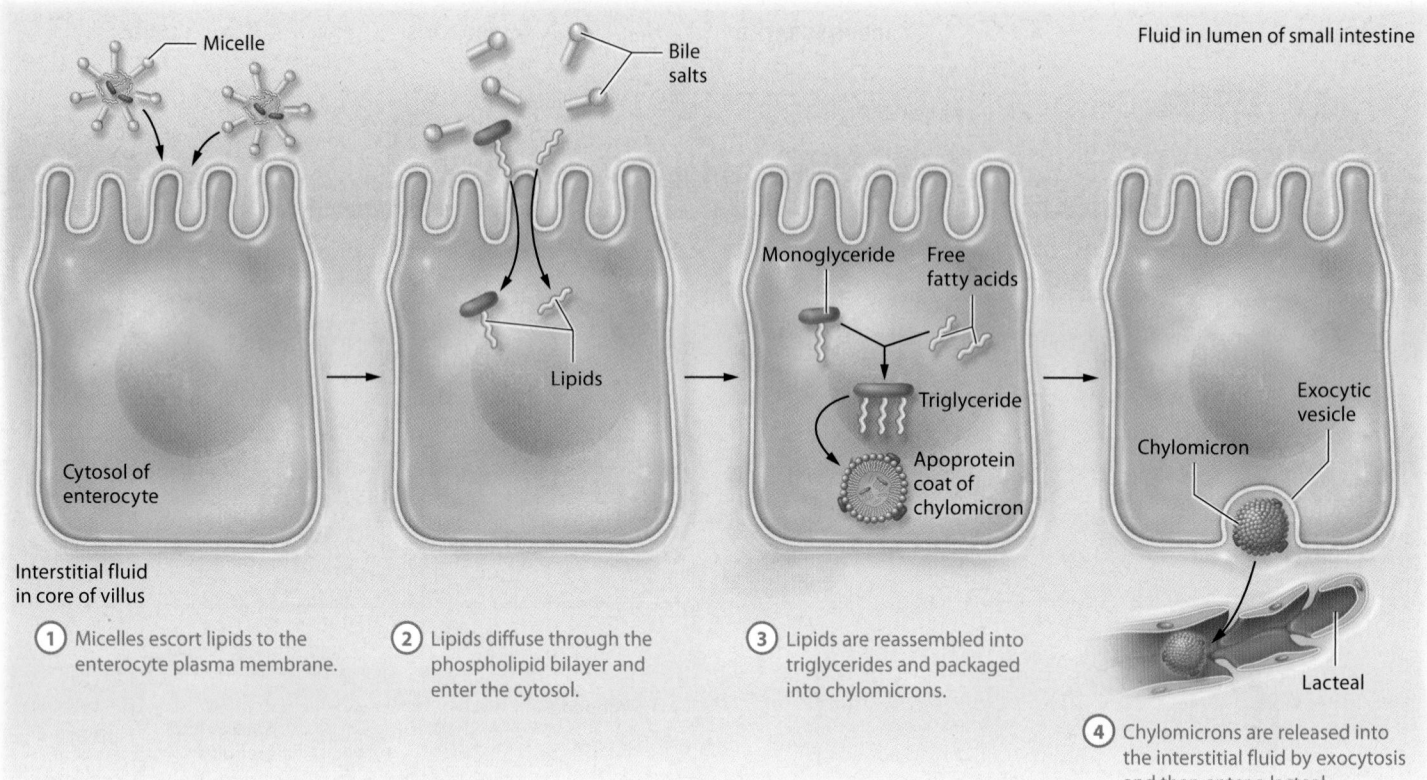

Figure 22.33 Lipid absorption in the small intestine.

Let's walk through the steps of lipid absorption (**Figure 22.33**):

① **Micelles escort lipids to the enterocyte plasma membrane.** Digested triglycerides and other lipids such as cholesterol and *fat-soluble vitamins* (such as vitamin D) remain associated with bile salts in micelles. Notice that these bile salts have their polar portions on the outside, facing the water-based environment, and their nonpolar portions on the inside, facing the digested lipids. The polar outer portion of the micelle allows it to approach the polar mucus layer and phosphate heads of the plasma membrane.

② **Lipids diffuse through the phospholipid bilayer and enter the cytosol.** Once the lipids in the micelles are at the plasma membrane, the attraction of the fatty acid tails of the phospholipid bilayer draws the lipids into the cytosol. This generally occurs by simple diffusion, although some lipids use facilitated diffusion by carrier-mediated transport. Note that the leftover bile salts in the micelles remain in the lumen of the small intestine and are reabsorbed by active transport mechanisms in the ileum.

③ **Lipids are reassembled into triglycerides and packaged into chylomicrons.** Within the enterocyte, enzymes catalyze reactions that turn the free fatty acids and monoglycerides back into triglycerides. The triglycerides are packaged with cholesterol, other dietary lipids, phospholipids, and lipid-binding protein molecules known as *apoproteins* into structures called **chylomicrons** (ky-loh-MY-krahnz).

Chylomicrons are similar to micelles in that their nonpolar lipids face the inside and their polar portions face the outside. This allows chylomicrons to travel with the polar water molecules in blood.

④ **Chylomicrons are released into the interstitial fluid by exocytosis and then enter a lacteal.** The newly formed chylomicrons are packaged into vesicles by the Golgi apparatus and released by exocytosis from the basal enterocyte membrane into the interstitial fluid in the core of the villus. Chylomicrons vary in size but are generally large—in fact, too large to enter the capillaries in the villus. However, lacteals, which are also found in the core of the villus, have valves in their walls that allow large substances to enter and exit. Chylomicrons therefore enter a lacteal, where they join the lymph.

Unlike proteins and carbohydrates, lipids are not delivered directly to the liver via the hepatic portal vein after absorption. However, the hepatic portal vein delivers the leftover bile salts to the liver, where they are used to make new bile. The lipids travel within chylomicrons through the lymphatic vessels and eventually to the thoracic duct, where they join the blood with the rest of the lymph as the thoracic duct meets the junction of the left internal jugular and left subclavian veins. As chylomicrons travel through blood capillaries, lipids are progressively removed and enter cells. Lipids removed from chylomicrons in the liver are processed by a number of metabolic processes. Chylomicrons that have been depleted of all of their lipids are taken in by hepatocytes and dismantled.

Quick Check

- ☐ 7. Where does lipid digestion begin and with which enzyme?
- ☐ 8. What is the role of bile salts in lipid digestion?
- ☐ 9. Why is lipid absorption more complex than carbohydrate or protein absorption?

Digestion and Absorption of Nucleic Acids

The nucleic acids in the food we eat begin chemical digestion in the small intestine with the help of pancreatic enzymes called *nucleases*. These enzymes catalyze the reactions that break nucleic acids into individual nucleotides. Further digestion occurs via brush border enzymes, which remove the phosphate group and the sugar from the nucleotide to leave a phosphate ion, ribose or deoxyribose, and a nitrogenous base. These three molecules are absorbed via primary and secondary active transport mechanisms into capillaries in the villi. The hepatic portal system then takes them via the blood to the liver to be metabolized.

Quick Check

- ☐10. Which enzymes catalyze reactions that digest nucleic acids?

Absorption of Water, Electrolytes, and Vitamins

In addition to nutrients, the alimentary canal absorbs large quantities of water, electrolytes (such as sodium, potassium, and calcium ions), and vitamins. Much of this absorption occurs in the small intestine, but a significant quantity occurs in the large intestine as well. The next subsections examine the absorption of each of these substances.

Absorption of Water

On average, over 9 liters of water enter the small intestine each day. About 2 liters of water are ingested, and the remaining liters are secreted into the alimentary canal by the alimentary canal itself and accessory organs. Of this 9 liters, about 8 liters are absorbed into the enterocytes of the small intestine, and most of the remaining water is absorbed into the enterocytes of the large intestine. This leaves only about 0.1 liter of water to be excreted in feces.

Water absorption occurs exclusively by osmosis. This requires a concentration gradient—both the cytosol of the enterocytes

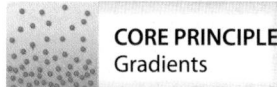
CORE PRINCIPLE
Gradients

and the extracellular fluid in the villi must be *more* concentrated than the fluid in the lumen of the intestine to drive water movement, which is another example of the Gradients Core Principle (p. 26). For this reason, anything that causes excess solutes to remain in the lumen of the intestines will hold onto water and prevent its reabsorption. This fact can be exploited in the use of drugs known as *osmotic laxatives,* which soften the feces. These drugs consist of

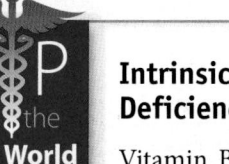
Intrinsic Factor and Vitamin B12 Deficiency

Vitamin B12 is part of the metabolic functions of every cell in the body; it is particularly important to the formation of erythrocytes and the normal functioning of the nervous system. Its absorption requires the presence of intrinsic factor (IF). An IF-B12 complex forms and then binds to the enterocyte plasma membrane, which enables its absorption.

Multiple conditions can impair production of intrinsic factor, including an autoimmune condition that attacks the stomach, atrophy of the gastric mucosa from ulcers or infections, and surgical removal of the stomach for weight loss or cancer treatment. Without intrinsic factor, vitamin B12 deficiency results, even if adequate B12 is present in the diet. This deficency can lead to pernicious anemia and nervous system problems and can only be treated with B12 injections.

solutes that either are not absorbed or are given in amounts in excess of what can be absorbed. As a result, water is retained in the feces, softening them and making them easier to eliminate.

Absorption of Electrolytes

Electrolytes are both taken in from the diet and present in secretions from digestive organs. Most of these electrolytes are absorbed in the small intestine, although significant amounts are absorbed in the large intestine as well. You have already seen mechanisms for absorption of sodium ions (cotransport with monosaccharides and amino acids). This sodium ion absorption is also key in the absorption of anions—as sodium ions enter the enterocytes, an electrical gradient is created that drives the absorption of anions such as chloride and bicarbonate ions.

Other electrolytes have specialized mechanisms for absorption. For example, as you've discovered, calcium ions are absorbed with the help of vitamin D (see Chapter 16). Certain electrolytes that are present in smaller amounts, such as iron and magnesium ions, are absorbed by active transport mechanisms.

Absorption of Vitamins

Vitamins are chemicals generally provided by the diet that are involved in a host of metabolic reactions. There are two types of vitamins: **water-soluble vitamins,** which are polar molecules; and **fat-soluble vitamins,** which are lipid-based and mostly nonpolar. Most of the water-soluble vitamins are absorbed in the small intestine by diffusing through the enterocytes' plasma membranes. One important exception is vitamin B12, which must bind to *intrinsic factor*—a molecule produced by the parietal cells of the stomach—to be absorbed in the ileum. A shortage of intrinsic factor can impair vitamin B12 absorption, as discussed in *A&P in the Real World.* Fat-soluble vitamins are packaged into micelles with fats and other lipids and are absorbed

TABLE 22.2 DIGESTIVE ENZYMES

Enzyme(s)	Source	Reaction Catalyzed
Carbohydrates		
Salivary amylase	Salivary glands	Polysaccharides into smaller polysaccharides and oligosaccharides
Pancreatic amylase	Pancreatic juice	Polysaccharides into oligosaccharides
Maltase, sucrase, lactase	Intestinal brush border	Oligosaccharides into monosaccharides
Proteins		
Pepsin	Chief cells of gastric glands (secreted as precursor pepsinogen)	Proteins into polypeptides and oligopeptides
Trypsin	Pancreatic juice	Oligopeptides into small peptides; activates itself and other pancreatic enzymes
Chymotrypsin	Pancreatic juice	Oligopeptides into small peptides
Carboxypeptidase	Pancreatic juice	Oligopeptides into small peptides
Dipeptidase and tripeptidase	Intestinal brush border	Dipeptides and tripeptides into amino acids
Lipids		
Gastric lipase	Gastric glands	Triglycerides into free fatty acids and diglycerides
Pancreatic lipase	Pancreatic juice	Triglycerides into free fatty acids and monoglycerides
Nucleic Acids		
Nucleases	Pancreatic juice	Nucleic acids into nitrogenous bases and simple sugars

with them. The fat-soluble vitamins include A, D, E, and K. Vitamins are covered in more depth in the metabolism and nutrition chapter (see Chapter 23).

To help you study what's been covered in this module, look to Table 22.2, which summarizes the digestive enzymes.

Quick Check

☐ 11. How is water absorbed in the small and large intestines?
☐ 12. How is sodium ion absorption achieved? How does it affect the absorption of other electrolytes?

Apply What You Learned

☐ 1. The weight-loss drug orlistat, available over the counter under the name Alli, blocks the activity of pancreatic lipase. Predict the effects of this drug. Why might it help a person lose weight?

☐ 2. Many dietary supplements contain supplemental digestive enzymes that manufacturers claim are needed to digest food properly.
   a. What do you think will happen to these enzymes in the stomach after they are ingested? (*Hint:* Remember that enzymes are proteins.)
   b. Do you think that these supplements are likely to be useful? Why or why not?

☐ 3. Would bile salts be effective if they were polar molecules instead of amphiphilic molecules? What if they were nonpolar molecules? Why or why not?

See answers in Appendix A.

MODULE **22.8**
Putting It All Together: The Big Picture of Digestion

Learning Outcomes

1. Describe the overall big picture of digestion and digestive processes.

Now let's put together the information we covered in previous modules to create a big picture of digestion. **Figure 22.34** traces the digestion pathway, starting with the ingestion of food and ending with the defecation of the remaining, indigestible portion.

Quick Check

☐ 1. In which parts of the alimentary canal does chemical digestion take place?
☐ 2. How do the accessory organs assist digestion?

Apply What You Learned

☐ 1. Patients who are unable to eat by mouth often have a feeding tube placed into the stomach. Which digestive processes are missed with this type of feeding?

See answers in Appendix A.

The Big Picture of Digestion

Figure 22.34

Bolus

1 Food is ingested and chemically and mechanically digested in the mouth into a bolus.

Bolus

2 The swallowing reflex propels the bolus to the stomach (see Figure 22.9).

Stomach

3 The bolus is churned into chyme and chemically digested in the stomach (see Figures 22.13 and 22.32).

Gallbladder

Pancreas

Duodenum

Chyme

Pancreatic juice and bile

4 Chyme enters the duodenum and mixes with the pancreatic juice and bile (see Figures 22.21 and 22.26).

Small intestine

5 Nutrients are chemically digested and absorbed as chyme passes through the small intestine (see Figures 22.32 and 22.33).

Large intestine

Electrolytes

H₂O   Vitamins

6 Material passes into the large intestine where water, electrolytes, and vitamins are absorbed.

7 Indigestable materials are excreted in the feces via defecation (see Figure 22.18).

CHAPTER SUMMARY

Take full advantage of the invaluable online practice and study tools for this chapter by going to MasteringA&P®. There you will find:

 Practice tests, module quizzes, art-labeling and sequence-the-steps activities

 Animated figures with quizzes

 Practicing A&P worksheets

 Author-narrated podcast

 Dynamic Study Modules

iP *Interactive Physiology*

MODULE **22.1**
Overview of the Digestive System 846

- The digestive system consists of the **alimentary canal** and the **accessory organs.** (Figure 22.1, p. 847)

  Practice art-labeling exercise: Figure 22.1.

- The functions of the digestive system include food, water, electrolyte, and vitamin intake; hormone production; and waste excretion.

- There are six basic digestive processes: **ingestion, secretion, propulsion, digestion, absorption,** and **defecation.**

- All regions of the alimentary canal are **motile** due to skeletal and/or smooth muscle. The motility of the alimentary canal is controlled largely by the **enteric nervous system (ENS),** the parasympathetic nervous system, and hormones from the endocrine system.

- The organs of the alimentary canal have four tissue layers: the inner **mucosa,** the **submucosa,** the **muscularis externa,** and either the **serosa** or the **adventitia,** depending on the location of the organ. (Figure 22.2, p. 849)

- The **peritoneum** is the largest serous membrane in the body. It consists of the **parietal peritoneum** and the **visceral peritoneum.** Between these two layers is the **peritoneal cavity. Mesenteries** are folds of the visceral peritoneum that bind and support abdominal organs. (Figure 22.3, p. 851)

  Practice art-labeling exercise: Figure 22.3.

- The abdominal digestive organs are supplied and drained by the **splanchnic circulation.**

- The nerve supply of the abdominal digestive organs comes from the enteric nervous system and the two branches of the autonomic nervous system.

iP Play Interactive Physiology tutorials on Digestive System: Orientation and Anatomy Review.

MODULE **22.2**
The Oral Cavity, Pharynx, and Esophagus 852

- The mouth, or *oral cavity,* is bounded laterally by the cheeks, anteriorly by the lips, and superiorly by the **hard** and **soft palate.** Within the mouth, and with the help of accessory organs, food is ground into a **bolus.** (Figure 22.4, p. 853)

  Practice art-labeling exercise: Figure 22.4.

- The **teeth masticate** and grind food into a bolus. They are located in *alveoli* within the mandible and maxilla. (Figure 22.5, p. 854; Figure 22.6, p. 855)
  - A tooth consists of a **crown,** the portion above the gum line, and the **root,** the portion embedded in the alveolus. The three classes of teeth are **incisors, canines,** and **molars.**
  - The **primary dentition,** or **deciduous teeth,** consists of 20 teeth that erupt from the age of 6 months to 24 months. The **secondary dentition,** or **permanent teeth,** consists of 32 teeth that replace the deciduous teeth.
  - The crown of a tooth is covered by an outer layer of **enamel,** and its root is covered by an outer layer of **cementum.** Deep to these substances is **dentin,** which encases the gelatinous **pulp.**

- The **tongue** consists of skeletal muscle covered with stratified squamous epithelium. The epithelium is arranged into **papillae.** The tongue is involved in mechanical digestion, and it pushes the bolus posteriorly during swallowing.

- The **salivary glands** are located around the oral cavity, into which they secrete **saliva.** The three pairs of salivary glands are the **parotid glands** over the masseter muscle, the **submandibular glands** medial to the mandible, and the **sublingual glands** under the tongue. (Figure 22.7, p. 856)
  - Saliva consists primarily of water, with mucus, **salivary amylase, lysozyme,** and secretory IgA. It moistens food and assists in mechanical and chemical digestion.
  - **Salivation** is controlled by a parasympathetic nervous system reflex arc mediated by the facial and glossopharyngeal nerves. These neurons release ACh onto acinar cells, which triggers the release of saliva.

- Food passes from the oral cavity to the **pharynx,** which is surrounded by skeletal muscles. These muscles contract during swallowing and propel the bolus inferiorly.

- Food next passes through the *upper esophageal sphincter* to enter the **esophagus,** where waves of peristalsis massage it inferiorly to the **gastroesophageal sphincter,** where it meets the stomach. (Figure 22.8, p. 858)

- **Swallowing,** or **deglutition,** is a type of propulsion with three phases. The **voluntary phase** is mediated by the contraction of the tongue and soft palate and is under conscious control. The **pharyngeal phase** involves reflex-mediated contraction of the pharyngeal constrictor muscles and is controlled by the *swallowing reflex.* The **esophageal phase** is mediated by **peristalsis,** also controlled by the swallowing reflex. (Figure 22.9, p. 859) iP Play Interactive Physiology tutorial on Digestive System: Anatomy Review.

MODULE 22.3
The Stomach 860

- The **stomach** is in the left upper quadrant. It has five regions: the **cardia, fundus, body, pyloric antrum,** and **pylorus.** The stomach controls emptying into the duodenum via the **pyloric sphincter.** (Figure 22.10, p. 861; Figure 22.11, p. 862)
  - The stomach's mucosa is folded into **gastric pits** that house **gastric glands.** There are four main cell types in gastric glands: **enteroendocrine cells,** which produce hormones; **chief cells,** which secrete the inactive precursor **pepsinogen; parietal cells,** which secrete hydrochloric acid and **intrinsic factor;** and **mucous neck cells,** which secrete acidic mucus.
  - The muscularis externa of the stomach has a third oblique layer that performs **churning,** which mechanically digests food into **chyme.**
  - The processes that take place in the stomach include secretion, chemical digestion, mechanical digestion, and propulsion.

  🖑 Practice art-labeling exercise: Figure 22.10.

- Gastric acid is secreted by H^+/K^+ pumps in the membranes of parietal cells. Secretion occurs in three phases. (Figure 22.12, p. 864)
  - During the **cephalic phase,** the vagus nerve directly stimulates parietal cells, stimulates release of **histamine** and **gastrin,** and inhibits the release of **somatostatin.** These four effects increase acid secretion.
  - The **gastric phase** begins when food enters the stomach. Distention of the stomach wall triggers the ENS and vagus nerve to increase acid secretion. Partially digested proteins trigger the release of gastrin, which also increases acid secretion. A pH below 2 stimulates release of somatostatin, which decreases acid secretion.
  - The **intestinal phase** begins when chyme enters the duodenum. This briefly stimulates acid secretion, but the declining pH and lipid presence in the duodenum trigger the **enterogastric reflex.** This results in decreased acid secretion.
- The stomach's motility has three functions: (1) receptive function, (2) churning function, and (3) emptying function. (Figure 22.13, p. 866)
  - When the stomach fills, it reflexively relaxes to prevent gastric pressure from rising.
  - Churning is accomplished by waves of peristalsis that are initiated by the ENS. Serotonin, intestinal gastrin, and the vagus nerve control the strength of contractions.
  - The stomach controls emptying into the duodenum to give the duodenum adequate time to process the incoming chyme. The rate of gastric emptying is controlled by the composition of chyme in the duodenum and hormones such as **secretin, cholecystokinin,** and **gastric inhibitory peptide.**

iP Play Interactive Physiology tutorials on Digestive System: Anatomy Review, Motility, and Control of the Digestive System.

MODULE 22.4
The Small Intestine 867

- The **small intestine** has three regions: the **duodenum, jejunum,** and **ileum.** (Figure 22.14, p. 867)
  - The internal surface of the small intestine is arranged into progressively smaller folds: **circular folds,** folds of the mucosa and submucosa; **villi,** folds of the mucosa; and **microvilli,** folds of the enterocyte plasma membrane.
  - Villi surround a central core of capillaries and a *lacteal*. Between villi are **intestinal crypts** that house glands. (Figure 22.15, p. 868)
  - The primary processes occurring in the small intestine are absorption, secretion, propulsion, and mechanical and chemical digestion.
- The small intestine's motility between meals is characterized by slow, rhythmic contractions of the **migrating motor complex.** It is controlled by the ENS and the hormone *motilin*. During eating, the small intestine exhibits two movements: **segmentation,** which is a churning movement, and peristalsis, which is a propulsive movement. Both are controlled by the vagus nerve. (Figure 22.16, p. 869)

iP Play Interactive Physiology tutorials on Digestive System: Anatomy Review, Motility, Segmentation, and Control of the Digestive System.

MODULE 22.5
The Large Intestine 870

- The **large intestine** frames the abdominopelvic cavity.
- The large intestine begins with the **cecum,** which receives material from the ileum and houses the **vermiform appendix.** The next segment is the **colon,** which has four segments: the ascending, transverse, descending, and sigmoid segments. The sigmoid colon becomes the **rectum,** which then becomes the **anal canal.** (Figure 22.17, p. 871)
- The large intestine houses a great number of bacteria that produce vitamins, metabolize undigested materials, deter the growth of harmful bacteria, and stimulate the immune system.
- Processes occurring in the large intestine include absorption of water, electrolytes, vitamins, and small amounts of nutrients; secretion; propulsion; and defecation.
- The proximal large intestine undergoes a churning or segmentation motion to assist with water and electrolyte absorption. It also undergoes propulsive **mass movements.** Both are controlled by the ENS. When mass movements propel fecal material into the distal colon and rectum, the **defecation reflex** is initiated. This reflex is mediated by the parasympathetic nervous system and the cerebral cortex, and mediates the release of fecal material from the large intestine. (Figure 22.18, p. 873)

iP Play Interactive Physiology tutorial on Digestive System: Anatomy Review.

MODULE 22.6
The Pancreas, Liver, and Gallbladder 874

- The **pancreas** is an endocrine and exocrine gland. Its exocrine portion, which consists of **acinar cells,** is involved in digestion through the secretion of **pancreatic juice** into the **main pancreatic duct.** Pancreatic juice consists of water, bicarbonate, and multiple digestive enzymes, and it exits the main pancreatic duct into the duodenum. Pancreatic secretion during meals is mediated by the parasympathetic nervous system and hormones such as cholecystokinin and secretin. (Figure 22.20, p. 875; Figure 22.21, p. 876)

- The liver is in the right upper quadrant and consists of four lobes. (Figure 22.22, p. 877; Figure 22.23, p. 878)
  - Microscopically, the liver consists of hexagonal **liver lobules** with hepatocytes surrounding a **central vein** and **portal triads** at each corner. The portal triad consists of a **bile duct,** a branch of the hepatic artery, and a branch of the hepatic portal vein.
  - The liver's main digestive function is the production of **bile,** which is required for lipid digestion. Bile contains **bile salts,** water, and metabolic wastes such as bilirubin.
  - Other functions of the liver include nutrient metabolism, detoxification, excretion, production of plasma proteins, and storage of nutrients.

- The **gallbladder** sits on the posterior liver. It receives bile from the common hepatic duct and stores it for later release. Bile leaves the gallbladder through the **cystic duct,** which merges with the common hepatic duct to form the **common bile duct.** This duct then empties into the duodenum. (Figure 22.24, p. 879)

  📋 Practice art-labeling exercise: Figure 22.24.

- The main trigger for bile release from the gallbladder and liver is cholecystokinin, which causes contraction of the smooth muscle of the gallbladder. Another stimulus for bile secretion is bile salts themselves, which are reabsorbed in the ileum. (Figure 22.26, p. 880)

  iP Play Interactive Physiology tutorial on Digestive System: Anatomy Review.

MODULE 22.7
Nutrient Digestion and Absorption 881

- Carbohydrate digestion begins in the mouth with salivary amylase. It continues in the small intestine with the help of **pancreatic amylase** and the *brush border enzymes* **lactase, maltase,** and **sucrase.** Carbohydrates are absorbed when they are broken down into their monomer subunits. Glucose and galactose are transported across an enterocyte's apical membrane using the secondary active transport **Na⁺/glucose cotransporter.** Fructose crosses the apical membrane by facilitated diffusion. All three monomers cross the enterocyte's basal membrane by facilitated diffusion. (Figure 22.29, p. 883)

- Protein digestion begins in the stomach with the enzyme **pepsin.** Pepsin is secreted as its inactive form pepsinogen, which is converted to pepsin in the presence of an acid environment. Protein digestion continues in the small intestine with pancreatic enzymes such as **trypsin** and brush border enzymes. Proteins must be broken down into oligopeptides and free amino acids to be absorbed. They cross the enterocyte's apical and basal membranes primarily by facilitated diffusion.

- Both absorbed amino acids and carbohydrates enter the capillaries in the villus, and are delivered to the hepatic portal system for processing. (Figure 22.30, p. 885)

- Lipid digestion is more complicated than protein and carbohydrate digestion because lipids are nonpolar and clump together. (Figure 22.31, p. 886)
  - Lipid globules are broken apart physically by chewing and churning and by **emulsification** by bile salts. During emulsification, the nonpolar parts of the bile salts interact with the lipids and the polar parts interact with the water molecules in the small intestine.
  - Emulsified lipids are broken down in reactions catalyzed by **pancreatic lipase** into two free fatty acids and a monoglyceride.
  - Digested lipids remain associated with other lipids and bile salts to form **micelles.**

- Micelles escort lipids to the enterocyte plasma membrane, where the lipids enter the cytosol. Once inside the enterocyte, fatty acids are reassembled into triglycerides and packaged with other lipid-based chemicals into **chylomicrons.** Chylomicrons exit the enterocyte by exocytosis and enter a lacteal, where they travel through the lymph and are delivered to the blood at the junction of the left internal jugular and subclavian veins. (Figure 22.33, p. 888)

- Water is absorbed across the small and large intestines by osmosis. Electrolytes are absorbed across the small and large intestines as well by passive and active transport mechanisms.

  iP Play Interactive Physiology tutorial on Digestive System: Digestion and Absorption.

MODULE 22.8
Putting It All Together: The Big Picture of Digestion 890

- The big picture of digestion is shown in Figure 22.34 (p. 891).

ASSESS WHAT YOU LEARNED

See answers in Appendix A.

LEVEL 1 Check Your Recall

1. Which of the following is not one of the six basic processes carried out by the digestive system?

   a. Propulsion   c. Secretion

   b. Gas exchange   d. Absorption

2. The peritoneal cavity is located:

   a. around each of the digestive organs.

   b. between layers of mesentery.

   c. superior to the greater omentum.

   d. between the parietal and visceral peritoneum.

3. Mark the following statements as true or false. If a statement is false, correct it to make a true statement.

   a. The mucosa from the stomach to the anus consists of an inner layer of stratified columnar epithelium.

   b. The muscularis externa of most of the alimentary canal consists of inner circular and outer longitudinal layers of smooth muscle.

   c. The soft palate and uvula move posteriorly during swallowing to prevent food from entering the laryngopharynx.

   d. The mucosa of the esophagus, pharynx, and oral cavity contains simple squamous epithelium to protect it from abrasion.

4. Match the following terms with the correct definition.

   _____ Chief cells
   _____ Parietal cells
   _____ Gastrin
   _____ Pyloric sphincter
   _____ Enteroendocrine cells
   _____ Gastroesophageal sphincter
   _____ Pepsin
   _____ Chyme

   a. Hormone that stimulates multiple digestive processes
   b. Enzyme that begins protein digestion
   c. Produce acid and intrinsic factor
   d. Liquid produced as a result of stomach churning
   e. Produce pepsinogen
   f. Cells in gastric glands that produce hormones
   g. Controls passage of bolus from esophagus to stomach
   h. Controls passage of stomach contents to duodenum

5. The common hepatic duct and main pancreatic duct enter into the:

   a. pyloric sphincter.
   b. duodenum.
   c. ileum.
   d. cecum.

6. What are the three folds of the small intestine called? What is their purpose?

7. Which of the following is *not* one of the functions of the bacterial flora of the large intestine?

   a. Metabolism of undigested carbohydrates
   b. Deterring the growth of harmful bacteria
   c. Chemical digestion of dietary proteins
   d. Stimulation of the immune system
   e. Vitamin production

8. Trace the pathway that food takes through the entire alimentary canal, from the oral cavity to the anal canal.

9. Mark the following statements as true or false. If a statement is false, correct it to make a true statement.

   a. The crown of a tooth is surrounded by an outer layer of pulp, which is the hardest substance in the body.

   b. The filiform papillae of the tongue are keratinized, which makes them rough and able to assist in mechanical digestion.

   c. The three pairs of salivary glands are the parotid, submandibular, and sublingual glands.

   d. The exocrine cells of the pancreas and salivary glands are islet cells.

10. Which of the following statements about accessory organ secretions is *not* true?

    a. Hepatocytes produce bile, which drains out of the liver via the common hepatic ducts.

    b. Saliva contains secretory IgA and lysozyme, which play an important role in preventing the growth of pathogenic bacteria in the oral cavity.

    c. Pancreatic juice contains digestive enzymes and bicarbonate to neutralize the acidic chyme.

    d. The gallbladder produces bile, which drains out of the gallbladder via the cystic duct.

11. Which of the following best describes the microscopic structure of the liver?

    a. Hexagonal plates of hepatocytes surrounding a central vein with portal triads at each corner

    b. Octagonal plates of hepatocytes surrounding a central artery with portal triads at each corner

    c. Irregular groups of hepatocytes surrounding a central vein with bile ducts at each corner

    d. Hexagonal plates of hepatocytes surrounding a portal vein with bile ducts at each corner

12. Match the following digestive enzymes with the digestive reaction they catalyze.

    _____ Salivary amylase
    _____ Sucrase
    _____ Trypsin
    _____ Pancreatic lipase
    _____ Pancreatic amylase
    _____ Pepsin

    a. Breaks down sucrose into glucose and fructose in small intestine
    b. Digests lipids in small intestine
    c. Breaks down proteins into polypeptides in stomach
    d. Breaks down polysaccharides into oligosaccharides in small intestine
    e. Assists in protein digestion in small intestine and activates other protein-digesting enzymes
    f. Breaks down polysaccharides into oligosaccharides in the mouth

13. Which of the following best describes the role of bile salts in lipid digestion?

    a. Bile salts begin chemical digestion of lipids.

    b. Bile salts bind to lipids and physically break them apart into monoglycerides and free fatty acids.

    c. Bile salts bind to lipids and physically break them apart into smaller pieces.

    d. Bile salts interact with pancreatic lipase and catalyze lipid breakdown.

14. Which of the following are absorbed in the small intestine by the Na^+/glucose cotransporter?

    a. Glucose
    b. Galactose
    c. Fructose
    d. Both a and b are correct.
    e. All of the above are correct.

15. The primary hormone that triggers the secretion of pancreatic juice and bile is:

    a. gastric inhibitory peptide.
    b. cholecystokinin.
    c. motilin.
    d. somatostatin.

16. Fill in the blanks: Lipids associate with bile salts in the small intestine to form _____. In the enterocyte, lipids are packaged with other lipids into _____.

17. Which of the following statements about absorption in the small and large intestines is *false?*

    a. Absorption of water occurs by secondary active transport.
    b. Absorption of sodium ions creates an electrical gradient that drives absorption of chloride and bicarbonate anions.
    c. Fat-soluble vitamins are absorbed with other lipids.
    d. Vitamin B12 requires intrinsic factor to be absorbed in the small intestine.

18. Mark the following statements as true or false. If a statement is false, correct it to make a true statement.

    a. Motility of the alimentary canal is regulated by the enteric nervous system, the parasympathetic nervous system, and hormones.
    b. The esophageal and pharyngeal phases of swallowing are mediated by the swallowing reflex of the brainstem and the enteric nervous system.
    c. Smooth muscle of the stomach contracts when food enters from the esophagus.
    d. Control of gastric emptying into the duodenum is important to allow the chyme to be neutralized and diluted by pancreatic secretions.
    e. During eating, the small intestine exhibits motion in the migrating motor complex pattern.

19. Defecation is stimulated by:

    a. stretch receptors in the rectum, initiating the defecation reflex in the spinal cord.
    b. segmentation of haustra in the proximal large intestine.
    c. contraction of the external anal sphincter.
    d. activation of the defecation reflex by the presence of fecal material in the proximal large intestine.

20. Which of the following stimuli mediates salivation?

    a. Neurons of the ENS
    b. ACh from parasympathetic neurons
    c. The hormone secretin
    d. Feedback from the cerebral cortex

21. Which of the following occur during the cephalic phase of gastric acid secretion?

    a. Stimulation of gastrin and histamine release
    b. Direct stimulation of parietal cells
    c. Inhibition of somatostatin release
    d. Both a and b are correct.
    e. All of the above are correct.

22. How does absorption of lipids differ from absorption of carbohydrates and proteins in the small intestine?

    a. Lipids are absorbed into a capillary; carbohydrates and proteins are absorbed into a lacteal.
    b. Lipids are not absorbed in the small intestine.
    c. Lipids are absorbed into a lacteal; carbohydrates and proteins are absorbed into a capillary.
    d. They are all absorbed into the same structure.

LEVEL 2 Check Your Understanding

1. You and your friend are having dinner and you tell a funny joke. Your friend laughs at the same time he tries to swallow some water, and the water comes out his nose. Why has this happened?

2. General anesthesia eliminates the swallowing reflex but does not eliminate the feedback loops that cause vomiting. Considering this, explain why patients are advised not to eat or drink anything for at least 12 hours prior to undergoing surgery.

3. Drugs with anticholinergic side effects block ACh receptors in the peripheral nervous system, including those on digestive organs. Predict the effects such drugs would have on motility and secretion on the specific organs in the digestive system.

4. Your friend insists that absorption is the most important process carried out by the digestive system. What do you think of this claim? How would you respond?

LEVEL 3 Apply Your Knowledge

PART A: Application and Analysis

1. Mr. Williams presents to your clinic with a complaint of abdominal pain in the right upper quadrant. He says that the pain worsens when he eats, particularly when he eats fatty meals. He has noticed that his stool has been an unusual clay color recently. You perform an ultrasound of his abdomen and find that gallstones are blocking his common bile duct, preventing bile from entering the duodenum.

    a. Why are his symptoms worse when he consumes a high-fat meal? Would you expect his symptoms to worsen or lessen when he consumes a meal consisting only of carbohydrates? Explain.
    b. You test Mr. Williams's stool and find high amounts of undigested fats. Explain this finding.
    c. Explain why his feces have become a clay color instead of a normal brown color.

2. A surgical procedure known as *gastric bypass* involves removing all or part of the stomach.

    a. Ms. Anthony has undergone gastric bypass and her stomach has been removed entirely. She goes out to dinner and eats an extremely large meal. Explain what will happen in her small and large intestine following this meal.
    b. How would stomach removal affect absorption of vitamin B12? Explain.

PART B: Make the Connection

3. Individuals experiencing prolonged vomiting lose a great deal of hydrochloric acid with the vomitus. Predict the effect this loss of acid will have on the pH of the blood. How will the respiratory system respond to the change in pH? *(Connects to Chapters 2 and 21)*

4. Irritant laxatives increase the amount of water in the feces by triggering an inflammatory response in the intestinal mucosa. Explain why an inflammatory response causes this effect. *(Connects to Chapter 20)*

5. Predict what types of nervous system damage might lead to a loss of the defecation reflex. *(Connects to Chapters 12 and 13)*

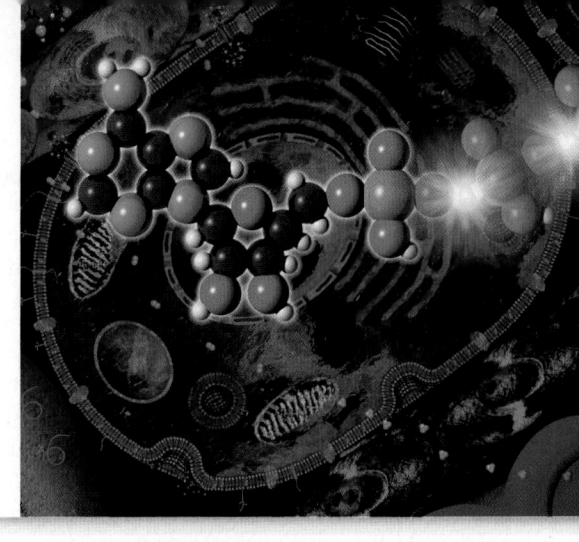

23

Metabolism and Nutrition

Have you ever noticed that when you're watching television, reading a magazine, or surfing the Internet, hardly a moment passes that you aren't faced with advertisements for products and diets promising to "rev your metabolism," "stimulate fat metabolism," or "cut calories"? Many of these ads make outrageous promises along the lines of, "Eat all you want and still lose 30 pounds in 2 weeks!" Although it is easy to spot exaggerated claims like these, others are more difficult to see. Most ads use terms that sound scientific and seem to be based on good physiological reasoning, which often makes it difficult to separate fact from fiction.

In our everyday lives, we generally use the term "calorie" to refer to the energy content of the food we eat. But *calorie* actually refers to a measure of heat: One calorie is the amount of heat required to raise the temperature of one gram of water by one degree Celsius. In terms of the human diet, the unit is actually the **Calorie** (with a capital C), which is really a **kilocalorie.** The Calorie is the amount of heat required to raise the temperature of one *kilogram* of water by one degree Celsius. So, we use heat units to measure the amount of energy in food. The references to heat production and energy don't end with the Calorie; you've likely heard that certain activities or dietary supplements will help you to "burn calories," or "burn fat," or even "turn your body into a fat-melting furnace." But why are these heat-related terms associated with food, diets, and metabolism? And can you really "burn calories" and "rev" your metabolism? This chapter explores these topics as part of a discussion of human metabolism and nutrition.

MODULE **23.1**
Overview of Metabolism and Nutrition

Learning Outcomes

1. Define the terms metabolism, catabolism, and anabolism, and identify the nutrients the body is able to use for fuel.
2. Compare and contrast endergonic and exergonic reactions.
3. Describe the process of phosphorylation.

Illustration: The adenosine triphosphate (ATP) molecule is the cell's main source of chemical energy to fuel its processes.

897

4. Describe the hydrolysis of ATP, and explain why this reaction is exergonic.
5. Explain what happens in an oxidation-reduction reaction and how electrons are transferred between reactants, including NADH and $FADH_2$.

Metabolism is defined as the sum of the body's chemical reactions. Metabolism includes four basic processes: (1) harnessing the energy in the chemical bonds of molecules obtained from the diet, called *nutrients,* that may be used to make **adenosine triphosphate,** or **ATP;** (2) converting one type of molecule into another for the cell's synthesis reactions; (3) carrying out synthesis reactions and assembling macromolecules such as proteins, polysaccharides, nucleic acids, and lipids; and (4) breaking down macromolecules into their monomers or other smaller molecules. Metabolism occurs through many series of enzyme-catalyzed reactions that are referred to as *metabolic pathways.* Each reaction in a series is characterized by a small chemical change in a reactant as the product is produced, and generally involves the addition or removal of one or more atoms or electron pairs.

In this module we explore the introductory concepts of metabolism, beginning with the phases of metabolism and different types of nutrients. Then we examine how the cell uses energy to carry out reactions, and the sources of energy for cellular metabolism, including ATP.

Phases of Metabolism: Catabolism and Anabolism

◄◄ **Flashback**

1. What are catabolic and anabolic reactions? (p. 42)
2. Define the monomers of carbohydrate, protein, and lipid, and describe their characteristic features. (p. 50)

Metabolism is generally considered to have two phases. The first is a series of reactions collectively known as **catabolism** (kah-TAH-bol-izm; *cata-* = "down"). Recall that a *catabolic reaction* is one in which a substance is broken down into smaller parts (see Chapter 2). Although certain reactions in a catabolic pathway may require the input of energy, the process as a whole releases energy that the cell can harness to drive processes such as ATP production. To provide materials for this process, macromolecules that we ingest are broken down into nutrient monomers by our digestive processes. The body uses three types of nutrient monomers for its catabolism to generate ATP:

- **Glucose.** Both dietary carbohydrates and those stored by the body are generally degraded by catabolic pathways to the monosaccharide glucose. Glucose catabolism is carried out by every cell in the body, and it is the preferred fuel for many cells, including those of the brain and the liver.
- **Fatty acids.** Catabolic pathways degrade lipids (mostly triglycerides) into fatty acids and glycerol. These molecules then enter their own catabolic pathways and are used by the cell.

- **Amino acids.** Proteins are degraded by catabolic pathways into amino acids. Like lipids, amino acids then enter their own catabolic pathways, during which they are converted into other molecules for use by the cell.

The second phase of metabolism is known as **anabolism** (ah-NAH-bol-izm; *ana-* = "up"). As we discussed previously, an *anabolic reaction* is one in which smaller molecules are combined to make a larger molecule (see Chapter 2). Via anabolism, cells take smaller molecules and build them into proteins, nucleic acids, carbohydrates, lipids, and all other macromolecules. Anabolic reactions require energy, and anabolism is driven by the energy liberated from catabolism.

Quick Check

☐ 1. How do catabolism and anabolism differ?
☐ 2. What three types of nutrients can the cell use to generate ATP?

Energy Requirements of Metabolic Reactions

◄◄ **Flashback**

1. How do endergonic and exergonic reactions differ? (p. 42)

In the preceding section, we spoke of energy being released from catabolic reactions and required by anabolic reactions. In other words, catabolic reactions are generally *exergonic,* and anabolic reactions are generally *endergonic.* Recall from the chemistry chapter (see Chapter 2) that we define these terms as follows:

- **Exergonic reactions.** Reactions that release energy are known as *exergonic reactions.* The products of these reactions have *less* energy than their reactants, and the remaining energy is released during the reaction. To understand why energy is released, we need to discuss the *law of conservation of energy,* which states that energy cannot be created or destroyed; it can only change forms. This means that the energy present before the reaction must equal the energy present after the reaction. Since the products of an exergonic reaction have less energy than the reactants, energy must be released so that the total energy is the same before and after the reaction. Most catabolic reactions are exergonic.
- **Endergonic reactions.** An *endergonic reaction* is one that requires the input of energy to proceed. This is because the products of the reaction have *more* energy than the reactants. To be in line with the law of conservation of energy, energy must be added to the reactant side so that the energy on both sides of the equation is equal. Anabolic reactions tend to be endergonic.

Endergonic and exergonic reactions in the cell are *coupled*—essentially, the cell "pays for" an endergonic reaction with the energy released by an exergonic reaction. To do so, the cell needs

① Energy is released from exergonic catabolic reactions.

Nutrients from food:
• Glucose
• Fatty acids
• Amino acids

A—B → *Exergonic catabolic reactions* → Energy + A + B

② This energy is used to fuel the endergonic anabolic reaction of ATP synthesis.

③ ATP is broken down in an exergonic catabolic reaction.

ADP + Phosphate group → *Endergonic anabolic reaction* → ATP → *Exergonic catabolic reactions* → Energy + ADP + Phosphate group

④ The energy from ATP breakdown fuels other endergonic anabolic reactions in the cell.

C + D → *Endergonic anabolic reaction* → C—D

Examples:
• Glycogen
• Phospholipids
• Proteins

Figure 23.1 Coupling of endergonic and exergonic reactions.

some sort of "money" with which to pay, that is, some way to store and transfer the released energy. This comes in the form of molecules such as ATP. An overview of how this process works is shown in **Figure 23.1**:

① **Energy is released from exergonic catabolic reactions.** Nutrients such as glucose, fatty acids, and amino acids are broken down via catabolic reactions that eventually release energy.

② **This energy is used to fuel the endergonic anabolic reaction of ATP synthesis.** Synthesis of molecules such as ATP is an endergonic process. The energy for this reaction comes from the exergonic reactions in the previous step.

③ **ATP is broken down in an exergonic catabolic reaction.** In another exergonic catabolic reaction, the third phosphate group from ATP is removed and energy is released.

④ **The energy from ATP breakdown fuels other endergonic anabolic reactions in the cell.** ATP breakdown is a highly exergonic process that can fuel a number of endergonic reactions in the cell.

Since ATP is a critical component of these reactions, we look more closely at this molecule and its reactions next.

Quick Check

☐ 3. How do endergonic and exergonic reactions differ?
☐ 4. Why are endergonic and exergonic reactions coupled?

Adenosine Triphosphate (ATP) and Phosphorylation

◄◄ **Flashback**

1. What is chemical energy? (p. 42)
2. What happens during a hydrolysis reaction? (p. 50)

ATP is the cell's main source of energy "money," but how can such a simple molecule drive nearly all cellular processes? To begin, recall what happened when we combined two unstable atoms, such as sodium and chlorine—they engaged in an exergonic reaction and formed products that have less energy and were therefore more stable (see Figure 2.4). In the process, they released their excess energy in a small explosion. The same principle is at work with ATP, because ATP, like lone sodium and chlorine atoms, is an unstable molecule. The phosphate groups of ATP have negative charges, and as a result they repel one another strongly. The repelling charges put a strain on the molecule, making these bonds highly unstable and the molecule reactive.

The bond between the second and third phosphate groups of ATP is particularly unstable because these groups repel each other due to their negative charges, and a great deal of energy is released when this bond is broken. Some sources call this bond a "high-energy bond." This term is not correct, as there is no more energy contained in this bond than in any other chemical bond. This reaction is highly exergonic only because the bond is the most unstable one in ATP due to the repelling forces of the first two phosphate groups.

The cell is able to harness energy from ATP by removing its third phosphate group in a *hydrolysis* reaction. In this reaction, the bond between the second and third phosphate groups is broken by a water molecule, resulting in a free phosphate molecule (P_i) and a molecule of *adenosine diphosphate,* or *ADP.* As we have discussed previously, these reactions are catalyzed by enzymes called *ATPases.*

Although ATP hydrolysis is highly exergonic, the cell is able to harness only about 40% of this energy to perform work. The low number reflects the fact that exergonic reactions are not completely efficient, and much of the released energy is lost as heat. In the cell, about 60% of the energy from ATP hydrolysis is lost as heat. This may seem wasteful, but bear in mind that most processes that burn fuel, such as a wooden log on a fire, lose about 70–90% of their energy as heat—which explains why fires and running engines feel hot to the touch. So ATP hydrolysis is pretty efficient, relatively speaking.

The energy that the cell harnesses from ATP drives cellular processes in two ways. First, it can be used by the cell to directly "pay for" certain reactions. ATP is used in this manner to fuel processes such as muscle contraction or the movement of a ribosome along a strand of messenger RNA.

The second mechanism by which ATP "pays for" most cellular processes is slightly more complex. In this process, called **phosphorylation** (fos-for′-ah-LAY-shun), ATP donates a phosphate group to a reactant, with the help of an enzyme. Phosphorylation energizes the reactant in much the same way it energizes ADP—by making the resulting product unstable. This instability favors the conversion of the reactant to the product.

A simple example of this is the conversion of the amino acid glutamate to a different amino acid, glutamine. This reaction requires the addition of an ammonia molecule to glutamate. ATP assists this reaction as follows: (1) ATP is hydrolyzed, yielding ADP and P_i. (2) The energy released in the first step is used in an endergonic reaction to add the P_i to glutamate, which produces an unstable molecule:

$$\text{Glutamate} + P_i + \text{Energy} \rightarrow \text{Glutamate-P}$$

The glutamate-P molecule is unstable for the same reason that ATP is unstable—the glutamate and P repel one another, which places strain on the molecule. (3) The P is removed from glutamate and replaced with an ammonia group, forming glutamine, the product:

$$\text{Glutamate-P} + NH_3 \rightarrow \text{Glutamine} + P_i + \text{Energy}$$

This reaction is exergonic because the product, glutamine, has less energy than the glutamate-P molecule.

Quick Check

☐ 5. Why is the removal of the third phosphate group from ATP highly exergonic?

☐ 6. How does ATP fuel cellular processes?

Nutrients and ATP Generation

◀◀ **Flashback**

1. What is an oxidation-reduction reaction? (p. 43)

Before we discuss ATP production, we need to take a look at how a cell uses the chemical energy in nutrients to drive ATP synthesis. This involves a type of reaction discussed in the chemistry chapter called an *oxidation-reduction reaction.* In this reaction, electrons are transferred from one reactant to another (see Chapter 2). The next subsections examine oxidation-reduction reactions and the molecules to which electrons are transferred: *electron carriers.*

Oxidation-Reduction Reactions

People often speak of "burning fat" when they exercise, and you may assume that this type of burning differs from burning wood in a fire. Surprisingly, however, this is not really the case. Although we obviously don't have mini fires raging inside our cells, the same basic reaction that takes place when wood burns in a fire also occurs within our cells. When something burns, some of its electrons are transferred from one molecule to another; this is an exergonic process that generates a great deal of heat. You can see this with the explosive sodium-chlorine reaction shown in **Figure 23.2**, in which heat is generated as the elements burn. As we discussed, such reactions are called **oxidation-reduction reactions** (see Chapter 2). The substance that loses electrons is **oxidized,** and the substance that gains electrons is **reduced** (a useful mnemonic is "OIL RIG," which stands for "Oxidation Is Loss; Reduction Is Gain").

When a wooden log burns, the molecules in the wood are oxidized—they lose electrons to the oxygen in the air, which is reduced. As electrons are stripped from the wood molecules, energy is released as heat, which keeps the fire hot and burning. Fires will not burn in the absence of oxygen because the oxygen must be present to accept electrons from the wood. Similarly, fires will not burn when the wood has been oxidized to ashes, because electrons have been removed and none are available to give to oxygen.

Essentially the same process takes place when our cells "burn" nutrients—only the oxidation-reduction reactions in our cells occur over a series of steps rather than all at once, as with a log in a fire. During catabolism, nutrient molecules such as glucose are oxidized and their electrons are passed to other molecules, which are reduced. The flow of electrons from one molecule to another is then used to perform cellular work, including the synthesis of ATP.

Figure 23.2 Energy release during an oxidation-reduction reaction.

Electron Carriers

Nutrient oxidation releases a great deal of chemical energy, but it produces very little ATP directly. Instead, it is the electrons that are stripped off the nutrients that generate most of the body's ATP.

ConceptBOOST 》》》

How Electron Movement Can Be Harnessed to Do Work

Using the movement of electrons to power the work of producing ATP is done in a way very similar to the function of a simple battery. A battery contains separate compartments with two metals (common examples are zinc and copper) that have differing *electron affinities*—that is, they attract electrons with unequal strength. As you can see here, when the two compartments are connected via a metal wire, electrons leave the metal with the lower electron affinity (zinc) and flow toward the metal with the higher electron affinity (copper):

The flow of electrons illuminates the light bulb.

Electrons flow away from zinc, which is being oxidized.

Electrons flow toward copper, which is being reduced.

Electron

Zinc metal (low affinity for electrons)

Copper metal (high affinity for electrons)

In the process, zinc is oxidized and copper is reduced. If this metal wire is placed within a device such as a light bulb or a motor, the spontaneous flow of electrons through the motor provides a force, the **electromotive force,** which can perform work. A battery "dies" when the metal with low electron affinity is completely oxidized and so can no longer supply electrons to the metal in the other compartment. Our cells contain "biological batteries" that work by this same principle. ■

During nutrient catabolism, nutrients are oxidized, and their electrons are removed (like the electrons of zinc in our battery example) and passed to molecules with higher electron affinities, called **electron carriers** (which would be analogous to the copper in the battery). This electron transfer generates an electromotive force that is used by the cell to perform work, the same way electricity from a battery can power a light bulb or a motor. The majority of the cell's electron carriers are integral and peripheral proteins found within the mitochondrial membranes. However, two electron carriers, **nicotinamide adenine dinucleotide (NAD$^+$)** and **flavin adenine dinucleotide (FAD),** are located within the cytosol and the mitochondrial matrix. These two electron carriers are the first to accept electrons from nutrients during catabolism. NAD$^+$ does this as follows:

$$NAD^+ + H^+ + 2e^- \longrightarrow NADH$$

Note that the electrons (purple dots) are carried by and added with hydrogen atoms (not hydrogen ions, which have no electrons). Therefore, when reduced, NAD$^+$ becomes NADH (and FAD becomes FADH$_2$). These electrons will eventually be passed to other electron carriers to generate an electromotive force linked to ATP production. The specifics of these processes are discussed in the next two modules.

Quick Check

☐ 7. In an oxidation-reduction reaction, what happens to the chemical that is oxidized? What happens to the chemical that is reduced?

☐ 8. What is electron affinity? Do electrons flow toward a chemical with a higher or lower electron affinity?

☐ 9. What is the electromotive force?

Apply What You Learned

1. Would the reaction involving two stable molecules likely be endergonic or exergonic? Explain. How would this change if the two molecules were unstable? Why?

2. Explain how a fire that burns coal is similar to the reactions that take place in your cells.

3. Riboflavin (ry-boh-FLAY-vin), also called vitamin B2, is required for synthesis of FAD. Predict the effects of a diet that is deficient in riboflavin.

See answers in Appendix A.

MODULE **23.2**

Glucose Catabolism and ATP Synthesis

Learning Outcomes

1. Explain the overall reaction for glucose catabolism.
2. Describe the processes of glycolysis, formation of acetyl CoA, and the citric acid cycle.
3. Describe the process of the electron transport chain.
4. Discuss the process of chemiosmosis and its role in ATP production.
5. Give the energy yield of each part of glucose catabolism and the overall energy yield of glucose catabolism.

Glucose catabolism refers to the reactions that involve the breakdown of the monosaccharide glucose and the use of the chemical energy in its bonds to drive ATP synthesis. These reactions are critical, as glucose is the preferred carbohydrate fuel for essentially every cell in the body. Many cells can use the monosaccharides fructose and galactose (which are isomers of glucose), as well, but they are almost always converted to glucose before they are catabolized.

Glucose catabolism has two main parts, each of which generates ATP: (1) *glycolysis* (gly-KOL-uh-sis), a series of reactions in the cytosol that split glucose; and (2) the *citric acid cycle,* a series of reactions in the mitochondrial matrix that break down glucose further. Then, to use the energy liberated by glucose catabolism, a series of oxidation-reduction reactions known as *oxidative phosphorylation* takes place in the inner mitochondrial membrane. This series of reactions involves the transfer of electrons between electron carriers known collectively as the *electron transport chain* (*ETC*), and this leads to ATP synthesis. You can see how these processes are related in **Figure 23.3**. The upcoming sections present each in greater depth.

Overview of Glucose Catabolism and ATP Synthesis

◀◀ **Flashback**

1. How does creatine phosphate work in a skeletal muscle fiber? (p. 361)
2. How do oxidative catabolism and glycolytic catabolism differ? (p. 361)

Glucose contains a great deal of potential energy in its bonds, and that energy is released when it is catabolized. The energy released from glucose catabolism can be used to produce ATP in two ways:

- **Substrate-level phosphorylation. Substrate-level phosphorylation** involves the transfer of a phosphate group directly from a phosphate-containing molecule (the substrate) to ADP, to form ATP. A simple example is one that you saw in

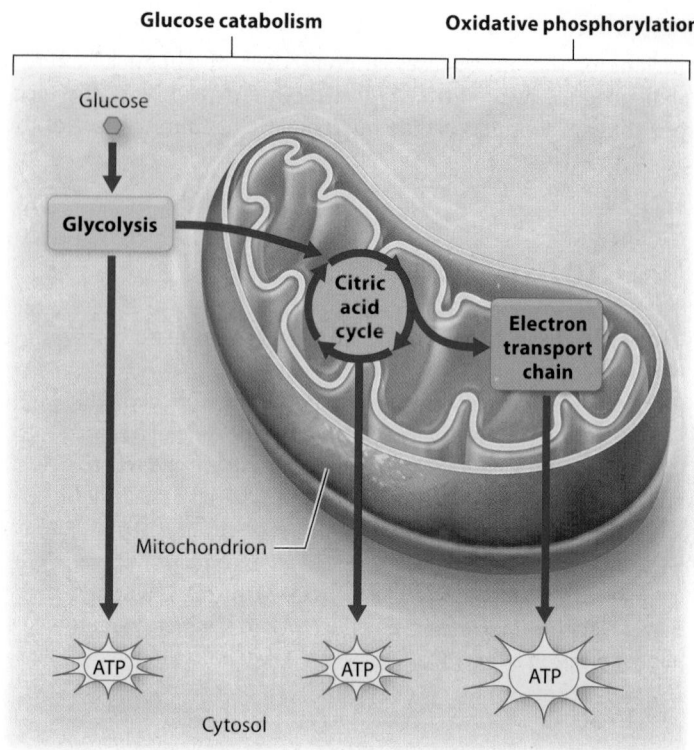

Figure 23.3 Overview of glucose catabolism and oxidative phosphorylation.

the muscle tissue chapter—that of creatine phosphate (see Chapter 10). Recall that creatine phosphate donates its phosphate group to ADP, producing ATP:

$$\text{Creatine phosphate} + \text{ADP} \rightarrow \text{ATP} + \text{Creatine}$$

In this reaction, creatine phosphate is the substrate that is phosphorylating ADP to yield ATP. Similar reactions occur during glycolysis and the citric acid cycle, as energy released from glucose catabolism drives the transfer of a phosphate group from a substrate to ADP.

- **Oxidative phosphorylation.** During **oxidative phosphorylation,** the energy from the flow of electrons is harnessed in a process that generates ATP (recall our earlier battery example). Oxidative phosphorylation produces a great deal more ATP for the cell than does substrate-level phosphorylation. Note that oxidative phosphorylation occurs in the mitochondria as part of the ETC. The reactions of glucose catabolism in glycolysis and the citric acid cycle reduce electron carriers, which are then fed into the ETC and participate in oxidative phosphorylation.

The reactions of glucose catabolism can be divided broadly into two classes. The first class is known as **glycolytic catabolism,** or **anaerobic catabolism,** because the reactions can take place in the absence of oxygen (an-air-OH-bik; *an-* = "without," *aero-* = "air"). As implied by the name, the process of glycolysis involves glycolytic catabolism. The second class is called **oxidative catabolism,** or **aerobic catabolism,** because the reactions do require oxygen. The citric acid cycle and the electron transport chain are both types of oxidative catabolism. Oxidative catabolism is also sometimes called *cellular respiration*. This should not be confused with pulmonary ventilation, which refers to the movement of air in and out of the lungs. Rather, cellular respiration refers to the cell's consumption of oxygen and production of carbon dioxide and water.

Glucose Catabolism Part 1: Glycolysis

Glycolysis (*glyco-* = "sugar," *lysis* = "splitting") is a series of 10 reactions that takes place in the cytosol (**Figure 23.4**). As with all human metabolism, all glycolytic reactions are catalyzed by enzymes. Its name suggests what happens during glycolysis: Glucose is split from a six-carbon molecule into two three-carbon sugars called **pyruvate** (PY-roo-vayt; also referred to as *pyruvic acid*). Glycolysis occurs in every cell in the body, even cells that lack mitochondria and other organelles, such as erythrocytes. The reactions of glycolysis can be broadly separated into two phases: the *energy investment phase*, during which a small amount of ATP is actually consumed, and the *energy payoff phase*, during which NADH and ATP are made. Although glycolysis consists of ten separate reactions (five in each phase), we have combined the reactions into five steps for simplicity. Let's look at what happens in each phase.

Energy Investment Phase

The energy investment phase of glycolysis consists of five reactions that have been condensed here into three major steps. They proceed as follows:

① **First phosphorylation: Glucose is phosphorylated by ATP, yielding glucose-6-phosphate and ADP.** The six-carbon glucose is phosphorylated in the first reaction of this phase, "spending" an ATP molecule. As you can see in Figure 23.4, a phosphate group is removed from ATP and attached to glucose, producing a molecule called *glucose-6-phosphate* (and ADP). This reaction is critical to glucose catabolism because it effectively traps the glucose in the cytosol and prevents it from leaving the cell.

② **Second phosphorylation: The carbon atoms in glucose-6-phosphate are rearranged, and the molecule is then phosphorylated by another ATP, yielding fructose-1,6-bisphosphate and ADP.** In reactions two and three, the carbons of glucose-6-phosphate are rearranged and a second phosphorylation reaction takes place, which consumes

another molecule of ATP. This reaction yields the six-carbon sugar fructose-1,6-bisphosphate (and ADP).

③ **Cleavage: The six-carbon fructose-1,6-bisphosphate is split, and two three-carbon molecules are formed.** During reactions four and five, the six-carbon sugar fructose is split and two three-carbon molecules are produced: glyceraldehyde-3-phosphate and dihydroxyacetone-3-phosphate. The latter molecule undergoes another reaction and is converted into a second molecule of glyceraldehyde-3-phosphate.

At the end of the energy investment phase of glycolysis, we have "spent" two molecules of ATP and are left with two three-carbon compounds, each of which contains a phosphate group. These compounds are now ready to enter the energy payoff phase.

Energy Payoff Phase

The energy payoff phase of glycolysis consists of five reactions that we have condensed into two major steps. During these reactions, the phosphate groups of the three-carbon sugars are transferred to ADP, to yield ATP, and the compounds are oxidized to produce NADH. Remember that each three-carbon compound produced in the energy investment phase goes through these reactions, so there are two rounds of energy payoff reactions for each molecule of glucose. The reactions proceed as follows:

④ **Oxidation: Glyceraldehyde-3-phosphate is phosphorylated and oxidized by NAD⁺ to yield NADH and 1,3-bisphosphoglycerate, which then donates a P_i to ADP, yielding ATP.** During reaction six, the glyceraldehyde-3-phosphate molecules are each given second phosphate groups, after which both are oxidized and their electrons transferred to NAD⁺, which produces two molecules of the reduced NADH. In reaction seven, one of the phosphate groups of each molecule of 1,3-bisphosphoglycerate is transferred to ADP, yielding two molecules of 3-phosphoglycerate and two molecules of ATP by substrate-level phosphorylation.

⑤ **ATP synthesis: The carbon atoms in 3-phosphoglycerate are rearranged to form phosphoenolpyruvate, which donates a P_i to ADP, yielding ATP and pyruvate.** After rearranging the carbon atoms in the two molecules of 3-phosphoglycerate to produce two molecules of phosphoenolpyruvate in reactions eight and nine, the remaining phosphate groups on the phosphoenolphyruvate molecules are transferred to ADP via substrate-level phosphorylation in reaction ten. This produces two molecules of ATP and two three-carbon pyruvate molecules.

Overall Yield of Glycolysis

So, in a nutshell, what have we accomplished during glycolysis? At the conclusion of the energy payoff phase, we have:

- spent two molecules of ATP,
- synthesized four molecules of ATP,
- synthesized two molecules of NADH, and
- split glucose into two three-carbon pyruvate molecules.

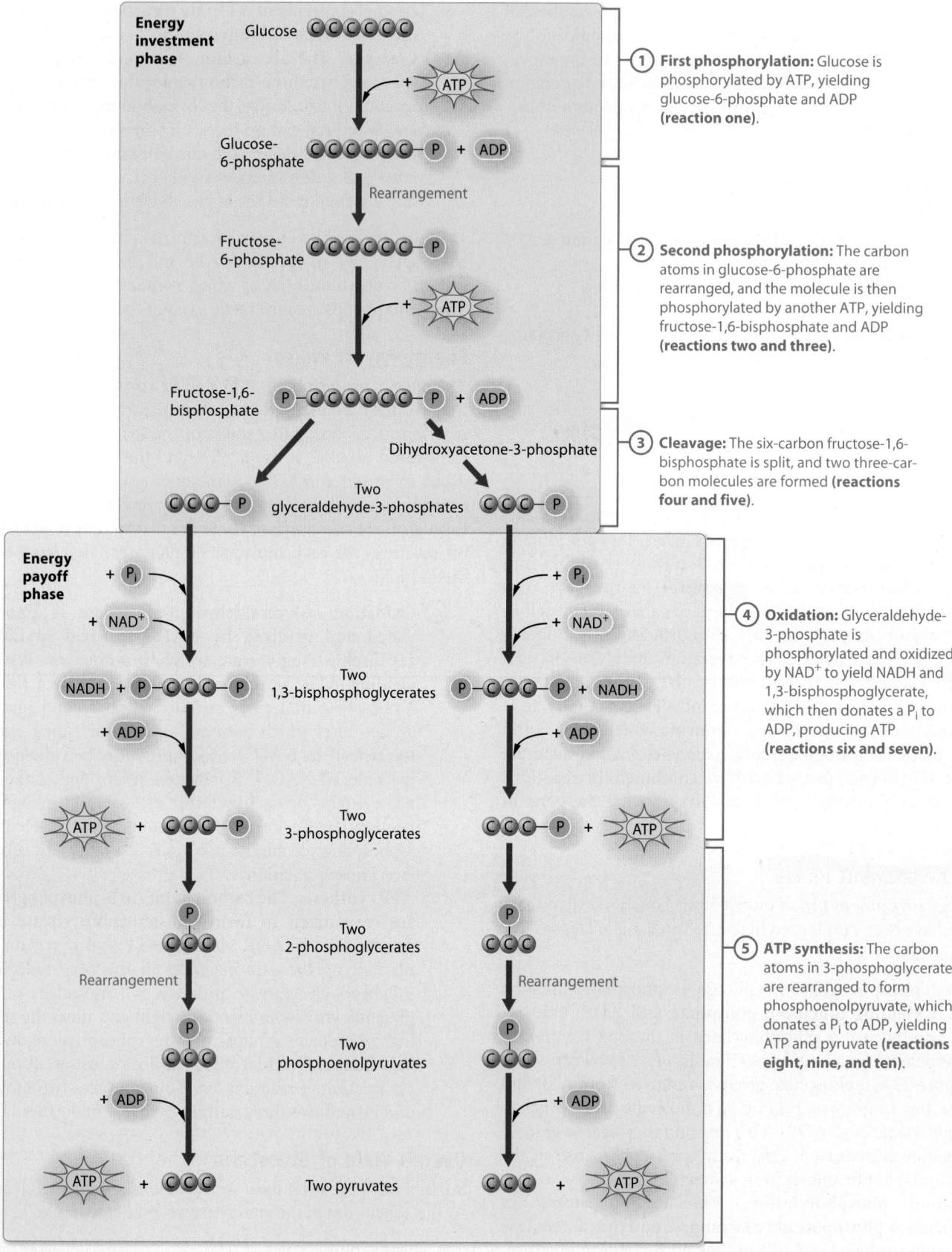

Figure 23.4 The reactions of glycolysis.

Therefore, our *net* energy yield from the splitting of a single glucose molecule by glycolysis is two molecules of ATP and two molecules of NADH.

Intermediate Step: The Fate of Pyruvate

The fate of the two pyruvate molecules produced at the end of glycolysis depends on the amount of oxygen available to the cell (**Figure 23.5**):

- **Under anaerobic conditions, pyruvate is reduced to lactate.** When oxygen is not present in significant quantities,

(a) **Lactate formation:** Under anaerobic conditions, pyruvate is reduced to lactate and NADH is oxidized to NAD⁺.

① Pyruvate loses a carbon atom to yield acetate and CO_2.

② Acetate is oxidized by NAD⁺ and combined with coenzyme A to yield acetyl-CoA and NADH.

(b) **Pyruvate oxidation:** Under aerobic conditions, pyruvate moves into the mitochondria, is oxidized, and then enters the citric acid cycle.

Sequence the steps @MasteringA&P®

Figure 23.5 Intermediate step: the fate of pyruvate after glycolysis. Note that two molecules of pyruvate are produced for every molecule of glucose broken down.

the NADH produced in the energy payoff phase of glycolysis gives its electrons to the two pyruvate molecules (see Figure 23.5a). This reduces them to two three-carbon molecules called **lactate.** (Lactate was long believed to be a largely dead-end product of glycolysis, but research in the past decade has indicated that lactate is able to directly enter the mitochondria to be oxidized.) As pyruvate is reduced, NADH is oxidized to NAD⁺. This reaction is critical because it regenerates NAD⁺, which is needed to oxidize glyceraldehyde-3-phosphate to 1,3-bisphosphoglycerate in step four. Without this NAD⁺, no ATP would be generated in glycolysis at all.

- **Under aerobic conditions, pyruvate moves into the mitochondria, is oxidized, and then enters the second step of glucose catabolism, the citric acid cycle.** When oxygen concentrations are adequate, the two pyruvate molecules enter the mitochondrial matrix for pyruvate oxidation (see Figure 23.5b). ① There the pyruvate loses a carbon atom to yield a molecule of acetate and a molecule of carbon dioxide, a process called *decarboxylation.* The carbon dioxide molecule exits the cell and enters the bloodstream, where it is released from the lungs during expiration. ② The acetate molecule is immediately oxidized by NAD⁺ and combined with a large molecule called **coenzyme A (CoA).** The result of this reaction is the formation of the molecule **acetyl-CoA** (ah-SEE-til) and a molecule of NADH. The two acetyl-CoA molecules then enter the citric acid cycle, discussed next.

Quick Check

☐ 4. How much ATP is expended during the energy investment phase of glycolysis? How much is generated during the energy payoff phase?
☐ 5. What is the fate of pyruvate molecules under anaerobic conditions?
☐ 6. Why is it important to regenerate NAD⁺ under anaerobic conditions?

Glucose Catabolism Part 2: The Citric Acid Cycle

◀◀ Flashback

1. What is the primary function of mitochondria? (p. 88)
2. Where is the mitochondrial matrix located? (p. 89)

The second part of glucose catabolism is a series of eight reactions that takes place in the mitochondrial matrix; this series is the **citric acid cycle** (it is also often called the *tricarboxylic acid cycle,* or the *Krebs cycle* after its discoverer, Sir Hans Krebs). The citric acid cycle represents the beginning of the cell's oxidative glucose catabolism. The reactions do not require oxygen directly, but they will not proceed unless sufficient oxygen is present. These reactions produce a relatively small amount of ATP by substrate-level phosphorylation.

When acetyl-CoA from pyruvate oxidation enters the citric acid cycle, it proceeds as follows (**Figure 23.6**; note that we have condensed the eight reactions into four steps):

① **Citrate synthesis: Acetyl-CoA combines with oxaloacetate to form citrate and CoA.** In reaction one, the acetyl-CoA molecule "donates" its two-carbon acetyl group to a four-carbon compound called oxaloacetate, forming the six-carbon compound **citrate** (for which the cycle was named) and leaving CoA. Acetyl-CoA in this reaction acts a little like an ATP molecule—acetyl-CoA has a great deal of potential energy, as its bond is unstable. When the bond is broken, the resulting CoA and acetate molecules have less energy, and the released energy helps to unite acetate and oxaloacetate.

② **First oxidation: Citrate is rearranged, then oxidized by NAD^+, generating CO_2 and NADH.** In reactions two through four, the atoms in the six-carbon citrate are rearranged and oxidized, eventually forming the four-carbon

compound *succinyl-CoA* (containing the same CoA that was part of acetyl-CoA). The two removed carbons are lost as two molecules of carbon dioxide, and the removed electrons are accepted by two molecules of NAD^+, producing two molecules of NADH.

③ **ATP synthesis: Succinyl-CoA is converted to succinate and CoA, while forming ATP.** In the fifth reaction, the CoA molecule is removed from succinyl-CoA. In this reaction we again see an unstable bond between CoA and its partner. The energy released when CoA is removed from succinate is used to drive the formation of ATP by substrate-level phosphorylation (sometimes the very similar molecule *GTP*, or *guanosine triphosphate*, is produced instead).

④ **Second oxidation: Succinate is oxidized by FAD and NAD^+, generating $FADH_2$ and NADH, and is converted back to oxaloacetate.** Reactions six through eight of the citric acid cycle see succinate oxidized to other four-carbon compounds until it is converted back into oxaloacetate. As the compounds are oxidized, their electrons are

From glycolysis and pyruvate oxidation

Oxaloacetate

① **Citrate synthesis:** Acetyl-CoA combines with oxaloacetate to form citrate and CoA **(reaction one).**

Acetyl-CoA

Citrate

④ **Second oxidation:** Succinate is oxidized by FAD and NAD^+ (removed electrons are in purple), generating $FADH_2$ and NADH, and is converted back to oxaloacetate **(reactions six, seven, and eight).**

Malate

Fumarate

Citric Acid Cycle (in mitochondrial matrix)

Isocitrate

Succinate

③ **ATP synthesis:** Succinyl-CoA is converted to succinate and CoA, while forming ATP **(reaction five).**

Succinyl-CoA

α-Ketoglutarate

② **First oxidation:** Citrate is rearranged, then oxidized by NAD^+ (the removed electrons are shown in purple), generating CO_2 and NADH **(reactions two, three, and four).**

Figure 23.6 The citric acid cycle. For each glucose molecule broken down, this cycle "turns" twice.

Sequence the steps
@MasteringA&P®

transferred to FAD and NAD$^+$, generating one molecule each of FADH$_2$ and NADH.

By the end of the citric acid cycle, we have regenerated oxaloacetate, the compound with which we started in step one. This is why this process is called a *cycle*—it begins and ends with the same molecule.

Note that each glucose molecule yields two acetyl-CoA molecules, so for each glucose molecule, this cycle "turns" twice. Keeping this in mind, we can tally up the overall yield per glucose molecule from the start of glucose catabolism:

- ten NADH molecules (two from glycolysis, two from pyruvate oxidation, and six from the citric acid cycle),
- two FADH$_2$ molecules, and
- four ATP molecules (two from glycolysis and two from the citric acid cycle).

As you can see, we have generated very little energy in the form of ATP from these reactions. However, we have conserved the chemical energy in the original glucose molecule by passing off its electrons to NADH and FADH$_2$. Both of these electron carriers will now move on to the process of ATP synthesis with the electron transport chain.

ATP Synthesis: The Electron Transport Chain and Oxidative Phosphorylation

◀◀ Flashback

1. How is potential energy present in a concentration gradient? (p. 75)

We are finally at the stage where most of the potential energy of glucose is used to make ATP: oxidative phosphorylation. Oxidative phosphorylation involves three processes: (1) the transfer of electrons between electron carriers, (2) the generation and maintenance of a hydrogen ion concentration gradient, and (3) the use of this gradient to drive the release of ATP. This section examines these processes.

Electron Transfer and Proton Pumping

The energy stored in the electrons generated in glucose catabolism and accepted by NADH and FADH$_2$ is released by a group of more than 15 different electron carriers collectively called the **electron transport chain (ETC)**. As you can see in **Figure 23.7a**, the electron carriers of the ETC are contained within four large enzyme complexes (numbered I–IV) embedded in the inner mitochondrial membrane.

These carriers are also sometimes called the *electron transfer chain*, in recognition of the fact that electrons are transferred between them. Remember that these electrons are usually passed between carriers as part of hydrogen atoms. The events of oxidative phosphorylation unfold as follows (see Figure 23.7a):

① **NADH and FADH$_2$ are oxidized; the flow of their electrons generates an electromotive force in the electron transport chain.** NADH is oxidized to NAD$^+$, its electrons are passed to the first electron carrier in enzyme complex I, and the electron carrier is reduced. As this process continues, the flow of electrons acts as a "cellular battery" and generates an electromotive force in the ETC. (FADH$_2$ is oxidized in a similar manner using a different electron carrier in complex II.)

② **Complex I uses this force to pump hydrogen ions into the intermembrane space against the concentration gradient.** Complex I harnesses the energy of this electromotive force, just as we saw with our earlier battery example, to pump hydrogen ions against the concentration gradient. Hydrogen ions are pumped from the mitochondrial matrix, where their concentration is relatively low, into the intermembrane space, where their concentration is higher. This first establishes and then maintains a steep hydrogen ion concentration gradient. (Hydrogen ions are also often called *protons* because that is their sole subatomic particle, so the movement of hydrogen ions is sometimes referred to as *proton pumping*.) Note that some of the electrons may also be transferred through complex II, which does not contribute to hydrogen ion pumping.

③ **Electron flow through complexes III and IV continues to drive the pumping of hydrogen ions into the intermembrane space.** As you can see in the figure, electrons continue to be transferred, first to complex III and then complex IV. As they gain electrons, the electron carriers in these complexes are reduced and hydrogen ions continue to be pumped into the intermembrane space.

④ **Complex IV transfers the electrons to oxygen—the final electron acceptor—to form H$_2$O.** Complex IV transfers electrons to the final electron acceptor, oxygen. In this reaction, oxygen is reduced and accepts the electrons while combining with two hydrogen ions. The end result of this reduction is the following reaction:

$$2\,H^+ + 2\,e^- + \tfrac{1}{2}\,O_2 \rightarrow H_2O$$

In other words, one oxygen atom forms a covalent bond with two hydrogen atoms to produce water. This step is absolutely critical—without oxygen acting as the final electron acceptor, glucose catabolism stops at glycolysis. In fact, this is the main reason why we breathe oxygen, and why we cannot live without it. For information on what happens when this step is blocked, see *A&P in the Real World: Cyanide and the ETC* on page 910.

At this point, you may be wondering why electrons have to pass through so many electron carriers before they get to oxygen. Although it would be simpler to transfer the electrons directly from NADH and FADH$_2$ to oxygen, doing so would cause most of the energy of the reaction to be lost as heat. The fact that this process proceeds by several small steps allows more of the released energy to be harnessed to do work, pumping hydrogen ions into the intermembrane space against the concentration gradient.

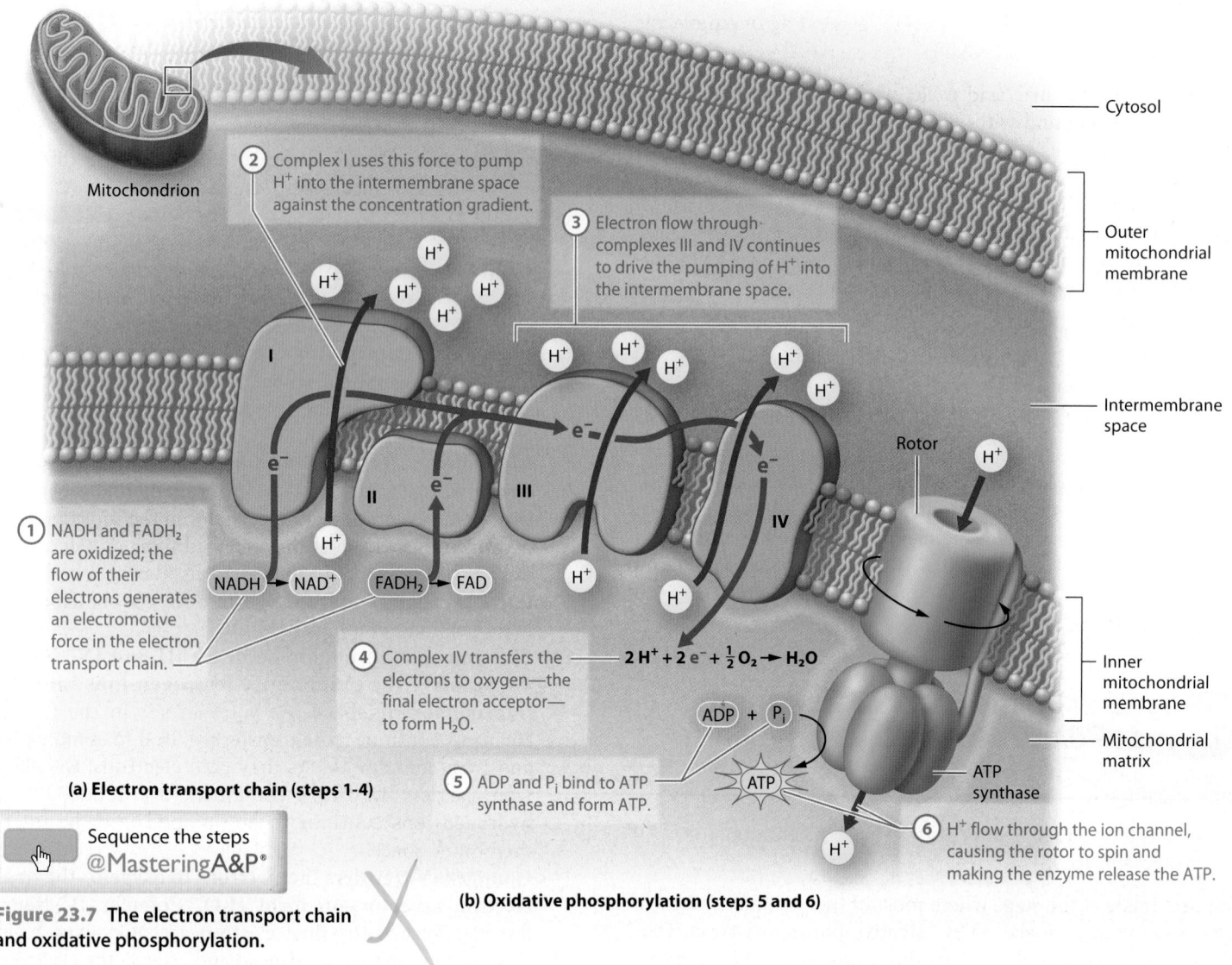

Mitochondrion

Cytosol

Outer mitochondrial membrane

2 Complex I uses this force to pump H^+ into the intermembrane space against the concentration gradient.

3 Electron flow through complexes III and IV continues to drive the pumping of H^+ into the intermembrane space.

Intermembrane space

Rotor

1 NADH and $FADH_2$ are oxidized; the flow of their electrons generates an electromotive force in the electron transport chain.

NADH → NAD⁺ $FADH_2$ → FAD

4 Complex IV transfers the electrons to oxygen—the final electron acceptor—to form H_2O.

$$2\,H^+ + 2\,e^- + \tfrac{1}{2}O_2 \rightarrow H_2O$$

Inner mitochondrial membrane

Mitochondrial matrix

ADP + P_i

ATP

ATP synthase

(a) Electron transport chain (steps 1-4)

5 ADP and P_i bind to ATP synthase and form ATP.

6 H^+ flow through the ion channel, causing the rotor to spin and making the enzyme release the ATP.

Sequence the steps
@MasteringA&P®

(b) Oxidative phosphorylation (steps 5 and 6)

Figure 23.7 The electron transport chain and oxidative phosphorylation.

ConceptBOOST »»»

Why Do We Breathe?

As you likely have already discovered, we take in oxygen-rich air when we inhale, and we breathe out carbon dioxide–rich air when we exhale (see Chapter 21). This has led to the unfortunate misconception that oxygen is somehow converted into carbon dioxide in the body's cells. Although it is true that both carbon dioxide and oxygen have to do with our metabolic processes, the two do not directly interact in any way, nor does one ever become the other. So let's clarify the roles and fates of oxygen and carbon dioxide:

- **Oxygen accepts the final electrons in the ETC to become a molecule of water.** The oxygen that we inhale is picked up by the bloodstream and delivered to our cells. In the cell, it ends up in the inner mitochondrial membrane, where it accepts the final electrons in the ETC. When oxygen accepts the electrons, it becomes part of a molecule of water. Thus, the oxidation-reduction reactions in the mitochondria generate

water as a final end product (sometimes called metabolic water).

- **Carbon dioxide comes from carbons lost during carbohydrate catabolism.** Before pyruvate enters the citric acid cycle, it loses one of its carbons as a molecule of carbon dioxide. Similarly, carbons are removed during the citric acid cycle and are also lost as carbon dioxide. As we've mentioned, these carbon dioxide molecules diffuse through the cytosol, exit the cell, and enter the bloodstream; when they reach the lungs, they are exhaled.

To summarize, oxygen is never transformed into carbon dioxide or vice versa. These two molecules are parts of two different processes of carbohydrate metabolism—inhaled O_2 becomes H_2O, whereas ingested CHO (part of glucose) becomes CO_2. Note, however, that if the blood oxygen level is low, less carbon dioxide will be formed. This is because under anaerobic conditions, pyruvate will not enter the citric acid cycle and so fewer carbons will be removed. ■

ATP Synthesis

At this point, the energy generated by the "cellular battery" has pumped hydrogen ions against the concentration gradient from the mitochondrial matrix into the intermembrane space. The result of this pumping is that we now have a steep concentration

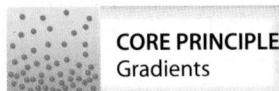 **CORE PRINCIPLE** Gradients

gradient, with more hydrogen ions in the intermembrane space than in the mitochondrial matrix. We also have an electrical gradient, with more positive charges in the intermembrane space than in the matrix, bringing us back to the Gradients Core Principle (p. 26).

The two gradients combined make up an electrochemical gradient, which is a source of potential energy (see Chapter 11). Overall, the potential energy in the bonds of the glucose molecule has been transformed into the potential energy of this electrochemical gradient. This potential energy becomes kinetic energy when hydrogen ions flow through an ion channel across the membrane. The kinetic energy is then used to perform work.

Using an electrochemical gradient to do work is a type of mechanism known as *chemiosmosis*. You can visualize how chemiosmosis works with the simple analogy of a hydroelectric dam—when the dam is opened, the potential energy of the water becomes kinetic energy as the water flows, and the flow of water generates a force that can be used to perform work, such as turning a turbine to produce electricity.

The flow of hydrogen ions back into the matrix down the electrochemical gradient drives the work of ATP synthesis by a mechanism surprisingly similar to our dam example. In this case, the "turbine" is part of a large protein in the inner mitochondrial membrane, called **ATP synthase.** This protein functions as both an ion channel through the membrane and an enzyme that catalyzes the production of ATP, as you can see in **Figure 23.7b.** The process continues as follows:

⑤ **ADP and P$_i$ bind to ATP synthase and form ATP.** ATP synthesis begins when ADP and P$_i$ bind to the enzyme portion of ATP synthase, uniting them and forming ATP. Since this enzyme is most stable when bound to ATP, it holds on to it tightly. ATP cannot be used by the cell until it is released from this enzyme.

⑥ **Hydrogen ions flow through the ion channel, causing the rotor to spin and making the enzyme release the ATP.** Now the energy from glucose catabolism finally comes into play. The electrochemical gradient causes hydrogen ions to flow into the matrix through the ion channel of ATP synthase, which generates a force called the *proton motive force*. The electrochemical energy of the proton motive force causes a portion of the enzyme called the *rotor* to start spinning just like a turbine. The spinning rotor causes the enzyme to release its ATP. This process is repeated, with a new ADP and P$_i$ binding and joining into ATP, and the spinning rotor causing ATP to be released. So the majority of the energy harnessed from the oxidation of glucose is used to create a hydrogen ion gradient. The energy of the hydrogen ion gradient then causes ATP synthase to release the newly formed ATP for the cell to use.

Putting It All Together: The Big Picture of Glucose Catabolism and ATP Synthesis

Now that we have progressed through all the parts of glucose catabolism and ATP synthesis, we can summarize the events and figure out how much ATP we have formed per molecule of glucose (**Figure 23.8**). The overall reaction of glucose catabolism is as follows:

$$C_6H_{12}O_6 + 6\,O_2 \rightarrow 6\,H_2O + 6\,CO_2 + 38\,ATP + heat$$

Here we are assuming that adequate oxygen is available and oxidative catabolism can take place.

↑ STUDY BOOST

ATP Yield from Glucose Catabolism and ATP Synthesis

Let's calculate how many total ATP are produced per glucose molecule. The energy released from the oxidation-reduction reactions of one NADH molecule pumps enough hydrogen ions into the intermembrane space to drive the release of three ATP molecules (we're rounding up a bit to use whole numbers here). The yield for FADH$_2$ is slightly lower at two ATP molecules. This brings us to the following total:

Process	Yield of ATP from glucose catabolism	Yield of electron carriers from glucose catabolism	Yield of ATP from electron carriers going through ETC and oxidative phosphorylation	Final yield of ATP
Glycolysis	2 ATP			2
		2 NADH	2 × 3 ATP	6*
Pyruvate oxidation		2 NADH	2 × 3 ATP	6
Citric acid cycle	2 ATP			2
		6 NADH	6 × 3 ATP	18
		2 FADH$_2$	2 × 2 ATP	4

TOTAL = 38 ATP **

*The NADH generated in the cytosol from glycolysis may yield 4 or 6 ATP depending on the process used to transport its electrons into the mitochondria.

**If only 4 ATP are generated from the cytosolic NADH, the total number of ATP produced is 36.

When one molecule of glucose is completely oxidized, then the overall energy gain by the cell is 38 ATP. ■

(1) Glycolysis: Glucose is split and oxidized during glycolysis, yielding two pyruvate, two NADH, and two ATP (see Figure 23.4).

Cytosol

GLYCOLYSIS

Glucose

2 NADH
+
2 Pyruvate

Intermembrane space

(4) Electron transport chain: Electrons from NADH and FADH$_2$ enter the electron transport chain (ETC), powering the pumping of H$^+$ into the intermembrane space (see Figure 23.7a).

(5) Oxidative phosphorylation: H$^+$ re-enter the matrix through ATP synthase, which releases ATP from the synthase enzyme (see Figure 23.7b).

2 Acetyl-CoA
+
2 NADH

(2) Intermediate step: The two pyruvates are oxidized to two acetyl-CoA, producing two NADH (see Figure 23.5).

CITRIC ACID CYCLE

H$^+$ H$^+$ H$^+$ H$^+$
H$^+$

ETC

6 NADH
+
2 FADH$_2$

(3) Citric acid cycle: The two acetyl-CoA are further oxidized in the citric acid cycle, yielding six NADH, two FADH$_2$, and two ATP (see Figure 23.6).

2 ATP

34 ATP

Mitochondrial matrix

Mitochondrion

2 ATP

2 ATP

34 ATP

TOTAL = 38 ATP

The advantages of continuing the catabolism of glucose beyond glycolysis are pretty obvious—if we stop at glycolysis, nearly all of the potential energy in the glucose molecule is wasted. In fact, glycolysis by itself cannot produce enough ATP to sustain most cells in the body. This is why lack of oxygen is so dangerous; without adequate oxygen, the cells are forced to use glycolytic catabolism exclusively, which does not generate enough ATP to keep the cells, and so the body, alive.

Apply What You Learned

☐ 1. In the absence of oxygen, glucose is consumed extremely rapidly by the cell. However, glucose consumption decreases significantly when the cell is exposed to oxygen. Explain this finding. (*Hint:* Think

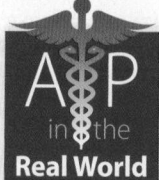

Cyanide and the ETC

The poison *cyanide* (SY-ah-ny'd) binds to a key enzyme of the ETC and inactivates it. This makes the cells unable to produce adequate ATP, even when the blood oxygen level is normal or elevated. Cyanide exposure is fatal only seconds after inhalation or injection; however, it may take several minutes to several hours for the poison to be fatal after ingestion or skin exposure. Several agents can be given as antidotes to cyanide poisoning, most of which either reverse its actions or bind to it and render it less harmful.

about the ATP yield of anaerobic processes versus aerobic processes.)

☐ 2. Why do you think that the amount of carbon dioxide you exhale increases when you exercise?

☐ 3. Predict what would happen to a cell if a mutation resulted in the formation of insufficient oxaloacetate in the citric acid cycle.

See answers in Appendix A.

MODULE **23.3**
Fatty Acid and Amino Acid Catabolism

Learning Outcomes

1. Describe lipolysis, deamination, and transamination.
2. Summarize the β-oxidation of fatty acids, and explain how it leads to ATP production and relates to ketogenesis.
3. Explain how amino acid catabolism leads to ATP production.
4. Describe the effect of amino acid catabolism on ammonia and urea production.

The cell is able to catabolize fatty acids and amino acids in addition to glucose. As we progress through fatty acid and amino acid catabolism, you will see that their pathways share many of the same products and pathways as in glucose catabolism. This is because the same basic principles are at work with the catabolism of all three nutrients—each is converted into chemicals that the cell can oxidize, and its electrons are removed and passed to electron carriers such as NADH and FADH₂. Just as with glucose, the electrons removed from fatty acids and amino acids are sent to the electron transport chain for oxidative phosphorylation, which is the final common pathway for each nutrient. We look at these pathways and processes in this section.

Fatty Acid Catabolism

◄◄ **Flashback**

1. What are the chemical components and properties of fatty acids and triglycerides? (pp. 52–53)
2. When are ketone bodies produced? (p. 615)

Recall that most fats in the body are found as *triglycerides,* which contain three long hydrocarbon chains called **fatty acids** bound to the modified sugar **glycerol** (see Figure 2.18). The fatty acids and glycerol are liberated by an enzyme-catalyzed process known as *lipolysis* (ly-POL-ih-sis). Both fatty acids and glycerol are used for fuel, but 95% of the energy generated by triglycerides comes from fatty acid oxidation. The catabolism of glycerol is fairly simple: It is converted to the compound glyceraldehyde-3-phosphate,

and it enters glycolysis (see Figure 23.4). However, the catabolism of fatty acids, which occurs by a process called *β-oxidation,* is slightly more complicated. Fatty acid catabolism generates acetyl-CoA molecules via β-oxidation, as well as molecules called *ketone bodies* by the process of *ketogenesis.*

β-oxidation

When fatty acids are released from adipocytes, they are transported through the bloodstream bound to protein carriers and delivered to tissues that can oxidize fatty acids, such as skeletal muscle and the heart. Inside these cells, the fatty acids enter the mitochondrial matrix. As shown in **Figure 23.9**, each fatty acid ① is bound to coenzyme A. From here, fatty acid oxidation proceeds by a series of reactions known as **β-oxidation.** ② In the reactions of β-oxidation, the fatty acid chain is oxidized by NAD⁺ and FAD, and two carbons are removed. ③ The products of this reaction are FADH₂, NADH, acetyl-CoA, and a fatty acid with two fewer carbons. FADH₂ and NADH go to the electron transport chain, and acetyl-CoA goes to the citric acid cycle. ④ Another CoA molecule binds to the shorter fatty acid, and rounds of β-oxidation are repeated until the entire fatty acid has been converted into acetyl-CoA subunits—fatty acids with six carbons yield three acetyl-CoA molecules, fatty acids with four carbons yield two acetyl-CoA molecules, and so forth. The released acetyl-CoA molecules are then fed into the citric acid cycle and further oxidized by the same pathway that we discussed with glucose catabolism. The NADH and FADH₂ from β-oxidation and from the citric acid cycle donate their electrons to the electron transport chain, and ATP is generated.

You may have heard that fat has more than twice the "calories" of carbohydrate or protein. This means that the oxidation of fats releases at least twice as much energy (or heat) as the other two classes of nutrients. Why, though, does fatty acid catabolism generate so much more energy? Let's examine this using the 16-carbon fatty acid palmitic acid as an example: With each round of β-oxidation, one molecule of NADH and one molecule of FADH₂ are produced as the oxidation-reduction reactions progress. For palmitic acid, β-oxidation will repeat seven times to break the whole fatty acid into two-carbon units. Once β-oxidation of palmitic acid is complete, we have generated seven NADH and FADH₂ molecules each, which will then enter the electron transport chain to generate 35 ATP.

But we haven't finished with our ATP-generating potential, as we have also produced eight acetyl-CoA molecules that will be fed into the citric acid cycle and generate a total of 96 more ATP. Clearly, then, the energy yield from fatty acids is much higher than that from a molecule of glucose, whose oxidation generates a maximum of 38 ATP. Note, though, that the energy yield from fatty acid oxidation will vary depending on the length of the fatty acid chain; those with longer chains yield more ATP, whereas those with shorter chains yield fewer.

Ketogenesis

Much of the β-oxidation in the body takes place in the liver, and under certain conditions the acetyl-CoA produced by β-oxidation can be converted into molecules called **ketone bodies** (KEE-tohn). This process, **ketogenesis,** occurs in the mitochondrial matrix and involves the combination of two acetyl-CoA molecules into a four-carbon ketone body. The liver cannot metabolize ketone bodies in any appreciable amount under normal conditions, so they are released into the bloodstream and delivered to other cells. In the cells, ketone bodies are split into two molecules of acetyl-CoA, after which they enter the citric acid cycle and are oxidized. Some tissues, such as those of the heart and skeletal muscle, take up and oxidize ketone bodies quite readily. Other tissues, such as the brain, are not as able to oxidize ketone bodies except under starvation conditions.

During extreme caloric restriction, carbohydrate restriction, or full starvation, the liver begins to rapidly oxidize fatty acids and produce large amounts of ketone bodies. However, the cells are able to use only so many of these ketone bodies because the lack of glucose decreases the amounts of oxaloacetate and other components of the citric acid cycle that are derived from glucose breakdown. As a result, ketone bodies accumulate in the blood, leading to a condition called *ketosis* (kee-TOH-sis). The body excretes many excess ketone bodies in the urine; however, if ketosis is severe, they accumulate and lower the pH of the blood, producing the dangerous condition *ketoacidosis.* Recall that this condition may also result from uncontrolled diabetes mellitus, in which the cells are unable to take in glucose and thus must rely on fatty acid oxidation for survival (see Chapter 16).

Quick Check

☐ 1. What is the basic process of β-oxidation? What happens to the resulting acetyl-CoA groups?
☐ 2. What happens to the ketone bodies often generated during fatty acid catabolism?

Amino Acid Catabolism

◄◄ **Flashback**

1. What is the basic chemical composition and structure of an amino acid? (p. 55)

Proteins used for catabolism are derived either from the diet or from proteins already in the cytosol. Dietary proteins are broken down into amino acid subunits in the digestive tract; these subunits then enter the bloodstream. Amino acid catabolism is important not just for oxidative catabolism and anabolism of new amino acids, but also for keeping the levels of amino acids in the blood within normal limits. Read *A&P in the Real World: Phenylketonuria* to find out what happens when the body is unable to catabolize the amino acid phenylalanine and its concentration in the blood increases dramatically.

Newly released amino acid subunits are taken up by hepatocytes (cells of the liver) to be catabolized. Recall from the chemistry chapter (see Figure 2.21) that an amino acid consists of carbon atoms, sometimes called the "carbon skeleton," bound to a nitrogen-containing amino group. Nitrogen is a relatively inert element and doesn't oxidize readily, so before amino acids can be

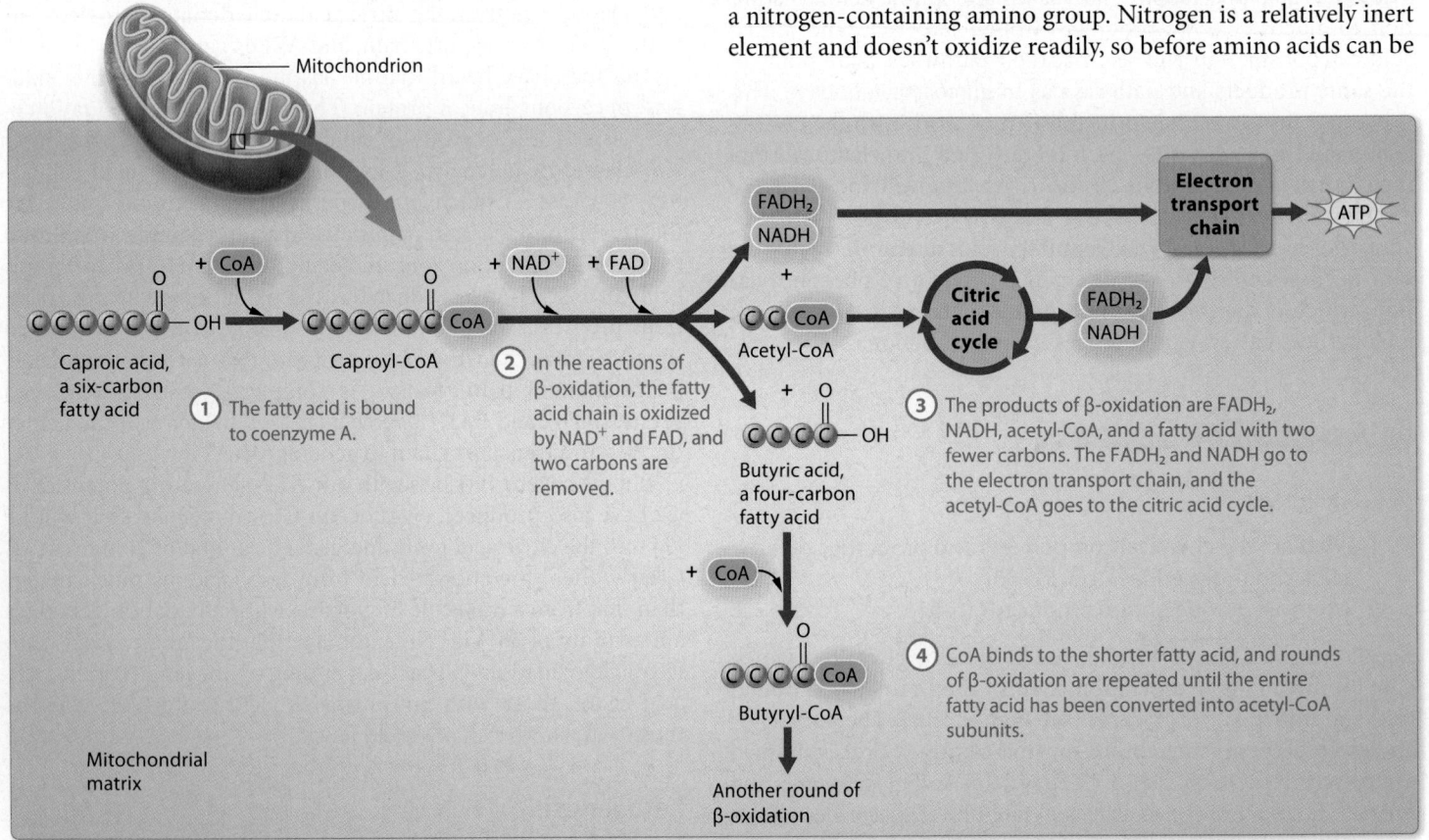

Figure 23.9 Fatty acid catabolism and β-oxidation.

oxidized for fuel, the hepatocytes first have to remove the amino group by **transamination** (*trans-* = "across").

The events of amino acid catabolism occur by the following steps, shown in **Figure 23.10**:

(1) **In transamination, the amino group is removed from the amino acid and transferred to α-ketoglutarate.** The removed amino group is transferred to the five-carbon compound α-ketoglutarate, one of the components of the citric acid cycle.

(2) **This generates two products: a carbon skeleton and the amino acid glutamate.**

(3) **The carbon skeleton can be converted into a variety of compounds, which can then be oxidized.** The carbon skeleton itself may become certain compounds, including pyruvate, acetyl-CoA, other citric acid cycle compounds, and even glucose. These compounds can then be oxidized via the reactions of glycolysis or the citric acid cycle.

(4) **In the mitochondria of the hepatocyte, glutamate undergoes oxidative deamination, producing NH₃ and α-ketoglutarate.** Glutamate "carries" the amino group to the mitochondria of the hepatocyte. In the mitochondria, glutamate undergoes a process called *oxidative*

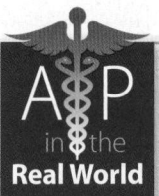

Phenylketonuria

Phenylketonuria (fee′-nul-kee-tohn-YOOR-ee-ah), or PKU, was one of the first genetic defects of metabolism detected in humans. PKU results from a deficiency in one of the enzymes involved in the catabolism of the amino acid phenylalanine. Normally, phenylalanine is degraded to acetyl-CoA; however, when the necessary enzyme is deficient, phenylalanine accumulates in the blood and is excreted in the urine (hence the name of the disease). A high level of phenylalanine interferes with normal development of the nervous system, and severe mental retardation may result.

The effects of PKU can be prevented by early detection and strict dietary modifications. For this reason, newborns are screened for the disease shortly after birth. Those diagnosed with PKU must limit their intake of protein-rich foods and foods with added phenylalanine, such as the artificial sweetener aspartame. Foods with aspartame have a warning label to alert those with PKU of the phenylalanine content.

Figure 23.10 Amino acid catabolism.

The Big Picture of Nutrient Catabolism

Figure 23.11

Practice art labeling
@MasteringA&P®

β-oxidation

Fatty acids

NADH + FADH₂

Lipids
(See Figure 23.9.)

Glycerol

Glycolysis
(See Figure 23.4.)

Citric acid cycle
(See Figure 23.6.)

NADH

Electron transport chain and oxidative phosphorylation
(See Figure 23.7.)

Carbohydrates
(See Figure 23.8.)

Glucose → Glyceraldehyde-3-phosphate → Pyruvate
(See Figure 23.5.)

Acetyl-CoA

FADH₂

Transamination

Proteins
(See Figure 23.10.)

Amino acids

Urea

Mitochondrion

Cytosol

---

deamination, during which the amino group is removed, forming ammonia (NH₃) and re-forming α-ketoglutarate.

⑤ **Some of the amino groups removed as ammonia are used in the synthesis of new amino acids.**

⑥ **The remaining ammonia molecules are removed by the urea cycle, which forms urea that is then eliminated by the kidneys in urine.** The remaining ammonia must be removed from the body, because it is toxic to cells. This occurs via the **urea cycle** (yoo-REE-ah), which takes place in hepatocytes. During the urea cycle, two ammonia groups are combined with carbon dioxide to form the molecule **urea.** Most of the urea resulting from this cycle is then eliminated in the urine by the kidneys.

Quick Check

☐ 3. Why is it necessary to remove the amino group from amino acids during their catabolism?

☐ 4. What is the fate of the carbon skeletons of amino acids?

☐ 5. Why must the body eliminate the ammonia that results from transamination?

Putting It All Together: The Big Picture of Nutrient Catabolism

The large numbers of catabolic reactions can seem overwhelming. To help you study them, the pathways of fatty acid, glucose, and amino acid catabolism are summarized in **Figure 23.11.**

Apply What You Learned

☐ 1. Explain how the pathways of protein, lipid, and carbohydrate catabolism are similar. How are they different?

☐ 2. Explain why kidney failure could result in the accumulation of nitrogen-containing waste products in the blood.

☐ 3. What does it mean when we say that fats have more "calories" than carbohydrates or proteins? Why is this true?

See answers in Appendix A.

Figure 23.13 Gluconeogenesis.

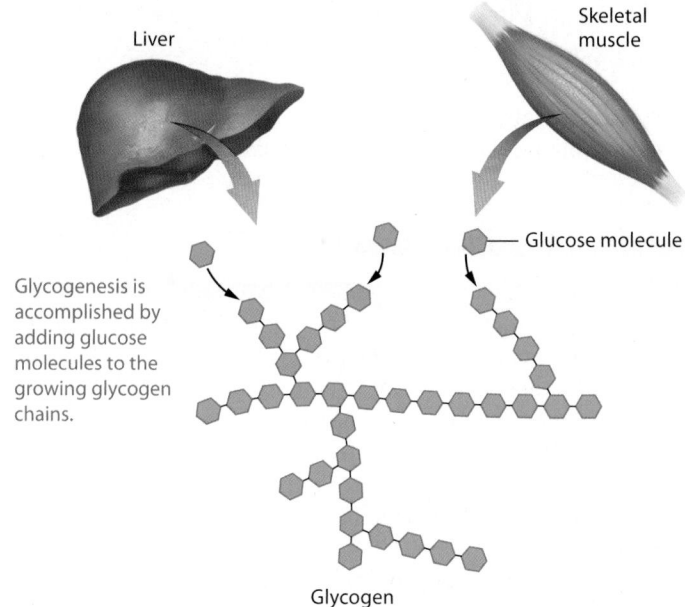

MODULE 23.4
Anabolic Pathways

Learning Outcomes

1. Describe the processes of glycogenesis, glycogenolysis, and gluconeogenesis.
2. Describe the process by which fatty acids are synthesized and stored as triglycerides in adipose tissue.
3. Explain how nutrients may be converted to amino acids and lipids if needed.
4. Explain the fate of excess dietary proteins and carbohydrates.

As we discussed in Module 23.1, catabolism tells only part of the story of metabolism. To get the full picture, we must also consider the other side of metabolism: anabolism, also known as *biosynthesis.* The body's anabolic reactions serve many purposes, including nutrient storage; synthesis of nutrient molecules; synthesis of structural elements such as phospholipids, proteins, and nucleic acids; and synthesis of special molecules such as FAD and NAD$^+$. This module focuses primarily on the anabolic reactions that store nutrients and build nutrient molecules.

The body stores nutrients for two reasons: (1) When we eat, we take in more energy than we can immediately use; and (2) the body needs a ready supply of nutrients to oxidize for maintaining metabolic homeostasis between the times we eat. The two main nutrient energy storage forms we find in the body are *glycogen* (GLY-koh-jen), which consists of linked glucose units, and adipose (fat), which is made up mostly of triglycerides. Interestingly, often the glucose in glycogen is derived from amino acids or glycerol; similarly, often the fatty acids of triglycerides are derived from glucose and amino acids. Let's examine how the body stores and manufactures nutrient molecules.

Glycogenesis is accomplished by adding glucose molecules to the growing glycogen chains.

Figure 23.12 Glycogenesis.

Glucose Anabolism

◀◀ **Flashback**

1. What is a polysaccharide? (p. 52)
2. How does the body store glucose? (p. 52)

There are two anabolic processes involving glucose: *glycogenesis,* which is the storage of excess glucose taken in from the diet; and *gluconeogenesis,* which is the formation of new glucose from noncarbohydrate molecules. The next two subsections explore these processes.

Glycogenesis

Glucose is stored in the form of **glycogen,** a large, branching molecule composed of thousands of glucose units. The process of glycogen synthesis, which is called **glycogenesis,** involves a series of enzyme-catalyzed reactions that add glucose units to the growing glycogen molecule, as shown in **Figure 23.12.** Although nearly all cells are able to store glycogen, the majority of the body's glycogen is stored in the liver and skeletal muscle tissue.

When the concentration of glucose in the bloodstream decreases, hormones such as glucagon and epinephrine trigger the breakdown of glycogen, a catabolic process called **glycogenolysis** (gly′-koh-jen-AWL-ih-sis). This releases glucose molecules into the bloodstream, which raises the concentration of blood glucose. Glycogen stores do not last long, however, and are depleted after about half a day (much faster when a person is engaged in strenuous activity).

Gluconeogenesis

When glycogen supplies are exhausted, the body needs to turn to alternative methods to keep up with the needs of cells that require glucose. The body's main method of doing so is **gluconeogenesis** (glew′-koh-nee-oh-JEN-eh-sis; literally, "making new glucose"), or synthesis of glucose from noncarbohydrate molecules (**Figure 23.13**). During gluconeogenesis, the hepatocytes and certain cells of the kidney convert three- and four-carbon compounds into glucose. Sources of these compounds include the following:

- glycerol from triglyceride catabolism,
- pyruvate and lactate from glycolysis,
- intermediate compounds from the citric acid cycle, and
- certain amino acids called *glucogenic* amino acids.

The one major nutrient molecule that cannot be converted into new glucose is a fatty acid. This is because fatty acid oxidation can produce only acetyl-CoA, and acetyl-CoA cannot be used by mammals to generate new glucose. The glucose synthesized by gluconeogenesis is either released into the bloodstream and delivered to cells or made into glycogen.

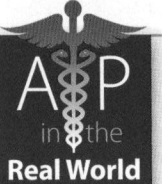

Fatty Liver Disease

The role of the liver in lipid metabolism becomes obvious in the condition called *steatohepatitis* (stee'-ah-toh-hep-ah-TY-tis; *steat-* = "fat"), commonly known as "fatty liver disease" (FLD). FLD has many causes, including alcohol abuse, obesity, diabetes, and many drugs and toxins. It is characterized by the accumulation of triglycerides and other fats in hepatocytes, as shown in the micrograph on the right here, next to one of a normal liver:

Hepatocytes   Fat deposits

LM (225×)   LM (225×)
Normal liver   Fatty liver

The exact mechanism by which FLD develops isn't precisely known, but may include decreased capacity for mitochondrial β-oxidation, increased fatty acid synthesis, increased delivery of fatty acids to the liver, and deficient export of triglycerides from the liver to other cells.

Untreated, FLD may progress to cirrhosis and hepatic failure. However, FLD is generally reversible if the underlying cause is addressed. Treatments include dietary modifications with weight loss, abstinence from alcohol, and certain medications.

Quick Check

☐ 1. How is glucose stored in the body?
☐ 2. Which molecules can be used by the body for gluconeogenesis?

Fatty Acid Anabolism

The synthesis of fatty acids somewhat resembles β-oxidation in reverse; however, they are still separate pathways that require different enzymes and take place in different locations. Fatty acid synthesis occurs by a process called **lipogenesis** (ly'-poh-JEN-eh-sis). Lipogenesis takes place in the cytosol, and involves a large enzyme complex called *fatty acid synthase*, which catalyzes reactions that progressively lengthen the fatty acid chain by two-carbon units. Most of the fatty acids are attached to glycerol and assembled into triglycerides in the endoplasmic reticulum, after which they are stored in adipocytes. The majority of the body's energy is stored as triglycerides in adipose tissue, although some triglycerides are also stored in the liver. See *A&P in the Real World: Fatty Liver Disease* to discover what happens when too many fatty acids are stored in the liver.

Not all fatty acids and glycerol in triglycerides begin as dietary fats; often they start out as glucose or amino acids. After eating, the body takes the glucose obtained from food and uses it for its immediate needs. Excess glucose is stored as glycogen; however, there is an upper limit to the amount of glycogen that cells can make. When this limit has been reached, the glucose molecules undergo the beginning of catabolism via glycolysis, but the products of glycolysis do not enter the citric acid cycle. Rather, the resulting acetyl-CoA molecules are combined to form fatty acids, and other products of the glycolytic reactions are converted into glycerol. Excess dietary amino acids can also undergo similar reactions to be converted into triglycerides. The truth of the matter is that any type of nutrient can be converted into fat.

Quick Check

☐ 3. How are fatty acids synthesized?
☐ 4. How are nonlipids converted into fats?

Amino Acid Anabolism

We have already discussed protein synthesis and seen how strings of amino acids are joined by peptide bonds in a sequence determined by the DNA of one gene (see Chapter 3). The body is able to synthesize 11 of the 20 amino acids found in human proteins. These amino acids are primarily synthesized by reactions that add an amino group to carbon skeleton molecules such as α-ketoglutarate, pyruvate, and oxaloacetate. The remaining amino acids must be supplied by the diet, which we discuss in Module 23.7.

Unlike carbohydrates and lipids, amino acids are not stored to any significant extent in the body's proteins. A cell needs to synthesize only so many proteins; once this limit has been reached,

The Big Picture of Nutrient Anabolism

Figure 23.14

any excess amino acids obtained by the diet are converted to other molecules for storage. Some amino acids are converted into glucose by gluconeogenesis and incorporated into glycogen. Others are converted to fatty acids and stored in the body's adipose tissue. You may have heard the claim by individuals promoting high-protein diets that proteins cannot be stored as fat. This is, of course, untrue.

Putting It All Together: The Big Picture of Nutrient Anabolism

The anabolic reactions of glucose, fatty acids, and amino acids are summarized in **Figure 23.14**. To help you study both catabolism and anabolism of nutrients, compare this figure to Figure 23.11.

Quick Check

☐ 5. How are amino acids produced by the body?
☐ 6. What happens to amino acids consumed in excess?

Apply What You Learned

☐ 1. Explain why diets extremely low in carbohydrates often lead to mental "fogginess" and decreased mental stamina.
☐ 2. The fat in low-fat foods is frequently replaced by large amounts of sugar. Will the intake of such products likely lead to fat loss? Explain.

See answers in Appendix A.

MODULE **23.5**

Metabolic States and Regulation of Feeding

Learning Outcomes

1. Compare and contrast the processes that occur in the absorptive and postabsorptive states.
2. Explain the roles of insulin and glucagon in the absorptive and postabsorptive states.
3. Describe the role of the hormones insulin and glucagon in regulating glucose and amino acid catabolism and anabolism.
4. Explain the significance of glucose sparing for neural tissue in the postabsorptive state.
5. Describe how feeding behaviors are regulated.

We have established that the body oxidizes the nutrients in food to generate energy. However, we do not eat continuously throughout the day; rather, nutrients are generally taken in when we are hungry and food is available. Yet even when food is not being ingested, the activities of the body continue, and the cells constantly require energy to carry out processes. The body has adapted to this fact by entering different metabolic "states" that ensure that the cells are provided with energy at all times. This module focuses on these metabolic states and how feeding behaviors are regulated.

Metabolic States

◄◄ Flashback

1. What are the actions and target tissues of the hormone insulin? (p. 616)
2. What are the actions and target tissues of the hormone glucagon? (p. 615)

The body has two basic metabolic states: the *absorptive state* and the *postabsorptive state*. They differ in terms of their proximity to feeding and the metabolic reactions that predominate. Let's take a closer look at the metabolic changes and events that accompany each state and the hormones that regulate them.

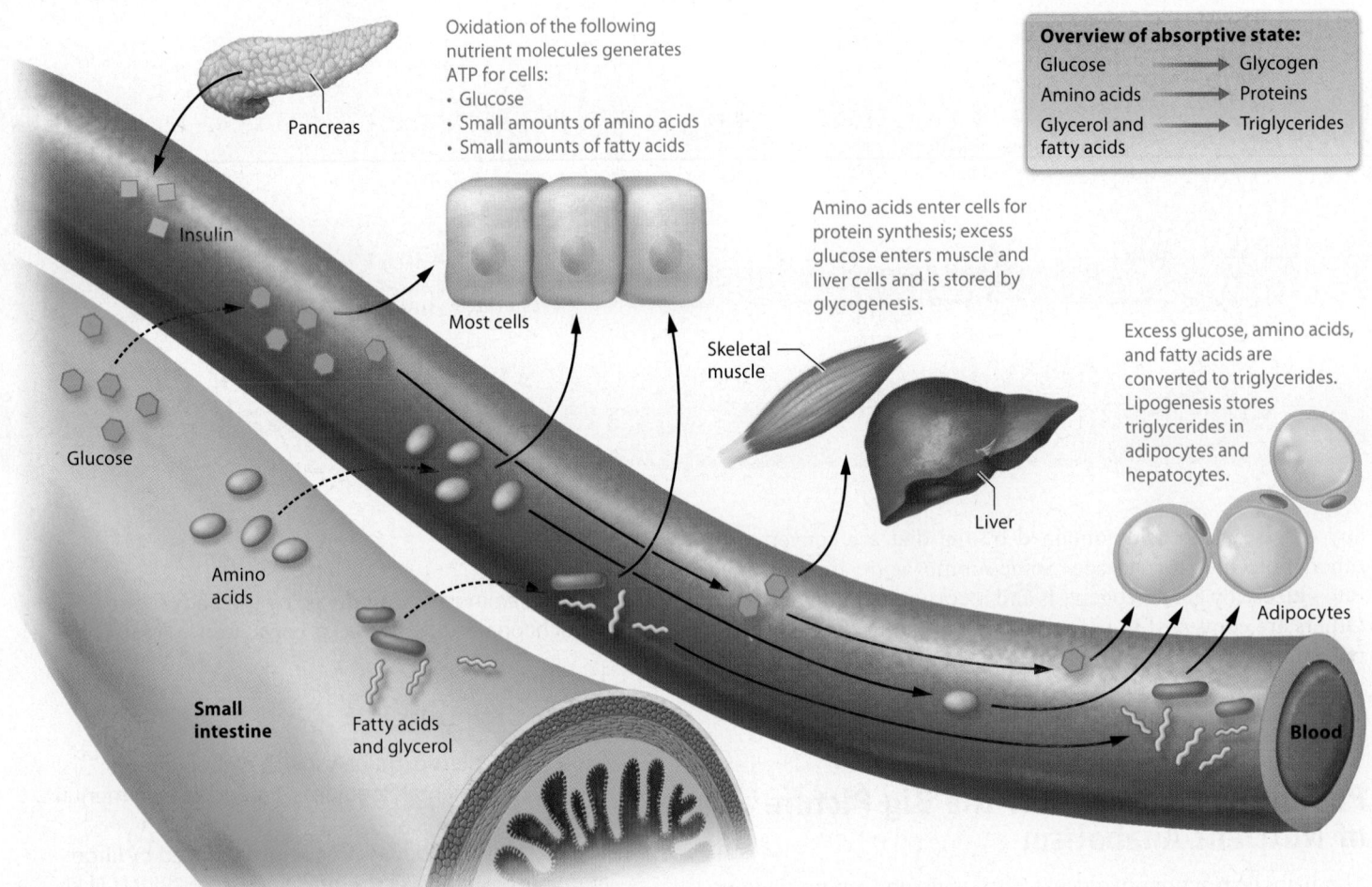

(a) **Changes in the absorptive state triggered by insulin.**

Figure 23.15 Comparison of the absorptive and postabsorptive states.

Absorptive State

The **absorptive state** occurs right after feeding, from the time that ingested nutrients enter the bloodstream to as long as 4 hours after feeding (**Figure 23.15a**). It is called the absorptive state because it is the period during which the nutrients from food are absorbed from the small intestine into the blood. During the absorptive state, the following processes occur:

- **Oxidation of nutrient molecules, primarily glucose, provides fuel to cells.** Glucose is the main fuel used by cells. Glucose is usually readily available during the absorptive state, and it enters cells rapidly to be oxidized as the main fuel. A small fraction of the absorbed fatty acids and glycerol are oxidized by different tissues for fuel; similarly, a small percentage of amino acids are deaminated and oxidized for fuel.

- **Synthesis of amino acids into proteins provides structural materials to cells.** Many ingested amino acids enter cells, where they are used for protein synthesis.

- **Glycogenesis stores excess glucose in skeletal muscle cells and hepatocytes.** Glucose that is not oxidized for energy is stored as glycogen.

- **Lipogenesis stores triglycerides in adipocytes and hepatocytes.** The majority of triglycerides are simply stored in adipose tissue or the liver. Excess glucose is also converted into fatty acids or glycerol and stored as fat, and some amino acids are converted to fatty acids and stored as well.

These changes are largely controlled by the pancreatic hormone *insulin*. As we discussed, insulin release is triggered by a rising level of glucose in the blood (see Chapter 16). This hormone targets most cells in the body and stimulates the uptake of glucose by cells, lowering the concentration of glucose in the blood. It also acts on the liver and several other tissues, initiating the anabolic processes we just covered. In the absorptive state, then, cells take in fuels and, under the control of insulin, use or store them.

Postabsorptive State

Once nutrient absorption is complete, the body enters the **postabsorptive state** (**Figure 23.15b**). The postabsorptive state generally begins about 4 hours after feeding, although this varies depending on the amount and types of nutrients ingested. For most individuals, the body is generally in the postabsorptive

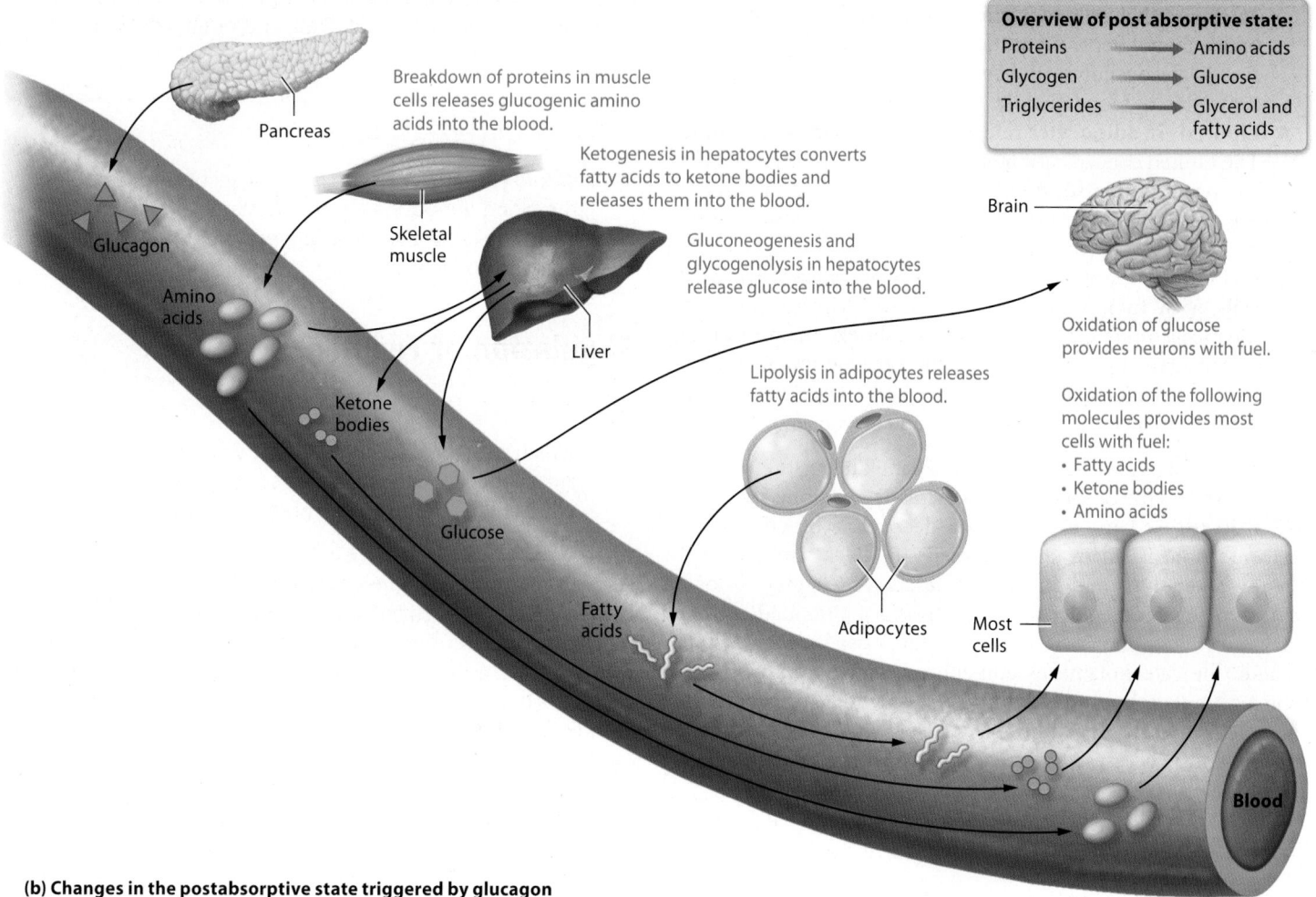

Overview of post absorptive state:

Proteins	→	Amino acids
Glycogen	→	Glucose
Triglycerides	→	Glycerol and fatty acids

Pancreas

Glucagon

Breakdown of proteins in muscle cells releases glucogenic amino acids into the blood.

Amino acids

Skeletal muscle

Ketogenesis in hepatocytes converts fatty acids to ketone bodies and releases them into the blood.

Liver

Gluconeogenesis and glycogenolysis in hepatocytes release glucose into the blood.

Brain

Ketone bodies

Glucose

Lipolysis in adipocytes releases fatty acids into the blood.

Fatty acids

Adipocytes

Most cells

Oxidation of glucose provides neurons with fuel.

Oxidation of the following molecules provides most cells with fuel:
- Fatty acids
- Ketone bodies
- Amino acids

Blood

(b) Changes in the postabsorptive state triggered by glucagon

state in late morning, late afternoon, and most of the night. During this state, the body slows and then ceases anabolic processes such as glycogenesis, lipogenesis, and protein synthesis because the steady supply of nutrients ends during the postabsorptive period. In their place, the following processes occur:

- **Breakdown of proteins in muscle cells releases glucogenic amino acids into the blood.** If necessary, muscle proteins are broken down to release glucogenic amino acids that can be taken up by cells and converted to glucose. See *A&P in the Real World: Fasting and Protein Wasting* to find out more about protein breakdown during the postabsorptive state.

- **Ketogenesis in hepatocytes converts fatty acids to ketone bodies and releases them into the blood.** Many of the fatty acids oxidized in the liver are converted to ketone bodies, which are released into the blood. The advantage to releasing ketones is that cells of the nervous system are able to oxidize them for fuel when glucose is unavailable, whereas these cells are not capable of oxidizing fatty acids to any significant extent.

- **Gluconeogenesis and glycogenolysis in hepatocytes release glucose into the blood.** The blood glucose concentration begins to decline during the postabsorptive state. To ensure that cells requiring glucose, such as those of the brain, are not deprived of their primary fuel, gluconeogenesis and glycogenolysis in hepatocytes begin to raise the concentration of blood glucose.

- **Lipolysis in adipocytes releases fatty acids into the blood.** The limited concentration of blood glucose means that cells will need another source of fuel, namely, the fuel in fats. For this reason, lipolysis begins and fatty acids are released into the bloodstream from adipocytes.

- **Oxidation of molecules such as fatty acids provides most cells with fuel.** When fatty acids are delivered to cells, most cells preferentially begin to catabolize them in order to conserve glucose for cells of the nervous system, a phenomenon called **glucose sparing.** Cells outside the nervous system can also use ketone bodies and amino acids for fuel.

Several hormones help the body adapt during the postabsorptive state. One of these is the pancreatic hormone *glucagon,* which is released in response to a decreasing blood glucose concentration. Its main target tissue is the liver, where it triggers glycogenolysis and gluconeogenesis (see Chapter 16 for a review of how blood glucose homeostasis is maintained). Other hormones that play a role during the postabsorptive state include the catecholamines epinephrine and norepinephrine, which stimulate lipolysis in adipose tissue and glycogenolysis in skeletal muscle. In addition, cortisol stimulates gluconeogenesis and processes that release glucogenic precursors into the blood.

Quick Check

☐ 1. Why does anabolism dominate in the absorptive state?

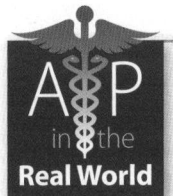

Fasting and Protein Wasting

Fasting, or a prolonged period during which little to no food is consumed, may be intentional, or could be due to limited availability of nutrient-dense foods or to a disease process that prevents ingestion or processing of nutrients. One consequence of fasting is a condition called *protein wasting.* Normally, after a few hours in the postabsorptive state, proteins are degraded and their amino acids are used as the primary source of carbon for gluconeogenesis. Some degree of protein degeneration is normal and does not harm the body. However, this cannot go on for long before problems begin to arise.

Protein wasting is generally first noticeable in muscle tissue, as those cells have a much higher protein content than other cells. Degradation of muscle proteins results in weakness and fatigue, along with a significant decrease in muscle mass. However, the body does not just rob the muscles of protein—the body does not discriminate among sources of protein or among cell populations. This can lead to a loss of critical structural and functional proteins in a multitude of tissues, including the heart, liver, and brain, which in severe cases may prove fatal.

☐ 2. Would you expect to find a high level of ketone bodies in the blood during the absorptive state?

☐ 3. Which metabolic activities are prevalent during the postabsorptive state?

Regulation of Feeding

◀◀ **Flashback**

1. Which structures are innervated by the vagus nerve? What are its effects on the digestive system? (pp. 488, 531)
2. Which tissue produces the hormone leptin, and what is its effect? (p. 620)

As with many other homeostatic variables, feeding is controlled by nuclei within the hypothalamus. Two nuclei directly involved in feeding are the *satiety center* (sah-TY-eh-tee), which elicits feelings of fullness and inhibits the desire to eat, and the *hunger center* (or the *feeding center*), which elicits feelings of hunger and stimulates the desire to eat. A variety of hormonal and neural signals stimulates and/or inhibits these centers.

Long-term regulation of feeding is primarily hormonal. The satiety center is stimulated by the hormone **leptin,** which is produced by adipocytes. Leptin inhibits neurons in the hunger center while stimulating those in the satiety center to produce neurotransmitters such as *pre-opiomelanocortin* (pree-oh'-pee-oh-meh-lan-oh-KOR-tin; POMC). These neurotransmitters have an overall *anorexigenic* effect (an'-oh-rek-sih-JEN-ik; *a-* = "without," *-orex* = "appetite"), depressing the appetite. This profound anorexigenic effect is evident when the receptors to which the neurotransmitters bind are defective, resulting in severe, early-onset obesity caused by uncontrolled overeating. The discovery of leptin generated a great deal of excitement due to its potential as a novel agent to suppress appetite and cause weight loss. Unfortunately, studies of leptin have shown that administering excess leptin does not result in decreased appetite or weight loss in humans.

The hunger center is stimulated by a decreasing level of leptin secretion and a rising level of secretion of the hormone **ghrelin** (GRAY-lin) from the cells of the stomach mucosa. Ghrelin stimulates neurons in the hunger center to produce neurotransmitters such as *neuropeptide Y* and *orexins,* both of which are *orexigenic,* promoting hunger. Interestingly, ghrelin levels decrease in obese individuals after gastric bypass surgery (removal of part of the stomach), which is hypothesized to partially account for their reduced food intake and subsequent weight loss.

A variety of short-term signals can also inhibit or stimulate feeding. Feeding triggers the release of several hormones, such as insulin, that have actions similar to those of leptin, which in turn decrease food intake. In addition, feeding stretches the stomach walls and initiates the release of gastrointestinal hormones such as cholecystokinin. Both stimuli cause the vagus nerve to indirectly suppress the hunger center and decrease the release of orexins and neuropeptide Y. The concentration of certain molecules in the blood can also directly stimulate or inhibit centers in the hypothalamus. For example, a low level of glucose in the blood (hypoglycemia) stimulates the hunger center and the release of orexins.

Of course, feeding behaviors are more complex than the signals we have discussed. A number of psychological and environmental factors affect feeding, as well, which has hampered efforts to find a "cure-all" medication that suppresses appetite.

Quick Check

☐ 4. Which centers in the hypothalamus regulate feeding?
☐ 5. What are the actions of leptin?
☐ 6. How do the actions of leptin differ from those of ghrelin?

Apply What You Learned

☐ 1. How would the absorptive state differ in a person who has completely eliminated carbohydrates from his or her diet?
☐ 2. Type 1 diabetes mellitus results from the inability to produce the hormone insulin. Predict how this condition changes the normal events of the absorptive state.
☐ 3. Certain antidepressants trigger the production of neurotransmitters by the neurons of the hypothalamic feeding center. Predict the effects these medications have on appetite and feeding.

See answers in Appendix A.

The Metabolic Rate and Thermoregulation

Learning Outcomes

1. Define metabolic rate and basal metabolic rate.
2. Describe factors that affect metabolic rate.
3. Differentiate between radiation, conduction, convection, and evaporation.
4. Explain the importance of thermoregulation in the body.

The metabolic modifications that occur during the absorptive and postabsorptive states ensure that the body has a ready supply of fuels that can be oxidized to drive its processes. The total amount of energy expended by the body to power all of its processes is called the **metabolic rate.** A consequence of this energy expenditure is a great deal of heat production. Indeed, metabolism is the primary source of body heat, and metabolic processes are an important part of body temperature regulation, or *thermoregulation.* In this module, we examine the metabolic rate and how it is measured. Then our attention shifts to heat production, the process of thermoregulation, and what happens when thermoregulation is impaired.

Metabolic Rate

The metabolic rate is expressed in a unit we encountered in the chapter opener: the Calorie or, more accurately, the kilocalorie. Recall that a kilocalorie is defined as the amount of heat required to raise the temperature of one kilogram of water by one degree Celsius. Why would we use a measure of heat to express the body's metabolic rate? The answer is because most of the body's energy is eventually lost to the environment as heat. Consider the following facts:

- The liberation of the chemical energy in foods is inefficient, and much of the energy liberated by catabolism is lost as heat.
- The use of ATP to drive cellular processes is similarly inefficient, and much of the chemical energy of ATP use is lost as heat.
- Much of the energy from the processes driven by ATP, such as muscle contraction, is also lost as heat.

Because of this conversion of energy to heat, a process called *thermogenesis,* heat production is a good way to measure energy expenditure. Heat production is determined by **calorimetry** (kal'-or-IM-eh-tree), in which the actual amount of heat an individual's body produces is measured. The value obtained from calorimetry is then used to directly determine the person's metabolic rate. However, because this process is physically difficult to perform, most often the metabolic rate is determined indirectly

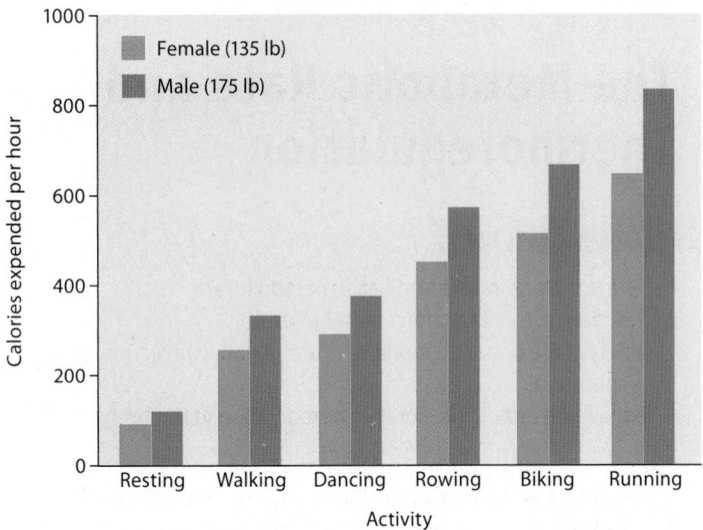

Figure 23.16 Approximate energy expenditures in adults during various activities.

by measuring the amount of oxygen a person consumes, as this is directly related to the amount of energy generated and so heat produced.

One of the values that calorimetry helps to measure is the **basal metabolic rate,** or **BMR,** which is defined as the minimal rate of metabolism for an awake individual, or the "energy cost of living." Even when the body is completely at rest, it is still using a considerable amount of energy to perform the processes necessary to maintain life, such as keeping the heart beating and filtering the blood through the kidneys. A person is said to be functioning at his or her BMR when the following conditions are met: (1) The person has not eaten for the past 12 hours and has had a restful night of sleep; (2) the person is not performing physical activity and has not performed strenuous physical activity for at least 1 hour; (3) the person is not under physical or emotional stress; and (4) the temperature surrounding the person is constant and comfortable.

Of course, the body's energy expenditure is not going to precisely equal the BMR most of the time. Rather, it will vary at different times of the day and with different activities (**Figure 23.16**). A variety of factors influence the metabolic rate, including hormones such as thyroid hormone and growth hormone, fever, nutritional status, physical activity, and certain drugs. For example, during exercise the metabolic rate can increase tenfold or more above the BMR, and during sleep it decreases to about 10–15% below the BMR. Additionally, the BMR varies from person to person, depending on factors such as gender and muscle mass (as skeletal muscle is a very metabolically active tissue). Is it possible to change your BMR? Find out in *A&P in the Real World: "Rev" Your Metabolism.*

Quick Check

☐ 1. What is a Calorie?
☐ 2. What is the basal metabolic rate (BMR)? When is a person functioning at his or her BMR?

Heat Exchange between the Body and the Environment

◀◀ **Flashback**

1. How does water take heat with it when it evaporates? (p. 46)

Heat, like solutes, water, and pressure, will move with a gradient, again reminding us of the Gradients Core Principle (p. 26). Specifically, heat moves from an area of higher temperature to one of lower temperature. The heat produced by the body's metabolic reactions is exchanged with the environment in this manner, mostly through the skin. Heat is exchanged by four mechanisms: *radiation, conduction, convection,* and *evaporation* (**Figure 23.17**).

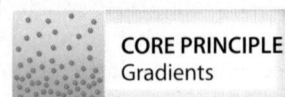

CORE PRINCIPLE
Gradients

Radiation

The transfer of heat from one object to another through electromagnetic waves is known as **radiation.** The body loses heat to the environment when the body temperature is higher than the surrounding air, which is generally the case. Radiation is the reason why you feel warmth from the sun. It's also why you might feel cold when you initially lie under a blanket but gradually feel warmer—your body heat is transferred to the air, and the heated air is trapped by the blanket. Note that this also works in reverse when the air temperature is higher than body temperature—the body will gain heat. This is why exposure to extreme heat for a long time can be dangerous. Radiation is responsible for about 50% of the body's total heat exchange with the environment.

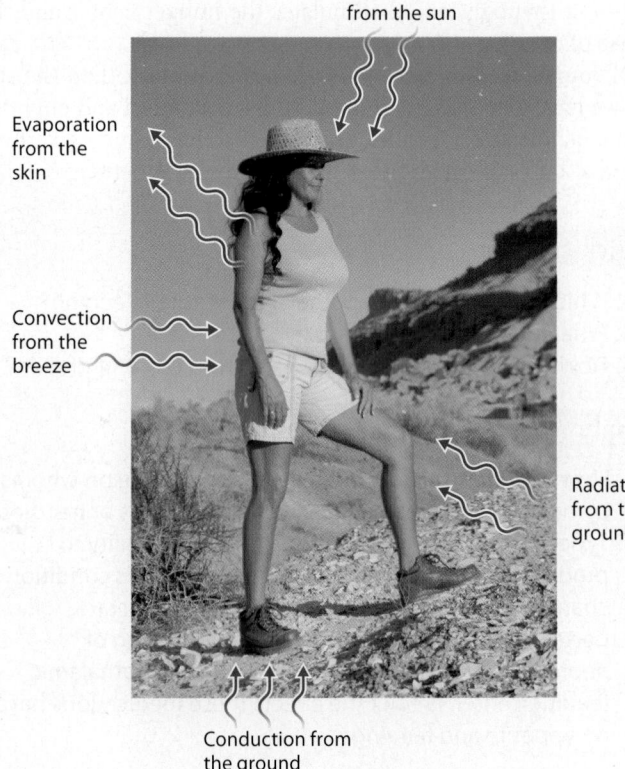

Figure 23.17 Mechanisms of heat transfer.

"Rev" Your Metabolism

You have probably heard about a lot of things that claim to speed up, or "rev," your metabolism, or increase your BMR. It generally is possible to increase your BMR. For example, you've already seen that exercise can increase the metabolic rate by up to 10 times. But this is just a temporary effect—after exercise stops, the metabolic rate gradually returns to its basal level. However, consistent exercise, particularly weight training, can affect the BMR. Adipose tissue is very energy efficient, and so has a low metabolic rate. By contrast, skeletal muscle tissue is energy inefficient and has a much higher metabolic rate. For this reason, an individual who loses adipose tissue and gains muscle tissue will generally increase his or her BMR.

But what about those over-the-counter diet pills that claim to "rev" your metabolism? The first thing you need to understand about dietary supplements in the United States is that manufacturers are not required to demonstrate safety or efficacy in order to sell their products. Also, quality testing is not required, so these products might not even contain the ingredients they claim. Most of these dietary supplements claim to contain varying amounts of stimulants, such as the following:

- *Caffeine.* Caffeine increases the metabolic rate temporarily, but over the long term, body cells become insensitive to its effects.
- *Green tea.* The active components in green tea do appear to increase the BMR. However, a certain dosage and a specific part of the plant are required to achieve effectiveness. There is no way to be certain that any given dietary supplement contains adequate amounts of green tea or the correct components.
- *Capsaicin.* A component of peppers, capsaicin (kap-SAY-sin) has been shown to increase the metabolic rate. However, its use is limited by adverse effects and stringent dosage requirements.

When dealing with over-the-counter dietary supplements, it is best to be cautious. They are often quite expensive and may not prove to be safe or effective. Besides, the best way to "rev" your metabolism can be free: a regular exercise program.

Conduction and Convection

Similar to radiation is **conduction,** in which heat is transferred from one object to another through direct contact. This can work through solids, liquids, and gases such as air, although it occurs much faster through contact with solids and liquids than with air. For example, conduction is the reason why you feel cold when you jump in a cold pool—your body heat is directly transferred to the water, lowering your body temperature—or why your feet feel hot when you walk on hot pavement.

Conduction works hand in hand with another mechanism of heat exchange: **convection,** in which heat is transferred through a liquid or gaseous medium. The simplest example of convection can be seen in many houses. Have you ever noticed that heating vents in a room are located near the floor and air conditioning vents are located near the ceiling? The reason for this is because warm air expands and rises from the floor, whereas cool air contracts, making it denser and causing it to fall into the room. Convection works the same way in body heat transfer; body heat rises away from the body and is replaced by falling cooler, denser air. Convection is also responsible for the cooling effect of a fan or breeze; the cooler, denser air in the breeze contracts and displaces the warmer air surrounding you.

Convection aids conduction because the cooler air around the body provides a steeper temperature gradient for heat transfer between the body and the surrounding air. Together these two processes account for about 15–20% of the body's heat transfer with the environment.

Evaporation

The final heat transfer mechanism is **evaporation,** during which water changes from a liquid to a gas. Water in liquid form absorbs a great deal of heat from its surroundings, and this heat enables liquid water to change into a gas. Water in and on the body absorbs body heat, removing the heat as it evaporates.

Evaporation occurs from the skin, lungs, and oral and nasal mucosae continuously without stimulation from the nervous system, producing what's known as *insensible heat loss.* This accounts for about 10% of the body's total heat transfer under resting conditions. When body temperature increases, the hypothalamus triggers *sweating,* which can result in the remaining 20–25% of heat loss, in addition to water loss. We discuss sweating in the next section.

Quick Check

☐ 3. What is the main heat transfer mechanism between the environment and the body? How does this mechanism transfer heat?

☐ 4. How do convection and conduction work together to transfer heat?

☐ 5. Why does evaporation have a cooling effect?

Thermoregulation: Body Temperature Regulation

 Flashback

1. How does the integumentary system contribute to thermoregulation? (p. 162)

2. How does thyroid hormone assist in thermoregulation? (p. 602)
3. What is the mechanism behind a fever? (p. 922)

As you can tell from the preceding Flashback questions, we have discussed the regulation of body temperature, or **thermoregulation,** several times in this book. The amount of heat in the body is a key homeostatic variable that must be maintained within a narrow range to ensure that all systems function properly. You have already learned about how metabolism produces heat, so let's find out how the body handles this heat and regulates its internal temperature.

Body Temperature

In spite of often significant swings in air temperature, body temperature remains relatively constant in the range of 35.8–38.2 °C (96–101 °F), averaging about 37.5 °C (99.5 °F). Note that this refers to the **core body temperature,** which is that of deep structures in the body, such as the liver and brain. It might strike you that this measurement differs from what is called "normal" on most oral and skin thermometers. This difference exists because measurements in these locations do not accurately reflect the core temperature due to variations with activity and air temperature, averaging instead about 35.6–37 °C (97.7–98.6 °F). Historically, the most accurate way to determine core body temperature is via a thermometer inserted into the rectum, but the inherent discomfort of this method now precludes its regular use. Certain newer thermometers have the ability to measure the temperature of the blood in the temporal artery near the surface of the skin, and this provides a noninvasive and generally accurate reading of core temperature.

Core body temperature fluctuates throughout the day. You may have noticed that you feel colder in the morning than in the evening. This reflects the fact that core body temperature is highest in the evening and decreases by about 1 °C in the morning. In women, core body temperature also fluctuates with the ovarian cycle, which proves useful for fertility planning.

Regulation of Body Temperature

Like most other homeostatic variables, body temperature is regulated by a negative feedback loop, another example of the Feedback Loops Core Principle (p. 21). The feedback loop that responds to a rising body temperature, illustrated in **Figure 23.18a,** proceeds as follows:

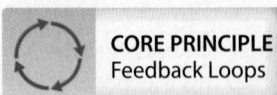
CORE PRINCIPLE
Feedback Loops

- **Stimulus: Body temperature increases above normal range.**
- **Receptor: Thermoreceptors in the skin and the hypothalamus detect the increased body temperature.** Many factors can cause body temperature to rise, including stimulation from the sympathetic nervous system, muscle activity, inflammation, and elevated environmental temperature. Temperature-sensitive *thermoreceptors* are located throughout the body, particularly in the skin and hypothalamus. These receptors detect the current higher temperature, and then send this information to the hypothalamus.

- **Control center: The heat-loss center in the hypothalamus is activated.** The hypothalamic thermoregulatory centers, which act as the control center, compare the current temperature with the temperature set point, and find that they are too far apart. This stimulates a hypothalamic nucleus, the **heat-loss center,** which sends signals to effectors.
- **Effector/response: Blood vessels in the skin dilate, and sweat glands release sweat.** Several effectors then produce responses that reduce body temperature. One response is vasodilation of the blood vessels in the skin. Recall that this increases blood flow through the skin and promotes heat loss (see Chapter 5). Another response is increased sweating from the skin's sweat glands. However, sweating only produces significant heat loss if the humidity in the air isn't too high. In high-humidity conditions, the rate of evaporation decreases, which is why humid heat feels hotter than "dry heat."
- **Homeostatic range and negative feedback: As body temperature returns to the homeostatic range, the hypothalamus stops stimulating the effectors, which decreases the responses.** When the thermoreceptors detect a return to the homeostatic body temperature range, they stop stimulating the hypothalamus through negative feedback, which in turn ends stimulation of the effector organs, decreasing or shutting down the responses.

A similar feedback loop is triggered by a falling body temperature, commonly caused by a low environmental air or water temperature, and proceeds as follows (**Figure 23.18b**):

- **Stimulus: Body temperature decreases below normal range.**
- **Receptor: Thermoreceptors in the skin and the hypothalamus detect the decreased body temperature.** As with higher body temperature, thermoreceptors in the skin and hypothalamus detect the lower body temperature. The thermoreceptors then send this information to the hypothalamus.
- **Control center: The heat-promoting center in the hypothalamus is activated.** Again the thermoregulatory centers of the hypothalamus act as the control center. These centers compare the current temperature with the temperature set point, and find that they are too far apart. They then activate the **heat-promoting center,** which sends signals to effectors.
- **Effector/response: Blood vessels in the skin constrict, shivering in skeletal muscles is triggered, and the metabolic rate increases.** These signals trigger several effectors to produce responses that raise the body's temperature. For one, it causes constriction of blood vessels in the skin. This diverts blood from the skin to deeper tissues and minimizes heat loss through radiation, conduction, and convection. It also increases heat production through promotion of involuntary muscle contractions known as **shivering.** Finally, it stimulates the release of thyroid hormones from the thyroid gland and the release of the catecholamines epinephrine and norepinephrine from the adrenal medulla. Both effects increase the metabolic rate of their target cells (an example of the Cell-Cell Communication Core Principle, p. 27), and therefore heat production.

CORE PRINCIPLE
Cell-Cell Communication

STIMULUS

Body temperature increases above normal range.

Normal range

RECEPTOR

Thermoreceptors in the skin and the hypothalamus detect increased body temperature.

Sensory receptors   Hypothalamus

IN HOMEOSTATIC RANGE

As body temperature returns to normal, feedback decreases effector responses.

Normal range

EFFECTOR/RESPONSE

Blood vessels in the skin dilate.

Sweat glands release sweat.

Blood vessels   Sweat gland

CONTROL CENTER

Heat-loss center in the hypothalamus is activated.

Heat-loss center in hypothalamus

(a) Negative feedback loop in response to rising body temperature

STIMULUS

Body temperature decreases below normal range.

Normal range

RECEPTOR

Thermoreceptors in the skin and the hypothalamus detect decreased body temperature.

Sensory receptors   Hypothalamus

IN HOMEOSTATIC RANGE

As body temperature returns to normal, feedback decreases effector responses.

Normal range

EFFECTOR/RESPONSE

Blood vessels in skin constrict.

Shivering is triggered.

Thyroid hormones and catecholamines increase the metabolic rate.

Blood vessels   Skeletal muscle   ATP

CONTROL CENTER

Heat-promoting center in the hypothalamus is activated.

Heat-promoting center in hypothalamus

(b) Negative feedback loop in response to falling body temperature

Figure 23.18 Maintaining homeostasis: regulation of core body temperature by negative feedback loops.

- **Homeostatic range and negative feedback: As body temperature returns to the homeostatic range, the hypothalamus stops stimulating the effectors, which decreases the responses.** The negative feedback mechanisms work the same way as they did with increased temperature—the thermoreceptors detect the return to the normal range and stop stimulating the hypothalamus, which then stops stimulating the effectors.

Impaired Temperature Regulation

As you have seen, the human body has a fairly narrow homeostatic temperature range. When the core temperature moves outside this range, significant disruptions in all body systems can result. An abnormal decrease in body temperature is known as **hypothermia,** commonly due to prolonged exposure to decreased temperatures. Hypothermia generally slows down the body's enzymes, which slows the metabolic rate and most of its processes, including heart and breathing rate. When body temperature falls below about 32 °C (89.6 °F), heat-promoting mechanisms cease, mental function becomes sluggish, and coma and death are possible. An individual who survives hypothermia may suffer from *frostbite,* or the death of peripheral structures such as the fingers, toes, and tips of the ears and nose. This is due to the constriction of blood vessels to these peripheral locations as part of the body's heat-conserving mechanisms.

Humans are more tolerant in general of hypothermia than of an abnormally elevated temperature, which is called **hyperthermia.** Technically, any temperature higher than the normal core temperature range could be considered hyperthermia, but the defining temperature is usually above 38.3 °C (101 °F). Humans can generally survive internal temperatures up to about 40.6–41.1 °C (105–106 °F), depending on their age and general health. Temperatures at or above 42.8 °C (109 °F) are generally incompatible with life.

Some cases of hyperthermia, known collectively as *heat illness,* are caused by prolonged exposure to elevated temperatures in which the body's ability to cool itself is insufficient. The elevated temperatures increase enzyme activity and the rate of the body's reactions. This actually produces more heat, which further increases enzyme activity in a positive feedback loop. This loop continues to operate until the body's temperature rises to about 41.1 °C (106 °F), at which point enzymes begin to denature and organs start to fail. The resulting condition, known as **heat stroke,** is characterized by hot, dry skin; rapid breathing; and a rapid, sometimes irregular pulse. Heat stroke can be quickly fatal if not immediately recognized and treated.

Another type of hyperthermia is **fever,** which you learned about with the immune system (see Chapter 20). Unlike heat illness, which is caused by elevated external temperatures and humidity, fever is due to an increase in the set point for body temperature. Recall that a fever is part of the nonspecific inflammatory response that generally occurs when chemicals called *pyrogens* are secreted by leukocytes such as macrophages. Pyrogens act on the thermoregulatory centers of the hypothalamus and reset the body's "thermostat" so that the set point is higher. Thermoreceptors then detect that the body temperature is too low and trigger the heat-promoting center in the hypothalamus, which in turn triggers

mechanisms such as shivering (the "chills" you get when you have a fever). When the body temperature rises to the new set point, the hypothalamus normally shuts off its heat-promoting mechanisms.

Quick Check

- [ ] 6. What is the normal core body temperature?
- [ ] 7. How does the body respond to a decrease in core temperature?
- [ ] 8. How does the response differ with an increase in core temperature?

Apply What You Learned

- [ ] 1. A condition for measurement of the BMR is that the individual is not under physical or emotional stress. How might physical or emotional stress affect the metabolic rate?
- [ ] 2. If a product increased lipolysis without increasing the metabolic rate, would this likely result in weight loss? Explain why or why not.
- [ ] 3. What happens to the metabolic rate during exercise? Explain why you feel warm when you exercise, even if the environmental temperature is cold.

See answers in Appendix A.

MODULE **23.7**

Nutrition and Body Mass

Learning Outcomes

1. Define the terms nutrient, essential nutrient, macronutrient, and micronutrient.
2. Describe the dietary sources, relative energy yields, and common uses in the body for carbohydrates, lipids, and proteins.
3. Discuss the major roles and sources of fat- and water-soluble vitamins and dietary minerals.
4. Describe the components of a balanced diet, including the concept of recommended dietary allowances and energy balance.
5. Explain how the liver processes cholesterol, and identify the roles of the different lipoproteins.
6. Describe how body mass is determined.

We've seen how nutrients are used by the body, so now let's examine how they are obtained from the diet. As we mentioned earlier, a **nutrient** is a molecule obtained from food that the body requires for its metabolic processes. The study of nutrients includes the fuel molecules carbohydrates, proteins, and lipids; molecules called *vitamins* and *minerals;* and other structural molecules taken in from the diet, such as cholesterol.

The upcoming sections take a look at the dietary sources of these nutrients and the amounts of each nutrient required for optimal health. The end of the module concentrates on how nutrient intake contributes to body mass and the problems that may arise when excessive intake leads to obesity. First, however, let's get acquainted with some basics of nutrition science.

Overview of Nutrients

There are five main categories of nutrients. The first three—carbohydrates, proteins, and lipids—are known as **macronutrients** because they make up the bulk of our diet and are required in relatively large amounts. The other two categories, vitamins and minerals, are called **micronutrients** because they are required from the diet in much smaller amounts. Note that water is often considered a sixth class of nutrient, but we discuss its many roles in the body in the fluids and electrolytes chapter (see Chapter 25).

For most types of nutrients, the amount required from the diet to maintain health has been calculated by the U.S. Department of Agriculture (USDA) and is referred to as the *recommended dietary allowance,* or *RDA.* The RDA for each nutrient depends on several factors, including a person's age and gender, the number and types of processes in which the nutrient is involved, and whether or not the body can produce the nutrient. Many molecules, such as glucose, several types of amino acids and fatty acids, as well as some vitamins, can be made by the body. However, the body is unable to produce certain molecules called **essential nutrients** (so named because they are an essential part of the diet). There are as many as 50 essential nutrients, including certain amino acids and fatty acids, most vitamins, and all minerals.

Quick Check

☐ 1. How do macronutrients and micronutrients differ?
☐ 2. What is an essential nutrient?

Macronutrients

◄◄ **Flashback**

1. How do saturated and unsaturated fats differ? (p. 53)

We take in a wide variety of carbohydrates, lipids, and proteins through the diet, some of which are used for fuel, and others for structural molecules; still others cannot be digested and so simply pass through the body unaltered. The following sections examine the dietary sources, the recommended dietary allowances, and the functions for each of these macronutrients (**Table 23.1**).

Carbohydrates

Dietary carbohydrates consist of monosaccharides such as glucose, disaccharides such as sucrose, and polysaccharides such as starch. Most of these forms of carbohydrates come from plant sources, although there are animal sources as well, notably the disaccharide lactose in milk. We find monosaccharides and disaccharides (nutritionally simple sugars) naturally in fruits, honey, milk, and sugar cane, as well as added to many foods and drinks, including breakfast cereals, candy, juices, and sodas. Polysaccharides, sometimes called *complex carbohydrates,* are found in foods such as grains and vegetables.

Certain polysaccharides known collectively as **fiber** are not fully digestible by humans because we lack the digestive enzymes for their catabolic reactions. *Insoluble fiber,* which is found in whole grains, many fruit skins, bran, and some vegetables, is not digestible at all and passes through the digestive tract and into the feces virtually unaltered. *Soluble fiber,* however, is partially digestible by bacteria in the colon into compounds that can be absorbed into the bloodstream. This type of fiber is found in the pulp of many fruits, grains such as oats, and many vegetables. Soluble fiber has been the topic of much research, as some of the absorbed compounds have been found to have beneficial health effects, such as lowering the level of cholesterol in the blood.

Nutritionists generally recommend that an individual's daily intake of Calories consist of about 45–65% carbohydrates. These carbohydrates should be mostly polysaccharides, rather than monosaccharides or disaccharides, because polysaccharides tend to be more nutrient-dense. Many foods with primarily simple sugars are highly processed and nutrient-poor, and for this reason are sometimes said to have "empty calories."

TABLE 23.1 SUMMARY OF MACRONUTRIENTS

Macronutrient Class	Dietary Sources	Daily Requirement (for Adults)	Functions
Carbohydrates • Monosaccharides and disaccharides (simple sugars) • Polysaccharides (complex carbohydrates)	• Fruits, dairy products, sugar cane, sugar beets, candy, additives to many foods and drinks • Rice, pasta, bread, potatoes, vegetables, glycogen in meat	45–65% of total caloric intake	• Oxidized for energy • Used in production of other molecules • Incorporated into multiple structural molecules
Lipids • Saturated • Unsaturated	Meats, eggs, dairy products, plant oils	< 30% of total caloric intake	• Oxidized for fuel • Used to form structural molecules • Glycerol used for gluconeogenesis
Proteins	Meats, legumes, nuts, seeds, dairy products, certain grains, eggs	10–35% of total caloric intake	• Used to form structural molecules • Used to manufacture enzymes, antibodies, and hormones • Oxidized for fuel • Used for gluconeogenesis

In the past decade, carbohydrates have gotten a lot of negative attention in the media. A number of diet plans say that carbohydrates are the main reason for most people's expanding waistlines. Nevertheless, carbohydrates remain an essential part of the diet. Glucose is the preferred fuel source for the brain and other tissues such as the liver, and the body will go to great lengths to ensure that the glucose level in blood remains constant. Diet plans that severely restrict or even eliminate carbohydrates may lead to some fat loss as the body initiates processes like β-oxidation and ketogenesis, but they also cause protein breakdown to release amino acids for gluconeogenesis. This can lead to wasting of skeletal muscle tissue and other vital organs such as the heart. In addition, the ketone bodies can accumulate in the blood and potentially affect the blood's pH, as you discovered earlier.

Lipids

Several types of lipids are ingested in the diet, including triglycerides, cholesterol, and certain vitamins. This discussion is restricted to those lipids used as a fuel source, namely, the triglycerides or neutral fats (cholesterol, a structural lipid, is discussed later in this module), which consist of three fatty acids bound to glycerol. As we discussed previously, saturated fatty acids have hydrocarbon chains with no double bonds between carbon molecules (see Figure 2.17). Saturated fats are found primarily in animal fats, including butter, cream, cheese, fatty meats, and lard. Certain plant products have saturated fats as well, including coconut oil, cottonseed oil, and cacao (the main ingredient in chocolate).

Unsaturated fatty acids have hydrocarbon chains with one or more double bonds between carbon molecules. An unsaturated fatty acid with a single double bond is a *monounsaturated fat,* and one with multiple double bonds is a *polyunsaturated fat.* Unsaturated fats can be classified according to the location of their final double bond in the fatty acid chain. This is indicated by the Greek letter omega, followed by a number indicating the location of the carbon (e.g., omega-3 fatty acid). They may also be classified according to the orientation of hydrogen atoms around the carbon molecules. Those with hydrogen atoms on opposite sides are known as *trans* fats; those with hydrogen atoms on the same side are *cis* fats. *Trans* fats have gotten a great deal of publicity in recent years, as they have demonstrated many negative health consequences. (For more information, look back at *A&P in the Real World: The Good, the Bad, and the Ugly of Fatty Acids,* in Chapter 2.)

Fat is an important nutrient; in fact, there are two essential fatty acids, those that the human body can't make and must be supplied in the diet. The first, the polyunsaturated omega-6 fatty acid *linoleic acid,* is an important building block for many lipids in the body. The second is the polyunsaturated omega-3 fatty acid *linolenic acid.* Both are present in most plant oils, and linolenic acid is found in abundance in fish oils.

Present USDA guidelines recommend that fats make up no more than 30% of an individual's total caloric intake. The majority of these fats should be unsaturated fats, some of which may have potential health benefits. For example, studies have demonstrated possible positive effects on cardiovascular health from ingestion of omega-3 fatty acids. Saturated fat intake should be no more than 10% of the total Caloric intake, as many correlations have been found between the consumption of large amounts of saturated fats and certain diseases, particularly coronary artery disease and certain cancers.

Proteins

As we have discussed throughout this text, proteins and amino acids are important molecular fuels, structural molecules, and enzymes. They are also used in the synthesis of other molecules, such as nucleic acids, glycoproteins, and glucose. Dietary sources of proteins include animal proteins from meats, dairy products, and eggs, as well as plant proteins from legumes, nuts, seeds, certain grains, and soy. Current guidelines recommend that about 10–35% of your total caloric intake come from proteins. This number varies with each individual; for example, children and athletes typically have higher protein requirements than do older, sedentary individuals.

Twenty amino acids make up the body's proteins, 11 of which, called **nonessential amino acids,** can be synthesized from carbon skeletons. The other nine **essential amino acids** must be supplied by the diet (Table 23.2). A dietary protein is considered a **complete protein** when it supplies all of the essential amino acids; an **incomplete protein** lacks one or more of them. Most animal proteins are complete proteins, but of the plant proteins, only soy is complete. For this reason, there is a common misconception that individuals eating vegetarian diets cannot obtain all of the essential amino acids. However, vegetarians can readily obtain them by combining incomplete proteins. The classic example of a good combination is rice and beans.

A dietary protein that has been getting a great deal of media attention in recent years is *gluten.* Gluten is actually a protein composite, a combination of two proteins occurring naturally in wheat and certain other grass-based grains. It is found in foods prepared directly from wheat as well as in foods in which it is used as an additive to improve the food's texture and increase its protein content. An indeterminate percentage of the population, variably given as between 4% and 7%, has a sensitivity to gluten. The symptoms range from mild digestive irritation to small intestine damage in the 1% who have an autoimmune disease known as *celiac disease* (SEE-lee-ak). Although the percentage of individuals with gluten sensitivity is relatively small, gluten has been incorrectly reported by the media to be potentially harmful

TABLE 23.2 ESSENTIAL AND NONESSENTIAL AMINO ACIDS	
Essential Amino Acids	**Nonessential Amino Acids**
Histidine	Alanine
Isoleucine	Arginine
Leucine	Asparagine
Lysine	Aspartate
Methionine	Cysteine
Phenylalanine	Glutamate
Threonine	Glutamine
Tryptophan	Glycine
Valine	Proline
	Serine
	Tyrosine

to nearly everyone and the cause of a host of diseases. However, the only people who truly need to avoid gluten are those with gluten sensitivity; there is no evidence that it is unhealthy for the remaining 93% of the population.

Quick Check

☐ 3. What are dietary sources of monosaccharides, disaccharides, and polysaccharides?
☐ 4. Which fatty acids are essential? What are the main sources of these fatty acids?
☐ 5. How does a complete protein differ from an incomplete protein?

Micronutrients

◀◀ Flashback

1. How do hydrophobic and hydrophilic substances differ? (p. 47)
2. What are some minerals important to human physiology? (p. 33)

The micronutrients—vitamins and minerals—are molecules that are not used for fuel but that play critical roles in nearly all of our physiological processes. The vitamins and minerals required in our diets, their sources, and their roles are examined next.

Vitamins

Some media reports tend to associate the term vitamin with anything that's deemed "good for you." But what are vitamins, and what do they do in the body? A **vitamin** (from its historical name, *vital amine*) is an organic compound required for the body's functions. There are 13 known vitamins, and their letter names indicate the order in which they were discovered. Some vitamins, such as vitamins A and D, act as hormones and influence the functions of other cells. Others, particularly the B vitamins, act as enzyme *cofactors,* or nonprotein molecules required for the enzyme to function properly. Still others, including vitamins E and C, are important antioxidants that detoxify free radicals formed as a normal part of the body's metabolism.

Vitamins are classified by their structure, which determines their water or lipid solubility. The **fat-soluble vitamins** have a structure similar to that of cholesterol with one or more hydrocarbon rings and long hydrocarbon chains. For this reason, they are nonpolar and are therefore hydrophobic and soluble in other lipids. The four fat-soluble vitamins are vitamins A, D, E, and K. As we have seen, fat-soluble vitamins bind bile salts in the small intestine and are absorbed via micelles with other lipids (see Chapter 22). In the body, excess fat-soluble vitamins can be stored in adipose tissue. This means that excess stores of fat-soluble vitamins have the potential to reach toxic levels.

Vitamin C and the B vitamins are **water-soluble vitamins,** meaning that they have predominantly polar bonds and are therefore hydrophilic and soluble in water. With the exception

of vitamin B12, they are absorbed from the gastrointestinal tract with water and polar solutes. Recall that vitamin B12 must bind to the molecule intrinsic factor, produced by the stomach, to be absorbed (see Chapter 22). Water-soluble vitamins are not stored in the body in any appreciable amount, and any ingested in excess are generally excreted by the kidneys into the urine.

The 13 vitamins are summarized in Table 23.3. It covers their roles, their sources, and the effects of their deficiency and excess.

Minerals

As we discussed, a **mineral** is any element other than carbon, hydrogen, oxygen, and nitrogen that is required by living organisms (see Chapter 2). Many minerals in the human body exist in their ionized forms, as this is the form in which they are most stable. Like vitamins, minerals are not used for fuel, but they are an important component of many physiological processes. There are seven minerals called *major minerals* that are required in moderate amounts for optimal health: calcium ions, chloride, magnesium ions, phosphorus, potassium ions, sodium ions, and sulfur. In addition to these seven, very small amounts of many other minerals, which are called *trace elements,* are necessary, including iodine, iron, selenium, and zinc. The biochemical roles of 10 trace elements have been established; many others are hypothesized, but research hasn't yet defined their roles.

Mineral-rich foods include fruits, vegetables, dairy products, and certain meats. Foods that are highly refined are typically very mineral-poor, as are high-fat foods. Many very low-carbohydrate diets prohibit intake of foods rich in minerals, which could lead to mineral deficiencies. Are minerals consumed in excess quantities beneficial? What about vitamins? Find out in *A&P in the Real World: Vitamin and Mineral Megadoses* on page 932.

The functions, sources, and symptoms of deficiencies of minerals are summarized in Table 23.4.

Quick Check

☐ 6. What roles do vitamins play in the body?
☐ 7. What are the major minerals required from the diet?

Structural Lipid: Cholesterol

◀◀ Flashback

1. What is the function of cholesterol in the body? (p. 55)
2. What are some functions of the liver? (p. 878)

A discussion of nutrition would be incomplete without mention of the steroid molecule **cholesterol.** Although cholesterol isn't oxidized as a fuel source, it is important in many of the body's anabolic reactions. As you've read throughout this book, cholesterol is modified to produce a variety of molecules, including steroid hormones, vitamin D, and bile salts. In addition, cholesterol is an important structural molecule, as it forms part of plasma membranes. This section discusses how cholesterol is processed

TABLE 23.3 SUMMARY OF VITAMINS

Vitamin	Dietary Sources and RDA	Functions	Symptoms of Deficiency	Symptoms of Excess
Fat-Soluble				
Vitamin A (retinol)	Leafy green vegetables, yellow vegetables; 0.7–0.9 mg	• Required for low-light and color vision • Functions as growth factor for epithelium • Needed for growth and development	• Delayed growth • Night blindness and potential total blindness • Immune dysfunction	• Liver damage • Nausea • Decreased appetite • Osteoporosis • CNS dysfunction • Death from extreme doses
Vitamin D	Synthesized in skin when exposed to certain wavelengths of UV light, supplemented in milk and other foods; 15–20 µg (unless exposure to sunlight is inadequate)	• Required for calcium ion homeostasis and bone growth	• Rickets (in children) • Osteoporosis	• Multiple organ dysfunction due to hypercalcemia
Vitamin E (tocopherol)	Vegetable oils, nuts, seeds, dairy products; 15 mg	• Antioxidant	• Anemia • CNS dysfunction	• Nausea • Fatigue • Vision disturbances
Vitamin K	Leafy green vegetables, produced by intestinal bacteria; 90–120 µg	• Required for synthesis of clotting factors II, VII, IX, and X	• Bleeding disorders	• Liver dysfunction • Excessive clotting in individuals taking the drug warfarin
Water-Soluble				
Vitamin B1 (thiamine)	Legumes, whole grains, certain meats; 1.1–1.2 mg	• Coenyzme in many catabolic pathways	• Disease beriberi—peripheral nerve dysfunction, heart disease	• Decreased blood pressure • Nausea
Vitamin B2 (riboflavin)	Dairy products, eggs, certain vegetables, added to many grains; 0.9–1.1 mg	• Component of FAD and so critical component of catabolic pathways	• Skin disorders • Dysfunction of other epithelial membranes	• Skin dryness and itching
Vitamin B3 (niacin)	Nuts, meats, grains, potatoes; 11–12 mg	• Component of NAD and so critical component of catabolic pathways	• Disease pellagra—CNS, gastrointestinal, and skin dysfunction	• Vasodilation, leading to skin flushing and burning • Death from extreme doses
Vitamin B5 (pantothenic acid)	Meats, dairy products, grains; 5 mg	• Required to synthesize coenzyme A	• Delayed growth • CNS dysfunction	None known
Vitamin B6 (pyridoxine)	Meat, whole grains, citrus fruits, vegetables; 1.3–1.7 mg	• Coenzyme in many metabolic pathways	• Peripheral nerve dysfunction • CNS dysfunction • Anemia • Delayed growth	• CNS dysfunction, including coma • Possible death from extreme doses
Vitamin B7 (biotin)	Red meat, organ meats, eggs, legumes, nuts, dairy products; 25–30 µg	• Coenzyme in many anabolic pathways	• Skin disorders • Neuromuscular dysfunction • Fatigue	None known
Vitamin B9 (folic acid)	Leafy vegetables, fruits, grains; 400 µg, 600 µg in pregnancy	• Coenzyme in many metabolic pathways	• Delayed growth • Anemia • Birth defects if deficient during pregnancy	• Can mask symptoms of vitamin B12 deficiency
Vitamin B12	Dairy products, meat, added to many other foods; 2.4 µg	• Coenzyme in nucleic acid metabolism • Required for development of erythrocytes	• Disease pernicious anemia	• Polycythemia
Vitamin C	Citrus fruits, red and green peppers; 75–90 mg	• Antioxidant • Coenzyme in collagen synthesis	• Disease scurvy—deterioration of skin and epithelial membranes	• Nausea and vomiting • Kidney stones

TABLE 23.4 SUMMARY OF MINERALS

Mineral	Dietary Sources	Functions	Symptoms of Deficiency	Symptoms of Excess
Major Minerals				
Calcium ions	Dairy products, legumes, dark green vegetables	• Required for bone and tooth synthesis and maintenance • Important for nerve function • Important in muscle contraction	• Osteoporosis • Muscle spasms • Nerve dysfunction	• Kidney stones • Disturbances in heart rhythm • Muscle weakness
Chloride	Table salt	• Osmotic particle • Required for synthesis of stomach acid • Important in acid-base balance	• Muscle cramps	• Fluid retention
Magnesium ions	Green vegetables, whole grains, dairy products	• Cofactor in many enzymes • Important in bone formation • Extracellular cation	• Muscle cramps	• Diarrhea
Phosphorus	Meat, dairy, grains, legumes	• Important component of nucleotides • Required for bone and tooth synthesis • Component of buffer system	• Osteoporosis • Fatigue • Muscle weakness	None known
Potassium ions	Sea salt, many fruits such as bananas, meats, dairy products, vegetables	• Important cation in electro-physiological processes • Osmotic particle	• Disturbances in heart rhythm • Heart failure • Muscle weakness	• Disturbances in heart rhythm • Death if severe excess
Sodium ions	Table salt	• Important cation in electro-physiological processes • Osmotic particle	• Disturbances in heart rhythm • Muscle weakness	• Disturbances in heart rhythm • Water retention • Hypertension
Sulfur	Many different proteins	• Component of certain amino acids	• Inability to produce certain amino acids and the proteins in which they occur	None known
Trace Elements				
Chromium	Yeast, meats, some vegetables	• Required for insulin activity	• Decreased effectiveness of insulin	None known
Cobalt	Meats, dairy products	• Component of vitamin B12	• Same symptoms as vitamin B12 deficiency	None known
Copper	Sea animals, nuts, legumes	• Component of coenzymes of the electron transport chain • Required for hemoglobin synthesis	• Anemia	• Liver failure • CNS dysfunction
Fluorine	Sea animals, added to most drinking water	• Prevention of tooth decay	• Higher incidence of tooth decay	None known
Iodine	Sea animals, dairy products, added to salt	• Required for synthesis of thyroid hormone	• Hypothyroidism	None known
Iron	Meats, eggs, legumes, certain grains, certain vegetables	• Required for synthesis of hemoglobin	• Iron-deficiency anemia	• Nausea and vomiting • Liver damage
Manganese	Nuts, grains, fruits, vegetables	• Cofactor in several enzymes • Potential multiple roles in physiology	• Abnormalities in bone and cartilage • Possible anemia	None known
Molybdenum	Legumes, grains	• Cofactor in many enzymes • Involved in urea processing	• Decreased ability to process amino acids	None known
Selenium	Meats, whole grains, sea animals	• Antioxidant • Enzyme cofactor	• Muscle cramps	• Nausea and vomiting • Fatigue
Zinc	Meats, grains	• Component of digestive enzymes • Component of carbonic anhydrase • Required for normal sperm production in males	• Impaired immunity • Low sperm count in males • Diminished reproductive drive	• CNS dysfunction • Difficulty walking

by the body, particularly the liver, and what can happen when the level of cholesterol in the blood is elevated.

Cholesterol Processing by the Liver

Cholesterol is found in the diet in animal-based foods such as meats and dairy products. The body generally has little to no need for dietary cholesterol, as it can be synthesized by the liver. Normally, about 15% of the body's cholesterol comes from the diet; the remaining 85% is made in the liver. Recall that cholesterol is packaged in the small intestine with other lipids as micelles, which escort these nonpolar molecules to the enterocyte membrane, where they are absorbed into the cell. In the enterocytes, cholesterol and other lipids are packaged into chylomicrons, which are then absorbed into a lacteal (see Chapter 22) and eventually delivered to the liver for processing.

The liver also plays a key role in distributing cholesterol throughout the body. As we noted previously, cholesterol is a steroid molecule and so is nonpolar and hydrophobic. For this reason, it must be packaged along with other lipids onto carrier proteins called **lipoproteins** for safe transport through the blood. The structure of a typical lipoprotein and its lipids is shown here:

Protein

Phospholipids:
- Polar heads
- Fatty acid tails

Lipids:
- Cholesterol
- Triglycerides

There are several types of lipoproteins that differ in their ratios of lipid to protein, which affects their density. Those carrying more lipids than proteins are larger and have a lower density, whereas those with fewer lipids and more proteins are smaller and have a higher density.

Let's look at these types in more detail (**Figure 23.19**):

- **Very low-density lipoproteins.** The **very low-density lipoproteins (VLDLs)** contain a higher percentage of lipids than proteins, and so have a relatively low density. VLDLs are triglyceride-rich particles that deliver triglycerides from the liver to other tissues, primarily adipose tissue for storage and muscle tissue for immediate use. Triglycerides are removed from VLDL particles via reactions that convert them into free fatty acids and monoglycerides. These reactions are catalyzed by the enzyme **lipoprotein lipase,** which is located on the surface of endothelial cells. The free fatty acids and monoglycerides then diffuse into cells, where they are either oxidized for fuel or reassembled into triglycerides.

- **Low-density lipoproteins.** After VLDLs unload triglycerides, they have a slightly higher density and become **low-density lipoproteins (LDLs).** LDLs are the main mechanism by which cholesterol is delivered to cells in peripheral tissues. When an LDL binds to a cellular receptor, it triggers receptor-mediated endocytosis and the uptake of cholesterol.

- **High-density lipoproteins. High-density lipoproteins (HDLs)** are synthesized by the liver and released into the circulation or the small intestine. Whereas LDLs and VLDLs transfer lipids to the body's cells, HDLs transfer lipids from the cells in peripheral tissues to the liver. In the liver, cholesterol is taken up by hepatocytes and excreted from the body via the bile.

You may have heard VLDL and LDL particles described as "bad" cholesterol. This refers to the fact that these particles deliver cholesterol to tissues and can contribute to disease processes such as atherosclerosis (the formation of fatty plaques in arteries). In contrast, HDL particles are described as "good" cholesterol because a higher level of HDL particles will lead to less total cholesterol deposition in peripheral tissues.

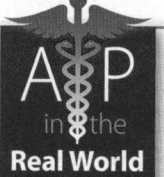

A&P in the Real World

Vitamin and Mineral Megadoses

If something is good for you in small doses, it must be even better for you in large doses, right? That certainly seems to be the thinking of many dietary supplement manufacturers. In the United States, vitamins and minerals are regulated as dietary supplements, not as drugs, and so manufacturers are not required to demonstrate their safety or efficacy. It's not uncommon to see supplements with extreme excesses of the RDAs of certain vitamins and minerals. So, are these megadoses of vitamins and minerals safe and/or effective?

Research has shown that *small* dose supplementation (e.g., 100% or less of the RDA) may be of benefit in certain individuals. For example, supplementation of calcium salts in some people may reduce their risk of osteoporosis. However, megadoses are generally of no benefit and can actually be harmful. High doses of water-soluble vitamins are simply excreted into the urine by the kidneys, but toxicity can result if the dose exceeds the kidneys' capacity for excretion. Fat-soluble vitamins are excreted less efficiently by the kidneys and so the potential for toxicity is much higher. The toxic effects of excessive doses of vitamins and minerals are listed in Tables 23.3 and 23.4—notice that these effects range from liver damage to death.

So, should you take those megadoses of vitamins and minerals? Unless you're told to do so by a medical doctor to correct a significant deficit, the answer is no. At best, they lead to the production of very expensive urine, and at worst, they could cause significant toxicity.

LIPOPROTEIN	FUNCTION	COMPOSITION
VLDL	Deliver triglycerides to adipocytes, which are removed by lipoprotein lipase; VLDLs then become LDL particles.	55–60% Triglyceride 15–20% Phospholipid 10–15% Cholesterol 5–10% Protein
LDL	Deliver cholesterol and triglycerides to the cells. Some of the lipids are used by the cell; others are transferred to empty HDL particles for delivery to the liver.	10% Triglyceride 20% Phospholipid 45% Cholesterol 25% Protein
HDL	Deliver cholesterol to the liver for excretion in the bile.	5% Triglyceride 30% Phospholipid 20% Cholesterol 45–50% Protein

Labels in figure: VLDL, VLDL, Fatty acids, Cholesterol, LDL, Hepatocyte, "Empty" HDL, "Full" HDL, Bile duct, Adipocyte, LDL receptor, Body cell, Used by cell, Exported from the cell, packaged into HDL, Blood vesssel

VLDL = Very low-density lipoprotein

LDL = Low-density lipoprotein

HDL = High-density lipoprotein

Figure 23.19 Cholesterol distribution by lipoproteins.

The ideal value for total cholesterol is less than 200 mg/dl, but just as important as total cholesterol is the ratio of HDL to LDL cholesterol. Current guidelines from the American Heart Association recommend that LDL cholesterol be less than 129 mg/dl in most individuals, and less than 70 mg/dl in those with a very high risk for cardiovascular disease. By contrast, it is recommended that HDL be *at least* 50 mg/dl and ideally over 60 mg/dl.

Disorders of Cholesterol Processing: Hypercholesterolemia

Cholesterol production is partially regulated by a negative feedback loop so that increases in intake will result in a limited decline in production by the liver, which brings us again to the Feedback Loops Core Principle (p. 21). However, if intake exceeds about 15% of the total requirement for cholesterol, then the total blood concentration of cholesterol is likely to increase, a condition

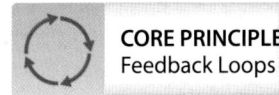

CORE PRINCIPLE
Feedback Loops

called **hypercholesterolemia** (hy'-per-koh-less-ter-oh-LEE-me-ah). Individuals with hypercholesterolemia typically have high total cholesterol and LDL levels, along with a low HDL level. Another factor that can increase cholesterol is increased production by the liver. This can also be related to the diet—a high intake of saturated fats increases cholesterol production while inhibiting its excretion. *Trans* fats have a similar effect, increasing the production of LDLs and decreasing that of HDLs. Hypercholesterolemia can also be due to genetic defects that cause the liver to produce excess cholesterol (a condition known as *familial hypercholesterolemia*).

Hypercholesterolemia is a health concern largely because it greatly increases the risk of developing atherosclerosis. As we mentioned, atherosclerosis is associated with conditions such as coronary artery disease, myocardial infarction (heart attack), and stroke (see Chapter 18). For this reason, treatment is generally recommended in individuals with a total cholesterol level over 200 mg/dl and/or an LDL level over 130 mg/dl. Treatment consists of dietary modification in combination with exercise and several different types of medications. Some medications, such as the group known as the *statins*, prevent the formation of cholesterol by the liver. Others decrease its absorption from the small intestines and enhance its excretion in the bile.

Quick Check

☐ 8. Why is HDL considered "good" cholesterol, whereas LDL and VLDL are considered "bad" cholesterol?

☐ 9. What role does the liver play in cholesterol metabolism?

Diet and Body Mass

As mentioned in the introduction to this chapter, diet products, supplements, and books are everywhere, and they form the basis of a billion-dollar industry. Clearly, food issues and weight loss are on the minds of many, many people in the United States. And there is a good reason for this—a significant portion of the American population weighs more than what is generally considered healthy for their height.

The term *body mass* refers to the amount of matter in the body; the more commonly used term *body weight* refers to the resulting force exerted on this mass by gravity. An individual's weight in and of itself often doesn't give a clinician enough information, as it varies with several factors, particularly an individual's height. For this reason, many healthcare providers and nutritionists instead calculate a number called the **body mass index (BMI),** which is obtained with the following equation:

$$ \text{BMI} = \frac{\text{Weight (kg)}}{[\text{Height (m)}]^2} $$

As you can see from the equation, BMI takes height into account, and so provides a somewhat more reliable index from which to determine relative body mass. From the BMI, a person can be classified as underweight (BMI < 18.5), normal weight (BMI = 18.5–24.9), overweight (BMI = 25.0–29.9), or obese (BMI ≥ 30.0). Note that these numbers are only reference

ranges, not absolutes, because the BMI does not consider factors such as muscle mass, body type, or ethnicity. For example, many highly trained athletes have BMIs in the overweight range due to their larger than average muscle mass.

This section first looks at factors that affect an individual's body mass. Then we turn to how this applies to a diet that meets all nutritional needs and is generally considered healthy. Finally, we examine obesity, the problems that accompany it, and the many ways in which people try to address it.

Energy Balance

Multiple factors determine an individual's recommended body mass, including genetics, age, gender, body type, and degree of physical training. However, the rate at which an individual gains or loses body mass is largely dependent on a principle known as **energy balance,** or **nitrogen balance,** the difference between energy intake and energy expenditure. The total amount of energy intake an individual requires to meet his or her daily needs depends on a number of factors, including age, gender, muscle mass, activity level, and genetic factors. The precise number of Calories an individual requires can be estimated by calculating the BMR and factoring in the activity level. For most adults, this ranges from 2000 to 2500 Calories.

So, how does an individual gain, lose, or maintain body mass? The answer is pretty simple:

- If the total number of Calories taken in equals the total number expended, body mass remains stable.
- If the total number of Calories taken in is greater than the total number expended, a condition known as *positive energy balance,* body mass increases.
- If the total number of Calories taken in is less than the total number expended, a condition known as *negative energy balance,* body mass decreases.

Essentially, if you eat more Calories than you burn, you will gain weight. This principle is the basic idea behind all varieties of diets—those intended to help an individual gain weight, lose weight, or maintain a given weight.

Healthy Diet

An individual's diet is considered healthy when it meets his or her nutritional needs in terms of energy requirement; types and amounts of macronutrients, vitamins, minerals, essential amino acids, and fatty acids; and no more than the recommended amounts of other molecules such as cholesterol. Healthy diets are known to promote health, as their name implies, and reduce the risk for illnesses such as heart disease, stroke, cancer, and diabetes.

But what is a healthy diet? In 2011, the USDA released guidelines called **MyPlate,** seen in **Figure 23.20.** MyPlate shows the proportions of the types of foods that should make up meals. As you can see, the guidelines recommend filling your plate half with vegetables and fruits, and half with grains and proteins. In addition, the small circle represents a serving of a low-fat dairy product such as milk or yogurt.

Since the release of MyPlate, other institutions have issued similar guides that include healthy oils and replace the dairy recommendation with water. These different nutritional

Figure 23.20 Current USDA healthy diet guidelines.

recommendations from so many sources can make it confusing to figure out exactly what is healthy for you. The best thing to do in the face of multiple, ever-changing guidelines is use good sense—you really can't go wrong with a diet rich in vegetables, fruits, whole grains, and lean protein in amounts that meet, but don't exceed, your body's energy requirements.

Obesity

The prevalence of adults who are overweight and **obese** in the United States is alarming—according to the Centers for Disease Control (CDC), approximately 34% are overweight, 33.8% obese (have a BMI > 30), and 5.7% morbidly obese (BMI > 40). Even more alarming is the trend that shows a relatively rapid increase in obesity rates—in 1960, only 13.4% of adults were considered obese, and 0.9% morbidly obese. Despite this increase in obesity, however, the percentage of adults meeting the BMI guidelines for overweight has remained nearly unchanged over the past 50 years. Similarly alarming is the obesity rate of 16.9% for children aged 2–19 years. This is an increase from only 5.0% in 1971.

Why is obesity such a concern? The answer has many factors. Chief among them are the health concerns that disproportionately face obese individuals. Obesity is one of the leading preventable causes of death worldwide, and decreases a person's lifespan by an average of 6–7 years. It increases the risk of type 2 diabetes, coronary artery disease, osteoarthritis, certain types of cancer, hypertension, and obstructive sleep apnea (periods in which breathing stops during sleep). Another concern with obesity is social in nature. Obese individuals, particularly children, can face social stigmatization. Finally, obesity has an economic cost, as well, both to the individual and to society. Obesity and its related conditions cost the U.S. government an estimated 100 billion dollars per year. In addition, an obese person often faces employment discrimination and pays more for health insurance premiums.

Obesity has many potential causes, but the number one cause is positive energy balance—that is, a caloric intake that exceeds the number of Calories that are burned. In the United States, this

is partly due to the widespread availability of food, particularly high-Calorie, high-fat, processed foods with few nutrients (e.g., fast food). This fact, combined with a sedentary lifestyle, can lead to an energy imbalance. In addition, a genetic connection to obesity exists in many individuals. For example, abnormalities in genes controlling appetite and the metabolic rate can contribute to the development of obesity. Other less common factors that predispose individuals to becoming obese include endocrine abnormalities such as hypothyroidism and Cushing's syndrome, certain medications, psychiatric conditions, and lower socioeconomic status.

The mainstay of treatment for obesity is reduced caloric intake combined with an exercise program. There are a host of diet programs out there, ranging from low-carbohydrate to low-fat to simply counting Calories. Each type of plan has its advantages and disadvantages, and anyone embarking on a weight loss journey should consult a dietician or physician to find the best plan for his or her needs. Sometimes, in conjunction with diet and exercise, diet medications may be used. As of 2014, two types of agents for weight loss have been approved by the Food and Drug Administration (FDA), one of which (orlistat) blocks fat digestion and absorption in the small intestine and several that act on the central nervous system to decrease appetite. In cases of morbid obesity, surgery may be considered. Surgical procedures, such as the "lap band," which constricts the stomach and prevents it from overfilling, or gastric bypass, which removes a portion of the stomach, can yield significant weight loss results, but only when combined with reduced caloric intake and exercise.

In addition, an ever-increasing number of diet plans and over-the-counter dietary supplements claim to produce miraculous weight loss results. Here we definitely have a case of buyer beware—many of these plans and products are the modern-day equivalents of snake oil, as they promise something they cannot possibly deliver. Most of these plans and products use scientific-sounding terms in their advertisements, which can make it quite difficult to determine if they are legitimate or a waste of money. Some warning signs to look for include promises of unreasonable results ("Eat all you want and still lose weight!"), diet plans that nearly or fully omit entire food groups or severely restrict caloric intake, and any product that uses fear or claims of "conspiracies" to induce you to buy. Remember also that manufacturers of weight loss dietary supplements are not required to test their ingredients for safety or efficacy, which allows them to make any claims about their product and weight loss without having to actually prove that the claims are true.

The best defense for your health (and your bank account) is knowledge and good sense. If something sounds too good to be true, trust your instincts. There are no miraculous instant solutions for weight loss. If you want to lose weight, you have to burn more energy than you take in; in other words, eat less and exercise more—simple, but usually not easy.

Quick Check

☐ 10. What conditions create a positive or negative energy balance? What effect will each have on body mass?

☐ 11. What are the causes and consequences of obesity?

☐ 1. Some low-carbohydrate diets prohibit almost all consumption of carbohydrates, including fruits and vegetables. Predict the potential consequences of these diets.

☐ 2. A patient comes into a clinic for a follow-up appointment after having been diagnosed with high cholesterol. The nurse notes that his total cholesterol has decreased from 250 mg/dl to 230 mg/dl, his LDL cholesterol has remained the same at 200 mg/dl, and his HDL cholesterol has decreased from 50 mg/dl to 30 mg/dl. Do these numbers represent an overall improvement in his condition? Why or why not?

☐ 3. A friend has embarked on a diet plan whose advertisements say that he can eat as much of anything as he wants and still lose weight. Explain to your friend, in physiological terms, whether or not this claim is likely to be true and why.

See answers in Appendix A.

CHAPTER SUMMARY

Take full advantage of the invaluable online practice and study tools for this chapter by going to MasteringA&P®. There you will find:

- 🖱 Practice tests, module quizzes, art-labeling and sequence-the-steps activities

- ▶ Animated figures with quizzes

- ✎ *Practicing A&P* worksheets

- ⑴ Author-narrated podcast

- ♨ Dynamic Study Modules

- iP *Interactive Physiology*

MODULE 23.1
Overview of Metabolism and Nutrition 897

- **Metabolism** is the sum of the body's chemical reactions, and includes harnessing the chemical energy in *nutrients* to make **ATP**, converting one type of molecule into another, carrying out synthesis reactions, and degrading macromolecules.

- **Catabolism** consists of the reactions that break down carbohydrates, proteins, and lipids into smaller substances and release energy.

- **Anabolism** consists of the anabolic reactions in which small molecules are combined to make larger ones; it is driven by the energy released by catabolism.

- *Exergonic reactions* release energy because the products have less energy than the reactants. *Endergonic reactions* require energy because the energy of the products is greater than that of the reactants. Endergonic and exergonic reactions are often *coupled.* (Figure 23.1, p. 899)

- ATP is an unstable molecule because its three phosphate groups are negatively charged and repel one another. When a phosphate group is removed in a hydrolysis reaction catalyzed by an *ATPase*, the reaction is highly exergonic.

- ATP drives reactions either directly by providing energy from its hydrolysis or, more commonly, by **phosphorylating** a reactant, which makes formation of the product of the reaction more energetically favorable.

- During an **oxidation-reduction reaction,** electrons are transferred from one molecule, which is oxidized, to another, which is reduced. This process can be highly exergonic. (Figure 23.2, p. 900)

- Electrons in the cell are passed from nutrient molecules to different **electron carriers** with increasing electron affinities, including **FAD** and **NAD$^+$**. This generates an **electromotive force** that provides the energy to generate ATP.

MODULE 23.2
Glucose Catabolism and ATP Synthesis 902

- Glucose catabolism proceeds by **glycolytic catabolism** and the **citric acid cycle,** which generate ATP, NADH, and FADH$_2$. The electron carriers are fed into the **electron transport chain,** where more ATP is generated. (Figure 23.3, p. 902)

- The energy released from glucose catabolism generates ATP via either **substrate-level phosphorylation** or **oxidative phosphorylation.**

- Oxygen is neither used nor required for **anaerobic catabolism;** however, oxygen is required for **aerobic catabolism.**

- **Glycolysis** is a series of 10 anaerobic reactions that occur in the cytosol during which glucose is split into two three-carbon **pyruvate** molecules. During the energy investment phase of glycolysis, two molecules of ATP are consumed. During the energy payoff phase, four molecules of ATP and one molecule of NADH are generated. (Figure 23.4, p. 904)

- 🖱 Practice sequencing the steps: Figure 23.4.

- The fate of pyruvate depends on the availability of oxygen (Figure 23.5, p. 905):
  - Under anaerobic conditions, pyruvate is reduced to **lactate** and NADH is oxidized back into NAD$^+$.
  - Under aerobic conditions, pyruvate moves into the mitochondria and is oxidized to yield **acetyl-CoA** and carbon dioxide.

- 🖱 Practice sequencing the steps: Figure 23.5.

- The **citric acid cycle** is a series of eight aerobic reactions that take place in the mitochondrial matrix. Acetyl-CoA enters the cycle and is combined with oxaloacetate to form **citrate.** As the cycle progresses, oxidation-reduction reactions generate NADH and FADH$_2$. At the completion of the citric acid cycle, oxaloacetate is regenerated and eight NADH, two FADH$_2$, and two ATP have been formed per glucose. (Figure 23.6, p. 906)

👆 **Practice sequencing the steps: Figure 23.6.**

- Oxidative phosphorylation consists of three processes (Figure 23.7, p. 907):
  - Electrons are transferred from NADH and FADH$_2$ to the **electron transport chain (ETC)**, a group of electron carriers in the inner mitochondrial membrane. The final electron acceptor is oxygen, which is reduced to become water.
  - The electromotive force that is generated as electrons pass from one electron carrier to another drives the pumping of hydrogen ions into the mitochondrial intermembrane space, which creates an electrochemical gradient.
  - Hydrogen ions flow back into the matrix through a pore that is part of the enzyme **ATP synthase.** The flow of hydrogen ions creates a *proton motive force* that causes ATP synthase to release its newly formed ATP and bind new ADP and P$_i$.

👆 **Practice sequencing the steps: Figure 23.7.**

- The overall yield from all parts of glucose catabolism and ATP synthesis is about 38 ATP per glucose molecule. (Figure 23.8, p. 910)

MODULE 23.3
Fatty Acid and Amino Acid Catabolism 911

- Triglycerides are catabolized in the following way: **Glycerol** is converted into glyceraldehyde-3-phosphate and fed into glycolysis. **Fatty acids** are oxidized in two-carbon units by the process **β-oxidation.** Each round of β-oxidation yields seven each of NADH and FADH$_2$ and a molecule of acetyl-CoA that is then fed into the citric acid cycle. (Figure 23.9, p. 911)

- Under certain conditions, the acetyl-CoA molecules produced in hepatocytes by β-oxidation can combine to form **ketone bodies.** These molecules enter cells and are converted back into two molecules of acetyl-CoA and then fed into the citric acid cycle. During starvation or similar conditions, ketone bodies are produced faster than the cells can use them, causing *ketosis.*

- Amino acid catabolism begins in the liver with *deamination* by the process of **transamination,** which yields a carbon skeleton and the amino acid glutamate. The carbon skeleton is converted into a variety of compounds that are fed into glycolysis or the citric acid cycle. Glutamate is deaminated and the ammonia enters the **urea cycle** to produce **urea,** which is eliminated from the body by the kidneys. (Figure 23.10, p. 913)

- The big picture of nutrient catabolism is shown in Figure 23.11 (p. 914).

👆 **Practice art-labeling exercise: Figure 23.11.**

MODULE 23.4
Anabolic Pathways 915

- Anabolism is responsible for nutrient synthesis and storage, as well as synthesis of molecules such as proteins, nucleic acids, NAD$^+$, and FAD.

- Glucose is stored as **glycogen,** primarily in the liver and in skeletal muscle tissue. Glycogen is synthesized by the process **glycogenesis** and broken down by **glycogenolysis.** (Figure 23.12, p. 915)

- **Gluconeogenesis** is the synthesis of glucose from noncarbohydrate molecules, including glycerol, amino acids, compounds from the citric acid cycle, lactic acid, and pyruvate. (Figure 23.13, p. 915)

- Most of the body's triglycerides are stored in adipose tissue and in the liver. Fatty acid synthesis, or **lipogenesis,** takes place in the cytosol, where a large enzyme complex builds fatty acids in two-carbon units. They are assembled into triglycerides by being combined with glycerol.

- When more glucose is consumed than can be deposited in glycogen, the excess is converted into fatty acids and glycerol.

- Cells can synthesize 11 of 20 amino acids from carbon skeletons. Excess amino acids are converted to glucose or fatty acids.

- Figure 23.14 (p. 917) summarizes the anabolic reactions of nutrients.

MODULE 23.5
Metabolic States and Regulation of Feeding 918

- The body can enter one of two metabolic states, depending on proximity to the consumption of food:
  - The **absorptive state** takes place from the time food is ingested to the time it is fully absorbed into the bloodstream. In this state, glucose is the main fuel used by the cells, glycogenesis and lipogenesis increase, and protein synthesis takes place. (Figure 23.15a, p. 918)
  - The **postabsorptive state** begins after ingested nutrients are fully absorbed into the bloodstream. In this state, glycogenesis, lipogenesis, and protein synthesis decrease, and gluconeogenesis, glycogenolysis, lipolysis, and ketogenesis increase. (Figure 23.15b, p. 919)

- Feeding is regulated by two hypothalamic nuclei: the *satiety center* and the *hunger center.*

- Long-term regulation of feeding is done by **leptin,** which is *anorexigenic,* and **ghrelin,** which is *orexigenic.*

- Short-term regulation of feeding involves multiple mechanisms. Insulin, digestive hormones, and the vagus nerve stimulate the satiety center. Hypoglycemia stimulates the hunger center.

MODULE 23.6
The Metabolic Rate and Thermoregulation 921

- The total amount of energy expended by the body to drive all of its processes, measured in kilocalories, is the **metabolic rate.** It is measured by **calorimetry** because nearly all of the energy spent by the body is eventually released as heat in the process of *thermogenesis.*

- The **basal metabolic rate (BMR)** is the "energy cost of living." Many conditions cause the body to deviate from the BMR, including certain hormones, drugs, nutritional status, activity, and fever. (Figure 23.16, p. 922)

- Heat may be exchanged between the body and the environment through **radiation, conduction, convection,** and **evaporation.** (Figure 23.17, p. 922)

- **Thermoregulation** is the mechanism by which the body maintains a stable internal temperature. Normal **core body temperature** averages about 37.5 °C (99.5 °F).
  - When body temperature increases, the hypothalamic **heat-loss center** promotes vasodilation of skin blood vessels and sweating. (Figure 23.18a, p. 925)
  - When body temperature decreases, the hypothalamic **heat-promoting center** triggers constriction of skin blood vessels, shivering, and an increased metabolic rate. (Figure 23.18b, p. 925)

- Impaired thermoregulation can lead to **hypothermia,** which is a decreased body temperature, or **hyperthermia,** which is an increased body temperature.

MODULE **23.7**
Nutrition and Body Mass 926

- Carbohydrates, proteins, and lipids are **macronutrients,** and vitamins and minerals are **micronutrients.** Nutrients that cannot be made by the body are **essential nutrients.** (Table 23.1, p. 927)

- Monosaccharides and disaccharides are found in foods such as fruits, milk, honey, and sugar cane. Polysaccharides are found in grains and vegetables. **Fiber** is a polysaccharide not digestible by humans. Carbohydrates should make up about 45–65% of the diet, mostly as polysaccharides.

- Triglycerides are the main dietary lipids used as a fuel source. Fatty acids bound to glycerol in triglycerides may be **saturated** or **unsaturated.** Fats should constitute 30% or less of the total caloric intake; saturated fats should constitute less than 10%.

- There are 20 amino acids in the body's proteins. Eleven amino acids can be produced by the body; the other nine are **essential.** A **complete protein** contains all of the essential amino acids. Proteins should make up between 10% and 35% of the total caloric intake. (Table 23.2, p. 928)

- There are 13 known **vitamins,** organic compounds that act as hormones, cofactors for enzymes, and antioxidants. **Fat-soluble vitamins** include vitamins A, D, E, and K, and are cholesterol-based and hydrophobic. **Water-soluble vitamins** include vitamin C and the B vitamins, and are hydrophilic. (Table 23.3, p. 930)

- A **mineral** is any element other than carbon, hydrogen, oxygen, and nitrogen that is required by living organisms. Functions, sources, and symptoms of deficiencies of minerals are summarized in Table 23.4 (p. 931).

- **Cholesterol** is a steroid molecule obtained from animal products in the diet and produced by the liver. Cholesterol, along with other lipids, is attached to **lipoproteins** by reactions that take place in the liver. **Very low-density lipoproteins (VLDLs)** deliver triglycerides to peripheral tissues, particularly adipose tissue. **Low-density lipoproteins (LDLs)** deliver cholesterol to peripheral tissues. **High-density lipoproteins (HDLs)** remove cholesterol from peripheral tissues and deliver it to the liver for excretion in the bile. (Figure 23.19, p. 933)

- *Body mass* refers to the amount of matter in the body. The **body mass index (BMI)** is a measurement of body weight relative to height that provides a range for normal body mass.

- The difference between energy intake and energy expenditure is known as **energy balance.** If there is a *positive energy balance,* body mass generally increases. If there is a *negative energy balance,* body mass generally decreases.

- A healthy diet meets an individual's needs in terms of micronutrients, macronutrients, essential nutrients, and other molecules such as cholesterol. (Figure 23.20, p. 935)

- Individuals who have a body mass index over 30.0 are defined medically as **obese.** Obesity increases the risk for many diseases, such as type 2 diabetes, coronary artery disease, cancer, hypertension, and osteoarthritis.

ASSESS WHAT YOU LEARNED

See answers in Appendix A.

LEVEL 1 Check Your Recall

1. Which of the following statements is false?

   a. Metabolism is the sum of all the reactions in the body.
   b. Anabolic pathways are exergonic and release energy.
   c. Catabolic pathways break larger substances into smaller substances.
   d. The body "pays for" anabolic reactions with the energy from catabolic reactions.

2. Which nutrient is the preferred fuel for essentially every cell in the body?

   a. Glucose
   b. Fatty acids
   c. Glycerol
   d. Amino acids

3. Fill in the blanks: A/an _____ reaction releases energy because the products have less energy than the reactants, whereas a/an _____ reaction consumes energy because the products have more energy than the reactants.

4. Mark the following statements as true or false. If a statement is false, correct it to make a true statement.

   a. During an oxidation-reduction reaction, the compound with the greatest electron affinity accepts electrons and is oxidized.
   b. The electromotive force is generated by the flow of electrons from one substance that is oxidized to another substance that is reduced.
   c. NAD^+ and FAD are electron carriers in the cell.
   d. ATP is the "money" that the cell uses to "pay for" the cell's exergonic reactions.

5. Does ATP possess a "high-energy bond"? Explain. How does ATP provide energy to drive a reaction or process?

6. Place the following events of glucose catabolism in the correct order, placing a 1 by the first event, a 2 by the second, and so forth.

_____ a. Electrons are passed between electron carriers in the inner mitochondrial membrane, and hydrogen ions are pumped into the intermembrane space.

_____ b. ATP is generated in the cytosol by substrate-level phosphorylation; NADH is generated as well.

_____ c. Hydrogen ions pass through the channel of ATP synthase and ATP is released from the enzyme.

_____ d. Acetyl-CoA is combined with oxaloacetate to form citrate.

_____ e. ATP is consumed to split glucose into two three-carbon compounds.

_____ f. Pyruvate loses a carbon forming acetyl-CoA, carbon dioxide, and NADH.

_____ g. Citrate undergoes a series of oxidation-reduction reactions that generate ATP, NADH, and $FADH_2$.

7. Which of the following statements is false?

a. Oxaloacetate is regenerated at the completion of the citric acid cycle.

b. Glycolysis takes place in the cytosol.

c. The energy yield from oxidative phosphorylation is much greater than the yield from substrate-level phosphorylation.

d. The electromotive force directly generates ATP.

8. What is the role of oxygen in oxidative phosphorylation?

9. The amino groups from amino acid catabolism are eliminated as which of the following molecules?

a. Uric acid   c. Ammonia

b. Ammonium ions   d. Urea

10. Mark the following as true or false. If a statement is false, correct it to make a true statement.

a. The reactions of β-oxidation crop fatty acids into three-carbon pyruvate units.

b. Most amino acids are converted into products that are oxidized by β-oxidation.

c. The final common pathway for carbohydrate, protein, and lipid catabolism is the same.

d. Fat oxidation generates more ATP than protein or carbohydrate catabolism.

e. Fat and protein catabolism takes place primarily in the cytosol.

11. When fats are catabolized by the liver, some of the acetyl-CoA is converted into:

a. amino acids.

b. ketone bodies.

c. glucose.

d. glycerol.

12. Fill in the blanks: Glucose is stored by the process of _____ and new glucose is generated by the process of _____.

13. Which of the following can be used to synthesize fatty acids?

a. Glucose   c. Both a and b

b. Amino acids   d. Neither a nor b

14. Identify the following characteristics as belonging to the absorptive state or the postabsorptive state.

a. Ketogenesis occurs. _____

b. Gluconeogenesis and glucose sparing take place. _____

c. Glycogenesis and lipogenesis take place. _____

d. Proteins synthesis decreases or stops. _____

e. Glucose is readily available and is used as the primary fuel for most cells. _____

f. Lipolysis releases fatty acids into the blood. _____

15. The hormone that directly stimulates the hunger center is:

a. ghrelin.

b. cholecystokinin.

c. insulin.

d. leptin.

16. Why is heat production a good way to measure the body's metabolic rate?

17. Which of the following best defines the basal metabolic rate?

a. The rate of heat production during exercise

b. The rate at which glucose is catabolized

c. The amount of energy needed for the body's anabolic reactions

d. The energy cost of living

18. Mark the following statements as true or false. If a statement is false, correct it to make a true statement.

a. Conduction is energy transfer between the body and the environment via electromagnetic waves.

b. Thermoreceptors send temperature information to centers in the hypothalamus.

c. Heat-loss mechanisms include sweating and vasodilation of the blood vessels in the skin.

d. The body temperature measured on the skin or in the mouth is generally higher than the core body temperature.

19. Fill in the blanks: A/an _____ is a nutrient that the body requires in large quantities, whereas a/an _____ is one that the body requires in much smaller quantities. A/an _____ is a nutrient that the body cannot produce and so must be supplied from the diet.

20. Which of the following statements is false?

a. Fiber is a polysaccharide that is completely or partially indigestible by the human body.

b. A complete protein contains all of the essential amino acids.

c. Saturated fats contain hydrogen atoms arranged on opposite sides of the carbon-carbon double bond.

d. Linoleic and linolenic acid are essential fatty acids.

21. How do fat-soluble and water-soluble vitamins differ? Which vitamins are fat-soluble?

22. Mark the following statements as true or false. If a statement is false, correct it to make a true statement.

a. Vitamins and minerals are used as fuel sources for cells.

b. High-density lipoproteins deliver cholesterol from the liver to peripheral tissues.

c. Negative energy balance refers to the condition in which fewer Calories are taken in than are expended.

d. Obesity is defined as a body mass index of 30.0 or greater.

LEVEL 2 Check Your Understanding

1. Exercise is said to "burn" fat. In what ways is this analogy true?

2. Certain diet plans claim that a person may eat all the protein he or she wants, as protein is not stored by the body. Is this claim completely accurate? Explain.

3. Erythrocytes lack organelles, including mitochondria. What sort of carbohydrate catabolism can erythrocytes undergo? Explain.

4. You are performing a lab experiment to measure your lab partner's metabolic rate. To perform the experiment, your lab partner is hooked up to a machine that measures her oxygen consumption.

   a. Why is the amount of oxygen that your lab partner consumes a good way to measure her metabolic rate?

   b. Your lab partner just ran up three flights of stairs to get to lab on time. Will the measured metabolic rate be her basal metabolic rate? Why or why not?

5. People with extremely restricted caloric intake often complain of being cold. What does this signify about their metabolic rate?

LEVEL 3 Apply Your Knowledge

PART A: Application and Analysis

1. A patient presents to a clinic with acute ingestion of the poison cyanide, which blocks the functioning of the electron transport chain. He is exhibiting many of the symptoms seen with suffocation (lack of breathing). Explain this finding. Would the level of oxygen in the body be normal, elevated, or decreased in cyanide poisoning? Explain.

2. Your friend hears about the latest fad diet that consists entirely of fat. Predict the initial effects such a diet might have on your friend's concentration, memory, and other nervous system functions. Would you recommend this diet to anyone? Why or why not?

3. Certain dietary supplements for weight loss contain drugs that are claimed to block the absorption of polysaccharides such as starch. If a "starch blocker" actually worked and fully prevented the absorption of carbohydrates into the bloodstream, what effects would this have on protein and lipid metabolism?

PART B: Make the Connection

4. The condition hyperthyroidism is characterized by an increase in the level of thyroid hormone. People with untreated hyperthyroidism often complain of feeling hot. Explain this finding. *(Connects to Chapter 16)*

5. When you are exercising, your respiratory rate increases and you exhale more carbon dioxide. Explain this. What will happen to the pH of the blood if the respiratory rate does not increase? Why? *(Connects to Chapter 21)*

24

The Urinary System

A nimals living in an aquatic environment face little risk of becoming dehydrated. However, animals that started to spend more time on dry land millions of years ago needed mechanisms to conserve water and prevent dehydration. The organ system that performs this function in humans—the **urinary system**—is the topic of this chapter. The organs of the urinary system are organs of **excretion**—they remove wastes and water from the body. Specifically, the urinary system "cleans the blood" of *metabolic wastes,* which are substances produced by the body that it cannot use for any purpose. However, as you will learn in this chapter, the urinary system does far more: This system is also essential for removing toxins, maintaining homeostasis of many factors (including blood pH and blood pressure), and producing erythrocytes. Read on to discover how the urinary system is vital to your body's homeostasis.

MODULE **24.1**

Overview of the Urinary System

Learning Outcomes

1. List and describe the organs of the urinary system.
2. Describe the major functions of the kidneys.

The urinary system is composed of the paired *kidneys* and the *urinary tract*. The **kidneys** filter the blood to remove metabolic wastes and then modify the resulting fluid, which allows these organs to maintain fluid, electrolyte, acid-base, and blood pressure homeostasis. This process produces **urine,** a fluid that consists of water, electrolytes, and metabolic wastes. Then the remaining organs of the urinary system—those of the urinary tract—transport, store, and eventually eliminate urine from the body. In this module, we first examine the basic structures of the urinary system, and then turn to the functional roles of the kidneys.

Overview of Urinary System Structures

As you can see in **Figure 24.1**, the kidneys resemble their namesake, the kidney bean, in both shape and color. The kidneys are situated against the posterior abdominal wall and are *retroperitoneal* (reh'-troh-pair-ih-ton-EE-ul; *retro-* = "behind") organs, meaning they are located posterior to the peritoneal membranes. Note, however, that the two kidneys differ slightly in position—the left kidney extends from about T12 to L3, whereas the right kidney sits slightly lower on the abdominal wall because of the position of the liver. The superior portions

Photo: This scanning electron micrograph shows glomeruli, the filtering units of the kidneys.

(a) Anterior view

(b) Posterior view

Figure 24.1 Organs of the urinary system in a female.

Practice art labeling
@MasteringA&P®

of both kidneys are partially protected by the 11th and 12th pairs of ribs. Each kidney is capped by an adrenal gland (*ad-* = "near," *ren-* = "kidney"); these glands perform endocrine functions and secrete a variety of hormones (see Chapter 16).

The **urinary tract** is composed of the paired *ureters* (YOOR-eh-terz), the *urinary bladder,* and the *urethra* (yoo-REE-thrah). Urine leaves each kidney through one of the two ureters, tubes that run along the posterior body wall, connecting the kidneys with the hollow urinary bladder. The bladder, which stores the urine, sits on the floor of the pelvic cavity. Urine is expelled from the body through the tube called the urethra, which connects the urinary bladder with the outside of the body.

Quick Check

☐ 1. What are the organs of the urinary system?

Overview of Kidney Function

◄◄ **Flashback**

1. What are three factors that determine blood pressure? (p. 675)
2. What is erythropoietin, and what is its main function? (p. 727)

The regulation of homeostasis by the urinary system takes place in the kidneys, so let's take a quick look at what the kidneys do and how they do it. The kidneys perform the following functions:

- **Removal of metabolic wastes.** As we have discussed, the kidneys filter the blood, removing metabolic wastes. These wastes are eliminated from the body via the urine.
- **Regulation of fluid and electrolyte balance.** The kidneys regulate blood solute concentration, or osmolarity, by conserving or eliminating water and electrolytes such as sodium, potassium, and calcium ions.
- **Regulation of acid-base balance.** The kidneys assist in the long-term regulation of blood pH by conserving or eliminating hydrogen (H^+) and bicarbonate (HCO_3^-) ions.
- **Maintenance of blood pressure.** The kidneys directly influence systemic blood pressure through their control of blood volume. Additionally, they secrete an enzyme that influences both blood volume and peripheral resistance.
- **Regulation of erythropoiesis.** The kidneys regulate red blood cell production in the bone marrow by releasing the hormone *erythropoietin* (eh-rith′-row-POY-eh-tin; see Chapter 19).
- **Performing other metabolic functions.** The kidneys play many important metabolic roles, including detoxifying substances in the blood, activating vitamin D, and making new glucose through the process of *gluconeogenesis* (glew′-koh-nee-oh-JEN-eh-sis).

□ 2. List the basic functions of the kidneys.

Apply What You Learned

□ 1. Inflammation of the peritoneal membranes, or *peritonitis,* can cause dysfunction of multiple organs in the abdominal cavity. Would you expect peritonitis to impact the kidneys? Explain.

□ 2. Explain why a patient with long-term renal failure might have a decreased number of erythrocytes in his or her blood.

See answers in Appendix A.

MODULE **24.2**

Anatomy of the Kidneys

Learning Outcomes

1. Describe the external structure of the kidney, including its location, support structures, and covering.
2. Trace the path of blood flow through the kidneys.
3. Identify the major structures and subdivisions of the renal corpuscle, renal tubules, and renal capillaries.
4. Describe the histological structure of the proximal tubule, nephron loop, distal tubule, and collecting system, and trace the pathway of filtrate flow through these tubules.
5. Compare and contrast cortical and juxtamedullary nephrons.

Let's now take a closer look at the anatomy of the kidneys. In this module, we explore first the external and internal anatomy of the kidneys. We then turn our attention to the structure and basic roles of the kidneys' functional units: the *nephrons* (NEF-ronz; *nephro-* = "kidney"). We conclude the module with an examination of the two main types of nephrons.

External Anatomy of the Kidneys

The kidneys are held in place on the posterior body wall and protected by three external layers of connective tissue (**Figure 24.2**). From superficial to deep, these layers are as follows:

- **Renal fascia.** The **renal fascia** (FASH-ee-ah) is a layer of dense irregular connective tissue that anchors each kidney to the peritoneum and to the fascia covering the muscles of the posterior abdominal wall.
- **Adipose capsule.** The middle and thickest layer, called the **adipose capsule,** consists of adipose tissue that wedges each kidney in place and shields it from physical shock. During prolonged starvation, the body uses the fatty acids in the adipose capsule of the kidney for fuel. This causes the kidney to droop, a condition called *nephroptosis* (nef-rop-TOH-sis; *-ptosis* = "drooping").
- **Renal capsule.** The **renal capsule** is an extremely thin layer of dense irregular connective tissue that covers the exterior of each kidney like plastic wrap. It protects the kidney from infection and physical trauma.

Without its connective tissue coverings, a typical adult kidney is about the size of a large bar of soap (11 cm long, 6 cm wide,

Anterior

Liver
Renal artery
Renal vein
Stomach
Pancreas
Large intestine
Renal fascia
Peritoneal cavity
Spleen
Left kidney
Inferior view
Adipose capsule
Renal capsule
Right kidney
Vertebra (L1)
Hilum
Renal pelvis

Posterior

Figure 24.2 Position and external structure of the kidneys. Note that you are looking at an inferior view of this transverse section of the trunk.

and 3 cm thick), and weighs about 150 grams. On the medial surface we find an opening called the **hilum** (HY-lum), through which the *renal artery, renal vein,* renal nerves, and ureter enter and exit the kidney. The hilum opens to a central cavity called the **renal sinus,** which is lined by the renal capsule and filled with urine-draining structures and adipose tissue. This connective tissue anchors the ureter, blood vessels, and nerves in place.

Quick Check

☐ 1. What are the three connective tissue coverings of the kidney?

Internal Anatomy of the Kidneys

A frontal section of the kidney reveals the three distinct regions of this organ: the outermost **renal cortex,** the middle **renal medulla,** and the inner **renal pelvis** (**Figure 24.3a**). Together, the renal cortex and the renal medulla make up the urine-forming portion of the kidney. The renal pelvis and associated structures drain urine that the cortex and medulla have formed.

Notice in Figure 24.3a that the renal cortex is reddish-brown. This is due to its rich blood supply—it houses 90–95% of the kidney's blood vessels. At specific points, extensions of the renal cortex called **renal columns** pass through the renal medulla toward the renal pelvis. The renal columns house blood vessels

that branch from the renal artery as they travel to the outer portion of the cortex.

Within the renal medulla we also find cone-shaped **renal pyramids** (or *medullary pyramids*), which are separated from one another by a renal column on each side. Notice that the renal pyramids are darker in color and appear striped, reflecting that they are made up of parallel bundles of small tubes, with fewer blood vessels than in the renal cortex.

The renal cortex and renal medulla of each kidney contain over one million microscopic filtering structures called **nephrons.** Nephrons are the functional units of the kidney—each one is capable of filtering the blood and producing urine. **Figure 24.3b** shows the basic structure of a nephron, which consists of two main components: the globe-shaped *renal corpuscle,* and a long, snaking tube of epithelium called the *renal tubule.* Notice how these components are arranged in the kidney. The renal corpuscle and the majority of the renal tubule reside in the renal cortex, whereas varying amounts of the renal tubule dip into the renal medulla.

The tip of each renal pyramid tapers into a slender **papilla** (pah-PIL-ah; *papil-* = "nipple"), which borders on the first urine-draining structure, a cup-shaped tube called a **minor calyx** (KAY-liks; plural *calyces,* KAL-ih-seez). Urine from three to four minor calyces drains into a larger **major calyx** (shown in Figure 24.3a). Two to three major calyces, in turn, drain urine into the large collecting chamber that is the renal pelvis, which leads into the ureter. Smooth muscle tissue in the walls of the calyces and renal pelvis contracts to help propel urine toward the ureter. Both the calyces and the renal pelvis reside in the renal sinus.

(a) **Kidney, frontal section**

(b) **Section of the renal cortex and renal pyramid showing nephrons**

Practice art labeling
@MasteringA&P®

Figure 24.3 Internal anatomy of the kidney, including the nephron.

☐ 2. What are the three regions of the kidney, and how do they differ structurally?

☐ 3. What is the functional unit of the kidney?

Blood Supply of the Kidneys

◀◀ **Flashback**

1. How is blood delivered to and drained from a typical capillary bed? (p. 670)
2. What are the major arteries and veins that supply and drain the kidneys? (p. 683)

The kidneys receive approximately one-fourth of the total cardiac output—about 1200 ml per minute—from the right and left **renal arteries,** which branch from the abdominal aorta. The renal arteries fan out into ever-smaller vessels as they pass through the renal sinus to the renal columns and cortex. **Figure 24.4** traces the path of blood flow through the kidney. From largest to smallest, the arteries shown in Figure 24.4 are as follows: ❶ Renal artery → ❷ Segmental artery → ❸ Interlobar artery → ❹ Arcuate artery → ❺ Interlobular (cortical radiate) artery.

You learned in the blood vessels chapter that systemic capillary beds are fed by arterioles and drained by venules (see Chapter 18). But in the kidneys, we find an unusual capillary bed that is both fed and drained by arterioles (Figure 24.4b). In the kidney's renal cortex, the interlobular arteries branch into tiny **afferent arterioles** (*affer-* = "to carry toward"), which feed a ball-shaped capillary bed called the **glomerulus** (gloh-MAIR-yoo-lus; *glom-* = "ball"). The glomerulus, which consists of glomerular capillaries and supporting cells, is part of the renal corpuscle of the nephron. The glomerular capillaries are then drained by a second arteriole called the **efferent arteriole** (*effer-* = "to carry away"). The efferent arteriole feeds a second capillary bed known as the **peritubular capillaries,** which together form a plexus or network around the renal tubule of each nephron. The order of this part of the system is as follows: ❻ Afferent arteriole → ❼ Glomerulus → ❽ Efferent arteriole → ❾ Peritubular capillaries.

The venous route of blood out of the kidney parallels the arterial path. Groups of peritubular venules unite into a series of progressively larger venous vessels. The sequence, from smallest to largest (see Figure 24.4), is as follows: ❿ Interlobular vein → ⓫ Arcuate vein → ⓬ Interlobar vein → ⓭ Renal vein. Note that no segmental veins are present in the kidneys; the interlobar veins merge in the renal sinus to form the large **renal vein.** The renal vein then exits the kidney through the hilum and empties into the inferior vena cava.

(a) Blood flow through kidney

(b) Blood flow around nephron

Figure 24.4 Blood flow through the kidney.

☐ 4. Trace the sequence of blood flow through the kidneys from the renal artery to the renal vein.

☐ 5. How does the arrangement of capillary beds in the kidneys differ from those elsewhere in the body?

Microanatomy of the Kidney: The Nephron and Collecting System

◀◀ **Flashback**

1. What are fenestrated capillaries? (p. 686)

As you learned in this module, most of the functions of the kidneys occur in their tiny nephrons (**Figure 24.5**). Nephrons filter the blood and modify the filtered fluid as it passes through the renal tubules. This fluid then leaves the nephron and drains into the tubules of the *collecting system* (which are not considered part of the nephron), where it is further modified until it finally becomes urine. Let's take a closer look at each of these components.

The Nephron

As we introduced earlier, the nephron has two main divisions: the renal corpuscle and the renal tubule. Both structures are composed of multiple parts.

The Renal Corpuscle The **renal corpuscle** is responsible for filtering the blood. Each globe-shaped renal corpuscle consists of two parts: the **glomerulus** and an outer sheath of epithelial tissue called the **glomerular capsule** (or Bowman's capsule; **Figure 24.6**). The glomerulus is a group of looping fenestrated capillaries. These capillaries are called fenestrated because of the large pores, or *fenestrations* (*fenestre-* = "window"), present within their plasma membranes and between their endothelial cells. These fenestrations, which make the capillaries extremely "leaky," or permeable, form a main part of the filtering structure of the renal corpuscle.

Surrounding the glomerulus is the double-layered glomerular capsule, which consists of an outer *parietal layer* (pah-RY-eh-tal) and an inner *visceral layer*. The parietal layer is a globelike extension of the renal tubule consisting of simple squamous epithelium. The visceral layer consists of modified epithelial cells called **podocytes** (POH-doh-sytz; *podo-* = "foot") that wrap around the glomerular capillaries. Extending from each podocyte are extensions called *foot processes,* or *pedicels* (PED-ih-selz). Pedicels weave together to form **filtration slits,** which make up another part of the renal corpuscle's filtering structure. Between the parietal and visceral layers we find a hollow region, or lumen, called the **capsular space,** which is continuous with the beginning of the renal tubule lumen.

The podocytes and fenestrated glomerular capillaries form part of a complex membrane (the filtering structure we mentioned earlier) that filters blood flowing through the glomerulus. This structure allows a large volume of fluid to be filtered from the blood (which we discuss in the next module). The fluid

Figure 24.5 A generalized nephron and collecting system.

Practice art labeling
@MasteringA&P®

that passes through the filter to leave the glomerular capillaries, which is known as *filtrate,* first enters the capsular space, then flows into the renal tubule lumen.

The Renal Tubule Newly formed filtrate flows from the capsular space into the "pipes" of the nephron: the renal tubule. The renal tubule is a winding tube responsible for modifying the filtrate. It has three regions: the proximal tubule, nephron loop, and distal tubule, each of which differs in structure and function (**Figure 24.7**).

1. **Proximal tubule.** The initial and longest segment of the renal tubule through which filtrate flows is the **proximal tubule;** it consists of simple cuboidal epithelial cells with prominent microvilli. This part of the tubule has both convoluted (coiled) and straight sections, which is why this text uses the broader term proximal tubule (rather than proximal convoluted tubule). The many microvilli projecting into the lumen of the proximal tubule form a *brush border,* so named because the fine projections resemble the bristles on a brush. This border greatly increases surface area.

(a) The renal corpuscle

Glomerular capillaries (covered by podocytes)

Glomerular capsule:
Squamous epithelium (parietal layer)
Capsular space
Podocyte (visceral layer)

Efferent arteriole
Afferent arteriole

Proximal tubule

(b) SEM of capillary surrounded by podocytes

Podocyte   Filtration slits   Capillary   Pedicels

SEM (92,000×)

Figure 24.6 The renal corpuscle.

2. **Nephron loop (loop of Henle).** The remaining filtrate flows on to the **nephron loop,** also known as the **loop of Henle,** which is the only part of the renal tubule that dips into the renal medulla. The nephron loop has two limbs: The **descending limb** travels toward the renal medulla, turns 180°, and becomes the **ascending limb,** which climbs back toward the renal cortex. The descending limb is composed of simple squamous epithelium, and thus is often called the **thin descending limb.** In some nephrons this thin segment also forms the bend region and part of the ascending limb, so there is also a **thin ascending limb.** However, the majority of the ascending limb is composed of thicker simple cuboidal epithelium, and is therefore referred to as the **thick ascending limb.**

3. **Distal tubule.** The final segment through which the filtrate passes in the renal tubule is the **distal tubule.** (Again, this text uses the broader term, as this segment of the tubule has both convoluted and straight sections.) Like the proximal tubule, the distal tubule is composed of simple cuboidal epithelium. However, the distal tubule lacks a brush border.

The Juxtaglomerular Apparatus

At the transition point between the ascending limb of the nephron loop and the distal tubule, we find a tightly packed group of cells called the **macula densa** (MAK-yoo-lah DEN-sah; "dense spot"). The macula densa comes into contact with modified smooth muscle cells in the afferent and efferent arterioles, known

Brush border   Microvilli

Lumen

Simple cuboidal epithelium with microvilli

Lumen

Simple squamous epithelium

Distal tubule

Microvilli

Lumen

Simple cuboidal epithelium with very few microvilli

Proximal tubule

Thin descending limb of nephron loop

Thick ascending limb of nephron loop

Lumen

Simple cuboidal epithelium with no microvilli

Figure 24.7 Structural characteristics of the renal tubule.

as **juxtaglomerular (JG) cells** (juks'-tah-gloh-MAIR-yoo-lar; *juxta-* = "next to"). Together, the macula densa and JG cells form a structure called the **juxtaglomerular apparatus (JGA),** which regulates blood pressure and glomerular filtration rate (**Figure 24.8**). We will revisit the JGA in Module 24.4.

The Collecting System

The nephron ends where filtrate in the distal tubule empties into the **collecting system,** another series of structurally and functionally distinct tubules that further modify the filtrate as it passes through them. Most of the collecting system consists of simple cuboidal or columnar epithelium with few microvilli (**Figure 24.9**).

The collecting system consists of the cortical collecting duct and the medullary collecting system. The distal tubule empties filtrate into the first part of the collecting system, the *cortical collecting duct,* which is found within the renal cortex. The cortical collecting duct is made up of simple cuboidal epithelial cells. Note that each cortical collecting duct drains several distal tubules. As the cortical collecting duct passes into the renal medulla, it becomes the *medullary collecting duct.* Deep within the renal medulla, several medullary collecting ducts empty filtrate into a larger **papillary duct.** The papillary duct contains low columnar epithelial cells. The *medullary collecting system* includes the medullary collecting ducts and the papillary ducts. When the filtrate reaches the end of the papillary duct, it is urine. The urine exits at the papilla of the renal pyramid into a minor calyx. The formation of crystals within the tubules of the collecting system can block the flow of filtrate and lead to intense pain. You can find out more about this condition in *A&P in the Real World: Nephrolithiasis.*

Quick Check

☐ 6. What are the two components of the renal corpuscle?
☐ 7. Trace the pathway filtrate takes through the nephron and collecting system from the capsular space to the papillary duct.

Types of Nephrons

The previous discussion simplified the nephron by presenting a "generalized" version of it. However, the kidneys actually contain two types of nephrons that are distinguished by both the structure and function of their nephron loops and the organization of the peritubular capillaries. These two types are labeled *cortical* and *juxtamedullary* (juk'-stah-MED-yoo-lair-ee).

About 80% of nephrons are **cortical nephrons,** so named because they are located primarily in the renal cortex (**Figure 24.10a**). The renal corpuscles of cortical nephrons are situated in the *outer* renal cortex, and they have very short nephron loops that either just dip into the superficial part of the renal medulla or never leave the cortex. Peritubular capillaries supply blood to the nephron loops of cortical nephrons (although they exchange materials through the interstitial fluid rather than directly).

The other, less numerous, type of nephron in the kidney is the **juxtamedullary nephron.** Notice in **Figure 24.10b** that the renal corpuscle of the juxtamedullary nephron sits close to the boundary between the renal cortex and the renal medulla. In addition, it has a long nephron loop that burrows deeply into the renal medulla. Here the loop is surrounded by a ladder-like network of capillaries called the **vasa recta** (VAY-zah REK-tah; "straight vessels") which arises from the efferent arteriole. Like the

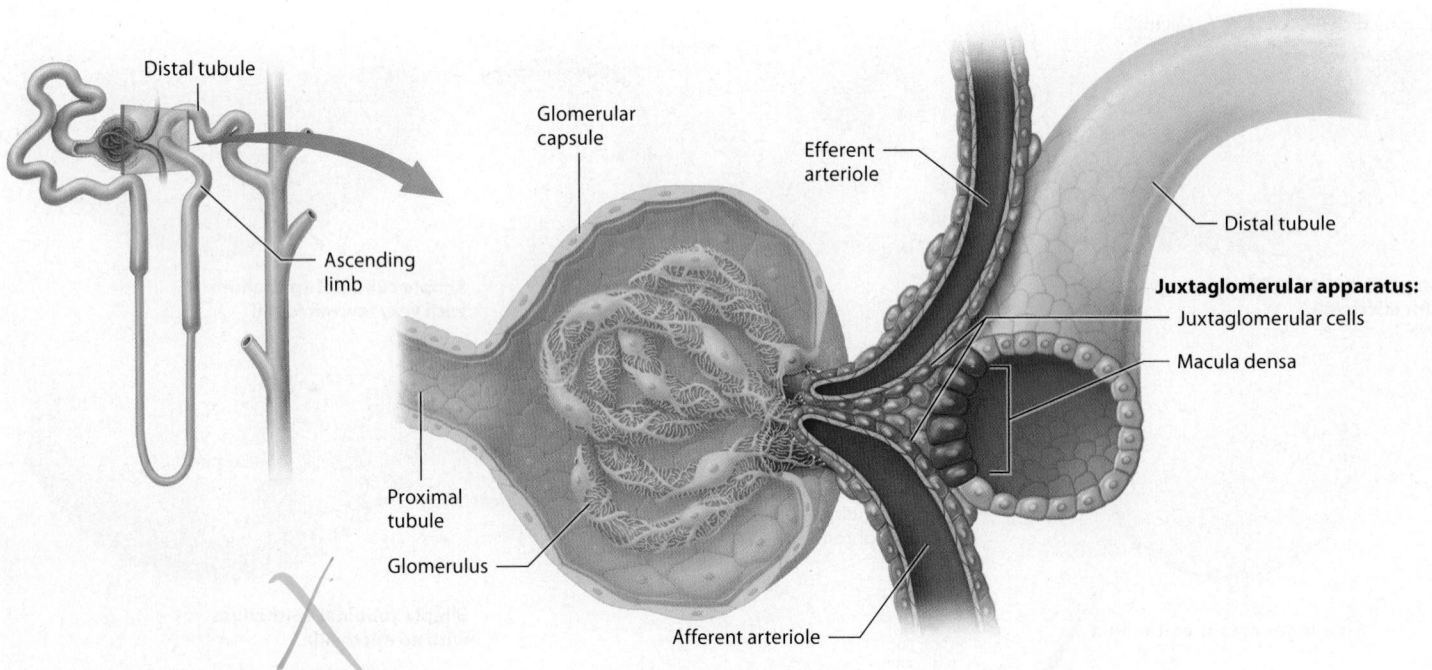

Figure 24.8 The juxtaglomerular apparatus.

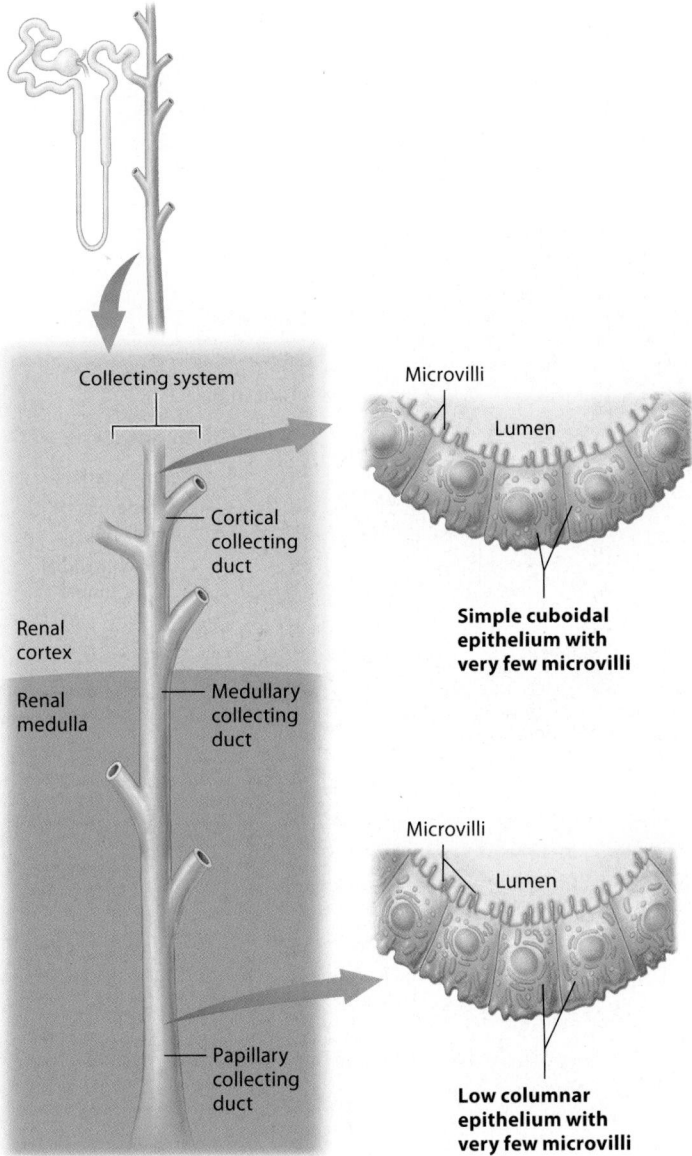

Figure 24.9 Structural characteristics of the collecting system.

peritubular capillaries, the vasa recta empty into interlobular veins. Notice that the portion of the nephron that lies in the cortex is surrounded by branches of the peritubular capillaries from the neighboring cortical nephrons. This unique structure allows the juxtamedullary nephrons to function as part of a system that controls the volume and concentration of urine, concepts we will discuss in Module 24.6.

Apply What You Learned

☐ 1. Predict the effects of a condition that results in gradual loss of microvilli from the proximal tubule.

☐ 2. Predict the potential effects of abnormally narrow renal arteries, a condition called *renal artery stenosis,* on the ability of the kidneys to carry out their functions.

See answers in Appendix A.

Nephrolithiasis

The condition known as **nephrolithiasis** (*nef'*-roh-lith-EYE-ah-sis; *lith-* = "stone") is characterized by the formation of **renal calculi** (KAL-kyoo-lye), commonly known as *kidney stones.* Renal calculi are crystalline structures composed most commonly of calcium oxalate salts. They form when the concentrations of these ions, as well as solutes such as sodium ions, hydrogen ions, and uric acid, are present in the filtrate in higher than normal amounts. This condition is known as *supersaturation,* and supersaturated ions are more likely to come out of solution and crystallize. Risk factors for supersaturation include dehydration; a diet high in fat, animal protein, and salt; and obesity.

Typically, the crystals form in the nephron loop, distal tubule, and/or collecting system. Most crystals simply pass unnoticed into the urine. However, sometimes the crystals adhere to the epithelium of the tubules, particularly in the collecting system, and form *seed crystals* that lead to the formation of stones. The stones may remain in the collecting system or may break off and lodge in the calyces, renal pelvis, and ureter. Stones lodged within the urinary system cause the most common symptom of nephrolithiasis: severe pain, known as *renal colic,* that radiates from the lumbar region to the pubic region. Other symptoms include blood in the urine, sweating, nausea, and vomiting.

Nephrolithiasis can be diagnosed in several ways, including computed tomography scans and an **intravenous pyelogram** (PY-loh-gram; *pyelo-* = "pelvis"), or **IVP.** An IVP is a radiograph of the urinary system that uses a contrast medium such as iodine to reveal the structure of the renal pelvis, the major and minor calyces, and the ureters and urinary bladder. An example of an IVP with a renal calculus blocking the left ureter is shown here:

Efferent arteriole

Afferent arteriole

Glomerular capillaries

Proximal tubule

Distal tubule

Interlobular vein

Interlobular artery

Peritubular capillary

Distal tubule

Efferent arteriole

Afferent arteriole

Glomerular capillaries

Proximal tubule

Renal cortex

Peritubular capillaries

Nephron loop

Arcuate artery

Arcuate vein

(a) Cortical nephron

Nephron loop

Vasa recta

Renal medulla

Medullary collecting duct

(b) Juxtamedullary nephron

Figure 24.10 Cortical and juxtamedullary nephrons.

MODULE 24.3
Overview of Renal Physiology

Learning Outcomes

1. List the three major processes in urine formation and describe where each occurs in the nephron and collecting system.

Nephrons carry out three basic physiologic processes that allow the kidneys to perform their homeostatic functions: *filtration, reabsorption,* and *secretion.* These functions, illustrated in Figure 24.11, are as follows:

- **Glomerular filtration.** The first process performed by the nephron is to filter the blood, a process known as **glomerular filtration.** This takes place as blood passes through the membrane of the glomerular capillaries and some of the plasma is filtered into the surrounding glomerular space. This filter is selective based on size, so it holds back cells and most proteins, which remain in the blood, but allows some of the smaller substances—including water, electrolytes (such as sodium and potassium ions), acids and bases (such as hydrogen and bicarbonate ions), organic molecules, and metabolic wastes—to exit the blood and enter the glomerular capsule. The fluid and solutes that enter the capsular space of the nephron form the **filtrate,** also known as *tubular fluid.*
- **Tubular reabsorption.** The next process the nephron performs is to modify the filtrate as it flows through the tubules. Much of this modification involves reclaiming substances from the filtrate, such as water, glucose, amino acids, and electrolytes, and returning them to the blood. This process is known as **tubular reabsorption.** The nephron is able to reabsorb the majority of water and solutes filtered by the glomerular membrane. Most reabsorption takes place in the proximal tubule and nephron loop. However, some more precisely controlled reabsorption also occurs in the distal tubule and collecting ducts; this process allows the nephron to vary the amounts of different substances reabsorbed to maintain homeostasis as the needs of the body change.
- **Tubular secretion.** The filtrate is also modified by a process known as **tubular secretion,** which is essentially tubular reabsorption in the reverse direction. Notice in Figure 24.11 that during tubular secretion substances are moved from the peritubular capillary blood into the filtrate to eventually be excreted. Secretion happens all along the tubule, although different substances may be secreted more in one area than another. Tubular secretion helps maintain electrolyte and acid-base homeostasis and removes toxins from the blood that did not enter the filtrate via filtration.

In the next two modules, Modules 24.4 and 24.5, we examine each of these processes and how they contribute to the overall function of the kidneys and urinary system. In Module 24.6 we see how these processes allow the kidneys to control the volume and concentration of urine produced.

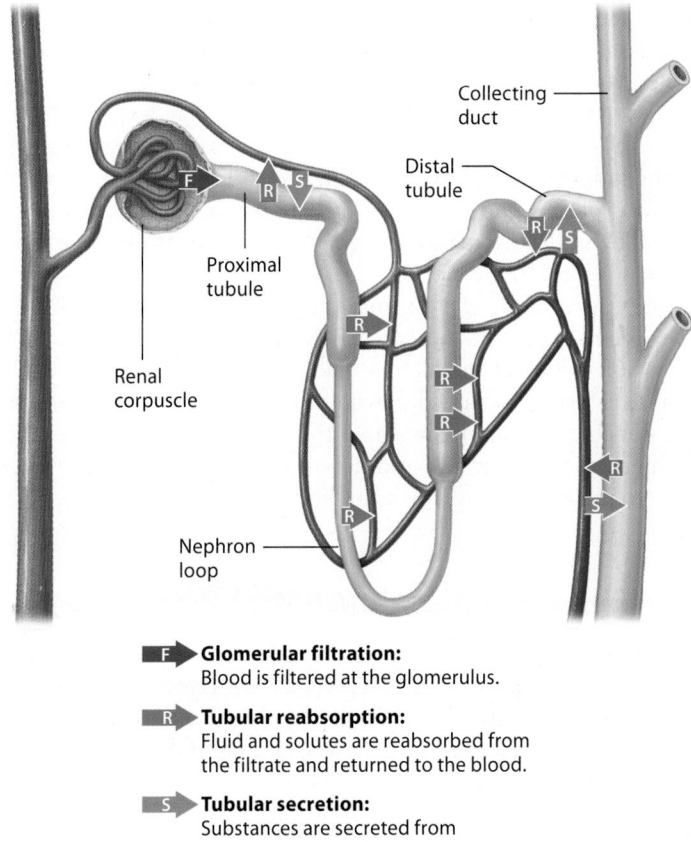

- **F** **Glomerular filtration:** Blood is filtered at the glomerulus.
- **R** **Tubular reabsorption:** Fluid and solutes are reabsorbed from the filtrate and returned to the blood.
- **S** **Tubular secretion:** Substances are secreted from the blood into the filtrate.

Figure 24.11 Three physiological processes carried out by the kidneys.

Quick Check

☐ 1. What happens during glomerular filtration?
☐ 2. How do tubular reabsorption and tubular secretion differ?

MODULE 24.4
Renal Physiology I: Glomerular Filtration

Learning Outcomes

1. Describe the structure of the filtration membrane.
2. Define the glomerular filtration rate (GFR) and state its average value.
3. Describe the hydrostatic and colloid osmotic forces that favor and oppose filtration.
4. Predict specific factors that will increase or decrease GFR.
5. Describe the myogenic and tubuloglomerular feedback mechanisms.
6. Describe the role of each of the following in the control of GFR: renin-angiotensin-aldosterone system, atrial natriuretic peptide, and sympathetic nervous system activity.

The mechanical process of filtration is easy to picture—you witness it every time you drain cooked spaghetti into a colander. When you "filter" the spaghetti, you cause the filtrate (water and small solutes) to pass through a filter (the colander) and leave large substances (the noodles) behind. A similar process occurs in the kidneys with glomerular filtration—the membrane in the renal corpuscle, known as the *filtration membrane,* forms a filter that allows filtrate to pass through, but leaves relatively large substances behind in the blood.

In this module we examine the structure of the filtration membrane and the mechanical process of filtration. We then take a look at the factors that determine the rate at which filtrate is formed, a value known as the *glomerular filtration rate.* We conclude by discussing how the body tightly regulates the glomerular filtration rate and how this in turn enables the kidneys to maintain many aspects of homeostasis.

The Filtration Membrane and the Filtrate

The inner part of the glomerular capsule consists of three layers that act as barriers, or filters; together these layers are called the **filtration membrane.** This membrane is made up of the glomerular capillary endothelial cells, a basal lamina, and podocytes (the visceral layer of the glomerular capsule) (**Figure 24.12**). Let's look at each of these layers, from deep to superficial (the direction of filtrate flow), more closely:

1. **Fenestrated glomerular capillary endothelial cells.** Like all capillaries, the glomeruli are composed of endothelial cells. Recall, however, that the glomerular endothelial cells are **fenestrated;** they have pores that make them leakier than most capillaries. The gaps between the glomerular endothelial cells are relatively large (about 70–100 nm) but are still small enough to prevent blood cells and platelets from exiting the capillaries.

2. **Basal lamina.** This thin layer of extracellular matrix gel separates the glomerular endothelial cells from the podocytes. Collagen fibers within the basal lamina form a meshwork that acts like a colander, preventing substances with a diameter greater than 8 nm from entering the capsular space. This effectively blocks the passage of most plasma proteins. In addition, the collagen fibers have negative charges that repel negatively charged plasma proteins, even those smaller than 8 nm in diameter.

3. **Podocytes (visceral layer of the glomerular capsule).** The podocytes composing the visceral layer of the glomerular capsule make up the third and finest filter in the filtration membrane. Note in Figure 24.12 that their finger-like pedicels wrap around the glomerular capillaries and interlace to form filtration slits. These narrow slits allow only substances with a diameter less than 6–7 nm to enter the capsular space. Albumin, the most prevalent plasma protein, has a diameter of 7.1 nm, so it is normally prevented from entering the filtrate.

The fluid and solutes that pass through the filtration membrane and enter the capsular space make up the filtrate. The size of the pores in this membrane determines the composition of

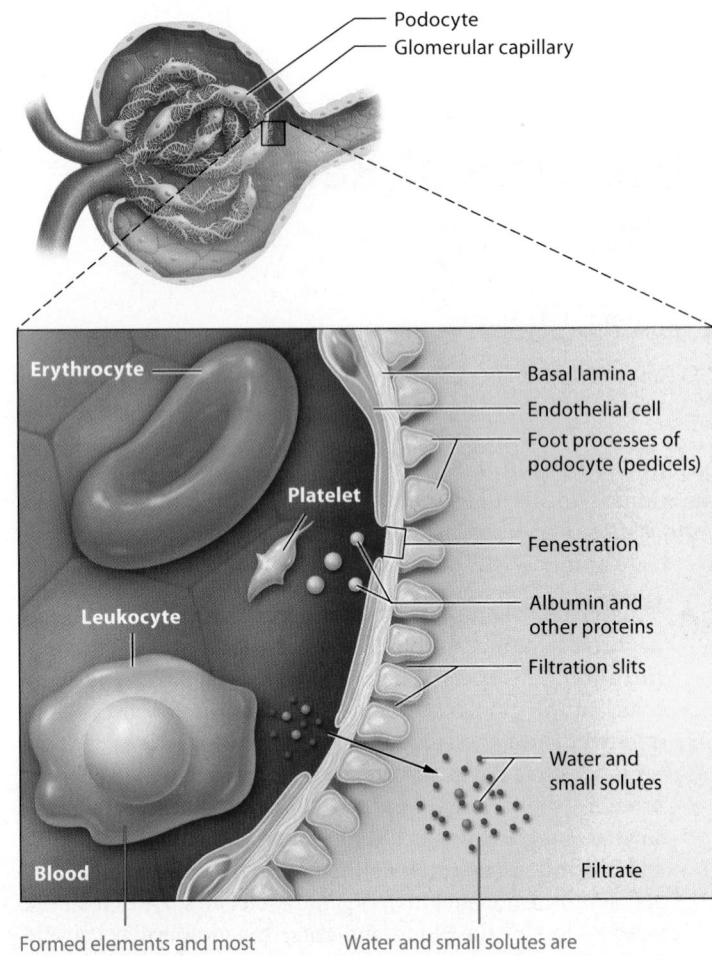

Formed elements and most proteins in blood are too large to fit through the filtration slits.

Water and small solutes are small enough to pass through the filtration slits and enter the filtrate.

Figure 24.12 Filtration and the filtration membrane.

Play animation
@MasteringA&P®

the filtrate. Observe in Figure 24.12 that the filtration membrane allows water and small dissolved solutes, including glucose, electrolytes, very small proteins, and amino acids, to leave the blood and enter the capsular space, but prevents the exit of formed elements, such as cells and platelets, and most proteins. Another important substance filtered from the blood is a group of molecules known as *nitrogenous wastes.* These molecules include *urea* and *ammonium ions* (NH_4^+), waste products of protein metabolism; *creatinine,* a waste product of the creatine kinase reaction that occurs in muscle cells and other body cells; and *uric acid,* a waste product of nucleic acid metabolism. All are relatively small molecules that pass easily through the pores of the filtration membrane.

The percentage of plasma that passes through the filtration membrane to enter the capsular space and become filtrate is called the *filtration fraction.* An average filtration fraction is 20%, which means that about one-fifth of the plasma that flows through the glomerulus exits the blood and enters the capsular space. The filtration fraction is so large because of the looping arrangement of the glomerular capillaries. This looping increases their surface area dramatically—if you were to stitch

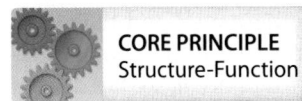

CORE PRINCIPLE
Structure-Function

all of the surfaces of the filtration membranes of one kidney together, their combined area would be about 6 m² (about the size of a small bedroom)! In an example of the Structure-Function Core Principle (p. 25), this large surface area makes filtration through this stack of ever-finer filters a very efficient process.

Quick Check

☐ 1. What are the three components of the filtration membrane?
☐ 2. What is the function of the filtration membrane?
☐ 3. Which substances are found in the filtrate?

The Glomerular Filtration Rate (GFR)

◀◀ **Flashback**

1. Define osmosis. (p. 75)
2. What is colloid osmotic pressure? (p. 690)
3. What is vascular resistance? (p. 689)

Filtrate is formed at the remarkably rapid rate of about 125 ml/min. This value, the amount of filtrate formed by both kidneys in 1 minute, is known as the **glomerular filtration rate (GFR).** Over the course of a day, the kidneys produce about 180 liters of filtrate (to put this into perspective, imagine filling your gastrointestinal tract with a similar volume by drinking 90 2-liter bottles of soda in one day!). This is an impressive feat, considering that we have only about 3 liters of plasma. Therefore, your entire plasma volume is filtered by your kidneys about 60 times per day.

The kidneys are able to filter blood so efficiently in part because the glomerular capillaries are remarkably permeable.

CORE PRINCIPLE
Gradients

However, even with fenestrated capillaries, filtration will happen only if a pressure gradient is present to push water and solutes through the filtration membrane (an example of the Gradients Core Principle, p. 26). In this section we discuss the forces that allow this process to occur.

Filtration Pressures

Let's first review the two forces that drive fluid movement in a typical capillary bed:

- **Hydrostatic pressure.** *Hydrostatic pressure* is the force of a fluid on the wall of its container. In the case of blood capillaries, hydrostatic pressure is equal to the blood pressure, and it tends to push water *out of* the capillary and into the interstitial space.
- **Colloid osmotic pressure.** *Colloid osmotic pressure* (*COP*) is the pressure created by proteins (primarily albumin) in the plasma. The osmotic gradient created by these proteins pulls water *into* the capillaries by osmosis.

Recall that these two forces work together in a capillary bed to determine the *net filtration pressure (NFP)* of the bed (see Chapter 18). The net filtration pressure determines the direction of water movement between the capillaries and the interstitial fluid. Simply stated, water moves out of the capillary if hydrostatic pressure is higher than COP, or into the capillary if COP is higher than hydrostatic pressure.

Net Filtration Pressure at the Glomerulus

We can apply these same two principles to the glomerular capillaries. However, the situation is slightly more complex in the glomerulus, because we have a third force to factor in: the hydrostatic pressure of the fluid in the capsular space. Let's examine each of these forces, which are shown in **Figure 24.13**:

- **Glomerular hydrostatic pressure.** The **glomerular hydrostatic pressure (GHP)**, which is largely determined by the systemic blood pressure, measures about 50 mm Hg. This pressure is considerably higher than that of a typical capillary bed (which ranges from 17 to 35 mm Hg). This is because blood leaving such a capillary bed enters a low-resistance venule, whereas blood leaving the glomerulus enters a high-resistance efferent arteriole. This arteriole's high resistance is due to its smooth muscle and its small diameter—notice in Figure 24.13 that the diameter of the efferent arteriole is smaller than that of the afferent arteriole. This causes blood to back up and push against the walls of the glomerular capillaries, which *favors* its movement through the filtration membrane.
- **Glomerular colloid osmotic pressure.** Like the COP in typical capillaries, the **glomerular colloid osmotic pressure (GCOP)** is created by the presence of proteins such as albumin in the plasma. The GCOP averages about 30 mm Hg, slightly higher than the COP in a typical capillary bed, because the blood in the glomerulus is a bit more concentrated. The reason for this is that water leaves the glomerular blood rapidly through the filtration membrane, which causes any solutes left in the blood to increase in concentration. The GCOP *opposes* filtration, "pulling" on water to hold it in the glomerular capillaries.

The three forces in the renal corpuscle = NFP

Figure 24.13 Net filtration pressure in the glomerular capillaries.

Play animation
@MasteringA&P®

- **Capsular hydrostatic pressure.** Like an emptying kitchen sink with the faucet turned on, the water in the capsular space can only drain into the renal tubule so quickly. The rapidly accumulating filtrate inside the capsular space of a nephron builds up a hydrostatic pressure of its own, called the **capsular hydrostatic pressure (CHP).** This pressure (about 10 mm Hg) tries to push water into the glomerular capillaries and so *opposes* filtration.

We can combine these three forces to yield the glomerular **net filtration pressure (NFP),** the total pressure gradient available to drive water across the filtration membrane and into the capsular space. To find the glomerular NFP, we subtract the two forces that oppose filtration (GCOP and CHP) from the one that favors filtration (GHP):

$$NFP = GHP - (GCOP + CHP)$$
$$= 50 \text{ mm Hg} - (30 \text{ mm Hg} + 10 \text{ mm Hg})$$
$$= 10 \text{ mm Hg}$$

So, we find a net filtration pressure of about 10 mm Hg in the glomerular capillaries. This pressure, combined with the leakiness of the glomerular capillaries and their large surface area, yields the GFR of about 125 ml/min. To gain perspective on this, compare the GFR to the rate at which systemic capillary beds lose water—about 1.5 ml/min, quite a significant difference.

Many conditions can impact the GFR. One common condition is discussed in *A&P in the Real World: Glomerulonephritis.*

Quick Check

- ☐ 4. What is the GFR?
- ☐ 5. Which three pressures combine to determine the net filtration pressure? Which pressure(s) promote filtration? Which pressure(s) oppose filtration?

Factors That Affect the Glomerular Filtration Rate

◀◀ **Flashback**

1. What are the primary functions of the hormones angiotensin-II and atrial natriuretic peptide? (p. 610)
2. What happens to a blood vessel when blood pressure increases and the vessel is stretched? (p. 671)

Glomerular filtration is essentially the "gatekeeper" of renal physiology because it begins the process of waste removal. The GFR determines how rapidly the blood is cleansed of metabolic wastes, how effectively the kidneys can carry out both tubular reabsorption and secretion, and how well the kidneys are able to maintain homeostasis in the body. For these reasons, factors that impact the GFR influence all functions of the kidney.

Several factors regulate and impact the GFR. These factors may originate within the kidney itself or may be due to the function of systems outside the urinary system. The internal mechanisms are called *autoregulation,* and have to do with maintaining

the GFR. The external factors, including neural and hormonal factors, are part of broader systems that work to maintain systemic blood pressure and affect the GFR in doing so. We discuss each of these factors in the upcoming sections. But first let's look at the physical changes that are used by all of these mechanisms to control the GFR.

ConceptBOOST »»»

How Changes in Arteriolar Diameter Influence the GFR

As we discussed earlier, filtration will occur only when a net pressure gradient in the glomerulus drives fluid out of the blood and into the capsular space. The size of this gradient determines how much filtration takes place—a small gradient will lead to only minimal filtration, whereas a large gradient leads to heavy filtration. Several factors determine the size of the pressure gradient in the glomerulus, but one of the most easily adjustable factors is the diameter of the afferent (entering) and efferent (leaving) arterioles. When either arteriole constricts or dilates, this changes the glomerular hydrostatic pressure (GHP), and therefore the entire net pressure gradient also changes.

You can think of blood flowing in and out of the glomerulus as being similar to water flowing in and out of a sink, where the afferent arteriole is the faucet, the basin is the glomerulus, and the efferent arteriole is the drainpipe. Keep this analogy in mind as we explore how this mechanism works:

- Vasoconstriction of the afferent arteriole "turns down the faucet." This allows less blood to flow into the glomerulus, which decreases the GHP and the GFR:

- Vasoconstriction of the efferent arteriole "clogs the drain." This causes blood to back up within the glomerulus, which increases the GHP and thus increases the GFR:

Vasodilation has the opposite effects:

- Vasodilation of the afferent arteriole "turns up the faucet." This increases the GHP and the GFR:

- Vasodilation of the efferent arteriole "unclogs the drain," allowing increased flow out of the glomerulus. This decreases the GHP and the GFR:

In the upcoming sections, you'll see that the main mechanisms the body uses to control or maintain GFR work by causing one or more of these physical changes. ∎

Autoregulation of the GFR

Because the GHP is mostly the result of systemic blood pressure, it seems logical that a change in systemic blood pressure would alter the GHP and so the GFR. However, this is *not* the case. The GFR remains relatively constant over wide ranges of systemic 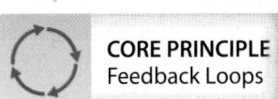 blood pressure because of the process known as **autoregulation,** local responses initiated and maintained by the kidneys. In an example of the Feedback Loops Core Principle (p. 21), autoregulation consists of two negative feedback processes: the *myogenic mechanism* and *tubuloglomerular feedback.*

CORE PRINCIPLE Feedback Loops

The Myogenic Mechanism Recall that an increase in blood pressure stretches a blood vessel, which triggers its smooth muscle cells to vasoconstrict (see Chapter 18). This decreases the amount of blood flowing through the vessel, which in turn minimizes the stretch imposed on the vessel. We find a similar phenomenon in the blood vessels of the kidney, which is termed the **myogenic mechanism** (my-oh-JEN-ik; *myo-* = "muscle," *-genic* = "producing"). The myogenic mechanism works by using the same

mechanisms to control GFR that we discussed in the Concept Boost. It occurs in the following way:

- An *increase* in systemic blood pressure stretches the afferent arteriole and leads to an increase in GFR. This leads the muscle cells to contract, constricting the arteriole. Vasoconstriction of the afferent arteriole decreases the blood flow through the glomerulus ("turns down the faucet"), which decreases glomerular hydrostatic pressure and the GFR back to normal levels.
- A *decrease* in the systemic blood pressure causes the afferent arteriole to be less stretched, decreasing the GFR and making its smooth muscle cells relax. The resulting vasodilation of the arteriole increases the blood flow and the glomerular hydrostatic pressure ("turns up the faucet"), and the GFR increases back to normal levels.

The myogenic mechanism acts rapidly—it can restore the normal GFR within seconds of even a significant change in blood pressure. Note, however, that these mechanisms can restore the GFR over a systolic blood pressure range of about 80–180 mm Hg (a normal range for systolic pressure is 100–120 mm Hg). If the systolic blood pressure goes higher than 180 mm Hg, the afferent arteriole can't constrict any further to restore the GFR. Similarly, if systolic blood pressure drops below this range, the afferent arteriole is unable to dilate enough to restore the GFR to normal levels.

Tubuloglomerular Feedback The second autoregulatory mechanism of the kidneys is *tubuloglomerular feedback* (too′-byoo-low-gloh-MAIR-yoo-lar), so named because the macula densa of the distal renal tubule (see Figure 24.8) is part of a negative feedback loop that controls pressure in the glomerulus. As it's currently understood, the mechanism of tubuloglomerular feedback is related to the concentration of NaCl in the

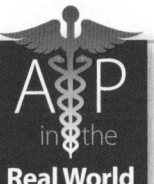

Glomerulonephritis

Glomerulonephritis (gloh-mair-yoo-loh-nef-RY-tiss) is a common condition that involves damage to and destruction of the glomeruli, with resulting inflammation of the glomerular capillaries and the glomerular basement membrane. Recall that the major hallmarks of inflammation are both increased blood flow and increased capillary permeability. This increases the GHP and causes the filtration membrane to become excessively leaky, leading to the loss of blood cells and proteins to the urine. As nephrons are further damaged and destroyed by the inflammation, the GFR decreases, which triggers compensatory mechanisms that further increase the GHP to maintain the GFR. This causes additional loss of cells and proteins to the urine. Eventually, these mechanisms are insufficient to restore the GFR, and toxic substances can accumulate in the blood. This condition may eventually lead to complete renal failure, in which the GFR is insufficient for the kidneys to perform their functions.

filtrate. The basic feedback loop proceeds by the following steps:

1. If the GFR increases, the volume of filtrate flowing through the renal tubule increases.
2. The increased filtrate volume makes the filtrate flow more rapidly through the tubules. This leads to an increased delivery of sodium and chloride ions to the macula densa cells in the distal tubule, causing them to absorb more of these ions from the filtrate.
3. In an example of the Cell-Cell Communication Core Principle (p. 27), the macula densa cells release chemicals (which researchers are currently trying to identify) that diffuse through the interstitial fluid to the afferent arteriole, which they trigger to constrict.

CORE PRINCIPLE
Cell-Cell Communication

4. The macula densa cells signal the renin-containing JG cells to reduce their release of renin, leading to a decreased renal production of angiotensin-II and dilation of the efferent arteriole.
5. The GFR decreases back toward normal.

Conversely, a decrease in the GFR is thought to have the opposite effect—fewer sodium and chloride ions are delivered to the macula densa cells, causing the afferent arteriole to dilate and the efferent arterioles to constrict. This increases the glomerular hydrostatic pressure and restores the GFR.

With both the myogenic mechanism and tubuloglomerular feedback at work, the GFR remains remarkably consistent through changes in blood pressure, which happen with some frequency. For example, every time you sneeze or get up off the couch, your blood pressure changes momentarily. However, these mechanisms are insufficient to maintain GFR when dealing with very large changes in blood pressure. If systolic blood pressure rises above 180 mm Hg, the GFR and urine output will increase dramatically. Conversely, if systolic blood pressure drops below about 70 mm Hg, glomerular filtration will cease because the compensatory mechanisms cannot create an NFP that favors filtration. At this point, no urine will be produced, a life-threatening condition called *anuria* (an-YOO-ree-ah; "absence of urine formation").

Quick Check

☐ 6. What happens to the GFR when the afferent arteriole constricts? What happens to the GFR when it dilates?
☐ 7. How does tubuloglomerular feedback affect the GFR?

Hormonal Effects on the GFR

Recall that *hormones* are chemical messengers released into the bloodstream that regulate the functions of other cells (see Chapter 16). Like the two autoregulatory mechanisms, hormones that affect the GFR do so by adjusting the glomerular hydrostatic pressure. However, hormones affect the GFR as part of a larger system that regulates systemic blood pressure. One hormone that regulates the GFR is *angiotensin-II* (an′-jee-oh-TEN-sin), a component of the *renin-angiotensin-aldosterone system*. Another

set of regulating hormones consists of the *natriuretic peptides* (nay′-tree-yer-ET-ik; *natri-* = "sodium") produced by the heart. First let's examine the former hormone system.

The Renin-Angiotensin-Aldosterone System The **renin-angiotensin-aldosterone system (RAAS)** is a complex system whose primary function is to maintain systemic blood pressure; it preserves the GFR as a secondary effect. Therefore, it acts on much more than just the afferent and/or efferent arterioles of the glomeruli. The RAAS also significantly impacts tubular reabsorption in the nephron and collecting system in order to influence electrolyte balance and blood volume, in addition to causing changes within the body as a whole.

This system may be triggered into action by three conditions: (1) stimulation from neurons of the sympathetic nervous system, (2) low glomerular hydrostatic pressure, or (3) stimulation from the macula densa cells as part of tubuloglomerular feedback. Note that often the system is "turned on" by a combination of these three factors.

The RAAS proceeds by the following steps, illustrated in **Figure 24.14:**

① **Systemic blood pressure decreases, causing a decrease in the GFR.**

② **JG cells release renin.** Decreased blood flow through the afferent arteriole triggers the JG cells to release the enzyme **renin** (REE-nin) into the bloodstream.

③ **Renin converts angiotensinogen to angiotensin-I.** Renin circulates until it encounters an inactive protein produced by the liver called **angiotensinogen** (an′-jee-oh-ten-SIN-oh-jen). Renin catalyzes the conversion of angiotensinogen to a product with minimal activity, **angiotensin-I (A-I).**

④ **ACE converts angiotensin-I to the active angiotensin-II.** A-I circulates through the blood until it encounters an enzyme called **angiotensin-converting enzyme (ACE),** which is made by cells such as the endothelial cells in the lungs. ACE converts A-I to its active form, **angiotensin-II (A-II).**

A-II has several different, simultaneous effects that influence both systemic blood pressure and the GFR, including:

⑤ₐ **Promotes vasoconstriction of efferent arterioles.** A-II constricts all of the renal blood vessels, but it has a greater effect on the efferent arterioles than on the afferent arterioles. This "clogs the drain," raising glomerular hydrostatic pressure. This increased pressure, in turn, increases the GFR to maintain the filtration rate despite the reduced blood flow.

⑤ᵦ **Promotes vasoconstriction of systemic blood vessels.** A-II is a powerful vasoconstrictor of nearly all systemic vessels (it causes effects more than 40 times stronger than those of norepinephrine). This vasoconstriction increases peripheral resistance, which in turn increases systemic blood pressure.

⑤ᵪ **Promotes reabsorption of sodium and chloride ions from the proximal tubule, and water follows.** One of the most powerful effects of A-II on blood pressure comes from its role in renal tubule reabsorption (covered in Module 24.5). In the proximal tubule, A-II promotes reabsorption of both sodium and chloride ions from the filtrate, which in turn causes reabsorption of water by osmosis. This increases blood volume, which raises blood pressure.

5d **Promotes aldosterone release, leading to increased sodium ion and water reabsorption.** A-II stimulates the adrenal glands to release the hormone **aldosterone** (al-DOHS-ter-own). Aldosterone acts on the distal tubule and parts of the collecting system to increase sodium ion reabsorption from the filtrate. If ADH is present, water follows by osmosis, so this action increases blood volume, and therefore blood pressure.

5e **Stimulates thirst.** Another effect of A-II is to stimulate the thirst centers in the hypothalamus. This can increase fluid intake, which would increase blood volume.

The RAAS maintains blood pressure over both the short and long term. The vasoconstriction produced by A-II is nearly immediate—blood pressure rises within seconds of exposure to A-II. However, this vasoconstriction lasts only 2–3 minutes. Long-term blood pressure control comes from the retention of sodium ions and water from the filtrate. Notice that the effects of

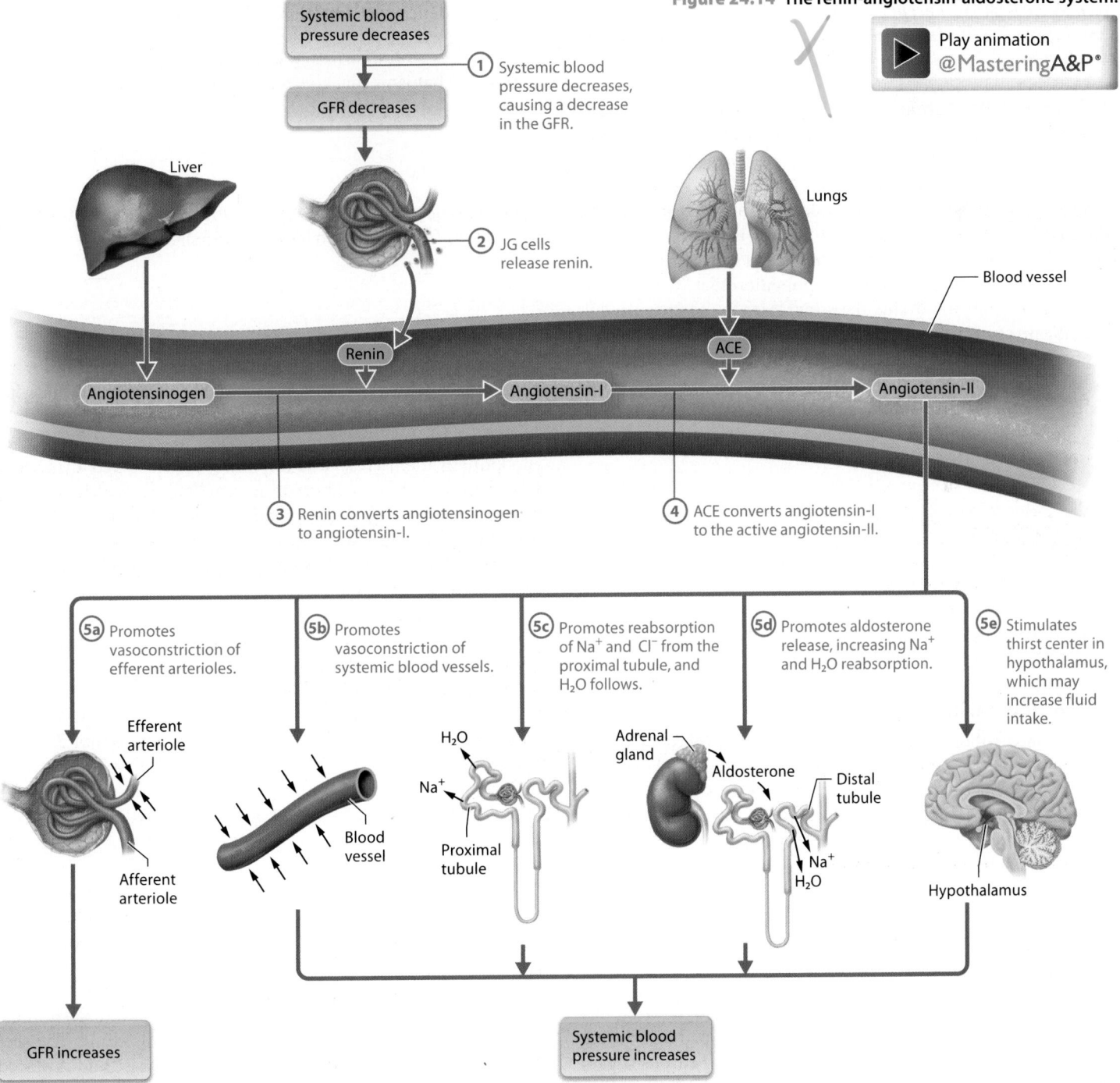

Figure 24.14 The renin-angiotensin-aldosterone system.

Play animation
@MasteringA&P®

Systemic blood pressure decreases

GFR decreases

① Systemic blood pressure decreases, causing a decrease in the GFR.

Liver

② JG cells release renin.

Lungs

Blood vessel

Renin

ACE

Angiotensinogen → Angiotensin-I → Angiotensin-II

③ Renin converts angiotensinogen to angiotensin-I.

④ ACE converts angiotensin-I to the active angiotensin-II.

5a Promotes vasoconstriction of efferent arterioles.

5b Promotes vasoconstriction of systemic blood vessels.

5c Promotes reabsorption of Na^+ and Cl^- from the proximal tubule, and H_2O follows.

5d Promotes aldosterone release, increasing Na^+ and H_2O reabsorption.

5e Stimulates thirst center in hypothalamus, which may increase fluid intake.

Efferent arteriole

Afferent arteriole

Blood vessel

H_2O

Na^+

Proximal tubule

Adrenal gland

Aldosterone

Distal tubule

Na^+

H_2O

Hypothalamus

GFR increases

Systemic blood pressure increases

the RAAS on both vasoconstriction and sodium ion reabsorption allow the body to increase systemic blood pressure while preserving the GFR.

In addition to its effects on blood pressure, the RAAS is also critical to maintaining sodium ion balance. When the sodium ion level decreases in the plasma, and so in the filtrate, the macula densa cells stimulate renin secretion from the JGA cells, and more sodium ions are reabsorbed into the blood from the filtrate in the renal tubules. When the plasma sodium ion level increases, renin secretion decreases, and the kidneys excrete more sodium ions. The RAAS is so effective at this task that sodium ion intake can increase up to 50 times and the level of sodium ions in the plasma will remain relatively unchanged. See *A&P in the Real World: The RAAS and Hypertension* for information on how the effects of the RAAS can be modified to treat high blood pressure.

Atrial Natriuretic Peptide Recall that certain cells of the atria in the heart produce the hormone **atrial natriuretic peptide (ANP)** (see Chapter 16). ANP, which is released when the volume of blood in the atria increases, acts to lower blood volume and blood pressure and thereby reduce the workload of the heart. One of the ways that ANP accomplishes this task is by increasing the GFR. ANP dilates the afferent arterioles and constricts the efferent arterioles of the glomeruli, a combination that "turns up the faucet" and "clogs the drain." This increases the glomerular hydrostatic pressure, raising the GFR. High GFR leads to more fluid loss and therefore decreases blood volume and effectively lowers blood pressure. Note that ANP is part of a system that works to decrease the systemic blood pressure; it simply happens to do so by impacting the GFR. We consider the other effects of ANP on the kidney in Module 24.5.

Quick Check

☐ 8. Describe the basic steps of the renin-angiotensin-aldosterone system.

☐ 9. What are the effects of the RAAS on the GFR?

☐ 10. What are the effects of ANP on the GFR?

Neural Regulation of the GFR

Neural control of the GFR is chiefly mediated by the sympathetic division of the autonomic nervous system. As with the hormonal systems just described, the sympathetic nervous system affects the GFR as part of a larger system to control systemic blood pressure. Sympathetic neurons release norepinephrine during times of increased sympathetic activity, which causes constriction of most systemic blood vessels, including the afferent arterioles, and so elevates systemic blood pressure. You might think this should decrease glomerular hydrostatic pressure and the GFR, but the sympathetic effect on the GFR isn't quite that simple. Instead, its impact depends on how much it is stimulated:

- If the level of sympathetic stimulation is low (e.g., during mild exercise such as walking), sympathetic neurons trigger

the JG cells to release renin. This leads to the formation of a low level of A-II, which raises systemic blood pressure and increases the GFR.

- If the level of sympathetic stimulation is high (e.g., during strenuous exercise or severe blood loss), a large amount of renin is secreted, and the blood concentration of A-II increases dramatically. A high level of A-II will actually *constrict* both the afferent and efferent arterioles of the glomeruli, decreasing the filtration rate. This is especially important in cases of severe hypotension and dehydration, as it helps the body to minimize fluid loss and preserve blood volume and blood pressure. In these cases, perfusion of muscle tissue and vital organs (brain, heart, etc.) takes precedence over filtration in the kidneys.

Table 24.1 summarizes the autoregulatory, hormonal, and neural mechanisms that control the GFR.

Quick Check

☐ 11. How does the sympathetic nervous system affect the GFR at both low and high levels of stimulation?

Renal Failure

If a person's GFR decreases, he or she may enter **renal failure,** a condition in which the kidneys are unable to carry out their vital functions. Many conditions may lead to renal failure, including factors that decrease blood flow to the kidneys, diseases of the kidneys themselves, and anything that obstructs urine outflow from the kidneys. Short-term renal failure, which is known as *acute renal failure* or *acute kidney injury,* is common among hospitalized patients and may resolve completely with treatment of the underlying cause. Some people develop *chronic* (long-term) *renal failure,* which is defined as a decrease in GFR lasting 3 months or longer.

The biggest risk factors for the development of chronic renal failure are diabetes mellitus and hypertension. Symptoms depend on the severity of the renal failure, and those with mild renal failure might not notice any symptoms. As renal function declines, patients experience fatigue, edema, nausea, and loss of appetite. Severe renal failure, in which the GFR is less than 50% of normal, results in a condition known as **uremia** (yoo-REEM-ee-ah). Uremia is characterized by a buildup of waste products and fluid, electrolyte, and acid-base imbalances. Untreated, it can lead to coma, seizures, and death.

When a patient develops the signs and symptoms of uremia, **dialysis** (dy-AL-ah-sis) treatment may be initiated. There are two types of dialysis. The first is **hemodialysis,** which temporarily removes an individual's blood and passes it through a filter that removes metabolic wastes and extra fluid, and normalizes electrolyte and acid-base balance. Hemodialysis must be performed three times per week at a dialysis clinic. The second type is **peritoneal dialysis,** in which dialysis fluid is placed

TABLE 24.1 SUMMARY OF CONTROL OF THE GLOMERULAR FILTRATION RATE

Stimulus	Responding Mechanism	Main Effect(s)	Effect on GFR
Autoregulation			
Increased stretching of the afferent arteriole (due to increased blood pressure)	Myogenic mechanism	• Vasoconstriction of the afferent arteriole	Decrease
Increased sodium ion delivery to the macula densa cells (due to increased GFR)	Tubuloglomerular feedback	• Vasoconstriction of the afferent arteriole • Vasodilation of the efferent arteriole	Decrease
Hormonal Mechanisms			
Sympathetic nervous system activated; decreased GFR; decreased systemic blood pressure	Renin-angiotensin-aldosterone system	• Vasoconstriction of the efferent arteriole • Systemic vasoconstriction • Reabsorption of sodium ions from the filtrate in the proximal tubule • Release of aldosterone • Reabsorption of water	Increase (decrease at high levels of activation)
Increased systemic blood pressure	Atrial natriuretic peptide	• Vasodilation of the afferent arteriole • Vasoconstriction of the efferent arteriole • Decreased reabsorption of sodium ions • Increased water loss	Increase
Neural Mechanisms			
Multiple	Sympathetic nervous system	• Constriction of all vessels, including the afferent and efferent arterioles • Stimulates RAAS	Decrease (increase at low levels of activation)

into the peritoneal cavity, allowed to circulate for several hours, and then drained. A patient is able to undergo peritoneal dialysis nightly at home, so this is often the preferred treatment for individuals needing long-term dialysis.

Apply What You Learned

☐ 1. Ms. Douglas has advanced liver disease; because her liver is no longer able to produce plasma proteins, her colloid osmotic pressure has decreased. Predict the effects that this loss of pressure will have on the net filtration pressure and the GFR in her nephrons.

☐ 2. Certain drugs that treat high blood pressure cause vasodilation of systemic arteries and arterioles, including those in the kidneys. What effect would these drugs have on the GFR? How would the myogenic mechanism and tubuloglomerular feedback respond to this change in the GFR?

☐ 3. Mr. Adams is taking an ACE inhibitor and an angiotensin-receptor blocker, both drugs that block the RAAS, for his high blood pressure. He complains that when he tries to engage in physical activity, he feels faint. He is asked to exercise on a treadmill, and his blood pressure remains very low when he exercises, rather than rising with his level of physical activity. Explain how his medications could be causing his current problem.

See answers in Appendix A.

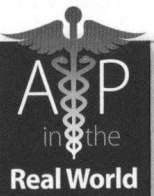

The RAAS and Hypertension

Because the effects of the RAAS on systemic blood pressure are so potent, three classes of drugs have been developed that act on this system to reduce blood pressure:

● **ACE inhibitors**, a class of drugs developed from snake venom, block angiotensin-converting enzyme, and therefore inhibit the conversion of angiotensin-I to angiotensin-II.

● **Angiotensin-receptor blockers** block the angiotensin receptors on the cells of blood vessels and the proximal tubule, preventing vasoconstriction and sodium ion and water reabsorption, respectively.

● **Aldosterone antagonists** block the effects of aldosterone on the distal tubule and decrease reabsorption of sodium ions and water; this leads to a diuretic ("water-losing") effect.

Although these drugs are useful in the treatment of hypertension, blocking the formation or actions of angiotensin-II can decrease the GFR. Typically, the GFR will noticeably decrease only in those patients with pre-existing renal disease, but all patients should be monitored for this potential adverse effect.

Renal Physiology II: Tubular Reabsorption and Secretion

Learning Outcomes

1. Describe how and where water, organic compounds, and ions are reabsorbed in the nephron by both passive and active processes.
2. Describe the location(s) in the nephron where tubular secretion occurs.
3. Compare and contrast tubular reabsorption and secretion, with respect to direction of solute movement, strength of concentration gradients, and energy required.
4. Describe how the renin-angiotensin-aldosterone system, antidiuretic hormone, and atrial natriuretic peptide each work to regulate reabsorption and secretion.

Because of the high rate of filtrate formation, the kidneys have to work to reclaim most of the filtrate and return it to the bloodstream. Recall that this occurs by the process of *tubular reabsorption* in the nephrons and collecting ducts. Of the 180 liters of water filtered by the kidneys every day, only about 1.8 liters (1%) ultimately leave the body as urine, which means that the kidneys are very efficient at reclaiming water. In addition, about 99% of the solutes filtered are subsequently reabsorbed. Given the magnitudes of these quantities, it's not surprising that the kidneys use approximately 20% of the body's ATP at rest.

Since our kidneys reabsorb so much of what they filter, you may wonder why our bodies need to filter blood through the kidneys at all. First of all, certain waste products are not reabsorbed. In addition, in spite of the high GFR, some waste substances do not pass through the filtration membranes of the glomeruli and therefore don't enter the filtrate. This necessitates the process of *tubular secretion*, which occurs in the renal tubules and collecting ducts at the same time as tubular reabsorption. Certain substances are secreted into the filtrate; these nonreabsorbed and secreted substances are often toxic when present in high concentrations, and the high GFR and process of secretion ensure their rapid clearance from the body. In addition, the presence of a large volume of filtrate in the nephrons gives the kidneys precise control over the homeostasis of fluid, electrolyte, and acid-base balance. If substances are not needed, they are excreted with the urine. If they are needed, they will be reabsorbed into the blood.

In the upcoming sections, we examine the processes of tubular reabsorption and secretion more closely. We also trace the filtrate as it flows through the different parts of the nephron and collecting system, and discuss how it changes in both composition and concentration to finally become urine.

Principles of Tubular Reabsorption and Secretion

◀◀ **Flashback**

1. What are tight junctions, and what are their functions? (p. 127)
2. What are facilitated diffusion, primary active transport, and secondary active transport? (p. 76)

This section looks at how substances move between the inside of the renal tubule and the surrounding peritubular capillaries. We also examine how solute and water movement are driven by either active or passive transport processes. Let's start by reviewing some of the principles we've learned that we can now apply to tubular reabsorption and secretion.

Paracellular and Transcellular Transport Routes

When discussing the movement of solutes and water, we first need to discuss exactly where and how the substances are moved. As you can see in **Figure 24.15**, in tubular reabsorption, substances must pass from the filtrate in the lumen (inside) of the tubule across or between the tubule cells, into the interstitial fluid, and finally across or between the endothelial cells of the peritubular capillaries to re-enter the blood. In tubular secretion, substances move in the opposite direction. But exactly how do these substances cross the tubule cells?

As you learned in the histology chapter, substances can be transported by two different routes across epithelia: paracellular or transcellular (see Figure 4.6). For a quick review, these processes are described as follows:

- **Paracellular route.** On the paracellular route (*para-* = "beside"), substances pass *between* adjacent tubule cells. The tight junctions between the tubule cells are just leaky enough to allow some substances such as small ions and water to move passively between them, particularly in the proximal tubule.
- **Transcellular route.** On the transcellular route (*trans-* = "across"), substances such as glucose and amino acids must move *through* the tubule cells. A reabsorbed substance first crosses the *apical* membrane of the tubule cell (the membrane facing the tubule lumen), then travels through the cytosol, and finally exits the cell through the *basolateral* membrane (the side of the membrane facing the interstitial fluid).

Transport along the paracellular route is passive, requiring no energy in the form of ATP, because substances move with their concentration gradients via diffusion or osmosis. The same is true for certain substances taking the transcellular route. However, other substances moving along the transcellular route travel against their concentration gradients and thus require energy, either directly or indirectly, from ATP. Secretion is an active process, so it must occur via the transcellular route across the tubule cell membrane.

Reabsorbed substances that have entered into the interstitial fluid may then cross the endothelial cells of the blood vessel and enter the blood. These substances can follow the same routes into the capillary

Figure 24.15 **Barriers to tubular reabsorption and secretion.**

that they followed to exit the tubule—they may take the paracellular route or the transcellular route. Generally, these processes are passive, and solutes move by diffusion and water by osmosis.

Carrier-Mediated Transport and the Transport Maximum

Most of the substances that are reabsorbed and secreted via the transcellular route require the use of a carrier protein in the tubule cell plasma membrane. Recall that there are three ways in which cells use carrier proteins to transport substances (see Chapter 3):

- *facilitated diffusion*, in which a carrier protein passively transports a substance with its concentration gradient, without using energy from ATP (an example of the Gradients Core Principle, p. 26);

CORE PRINCIPLE
Gradients

- *primary active transport*, in which a carrier protein "pump" directly uses ATP to move a substance against its concentration gradient; and
- *secondary active transport*, in which a concentration gradient set up by a primary active transport pump is used to drive the transport of a second substance against its concentration gradient via another carrier protein.

There are two types of active transport carrier proteins: *Antiport pumps* (or *antiporters*) move two or more substances in opposite directions, and *symport pumps* (or *symporters*) move two or more substances in the same direction (see Chapter 3). Both have a limited number of sites on which they can transport substances, much as a train has only a certain number of seats for passengers. If all of their sites become filled, the carrier proteins are said to be **saturated,** as they have reached their

transport maximum (T$_M$). Any substances unable to bind to their carrier proteins will likely not be transported and will end up in the urine. This is what happens to glucose in diabetes mellitus, as discussed in *A&P in the Real World: Glycosuria.*

Quick Check

☐ 1. What are the two routes of reabsorption and secretion?
☐ 2. List the three types of transport processes that involve carrier proteins in the renal tubule and collecting system.

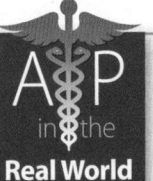

Glycosuria

The transport maximum (T$_M$) becomes especially important with substances such as glucose that "fill seats" fairly rapidly in the carriers of the proximal tubule cells. If too much glucose is in the filtrate, the T$_M$ will be reached before all of the glucose is reabsorbed and glucose will appear in the urine, a condition called *glycosuria* (gly'-koh-SOOR-ee-ah). Glycosuria is commonly seen with the disorder **diabetes mellitus,** a condition characterized by defects in the production of or response to the pancreatic hormone *insulin.* Insulin causes most cells to take in glucose; in its absence, these cells are unable to bring glucose into their cytosol. This leads to a high level of circulating blood glucose, or *hyperglycemia,* which causes excessive amounts of glucose to be present in the filtrate and therefore ultimately in the urine.

Reabsorption and Secretion in the Proximal Tubule

◀◀ **Flashback**

1. What is the function of the Na⁺/K⁺ pump? In which direction(s) does it move sodium and potassium ions? (p. 79)
2. What is the function of the carbonic acid–bicarbonate ion buffer system? (p. 49)

The remainder of this module follows the filtrate from the capsular space through the nephron as it is modified by tubular reabsorption and secretion. We start, of course, with the proximal tubule. Recall that the cells of the proximal tubule have prominent microvilli that provide these cells with a large surface area. This facilitates the remarkably rapid reabsorption that occurs in this very active segment of the renal tubule. In fact, the proximal tubule is the most metabolically active part of the nephron, as most of the filtrate is reabsorbed here—its Na⁺/K⁺ pumps alone consume about 6% of the body's ATP at rest. In addition to all of this reabsorption, a great deal of secretion takes place in the proximal tubule as well. The following sections examine the changes that the filtrate undergoes in the proximal tubule; we discuss first reabsorption and then secretion.

The main roles of the proximal tubule in reabsorption from the filtrate back to the blood are as follows:

- reabsorption of a large percentage of electrolytes, including sodium, chloride, potassium, sulfate, and phosphate ions, an activity that is vital for electrolyte homeostasis;

- reabsorption of nearly 100% of nutrients such as glucose, amino acids, and other organic substances (e.g., lactic acid, water-soluble vitamins);
- reabsorption of many of the bicarbonate ions, which is critical for acid-base homeostasis; and
- reabsorption of about 65% of the water, which is required for maintenance of the body's fluid homeostasis.

Let's look more closely at each of these roles.

Sodium Ion Reabsorption

We begin with the reabsorption of sodium ions, because this process turns out to be the key to reabsorbing many other substances in the proximal tubule. First, the majority of sodium ion reabsorption occurs through sodium ion leak channels on the apical surface of the proximal tubule cell, driven by its concentration gradient. Then, for active transport by the transcellular route, the cells of the proximal tubule have three types of carrier proteins for sodium ions in their apical membranes:

- carrier proteins specific for sodium ions that enable facilitated diffusion of sodium ions from the filtrate into the tubule cells,
- Na⁺ symporters that bring sodium ions from the filtrate into the cells with other solutes (such as glucose), and
- Na⁺/H⁺ antiporters that bring sodium ions into the cells while secreting hydrogen ions into the filtrate.

Each of these three carrier proteins transports sodium ions down its concentration gradient into the tubule cell. This concentration gradient isn't present naturally—it is created by

Figure 24.16
Glucose reabsorption in the proximal tubule.

Filtrate in tubule lumen | Cytosol in proximal tubule cell | Interstitial fluid | Blood in peritubular capillary

Na⁺/glucose symporter

Na⁺

Na⁺

K⁺

Na⁺/K⁺ pump

Na⁺

Glucose

High Na⁺ | Low Na⁺ | High Na⁺

Carrier protein

Glucose

① Na⁺/K⁺ pumps move Na⁺ out of the proximal tubule cell into the interstitial fluid, creating a Na⁺ concentration gradient via primary active transport.

② Na⁺ and glucose are moved into the cell from the filtrate by Na⁺/glucose symporters, using the energy of the Na⁺ gradient.

③ Glucose is transported from the proximal tubule cell to the interstitial fluid via facilitated diffusion, and then diffuses into the peritubular capillary.

Na^+/K^+ pumps in the basolateral membrane that continually pump sodium ions out of the tubule cells and into the interstitial fluid. These pumps create a relatively low sodium ion concentration in the proximal tubule cells (about 12 milliosmoles, or mOsm), whereas the sodium ion concentration in the filtrate is higher (about 142 mOsm). This concentration gradient is critical for the secondary active transport of many other solutes.

Most activity in the proximal tubule occurs without any outside control. Exceptions occur, such as parathyroid hormone decreasing phosphate ion reabsorption. In addition, the Na^+/K^+ pumps in the basolateral membrane become more active in the presence of angiotensin-II, which explains how this hormone increases sodium ion reabsorption from proximal tubule cells.

Reabsorption of Organic Solutes and Ions

The cells of the first half of the proximal tubule contain both Na^+/K^+ pumps and **Na^+/glucose symporters.** The symporters use the sodium ion gradient created by the pumps to carry both glucose and sodium ions from the filtrate into the tubule cell, an example of secondary active transport (**Figure 24.16**). Once in the cell, glucose is transported via facilitated diffusion into the interstitial fluid, where it diffuses into the peritubular capillaries. Other symporters in the apical membrane of the proximal tubule cells function in a similar fashion, allowing the secondary active transport of sodium ions and another solute, such as an ion (e.g., SO_4^{-2} or HPO_4^{-2}) or an organic solute (e.g., amino acids or lactic acid).

The reabsorption of sodium ions also leads to the reabsorption of anions such as chloride ions in another way, too. As sodium ions are passively transported out of the tubule lumen, the lumen accumulates a net negative charge. This creates an electrical gradient that pushes the negatively charged chloride ions across the epithelium through the paracellular route. These chloride ions similarly follow sodium ions into the interstitial fluid and into the plasma, as well.

Bicarbonate Ion Reabsorption

Bicarbonate ion reabsorption from the proximal tubule involves principles that you learned in the chemistry chapter about the carbonic acid–bicarbonate buffer system (see Chapter 2). Recall that carbon dioxide (CO_2) in the blood reacts with water (H_2O) to produce carbonic acid (H_2CO_3); this reaction is catalyzed by the enzyme *carbonic anhydrase (CA)*. Newly formed carbonic acid spontaneously dissociates into bicarbonate (HCO_3^-) and hydrogen (H^+) ions. The complete reaction is as follows:

$$CO_2 + H_2O \xrightleftharpoons{\substack{Carbonic \\ anhydrase}} H_2CO_3 \rightleftharpoons H^+ + HCO_3^-$$

Remember that carbonic anhydrase also catalyzes the reverse reaction, turning carbonic acid into carbon dioxide and water.

Bicarbonate reabsorption from the proximal tubule occurs in a roundabout way (**Figure 24.17**). The process somewhat resembles a game of "fetch": The cell "tosses" hydrogen ions into the

Figure 24.17 Bicarbonate ion reabsorption in the proximal tubule.

filtrate, and the hydrogen ion "fetches" a bicarbonate ion, which is then brought back into the cell. This toss is accomplished with the help of another carrier protein: the **Na^+/H^+ antiporter.** This carrier protein transports sodium ions into the cell while secreting hydrogen ions from the cell into the filtrate. Here's how the process unfolds:

① **Hydrogen ions secreted into the filtrate combine with bicarbonate ions to form carbonic acid.** Hydrogen ions are secreted from the cytosol of the proximal tubule cell into the filtrate by the Na^+/H^+ antiporter. The H^+ react with HCO_3^- in the filtrate to form H_2CO_3.

② **Carbonic acid is converted, via carbonic anhydrase, to carbon dioxide and water.** Carbonic anhydrase on the apical plasma membrane of the tubule cell catalyzes the conversion of H_2CO_3 into CO_2 and H_2O.

③ **Carbon dioxide diffuses into the tubule cell cytosol and combines with water to become bicarbonate and hydrogen ions.** CO_2 diffuses into the cytosol of the tubule cell with its concentration gradient. Carbonic anhydrase within the cell catalyzes the conversion of CO_2 and H_2O into H_2CO_3, which then dissociates into HCO_3^- and H^+.

④ **Bicarbonate ions are transported into the interstitial fluid and then move into the blood.** The HCO_3^- is transported across the basolateral membrane to enter the interstitial fluid and then diffuses into the peritubular capillaries.

⑤ **The process repeats as hydrogen ions are again secreted into the filtrate.** Hydrogen ions are again secreted into the filtrate by the Na^+/H^+ antiporter, and the process is repeated.

Though this process may seem complicated, it is effective and allows the cells of the proximal tubule to reabsorb about 90% of the bicarbonate ions from the filtrate. This is a key component of the body's ability to maintain the pH of the blood within a very specific range, 7.35–7.45 (see Chapter 25).

Obligatory Water Reabsorption and Its Effect on Other Electrolytes

By the time the filtrate has reached the second half of the proximal tubule, many of the sodium ions as well as glucose and other organic molecules have been reabsorbed. This creates a gradient favoring the passive movement of water by osmosis out of the filtrate along both the paracellular and transcellular routes. Remember that in osmosis, water moves to the solution with a higher solute concentration.

This type of water reabsorption is called **obligatory water reabsorption,** because water is "obliged" to follow solute movement (**Figure 24.18**). A kind of water channel in the plasma membrane called an **aquaporin** (ah-kwah-POUR-in) greatly enhances rapid water reabsorption. These channels, which are located in both the apical and basolateral membranes of proximal tubule cells, allow water to move through these cells via the transcellular route.

As obligatory water reabsorption continues, the concentration of solutes, such as potassium, calcium, and magnesium ions, rises in the filtrate. This creates a concentration gradient that favors their diffusion into or between the proximal tubule cells. Notice that in this process, active reabsorption of solutes stimulates further reabsorption of water by osmosis. This in turn stimulates passive reabsorption of other solutes.

Figure 24.18 Obligatory water reabsorption in the proximal tubule.

Secretion in the Proximal Tubule

In addition to the hydrogen ion secretion we discussed earlier, other substances are secreted into the filtrate by the proximal tubule cells, including many nitrogenous waste products and drugs. In the first half of the proximal tubule, most of the uric acid in the filtrate is reabsorbed, but nearly all of it is secreted back into the filtrate in the second half of the tubule. Additionally, ammonium ions (NH_4^+), creatinine, and small amounts of urea are also secreted. Drugs such as penicillin and morphine have significant renal secretion. These drugs must be taken often (typically 3–5 times per day), because medicine lost through renal secretion must be replaced in order to maintain relatively consistent blood levels.

Quick Check

☐ 3. Which substances are reabsorbed from the proximal tubule? Which of these substances are reabsorbed using the sodium ion gradient?

☐ 4. What is obligatory water reabsorption?

Reabsorption in the Nephron Loop

By the time the filtrate reaches the nephron loop, it barely resembles the original filtrate—about 60–70% of the electrolytes and water have been reabsorbed from the filtrate and returned to the blood, in addition to most of the organic solutes such as glucose and amino acids. As the filtrate flows through the nephron loop, it undergoes further losses: Approximately 20% of the total water, 25% of the total sodium and chloride ions, and a significant portion of the remaining ions are reabsorbed and returned to the blood.

In the proximal tubule, you saw that water reabsorption is proportional to solute reabsorption. For this reason, the filtrate in the proximal tubule has the same concentration, or *osmolarity*, as the interstitial fluid, about 300 mOsm. In the nephron loop, however, the filtrate's osmolarity changes as it flows through the loop. This is due to the differing permeabilities in the ascending and descending limbs of the nephron loop.

The thin descending limb of the nephron loop is freely permeable to water, but much less permeable to solutes such as sodium and chloride ions. So water can move out of the thin descending limb cells by osmosis, but few solutes follow. This causes the osmolarity of the filtrate to increase as it passes down the descending limb and rounds the bend of the loop.

The cells of the thick ascending limb are impermeable to water, but they transport NaCl into the tubule cells with the use of $Na^+/K^+/2Cl^-$ **symporters.** This secondary active transport system brings one sodium, one potassium, and two chloride ions into the tubule cell, relying on the Na^+/K^+ pump on the basolateral membrane to create a favorable sodium ion gradient. Remember that the Na^+/K^+ pumps drive potassium ions back *into* the tubule cell, so there isn't much net reabsorption of potassium ions in this process. As filtrate passes through the ascending limb, it loses solutes and gradually becomes less concentrated as ions are pumped into the interstitial fluid. These differing permeabilities in the two limbs and the changing concentration of the filtrate are part of a larger system that allows for extensive water reabsorption from the filtrate in both the loop and the collecting system, which we cover in Module 24.6.

Quick Check

☐ 5. How do the permeabilities of the two limbs of the nephron loop differ?

☐ 6. What is reabsorbed from the filtrate in the nephron loop?

Reabsorption and Secretion in the Distal Tubule and Collecting System

By the time the filtrate enters the first part of the distal tubule, about 85% of the water and 90% of the sodium ions have been reabsorbed. For this reason, the rate of filtrate flow in this part of the tubule is significantly lower (about 20 ml/min) than it was in the early proximal tubule (about 120 ml/min). As you learned earlier, the cells of the distal tubule lack microvilli. This reflects the

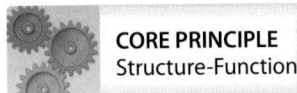
CORE PRINCIPLE
Structure-Function

Structure-Function Core Principle (p. 25)—most of the water reabsorption has already taken place, and so the cells do not require microvilli to perform their functions of reabsorbing much of the remaining water and solutes. Even though so much reabsorption has taken place, if we excreted the remaining water and sodium ions in the urine, we would still lose about 29 liters of water and a significant portion of our sodium ions every day. This situation would be incompatible with life. Reabsorption of the remaining water and sodium ions, then, is still critical.

The early distal tubule is structurally and functionally similar to the ascending limb of the nephron loop. However, the latter portion of the distal tubule is very similar to the cortical collecting duct, so we discuss them together here. Then we examine the medullary collecting system, which differs structurally and functionally from these other areas.

The Late Distal Tubule and Cortical Collecting Duct: Hormone Regulation

Cells of the late distal tubule and cortical collecting duct have hormone receptors that determine their function; the number and types of these receptors vary with hormone stimulation. The majority of their activity is therefore regulated by hormones in order to fine-tune water, electrolyte, and acid-base balance. For this reason, the water reabsorption here is called **facultative water reabsorption,** because water is reabsorbed in accordance with the body's needs (*facultative* means "able to adapt to a need"). The hormones involved in facultative water reabsorption as well as water and electrolyte balance include the following:

- **Aldosterone.** Aldosterone is a steroid hormone synthesized and released by the adrenal cortex that interacts with the DNA of cells to increase their permeability to sodium ions and the number of their Na^+/K^+ pumps. Both actions increase the reabsorption of sodium ions from the filtrate and the secretion of potassium ions into the filtrate. Note that these actions also indirectly cause reabsorption of water and chloride ions, because as sodium ions are reabsorbed, water and chloride ions passively follow (if antidiuretic hormone is present, as discussed next). Aldosterone also stimulates secretion of hydrogen ions into the filtrate by specialized cells called *intercalated cells*.

- **Antidiuretic hormone.** Recall that **antidiuretic hormone (ADH)** is made by the hypothalamus and released from the posterior pituitary gland (see Chapter 16). Note that *diuresis* (dy-yoo-REE-sis) refers to losing body water to the urine, and a *diuretic* is an agent that promotes diuresis. An *anti*-diuretic, therefore, refers to an agent that causes water retention and reduces urine output. ADH exerts these effects by causing aquaporins to be inserted into the cells' apical membranes, permitting rapid water reabsorption. In the absence of ADH, the cells of the late distal tubule and cortical collecting duct are barely permeable to water, and a larger volume of water is lost in the urine.
- **Atrial natriuretic peptide (ANP).** ANP triggers *natriuresis* (nay′-tree-yoo-REE-sis), or urinary excretion of sodium ions. It also appears to inhibit release of ADH and aldosterone, causing fewer sodium ions (and also less water) to be reabsorbed, and so more sodium ions and water to appear in the urine.

The Medullary Collecting System

As filtrate flows through the medullary collecting ducts and the papillary ducts, this is the kidney's last chance to regulate fluid, electrolyte, and acid-base balance before the filtrate becomes urine. The cells of the medullary collecting system have the following properties:

- They are impermeable to water in the absence of ADH. However, in the presence of ADH, they reabsorb large volumes of water.
- They are permeable to urea, which allows some urea to move down its concentration gradient into the interstitial fluid.
- Intercalated cells in this region actively secrete hydrogen ions from the interstitial fluid into the filtrate against a very high concentration gradient; they can increase the concentration of hydrogen ions in the filtrate about 900 times.

In addition, the cells of the medullary collecting system continue to reabsorb ions such as sodium and chloride from the filtrate. We revisit the medullary collecting system in Module 24.6, as it is part of a mechanism that allows the kidneys to concentrate urine and conserve water.

Quick Check

☐ 7. How does aldosterone impact reabsorption and secretion in the late distal tubule and cortical collecting duct?

☐ 8. How does ADH influence water reabsorption? In which parts of the nephron and collecting system does ADH act?

How Tubular Reabsorption and Secretion Maintain Acid-Base Balance

Variable reabsorption and secretion of hydrogen and bicarbonate ions contribute to pH homeostasis of the extracellular fluids, including blood. Thus far, you have seen two examples of acid-base regulation throughout the renal tubule and collecting system: The cells of the proximal tubule secrete hydrogen ions as a way to reabsorb bicarbonate ions, and the intercalated cells of the late distal tubule and collecting system actively secrete hydrogen ions under the influence of aldosterone. Another mechanism in renal tubule cells is stimulated when the pH of the blood becomes abnormal. If the pH of the blood *decreases,* making it too acidic, enzymes in the tubule cells will remove the amino group ($-NH_2$) from the amino acid glutamine in the cytosol. In doing this, the cells generate two ammonia molecules (NH_3) and two bicarbonate ions. The ammonia is then secreted, and the bicarbonate ions are reabsorbed. In the filtrate, the ammonia molecules bind and buffer hydrogen ions to form the ammonium ion (NH_4^+). Both ammonia buffering and bicarbonate ion reabsorption help to raise the pH of the blood back to normal. If the pH of the blood *increases,* making it too alkaline, the tubule cells will reabsorb fewer bicarbonate ions from the filtrate, excreting them in the urine and lowering the pH of the blood.

Putting It All Together: The Big Picture of Tubular Reabsorption and Secretion

As you've seen, both tubular reabsorption and secretion are vital functions, and enormous quantities of water and solutes cross the tubule cells every day. Filtrate entering the proximal tubule contains water; ions such as sodium, potassium, chloride, calcium, and bicarbonate; and organic solutes such as glucose, amino acids, and metabolic wastes. By the time the filtrate leaves the papillary ducts to become urine, most of the water and solutes have been reclaimed. In **Figure 24.19** we summarize the substances that are filtered, reabsorbed, and secreted by the nephrons and collecting system of the kidneys.

The final product that exits the papillary ducts is urine; the properties of this fluid are discussed in Module 24.7.

Apply What You Learned

☐ 1. You discover a new toxin that blocks the reabsorption of all sodium ions from the proximal tubule. What effect would this drug have on the reabsorption of water and other electrolytes from this tubule?

☐ 2. What effect would this toxin have on the reabsorption of glucose and bicarbonate?

☐ 3. Respiratory conditions can cause chronic *hypoventilation* that leads to a decreased blood pH. Predict how the kidneys will respond to this change in pH.

See answers in Appendix A.

The Big Picture of Tubular Reabsorption and Secretion

Figure 24.19

Play animation
@MasteringA&P®

Proximal tubule

Thick ascending limb

Thin descending limb

Distal tubule

Collecting duct

R REABSORPTION

S SECRETION

TUBULAR SEGMENTS	PROXIMAL TUBULE	NEPHRON LOOP		DISTAL TUBULE AND COLLECTING DUCT
		Thin descending limb	Thick ascending limb	
SUBSTANCES REABSORBED R→	• 65% of H_2O in the filtrate • Nearly 100% of glucose, amino acids, and other organic solutes • About 90% of bicarbonate ions (HCO_3^-) • 65% or more of Na^+, K^+, Ca^{2+}, Cl^-, and Mg^{2+}	• 20% of H_2O in the filtrate	• 25% of Na^+ and Cl^-	• Most of remaining H_2O • Nearly all of the remaining Na^+, Cl^-, and Ca^{2+} • Bicarbonate ions (HCO_3^-)
SUBSTANCES SECRETED ←S	• Hydrogen ions (H^+) • Nitrogenous wastes such as uric acid • Some drugs			• K^+ and H^+ (regulated by hormones)

Renal Physiology III: Regulation of Urine Concentration and Volume

Learning Outcomes

1. Explain why the differential permeability of specific sections of the renal tubule is necessary to produce concentrated urine.
2. Predict specific conditions that cause the kidneys to produce dilute versus concentrated urine.
3. Explain the role of the nephron loop, the vasa recta, and the countercurrent mechanism in the concentration of urine.

In the previous two modules on renal physiology, we followed the path of the filtrate from its formation via glomerular filtration to its modification in the renal tubules and collecting system by re-absorption and secretion. Now we are moving on to discuss the variable water reabsorption in the late distal tubule and collecting system, and the impact of this on urine concentration and volume.

As you learned in the previous module, about 85% of water reabsorption in the kidney is obligatory—water is "obliged" by osmosis to follow solutes that have been reabsorbed. The last 15% of water reabsorption is facultative water reabsorption, which is adjusted by hormones to meet the body's needs and maintain fluid homeostasis. Facultative water reabsorption is what determines final urine concentration and volume. In this module, we examine how the kidneys adjust facultative water reabsorption to produce either dilute or concentrated urine, and the role that hormones, particularly ADH, play in this process.

Osmolarity of the Filtrate

Let's first trace the filtrate through the nephron and follow how its osmolarity can change on its way to becoming urine. The filtrate that exits the blood and enters the renal tubule initially is *iso-osmotic*, or equally osmotic, to the plasma at 300 mOsm. Recall that in the nephron loop, however, the filtrate's osmolarity changes because of the differing permeabilities of its ascending and descending limbs.

The thin descending limb of the nephron loop is permeable to water but not solutes, as you learned, so water flows from the filtrate to the interstitial fluid by osmosis, but very few solutes follow. This causes the filtrate to become progressively more concentrated as it travels down the loop. By the time it reaches the bottom of the loop, on average it is approximately 900 mOsm, about three times more concentrated than plasma.

As the filtrate enters the thick ascending limb, sodium and other ions are pumped out of the filtrate and into the interstitial fluid. However, because this part of the limb is virtually imperme-able to water, water can't follow these solutes. So the concentration

of the filtrate decreases as it moves up the ascending limb of the loop. By the time the filtrate exits the thick ascending limb and moves into the distal tubule, its osmolarity is generally less than that of filtrate at the same level in the thin descending limb. In the early distal tubule, ions continue to leave the filtrate while water stays behind, and the concentration of the filtrate decreases even further, to about 100 mOsm, or three times *less* concentrated than plasma.

Once the filtrate enters the late distal tubule and collecting system, facultative water reabsorption may begin, so the concentration of the filtrate varies with the amount of water reabsorbed. If less water is reabsorbed, the concentration of the filtrate remains low as it passes through the late distal tubule and collecting system. The end result is the production of *dilute urine*, or urine with a concentration less than 300 mOsm. If, however, more water is reabsorbed, the concentration of the filtrate progressively increases as it passes through the late distal tubule and collecting system. The end result is the production of *concentrated urine*, or urine with a concentration greater than 300 mOsm. The upcoming sections examine the means by which both dilute urine and concentrated urine are produced.

Quick Check

☐ 1. How does the concentration of the filtrate change as it passes through the renal tubule and collecting system?
☐ 2. What are concentrated urine and dilute urine?

Production of Dilute Urine

The kidneys produce dilute urine when the solute concentration of the body's extracellular fluid is too low (which means the extracellular fluid contains excessive water) (**Figure 24.20**). The filtrate entering the late distal tubule is already less concentrated than the surrounding interstitial fluid. For this reason, the kidneys simply have to not reabsorb any additional water from the filtrate, or "turn off" facultative water reabsorption, to produce dilute urine. This is accomplished through a reduction in ADH release, which renders the late distal tubule and collecting system essentially impermeable to water.

Note that the distal tubule and collecting system continue to reabsorb sodium and chloride ions from the filtrate, so the number of solutes in the filtrate decreases while the amount of water remains the same. This causes the concentration of the filtrate to progressively decrease as it passes through the collecting system. When the amount of water in the extracellular fluid is greatly in excess, the osmolarity of the urine can fall as low as 50 mOsm, which is only one-sixth the osmolarity of plasma.

Typically, the volume of urine produced also increases when the urine is very dilute. As you learned, the normal volume of urine produced is about 1.8 liters per day. However, this value can increase dramatically when ADH secretion is low. Note that urine volume is also influenced by many factors, including fluid intake, general health, diet, and blood pressure. It is also impacted by **diuretics,** which act on various parts of the renal tubule to block reabsorption of water or solutes and promote diuresis.

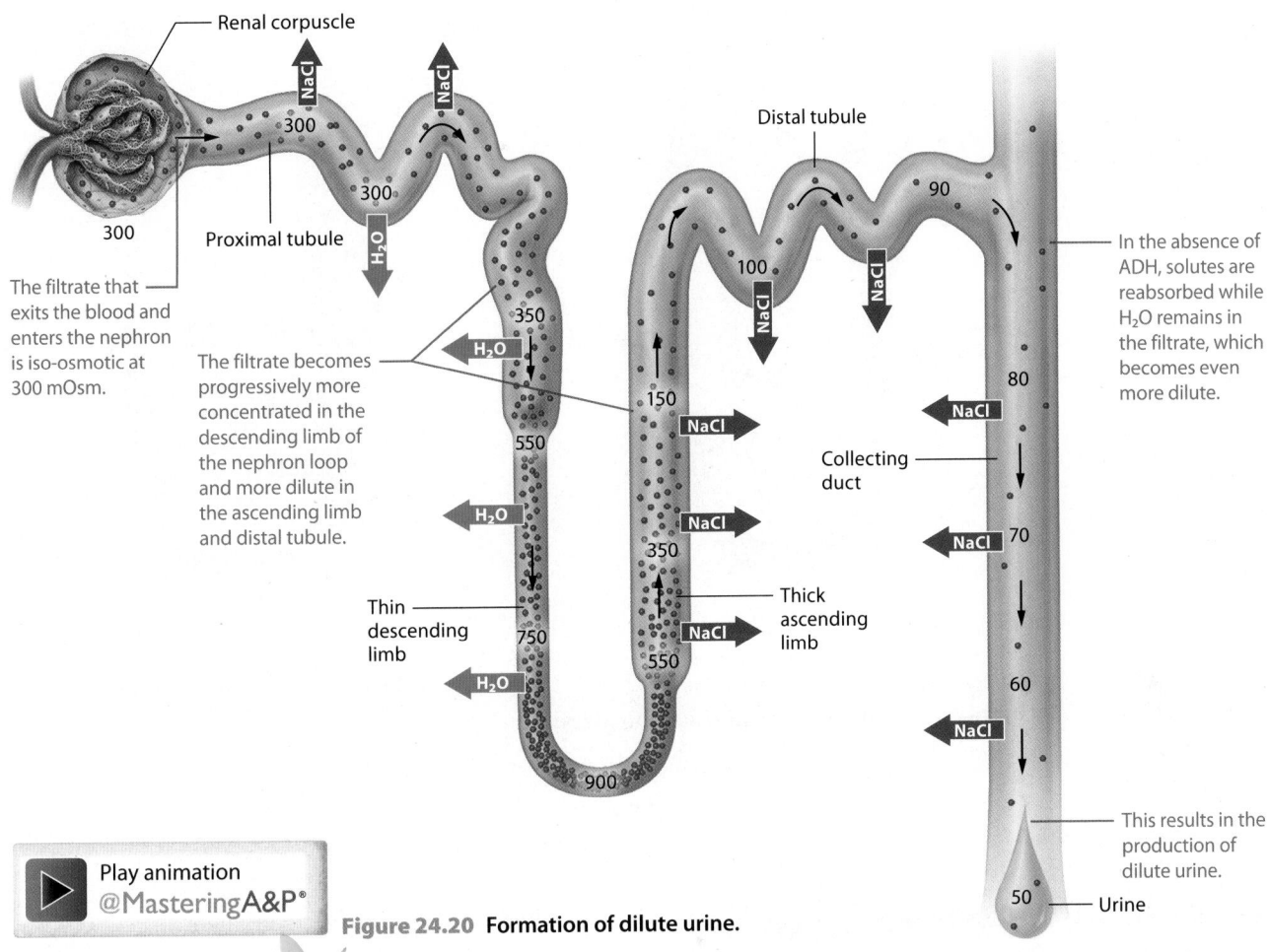

Figure 24.20 Formation of dilute urine.

Play animation @MasteringA&P®

The filtrate that exits the blood and enters the nephron is iso-osmotic at 300 mOsm.

Renal corpuscle

Proximal tubule

The filtrate becomes progressively more concentrated in the descending limb of the nephron loop and more dilute in the ascending limb and distal tubule.

Thin descending limb

Distal tubule

Thick ascending limb

Collecting duct

In the absence of ADH, solutes are reabsorbed while H_2O remains in the filtrate, which becomes even more dilute.

This results in the production of dilute urine.

Urine

Quick Check

☐ 3. Under what condition do the kidneys produce dilute urine?
☐ 4. How do the kidneys produce dilute urine?

The Countercurrent Mechanism and the Production of Concentrated Urine

Our kidneys are quite effective at conserving water and can produce urine with a concentration up to about 1200 mOsm. (This is impressive, but the kidneys of the Australian desert hopping mouse can produce urine with a concentration as high as 10,000 mOsm!) Concentrated urine results from "turning on" facultative water reabsorption in the late distal tubule and collecting system by the release of ADH. However, facultative water reabsorption requires more than simply the presence of ADH. Recall that water reabsorption happens only by osmosis, and osmosis, being a passive process, will occur only if a concentration gradient is present to drive it. This means that facultative water reabsorption takes place only if the interstitial fluid surrounding the nephron is *more* concentrated than the filtrate (an example of the Gradients Core Principle, p. 26).

CORE PRINCIPLE
Gradients

The interstitial fluid within the renal cortex has about the same osmolarity as the interstitial fluid elsewhere in the body, approximately 300 mOsm. The filtrate entering the late distal tubule and cortical collecting duct has an osmolarity of about 100 mOsm, so it is less concentrated than the interstitial fluid. This creates a small but significant gradient to drive water reabsorption. However, by the time the filtrate enters the medullary collecting duct, its osmolarity has risen to a value about equal to that of interstitial fluid, and the gradient disappears. Therefore osmosis will not take place unless the nephrons work to create a concentration gradient within the renal medulla. This gradient, known as the **medullary osmotic gradient,** starts at 300 mOsm at the cortex/medulla border, and increases as we go deeper into the medulla. Within the deepest regions of the renal medulla, it reaches a concentration of about 1200 mOsm, four times more concentrated than plasma.

The medullary osmotic gradient is created and maintained by a system called the **countercurrent mechanism,** which is a type of mechanism that involves the exchange of materials or heat between fluids flowing in opposite directions. In the kidneys, this mechanism consists of three factors: (1) a *countercurrent multiplier* system in the nephron loops of juxtamedullary nephrons, (2) the recycling of urea in the medullary collecting ducts, and (3) a *countercurrent exchanger* in the vasa recta.

(1) NaCl is actively transported from the filtrate in the thick ascending limb into the interstitial fluid, raising its NaCl concentration.

(2) The NaCl pumped into the interstitial fluid draws water out of the filtrate in the thin descending limb into the interstitial fluid by osmosis.

(a) The process of the countercurrent multiplier system

Figure 24.21 The countercurrent multiplier in the nephron loop.

▶ Play animation
@MasteringA&P®

The Nephron Loop and the Countercurrent Multiplier

Recall that there are two types of nephrons: cortical and juxtamedullary. Up to this point in the chapter, we have been discussing the physiology of both types of nephron. Now, however, we shift our attention specifically to the physiology of juxtamedullary nephrons—those with long nephron loops that descend deeply into the renal medulla. Within these long nephron loops we find a system called the **countercurrent multiplier,** which helps to create the medullary osmotic gradient.

In this system, the term *countercurrent* refers to the fact that the filtrate in the two limbs of the nephron loop flows in opposite directions—the filtrate in the descending limb flows toward the renal pelvis, and the filtrate in the ascending limb flows back up toward the renal cortex. Although the two limbs do not directly touch, they are close enough to influence each other. The term *multiplier* means that the effect increases, or multiplies, as the process proceeds due to the input of energy from ATP.

As depicted in **Figure 24.21a**, the system works like this:

(1) **NaCl is actively transported from the filtrate in the thick ascending limb into the interstitial fluid, raising its NaCl concentration.** $Na^+/K^+/2Cl^-$ symporters pump NaCl from the cells of the thick ascending limb into the interstitial fluid. This increases the interstitial fluid's osmolarity.

(2) **The NaCl pumped into the interstitial fluid draws water out of the filtrate in the thin descending limb into the interstitial fluid by osmosis.** The concentrated interstitial

fluid creates an osmotic gradient that draws water from the filtrate in the thin descending limb into the interstitial fluid.

(3) **Due to the continuing loss of water, the NaCl concentration of the filtrate increases as it approaches the bottom of the loop.** As water leaves the filtrate in the thin descending limb, NaCl remains, so the filtrate becomes progressively more concentrated as it reaches the bottom of the nephron loop.

(4) **The high NaCl concentration of the filtrate that reaches the thick ascending limb allows the NaCl reabsorption to continue.** The filtrate reaches the thick ascending limb with a very high NaCl concentration. This is key because the symporters in the thick ascending limb will work only if the filtrate has a high NaCl concentration. The symporters then begin pumping NaCl into the interstitial fluid, and we return to step (1).

All of these steps are occurring constantly; we separated them here for simplicity. You can see how the process works in **Figure 24.21b**.

The interstitial fluid is most concentrated in the deepest part of the renal medulla because the amount of NaCl pumped out of the thick ascending limb is proportional to its concentration in the filtrate, and its concentration is highest at the base of the loop. As the filtrate moves up the thick ascending limb and NaCl is pumped out, the NaCl concentration in the filtrate decreases and less NaCl can be pumped into the interstitial fluid.

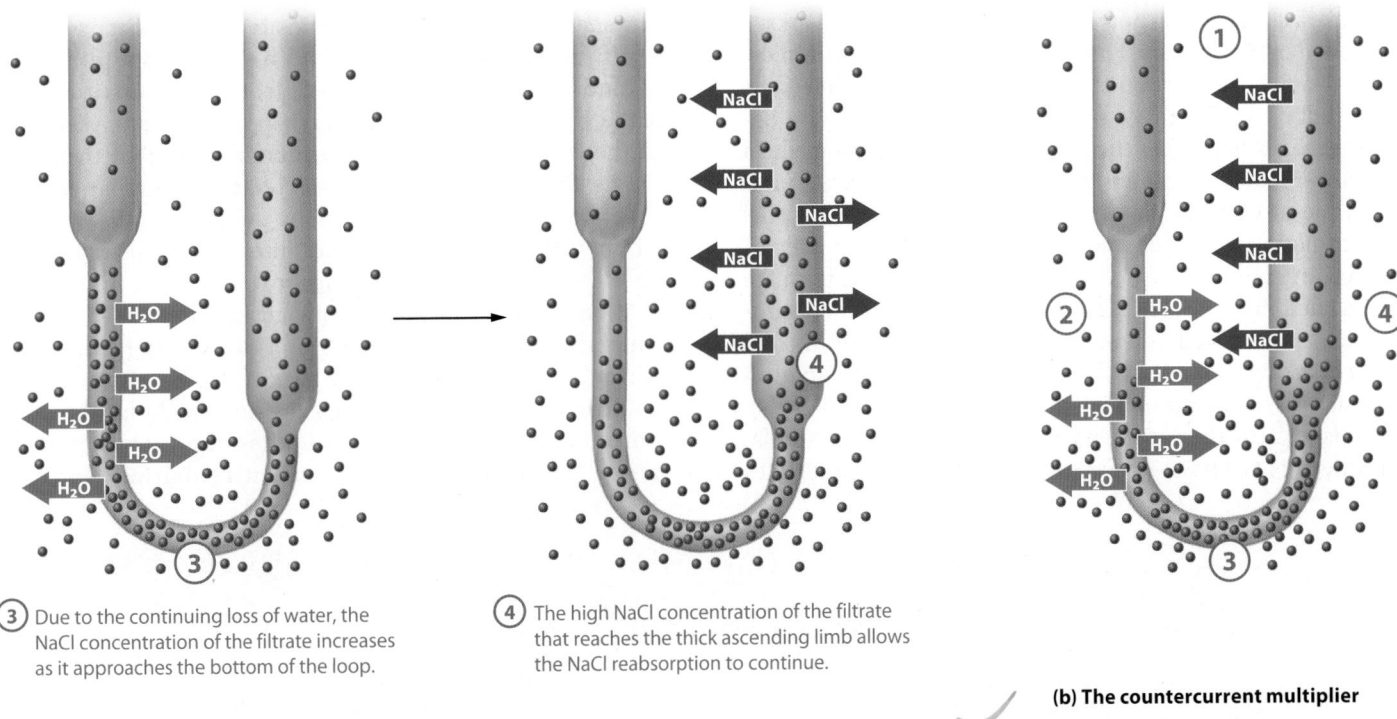

③ Due to the continuing loss of water, the NaCl concentration of the filtrate increases as it approaches the bottom of the loop.

④ The high NaCl concentration of the filtrate that reaches the thick ascending limb allows the NaCl reabsorption to continue.

(b) The countercurrent multiplier

ConceptBOOST ≫≫≫

Demystifying the Countercurrent Multiplier

Let's be honest: The countercurrent multiplier is a difficult concept for everyone to understand at first. To demystify it, first remember that this process has two main functions:

- to reabsorb about 15–25% of the remaining water and electrolytes in the filtrate, and
- to create a steep concentration gradient in the deepest parts of the renal medulla that allows water reabsorption from the medullary collecting ducts by osmosis.

The second thing to remember is that this is a *positive feedback loop*, meaning that its effects amplify. The countercurrent multiplier is actually a two-part positive feedback loop. In the first part, sodium and chloride ions that are reabsorbed from the ascending limb increase the amount of water that is reabsorbed from the descending limb. In the second part, as more water is reabsorbed from the descending limb, the concentration of sodium and chloride ions in the ascending limb increases. The higher the concentration of sodium and chloride ions in the ascending limb, the more that are pumped out in the first part of the loop. So, both parts amplify one another, allowing the system to perform its two main functions. ■

The Medullary Collecting System and Urea Recycling

A second factor that helps to create the medullary osmotic gradient is the permeability of the medullary collecting system to urea. As water is reabsorbed from the filtrate, urea becomes more concentrated in the remaining fluid. In the medullary collecting ducts and papillary ducts, urea follows its concentration gradient and passively diffuses out of the filtrate and into the interstitial fluid, further concentrating the medullary interstitial fluid. Some urea then enters the thin descending limb, so it continuously recycles. Note that the urea diffusing out of the medullary collecting duct constitutes only a small amount of the total urea; much of the urea remains in the filtrate and is excreted in the urine.

The Vasa Recta and the Countercurrent Exchanger

The steep medullary osmotic gradient created by the countercurrent multiplier and urea recycling is maintained by the vasa recta, the capillaries surrounding the nephron loops of juxtamedullary nephrons. The vasa recta act as a special kind of vascular system called a **countercurrent exchanger.** Like the limbs of the nephron loop, the vasa recta descend into the renal medulla, and then, following a hairpin turn, ascend toward the renal cortex. This arrangement of countercurrent flow, the blood flowing in the opposite direction from the filtrate, enables them to exchange substances.

Let's take a closer look at how countercurrent exchange works—it's illustrated in **Figure 24.22**. First notice that the blood within the vasa recta has a concentration of about 300 mOsm as it enters the renal medulla (to the right of the nephron, as shown in the figure). This means that ① as the blood descends into the medulla it is hypo-osmotic to the interstitial fluid. As we saw earlier, this situation causes water to leave the blood and enter the interstitial fluid by osmosis. In addition, more NaCl is present in the interstitial fluid than in the blood, which causes NaCl to diffuse from the interstitial fluid into the blood. The blood in the vasa recta continues to pick up NaCl and lose water as it descends deeper into the renal medulla. By the time the blood reaches the deepest part of the medulla, it has a concentration of about 1200 mOsm.

However, ② as the vasa recta ascend through the medulla, the gradient is reversed—the blood is now hyperosmotic to the interstitial fluid, and the opposite process occurs. NaCl now diffuses out of the blood and back into the interstitial fluid, and water moves by osmosis from the interstitial fluid into the blood. Notice what has happened here: All of the NaCl that was removed from the interstitial fluid by the blood in the descending limb of the vasa recta was "exchanged," or put back into the interstitial fluid by the blood in the ascending limb of the vasa recta. By the time the vasa recta exit the renal medulla, the blood has approximately the same concentration (about 300 mOsm) it had upon entering the renal medulla. The return of the blood to its initial osmolarity is critical, because it allows the vasa recta to deliver oxygen and nutrients to the cells of the medulla without depleting the medullary osmotic gradient necessary for water reabsorption and the production of concentrated urine. Note that

Figure 24.22 The countercurrent exchanger in the vasa recta.

Play animation
@MasteringA&P®

Figure 24.22 is a simplified model, as numerous vasa recta and nephrons are actually in proximity to each other in the kidney.

How the Countercurrent Mechanism Produces Concentrated Urine

The countercurrent system is complicated, so let's take the time to summarize its function. First, keep in mind that the entire homeostatic function of this system is to conserve water for the body when needed. Water can be reabsorbed from collecting ducts only if ADH is present and a concentration gradient is there to drive osmosis. This medullary osmotic gradient doesn't exist on its own, so the kidneys must create and maintain it in the following ways:

- The countercurrent multiplier of the thick ascending limb establishes the medullary interstitial gradient by pumping NaCl into the interstitial fluid.
- Continued solute reabsorption, including urea recycling, from the filtrate in the medullary collecting duct adds to the gradient.
- The countercurrent exchanger of the vasa recta allows perfusion of the inner medulla while maintaining the medullary interstitial gradient.

Let's look at **Figure 24.23** to see how these combine to produce concentrated urine:

1. **When filtrate enters the cortical collecting duct in the renal medulla, there is no osmotic gradient between the filtrate and the interstitial fluid, so no water is reabsorbed.** Filtrate entering the renal medulla has the same concentration as the interstitial fluid, so no gradient is present to drive water reabsorption.

2. **In the presence of ADH, the concentrated medullary interstitial fluid creates a gradient for water reabsorption from the filtrate in the medullary collecting duct.** The interstitial fluid in the renal medulla is more concentrated than the filtrate. So as the filtrate passes deeper into the renal medulla through the medullary collecting duct, if ADH is present, water is drawn into the interstitial fluid by osmosis, thanks to the aquaporins and the medullary osmotic gradient. Water is reabsorbed until the filtrate and interstitial fluid are iso-osmotic.

3. **Deeper into the medulla, interstitial fluid is more concentrated, so water reabsorption continues from the medullary collecting duct.** The process continues because the interstitial fluid becomes progressively more concentrated in the deep renal medulla, allowing continued water reabsorption.

4. **Concentrated urine is produced.** This process produces urine with a concentration up to 1200 mOsm. We cannot make more concentrated urine, because after this point there is no longer a gradient to drive osmosis.

To find out what happens to the concentration of urine when too much ADH is secreted, see *A&P in the Real World: SIADH* on page 974.

Figure 24.23 Formation of concentrated urine.

Play animation
@MasteringA&P®

1. When filtrate enters the cortical collecting duct in the renal medulla, there is no osmotic gradient between the filtrate and the interstitial fluid, so no water is reabsorbed.

2. In the presence of ADH, the concentrated medullary interstitial fluid creates a gradient for water reabsorption from the filtrate in the medullary collecting duct.

3. Deeper into the medulla, interstitial fluid is more concentrated, so water reabsorption continues from the medullary collecting duct.

4. Concentrated urine is produced.

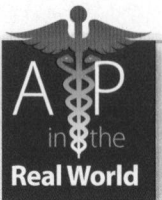

SIADH

The condition known as **syndrome of inappropriate ADH secretion,** or **SIADH,** is characterized by the secretion of excess antidiuretic hormone. SIADH has multiple causes, including ADH-secreting tumors (certain lung tumors are especially prone to this). Regardless of the cause, the result is the same: excessive fluid retention leading to a decreased plasma osmolarity and a high urine osmolarity.

Apply What You Learned

☐ 1. Glomerulonephritis, or inflammation of the glomerulus, results in excessively leaky glomerular capillaries and damaged glomeruli. The damaged and destroyed glomeruli cause the GFR to decrease. Which compensatory mechanisms would you expect to be triggered, and what effects would they have?
☐ 2. Certain diuretics block the effects of carbonic anhydrase in the proximal tubule. Predict the effects these drugs would have on the pH of the blood. How might the kidneys compensate for this?

See answers in Appendix A.

Quick Check

☐ 5. What three factors allow the kidney to produce and maintain the medullary osmotic gradient?
☐ 6. How is concentrated urine produced?

Apply What You Learned

☐ 1. Alcohol inhibits the release of ADH. Predict how this inhibition will influence urine volume and concentration.
☐ 2. Drugs called *loop diuretics* block the $Na^+/K^+/2Cl^-$ symporters in the thick ascending limb of the nephron loop.
  a. Predict the effect these drugs will have on water reabsorption from the thin descending limb of the nephron loop.
  b. Predict the effect these drugs will have on the gradient in the renal medulla, and the resulting effect on water reabsorption from the medullary collecting system. How will this impact urine volume?

See answers in Appendix A.

MODULE **24.7**

Putting It All Together: The Big Picture of Renal Physiology

Learning Outcomes

1. Describe the overall process by which blood is filtered and filtrate is modified to produce urine.

Let's now take a big picture look at all of the processes involved in renal physiology. We can break these processes down into six broad steps that are outlined in **Figure 24.24**. Note that Figure 24.24 shows the production of concentrated urine only.

MODULE **24.8**

Urine and Renal Clearance

Learning Outcomes

1. Explain how the physical and chemical properties of a urine sample are determined and relate these properties to normal urine composition.
2. Explain how filtration, reabsorption, and secretion determine the rate of excretion of any solute.
3. Explain how renal clearance rate can be used to measure GFR.

In the previous modules, we discussed how blood is filtered and how filtrate is modified as it flows through the nephron and collecting system. Now we take a look at the final product that exits the papillary ducts and is excreted by the body: the urine. In this module, we examine the components of normal urine and how abnormalities in the urine can signify problems in the kidneys and elsewhere in the body. We also explore another way in which the health of the kidneys may be assessed, through a value known as *renal clearance.*

Urine Composition and Urinalysis

As you learned in earlier modules, urine is the fluid that remains after tubular reabsorption and secretion have taken place. It normally contains water; sodium, potassium, chloride, and hydrogen ions; phosphates; sulfates; and metabolic waste products such as urea, creatinine, ammonia, and uric acid. It may also contain trace amounts of bicarbonate, calcium, and magnesium ions.

Because certain conditions cause abnormalities in the urine, an analysis of the urine, or **urinalysis,** can be a valuable tool in the diagnosis of disease. The factors examined in a typical urinalysis are as follows:

● **Color.** Urine is colored by a yellow pigment called *urochrome* (YER-oh-krohm), a breakdown product of hemoglobin. The more concentrated the urine, the less water is present, so the

The Big Picture of Renal Physiology

Figure 24.24

  Play animation @MasteringA&P®

① Glomerular filtration: In the renal corpuscle, filtrate is formed as blood is filtered through the filtration membrane (see Figure 24.12).

Filtrate

Filtration membrane

Blood

② GFR and its regulation: The GFR is determined by the net filtration pressure in the renal corpuscle, which is influenced by many factors, such as angiotensin-II (see Figures 24.13 and 24.14).

Renal corpuscle

NFP

③ Reabsorption and secretion in proximal tubule: The proximal tubule is the site of extensive tubular reabsorption and select secretion (see Figure 24.19).

Proximal tubule

Distal tubule

Aldosterone

Na^+ K^+ H_2O

A

A H^+ ADH

⑤ Reabsorption and secretion in distal tubule: In the late distal tubule and cortical collecting duct, reabsorption and secretion are controlled by hormones (see Figures 24.20 and 24.23).

Cortical collecting duct

Renal cortex
Renal medulla

Antidiuretic hormone

ADH

Medullary collecting duct

H_2O

Interstitial fluid (mOsm)

300

400

H_2O

H_2O

600

H_2O

900

H_2O

1200

NaCl

NaCl

NaCl

NaCl

ADH

H_2O

ADH

ADH

H_2O

ADH

H_2O

⑥ Production of dilute or concentrated urine: Water is reabsorbed in the medullary collecting duct in the presence of ADH and the medullary concentration gradient (shown here). Water is not reabsorbed in the absence of ADH. The amount of water reabsorbed determines whether dilute or concentrated urine will be produced (see Figure 24.23).

④ Countercurrent multiplication and exchange: In the nephron loop and vasa recta, countercurrent multiplication and exchange occur (see Figures 24.21 and 24.22).

Vasa recta

Nephron loop

Urea

Urine

same amount of urochrome makes the urine darker yellow. The color of urine may also be altered by certain foods, vitamins, drugs, and food dyes, or by the presence of blood.

- **Translucency.** Regardless of color, urine should always be translucent (you should be able to see light through it). Cloudy urine typically indicates an infection but may also indicate that the urine contains large quantities of protein.
- **Odor.** Recently voided urine should have a mild odor. If urine is allowed to sit out, however, bacteria metabolize the urea in the urine to produce ammonia, giving it a stronger odor. The odor of urine may also be altered by certain disease states, such as diabetes mellitus or infection, or by eating certain foods, such as asparagus.
- **pH.** The pH of urine is normally around 6.0—slightly acidic—but it can range from 4.5 to 8.0. The reason for the minimum pH of 4.5 is that the H^+ transport pumps in the distal tubule and collecting system cannot pump against a higher hydrogen ion gradient than this.
- **Specific gravity.** Specific gravity compares the amount of solutes in a solution to the amount in deionized water. Deionized water has no solutes and is assigned a specific gravity of 1.0; because urine has solutes, its specific gravity will be higher than that of water. This value typically ranges from 1.001 (very dilute urine) to 1.035 (very concentrated urine).

Other properties of urine, such as the presence and relative levels of certain solutes, are analyzed with urinalysis test strips. These strips are dipped in the urine, and their chemical indicators change color in the presence of specific substances. Substances commonly tested for in this way include blood, protein, leukocytes, and glucose. If the kidneys are functioning properly, these substances should not be present in urine in significant amounts.

Quick Check

☐ 1. What is the normal composition of urine?
☐ 2. What are its normal characteristics?

Renal Clearance

Evaluation of renal function is very important in clinical settings to monitor the health of the kidneys. Although urinalysis may provide valuable clues about renal function, a more complete assessment is provided by measuring the rate at which the kidneys remove a substance from the blood, a process known as **renal clearance.** The renal clearance of the chemical is then used to estimate the GFR. Renal clearance and GFR both are measured in the same units: milliliters of plasma per minute.

For a substance to provide an accurate measure of renal clearance and the GFR, the substance should be completely filtered and neither reabsorbed nor secreted. Substances secreted by renal tubules have a renal clearance greater than their GFR, whereas those that are reabsorbed have a renal clearance less than their GFR. To measure renal clearance, we therefore have a limited group of substances from which to choose. Two such commonly used substances are creatinine and inulin.

Creatinine, as you learned earlier, is a waste product of the metabolism of muscle and other cells. Nearly all of the creatinine produced is excreted by the kidneys. When the kidneys are impaired, the level of creatinine in the blood tends to rise. Generally, a plasma creatinine level above 1.2 mg/dl (milligrams per deciliter) is considered abnormal, but this varies with age, gender, and body mass. Creatinine excretion may be used to estimate the GFR by comparing the amount of creatinine excreted in the urine to the plasma concentration of creatinine. The main difficulty with using creatinine as an indicator of glomerular filtration is that between 15% and 50% of creatinine in the urine arrived there via secretion, not filtration. So a patient with a very low GFR may still have a nearly normal result from this test.

A more accurate assessment of the GFR can be obtained using the substance *inulin* (IN-yoo-lin). Inulin (not to be confused with the hormone insulin) is a complex carbohydrate found in plants such as garlic and artichokes that is filtered by the glomerulus, but is neither reabsorbed nor secreted by the renal tubule or collecting system. The GFR may be measured by injecting inulin and comparing its excretion in the urine to its plasma concentration.

Quick Check

☐ 3. What is renal clearance, and what is it used to estimate?

Apply What You Learned

☐ 1. The condition known as metabolic acidosis is characterized by a decreased blood pH from the accumulation of metabolic acids. Predict the effects this condition will have on the pH of the urine. What effect would you expect the opposite condition, metabolic alkalosis, to have on the urine pH?
☐ 2. Dietary supplementation with creatine phosphate is popular among athletes for its supposed performance-enhancing effects. What effect would creatine phosphate supplementation have on the amount of creatinine in the blood, and therefore the amount of creatinine that the kidneys must excrete?

See answers in Appendix A.

MODULE **24.9**

Urine Transport, Storage, and Elimination

Learning Outcomes

1. Identify and describe the structure and functions of the ureters, urinary bladder, and urethra.
2. Relate the anatomy and histology of the bladder to its function.
3. Compare and contrast the male and female urinary tracts.
4. Describe the micturition reflex.
5. Describe voluntary and involuntary neural control of micturition.

Newly formed urine drains from the papillary ducts to minor ca-lyces, then to major calyces, and finally to a renal pelvis. The renal pelves then drain urine into the organs of the urinary tract. In this module, we examine the structure and function of the organs of the urinary tract. We conclude with a look at the process by which urine is expelled from the body, *micturition* (mik-choo-RISH-un).

Anatomy of the Urinary Tract

◀◀ Flashback

1. What are the properties of transitional epithelium? (p. 135)

The urinary tract consists of the two ureters, the urinary bladder, and the urethra (**Figure 24.25a**). The upcoming sections exam-ine their anatomy and histology and the differences between the male and female urinary tracts.

Ureters

The **ureters** transport urine from the kidneys to the urinary blad-der. The ureters are generally about 25–30 cm long and 3–4 mm in diameter in an adult. They begin at roughly the level of the sec-ond lumbar vertebra, travel behind the peritoneum, and empty into the urinary bladder.

Like any hollow organ, the ureters have a multilayered wall. The three layers are named as follows, from superficial to deep (**Figure 24.25b**):

1. **Adventitia.** This superficial layer, known as the **adventitia** (ad-ven-TISH-uh), is fibrous connective tissue that supports the ureters.
2. **Muscularis.** The middle layer is the **muscularis,** and it consists of smooth muscle. Like the smooth muscle of the alimentary canal, the smooth muscle in the muscularis contracts rhythmi-cally via *peristalsis* to propel urine toward the urinary bladder. Waves of peristalsis course through the muscularis as often as five times per minute, depending on the rate of urine production.
3. **Mucosa.** The innermost layer is the **mucosa,** a mucous mem-brane composed of transitional epithelium and its underlying basal lamina. As you learned in Chapter 4, transitional epi-thelium is stratified with cells that can change from having a dome shape to being squamous. This property allows the epi-thelium to expand and recoil.

The ureters drain posteriorly into the inferior urinary blad-der. In this region a mechanism prevents urine from flowing backward through the ureter. As each ureter passes along the posterior urinary bladder, it travels obliquely through a "tunnel" in the bladder wall. As urine collects in the bladder, the pressure rises and compresses this tunnel, pinching the ureter closed and preventing backflow of urine.

Urinary Bladder

The **urinary bladder** is a hollow, distensible organ that sits on the floor of the pelvic cavity, suspended by a fold of parietal peritoneum.

(a) Organs of the urinary tract

(b) Cross-sectional histology of ureter — LM (8×)

(c) Histology of urinary bladder wall — LM (50×)

Practice art labeling @MasteringA&P®

Figure 24.25 Anatomy of the urinary tract.

It collapses when empty, but when distended, it becomes pear-shaped, and can hold up to about 700–800 ml of urine in males (slightly less in females because of the position of the uterus).

Like the ureters, the wall of the urinary bladder has three tissue layers (**Figure 24.25c**). From superficial to deep, these layers are as follows:

1. **Adventitia.** The adventitia, the most superficial layer, is composed of areolar connective tissue. On the superior surface of the urinary bladder is an additional serosa, which is a fold of the parietal peritoneum.
2. **Detrusor muscle.** The middle tissue layer is composed of smooth muscle known as the **detrusor muscle** (dee-TROO-sor; "to push down"). The muscle fibers are arranged into inner longitudinal, middle circular, and outer longitudinal layers. The detrusor muscle forms a circular band around the opening of the urethra, called the **internal urethral sphincter,** shown in Figure 24.25a.
3. **Mucosa.** The mucosa is composed of transitional epithelium with an underlying basal lamina. It is a mucous membrane that produces mucus to protect the bladder epithelium from urine. When the bladder is not full, folds of mucosa called *rugae* (ROO-gee) are visible. What would happen if this mucus was not present? See *A&P in the Real World: Interstitial Cystitis* to find out.

Notice that the floor of the urinary bladder contains a triangular area called the **trigone** (TRY-gohn; "triangle"). The trigone lacks rugae and appears smooth because its mucosa is tightly bound to the underlying muscularis. The two posterior corners of the trigone are formed by the two ureteral orifices (openings). These orifices have mucosal flaps that act as valves to prevent backflow of urine during elimination. The apex of the trigone is formed by the opening to the urethra, the **internal urethral orifice.**

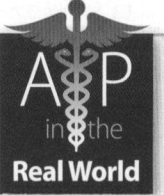

Interstitial Cystitis

Inadequate mucus production by the mucosa of the urinary bladder can cause the disease *interstitial cystitis* (sis-TY-tis; IC). This condition allows the acid and other toxic substances in the urine to damage the underlying mucosa and other tissues. IC is characterized by frequent urination and pelvic pain resulting from mucosal ulcerations. The underlying cause of IC is unknown, and unfortunately it responds poorly to treatment. Over the long term, IC can lead to bladder scarring and contraction.

Notice in **Figure 24.26** that the position of the urinary bladder differs in males and females. In males, it is anterior to the rectum; in females, it is anterior to the vagina and inferior to the uterus.

Urethra

The **urethra** is the terminal portion of the urinary tract; it drains urine from the urinary bladder to the outside of the body. Like the rest of the urinary tract, the urethra has an outer adventitia, middle muscularis, and inner mucosa. It begins at the internal urethral orifice in the urinary bladder, which is surrounded by the internal urethral sphincter. This sphincter remains closed unless urine is being eliminated. A second urethral sphincter, the **external urethral sphincter,** is formed from the *levator ani muscle* (leh-VAY-ter AY-nai; sometimes called the *urogenital diaphragm*), the muscular floor of the pelvic cavity. This sphincter is composed of skeletal muscle and is under voluntary control. There is no sphincter at the external orifice to the urethra in either males or females.

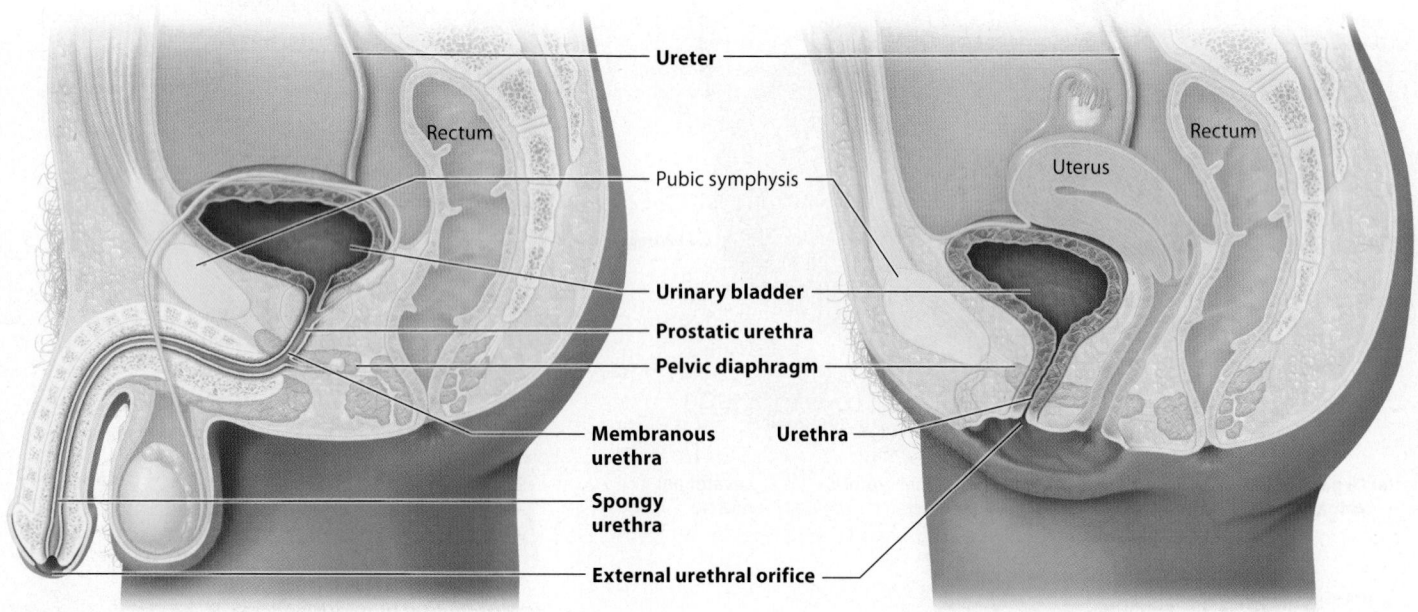

(a) Sagittal section through male pelvis

(b) Sagittal section through female pelvis

Figure 24.26 Comparison of urinary tract anatomy in the male and female.

The urethra differs in males and females structurally, as shown in Figure 24.26, which means that functional differences must exist as well. The female urethra is shorter (about 4 cm in length) and opens at the **external urethral orifice** between the vagina and the clitoris. It serves exclusively as a passage for urine.

The male urethra is considerably longer (about 20 cm in length) and consists of three regions:

1. **Prostatic urethra.** As the urethra exits the urinary bladder, it passes through the *prostate gland,* which sits inferior to the urinary bladder. This section of the urethra is called the **prostatic urethra.**
2. **Membranous urethra.** The shortest segment, known as the **membranous urethra,** passes through the levator ani muscle.
3. **Spongy urethra.** The longest segment of the male urethra, the **spongy urethra,** passes through the penis to the external urethral orifice. It is called the spongy urethra because it passes through an erectile body of the penis called the *corpus spongiosum.*

The location of the male urethra allows it to serve a dual purpose: It transports both urine and semen (see Chapter 26).

Quick Check

- [ ] 1. What are the three tissue layers of the organs of the urinary tract?
- [ ] 2. What are the functions of the ureters and urinary bladder?
- [ ] 3. How does the urethra differ structurally and functionally in males and females?

Micturition

◄◄ **Flashback**

1. What are afferent and efferent neurons? (p. 478)
2. What is a reflex arc? (p. 508)

Micturition, also called urination or voiding, is the discharge of urine from the urinary bladder to the outside of the body. It is activated by a reflex arc called the **micturition reflex,** which is mediated by the parasympathetic nervous system. This reflex arc is carried out by three components: (1) stretch receptors in the wall of the urinary bladder, (2) sensory afferent nerve fibers that convey this information to the sacral portion of the spinal cord (S2 and S3), and (3) parasympathetic efferent fibers that travel to the detrusor muscle and internal sphincter of the urinary bladder.

Micturition is shown in **Figure 24.27.** When urine fills the bladder and stretches its walls ①, the micturition reflex occurs. In infants and young children, this involuntary reflex is the primary mechanism by which the urinary bladder is emptied. This reflex arc has two basic steps:

② Stretch receptors send a signal via sensory afferent fibers to the sacral portion of the spinal cord.
③ Parasympathetic efferent fibers stimulate the detrusor muscle to contract and the internal urethral sphincter to relax, causing micturition.

As children mature, however, pathways develop between these parasympathetic neurons and the brain that allow control over the external urethral sphincter. At this point, micturition is predominantly controlled by the **micturition center** in the pons,

Involuntary reflex:

① Urine fills the bladder and stretches its wall.

② Stretch receptors send a signal via sensory afferent fibers to the sacral portion of the spinal cord.

③ Parasympathetic efferent fibers stimulate the detrusor muscle to contract and the internal urethral sphincter to relax, causing micturition.

Voluntary control:

To pons   From brain centers

④ Interneurons in the spinal cord communicate the "full bladder" signal to the micturition center in the pons.

Urinary bladder

Internal urethral sphincter

External urethral sphincter

Spinal cord

⑤ If micturition is appropriate, the cerebral cortex facilitates this process by allowing the external urethral sphincter to relax, and urine is voided.

Figure 24.27 Micturition.

making the process voluntary. When the bladder is full, two additional steps then occur simultaneously with the involuntary process (see Figure 24.27):

④ Interneurons in the spinal cord communicate the "full bladder" signal to the micturition center in the pons.

⑤ If micturition is appropriate, the cerebral cortex facilitates this process by allowing the external urethral sphincter to relax, and urine is voided.

If micturition is not appropriate, then the detrusor muscle relaxes, the internal and external urethral sphincters remain closed, and the urge to urinate passes. The reflex generally initiates again within about an hour. This cycle repeats until the sensation of having to urinate becomes more acute. By the time about 500–600 ml has accumulated in the urinary bladder, the urge to urinate becomes too strong, voluntary control over the external urethral sphincter is lost, and micturition occurs. After micturition, the bladder contains only about 10 ml of urine.

Quick Check

☐ 4. What are the steps of the micturition reflex?
☐ 5. How is micturition consciously controlled?

Apply What You Learned

☐ 1. Predict what would happen if the epithelium of the urinary tract were made of simple squamous epithelium instead of transitional epithelium.
☐ 2. How would a spinal cord injury above the level of S1–S2 affect micturition in that patient? How would the situation change if the injury were below S1–S2?

See answers in Appendix A.

CHAPTER SUMMARY

Take full advantage of the invaluable online practice and study tools for this chapter by going to MasteringA&P®. There you will find:

- 👆 Practice tests, module quizzes, art-labeling and sequence-the-steps activities

- ▶ Animated figures with quizzes

- ✎ *Practicing A&P* worksheets

- 🔊 Author-narrated podcast

- 🔥 Dynamic Study Modules

- iP *Interactive Physiology*

MODULE 24.1
Overview of the Urinary System 941

- The organs of the **urinary system** are the **kidneys** and the ureters, urinary bladder, and urethra. (Figure 24.1, p. 942)

  👆 Practice art-labeling exercise: Figure 24.1.

- The kidneys remove waste products from the blood and produce and expel **urine;** regulate fluid, electrolyte, and acid-base balance, blood pressure, and erythrocyte production; and perform other metabolic functions.

MODULE 24.2
Anatomy of the Kidneys 943

- The kidneys are paired retroperitoneal organs with three layers of protective coverings. (Figure 24.2, p. 943)

- Each kidney consists of over one million functional units called **nephrons.**

- Internally, the kidney consists of three regions: the superficial **renal cortex,** which houses blood vessels and most parts of the nephrons; the middle **renal medulla,** which consists of triangular **renal pyramids** that are separated by **renal columns;** and the innermost **renal pelvis** and **major** and **minor calyces,** which form drainage and collection areas for urine. (Figure 24.3, p. 944)

  👆 Practice art-labeling exercise: Figure 24.3.

- The kidney is supplied by the large **renal artery,** which branches multiple times as it passes from the renal pelvis to the renal cortex. In the renal cortex, each tiny **afferent arteriole** feeds a **glomerulus,** which is then drained by an **efferent arteriole.** The efferent arteriole feeds a second capillary bed, the **peritubular capillaries.** These vessels lead into a series of veins that drain into the **renal vein.** (Figure 24.4, p. 945)

  👆 Practice art-labeling exercise: Figure 24.4.

- Nephrons are made up of two main parts: (Figure 24.5, p. 946)
  ○ The first part of the nephron is the **renal corpuscle,** which consists of the **glomerulus** and the surrounding **glomerular capsule.** The glomerular capsule has a parietal layer consisting of simple squamous epithelium and an inner visceral layer consisting of **podocytes.** (Figure 24.6, p. 947)
  ○ The second part of the nephron is the renal tubule, which consists of the **proximal tubule,** the **nephron loop,** and the **distal tubule.** (Figure 24.7, p. 947)

  👆 Practice art-labeling exercise: Figure 24.5.

- The **juxtaglomerular apparatus** is a specialized region of the nephron where a group of cells located at the transition point between the nephron loop and the distal tubule **macula densa** come into contact with the **JG cells** of the afferent arteriole. (Figure 24.8, p. 948)

- Nephrons drain into the **collecting system,** which consists of the *cortical collecting ducts* and the *medullary collecting system*; the latter includes the *medullary collecting ducts* and **papillary ducts.** A papillary duct drains into a minor calyx. (Figure 24.9, p. 949)

- There are two types of nephrons: **cortical nephrons,** which have short nephron loops, and **juxtamedullary nephrons,** which have long nephron loops that dip deeply into the renal medulla and are surrounded by capillaries called the **vasa recta.** (Figure 24.10, p. 950)

iP Play Interactive Physiology tutorial on Urinary System: Anatomy Review.

MODULE **24.3**
Overview of Renal Physiology 951

- Nephrons carry out three basic physiological processes: (Figure 24.11, p. 951)
  - **Glomerular filtration** is performed by the renal corpuscles as blood passes through the glomerulus, resulting in the production of **filtrate.**
  - **Tubular reabsorption** is the process by which solutes and water are reclaimed from the filtrate and re-enter the blood.
  - **Tubular secretion** is the process by which substances are removed from the blood and transported into the filtrate.

MODULE **24.4**
Renal Physiology I: Glomerular Filtration 951

- Three barriers to filtration constitute the **filtration membrane:** the glomerular capillary endothelial cells, the basal lamina of the glomerulus, and the podocytes (visceral layer of the glomerular capsule). The filtration membrane permits the passage of small molecules into the filtrate, but does not permit larger molecules, such as proteins and the formed elements of blood, to pass. (Figure 24.12, p. 952)

▶ Play animation on filtration: Figure 24.12.

- The rate at which filtrate is formed is called the **glomerular filtration rate (GFR);** it averages about 125 ml/min. The GFR is determined by the net filtration pressure (NFP) in the glomerulus, which is the sum of the forces that favor filtration (glomerular hydrostatic pressure) and those that oppose filtration (colloid osmotic pressure and capsular hydrostatic pressure). The NFP averages about 10 mm Hg. (Figure 24.13, p. 953)

▶ Play animation on net filtration pressure: Figure 24.13.

- Maintenance of the GFR is crucial to homeostasis of fluid, electrolyte, and acid-base balance.
  - The kidneys can autoregulate to maintain the GFR by the **myogenic mechanism** and *tubuloglomerular feedback.* Both mechanisms maintain the GFR by modifying the diameter of the glomerular afferent arterioles.
  - The **renin-angiotensin-aldosterone system (RAAS)** works to maintain the GFR as part of the larger process of maintaining systemic blood pressure. The RAAS culminates in the formation of the protein **angiotensin-II,** which increases glomerular hydrostatic pressure through greater constriction of the efferent arterioles to preserve the GFR, stimulates sodium ion and water reabsorption from the proximal tubule, causes systemic vasoconstriction, and triggers aldosterone release to increase systemic blood pressure. (Figure 24.14, p. 957)

▶ Play animation on the renin-angiotensin-aldosterone system: Figure 24.14.

  - **Atrial natriuretic peptide** increases the GFR by dilating the afferent arteriole.
  - The sympathetic nervous system decreases the GFR by causing vasoconstriction of the afferent arterioles.
- The ways in which the GFR is controlled are summarized in Table 24.1 (p. 959).

iP Play Interactive Physiology tutorial on Urinary System: Glomerular Filtration.

MODULE **24.5**
Renal Physiology II: Tubular Reabsorption and Secretion 960

- To be reabsorbed, substances must pass from the filtrate in the tubule lumen, past the tubule cells into the interstitial space, where they then diffuse through the cells of the capillary walls into the blood. Substances may cross the tubule cells by passing *between* the cells (the **paracellular route**) or *through* them (the **transcellular route**). Substances are secreted only through the transcellular route. (Figure 24.15, p. 961)
- Reabsorption and secretion occur in the renal tubule and collecting system:
  - In the proximal tubule, sodium ion reabsorption drives **obligatory water reabsorption,** reabsorption of many solutes, and hydrogen ion secretion. About 65% of the water, much of the HCO_3^-, nearly all of the amino acids and glucose, and variable amounts of electrolytes are reabsorbed here. (Figure 24.16, p. 962; Figure 24.17, p. 963; Figure 24.18, p. 964)
  - In the nephron loop, another 15–20% of the water and 20–25% of the NaCl are reabsorbed.

iP Play Interactive Physiology tutorial on Urinary System: Early Filtrate Processing.

  - The late distal tubule and cortical collecting ducts are relatively impermeable to water and solutes unless influenced by hormones.
    - **Aldosterone** causes reabsorption of sodium ions and secretion of potassium and hydrogen ions.
    - **Antidiuretic hormone** (ADH) stimulates the reabsorption of water through the formation of **aquaporin** channels.
  - The medullary collecting system reabsorbs most of the remaining water from the filtrate under the influence of ADH. These ducts also reabsorb many of the ions by diffusion and secrete hydrogen ions by primary active transport.
  - The pH of the blood is regulated by adjusting the secretion of hydrogen ions and the reabsorption of bicarbonate ions.

iP Play Interactive Physiology tutorial on Urinary System: Late Filtrate Processing.

- The big picture of tubular reabsorption and secretion is shown in Figure 24.19 (p. 967).

▷ **Play animation on tubular reabsorption and secretion: Figure 24.19.**

Renal Physiology III: Regulation of Urine Concentration and Volume 968

- **Facultative water reabsorption** in the late distal tubule and collecting system determines the concentration and volume of urine that the kidneys produce. If less water is reabsorbed, dilute urine is produced. If more water is reabsorbed, concentrated urine is produced.

- Dilute urine is produced by decreasing the release of ADH, which causes less water to be reabsorbed. (Figure 24.20, p. 969)

▷ **Play animation on the formation of dilute urine: Figure 24.20.**

- The ability to produce concentrated urine relies on two factors: ADH acting on the late distal tubule and collecting system to cause reabsorption of water and a **medullary osmotic gradient** in the interstitial fluid of the renal medulla that drives the reabsorption of water by osmosis.

  ○ The medullary osmotic gradient is created by the **countercurrent multiplier** in the nephron loops of juxtamedullary nephrons, whereby NaCl is pumped out of the thick ascending limb of the loop into the interstitial fluid and water moves out of the thin descending limb into the interstitial fluid by osmosis. (Figure 24.21, pp. 970–971)

▷ **Play animation on the countercurrent multiplier: Figure 24.21.**

  ○ Urea passively diffuses into the interstitial fluid from the medullary collecting system, which contributes to the medullary osmotic gradient.

  ○ The vasa recta maintain the gradient by acting as a **countercurrent exchanger.** These vessels remove NaCl from the interstitial fluid as they descend into the renal medulla and then redeposit the NaCl and absorb water as they ascend into the renal cortex. (Figure 24.22, p. 972)

▷ **Play animation on the countercurrent exchanger: Figure 24.22.**

  ○ The result of the medullary osmotic gradient in the presence of ADH is continued water reabsorption and a low volume of concentrated urine. (Figure 24.23, p. 973)

▷ **Play animation on the formation of concentrated urine: Figure 24.23.**

Putting It All Together: The Big Picture of Renal Physiology 974

- The big picture of renal physiology is shown in Figure 24.24 (p. 975).

▷ **Play animation on the big picture of renal physiology: Figure 24.24.**

Urine and Renal Clearance 974

- Urine is composed of mostly water with solutes such as electrolytes and metabolic wastes.

- Renal function may be assessed by measuring the rate at which certain substances are removed from the blood by the kidney, known as **renal clearance,** as a way to estimate GFR. Commonly measured substances include creatinine and inulin.

Urine Transport, Storage, and Elimination 976

- The ureters, urinary bladder, and urethra comprise the **urinary tract.** (Figure 24.25, p. 977)

🖑 **Practice art-labeling exercise: Figure 24.25.**

- The **ureters** are muscular tubes that undergo peristalsis to propel urine toward the urinary bladder.

- The wall of the **urinary bladder** consists of three layers: the inner **mucosa,** made of transitional epithelium; the middle **detrusor muscle;** and the outer **adventitia.** The inferior portion of the urinary bladder is called the **trigone.**

- The **urethra** contains two sphincters: the involuntary **internal urethral sphincter** and the voluntary **external urethral sphincter.** The longer male urethra contains three regions and serves as a conduit for both urine and semen. The shorter female urethra serves as a passageway only for urine. (Figure 24.26, p. 978)

- **Micturition** is the process whereby urine is voided from the urinary bladder through the urethra. It occurs via the **micturition reflex,** mediated by the parasympathetic nervous system, though typically under voluntary control. (Figure 24.27, p. 979)

ASSESS WHAT YOU LEARNED

See answers in Appendix A.

LEVEL 1 Check Your Recall

1. What are the four main organs of the urinary system?

2. Which of the following is not a physiological process carried out by the kidneys?
   a. Blood pressure regulation
   b. Tubular reabsorption
   c. Tubular secretion
   d. Glomerular filtration
   e. All of the above are physiological processes carried out by the kidneys.

3. Mark the following statements as true or false. If a statement is false, correct it to make a true statement.
   a. The kidneys are retroperitoneal and covered by three layers of connective tissue.
   b. Internally, the kidneys consist of an outer renal medulla, a middle renal pelvis, and an inner renal cortex.

c. The first capillary bed of the kidneys is the peritubular capillaries, which are fed by the afferent arteriole and drained by the efferent arteriole.

d. Filtrate flows from the renal corpuscle to the distal tubule, the nephron loop, the proximal tubule, and into the collecting system.

4. Cortical and juxtamedullary nephrons differ in the:

a. lengths of their nephron loops.

b. structure of the capillaries surrounding them.

c. structure of their renal corpuscles.

d. Both a and b are correct.

e. Both b and c are correct.

5. Describe the structure of the filtration membrane.

6. Which of the following substances would pass through the filtration membrane to become part of the filtrate under normal circumstances? (Circle all that apply.)

a. Sodium ions   e. Leukocytes

b. Albumin   f. Amino acids

c. Glucose   g. Urea

d. Erythrocytes

7. Fill in the blanks: Glomerular hydrostatic pressure _____ filtration; colloid osmotic pressure and capsular hydrostatic pressure _____ filtration.

a. favors; favor   c. favors; oppose

b. opposes; oppose   d. opposes; favor

8. Fill in the blanks for the following statements:

a. When the GFR decreases, the macula densa releases chemicals to _____ the afferent arteriole.

b. The sympathetic nervous system _____ the blood vessels supplying the kidney to _____ the glomerular filtration rate.

c. The enzyme _____ is released by JG cells in response to a decrease in the GFR.

d. The enzyme _____ converts angiotensin-I to angiotensin-II.

e. Generally, angiotensin-II _____ systemic blood pressure while _____ the GFR.

9. Which of the following is *false* about the GFR?

a. The GFR averages about 120 ml/min.

b. The GFR increases when the afferent arteriole dilates.

c. The GFR decreases when the efferent arteriole constricts.

d. The GFR decreases when the afferent arteriole constricts.

10. The route by which substances are reabsorbed by crossing through the cells of the renal tubule and collecting system is known as the:

a. paracellular route.

b. transcellular route.

c. primary active transport route.

d. secondary active transport route.

11. Mark the following statements as true or false. If a statement is false, correct it to make a true statement.

a. Sodium ions and glucose are cotransported into the proximal tubule cell by secondary active transport.

b. The distal tubule reabsorbs sodium ions and secretes potassium and hydrogen ions in response to ADH.

c. Sodium ion reabsorption creates a gradient that helps drive the reabsorption of water and many other solutes from the proximal tubule.

d. ADH triggers water reabsorption from the nephron loop.

e. Obligatory water reabsorption occurs in the distal tubule and collecting system.

12. Dilute urine is produced when decreased levels of _____ are secreted:

a. aldosterone

b. atrial natriuretic peptide

c. ADH

d. none of the above

13. Which of the following conditions does not contribute to the creation and/or maintenance of the medullary osmotic gradient?

a. The countercurrent exchanger of the vasa recta

b. The countercurrent multiplier of the nephron loops of cortical nephrons

c. The countercurrent multiplier of the nephron loops of juxtamedullary nephrons

d. The permeability of the medullary collecting system to urea and other ions

14. Fill in the blanks: The kidneys produce _____ urine when the osmolarity of the body's fluids increases. They produce _____ urine when the osmolarity of the body's fluids decreases.

15. Normal urine should have which of the following properties? Circle all that apply.

a. Translucency

b. Yellowish pigment

c. Cloudy appearance

d. pH less than 4.5

16. The GFR may be estimated by measuring the rate at which certain substances are removed from the blood, which is known as:

a. renal clearance.

b. plasma creatinine.

c. glomerular hydrostatic pressure.

d. inulin estimation.

17. Fill in the blanks for each of the following statements:

a. The process by which urine is eliminated is called _____, and it is mediated by reflexes involving the _____ nervous system.

b. The mucosa of the organs of the urinary tract is lined with _____ epithelium.

c. The three layers of smooth muscle in the urinary bladder are known as the _____ muscle.

d. The female urethra provides a passageway for _____, whereas the male urethra provides a passageway for _____ and _____.

LEVEL 2 Check Your Understanding

1. Predict the effects the following scenarios would have on glomerular filtration:

   a. Having excess proteins in the blood, increasing colloid osmotic pressure
   b. Having low arterial blood pressure (hypotension)
   c. Having high arterial blood pressure (hypertension)

2. Trace the pathway taken by a molecule of urea through the kidney from the glomerulus to the renal pelvis if the urea is recycled.

3. Explain why urinary tract infections, which involve the urethra and urinary bladder, are much more common in females than males.

4. Why must the kidneys establish a concentration gradient in the interstitial fluid of the renal medulla in order to produce concentrated urine?

LEVEL 3 Apply Your Knowledge

PART A: Application and Analysis

1. Drugs that treat hypertension, or high blood pressure, have the following actions. Discuss the potential effects of each of these drug actions on the functions of the kidneys. Might any of these drugs cause adverse effects? If so, what are the potential adverse effects of each drug type?

   a. Blocking the action of aldosterone on the kidneys
   b. Blocking the receptor for angiotensin-II on blood vessels, including the vessels of the kidney
   c. Blocking the $Na^+/Cl^-/2K^+$ transport pumps in the thick ascending limb of the nephron loop

2. Mr. Wu is a patient with kidney disease who presents to your clinic for monitoring. You notice on his chart that his GFR was estimated through inulin administration to be about 35 ml/min. What does this tell you about the health of his kidneys? Mr. Wu is taking a medication that is normally excreted from the body in the urine. You order blood work and find that the concentration of this medication in his plasma is much higher than normal. How does his decreased GFR explain the elevated level of medication in his plasma?

3. Deana is a 4-year-old girl with a rare genetic defect that causes the $Na^+/$glucose symporters in the proximal tubule to reabsorb less glucose and sodium ions than normal. Predict the effects this defect will have on the composition and volume of Deana's urine. Explain why you would expect to see increased activity of the tubuloglomerular feedback and the renin-angiotensin-aldosterone systems in Deana's kidneys.

PART B: Make the Connection

4. Explain how each of the drugs in question 1 from this section would lower blood pressure. *(Connects to Chapter 17)*

5. What might it mean if you found a high concentration of urobilinogen in your patient's urine? (*Hint:* Consider the source of urobilinogen.) *(Connects to Chapter 19)*

25

Fluid, Electrolyte, and Acid-Base Homeostasis

You have read about the topics of this chapter—body fluids, electrolytes, and acids and bases—multiple times already in this book. We looked at the roles of water in the body in the chemistry chapter (see Chapter 2), the roles of electrolytes with muscle and nervous tissue (see Chapters 10, 11, and 17), acid-base homeostasis with the respiratory system (see Chapter 21), and all three with the digestive and urinary systems (see Chapters 22 and 24). Now it's time to examine these three critical homeostatic variables on their own so that you can see the overall picture of how they affect the body and how the body works to maintain them within their normal ranges. The upcoming modules survey body fluids, electrolytes, and acid-base balance in detail, including their effects on homeostasis in the presence of an imbalance. The chapter concludes with an example of how all three variables are regulated together.

MODULE **25.1**
Overview of Fluid, Electrolyte, and Acid-Base Homeostasis

Learning Outcomes

1. Explain the concept of balance with respect to fluids and electrolytes and acids and bases.
2. Define the terms body fluid, electrolyte, acid, base, pH scale, and buffer.

Fluids, electrolytes, and acids and bases are not exactly new to you at this point in your study of anatomy and physiology. Still, it's helpful to quickly review the basic principles of these three variables. This module provides such a review before we delve into the details of their regulation. You may also wish to refer back to the chemistry chapter for more background information about these concepts (see Chapter 2).

Body Fluids

The term body fluids is used to encompass all of the body's water-based liquids, including blood plasma, interstitial fluid, cytosol, cerebrospinal fluid, lymph, exocrine secretions, and other specialized fluids. Each of these fluids has a slightly different composition, but the main component of each is water. As water is their main component, the study of **fluid balance**—maintaining the appropriate volume and concentration of the body's intracellular and extracellular fluids—is largely the study of water balance.

Photo: Colored scanning electron micrograph (SEM) showing a calcium phosphate crystal, which can form in the body when phosphate ions are present in large numbers.

The functions of water in the body's fluids are diverse. As we've described, water is a polar solvent, meaning that compounds with ionic and polar covalent bonds are able to dissolve in it. This allows water to transport and distribute a large number of solutes throughout the body. Water also distributes body heat, cushions organs and tissues, and lubricates organs and tissues as they move.

Water obeys the *principle of mass balance*—in other words, what is gained by the body must equal what is lost by the body. Multiple factors impact fluid balance, such as the amount of water ingested, the amount of physical activity, kidney function, medications, and digestive activities. When an imbalance results in inadequate or excess water in the body, the body generally compensates, but if it cannot, serious disturbances to homeostasis can occur. As you will see in the modules that follow, such disturbances affect more than just body fluids—they impact electrolyte and acid-base balance, too, so their effects are wide-ranging.

Quick Check

☐ 1. What is a body fluid?
☐ 2. What is balance with respect to body fluids?

Electrolytes

◀◀ **Flashback**

1. How is an electrolyte defined? (p. 49)
2. What are the main electrolytes in the body? (p. 49)

As we've discussed, an **electrolyte** is a substance that dissociates into ions when placed in water, whereas a **nonelectrolyte** is a substance that generally has covalent bonds and does not dissociate into charged particles in water. Electrolytes are named for the fact that the resulting charged particles conduct electricity in a solution of water (**Figure 25.1**). This is, of course, why you are warned to never allow electrical devices to come into contact with tap water when you're in it. Tap water contains electrolytes, and will therefore conduct an electrical current from the device. Your body also contains a large number of electrolytes, and so this electrical current will flow from the device to the tap water and finally into your body, possibly inflicting a fatal injury.

Throughout this book, you have seen the importance of several electrolytes, particularly sodium, potassium, and calcium ions. Note that we have to call them "ions" because if we didn't include that term, we would be referring to the atomic, metal form. The difference is significant—in their atomic forms, sodium and potassium metals both explode when placed in water, and calcium metal causes a reaction that emits hydrogen gas. In their ionic forms, they simply dissolve into the water of the body's fluids, where they have a number of important roles. Other examples of electrolytes include chloride ions, acids, bases, and some proteins.

Like water balance, *electrolyte balance* in the body is maintained through the principle of mass balance, by which the amount of electrolytes gained through the diet is equal to that lost from the body via various routes. There are numerous mechanisms, most of which are hormonal, that act to maintain a state

Electrolyte solution:

$$NaCl \xrightarrow{\text{in water}} Na^+ + Cl^-$$

The bulb is lit with the NaCl solution, indicating that this solution conducts an electric current.

Nonelectrolyte solution:

$$C_6H_{12}O_6 \xrightarrow{\text{in water}} C_6H_{12}O_6$$

The bulb is not lit with the glucose solution, indicating that this solution does not conduct an electric current.

Figure 25.1 Comparison of electrical conduction in electrolytes and nonelectrolytes.

of electrolyte balance. Consequently, the numbers of these ions in the blood stay relatively stable. But equally important to the total number of ions in the body is their concentration, which depends not only on the number of ions but also on the amount of water in the body. For this reason, fluid balance is a critical factor that helps to determine electrolyte balance. We will examine how this works, and the effects of imbalances, in Module 25.3.

Quick Check

☐ 3. How does an electrolyte differ from a nonelectrolyte?
☐ 4. What is electrolyte balance?

Acids, Bases, and pH

◀◀ **Flashback**

1. How are acids and bases defined? (p. 48)
2. What is the pH scale? (p. 48)

As we've covered, an **acid** is a chemical that dissociates in water to release a hydrogen ion (H^+), as in the following reaction that uses HCl as an example:

$$HCl \rightarrow H^+ + Cl^-$$

Hydrochloric acid   *Hydrogen ion*   *Chloride ion*

Common acids in the human body include hydrochloric acid (HCl) in the stomach and carbonic acid (H_2CO_3) in the blood. Hydrogen ions released from acids play many important physiological roles, such as digestion of ingested food, inactivation of bacteria and other pathogens, and intracellular digestion by lysosomes. In addition, recall that creation of a hydrogen ion gradient is a critical component of ATP synthesis by mitochondria (see Chapter 23).

A **base,** or **alkali,** however, is a chemical that accepts a hydrogen ion in a solution, such as the following:

$$NaOH \quad + \quad HCl \quad \rightarrow \quad NaCl \quad + \quad H_2O$$

Sodium   Hydrochloric   Salt   Water
hydroxide (base)   acid

The result of this reaction is generally a salt and a molecule of water. Most bases that you encounter in the laboratory have an electrolyte bound to OH^-, the *hydroxide ion.* However, in the human body, the most common base is the *bicarbonate ion,* or HCO_3^-. Bicarbonate and certain other bases are part of *buffer systems,* meaning that they bind hydrogen ions released from acids to resist dramatic swings in the hydrogen ion concentration of body fluids.

Recall that the hydrogen ion concentration of a solution is measured by the **pH scale.** As you've read, an increased hydrogen ion concentration results in a solution with a lower pH. Conversely, a solution with a lower hydrogen ion concentration has a higher pH. Solutions with a pH less than 7 are considered acidic, those with a pH greater than 7 are considered basic, and those with a pH of exactly 7 are considered neutral (see Chapter 2).

Quick Check

☐ 5. How do acids and bases differ?
☐ 6. Which pH values are considered acidic, basic, and neutral?

Apply What You Learned

☐ 1. A patient suffers from renal failure and so produces less urine. If this patient increased his water intake, what state of fluid balance or imbalance would exist? Why?
☐ 2. Would you expect this same patient's electrolytes to be balanced or imbalanced? Explain.

See answers in Appendix A.

MODULE **25.2**
Fluid Homeostasis

Learning Outcomes

1. Describe the fluid compartments, and explain how each contributes to the total body water.
2. Compare and contrast the relative concentrations of major electrolytes in intracellular and extracellular fluids.
3. Explain how osmotic pressure is generated, and compare and contrast the roles that hydrostatic and osmotic pressures play in the movement of water between fluid compartments.
4. Describe the routes of water gain in and loss from the body.
5. Describe the mechanisms that regulate water intake and output, and explain how dehydration and overhydration develop.

You've probably heard the oft-repeated advice that the average person should consume at least eight glasses of water (at 8 fluid ounces each) per day. This is widely reported as fact by many media outlets, websites, and diet and health books. However, this value equates to nearly 2 liters of pure water per day, which is quite a lot of water. Other claims about water and hydration have been made, such as its purported health benefits, its ability to promote weight loss, and the effects of caffeine on hydration. But are any of these claims true? And just how much water does a person really need to consume every day?

This module answers these questions as it examines the most abundant chemical compound in the human body. We first discuss the amount of water in the body and where it is located. Next, we revisit a basic principle we have seen many times in this book: water movement due to hydrostatic pressure and osmosis. Finally, we address water gains and losses and the mechanisms by which fluid balance is regulated.

Total Body Water

You may have heard or read that the human body is composed of about 60% water. However, this value refers to the amount of water in the "standard man," a reference point that considers the standard values for a 20- to 30-year-old 70-kg (154-lb) male of European descent. For this standard man, the amount of water in the body is about 42 kg (92.5 lb). One kilogram of water is equal to one liter, so this equates to 42 liters (11 gal) of water. This volume represents a value known as the **total body water.** To visualize how much water this is, collect 21 two-liter bottles of soda. The amount of liquid in those bottles is equal to the total body water for this hypothetical individual.

Of course, most people don't fit the profile of the standard man, and so total body water differs from this value. An individual's body mass, age, gender, and amount of adipose tissue all affect total body water. For example, women tend to have a lower percentage of total body water than men because they have a higher percentage of body fat, which contains less water than other tissues (and because men generally have more muscle, a tissue that contains a lot of water). In addition, infants of both genders have a higher total body water, averaging about 65%, whereas individuals over 60 years of age average a total body water of about 52% for men and 46% for women.

Quick Check

☐ 1. What is total body water?
☐ 2. What factors affect total body water?

Fluid Compartments: Intracellular and Extracellular Fluids

There are two basic places, or *compartments,* where we find body fluids: inside the cells, the **intracellular compartment,** and the area outside the cells, the **extracellular compartment.** The intracellular compartment isn't a true single compartment, but is instead trillions of microscopic compartments inside the plasma membranes of our cells. The fluid within the intracellular compartment is, of course, **cytosol** (or *intracellular fluid* [*ICF*]), which accounts for about 60% of the body's fluids, or about 26 liters. The extracellular compartment is filled with

extracellular fluid (ECF), made up of a variety of fluids that you have studied throughout this book. Two of the fluids that we have discussed at length are *plasma,* the fluid portion of the blood, and *interstitial fluid,* the fluid in the spaces between the cells. Plasma accounts for approximately 8% (3 liters) of the total body water, and interstitial fluid for about 32% (13 liters) (**Figure 25.2**).

The solute composition of plasma and interstitial fluid is similar—both contain various types of ions in similar amounts. The main difference in the composition of the two fluids is in protein content: Plasma has a much higher protein content than interstitial fluid, which has virtually no proteins. As you can see in **Figure 25.3**, however, the solute composition of the ECF and the cytosol varies starkly. The concentrations of sodium, chloride, calcium, and bicarbonate ions are much higher in the ECF than in the cytosol. In contrast, proteins and potassium, magnesium, sulfate, and monohydrogen phosphate (HPO_4^{2-}) ions have a much higher concentration in the cytosol than in the ECF. The concentration gradients of these solutes are due to factors such as particle size and the reliance of certain solutes on transport proteins or pumps to cross the membrane. Many of these concentration gradients, particularly those of sodium, potassium, and calcium ions, play important physiological roles.

Quick Check

☐ 3. Where are the intracellular and extracellular compartments located?
☐ 4. How do the compositions of cytosol, interstitial fluid, and plasma differ?

Figure 25.2 The distribution of water in the body.

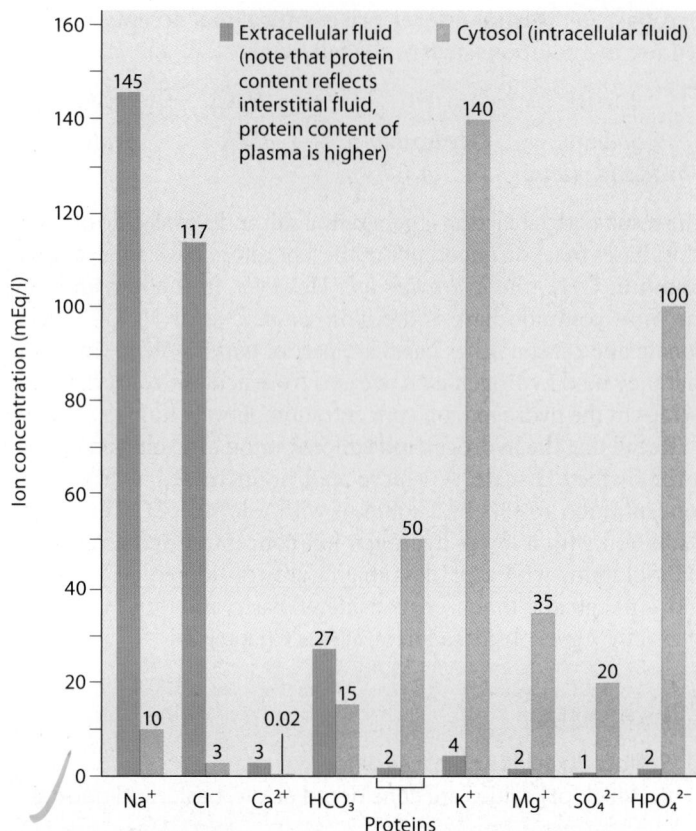

Figure 25.3 The solute composition of extracellular fluid and cytosol.

Osmotic Movement of Water between Compartments

◀◀ **Flashback**

1. What are osmotic and hydrostatic pressures? In which directions do they drive water movement? (p. 690)
2. In which direction does water move during osmosis? (p. 76)
3. What is tonicity? (p. 77)

Although solute movement between fluid compartments is restricted, water generally moves freely between them. We have discussed this before in this book, so let's just quickly review. Recall from the blood vessels chapter (see Chapter 18) that the direction in which water moves between compartments is influenced by gradients involving two pressures: hydrostatic pressure and osmotic pressure, another example of the Gradients Core Principle (p. 26). The hydrostatic pressure gradient, or the force that the fluid exerts on cells, 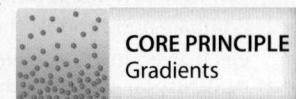 **CORE PRINCIPLE** Gradients tends to push water away from an area of higher hydrostatic pressure to one with lower hydrostatic pressure. The osmotic pressure gradient, or the force of solutes in a solution, tends to pull water toward the solution with the higher osmotic pressure, or the more concentrated solution, by osmosis. A solution's osmotic pressure is determined by its *osmolarity,* or the number of solute particles present in the solution. Now let's see how these two pressures affect fluid movement between compartments.

First, consider water movement between the plasma and interstitial fluid. In most blood vessels, the high hydrostatic pressure at the arteriolar end dominates, pushing water out of the capillary and into the interstitial fluid. At the venular end, the high osmotic pressure dominates, pulling most of this water back into the vessel by osmosis. The small amount of water lost to the interstitial fluid is returned to the plasma via the lymphatic system (look again at Figure 20.3 for a review).

Now consider water movement between the plasma and interstitial fluid: (ECF) and the cytosol (ICF), which is illustrated in **Figure 25.4**. Notice that under normal conditions, the ECF's hydrostatic and osmotic pressures are very close to those of the ICF, so not much of a gradient exists between the two compartments and there is no significant net movement of water between them. However, you can see that if water is gained by the plasma (see Figure 25.4a) or lost from it (see Figure 25.4b), its hydrostatic and osmotic pressures change. This in turn influences how much water leaves the plasma and enters the interstitial fluid, and so the hydrostatic and osmotic pressures of interstitial fluid change, too. The change in the osmotic pressure of interstitial fluid creates a gradient and has a direct influence on the volume of the cytosol. Let's see how this works.

In the cell chapter, you read about **tonicity**—the measurement of the osmotic pressure gradient between two fluid compartments (see Figure 3.9). As we just described, the ECF and cytosol generally have the same osmotic pressure and are *isotonic*, leading to no net movement of water between the two compartments. However, when the ECF's osmotic pressure changes, water movement does occur, as follows:

- **Hypotonic ECF causes a cell to gain water.** The ECF is *hypotonic* to the cytosol when its osmotic pressure is lower than that of the cytosol, as you can see in Figure 25.4a. The cytosol now has the higher osmotic pressure, which pulls water molecules from the interstitial fluid and plasma into the cell. As a result, the cell swells.
- **Hypertonic ECF causes a cell to lose water.** The ECF is *hypertonic* to the cytosol when its osmotic pressure is higher than that of the cytosol, as shown in Figure 25.4b. This exerts a "pull" on the water molecules in the cytosol, and water moves out of the cytosol and into the interstitial fluid and then plasma. As a result, the cell shrinks.

A variety of conditions can cause cells to be exposed to hypertonic and/or hypotonic solutions. The most common cause is simply varied fluid intake; drinking a large volume of water makes the plasma temporarily hypotonic to the cytosol, and losing water via sweating without replacing it makes the plasma temporarily hypertonic to the cytosol.

Quick Check

☐ 5. In which direction(s) do hydrostatic pressure and osmotic pressure move water?
☐ 6. If a cell is in interstitial fluid that is hypertonic to its cytosol, how does this affect the cell? Why?
☐ 7. What happens if the interstitial fluid is hypotonic to the cytosol of the cell?

Figure 25.4 Fluid movements between compartments.

Water Losses and Gains

◀◀ **Flashback**

1. What does oxygen become after it accepts the final electrons in the electron transport chain? (p. 907)
2. What effect does angiotensin-II have on thirst and the amount of water in the body? (pp. 956–957)
3. What are baroreceptors? (p. 672)

In the module introduction, we posed a common question: How much water do you need to consume every day? To answer this, we must first find out how water is lost and gained from the body. The next two subsections focus on these topics, and the third answers the question.

Factors That Influence Water Loss

Water is lost from the body in several ways, but most is lost via the kidneys in urine (**Figure 25.5a**). A certain amount of urine, called the **obligatory water loss,** must be produced each day irrespective of fluid intake. This is explained by the fact that the body must excrete a minimum amount of solutes to prevent buildup of toxic molecules and electrolyte imbalances. The kidneys are not capable of producing urine with an osmolarity greater than 1200 mOsm, so a certain amount of water must be excreted with the solutes. The obligatory water loss for most people is about 500 ml, even if a person has not taken in any water at all. This water loss is why a human stranded without water is in trouble—the kidneys must continue to produce this volume of urine, and without water to replace the loss, a human can succumb to dehydration in a matter of days.

Normally, of course, we do consume liquids throughout the day, and so urine output is about 1500 ml, greater than obligatory water loss. In addition to urine, a small amount of water, about 100 ml, is also lost in the feces. Together, the water loss from these two sources is known as **sensible water loss** because we normally notice or "sense" it happening. About 600 ml is lost from the skin due to sweating and evaporation, and another 300 ml is lost during expiration of humidified air. We normally don't notice when this water loss occurs, and for this reason it is called **insensible water loss** (the degree to which sweating is insensible, however, depends on the relative humidity—it can be considered sensible at higher humidity levels). Between the sensible and insensible water losses, most people will lose about 2.5 liters of water per day. This value fluctuates with water intake, physical activity, food intake, and other factors.

Factors That Influence Water Gain

As you can see in **Figure 25.5b**, the body gains water from three main sources. The first source is our own catabolic reactions. Recall that when oxygen molecules accept the final pair of electrons in the electron transport chain, a molecule of water is formed. Water is also formed as a byproduct of several other catabolic reactions. Together, water from all these reactions is called **metabolic water** (or *water of oxidation*), and it amounts to a total gain of about 250 ml per day. The other two sources are the intake of water via food and liquid in the diet. Water from food totals about 750 ml per day, and water from liquids about 1500

Figure 25.5 Water losses and gains.

ml per day. All three sources combined generally equal 2.5 liters, or about the same amount of water lost every day.

The mechanism that drives water intake from liquids is known as the **thirst mechanism,** which is controlled by the hypothalamus. Many stimuli can trigger thirst, but by far the most potent is an increase in the osmolarity of the plasma. As you can see in **Figure 25.6a**, changes in plasma osmolarity are detected by neurons called **osmoreceptors,** located in the *thirst center* of the hypothalamus. An increase in the solute concentration of the ECF causes water to move out of the osmoreceptors, which triggers the sensation of thirst. This mechanism is very sensitive, as an increase in plasma osmolarity of even 2–3% is enough to initiate thirst.

When plasma osmolarity decreases, water moves back into the osmoreceptors and thirst is inhibited. Interestingly, however, we feel relieved of thirst as soon as we ingest liquids, well before they are absorbed and have the chance to affect osmoreceptors. This is believed to be due to receptors in the pharynx that detect the ingestion of liquid. Such a response is critical to preventing excessive liquid intake. If the liquid were not detected until ECF osmolarity decreased to the normal range—which takes about 30–60 minutes—one could continue drinking and overhydration could result.

Other stimuli that can trigger thirst include a decreased plasma volume and the hormone angiotensin-II (**Figure 25.6b**). A decrease in plasma volume, which is generally accompanied by a reduction in blood pressure, is detected by baroreceptors in several arteries. It is a much less potent stimulus for thirst, as a decrease of 10–15 mm Hg is required. Decreased plasma volume

STIMULUS

Plasma osmolarity increases above its normal range.

Normal range

RECEPTOR

Osmoreceptors in the hypothalamus detect the change in osmolarity.

Hypothalamus

EFFECTOR/RESPONSE

Hypothalamus triggers thirst—fluid intake increases.

IN HOMEOSTATIC RANGE

Receptors detect fluid intake; plasma osmolarity returns to normal; thirst is relieved.

Normal range

CONTROL CENTER

Osmoreceptors stimulate the thirst center in the hypothalamus.

Thirst center in hypothalamus

(a) Control of thirst response due to increased plasma osmolarity by a negative feedback loop

STIMULUS

Plasma volume decreases below its normal range.

Normal range

RECEPTOR

Baroreceptors in the arteries detect the change in blood pressure due to decreased plasma volume.

Juxtaglomerular (JG) cells in nephrons detect the change as well, activating the renin-angiotensin-aldosterone system (RAAS).

Baroreceptors

Heart

JG cells

EFFECTOR/RESPONSE

Hypothalamus triggers thirst—fluid intake increases.

IN HOMEOSTATIC RANGE

Receptors detect fluid intake; plasma volume returns to normal; thirst is relieved.

Normal range

CONTROL CENTER

Baroreceptors and angiotensin-II from RAAS stimulate the thirst center in the hypothalamus.

Thirst center in hypothalamus

(b) Control of the thirst response due to decreased plasma volume by a negative feedback loop

Sequence the steps
@MasteringA&P®

Figure 25.6 Maintaining homeostasis: regulation of thirst by negative feedback loops.

also stimulates the juxtaglomerular cells in the kidneys and initiates the *renin-angiotensin-aldosterone system* (*RAAS*), which culminates in the formation of angiotensin-II. Both mechanisms further enhance the thirst sensation.

There is a common claim that by the time the thirst mechanism is activated, you are already dehydrated. Proponents therefore advise that an individual consume water nearly continuously throughout the day in order to prevent dehydration. This is, of course, not scientifically accurate; we have said that it takes only a 2–3% change in the osmolarity of the ECF to stimulate thirst. However, there are occasions when we need to continuously administer fluids. In such cases in the medical field, it is common to use an *intravenous line*, discussed in *A&P in the Real World: Intravenous Fluids.*

How Much Water Do We Need to Drink?

Now we have enough information to answer the question of how much water a person should drink. Simply put, water intake should balance water loss. Since on average a person loses about 2.5 liters per day, then water gains should equal about 2.5 liters per day. On the surface, this might seem to support the "eight 8-oz glasses of water" advice. However, recall from Figure 25.5b that we usually obtain about 750 ml of water from the foods that we eat, and we produce about 250 ml of water from oxidative catabolism. So this leaves only about 1.5 liters of water that we need to drink. Of course, this amount will vary with activity. A person engaged in strenuous physical activity can lose 4–5 liters of fluid through sweating, and naturally this volume needs to be replaced to prevent serious imbalances.

Many claim that there are health benefits to consuming more water than we need. The claims seem logical on the surface, but are they true? Several studies have examined this question

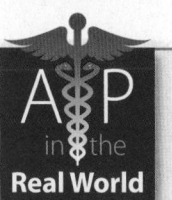

Intravenous Fluids

Sometimes it's necessary to bypass the gastrointestinal system to get fluids into an individual rapidly. In these cases, a medical professional generally starts an **intravenous line,** often just called an **IV.** With this line, fluids can be administered directly into a vein, allowing them to have a nearly immediate effect on the body.

An IV consists of a flexible catheter that is inserted into a patient's vein, with an external port onto which bags of different fluids can be attached. Fluid from the bag drips into a chamber and then into a line at a specified rate, allowing the rate of fluid administration to be controlled. Most often, an IV is placed in a vein in the upper limb, which is known as a *peripheral line.* However, when intravenous access is required over the long term, a larger port may be inserted in a vein such as the subclavian or internal jugular vein. These lines are known as *central lines.*

and found that there were no noticeable benefits in individuals who consume more water than needed to replace normal losses. The individuals weren't harmed by the increased water intake but were not benefited by it, either. Many also claim that consuming excess water will help a person lose weight. This is also unfortunately not scientifically accurate—water intake does stretch the stomach, but it does not trigger the signals that make us feel full.

Quick Check

☐ 8. How is thirst stimulated?
☐ 9. How are fluids lost from the body?
☐ 10. What are the water requirements for an average person?

Hormonal Regulation of Fluid Balance

◄◄ **Flashback**

1. What is antidiuretic hormone, and what are its actions on the kidneys? (p. 594)

The mechanisms that lead to water loss and gain are imprecise. You can't always take a drink when you're thirsty, and sometimes you might drink liquid when you're not thirsty. For these reasons, the body needs ways to ensure that water input matches water output. There are four main hormones that perform this task. By far the most important of these hormones is **antidiuretic hormone (ADH).** We have discussed ADH and the other three hormones that affect fluid balance—angiotensin-II, aldosterone, and atrial natriuretic peptide (ANP)—several times in this book (see Chapters 16 and 24).

ADH is produced by the hypothalamus and released in the posterior pituitary gland. Its main target tissues are the cells of the distal tubule and collecting system of the kidney. It triggers the insertion of water channels called *aquaporins* into their plasma membranes, which allows for the reabsorption of water from the kidneys back into the ECF by osmosis. When the ADH level rises, more water is reabsorbed, the ECF volume increases, and urine production decreases. Conversely, when the ADH level falls, less water is reabsorbed, the ECF volume decreases, and urine production increases. When ADH is at its maximal secretion, the only water lost to the urine is that due to obligatory water loss.

Of course, ADH secretion by itself isn't the whole story of water reabsorption in the collecting system. Recall from the urinary system chapter that even with large numbers of aquaporins in the plasma membranes of tubule cells, osmosis will not occur unless a concentration gradient is present to drive the process. This gradient is set up and maintained by the countercurrent system within the nephron loops of juxtamedullary nephrons (look at Figure 24.21 for a review of this concept).

Quick Check

☐ 11. What is the role of ADH in fluid balance?
☐ 12. How is ADH secretion stimulated?

Imbalances of Fluid Homeostasis

The amount of water in your body may increase and decrease throughout the day. Generally, the body compensates for these changes by altering urine output and there is no noticeable disturbance in fluid balance. However, some conditions can either overwhelm the compensatory mechanisms or prevent the body from compensating at all. Certain imbalances can lead to a change in the osmolarity of the ECF, whereas others produce an *isosmotic* state, meaning that the ECF has the same tonicity as the cytosol. In this section, we look at both types of imbalances more closely.

Dehydration: Decreased Volume and Increased Osmolarity

Dehydration is defined as decreased volume and increased concentration of the ECF. Common causes for dehydration include profuse sweating, prolonged diarrhea and/or vomiting, certain endocrine conditions, and overuse of diuretics. In such cases, the water loss decreases the plasma volume, increasing its solute concentration and osmotic pressure. As we saw earlier in Figure 25.4, if water is lost from one compartment, then water from another compartment will move by osmosis into it, and the osmotic pressure of the two compartments will be equalized. For this reason, as the ECF osmolarity increases, water moves out of the cells by osmosis and the cells crenate. This not only causes a water imbalance within the cells but also leads to electrolyte imbalances, which are covered in the next module.

Initially, a person becoming dehydrated will experience dryness of the oral mucosa and thirst. This situation progresses to dryness of the skin and decreased urine production (called *oliguria;* ohl´-ih-GYU-ree-ah); eventually, fever may develop and organ function is impaired. The condition, and its associated electrolyte imbalances, is fatal if hydration is not restored.

Dehydration may be treated by consuming pure water, but this should be approached with caution. If water is consumed in moderate amounts, it will mildly dilute the ECF, making it slightly hypotonic to the cytosol. This will trigger a shift of water back into the cells, rehydrating them. However, if large amounts of water are consumed in a short period of time, the ECF can become overly dilute, causing too much water to enter the cells and resulting in cellular swelling. For this reason, treating dehydration with sports drinks is preferable in many cases, as sports drinks contain glucose and electrolytes and so are hypertonic to pure water.

Overhydration: Increased Volume and Decreased Osmolarity

Overhydration, also known as **hypotonic hydration,** results when the volume of the ECF increases and its osmotic pressure falls. Cells exposed to the hypotonic ECF swell as water enters the cells by osmosis. Normally, when water is consumed, hypotonic hydration is prevented by a decline in ADH production, which leads to loss of the excess water in the urine. However, this mechanism can fail when renal function is impaired, ADH is secreted abnormally, or extreme amounts of water are consumed in a short time period (a condition called *water toxicity*).

Hypotonic hydration results in electrolyte imbalances due to dilution of the ECF. Particularly affected is the sodium ion concentration, which can fall dramatically, leading to *hyponatremia.* Together, the hyponatremia and cellular swelling can cause severe disturbances to homeostasis, particularly to the central nervous system. The swelling of neurons, or *cerebral edema,* affects their ability to function and can lead to mental status changes, seizures, coma, and even death. Treatment consists of administering hypertonic fluids to draw the water back out of the cell and into the interstitial fluid.

Isosmotic Fluid Imbalances

Disruptions in homeostasis can result even when the ECF remains isotonic to the cytosol. For example, fluid depletion occurs in cases of blood loss, but the amount of solutes and water lost are the same. For this reason, the plasma's osmotic pressure is unchanged and there is no osmotic gradient to pull water into or out of a cell. This type of fluid imbalance is known as *hypovolemia,* and, if untreated, it can lead to shock, organ failure, and death due to extremely low blood pressure. Generally, if the blood products cannot be immediately replaced, isotonic fluids are administered to help restore ECF volume.

The same situation can exist with fluid excess, or *hypervolemia*—the ECF volume can increase without a significant change in osmotic pressure. A common example is **edema** (eh-DEE-mah), in which fluid accumulates in the interstitial fluid, causing swelling. As with hypovolemia, the fluid is isotonic to the cytosol, so the volume of the cells doesn't change. Edema can result from a number of causes, including high hydrostatic pressure, as seen with hypertension; a decrease in the return of venous blood to the heart; low colloid osmotic pressure; removal of lymphatic vessels; excess sodium ion retention; or release of inflammatory mediators. The excess interstitial fluid can make basic movements such as walking difficult and can also cause organ dysfunction. Edema may be managed by treating the underlying cause and by removing excess fluid through drugs such as diuretics.

Quick Check

☐ 13. How does dehydration affect the volume of the cytosol?
☐ 14. What are some causes of overhydration, and how does it affect the cells?
☐ 15. How do dehydration and overhydration differ from isosmotic fluid imbalances?

Apply What You Learned

☐ 1. Would you expect an organ such as the liver, whose cells are protein-rich, to contain more or less water than lipid-rich adipose tissue? Explain your answer.
☐ 2. A nurse has been working for 11 hours on an overnight shift. He is asked to start an intravenous line and hang a bag of isotonic 0.9% sodium chloride. He accidentally grabs a bag of hypertonic 9% sodium chloride instead. What is likely to happen to his patient? Why?
☐ 3. Under what condition might it be beneficial to administer hypertonic sodium chloride to a patient?

See answers in Appendix A.

MODULE **25.3**
Electrolyte Homeostasis

Learning Outcomes

1. Describe the function of the most prevalent electrolytes found in body fluids.
2. Describe the hormonal regulation of electrolyte levels in the plasma.
3. Explain how calcium ion regulation is related to phosphate ions.

Technically, an electrolyte is any molecule that ionizes in solution, but in the body we generally refer to electrolytes as salts that dissociate into ions such as sodium, potassium, calcium, magnesium, and chloride ions. The most critical systems for maintenance of electrolyte imbalance are the endocrine and urinary systems. The importance of preserving electrolyte balance cannot be overstated, as imbalances can very easily be fatal. In this module, we examine the roles and regulation of sodium, potassium, calcium, phosphate, chloride, and magnesium ions. We also look at the most commonly encountered electrolyte imbalances—those of sodium, potassium, and calcium ions.

Sodium Ions

◀◀ Flashback

1. What is the function of the Na^+/K^+ ATPase pump? (p. 79)
2. What role do sodium ions play in the action potential? (p. 399)

Sodium ions are the most abundant extracellular cation, measuring about 135–145 mEq/l (milliequivalents per liter, a common way to express electrolyte concentration); in contrast, the levels in the cytosol are very low. This is due in large part to two factors: (1) the Na^+/K^+ ATPase pump, which uses ATP to pump sodium ions out of the cytosol against their concentration gradient; and (2) the relatively low permeability of the plasma membrane to sodium ions, which prevents these ions from leaking back into the cytosol in large numbers. This high extracellular concentration of sodium ions makes them one of the most important types of osmotic particles in the ECF.

The steep sodium ion gradient between the cytosol and ECF also makes this ion critical for all of our electrophysiological processes. Recall that cells have a negative resting membrane potential—the inside of the cell is negative compared to the outside. When sodium ion channels open, the sodium ions follow their

Sequence the steps
@MasteringA&P®

Figure 25.7 Maintaining homeostasis: regulation of the extracellular concentration of sodium ions by a negative feedback loop.

concentration gradient and rush into the cell. This causes the cell to gain positive charges and *depolarize:*

Na⁺ inflow is responsible for depolarization.

Depolarization is a key event in the functioning of all excitable cells, including neurons and the three different types of muscle cells.

Due to the critical roles of sodium ions in many processes, an imbalance can have a significant impact on overall homeostasis. This section looks at how sodium ion balance is maintained and then what happens when the sodium ion concentration increases or decreases abnormally.

Regulation of Sodium Ion Concentration

You might be surprised to learn how little salt a person would need to ingest in one sitting to prove fatal. As little as 2 teaspoons can raise plasma sodium ion concentration by 10 mEq/l, and 2 tablespoons by 30 mEq/l, which can easily cause death. In the typical American diet, we don't usually ingest that much salt at one time, but we tend to consume a lot of it over the course of a day. Most people consume about 9 grams of salt in a day, which equals about 2 teaspoons. Yet in spite of this significant sodium ion intake, the concentration of sodium ions in the ECF remains stable, thanks to several mechanisms. This is critical not only to sodium ion balance but also to fluid balance, as water reabsorption in the kidneys depends on a gradient consisting largely of sodium ions in the interstitial fluid (see Chapter 24).

The two main hormones that increase sodium ion retention are *angiotensin-II* and *aldosterone* (**Figure 25.7**). A low extracellular sodium ion concentration is one of the major stimuli that trigger the RAAS. The low sodium ion concentration is detected by receptors in the macula densa of the nephron, which leads to the release of renin and the eventual formation of angiotensin-II. As we've covered, angiotensin-II increases the reabsorption of sodium ions from the proximal tubule. Angiotensin-II also stimulates aldosterone secretion, which causes sodium ion reabsorption from the distal tubule. Note that both actions also indirectly increase water reabsorption, as water follows sodium ions by osmosis. When secretion of both hormones decreases, the number of sodium ions (as well as the amount of water) reabsorbed decreases. Another hormone that influences sodium ion balance is *atrial natriuretic peptide* (*ANP*), which decreases sodium ion (and water) reabsorption. Together, these three hormones maintain a relatively stable plasma sodium ion concentration in the face of sometimes significant swings in sodium ion intake.

Sodium Ion Imbalances

The concentration of sodium ions in the body's fluids is determined by both the absolute number of sodium ions in the body fluids and the number of water molecules. So anything that abnormally increases or decreases the number of sodium ions and/or the number of water molecules can cause a sodium ion imbalance. This potential imbalance comes in two forms: *hypernatremia,* an elevated sodium ion concentration, and *hyponatremia,* a decreased sodium ion concentration.

ConceptBOOST »»»

Why Does the Amount of Water in the Body Affect the Sodium Ion Concentration?

You might be wondering how the amount of water in the body can affect the number of sodium ions in the ECF. The answer is that it actually doesn't have any impact on the absolute number of sodium ions. Instead, it changes the *concentration* of sodium ions in body fluids, meaning the percentage of sodium ions relative to water molecules. Let's look at a simplified example here. Say the following three beakers each contain solutions of NaCl with 10 sodium ions (for simplicity we'll omit the chloride ions) and a different number of water molecules:

Beaker A: 10% NaCl solution | Beaker B: 20% NaCl solution | Beaker C: 5% NaCl solution

Beaker A contains 90 water molecules, and so its 10 sodium ions give it a 10% sodium ion concentration. Beaker B, however, has only 40 water molecules. Now those same 10 sodium ions create a solution with a 20% sodium ion concentration. Finally, turn to beaker C—it has 190 water molecules, and so it has a 5% sodium ion concentration.

Now let's see how this applies to the human body. In this example, an individual with normal hydration is represented by the solution in beaker A. If that person were dehydrated, the blood would most closely resemble the solution in beaker B. So the individual would have hypernatremia without the number of sodium ions changing at all. Similarly, if the person were overhydrated, as in beaker C, he or she would develop hyponatremia, with no change in the number of sodium ions.

So, as you can see, the number of water molecules in the body can have a profound effect on the sodium ion concentration of the blood and body fluids without any change in the number of sodium ions. For this reason, sodium ion imbalances due to dehydration or overhydration can be corrected by simply adjusting the amount of water in the body. ■

Hypernatremia **Hypernatremia** (hy′-per-nah-TREE-mee-ah; *natri-* = "sodium") is defined as a sodium ion concentration in the blood above 145 mEq/l. The most common cause of hypernatremia is actually dehydration. Although hypernatremia can impact the depolarization of excitable cells, its most immediate, obvious effects are due to the osmotic imbalances that it causes. As occurs in most cases of dehydration, water moves out of the cell into the hypertonic ECF, leading to a reduced volume of cytosol and to cell crenation. This affects cells throughout the body, and can lead to the same symptoms we listed for dehydration. Treatment revolves around restoring hydration, which will normalize the sodium ion concentration.

Note that sometimes hypernatremia can be accompanied by overhydration. This occurs when the hypernatremia is due to rapid ingestion of excess table salt or increased secretion of aldosterone, both of which increase water retention by the kidneys. Rehydration will be little help in these cases; for this reason, it's critical for a clinician to determine the primary cause of the hypernatremia so the proper course of treatment can be initiated.

Hyponatremia **Hyponatremia** is defined as a plasma sodium ion concentration less than 135 mEq/l. Like hypernatremia, hyponatremia is often not related to the absolute number of sodium ions but, rather, is due to overhydration. One cause of hyponatremia with overhydration is hypersecretion of ADH, a condition known as *syndrome of inappropriate ADH secretion (SIADH)*. As with overhydration, hyponatremia can result in cellular swelling due to movement of water into the cytosol. It can also lead to problems with electrophysiology; specifically, it reduces the sodium ion gradient across the plasma membrane and for this reason can slow depolarization and so the rate of action potential generation. Hyponatremia due to overhydration is generally treated with hypertonic saline to restore the plasma sodium ion concentration and draw water back out of the cells.

Sometimes hyponatremia can exist with dehydration, which occurs with a primary loss of sodium ions and water molecules. This can be caused by prolonged vomiting and/or diarrhea, decreased aldosterone secretion, and diuretic overuse. In such cases, the osmolarity of the ECF is generally nearly isotonic to the cytosol, so isotonic saline is used instead of hypertonic saline.

Quick Check

☐ 1. What are the main roles of sodium ions in the body?
☐ 2. How is sodium ion concentration regulated?
☐ 3. How are sodium ions and fluid balance related?

Potassium Ions

◀◀ **Flashback**

1. What is the role of potassium ions in an action potential? (p. 399)
2. How do potassium ions contribute to generation of the resting membrane potential? (p. 395)

Whereas sodium ions are the most abundant extracellular cation, potassium ions are the most abundant intracellular cation. Indeed, the extracellular potassium ion concentration averages from 3.9 to 4.5 mEq/l, far lower than the average of 139 mEq/l in the cytosol. The concentration of potassium ions is higher in the cytosol of cells than in any other body fluids due primarily to the actions of the Na^+/K^+ pump. This steep concentration gradient, like that of sodium ions, is critical to the functioning of neurons and the three types of muscle cells in the body. The movement of potassium ions out of the cell down their concentration gradient through potassium ion channels is responsible for the repolarization phase of the action potential, shown here:

The repolarization phase is critical to the functioning of excitable cells because it returns cells to their resting membrane potential. A cell cannot be re-excited until it returns to this resting value. This means that unless repolarization occurs, a neuron cannot fire another action potential and muscle cells cannot contract. In short, without repolarization, the nervous system shuts down, the heart stops, and the muscles for breathing cease functioning.

Not only are potassium ions critical for action potentials but they are also the main factor responsible for generating the resting membrane potential. Recall that small numbers of potassium ions are constantly being lost from the cell to the ECF through leak channels. This loss of positive charge from the cytosol to the ECF helps to generate the negative resting membrane potential. As you've seen throughout this book, the negative resting membrane potential is the basis for essentially all of the body's electrophysiology.

We briefly discussed the effects of potassium ion imbalances on the action potentials of the heart (see Chapter 17). Let's take a more detailed look at how the potassium ion concentration of the body fluids is regulated and at the wide-ranging effects that can result from imbalances.

Regulation of Potassium Ion Concentration

Significant numbers of potassium ions can be added to the body fluid from the diet, particularly when you eat a meal rich in fruits and vegetables. Such foods could raise the potassium ion concentration to a lethal level if all the ions remained in the ECF. Fortunately, several mechanisms within the endocrine and urinary systems help to prevent this from happening. For one, hormones such as insulin, aldosterone, and epinephrine stimulate the uptake of potassium ions by cells. This shifts the newly ingested potassium ions into the intracellular compartment.

The remaining excess potassium ions must be excreted, which is mostly handled by the kidneys via the process of secretion. Recall that potassium ions in the filtrate are reabsorbed, not secreted, in the proximal tubule and the nephron loop (see Chapter 24). This makes potassium ion secretion possible only in the distal tubule and collecting system. As you have discovered, this is under the control of aldosterone, which triggers secretion of potassium ions (and hydrogen ions) and reabsorption of sodium ions. One of the more important stimuli for aldosterone release is a rising extracellular concentration of potassium ions. Conversely, when the extracellular potassium ion concentration falls, the secretion of aldosterone also generally falls, and more potassium ions are conserved. Note that aldosterone secretion is stimulated by several other factors that are unrelated to potassium ions, including decreased blood volume and blood pressure (and any other stimulus that triggers secretion of angiotensin-II) and a decreased blood pH. Therefore, potassium ion imbalances can result when these conditions arise.

Potassium Ion Imbalances

When the mechanisms we just described fail or when excess potassium ions are taken in or lost, potassium ion imbalances can result. Keep in mind that water imbalances can impact potassium ion concentration as well, but the effect is not as noticeable as with sodium ion concentration because fewer potassium ions are present in the ECF. The effects of excess potassium ions, or *hyperkalemia,* and insufficient potassium ions, or *hypokalemia,* are primarily electrophysiological in nature. Let's turn our attention to these imbalances.

Hyperkalemia We define **hyperkalemia** (*kalem-* = "potassium") as a plasma potassium ion concentration above 4.5 mEq/l. Hyperkalemia is arguably one of the most dangerous electrolyte imbalances; indeed, a high dose of potassium ions can cause death almost immediately. Mild hyperkalemia alters the resting membrane potential—the excess of potassium ions in the ECF causes fewer potassium ions to leave the cell through leak channels. This means that the cell retains more positive charges, and as a result the resting membrane potential is more positive at rest than is normally the case (i.e., the membrane is slightly depolarized). For this reason, an action potential is more easily generated and the cells are more excitable with mild to moderate hyperkalemia. However, with severe hyperkalemia, the membrane can become so depolarized at rest that the cells are no longer excitable. In effect, this shuts down every excitable cell in the body, including neurons and muscle cells, and the result can be nearly instantaneous death.

Mild hyperkalemia is generally the result of renal failure or aldosterone insufficiency. It may also be due to widespread tissue damage that occurs with severe burns or trauma, as the damaged cells rupture and release their potassium ions into the ECF. Other potential causes include drugs such as aldosterone blockers and one commonly used for heart failure (find out more about this in *A&P in the Real World: Digoxin Toxicity and Hyperkalemia* on page 998). Severe hyperkalemia is almost always due to ingestion or administration of excess potassium ions. This can be intentional, as with a lethal injection, or accidental, as occurs when a medical professional mistakenly gives the wrong dose of potassium salts.

Hypokalemia A plasma potassium ion concentration below 3.9 mEq/l is known as **hypokalemia.** The most common cause of hypokalemia is the use of diuretics, many of which lead to excess potassium ion loss in the urine. As with hyperkalemia, the effects of hypokalemia are primarily electrophysiological. However, whereas hyperkalemia makes the resting membrane potential more positive, hypokalemia makes it more negative, meaning that the cells become hyperpolarized. This change happens because more potassium ions exit the cell through leak channels, and the loss of positive charges makes the cell more negative. As a result, the cell becomes more difficult to stimulate, which leads to muscle weakness, mental status changes, and a slowed heart rate.

Quick Check

☐ 4. What is the main role of potassium ions in the body?
☐ 5. How is the concentration of potassium ions in the ECF regulated?
☐ 6. What happens to the resting membrane potential of excitable cells and their generation of action potentials in hyperkalemia and hypokalemia?

Calcium and Phosphate Ions

◀◀ Flashback

1. What are the effects of parathyroid hormone? (p. 606)
2. How does vitamin D influence calcium ion homeostasis? (p. 606)

Calcium ions play many critical roles in the body. As you may recall, they are one of the main components of osseous tissue, giving bone its hardness and resistance to compression. In fact, about 99% of the total calcium ions in the body are located in bone as part of hydroxyapatite salts (see Chapter 6). Calcium ions are also required for muscle contraction, the plateau phase of the cardiac action potential, intracellular signaling, blood clotting, neuronal synaptic transmission, and more. In addition, they have an important role in the excitation of neurons—when calcium ion levels rise, neurons become less permeable to sodium ions, and the reverse is true as well.

We cannot discuss calcium ions without also discussing phosphate ions (PO_4^{3-}) because they are found together in hydroxyapatite crystals. The inorganic portion of bone cannot be built unless both ions are present. For this reason, they are regulated in tandem by the same hormones. Phosphate ions have other important functions in the body, for example, as an integral part of the ATP molecule. This section first examines the hormones and mechanisms that control phosphate and calcium ion levels in the ECF. The discussion then focuses on calcium ions, exploring what happens when these mechanisms are inadequate and imbalances result.

Regulation of Calcium and Phosphate Ion Homeostasis

The calcium ion concentration of the ECF measures about 8.7–10.4 mg/dl and rarely varies from this range by more than a few percentage points in either direction. This strict control is thanks to compensatory mechanisms that occur at three main sites: bone tissue, the distal tubules and collecting systems of the kidneys, and the small intestine. Simply put, when the level of calcium ions in the ECF falls, calcium ions are released from the bone by osteoclasts and more are reabsorbed from the filtrate in the kidneys and absorbed from ingested food and liquids in the small intestine. Conversely, when the calcium ion level rises, calcium ions are deposited into bone by osteoblasts, and fewer are reabsorbed from the filtrate and absorbed from the small intestine.

The level of phosphate ions, however, is not nearly as tightly regulated, and the extracellular phosphate ion concentration varies throughout the day. Generally, the body maintains a slightly lower phosphate ion concentration than calcium ion concentration. This is because cells require a certain number of free calcium ions in their cytosol. When phosphate ions are present in large numbers, they bind calcium ions and form solid calcium phosphate salts, reducing the number of free calcium ions. Therefore, the body must excrete more phosphate ions and retain more calcium ions.

The two main hormones involved in calcium and phosphate ion regulation are *parathyroid hormone (PTH)* and *vitamin D3 (calcitriol)*. However, many other hormones influence this balance as well, including calcitonin, cortisol, and estrogens. Recall that PTH is released from the parathyroid glands in response to a decrease in the calcium ion concentration of the blood, and it triggers osteoclast activity and calcium ion reabsorption in the kidneys (see Chapter 16). In addition, it decreases the reabsorption of phosphate ions.

Another important effect of PTH is the activation of vitamin D3, which is a potent stimulator of calcium ion absorption by the small intestine. Vitamin D3 also assists PTH in increasing osteoclast activity and calcium ion reabsorption from the kidneys. With regard to phosphate ions, vitamin D3 triggers their absorption from the small intestine. This increases the concentration of phosphate ions in the ECF, in contrast to the effects of PTH, which decreases their level. This helps to ensure that the level of phosphate ions in the ECF does not fall too low.

Calcium Ion Imbalances

Unlike changes in phosphate ion concentration, small changes in calcium ion concentration can have immediate, dramatic impacts on many different body functions. Let's look at the effects of an abnormal increase and decrease in blood calcium ion concentration, known as *hypercalcemia* and *hypocalcemia*, respectively.

Hypercalcemia Hypercalcemia is defined as a plasma calcium ion concentration above 10.5 mg/dl. This condition is usually due to hyperparathyroidism but may also be caused by certain cancers, excess vitamin D, certain bone disorders, and renal failure. Hypercalcemia has wide-ranging effects on multiple organ systems. The most immediately noticeable effects are on the nervous system. As we discussed earlier, an elevated calcium ion concentration in the ECF makes neurons less permeable to sodium ions, which diminishes the ability of the neurons to depolarize. This in turn decreases the neurons' activity, leading to mental sluggishness, reduced reflex activity, and weak muscle activity (due to lack of stimulation from motor neurons).

Hypercalcemia affects other tissues, as well. Recall that calcium ions are responsible for the plateau phase of the cardiac action potential, which lengthens and strengthens the heart's contraction. When excess calcium ions are present, the plateau phase is shortened, making the contraction shorter and weaker. Hypercalcemia can cause decreased appetite and constipation due to decreased activity of gastrointestinal smooth muscle. It can also cause kidney stones, bone pain, and frequent urination.

Hypocalcemia A plasma calcium ion concentration lower than 8.7 mg/dl is considered to be **hypocalcemia.** The condition is generally due to a decrease in PTH, ineffective PTH (as occurs with renal failure), and vitamin D deficiency, although many other factors can contribute to it. Hypocalcemia has the opposite effect on neurons from hypercalcemia—the neurons become hyperexcitable, as more sodium ions enter and depolarize the neurons, stimulating action potentials. When hypocalcemia is severe, neurons begin firing spontaneously, leading to repetitive stimulation of skeletal muscles and tetanic (sustained) contractions. For example, the characteristic *carpopedal spasm* of hypocalcemia is shown in **Figure 25.8**. In addition, the plateau phase of the cardiac muscle cell action potential is lengthened, leading to excessively long and strong contractions. A calcium ion

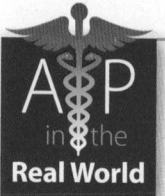

A&P in the Real World

Digoxin Toxicity and Hyperkalemia

Digoxin (dih-JOKS-in) is a drug derived from the foxglove plant that is used to treat heart failure. Its effectiveness as a heart failure agent comes from the fact that it blocks the functioning of the Na^+/K^+ pump in cardiac muscle cells. This blocking activity effectively lengthens the action potentials of the cardiac muscle cells, enabling a longer, stronger contraction.

Unfortunately, as with many drugs, digoxin comes with potential adverse side effects, including abdominal pain and dizziness. In addition, the range between a therapeutic blood concentration of the drug and a toxic concentration is very small. As a result, it is relatively easy to develop *digoxin toxicity.* One of the main symptoms of digoxin toxicity is hyperkalemia. This is not true hyperkalemia in that the total number of potassium ions in the body is not elevated. Instead, these potassium ions are concentrated in the wrong compartment—the Na^+/K^+ pump is unable to drive them into the cytosol, and so they are not shifted from the interstitial fluid to the cytosol. Instead, they remain in the interstitial fluid around cardiac muscle cells. From here they diffuse into the blood, leading to the elevated blood concentration of potassium ions.

Figure 25.8 Carpopedal spasm of hypocalcemia.

concentration below about 4 mg/dl is generally lethal because of these and other cardiac effects and tetanic spasm of the respiratory muscles.

Other Ions Critical to Human Physiology

Many other ions play a role in human physiology, including the anion chloride and the cation magnesium. Let's briefly discuss their roles here.

Chloride Ions

Chloride ions have a variety of functions in the body. Their high concentration in the ECF makes them an important osmotic particle, along with sodium ions. Chloride ions are also a critical component of some other processes, including the production of hydrochloric acid secreted by the stomach. In addition, chloride ions are involved in the secretion of newly formed bicarbonate ions from erythrocytes. The concentration of chloride ions in the ECF is mostly determined by the kidneys, where the reabsorption of these ions is largely coupled to that of sodium ions. Chloride ions follow sodium ions passively from the filtrate to the ECF due to the electrical gradient created by sodium ion reabsorption. In addition, several transporters move the two ions together.

Magnesium Ions

Magnesium ions are critical to multiple cellular processes, as they are involved in the activation of many enzymes. In addition, magnesium salts are an important component of bone tissue. In fact, about half of the body's magnesium ions are located in bone; most of the rest reside in the cytosol, with only about 1% in the ECF. For this reason, the kidneys must excrete the majority of the magnesium ions that are ingested. Maintaining this low extracellular concentration of magnesium ions is critical because if the level in the ECF rises too high, it blocks the entry of calcium ions into cells, which can in turn lead to hypercalcemia.

MODULE **25.4**

Acid-Base Homeostasis

We now turn our attention toward ions with a different role in the body: hydrogen ions and various base ions such as bicarbonate. The normal hydrogen ion concentration of body fluids is equal to a pH range of about 7.35–7.45. The body must maintain the pH within this narrow, slightly alkaline, range because even a slight deviation can have disastrous consequences for our cells. Blood pH is maintained through a number of mechanisms, the most important of which are the respiratory system, the urinary system, and the body's **buffer systems.** There are two major types of buffer systems: *chemical buffer systems* and *physiological buffer systems.* Chemical buffers are, as the name suggests, chemical systems that work to buffer fluids in the body, whereas physiological systems are functions of organ systems that work to buffer fluids. In this module, we examine how the respiratory, urinary, and buffer systems maintain the pH of the blood and what happens when the pH moves out of its normal range. First, let's look at the sources of acids and bases in body fluids.

Sources of Acids and Bases in the Body

◀◀ **Flashback**

1. How is carbon dioxide converted into carbonic acid? (p. 830)

Acids and bases in the body come from two sources: (1) those that are formed as a normal part of metabolic processes, and (2) those that are ingested in the diet. Throughout this book, you have seen how the body produces multiple metabolic acids. The biggest source of metabolic acids is carbon dioxide, which is a byproduct of glucose catabolism (see Chapter 23). Recall that carbon dioxide isn't an acid, but when it reacts with water, it forms carbonic acid (H_2CO_3), which can then dissociate to a hydrogen ion and a bicarbonate ion (HCO_3^-). Carbon dioxide is known as a **volatile acid** because in solution it is able to change states from carbonic acid back to carbon dioxide gas. This allows carbon dioxide to be eliminated from the body through both the lungs and the kidneys.

The body's reactions produce many other metabolic acids. For example, lactic acid is a byproduct of glycolysis, uric acid is produced from the breakdown of nucleic acids, the intermediates of the citric acid cycle are acids, and acidic ketone bodies are produced from fat breakdown. Such acids are known as **fixed acids** (or *nonvolatile acids*) because they cannot change to a gaseous state and so must be eliminated by the kidneys.

In addition, the diet is a source of both acids and bases. Protein-rich foods contain amino acids, fat-rich foods contain fatty acids, and some fruits contain acids such as citric acid. Some foods, such as certain fruits and vegetables, also contribute small numbers of base ions, particularly bicarbonate. Note that although these foods do contribute some amounts of acids and bases to the body, they represent an exceedingly small minority that have essentially no effect on overall pH—the vast majority of acids and bases come from the body's own metabolic reactions, not from outside sources.

You may have noticed that there are a lot of sources of hydrogen ions, but relatively few for base ions. For this reason, the body's acid-base balance requires mechanisms that deal primarily with buffering hydrogen ions, excreting hydrogen ions, and conserving base ions. These mechanisms are examined next.

Quick Check

☐ 1. What are the major sources of acids for the body?
☐ 2. Why are acids the main focus of the body's acid-base balance mechanisms?

Chemical Buffer Systems

◀◀ **Flashback**

1. How is a base related to an acid? (p. 48)
2. Which body system controls the carbonic acid–bicarbonate ion buffer system? How does this system control the amount of carbonic acid in the body? (p. 830)

Before discussing buffer systems, let's review a few definitions from chemistry (see Chapter 2). Recall that a *strong acid* is one that releases most of its hydrogen ions when placed in water, which lowers the pH of the solution. A *weak acid,* however, releases relatively few hydrogen ions in solution, and so has a much smaller impact on pH. Conversely, a *strong base* is one that binds and removes a great number of hydrogen ions from the solution, increasing the solution's pH. A *weak base* binds relatively few hydrogen ions in a solution and so has less impact on its pH.

The body's **chemical buffer systems,** which consist of a weak acid and its conjugate weak base, function to resist large swings in pH. When a strong acid is added to a solution, the weak base of the buffer system binds the released hydrogen ions and removes them from solution. When a strong base is added to a solution, the buffer system's weak acid releases its hydrogen ions to bind the base ions. Both situations minimize pH changes and help maintain acid-base homeostasis. The buffering capacity of the body's fluids is known as the *alkaline reserve.* When the alkaline reserve is depleted, acid-base imbalances can result.

There are three main chemical buffer systems in the body: the *carbonic acid–bicarbonate ion buffer system,* the *phosphate buffer system,* and the *protein buffer system.* These three systems are explored in this section.

Carbonic Acid–Bicarbonate Ion Buffer System

The most important chemical buffer system in the blood is the **carbonic acid–bicarbonate ion buffer system,** which consists of the weak acid carbonic acid and its weak conjugate base, the bicarbonate ion. One source of carbonic acid is the reaction that we described earlier in which carbon dioxide reacts with water in erythrocytes, producing carbonic acid. When carbonic acid dissociates, it yields a hydrogen ion and a bicarbonate ion:

$$CO_2 + H_2O \leftrightarrow H_2CO_3 \leftrightarrow HCO_3^- + H^+$$

Carbon dioxide   Water   Carbonic acid   Bicarbonate ion   Hydrogen ion

Bicarbonate ions are also found as salts, primarily sodium bicarbonate ($NaHCO_3$). Normally, sodium bicarbonate completely dissociates in solution to form sodium and bicarbonate ions:

$$NaHCO_3 \leftrightarrow Na^+ + HCO_3^-$$

Sodium bicarbonate   *Sodium ion*   *Bicarbonate ion*

How do carbonic acid and bicarbonate ion buffer a solution? First, remember that a beaker of pure water contains an equal number of hydrogen and hydroxide ions, as shown at the top of Figure 25.9. Next, let's look at this process in an acidic solution. The upper left illustration in Figure 25.9a shows a strong acid, such as hydrochloric acid, being added to a beaker of water. Notice that the acid immediately donates its hydrogen ions to the solution, which increases its hydrogen ion concentration and therefore reduces its pH.

Now see what happens when we add the same acid to a solution buffered by sodium bicarbonate, shown at the bottom left in Figure 25.9a. Notice that here (after sodium bicarbonate

Figure 25.9 The effects of buffers.

dissociates), the hydrogen ion from the strong acid is accepted by the bicarbonate ion, forming carbonic acid. Since carbonic acid is a weak acid, it doesn't release its hydrogen ions as readily, minimizing any pH changes in the solution.

Carbonic acid can similarly buffer a very basic solution. For example, see what happens when a strong base such as sodium hydroxide (NaOH) is added to water, as shown in the upper right illustration in Figure 25.9b. In this unbuffered solution, the released hydroxide ions remove many of the hydrogen ions from the solution and raise its pH, making it very basic. When carbonic acid is present, however, as shown at the bottom right in Figure 25.9b, it releases bicarbonate and hydrogen ions into the solution. The hydrogen ion binds to the hydroxide ion released from sodium hydroxide, forming water and a molecule of sodium

bicarbonate. The bicarbonate ion is a weak base and will not remove significant numbers of hydrogen ions from the solution.

Phosphate Buffer System

The **phosphate buffer system** consists of the weak acid dihydrogen phosphate ($H_2PO_4^-$) and its conjugate base, hydrogen phosphate (HPO_4^{2-}). As in the carbonic acid–bicarbonate ion buffer system, the acid and conjugate base balance excess hydroxide and hydrogen ions, respectively, keeping the pH of a solution stable. The concentration of hydrogen phosphate is too low in the ECF—only about 8% of the bicarbonate concentration—to make it an important buffer here. However, it has a higher concentration in the cytosol and in the filtrate within the kidney tubules. For this reason, it is an important buffer in these locations.

Protein Buffer System

A huge variety of proteins are found in the body, particularly in the cytosol of our cells and in plasma. As you've discovered, proteins are composed of strings of amino acids, each containing a carboxylic acid group (—COOH). In solution, some of these carboxylic acid groups ionize to their weak conjugate base form, allowing them to act as a **protein buffer system.** When hydrogen ions are added to the cytosol, they can bind to these bases and be removed from the cytosol, preventing large swings in pH. There are many important intracellular protein buffers; in fact, about 60–70% of the body's total buffering capacity takes the form of protein buffers in the cytosol. One example of an intracellular protein buffer is the hemoglobin inside erythrocytes, which acts as a buffer for hydrogen ions formed from the dissociation of carbonic acid.

Quick Check

☐ 3. How can carbon dioxide become an acid?
☐ 4. How do carbonic acid and bicarbonate ions buffer the plasma?
☐ 5. What are the components of the phosphate and protein buffer systems?

Physiological Buffer Systems: Respiratory and Renal Regulation of Blood pH

◀◀ **Flashback**

1. What happens to the pH of the blood when a person hypoventilates or hyperventilates? (p. 832)
2. How is bicarbonate ion absorption coupled with hydrogen ion secretion? (p. 962)
3. What effect does aldosterone have on hydrogen ion secretion in the kidneys? (p. 965)

The chemical buffers that we just discussed respond to *changes* in the hydrogen ion concentration of the body fluids in order to maintain it within its normal range. However, for these systems to work, the body fluids must contain adequate numbers of base ions, particularly bicarbonate ions. The body must also have a way to rid itself of hydrogen ions in order to maintain the normal, slightly alkaline pH. Two systems perform these tasks: the respiratory system, which controls the amount of volatile acids in the ECF, and the urinary system, which controls the amount of fixed acids and bicarbonate in the ECF. These systems can also work to correct hydrogen ion imbalances. For this reason, these systems are sometimes referred to as **physiological buffer systems.** You have read about these mechanisms previously in this book (see Chapters 21 and 24), but let's review them now.

Respiratory System Effects on Blood pH

In the previous section, we saw that the carbonic acid in body fluids is formed from the reaction of carbon dioxide with water.

The lungs directly control the amount of carbon dioxide in the blood, so it follows that the lungs also control, to a significant extent, the levels of carbonic acid and hydrogen ions. Under normal conditions, the amount of carbon dioxide expired by the lungs matches the amount of carbon dioxide produced by metabolic reactions, which is why your respiratory rate doesn't remain constant. When you are physically active, glycolytic catabolism and oxidative catabolism occur more rapidly and more carbon dioxide is generated. As a result, your respiratory rate increases to match the higher level of carbon dioxide production. The opposite is also true—when you are sedentary, you generate less carbon dioxide and so your respiratory rate decreases.

The respiratory system also influences the number of bicarbonate ions in the plasma. Remember that the reaction between carbon dioxide and water occurs rapidly in erythrocytes due to the presence of the enzyme *carbonic anhydrase* (refer back to Figure 21.25). When carbonic acid is formed in erythrocytes, much of it dissociates into bicarbonate and hydrogen ions. The hydrogen ions that form bind to hemoglobin molecules, which buffers the cytosol of erythrocytes. The bicarbonate ions, however, are transported into the plasma, where they buffer fixed acids.

Urinary System Effects on Blood pH

The kidneys work with the lungs in two ways to determine the number of hydrogen ions and base ions in the blood. First, the kidneys can excrete fixed acids that the lungs cannot excrete, such as lactic acid, ketones, phosphoric acid, uric acid, and ammonium ions (NH_4^+). About 25% of the body's hydrogen ions are eliminated in this manner.

Second, the kidneys contribute to acid-base balance by controlling the concentration of bicarbonate ions in the blood. Generally, the kidneys reabsorb all of the bicarbonate ions from the filtrate. Recall, though, that there are no true transport mechanisms to move bicarbonate ions from the filtrate into the tubule cells. For this reason, other, more roundabout methods for reabsorption are needed. In addition, the kidneys can actually manufacture new bicarbonate ions when the pH of the blood falls, and secrete bicarbonate ions when the pH of the blood rises. We examine these mechanisms next.

Hydrogen Ion Secretion and Bicarbonate Ion Reabsorption in the Proximal Tubule The cells of the proximal tubule secrete hydrogen ions from fixed acids by a secondary active transport process. Recall that this process is coupled to bicarbonate ion reabsorption—the secreted hydrogen ions bind to bicarbonate ions in the filtrate to form carbonic acid. Carbonic anhydrase then catalyzes the conversion of the carbonic acid to carbon dioxide and water, and these two molecules enter the kidney tubule cell. Inside the cell, they react again with carbonic anhydrase to form carbonic acid, which dissociates into bicarbonate and hydrogen ions. The bicarbonate ion leaves the cell and enters the blood, whereas the hydrogen ion is recycled, as it is re-secreted to bind another bicarbonate ion in the filtrate (Figure 24.17 provides a diagram of this process). This process accounts for reabsorption of nearly 100% of the bicarbonate ions in the filtrate and secretion of large numbers of hydrogen ions.

Another process that occurs in the proximal tubule secretes hydrogen ions while actually forming *new* bicarbonate ions. As you can see in **Figure 25.10**:

(1) **Glutamine enters a proximal tubule cell and undergoes reactions that yield two ammonia molecules and two bicarbonate ions.**

(2) **Ammonia molecules combine with hydrogen ions, producing acidic ammonium ions. The ammonium ions are secreted into the filtrate for excretion in urine.**

(3) **Bicarbonate ions are transported into the interstitial fluid and then into the blood.**

Although a huge number of hydrogen ions are secreted in the proximal tubule, the pH of the filtrate doesn't change all that much. The explanation for this is that many of the secreted hydrogen ions are recycled during the reabsorption of bicarbonate ions, and so the concentration of hydrogen ions in the filtrate increases only about three or four times. In addition, the hydrogen ions that are not recycled are either buffered as ammonium ions or bind to other buffers in the filtrate, particularly phosphate buffers. These two facts combined mean that the filtrate pH reaches a minimum value of only about 6.7 in the proximal tubule.

Hydrogen Ion Secretion in the Distal Tubule and Collecting System

The major acidification of the urine takes place in the late distal tubule and collecting system by cells called *intercalated cells* (in-TER-kah-lay'-ted). These cells have primary active transport pumps in their plasma membranes that pump hydrogen ions into the filtrate. As you can see in **Figure 25.11**, the source of the hydrogen ions is actually carbon dioxide. The process proceeds by the following steps:

(1) **Carbon dioxide diffuses from the blood into the distal tubule cell.** Carbon dioxide from the blood

Filtrate in tubule lumen   Cytosol in proximal tubule cell   Interstitial fluid   Blood in peritubular capillary

(1) Glutamine enters a proximal tubule cell and undergoes reactions that yield 2 NH_3 and 2 HCO_3^-.

(2) NH_3 molecules combine with H^+, producing acidic NH_4^+. The NH_4^+ are secreted into the filtrate for excretion in urine.

(3) HCO_3^- are transported into the interstitial fluid and then into the blood.

Figure 25.10 Production of bicarbonate ions from glutamine in the proximal tubule.

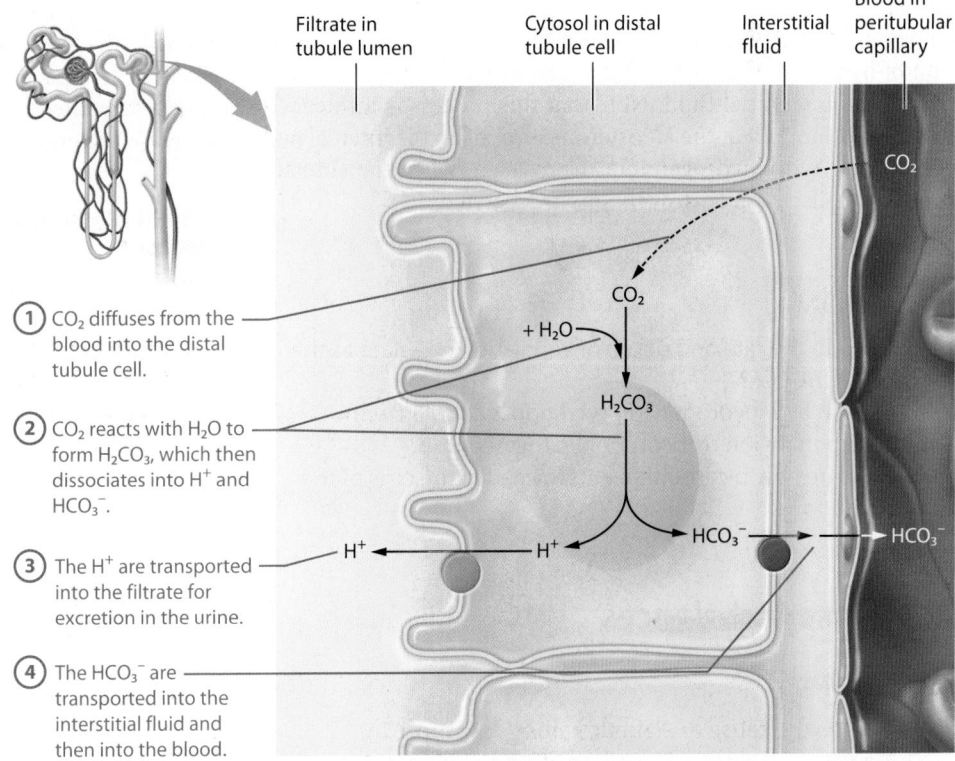

Filtrate in tubule lumen   Cytosol in distal tubule cell   Interstitial fluid   Blood in peritubular capillary

(1) CO_2 diffuses from the blood into the distal tubule cell.

(2) CO_2 reacts with H_2O to form H_2CO_3, which then dissociates into H^+ and HCO_3^-.

(3) The H^+ are transported into the filtrate for excretion in the urine.

(4) The HCO_3^- are transported into the interstitial fluid and then into the blood.

Figure 25.11 Secretion of hydrogen ions in the distal tubule and collecting system.

and interstitial fluid diffuses freely into the tubule cells when a concentration gradient is present.

(2) **Carbon dioxide reacts with water to form carbonic acid, which then dissociates into hydrogen and bicarbonate ions.** In the distal tubule cell, carbon dioxide reacts with water in a reaction catalyzed by carbonic anhydrase to form carbonic acid. The carbonic acid dissociates into hydrogen and bicarbonate ions.

(3) **The hydrogen ions are transported into the filtrate for excretion in the urine.** The newly formed hydrogen ions are transported into the filtrate by active transport pumps, including one that exchanges hydrogen ions for potassium ions. Note that these hydrogen ions, unlike the ones in the proximal tubule, are not recycled.

(4) **The bicarbonate ions are transported into the interstitial fluid and then into the blood.** The newly formed bicarbonate is transported across the tubule cell plasma membrane into the interstitial fluid and blood.

Although the body needs to be able to eliminate excess hydrogen ions in the urine to maintain acid-base homeostasis, a pH in the filtrate that is too acidic inhibits hydrogen ion pumps. If the pH of the filtrate reaches about 4.5, the pumps are unable to secrete any more hydrogen ions. To avoid this, the secreted hydrogen ions in the distal tubule filtrate bind to hydrogen phosphate ions (HPO_4^{2-}) of the phosphate buffer system, forming dihydrogen phosphate ($H_2PO_4^-$). This allows the hydrogen ions to be excreted in urine without contributing to its acidity.

The process in the distal tubule and collecting system accounts for only about 5% of the total hydrogen ion secretion, but it's the major factor that determines the pH of the urine. This is largely because the hydrogen ions secreted from the distal tubule and collecting system are not recycled. As a result, the concentration of hydrogen ions in the filtrate can increase up to 900 times that of the interstitial fluid. Note that this process is influenced by aldosterone. When the aldosterone level rises, the intercalated cells secrete more hydrogen ions. Conversely, when the aldosterone level falls, fewer hydrogen ions are secreted.

Quick Check

☐ 6. How do the rate and depth of ventilation correlate to the pH of the ECF?

☐ 7. Explain how hydrogen ion secretion is coupled with bicarbonate ion reabsorption in the nephron.

☐ 8. How are new bicarbonate ions formed in the cells of the nephron?

Acid-Base Imbalances

◀◀ **Flashback**

1. What is the cause of diabetic ketoacidosis? (p. 616)
2. How does prolonged vomiting affect the pH of the blood? (p. 845)

Normally, the body's buffer systems ensure that the pH of body fluids remains within a narrow range, even when acids or bases are added or lost. But when these buffer systems are overwhelmed by a significant loss or gain of acids or bases, they can fail, resulting in a pH imbalance. The two basic types of pH imbalance are *acidosis* and *alkalosis*. The body attempts to correct both types of imbalances through a process called *compensation*, which involves **respiratory compensation** and/or **renal compensation**.

Although compensation can bring the pH closer to the normal range, it is not a permanent solution to the imbalance. Rather, the permanent solution lies in correcting the root cause of the imbalance. Most pH imbalances are secondary to other causes, such as respiratory disorders, kidney failure, or gastrointestinal problems. For this reason, treatment is generally aimed at the primary cause, and when this is resolved, the pH returns to the normal range. The upcoming subsections explore the types and causes of pH imbalances and how compensation occurs.

Acidosis

Acidosis is defined as a body fluid pH of less than 7.35 (note that *acidemia* refers to low blood pH specifically). The pH of a patient with acidosis generally remains in the alkaline range (it is rare to see a pH less than 7.0), but it is still more acidic than the normal range. Acidosis develops when more hydrogen ions are added than the body's buffers can bind or when the number of buffer molecules such as bicarbonate ions decreases. This causes neurons to become less excitable, leading to signs and symptoms of nervous system depression, such as slurred speech, drowsiness, and unresponsiveness. Left untreated, acidosis can lead to coma and, in severe cases, death.

There are many different causes of acidosis, which can be broadly divided into two classes: *respiratory acidosis* and *metabolic acidosis*. Irrespective of the cause of acidosis, the body's response is generally the same: decrease the number of hydrogen ions and increase the number of base ions, particularly bicarbonate.

Respiratory Acidosis and Compensation We define **respiratory acidosis** as a decrease in the pH of body fluids due to excess carbon dioxide. The carbon dioxide excess happens when ventilation decreases, which is known as *hypoventilation* (see Chapter 21). Hypoventilation leads to the accumulation of carbon dioxide because less of it is being exhaled. This interferes with the carbonic acid–bicarbonate ion buffer system and causes a shift toward excessive carbonic acid, lowering the pH of the blood.

There are three general causes of respiratory acidosis: suppressed ventilation from brainstem dysfunction, blockage of air passages in the lungs, and decreased gas exchange in the alveoli. Whatever the cause, the respiratory and urinary systems attempt to compensate for the acidosis.

Respiratory compensation begins within minutes of the pH disturbance, as excess hydrogen ions in the ECF stimulate chemoreceptors in the brain. These neurons then trigger an increase in the rate and depth of ventilation in an attempt to rid the body of excess carbon dioxide (and indirectly hydrogen ions). This is why an individual in respiratory distress, which initially causes hypoventilation, will generally have an increased respiratory rate. It might seem that the individual is breathing rapidly because of an inability to "catch" his or her breath, but the elevated respiratory rate is generally due to elevated carbon dioxide not low oxygen.

Respiratory compensation for respiratory acidosis is often not terribly effective because disordered ventilation was the initial cause of the problem. Such cases therefore rely more on renal compensation: The kidney tubule cells absorb all of the available bicarbonate ions from the filtrate, secrete hydrogen ions, and produce new bicarbonate ions from glutamine. Renal compensation takes several hours to days to have a significant effect on overall blood pH, so these actions by the kidneys won't be immediately noticeable.

Metabolic Acidosis and Compensation Metabolic acidosis is defined as the addition of hydrogen ions to the ECF (from acids other than carbon dioxide) or loss of bicarbonate ions. The term refers to the fact that the condition can result from the addition of metabolic acids such as lactic acid, uric acid, and ketones to the blood. There are many different causes of metabolic acidosis, including prolonged diarrhea (which results in the loss of bicarbonate ions to the feces), the production of excess metabolic acids (such as occurs with diabetic ketoacidosis), failure of the kidneys to reabsorb bicarbonate ions or secrete hydrogen ions, and the ingestion of acidic drugs and toxins such as methanol.

The first compensatory response to metabolic acidosis is respiratory compensation, which occurs in the form of hyperventilation. In fact, hyperventilation in the absence of overt respiratory distress is one of the main signs that can point a clinician toward a diagnosis of metabolic acidosis.

The extent to which renal compensation occurs depends on the health of the kidneys. Obviously, if renal failure is the cause of acidosis, the kidneys aren't going to be able to contribute to the compensation. However, if the cause is unrelated to the kidneys, they will work over time to eliminate hydrogen ions and retain bicarbonate ions, as we saw with respiratory acidosis.

ConceptBOOST ⟫⟫⟫

How Can Respiratory Changes Compensate for Metabolic Acidosis?

We have just established that the respiratory rate increases in metabolic acidosis as part of respiratory compensation. But how does changing the level of carbon dioxide in the blood help when the problem is metabolic in nature? The answer is that respiratory compensation cannot correct the actual cause of the metabolic imbalance. It can, however, correct the resulting pH imbalance. Remember that the pH of a solution depends only on the *number* of hydrogen ions present, not on the source of the hydrogen ions. Take a look at the following simple graphic, which shows two hypothetical samples of blood, one from a person with normal pH (on the left) and one from a person suffering from metabolic acidosis (on the right). The hydrogen ions from carbonic acid are represented in purple, and those from the metabolic acids are represented in blue:

H^+ from carbonic acid (H_2CO_3)

H^+ from metabolic acids

Normal sample:
pH = 7.4

Metabolic acidosis
sample: pH = 7.0

Notice that the normal sample of blood has a total of 20 hydrogen ions (15 from carbonic acid and 5 from metabolic acids) and a pH of 7.4. The metabolic acidosis sample, however, has a total of 30 hydrogen ions as a result of the excess metabolic acids and a pH of 7.0.

Now let's see what happens when the person with metabolic acidosis starts hyperventilating and so exhaling more carbon dioxide. As carbon dioxide molecules are exhaled, the number of carbonic acid molecules and thus of hydrogen ions in the blood decreases. You can see this in the blood sample on the right.

Notice that this sample of blood has 20 hydrogen ions (15 from metabolic acids, 5 from carbonic acid), the same number as in the initial tube, with a normal pH of 7.4. At this point, we have a compensated metabolic acidosis. The condition that caused the acidosis in the first place has not been resolved, as shown by the fact that there is still an elevated number of metabolic hydrogen ions. But the total number of hydrogen ions has gone down, thanks to respiratory compensation.

Of course, our diagrams here are very simplified and the pH values in the figures aren't realistic given the numbers of hydrogen ions illustrated. But the basic principle still stands—respiratory compensation for metabolic acidosis works by decreasing the total number of hydrogen ions in the blood. As you will see in the next section, the same situation exists for metabolic alkalosis but in reverse—carbon dioxide is retained to *increase* the total number of hydrogen ions in the body and return the pH to normal. ∎

CO_2

CO_2

Hyperventilation begins to compensate for the acidosis.

Compensated metabolic acidosis has moved the pH back toward 7.4.

Alkalosis

Alkalosis is defined as a body fluid pH greater than 7.45 (*alkalemia* refers specifically to an elevated blood pH). This occurs when more base ions are added than the buffers can handle or when the number of hydrogen ions in the body fluids decreases. Alkalosis increases the excitability of neurons, and so they fire action potentials inappropriately. This results in sensory symptoms such as numbness and tingling and motor symptoms such as muscle twitches or tetanic contractions. Untreated, alkalosis can lead to seizures and death due to sustained contractions of respiratory muscles.

As with acidosis, there are two types of alkalosis: *respiratory alkalosis* and *metabolic alkalosis*. The compensatory response for both involves excretion of bicarbonate ions and retention of hydrogen ions.

Respiratory Alkalosis and Compensation Respiratory alkalosis is caused by a loss of carbon dioxide through the lungs due to hyperventilation. Generally, this results from psychological states that increase the rate of breathing. The condition may also occur when an individual is at extremely high altitudes and the low P_{O_2} of the atmosphere triggers hyperventilation.

If hyperventilation is psychological in nature, there generally isn't much of a metabolic compensatory response. Usually, if the pH of the blood becomes too high, the individual quickly loses consciousness and the respiratory centers in the brainstem slow the respiratory rate and retain carbon dioxide to try to return the fluid pH to the normal range. If the hyperventilation is physiological in nature, however, the compensatory response primarily involves the kidneys excreting bicarbonate ions and retaining hydrogen ions.

Metabolic Alkalosis and Compensation Metabolic alkalosis results from a loss of hydrogen ions or an excess of bicarbonate ions. Hydrogen ion loss is commonly the result of prolonged vomiting and the loss of acidic stomach contents. It may also be produced by certain diuretics or excess aldosterone, both of which lead to increased sodium ion reabsorption with concurrent hydrogen ion secretion. An increase in the bicarbonate ion level can happen when excessive bicarbonate salts have been ingested; the "culprit" is usually oral antacids taken for stomach ulcers or severe acid reflux.

Compensation for metabolic alkalosis is both respiratory and renal. The respiratory response to a decreased hydrogen ion concentration is hypoventilation. This retains carbon dioxide and increases the number of hydrogen ions in the body. The renal response to metabolic alkalosis is the same as for respiratory alkalosis—the kidneys retain hydrogen ions and secrete bicarbonate ions into the urine.

Arterial Blood Gases

Acid-base imbalances can be difficult to distinguish from one another with standard blood tests, which examine venous blood. To get an accurate picture of a patient's acid-base balance, we need instead to look at the arterial blood. In a procedure to determine *arterial blood gases* (ABGs), blood is drawn from the radial or brachial artery, and the pH, P_{CO_2}, and concentration of bicarbonate ions in the blood are measured. From these values, it is possible in many cases to surmise the type of acid-base imbalance and the degree of compensation that has occurred. If little to no compensation has occurred, the imbalance is called *uncompensated*; if some compensation has occurred, but the pH of the blood is still abnormal, it is called *partially compensated*; and if the pH of the blood is normal, but the P_{CO_2} and the bicarbonate ion level are abnormal, it is called *fully compensated*.

TABLE 25.1 SUMMARY OF ACID-BASE DISORDERS

Type of Disorder	Characteristic Blood Gas Findings (without Compensation)	Causes	Main Methods of Compensation
Respiratory acidosis	• Decreased pH • Increased P_{CO_2}	• Decreased ventilation from brainstem dysfunction • Blockage of air passages in the lungs • Decreased gas exchange in the alveoli	• Hyperventilation to remove CO_2 • Retention of HCO_3^- by the kidneys • Excretion of H^+ by the kidneys
Respiratory alkalosis	• Increased pH • Decreased P_{CO_2}	• Psychological states that cause hyperventilation • Brain tumor or other damage to respiratory centers • Hyperventilation due to high altitudes and low atmospheric P_{O_2}	• Slowed respiratory rate • Excretion of HCO_3^- and retention of H^+ by the kidneys (if alkalosis is physiological rather than psychological)
Metabolic acidosis	• Decreased pH • Decreased HCO_3^-	• Prolonged diarrhea • Excessive production of metabolic acids • Renal failure • Ingestion of acidic drugs or toxins	• Hyperventilation • Retention of HCO_3^- from the kidneys • Excretion of H^+ by the kidneys (The second and third responses are only possible if renal failure is not the cause of metabolic acidosis.)
Metabolic alkalosis	• Increased pH • Increased HCO_3^-	• Prolonged vomiting • Ingestion of certain diuretics or oral antacids • Excess aldosterone	• Slowed respiratory rate • Retention of H^+ by the kidneys • Excretion of HCO_3^- by the kidneys

Each type of acid-base disturbance has a characteristic set of ABG findings, which are as follows:

- *Respiratory acidosis.* The primary abnormality seen with respiratory acidosis is an elevated arterial P_{CO_2} and lower than normal blood pH. If the bicarbonate ion level is elevated, then renal compensation has begun and the acidosis is partially compensated.
- *Metabolic acidosis.* The values for metabolic acidosis are slightly more complex, but the general rule of thumb is that you will have a decreased pH and lower bicarbonate ion level in the blood, although this might not always be the case. The P_{CO_2} is also generally lower than normal because of respiratory compensation. Due to the rapid effects of respiratory compensation, most cases of metabolic acidosis are partially compensated.
- *Respiratory alkalosis.* The main findings with respiratory alkalosis are fairly straightforward—an elevated blood pH and a low P_{CO_2}. If the bicarbonate ion level is also low, this signifies partial compensation by the kidneys.
- *Metabolic alkalosis.* Metabolic alkalosis shows an elevated blood pH and a high bicarbonate ion level. If respiratory compensation has occurred, the P_{CO_2} will be elevated.

Keep in mind that these are just general guidelines; many cases of acid-base imbalance are quite complex and do not fit these profiles.

Table 25.1 will help you study and remember the causes and compensations of acidosis and alkalosis.

Quick Check

☐ 9. How do metabolic acidosis and respiratory acidosis differ?

☐ 10. How does the body compensate for metabolic acidosis and respiratory alkalosis?

☐ 11. What are the main differences between metabolic alkalosis and respiratory alkalosis?

Apply What You Learned

☐ 1. Mr. Wong presents with pneumonia and asthma that has progressively worsened over several days, and has led to obstruction of his airways. The physician in the emergency department orders arterial blood gases. Predict what you are likely to find in the results of this test.

☐ 2. You note that Mr. Wong's respiratory rate is elevated. Explain why he is hyperventilating. Will this likely help his pH imbalance?

☐ 3. A urinalysis shows that the pH of Mr. Wong's urine is 4.5. Are his kidneys going to be able to compensate any further for his acidosis? Why or why not?

☐ 4. Mr. Wong's wife has been giving him large doses of vitamin C at home because she thinks it might help him fight off his infection. Vitamin C is a chemical also known as ascorbic acid. What effect might this have on the pH of his body fluids?

See answers in Appendix A.

An Example of Fluid, Electrolyte, and Acid-Base Homeostasis

Learning Outcomes

1. Provide specific examples to demonstrate how the cardiovascular, endocrine, and urinary systems respond to maintain homeostasis of fluid volume, electrolyte concentration, and pH in the body.

You've seen in this chapter that the maintenance of fluid, electrolyte, and acid-base homeostasis requires the input of the two systems in the body whose primary responsibilities are the maintenance of homeostasis: the endocrine and nervous systems. You've also discovered that control of all three variables is intertwined. All three must be balanced together, as imbalances of one variable cause imbalances of the others. To illustrate this, we examine the effects of and response to a common imbalance: dehydration from inadequate fluid intake.

The process shown in **Figure 25.12** first deals with the homeostatic imbalances that occur with dehydration, and then with how all of these imbalances are corrected together. The general sequence of events is another example of the Feedback Loops Core Principle (p. 21), as it follows a multistep feedback loop. The steps are as follows:

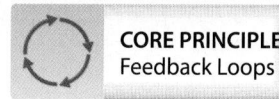

CORE PRINCIPLE
Feedback Loops

① **Total body water decreases, leading to the following imbalances of electrolytes and extracellular fluid pH:**
- decreased blood volume and possibly blood pressure;
- increased extracellular electrolyte concentrations, particularly that of sodium ions;
- increased osmolarity of the ECF, which draws water out of the cytosol of cells; and
- increased concentration of metabolic acids in the ECF (due to fewer water molecules)—metabolic acidosis may develop.

② **Juxtaglomerular cells release renin, which leads to formation of angiotensin-II.** The decreased blood pressure and blood volume trigger renin release from the juxtaglomerular cells of the kidneys, leading to the production of more angiotensin-II.

③ **Angiotensin-II has multiple effects:**

③a **It causes vasoconstriction, increases sodium ion (and so water) reabsorption, and stimulates thirst.** One of the most immediate effects of angiotensin-II is to increase blood pressure through vasoconstriction. It also acts on the proximal tubule of the nephron to increase fluid retention from the kidneys through stimulation of sodium

(1) **Dehydration:** total body water decreases, leading to:

- Decreased blood volume/blood pressure
- Increased Na⁺ concentration
- Increased osmolarity of ECF, which draws water out of the cytosol of cells
- Increased concentration of metabolic acids in the ECF—metabolic acidosis may develop

Renin

JG cells

(2) JG cells release renin, which leads to formation of angiotensin-II.

Angiotensin-II

(3a)
Causes systemic vasoconstriction

Blood vessel

Increases Na⁺ (and so H₂O) reabsorption

H_2O

Na^+

Proximal tubule

Stimulates thirst

Hypothalamus

Increased water intake

(3b)
Stimulates antidiuretic hormone secretion

Increased H_2O reabsorption

Distal tubule

Collecting duct

H_2O

H_2O

(3c)
Stimulates aldosterone secretion

- Increased Na⁺ and H₂O reabsorption
- Increased H⁺ secretion
- Increased K⁺ secretion

H^+ K^+

Distal tubule

Na^+

H_2O

(4) Fluid homeostasis is restored:

- Body water restored
- Blood volume/pressure restored
- Na⁺ concentration decreases to normal
- Osmolarity of ECF decreases to normal
- pH restored

Sequence the steps
@MasteringA&P®

Figure 25.12 Physiological responses to dehydration.

ion reabsorption. Finally, it also stimulates thirst, which will lead to increased water intake if conditions permit.

(3b) **It stimulates antidiuretic hormone (ADH) secretion.** Several stimuli, primarily angiotensin-II and the increased osmolarity of the ECF, stimulate increased release of ADH. This dramatically increases water reabsorption from the distal tubule and collecting system.

(3c) **It stimulates aldosterone secretion.** Angiotensin-II also stimulates aldosterone release. Aldosterone triggers sodium ion (and so water) reabsorption from the distal tubule and collecting system, while also stimulating potassium and hydrogen ion secretion. This hydrogen ion secretion helps to restore the normal pH of the blood.

(4) **Fluid homeostasis is restored.** All of the actions of angiotensin-II, ADH, and aldosterone combined increase the amount of water in the body. This restores the total body water, blood volume, and blood pressure, and decreases the osmolarity of the ECF back to normal. The increased amount of water also restores the normal sodium ion concentration. Note that although these mechanisms caused

sodium ion reabsorption, more water molecules are reabsorbed than sodium ions, so the concentration still returns to normal. Finally, the excretion of hydrogen ions restored the normal pH of the body fluids.

When the imbalanced variables return to their normal ranges, the secretion of angiotensin-II, ADH, and aldosterone is restored to the previous lower levels. Remember that this process shows how correction occurs under ideal conditions and that some cases of dehydration are either too severe or too complicated for these mechanisms to resolve fully.

Notice from this discussion that the mechanisms for correcting water imbalance in the body also correct electrolyte and pH imbalances. So, not only are the imbalances of each variable tied together, but the mechanisms that correct them are also inextricably linked.

Quick Check

☐ 1. What imbalances result from dehydration?
☐ 2. What hormonal mechanisms correct for these imbalances?

Apply What You Learned

☐ 1. How could the response to dehydration lead to a potassium ion imbalance? Explain.
☐ 2. Drugs known as ACE (angiotensin-converting enzyme) inhibitors block the production of active angiotensin-II. How would this blockage affect the response to dehydration?
☐ 3. Predict what would happen to overall homeostasis if this response operated as a positive feedback loop rather than a negative feedback loop.

See answers in Appendix A.

CHAPTER SUMMARY

Take full advantage of the invaluable online practice and study tools for this chapter by going to MasteringA&P®. There you will find:

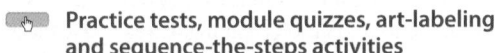 Practice tests, module quizzes, art-labeling and sequence-the-steps activities

 Animated figures with quizzes

 Practicing A&P worksheets

 Author-narrated podcast

 Dynamic Study Modules

iP *Interactive Physiology*

MODULE 25.1
Overview of Fluid, Electrolyte, and Acid-Base Homeostasis 985

- Body fluids include all of the body's water-based liquids. They function in transport, regulation of body temperature, cushioning, and lubrication.

- Water obeys the *principle of mass balance*—what is gained by the body must equal what is lost by the body.

- An **electrolyte** is a substance that dissociates into ions in water; the resulting charged particles conduct an electric current. Electrolytes in the body obey the principle of mass balance. (Figure 25.1, p. 986)

- An **acid** is a chemical that dissociates in water to release a hydrogen ion. A **base** is a chemical that dissociates in water to form an ion that can accept a hydrogen ion.

iP Play Interactive Physiology tutorial on Fluids, Electrolytes, and Acid/Base Balance: Introduction to Body Fluids.

MODULE 25.2
Fluid Homeostasis 987

- The **total body water** of the standard man is about 42 kg, or about 60% of the total body weight. Total body water is generally higher in men than in women, and higher in infants and young children than in adults.

- Body fluids are in two general compartments: the **intracellular compartment** and the **extracellular compartment.** (Figure 25.2, p. 988; Figure 25.3, p. 989)

- Water movement between the plasma and cytosol is determined primarily by osmotic pressure. The plasma is generally *isotonic* to the cytosol, so no net movement occurs between the two compartments. If the ECF is *hypertonic,* water is drawn out of the cytosol; if it is *hypotonic,* water is drawn into the cytosol. (Figure 25.4, p. 989)

- **Obligatory water loss** is the amount of water, about 500 ml, that must be lost each day in the urine in order to excrete a minimum amount of solutes. (Figure 25.5, p. 990)
  - Generally, 1500 ml of water is lost from the kidneys and 100 ml from the feces each day, which together make up the body's **sensible water loss.**
  - Another 900 ml of water is lost via evaporation from the skin and during expiration; losses from both of these routes make up the body's **insensible water loss.**

- The main factor that influences water gain is water intake, which is driven by the **thirst mechanism.** Changes in plasma osmolarity are detected by **osmoreceptors** in the thirst center of the hypothalamus. When plasma osmolarity increases, osmoreceptors trigger thirst; when it decreases, thirst is inhibited. Decreased plasma volume and angiotensin-II can also trigger thirst. (Figure 25.6, p. 991)

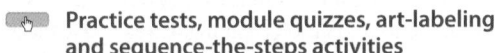 Practice sequencing the steps: Figure 25.6.

- On average, a person needs to consume about 1.5 liters of water per day, although this amount varies with activity.

- **Antidiuretic hormone (ADH)** plays a role in fluid balance by causing the insertion of water channels into the distal tubules and collecting systems of the kidneys, increasing water reabsorption. ADH secretion is stimulated by osmoreceptors, angiotensin-II, decreased blood volume, and decreased blood pressure.

- **Dehydration** is characterized by decreased volume and increased osmolarity of the ECF, and tends to cause cellular crenation. **Over-hydration** features an increased volume and decreased osmolarity of the ECF, and tends to cause cellular swelling. Isosmotic fluid imbalances include *hypovolemia* and **edema.**

iP Play Interactive Physiology tutorial on Fluids, Electrolytes, and Acid/Base Balance: Water Homeostasis.

MODULE 25.3
Electrolyte Homeostasis 994

- Sodium ions are more concentrated in the ECF. This concentration gradient for sodium ions is critical for depolarization of excitable cells and for the maintenance of ECF osmolarity. (Figure 25.7, p. 994)
  - **Hypernatremia** is an abnormal increase in the plasma sodium ion concentration. The most common cause of hypernatremia is dehydration, which tends to cause cellular crenation.
  - **Hyponatremia** is an abnormal decrease in the plasma sodium ion concentration. The most common cause is overhydration, and it tends to cause cellular swelling.

  🖑 Practice sequencing the steps: Figure 25.7.

- Potassium ions are more concentrated in the cytosol than the ECF. They are largely responsible for generating the negative resting membrane potential, and they cause the repolarization phase of the action potential.
  - When potassium ions are ingested, aldosterone, insulin, and epinephrine trigger their uptake by cells.
  - **Hyperkalemia** can be caused by renal failure, aldosterone insufficiency, cellular destruction from severe burns or trauma, and accidental administration or ingestion of excess potassium salts. It causes the resting membrane potential to become more positive than normal, and so cells are more easily excitable.
  - **Hypokalemia** is most commonly caused by diuretics. It causes the resting membrane to hyperpolarize, and so cells are more difficult to stimulate.

- Calcium ions are required for bone tissue, synaptic transmission, contraction of all three types of muscle tissue, blood clotting, and as intracellular messengers. They are regulated in tandem with phosphate ions.
  - PTH increases calcium ion reabsorption from the kidneys, inhibits phosphate reabsorption, and stimulates osteoclasts. It also activates vitamin D3, which assists PTH and stimulates absorption of phosphate and calcium ions from the small intestine.
  - **Hypercalcemia** is generally caused by hyperparathyroidism. It decreases the excitability of neurons and shortens the plateau phase of the cardiac action potential.
  - **Hypocalcemia** is generally due to decreased or ineffective PTH or inadequate vitamin D. Neurons become more excitable,

which can lead to tetanic muscle contractions. (Figure 25.8, p. 999)

- Chloride ions are important osmotic particles in the ECF. They, along with sodium ions, are generally reabsorbed from the kidneys.

- Magnesium ions are found in bone and are important in many biochemical reactions in the cells. Their low concentration in the ECF is maintained by renal excretion.

iP Play Interactive Physiology tutorial on Fluids, Electrolytes, and Acid/Base Balance: Electrolyte Homeostasis.

MODULE 25.4
Acid-Base Homeostasis 999

- The pH of the blood is maintained between 7.35 and 7.45 by the urinary system, respiratory system, and **buffer systems.**

- Sources of acids include those formed during metabolism, including the **volatile acids,** formed from carbonic acid, and **fixed acids,** such as lactic acid, amino acids, and fatty acids.

- **Chemical buffer systems** consist of a weak acid and its conjugate weak base. Their function is to resist large swings in pH. The *alkaline reserve* is the buffering capacity of the body's fluids. (Figure 25.9, p. 1001)
  - The **carbonic acid–bicarbonate ion buffer system** consists of the weak acid carbonic acid and its conjugate weak base bicarbonate ion. It is the most important buffer system for the blood.
  - The **phosphate buffer system** consists of the weak acid dihydrogen phosphate and its conjugate base, hydrogen phosphate. It is an important buffer system for the cytosol and the filtrate within the kidney tubules.
  - In **protein buffer systems,** many proteins such as hemoglobin can bind hydrogen ions and act as buffers.

- The respiratory system and urinary system are the body's **physiological buffer systems.** The respiratory system controls the amount of carbon dioxide in the body, and so it controls, to a large degree, the number of hydrogen and bicarbonate ions in the ECF.

- The kidneys contribute to pH balance in the following ways:
  - Proximal tubule cells secrete hydrogen ions and reabsorb bicarbonate ions.
  - New bicarbonate ions are formed in proximal tubule cells from the amino acid glutamine. (Figure 25.10, p. 1003)
  - New bicarbonate and hydrogen ions are formed in distal tubule cells from carbon dioxide in the interstitial fluid. The bicarbonate ions are reabsorbed into the interstitial fluid, and the hydrogen ions are secreted into the filtrate. (Figure 25.11, p. 1003)
  - Bicarbonate ions can be secreted into the late distal tubule and collecting system if the blood pH rises to an abnormally high level.

- Acid-base imbalances include **acidosis** and **alkalosis.** They are corrected by the respiratory and urinary systems, a process called **compensation.** (Table 25.1, p. 1006)
  - Acidosis is an ECF pH less than 7.35. It can lead to coma and death. **Respiratory acidosis** is caused by an abnormal decrease in ventilation and retention of carbon dioxide. It is

compensated for by the urinary system excreting more hydrogen ions and retaining more bicarbonate ions. **Metabolic acidosis** is caused by the addition of hydrogen ions to the ECF from a source other than carbon dioxide or by the loss of bicarbonate ions. The main compensatory mechanism is hyperventilation.

○ Alkalosis is an ECF pH greater than 7.45. It can lead to an increase in the excitability of neurons. **Respiratory alkalosis** is due to abnormal hyperventilation. If the hyperventilation has a physiological cause, the kidneys compensate by secreting bicarbonate ions and reabsorbing hydrogen ions. **Metabolic alkalosis** results from a loss of hydrogen ions or an excess of bicarbonate ions. Respiratory compensation occurs via hypoventilation, and the kidneys respond the same way as for respiratory alkalosis.

○ **Arterial blood gases** measure the pH, P_{CO_2}, and bicarbonate ion content of the arterial blood. Each type of acid-base imbalance has a characteristic finding on arterial blood gas analysis.

iP Play Interactive Physiology tutorial on Fluids, Electrolytes, and Acid/Base Balance: Acid/Base Homeostasis.

MODULE **25.5**
An Example of Fluid, Electrolyte, and Acid-Base Homeostasis 1007

● An imbalance in one variable will lead to imbalances in other variables, as evidenced by dehydration. In dehydration, the decreased amount of water in the body leads to increased levels of sodium ions (hypernatremia) and acidosis. (Figure 25.12, p. 1008)

Practice sequencing the steps: Figure 25.12.

● Mechanisms that correct dehydration also correct the accompanying hypernatremia and acidosis. These mechanisms include angiotensin-II, ADH, and aldosterone, which operate via negative feedback loops to restore homeostasis.

ASSESS WHAT YOU LEARNED

See answers in Appendix A.

LEVEL 1 Check Your Recall

1. Which of the following statements best describes the principle of mass balance?
   a. The amount of a variable that is gained by the body through ingestion equals the amount that is lost from the body.
   b. The body maintains a stable mass at all times.
   c. The amount of each variable brought into the body must be balanced by all other variables.
   d. The amount of a variable ingested is regulated by a positive feedback loop.

2. How does an electrolyte differ from a nonelectrolyte?

3. What happens to the pH of a solution when hydrogen ions are added?
   a. The pH increases.
   b. The pH decreases.
   c. The pH does not change.
   d. The pH does not measure hydrogen ion concentration.

4. As a percentage of body weight, the total body water tends to be higher in _____ and lower in _____.
   a. infants; men
   b. women; men
   c. men; infants
   d. infants; women

5. Fill in the blanks. A cell in a/an _____ fluid will lose water, and a cell in a/an _____ fluid will gain water.

6. Mark the following statements as true or false. If a statement is false, correct it to make a true statement.
   a. The thirst mechanism is mediated by osmoreceptors located in the cerebral cortex.
   b. The sensible water loss includes the water lost from the body via the skin and the respiratory system.
   c. The main hormone that regulates fluid balance is antidiuretic hormone (ADH).
   d. Atrial natriuretic peptide promotes water retention in the kidneys and increases the amount of water in the body.
   e. Dehydration is characterized by a decreased volume and increased osmolarity of the ECF.

7. Which of the following is *false* with respect to sodium ions in human physiology?
   a. Sodium ions are the most abundant extracellular cation.
   b. Sodium ions are an important osmotic particle in the ECF.
   c. The entry of sodium ions into a cell causes depolarization.
   d. Sodium ions are more concentrated in the cytosol than in the ECF.

8. List the effects of each of the following hormones on electrolyte balance. Note that some hormones affect more than one electrolyte.
   a. Angiotensin-II
   b. Aldosterone
   c. Parathyroid hormone
   d. Vitamin D
   e. Atrial natriuretic peptide

9. Which of the following effects tend to be caused by hypernatremia?
   a. Inhibition of ADH secretion
   b. Cellular crenation
   c. Cellular swelling
   d. Increased urine production

10. Mark the following statements as true or false. If a statement is false, correct it to make a true statement.
    a. Potassium ions are responsible for the repolarization phase of the action potential.
    b. Insulin, aldosterone, and epinephrine stimulate the uptake of potassium ions into cells.
    c. Hyperkalemia tends to decrease the resting membrane potential and hyperpolarize excitable cells.
    d. Hypocalcemia causes neurons to become hyperexcitable, leading to potential tetanic contractions.
    e. Chloride ions are generally reabsorbed from the kidneys, along with bicarbonate ions.

11. The biggest source of metabolic acids in the body is:

    a. lactic acid.
    b. ketone bodies.
    c. carbon dioxide.
    d. uric acid.

12. What is the main buffer system of the ECF?

    a. Protein buffer system
    b. Carbonic acid–bicarbonate ion buffer system
    c. Phosphate buffer system
    d. None of the above

13. Explain what happens to the pH of a buffered solution when hydrogen ions are added. Why does this happen?

14. Fill in the blanks: An increase in ventilation _____ the pH of the blood due to a/an _____ of carbon dioxide in the blood. A decrease in ventilation _____ the pH of the blood due to a/an _____ of carbon dioxide in the blood.

15. Which of the following mechanisms is/are used by the kidneys to regulate the pH of the blood? Circle all that apply.

    a. Hydrogen ions are secreted from the proximal and distal tubules and the collecting system.
    b. Hydrogen ions are reabsorbed from the nephron loop.
    c. New bicarbonate ions are formed from glutamine and carbon dioxide in the interstitial fluid that enters proximal tubule cells.
    d. Bicarbonate ions can be secreted.
    e. Bicarbonate ions are reabsorbed directly from the filtrate.

16. Mark the following statements as true or false. If a statement is false, correct it to make a true statement.

    a. Respiratory acidosis is caused by hypoventilation.
    b. Renal compensation for acid-base disturbances begins within minutes, whereas respiratory compensation begins within several hours.
    c. Respiratory compensation for metabolic acidosis consists of hypoventilation.
    d. Renal compensation for metabolic alkalosis consists of reabsorption of hydrogen ions and secretion of bicarbonate ions.
    e. You would expect to find an elevated P_{CO_2} in arterial blood gas analysis of a patient with respiratory acidosis.

17. Which of the following statements correctly describes the role of aldosterone?

    a. Aldosterone triggers sodium ion and so water reabsorption.
    b. Aldosterone triggers potassium ion secretion.
    c. Aldosterone triggers hydrogen ion secretion.
    d. Both a and b are correct.
    e. All of the above are correct.

18. How does angiotensin-II help to restore fluid balance when a person is dehydrated?

LEVEL 2 Check Your Understanding

1. Your friend argues that *all* water conducts electricity, regardless of what it contains. You prepare three liquids to test this hypothesis: one with deionized water (with no solutes, only water molecules), one with 5% glucose in water, and one with 5% sodium chloride in water. Which of these solutions, if any, will conduct an electric current? Explain.

2. A woman begins a diet and exercise regimen and loses 30 lb. Will her total body water increase, decrease, or stay the same? Why?

3. Explain how the amount of water in the body affects the concentration of ions and solutes in the ECF.

4. Diabetic ketoacidosis is characterized by an increased level of ketone bodies, which causes metabolic acidosis. A patient in diabetic ketoacidosis will have an altered rate of ventilation. Will the patient be hyperventilating or hypoventilating? How will a change in the rate and depth of ventilation compensate for an acidosis that is metabolic in nature?

LEVEL 3 Apply What You Learned

PART A: Application and Analysis

1. Elise Anderson is a 6-year-old girl who presents to the emergency department with a history of vomiting for the past 3 days. The nurse notices that her respiratory rate is abnormally low. What is the likely reason for this change in ventilation? Predict what Elise's arterial blood gas values would show.

2. What do you think has happened to Elise's ECF volume and osmolarity over the past 3 days? Will this lead to a change in the volume of water in the cytosol of her cells? Explain.

3. Ms. Johanssen is a patient in the hospital. The nurse examines her laboratory reports and notices that she has developed hyperkalemia and acidosis over the past several days. On closer examination of her medical chart, the nurse also sees that her physician recently doubled her dose of *spironolactone*, an aldosterone-blocking diuretic. How does this explain her laboratory findings?

4. A laboratory printout of arterial blood gases indicates that a patient has an increased P_{CO_2}, decreased pH, and normal bicarbonate ion concentration. Is this patient in acidosis or alkalosis? Is the pH disturbance respiratory or metabolic in nature? Explain your reasoning. How long do you think the patient has had this pH disturbance? (*Hint:* Look at the bicarbonate ion concentration. What system controls the concentration of bicarbonate ions, and how quickly does it compensate for pH disturbances?)

PART B: Make the Connection

5. What happens to the concentration of sodium ions in the ECF if you consume a large amount of salt without consuming any water? How will this impact the osmotic pressure of the ECF? Why could this lead to an elevation in blood pressure? (*Connects to Chapter 19*)

26

The Reproductive System

S o far, we have discussed organ systems that are absolutely required for survival. For example, how long could we live without a functional heart or brain? However, there is one body system that—although it does produce important hormones that affect other systems—is not technically necessary for survival: the **reproductive system.** In fact, this system appears to almost hibernate throughout most of an individual's childhood until **puberty**—the period of life when reproductive organs mature and become functional.

This chapter will explore the male and female reproductive organs. We will also look at how the reproductive systems are regulated and how the sex hormones influence other body systems. (We will save discussion of pregnancy and the functions of the female reproductive tract in this process until Chapter 27.) Let's begin by looking at the main parts of the reproductive system and the process of cell division that produces the *gametes,* or sex cells.

MODULE | **26.1**

Overview of the Reproductive System and Meiosis

Learning Outcomes

1. Compare and contrast the basic structure and function of the male and female reproductive systems.
2. Contrast the overall processes of mitosis and meiosis.
3. Describe the stages of meiosis.

In all animals, two forms of cell division take place: *mitosis* and *meiosis*. **Mitosis** (my-TOH-sis) is the cell division undergone by all *somatic*, or body, cells that are capable of division (see Chapter 3). (Recall that some cells are *amitotic*, or unable to divide once they have differentiated.) Mitosis produces "daughter" cells that are genetically identical to the "mother" cell. *Meiosis* (my-OH-sis), on the other hand, is a special type of cell division that occurs only in cells destined to become gametes. This module explores the process of meiosis and compares it to mitosis. Before we begin, however, let's introduce some basic principles and terms associated with the reproductive system.

Computer-generated image: Meiosis occurs only in the sex cells of testes and ovaries. This image shows pairs of chromosomes (orange) being pulled to opposite ends of a cell by spindles (blue).

Introduction to the Male and Female Reproductive Systems

◄◄ Flashback

1. What are the male and female gonads? (p. 619)
2. What hormones do the gonads produce? (p. 619)

As you are likely aware, the male and female reproductive systems differ in many ways. However, they also share a number of similarities, including the following:

- In both males and females, the most important organs in the reproductive system are the **gonads,** or **primary sex organs.** These are the *testes* in males and the *ovaries* in females.
- The gonads secrete **sex hormones,** including *testosterone* and *estrogens.*
- The gonads also produce **gametes,** or sex cells, through the process of meiosis. In males the gametes are **sperm;** in females the gametes are **ova,** also called *egg cells.*
- Both males and females have additional organs and structures that contribute to the functioning of the reproductive system. These are known as **accessory reproductive organs.**

The similarities between the male and female reproductive systems reflect their similar function. Keeping this in mind will help you as you proceed through this chapter.

Quick Check

☐ 1. What are the male and female gonads? What are their functions?

Overview of Meiosis

During **fertilization,** the process by which a sperm and egg cell fuse, a new single cell called a **zygote** is formed. The zygote is the cell that will eventually divide to produce all of the cells in a new individual. For this reason, it must contain the correct number of chromosomes. Thus, each sperm and ovum must contribute only half of the chromosomes that we find in a somatic cell so the zygote does not end up with double the normal number. This is made possible by the process of **meiosis,** during which a cell divides to form daughter cells with half the number of chromosomes.

All human somatic cells with a nucleus have 46 chromosomes, and these chromosomes are paired so that there are 23 pairs of chromosomes in each nucleus (see Chapter 3). Somatic cells are called **diploid** (DIP-loyd; *di* = "two," *ploid* = "having chromosome sets"), or **2n,** because the chromosomes are paired. One set, or 23, of these chromosomes consists of the **paternal chromosomes** that came from the father of the individual. The other set, the **maternal chromosomes,** came from the individual's mother.

Each member of a pair of chromosomes has the same **genes,** which are segments of the DNA that code for a specific protein. For this reason, the pairs are called **homologous chromosomes** (hah-MAH-luh-gus; "sharing a similar structure"), shown here:

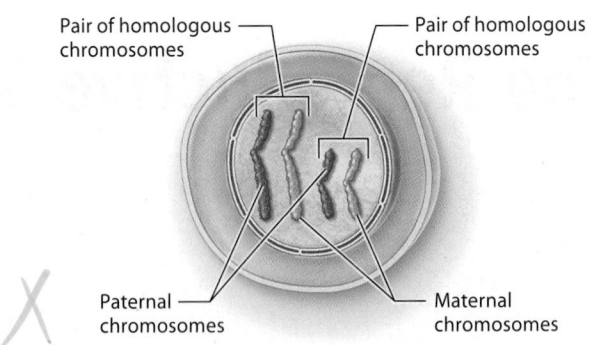

However, the matching genes on the homologous chromosomes may have different variants, or **alleles** (uh-LEELZ). For example, let's say that one homologous pair of chromosomes includes the gene for hair color. The paternal chromosome of this homologous pair might have an allele for brown hair, whereas the maternal chromosome of this pair might have an allele for blond hair. (For an in-depth discussion of this topic, see Chapter 27.)

As with mitosis, just before meiosis begins, all of the cell's DNA is replicated so that each new cell will get a complete copy. The result of DNA replication is a cell with two identical copies of its chromosomes, which are called **sister chromatids** (KROH-mah-tidz). Each pair of sister chromatids is connected in the center by a region known as a *centromere,* which you can see here:

Recall from the cell chapter (Chapter 3) that when a cell is not dividing, its DNA is in the form of **chromatin,** which is one very long piece of DNA with its associated proteins. After replication, when cell division begins, chromatin threads coil tightly and condense into the barlike chromosomes.

Also like mitosis, meiosis has four basic phases: prophase, metaphase, anaphase, and telophase. However, in meiosis each of these phases occurs twice. Meiosis includes two successive divisions, **meiosis I,** or the *first meiotic division,* and **meiosis II,** or the *second meiotic division.*

Meiosis I (First Meiotic Division)

◄◄ Flashback

1. What are spindle fibers? (p. 110)
2. What are centrioles? (p. 97)

The first meiotic division separates the homologous pairs to produce **haploid** (or **1n**) cells, which have only half the number of chromosomes of the original cell. For this reason, this process is sometimes referred to as *reduction division*.

ConceptBOOST »»»

Understanding Diploid and Haploid Cells

There is often confusion as to why the daughter cells at the end of meiosis I are haploid because they might *appear* to have the same absolute number of chromosomes as the starting mother cell (four in our simplified example in Figure 26.1). However, remember that you are not seeing four individual chromosomes—you are seeing two sets of sister chromatids. A set of sister chromatids is simply two identical copies of a *single* homologous chromosome.

Here's a simple analogy: Say your class requires two textbooks, and each book contains similar but slightly different information. The two books represent a pair of homologous chromosomes found in a diploid cell—each homologous chromosome contains similar but slightly different information in its DNA. Now imagine that the bookstore accidentally ships you two copies of the *same* book instead of two different ones. You now have two identical copies of a single book, not two individual books. This represents a pair of sister chromatids—they are simply two copies of the exact same genetic information. You can see a simple progression here, using an example of a cell with two homologous chromosomes in the mother cell:

2 homologous chromosomes– Diploid

DNA replication

2 copies of 2 homologous chromosomes– Diploid

Meiosis I

2 copies of one homologous chromosome– Haploid

2 copies of one homologous chromosome– Haploid

So, the diploid mother cell here has two homologous chromosomes, or two individual books. But by the end of meiosis I, the resulting cells have only one replicated homologous chromosome, or two copies each of a single book. The number of distinct chromosomes (or books) has been reduced by half, from two to one, making the cell now haploid. Note that we have

simplified this somewhat because, as you will see, sister chromatids are not necessarily identical at the end of meiosis I. However, the basic principle remains: A set of sister chromatids represents a single homologous chromosome. ∎

Even though the DNA is in its noncondensed form (chromatin) while undergoing replication, the first cell diagram in **Figure 26.1** shows what the chromosomes would look like if they had already assumed their characteristic shape. Notice that one chromosome in each pair is labeled "paternal" and the other is labeled "maternal."

Early Prophase I

At the beginning of prophase I, the DNA of the chromosomes has already been replicated. The chromatin then condenses so that individual sister chromatids become visible through a microscope as thin lines. At the same time, nucleoli disappear, the nuclear membrane temporarily breaks down, and spindle fibers and centrioles form and align at opposite poles.

Middle to Late Prophase I

During middle to late prophase I, the homologous chromosomes move together and pair up tightly side by side in a process called **synapsis** (sih-NAP-sis). When the two homologous chromosomes are touching, they are called **tetrads** (TEH-tradz). While touching, some sister chromatids break in one or more places and homologous chromosomes exchange segments of DNA, forming chromatids with new combinations of genetic information in a step referred to as **crossing over.** This is the first process that shuffles the alleles, and changes the chromosomes so that the sister chromatids are *no longer identical.* This will result in genetically unique cells at the end of meiosis.

Metaphase I

At the time of metaphase I, the sister chromatids of the homologous pairs line up on either side of the equator, or midline of the cell, forming a *double line* of chromosomes. Spindle fibers then branch out from the centrioles at opposite ends of the poles of the cell and attach to each set of sister chromatids at the centromere. Note that the orientation of paired chromosomes is random with respect to which chromosome from the pair is on one side of the equator or the other. For example, any combination of maternal chromosomes may line up on one side of the equator, with the remainder on the other side. The same random combinations occur with paternal chromosomes.

Anaphase I

In anaphase I, one set of sister chromatids from each homologous pair separates from its partner as they are pulled to opposite poles of the cell by the spindle fibers. Due to the random orientation of homologous pairs that occurred during metaphase I, separation of the maternal and paternal chromosomes is random as well. This process is called **independent assortment** and

BEFORE BIRTH

Cells before DNA Replication
- This is what the cell would look like if the chromatin condensed into chromosomes before the DNA replicated.

"Mother cell" — Maternal chromosomes

Paternal chromosomes

MEIOSIS I

Early Prophase I
- Chromosomes form with two sister chromatids.

Chromosomes

Centrioles

Mid- to Late Prophase I
- During synapsis, homologous chromosomes form tetrads and **crossing over** occurs.

Tetrad

Crossing over

Spindle fibers

Metaphase I
- Tetrads align randomly at equator (random orientation).

Paired homologous chromosomes

Equator

Random orientation

Anaphase I
- Random orientation in metaphase I leads to **independent assortment.**

Independent assortment

Telophase I
- Cytokinesis may follow, resulting in two genetically different haploid cells with sister chromatids still attached.

MEIOSIS II

Prophase II
- Chromosomes remain condensed.

Metaphase II
- Chromosomes line up along equator.

Anaphase II
- Sister chromatids separate.

Telophase II
- Cytokinesis follows.

Meiosis produces four genetically unique, haploid daughter cells.

Figure 26.1 Stages of meiosis. For simplicity, the cell is shown with only two pairs of homologous chromosomes.

is the second shuffling of the alleles, further increasing genetic variability of the daughter cells.

Telophase I

By the time we reach telophase I, the sets of sister chromatids arrive at opposite ends of the cell. A nuclear membrane may re-form around them, but this does not always occur. **Cytokinesis** (sy′-toh-kuh-NEE-sis), or division of the cytoplasm, then takes place, separating the cells.

At the end of this phase, we have two haploid cells, each containing 23 individual homologous chromosomes, with each chromosome having two sister chromatids. Also, because of crossing over and independent assortment, these two cells are genetically different from each other.

Quick Check

- ☐ 2. Which mechanisms increase the genetic variability of daughter cells?
- ☐ 3. Why are cells haploid at the end of meiosis I?

Meiosis II (Second Meiotic Division)

The second meiotic division separates the chromatids of each chromosome. The resulting cells will still be haploid, but each cell will have only one chromatid from the original four in a pair of homologous chromosomes.

Prophase II

The second prophase event of meiosis is the same as prophase of mitosis, and it occurs in both cells produced from meiosis I. The nuclear membrane, if re-formed, breaks down and the sister chromatids remain condensed.

Metaphase II

During metaphase II, each pair of sister chromatids aligns to form a single line along the equator of the cell. Spindle fibers extend from the centrioles to each sister chromatid, attaching at the centromere.

Anaphase II

The sister chromatids of each chromosome separate during ana-phase II and are pulled to opposite ends of the cell by spindle fibers. Each sister chromatid is now called a chromosome or a nonreplicated chromosome.

Telophase II

Finally, during telophase II, the nonreplicated chromosomes arrive at opposite ends of the cell. The nuclear membrane re-forms, cytokinesis occurs, and the spindle fibers and centri-oles disassemble. One original cell has become four genetically unique daughter cells, each with 23 nonreplicated chromosomes. In males, these four daughter cells mature into four sperm cells. In females, only one daughter cell will go on to poten-tially become an ovum. However, the basic steps of meiosis are the same for sperm and ovum production, as they are necessary to achieve the correct chromosome number and ensure genetic variability in the cells. The specifics of sperm and ovum produc-tion are discussed in later modules.

To compare the processes of mitosis and meiosis, see **Figure 26.2.**

ConceptBOOST »»»

Comparing Mitosis and Meiosis

Mitosis occurs because new cells are needed for tissue growth or repair, and these new cells must be genetically identical to the original cell. So, the process of mitosis occurs by first making a complete copy of the DNA and then dividing the cells so that each cell gets one copy of the DNA (see Figure 26.2, left side). The result is two genetically identical daughter cells.

In contrast, meiosis produces sperm and ova for reproduction. These cells need to have half the chromosome number of the original cell so that when they combine to form a new cell, the chromosome number will not be doubled. Also, meiosis mixes up the possible combinations of genes that each sperm or ovum will receive, to produce a genetically unique individual. Meiosis I includes the unique set of steps that reduce the chromosome number *and* shuffle the alleles (see Figure 26.2, right side). Meiosis II then completes the process to potentially form four genetically unique daughter cells. ∎

Figure 26.2 Comparing mitosis and meiosis.

Quick Check

☐ 4. How many daughter cells can potentially be produced from one mother cell in meiosis?
☐ 5. How do daughter cells differ from the mother cell?
☐ 6. How do mitosis and meiosis differ?

Apply What You Learned

☐ 1. Even though we pass on the genes we inherit from our own parents to our children, we do not pass on the same alleles to each child. How can you explain this statement?
☐ 2. What do you think would happen if crossing over and independent assortment no longer occurred during meiosis? (*Hint:* Compare the final cells of mitosis to those of meiosis.)
☐ 3. Suppose that a homologous chromosome pair fails to separate during the first meiotic division, a problem known as nondisjunction. Predict the effects on potential offspring.

See answers in Appendix A.

MODULE **26.2**

Anatomy of the Male Reproductive System

Learning Outcomes

1. Describe the structure and functions of the male reproductive system.
2. Trace the pathway of sperm from the seminiferous tubules to the external urethral orifice of the penis.

3. Describe the organs involved in semen production.
4. Discuss the composition of semen and its role in sperm function.

We turn first to the anatomy of male reproductive structures (**Figure 26.3**). To begin, the paired *testes,* or male gonads, are located in the *scrotum.* The testes produce sperm, which travel through a series of tubes known as *ducts* opening to the outside of the body through the *penis. Accessory glands* secrete fluid as part of *semen,* the fluid that accompanies sperm, into the ducts. Figure 26.3 shows a midsagittal section of the internal organs of the male reproductive system. Note the location of the reproductive organs with respect to the urinary bladder and rectum.

Testes

◀◀ **Flashback**

1. Where is the abdominopelvic cavity? (p. 17)
2. What is the peritoneum? (p. 20)

The paired **testes** (or *testicles;* singular, *testis*) are ovoid structures approximately 4 cm (1.6 in.) long and 2.5 cm (1 in.) wide. They are located outside the abdominopelvic cavity in a saclike structure, the **scrotum,** that is composed of skin, connective tissue, and smooth muscle. Connective tissue surrounds each testis and divides it internally into about 250 internal segments known as **lobules** (**Figure 26.4a**), which contain one to four tightly coiled looped **seminiferous tubules** (seh'-men-IF-er-us). These tiny tubules are the sites of sperm production.

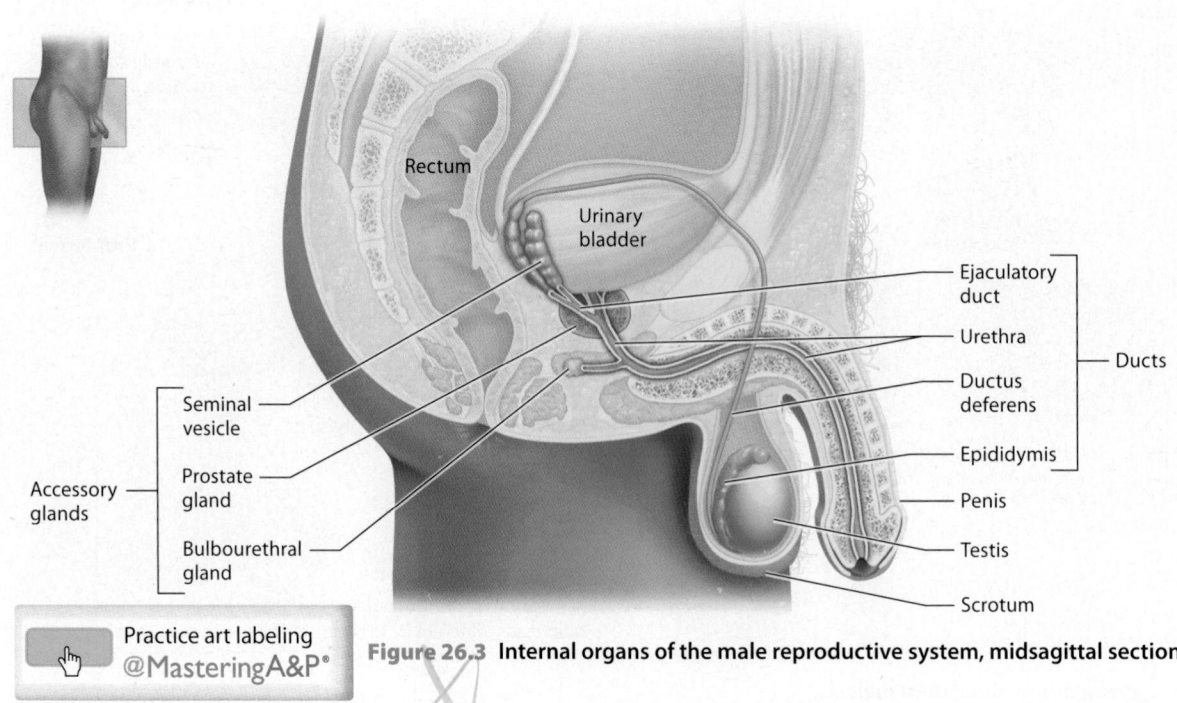

Rectum

Urinary bladder

Accessory glands
— Seminal vesicle
— Prostate gland
— Bulbourethral gland

Ejaculatory duct
Urethra
Ductus deferens
Epididymis
Penis
Testis
Scrotum

Ducts

Practice art labeling
@MasteringA&P®

Figure 26.3 Internal organs of the male reproductive system, midsagittal section.

The testes perform two very important functions. They (1) produce sperm cells and (2) secrete androgen hormones, most importantly *testosterone.* The seminiferous tubules contain two types of cells: *spermatogenic,* or sperm-forming cells, and *sustentacular* cells, which have several functions that support the production of sperm. One such function includes the production of *testicular fluid,* which contains nutrients for sperm cells. Between the seminiferous tubules lie **interstitial cells,** or *Leydig cells* (**Figure 26.4b and c**), which produce and secrete androgens, mostly testosterone, into the surrounding interstitial fluid.

Seminiferous tubules are surrounded by muscle-like **myoid cells,** which contract to push sperm and testicular fluids through the tubules and out of the testes. The seminiferous tubules of each lobule merge into a single **straight tubule** that moves sperm into the **rete testis** (REE-tee TES-tus), a network of tubules on the posterior side of the testis. Sperm then pass through **efferent ductules** and enter the first portion of the duct system, the *epididymis.*

Each testis receives blood from a **testicular artery** that branches from the abdominal aorta above the pelvis. A network of veins called the **pampiniform venous plexus** drains blood from the testes into *testicular veins.* The testes are innervated by both the sympathetic and parasympathetic divisions of the autonomic nervous system. In addition, a high number of pain receptors, or *nociceptors,* and thermoreceptors are concentrated in the testes and surrounding connective tissues.

Quick Check

☐ 1. Which cell type in the testes produces sperm? Which cell type produces testosterone?
☐ 2. What is the specific site of sperm production in the testes?

Duct System

◄◄ **Flashback**

1. What are microvilli? (p. 97)
2. What other system includes the urethra as an important structure? (p. 942)

After leaving the testis, the sperm travel through a system of ducts: the *epididymis, ductus deferens, ejaculatory duct,* and *urethra.* As sperm move through these ducts, they are supported and nourished by secretions, collectively known as *semen,* from accessory structures (discussed in the next section). The duct system transports sperm and is also involved in **ejaculation** (ee-jak′-yu-LAY-shun), the process by which semen is expelled from the penis.

(a) Partial midsagittal section through testis attached at spermatic cord

(b) Cross-section of seminiferous tubule

LM (200×)

(c) Light micrograph of seminiferous tubule

Practice art labeling @MasteringA&P®

Figure 26.4 Internal structures of the testis and epididymis.

Epididymis

The **epididymis** (ep′-uh-DID-uh-miss) is the site of sperm maturation and storage and is filled with many ductules. As you can see in Figure 26.4a, each epididymis sits on the superior and posterior surface of a testis and includes head, body, and tail regions. The *head* of the epididymis contains the efferent ductules of the testis and sits on its superior surface. Sperm move from the head to the *body* of the epididymis, which is its narrow midportion, and finally into the *tail*, its smaller, inferior portion.

Sperm move from the efferent ductules into a single tube called the **ductus epididymis** that passes through all regions of the epididymis. The ductus epididymis is quite long – about 6 m (19.7 ft) – but is so tightly coiled and twisted that it fits inside the 3.8 cm (1.5 in.) long epididymis. The mucosa lining the hollow lumen of the ductus epididymis includes pseudostratified epithelial cells with long, nonmotile microvilli (or *stereocilia*) that absorb excess testicular fluid, pass nutrients to the sperm in the lumen, and complete the maturation process of the sperm. The long length

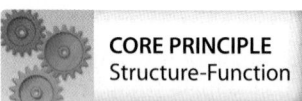

CORE PRINCIPLE
Structure-Function

of the ductus epididymis allows ample time for sperm to mature, and also permits sperm to be stored for several months if ejaculation

does not occur. Here we see an example of the Structure-Function Core Principle (p. 25). Sperm that are not ejaculated are resorbed.

Ductus Deferens

The ductus epididymis widens within the tail of the epididymis and becomes the **ductus deferens** (or *vas deferens*). The ductus deferens is very long (40–45 cm, or 15.7–17.7 in.) and thin (about 2.5 mm, or 0.1 in., in diameter). Notice in Figure 26.4a that it extends along the posterior border of the epididymis inside the *spermatic cord* with the testicular arteries, veins, and nerves. It enters the pelvic cavity through a mostly fibrous tunnel called the *inguinal canal* (see Figure 26.6), after which it passes along the lateral side of the bladder, over the ureter, and then along the posterior bladder. Here it ends in a saclike wider area called the **ampulla** (**Figure 26.5a**).

The organization of the tissue layers of the ductus deferens is the same as you have seen for other organs: an inner mucosa, a middle layer of smooth muscle, and an outer adventitia layer. The mucosa of the ductus deferens consists of pseudostratified columnar epithelium resting on loose connective tissue connected. The muscularis consists of three layers of smooth muscle; the inner and outer layers are longitudinal, and the middle layer is circular. During ejaculation,

Ureters

Urinary bladder

Ductus deferens

Ampulla of ductus deferens

Seminal vesicle

Ejaculatory duct

Prostate gland

Bulbourethral gland

Root of penis — Crus of penis

Bulb of penis

Body of penis — Corpora cavernosa

Corpus spongiosum

Glans penis — Prepuce

External urethral orifice

Prostatic urethra

Membranous urethra

Urethra

Spongy urethra

Epididymis

Testis

(a) Posterior view of the duct system showing a frontal section of the penis

Dorsal arteries

Corpora cavernosa

Deep arteries

Corpus spongiosum

Spongy urethra

(b) Cross-section of the penis with dorsal surface on top

Figure 26.5 Male reproductive duct system and penis.

the smooth muscle creates strong contractions that rapidly squeeze the sperm forward along the tract. Like the epididymis, the ductus deferens can store sperm for several months and resorb any sperm not ejaculated by that time.

Ejaculatory Duct

Sperm next move into the short **ejaculatory duct,** which is approximately 2 cm (0.8 in.) long. It is located at the junction of the ampulla of the ductus deferens and the duct of an accessory gland called the *seminal vesicle.* Each ejaculatory duct then travels through another accessory gland, the *prostate gland,* and empties into the *urethra* (Figure 26.5a).

Urethra

Remember from your study of the urinary system that the **urethra,** which is about 18–20 cm (7.1–7.9 in.) long, connects the urinary bladder to the external body surface. In males, it moves both urine and semen (not at the same time, however), so it belongs to both urinary and reproductive systems. You can see in Figure 26.5a that the male urethra has three regions:

- the *prostatic urethra* is surrounded by the prostate gland;
- the *membranous urethra* passes through the external urethral sphincter; and
- the *spongy urethra* extends through the *penis* and terminates at the *external urethral orifice.*

As with the ductus deferens, smooth muscle in the wall of the urethra contracts during ejaculation to help expel semen through the external urethral orifice, The mucosa of the urethra contains several small, mucus-secreting *urethral glands* empty into the urethra and contribute to semen.

Quick Check

☐ 3. What is the function of the epididymis? How does it perform this function?
☐ 4. Trace the pathway that sperm take from the seminiferous tubules to the outside of the body.

Penis

◀◀ **Flashback**

1. Where are the iliac arteries and veins? (p. 694)
2. What muscles are found in the floor of the perineum? (p. 310)

The **penis** delivers sperm into the female reproductive tract and is therefore called the male *copulatory* organ. At the base of the penis is the **root.** The **body,** or *shaft,* enlarges into the end of the penis, called the **glans penis,** which is where the external urethral orifice is located (Figure 26.5a). The skin covering the penis is loose and forms a circular fold called the **prepuce** (PREE-pyus), or *foreskin,* around the glans. The prepuce may be removed shortly after birth in a procedure called *circumcision.*

There is a great deal of controversy about the medical necessity of circumcision, with many studies showing no measurable benefit.

Internally, the penis contains three cylindrical erectile bodies, or *corpora* (singular, *corpus*), covered by sheaths of dense fibrous connective tissue, which you can see in the transverse section in Figure 26.5b. Each erectile body is a spongy network of connective tissue and smooth muscle filled with vascular spaces. The first two erectile bodies are the paired dorsal **corpora cavernosa** (KOR-por-uh ka-ver-NOH-sa). At the base of the penis, the two corpora cavernosa split to form the two **crura** (KROO-ra; singular, *crus*) of the penis that attach to the ischial rami. The other erectile body is the ventral **corpus spongiosum** (KOR-pus spun-jee-OH-sum), which expands to form the **bulb** of the penis at the base. You can see in Figure 26.5a how the bulb sits between the two crura; together these structures make up the root of the penis and connect the penis to the pelvic bones. The corpus spongiosum also expands at its distal end to form the glans. During sexual excitement, the vascular spaces fill with blood, which causes the penis to enlarge and become rigid. This event, called *erection,* is discussed in Module 26.3.

The blood supply to the penis comes from the internal iliac arteries, which give rise to the internal penile arteries; these divide to form the *dorsal arteries* and the *deep arteries.* The **dorsal arteries** supply the penile skin, fascia, and corpus spongiosum, and the **deep arteries** feed the corpora cavernosa.

The penis is innervated by many sensory and motor fibers, including both sympathetic and parasympathetic nerves. Tactile, pressure, and temperature receptors are abundant here. These are important for sexual stimulation as well as control of muscles that regulate the distance of the scrotum from the body wall.

Together, the penis and scrotum make up the **external male genitalia.** They are part of the male **perineum,** a diamond-shaped area between the thighs bordered by the pubic symphysis anteriorly, the ischial tuberosities laterally, and the coccyx posteriorly. The perineum is separated into the **urogenital** (yoo′-roh-JEN-uh-tul) **triangle,** which contains the base of the penis and the scrotum, and the **anal triangle,** which contains the anus. The floor of the perineum is formed by muscles (described in Chapter 9).

Quick Check

☐ 5. What happens to the corpora cavernosa and the corpus spongiosum during sexual excitement?
☐ 6. Why are sensory receptors abundant in penile tissue?

Accessory Sex Glands

◀◀ **Flashback**

1. What kinds of tissues make up an exocrine gland? (p. 135)
2. How do exocrine glands secrete their products? (p. 135)

The accessory sex glands of the male reproductive system are exocrine glands. They produce most of the liquid portion of semen, the mixture of secretions and sperm expelled during ejaculation, and help lubricate the outside of the penis during intercourse. The accessory glands include the *seminal vesicles, prostate, and bulbourethral glands* (see Figure 26.5a).

Seminal Vesicles

The paired **seminal vesicles** (SEH-min-ul) are found on the posterior surface of the urinary bladder near the ampullae of the ductus deferens. Each seminal vesicle is about 15 cm (5.9 in.) long, but is coiled and folded until it is about 5 cm (2.0 in.) long. The duct of each seminal vesicle converges with the ductus deferens to form the ejaculatory ducts. The exterior of each gland has a fibrous capsule that covers a thick layer of smooth muscle. Deep to this is the mucosa layer, which is made of a pseudostratified columnar epithelium that secretes *seminal fluid.*

Seminal fluid, a yellowish secretion that makes up approximately 60–70% of total semen volume, contains:

- fructose, a sugar that sperm use as a nutrient to make ATP;
- prostaglandins, which stimulate smooth muscle contractions in both the male and female reproductive tracts and increase sperm viability; and
- coagulating proteins and enzymes (a coagulating protein, together with an enzyme from the prostate, forms a temporary clot of semen in the female reproductive tract).

The pH of the seminal fluid is alkaline to help neutralize acids from the male urethra and the female reproductive tract.

Prostate

The **prostate** (PRAHS-tayt; be careful not to call it the "pros*trate*") is an egg-sized gland that sits just inferior to the urinary bladder and surrounds the urethra and ejaculatory ducts. Covered by an outer fibrous capsule, it is made up of 20–30 tubular glands embedded in a mass of smooth muscle and dense connective tissue. During ejaculation, prostatic smooth muscle contracts and squeezes the *prostatic secretions* into the prostatic urethra via several ducts. The prostate gland tends to enlarge with age; you can find out more about this in *A&P in the Real World: Benign Prostatic Hyperplasia (BPH) and Prostate Cancer.*

Prostatic secretions (prah-STAT-ik) are milky in appearance and make up about 20–30% of semen volume. They contain:

- citrate, a sugar that sperm use as a nutrient to make ATP;
- prostate-specific antigen (PSA) and other enzymes, which dissolve the clot of semen that initially forms in the female reproductive tract so that the sperm can move deeper into the tract; and
- antimicrobial chemicals, which inhibit the growth of some bacteria to decrease the risk of infection in the female reproductive system.

The pH of the prostatic fluid is also alkaline to help neutralize acids from the male urethra and the female reproductive tract.

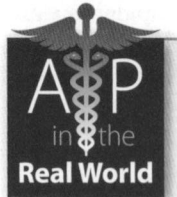

A&P in the Real World

Benign Prostatic Hyperplasia (BPH) and Prostate Cancer

The prostate slowly enlarges from birth to puberty and then expands rapidly until about age 30. Further enlargement often occurs after age 45. If the prostate is noncancerous but expands to the point of pushing on the urethra, the condition is called **benign prostatic hyperplasia,** or **BPH.**

Sometimes an enlarged prostate is secondary to prostate cancer, which is the second most common cancer in U.S. men (after non-melanoma skin cancer). Prostate cancer is one of the leading causes of cancer death among men of all races. Risk factors for the development of prostate cancer include genetic predisposition, African ethnic heritage, age over 50 years, obesity, and poor diet. Screening for prostate cancer usually includes a digital rectal examination—in which a medical professional palpates the prostate through the anterior rectal wall—and assessment of blood prostate-specific antigen (PSA) levels.

PSA blood levels below 2.5 ng/ml are considered normal. When there is an increase in cell number, as in either BPH or prostate cancer, the PSA rises. If there are any suspicious areas, such as lumps, or if PSA levels are over 4.0 ng/ml, tissue biopsy specimens from different sites in the prostate are taken to look for cancerous cells.

Prostate cancer may be treated with surgery, radiation therapy, chemotherapy, or hormone therapy. In some men, the tumors grow very slowly and have a low risk of metastasis. In these cases, regular monitoring of the disease may be all that is required. However, some tumors are quite aggressive and metastasize to other tissues rapidly. For this reason, screening and early detection are critically important.

Bulbourethral Glands

The paired **bulbourethral glands** (bul'-boh-yoo-REETH-ruhl), or *Cowper's glands,* are located at the base of the penis, on either side of the membranous urethra. The size of a small marble, each round gland is about 1 cm (0.4 in.) in diameter with a short duct that connects to the urethra.

In response to sexual stimulation, the bulbourethral glands secrete a thick alkaline mucus-like fluid that helps neutralize any acidic urine remaining in the urethra prior to ejaculation. It also lubricates the urethra for the passage of ejaculate. Unlike the prostate gland, bulbourethral glands grow smaller with age and are very small in older men.

Semen

◀◀ **Flashback**

1. What is coagulation? What are clotting proteins? (p. 725)
2. What is a flagellum? (p. 98)

Semen is a milky white, somewhat sticky mixture of sperm and fluids from the testes, seminal vesicles, prostate, bulbourethral glands, and urethral glands. Sperm contribute only about 5% to semen's total volume. Each typical **ejaculate,** which is the amount of semen expelled during one ejaculation, measures 2.5–5 ml in volume and contains 40–750 million sperm cells. This may seem like a large number of sperm; however, only a small fraction will ever reach the ovum.

About 5 minutes after semen is ejaculated, it coagulates because of the combination of a clotting protein from the seminal vesicles and a clotting enzyme from the prostate. This coagulation prevents the sperm from leaking back out of the female reproductive tract. Fifteen to 30 minutes later, the clot breaks down because of PSA and other anticoagulant enzymes from the prostate. As it dissolves, the sperm are then gradually released from the clot, are activated, and begin their ascent up the female reproductive tract.

The alkaline pH of semen helps make the sperm fully motile, and the sperm begin the process of **capacitation** (kuh-pa′-suh-TAY-shun). By this process, sperm undergo changes that enable them to penetrate and fertilize an immature female gamete. They will finish capacitation in the female reproductive tract, where secretions will prepare the sperm cell's plasma membrane to fuse with that of the female gamete.

Prostaglandins from the prostate are thought to thin the mucus in the female reproductive tract and may stimulate contractions, which pull the semen farther into the tract, facilitating fertilization. Antimicrobial chemicals from the prostate kill some bacteria, including *Escherichia coli,* which live in the large intestine and may cause illness if allowed to colonize the reproductive tract.

Quick Check

☐ 7. How many sperm are in each milliliter of semen?
☐ 8. What are the functions of semen?

Support Structures: Scrotum and Spermatic Cord

◄◄ Flashback

1. What is fascia? (p. 161)
2. What are thermoreceptors? (p. 162)

The *scrotum,* the sac that contains the testes, is located between the thighs (**Figure 26.6**). A midline **septum** divides the scrotum into two internal compartments, one for each testis. The location of the septum is marked externally by a ridgelike seam at the midline, called the **raphe** (RAY-fee), which continues anteriorly along the ventral side of the penis and posteriorly to the anus. The wall of the scrotum includes an outer layer of skin, a layer of superficial fascia, and a layer of smooth muscle called the **dartos muscle.**

Extending from the scrotum is the **spermatic cord,** a tube that is made up largely of layers of fascia and contains the ductus deferens, which passes into the pelvic cavity. The spermatic cord also contains blood vessels, nerves, and lymphatic vessels, as well as smooth muscle tissue known collectively as the **cremaster muscle** (kreh-MASS-ter), which controls the height of the testes. It enters the pelvic cavity by passing through the *external inguinal ring,* and ends at the **inguinal canal,** a passageway into the abdomen. Note that sometimes the abdominal contents can protrude through the inguinal canal, leading to an *inguinal hernia.* This condition is about 25 times more common in men than in women, due mostly to the large size of the inguinal canal and external ring in males, necessary to accommodate the spermatic cord.

The scrotum and spermatic cord together help to support sperm production through temperature regulation. The core body temperature of 37° C is too warm for producing large numbers of viable sperm. Under normal circumstances, the temperature of the scrotum is about 3° C lower, which is the optimal temperature for developing sperm. This temperature difference is controlled in two ways.

First, remember that the spermatic cord contains the testicular artery and an extensive network of veins, the pampiniform venous plexus, which surrounds the testicular artery before draining into the testicular veins. In another example of the Structure-Function Core Principle (p. 25), this unique anatomical arrangement of blood vessels results in a *countercurrent heat exchange.* When warm arterial blood passes into the testis, the cooler blood from the pampiniform plexus passes close by, moving in the opposite direction. Some of the heat from the arterial blood transfers to the cooler blood, making the blood that enters the testis a few degrees cooler.

CORE PRINCIPLE
Structure-Function

External inguinal ring (end of inguinal canal)

Ductus deferens

Spermatic cord

Penis

Cremaster muscle

Dartos muscle

Scrotum

Urinary bladder

Ductus deferens

Pampiniform venous plexus

Testicular artery

Septum of scrotum

Epididymis

Testes

Raphe

Figure 26.6 The scrotum and spermatic cord.

Second, the scrotum can be cooled or heated slightly by controlling its proximity to the body cavity. For example, when the external temperature is cold, the testes are pulled closer to the pelvic floor and the warmth of the body wall, and the scrotum becomes shorter and heavily wrinkled. This decreases the surface area of the scrotum and reduces heat loss. When the external temperature is hot, the scrotal sac relaxes and becomes *flaccid*, or limp. This increases the surface area for cooling, which may involve sweating, and moves the testes farther away from the warm body trunk. These changes in scrotal surface area are controlled by the dartos muscle, which contracts to wrinkle the scrotal skin, and the cremaster muscle, which contracts to elevate the testes.

The scrotum is innervated by many sensory and motor fibers and includes both sympathetic and parasympathetic innervation. In addition to the thermoreceptors listed above, the scrotum has a high number of tactile receptors important in sexual stimulation. The cremaster muscle also often contracts when a man is sexually aroused to protect the area during sexual intercourse.

Table 26.1 lists functions of the male reproductive organs.

TABLE 26.1 FUNCTIONS OF THE MALE REPRODUCTIVE STRUCTURES

Structure	Function
Internal Genitalia	
Testis	Produces sperm cells and testosterone and inhibin
Duct System	
Epididymis	Promotes sperm cell maturation; stores sperm; moves sperm cells to ductus deferens
Ductus deferens	Stores sperm; moves sperm to ejaculatory duct
Ejaculatory duct	Transports sperm from ductus deferens into urethra
Urethra	Transports semen out of penis
Accessory Glands	
Seminal vesicle	Secretes alkaline fluid with nutrients, prostaglandins, and a coagulating enzyme that support sperm, enhance their motility, and maintain semen in the female reproductive tract after ejaculation
Prostate gland	Secretes slightly alkaline fluid with nutrients and several components that support sperm, help activate sperm for fertilization, and function as an anticoagulant; immunological functions
Bulbourethral gland	Secretes mucus to lubricate glans penis and neutralize acidic traces of urine in urethra
External Genitalia	
Scrotum	Encloses, protects, and regulates temperature of testes
Penis	Moves urine and semen out of body; inserts into vagina during intercourse to deposit semen

☐ 9. Why are the testes located in the scrotum rather than the abdominopelvic cavity?
☐ 10. Which part of the duct system passes through the inguinal canal inside the spermatic cord?

Apply What You Learned

☐ 1. A patient has a blockage that prevents secretions from the seminal vesicles. How might this affect the amount of semen produced?
☐ 2. A man is taking a medication that triggers involuntary muscle contractions as a side effect. In addition, the weather is very hot. Predict how this combination of factors might affect his production of viable sperm. (*Hint:* What is one function of the muscles in the scrotum?)

See answers in Appendix A.

MODULE **26.3**

Physiology of the Male Reproductive System

Learning Outcomes

1. Relate the general stages of meiosis to the specific process of spermatogenesis.
2. Discuss endocrine regulation of spermatogenesis.
3. Discuss the events and endocrine regulation of male puberty.
4. Describe the male sexual response.
5. Describe male secondary sex characteristics and their role in reproductive system function.
6. Explain the effects of aging on reproductive function in males.

Now that we have examined the anatomy of the male reproductive system, we turn our attention to its physiology. In this module, we first discuss how sperm cells develop via a process known as *spermatogenesis*. Next we focus on how the hypothalamus and anterior pituitary regulate the production of hormones responsible for reproduction in males. The module closes with a look at the male sexual response, the changes that take place in males at puberty, and the changes in male reproductive function that occur with age.

Spermatogenesis

◄◄ **Flashback**

1. What is the function of mitochondria? What is ATP? (p. 88)
2. What is peristalsis? (p. 848)

Spermatogenesis (sper-mat'-oh-JEN-uh-sis) is the process of male gamete, or sperm cell, development. It begins at puberty and continues throughout life. Every day, a healthy young adult male makes about 400 million sperm.

Spermatogenesis occurs in the seminiferous tubules of the testes by *spermatogenic,* or sperm-forming, cells. **Spermatogonia** (sper-mat'-oh-GOHN-ee-ah; singular, *spermatogonium*), the stem cells that begin the process, have 46 chromosomes and are found along the basement membrane of the seminiferous tubules, as you can see in **Figure 26.7**.

Before puberty, spermatogonia divide by mitosis, but only to produce more spermatogonia. When puberty begins, they become functional stem cells and divide, again by mitosis, to maintain the stem cell line and create more spermatogonia. Some of the cells resulting from mitosis go on to become diploid **primary spermatocytes** (sper-MAT-oh-sytz; step ①). As spermatogonia divide into these two cell types, the cell destined to become a primary spermatocyte is pushed toward the lumen of the seminiferous tubule while the spermatogonium takes its place along the basement membrane to replace its parent cell. Each primary

spermatocyte then undergoes the first meiotic division, forming two smaller joined haploid cells called **secondary spermatocytes** (step ②). The two secondary spermatocytes rapidly progress to finish the second meiotic division and produce four small, round, haploid **spermatids** (step ③). Notice in Figure 26.7 that the spermatids are moved toward the lumen of the tubule.

Quick Check

☐ 1. What are the steps of spermatogenesis?

Sustentacular Cells

◄◄ **Flashback**

1. What is the function of tight junctions? (p. 127)
2. What are microtubules? (p. 96)

Figure 26.7 Spermatogenesis in the seminiferous tubules. Spermatogenesis begins with a spermatogonium that has 46 chromosomes, and produces 4 spermatids that have 23 chromosomes each.

Golgi apparatus Acrosomal vesicle

Mitochondria

Flagellum extends upward, toward nucleus

Acrosome

Head

Nucleus

Mitochondria

Midpiece

Excess cytoplasm

Shedding of excess cytoplasm

Nucleus Centrioles

Microtubules

Tail

Spermatid

Flagellum

Sperm cell

(a) Maturation from a spermatid to a sperm cell

(b) SEM of mature sperm cells   SEM (1650×)

Figure 26.8 Spermiogenesis.

The yellow cells in Figure 26.7 that you can see surrounding the spermatogenic cells are supporting cells called **sustentacular cells** (sus′-ten-TAK-yoo-luhr), also referred to as *nurse* or *Sertoli cells*. These cells extend from the basement membrane to the lumen of the tubule and are bound to each other by tight junctions that separate the forming sperm cells from the immune system. This **blood-testis barrier** is important because it prevents immune cells in the bloodstream from encountering the new antigens formed on the genetically unique sperm cells. Haploid sperm cells are not present in the body until puberty, long after the immune system has determined how to identify cells with "self" antigens (those that do not require an immune response). The spermatogonia, however, can be outside the blood-testis barrier because they are genetically identical to other body cells and so are recognized as self. Indeed, they need to be outside of the blood-testis barrier for hormones and paracrines to reach them and stimulate spermatogenesis.

In addition to forming the blood-testis barrier, sustentacular cells perform many other functions important to spermatogenesis, which explains why they are nicknamed nurse cells. These functions include:

- providing structural support for stem cell development;
- secreting testicular fluid, which helps transport sperm in the lumen of the seminiferous tubules;
- providing nutrients to the dividing cells;

- phagocytizing damaged spermatogenic cells and excess cytoplasm released from maturing spermatids; and
- producing *androgen-binding protein (ABP)* and *inhibin* (see the next module), which help regulate spermatogenesis.

Quick Check

☐ 2. How do sustentacular cells support developing sperm?

Spermiogenesis

The maturation process of sperm is called **spermiogenesis** (sper-mee′-oh-JEN-uh-sis) and involves changing the size and shape of the cell. Spermiogenesis begins in the testes as the spermatid elongates and sheds excess cytoplasm (step (4) in Figure 26.7). As it continues, the following occur (step (5)): A structure known as the **acrosome** (AK-roh-sohm), which helps with fertilizing the female gamete, forms over the nucleus; more mitochondria are produced and move to their positions around the beginning part of the flagellum; microtubules in the cell form the flagellum, or tail; and cells lose their connections to sustentacular cells and are released into the lumen of the seminiferous tubule. The changing appearance of the cell as it matures from a spermatid to a sperm cell is shown in **Figure 26.8**. Each sperm cell has **head, midpiece,** and **tail** regions and is about 55–60 μm (0.0021–0.0024 in.) long and about 3 μm (0.0001 in.) wide (Figure 26.8a). The head includes the nucleus and the acrosome and very little cytoplasm; the midpiece contains the mitochondria; and the tail is basically a flagellum surrounded by the plasma membrane. The tail moves in a mature sperm cell because of ATP provided by the mitochondria of the midpiece.

Even though the sperm now has a flagellum, it is still nonmotile and will migrate to the epididymis to finish maturing. Sperm move slowly, taking about 12 days to reach the tail of the epididymis, where they can remain viable for several months. If the sperm are not ejaculated by this time, the body resorbs them. Together, the processes of spermatogenesis and spermiogenesis take about 60–70 days to complete.

Quick Check

☐ 3. How do spermatogenesis and spermiogenesis differ?
☐ 4. At what point is a sperm cell mature?

Hormonal Control of Male Reproduction

One of the main hormones that regulates spermatogenesis and other aspects of male reproductive physiology is tes-

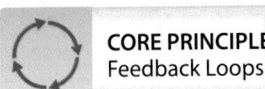

CORE PRINCIPLE
Feedback Loops

tosterone. Secretion of testosterone is controlled by a multi-tiered negative feedback loop involving the hypothalamus, anterior pituitary, and testes, in an example of the Feedback Loops Core Principle (p. 21).

These hormones interact in a relationship called the **hypothalamic-pituitary-gonadal (HPG) axis.** Let's look at the events that produce and regulate testosterone and testicular function (**Figure 26.9**):

- **First-tier control.** The hypothalamus secretes **gonadotropin-releasing hormone (GnRH).** (Remember from Chapter 16 that GnRH stimulates the secretion of *gonadotropins*, which stimulate the gonads to release hormones.)

FIRST-TIER CONTROL

Hypothalamus releases GnRH.

Hypothalamus

GnRH

As levels of testosterone and inhibin rise, secretion of GnRH, FSH, and LH decrease.

SECOND-TIER CONTROL

Anterior pituitary detects GnRH and secretes FSH and LH.

Anterior pituitary

FSH   LH

Testis

THIRD-TIER CONTROL

LH stimulates testosterone production from interstitial cells; FSH stimulates sustentacular cells to secrete ABP.

Sustentacular cells

Interstitial cells

Testosterone   ABP

EFFECTS

ABP binds testosterone.

Testosterone stimulates spermatogenesis and development of male characteristics.

Negative feedback

Figure 26.9 Hormonal regulation of testicular function via the hypothalamic-pituitary-gonadal (HPG) axis.

- **Second-tier control.** The anterior pituitary detects the presence of GnRH, which stimulates it to secrete **follicle-stimulating hormone (FSH)** and **luteinizing hormone** (LOO-tee′-uh-nyz-ing) **(LH)**, also called *interstitial cell–stimulating hormone (ICSH)* in males.
- **Third-tier control.** The testes are the target organs of FSH and LH. LH stimulates the interstitial cells to produce testosterone. FSH stimulates the sustentacular cells to secrete **androgen-binding protein (ABP)**, which keeps some of the testosterone bound near the spermatogenic cells, and the hormone *inhibin*.
- **Effects:** Testosterone stimulates spermatogenesis and the development of male characteristics.

Testosterone and inhibin both exert negative feedback controls on the hypothalamus and anterior pituitary. Inhibin decreases the release of FSH, while testosterone reduces GnRH secretion.

It usually takes a few years to establish the HPG axis, but then testosterone levels and spermatogenesis remain fairly stable throughout life, with only gradual decreases in both as men age. Sometimes certain factors disrupt testicular function, leading to infertility (see *A&P in the Real World: Male Infertility*).

Table 26.3 in Module 26.5 summarizes the functions of the major reproductive hormones in both males and females.

Quick Check

☐ 5. What is the relationship between the hypothalamus, anterior pituitary gland, and testes in the HPG axis?

☐ 6. On what type of cell do FSH and LH act in males, and what do they stimulate in their target cells?

Male Sexual Response

◀◀ **Flashback**

1. What is the basic function of the parasympathetic division of the autonomic nervous system? What is the basic function of the sympathetic division of the autonomic nervous system? (pp. 521, 528)

For fertilization to occur, sperm cells must reach female egg cells deep inside the female reproductive tract. To optimize the chances that fertilization will occur, sperm should enter into the female reproductive tract as far as possible. This is usually accomplished by **sexual intercourse,** also called *copulation* or *coitus* (KOY-tus), most commonly defined as the insertion of the male penis into the copulatory organ of the female, the *vagina.*

The basic phases of the male sexual response include *erection* and *ejaculation.* **Erection** is the enlargement and stiffening of the penis, which allows the penis to enter the vagina. An erection results when the erectile tissue in the penis becomes engorged with blood. Normally, blood vessels in the penis that supply erectile tissue are constricted, and the penis is in a **flaccid,** or relaxed, state. During sexual arousal, a parasympathetic reflex triggers the release of nitric oxide (NO) from blood vessel endothelial cells. This relaxes smooth muscle in the walls of the arterioles of the penis, dilating them, and within the blood sinuses in erectile tissue, which allows large amounts of

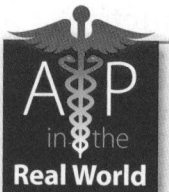

Male Infertility

Infertility is the inability to produce a pregnancy after 1 year of unprotected intercourse. Approximately 40 percent of all infertility cases result from male infertility, which is usually due to a low sperm count. Normal levels are 40–750 million sperm per ejaculate; a count of less than 15 million sperm cells per milliliter of semen usually indicates infertility. Low sperm count can result from any sort of damage to the testis, such as physical trauma, exposure to radiation, or disease. It could also be due to developmental defects. During normal development, the testes begin forming inside the abdominopelvic cavity and then descend into the scrotum. If a testis does not descend into the scrotum, a disorder called *cryptorchidism,* sperm cells will not be produced. In addition, inadequate secretion of GnRH, FSH, LH, or testosterone for any reason will also lower sperm count.

blood to enter the tissue. The combination of increased blood flow and widening blood sinuses results in an erection. The erection is maintained, at least partially, by the compression of the veins that drain the penis. Note that the corpus spongiosum does not become as rigid as the corpora cavernosa, which allows the urethra to remain open during ejaculation so that sperm and semen can pass.

Erection can occur because of a variety of sexual stimuli, including touch; mechanical stimulation of the penis; and erotic sights, sounds, and smells. Emotions and thoughts, controlled by the higher thought centers of the brain, may cause or prevent erections. **Impotence,** or *erectile dysfunction (ED),* is the inability to maintain an erection long enough to participate in sexual intercourse (you can read more about this in *A&P in the Real World: Erectile Dysfunction*).

Ejaculation is the process by which semen is expelled from the penis. Although erection is under parasympathetic control, ejaculation is under *sympathetic* control. It actually occurs in two stages, *emission* and *expulsion.* **Emission** is the movement of sperm and testicular fluid, as well as secretions from the prostate gland and seminal vesicle, into the urethra. As semen accumulates in the urethra with the increase in sexual tension, sensory impulses are stimulated and pass into the sacral portion of the spinal cord. Motor neurons from the spinal cord stimulate skeletal muscles at the base of the erectile columns of the penis to rhythmically contract, resulting in **expulsion** of the semen from the urethra. **Orgasm** refers to the time period during which feelings of pleasure are experienced. In males, this coincides with ejaculation.

Orgasm and ejaculation are quickly followed by **resolution,** or *relaxation.* Blood vessels in the erectile tissues and blood sinuses constrict, forcing blood out of the penis, which soon becomes flaccid again. After ejaculation, a *latent period,* or **refractory period,** follows; during this time, ranging from minutes to hours, a man is unable to achieve another orgasm. The refractory period tends to get longer as men age.

Quick Check

- [ ] 7. What is the role of the parasympathetic nervous system in the male sexual response? What is the role of the sympathetic nervous system?
- [ ] 8. What are the two stages of ejaculation? What occurs during these stages?

Effects of Testosterone on Other Body Systems

◀◀ Flashback

1. What are the androgenic and anabolic effects of testosterone? (p. 619)
2. Which endocrine organs secrete androgens? (p. 612)

Testosterone is known as the male steroid hormone for more reasons than its importance in stimulating spermatogenesis. Without testosterone, many characteristics we identify as "male" or "masculine" would not be distinguishable. Levels of this hormone increase dramatically at puberty and are responsible for the growth, maturation, and maintenance of male reproductive organs, including the testes, scrotum, penis, and duct system. Testosterone also triggers the development of **male secondary sex characteristics,** which are features induced in nonreproductive organs that are associated with adult males.

Puberty

Throughout a boy's childhood, testosterone from the testes and androgens from the adrenal cortex are secreted in low, but sufficient, amounts to inhibit GnRH release from the hypothalamus. This continues until puberty, when the hypothalamus becomes much less sensitive to inhibition by the androgens, causing GnRH levels to increase. GnRH stimulates the HPG axis, and

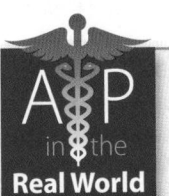

Erectile Dysfunction

Various psychological and physical factors may cause erectile dysfunction (ED). Psychological influences include stress, depression, and anxiety, while common physical causes include cardiovascular disease and diabetes mellitus. Obesity, tobacco and alcohol use, and certain prescription medications can also play a role. Older men have a greater risk of erectile dysfunction because they are more likely to have physical conditions associated with it. Also, the amount of connective tissue in the erectile tissue of the penis increases with age, reducing blood flow to the penis.

Treatment for ED includes medications that permit blood vessels supplying the penis to remain dilated, thereby maintaining an erection. Other treatments may involve surgery to repair damaged blood vessels or to insert penile implants.

blood testosterone levels begin to rise, triggering spermatogenesis and the appearance of male secondary sex characteristics.

In most boys, puberty begins between 12 and 14 years of age, but this age varies. Usually within a few years after the onset of puberty, the hormone levels stabilize. As soon as spermatogenesis begins, males are capable of sexual reproduction, even if hormone levels are still fluctuating.

Secondary Sex Characteristics

Male secondary sex characteristics develop primarily at puberty, but may continue to develop into early adulthood in some males. They include:

- growth of pubic, axillary, and facial hair;
- enhanced hair growth on the chest or other body areas in some men;
- an enlarged larynx and thicker vocal cords, resulting in a noticeable "Adam's apple" and deepening voice; and
- thickened skin and increased secretions of sebaceous glands (the oilier skin often results in acne).

Testosterone also causes **somatic** (body) **effects,** many of which are anabolic. Under the influence of testosterone, more calcium salts are deposited in bone, making it denser. In addition, skeletal muscle mass increases, accounting for the higher muscle mass in males compared to females. The number of erythrocytes also increases because testosterone has a direct effect on the production of erythropoietin. This is why males have a higher number of erythrocytes than females.

Testosterone also influences behavior. It is the basis for the male **libido,** which is the desire for sexual activity.

Quick Check

- [ ] 9. Why do GnRH levels increase at the beginning of puberty?
- [ ] 10. What are the most noticeable secondary sex characteristics caused by testosterone?
- [ ] 11. What are the anabolic effects of testosterone?

Effects of Aging: Male Climacteric

Male climacteric (kly-MAK-ter-ik), or *andropause*, is the period in which reproductive functions begin to decline in men. These changes are usually gradual and typically begin in the early fifties, but may or may not produce noticeable changes in all men. The size and weight of the testes may decrease and the number of sustentacular and interstitial cells may begin to decline. When sustentacular cells produce less inhibin and interstitial cells produce less testosterone, GnRH increases the production of FSH and LH because negative feedback decreases. If FSH and LH levels remain high, it indicates that the testes are not producing enough testosterone to suppress their production. Reduced levels of testosterone result in the gradual production of fewer sperm cells, although this is not absolute, as many men in their seventies or eighties have fathered children. In addition, low levels of testosterone are associated with depressed mood and fatigue in some men and lead to a decrease in muscle strength and some decrease in bone density. However, bone density typically declines more slowly in aging men than in women, because the loss of testosterone is gradual.

Probably the most noticeable change for most men during climacteric is enlargement of the prostate gland. This compresses the prostatic urethra, which often causes difficulty in urination.

□12. Why does sperm cell production decrease as men age?
□13. Which organs not directly associated with the reproductive system are affected by decreased testosterone levels?

Apply What You Learned

□ 1. If a man's anterior pituitary did not secrete LH, how would this affect spermatogenesis?
□ 2. Predict how a drug that inhibits nitric oxide production might affect a man's sexual response.
□ 3. Explain why a man experiencing emotional or physical stress is often unable to develop or maintain an erection. (*Hint:* Which branch of the autonomic nervous system is involved in the stress response?)

See answers in Appendix A.

(a) Midsagittal section of female pelvic cavity

Labels: Suspensory ligament, Uterine tube, Ovary, Round ligament, Uterus, Vesicouterine pouch, Vagina, Urethra, Peritoneum, Rectum, Urinary bladder, Rectouterine pouch, *Posterior*, *Anterior*

(b) Superior view of female pelvic cavity

Labels: View, Female reproductive ligaments, Suspensory ligament, Broad ligament, Ovarian ligament, Uterine tube, Uterus, Vagina, *Posterior*, Ovaries, Urinary bladder, *Anterior*

Practice art labeling @MasteringA&P®

Figure 26.10 Internal organs of the female reproductive system.

MODULE **26.4**

Anatomy of the Female Reproductive System

Learning Outcomes

1. Describe the structure and functions of the female reproductive organs.
2. Describe the histology of the uterine wall.
3. Trace the pathway of the female gamete from the ovary to the uterus.

The male and female reproductive systems share a number of similarities. Both systems produce sex hormones and gametes, and both have a series of ducts to move the gametes. However, the female reproductive system must also support a developing embryo and nourish an infant, and these functional differences lead to structural differences.

The internal female genitalia include the *ovaries, uterine tubes, uterus,* and *vagina.* See **Figure 26.10a** for a midsagittal view of these organs and some of the ligaments that hold them in place. In this view you can also see how these organs are related to the urinary bladder and the rectum. Notice that the peritoneum dips down on either side of the uterus. This creates two "pouches" between the uterus and the bladder and rectum—the *vesicouterine pouch* (vez'-uh-koh-YOO-ter-in) and the *rectouterine pouch* (rek'-toh-YOO-ter-in), respectively.

Unlike the male reproductive organs, the female reproductive organs are located mostly in the pelvic cavity. This can be appreciated by looking at Figure 26.10b, which shows how the organs are arranged from a superior view (imagine you are holding a pelvis and looking down into it). This module first examines the internal female genitalia and then discusses the *external genitalia* and *mammary glands*.

Ovaries

◀◀ Flashback

1. What are the main functions of estrogens and progesterone? (p. 619)

The paired **ovaries** are the main female gonads and are found on the lateral walls of the pelvic cavity (Figure 26.10b). Each ovary weighs only about 6–8 g and is about the size of a walnut in women who are actively releasing ova on a monthly cycle. These organs appear relatively smooth before puberty, but as a woman matures, they take on a bumpy appearance due to scars left behind when ova are released approximately once each month.

The ovaries perform two main functions: They (1) produce ova and (2) secrete several hormones. Among these hormones are the *estrogens*, which include estradiol, estrone, and estriol,

with estradiol being the most abundant. We will use the term estrogens to refer to all of these hormones collectively. The ovaries also produce *progesterone* (proh-GES-ter-ohn), inhibin, and relaxin. (The function of relaxin is discussed in Chapter 27.)

As in many other organs, the superficial portion of the ovary is the cortex and the deeper portion is the medulla (Figure 26.11). The **ovarian cortex** is the site of *oogenesis*, the production of female gametes. In females, gametes are produced inside saclike structures called *follicles*, which develop and mature along with the gametes. The **ovarian medulla** houses blood vessels, lymphatic vessels, and nerves.

Each ovary is held in place by three ligaments. First is the **broad ligament**, a large, flat ligament that attaches to the ovaries, uterine tubes, and uterus and connects them to the bony pelvis (you can see this attachment to the pelvis in Figure 26.10b). Next is the **ovarian ligament**, which connects the medial surface of the ovary to the uterus, near the attachment of the uterine tube. Finally, the **suspensory ligament** (or *infundibulopelvic ligament*; in-fun-DIB-yoo-loh-pel'-vik) connects the lateral surface of the ovary to the pelvic wall. Notice in Figure 26.11 that this ligament carries the **ovarian artery** and **ovarian vein.**

Quick Check

☐ 1. What are the main functions of the ovaries?
☐ 2. Which three ligaments support the ovary, and to which structures are they attached?

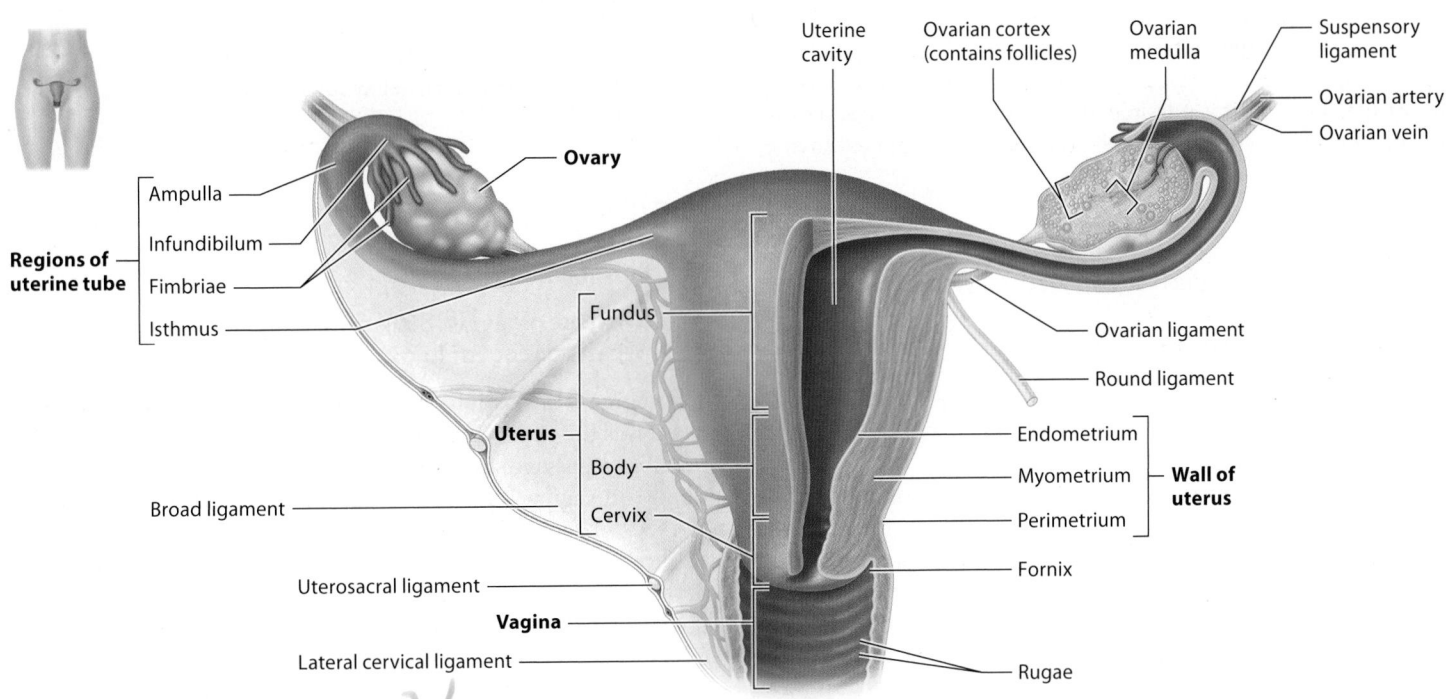

Figure 26.11 Internal organs of the female reproductive system, anterior view. The left ovary, uterus, and uterine tube and the vagina are cut in a frontal section to show the internal anatomy. The left side of the broad ligament has been removed.

Practice art labeling
@MasteringA&P®

Uterine Tubes

◀◀ **Flashback**

1. What is the function of cilia? (p. 98)

The **uterine tubes** (YOO-ter-ihn), or *fallopian tubes* or *oviducts,* form the first part of the duct system of the female reproductive system (see Figure 26.11). Externally, they are covered by the peritoneum and are supported along their length by part of the broad ligament. Each uterine tube is approximately 10 cm (3.9 in.) long and extends medially from the ovary to the superior and lateral region of the uterus. At its proximal end, the uterine tube connects to the uterus in a constricted region called the **isthmus.** The distal end of each uterine tube expands as it curves around the ovary, forming the **ampulla,** which connects to the **infundibulum,** an open funnel-shaped structure. At the end of the infundibulum, we find ciliated, finger-like projections known as **fimbriae** (FIM-bree-eh), which drape over the ovary.

During the process of *ovulation,* when an immature female gamete called an *oocyte* (OH-uh-sy't) is expelled from the ovary, the fimbriae sweep the ovarian surface and "catch" the oocyte. The oocyte is then carried into the uterine tube, which contains sheets of smooth muscle and folded mucosa with both ciliated and nonciliated cells. A combination of peristalsis and the beating of the cilia moves the oocyte toward the uterus. Nonciliated cells produce a mucus-like secretion that keeps the oocyte moist and nourished.

Fertilization, which begins when a sperm cell binds to and penetrates the plasma membrane of an oocyte, occurs in the uterine tubes, usually in the ampulla. If fertilization occurs, it must do so within 24 hours of ovulation. However, sperm can survive in the female reproductive tract for up to 5 days, so they often end up waiting for the oocyte. Since the ovary is not directly attached to the uterine tube, it is possible that the fimbriae will not pick up the oocyte, in which case it is released into the peritoneal cavity.

Quick Check

☐ 3. What structures catch an ovulated oocyte and move it toward the uterus?
☐ 4. What is the usual site of fertilization?

Uterus

The **uterus** (YOO-ter-uhs), or womb, functions as part of the pathway by which sperm reach the uterine tubes. It is also the site where a fertilized ovum implants, and it provides protection and nutritional support for a developing fetus. In addition, contractions of the uterus are responsible for ejecting the fetus at birth. When pregnancy does not occur, the uterus is the source of menstrual flow.

The uterus is located in the pelvis anterior to the rectum and posterior to the urinary bladder. A hollow, thick-walled, muscular organ, it is similar in size and shape to an inverted pear. It weighs 30–40 g, but it is usually a little larger in women who have borne children. Like the ovaries, the uterus is supported by several ligaments: inferiorly by the **lateral cervical ligaments,** which extend from the cervix and superior vagina to the lateral walls of the pelvis; posteriorly by the paired **uterosacral ligaments** (yoo'-ter-oh-SAYK-ruhl), which connect the uterus to the sacrum; anteriorly by the **round ligaments,** which connect the uterus to the anterior body wall (you can see this attachment in Figure 26.10); and laterally by the broad ligaments. In addition to these ligaments, the uterus also receives a great deal of support from the muscles of the pelvic floor.

There are three regions of the uterus: the main portion, or **body;** the rounded region superior to the entrance of the uterine tubes, called the **fundus;** and the **cervix,** the narrow neck that projects into the vagina inferiorly. The most inferior portion of the cervix, which serves as the entrance into the uterus, is known as the *external os.* It is this portion of the cervix that is generally observed during a pelvic examination. Glands in the mucosa of the cervical canal secrete mucus that covers the external os, acting as a protective plug.

The wall of the uterus is composed of three layers, which are labeled in Figure 26.11. The **perimetrium** is the outermost serous layer and is an extension of the parietal peritoneum. The **myometrium** is the thick middle layer, composed of bundles of smooth muscle. It contracts rhythmically during orgasm and during childbirth to expel the baby from the mother's body and in some women, to a lesser degree, during the menstrual cycle. The innermost **endometrium** lines the uterine cavity and is a mucous membrane composed of simple columnar epithelium on a layer of connective tissue called the lamina propria.

Quick Check

☐ 5. What are the ligaments that support the uterus?
☐ 6. From deep to superficial, what are the layers of the uterine wall?

Vagina

◀◀ **Flashback**

1. What are rugae? (p. 860)
2. What is glycogen? (p. 52)

The **vagina** (vuh-JY-nuh) is the female organ of copulation because it receives the penis (and semen) during sexual intercourse. It also provides a passageway for delivery of an infant and for menstrual flow. The vagina is a thick-walled tube, approximately 8–10 cm (3.1–3.9 in.) long that extends from the cervix to the exterior of the body (see Figure 26.10). It parallels the anterior urethra, lying between the urinary bladder and the rectum. The superior end of the vaginal canal surrounds the external os of the cervix and produces a recess called the vaginal **fornix** (shown in Figure 26.11). The wall of the vagina is covered with transverse ridges, or **rugae,** which stimulate the penis during intercourse. Generally, the lumen of the vagina is quite small and, except where it is held open by the cervix, its walls touch one another. However, the vagina stretches considerably during intercourse and childbirth.

The vaginal mucosa is a stratified squamous epithelium, allowing it to withstand friction. It has no glands, but is lubricated by cervical mucus and secretions from the epithelial cells, which increase during sexual excitement. The epithelial cells also secrete glycogen, which resident bacteria break down to produce ATP and lactic acid. This helps maintain the acidic pH that keeps the vagina healthy and free of infection. Earlier in the chapter, we mentioned that the acidic pH of the vagina is also hostile to sperm. This is one reason why the alkaline semen is so important for maintaining sperm health in the female reproductive tract.

In most women, the mucosa near the distal vaginal orifice forms an incomplete partition called the *hymen*. The hymen is very vascular and may bleed when it ruptures, which happens in about 50% of women during their first experience of sexual intercourse. However, it may also rupture from an injury or during activities such as sports.

Figure 26.12 The female perineum.

Quick Check

- [ ] 7. What are the functions of the vagina?
- [ ] 8. Which cells produce the lactic acid that gives the vagina an acidic pH? What is the importance of maintaining this acidic pH?

Female External Genitalia

The **external female genitalia,** or **vulva,** are the female reproductive structures that lie external to the vagina (**Figure 26.12**). The external genitalia include the *mons pubis, labia majora, labia minora, clitoris,* and structures associated with the *vestibule.* Although the external genitalia are often referred to as the "vagina," the vagina is located internally and is not part of the external genitalia.

The **mons pubis** (mahnz PYOO-bis) is a fatty, rounded area overlying the pubic symphysis that is covered with hair after puberty. Posterior to the mons pubis are two elongated, fatty protective skin folds, the **labia majora** (LAY-bee-uh), which enclose two thinner skin folds called the **labia minora.** Enclosed within the labia minora is a recess known as the **vestibule.** The vestibule contains the **urethral orifice** and **vaginal orifice;** the **paraurethral glands** (payr-uh-yoo-REETH-ruhl), which discharge mucus into the urethra; and the openings of the **greater vestibular glands** (*Bartholin's glands*), which release a small amount of mucus into the vestibule, helping to lubricate the opening of the vagina during intercourse.

Anterior to the vestibule is the **clitoris** (KLIT-ur-us), a small, protruding structure composed of erectile tissue. It corresponds to the corpora cavernosa of the penis and is important in the female sexual response. Just as in the penis, erectile tissue of the clitoris becomes engorged with blood and is innervated by many sensory, motor and autonomic fibers. Tactile, pressure, and temperature receptors are abundant here as well, and are also important for sexual stimulation.

The female **perineum** is a diamond-shaped region located between the pubic arch anteriorly, the coccyx posteriorly, and the ischial tuberosities laterally (see Figure 26.12). The tissues of the perineum overlie the muscles of the pelvic outlet, and the posterior ends of the labia majora overlie the central tendon, which is the insertion of most muscles supporting the pelvic floor.

The region between the vagina and anus is the *clinical perineum.* During childbirth some doctors will cut a straight incision, called an **episiotomy** (ep-EE-zee-ah'-tuh-mee), through the skin and muscle of this area if a tear appears likely. Episiotomies are done because straight incisions usually heal more quickly than jagged tears.

Quick Check

- [ ] 9. What are two functions of the female external genitalia?
- [ ] 10. How are the external genitalia of the female similar to those of the male? How are they different?

Mammary Glands

◀◀ **Flashback**

1. What are the layers of the skin? (p. 165)
2. What is the basic structure of a sweat gland? (p. 176)

The **mammary glands** are part of the integumentary system and are actually modified sweat glands. Their major biological role is to produce milk to nourish a newborn baby, so they are important only when reproduction has already been accomplished. They are present in both sexes but normally function only in females.

Each mammary gland is contained within a rounded skin-covered breast in the hypodermis region, superficial to the pectoralis

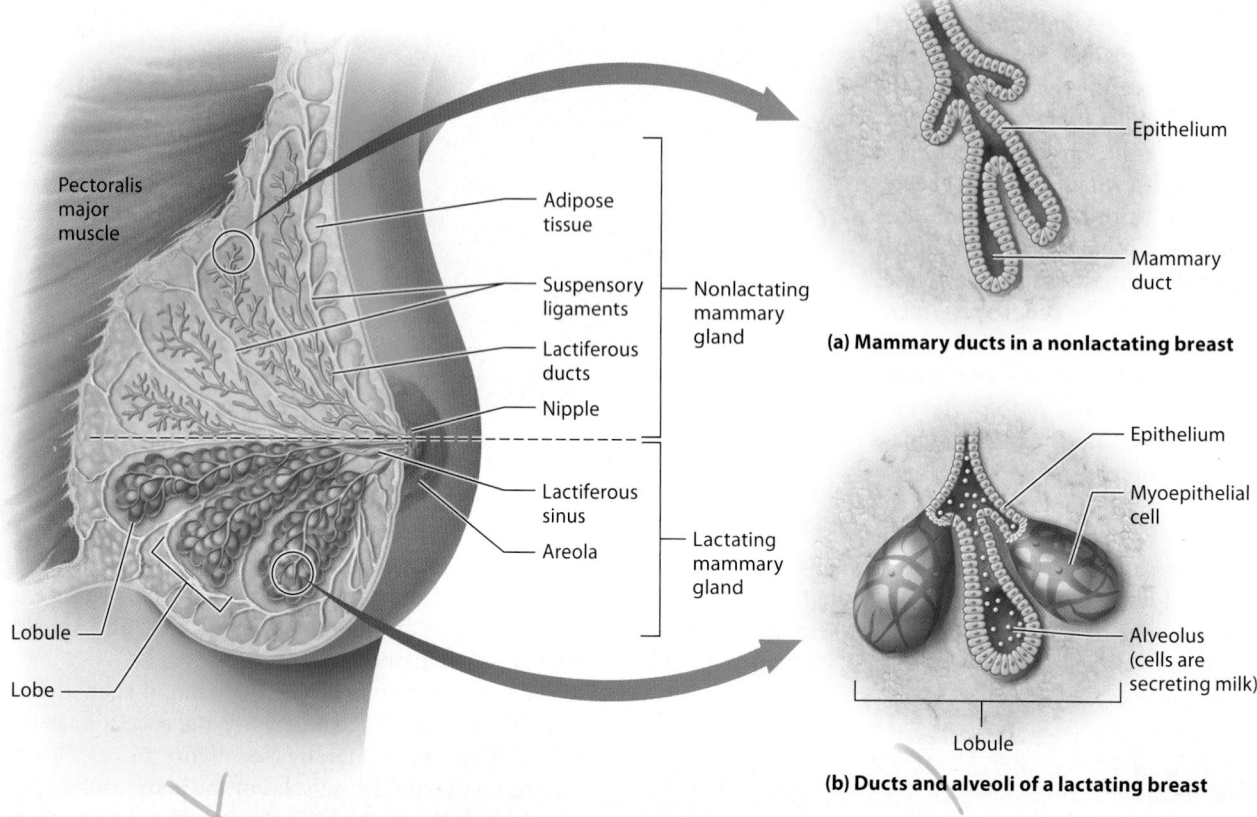

Figure 26.13 Internal anatomy of the female breast.

(a) Mammary ducts in a nonlactating breast

(b) Ducts and alveoli of a lactating breast

major muscle of the chest (**Figure 26.13**). Just below the center of each breast is an area of pigmented skin, the **areola,** which surrounds a central protruding **nipple** through which milk exits the gland.

Internally, each mammary gland consists of 15–25 **lobes** that radiate around and open at the nipple. The lobes are padded and separated from each other by fibrous connective tissue and fat, and they receive support from **suspensory ligaments** that attach the breast to the underlying muscle fascia and the overlying dermis. Within each lobe are smaller units called **lobules,** which contain glandular **alveoli** that produce milk when a woman is lactating (compare Figures 26.13a and b). Contraction of **myoepithelial cells** that surround the alveoli helps propel milk toward the nipples.

When milk is being produced, it passes from the alveoli into a series of **mammary ducts,** which flow into **lactiferous ducts** that open to the outside at the nipple. Just deep to the areola, each lactiferous duct has a dilated region called a **lactiferous sinus,** where milk accumulates during nursing. (Chapter 27 will discuss the process and regulation of lactation in more detail.)

In nonpregnant women, the glandular structure of the breast is mostly undeveloped and alveoli are not present. For this reason, breast size is largely due to the amount of *adipose tissue* in the breast. One of the most common sites of cancer in women is the breast (see *A&P in the Real World: Breast Cancer*).

Table 26.2 summarizes the functions of the female reproductive organs.

TABLE 26.2 FUNCTIONS OF THE FEMALE REPRODUCTIVE STRUCTURES	
Structure	**Function**
Internal Genitalia	
Ovary	Produces oocytes and estrogens, progesterone, and inhibin
Uterine tube	Moves oocytes or fertilized ovum toward uterus; site of fertilization and early stages of development
Uterus	Protects and sustains embryo/fetus during pregnancy; cyclic shedding results in menstrual flow
Vagina	Receives penis during sexual intercourse; provides passageway for menstrual flow, for sperm, and for infant during birth
External Genitalia	
Labia majora	Enclose and protect other external reproductive structures
Labia minora	Enclose vestibule and protect openings of vagina and urethra
Clitoris	Erectile tissue innervated with sensory nerve endings
Vestibule	Recess that contains external openings of vagina and urethra
Greater vestibular gland	Secretes mucus, which helps lubricate the opening of the vagina

Quick Check

☐ 11. Which structures do not fully develop in the nonlactating breast?
☐ 12. What is the pathway of milk through the internal structures of the lactating breast?

Breast Cancer

Real World

Breast cancer is the second most common type of cancer in women (after non-melanoma skin cancer) in the United States. Over 200,000 cases a year are diagnosed, a few hundred of which occur in men. Risk factors for breast cancer include maternal relatives with breast cancer, longer reproductive span (early first menstrual cycle coupled with menstruation continuing until a later age), obesity, no pregnancies or first pregnancy at or after the age of 35, and the presence of breast cancer genes. In particular, two genes that increase susceptibility to breast cancer have been identified: *BRCA1* and *BRCA2*.

Unfortunately, most breast cancers are painless until they become advanced and have metastasized. Therefore, the best way to prevent the cancer from spreading is to detect it early. Monthly self-examinations can locate lumps that may be malignant. Mammograms (x-rays of the breast) can detect small areas of increased tissue density:

— Tumor

(a) Normal breast tissue   **(b) Breast tissue with cancerous tumor**

A mammogram can detect many small malignancies that are not yet palpable in a self-examination. Most physicians recommend that women over age 50 have a mammogram every 2 years and women with a family history of breast cancer have regular mammograms before age 40.

Treatment for breast cancer may involve chemotherapy, radiation therapy, lumpectomy (surgical removal of the tumor and immediate surrounding tissue), or mastectomy (surgical removal of the breast and possibly the underlying pectoral muscles and axillary lymph nodes). Several chemotherapeutic drugs may decrease the risk of recurrence or disease progression.

Apply What You Learned

☐ 1. A tumor is growing in the isthmus of a uterine tube. It is still small enough to allow sperm to enter, but too large to allow an ovulated oocyte to pass. Predict what may happen if sperm enter this uterine tube following ovulation.
☐ 2. A woman has been trying to become pregnant for several months without success. A scan of her abdomen and pelvis reveals that her uterine tubes lack fimbriae. How might this finding help explain her lack of success in becoming pregnant?

See answers in Appendix A.

MODULE **26.5**

Physiology of the Female Reproductive System

Learning Outcomes

1. Relate the general stages of meiosis to the specific process of oogenesis.
2. Describe the events of the ovarian cycle and the uterine cycle.
3. Discuss endocrine regulation of oogenesis.
4. Analyze graphs depicting the typical female monthly sexual cycle and correlate ovarian activity, hormonal changes, and uterine events.
5. Describe female secondary sex characteristics and their role in reproductive system function.
6. Describe the physiological changes associated with menopause, and explain the fertility changes that precede menopause.

Female gametes are produced by the process of *oogenesis*. This module examines oogenesis and how it differs from spermatogenesis. Attention then shifts to the female's ovarian and uterine cycles and how the cyclic fluctuation of hormones controls these cycles. The module concludes with an examination of the female sexual response and the changes that occur at puberty, as well as how the reproductive cycle changes as a woman ages.

Oogenesis

◀◀ **Flashback**

1. What is apoptosis? (p. 776)

Recall that in males the process of gamete development is spermatogenesis, which yields sperm cells. In females, this process is **oogenesis** (oh′-uh-JEN-uh-sis), which produces **ova** (singular, *ovum*). Oogenesis differs from spermatogenesis in many ways. For one, oogenesis actually begins before a female infant is born but is then arrested until puberty. At that time, it resumes and remains active until ending at *menopause*, which usually occurs when a woman is 45–55 years old. Between puberty and menopause, oogenesis continues approximately once per month as part of the *ovarian cycle.* The left side of **Figure 26.14** summarizes the stages of oogenesis:

Before Birth

- Unlike the spermatogonia that function in spermatogenesis, the stem cells of females, called **oogonia** (oh′-uh-GOH-nee-ah;

singular, *oogonium*), complete their mitotic divisions before birth. This means that all oogonia that will ever be present in a female are formed before she has been born.

- Between the third and seventh months of fetal development, the oogonia begin to undergo meiosis I, at which point they are called **primary oocytes.** The primary oocytes proceed as far as prophase I, after which their development is arrested.

OOGENESIS (development of an ovum)   DEVELOPMENT OF A FOLLICLE

Figure 26.14 Comparison of the stages of oogenesis and follicular development.

Childhood

- The primary oocytes remain in prophase I and do not develop further meiotically until a girl reaches puberty. At that time, rising hormone levels trigger the start of the ovarian cycle.

Puberty to Menopause

- About once a month after puberty, 20–30 primary oocytes are stimulated to continue development. Usually one primary oocyte (or more in some women) will complete the first meiotic division to produce two haploid cells that are very different in size. The smaller cell, which contains DNA but very little cytoplasm, is the first **polar body,** which often degenerates. The larger cell, which contains DNA and most of the cytoplasm plus fluid absorbed from the extracellular environment, is the **secondary oocyte.** Only the secondary oocyte has the potential to become an ovum. The extra cytoplasm present in the secondary oocyte ensures that if it is fertilized, it will have adequate nutrients for its 4- to 7-day journey toward the uterus.
- The secondary oocyte proceeds to metaphase II and its development is again arrested.
- If the secondary oocyte is fertilized, it will complete the second meiotic division and divide to form a second polar body and the fertilized ovum. Occasionally, the first polar body proceeds through meiosis II, and will divide and produce two polar bodies. However, the fate of all polar bodies is the same: They degenerate.

Each ovary of a newborn baby girl has 2–4 million primary oocytes. During the female's lifetime, however, this number decreases as the oocytes degenerate. By the time a girl reaches puberty, the number has dropped to between 80,000 and 400,000. Only about 400–500 primary oocytes will reach maturity and ovulate during a woman's reproductive lifetime.

ConceptBOOST »»»

Spermatogenesis versus Oogenesis

Both spermatogenesis and oogenesis produce gametes and provide the same amount of genetic material. However, their gametes play very different roles in reproduction.

Sperm are produced in high numbers for two important reasons. First, sperm have to travel through the inhospitable female reproductive tract to reach the secondary oocyte and most will not survive. The more sperm that enter the reproductive tract, the more sperm that will probably make it to the secondary oocyte. Second, sperm cells produce an enzyme that allows them to penetrate the secondary oocyte, but one sperm cell by itself cannot produce enough of this enzyme—hundreds of sperm are needed.

In contrast, there is no reason for the female reproductive system to produce multiple ova at one time. In fact, the female system has a difficult time supporting more than two or three embryos during one pregnancy. When an ovum forms, it receives

most of the nutrient-rich cytoplasm, whereas the polar body receives very little of it. The nutrients from the ovum help support the developing embryo until it embeds in the uterus and begins obtaining nutrients from the mother. The situation differs for a sperm cell: It only needs to deliver genetic material to the developing oocyte, so excess cytoplasm is removed during spermiogenesis to keep the sperm cell small.

Now let's relate this information to the processes of spermatogenesis and oogenesis. Spermatogenesis produces millions of sperm every day (in case they are needed), whereas oogenesis produces one viable secondary oocyte approximately once a month. Here is a summary of the key differences between spermatogenesis and oogenesis:

Characteristic	Spermatogenesis	Oogenesis
Time of onset	Begins at puberty	Begins before birth
What happens at puberty	Begins and continually produces sperm cells	Continues meiosis and releases an immature gamete (the secondary oocyte) each month
Number of cells produced	Produces millions of small, motile sperm each day	Produces one large oocyte each month
When process ends	Continues until death	Continues until age 45–55
Duration of meiosis	Continuous throughout life	Pauses twice: at prophase I from prenatal development to puberty, and at metaphase II from ovulation to fertilization

Quick Check

- ☐ 1. When in the life cycle of a female does oogenesis begin?
- ☐ 2. When is development of an oocyte arrested, and why?
- ☐ 3. How many ova are produced at the end of oogenesis? Why?

Hormonal Control of Female Reproduction

As in a male, secretion of sex hormones in a female—in this case estrogens and progesterone—is controlled by a multi-tiered negative feedback loop involving the hypothalamus, anterior pituitary, and ovaries. However, hormonal regulation of the female reproductive system is more complicated than that of the male because it requires a rhythmic fluctuation of hormones from the pituitary and ovaries. For example, hormones from the anterior pituitary regulate the maturation of oocytes and follicles. Ovarian follicles then produce fluctuating levels of ovarian hormones that cause monthly cyclic changes to the uterine endometrium.

Comparison of Oogenesis and Follicle Development

An oocyte grows and develops within an **ovarian follicle,** which is a small blister-like structure inside the ovarian cortex.

As you can see in Figure 26.14, the stages of oogenesis (left side of the figure) are related to the development of a follicle (right side).

Before Birth

- The first follicles formed in the female fetus are called **primordial follicles.** Each one contains one primary oocyte.

Childhood to Puberty

- Some primordial follicles grow into primary follicles as primary oocytes enlarge. However, many primordial follicles die by a process called *apoptosis,* or programmed cellular death, before puberty due to the lack of gonadotropins. By the time puberty is reached, the number of primordial follicles has fallen by 50–90%.
- Primary follicles develop into secondary follicles, each still containing an arrested primary oocyte.

Puberty to Menopause

- Approximately once per month, one secondary follicle grows into a mature follicle called a **vesicular follicle,** or *tertiary follicle,* which contains a secondary oocyte arrested in metaphase II.
- The vesicular follicle ruptures, resulting in ovulation of the secondary oocyte.
- The ruptured follicle becomes a hormone-secreting structure called the corpus luteum, which persists if the oocyte is fertilized. If fertilization does not occur, the corpus luteum degrades into scar tissue.

Beginning at menopause, follicle development ceases and ova no longer develop.

The Ovarian Cycle

The **ovarian cycle** includes the monthly series of events associated with the maturation of an oocyte and its follicle in an ovary. Keep in mind that the maturation of any *one* individual oocyte takes several cycles to complete and requires more than 350 days to progress from primordial follicle to ovulation. At any time in an ovary, several follicles are in various stages of development and decline. Let's examine the stages of a follicle (**Figure 26.15**):

Follicular Phase (Stages 1–4) During the **follicular phase,** the follicle grows and develops, as you can see in the different follicular stages in Figures 26.14 and 26.15. The maturation of a follicle proceeds as follows:

1. **Primordial follicle.** A single layer of squamous **follicular cells** surrounds the primary oocyte. At birth, each ovary has approximately 2–4 million primordial follicles, each follicle containing a primary oocyte. During childhood, many of the primordial follicles will begin developing into primary follicles.
2. **Primary follicle.** The follicular cells change in shape from flattened to cuboidal, at which point they are referred to as **granulosa cells.** Two important structures appear in the primary follicles: microvilli and thecal cells. *Microvilli* develop in the oocyte and grow into the surrounding granulosa cells and form a connection. Part of this

connection is a glycoprotein-rich area called the *zona pellucida,* which becomes important if the oocyte is later fertilized (see Chapter 27). Microvilli also increase the surface area available for the granulosa cells to transfer materials to the growing primary oocyte through gap junctions. Only a small number of the primary follicles will continue development.

3. **Secondary follicle.** The wall of the follicle thickens, and the deeper granulosa cells secrete **follicular fluid,** which forms small pockets and expands the overall size of the follicle. The granulosa cells grow, enlarge, and stimulate nearby cells in the ovary to form a layer of **thecal cells** around the follicle. Granulosa cells produce estrogens from the secretions of the thecal cells. The primary oocyte is still growing slowly at this stage. Usually only one secondary follicle per cycle will complete development.
4. **Vesicular (tertiary) follicle.** The smaller pockets of follicular fluid merge to form a single large cavity called the **antrum.** The primary oocyte and its surrounding capsule of granulosa cells, known as *cumulus cells,* project into the antrum on a stalk. As the vesicular follicle continues to enlarge, it may reach 15–25 mm (0.6–1.0 in.) in diameter and create a bulge in the surface of the ovary. At this point, the primary oocyte, which has been suspended in prophase I, completes meiosis I to form the secondary oocyte and the first polar body. However, it is then suspended in metaphase II and will not complete meiosis unless fertilization occurs.

Ovulation Phase (Stage 5) In **ovulation** the ovary expels the secondary oocyte and associated granulosa cells. Ovulation occurs at the end of the follicular phase.

5. **Ruptured follicle.** The bulging ovary wall and vesicular follicle rupture and expel the secondary oocyte, surrounded by a layer of granulosa cells called the **corona radiata,** into the peritoneal cavity. The vesicular follicle collapses and ruptured vessels bleed into the antrum.

Usually only one vesicular follicle is at the peak of maturation when ovulation occurs; however, sometimes more than one vesicular follicle releases a secondary oocyte. This phenomenon, which is more common in some families and also increases with maternal age, may result in **fraternal twins,** or *nonidentical twins.* **Identical twins** result from the fertilization of a single oocyte by a single sperm, followed by separation of the dividing cells in early development.

The Luteal Phase (Stages 6–7) The **luteal phase** (LOO-tee-uhl) is the period in which the remnants of the ruptured follicle become an endocrine organ called the **corpus luteum.** The steps of this phase proceed as follows:

6. **Corpus luteum.** The remaining granulosa cells from the ruptured follicle enlarge and, along with the thecal cells, form the yellowish corpus luteum. The corpus luteum secretes progesterone and some estrogen. As we will discuss shortly, progesterone is necessary to maintain a pregnancy. So, if pregnancy does occur, the corpus luteum will

Ovarian ligament

Ovarian artery

Ovarian vein

STAGES OF THE OVARIAN CYCLE		DESCRIPTION
① **Primordial follicle**	Primary oocyte — Follicular cells — 215×	• Primary oocyte surrounded by a single layer of squamous follicle cells
② **Primary follicle**	Granulosa cells — Thecal cells — 260×	• Follicle cells become cuboidal granulosa cells • Microvilli and thecal cells develop
③ **Secondary follicle**	Granulosa cells — Pockets of follicular fluid — 125×	• Primary oocyte surrounded by multiple layers of granulosa cells • Follicular fluid found in small cavities around the oocyte
④ **Vesicular follicle**	Secondary oocyte — Granulosa cells — Antrum — Thecal cells — 90×	• Primary oocyte finished meiosis I to become secondary oocyte • Surrounded by granulosa cells and a fluid-filled antrum • Maturation of a primordial follicle to this stage takes around 350 days
⑤ **Ruptured follicle**	Ovulated secondary oocyte — Corona radiata — 90×	• Secondary oocyte is released from the vesicular follicle
⑥ **Corpus luteum**	Corpus luteum — 3×	• Remnant of the ruptured follicle • Secretes progesterone and some estrogen
⑦ **Corpus albicans**	Corpus albicans — 20×	• Remnant of the corpus luteum after it ceases hormone secretion

Sequence the steps
@MasteringA&P®

Figure 26.15 The ovarian cycle.

persist for approximately 3 months to produce hormones. If pregnancy does not occur, the corpus luteum begins to degenerate in approximately 10 days and stops producing hormones.

(7) **Corpus albicans.** When levels of hormones secreted by the corpus luteum decline, macrophages within the ovary invade the corpus luteum and degrade it while fibroblasts lay down collagen. These actions reduce the corpus luteum to the **corpus albicans,** a whitish knot of scar tissue (*albus* = "white"). This event marks the end of one ovarian cycle.

Assuming an ovarian cycle of 28 days, the follicular phase extends from the first to the 14th day of the cycle, ovulation typically occurs on day 14, and the luteal phase extends from the 14th to the 28th day. However, fewer than 25% of women normally have 28-day cycles, and cycles as short as 21 days or as long as 40 days are relatively common. In addition, it is also common for the same woman to experience 1- or 2-day fluctuations from one cycle to the next. In these cases, the length of the follicular phase varies, but the luteal phase normally remains constant: It is 14 days from the time of ovulation to the end of the cycle.

Quick Check

☐ 4. What are the seven stages of the ovarian cycle?
☐ 5. How do these stages correspond to oogenesis?
☐ 6. How do these stages correspond to the phases of the ovarian cycle?

Hormonal Interactions during the Ovarian Cycle

Regulation of ovarian function requires a rhythmic fluctuation of hormones from the hypothalamus, the anterior pituitary, and the ovaries themselves. These hormones interact via negative feedback in the same hypothalamic-pituitary-gonadal (HPG) axis that operates in males, with the exception of a positive feedback mechanism that triggers ovulation. Here we see another example of the Feedback Loops Core Principle (p. 21). During a 28-day cycle, levels of hypothalamic, anterior pituitary, and ovarian hormones fluctuate as follows (**Figure 26.16**):

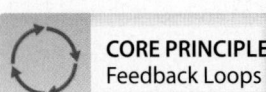
CORE PRINCIPLE
Feedback Loops

- **First-tier control.** The hypothalamus releases **gonadotropin-releasing hormone (GnRH).**
- **Second-tier control.** The anterior pituitary releases follicle stimulating hormone (FSH) and luteinizing hormone (LH) in response to GnRH from the hypothalamus.
- **Third-tier control.** The ovaries are the target organ of FSH and LH. LH stimulates thecal cells in the follicle to secrete androgens, which diffuse through the basement membrane to the granulosa cells. FSH stimulates granulosa cells in follicles to secrete estrogens, convert the androgens to additional estrogens, and secrete inhibin.
- **Effects:**
  - ○ Estrogens stimulate a follicle (called the *dominant follicle*) to continue developing into a vesicular follicle. Usually only one dominant follicle continues development.

- ○ The new vesicular follicle eventually produces enough estrogens to exert *positive feedback* on the anterior pituitary to trigger a large increase in LH release known as the *LH surge.*
- ○ The LH surge, accompanied by increased levels of FSH, triggers ovulation. The resulting corpus luteum produces additional progesterone, estrogens, and inhibin.

Increasing levels of estrogens and inhibin exert negative feedback controls on the hypothalamus and pituitary. Estrogens inhibit GnRH and LH secretion, and inhibin inhibits FSH secretion. This negative feedback prevents other primary follicles from beginning to develop too early.

If fertilization occurs, progesterone from the corpus luteum inhibits the release of gonadotropins and will continue to do this. If fertilization does not occur, the corpus luteum degenerates into the corpus albicans and estrogen, progesterone, and inhibin levels drop. With ovarian hormones no longer inhibiting secretion of GnRH, its levels rise and the cycle begins again.

Keep in mind that more than one follicle develops at a time. Each rise in FSH at the beginning of the ovarian cycle activates several follicles. Each midcycle LH surge triggers ovulation of a follicle that was originally activated several months earlier. An imbalance in any part of this loop can interfere with ovulation. For more information, see *A&P in the Real World: Female Infertility.*

A&P in the **Real World**

Female Infertility

The most common reason for female infertility is a problem with ovulation. Without ovulation, there are no secondary oocytes to fertilize. Possible causes include:

- polycystic ovarian syndrome, a hormone imbalance that is due to excessive androgens and interferes with normal ovulation;
- primary ovarian insufficiency, which occurs when a woman's ovaries stop working normally before age 40 (this is not the same as early menopause); and
- blockage of uterine tubes due to pelvic inflammatory disease (discussed in Module 26.7), endometriosis, or uterine fibroids (noncancerous clumps of tissue and muscle on the internal walls of the uterus).

In addition, poor diet, being overweight or underweight, smoking, excessive alcohol consumption, stress, excessive athletic training, and age can all affect a woman's ability to conceive.

Treatment for infertility in women depends on the cause. Medications, including synthetic gonadotropins, may stimulate ovulation. Abdominal surgery may be necessary to clear blocked uterine tubes or remove endometriosis or fibroids. If no other treatments are successful, in vitro fertilization (IVF) is an option. In this procedure, oocytes and sperm, from the mother and father or from donors, are fertilized in a laboratory and then implanted in the mother's uterus.

Other Effects of Female Hormones

Estrogen and, to a lesser degree, progesterone also stimulate the development of female sex characteristics:

- maturation of sex organs,
- development of external genitalia, and
- maintenance of anatomical features unique to adult females, such as breast development and fat accumulation around the hips and thighs.

Progesterone is also responsible for maintaining a pregnancy if fertilization occurs and is discussed in more detail in Chapter 27.

Estrogens also have far-reaching effects on other tissues, many of which are beneficial. For example, estrogens reduce the breakdown of bone, helping to prevent osteoporosis, and increase the level of HDLs ("good" cholesterol) in the blood while decreasing the level of LDLs ("bad" cholesterol), helping to protect the cardiovascular system. However, estrogens also promote blood coagulation, which can lead to the formation of inappropriate blood clots under certain conditions.

Quick Check

☐ 7. What is the function of FSH in the ovarian cycle?
☐ 8. What is the function of LH in the ovarian cycle?
☐ 9. Which processes are stimulated by estrogens and progesterone?

Figure 26.16 Hormonal regulation of ovarian function via the hypothalamic-pituitary-gonadal (HPG) axis.

The Uterine Cycle

The **uterine cycle,** or *menstrual cycle,* is the series of cyclic changes that the uterine endometrium goes through each month as it responds to the fluctuating levels of ovarian hormones. These uterine changes are coordinated with the levels of estrogen and progesterone released during the phases of the ovarian cycle, which are controlled by gonadotropins released from the anterior pituitary. Let's first look at the structures of the endometrium that are affected by the changing hormone levels.

The endometrium has two main strata, or layers. The **stratum functionalis,** or *functional layer,* undergoes cyclic changes in response to ovarian hormones. It detaches from the uterine wall and is shed as a discharge of roughly 35–50 ml of blood and other materials from the vagina approximately once per month during **menstruation.** The thinner, deeper **stratum basalis,** or *basal layer,* does not thicken in response to ovarian hormones, but forms a new stratum functionalis after menstruation ends.

Numerous **endometrial (uterine) glands** change in length as endometrial thickness changes. The main vascular supply of the uterus comes from the *uterine arteries,* which arise from the internal iliac arteries in the pelvis. After branching several times, they eventually form the **spiral (coiled) arteries** of the stratum functionalis.

Now let's examine the three phases of the uterine cycle. For simplicity, we will assume a 28-day cycle (**Figure 26.17**):

1. **Menstrual phase, days 1–5.** During the menstrual phase, the uterus sheds the stratum functionalis, resulting in menstruation.

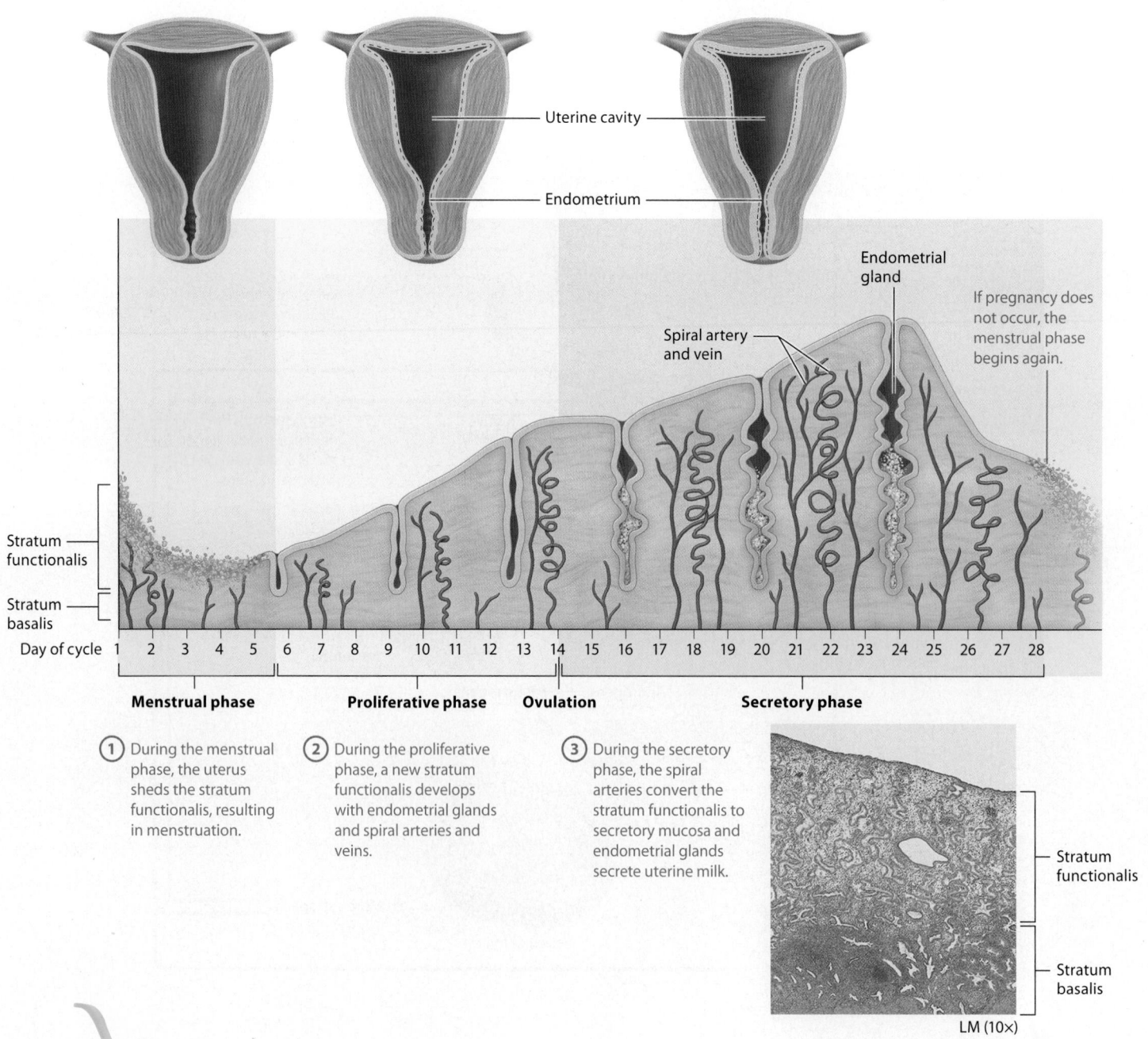

(1) During the menstrual phase, the uterus sheds the stratum functionalis, resulting in menstruation.

(2) During the proliferative phase, a new stratum functionalis develops with endometrial glands and spiral arteries and veins.

(3) During the secretory phase, the spiral arteries convert the stratum functionalis to secretory mucosa and endometrial glands secrete uterine milk.

LM (10×)

Endometrium during secretory phase

Figure 26.17 Endometrial changes during the uterine cycle.

2. **Proliferative (preovulatory) phase, days 6–14.** During the proliferative phase, a new stratum functionalis develops with endometrial glands and spiral arteries and veins. As this new layer thickens (proliferates), its endometrial glands enlarge and its spiral arteries and veins increase in number. Cervical mucus is thick and sticky until just before ovulation, when it thins to facilitate the passage of sperm into the uterus. At the end of the proliferative phase around day 14, ovulation occurs.

3. **Secretory (postovulatory) phase, days 15–28.** Regardless of the normal length of a woman's cycle, this phase usually lasts 14 days. During the secretory phase, the spiral arteries convert the stratum functionalis to secretory mucosa and endometrial glands secrete a nutritious glycogen-rich fluid, also known as *uterine milk,* into the uterine cavity to sustain the embryo.

If pregnancy does not occur, the cells of the stratum functionalis die and, on day 28, the menstrual phase begins again. If pregnancy does occur, the secretory phase will continue and the uterus will continue to develop.

Hormonal Interactions during the Uterine Cycle

Regulation of uterine function also requires a rhythmic fluctuation of hormones from the hypothalamus, pituitary, and ovaries. Again, for simplicity we will assume a 28-day cycle as we examine the changes that occur during the uterine cycle.

Figure 26.18 presents two graphs: one showing the changing levels of gonadotropins during the phases of the uterine cycle, and another showing the changing levels of ovarian hormones. As you can see, at the beginning of the menstrual phase, FSH and LH levels are beginning to rise. Ovarian hormones, however,

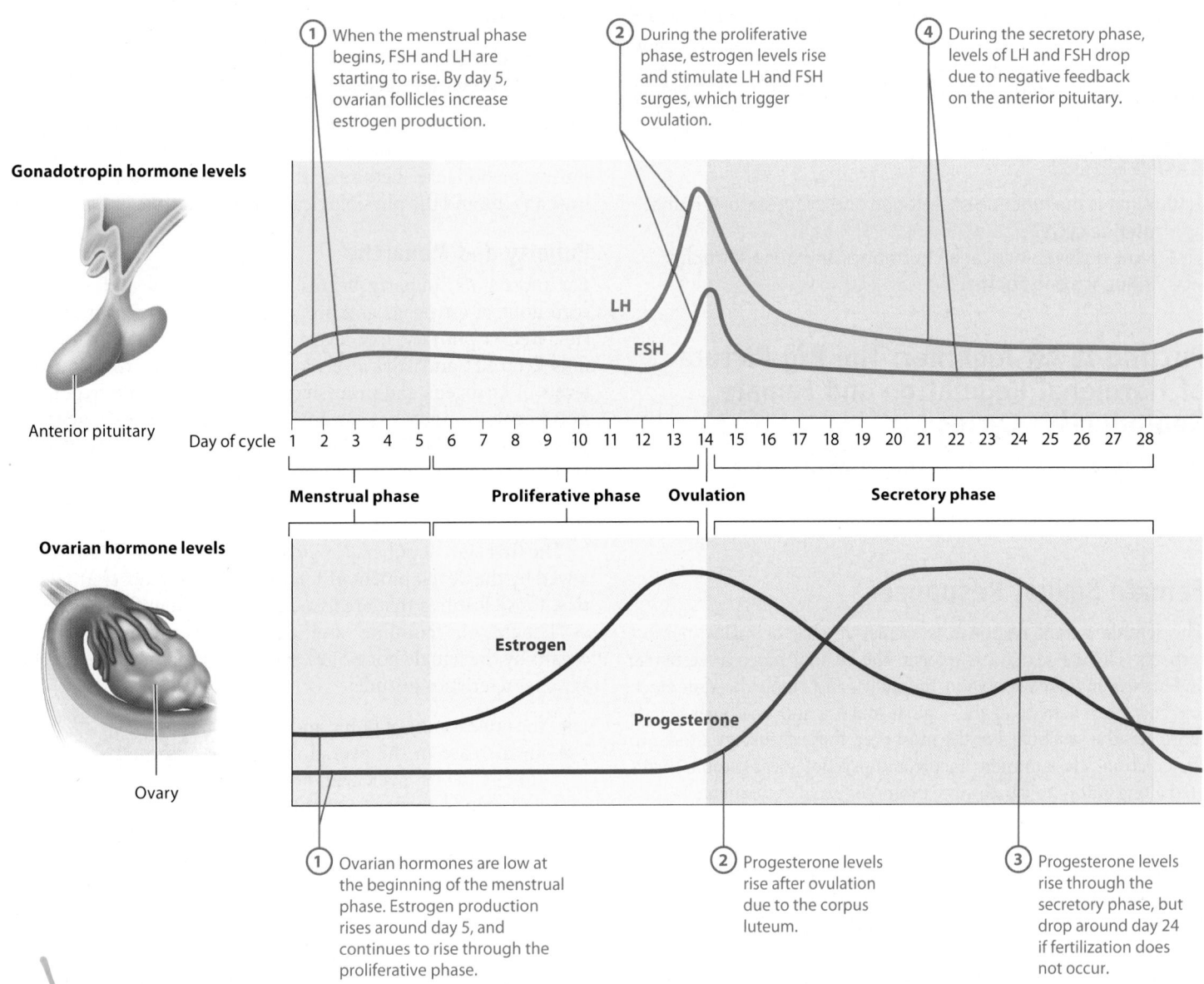

Figure 26.18 Pituitary and ovarian hormone interactions during the uterine cycle.

are at their lowest normal levels until day 5, at which point the growing ovarian follicles start to produce more estrogens. As the proliferative phase begins, the rising estrogen levels cause a new stratum functionalis to generate and trigger endometrial cells to synthesize progesterone receptors. The high estrogen levels trigger release of more FSH and LH, and their levels peak around day 14, which causes ovulation. At the start of the secretory phase, the levels of progesterone from the corpus luteum begin to rise, while the amount of estrogen secreted decreases. The high progesterone level and falling estrogen levels inhibit the anterior pituitary, which, as you can see in the graph, leads to low amounts of LH and FSH.

If the ovulated oocyte is not fertilized, the corpus luteum degenerates toward the end of the secretory phase and progesterone levels fall. Without progesterone, the stratum functionalis layer loses its blood supply and breaks down. This triggers the beginning of a new menstrual phase. If fertilization does take place, the outer layer of the developing embryo begins to secrete the LH-like substance **human chorionic gonadotropin (hCG),** which keeps the corpus luteum from degenerating and the progesterone level high. Since hCG is unique to the developing embryo, it is the target molecule of home pregnancy tests.

Quick Check

☐ 10. What is the function of estrogen and progesterone in the uterine cycle?
☐ 11. How do levels of ovarian hormones and gonadotropins influence one another?

Putting It All Together: The Big Picture of Hormonal Regulation and Female Reproductive Cycles

To understand how hormones coordinate the ovarian and uterine cycles, let's look at both cycles at the same time. **Figure 26.19** allows the two cycles to be easily compared.

Female Sexual Response

The **female sexual response** is similar to that of males in most respects. During sexual excitement, the vaginal mucosa, vestibule, and breasts engorge with blood; the clitoris and nipples become erect; and increased activity of the vaginal mucosa and vestibular glands lubricates the vestibule. For the most part, these events are similar to the erection phase in men. Tactile and psychological stimuli, mediated along autonomic pathways, promote sexual excitement.

As in males, orgasm refers to the period during which women experience intense feelings of pleasure. However, unlike in males, no ejaculation takes place. The uterus does exhibit peristaltic waves of contraction, and the cervix pushes down somewhat into the vagina. If semen is present, these actions may draw semen farther into the reproductive tract, but female orgasm is not required for conception. Females do not experience a refractory period, so they may have multiple orgasms during a single sexual experience.

The female libido, or sex drive, may be related to the production of the androgen *dehydroepiandrosterone* (*DHEA*), produced by the adrenal cortex in women. Androgens and estrogens both appear to influence sexual behavior because of their interactions in the hypothalamus. In addition, many women experience fluctuations in libido in response to monthly fluctuations of estrogens and progesterone.

Quick Check

☐ 12. What are the similarities between the male and female sexual responses?
☐ 13. What are the differences between the male and female sexual responses?

Puberty and Menopause

◄◄ **Flashback**

1. What is adipose tissue? (p. 139)
2. What is leptin? (p. 620)

As we have discussed previously, levels of estrogens and progesterone increase dramatically at puberty and begin to decline during menopause. Let's take a look at these two periods in a woman's life and the physiological changes that accompany them.

Puberty and Menarche

For most girls, puberty begins between ages 9 and 11, when secretions of estrogens and progesterone increase from the ovaries. Before puberty, low levels of GnRH from the hypothalamus keep LH and FSH levels low, which in turn results in low levels of estrogens and progesterone. At puberty, the hypothalamus becomes much less sensitive to inhibition by estrogens and progesterone, and GnRH levels increase. GnRH stimulates the HPG axis, and blood estrogens and progesterone levels begin to rise, resulting in the appearance of female secondary sex characteristics.

The first sign of puberty in girls is budding breasts. This is followed by the development of **female secondary sex characteristics,** those features that are associated with identifying someone as "female" or "feminine" and are induced in nonreproductive organs by the female hormones, mostly estrogens. These secondary characteristics include:

- the appearance of pubic and axillary hair;
- an increase in the overall amount of adipose tissue in the subcutaneous layer, with additional deposits in the breasts and around the hips and thighs;
- increased secretions of sebaceous glands (the oilier skin often results in acne); and
- skeletal changes, including increased height and widening pelvis.

Approximately 2 years after the onset of puberty, females experience their first episode of menstrual bleeding, called **menarche** (men-AR-kee). Perhaps because of the physical toll exerted by

The Big Picture

Figure 26.19

of Hormonal Regulation of the Ovarian and Uterine Cycles

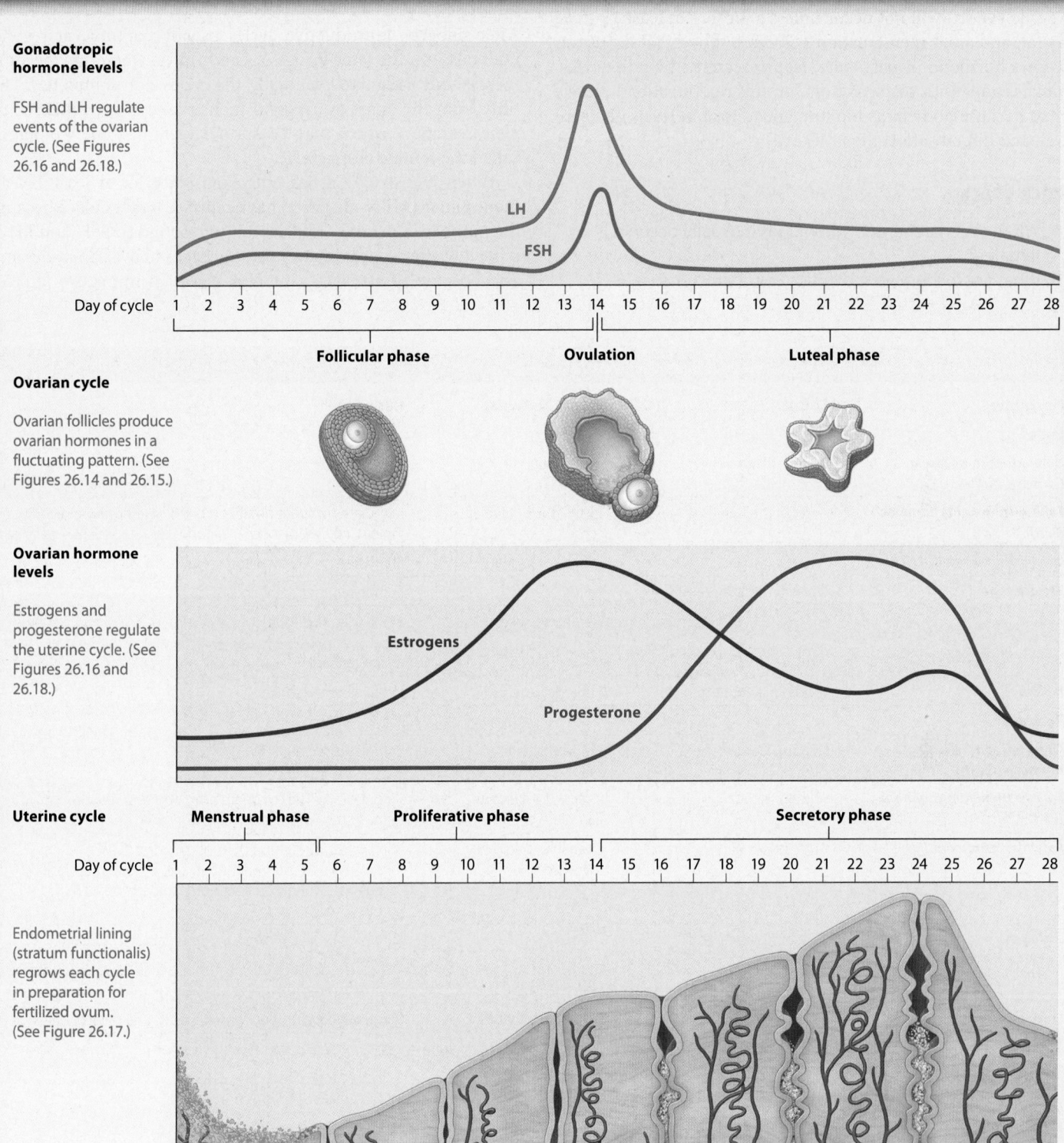

Gonadotropic hormone levels

FSH and LH regulate events of the ovarian cycle. (See Figures 26.16 and 26.18.)

LH

FSH

Day of cycle 1 2 3 4 5 6 7 8 9 10 11 12 13 14 15 16 17 18 19 20 21 22 23 24 25 26 27 28

Follicular phase | **Ovulation** | **Luteal phase**

Ovarian cycle

Ovarian follicles produce ovarian hormones in a fluctuating pattern. (See Figures 26.14 and 26.15.)

Ovarian hormone levels

Estrogens and progesterone regulate the uterine cycle. (See Figures 26.16 and 26.18.)

Estrogens

Progesterone

Uterine cycle

Menstrual phase | **Proliferative phase** | **Secretory phase**

Day of cycle 1 2 3 4 5 6 7 8 9 10 11 12 13 14 15 16 17 18 19 20 21 22 23 24 25 26 27 28

Endometrial lining (stratum functionalis) regrows each cycle in preparation for fertilized ovum. (See Figure 26.17.)

pregnancy on the female body, menarche (and the accompanying cyclic events) will not occur unless a girl has at least 15–17% body fat, and adult menstruation will cease if body fat drops too low. One hormone in particular, *leptin*, secreted by adipocytes, stimulates gonadotropin secretion. For this reason, if leptin levels decline because body fat is too low, gonadotropin levels decrease as well and directly affect the HPG axis.

Quick Check

☐ 14. What is menarche, and when does it usually occur in females?

☐ 15. What are the female secondary sex characteristics?

Effects of Aging: Menopause

After puberty, reproductive cycles usually continue at regular intervals into the later forties or early fifties, when they become increasingly irregular. Eventually, the cycles stop completely. The time from the onset of irregular menstrual cycles to their complete cessation, which may be up to 5 years in some women, is called the **female climacteric.**

When menstruation has not taken place for at least 1 year, **menopause** (MEN-ah-pawz) has occurred. It takes place because the number of primary follicles left to respond to FSH and LH is quite low after approximately three decades of ovarian cycles and also because of **atresia**—a process in which immature follicles

---

TABLE 26.3 MAJOR REPRODUCTIVE HORMONES IN MALES AND FEMALES

Hormone	Source	Target Tissue(s)	Response
Males			
Gonadotropin-releasing hormone (GnRH)	Hypothalamus	Anterior pituitary	Stimulates secretion of LH and FSH
Follicle-stimulating hormone (FSH)	Anterior pituitary	Testes (sustentacular cells)	Stimulates production of androgen-binding protein (ABP), inhibin; necessary for spermatogenesis
Luteinizing hormone (LH)	Anterior pituitary	Testes (interstitial cells)	Stimulates production and secretion of testosterone
Testosterone	Testes (interstitial cells)	Testes	Stimulates spermatogenesis
		Other body tissues	Stimulates development and maintenance of reproductive organs; responsible for development of secondary sex characteristics
Inhibin	Testes (sustentacular cells)	Anterior pituitary	Inhibits FSH secretion
Females*			
Gonadotropin-releasing hormone (GnRH)	Hypothalamus	Anterior pituitary	Stimulates secretion of LH and FSH
Follicle-stimulating hormone (FSH)	Anterior pituitary	Ovarian follicles and granulosa cells	Stimulates follicles to begin development; stimulates granulosa cells to convert androgens to estrogens
Luteinizing hormone (LH)	Anterior pituitary	Ovarian follicles and thecal cells	Stimulates production of gonadal hormones (e.g., estrogens); stimulates follicles to finish maturation; stimulates ovulation; contributes to development of corpus luteum
Estrogens	Ovaries (follicles) and granulosa cells	Uterus	Stimulate proliferation of endometrial cells; stimulate uterine contractions
		Mammary glands	Stimulate development of ducts of mammary glands
		Other body tissues	Stimulate development of secondary sex characteristics
		Hypothalamus and anterior pituitary	Before ovulation, cause positive feedback to increase LH and FSH secretion
		Ovaries	Inhibit adjacent follicles
Progesterone	Ovaries (corpus luteum)	Uterus	Causes hypertrophy of endometrial cells; stimulates secretion of mucus from uterine glands; helps maintain pregnancy and prevents uterine contractions
		Mammary glands	Stimulates alveoli of mammary glands to develop during pregnancy
		Other body tissues	Stimulates development of secondary sex characteristics
Inhibin	Ovaries (follicles) and granulosa cells	Anterior pituitary	Inhibits FSH secretion

*Other hormones secreted in females, primarily during pregnancy, include relaxin, prolactin, oxytocin, and human chorionic gonadotropin. (These hormones are covered in Chapter 27.)

degenerate and are resorbed during the follicular phase of the menstrual cycle. Those follicles that remain in the ovaries are less responsive to gonadotropins, and so less estrogen and progesterone are secreted. As a result, ovulation does not occur and the blood concentration of estrogens declines sharply, although many women continue to synthesize some estrogen from adrenal androgens. However, secretions of FSH and LH from the pituitary are no longer inhibited, so these hormones may be continuously released for some time.

The lower concentration of estrogens and progesterone may change the female secondary sex characteristics. The breasts, vagina, uterus, and uterine tubes may shrink, and the pubic and axillary hair may thin. Often bone density decreases, which may lead to osteoporosis, and the skin becomes thinner.

Some women experience unpleasant vasomotor signs, including sensations of heat called "hot flashes." A hot flash may last up to 5 minutes and may be accompanied by chills and/or sweating. These vasomotor symptoms may result from changes in the rhythmic secretion of GnRH by the hypothalamus in response to declining concentrations of sex hormones. Women may also experience migraine headaches, backaches, fatigue, and mood swings during menopause.

Some women experience menopause before age 45, which is called **early menopause.** If it occurs before age 40, it is **premature menopause.** Early or premature menopause can happen for several reasons, including chemotherapy or radiation treatments for cancer, surgery to remove the ovaries or uterus, and autoimmune diseases. In some cases, the reason may simply be genetics: Women with a family history of early menopause are more likely to experience early menopause.

Table 26.3 summarizes the functions of the major reproductive hormones in both males and females, identifying the source, the target tissue(s), and the body's response to each hormone.

Quick Check

☐ 16. What is menopause?
☐ 17. Why does menopause occur?

Apply What You Learned

☐ 1. A new drug is able to stop the atresia of primary oocytes *without* increasing the number of secondary oocytes ovulated each cycle. Predict the effect this drug would have on the reproductive lifespan of an adult female.

☐ 2. A young healthy woman begins to take a daily herbal supplement that mimics the effects of progesterone. Predict how this might affect her ovarian and uterine cycles.

☐ 3. A 38-year-old woman has decided not to have children. She is considering a form of birth control called tubal ligation, which involves closing off the uterine tubes. However, she is concerned that this procedure will push her into early menopause. How would you address her concerns?

See answers in Appendix A.

MODULE **26.6**
Methods of Birth Control

Learning Outcomes

1. Explain why changes in cervical mucus can predict a woman's monthly fertility.
2. Provide examples of how birth control methods relate to normal reproductive function.

Birth control, or contraception, refers to any procedure or device that is intended to prevent pregnancy. In this module, we explore common methods of contraception.

Behavioral Methods

Abstinence—the avoidance of intercourse—is the only 100% reliable method of avoiding pregnancy. Periodic abstinence, also called the *rhythm method,* is based on avoiding intercourse near the time of ovulation. The rhythm method has a high failure rate partly because it is often difficult to predict the exact date of ovulation. Also, intercourse must be avoided at least 7 days before ovulation and at least 2 days after ovulation, so there are no surviving sperm and fertile oocytes in the reproductive tract at the same time.

Withdrawal, also called *coitus interruptus,* involves withdrawing the penis before ejaculation. This method also has a high failure rate because it is difficult to time withdrawal accurately, and because sperm may be present in pre-ejaculatory fluid. In addition, sperm ejaculated anywhere in the vulva can potentially enter the female reproductive tract.

Quick Check

☐ 1. Why do most behavioral methods of birth control have high failure rates?

Barrier Methods

Barrier methods of birth control prevent sperm from moving beyond the vagina. These methods include condoms, diaphragms, cervical caps, and the sponge (**Figure 26.20**).

Most *condoms* (Figure 26.20a) are made of synthetic materials or latex, although some are made from animal membranes (lamb intestine). Male condoms, worn on the penis, are the most common type of condom in use. Female condoms are inserted into the vagina and include a flexible ring that covers the cervix and an outer ring that covers the external genitalia. In addition to reducing the risk of pregnancy, synthetic and latex condoms lower the risk of sexually transmitted infections (discussed in Module 26.7).

Diaphragms or *cervical caps* are placed inside the vagina and cover the cervix to block sperm (Figure 26.20b). Both of these barriers require a prescription and a physical exam by a health

(a) Male and female condoms

(b) Diaphragm, cervical cap, and sponge

(c) Spermicide

(d) Oral contraceptives

(e) Vaginal ring and skin patch

(f) Progesterone injections and implants

(g) IUD

Figure 26.20 Temporary methods of birth control.

care provider for proper fit. A similar form of birth control that requires no fitting is the *sponge,* a foam disk that is inserted before intercourse to cover the cervix.

Many barrier methods, particularly the diaphragm, cervical cap, and sponge, are made more effective by using a chemical **spermicide,** which kills or damages sperm cells it contacts. Spermicides are available without a prescription and may be purchased as foams, creams, or jellies (Figure 26.20c).

Quick Check

☐ 2. What are some common barrier methods of contraception?

Hormonal Methods

Oral contraceptives, or *birth control pills,* contain synthetic estrogens and progesterone (Figure 26.20d). The artificially high levels of estrogens and progesterone have a negative feedback effect on the release of FSH and LH. Low levels of FSH and LH prevent follicle maturation and, subsequently, ovulation. In addition, they interfere with the buildup of the endometrial lining necessary for implantation of an embryo. A side benefit is that oral contraceptives often reduce the amount of monthly bleeding during menstruation. In fact, some reduce the number of menstrual cycles to only about four per year.

Oral contraceptives are very effective at preventing pregnancy if taken as prescribed. However, they may cause side effects.

Some of these are relatively minor, such as nausea, fluid retention, and breast tenderness, whereas others are potentially more severe. The risk of serious side effects, such as cardiovascular or liver problems, increases in women over 35, particularly those who smoke or have a history of hypertension or bleeding disorders.

One alternative to the combined estrogen-progesterone pill is the *minipill,* which contains only synthetic progesterone. High levels of progesterone thicken the cervical mucus and make it difficult for sperm to reach the egg. The minipill is slightly less effective than the combined pill.

Hormonal alternatives to a daily pill include:

- vaginal rings containing estrogen and progesterone, which are worn for the 3 weeks or so between menstrual cycles (NuvaRing) (Figure 26.20e);
- skin patches containing estrogen and progesterone that are replaced weekly (ORTHO EVRA) (Figure 26.20e);
- intramuscular injections of progesterone every 3 months (Depo-Provera) (Figure 26.20f); and
- thin rods containing progesterone, which are implanted under the skin of the arm and remain in the body for approximately 3 years (Nexplanon) (Figure 26.20f)

Quick Check

☐ 3. How do oral contraceptive pills prevent pregnancy?
☐ 4. Why is a progesterone-only contraceptive less effective than one containing estrogen and progesterone?

Intrauterine Methods

An **intrauterine device (IUD)** is a small T-shaped device made of plastic and copper that is placed into the uterus by a health provider, where it can remain for approximately 10 years. It prevents pregnancy by not allowing a fertilized ovum to implant into the uterine wall. In addition, copper is released inside the womb and spreads to the uterine tubes, where it poisons the sperm and oocyte (Figure 26.20g).

An **intrauterine system (IUS)** is a small T-shaped device similar to the IUD, but it is made of plastic and releases a small amount of progesterone each day to prevent pregnancy. The IUS can remain in the uterus for approximately five years.

Quick Check

☐ 5. How do intrauterine devices prevent pregnancy?

Permanent Methods

Permanent methods of birth control are referred to as "sterilization" because they render the individual sterile, or incapable of reproduction. Most of these techniques require a surgical procedure. Although they can sometimes be reversed with additional surgery, the second surgery is successful less than 45% of the time. Therefore, permanent methods are best for people who do not wish to have children at any point in the future.

Male sterilization is accomplished by a surgical procedure called a **vasectomy** (va-SEK-tuh-mee) in which the physician cuts through and then either *ligates* (ties off) or *cauterizes* (burns) the ends of the ductus deferens (**Figure 26.21a**). This prevents sperm from leaving the man's body with the semen during ejaculation. After a vasectomy, sperm are still produced, but they degenerate and are destroyed by phagocytosis. Semen is also still produced because the vas deferens is cut below the area where the accessory structures secrete the other components of semen. Also, because the blood vessels are unaffected, blood testosterone levels remain normal. For these reasons, a vasectomy has no effect on sexual desire or performance. Since some sperm remain in the vas deferens, it is necessary to postpone unprotected intercourse until all sperm have been phagocytized, which may take up to 12 weeks.

A similar surgical procedure, called **tubal ligation** (lye-GAY-shun), is performed in women (**Figure 26.21b**). During this procedure, a surgeon makes an insertion into the wall of the body cavity, cuts through the uterine tubes, and either ligates or cauterizes their ends. The oocyte is thus prevented from passing through the uterine tubes and the sperm from reaching the oocyte.

Nonsurgical permanent methods of birth control include *tubal implants*. These small metal springs are placed in each uterine tube and do not require incisions in the abdomen. Over time, scar tissue grows around each implant and permanently blocks the tubes. Three months after the procedure, an x-ray is taken of the uterine tubes, which have been filled with contrast dye, to determine if they are completely blocked.

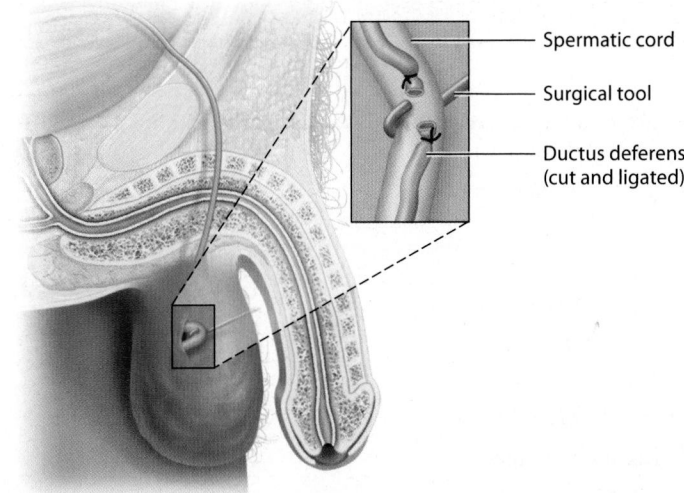

(a) Vasectomy

Spermatic cord
Surgical tool
Ductus deferens (cut and ligated)

(b) Tubal ligation

Uterine tubes (cut and ligated)

Figure 26.21 Permanent methods of birth control.

Unfortunately, no perfect form of birth control exists. Table 26.4 shows the failure rates of the available contraceptive methods.

Quick Check

☐ 6. Which methods of birth control are also called sterilization? Why?

Apply What You Learned

☐ 1. Choose an appropriate method of contraception for each of the following individuals and briefly explain your reasoning:
   a. A woman is 25 years old. She is married and is in a committed relationship. She currently has no children but wants children "someday." She has experienced blood clots due to a coagulation disorder in the past.
   b. A man is 25 years old. He is single and has multiple sexual partners. He may want children at some point in his life, but he is not sure.

See answers in Appendix A.

TABLE 26.4 FAILURE RATES FOR COMMON METHODS OF BIRTH CONTROL

Method (most successful to least successful within each category)	Rate of Failure (pregnancies per 100 users)	
	Perfect Use (method used correctly and consistently)	Typical Use (method may be used incorrectly or inconsistently)
Behavioral Methods		
Complete abstinence	0	Unknown
Withdrawal method	4	27
Rhythm method	3–5	25
Permanent Methods		
Vasectomy	0.10	0.15
Tubal ligation and implants	0.5	0.5
Hormonal Methods		
Skin patch	0.1	1–2
Vaginal contraceptive ring	0.1	1–2
Birth control pill	0.3	1–2
Hormone injections	0.3	1–2
Implants	0.3	1–2
Minipill	0.5	2
Intrauterine Devices and Systems	0.2–0.6	0.2–0.8
Barrier Methods		
Male and female condoms	2–5	15–21
Diaphragm, cervical cap, sponge (with spermicide)	6–9	16–32
Spermicides (foams, creams, jellies) used alone	18	29
No Birth Control	85	85

MODULE **26.7**

Sexually Transmitted Infections (STIs)

Learning Outcomes

1. Predict factors or situations affecting the reproductive system that could disrupt homeostasis.
2. Describe the causes, symptoms, and potential complications of common sexually transmitted infections.

A **sexually transmitted infection (STI),** also called a *sexually transmitted disease* (*STD*) or *venereal disease,* is an infectious disease spread through sexual contact. The most important cause of reproductive disorders, STIs often cause infertility in adults and deformities or even death in a fetus.

We have already discussed the most notorious STI, HIV and AIDS, in the chapter on the lymphatic system and immunity (see Chapter 20). This module presents several of the most common STIs, and distinguishes between those caused by bacteria and parasites and those caused by viruses.

Bacterial and Parasitic STIs

 Flashback

1. What is a pathogen? (p. 761)
2. What is the conjunctiva? (p. 547)

The first group of STIs we discuss are those caused by pathogenic, or disease-causing, bacteria and parasites. Both types of pathogens are foreign cells that invade tissues and create an inflammatory reaction. The common bacterial and parasitic STIs we discuss in this section are chlamydia, gonorrhea, syphilis, and trichomoniasis.

Chlamydia

Chlamydia (klah-MID-ee-ah) is a common STI caused by the bacterium *Chlamydia trachomatis.* Even though it is easily treated with antibiotics, an estimated 3 million chlamydial infections occur annually in the United States. This may be because chlamydia produces no symptoms in most people. If symptoms do occur, they may not appear until several weeks after exposure and most commonly include painful urination, pain in the pelvic region and testes, and vaginal or penile discharge. Newborns may become infected in the birth canal, and as a result develop respiratory tract infections and a form of conjunctivitis (infection of the conjunctiva) called *trachoma,* which can lead to blindness.

In women, untreated chlamydia can spread into the uterus and uterine tubes and cause *pelvic inflammatory disease* (PID), which scars the uterine tubes. Scar tissue may prevent fertilization by completely blocking sperm, resulting in infertility, or it may partially block the uterine tube, preventing a fertilized ovum from moving into the uterus and causing it to implant in the uterine tube (see Chapter 27).

Gonorrhea

Gonorrhea (gon-oh-REE-uh), commonly called "the clap," is caused by the bacterium *Neisseria gonorrhoeae*. Gonorrhea infects an estimated 800,000 to 1 million people each year in the United States. It is treatable with antibiotics, but antibiotic-resistant strains of the bacteria are becoming more prevalent. Although some men have no symptoms, most exhibit clinical features that include painful urination and penile discharge, which ranges in color from white to yellow to green. Most women are asymptomatic, but painful urination and vaginal bleeding may occur. As with chlamydia, untreated gonorrhea can lead to PID in women and blindness in newborns. Gonorrhea and chlamydia often occur together.

Syphilis

Syphilis (SIFF-ih-liss) is caused by the spiral-shaped bacterium *Treponema pallidum*. Approximately 50,000 individuals are infected with syphilis each year in the United States. The bacterium incubates in the body for up to 6 weeks and then produces a small painless lesion called a *chancre* (SHAN-kur), at which point the disease is transmissible to other people via contact with the chancre. In males, the chancre usually forms on the penis, but in females, the chancre may form inside the vagina and go undetected. The chancre generally heals within 6 weeks, and the disease seems to have run its course. However, the second stage of the disease appears a few weeks later, when a pink skin rash appears all over the body. Fever and joint pain are common, and hair loss sometimes occurs. Symptoms again subside, this time within about 3 months, but the disease has not been cured—it has entered the latent stage.

Symptoms of syphilis can come and go for years, although the immune system may eventually kill the bacteria. It is also possible for the disease to progress to *tertiary syphilis*, which produces destructive lesions in the brain and damages the cardiovascular organs, bones, and joints, damage that can be fatal.

Antibiotics can treat the infection but cannot repair damage done by the bacterium. Infected newborns may exhibit no symptoms at first, but can have many health problems, including impaired vision and hearing, if not treated with antibiotics.

Trichomoniasis

Trichomoniasis, or "trich," is an STI caused by the protozoan parasite *Trichomonas vaginalis*. An estimated 3–4 million people have the infection, which can be easily cured with antibiotics. Trichomoniasis may cause itching, burning, or redness of the genitals and, in women, may result in a white to greenish vaginal discharge with a strong odor. However, as with several other STIs, many individuals exhibit no symptoms. Infection during pregnancy can lead to early birth of the baby and low birth weight.

Quick Check

☐ 1. Which of the four STIs discussed in this section are bacterial? Which are parasitic?
☐ 2. What are the stages of untreated syphilis?

Viral STIs

◀◀ **Flashback**

1. What is interferon? (p. 766)

Unlike bacterial and parasitic infections, viral infections are caused by small nonliving viruses taking over the cellular organelles responsible for protein synthesis and using them to make copies of themselves. The new viral particles are then released from the cell and go on to infect other cells. Common viral STIs discussed in this section are human papillomavirus and herpes simplex virus 2.

Genital Warts: Human Papillomavirus (HPV)

Genital warts, caused by over 40 types of the **human papillomavirus (HPV),** are the most common STI in the United States. Nearly 80 million people are currently infected with HPV, and about 14 million new infections occur each year.

Genital warts associated with HPV occur in just under 400,000 people a year. Warts may appear on the penis, perineum, or anus in males and on the cervix, vaginal wall, perineum, or anus in females. The most common risk to newborns is warts

A&P in the Real World

Cervical Cancer

Cervical cancer occurs most often in women between the ages of 30 and 50, particularly those with a history of sexually transmitted infections, cervical inflammation, or multiple pregnancies. It is frequently caused by the human papillomavirus (HPV), which is transmitted sexually.

Cervical cancer used to be a leading cause of cancer death for women in the United States. However, in the past few decades, both the number of cases and the number of deaths from cervical cancer have decreased significantly, and are projected to decrease further as the HPV vaccine becomes more widespread. This decline is due in large part to the Pap (Papanicolaou) smear test, which detects precancerous cells and early-stage cancers before symptoms are noticeable. The Pap test involves scraping loose cells from the external os of the cervix and examining them microscopically. Cells showing signs of abnormal development, called *dysplasia,* warrant further investigation, including visual examination of the cervix or a biopsy to determine if cancerous cells are present. If cancerous cells are detected, surgical removal of the tumor or even a hysterectomy (removal of the uterus) may be required. In addition, radiation therapy or chemotherapy may be needed.

in the throat, called *recurrent respiratory papillomatosis,* a rare condition that may result from the mother having genital warts during late pregnancy. Genital warts can be treated with cryosurgery, laser surgery, or *interferon,* a cytokine released by infected cells that helps combat intracellular pathogens.

Many cases of cervical cancer are caused by HPV (see *A&P in the Real World: Cervical Cancer*), although most of the strains that cause genital warts do not lead to cancer. A vaccine against HPV can protect against some of the most common disease-causing strains, including those that cause 70% of cervical cancers, most cases of HPV-induced penile, vulvar, vaginal, and anal cancers, and 90% of genital warts. No medication is available to treat HPV after it has been contracted, which underlies the importance of preventing it.

Genital Herpes

Genital herpes is usually caused by the herpes simplex virus type 2. Approximately 750,000 people in the United States are infected each year, but many are unaware that they harbor the virus. The disease causes blisters that are often extremely painful on the external genitalia of both sexes and sometimes on the thighs and buttocks as well. The virus is most easily transmitted when such blisters occur, if there is direct skin-to-skin contact. However, a person is contagious even during the dormant stage, which can last for weeks to years. Herpes infections can be passed to a fetus, resulting in neonatal herpes, a potentially fatal

infection. Genital herpes may also increase the risk of HIV infections and some cancers.

The only treatment for genital herpes is antiviral medication, which shortens or sometimes prevents outbreaks. Daily use of antiviral medication may reduce the risk of transmission to sexual partners.

Quick Check

☐ 3. What is the relationship between human papillomavirus and certain cancers?

☐ 4. When is an individual with genital herpes contagious?

Apply What You Learned

☐ 1. Jim is presenting for a physical exam that is required to play college baseball. He says he has multiple sexual partners and practices safe sex occasionally. However, he states that he is sure he does not have any STIs because he has never had any lesions or other symptoms. What do you tell him, and why?

☐ 2. Why does the Centers for Disease Control (CDC) recommend that girls as young as 11 or 12 be vaccinated against the human papillomavirus? Why does the CDC recommend that boys also receive the vaccination?

See answers in Appendix A.

CHAPTER SUMMARY

Take full advantage of the invaluable online practice and study tools for this chapter by going to MasteringA&P®**. There you will find:**

- 🖱 Practice tests, module quizzes, and art-labeling and sequence-the-steps activities

- ▶ Animated figures with quizzes

- ✏ Practicing A&P worksheets

- 🔊 Author-narrated podcast

- 🔦 Dynamic Study Modules

- **iP** *Interactive Physiology*

MODULE **26.1**
Overview of the Reproductive System and Meiosis 1013

- The **gonads** produce **gametes** and **sex hormones.**

- Gametes, or **sperm** and **ova,** are produced by meiosis. **Meiosis** includes two successive divisions: first meiotic division (**meiosis I**) and second meiotic division (**meiosis II**). (Figure 26.1, p. 1016)

- These events occur during the first meiotic division:
  - Pairs of **homologous chromosomes,** each consisting of two **sister chromatids,** separate to produce **haploid (1n)** cells with 23 chromosomes.

  - **Crossing over** and **independent assortment** randomly "shuffle" the alleles and align the maternal and paternal chromosomes, which contributes to the genetic diversity of the offspring.

- These events occur during the second meiotic division:
  - Homologous chromosomes separate into individual chromatids.
  - Potentially there are four haploid cells at the end of telophase II.

- **Mitosis** and **meiosis** differ in important ways:
  - Mitosis occurs in almost every cell type in the body, but meiosis occurs only in the testes and ovaries, to produce gametes.
  - Mitosis produces two cells that are genetically identical to the original cell, but meiosis produces up to four cells that are genetically unique with half the number of chromosomes as the original cell. (Figure 26.2, p. 1017)

MODULE **26.2**
Anatomy of the Male Reproductive System 1018

- The paired **testes,** or male gonads, produce **sperm cells** and secrete the male hormone **testosterone.** (Figure 26.3, p. 1018)

- 🖱 Practice art-labeling exercise: Figure 26.3.

- **Seminiferous tubules** in the testes contain *spermatogenic cells* that form sperm and **sustentacular cells,** which support sperm cells and form the **blood-testis barrier. Myoid cells** contract to push sperm and testicular fluids through the tubules and out of the testes. The tubules are surrounded by **interstitial cells,** which produce testosterone. (Figure 26.4, p. 1019)

Practice art-labeling exercise: Figure 26.4.

- After leaving the testis, sperm travel through ducts that move the sperm into the abdominopelvic cavity and past accessory structures that produce **semen.** The ducts are as follows:
  - The **epididymis** sits on the superior and posterior surface of the testis and stores sperm. It is the site of sperm maturation.
  - The **ductus deferens** is a muscular tube that propels sperm. It travels through the **spermatic cord** and **inguinal canal** into the abdominopelvic cavity. It merges with the duct of the seminal vesicle to form the ejaculatory duct on the posterior bladder.
  - The **ejaculatory duct** extends into the prostate and empties into the urethra.
  - The **urethra** carries sperm out of the body through the penis.
- Accessory structures that produce semen include the following (Figure 26.5, p. 1020):
  - **Seminal vesicles** produce an alkaline secretion that contains nutrients, prostaglandins, and a coagulating enzyme. It supports sperm, enhances their motility, and maintains semen in the female reproductive tract after ejaculation.
  - The **prostate** gland produces a slightly alkaline fluid with nutrients and several components that support and activate sperm and function as an anticoagulant.
  - **Bulbourethral glands** produce mucus to lubricate the penile urethra.
- The external genitalia of the male include the following (Figure 26.6, p. 1023):
  - The **penis** is the male copulatory organ and delivers sperm into the female reproductive tract through the urethra.
  - The testes are located in the **scrotum,** outside the body cavity just below the abdominopelvic cavity. Each side of the scrotum contains a testis and epididymis. The scrotum keeps the developing sperm approximately 3° C below body temperature, which is necessary for their development.
- Table 26.1 summarizes the male reproductive organs. (p. 1024)

MODULE 26.3
Physiology of the Male Reproductive System 1024

- **Spermatogenesis** is the process of sperm cell development. It includes all phases of meiosis and *spermiogenesis,* the cytoplasmic changes required to form a mature sperm cell. The process begins at puberty and continues throughout life.
  - Spermatogenesis occurs in the seminiferous tubules of the testes by spermatogenic cells. **Spermatogonia** divide to produce **primary spermatocytes,** which undergo the first meiotic division to form two **secondary spermatocytes.** Each secondary spermatocyte undergoes the second meiotic division to produce four total **spermatids.** (Figure 26.7, p. 1025)
  - Spermiogenesis is the maturation of spermatids into motile sperm cells. (Figure 26.8, p. 1026)
- Testosterone secretion and spermatogenesis are controlled by the **hypothalamic-pituitary-gonadal (HPG) axis.** (Figure 26.9, p. 1027)
  - The hypothalamus releases **gonadotropin-releasing hormone (GnRH),** which stimulates the anterior pituitary to secrete **follicle-stimulating hormone (FSH)** and **luteinizing hormone (LH).**
  - LH stimulates interstitial cells to produce **testosterone.**

  - FSH stimulates sustentacular cells to secrete **androgen-binding protein (ABP)** and inhibin.
  - Testosterone and inhibin exert negative feedback controls on the hypothalamus and anterior pituitary.
- Testosterone stimulates spermatogenesis; maturation of sex organs; development and maintenance of secondary sex characteristics and libido; and anabolic effects, including bone growth and increased muscle mass.
- The male sexual response begins with **erection** of the penis. This is followed by **ejaculation,** which is the expulsion of semen.
- **Puberty** in boys usually begins between the ages of 12 and 14, when GnRH levels begin to rise. This stimulates the HPG axis and testosterone levels increase, causing spermatogenesis to begin and male secondary sex characteristics to develop. Reproductive function begins to decline in men during **male climacteric.**
- **Male secondary sex characteristics** include pubic, axillary, and facial hair; deepening voice; thicker skin; increased bone density; larger skeletal muscles; increased basal metabolic rate and red blood cell count; increased volume of body fluids; and male **libido.**

MODULE 26.4
Anatomy of the Female Reproductive System 1030

- The paired **ovaries,** or female gonads, are located inside the pelvic cavity close to the lateral walls and produce ova, and secrete the hormones *estrogens, progesterone,* inhibin, and relaxin. (Figure 26.10, p. 1030)

Practice art-labeling exercise: Figure 26.10.

- Oocytes, or immature ova, mature in hormone-producing **follicles,** which develop and mature along with the oocytes. After an oocyte is ovulated from an ovary, it is picked up by the **fimbriae** of the **uterine tube** and moves toward the uterus. The uterine tube is the site of fertilization. (Figure 26.11, p. 1031)

Practice art-labeling exercise: Figure 26.11.

- The **uterus** sits in the pelvis anterior to the rectum and posterosuperior to the urinary bladder. The **endometrium** lines the uterus and is the site of implantation of the fertilized ovum; the other layers are the **myometrium** and **perimetrium.**
- The **vagina** is the female copulatory organ and receives the penis during intercourse. In addition, the vagina provides a passageway for delivery of an infant and for menstrual flow.
- The **external genitalia** of the female include the **mons pubis, labia majora** and **labia minora, clitoris,** and **vestibule.** The vestibule contains the external openings of the urethra and the vagina. (Figure 26.12, p. 1033)
- The **mammary glands** are modified sweat glands that sit over the pectoralis major muscles of the chest. They are composed of adipose and fibrous connective tissue filled with glandular **lobes** that contain lobules with **alveoli.** If pregnancy occurs, the alveoli produce milk. (Figure 26.13, p. 1034)
- Table 26.2 summarizes the female reproductive organs. (p. 1034)

- **Oogenesis** is the process by which an **ovum** develops. The phases of meiosis are not continuous, but are arrested at different points in the female's life cycle. (Figure 26.14, p. 1036)
  - Oogenesis occurs in the ovarian cortex within follicles.
  - Before a female is born, **oogonia** divide and differentiate to produce **primary oocytes** arrested in prophase I until puberty.
  - About once per month after puberty, a primary oocyte completes the first meiotic division to produce one large **secondary oocyte** and one small **polar body.** The secondary oocyte proceeds to metaphase II, where its development is arrested unless it is fertilized. If fertilized, it completes the second meiotic division and becomes the ovum and expels the last polar body. Polar bodies are resorbed by the body.

- The **ovarian cycle** is the monthly series of events associated with maturation of oocytes and their follicles. It includes the following phases: (Figure 26.15, p. 1039)
  - In the **follicular phase,** several **primary follicles** (with primary oocytes) begin to mature into **secondary follicles.** Usually only one secondary follicle with a primary oocyte will develop into a **vesicular follicle,** and the primary oocyte inside will become a secondary oocyte after the LH surge.
  - **Ovulation** releases the secondary oocyte into the peritoneal cavity.
  - In the **luteal phase,** the ruptured follicle forms the **corpus luteum,** which secretes progesterone and estrogens for the rest of the cycle. If fertilization does not occur, the corpus luteum degenerates into the **corpus albicans.**

  👆 **Practice sequencing the steps: Figure 26.15.**

- The HPG axis controls the ovarian cycle. (Figure 26.16, p. 1041)
  - The hypothalamus releases **GnRH,** which stimulates the anterior pituitary to secrete **FSH** and **LH.**
  - FSH and LH stimulate follicles to grow and secrete estrogens and inhibin, which inhibit the release of more FSH and LH.
  - One follicle eventually produces enough estrogens to exert positive feedback on the HPG axis, resulting in an LH surge and a rise in FSH. This triggers ovulation and transforms the ruptured follicle into the corpus luteum.
  - The corpus luteum produces progesterone and estrogens and will continue to do so if fertilization occurs. If fertilization does not occur, progesterone levels drop, and the cycle repeats.

- Estrogens promote maturation of sex organs, development of external genitalia, development of breast tissue, preservation of bone mass, and maintenance of vascular health.

- The **uterine cycle,** the series of cyclic changes in the endometrium that occur each month in response to fluctuating ovarian hormones, includes the following phases, regulated by the HPG axis: (Figure 26.17, p. 1042; Figure 26.18, p. 1043)
  - In the **menstrual phase,** the uterus sheds the **stratum functionalis** of the endometrium through the vagina. This phase results from rising FSH and LH levels and normal/low levels of estrogens and progesterone.
  - In the **proliferative phase,** a new stratum functionalis grows and thickens as estrogen levels rise. Ovulation occurs in the ovary at the end of this phase.
  - In the **secretory phase,** the endometrium prepares for implantation of an embryo by continuing to grow, increasing vasculature, and secreting nutrients into the uterine cavity. This phase occurs as rising progesterone levels from the corpus luteum stimulate continued growth of the endometrium and inhibit LH release. If fertilization does not occur, the menstrual phase will follow.

- The big picture of hormonal regulation of the ovarian and uterine cycles is illustrated in Figure 26.19 (p. 1045).

- Table 26.3 (p. 1046) summarizes the functions of major reproductive hormones in both males and females.

- The **female sexual response** is similar to that of males, and erectile tissues also engorge with blood. However, orgasm in females is not accompanied by ejaculation.

- Puberty usually begins when girls are 9–11 years old, when GnRH levels increase. GnRH stimulates the HPG axis, and estrogens and progesterone levels rise, which stimulates oogenesis to resume and female sex characteristics to develop.

- **Female secondary sex characteristics** include changes in bone structure, such as widening pelvis; an increase in adipose tissue in general and particularly in the breasts and around the hips and thighs; hair in the pubic and axillary areas; and development of female libido.

- Birth control, or contraception, refers to any procedure or device intended to prevent pregnancy. (Figure 26.20, p. 1048)

- Methods of birth control include behavioral modification such as *abstinence;* **barrier methods,** such as condoms; hormonal methods, such as **oral contraceptives;** intrauterine methods; and permanent methods, such as **vasectomy, tubal ligation,** and tubal implants. (Figure 26.21, p. 1049; Table 26.4, p. 1050)

- **Sexually transmitted infections (STIs)** are the most important cause of reproductive disorders and often result in infertility, fetal deformities, and even fetal death.

- Common STIs include HIV and AIDS, **chlamydia, gonorrhea, syphilis, genital warts (HPV), genital herpes,** and **trichomoniasis.**

ASSESS WHAT YOU LEARNED

See answers in Appendix A.

LEVEL 1 Check Your Recall

1. Fill in the blanks: At the end of meiosis, each cell has
   _____ chromosomes and they are genetically
   _____ from the original cell.

2. Match the specific phase of meiosis with the correct description.

_____ Prophase I
_____ Metaphase I
_____ Anaphase I
_____ Telophase I
_____ Prophase II
_____ Metaphase II
_____ Anaphase II
_____ Telophase II

a. Homologous chromosomes arrive at opposite poles.
b. Homologous chromosomes line up.
c. Chromosomes line up at the equator.
d. Independent assortment occurs.
e. Separated sister chromatids arrive at opposite poles.
f. Sister chromatids remain condensed.
g. Crossing over occurs.
h. Sister chromatids pull apart.

3. Mark the following statements about spermatogenesis as true or false. If a statement is false, correct it to make a true statement.

a. Spermatogenesis is the process of sperm cell development. It begins at puberty and ends when a man reaches approximately 45–55 years of age.
b. Spermatogonia are the stem cells that begin the process of spermatogenesis and divide by mitosis.
c. Primary spermatocytes undergo the first meiotic division and form two smaller diploid cells called secondary spermatocytes.
d. Spermiogenesis is the maturation process of sperm cells and involves changing the size and shape of the cell.

4. Which of the following structures is the site of sperm maturation?

a. Epididymis
b. Ductus deferens
c. Ejaculatory duct
d. Urethra
e. Scrotum

5. Which of the following occurs in the testes?

a. Production of gametes
b. Secretion of testosterone
c. Production of semen
d. Both a and b
e. Both b and c

6. What are the three regions of the male urethra?

7. Match the component of the glandular secretions with the correct accessory sex gland. A gland may match with more than one component.

_____ Seminal vesicle
_____ Prostate
_____ Bulbourethral gland

a. Citrate
b. Fructose
c. Prostaglandins
d. Coagulating enzyme
e. Mucus
f. Proteolytic enzymes

8. Fill in the blanks: Internally, the penis contains three cylindrical erectile bodies, the paired _____ and the _____.

9. Which of the following is the main function of the scrotum?

a. The scrotum contains the testes and provides additional blood supply to them because of the location outside the abdominopelvic cavity.
b. The scrotum contains the testes and maintains the temperature of the testes about 3° C lower than body temperature, which is the optimal temperature for developing sperm.
c. The scrotum contains the testes and maintains the temperature of the testes about 3° C higher than body temperature, which is the optimal temperature for developing sperm.
d. The scrotum contains the testes and contains the blood-testis barrier to separate the seminiferous tubules from the bloodstream.

10. Number the sequence of events in the hormonal regulation of male reproduction, placing a 1 by the first event, a 2 by the second, and so forth.

_____ a. Luteinizing hormone (LH) stimulates interstitial cells found in the connective tissue around the seminiferous tubules to produce testosterone.
_____ b. Testosterone and inhibin both exert negative feedback controls on the hypothalamus and anterior pituitary.
_____ c. Follicle-stimulating hormone (FSH) stimulates sustentacular cells to secrete inhibin and androgen-binding protein, which keeps some of the testosterone bound near the spermatogenic cells.
_____ d. Gonadotropin-releasing hormone (GnRH) stimulates the anterior pituitary to secrete FSH and LH.
_____ e. The hypothalamus releases GnRH.

11. Fill in the blanks: _____ is the enlargement and stiffening of the penis, which results when the tissue in the penis becomes engorged with blood, and _____ is the process by which semen is expelled from the penis.

12. Which of the following hormones is/are *not* produced by the ovaries?

a. Relaxin
b. Inhibin
c. Estrogens
d. Oxytocin
e. Progesterone

13. Which of the following ligaments does *not* attach to the ovaries?

a. Lateral cervical ligaments
b. Broad ligament
c. Ovarian ligament
d. Suspensory ligament
e. All of these attach to the ovaries.

14. Fill in the blanks: The _____ is the site of fertilization in the female reproductive system, and the _____ is the site of implantation of the fertilized ovum.

15. Fill in the blanks: Within the lobes of the mammary glands are smaller units called _____, which contain glandular _____ that produce milk when a woman is lactating.

16. Mark the following statements about oogenesis as true or false. If a statement is false, correct it to make a true statement.

a. Oogenesis begins before a female is born and continues for the remainder of her life.
b. About once a month after puberty begins, some primary oocytes are stimulated to continue development. Usually one will complete the first meiotic division to produce two haploid cells that are the same size.
c. The fate of all polar bodies is the same; they all degenerate and are reabsorbed by the body.
d. Only if a secondary oocyte is fertilized will it complete the second meiotic division and expel the last polar body.

17. Match the follicle stage with the correct description.

_____ Primordial follicle

_____ Primary follicle

_____ Secondary follicle

_____ Vesicular follicle

_____ Corpus luteum

_____ Corpus albicans

a. Several layers of granulosa cells with pockets of follicular fluid surround the primary oocyte.

b. Knot of whitish scar tissue

c. Yellow glandular-looking endocrine structure

d. A single layer of squamous follicular cells surrounds the primary oocyte.

e. One or two layers of cuboidal granulosa cells surround the primary oocyte.

f. Several layers of granulosa cells surround the secondary oocyte and project into the antrum.

18. Number the sequence of events in the hormonal regulation of female reproduction, placing a 1 by the first event, and so forth.

_____ a. Increasing levels of estrogens and inhibin exert negative feedback controls on the hypothalamus and pituitary to inhibit the release of more FSH and LH, leaving only one follicle to survive.

_____ b. LH and FSH surge stimulates the primary oocyte of one follicle to complete meiosis to metaphase II, which triggers ovulation.

_____ c. FSH and LH stimulate follicles to grow and secrete estrogens and inhibin.

_____ d. Progesterone from the corpus luteum inhibits the release of gonadotropins and will continue to do this if pregnancy is achieved.

_____ e. The hypothalamus releases GnRH.

_____ f. A single follicle produces enough estrogens to exert positive feedback on the hypothalamus and anterior pituitary to trigger an LH surge.

_____ g. GnRH levels rise to begin the cycle over.

_____ h. LH transforms the ruptured follicle into the corpus luteum, which produces progesterone and some estrogens.

_____ i. If pregnancy is not achieved, the corpus luteum degenerates into the corpus albicans and progesterone levels drop.

_____ j. GnRH stimulates the anterior pituitary to secrete FSH and LH.

19. Mark the following statements about the uterine cycle as true or false. If a statement is false, correct it to make a true statement.

a. During the proliferative phase, the endometrium prepares for the implantation of the embryo by enlarging the endometrial glands and secreting glycogen into the uterine cavity.

b. During the secretory phase, the endometrium generates a new stratum functionalis.

c. During the menstrual phase, the uterus sheds the stratum basalis of the endometrium, which results in the discharge of blood and tissue from the vagina.

20. Which of the following hormones stimulates the generation of a new stratum functionalis in the endometrium?

a. Estrogens   c. Inhibin   e. FSH
b. Progesterone   d. LH

21. Which of the following methods of birth control is classified as "sterilization"? Mark all that apply.

a. Oral contraceptive pill   c. Intrauterine device
b. Tubal ligation   d. The patch

22. Which of the following sexually transmitted infections is/are caused by a virus? Mark all that apply.

a. Chlamydia   c. Genital herpes   e. Gonorrhea
b. Genital warts   d. Syphilis

LEVEL 2 Check Your Understanding

1. The second meiotic division is very similar to mitosis. How would you explain the differences between the first meiotic division and mitosis?

2. Sustentacular cells in the testes are nicknamed "nurse cells." Explain why the nickname is appropriate.

3. How might the female orgasm increase the chance for fertilization to occur?

4. Explain why oral contraceptives, which artificially raise levels of estrogens and progesterone, prevent pregnancy.

LEVEL 3 Apply Your Knowledge

PART A: Application and Analysis

1. Mr. Dell has recently started getting up during the night to urinate and has also noticed that he is having trouble emptying his bladder. He has made a doctor's appointment, but his wife has already told him that the doctor will want to check his prostate. Do you think his wife is correct? Explain your answer.

2. A professional male athlete has been taking a synthetic testosterone supplement for several months. He and his wife have been trying to have a baby, with no success, and have decided to begin treatment for infertility. Before testing the man's sperm count, the doctor discovers that the man's LH levels are very low. Why do you think his LH levels are low? Would low LH levels affect sperm production?

3. Fabiola, a 17-year-old college student, suffers from an eating disorder. Her percentage of body fat has dropped to 14%. Her menstrual cycles were irregular, and now they have stopped altogether. Why do you think this has happened? Could the lack of menstrual periods be beneficial in this situation? Explain.

PART B: Make the Connection

4. A 60-year-old man has been taking an herbal medication for impotence. Recently, he has been having periods of dizziness when he is exercising and has started to check his blood pressure, which drops after he takes the medication. Why do you think a medication that prevents impotence also affects blood pressure? *(Connects to Chapter 18)*

27

Development and Heredity

Every day, the population of the earth increases by over 200,000 people. Perhaps because births are so common, it's easy to forget that the development of a new human life is an incredible event. That a single, fertilized egg can grow into a complex, independent individual made of trillions of cells is truly astounding.

The reproductive system chapter examined the anatomy and physiology of the male and female reproductive systems and described the formation of *gametes* (GAM-eetz)—the *sperm* and *eggs* (see Chapter 26). Let's now explore the developmental processes that occur when a fertilized egg develops into a new individual. This chapter covers the major developmental changes that occur before birth and ultimately result in labor and delivery of the child. Then, our focus shifts to what happens to both baby and mother in the first few months after birth. The chapter ends with a brief description of how genetic characteristics are transmitted from parent to offspring—a mechanism called *heredity* (heh-RED-ih-tee).

MODULE **27.1**

Overview of Human Development

Learning Outcomes

1. Explain the difference between prenatal and postnatal development.
2. Describe the periods of prenatal development.

Human development begins at *fertilization,* with the union of gametes, and continues throughout life. **Developmental biology** is the science that studies the changes in form and function from the fertilized egg through old age. Within the broad field of developmental biology is **embryology** (em'-bree-AH-luh-jee), the study of the events that occur during the approximately 38 weeks of development inside the mother's womb. This period is referred to as the **prenatal period** (pree-NAY-tuhl; *nata-* = "birth"). Developmental changes occurring from birth through all life stages are part of the **postnatal period**, although "postnatal" may also refer to the short period following birth. Let's look a bit deeper into these periods.

Photo: Image of a human fetus during the fourth month of development.

The Process of Prenatal Development

◀◀ **Flashback**

1. What is the function of mitosis? (p. 110)
2. What layers make up the wall of the uterus? (p. 1031)

The prenatal period is what people usually refer to as pregnancy. The term **pregnancy** embraces all of the events from **conception**—when an egg is fertilized and begins to develop—until birth, and includes the anatomical and physiological changes to the mother during this time. The growing offspring is generally referred to as the **conceptus** (kon-SEP-tuss) and develops during the **gestation period** (jes-TAY-shun) of pregnancy, which extends from the mother's last menstrual period until birth, or for approximately 280 days, or 40 weeks. The gestation period is longer than the prenatal period because it includes the 2 weeks after the mother's last menstrual cycle—a date that women can usually recall. We will focus on the approximately 38 weeks of prenatal development, which is divided into three stages (**Figure 27.1**):

	PRE-EMBRYONIC PERIOD	EMBRYONIC PERIOD	FETAL PERIOD
Developmental stage	Blastocyst	Embryo	Fetus
Events	**Weeks 1 and 2:** • Zygote divides mitotically many times to produce a multicellular blastocyst that implants in the uterus.	**Weeks 3 through 8:** • Blastocyst grows, folds, and forms rudimentary organ systems. • It is now called an embryo.	**Weeks 9 through 38 (until birth):** • Embryo is now called a fetus. • It grows larger and develops until its organ systems can function without assistance from the mother.

Figure 27.1 Summary of developmental changes during the prenatal period. Note that these illustrations are not to scale.

- **Pre-embryonic.** The **pre-embryonic period** lasts for the first 2 weeks after fertilization. Recall from the reproductive system chapter that the female gamete is the *secondary oocyte* (see Chapter 26). When a secondary oocyte is fertilized, it divides to become a single-celled *zygote*. During the pre-embryonic stage, the zygote divides mitotically many times to produce a multicellular structure, the **blastocyst** (BLAS-toh-sist), which *implants* in, or attaches to, the endometrium of the uterus.
- **Embryonic.** The **embryonic period** extends from week 3 through week 8 of gestation. During this stage, the conceptus grows, folds, and forms rudimentary organ systems. As you might expect from the name of the period, it is now called an **embryo** (EM-bree-oh).
- **Fetal.** The final stage, the **fetal period**, lasts from week 9 through week 38 of gestation, or until birth. In this stage, the conceptus is now called a **fetus** (FEE-tus); it grows larger and continues to develop until its organ systems can function without assistance from the mother.

Several intermediate stages, particularly in the embryonic period, are responsible for the progressive tissue and organ development that occurs during the prenatal period. We discuss these intermediate stages in more detail in the next three modules. After birth, the fetus becomes a **newborn,** and the postnatal period of life begins.

The Postnatal Period

In the postnatal period, rapid developmental changes continue to take place, particularly in the first few months, although they are not as dramatic as in the prenatal period. This period can be broadly divided into five stages: the *neonatal period* (nee-oh-NAYT-uhl), *infancy, childhood, adolescence* and *adulthood* (*maturity*). The neonatal period extends from birth to one month of age and will be the focus of our study of the postnatal period in Module 27.6. Infancy continues development through two years of age, and childhood lasts until the period of sexual maturation, or puberty, begins. At this point, adolescence begins, and finishes when the individual is capable of sexual reproduction. Adulthood extends from the end of adolescence until death, and is the stage at which development ends and the degeneration of tissues and organs, called **senescence** (seh-NESS-ens), begins. Eventually senescence will lead to death, even without a specific disease state.

MODULE **27.2**

Pre-embryonic Period: Fertilization through Implantation (Weeks 1 and 2)

Learning Outcomes

1. Describe the process of fertilization.
2. Describe the events of sperm capacitation, acrosomal reaction, sperm penetration, cortical reaction, and fusion of pronuclei.
3. Explain the formation and function of the extraembryonic membranes.

The pre-embryonic period begins with fertilization of the **secondary oocyte** by a **sperm cell,** and continues until the blastocyst moves into the uterus and *implants* in the endometrium of the uterine wall. During this period, some cells of the blastocyst begin to form *extraembryonic membranes* (EK-strah-em′-bree-ahn-ik; "outside the embryo"), which help support the developing embryo.

The upcoming sections examine fertilization, the formation of the blastocyst, and implantation. We then look at the structure and function of the extraembryonic membranes that develop during these processes.

Fertilization

◀◀ Flashback

1. What are gametes? (p. 1014)
2. How do spermatogenesis and oogenesis differ? (p. 1035)
3. What are the components of semen? (p. 1022)

As discussed in the reproductive system chapter, the process of **meiosis** (my-OH-sis) produces **gametes,** or sex cells (see Figure 26.1). Meiosis results in genetically unique *haploid* cells (*n*), with half the number of chromosomes normally found in body cells. Each gamete contains 23 chromosomes; when gametes combine, they restore the correct *diploid* number (2*n*) of 46 to future cells of the new offspring.

Sperm cells are ejaculated in **semen** (SEE-men; *semen* = "seed"), which is a mixture of sperm and fluid secretions from the male reproductive glands. The fluid secretions in semen contain chemicals that make the female reproductive tract more hospitable to sperm, thus giving more sperm an opportunity to compete in fertilizing the secondary oocyte.

Each ejaculate contains 40–750 million sperm cells. However, only a small fraction of this number ever reaches the ovulated secondary oocyte. Many sperm are destroyed by acidic vaginal secretions. Still other sperm fail to penetrate the mucus of the cervical canal or, if they do, are destroyed by immune cells in the uterus. Sperm motility is also influenced by a woman's stage in her reproductive cycle. When a woman is in the secretory phase of her reproductive cycle, the level of estrogen is high and the cervical mucus is thin and watery, which promotes sperm motility. However, if the female is in the proliferative phase of her reproductive cycle, the progesterone level is high and the cervical mucus is viscous, which inhibits sperm motility. Also, many sperm survive these obstacles only to go up the wrong uterine tube (the one that currently has no ovulated oocyte).

In fact, no oocyte may be available at all when the sperm enter the female reproductive tract. An ovulated secondary oocyte is viable for only 24 hours after it is released from the ovary. Because oocytes travel slowly through the uterine tubes, they are usually in the distal end of the uterine tube, called the **ampulla** (am-POOL-ah), when they are fertilized. Pregnancy most frequently occurs when sperm enter the female reproductive tract during a 3-day window—from 2 days before ovulation to 1 day after—but sperm can remain viable in the female's reproductive tract for up to 5 days.

Fertilization is the fusion of a sperm cell and secondary oocyte to form a zygote. Remember that a secondary oocyte is released during ovulation because meiosis II has been suspended in metaphase (see Figure 26.14). The secondary oocyte divides to form an *ovum* only if it is fertilized.

Even though both gametes contribute equally to the chromosome number, and therefore the genetic makeup of the new offspring, other functional roles of the gametes are unique. A sperm cell must travel a long distance to reach the oocyte, so it needs to be small and motile. In contrast, a secondary oocyte is relatively large because it must provide the cellular machinery necessary for the first week of development. Cilia lining the uterine tubes and gentle peristaltic contractions propel the oocyte toward the uterus, so the oocyte is not motile. These structural differences between male and female gametes demonstrate the Structure-Function Core Principle (p. 25).

CORE PRINCIPLE
Structure-Function

① Sperm undergo capacitation as they migrate to the oocyte.

Migrating sperm

② The acrosomal reaction releases enzymes from the head of the sperm.

③ The sperm binds to the plasma membrane of the oocyte.

④ Sperm entry stimulates the cortical reaction, which destroys sperm-binding receptors.

⑤ Additional sperm cells are rejected.

Sperm-binding receptors

Oocyte plasma membrane

Oocyte plasma membrane receptors

Cortical granules

Zona pellucida

Corona radiata

Sperm nucleus

Male pronucleus

Polar body

⑥ The sperm nucleus swells to form a male pronucleus.

Female pronucleus

⑦ The oocyte completes meiosis II, and the ovum's nucleus swells to form a female pronucleus.

⑧ A spindle forms between male and female pronuclei, and the chromosomes intermix.

Sequence the steps
@MasteringA&P®

Figure 27.2 Events leading to fertilization.

Several events must occur before fertilization can occur between the sperm cell and the oocyte (**Figure 27.2**):

1. **Sperm undergo capacitation as they migrate to the oocyte.** A sperm cell moves by a whiplike motion of its *flagellum*, or "tail." Prostaglandins in semen are also thought to stimulate uterine contractions, which assist with moving the sperm farther into the female reproductive tract. Although female orgasm is not required for fertilization, it also results in uterine contractions and, again, may help in moving sperm. All of these processes together are called **sperm migration.**

   Before it can fertilize an oocyte, a sperm must undergo **capacitation** (kah-pah′-sih-TAY-shun), a series of functional changes that make it fully motile, and modify its plasma membrane so it can fuse with the plasma membrane of the oocyte. The alkaline composition of semen causes the sperm's flagellum to beat more quickly, and fluids produced by the female reproductive tract alter the composition of the sperm's plasma membrane. In addition, the oocyte itself attracts the sperm to its location, using chemical signals (a process called *chemotaxis*). Researchers have identified a receptor on human sperm cells that responds to a chemical secreted by oocytes, which appears to be progesterone.

2. **The acrosomal reaction releases enzymes from the head of the sperm.** An ovulated secondary oocyte is surrounded by a layer of *granulosa* cells called the **corona radiata** (koh-ROH-nah ray-dee-AH-tah). To reach the secondary oocyte, sperm must penetrate this cell layer. This is accomplished through a combination of sperm movements and *hyaluronidase* (hy′-ah-loo-RON-ih-days), an enzyme released from the membrane-bound sac on the head of the sperm called the **acrosome** (AK-row-sohm). This enzyme catalyzes reactions that break down the protein layer that serves as a barrier between the granulosa cells.

   Immediately deep to the oocyte's corona radiata is the **zona pellucida** (ZOH-nah peh-LOO-sih-dah), a layer of extracellular matrix secreted by the oocyte that contains **sperm-binding receptors.** These receptors are an important site for communication between cells, one example

   CORE PRINCIPLE
   Cell-Cell Communication

   of the Cell-Cell Communication Core Principle (p. 27). When a sperm cell binds to a sperm-binding receptor in the zona pellucida, calcium ion channels in the sperm's plasma membrane open, allowing calcium ions to flow into the cell. The resulting increase in the intracellular calcium ion level triggers the **acrosomal reaction,** which releases acrosomal enzymes, including additional hyaluronidase and *acrosin*. These enzymes catalyze reactions that digest holes in the zona pellucida. Numerous sperm cells are needed to release enough hyaluronidase and acrosin to break through the corona radiata and zona

pellucida, so a single sperm cell is not capable of fertilizing an oocyte alone. This is why males with sperm counts below 20 million have a lowered ability to produce offspring.

3. **The sperm binds to the plasma membrane of the oocyte.** When the first sperm cell penetrates the zona pellucida and binds to the oocyte's plasma membrane, changes in the oocyte and zona pellucida occur so that no more sperm can enter.

4. **Sperm entry stimulates the cortical reaction, which destroys sperm-binding receptors.** Specialized secretory vesicles called *cortical granules* are present in the oocyte that contain enzymes that catalyze reactions to destroy the sperm-binding receptors. Sperm entry triggers the release of the enzymes from the cortical granules, a process termed the **cortical reaction.**

5. **Additional sperm cells are rejected.** The cortical reaction prevents additional sperm cells from binding. On rare occasions, more than one sperm cell enters the secondary oocyte, a phenomenon known as **polyspermy** (pol-ee-SPER-mee). In this case development cannot proceed because the fertilized oocyte is triploid (3*n*)—has three sets of chromosomes—which is fatal to the zygote.

6. **The sperm nucleus swells to form a male pronucleus.** The plasma membranes of the sperm and oocyte bind and fuse together; the cytoplasmic contents of the sperm cell, including the nucleus, are then released into the cytoplasm of the oocyte. The sperm nucleus swells to form the **male pronucleus.**

7. **The oocyte completes meiosis II, and the ovum's nucleus swells to form a female pronucleus.** At the same time, the fusion of the plasma membranes signals the secondary oocyte to complete meiosis II and become an ovum. The released polar body is expelled but sometimes remains visible at the edge of the ovum. The ovum's nucleus also swells and becomes the **female pronucleus.**

8. **A spindle forms between male and female pronuclei, and the chromosomes intermix.** Spindle fibers form between the male and female pronuclei, and the pronuclear membranes break down, allowing the chromosomes to mingle.

If any part of this process is interrupted or defective, reproduction will not be successful. **Infertility** is the inability to produce a pregnancy after 1 year of unprotected sexual intercourse. Individuals who are infertile can benefit from techniques that increase their chances of having children (see *A&P in the Real World: Assisted Reproductive Technology* on page 1063).

Quick Check

☐ 1. What is capacitation, and where does this process occur?
☐ 2. What are the steps of fertilization?

Cleavage and Blastocyst Formation

◀◀ **Flashback**

1. What are the areas of the uterine tube? (p. 1031)
2. What happens to the oocyte during ovulation? (p. 1032)

The next steps of development are shown in **Figure 27.3**. When the secondary oocyte is fertilized, chromosomes from the male and female pronuclei combine in the process of *amphimixis* (am-fih-MIK-sis), and the diploid zygote is formed. Approximately 30 hours after fertilization occurs, the zygote begins **cleavage**— a series of rapid mitotic divisions that produce small, genetically identical cells called **blastomeres** (BLAS-tow-meerz). The divisions occur too quickly for cells to grow; for this reason, the cell number increases, but the cell size becomes progressively smaller. Each subsequent division takes less time than the previous one,

so that by the second day there are 4 cells and by the end of the third day there are 16 cells. At this stage, the cells start to differentiate and will be the building blocks of all future tissues of the developing conceptus. At this point, the conceptus is known as a **morula** (MOR-yoo-lah; "little raspberry") and remains covered by the zona pellucida.

Notice in the figure that while these mitotic divisions are taking place, the conceptus has moved through the uterine tube and into the uterine cavity. Around day four, a glycogen-rich secretion from endometrial glands in the lumen of the uterus, called *uterine milk*, nourishes the conceptus. The conceptus, now called a blastocyst, hatches from the zona pellucida and the blastomeres reorganize to surround an internal fluid-filled cavity.

By the time the blastocyst is ready for implantation, it has two distinct cell populations: an outer layer of large, flattened cells, called **trophoblast cells** (TROF-oh-blast; *troph-* = "nourish")

Practice art labeling @MasteringA&P®

Figure 27.3 **Steps of the pre-embryonic period: fertilization through implantation.**

that surround the fluid-filled cavity; and an inner cluster of rounded cells, the **inner cell mass,** or *embryoblasts* (EM-bree-oh-blastz). The trophoblast cells will participate in forming the *placenta* (plah-SEN-tah), the temporary organ that provides nutrients and oxygen to the conceptus. They also suppress the mother's immune system, preventing it from attacking the conceptus. The inner cell mass will form the *embryo proper*—the developing body.

Assisted Reproductive Technology

According to the Centers for Disease Control, in 2014 1 in 8 couples in the United States have had trouble getting pregnant or sustaining a pregnancy. In males, the most common cause of infertility is a low sperm count. Infertility in females may be caused by hormonal problems associated with ovulation, diseases of the ovary, obstruction of the uterine tubes, or conditions in which the uterus is not able to receive or maintain a fertilized ovum. Lifestyle changes or hormone therapy can resolve some cases. Others may require surgery, such as repairing an obstructed uterine tube. Still others involve direct manipulation of the reproductive process using **assisted reproductive technology,** or **ART.**

One type of ART is *intrauterine insemination* (IUI), also called *artificial insemination,* in which sperm that have been washed and concentrated are placed directly into the woman's uterus at the time of ovulation. Sperm must then swim up the uterine tube and fertilize the oocyte naturally. A similar procedure is *gamete intrafallopian transfer* (GIFT), in which sperm and oocytes are placed into the woman's uterine tube. Fertilization and all developmental steps then must occur naturally in both procedures.

For other procedures, oocytes are fertilized outside the female's reproductive tract. For example, in *zygote intrafallopian transfer* (ZIFT), a new zygote is immediately placed into the woman's uterine tube. The zygote will then need to develop into a blastocyst and implant. In *in vitro fertilization* (IVF), most or all of the steps of the pre-embryonic period are completed outside the female reproductive tract. Oocytes are incubated with sperm cells to allow fertilization and the first stages of development to occur before transferring the embryos into the woman's uterus, where they must implant naturally. In some cases, to make the embryo, the oocytes are directly injected with a sperm cell or a sperm cell nucleus.

In many ART procedures, it is possible to use sperm and/or oocytes donated by a third person to enable an infertile individual or couple to conceive. The donor may be known or anonymous to the intended recipient. Recent statistics (from the Society for Assisted Reproductive Technology in 2012) indicate the birth rate per embryo transfer is 47% if the woman is under 35 years of age and 38% if the woman is age 35–37.

Sometimes during these early cell divisions, within the first 8 days after fertilization, the cells separate into two groups and develop into two individuals, called **monozygotic** (mahn′-oh-zy-GOT-ik; *mono-* = "single"), or **identical,** twins. Monozygotic twins are genetically identical, or practically so, and have a nearly identical appearance. They usually share the same placenta. Very rarely, identical triplets or even quadruplets result from separations during these early cell divisions. A more common type of twinning results from the ovulation of two secondary oocytes at the same time, which are then fertilized by two separate sperm cells. These individuals are **dizygotic** (dy-zy-GOT-ik; *di-* = "double"), or **fraternal,** twins, and may be different sexes and as dissimilar in appearance as any siblings. Multiple ovulations can result in triplets, quadruplets, or even greater numbers of offspring. However, this is extremely rare, except with the use of assisted reproductive technology that artificially stimulates the release of multiple oocytes.

Quick Check

☐ 3. What is the function of cleavage of the zygote?
☐ 4. How does a morula differ from a blastocyst?

Implantation

◀◀ Flashback

1. What is a corpus luteum, and what does it secrete? (p. 1038)
2. Which hormones are responsible for regulating the uterine cycle? (p. 1042)

Approximately 4–7 days after fertilization, the blastocyst begins to attach to the endometrium of the uterus in the process of *implantation.* The trophoblast invades the **stratum functionalis** (funk-shun-AL-iss) layer of the endometrium by secreting digestive enzymes that catalyze reactions to degrade the endometrial lining (**Figure 27.4**).

Trophoblast cells divide into two layers: (1) the inner **cytotrophoblast** (sy′-toh-TROF-oh-blast), or **cellular trophoblast,** where cells remain intact; and (2) the outer **syncytiotrophoblast** (sin-SISH-ee-oh-trof′-oh-blast; *syncyt-* = "cells together"), or *syncytial trophoblast* (sin-SISH-ee-al), where the cells' plasma membranes disappear and their cytoplasmic contents merge. This makes the syncytiotrophoblast functionally a large, single, multinucleated cell. Almost immediately after formation, the syncytiotrophoblast begins to secrete enzymes to digest uterine cells. Interestingly, it is nonantigenic, which means that it does not trigger an immune reaction from the maternal tissue it is invading.

The syncytiotrophoblast progressively invades the uterine lining over the next several days. It reaches maternal blood vessels on or before day 12, at which point it surrounds them and digests the walls of the vessels. This causes the maternal blood to pool and fill cavities called **lacunae** (lah-KOO-nee; *lacun-* = "cave"), or *intervillous spaces.* The syncytiotrophoblast also makes contact with nutrients in the uterine glands. By day 16, the blastocyst moves farther into the lining and becomes covered by regrowth of the maternal epithelial cells on the

Day 4-7

Stratum functionalis

Trophoblast:
Syncytiotrophoblast
Cytotrophoblast

Lumen of uterus

Inner cell mass

Days 4–7:
The blastocyst attaches to the stratum functionalis and the trophoblast differentiates into the syncytiotrophoblast and the cytotrophoblast.

Day 8

Hypoblast

Epiblast

Day 8:
The syncytiotrophoblast digests the uterine wall as the inner cell mass differentiates into the hypoblast and epiblast.

Lacunae filled with maternal blood

Day 12

Hypoblast

Epiblast

Amniotic cavity

Day 12:
The syncytiotrophoblast reaches maternal blood vessels and degrades them, creating lacunae; the amniotic cavity forms within the epiblast.

Uterine glands

Day 16

Lacunae

Ectoderm

Mesoderm

Endoderm

Amniotic cavity

Day 16:
The blastocyst is fully implanted when it is completely covered by uterine tissue.

Practice art labeling
@MasteringA&P®

Figure 27.4 Implantation of the blastocyst in the uterine lining.

surface. When the blastocyst is embedded in the uterine wall and completely covered by uterine epithelial cells, it has fully implanted.

The syncytiotrophoblast is also responsible for secreting the hormone **human chorionic gonadotropin (hCG)** (koh-ree-ON-ik goh-nad′-oh-TROP-in), which stimulates the corpus luteum in the ovary (a structure formed after an ovarian follicle ruptures) to secrete the hormones *estrogen* and *progesterone* and helps promote placental development. Progesterone suppresses menstruation by maintaining the endometrium of the uterus and prolonging the secretory phase of the uterine cycle. Some hCG is secreted into the urine; because hCG is produced only during pregnancy, it is the basis for pregnancy tests. The hCG level rises in the mother's blood until the end of the second month of development, when a membrane called the *chorion*

(KOR-ee-on) takes over the secretion of estrogen and progesterone. At that point, hCG is no longer needed, and its level declines and remains low for the remainder of the pregnancy.

Notice in Figure 27.4 that during the same time period, the *inner cell mass,* or developing offspring, separates from the trophoblast and differentiates into two layers: the superior **epiblast** and the inferior **hypoblast.** Together, these layers form a flat **bilaminar embryonic disc** (*bilaminar* = "having two layers"). A small cavity appears within the epiblast and will enlarge to become the *amniotic cavity* (am-nee-OT-ik), which will surround the embryo and fill with fluid. The bilaminar embryonic disc will eventually become the three primary germ layers that differentiate to produce all tissues in the body: *ectoderm, mesoderm,* and *endoderm.* These three layers are the blueprint for future organ development and are discussed in Module 27.3.

If for some reason the conceptus is not viable—for example, a chromosomal abnormality is present—it may die. This is called a **spontaneous abortion,** or **miscarriage,** and the pregnancy will end. On rare occasions, the blastocyst implants in a site different from the uterus. If the pregnancy continues, the result is an *ectopic pregnancy* (*ectop-* = "displaced"; see *A&P in the Real World: Ectopic Pregnancy* on page 1066).

Quick Check

☐ 5. What is the ultimate fate of the inner cell mass?
☐ 6. Which cell type secretes hCG?

Development of Extraembryonic Membranes

◀◀ **Flashback**

1. What are the structures of the urinary system? (p. 941)

The bilaminar embryonic disc and trophoblast together produce the **extraembryonic membranes,** which first appear during the second week of development but continue to develop during the embryonic and fetal periods. Collectively, these membranes protect the embryo and assist with vital functions, including nutrition, gas exchange, and storage and removal of waste. The extraembryonic membranes we discuss in the upcoming subsections include the *yolk sac, amnion, allantois,* and the already mentioned chorion. **Figure 27.5** illustrates these membranes in week 3 of development (part a), and early (part b) and late (part c) in week 4.

Yolk Sac

The **yolk sac** arises from hypoblast cells and is the first extraembryonic membrane to develop. In some animals such as birds, the yolk sac is the main nutritional source for the developing offspring, but in humans, cells from the yolk sac form part of the digestive tract and are the source of the first blood cells and blood vessels. In addition, they produce the first *germ cells,* the precursors of the gametes, that will eventually migrate into the developing ovaries and testes.

(a) **Week 3 of development**

Chorionic villi are found around the entire surface of the chorion.

(b) **Early week 4 of development**

Chorionic villi grow more in the direction of the stratum functionalis.

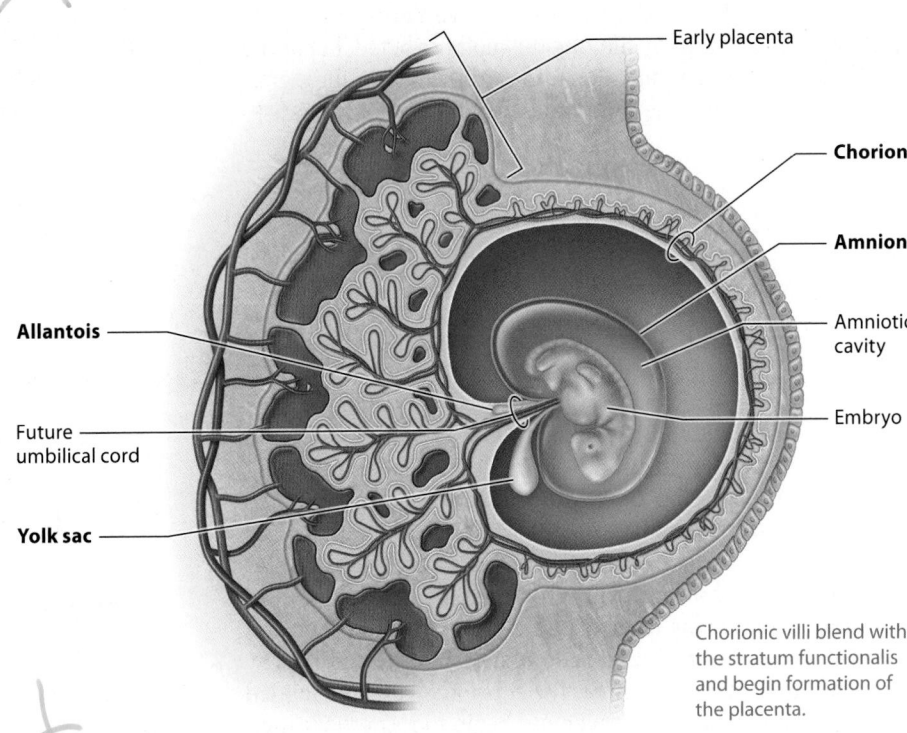

(c) **Late in week 4 of development**

Chorionic villi blend with the stratum functionalis and begin formation of the placenta.

Figure 27.5 Formation of the extraembryonic membranes.

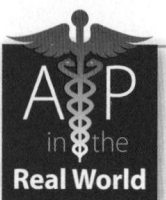

Ectopic Pregnancy

In an ectopic pregnancy, a conceptus implants and grows in any location other than the endometrium of the uterus. In the United States, 1–2% of all pregnancies are ectopic. Almost all of these are "tubal pregnancies," meaning that they are in the uterine tubes; however, they can occur in other locations, including the abdominal cavity, ovary, or cervix. An ectopic pregnancy presents a large risk to the mother, as only the uterus is able to expand and sustain a growing conceptus. If an ectopic pregnancy does not terminate on its own in a spontaneous abortion, medical intervention is necessary to remove the conceptus. Otherwise, the region in which it implants can eventually rupture and result in serious bleeding and possibly even death of the mother.

Amnion

The **amnion** (AM-nee-ahn) is a transparent membrane that develops from the epiblast. It grows to completely enclose the embryo in the fluid-filled **amniotic cavity**, which is penetrated only by the *umbilical cord* (um-BIL-ih-kuhl), whose vessels will connect the growing embryo or fetus to the mother's blood supply (see Figure 27.5c).

The amnion secretes **amniotic fluid** into the cavity. Amniotic fluid protects the embryo from trauma and helps maintain a constant temperature. It also allows freedom of movement, which is important for symmetrical muscle development, and prevents body parts from drying out and adhering to each other during growth. At first, all of the fluid forms by filtering the mother's blood, but as the embryo's kidneys become functional toward the end of the eighth or ninth week, its own urine contributes to amniotic fluid volume.

Allantois

The **allantois** (ah-LAHN-tow-is) is a small outpocketing of the caudal wall of the yolk sac. It forms the base for the umbilical cord and ultimately will become part of the urinary bladder.

Chorion

The **chorion,** the outermost extraembryonic membrane that develops from the cytotrophoblast and synctiotrophoblast, encloses all the other membranes and the embryo (see Figure 27.5). As the chorion develops, it forms outgrowths called the **chorionic villi** (singular, *villus*), which are initially found around the entire surface of the chorion. However, as the pregnancy progresses, chorionic villi grow only in the direction of the stratum functionalis layer of the endometrium and do not continue to surround the other membranes and the embryo. By late in week four of development, the chorionic villi are blended with the stratum functionalis layer and begin to form the principal embryonic part of the placenta.

TABLE 27.1 SUMMARY OF THE FOUR EXTRAEMBRYONIC MEMBRANES

Extraembryonic Membrane	Primary Functions
Yolk sac	• Contributes to the formation of the digestive tract • Source of the first blood cells, blood vessels, and germ cells
Amnion and amniotic cavity	Amnion: • Surrounds the embryo • Produces amniotic fluid Amniotic cavity contains amniotic fluid, which: • Protects the embryo from trauma • Helps maintain a constant temperature • Allows symmetrical muscle development • Prevents drying out and adhesion of body parts during growth
Allantois	• Forms the base for the umbilical cord that links the embryo to the placenta • Ultimately becomes part of the urinary bladder
Chorion	• Encloses all other extraembryonic membranes • Forms the chorionic villi • Forms the main embryonic part of the placenta

Table 27.1 summarizes the functions of the four extraembryonic membranes.

Quick Check

☐ 7. What are the functions of amniotic fluid?
☐ 8. What are the main functions of each of the extraembryonic membranes?

Apply What You Learned

☐ 1. What would happen if the mother's immune system recognized the syncytiotrophoblast as foreign tissue? (*Hint:* Remember that the syncytiotrophoblast dissolves the walls of maternal blood vessels.)
☐ 2. What would happen to progesterone and estrogen levels if the trophoblast in an implanted embryo failed to secrete hCG?

See answers in Appendix A.

MODULE **27.3**

Embryonic Period: Week 3 through Week 8

Learning Outcomes

1. Describe the major events of embryonic development.
2. Discuss the process of gastrulation and embryonic folding.
3. Describe organogenesis, focusing on ectoderm, mesoderm, and endoderm differentiation.

At the end of the pre-embryonic period, the blastocyst is fully implanted, the inner cell mass has differentiated into the bilaminar embryonic disc, and the extraembryonic membranes have developed. We then enter the **embryonic period,** which begins at the third week of development and continues through the eighth week. During this stage, the conceptus is referred to as an embryo.

The embryonic period starts with a process called **gastrulation** (gas-troo-LAY-shun), which is the rearrangement and migration of the cells of the bilaminar embryonic disc to form the **trilaminar embryonic disc.** Trilaminar means "having three layers," so this is the stage at which the three germ layers develop. The three germ layers will become all of the major organ systems in the process of *organogenesis* by the end of week 8, although for the most part the organs are not functional until later in development. Also during the embryonic period, the placenta forms and begins to provide nutrition and oxygen to the embryo and remove wastes. Let's now walk through the processes of gastrulation and organogenesis.

Gastrulation and Formation of Germ Layers

In week 3, we start with a bilaminar embryonic disc containing two cell layers: the epiblast and hypoblast. At this time, a thin groove on the dorsal surface of the epiblast develops called the **primitive streak**. The primitive streak elongates along the cephalic-caudal (head-tail) line of the embryo (**Figure 27.6a**), establishing the head and tail regions, the right and left sides, and the dorsal and ventral (posterior and anterior) surfaces of the embryo.

When the primitive streak has developed, gastrulation begins. It starts as some cells detach from the epiblast layer and move into and then underneath the primitive streak in a process called *ingression*. The first cells that migrate in this way become the inner germ layer, or **endoderm,** which replaces the hypoblast (**Figure 27.6b**). The next cells migrate to a position between the epiblast and endoderm, and this becomes the middle layer, the **mesoderm.** The remaining cells of the epiblast form the outer layer, the **ectoderm** (**Figure 27.6c**). By the end of week 3, all three primary germ layers have formed, making the embryo a trilaminar embryonic disc. At this point, the ectoderm and endoderm are tightly packed and surround the loosely organized mesoderm.

As you can see in Figure 27.6, the trilaminar embryonic disc is fairly flat. This begins to change during the fourth week of development, when two types of folding cause the flat embryonic disc to become cylindrical. The first, **cephalocaudal folding** (sef′-ah-loh-KAW-dul; *cephalo-* = "head," *caud-* = "tail"), occurs in the head and tail regions of the embryo, as its name implies

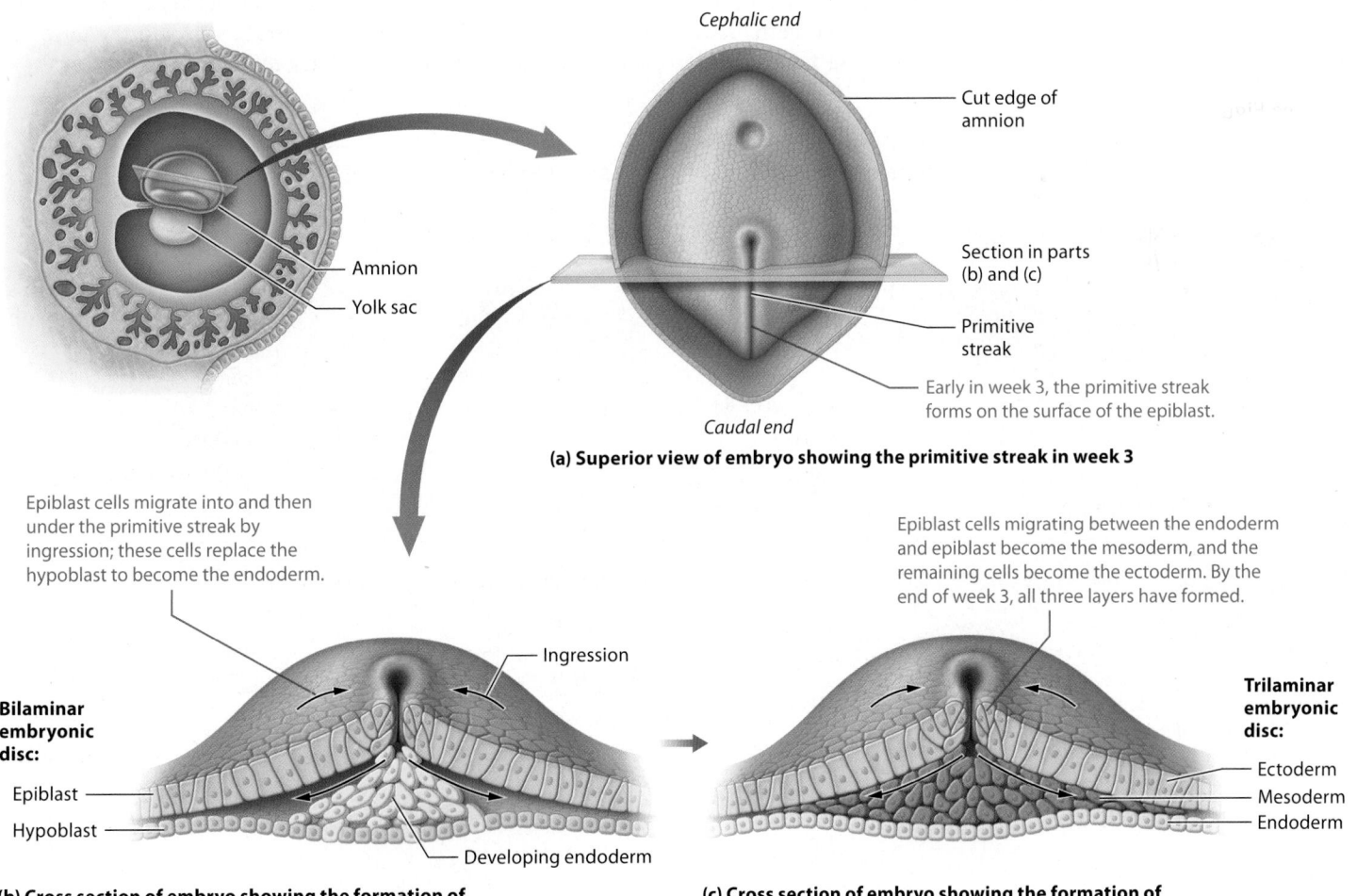

Cephalic end

Cut edge of amnion

Section in parts (b) and (c)

Primitive streak

Early in week 3, the primitive streak forms on the surface of the epiblast.

Caudal end

(a) Superior view of embryo showing the primitive streak in week 3

Amnion

Yolk sac

Epiblast cells migrate into and then under the primitive streak by ingression; these cells replace the hypoblast to become the endoderm.

Ingression

Bilaminar embryonic disc:

Epiblast

Hypoblast

Developing endoderm

(b) Cross section of embryo showing the formation of endoderm *early* in week 3

Epiblast cells migrating between the endoderm and epiblast become the mesoderm, and the remaining cells become the ectoderm. By the end of week 3, all three layers have formed.

Trilaminar embryonic disc:

Ectoderm

Mesoderm

Endoderm

(c) Cross section of embryo showing the formation of mesoderm *late* in week 3

Figure 27.6 Gastrulation and formation of the three primary germ layers.

(**Figure 27.7a**). The second type of folding, **transverse, or lateral, folding,** occurs when the left and right sides of the embryo curve and fold toward the midline (**Figure 27.7b**). As the sides move closer together, the yolk sac is pinched off, leaving only a small area of its endoderm called the **vitelline duct** (vy-TEL-een).

Cephalocaudal folding creates the future head and buttocks regions of the embryo. Transverse folding creates the future trunk region and almost immediately forms the **primitive gut,** which later becomes the digestive tract inside the abdominal cavity.

Quick Check

☐ 1. What is the primitive streak, and what does it establish?
☐ 2. Which cells form the three germ layers?
☐ 3. What are the two types of embryonic folding?

Organogenesis

◀◀ **Flashback**

1. What are the layers of the skin? (p. 161)
2. What are the divisions of the nervous system? (p. 382)

In the process of **organogenesis** (or-gan′-oh-JEN-ih-sis), the three primary germ layers differentiate into organs and organ systems. When organogenesis begins, the primary germ layers have undergone cephalocaudal and transverse folding and

are beginning to differentiate, and the embryo is approximately 1.5 mm (0.6 in.) long. When the embryonic period ends at the end of week 8, the embryo will be approximately 25 mm (1.0 in.) long and will have recognizable organ systems. Although not all the organs will be functional, some will begin to work.

Let's consider each of the major structures that will form by examining the differentiation of each primary germ layer. We'll start with the structures formed by the most superficial layer and end with those of the deepest.

Differentiation of the Ectoderm

After the embryo undergoes cephalocaudal and transverse folding, the ectoderm is on the external surface of the cylindrical embryo. This arrangement is appropriate because the ectoderm will form the epidermis of the skin. The ectoderm is also responsible for forming the majority of the nervous system and sense organs. In fact, the first major event of organogenesis is **neurulation** (NOOR-uh-lay-shun), which produces the brain and spinal cord. Cells of the ectoderm thicken and form the *neural plate,* which folds inward and deepens. The edges fuse to form a *neural tube,* which will pinch off from the ectodermal layer. The anterior end of the neural tube will become the brain, and the remainder will become the spinal cord. By the end of the fourth week, the anterior end of the neural tube will develop into three enlarged areas called the **primary brain vesicles:** the *forebrain, midbrain,* and *hindbrain* (refer back to Figure 12.3).

Figure 27.7 Embryonic folding and developmental changes during weeks 3 and 4 of development.

Between the layers of the ectoderm that form the epidermis and the neural tube are special cells called **neural crest cells.** These cells will migrate throughout the embryo and give rise to various structures, including cranial, spinal, and sympathetic ganglia and associated nerves; pigment cells of the skin; cells of the adrenal medulla; and connective tissue cells.

Differentiation of the Mesoderm

Immediately deep to the primitive streak, some mesodermal cells at the midline form a structure called the **notochord** (NOH-toh-kord), which serves to support and organize the embryo around a central axis. Remnants of the notochord will remain in the intervertebral discs as the notochord is surrounded by the vertebrae during development.

Mesoderm on either side of the notochord forms **somites** (SOH-mytz), or blocklike structures. Each somite has three regions:

- The *sclerotome* (SKLER-oh-tohm) develops into the vertebrae and ribs.
- The *dermatome* develops into the dermis layer of the skin.
- The *myotome* develops into most of the skeletal muscles.

On the lateral sides of the somites are clusters of mesoderm called **intermediate mesoderm** and sheets of **lateral plate mesoderm.** The intermediate mesoderm forms the gonads and kidneys, and the lateral plate mesoderm develops into the spleen, adrenal cortex, most of the cardiovascular system, serous membranes of the body cavities, and the connective tissue components of the limbs.

Differentiation of the Endoderm

The endoderm becomes the internal epithelial layers of the digestive, respiratory, urinary, and reproductive systems. It forms the cavity of the middle ear and the auditory tube. Several glands and accessory digestive organs—including the thyroid gland; parathyroid glands; thymus; parts of the palatine tonsils; and the majority of the liver, gallbladder, and pancreas—also form from the endoderm.

Summary of Organogenesis

All body structures derive from the three primary germ layers. **Table 27.2** summarizes additional structures produced by each germ layer.

By the end of the embryonic period, organogenesis has occurred, although the organs are rudimentary. **Figure 27.8** compares fiber optic views of embryos at 4 and 8 weeks of development. Notice that at 4 weeks the embryo has a prominent forebrain and somites, while at 8 weeks the embryo looks essentially human, albeit with different body proportions.

TABLE 27.2 BODY STRUCTURES PRODUCED BY THE THREE PRIMARY GERM LAYERS

Layer	Major Structures
Ectoderm	Nervous system; epidermis; hair follicles; arrector pili muscles; nails; cutaneous glands; epithelium of nasal, oral, and anal canals; lens and cornea of eye; internal eye muscles; internal and external ear; salivary glands; adrenal medulla; pituitary and pineal glands; melanocytes
Mesoderm	Skeleton; cartilage and connective tissues; skeletal, cardiac, and most smooth muscle; adrenal cortex; middle ear; dermis; blood; blood and lymph vessels; bone marrow; lymphatic tissues; kidneys; ureters; gonads; genital ducts; mesothelium of abdominopelvic and thoracic cavities
Endoderm	Epithelium of the lining of digestive tract (except oral, nasal, and anal canals) and respiratory tract; epithelium of urinary bladder and urethra; epithelium of accessory reproductive and digestive glands (except salivary glands); thyroid, parathyroid, and thymus glands

During this period (weeks 3–8), the embryo is especially sensitive to **teratogens** (teh-RAT-oh-jenz; *terato-* = "monster"), or substances that can cause birth defects or even death. Some infectious agents that produce maternal illness, particularly German measles (rubella), are teratogenic. Other teratogens include alcohol, nicotine, and many over-the-counter and prescription drugs, such as aspirin, anticoagulants, and some antibiotics. For example, the acne medication Accutane® has resulted in life-threatening birth defects. Possibly the most notorious teratogen is the sedative thalidomide, which was prescribed to thousands of pregnant women in the 1960s to alleviate *morning sickness*

(a) **Mini MRI of embryo at week 4 (reconstructed with 3D imagery, 6×)**

(b) **Fiber optic view of embryo at week 8 (1.5×)**

Figure 27.8 Development during the embryonic period.

(nausea), but resulted in many severely deformed infants with short flipper-like legs and arms.

Quick Check

☐ 4. What is the first major event that occurs during organogenesis?
☐ 5. List two structures formed by each of the primary germ layers.

Apply What You Learned

☐ 1. Why does exposure to teratogens during organogenesis increase the risk of birth defects?
☐ 2. Why do you think the first major event in organogenesis is neurulation? (*Hint:* Think about the function of the brain and spinal cord.)

See answers in Appendix A.

MODULE **27.4**
Fetal Period: Week 9 until Birth (about Week 38)

Learning Outcomes

1. Describe the formation and function of the placenta.
2. Identify the major events of fetal development.
3. Describe the specific structures unique to the fetal circulation.

The **fetal period** of development extends from the beginning of week 9 until birth, which usually occurs near week 38. By the 12th week, the placenta is fully functional, providing nutrition to the fetus for the rest of the gestation period. Let's look at the development of both the placenta and the fetus during this period, and finish with a big picture look at prenatal development.

Placentation

◀◀ **Flashback**

1. What is diffusion? (p. 75)
2. How does hemoglobin transport oxygen? (p. 727)

During the pre-embryonic period, the developing embryo receives nutrients from two sources: uterine milk secreted from endometrial glands, and digested endometrial cells from the uterine wall. The embryo will continue to receive nutrition from digested endometrial cells until the placenta is fully formed at about week 12, during the fetal period.

The **placenta,** a temporary organ that is shed after the infant is born, is the site of exchange of oxygen, nutrients, and waste between the mother and fetus. In addition, the placenta produces hormones to support the pregnancy. **Placentation**

(plah-sen-TAY-shun) is the process of forming the disc-shaped placenta, which attaches to the uterine wall and to the embryo or fetus through the umbilical cord (**Figure 27.9a**). Placentation begins during implantation, but the majority of placental growth occurs in the fetal period. If the placenta grows in the wrong place, it may block the cervix of the mother's uterus, resulting in a potentially serious condition during birth (see *A&P in the Real World: Placenta Previa* on page 1075).

The **umbilical cord** connects the center of the placenta to the fetus' *umbilicus* (belly button). The cord normally contains two **umbilical arteries** that carry deoxygenated fetal blood away from the fetal heart to the placenta, and one **umbilical vein** that carries oxygen and nutrients toward the fetal heart. The umbilical cord also contains a soft connective tissue with a jelly-like consistency, called *Wharton's jelly,* that insulates and protects the umbilical arteries and vein; its external surface is covered by the amniotic membrane.

Recall that when the embryonic period begins, the blastocyst has already implanted in the uterus and chorionic villi have penetrated uterine blood vessels to form lacunae filled with maternal blood. Eventually, these lacunae merge into a single blood-filled cavity called the **placental sinus** (**Figure 27.9b**). Once placentation has begun, the stratum functionalis of the uterus is known as the **decidua** (deh-SID-yoo-uh). The region of the endometrium that lies beneath the fetus becomes the **decidua basalis,** and the region that surrounds the uterine cavity forms the **decidua capsularis** (see Figure 27.9a).

The placenta is a unique organ because it develops from both fetal and maternal structures. The fetal portion is formed by the chorionic villi, and the maternal portion by the decidua basalis. The chorionic villi contain fetal blood vessels that are surrounded by maternal blood in the placental sinus; however, maternal and fetal blood do not mix because they are separated by the **placental barrier.** As you can see in **Figure 27.9c**, the barrier consists of the maternal and fetal basal laminae, fetal connective tissue, the cytotrophoblast, and the synctiotrophoblast.

Some substances are exchanged across the placental barrier between the fetal and maternal blood supplies. This movement of substances occurs through diffusion—an important example of the Gradients Core Principle (p. 26). Since the maternal and fetal blood supplies are so close, oxygen and nutrients diffuse from the maternal blood into the fetus for transport to tissues, and

CORE PRINCIPLE
Gradients

waste diffuses from the fetus to the maternal blood for removal (see Figure 27.9c). Let's focus in detail on the maternal and fetal blood flow through the placenta.

Maternal blood flows to the decidua basalis layer of the uterus and then through maternal arteries into the placental sinus. Maternal blood touches the placental barrier, which allows diffusion of substances, but not passage of blood cells. Think of the placental barrier as a "screen" that allows certain gases, nutrients, and waste to get through because they are small enough to fit through the "holes." Once diffusion has taken place, the maternal blood, now deoxygenated and carrying waste, flows into the maternal veins and so back into the mother's cardiovascular system.

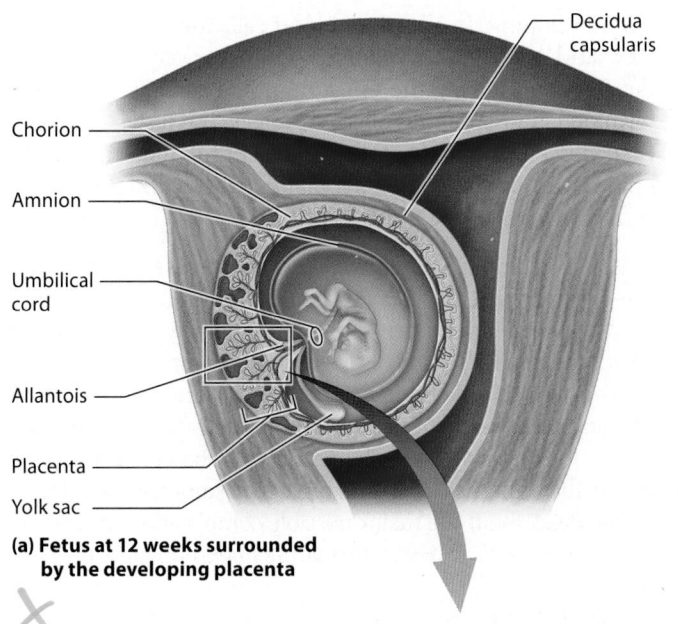

(a) Fetus at 12 weeks surrounded by the developing placenta

Decidua capsularis
Chorion
Amnion
Umbilical cord
Allantois
Placenta
Yolk sac

Decidua basalis Chorion Amnion
Umbilical arteries
Umbilical vein
Maternal vein
Umbilical cord
Maternal artery
Maternal blood in placental sinus
Placental barrier

(b) Section of the placenta and umbilical cord

Fetal blood in capillary
Fetal endothelial cell
Fetal basal lamina
Fetal connective tissue
O_2 Nutrients   Waste
Cytotrophoblast
Placental barrier
Syncytiotrophoblast
Maternal basal lamina
Maternal blood in placental sinus

(c) Schematic showing movement of substances across the placental barrier between maternal and fetal blood

Practice art labeling
@MasteringA&P®

Figure 27.9 Development and nutritive functions of the placenta.

Fetal blood leaves the fetus and enters the two umbilical arteries (shown as blue in Figure 27.9 because the blood is deoxygenated). Blood flows into the placenta and moves into fetal capillaries in the chorionic villi. These capillaries are on the fetal side of the placental barrier. Fetal blood then picks up oxygen and nutrients and delivers waste by diffusion, but again, no blood cells move through the barrier. Fetal hemoglobin has a greater affinity for oxygen than does maternal hemoglobin, so it picks up oxygen from the mother's blood at the placental barrier. Next, the oxygenated blood leaves the placenta and flows back to the fetus through the single umbilical vein (notice it is red because the blood is now oxygenated).

If fetal and maternal blood were to mix, it could be very dangerous for both the mother and fetus. For example, if the mother is blood type A and the fetus is blood type B, both could die because of the agglutination of erythrocytes caused by the incompatibilities between their blood types (see Chapter 19).

In addition to its nutritive functions, the placenta also functions as an endocrine organ. The placenta assumes the production of hCG as it grows and results in the subsequent reduction of the syncytiotrophoblast. In addition, the corpus luteum relinquishes production of progesterone and estrogens to the placenta by the end of the third month of gestation. The placenta also produces *human placental lactogen* and *placental prolactin*, which help prepare the mammary glands for milk production, and *relaxin*, which relaxes the body's muscles, joints, and ligaments, presumably to facilitate stretching during delivery of the newborn.

Quick Check

☐ 1. What are the functions of the placenta?
☐ 2. Which structure connects the embryo to the placenta?

Fetal Development

◀◀ Flashback

1. What is surfactant? (p. 810)
2. What are the four chambers of the heart? (p. 631)

During fetal development, the growth of the fetus is characterized by the maturation of tissues and organs. In addition, fetal growth is rapid and fetal size increases dramatically. Measurements of length are usually taken as crown-rump length (CRL) or crown-heel length (CHL). An embryo 25 mm (1.0 in.) long can grow to be as large as 50 cm (19.7 in.; CHL) during the fetal period. Weight increases as well, most obviously in the last 8 weeks of development. The average weight of a full-term fetus ranges from 3.2 to 3.5 kg (7 to 8 lbs).

One way that a fetus and newborn infant differ noticeably is in their body proportions. At the beginning of the fetal stage, the head is disproportionately large and the lower limbs are short. During the fetal period leading up to birth, the body proportions become more like those of an infant. After birth, body proportions will continue to change until adulthood, when the head is the smallest proportionally to the rest of the body.

Major Events in Fetal Development

Let's summarize the most important changes that occur during the 6 months of the fetal period (months 3–9 of the pregnancy).

- **Month 3:** The fetal body lengthens as its head growth slows, while the upper limbs grow to their birth length. Ossification begins in most bones. In the fetus' face, the eyes are well developed, the eyelids are fused, the bridge of the nose develops, and the external ears are present but not prominent. During this period, the genitals are distinguishable as male or female. By the end of the third month, the approximate CRL is 9 cm (3.5 in.).

- **Month 4:** Body growth is rapid during the fourth month. The lower limbs lengthen and joints are forming as the skeleton continues to ossify. The kidneys are well formed, the digestive glands are forming, and the heartbeat can be heard with a stethoscope. The fetus begins to develop reflexes, and will startle and turn away from loud noises and bright lights. The CRL is about 14 cm (5.5 in.) by the end of the fourth month.

- **Month 5:** Growth slows during the fifth month and the lower limbs achieve their correct proportions relative to the body. Hair grows on the head, and the skin is covered by a downy hair called **lanugo** (lah-NEW-goh) and a white-to-gray secretion composed of shed epithelial cells and sebum known as **vernix caseosa** (kay-see-OH-suh). Both lanugo and vernix caseosa cover the skin to protect it from amniotic fluid. Brown fat forms and helps with heat production after the fetus is born. The mother can feel movements—a phenomenon referred to as *quickening*—as skeletal muscles begin to contract. The approximate CRL is 19 cm (7.5 in.) by the end of the fifth month.

- **Month 6:** The fetus gains significant weight during the sixth month. The eyebrows and eyelashes appear, and the eyelids, which have been fused since the third month, are partially open. The fetus' skin is wrinkled and translucent. Importantly, the fetal lungs begin to produce surfactant, which is key to survival if the infant is born prematurely. The approximate CRL is 23 cm (9.1 in.) by the end of the sixth month.

- **Month 7:** During the seventh month, the eyelids open completely, and fat is deposited in subcutaneous tissue so the skin is slightly smoother, but still wrinkled and red. The fetus usually turns upside down, assuming the **vertex** position. In males, the testes begin to descend through the inguinal canal and into the scrotum. By the end of the seventh month, the approximate CRL is 28 cm (11.0 in.).

- **Months 8 and 9:** The two final months see the fetal neurons forming networks, and organs growing and developing. In addition, blood cells form in bone marrow, and the digestive and respiratory systems complete development during the ninth month. The fetal skin is less wrinkled, and lanugo is shed. In males, the testes complete their descent into the scrotum. The approximate CRL is 30 cm (12 in.) by the end of the eighth month and 36 cm (14 in.) by the end of the ninth month.

Figure 27.10 compares fiber optic views of fetuses at months 3 and 5 of development. As you can see, at 3 months the limbs are short and the head is very large. At 6 months the limbs are much longer and the head is closer to the postnatal proportion.

Approximately 266 days after fertilization, a fetus is considered full-term and ready to be born. The scalp usually has hair, and the fingers and toes have well-developed nails. If, however, a fetus is born before that time, its tissues and organs may not have developed completely. Read about the major possible complications of prematurity in *A&P in the Real World: Premature Infants* on page 1075.

Fetal Circulation

All organ systems grow and develop after birth, but one system must make rapid changes soon after birth to adjust to life without a placenta—the fetal cardiovascular system. The greatest anatomical changes between the prenatal and postnatal states occur in this system. The fetus depends on the placenta and umbilical cord for its blood supply. Cardiovascular structures seen only during prenatal development include the umbilical arteries and vein, and three circulatory "shortcuts" called **vascular shunts**: the *ductus venosus, foramen ovale,* and *ductus arteriosus* (**Figure 27.11a**).

As we know from our discussion of the placenta, the umbilical arteries take fetal blood to the placenta for exchange with maternal blood, and the umbilical vein returns that blood to the

(a) Fiber optic view of embryo at month 3

(b) Fiber optic view of embryo at month 5

Figure 27.10 Development during the fetal period (images are approximately actual size).

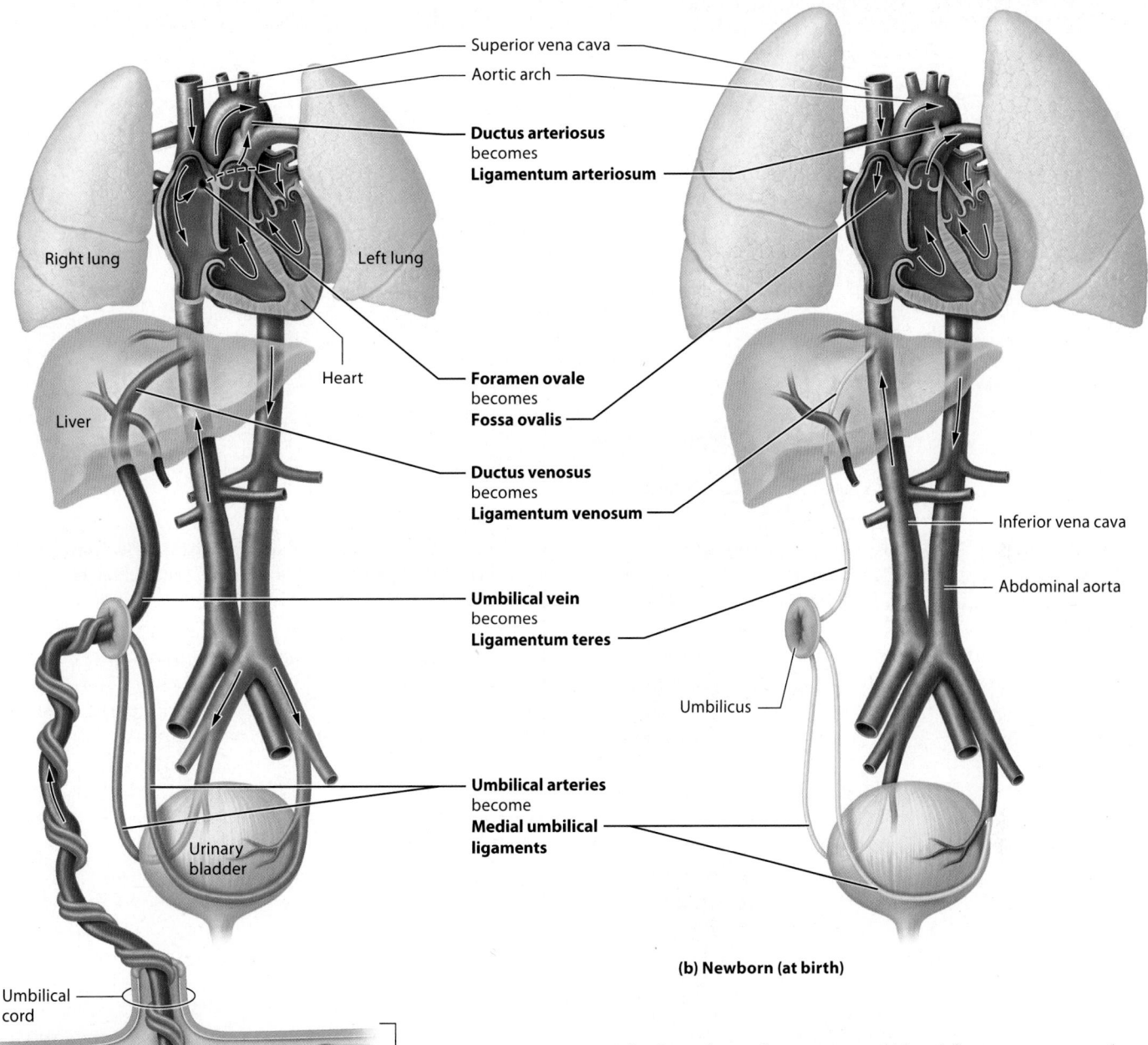

Superior vena cava

Aortic arch

Ductus arteriosus
becomes
Ligamentum arteriosum

Right lung

Left lung

Heart

Foramen ovale
becomes
Fossa ovalis

Liver

Ductus venosus
becomes
Ligamentum venosum

Inferior vena cava

Abdominal aorta

Umbilical vein
becomes
Ligamentum teres

Umbilicus

Umbilical arteries
become
**Medial umbilical
ligaments**

Urinary
bladder

(b) Newborn (at birth)

Umbilical
cord

Placenta

(a) Fetus (38 weeks)

Figure 27.11 **Comparison of fetal and newborn cardiovascular systems.**

fetus. Blood that enters the fetus from the umbilical vein travels to the liver first. However, the immature liver cannot perform many of its filtering or storage functions (a job done by the maternal liver), so complete perfusion is not necessary before birth. Therefore, most blood from the umbilical vein bypasses the liver via the **ductus venosus** (veh-NOH-suhs), which connects to the inferior vena cava and flows to the right atrium of the heart.

The fetus depends on maternal blood for its oxygen, so there is little need for all of the blood that enters the right side of the heart to travel to the lungs, which are not fully functional until just after birth. For this reason, two vascular shunts bypass the fetal pulmonary circulation:

- The **foramen ovale** (foh-RAY-men oh-VAH-lay) is a hole in the interatrial septum that directly connects the right and left atria. The foramen ovale allows blood to "skip" ahead and avoid the pulmonary circulation by moving directly from the right to the left side of the heart and the systemic circulation.
- The **ductus arteriosus** (ar-TEER-ee-oh′-sis) is a short passage that connects the pulmonary trunk to the aorta. The ductus arteriosus allows blood to move from the pulmonary trunk directly into the aorta.

After birth, these shunts and the umbilical vessels close, and the regular circulatory pattern begins. Within a year in most

The Big Picture of Prenatal Development

Figure 27.12

Fertilization Day 1

(1) Fertilization:
Sperm cell and secondary oocyte unite in the uterine tube (see Figure 27.2).

Zygote Day 1

(2) Pre-embryonic period:
Chromosomes from the male and female pronuclei combine and form the zygote. Cleavage produces a blastocyst of about 100 cells that surround a fluid-filled cavity. The blastocyst implants in the uterine endometrium (see Figures 27.3 and 27.4).

Blastocyst Days 4–7

Embryo Week 3

(3) Embryonic period:
Cells in the blastocyst migrate to form the three germ layers (endoderm, mesoderm, and ectoderm). Other cells of the blastocyst form the extraembryonic membranes that support the developing embryo. Differentiation of the three germ layers results in organogenesis—rudimentary formation of organs and organ systems (see Figures 27.5, 27.6, and 27.7).

Embryo Week 8

Fetus Week 9

(4) Fetal period:
The placenta completes its development and, through the umbilical cord, provides nutrients to the fetus for the remainder of the gestation period. The fetus grows rapidly. Tissues and organs continue to mature until gestation is complete at 38 weeks (see Figure 27.9).

Fetus Week 38

Placenta Previa

In **placenta previa** (PREE-vee-uh), a part of or all of the placenta attaches to the inferior portion of the uterus, near or even covering the cervix. Although most cases result in live births, placenta previa may cause spontaneous abortion or premature birth. The condition is also dangerous to the mother because as the uterus stretches to accommodate fetal and placental growth, the placenta may hemorrhage or become infected, which increases the risk of maternal mortality. Typically, caesarean section—surgical removal of the fetus (described in Module 27.5)—is the preferred method of delivery in cases of placenta previa.

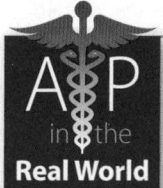

Premature Infants

An infant is considered premature if it is born more than 3 weeks before full-term, which is 38 weeks. More than 12% of the babies born in the United States each year are premature. The earlier the birth, the more complications the infant is likely to experience. Most commonly, premature infants suffer from respiratory, digestive, and thermoregulatory difficulties.

The most common respiratory problem is infant respiratory distress syndrome (IRDS). IRDS develops in infants born before 24 weeks because surfactant is not produced in the alveoli of the lungs until this time. Without surfactant, alveoli collapse during exhalation and are difficult to re-inflate due to alveolar surface tension. Synthetic surfactants can help premature infants breathe until their lungs develop sufficiently.

Premature infants face many other difficulties. For example, they often have minimal sucking and swallowing reflexes and require feeding tubes. In addition, the underdeveloped liver cannot produce enough clotting factors or albumin, so they bleed easily and may develop swelling, or edema. Also, their hypothalamus has not completely developed, so they cannot maintain a constant body temperature and must be kept warm in an incubator.

Premature birth is the most common cause of infant death in the United States. However, due to advances in neonatal care, survival rates have increased to almost 30% for infants born at 23 weeks of pregnancy, about 50 to 60% of those born at 24 weeks, about 75% born at 25 weeks, and more than 90% born at 27 to 28 weeks. Note, though, that infants who survive often face lifelong issues such as cerebral palsy, learning disabilities, and hearing and vision loss.

infants, the flaps in the interatrial septum seal and the foramen ovale becomes a depression called the **fossa ovalis** (**Figure 27.11b**). In some cases, however, the flap does not fuse shut, but remains closed anyway because of the high pressure in the left atrium. Resistance and pressure changes in the pulmonary trunk and aorta cause the ductus arteriosus to close approximately 3 months after birth. The remnant of the ductus arteriosus is a permanent cord, the **ligamentum arteriosum.** The ductus venosus similarly degenerates and becomes the fibrous **ligamentum venosum.** The umbilical arteries become the **medial umbilical ligaments**, and the umbilical vein becomes the **ligamentum teres** (TEHR-eez), or **round ligament of the liver.**

Putting It All Together: The Big Picture of Prenatal Development

Figure 27.12 summarizes key events of prenatal development. This figure starts with fertilization and ends with week 38, just before birth usually takes place.

Quick Check

☐ 3. What specific changes occur in each month of fetal development?
☐ 4. What are the three vascular shunts, and what are their functions?

Apply What You Learned

☐ 1. If a pregnant woman wants to know the sex of her baby as early as possible, when should she schedule an ultrasound exam? (*Hint:* Ultrasound allows visualization of the fetus by applying a probe to the skin of the mother's abdomen.)
☐ 2. If the ductus venosus of an infant does not close after birth, what consequences might this have?

See answers in Appendix A.

MODULE **27.5**
Pregnancy and Childbirth

Learning Outcomes

1. Describe hormonal changes during pregnancy and the effects of the hormones.
2. Describe the functional changes in the maternal reproductive, endocrine, cardiovascular, respiratory, digestive, urinary, and integumentary systems during pregnancy.
3. Explain the hormonal events that initiate and regulate labor.
4. Describe the three stages of labor.

So far, we have examined the changes that occur in the developing embryo and fetus. Now, let's turn our attention to the dramatic anatomical and physiological changes that occur in the mother's body during this 9-month process of pregnancy. It is important to remember that each woman, and in fact each pregnancy, is different. Some pregnancies are uneventful, whereas others result in significant medical complications. Some women feel ill while others feel wonderful. As it is not possible to cover all the potential complications and changes that might happen during pregnancy, we concentrate on the most important events. We conclude this module with the most dramatic event in this process—childbirth.

Maternal Changes during Pregnancy

◄◄ **Flashback**

1. How do hormone levels fluctuate during a female's ovarian and menstrual cycles? (p. 1040)
2. What are the functions of cortisol and aldosterone? (p. 609)
3. What are the normal curvatures of the spine? (p. 231)

From a clinical standpoint, pregnancy is subdivided into three **trimesters**—each lasting 3 months—which differ slightly from the subdivisions of embryonic development.

- **First trimester: months 1–3 of pregnancy.** Recall that within the first 12 weeks of development, all pre-embryonic and embryonic development is completed and fetal development begins. Therefore, by the end of the first trimester, the rudiments of all of the major organ systems are present. This fact makes the first trimester the most critical stage of development. This is also the reason why most spontaneous abortions occur during the first trimester, because of chromosomal abnormalities that prevent the organ systems from developing correctly. Some women are unaware that they are pregnant until a good portion of this trimester is over, while others realize they are pregnant more quickly. Many women find the first trimester to be plagued with morning sickness, fatigue, and other symptoms such as breast tenderness. There are some women who do not suffer from the typical symptoms of early pregnancy and still others who experience increased energy levels.

- **Second trimester: months 4–6 of pregnancy.** During the second trimester, the fetus continues to grow and develop, and the pregnancy usually becomes obvious as the uterus and abdomen expand. Many women feel relief from morning sickness during this trimester and may feel more energetic, although some women continue to experience these or other symptoms. Reaching the second trimester makes many women begin to feel comfortable that the pregnancy is "safe" because spontaneous abortions are less common during the fetal period of development. Additionally, this is the time period in which many women feel the movements of the growing fetus for the first time.

- **Third trimester: months 7–9 of pregnancy.** The fetus grows rapidly during the third trimester and gains a significant amount of weight. The woman's uterus and abdomen enlarge further, and many women experience new symptoms related to the size of the fetus, such as backaches and pressure on internal organs, which are now competing for space in the abdominal cavity. In addition, the weight of the fetus rests on the urinary bladder, increasing the urge to urinate. As the delivery nears, many women grow more uncomfortable as the fetus settles into a lower position in the abdomen.

The rest of this section takes a look at the hormonal, anatomical, and physiological changes that occur during these three trimesters.

Hormonal Changes

During a typical ovarian cycle, the corpus luteum degenerates about 2 weeks after ovulation. Consequently, the concentrations of estrogen and progesterone decline rapidly, the uterine lining breaks down, and the endometrium sloughs off as menstrual flow. If this were to occur following implantation, the blastocyst would be lost in a spontaneous abortion.

Remember that when the blastocyst implants in the uterine lining, the syncytiotrophoblast begins secreting human chorionic gonadotropin, or hCG, which maintains the corpus luteum so that it will continue secreting estrogen and progesterone. For this reason, the inner lining of the uterine wall, the stratum functionalis, continues to grow and develop. At the same time, the hormones stimulated by hCG inhibit the release of *follicle-stimulating hormone* (*FSH*) and *luteinizing hormone* (*LH*) by the anterior pituitary gland, which stops the normal ovarian cycle. Secretion of hCG continues at a high level for about 2 months, then declines by the end of the fourth month, but by this time, the corpus luteum is no longer needed, because the placenta secretes sufficient estrogen and progesterone. For the remainder of the pregnancy, placental estrogen and placental progesterone maintain the uterine lining.

In addition to estrogen and progesterone, several other hormones have important roles during pregnancy. Along with progesterone, **relaxin** from the placenta helps suppress uterine contractions until the birth process begins. When labor nears, relaxin loosens the connective tissue of the pubic symphysis and sacroiliac joints, which allows greater movement of the joints to help the fetus pass through the birth canal.

The placenta also secretes a variety of other hormones. For example, the hormone called **human placental lactogen** (**hPL;** LAK-toh-jin), or *human chorionic somatomammotropin* (hCS; soh-MAH-toh-mam′-oh-troh-pin), helps stimulate breast development and prepares the mammary glands to secrete milk. It may also affect a pregnant woman's metabolism of glucose by inhibiting the release of insulin, which in turn raises the glucose level in the blood supply, allowing more to go to the fetus. A decreased maternal blood glucose level results in increased fatty acid metabolism by the mother. Another placental hormone is **melanocyte-stimulating hormone** (**MSH**), which is partially responsible for darkening of the areolae and nipples of the breast as well as changes in skin tone in some women. Finally, the placenta also produces **corticotropin-releasing hormone (CRH),** which current evidence indicates is important in determining the length of pregnancy and the timing of childbirth. In addition, CRH increases secretion of *cortisol* from the adrenal cortex, which is needed for maturation of the fetal lungs and production of surfactant. The placenta's secretion of CRH begins about the 12th week of development and rises significantly toward the end of pregnancy.

Other endocrine glands increase secretion during pregnancy as well. Aldosterone from the adrenal cortex promotes reabsorption of sodium ions from kidney tubules, which causes water to follow by osmosis. This leads to an overall increase in blood volume during pregnancy. Parathyroid hormone from the parathyroid glands helps to maintain a high concentration of maternal blood calcium ions because fetal demand for calcium ions is high to support bone development. **Prolactin** secreted from the anterior pituitary stimulates milk production by the mammary glands. Finally, **oxytocin** from the fetal and maternal hypothalamus is secreted during the second and third trimesters and peaks during labor to stimulate uterine contractions and allow milk release from the mammary glands.

Anatomical and Physiological Changes

Many anatomical and physiological changes take place to support pregnancy and prepare a woman's body for *childbirth* and *lactation* (lak-TAY-shun), or breastfeeding. These changes affect multiple organ systems.

Reproductive System The uterus enlarges greatly over the 9-month pregnancy and eventually extends upward from its normal location in the pelvic cavity to reach the level of the xiphoid process of the sternum. Before pregnancy, the uterus is approximately the size of a fist, but by the fourth month, the uterus fills the pelvic cavity and has already begun to push into the abdominal cavity (**Figure 27.13a** and **b**). The bulk of uterine enlargement is not because of the developing fetus—which is only about 90 mm (3.5 in.) long from crown to rump at this time—but because of hypertrophy of the myometrium, placental growth, and accumulation of amniotic fluid. In the last few weeks of pregnancy, estrogens promote the formation of gap junctions between the uterine muscle cells to facilitate coordinated contractions during childbirth (see Chapter 10).

By about 28 weeks, or 7 months, the uterus reaches above the umbilicus (**Figure 27.13c**). As pregnancy progresses, the uterus continues to expand, displacing abdominal organs upward, compressing the diaphragm, and pressing on the urinary bladder. As a result, a pregnant woman may be unable to eat large meals, may develop heartburn, and may have to urinate frequently. The increased size of the anterior abdomen changes the center of gravity and some women develop *lordosis,* which is accentuated lumbar curvature, and backaches during the last few months of pregnancy (**Figure 27.13d**).

Notice in the figure that the mammary glands also undergo significant changes in preparation for lactation after the newborn is born. They are typically sore during the first trimester because of enlargement due to increased levels of estrogen and progesterone. Glandular tissue grows and additional **alveoli**—the structures of the breast that produce milk—develop under the influence of prolactin. Typically, the breasts enlarge because of the increase in glandular tissue and additional adipose tissue that accumulates during the pregnancy (refer back to Figure 26.13).

Cardiovascular System The growing placenta requires increased blood volume, and as the offspring grows, it needs more oxygen

(a) **Nonpregnant female**   (b) **4 months pregnant**   (c) **7 months pregnant**   (d) **9 months pregnant**

Figure 27.13   Changes in a woman's body during pregnancy.

and produces more waste that must be removed. As a result of these changes, the woman's blood volume and, subsequently, cardiac output increase. Blood pressure may rise to some extent, but typically, pregnant women have a lower hematocrit (number of erythrocytes) than nonpregnant women. This makes the blood slightly thinner, which decreases resistance and tends to prevent a significant rise in blood pressure. As the uterus grows and presses on pelvic blood vessels, venous return from the lower limbs may be impaired and sometimes results in *varicose veins* and leg swelling, or *edema.*

Perhaps because of the close association between maternal and fetal circulations, consequences arise when these systems do not work normally. See *A&P in the Real World: Pre-eclampsia* to read about a potentially serious condition that highlights this association.

Respiratory System As previously mentioned, more oxygen is needed during pregnancy, and more carbon dioxide is produced to be exhaled. Progesterone makes the medullary respiratory center more sensitive to carbon dioxide, which stimulates removal of more carbon dioxide from the blood by increasing the rate and depth of ventilation. However, during the latter part of pregnancy, the uterus may prevent the diaphragm from contracting completely and the lungs from expanding completely (see Figure 27.13d). Other effects on the respiratory system are mediated by estrogen, which may cause the nasal mucosa to swell, resulting in nasal congestion and occasional *epistaxes* (ep-ih-STAK-seez), or nosebleeds. These changes can contribute to shortness of breath in some women.

Digestive System For normal development of the fetus, good nutrition is necessary all during pregnancy. A pregnant woman typically needs about 300 extra calories each day to supply both herself and the fetus. Specifically, protein, calcium salt, iron salt, and folic acid requirements increase from the recommended daily levels required for nonpregnant women. The

increased intake of food, water retention, and the growing fetus lead to weight gain. The average weight gain is about 13 kg (28.7 lb), but there are significant differences among women.

During the first trimester, many pregnant women suffer from **morning sickness,** nausea and sometimes vomiting (that do not necessarily occur only in the morning). These symptoms are thought to be related to elevated hCG, estrogen, and progesterone levels, but no definitive cause has been identified. Some women suffer from an extreme form of morning sickness called **hyperemesis gravidarum** (hy'-per-EM-uh-sis grav-ih-DAH-rem). This results in excessive vomiting leading to dehydration, electrolyte imbalances, and weight loss in the mother. Severe cases require hospitalization and administration of fluids intravenously.

Some women experience food cravings and/or food aversions. Although no specific reason has been identified for this phenomenon, it could relate to the fact that some foods, such as bland carbohydrates, might contain fewer associated toxins and therefore be "safer" for the fetus. For example, many women crave fruits and vegetables and tend to lose interest, at least temporarily, in caffeinated or spicy foods.

Another digestive system effect is linked to progesterone. This hormone relaxes the smooth muscle of the gastrointestinal tract and slows peristalsis, resulting in constipation.

Urinary System The fetus produces metabolic wastes that the mother's kidneys must excrete. This, combined with the mother's higher blood volume, cause the glomerular filtration rate to rise by up to 50% during pregnancy. However, even as more urine is being produced, progesterone signals relaxation of the smooth muscle of the ureters and may slow the movement of urine out of the urinary tract. The uterus tends to compress the urinary bladder as it expands (see Figure 27.13), sometimes leading to *stress incontinence,* or unintentional loss of urine.

Integumentary System Changes in the skin during pregnancy are more apparent in some women than in others. Some experience increased pigmentation around the eyes and cheekbones in a masklike pattern called **chloasma** (kloh-AZ-muh), in the areolae of the breasts, and along the midline of the lower abdomen. *Stretch marks,* or *striae,* over the abdomen can occur as the uterus enlarges rapidly and the elastic fibers in the dermis break. In addition, progesterone and estrogen decrease the normal hair loss of some women, who subsequently experience thicker hair during pregnancy. On the other hand, nails may become stronger and thicker or more brittle and fragile during pregnancy. A woman's hair and nails typically return to normal within six months of childbirth.

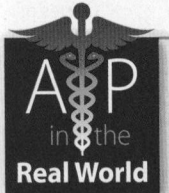

Pre-eclampsia

Approximately 10% of pregnant women in the United States experience a dangerous complication known as *pre-eclampsia* (pree'-eh-KLAMP-see-uh). The causes are not known, but may involve insufficient blood flow to the uterus or a problem with the mother's immune system. Its occurrence seems to be related to the number of fetal cells that enter the maternal circulation. It leads to the fetus being deprived of oxygen and the mother developing edema, hypertension, and high protein levels in urine (proteinuria).

Undiagnosed pre-eclampsia can lead to *eclampsia,* a potentially fatal condition that can result in seizures and put both mother and fetus at risk. Unfortunately, the only treatment for pre-eclampsia is to deliver the baby, but once that is done the condition usually resolves within a few weeks.

Quick Check

☐ 1. What is the first hormone secreted that prevents a menstrual cycle, thereby blocking removal of an implanted blastocyst?

☐ 2. How do a woman's nutritional requirements change during pregnancy?

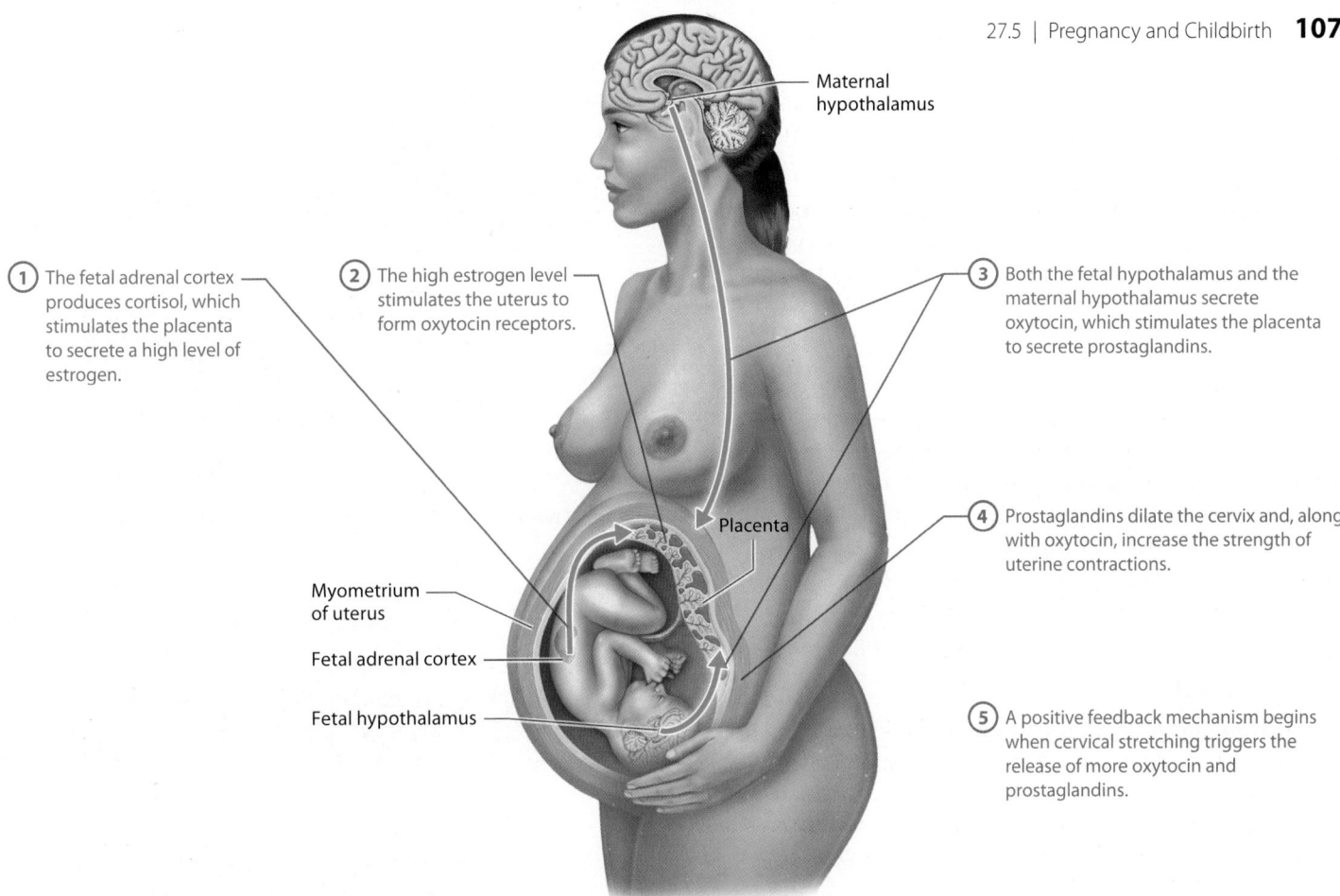

① The fetal adrenal cortex produces cortisol, which stimulates the placenta to secrete a high level of estrogen.

② The high estrogen level stimulates the uterus to form oxytocin receptors.

③ Both the fetal hypothalamus and the maternal hypothalamus secrete oxytocin, which stimulates the placenta to secrete prostaglandins.

④ Prostaglandins dilate the cervix and, along with oxytocin, increase the strength of uterine contractions.

⑤ A positive feedback mechanism begins when cervical stretching triggers the release of more oxytocin and prostaglandins.

Maternal hypothalamus

Placenta

Myometrium of uterus

Fetal adrenal cortex

Fetal hypothalamus

Figure 27.14 Initiation and regulation of childbirth by a positive feedback mechanism.

Childbirth (Parturition)

◀◀ **Flashback**

1. What is the "true pelvis"? (p. 246)
2. What is the function of oxytocin? (p. 596)

Pregnancy ends with **childbirth,** or **parturition** (par-too-RIH-shun), when the fetus is expelled from the uterus through the vagina. The series of events that expel the fetus are collectively called **labor.** Although birth is estimated to take place during week 38 of pregnancy, many occur 1–2 weeks earlier or later.

The series of events involved in initiating and regulating childbirth are as follows (**Figure 27.14**):

① **The fetal adrenal cortex produces cortisol, which stimulates the placenta to secrete a high level of estrogen.** The initial stimulus for labor comes from the fetus. As the time of childbirth nears, the fetus begins to secrete a high level of cortisol. The cortisol binds to receptors on the placenta and stimulates it to produce large amounts of estrogen.

② **The high estrogen level stimulates the uterus to form oxytocin receptors.** The rising estrogen level stimulates the formation of additional oxytocin receptors on the myometrium. It also overrides the quieting influence of progesterone on the myometrium. The myometrium becomes "irritable," leading to irregular contractions. These **Braxton-Hicks contractions** are referred to as

false labor because they do not result in the three stages of labor. Some women experience strong Braxton-Hicks contractions and because of this may think they have begun the process of labor. However, true labor results in uterine contractions that increase in intensity and regularity and cause changes in the cervix.

③ **Both the fetal hypothalamus and maternal hypothalamus secrete oxytocin, which stimulates the placenta to secrete prostaglandins.** Now that the uterus has more receptors for oxytocin, the fetus and mother both increase oxytocin production. This causes the placenta to produce chemicals called *prostaglandins.* It also leads to early, usually weak, contractions of the myometrium.

④ **Prostaglandins dilate the cervix and, along with oxytocin, increase the strength of uterine contractions.** These prostaglandins cause the cervix to open or *dilate,* which is necessary for the fetus to exit the uterus. Both prostaglandins and oxytocin cause the uterine contractions to progressively increase in strength and frequency.

⑤ **A positive feedback mechanism begins when cervical stretching triggers the release of more oxytocin and prostaglandins.** As the head of the fetus pushes on and stretches the cervix, more oxytocin is released. As more oxytocin is released, the myometrium contracts more forcefully and the placenta secretes more prostaglandins. Both effects cause the cervix to stretch more, which stimulates the release of more oxytocin in a

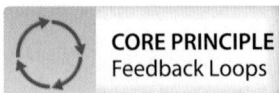

CORE PRINCIPLE
Feedback Loops

positive feedback loop. This brings us back to the Feedback Loops Core Principle.

During pregnancy, the rising level of progesterone leads to the formation of a *mucus plug* at the cervix. When the cervix begins to dilate slightly, the mucus (which may be clear or bloody) is discharged into the vagina. Labor may begin within hours after the mucous plug is discharged, or it may not occur for up to 2 weeks.

In some pregnancies, labor does not begin as it should or, for some reason, the mother wants to choose the time she gives birth. In these situations, labor can be initiated, or **induced**, in a clinical setting. Prostaglandin gels may be applied to the cervix to stimulate dilation and an intravenous solution of synthetic oxytocin, called *Pitocin,* is administered.

Labor can be subdivided into three stages—the *dilation, expulsion,* and *placental stages.* Let's explore what happens in each of these stages.

Dilation Stage

The **dilation stage** is the time from the onset of labor until the cervix is fully dilated to 10 cm (3.9 in.) in diameter (**Figure 27.15a and b**). When labor begins, contractions originate in the upper part of the uterus and move toward the vagina. As labor progresses, contractions become stronger and the lower part of the uterus becomes involved.

Dilation is the longest stage of labor, but its length varies. The range is usually longer (8–24 hours) in women who are having their first child and shorter (4–12 hours) in those who have previously given birth. As long as the fetus is in the *vertex,* or head-first, presentation, the fetus' head is forced against the cervix, helping it to thin and dilate. At some point during this stage, the amnion ruptures and releases the amniotic fluid, which is the event known as the "water breaking" or "rupture of membranes." If the amnion does not rupture on its own by the end of this stage, it may need to be ruptured manually. Toward the end of the dilation stage, the head of the fetus enters the true pelvis and the fetus rotates so that its head can pass through the pelvic outlet.

Expulsion Stage

The **expulsion stage** is the time from full dilation to delivery of the newborn (**Figure 27.15c**). Once the cervix is fully dilated, strong contractions occur frequently, often assisted by the mother voluntarily contracting her abdominal muscles, called "pushing" or "bearing down." This stage may take up to a few hours, but typically ranges from 30 minutes to 1 hour.

When the first part of the fetus's head distends the vagina, **crowning** has occurred. An incision, called an **episiotomy** (eh-piz'-ee-OT-ah-mee), may be done at this time to widen the vaginal orifice and reduce tearing of the surrounding tissues. The fetus' head is then born, followed by the rest of the body, which is delivered more easily. The umbilical cord is then clamped and cut.

(a) Normal position of fully developed fetus before labor begins

Fetus · Umbilical cord · Amnion · Cervical canal · Placenta · Myometrium of uterus · Vagina

(b) Dilation stage

Ruptured amnion · Amniotic fluid · Cervical canal dilated to 10 cm

(c) Expulsion stage

(d) Placental stage

Placenta

Figure 27.15 **The stages of labor.**

Under certain circumstances, a vaginal delivery is not possible or is unsafe for the mother and fetus. For instance, in a **breech birth,** or *buttocks-first,* or other *nonvertex* presentation, labor often lasts longer. Complications may arise, such as the umbilical cord wrapping around the fetus or becoming compressed.

In these cases, a **caesarean section,** or **C-section,** may be performed. A C-section is a surgical procedure in which an incision is made through the abdominal and uterine walls in order to remove the fetus and placenta. In most cases delivery in subsequent pregnancies will also require C-section, although vaginal delivery after a prior C-section is sometimes an option.

Placental Stage

The **placental stage** is the time after delivery of the newborn when the placenta and the attached fetal extraembryonic membranes, collectively called the **afterbirth,** are delivered (**Figure 27.15d**). The uterus continues to contract, compressing uterine blood vessels to limit bleeding and pulling the placenta away from the uterine wall. This stage typically takes about 30 minutes, although often a healthcare provider will simply tug on the umbilical cord to manually remove the afterbirth more quickly. Complete removal of the placenta is important to stop uterine bleeding. A retained placenta can lead to significant blood loss for the mother and an increased risk for infection.

Quick Check

☐ 3. Which hormone stimulates the beginning of childbirth, and where does this hormone originate?

☐ 4. Summarize what happens during each stage of labor.

Apply What You Learned

☐ 1. How could a breech birth result in compression of the umbilical cord? (*Hint:* Review Figure 27.15.)

☐ 2. Pregnancies involving two or more fetuses usually result in the infants being smaller at birth than babies born individually. Why do you think this is the case? (*Hint:* Review Figure 27.13.)

See answers in Appendix A.

Postnatal Changes in the Newborn and Mother

Learning Outcomes

1. Describe the major changes that occur to the newborn during the neonatal period.
2. Describe the major anatomical and physiological changes to the mother after parturition.
3. Describe the structure and function of the mammary glands.
4. Explain the hormonal regulation of lactation.

Recall that the *postnatal period* of development lasts from birth until death. In this section we focus on the neonatal period and the immediate changes the newborn and mother must make to regain homeostasis after birth.

Changes in the Newborn

◄◄ Flashback

1. What is brown fat? (p. 139)
2. What are the functions of IgG and IgA antibodies? (p. 779)

The first 4 weeks after birth constitute the **neonatal period** for the newborn, or **neonate.** Immediately after birth, the newborn must make significant physiological adjustments to adapt to life outside the uterus. It must now do for itself all of the physiological functions that the mother had been doing, including breathing, obtaining nutrients, excreting waste, and maintaining body temperature.

To determine whether a newborn is adjusting appropriately, its physical status is evaluated at 1 minute and 5 minutes after birth, based on five criteria: skin color, pulse rate, respiratory level, muscle activity, and grimace (which assesses reflexes). Each criterion is given a score of 0–2, and the total is called the **Apgar score** (**Table 27.3**) An Apgar score of 8–10 indicates a healthy

TABLE 27.3 APGAR SCORING FOR NEWBORNS

Physiological Criteria	Score 0	Score 1	Score 2
Skin color	Pale to blue	Limbs blue, body pink	Pink; in dark-skinned babies, the color inside the mouth, the lips, and the palms of the hands and soles of the feet are examined.
Pulse rate	No pulse	100 beats per minute or lower	Over 100 beats per minute
Respiratory level	No breathing	Slow and/or irregular breathing; weak cry	Normal regular breathing; strong cry
Muscle activity	No movement; muscles flaccid	Some movement; poor muscle tone	Active movement; good muscle tone
Grimace (reflex involving face and body, initiated by mechanically stimulating plantar surface of foot)	No response	Facial grimace; some body movement	Facial grimace; significant body movement

baby. Lower scores signify a problem in one or more of the physiological functions assessed.

The most dramatic event after the birth is the first breath. Remember that before birth, most blood bypassed the lungs, so in addition to inflating the alveoli of the lungs for the first time, the first breath also causes the blood pressure in the lungs to drop and decreases pressure in the right side of the heart. For this reason, blood flows from the left side of the heart to push flaps over the foramen ovale, which closes the opening. Due to this initial change in respiration, fetal circulatory routes are converted to the newborn circulatory system (look back at Figure 27.11). The neonatal respiratory rate (30-60 breaths per minute) and heart rate (120–140 beats per minute) are significantly higher than those of adults. These higher rates result because compared to adults, neonates have smaller lung capacity and the physical size of their heart in relation to their body is reduced.

Newborns lose heat more rapidly than adults because they have a larger ratio of surface area to volume. However, keep in mind that newborns are born with brown fat, which is a special adipose tissue deposited during the fifth month of fetal development that produces heat rather than storing energy. Eventually, most of this brown fat is metabolized and is not replaced. However, as an infant grows, its metabolic rate increases, and it accumulates more subcutaneous white (or adult) fat, which helps insulate against heat loss.

Although the kidneys are almost fully developed at birth, they are less capable than adult kidneys of concentrating urine. Consequently, newborns have a higher rate of water loss and require more fluid intake, relative to body weight, than adults do. It helps that their food source is liquid during the first few months of life.

Babies are born with a near-adult level of IgG antibodies acquired passively from the mother through the placenta. However, maternal IgG breaks down rapidly and is virtually gone by 1 year after birth. Depending on the disease, maternal IgG may be sufficient to prevent infection after exposure until the infant's immune system is fully developed. In addition, breastfed infants receive added protection from gastroenteritis thanks to IgA antibodies in the *first milk,* or *colostrum* (kohl-OSS-trum).

Other notable changes occur in the nervous and skeletal systems. During the first year of life, the nervous system grows and many neurons develop a myelin sheath. As for the skeleton, various areas of it are cartilaginous rather than ossified at birth; these will be converted to bone over time.

In some cases, infants are born with anatomical malformations, called *congenital disorders* because the deformity exists from birth. Two congenital disorders involve the vascular shunts in the newborn cardiovascular system (see *A&P in the Real World: Patent Ductus Arteriosus and Patent Foramen Ovale*).

Quick Check

☐ 1. What are the criteria used to determine a newborn's Apgar score?
☐ 2. What is the benefit to the newborn of acquiring IgG antibodies from the mother?
☐ 3. Explain what happens to each structure in the fetal circulation after the infant is born.

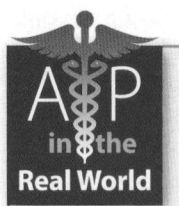

Patent Ductus Arteriosus and Patent Foramen Ovale

Failure of an infant's ductus arteriosus to close after birth results in **patent ductus arteriosus** (PAY-tent). This condition causes difficulties because it allows oxygen-poor blood being carried by the pulmonary artery to mix with oxygen-rich blood being carried by the aorta. In addition, this condition causes an increase in pulmonary blood pressure that can damage the heart and lungs. If not corrected, patent ductus arteriosus leads to cardiac dysrhythmias and possibly heart failure, because the heart must work harder to supply blood to body tissues.

Similarly, if an infant's foramen ovale does not close because the septal flaps fail to seal after birth, **patent foramen ovale** results. As the heart grows larger, the patent foramen ovale also enlarges. In most cases, there are no symptoms and no treatment is prescribed. However, evidence increasingly shows that in adults, patent foramen ovale is associated with strokes because it increases the formation of inappropriate clots. Patent foramen ovale may also increase the risk of migraine headaches.

Changes in the Mother

◀◀ **Flashback**

1. What is the function of endorphins? (p. 456)
2. What are the functions of complement proteins and interferon? (p. 776)

Now let's shift our focus to the mother. For her, the first 6 weeks following birth make up the **postpartum period.** The woman's body undergoes several changes to return to pre-pregnancy homeostasis and to begin and then maintain the production of breast milk—the process of lactation. Most of these changes are completed within 6 weeks, but lactation can continue up to a year or more.

Anatomical and Physiological Changes

Just as the mother's body changed during pregnancy, many adjustments occur after birth to reverse those changes and restore homeostasis to the mother's system. Immediately after birth, endorphins released by the pituitary gland allow the mother to ignore fatigue and pain to take care of the new infant. However, endorphin production is short-lived, and these positive feelings soon dissipate. In addition, within a few days after birth, the levels of estrogen and progesterone fall significantly because they are no longer needed to maintain the uterine lining. The level of corticotropin-releasing hormone (CRH) also declines because the placenta is no longer present. Sudden reductions in the levels of estrogen, progesterone, and CRH have been linked to a severe form of depression called

postpartum depression. Women who suffer from this condition may require medical treatment.

For most women, the respiratory and digestive systems resume normal function soon after birth because the need for oxygen and nutrients returns to pre-pregnancy levels and the respiratory and digestive organs are no longer compressed. The digestive symptoms associated with pregnancy—such as nausea, vomiting, heartburn, and constipation—usually stop after childbirth.

Some of the excess fluid volume retained in the mother during pregnancy is removed during birth through blood loss and expulsion of the amniotic fluid. More of the fluid is removed as **lochia** (LOW-kee-uh), a bloody, yellowish discharge that is expelled from the uterus for up to 6 weeks after birth. Lochia is usually heavy for the first week or so and then lightens for the remainder of the time. The urinary system removes still more excess fluid. When the CRH level declines, the level of aldosterone also declines and urine production increases. When the fluid volume returns to normal, cardiac output also returns to the pre-pregnancy level.

Other changes, such as altered pigmentation in the skin, usually return to the pre-pregnancy state. The thickened hair is also usually temporary, as many women experience significant hair loss after pregnancy. However, some changes, such as stretch marks and varicose veins, are often irreversible.

Lactation

Lactation refers to the production and release of **breast milk** from the mammary glands. In the last month of pregnancy, increasing levels of placental estrogen, progesterone, and human placental lactogen stimulate the production of prolactin from the anterior pituitary. Estrogen and prolactin stimulate alveoli in the mammary gland to grow and lactiferous ducts to branch. However, high levels of estrogen and progesterone inhibit the actual production of milk.

For the first few days after birth, the mammary glands secrete a thick yellowish fluid called **colostrum.** Colostrum has very little fat and less lactose than milk, but it contains protein and is rich in IgA antibodies. These antibodies from the mother provide passive immunity and protect against certain infections, such as bacterial ear infections, respiratory syncytial virus, and gastrointestinal viruses that cause diarrhea.

A few days after birth, true breast milk starts to be produced, because with the placenta gone, the levels of estrogen and progesterone decline and prolactin is not inhibited. Breast milk has a high fat content and contains substances such as essential fatty acids, growth factors, enzymes to help digest the milk, IgA, and immune chemicals such as complement proteins and interferon. Infants can easily digest and absorb it. Breast milk also acts as a mild laxative and helps cleanse **meconium** (mih-KOH-nee-um), a tarry, green fecal material made up of bile and other wastes that is first excreted by the newborn. Removing meconium quickly is important because fetal red blood cells break down rapidly after birth and, as a result, bilirubin can accumulate in the infant's tissues, causing jaundice.

Oxytocin is responsible for the actual *ejection* of milk from the alveoli of the mammary glands, a process called the **let-down reflex** (Figure 27.16). Let-down occurs when oxytocin binds to *myoepithelial cells* surrounding the glands and causes them to contract and eject the milk. During nursing, oxytocin also stimulates the recently emptied uterus to contract, which helps return it to pre-pregnancy size.

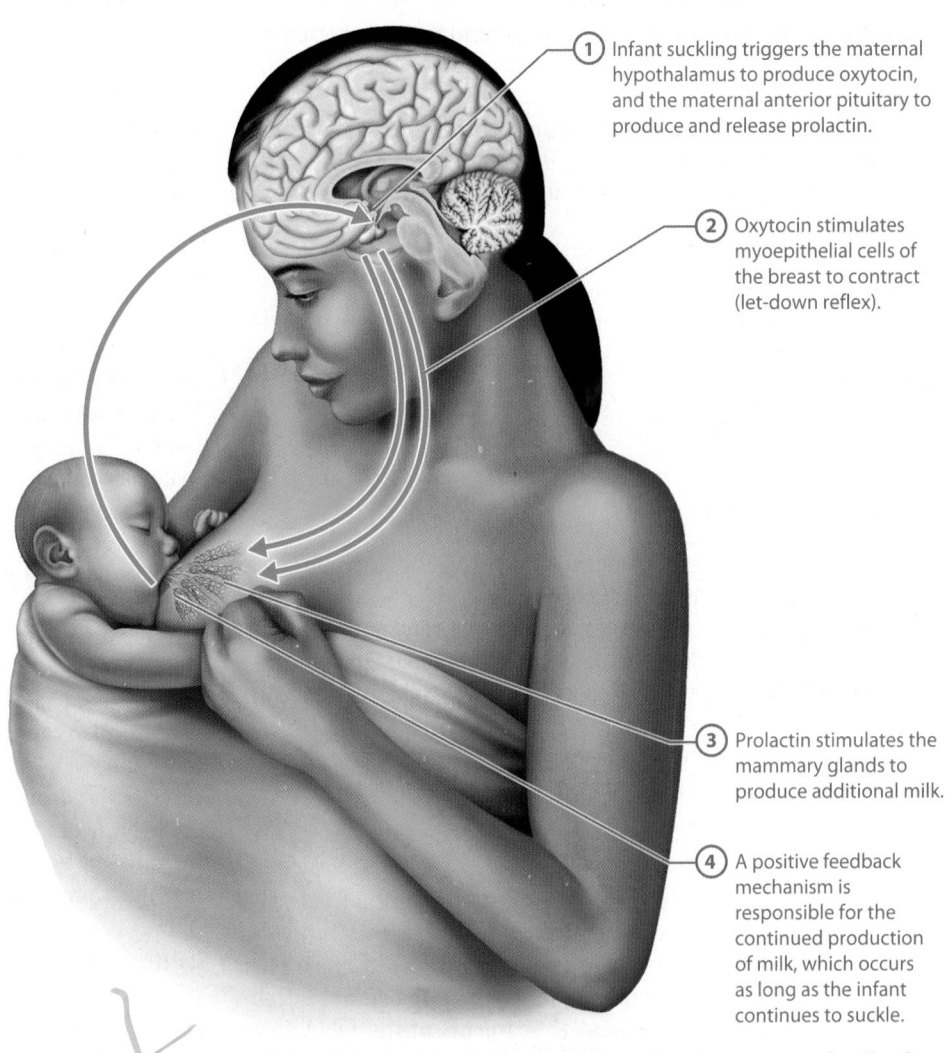

1 Infant suckling triggers the maternal hypothalamus to produce oxytocin, and the maternal anterior pituitary to produce and release prolactin.

2 Oxytocin stimulates myoepithelial cells of the breast to contract (let-down reflex).

3 Prolactin stimulates the mammary glands to produce additional milk.

4 A positive feedback mechanism is responsible for the continued production of milk, which occurs as long as the infant continues to suckle.

Figure 27.16 Hormonal regulation of lactation (the let-down reflex) by a positive feedback mechanism.

Continual milk production is an example of a positive feedback mechanism. In an example of the Feedback Loops Core Principle (p. 21), continuous milk production depends on mechanical stimulation of the nipples, which is provided by the infant's suckling. Mechanoreceptors in the nipple trigger secretion of bursts of prolactin, which stimulate the production of milk needed for the next feeding. Milk production recedes only when the infant no longer suckles. At this point, the mammary glands return to their pre-pregnancy state.

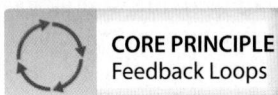

CORE PRINCIPLE
Feedback Loops

Quick Check

☐ 4. Describe the benefits of colostrum for an infant.
☐ 5. What role do hormones play in the let-down reflex and continual milk production?

Apply What You Learned

☐ 1. Why do infants begin receiving vaccinations for certain diseases immediately after birth?
☐ 2. As a newborn grows, his or her milk requirements increase. How does breast milk production increase to keep up with the demand?
☐ 3. Why does continued breastfeeding decrease the chance of pregnancy? (*Hint:* Think about the effect the breastfeeding hormones have on estrogens and progesterone.)

See answers in Appendix A.

MODULE **27.7**

Heredity

Learning Outcomes

1. Define chromosome, gene, allele, homologous, homozygous, heterozygous, genotype, and phenotype.
2. Analyze genetics problems involving dominant and recessive alleles, incomplete dominance, codominance, and multiple alleles.
3. Explain how polygenic inheritance differs from inheritance that is controlled by only one gene.
4. Explain how environmental factors can modify gene expression.
5. Discuss the role of sex chromosomes in sex determination and sex-linked inheritance.

How does a single-celled zygote become a human made of trillions of cells? And how does it "know" whether that individual should have brown or blue eyes, or brown or red hair? The answers to these questions are the same—it's in the genes. Recall from the cell chapter that **genes** are segments of the DNA that code for a specific protein (see Chapter 3). Proteins make up the bulk of body structures, so genes are basically the blueprints for tissues and organs. The genes are already present in that first

small cell, and they are in every nucleated cell that is produced thereafter by mitotic divisions.

Although all healthy individuals have the same tissues and organs, the exact blueprint for each person is unique. Genes have different variants, or *alleles,* that allow for a variety of combinations. Most inherited traits are actually determined by multiple alleles or the interaction of more than one gene. In this module, we discuss the principles of heredity and examine the interactions of various genes and their alleles.

Introduction to Heredity

◀◀ **Flashback**

1. What are homologous chromosomes? (p. 102)

Heredity is the transmission of genetic characteristics from parent to offspring through the genes. The study of heredity and how it is transmitted is the field of **genetics.**

Collectively, all of an individual's DNA is referred to as the **genome;** each genome contains around 20,000 genes with about 3.2 billion nucleotides. These nucleotides are folded and compressed to form the 46 **chromosomes** (*chromo-* = "colored," *soma-* = "body," so named because they stain darkly) that are found in all somatic cells with a nucleus. These chromosomes are paired so that there are 23 pairs of **homologous chromosomes** in each nucleus (see Figure 3.27). Each pair of homologous chromosomes contains one from each parent, so in total our cells have 23 maternal chromosomes and 23 paternal chromosomes. For this reason, the cells are called *diploid* (*di-* = "two"), meaning that they have two copies of each chromosome.

Chromosomes 1 through 22 are known as **autosomes** (AW-toh-sohmz). Each member of a pair of autosomes carries the same genes. The twenty-third pair are the **sex chromosomes**, the X and Y chromosomes, which determine the sex of the individual. In humans, a female normally has two X chromosomes, whereas a male has one X and one Y chromosome (in this case, the X and Y chromosomes are not truly homologous, as they contain different genes). A full set of the 23 chromosome pairs can be arranged and displayed in what is called a **karyotype** (KEHR-ee-oh-ty'p). To produce a karyotype, cells are chemically induced to divide, and then arrested in metaphase. A computer then arranges the chromosomes into pairs according to size and structure (see Figure 3.27).

The presence of abnormalities in the genome results in a **genetic disorder.** There are many possible abnormalities in an individual's DNA, ranging from a single incorrect nucleotide in a gene to the addition or subtraction of an entire chromosome or set of chromosomes. Let's now take a closer look at genes and examine how the genetic makeup, called the *genotype,* relates to how it manifests physically as a trait, which is known as the *phenotype.*

Genes and Alleles

Recall that homologous chromosomes have pairs of genes that code for the same trait. These genes are found at the same location, or **locus,** of that chromosome (see Chapter 26). However, these genes are not necessarily identical—they may have slightly different sequences of nucleotides. These variants of a gene are called **alleles** (uh-LEELZ). For example, both maternal and paternal chromosomes have genes that code for eye color. However, one gene may have an allele that codes for brown eyes, whereas the other gene may have an allele coding for blue eyes. If two alleles code for the same trait (such as blue eyes), they are said to be **homozygous** (HOH-moh-zy′-gus; *homo-* = "same") for that trait. If the two alleles are different (one codes for blue eyes, whereas the other codes for brown), they are **heterozygous** (HET-er-oh-zy′-gus; *hetero-* = "different").

Some alleles mask or suppress the expression of another allele for the same gene. An allele that has the ability to mask another is called **dominant,** and the allele that is masked is known as **recessive.** Alleles are generally represented with letters: Dominant alleles are shown as capital letters and recessive alleles as lowercase letters. Let's use the example of freckles to further explain these concepts. Freckles are the result of a dominant trait, which we represent with an "*F.*" A lack of freckles is the result of a recessive trait, so we represent it with an "*f.*" Individuals with the genetic makeup *FF* are said to be homozygous dominant, those with *Ff* are heterozygous, and those with *ff* are homozygous recessive. Those who are homozygous dominant (*FF*) will have freckles. In addition, those who are heterozygous (*Ff*) will also have freckles, because the dominant allele masks the recessive one. Only individuals who are homozygous recessive, or *ff,* will have no freckles.

Note that not all alleles are dominant or recessive, and many factors often influence the inheritance of a trait. Also, most genes have more than two different alleles in the population (such as multiple choices for eye color), but each individual receives only two alleles—one from each parent.

Genotype versus Phenotype

An individual's genetic makeup is called the **genotype** (JEEN-oh-ty′p). The physical expression of an individual's genotype in the form of a trait is referred to as the **phenotype** (FEE-noh-ty′p). In the freckles example just described, *FF* and *Ff* are different genotypes, but they produce the same phenotype: the appearance of freckles. The genotype *ff* would yield the phenotype of no freckles. Although a person with a heterozygous genotype displays only the dominant phenotype, the recessive allele remains in his or her genotype. If this individual reproduces, he or she can pass on the recessive trait, which may appear as the phenotype in the offspring or other future generations.

Quick Check

☐ 1. What is the genome? How many chromosomes are in the human genome?
☐ 2. What is the difference between homozygous and heterozygous alleles?
☐ 3. How does a genotype differ from a phenotype?

Patterns of Inheritance

◀◀ Flashback

1. What is sickle cell anemia? (p. 731)
2. What are the ABO blood groups? (p. 744)

3. Which photoreceptors produce high-acuity color vision? (p. 530)

Some human phenotypes can be linked to a single gene pair, but most inherited traits are determined by multiple alleles or by the interaction of several genes. Environmental influences may also affect gene expression and ultimately a person's phenotype. The upcoming sections look at some of the different patterns of inheritance, and the resulting genotypes and phenotypes.

Dominant and Recessive Traits

The pattern of inheritance known as **autosomal dominant-recessive inheritance** refers to the interaction of alleles that are strictly dominant or recessive. We already discussed this type of inheritance with the example of freckles in the previous section. The dominant allele (*F*) is always expressed in the phenotype, whether the individual is homozygous (*FF*) or heterozygous (*Ff*). Recessive alleles are expressed only if two of them are present (*ff*), because there is no dominant allele to mask the recessive one.

Other examples of autosomal dominant traits include dimples, astigmatism (irregularly shaped cornea or lens of the eye that causes blurred vision), and free (unattached) earlobes. The corresponding recessive traits are lack of dimples, nondistorted vision, and attached earlobes.

A **Punnett square** is a simple diagram used to predict the possible offspring genotypes that might occur if a man and woman have a child. **Figure 27.17** shows a Punnett square for the possible genotypes of offspring from a man and woman who are both heterozygous for dimples, or *Dd.* Remember that when sperm cells and oocytes form during meiosis, each receives only one allele. For this reason, one oocyte might have the dominant "*D*" allele, whereas the other has the recessive "*d*" allele, and the same options exist for the sperm cell. We can see in the figure that when both parents are heterozygous for dimples, the probabilities of offspring with each possible genotype are the following: *DD* (1/4, or 25%), *Dd* (1/2, or 50%), and *dd* (1/4, or 25%).

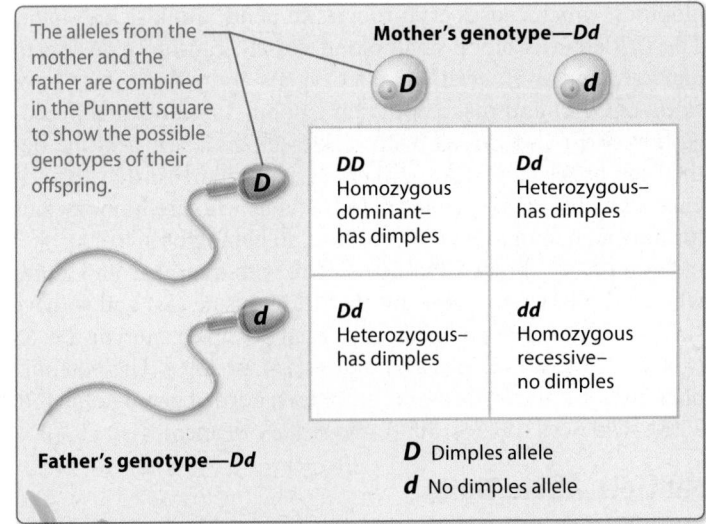

Figure 27.17 Possible offspring with a dominant-recessive trait: dimples.

Mother's genotype—Ss

	S	**s**
S	**SS** Normal hemoglobin	**Ss** Sickle-cell trait
s	**Ss** Sickle-cell trait	**ss** Sickle-cell anemia

Father's genotype—Ss

S Normal hemoglobin allele
s Sickling allele

Figure 27.18 Possible offspring with an incomplete dominant trait: sickle-cell anemia.

The phenotypes of these offspring are as follows: dimples (*DD* or *Dd*, totaling 3/4, or 75%), and no dimples (*dd*, 1/4, or 25%).

Each time fertilization occurs, a new sperm and oocyte are involved, so the probabilities remain the same for each child. In other words, if the first child has no dimples, this does not mean the second child will have dimples. Each child has a 25% chance of lacking dimples.

Traits with Incomplete Dominance

Relatively few human traits follow a strict dominant-recessive inheritance pattern. In the pattern known as **incomplete dominance,** the heterozygous genotype results in a phenotype that is "between" the phenotypes of individuals who are homozygous dominant and homozygous recessive.

An example of incomplete dominance in humans involves the sickling gene that causes sickle-cell anemia. Individuals who are homozygous for the recessive sickling gene (*ss*) have sickle-cell anemia and will produce an abnormal hemoglobin, called hemoglobin S, which causes erythrocytes to bend into a sickle shape. The sickled cells block small blood vessels, causing pain, organ damage, and even death. Heterozygous individuals (*Ss*) have *sickle-cell trait* and make both hemoglobin S and normal hemoglobin. Most individuals with sickle-cell trait are healthy, but they can experience sickle-cell crises, particularly if they experience a low blood oxygen level. Individuals who are homozygous for normal hemoglobin (*SS*) produce no hemoglobin S.

Figure 27.18 shows a Punnett square with a mother and father who are both heterozygous for the sickling gene (*Ss*) and so have sickle-cell trait. The probabilities of each possible genotype are: *SS* (1/4, or 25%), *Ss* (1/2, or 50%), and *ss* (1/4, or 25%). The potential phenotypes of the offspring are as follows: normal hemoglobin (*SS;* 25%), sickle-cell trait (*Ss;* 50%), and sickle-cell anemia (*ss;* 25%).

Multiple-Allele Traits

Many human traits involve the interaction of multiple alleles. A good example of one such trait is the ABO blood group. Recall

from the blood chapter that your blood type is determined by the presence or absence of the A and B antigens on erythrocytes (see Chapter 19). The allele for antigen A is represented as I^A, while the allele for antigen B is I^B. The allele that codes for a lack of either antigen (resulting in blood type O) is represented with an *i*.

ABO blood types exhibit the phenomenon of **codominance,** meaning that some alleles are equally dominant and are therefore equally expressed. I^A and I^B are codominant alleles, while *i* is recessive. Possible genotypes and phenotypes of the ABO blood group are shown here:

Genotype	Phenotype
$I^A I^A$	Blood type A
$I^A i$	Blood type A
$I^B I^B$	Blood type B
$I^B i$	Blood type B
$I^A I^B$	Blood type AB
ii	Blood type O

Figure 27.19a shows a Punnett square with a mother who has type A blood and is heterozygous for the trait ($I^A i$), and a father who has type B blood and is also heterozygous for the trait ($I^B i$). There are four possible genotypes for their offspring, with equal probabilities: $I^A I^B$ (1/4, or 25%), $I^A i$ (1/4, or 25%), $I^B i$ (1/4, or 25%), and *ii* (1/4, or 25%). The phenotypes of these individuals would be AB, A, B, and O, respectively.

Sex-Linked Traits

Sex-linked traits are those that are specifically expressed on the X chromosome or Y chromosome. The Y chromosome contains genes that determine male phenotypic traits, which explains why females have the genotype *XX* and males have the genotype *XY*. All oocytes carry an X chromosome, so female gametes do not determine the sex of an offspring. Sperm cells, however, carry either an X chromosome or a Y chromosome, so the male gamete determines the offspring's sex.

As you can see from this scanning electron micrograph, the X chromosome is significantly larger than the Y chromosome:

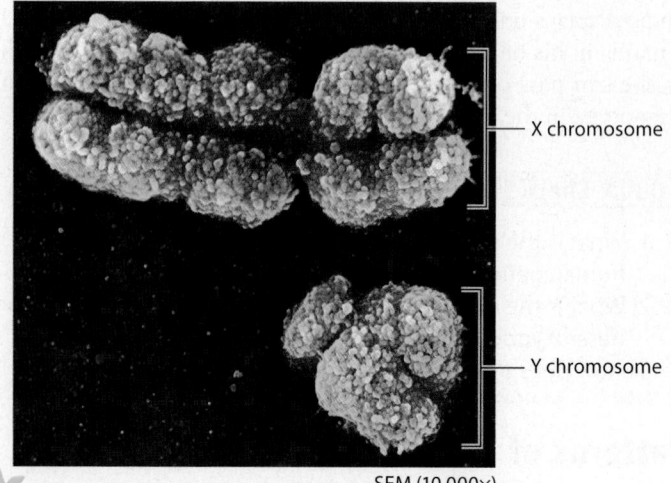

X chromosome

Y chromosome

SEM (10,000×)

For this reason, the X chromosome holds approximately five times more genes than the Y chromosome, including genes that

Mother's genotype—$I^A i$

	I^A	i
I^B	$I^A I^B$ Type AB	$I^B i$ Type B
i	$I^A i$ Type A	ii Type O

Father's genotype—$I^B i$

I^A Antigen A allele
I^B Antigen B allele
i Neither antigen A nor antigen B alleles

If the parents are heterozygous type A ($I^A i$) and type B ($I^B i$) for the codominant blood type trait, any child of theirs has an equal chance (¼, or 25%) of having blood type AB, type A, type B, or type O.

(a) Possible offspring with a multiple-allele trait: blood type

Mother's genotype—$X^R X^r$

	X^R	X^r
X^R	$X^R X^R$ Normal vision female	$X^R X^r$ Normal vision female (carrier)
Y	$X^R Y$ Normal vision male	$X^r Y$ Color-blind male

Father's genotype—$X^R Y$

X^R Normal red-green vision allele (on X chromosome)
X^r Red-green color blindness allele (on X chromosome)
Y No corresponding allele (on Y chromosome)

(b) Possible offspring with X-linked trait: red-green color blindness

Figure 27.19 Possible offspring with multiple-allele traits and X-linked traits.

code for traits other than those that make an individual a female. Unfortunately, there can be alleles for genes on the X chromosome that encode harmful traits, producing what are called **sex-linked,** or **X-linked disorders.** (*Y-linked disorders* do occur, but they are extremely rare.)

If a male inherits a trait for an X-linked disorder, he will likely express the trait because the Y chromosome does not have a corresponding allele to influence its expression. However, a female will express the trait for an X-linked disorder only if

she is homozygous for the recessive allele. A female who is heterozygous for a recessive allele does not show the phenotype of the trait, but she is a *carrier* of that allele. This means that she can pass the allele on to her offspring. If she passes this on to a female child, her daughter will also be a carrier. If she passes it on to a male child, her son will express the trait. Note that a recessive allele is never passed from a father to a son because the son receives only the father's Y chromosome. A father can, however, pass on a recessive allele to a daughter, because he contributes one of her X chromosomes.

Examples of X-linked disorders include Duchenne muscular dystrophy (a muscle-destroying disease; see Chapter 10), certain forms of baldness, color blindness, and hemophilia (a disorder of blood clotting; see Chapter 19). Let's look further into what happens with color blindness. In red-green color blindness, the recessive allele results in defective cones that do not allow the affected individual to distinguish between red and green. **Figure 27.19b** shows a Punnett square with a mother who is heterozygous for red-green color blindness ($X^R X^r$), and a father with one allele for normal red-green vision ($X^R Y$). The four possible genotypes of offspring are as follows, with equal probabilities: $X^R X^R$ (1/4, or 25%), $X^R X^r$ (1/4, or 25%), $X^R Y$ (1/4, or 25%), and $X^r Y$ (1/4, or 25%). The phenotypes of these individuals are, respectively: female with normal vision, female carrier with normal vision, male with normal vision, male with red-green color blindness. So while a female child of these parents has no risk of red-green color blindness, a male child has a 50% chance of being red-green color blind.

Polygenic Inheritance Patterns

Most inherited traits are controlled by the combined effects of two or more genes, a condition called **polygenic inheritance** (pol-ee-JEN-ik). Examples of traits with polygenic inheritance include height, skin color, and eye color. **Figure 27.20** shows a simplified example of polygenic inheritance in which two genes regulate height. The first gene exhibits three genotypes: *TT* is very tall, *Tt* is moderately tall, and *tt* is short. The second gene also exhibits three genotypes: *AA* reduces height significantly, *Aa* reduces height slightly, and *aa* has no effect on height. Notice that this Punnett square is somewhat more complicated than our previous examples—it crosses two traits at the same time, and so has 16 possible outcomes for the offspring.

When the combined effects of many genes are further influenced by environmental factors, this is referred to as **multifactorial inheritance.** In multifactorial inheritance, one genotype can have many possible phenotypes, depending on the environment, or one phenotype can result from many possible genotypes. For example, even though a person inherits several genes for tallness, the individual may not reach full height due to environmental factors such as disease or malnutrition during childhood.

Thanks to modern technology, it is now possible to test for genetic traits in a fetus during prenatal development. Some of these techniques are discussed in *A&P in the Real World: Prenatal and Newborn Genetic Screening* on page 1088.

Figure 27.20 **Possible offspring with polygenic inheritance: height.**

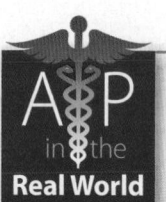
A P in the **Real World**

Prenatal and Newborn Genetic Screening

Cells and amniotic fluid may be withdrawn and analyzed to test for chromosomal abnormalities. Many genetic disorders can be detected by such procedures, including Down syndrome, sickle-cell anemia, and hemophilia. These tests are usually recommended for women who will be 35 years or older at delivery, as their oocytes are older, which increases the risk of chromosomal abnormalities. They are also recommended for women who know they or the father are carriers of inherited diseases, or when possible fetal abnormalities are discovered on ultrasound.

In **amniocentesis** (am′-nee-oh-sen-TEE-sis), which is done between 14 and 20 weeks of pregnancy, amniotic fluid is withdrawn using a needle inserted into the amniotic cavity, as shown to the right.

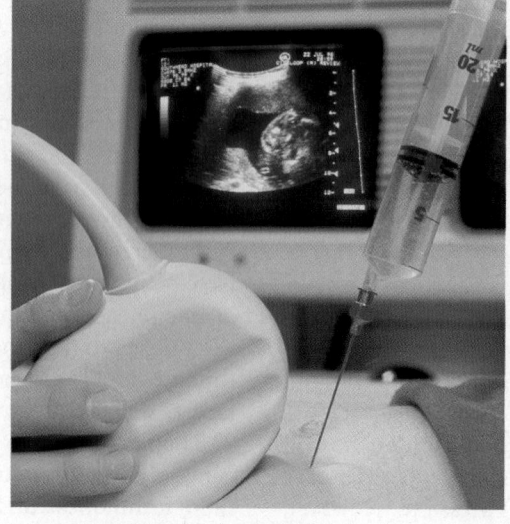

Similarly, **chorionic villi sampling,** which is usually performed between 10 and 12 weeks of pregnancy, withdraws chorionic villi tissue using a suction catheter guided through the vagina and into the uterus. It is also possible to remove chorionic villi by inserting a needle through the abdominal cavity. Both of these procedures require an ultrasound transducer, which uses sound waves to monitor the location and position of the fetus and the movement of the needle or catheter. Although the risk of amniocentesis or chorionic villi sampling is low, there is a possibility that the test could harm the fetus.

After birth, newborn screening procedures involve the analysis of blood samples to determine if the baby has a genetic or metabolic disorder. Prompt identification of any disorders can allow for early treatment and prevent long-term complications.

Quick Check

☐ 4. If a trait is autosomal dominant, how many alleles are necessary for the trait to be seen in the phenotype?

☐ 5. What is codominance?

☐ 6. What is an X-linked disorder? How can a female be a carrier of such a disorder?

Apply What You Learned

☐ 1. Draw a Punnett square to determine if it is possible for a man and a woman who both have attached earlobes to have a child with free earlobes. (*Hint:* Free earlobes

are autosomal dominant, and the trait follows the strict dominant-recessive inheritance pattern.)

☐ 2. A couple who are expecting a baby come to you for genetic counseling. The father is red-green color blind. The mother has normal color vision, but her father had red-green color blindness. What are the chances that the baby will be color blind? Are the probabilities different for male and female offspring? If so, why?

☐ 3. Determine the possible genotypes and phenotypes of offspring from a mother with the blood type AB and a father with the blood type O.

CHAPTER SUMMARY

Take full advantage of the invaluable online practice and study tools for this chapter by going to MasteringA&P®. There you will find:

- Practice tests, module quizzes, art-labeling and sequence-the-steps activities
- Animated figures with quizzes
- *Practicing A&P* worksheets
- Author-narrated podcast
- Dynamic Study Modules
- iP *Interactive Physiology*

MODULE 27.1
Overview of Human Development 1057

- **Developmental biology** is the study of changes in form and function from the fertilized egg through old age, and **embryology** is the study of the period of development inside the mother's uterus.
- A pregnancy consists of three **trimesters,** each lasting 3 months. The **pre-embryonic period** includes the first 2 weeks, the **embryonic period** includes weeks 3–8, and the **fetal period** includes weeks 9–38, or until birth. (Figure 27.1, p. 1058)
- The **postnatal period** includes the changes from newborn to toddler, childhood, and adolescence into adulthood.

MODULE 27.2
Pre-embryonic Period: Fertilization through Implantation (Weeks 1 and 2) 1059

- **Fertilization** is the fusion of a **sperm cell** and **secondary oocyte** to form a single-celled zygote. Sperm must migrate to the oocyte, undergo **capacitation,** and penetrate the **corona radiata** and **zona pellucida** to successfully fertilize an oocyte. (Figure 27.2, p. 1060)
- Once the sperm cell nucleus enters the oocyte, it swells to form the **male pronucleus.** The oocyte completes meiosis and forms

the **female pronucleus,** and *amphimixis* fuses the two pronuclei to form the diploid zygote. (Figure 27.2, p. 1060)

Practice sequencing the steps: Figure 27.2.

- The zygote undergoes **cleavage** to become a **blastocyst** that implants in the endometrium. The blastocyst differentiates into the **trophoblast** and an **inner cell mass.** The trophoblast produces the hormone **human chorionic gonadotropin (hCG)** and the **syncytiotrophoblast,** which digests the uterine lining as the blastocyst implants. The inner cell mass becomes the embryo proper. (Figure 27.3, p. 1062; Figure 27.4, p. 1064)

Practice art-labeling exercise: Figure 27.3.

Practice art-labeling exercise: Figure 27.4.

- hCG stimulates the corpus luteum to secrete estrogen and progesterone and promotes placental development.
- The inner cell mass differentiates into **epiblast** and **hypoblast** layers and forms the embryo and the **extraembryonic membranes,** including the **amnion, yolk sac, allantois,** and **chorion.** (Figure 27.5, p. 1065)

MODULE 27.3
Embryonic Period: Week 3 through Week 8 1066

- At the start of the embryonic period, **cephalocaudal folding** and **lateral folding** transform the embryo to its cylindrical shape. **Gastrulation** leads to the development of the three primary germ layers. (Figure 27.6, p. 1067; Figure 27.7, p. 1068)
- **Organogenesis** involves differentiation of the germ layers into rudimentary organ systems. The basic structures formed by each germ layer include the following (Table 27.2, p. 1069):
  ○ The **ectoderm** forms the epidermis of the skin, the nervous system and the sense organs.
  ○ The **mesoderm** forms the skeletal structures, muscles, most organs (except the epithelial linings), mesenteries, and dermis.
  ○ The **endoderm** forms the linings and glands of the digestive and respiratory tracts and several endocrine glands.

MODULE **27.4**

Fetal Period: Week 9 until Birth (about Week 38) 1070

- The **placenta** forms when the syncytiotrophoblast produces **chorionic villi** that extend into the **stratum functionalis** of the endometrium (Figure 27.9, p. 1071). The placenta protects the fetus and provides oxygen and nutrients during development, and secretes estrogen and progesterone to maintain the uterine lining.

  🖑 Practice art-labeling exercise: Figure 27.9.

- The **fetal period** is a time of rapid growth and maturation of tissues and organs. Fetal length and weight increase throughout the fetal period, although weight gain is most obvious in the last 8 weeks. (Figure 27.10, p. 1072)

- The fetal cardiovascular system includes **shunts** that bypass the pulmonary circuit and the liver to expedite oxygen and nutrient delivery to the developing tissues. In the newborn, these shunts close. (Figure 27.11, p. 1073)

MODULE **27.5**

Pregnancy and Childbirth 1075

- In addition to estrogen and progesterone, several hormones maintain the pregnancy and prepare for delivery and lactation. Examples include **relaxin, placental lactogen,** and **prolactin.**

- During pregnancy, the mother's uterus eventually fills most of her abdominopelvic cavity and compresses her abdominal organs. Glandular tissue in her mammary glands grows in preparation for **lactation.** (Figure 27.13, p. 1077)

- Important physiological changes in the mother during pregnancy include the following:
  ○ Blood volume and cardiac output increase.
  ○ Weight is gained, with the average being 13 kg.
  ○ The glomerular filtration rate increases.

- Pregnancy terminates with **parturition,** or **childbirth,** when the fetus is expelled from the uterus through the vagina.

- Labor is initiated by a positive feedback mechanism that requires the secretion of **prostaglandins** and **oxytocin** to dilate the cervix and contract the uterus. Physical distention of the cervix signals the release of more oxytocin, and the process continues until birth. (Figure 27.14, p. 1079)

- The stages of labor include the **dilation, expulsion,** and **placental stages.** (Figure 27.15, p. 1080)

MODULE **27.6**

Postnatal Changes in the Newborn and Mother 1081

- The first 4 weeks after birth constitute the **neonatal period** of life for the newborn, or **neonate.** The time period after birth for the mother is called the **postpartum period.**

- At 1 minute and 5 minutes after birth, an **Apgar score** is recorded to evaluate a newborn's physical status based on five criteria: skin color, pulse rate, respiratory level, muscle activity, and grimace. (Table 27.3, p. 1081)

- The newborn's first breath changes the blood pressure in the pulmonary circulation, closing fetal vascular shunts.

- The mother's system returns to homeostasis after birth. Several events occur, including the following:
  ○ Estrogen and progesterone levels decline.
  ○ The digestive, respiratory, and urinary system functions return to pre-pregnancy levels.
  ○ The uterus returns to its pre-pregnancy size and secretes **lochia** for up to 6 weeks.

- **Lactation** refers to the production and release of breast milk from the mammary glands.
  ○ The first few days after birth, a thick, yellowish fluid called **colostrum** is secreted. It contains protein and is rich in IgA antibodies.
  ○ A few days after birth, true breast milk is produced. A positive feedback mechanism involving mechanical stimulation of the mammary glands is responsible for continued milk production. (Figure 27.16, p. 1083)

MODULE **27.7**

Heredity 1084

- The transmission of genetic characteristics from parent to offspring through the **genes** is called **heredity; genetics** is the science that studies heredity.

- Genes are segments of DNA that code for a specific protein, and **alleles** are different variants of genes. The genetic makeup of an individual is referred to as the **genotype,** and the physical expression of an individual's genotype is the **phenotype.**

- **Autosomal dominant-recessive inheritance** refers to the interaction of alleles that are strictly **dominant** or **recessive,** which means that a dominant trait masks a recessive trait. (Figure 27.17, p. 1085)

- Few human traits actually follow a strict dominant and recessive inheritance pattern. Many involve **incomplete dominance,** or the interaction of multiple alleles. (Figure 27.18, p. 1086; Figure 27.19, p. 1087)

- **Sex-linked traits** are those that are specifically expressed on the X or Y chromosome, called the **sex chromosomes.** Most **sex-linked disorders** are X linked.

- **Polygenic inheritance** occurs when the combined effects of two or more genes produce a trait. (Figure 27.20, p. 1088)

- **Multifactorial inheritance** involves the influence of environmental factors in addition to the combined effects of two or more genes to produce a trait.

ASSESS WHAT YOU LEARNED

See answers in Appendix A.

LEVEL 1 Check Your Recall

1. Match the correct time period of gestation with the prenatal period of development:

_____ Pre-embryonic period	a. Weeks 3–8
_____ Fetal period	b. Weeks 1–2
_____ Embryonic period	c. Weeks 9–38

2. Fill in the blanks: During the pre-embryonic period, the conceptus divides into hundreds of cells by a process called _____ and becomes a/an _____, which implants in the uterine wall.

3. Mark the following statements about fertilization as true or false. If a statement is false, correct it to make a true statement.

 a. Fertilization is the fusion of a sperm and a secondary oocyte to form a new cell called a zygote.
 b. Prostaglandins in semen may assist with the movement of sperm farther into the female reproductive tract by stimulating uterine contractions.
 c. Secondary oocytes are covered with a layer of cells called the zona pellucida.
 d. The cortical reaction allows additional sperm cells to bind to the oocyte.

4. Which of the following areas in the female reproductive tract is the site of fertilization?

 a. Uterus
 b. Ampulla of the uterine tube
 c. Vagina
 d. Cervix of the uterus
 e. Any of these areas are possible sites of fertilization.

5. Number the sequence of events in fertilization.

 _____ a. Formation of male and female pronuclei
 _____ b. Acrosomal reaction
 _____ c. Rejection of sperm
 _____ d. Binding of sperm cells
 _____ e. Cortical reaction
 _____ f. Fusion of pronuclei

6. Fill in the blanks. The blastocyst has two distinct cell populations: large, flat cells called the _____, and a cluster of round cells called the _____.

7. Match the extraembryonic membrane with the correct statement:

 _____ Amnion
 _____ Yolk sac
 _____ Allantois
 _____ Chorion

 a. Forms part of the digestive tract and is the first extraembryonic membrane to form
 b. Forms the base for the umbilical cord and later becomes part of the urinary bladder
 c. Encloses the embryo and secretes fluid to fill the sac
 d. Forms outgrowths that blend with the stratum functionalis layer of the uterus

8. Fill in the blanks: _____ is the process of forming the placenta, which is a disc-shaped organ that attaches to the fetus via a/an _____.

9. Match the structures of the embryo with the primary germ layer from which each developed. Each germ layer will have more than one structure.

 _____ Endoderm
 _____ Mesoderm
 _____ Ectoderm

 a. Nervous system
 b. Linings of the digestive tract
 c. Skeletal structures
 d. Epidermis of the skin
 e. Thyroid and parathyroid glands
 f. Skeletal muscles

10. What are the two types of embryonic folding?

11. Match the fetal developments to the correct month of gestation.

 _____ Month 3
 _____ Month 4
 _____ Month 5
 _____ Month 6
 _____ Month 7
 _____ Months 8 and 9

 a. A downy hair called lanugo develops.
 b. Organs specialize and grow.
 c. Lungs begin to produce surfactant.
 d. The heartbeat can be heard with a stethoscope.
 e. Fetus turns to the vertex position.
 f. Ossification begins in most bones.

12. Match the structures found in the fetal circulation with what each fetal structure becomes after birth:

 _____ Foramen ovale
 _____ Ductus arteriosus
 _____ Ductus venosus

 a. Ligamentum venosum
 b. Fossa ovalis
 c. Ligamentum arteriosum

13. The _____ is responsible for secreting the hormone human chorionic gonadotropin (hCG).

 a. Trophoblast
 b. Ovary
 c. Endometrium
 d. Morula

14. Which of the following hormones secreted during pregnancy assists with increasing the mother's blood volume?

 a. Estrogen
 b. Progesterone
 c. Cortisol
 d. Aldosterone
 e. Parathyroid hormone

15. Number the sequence of events that are responsible for initiating labor.

 _____ a. Hypothalamus of the mother and fetus stimulate the placenta to secrete prostaglandins.
 _____ b. Fetal cortisol levels increase and stimulate the placenta to release estrogen.
 _____ c. Prostaglandins and oxytocin stimulate the uterus to contract.
 _____ d. Estrogen stimulates the myometrium to form oxytocin receptors.

16. Mark the following statements about the stages of labor as true or false. If a statement is false, correct it to make a true statement.

 a. The dilation stage is the time from the onset of labor until the cervix is fully expanded.
 b. The placental stage is the time after delivery of the infant when the placenta and the attached fetal extraembryonic membranes are delivered.
 c. The dilation stage of labor is usually longer than the expulsion or placental stages.

17. Fill in the blanks: The first _____ (time frame) after birth constitute(s) the _____ period of life for the newborn.

18. Number the sequence of events that initiate lactation.

    _____ a. Estrogen and progesterone levels decrease when the placenta is expelled and prolactin stimulates milk production.

    _____ b. In the last month of pregnancy, placental estrogen, progesterone, and placental lactogen stimulate the production of prolactin.

    _____ c. Estrogen and prolactin stimulate acini in the mammary gland to grow and lactiferous ducts to branch.

    _____ d. Oxytocin binds to myoepithelial cells in the mammary glands and causes them to contract and eject milk.

19. Fill in the blanks: _____ are segments of DNA that code for a specific protein, and the human _____ includes over 20,000 of these segments.

20. Which pattern of inheritance includes examples of codominance?

    a. Autosomal dominant and recessive traits
    b. Multiple-allele traits
    c. Sex-linked traits
    d. Polygenic inheritance

LEVEL 2 Check Your Understanding

1. The prenatal period typically lasts 38 weeks. From the clinical viewpoint, a pregnancy typically lasts 40 weeks. What is the reason for this difference?

2. Why is the single-celled zygote approximately the same size as the multicellular blastocyst before it implants in the uterine wall?

3. Why do most spontaneous abortions occur in the first trimester?

4. Explain why many women frequently feel the urge to urinate during their last trimester.

LEVEL 3 Apply Your Knowledge

PART A: Application and Analysis

1. Ms. Noble is in the first trimester of her pregnancy when she suddenly feels severe pain in her lower abdomen. She rushes to the hospital where it is discovered that she has an ectopic pregnancy. The physician wants to do immediate surgery, but Ms. Noble argues that the embryo could simply move to the uterus if they wait. What would you tell this patient?

2. A new allergy medicine has just been approved by the Food and Drug Administration (FDA); however, it is not approved for pregnant women. What might be the reason for this exclusion?

3. Determine the probability of a couple having a child with a widow's peak if the conditions are as follows: The gene for widow's peak follows strict autosomal dominant-recessive inheritance. Widow's peak (W) is dominant, and no widow's peak (w) is recessive. The genotype of the mother is Ww, and the genotype of the father is ww.

4. Elise is a freshman college student taking Introduction to Biology. In lab, she is asked her parents' blood types to complete a Punnett square. She knows that she has type A blood and that her mother has type B, but she does not know her father's blood type. Elise's professor assures her that she can figure out her father's blood type just by knowing her own and her mother's. How can this be determined, and what is her father's blood type?

PART B: Make the Connection

5. Laura has diabetes and is 6 months pregnant. Lately, her blood pressure and blood glucose levels have been higher than normal. What factors related to the pregnancy might cause Laura's blood pressure and blood glucose to rise? (*Connects to Chapter 18*)

Appendix A
Answers to Apply What You Learned and Assess What You Learned

CHAPTER 1
Apply What You Learned

MODULE 1.1

1. and 2. Answers will be unique for each student, so none are provided here

MODULE 1.2

1. It involves the *chemical* thyroid hormone made by *cells* in the *tissues* of the *organ* the thyroid gland. This gland is part of the *endocrine system*. The entire *organism* can be affected by this condition. 2. As living organisms, a human and a rose plant share the features common to all life forms: composed of cells, metabolism, responsiveness, movement (the movement of the rose plant is internal), growth, excretion, and reproduction.

MODULE 1.3

1. The answer to this question will, of course, be different for each student. Here is one example: The mole is located in the left anterior buccal region, 1 centimeter lateral to the oral region, and 4 centimeters superior to the mental region. 2. Either a frontal section or a transverse section would allow you to view both lungs at the same time.

MODULE 1.4

1. The fragment entered the pelvic cavity, then passed into the abdominal cavity, where it likely encountered the peritoneal cavity (and the parietal and visceral peritoneal membranes). The fragment then pierced the diaphragm to enter the thoracic cavity, and finally the pleural cavity (passing through the parietal and visceral pleural membranes). 2. Inadequate serous fluid impairs an organ's ability to move within the cavity without friction. This could lead to pain as the two layers of the serous membrane rub together. 3. This condition will impair an organ's ability to expand, as the excess serous fluid will compress the organ.

MODULE 1.5

1. Such cells will likely be unable to perform most of their original functions because they will have lost the cellular structures required to perform them. 2. With a positive feedback loop, sweating would trigger more sweating, which would in turn trigger even more sweating, and the response would continue to amplify. This would lead to a lower than normal body temperature and dehydration. 3. A gradient is present any time there is more of a substance in one area than another. In this case, there are a large number of molecules of air freshener producing the scent near the outlet, and the number of molecules decreases as you move farther away from the outlet.

Assess What You Learned

Level 1

1. anatomy; physiology
2. a. False: *Visual/nonverbal* learners tend to learn best from diagrams and figures.
   b. False: You should budget your time for studying at least *one week* before an exam.
   c. True
   d. False: It is better to *get enough sleep* before an exam.
3. It is easier to navigate through familiar material; your professor prepares lectures with the assumption that you have read the assignment; your brain is best able to perceive concepts, ideas, and facts that it expects to see and are familiar.

4. e   5. d   6. b
7. a. False: Histology is the division of microscopic anatomy that studies the *tissue* level of organization.
   b. True
   c. True
   d. False: The anatomical position features the person facing forward, feet shoulder width apart, and palms facing *anteriorly*.
8. _d_ Anterior   _h_ Inferior
   _g_ Lateral   _c_ Distal
   _e_ Proximal   _b_ Medial
   _a_ Posterior   _f_ Superior
9. a   10. c   11. d   12. cranial, vertebral
13. thoracic, abdominopelvic   14. b, c, f   15. b   16. c
17. a. False: Structure and function are closely related at *all levels of organization*.
    b. True
    c. False: *Negative* feedback loops are triggered by a deviation from a homeostatic set point and are shut down when conditions return to the set point.
    d. False: The effects of *positive* feedback loops are amplified to create an escalating response.
    e. True
18. d

Level 2

1. The skull is hard and able to withstand tension (pulling) and compression (pushing) forces, so it is well suited to providing protection to underlying structures. The skull also has a number of cavities for housing delicate structures, and a number of prominences (bulges) to which muscles can attach.
2. a. The esophagus is located in the mediastinum of the thoracic cavity, approximately along the body's midline. It begins inferior to the oral cavity in the cervical region, and ends at the stomach in the epigastric region. It is located posterior to the trachea and great blood vessels and anterior to the spinal column.
   b. The brain is located within the cranial cavity, deep to the skull. It is superior to the spinal cord, and extends from the frontal region to the occipital region.
   c. The urinary bladder in a female is located in the inferior pelvic cavity, inferior to the uterus, and approximately along the midline. It is located deep to the skin, muscle, and bone of the pelvis.

Level 3

1. The organs that could be involved in causing the pain include parts of the large intestine, the liver, the gallbladder, the right kidney, and portions of the small intestine. The cause of the pain could also involve the peritoneal cavity.
2. The surgeon will first have to cut through the skin and muscles of the anterior abdomen. Next, the surgeon will enter the peritoneal cavity by cutting through the parietal peritoneum. Finally, the pancreas is deep to the stomach, so the surgeon will have to move the stomach out of the way in order to reach it.
3. The incisions will be unlikely to pass through the parietal peritoneum or enter the peritoneal cavity because the kidneys are retroperitoneal—behind the peritoneum. If the incision were made on the anterior lumbar region, then both sides of the parietal peritoneum would need to be incised to reach the kidneys. First, the parietal peritoneum on the anterior body wall would be cut, then the parietal peritoneum on the posterior body wall would be cut.

4. A negative feedback loop would restore the blood pressure to normal levels and then shut off; a positive feedback loop would restore the blood pressure to normal levels and then cause it to continue to drop below its set point. The baroreceptor reflex is therefore likely to be a negative feedback loop, as a positive feedback loop would cause another homeostatic imbalance: low blood pressure.

CHAPTER 2
Apply What You Learned

MODULE 2.1

1. One sample is likely a different isotope of carbon, meaning that it has a different number of neutrons than the other sample. The added neutrons gave one of the samples a greater mass. **2.** Lithium has three protons, indicated by its atomic number, and three electrons. Its mass number of 7 tells us that it has four neutrons.

MODULE 2.2

1. Fluorine is much more reactive than neon. This is due to the valence electrons—fluorine has seven in its outer shell, and neon has eight. Therefore, neon obeys the octet rule and is relatively stable, and fluorine does not obey the octet rule and is reactive. **2.** No, it would not. Hydrogen bonds only form between two molecules with partial charges. A molecule of H_2 has a nonpolar covalent bond in which electrons are shared equally and neither hydrogen atom has a partial charge. **3.** Sodium atoms have a single valence electron, and so each of these atoms has only one electron to share. If two of them were to form a nonpolar covalent bond, each atom would still have only two valence electrons and would not obey the octet rule. For this reason, the Na_2 molecule would be very unstable and would degrade very quickly.

MODULE 2.3

1. You feel hot during and after exercise due to the production of heat, which is a byproduct of the conversion of chemical energy in the cells to mechanical energy of movement. **2.** An unstable molecule is one that tends to react with other unstable molecules in order to form stable ones. If most of our biological molecules were unstable, they would not maintain their structure or function in the body. **3.** Without enzymes, reactions in our cells would proceed at an extremely slow pace—too slow a pace to maintain life.

MODULE 2.4

1. The water around a fetus keeps it buoyant within the uterus and cushions it from mechanical trauma. The water also helps to maintain a stable temperature around the fetus. **2.** Heat produced by the body is normally absorbed by the surrounding water, which changes very little in temperature. When the amount of water is depleted, the heat is absorbed less efficiently, which can lead to an overall increase in body temperature.

MODULE 2.5

1. Glucose would be the most soluble. A triglyceride would not be soluble at all due to its nonpolar structure. A large protein might have polar or even ionic bonds, but is likely too large to be fully soluble in water. **2.** The body's reactions are catalyzed by enzymes, most of which are proteins. At high temperatures, enzymes lose their structure or are denatured. A denatured enzyme is no longer able to function, and so the rates of the body's reactions would fall precipitously. **3.** Most enzymes are proteins. The structure of every protein in the body is coded for by the DNA. If the code of DNA for an enzyme is changed via a gene defect, then the structure of the enzyme will change. An enzyme's structure is very specific for its function, so the change in structure could render it nonfunctional.

Assess What You Learned
Level 1

1. a. True
   b. False: Electrons travel in *general areas called electron shells* around the nucleus.
   c. False: Protons have a positive charge, *neutrons have no charge,* and electrons have a negative charge.
   d. True
   e. False: Every element has a characteristic number of protons, which is called the element's *atomic* number.

2. atomic number; mass number   **3.** d

4. Electrons are transferred from a metal to a nonmetal in an ionic bond. In a covalent bond, however, electrons are shared between two nonmetals.

5. a. polar covalent   d. nonpolar covalent
   b. ionic   e. ionic
   c. nonpolar covalent   f. polar covalent

6. Hydrogen bonds are weak attractions between partially negative atoms and partially positive hydrogen atoms in one or more molecules with polar covalent bonds.

7. During a chemical reaction, electrons may be transferred from one atom to another, or chemical bonds may be broken, rearranged, and/or formed.

8. __c__ Endergonic reaction   __h__ Oxidation-reduction reaction
   __f__ Potential energy   __b__ Chemical energy
   __e__ Electrical energy   __d__ Catabolic reaction
   __g__ Anabolic reaction   __a__ Kinetic energy

9. b   **10.** c   **11.** b   **12.** hydrophobic; hydrophilic

13. a. True
   b. False: *Acids* are defined as hydrogen ion donors.
   c. False: As a solution's pH rises, it becomes more *basic.*
   d. True
   e. False: A solution with a pH of 10 has *fewer* hydrogen ions than a solution with a pH of 2.

14. d

15. Triglycerides consist of three fatty acids bound to glycerol. They are highly nonpolar molecules. Phospholipids consist of two fatty acids and a phosphate group bound to glycerol. The two fatty acids are highly nonpolar, but the phosphate group is highly polar, making the overall molecule amphiphilic.

16. a. True
   b. False: The main storage form of glucose in the human body is *glycogen.*
   c. False: Lipids contain *hydrocarbon chains* and are therefore *nonpolar* covalent molecules.
   d. True
   e. False: Proteins are composed of strings of *amino* acids linked by *peptide* bonds.
   f. True

17. The structure of a protein is closely related to its function. If a protein loses its structure, which is known as denaturation, then the protein is unable to function.

18. a

19. a. RNA   d. DNA
   b. DNA   e. DNA
   c. RNA   f. RNA

20. d

Level 2

1. The product of the radioactive decay will not be the same element. An element is identified by its atomic number, or number of protons. If the number of protons changes, then the atomic number and the identity of the element also change.

2. Lipids are nonpolar and therefore hydrophobic, which causes them to clump together in the water-based environment of the digestive tract. Carbohydrates and proteins are polar, hydrophilic molecules, and so they mix and interact with the contents of the digestive tract more easily.

3. The difference in polarity is due to the ratio of oxygen atoms to carbon and hydrogen atoms in the two molecules. Monosaccharides have proportionally more oxygen atoms, and they are arranged in such a way that the overall molecule is polar covalent. Fatty acids have fewer oxygen molecules and long hydrocarbon chains that make the overall molecule nonpolar.

Level 3

1. Cellulose is composed of glucose molecules linked by glycosidic bonds. If humans were to break down the cellulose, it would lead to the release of glucose molecules that would be absorbed into the bloodstream. This would actually increase the calories consumed, and not lead to weight loss.

2. Extremely acidic foods could transiently affect the pH of the blood, but the extra hydrogen ions would simply bind the conjugate bases of buffers, which would minimize the pH change. The overall effect of the acidic foods would be negligible.

3. Enzymes are necessary for essentially all of our chemical reactions, including the production of ATP and synthesis of proteins. With these processes inhibited, the cell ceases functioning and dies.

4. The phospholipids will line up so that the polar phosphate heads face the polar water and the nonpolar fatty acid tails face the nonpolar air. This allows the molecules to follow the "like interacts with like" rule.

5. When the pH of a solution increases, a buffer releases hydrogen ions to decrease the solution's pH. When the pH of the solution returns to its normal value, the buffer stops releasing hydrogen ions. This is an example of a negative feedback loop. If this were a positive feedback loop, the buffer would release progressively more hydrogen ions as the pH decreased.

CHAPTER 3
Apply What You Learned

MODULE **3.1**

1. Bacteria must have different mechanisms for modifying proteins since they lack much of the cellular machinery for protein modification (Golgi and ER). They may also be limited in the amount of ATP they can produce since they lack mitochondria, and they have to carry out their cellular reactions in the cytosol since they lack membrane-bounded organelles. 2. Cells are specialized so that their structure enables them to carry out their functions. If a cell loses its specialized structure and becomes more generalized, it will also lose its ability to carry out these specific functions.

MODULE **3.2**

1. The nonpolar parts of the detergent would interact with the fatty acid tails of the phospholipids, and the polar parts of the detergent would interact with the phosphate heads. These interactions would pull the phospholipid bilayer apart and rupture the cell. 2. Structural proteins of the plasma membrane help to maintain the membrane's stability, and when this is disrupted, the plasma membrane in essence falls apart. When the plasma membrane falls apart, the cell dies.

MODULE **3.3**

1. a. The hypotonic solution will cause an individual's red blood cells to swell, as water will move into the more concentrated cytosol by osmosis.

b. Under normal conditions, this solution will cause no net change in cell volume by osmosis.

c. In a hypertonic solution, the red blood cells will crenate because water will move out of the cells into the more concentrated extracellular solution by osmosis.

2. The energy for secondary active transport relies on concentration gradients created by primary active transport pumps, especially the Na^+ gradient produced by the Na^+/K^+ pump. When this pump is disrupted, the cell is not able to establish this gradient, preventing normal secondary active transport.

MODULE **3.4**

1. It would block the production of the bulk of the cell's ATP, resulting in insufficient ATP to fuel the cell's many energy-requiring processes. 2. Phagocytes are cells that ingest large particles, which are then broken down inside the cells by products released from organelles such as lysosomes. Defective lysosomal enzymes would inhibit the phagocytes' ability to adequately digest these ingested particles, particularly bacteria. 3. This poison inhibits protein synthesis. Proteins in the cell may act as enzymes, structural molecules, transport molecules, and more. Without the ongoing synthesis of these proteins, a cell will rapidly die.

MODULE **3.5**

1. Microtubules are assembled and disassembled as needed by a cell—when a microtubule is no longer needed, it is disassembled and then a new one is reassembled to suit a different need. If microtubules could not be assembled, then essentially all cellular functions and elements that require them would be hindered. The cell's shape would become abnormal; organelles would not be held in position; the endoplasmic reticulum would collapse; and the chromosomes would not separate into two sets during mitosis. 2. Both cilia and flagella have a core of microtubules that are critical to their structure and their movement. Blocking microtubule assembly would inhibit their motility and therefore their functions. 3. Intermediate filaments consist of protein filaments wrapped around one another. This creates a ropelike filament that is quite strong and much harder to disassemble than either microtubules or actin filaments, both of which are assembled out of globular proteins.

MODULE **3.6**

1. Ribosomes are made in the nucleolus, and they include rRNA encoded by genes in that structure. The nucleolus is a component of the nucleus. Anucleate cells would therefore be unable to produce new ribosomes. 2. The DNA in a cell is packaged into long chromatin threads, which consists of the DNA molecule itself wrapped around various proteins such as histones. To isolate the DNA, the proteins must first be degraded. 3. Cells that are very active in protein synthesis need more ribosomes and larger nucleoli due to the prominent role ribosomes play in this process.

MODULE **3.7**

1. The TGA triplet corresponds to the ACU codon, which codes for the amino acid threonine. The mutated triplet corresponds to the ACG codon, which also codes for the amino acid threonine. Therefore, this particular mutation will not change the structure of the protein. The second mutation corresponds to the codons CGC mutating to CCC. This will result in the change of amino acid from arginine to proline, which will change the protein's structure. 2. In inhibiting ribosomes, the poison ricin specifically inhibits translation. Protein transcription will not be impacted, but the ribosomes will be unable to bind the mRNA transcript and begin translation.

MODULE **3.8**

1. The process of cytokinesis divides the cytoplasm and other organelles between the two daughter cells. If cytokinesis did not take place, the result would be a huge cell with several times the normal cytoplasmic volume. 2. Periwinkle will inhibit mitosis by preventing the formation of the mitotic spindle from microtubules and so preventing the

separation of chromosomes. Yew will inhibit it by preventing the disassembly of the spindle, which will block the cell's ability to re-form the nuclear membrane and chromatin.

Assess What You Learned

Level 1

1. d   2. plasma membrane; cytoplasm; nucleus

3. The cytosol (intracellular fluid) is the fluid compartment in the cells, and the extracellular fluid is the fluid outside the cells. The two are kept separate by the cells' plasma membranes.

4. b

5. a. True
   b. False: Cholesterol provides the plasma membrane with stability in the face of changing *temperatures.*
   c. False: Membrane *glycoproteins and glycolipids* are vitally important for cell-cell recognition.
   d. True
   e. False: The overall structure of the plasma membrane is a mosaic with the components *moving fluidly within the membrane.*

6. Passive transport processes require no expenditure of energy by the cell and are driven by the concentration gradient. Active transport processes require energy expenditure in the form of ATP.

7. d Osmosis   c Simple diffusion
   h Secondary active transport b Primary active transport
   g Exocytosis   f Pinocytosis
   a Phagocytosis   e Facilitated diffusion

8. into, swell; out of, crenate

9. f Peroxisome   b Golgi apparatus
   d Ribosome   c Lysosome
   h Smooth endoplasmic   a Rough endoplasmic reticulum
     reticulum
   e Mitochondrion   g Vesicle

10. The smooth and rough endoplasmic reticulum completes the synthesis of lipids and proteins. They are then sent by vesicles to the Golgi apparatus, which modifies and sorts the products, and then packages them into another vesicle. The vesicles are sent to lysosomes, other organelles, or the plasma membrane.

11. a. False: Actin filaments combine with myosin motor proteins to provide the cell with *movement.*
    b. False: *The flagella* found on sperm cells propel the cells through a liquid medium.
    c. True
    d. False: Intermediate filaments are ropelike structures composed of *fibrous* proteins.
    e. True
    f. False: Endocytosis, exocytosis, muscle cell contraction, and cellular "crawling" are all mediated by *actin* filaments.

12. d

13. Chromatin is condensed during cell division by being coiled repeatedly around histone proteins. This enables precise division of the chromosomes into two complete sets when the cell divides.

14. a

15. a. In a gene, each *triplet* specifies one amino acid in a protein sequence.
    b. A transcription factor must bind to the promoter region of a gene before the enzyme *RNA polymerase* is able to bind and begin transcription.
    c. The enzyme RNA polymerase builds a strand of *messenger* RNA, whose codons are complementary to DNA's triplets.
    d. Proteins destined for secretion from the cell enter the *rough ER* after translaterion, to be folded and modified.
    e. During translation, amino acids are delivered by the *tRNA molecule.*

16. a. 8   f. 2
    b. 6   g. 5
    c. 3   h. 7
    d. 1   i. 4
    e. 9

17. a

18. The cell cycle must be tightly regulated in order to balance cell division with cell death. Improperly regulated cell division may lead to a progressive deterioration of the tissue (if cells do not divide enough) or an increase in the number of cells in a tissue (if the cells divide too often).

19. a. False: During the G_0 phase of the cell cycle, the cell stalls until conditions for division are more favorable.
    b. False: During *anaphase,* the sister chromatids are pulled apart and the chromosomes move to the opposite poles of the cell.
    c. False: The main enzyme that builds the new DNA strands during DNA synthesis is *DNA polymerase.*
    d. True

20. g G_1 h Telophase
    d Metaphase   c M phase
    e S phase   b G_2
    a Cytokinesis   f Anaphase

Level 2

1. a. Diffusion is movement of solute molecules from a high concentration to a low concentration.
   b. Osmosis is movement of a solvent from a solution with a lower solute concentration to one with a higher solute concentration.
   c. Primary active transport is transport of a substance across the plasma membrane against its concentration gradient with the use of energy from ATP.
   d. Secondary active transport is transport of a substance across the plasma membrane against its concentration gradient using the energy from another substance's concentration gradient.
   e. Transcription means copying the code of a gene by creating a complementary strand of mRNA.
   f. Translation means translating the code in the mRNA into a protein with the help of tRNA and ribosomes.
   g. DNA synthesis is the replication of a cell's DNA in preparation for cell division.
   h. Mitosis is division of a cell's nuclear material between two daughter cells.

2. Cells with a very high demand for ATP will be disproportionately affected by mitochondrial diseases. Such cell types include those of muscles, the nervous system, the liver, and the kidneys.

3. a. Cilia are short, motile extensions from the cell. Their structure enables them to beat rhythmically to sweep substances past the cell.
   b. Microvilli are folds of the plasma membrane that increase the surface area for absorption, which increases the efficiency of absorption in cells for which absorption is a primary function.
   c. The ropelike structure of intermediate filaments allows them to be flexible but very strong and therefore provide the cell with structural support.
   d. Lysosomes are composed of a phospholipid bilayer surrounding digestive enzymes in an acidic environment. This structure allows them to digest particles within the cell while preventing the enzymes in acid from damaging the rest of the cell.
   e. The nuclear envelope is composed of a double phospholipid bilayer with large nuclear pores. It contains the chromatin and other nuclear structures within the nucleus while allowing fairly large molecules such as mRNA to move between the nucleoplasm and cytosol.

4. Healthy cells produce only specific proteins in specific amounts due to mechanisms that regulate gene expression. In cancer cells, these mechanisms may be lost due to changes in the DNA, causing them to produce inappropriate proteins.

5. Some cell types are subject to a high degree of wear and tear because of their location and/or function, including the cells of the skin and the gastrointestinal tract. Such cells must have a rapid rate of mitosis to replace dead or damaged cells. Cells that have slow or absent cell division include those that have complex structure and/ or do not have the cellular machinery for cell divisions. These cells include neurons and muscle cells.

Level 3

1. The fluids were likely hypertonic instead of isotonic. The hypertonic fluids caused the water to leave the patient's cells by osmosis, which led to his cells becoming dehydrated.

2. Introns are noncoding DNA, meaning that they do not code for any protein in the human body. Even if a virus were to "activate" them, it wouldn't result in the production of a protein, as they do not code for anything. In addition, introns are removed from mRNA transcripts before they are translated.

3. This poison would prevent the synthesis of tubulin proteins and therefore microtubules. This would inhibit the development of the mitotic spindle, and mitosis would be unable to proceed.

4. Microtubules are responsible for maintaining cell shape and the positions of organelles. Without microtubules, the organelles would disperse, particularly the members of the endomembrane system. In addition, vesicles and other cell structures would not be transported from place to place, disrupting vesicular transport within the cell and between the cell and the environment. Finally, if the cell has cilia or a flagellum, they would be unable to function. These conditions would result in the death of the cell.

5. Polar solutes are soluble in polar solvents, and nonpolar solutes are soluble in nonpolar solvents. Since phospholipids have both polar and nonpolar parts, the phospholipid bilayer will be soluble in a solvent with polar and nonpolar parts, such as a detergent, and thus its structure will be disrupted.

CHAPTER 4
Apply What You Learned

MODULE 4.1

1. This would cause the tissues to be unable to resume their original shape and size once the stretching forces were removed. 2. Tight junctions make the spaces between the cells relatively impermeable to water and many solutes. Nonfunctional tight junctions in blood vessels would allow much of the fluid and solutes in blood to leak out of the vessel and into the extracellular fluid. 3. Desmosomes help to increase a tissue's mechanical strength. This is particularly important in tissues subject to high amounts of stress, like the skin. If the skin's desmosomes were not working properly, the skin would not be as resistant to mechanical stresses and could break down more easily.

MODULE 4.2

1. The thinness of the epithelium of the lungs allows gases to diffuse across this barrier rapidly and efficiently. If the epithelium thickens, the distance over which the gases must diffuse will increase, making the process slower and less efficient. 2. The thickness of the stratified squamous epithelium in the skin helps to resist mechanical trauma and other stresses. As the epithelium gets progressively thinner, it is less able to resist these mechanical stresses, resulting in more skin lacerations. 3. Like the epithelium in the lung, the simple columnar epithelium in the digestive tract allows substances to diffuse across it. If the digestive tract had stratified columnar epithelium, it would be far too thick to allow efficient diffusion of water and nutrients into the bloodstream.

MODULE 4.3

1. Collagen fibers are very tough but not very elastic or extensible. Replacement of elastic fibers with collagen fibers would therefore reduce a tissue's

elasticity and extensibility and instead make it quite tough. 2. Hyaline cartilage has far more ground substance than fibrocartilage, which has far more protein fibers. This makes fibrocartilage much less smooth as a surface for articulation. 3. Osteoblasts are bone-building cells, and osteoclasts are bone-destroying cells. If osteoclasts were more active, there would be excessive destruction of bone and as a result the bone would be much weaker.

MODULE 4.4

1. a. Skeletal muscle tissue is composed of contractile muscle fibers. Dense irregular collagenous connective tissue is composed of fibroblasts, bundles of collagen fibers, and ground substance.
   b. Skeletal muscle fibers generate tension, whereas the components of dense irregular collagenous connective tissue provide an organ with strength and resistance to mechanical stresses.
   c. None of the components of dense irregular collagenous connective tissue are contractile, and so the functional abilities of the tissue would be completely altered.
2. The intercalated discs allow the cardiac muscle fibers to contract in unison, producing a uniform contraction of the heart. If they were not functional, the cardiac muscle cells would not contract in unison and the heart would not have uniform, functional contractions.

MODULE 4.5

1. Cells of brain tumors are generally neuroglial cells, as they have the capability for mitosis. 2. An axon sends messages to target cells, including other neurons, muscle cells, and glands. If the axon is damaged, a neuron can still receive messages from other neurons, but can no longer send messages to target cells.

MODULE 4.7

1. Serous fluid is thin and watery, whereas mucus is thick and sticky. Serous fluid in the respiratory tract would not be sticky enough to trap inhaled debris, and serous fluid in the digestive tract would not be thick enough to protect the epithelium from damaging substances like stomach acid. 2. Synovial fluid lubricates a joint so that two articulating bones can move smoothly against one another. A lack of synovial fluid could lead to a "dry joint" that doesn't move smoothly and causes pain on movement.

MODULE 4.8

1. Most of the damaged cardiac muscle tissue will not regenerate; instead, it will undergo fibrosis. This results in the formation of collagenous connective tissue, which lacks the contractile properties of cardiac muscle tissue. For this reason, function of the heart muscle rarely returns to normal. 2. Nutrition is a key factor in wound healing by both fibrosis and regeneration. Individuals who are homeless or alcoholic generally have poor nutritional status, and their diets often lack factors that are essential for healing a wound.

Assess What You Learned

Level 1

1. b

2. a. connective   d. nervous
   b. muscle   e. muscle
   c. epithelial   f. connective

3. a. Collagen fibers are very tough fibers that make a tissue resistant to tension and pressure.
   b. Glycosaminoglycans create a concentration gradient to draw water into the ECM by osmosis and trap it there; this makes a tissue more resistant to compression.
   c. Reticular fibers are smaller fibers that interweave to form supportive meshworks around the cells of many tissues and "webs" in certain organs to trap foreign cells.
   d. Proteoglycans are large molecules that make a tissue firmer and more resistant to compression and also act as a barrier to substances diffusing through the ECM.

e. Glycoproteins hold cells in their places in a tissue and connect cells to one another and to components of the ECM.

f. Elastic fibers have the ability to be stretched and return to their original length, and so give a tissue the properties of extensibility and elasticity.

4. impermeable; mechanical strength

5. a. True
b. True
c. False: Epithelial tissue is *avascular*.
d. True
e. False: Pseudostratified epithelium appears to be *stratified* but is actually *simple epithelium*.
f. False: *Simple* epithelia are specialized to allow substances to cross their cells rapidly.

6. _c_ Simple squamous   _f_ Simple columnar
_e_ Pseudostratified columnar   _b_ Transitional
_a_ Keratinized stratified squamous   _d_ Simple cuboidal

7. a. Endocrine glands release hormones into the bloodstream and contain no ducts. Exocrine glands release substances into a body cavity or the outside of the body through a duct.
b. A unicellular gland consists of a single cell that produces and secretes its product. A multicellular gland consists of a group of cells that function together to produce and release a product.
c. During holocrine secretion, product accumulates until the cell ruptures and releases its product. During merocrine secretion, the cell secretes its product by exocytosis.

8. Connective tissues have ECM as their main component, which allows them to serve their functions of binding and supporting. Epithelial tissue is composed of sheets of cells with little ECM, which enables it to form a barrier between the body and the external environment and between the organs and fluid-filled cavities.

9. a. False: *Adipocytes* store lipids in a large inclusion in their cytoplasm.
b. False: Loose connective tissue features *ground substance* as its primary component.
c. True
d. False: *Hyaline cartilage* provides a smooth surface on which bones may articulate with little friction.
e. False: The ECM of bone tissue consists of calcium phosphate crystals *and an organic component*.

10. _f_ Hyaline cartilage   _g_ Fibrocartilage
_d_ Elastic cartilage   _h_ Cardiac muscle
_b_ Smooth muscle   _c_ Dense regular elastic tissue
_a_ Dense irregular   _e_ Reticular connective tissue
connective tissue

11. c   12. Neurons; neuroglia; dendrite; axon

13. a. Mucous membranes are composed of *epithelium* and the underlying loose connective tissue.
b. *Goblet* cells secrete mucus.
c. Synoviocytes are the secretory cells of *synovial membranes*.
d. The cutaneous membrane is composed of *stratified* squamous epithelium, loose connective tissue, and dense irregular connective tissue.
e. Serous membranes line all *body cavities*.

14. Epithelial tissues, many connective tissues, and smooth muscle tissue often heal by regeneration because their cells retain the capacity to divide by mitosis. Nervous tissue and other types of muscle tissue generally repair by fibrosis because their cells usually are unable to undergo mitosis.

Level 2

1. You would not expect bleeding to occur, because epithelial tissue is avascular, or lacking blood vessels.

2. Collagen fibers are an important component of all tissues listed, giving them strength and the ability to resist tension. In vitamin C deficiency,

collagen synthesis is impaired, which leads to bone pain and weakness, skin fragility, tooth loss, joint pain, and myriad other symptoms.

3. Mucus protects many tissues, including the airways and the gastrointestinal tract. Inadequate mucus would allow bacteria and other irritants to have easier access to these tissues and would put the digestive tract at risk from its own acid and enzymes. Too much mucus could interfere with both breathing and digestion, processes that depend on transporting gases or nutrients across the epithelium.

Level 3

1. Desmosomes hold the epithelial cells of the skin together so that the skin functions as a unit. When the desmosomes are attacked and rendered nonfunctional, the cells will not be held to one another or to the underlying basement membrane. This leads to skin fragility and separation of the skin layers, which manifests as large blisters.

2. Each of these components of the ECM makes a tissue firm. When the ECM of all tissues has more collagen fibers, GAGs, and glycoproteins, the tissues will become firmer than normal, eventually reaching a point where they are stiff and immoveable.

3. The lungs contain elastic fibers, which enable the lungs to be both extensible and elastic. When elastic fibers are replaced with nonextensible and nonelastic collagen fibers, the lungs progressively lose their ability to stretch and then return to their original size after the stretching force is removed. This can severely limit one's ability to breathe.

4. The simple epithelia in these organs enable substances to rapidly cross the cells and move from one place to another. If the simple epithelia were replaced with keratinized stratified squamous epithelia, substances would either cross the epithelia too slowly or be unable to cross them at all.

5. a. The epithelium is likely to be simple, as the single layer of cells in simple epithelia allow substances such as ions to cross more efficiently and rapidly.
b. When sodium ions move from one compartment to another, a concentration gradient is created that causes water to follow by osmosis.

6. When a cell undergoes mitosis, it must replicate its DNA and then divide the DNA and the cytoplasm equally between the two daughter cells. For a cell such as a skeletal muscle fiber that is multinucleated, it would be extremely difficult to accurately replicate and divide the DNA from many separate nuclei into two daughter cells.

7. Cells such as fibroblasts should have large numbers of protein-making organelles, including ribosomes, rough ER, and the Golgi apparatus. High rates of protein synthesis also require a great deal of ATP, so one would expect to find large numbers of mitochondria in these cells as well.

CHAPTER 5
Apply What You Learned

MODULE 5.1

1. The cells of the epidermis receive oxygen and nutrients from blood vessels located in the dermis. If these vessels were constricted for several hours, very little oxygen and very few nutrients would reach the epidermal cells, which could damage them and lead to their subsequent death. 2. The skin would feel cold, as the heat from the blood would be directed to deeper tissues. 3. Complications experienced by burn victims include difficulty regulating body temperature and fluid homeostasis, increased susceptibility to infection and harm from the environment, and decreased or abnormal sensation due to damaged nerves.

MODULE 5.2

1. There are touch receptors in the epidermis (Merkel cells), which is why you are able to feel the cut. However, there are no blood vessels in

the epidermis, which is why the cut does not bleed. **2.** It could potentially affect these superficial cells because they would have increased access to blood vessels. In addition, the thickness of the skin might increase, as the cells in the stratum granulosum might continue dividing if they gained access to high amounts of oxygen and nutrients.

MODULE **5.3**

1. Her wound likely extends down to the hypodermis, which is where we find adipose tissue. **2.** The wound runs perpendicular to the tension lines in the anterior forearm. Wounds that run perpendicular to skin tension lines are more likely to scar, as they tend to gape.

MODULE **5.4**

1. The skin of an individual with vitiligo tends to appear mottled, with alternating patches of darker-pigmented and lighter-pigmented skin.
2. Breanna might want to tell Max that he should take an anatomy and physiology course! Melanin does naturally shield the nuclei of keratinocytes from the effects of UV rays, but only to a certain point. Excess exposure to this radiation will mutate the DNA and potentially burn the skin of anyone, irrespective of skin pigmentation. **3.** The reddish hue imparted to fair skin by hemoglobin depends on the amount of oxygen to which the hemoglobin is bound. If the blood is well oxygenated, it will be bright red and give the skin a pink hue. If the blood is poorly oxygenated, it will be much less red and could even develop a faint purplish hue.

MODULE **5.5**

1. The amount that a hair grows depends on the rate of cell division and the amount of time the cells spend in the growth phase. Increasing the rate of cell division and/or increasing the amount of time that hair is in the growth phase could increase the length of the eyelashes. **2.** It is true that hair and nails are composed largely of the protein keratin. However, protein and amino acids are needed by dividing cells in the nail and hair matrices, not by the cells in the hair shaft and nail plate, which are dead. The products could potentially affect future hair and nail growth, but they will not change the present state of hair or nails.

MODULE **5.6**

1. She has suffered a second-degree burn, which has involved all or part of the dermis. She is unlikely to develop significant scarring from the burn, but the UV radiation could have damaged her DNA. **2.** This carcinogen in all likelihood will not lead to any type of cancer unless the skin is damaged or the exposure is repeated. Because hydrophobic lipid-like substances coat the keratinocytes of the epidermis, the hydrophilic carcinogen will have difficulty penetrating the skin to the deeper layers where dividing cells are located.

Assess What You Learned

Level 1

1. An organ by definition is two or more tissues that work together to achieve common functions. The skin consists of keratinized stratified squamous epithelium, loose connective tissue, and dense irregular connective tissue. These three tissues form a barrier between the environment and the internal body and perform a host of other functions.
2. b   3. d
4. Dermal blood vessels constrict when in a cold environment, to direct heat to deeper tissues and minimize heat loss to the environment.
5. __2__ Stratum spinosum
   __5__ Stratum corneum
   __1__ Stratum basale
   __4__ Stratum lucidum
   __3__ Stratum granulosum

6. a
7. a. False: *Keratinocytes* account for the bulk of the epidermis.
   b. False: Keratinocytes begin life in the stratum *basale* and gradually are pushed into the stratum *corneum*.
   c. True
   d. False: *Melanocytes* produce the pigment melanin.
8. c
9. Dermal papillae house capillary loops that provide oxygenated blood to the epidermal cells. They also house sensory receptors that detect changes in the environment.
10. b
11. a. True
    b. False: Melanin is produced by melanocytes and covers the nuclei of neighboring *keratinocytes*.
    c. False: Carotene is a *yellow-orange* pigment that accumulates in the stratum corneum.
    d. False: *Decreased* amounts of blood flowing through the dermis lead to pallor in the skin.
    e. True
12. d   13. hair shaft, hair root, hair follicle; matrix
14. b   15. holocrine, merocrine
16. __c__ Eccrine sweat gland
    __a__ Sebaceous gland
    __d__ Ceruminous gland
    __e__ Mammary gland
    __b__ Apocrine sweat gland
17. Sweat is a water-based secretion that functions primarily in thermoregulation and is secreted via merocrine secretion. Sebum is a lipid-based secretion that provides a hydrophobic barrier and is secreted via holocrine secretion.
18. b   19. b

Level 2

1. Both hair and nails are derived from the epidermis, which is avascular. Since these structures have no blood vessels, there is no bleeding when they are damaged unless the injury involves the underlying dermis.
2. Hair shafts consist of dead keratinocytes. Dead cells do not carry out chemical reactions, and so they have no need of vitamins. For shampoos and conditioners to have any effect on the hair, the vitamins would have to reach the cells in the hair matrix, and this is unlikely to happen. The claims are therefore very likely to be invalid.
3. An organ consists of two or more tissues. The hair and nails consist only of epidermal cells, and are therefore not organs. However, if the surrounding connective tissue is considered, they could be called organs.

Level 3

1. The patient has likely suffered heat stroke. The bright red color of the skin is due to the body's attempt to release heat via dilation of the vessels in the dermis.
2. The lipid-based toxins in poison ivy easily penetrate the epidermis and reach the cells of the deep epidermis and the dermis. However, the water-soluble toxins in venom cannot pass through the lipid-based substance that coats keratinocytes in significant amounts, and so the venom does not reach the deep epidermis and the dermis.
3. a. The immediate concern is malignant melanoma, because her lesion has the characteristics of this particular cancer.
   b. Exposure to UV radiation mutates the DNA of cells in the skin. This DNA mutation is what triggers the production of more melanin by melanocytes. For this reason, any tanning, irrespective of the source, comes from the initial DNA damage and is therefore not completely safe.

4. Sebum helps to repel water from the skin, along with the lipid-based substance produced by the keratinocytes. If sebum were water-soluble, the skin would be less water-resistant and more prone to gain or lose water to the external environment.

5. a. Both chemicals are components of the ECM. Collagen is a protein fiber that gives the skin tensile strength, and hyaluronic acid is a glycosaminoglycan that traps water in the ECM and increases its firmness.
   b. Most proteins are large polar molecules.
   c. Since these molecules are large and polar, they are unlikely to penetrate the water-resistant lipid coating of the epidermis in significant amounts.
   d. The collagen and hyaluronic acid would need to reach the dermis in order to have any effect. Since they are unlikely to penetrate the epidermis in any significant amounts, they are not likely to be effective.

6. A second-degree burn involves epithelium and connective tissue, which generally heal by regeneration unless the damage is extensive. However, muscle tissue generally heals by fibrosis, so healing of a severe third-degree burn will involve some degree of fibrosis.

CHAPTER 6
Apply What You Learned

MODULE 6.1

1. If the spongy bone were on the outside, the thin trabeculae would be subjected to more mechanical trauma than usual, which would damage their structure. In addition, the bone marrow housed within the trabeculae could also sustain damage without the outer shell of compact bone to protect it. **2.** The epiphyseal plate is the structure from which a bone grows in length. If this structure is damaged, longitudinal growth could be impacted, and the bone on one side of the body could be shorter than the same bone on the other side of the body. **3.** The nutrient artery supplies a good portion of the blood flow to the bone. If this artery is damaged, portions of the bone could be cut off from the normal blood supply, which could lead to further bone injury. A bone with a damaged nutrient artery will take longer to heal because it will lack sufficient blood flow to deliver adequate oxygen, nutrients, and other components vital to healing.

MODULE 6.2

1. Bones affected by osteogenesis imperfecta are extremely brittle, as they lack the collagen fibers to give them tensile strength. This makes them much more likely to fracture when subjected to even normal physical forces. **2.** Excess osteoclast activity would mean that more ECM would be broken down than built, which would lead to progressive weakening of the bone. **3.** The function of compact bone is to provide a hard outer shell that resists all types of forces and protects the inner spongy bone and marrow. The concentric rings of bone matrix that characterize compact bone make it extremely strong and hard and enable it to perform these functions.

MODULE 6.3

1. Osteoclasts function in both types of ossification to resorb the initial primary bone so that it can be replaced with secondary bone. If osteoclasts failed to function, the weaker, less organized primary bone would remain instead of being replaced. In addition, nonfunctional osteoclasts would result in the failure of the medullary cavity to form and enlarge during endochondral ossification. **2.** Primary bone is much weaker than secondary bone because it has little inorganic matrix, which makes it much less able to resist compressive forces. In addition, the irregularly arranged collagen bundles make primary bone less resistant to tensile forces, which further weakens it.

MODULE 6.4

1. The decreased rate of mitosis would reduce the number of chondroblasts at the epiphyseal plate, and therefore slow the production of hyaline cartilage and the longitudinal growth of the bones. In addition, malnutrition would lead to an insufficient intake of calcium salts, vitamin D, vitamin C, and other nutrients needed for ossification, which could weaken the bones and further delay their growth. **2.** Growth hormone also promotes appositional bone growth, so if it is secreted in excess, the bones will grow in width, even if they can no longer grow in length. **3.** A young girl with excessive estrogen secretion is likely to have a significant increase in longitudinal bone growth, but is also likely to suffer from premature closure of the epiphyseal plates. Her height would increase in the short term, but her adult height might be less than average.

MODULE 6.5

1. Load-bearing exercise exerts compression and tension on bones, which increases osteoblast activity. Astronauts in space for a long period are not able to engage in load-bearing activities, and so their osteoblasts are less stimulated than normal, leading to reduced bone mass. **2.** A tumor that secretes parathyroid hormone would trigger excessive osteoclast activity, and the resulting breakdown of bone would lead to more frequent bone fractures. In addition, it would cause the concentration of calcium ions in the blood to be much higher than normal because the normal negative feedback loop would be ineffective.

Assess What You Learned

Level 1

1. a

2. __e__ Long bone   __c__ Short bone
   __g__ Epiphysis   __b__ Periosteum
   __a__ Nutrient artery __d__ Diaphysis
   __h__ Flat bone   __f__ Medullary cavity

3. Yellow bone marrow consists largely of adipose tissue, whereas red bone marrow contains the hematopoietic stem cells that divide and differentiate to form all formed elements of the blood.

4. a. True
   b. False: *Both the organic and inorganic matrices of bone are* responsible for the strength of bone tissue.
   c. False: Collagen fibers are one of the predominant parts of the *organic* matrix.
   d. True
   e. False: *Osteoclasts* are responsible for bone resorption, and *osteoblasts* are responsible for bone deposition.
   f. True

5. osteon; lamellae, central canal; perforating canals; lacunae, canaliculi

6. d   7. b

8. a. Endochondral   d. Intramembranous
   b. Both   e. Both
   c. Intramembranous   f. Endochondral

9. The primary ossification center is generally in the diaphysis of a long bone, whereas the secondary ossification centers are generally in the epiphyses.

10. a   11. c

12. a. __4__
    b. __1__
    c. __3__
    d. __2__

13. a. Growth hormone increases both longitudinal and appositional growth by increasing the rate of mitosis of chondrocytes in the epiphyseal plate, increasing the activity of osteogenic cells, and directly stimulating the osteoblasts in the periosteum.
    b. Testosterone increases the rate of appositional bone growth, the rate of calcium ion deposition, and the rate of mitosis at the epiphyseal plate. It also accelerates closure of the epiphyseal plate.
    c. Estrogen increases the rate of mitosis of chondrocytes at the epiphyseal plate, inhibits osteoclasts, and accelerates closure of the epiphyseal plate.

14. osteoblasts, organic matrix; inorganic matrix, vesicles; osteoclasts, hydrogen ions, enzymes

15. a. True
    b. True
    c. False: The greater the load the bone must carry, the *less* bone that is resorbed by osteoclasts.
    d. True
    e. False: Parathyroid hormone increases the blood calcium ion concentration by increasing the activity of *osteoclasts*.
    f. True

16. d

17. a. 3
    b. 1
    c. 4
    d. 2

Level 2

1. An individual who is not bearing weight regularly is not placing an adequate load on his or her bones. This leads to insufficient stimulation of bone deposition, and so bone resorption predominates. This leads to progressive loss of bone mass.

2. The skin is involved in the production of vitamin D, and the kidneys and intestines must be able to absorb calcium ions into the blood. If any of these organs are diseased and unable to perform their functions, levels of vitamin D and/or calcium ions in the blood will be inadequate to maintain bone mass.

3. The epiphyseal plate is composed of hyaline cartilage. Any disease that affects hyaline cartilage will impair the function of the epiphyseal plate, and therefore the structure and function of the bone as it grows.

Level 3

1. The periosteum is where the majority of the osteoblasts reside. Osteoblasts are needed in fracture repair to rebuild the bone and bridge the gap between bone fragments. If the periosteum is stripped, bone healing will occur much more slowly because the population of osteoblasts will be dramatically reduced.

2. Testosterone and other anabolic steroids accelerate the closure of the epiphyseal plates. If anabolic steroids are taken during adolescence, the elevated levels of androgens will cause the plates to close even more rapidly, which will lead to shorter stature.

3. Lucy lives in an environment in northern Canada where she likely receives inadequate amounts of ultraviolet radiation. This has decreased the amount of vitamin D synthesis taking place in her skin. This decrease, combined with the lack of vitamin D supplementation, has resulted in a severe lack of vitamin D overall and the consequent leaching of calcium ions out of her bones. Such a condition has led to her stunted growth and weakened bones.

4. Lysosomes are organelles that generally perform digestive functions. Their main enzymes for catalyzing these breakdown reactions are functional only at an acidic pH. The lysosomes of osteoclasts contain enzymes responsible for catalyzing reactions that break down the organic component of bone. If the lysosomes are unable to maintain an acidic pH, these enzymes cannot function. This will lead to an imbalance in which bone deposition abnormally dominates bone resorption.

CHAPTER 7
Apply What You Learned

MODULE 7.1

1. The bones of the axial skeleton protect delicate vital structures, such as the spinal cord, brain, heart, and lungs. The bones of the appendicular skeleton are more important for movement than protection, and so damage to these bones is likely to be less damaging to overall homeostasis. 2. Openings typically allow the passage of important structures, such as nerves, blood vessels, and the spinal cord. Openings that are too small could compress these important structures and impair their functions. Those that are too large could lead to instability and inadequate support of the structures.

MODULE 7.2

1. a. The ethmoid bone has likely been fractured. The superior part of the ethmoid forms both the roof of the nasal cavity and part of the floor of the cranial cavity. A fracture of this bone could connect the two cavities, allowing the fluid to leak out from around the brain through the nose.
   b. This portion of the sphenoid bone forms the posterior portion of the orbit through which the nerve for vision passes. A fracture in this location could damage the nerve, leading to vision problems.
2. Ms. Midna likely has an infection in her frontal and maxillary sinuses, leading to her pain. The paranasal sinuses are connected to the nasal cavity via small bony canals, which allow infectious agents to pass from one location to the other.

MODULE 7.3

1. Ms. Cho has also likely fractured her fourth rib, which articulates with the transverse process. The ribs are attachment points for muscles of breathing, and so each breath pulls on the fractured rib, causing pain.
2. a. The abnormal curvature is scoliosis.
   b. The muscle weakness has allowed the curvature to develop. The curvature, in turn, compressed the discs in the lower thoracic region, leading to their herniation.

MODULE 7.4

1. The trochlear notch of the ulna forms a joint with the distal humerus. Normally, the notch is fairly deep, so the two bones fit together well and are stable. However, if the notch were abnormally shallow, the bones would not fit together as snugly and the stability of the joint would be compromised. 2. Mr. Heller has likely fractured his clavicle. The clavicle normally braces the upper limb in its proper position. When it is fractured, this function is disrupted, and the upper limb hangs in a more medial position than normal.

MODULE 7.5

1. Many characteristics can be examined, including the overall shape, the angle of the pubic arch, the shape of the pelvic inlet, the orientation of the ischial tuberosities, and the shape and size of the sacrum. 2. The head of the femur fits into the acetabulum, which is normally fairly deep. This gives the hip joint a fair amount of stability, which would be lacking were the acetabulum too shallow. The lack of stability is likely putting abnormal strain on her pelvic bones, femur, and surrounding structures, leading to her pain.

Assess What You Learned

Level 1

1. c, d, and e   2. b   3. sagittal; coronal, squamous, lambdoid
4. a. False: The four paranasal sinuses include the frontal, *ethmoidal*, sphenoidal, and *maxillary* sinuses.
   b. True
   c. True
   d. False: The *mastoid* process of the temporal bone is a thick, posterior projection.
   e. False: The most conspicuous feature of the *occipital* bone is the foramen magnum.
5. c   6. e   7. a
8. a. False: The thoracic and sacral curvatures are the vertebral column's *concave* curvatures.
   b. False: A vertebral disc is composed of an *outer* anulus fibrosus and an *inner* nucleus pulposus.

c. True

d. False: The *sacral and coccygeal* vertebrae are fused in an adult.

9. d   **10.** xiphoid process; manubrium, clavicle

11. True ribs (1–7) attach to the sternum by their own costal cartilage; false ribs (8–12) do not attach directly to the sternum. Ribs 8–10 attach to the costal cartilage of true ribs; floating ribs (11–12) do not attach to the sternum at all.

12. b   **13.** humerus; ulna, radius   **14.** c   **15.** a

16. a. False: The *acetabulum* articulates with the head of the femur at the hip joint.

b. True

c. True

d. False: The two pubic bones articulate at the *pubic symphysis*.

e. True

17. c   **18.** tibia, fibula; patella   **19.** d   **20.** e

Level 2

1. The atlas lacks a body and has large superior articular facets that articulate with the occipital bone. The axis has a superior projection called the dens that fits into the large vertebral foramen of the atlas. This allows the two bones to move around one another at the atlantoaxial joint.

2. Bones serve as levers for muscles, and so of course abnormal bone structure can cause muscles to function improperly. But even minor bone abnormalities can disrupt muscle function, because muscles attach to bone prominences such as tuberosities, trochanters, and tubercles. If these projections are abnormal in shape or size, then the muscle attachment may be affected.

3. The knee joint is formed by the medial and lateral femoral and tibial condyles; the elbow joint is formed by the capitulum and trochlea of the humerus, the head of the radius, and the trochlear notch of the ulna. The elbow joint is more stable because of the shapes of the articulating bones, which allow a tighter fit than in the knee joint.

Level 3

1. Generally, the bones involved are the perpendicular plate of the ethmoid bone and/or the vomer. When the nasal septum shifts to one side, it obstructs part of the nasal cavity, which is the first portion of the respiratory tract. This can make breathing through the nose difficult and necessitate breathing through the mouth instead.

2. Kyphosis is an abnormal curvature of the spine in the anterior direction. This places stress on the anterior portion of the vertebrae, specifically the vertebral bodies, which are the part of the vertebrae most likely fractured. (Such a fracture is called a wedge fracture.)

3. The person who performed CPR likely performed it incorrectly, doing compressions over the xiphoid process or over the ribs instead of over the body of the sternum.

4. a. Anterior portion of the head (frontal region of the head)

b. Superior, anterior cervical region

c. Medial side of the forearm (medial antebrachial region)

d. Medial leg (medial crural region)

e. Located between the ribs

f. Anterior pelvic region

CHAPTER 8
Apply What You Learned

MODULE 8.1

1. The bones of the mature skull are joined by joints that are functionally classified as synarthroses. The sutures between skull bones are interlocked and become fused once an individual reaches skeletal maturity. This is important because these joints have evolved for protection of the underlying brain. If these joints were freely moveable or even slightly moveable, this protective function would be compromised. **2.** The joint resulting from this surgical procedure resembles a synarthrosis. The fused joint would no longer have a synovial cavity or hyaline articular cartilage and would be more stable than mobile.

MODULE 8.2

1. Such diseases can completely disrupt the structure of the gomphoses because the periodontal ligament that locks a tooth into the alveolus of the mandible or maxilla would not be strong enough to withstand the demands of chewing. The teeth would be at risk for falling out or being damaged, as they would no longer be structurally sound. When the structure is compromised, the function may be lost. **2.** If the sutures in the skull were fully fused at birth, the brain tissues could not grow within the skull. The sutures do not fuse until later in life to allow for brain growth.

MODULE 8.3

1. Synchrondroses are classified functionally as immoveable, and immobility is vital to their function. Usually these joints are involved in protection of and stability for the surrounding anatomical structures. Any motion at these joints would compromise these vital functions. **2.** Symphyses are classified functionally as amphiarthroses. Some degree of motion has to be allowed for these joints to function normally.

MODULE 8.4

1. Ligaments are important in supporting synovial joints. If the ligaments are too loose, they will not fulfill their supportive role, and the synovial joint will be more likely to allow abnormal motion, leading to injury such as dislocation. On the other hand, cartilaginous and fibrous joints do not rely on ligaments for stability as these two joint types are inherently more stable. **2.** Tendons generally pass over or around synovial joints on their way to their insertions on bones. When the muscle associated with an injured tendon is contracted, the tendon is pulled taut over the joint and this causes pain. As the joint is moved over and over again, the injured tendon will constantly be a source of pain.

MODULE 8.5

1. The condition of drop foot leaves the foot in a plantarflexed position. Individuals with this condition drag their toes as they walk because they are unable to flatten their feet. As you might expect, this makes the knee and hip work harder, as they have to pull the foot up farther so that the pointed toes clear the ground with each step. **2.** Like any disorder or disease that affects the function of the ligaments, Marfan syndrome can cause hypermobility at any number of joints. In a person with the syndrome, the elbow will have a significantly greater range of motion, allowing some amount of hyperextension. The joint will be at risk for dislocation, as well, because the increased range of motion reduces its stability.

MODULE 8.6

1. a. The anterior cruciate ligament runs from an anterior insertion site on the tibia to the posterior aspect of the femur. When the knee is extended, this ligament is pulled taut to prevent hyperextension. The injured ligament would permit an increase in the amount of extension or hyperextension, allowing the tibia to move too far anteriorly on the femur.

b. The tibial collateral ligament supports the medial aspect of the knee joint and prevents the lower leg from moving too far laterally relative to the thigh at the knee joint. The injured ligament would allow too much lateral movement of the lower leg relative to the thigh at the knee joint.

c. The posterior cruciate ligament is found just posterior to the anterior cruciate ligament and runs from the anteroinferior femur to the posterior side of the tibia. This ligament is pulled tightly when the knee is flexed to prevent hyperflexion and dislocation of the tibia from the femur. The injured ligament would allow an increased amount of flexion at the knee, and the tibia would be at risk for being dislocated from the femur posteriorly.

2. This change to the structure of the thumb joint would completely alter the functional capability of the hand. The special movements of opposition would be lost, as a uniaxial hinge joint would not allow that type of movement. The thumb would move only around a single axis, allowing for a degree of flexion and extension similar to that of the other digits of the hand, but not allowing abduction and adduction. Grasping and holding objects, which is taken for granted, would be severely limited if the thumb joint were uniaxial.

Assess What You Learned

Level 1

1. b   **2.** c

3. a. cartilaginous
b. synovial
c. cartilaginous
d. fibrous
e. fibrous

4. d

5. a. True
b. False: A syndesmosis is a type of *fibrous* joint.
c. False: Cartilaginous joints are synarthroses *or amphiarthroses.*
d. True

6. articular cartilage; synovial membrane; articular capsule

7. Synovial fluid lubricates a joint to reduce friction, absorbs shock, and helps deliver nutrients to cartilage cells.

8. c   **9.** b

10. _c_ Plane joint   _c_ Condylar joint
_f_ Saddle joint   _a_ Pivot joint
_b_ Ball-and-socket joint   _d_ Hinge joint

11. a. Flexion decreases the angle between two bones, and extension increases the angle between two bones.
b. Adduction moves a body part toward the body's midline, and abduction moves it away from the midline.
c. Rotation is the turning of a bone along its own longitudinal axis.
d. Circumduction is movement of a bone in a cone-shaped radius.
e. Dorsiflexion is the movement of the foot and ankle toward the leg, and plantarflexion is the movement of the foot and ankle away from the leg.
f. Elevation is movement of a body part in a superior direction, and depression is movement in an inferior direction.

12. a. False: The knee and the elbow are *uniaxial* joints.
b. True
c. True
d. False: The *knee* joint is stabilized by the medial and lateral menisci.
e. False: The *shoulder* joint is less stable than the *hip* joint, but it allows more motion.

13. c   **14.** a

15. The acetabulum is relatively deep, which allows the femoral head to fit inside it snugly. The fit is enhanced by the acetabular labrum, the articular capsule, and multiple ligaments.

Level 2

1. a. The primary function of fibrous joints is stability, or maintaining bones in certain positions relative to other bones. These joints consist of bones united by collagen fibers, which prevent them from moving freely and give them stability.
b. Most cartilaginous joints (except the epiphyseal plate) must be both stable and allow for motion. The fibrocartilage that joins

the bones in these joints is flexible, which allows some motion, but it is also tough, which provides stability.
c. Synovial joints must allow for a significant amount of motion between two bones. This is made possible by the fact that the two articulating bones are separated by a fluid-filled joint cavity.

2. The tendon of the biceps brachii passes through the articular capsule of the shoulder joint as it attaches to its origin point. For this reason, inflammation of the tendon in this area leads to inflammation in the articular capsule of the shoulder joint. This causes pain on movement of the joint.

3. The head of the humerus doesn't fit very snugly into the glenoid cavity, as the cavity is too shallow. The glenoid labrum helps improve this fit and provides the shoulder with extra stability. If the glenoid labrum is smaller than normal, the degree of fit decreases, leading to instability and possibly chronic dislocations of the shoulder joint.

Level 3

1. The sutures are fused fibrous joints that are synarthroses. No motion is allowed at these joints, and therefore the claims are unlikely to be true.

2. Lauren has likely injured her anterior cruciate ligament, which normally prevents hyperextension of the knee and anterior displacement of the tibia. She has also likely injured her tibial collateral ligament, as it normally prevents the abnormal lateral tibial motion seen in the exam.

3. Fibrocartilage is very tough and functions as a good shock absorber in a joint. However, it contains large numbers of protein fibers, particularly collagen, and so it isn't as smooth as hyaline cartilage and is not as effective at reducing friction. For this reason, a synovial joint with fibrocartilage will not function as well as one with hyaline cartilage.

CHAPTER **9**
Apply What You Learned

MODULE **9.1**

1. The name would be sternohyoid muscle. **2.** The temporal bone serves as the origin, as it is the more fixed end of the muscle attachment. The insertion is on the mandible, which is moved toward the origin during contraction of the temporalis. **3.** A first-class lever system is used to dislodge the large rock, and the work was accomplished at a mechanical advantage. In this case, the effort arm moved a greater distance than the load arm.

MODULE **9.2**

1. Any muscle that is not a muscle of facial expression or supplied by the facial nerve would be unaffected by Bell's palsy. Examples include the muscles of chewing, the muscles of the tongue, and the muscles of the pharynx. **2.** Damage to the oculomotor nerve would be a greater problem because it supplies most of the muscles that move the eye, in addition to the upper eyelid. The trochlear nerve supplies only one muscle, the superior oblique. **3.** Severe injury to muscles during childhood can have a negative effect on the growth of bones to which the muscles attach. Severe injury to the erector spinae muscles can alter the curvature of the spine and affect posture.

MODULE **9.3**

1. The diaphragm contracts during inspiration to increase the volume inside the thoracic cavity and draw air into the lungs. This action is aided by the external intercostals, which raise and spread the ribs. During expiration, air is expelled by relaxation of the diaphragm (which allows the lungs to compress by elastic recoil), aided by the internal intercostal muscles, which depress the ribs and rib cage and further reduce thoracic volume. **2.** Abdominal muscles flex and twist the trunk, so they are strengthened by exercises that force the trunk to flex or twist against resistance (such as gravity). These exercises include various kinds of sit-ups, crunches, and leg lifts.

MODULE **9.4**

1. The origins of these two muscles are anterior and posterior, respectively, to their insertions, and so the pectoralis major muscle is able to flex the arm and the latissimus dorsi muscle is able to extend the arm. These two muscles work together in adduction because both of them have origins that are medial to their insertions. **2.** The brachialis muscle crosses only one joint, the elbow joint, and has flexion of this joint as its sole action. The biceps brachii muscle crosses two joints, the shoulder and the elbow, and has actions of flexion at each of these. In addition, its attachment to the radius enables it to be a powerful supinator of the forearm. **3.** Finger movements in plucking guitar strings involve use of the flexor digitorum profundus muscles to flex the distal interphalangeal joints and the flexor pollicis longus to flex the interphalangeal joint of the thumb. These actions would be augmented by the lumbrical muscles that flex the metacarpophalangeal joints and extend the interphalangeal joints. In addition, palmar and dorsal interossei enable the fingers to make short, subtle, but precise movements toward or away from one another.

MODULE **9.5**

1. The rectus femoris crosses two joints: the hip and the knee. It has proximal attachments to the ilium that enable it to flex the thigh and a distal attachment to the tibia enables it to extend the knee. **2.** Plantar flexion allows us to overcome gravity and lift our body weight off the ground to move forward in walking or running. This takes more muscle power and energy to accomplish than raising the toes of the foot to clear it from the ground when taking the next step. **3.** These sets of muscles arise from the metatarsal and metacarpal bones, respectively, of the foot and hand, but the plantar interossei arise from beneath the metatarsals. The dorsal interossei spread the toes and fingers apart (abduction), and the plantar and palmar interossei draw the toes and fingers closer to each other (adduction).

Assess What You Learned

Level 1

1. d   **2.** e   **3.** b   **4.** a

5. _c_ Brevis   _b_ Rectus
       _a_ Digitorum   _d_ Flexor
       _f_ Hallucis   _e_ Pronator

6. a   **7.** d   **8.** c   **9.** d   **10.** a   **11.** c

12. d   **13.** c   **14.** a   **15.** c   **16.** d   **17.** c   **18.** True

Level 2

1. Lateral movement of the eye would be impaired.

2. The scalenes and the sternocleidomastoid muscles are involved in movement of the head/neck and breathing. These muscles attach to the rib cage and can raise the ribs to assist breathing when the head/neck is fixed. They move the head/neck when the rib cage is fixed. Damage to these muscles would affect both of these actions.

3. You might explain that this muscle is part of the erector spinae group that runs along the sides of the vertebral column and serves to extend that column or bend it laterally. The longissimus thoracis is the middle muscle of the group, from side to side, and spans the posterior rib cage area near the vertebrae.

4. The serratus anterior, pectoralis minor, trapezius, rhomboids, and levator scapulae have actions mainly limited to fixing/moving the scapula, while the infraspinatus, supraspinatus, and subscapularis both stabilize the shoulder joint and contribute to arm movement. The deltoid, teres major and teres minor, and coracobrachialis contribute to arm movements as their primary actions.

5. The name tells you the origin is the area of the scapula above the spine (the supraspinous fossa), and the action implies that the insertion is on the superior lateral surface of the humerus (specifically, the greater tubercle).

6. The extensor digitorum and extensor indicis extend and hold up the fingers. Because this movement requires moving the thumb and little finger toward each other (opposition) and flexing, the opponens pollicis, opponens digiti minimi, flexor pollicis longus, and portions of the flexor digitorum muscles also come into play.

7. The pectineus and gracilis also act in adducting the leg.

8. To grip, hold, and manipulate an item requires flexion of the hand and fingers. Once that grip is relaxed, minimal muscle strength is required to extend the hand back to its anatomical position.

9. Plantar flexion is the action needed for pushing off when walking, running, or jumping. The muscles of plantar flexion are pushing the entire weight of the body, but very little weight is opposing the muscles of dorsiflexion. Both sets of muscles do have to work together to ensure that the foot is in the correct position during walking or running.

10. The levator ani muscle is an important component of the pelvic diaphragm, which supports the pelvic floor, anal canal, and genitourinary organs. It consists of two paired muscles: the pubococcygeus and ileococcygeus. These muscles attach, respectively, to the pubis and ilium at one end and the coccyx and anococcygeal ligament at the other end. The coccygeus muscle is the other component of the pelvic diaphragm.

11. The three-headed triceps brachii muscle narrows into a single tendon that attaches to the olecranon process of the ulna. It is the principal extensor of the forearm at the elbow joint. Forearm extension is assisted by a small muscle, the anconeus, which attaches from the lateral epicondyle of the humerus to the olecranon process.

Level 3

1. Malfunction of the abducens nerve will cause subsequent malfunction of the lateral rectus muscle of the eye (see Table 9.3). This will result in impaired ability to move the eye laterally and reduced ability to hold the eye from moving medially.

2. Because humans use their arms for a wide variety of actions, and the legs are primarily used for standing, walking, and running, the shoulder joint is much more moveable than the hip joint. Therefore, there are far more muscles in the shoulder area to stabilize the scapula and the shoulder joint. Also, the dexterity required in hand/wrist movements requires muscles like the pronators and the supinator to efficiently move the forearm and hand to grasp and manipulate objects.

3. The man has likely experienced damage to the facial nerve that has weakened the muscles on the right side of his face. The important muscles that relate to his problem are the orbicularis oculi, which closes the eye and keeps the cornea moist, and the buccinator and orbicularis oris muscles, which maintain the tone of the lips and cheek wall and prevent fluids from leaking out of the mouth.

4. The skateboarder suffered damage to the radial nerve during the fall. This nerve supplies the muscles that extend the elbow, wrist, and fingers. Because he is still able to extend his elbow, nerves to the triceps brachii were not affected—just those to the other extensor muscles.

5. The bone involved in a hip fracture is the femur; it was likely the femoral neck that was fractured. Muscles that attach to the proximal femur include the gluteus medius (the greater trochanter), the iliopsoas (the lesser trochanter), the vastus muscles of the quadriceps femoris group (the proximal femoral shaft), and the adductor group (the proximal femoral shaft). Exercises to strengthen these muscles involve hip abduction for the gluteus medius, hip extension for the gluteus maximus, hip adduction for the adductors, and knee extension for the vastus muscles.

CHAPTER **10**
Apply What You Learned

MODULE **10.1**

1. We would expect the new type of muscle tissue to have structural elements similar to those of the other three types of muscle tissue. This is due to the structure-function relationship—the new muscle tissue type has the same function as the other types, so it would likely have a similar structure. **2.** A muscle cell that is extensible but not elastic would be able to stretch but would not be able to return to its original resting length. A cell that is not extensible would not be able to stretch at all, irrespective of its elasticity.

MODULE **10.2**

1. The spring shape of titin allows it to perform its primary functions, which are to hold thick filaments in place, help return the thick filaments to their resting positions after a contraction, and provide the muscle fiber with elasticity. Were titin to lose this structure, it would not be able to adequately perform its function. The thick filaments would be less likely to return to their normal resting positions after a contraction and the sarcomere might not return to its normal resting length. **2.** The basic mechanism of contraction involves the sliding of thick and thin filaments past one another. If they were stationary in a myofibril, no contraction would occur and no tension would be generated.

MODULE **10.3**

1. Blocking the Na^+/K^+ pump would disrupt the resting membrane potential and eventually deplete the Na^+ and K^+ gradients. This would diminish the amount of Na^+ in the extracellular fluid available for depolarization and the amount of K^+ in the cytosol available for repolarization. Eventually, action potentials would cease. **2.** Tetrodotoxin will prevent Na^+ from entering the cytosol when the voltage-gated channels are triggered to open by the end-plate potential. This will inhibit the depolarization phase of the action potential and prevent an action potential from propagating along the sarcolemma. **3.** If the voltage-gated potassium ion channels in the sarcolemma were blocked, it would prevent the cell from repolarizing. A cell that cannot repolarize will never be able to depolarize again, and so no further action potentials will be generated or propagated.

MODULE **10.4**

1. The progressive destruction of ACh receptors leads to a progressive decline in the strength of end-plate potentials. This will make it harder for a neuron to elicit action potentials, resulting in muscle weakness and partial or full paralysis. **2.** The troponin-tropomyosin interaction is necessary to regulate contraction and relaxation. Were troponin unable to bind tropomyosin, it would not be able to hold tropomyosin in place when the muscle is not contracting, and it would not be able to move it away from the active sites on actin to initiate a contraction. **3.** The action potential would be unable to reach and open calcium ion channels in the terminal cisternae without T-tubules. This would prevent calcium ions from being released from the SR, which would in turn prevent contraction from occurring.

MODULE **10.5**

1. Creatine phosphate, even when provided in excess, provides only enough energy to fuel activities lasting a few seconds. Activities lasting longer than this will be fueled by glycolytic and oxidative catabolism, and extra creatine phosphate will be of no benefit. **2.** Long-lasting activities typically use all of the available glucose and require the use of other fuels such as fatty acids. This is why aerobics are said to "burn fat." Short sprints will not produce this same effect because they use primarily glycolytic catabolism and glucose as a fuel.

MODULE **10.6**

1. When you lift a heavy box, the muscle fibers will likely be stimulated at a very high rate, about 80–100 times per second. This will result in the production of wave summation and fused tetanus, which are needed to produce a strong enough contraction. **2.** The difference in strength has to do with the difference in starting length of the muscle fibers. When your friend's arm was flexed at the elbow, her muscle fibers were shortened and therefore able to produce less tension due to increased overlap of the thin and thick filaments. However, when she picked it up from a relaxed or neutral position, the overlap of the filaments was ideal for maximal tension production. **3.** This new fiber type would likely use primarily oxidative catabolism, as it has a ready supply of oxygen from myoglobin and blood vessels and a large number of mitochondria in which to carry out oxidative catabolism. Such fibers tend to be fatigue-resistant because they continue to produce adequate quantities of ATP for as long as chemical fuels are available.

MODULE **10.7**

1. When more tension is needed, fast motor units are recruited in addition to slow motor units. These fast motor units are less fatigue-resistant than slow motor units, which is why you are likely to feel fatigued more quickly. **2.** Lowering the weight is accomplished by an isotonic eccentric contraction, which involves lengthening of the muscle fibers. Such contractions can cause injury due to overstretching if they are not done in a controlled manner.

MODULE **10.8**

1. An athlete who runs marathons likely undergoes endurance training. This individual probably has a higher number of type I fibers, and the training would likely increase the number of mitochondria, the amount of oxidative enzymes, and the blood flow. Each of these adaptations would make the athlete's muscle fibers more resistant to fatigue. **2.** This wonder drug would most benefit athletes engaged in endurance activities, as those activities rely primarily on oxidative catabolism. Resistance activities rely more heavily on immediate and glycolytic energy sources, and so the extra myoglobin would be less beneficial to them. **3.** If your friend had truly "used up all of his oxygen," it would be because he was no longer breathing and would not be long for this world. Instead, the rapid rate of breathing, or excess postexercise oxygen consumption, is due to the body's attempt to correct homeostatic imbalances that result from exercise.

MODULE **10.9**

1. No, you would not find recruitment in single-unit smooth muscle tissue and cardiac muscle tissue, as both types contain muscle cells that are connected by gap junctions and so function as a unit. When the cells contract, they all contract together, so there are no cells left to recruit. **2.** The excess calcium ions in the extracellular fluid could enter the smooth muscle fiber and lead to a longer and stronger contraction—perhaps longer and stronger than is needed. **3.** Decreased ATP production would have a more detrimental effect on cardiac muscle tissue, which generally uses far more ATP than smooth muscle tissue. This is because smooth muscle cells can enter the latch state, during which they are able to sustain a contraction while expending very little ATP. Cardiac muscle cells are unable to enter such a state, and must repeatedly use ATP to contract and relax cyclically.

Assess What You Learned

Level 1

1. a. False: a property of all muscle cells is *extensibility,* which means that the tissue is able to stretch.
   b. True
   c. False: a muscle cell's plasma membrane is called the *sarcolemma.*
   d. True
   e. True

2. Skeletal muscle fibers are typically long, thin cells. Cylindrical myofibrils make up about 80% of their sarcoplasm, along with a large number of mitochondria and multiple nuclei. The myofibrils are surrounded by sarcoplasmic reticulum.

3. b

4. <u>c</u> Z-disc   <u>e</u> H zone
   <u>d</u> Sarcomere   <u>f</u> I band
   <u>a</u> A band   <u>b</u> M line

5. The thick and thin filaments slide past one another, which generates tension, as described by the sliding-filament mechanism of contraction. As multiple sarcomeres generate tension, the entire muscle fiber contracts.

6. d

7. The three components of the neuromuscular junction are (1) the axon terminal of the motor neuron, which contains synaptic vesicles with the neurotransmitter acetylcholine; (2) the narrow synaptic cleft; and (3) the motor end plate of the muscle fiber, which contains acetylcholine receptors.

8. <u>2</u> The motor end plate generates an end-plate potential.
   <u>4</u> The action potential spreads past the T-tubules, SR calcium ion channels are pulled open, and calcium ions flood the cytosol.
   <u>1</u> Acetylcholine binds to receptors on the motor end plate, and ligand-gated ion channels open.
   <u>5</u> Calcium binds troponin, which allows tropomyosin to move away from the actin active site, initiating a contraction cycle.
   <u>3</u> The action potential propagates through the sarcolemma and dives into the cell along the T-tubules.

9. c   10. a

11. a. True
    b. False: glycolytic catabolism occurs in the *cytosol* and requires *no oxygen*.
    c. False: *oxidative* catabolism provides enough ATP to fuel from several minutes to hours of activity.
    d. True

12. a. False: muscle fibers generate more tension if the resting length of their sarcomeres is *100–120% of their optimal length*.
    b. True
    c. False: muscles that require a great deal of precise control will have *small* motor units.
    d. True

13. a   14. b   15. glycolytic; oxidative   16. c   17. a, b, c, d

18. Excessive postexercise oxygen consumption (EPOC) is due to the body's efforts to correct the disturbances to homeostasis that were brought on by exercise, including dissipating heat, restoring ion concentrations, and correcting blood pH.

19. Functions of smooth muscle include peristalsis, regulating the flow through organs such as blood vessels, and the formation of sphincters.

20. c

21. a. smooth   e. cardiac, some smooth
    b. cardiac   f. skeletal, cardiac
    c. skeletal, cardiac   g. smooth
    d. smooth   h. skeletal

Level 2

1. You would expect to find larger motor units in the back. The hand requires precise motor control, which necessitates multiple small motor units.

2. This condition will cause depolarization to be weak or even absent because it diminishes the size of the sodium ion gradient between the extracellular fluid and the cytosol.

3. The result of hyperkalemia depends on its severity. In mild to moderate hyperkalemia, the gradient is decreased and the resting membrane is closer to threshold. This makes the fiber more excitable and slows repolarization. In severe hyperkalemia, the resting membrane potential becomes higher than threshold, and the fiber is unable to become excited because it cannot recover properly from previous activation.

4. Cardiac muscle cells and some smooth muscle cells are stimulated by pacemaker cells that stimulate the cells to have action potentials. Such cells do not require stimulation from an external nerve supply.

Level 3

1. Curare would prevent acetylcholine from binding to the motor end plate and initiating contraction, effectively paralyzing the recipient. This could be useful in the treatment of muscle spasms and in cases where a patient has to be paralyzed, such as during surgery. An overdose would be lethal because it would paralyze the respiratory muscles and prevent breathing.

2. Neostigmine prolongs the activity of ACh in the synaptic cleft. This results in stimulation of the muscle fiber for a longer period of time.

3. Glycogen stores are one of the main sources of energy for short, powerful bursts of activity. Therefore, protein loading in an athlete engaging in these types of activities would actually decrease his or her performance. An athlete engaging in light endurance-type activities would not likely suffer a great decrease in performance, as glycogen is not the primary fuel for these activities.

4. Damage to the nerves has impaired their ability to stimulate the muscle fibers, which means that the fibers will not receive the signal from the neuron to have an action potential or contract. However, the muscle fibers are still able to contract when stimulated by an electrode because the electrode triggers the initiation of an action potential along their sarcolemmas, which then leads to a contraction.

5. Blocking the release of calcium ions from the SR will prevent calcium ions from flooding the cytosol and binding to troponin. This will prevent actin and myosin from binding and the initiation of contraction.

6. The segment of DNA that codes for a single protein is a gene. Each gene is a unique sequence of nucleic acids in the DNA that specify a unique sequence of amino acids. When this sequence is modified, the resulting protein and its amino acid sequence are modified, often leading to a nonfunctional protein. Jesse's genetic defect prevents normal muscle relaxation by preventing the removal of calcium ions from the cytosol after a contraction.

7. ATP is required to fuel multiple aspects of muscle contraction, including cocking of the myosin head, running the pumps in the SR, and running the Na^+/K^+ pumps in the sarcolemma. A decrease in ATP production would therefore interfere with all aspects of muscle tissue function. Other organs and tissues with high metabolic rates would also be affected, including the kidneys, the heart, the brain, and the liver.

CHAPTER 11
Apply What You Learned

MODULE 11.1

1. PNS neurons of the somatic sensory division deliver information on the temperature, smell, and sight of the coffee to the CNS. In the CNS, different neurons integrate these stimuli. Then, PNS neurons of the somatic motor division coordinate lifting the cup of coffee to your mouth and taking a sip. 2. An injury to the visceral motor division would likely be the most threatening to survival. This division innervates the viscera, including the heart, the smooth muscle of hollow

organs, and the glands. Damage to this division would therefore produce disturbances in variables such as heart rate and blood pressure, which could prove fatal.

MODULE **11.2**

1. Microglia are phagocytes that ingest dead or damaged cells in the brain, including neurons and other neuroglia. A brain injury results in damaged cells in the brain, and as a result, more microglia will be present. **2.** In a patient with this syndrome, conduction will slow significantly in PNS neurons due to the destruction of myelin. This will decrease the delivery speed of motor impulses to muscles, leading to partial or full paralysis. It will also slow the delivery of sensory impulses to the CNS, leading to decreased sensation. **3.** Ms. Karabekian might regain some sensation, but she is unlikely to recover fully. Ganglia contain the cell bodies of PNS neurons. These neurons have some limited ability to regenerate if the cell body is intact. However, since her cell bodies were damaged in the injury, her neurons are unlikely to regenerate.

MODULE **11.3**

1. The gradients for Na$^+$ and K$^+$ change very little with a single action potential. However, the gradients will start to dissipate with repeated action potentials, and the poison will prevent the pumps from restoring them, reducing the ability of the neuron to have an action potential. **2.** In this hypothetical situation, the opening of sodium ion channels would lead to Na$^+$ leaving the cytosol and entering the extracellular fluid. This would cause the cell to lose positive charges, and the cell would hyperpolarize instead of depolarize. **3.** The impulses are conducted relatively slowly because temperature sensations are carried on type C fibers, which are small and unmyelinated.

MODULE **11.4**

1. The influx of calcium ions in the axon terminal is what triggers exocytosis of the synaptic vesicles. If these calcium ion channels are blocked, exocytosis will not be triggered, and synaptic transmission will effectively be halted. **2.** Calcium ions are more concentrated in the extracellular fluid, so when calcium ion channels are opened, calcium ions follow their gradient and flow into the cytosol of the neuron. The influx of positive charges produces an EPSP. Repeated EPSPs make an action potential more likely to occur. **3.** To make an action potential more likely, the number of EPSPs needs to be increased. This can occur via temporal summation, in which one or more neurons fires repeated action potentials and triggers multiple EPSPs in rapid succession. It can also occur via spatial summation, in which multiple neurons fire action potentials and trigger EPSPs all at the same time.

MODULE **11.5**

1. Glutamate binds ionotropic postsynaptic receptors that trigger EPSPs in the postsynaptic neurons. Blocking glutamate receptors will prevent the generation of these EPSPs and reduce the number of action potentials fired in the CNS. **2.** Glycine typically triggers IPSPs and so has an inhibitory effect in the CNS. Blocking glycine receptors will increase the amount of inappropriate excitation in the CNS, leading to abnormal motor, sensory, and mental processes. **3.** Such blockage would prolong the actions of the neurotransmitters in the synaptic cleft, increasing excitation of the postsynaptic neurons. If you prolonged the actions of inhibitory neurotransmitters, you would decrease the excitation of postsynaptic neurons.

MODULE **11.6**

1. Such circuits are required for certain types of muscle contractions, as these often involve both agonist and antagonist muscle groups, of which one group must be stimulated while the other must be inhibited. **2.** Prolonged abuse of the drug leads to synaptic fatigue; the number of dopamine receptors declines over time, and the remaining receptors become less sensitive to dopamine.

Assess What You Learned

Level 1

1. c   **2.** d

3. __c__ Schwann cells   __e__ Oligodendrocytes
__f__ Ependymal cells   __b__ Satellite cells
__a__ Microglial cells   __d__ Astrocytes

4. a   **5.** b   **6.** internode; nodes of Ranvier

7. absolute refractory period; relative refractory period   **8.** a

9. a. ES
b. CS
c. B
d. CS
e. ES

10. a

11. __d__ GABA   __c__ Glutamate
__g__ Dopamine   __e__ Endorphins
__a__ Substance P   __f__ Norepinephrine
__b__ Acetylcholine

12. c   **13.** d

14. a. True
b. False: Dendrites are highly branched receptive processes that generate and conduct *local* potentials.
c. True
d. False: Multipolar neurons have multiple *dendrites* and a single *axon*.
e. False: Areas of unmyelinated processes and cell bodies appear *gray* and are called *gray* matter.
f. True

15. a. True
b. False: The concentration of Na$^+$ is highest in the *extracellular fluid*, and the concentration of K$^+$ is highest in the *cytosol*.
c. True
d. True
e. False: *An action* potential is a change in membrane potential that conducts the long-distance signals of the nervous system.

16. a. 2
b. 5
c. 1
d. 3
e. 4

17. a. False: An *inhibitory* postsynaptic potential is caused by K$^+$ or Cl$^-$ channels opening in the membrane of the postsynaptic neuron.
b. True
c. False: An *excitatory* postsynaptic potential causes the membrane potential of the postsynaptic neuron to approach threshold.
d. True

Level 2

1. Such a drug would not only prevent the conduction of action potentials but also block the generation of local potentials. This would prevent the depolarization of the axon hillock to threshold, and thus the generation of action potentials.

2. In this case, the drug would prevent the neuron from repolarizing. The neuron would remain depolarized and unable to generate another action potential.

3. The axon generally must be in the PNS and its cell body must be intact in order for regeneration to take place. The process of regeneration begins when the axon and myelin sheath distal to the injury degenerate as growth processes stem from its proximal end. Schwann cells and the basal lamina then form a regeneration tube, into which a single growth process extends. Regeneration is complete when the axon re-establishes connections with its target cell.

4. In continuous connection, the initial segment of the axolemma depolarizes to threshold and has an action potential, which depolarizes the next segment of the axolemma to threshold. This continues down the length of the axon. Saltatory conduction is faster than continuous conduction. In saltatory conduction, only the nodes of Ranvier must be depolarized to threshold, allowing action potentials to "leap" down the length of the axon.

Level 3

1. Initially, synaptic transmission is increased as the actions of acetylcholine are prolonged. However, eventually the postsynaptic cell becomes refractory to stimulation because its membrane is unable to repolarize, and so it cannot be depolarized again or have another action potential. This will decrease the excitation at synapses that use acetylcholine, including the neuromuscular junction.

2. Opening Cl⁻ channels will cause the patient's neurons to hyperpolarize, making it very difficult to stimulate them to have an action potential. This slows neuronal activity in the brain, which puts the patient "to sleep" for the surgical procedure

3. Albert's nervous system will partially or completely shut down, as action potentials will be unable to propagate along his axons. This will result in paralysis, loss of autonomic functions, and mental impairment. The severity of the symptoms will depend on the amount of poison he ingested, but could include a decrease in all functions of the brain and nervous system. Death results from moderate to severe poisoning.

4. Lithium would simply make the tetrodotoxin poisoning worse, as it would further decrease the ability of the neurons to generate and conduct action potentials. The drug should not be administered while the patient recovers from the effects of the poison.

5. The poison would prevent the spread of action potentials along the sarcolemma of muscle fibers. However, the end-plate potentials at the motor end plate would not be directly affected, as those are generated by ligand-gated ion channels. However, because of the effect on action potentials, the neurons supplying the muscle fibers would be unable to stimulate those fibers.

6. If depolarization triggered a negative feedback loop, the initial depolarization to threshold would not lead to further depolarization. Instead, it would lead to repolarization and the cell would return to its resting membrane potential.

CHAPTER 12
Apply What You Learned

MODULE 12.1

1. An injury to the brainstem would be most damaging in terms of survival because the brainstem is responsible for maintaining basic aspects of homeostasis, such as breathing and blood pressure. 2. Damage to the cerebrum, specifically the frontal lobe, could lead to changes in personality. This is because of the association areas such as the prefrontal cortex, which house neurons involved in higher functions like personality and conscience.

MODULE 12.2

1. The basic homeostatic functions that enable life are controlled primarily by the hypothalamus and brainstem. For this reason, such functions can continue, even in the absence of cortical activity. However, an individual in a vegetative state does not have any true sensory awareness of pain or other stimuli and lacks the ability to initiate voluntary movement, since that requires upper motor neurons of the motor cortices.
2. Damage to the basal nuclei will cause a patient to have difficulty initiating movement. Damage to the cerebellum will cause a patient to have difficulty controlling ongoing movement. 3. The gyri and sulci increase the surface area of the cerebral cortex, which in essence allows more

neurons to "fit" in a small area. A lack of gyri and sulci severely reduces the surface area of the cerebral cortex and thus the number of neurons located within it. 4. Effects of damage to the thalamus depend on which portions are injured. Because the thalamus controls the inflow of information to the cerebral cortex, thalamic injury can lead to deficits in perception and awareness of sensory stimuli. Such injury can also affect an individual's ability to initiate voluntary movement. Severe damage to the thalamus will essentially cut off the cerebral cortex from the rest of the nervous system, leading to permanent coma or a persistent vegetative state.

MODULE 12.3

1. CSF makes the brain buoyant, preventing it from crushing itself under its own weight. A decreased amount of CSF will put pressure on the brain, leading to headaches and possible impairment of various sensory and motor functions, depending on the degree of CSF loss. 2. Normally, the capillaries of the brain are highly impermeable to most large molecules due to the blood-brain barrier. Disrupting the tight junctions of the blood-brain barrier with mannitol will increase the permeability of the capillaries and allow more molecules of a drug to exit the blood and enter the ECF.

MODULE 12.4

1. The dural sinuses are located between the periosteal and meningeal layers of the cranial dura. There is no periosteal dura in the spinal cord; instead, the dura there is composed of a single layer. For this reason, there is no place for a dural sinus in the spinal cord. 2. If the damage is on one side only, the individual will still have sensation carried by the posterior columns on the other side. In addition, other types of sensory stimuli travel through tracts in the anterior and lateral spinal cord, such as those of the anterolateral system, and those tracts are likely to be unaffected by damage to the posterior columns. However, the sense of fine touch and conscious proprioception will be lost from the limb served by the damaged posterior column.

MODULE 12.5

1. Areas of the body that have a higher density of sensory receptors have a larger proportional area of the primary somatosensory cortex dedicated to them. The fingertips have a higher density of sensory receptors than the back does, so their area of primary somatosensory cortex is larger than that for the back. 2. The pathways of the anterolateral system, which carry pain stimuli from the left side of the body, decussate in the posterior horn and travel up the right side of the spinal cord. The pathways of the right posterior columns, which carry tactile stimuli, travel in the right side of the spinal cord and do not decussate until the brainstem. Therefore, the injury is on the right side of the spinal cord. 3. These drugs bind to receptors in the brainstem and spinal cord, and block the transmission and perception of pain stimuli. They do not treat the cause of the pain, so when their effects wear off, the perception of pain will return.

MODULE 12.6

1. The globus pallidus sends inhibitory impulses to the thalamus, blocking the initiation of inappropriate movement. Degeneration of the globus pallidus would remove this inhibition, and the individual would experience inappropriate, involuntary movements. 2. The caudate nucleus, putamen, and substantia nigra are instrumental in initiating movement by inhibiting the globus pallidus. With their degeneration, an individual would have difficulty in initiating movement, as the globus pallidus would inhibit movement continuously. 3. Cerebellar damage will cause an individual to appear clumsy and uncoordinated, as the cerebellum will be unable to detect motor error and/or unable to send feedback to the primary motor cortex to correct for motor error. 4. The stroke likely occurred on the left side of the brain. Motor pathways initiate in one cerebral hemisphere, and then most of the fibers decussate in the medulla to serve the opposite side of the body. Therefore, damage to the left hemisphere will mostly affect the right side of the body.

MODULE **12.7**

1. The structures mainly responsible for these changes are the hypothalamus and the reticular formation. The cerebral cortex also plays a role, as it is the cortical mediation of "nervous" feelings that trigger the hypothalamus to stimulate the reticular formation nuclei. (But don't be nervous or anxious. Have confidence that you're going to ace the exam!) **2.** Orexins stimulate appetite and wakefulness. Therefore, an individual with such a tumor is likely to experience an increase in appetite and difficulty falling asleep at night. **3.** Under most conditions, muscle groups are paralyzed during REM sleep, which prevents sleeping individuals from acting out their dreams. These muscle groups are not fully paralyzed during non-REM sleep, and thus a sleeper can potentially engage in sleeptalking or sleepwalking.

MODULE **12.8**

1. The lateralization of certain functions means that some functions are performed primarily by the right or left side of the cerebrum. Mr. Jacobs can expect difficulties with attention, facial recognition, "negative" emotions such as anger, and interpretation of the emotional aspects of language. Furthermore, damage to sensory and motor regions of the right hemisphere will impair his ability to sense stimuli from the left or command muscles on the left side of his body. **2.** Ms. Mah will have deficits in the formation of new declarative memories, as immediate and short-term declarative memories are processed by the hippocampus before becoming long-term memories. Existing long-term memories and new and old nondeclarative memories will be largely unaffected by the surgery. **3.** The amygdala is a key player (although not the only one) in the highly subjective feelings that we experience. Thus, Ms. Mah can expect to have some emotional abnormalities, although the degree to which this will be a problem for her is difficult to predict.

Assess What You Learned

Level 1

1. a

2. a. False: Humans use *100%* of their brains.
  b. True
  c. False: The right and left lateral ventricles are the largest of the ventricles in the brain and are located in the *cerebrum.*
  d. False: The *brainstem* is responsible for our basic, involuntary homeostatic functions and reflexes.

3. d **4.** b

5. a. True
  b. False: *Multimodal* association areas integrate information from many different sources.
  c. True
  d. False: The prefrontal cortex is located in the frontal lobe and is concerned with *behavior, personality, learning, and memory.*

6. c

7. _c_ Amygdala   _b_ Medulla oblongata
  _a_ Thalamus   _d_ Hypothalamus
  _f_ Hippocampus   _e_ Cerebellum
  _h_ Midbrain   _g_ Pons

8. a

9. Cerebrospinal fluid is formed by the choroid plexuses. The capillaries of the choroid plexuses are fenestrated, and so fluid and electrolytes leak out of them and into the space under ependymal cells, which then secrete the CSF into the ventricles of the brain.

10. The blood-brain barrier consists of the endothelial cells of the capillaries in the brain, which are joined by more tight junctions than most other capillaries and have limited capacity for endocytosis and exocytosis, as well as the ependymal cells that surround the capillaries of the choroid plexuses. This barrier functions to keep many

chemicals out of the ECF of the brain and the CSF, protecting the brain from many different types of toxins.

11. a. True
  b. True
  c. False: The epidural space around the spinal cord is *a true* space.
  d. False: The posterior horn of spinal gray matter contains the cell bodies of *sensory* neurons.
  e. False: The corticospinal tracts are the main *motor* tracts in the spinal cord.

12. b   **13.** medulla, posterior horn of the spinal cord

14. The fingertips, lips, and hands have the greatest representation within the primary somatosensory cortex because of the importance of manual dexterity and facial expression to human activities.

15. d

16. cerebral cortex, plan and initiate voluntary movement; anterior horn of the spinal cord, directly innervate skeletal muscles

17. a. _4_   d. _1_
  b. _6_   e. _2_
  c. _3_   f. _5_

18. a. True
  b. False: The cerebellum *does not* monitor the initiation of movement but *does* monitor ongoing movements.
  c. True
  d. True

19. hypothalamus, reticular formation

20. b

21. _f_ REM sleep   _c_ Beta waves
  _d_ Stage I sleep   _e_ Theta waves
  _a_ Delta waves   _b_ Stage IV sleep

22. d   **23.** long-term potentiation, hippocampus, cerebral cortex

24. c

Level 2

1. The basal nuclei inhibit abnormal or involuntary movements. In a person with Huntington's disease, you will therefore see involuntary movements and vocalizations due to loss of inhibition.

2. The basal nuclei are responsible for the initiation of movement, and the cerebellum is responsible for monitoring ongoing movement. An individual with damage to the basal nuclei will have difficulty in initiating a movement. An individual with damage to the cerebellum, however, will have difficulty in continuing a movement in a smooth, fluid, balanced fashion after it has been initiated.

3. The hippocampus is involved in the formation of declarative memories, but not their storage (they are stored in the cerebral cortex). Nondeclarative memories are not formed by the hippocampus, but rather by the motor cortices, the cerebellum, and the basal nuclei, so damage to the hippocampus will not affect these memories.

Level 3

1. Damage to the posterior of the head is most likely to affect the occipital lobe and perhaps parts of the parietal and temporal lobes. Occipital lobe damage could interfere with interpretation of visual stimuli. Temporal lobe damage could interfere with many different functions, such as memory, hearing, and language. Parietal lobe damage could disrupt interpretation of various types of somatic sensory stimuli.

2. Ms. Norris likely has sustained damage to her reticular formation, which is the part of the brainstem responsible for mediating the sleep/wake cycle. Damage to the reticular formation can prevent the normal arousal of the cerebral cortex seen in an awake individual, and the patient therefore remains unarousable.

3. Arlene will lose tactile sensation and muscle control on the left side of her body. These stimuli are carried on the posterior columns and corticospinal tracts, respectively, and these tracts do not decussate until they reach the medulla. Therefore, at the level of the thoracic spinal cord, they carry information from the same side. Pain perception, however, could be lost from the right side of her body. Pain and other touch stimuli are carried by the anterolateral system, which decussates in the spinal cord. Damage to the left side of the spinal cord will therefore affect the second-order neurons serving the right side of the body.

4. Orexins are neurotransmitters that induce feeding. Blocking orexin release could potentially decrease appetite and lead to weight loss. However, orexins are also involved in the sleep/wake cycle. Normally, they are released on waking and help to maintain a state of wakefulness throughout the day via stimulation of the reticular formation. Blocking orexin release could therefore potentially lead to daytime drowsiness.

5. Endorphin release does not stop the transmission of pain impulses—they are detected normally by peripheral nerves. However, endorphins block the perception of pain by inhibiting the delivery of these impulses to the cerebral cortex. A decrease in the perception of pain allows an athlete to continue strenuous exercise with less discomfort.

CHAPTER 13
Apply What You Learned

MODULE 13.1

1. A nerve is a group of axons and small blood vessels surrounded by connective tissue. Damage to a single motor or sensory neuron would produce far less functional impairment than damage to an entire nerve, which contains multiple sensory and motor neurons. 2. Like a muscle, a nerve consists of multiple cells bound together in fascicles by connective tissue sheaths. The fascicles, in turn, are surrounded and held together by another connective tissue sheath. 3. Damage to the visceral sensory or motor division would be the most life-threatening, because these divisions innervate vital organs such as the heart, blood vessels, and lungs and maintain homeostasis.

MODULE 13.2

1. a. CN VII, CN IX, and CN X detect taste.
   b. CN XII provides motor innervation to most of the muscles of the tongue.
   c. CN III, CN IV, and CN VI provide motor innervation to the extra-ocular muscles that move the eyes.
   d. CN VIII, specifically the vestibular portion, is responsible for balance and equilibrium.
   e. CN V innervates the muscles of mastication that close the jaw.
2. The cranial nerves that innervate the salivary glands are CN VII and CN IX. If the activity of these nerves is inhibited, less saliva will be produced, and the individual will experience a dry mouth.

MODULE 13.3

1. The phrenic nerve, which innervates the diaphragm and so is the main nerve that drives breathing, is formed of axons from C_3, C_4, and C_5. Damage at or above the level of C_4 would render this nerve unable to function.
2. a. Inability to flex the wrist and fingers; numbness or pain on the anterior thumb, the anterior second and third digits, and the lateral side of the fourth digit
   b. Inability to dorsiflex and evert the ankle; numbness or pain around the knee, the anterior and distal leg, and the dorsum of the foot
   c. Inability to flex the hip and extend the knee; numbness or pain along the anterior thigh and the medial thigh, leg, and foot

   d. Significant impairment of most muscles of the lower limb except the anterior thigh muscles; numbness or pain on the posterior lower limb and part of the anterior leg
   e. Reduced ability to flex the forearm and difficulty moving the hand; numbness or pain along the fifth digit and the medial side of the fourth digit

MODULE 13.4

1. Loss of their myelin sheaths would cause these neurons to conduct action potentials much more slowly than normal. This would delay the arrival at the cerebellum of critical feedback on motor error. 2. Reading Braille relies largely on Merkel cell fibers, as they have the finest spatial resolution and are numerous in the fingertips. 3. Extremes in temperature and painful stimuli can cause serious homeostatic imbalances. If our thermoreceptors and nocioceptors were rapidly adapting, we would not be adequately sensitive to these ongoing stimuli.

MODULE 13.5

1. a. Upper motor neurons of the premotor cortex are involved in planning movement and selecting a motor program, and so damage to them would lead to a loss of specific motor skills.
   b. Damaged upper motor neurons of the primary motor cortex cannot stimulate lower motor neurons, leading to uncontrolled muscle contractions mediated by those lower motor neurons and the spinal cord.
   c. A lower motor neuron is the only path for stimulation of the muscle fibers of its motor unit, so damage to lower motor neurons produces paralysis or paresis.
2. Information from proprioceptors allows the cerebellum to detect and correct motor error. Damaged proprioceptors would not effectively communicate motor error to the cerebellum, leading to dyskinesia (difficulty executing smooth movements).

MODULE 13.6

1. The feedback from muscle spindles helps ensure that movement is fluid and prevents damage from excessive stretching of a muscle. Inhibition of muscle spindles would result in jerky movements, difficulty maintaining proper posture and balance, and potential muscle damage from excessive stretching. 2. The damage to the facial nerve would impair the motor arm of the corneal blink reflex, so the eyelid wouldn't close even though the stimulus would be detected. 3. Lower motor neurons directly trigger muscle contraction, so if they are not functioning, muscles will not contract after the stretch stimulus. Upper motor neurons do not directly contact muscle, so lower motor neurons may continue to stimulate muscles to contract when the stretch receptors are triggered.

Assess What You Learned

Level 1

1. a. False: The *visceral* sensory division of the PNS detects sensory information from the organs in the thoracic and abdominopelvic cavities.
   b. True
   c. True
   d. False: The term nerve is *not* the equivalent of the term neuron. (*Nerves are composed of the axons of many neurons.*)
   e. True

2. anterior ramus, posterior ramus

3. a. Peripheral nerve: a collection of axons, connective tissue sheaths, and blood vessels
   b. Nerve plexus: a network of nerves that innervates a specific body region
   c. Posterior root ganglion: a collection of cell bodies of sensory neurons located in the posterior root outside the spinal cord

4. a. CN VIII   Vestibulocochlear nerve __e__
   b. CN V   Trigeminal nerve __k__
   c. CN XII   Hypoglossal nerve __g__
   d. CN VI   Abducens nerve __a__
   e. CN X   Vagus nerve __j__
   f. CN I   Olfactory nerve __c__
   g. CN XI   Accessory nerve __h__
   h. CN III   Oculomotor nerve __l__
   i. CN VII   Facial nerve __b__
   j. CN II   Optic nerve __i__
   k. CN IX   Glossopharyngeal nerve __d__
   l. CN IV   Trochlear nerve __f__

5. CN I, II, and VIII are sensory; CN III, IV, VI, XI, and XII are motor; CN V, VII, IX, and X are mixed.

6. __d__ Phrenic nerve   __a__ Radial nerve
   __h__ Median nerve   __g__ Intercostal nerves
   __e__ Femoral nerve   __c__ Common fibular nerve
   __f__ Tibial nerve   __b__ Musculocutaneous nerve

7. b   8. d

9. Many spinal nerves carry both somatic and visceral neurons, so visceral sensations travel along the same pathways as somatic sensations. This causes referred pain.

10. b

11. a. __4__
    b. __3__
    c. __1__
    d. __5__
    e. __2__

12. Upper motor neurons are located in the motor areas of the cerebral cortex; they do not directly contact skeletal muscle fibers. They are involved in making the "decision" to move, as well as planning and monitoring movement. Lower motor neurons directly stimulate skeletal muscle fibers to contract, but they are not involved in the planning of movement.

13. Upper motor neurons select a motor program. They stimulate the caudate nucleus and putamen, which inhibit the globus pallidus. This allows the thalamus to stimulate the upper motor neurons, which in turn stimulate lower motor neurons. The lower motor neurons then stimulate muscle fibers to contract.

14. a   15. Muscle spindles, Golgi tendon organs   16. c

17. a. True
    b. False: *Muscle spindles* detect stretch in a simple stretch reflex.
    c. False: A *simple stretch* reflex is a monosynaptic reflex with only one synapse in the spinal cord.
    d. False: The crossed extension reflex occurs simultaneously with the *flexion* reflex.

18. The sensory arm is mediated by the glossopharyngeal nerve, and the motor arm is mediated by the vagus nerve.

Level 2

1. CN I: identify smells; CN II: test vision; CN III, IV, and VI: test ability to follow a moving finger with the eyes; CN V: test ability to open and close the jaw or test sensation on the face; CN VII: test ability to make different facial expressions or test ability to taste; CN VIII: test hearing and have patient stand on one foot to assess balance; CN IX and X: test ability to swallow and/or taste; CN XI: test ability to shrug shoulders and rotate neck; CN XII: test ability to move the tongue.

2. The change would be potentially harmful. If the nocioreceptors were rapidly adapting, we would be unable to detect the continuing presence of painful stimuli, which would be likely to lead to tissue damage.

3. All spinal nerves result from the fusion of the anterior and posterior roots, and, as a result, all carry both sensory and motor signals.

Level 3

1. One of the simplest tests to do is to test Ms. McKenna's reflexes. Reflexes are diminished to absent with lower motor neuron disorders and are normal to exaggerated with upper motor neuron disorders. It would also help to look at her muscle tone, which is decreased with lower motor neuron disorders and generally increased with upper motor neuron disorders.

2. Jason is likely to experience weakness or paralysis of the muscles of his arm, forearm, wrist, hand, and digits. He will also experience sensory deficits in the arm of varying severity, depending on how severe the damage to the nerves was. He may also experience pain associated with peripheral neuropathy from the injury.

3. The pain could be related to the heart, but is more likely to be referred pain from the stomach or the pancreas, which tend to refer to the area along the diaphragm because of their dermatomal distribution.

4. CIPA affects all somatic and visceral sensory neurons, and patients are unable to feel pain or distinguish temperatures or textures. This condition is dangerous because Maria will not be able to feel pain that leads to tissue damage, and may suffer burns or severe injuries.

5. Sweating is an important aspect of thermoregulation, as it helps to cool the body. Maria's inability to produce sweat will make her prone to overheating from normal physical activities, leading to hyperthermia, which can prove fatal.

CHAPTER **14**
Apply What You Learned

MODULE **14.1**

1. An individual with autonomic neuropathy would have difficulty maintaining homeostasis of many vital functions, including blood pressure, digestion, and urine production. An individual with somatic motor neuron neuropathy would have difficulty controlling voluntary muscle movements, and could eventually exhibit paralysis of affected muscles. 2. The sympathetic nervous system is the "fight or flight" division, meaning that it prepares the body for emergencies. Blocking many of the effects of this system would make Mrs. Williams less able to regulate her blood pressure and make other necessary adjustments during an emergency.

MODULE **14.2**

1. Activating the receptors will dampen the sympathetic response, as these receptors inhibit preganglionic sympathetic neurons. Such a drug could be used to lower blood pressure and heart rate. 2. When preganglionic sympathetic neurons function normally, they stimulate postganglionic cell bodies by releasing ACh onto nicotinic receptors. Blocking these receptors would prevent stimulation of postganglionic sympathetic neurons, which is the same as stimulating receptors on the preganglionic neurons. 3. Epinephrine and norepinephrine from the adrenal medulla will still stimulate this patient's heart and blood vessels to increase his heart rate and blood pressure during exercise.

MODULE **14.3**

1. Atropine will block the muscarinic ACh receptors on all parasympathetic target cells (and on some sympathetic target cells, too). This leaves the sympathetic nervous system unopposed, and all of its effects are amplified, including elevated heart rate, increased blood pressure, and pupil dilation. 2. Stimulation of ACh release will trigger many of the effects that the parasympathetic nervous system has on its target cells. These include increasing saliva release (drooling), pupil constriction and adjustment of the lens for near vision (poor distance vision), and excess digestive activity (diarrhea).

MODULE **14.4**

1. Autonomic centers and the cell bodies of many parasympathetic neurons are located in the brainstem, and so damage to it causes autonomic dysfunction. **2.** Dominance of parasympathetic tone would lead to bronchoconstriction because parasympathetic input would outweigh sympathetic input to the smooth muscle of the airways.

Assess What You Learned

Level 1

1. b

2. thoracolumbar, thoracic and lumbar regions of the spinal cord

3. c

4. Adrenergic receptors bind to norepinephrine and epinephrine and are located on many target cells of the sympathetic nervous system. Cholinergic receptors bind acetylcholine. They are found on sympathetic ganglia neurons and certain sympathetic target cells, as well as on all parasympathetic ganglia neurons and target cells.

5. e

6. Sympathetic neurons stimulate the adrenal medulla to secrete epinephrine and norepinephrine, which function as hormones, into the bloodstream. This allows epinephrine and norepinephrine to reach target cells that are not innervated by sympathetic neurons and prolongs the effects of sympathetic activation.

7. d

8. a. sympathetic   e. sympathetic
   b. parasympathetic   f. sympathetic
   c. parasympathetic   g. parasympathetic
   d. sympathetic

9. a. False: The parasympathetic nervous system generally *increases* the secretion from digestive glands.
   b. True
   c. True
   d. False: *Parasympathetic* tone controls the resting rate of the heart.

10. Nicotinic, muscarinic   **11.** b   **12.** d

Level 2

1. a. Sympathetic nervous system: the system that maintains homeostasis in times of exercise, emergency, and emotion
   b. Parasympathetic nervous system: the system that maintains homeostasis in times of rest; promotes digestion, diuresis, and defecation

2. The sympathetic nervous system increases the rate and force of the contraction of the heart and constricts blood vessels that supply the digestive, integumentary, and urinary systems while dilating blood vessels that supply skeletal muscles. This increases blood pressure and routes blood to the skeletal muscles so that they can continue to function as you run. Sympathetic neurons also dilate the bronchioles so that more oxygen is available to be delivered to skeletal muscles and the heart, and they increase the metabolic rate so that there is more ATP for the heart and skeletal muscles to consume. These neurons also trigger the release of metabolic fuels so that the cells can continue to produce ATP at a higher rate. Finally, they increase sweat secretion to remove heat from the body.

3. Once the race is over, the parasympathetic nervous system restores the heart rate and blood pressure to resting levels, which ensures that all organs receive the proper amount of blood flow. The absence of sympathetic stimulation causes the blood vessels serving the digestive, integumentary, and urinary systems to dilate, which promotes their functions. Parasympathetic neurons trigger secretions by digestive glands so that digestive processes may take place. They also lower the metabolic rate and promote the storage of metabolic fuels to save them for the next time the sympathetic nervous system is activated.

Level 3

1. Cranial nerves III, VII, IX, and X would be affected by this drug. Potential adverse effects include excessive salivation, constriction of the pupils, increased blood flow to the skin (flushing), blurry distance vision, low heart rate and blood pressure, bronchoconstriction, and excessive digestive activity.

2. Yes, the pairing of these two drugs will cause problems for the patient. The albuterol will be unable to bind and activate the β_2 receptors because the propranolol will block these receptors. The albuterol is therefore unlikely to be effective, and her asthma symptoms will not be alleviated.

3. The symptoms of Horner syndrome mimic the symptoms of excessive parasympathetic activity because there is no sympathetic activity to balance it. Mr. Chevalier is likely experience constricted pupils, blurry distance vision, decreased sweating on the face, flushing of the skin due to vasodilation, and droopy eyelids due to reduced sympathetic stimulation of the superior tarsal muscle that opens the eye.

4. This poison results in excessive stimulation of skeletal muscles as a result of prolonged action at the motor end plate. Eventually, the skeletal muscle fibers become so overstimulated that they are unable to respond to further stimulation, which leads to paralysis. The ANS effects mimic those of cholinergic drugs, causing a low heart rate and blood pressure, drooling, increased blood flow to the skin (flushing), bronchoconstriction, and constriction of the pupils.

CHAPTER **15**
Apply What You Learned

MODULE **15.1**

1. Stimuli for the special senses (with the exception of olfaction) are detected by specialized receptor cells instead of neurons. For example, taste chemicals are detected by gustatory cells, which then transmit the stimuli to a sensory neuron. The receptor cells must transduce, or convert, the stimuli from a chemical signal to an electrical signal.

MODULE **15.2**

1. Blockage of the nasal cavity will prevent air (and odorants) from reaching the olfactory epithelium. **2.** The olfactory bulb and tracts lie just inferior to the frontal lobe and would be compressed by such a tumor. This would interfere with the delivery of olfactory stimuli to the inferomedial temporal lobe, decreasing the ability to identify and perceive odorants.

MODULE **15.3**

1. Such a tumor will prevent production of saliva. Without saliva, the taste chemicals will not dissolve and bind to receptors on the gustatory cells, so taste sensation will be impaired. **2.** If the facial, glossopharyngeal, and vagus nerves are intact, taste sensation is preserved, and the individual will be able to distinguish the tastes of hot peppers and French fries. However, he or she will not detect a difference in the spiciness, or "heat," of the two foods, because this is detected by the trigeminal nerve. **3.** The limbic system stores emotional reactions to food, and it created an association between eating coleslaw and severe vomiting. For this reason, coleslaw remains something you have no desire to eat.

MODULE **15.4**

1. Pathogens may travel from the nasal cavity through the nasolacrimal duct, lacrimal sac, and canaliculi to reach the conjunctiva and cause infection and inflammation. **2.** The lateral rectus muscle in the left eye is not functioning properly. **3.** The ciliary body or suspensory ligaments are no longer intact, and so the lens is not held in place.

MODULE **15.5**

1. Lenses that correct hyperopia increase the convergence of rays. Therefore, the focal point will fall in front of your retina. **2.** cGMP wouldn't be converted to GMP and would continue to keep sodium ion channels open, depolarizing the rod. It would remain depolarized and would be unable to respond to light. **3.** Blindness would occur in the right half of the visual field (lateral side of the left eye and medial side of the right eye) because all of the visual signals from that half are carried in the left optic tract.

MODULE **15.6**

1. This person would experience dryness and a buildup of debris in the external auditory canal. **2.** Contraction of the tensor tympani muscle reduces vibration of the tympanic membrane, and the energy of sound waves is less likely to reach the inner ear. A persistent contraction of the muscle would reduce the amount of sound energy reaching the inner ear, and hearing would be impaired.

MODULE **15.7**

1. The sound is loud and has a high pitch. High-pitched sounds produce vibration of the basilar membrane at the base of the cochlea, and loud sounds produce greater vibration.

MODULE **15.8**

1. The utricle detects tilting of the head, and dysfunction of it could lead to dizziness and loss of balance. **2.** The vestibulo-ocular reflex produces eye movement in the direction opposite to that of head movement, so your eyes move to the left. The oculomotor, trochlear, and abducens nuclei receive input from the vestibular nuclei and produce a motor response to this input. **3.** The loss of balance could be due to dysfunction of one of the components of the vestibular system in the inner ear (the semicircular ducts and ampulla and/or the utricle and saccule). An ear, nose, and throat specialist will be able to test the structures of the inner ear to determine if this is indeed the source of her symptoms.

MODULE **15.9**

1. Any of the structures in the visual pathway may be nonfunctional: the cells in the retina, the axons of the optic nerve or tract, the lateral geniculate nucleus of the thalamus, the optic radiations, or the primary visual cortex. **2.** Olfaction will not be affected because the olfactory pathway does not travel through the thalamus. The pathways for gustation, vision, hearing, and vestibular sensation pass through the thalamus to reach various areas of cortex where conscious awareness occurs. Damage to the thalamus prevents this information from reaching the cortex, and the patient will lack awareness of these sensations.

Assess What You Learned

Level 1

1. __c__ Rod or cone
   __e__ Hair cell in cochlea
   __d__ Gustatory cell
   __b__ Olfactory neuron
   __a__ Hair cell in vestibule

2. b   3. odorant, action, olfactory   4. d   5. a

6. __b__ Sweet
   __e__ Sour
   __d__ Salty
   __a__ Bitter
   __c__ Umami

7. b   8. d   9. a

10. If you look directly at an object, its image will be focused on the fovea of the retina. There are only cones in the fovea, which do not function in dim light. To see the object, you must look to the side of it, so that the image falls adjacent to the fovea, where rods are located.

11. a. Photons are absorbed by rhodopsin in *rods*.
    b. *Only cones* are found in the fovea centralis.
    c. In the dark, rods and cones *do not* produce action potentials. (*Rods and cones produce only graded potentials.*)

12. a

13. a. False: The *malleus* is connected to the tympanic membrane.
    b. True
    c. False: The auditory canal is separated from the middle ear by the *tympanic membrane*.
    d. False: The cochlear duct is filled with *endolymph*.
    e. True
    f. False: The spiral organ is located in the scala *media* (*cochlear duct*).

14. High-frequency sounds cause the basilar membrane to vibrate near the base, where it is narrow and stiff; low-frequency sounds cause it to vibrate near the tip, where it is wide and flexible. Different populations of neurons are activated in the cochlear nerve, which the brain interprets as different frequencies.

15. Turning the head causes the endolymph of the semicircular duct to lag behind and push the cupula to one side. This bends the stereocilia and kinocilium on each hair cell, causing it to depolarize or hyperpolarize, which changes the amount of neurotransmitters released onto dendrites in the neurons of the vestibular nerve.

16. c   17. basilar, tectorial   18. True

19. b   20. crista ampullaris, cupula   21. e

Level 2

1. You would gradually lose your sense of smell because there would be no replacement of the olfactory neurons as they died off.

2. You would be unable to move your right eye to the right, since the lateral rectus muscle would be paralyzed.

3. The structures after the chiasma carry information from both eyes, so damage there would cause impairment in both. If only one eye is affected, the damage must be in the retina or the optic nerve.

4. The following structures on the right side may be damaged: optic tract, lateral geniculate nucleus, optic radiations, and/or primary visual cortex.

5. Equilibrium relies on input from the inner ear, the visual system, and proprioceptors. With your eyes closed, your brain has lost one of the three inputs and must rely on the other two. However, if the vestibular system and proprioception are functioning normally, they are adequate to prevent you from losing your balance.

6. The endolymph in the semicircular canals keeps moving after you stop, due to inertia. This bends the stereocilia and kinocilia on the hair cells, which the brain interprets as movement.

Level 3

1. The facial, glossopharyngeal, or vagus nerve may be damaged, or there may be impairment of the nuclei for taste in the medulla oblongata or the primary gustatory cortex in the parietal lobe. More commonly, there is some type of deficit in olfaction, since it is highly involved in our sense of taste.

2. If the pharyngotympanic tube didn't open, the pressure on each side of the tympanic membrane would not be able to equalize. If the pressure difference increased, the membrane could rupture and hearing would be impaired.

3. As the stapes vibrates, it pushes the oval window into the perilymph of the vestibule and the scala vestibuli. Sound waves in the scala vestibuli cross the scala media to the scala tympani partway down the cochlea, so any fluid movement in the scala vestibuli causes fluid movement in the scala tympani, which pushes on the round window. If the round window couldn't move, there would be no fluid movement in the scala vestibuli or the scala tympani, since fluid cannot be compressed.

4. The axons of the olfactory nerve run through the cribriform plate of the ethmoid and may be torn if the plate is fractured. Much of what we perceive as taste is actually stimulation of the olfactory receptors, so impairment of olfaction leads to a reduction in the sense of taste.

5. The orbicularis oculi muscle may be paralyzed, preventing the eyelid from closing completely. If the eye cannot close, tears cannot be spread over the sclera and cornea, leading to dryness of the eye, and potential damage.

CHAPTER 16
Apply What You Learned

MODULE 16.1

1. Its structural similarity to cholesterol indicates that it is likely hydrophobic, which means it may bind to receptors in the plasma membrane, in the cytosol, or in the nucleus. **2. a.** Secretion of hormone X is likely to increase. **b.** Hormone X should increase the level of chemical Y back to its normal range. **c.** The secretion of hormone X should decrease by negative feedback.

MODULE 16.2

1. With low TRH, both TSH secretion and thyroid hormone secretion will decrease. **2.** The extra ADH will abnormally increase the amount of water in the body and decrease the osmolarity of the extracellular fluid. **3.** Such a tumor would increase the secretion of GH and IGF, increasing their levels in the blood. In a child, this could lead to pituitary gigantism; in an adult, it could lead to acromegaly.

MODULE 16.3

1. Thyroid hormone increases the metabolic rate, which could promote weight loss. Unfortunately, such a supplement could mimic the signs and symptoms of hyperthyroidism. It could also decrease the secretion of a person's own thyroid hormone through negative feedback, resulting, at best, in no effect. **2.** The patient's levels of T_3 and T_4 are likely low. The low levels are causing the anterior pituitary to secrete extra TSH in an attempt to stimulate T_3 and T_4 production and secretion from the thyroid gland. However, the extra TSH is not effective, because of either iodine deficiency or Hashimoto's thyroiditis. **3.** The result would likely be hypercalcemia and weakened bones due to excessive osteoclast activity.

MODULE 16.4

1. An ACTH-secreting tumor will increase the cortisol level in the blood, but the elevated cortisol level will decrease CRH secretion through negative feedback. **2. a.** The sodium ion concentration could potentially decrease, and the potassium ion concentration could increase. **b.** Blocking aldosterone could increase the hydrogen ion concentration of the blood, decreasing the pH. **c.** By decreasing water reabsorption, these drugs could cause the blood concentration of solutes other than sodium to increase slightly. **3.** Large amounts of epinephrine and norepinephrine will mimic a prolonged sympathetic response, leading to elevated heart rate and blood pressure, decreased digestive and urinary activity, elevated metabolic rate, and more.

MODULE 16.5

1. A glucagon-secreting tumor would cause hyperglycemia and ketosis, and thus resemble diabetes mellitus. **2.** The signs and symptoms would be almost the exact opposite of those in question 1; Hypoglycemia would result from excess glucose uptake by target cells. **3.** The sympathetic nervous system triggers the release of glucagon to increase the release of metabolic fuels. Drugs that block sympathetic neurons could hypothetically decrease the release of glucagon, although the effect would likely be minor due to other stimuli that trigger glucagon release.

MODULE 16.6

1. Melatonin secretion is stimulated by a decreasing light level. So when the light level is low all day, secretion will increase and individuals might feel more tired than normal. When the light level is high all day, secretion will decrease and individuals might feel less able to fall asleep. **2.** Leptin promotes satiety, and so the mice with this mutation are likely to always feel hungry and therefore to overeat. **3.** The consequences will be wide-ranging and include a decrease in erythrocyte production due to decreased erythropoietin, a loss of bone mass due to lower levels of vitamin D (and so a decreased blood calcium ion concentration), and disrupted blood pressure regulation due to abnormal renin secretion.

MODULE 16.7

1. The ADH-secreting tumor will increase the blood volume and decrease the blood osmolarity. This will reduce secretion of aldosterone and stimulate the secretion of ANP in order to lower the blood volume (and blood pressure). **2.** The amounts of adrenal steroids and catecholamines might decrease slightly, but many other stimuli trigger secretion of these hormones. For this reason, the amounts secreted will not likely decrease significantly. **3.** Glucagon secretion will increase in order to raise the concentration of glucose in the blood.

Assess What You Learned

Level 1

1. **a.** True
   b. True
   c. False: The pancreas, thyroid gland, and parathyroid glands secrete *endocrine hormones*.
   d. False: *Most amino acid–based hormones* are hydrophilic molecules that bind to plasma membrane proteins as part of a second messenger system.
   e. True

2. d **3.** d

4. ADH increases the amount of water in the body by increasing water retention from the fluid in the kidneys and so decreasing the amount of water lost from the body as urine. This reduces the osmolarity of the blood.

5. hypothalamic-hypophyseal portal, anterior pituitary

6. **a.** Thyroid-stimulating hormone: target tissue, thyroid gland; stimulates production and release of thyroid hormones, growth and development of the thyroid gland
   b. Adrenocorticotropic hormone: target tissue, adrenal cortex and medulla; stimulation of secretion from the adrenal gland
   c. Prolactin: target tissue, mammary gland; stimulation of milk production
   d. Gonadotropins: target tissues, ovaries and testes; FSH stimulates production of chemicals that bind and concentrate testosterone by the testes in males and triggers the production of estrogens and maturation of ovarian follicles in females; LH promotes the production of testosterone in males and the production of estrogens and progesterone and the release of an oocyte in females.
   e. Growth hormone: multiple target tissues, including bone, muscle, adipose, liver, and cartilage; promotes fat breakdown, gluconeogenesis, protein synthesis, and cell division

7. b **8.** d

9. **a.** False: About 90% of the thyroid hormone produced is *thyroxine* (T_4).
   b. False: *Triiodothyronine* (T_3) is the more active of the two thyroid hormones.
   c. True
   d. False: Iodine molecules are a key component of *thyroid hormone*.
   e. False: *Decreased* secretion of thyroid hormones produces weight gain, cold intolerance, and slow heart rate.

10. decrease; increase   11. b

12. adrenal cortex, steroid hormones; adrenal medulla, catecholamines

13. c   14. c

15. The hypothalamus produces CRH, which stimulates secretion of ACTH from the anterior pituitary. ACTH, in turn, stimulates secretion from the adrenal cortex. When levels of adrenal hormones rise, production of CRH and ACTH decreases through negative feedback.

16. b

17. Glucagon triggers changes that increase the concentration of glucose in the blood, such as glycogen breakdown and gluconeogenesis. Insulin triggers changes that decrease the concentration of glucose in the blood, including the formation of glycogen; uptake of glucose, fatty acids, and amino acids; and storage of these molecules.

18. d   19. d

20. _e_ Leptin   _f_ Estrogens
    _c_ Atrial natriuretic peptide   _b_ Erythropoietin
    _a_ Melatonin   _d_ Testosterone

21. a. True
    b. False: Thyroid hormones and *glucagon* maintain blood glucose concentration during fasting.
    c. False: During exercise, the *sympathetic* nervous system increases the metabolic rate.
    d. True

Level 2

1. An insulin-secreting tumor would result in extreme hypoglycemia, as the excess insulin would trigger glucose uptake and storage and removal from the blood. A glucagon-secreting tumor would have the opposite effect, and lead to hyperglycemia due to excess glucose release and formation.

2. These tumors may produce androgenic steroids, which, when present in excess, promote the development of male secondary sex characteristics in females.

3. The hormones of the anterior pituitary will need to be replaced, because most of them are required for the function of other glands in the body, including the adrenal and thyroid glands. ADH and oxytocin will also need to be replaced, because although they are produced in the hypothalamus, they are released in the posterior pituitary.

Level 3

1. Her parathyroid glands were removed with her thyroid gland, and her symptoms are the result of a parathyroid hormone deficiency. Simply replacing her parathyroid hormone should be enough to restore calcium ion homeostasis.

2. Abnormal hypersecretion of cortisol does cause weight gain by increasing appetite and fluid retention and decreasing the metabolic rate. However, blocking *normal* cortisol secretion will not lead to weight loss because normal levels of cortisol do not cause weight gain.

3. If the supplement actually blocked normal cortisol secretion, it could lead to significant problems with glucose and other aspects of homeostasis, resulting in symptoms similar to those of Addison's disease.

4. Mr. Montez's blood glucose level will be elevated (hyperglycemia) because glucose is not entering the cells. Ingesting sugar will not help because his cells cannot take in the glucose, and will simply lead to even more profound hyperglycemia.

5. He is likely in diabetic ketoacidosis, meaning that his blood will have a high concentration of ketone bodies, increasing hydrogen ion concentration and lowering the pH of the blood. His buffer systems will respond by binding and removing the excess hydrogen ions from the blood, restoring the pH to normal.

6. Blocking aldosterone will decrease the retention of sodium ions from the fluid in the kidneys. This will increase the solute concentration of this fluid, which will cause water to remain in the fluid due to osmosis. Overall, the amount of water retained by the kidneys will decrease.

CHAPTER 17
Apply What You Learned

MODULE 17.1

1. The right side of the heart pumps deoxygenated blood into the pulmonary circuit. Failure of the right ventricle to pump adequately would reduce the oxygenation of the blood that normally occurs in the lungs.
2. In left-sided heart failure, oxygenated blood is not delivered to tissues adequately, so cells potentially won't receive enough oxygen and nutrients.

MODULE 17.2

1. Dysfunction of the papillary muscles and/or chordae tendineae would affect only the atrioventicular valves, as these muscles and cords do not attach to the semilunar valves. Such dysfunction could allow the AV valves to evert, letting blood leak back into the atria from the ventricles. 2. Death of cells in the left ventricle leads to the worst prognosis because it must pump blood to the entire systemic circuit. If the left ventricle fails, none of the body's cells are supplied with adequate oxygenated blood. The right ventricle does not have to contract as strongly to pump blood through the lower-resistance pulmonary circuit. 3. Coronary artery anastomoses provide alternate routes of blood flow when the flow through a vessel is blocked.

MODULE 17.3

1. Slowing calcium ion entry into the cell lengthens the plateau phase and therefore lengthens the contraction. These drugs also reduce the calcium ion concentration in the sarcoplasm, which leads to weaker contractions. 2. Blocking sodium ion channels in contractile cardiac cells would interfere with their ability to depolarize and have an action potential. These types of sodium ion channels are not present in the SA or AV nodal cells, so a local anesthetic would not directly affect their action potentials.

MODULE 17.4

1. Most of the blood drains from the atria to the ventricles passively, without any need for the atria to contract. For this reason, atrial fibrillation may not significantly affect end-diastolic volume in certain patients. 2. Heart failure generally causes the end-systolic volume to increase because the heart is unable to pump as much of the blood with each beat. The end-diastolic volume could potentially decrease if venous return decreases due to blood backing up in the pulmonary and systemic circuits. 3. If the Q-T interval is prolonged, the ventricles take longer than normal to complete an action potential. Under some circumstances, a long Q-T interval can lower the heart rate and can lengthen and strengthen the ventricular contraction, or ventricular ejection.

MODULE 17.5

1. Digoxin increases contractility, which will raise stroke volume and so cardiac output. 2. In this case, the heart would beat faster but not contract more strongly. This could actually lower cardiac output by lowering stroke volume.
3. a. The stroke volume is 40 ml.
   b. Her ejection fraction is about 36%.
   c. Her cardiac output is about 2.8 liters/min.
   d. A normal resting cardiac output is about 5 liters/min, so this patient's is lower than normal.

Assess What You Learned

Level 1

1. a. True
   b. False: The heart consists of two *inferior* ventricles and two *superior* atria.
   c. False: Arteries *generally* carry oxygenated blood away from the heart, and veins *generally* carry deoxygenated blood toward the heart.
   d. True
   e. True

2. c

3.
   e Auricle   j Pectinate muscle
   i Aorta   d Venae cavae
   a Coronary sinus   g Pulmonary trunk
   f Papillary muscle   b Chordae tendineae
   c Fossa ovalis   h Pulmonary veins

4. a

5. ascending aorta; posterior interventricular artery; anterior interventricular artery, circumflex artery

6. Pacemaker cells are autorhythmic, meaning that they spontaneously generate rhythmic action potentials. They contain fewer contractile elements and lack Na⁺ channels.

7. d

8. a. False: The rapid depolarization phase of the contractile cell action potential is due to the opening of voltage-gated *sodium* ion channels.
   b. True
   c. True
   d. False: The repolarization phase of the contractile cell is due to the potassium ions rushing *out of* the cell through potassium channels.
   e. False: Open *potassium* ion channels cause hyperpolarization in pacemaker cells, which triggers nonselective cation channels to open and begins a new action potential.

9. The plateau lengthens and strengthens the contraction, lengthens the action potential, and lengthens the cardiac refractory period, thus preventing tetanus.

10. b   11. c

12. a. P wave: depolarization of the atria
    b. QRS complex: depolarization of the ventricles
    c. T wave: repolarization of the ventricles
    d. P-R interval: the delay at the AV node
    e. S-T segment: the plateau phase of ventricular action potentials

13. a. True
    b. False: Atrial systole is responsible for ejecting *about 20%* of the blood into the ventricles during the inflow phase of the cardiac cycle.
    c. False: The amount of blood in the ventricles at the end of the inflow phase is the end-*diastolic* volume.
    d. True

14. d

15. S1, atrioventricular; isovolumetric contraction; S2, semilunar; isovolumetric relaxation

16. b   17. increase, Frank-Starling; decrease; increase   18. c

Level 2

1. a. The left ventricle is thicker-walled because it pumps against a greater resistance and therefore needs to be stronger and have greater muscle mass.
   b. This birth defect will cause the right ventricle to grow larger, as it has to pump against much greater resistance in the systemic circuit. Also, the left ventricle will atrophy because it is pumping against the lower resistance of the pulmonary circuit.

2. Both would cause significant problems and potential death of cardiac muscle and of the individual. However, blockage of the left coronary artery would be more damaging to long-term survival because that artery supplies the left ventricle. The left ventricle pumps blood into the systemic circuit, and so if its functions are compromised, body cells are not supplied with oxygen and nutrients.

3. A junctional rhythm is bradycardic, meaning that the heart rate is below 60 beats per minute, averaging about 40 beats per minute. The decrease in heart rate also lowers the cardiac output. We do not see P waves on the ECG because the atria are not depolarizing.

4. Heart failure results in a decrease in cardiac output. To compensate and increase the cardiac output, both stroke volume and heart rate increase. Fluid retention increases preload, which increases stroke volume. Sympathetic nervous system stimulation increases contractility through positive inotropic effects, which increases stroke volume and also elevates heart rate through positive chronotropic effects.

Level 3

1. The trainee should be informed that increasing his heart rate up to a certain point will increase his cardiac output. However, too much of an increase in heart rate will actually decrease cardiac output because the heart will have inadequate time to fill with and eject blood.

2. The nursing student should absolutely discard both syringes and fill new ones with proper labels. A syringe full of isotonic sodium chloride is unlikely to have any effect on the body's electrophysiology. However, a syringe full of potassium chloride could have a significant negative impact on the electrical processes of the cells of the heart, the neurons, and skeletal muscle cells. The potassium could potentially disrupt the resting membrane potentials of the cells, making the membrane potential more positive and the cells more easily excitable. Eventually the cells would become refractory and unable to have further action potentials.

3. The signs and symptoms of this condition include easy fatigability and weakness. This is because the condition allows blood to flow backward into the left atrium instead of the aorta, which reduces stroke volume and cardiac output. Mr. Watson's cardiac output might be improved by first fixing the defective valve. He might also be helped by drugs that increase contractility or decrease afterload, both of which increase stroke volume.

4. The refractory period of skeletal muscle fibers is quite short, which allows the fibers to be repeatedly stimulated and generate a sustained contraction. If the cardiac muscle cells had a similarly short refractory period, they would be able to maintain a sustained contraction if repeatedly stimulated. This could potentially result in death, as the heart chambers would not be able to go through diastole, as is usual in the cardiac cycle.

CHAPTER **18**
Apply What You Learned

MODULE **18.1**

1. The elastic fibers in an artery allow the vessel to stretch and recoil to its original size after being stretched. Defective elastic fibers would make a vessel either noncompliant (not able to stretch) or unable to recoil to its original size after stretching. 2. Veins would also be affected, as they are very compliant. However, they are under lower pressure than arteries and so the effects would be less obvious. 3. Atherosclerosis makes an artery narrow and hard through damage to the endothelium. This interferes with its function of delivering blood to capillary beds by decreasing the amount of blood that can flow through the artery.

MODULE **18.2**

1. Vasoconstriction increases resistance, which raises blood pressure. For this reason, regular consumption of a vasoconstrictor such as nicotine will

increase the risk of developing hypertension. **2.** Arteriosclerosis increases blood pressure because the vessels are no longer able to stretch to accommodate small increases in blood volume. **3.** Severe blood loss causes a significant decrease in blood volume, which decreases blood pressure.

MODULE **18.3**

1. a. Diuretics decrease blood volume, which decreases blood pressure.
  b. Angiotensin receptor-blockers block the effects of angiotensin-II on blood vessels, decreasing resistance and therefore blood pressure.
  c. Beta-blockers decrease the rate and force of heart contraction, decreasing cardiac output and blood pressure.
2. Extremely low blood pressure decreases blood flow to the kidneys, which in turn decreases filtration of fluid into the tubules of the kidneys. In addition, hormones such as ADH will trigger increased retention of water from the kidneys, further decreasing urine output. Reduced urine output decreases loss of blood volume. **3.** Salt increases the osmolarity of the extracellular fluid, which draws water out of cells by osmosis, increasing blood volume and blood pressure.

MODULE **18.4**

1. Thicker epithelium would slow transport of substances across the epithelium, which would considerably decrease the efficiency of nutrient and gas exchange between capillaries and tissues. **2.** This poison would partially destroy the blood-brain barrier, allowing more potentially toxic substances and pathogens into the brain's interstitial fluid than would normally occur. **3.** Perfusion to the heart muscle decreases during systole, so if the heart momentarily isn't contracting, perfusion will actually increase (this will stop, though, once the pressure gradient driving blood flow has ceased).

MODULE **18.5**

1. Hypotension would reduce the force driving water out of the capillaries, potentially causing the net filtration pressure to favor absorption instead of filtration. **2.** Osmotic pressure is normally higher in the blood than in the interstitial fluid because of blood proteins. However, if blood proteins entered the interstitial fluid, the osmotic pressure there would rise, potentially eliminating the gradient for osmosis. **3.** Poison X would cause the overall net filtration pressure to strongly favor filtration, resulting in potentially massive fluid loss from the blood to the interstitial fluid.

MODULE **18.6**

1. Although blockage of the right internal carotid artery would decrease blood flow to the brain, it wouldn't cut it off entirely due to the structure of the cerebral arterial circle, which provides alternate routes of blood flow to the brain. **2.** The blood supply to the hand is provided by the superficial and deep palmar arches, which form from an anastomosis of the ulnar and radial arteries. So although the blood flow to the hand and digits would be reduced if the ulnar artery were severed, it would not be completely disrupted. **3.** The femoral artery is the only major artery that supplies the entire lower limb. If it is irreparably damaged, little can be done to restore blood flow to the lower limb, resulting in loss of the limb.

MODULE **18.7**

1. The dural sinuses are not formed by regular veins, but rather are venous channels between layers of the dura mater. The dura mater is not very distensible. Therefore, compliance of the dural sinuses is low compared to that of other veins. **2.** A drug such as ticarcillin is ineffective when administered by mouth, because most of never enters the capillary beds of the regular systemic circuit. However, if such a drug is injected into a vein, it will go to the heart, then the arteries, and then the capillary beds to be delivered to the tissues, where it will have an effect. **3.** Removal of the great saphenous vein does decrease venous drainage slightly. However, the venous system is characterized by a large number of anastomoses, so generally the venous blood is simply diverted to another vein.

MODULE **18.8**

1. Blood always flows down a pressure gradient. The pressure in the arteries is higher than the pressure in the veins, so blood can only flow from arteries to veins. **2.** The internal jugular vein is adjacent to the carotid artery, whereas the external jugular vein is more superficial and is separated from the carotid artery by muscles in the neck.

Assess What You Learned

Level 1

1. a. False: *Capillaries* are the exchange vessels of the cardiovascular system.
  b. True
  c. True
  d. False: Muscular arteries control blood flow at the *organ* level.
  e. False: *Arteries* have smaller lumens, more elastic fibers, and more smooth muscle than *veins*.

2. c

3. a. A pressure gradient is a situation in which pressure is higher in one area than in another.
  b. Blood pressure is the force exerted on a blood vessel wall by the blood.
  c. Blood flow is the volume of blood that flows per minute.
  d. Resistance is any impedance to blood flow.

4. a   **5.** b   **6.** systolic, diastolic; pulse   **7.** a

8. Venous valves prevent blood from flowing backward; smooth muscle in the walls of veins can contract to reduce the volume; the skeletal muscle pump and the respiratory pump move blood toward the heart.

9. a. True
  b. True
  c. False: Hormones that *increase* blood volume and blood pressure include antidiuretic hormone and aldosterone.
  d. True
  e. True

10. a   **11.** c

12. Substances may cross the wall (1) through gaps or pores in the endothelial layer, (2) through the membranes of endothelial cells by diffusion, or (3) through the endothelial cells by transcytosis.

13. b

14. a. False: Arterioles reflexively *constrict* when blood pressure increases.
  b. False: Increased concentrations of carbon dioxide and hydrogen ions and a decreased concentration of oxygen in the interstitial fluid cause local arteriolar *dilation*.
  c. True
  d. True
  e. False: The sympathetic nervous system causes *vasoconstriction* in the skin when body temperature decreases.

15. b   **16.** c

17. __d__ Radial artery   __e__ Femoral artery
     __g__ Celiac trunk   __b__ Internal iliac artery
     __j__ Basilar artery   __i__ Renal artery
     __a__ Superior mesenteric artery   __f__ Internal carotid artery
     __h__ Dorsalis pedis artery   __c__ Subclavian artery

18. b   **19.** d

20. __g__ Cephalic vein   __i__ Splenic vein
     __d__ Great saphenous vein   __e__ Internal jugular vein
     __j__ Dural sinus   __b__ Brachiocephalic veins
     __a__ Azygos vein   __h__ Brachial vein
     __c__ Hepatic portal vein   __f__ Inferior mesenteric vein

Level 2

1. The blood in an artery is under much higher pressure than the blood in a vein. For this reason, blood is propelled out of a severed artery with much greater force than it is out of a severed vein.

2. A very tall person simply has more blood vessels that are longer. This increases the resistance of the circuit, and the increase in resistance leads to an increase in blood pressure.

3. The collagen in the tunica externa prevents blood vessels from stretching excessively when the blood pressure increases during systole. Dysfunctional collagen fibers would weaken the vessels, and they could abnormally expand, potentially causing an aneurysm.

Level 3

1. A person hanging upside-down increases the blood flow to the head, which increases the pressure detected by the baroreceptors in the carotid sinus. This will trigger the baroreceptor reflex, which will lead to an overall decrease in systemic blood pressure. When a person in this situation returns to an upright position, his or her blood pressure will be quite low, leading to dizziness and temporary feelings of being ill.

2. The loss of albumin will decrease colloid osmotic pressure and cause the net filtration pressure to further favor filtration. Eventually, this will cause an excessive loss of fluid from the blood to the interstitial fluid, leading to edema.

3. a. An increase in the number of erythrocytes will increase the viscosity of the blood, so resistance and blood pressure will increase.
   b. After ingestion of caffeine, both vasoconstriction and increased cardiac output will increase the blood pressure.
   c. A bleeding ulcer will decrease blood volume, which will in turn decrease blood pressure.

4. The precapillary sphincters in the skin cut off blood flow to the skin to divert blood to deeper tissues as a heat conservation mechanism. The lack of oxygenated blood in the capillaries of the skin causes the skin to take on a bluish hue, particularly in the lips, where the epidermis is thin.

5. SIADH will increase the amount of water retained from the kidneys and therefore the blood volume. Increased blood volume will elevate blood pressure and raise the hydrostatic pressure in capillary beds, causing the net filtration pressure to further favor filtration.

CHAPTER 19
Apply What You Learned

MODULE 19.1

1. A decreased amount of albumin in the plasma will decrease its colloid osmotic pressure. As a result, less water will be drawn into capillaries at the venous end of a capillary bed, leading to a net loss of more water than normal from the plasma. The water lost to the interstitial fluid will accumulate there, leading to edema. 2. Proteins are composed of largely polar amino acids (although some have nonpolar R groups) and therefore mix with polar water. Lipids are composed largely of nonpolar hydrocarbons and therefore do not mix with polar water. Instead, lipids would congregate, forming a clump in the blood, which could block blood flow. For this reason, they must be transported through the blood on plasma proteins. 3. Extensive blood loss impairs the ability of blood to carry out its homeostatic functions. This will disrupt gas exchange, solute distribution, the volume of extracellular fluid and cytosol, the body's acid-base balance, and blood pressure.

MODULE 19.2

1. An erythrocyte is essentially a "bag" of cytosol with hemoglobin molecules. This structure allows an erythrocyte to carry a great number of oxygen molecules (and some carbon dioxide), and its disc shape gives it a large surface area for rapid diffusion of gases. In addition, the lack of most organelles and the disc shape make the erythrocyte flexible so that it can fit through small vessels without sustaining damage. 2. Erythropoietin increases a person's hematocrit and therefore the oxygen-carrying capacity of the blood. An athlete who can carry more oxygen in his or her blood will likely have greater endurance and resistance to muscular fatigue. However, an abnormally elevated hematocrit increases the viscosity of the blood and makes blood flow sluggish and could potentially block the flow through blood vessels.

MODULE 19.3

1. a. The primary leukocyte that responds to bacterial infections is the neutrophil, so one would expect to see elevated levels of neutrophils in this case.
   b. Virally infected cells are destroyed by T lymphocytes, so we see elevated levels of T lymphocytes with viral infections.
   c. The main leukocyte that responds to parasitic infections is the eosinophil, so elevated eosinophil levels are seen with such infections.
   d. Both macrophages and neutrophils ingest dead and dying cells after tissue injury, so one may see elevated levels of both monocytes and neutrophils in the blood after trauma.

MODULE 19.4

1. This will not affect already formed platelets, which lack nuclei and other organelles required for protein synthesis. It will, however, affect the formation of new platelets via thrombopoiesis. 2. Megakaryocytes undergo repeated rounds of mitosis to become cells with a large cytoplasmic mass. If their mitosis is inhibited, they will be much smaller cells (and fewer in number), resulting in decreased levels of platelets.

MODULE 19.5

1. The activation of these three factors will stimulate coagulation, causing the blood to begin to clot abnormally. 2. These molecules will have the opposite effects of those in question 1—they will inhibit coagulation and lead to abnormal bleeding. 3. tPA activates plasmin, which catalyzes the reactions that degrade fibrin and break up clots. After CPR, there is generally some amount of internal injury to blood vessels, and clots prevent blood loss. If tPA is administered, these clots will dissolve and the patient will bleed internally.

MODULE 19.6

1. Mr. Ramirez cannot donate blood safely to Mr. Jones. With type A− blood, Mr. Jones has anti-B antibodies and possibly anti-Rh antibodies in his plasma (if he has been exposed to Rh+ blood previously). These antibodies will bind to the B and Rh antigens of Mr. Ramirez's blood. 2. Ms. Wright can donate blood to Mr. Jones safely only if he has not been exposed to Rh+ blood previously. If he has had such prior exposure, his anti-Rh antibodies will bind to her Rh antigens. 3. A person with type O− blood has no A, B, or Rh antigens on the surface of the erythrocytes, and so there can be no reaction with a recipient's anti-A, anti-B, or anti-Rh antibodies. However, this person's plasma contains anti-A, anti-B, and possibly anti-Rh antibodies, which would react with any blood type other than O−.

Assess What You Learned

Level 1

1. c   2. b

3. Gas exchange, solute distribution, immune functions, maintaining body temperature, blood clotting, maintaining acid-base homeostasis, stabilizing blood pressure

4. a. False: Erythrocytes are biconcave discs with *no nuclei as mature cells*.
   b. True
   c. True
   d. False: Hemoglobin consists of *four* polypeptide chains *each* bound to a heme group.

5. d 6. hematopoietic stem cells, 100–120, spleen 7. a 8. b

9. _d_ Basophil
_f_ B lymphocyte
_a_ Neutrophil
_e_ Monocyte
_c_ T lymphocyte
_b_ Eosinophil

10. lymphoid, myeloid 11. c

12. _3_ The intrinsic and extrinsic pathways produce factor Xa.
_6_ The clot retracts.
_5_ Thrombin converts fibrinogen to fibrin and fibrin glues the plug together.
_2_ Platelets are activated, and the platelet plug forms.
_1_ Vasoconstriction and increased tissue pressure decrease blood flow through the vessel.
_7_ Tissue plasminogen activator activates plasmin, which degrades fibrin.
_4_ The common pathway produces thrombin.

13. The intrinsic pathway begins when inactive factor XII comes into contact with exposed collagen, whereas the extrinsic pathway begins when subendothelial cells display tissue factor, which activates factor VII. In addition, the extrinsic pathway involves fewer steps to the activation of factor X (thrombin) than does the intrinsic pathway. The pathways are similar in that they both culminate in the activation of factor X, which in turn leads to the common pathway and the activation of thrombin and the production of fibrin.

14. The common pathway converts fibrinogen to fibrin. Fibrin then "glues" the platelet plug and other formed elements together to form a blood clot.

15. d 16. b 17. ABO, Rh 18. a 19. d 20. b, c, and e

Level 2

1. The formed elements of blood must be distributed throughout the entire body to perform their functions—deliver oxygen to the tissues, perform immune functions, and clot damaged vessels—which is possible only in a liquid connective tissue. In addition, the plasma proteins and dissolved solutes must also be distributed throughout the body, which again is possible only in a liquid medium.

2. Oxygen molecules bind to the iron-containing heme groups, of which there are normally four. If only two iron ions are present, there are half as many binding sites for oxygen molecules, which means the blood is capable of carrying half as much oxygen. This will lead to a reduction in oxygenation of the tissues, which will trigger erythropoietin production and increase the number of erythrocytes in the blood. The excess erythrocytes will increase the oxygen-carrying capacity of the blood but will also increase the risk for blood clots, as an increased number of erythrocytes will make blood flow sluggish.

3. Disrupting either the intrinsic or extrinsic pathway leads to disruption of the entire process because both pathways rely heavily on one another to proceed to completion. In addition, the extrinsic pathway is the most important path to the production of factor Xa, which is required for the common pathway.

4. A decrease in albumin will result in a decrease in osmotic pressure and a resulting loss of water from blood vessels into the interstitial fluid. A decrease in clotting proteins will interfere with the blood's ability to clot and may prolong bleeding time when an injury happens. Cirrhosis will not affect the production of γ-globulins, as they are produced by B lymphocytes, not the liver.

Level 3

1. Neutrophils are very active phagocytes, particularly in response to a bacterial infection. A deficiency in neutrophils will make a patient especially at risk for bacterial infections. T lymphocytes are involved in directly killing virally infected cells and cancer cells, but they are also responsible for activating other cells and components of the immune system. With a decrease in T lymphocytes, a patient is at risk for all types of infectious agents and certain cancers.

2. Platelets will not help Mr. Jackson, as he is able to form a platelet plug. He has a deficiency in a clotting factor, which prevents coagulation. This means that fibrin will not be produced and the platelet plug will not be "glued" together. Giving him platelets will only increase the size of the ineffectual platelet plug.

3. None of her family members could safely donate blood to Ms. Wu. Her blood type is O−, meaning she lacks A, B, and Rh antigens and has anti-A, anti-B, and anti-Rh antibodies (we know that she has anti-Rh antibodies because her daughter is type O+, so she has been exposed to the Rh antigen). All of the individuals who wish to donate to Ms. Wu have B and/or Rh antigens, so they cannot donate.

4. Since Elise is an adult, she is unlikely to have red bone marrow in her tibia, as much of the red bone marrow in an adult is replaced with fat-filled yellow bone marrow. A better place to withdraw red bone marrow from Elise would be the iliac crest, which does contain red bone marrow in an adult and is relatively superficial and easily accessed surgically.

CHAPTER 20
Apply What You Learned

MODULE 20.1

1. Removal of the lymph nodes and vessels will increase the patient's risk of infection and also decrease the return of excess interstitial fluid back to the cardiovascular system, leading to edema. 2. Bacteria or other pathogens were likely introduced into the wound, after which they traveled through the lymphatic vessels to the inguinal lymph nodes, where they were trapped.

3. a. Removal of the spleen will increase the risk of infection slightly. In addition, destruction of old erythrocytes will have to take place elsewhere, such as in the plasma and liver.
b. Removal of the thymus in an adult will likely have few consequences, as T cell maturation has already taken place.
c. Removal of the thymus in a newborn will stop T cell maturation, leading to severe dysfunction in the entire immune system.
d. Removal of the tonsils will increase the risk of infection slightly, particularly around the oral and nasal cavities.

MODULE 20.2

1. Erosion of this mucous membrane compromises the first line of defense against pathogens and tissue damage. As a result, the urinary bladder is at higher risk of infection and damage from chemicals in urine. 2. All three lines of defense rely on one another to function properly. If one line fails, the other two have great difficulty in protecting the body.

MODULE 20.3

1. A deficiency in neutrophils puts one at great risk of bacterial infection, as neutrophils are one of the main types of cells responsible for killing bacteria.
2. Although it is possible that the patient has an infection, the fever could also be caused by inflammation due to tissue damage from a heart attack.
3. A fever is a sign that inflammation is occurring in the body. Although aspirin might make the fever decrease, it will not treat the underlying cause of the inflammation. For this reason, further examination is warranted any time a patient in such a setting develops an unexpected fever.

MODULE 20.4

1. T_H cells secrete interleukin-2, which is necessary for T_C cells to become fully activated. If T_H cells are inhibited, T_C cells will do less damage to the cells of the donated organ. 2. The drug would decrease

many other components of the immune response, including innate and antibody-mediated immunity. **3.** The change in the self antigens would likely cause the cells to be recognized by T_C cells as foreign, and T_C cells would attack and destroy the kidney cells.

MODULE **20.5**

1. The lack of memory cells would effectively prevent the development of immunological memory, and a secondary immune response would be impossible. For this reason, every encounter with an antigen would initiate a primary immune response.

2. a. Decreased secretion of IgA will decrease the effectiveness of surface barriers that release IgA with their secretions.

b. IgG is the main antibody secreted in the secondary immune response. For this reason, a deficiency in IgG will prolong the duration and increase the severity of infection with many different types of pathogens.

c. A deficiency of IgG in a pregnant female puts both the mother and fetus at higher risk for developing prolonged and severe infections. It is particularly dangerous for the fetus, as maternal IgG is the only source of immunity before and immediately after birth.

d. A decrease in IgM will prolong the primary immune response and potentially increase the severity of an infection, as IgM is generally the first antibody to arrive and contain an infection.

3. The tetanus antitoxin provided passive immunity, as it contains pre-formed antibodies that will last only about 3 months. The tetanus antigens stimulates your immune system to produce its own antibodies, and so provides active immunity.

MODULE **20.6**

1. An inability to activate T_C cells would decrease the ability of the body to fight off the viral infection, as T_C cells would not lyse and kill infected cells. Essentially every other component of the immune response, particularly T_H, NK, and B cells, would need to compensate for the lack of T_C cells. **2.** The result would be similar—T_C cells would be unable to detect and destroy the cancer cells. **3.** The fever indicates that an inflammatory response is occurring in the patient's body. **4.** The elevated level of neutrophils points to a bacterial infection as the possible cause of the fever, but this can't be determined definitively without more information.

MODULE **20.7**

1. Such a drug will decrease allergen-mediated inflammation by preventing the release of mast cell granules. It would primarily affect type I hypersensitivity. **2.** The antibodies secreted by B cells would bind to a specific HIV-1 antigen, which could temporarily slow the infection. However, as the virus mutates, the antibodies would no longer bind the viral antigens, and a new B cell clone would be required. Note that the B cell response also depends on the levels of T_H cells—if there is a very low number of T_H cells, B cells won't activate fully or produce the higher levels of antibodies. **3.** Suppressing the immune system will decrease the production of autoantibodies, reducing some of the symptoms of an autoimmune disorder. However, this suppression will also raise the risk of infection.

Assess What You Learned

Level 1

1. b

2. a. False: Lymphatic vessels constitute a one-way system that delivers fluid from the *extracellular space to the blood vessels.*

b. False: Lymph from the lower limbs drains into the *thoracic* duct.

c. True

d. False: The thymus is the site of maturation of *T* cells.

e. False: Clusters of MALT located around the oral and nasal cavities are known as *tonsils.*

3. spleen, lymph node   **4.** innate immunity; acquired immunity; memory

5. a   **6.** d

7. a. True

b. False: Phagocytic cells of innate immunity include *macrophages* and *neutrophils.*

c. True

d. False: *Interferon* is a cytokine that prevents viral replication in infected cells.

e. False: Fever is generated by pyrogens that reset the temperature set point of the hypothalamus to a *higher* value.

8. e   **9.** inflammatory mediators, edema, redness, pain, heat   **10.** c

11.   _e_ IgD
   _d_ IgM
   _b_ IgG
   _a_ IgA
   _c_ IgE

12. a. True

b. False: The secondary immune response is mediated by *memory* cells.

c. False: *Attenuated* vaccines consist of pathogens that are alive but unable to cause disease.

d. False: Vaccinations are given to induce the production of *memory cells.*

13. Active immunity involves exposure to an antigen and a subsequent primary immune response in which antibodies and memory cells are generated. Passive immunity involves the transfer of preformed antibodies into an individual. Passive immunity does not result in the production of memory cells or a secondary immune response.

14. c   **15.** endogenous, cytotoxic T; exogenous, helper T   **16.** d

17. a. False: The cells involved in organ and transplant rejection are primarily T_C cells.

b. True

c. False: Neutrophils are a critical component of the response to a *bacterial* infection.

d. True

18. d

19. HIV mediates the destruction of T_H cells, and these cells are responsible for activating most elements of the specific and nonspecific immune responses. Without these cells, a patient is at risk for bacterial, viral, and parasitic infections. In addition, immune surveillance is compromised, leading to an increased risk for certain cancers.

20. e

Level 2

1. a. The body aches are a result of inflammation, and the fever and chills are a result of pyrogens resetting the hypothalamic temperature set point. Both inflammation and fever are part of the innate immune response.

b. As neutrophils are the body's main response to bacteria, the infection is likely bacterial in nature. The elevated number of neutrophils means that the body is producing more neutrophils to deal with the increased number of bacteria.

c. If lymphocytes were elevated instead, this would indicate a possible viral infection, as lymphocytes are the main immune system response to viruses. It could also indicate other generalized causes of systemic inflammation such as cancer or a severe injury.

2. A complement deficiency could result in a type III hypersensitivity disorder, as the uncleared immune complexes could deposit in tissues and create inflammatory reactions. It could also depress other processes that complement proteins support, including lysing bacteria, enhancing inflammation, neutralizing viruses, and enhancing phagocytosis through opsonization.

3. IgE is the main antibody that responds to allergens. Individuals with type I hypersensitivity disorders tend to produce excess IgE that responds abnormally to allergens. Blocking the functioning of

IgE would prevent it from activating basophils and mast cells, and therefore reduce Terrence's symptoms.

Level 3

1. This is an example of passive immunity, as the antibodies are pre-formed and not made by your friend's plasma cells. The antivenin will not confer lasting immunity; however, your friend will produce her own antibodies after exposure to the snake venom via an active response, which will have a longer-lasting effect.

2. a. Cancer cells likely broke off from the tumor at the start of metastasis. Once in the interstitial fluid, they entered a lymph vessel and were delivered to a lymph node. Here, they encountered T cells, macrophages, and other immune cells that mounted an inflammatory response, which resulted in the swelling of the nodes.
   b. Removal of lymphatic vessels will inhibit the drainage of excess interstitial fluid, causing it to accumulate in the extracellular space. This results in sometimes significant edema. Removal of the lymph nodes reduces the body's ability to respond to pathogens, although it is likely that the pathogens would simply be caught by other, neighboring lymph nodes.

3. Neutrophils are the body's main defense against bacteria, so when the number of neutrophils is low, the body is unable to respond as it normally would to bacterial pathogens. This is particularly dangerous because the normal response to a bacterial infection is an *increase* in neutrophil numbers; the fact that the neutrophil count is already too low combined with the inability to increase it could make the bacterial infection rapidly lethal.

4. One would expect to see a decrease in the numbers of all types of leukocytes, as they are all produced by the bone marrow. The change would also cause a decrease in the hematocrit, as erythrocytes are also produced by the bone marrow. The decreased hematocrit would result in symptoms of anemia: fatigue, weakness, and pallor.

5. The cells of the epidermis normally undergo rapid mitosis to replace cells that die and are sloughed. If the epidermis is unable to undergo rapid mitosis, the skin will thin and be more prone to injury. This reduces its ability to function as the first line of defense of the immune system.

CHAPTER 21
Apply What You Learned

MODULE 21.1

1. Blocking the structures of the conducting zone would prevent delivery of air to and from the structures where gas exchange occurs. Blocking structures of the respiratory zone would directly impair gas exchange, as this is where it takes place. 2. Of course, the number one homeostatic disturbance would be a lack of oxygenation of the cells. Respiratory diseases also lead to imbalances in the pH of the body's fluids and potential abnormalities in blood pressure and fluid balance.

MODULE 21.2

1. The pseudostratified columar epithelium with goblet cells is a component of the mucous membrane lining much of the respiratory tract. It secretes mucus, which traps the inhaled debris, and the cilia projecting from the apical surfaces of the epithelial cells sweep the debris out of the respiratory tract. 2. Inflammation of the larynx and vocal folds makes the folds much less elastic. As a result, they don't vibrate when air passes over them, and thus the person "loses" his or her voice. 3. The tracheal rings keep the trachea patent, and if they were absent, the trachea could potentially collapse and block ventilation. The flexible nature of the cartilage rings allows the trachea to remain firmly open yet able to move with the body and surrounding organs. Bone is not flexible, and if the rings were composed of bone, the trachea would be stiff and inflexible. 4. The type I alveolar cells are simple squamous epithelial cells. They are extremely thin, a feature that allows gases to cross their apical and basal membranes very rapidly.

MODULE 21.3

1. The blood in the left pleural cavity will increase the intrapleural pressure. If the intrapleural pressure rises above the intrapulmonary pressure, the patient's left lung will collapse.
2. a. The increased amount of mucus in the respiratory tract will increase airway resistance, which will decrease the effectiveness of pulmonary ventilation.
   b. Pulmonary edema increases surface tension on the cells of the alveoli. This causes atelectasis, which also decreases the effectiveness of pulmonary ventilation.
   c. Inspiratory capacity = 3450 ml; vital capacity = 4100 ml. Inspiratory capacity is slightly lower than normal, and vital capacity is significantly lower than normal.

MODULE 21.4

1. An increase in the thickness of the respiratory membrane increases the distance over which gases must diffuse to enter and exit the blood. This will decrease the effectiveness of pulmonary gas exchange. 2. Low partial pressures of oxygen in the alveoli cause the pulmonary arterioles to constrict. If this happens over the entire pulmonary circuit, blood pressure in this circuit will increase. 3. This will decrease the amount of oxygen and nutrients delivered to the tissues. Not only will blood flow be reduced by the vasoconstriction, but the damage to the capillaries will cause them to thicken, which will decrease the efficiency of gas exchange.

MODULE 21.5

1. Anything that increases hemoglobin's affinity for oxygen will decrease its ability to release oxygen to the tissues. As a result, the amount of oxygen in the venous blood will remain nearly as high as in the arterial circuit, and the tissue cells will be starved of oxygen.
2. a. A fever increases the temperature of the tissues, which will increase the unloading of oxygen from hemoglobin.
   b. A lower number of hydrogen ions in the blood will decrease the unloading of oxygen from hemoglobin.
   c. An elevated level of carbon dioxide in the blood will increase the unloading of oxygen from hemoglobin.
3. The forward reaction must take place in the capillary beds of the systemic circuit in order to convert carbon dioxide to bicarbonate. In the pulmonary circuit, the reverse reaction must take place to convert the bicarbonate back into carbon dioxide so that it may be exhaled.

MODULE 21.7

1. The neurons in the VRG directly stimulate the neurons that stimulate the muscles of inspiration. Without a functioning VRG, the patient will be unable to breathe on her own and will have a poor prognosis. 2. The central and peripheral chemoreceptors lose their sensitivity over time when exposed to continually high partial pressures of carbon dioxide. This patient has chronic pulmonary disease, and the lab results indicate that this is what has happened.

MODULE 21.8

1. The residual volume is the amount of air that remains in the lungs after expiration, or the air that is not exchangeable. The vital capacity includes all of the exchangeable volumes of air. If the residual volume increases, the volumes of exchangeable air all decrease, which in turn decreases vital capacity. 2. Without functional cilia, mucus and inhaled debris are not cleared from the respiratory tract in the normal way, and the only way to clear the tract is with a deep cough. 3. During an asthma attack, a person hypoventilates as a result of the obstruction of the airways. This increases the retention of carbon dioxide, which in turn increases the amount of hydrogen ions in the blood. The elevated level of hydrogen ions stimulates the central chemoreceptors, which in turn increase the rate of ventilation to blow off the excess carbon dioxide.

Assess What You Learned

Level 1

1. a, c, e, f   2. respiratory tract, alveoli; pleural

3. a. False: The trachea contains *C*-shaped rings of hyaline cartilage.
   b. False: Goblet cells secrete *mucus* into the respiratory tract.
   c. True
   d. True
   e. False: The epithelium of the oropharynx changes from *pseudostratified ciliated* columnar epithelium to *stratified* squamous epithelium to enable it to resist abrasion from food.

4. d   5. true vocal cords (or vocal folds); adducted, abducted   6. b

7. a. Type *II* alveolar cells secrete a chemical called surfactant.
   b. *Type I alveolar cells* are squamous epithelial cells that constitute 90% of the total cells in the alveoli.
   c. The respiratory membrane consists of the type *I* alveolar cells, the pulmonary capillaries' endothelial cells, and their shared basal lamina.
   d. The respiratory membrane must be *thin* to function effectively in gas exchange.

8. a

9. __c__ Airway resistance
   __d__ Surface tension
   __a__ Surfactant
   __e__ Pulmonary compliance
   __b__ V/Q ratio

10. a. True
    b. False: The *residual* volume is the amount of air left in the lungs after maximal expiration.
    c. False: The inspiratory capacity is equal to the *inspiratory reserve volume* plus the tidal volume.
    d. True

11. c   12. d   13. constrict; dilate

14. a. False: Oxygen is transported primarily by *binding to hemoglobin*.
    b. True
    c. True
    d. False: When the P_{O_2} of arterial blood drops slightly, the percent saturation of Hb drops *only slightly*.
    e. False: Increased P_{CO_2}, temperature, and hydrogen ion concentration all *decrease* Hb's affinity for oxygen.

15. __d__ Bicarbonate   __c__ Carbonic anhydrase
    __e__ Hemoglobin   __a__ Chloride shift
    __f__ BPG   __b__ Carbaminohemoglobin

16. decrease, increase; increase, decrease   17. d

18. a. False: The *respiratory rhythm generator* maintains the respiratory rhythm.
    b. True
    c. True
    d. False: The respiratory groups in the *medullary reticular formation* maintain eupnea.

19. arterial P_{CO_2} and hydrogen ions within the extracellular fluid of the brain; arterial P_{O_2}.

20. d

Level 2

1. Boyle's law states that pressure and volume are inversely related. This means that as the volume of a container increases, the pressure decreases, and as the volume of a container decreases, the pressure increases. The pressure inside the cylinder in this case would therefore increase as its volume decreased.

2. The esophagus is located on the posterior side of the trachea, which lacks cartilage rings of support. When the esophagus expands, it pushes into the trachea's soft posterior side, giving the feeling of being less able to breathe in spite of the fact that the food is in the esophagus, not the respiratory tract.

3. Both volumes would decrease dramatically, as the phrenic nerve is responsible for innervating the diaphragm and stimulating it to contract. Without the phrenic nerve, the main muscle of inspiration would be unable to contract, and inspiration would be severely decreased. The accessory muscles of inspiration and the external intercostals would try to increase the volume of the thoracic cavity in an effort to compensate.

4. The epiglottis covers the larynx during swallowing to prevent food and liquid from entering the respiratory tract. If the epiglottis is swollen, it can potentially seal off the larynx during ventilation as well, and cause the person to suffocate.

Level 3

1. A person loses carbon dioxide during hyperventilation, which increases the pH of his or her blood. Breathing into a paper bag can help because the bag contains exhaled carbon dioxide, which the person will rebreathe. This returns the partial pressure of carbon dioxide in the blood to normal and reduces the pH.

2. An individual with more erythrocytes is able to carry more oxygen in the blood, which increases the capacity for endurance. Elevated levels of BPG decrease hemoglobin's affinity for oxygen, making it easier to unload in the tissues. This also increases the capacity for endurance.

3. When your friend hyperventilated, he decreased the amount of carbon dioxide in his blood. This made the accumulation of carbon dioxide as he held his breath have less effect on the overall pH of his blood, resulting in less stimulation of the central chemoreceptors. This enabled him to hold his breath for a longer time.

4. Such a person would hypoventilate. The decrease in hydrogen ions in his or her blood would reduce the conversion of bicarbonate ions to carbonic acid, which could dissociate into CO_2 and water. Central chemoreceptors would detect the resulting decline in P_{CO_2} and trigger hypoventilation. That in turn would retain carbon dioxide and decrease the blood pH to a more normal range.

5. a. Diabetic ketoacidosis is the accumulation of ketones in the blood as a result of excessive catabolism of fatty acids. It results from insufficient insulin, which makes the individual unable to take glucose into the cells.
   b. Mrs. Jordan likely has type I diabetes mellitus, the type that is more prone to lead to ketoacidosis.
   c. As the name implies, ketoacidosis results in a decrease in blood pH. This is due to the accumulation of ketones, which are organic acids.
   d. Her hyperventilation was triggered to correct the pH of her blood. As she hyperventilates, she blows off carbon dioxide, which decreases the production of carbonic acid and therefore reduces the total number of hydrogen ions in her blood.

6. The rate of ATP production increases during exercise increases to meet the demands of the contracting muscle fibers. A waste product of this ATP production is carbon dioxide. The accumulated carbon dioxide triggers the central chemoreceptors to increase the rate of ventilation in order to rid the body of it.

CHAPTER 22
Apply What You Learned

MODULE 22.1

1. The peritoneal membranes surround a number of organs within the abdominal cavity. For this reason, it is quite easy for an infection to spread rapidly from one organ to another. 2. The digestive system plays a role in many homeostatic processes, including regulation of

fluid balance, acid-base homeostasis, and metabolic homeostasis. A disorder of the digestive system can therefore interfere with any of these processes. **3.** The epithelium of the alimentary canal contains regenerative cells that have a rapid rate of mitosis to replace the cells that are sloughed off or damaged. Chemotherapy drugs that cause the death of rapidly dividing cells will kill many of these cells, reducing the ability of the mucosa of the alimentary canal to maintain its epithelial lining.

MODULE **22.2**

1. Saliva washes away many of the acids and other material produced by the normal flora in the mouth that cause tooth decay. In addition, it contains substances such as lysozyme and IgA that kill pathogenic bacteria. Any condition that decreases the production of saliva will therefore create favorable conditions for bacterial activity and tooth decay. **2.** Secretion of saliva is stimulated by parasympathetic neurons that release acetylcholine (ACh) onto target cells. Anything that increases the amount of ACh will also increase the production and release of saliva, which results in drooling. **3.** Mr. Kaku's difficulty in swallowing can lead to aspiration of food because the swallowing reflex doesn't properly trigger elevation of the larynx. In addition, the decrease in activity of the muscularis externa of the esophagus will decrease peristaltic activity, potentially allowing the bolus to become lodged in the esophagus.

MODULE **22.3**

1. When the stomach cannot empty normally, the pressure inside it rises. This distends the stomach, which can trigger a vomiting reflex. Thus, individuals with pyloric stenosis can suffer from frequent vomiting, resulting in malnutrition and fluid, electrolyte, and acid-base imbalances. **2.** The vagus nerve stimulates acid secretion from parietal cells, so severing this nerve will decrease acid secretion. However, it will also decrease the activation of pepsinogen and so affect protein digestion in the stomach.

MODULE **22.4**

1. Celiac disease can severely decrease the surface area available for digestion and absorption of nutrients in the small intestine. This leads to the presence of undigested nutrients and excess unabsorbed water and electrolytes in the feces, which will likely cause chronic diarrhea. In addition, it can lead to malnutrition and fluid, electrolyte, and acid-base imbalances. **2.** A severed vagus nerve would decrease the motility of the stomach; however, it would not eliminate it. Much of the churning motion in the stomach is controlled by ENS neurons and hormones such as serotonin and intestinal gastrin. In the small intestine, the migrating motor complex would continue to function under the influence of the ENS and motilin, but peristalsis and segmentation would be significantly diminished.

MODULE **22.5**

1. Oral antibiotics often kill off the normal flora in the large intestine. The main danger from this is that pathogenic bacteria that are not killed by the antibiotic may take over and cause infection. Ingesting supplements containing the normal flora helps to replace flora killed by the antibiotic, preventing pathogenic bacteria from causing infection. **2.** Cutting of the spinal cord would prevent conscious control of defecation, as the spinal cord would no longer be able to relay impulses from the cerebral cortex to voluntarily control the external anal sphincter and levator ani muscle.

MODULE **22.6**

1. Pancreatic acini produce pancreatic juice, which contains bicarbonate ions, water, and multiple digestive enzymes needed to digest all types of nutrients. Removal of the pancreas would therefore severely impact chemical digestion, and supplementary digestive enzymes would need to be administered. **2.** Under normal conditions, the liver will metabolize the drug and steadily decrease its concentration in the blood over a period of several hours. However, in a patient with liver disease, this function will be performed much less efficiently, prolonging the life of the drug in the blood. As a result, a lower dose is needed.

MODULE **22.7**

1. By preventing the activity of pancreatic lipase, orlistat blocks chemical digestion of lipids, and therefore also their absorption. This decreases the amount of dietary fat that the body absorbs, which could potentially lead to weight loss. Unfortunately, it can also lead to a deficiency of fat-soluble vitamins. **2. a.** Being proteins, the supplemental digestive enzymes are likely to be damaged and possibly denatured by the acid in the stomach, and pepsin will begin digesting them into smaller polypeptides. **b.** Such supplements are fairly useless, as the enzymes will be rendered nonfunctional in the stomach. A healthy person probably doesn't need supplemental digestive enzymes unless a specific deficiency is noted. **3.** If bile salts were polar molecules, they would be unable to interact with lipids. If they were nonpolar, they would be unable to interact with water molecules.

MODULE **22.8**

1. The parts of the digestive process that will be missed when a patient is fed directly into the stomach are mechanical digestion by mastication and a small amount of chemical digestion by salivary amylase.

Assess What You Learned

Level 1

1. b   **2.** d

3. a. False: The mucosa from the stomach to the anus consists of an inner layer of *simple* columnar epithelium.
b. True
c. True
d. False: The mucosa of the esophagus, pharynx, and oral cavity contains *stratified* squamous epithelium to protect it from abrasion.

4. _e_ Chief cells _f_ Enteroendocrine cells
_c_ Parietal cells _g_ Gastroesophageal sphincter
_a_ Gastrin _b_ Pepsin
_h_ Pyloric sphincter _d_ Chyme

5. b

6. The three folds of the small intestine are microvilli, folds of the enterocyte plasma membrane; villi, folds of the mucosa; and circular folds (plicae circulares), folds of the mucosa and submucosa. These folds increase the surface area of the small intestine for absorption and chemical digestion.

7. c

8. Oral cavity → oropharynx → laryngopharynx → esophagus → stomach → duodenum → jejunum → ileum → cecum → ascending colon → transverse colon → descending colon → sigmoid colon → rectum → anal canal

9. a. False: The crown of a tooth is surrounded by an outer layer of *enamel*, which is the hardest substance in the body.
b. True
c. True
d. False: The secretory cells of the pancreas and salivary glands are *acinar* cells.

10. d   **11.** a

12. _f_ Salivary amylase _b_ Pancreatic lipase
_a_ Sucrase _d_ Pancreatic amylase
_e_ Trypsin _c_ Pepsin

13. c   **14.** d   **15.** b   **16.** micelles; chylomicrons   **17.** a

18. a. True
    b. True
    c. False: Smooth muscle of the stomach *relaxes* when food enters from the esophagus.
    d. True
    e. False: During *fasting*, the small intestine exhibits motion in the migrating motor complex pattern.

19. a   20. b   21. e   22. c

Level 2

1. Normally during swallowing, the uvula and soft palate move posteriorly to close off the nasopharynx and direct food down into the laryngopharynx. However, when your friend laughs, he exhales through his nose and this mechanism fails, causing the water to go up into the nasopharynx and nasal cavity instead of down into the laryngopharynx.

2. A patient who has food or liquid in his or her gastrointestinal tract may vomit during surgery as a result of the effects of the general anesthesia. With no reflex for swallowing in place (and no protection for the respiratory tract), the vomit can potentially go into the larynx, causing a patient to choke.

3. In general, such drugs inhibit the motility and secretion of most organs of the digestive system. They decrease salivation and secretion from the stomach, and they slow the movement of the stomach, small intestine, and large intestine. This can lead to dry mouth and constipation.

4. Absorption is indeed an important process of the digestive system, but each of the processes of the system is required for the whole to work. For example, nutrients cannot be absorbed unless they are physically and chemically digested, they cannot be digested if they are not propelled through the alimentary canal, and they cannot be propelled if they are not first ingested. All of the processes are connected, and one cannot be deemed "more important" than any other.

Level 3

1. a. Bile release increases when fatty foods are ingested, and the contraction of the gallbladder onto the stones causes Mr. Williams's pain. If he ate a carbohydrate-only meal, his symptoms would likely be much less severe, as bile is not required for carbohydrate digestion.
   b. His bile duct is likely blocked, preventing the release of bile into his small intestine. Bile is needed to emulsify fats so that they can be chemically digested by pancreatic lipase. In the absence of bile, pancreatic lipase will be unable to digest significant amounts of fats, leading to the presence of these fats in the feces.
   c. Bile contains the pigment stercobilin, which is responsible for the brown color of feces. Since bile is not being released into the small intestine, stercobilin isn't either, resulting in the production of nonpigmented feces.

2. a. Total removal of the stomach means that it is not present to control emptying into the duodenum. Instead, the bolus will be "dumped" straight into the duodenum, which will be unable to mix the bolus with pancreatic juice. As a result, nutrients will not be absorbed fully and the concentration of the chyme in the small intestine will be too high. This will cause water to be drawn out of the large and small intestine, resulting in diarrhea.
   b. Absorption of vitamin B12 is dependent on production of intrinsic factor by the stomach. With the stomach gone, no intrinsic factor will be produced, preventing the absorption of vitamin B12 in the small intestine.

3. When acid is lost through vomiting, more acid must be secreted from the parietal cells of the stomach. This decreases the hydrogen ion concentration of the blood, leading to an increase in its pH. This increase in pH will cause a decrease in respiratory rate as the body attempts to retain carbon dioxide to lower the pH to a normal level.

4. The inflammatory response leads to vasodilation, along with increased capillary permeability. Both of these factors result in water loss, which is the desired effect of the laxative.

5. Three types of nervous system damage can lead to a loss of the defecation reflex (or a loss of conscious control over the reflex): Damage to the peripheral nerves that innervate stretch receptors, the internal anal sphincter, and/or the external anal sphincter can lead to loss of the reflex. Damage to the spinal cord can impair the ability of these neurons to communicate with their target cells. Finally, damage to the cerebral cortex can impair a person's ability to determine the appropriate time for defecation.

CHAPTER 23
Apply What You Learned

MODULE 23.1

1. The reaction of two stable molecules would likely be endergonic, as it requires energy to break bonds in a stable molecule. If the two molecules were unstable, the reaction could be exergonic, as the molecules would release energy as they became more stable. 2. A fire that burns coal involves oxidation-reduction reactions, in which the coal is oxidized, and the fire results from the energy released by the oxidation. In your body cells, fuels such as glucose are oxidized, and the released energy is harnessed in the synthesis of ATP. 3. A riboflavin deficiency will affect almost all cells and decrease their ability to make ATP. The first cells affected will be those that are highly active metabolically, such as neurons, muscle fibers, and liver cells.

MODULE 23.2

1. In the absence of oxygen, glucose is catabolized by glycolysis, which yields only 2 molecules of ATP. For this reason, large amounts of glucose must be catabolized in order to meet the needs of the cell. When oxygen is available, glucose is catabolized by oxidative catabolism, which generates more ATP (38 molecules total). So, less glucose must be broken down to produce the same amount of ATP. 2. The rate at which oxidative catabolism takes place must increase during exercise to meet the body's higher needs. Because carbon dioxide is produced as a waste product from these processes, a higher rate of oxidative catabolism also leads to a higher rate of carbon dioxide production. 3. Oxaloacetate is required in the first step of the citric acid cycle, where it combines with acetyl-CoA. This mutation would cause less oxaloacetate to be available to react with acetyl-CoA, and therefore fewer electron carriers would be reduced and fewer ATP molecules produced.

MODULE 23.3

1. The pathways of catabolism for all three nutrients finish by a common pathway in the citric acid cycle, and all three generate reduced electron carriers for the electron transport chain. They differ in the way in which they are oxidized—lipids are oxidized via β-oxidation, amino acids must be deaminated before they are converted to citric acid cycle intermediates or glucose, and glucose is split into two molecules of pyruvate. 2. Urea, a nitrogen-containing waste product, formed in the deamination of amino acids, is eliminated by the kidneys. Kidney failure decreases the rate at which urea can be excreted, causing it to build up to toxic levels. 3. Lipids generally have more carbons than glucose or amino acids. For this reason, the catabolism of lipids generates more reduced electron carriers, and so more ATP, than does the catabolism of glucose or amino acids.

MODULE 23.4

1. Glucose is the fuel that is preferred by neurons, and depriving neurons of glucose can decrease the efficiency of their activity. 2. If the total caloric content of the food is unchanged, then eating it will not lead to fat loss. The extra sugar will simply be converted to lipids and stored in adipose tissues.

MODULE **23.5**

1. Normally glucose is the main fuel used by cells during the absorptive state, but if carbohydrates are unavailable, cells will instead oxidize lipids and amino acids. In addition, glycerol and glucogenic amino acids will be converted to glucose for those cells that require it as their main fuel source. **2.** In the absence of insulin, most body cells will be unable to take up glucose, and the processes that normally take place during the absorptive state will decrease or stop all together. Blood glucose levels will climb, the body will catabolize lipids and amino acids for fuel, and ketone production will be high. **3.** Stimulating the neurons of the feeding center in the hypothalamus in this way will cause hunger and likely trigger feeding.

MODULE **23.6**

1. Stressors activate the sympathetic nervous system and trigger the release of epinephrine and norepinephrine from sympathetic neurons and the adrenal medulla. This will elevate the metabolic rate above its basal level. **2.** Stimulating lipolysis without increasing the metabolic rate will not cause weight loss because the released fatty acids will not be catabolized. Instead, they will simply be stored in adipocytes in a different location. **3.** The metabolic rate increases during exercise in order to provide more ATP for contracting skeletal muscles and the heart. The reactions involved in ATP synthesis are not completely efficient, and much of the energy is lost as heat. This heat makes you feel warm, regardless of the environmental temperature.

MODULE **23.7**

1. There are many potential consequences to such diets, depending on which low-carbohydrate foods are consumed. Decreased intake of fruits and vegetables will decrease the amount of certain vitamins and minerals, as well as fiber, that is consumed. This could lead to changes in digestive habits, such as constipation. In addition, the lack of glucose in the diet will increase the rate at which gluconeogenesis takes place, which could cause protein breakdown to liberate glucogenic amino acids. **2.** The patient's condition has not improved—if anything, it is actually worse. Ideally, his HDL should have increased, as HDL particles take cholesterol to the liver for excretion. Instead, his HDL decreased. In addition, his LDL should have decreased, as the LDL particles deposit cholesterol in tissues, but his LDL remained unchanged. **3.** It's a matter of energy balance—if he takes in more than he oxidizes for fuel, then those excess nutrients will be stored. It doesn't matter what type of nutrient is consumed—consuming it in excess will not allow for weight loss unless energy expenditure is also increased.

Assess What You Learned

Level 1

1. b   **2.** a   **3.** exergonic, endergonic

4. a. False: During an oxidation-reduction reaction, the compound with the greatest electron affinity accepts electrons and is *reduced*.
   b. True
   c. True
   d. False: ATP is the "money" that the cell uses to "pay for" the cell's *endergonic* reactions.

5. ATP does not contain a "high-energy bond"; rather, its terminal bond is very unstable. When this bond is broken, the released energy can either directly fuel a process or trigger the phosphorylation of another molecule to make a reaction more energetically favorable.

6. a. _6_   **e.** _1_
   b. _2_   **f.** _3_
   c. _7_   **g.** _5_
   d. _4_

7. d

8. Oxygen is the final electron acceptor in the electron transport chain.

9. d

10. a. False: The reactions of β-oxidation crop fatty acids into *two-carbon acetyl-CoA* units.
   b. False: Most amino acids are converted into products that are oxidized by *the citric acid cycle*.
   c. True
   d. True
   e. False: Fat and protein catabolism takes place primarily in the *mitochondria*.

11. b   **12.** glycogenesis, gluconeogenesis   **13.** c

14. a. postabsorptive   **d.** postabsorptive
   b. postabsorptive   **e.** absorptive
   c. absorptive   **f.** postabsorptive

15. a

16. Heat is a byproduct of every metabolic reaction, and nearly all of the energy in the human body is eventually lost to the environment as heat.

17. d

18. a. False: *Radiation* is energy transfer between the body and the environment via electromagnetic waves.
   b. True
   c. True
   d. False: The body temperature measured on the skin or in the mouth is generally *lower* than the core body temperature.

19. macronutrient, micronutrient; essential nutrient   **20.** c

21. Fat-soluble vitamins have a cholesterol-based structure and are hydrophobic, so the body can store these compounds in adipose tissue when present in excess. Water-soluble vitamins have a polar structure and are hydrophilic, and so are excreted by the kidneys in the urine when present in excess. The fat-soluble vitamins include vitamins A, D, E, and K.

22. a. False: *Carbohydrates, lipids, and proteins* are used as fuel sources for cells.
   b. False: *Low-density* lipoproteins deliver cholesterol from the liver to peripheral tissues.
   c. True
   d. True

Level 2

1. When the body has exhausted its supply of glycogen during exercise, it begins to catabolize fatty acids. The fatty acids are broken down by β-oxidation, and the resulting acetyl-CoA molecules are sent to the citric acid cycle. Here they are oxidized, which is the same process that occurs when wood is burned in a fire.

2. Amino acids from proteins may be converted to glucose via gluconeogenesis, and any excess glucose could be stored as fat.

3. Erythrocytes are only able to undergo glycolysis, and so all NADH is reoxidized to NAD^+ and lactic acid is produced instead of pyruvate.

4. a. Oxygen is the final electron acceptor in the final common pathway (the electron transport chain) for carbohydrates, lipids, and proteins. The amount of oxygen consumed is therefore a good indicator of the rate at which the body's many metabolic reactions are occurring.
   b. The criteria for measuring the basal metabolic rate include no engagement in strenuous physical activity within the past hour. For this reason, the metabolic rate measured will not be your lab partner's basal rate.

5. Heat is a byproduct of all metabolic reactions. A person with extremely restricted caloric intake feels cold because his or her metabolic rate has slowed and less heat is being generated.

Level 3

1. A person who is suffocating is unable to produce adequate ATP via oxidative phosphorylation to sustain life. The same situation exists with cyanide poisoning—even though the oxygen level is normal to elevated, electrons cannot be transferred to the oxygen molecules, resulting in insufficient ATP production.

2. A diet that consists of nothing but fat would, of course, contain inadequate protein and carbohydrates. Glucose is the preferred fuel of the brain under all but near-starvation conditions, and so this diet would have a negative impact on concentration, memory, and other nervous system functions. In addition, it would lack the essential amino acids needed to make body proteins, which would accelerate muscle wasting and other forms of protein breakdown. A diet like this would not be advisable for anyone.

3. If starches were not absorbed, this would likely increase the body's reliance on fats and proteins for fuel and to make new glucose. This would lead to protein breakdown to release amino acids for gluconeogenesis and accelerated lipolysis, resulting in potential ketosis.

4. Thyroid hormone is a major determinant of the metabolic rate. Excess thyroid hormone, as seen with hyperthyroidism, will elevate the metabolic rate above normal levels, which means that more heat is produced as a byproduct.

5. As you exercise, the rate at which you catabolize metabolic fuels increases, which results in the production of more carbon dioxide. This elevated CO_2 level triggers the increased respiratory rate to remove the extra CO_2 from the blood. If the respiratory rate does not increase, the CO_2 can accumulate and decrease the pH of the blood.

CHAPTER 24
Apply What You Learned

MODULE **24.1**

1. Peritonitis would be unlikely to impact the kidneys, as they are retroperitoneal, or located posterior to the peritoneal membranes. 2. The kidneys produce the hormone erythropoietin, which stimulates erythrocyte production. In renal failure, all aspects of kidney function decrease, including production of erythropoietin. The lower levels of the hormone cause decreased rates of erythropoiesis.

MODULE **24.2**

1. The microvilli of the proximal tubule increase the surface area available for absorption of water and other solutes. If the microvilli decrease in numbers, the cells will be unable to reabsorb the same amount of water and solutes. As a result, urine volume would increase and solutes such as glucose and amino acids would be found in the urine. 2. Anything that decreases the blood flow to the kidneys will decrease the rate of filtration of fluid from the blood into the nephrons and therefore decrease the regulation of fluid, electrolyte, and acid-base balance performed by the kidneys.

MODULE **24.4**

1. A decrease in colloid osmotic pressure will decrease the force that retains water in the blood in the glomerulus. As a result, the net filtration pressure will increase in favor of filtration, and more filtrate will be produced. 2. Such drugs would increase the GFR temporarily by increasing the glomerular hydrostatic pressure due to the dilation of renal arteries and afferent arterioles. However, these changes would be countered very rapidly by the myogenic mechanisms and tubuloglomerular feedback, both of which would trigger constriction of the afferent arterioles to decrease the GFR. 3. Drugs that block the RAAS will inhibit overall vasoconstriction, and this can reduce the vasoconstriction that normally occurs with physical activity. Without this normal vasoconstriction, the blood pressure remains low, even when it would normally rise.

MODULE **24.5**

1. The drug would significantly decrease the amount of water and certain electrolytes reabsorbed from the proximal tubule. If fewer sodium

ions are reabsorbed, there will be a smaller electrical gradient for reabsorbing chloride, a smaller osmotic gradient for reabsorbing water, and therefore a smaller concentration gradient for reabsorption of other electrolytes. 2. Glucose molecules are reabsorbed with sodium ions as a result of secondary active transport. Bicarbonate ions are reabsorbed as a result of the Na^+/H^+ antiporter in the proximal tubule. If all sodium ion reabsorption were blocked, the glucose/sodium ion cotransporter and Na^+/H^+ antiporter would not function, and high amounts of glucose and bicarbonate ions would appear in the urine. 3. Hypoventilation will lead to acidosis, which causes the kidneys to increase their secretion of hydrogen ions and increase the reabsorption of bicarbonate ions.

MODULE **24.6**

1. With lower levels of ADH, less water will be reabsorbed from the distal tubule and collecting system. This will increase urine volume and decrease urine concentration.
2. a. When the pumps in the thick ascending limb are blocked, the osmotic gradient for water reabsorption from the descending limb is decreased. As a result, less water will be reabsorbed, which will further decrease the amount of sodium chloride pumped into the interstitial fluid.
   b. Since less sodium chloride is pumped into the interstitial fluid, the gradient for water reabsorption from the medullary collecting system will be decreased. This will reduce the volume of water reabsorbed and thus increase urine volume.

MODULE **24.7**

1. The decreased GFR would trigger the RAAS and tubuloglomerular feedback, both of which would lead to increased glomerular hydrostatic pressure and a slight increase in GFR. 2. Blocking carbonic anhydrase will result in the reabsorption of fewer sodium and bicarbonate ions. This will lead to a decreased pH of the blood, as fewer bicarbonate ions will be present to buffer the blood. The resulting acidosis will cause the kidneys to increase secretion of hydrogen ions and retain bicarbonate ions from the distal tubule.

MODULE **24.8**

1. With metabolic acidosis, you would expect the pH of the urine to be low; however, it will not likely drop below 4.5 irrespective of the degree of acidosis. With metabolic alkalosis, the urine pH will rise as more hydrogen ions are retained and bicarbonate ions are excreted. 2. Creatine phosphate is metabolized to creatinine, and so an increase in consumption of creatine will lead to an increased amount of creatinine in the blood. This will increase the workload of the kidneys, as they must excrete the excess creatinine.

MODULE **24.9**

1. Simple squamous epithelium is very thin and would not provide adequate protection to the underlying connective and muscle tissue of the ureters and urinary bladder. In addition, the simple squamous cells would not be as able to change shape with the changing size of these structures, and would be more prone to tears and other structural disruptions. 2. A spinal cord injury above S1-S2 would prevent voluntary control of micturition, as the impulses would be unable to reach the micturition center in the pons. An injury below S1-S2 could damage the neurons in the involuntary pathway of the micturition reflex. This could lead to urinary retention, because the signals that trigger the emptying of the urinary bladder would be interrupted.

Assess What You Learned

Level 1

1. The four main organs of the urinary system are the kidneys, ureters, urinary bladder, and urethra.
2. e

3. a. True
   b. False: Internally, the kidneys consist of an outer renal *cortex*, a middle renal *medulla*, and an inner renal *pelvis*.
   c. False: The kidneys' first capillary bed is the *glomerulus*, which is fed by the afferent arteriole and drained by the efferent arteriole.
   d. False: Filtrate flows from the renal corpuscle to the *proximal* tubule, the nephron loop, the *distal* tubule, and into the collecting system.

4. d

5. The filtration membrane consists of a series of three progressively finer filters, including the glomerular endothelial cells, the basal lamina, and the filtration slits of the podocytes.

6. a, c, f, g   7. c

8. a. dilate
   b. constricts; reduce
   c. renin
   d. angiotensin-converting enzyme
   e. increases; maintaining

9. c   10. b

11. a. True
    b. False: The distal tubule reabsorbs sodium ions and secretes potassium and hydrogen ions in response to *aldosterone*.
    c. True
    d. False: ADH triggers water reabsorption from the *late distal tubule and collecting system*.
    e. False: *Facultative* water reabsorption occurs in the distal tubule and collecting system.

12. c   13. b   14. concentrated; dilute   15. a, b   16. a

17. a. micturition; parasympathetic
    b. transitional
    c. detrusor
    d. urine; urine and semen

Level 2

1. a. An increase in colloid osmotic pressure will increase the forces opposing filtration in the glomerulus, so the GFR will decrease.
   b. This condition will decrease the glomerular hydrostatic pressure, and therefore decrease the force favoring filtration, leading to a reduced GFR. Note that the renin-angiotensin-aldosterone system would compensate somewhat for this, as would tubuloglomerular feedback.
   c. Hypertension will increase the glomerular hydrostatic pressure and increase the force favoring filtration, leading to an increase in GFR. Note that tubuloglomerular feedback and the myogenic mechanism will compensate for this.

2. Glomerulus → proximal tubule → nephron loop → distal tubule → cortical collecting duct → medullary collecting duct → reabsorbed into the interstitial fluid → enters thin descending limb of nephron loop → distal tubule → cortical collecting duct → medullary collecting duct → papillary duct → minor calyx → major calyx → renal pelvis.

3. The urethra in females is much shorter and only a short distance from the anus and vagina, where large numbers of bacteria normally reside. The male urethra is much longer and not anatomically close to the anus, making urinary tract infections less common.

4. Water reabsorption, which is needed to produce concentrated urine, occurs by osmosis. Osmosis is a passive process that relies on a concentration gradient. The kidneys must therefore establish a steep concentration gradient in the interstitial fluid so that osmosis may occur.

Level 3

1. a. Aldosterone promotes the retention of Na$^+$ and water in the late distal tubule and cortical collecting duct. Adverse effects of blocking aldosterone include high plasma K$^+$ and acidosis, as aldosterone normally promotes the secretion of K$^+$ and H$^+$.
   b. A-II normally causes constriction of the efferent arteriole and systemic blood vessels. If you block the receptors for A-II on the vessels of the kidney, you could potentially dilate the efferent arteriole and lower the GFR.
   c. These pumps are largely responsible for creating the gradient needed for countercurrent multiplication and the production of concentrated urine. Blocking these pumps leads to a large volume of dilute urine. However, these drugs also block the reabsorption of K$^+$, which can lead to low plasma K$^+$ concentrations.

2. Mr. Wu's GFR, at 35 ml/min, is much lower than normal, which means he is in renal failure and inadequate amounts of blood are being filtered each minute. The plasma concentrations of the drug are higher than expected, therefore, because the drug is not being eliminated as it should by the kidneys.

3. Sodium ion reabsorption in the proximal tubule drives the reabsorption of many solutes, so the concentration of these solutes, including glucose and many electrolytes, would increase in the urine. Sodium ions also drive the reabsorption of water by osmosis, and therefore the volume of urine would increase. The presence of excess sodium ions in the filtrate would trigger the macula densa to release chemicals that cause constriction of the afferent arteriole and a decrease in the GFR. The macula densa will also trigger the release of renin from JG cells, which activates the RAAS and culminates in systemic vasoconstriction and other effects of A-II.

4. If you block aldosterone, you also block sodium ion and water reabsorption, which reduces blood volume and blood pressure. Drugs that block angiotensin-II receptors on blood vessels will lower blood pressure by allowing for systemic vasodilation. The result of blocking pumps in the nephron loop and reducing the medullary osmotic gradient will be the production of a large volume of dilute urine, which decreases blood volume and blood pressure.

5. Urobilinogen is a breakdown product of hemoglobin that is excreted by the kidneys. Excess urobilinogen in the urine could indicate an increased plasma concentration of urobilinogen, which could be caused by an elevated rate of erythrocyte destruction.

CHAPTER 25
Apply What You Learned

MODULE 25.1

1. The kidneys produce urine, which aids in fluid balance by excreting excess water. If the kidneys are unable to do this and fluid intake increases, water will not obey the principle of mass balance—a water imbalance will exist. **2.** The excess water in the patient's body will dilute the body fluids and change the concentration of extracellular electrolytes, leading to an imbalance.

MODULE 25.2

1. Lipids repel water and are hydrophobic, so lipid-rich cells will contain less water than cells such as hepatocytes that are rich in other molecules. **2.** A 9% sodium chloride solution is very hypertonic to the cytosol and will draw water out of the patient's cells by osmosis. As a result, the patient's cells will shrink, or crenate. This solution could also affect the patient's sodium ion balance. **3.** If a patient is overhydrated, meaning that too much water is in the plasma and cytosol, hypertonic saline could be used to draw the excess water out of the cells.

MODULE 25.3

1. The levels of sodium and chloride ions in the ECF will decrease because their reabsorption isn't stimulated by aldosterone. In addition, the level of potassium ions will increase because their secretion

isn't stimulated by aldosterone. **2.** Water normally follows reabsorbed sodium and chloride ions by osmosis. With the reabsorption of these ions decreased, water reabsorption will decrease as well. **3.** Excess aldosterone would lead to reabsorption of large numbers of sodium and chloride ions, and also water by osmosis. It would also lead to a low blood potassium ion concentration as a result of excessive secretion of potassium ions.

MODULE **25.4**

1. Mr. Wong is likely to have developed respiratory acidosis. Due to the long time period over which the acidosis developed, it will likely be partially compensated. The arterial blood gases will probably show a low pH, high P_{CO_2}, and high HCO_3^-. **2.** Hyperventilation is a response to an acidotic state, as it rids the body of excess carbon dioxide. However, it is unlikely to be effective in this case due to Mr. Wong's state of impaired gas exchange. **3.** The hydrogen ion pumps of the kidneys are unable to pump against a gradient greater than a pH of 4.5, so they will be unable to secrete further hydrogen ions (unless some of the hydrogen ions in the filtrate are buffered). **4.** Normally, large doses of vitamin C or other acidic compounds will impact the pH of body fluids very little because the extra hydrogen ions are eliminated by the kidneys. However, Mr. Wong's kidneys are already producing urine at a pH of 4.5, and cannot excrete any further hydrogen ions. So the excess vitamin C may indeed be contributing to his acidosis by creating a metabolic acidosis in addition to his respiratory acidosis.

MODULE **25.5**

1. Dehydration triggers aldosterone secretion, which stimulates potassium ion secretion. This could lead to hypokalemia if large amounts of aldosterone are released. **2.** The blockage would hamper the ability of the body to adjust to dehydration, as the response begins with angiotensin-II. Note that hormones such as ADH would still be secreted due to stimulation of osmoreceptors, so the body would be able to respond at least somewhat to dehydration. **3.** As a positive feedback loop, the response would continue even after fluid balance was restored, leading to overhydration and other electrolyte and pH imbalances.

Assess What You Learned

Level 1

1. a

2. An electrolyte dissociates into ions in a solution, and these ions can conduct an electric current. A nonelectrolyte does not dissociate into ions in a solution, and the solution cannot conduct an electric current.

3. b   **4.** d   **5.** hypertonic, hypotonic

6. a. False: The thirst mechanism is mediated by osmoreceptors located in the *hypothalamus.*
b. False: The *insensible* water loss includes the water lost from the body via the skin and the respiratory system.
c. True
d. False: Atrial natriuretic peptide promotes water *loss from* the kidneys and *decreases* the amount of water in the body.
e. True

7. d

8. a. Angiotensin-II: promotes sodium ion reabsorption from the proximal tubule
b. Aldosterone: promotes sodium ion reabsorption and potassium ion secretion from the distal tubule and collecting system
c. Parathyroid hormone: promotes calcium ion reabsorption and inhibits phosphate ion reabsorption from the kidneys, as well as stimulating osteoclast activity
d. Vitamin D: promotes calcium ion and phosphate ion absorption from the small intestine
e. Atrial natriuretic peptide: inhibits sodium ion reabsorption in the distal tubule and collecting system

9. b

10. a. True
b. True
c. False: *Hypokalemia* tends to decrease the resting membrane potential and hyperpolarize excitable cells.
d. True
e. False: Chloride ions are generally reabsorbed from the kidneys, along with *sodium* ions.

11. c   **12.** b

13. The pH changes very little when hydrogen ions are added to a buffered solution because the buffer's weak conjugate base binds the hydrogen ions, taking them out of solution.

14. increases, decrease; decreases, increase   **15.** a, c, d

16. a. True
b. False: *Respiratory* compensation for acid-base disturbances begins within minutes, whereas *renal* compensation begins within several hours.
c. False: Respiratory compensation for metabolic acidosis consists of *hyperventilation.*
d. True
e. True

17. e

18. Angiotensin-II has several effects that result in water reabsorption: It promotes sodium ion reabsorption, which creates an osmotic gradient for water reabsorption; it stimulates thirst; it stimulates release of ADH; and it triggers the release of aldosterone.

Level 2

1. Only the solution with 5% sodium chloride will conduct electricity. Deionized water contains no electrolytes, and glucose does not dissociate into ions in water and is therefore a nonelectrolyte.

2. Her body water will increase with the loss of adipose tissue, as adipose tissue contains very little water.

3. The amount of water does not affect the absolute number of ions and other solutes, but it does affect the percentage of ions and solutes relative to water. So if the total amount of water decreases, the relative proportion of solutes increases, and their concentration rises.⁺

4. The patient is likely to be hyperventilating. This removes carbon dioxide from the ECF, which reduces the hydrogen ion content. The source of the hydrogen ions doesn't matter, only that their number decreases.

Level 3

1. Elise is hypoventilating because she likely has metabolic alkalosis due to the loss of hydrogen ions from vomiting. The hypoventilation will help her retain carbon dioxide and reduce the pH of her ECF. Her arterial blood gases will likely show an elevated pH, high bicarbonate ion concentration, and elevated P_{CO_2}.

2. The volume of her ECF has likely decreased due to fluid loss from the vomiting, and the osmolarity has likely increased. This could draw fluid out of her cells and into the ECF, resulting in cell crenation.

3. By blocking aldosterone, the drug has inhibited the secretion of both potassium and hydrogen ions from the kidneys. This has caused retention of excess potassium and hydrogen ions, leading to hyperkalemia and acidosis.

4. The patient likely has respiratory acidosis. Renal compensation has not yet occurred, as indicated by the normal bicarbonate ion concentration, so the acidosis has likely not existed for longer than a few hours.

5. The sodium ion concentration will increase, resulting in an increase in the osmolarity and osmotic pressure of the ECF. This will draw water out of the cells and into the ECF, which leads to an increased blood volume. The higher blood volume will tend to elevate the blood pressure.

CHAPTER 26
Apply What You Learned

MODULE **26.1**

1. Alleles are separated during meiosis and "shuffled" because of crossing over and independent assortment. Each sperm or ovum receives only one allele for each gene, so the combinations children receive will differ. **2.** Without crossing over and independent assortment, chromosomes would remain intact and would separate consistently, resulting in ova or sperm with the same alleles. **3.** One kind of gamete would get both homologous chromosomes, and the other kind of gamete would get neither one. Therefore, the offspring would either receive an extra chromosome or have a missing chromosome. The effects on the offspring would vary, depending on the chromosome involved.

MODULE **26.2**

1. The seminal vesicles contribute approximately 60–70% of the volume of semen, so the amount of semen produced would decrease. **2.** If the dartos and cremaster muscles of the scrotum contract, the scrotum will be pulled close to the abdominopelvic cavity. If the weather is hot, this will raise the temperature of the scrotum to body temperature and will decrease the production of sperm.

MODULE **26.3**

1. LH stimulates interstitial cells in the connective tissue around the seminiferous tubules to produce testosterone, which is necessary for sperm production. **2.** Nitric oxide relaxes and dilates smooth muscle in the walls of the arterioles of the penis and within the blood sinuses in erectile tissue. This allows large amounts of blood to enter the tissue and results in an erection. Without nitric oxide, an erection would not occur. **3.** Stress activates the sympathetic nervous system, which suppresses the parasympathetic response needed for an erection.

MODULE **26.4**

1. If sperm enter the tube and fertilize an oocyte, the fertilized ovum will not be able to pass through the tube to the uterus, causing it to remain in the much smaller uterine tube. **2.** Fimbriae "catch" an ovulated oocyte and draw it into the uterine tube. A lack of fimbriae would cause the oocyte to be released into the peritoneal cavity instead, where sperm likely cannot reach it.

MODULE **26.5**

1. If the primary oocytes survive and are permitted to continue to cycle monthly, the female will continue to ovulate for as long as she lives. **2.** Taking a supplement that produces a continuously high level of progesterone in a woman's body will inhibit the release of gonadotropins, which will prevent the ovarian cycle from beginning again. Also, mimicking the effects of progesterone will maintain the endometrial lining of the uterus, preventing menstruation. **3.** The uterine tubes are not involved in the production of hormones responsible for the ovarian and uterine cycles. Therefore, the surgery will have no effect on the onset of menopause.

MODULE **26.6**

1. a. This woman could use any of the barrier methods (condoms, cervical cap, or sponges) and spermicides. In addition, she could use an IUD, but should not consider an IUS or any hormonal contraceptives because of her bleeding disorder. Tubal ligation is not an option because of her age and unknown future plans.
b. The man should be using condoms for each sexual encounter and should include spermicides if possible. A vasectomy is not an option because of his age and unknown future plans.

MODULE **26.7**

1. Many cases of STIs are asymptomatic, and some individuals with an infection never develop symptoms or lesions. However, even an asymptomatic individual can pass an STI on to other partners. Furthermore, some diseases enter a latent period when they are inactive, and then manifest at a later date. **2.** Girls should be vaccinated for HPV before they become sexually active to avoid contracting the virus in the future. This will significantly minimize their risk of developing cervical and other cancers and genital warts. Boys should receive the vaccine to prevent them from spreading the virus to sexual partners and also to reduce their risk of contracting certain cancers and genital warts.

Assess What You Learned

Level 1

1. 23, different

2. _g_ Prophase I   _f_ Prophase II
    _b_ Metaphase I   _c_ Metaphase II
    _d_ Anaphase I   _h_ Anaphase II
    _a_ Telophase I   _e_ Telophase II

3. a. False: Spermatogenesis is the process of sperm cell development. It begins at puberty and *continues throughout life*.
b. True
c. False: Primary spermatocytes undergo the first meiotic division and form two smaller *haploid* cells called secondary spermatocytes.
d. True

4. a   **5.** d

6. The prostatic urethra, the membranous urethra, and the spongy urethra

7. _b, c, d_ Seminal vesicle
    _a, f_ Prostate
    _e_ Bulbourethral gland

8. corpora cavernosa, corpus spongiosum   **9.** b

10. _3 or 4_ a.
    _5_ b.
    _3 or 4_ c.
    _2_ d.
    _1_ e.

11. Erection, ejaculation   **12.** d   **13.** a

14. uterine tube, uterus   **15.** lobules, alveoli

16. a. False: Oogenesis begins before a female is born and continues *until she reaches menopause (usually between 45 and 55 years of age)*.
b. False: About once a month after puberty begins, some primary oocytes are stimulated to continue development. Usually one will complete the first meiotic division to produce two haploid cells, *one of which is large and the other small*.
c. True
d. True

17. _d_ Primordial follicle   _f_ Vesicular follicle
    _e_ Primary follicle   _c_ Corpus luteum
    _a_ Secondary follicle   _b_ Corpus albicans

18. _4_ a.   _5_ f.
    _6_ b.   _10_ g.
    _3_ c.   _7_ h.
    _8_ d.   _9_ i.
    _1_ e.   _2_ j.

19. a. False: During the *secretory* phase, the endometrium prepares for the implantation of the embryo by enlarging the endometrial glands and secreting glycogen into the uterine cavity.
b. False: During the *proliferative* phase, the endometrium generates a new stratum functionalis.
c. False: During the menstrual phase, the uterus sheds the stratum *functionalis* of the endometrium, which results in the discharge of blood and tissue from the vagina.

20. a   **21.** b and e   **22.** b and c

Level 2

1. The main purpose of the first meiotic division is to create gametes with half the number of chromosomes that were in the original cell. To accomplish this, homologous pairs of chromosomes are separated, rather than the sister chromatids of each chromosome. Also, during the first meiotic division, genetic variability is increased by two processes—crossing over and independent assortment—which randomly "shuffle" the paternal and maternal genes. The purpose of mitosis is to create two cells that are genetically identical to the original cell.

2. In addition to producing androgen-binding protein and inhibin, sustentacular cells create the blood-testis barrier by forming tight junctions with each other to separate the forming sperm cells from the immune system. This prevents the immune system from attacking the genetically unique sperm cells. Also, these cells secrete testicular fluid, which helps transport sperm through the lumen of the seminiferous tubules, and they provide nutrients to the dividing cells. They phagocytize damaged spermatogenic cells and the excess cytoplasm released from maturing spermatids.

3. During the female orgasm, the uterus exhibits peristaltic waves of contraction and the cervix pushes down somewhat into the vagina. If semen is present, it is possible that these actions may draw the semen farther into the female reproductive tract, which could increase the chances that fertilization will occur.

4. The artificially high levels of estrogens and progesterone have a negative feedback effect on the release of FSH and LH. Low levels of FSH and LH prevent follicle maturation and subsequently ovulation and also interfere with the buildup of the endometrial lining of the uterus necessary for implantation of an embryo.

Level 3

1. His wife is correct. The prostate gland surrounds the prostatic urethra and enlarges as men age. As it enlarges, it pushes in on the urethra and frequently affects urination.

2. High blood levels of testosterone will inhibit secretion of FSH and LH from the anterior pituitary, which is needed for testosterone production in the testes. LH stimulates interstitial cells to produce testosterone, and this local production of testosterone is required for spermatogenesis.

3. Body fat is essential to support a pregnancy. The brain monitors body fat using hormones such as leptin. Leptin stimulates gonadotropin secretion, so if leptin levels drop because the percent of body fat is low, gonadotropin secretion will decrease as well. This will affect the HPG axis and cause amenorrhea. For a female with such a low percentage of body fat, a pregnancy would be particularly stressful, so the fact that she is not having her menstrual cycle is actually beneficial to her at this point.

4. Erections occur when the corpora of the penis are engorged with blood. Most herbal medications for impotence increase the effects of the nitric oxide that is already present, which results in vasodilation of blood vessels to the penis. However, other blood vessels are often also affected, which can lower vascular resistance and decrease systemic blood pressure.

CHAPTER 27
Apply What You Learned

MODULE 27.1

1. The most fundamental events, such as formation of the organs, occur during the first trimester (first 3 months). 2. The last 10 weeks of fetal development are important to complete the growth of the organs.

MODULE 27.2

1. The syncytiotrophoblast has to touch maternal cells and be able to dissolve maternal blood vessels without being attacked by the maternal immune system. If this cell were antigenic, the pregnancy would probably not progress beyond implantation because the mother's immune system would destroy the conceptus. 2. Without secretion of hCG at the beginning of a pregnancy, the corpus luteum would degenerate and maternal levels of progesterone and estrogen would decrease.

MODULE 27.3

1. Teratogens are chemicals or biological agents, such as bacteria, that can cross the placenta and harm the conceptus. During the pre-embryonic and embryonic periods of development, tissues and organs are forming and are most susceptible to these agents. 2. The nervous system integrates and allows communication between all of the developing tissues.

MODULE 27.4

1. She should schedule an ultrasound exam by the end of the third month of development. 2. The ductus venosus acts as a shunt between the hepatic portal vein and the inferior vena cava. If it remains open after birth, some of the blood from the gastrointestinal tract will bypass the liver and not be filtered. Toxins will be allowed to pass to other organs, and nutrients that should be stored in the liver will also bypass the organ.

MODULE 27.5

1. The umbilical cord is usually not below the fetus's head in the vertex birth, so it stays behind the fetus. If the legs are touching the cervix, the umbilical cord can collect under the fetus, and during labor, the fetus may press on the cord. 2. Even with one fetus, the uterus takes up a large space in the abdominal cavity when the fetus is full term. If there are two or more fetuses, the space is even more limited, and each fetus has less space to grow. Also, the fetuses may be born earlier because crowding inside the uterus might trigger contractions.

MODULE 27.6

1. Newborns receive some antibody protection from their mothers, but they need protection from diseases that might make them seriously ill or even be fatal. 2. Milk production is controlled by a positive feedback mechanism. Prolactin stimulates the production of milk, and suckling by the infant increases prolactin production. The more the infant suckles, the more prolactin is produced, and, subsequently, the more milk is produced. 3. Breastfeeding stimulates oxytocin, which causes contractions in the uterus.

MODULE 27.7

1. Because both parents have attached earlobes, they must each be homozygous for the recessive allele. Using E for the dominant allele and e for the recessive allele, the Punnett square is as follows:

	e	e
e	ee	ee
e	ee	ee

Because none of the offspring genotypes include a dominant E allele, it is not possible for these parents to produce a child with free earlobes.

2. The mother's father had only one X chromosome to pass on, so she must have inherited his X^r allele. Because she is not color blind, she must have inherited X^R from her mother. Therefore, her genotype is $X^R X^r$ and she is a carrier. Her partner's genotype is $X^r Y$ because he is color blind. This is the Punnett square:

	X^r	Y
X^R	$X^R X^r$	$X^R Y$
X^r	$X^r X^r$	$X^r Y$

The offspring has a 50% chance of being color blind, regardless of sex.

3. The parents' genotypes produce the following offspring genotypes: $I^A i$ (1/2, or 50%) and $I^B i$ (1/2, or 50%). The phenotypes are blood type A (1/2, or 50%) and blood type B (1/2, or 50%).

Assess What You Learned

Level 1

1. b Pre-embryonic period
    c Fetal period
    a Embryonic period

2. cleavage, blastocyst

3. a. True
   b. True
   c. False: Secondary oocytes are covered with a layer of cells called the *corona radiata.*
   d. False: The cortical reaction *prevents* additional sperm cells *from binding* to the oocyte.

4. b

5. 5 a.   2 d.
    1 b.   3 e.
    4 c.   6 f.

6. trophoblast, inner cell mass

7. c Amnion
    a Yolk sac
    b Allantois
    d Chorion

8. Placentation, umbilical cord

9. b, e Endoderm
    c, f Mesoderm
    a, d Ectoderm

10. Cephalocaudal and lateral

11. f Month 3   c Month 6
     d Month 4   e Month 7
     a Month 5   b Months 8 and 9

12. b Foramen ovale
     c Ductus arteriosus
     a Ductus venosus

13. a   14. d

15. 3 a.
     1 b.
     4 c.
     2 d.

16. a. True
    b. True
    c. True

17. 4 weeks, neonatal

18. 3 a.
     1 b.
     2 c.
     4 d.

19. Genes, genome   20. b

Level 2

1. From the mother's perspective, a pregnancy begins at the last menstrual cycle rather than at fertilization. This is typically 2 weeks longer because ovulation and fertilization occur approximately 14 days after menstruation.

2. The zygote is a very large cell because the ovum is very large. Cleavage occurs quickly, and cells produced from these mitotic divisions do not have time to grow before they divide again.

3. If there are significant chromosomal abnormalities, with a few exceptions, the embryo cannot survive past the embryonic period.

4. The weight of the rapidly growing fetus rests on the urinary bladder, increasing the urge to urinate.

Level 3

1. Once the blastocyst implants, it embeds in the lining and is covered by endometrial cells. Detachment and movement would kill the embryo, so the only choice is to perform surgery to physically remove the embryo and try to prevent damage to the uterine tube.

2. The drug may cross the placenta and injure the fetus. The embryonic stage is particularly vulnerable to such effects because the tissues and organs are still forming.

3. The Punnett square is as follows:

	w	w
W	Ww	Ww
w	ww	ww

The probability of a child having a widow's peak is 50%.

4. Her professor is wrong. Elise can only narrow her father's blood type to two possibilities. Her mother has type B, so her genotype could be $I^B I^B$ or $I^B i$. However, Elise doesn't have type B or AB, so she must have inherited an i allele from her mother instead of I^B. Therefore, Elise's genotype must be $I^A i$, and her mother's must be $I^B i$. Elise must have inherited the I^A allele from her father. Therefore, his genotype could be $I^A I^A$, $I^A i$, or $I^A I^B$, and his blood type could be A or AB.

5. The growing placenta requires increased blood volume, and as the fetus grows, it needs more oxygen and produces more waste that must be removed. Because of these changes, Laura's blood volume and, subsequently, her cardiac output have increased. Blood pressure may rise because of increased blood volume. Also, some women develop preeclampsia, which results in high blood pressure. The placenta secretes human placental lactogen, which inhibits maternal insulin and increases blood glucose. The mother's cortisol levels also increase to promote maturation of fetal lungs, and this too can increase blood glucose.

Appendix B
The Metric System

Measurements, Units, and Equivalents

Measurement	Metric Unit	Metric-to-English Equivalent	English-to-Metric Equivalent
Length *SI unit* = meter*	Kilometer (km)	1 km = 0.62 mile	1 mile = 1.61 km
	Meter (m)	1 m = 1.09 yards or 3.28 feet	1 yard = 0.914 m 1 foot = 0.305 m
	Centimeter (cm)	1 cm = 0.394 inch	1 inch = 2.54 cm
	Millimeter (mm)	1 mm = 0.039 inch	1 inch = 25.4 mm
	Micrometer (μ or μm)	1 μm = 0.0039 inch	1 inch = 2540 μm
	Nanometer (nm)	1 nm = 0.0000039 inch	1 inch = 2,540,000 nm
Mass *SI unit = kilogram*	Kilogram (kg)	1 kg = 2.205 pounds	1 pound = 0.4536 kg
	Gram (g)	1 g = 0.0353 ounce	1 ounce = 28.35 g
	Milligram (mg)	1 mg = 0.015 grain	1 grain = 64.799 mg
	Microgram (μg)	1 μg = 0.000015 grain	1 grain = 64,799 μg
Area *SI unit = square meter*	Square meter (m^2)	1 m^2 = 1.1960 square yards or 10.764 square feet	1 square yard = 0.8361 m^2 1 square foot = 0.0929 m^2
	Square centimeter (cm^2)	1 cm^2 = 0.155 square inch	1 square inch = 6.4516 cm^2
Volume: solids *SI unit = cubic meter*	Cubic meter (m^3)	1 m^3 = 1.3080 cubic yards or 35.315 cubic feet	1 cubic yard = 0.7646 m^3 1 cubic foot = 0.0283 m^3
	Cubic centimeter (cm^3)	1 cm^3 = 0.0610 cubic inch	1 cubic inch = 16.387 cm^3
	Cubic millimeter (mm^3)	1 mm^3 = 0.000610 cubic inch	1 cubic inch = 1638.7 mm^3
Volume: liquids and gases *SI unit = liter*	Kiloliter (kl)	1 kl = 264.17 gallons	1 gallon = 0.003785 kl or 3.785 l
	Liter (l)	1 l = 0.264 gallon or 1.057 quarts	1 quart = 0.946 l 1 pint = 0.473 ml
	Milliliter (ml)	1 ml = 0.034 fluid ounce	1 fluid ounce = 29.57 ml
	Microliter (μl)	1 μl = 0.000034 fluid ounce	1 fluid ounce = 29,570 μl
Temperature *SI unit = Celsius degree or Kelvin*	Degrees Celsius (°C) Degrees Kelvin (K = °C + 273)	°F = 9/5(°C) + 32	°C = 5/9(°F − 32)

*SI units are the internationally recognized standard metric system measurements that are used in science and commerce. The abbreviation SI is derived from the French *Système International* (international system). Note that the liter is not an official SI unit, but its use is accepted.

Laboratory Reference Values

Blood Chemistry (Comprehensive Metabolic Panel)

Substance	Normal Range: Adult*	Potential Causes of Abnormal Values
Albumin	3.9–5.0 g/dl	• Low albumin is seen with malnutrition, renal failure, hepatic disease. • Elevated albumin is most commonly seen with dehydration.
Alkaline phosphatase	44–147 IU/l	Elevated levels are seen with certain cancers or diseases of the bone or liver, bile duct obstruction.
Alanine aminotransferase (ALT)	8–37 IU/l	Elevated level is seen with hepatic disease or failure.
Aspartate aminotransferase (AST)	10–34 IU/l	Elevated level is seen with hepatic disease or failure, severe muscle injury.
Blood-urea-nitrogen (BUN)	7–20 mg/dl	• Low BUN typically indicates overhydration or malnutrition. • Elevated BUN typically indicates dehydration or renal, heart, or hepatic failure.
Calcium ions	8.5–10.9 mg/dl	• Low calcium ion concentration is seen with malnutrition; nutritional deficiency of vitamin D, vitamin K, or magnesium; pancreatitis; low parathyroid hormone levels. • Elevated calcium ion concentration is seen with hyperparathyroidism, renal disease, certain cancers.
Carbon dioxide	20–29 mmol/l	• Low carbon dioxide levels may indicate compensation for metabolic acidosis, renal failure. • Elevated levels generally indicate respiratory disease; may also indicate compensation for metabolic alkalosis.
Chloride ions	96–106 mmol/l	• Level tends to be decreased with respiratory diseases and overhydration. • Level tends to be increased with dehydration, excess cortisol secretion, renal failure.
Creatinine	0.8–1.4 mg/dl	Levels are elevated in renal failure, severe dehydration.
Glucose	70–100 mg/dl (fasting) 70–125 mg/dl (nonfasting)	• Low blood glucose may indicate excess insulin, hypoglycemia, hepatic disease. • High blood glucose may indicate diabetes mellitus, pancreatitis, hyperthyroidism.
Potassium ions	3.7–5.2 mEq/l	• Low levels of potassium ions may indicate overuse of certain diuretics, excess aldosterone secretion. • Elevated levels of potassium ions may indicate renal failure, dehydration, adrenal insufficiency.
Sodium ions	136–144 mEq/l	• Low levels of sodium ions may indicate overhydration, use of certain diuretics, diarrhea, adrenal insufficiency. • Elevated levels of sodium ions may indicate dehydration, renal failure, elevated cortisol levels.
Total bilirubin	0.2–1.9 mg/dl	Elevated levels may indicate liver disease, bile duct obstruction, hemolysis.
Total protein	6.3–7.9 g/dl	• Low protein is seen with malnutrition, renal failure, hepatic disease. • Elevated protein is seen with certain cancers such as multiple myeloma, renal or hepatic failure.

*Source: National Institutes of Health.

Arterial Blood Gases

Measurement	Normal Range: Adult at sea level*	Potential Causes of Abnormal Values
Partial pressure of oxygen (PaO$_2$)	75–100 mmHg	Low PaO$_2$ usually indicates respiratory dysfunction; may be seen at high altitudes.
Partial pressure of carbon dioxide (PaCO$_2$)	38–42 mmHg	• Low PaCO$_2$ indicates compensated metabolic acidosis or respiratory alkalosis. • Elevated PaCO$_2$ indicates compensated metabolic alkalosis or respiratory acidosis.
Blood pH	7.38–7.42	• Low pH means acidemia due to metabolic or respiratory causes. • Elevated pH means alkalemia due to metabolic or respiratory causes.
Oxygen saturation (SaO$_2$)	94–100%	Low oxygen saturation usually indicates respiratory dysfunction; may be seen at high altitudes.
Bicarbonate (HCO$_3^-$)	22–28 mEq/l	• Low HCO$_3^-$ indicates metabolic acidosis or compensated respiratory alkalosis. • Elevated HCO$_3^-$ indicates metabolic alkalosis or compensated respiratory acidosis.

*Source: National Institutes of Health.

Complete Blood Count with Differential

Component	Normal Range: Adult Male*	Normal Range: Adult Female*	Potential Causes of Abnormal Values
Red blood cell (erythrocyte) count	4.1–5.8 million cells/µl	3.7–5.2 million cells/µl	• Low total erythrocyte count may indicate bleeding, destruction of erythrocytes, vitamin or mineral deficiency (iron, vitamin B12, folate), bone marrow disorder or damage, renal failure. • High total erythrocyte count may indicate excess erythropoietin, lung disease, high altitude, or dehydration.
Hemoglobin	12.7–17.1 g/dl	10.4–15.2 g/dl	Causes of elevated or depressed hemoglobin levels generally parallel the causes of abnormal erythrocyte count.
Hematocrit (%)	38.0–50.3%	32.0–45.0%	Causes of elevated or depressed hematocrit generally parallel the causes of abnormal erythrocyte count.
White blood cell (leukocyte) count	3.9–12.1 thousand cells/µl	3.9–12.5 thousand cells/µl	• Low total white cell count may indicate cancer or diseases of the bone marrow, immunosuppression due to drugs or infection (HIV). • High total white cell count may indicate infection, inflammation, allergic disorders, trauma, leukemia.
· **Neutrophils (%)**	39.7–77.3%	39.6–77.8%	Elevated neutrophil count most commonly indicates bacterial infection or severe systemic inflammation.
· **Lymphocytes (%)**	17.8–51.8%	17.8–52.8%	Elevated lymphocytes may indicate viral infection, certain bacterial infections, leukemia.
· **Monocytes (%)**	0–12%	0–12%	Elevated monocytes may indicate chronic infection or inflammation, certain types of leukemia.
· **Eosinophils (%)**	0–8%	0–8%	High eosinophil levels most commonly indicate allergic disease or parasitic infection.
· **Basophils (%)**	0–2%	0–2%	High numbers of basophils may indicate certain types of allergic reactions or inflammation.
Platelets	157–441 thousand platelets/µl	172–453 thousand platelets/µl	• Low platelet levels are associated with certain infections, certain drugs, hepatic disease, bone marrow disease or destruction. • Elevated platelet count is associated with certain cancers, autoimmune diseases, certain types of anemia.

*Source: Centers for Disease Control and Prevention.

Urinalysis

Characteristic or Component	Normal Range or Value: Adult*	Potential Causes of Abnormal Values
Color	Light to dark yellow, slight amber hue	• Hydration status: urine is darker in dehydration, lighter in overhydration. • Certain drugs and foods may change urine color to red, blue, brown, or another color. • Blood in the urine from bacterial infections, kidney stones, renal cancer, etc., may give it a red hue.
Clarity	Clear	Urine appears cloudy during infection, renal failure, and with kidney or bladder stones.
pH	4.5–8.0	pH impacted by diet (high-protein diets lead to acidic urine; high-fat diets lead to alkaline urine), kidney stones, bacterial infection, certain drugs, and renal failure.
Specific gravity	1.005–1.025	Low specific gravity indicates overhydration, renal disease, diabetes insipidus; high specific gravity generally indicates dehydration.
Urobilinogen	0.5–1.0 mg/dl	Elevated urobilinogen may indicate hemolysis, hepatic disease, and constipation. Low urobilinogen may indicate obstructive biliary disease.
Bilirubin	Negative	Presence often indicates hepatic failure.
Blood (hemoglobin)	< 3 erythrocytes/ml (read as negative on urinalysis test strip)	Blood in the urine may indicate renal failure, infection, kidney stones.
Protein	< 150 mg/dl (read as negative on urinalysis test strip)	Proteinuria may indicate heart failure, fever, muscle damage, renal failure, glomerulonephritis, diabetic nephropathy.
Glucose	< 130 mg/dl	Elevated urine glucose is seen with diabetes mellitus, pregnancy.
Ketones	Negative	Ketones in the urine may be present with uncontrolled diabetes mellitus, low-carbohydrate diets, starvation, severe exercise, vomiting, and pregnancy.
Nitrite	Negative	Nitrites generally indicate urinary tract infection.
Leukocyte esterase	Negative	Positive leukocyte esterase generally indicates a urinary tract infection.

*Source: Medscape.

Credits

Front matter   iii top: Courtesy of Erin Amerman; iii bottom: Courtesy of Virginia Irintcheva; ix (image of author): Winston Charles Poulton, Pearson Education; x (Concept Boost): left and right, William Karkow, Pearson Education; middle, Nina Zanetti, Pearson, Education; xv (brain): Winston Charles Poulton, Pearson Education; xvii (smartphone): S.V. Art/Shutterstock; xvii (tablet): fotoscool/Fotolia.

Chapter 1   Opener: Simon Fraser/Science Source; 1.1 left: Lucky Business/Shutterstock; 1.1 right: Adam Lubroth/Digital Vision/Getty Images; 1.8a: Roxana Wegner/Flickr/Getty Images; 1.8b: Medical Body Scans/Science Source; 1.8c: Stephen Rees/Shutterstock; APRW top: Yen Teoh/Thinkstock/Getty Images; APRW middle: CNRI/Science Source; APRW bottom left: Scott Camazine/Science Source; APRW bottom right: Ramare/BSIP SA/Alamy.

Chapter 2   Opener: Jim Dowdalls/Science Source; APRW: Simon Fraser/Medical Physics, RVI, Newcastle upon Tyne/Science Source; 2.7b: Tony Watson/Alamy.

Chapter 3   Opener: Colin Anderson/Photographer's Choice/Getty Images; UN3.1 (dye in water): GIPhotoStock/Science Source; 3.9a, b, c: David M. Phillips/Science Source; 3.14b: Don W. Fawcett/Science Source; 3.16: Pietro M. Motta & Tomonori Naguro/Science Source; 3.19 top and bottom: Don W. Fawcett/Science Source; 3.20: Biophoto Associates/Science Source; Table 3.3 left: Gopal Murti/Science Source; Table 3.3 middle: Alvin Telser, Cultura Science/Oxford Scientific/Getty Images; Table 3.3 right: Torsten Wittmann/Science Source; 3.22: Don W. Fawcett/Science Source; 3.23: Steve Gschmeissner/Science Source; 3.24 top: Kallista Images/Getty Images; 3.24 bottom: Biophoto Associates/Science Source; Table 3.4 left: SPL/Science Source; Table 3.4 right: Juergen Berger/Science Source; 3.25: Don W. Fawcett/Science Source; 3.27c: Biophoto Associates/Science Source; 3.35 all: Ed Reschke/Photolibrary/Getty Images; 3.36: E. Walker/Science Source.

Chapter 4   Opener: Steve Gschmeissner/SPL/Science Source; CB left and right: William Karkow, Pearson Education; APRW: William Karkow, Pearson Education; 4.5a, b, c: William Karkow, Pearson Education; 4.5d: Stephen W. Downing, Pearson Education; 4.7a, b: Stephen W. Downing, Pearson Education; 4.7c, 4.7d: William Karkow, Pearson Education; 4.9: Nina Zanetti, Pearson Education; 4.12a: William Karkow, Pearson Education; 4.12b: Nina Zanetti, Pearson Education; 4.12c: Don W. Fawcett/Science Source; 4.12d: Eye of Science/Science Source; 4.13 left: Karen Krabbenhoft, Pearson Education; 4.13 right: William Karkow, Pearson Education; 4.14a, b, c left: Karen Krabbenhoft, Pearson Education, 4.14a right: Lisa Lee, Pearson Education; 4.14b, c right: Stephen W. Downing, Pearson Education; 4.15: Lisa Lee, Pearson Education; 4.16 left: Karen Krabbenhoft, Pearson Education; 4.16 right, 4.17a right: Lisa Lee, Pearson Education, 4.17b right: William Karkow, Pearson Education, 4.17c right: Stephen W. Downing, Pearson Education, APRW left: P.M. Motta/Dept. of Anatomy/University La Sapienza, Rome/Science Source; APRW right: SPL/Science Source; 4.18a, b, 4.19: William Karkow, Pearson Education, 4.21a: Nina Zanetti, Pearson Education, 4.21b: Stephen W. Downing, Pearson Education, 4.21c: Lisa Lee, Pearson Education, 4.22, William Karkow, Pearson Education; CB left and right: William Karkow, Pearson Education; CB middle: Nina Zanetti.

Chapter 5   Opener: NASA/Science Source; 5.3: Lisa Lee, Pearson Education; 5.4a, b, 5.5: William Karkow, Pearson Education; 5.6a: Lisa Lee, Pearson Education; 5.6b: Clouds Hill Imaging/Science Source; 5.9b: Stephen W. Downing, Pearson Education; 5.12a, b: P. Marazzi/Science Source; 5.12c: John Radcliffe Hospital/Science Source; 5.14a, b: P. Marazzi/Science Source; 5.14c: Medical-on-Line/Alamy.

Chapter 6   Opener: Steve Gschmeissner/Science Source; 6.5 left: Creative Digital Visions, LLC, Pearson Education; 6.5.2a, b: Karen Krabbenhoft and Andrew Lokuta, Pearson Science; 6.6: Stephen W. Downing, Pearson Education; 6.9: William Karkow, Pearson Education; APRW left: Steve Gschmeissner/Science Source; APRW right: Pietro M. Motta/Science Source; UN6.2 (x-ray of hand): Steven Needell/Custom Medical Stock Photo; 6.13: Lisa Lee, Pearson Education; Table 6.1 top left: P. Marazzi/Science Source; Table 6.1 middle left: Custom Medical Stock Photo; Table 6.1 bottom left: skyhawk x/Shutterstock; Table 6.1 top right: Zephyr/Science Source; Table 6.1 middle right: kuehdi/Shutterstock; Table 6.1 bottom right: CC-BY-3.0: Gilo1969.

Chapter 7   Opener: Riccardo Cova/iStock/Getty; 7.5b, 7.6b, 7.8b: Larry DeLay, Pearson Education; 7.9b: Creative Digital Visions, LLC, Pearson Education; 7.14c: Larry DeLay, Pearson Education; APRW: Bone Clones, www.boneclones.com; 7.17a: Apogee/Science Source; 7.17b: Zephyr/Science Source; 7.17c: Scoliosis Research Society (SRS.org); Table 7.3 all: Creative Digital Visions, LLC, Pearson Education; 7.31a, b: Eksel Perez and Peter Westra, Pearson Education; APRW: CC-BY-SA 3.0: Lucien Monfils; 7.40b: Media for Medical/Universal Images Group User/age footstock.

Chapter 8   Opener: Zephyr/Science Source; 8.1a: Larry DeLay, Pearson Education; 8.6, 8.7, 8.8, 8.9, 8.10: John Wilson White, Studio Phocasso, Pearson Education; 8.14b: Karen Krabbenhoft, Pearson Education; APRW left: Jordan B. Renner, M.D., University of North Carolina at Chapel Hill; APRW right: Du Cane Medical Imaging Ltd./Science Source; 8.15d: Karen Krabbenhoft, Pearson Education; 8.16d: Winston Charles Poulton, Pearson Education.

Chapter 9   Opener: cosmin4000/IStock/Thinkstock; 9.8b, 9.13c: Karen Krabbenhoft, Pearson Education; 9.15b, 9.21d: Winston Charles Poulton, Pearson Education; 9.22b: Karen Krabbenhoft, Pearson Education.

Chapter 10   Opener: Animated Biomedical Productions, Pearson Education; APRW left: Lisa Lee, Pearson Education; APRW right: William Karkow, Pearson Education; 10.7: Don W. Fawcett/Science Source; 10.24: National Institutes of Health.

Chapter 11   Opener: Medi-Mation Ltd/Science Source; 11.4: William Karkow, Pearson Education; 11.8a: David Furness, Keele University/Science Source; APRW: Living Art Enterprises/Science Source.

Chapter 12   Opener: Tom Barrick, Chris Clark, SGHMS/Science Source; 12.2a: Winston Charles Poulton, Pearson Education; 12.2b: Lisa Lee, Pearson Education; 12.4a, 12.11d, 12.12b: Karen Krabbenhoft, Pearson Education; 12.36: Stegerphoto/Photolibrary/Getty Images.

Chapter 13   Opener: Eraxion/iStock/Getty Images; 13.2b: Thomas Deerinck, NCMIR/Science Source; APRW: P. Marazzi/Science Source; 13.7a, 13.8a: Winston Charles Poulton, Pearson Education; 13.9a: Karen Krabbenhoft, Pearson Education; 13.21a, b: Maureen Spuhler, Pearson Education.

Chapter 14   Opener: 4X-image/E+/Getty Images.

Chapter 15   Opener: Paul Whitten/Science Source; 15.2c: Steve Gshmeissner/SPL/Science Source; 15.5c: William Karkow, Pearson Education; 15.8a: nrt/Shutterstock; 15.9b: From: Balloon catheter dilation in common canalicular obstruction of the lacrimal system: safety and long-term effectiveness. Ko GY, Lee DH, Ahn HS, Yoon HK, Sung KB, Song HY. Radiology. 2000 Mar; 214(3):781-6. Figure 1e; 15.12 left: cristi180884/Fotolia; 15.12 middle: samsonovs/Fotolia; 15.12 right: johanna goodyear/Fotolia; 15.13: Mark Miller/Science Source; 15.21b: Nina Zanetti, Pearson Education; 15.22c: Ralph C. Eagle, Jr./Science Source; APRW1: Portokalis/Shutterstock; 15.26, Dennis MacDonald/Alamy; APRW left: G. Lacher/Science Source; APRW middle: Medical-on-Line/Alamy; APRW right: SPL/Science Source; 15.30a: Bobkeenan Photography/Shutterstock; 15.32b: Lisa Lee, Pearson Education; 15.33c: David Furness, Keele University/Science Source; 15.35, 15.40: Dennis MacDonald/Alamy.

Chapter 16   Opener: UGREEN 3S/Shutterstock; 16.11b: Stephen W. Downing, Pearson Education; 16.15: Scott Camazine/Science Source; 16.17b: Lisa Lee, Pearson Education; 16.20a, b: AR Waltner; 16.21b: Stephen W. Downing, Pearson Education; 16.24: bikeriderlondon/Shutterstock.

Chapter 17   Opener: Kozorez Vladislav/Shutterstock; UN17.2 (angiogram): Zephyr/Science Source; 17.5b: Karen Krabbenhoft, Pearson Education; 17.7.1b left and right: SPL/Science Source.

Chapter 18   Opener: MedicalRF.com/GettyImages; 18.2: Lisa Lee, Pearson Education; APRW: Richard Kirby, David Spears Ltd./Science Source; 18.10b: Stephen W. Downing, Pearson Education.

Chapter 19   Opener: National Cancer Institute/Science Source; 19.1a: William Karkow, Pearson Education; 19.2b: Susumu Nishinaga/Science Source,; 19.7a: Eye of Science/Science Source; 19.8: Lisa Lee, Pearson Education; 19.13: CNRI/Science Source; 19.18: Holyoke College.

Chapter 20   Opener: Steve Gschmeissner/Science Source; APRW: Sergio Azenha/Alamy; 20.6b: Lisa Lee, Pearson Science; 20.7b: Winston Charles Poulton, Pearson Science; 20.8b: Lisa Lee, Pearson Science; UN20.2 (macrophage): Eye of Science/Science Source; 20.28: National Cancer Institute.

Chapter 21   Opener: Anatomical Travelogue/Science Source; 21.3b: Jupiterimages/Stockbyte/Getty Images; 21.4b: Living Art Enterprises/Science Source; 21.8c: Nina Zanetti, Pearson Education; 21.9b: SPL/Science Source; 21.10b: David M. Phillips/Science Source; 21.32: Biophoto Associates/Science Source.

Chapter 22   Opener: Susumu Nishinaga/Science Source; 22.3c: Karen Krabbenhoft, Pearson Science; 22.8a: Steve Downing, Pearson Science; 22.8b: William Karkow, Pearson Education; 22.10b: Karen Krabbenhoft, Pearson Education; 22.15b: William Karkow, Pearson Education; 22.23b: Nina Zanetti, Pearson Science; 22.25: BSIP/Science Source.

Chapter 23   Opener: Carol and Mike Werner/Science Source; 23.2: Charles D. Winters/Science Source; APRW left: Lisa Lee, Pearson Education; APRW right: Biophoto Associates/Science Source; 23.17: Kriss Russell/iStock/Getty Images, 23.20: http://www.choosemyplate.gov/print-materials-ordering/graphic-resources.html.

Chapter 24   Opener: Susumu Nishinaga/Science Source; 24.6b: Thomas Deerinck, NCMIR/Science Source; APRW: SPL/Science Source; 24.25b, c: Lisa Lee, Pearson Education.

Chapter 25   Opener: Steve Gschmeissner/Science Source; 25.8: John Lynn, www.endocrinesurgeon.co.uk.

Chapter 26   Opener: Tim Vernon/Science Source; 26.4c: Nina Zanetti, Pearson Education; 26.8b: Juergen Berger/Science Source; APRW: UHB Trust/The Image Bank/GettyImages; 26.15 top: Stephen W. Downing, Pearson Education; 26.15 second and third from top: William Karkow, Pearson Education; 26.15 middle: Ed Reschke/Photolibrary/GettyImages; 26.15 third from bottom: Claude Edelmann/Science Source; 26.15 second from bottom: Biophoto Associates/Science Source; 26.15 bottom: Stephen W. Downing, Pearson Education; 26.17: William Karkow, Pearson Education; 26.20a left: Mega Pixel/Shutterstock; 26.20a right: Image Point Fr/Shutterstock; 26.20b left and middle: Gary Parker/Science Source; 26.20b right, 26.20c: Garo/Phanie/age fotostock; 26.20d: Voisin/Phanie/age fotostock; 26.20e left: GmbH & Co. KG/Alamy; 26.20e right: Gusto/Science Source; 26.20f left: P. Marazzi/Science Source; 26.20f right: Hattie Young/Science Source; 26.20g: Image Point Fr/Shutterstock.

Chapter 27   Opener: Nestle/Petit Format/Science Source; 27.2: David M. Phillips/Science Source; 27.8a: Anatomical Travelogue/Science Source; 27.8b: G. Moscoso/Science Source; 27.10a: Petit Format/Science Source; 27.10b: Neil Bromhall/Science Source; UN27.1: Biophoto Associates/Science Source; APRW: Saturn Stills/Science Source.

Glossary

A band The component of the sarcomere that appears darker because it contains thick filaments and areas in which the thick and thin filaments overlap.

Abdominal cavity The portion of the abdominopelvic cavity located between the bony pelvis and the diaphragm.

Abdominopelvic cavity (ab-dom´-ih-noh-PEL-vik) A ventral body cavity located within the area inferior to the diaphragm that encompasses the area of the abdomen and the pelvis.

Abduction (ab-DUHK-shuhn) An angular movement whereby the body part moves away from the midline of the body or another reference point.

ABO blood group A blood group that classifies blood type based on the presence or absence of two antigens: the A and B antigens.

Absorption The transport of a substance across an epithelial lining into the bloodstream.

Absorptive state The time period from the point when ingested nutrients enter the bloodstream to about 4 hours after feeding.

Accommodation The rounding of the lens to focus light from near objects on the retina.

Acetabulum (ah´-suh-TAB-yoo-lum) The bony socket formed by the ilium, ischium, and pubis into which the femur fits to form the hip joint.

Acetylcholine (Ach) (ah-SEE-til-koh´-leen) A neurotransmitter involved in a wide variety of processes, including those of the autonomic nervous system and muscle contraction.

Acetylcholinesterase (AChE) (ah-SEET-´l-kohl-in-ess´-teh-rayz) An enzyme located in the synaptic cleft that degrades acetylcholine.

Acid A compound that releases one or more hydrogen ions when placed in water; has a pH less than 7.0.

Acromioclavicular joint (AC joint) The articulation between the acromion of the scapula and the lateral portion of the clavicle.

Acromion (ah-KROH-mee-ahn) The anterosuperior projection of the scapula that articulates with the clavicle.

Acrosome (AK-row-sohm) A membrane-bounded sac on the head of the sperm cell that contains digestive enzymes.

Actin (AK-tin) A bead-shaped contractile protein found in muscle fibers and motile cells.

Actin filaments (microfilaments) Thin protein filaments of the cytoskeleton composed of two intertwining actin strands that provide a cell with structural support, bear tension, help to maintain its shape, and function in cellular motion.

Action potential A quick, temporary change in the membrane potential of a cell in a single region of the plasma membrane.

Activation energy (E_a) The energy required for an adequately strong collision to occur between atoms for a reaction to take place.

Active immunity A type of immunity mediated by an individual's own immune system after exposure to an antigen.

Active transport A type of membrane transport against a gradient that requires energy expenditure in the form of ATP.

Adaptive immunity (specific immunity) The arm of the immune system that consists of cells and proteins that must bind to specific antigens to become activated.

Adduction (uh-DUHK-shuhn) An angular movement whereby the body part moves toward the midline of the body or another reference point.

Adductor A muscle whose function is to move a structure or the body part it is attached to closer to the midline of the body.

Adenine A purine nitrogenous base found in nucleic acids and molecules such as ATP.

Adenosine triphosphate (ATP) A nucleotide consisting of adenine, ribose, and three phosphate groups; the main energy currency of human cells.

Adipocyte (AD-ih-poh-sy´t) The main cell type in adipose tissue; each cell contains a large lipid droplet in its cytoplasm.

Adipose tissue A type of connective tissue proper that contains adipocytes; found in locations such as the hypodermis and around organs.

Adrenal cortex The outer region of the adrenal gland that produces steroid hormones.

Adrenal gland An endocrine gland located at the superior pole of each kidney; consists of an outer cortex and an inner medulla.

Adrenal medulla The inner region of the adrenal gland; consists of chromaffin cells that release epinephrine and norepinephrine when stimulated by the sympathetic nervous system.

Adrenergic receptors (ad-ren-ER-jik) Receptors in the plasma membranes of the target cells of sympathetic neurons that bind to catecholamines.

Adrenocorticotropic hormone (ACTH) A hormone released by the adrenal cortex that stimulates the production of mostly glucocorticoids from the adrenal cortex.

Afferent arterioles Arterioles that deliver blood to the glomeruli in the kidney.

Afterload The force that the right and left ventricles must overcome in order to eject blood into their respective arteries.

Agonist A muscle that provides the principal force required in a movement; also known as the *prime mover*.

Agranulocyte (ay-GRAN-yoo-low-sy´t) A class of leukocyte without visible cytosolic granules; includes lymphocytes and monocytes.

Airway resistance The impedance to air flow within the respiratory tract.

Albumin (al-BYOO-mun) The most abundant plasma protein; largely responsible for the plasma's colloid osmotic pressure.

Aldosterone (al-DOHS-ter-ohn) A hormone produced by the adrenal cortex that increases reabsorption of sodium ions (and indirectly) water from the distal tubule, as well as the secretion of potassium and hydrogen ions.

Alimentary canal (gastrointestinal, or GI, tract) A series of hollow digestive organs through which food passes as it is ingested, digested, and absorbed; also carries out defecation.

Alleles (uh-LEELZ) Different variants of the same gene.

Alpha cells (α cells) Cells of the pancreatic islets that secrete the hormone glugacon.

Alpha motor neurons (α motor neurons) Lower motor neurons that innervate extrafusal muscle fibers.

Alveoli (al-vee-OHL-eye; singular, *alveolus*) Thin-walled structures in the lungs across which gases are exchanged.

Amino acid A monomer consisting of carbon, hydrogen, oxygen, and nitrogen atoms.

Amino acid–based hormone A chemical messenger consisting of one or more amino acids; most amino acid–based hormones are hydrophilic and so bind plasma membrane receptors.

Amnion (AM-nee-ahn) A membrane that develops from embryonic tissues; completely encloses the embryo in the fluid-filled amniotic cavity.

Amniotic fluid The fluid in the amniotic cavity that protects the developing conceptus.

Amphiarthrosis (am´-fee-ahr-THROH-sis) A functional classification of articulations in which the joint allows for only a small amount of motion between articulating bones.

Ampulla (am-PEWL-ah) An enlarged area at the base of each semicircular duct in the ear that contains receptor cells that detect rotation of the head.

Amygdala (ah-MIG-duh-lah) A component of the limbic system in the brain that functions in the behavioral expression of emotion.

Anabolic reaction (an-ah-BOL-ik) A reaction in which smaller substances bond to form larger substances.

Anal canal The terminal portion of the large intestine that contains sphincters to regulate the passage of feces during defecation.

Anaphase The third phase of mitosis during which the spindle fibers begin to pull chromatids to opposite poles of the cell.

Anatomical position The standard position in which an anatomical specimen is presented, in which a subject is facing forward, arms at the side with the palms facing outward, and the feet are shoulder-width apart.

Androgenic steroids Hormones produced by the adrenal cortex that have androgenic effects on a variety of tissues.

Anemia (ah-NEE-mee-ah) A condition in which the oxygen-carrying capacity of the blood is reduced due to a decreased amount of hemoglobin, a decreased number of erythrocytes, and/or abnormal hemoglobin.

Angiotensin-converting enzyme (ACE) An enzyme produced by the lungs and other tissues that converts angiotensin-I to the more active angiotensin-II.

Angiotensin-II (A-II) A hormone that has widespread effects on the body, which include causing vasoconstriction of all vessels, restoring the glomerular filtration rate, promoting ADH and aldosterone release, and increasing sodium ion and water reabsorption from the kidneys.

Anion (AN-eye-on) A negative ion with more electrons than protons.

Antagonist A muscle generally located on the opposite side of a joint from its agonist that opposes or slows the action of the agonist.

Anterior (an-TEER-ee-ur) Toward the front; also known as *ventral*.

Anterior cruciate ligament (ACL) A ligament found within the synovial cavity of the knee that, when the knee joint is extended, tightens to prevent hyperextension.

Anterior horn The anterior region of gray matter in the spinal cord that contains the cell bodies of motor neurons.

Anterior pituitary gland The anterior portion of the pituitary gland that is derived from glandular epithelium; produces many tropic hormones and growth hormone.

Anterior ramus The division of a spinal nerve that travels to the anterior side of the body.

Anterior root An extension from the anterior horn of the spinal cord containing the axons of lower motor neurons; some also contain the axons of visceral motor neurons.

Anterolateral system A set of ascending tracts in the spinal cord that convey sensations relating to pain, temperature, and nondiscriminative touch.

Antibody A protein produced by activated B lymphocytes that binds to a specific antigen and facilitates its removal from a tissue.

Antibody-mediated immunity (humoral immunity) The arm of adaptive immunity that is mediated by B lymphocytes and antibodies.

Anticoagulant (an'-tee-ko-AG-yoo-lunt) A molecule that inhibits some portion of the coagulation cascade.

Anticodon A three-nucleotide portion of transfer RNA that binds a specific messenger RNA codon.

Antidiuretic hormone (ADH) A hormone produced by the hypothalamus and stored in the posterior pituitary that causes the insertion of aquaporin channels in the cells of the distal tubule and collecting system to allow water reabsorption.

Antigen (AN-ti-jen) A unique glycoprotein marker found on the surface of almost all cells and biological molecules.

Antigen-presenting cells Immune cells that process and display antigen fragments on their class II major histocompatibility molecules to activate other immune cells such as helper T cells.

Antithrombin-III (AT-III) An anticoagulant that inhibits the activity of factor Xa and thrombin.

Anulus fibrosus (AN-yoo-lus fy-BRO-sus) The outer, fibrous, supportive tissue of an intervertebral disc.

Aorta (ay-OR-tah) The largest artery in the body; arises from the left ventricle and delivers blood to the systemic circuit.

Aortic sinus An area within the aortic arch that houses baroreceptors.

Aortic valve The semilunar valve located between the left ventricle and the aorta.

Apocrine glands (AP-oh-krin) Sweat glands found in the axillae, the anal area, and the areola that secrete a thick sweat rich in proteins.

Aponeurosis (ap'-oh-noo-ROW-sis) A thin tendon sheet that attaches a muscle to another structure.

Appendicular region (ap'-en-DIK-yoo-lar) The portion of the body consisting of the upper and lower limbs.

Appositional growth The process by which bones grow in width.

Aquaporin (ah-kwah-POUR-in) A protein channel in the plasma membrane of renal tubule and collecting duct cells that allows the passage of water from the filtrate to the interstitial fluid.

Aqueous humor (AY-kwee-us) The watery fluid that fills the anterior cavity of the eyeball.

Arachnoid mater (ah-RAK-noyd) The middle meninx that surrounds the brain and spinal cord.

Arbor vitae (AHR-bor VEE-tay) Branching tracts of white matter located within the cerebellum.

Arm The proximal portion of the upper limb from the elbow to the shoulder that consists of the humerus.

Arrector pili A small band of smooth muscle cells that causes a hair to stand erect when contracted.

Arteries The vasculature's distribution system; pulmonary arteries carry deoxygenated blood to the lungs, and systemic arteries carry oxygenated blood to the body's cells.

Arterioles (ar-TEER-ee-olz) The smallest arteries that control blood flow to individual tissues.

Articular capsule (ar-TIK-you-luhr) A double-layered structure surrounding synovial joints that holds the articulating bones together.

Articulation The location where two or more bones come together; also known as a *joint*.

Association areas Areas of the cerebral cortex that integrate different types of information.

Association fibers White matter fibers in the brain that connect the cerebral cortex in adjacent gyri.

Astigmatism (ah-STIG-mah-tizm) An irregular curvature of the lens or cornea that produces uneven refraction and blurred vision.

Astrocyte (ASS-troh-sy't) A neuroglial cell of the central nervous system that facilitates the formation of the blood-brain barrier, regulates the extracellular environment of the brain, anchors neurons and blood vessels in place, and repairs damaged brain tissue.

Asystole (ay-SIS-toh-lee) A condition in which there is no electrical activity in the heart, resulting in a flattened line on the ECG.

Atlas The first cervical vertebra.

Atmospheric pressure The pressure exerted by the air of the surrounding atmosphere.

Atom The smallest unit of matter that still retains its original properties.

Atomic nucleus (NEW-klee-us) The central core of an atom where the protons and neutrons reside.

Atomic number The number of protons found in an atom; the atomic number is unique to each element.

ATP synthase A large protein in the inner mitochondrial membrane that functions as an ion channel and an enzyme that catalyzes the production of ATP.

Atria (AY-tree-ah) The superior chambers of the heart that receive blood from veins and pump blood into the ventricles.

Atrial natriuretic peptide (ANP) A hormone produced by cells of the heart that triggers the excretion of sodium ions (and indirectly water) from the renal tubule of the kidneys.

Atrioventricular bundle (AV bundle) The first component of the heart's Purkinje fiber system that propagates action potentials from the atria to the ventricles.

Atrioventricular node (AV node) A cluster of pacemaker cells in the inferior right atrium that delays the propagation of the cardiac action potential from the atria to the ventricles.

Atrioventricular valves (AV valves) Valves located between the atria and ventricles of the heart.

Atrophy (AH-troh-fee) A decrease in cell size.

Auditory ossicles (OSS-ih-kulz) A chain of three bones (malleus, incus, and stapes) that connect the tympanic membrane to the oval window.

Auricle (heart) (OR-ih-kul) A muscular pouch extending from an atrium of the heart that allows the atrium to hold more blood.

Autocrine (AW-toh-krin) A chemical messenger that influences the function of the same cell or cell type that produced and secreted it.

Autoimmune disorder A condition characterized by the production of autoantibodies, or antibodies that bind and react to self-antigens.

Autonomic ganglion A collection of cell bodies of postganglionic autonomic neurons.

Autonomic nervous system (ANS) (aw-toh-NOM-ik) The division of the peripheral nervous system that controls homeostatic responses of the organs of many other systems; also known as the *visceral motor division of the PNS*.

Autonomic tone The constant amount of activity exhibited by each branch of the ANS on specific target cells.

Autoregulation A process in which a tissue regulates aspects of its own structural and functional properties.

Autorhythmicity (aw′-toh-rith-MISS-ih-tee) A property whereby a tissue sets its own rhythm of action potential generation without external influences.

Autosomal dominant-recessive inheritance Pattern of inheritance involving autosomal alleles that are strictly dominant or recessive.

Axial region (AKS-ee-ul) The portion of the body consisting of the head, neck, and trunk.

Axis The second cervical vertebra.

Axolemma (aks-oh-LEM-ah) The plasma membrane of an axon.

Axon A single extension of a neuron that can generate action potentials; generally carries information away from the cell body.

Axon hillock The initial portion of an axon where an action potential is generated.

Axon terminal A knoblike structure at the end of an axon that contains synaptic vesicles with neurotransmitters.

Azygos system (AY-zye-gohs) The group of veins—including the azygos vein, hemiazygos vein, and accessory hemiazygos vein—that drain structures of the posterior thorax.

B lymphocyte The class of immune cell responsible for the antibody-mediated arm of adaptive immunity.

Ball-and-socket joint A multiaxial joint in which the articulating surface of one bone is ball-shaped and fits into a cup-shaped socket in the other bone, allowing for movement around three axes.

Baroreceptor reflex A negative feedback loop that responds to an increase in blood pressure by triggering the parasympathetic nervous system to decrease blood pressure.

Baroreceptors Internal sensory receptors that detect pressure.

Basal metabolic rate (BMR) The minimal rate of caloric expenditure for an awake individual.

Basal nuclei Nuclei within the cerebral hemispheres that function in the initiation of movement and the prevention of inappropriate movements.

Base (alkali) A compound that accepts a hydrogen ion from an acid; has a pH greater than 7.0.

Basement membrane Two layers of extracellular matrix—the basal lamia and the

reticular lamina—that anchor an epithelial tissue to the underlying connective tissue.

Basophil (BEH-zuh-fill) A class of granulocyte that is involved in mediating inflammation.

Beta cells (β cells) Cells of the pancreatic islets that secrete the hormone insulin.

Beta oxidation (β oxidation) The process of fatty acid oxidation, during which two-carbon molecules are removed as acetyl-CoA and $FADH_2$ and NADH are generated.

Beta waves Low-amplitude, high-frequency brain waves that are characteristic of an awake adult.

Biaxial joints A functional classification of synovial joints in which motion between articulating bones occurs around two axes.

Bile A liquid that contains water, electrolytes, and organic compounds such as bile salts; it is required for the digestion and absorption of lipids, and it is a mechanism by which the liver excretes wastes.

Biochemistry The study of the chemistry of living beings.

Bipolar neuron A neuron with one axon and one dendrite.

Blastocyst (BLAS-toh-sist) The early stage of development during which implantation occurs.

Blood The fluid connective tissue consisting of formed elements and plasma.

Blood-brain barrier The collective term for the tight junctions between the endothelial cells of the brain capillaries, their basal laminae, and astrocytes; keeps cerebrospinal fluid and brain extracellular fluid separate from the blood, protecting the brain from certain substances in the blood.

Blood clot A collection of platelets, erythrocytes, fibrin, and other clotting proteins that prevents further loss of blood from an injured blood vessel.

Blood flow The volume of blood that flows through the systemic or pulmonary circuit in one minute.

Blood pressure The hydrostatic pressure exerted on a blood vessel wall by the blood.

Blood-testis barrier A barrier formed by sustentacular cells that prevents immune cells in the bloodstream from encountering the new antigens formed on the genetically unique sperm cells.

Body mass index (BMI) A calculation of body mass that takes into account an individual's height; allows the classification of an individual as underweight, normal weight, overweight, or obese.

Bolus (BOW-lus) A moistened, partially digested ball of ingested food that is swallowed and transported to the stomach via the pharynx and esophagus.

Bone deposition The formation of new bone by osteoblasts.

Bone remodeling The continual process of bone growth and bone loss that takes place within healthy bone.

Bone resorption The destruction of bone tissue by osteoclasts.

Boyle's law A gas law that states that pressure and volume are inversely proportional.

Brachial plexus A group of nerves consisting of the anterior rami from C5–T1 that innervates structures within and around the upper limb.

Brain The organ of the central nervous system that performs multiple integrative functions.

Brain waves Characteristic patterns of electrical activity in the brain that can be measured on an electroencephalogram (EEG).

Brainstem The most inferior part of the brain; responsible for automatic, subconscious functions.

Broca's area A premotor multimodal association area for speech located in the anterolateral frontal lobe; responsible for the ability to produce the motor actions necessary for language.

Bronchi (BRONG-kye; singular, *bronchus*) Larger passages of the bronchial tree.

Bronchial tree (BRONG-kee-uhl) Branching series of passageways that conducts air to and from the respiratory zone of the respiratory tract.

Bronchioles (BRONG-kee-olz) Passages of the bronchial tree that are less than one millimeter in diameter.

Bronchoconstriction The process by which the diameter of the bronchioles decreases, generally due to smooth muscle contraction.

Bronchodilation The process by which the diameter of the bronchioles increases, generally due to smooth muscle relaxation.

Buffer system A chemical system that resists changes in pH and prevents large swings in the pH when acid or base is added to the solution.

Bulbourethral gland (bul′-boh-yoo-REETH-ruhl) An accessory gland in the male reproductive system that secretes thick alkaline fluid that helps neutralize acidic urine remaining in the urethra prior to ejaculation.

Bundle branches The second component of the heart's Purkinje fiber system that propagates action potentials along the right and left sides of the interventricular septum.

Bursa (BER-suh) A synovial fluid–filled structure located between bones, tendons, and muscles of certain synovial joints; minimizes friction between moving parts.

Calcaneus (kal-KEHN-ee-us) The posterior tarsal bone; "heel bone."

Calcitonin (kal-sih-TOW-nin) A hormone produced by the parafollicular cells of the thyroid gland; decreases blood calcium ion concentration by stimulating osteoblasts to build bone.

Calcitriol (kal-sih-TRY-ohl) The active form of vitamin D; increases the number of calcium ions absorbed from the small intestine.

Calmodulin (Cam) (kal-MOD-yoo-lin) A protein in the cytosol of smooth muscle cells that binds to calcium ions during a contraction.

Calorie The amount of heat required to raise the temperature of one kilogram of water by one degree Celsius; also known as the *kilocalorie*.

Calorimetry (kal'-or-IM-eh-tree) The process by which the amount of body heat an individual produces is measured.

Calvarium The superior portion of the cranial cavity; also called the *cranial vault*.

Capillaries The vasculature's exchange system through which oxygen, nutrients, water, and wastes are exchanged between the blood and the cells; tiny vessels with only an endothelium and basal lamina.

Carbohydrate An organic molecule composed of carbon, hydrogen, and oxygen atoms, often in the ratio of 1:2:1.

Carbonic acid–bicarbonate buffer system The major buffer system in the blood; consists of the weak acid carbonic acid and its conjugate base, the bicarbonate ion.

Carbonic anhydrase (CA) An enzyme that catalyzes the reversible reaction in which carbon dioxide and water are combined to become carbonic acid.

Carcinoma (kar-sih-NOH-mah) A cancer of epithelial tissue.

Cardiac conduction system The collective term for the three populations of pacemaker cells in the heart: the sinoatrial node, the atrioventricular node, and the Purkinje fiber system.

Cardiac cycle The sequence of mechanical events in the heart that takes place from one heartbeat to the next.

Cardiac muscle A tissue type composed of wide, branching cardiac muscle cells and the surrounding endomysium.

Cardiac muscle cells (myocytes) Striated, branched muscle cells joined by intercalated discs; make up the majority of cardiac muscle tissue.

Cardiac output (CO) The volume of blood pumped into the pulmonary and systemic circuits in one minute.

Cardiovascular system The organ system made up of the heart and the blood vessels; delivers and drains blood to and from tissues.

Carina (kar-EYE-nah) The final cartilage ring of the trachea where cough receptors are housed.

Carotid sinus Area in the common carotid artery near its bifurcation into the internal and external carotid arteries that houses baroreceptors.

Carpals The short bones of the wrist.

Cartilaginous joints (kar'-tih-LAJ-ih-nuhs) A structural classification of articulations in which the joints are held together by cartilage and have neither a synovial cavity nor an articular capsule.

Catabolic reaction (kat-ah-BOL-ik) A reaction in which larger substances are broken down into smaller ones.

Cataract (KAT-ah-rakt) A condition in which clouding of the lens interferes with vision.

Catecholamines (kat'-eh-KOHL-ah-meenz) A group of neurotransmitters derived from the amino acid tyrosine; includes norepinephrine, epinephrine, and dopamine.

Cation (KAT-eye-on) A positive ion with more protons than electrons.

Cauda equina (KOW-dah eh-KWY-nah) A large bundle of nerve roots given off at the terminal end of the spinal cord.

Caudate nuclei (KOW-dayt) Basal nuclei consisting of two C-shaped rings of gray matter; work with the putamen to inhibit the globus pallidus and initiate movement.

Cecum (SEE-kum) A blind pouch that is the first portion of the large intestine.

Celiac trunk (SEE-lee-ak) The single large vessel that branches off the superior descending abdominal aorta; branches into vessels that supply many abdominal organs.

Cell The smallest unit capable of carrying out the functions of life.

Cell-adhesion molecules (CAMs) Glycoproteins in the extracellular matrix that hold cells in a tissue together.

Cell body The central portion of a neuron that contains the nucleus and the bulk of the organelles.

Cell cycle The ordered series of events from cell formation to cell division.

Cell-mediated immunity The arm of adaptive immunity that is mediated by T lymphocytes.

Cell metabolism The collection of reactions that a cell carries out to maintain life.

Central canal A small, cerebrospinal fluid–filled canal located in the center of the spinal cord.

Central nervous system (CNS) The division of the nervous system consisting of the brain and the spinal cord.

Central sulcus A sulcus of each cerebral hemisphere that separates the frontal lobes from the parietal lobes.

Centrioles A pair of organelles within the centrosome that consist of modified microtubules and other proteins; play an important role in cell division.

Centrosome The central area of the cell from which microtubules emanate.

Cerebellum (sair-ah-BEL-um) The posteroinferior portion of the brain; responsible for regulating ongoing movement and the correction of motor error.

Cerebral aqueduct A hollow, cerebrospinal fluid–filled channel that connects the third ventricle in the diencephalon with the fourth ventricle in the brainstem.

Cerebral arterial circle A vascular anastomosis on the inferior surface of the brain that provides alternate routes of blood flow to the regions of the brain.

Cerebral cortex The outer two millimeters of gray matter of the cerebrum that function in conscious processes.

Cerebral hemispheres (cerebrum) The two large, superior structures of the brain that are responsible for higher mental functions, control of movement, and interpretation of sensations.

Cerebral lateralization The unequal distribution of cognitive functions in the right and left cerebral hemispheres.

Cerebrospinal fluid (CSF) A fluid similar in composition to blood plasma (but with fewer proteins) that bathes and surrounds the brain and spinal cord.

Cerumen (si-ROO-min) A modified sweat secreted by ceruminous glands in the external ear.

Ceruminous glands (si-ROO-min-us) Modified sweat glands located in the external ear.

Cervical curvature A convex curvature of the cervical portion of the vertebral column.

Cervical plexus A group of nerves consisting of the anterior rami from C1–C4 that innervates primarily structures around the head and the neck.

Cervical vertebrae Seven bones of the vertebral column located in the neck.

Cervix The narrow neck, or outlet, of the uterus that projects into the vagina inferiorly.

Chemical A substance with a unique molecular composition that is used in or produced by chemical processes.

Chemical bond An energy relationship between atoms or ions caused by the sharing or transfer of valence electrons.

Chemical buffer system A system that resists changes in the pH of the body's fluids; consists of a weak acid and its conjugate base.

Chemical energy Energy stored in the chemical bonds of all molecules.

Chemical reaction The process that occurs when chemical bonds are formed, broken, or rearranged, or electrons are transferred between reactants.

Chemical synapse A type of synapse in which a presynaptic neuron releases neurotransmitters to trigger a change in a postsynaptic neuron.

Chemistry The study of matter and its interactions.

Chemoreceptor (KEE-moh-ree-sep'-ter) A sensory receptor that responds to a change in the concentration of a specific chemical in the air or the body's fluids.

Chemotaxis (kee-moh-TAK-sis) The process by which cells are attracted to a particular location in the body via chemicals released by other cells.

Cholecystokinin (CCK) (kohl'-eh-sih-stoh-KY-nin) A hormone produced by duodenal enteroendocrine cells; triggers the secretion of digestive enzymes and other proteins from the pancreas.

Cholesterol A lipid that is used as a precursor for many molecules in the body, including bile salts and steroid hormones.

Cholinergic receptors (kohl-in-ER-jik) Receptors in the plasma membranes of the target cells of parasympathetic neurons and certain sympathetic neurons that bind acetylcholine.

Chondrocyte (KON-droh-sy't) The main cell type in cartilage that maintains the extracellular matrix of the tissue.

Chordae tendineae Fibrous cords that attach papillary muscles to cusps of the atrioventricular valves.

Chorion (KOR-ee-on) The outermost extraembryonic membrane that forms from maternal and embryonic tissues.

Chorionic villi (kohr-ee-ON-ik VIL-eye; singular, *villus*) Outgrowths of the chorion that grow in the direction of the stratum functionalis layer of the endometrium.

Choroid (KOR-oyd) The pigmented portion of the vascular layer of the eyeball that contains many blood vessels to nourish the retina.

Choroid plexuses (KOR-oyd) Groups of brain capillaries that come into contact with ependymal cells and produce cerebrospinal fluid.

Chromaffin cells (KROH-maf-in) Cells found in the adrenal medulla that secrete epinephrine and norepinephrine when stimulated by the sympathetic nervous system.

Chromatin (KROH-mah-tin) A long, thin strand of DNA and its associated proteins.

Chromosome (KROH-moh-sohm) The tightly coiled and condensed chromatin molecule that forms in preparation for cell division.

Chylomicron (ky-loh-MY-krahn) A large molecule composed of triglycerides packaged with cholesterol, other dietary lipids, phospholipids, and lipid-binding apoproteins that delivers dietary lipids to cells.

Chyme (KYME) A liquid mixture that results from the stomach churning a bolus and mixing it with gastric juice.

Cilia (SILL-ee-uh) Microtubule-containing extensions from the plasma membrane that sweep substances past the cell.

Ciliary body (SILL-ee-air-ee) A ring of smooth muscle that alters the shape of the lens to focus light on the retina.

Circadian rhythm (sir-KAY-dee-an) A bio-logical process that follows a set pattern over the 24-hour day.

Circumduction (sir-kuhm-DUHK-shuhn) An angular movement that is the sum total of flexion-extension and abduction-adduction; results in a conelike motion best seen in hip and shoulder joints.

Citric acid cycle The second part of glucose catabolism; a series of eight reactions that takes place in the mitochondrial matrix during which sugars are oxidized and ATP and reduced electron carriers are generated.

Clavicle The anterosuperior bone of the pectoral girdle that spans from the scapula to the sternum.

Cleavage An early series of rapid mitotic divisions of a conceptus in the pre-embryonic period.

Clitoris (KLIT-ur-us) A small, protruding structure of the female vulva composed of erectile tissue; important in the female sexual response.

Clonal selection The process by which an antigen selects and activates a specific T or B lymphocyte clone.

Clotting factors Proteins found in the blood in their inactive forms that are involved in the coagulation cascade.

Coagulation (koh-ah-gyoo-LAY-shun) The process of formation of fibrin, which "glues" the components of a blood clot together.

Coagulation cascade A series of enzyme-catalyzed reactions involving clotting factors that culminates in the activation of fibrinogen to fibrin.

Coccygeal vertebrae Three to five small, fused vertebrae of the inferior vertebral column.

Cochlea (COKE-lee-ah) The bony spiral-shaped portion of the inner ear important in hearing.

Codon (KOH-don) A three-nucleotide portion of messenger RNA that specifies a single amino acid.

Cognition The mental functions that include processing and responding to complex external stimuli, recognizing related stimuli, processing internal stimuli, and planning appropriate responses to stimuli.

Collagen fibers Protein fibers in the extracellular matrix of a tissue that are composed of the protein collagen and give a tissue tensile strength.

Collateral circulation Alternate routes of blood flow within a tissue.

Collateral ganglia Autonomic sympathetic ganglia located near the neurons' target organs.

Collecting system The series of tubules that receive filtrate from the distal tubules in the kidneys; responsible for fine-tuning fluid, electrolyte, and acid-base balance.

Colloid (KOL-oyd) A mixture in which the molecules are in different phases but the particles will not settle out when left undisturbed; the gelatinous material in thyroid follicles in which iodine and thyroid hormones are stored.

Colloid osmotic pressure (COP) An osmotic pressure gradient created by proteins within the plasma; draws water into the plasma from the interstitial fluid.

Colon (KOH-lun) The longest portion of the small intestine that consists of the ascending, transverse, descending, and sigmoid colon.

Columnar cell An epithelial cell with a column shape that is taller than it is wide.

Commissural fibers (kom-ih-SHUR-al) White matter fibers that connect the right and left cerebral hemispheres.

Compact bone The hard, dense bone tissue located on the exterior of a bone; composed of repeating units called osteons.

Complement system A group of proteins that function in innate immunity.

Complementary base pairing The formation of hydrogen bonds between the nucleic acid bases adenine and thymine (or uracil) and between guanine and cytosine.

Compound A molecule composed of two or more atoms of different elements.

Concentration gradient A situation in which a differential concentration of molecules exists, meaning that a high concentration of the substance is present in one location and a low concentration in a neighboring area.

Conducting zone The portion of the respiratory tract that conducts air toward and away from the alveoli.

Conduction The transfer of heat from one object to another through direct contact.

Condylar joint (KAHN-duh-luhr) A biaxial joint in which the oval-convex surface of one bone fits into a shallow concave depression on the other bone, allowing for movement around two axes.

Cones Photoreceptors that have a cone-shaped outer segment and respond to wavelengths of light that we perceive as color; responsible for high-acuity color vision.

Conjunctiva (kon-junk-TY-vah) A transparent membrane that covers the inside of the eyelids and the anterior sclera.

Connective tissue A type of tissue characterized by extensive extracellular matrix; functions in support, protection, and transport.

Connective tissue proper A type of connective tissue that generally functions to connect and support other tissues.

Consolidation The transfer of information from immediate and working memory into long-term memory.

Continuous capillaries Capillaries whose endothelial cells are joined by tight junctions; the least leaky type of capillary.

Continuous conduction A relatively slow type of neuronal action potential conduction that occurs in unmyelinated axons in which every segment of the axolemma is depolarized.

Contractility The intrinsic pumping ability of the heart.

Contraction The generation of tension by a muscle cell.

Control center The component of a feedback loop that receives a stimulus from a receptor and regulates the output of an effector.

Conus medullaris (KOHN-us med-yoo-LAIR-us) The cone-shaped terminal portion of the spinal cord between L1 and L2.

Convection The transfer of heat through a liquid or gaseous medium.

Convergence The medial movement of the eyes to focus on a near object.

Converging circuit A type of neural circuit in which several input neurons converge on a single output neuron.

Core body temperature The body temperature of deeper structures in the body such as the liver and brain; averages 35.8–38.2 °C (96–101 °F).

Cornea (KOR-nee-ah) A transparent structure on the anterior eye that refracts light as it enters the eye.

Coronary circulation The set of arteries, capillaries, and veins that supply blood to and drain blood from the myocardium of the heart.

Corpora cavernosa (KOR-por-uh ka-ver-NOH-sa) Paired dorsal erectile bodies of spongy connective tissue and smooth muscle filled with vascular spaces in the penis.

Corpus albicans A whitish knot of scar tissue that forms as the corpus luteum degrades.

Corpus callosum The largest group of commissural fibers in the brain; connects the right and left cerebral hemispheres.

Corpus luteum An endocrine gland that forms from a ruptured ovarian follicle; secretes progesterone and some estrogens.

Corpus spongiosum (KOR-pus spun-jee-OH-sum) Ventral erectile body of spongy connective tissue and smooth muscle filled with vascular spaces in the penis.

Cortical nephrons Nephrons with short nephron loops that are either confined to the renal cortex or extend only into the superficial renal medulla.

Cortical reaction The process by which cortical granules of the secondary oocyte release enzymes to destroy the sperm-binding receptors on the oocyte plasma membrane.

Corticonuclear tracts Motor tracts that control the muscles of the head and neck via the lower motor neurons of the cranial nerves.

Corticospinal tracts The largest descending spinal tracts; consist of the axons of upper motor neurons that control movement of muscles below the head and neck via stimulation of lower motor neurons of spinal nerves.

Cortisol (KOR-tih-zohl) The main glucocorticoid produced by the adrenal cortex; stimulates gluconeogenesis as well as fat and protein breakdown and inhibits the inflammatory response.

Costal cartilage The cartilage that attaches the ribs to the sternum.

Costal margin The palpable margin of the costal cartilage along the medial rib cage.

Countercurrent mechanism The mechanism in the nephron loops of juxtamedullary nephrons of the kidneys that establishes a concentration gradient in the medullary interstitial fluid in order to produce concentrated urine.

Covalent bonding Chemical bonding resulting from the sharing of electrons between two nonmetals or a nonmetal and hydrogen.

Coxal bones The two bones that, together with the sacrum, form the pelvis; consist of three fused bones, the ilium, ischium, and pubis.

Cranial base The inferior portion of the cranial cavity that cradles the brain.

Cranial bones The eight bones of the skull that encase the brain.

Cranial cavity A dorsal body cavity located within the skull that houses the brain.

Cranial nerve A nerve containing axons whose origin or destination is the brain.

Cranial vault The superior portion of the cranial cavity; also called the *calvarium*.

Creatine phosphate (CP) (KREH-ah-tin FOS-fayt) An immediate energy source for certain cell types that donates a phosphate group to adenosine diphosphate (ADP).

Creatinine The metabolic waste product of creatine phosphate catabolism.

Crista ampullaris (KRIS-tah am-pew-LAIR-is) A cluster of receptor cells and supporting cells in the ampulla that detect rotation of the head.

Crossbridge cycle A series of events in a muscle fiber during which a myosin head grabs onto a series of actin molecules in the thin filament and pulls the thin filament progressively closer to the M line of the sarcomere.

Crossed-extension reflex A polysynaptic reflex that occurs concurrently with the flexion reflex; triggers extension of the limb on the opposite side of the body from a painful stimulus.

Crossing over An exchange of segments of DNA between homologous chromosomes.

Cuboidal cell An epithelial cell with a shape that is about as wide as it is tall.

Cutaneous membrane (kyoo-TAYN-ee-us) Another name for the skin or integument.

Cyclic adenosine monophosphate (cAMP) A common second messenger molecule derived from ATP.

Cytokines (SY-toh-kynz) Proteins produced by several types of leukocytes that enhance the immune response in some way.

Cytokinesis (sy'-tow-kin-EE-sis) The division of the cytoplasm among two daughter cells that occurs near the end of cell division.

Cytology (sy-TOL-uh-jee) The study of cells.

Cytoplasm The part of a cell located inside the plasma membrane that consists of the cytosol and the organelles.

Cytosine A pyrimidine nitrogenous base found in nucleic acids.

Cytoskeleton A network of protein filaments within a cell that supports the cell, maintains its shape, holds its organelles in place, and functions in cellular motion.

Cytosol (intracellular fluid) (SY-tow-sol) The fluid located inside a cell that contains water and a variety of solutes.

Dalton's law of partial pressures A gas law stating that the total pressure of a container is equal to the sum of the partial pressures of the gases in the mixture.

Declarative memory (fact memory) The memory of things that are readily available to consciousness that could in principle be expressed aloud.

Decussation The crossing-over of tracts of white matter from one side of the brain or spinal cord to the other side.

Deep A position that is farther away from the surface of the body or more within the body's interior.

Defecation reflex Reflex arc triggered by the presence of feces in the anus; mediated by the parasympathetic nervous system and controlled by the cerebral cortex.

Deglutition (swallowing) (dee-gloo-TISH-un) A specialized type of propulsion that pushes a bolus of food from the oral cavity through the pharynx and esophagus to the stomach.

Dehydration A condition characterized by a decreased volume and increased concentration of the extracellular fluid.

Dehydration synthesis A reaction in which two monomers are linked by a covalent bond and that results in the formation of a water molecule.

Delta waves Highest-amplitude, lowest-frequency brain waves characteristic of stage IV of sleep.

Denaturation The destruction of the three-dimensional structure of a protein and the subsequent loss of its function.

Dendrite (DEN-dryt) A process extending from a neuron that receives input from other neurons.

Dendritic cells A class of leukocytes that often function as phagocytes and antigen-presenting cells and are located in places such as the epidermis of the skin.

Dens The superior projection from the axis that fits inside the atlas to form the atlantoaxial joint.

Dense connective tissue A type of connective tissue proper in which protein fibers are the dominant element.

Dense irregular connective tissue A type of connective tissue proper that contains collagen fibers arranged in irregular bundles; found in locations such as joint and organ capsules and the dermis.

Dense regular collagenous connective tissue A type of connective tissue proper that contains collagen fibers arranged in regular bundles; found in locations such as tendons.

Dense regular elastic connective tissue A type of connective tissue proper that contains elastic fibers arranged in regular bundles; found in locations such as large blood vessels.

Dentin The inner mineralized material of a tooth composed largely of calcium hydroxyapatite crystals.

Deoxyribonucleic acid (DNA) A double-stranded nucleic acid that is found in the cell nucleus and contains the code for every protein in the body.

Depolarization A temporary increase in a cell's membrane potential.

Depression The movement of a joint in which the corresponding body part moves in an inferior direction.

Dermal papillae (DER-mal pah-PILL-ee) Projections of the papillary layer of the dermis that indent the overlying epidermis and house capillary loops and sensory receptors.

Dermatome (DER-mah-tohm) A region of the skin supplied by a specific spinal nerve.

Dermis The deeper layer of the skin consisting of loose and dense irregular collagenous connective tissue.

Desmosomes (DEZ-moh-sohmz) Intercellular junctions that hold cells together in a tissue to increase the tissue's resistance to mechanical stresses.

Diaphragm (DY-ah-fram) The dome-shaped muscle located between the thoracic and abdominal cavities; the main muscle for breathing.

Diaphysis (dy-AF-eh-sis) The shaft of a long bone.

Diarthrosis (die-ahr-THROH-sis) A functional classification of articulations in which the joint allows for a wide range of motion between articulating bones.

Diastole (dy-ASS-toh-lee) The period during which the heart is relaxing.

Diastolic pressure The blood pressure in the arteries when the ventricles are in diastole.

Diencephalon (dy′-en-SEF-ah-lon) The central core of the brain; made up of the thalamus, hypothalamus, epithalamus, and subthalamus.

Diffusion A type of passive transport in which solute molecules move with their concentration gradients from an area of high solute concentration to one of low solute concentration.

Digestion The process of breaking a substance into smaller subunits; may be physical or chemical.

Digestive system The collection of organs concerned with ingestion and digestion of food; absorption of nutrients; propulsion of ingested food through the digestive tract; and defecation of indigestible substances.

Diploë (dip-LOW-ee) The spongy bone tissue of flat bones.

Diploid (2n) A cell with 23 pairs of homologous chromosomes.

Dipole (DY-pohl) A partially positive and partially negative pole that forms as a result of a polar covalent bond.

Disaccharide (dy-SAK-ah-ryd) A carbohydrate with two monosaccharides joined by dehydration synthesis.

Distal A position that is farther away from the point of origin, usually the trunk.

Distal tubule The final segment of the renal tubule in which filtrate modification is controlled by hormones to fine-tune fluid, electrolyte, and acid-base balance.

Diverging circuit A type of neural circuit in which a single input neuron contacts several output neurons.

DNA synthesis (replication) The process during which a cell's DNA is duplicated prior to cell division.

Dominant trait A trait for which one allele is expressed in the phenotype over another allele.

Dorsal body cavity The major body cavity located mostly on the posterior side of the body.

Dorsiflexion (dawr-suh-FLEK-shuhn) An angular movement of the foot at the ankle in which the angle between the foot and the leg decreases.

Ductus deferens (vas deferens) A tube that carries semen from the epididymis through the inguinal canal into the pelvic cavity.

Duodenum (doo-AH-den-um) The initial 25-cm-long segment of the small intestine into which bile and pancreatic juice are secreted.

Dura mater (DOO-rah MAH-ter) The tough outermost meninx that surrounds the brain and the spinal cord.

Dural sinuses The set of venous channels that are located between two layers of dura mater and drain the cerebral veins of the brain.

Dynamic equilibrium The ability to maintain balance during linear or rotational acceleration of the head or body.

Dyspnea (DISP-nee-uh) A condition of abnormal breathing.

Dysrhythmia (DIS-rith-mee-ah) Any disturbance in the electrical rhythm of the heart.

Eccrine sweat glands (EK-rin) Glands in the skin that produce a watery sweat as part of the body's thermoregulatory mechanisms.

Ectoderm The outermost germ layer of an embryo.

Edema (eh-DEE-mah) The collection of excess water in the interstitial fluid that results in swelling.

Effector Cell or organ that causes a physiological response.

Efferent arterioles Arterioles that drain blood from the glomeruli in the kidney.

Ejaculate The semen expelled during one ejaculation; typically measures 2.5–5 ml in volume and contains 40–750 million sperm cells.

Elastic arteries The largest arteries that are near the heart and so under the highest pressure; have a great deal of elastic fibers.

Elastic cartilage A type of cartilage with elastic fibers as the predominant element; found in locations such as the external ear.

Elastic fibers Protein fibers in the extracellular matrix of a tissue that give a tissue extensibility and elasticity.

Elastic filament A myofilament that consists of the structural protein titin.

Elasticity A property of a cell by which it will return to its resting length after it has been stretched.

Electrical synapse A type of synapse in which two cells are joined by gap junctions that allow the bidirectional flow of ions.

Electrocardiogram (ECG) A graphic representation of all of the action potentials occurring in the heart.

Electrochemical gradient The sum of the electrical gradient and chemical (concentration) gradient for an ion.

Electroencephalogram (EEG) A recording of the electrical activity of the brain.

Electrolyte (ee-LEK-troh-lyt) Cations and anions that result when ionic compounds are placed in a solution; these ions will conduct an electric current.

Electromotive force A force generated by the flow of electrons or ions that can perform work.

Electron A negatively charged subatomic particle located in electron shells around the atomic nucleus.

Electron transport chain (ETC) A group of electron carriers in the inner mitochondrial membrane through which electrons are passed as part of oxidative phosphorylation.

Electrophysiology The study of electrical changes in the body's cells and the physiological processes that accompany these changes.

Element A substance that cannot be broken into simpler substances by chemical means; consists of one or more atoms with the same atomic number.

Elevation The movement of a joint in which the corresponding body part moves in a superior direction.

Embryo (EM-bree-oh) The stage of development from the first to the eighth week of gestation during which the rudimentary organ systems are formed.

Emesis (vomiting) (EM-uh-sis) The process during which stomach contents move backward and are expelled from the mouth.

Emulsification (ee-mul′-sih-fih-KAY-shun) A process in which an amphiphilic compound physically breaks lipids into smaller clusters through polar and nonpolar interactions.

End-diastolic volume (EDV) The volume of blood present in the ventricles at the end of ventricular filling.

End-plate potential A depolarization at the motor end plate of a skeletal muscle fiber that can trigger an action potential.

End-systolic volume (ESV) The volume of blood present in the ventricles at the end of the ventricular ejection phase.

Endergonic reaction (en-der-GON-ik) A reaction in which the products have more energy than the reactants; requires the input of energy to proceed.

Endocardium (en′-doh-KAR-dee-um) The inner simple squamous epithelial layer of the heart wall; continuous with the endothelium of the blood vessels.

Endochondral ossification (en-doh-KON-dral) The process by which most bones form from a hyaline cartilage model.

Endocrine gland (EN-doh-krin) A gland that secretes a hormone or hormones directly into the bloodstream to influence the functions of distant target cells.

Endocrine system The group of organs that secretes hormones, which influence the functions of other cells and tissues.

Endocytosis (en′-doh-sy-TOH-sis) A type of vesicular transport in which substances are taken into the cell via an infolding of the plasma membrane.

Endoderm The innermost germ layer of an embryo.

Endomembrane system A group of organelles consisting of the endoplasmic reticulum, Golgi apparatus, and lysosomes that together work to process substances imported to and exported from the cell.

Endometrium The innermost tissue layer of the uterus composed of simple columnar epithelium.

Endoneurium (en′-doh-NOOR-ee-um) The connective tissue that surrounds an axon within a peripheral nerve.

Endosome (EN-doh-sohm) The vesicle into which substances are ingested in endocytosis; it forms from a pinched-off portion of the plasma membrane.

Endosteum (en-DOS-tee-um) The connective tissue membrane lining the internal surfaces of a bone.

Endothelium A modified simple squamous epithelium that lines the inner surface of blood vessels and the heart.

Energy The capacity to do work.

Energy balance The difference between energy intake and energy expenditure; also known as *nitrogen balance*.

Enteric nervous system (ENS) (en-TEHR-ik) A self-contained branch of the autonomic nervous system that extends from the esophagus to the anus and regulates secretion and motility of the digestive organs.

Enzyme A molecule (almost always a protein) that acts as a biological catalyst, speeding up a reaction without changing its products or reactants.

Eosinophil (ee′-uh-SIN-uh-fill) A class of granulocyte that is involved in the immune response to parasites and in allergic reactions.

Ependymal cell (eh-PEN-dih-mal) A ciliated neuroglial cell of the central nervous system that lines the hollow cavities of the brain and spinal cord; cilia beat to circulate cerebrospinal fluid.

Epidermal ridges Folded patterns of the epidermis in thick skin that enhance gripping ability.

Epidermis (eh-puh-DER-mis) The superficial layer of the skin consisting of stratified squamous keratinized epithelium.

Epididymis (ep′-uh-DID-uh-miss) Portion of the male duct system that is the site of sperm maturation and storage.

Epidural space The fat-filled space between the spinal dura mater and the internal surface of the vertebral cavity.

Epiglottis A flap of elastic and hyaline cartilage that covers the larynx during swallowing to prevent food and liquid from entering the lungs.

Epinephrine A catecholamine that functions largely as a hormone released by the adrenal medulla and that is also released as a neurotransmitter by certain neurons in the sympathetic nervous system.

Epineurium (ep′-ih-NOOR-ee-um) The connective tissue sheath that surrounds a peripheral nerve.

Epiphyseal line (eh-PIF-eh-see-uhl) A calcified remnant of the epiphyseal plate.

Epiphyseal plate A structure composed of hyaline cartilage from which long bones grow in length.

Epiphysis The end of a long bone.

Epithelial tissue (ep′-ih-THEE-lee-ul) A tissue type composed of epithelial cells and a thin basal layer of extracellular matrix; functions to cover and line all body surfaces and hollow organs.

Eponychium (ep-oh-NIK-ee-um) The most distal part of the proximal nail fold; consists of stratum corneum and covers the proximal part of the nail body.

Erythrocyte (red blood cell, or RBC) The primary formed element of the blood; biconcave disc-shaped cell that transports oxygen and carbon dioxide on its hemoglobin molecules.

Erythropoiesis (ee-rith′-row-poh-EE-sis) The process of differentiation and maturation of erythrocytes.

Erythropoietin (ee-rith′-row-poh-EE-tin) The hormone produced by the kidneys that increases the rate of erythrocyte production and decreases the time it takes for them to mature.

Esophagus A tubular organ in the posterior thoracic cavity that transmits food from the pharynx to the stomach.

Essential nutrient A nutrient that cannot be manufactured by the body and so must be obtained from the diet.

Eupnea (YOOP-nee-uh) The condition of having a normal rate and pattern of ventilation.

Evaporation The transformation of water from a liquid to a gas; in the process, water must absorb a great deal of heat from its surroundings.

Eversion (ee-VER-zhuhn) A rotational movement of the foot in which the plantar surface rotates laterally away from the midline of the body.

Excess postexercise oxygen consumption (EPOC) The persisting increased rate of breathing during the recovery period after completing exercise.

Exchange reaction A reaction in which one or more atoms or electrons from the reactants are exchanged.

Excitability A property of a cell in which it is responsive in the presence of various stimuli.

Excitation-contraction coupling The linking of muscle fiber excitation via an action potential with contraction via the release of calcium ions from the sarcoplasmic reticulum.

Excitatory postsynaptic potential (EPSP) A change in the membrane potential of a neuron in which it becomes less negative due to the influx of positive ions.

Excretion The process by which waste products are removed from the body.

Exergonic reaction (eks-er-GON-ik) A reaction in which the products have less energy than the reactants; releases energy once completed.

Exocrine gland (EKS-oh-krin) A gland that secretes a product through a duct to the external surface of the body or into the respiratory, gastrointestinal, and/or genitourinary tract.

Exocytosis (eks′-oh-sy-TOH-sis) A type of vesicular transport in which substances are released from the cell via the fusion of a vesicle with the plasma membrane.

Expiration The movement of air out of the lungs.

Extensibility A property of a cell by which it can be stretched without sustaining damage.

Extension (eks-TEN-shuhn) An angular movement whereby the angle increases between articulating bones, moving the bones away from each other.

Extensor A muscle that contracts to increase the angle of the joint on which it operates.

Exteroceptor (EK-ster-oh-sep′-ter) A sensory receptor that responds to a stimulus exterior to the body.

Extracellular compartment The fluid compartment located outside cells that contains extracellular fluid.

Extracellular fluid (ECF) The fluid located in the spaces between cells.

Extracellular matrix The substance in a tissue outside of the cells that consists of extracellular fluid, ground substance, and protein fibers.

Extrafusal muscle fibers (ek-strah-FYOO-zul) Regular contractile skeletal muscle fibers.

Facial bones The 14 bones of the skull that form the framework of the face.

Facilitated diffusion A type of diffusion in which molecules move through a membrane using a protein channel or carrier.

Facultative water reabsorption Water reabsorption that is controlled by hormones to maintain a constant extracellular fluid osmolarity.

False rib A rib that attaches to the costal cartilage of another rib rather than attaching directly to the sternum.

Falx cerebri (seh-REE-bry) A double fold of meningeal dura that dives into the longitudinal fissure and separates the two cerebral hemispheres.

Fascicle (FASS-ih-kul) A bundle of axons in a nerve surrounded by the perineurium.

Fast-twitch fibers Skeletal muscle fibers with high myosin ATPase activity that proceed more rapidly through their crossbridge cycles; generate rapid but generally short-duration contractions.

Fatty acid A lipid molecule with a hydrocarbon chain bound to a carboxylic acid group.

Feces The collection of indigestible or undigested food, bacteria, and water in the large intestine that is eliminated via defecation.

Feedback loop A homeostatic control mechanism in which a change in a regulated variable causes effects that feed back and affect that same variable.

Femur The only bone of the thigh.

Fenestrated capillaries Capillaries whose cells have fenestrations to allow large volumes of fluids and solutes to be exchanged.

Fenestrations (fen-eh-STRAY-shunz) Small pores within endothelial cells of capillaries.

Fertilization The union of a sperm and ovum to form a zygote.

Fetus (FEE-tus) The stage of development from weeks 9 through 38 of gestation.

Fever A body temperature above the normal range of 36–38 °C (97–99 °F).

Fiber Polysaccharides that are not fully digestible by humans due to a lack of the enzymes required to catabolize them.

Fibrillation A condition in which muscle cells are contracting individually due to chaotic electrical activity.

Fibrin (FY-brin) A sticky, threadlike protein that "glues" the formed elements and clotting factors together in a blood clot.

Fibrinolysis (fy-brin-AH-luh-sis) The process by which fibrin is degraded during thrombolysis.

Fibroblast A cell type within connective tissue proper that produces components of the extracellular matrix.

Fibrocartilage A type of cartilage with collagen fibers as the predominant element; found in locations such as the intervertebral discs.

Fibrosis (fy-BROH-sis) The process by which a damaged tissue is replaced with fibrous connective tissue during healing.

Fibrous joints A structural classification of articulations in which the joints are held together by dense regular collagenous connective tissue and have neither a synovial cavity nor an articular capsule.

Fibula (FIB-yoo-luh) The lateral leg bone.

Filtrate The fluid resulting from glomerular filtration that flows through the renal tubule and collecting system.

Filtration The movement of a fluid by a force such as hydrostatic pressure or gravity.

Filtration membrane A structure consisting of the glomerular endothelial cells, basal lamina, and podocytes through which blood is filtered to produce filtrate in the kidneys.

Filtration slits Tiny spaces between the foot processes of podocytes surrounding the glomerular capillaries.

Filum terminale (FY-lum ter-mee-NAL-ay) The inferior continuation of the spinal pia mater; continues after the end of the spinal cord through the vertebral cavity and anchors into the first coccygeal vertebra.

First-class lever A lever system in which the fulcrum is located between the applied force and the load to be moved.

Fissure A deep groove between major brain structures or lobes of the cerebrum.

Fixator A muscle that holds a bone in place, allowing other muscles to move the bone and joint more effectively.

Fixed acid An acid that cannot be converted to a gaseous form and so must be excreted by the kidneys.

Flagellum (flah-JEL-um) A single motile extension from a cell that propels the cell forward; found in humans only on sperm cells.

Flat bone A bone that is flat in shape.

Flavin adenine dinucleotide (FAD) One of the main electron carriers involved in nutrient catabolism; when reduced, it becomes $FADH_2$.

Flexion (FLEK-shuhn) An angular movement whereby the angle decreases between articulating bones, bringing the bones closer together.

Flexion reflex (withdrawal reflex) A polysynaptic reflex initiated by painful stimuli that triggers withdrawal of the affected body part.

Flexor A muscle that contracts to decrease the angle of the joint on which it operates.

Floating rib A rib that lacks an attachment to the sternum.

Fluid mosaic model The model that describes the plasma membrane as a dynamic structure consisting of a phospholipid bilayer with multiple components interspersed throughout.

Follicle-stimulating hormone (FSH) Hormone produced by the anterior pituitary gland that stimulates reproductive functions in males and females.

Fontanel A "soft spot" in the skull of a fetus and infant in which the cranial sutures have not yet fused.

Foramen ovale (for-AY-men oh-VAH-lay) An opening in the interatrial septum in the fetal heart that allows blood to bypass the pulmonary circuit.

Forearm The distal portion of the upper limb that consists of the radius and the ulna.

Fossa ovalis An indentation in the interatrial septum that marks the former location of the foramen ovale in the fetal heart.

Fourth ventricle A hollow, cerebrospinal fluid–filled cavity located within the brainstem.

Fovea centralis (FOE-vee-ah sen-TRA-liss) An area of the retina with the highest density of cones that produces high-acuity vision.

Frank-Starling law A mechanism by which a ventricular muscle cell of the heart contracts more forcefully when it is stretched.

Frontal lobes The two anterior lobes of the cerebral hemispheres; responsible for planning and executing movement and complex mental functions.

Frontal plane A plane of section that divides the body or body part into anterior and posterior portions; also known as the *coronal plane*.

Fulcrum The pivot or hinge point around which movement is produced when a muscle or group of muscles contracts.

Funiculus (foo-NIK-yoo-luss) A general region of white matter in the spinal cord that is organized into tracts.

Fused tetanus (complete tetanus) A type of wave summation in which a muscle fiber is stimulated rapidly and the muscle fiber is not allowed to relax between contractions.

G-proteins Intracellular enzyme complexes involved in a variety of second messenger systems.

Gallbladder The small, hollow organ on the posterior side of the liver that stores and releases bile.

Gametes Sex cells (sperm and ova) that are produced in the gonads.

Gamma motor neurons (γ motor neurons) Lower motor neurons that innervate intrafusal muscle fibers.

Ganglion A cluster of neuronal cell bodies in the peripheral nervous system.

Gap junctions Intercellular junctions that connect the cytosol of neighboring cells and allow water and solutes to pass between them.

Gastrin A hormone secreted by G cells of gastric glands that triggers acid secretion from parietal cells.

Gastrulation (gas-troo-LAY-shun) The rearrangement and migration of the cells of the bilaminar embryonic disc to form the three germ layers.

Gene A segment of DNA that specifies the amino acid sequence for a single protein.

Genetic code A list of the amino acids coded for by a specific DNA triplet or mRNA codon.

Genome (JEE-nohm) Collectively, all of an individual's DNA.

Genotype (JEEN-oh-ty'p) The genetic makeup of an individual.

Gestation period (jeh-STAY-shun) The period that extends from the mother's last menstrual period until birth, which is approximately 280 days, or 40 weeks.

Glaucoma (glaw-KOH-mah) A condition in which the aqueous humor builds up in the eye and puts pressure on the retina and optic nerve.

Glenoid cavity (GLEN-oyd KA-vuh-tee) The lateral depression of the scapula that articulates with the humerus at the shoulder joint.

Globus pallidus Two basal nuclei that sit medial to the putamen; inhibit the initiation of movement by inhibiting the thalamus.

Glomerular capsule The double-layered epithelium that surrounds the glomeruli of the kidneys; encloses the capsular space.

Glomerular filtration The process by which blood is filtered in the glomeruli of the kidneys.

Glomerular filtration rate (GFR) The rate at which filtrate is produced in the glomeruli of the kidneys.

Glomerular hydrostatic pressure (GHP) The blood pressure within the glomerular capillaries; favors filtration into the capsular space.

Glomerulus (gloh-MAIR-yoo-lus) Ball-shaped capillary bed where the blood is filtered in the kidney.

Glottis The opening between the vocal cords of the larynx through which air passes.

Glucagon A hormone produced by the pancreas that triggers changes to raise the concentration of glucose in the blood.

Glucocorticoids (glew′-koh-KOR-tih-koydz) A class of hormones produced by the adrenal cortex that are part of the body's response to a stressor.

Gluconeogenesis (glew′-koh-nee-oh-JEN-eh-sis) The formation of new glucose from

GLOSSARY

noncarbohydrate precursors, including glucogenic amino acids and glycerol.

Glucose (GLUW-kohs) The monosaccharide used as the primary fuel by most cells in the body.

Glycerol A modified sugar alcohol that is found in triglycerides and can be used as a glucogenic precursor.

Glycogen (GLY-koh-jen) A polysaccharide that is the storage form of glucose in animals.

Glycogenesis The formation of glycogen from glucose monomers.

Glycogenolysis (gly′-koh-jen-AWL-ih-sis) The breakdown of glycogen to release glucose monomers into the blood.

Glycolysis (gly-KOL-uh-sis) A series of 10 reactions that takes place in the cytosol during which glucose is split and small amounts of ATP and NADH are generated.

Glycolytic catabolism (anaerobic catabolism) A series of ATP-producing reactions that occur in the cytosol of cells in which glucose is broken down into two molecules of pyruvate; these reactions do not require oxygen to proceed.

Goblet cell A unicellular exocrine gland that secretes mucus.

Golgi apparatus (GOHL-jee) A group of membrane-bounded sacs that receive and modify products from the endoplasmic reticulum for export or use in the cell.

Golgi tendon organs Mechanoreceptors located within tendons near the muscle-tendon junction; monitor the tension generated by a muscle contraction.

Golgi tendon reflex A reflex mediated by Golgi tendon organs that causes relaxation in response to increased tension within a muscle.

Gomphosis (gahm-FOH-sis) A fibrous joint structurally and a synarthrosis functionally that is found between each tooth and the bony socket in either the mandible or the maxilla.

Gonads The primary organs of the reproductive system that produce gametes and hormones; the male gonads are testes and the female gonads are ovaries.

Gradient A condition in which more of something exists in one area than in another, and the two areas are connected.

Granulocyte (GRAN-yoo-low-syt) A class of leukocyte with visible cytosolic granules; includes neutrophils, eosinophils, and basophils.

Gray matter Collections of unmyelinated axons, dendrites, and cell bodies that appear gray.

Great saphenous veins Large superficial veins of the medial lower limb.

Greater pelvis The portion of the pelvis superior to the pelvic brim.

Greater trochanter (tro-KAN-tuhr) Large, lateral projection from the proximal femur.

Gross anatomy The study of structures of the human body that can be seen with the unaided eye.

Ground substance An amorphous component of the extracellular matrix containing water, proteoglycans, glycosaminoglycans, and other molecules.

Growth An increase in the size of an individual cell or an increase in the number of cells; occurs when anabolic processes outweigh catabolic processes.

Growth hormone (GH) A hormone produced by the anterior pituitary gland that promotes fat breakdown and gluconeogenesis; also stimulates the release of insulin-like growth factor, which promotes growth and protein synthesis.

Guanine A purine nitrogenous base found in nucleic acids.

Gustation (gus-TAY-shun) The sense of taste.

Gyrus An elevated ridge of the superficial cerebrum.

Hair Small, filamentous structures that project from the surface of the skin and are composed of dead, keratinized cells.

Hair follicle An infolding of the epidermis in which a hair is embedded in the skin.

Hair follicle receptors Free nerve endings wrapped around the base of a hair follicle in the dermis or hypodermis.

Hallux The anatomical term for the big toe.

Haploid (1n) A cell with 23 chromosomes.

Heart The muscular organ that pumps blood to the pulmonary and systemic circuits.

Heart rate (HR) The number of heartbeats that occurs in a set period of time.

Heart sounds Noises heard during auscultation of the heart that are caused by vibrations of the ventricular wall when the valves close.

Heartbeat A single, coordinated contraction of cardiac muscle cells.

Hematocrit (heh-MAH-toh-krit) The portion of the blood that consists of erythrocytes.

Hematopoiesis (he′-mah-toe-poh-EE-sis) The process of differentiation and maturation of the formed elements of blood.

Hematopoietic stem cells Cells located in red bone marrow that can become any type of formed element of the blood.

Hemoglobin The iron-containing protein in erythrocytes that binds and carries oxygen throughout the blood.

Hemolysis (hee-MAH-luh-sis) The rupture of erythrocytes.

Hemostasis (he-mo-STAY-sis) The process by which blood loss is stopped from a damaged blood vessel.

Henry's law A gas law stating that the degree to which a gas dissolves in a liquid is proportional to both its partial pressure and its solubility in the liquid.

Hepatic portal system A system that delivers blood from the digestive organs and the spleen to hepatocytes via the hepatic portal vein for processing of nutrients, wastes, and toxins.

Heredity (heh-RED-ih-tee) The transmission of genetic characteristics from parent to offspring through the genes.

Heterozygous (HET-er-oh-zy′-gus) Two alleles of a gene that code for different traits.

High-density lipoproteins (HDLs) Lipoproteins that transfer lipids from cells in peripheral tissues to the liver for excretion.

Hinge joint A uniaxial joint in which the convex articular surface of one bone fits into the concave depression of the other bone, creating a hinge that allows for motion around one axis.

Hippocampus (hip-poh-KAM-pus) A component of the limbic system in the brain that functions in memory and learning.

Histology (hiss-TOL-uh-jee) The study of tissues.

Holocrine secretion (HOH-loh-krin) A type of exocrine secretion in which the product accumulates in the cell until the cell ruptures and dies.

Homeostasis (hoh′-mee-oh-STAY-sis) The maintenance of the body's stable internal environment.

Homologous chromosomes (huh-MAH-luh-gus) A set of maternal and paternal chromosomes with the same genes.

Homozygous (HOH-moh-zy′-gus) Two alleles of a gene that code for the same trait.

Hormone A chemical messenger secreted into the blood that triggers changes within its target cells.

Human anatomy The study of the structure of the human body.

Human chorionic gonadotropin (kor-ee-ON-ik goh-NAH-doh-troh′-pin) A hormone produced by the outer layer of a developing embryo that keeps the corpus luteum from degenerating.

Human physiology (fiz′-ee-AWL-oh-gee) The study of the functions of the human body.

Humerus The only bone of the upper arm.

Hyaline cartilage (HY-ah-lin) A type of cartilage with ground substance as the predominant element; found in locations such as the articulating ends of bones, between the sternum and ribs, and in the nose.

Hydrogen bonds Weak attractions between partially positive hydrogen atoms of one molecule or functional group and partially negative atoms of another molecule or functional group.

Hydrolysis (hy-DROL-uh-sis) A reaction in which a molecule of water is added to a polymer, splitting it into component monomers.

Hydrophilic (hy-droh-FIL-ik) A property of substances with partially or fully charged ends (i.e., polar covalent and ionic molecules) that will dissolve when placed in water.

Hydrophobic (hy-droh-FOH-bik) A property of substances with no dipole (nonpolar covalent bonds) that will not dissolve when placed in water.

Hydrostatic pressure (HP) The pressure that a fluid exerts on the wall of its container.

Hydroxyapatite A calcium phosphate–based mineral that makes up the inorganic portion of the osseous tissue extracellular matrix.

Hyoid bone A bone in the superior neck to which muscles and ligaments attach.

Hyperextension (hy′-per-eks-TEN-shuhn) An angular movement whereby the joint allows for motion beyond the anatomical position of the articulation.

Hyperopia (hy′-per-OH-pee-ah) A condition in which the eye is too short or the cornea is too flat and the lens cannot round up enough for near vision.

Hyperpolarization A change in the membrane potential of an excitable cell to a value more negative than its resting membrane potential.

Hypersensitivity disorder A condition characterized by an exaggerated response of the immune system to a foreign antigen.

Hypertension High blood pressure; generally defined as a systolic pressure of 140 mm Hg or higher and/or a diastolic pressure of 90 mm Hg or higher.

Hypertonic solution A solution that has a higher solute concentration (and so osmotic pressure) than the reference solution; a hypertonic solution will cause water loss from the reference solution by osmosis.

Hypertrophy (hy-PER-troh-fee) An increase in cell size.

Hyperventilation An increased rate of ventilation.

Hypodermis (hy-poh-DER-mis) The layer of adipose tissue deep to the skin; not considered part of the integument.

Hyponychium The fold of stratum corneum that anchors the distal end of the nail plate to the underlying epithelium.

Hypotension Any abnormally low blood pressure, defined by a systolic pressure lower than 90 mm Hg and/or a diastolic pressure lower than 60 mm Hg.

Hypothalamic-pituitary-gonadal (HPG) axis The multi-tiered feedback loops of the hormones of the hypothalamus, anterior pituitary, and gonads.

Hypothalamus The small anterior and inferior component of the diencephalon of the brain; responsible for many homeostatic functions, producing the hormones oxytocin and antidiuretic hormone, and producing releasing and inhibiting hormones that affect the anterior pituitary gland.

Hypotonic solution A solution that has a lower concentration (and so osmotic pressure) than the reference solution; a hypotonic solution will cause the reference solution to gain water by osmosis.

Hypoventilation A decreased rate of ventilation.

I band The component of the sarcomere that appears lighter because it contains only thin filaments.

Ileum (ILL-ee-um) The final and longest segment of the small intestine, measuring about 3.6 meters in length.

Ilium (ILL-ee-um; plural, *ilia*) The largest bone of the pelvis that forms its superior, lateral, and posterior walls.

Immune system A group of cells and proteins that works to defend the body against internal and external threats.

Immunodeficiency disorder A condition characterized by a deficit in one or more components of the immune response.

Immunological memory The phenomenon whereby cells of adaptive immunity respond more quickly and efficiently to subsequent exposures to an antigen.

Inclusions Clusters of identical storage molecules in the cytosol of certain cell types.

Inferior Away from the head or toward the tail; also known as *caudal* (KAW-d'l).

Inferior vena cava The large vein that drains mostly structures inferior to the diaphragm; empties into the right atrium.

Inflammatory response The series of events that occurs in reaction to any cellular injury; involves vasodilation, increased capillary permeability, pain, swelling, and decreased function.

Inhibin A hormone secreted by the testes and ovaries that inhibits FSH secretion.

Inhibitory postsynaptic potential (IPSP) A change in the membrane potential of a neuron in which it becomes more negative due to the outflow of positive ions or the influx of negative ions.

Innate immunity (nonspecific immunity) The arm of the immune system that consists of cells and proteins that are not specific for individual antigens and that always respond to cell injury and pathogens in the same manner.

Inorganic compounds Compounds that do not contain carbon and hydrogen atoms bound to one another.

Insensible water loss A water loss that is not "sensed" or noticed; includes water lost to sweat and ventilation.

Insertion (muscle) The end of a muscle attached to the structure that will be moved when the muscle contracts.

Inspiration The movement of air into the lungs.

Inspiratory muscles Muscles that increase the volume of the lungs, which decreases the intrapulmonary pressure below that of atmospheric pressure, causing inspiration.

Insulas (IN-syoo-lahz) Deep lobes of the cerebrum; have functions relating to taste and to viscera.

Insulin A hormone produced by the pancreas that triggers changes to lower the concentration of glucose in the blood.

Insulin-like growth factor (IGF) A hormone produced largely by the liver in response to growth hormone; stimulates protein synthesis and cell division.

Integral protein A protein within a membrane that spans the width of the membrane.

Integument (in-teg′-you-MENT) The skin and its accessory structures.

Integumentary system (in-teg′-you-MEN-tuh-ree) The organ system consisting of the skin and its accessory structures.

Intercalated disc (in-TER-kuh-lay′-t'd) Specialized structure that connects adjacent cardiac muscle cells and contains gap junctions and desmosomes.

Intercostal space The space located between two ribs.

Interferon (INF) (in-ter-FEER-on) A cytokine family that is generally produced in response to infection by intracellular pathogens; they inhibit viral replication inside host cells and activate components of innate and adaptive immunity.

Intermediate filaments Ropelike protein filaments of the cytoskeleton that hold organelles in place, form much of the framework of the cell, and help a cell withstand mechanical stresses.

Internal carotid arteries Branches of the common carotid arteries that supply the brain.

Internal jugular veins Veins that drain the brain and parts of the head and neck.

Interneurons Neurons between sensory and motor neurons that perform integrative functions.

Internodes The myelinated segment of an axon.

Interoceptor (IN-ter-oh-sep′-ter) A sensory receptor that responds to a stimulus originating within the body.

Interosseous membrane A supportive band of dense irregular collagenous connective tissue between the radius and ulna and the tibia and fibula.

Interphase The phase of the cell cycle during which the cell grows, replicates its DNA, and prepares for cell division.

Interstitial (cells) Cells of the testes that produce androgens, mostly testosterone.

Intervertebral disc The fibrocartilage pad between two vertebrae that absorbs shock and supports the vertebral column.

Intracellular compartment The fluid compartment located within cells that contains cytosol or intracellular fluid.

Intrafusal muscle fibers Specialized muscle fibers that are contractile only at the poles and are found in muscle spindles; innervated by sensory neurons that detect stretch.

Intramembranous ossification (in′-tra-MEM-brah-nus) The process by which certain flat bones form from a mesenchymal model.

Intrapleural pressure The pressure within the pleural cavities.

Intrapulmonary pressure The pressure exerted by the air within the lungs.

Inversion (in-VUR-zhun) A rotational movement of the foot in which the plantar surface rotates medially toward the midline of the body.

Ion (EYE-on) A charged particle that has lost or gained one or more electrons.

Ionic bond (eye-ON-ik) A chemical bond resulting from the transfer of electrons and resulting attraction between a positive metal cation and negative nonmetal anion.

Iris A pigmented ring of smooth muscle cells that surrounds the pupil and controls the amount of light that enters the eye.

Irregular bone A bone that has an irregular shape and that does not fit into any other category.

Ischium (ISS-kee-um) The posteroinferior bone of the pelvis.

Isomers (EYE-soh-merz) Two or more molecules with the same molecular formula but different three-dimensional arrangement of their atoms.

Isometric contraction (eye-soh-MET-rik) A type of muscle contraction in which the tension generated is equal to that of the external load, and so the muscle cell remains at a constant length.

Isotonic concentric contraction A type of muscle contraction in which the tension generated is greater than that of the external load, and so the muscle cell shortens with the contraction.

Isotonic eccentric contraction A type of muscle contraction in which the tension generated is less than that of the external load, and so the muscle cell lengthens with the contraction.

Isotonic solution A solution that has the same concentration (and so osmotic pressure) as the reference solution; two isotonic solutions will result in no net movement of water molecules by osmosis.

Isotope (EYE-soh-towp) An element that has the same atomic number as another element but a different mass number (due to a different number of neutrons).

Isovolumetric contraction (eye′-soh-vol-yoo-MET-rik) The period of the cardiac cycle during which the ventricles are contracting but their volume does not change because the pressure is not high enough to open the semilunar valves.

Isovolumetric relaxation The period of the cardiac cycle during which the ventricles are relaxing but their volume does not change because the pressure is not high enough to open the atrioventricular valves.

Jejunum (jeh-JOO-num) The second segment of the small intestine, measuring about 2.5 meters in length.

Joint The location where two or more bones come together; also known as an *articulation*.

Juxtaglomerular apparatus (JGA) The location where the macula densa contacts the juxtaglomerular cells of the afferent arteriole; monitors and maintains the glomerular filtration rate.

Juxtamedullary nephrons Nephrons with long nephron loops that dip deeply into the renal medulla; play a role in the production of concentrated urine.

Keratin (KEHR-uh-tin) A hard, fibrous protein that increases a tissue's resistance to mechanical stresses.

Keratinocyte (kehr′-ah-TIN-oh-syt) The main cell type found in the epidermis.

Ketone bodies (KEE-tohn) Four-carbon molecules formed during lipid catabolism that a cell can catabolize to form two molecules of acetyl-CoA.

Kidneys Paired retroperitoneal organs that filter the blood to remove wastes and maintain fluid, electrolyte, and acid-base homeostasis.

Kinetic energy Energy in motion or action.

Kyphosis (ky-FOH-sis) A potentially abnormal exaggeration of the thoracic curvature of the vertebral column.

Labia (LAY-bee-uh) Two sets of fatty protective skin folds that are posterior to the mons pubis; includes the outer labia majora and smaller inner labia minora.

Lacrimal apparatus (LACK-rah-mal) A collection of structures involved in the production and drainage of tears.

Lactation (lak-TAY-shun) The production and release of breast milk from the mammary glands.

Lacteal (lak-TEE-ul) A specialized lymphatic capillary in an intestinal villus that receives absorbed fats in addition to lymph.

Lamellated corpuscle (Pacinian corpuscle) (pah-SIN-ee-an KOR-pus-ul) A sensory receptor located in the reticular layer of the dermis that detects pressure and vibration stimuli.

Language The ability to comprehend and produce words through speaking, writing, and/or signing, and to assign and recognize the symbolic meaning of a word correctly.

Large intestine A tubular organ that is the last part of the alimentary canal; responsible for the absorption of substances such as water and electrolytes and the propulsion of feces out of the body.

Laryngopharynx (lah-ring′-goh-FAIR-inks) The portion of the pharynx that is posterior to the larynx; serves as a passageway for food, liquids, and air.

Larynx (LAIR-inks) The segment of the respiratory tract in the anterior neck that houses the vocal cords.

Lateral A position that is farther away from the midline of the body or a body part.

Lateral fissure A fissure on the lateral side of each cerebral hemisphere that separates the temporal lobes from the frontal and parietal lobes.

Lateral horn The lateral region of gray matter in the thoracic and lumbar spinal cord that contains the cell bodies of autonomic neurons.

Lateral malleolus The distal end of the fibula; lateral ankle bone.

Lateral ventricles Hollow, cerebrospinal fluid–filled cavities located within the two cerebral hemispheres.

Leak channel An ion channel that is always open to allow specific ions to cross the plasma membrane with their concentration gradients.

Leg The distal portion of the lower limb that consists of the tibia and the fibula.

Length-tension relationship The relationship between the length of the sarcomeres of a muscle fiber while at rest and the amount of tension that can be generated by a contraction.

Lens A transparent structure posterior to the pupil of the eye that refracts light rays to focus them on the retina.

Leptin A hormone produced by adipocytes that promotes satiety (feelings of fullness after eating).

Lesser pelvis The portion of the pelvis inferior to the pelvic brim.

Leukocyte (white blood cell, or WBC) (LEWK-oh-syt) A class of formed elements of the blood that perform functions relating to the immune system.

Leukopoiesis (loo′-koh-poh-EE-sis) The process of differentiation and maturation of leukocytes.

Levator (LEE-vay-tore) A muscle whose function is to elevate a body part.

Ligament A strand of dense regular collagenous connective tissue that connects one bone to another to strengthen and reinforce an articulation.

Ligand (LY-gand) A chemical that binds to a receptor in the plasma membrane of a cell.

Ligand-gated channel An ion channel in the plasma membrane of a cell that opens or closes in response to the binding of a specific ligand to a receptor or to the channel.

Limbic system A functional brain system that participates in learning, memory, emotion, and behavior; consists of the cingulate gyri, the parahippocampal gyri, the hippocampi, and the amygdalae.

Linea alba (LINN-ee-ah AHL-bah) The midline connective tissue seam that separates the right and left rectus abdominis muscles.

Lipid (LIP-id) A nonpolar organic molecule that contains primarily carbon and hydrogen atoms; examples include fats and oils.

Lipogenesis (ly′-poh-JEN-eh-sis) The process by which fatty acid and triglyceride synthesis occurs.

Liver The large organ in the right upper quadrant of the abdominopelvic cavity; functions include diverse metabolic activities, filtering blood from most abdominal organs, and bile production.

Local potential A small change in the membrane potential in a specific region of a cell's plasma membrane.

Long bone A bone that is longer than it is wide.

Long-term potentiation (LTP) An increase in synaptic activity between associated neurons; the mechanism by which hippocampal neurons encode long-term declarative memories.

Longitudinal fissure The deep groove that separates the two cerebral hemispheres.

Longitudinal growth The process by which long bones grow in length.

Loose connective tissue A type of connective tissue proper in which ground substance is the dominant element; also known as *areolar connective tissue*.

Lordosis (lor-DOH-sis) A potentially abnormal exaggeration of the lumbar curvature of the vertebral column.

Low-density lipoproteins (LDLs) Cholesterol-rich particles that deliver cholesterol to cells in peripheral tissues.

Lower limb The portion of the appendicular skeleton that consists of the thigh, leg, ankle, and foot.

Lower motor neuron Motor neurons of the PNS that directly innervate skeletal muscles.

Lumbar curvature A concave curvature of the lumbar portion of the vertebral column.

Lumbar plexus A group of nerves consisting of the anterior rami from L1–L4 that innervates structures within the pelvis and lower limb.

Lumbar vertebrae Five large, blocky bones of the vertebral column located inferior to the thoracic vertebrae and superior to the sacrum.

Lumen The space enclosed by a hollow organ.

Lungs Paired spongy organs of the respiratory system that consist of alveoli and surrounding elastic connective tissue.

Luteinizing hormone (LH) Hormone produced by the anterior pituitary gland that stimulates reproductive functions in males and females.

Lymph (LIMF) The watery liquid located within lymphatic vessels; composition is similar to that of interstitial fluid but with fewer proteins.

Lymph nodes Clusters of lymphatic tissue located along lymphatic vessels throughout the body; they filter lymph.

Lymphatic system (lim-FAH-tik) A group of organs and tissues with functions including immunity and fluid homeostasis.

Lymphatic vessels Blind-ended vessels that transport lymph from the interstitial fluid to the blood.

Lymphocyte (LIM-fo-syt) A class of agranulocyte that has diverse functions in the immune system; includes B and T lymphocytes.

Lysosome (LY-soh-sohm) A spherical membrane-bounded organelle that contains digestive enzymes.

Macronutrients Nutrients required in relatively large amounts in the diet; include carbohydrates, proteins, and lipids.

Macrophage (MAK-roh-fehj) A phagocyte that ingests damaged cells, bacteria, and cellular debris.

Macula (MACK-you-lah) A cluster of receptor cells (hair cells) in the lining of the utricle and saccule responsible for detection of head tilting and linear movement.

Macula densa (MAK-yoo-lah DEN-sah) A group of cells in the segment of the renal tubule between the ascending limb of the nephron loop and the distal tubule whose function is to monitor flow through the nephron.

Major histocompatibility complex (MHC) molecules (hiss'-toh-kom-pat-ih-BIL-ih-tee) Glycoproteins found on the surface of cells that bind and display antigen fragments.

Mammary gland Exocrine gland contained within a rounded skin-covered breast; consists of lobules with glandular alveoli that produce milk.

Manubrium (muh-NOO-bree-um) The superior portion of the sternum.

Mass number The number of protons plus the number of neutrons found in an atom.

Mast cell A tissue-bound immune cell found in many types of connective tissue that secretes inflammatory mediators.

Mastication The process of chewing during which food is physically broken into smaller particles by the teeth, the tongue, and the hard palate.

Matrix The group of cells at the base of a hair that are actively dividing; the area from which a hair grows.

Matter The material of our universe; anything that has mass and occupies space.

Mean arterial pressure (MAP) The average pressure in the systemic circuit during a complete cardiac cycle.

Mechanically gated ion channel An ion channel that opens or closes in response to mechanical deformation.

Mechanoreceptor (mek'-ah-noh-ree-SEP-ter) A sensory receptor that responds to a mechanical deformation of the cell.

Medial A position that is closer to the midline of the body or a body part.

Medial malleolus (mal-lee-OH-lus; plural, *malleoli*) The distal end of the tibia; medial ankle bone.

Mediastinum (meh'-dee-ass-TY-num) The portion of the thoracic cavity located between the lungs.

Medulla oblongata The final portion of the brainstem.

Medullary cavity (MED-yoo-lehr-ee) The largely hollow interior portion of the diaphysis of a long bone.

Megakaryocyte Cell located in the bone marrow from which platelets are derived.

Meiosis (my-OH-sis) A type of cell division that produces gametes.

Melanin (MEL-ah-nin) A pigment produced by melanocytes in the epidermis that ranges in color from brown-black to orange-red; responsible for much of the pigmentation of the skin, hair, and irises of the eyes.

Melanocyte A cell located in the epidermis of the skin that produces the pigment melanin.

Melatonin (mel-ah-TOH-nin) A hormone produced by the pineal gland that regulates the sleep/wake cycle.

Membrane A sheet of one or more tissues that lines a body surface or body cavity.

Membrane potential The difference in voltage between the extracellular fluid and the cytosol in the area near the plasma membrane.

Membranous urethra The short middle portion of the male urethra.

Menopause (MEN-ah-pawz) The cessation of regular menstrual cycles that normally occurs between the ages of 45 and 55 years.

Menstruation The discharge of blood and other materials from the vagina that occurs when a secondary oocyte is not fertilized.

Merkel cell A sensory receptor located in the stratum basale of the epidermis.

Merocrine secretion (MAIR-oh-krin) A type of exocrine secretion in which the product is secreted by exocytosis.

Mesentery Double folds of visceral peritoneum located around certain abdominal organs, such as the small and large intestines.

Mesoderm The middle germ layer of an embryo.

Messenger RNA (mRNA) A strand of RNA that contains the transcript for a single gene.

Metabolic water Water produced as a byproduct of metabolic reactions.

Metabolism The sum of the body's chemical reactions.

Metacarpals The long bones of the hand.

Metaphase The second phase of mitosis during which sister chromatids line up along the equator of the cell.

Metarterioles The smallest arterioles that directly feed capillaries in many tissues.

Metatarsals The long bones of the foot.

Micelle (my-SEL) A small vesicle composed of an outer layer of bile salts surrounding an inner core of chemically digested lipids.

Microcirculation The total blood flow that takes place within the capillary beds.

Microglial cell A neuroglial cell of the central nervous system that acts as a phagocyte.

Micronutrients Nutrients required in relatively small amounts in the diet; include vitamins and minerals.

Microscopic anatomy The study of structures of the human body that require the use of a microscope for observation.

Microtubules Hollow rods of the cytoskeleton composed of tubulin subunits that help a cell maintain its size and shape, position organelles, and play a role in motion within the cell.

Microvilli Highly folded extensions of the plasma membrane that increase its surface area for absorption.

Micturition (mik-choo-RISH-un) The process of voiding urine from the urinary bladder through the urethra.

Midbrain The first portion of the brainstem.

Midsagittal plane A plane of section that divides the body or body part into equal right and left portions; also known as the *median plane*.

Mineral Any element other than carbon, hydrogen, oxygen, and nitrogen that is required by living organisms.

Mineralocorticoids (min'-er-ah-loh-KOR-tih-koydz) A class of hormones produced by the adrenal cortex that regulates fluid, electrolyte, and acid-base balance.

Mitochondria (my'-toh-KON-dree-ah; singular, *mitochondrion*) Organelles surrounded by a double membrane that produce the bulk of the cell's ATP by oxidative catabolism.

Mitosis (my-TOH-sis) The division of somatic cells that results in two identical daughter cells.

Mitotic spindle A structure that emanates from centrioles and consists of microtubules that attach to sister chromatids at the centromere; pulls chromatids to opposite poles of the cell during mitosis.

Mitral valve (MY-tral) Atrioventricular valve located between the left atrium and left ventricle; also known as the bicuspid valve.

Mixture Two or more substances that are physically combined such that their chemical properties remain unaltered.

Molecule (MOL-eh-kyul) Two or more atoms joined by a chemical bond.

Monocyte (MAH-no-syt) A class of agranulocyte that matures into macrophages, very active phagocytes.

Monomer (MON-oh-mer) A single molecular subunit that can be combined with another monomer to build a larger molecule.

Monosaccharide (mon′-oh-SAK-ah-ryd) A carbohydrate monomer.

Monosynaptic reflex A reflex arc consisting of a single synapse between a sensory and motor neuron within the spinal cord or brainstem.

Morula (MOR-yoo-lah) The early developing conceptus composed of 16 or more cells.

Motility The movement of a cell or tissue.

Motor division (efferent division) The division of the peripheral nervous system that relays motor information from the central nervous system to muscle and glandular tissues.

Motor end plate The specialized region of the skeletal muscle fiber plasma membrane that contains receptors for acetylcholine.

Motor learning The correction of motor error by the cerebellum.

Motor neuron (efferent neuron) A neuron that transmits motor impulses from the central nervous system to a muscle or gland cell.

Motor program The group of neurons selected to perform a given action.

Motor unit The group of muscle fibers innervated by a single motor neuron.

Mucosa-associated lymphatic tissue (MALT) Loosely organized clusters of lymphoid tissue located along mucous membranes.

Mucous membrane A sheet of epithelium and the underlying basement membrane that produces mucus; lines internal hollow organs that open to the outside.

Mucus A thick, sticky secretion from goblet cells and other mucous membranes; protects underlying epithelial cells and traps foreign debris.

Multiaxial joints (triaxial joints) (muhl′-tee-AK-see-uhl; try-AK-see-uhl) A functional classification of synovial joints in which motion between articulating bones occurs around three axes.

Multipolar neuron A neuron with one axon and two or more dendrites.

Multi-unit smooth muscle Smooth muscle cells that are able to contract individually.

Murmur An abnormal sound heard in the heart caused by the turbulent flow of blood through a valve or other heart opening.

Muscle fiber An alternate name for a skeletal muscle cell.

Muscle relaxation The return of a muscle cell to its resting length due to the decreasing concentration of calcium ions in the cytosol.

Muscle spindle A specialized bundle of intrafusal muscle fibers innervated by sensory neurons that detect stretch.

Muscle tone The small amount of tension produced by a muscle at rest due to the involuntary activation of motor units by the brain and spinal cord.

Muscle twitch A single cycle of contraction and relaxation of a muscle fiber generated by a single action potential.

Muscular arteries Medium-sized arteries that control blood flow to organs; have a thick tunica media with a large amount of smooth muscle.

Muscular fatigue An inability to maintain a given level of intensity of a particular exercise.

Muscular system The body system consisting of the skeletal muscles and the associated connective tissues.

Mutation An alteration in the DNA.

Myelin (MY-eh-lin) A fatty substance that envelops and insulates the axons of certain neurons, increasing the speed of action potential conduction; formed from the plasma membranes of oligodendrocytes and Schwann cells.

Myelination (my′-eh-lin-AY-shun) The process of formation of the myelin sheath by Schwann cells or oligodendrocytes.

Myocardium (my′-oh-KAR-dee-um) The layer of the heart wall that consists of cardiac muscle cells and the heart's fibrous skeleton.

Myofibrils (my-oh-FY-brilz) Long, cylindrical organelles composed of muscle proteins in a muscle fiber.

Myofilaments Muscle proteins that make up a myofibril in a muscle fiber.

Myogenic mechanism (my-oh-JEN-ik) An autoregulatory mechanism in blood vessels by which the degree of stretch of the vessel wall triggers a reflex that maintains blood flow to a tissue.

Myoglobin (MY-oh-glohb-in) An oxygen-binding protein in muscle cells that increases the amount of oxygen immediately available to the cell.

Myometrium The thick middle layer of the uterus composed of bundles of smooth muscle.

Myopia (my-OH-pee-ah) A condition in which the eye is too long or the cornea is too curved and the lens cannot flatten enough for far vision.

Myosin (MY-oh-sin) A club-shaped contractile protein found in muscle fibers and motile cells.

Nail A hard structure located at the distal end of a digit that consists of dead, keratinized cells.

Nail matrix The proximal portion of a nail that consists of living, dividing cells from which a nail grows.

Nail plate The portion of a nail that rests on the superficial epidermal nail bed.

Nasal cavity The two-sided cavity within the anterior skull that houses the sensory receptors for olfaction and serves as the first portion of the respiratory tract.

Nasal conchae (KON-kee; singular, *concha*) Ridges of bone projecting from the lateral walls of the nasal cavity that make the air flow within the nasal cavity turbulent.

Nasal septum The bony and cartilaginous structure that separates the two sides of the nasal cavity.

Nasopharynx (nayz-oh-FAIR-inks) The portion of the pharynx that is posterior to the nasal cavity; normally serves as a passageway for air only.

Negative feedback loop The response in which the change in a regulated variable in one direction results in actions that cause changes in the variable in the opposite direction.

Neocortex The most recent part of our brains to evolve; makes up the majority of the cerebral cortex.

Neonatal period (nee-oh-NAYT-uhl) The first 4 weeks after birth.

Nephron (NEF-ron) The functional unit of the kidney where blood is filtered and the resulting filtrate is modified.

Nephron loop The second segment of the renal tubule in which water and electrolytes are reabsorbed; consists of a descending limb and an ascending limb.

Nerves Organs of the peripheral nervous system that consist of bundles of axons, connective tissue sheaths, and blood vessels.

Nervous system The body system consisting of the brain, spinal cord, and nerves; one of the chief homeostatic systems in the body.

Net filtration pressure (NFP) The difference between the opposing forces of the hydrostatic pressure gradient and colloid osmotic pressure.

Net filtration pressure of the glomerulus (NFP) The overall filtration pressure in the glomerulus that consists of the sum of the pressure that favors filtration (glomerular hydrostatic pressure) and those that oppose filtration (glomerular colloid osmotic pressure and capsular hydrostatic pressure).

Neural circuits Specific patterns of connections between neuronal pools.

Neural integration The process by which a neuron integrates all of the postsynaptic potentials from multiple presynaptic neurons.

Neural tube A structure formed by the fourth week of development from which nervous tissue, the brain, and the spinal cord arise.

Neuroendocrine organ An organ that consists of nervous tissue and secretes hormones.

Neuroglial cell (noor-oh-GLEE-ahl) A supporting cell of nervous tissue.

Neuromuscular junction (NMJ) The location where a neuron communicates with a muscle fiber.

Neuron (NOOR-on) A nerve cell capable of sending and receiving messages in the form of local and action potentials.

Neuronal pools Groups of interneurons in the central nervous system that process specific types of information.

Neuropeptides A group of peptide neurotransmitters produced by the cell body of a neuron.

Neurotransmitters Chemical messengers produced by neurons that communicate with target cells.

Neutral A solution with a pH of exactly 7.0.

Neutral fat (triglyceride) (try-GLISS-er-iyd) A lipid consisting of three fatty acids bound to a glycerol molecule.

Neutron (NEW-tron) A subatomic particle with no charge that resides in the atomic nucleus.

Neutrophil (noo-TROH-fill) A class of granulocyte and very active phagocyte that ingests damaged cells and bacteria.

Nicotinamide adenine dinucleotide (NAD⁺) One of the main electron carriers involved in nutrient catabolism; when reduced, it becomes NADH.

Nociception (NOH-sih-sep-shun) The detection and perception of painful stimuli.

Nociceptor (NOH-sih-sep-ter) A sensory receptor that responds to painful stimuli.

Node of Ranvier (RAHN-vee-ay) The unmyelinated segment of an axon between two internodes.

Nonaxial joints (nahn-AX-ee-uhl) A functional classification of synovial joints in which motion between articulating bones occurs in one or more planes but does not move around an axis.

Nondeclarative memory The memory of things that are largely procedural or skill-based.

Nondiscriminative touch The senses such as deep pressure that lack the fine spatial resolution of the tactile senses.

Nonpolar covalent bond A covalent bond in which electrons are shared equally among the atoms in the bond.

Non–rapid eye movement sleep (non-REM sleep) The name for stages I–IV of the sleep cycle.

Norepinephrine A catecholamine that functions largely as a neurotransmitter released by neurons of the sympathetic nervous system and that is also released in small amounts by the adrenal medulla.

Nostrils The paired anterior openings to the nasal cavity.

Nuclear envelope The double-layered phospholipid membrane surrounding the nucleus.

Nucleolus (new-klee-OH-luss) The structure located in the nucleus where ribosomes are assembled.

Nucleotide (NEW-klee-oh-tyd) An organic molecule consisting of a nucleic acid base, a monosaccharide, and one or more phosphate groups.

Nucleus (cellular) The biosynthetic center of the cell that contains the cell's DNA and much of its RNA.

Nucleus (in CNS) A cluster of neuronal cell bodies in the central nervous system.

Nucleus pulposus (NOO-klee-us pull-POH-sis) The inner, gelatinous, shock-absorbing portion of an intervertebral disc.

Nutrient A molecule obtained from food that the body requires for its metabolism.

Obligatory water loss The amount of urine that must be produced daily irrespective of fluid intake to excrete wastes and other solutes.

Obligatory water reabsorption Water reabsorption in the kidneys that occurs without the regulation of hormones and irrespective of the medullary concentration gradient.

Obstructive lung disease A type of pulmonary disease in which the airway resistance increases.

Obturator foramen (AHB-too-ray-tor for-AY-mun) The large hole in the anterior pelvis formed by the ilium and pubis.

Occipital lobes The posterior lobes of the cerebrum; process all information relating to vision.

Octet rule The principle that states that a molecule is most stable with eight electrons in its valence shell.

Olecranon (oh-LEK-ruh-nahn) The posterior, proximal projection from the ulna; commonly known as the "elbow."

Olfaction (ohl-FAK-shun) The sense of smell.

Olfactory mucosa A small patch of specialized epithelium on the superior surface of the nasal cavity that contains olfactory receptor cells.

Oligodendrocyte (oh-lig′-oh-DEN-droh-syt) A neuroglial cell of the central nervous system that myelinates certain axons.

Omentum (oh-MEN-tum) A specialized mesentery composed of four layers of folded visceral peritoneum that covers the digestive organs in the anterior abdominal cavity.

Oogenesis (oh′-oh-JEN-uh-sis) The process of female gamete, or ova, production in females.

Opioids (OH-pee-oydz) A group of neuropeptides that block pain perception and have various effects on behavior.

Opposition The special movement that occurs only at the first carpometacarpal joint, in which the thumb travels across the palmar surface of the hand.

Optic chiasma (KYE-azma) An X-shaped structure formed by the meeting of the optic nerves at the midline; location where some of the axons of the optic nerve cross to the other side of the brain.

Oral cavity The space enclosed by the teeth anteriorly and laterally, the palate superiorly, the tongue inferiorly, and the oropharynx posteriorly; the first portion of the alimentary canal.

Orbit The cavity in the skull that houses the eyeball and its associated structures.

Organ Two or more tissues combined to produce a structure that has a recognizable shape and that performs a specialized task.

Organ system Two or more organs that work together to carry out a broad function in the body.

Organelles (or-gan-ELZ) Structures inside the cell that carry out specific, compartmentalized functions.

Organic compounds Compounds that contain carbon atoms bonded to hydrogen atoms.

Orgasm The time period during which feelings of pleasure are experienced during sexual intercourse; in males, this coincides with ejaculation.

Origin (muscle) The less moveable attachment point of a muscle on a bone.

Oropharynx (OR-oh-fair-inks) The portion of the pharynx that is posterior to the oral cavity; serves as a passageway for food, liquids, and air.

Osmoreceptors Neurons in the thirst center of the hypothalamus that detect the osmolarity of the extracellular fluid.

Osmosis (oz-MOH-sis) A type of passive transport in which solvent molecules, usually water, move through a selectively permeable membrane from an area of lower solute concentration to one of higher solute concentration; driven by osmotic pressure gradients.

Osmotic pressure (OP) The force exerted by solute particles in a solution that "pulls" water molecules toward a solution of higher osmotic pressure.

Osseous tissue (OSS-ee-us) The hard, dense tissue that forms the majority of a bone.

Osteoblast (OSS-tee-oh-blast) An immature bone cell that secretes osteoid and aids in the deposition of bone's inorganic matrix.

Osteoclast (OSS-tee-oh-klast) A bone-destroying cell that secretes chemicals to dissolve the organic and inorganic matrices of bone tissue.

Osteocyte (OSS-tee-oh-sy't) A mature bone cell that monitors and maintains the surrounding bone matrix.

Osteoid (OSS-tee-oyd) The organic component of the extracellular matrix of osseous tissue; consists of collagen fibers, proteoglycans, glycosaminoglycans, glycoproteins, and bone-specific proteins.

Osteon (OSS-tee-on) The functional unit of compact bone; consists of concentric rings of bone matrix called lamellae that surround a central canal.

Ova (singular, *ovum*) Female gametes.

Ovarian cycle The monthly series of events associated with the maturation of an oocyte and its follicle in an ovary.

Ovarian follicle A small structure inside the ovarian cortex that contains a developing oocyte.

Ovaries The female gonads; the site of ova production and hormone secretion.

Overhydration A condition characterized by an increased volume and decreased concentration of the extracellular fluid; also known as hypotonic hydration.

Ovulation (ah-vyoo-LAY-shun) The process by which the ovary expels a secondary oocyte.

Oxidation-reduction reaction A type of exchange reaction in which electrons are exchanged among reactants.

Oxidative catabolism (aerobic catabolism) A series of reactions that occur in the mitochondria in the presence of oxygen during which electrons are removed from carbon-based molecules and the energy released is used to fuel the synthesis of ATP.

Oxidative phosphorylation The formation of ATP by harnessing the energy from the flow of electrons to create a hydrogen ion gradient in mitochondria.

Oxygen-hemoglobin dissociation curve A graphic representation of the relationship between the partial pressure of oxygen in the blood and the degree of saturation of hemoglobin.

P wave The first wave on an ECG; represents the depolarization of the atria.

Pacemaker cell A cell that depolarizes spontaneously and triggers action potentials in neighboring cells.

Pacemaker potential An action potential spontaneously generated by a pacemaker cell in cardiac or smooth muscle.

Palate (PAL-it) The roof of the oral cavity; consists of the bony hard palate anteriorly and the muscular soft palate posteriorly.

Pancreas An endocrine and exocrine organ located inferior and posterior to the stomach; secretes enzymes and other products for digestion and the hormones insulin, glucagon, and somatostatin.

Pancreatic islet A cluster of endocrine cells in the pancreas that produces and secretes the hormones insulin, glucagon, and somatostatin.

Pancreatic juice A liquid consisting of water, bicarbonate ions, and enzymes produced by pancreatic acinar cells and released into the small intestine during digestion.

Papillae (pah-PILL-ee) Finger-like protrusions on the surface of the tongue, some of which contain taste buds.

Papillary layer (PAP-uh-lar-ee) The superficial layer of the dermis composed of loose connective tissue.

Papillary muscles (PAP-ih-lair-ee) Projections of muscle from the right and left ventricular walls that attach to the atrioventricular valves via chordae tendineae.

Paracrine (PAR-ah-krin) A chemical messenger secreted into the extracellular fluid to influence nearby target cells.

Paranasal sinuses A group of hollow cavities that are continuous with the nasal cavity; located within the frontal, maxillary, sphenoid, and ethmoid bones.

Parasagittal plane (pair′-ah-SAJ-ih-tul) A plane of section that divides the body or body part into unequal right and left portions.

Parasympathetic nervous system "Rest and digest" division of the ANS that promotes maintenance functions such as defecation, diuresis, and digestion.

Parathyroid gland A group of three to five small endocrine glands located on the posterior thyroid gland; produces parathyroid hormone.

Parathyroid hormone (PTH) A hormone produced by the parathyroid gland that increases the concentration of calcium ions in the extracellular fluid.

Parietal lobes (pah-RY-eh-tal) Lobes of the cerebral hemispheres located posterior to the frontal lobes; responsible for processing and integrating sensory information, and also function in attention.

Parotid glands (puh-ROT-id) Large salivary glands located over the lateral mandible that secrete watery saliva.

Parturition (childbirth) The end of pregnancy when the fetus is expelled from the uterus through the vagina.

Passive immunity A temporary type of immunity mediated by preformed antibodies that are passed from one organism to another.

Passive transport A type of membrane transport with a gradient that does not require energy expenditure in the form of ATP.

Patella The sesamoid bone located within the tendon of the quadriceps femoris muscle.

Pectinate muscles (PEK-tin-et) Muscular ridges on the interior surface of the anterior atria, particularly the right atrium.

Pectoral girdle Bones that support the upper limb and hold it in its proper position; consists of the clavicle and scapula.

Pelvic brim The bony rim that forms the pelvic inlet.

Pelvic cavity (PEL-vik) The portion of the abdominopelvic cavity located within the bony pelvis.

Pelvic girdle The bony structure that attaches the lower limbs to the axial skeleton and that supports the weight of the upper body.

Pelvic inlet The superior opening of the bony pelvis.

Pelvic outlet The opening located at the inferior boundary of the lesser pelvis.

Pelvis The bowl-shaped bony structure formed by the two coxal bones and the sacrum.

Penis Male copulatory organ that elivers sperm into the female reproductive tract.

Peptide A polymer consisting of two or more amino acids joined by a peptide bond via dehydration synthesis.

Peptide bond A covalent bond between the amino group of one amino acid and the carboxyl group of a second amino acid.

Pericardial cavity The very narrow potential space between the parietal and visceral pericardial membranes; contains a thin layer of serous fluid.

Pericardium (pehr′-ih-KAR-dee-um) The outer layer of the heart wall; consists of

the fibrous pericardium and the serous pericardium.

Perineum (payr-uh-NEE-um) The diamond-shaped area on the pelvic floor bordered by the pubic symphysis anteriorly, the ischial tuberosities laterally, and the coccyx posteriorly.

Perineurium (pehr′-ih-NOOR-ee-um) The connective tissue that surrounds a fascicle of a peripheral nerve.

Periosteum (pear′-ee-OSS-tee-um) A membrane surrounding bones; composed of dense irregular connective tissue that houses osteogenic cells and osteoblasts.

Peripheral nervous system (PNS) The division of the nervous system consisting of the cranial and spinal nerves.

Peripheral protein A protein associated with either the intracellular or extracellular face of the plasma membrane.

Peripheral resistance Any impedance to blood flow found in the periphery of the vasculature.

Peristalsis (pehr-uh-STAL-sis) Rhythmic contractions of layers of smooth muscle that move material through a hollow organ.

Peritoneal cavity The narrow space between the visceral and parietal peritoneal membranes that is filled with serous fluid.

Peritoneum A set of double serous membranes around several abdominal organs; consists of the inner visceral peritoneum and the outer parietal peritoneum.

Peritubular capillaries Capillary beds that are fed by the efferent arterioles and surround the renal tubule.

Peroxisome (puh-ROKS-ih-sohm) An organelle containing enzymes that detoxify certain substances, metabolize fatty acids, and produce certain phospholipids.

pH The negative logarithm of the hydrogen ion concentration of a solution.

Phagocyte (FAYG-oh-syt) A cell that ingests foreign cells, dead cells, and other cellular debris by phagocytosis.

Phagocytosis (fayg′-oh-sy-TOH-sis) "Cellular eating"; a type of endocytosis that involves the ingestion of large particles.

Phalanx (FAY-langks) A long bone of the fingers or toes.

Pharyngotympanic tube (fah-ring′-go-tim-PAN-ick) The passageway that connects the middle ear and the nasopharynx.

Pharynx (FAIR-inks) The region of the respiratory tract that is located posterior to the nasal cavity, oral cavity, and larynx.

Phenotype (FEE-noh-ty′p) The way an individual's genotype is expressed physically.

Phosphate buffer system A buffer that consists of the weak acid dihydrogen phosphate ($H_2PO_4^-$) and its conjugate base, hydrogen phosphate (HPO_4^{2-}).

Phospholipid (FOSS-foh-lip-id) A lipid consisting of two fatty acids and a phosphate group bound to a glycerol molecule.

Phospholipid bilayer A double layer of phospholipids in which the fatty acid tails face one another and the phosphate heads face the surrounding aqueous environment; forms the plasma membrane and the membranes around organelles.

Photoreceptor A sensory receptor that responds to a change in light stimuli.

Phrenic nerve (FREN-ik) The main nerve of the cervical plexus that innervates the diaphragm, the main muscle of breathing; consists of the anterior rami from C3–C5.

Physiological buffer systems Actions taken by the respiratory and urinary systems to excrete hydrogen ions and/or retain bicarbonate ions in order to maintain homeostasis of the pH of the body's fluids.

Pia mater (PEE-ah) The inner, delicate meninx that surrounds the brain and spinal cord.

Pineal gland (PIN-ee-al) The part of the diencephalon that functions as a neuroendocrine organ by secreting melatonin.

Pinocytosis (peen´-oh-sy-TOH-sis) "Cellular drinking"; a type of endocytosis that involves the ingestion of smaller amounts of material through indentations in the plasma membrane coated with specific proteins.

Pivot joint A uniaxial joint in which the rounded surface of one bone fits into a groove on the surface of another bone, allowing for rotational movement around one axis.

Placenta (plah-SEN-tah) A temporary organ in the lining of the uterus through which oxygen, nutrients, and wastes are exchanged between the mother and conceptus.

Plane joint A synovial joint in which the flat surfaces of each articulating bone allow for nonaxial gliding motions.

Plantarflexion (plant-uhr-FLEK-shuhn) An angular movement of the foot at the ankle in which the angle between the foot and the leg increases.

Plasma The fluid extracellular matrix of blood; consists of water, proteins, and dissolved solutes.

Plasma cells Activated B cells that secrete antibodies.

Plasma membrane The selectively permeable barrier surrounding a cell that is composed of a bilayer of phospholipids and other components such as proteins and cholesterol.

Plasmin An enzyme that catalyzes the reaction that degrades fibrin and so removes a blood clot.

Plateau phase The portion of the cardiac action potential during which calcium ions enter the cardiac muscle cell as potassium ions exit the cell; lengthens and strengthens the resulting contraction of the cell.

Platelet (PLAYT-let) The cellular fragment of a megakaryocyte that participates in blood clotting.

Platelet plug A group of aggregated platelets that collect over an injured blood vessel to prevent blood loss.

Pleural cavity (PLOOR-al) A thin potential space between the parietal and visceral pleural membranes surrounding the lungs; contains a thin layer of serous fluid.

Podocytes (POH-doh-sytz) Cells that make up the inner visceral layer of the glomerular capsule and surround the glomerular capillaries.

Polar body The smaller cell that is produced during meiosis of oogenesis; contains DNA but very little cytoplasm and generally does not divide further.

Polar covalent bond A covalent bond in which electrons spend more time around the more electronegative atom(s), which results in the formation of a dipole.

Pollex The anatomical name of the thumb.

Polymer (POL-ih-mer) A molecule formed by joining two or more monomers chemically.

Polysaccharide (pol´-ee-SAK-ah-ryd) A large carbohydrate consisting of multiple monosaccharides joined by dehydration synthesis.

Polysynaptic reflex A reflex arc involving multiple synapses between the PNS and CNS.

Pons The middle portion of the brainstem.

Positive feedback loop A type of feedback loop in which the effector's activity increases, reinforcing the initial stimulus and amplifying the response of the effector.

Postabsorptive state The period of time from about 4 hours after feeding until the next meal.

Postcentral gyrus A gyrus located posterior to the central sulcus of each cerebral hemisphere in the parietal lobes; contains the primary somatosensory cortex.

Posterior Toward the back; also known as *dorsal*.

Posterior columns/medial lemniscal system A set of ascending tracts in the spinal cord that convey sensations relating to discriminative touch and proprioception; the posterior columns are located in the spinal cord, and they become the medial lemniscus after they decussate in the medulla.

Posterior cruciate ligament (PCL) A ligament found within the synovial cavity of the knee joint that, when the knee is flexed, tightens to prevent the tibia from being displaced posteriorly on the femur.

Posterior horn The posterior region of gray matter in the spinal cord that contains the cell bodies of second-order sensory neurons.

Posterior pituitary gland The posterior portion of the pituitary gland that is derived from nervous tissue; stores ADH and oxytocin produced by the hypothalamus.

Posterior ramus (RAY-muss; plural, *rami*) The division of a spinal nerve that travels to the posterior side of the body.

Posterior root The extension from the posterior horn of the spinal cord containing the axons of somatic sensory neurons.

Posterior root ganglion The swollen area of the posterior root where the cell bodies of somatic sensory neurons are housed.

Postganglionic neuron An autonomic neuron whose cell body is located in an autonomic ganglion in the PNS.

Postsynaptic neuron A neuron in a synapse that receives a message from a presynaptic neuron.

Postsynaptic potential A positive or negative change in the membrane potential of a neuron as a result of synaptic transmission.

Potential energy Energy that is stored, ready to be released and used to do work.

Power stroke A pivoting motion of a myosin molecule in which it moves from its cocked position to a relaxed position, pulling actin with it as it relaxes.

Precentral gyrus A gyrus located anterior to the central sulcus of each cerebral hemisphere in the frontal lobes; contains the primary motor cortex.

Pre-embryonic period The period of development from fertilization through implantation.

Prefrontal cortex A multimodal association area in the frontal lobe responsible for many higher mental functions.

Preganglionic neuron (PREE-gang-glee-on´-ik) An autonomic neuron whose cell body is located in the CNS in the spinal cord or brainstem.

Pregnancy The time period that encompasses the events from conception until birth.

Preload The length or degree of stretch of the sarcomeres in the ventricular cells of the heart before they contract.

Presbyopia (prez´-bee-OH-pee-ah) A condition in which the lens stiffens with age and cannot focus on near objects.

Pressure gradient A condition in which there are two connected areas with different gas, hydrostatic, or osmotic pressures.

Presynaptic neuron A neuron in a synapse that delivers a message to a target cell.

Primary active transport A type of active transport in which ATP is expended to move a molecule against its concentration gradient using a protein pump.

Primary motor cortices Two motor areas in the frontal lobes of the cerebrum (in the precentral gyri) that house upper motor neurons involved in the conscious planning and initiation of movement.

Principle of complementarity of structure and function A core principle of anatomy and physiology; states that the structure of a chemical, cell, tissue, or organ is always such that it best suits its function.

Principle of myoplasticity The principle stating that the structure of a muscle will change in accordance with its functional use.

Projection fibers White matter fibers that connect two areas of the cerebral cortex with one another, with other parts of the brain, and/or with the spinal cord.

Propagation The spread of an action potential along the length of a cell's plasma membrane.

Prophase The first phase of mitosis during which the chromatin condenses into chromosomes and the nuclear envelope disperses.

Proprioceptor A sensory receptor that responds to a change in position of a body part, particularly a ligament or tendon.

Propulsion The movement of a substance through a hollow organ via peristaltic contractions of smooth muscle layers.

Prostate (PRAHS-tayt) An accessory gland in the male reproductive system that produces prostatic secretions—a component of semen.

Prostatic urethra The initial segment of the male urethra that passes through the prostate gland.

Protein (PROH-teen) An organic molecule consisting of one or more polypeptide chains folded into a specific structure.

Protein buffer system A buffer that consists of proteins whose amino acids contain carboxylic acid groups that ionize to a weak base capable of binding released hydrogen ions.

Protein C An anticoagulant that catalyzes the reactions that degrade clotting factors Va and VIIIa.

Protein synthesis The process of making a protein from DNA to RNA to assembled amino acids.

Proton A positively charged subatomic particle located in the atomic nucleus.

Protraction (proh-TRAK-shuhn) The movement of a joint in which the corresponding body part moves in an anterior direction.

Proximal (PROKS-ih-mul) A position that is closer to the point of origin, usually the trunk.

Proximal tubule The first segment of the renal tubule in which water, electrolytes, and organic nutrient molecules are reabsorbed.

Pseudostratified columnar epithelium (soo′-doh-STRAT-ih-fyd) A single layer of columnar epithelial cells and the underlying basal lamina; the nuclei of the cells are uneven, giving the tissue the appearance of being stratified; found in locations such as the lining of the nasal cavity and much of the respiratory tract.

Pseudounipolar neuron (soo′-doh-yoo-nih-POH-lar) A neuron with two axons—a peripheral process that brings input to the cell body and a central process that brings input to a target cell; formerly referred to as *unipolar neurons*.

Puberty The period of life when reproductive organs mature and become functional.

Pubic symphysis The pad of fibrocartilage between the two pubic bones.

Pubis (PYOO-biss) The anteroinferior bone of the pelvis.

Pulmonary compliance The extensibility of the lungs and chest wall, or the ability of the lungs and chest wall to stretch.

Pulmonary gas exchange The exchange of oxygen and carbon dioxide across the respiratory membrane between the air in the alveoli and the blood in the pulmonary capillaries.

Pulmonary valve A semilunar valve located between the right ventricle and the pulmonary trunk.

Pulmonary veins Veins that transport oxygenated blood from the pulmonary circuit to the left atrium of the heart.

Pulmonary ventilation The physical movement of air in and out of the lungs.

Pulse points Locations where pulses may be palpated through the skin.

Pulse pressure The difference between the systolic and diastolic pressures.

Punnett square A simple diagram used to predict the possible offspring genotypes that might occur, given the genotypes of the parents.

Pupil An opening in the iris through which light enters the eye.

Purkinje fiber system (per-KIN-jee) A group of atypical pacemaker cells located in the ventricles.

Putamen (poo-TAY-men) Two basal nuclei associated with the caudate nuclei; the two structures work together to inhibit the globus pallidus and initiate movement.

Pyloric sphincter A ring of smooth muscle between the stomach and duodenum that controls the flow of chyme between the two organs.

QRS complex The second wave complex on an ECG; represents depolarization of the ventricles.

Radiation The transfer of heat from one object to another through electromagnetic waves.

Radius The lateral bone of the forearm.

Range of motion A measurement of the amount of movement that a particular joint is capable of under normal circumstances.

Rapid eye movement sleep (REM sleep) The stage of the sleep cycle characterized by rapid, back-and-forth eye movements during which most dreaming occurs.

Receptor A protein within a membrane that binds to a ligand.

Receptor-mediated endocytosis A type of endocytosis in which specific substances are taken into the cell after binding with a receptor on the plasma membrane surface.

Receptor potential A temporary depolarization caused by stimulation of a sensory receptor.

Recessive trait A trait for which an allele is masked in the phenotype by a dominant allele.

Recoil The return of an elastic tissue to its original shape and size.

Recovery period The period of time after exercise that is required to return internal conditions to the pre-exercise state.

Recruitment An increase in the number of motor units of a skeletal muscle that are stimulated in order to produce a contraction with greater tension.

Rectum (REK-tum) The portion of the large intestine between the sigmoid colon and the anal canal; located along the sacrum.

Rectus sheath The connective tissue that encloses the rectus abdominis muscle.

Red bone marrow Hematopoietic tissue located within certain bones that produces all of the formed elements of the blood.

Referred pain Painful stimuli that are perceived as cutaneous pain along a specific dermatome.

Reflex A programmed, automatic motor response to a sensory stimulus.

Refraction Bending of light rays when they pass through a translucent object.

Refractory period The period during which an excitable cell either cannot respond to another stimulus (the absolute refractory period) or requires a stronger stimulus to respond (the relative refractory period).

Regeneration The process by which a damaged tissue is replaced with the same tissue during healing.

Regional anatomy The study of the human body taken from the perspective of specific body regions.

Renal arteries Branches of the descending abdominal aorta that supply the kidneys.

Renal clearance The removal of a solute from the body by the kidneys.

Renal compensation An adjustment of the secretion or retention of hydrogen and/or bicarbonate ions to compensate for an increase or decrease in the pH of the body's fluids.

Renal corpuscle A structure in the kidneys consisting of the glomerulus and the surrounding glomerular capsule.

Renal cortex The outer region of the kidneys that consists of blood vessels and most components of a kidney's nephrons.

Renal medulla The middle region of the kidneys that consists of renal pyramids and renal columns.

Renal pelvis The inner region of the kidneys that receives urine drained from the major and minor calyces.

Renal pyramid The triangular component of the renal medulla that consists of parallel bundles of tubules.

Renal veins Veins that drain the kidneys into the inferior vena cava.

Renin (REE-nin) An enzyme produced by JG cells in response to a decline in systemic blood pressure or sympathetic nervous system stimulation; converts angiotensinogen to angiotensin-I.

Renin-angiotensin-aldosterone system (RAAS) A system that results in the formation of angiotensin-II and aldosterone; the system raises systemic blood pressure and preserves blood flow to the kidneys.

Repolarization Movement of a cell's membrane potential back toward resting level after a depolarization has taken place.

Reposition The movement of the first carpometacarpal joint back to anatomical position from an opposed position.

Reproduction The production of new cells within an organism or the production of offspring.

Reproductive system An organ system concerned with the production of gametes, hormones, and offspring.

Respiration The group of processes that combine to provide the body's cells with oxygen and remove the waste product carbon dioxide.

Respiratory acidosis A condition characterized by a decrease in the pH of the body's fluids due to hypoventilation.

Respiratory alkalosis A condition characterized by an increase in the pH of the body's fluids due to hyperventilation.

Respiratory compensation An adjustment of the respiratory rate to compensate for an increase or decrease in the pH of the body's fluids.

Respiratory membrane A thin structure across which gases are exchanged, consisting of the alveolar epithelium, basal lamina, and the endothelial cells of pulmonary capillaries.

Respiratory mucosa The pseudostratified ciliated columnar epithelium that makes up much of the respiratory tract.

Respiratory rhythm generator The collection of neurons in the ventral respiratory column of the reticular formation that generates the basic rhythm for breathing.

Respiratory system The group of organs consisting of the lungs and the respiratory tract that is responsible for gas exchange and other functions.

Respiratory tract A series of hollow organs, beginning with the nasal cavity and ending with the alveolar sacs, through which air is conducted and exchanged in the lungs.

Respiratory zone The portion of the respiratory tract that contains alveoli and so participates in gas exchange.

Responsiveness The property of living organisms by which they sense and react to changes in their environment.

Resting membrane potential The voltage difference across the plasma membrane of a cell when it is not being stimulated.

Restrictive lung disease A type of pulmonary disease in which the compliance of the lungs is decreased.

Reticular fibers Thin protein fibers in the extracellular matrix of a tissue that form "nets" within organs such as the spleen and lymph nodes and form supportive networks around blood vessels and nerves.

Reticular formation A group of more than 100 nuclei in the brainstem that are involved in a variety of functions, such as the maintenance of homeostasis and the sleep/wake cycle.

Reticular layer The deeper layer of the dermis that consists of dense irregular collagenous connective tissue.

Reticular tissue A type of connective tissue proper that contains reticular fibers; found in locations such as the spleen and lymph nodes.

Reticulocyte (reh-TI-kyu-loh-syt) An immature erythrocyte released into the circulation that still has a nucleus and some organelles.

Retina (RET-ih-nah) A thin membrane lining the inside of the posterior eye that contains cells specialized to detect light as well as neurons that give rise to the optic nerve.

Retraction (ree-TRAK-shuhn) The movement of a joint in which the corresponding body part moves in a posterior direction.

Rh blood group A blood group that classifies blood type based on the presence or absence of the Rh antigen.

Rib cage Twelve pairs of ribs that form much of the thoracic cage via posterior attachments to the thoracic vertebrae and anterior attachment to the sternum or to costal cartilage (excepting ribs 11 and 12).

Ribonucleic acid (RNA) A single-stranded nucleic acid that is involved in the transcription and translation of the code in the DNA; also forms the bulk of ribosomes.

Ribosomal RNA (rRNA) A type of RNA produced in the nucleolus that serves as a main component of ribosomes.

Ribosomes (RY-boh-sohmz) Granular, non–membrane-bounded organelles that are the site of protein synthesis.

Rods Photoreceptors that have a rod-shaped outer segment and that are sensitive to low-light conditions; responsible for low-acuity, black and white vision in dim light.

Root The portion of a hair that is embedded in the dermis.

Rotation A nonangular pivoting motion in which one bone rotates or turns on a longitudinal axis that runs down the shaft of the bone.

Rotator cuff Supportive structure around the shoulder joint formed by the tendons of four muscles: subscapularis, supraspinatus, infraspinatus, and teres minor; provides the shoulder with most of its structural support, strength, and stability.

Rough endoplasmic reticulum (RER) A series of winding membranes whose surfaces are studded with ribosomes; functions in protein synthesis and modification.

Round ligaments Fibrous cords that connect the uterus to the anterior body wall.

Ruffini endings Slowly adapting receptors that respond to stretch and movement.

S1 The first heart sound auscultated during the cardiac cycle; caused by the closing of the atrioventricular valves.

S2 The second heart sound auscultated during the cardiac cycle; caused by the closing of the semilunar valves.

Sacral curvature A convex curvature of the sacral portion of the vertebral column.

Sacral plexus A group of nerves consisting of the anterior rami from L4–S4 that mostly innervates structures of the pelvis and lower limb.

Sacrum Five fused vertebrae that articulate laterally with the pelvic bones and form the posterior boundary of the pelvis.

Saddle joint A biaxial joint in which the concave surface of one bone fits into the convex surface of the other bone, allowing for movement around two axes.

Sagittal plane (SAJ-ih-tul) A plane of section that divides the body or body part into right and left portions.

Saliva (suh-LY-vah) A fluid secreted by salivary glands into the oral cavity that consists of water, mucus, salivary amylase, lysozyme, secretory IgA, and other solutes.

Salivary glands (SAL-uh-vehr-ee) A set of three pairs of glands around the oral cavity that secrete saliva into it.

Salt An inorganic compound consisting of a nonmetal anion and a metal cation joined by an ionic bond.

Saltatory conduction A rapid type of neuronal action potential conduction that occurs in myelinated axons in which only the nodes of Ranvier are depolarized.

Sarcolemma (sar-koh-LEM-ah) The plasma membrane of a muscle fiber.

Sarcomere (SAR-koh-meer) The functional unit of muscle contraction; consists of the area of the myofibril from one Z-disc to the next Z-disc.

Sarcoplasm (SAR-koh-plazm) The cytoplasm of a muscle fiber.

Sarcoplasmic reticulum (SR) (sar-koh-PLAZ-mik reh-TIK-yoo-lum) The specialized smooth endoplasmic reticulum of a muscle fiber that stores calcium ions.

Satellite cell A neuroglial cell that surrounds cell bodies of neurons in the peripheral nervous system.

Saturated fatty acid A fatty acid with no double bonds between any of its carbon atoms.

Scapula The posterolateral bone of the pectoral girdle that articulates with the humerus and the clavicle.

Scar tissue Fibrous connective tissue that forms during fibrosis.

Schwann cell A neuroglial cell of the peripheral nervous system that myelinates the axons of certain neurons.

Sclera (SKLER-ah) The "white" part of the eyeball; thick collagenous structure that helps to maintain the shape of the eye.

Scoliosis (skoh-lee-OH-sis) An abnormal lateral curvature of the vertebral column.

Scrotum (SKROH-tum) A saclike structure located outside the abdominopelvic cavity of the male that contains the paired testes.

Sebaceous glands (seh-BEH-shus) Glands associated with hair follicles that secrete sebum via holocrine secretion.

Sebum (SEE-bum) An oily secretion from sebaceous glands that lubricates and moisturizes the surface of the skin.

Second-class lever A lever system in which the fulcrum is located farther from the applied force, with the load moved in between.

Second messenger A chemical formed inside a cell that triggers some sort of change within the cell.

Secondary active transport A type of active transport in which a primary active transport pump establishes a concentration gradient; the potential energy of this gradient is then used to fuel the transport of a second molecule against its concentration gradient.

Semen (SEE-muhn) A mixture of sperm and fluids from the testes, seminal vesicles, prostate, and bulbourethral and urethral glands.

Semiconservative replication A type of DNA replication in which each daughter cell receives one original strand and one newly replicated strand of DNA.

Semilunar valves Valves located between the ventricles and the arteries into which they pump blood.

Seminal vesicles (SEH-min-ul) Accessory glands in the male reproductive system that produce seminal fluid—a component of semen.

Seminiferous tubules (seh′-men-IF-er-us) Tightly coiled tubes inside the testes that are the site of sperm production.

Senescence (seh-NESS-ens) The period of life beginning in adulthood that involves degeneration of tissues and organs.

Sensible water loss A water loss that is "sensed" or noticed; includes water lost to the urine and the feces.

Sensory division (afferent division) The division of the peripheral nervous system that detects changes in the environment and transmits them to the central nervous system.

Sensory neuron (afferent neuron) A neuron that detects changes in the environment and transmits this information to the central nervous system.

Sensory receptor The specialized part of a neuron that detects changes in the internal and/or external environment.

Sensory transduction The conversion of a sensory stimulus to an electrical signal.

Serous fluid A thin, watery secretion from a serous membrane that lubricates an organ in a cavity within the serous membrane.

Serous membranes (SEER-us) Thin sheets of tissue that envelop certain organs and produce serous fluid.

Serum (SEAR-um) A liquid consisting of plasma without clotting factors.

Sesamoid bone (SEH-sah-moyd) A bone located within a tendon.

Set point The normal range of values of a regulated variable.

Sex-linked traits Traits coded for by alleles that are located on the X or Y chromosome.

Shaft The portion of a hair that projects above the surface of the epidermis.

Short bone A bone that is about as long as it is wide.

Simple columnar epithelium A single layer of columnar epithelial cells and the underlying basal lamina; found in locations such as lining the digestive tract and many other hollow organs.

Simple cuboidal epithelium A single layer of cuboidal epithelial cells and the underlying basal lamina; found in locations such as the kidney tubules, many endocrine glands, and the ducts of many exocrine glands.

Simple diffusion A type of diffusion in which molecules move through a membrane or fluid without the aid of a protein channel or carrier.

Simple epithelia A type of epithelial tissue composed of a single layer of epithelial cells and the underlying basal lamina.

Simple squamous epithelium A single layer of squamous epithelial cells and the underlying basal lamina; found in locations such as the alveoli of the lungs, the kidneys, and lining the inner surface of blood vessels.

Simple stretch reflex A monosynaptic reflex triggered by muscle stretch, which produces an automatic contraction of the muscle to counter the stretch.

Single-unit smooth muscle (unitary smooth muscle) Smooth muscle cells that contract together as a single unit.

Sinoatrial node (SA node) (sy′-noh-AY-tree-ul) A cluster of pacemaker cells in the superior right atrium that normally sets the pace for the heart.

Sinus rhythm The normal electrical rhythm of the heart generated by the sinoatrial node.

Sinusoidal capillaries Capillaries whose cells have large pores and gaps between them to permit large volumes of fluids, solutes, and cells to be exchanged.

Sister chromatids (KROH-mah-tidz) The two identical copies of a replicated chromosome.

Skeletal muscle A tissue type composed of multinucleate skeletal muscle cells and the surrounding endomysium.

Skeletal system The body system that consists of the bones and their associated skeletal cartilages.

Skeleton The complete set of bones of the skeletal system; perform the common functions of protection, providing mechanical levers for movement, support, and housing hematopoietic tissue.

Skin An organ consisting of the superficial epidermis and deep dermis.

Skull The superior portion of the axial skeleton that houses the brain and special sense organs; consists of cranial and facial bones.

Sleep A reversible and normal suspension of consciousness.

Sliding-filament mechanism The mechanism of contraction of a muscle cell in which the thin and thick filaments slide past one another while generating tension.

Slow-twitch fibers Skeletal muscle fibers with low myosin ATPase activity that proceed relatively slowly through their crossbridge cycles; generate slower but generally longer-lasting contractions.

Small intestine A tubular organ in the abdominopelvic cavity in which ingested food is digested, absorbed, and propelled through the remainder of the alimentary canal.

Smooth endoplasmic reticulum (SER) A series of winding membranes whose surfaces are not studded with ribosomes; functions in detoxification reactions, calcium ion storage, and lipid synthesis.

Smooth muscle A tissue type composed of flattened, spindle-shaped, uninucleate smooth muscle cells and the surrounding endomysium.

Sodium-potassium pump (Na⁺/K⁺ pump) A primary active transport pump consisting of a protein ATPase enzyme in the plasma membrane that pumps three Na^+ from the cytosol into the extracellular fluid and two K^+ from the extracellular fluid into the cytosol.

Solute (SOL-yoot) The component present in a solution in lower amounts that is dissolved in the solvent; often a solid substance.

Solution A mixture in which the molecules are in the same phase and the particles will not settle out when left undisturbed.

Solvent The component present in a solution in higher amounts; generally a liquid that dissolves a solute.

Somatic motor division A subdivision of the PNS that provides motor innervation to the skeletal muscles.

Somatic sensory division A subdivision of the PNS that provides sensory innervation to the skin, muscles, and joints.

Somatostatin (soh′-mah-toh-STAH-tin) Hormone produced by the hypothalamus that inhibits the release of growth hormone from the anterior pituitary gland.

Somatotopy A term that refers to the organization of the primary somatosensory cortex in which each part of the body is represented by a specific region in the cortex.

Somites (SOH-mytz) Blocklike structures of the mesoderm that develop into specific structures.

Spatial summation The additive effect of excitatory postsynaptic potentials triggered by multiple presynaptic neurons firing action potentials simultaneously.

Sperm Male gametes.

Spermatic cord Supporting structure of the male reproductive system extending from the scrotum to the inguinal canal.

Spermatogenesis (sper-mat′-oh-JEN-uh-sis) The process of sperm cell development.

Spermiogenesis (sper-mee′-oh-JEN-uh-sis) The maturation process of sperm.

Sphincter (SFINGK-ter) Ring of smooth muscle that generally remains in a state of contraction to control the flow of material through a hollow organ.

Sphincter muscle A muscle (either skeletal or smooth) whose circular arrangement around an opening controls passage of materials or protects structures found at that opening.

Spinal cord The organ of the central nervous system that connects the brain with the peripheral nervous system and performs certain integrative functions.

Spinal nerve A short nerve that results from the fusion of the anterior and posterior roots of the spinal cord.

Spinothalamic tracts The largest white matter tracts of the anterolateral system in the spinal cord.

Spinous process A posterior projection from a cervical, thoracic, or lumbar vertebra.

Spiral organ (organ of Corti) A cluster of receptor cells and supporting cells at the boundary between the scala tympani and the cochlear duct that transduce the energy from sound waves into action potentials.

Spirometer (spy-ROM-eh-ter) A laboratory instrument that measures volumes of air exchanged with ventilation.

Splanchnic circulation (SPLANK-nik) The collection of blood vessels that supplies and drains the digestive organs in the abdominopelvic cavity.

Splanchnic nerves (SPLANK-nik) Sympathetic nerves that synapse on collateral ganglia in the abdominopelvic cavity.

Spleen The large lymphoid organ in the left upper quadrant of the abdominopelvic cavity that filters blood and processes old erythrocytes.

Spongy bone The bone tissue located inside a bone; composed of small spicules of bone called trabeculae.

Spongy urethra The portion of the male urethra that passes through the corpus spongiosum of the penis.

Squamous cell (SKWAY-muss) An epithelial cell with a flat shape.

Static equilibrium The ability to maintain balance when the head and body are not moving.

Stem cell A cell that is undifferentiated and may develop into other cell types.

Sternum The central bone of the thoracic cage to which ribs attach.

Steroid A lipid consisting of a core of four hydrocarbon rings attached to other chemical groups.

Steroid hormones Chemical messengers derived from cholesterol, with a core of hydrocarbon rings; steroid hormones are hydrophobic and so interact with plasma membrane or intracellular receptors.

Stomach The J-shaped organ in the abdominopelvic cavity that is responsible for chemical and mechanical digestion and propulsion of ingested food.

Stratified columnar epithelium Several layers of cuboidal epithelial cells and the underlying basal lamina; lines the ducts of certain exocrine glands.

Stratified cuboidal epithelium Several layers of cuboidal epithelial cells and the underlying basal lamina; lines the ducts of certain exocrine glands.

Stratified epithelia A type of epithelial tissue composed of two or more layers of epithelial cells and the underlying basal lamina.

Stratified squamous epithelium Several layers of squamous epithelial cells and the underlying basal lamina; located in the oral cavity and esophagus, the anus, and vagina.

Stress response The series of physiological changes that maintains homeostasis when the body is faced with a stressor; mediated largely by the sympathetic nervous system and cortisol.

Striations (stry-AY-shunz) Alternating light and dark bands seen in skeletal and cardiac muscle cells.

Stroke volume (SV) The volume of blood ejected by a ventricle in one heartbeat.

Subarachnoid space The space between the arachnoid mater and the pia mater.

Subcutaneous tissue The adipose and other tissue deep to the skin.

Subdural space The very narrow space between the dura mater and the arachnoid mater.

Sublingual glands (sub-LING-gwuhl) Salivary glands located under the tongue that secrete mucus-rich saliva.

Submandibular glands Small salivary glands located on the medial side of the mandible.

Substance P A neuropeptide involved in the perception of pain.

Substantia nigra (sub-STAN-chah NIH-grah) A nucleus in the midbrain of the brainstem that works with the basal nuclei to control and initiate movement.

Substrate-level phosphorylation The formation of ATP by transferring a phosphate group directly from a phosphate-containing molecule to ADP.

Sulcus A shallow groove on the surface of the brain that separates two gyri.

Superficial A position that is closer to the surface of the body.

Superior Toward the head; also known as *cranial.*

Superior vena cava The large vein that drains most structures superior to the diaphragm; empties into the right atrium.

Surface anatomy The study of the surface markings of the human body.

Surface tension A visible film on the top of a water-based solution resulting from the formation of hydrogen bonds between adjacent water molecules.

Surfactant An amphiphilic chemical produced by type II alveolar cells that disrupts hydrogen bonds between water molecules and thus reduces surface tension.

Suspension A mixture in which the molecules are in different phases and the particles settle out when left undisturbed.

Sustentacular cells (sus-ten-TAK-yoo-luhr) Supporting cells that separate the forming sperm cells from the immune system.

Suture (SOO-cher) A fibrous joint structurally and a synarthrosis functionally that is found between the bones of the skull.

Sweat glands (sudoriferous glands) (soo'-doh-RIF-er-us) Glands in the skin that produce the watery secretion known as sweat.

Sympathetic chain ganglia A collection of autonomic ganglia located in a chain along the vertebral column.

Sympathetic nervous system "Fight or flight" division of the ANS that prepares the body for exercise, emergency, and emotion.

Symphysis (SIM-fuh-sis) A cartilaginous joint structurally and an amphiarthrosis

functionally in which the articulating bones are connected by a fibrocartilaginous pad.

Synapse (SIN-aps) The location where a presynaptic neuron communicates with its target cell.

Synapsis (sih-NAP-sis) The pairing up of homologous chromosomes to form tetrads during meiosis.

Synaptic cleft The small space between the axon terminal of a presynaptic neuron and its target cell.

Synaptic delay The short delay between the arrival of the action potential at the axon terminal of a presynaptic neuron and the postsynaptic potential of a postsynaptic neuron.

Synaptic transmission The process by which a presynaptic neuron communicates with a postsynaptic neuron through either chemical or electrical synapses.

Synaptic vesicle A membrane-bounded structure in an axon terminal that contains neurotransmitters.

Synarthrosis (sin-ar-THROH-sis) A functional classification of articulations in which the joint allows no motion between bones.

Synchondrosis (sin-kahn-DROH-sis) A cartilaginous joint structurally and a synarthrosis functionally in which the articulating bones are held together by hyaline cartilage.

Syndesmosis (sin-dez-MOH-sis) A fibrous joint structurally and an amphiarthrosis functionally in which the articulating bones are joined by an interosseous membrane or ligament.

Synergist A muscle that works together with the agonist to make the movement more efficient and smooth.

Synovial cavity A unique structural element of a synovial joint characterized by a fluid-filled space between the articulating bones of a synovial joint.

Synovial fluid The fluid secreted by the synovial membrane to provide lubrication, shock absorption, and nutrients to the articulating bones and articular cartilage as well as providing a medium in which metabolic functions can occur.

Synovial joints (sih-NO-vee-uhl) A structural classification of articulations in which the joints have a fluid-filled synovial cavity between articulating bones that are enclosed within an articular capsule.

Synovial membrane The inner layer of the articular capsule that secretes synovial fluid into the synovial cavity.

Systemic anatomy The study of the human body taken from the perspective of individual organ systems.

Systole (SIS-toh-lee) The period during which the heart is contracting.

Systolic pressure The blood pressure in the arteries when the ventricles are in systole.

T lymphocyte The leukocyte responsible for the cell-mediated arm of adaptive immunity.

T wave The third wave on an ECG; represents repolarization of the ventricles.

Tactile corpuscle (Meissner corpuscle) (MYSS-ner KOR-pus-ul) A sensory receptor located in the dermal papillae of the skin that detects light touch stimuli.

Tactile senses The senses that pertain to fine or discriminative touch, including vibration, two-point discrimination, and light touch.

Talus (TAYL-us) The tarsal bone that articulates with the tibia and fibula at the ankle joint.

Tarsals The short bones of the ankle and foot.

Taste buds Clusters of receptor cells and supporting cells on the tongue that detect taste sensations.

Telodendria (tee'-loh-DEN-dree-ah) Small branches at the end of an axon.

Telophase (TEEL-uh-fayz) The final phase of mitosis during which the cell is cleaved into two daughter cells and the nuclear membranes re-form.

Temporal lobes Cerebral lobes located on the lateral surface of the cerebrum; perform functions related to hearing, language, memory, and emotions.

Temporal summation The additive effect of excitatory postsynaptic potentials triggered by a single presynaptic neuron that fires action potentials in rapid succession.

Terminal cisternae (SIS-ter-nee) Enlarged regions of the sarcoplasmic reticulum where it contacts T-tubules.

Testes Male gonads that are the site of sperm and testosterone production.

Testosterone Hormone that stimulates spermatogenesis and the development of male sex characteristics.

Thalamus (THAL-ah-muss) The central and largest component of the diencephalon of the brain, consisting of two egg-shaped masses of gray matter; edits and sorts information entering the cerebrum.

Thermoreceptor A sensory receptor that responds to thermal stimuli.

Thermoregulation The process by which the body maintains a relatively constant internal temperature using several negative feedback loops.

Theta waves Higher-amplitude, lower-frequency brain waves characteristic of stages I–III of sleep.

Thick filament A myofilament composed of many molecules of the protein myosin.

Thick skin Skin that has five epidermal layers and in which the epidermal layers, particularly the stratum corneum, are generally quite thick; located on the palms of the hands and the soles of the feet.

Thigh The proximal portion of the lower limb that consists of the femur.

Thin filament A myofilament composed of molecules of the proteins actin, troponin, and tropomyosin.

Thin skin Skin that lacks a stratum lucidum and has thinner epidermal layers; located

everywhere on the body except the palms of the hands and the soles of the feet.

Third-class lever A lever system in which the applied force is closer to the fulcrum, with the load moved farther away from the fulcrum.

Third ventricle A hollow, cerebrospinal fluid–filled cavity located in the center of the diencephalon.

Thirst mechanism The main mechanism that drives fluid intake; controlled by the hypothalamus.

Thoracic cage A bony "cage" that encases the thoracic cavity; consists of the sternum, twelve pairs of ribs, and the thoracic vertebrae.

Thoracic cavity (thoh-RAH-sik) A ventral body cavity located within the area superior to the diaphragm that encompasses the area of the thorax.

Thoracic curvature A concave curvature of the thoracic portion of the vertebral column.

Thoracic vertebrae Twelve bones of the vertebral column that articulate laterally with the ribs and are located inferior to the cervical vertebrae and superior to the lumbar vertebrae.

Thrombin The active form of an enzyme that catalyzes the reaction converting fibrinogen to its active form, fibrin.

Thrombolysis (thrahm-BAH-luh-sis) The process by which a blood clot or thrombus is degraded.

Thrombus A blood clot; a collection of platelets, erythrocytes, fibrin, and other clotting proteins that prevents further loss of blood from a blood vessel.

Thymine A pyrimidine nitrogenous base found in DNA.

Thymus An endocrine gland in the anterior mediastinum that houses developing T lymphocytes and secretes paracrines that stimulate their maturation.

Thyroid cartilage The broad, shield-shaped cartilage framing the anterosuperior portion of the larynx.

Thyroid follicle The spherical subunit of the thyroid gland in which thyroid hormones are produced and stored.

Thyroid gland An endocrine gland located in the anterior neck that produces thyroid hormones and calcitonin.

Thyroid-stimulating hormone (TSH) The anterior pituitary hormone that stimulates the production and release of thyroid hormone and growth of the thyroid gland.

Thyroxine (T_4) (thy-ROKS-een) A hormone produced by the thyroid gland that contains four iodine molecules; less active form of the hormone, which can be converted into T_3 by cells when needed.

Tibia (TIB-ee-uh) The medial leg bone.

Tight junctions Intercellular junctions that hold cells tightly together in a tissue and make the space between them relatively impermeable.

Tissue A group of structurally and functionally related cells and their extracellular matrix.

Tissue gas exchange The exchange of oxygen and carbon dioxide between the blood in the systemic capillaries and the tissue cells.

Tissue perfusion The amount of blood flowing to a tissue through a capillary bed.

Tissue plasminogen activator (tPA) (plaz-MIN-no-jen) An enzyme produced by healed endothelial cells that catalyzes the reaction that activates plasminogen to plasmin.

Tissue repair The process by which tissue damage is repaired.

Tonicity (toh-NISS-ih-tee) A comparison between the osmotic pressure of two solutions separated by a selectively permeable membrane.

Tonsils Specialized clusters of lymphatic tissue located around the oral and nasal cavities.

Total body water The total amount of water in the intracellular and extracellular compartments of the body; averages about 42 liters.

Trabeculae (trah-BEK-yoo-lee) Spicules of bone that make up spongy bone tissue.

Trachea (TRAY-kee-uh) The hollow organ that conducts air to and from the larynx and bronchial tree of the respiratory tract.

Tract A bundle of white matter (myelinated axons) in the central nervous system.

Transcription The process by which a specific gene is copied onto a complementary strand of messenger RNA during protein synthesis.

Transcytosis A type of vesicular transport in which a substance is taken into one side of the cell by endocytosis and released from the other side of the cell by exocytosis.

Transduction The process of converting energy in the form of light, sound, movement, or touch into a neural signal.

Transfer RNA (tRNA) RNA molecules that contain an amino acid and an anticodon that binds to a specific mRNA codon.

Transitional epithelium Several layers of epithelial cells and the underlying basal lamina found in the organs of the urinary system; cells can change shape from dome-shaped to squamous when stretched.

Translation The process by which a messenger RNA transcript is read by a ribosome and an amino acid string is synthesized.

Transverse plane A plane of section that divides the body or body part into superior and inferior or proximal and distal portions; also known as the *horizontal plane* or *cross-section*.

Transverse tubules (T-tubules) Hollow inward extensions of the muscle fiber sarcolemma that surround myofibrils; filled with extracellular fluid.

Triad The combination of a T-tubule and two adjacent terminal cisternae in a muscle fiber.

Tricuspid valve Atrioventricular valve located between the right atrium and right ventricle.

Triiodothyronine (T_3) (try'-eye-oh-doh-THY-roh-neen) A hormone produced by the thyroid gland that contains three iodine atoms; form of the hormone that has higher physiological activity in promoting growth and development, regulating the metabolic rate and heat production, and promoting the effects of the sympathetic nervous system on its target cells.

Trophoblast cells (TROF-oh-blast) Large flattened cells of the blastocyst that participate in forming the placenta.

Tropomyosin (trohp′-oh-MY-oh-sin) A filamentous regulatory protein that covers the active sites of actin molecules in a thin filament.

Troponin (TROH-poh-nin) A regulatory protein with three subunits that binds tropomyosin and calcium ions in a thin filament.

True rib A rib that attaches to the sternum by its own costal cartilage.

Tubular reabsorption The process by which water and solutes are reclaimed from the filtrate in the renal tubules and collecting systems of the kidneys and returned to the blood.

Tubular secretion The process by which solutes are removed from the blood and placed into the filtrate in the renal tubules of the kidneys for excretion from the body.

Tubuloglomerular feedback (too′-byoo-low-gloh-MAIR-yoo-lar) A negative feedback mechanism in which the macula densa and JG cells regulate the GFR.

Tunica externa The outer layer of a blood vessel's wall; composed of fibrous connective tissue to support the vessel.

Tunica intima The inner layer of a blood vessel's wall; consists of endothelium, a basal lamina, and a varying number of elastic fibers.

Tunica media The middle layer of a blood vessel's wall that consists largely of smooth muscle.

Tympanic membrane (tim-PAN-ick) The membrane that separates the external auditory canal from the middle ear; vibrates when struck by sound waves and transmits the vibrations to the auditory ossicles.

Ulna The medial bone of the forearm.

Umbilical cord Structure that connects the center of the placenta to the umbilicus of the embryo or fetus; contains two umbilical arteries and one umbilical vein.

Unfused tetanus (incomplete tetanus) A type of wave summation in which a muscle fiber is stimulated rapidly and only allowed to partially relax between contractions.

Uniaxial joints (you′-nee-AK-see-uhl) A functional classification of synovial joints in which motion between articulating bones occurs around only one axis.

Universal donor An individual with blood type O– who is able to donate blood to any blood type due to the absence of A, B, and Rh antigens on his or her erythrocytes.

Universal recipient An individual with blood type AB+ who is able to receive donated blood from any blood type due to the absence of anti-A, anti-B, and anti-Rh antibodies in his or her plasma.

Unsaturated fatty acid A fatty acid with one or more double bonds between its carbon atoms.

Upper limb The portion of the appendicular skeleton that consists of the arm, forearm, wrist, and hand.

Upper motor neurons Interneurons located in the frontal lobe of the cerebral cortex that are involved in the conscious planning of movement; initiate movement via lower motor neurons of the PNS.

Uracil A pyrimidine nitrogenous base found in RNA in place of thymine.

Urea cycle The process that combines two molecules of ammonia with carbon dioxide to produce urea, which is eliminated by the kidneys.

Ureters (YOOR-eh-terz) Muscular tubes that transport urine from the kidneys to the urinary bladder.

Urethra (yoo-REE-thrah) The final segment of the urinary tract through which urine is voided.

Urinalysis A test performed on the urine to analyze its volume, color, and chemical characteristics.

Urinary bladder The hollow organ of the urinary tract that stores urine and contracts to expel it from the body.

Urinary system The body system consisting of the kidneys, ureters, urinary bladder, and urethra; filters the blood of wastes and works to maintain fluid, electrolyte, and acid-base homeostasis.

Urinary tract The organs of the urinary system (ureters, urinary bladder, and urethra) that transport, store, and eliminate urine.

Urine A liquid produced by the kidneys that consists of water, electrolytes, and metabolic wastes.

Urokinase (yoo-roh-KI-naz) An enzyme produced by cells of the kidney that catalyzes the reaction that activates plasminogen to plasmin.

Uterine cycle (menstrual cycle) The series of cyclic changes that the uterine endometrium goes through each month as it responds to the fluctuating levels of ovarian hormones.

Uterine tubes (fallopian tubes) Tubes that transmit an oocyte from the ovaries to the uterus.

Uterus (YOO-ter-uhs) The organ in which a conceptus implants and develops until birth.

Uvula (YOO-vyoo-luh) A portion of the soft palate suspended in the posterior oral cavity; seals off the nasopharynx during swallowing to prevent food and liquid from entering the nasal cavity.

Vaccination The administration of an antigen to elicit a primary immune response and generate memory cells.

Vagina (vuh-JY-nuh) The female organ of copulation that receives the penis and semen during sexual intercourse; it is also the passageway for delivery of an infant and for menstrual flow.

Valence electrons The electrons located in an atom's outer shell; involved in chemical reactions and the formation of chemical bonds.

Vasa recta (VAY-zah REK-tah) Capillary beds that surround the nephron loops of juxtamedullary nephrons.

Vasa vasora (VAY-zah vay-ZOH-rah) Small blood vessels that supply the walls of larger blood vessels.

Vascular anastomoses (an-ass′-toh-MOH-seez) Locations where blood vessels connect by collateral vessels.

Vasculature (VASS-kyoo-lah-chur) The collective term for the body's blood vessels.

Vasoconstriction The narrowing of a blood vessel due to contraction of vascular smooth muscle.

Vasodilation The opening of a blood vessel due to relaxation of vascular smooth muscle.

Veins The vasculature's collection system; systemic veins collect deoxygenated blood from systemic capillaries and return it to the right side of the heart, and pulmonary veins collect oxygenated blood from pulmonary capillaries and transport it to the left side of the heart.

Venae cavae (superior and inferior) (VEE-nee KAY-vee; singular, *vena cava*) The two largest systemic veins in the body; deliver deoxygenated blood from the systemic circuit to the right atrium.

Venous valves Flaps of endothelium in certain veins that prevent blood from flowing backward.

Ventilation/perfusion (V/Q) ratio The ratio of pulmonary ventilation of a segment of the lung to its blood flow (perfusion).

Ventral body cavity The major body cavity located mostly on the anterior side of the body.

Ventricles (of brain) (VEN-trih-kulz) Cerebrospinal fluid–filled cavities located within the brain.

Ventricles (of heart) The inferior chambers of the heart that receive blood from the atria and pump blood into arteries.

Ventricular ejection phase The period of the cardiac cycle during which blood is ejected from the ventricles into the pulmonary trunk and aorta.

Ventricular filling phase The period of the cardiac cycle during which blood drains from the atria into the ventricles.

Venules Small veins that drain blood from capillary beds.

Vermiform appendix (VER-muh-form uh-PEN-diks) A blind-ended extension from the cecum of the large intestine that contains lymphatic nodules.

Vermis (VER-miss) The structure that connects the two lobes of the cerebellum.

Vertebral arteries Arteries supplying the brain that travel through the transverse foramina of the cervical vertebrae.

Vertebral cavity A dorsal body cavity located within the vertebral column that houses the spinal cord.

Vertebral column A skeletal structure consisting of 24 individual vertebrae, 5 fused vertebrae of the sacrum, and 3–5 fused vertebrae of the coccyx; together encase the spinal cord and nerves within the vertebral cavity.

Very low-density lipoproteins (VLDLs) Triglyceride-rich particles that deliver triglycerides from the liver to other tissues.

Vesicle (VESS-ih-kul) A small, membrane-enclosed body in the cytosol that generally functions in transport or storage.

Vesicular transport (vess-IK-yoo-lar) A type of active transport in which materials are moved into and out of the cell by vesicles.

Vestibule The bony portion of the inner ear involved in detecting head position and linear movement.

Villi (VILL-ahy; singular, *villus*) Folds of the mucosa and submucosa of the small intestine that increase surface area for absorption.

Visceral reflex arc A series of events in which a sensory stimulus in an organ leads to a predictable visceral motor response mediated by the ANS.

Viscosity The thickness of a liquid or the inherent resistance all liquids have to flow.

Visual field The area one can observe with one or both eyes when focusing on a central point.

Vital capacity The total volume of exchangeable air within the lungs; equal to the tidal volume plus the expiratory and inspiratory reserve volumes.

Vitamin An organic compound required in small amounts for specific functions within the body.

Vitreous humor The thick fluid located in the posterior cavity of the eyeball.

Vocal folds A set of elastic mucosal folds in the larynx that vibrate to produce sound.

Volatile acid An acid that can be converted to a gaseous form that may be excreted through the lungs.

Voltage The difference in electrical potential between two points.

Voltage-gated channel An ion channel in the plasma membrane of a cell that opens or closes in response to a change in the membrane potential.

Wave summation A phenomenon in which repeated stimulation of a muscle fiber by a motor neuron results in muscle twitches with progressively greater tension.

Wernicke's area An integrative multimodal association area for speech located in the temporal and parietal lobes; responsible for the ability to understand and produce intelligible language.

White matter Collections of myelinated axons that appear white.

Wrist The portion of the upper limb that articulates proximally with the forearm and distally with the hand.

Xiphoid process (ZIH-foyd) The inferior portion of the sternum.

Yellow bone marrow Adipose tissue located within most bones of an adult.

Yolk sac The first extraembryonic membrane to develop; forms part of the digestive tract and is the source of the first blood cells, blood vessels, and the first germ cells.

Z-disc The component of the sarcomere that contains structural proteins to which thin and elastic filaments attach.

Zygote The earliest stage of human development; consists of a single cell formed from the fusion of an ovum and a sperm cell.

Index

Page numbers in **boldface** indicate a definition. A *t* following a page number indicates tabular material, an *f* indicates an illustration, and a *b* indicates boxed material.

Prefixes, Suffixes, Word Roots, and Combining Forms

Many of the words we use in everyday English have their roots in other languages, particularly Greek and Latin. This is especially true for anatomical terms. The lists that follow include some of the foreign prefixes, suffixes, word roots, and combining forms that are part of many of the biological and anatomical terms used in this text.

PREFIXES

a-, an-, without or lacking: *anesthesia* = without feeling, particularly pain

ab-, away from, departing: *abduct* = move away from midline

ad-, to or toward: *adduct* = move toward midline

af-, toward: *afferent neurons* = neurons that send stimuli toward the CNS

ante-, before: *antebrachium* = the forearm (or the region before the arm)

anti-, against: *antidiuretic hormone* = decreases urine output, working against diuresis

auto-, self: *autoimmune disease* = inappropriate immune response against the body's own tissues

bi-, two: *biceps* = a muscle with two heads or attachment sections

brady-, slow: *bradycardia* = slower than normal heart rate

circum-, around: *circumduction* = circular movement around a point

co-, con-, with, together: *coenzyme* = molecule that temporarily combines with an enzyme to help it function

contra-, against, opposite: *contralateral* = on the opposite side of the body from the point of reference

cyan-, blue: *cyanotic* = bluish color of skin due to hypoxia

de-, reversal, loss, undoing: *depolarization* = reversal of membrane polarity

di-, two: *disaccharide* = molecule made of two joined simple sugars

dia-, across, through: *dialysis* = diffusion across a membrane

dys-, difficult, painful: *dysmenorrhea* = painful menstruation

ecto-, out, away from: *ectoderm* = outer layer of embryonic tissue

ectop-, displaced: *ectopic pregnancy* = embryo implants outside the uterus

ef-, away from: *efferent neurons* = neurons that send stimuli away from the CNS

em-, en-, in, inside: *embolus* = abnormal object moving within blood vessels

endo-, within, inner: *endocytosis* = taking larger particles into the cell

epi-, over, upon: *epidermis* = outermost layer of the skin over the dermis

eryth-, red: *erythrocyte* = red blood cell

eu-, well, true: *eukaryotic cell* = cell with a true membrane-bounded nucleus

ex-, exo-, outer, outside: *exocytosis* = moving larger particles out of the cell

extra-, beyond, outside: *extracellular fluid* = fluid outside of most body cells

hemi-, half: *hemisphere* = left or right half of the brain

hetero-, different: *heterozygous* = having two different alleles for the same trait

homo-, same, equal: *homozygous* = having two identical alleles for the same trait

hyal-, clear: *hyaline cartilage* = type of cartilage with a fairly clear matrix

hyper-, over, excessive: *hypertension* = higher than normal blood pressure

hypo-, below, under: *hypotension* = lower than normal blood pressure

juxta-, next to, near: *juxtaglomerular* = near the glomerulus of the kidney

in-, im-, not, unable: *involuntary* = not under the conscious control of the brain

infra-, beneath: *infraorbital* = below the eye socket

inter-, between: *intercostal* = between the ribs

intra-, within: *intravenous* = within a vein

iso-, same: *isotonic* = having the same osmotic pressure

leuk-, white: *leukocyte* = white blood cell

macro-, large: *macrophage* = large phagocytic cell

mal-, bad: *malnutrition* = poor nutrition

medi-, middle: *mediastinum* = middle portion of the thoracic cavity

mega-, large: *megakaryocyte* = large cell that produces platelets

melan-, black: *melanocyte* = cell that produces dark pigment molecules

mes-, middle: *mesoderm* = middle layer of embryonic tissue

meta-, after, beyond: *metastasis* = spread of cancer beyond its point of origin

micro-, small: *microvilli* = small projections of the cell's plasma membrane

milli-, thousandth: *millimeter* = one thousandth of a meter

mono-, single: *monosaccharide* = a single simple sugar molecule

neo-, new: *neonatal* = referring to a newborn infant up to 4 weeks of age

non-, not: *nonpolar covalent bond* = bond with no poles or charged areas

oligo-, few: *oligosaccharide* = smaller carbohydrates made of three (or a few more) simple sugars

ortho-, straight, correct: *orthopedics* = medical field related to correcting musculoskeletal conditions

para-, near, at the side: *parathyroid glands* = glands on the back side of the thyroid

per-, through: *perforin* = protein that produces a small hole through the plasma membrane of a cell targeted for destruction

peri-, surrounding, around: *perimysium* = connective tissue wrapping around a fascicle (bundle) of muscle cells

poly-, many: *polysaccharide* = molecule made of many joined simple sugars

post-, after: *postpartum* = after childbirth

pre-, before: *presynaptic* = before the synapse of two neurons

presby-, old: *presbyopia* = age-related farsightedness due to loss of lens elasticity

pro-, first, promoting: *prolactin* = hormone that promotes mammary gland milk production

quad-, four: *quadriceps* = a muscle with four heads or attachment sections

re-, returning, doing again: *reflux* = return flow of fluid (e.g., esophageal)

retro-, behind: *retroperitoneal* = organs behind the abdominal cavity membranes

semi-, half: *semilunar valve* = valve with a half-moon shape

sub-, below, under: *submandibular* = below the mandible

super-, supra-, above, excessive: *supraspinatus* = muscle above the spine of the scapula

syn-, together: *synapse* = area where two neurons come together

tachy-, fast, rapid: *tachycardia* = faster than normal heart rate

trans-, through, across: *transversus abdominis* = muscle with fibers that run across the abdominal cavity

tri-, three: *triglyceride* = a lipid containing three fatty acid molecules

ultra-, beyond: *ultraviolet* = wavelength of light beyond the range of normal human vision

xanth-, yellow: *xanthosis* = yellowish discoloration of tissue (not the same as jaundice)

SUFFIXES

-al, -ac, pertaining to: *menopausal* = pertaining to menopause in women

-algia, pain: *neuralgia* = sharp pain caused by nerve damage

-aps, juncture: *synapse* = junction of two neurons

-arche, beginning: *menarche* = start of menstrual cycles in girls

-ary, associated with, referring to: *dietary* = associated with food and nutrition

-ase, suffix on names of most enzymes: *acetylcholinesterase* = enzyme that catalyzes the breakdown of acetylcholine

-asthenia, weakness: *myasthenia* = disease that produces muscle weakness

-blast, sprout, immature cell: *osteoblast* = bone cell that builds bone tissue

-centesis, draining: *amniocentesis* = removal of a sample of amniotic fluid

-cide, kill: *bactericide* = a chemical that kills bacteria

-clast, -clasia, breaking down: *osteoclast* = bone cell that breaks down bone tissue

-crine, secrete, release: *endocrine* = gland that releases hormones into the blood

-ectomy, -tomy, to excise surgically: *nephrectomy* = removal of the kidney

-ell, -elle, small: *organelles* = small structures in cells performing specific functions

-emia, condition of the blood: *leukemia* = abnormal development of leukocyte-producing bone marrow tissue

-fferent, carry: *efferent arteriole* = carries blood away from the glomerulus of the kidney

-form, -forma, shape: *lentiform nucleus* = lens-shaped area of gray matter in the brain

-gen, initiating or causing agent: *allergen* = substance capable of evoking an allergic response

-genesis, production, creation: *gluconeogenesis* = production of glucose molecules

-gram, recorded data: *electrocardiogram* = a graphic report of impulse transmission through the heart's conduction system

-graph, instrument that records data: *electrocardiograph* = machine that records impulse transmission through the heart's conduction system

-ia, -sia, condition of, process of: *hypoxia* = delivery of too little oxygen to the body or a tissue

-iasis, abnormal condition of: *mydriasis* = potentially abnormal dilation of the pupil

-ic, -ac, pertaining to: *pancreatic* = pertaining to the pancreas

-ism, condition of: *hypothyroidism* = condition in which one produces too little thyroid hormone

-itis, inflammation of: *arthritis* = inflammation of the joints

-lemma, husk, shell, covering: *neurilemma* = covering around axons of neurons

-lith, stone: *otoliths* = crystalline structures resembling microscopic pebbles in the inner ear

-lysis, splitting: *hemolysis* = rupture of erythrocytes

-malacia, softening: *osteomalacia* = a disorder such as rickets that produces abnormal bone softening

-megaly, growing, enlarged: *acromegaly* = excessive thickening of bones

-metric, -metry, measurement, length: *spirometry* = measuring lung volumes

-odyn, pain, painful: *ophthalmodynia* = pain centered around the eyes

-oid, like or resembling: *lipoid* = a chemical resembling a lipid in composition

-ology, study of: *histology* = study of tissues

-oma, tumor: *melanoma* = skin tumor originating in melanocytes

-opia, vision: *myopia* = nearsightedness

-ory, pertaining to: *inflammatory* = pertaining to inflammation

-osis, condition of: *osteoporosis* = condition in which bones lose density and weaken

-pathy, disease state: *cardiomyopathy* = disease of heart muscle

-penia, lack of: *leukopenia* = low white blood cell count

-phagy, eating: *geophagy* = compulsion to eat soil, possibly driven by mineral deficiencies

-phasia, speech: *aphasia* = loss of the ability to speak

-philic, liking, loving: *hydrophilic* = describes molecules that are attracted to and freely dissolve in water

-phob, fearing, disliking: *acrophobia* = fear of heights

-plasia, growth: *neoplasia* = new abnormal cell growth

-plasm, matter or substance: *cytoplasm* = liquid substance inside the cell

-plasty, reconstruction of a part: *mammoplasty* = breast reshaping or reconstruction after surgery

-plegia, paralysis: *paraplegia* = paralysis of both legs

-pnea, breathing: *apnea* = absence of respiration

-rrhagia, -rrhage, discharge, break out: *hemorrhage* = excessive bleeding

-rrhea, flow: *diarrhea* = frequent excess flow of watery loose stools

-some, -soma, body: *phagosome* = vesicle containing a particle that a cell has "eaten," or brought in via endocytosis

-spire, breathing: *inspire* = to breathe in

-stalsis, compression: *peristalsis* = wavelike squeezing of tubular organs

-stasis, to stop, arrest: *hemostasis* = process of stopping blood loss

-tensin, -tension, pressure: *hypertension* = high blood pressure

-tomy, to cut: *mastectomy* = surgical removal of the breast

-tonic, pressure, tension: *isotonic* = same amount of osmotic pressure on both sides of a membrane

-trophy, growth: *hypertrophy* = excess growth of a tissue

-ty, state or condition of: *obesity* = having an abnormal amount of body fat

-ule, tiny, small: *tubule* = small tube (e.g., microscopic collecting tubules in the kidneys)

-uria, urine: *hematuria* = blood in the urine

WORD ROOTS WITH COMMON COMBINING VOWELS

acr/o-, limb: *acromegaly* = enlargement of the limbs

aden/o-, gland: *adenoma* = tumor of a gland

alve/o-, small cavity: *alveoloplasty* = surgical reconstruction of a dental alveolus (tooth socket)

ang/i-, vessel: *angiogram* = examination of a vessel

arter/i-, artery: *polyarteritis* = inflammation of many arteries

arthr/o-, joint, articulation: *arthralgia* = joint pain

bar/o-, pressure: *baroreceptor* = nerve cell that detects pressure

bil/i-, bile: *endobiliary* = within the bile duct

brach/i-, arm: *antebrachial* = relating to the forearm

bronch/o-, bronchus (air passage): *intrabronchial* = within the bronchus

carcin/o-, cancer: *carcinogen* = chemical or condition that can cause cancer

card/i-, heart: *pericardiocentesis* = draining of the space around the heart

cephal/o-, relating to the head: *cephalodynia* = head pain or headache

cerv/i-, neck: *cervicalgia* = neck pain

chol/e-, bile: *cholecystectomy* = removal of the bile-storing gallbladder

chondr/o-, cartilage: *chondrosarcoma* = tumor of the cartilaginous soft tissue

chrom/o-, color: *hypochromasia* = the condition of decreased color

corn/e-, horn: *corneocyte* = hard cell in the outer layer of the skin or a horn in a nonhuman animal

corp/u-, body: *corpuscle* = small body

cortic/o-, bark (outer part of a structure): *corticosteroid* = steroid hormone produced by the outer part of the adrenal gland

cran/i-, head, skull: *intracranial* = within the head or skull

cusp/i-, point: *tricuspid* = having three points or cusps

cutane/o-, skin: *percutaneous* = through the skin

cyst/o-, bladder: *cystolith* = stone in the urinary bladder

cyt/o-, cell: *cytology* = study of cells

dactyl/o-, digits (fingers, toes): *polydactyly* = condition of having more than the normal number of fingers and/or toes

dendr/o-, branch: *dendritic cell* = cell with extensions resembling the branches of a tree

dent/o-, dont/o-, tooth: *endodontic* = pertaining to the inner structures of the tooth

derm/o-, dermat/o-, skin: *intradermal* = within the skin

diastol/i-, standing between: *diastolic pressure* = blood pressure between beats of the heart

dips/i-, thirst: *polydipsia* = condition of elevated level of thirst

encephal/o, brain: *encephalitis* = inflammation within the brain

enter/o-, intestine: *enterohepatic* = pertaining to the intestines and the liver

esth/e-, feeling, sensation: *hyperesthesia* = excessive feeling or sensitivity

gastr/o-, stomach: *gastroesophageal* = pertaining to the stomach and esophagus

gest/u-, carried: *gestational* = pertaining to the period during which a fetus is carried by a pregnant woman

gingiv/o-, gums: *supragingival* = pertaining to the area superior to the gums

glom/o, glomerul/o-, ball: *glomerulonephritis* = inflammation of the glomerular capillaries of the kidney

gloss/o-, tongue: *microglossia* = abnormal smallness of the tongue

gluc/o-, glyc/o-, sugar: *gluconeogenesis* = formation of glucose molecules

hepat/o-, liver: *perihepatic* = around the liver

hist/o-, tissue: *histopathology* = study of disease states of tissues

hydr/o-, water: *dehydrate* = to decrease the amount of water in a fluid

hyster/o-, uterus: *hysterectomy* = surgical removal of the uterus

iatr/o-, treatment: *iatrogenic* = caused by treatment

inguin/o, groin: *retroinguinal* = behind the inguinal ligament in the groin

jugul/o-, throat: *jugulodigastric* = location in the throat near the digastric muscles

kal/e-, potassium: *hyperkalemia* = condition of elevated potassium ions in the blood

kary/o-, nucleus: *eukaryote* = cell with a true nucleus

kerat/o-, horn: *hyperkeratosis* = condition in which the outermost layer of the skin develops horny deposits

kin/e-, movement: *kinesiology* = study of movement

lab/i-, lip: *supralabial* = above the lip

lact/o-, milk: *lactorrhea* = excess or abnormal flow of milk

lacun/o-, space: *lacunae* = small spaces within bone tissue

lamell/o-, plate or thin sheet: *circumferential lamellae* = thin, circular plates of bone tissue

lamin/o-, layer: *laminar flow* = condition in which fluid flows in parallel layers

lapar/o-, abdomen: *laparotomy* = incision into the abdomen

lig/a-, tie, binding: *ligament* = structure that binds two bones together

lip/o-, fat (adipose): *lipolysis* = breakdown of fat

men/u, menstr/u-, monthly: *dysmenorrhea* = painful monthly discharge

metr/i-, uterus: *endometrium* = inner lining of the uterus

mut/a-, change: *mutagen* = agent generating a change in DNA

my/o-, muscle: *cardiomyopathy* = disease of heart muscle cells

myel/o-, marrow: *myeloblast* = immature cell of the bone marrow

nat/a-, birth: *prenatal* = before birth

natr/i-, sodium: *hyponatremia* = condition of decreased sodium ions in the blood

necr/o-, death, dead: *necrosis* = condition of death

nephr/o-, kidney: *nephroptosis* = drooping of a kidney

neur/o-, nerve: *neurology* = study of the nervous system

noct/o, nyct/o-, night: *nocturia* = urination occurring at night

ocul/o-, eye: *monocular* = pertaining to one eye

onc/o-, cancer: *oncogene* = a gene involved in the pathogenesis of cancer

opt/i, ophthalm/o-, eye: *ophthalmoscope* = instrument for examining the eye

orchid/o-, testis: *orchidectomy* = removal of a testis

oste/o-, bone: *osteoblast* = immature bone cell

ot/o-, ear: *otoscope* = instrument for examining the ear

ov/u-, egg: *ovular* = pertaining to an egg, or ovum

path/o-, disease: *pathogenesis* = development of disease

ped/i-, child: *pediatrics* = medical specialty focusing on children